Reading's Sedimentary Environments

Reading's Sedimentary Environments

Processes, Facies and Stratigraphy

Fourth Edition

Edited by

Bruce Levell
University of Oxford, Oxford, UK

WILEY

This edition first published 2026

Edition History
Wiley-Blackwell (1978, 1e; 1986, 2e; 1996, 3e)

Registered Offices
John Wiley & Sons, Inc., 111 River Street, Hoboken, NJ 07030, USA
John Wiley & Sons Ltd, New Era House, 8 Oldlands Way, Bognor Regis, West Sussex, PO22 9NQ, UK

For details of our global editorial offices, customer services, and more information about Wiley products visit us at www.wiley.com.

The manufacturer's authorized representative according to the EU General Product Safety Regulation is Wiley-VCH GmbH, Boschstr. 12, 69469 Weinheim, Germany, e-mail: Product_Safety@wiley.com.

Wiley also publishes its books in a variety of electronic formats and by print-on-demand. Some content that appears in standard print versions of this book may not be available in other formats.

Library of Congress Cataloging-in-Publication Data Applied for:
Hardback ISBN: 9781119739838

Cover Design: Wiley
Cover Image: © Bruce Levell

Set in 9.5/12.5pt STIXTwoText by Straive, Pondicherry, India

SKY10150481_032426

Contents

List of Contributors

Kerrie L. Bann
Ichnofacies Analysis Inc., Calgary, Alberta
Canada

Clare Boston
School of the Environment and Life Sciences
University of Portsmouth
Portsmouth
UK

Luis A. Buatois
Department of Geological Sciences
University of Saskatchewan
Saskatoon
Canada

Daniel S. Collins
Shell International Ltd.
London
UK

Luca Colombera
Dipartimento di Scienze della Terra e dell'Ambiente
Università di Pavia
Pavia
Italy

Grace I. E. Cosgrove
School of Earth and Environment
University of Leeds
Leeds
UK

Bethan Davies
School of Geography
Politics and Sociology
Newcastle University
Newcastle upon Tyne
UK

Lauren Edgar
U.S. Geological Survey
Astrogeology Science Center
Flagstaff
AZ, USA

NASA Johnson Space Center, Houston, TX
USA

Massimiliano Ghinassi
Dipartimento di Geoscienze
Università degli Studi di Padova
Padova
Italy

Martin R. Gibling
Department of Earth and Environmental Sciences
Dalhousie University
Halifax
Canada

Murray K. Gingras
Department of Earth and Atmospheric Sciences
University of Alberta
Edmonton, Alberta
Canada

Chenglin Gong
College of Geosciences
China University of Petroleum (Beijing)
Beijing
China

Sanjeev Gupta
Department of Earth Science and Engineering
Imperial College London
London
UK

Gary J. Hampson
Department of Earth Science and Engineering
Imperial College
London
UK

David M. Hodgson
School of Earth and Environment
University of Leeds
Leeds, UK

Alessandro Ielpi
Earth, Environmental and Geographic Sciences
University of British Columbia
Okanagan Campus
Kelowna, BC
Canada

Christopher A.-L. Jackson
WSP UK Ltd, Manchester
UK

Howard D. Johnson
Department of Earth Science and Engineering
Imperial College
London
UK

Ian A. Kane
Department of Earth and Environmental Sciences
University of Manchester
Manchester
UK

Mathieu Lapôtre
Department of Earth & Planetary Sciences
Stanford University
Stanford
CA, USA

Daniel Paul Le Heron
Department of Geology
University of Vienna
Vienna
Austria

Bruce Levell
Department of Earth Sciences
University of Oxford
Oxford, UK

Shunli Li
School of Energy Resources
China University of Geosciences
Beijing, China

Stephen W. Lokier
School of Science
University of Derby
Derby, UK

James A. MacEachern
Department of Earth Sciences
Simon Fraser University
Burnaby, British Columbia
Canada

Mike Mayall
Department of Earth Science and Engineering
Imperial College
London
UK

Elda Miramontes
MARUM-Center for Marine Environmental Sciences
University of Bremen
Bremen
Germany
Faculty of Geosciences, University of Bremen, Bremen
Germany

Nigel P. Mountney
School of Earth and Environment
University of Leeds
Leeds
UK

Marcello Natalicchio
Department of Earth Sciences
University of Turin
Turin, Italy

Guy Plint
Department of Earth Sciences
The University of Western Ontario
London
Ontario
Canada

John J.G. Reijmer
Department of Earth Sciences, Faculty of Science
Vrije Universiteit Amsterdam
Amsterdam
The Netherlands

Stuart A. Robinson
Department of Earth Sciences
University of Oxford
Oxford
UK

Ronald J. Steel
Department of Geological Sciences
Jackson School of Geosciences
University of Texas, Austin
Texas
USA

Frans van Buchem
Physical Sciences and Engineering Division, King Abdullah University of Science and Technology, Thuwal
Saudi Arabia

Emmanuelle Vennin
Unité Mixte de Recherche Biogéosciences
University of Burgundy
Dijon
France

Xinghe Yu
School of Energy Resources
China University of Geosciences
Beijing
China

Xiaomin Zhu
College of Geosciences
China University of Petroleum (Beijing)
Beijing
China

Foreword

Harold Reading was an inspiring teacher and supervisor who left a lasting personal legacy. ...personal legacy, largely simply through the way he interacted with people. Many have fond memories of lectures and field trips, or learned from his discussions in the field, in the pub, or through his incisive, meticulous editing of thesis drafts. Harold's early career included a short but influential spell with Royal Dutch Shell. He then took an academic position at Oxford in 1957, where he stayed for the rest of his career. Here he started to develop, with his graduate students, the method of 'facies sedimentology', initially with attention to turbidites and later, deltaic, shallow-marine and fluvial deposits. These studies led to consideration of wider controls on sedimentation, in particular tectonism, as seen through the lens of the exciting, then new, plate tectonic paradigm.

As an educator, Harold's influence has been pervasive. Of his 35 graduate students supervised between 1961 and 1994, many entered academia, all over the world. Those students, and their students, have now trained as many as four generations of sedimentary geologists, totalling at least 780 people. Striking is the rise in female participation in sedimentary geology: 8% of his direct graduate students were women whereas amongst the fourth generation, the ratio is close to 50%.

Arguably his most lasting contribution to the discipline was his textbook Sedimentary Environments and Facies. The first edition (1978) was written largely by Harold's former graduate students and other Oxford colleagues. This volume became an essential gateway to the literature for young graduate students and young professionals (including many authors of this, 4th edition)! 'The Book' served to disseminate Harold's approach around the world. As the field (and the literature!) rapidly expanded and diversified, Harold was motivated to edit second (1984) and third (1996) editions, with an evolving cast of authors but always with his customary careful editing.

Harold played a leading role in the early development of the International Association of Sedimentologists, serving as Publications Secretary, General Secretary, and President. He was particularly concerned that education in sedimentary geology be available internationally and to as diverse an audience as possible. He promoted these objectives through the IAS Friendship Scheme, sponsored lecture tours, workshops and student grants as well as his personal hospitality at Oxford. He was a passionate supporter of lifelong learning including 'continuing professional development'.

It is now 29 years since the third edition of Sedimentary Environments was published, and the field has of course developed enormously. With rising public awareness of the causes and consequences of climate change, it is timely to show that 'the past is the key to the future', and that sedimentary geology provides 'deep time' insight that can help in current mitigation and adaptation efforts. It is very fitting that a fourth edition of 'The Book' should be available to assist and inspire a further generationof sedimentologists. The authors are drawn from around the world, and include a number perched on various branches of Harold's very extensive 'family tree'.

IAS as a society is proud to have contributed to this publication, which, with its new name "Reading's Sedimentary Environments" honours the memory of one of our pioneering supporters.

Daniel Ariztegui
President International Association of Sedimentologists,
May 2025

Preface

After 29 years, this 4th edition is, of necessity, an entirely new work.

Many of the authors when first approached about this book expressed great affection for its previous editions. Several said that it was this book that had attracted them to a career in sedimentology. More than one confessed to taking the book to bed to read. This loyalty bound many of us as a group during the long gestation of the volume. We hope we have done him proud. We are certain that he would welcome the increase in diversity that the past quarter-century has seen in the subject, its applications, and most of all in its practitioners. We thank the Reading family for their support and for granting permission to use his name in the title of this and subsequent editions.

How times have changed! The 1996 edition of this book had 15 authors 11 of whom were British and living in the United Kingdom, a further 2 were British living overseas, and just 2 were women. This edition has 37 authors, including 7 women and less than half, 16, are from the United Kingdom. Harold, as one of the founders of the International Association of Sedimentologists, would have strongly approved of this increase in diversity and internationalism, but would also have been disappointed that it did not go even further.

The authors have collectively agreed to donate all royalties from the sales of this book to the International Association of Sedimentologists for the purposes of funding support to sedimentologists, or would-be sedimentologists, from developing countries.

Each chapter has been extensively reviewed by the co-authors and a minimum of two external reviewers. Such reviews always contribute greatly to the readability, simplicity, and conciseness of the work and to removing errors in the text. We sincerely thank all 33 of our external referees for their careful work and time, which have contributed greatly to making this a better book. We could not adopt all of your suggestions, often because of space limitations, but all were carefully considered and deeply appreciated. A work of this scope can never be error-free, and the editor would welcome correspondence on any remaining errors, which we will then address in future updates. We thank sincerely all of the following who generously gave their time and expertise to improve this work through review, encouragement, or contributing material.

Bruce Ainsworth (University of South Australia, Australia), *Alessandro Amorosi* (Università di Bologna, Italy), *John Anderson* (Rice University, USA), *Maciej Bąbel* (University of Warsaw, Poland), *Giorgio Basilici* (University of Campinas, Brazil), *Sam Birch* (Brown University, USA), *Kevin Bohacs* (KM Bohacs Geoconsulting, USA), *Benjamin Brigaud* (Université Paris-Saclay, France), *Antonio Cattaneo* (Ifremer, Géosciences Marines, France), *Annamaria Correggiari* (ISMAR CNR Istituto di Scienze Marine, Consiglio Nazionale delle Ricerche, Italy), *Glen Cushing* (United States Geological Survey, USA), *Shahin Dashtgard* (Simon Fraser University, Canada), *Colin Dundas* (United States Geological Survey, USA), *Gregor Eberli* (University of Miami, USA), *Tracy Frank* (University of Connecticut, USA), *Marcello Gugliotta* (University of Bremen, Germany), *Mike Hambrey* (University of Aberystwyth, UK), *Adrian Hartley* (University of Aberdeen, UK), *Peter Homewood* (Retired), *John Howell* (University of Aberdeen), *Kevin Jones* (United States Geological Survey, USA), *Libby Ives* (NASA Jet Propulsion Laboratory, USA), *Charlie Kerans* (Jackson School, University of Texas, USA), *Colleen Kurcinka* (Ontario Geological Survey), *Kate Littler* (Camborne School of Mines, UK), *Sergio Longhitano* (Università degli Studi della Basilicata, Italy), *Tim Lowenstein* (Binghamton University, USA), *Joe Macquaker* (Exxon, USA), *Jason Muhlbauer* (Boise State University, USA), *Björn Nyberg* (University of Bergen, Norway), *Piret Plink-Bjorklund* (Colorado School of Mines, USA), *Miquel Poyatos More* (Universitat Autònoma de Barcelona, Spain), *Rodmar Ravnås* (Aker BP, Norway), *Janet Richie* (United States Geological Survey, USA), *Mike Simmons* (Halliburton, UK), *Katie Stack* (NASA Jet Propulsion Laboratory, USA), *Michael Steventon* (Royal Dutch Shell, UK), *Andreas Strasser* (University of Fribourg, Switzerland), *Peter Swart* (University of Miami, USA), *Maurice Tucker* (Bristol University, UK), *Anna Van Yperen* (University of Oslo, Norway), *Fernando Farias Vesely* (Universidade Federal do Paraná, Brazil), *Maxime Virolle* (Université Paris-Sud, France), and *Valentin Zuchuat* (RWTH Aachen Germany).

Dedications

Chapter 3 is dedicated to the memory of our friend and mentor Dr. S. George Pemberton, who was a champion of trace fossil applications to facies analysis and stratigraphy. George greatly admired H.G Reading and would have been proud to have contributed to this chapter, even if only indirectly.

Chapter 7 is dedicated to Professor Gary Kocurek, author of "*Desert aeolian systems*" in the 1996 edition of this book whose untimely passing occurred during the writing of the volume.

Chapter 13 is dedicated to Professor Hugh Jenkyns author of pelagic sediments in the very first edition of this book for his inspiration to many of us over very many years.

Special Thanks

A special word of thanks is needed for Didier Granjeon (IFP Énergies Nouvelles), who contributed the illuminating 3D stratigraphic simulations in Chapter 4, and especially to Kerrie Bann, whose fantastic scientific figures illustrating cores and ichnofacies grace Chapter 3, as well as finding a place in Chapters 2, 5, 8, and 14. Kerrie's work was, in part, funded by a generous donation from the International Association of Sedimentologists. Further draughting of figures in Chapters 2, 4 and 11 by graphic designer A. Egdane, was generously sponsored by the baseline funding of Frans van Buchem of the King Abdullah University of Science and Technology, Thuwal, Saudi Arabia.

We would also like to thank the team at Wiley who have seen this mammoth effort through to publication with great care and attention to detail Among the many involved those who were most closest to the project were: Andrew Harrison, Rosie Hayden, Frank Weinreich, Nandhini Karuppiah and Aswini Murugadass. Thank you all! Deep personal thanks are due to Howard Johnson who has not only now contributed to all four editions of this volume but also provided fantastically positive support throughout.

About the Companion Website

This book is accompanied by a companion website.

www.wiley.com/go/levell4e

This website contains

Instructors and Students

- Figure PPTs
- Supplemental Tables

1

Introduction

Bruce Levell

Department of Earth Sciences, University of Oxford, Oxford, UK

Summary

This short introduction sets out the purpose and structure of the book. In broad terms the theme is one of diversity: diversity of approaches and of emphasis, reflecting the diverse range of applications and questions which can be addressed using the twin subjects of sedimentology and stratigraphy.

1.1 Motivation for This Book

In 1996, when the third edition of this book was published, the most obvious application of sedimentology was to the exploration and production of hydrocarbons. The insatiable hunger for raw energy, food (fertiliser), steel and concrete of a global population (then 5.5 billion and growing at 80 million people per year) was driving demand for cheap and plentiful energy and food. The energy demand was, at that time, met for 84% by oil, gas and coal. Sedimentology, in particular the combination of a well-developed facies analysis methodology and insights into sedimentary architecture gained by sequence stratigraphy, was proving its worth in discovering and producing this energy. A main focus of sedimentological studies was the prediction of the properties and geometries of reservoirs, hydrocarbon source rocks and seals. At that time, a large amount of innovative academic research was either directly funded by the hydrocarbon industry, or inspired by the problems of subsurface prediction. Around 1996, this included a dramatic expansion in the collection of large amounts of quantitative analogue data from outcrops, wells, and high-resolution 3D seismic imaging of large swathes of the Earth's sedimentary basins.

Fast-forward to 2025 and sedimentology as both a field of research and as a profession finds itself in a very different environment. Global net population is now 8 billion and its growth is in only slightly down, at 75 million a year. Surprisingly hydrocarbons still underpin the energy demand of this 60% larger population, accounting, still, for around 84% of total energy supply. That energy demand has of course increased markedly, from ca 400 to nearly 700 EJ a^{-1} and the consequence is known to us all – a greenhouse gas related climate crisis.

Sedimentology now no longer has a single predominant driver as it did in 1996. The subject has multiple applications and is being applied to a much wider range of issues, by a much more diverse set of practitioners. Sedimentology is now well embedded both as a tool and in its own right as an integral part of Earth System approaches and is taught in broader Earth Science or Environmental Science courses rather than 'Geology' degrees. Sedimentological understanding, or at least awareness, is used to contextualise a vast range of scientific endeavours as indicated below.

At the same time, the 'schools of sedimentology' described in the introduction to the 3rd edition (namely the Anglo-Dutch school and two American schools) are no longer recognisable. The subject has gone global, and, at the same time, the male dominance of 1996 has weakened considerably.

Ironically, the diversity of applications and questions to which sedimentology is now applied occurs in the context of an apparently widespread view outside the subject that classical sedimentology and stratigraphy have been 'done'. In other words, a working toolkit that can address the main questions exists and just needs to be applied. This view (expressed in a relative reduction in pure

Reading's Sedimentary Environments: Processes, Facies and Stratigraphy, Fourth Edition. Edited by Bruce Levell.
 Published 2026 by John Wiley & Sons Ltd.
Companion website: www.wiley.com/go/levell4e

sedimentological research and teaching) is in a way a compliment to the success of facies analysis, but it is both unhelpful and wrong. The new applications of the subject are in fact doing rather a good job of exposing our lack of knowledge. For example: (i) the uncertainties and non-uniquenesses in determining causal explanations for the observed sedimentology and stratigraphy; (ii) the biases and gaps in the record that depositional and preservational processes inevitably introduce; (iii) the updating of classical facies models with new data from process measurements in a wider range of modern examples, such as sub-surface imaging, short-lived isotope dating methods and, probably, in the near future, artificial intelligence for unravelling concatenations of complex non-linear processes; (iv) the widespread integration of sedimentology with geochemical and palaeoecological data to broaden the scope of environmental reconstructions beyond the purely physical.

Direct applications of sedimentology are many and certainly much more diverse than in 1996, for example:

i) Predicting sediment architecture and properties for the subsurface extraction and storage of fluids (e.g. hydrocarbons, water (Taylor et al. 2022), carbon dioxide (Krevor et al. 2023) and hydrogen).
ii) Exploration for sediment-hosted ores and brines (e.g. copper, cobalt (Hitzman et al. 2010), lead, zinc, lithium and placer rare earth element deposits (Balaram 2023)).
iii) Geotechnical engineering, for example, ground models for infrastructure projects, roads, tunnels, dykes and dams (de Freitas 2021).
iv) Understanding the formation and protection of the critical zone (soil science (Hou et al. 2020)).
v) Engineering coastal defence against rising relative sea levels (Vousdoukas et al. 2020; Singhvi et al. 2022).
vi) Natural hazard prediction and understanding landscape evolution: for example, glacier, river and flood basin behaviour, slope stability and landslides in the context of increased temperatures, changes in precipitation and storminess (Ercilla et al. 2021; Syvitski et al. 2022).
vii) Tracing the dispersal of pollutants (including microplastics (Harris 2020; Waldschläger et al. 2022)).

Still central to many of these applications is the predictive power of being able to reconstruct the 3D geometry of sedimentary bodies through interpreting their depositional environment.

It was these changes in understanding, context and applications that prompted the thinking behind this completely new 4th edition of Harold Reading's book.

1.2 Structure of the Book

We decided that the diversity of end-uses and therefore of approaches to sedimentology would be reflected in the way the different chapters of this book have been presented. Rather than force the subject matter into a one-size-fits-all straight-jacket, we have tried to allow the authors to express their interests and personalities in the way they present each chapter.

The 'tools of the trade', an understanding of facies analysis, facies models and the larger-scale factors, such as climate, tectonics and changing base level, that control the nature of the depositional and preservational processes, are the subjects of Chapters 2–4. Chapters 2 and 3 explain and illustrate the work-flows of, respectively, physical and biogenic facies analysis using case studies from clastic and carbonate successions. Chapter 4 addresses stratigraphy and deliberately takes a more philosophical approach highlighting some of the uncertainties and debates that are still being addressed in our understanding of how the Earth system has constructed the 3D sedimentary record from the bed-scale to the basin-scale. It emphasises the importance of gaps in that record, differential preservation and how sedimentary architecture is controlled by the balance between rates of sediment supply and available depositional space (accommodation space). It concludes by presenting the conceptual framework of sequence stratigraphy as a unifying work flow for the analysis of sedimentary architecture and as a basis for understanding the terminology and methodology used in subsequent chapters.

The bulk of the book (Chapters 5–15) shows how the sedimentary rock record may be read by interpreting the process or processes which gave rise to the sediments, and thereby the environments in which these processes operated. We also explain how reconstructing the attributes of these environments can help us place them in the predictive context of a dynamic stratigraphy (Miall 2016) and a sediment routing system (Allen 2017).

The goal of reconstruction of past environments on the basis of understanding modern process forms a common structure for all chapters and requires the following:

1) A thorough **description** of the rocks, either in the field or in core, with additional laboratory data obtained from samples collected to answer specific questions. Since time is limited, rock description is inevitably selective, emphasising some features, underplaying others and rejecting yet others as unimportant. A useful basic check list is: 'colour, texture, lithology, bed geometry, sedimentary structures and fossils'. The selection depends on the judgement, experience and purpose of the investigator. Judgement and

experience take time to acquire and can only be gained by seeing lots of rocks. The absence of certain features is often as important as their presence. For example, the consistent absence of shallow-water features, rather than any positive evidence for great depth, led sedimentologists to infer that most turbidites were deposited in deep water. The utilisation of negative evidence requires a familiarity with a wide range of sedimentary rocks and environments. This text does not address the descriptive aspects in detail. The reader is referred to other texts, for example, Nichols (2023), Tucker and Jones (2023) and, for physical sedimentary structures, Collinson and Mountney (2019).

2) An understanding of **processes** so that, simultaneously with rock description, the strength or direction of the type of flow which carried and deposited each grain is considered. Such questions as 'What was the oxidation state, salinity or pH of the water?' or 'What forms of life were present?' can also be addressed. Later alteration weathering or diagenetic processes may have changed not only the colour of the rocks but also their grain size and composition. Particular processes are seldom totally diagnostic, that is, confined to just one environment and similar rocks may therefore form in different environments.
3) A knowledge of **present-day environments**, the processes which operate within them and the factors that control them. We need to know how environments evolve both under stable conditions through the intrinsic variability of processes and their responses to extrinsic change in a sea level, climate, tectonic activity or sediment supply.

In order to address these aspects systematically, the organisation of the main environmental chapters in the book is in general as follows:

1) *Environments,* reviewing modern environments, their appearance, context and their associated processes and products.
2) *Processes,* concentrating on those that occur in each environment and showing how they relate to the resultant sediment. We here emphasise those that are peculiar to, or diagnostic of, particular settings.
3) *Facies,* stressing field data, facies relationships, sequences and associations and including both modern and ancient examples.
4) *Sequence stratigraphy,* or analysis of facies successions using a chronostratigraphic framework defined by extensive correlative surfaces.
5) *Controls,* emphasising the interplay of sediment supply, tectonics and sea-level changes.

Within this framework, different chapters take a diverse and often individualistic approach: emphasising different aspects of the science of sedimentology. This is important as the approach adopted in any investigation will, at least in part, be environment-dependent. The conceptual framework chosen will also, rightly, reflect the questions that interest the authors of a particular chapter. As an example a geomorphological 'land system' approach is adopted in the Glacial chapter (Chapter 14), which consequently has less emphasis on facies. This is particularly appropriate for workers reconstructing more recent glacial systems and understanding their detailed response to climate change. It also provides a stimulating actualistic perspective with which to approach past glaciations or glaciation on other planets (Chapter 15). A landscape systems approach is also adopted in both the chapter on Rivers (Chapter 5) and in the Aeolian chapter (Chapter 7) which addresses linkages between environments (dunes, lakes and rivers) as records of climate change and explores their reflection in preserved stratigraphic architectures. The carbonate chapter (Chapter 11) is a variation on this approach in that it sets out the principles of carbonate sedimentology through the twin lenses of carbonate factories (sediment production sites) and depositional profiles related to the large-scale basin setting.

In contrast, the deep water clastic environment (Chapter 12) is one in which process understanding is both particularly important (as large-scale geometries are rarely visible at outcrop in such extensive systems) and also challenging (as modern catastrophic deep sea processes are rarely directly observed). This chapter therefore focuses on bed-scale processes and facies. In yet another different approach, Chapter 8, addressing coarse-grained shallow marine and coastal siliciclastics, focuses on mixed process systems, moving beyond a stereotypical characterisation of process in terms of end-members (wave, tide, fluvial) into the realm of more realistic and widespread "mixed" systems. It does this largely by focusing on case studies of facies successions with mixed characteristics highlighting the diagnostic observations.

In sum, each chapter has a different sedimentological emphasis. This is a deliberate feature of this book: a diversity of approach is simply a good thing. At the same time most authors have used as case studies examples that are personally familiar to them and have addressed them in some depth rather than summarising a global suite of case studies. The intent here is to illustrate how the subject is researched as well as what is researched. To explore this diversity in approach, it will repay the reader to read all the different chapters even if they are directly interested in just one or two of the specific depositional environments. In fact, several of our authors confessed to having taken the third edition of the book to bed with them! We hope to similarly inspire a new generation to enjoy learning and growing in this fascinating and rewarding subject as much as we, as an author group, self-evidently do!

Further Reading

Allen, P.A. and Allen, J.R. (2013). *Basin Analysis: Principles and Applications to Petroleum Play Assessment*, 3e. Oxford: Wiley Blackwell, 632p.

James, N.P. and Dalrymple, R.W. (2010). *Facies Models 4*. St. John's, NL, Canada: *Geological Association of Canada*, 586p. ISBN-13: 978-1-897095-50-8; ISSN: 1208-2260.

James, N.P. and Jones, B. (2016). *Origin of Carbonate Sedimentary Rocks*. Wiley, 446p. ISBN 978-1118-65273-2.

Leeder, M.R. (2011). *Sedimentology and Sedimentary Basins: From Turbulence to Tectonics*, 2e. Wiley. ISBN 978-1-4443-4992-4.

Miall, A.D. (2016). *Stratigraphy: A Modern Synthesis*, Springer, 454p. ISBN 978-3-319-24302-3.

Tucker, M. and Jones, S.J. (2023). *Sedimentary Petrology*, 448p. Wiley ISBN: 978-1-118-78648-2.

References

Allen, P.A. (2017). *Sediment Routing Systems: The Fate of Sediment from Source to Sink*, Cambridge University Press, 407p. ISBN 978-1-107-09199-3.

Balaram, V. (2023). Potential future alternative resources for rare earth elements: opportunities and challenges. *Minerals* 13 (3): 425.

Collinson, J. and Mountney, N. (2019). *Sedimentary Structures*, 4e. Edin Academic Press, 349p. ISBN 978-1-78046-002-0.

Ercilla, G., Casas, D., Alonso, B. et al. (2021). Offshore geological hazards: charting the course of progress and future directions. *Oceans* 2 (2): 393–428.

de Freitas, M.H. (2021). Future developments for ground models. *Quarterly Journal of Engineering Geology and Hydrogeology* 54 (2): qjegh2020-034.

Harris, P.T. (2020). The fate of microplastic in marine sedimentary environments: a review and synthesis. *Marine Pollution Bulletin* 158: 111398.

Hitzman, M.W., Selley, D., and Bull, S. (2010). Formation of sedimentary rock-hosted stratiform copper deposits through Earth history. *Economic Geology* 105 (3): 627–639.

Hou, D., Bolan, N.S., Tsang, D.C. et al. (2020). Sustainable soil use and management: an interdisciplinary and systematic approach. *Science of the Total Environment* 729: 138961.

Krevor, S., De Coninck, H., Gasda, S.E. et al. (2023). Subsurface carbon dioxide and hydrogen storage for a sustainable energy future. *Nature Reviews Earth & Environment* 4 (2): 102–118.

Nichols, G. (2023). *Sedimentology and Stratigraphy*, 3e. Wiley, 518p. ISBN: 978-1-119-41727-9.

Singhvi, A., Luijendijk, A.P., and van Oudenhoven, A.P. (2022). The grey–green spectrum: a review of coastal protection interventions. *Journal of Environmental Management* 311: 114824.

Syvitski, J., Ángel, J.R., Saito, Y. et al. (2022). Earth's sediment cycle during the Anthropocene. *Nature Reviews Earth & Environment* 3 (3): 179–196.

Taylor, R., Aureli, A., Allen, D., Banks, D., Villholth, K. and Stigter, T., 2022. Groundwater, aquifers and climate change, Chapter 7. In; *The United Nations World Water Development Report 2022: Groundwater: Making the Invisible Visible*. 101–114. UNESCO, Paris. Tucker, M., and Jones, S.J. 2023 Sedimentary petrology Wiley, 448p. ISBN: 978-1-118-78648-2

Vousdoukas, M.I., Ranasinghe, R., Mentaschi, L. et al. (2020). Sandy coastlines under threat of erosion. *Nature Climate Change* 10 (3): 260–263.

Waldschläger, K., Brückner, M.Z., Almroth, B.C. et al. (2022). Learning from natural sediments to tackle microplastics challenges: a multidisciplinary perspective. *Earth-Science Reviews* 228: 104021.

2

Facies Analysis – Principles and Practice

Howard D. Johnson[1], Mike Mayall[1], and John J.G. Reijmer[2]

[1] *Department of Earth Science and Engineering, Imperial College London, UK*
[2] *Department of Earth Sciences, Faculty of Science, Vrije Universiteit Amsterdam, Amsterdam, The Netherlands*

Summary

Facies analysis was once largely synonymous with sedimentology, as evident in the first three editions of this book. Today, facies analysis is an inseparable companion of sequence stratigraphy. Together, they provide a seamless hierarchical methodology for studying the processes that control the nature and distribution of sedimentary rocks at a wide range of periodicities and magnitudes. This chapter introduces the concepts of the facies-sequence stratigraphy workflow with three aims. First, to outline the basic principles and methods of facies analysis, including reviews of key developments, the scale hierarchy of facies organisation, and depositional models. Second, to provide a rock-focused, physical conceptual framework (biological aspects are covered in Chapter 3) as a basis for the subsequent chapters. Third, to share case studies of practical applications of both siliciclastic and carbonate facies analysis, covering diverse depositional environments and contrasting datasets. The basic premises of facies analysis that lie at the heart of this book, are summarised in the following workflow: (i) groups of genetically related rocks (facies) are systematically evaluated in terms of physical, biological and chemical processes at a range of scales; (ii) an hierarchical facies organisation (facies ➔ facies associations ➔ architectural elements ➔ facies successions) is established incorporating larger stratigraphic relationships and considerations of preservation, enabling diagnosis of sedimentary environments; and (iii) iterative comparison of analogous facies in modern, recent and ancient examples provides tests that support the choice of a final environmental interpretation. Crucial in this workflow are the recognition of substantial depositional breaks (e.g. unexpected and/or dislocated facies juxtapositions, unconformities, transgressive erosion surfaces) and/or the significance of volumetrically minor facies types (e.g. condensed facies, palaeosols and winnowed lags) that carry disproportionately large environmental significance. Through highlighting such breaks, facies analysis plays a key role in distinguishing the hierarchy of stratal surfaces, whether those intrinsic to an environment (autogenic) or those related to external forcing (allogenic). Depositional (or facies) models are one of the most important outcomes of facies analysis, providing a conceptual backdrop for deeper understanding of sedimentary systems. These models range from global syntheses of depositional systems, emphasising autogenic processes, through to more local models summarising specific rock units. Depositional models guide interpretation and prediction in basin- and/or reservoir-scale sedimentary systems, such as maps of palaeogeography, rock/facies types, thickness patterns and rock properties (e.g. porosity, permeability, etc.). Five case studies, based on outcrop and subsurface data, ranging in age from Mesoproterozoic to Quaternary and using approaches from regional geology to microfacies, demonstrate practical application of the facies approach.

2.1 Introduction

2.1.1 Background

The term 'facies' ('general appearance' from Latin[1]) is used in sedimentary geology to describe a rock unit characterised by distinctive and distinguishing features based on the following definition:

1 'Facies'. *Merriam-Webster.com Dictionary*, Merriam-Webster. https://www.merriam-webster.com/dictionary/facies (accessed 24 March 2022).

Reading's Sedimentary Environments: Processes, Facies and Stratigraphy, Fourth Edition. Edited by Bruce Levell.

Published 2026 by John Wiley & Sons Ltd.
Companion website: www.wiley.com/go/levell4e

> a rock unit with specified characteristics ... defined on the basis of [for example] colour, bedding, texture, fossils and sedimentary structures....that forms under certain conditions of sedimentation, reflecting a particular process or environment (Reading 1978a).

Figure 2.1 illustrates some siliciclastic facies and highlights the diagnostic features that allow a grouping of rocks with similar characteristics to be assigned a reasoned process interpretation in this case fluvial-tidal interactions.

Facies analysis is the methodology whereby groups of facies are interpreted in terms of depositional process, preservation, stratigraphic relationships and, thereby, sedimentary environment. This is achieved by defining several, ideally genetically related, facies types and deciphering for each the physical, biological and chemical processes responsible for their formation, including preservational aspects. This evaluation leads to a diagnosis of the most likely sedimentary environment(s), including an iterative comparison of analogous facies in Modern, Recent and

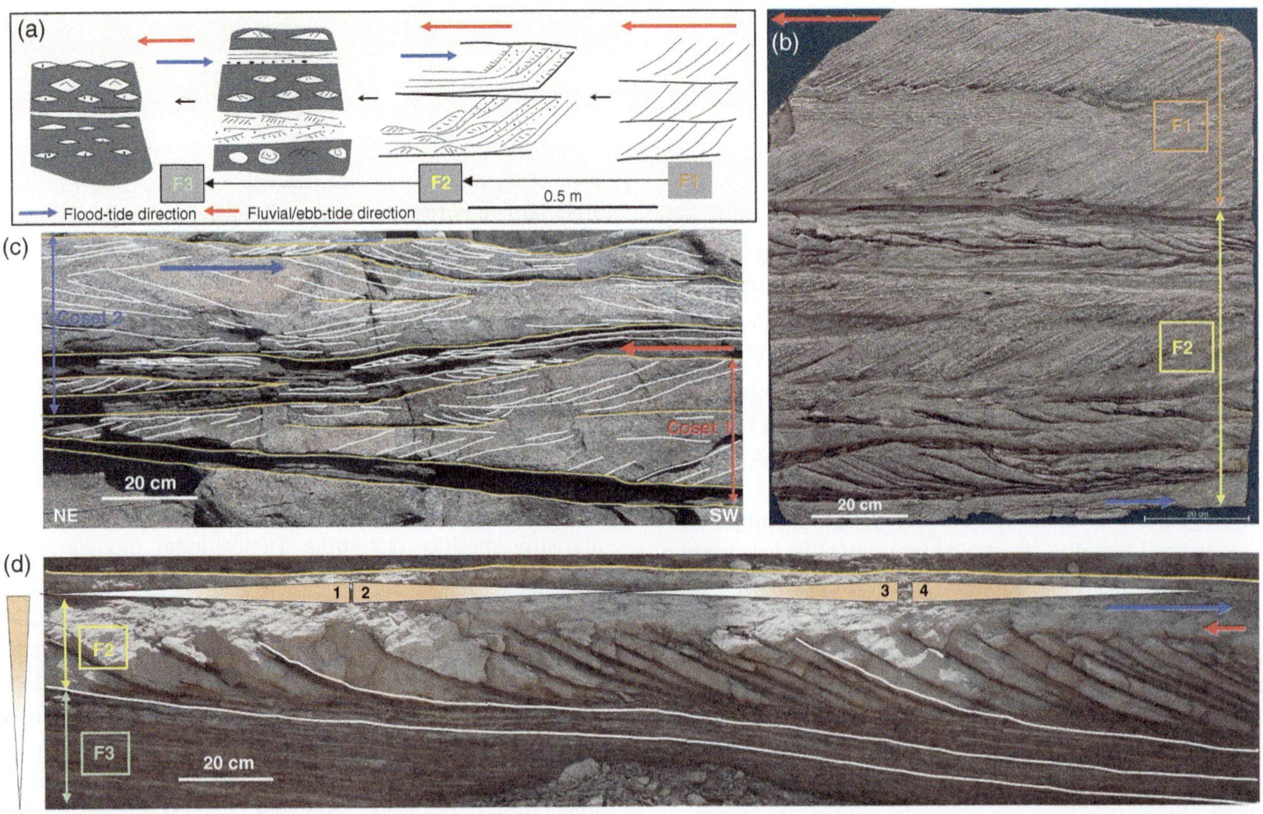

Figure 2.1 An illustration of facies analysis. (a) Graphic summary of diagnostic sedimentary structures and bedding styles of the fluvial–tidal transition zone of the modern Rhine-Meuse system, showing the characteristic rapid lateral facies change (F1 to F3; right to left) of laterally contiguous facies types: F1 – Cross-bedding (mainly fluvial) → F2 – Shovel-shaped cross-bedding with long muddy toesets (mainly tidal) → F3 – Heterolithic bedding (lenticular/wavy bedding) (mainly tidal). *Source:* Van den Berg et al. (2007) / Geologie en Mijnbouw/ Public Domain. (b) Lacquer peel from the Rhine-Meuse estuary/delta showing fluvial-dominated cross-bedding (F1) and flood-tide cross-bedding with shovel-shapes, extended muddy toesets and bidirectional foreset dip directions (F2) *Source:* Van den Berg et al. (2007) / Netherlands Journal of Geosciences. (c) Bidirectional cross-bedding in coarse-grained sandstones (compare with F2 in a and b) from the fluvio-tidal late Precambrian Jura Quartzite, Scotland (Levell et al. 2020): (i) Coset 1 is interpreted as a fluvial/ebb-tide transverse dune migrating towards the NE (red arrow); (ii) Coset 2 is interpreted as a flood-tide transverse dune migrating towards the SW (blue arrow). *Source:* Courtesy of Howard Johnson. (d) A c. 1 m-thick coarsening-upward succession comprising mudstone-dominated heterolithic facies (F3) at the base, which passes gradationally upwards into cross-bedded sandstones with regularly spaced mud-draped foresets (F2). This large-scale shovel-shaped cross-bedding demonstrates downdip bedform migration (towards the right = inferred flood-tide direction – blue arrow): (i) the deeper part of this structure is mud-dominated (F3 – comparable to F3 in a) with sandy laminae on very low-angle toesets (<5° dips), and (ii) cross-bedded sandstone facies dominates the shallow part of the structure (foresets dipping to the right at 25–30°) with regularly spaced mudstone drapes that thicken and flatten in the depositional dip direction – four distinctive bundles are highlighted with 1 and 3 thickening, and 2 and 4 thinning laterally. The facies relationships demonstrate Walther's law: a lateral facies change along discrete stratal surfaces (timelines) becomes a vertical facies succession with steeply dipping sandy foresets (F2) passing downdip into low-angle to horizontal surfaces in the muddy toeset unit (F3); note also how F2 itself becomes more mud-rich down the sediment transport pathway (left to right). From the fluvio-tidal deposits of the Eocene Dir Abu Lifa Member, Western Desert, Egypt (Legler et al. 2013). *Source:* Courtesy of Howard Johnson.

Ancient examples. This iterative comparative assessment is a key component of the facies analysis method.

Facies analysis is central to sedimentology and is a cornerstone of sequence stratigraphy; both topics combine to provide the most powerful means of deciphering the spatial and temporal evolution of sedimentary environments (Figure 2.1). The subject has been extensively reviewed in the past, not least in the three previous editions of *Sedimentary Environments* (Reading 1978b, 1986, 1996) and in the four versions of *Facies Models* (Walker 1979, 1984a; Walker and James 1992; James and Dalrymple 2010). Other notable reviews are those by Posamentier and Walker (2006), with emphasis on facies models, Miall (2010, 2016), with emphasis on integrating sedimentology and stratigraphy, and Anderton (1985) with a thought-provoking review of facies analysis itself. Historical aspects are comprehensively reviewed in these publications, and in Reading (2001) for siliciclastic facies, and are only very briefly touched upon here to provide context.

Analysis of sediments, both modern and ancient, can be undertaken at a range of scales, which reflects the magnitude, frequency and duration of the plethora of geological processes involved in their deposition and accumulation (Table 2.1; Figure 4.42). The stratigraphic record comprises a nested hierarchy of scales ranging from grains and pores (microfacies - level 1 in Table 2.1), through physical-, biological- or chemical-based facies analysis (facies, facies associations and successions – levels 2-4 in Table 2.1) up to entire depositional systems (level 5 in Table 2.1). This hierarchical structure also defines the basic facies/sequence stratigraphic analysis workflow for siliciclastic (Figure 2.2), carbonate (Figure 2.3) and mixed clastic-carbonate depositional systems (Figure 2.4).

Evidence from primary sedimentary structures, fossil remains of single-celled organisms and biologically related structures in carbonate rocks (e.g. stromatolites) confirm that identifiable sedimentary rocks on Earth date back to the Eoarchean (e.g. the c. 3700 Ma Isua supracrustal belt of southern Greenland; Nutman et al. 2019). For rocks that are Ediacaran (end of the Neoproterozoic Era) or younger, facies analysis is based on physical processes combined with biological analysis of the rock record, whether macrofauna, microfauna, palynology and/or ichnology (Chapter 3). This combination provides invaluable additional insight into salinity, substrate conditions and water depth, as well as crucial stratigraphic age-relationships. Carbonate deposits are also conditioned by additional life-related processes that include changes in physico-chemical conditions and their effect on organisms, as well as ecology, evolution and organic sediment production. Throughout the Phanerozoic carbonate-producing organisms in all carbonate depositional environments changed continuously. For example, the main reef-building organisms changed through time (e.g. successively: stromatolites, archaeocyathids, stromatoporoids, bryozoans, rudists, corals, coralline algae) and therefore so too have the mineralogy and texture (grain size, sorting, shape, etc.) of primary carbonate framework grains. Modern carbonate depositional systems are therefore not exact equivalents of ancient carbonate environments. Outcrop-based facies analysis of ancient carbonates is further complicated, to a much greater degree than for most siliciclastic rocks, by diagenetic modification of mineralogy and texture. Diagenesis is often initiated at the time of deposition and continues during burial (see Section 11.3.2).

Facies analysis provides a comprehensive approach for evaluating sedimentary deposits due to its flexible methodology and wide practical applications. The concepts of facies analysis have even been shown to be powerful in unravelling sedimentary processes on Mars (e.g. Grotzinger and Milliken 2012; Chapter 15). This reflects the following:

i) The basic concepts and workflow (Table 2.1) are applicable to all sedimentary environments, covering clastic, carbonate and evaporite depositional systems on Earth and other planets, and can be adjusted to accommodate a wide array of different project aims and objectives.
ii) The observational data used can be highly varied, including outcrops, cores, well logs and seismic data (Table 2.1), which extends the concept beyond its initial macroscopic lithology-based roots.
iii) The fundamental analytical approach is scale independent (Table 2.1),
iv) The results are directly suited for diverse problem-solving tasks (Table 2.1-IV): for example, research-focussed modern and palaeoenvironmental studies, geohazard assessment and the prediction and extraction of sedimentary rock-hosted natural resources (e.g. water, minerals, metals and hydrocarbons).

This chapter applies the methodology to a selection of diverse case studies (from the Mesoproterozoic to Quaternary). The case studies are central. They illustrate how the facies approach is used with different datasets and at different scales. They cover rock-based (outcrops and/or cores), well log-based and seismic-based approaches and their applications.

2.1.2 Key Conceptual Developments

Facies analysis has remained central through some of the most important historical developments in sedimentary geology, most notably: (i) linking facies and hence environments with stratigraphic aspects; (ii) developing a process-response understanding of sedimentology; (iii) constructing and applying facies models capable of making predictions

Table 2.1 Summary of five, scale-dependent stages in facies analysis (1–5; horizontal axis) and key descriptive and interpretational components (I–V; vertical axis). A fully integrated ('bottom-up') facies analysis may encompass all, or part, of these scales of investigation (1–5), but stage 2, 'facies', is the critical part. Aims of a project will vary and could be skewed towards any of the five stages. The 'top-down' approach typically focuses on stages 4 and 5.

		1 Microfacies	2 Facies	3 Facies associations and architectural elements	4 Facies successions	5 Depositional systems
I	Vertical scale	micron to cm	cm to m	m to 10 s m	10 s m to 100 s m	10 s m to km
	Horizontal scale	micron to cm	cm to m	m to 100 s m	100 s m to km	100 s m to 10 s/100 s km
II	Characteristics	Texture	Grainsize and mineralogy trends	Architectural Elements/Geobodies	Facies belts	Facies belt evolution
		Mineral composition	Bioturbation/trace fossils	Lateral facies changes (smaller scale)	Lateral facies changes (large scale)	Biotic associations evolution
		Micro-fossil content (fauna and flora)	Biotic associations (micro and macro)	Depositional profiles	Depositional profile evolution	
		Pore-space properties	Sedimentary structures	Depositional geometries (e.g. clinoforms, mounds)	Geobody stacking (e.g. channels)	Systems Tracts
			Stratigraphic surfaces	Stratigraphic surfaces	Stratigraphic surfaces (timelines)	Hierarchy of stratigraphic surfaces (timelines)
			Bedding patterns	Small-scale stacking patterns	Medium-scale stacking patterns	Orders of stacking patterns (small-medium-large)
		Diagenetic features	Geochemical features (e.g. nodules)	Geochemical feature trends	Geochemical feature trends	
III	Data types	Hand specimens from outcrop	Outcrop logs	Outcrop logs and photo panels		Large-scale outcrops, logs and photo panels
			Well logs	Well logs	Well logs and well log correlations	Regional well log correlations
		Plugs from cores, side-wall-samples and cuttings (well bore)	Core descriptions	Core descriptions	(Core)	(Core)
			Modern analogues	Modern analogues		
				High-resolution 3D seismic	2D and 3D seismic	2D and 3D seismic
IV	Analysis	Petrographic analysis (thin sections)	Facies analysis	Lateral and vertical facies relationship analysis	Lateral and vertical stacking analysis	Basin analysis (tectono-stratigraphy)
		Microfacies analysis	Integration of micro- and macro-scopic observations and analysis scales	Geometrical and architectural element analysis from outcrop and/or high-resolution seismic	Integration of geometries, architectural elements, profile evolution and stacking patterns	Sequence stratigraphy (systems tracts, sequence boundaries, maximum flooding surfaces, etc.)
		Biostratigraphic analysis				Biostratigraphy
		Geochemical analysis				Isotope stratigraphy
		Petrophysical analysis		Seismic stratigraphy and facies and analysis	Seismic stratigraphy and facies analysis	Seismic stratigraphy and facies and analysis

V	Interpretation	Microfacies classification	Facies classification and definition (schemes)	Depositional environment interpretation and facies models	Gross depositional environment (GDE) definition and maps	Depositional models, paleogeographical and thickness maps
		Provenance	Physical processes	Depositional profile definition	Regional facies trends	Tectono-sedimentary phases
		Diagenetic characteristics	Biological processes	Lateral facies trends and facies relationships	Facies model evolution	Tectonic, eustatic and climatic controls
		Reservoir properties	Diagenetic processes	Architectural Element definition	Allocyclic versus autocyclic control	
		Biota identification	Biostratigraphic age assignment	Local versus generic facies models		

VI Siliciclastic examples

VII Carbonate examples

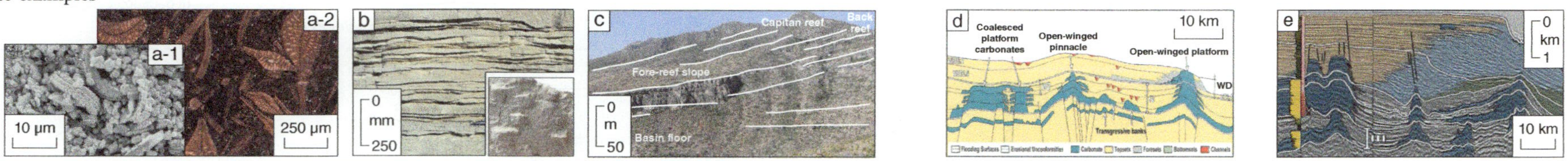

VI: Images of different scales of analysis in clastic-dominated successions:

VI a-1 and a-2, SEM and photomicrograph, showing well-sorted sandstone with minor quartz overgrowths and ferric oxide grain coatings (Permian aeolian sandstones; porosity c. 15-20%/permeability c. 500 md; North Sea Basin).

b: Ebb-dominated, estuarine tidal sandstones (large subaqueous dunes/'sand wave facies') with minor oppositely dipping sets (Early Cretaceous Woburn Sands, Leighton Buzzard, southern England) (based on Johnson and Levell 1995).

c: Lower coastal plain deposits with *in situ* mangrove-rich coal and floodplain mudstones overlain by two fluvial channel sandstones (red lines define channel bases) – architectural elements are lateral accretion surfaces (LAC – inclined white lines), indicating point bar deposition in a meandering channel (ML – muddy levee) (middle Miocene. Belait Formation, Labuan Island, NW Borneo).

d: Stratigraphic architecture comprising (bottom to top): (i) unstable basin margin with prograding deltas (abundant soft sediment deformation structures) ->(ii) sequence boundary (base of orange unit) and thin lowstand systems tract (braided fluvial channels in orange) ->(iii) transgressive systems tract consisting of meandering fluvial to tide-influenced channels with mangrove coals (pink-red colours) (see VIc for further details) – retrogradational coastal plain; (iv) maximum flooding surface (datum); and (v) highstand systems tract with stable prograding shoreface/delta front sands (yellow) – stable shelf.

e: Large-scale regressive-transgressive cycles defined by prograding seismic foresets (orange and yellow lines) and by retrograding seismic stratal patterns (blue); tectonically controlled accommodation between two active bounding fault systems defines the syndepositional Labuan-Berakas Syncline (c. 9 km of sediment accumulated in the basin axis during c. 7 Myr; relative sea level rise c. > 1 m/1000 yr in the middle Miocene) (2D seismic line length = 40 km (outcrop correlation panel in d equates to the lower regressive–transgressive cycle – the deeper red and blue triangles, respectively).

VII. Images of different scales of analysis in carbonate-dominated successions:

a-1: SEM of 2–10 μm-diameter, well-sorted calcite grains are from planktonic coccoliths in the Maastrichtian Tor Formation, North Sea Basin (porosity c. 25–30%/permeability c. 1–10 md).

a-2: photomicrograph of mud-rich limestone (wackestone) with abundant larger benthic foraminifera (*Lepidocyclina sp.*) (early Miocene, Subis Limestone, Niah, central Sarawak).

b: Peloidal to oolitic limestone with wavy to ripple cross-laminated, cm-thick beds alternating with mm-thick carbonate mudstones and with common halite pseudomorphs (inset) – interpreted as a lagoonal/tidal flat/sabkha environment (Late Jurassic, Purbeck Beds, Wessex Basin, southern England).

c: Shallowing-upward/regressive limestone reef succession, with large-scale foresets that preserve a regressive basin floor ➔ fore-reef ➔ reef ➔ back-reef succession (the 'Capitan Reef', Middle Permian, Delaware Basin, New Mexico, USA).

d: Carbonate platforms and pinnacle reefs, mostly developed during a period of drowning of a rifted succession of coastal plain clastic deposits – vertical facies successions were controlled by fluctuations in rates of carbonate production, relative sea-level change and clastic sediment supply from adjacent deltaic systems (Miocene, Central Luconia Province, Sarawak; WD = water depth) (Koša et al. 2015).

e: Regional seismic line across the Central Luconia Province, which was carbonate dominated in the middle Miocene (lower part of the section – carbonates in dark blue). The overlying clastic-dominated Plio-Pleistocene (c. 5 Myr duration) succession is preserved as a 100 s m-km thick regressive basin-margin succession and defined by large-scale foresets; the clastic influx terminated the earlier phase of widespread carbonate growth.

Photograph/image credits: VIa-1 = Nautilus Ltd.; VI-a2, VI-b, VI-c. VII-b and VIIc = Howard Johnson; VI-e seismic courtesy of PETRONAS; VIIa-2 – Yamin Ali 2014; VIId and VIIe = Koša et al. 2015.

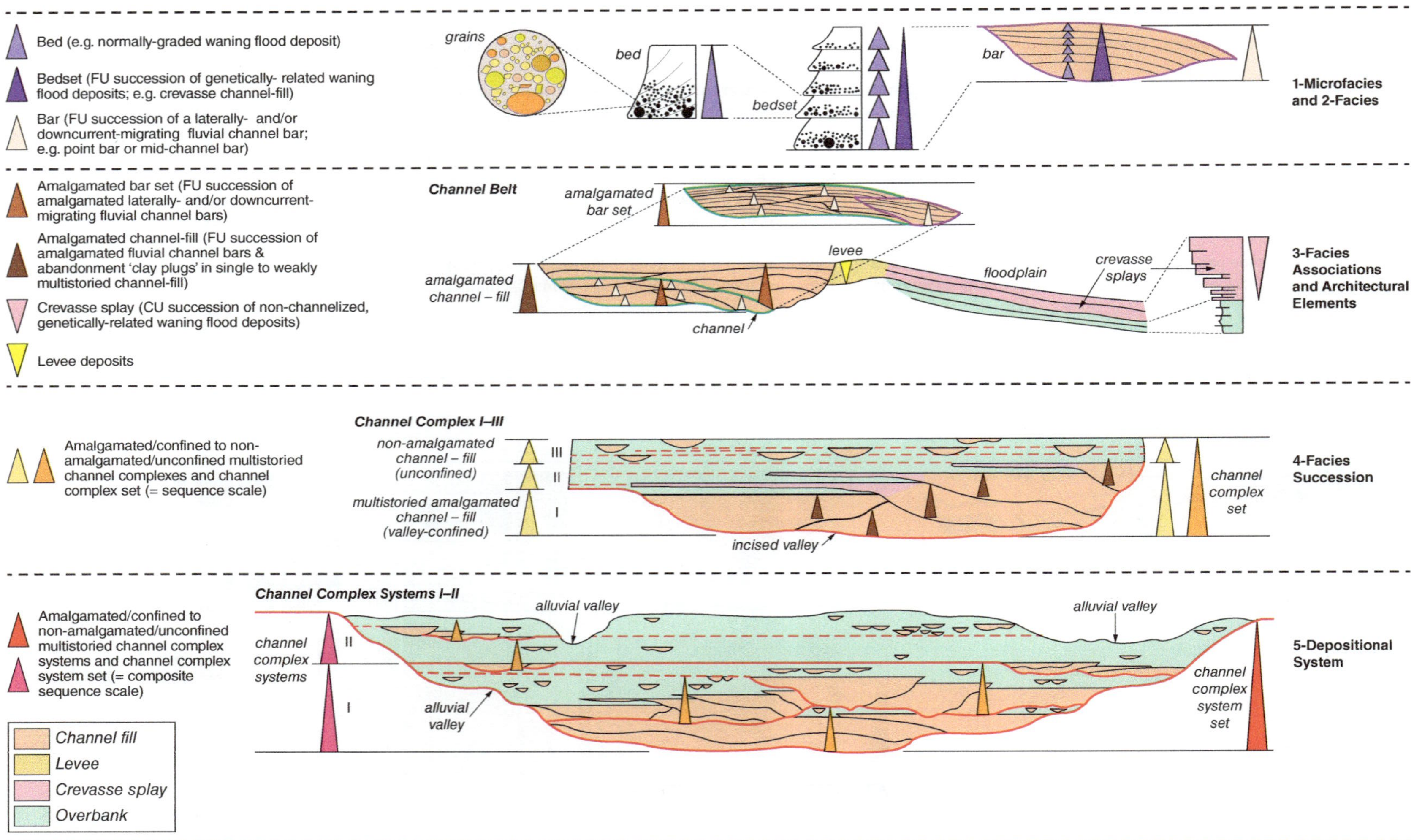

Figure 2.2 The five-step hierarchical nature of clastic facies and sequence stratigraphic architecture (Table 2.1), illustrated by the fluvial depositional system (Sprague et al. 2002, modified after Kendall 2014). 1 – Microfacies-scale (sub-mm/). 2 – Facies-scale (dm/m). 3 – Facies association-/architectural element-scale (m-10 s m). 4 – Facies succession-scale (10s–100 s m). 5 – Depositional system-scale (km). The vertical extent of the different stratigraphic units is shown by the coloured triangles, which emphasise the dominance of fining-upwards facies trends at a range of scales. Depositional strike sections.

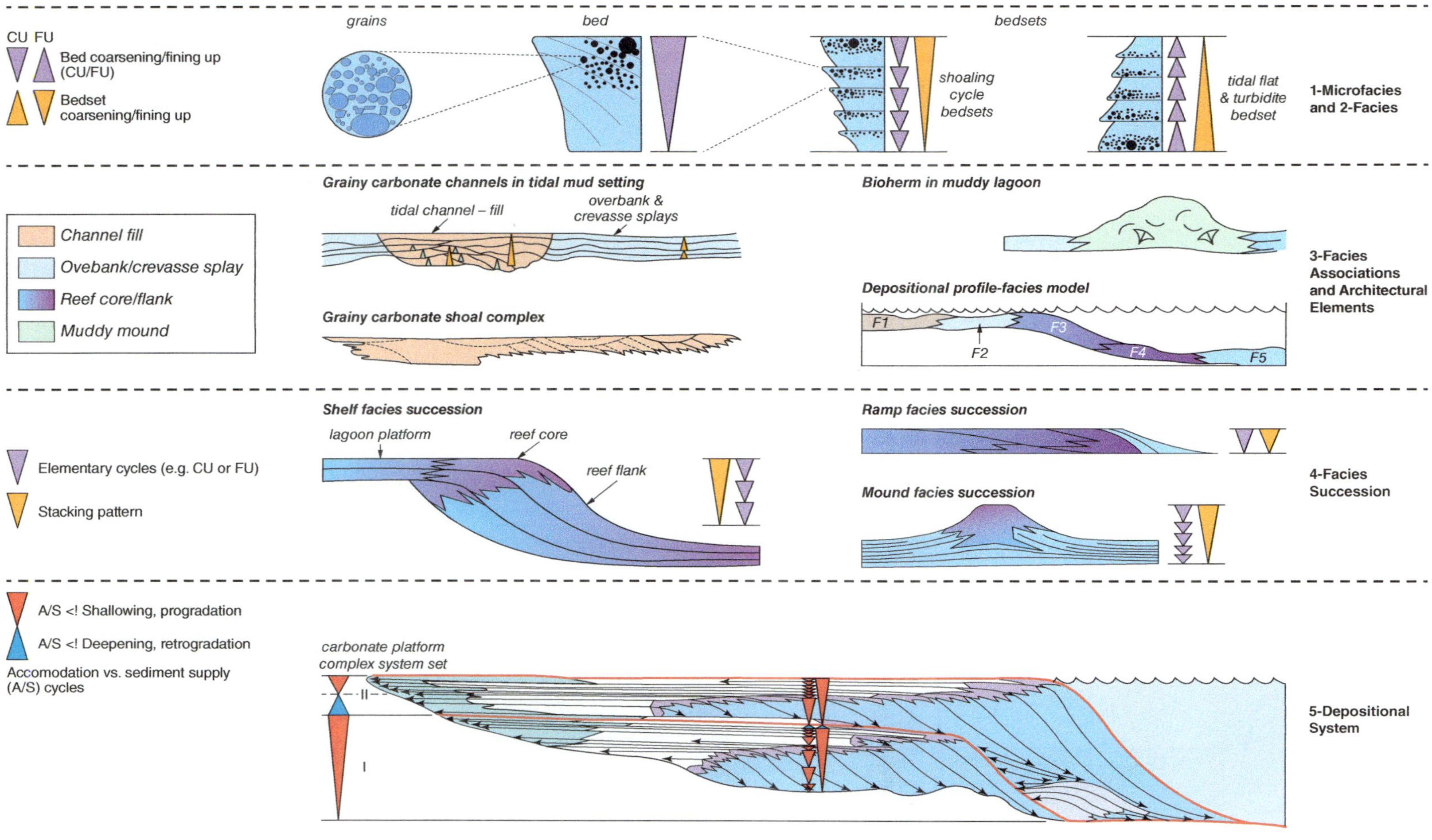

Figure 2.3 The hierarchical nature of carbonate facies and sequence stratigraphic architecture, illustrated by carbonate platform and shelf depositional systems (*Source:* Adapted from Kendall (2014): 1-Microfacies (e.g. grains, cements, pores, etc.). 2-Facies (e.g. beds, cycles, bedsets, etc.). 3-Facies associations and architectural elements (e.g. channels, shoals and bioherms). 4-Facies successions (e.g. shelf, ramp, and mound). 5-Depositional system (e.g. carbonatecomplex system sets). (Mainly depositional-dip sections).

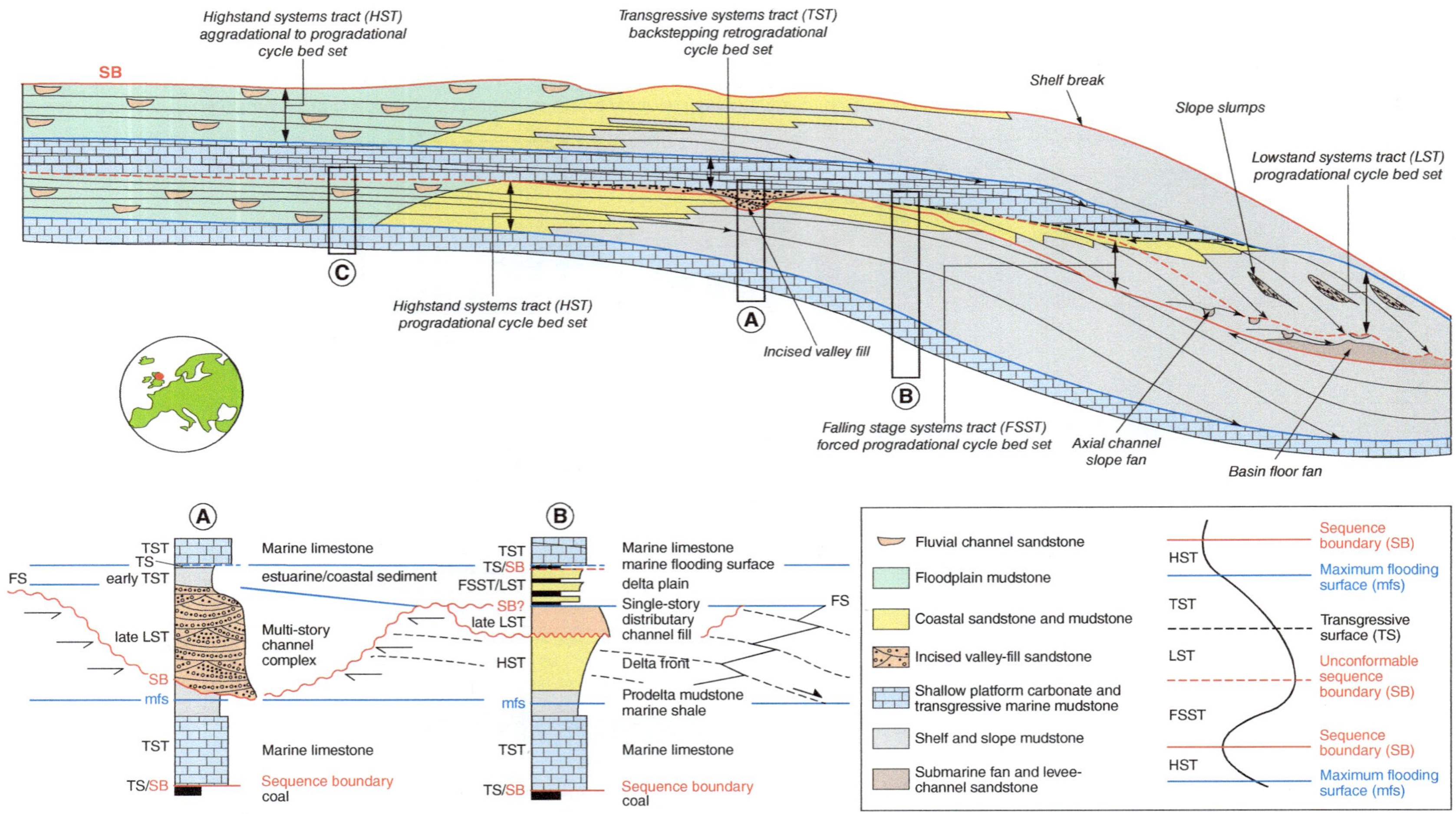

Figure 2.4 The hierarchies of facies and sequence stratigraphic architectures shown separately in Figures 2.2 and 2.3 are shown here in a single mixed clastic-carbonate succession. This simplified depositional-dip-oriented stratigraphic cross section is based on the Yoredale cycles in the mid-Carboniferous of northern England. The section shows two clastic-dominated, regressive deltaic wedges, and two broadly transgressive carbonate platform deposits (blue). The vertical facies distributions, their environmental and sequence stratigraphic interpretations of the lower clastic unit are illustrated in the two simplified logs. The first log (A) comprises a clastic interval with an incised valley-fill (IVF) sandstone cutting into delta plain-pro-delta mudstones – the IVF overlies a regionally extensive sub-aerial erosion surface (sequence boundary, in red). The second log (B) comprises coarsening-up, regressivedeltaic deposits. A third log (C) on the regional correlation panel is from a more updip depositional setting. Details of log C, including photopanoramas, are shown in Figure 2.7, which shows delta plain deposits, including distributary channels and interdistributary bay-fill facies successions, and bounding carbonate platform deposits. *Source:* Kendall (2014); Tucker et al. (2009).

beyond the observations; and (iv) enabling application of sequence stratigraphy for prediction and understanding of the dynamics of stratigraphic relationships in time and space.

The first concept, linking facies and environments with stratigraphy, was initially demonstrated in Gressly's work on Mesozoic carbonates in the Jura Mountains (Gressly 1837, 1838). This launched the far-reaching concept of linking a genetic understanding of stratigraphic relationships to palaeogeographic reconstructions (Cross and Homewood 1997). The facies to environment link was further cemented by Walther's (1894) Law of Correlation (or Succession) of Facies (Figure 4.18):

> "The various deposits of the same facies areas and similarly the sum of the rocks of different facies areas are formed beside each other in space, though in cross section we see them lying on top of each other. As with biotopes, it is a basic statement of far-reaching significance that only those facies and facies areas can be superimposed primarily, which can be observed beside each other at the present time." (Translation from original German in Middleton 1973).

The second, process-response concept, underpins all facies analysis work. This concept's strength was first demonstrated by the work of Kuenen and Migliorini (1950) on turbidity currents and graded bedding, which provided a link between a (still) rarely observed physical sedimentary process (turbidity currents), and a known response comprising a widely occurring rock type (turbidite facies) (Chapter 12). This phase was dominated by linking hydrodynamic processes and flow regimes, based on measurements in flume tanks and/or modern environments, to the interpretation of primary sedimentary structures and related physical bedding styles in the rock record (e.g. Allen 1970, 1982, 1985; Harms and Fahnestock 1965; Simons et al. 1965; Walker 1978). This provided the groundwork for many future examples of facies observed in the Modern (e.g. cross-bedded sands in tidal bars) that have closely linked process significance (e.g. lunar periodicities, tidal ellipses and measured tidal current velocities) that then guides interpretation of ancient facies (e.g. cosets of cross-stratified sandstones with systematically distributed mud drapes, sand–mud lamination bundles and/or bidirectional palaeocurrent patterns; Figure 2.1; Section 8.3.2).

Third, the facies model concept provides a spectrum of idealised representations of facies distribution in different depositional environments (Reading 2001). Building facies models involves the 'distillation' of key information from a range of representative recent sediments and ancient rocks, hence:

> the constant cross-comparison of the ancient and the modern/recent leads to simplifications (Walker 1984b, paraphrasing Walther 1894).

These distillations came from two main sources (Walker 1984b): (i) integration of process-based facies analysis with geometrical properties for a wide range of ancient depositional systems, initially with emphasis on fluvial and deltaic settings, and (ii) rapid expansion of work on modern sedimentary environments, driven partly by the need to understand the dynamic behaviour of such systems from an engineering perspective. Early examples of this approach are in the modern Mississippi River and delta in the Gulf Coast area of the United States (Figure 2.5) and the Rhine-Meuse delta, estuaries and seaways of NW Europe. A closely related driver was the need to generate depositional models to guide oil and gas exploration, which was motivated by the close link between depositional environment and key rock properties: (i) texture and pore space (e.g. porosity and permeability); (ii) reservoir heterogeneities (e.g. shale layer distribution); and (iii) reservoir body geometries (e.g. Figure 2.6). The powerful linkage of outcrop and subsurface facies analysis, integration with modern analogues (e.g. the modern Mississippi delta) and reservoir description applications was demonstrated in the cyclic, coal-bearing successions in the Carboniferous (e.g. N. America and NW Europe). This contributed to a surge in ancient facies analysis studies (e.g. Figure 2.7), integration of data from modern analogues (e.g. Figure 2.8) and the development of generic facies models, such as that of fluvial-dominated interdistributary bays (Figure 2.9). There exists now a set of 'global' facies models for all the major clastic and carbonate depositional systems, with examples covered in subsequent chapters of this book, and elsewhere (e.g. James and Dalrymple 2010; Nichols 2023).

The fourth conceptual development, sequence stratigraphy, uses facies analysis to determine environments and to identify significant breaks in stratigraphic successions. A common example of the latter would be where facies juxtapositions cannot be explained by a simple autogenic change but rather allogenic changes, such as a shift in sea level, need to be considered. This central role of facies analysis was recognised from the outset:

> Vertical facies analysis must be done within conformable stratal packages to accurately interpret coeval, lateral facies relationships along a single depositional surface (Van Wagoner et al. 1990; Catuneanu 2022).

The heightened focus on significant stratigraphic surfaces, representing key breaks in a facies succession, was

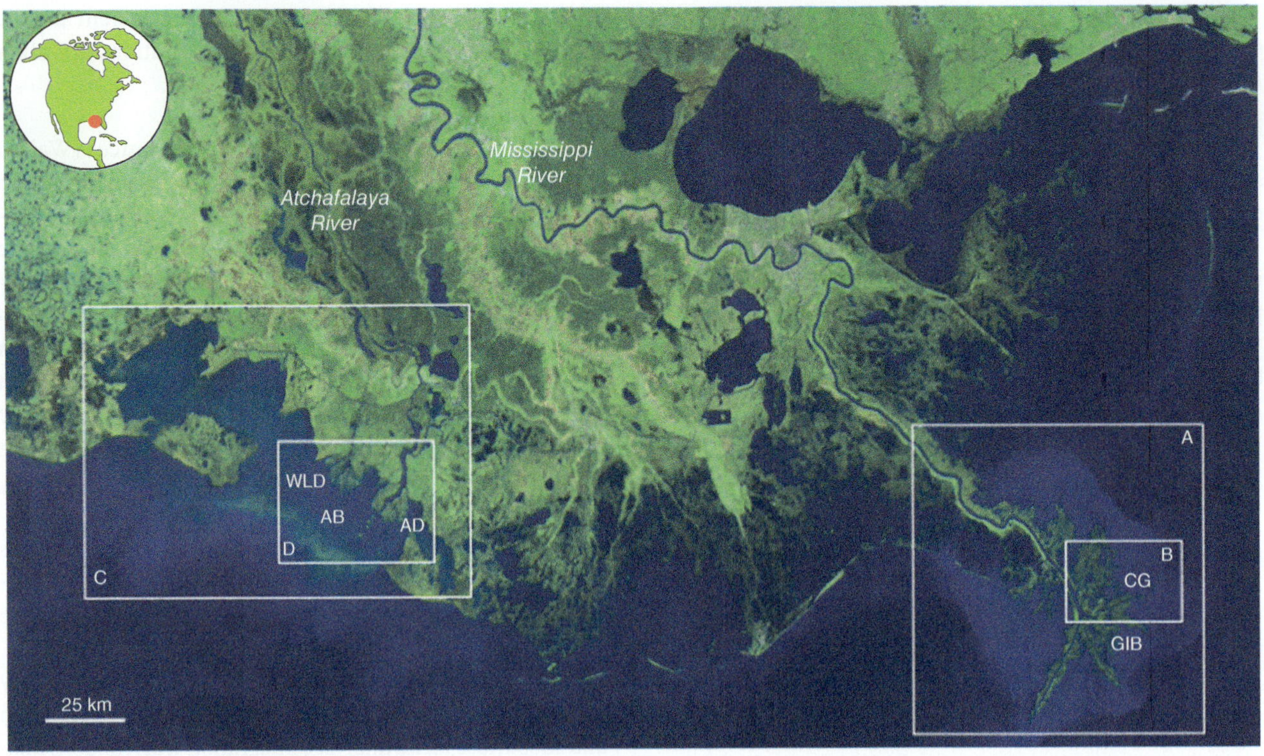

Figure 2.5 Satellite image of the Mississippi Delta (Balize & Modern lobes) and adjacent Gulf Coast of Louisiana (February 2014). White inset boxes (A–D) highlight the following locations: A. the Mississippi distributary channel and bar fingers (mouth bars) of the presently active shelfal delta front, and shallow water interdistributary bays, such as Garden Island Bay (GIB); B. Cubits Gap (CG) area, a modern example of an interdistributary bay-fill comprising a crevasse delta that evolved into a sub-delta lobe; C. Atchafalaya Bay (AB) and adjacent shallow-water bays, which coincide with previously abandoned and now drowned delta lobes; D. The Wax Lake Delta (WLD) and Atchafalaya Delta (AD), both are river-dominated bay-head deltas discharging into the Atchafalaya Bay – the Wax Lake Delta is an engineered flood discharge outlet, while the Atchafalaya Delta is the natural termination of the Atchafalaya River. Historical records document in detail the evolution of the WLD. *Source:* NASA / Public Domain.

a break-through that sharpened both the stratigraphic and process understanding of sedimentary successions. The allogenic controls cause more widely correlatable shifts between sets of environments. Notwithstanding this, distinguishing between autogenic processes, as emphasised in facies sedimentology, and allogenic factors (e.g. tectonics, climate and eustasy), as emphasised in sequence stratigraphy, remains a topic of active research with considerable uncertainty and debate (e.g. Hajek and Straub 2017).

Facies analysis of carbonates followed a similar pathway to siliciclastics. The first carbonate process-response models were based on studies of the facies distributions on Great Bahama Bank (Ginsburg and Lowenstam 1958) and the contemporaneous development of petrographic classification schemes for carbonate sediments (Folk 1959; Dunham 1962). These studies linked carbonate sand–mud distribution (in a simple textural sense) on a shallow-water carbonate platform to the impact of hydrodynamic processes. For example, grain-dominated sediment occurred at the high-energy margin of the platform and mud-dominated sediment accumulated in the protected inner platform environment (Figure 2.10). In the ancient, repeating facies patterns, the 'Lofer cyclothems', were first discussed for the Triassic Dachstein shallow-water carbonate platform in Austria (Figure 2.11; Schwarzacher 1947, 1954; Fischer 1964). The shifting of facies belts and associated sediment export in response to allogenic factors, notably climatic change and sea-level variation, was a further major step in the development of carbonate facies analysis (Schlager 2005).

Standard carbonate facies models were developed through the studies of Ginsburg and Lowenstam (1958), Bathurst (1971, 1975), Folk (1974), Milliman (1974) and Wilson (1975). Early examples were facies distribution patterns along recent, isolated, flat-topped, carbonate platforms like the Bahamas (Purdy 1963a, 1963b), and

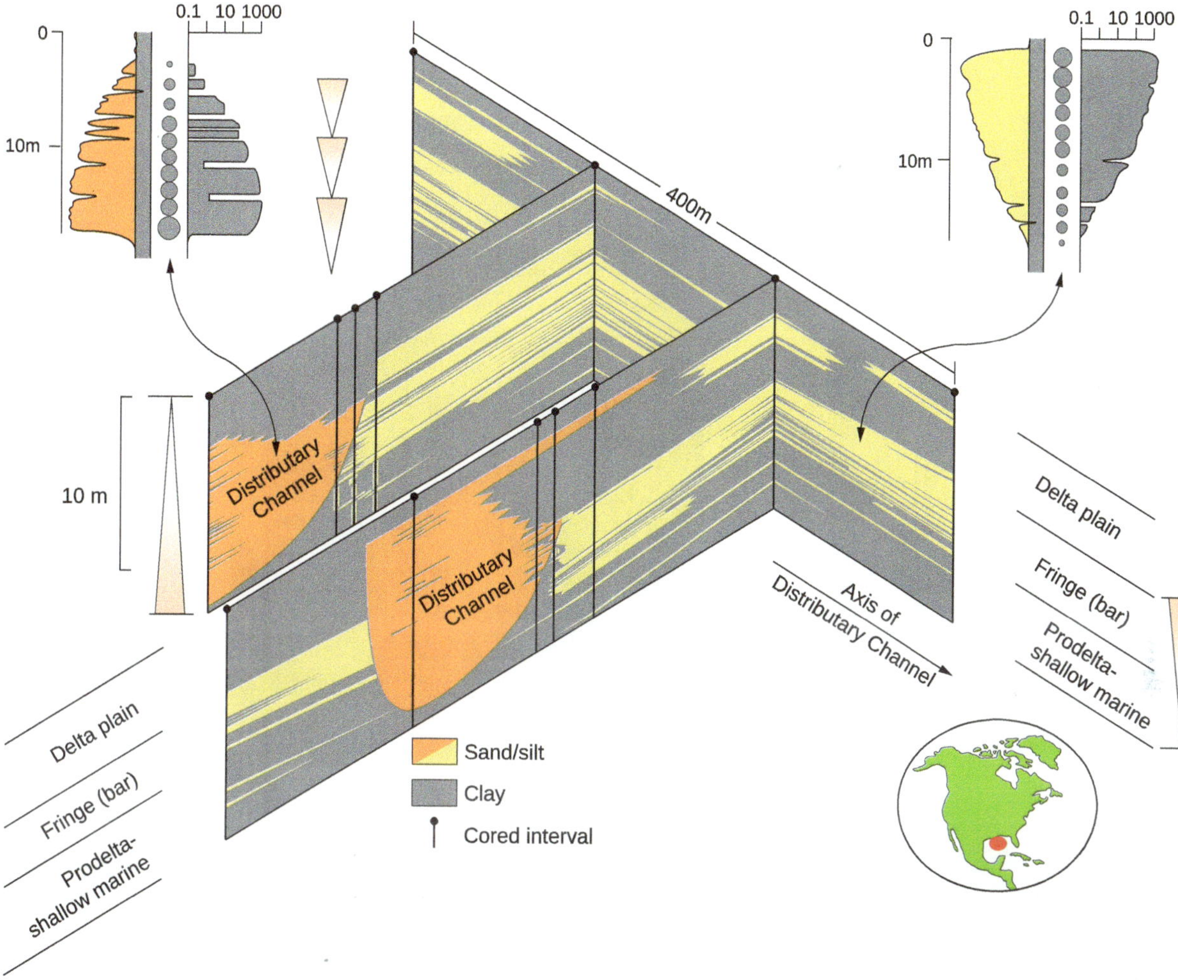

Figure 2.6 A fence diagram based on the facies interpretation of cores and well logs (grain size indicated schematically by size of grey circles, from fine sand to mud) from the pre-Modern deposits of the Mississippi Delta. This type of data provides excellent reservoir analogue information, in this case of distributary channel and mouth bar/delta fringe deposits, especially quantification of thickness, width and geometry of both sand bodies and shale layers. *Source:* Adapted from Sneider et al. (1978). The contrasting shale layer lengths are especially clearly highlighted between channel sand bodies (short, discontinuous and increasing frequency upwards) and mouth bars (longer, more continuous and decreasing frequency upwards).

shallow carbonate environments like the Persian Gulf (Purser 1973). These facies distribution patterns set the stage for the development of carbonate facies models for a variety of carbonate platform morphologies and introduced standard microfacies-type distribution and associated characteristics (Bathurst 1971, 1975; Wilson 1975). The microfacies concept was further detailed by Flügel (1982, 2010) with an emphasis on the diversity of skeletal and non-skeletal grains over geological time, their recognition in thin section, their occurrence within specific environmental settings as well as diagenetic processes operating in various depositional and burial settings. An important driver of such studies, especially microfacies, was to improve understanding of the diverse array of carbonate porosity types. Central to this was to improve the prediction of flow behaviour in oil- and gas-bearing reservoirs, including the interpretation of rock properties from wireline log responses, estimation of fluid saturation and effective pore-space distribution, and the prediction of fluid-flow behaviour in general. A major application of microfacies analysis in carbonates was directed towards rock-fabric classification schemes, which were initiated by Archie (1952) and extended by others (e.g. Lucia 1999, 2007).

Sequence stratigraphic expressions in carbonate systems differ significantly from those in siliciclastic-dominated environments (Chapter 4). Regrettably, even in recent textbooks, these differences are often reduced to a single additional sequence stratigraphic model, the 'highstand-shedding model' (Catuneanu 2022) that only applies to

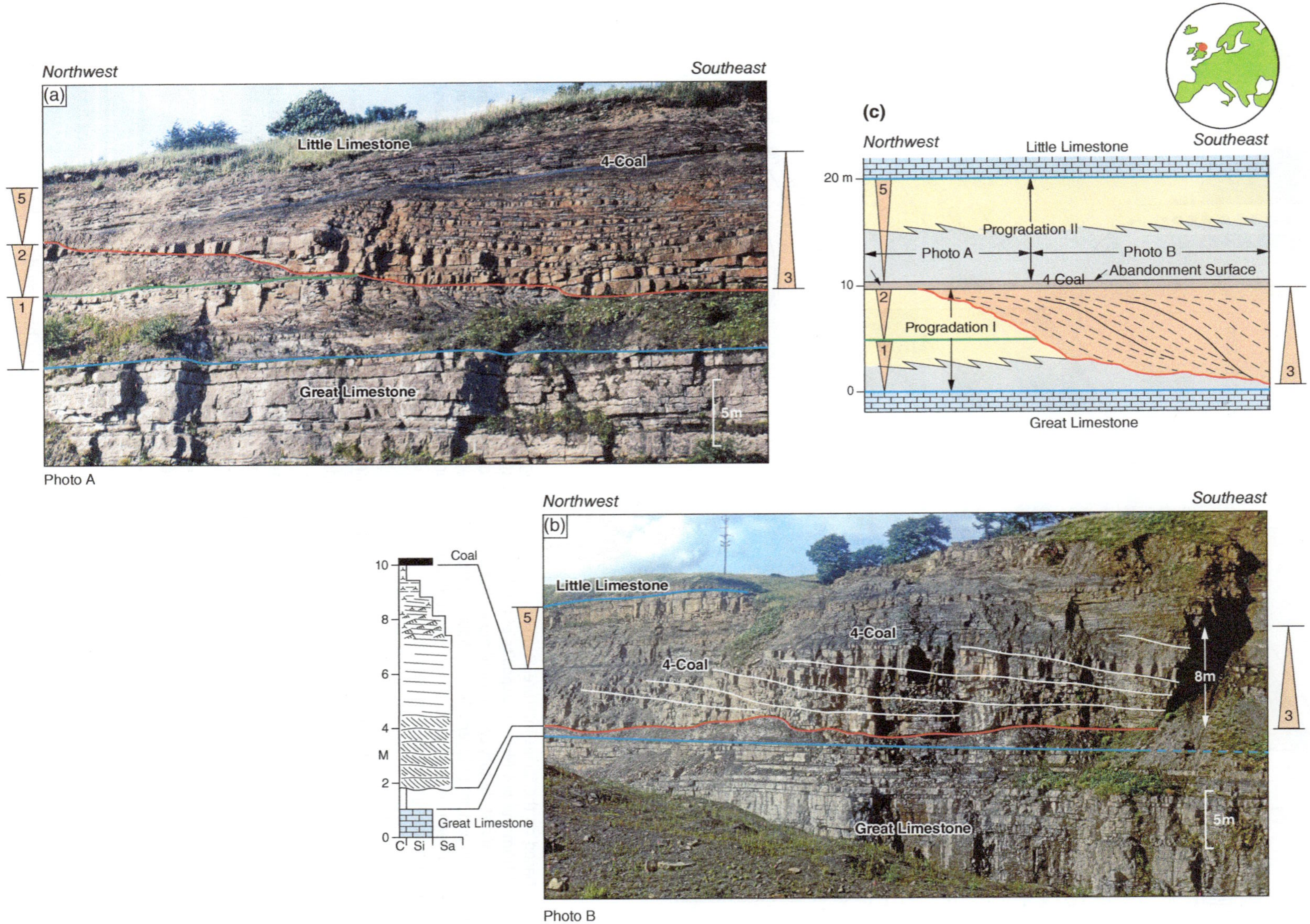

Figure 2.7 Example of vertical facies successions, architectural elements and contrasting sand body geometries in fluvio-deltaic deposits (Great Limestone cyclothem, Upper Limestone Group, Namurian E_1 (Pendleian), northern England - see Figure 2.7 for broader stratigraphic and depositional context Two shallow-marine platform carbonates bound a 20-m-thick deltaic clastic unit – centre of both photographs in a and b and line drawing in c. (a) Shows the pinch-out edge of a meandering distributary channel system, with numerous intercalated mudstones, thinner-bedded sandstones and lateral accretion surfaces (dipping to the right). The fining-upward distributary channel (right-side) cuts through two coarsening-upward successions (1 and 2), bounded by mudstones and interpreted as interdistributary bay infill facies successions (left side). A coal layer (4) represents a delta abandonment horizon, which overlies the both the two coarsening-upward bay-fill succession (1 and 2) and the fining-upward channel-fill succession (3). (b) This photo (c. 50–100 m along strike from A) shows the two transgressive shallow-marine carbonates (blue lines) bounding the clastic interval and the same meandering distributary channel sandstone body shown in A. Here the channel has an erosional base (red line), fining-upwards facies trend (facies log to left), low-angle lateral accretion surfaces (inclined white lines – dipping to the right/SE) and is capped by an in situ coal layer with underlying grey mudstones (seat earth; labelled '4-Coal'). This image is from the axial part of the inferred meanderbelt with southwards palaeoflow (directed towards the viewer). *Source:* Courtesy of Howard Johnson. (c) Simplified, strike-oriented cross section with sequence of events as follows: (1–2) fluvial-dominated, coarsening-upward interdistributary bay-fill successions; (3) avulsion, incision and lateral migration of a meanderbelt sandstone; (4) delta lobe abandonment and formation of a lateral extensive peat swamp (coal layer); (5) transgression followed by widespread coastal progradation of a shoreface-barrier island-lagoon system *Source:* Adapted from Elliott (1975). Rogerley Quarry, Stanhope, Weardale (UK). Sedimentary log and overall interpretation from Elliott (1975, 1976). Photo credits Howard Johnson.

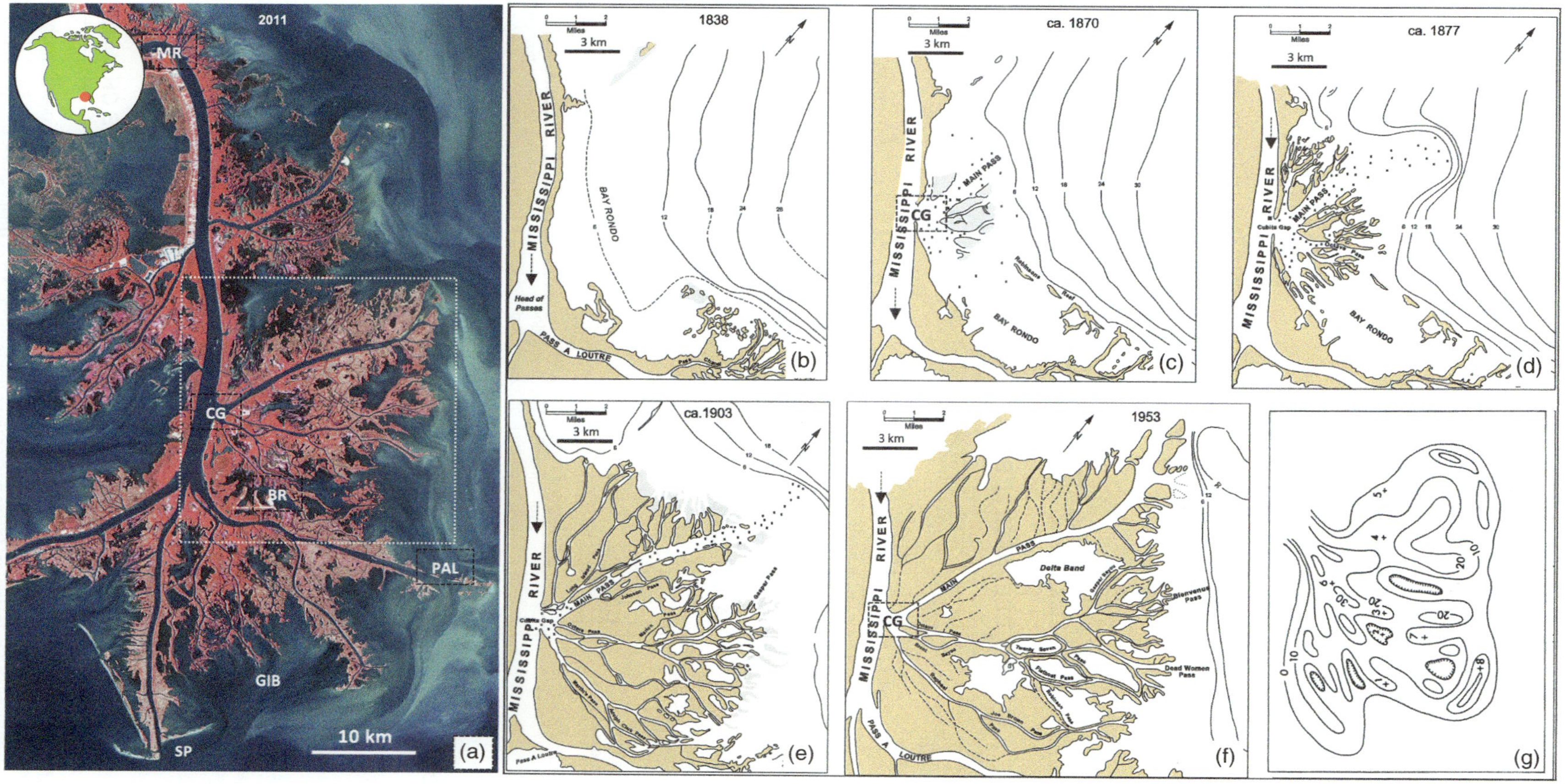

Figure 2.8 Evolution of the Cubits Gap crevasse delta. (a) Satellite image (3 October 2011) of the Mississippi River (MR) and south-east part of the present-day 'bird's foot' delta (false-colour: vegetated areas are pink-red and sediment-laden waters are blue/green). Dotted white box is the area of maps shown in b–g. Note the examples of crevasse channels and crevasse deltas stemming from flood-induced breaches in the natural levees of the Mississippi River. *Source:* European Space Agency (ESA). (b–f) Evolution of the Cubits Gap (CG) sub-delta, covering the pre-crevasse period (b 1838), through initiation of the crevasse channel at CG (c; c. 1870) and the infill of the Bay Rondo by the crevasse delta (d–f; c. 1877–1953). *Source:* Wells and Coleman (1987); Roberts (1997). (his Figure 16). (g) Conceptual isopach map of a crevasse sub-delta sand body: a lobate sediment body 3–10 m thick (*Source:* Adapted from Coleman and Prior (1982) – their Figure 9). Satellite image: **©ESA** Landsat image of 3 October 2011.

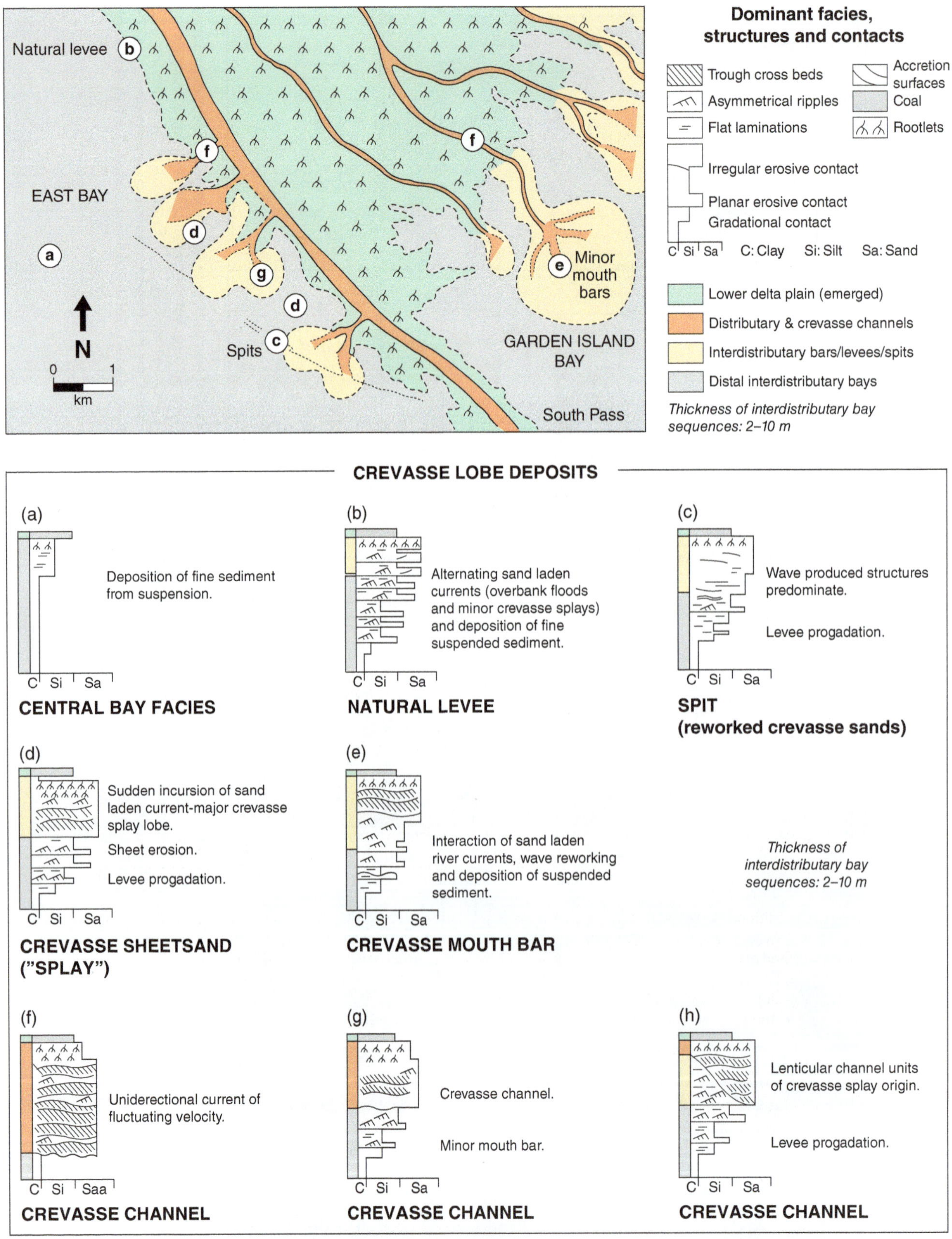

Figure 2.9 A simplified map of sedimentary environments in a fluvial-dominated interdistributary bay area, based on the modern Mississippi Delta (see Figures 2.5A and 2.8, for the location of Garden Island Bay - GIB). The facies model is captured in the eight logs (a–g, located in the map) and synthesises the variability in vertical facies successions (typically 2–10 m thick) and is based on (i) ancient examples mainly from late Carboniferous Yoredale deposits of northern England and (ii) modern Mississippi bay-fill successions (see Figure 2.8). The terms 'minor mouth bar' and 'crevasse mouth bar' are synonyms and equivalent to the terms 'bay-fill' or 'crevasse delta'. Idealised vertical facies logs are illustrated for the following depositional environments (see the map for geographic reference): (a) central bay facies; (b) natural levee; (c) spit; (d) crevasse splay; (e) crevasse mouth bar; and (f-h) crevasse channels. *Source:* Adapted from Elliott (1975).

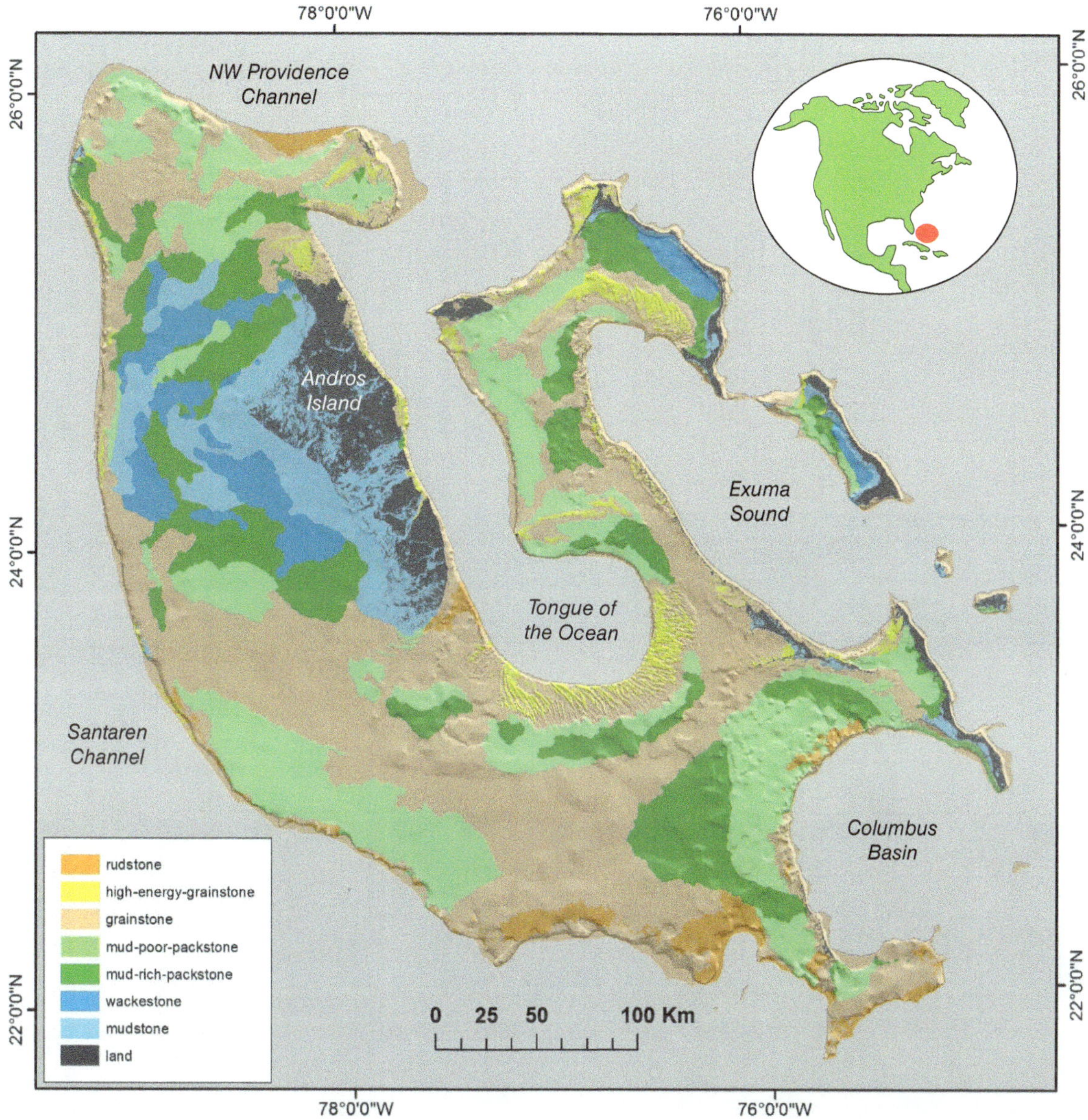

Figure 2.10 Map of depositional facies (sediment types) for the Great Bahama Bank (GBB). Basis of this map comes from data discussed in Reijmer et al. (2009) and Swart et al. (2009). From Harris et al. (2015). Map based on seafloor sample data compiled into ArcGIS, integrated with the Landsat mosaic and analysed with eCognition. The facies map is superimposed upon a simplified version of the bathymetric Digital Elevation Model for GBB. *Source:* Harris et al. (2015) / John Wiley & Sons. Note the location of the mud-free and mud-lean sediment types (rudstone, high-energy grainstone, grainstone, mud-poor packstone) at the edges of the bank and open parts of the platform (grainstone, mud-poor packstone) and the mud-rich sediment types (mud-rich packstone, wackestone, mudstone) on the leeward side of islands.

one carbonate factory: tropical carbonate systems (Schlager 2005; Reijmer 2021). However, it is now recognised that carbonates have different sediment production modes ('carbonate factories'), which result in significant variations in sequence stratigraphic facies stacking patterns depending on the carbonate production and distribution pattern of the particular carbonate factory (Schlager 2005; Pomar and Hallock 2008; Reijmer 2021) (Chapters 4 and 11). Examples of basin-scale correlation across complete carbonate systems using high-resolution sequence stratigraphic stacking patterns include those of the Cenozoic and Mesozoic of the Mediterranean and the Middle East (Razin et al. 2010; van Buchem et al. 2010) and the Permian Capitan Reef of the Delaware Basin (Table 2.1-VI-c; Saller et al. 1999).

2.1.3 Key Technological Developments

Allied to the conceptual advances discussed above, technology developments have also advanced facies analysis methodology, especially seismic imaging of stratigraphic

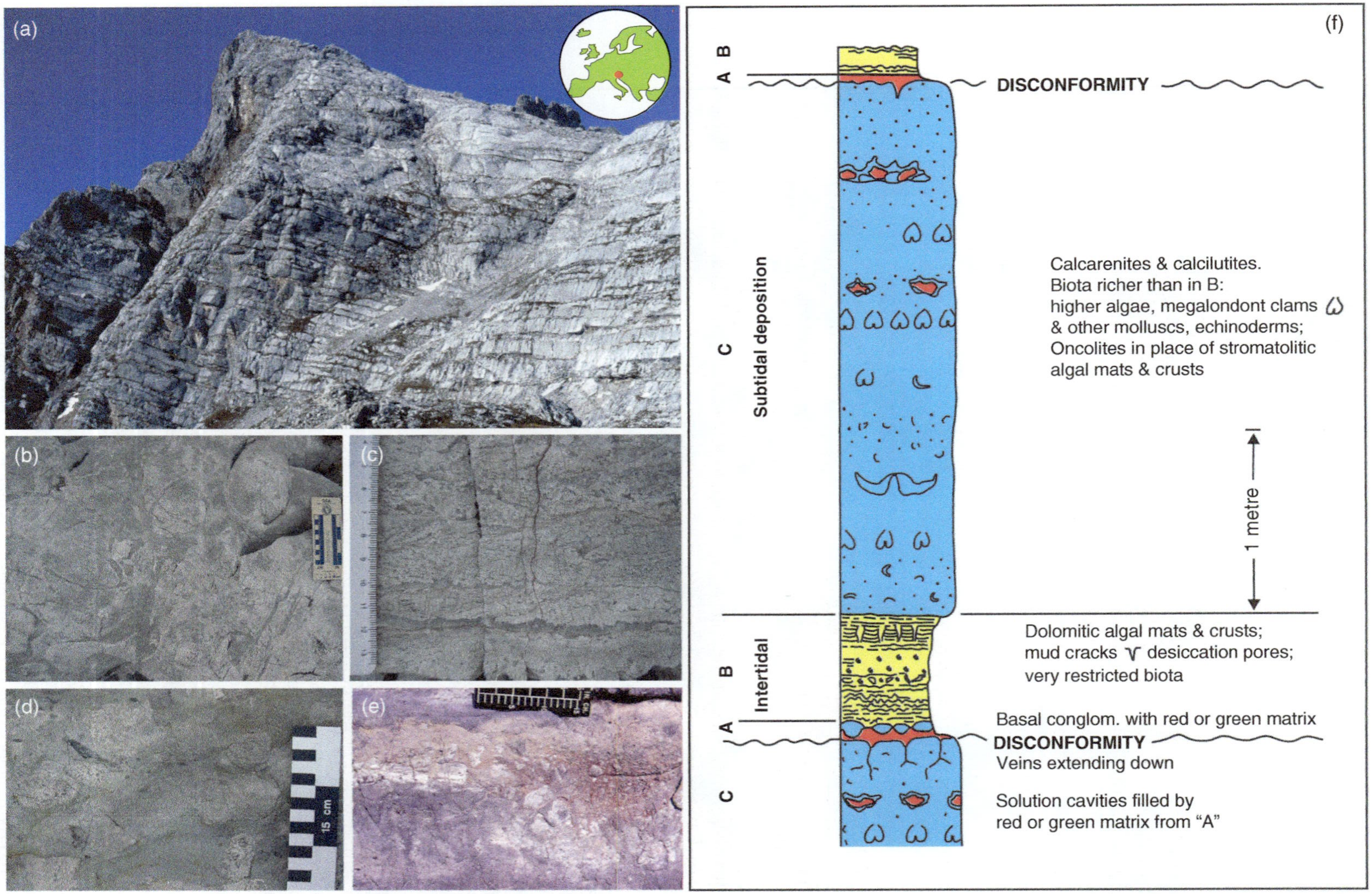

Figure 2.11 Lofer cyclothems, Triassic Dachstein shallow-water carbonate platform (Steinernes Meer, Austria). (a) Cyclic Dachstein Limestone. Part of the classic Steinernes Meer section, which also measured by Fischer (1964, his figs. 4 and 5B), Goldhammer et al. (1990, fig. 16) and the base of Satterley's (1996a, 1996b). See also Enos and Samankassou (1998). (b) Megalodont lime packstone; member C. Cement-filled megalodont mould at upper left of scale and elsewhere in image. (c) Dolomitic mudstone with vague wavy lamination, aligned fenestrae, sheet cracks and some auto-brecciation; typical member B. Scale in centimetres. (d) Soilstone with rounded clasts of lime wackestone and fenestral dolomudstone in lime mudstone matrix; member A. (e) Red palaeosol (member A) dominated by angular clasts of whitish dolomudstone with aligned fenestrae overlying and filling karstic depression within subtidal lime wackestone (member C). *Source:* Courtesy of John Reijmer. (f) Fischer's 'Lofer cyclothem' (modified slightly from Fischer 1964, p. 113). His member designations, A, B and C are used for the principle lithologies, as well as 'd' for disconformity. For more detailed lithologic descriptions, see Fischer (1964) and Enos and Samankassou (1998). Note that Fischer's cycle ABC(d) is deepening upward. (a–e) From Samankassou and Enos (2019). (f) *Source:* Enos and Samankassou (2021)/ John Wiley & Sons

geometries and depositional systems (e.g. Table 2.1-VIe and VIIe). Two-dimensional seismic data was sufficient to support the new concepts of seismic stratigraphy in the 1980s and provided guidelines, based on the type of stratal geometry and bounding erosion surfaces, to interpret regional stratigraphic architecture (Payton 1977). In particular, the revolutionary publications of Vail and co-workers (in Payton *op cit*) demonstrated that seismic reflection data preserved geological timelines. Subsequently (mid-1990s onwards), regional 3D seismic data fundamentally changed the perception of many aspects of stratigraphy, facies and environments. Specifically with the concept of seismic facies. For example, seismic amplitude data displayed on stratal surfaces, can be characterised in terms of amplitude value contrast and polarity, frequency, coherence and geometry into seismic facies groupings. These seismic facies, together with the exceptional imaging quality of modern 3D seismic, allow the definition of the dimensions, shapes and geometries of facies units and depositional elements (e.g. Figures 2.12 and 2.13a, c). At the basin-scale, large regional 3D datasets, especially in deep-marine settings, such as the Gulf of Mexico and the conjugate margins of the South Atlantic, have produced spectacular images of complete depositional systems (Figure 2.12). These images can be of such high quality that we have to remind ourselves that these are not from the present-day seafloor, rather in this case they show the palaeo-seafloor at c. 75 Ma! These types of images have also had a major impact in moving away from idealised global facies models and towards actually mapping different types of architectural elements. Gross depositional environment maps can be drawn directly from such images and reveal the key components of the depositional system. For example, in the Santos Basin, the following palaeoenvironments are clearly differentiated (Figure 2.12b,c): linear beach ridges, muddy shelf area, shelf edge, channel cutting through the slope with submarine fans at the base and slump scars in the upper slope.

In carbonates, a particular result of high-resolution 3D seismic data was to provide unprecedented insight into the external geometry and internal seismic facies characteristics of carbonate build-ups. The integration of stratal geometries of key surfaces, such as foresets, with the seismic facies of the deposits was especially powerful. For example, stratal geometries from the middle Miocene build-ups in the Central Luconia Province (offshore Sarawak) document their initiation, growth and abandonment (Table 2.1-VIId and e, and Figure 2.13a, c; Zampetti et al. 2004; Koša 2015; Koša et al. 2015; Rankey et al. 2019). Similarly, images of the Miocene Malampaya build-up of offshore Philippines (Fournier et al. 2005) enable reconstruction of its evolution and the interpretation of lithologic and diagenetic characteristics. Three-dimensional seismic imagery from the Gulf of Papua (Tcherepanov et al. 2008a, 2008b, 2010) reveals the interaction of different depositional styles in a mixed carbonate-siliciclastic system (Figure 2.14) (see also Case Study-5). A series of 2D seismic lines calibrated with continuous cores has been used to test the chronostratigraphic significance of seismic reflections (Bahamas Transect; Eberli et al. 2002).

A further technological development enabled by increased computational power was the growth of large 3D reservoir (or geocellular) models. Such models are major users of the rock facies-scale of analysis (Levels 1–3 in Table 2.1) and are required in all subsurface reservoir characterisation work, especially for oil and gas production projects (Ringrose and Bentley 2015). Most significant from a facies perspective is the ability of these modelling systems to capture fine-scale sedimentological and stratigraphic heterogeneity, based on extrapolation from high-resolution 3D seismic, core and outcrop datasets. The microfacies to facies scale properties are directly linked to sedimentary processes and exert the greatest influence on fluid flow. The facies-scale 3D rock property models are typically up-scaled via facies associations and architectural elements, thereby retaining a close sedimentological link. In many such examples, volumetrically small facies may have a major impact on flow behaviour, such as mudstone layers and/or thin, high-permeability coarse-grained layers, which need to be captured in reservoir models. These relatively local details normally override what may be predicted from distilled global facies models, where the detail has perforce been generalised. The upsurge in 3D reservoir modelling during the 2000s was accompanied by a huge amount of quantitative sedimentological work, using tools such as LIDAR and hand-held XRF and gamma ray (GR) logging devices, directed at measuring the dimensions and rock properties of outcrop analogues. Such work was applied across the full spectrum of clastic and carbonate depositional systems and reservoir types.

Finally, remote sensing techniques, including satellite imagery (Figures 2.5 and 2.8a), airborne surveys, ground penetrating radar (GPR), etc. (Strick et al. 2019), have all helped better characterise and visualise ancient and modern depositional elements. Modern satellite imagery has greatly improved the task of finding modern depositional environment analogues, sometimes with the benefit of a dynamic, 4D, perspective (see the Wax Lake and Atchafalya delta time lapse link in Figure 2.5). New geochemical methods, such as using short-lived isotopes, now provide vital information on ages and rates of deposition in modern environments.

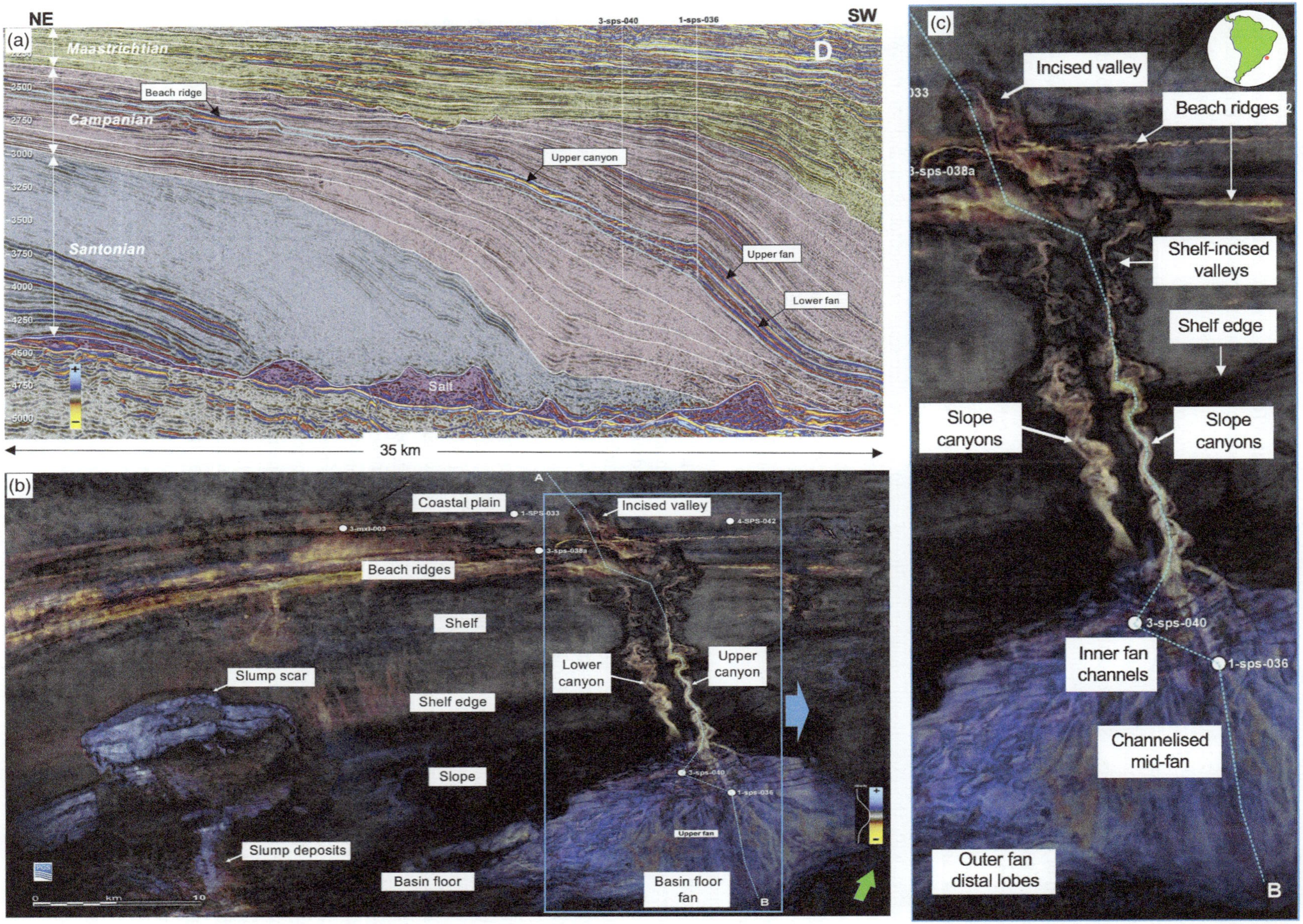

Figure 2.12 Seismic stratigraphic and amplitude mapping (Santos Basin, Brazil). (a) Interpreted dip-oriented seismic line showing the Late Cretaceous (Santonian-Maastrichtian) prograding succession in the Santos Basin. (b) Seismic amplitude map from a prograding surface within the middle part of the Campanian interval (pale blue seismic horizons in a), showing the wide range of gross depositional environments based on amplitude style and geometry. (c) Details of part of b, highlighting a source-to-sink profile, comprising the following depositional environments: beach ridges and shelf valleys, shelf edge, slope canyons and basin floor fan. *Source:* With permission of Hadler-Jacobsen et al. (2010) / GeoScienceWorld.

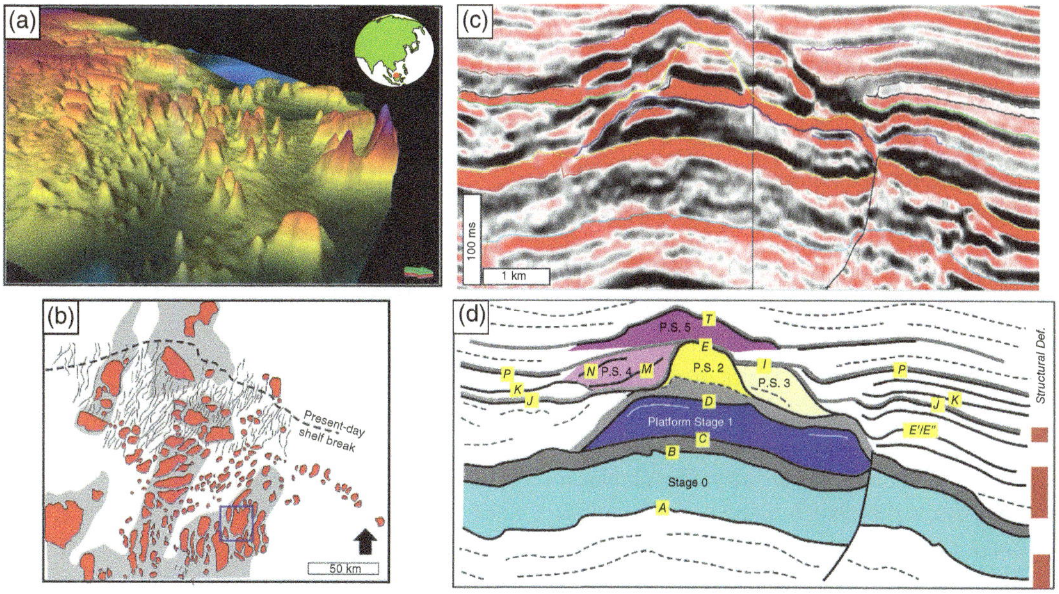

Figure 2.13 Miocene carbonates of the Central Luconia Province, offshore Sarawak, NW Borneo. (a) A 3D view of the seismic two-way travel time (TWT) surface that defines the top of the main middle Miocene carbonate build-up (looking westwards) – it shows the gross distribution of the main build-ups and their increasing relief in a basinwards direction (to the north/right) *Source:* (Koša et al. 2015) / GeoScienceWorld; Koša et al. (2015b) / American Association of Petroleum Geologists. (b) Middle Miocene paleogeographic map of Central Luconia. Structural highs (greys) and lows (unshaded) set the framework for carbonate platforms (red shades). Several faults (black lines), generally oriented NE–SW, have influenced the growth and development of many isolated platforms. Field EX lies within the blue-outlined square. (c) A 3D seismic line with interpreted seismic horizons from the EX Field. (d) Interpretation of the line in part c, illustrating the platform growth stages. Grey intervals are interpreted transgressive stages. Some of these are comparable to build-in phases of Epting (1980), others are defined based on downlap stratal termination. Red bars represent intervals of interpreted structural deformation. *Source:* Rankey et al. (2019) / with permission of Elsevier

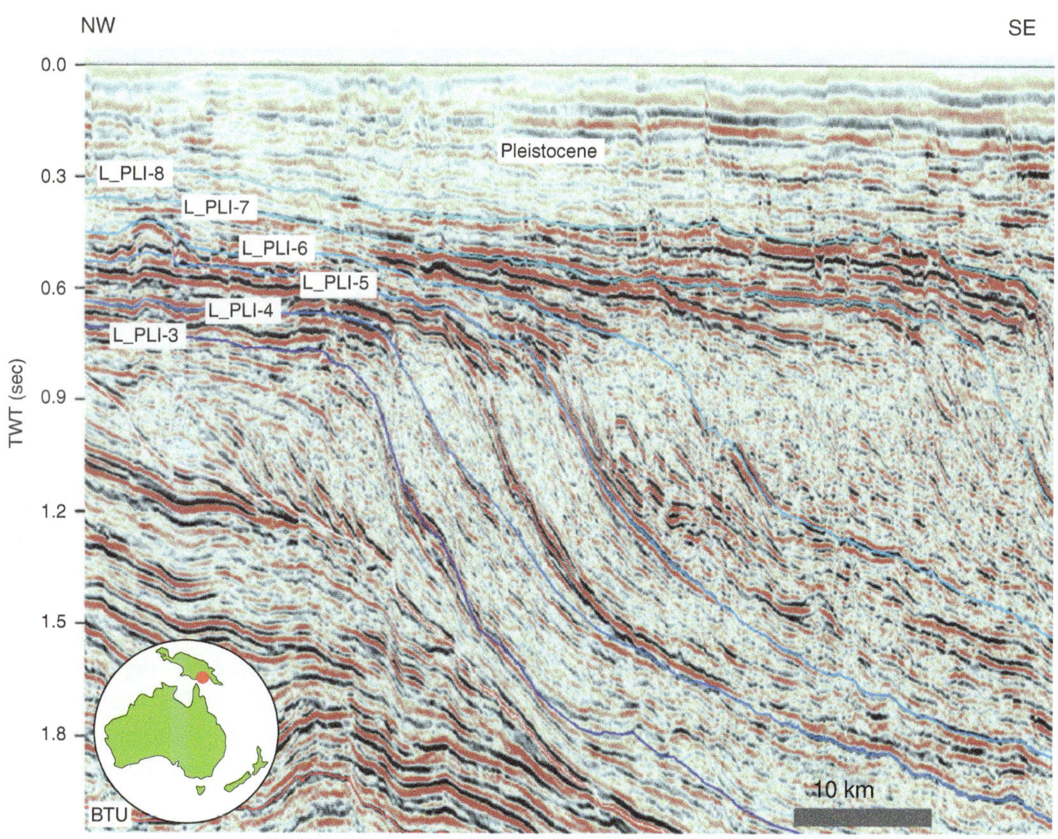

Figure 2.14 Seismic facies variations, Gulf of Papua. *Source:* Tcherepanov et al. (2010) / with permission of Elsevier. This example (see also Figure 9.11) shows significant seismic facies change between Pliocene outer shelf carbonates (high-amplitude reflections between 0.4 and 0.7 s) and Pleistocene siliciclastics (low amplitude reflections above 0.4 s). The late Pliocene sequences are characterised by extensive carbonates, while the Pleistocene sequences are predominantly siliciclastics. The Base Tertiary (Cenozoic) Unconformity (BTU) approximates to the base Eocene carbonates (lower left). 2.0 s corresponds roughly to 3.5 km, note the severe vertical exaggeration of the section.

2.1.4 Facies to Stratigraphic Synthesis

In principle, the facies analysis approach can systematically address the temporal and spatial hierarchies of process periodicities and magnitudes involved in the deposition of sedimentary rocks (Figures 2.2 and 4.42). This implies time scales of seconds to millions of years. Facies analysis achieves this feat by working successively through the spatial hierarchy discussed above, from facies to stratigraphic synthesis (Table 2.1 and Figures 2.2–2.4).

In this chapter, the focus is on the smaller-scale facies or sedimentology component of this hierarchy. The emphasis is on physical processes. It is complemented by Chapter 3, which focuses on biogenic processes, also at the facies level (scales 1–3 in Table 2.1). These two chapters address the following topics:

> Facies ➔ Facies associations/Architectural elements ➔ Facies successions ➔ The depositional system environments and depositional models

Chapter 4 then deals with larger-scale stratigraphic aspects (scales 4–5 in Figure 2.1), including the integration of sedimentology and sequence stratigraphy (Figure 4.1), addressing the following topics:

> Facies successions/sequences ➔ Stratigraphic concepts ➔ Sequence stratigraphy ➔ The preservation filter: preservational systems/models

Inevitably with such a continuum, the separations are arbitrary and there is some useful overlap in coverage. This is especially so in relation to the three clastic and two carbonate case studies (Section 2.6), which address different approaches to the complete sedimentology to stratigraphy continuum. Evaluating this broad range of scales requires the analysis of diverse datasets and methods (Figure 2.15).

2.2 Facies Analysis – Basic Concepts and Hierarchy

2.2.1 Facies

For most siliciclastic facies analysis, the facies scale (stage 2 in our hierarchy in Table 2.1) represents the smallest level of detail, covering the scale of grains, lamination, bedding styles and cm-dm lithology fluctuations. Individual facies characteristics are collated from relatively small rock volumes (e.g. dm-m thick and metres to kilometres wide). However, for rocks with complex and/or even smaller-scale fabrics (i.e. sub-mm), most notably in carbonates, evaporites and mudstones/siltstones, there is a need to move down to the microfacies-scale.

2.2.1.1 Siliciclastic Facies

Siliciclastic facies are described and distinguished based mainly on physical rock characteristics as follows:

i) grain size, sorting, and other textural and mineralogical properties;
ii) primary sedimentary structures;
iii) bedding scale and style (laminae, beds/bedsets, sets/cosets, beds/bedsets, etc.);
iv) palaeocurrent data;
v) sand/mud content;
vi) biogenic sedimentary structures (see Chapter 3);
vii) other biogenic components (macro and/or microfossils; palynological content);
viii) diagenetic aspects.

In subsurface studies, when core and well log data are available, the following information can be added to the description:

ix) porosity and permeability;
x) pore throat dimensions (McPhee et al. 2015);
xi) well log response (e.g. gamma ray, sonic and density); and
xii) high-resolution borehole image logs (useful for documenting bedding scale, style, and dip orientations for palaeocurrent analyses).

Finally, there is one element that is integral to all facies studies:

xiii) geological context, which should guide and inform, but not dictate, environmental diagnoses.

The majority of this work is captured by rock facies-scale evaluations (Section 2.3.2), although mudstones require microscopic methods to fully describe and understand particles, textures and sub-mm stratification styles (Figure 9.13; Section 2.3.3).

In siliciclastic facies descriptions, texture, primary sedimentary structures, bedding styles and gross sand/mud ratios are pre-eminent because they are closely tied to physical sedimentary processes, depositional energy and hydrodynamics. Biogenic sedimentary structures, when present, provide essential environmental insight, including substrate conditions, sedimentation patterns and water column aspects (e.g. salinity, oxygenation, see Chapter 3). Primary sedimentary structures (Collinson and Mountney

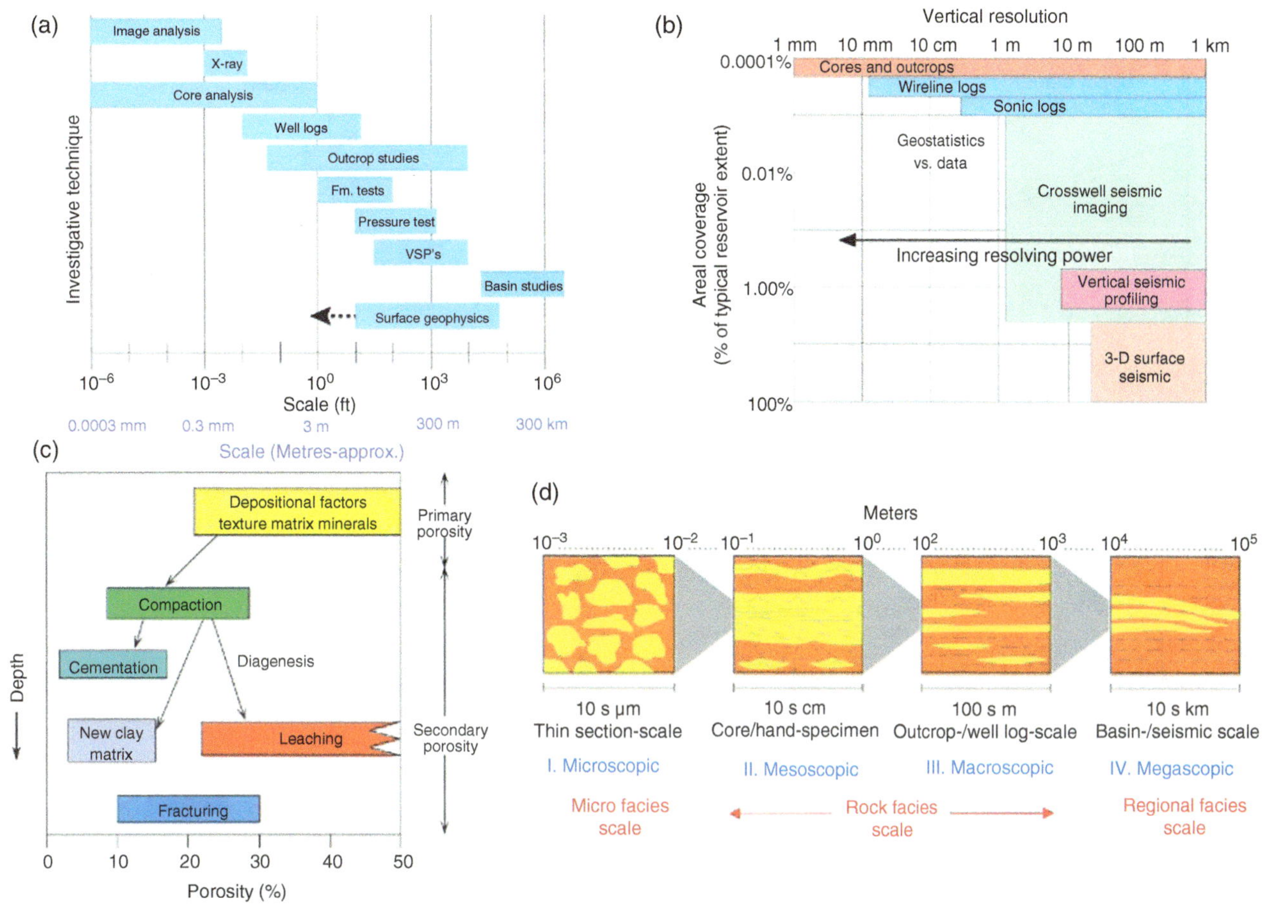

Figure 2.15 Types, scales and resolution of data used in facies analysis. (a) Range of scales typical for facies analysis, extending from pores to basins and involving the application of different analytical methods. (b) Typical reservoir-scale facies analysis, including those involving outcrop and/or subsurface reservoir characterisation, which combines data at a wide range of scales: for example, high vertical (mm-cm and m resolution) but low areal coverage is typical of cores, well logs and many outcrop studies, while low vertical (e.g. c. 10 s m) but high areal coverage is provided by 3D seismic surveys. (c) Understanding grain mineralogy, texture (grains/matrix) and pore-scale properties at the time of deposition is the starting point for assessing the impact of depth and/or compaction on porosity during burial, which requires microfacies-scale analysis. (d) The range of scales of different types of facies analysis (see Section 2.3). *Source:* Slatt (2006) / with permission of Elsevier.

2017) are major contributors to interpreting ancient flow conditions and inferring mechanisms of sediment transport. Our understanding of such process-response links is driven by a combination of measurements in modern environments and in more controlled conditions in flume tank experiments (Figure 2.16). The latter provide the basis for estimating flow conditions based on texture, bedform morphology and internal sedimentary structures (Allen 1982, 1985; Baas et al. 2016). Much of this information is synthesised in bedform stability ('phase') diagrams, which describe the movement of granular sediment in response to unidirectional, wave-generated oscillatory and combined (unidirectional and oscillatory) flows under varying flow velocity, stream power and sediment texture (Figure 2.16). However, direct observations are often hampered by accessibility, for example, measuring rare and extreme events. It is therefore no surprise that the measurement of processes and their facies responses advanced faster in terrestrial and coastal settings (e.g. fluvial, glacial, aeolian, deltaic and lacustrine environments), compared to the more challenging offshore-shelf ('shallow-marine') and deep-water (e.g. slope, submarine fan and abyssal plain) settings. Consequently, process and facies understanding from these latter environments relies relatively more on reasoned inference from the facies analysis of their rock records (Bouma 1962).

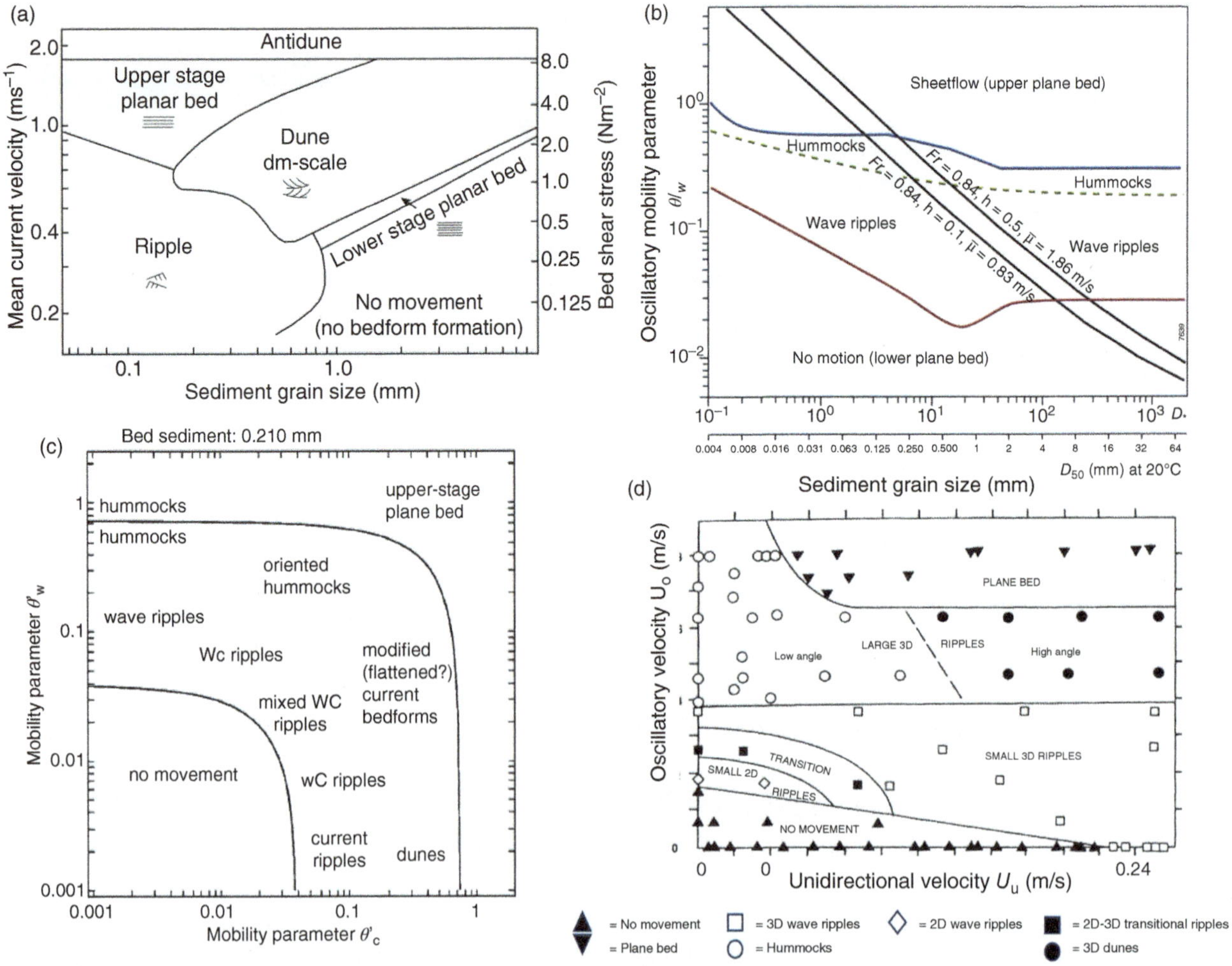

Figure 2.16 Bedform stability (or phase) diagrams with variable axes for different flow types. Most experimental flume tanks have water depths less than 1 m. (a) Bedform phase diagram for unidirectional currents for a flow depth of 0.25–0.40 m *Source:* Modified from Southard and Boguchwal (1990). (b) Bedform phase diagram for oscillatory currents (from van den Berg and Nio 2010, *Source:* Modified from Kleinhans (2005)). (c) Non-dimensional bedform phase diagram for combined flows, where the abscissa denoting an increase in oscillatory wave mobility and the ordinate an increase in unidirectional current mobility *Source:* Modified from Kleinhans (2005). (d) Experimental results of bedform stability fields in very fine sand for combined oscillatory and unidirectional flows; 'low-angle' refers to foreset dips of <10° and 'high-angle' >10°. *Source:* Arnott and Southard (1990) / Society for Sedimentary Geology.

Individual physical facies types are mostly indicative of either a single process (e.g. unidirectional flow) or a small set of genetically linked processes (e.g. waves and tidal currents), rather than to a particular environment are rare. For example, a facies composed of climbing current-ripple cross-laminated sandstone (description) can be interpreted to record the rapid migration of current-ripple bedforms in response to high sedimentation rates, including suspension fall-out, by unidirectional currents (process). However, this process can occur in a range of environments, including: (i) basin floor fan in bathyal water; (ii) lower coastal plain (levee deposits); (iii) prograding fluvial-dominated delta front. For further interpretation geological context is key. The alluvial levee interpretation, for example, would be indicated by adjacent *in situ* coal layers and/or palaeosols. The larger gross depositional environment context could also be provided by biological aspects, such as fauna, palynology and ichnofabrics.

Consequently, facies diagnostic of a single, specific, environment are relatively rare, but the link of facies to process is important. For example, combinations of several facies characteristics may be extremely diagnostic of process and therefore environment, as in the case of high-energy tidal sandstones (Figure 2.17).

As an illustration of the importance of context we take the example of the Early Triassic Budleigh Salterton Pebble Bed (Figure 2.18b; Figure 2.19). In this case, it is the additional evidence of associated pedogenesis plus an

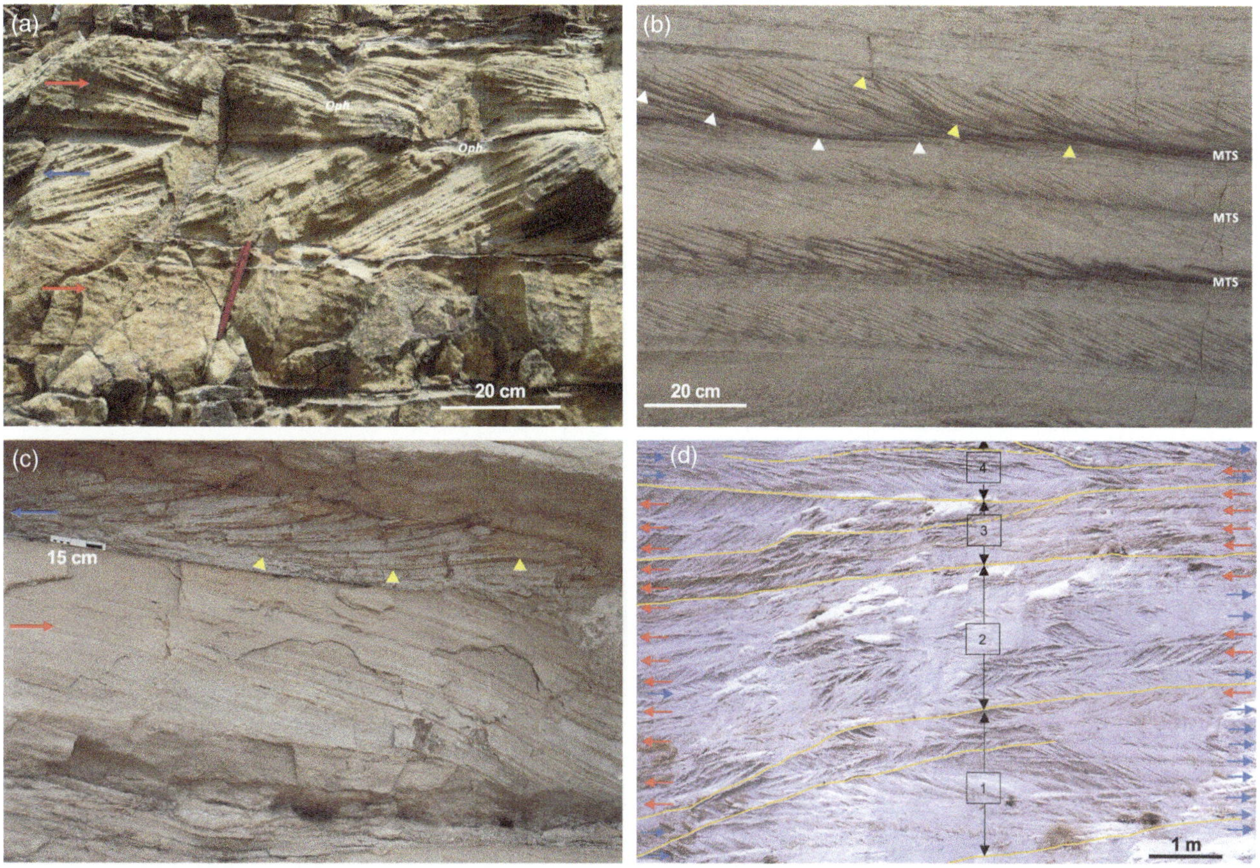

Figure 2.17 Facies-scale photographs illustrating typical features in tidal cross-bedding: (a) bidirectional cross-bedding ('herringbone' pattern: red arrow = fluvial/ebb-tide direction, blue arrow = flood-tide direction) in the middle Miocene Miri Formation (NW Borneo); note isolated *Ophiomorpha* burrows (Oph.). (b) Cosets of mud-draped cross-bed sets, with regular mud-drape spacing, shovel-shaped geometries (white and yellow arrow heads), mudstone-rich toe-sets (MTS) and reactivation surfaces (white block arrow) (Eocene Dir Abu Lifa Member, Western Desert, Egypt). (c) Bidirectional cross-bedding and muddy foresets in the Late Cretaceous Hosta Formation, San Juan Basin, New Mexico, USA (scale bar – 15 cm). (d) Stacked cross-bed cosets with bidirectional dips, including multiple herringbone sets and both upslope- and downslope-dipping cosets (Early Cretaceous Woburn Sands, southern England). *Source:* Courtesy of Howard Johnson.

understanding of the regional geological setting that confirms the environmental interpretation of a gravel-rich braided river. Sediment texture alone is a poor guide to environmental interpretation. For example, conglomeratic facies, result from deposition by a wide range of tractional, gravity-related and mass flow processes, and can form in many different environments (Figure 2.18). Often identification of primary sedimentary structures in very coarse-grained conglomerates (boulder-scale and larger) can be difficult but, when recognised, can confirm a particular depositional process. Large-scale cross-stratification in this example confirms tractional processes (Figure 2.19a). Context places this into a large, tectonically controlled, south-north-flowing, gravel-rich braided river system (Figure 2.19b, c).

The interpretation of siliciclastic facies in coastal-shallow-marine settings (e.g. delta front/shoreface/inshore/coast/shelf environments) is one of the most challenging. Sediments deposited in these settings reflect the variable interactions of fluvial, wave and tidal processes, which gives rise to a wide range of distinctive bedding styles (Reineck and Singh 1973). Early investigations resulted in criteria for distinguishing end-member environments (e.g. fluvial-/wave-/tide-dominated deltas), based on facies characteristics that supposedly identified the relative interaction of these three main process groupings (see Chapter 8). More recent work has shown much greater variability due to the diversity of processes and their interactions, especially those processes that dominate during peak energy levels, which are only rarely recorded in the modern (e.g. variable combinations of fluvial floods, spring tides and/or meteorological storms). Such combinations of processes can dominate in preserved records of sediment transport in preserved records of sediment transport

Figure 2.18 Contrasting types of conglomeratic facies illustrating the need for context in interpretation. *Source:* Courtesy of Howard Johnson. (a) Description: angular blocks of varied rock types (multi-mineralic and weakly metamorphosed sandstones, siltstone and slate, with limestone clasts shown in inset), extremely poorly sorted with widely varying clast size (1–20 cm) and shape, in a dark red mudstone/siltstone matrix; interpretation: rapidly deposited breccia resulting from flash floods in an alluvial fan/arid climate setting; context: Permian, basin-margin breccias (lateral, west to east sediment transport but only c. 10–20 km basinward extent), from the western margin of the Wessex Basin (SW England) – sediments are derived from local Carboniferous rocks from the adjacent, uplifted, Culm Basin. (b) Description: clast-supported conglomerate with well-rounded clasts of quartzite and a subordinate matrix of fine- to medium-grained sandstone in 1–3 m thick tabular cross-bed sets (see Figure 2.19a); interpretation: pebbly braided river deposits derived from Ordovician shallow-marine quartzites (monomineralic) from the Armorican Massif of NW France – long distance, bedload-dominated transport explains the well-rounded clasts with a significantly wetter semi-arid to temperate climate than for (a) context: Triassic axial rift valley-fill fluvial system forming part of a 100 s-km-long, south to north braid plain (extending from central-northern France to NW England/East Irish Sea Basin) – see Figure 2.19b, c; early Triassic, Wessex Basin. (c) Description: conglomerate with a fine-grained sandstone matrix and angular and often deformed, clasts of grey mudstone and silty mudstone, sometimes rich in carbonaceous debris and always extremely poorly sorted (evidence of auto-brecciation, loading and injection); interpretation: blocks of mudstone slumped into an unconsolidated sandy substrate with lateral transport of c. <10 m; context: collapse of a mudstone-rich cut-bank of a meandering river into point bar and channel floor sandstones of the Middle Jurassic Ravenscar Group of NE England (Whitby, East Cliff) – point bar sandstones and rooted levee deposits with dinoflagellates confirm the context of a meandering channel in a brackish water, lower delta plain setting (Figure 2.31). (d) Description: angular and variably sized clasts of dolomitic mudstone and subordinate/smaller extra-basinal clasts in a red-brown, silty mudstone-supported matrix that is lacking any internal sedimentary structures (massive); interpretation: rapidly emplaced mud-supported breccia, probably in sub-glacial depositional setting; context: part of the Neoproterozoic Smalfjord Tillite Formation in Finnmark, north Norway, where this facies is closely associated with a wide range of glacial features, including a striated pavement at the base of the formation (Photo credits Howard Johnson).

(e.g. Duke 1985; Duke et al. 1985). A related challenge is to distinguish the primary sedimentary structures resulting from unidirectional, oscillatory and combined (i.e. unidirectional plus oscillatory) flows (Dumas et al. 2005). We can begin with the results of flume experiments, which show hummocky bedforms resulting from high oscillatory wave conditions (Figure 2.16). But modern examples of the deposits of major oceanographic events, such as storms and hurricanes that can be tied to the events that deposited them, are still rare and, much like with turbidity currents and turbidites, much of our understanding stems from interpretation of the ancient record. For example, hummocky

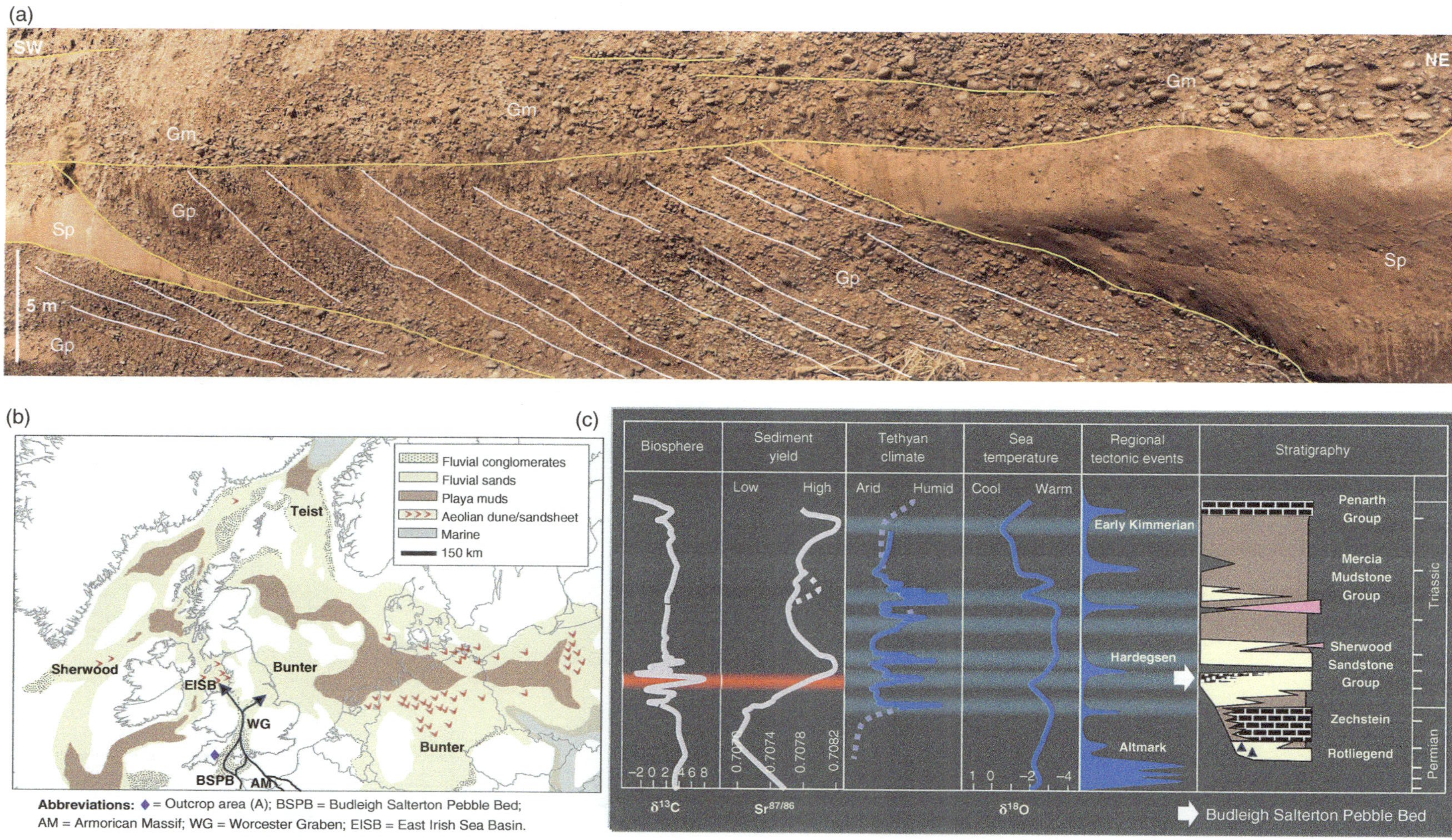

Figure 2.19 Pebbly/gravel-rich braided fluvial facies. (a) Description: a 3-m-thick, planar-tabular cross-bed set consisting of well-rounded, quartzite-dominated clasts (up to c. 25 cm diameter) with a fine- to medium-grained feldspathic sandstone matrix (Gp). This gravel-rich structure is bounded laterally by subordinate sandstone-dominated, planar-tabular cross-bedded facies (Sp). Palaeocurrents in both the Gp and Sp facies are towards the north-east. Above and below this unit are weakly stratified to cross-bedded gravel-dominated facies (Gm). Interpretation: deposition of large transverse, low-sinuosity dunes, probably in a gravel-dominated braid plain setting. Sediment transport was by bedload transport as indicated by the primary sedimentary structures and by the well-rounded nature of the clasts (Figure 2.18b). Context: Part of the Early Triassic Budleigh Salterton Pebble Bed (BSPB). The quartzite pebbles contain Ordovician-age brachiopods. (b) Early Triassic palaeogeography, highlighting the location of the outcrop in A (purple diamond) and the extent of the large BSPB river system (*Source:* McKie and Williams (2009) / John Wiley & Sons; Morton et al. 2016). Facies types are typical of a semi-arid, terrestrial setting. Deposition coincided with the onset of the break-up of Pangea, with the northward-flowing BSPB river draining a hinterland comprising Lower Palaeozoic rocks in the south (Variscan Massif – AM). The BSPB river system as at least 500 km-long and followed the newly created Worcester Graben (WG). (c) Summary of the main geological factors influencing Permo-Triassic deposition and its stratigraphic framework in NW Europe (*Source:* Based on McKie and Williams 2009; Gallois 2019); courtesy Tom McKie. Photo credit A Howard Johnson.

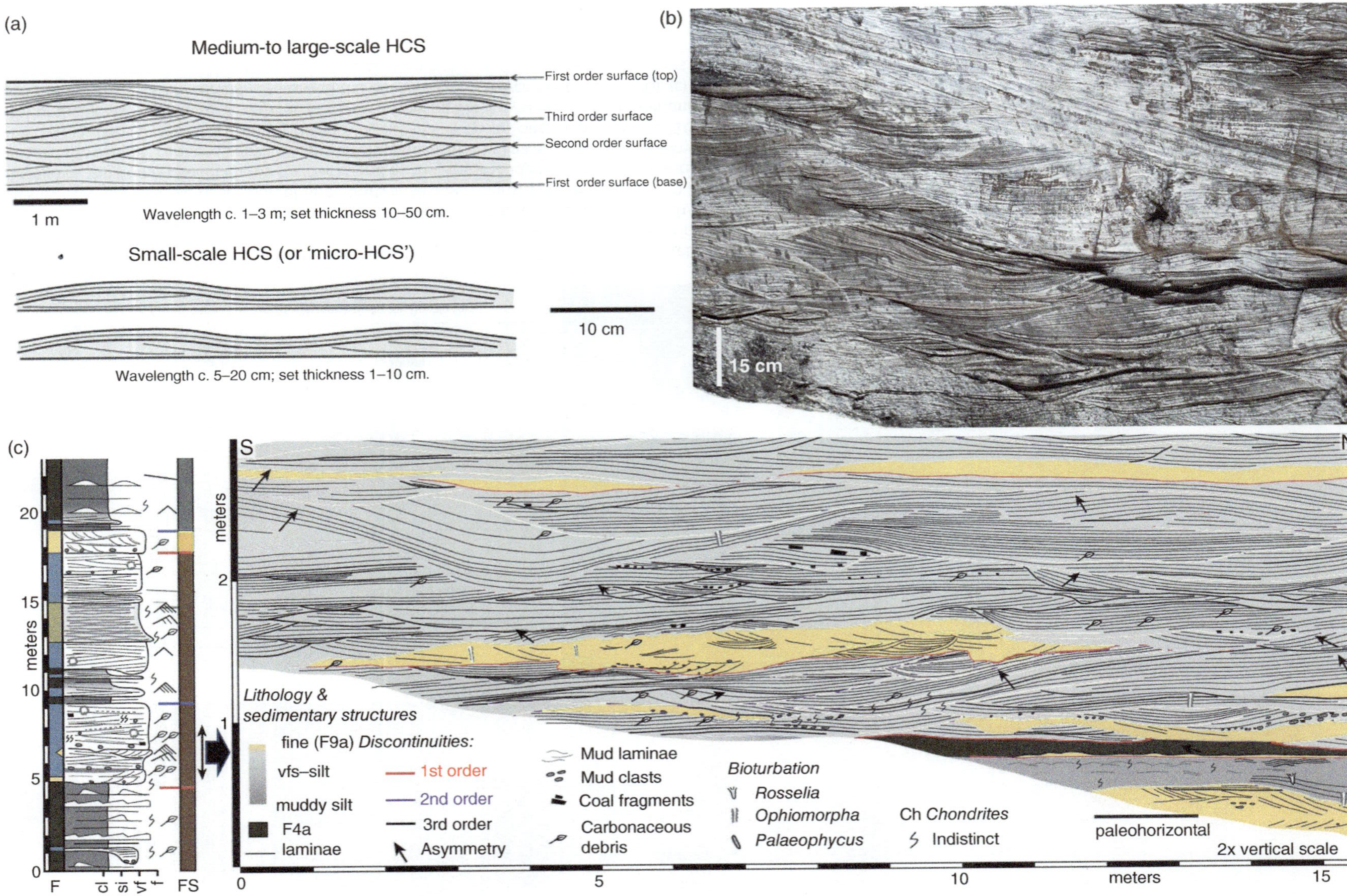

Figure 2.20 Facies evidence supporting wave-/storm-dominated, oscillatory to combined flow processes. (a) Sketches of hummocky cross-stratification (HCS), illustrating the lateral co-existence of mounds ('hummocks') and troughs ('swales'), plus distinctive mm-scale lamination, which defines this bedform and its internal structure. Individual laminasets typically have low-angle dips (<10°), they drape both mounds and troughs and have cross sections ranging from symmetrical (isotropic – with multiple, equally oriented dip directions = HCS *sensu stricto*) to asymmetrical (anisotropic – preferentially inclined in one direction, often with slightly steeper dips of c.10–15° and greater preservation of troughs = swaley cross-stratification or SCS *sensu stricto*). Isotropic forms are held to be typical of oscillatory-dominated flows, while anisotropic forms reflect combined flow-dominated regimes (*Source:* Adapted from Tinterri 2011). (b) Photograph of HCS-like structures. Note upward increase in set size (from smaller to larger wavelengths), asymmetrical dips, with infilling of troughs dominant relative to the draping of mounds – this is typical of the SCS variant and implies combined flows. From the Middle Jurassic Scarborough Formation, NE England, a transgressive interval that preserves the distal toes of prograding delta fronts (for depositional context see Gowland and Riding 1991). *Source:* Courtesy of Howard Johnson. (c) A sedimentary log and accompanying line drawing of hummocky to swaley cross-stratification (HCS to SCS) in a 15 m-long panel (up to 3 m-thick unit). The asymmetry of the scour-fills and the intercalated cross-bed sets support a combined flow regime setting. In this example, intense storm-related, oscillatory waves are interpreted to have been augmented by fluvial flood discharge, as indicated by abundant mangrove-derived, carbonaceous debris (Miocene proximal delta front environment from the Belait Formation of Brunei, NW Borneo). *Source:* Collins et al. (2018) / Society for Sedimentary Geology.

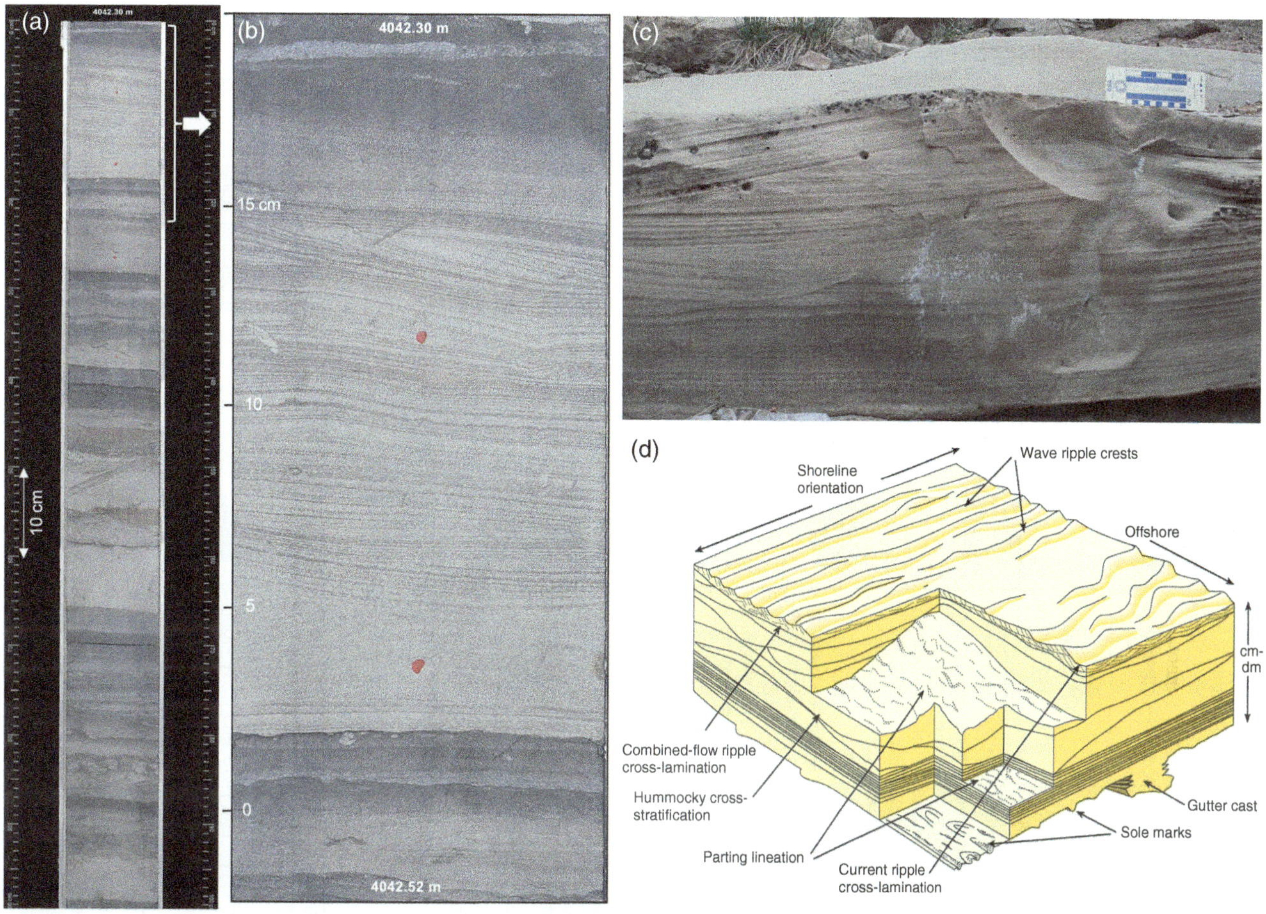

Figure 2.21 Examples of thin-bedded hummocky cross-stratification (HCS) and its depositional significance. (a) A one meter-thick cored interval of multiple, thin-bedded, HCS-dominated sandstones (pale grey colour) intercalated with low to moderately bioturbated mudstones in Middle Triassic deltaic deposits, Bedout sub-basin, NW Shelf, Australia. Arrow highlights close-up of one single storm-event bed shown in B. Courtesy Geoscience Australia. *Source:* National Offshore Petroleum Titles of Australia / https://www.nopta.gov.au/copyright.html / last accessed on April 07, 2025. (b) Description: example of a single storm-generated event bed comprising hummocky cross-stratification (HCS), passing gradationally upwards into wave-ripple/micro-HCS. The bed (c. 16 cm thick; 5 cm interval markers on left side of b) fines upwards, it has a scoured base and contains mm-scale low-angle and variably dipping lamination. Interpretation: a single storm-event bed. It reflects intense, storm-generated wave reworking at the seabed with deposition above the storm-wave base; evidence of combined flow towards the top. Context: facies association comprises the distal part of a fluvial/wave-storm-influenced delta front or a distal lower shoreface. For geological context, see: Allgöwer and Lignum 2019; Minken et al. 2019; Thompson et al. 2019. Courtesy Geoscience Australia. *Source:* National Offshore Petroleum Titles of Australia / https://www.nopta.gov.au/copyright.html / last accessed on April 07, 2025. (c) Description: an outcrop example displaying similar bedding styles, dominated by hummocky cross-stratification. Context: distal lower shoreface deposits, Late Cretaceous wave-dominated shoreface/delta front deposits (Spring Canyon Member, Gentile Wash, Book Cliffs, Utah; Hampson et al. 2001). *Source:* Photo by Howard Johnson; Courtesy of Howard Johnson. (d) Depositional concept: idealised geometry and bedding styles of a storm-generated sandstone bed (*Source:* Leckie and Krystinick (1989) / Society for Sedimentary Geology), but compare with inferred combined flow HCS beds in Figure 2.22.

cross-stratification (HCS) is indicative of storm-dominated processes based on a suite of distinctive features (Figures 2.20 and 2.21). However, recognition of the variability of 'HCS-like facies' plus its preservation context, based on facies analysis (e.g. Figures 2.20b,c, and 2.21), has resulted in a much-improved understanding of the relationship between HCS formation and fluvial flood events, especially in delta front settings (Figure 2.22).

The microfacies analysis approach to fine-grained (mudstone-/siltstone-dominated) facies has seen a revolution in recent years (Chapter 9). Many studies from diverse stratigraphic intervals have revealed that mudrocks are far from uniform, low-energy and simple suspension fall-out deposits. This is illustrated in our oldest case study, the Mesoproterozoic Roper Group (Figure 2.23, Section 2.6.1).

2.2.1.2 Carbonate Facies

The importance of biological, ecological and geochemical processes implies that there are additional criteria for carbonate facies schemes. For example, sedimentary structures and trace fossil types are complemented to a greater

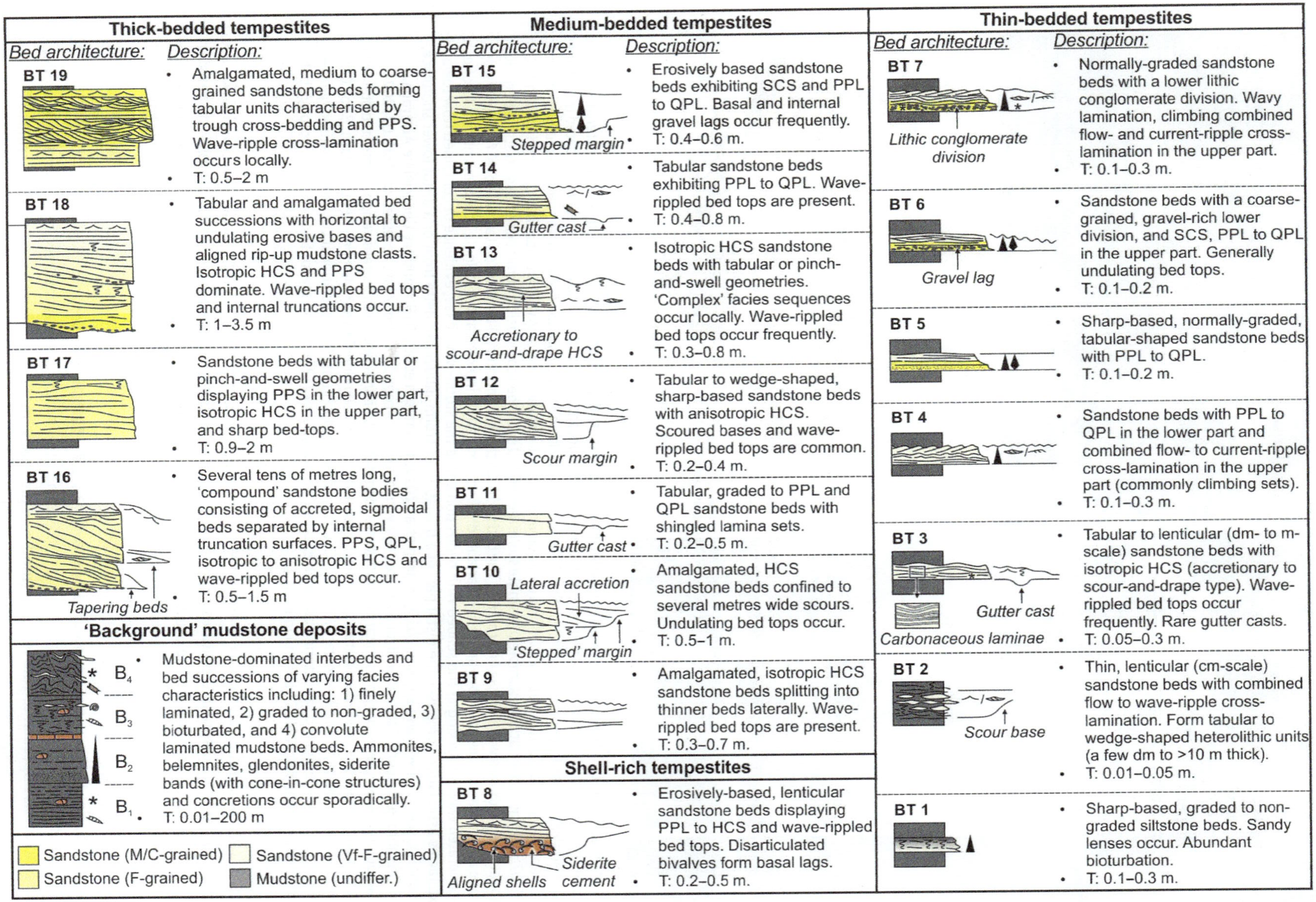

Figure 2.22 Example of a facies classification scheme for HCS-dominated storm deposits based on outcrop studies of Early Cretaceous shallow-marine successions in Svalbard (from Grundvåg et al. 2021). See also Jelby et al. (2020). Isotropic HCS is interpreted to reflect oscillatory flow-dominated storm processes. Anisotropic HCS, often with basal gutter casts, is interpreted to reflect combined flows, including hyperpycnal flows linked to river mouths and coupled storm-flood processes. Abbreviations: HCS = hummocky cross-stratification; PPS/PPL = planar parallel stratification/lamination; QPL = quasi-planar lamination; SCS = swaley cross-stratification. *Source:* Grundvåg et al. (2021) / John Wiley & Sons / CC BY 4.0.

degree by sedimentary fabrics, cement types, palaeoecological information from the fossils and geochemical data. Collectively, these allow diagnosis of more than just physical processes, while analysis of the (non-) skeletal grains and the matrix (micrite or sparite), can alone enable interpretation of the depositional environment. On the other hand, complexity arises from (i) the highly variable source-to-sink distances for particles (hundreds of metres to tens of km range), and (ii) the wide range in grain-size spectra due to particle production processes in addition to current strength. A detailed facies analysis helps decipher whether the grains and muds were deposited close to their sites of production or transported from more distant production sites (distances of 40 km or more; Andresen et al. 2003; Jorry et al. 2020).

Carbonate facies and their distribution patterns (e.g. patches or belts) are additionally controlled by the primary particle properties (such as particle shape and buoyancy as these interact with hydrodynamic processes (see de Kruijf et al. 2021 and Slootman et al. 2023, for summaries). Hence,

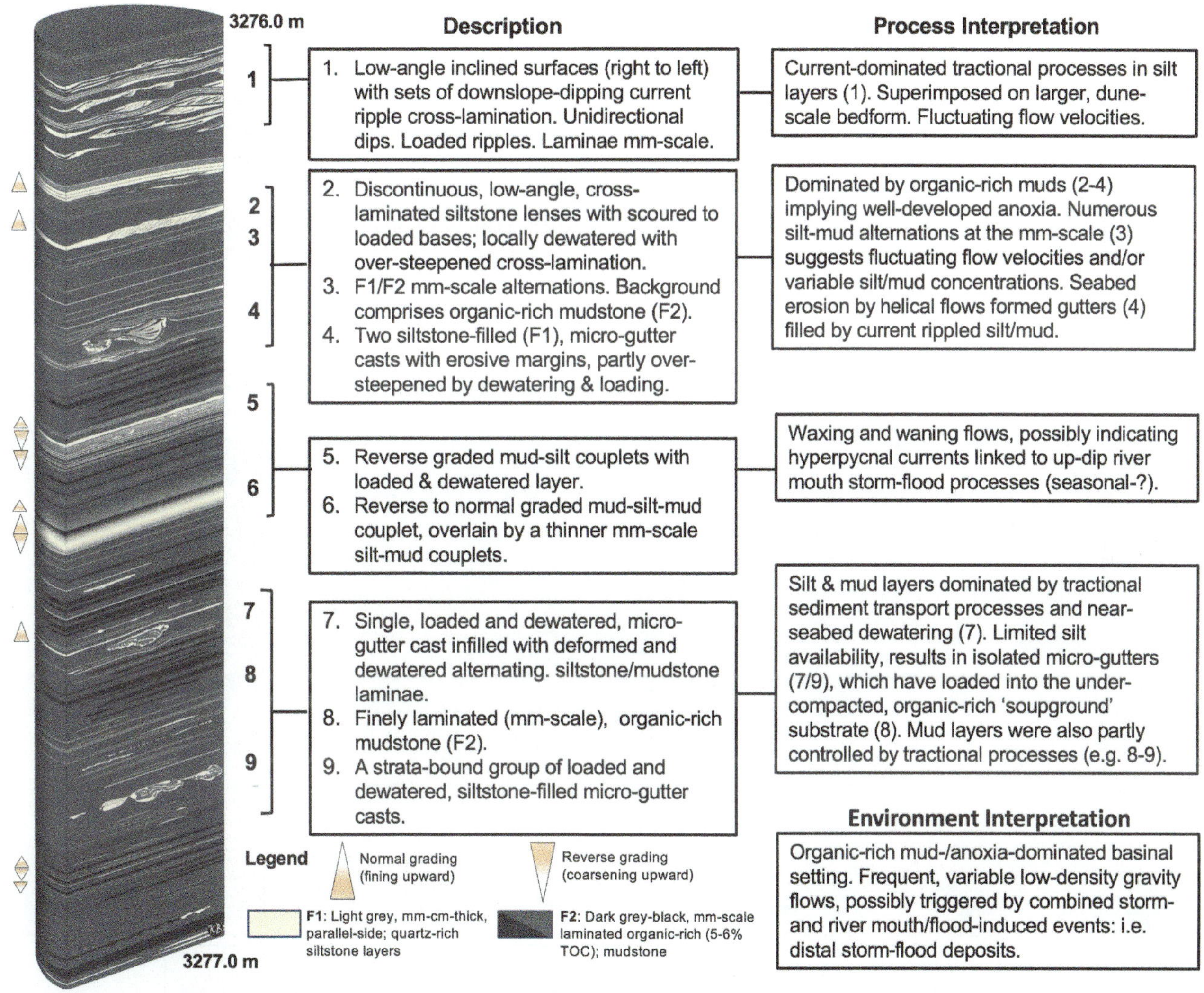

Figure 2.23 A detailed line drawing of a c.1 m-long section of slabbed core from the Mesoproterozoic Middle Velkerri Formation in well Tanumbirini-1 (courtesy of Kerrie Bann). The succession consists of mm-cm-scale alternations of pale grey quartz-rich siltstones (F1 – pale yellow) and dark grey organic-rich mudstones (F2 – black/grey shades) (part of FA4 in Case Study-1). The section shows examples of primary stratification styles, comprising normal- and reverse-graded siltstone/mudstone couplets (orange-white triangles), current-ripple cross-lamination and micro-gutter casts. Loading- and other dewatering-related soft sediment deformation structures are locally common. Highlighted laminae and laminasets are numbered (1–9). This interval is interpreted as being part of a basin floor environment (see 'Environmental Interpretation'). Core width is 90 mm. See Section 2.6.1 for geological context and for further details (*Source:* Johnson et al. (2024) / PESA.)

carbonate facies are less 'standardised' than siliciclastic facies, being subject to biotic change over time (on both evolutionary and shorter timescales).

The depositional profiles of carbonate platforms vary between flat-topped, shelf, atoll, mound and ramp (Read 1985; Handford and Loucks 1993). These profiles together with parameters such as (i) land-attached or fully marine setting, (ii) windward or leeward orientation, (iii) shallow- and deep-water hydrodynamics, (iv) tectonic setting as well as (iv) climate setting introduce a wide variety in facies distribution patterns (Bosence 2005). Present large-scale, tropical carbonate systems include carbonate ramps as present in the Persian Gulf (Arabian Gulf) (Purser 1973) and flat-topped carbonate platforms with varying topographies, such as the Bahamas, Maldives, Saya de Malha Bank and Great Barrier Reef (Laugié et al. 2019). Facies on carbonate ramps include coastal sabkhas, microbial mats, mixed flat, mud flat, channel deposits (mixed skeletal grain and pellet), oolitic sand and coral reef (Figure 2.24). Diverse facies zones occur within this depositional system (Figure 2.25). Facies on flat-topped carbonate platforms (Figure 2.26) are marked by reefs, sand patches (with both non-skeletal and skeletal grains) and islands at the edge of the platform together with tidal inlets. The interior facies may include mangroves, mud flats, patch reefs, various sediments (mud- to packstones), but also wash-over fans attached to the barrier reef and beach rocks along the shoreline depending on the setting of the barrier reef, windward or leeward (Figure 2.27). Facies on the slopes of these carbonate systems are also highly variable and include muddy slope wedges, microbial-influenced upper-slope sediments, channels and periplatform drift deposits (Wilber et al. 1990; Grammer et al. 1991; Betzler et al. 2014; Reolid et al. 2017; Fauquembergue et al. 2024).

A distinctive feature of carbonate sediments is that they are rapidly cemented (Grammer et al. 1993; Dravis 1996), not only when they become exposed to fresh water but also in shallow-water during early burial diagenesis; a process in which microbial organo-mineralisation also plays an important role. As a result, a wide variation in discontinuity surfaces can develop in shallow-marine carbonate platform deposits as shown, for example, in the Jurassic of France and Switzerland (Figure 2.28).

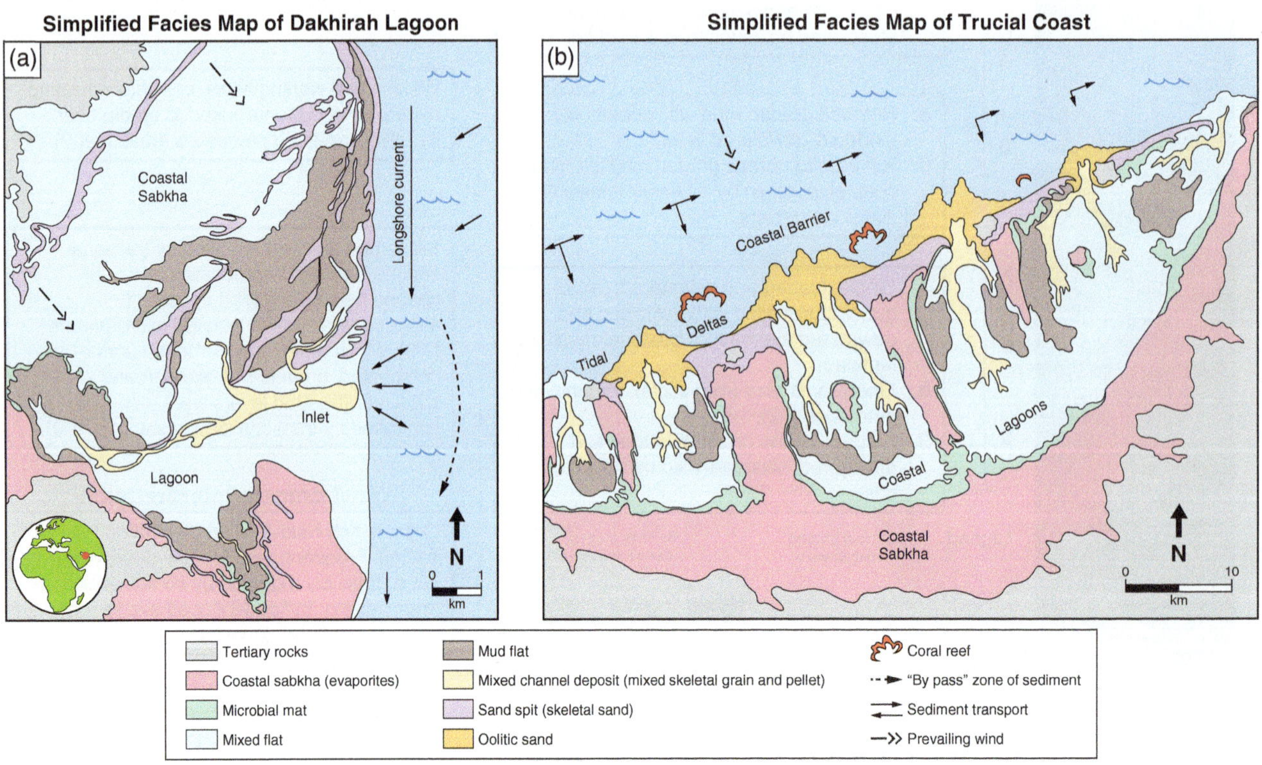

Figure 2.24 Facies models of a carbonate ramp lagoon, Persian/Arabian Gulf. *Source:* Billeaud et al. (2014) / GeoScienceWorld. This figure shows comparative facies models for lagoonal systems along the South Persian/Arabian Gulf coast represented by (a) the Al Dakhirah lagoon (Qatar), and (b) the UAE; after Purser and Evans 1973. The different facies distribution patterns relate to differences in hydrodynamic processes and the pre-existing coastal morphology. The Al Dakhirah lagoon is protected from the dominant NE winds, while the UAE example is fully exposed to these winds. The Al Dakhirah lagoon is also influenced by longshore drift currents whereas the UAE Coast is affected by direct wave action and cross-shore sediment transport. The combined effect of the differences between the two lagoonal systems is the development of completely different facies patterns. Note the different scales for both systems.

Figure 2.25 Facies zones in carbonate ramps of the Persian/Arabian Gulf. (a) Microbial mats with a distinctive 'cat ear' morphology, intertidal zone, backreef near Al-Ruwais (north of Qatar) surrounded by mangroves. Image width approximately 2.5 m. Photo credit: Ruqaiya Yousif. (b–g) Abu Dhabi sabkha system, near Al-Qanatir Island, UAE. From Strohmenger et al. (2010, 2011). (b) Outer, upper sabkha sub-environment. Close-up of buckled polygonal halite crust displaying desiccation polygons and teepee structures. (c) Transition of low upper sabkha to middle sabkha. Close-up of middle sabkha ('cottage-cheese-like') oxidised anhydrite polygons. (d) Upper intertidal microbial mat environment. Close-up of upper intertidal, thin, leathery, crinkled microbial mat. (e) Upper intertidal microbial mat environment. Trench showing from base to top: (i) hardground; (ii) crinkly laminated microbial mat (displaying minor erosive/burrowed upper surface and reworking); (iii) gypsum mush facies; and (iv) surface crinkled microbial mat. (f) Lowermost intertidal to shallow-subtidal, peloid-skeletal tidal-flat environment. Close-up of the transition between lower intertidal, tufted or cinder-like microbial mat and lowermost intertidal to shallow-subtidal, peloid-skeletal tidal-flat environment. (g) Partially exposed ooid tidal bar separating the lagoon (left) with oncolite(s) from a tidal channel (right). Hudairiat Island, UAE. (h) *Pirenella cingulata* gastropods (1–2 cm in size) commonly found in subtidal to intertidal settings. Tide-dominated coastal embayment, Khor Al Adaid, Southern Qatar. Photo credit: Ruqaiya Yousif (i) *Porites harrisoni* coral covered with *Hormophysa* brown algae and sea urchins (30 cm size). Offshore northeast Qatar, near Al Maroona Beach, water depth approximately 2–3 m. Photo credit: Ruqaiya Yousif.

2.2.2 Facies Associations and Architectural Elements

2.2.2.1 Siliciclastics

Facies typically occur in recognisable and repeatable combinations, which form larger sediment or rock volumes (m to tens m thick), and are referred to as facies associations (stage 3 in our hierarchy in Table 2.1). Whereas individual facies place emphasis on process interpretation, facies associations group genetically related facies and use their process significance to develop an environmental diagnosis. When the facies associations are bounded by gradational contacts, then Walther's law can be applied. There are no hard and fast rules, nor scale limits, for combining facies into facies associations.

Facies associations are here illustrated using outcrops from the c. 8 km-thick, strongly cyclical Neogene to present-day succession of the Baram Delta Province of NW Borneo. This succession is dominated by multiple, nested

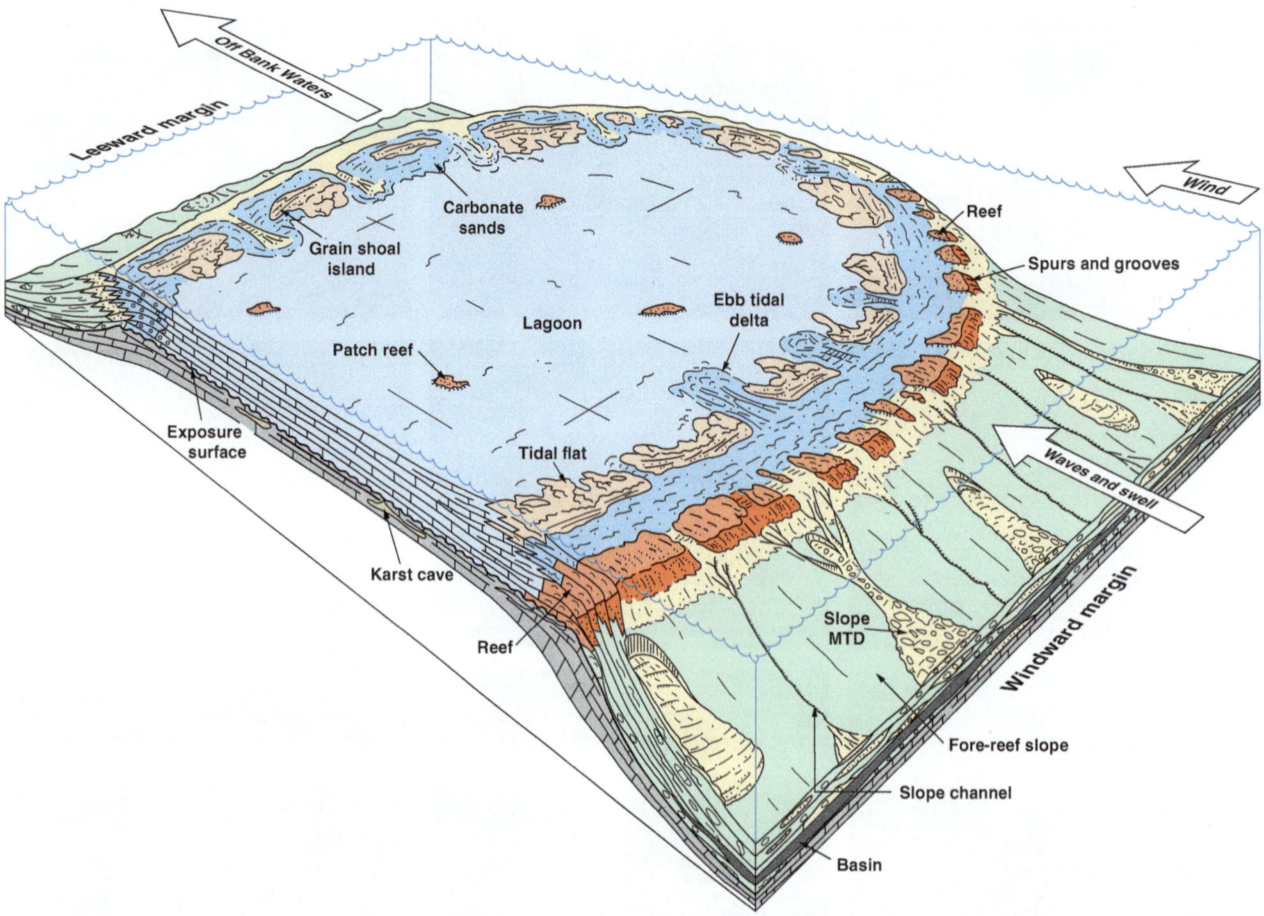

Figure 2.26 This idealised depositional model of a detached, humid carbonate-rimmed shelf (flat-topped platform) (modified from Handford and Loucks (1993) shows (i) lagoon with patch reefs and mud-dominated sediments; (ii) a barrier reef with grain-shoals, tidal flats and ebb tidal deltas; (iii) reef passages, some of which are connected to slope channels; (iv) a slope with slope channels and mass-transport deposits (MTD), reefs; and (v) onlapping basin sediments. Note the exposure surface and associated karst caves as well as the differences in sediment export patterns for the windward and leeward sides of the platform.

coarsening/sandier-upward units, which stack together to form km-thick progradational intervals bounded by regional marine flooding surfaces (see Section 2.6.2, Case Study-2 for an outcrop to subsurface example). The smallest-scale units comprise c. 10–30 m thick, coarsening/sandier-upward facies successions, which are bounded by marine mudstones. The example in Figure 2.29 consists of nine facies, ranging from mudstone-rich variants (F1–5) at the base through to increasingly sandstone-dominated examples (F6–9) towards the top (Figure 2.29a,b). Mudstone facies show evidence of high energy and rapid deposition, including intercalated thin (cm-scale) sandstone beds displaying evidence of oscillatory and combined waning flow events, while restricted trace fossil assemblages and benthic foraminifera support an environment with physico-chemical stress (Figure 2.29b). Sandstone facies indicate a high wave-/storm-dominated regime with combined flows (e.g. anisotropic hummock/swaley cross stratification), prominent erosion by seaward-directed combined flows (e.g. gutter casts) and abundant reworked terrestrial plant debris (Figure 2.29c). Based on sandstone/mudstone proportions, sandstone bed thickness and frequency, and degree of sandstone bed amalgamation, three distinctive facies associations are recognised (Figure 2.29d): (i) FA-1 is mudstone-dominated and comprises Facies 1–5, (ii) FA-2 alternating mudstones and sandstones, comprising Facies 3–5 and F6–8, and (iii) FA-3 sandstone-dominated, comprising Facies 2–3 and Facies 6–9. The gradational nature of changes in facies characteristics within an association (e.g. scale of cross-stratification, bed thickness, ichnofabrics and/or grain size) plus gradational facies boundaries between facies associations confirms the inference of 'genetic association' between them (Figure 2.29).

A critical assessment of potential sedimentary environments often involves several iterations. The gradational

Figure 2.27 Facies from flat-topped carbonate platforms. (a) Coral reef system of New Caledonia with lagoon (upper left), back reef, island, fore reef and spur and groove system marking the transition to the slope (lower right) at Huon Atoll situated NNW of Grande Terre. Photo by Khaled bin Sultan Living Oceans Foundation. (b) Coral reef and lagoon system (upper right) with forereef with spur and groove system of the fore reef. Clear facies zonation fore reef, barrier reef with breaking waves, back reef and lagoon (upper right) at Cook Reef, New Caledonia. Photo provided by the Khaled bin Sultan Living Oceans Foundation. (c) Photograph of a living *Halimeda* meadow near Lizard Island in the northern Great Barrier Reef (GBR). McNeil et al. (2021) / John Wiley & Sons / CC BY 4.0. (d) Detail *Halimeda* calcareous green macroalgae at same location (bottom). *Source:* Courtesy of E. Kennedy. From McNeil et al. (2020). (e) Scott Reef, Western Australia. Upper reef slope exposed to wave action. *Source:* Courtesy of Charlie Veron. (f) Southern shoal complex at the Crooked–Acklins Platform (Rankey and Reeder 2010). Details of representative bottom types shown in photos g–j (*Source:* Rankey and Reeder (2010) / John Wiley & Sons). (g) Underwater photograph of burrowed sand with scattered seagrass and calcareous algae from the platform interior. Seagrass is c. 10–15 cm tall. (h) Underwater image of *Acropora palmata* reef on the northern reef margin. Photo image is c. 20 m wide. (i) Broad supratidal plain that forms just south of Crooked Island, including areas of mangrove swamps and cemented crusts. The mangrove bushes in the foreground are c. 30 cm tall. (j) Oolitic shoal crest, exposed at spring low tide. Ripples in the foreground are <10 cm high. (k) Morphological characteristics of the Holocene uppermost marginal foreslope based on a sea-bed image of the northern Little Bahama Bank. Note that the slope includes updip bypass (rippled sands) or erosion (rocky surfaces, terraces and escarpments) and downdip accumulation (onlapping Holocene wedge). *Source:* Rankey and Doolittle (2012) / John Wiley & Sons.

Surface Groups	Characterization							Interpretation
	Geometry	**Lateral Extension**	**Morpho-logy**	**Biological Activity** at/below/ (above) surface	**Minera-lization**	**Facies and Contrast** across surface	**Early Diagenesis** below surface	
1 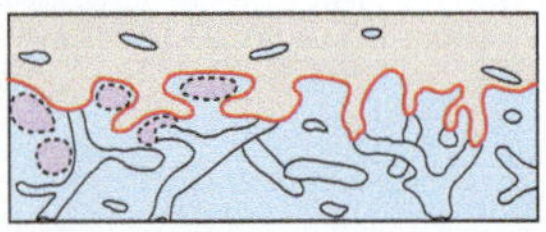	Erosional: cutting primary structures	100 m–1 km	Irregular: relief often < 5 cm	Bioturbation suites: intense preomission, moderate-intense omission; all *Thalassinoides*	Weak FeOx	Grain size composition microfacies	Beginning marine phreatic cementation fibrous aragonite	**Firmground to incipient hardground in subtidal environment**
2 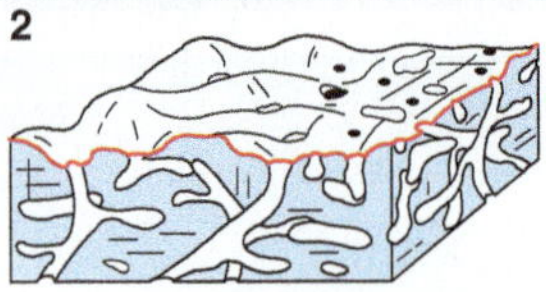	Weakly erosional	> 100 m	Flat to undulating	Bioturbation suites: intense preomission, intense omission, weak-moderate post-omission: all *Thalassinoides* boring, encrusting: *Lithophaga*, Oysters	Weak-strong FeOx MnOx possibly phosphates or sulfides	grain size composition microfacies	marine phreatic cementation fibrous aragonite microbial activity	**Hardground in subtidal low-energy environment**
3 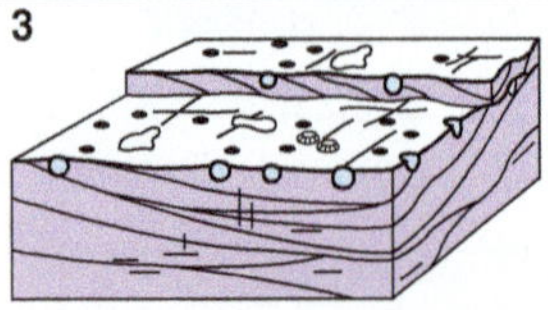	Weakly erosional	< 1 km	Flat	Bioturbation suites: occasionally weak omission, *Thalassinoides* boring, encrusting: *Lithophaga*, Oysters	Weak FeOx	Grain size	Marine phreatic cementation fibrous aragonite microbial activity	**Hardground in subtidal low-energy environment**
4 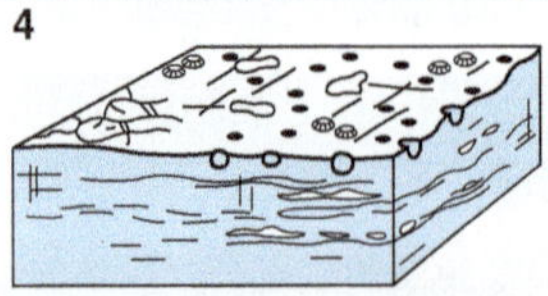	Weakly erosional	100 m–1 km	Flat	Bioturbation suites: weak preomission; (unidentified) boring, encrusting: *Lithophaga*, Oysters	Weak FeOx	Composition	Cementation microbial activity calcretization (desiccation) occasionaly relative depletion in ^{13}C and enrichment in ^{18}O	**Hardground in inter-supratidal environment**
5 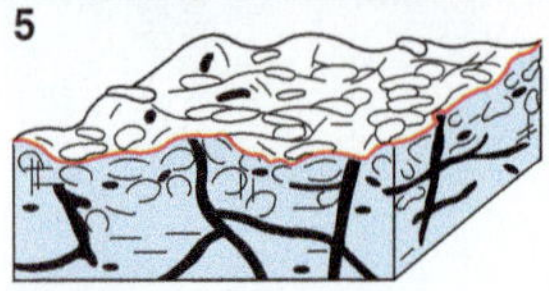	Weakly erosional	> 1 km	Undulating often brecciated	Bioturbation suites: weak-intense preomission: (unidentified) rhizoturbation: colonization by plants	Stron FeOx	Grain size composition microfacies	Pedogenesis microbial activity dolomitization calcretization silification relative depletion in ^{13}C and enrichment in ^{18}O	**Palaeosol commonly super-imposed on subtidal facies**
6 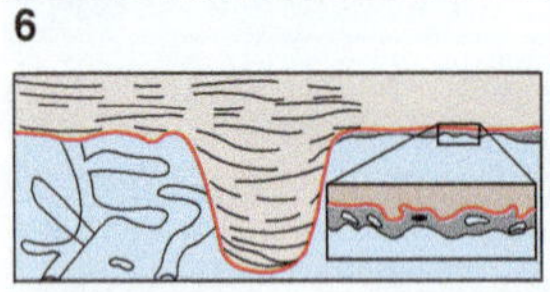	Erosional: cutting primary and secondary structures	> 1 km	Flat undulating jagged pitted etched cavities	Bioturbation: weak-intense	Strong FeOx penetrative staining	Grain size composition microfacies	Vadose cementation dissolution association with pedegenesis, gravitation cements and sediment infill in cavities	**Microkarst/ epikarst (karst)**
7 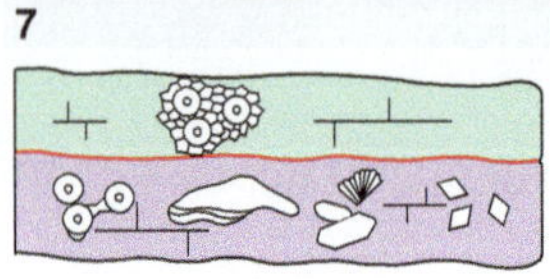	Depositional (erosional)	< 1 km	Flat	Bioturbation: weak-intense		Microfacies grain size	Cementation contrast across the surface often vadose cementation, dolomitization, silification below discontinuity	**Diagenetic discontinuity**
8 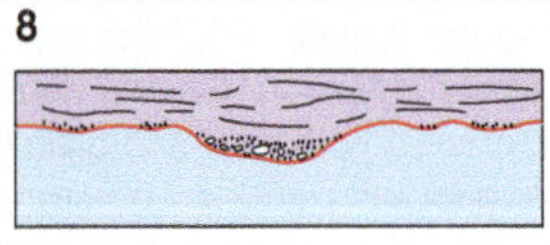	Erosional	< 1 km	Undulating irregular	Bioturbation: weak-intense		Grain size composition microfacies	Possible cementation contrast across the surface due to different rock properties	**Erosion surface in sub- intertidal environment**
9 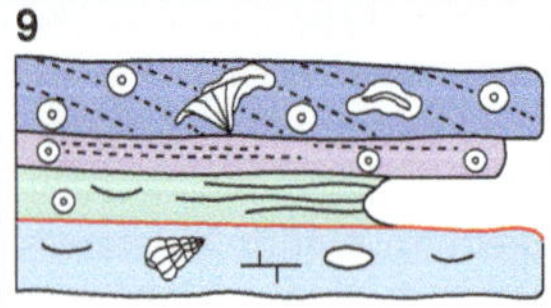	Depositional: preserving primary structures	< 1 km	Flat undulating	Bioturbation: weak-intense		Grain size composition microfacies	Possible cementation contrast across the surface due to different rock properties	**Simple discontinuity surface in sub-intertidal environment**

Figure 2.28 Groups of surfaces and their facies, biological and diagenetic characteristics based on Early Cretaceous shallow-marine carbonate platform deposits of the French and Swiss Jura. *Source:* Hillgärtner (1998) / SEPM Society for Sedimentary Geology.

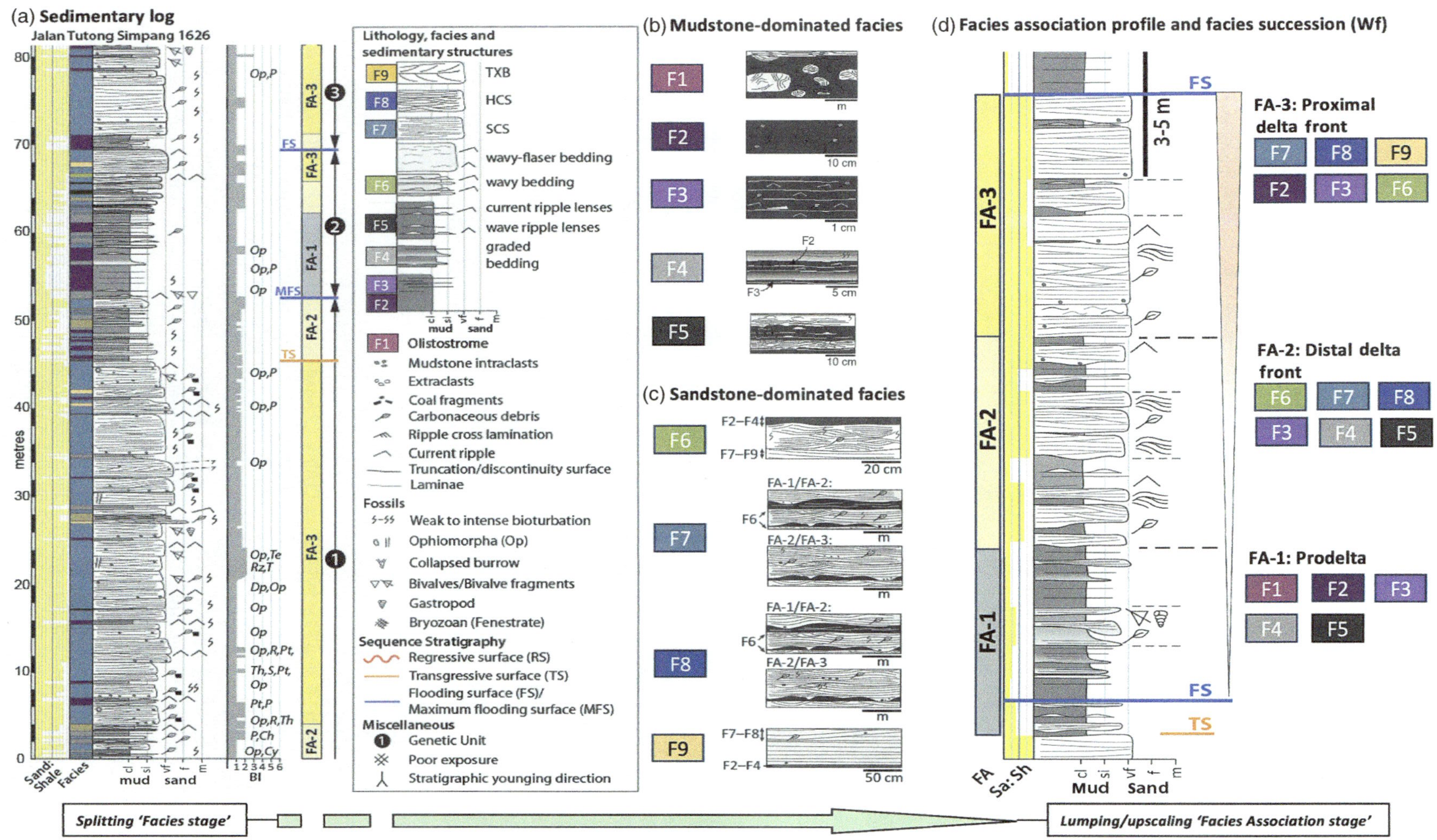

Figure 2.29 An example of the hierarchical facies analysis workflow (working from left to right): Facies description (sedimentary log) → Facies classification (F1–9) → Facies associations (FA1–3) → Facies successions → Key stratal surfaces. See Collins et al. 2017, for further details; from the middle Miocene Belait Formation outcrops, Brunei, NW Borneo. (a) Sedimentary log comprising (from left to right): (i) sand-shale ratio (sand % in yellow); (ii) facies (F1–9; see legend); (iii) grain size/structures column (hand-drawn) plus additional features (symbols) to right; (iv) bioturbation index (BI) with abbreviations of main trace fossil types; (v) facies association column (FA1–3) plus key stratal surfaces (MFS = maximum flooding surface and TS = transgressive surface); (vi) facies successions/genetic units; (vii) legend *Source:* Collins et al. (2017) / John Wiley & Sons / CC BY 4.0. (b) Thumbnail sketches of bedding styles in mudstone-dominated facies. (c) Thumbnail sketches of bedding styles in sandstone-dominated facies. (d) Typical coarsening/sandier-upward facies succession comprising (bottom to top) FA-1 → FA-2 → FA-3 (dominant facies types in coloured boxes), interpreted as a wave/storm-dominated, fluvially influenced delta (Wf) (Facies succession 3 in Figure 8.33a). Courtesy D.S. Collins.

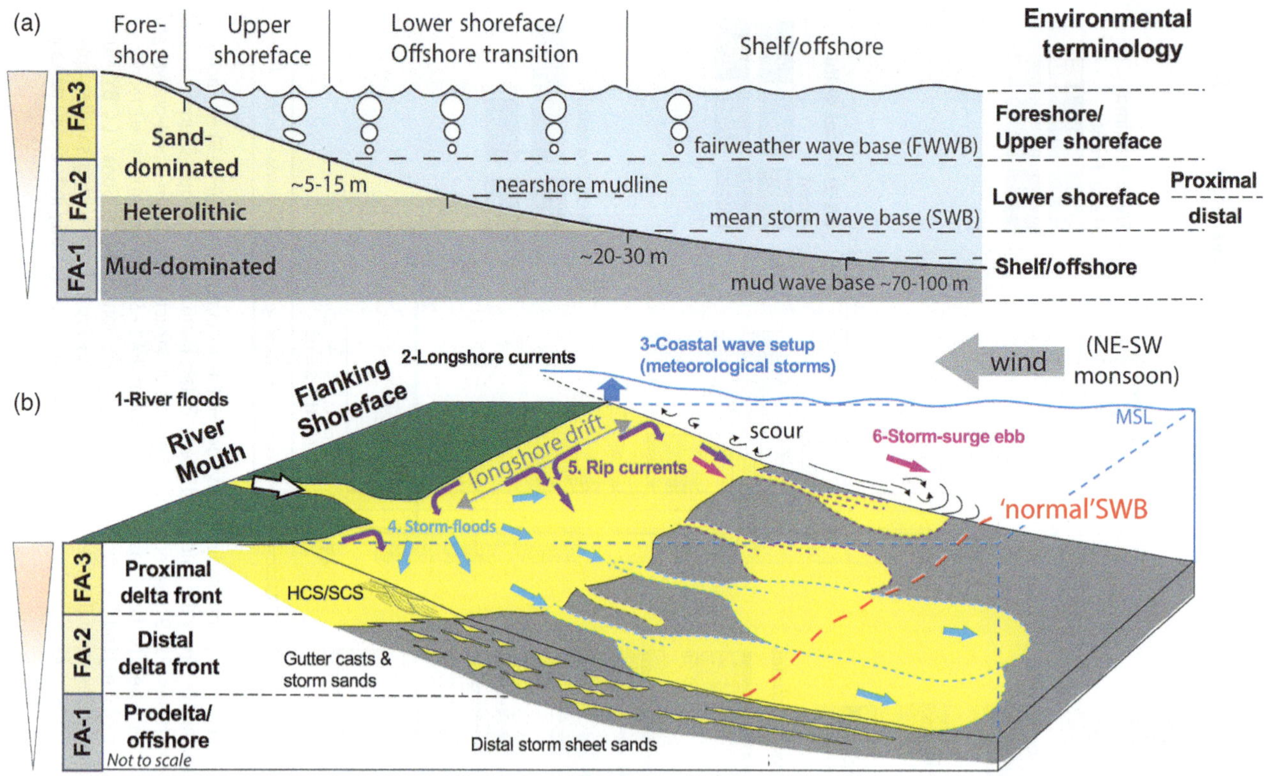

Figure 2.30 Two types of wave-influenced shoreline: shoreface (a) and delta front (b). (a) An idealised two-dimensional cross section through a wave-dominated shoreline, highlighting the relationship between wave energy, water depth and environmental terminology. This model assumes sediment supply predominantly from along-strike sources, most notably longshore drift. *Source:* Courtesy D.S. Collins / John Wiley & Sons / CC BY 4.0. (b) An idealised three-dimensional model of a storm-flood depositional setting, based on ancient (Neogene) and modern observations in NW Borneo. Note the greater diversity of sedimentary processes, including sediment supply from both immediate updip and along-strike sources. One consequence of this model is the greater variability in sand distribution in both dip and strike directions, as recognised in several subsurface Neogene examples (*Source:* Adapted from Collins et al. 2017).

upwards transition from FA-1 → FA-2 → FA-3 resembles the depositional profile of a shallowing-upward, prograding shoreline, probably wave-dominated (e.g. Figure 2.30a). However, closer inspection highlights significant fluvial influence, including combined flows, abundant terrestrial material and stressed conditions at the seabed. Hence, a delta to flanking shoreface setting, with both wave and fluvial interactions, is more likely (Figure 2.30b). This depositional model highlights the importance of peak energy periods, when rivers are in flood and when storm-wave activity is heightened. For example, present-day tropical storms are always accompanied simultaneously by intense rainfall, heightened fluvial discharge, strong winds and elevated wave conditions, which provides ideal conditions for 'storm-flood' processes (Figure 2.30b). This interpretation is further consistent with the present-day geological setting, which closely resembles the Neogene in terms of tectonics, palaeo-latitude, climate, hinterland vegetation (humid-tropical), fluvial hydrology and coastal-shelf hydrodynamics. The storm-flood model is further enhanced by the short length (300–400 km), short-spaced (10 s-km) and multiple and adjacent river systems, which are typical of tectonically active margins; this contrasts with >1000 km-long rivers of passive continental margins. This close link between the ancient and its modern equivalent narrows the range of environmental interpretations (Collins et al. 2017).

Architectural elements form one of the cornerstones of the nested siliciclastic facies hierarchy described earlier (Table 2.1). Groupings of facies in genetically related units, such as facies associations, display not only distinctive facies types and vertical facies trends but also discrete geometries with mappable bounding surfaces. The latter include relatively large-scale (tens to hundreds of metres) stratal surfaces that determine the geometries of sand bodies, such as channel and bar forms in channelised settings, and clinoforms that define the shape of prograding shoreface and/or delta front systems. In some cases thick mudstone units also form architectural elements when they preserve distinctive geometries (e.g. abandoned channel-fills, incised valleys and canyons).

Architectural elements in many fluvial successions are dominated by channel-fill, crevasse splay, levee and floodplain/palaeosol facies associations, reflecting fluctuating channel migration and avulsion, plus overbank accumulation (Figure 2.2). Architectural elements include channel forms (Figure 2.31b,c), gravel bars in pebbly braided rivers (Figure 2.19a) and lateral accretion surfaces from meandering rivers (Figure 2.7). Sets of lateral accretion surfaces represent one of the first, and best-known, examples of an architectural element, preserving the lateral migration of meandering channels and the deposition of point bars (e.g. Figures 2.2 and 2.7). The Middle Jurassic Ravenscar Group of NE England is a good example because point bars are preserved in both plan view (in wave-cut platforms) and cross section (in extensive cliffs – e.g. Figure 2.31). Some migrating channels are linked to regional unconformities and sequence boundaries (e.g. Figure 2.31a), whereas others appear to reflect random avulsion processes leading to a range of fluvial channel stacking patterns (Figure 2.31b).

Clinoforms are large-scale depositional surfaces that form in a wide range of settings with a depositional slope They define the migratory behaviour of various scales of deposit, ranging from large transverse and/or elongate sand bars/bedforms (e.g. tens m-thick tidal sand ridges, sand waves/very large subaqueous dunes in tidal estuaries and embayments, and huge dune complexes in deserts), through different types of deltaic and non-deltaic shorelines (e.g. 10 s-100 m-thick bayhead deltas, shelfal to shelf-edge deltas, mouth bars and shorefaces) and up to the scale of basin margins (e.g. 100 s-1000 s m-thick regressive shelf-slope-basin floor profiles (e.g. Figures 2.12a and 2.14). Clinoforms bound associations of facies (clinothems), which vary in a predictable way up and down the clinoform in response to variations in water depth and physical processes.

Facies associations and architectural elements are closely aligned and sometimes synonymous. For example, tide-dominated deltaic deposits in the Eocene Dir Abu Lifa Member (Western Desert, Egypt) comprise seven facies associations, each defining sediment bodies of different thickness, geometry and lateral extent (Figures 2.32 and 2.33) – defining such bodies is a key aim of much facies analysis work. The large-scale of some of these units (e.g. up to 25 m-thick channel-fills) reflects the interaction of fluvial and tidal energy. For example, a major consequence of high tidal range (e.g. >4 m) is to create multiple, laterally adjacent channel systems, often up to tens of m deep (Figure 2.32b,c), separated by elongate sand bars, which trend parallel to the dominant river/tidal current direction. Additionally, the superimposed bidirectional flows, between fluvial-/ebb-tidal currents and opposing flood tidal currents, causes widening of river mouths and creates mutually evasive (ebb versus flood tide) channels, while mixing of fresh river water and saline marine water promotes clay flocculation and deposition of fluid mud layers.

Facies associations and architectural elements form the building blocks for interpreting depositional environments (Figures 2.2–2.4). Trends within facies associations are also significant although there are few, if any, that are environmentally unique. For example, most channelised deposits display fining-upward trends due to a combination of lateral channel migration and/or abandonment due to channel switching, irrespective of whether the channel is on a floodplain, in a tide-influenced coastal zone, on a slope or within a deep-sea fan. In another example, large-scale accretion surfaces in shallow-marine sandstones (e.g. clinoforms) may reflect a relative shallowing-upward trend, but this can occur both by lateral (due to along-strike currents) and/or by forward (downdip current) migration and in a wide range of environments. Architectural elements in shallow-marine deposits may record non-channelised coastal or deltaic progradation, or the lateral migration of large elongate bars, such as within a deeply scoured tidal channel (e.g. Figure 2.33b–d) or on a shelf. Distinguishing between these and other environmental options will be guided by the autogenic process interpretations generated at the more detailed facies level (i.e. the facies model) as well as the higher-level depositional context. A further cautionary note concerns the use of proxy data such as the well log response. For example, a lowstand shoreface sand body overlain and underlain by shallow-marine mudstones (Section 8.5.5) will have the same blocky log shape as a channel-fill, but a completely different geometry.

The scale at which facies are grouped into facies associations is highly variable and there are no hard and fast rules (the onus lies with the practitioner to define and justify their facies classification scheme). Presenting the fine details of facies was once common and occupied much publication space. Today, it has become routine in sedimentological literature to tabulate basic facies, and sometimes facies association, information, thereby maximising text space for broader environmental diagnoses (see Dalrymple 2010, for further discussion). The main driver in the 'splitting' versus 'lumping' question is: 'what information is critical to answering the research question(s) under consideration'? This may legitimately result in contrasting formats, even in the same rock succession.

One constraint in subsurface sedimentological projects results from the vertical resolution of well logs and/or seismic data. In some projects, such as core logging and/or well log interpretation, facies analysis may even start at the facies association level. This flexibility is a strength of the approach, provided that pitfalls and non-uniqueness of

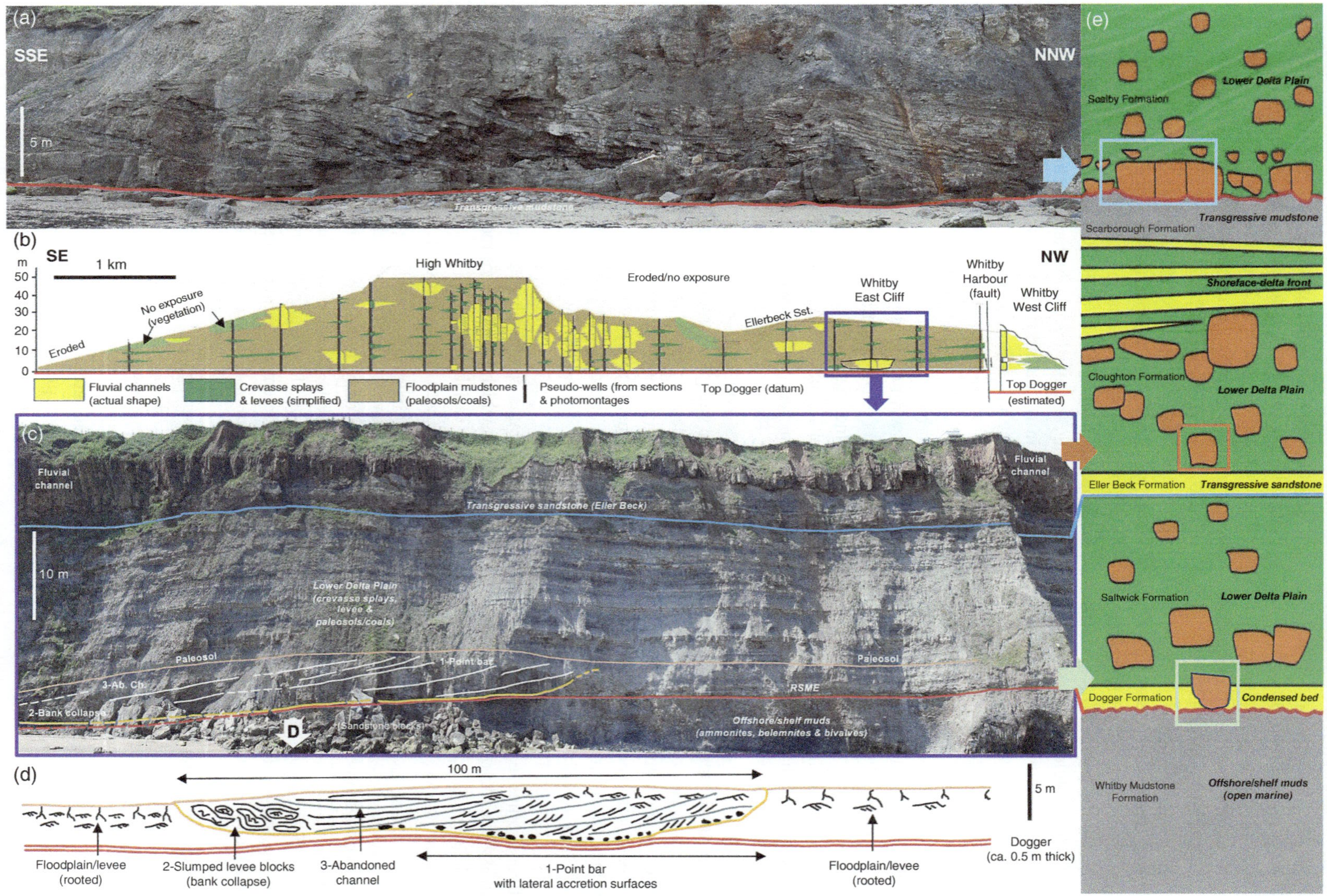

Figure 2.31 Fluvial architectural elements and stratal surfaces in the Middle Jurassic Ravenscar Group, NE England (Scarborough-Whitby area, a–d) and their sequence stratigraphic context (e). (a) Photo-panorama of laterally coalesced lateral accretion surfaces (dipping left to right) in a multilateral sandstone body comprising fluvial to tide-influenced (estuarine) channels. Note the sequence boundary (red line) at the base of the Scalby Formation (see e for stratigraphic context). (b) A depositional strike panel (c. 5 km-long) through the Saltwick Formation (see (e) for stratigraphic context). Fluvial channels show lateral accretion surfaces, implying high sinuosity channels, and occur as both single-story and multi-story sand bodies. The Saltwick Formation is bounded sharply by two regionally extensive stratal units: (i) the Dogger (below) is a reworked (palimpsest) deposit, which erosionally overlies marine mudstones of the Whitby Mudstone Formation (Toarcian), and (ii) the shallow-marine Eller Beck bed (above) is a transgressive sandstone. The Dogger unit represents a Regressive Surface of Marine Erosion (RSME) and is used as the datum (*Source:* Adapted from Kulpecz (2008).). (c) Panorama of Whitby East Cliff (purple box in b) showing the stratigraphic architecture of lower delta plain deposits (middle of cliff section = Saltwick Formation). Note the sheet-like crevasse splays, coal-bearing pedogenic horizons and at the base a single fluvial channel with lateral accretion surfaces (dipping right to left). Offshore marine mudstones form the base of the cliff (Whitby Mudstone Formation). (d) Simplified sketch of the basal single-story fluvial channel (see b and c for context). The channel base is in contact with the top of the Dogger, potentially a hard bed (0.5 m thick) due to early siderite cementation. The channel-fill comprises three architectural elements: (i) an active sand-dominated channel with lateral accretion surfaces/point bar, (ii) a collapsed channel bank with slumps, and (iii) mud-dominated abandoned channel fill. (e) Sequence stratigraphic framework (*Source:* Adapted from Mjøs and Prestholm (1993)). The regionally extensive deltaic deposition in the Middle Jurassic of NW Europe was initiated by pre-rift doming of the Mid-North Sea High (Underhill and Partington 1993), which caused abrupt shallowing and emergence of previously widespread muddy shelves (here marked by the base of the Dogger Formation – see red surface in b–d, above). (a, c) Courtesy of Howard Johnson.

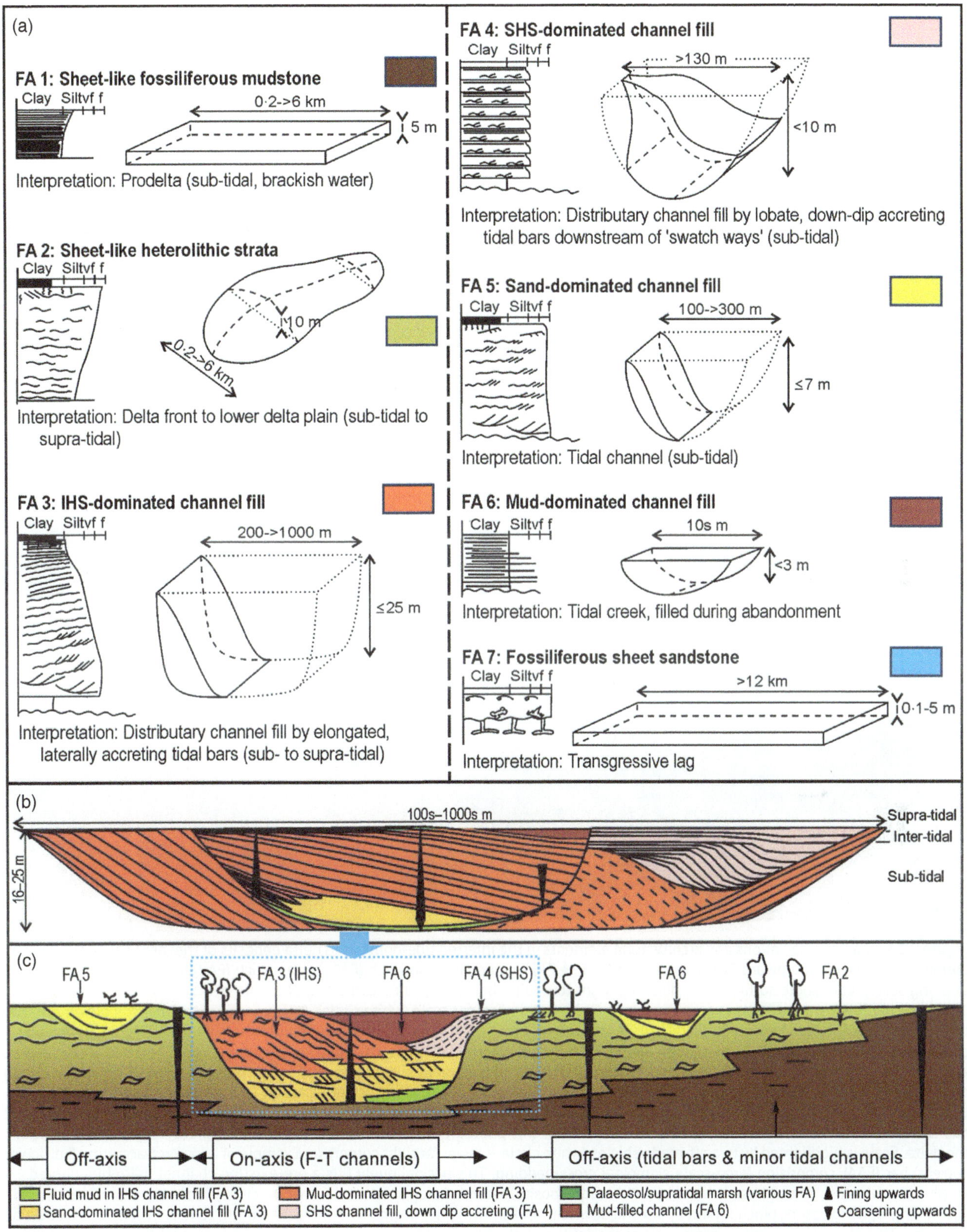

Figure 2.32 Summary of facies associations and large-scale architectural elements in the tide-dominated deltaic deposits of the late Eocene Dir Abu Lifa Member. (a) Summary of the vertical facies successions, external architectural element geometries and interpreted depositional environments (coloured boxes provide key to facies associations in b and c). IHS = inclined heteroilithic strata, SHS = sigmoidal heterolithic strata. (b) Simplified cross section through the large-scale, multi-storey channel belt (Gebel Sagha), illustrating the main channel-fill facies associations (FA3 to FA6), typical vertical facies successions and simplified internal architectural styles. (c) Schematic summary of the vertical and lateral relationships between the facies associations (FA1 to FA6; the transgressive sandstone deposits of FA7 are excluded). Context: the succession overlies the shallow-marine deposits of Birket Qarun Formation, which hosts the spectacular whale bone fossil remains in the UNESCO Whale Valley site at Wadi Al-Hitan: https://whc.unesco.org/en/list/1186/). Part of the Qasa El Sagha Formation, Fayoum Basin (Western Desert, Egypt). *Source:* Legler et al. (2013) / John Wiley & Sons.

Figure 2.33 Facies associations and architectural elements in the inferred tide-dominated delta deposits of the late Eocene Dir Abu Lifa Member. (a) Coarsening-upward facies succession (FA1 → FA2), comprising mud-rich (FA1) to sand-rich (FA2) heterolithic facies (cm-dm bedding scale). Low-angle surfaces (<5° dips) highlight low-relief bedforms, while current ripple-scale bedforms with variable dip directions and frequent mud drapes are widespread. Sedimentologists for scale (white circle). (b) Fluvio-tidal channel complex (FA3) with large, low-angle inclined heterolithic strata (IHS), comprising up to 25 m-thick multi-storey/multilateral channel-fill units defined by pronounced concave-upwards erosive bases (defined by different coloured arrowheads). Fluid mud layers and/or dune-scale cross-bed sets are common at the channel bases. Cross-bedding displays extensive mud-drapes, including mud couplets and lateral cyclicity (Figure 2.1d) plus subordinate bidirectional palaeocurrent indications. In this image, the palaeocurrent direction is away from the viewer, in a basinward direction, supporting a fluvial- to ebb-tide-dominated distributary channel interpretation. Lateral migration of the channel belt has created a multilateral sand body (parallel to the face of this photograph) that is around 6 km wide, representing the axial part of the delta. Scale is highlighted by two off-road vehicles (white circle – lower far right corner). (c) Example of a forward-accreting (i.e. down-current) bar (FA4), comprising large-scale sigmoidal cross-bedding. This bar obliquely fills a pre-existing channel (base highlighted by pink arrowheads), whose base is lined by fluid mud deposits (sedimentologist for scale – white circle). Note the sigmoidal-shaped, coarsening-upward cosets, pinching-out updip and downdip, and showing net downstream accretion (right to left) in an inferred ebb-tide direction. This channel-bar unit is interpreted as the infill of a channel swatchway (i.e. a secondary tidal channel that cuts across sand banks connecting larger tidal channel systems). (d) Example of a mud-filled channel (FA6) that cuts into the top of a unit of IHS (FA3). The mud-fill drapes and onlaps the basal channel surface (brown arrowheads) and is truncated by the base of the overlying transgressive surface (blue line/blue arrowhead). See Legler et al. 2013 for further details and broader context (see also Figure 2.32).

the dataset (e.g. seismic facies, well log facies) are recognised.

2.2.2.2 Carbonates

The strong environmental controls on carbonate sediment production and deposition are important. They mean that composition and texture as well as physical processes are clues to the environment. This extra dimension, as well as prevalent diagnostic palaeontological information, results in relatively well-defined vertical and lateral facies associations and relationships, which are in general predictable and follow tight depositional rules (although early diagenetic processes may blur certain trends).

Facies associations, architectural elements and stacking patterns in carbonates are, in addition to the normal controls of relative sea-level and climate changes, also constrained by the following: (i) platform morphology and depositional profile; (ii) hydrodynamic processes; and (iii) changes in atmospheric and oceanographic circulation. All these features are also highlighted in the variability in sediment export cycles due to excess carbonate production. For example, 'highstand shedding' is a consequence of increased shallow-water sediment production on low-latitude carbonate platforms related to rising sea level. It results in carbonate sediment export in both Cenozoic and Mesozoic carbonate systems (Schlager et al. 1994). Microbial dominated carbonate systems differ in their sediment production and export variations when compared to tropical highstand-shedding systems and develop distinctive slope stacking patterns related to 'slope shedding' processes (Kenter et al. 2005) as observed in the Carboniferous of NW-Spain (Della Porta et al. 2004, 2017; see also Section 2.6.4, Case Study-4), the Devonian of Australia (Playton and Kerans 2015a, 2015b) and the Triassic of Austria (Reijmer et al. 1994). This topic is developed much further in Chapter 11.

All these variations in sediment production, sediment export and depositional processes lead to a greater lithological variety, including the need to identify biotic particles, and ultimately result in a more laborious data collection phase for carbonates; this involves significant amounts of hand-lens, microscopic and geochemical analyses. The reward is that this work provides a wealth of data not only about the physical sedimentologic and stratigraphic conditions but also the biological, ecological, biogeochemical, palaeoceanographic and climatic circumstances. Hence, carbonate sedimentological logs start with texture, bedding patterns and sedimentary structures (Figure 2.34) and also need to be calibrated with microscopic observations, using either a hand lens in the field or thin sections and microscopy in the laboratory. The macro- and micro-fossil content of carbonate sediments provides valuable information about the environment of deposition and, sometimes, can provide biostratigraphic information for age dating. The sedimentological log of shallow-water carbonate facies of the Jurassic in Switzerland exemplifies what can be obtained through detailed logging and microfacies analysis (Figure 2.34). In addition, microscopic observations are also critical for the evaluation of the porosity network and the impact of diagenetic dissolution and precipitation processes, which may also reflect important palaeoenvironmental implications. Geochemical analyses may add information about the post-depositional chemical alterations (e.g. following karstification) or the chemical composition of palaeo-oceans. Carbon, oxygen, and strontium isotopes add further insight into age dating, atmospheric composition and climate.

Hence, the documentation of carbonate facies associations and architectural elements in sedimentological logs (e.g. Figure 2.34) share many of the features described earlier for siliciclastic deposits, but with adjustments for the following: (i) carbonate texture, most notably relative mud content (e.g. Dunham classification; Dunham 1962); (ii) the main biotic components (e.g. macro- and/or micro-fauna); (iii) abiotic components (e.g. ooids, peloids and intraclasts); and (iv) cementation/diagenetic fabrics. See Chapter 11 for more details.

The detailed analytical information from sedimentological logs often makes more sense when placed into a large-scale depositional geometrical context. This can be provided by seismic scale-outcrops, which allow beds to be traced laterally and depositional profiles and dimensions of geobodies to be determined (Figure 2.35). These detailed outcrop studies then provide the link to three-dimensional seismic data and may enable the complete geometry to be assessed, including restoration of initial depositional surfaces (Figure 2.13), where appropriate using horizon flattening. However, sometimes 3D seismic datasets lack detailed facies control or are dependent on sparse and widely spaced cored wells. Therefore, seismic scale-outcrop studies like those in Figure 2.35 will be essential to assess the internal structures that may not be recognised in time-equivalent carbonate build-ups in the subsurface, such as the Miocene examples in SE Asia (Figures 2.13 and 2.14).

2.2.3 Facies Successions

Facies successions arise through the stacking of facies associations and/or architectural elements and reflect the fourth, larger-scale level of facies organisation (Table 2.1). Typically, facies successions range from tens of m thick single genetic units up to hundreds of m thick, composite successions. However, the terminology is loose, which allows practitioners to apply facies succession

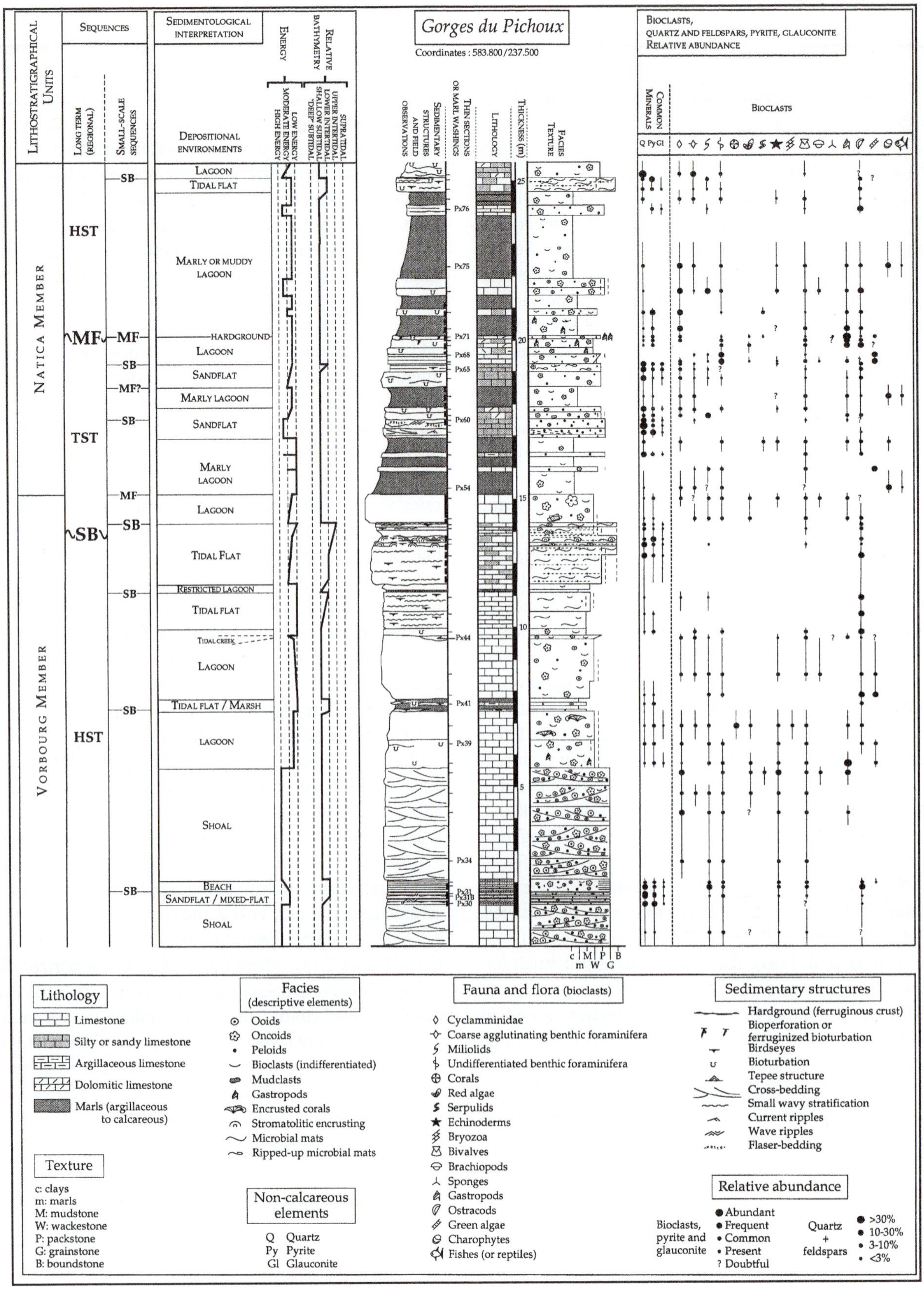

Figure 2.34 Sedimentological log of shallow-water carbonate facies of the Jurassic of France (Pittet and Gorin 1997). This exemplary log shows the lithostratigraphic units, sequence stratigraphic interpretation, sedimentologic interpretation (e.g. depositional environment, eustacy and water depth), as well as the erosional profile with sedimentary structures, lithology log, facies log and faunal distribution. Left-side bedding pattern and sedimentary structures, right-side texture (mudstone, etc.) and faunal components together define the carbonate sedimentary facies (see also Strasser et al. 1999). The importance of distinguishing between bedding pattern and texture is that a single bed can show variety in texture and sedimentary features that otherwise cannot be captured. The carbonate sedimentary facies are determined by the combination of the information of sedimentary structures and bedding, lithology, and texture and faunal content. Columns with other types of data are added (fossil counts, stratigraphic units and interpretation of the sedimentary environment, bathymetry and hydrodynamic energy).

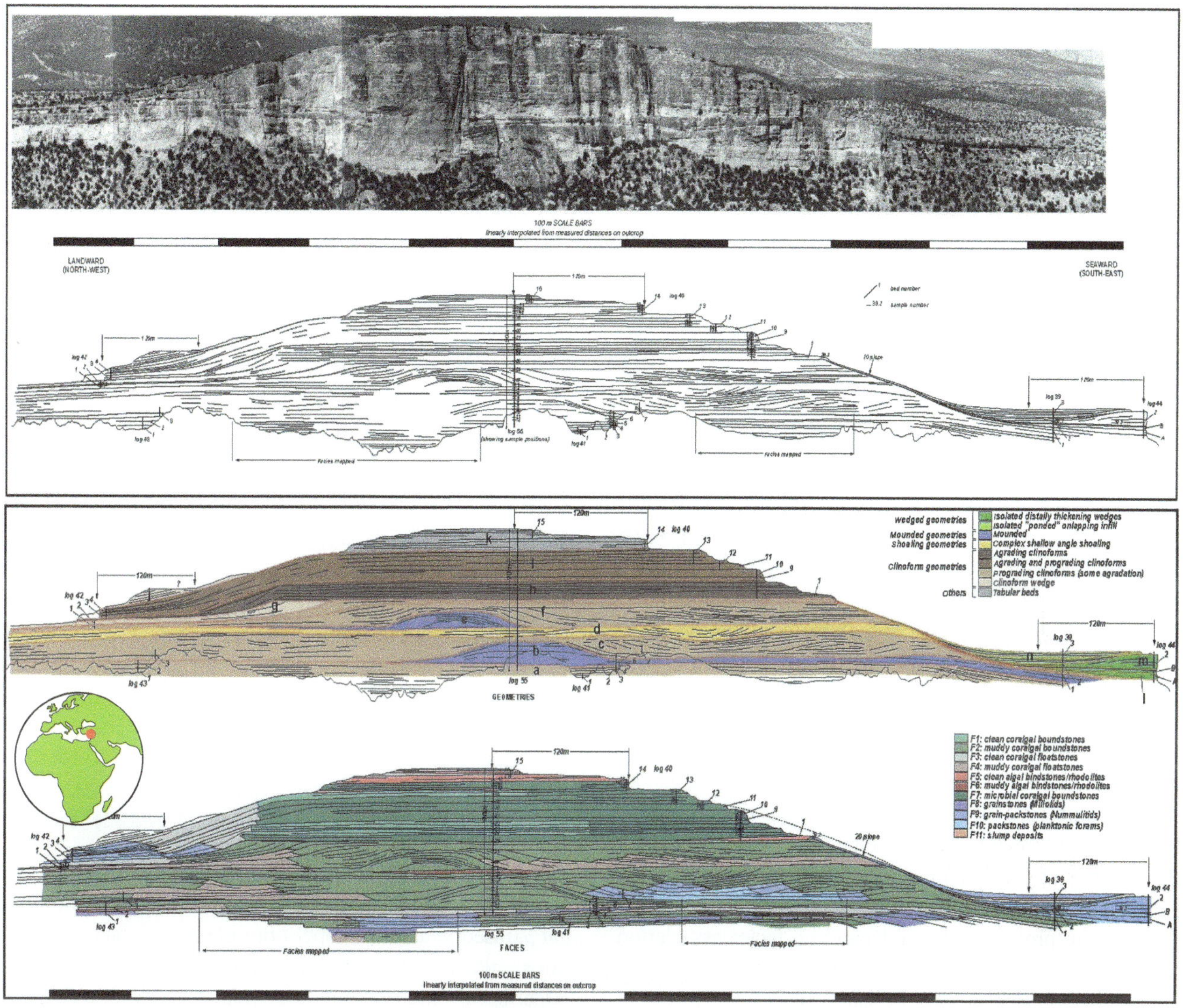

Figure 2.35 Isolated carbonate platform, Miocene, S. Turkey (Bassant et al. 2005, 2021), illustrating data integration at different scales, from top to bottom: (a) outcrop picture; (b) line drawing with bedding pattern interpretation; (c) colour-coded depositional geometries. Labels a-k mark the individual sequences with specific geometries (see legend with colours); (d) infill with carbonate sedimentary facies based on hand samples collected at the surface and along the cliff face. This carbonate platform is a good example of an outcrop analogue case study for time-equivalent carbonate build-ups in SE Asia (Bassant et al. 2004).

concepts at a range of different scales. The key point in defining a facies succession is to establish a genetic linkage between its component parts, from bottom to top. This involves an objective assessment of not only the relatedness of the deposits themselves (e.g. through facies, facies associations and architectural element similarities and/or repetition) but also, crucially, the nature and significance of bed boundaries. Specifically, whether a key stratigraphic surface has been crossed that separates genetically unrelated sediment packages. The decision that deposits are genetically related implies that some kind of facies model defining genetic linkages is under consideration. Conversely, a change in facies characteristics demands an assessment as to whether this is gradual or abrupt and, if the latter, what is the nature of the stratigraphic surface across which the facies change occurs.

For example, a change from a river-dominated deltaic facies association to a tide-dominated facies association in a candidate facies succession raises questions about the significance of the change. Is it due, for example, to a tectonic or climatic change in sediment input, a depositional break (e.g. across a thin transgressive surface), or more subtle changes in river flow strength, tidal amplification or coastal geometry? In another example, from Section 2.6.2/Case Study-2, multiple coarsening/shallowing-upward delta-front parasequences (100 s m thick succession) are succeeded by fluvial deposits with intercalated *in situ* coal layers. The simplest interpretation of this change in environments would be an autogenic response to delta progradation. However, a regional erosion surface separates the two environments, which is correlated with a basin-wide tectonic event. Here, the facies break was formed by extensive subaerial erosion in response to regional relative sea-level fall and represents a sequence boundary. Hence, defining the bounding surfaces that separate contrasting facies successions requires consideration of both autogenic and allogenic stratigraphic controls, which will include establishing the nature and lateral extent of key stratal surfaces (Chapter 4). In many cases, this can be aided by including environment-related biogenic aspects (Chapter 3).

The variablity of carbonate facies successions is best discussed in the conceptual framework of the carbonate factory and the depositional profile (Chapter 11).

2.2.3.1 Siliciclastics

Large-scale physical stratal geometries can offer compelling evidence of environmental linkage. For example, seismic clinoforms, calibrated with coarsening/sandier-upwards trends in coastal-deltaic deposits, define kilometre-thick facies successions of tectonically active basins and/or continental margin-scale regression (e.g. Figure 2.14). Such large-scale stratal geometries confirm a genetic linkage of multiple smaller-scale facies successions, including recognition of stacking patterns (e.g. aggrading, prograding and retrograding). Facies successions can also be confidently identified when outcrop quality is high (e.g. Figures 2.7 and 2.33) or when seismic data or other imaging techniques are available (e.g. Figures 2.12 and 2.14). In the absence of such geometrical data, the identification of vertical facies successions is increasingly reliant on the process interpretations at the facies and facies association levels. In subsurface situations, where both geometrical and facies information may be sparse (e.g. when core data is limited or absent), reliance on well log-only interpretations (log shapes) is sometimes unavoidable for this scale of analysis. However, well log calibration with cores is always preferable and, at a minimum, interpretation should be guided by consistency with the chosen facies model(s). In other subsurface situations, where high-quality 3D seismic data are available, insight into diagnostic stratal geometries can provide near-unequivocal environmental interpretations (e.g. Figure 2.12). For example, distinctive seismic amplitude variations calibrated with cores and logs can provide extremely high confidence levels to environmental interpretations in many depositional settings.

In summary, the facies succession terminology can be used in any depositional environment at a range of scales to be determined by the practitioner. The terminology is simple and flexible and avoids unintended links to sequence stratigraphic models and any implied (relative) sea-level control on bounding surface interpretation.

In contrast, parasequence concepts and terminology (Figure 2.36) are a particular type of facies succession, which is deeply embedded within sequence stratigraphy with the following definition (Van Wagoner et al. 1990):

> "relatively conformable succession of genetically related beds or bed sets bounded by marine flooding surfaces and their correlative surfaces".

The hard linkage to marine flooding surfaces largely restricts their occurrence to coastal-shelf deposits and results in parasequences having special significance in sequence stratigraphic methodology:

i) a single parasequence represents the accumulation of one individual genetically related facies unit, thus forming the single building blocks of idealised sequence stratigraphic models (Figure 2.36a);
ii) the bounding marine flooding surfaces approximate to local chronostratigraphic horizons, thereby providing a high-resolution chronostratigraphic correlation framework; and
iii) multiple parasequences form vertical facies successions (parasequence sets) that are genetically related and create larger-scale stacking patterns (at the systems tracts scale) that define long-term coastal-shelf behaviour, most notably in terms of progradation, aggradation and/or retrogradation (Figures 2.36 and 2.37; Section 2.6.1/2/Case Studies-1 and -2) (see also Chapter 4; Figures 4.11 and 4.30).

Sequence stratigraphic models typically simplify individual parasequences into repetitions of uniform facies successions, albeit with variable stacking patterns involving larger-scale parasequence sets (Figure 2.36b,c and 2.37). However, single parasequences can be highly variable, reflecting the facies-scale variability of coastal depositional systems (see Chapter 8). Each parasequence comprises a vertical stack of one or more facies associations representing a range of laterally contiguous sub-environments, thereby honouring Walther's Law (see Chapter 4 for further details). The default example of a parasequence is a coarsening/sandier-upward facies succession, interpreted as the result of coastal progradation (e.g. deltas and flanking shoreface environments; Figure 2.36a). Less commonly, they form fining/muddier-upwards facies successions, such as shallowing-upward facies profiles in inshore/back barrier settings (e.g. in tide-influenced channels, estuaries and embayments; Figure 2.33). Correlative fining-upwards facies successions in terrestrial settings may relate to the same base level fluctuations as those controlling contemporaneous coastal parasequences and parasequence sets, albeit landward of direct marine incursions (e.g. Figure 2.2). For example, the lateral equivalent of marine flooding surfaces are typically condensed sections in shelf and deep-water settings, but in terrestrial successions they

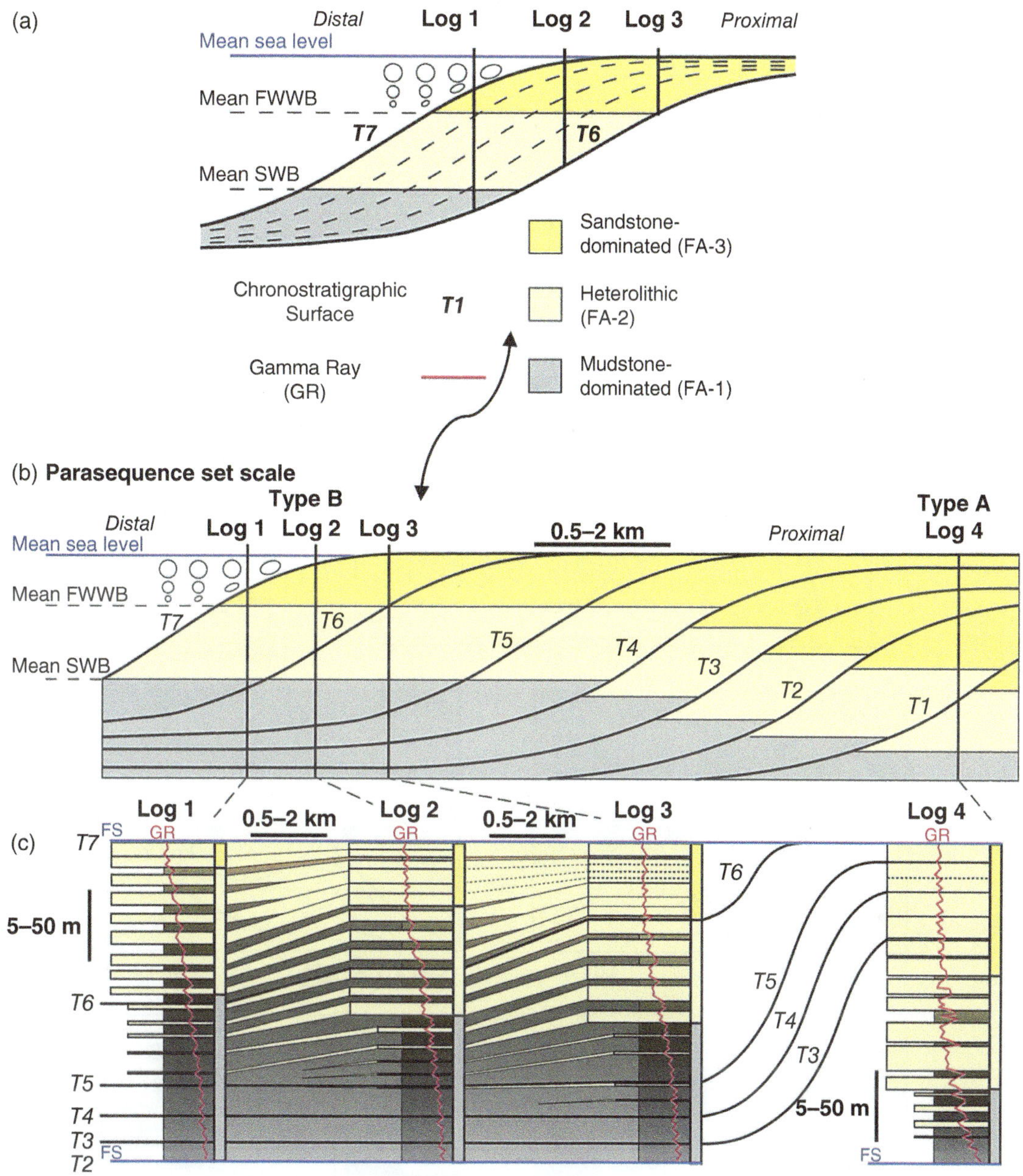

Figure 2.36 From process synthesis to environment interpretation: correlation implications of the facies association interpretation stage. (a) Conceptual cross section of a prograding wave-dominated sandy coast (delta front to shoreface), showing the inferred spatial and temporal relationships between the FA1→FA2-FA3 succession (Figure 2.29) – the two timelines (T6 and T7) represent clinoforms, which bound a single, shallowing upward clinothem (FA-1>FA-2>FA-3), which is equivalent to a parasequence (see Section 2.3 and Chapter 8 for further discussion). (b) A larger-scale (parasequence set scale), 2D panel showing (i) the inferred distribution of the three facies associations (shallowing-upwards and proximal to distal, right to left); and (ii) schematic chronostratigraphic surfaces (T1 to T7) – approximate scale derived from subsurface well log correlations (Baram Delta Province, NW Borneo). (c) Facies (simplified sandstone/mudstone proportions (blocky profile) and GR log (red line) correlation panel derived from the 2D panel in b; two end-member, sandier-upward parasequence types are shown ('A' and 'B'): type A is relatively sandy and displays sharp facies association transitions, whereas type B is more heterogeneous and displays more gradual transitions between facies associations. It is evident from b and c that (i) timelines cross-cut the gradational facies boundaries, and (ii) timelines (clinoforms) control the low-angle (c. 0.5–1° dip) 'shingled' geometry of the rock units. This also has an impact on reservoir zonation and correlation and is affected by the length of correlation panels and by the orientation of such panels relative to the dip and strike of the main facies belts. Based on Collins et al. 2017. *Source:* Adapted from Collins et al. (2017).

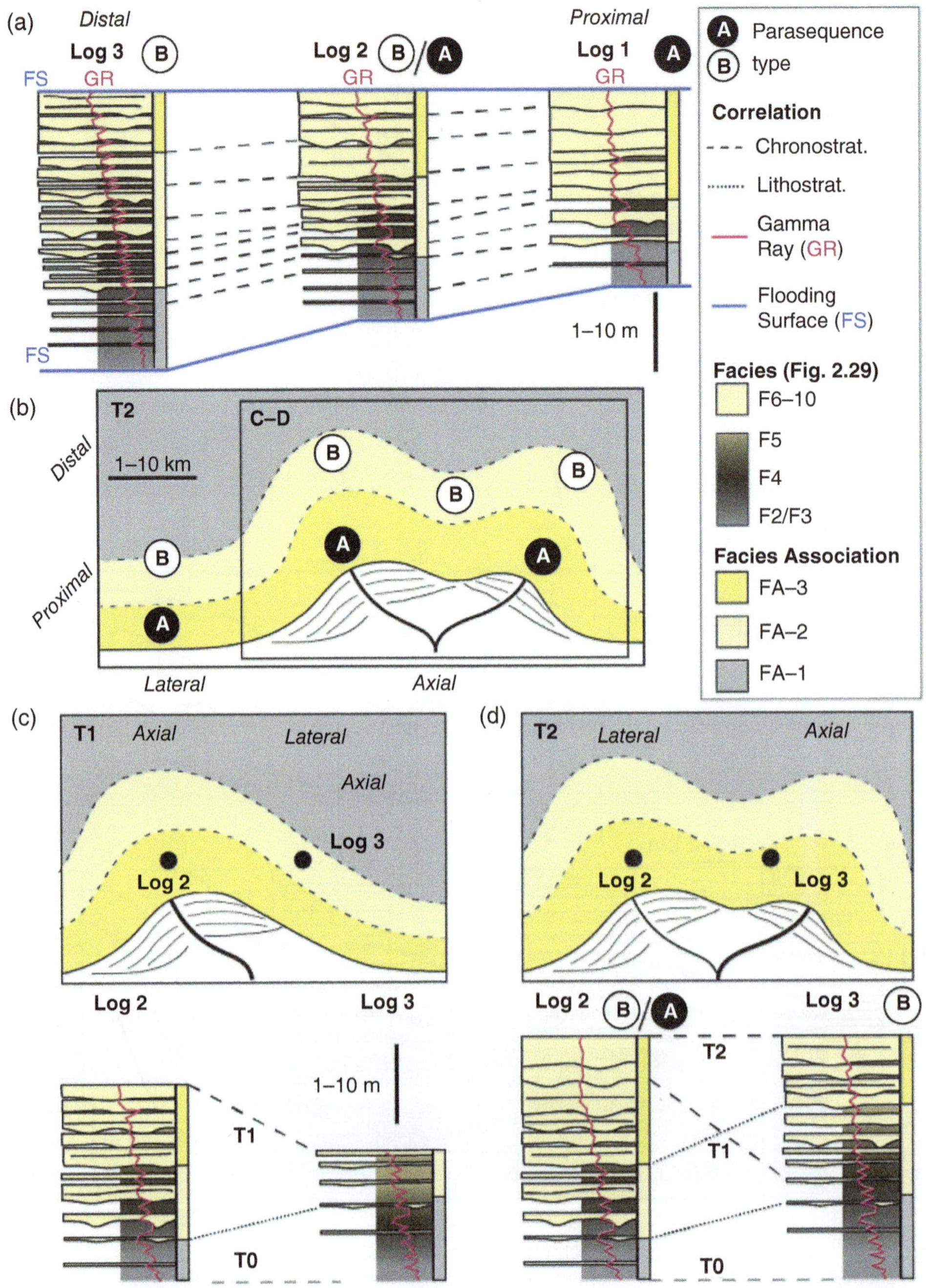

Figure 2.37 Vertical and plan-view variability of parasequences and possible interpretations. (a) A schematic proximal to distal (right to left), facies and gamma ray (GR) log correlation panel (Logs 1–3) (Log 1 is located onshore of Log 2 in the beach ridges), illustrating contrasting styles of sandier-upward parasequence (types 'A' and 'B'): (i) type A (proximal) is sandier, with thicker beds and relatively sharp facies association transitions, and (ii) type B (relatively more distal) is more heterogeneous with thinner beds, thicker and more frequent mudstone layers and with more gradational facies transitions. (b) A simplified plan-view model for facies association and parasequence relationships, based on the inferred wave-dominated, fluvial-influenced delta system and guided by modern equivalents; type A facies successions are interpreted as relatively proximal and axial, while type B is relatively distal and lateral. (c, d) These show time steps depicting one way in which the situation at T2 in b could have evolved, thereby illustrating issues in parasequence interpretation. For example, during T1 (c), a heterolithic to sandy delta lobe progrades north-westwards (Log 2), resulting in reduced sediment supply in the east (Log 3); differential compaction creates a depositional topographic high in the west. By T2 (d), nodal avulsion has occurred causing the axial part of the delta to shift eastwards (Log 3), while the western lobe is abandoned (Log 2). The combination of mainly dip-oriented progradation and strike-oriented lateral migration adds further complexity, including the crossing of facies boundaries by timelines. From Collins et al. 2017. *Source:* Collins et al. (2017) / John Wiley & Sons / CC BY 4.0.

may be represented by thick coal layers and mature soil horizons. The absence of genetically related facies successions unequivocally defined by sea level rise (i.e. an absence of marine flooding surfaces) precludes the use of the term parasequence in continental and deep-water settings.

In practice, there can be a model-driven element to the use of parasequence nomenclature, especially with regards to stacking patterns (e.g. Figure 4.30). This topic is discussed in depth by Colombera and Mountney (2020a, 2020b).

2.2.3.2 Carbonates

The definition of facies successions and parasequences in shallow-water carbonate environments is in principle the same as in siliciclastics with a similar range of scales (Figure 2.3). However, their carbonate equivalents display considerable variability as they reflect wider variation in sediment production (controlled by environmental parameters and especially sensitive to water depth), transportation (hydrodynamic controls with a wide variety of grain size and texture) and accommodation (a wide variety of depositional profiles). This difference (Catuneanu 2022) arises from the sediment production and delivery processes, with a source-to-sink sedimentation model in clastic environments and a 'source-within-sink' sedimentation model in carbonate settings. In addition, carbonates are more sensitive to early diagenetic modification processes. This sensitivity results in the formation of condensed surfaces, firmgrounds and hardgrounds (Christ et al. 2012, 2015). Rapid cementation and/or dissolution may accompany subaerial exposure (Dravis 1979, 1996). These factors can aid greatly in the recognition of the significant stratigraphic breaks that divide facies successions.

One consequence of this parasequence-scale variability in carbonate systems has been to hinder the collection of statistical datasets for reservoir analogue purposes for facies associations, architectural elements and facies successions. This in turn has prompted the use of 3D forwards stratigraphic models based on the variability of a wide series of parameters (e.g. light, temperature, currents, waves, base level changes, sea-level variations, sediment production potential, individual biota and grain-size variations). This modelling has been used to determine intrinsic carbonate sediment production and transport processes for the individual carbonate facies models. These studies aim to quantify and predict the stratigraphic architecture of individual carbonate platform environments. However, it remains difficult to predict the interaction between this broad range of parameters over time and their impact on the overall carbonate platform development (e.g. progradation, retrogradation, aggradation and drowning). For example, flat-topped platforms such as the Great Bahama Bank (Busson et al. 2019), isolated platforms such as in the Central Luconia Province (Figure 2.13; Granjeon and Gervais 2023) or the variety of other carbonate platform morphologies (Williams et al. 2011).

Large-scale physical stratal geometries and facies successions in carbonates depend upon the operating carbonate system (Figure 2.3). For example, Tropical and Microbial factories (T- and M-factories, respectively, *sensu* Schlager 2003; Reijmer 2021) create stratal geometries with an inclined slope (various clinoform morphologies), platform margin (various morphologies, sturdy structures, cemented) and protected lagoonal areas (flat-topped morphologies) and/or restricted environments (sabkhas and mangroves in carbonate ramp settings). On the other hand, Cold-water coral carbonates (CWC-factory) create different morphologies in response to fluctuating nutrient levels. Large-scale physical stratal geometries and hence facies successions in cool-water factory systems (C-factory) have large similarities to those found in clastic systems as both systems are characterised by loose sands and muds whose distribution is controlled mostly by hydrodynamic processes.

2.2.4 Stacking Patterns and Stratigraphic Architecture

2.2.4.1 Siliciclastics

Facies successions and parasequences occur as combinations in thicker (km-scale), vertically stacked and genetically related successions (level 5 in Table 2.1), including systems tracts and depositional systems (e.g. Figures 2.4 and 2.38). Their distinctive stacking patterns define a larger-scale stratigraphic architecture, especially parasequence sets (Figures 2.36 and 2.37). Stacking patterns at this scale are fundamentally controlled in most depositional environments by the shifting availability of accommodation space. The classical examples are provided by 10s–100 s m-thick, coastal sandier-upward units which pass upwards into either shallow-marine or coastal plain facies (Figure 2.38a). These sandier-upward stacks consist of smaller coarsening-upward facies successions or parasequences, which are initially mud-rich and relatively distal, becoming progressively sand-rich and more proximal upwards. Such successions are indicative of overall net progradation. They are commonly called 'progradational parasequence sets' (Figures 2.36b, c, 2.37a, and 2.38a).

Muddier-upward successions developed in comparable shallow-marine settings may either comprise (i) successively smaller coarsening-upward facies successions, which become more mud-rich and distal upwards, or (ii) progressively more mud-rich intervals but without discernible coarsening-upward elements. Such patterns typically indicate overall net retrogradation, or depositional transgression, when long-term relative sea level rise exceeded sediment supply. They are commonly called 'retrogradational parasequence sets' (Figure 2.38b).

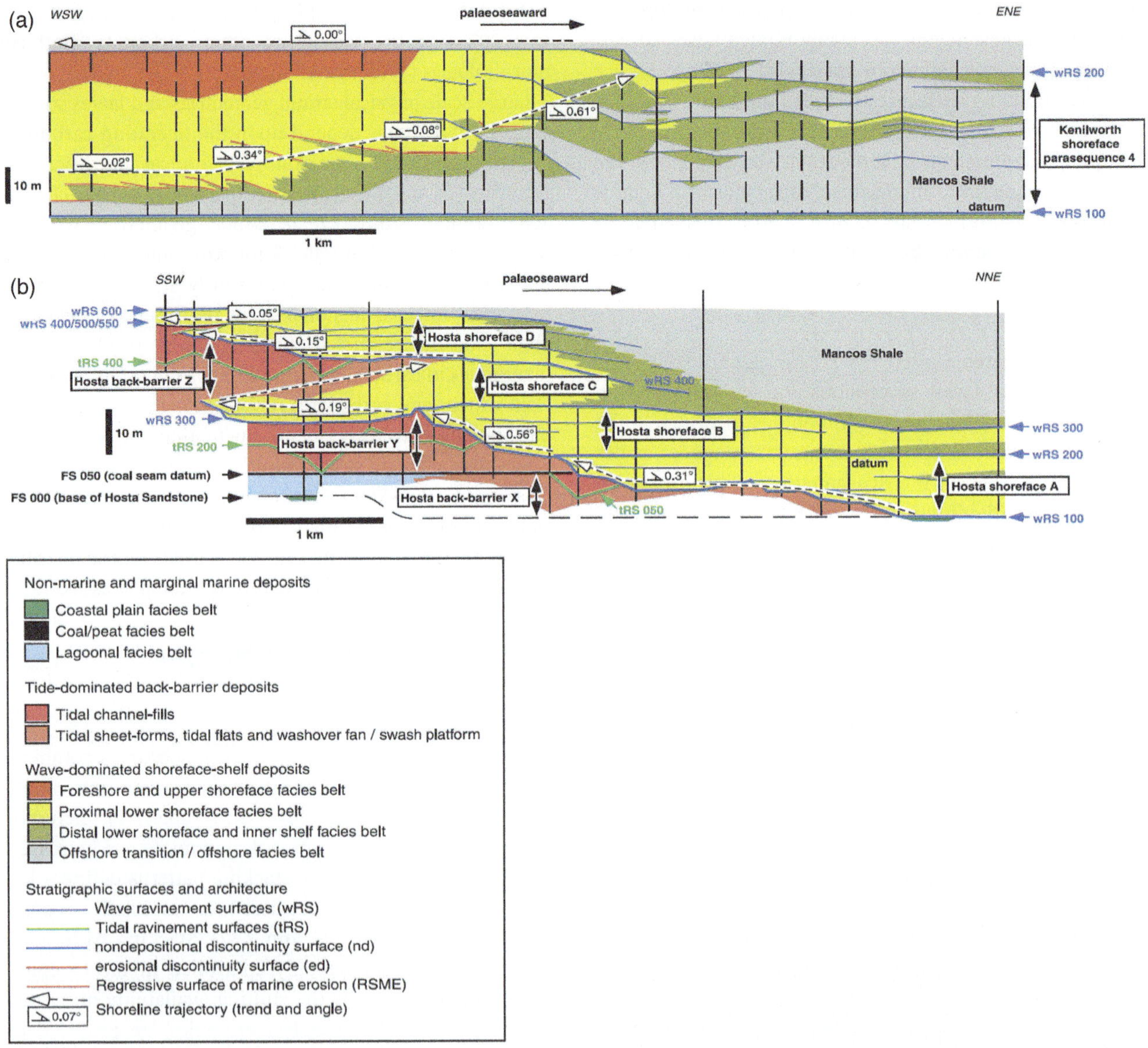

Figure 2.38 Two contrasting depositional dip-oriented facies panels of shoreline trajectories and stratigraphic architecture styles interpreted from fully exposed outcrop datasets. (a) A progradationally stacked succession of a sand-rich, wave-dominated deltaic shoreline-shelf system from the Late Cretaceous Kenilworth K4 tongue, Western Interior Basin, USA; facies characteristics illustrated in Figure 8.30 (*Source:* Hampson 2000 / John Wiley & Sons). (b) A transgressive, retrogradationally stacked sand-rich, mixed wave-/tide-dominated barrier shoreline system from the Late Cretaceous Hosta Sandstone, San Juan Basin, USA (after Sixsmith et al. 2008). *Source:* Helland-Hansen and Hampson (2009) / John Wiley & Sons.

The former reflects (Figure 2.38b) the landward retreat of the distal toes of deltas and lower shoreface deposits (Figure 4.34), whereas the latter accumulated beyond the distal delta/shoreline system (see Chapter 8) or in mud-dominated shoreline-shelf settings (see examples in Section 9.6).

Variations in the style of vertical and lateral organisation of progradational parasequence sets are embedded in sequence stratigraphic models (e.g. Figure 2.4 and Section 4.4.2).

High-quality outcrops enable different stacking patterns and architectural styles to be linked with physical depositional processes and environments and their associated key stratal surfaces (Figure 2.38). This type of work includes the calibration of seismic-scale and outcrop-/core/well-log-scale datasets and trajectory analysis (Figure 4.26) in monitoring longer-term shelf-shelf margin depositional patterns and evolution (e.g. Helland-Hansen and Martinsen 1996; Helland-Hansen and Hampson 2009; Helland-Hansen et al. 2012).

2.2.4.2 Carbonates

Facies successions and parasequence sets in carbonate systems respond differently to the sea-level fluctuations and to the tectonic processes that create and destroy accommodation space (Figure 2.3). This is because: (i) carbonate sediment *production* (rather than sediment *supply* in clastic systems) varies depending on the surface area over which carbonate sediment can be produced, and this area is in turn very sensitive to changes in depth with respect to the euphotic zone (Tropical-factory systems); and (ii) different carbonates respond differently to the same variations because they influence the production capacity of the carbonate system differently (Cool-water factory, Microbial factory). For example, during sea-level rise, carbonate sediment production realms may expand rapidly resulting in: (i) a rapid increase in carbonate growth and sediment production (Tropical-factory; highstand shedding *sensu* Droxler and Schlager 1985; Schlager et al. 1994); or (ii) a change in sediment production and export (Cool-water factory; Saxena and Betzler 2003); or (iii) just a slight increase in the continuous sediment production (Microbial factory; slope shedding *sensu* Kenter et al. 2005).

Hence, these changes affect the parasequence stacking patterns in different ways in different settings (Figure 2.3). For example, there are four possible outcomes resulting from rising relative sea level with sediment production differences, based on tropical, T-factory, systems alone (Schlager 2005): (i) progradation, with sediment production exceeding accommodation; (ii) backstepping with sediment production unable to fill all accommodation; (iii) aggradation with sediment production equalling accommodation; and (iv) drowning, a process in which sediment production is exceeded by the increase in accommodation.

In cool-water systems, the balance between accommodation and sediment supply is dominated by the hydrodynamic regime of the depositional realm: with a shaved shelf during highstands in sea level (James 1997) and reduced production and modified supply (different biota) during lowstands. In Microbial-factory systems, sediment supply (production) remains fairly constant with progradation (over long distances) and aggradation patterns, but no sharp contrast between sea-level highstand and lowstand periods (Kenter et al. 2005).

2.2.5 Facies Recognition of Key Stratal Surfaces

Some of the most environmentally diagnostic facies are volumetrically small, while major stratal surfaces may be highlighted more by what is missing rather than by what has been preserved. The abrupt juxtaposition of highly disparate, genetically unrelated facies types is one of the easier ways to recognise a key stratal surface. However, with more subtle or negligible facies changes across major stratal surfaces the regional context and/or biostratigraphic data will be needed.

In clastic successions, palaeosols, rootlets and *in situ* coals provide typical examples of relatively thin, but highly diagnostic facies. Two examples are given here: (i) temperate-humid climate setting example from the Middle Jurassic of NE England (Figure 2.39), and (ii) arid to semi-arid setting example from the Early Triassic of southern England (Figure 2.40). Additionally, these examples also include major stratal surfaces highlighted by significant facies dislocations, which are highlighted by thin and/or less common facies types: (i) condensed/palimpsest facies plus offshore marine/delta plain juxtaposition (Figure 2.39 – represented by the Dogger Formation in the Middle Jurassic of the Cleveland Basin, NE England), and (ii) an aeolian deflation surface/winnowed lag/palaeosol plus gravel braided river/aeolian juxtaposition (Figure 2.40 – represented by the Early Triassic Sherwood Sandstone Group in the Wessex Basin, southern England). In shallow-marine successions, winnowed lags are distinctive by virtue of their thin (<10 cm thick) and unusually coarse-grained texture compared with adjacent facies, and often abruptly separate contrasting facies (e.g. inshore tidal channel sandstones below and offshore shoreface sandstones above; Figure 2.38b).

In carbonate successions, a well-documented case study of the development of key stratal surfaces is the analysis of a series of continuous cores recovered during the Bahamas Drilling Project (BDP; Ginsburg 2001) and the Ocean Drilling Programme (ODP) Leg 166 (Eberli et al. 1997; Swart et al. 2000). The cores covered a transect from the top of the Great Bahama Bank towards the basin in the Straits of Florida. The BDP-ODP platform-to-basin transect enclosed shallow-water carbonates in the east, platform margin sediments, to slope and drift deposits in the Straits of Florida in the west and provided a unique dataset demonstrating that seismic reflections in the seismic profiles represented timelines with a chronostratigraphic significance (Figure 2.41) (Eberli et al. 2002). Seismic reflections arise from impedance contrasts (changes in density and/or sonic velocity) that are linked either to changes in sediment composition or diagenetic modifications or a combination of both. The variations in sediment production and sediment export combined with current-driven sediment redistribution at the Bahamas can be linked to sea-level changes that together with the platform morphology, carbonate ramp or flat-top platform, control the overall sediment production and the type of sediment produced, for example, skeletal, non-skeletal, mud–sand ratio. These variations are ultimately reflected in seismic velocity variations and thus impedance contrasts as shown in the seismic profile (Figure 2.41). See Chapter 11 for further details.

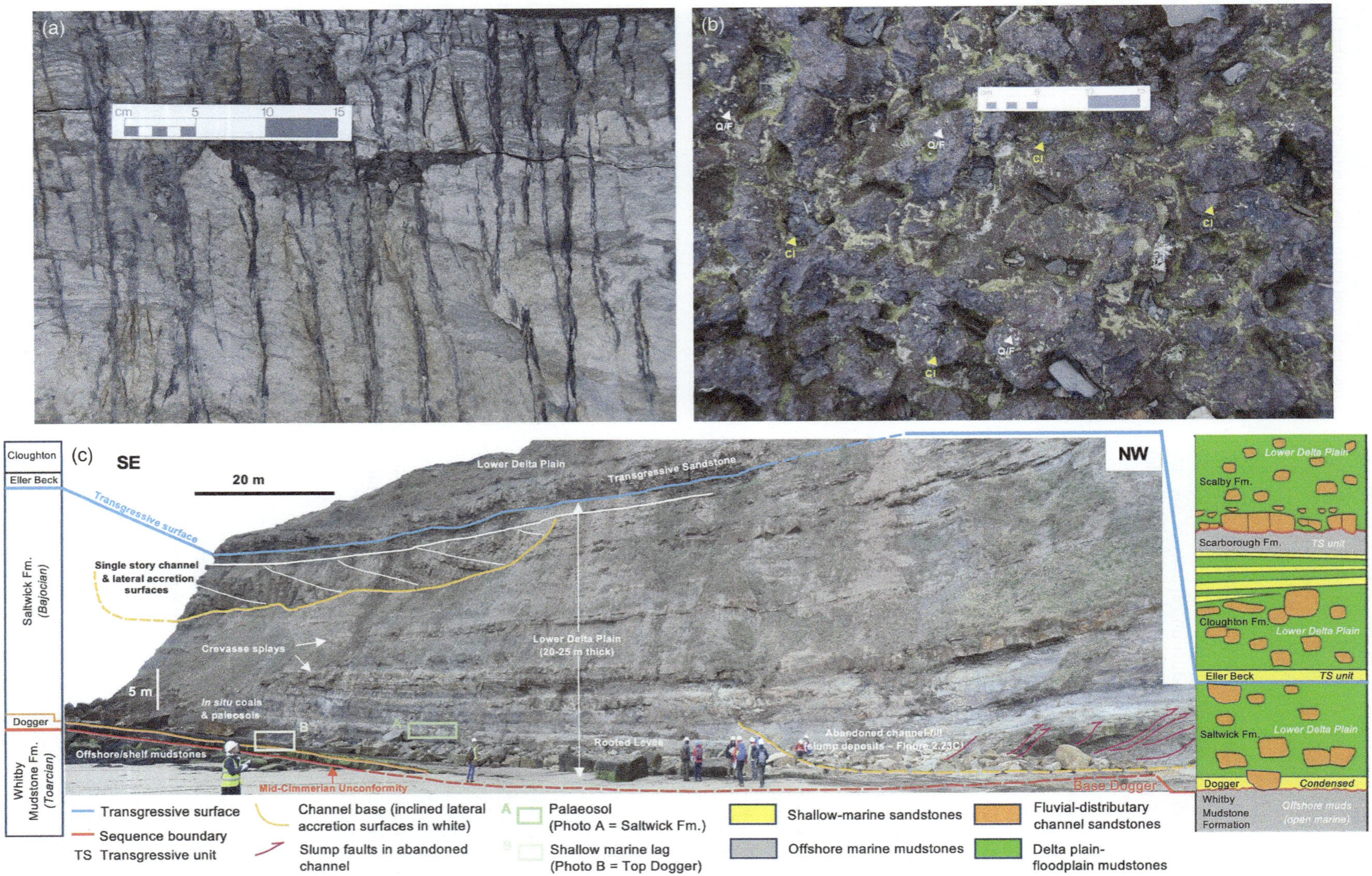

(a)
cm 5 10 15
(b)
Q/F
Cl
(c)
SE
NW
20 m
5 m
Cloughton
Eller Beck
Saltwick Fm. (Bajocian)
Dogger
Whitby Mudstone Fm. (Toarcian)
Lower Delta Plain
Transgressive Sandstone
Transgressive surface
Single story channel & lateral accretion surfaces
Crevasse splays
Lower Delta Plain (20-25 m thick)
In situ coals & paleosols
Offshore/shelf mudstones
Rooted Levee
Abandoned channel-fill (slump deposits – Figure 2.23C)
Mid-Cimmerian Unconformity
Base Dogger
Scalby Fm.
Lower Delta Plain
Scarborough Fm.
TS unit
Cloughton Fm.
Lower Delta Plain
Eller Beck
TS unit
Saltwick Fm.
Lower Delta Plain
Dogger
Condensed
Whitby Mudstone Formation
Offshore muds (open marine)
Transgressive surface
Sequence boundary
TS Transgressive unit
Channel base (inclined lateral accretion surfaces in white)
Slump faults in abandoned channel
Palaeosol (Photo A = Saltwick Fm.)
Shallow marine lag (Photo B = Top Dogger)
Shallow-marine sandstones
Offshore marine mudstones
Fluvial-distributary channel sandstones
Delta plain-floodplain mudstones

Figure 2.39 Contrasting stratal surfaces (autogenic and allogenic) in coastal/deltaic-shelf deposits from the Early-Middle Jurassic (Toarcian-Bajocian) of NE England (Whitby, East Cliff). (a) A palaeosol from the main cliff face dark (green box in (c)), which comprises the Bajocian Saltwick Formation. Description: massive, pale grey, silty mudstone to very fine-grained sandstone that is cut, and largely de-stratified by, vertical, mainly downward-tapering, carbonaceous-rich root traces, and overlain by a thin coal layer (above photograph); interpretation: a palaeosol, which formed the substrate of an *in situ* coal (seatearth) – adjacent point bars, levee deposits and floodplain mudstones with dinoflagellates, suggest a brackish water, lower delta plain setting. (b) Top surface of the c. 0.5 m-thick Dogger Formation (regionally up to 1–12 m thick) (light green box in (c)). Description: fine- to very coarse-grained, argillaceous sideritic sandstone with quartz and feldspar (white arrowheads with 'Q/F') and glauconite grains, locally conglomeratic (quartz pebbles, mudstone clasts – yellow arrowheads labelled 'Cl'), reworked ammonites with abraided exteriors, very poorly sorted (mud-rich matrix) and iron-cemented (red-brown, siderite cement); no internal structures visible, due to pervasive cementation and bioturbation (e.g. *Thalassinoides*). Interpretation: a winnowed lag mostly produced by marine processes. The diverse mixture of sediment and grain types makes further environmental diagnosis difficult – the sediment types reflect temporal variability: earlier (e.g. reworked ammonites), synchronous (phosphate nodules, burrows and iron-rich cementation) and later (carbonaceous root traces) processes and mixed depositional environments; these features identify this facies as a 'palimpsest' deposit. (c) Context: the base of the Dogger marks an unconformity, separating underlying mud-dominated shelf deposits (the Toarcian Whitby Mudstone Formation), from overlying paralic deposits (the Bajocian Saltwick Formation) (Mjøs and Prestholm 1993). The Dogger is a distinctive and regionally extensive unit across the Cleveland Basin, forming a condensed interval spanning around 3–4 Myr (*Opalinum* and *Murchisonae* zones of the Aalenian stage). It is a sequence boundary correlated with the Mid-Cimmerian Unconformity (red surface at the base of the Dogger), which represents a period of regionally extensive, tectonically induced uplift (the Mid-North Sea High 500 km to the north) and widespread relative sea-level fall. At this location, this unconformity separates underlying richly fossiliferous (e.g. abundant bivalves, ammonites and belemnites) offshore marine mudstones (Lias/Early Jurassic – Aalenian) from emergent lower delta plain deposits with fluvio-distributary channels, crevasse splays and palaeosols with thin coals (Saltwick Formation in the c. 25 m-high cliff). The palaeosols (a) form at numerous stratal surfaces throughout the paralic succession, are traceable laterally for 100 s m and mainly reflect autogenic processes, most notably fluvial channel avulsion and floodplain accumulation. The unconformity (b – base of the Dogger) marks a single stratal surface, it is traceable laterally for 10 s–100 s km and reflects mostly allogenic processes. The Dogger is a classic lag deposit formed mainly by shallow-marine processes during a period of relative sea-level fall. Stratal surfaces linked to the latter are sometimes termed 'Regressive Surfaces of Marine Erosion' (RSMEs) (Section 4.4). In this example, net deposition (up to 12 m) might suggest a superimposed (or slightly later) and short-lived period of marine transgression to provide accommodation for Dogger preservation – this is supported by evidence for onlap towards the west (Powell 2010). However, further east, detailed mapping revealed accommodation for Dogger preservation was provided by basal erosion: '...the Dogger was laid down on an irregular or undulating surface of Liassic mud, with localized shoals and deeps' (Tonks 1922). The first sign of unequivocal emergence is above the top of the Dogger, which is a very abrupt surface (orange line in c). This photo-panorama (c) continues laterally to the east (right) in Figure 2.31c. Courtesy of Howard Johnson.

Figure 2.40 Facies types associated with contrasting (autogenic and allogenic) stratal surfaces. A rhizocretion palaeosol horizon (a) and details of internal fabric (b), which are interbedded with braided fluvial sandstones from the Triassic Sherwood Sandstone of the Wessex Basin, southern England (Budleigh Salterton Promenade). Description: (a) trough cross-bedded, fine-grained red sandstone that is cut by bedding plane-parallel rhizocretion horizons (Rhz-h), and by vertical, more typical downward-tapering forms (e.g. b) of calcite-cemented, inferred tap roots (Rhz-tr); interpretation: these rhizoliths are probably the result of evaporative concentration, evapo-transpiration and/or CO_2 degassing of Ca-bearing ground water mixed with deeper saline, alkaline ground water. They form during tree growth (synchronous with deposition), especially around deep tap roots in semi-arid environments. This rhizocretion development probably formed in the abandoned part of a sandy braid plain, adjacent to ephemeral braided rivers. Context: co-existing rhizocretions and sandy braid plain deposits are a common constituent of semi-arid terrestrial environments, especially in the Triassic of NW Europe. Rhizocretion development on abandoned braid plains provided an intra-basinal source of calcite, often reworked into adjacent fluvial channel lags, which later provided localised sites of calcite cementation (an important form of heterogeneity in the Sherwood Sandstone reservoir). The distribution of rhizocretion horizons is mainly linked to autogenic processes, especially channel switching. (c) An impermeable palaeosol defines the contact between pebbly braided river facies (Budleigh Salterton Pebble Bed – BSPB – below lens cap) and highly porous and permeable aeolian sandstone facies (Otter Sandstone – above red line – the yellow colour at the base of the aeolian sandstones is a result of iron leaching by modern groundwater flow at the permeability barrier formed by the palaeosol). Description: the palaeosol is c. 30 cm-thick and comprises dark, red-brown silty mudstone facies (lens cap). At the top of the palaeosol are wind-faceted pebbles with distinctive desert varnish ('dreikanter' or ventifacts). Rounded pebbles of the BSPB increase downwards (base of photograph), while mud and silt content increase upwards. The aeolian sandstone facies is well-sorted and mainly flat-bedded with occasional low-angle cross-bedding. Interpretation: the massive (structureless) texture of the red-brown silty mudstone resembles mature palaeosols that form during prolonged periods of erosion and wind deflation. The fines are probably wind-blown and are absent in the BSPB. The location of this palaeosol between the c. 20 and 30 m-thick gravel-rich braided fluvial deposits (BSPB – below the red line) and aeolian sandstones (basal Otter Sandstone – with honeycomb weathering above the red line) defines a major change in environmental conditions. Context: see Figure 2.19b, c. The palaeosol has been correlated with the regionally extensive Hardegsen unconformity, which separates the Lower and Middle Triassic in NW Europe (red line in (c) and (d)). This surface is mainly of allogenic origin. *Source:* Courtesy of Howard Johnson.

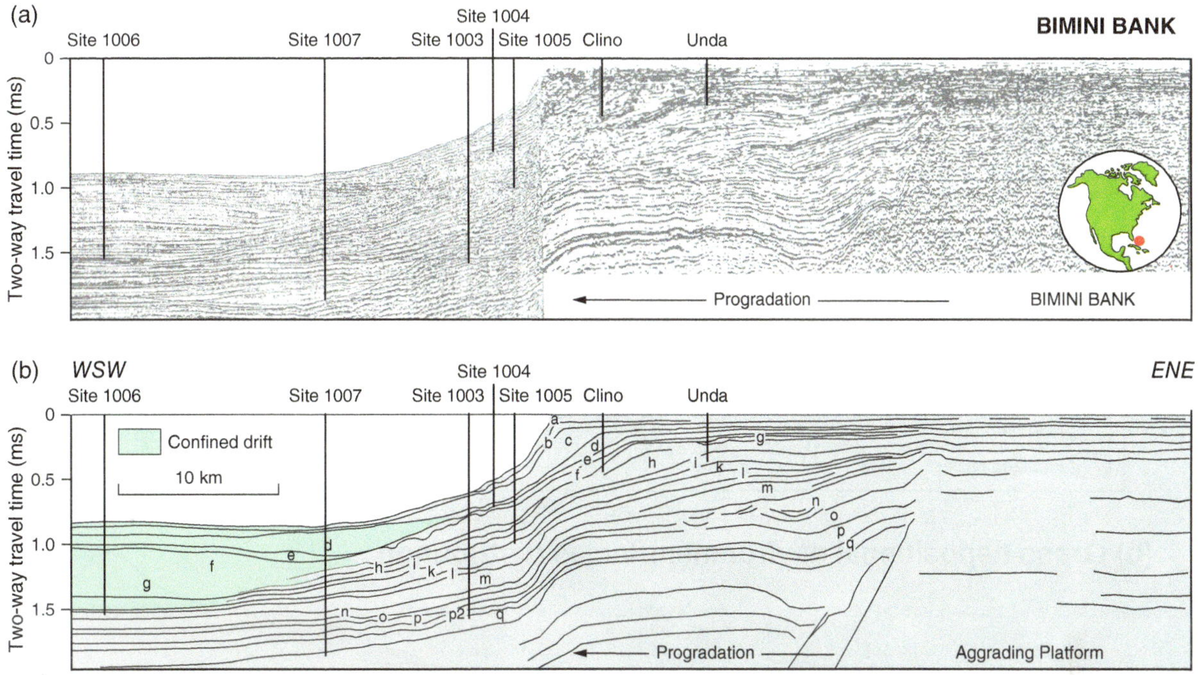

Figure 2.41 Seismic sequences and sequence interpretation along the western platform-to-basin transect of the Great Bahama Bank. (a) Seismic line and interpretation across the western margin of the Great Bahama Bank showing the progradation of the accretionary slope (Anselmetti et al. 2000; Eberli 2000) and toe of slope deposits (Mulder et al. 2017). (b) Sequence interpretation of the prograding slope (Eberli 2000). Note the differences in two-way-time-thickness for seismic sequences g-a when moving from the platform top to the basin as seismic sequences q-h display equal two-way-time-thicknesses. These changes reflect a transition from a carbonate ramp (h-q) towards a flat-top carbonate platform (g-a) in response to oceanographic changes (Betzler et al. 1999; Reijmer et al. 2002).

2.3 Methodologies and Applications

2.3.1 Introduction

Three end-member approaches to facies analysis span the spectrum of scales discussed earlier (levels 1–5 in Table 2.1): (i) regional-scale; (ii) rock facies-scale; and (iii)-microscopic-scale (Figure 2.15d). The type of investigation undertaken is determined by the nature of the data available, aims of the study and available time frame. These approaches may be conducted independently, in parallel, or sequentially, depending on the circumstances of the investigation. Sometimes they reflect different starting points in the context of larger investigations: (i) the regional approach ('top-down') progresses from larger to smaller hierarchical scales (e.g. systems tracts to architectural elements); (ii) the classical rock facies approach ('bottom-up') progresses upwards from smaller into successively larger hierarchical scales (e.g. from grains to facies and onto facies successions – see Section 2.2); and (iii) microscopic facies investigations start at a much smaller scale (i.e. pores to grains and onto laminae) and typically build-up to rock facies and maybe to flow units in a reservoir model (architectural elements or facies associations). In practice, the consistent integration of the observations and interpretations at all scales yields the greatest insight. However, in many cases (e.g. graduate and post-graduate projects), dataset limitations may restrict an investigation to a single level.

2.3.2 Regional-Scale

This approach can be adopted when a major aim is to link facies and stratigraphic interpretations with coeval tectonic processes and events, typically within a framework of megasequences (Hubbard 1988). It requires sufficient regional data, such as 2D or large 3D seismic surveys or extensive and/or previously studied wells or outcrops, to allow early definition of the tectono-stratigraphic framework, environments and the depositional profile of the basin (e.g. Figures 2.42, 4.22 and 4.23).

Megasequences define level 5 in the hierarchy of scales discussed earlier (Table 2.1) and typically link major sedimentary units to coeval tectonic settings, such as syn-rift, sag, passive margin and foreland basin-fills.

Each megasequence is typically hundreds to thousands of metres thick, displaying a different and characteristic stratigraphic pattern and is bounded by major unconformities (Hubbard et al. 1985). Within this framework, sequence stratigraphic approaches are utilised to define the next smaller level of stratigraphic hierarchy through recognising sequence boundaries, maximum flooding surfaces and systems tracts (see Section 4.4 for definitions). Incorporating data from wells and/or outcrops in a sequence stratigraphic context allows a further step in the depositional hierarchy to be evaluated. For example, validating 'bottom-up' interpretations of facies and gross depositional environments, such as sandy delta top, muddy slopes, carbonate build-ups or

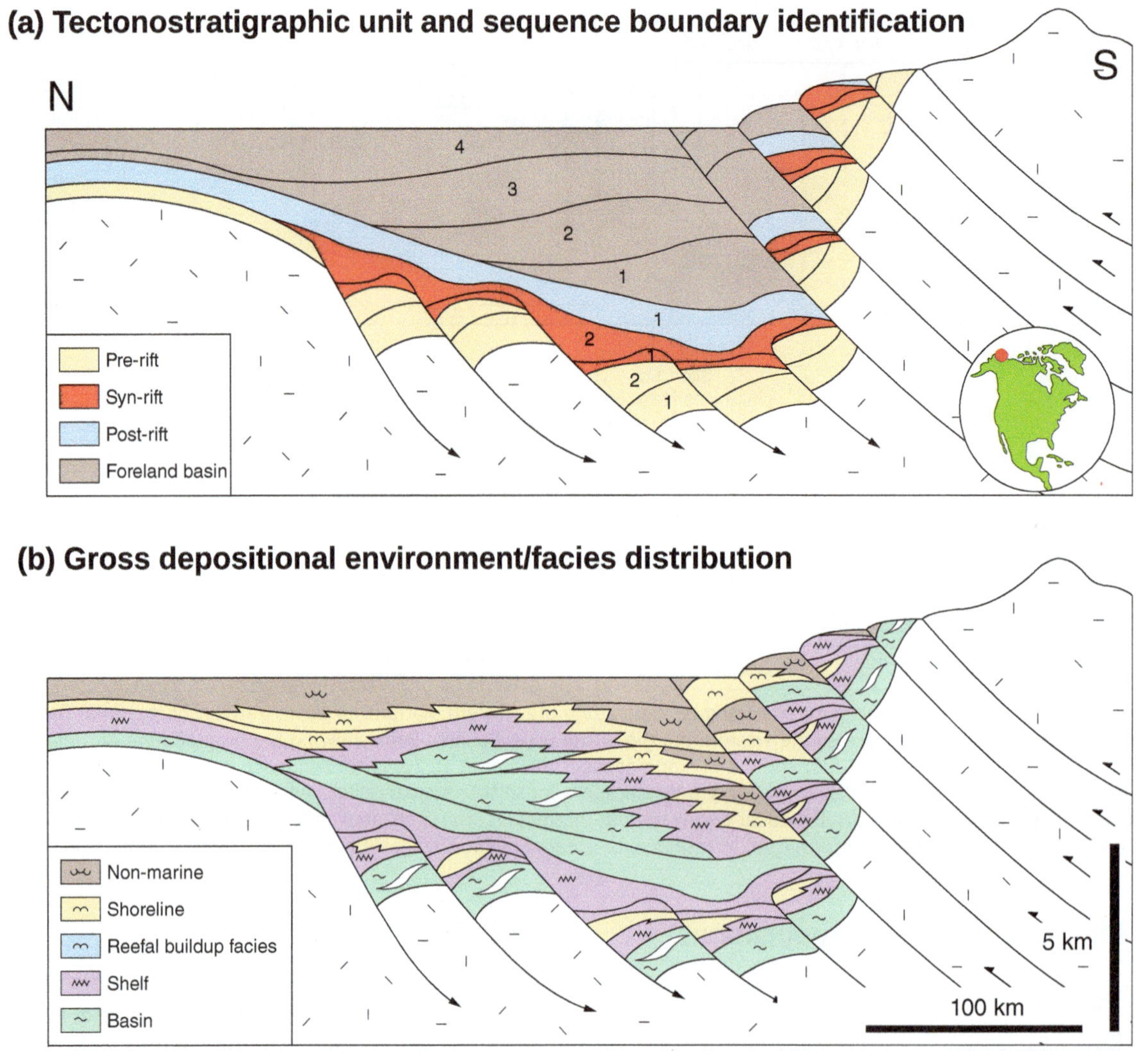

Figure 2.42 A schematic cross section based on the Colville basin Alaska, illustrating megasequences, showing the typical two-stage workflow (Hubbard et al. 1985 and Hubbard 1988, for interpretation methodology). (a) Sequence boundary identification, mainly from stratal terminations, stratal geometries and unconformity recognition. (b) Regional-scale facies interpretation (i.e. gross depositional environments), mainly from well log-calibrated seismic facies identification. A detailed chronostratigraphic chart and seismic cross section for this basin are also shown in Figures 4.22 and 23.

submarine fans. A major outcome is to construct sequence stratigraphic models and to assess the impact of temporal changes in allogenic processes, especially tectonics (e.g. expressed in tectono-stratigraphic charts and Wheeler diagrams; see also Williams 1993). Any subsequent rock-based facies analysis can be placed into this larger depositional context. Conversely, in some cases, a re-evaluation of previous rock facies-scale studies may be required to address any inconsistencies with a new regional interpretation.

The regional-scale typically focuses on several tens to hundreds of metre-thick successions that can be subdivided based on their distinct seismic expression, including both the seismic stratigraphic (rock) intervals and the nature and extent of their bounding stratal surfaces. The assumption is that individual seismic facies, displaying similar frequencies, amplitudes and reflector continuity characteristics, were deposited in similar gross depositional environments (Veeken and van Moerkerken 2013). However, seismic facies interpretation should always be guided by calibration with the relevant rock properties, including at the appropriate depth of burial (velocity and density), even if this means using global rather than local relationships (Bacon et al. 2007; Simm and Bacon 2014). When dense and high-quality data allow (e.g. regional 3D seismic), and where rock properties are favourable (e.g. high porosities, high sedimentation rates, km-thick intervals and moderate compaction and/or cementation), imaging of gross depositional environments can be of exceptionally high quality (e.g. Figure 2.14b,c and Section 2.6.3/Case Study-3).

Seismic facies analysis provides a systematic framework for describing the nature of stratigraphic units in terms of their internal seismic characteristics and the geometries of their bounding surfaces (Mitchum et al. 1977; Vail et al. 1977; Veeken and van Moerkerken 2013). Typically, this includes amplitude strength (bright, moderate, dim), amplitude continuity (continuous versus discontinuous), frequency (low, moderate, high) and geometrical patterns (parallel, divergent, mounded, channels). Many of these descriptors ('attributes') are generic and their interpretation requires an understanding of the broader geological context and/or well log calibration (e.g. Mitchum et al. 1991). Geometrical aspects can convey environmental significance when amplitude maps resemble geomorphological features, for example: meandering channels with point bars and lateral accretion surfaces (see Section 2.6.3/Case Study-3) or clinoforms defining the shape and geometry of carbonate build-ups (Figure 2.13a, c) or basin margin slopes (Figure 2.14).

The confidence level of a seismic facies interpretation is enhanced by rigorous rock facies-scale calibration of petrophysical data and/or facies associations derived from wells (e.g. wireline logs, cuttings, sidewall samples and/or cores) or outcrop data. During regional studies, the key aim is often to establish the regional stratigraphic and depositional setting, including the construction of palaeogeographic maps and/or tectono-stratigraphic charts. This is difficult to achieve with seismic data alone. Hence, a common starting point is to synthesise biostratigraphic and lithology data from well reports or from geological survey reports. Microfossils (e.g. foraminifera, palynomorphs and nannofossils) often provide initial insight into the stratigraphy and depositional environments: (i) biostratigraphic dating from pelagic microfauna (e.g. nannofossils can define stratigraphy with resolution down to 20–50 ky); and (ii) environmental interpretations from benthic foraminifera (e.g. often showing water depth and/or salinity controls) and palynological data (critical in coastal plain and coastal/deltaic successions). Abrupt changes in the type, abundance and/or concentration of biostratigraphic data often highlight regional (allogenic) changes in depositional conditions: (i) an abrupt upward change from abundant and diverse marine microfossils to rocks rich in pollen might reflect a sequence boundary; (ii) concentrations of marine microfauna with phosphate, glauconite and/or organic-rich sediments commonly reflect marine maximum flooding surfaces; and (iii) the sudden appearance of nannofossils, dinoflagellates or acritarchs in a predominantly non-marine or deltaic succession is a good indicator of a marine incursion (flooding surface). Micro-fossils are commonly used to develop a stratigraphic framework with key species or assemblages forming well-constrained marker horizons. Key biostratigraphic events must then be calibrated with their well log and seismic expressions. Biostratigraphic interpretations are invaluable at all scales of analysis, including reservoir studies.

Understanding the depositional profile (e.g. ramp versus rimmed shelf; Figure 2.3) is an important component of the large-scale approach in both clastic and carbonate systems (Table 2.1-IV5 and V5). This is particularly true for carbonates because of the variety and fundamental importance of depositional profiles as well as the identification of the correct factory type(s) (see Chapter 11). Seismic data provide one of the most reliable means of determining large-scale depositional profiles (e.g. Table 2.1-VIe and VIIe and -VIE). Other approaches include chronostratigraphic- and/or genetic facies-constrained correlations (Figure 2.36), which can be further facilitated by extensive, high-quality outcrops (Figure 2.31b) or dense grids of wells distributed at a regional scale (Figure 2.38a). Volcanic tuffs are relatively rare but can also provide direct insight into the depositional profiles of mud-dominated shallow- to deep-marine depositional systems, as first demonstrated by Asquith (1970) in the Mesozoic of the Western Interior Seaway (see Section 9.6). A more common practice is to use marine flooding surfaces, including condensed horizons, transgressive lags (Figures 2.32 and 2.33) and/or winnowed layers (e.g. wave ravinement surfaces in Figure 2.38b). These provide a high-resolution chronostratigraphic framework, which, when expressed in depth relative to the palaeo-horizontal at the time of deposition, approximates the geometry of the depositional profile and shoreline trajectory (Figure 2.38b).

The regional-scale approach has several advantages. First, it results in an evaluation of the impact of a spectrum of allogenic processes: (i) relative sea-level (eustasy and subsidence), (ii) tectonics, (iii) climate, (iv) hinterland geology, (v) source-to-sink and sediment flux, (vi) chronostratigraphic framework, (vii) depositional profile (Figure 2.42a); and, especially for carbonates, (vii) inherited morphology and (viii) variations in production and export relationships depending on the carbonate factory. Second, the larger sequence stratigraphic framework provides a context that can guide future detailed stratigraphic and/or rock facies analysis (Figure 2.42b). Third, it can also help focus on new or potential future problems, enabling short-cuts to be identified for future investigation and/or new data acquisition to be planned.

2.3.3 Rock Facies-Scale

This is the traditional approach to facies analysis, focusing on rocks, either from outcrops or cores (Section 2.2). In both clastic and carbonate studies, the primary tasks of the rock facies approach are often twofold:

i) to determine the relevant sedimentary processes enabling diagnosis of depositional environments and prediction of sedimentary geometries, and
ii) to evaluate the geological processes controlling rock properties and/or reservoir characteristics, starting at the time of deposition through to their subsequent modification during depth-related burial diagenesis.

Determination of depositional environment is nearly always an end point of the rock facies approach as this enables predictions that can be tested by further observation, thereby narrowing the range of hypotheses that fit the data (i.e. steps 1–3 in Table 2.1). It is also a pre-requisite for understanding the controls on reservoir and non-reservoir rock properties in both siliciclastic and carbonate deposits, even when diagenesis has had a major influence. This approach delivers the environmental 'building blocks' for a raft of subsequent applications, spanning sequence stratigraphic studies at the larger scale down to rock property predictions for various reservoir characterisation and modelling purposes (e.g. Sneider 1990; Weber and Van Geuns 1990, 2005; Lucia 1999, 2007).

2.3.4 Microfacies-Scale

There are two important drivers for a microfacies-scale analysis:

i) When petrographic microscopy and allied analytical techniques are used to characterise framework grain mineralogy and its environmental significance (always essential in carbonate rocks, but generally of subordinate significance for most siliciclastic rocks).

ii) When pore-space properties are a key part of a project, such as when facies need to be understood with respect to their storage, seal and/or flow potential (e.g. oil and gas reservoirs, aquifer studies, and waste disposal projects).

We start with a brief summary of the methods used for determining the geological controls on pore-space characteristics and rock quality in both carbonate and siliciclastic rocks. Petrophysics is beyond the scope of this review, however facies analysis has always been connected to petrophysical applications, not least because of the intimate link between facies characteristics (at the rock facies-scale), quantitative physical log responses, and pore-scale properties (at the microscopic-scale).

In subsurface projects, rock facies types form the starting point for core analysis (based on core plug measurements) and for identifying statistically significant groups of different rock types (Figures 2.43 and 2.44): Key parameters for storage and fluid-flow potential and seal capacity estimation are:

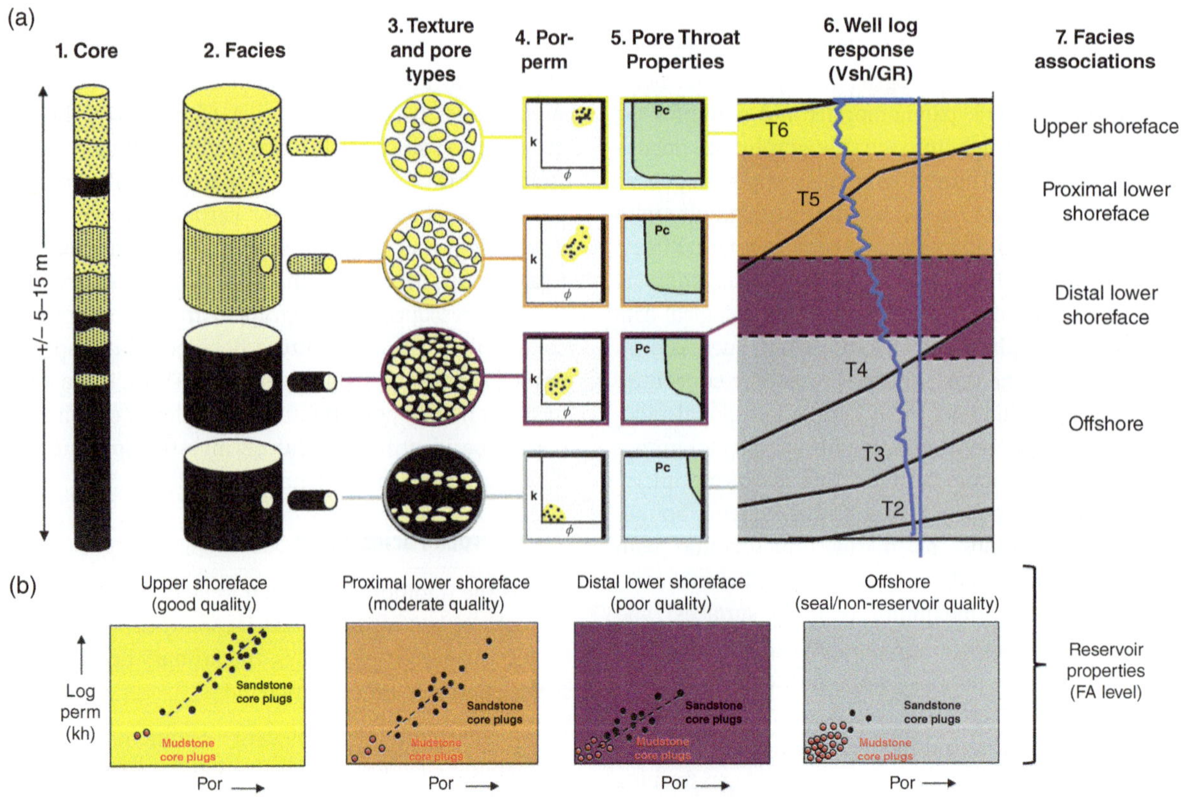

Figure 2.43 Schematic methodology for calibrating facies with core analysis data (i.e. pore-space properties) and well log response. (a) From left to right: (1) obtain representative core data for all facies types; (2) determine macroscopic facies types, conduct facies analysis and compile representative core analysis measurements from core plugs; (3) select representative core plug samples for each facies/facies association and evaluate microscopic rock properties (grain size, sorting and pore types); (4) collate representative routine core analysis data (porosity and permeability) for the facies/facies associations; (5) extract pore throat dimensions and sorting properties from capillary pressure (Pc) plots representative of each facies/rock type; (6) calibrate core analysis measurements to well log response using the main facies/facies associations and rock types defined earlier (T2-6 = relative time lines) (see well log 3 in Figure 2.36b for larger depositional context); (7) subdivide the reservoir interval into its facies associations/environments and/or flow units, usually guided mainly by texture, permeability and pore throat properties. (b) Representative porosity/permeability relationships for the main facies/facies associations, establish core porosity versus log porosity trends and use these to infer permeability estimations in uncored wells. This workflow provides a genetic link between depositional environment and reservoir characteristics and provides the basis for constructing 3D reservoir geological models *Source:* Ebanks et al. 1992 / Scientific Research Publishing Inc.; logs (a column 6) and cross plots. (b) Courtesy of J-P. Tromp.

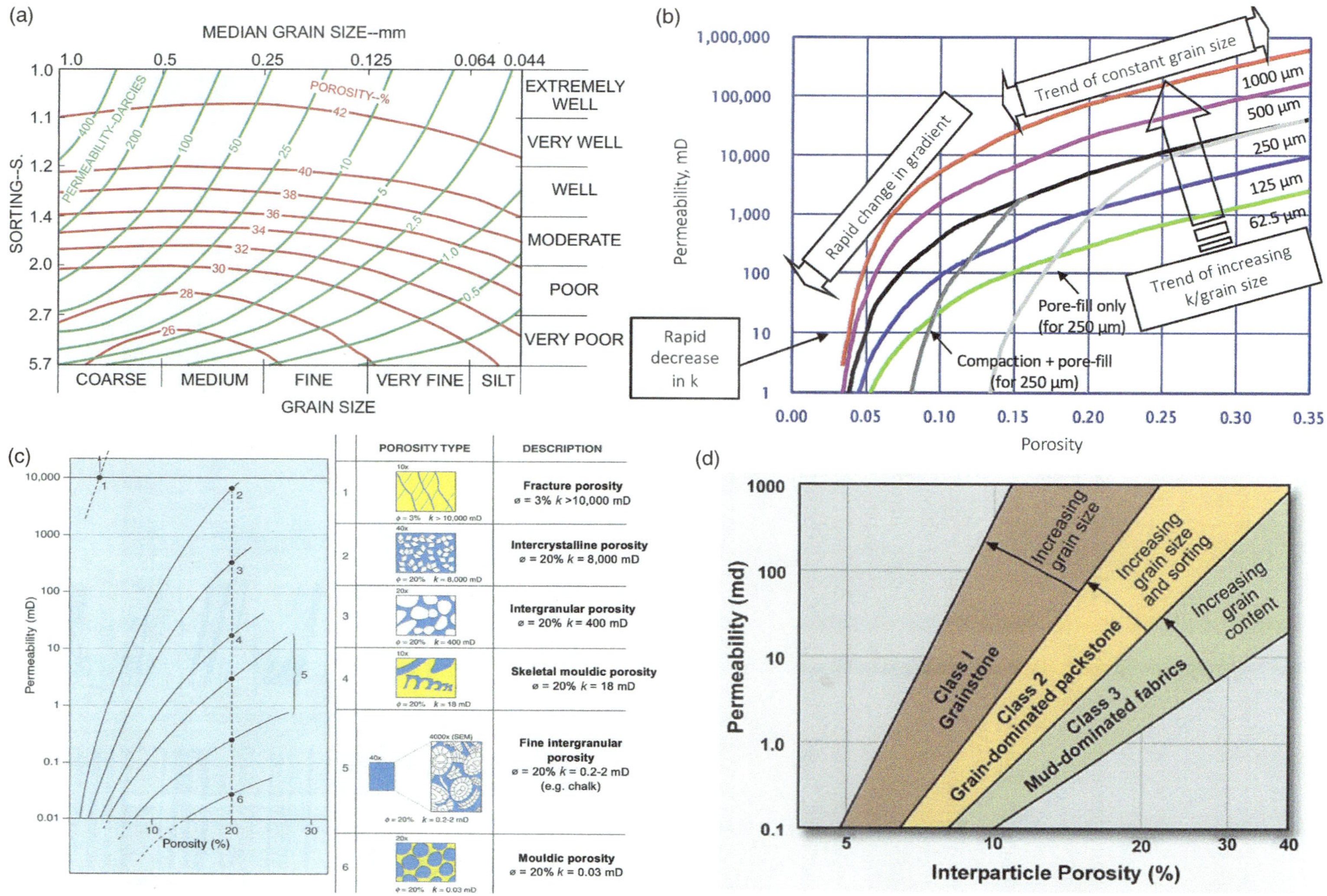

Figure 2.44 Geological controls on pore-space properties in siliciclastic (a, b) and carbonate (c, d) rocks. (a) Experimental work by Beard and Weyl (1973) using artificial sand packs to demonstrate the relationship between sediment grain size and grain sorting on porosity (red iso-porosity lines 26–42%), and permeability (k, green iso-permeability lines 0.1–400 D). Note how grain sorting controls porosity, irrespective of particle size, while grain size is the dominant control on permeability (from Sneider et al. 1984; reproduced in Slatt 2006). (b) Siliciclastic porosity–permeability trends calculated from a digitised random bead pack, which have been calibrated with those measured in actual reservoir rocks. The plot summarises the control exerted by grain size, highlighted by five grain size classes: 62.5 μm (silt/vfs), 125 μm (vfs), 250 μm (f/ms), 500 μm (m/cs) and 1000 μm (c/vcs). The additional impact of reduced porosity and permeability due to cementation/pore-fil l (i.e "fill") (pale grey line) and compaction plus pore-fill (dark grey line) is only shown for the 250 μm grain size class (black grain size line). Based on Bryant et al. 1993, and Cade et al. 1994; reproduced courtesy of Nautilus Ltd. (c) Summary of carbonate reservoir pore systems. To the left is a typical porosity–permeability cross-plot covering 6 orders of magnitude in permeability (0.01–10,000 mD) and a porosity range of 0–30%. Six schematic porosity–permeability regression lines correspond to six different carbonate rock types. Samples along the 20% porosity line are highlighted for five samples (numbers 2–6) and cover 6 orders of magnitude of permeability. The sixth sample (number 1) has a much lower porosity, but very large permeability, and so occupies a different space on this cross-plot (typical of a fractured reservoir rock). To the right is a summary of the six rock types in terms of (i) simplified graphical image of each of the six the pore networks and (ii) text summary of the pore network and lithology. Looking at the 20% porosity line highlights the problem facing all carbonate reservoir studies: which is the most appropriate porosity–permeability relationship for estimating permeability from porosity logs? Answer: you need to determine the nature and origin of the pore network(s) initially from cores and then calibrate the results with well log data (*Source:* (Weber and van Geuns 2005) / with permission of Elsevier). (d) Three examples of limestone rock classes (1–3) differentiated by their interparticle porosity versus permeability relationships (*Source:* from Lucia, after Lucia (1999, 2007) / American Geological Institute (AGI)).

i) porosity,
ii) permeability,
iii) pore throat properties.

Petrographic investigation of framework grains and any diagenetic modifications is typically made on the same core plug samples enabling calibration with rock facies.

2.3.4.1 Carbonates

Investigating the microfacies-scale of carbonate rocks involves a range of petrographic and geochemical analytical methods (e.g. standard microscopic, scanning electron optical techniques, X-ray fluorescence and isotope geochemistry). These methods generate unique data needed to investigate three critical topics: (i) particle type, (ii) depositional texture and (iii) diagenetic fabric. Pore-space properties of carbonates display a much wider range of porosity–permeability relationships compared to their sandstone counterparts (in Figure 2.44 compare a and b with c and d). A carbonate example is provided from Late Cretaceous reservoirs in the Miskar Field, Tunisia, showing different facies with contrasting pore systems and reservoir quality (Figure 2.45). The applications of microfacies analysis in carbonate rocks are wide ranging. For example they can lead to the prediction of palaeoceanographic conditions and climatic variations determining the overall depositional setting and thus the variation in the sediment production and export (shallow to deep) of both non-skeletal and skeletal grains. The microfacies analysis of carbonates is discussed further in Chapter 11.

2.3.4.2 Siliciclastics

The microfacies-scale of analysis has seen a recent resurgence in mudstone facies analysis related largely to shale gas plays. The approach (analytical techniques and acronyms in Figure 9.6) has been successfully utilised to determine a wide range of particle types, textures and primary stratification styles in mudstones that were previously largely unknown. This has resulted in much more refined interpretations of processes of erosion, transportation and deposition of mud-grade sediment (e.g. Schieber et al. 2007). Consequently, there is now a wider range of models

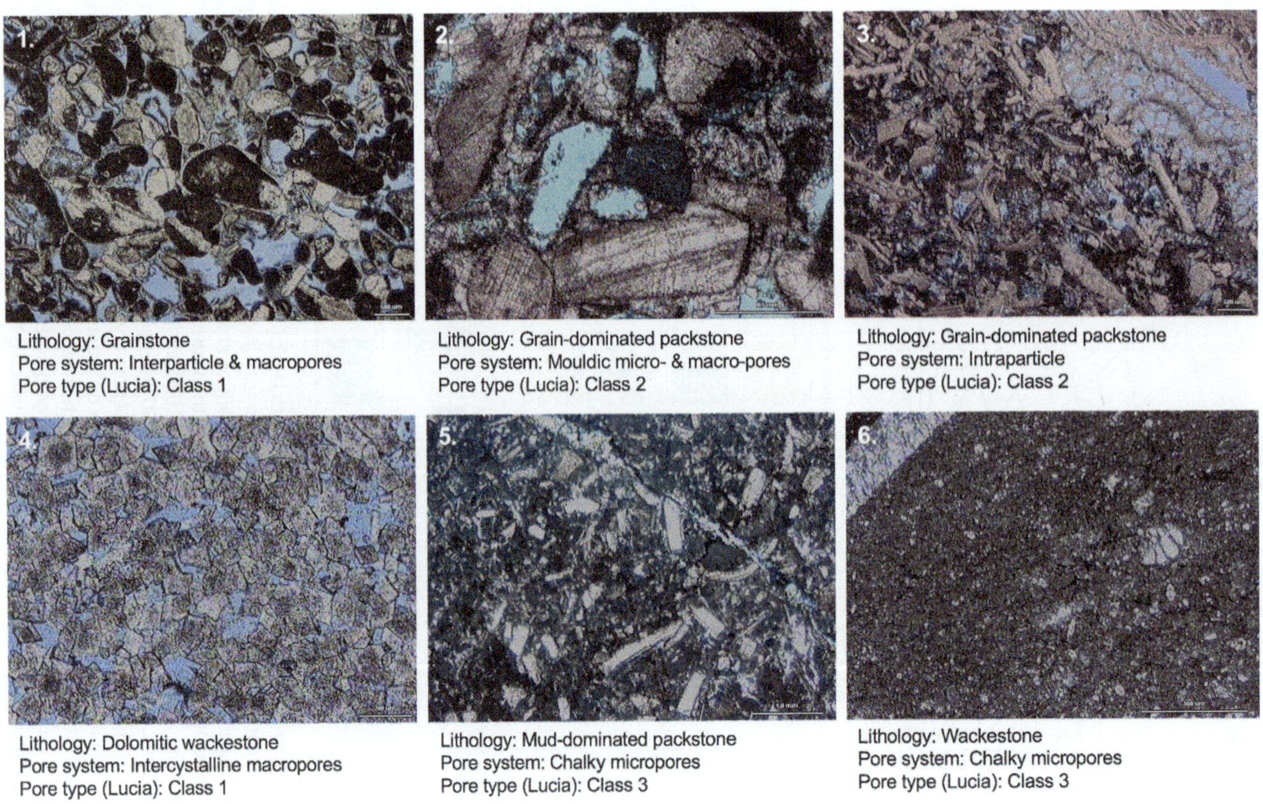

Figure 2.45 An example of pore-space variability in carbonate reservoirs in the Late Cretaceous carbonates of the Miskar Field (Tunisia), which shows a typically wide range of facies and pore types, comprising the following: (1) grainstone with interparticle macro-pores (top left); (2) grain-dominated packstone with mouldic micro- and macro-pores (top middle); (3) grain-dominated packstone with intraparticle pores (top right); (4) dolomitic wackestone with intercrystalline macro-pores (bottom left); (5) mud-dominated packstone with chalky micro-pores (bottom middle); and (6) wackestone with chalky micro-pores. Each pore type (Classes 1–3, after Lucia 1999) has a separate porosity–permeability relationship (Lucia 2007). Reservoir quality ranking (based on visual estimation) shows the following: **4 > 1 > 2 > 3 > 5 > 6.** *Source:* Kay Johnson et al. (2010) / AAPG.

for interpreting mud-rich depositional environments, especially in shallow- to deep-marine settings (see Section 9.3 and Section 2.6.1/Case Study-1).

Aspects of the microscopic analysis of coarser-grained siliciclastic rocks for the understanding of petrophysical rock properties are summarised in Figures 2.43 and 2.44. This approach enables facies analysis to be applied to reservoir characterisation studies, especially calibration of a genetically based facies scheme with its well log response, seismic expression and/or outcrop analogue data (e.g. Figure 2.46). Such analyses are the starting point of most 3D reservoir geological investigations, including the construction of 3D geocellular models. A 3D facies model of a reservoir can thus be converted into an equivalent numerical version, providing a genetically constrained representation of pore-scale rock and fluid properties (Weber and Van Geuns 1990; Ringrose and Bentley 2015).

Microscopic-scale methods are also required when pore-space properties, diagenesis and/or provenance aspects are being studied at a regional scale. For example, if the investigation is tasked with evaluating the quality of sandstone reservoirs, then understanding framework grain mineralogy and texture (Figure 2.47) together with depth-related diagenetic trends due to compaction and/or cementation (Figure 2.48) becomes essential. Mineralogical and textural maturity are linked to hinterland conditions (composition, climate and weathering) and to physical-chemical reworking, first, along the sediment transport pathway and, second, during preservation in the depositional environment (Figure 2.47). Thereafter, depth-related diagenesis progressively reduces primary reservoir properties, albeit at different rates (Figure 2.48).

Petrographic studies are needed in the identification of sediment provenance and sediment routing systems. Especially useful are distinctive heavy mineral suites or radiometrically dateable framework grains, such as detrital zircons or some feldspars. Changes in framework grain mineralogy, such as from a quartz-rich cratonic source to a mafic-rich, volcanic source, can result in major differences in reservoir quality and contrasting porosity-depth trends. Sandstones with low mineralogical and textural maturity are more susceptible to the alteration of unstable framework grains during depth-related burial diagenesis, including transformation into different authigenic clay minerals at different burial depths, often with rapid deterioration in reservoir quality. In contrast, sandstones with high mineralogical and textural maturity contain much lower proportions of mineralogically unstable material and, consequently, display slower rates of depth-related reduction in porosity and permeability.

2.4 Depositional Models

2.4.1 Introduction

'Depositional model' is a general term used to describe a graphical summary of the interpretation of sedimentary rocks in terms of a depositional system. The term is broad in scope, covers a wide range of approaches and forms a vital means of communicating sedimentary geology information and interpretations (e.g. Figure 2.49). Sometimes, but not always, the term is synonymous with 'facies model'. Here we follow Walker's (1984) definition of facies model (aimed mostly at levels 2–3 in Table 2.1), with its focus on autogenic processes, and use the term depositional model for higher levels (5–6 in Table 2.1), which extend into allogenic processes and spatial and temporal variations within a depositional system.

Depositional models range from those directed towards advancing our basic understanding of sedimentary processes and depositional systems, through to those summarising relatively local investigations directed at interpreting individual rock units or part of a modern environmental setting. Some depositional models may summarise one ancient rock succession (Figure 2.49a) or synthesise one modern depositional environment (Figure 2.49b). Others may be a synthesis of a wider range of modern and/or ancient examples (Figure 2.49c). All facies analysts need to be skilled in developing their own depositional models, since these are needed to guide the construction of maps and correlation sections and/or to convey interpretations and conceptual thinking. All applied aspects of sedimentary geology require depositional models to guide the prediction of an array of sedimentary features at basin- and/or reservoir-scale, such as palaeogeography, rock/facies types, thickness patterns and rock properties (e.g. porosity, permeability), among others. These data enable prediction and volumetric estimation of the raw materials hosted in sedimentary rocks, such as building materials, water, geothermal energy, oil, gas, minerals and/or other geological resources. They also guide other investigations, such as aquifer characterisation, waste disposal efficacy, geohazard risk mitigation and impact of recent-future environmental change, among others.

The first step in constructing a depositional model, after formulating the study's aims and objectives, often involves adopting an existing facies model that pertains to the inferred depositional setting of the rock unit under investigation. This also serves the second step, which is to consider the question: "How, and/or by how much, do these rocks conform to, or differ from, established facies models of the environment(s) being studied?" The third step is to carefully consider the type of data gathering needed to

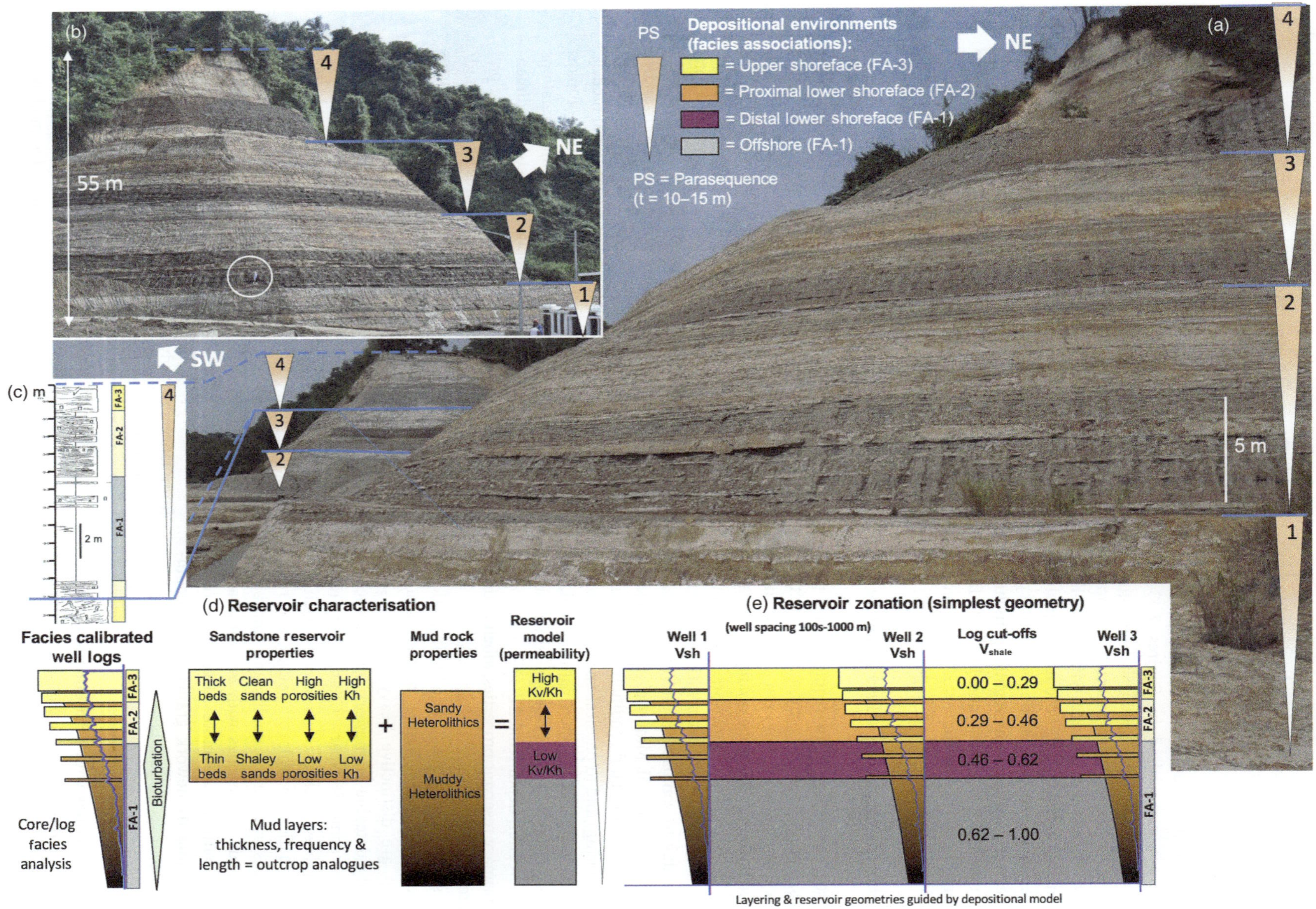

Figure 2.46 Example of a three-dimensional outcrop, which has been used as a reservoir analogue for nearby oil and gas reservoirs (Baram Delta Province, northern Sarawak and Brunei). The vertical facies successions 1–4 are parasequences, based on their shallowing-upward facies profiles, vertical scale and bounding flooding surfaces (blue lines). (a) A tiered outcrop (view to SW, with 'benches' cut at the base of major shale units), comprising four stacked parasequences interpreted as wave-dominated, fluvially influenced, progradational facies successions. Note the discontinuous, gutter-cast sandstones in the shale unit of parasequence-2. The main sandstone units are dominated by HCS/SCS facies. (b) The same outcrop with a different orientation (view towards the NE), enabled the three-dimensional geometry of mudstone layers and primary sedimentary structures to be determined. Geologist for scale (white circle). (c) Sedimentary facies log (parasequence-4), highlighting the three main facies associations (FA-1, -2 and -3) for subsurface well log calibration and reservoir characterisation (see Figure 2.29 for further details of a similar, nearby outcrop in Brunei). (d) Simplified reservoir characteristics, highlighting outcrop-well log calibration, sandstone reservoir property trends, mudstone layer properties and qualitative permeability trends kv and kh are vertical and horizontal pemeabilities (see Figure 2.43 for typical GR log and porosity/permeability relationships). (e) Conceptual reservoir zonation and correlation scheme based on schematic Vshale (volume of mudstone) logs with effective Vshale log cut-off values (from nearby subsurface examples). *Source:* Courtesy of Howard Johnson. d–e courtesy of J-P. Tromp.

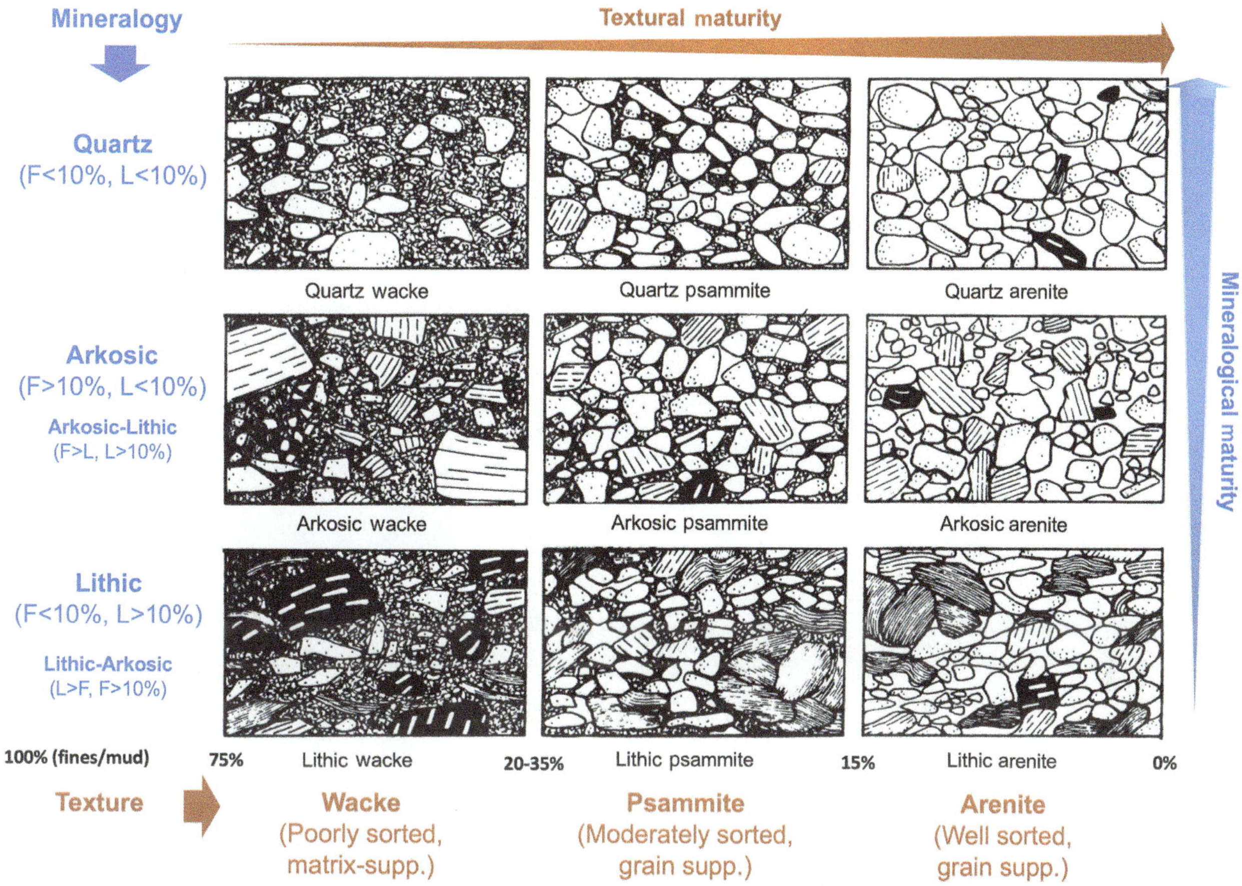

Figure 2.47 Textural and mineralogical maturity of different types of sandstone. Texture and mineralogy exert a primary control on pore-space properties and overall sandstone reservoir quality: (i) proportion of mineralogical stable/unstable grains ('mineralogical maturity') and (ii) the proportion of fine-grained matrix (0–100%) and rock fragments to stable grains ('textural maturity'). This has two important consequences. First, it determines the relative reservoir quality at the time of deposition: texturally and mineralogically mature sandstones (e.g. quartz arenites) have higher initial porosity and permeability compared to, for example, lithic- and/or arkosic-rich wackes. Second, it determines subsequent porosity–permeability trends during burial diagenesis: that is, the rate of reduction of porosity and permeability during burial is highest within immature sediments *Source:* (Nagtegaal 1978) / Geological Society of London.

support a new depositional model. Failure to undertake a careful facies analysis at an early stage carries the risk that the larger depositional model is unreliable, inappropriate, or even fanciful (e.g. Anderton 1985).

2.4.2 Types of Depositional Model

Depositional models are abundant, diverse in content and have evolved in response to changing concepts and applications. Early models (pre-1950) were descriptive and focused on classification schemes, while later models (post-1950s–1980s) evolved from an initial emphasis on process to those more focused on environmental facies models with emphasis on autogenic controls (Reading 2001). These progressively morphed into sequence stratigraphic models by incorporating time, sediment supply/accommodation relationships and allogenic controls (1990s to present-day). Here we provide an overview of the most common types of depositional model (Figure 2.49).

1) Local depositional models are based on specific recent to ancient systems, which may, or may not, have broader implications beyond the investigation. There are two main categories:
   i) Outcrop studies (Figure 2.49a) provide a very high level of detail on facies, facies associations and stratigraphic surfaces. Many are often limited to two dimensions but still provide useful information on dimensions and lateral relationships of sedimentary units for generalised facies models. When outcrop quality is exceptional, and ideally three-dimensional, some may feature prominently in global facies models (e.g. Figures 2.29 and 2.30b; the storm-flood model informed by the Miocene Belait Formation, Baram Delta Province, NW Borneo;

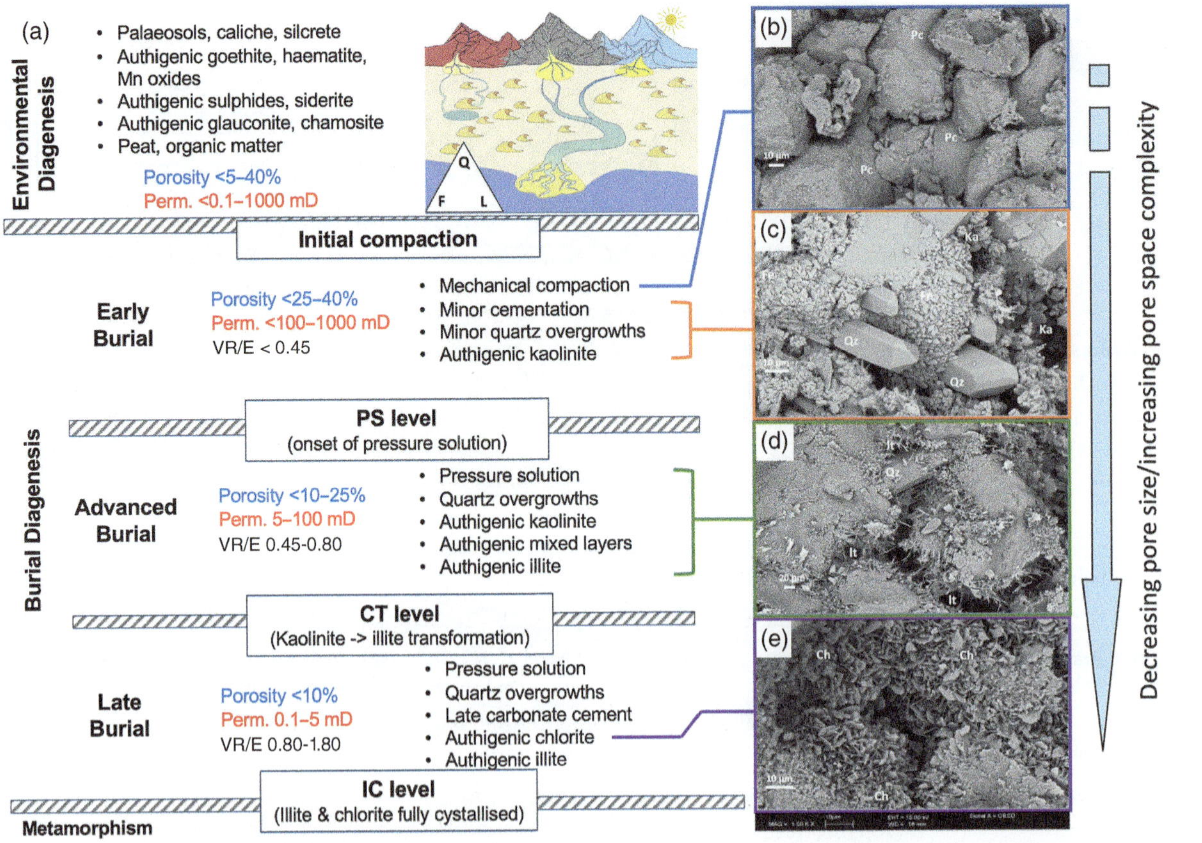

Figure 2.48 (a) Summary of the stages in burial diagenesis of sandstones and their impact on reservoir quality. Note the gradual loss of porosity and permeability. Abbreviations: PS = pressure solution; CT = chlorite transformation; IC = illite-chlorite; VR/E (equivalent vitrinite reflectance, in percent: a measure of source rock maturation) (Modified from Nagtegaal 1978). (b) Minor mechanical compaction typical of early burial with no apparent cementation, minor detrital clay particles and grain point-contacts (Pc); grain surfaces, pores and pore throats are all clearly visible and well connected. (c) Ferric oxide (Fe) grain coatings (micron-thick grain coatings), onset of quartz (Qz) overgrowths and minor clay mineral authigenesis (kaolinite – Ka). (d) Fibrous illite (It) crystals are growing across the pores ('pore-bridging'-type), reflecting increasing pressure and temperature during burial; porosity still just visible, albeit poorly connected, but permeability is greatly reduced due to illite fibres. (e) Grain coating chlorite (Ch) with almost complete occlusion of the pore network (isolated pores are visible). Environment sketch in a from Caracciolo 2020. Courtesy of Nautilus Ltd.

Collins et al. 2017). Such outcrop studies provide unrivalled information on the dimensions and geometries of architectural elements and reservoir heterogeneities, including, for example, dimensions of mudstone layers, pebble beds and/or cemented layers (e.g. Figures 2.32, 2.33, 2.34 and 2.38).

ii) Subsurface studies, especially those with high-resolution 3D seismic data, well logs and cores, can demonstrate fine detail of three-dimensional geometries (e.g. Figures 2.12 and 2.13a) and internal stratigraphy (e.g. Figure 2.13c,d), while still allowing process interpretation through classical facies analysis of core. 3D seismic amplitude or attribute displays in map view may generate almost unequivocal interpretations of depositional environments. sometimes even approximating satellite images (Section 2.6.3/Case Study-3).

2) Models based on the geomorphology and/or hydrology of specific recent to modern systems, such as fluvial and deltaic environments (Figures 2.49b, c). The integration of satellite imagery with other data types, including direct sampling and boreholes, provides a spectacular example of a modern depositional model for the bayhead and/or crevasse deltas in the US Gulf Coast area (Figures 2.8 and 2.9). This is the location of one of the world's largest, albeit unintended, sedimentological experiments in Atachafalaya Bay: the link in the caption to Figure 2.5 shows a period of 33 years (1984–2017) of evolution of the Wax Lake Delta, which has documented the growth of a bayhead delta in a prolonged period of far-sighted data collection (Wellner et al. 2005). This river-dominated delta, formed when flood waters from the Atchafalaya River were diverted through an engineered channel – the Wax Lake Outlet.

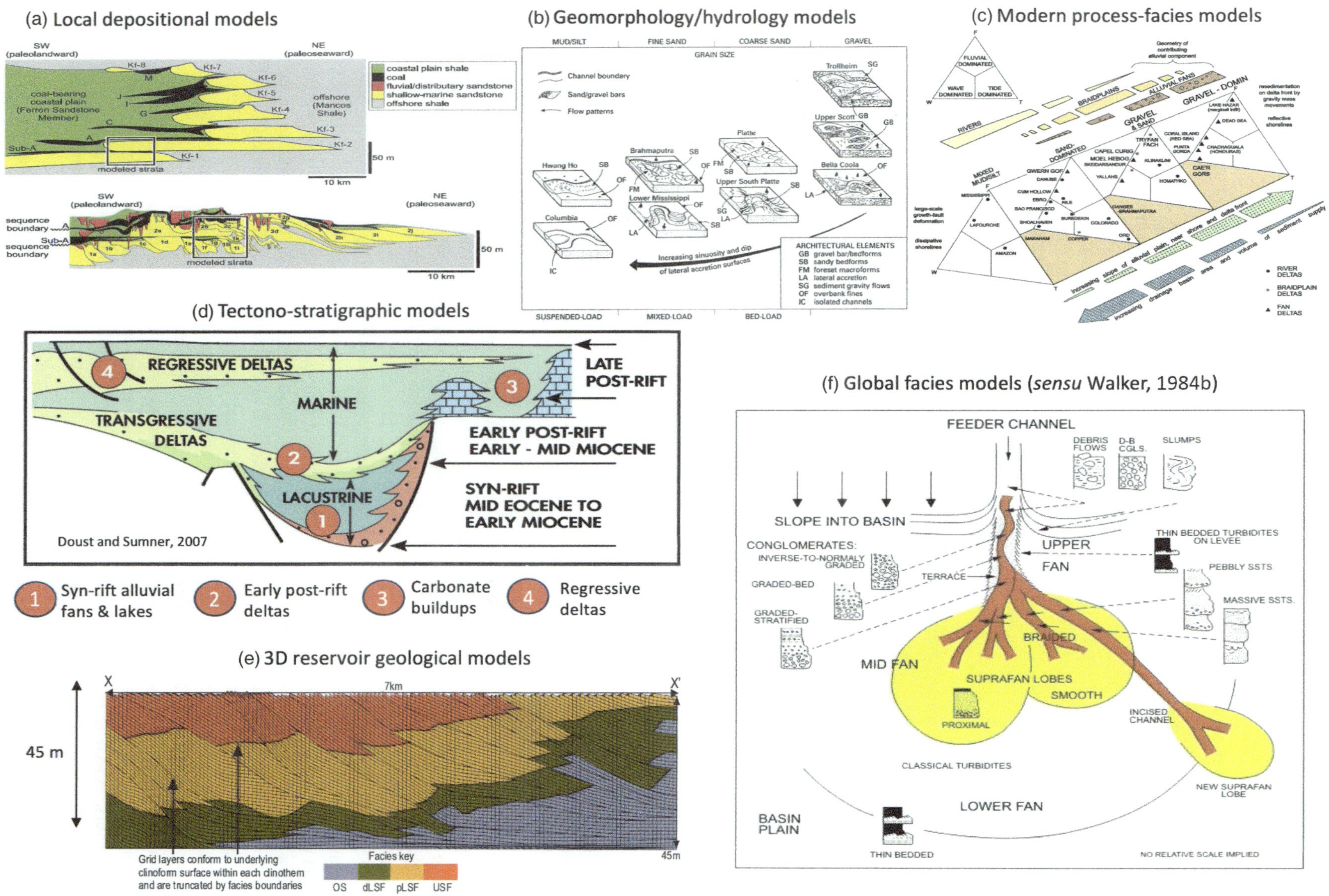

Figure 2.49 A selection of different types of depositional model: (a) local facies model – Late Cretaceous Ferron Sandstone, Utah, USA *Source:* Garrison and van den Bergh 2004 / GeoScienceWorld. (b) Modern geomorphological model – fluvial channel style, architectural elements and dimensions . *Source:* (Orton and Reading 1993) / John Wiley & Sons. (c) Modern process-facies model, combining the ternary process classification of deltas with variations in sediment type *Source:* Adapted from Orton and Reading (1993). (d) Tectono-stratigraphic models (*Source:* Doust and Sumner 2007; Henstra et al. 2016) are generated from combining regional stratigraphy from well data, structure/depth and thickness/isochore maps from well and seismic data with environmental information from biostratigraphic, seismic, well and/or outcrop data. (e) 3D geocellular facies models at variable scales capture fluid-flow behaviour (*Source:* (Sech et al. 2009) / American Association of Petroleum Geologists). The low-angle surfaces (dipping left to right) define clinoforms, guided by outcrop observations and the grid boxes are aligned parallel to the basal clinoform of each clinothem. Facies/environment abbreviations: OS = offshore; dLSF = distal lower shoreface; pLSF = proximal lower shoreface. USF = upper shoreface. (f) A vintage facies model of submarine fans (Walker 1978, 1984c) designed as a global synthesis of a particular depositional system *Source:* Walker (1978, 1984).

Since 1984, the Wax Lake Outlet has discharged into the shallow waters (c. up 10–20 m) of the Atchafalaya Bay and created the artificial Wax Lake Delta. Detailed shallow boreholes and cores, sea-bed sediment sampling and precise mapping of the evolution of the multiple and diverging channels and associated mouth bars have documented the Wax Lake Delta's history (Wellner et al. 2005). Both the dynamic behaviour and facies characteristics of this modern delta have been recorded, including its complex internal stratigraphic architecture. This results mostly from autogenic processes and is dominated by river floods and coastal hurricanes. This example provides abundant analogue information to guide interpretation of comparable ancient deltaic successions (e.g. Bhattacharya et al. 2020).

3) Models and classification schemes based on syntheses of modern environments from around the world (Figure 2.49c), such as deltas (e.g. Galloway 1975; Orton and Reading 1993), clinoforms (Patruno et al. 2015), shoreline-shelf processes classification (Boyd et al. 1992; Nyberg and Howell 2016) and submarine fans (e.g. Reading and Orton 1991; Reading and Richards 1994; Richards et al. 1998). These models summarise the main controlling depositional processes in a depositional system and relate these to planform geometries. Often classifications comprise a series of end members, and the variations between these are represented, for example, along a spectrum or by a triangular plot. Such schemes are commonly developed for modern depositional systems and feature prominently as a source of data and ideas in global facies models (see 6, below). They are a powerful means of demonstrating the range of variation that exists within natural depositional systems. They also provide a basis for classifications that may subsequently be extended into the ancient rock record via facies models.
4) Models based on a particular tectono-stratigraphic setting (e.g. Leeder and Gawthorpe 1987; Gawthorpe and Leeder 2000; Doust and Sumner 2007; Figure 2.49d). These models can be a product of any of the approaches described above but address a specific tectonic setting. Tectono-stratigraphic models are especially relevant where tectonic processes have a profound impact on the type and distribution of depositional environments, including the depositional profile and stratigraphic geometries, such as in rift basins or fold-thrust belts. The tectonic evolution of these settings can be the dominant control on a range of allogenic factors, especially elevation, slope, accommodation and sediment supply, resulting in extreme spatial and temporal stratigraphic variability (e.g. shallow- to deep-marine Upper Jurassic deposits in the North Sea Basin; Fraser et al. 2003). Typically such models are made for the main megasequence settings (e.g. Figure 2.42) but they can also include smaller elements such as syn-depositional topography, models for stepped slopes and ponded basins in deep-water depositional settings (e.g. Cronin et al. 2023). In these situations, the tectonic setting might be the primary control on stratigraphy and the specific depositional environment becomes secondary. This often occurs when the substrate is mobile (e.g. salt and/or under-compacted mud), resulting in especially close links between depositional elements and locally evolving topography.
5) Three-dimensional reservoir models ('geocellular models') are designed to capture 3D fluid-flow (Figure 2.49e). These models are used to simulate the movement of fluids, capturing how these might change due to fluctuations in pressure and/or production and injection rates (e.g. Weber and Van Geuns 1990). Reservoir models typically simulate flow processes on decadal timescales and are commonly used to predict the dynamic behaviour of oil, gas and water during the production history of petroleum accumulations (e.g. Ringrose and Bentley 2015). Such models are equally applicable to aquifer characterisation and to reservoir modelling for carbon storage and carbon dioxide injection and geothermal projects (e.g. Heinz and Aigner 2003; Soltanian and Ritzi 2014; Medici et al. 2016; Halaj et al. 2022). Reservoir characterisation is required at different scales to capture the impact of a range of geological heterogeneities on fluid flow, which are often linked to primary depositional controls (e.g. Weber and Van Geuns 1990; Hornung and Aigner 2002.). Hence, this work is underpinned by facies analysis because the range of reservoir heterogeneities follows the natural stratigraphic hierarchy of all depositional systems (Table 2.1). These models are guided by an understanding of how a depositional system and its facies influence the three-dimensional distribution of reservoir properties, including reservoir and non-reservoir (mudstone layers) body dimensions, geometries and lateral extent. A range of geostatistical methods is used that are designed to quantitatively capture primary depositional patterns. The resulting three-dimensional reservoir facies model is calibrated with petrophysical properties and converted into a numerical geocellular model where fluid flow simulations can be undertaken (Figure 2.49e).
6) Models based on syntheses of facies distributions in recent-modern and ancient examples (Figure 2.49f), with the results distilled into idealised graphic representations, are termed facies models *sensu* Walker (1979, 1984, 1992), a facies model is a general summary of a particular depositional system, based on many individual examples from recent sediments and ancient rocks.

The essence of this definition is the integration of a representative range of recent deposits (i.e. including modern sediments) and ancient examples, which are synthesised into a single model, stripping away local variability and distilling out only those features that are near-universal in occurrence (i.e. a 'global facies model'). The inclusion of 'ancient rock' examples in Walker's definition is not simply to plug the gaps in poorly observed modern processes (such as in deep water and shelf settings). It is only in the ancient record that the effect of the 'preservation filter' can be determined. As discussed in Chapter 4, not all parts of any depositional environment have equal preservation potential.

Facies models of carbonate depositional systems, comprising facies zonation with associated microfacies fabrics, have proven to be remarkably resilient, most notably in the following examples: (i) rimmed carbonate platforms (Burchette and Wright 1992); (ii) carbonate ramps (James 1997); (iii) shelves, epeiric platform and ramps (Lukasik et al. 2000). All models were summarised in Flügel (2010), as follows: (i) facies zones (pp. 661–671); (ii) standard microfacies types (Wilson 1975) (pp. 680–716); and (iii) generalised standard microfacies types for rimmed carbonate platforms and carbonate ramps (pp. 720–723). A detailed summary is also provided in Chapter 11 (this book). Other more general facies models are discussed in James and Dalrymple (2010) and Tucker and Wright (2009).

Facies models start with a thorough facies analysis, which provides the intellectual backdrop to the development of broader depositional models. Facies and depositional models are now embedded in a wide range of other geoscience activities, including sequence stratigraphy (Catuneanu 2022), basin analysis (Allen and Allen 2013), palaeogeographic mapping (Ziegler 1982, 1988, 1990), forward stratigraphic modelling (Burgess and Steel 2008; Burgess and Prince 2015; Granjeon and Gervais 2023), 3D seismic interpretation (Bacon et al. 2007; Rankey and Pöppelreiter 2023), seismic amplitude analysis (Røgen et al. 2005; Simm and Bacon 2014), reservoir characterisation (Sneider 1990; Lucia 2007) and aquifer reservoir studies (e.g. Heinz and Aigner 2003).

2.5 Debatable Depositional Models

There are many cases of debatable sedimentological interpretations, no depositional environment has been exempted, and it will happen again. Sometimes, such debates have been heated, even with negative consequences. However, an exemplar of such a debate that rises above all others is that of isolated shallow-marine sand bodies. This is not because of specific aspects of the different interpretations, interesting though they are, but rather in how the issue was addressed and documented. Here follows a brief description of the situation, then a summary of the three interpretations, followed by the 'solution'.

The term isolated shallow-marine sand bodies is used to refer to sandstone units with the following overall features:

i) 5–20 m-thick, coarsening/sandier-upward, shallow-marine, often glauconitic, sandstones, more rarely conglomeratic;
ii) narrow (c. 5–15 km wide) and elongate (20–50 km long) geometries;
iii) interfinger laterally with, and/or entirely encased in, shallow-marine mudstones;
iv) orientations mostly parallel with internal palaeocurrent directions, sandstone thickness patterns and regional shoreline trends;
v) deposition in shallow water depths based on micropalaeontological data (c. 30 m); and
vi) located some 10 s-km basinward of more continuous shoreline deposits (in some, but not all, interpretations).

Such sandstone bodies are common in Cretaceous successions of the Western Interior Seaway in North America in both outcrop (e.g. Shannon Sandstone in Wyoming) and subsurface settings (e.g. Cardium Sandstone in Alberta) and frequently form stratigraphic traps (e.g. Pembina Field in Alberta and Hartzog Draw Field in Wyoming). Three depositional models have been proposed, sometimes for the same sandstone unit, and fiercely defended (Figure 2.50): (i) shelf sand ridge; (ii) incised shoreface; and (iii) incised valley-fill.

The shelf sand ridge model (Figure 2.50a) was prompted primarily by the apparent location of the sandstone units deposited several 10 s km offshore from the inferred coeval shoreline (Tillman and Martinsen 1984, 1987). Facies evidence, including hummocky cross-stratification, supported storm deposition, while cross-bedding orientation suggested shore-parallel shelfal currents, most likely storm-generated. This placed the depositional setting above storm-wave base, but below fair weather wave base. This setting was consistent with the location of potential modern analogues, most notably the storm-dominated sand ridges located along the trailing edge of the transgressive shorelines of the eastern seaboard of the United States. This also provided a mechanism for sand supply via *in situ* erosion of the seabed during transgression, followed by along shelf transport (e.g. Snedden and Dalrymple 1999). In this model, sand body geometry and isolation are primary depositional features with minimal subsequent modification and largely preserved intact (Figure 2.50a).

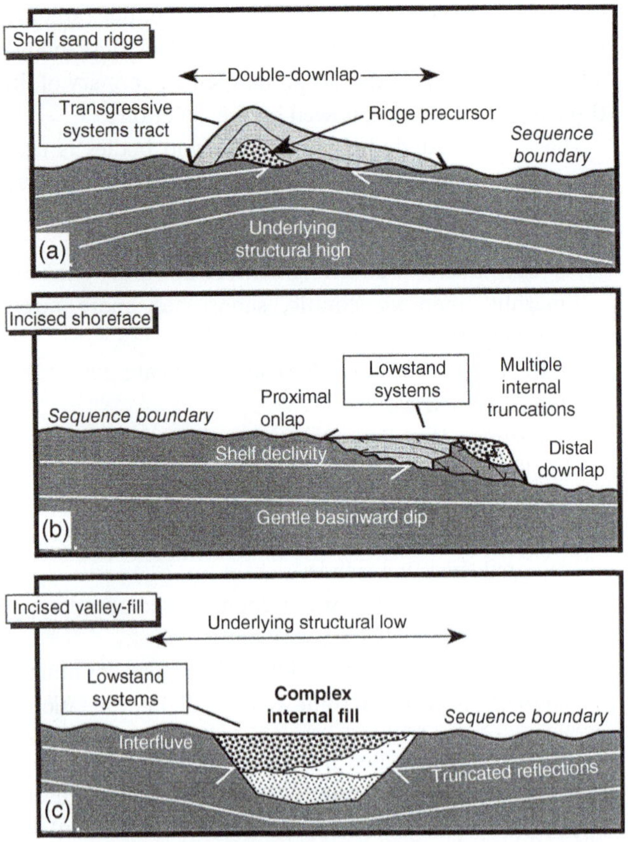

Figure 2.50 Three contrasting types of isolated shallow-marine sandstones. *Source:* (Snedden and Bergman 1999) / GeoScienceWorld.

The incised shoreface model came later and can be traced to sequence stratigraphy concepts, involving sea-level fluctuations, forced regression, marine erosion and variable stratigraphic preservation (Figure 2.50b). This model is supported by a regional correlation framework based on extensive marine erosion surfaces identified by winnowed pebble lags (Plint 1988; Walker and Plint 1992; Walker and Bergman 1993; Bergman 1994). The concept involves rapid basinward migration of the shoreline during periods of falling sea-level (forced regression), which causes the lowstand shoreline to be detached from its highstand equivalents. The basal surface, and its landward equivalent, is interpreted as a regressive surface of marine erosion. During subsequent transgression, a transgressive surface of erosion truncates and winnows the tops of these sands, which contributes to their physical separation from earlier highstand systems (Bergman and Walker 1987). Hence, two factors control sand body geometry in this second model (Figures 2.50, 2.91b): (i) wave and storm processes control shoreface facies characteristics and the orientation of the bounding marine erosion surfaces (in parallel with the shelf sand ridge model), and (ii) convergence of the basal regressive erosion surface and the overlying transgressive erosion surface contributes to landward sand body pinch-out and detachment (in contrast to the shelf sand ridge model).

The incised valley-fill model (Figure 2.50c) stems from the occurrence of some of these sand bodies within brackish water settings, based on their palynological characteristics, plus some facies display inferred tidal indicators (e.g. MacEachern and Pemberton 1994; Sullivan et al. 1997). This model is also supported by the recognition of regionally correlatable erosion surfaces, but in this case, the basal bounding surface, and valley incision, is interpreted to reflect fluvial erosion. The incised valley-fill results from rising sea-level and mixed fluvio-tidal interactions. Also, the elongate sand bodies are aligned parallel to the direction of the dominant fluvio-tidal currents, which are considered to dominate over wave processes. The tops of the incised valley-fill are also bounded by transgressive erosion surfaces (Figure 2.50c). In this model, sand body geometry and isolation reflect a primary depositional geometry, but now mainly defined by bidirectional, rectilinear tidal currents and the orientation of the large-scale valley/estuarine margins (Figure 2.50c).

All three models have been proposed for the Shannon Sandstone, which is widely considered typical of this type of sand body, by experienced and highly respected sedimentologists. So, how to reconcile the differences and answer the opening question? Answer: organise a SEPM-sponsored research conference (held in 1995: *'Tongues, Ridges, and Wedges'*) and get the key players associated with the different models to come face-to-face in the conference meeting room, core store and on the outcrops. First, create a respectful, open-minded and non-confrontational environment, and, second, publish the results with the honest intent that *'no preference for one model over another is given in the volume'* (Bergman and Snedden 1999, their Forward).

This is exquisitely achieved by Suter and Clifton (1999) in their final synthesis chapter, which provides an exemplary, impartial and objective review of the lingering controversies surrounding the Shannon Sandstone. It is a timeless model of sedimentological reasoning and synthesis, carefully weighing up the pros and cons of each element of the interpretation (e.g. facies, geometry, internal architecture, inferred process and modern analogues). By way of analogy, it is presented in the form of a judge-led public inquiry (i.e. primarily gathering and synthesising the facts), as opposed to a legal prosecution (i.e. finding, exposing and punishing the guilty parties!). It concludes, respectfully and honestly, that *'no interpretation proposed to date adequately explains all aspects of the Shannon Sandstone'.*

Suter and Clifton (1999, p. 353) also respectfully remind us all of a crucial limitation in facies analysis: *'We who decipher ancient depositional systems prefer to think that we apply rigorous scientific methods to our efforts. Yet we must acknowledge that* ***woefully few incontrovertible diagnostic tools lie at our disposal**'* (our emphasis). This serves as a valuable, and still current, reminder for the rest of this book and your next interpretation project! It also re-emphasises the importance of integrating all aspects of a rock succession when interpreting depositional environments, including the need to openly acknowledge limitations that might stem from weak or conflicting evidence. In other words, there are no 'silver bullets'!

2.6 Case Studies

2.6.1 Case Study-1: Facies Analysis of Mesoproterozoic Roper Group, Greater McArthur Basin, Northern Australia – An Integrated Core-Well Log Facies Analysis

2.6.1.1 Aim

This case study synthesises the facies, depositional environments and sequence stratigraphic relationships in the 1.4 Ga Mesoproterozoic Roper Group of the Greater McArthur Basin (GMB) in northern Australia (Figure 2.51). The facies analysis utilises extensive core and well data to reconstruct the sedimentary processes and environments with two main outcomes: (i) a facies-based depositional model and (ii) a genetic stratigraphic framework. The intent here is to provide an example of methodology rather than presenting a full description and interpretation (see Abbott and Sweet 2001, Crombez et al. 2023 and Johnson et al. 2024, for further details).

Using such an old part of the geological record places exclusive reliance on physical facies characteristics for process-environmental interpretations, since only the most primitive life-forms, most notably single-celled acritarchs, are evident (Javaux et al. 2001). Additionally, Mesoproterozoic landscapes were barren of much biological activity, while physico-chemical processes in the oceans, shallow seas and atmosphere differed from those of today.

2.6.1.2 Data

The study area is onshore, mostly in the subsurface but with adjacent outcrop equivalents (investigated separately), and has suffered minimal deformation. This study used 4 km of open file drill core from 10 wells, plus additional uncored wells with comprehensive and relatively modern well log suites. The cores, typically more than several hundreds of metres long, enable direct facies observation of not only all the main tens of metres-thick sandstone units but also the volumetrically dominant hundreds of metres thick mudstone intervals, including well-developed organic-rich shales (Figure 2.52).

Continuous core-calibrated well logs cover the stratigraphy and extend across c. 60,000 km^2 (c. 250 km along depositional strike, from NE to SW, and c. 250 km along depositional dip, from SE to NW).

2.6.1.3 Tectono-Stratigraphic Setting

The GMB is a huge (c. 180,000 km^2) Proterozoic intracratonic basin (extending c. 700 km N-S and up to c. 800 km E-W), located in the northwestern part of the Australian Craton. The craton is bounded to the south by northward-subducted oceanic crust (Allen et al. 2015). The Palaeo- to Mesoproterozoic tectono-stratigraphic history comprises four major phases of basin formation. From approximately 1800–1300 Ma, the GMB accumulated up to c. 15 km of mainly fluvial to shallow-marine clastic sediments and subordinate carbonates, in a series of fault-bounded sub-basins, adjacent to, and unconformably overlying, strongly deformed and metamorphosed basement. We focus on the youngest (Mesoproterozoic) part of the GMB fill, close to the basin centre.

The Roper Group (c. 1590–1324 Ma based on indirect geochronology) comprises a 2–6 km-thick, cyclical succession of shallow-marine clastic sediments that are dominated by mudrocks, including exceptionally organic-rich variants (up to 10% TOC) and significant (tens to hundreds of metres-thick) intercalated sandstone units (Figure 2.51b, d). The gross Roper Group succession is dominated by at least five relatively thick regressive (R) sandier-upward facies successions, which are capped by generally thinner transgressive (T) muddier-upward units (Figure 2.51e).

The Roper Group is bounded by regional unconformities and is interpreted as a 1st-order stratigraphic cycle. Its internal stratigraphic architecture comprises a series of nested higher order cycles (2nd, 3rd and 4th order) (Crombez et al. 2023).

The Roper Group displays significant vertical and lateral variability in terms of thickness and facies. The focus here is on the Velkerri-Moroak sequence: 600–1900 m thick and with variable facies development (Figure 2.52). This unit includes 3–20% gross sandstone. Organic-rich mudstone intervals are widespread and reach a maximum in the middle of the Velkerri Formation (A, B and C Shales; Figures 2.51 and 2.52).

2.6.1.4 Facies Analysis

Six facies associations (FA) are identified and summarised in Figure 2.53. Mudstones dominate FA4, 3 and 2.2 (c. 70% of the Roper Group), representing shallow- to deeper-marine environments. Sandstones occur mainly in FA2.1 and FA 1.1 and 1.2 (c. 30% of the Roper Group) and were deposited in fluvial-paralic to coastal environments.

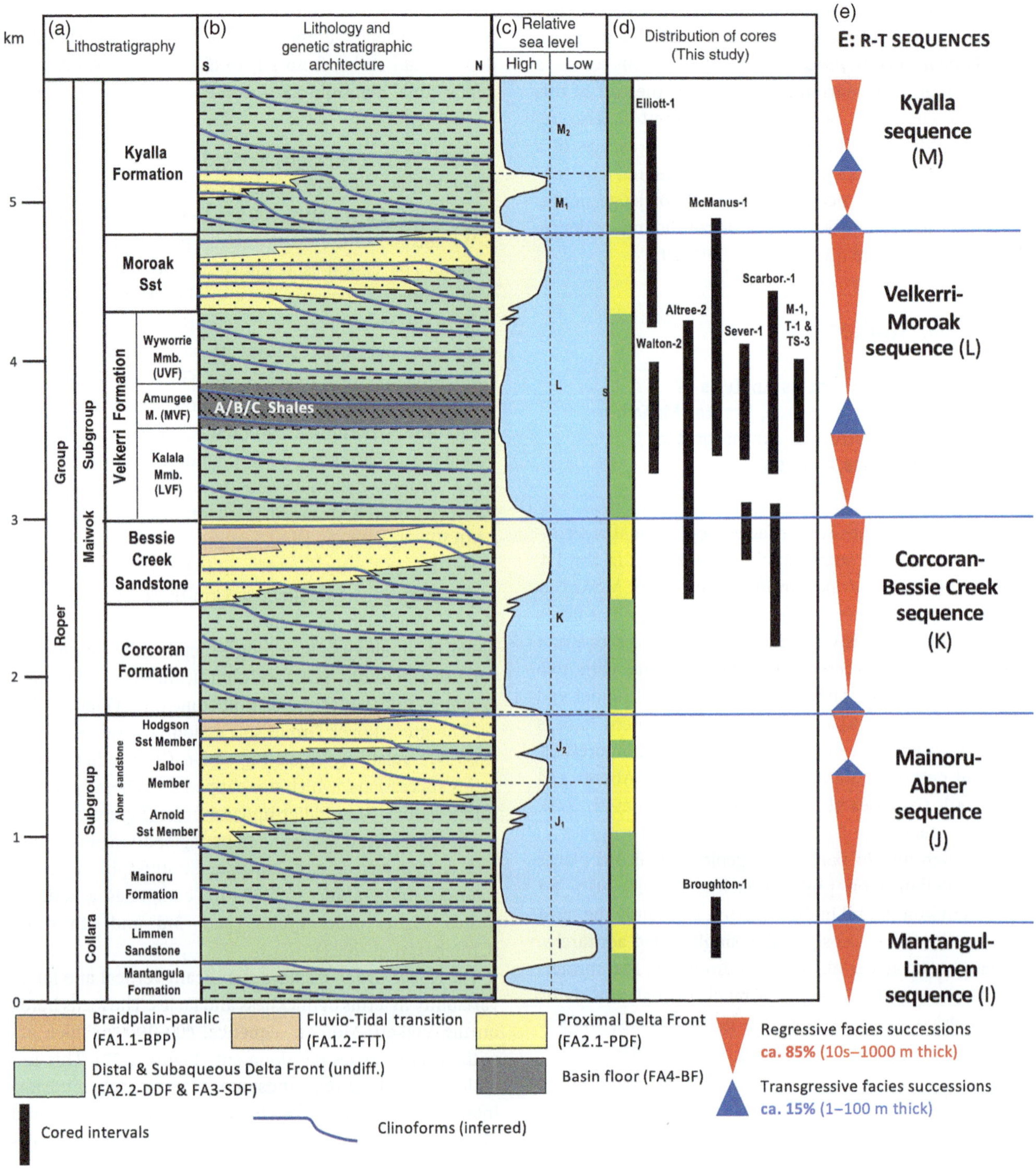

Figure 2.51 Stratigraphic and depositional framework of the Roper Group (from Johnson et al. 2024; modified after Jackson et al. 1988). (a) Lithostratigraphy derived from mapping outcrops north of the subsurface study area (*Source:* Jackson et al. (1988); Munson (2016); Munson and Revie 2018). (b) Simplified gross lithology and facies relationships, colour coded in relation to gross depositional environments, with inferred clinoforms (blue lines) based on facies relationships documented in outcrops, cores and well logs (e.g. Powell et al. 1987; Jackson et al. 1988; Abbott and Sweet 2000; Munson 2016). (c) Inferred relative sea-level changes (*Source:* Adapted from Jackson et al. (1988)). (d) Gross shallow-marine to fluvial sandstone (yellow/orange) and mudstone (green units), with the stratigraphic positions of the cored intervals used in this study. Well abbreviations are: M-1 = Marmbulligan-1; T-1 = Tanumbirini-1; TS-3 = Tarlee South-3. (e) Five major regressive–transgressive (R–T) sequences (I-M), which define the gross genetic stratigraphic framework of the Roper Group (*Source:* Adapted from Jackson et al. 1988). Regressive intervals defined by large-scale coarsening/sandier-upwards facies successions are dominant. Transgressive intervals are thinner and display greater facies variability.

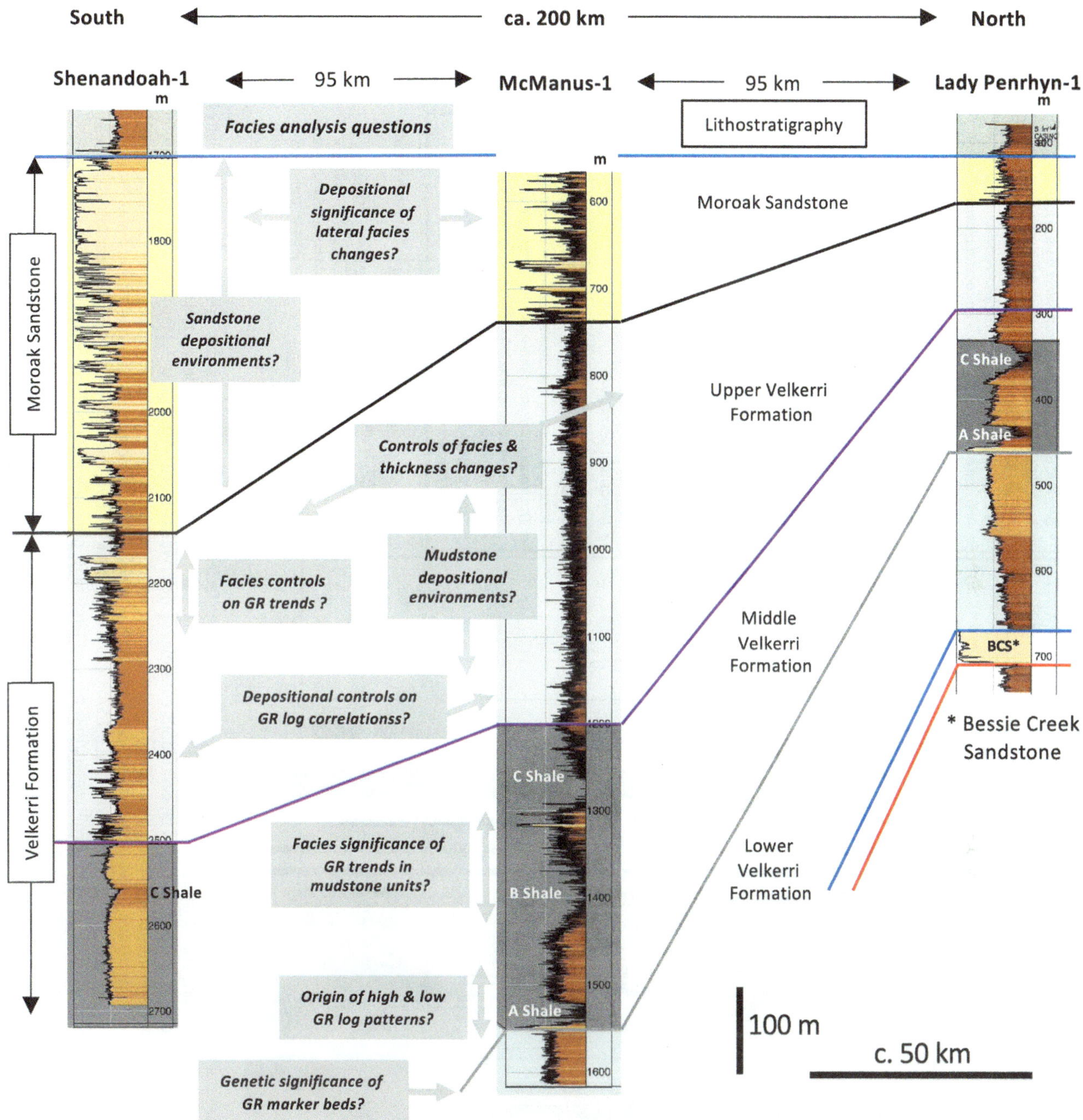

Figure 2.52 Facies analysis objectives are illustrated in this simplified N-S correlation panel of three gamma ray well log profiles (each ca. 95 km apart) through the upper part of the Velkerri-Moroak sequence. The facies analysis aims to explain this variability with some pertinent questions highlighted in the grey-shaded boxes (left side). Simplified lithostratigraphic terminology (right side).

The typical vertical distribution of facies associations is represented by the upper part of the Velkerri-Moroak unit, comprising a large-scale, regressive facies succession bounded by two regionally correlatable organic-rich mudstones (Figure 2.53). The maximum total organic carbon peaks within the middle of the organic-rich mudstones, which, supported by their large regional extent, are interpreted as maximum flooding surfaces (MFS). The intervening vertical facies succession becomes sandier upwards: FA4 → FA3 → FA2.2/2.1 → FA1.2/1.1 (Figure 2.53). The gradational vertical contacts between successive facies associations support the interpreted genetic relationship.

FA4 is mudstone-dominated, laterally extensive (at least c. 200–500 km) and formed in a basin floor environment (Figure 2.53e). It comprises variable intercalations of

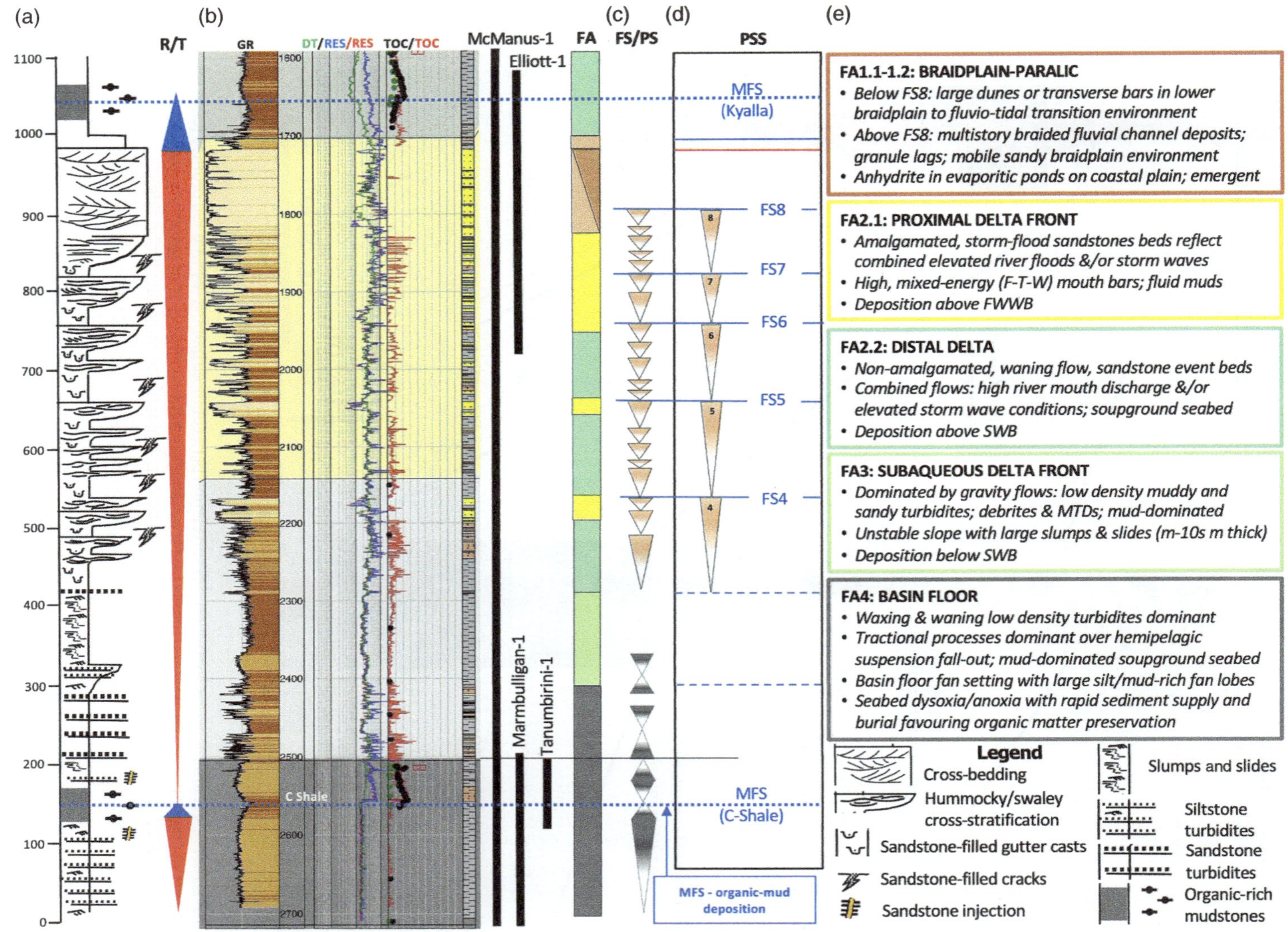

Figure 2.53 A composite vertical profile through the upper part of the Velkerri-Moroak succession summarising the facies-sequence stratigraphic analysis workflow sequentially from left to right: (a) core description and facies successions based on merged core descriptions and illustrated with a simplified log (thickness in metres) through the upper part of the Velkerri-Moroak R-T sequence, typical of the Beetaloo Sub-basin, and highlighting characteristic physical sedimentary features based on continuous core, mainly from wells McManus-1, Elliott-1, Marmbulligan-1 and Tanumbirini-1 (R/T – regressive/transgressive successions). (b) Calibrate facies with log response and interpret depositional environments and facies associations. Based on a representative well log through the same interval (Shenandoah-1): Track 1 = colour-shaded GR log (yellow/blocky = sandstone; yellow-brown/serrated = heterolithic (thin-bedded sandstone–mudstone alternations; brown-orange/finely serrated = mudstone-/siltstone-dominated); Track 2 = sonic log (DT/green), and two resistivity logs (RES/blue and red); Track 3 = two, log-derived, total organic carbon estimates (TOC/black and red); Track 4 = log-derived lithology estimations (yellow = clean/well-sorted sandstone; orange = argillaceous/thin-bedded sandstone; grey = mudstone). Thick black lines define the stratigraphic extent of continuous cores from the four mentioned wells. Final column: summary of the main facies associations (FA) – see colour code legend in Figure 2.51). (c) Define the individual vertical coarsening/sandier-upwards facies successions (i.e. single genetic units); facies successions (FS) that approximate to parasequences (PS) are illustrated by colour-shaded orange/white triangles denoting inferred deltaic facies successions. In contrast, the grey/black/white triangles denote siltier-upwards (SUS) and muddier-upwards (MUS) facies successions in inferred basinal settings. These are not parasequences. (d) Synthesis of the main FA into larger-scale facies successions (FS4-8) (approximating to parasequence sets – PSS); key stratal surfaces are highlighted: (i) bounding maximum flooding surfaces defined by two prominent organic-rich mudstones (C-Shale at the base and a near-base Kyalla mudstone defining the top), partly picked using the TOC logs in b; (ii) a surface of maximum regression (red line) near the top of the Moroak Sandstone – MS); (iii) initial flooding surface at the top of the MS, and (iv) intra-sequence flooding surfaces bounding larger-scale coarsening/sandier upward successions (FS4-8), which are candidate surfaces for further well log correlations. (e) Summary of the gross depositional environments combining the core- and log-based facies analysis, regional well log correlations and the broader geological context.

organic-rich mudstones and lean quartz-rich siltstones, most strikingly at the mm- to m-scale with a wide range of small-scale primary sedimentary structures and small-scale soft-sediment deformation features (facies characteristics are summarised in Figures 2.23, 2.53 and 2.54). This suggests a dominance of low-density, mud- and silt-rich turbidity currents flowing across an under-compacted soupground substrate. There is a larger-scale (10 s m) facies organisation comprising siltier-upward and muddier-upward facies successions (SUS and MUS, respectively; illustrated by the grey-white triangles in Figures 2.53c). These facies successions within the basin floor are

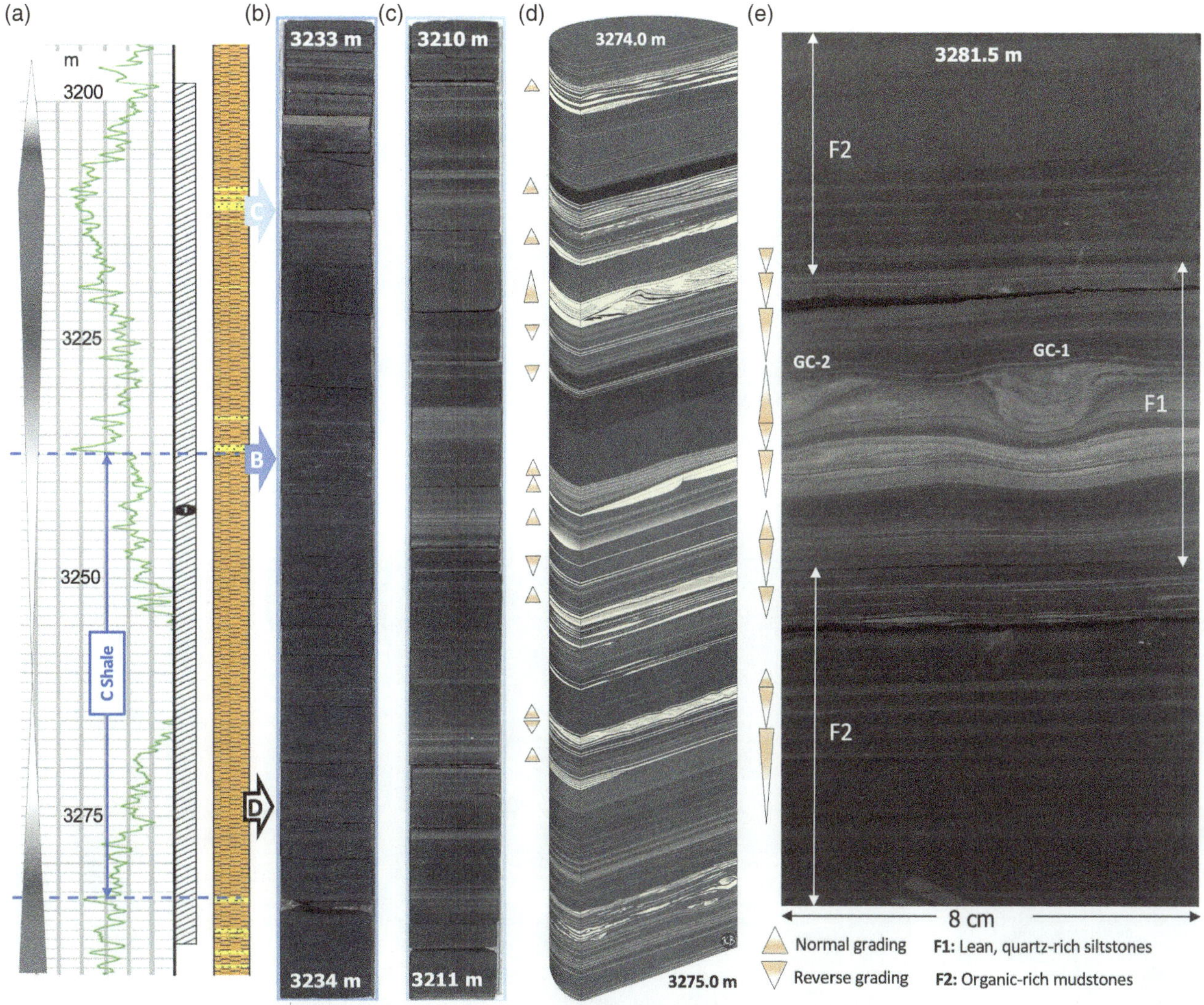

Figure 2.54 Facies Association 4, showing representative facies types and well log response (all from well Tanumbirini-1). (a) The 90 m long FA4 cored interval (cross-hatch shading), the gamma ray log (green), and simplified lithologies (mudstone/siltstone = orange; siltstone/very fine sandstone = yellow). The large-scale siltier-upward (SUS) and muddier-upward (MUS) facies successions are highlighted by the grey/white triangles to the left of the GR log. Locations of the two core segments are shown by the blue arrows and the 1 m-long core sketch by the black arrow. (b) Dark grey-black, organic-rich mudstone from near the top of the C-Shale, showing a reduction in organic-rich and an upward increase in lean mudstones. Note slight textural variations in the form of cm-dm scale, lighter/darker grey-scale colour banding. The TOC reduces from around 10% (base) to 2–4% (top). (c) Light grey, mm-cm-thick, parallel-sided, quartz-rich siltstone layers are more prominently intercalated with the darker organic-rich mudstone facies, TOC is reduced further and the interval corresponds to an overall lower GR log response (see (a) and the SUS). Individual siltstone layers show normal and reverse grading and some thicker (dm-scale) beds are massive. (d) Detailed line drawing of bedding styles in a 1 m-long core section in FA4 (lower part of the C-Shale – see (a) and black arrow with the letter 'D' for context) (courtesy of Kerrie Bann). Note the current-ripple cross-lamination, plus examples of loading- and dewatering-related soft-sediment deformation structures. Siltstones are coloured yellow, and mudstones are in various grey shades. Core width is 90 mm. (e) Details of mm-cm laminae and lamination styles in FA4, including and laminaset successions. Two small gutter casts (GC1 and 2) are highlighted. The background facies is dominated by organic-rich mudstone. Intercalated siltstone layers display current ripples, sometimes as low-relief, discontinuous lenses, and normal and reverse grading (mm-scale). The immediately underlying 1 m-long core (3276–3277m) is described and interpreted in Figure 2.23.

interpreted to reflect the lateral migration of mud-rich lobes (with <10% gross sand content). FA4 is interpreted to have accumulated at the distal end of a major sediment transport system, probably in relatively deep water (>100 m) and below maximum storm-wave base. In this setting, the waxing and waning flows may have been triggered by either river mouth (storm-flood)-induced currents, meteorological storms and/or wave-/storm-generated sediment gravity flows. Abundant well-preserved acritarchs confirm marine conditions.

FA3 is also mudstone-dominated, it gradationally overlies FA4 and is distinguished by its low organic carbon and siltstone-rich composition (lower overall GR log response; Figures 2.53 and 2.55). It contains abundant mm- to cm-scale siltstone to fine-grained sandstone intercalations (with scoured to loaded bases, current ripple cross-lamination and normal to reverse grading) and widespread soft-sediment deformation, including metre-scale slumps (Figure 2.55b–d). This supports rapid deposition with extensive seabed instability. The paucity of organic-rich

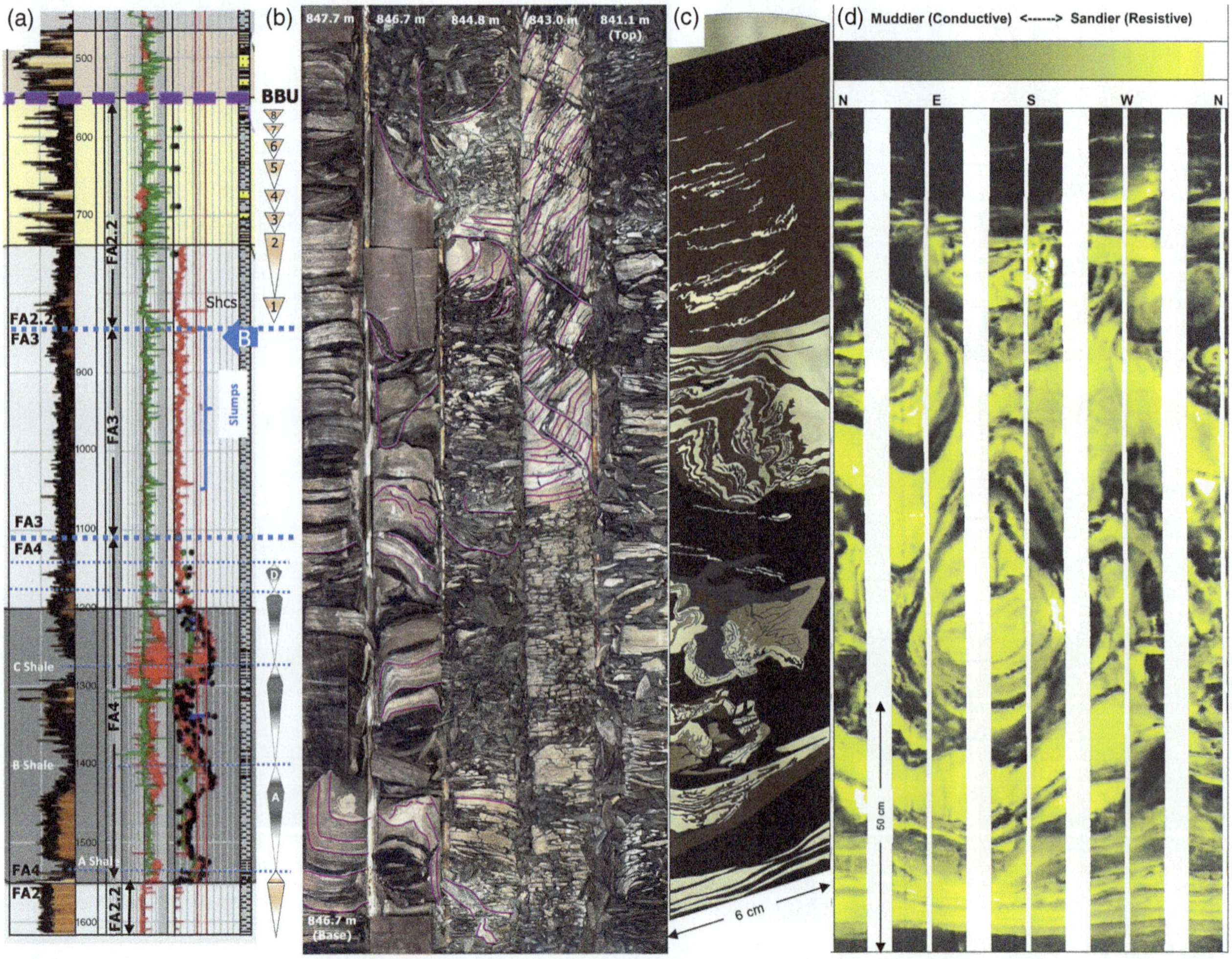

Figure 2.55 Facies Association 3. (a) Well log (McManus-1, depth in metres) expression of FA3 silty mudstones, which lack internal facies organisation illustrating typical examples of soft-sediment deformation structures (maximum frequency of slumping is defined by a blue bracket; c. 840–1040 m,); the underlying FA4 contains the main organic-rich mudstone units (A-, B- and C-Shale) and comprises correlatable SUS-MUS cycles (denoted by grey-white triangles and letters A-D); overlying deltaic cycles are also highlighted (coloured orange-white, and numbered 1–8, mainly in the Moroak Sandstone). (b) HyLogger Images of a 5 m-long, continuous core (in 5 × 1-m long sections) from the upper part of FA3 in McManus-1 (see blue arrow in (a) for stratigraphic location), displaying strongly deformed light-grey to brown-coloured mudstones and silty mudstones, with paler coloured (grey-yellow), cm-thick sandstone layers deformed by intense folding and faulting (highlighted by the purple lines). Johnson et al. (2024) / Petroleum Exploration Society of Australia (PESA). (c) Detailed line drawing of a c. 0.5 m-thick slumped interval (siltstone in yellow and mudstone in dark brown/black) in a laterally equivalent interval (Altree-2, c. 491.64–492.13 m depth), showing intensely folded sediment in the lower part, with some degree of left to right downslope translation, and an overlying unit of disrupted thin siltstone layers, caused mainly by vertical loading (courtesy of Kerrie Bann). (d) Example of similar soft-sediment deformation structures (static normalised FMI image), illustrating intensely folded FA3 deposits (from well Velkerri 76 S2-1). Courtesy of Andrew Wilson, ImageStrat and Tamboran Resources *Source:* Wilson et al. (2023) / Petroleum Exploration Society of Australia (PESA).

mudstone suggests either a reduced supply of organic carbon or lower preservation potential or greater dilution. Overall, deposition occurred on an unstable subaqueous slope, below storm-wave base and, based on the gradational upward change into more fluvially influenced deposits (FA1 and 2), in a delta-front setting.

FA2 is characterised by 10 s-m-thick coarsening/sandier-upward successions (Figures 2.53c, 2.55a and 2.56a;), comprising gradationally bounded mudstone- to sandstone-dominated units interpreted as distal (FA2.2) to proximal (FA2.1) delta-front deposits (Figure 2.56).

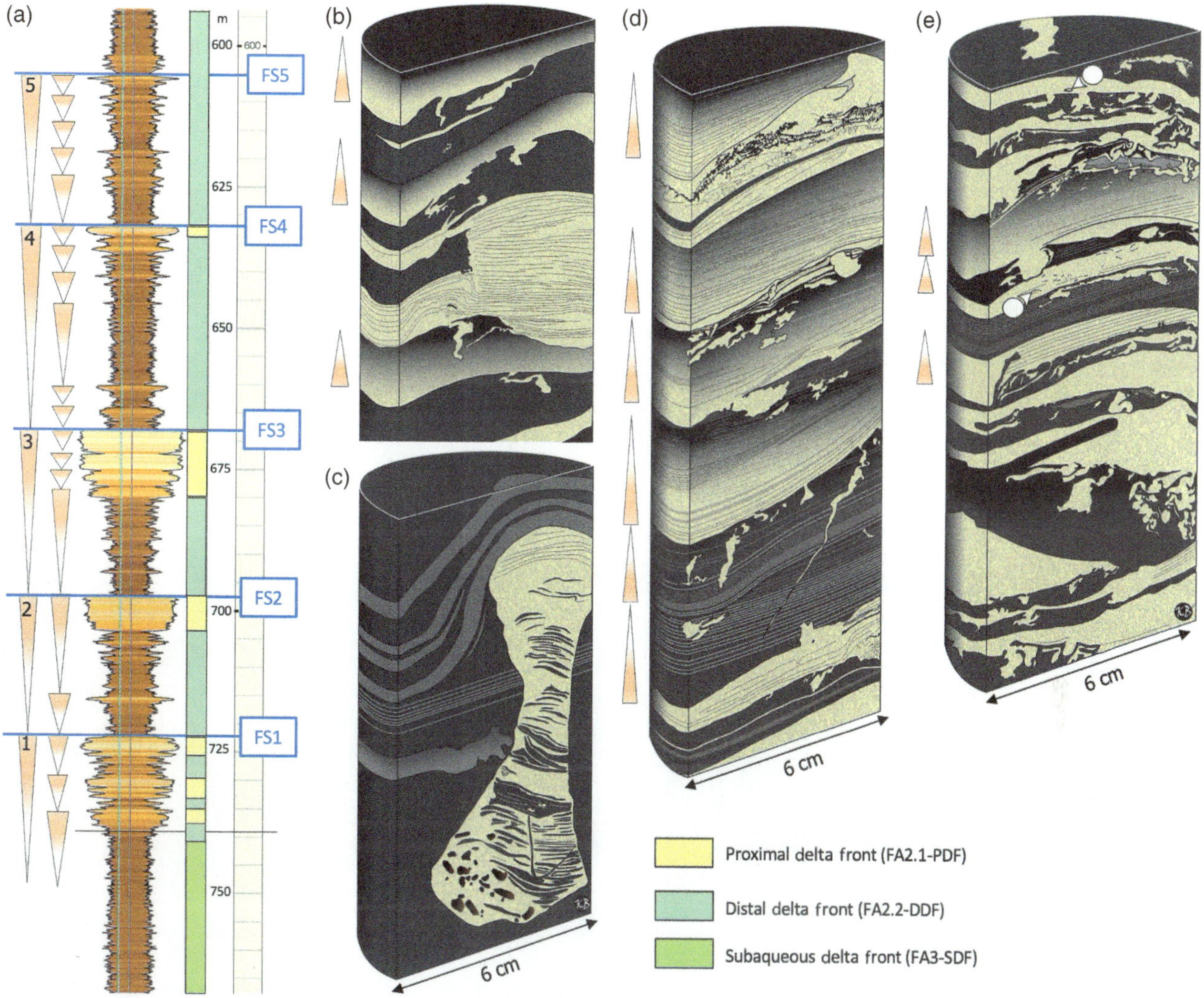

Figure 2.56 Facies Association 2.2 (distal delta front), illustrating bedding styles and well log expression from the Velkerri-Moroak sequence. (a) Colour-shaded GR well log profile in McManus-1, highlighting the upper part of the sequence, which is divided into five main coarsening/sandier-upward facies successions (parasequence sets, numbered 1–5) and their bounding flooding surfaces (FS1-5); this succession comprises a complex nesting of coarsening/sandier-upward facies successions, each dominated by FA2.2, and sometimes capped by subordinate FA2.1 (proximal delta front FA). Typical bedding styles in FA2.2 are illustrated in four detailed line drawings (b–e). (b) A single, sandstone-filled gutter cast with flat to low-angle/hummocky lamination (lower-middle part) and wave-ripple cross-lamination (top); interbedded with cm-scale, normally graded sandstone beds with erosive bases and downward-tapering sandstone-filled cracks (Elliott-1). (c) One compound gutter with multiple infill episodes; dark-grey/black mudstone intraclasts above basal erosion surface; downward 'necking' caused by vertical loading of the denser basal sandstone/intraclast layer and differential compaction at the top (CUS-3, Elliott-1, c. 1682 m). (d) Example of cm-scale, normally graded sandstone beds with wavy to low-angle combined flow ripple lamination, including micro-HCS; localised intense soft-sediment deformation intervals (e.g. upper 5 cm), comprising loading, injection and sand-filled cracks (Elliott-1, CUS-3, c. 1689 m). (e) Thin beds of erosively based sandstone and interbedded mudstone; sandstones display normal grading (orange to white shaded, upward-tapering triangles) and sandstone- and mudstone-filled gutter casts; note extensive soft-sediment deformation structures, mainly comprising sandstone-filled injection structures (mainly downward-tapering varieties) and loading (McManus-1, c. 712.5 m). b–e Sketches are courtesy of Kerrie Bann.

The distal delta-front (**FA2.2**) units are dominated by mudstone with interbedded, normally graded, c. 5–50 cm thick and mostly non-amalgamated, waning flow sandstone event beds with combined flow sedimentary structures, including frequent gutter casts (summarised in Figure 2.56). This implies rapid deposition from waning combined flows in a distal storm-flood environment, reflecting hyperpycnal flows generated at river mouths and/or storm reworking of proximal delta-front/shoreface sandstones.

The proximal delta front (**FA2.1**) is dominated (80–90%) by fine- to coarse-grained, medium- to thick-bedded sandstones, with planar-parallel, quasi-planar-parallel and low-angle (<10° dip) to undulose stratification, combined flow structures and dewatering features (summarised in Figure 2.57). Sandstone bed bases are erosional and downcutting, tops sharp to undulose, and dark grey-coloured mudstone intraclasts are common. This implies high-energy deposition by combined unidirectional and oscillatory currents, possibly in an axial river mouth to upper shoreface setting.

FA2.1 gradationally overlies FA2.2, forming coarsening/sandier upward successions (CSUS) at a range of scales and

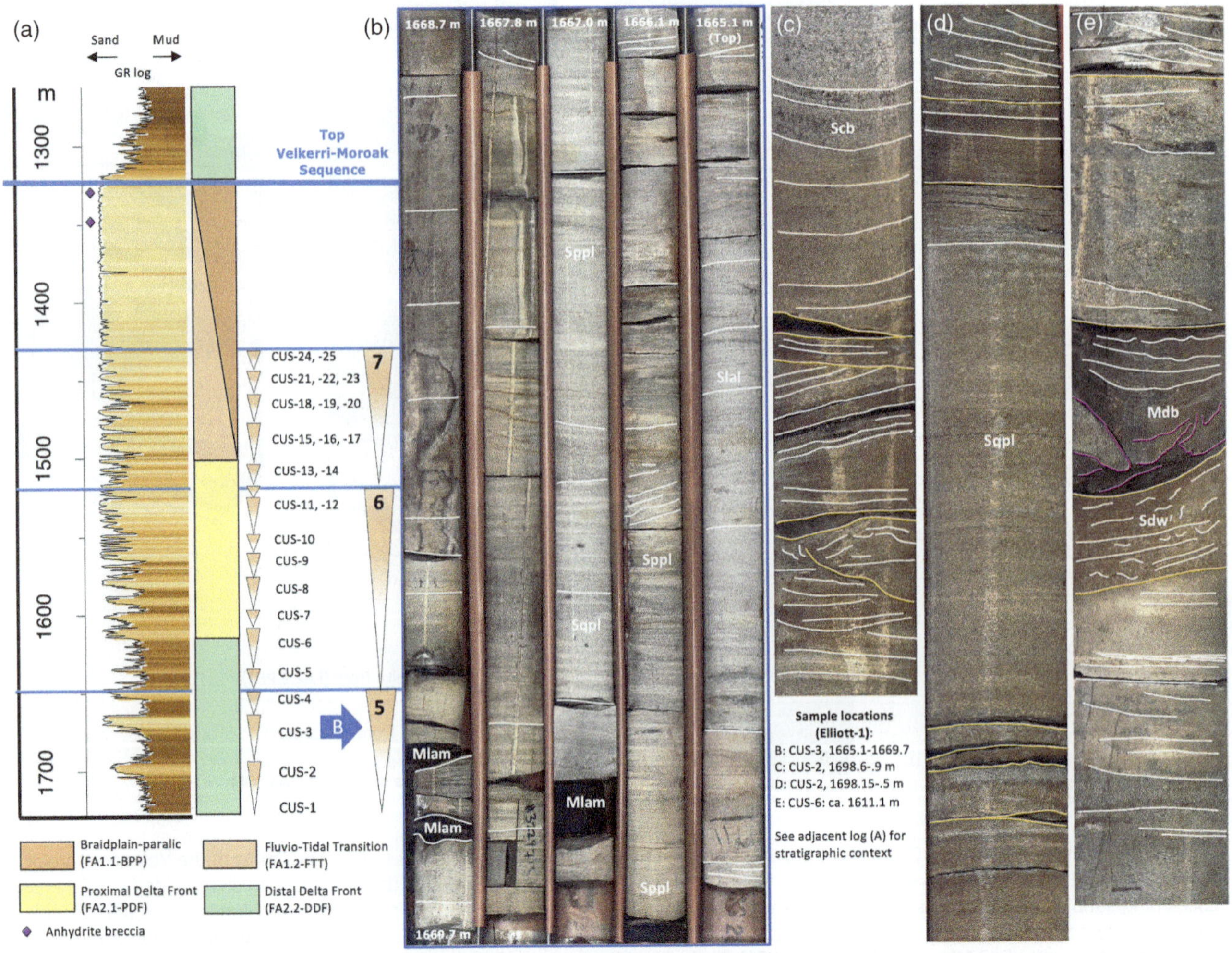

Figure 2.57 Facies Association 2.1 (proximal delta front), illustrating bedding styles and well log expression. (a) Colour-shaded GR well log profile through upper (c. 400 m) part of the Velkerri-Moroak sequence (mainly Moroak Sandstone) in the fully cored well Elliott-1; vertical distribution of facies associations, coarsening-upward facies successions/parasequences (CUS1-25) and interpreted parasequence sets (numbered 5–7) are highlighted to the right of the well log profile. (b) HyLogger images of a c. 5 m-thick continuous core from the upper part of CUS-3 vertical facies successions and facies associations are highlighted to the right of the well log profile (blue arrow/'B'). (c) Trough cross-bedding with localised soft-sediment deformation. Remnants of mudstone layers below erosive sandstone bed bases. (d) Quasi-parallel lamination with mm-cm-scale grain size fluctuations and combined flow to wave-rippled tops with low-angle cross-lamination. (e) Low-angle stratification, one internally deformed mudstone layers (Mdb) and localised dewatering (Sdw). Abbreviations: Mlam: laminated mudstone; Mdb: muddy debrite; Sqpl: quasi-planar laminated sandstone; Sppl: planar-parallel-laminated sandstone; Scb: cross-bedded sandstone; Sdw: sandstone with dewatering structures. (b-e) *Source:* Johnson et al. (2024) / Petroleum Exploration Society of Australia (PESA).

with varying facies proportions (compare Figures 2.56a and 2.57a) These are comparable to stacked parasequences, including progradational parasequence sets (e.g. parasequence sets 5-7 in Figure 2.57a). The facies and stratigraphic architecture suggests deposition in a mixed energy, storm-flood-dominated delta front to flanking shoreface setting.

FA1 is sandstone-dominated (>90%), it represents the shallowest part of the Velkerri-Moroak succession with a blocky gamma ray log response (Figure 2.57a). It comprises medium- to very coarse-grained, red-pink-coloured sandstones with dm- to m-scale angle-of-repose cross-bedding, thin and subordinate intercalations of red, grey and black mudstone layers (cm to dm thick) and, towards the top, rare anhydrite breccias with angular black mudstone clasts. The gradual increase in gross sand content and maximum grain size, gradual reduction in mudstone intercalations and the increase in red colouration with rare anhydrite supports a progressive upward shallowing. This comprises a sand-dominated mixed fluvio-tidal transition zone (FA1.2) at the base, which is gradationally overlain by a sandy, braided river-dominated lower delta plain (FA1.1) at the top (Figure 2.57a).

2.6.1.5 Depositional Model

The depositional model of the Velkerri-Moroak succession is underpinned by the interpretation that the FA4 ➔ FA3 ➔ FA2 ➔ FA1 facies succession is genetically related (Figure 2.58). A mixed energy, storm-flood-dominated delta model is proposed, which is guided by the facies analysis (from base to top):

i) Basinal mudstones (FA4), including organic-rich types, are dominated by tractional processes, especially low-density turbidites, plus other gravity flow deposits, while larger-scale facies organisation suggests mud-rich fan lobes, with a paucity of sand – this environment was highly energetic, subject to frequent dysoxia/anoxia and the seabed was under-compacted and mobile.
ii) Subaqueous delta front (FA3) comprises sub-storm-wave base heterolithic mudstones with larger and more frequent soft-sediment deformation structures, including loading, foundering and metre-scale slumping and sliding, which passes gradationally downwards into, and interfingers with, basin floor muds (FA4) – this environment was highly unstable and subject to gravity-flow related tractional processes.
iii) Proximal to distal (subaerial) delta front containing a plethora of combined flow sandstone event beds, including abundant sandstone-filled gutter casts, with deposition above storm-wave base and probably due to storm-flood processes directly connected to river mouths – this environment was dominated by mixed-energy storm-flood deltas in on-axis areas and flanking shoreface in off-axis areas.
iv) Braidplain to paralic (FA1.1) and fluvio-tidal transition (FA1.2), which together represent a sandstone-dominated, subaerial lower delta plain to braidplain setting – this environment was sandy, highly mobile, subject to highly erratic fluvial discharge and under the influence of tidal processes in downstream/river mouth regions.

2.6.1.6 Facies Successions, Correlation and Stratigraphic Architecture

This step uses the facies analysis and the core-defined facies associations, to help identify candidates for the key stratal surfaces for correlation. The well log correlation itself is an iterative process which must honour the facies analysis and depositional model (Figure 2.58).

The first challenge is the lack of a chronostratigraphic framework that could independently guide well log correlation in the 2 km thick Velkerri-Moroak sequence. In the absence of timelines, flooding surfaces have been adopted as a proxy for a chronostratigraphic framework. Transgressive surfaces with prominent winnowed lags are the most reliable, because they are significant events with distinctive facies (e.g. surfaces lined with pebbles and coarse sandstone, rich in glauconite, and carbonate-cemented). Those that cap individual parasequences (e.g. CUS 1-25 in Figure 2.57) are the least reliable because they likely reflect localised switching of delta lobes and may well have more limited lateral extents, even, most crucially, less than the tens of kilometres well spacing. Conversely, flooding surfaces that bound parasequence sets are more likely to extend beyond the well spacing in this setting (e.g. parasequence sets 1–5 in Figure 2.56 and 5–7 in Figure 2.57). However, there are two main limitations of marine flooding surfaces: (i) they cannot be recognised, and are therefore uncorrelatable, updip in coeval non-marine intervals, and (ii) downdip, in marine mudstone-dominated intervals, they lack unique distinguishing features.

The concentration of organic carbon is a common feature of Phanerozoic maximum flooding surfaces (or condensed sections), often in association with winnowing and submarine erosion in shallow water settings, and biostratigraphic hiatus and authigenic mineralisation in deeper water. There are three laterally extensive and correlatable organic-rich mudstones (A, B and C Shales) in this succession (Figure 2.58), which, based on the facies analysis, appear to define the turnaround between overall regressive and transgressive conditions. However, each represents a

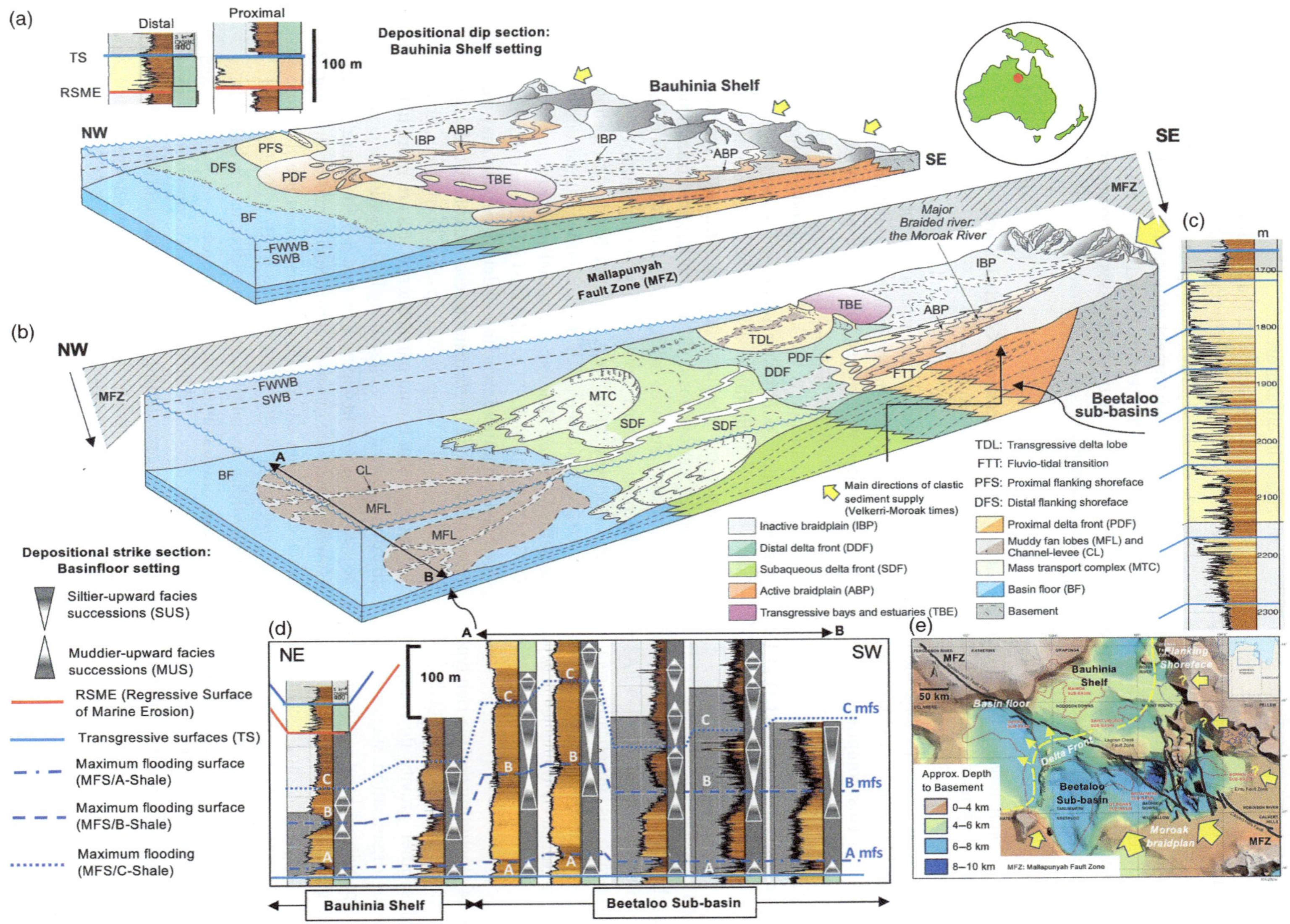

Figure 2.58 (a) Depositional model of the Upper Velkerri-Moroak sequence in the Bauhinia Shelf area (north of the strike-slip Mallapunyah Fault Zone – MFZ), showing overall condensation of the succession relative to that of the Beetaloo Sub-basin (south of the MPZ). The tens of metres-thick shallow-marine sandstone-dominated interval (Moroak Sandstone) is interpreted as a sharp-based flanking shoreface succession. It is characterised by a blocky GR log response, which reflects the sharp to erosional contacts with the underlying and overlying mudstone-dominated intervals (i.e. Upper Velkerri and Lower Kyalla units, respectively); the basal contact is interpreted as a regressive surface of marine erosion (RSME) and the top contact as a transgressive surface (TS). (b) Depositional model of the Velkerri-Moroak sequence in the Beetaloo Sub-basin area (south of the MFZ), showing overall expansion of the succession relative to Bauhinia Shelf (see GR log in c). Here the succession is dominated by a c. 1 km-thick regressive interval preserving the complete basin floor to braidplain facies succession (FA4→FA3→FA2→FA1). (c) Colour-shaded GR log showing the hundreds of metres-thick regressive coarsening/sandier-upward successions comprising multiple stacked distal- to proximal-dominated parasequences and overlying braidplain sandstones. The main transgressive surfaces are shown in blue (with a thicker blueline a maximum flooding surface at the top). A major fluvial system supplied sediment from the south to south-east, approximately sub-parallel to the MPZ, while basinal areas were mostly in the west and north-west (see e). (d) Colour-shaded GR logs along a NE-SW transect (see line A-B in b) within the deep-water-dominated part of the succession (underlying the GR log in c). The three organic-rich mudstone intervals are highlighted (A-, B- and C-Shales). The C-Shale defines the base of the Upper Velkerri-Moroak R-T sequence. Siltier-upward (SUS) and muddier-upward (MUS) facies successions are highlighted by the white triangles and interpreted as laterally migrating mud-rich basin floor fans. (e) Simplified depth-to-basement and tectonic framework map (*Source:* (from Munson 2016) / Northern Territory Government of Australia / CC BY 4.0). Interpreted outline of the maximum regression of the 'Moroak delta' is shown by the yellow dashed line, plus the inferred directions of fluvial sediment supply (yellow block arrows) and delta-front progradation (yellow normal arrows).

high-energy, albeit, mud-rich depositional setting, rather than simply a condensed section.

These considerations, together with the depositional model, provide one possible correlation of the upper part of the Velkerri-Moroak succession (Figure 2.59):

The absence of a unique chronostratigraphic framework for the Roper Group means that good practice is to apply a range of correlation approaches. These may be equally valid but of varying value. Four correlation scenarios are presented (Figure 2.60) and summarised in the figure caption.

2.6.1.7 Lessons Learnt

1) The core-based facies analysis provides the most reliable means of inferring sedimentary processes, interpreting facies successions and predicting depositional environments. This is an old lesson, but the unusually long length and continuous nature of the cores in the Roper Group re-emphasise this important point.
2) Mudrocks dominate the Velkerri-Moroak succession and core data were essential for extracting critical information on both depositional and soft sediment deformational processes. In particular (i) the mud-dominated basin floor (FA4) was high-energy and vertical facies organisation (tens of metres-thick siltier and muddier-upward cycles) suggest deposition in extensive mud-rich basin floor fans; and (ii) recognition of widespread slumping enabled definition of a rarely recognised subaqueous delta front (FA3).

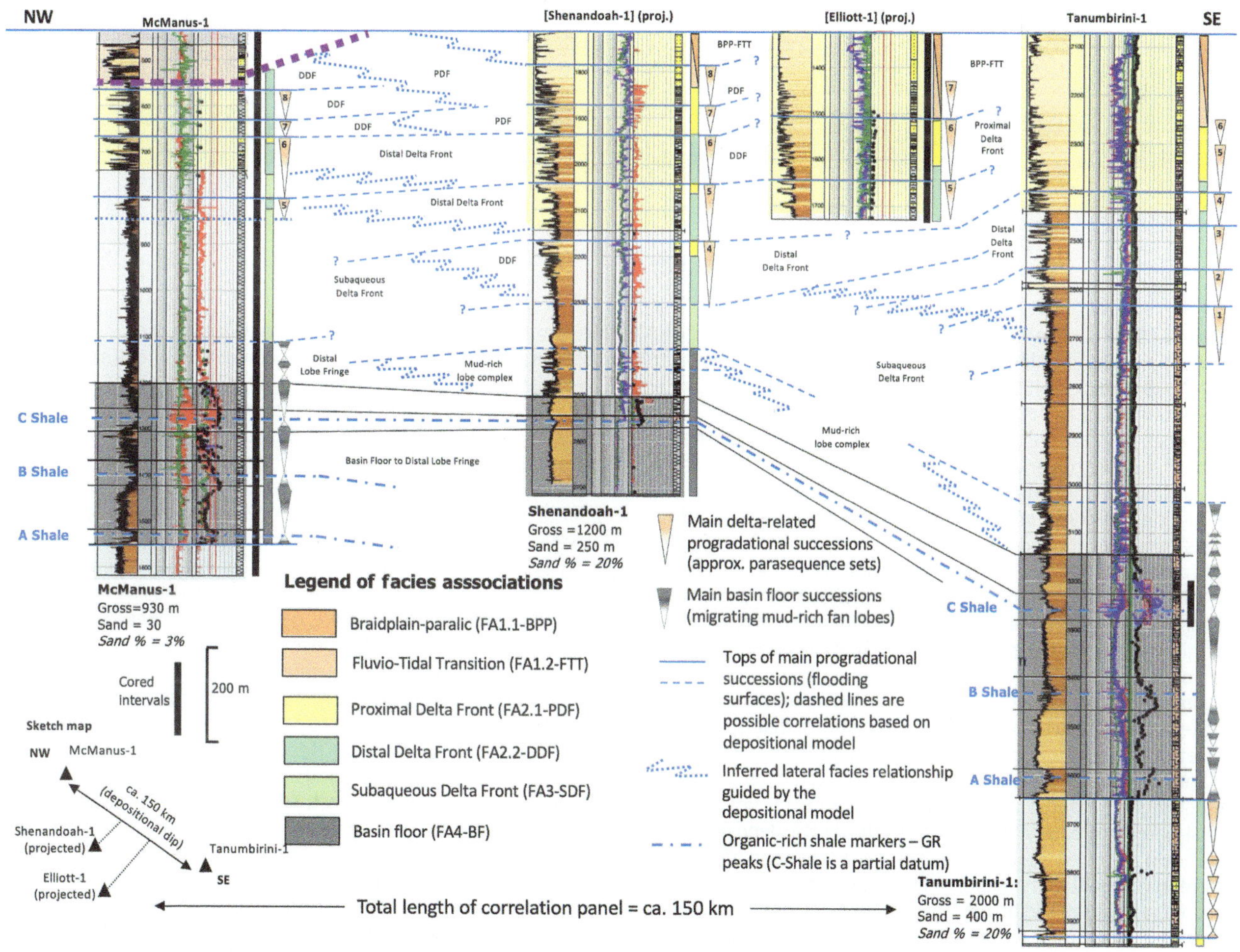

Figure 2.59 One possible interpretation of a NW-SE correlation panel for the Velkerri-Moroak sequence within the Beetaloo Sub-basin trending approximately parallel to depositional dip. In the absence of any reliable chronostratigraphic data, the correlations and facies relationships are guided by the facies analysis, most notably the inferred physical depositional process and its environmental significance. The section is c. 150 km long between wells McManus-1 and Tanumbirini-1 (wells Shenandoah-1 and Elliott-1 are projected into this section, so their positions are only approximate). The large well spacing means that lateral facies relationships can only be conceptual. However, they are consistent with the deltaic depositional model of Figure 2.58. The Top Moroak Sandstone is used as the datum in three wells: Tanumbirini-1, Elliott-1 and Shenandoah-1. However, in McManus-1, the top of Moroak Sandstone is truncated by the Base Bukalorkmi Unconformity (dashed purple line), hence, wells McManus-1 and Shenandoah-1 also use a secondary datum: the maximum GR peak in the centre of the C-Shale (dotted blue line).

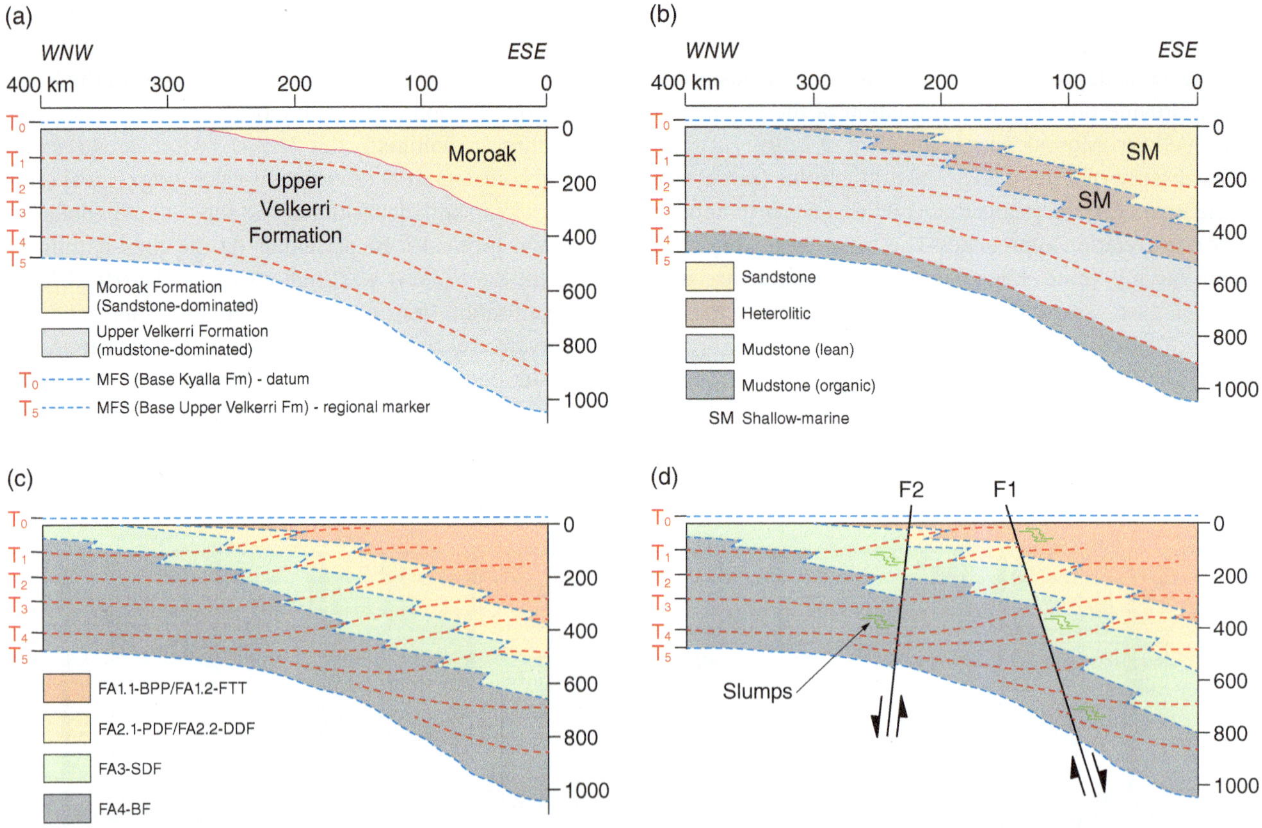

Figure 2.60 Four scenarios of correlation and stratigraphic architecture predictions along a simplified depositional dip section (landward/proximal to the right). Each panel is scaled to the actual stratigraphy and geography of the Velkerri-Moroak sequence: c. 500–1200 m thick; c. 400 km long (the Beetaloo Sub-basin is in the ESE parts of these panels). Conceptual timelines are in red (T0–T5); the horizontal datum at the top of the panel is the uppermost maximum flooding surface (organic-rich mudstone at the base of the Kyalla sequence). (a) A basic lithostratigraphic correlation panel; the main uncertainty is the gradational/transitional base between the Moroak Sandstone at the top and the underlying Upper Velkerri Formation; lithostratigraphic panels inevitably highlight lateral continuity: in this case a clear westward-tapering sandstone wedge. Timelines T0–T5 are entirely arbitrary, albeit sub-parallel to stratigraphic boundaries, and define five layers of approximately equal thickness (c. 100–200 m per layer). Such panels are easy to construct, require only sparse depositional insight and are of limited use in prediction. (b) This panel is still largely lithostratigraphic but differs from (a) a by capturing the transitional nature between the MS and UVF rock units by highlighting a zone of heterolithic facies. This interpretation incorporates slightly more insight into the depositional system. In this case, a westward-prograding deltaic sandstone wedge and a coeval mud-rich basin. The timelines are again arbitrary (as in (a)), so they offer no meaning with regards to, for example, geometrical shapes and relations. (c) This scenario is guided by the facies analysis, especially the depositional processes and environments described and interpreted in the cores, plus calibration of the vertical well log profiles with the main facies associations (in this case, with a relatively high degree of confidence). Most fundamental in this example is the depositional significance of vertical genetic stratigraphic relationships based on the gradationally bounded facies associations; that is, the basin floor (FA4) was partly coeval with the progradation of a large (hundreds of metres-thick) deltaic clastic wedge (FA1-3). From a correlation perspective, the interpretation of flooding surfaces at the top of parasequences and parasequence sets, which can be extended to each well, provides a possible chronostratigraphic framework based on sequence stratigraphic concepts, even though the time relationships are purely relative. Nonetheless, stratigraphic geometries can be investigated in greater depth, guided by the deltaic depositional system highlighted by the facies analysis. Hence, the timelines in this case are shown to bound westward-dipping delta lobes and offer the opportunity for more detailed stratigraphic analysis (e.g. thickness, facies and reservoir quality mapping). This scenario is highly deterministic and captures the thinking of the practitioner. It can also be readily updated and/or revised as both data and knowledge increase. The large well spacing inevitably requires a strong framework, but this provides the potential for greater prediction and testability. (d) This scenario follows the depositional and stratigraphic interpretation embedded in (c), but adds an additional uncertainty, namely the impact of intra-basinal faults that could have been active at the time of deposition. The MFZ was certainly a major, active fault during deposition of the upper Velkerri-Moroak succession based on basin-scale thickness changes. Secondary tectonic faults are likely but are poorly imaged on seismic data. The nature of the facies, supported by the high mud/sand ratio, is typical of basins that develop delta, or gravity-related, tectonics. This is most commonly manifested by growth faults, as seen in many Cenozoic deltas around the world. Hence, tectonic and gravity-driven faulting remains a major uncertainty and the correlation panels presented here are, consequently, highly simplified.

3) The widespread occurrence and diverse nature of soft-sediment deformation is consistent with other evidence of rapid mud deposition (soupground conditions with extensive loading, foundering, re-deposited mud-rich debrite beds, etc.), signs of sudden release of overpressures (e.g. sand and silt injection structures) and downslope translation (slumps and slides in FA3). Trigger mechanisms to initiate seabed instability (liquefaction/fluidisation and/or shear failure) were plentiful, including wave loading during storms (a dominant aspect of inferred delta-front/shelf depositional environment) and proximity to an active strike-slip fault (the Mallapunyah Fault Zone - Figure 2.58).
4) Many of the Roper Group regressive deltaic successions contain an abundance of sandstone-filled gutter casts in combined flow, waning-upward, event beds in the distal delta front (FA2.2) (Figure 2.56). This is increasingly seen as a common constituent of storm-flood-dominated deltaic depositional settings in many Phanerozoic successions. The abundance of gutter casts and closely related small-scale deformation features suggests that the upper Velkerri-Moroak succession is, at present, the oldest known example of this type of depositional setting in the world!
5) The lack of any independent high-resolution chronostratigraphic framework to confirm gross depositional morphologies requires alternative solutions to guide correlation. Here, flooding surfaces bounding parasequences and parasequence sets have been adopted as a proxy to provide a correlation framework; only flooding surfaces bounding the thicker parasequence sets have the lateral extent sufficient for correlation when well spacing is tens to hundreds of km. Well spacing always places critical constraints on all well correlation interpretations and a scenario approach to this task is always advisable.
6) Facies and depositional models underpin the construction of palaeogeographic maps (or gross depositional environments – GDEs) and well log correlations, which provide the cornerstone for all facies-related predictions, including those in energy, resource and waste disposal activities.

2.6.2 Case Study-2: Miocene Tectonically Active Margin of NW Borneo: Facies Analysis Integrating Outcrop, Core/Well Log and Seismic Stratigraphy

2.6.2.1 Aim

This case study investigates the sedimentary facies, depositional environments and basin-scale stratigraphy in the several km-thick, deltaic to deep-water Neogene succession in the tectonically active Baram-Balabac Basin of NW Borneo (Figure 2.61). It aims to demonstrate how linking results from even relatively sparse outcrop-based sedimentological studies with broader subsurface well and seismic data provides a more complete assessment of a depositional system. In this case, this is aided by the close affinities between the observable modern depositional environments, and those of the past 20 My or so (viz: the same tectono-stratigraphic framework, basin setting, humid-tropical climate and sediment routing systems (Figure 2.61b)).

2.6.2.2 Data

NW Borneo is a data-rich region, it contains a wealth of subsurface material and knowledge (reflecting >100 years of petroleum exploration), including 1000 s of wells and large (>100 km^2), high-quality 3D seismic datasets. Outcrops suitable for sedimentological analysis are scattered, often ephemeral, but of exceptional quality (e.g. Figure 2.46).

This facies analysis is based on c. 2.5 km of total logged stratigraphy from the early-middle Miocene, mainly from the exposed margins of the Labuan-Berakas Syncline sub-basin (Figure 2.62a).

2.6.2.3 Regional Tectonic and Palaeogeographic Setting

Plate Tectonics

The Neogene basin-fill (Figure 2.63) unconformably overlies an older fold-thrust belt comprising a deep-water Palaeogene accretionary wedge deposited during subduction of the proto-South China Sea (Figures 2.61a and 2.62c). The resulting tectonic framework controls the distribution of sub-basins along the active margin, each with distinctive structural and depositional timings. The bounding faults of the sub-basins controlled the gross depositional environments, thickness and stratigraphic architecture during Neogene deposition (Figure 2.62b,c). Tectonic inversion has occurred intermittently. However, what appear to be Pliocene to Recent synclines and anticlines mostly originated in the early-middle Miocene, when they were active depocentres and flanking, fault-cored highs, respectively (Figure 2.64).

Gravity and Regional Tectonics

The rapid infilling of structural sub-basins with up to 8 km of mostly coastal-deltaic siliciclastic sediment along an accretionary margin (Figure 2.63) is reflected in widespread gravity-related tectonics, including basinward- and landward-dipping growth faults (synthetic and antithetic, respectively; Figure 2.64b), gentle rollover anticlines and mud intrusions. However, in contrast to Cenozoic passive margin delta systems, such as the Mississippi, Niger and Orinoco, the NW Borneo margin was also tectonically active during deposition (Figure 2.62b,c). Specifically,

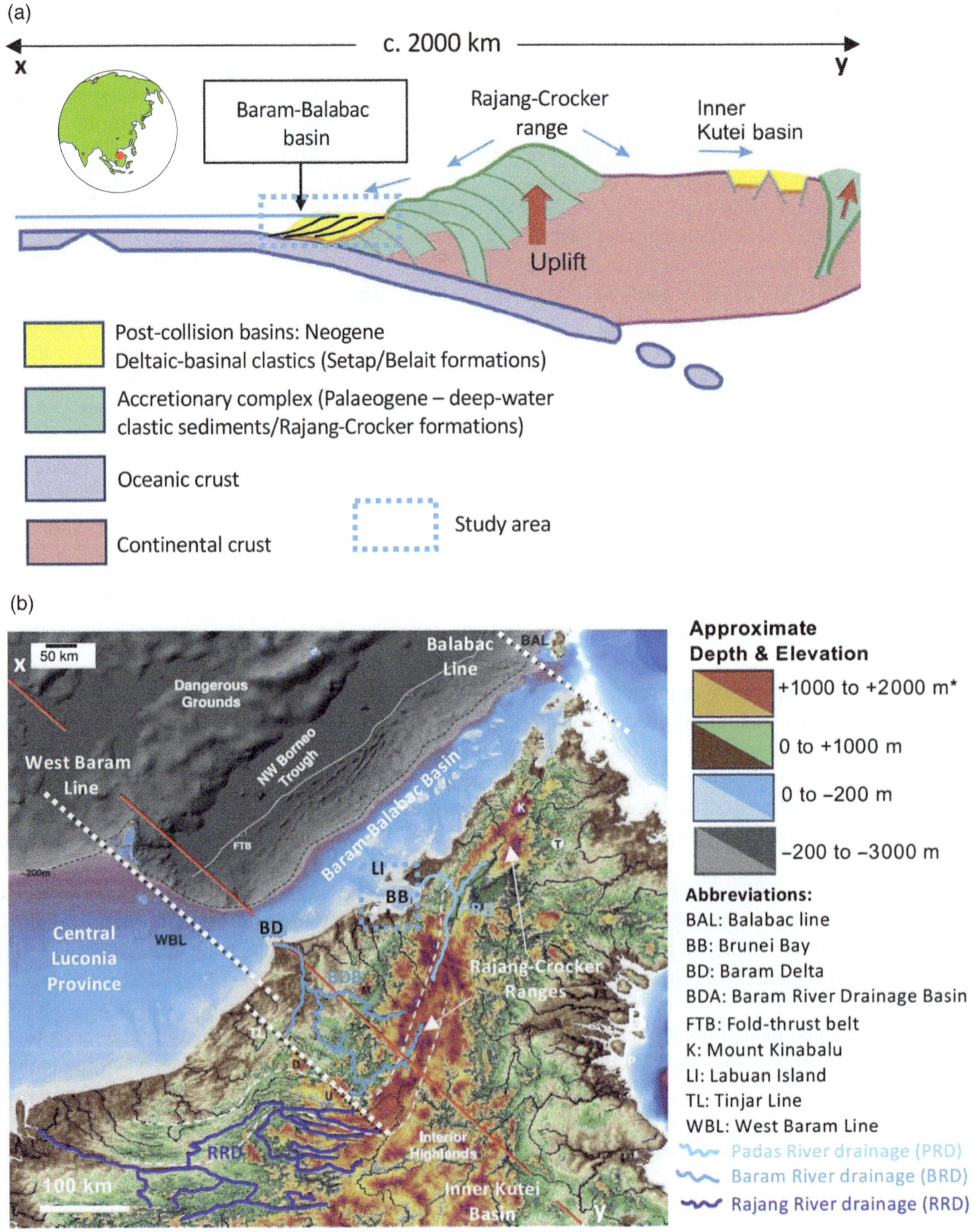

Figure 2.61 Regional geological setting of NW Borneo. (a) Schematic NW-SE geological cross section, extending from the South China Sea to eastern Borneo (Kalimantan). This section highlights the Neogene deltaic clastic wedge of the Baram-Balabac Basin (this case study) shown here with schematic NW-dipping clinoforms (*Source:* Modified from Darman and Handoyo (2013)). (b) Digital elevation model of topography and seabed bathymetry of the onshore to offshore Baram-Balabac Basin, located between the West Baram Line (WBL) and the Balabac Line (BAL). The red-shaded areas of the DEM are around 2 km elevation (maximum 4 km), blue colours define the shelf (0–200 m water depth) and the grey colours are deep water, passing from a 2 to 4°-dipping slope, through to a flat basin floor (NW Borneo Trough). The three largest river systems are highlighted (in decreasing size): Rajang, Baram and Padas. *Source:* Cullen et al. (2010) / EAGE. The dotted pale blue box in (a) and (b) defines the study area.

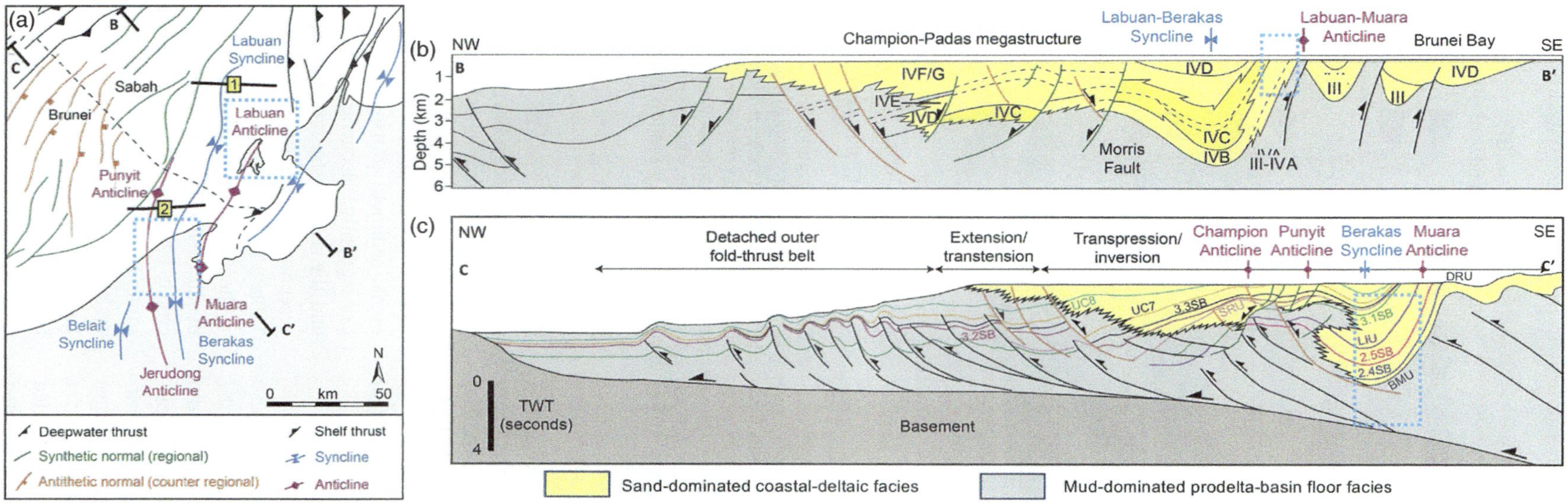

Figure 2.62 Location and geological setting of the study area. (a) Simplified structural map of the eastern Baram Delta Province (*Source:* Morley et al. (1998); Collins et al. (2017); courtesy D.S. Collins). Study areas highlighted by the pale blue dotted boxes: Labuan Island in the NE and onshore Brunei in the SW (courtesy D.S. Collins). (b) NW-SE regional geoseismic cross section (B-B′) through the Labuan-Berakas Syncline in SW Sabah (see (a) for location), aligned approximately parallel to depositional dip (towards the NW). This cross section traverses the outcrop belt around the Labuan Anticline (pale blue dotted box) and the adjacent Labuan Syncline (west of the island). It shows several major, basement-linked and partly strike-slip faults, which were active during the early Miocene (Stages III and IVA-D; see Figure 2.63) (*Source:* Adapted from Tan and Lamy 1990). (c) NW-SE regional geoseismic cross section (C-C′) through the Berakas Syncline in Brunei (see (a) for location), aligned approximately parallel to depositional dip (towards the NW). This section shows thrust faults within the Palaeogene fold-thrust belt and most dramatically the rapid thickening into the area of the Berakas Syncline (pale blue dotted box). The modern fold and thrust belt reaches the seabed, see Figure 2.61b) (courtesy D.S. Collins; after Gartrell et al. 2011, 2012; Torres et al. 2011).

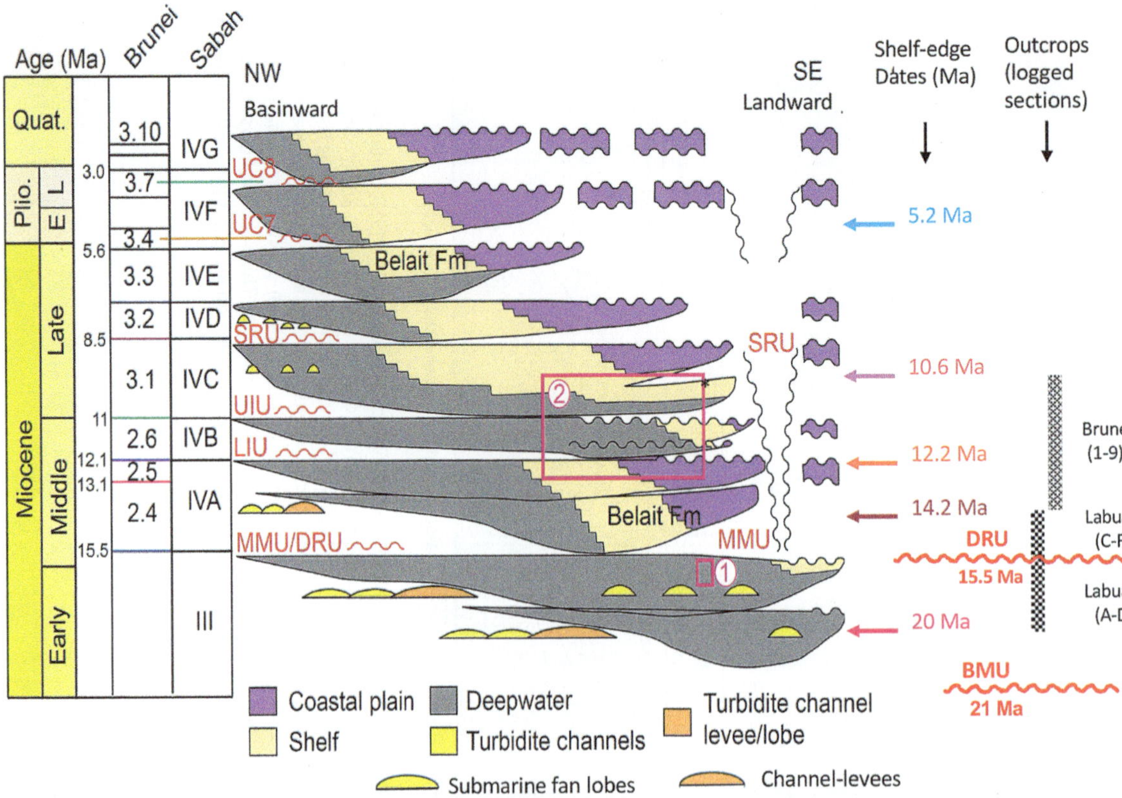

Figure 2.63 A simplified NW-SE depositional dip-oriented tectono-stratigraphic chart through the c. 8 km-thick Neogene of the study area. Note the approximately eight unconformity-bounded, regressive clastic wedges (or 'Stages'), comprising updip coastal plain to deltaic/shelf intervals and coeval, downdip deep-water units (*Source:* Bol and van Hoorn (1980); Levell (1987); Krebs (2011); Ballaguru and Lukie (2012); Courtesy D.S. Collins). The stratigraphic distribution of the outcrops used in this study is highlighted on the right, with emphasis on the Labuan area (Stages III and IVA). Shelf-edge dates are based on biostratigraphically constrained seismic surfaces. Brunei outcrops (1–9; red box number 2) are discussed in Chapter 8. Courtesy D.S. Collins. Abbreviations of unconformities: BMU = Base Miocene; DRU = deep regional; LIU = lower intermediate; MMU = middle Miocene; SRU = shallow regional; UC7 = Unconformity 7; UC8 = Unconformity 8; UIU = upper intermediate (see Levell 1987).

gravitationally unstable deltas were deposited above, and in front of, an active or partially reactivated accretionary wedge. Hence, gross stratigraphy thickens into synclinal areas and thins over intra-basinal structural highs (Figure 2.62), and there is regional truncation landwards towards the rotating orogenic belt along the proximal south-east margin (Figure 2.63).

2.6.2.4 Stratigraphic Framework and Gross Depositional Environments

The tectono-stratigraphic evolution of this margin over the past 20 Myr has involved intermittent hinterland uplift and repeated basinwards progression of subsidence and deposition in the sub-basins. Hence, there is an overall gross offlapping stacking pattern of successive clastic wedges (Figure 2.63). There is also marked internal cyclicity within individual clastic wedges, often accompanied by NW-dipping clinoforms, some of which are dated (Figure 2.64b). The combination of deep humid-tropical weathering and high fluvial discharge created an exceptionally efficient source-to-sink system with high sedimentation rates. The tectono-stratigraphic setting ensured high stratigraphic preservation due to rapidly subsiding basins and fluvial sediment transport lengths of only c. 300–500 km. Consequently, Neogene thicknesses are large c. 5–8 km and the seismic stratigraphic expression of the basin-fill is exceptionally clear and highly instructive (Figure 2.64 and Table 2.1-VIe).

The basinward progression of subsidence and deposition in sub-basins (Figure 2.63) enables rock facies types at outcrop to be calibrated with their seismic facies equivalents in some of the older and exposed large-scale depositional cycles. In particular, outcrops along the western side of Labuan Island can be compared with their lateral equivalents in the subsurface of the adjacent Labuan-Berakas Syncline offshore to the west (Figure 2.62a,b).

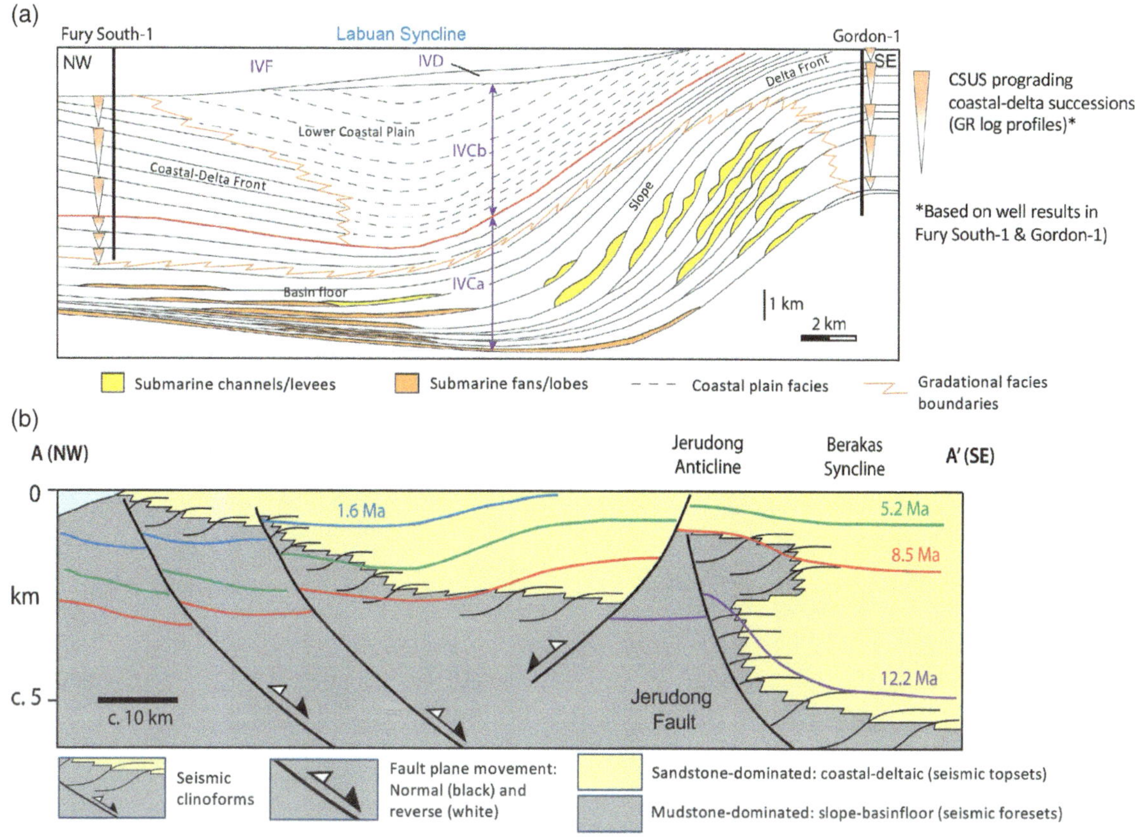

Figure 2.64 Simplified seismic stratigraphic profiles of the Neogene succession in the Labuan-Berakas Syncline area, presented in the form of geoseismic cross sections (see Figure 2.62a for line locations). (a) A SE to NW geoseismic section aligned approximately parallel to depositional dip (towards the NW), showing the stratigraphic architecture of the Stage IVC progradational clastic wedge (see Line 1 in Figure 2.62b, located c. 25 km north of Labuan) (*Source:* (Wong 1997) / Geological Society of Malaysia). Seismic data across the Labuan-Berakas Syncline and a chronostratigraphic chart are also shown in Figure 4.5. (b) NW-SE cross section from the western part of the Berakas Syncline (onshore Brunei), through the Jerudong Anticline and the present-day shelf edge, showing the main active faults and simplified seismic foresets. Four timelines are highlighted. *Source:* Modified from Gartrell et al. (2012) and Torres et al. (2012).

2.6.2.5 Facies Analysis

This facies analysis is based on a variable stratigraphic succession through Stages III and IVA in outcrops on Labuan Island (Figure 2.62a,b), which is divided into two gross depositional environments, separated by a regional unconformity (Figure 2.65): (i) deep-water to coastal-deltaic environments (Stage III), which are abruptly overlain by (ii) coastal plain to coastal-deltaic environments (Stage IVA).

Stage III is interpreted as a single, large-scale (c. 2.5–3 km-thick), mudstone-dominated, regressive cycle, which is comparable in scale to other unconformity-bound stages (e.g. Figure 2.63).

The deep-water interval comprises outer fan lobes and lower slope deposits (Figure 2.66). Outer fan lobes are interpreted to form part of the basin floor (Unit 1a/equivalent to seismic toesets) and are dominated by low-density turbidites (mostly Bouma Tb-c types; Figure 2.66a–d), organised mainly into 10 s m-thick coarsening/sandier-upward successions (Figure 2.66e) and with subordinate mass transport deposits (Figure 2.66f). Trace fossils and bathyal microfauna support a deep-water setting. Abundant disseminated mangrove-derived plant debris and westerly directed palaeocurrents (towards the basin axis) suggest derivation from a lower delta plain in the east/south-east, possibly supplied from hyperpycnal flows linked to river mouths.

Turbidite channels and mass transport deposits overlie the outer fan lobes and are interpreted to represent the lower part of a subaqueous delta slope (Unit 1b/equivalent to seismic foresets). This interval is dominated by 10 s m-thick debrites with m-scale externally derived blocks, scoured bases and irregular tops (Figure 2.67a,b). Intercalated coarse-grained high-density sandstones are occasionally conglomeratic and display parallel lamination, cross-bedding and abundant mangrove debris (Figure 2.67c–e). Sediment supply was closely linked to

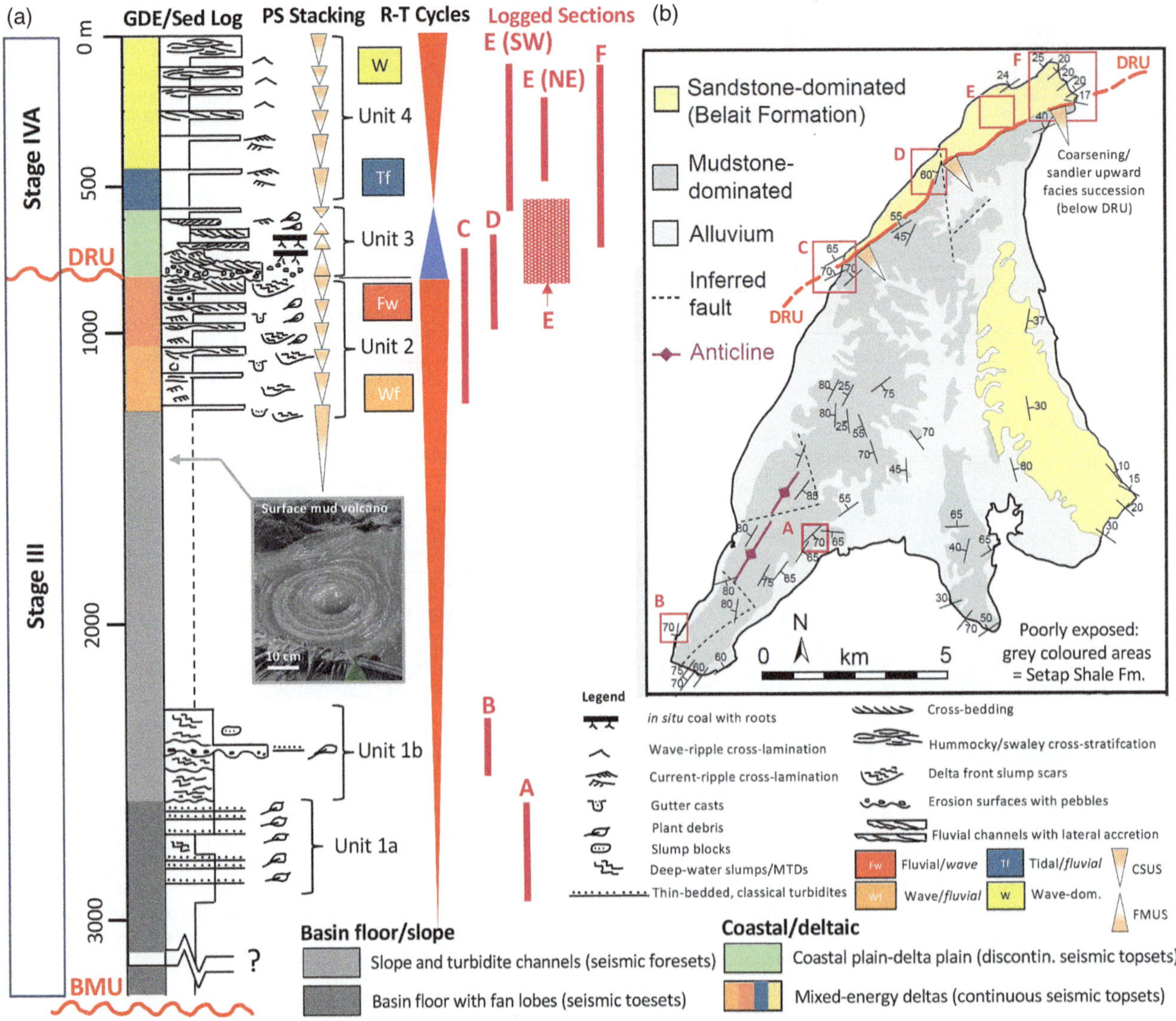

Figure 2.65 A depositional and stratigraphic synthesis of the early-middle Miocene on Labuan Island, representing a composite, large-scale regressive-transgressive-regressive cycle (red/blue triangles in A). Outcrops are mostly located on the steeper, western limb of the asymmetric Labuan Anticline. (a) Simplified vertical facies succession through Stage III (Units 1–2) and Stage IVA (Units 3–4), separated by the Deep Regional Unconformity (DRU). The fine-grained rocks have biostratigraphic ages indicating either early Miocene (below the DRU) or middle Miocene (above the DRU) (Lunt 2022). The sedimentary log is a compilation of measured sections from outcrops A-F (see column 'Logged Sections'); inferred thicknesses (dashed lines) have been derived from geological maps and satellite images. The poorly exposed centre of the island hosts modern mud volcanoes at the surface confirming overpressured mud at depth (see inset from Layang Layangan – immediately inland from outcrop C). Parasequence-scale facies successions are shown in the orange-white triangles (CSUS = coarsening/sandier-upward facies successions; FMUS = fining/muddier-upward facies successions). Logged sections shown by the black lines have been documented in detail (c. 1150 m stratigraphic thickness) the shaded box and dashed line are more generalised (exposure gaps). (b) Simplified geological map of Labuan Island (partly after Wilson 1964, but with chronostratigraphy based on Lunt 2022; from Puasa 2018), showing the locations of measured sections A–F. The map shows an older, early Miocene, mudstone-dominated, deep- to shallow-marine succession (shown in grey; the lithostratigraphic Setap Formation) and an overlying middle Miocene sandstone-dominated, coal-bearing, coastal plain/coastal succession (shown in yellow; Belait Formation). Outcrop names: A = Bebuloh; B = Kiam Sam (PETRONAS – formerly Shell – Oil Terminal); C = Layang Layangan; D = Ganggarak (near Pohon Batu); E = Manikar and adjacent Royal Malaysian Airforce Camp); and F = Tanjung Kubong to Bethune Head.

extensive slumping from collapsing delta front/mouth bars. Turbidite channels may have been connected updip to river mouths, with facies cyclicity linked to a combination of delta-front-related avulsion, mouth bar collapse and tectonics. Younger parts of the inferred slope are mud-dominated and poorly exposed, forming part of the widespread, often overpressured, mudstones of the Setap Formation (shaded grey in Figures 2.63, 2.64b and 2.65a; present-day mud volcanoes at surface – see inset photograph in Figure 2.65a).

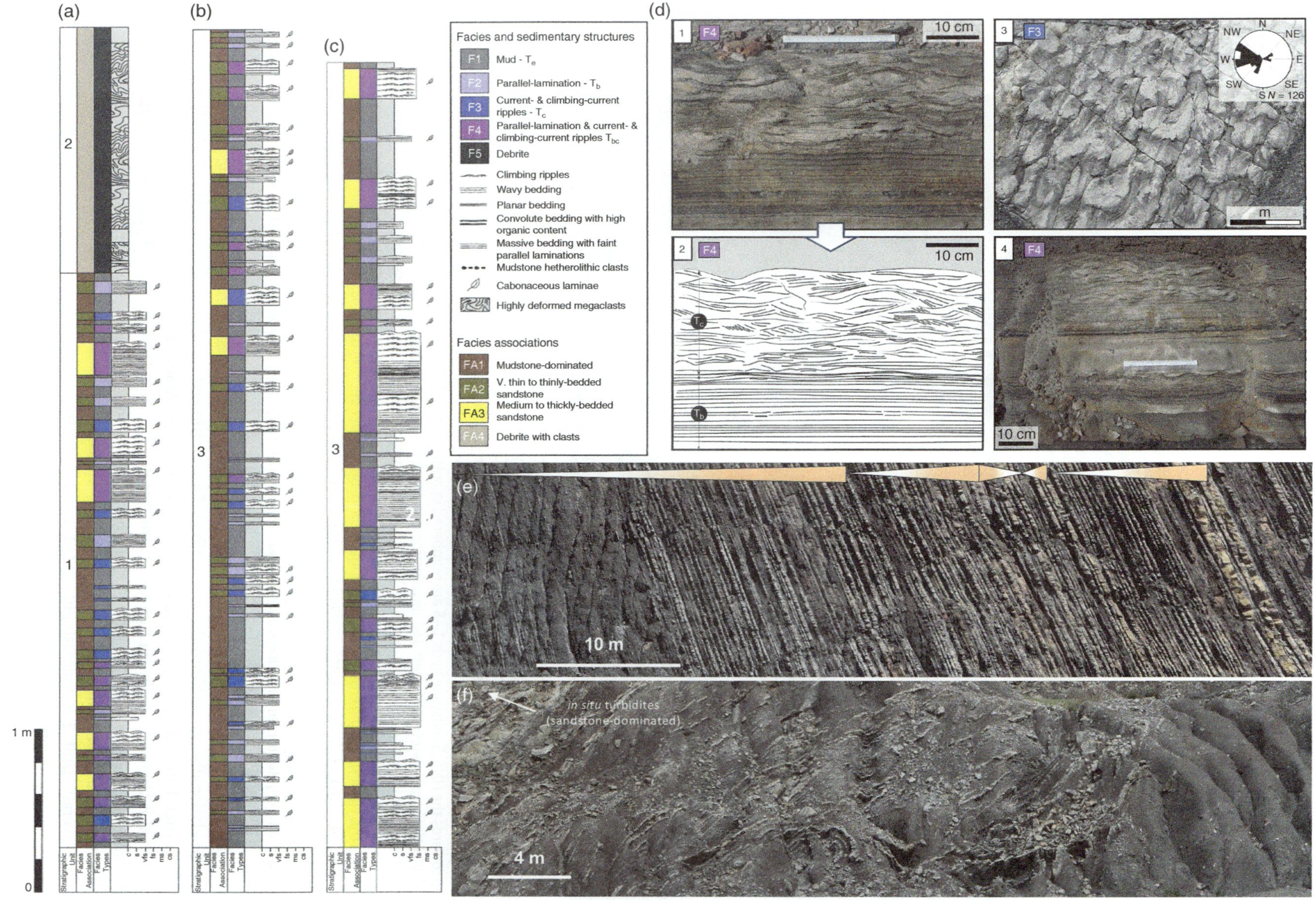

Figure 2.66 Facies characteristics of distal/outer fan lobe deposits (Stage III), Bebuloh outcrop, Labuan (stratigraphic Unit 1a in Figure 2.65a and outcrop location A in Figure 2.65b). (a) Sedimentary log summarising (from left to right) facies associations, facies and grainsize/lithology/sedimentary structures, including a mass transport deposits (labelled as 2 in the left column – see legend and (f) for further details). *Source:* Puasa (2018) / Imperial College London. (b) Sedimentary log of a mudstone-rich interval with subordinate intercalations of thin-bedded, classical (Bouma) low-density turbidites (c) Sedimentary log of a sandstone-rich interval comprising thin-to-medium-bedded, low-density turbidites. (d) Details of single low-density, classical turbidite beds (Tb→Tc). In the first bed (photograph 1 and line drawing 2 – same bed), the dark lines in the Tb division of (base of photograph 1) are organic-rich layers dominated by mangrove-derived material, which fine upwards into current-ripple cross-lamination (Tc – see line drawing 2). A bedding plane surface of a Tb–Tc turbidite bed top shows a series of lunate to linguoid current ripples. See inset rose diagram of palaeocurrent directions (showing a dominant westerly transport direction). (e) Photopanorama of a c. 60 m-thick, overall coarsening/sandier-upward facies succession dominated by thin-bedded, low-density classical turbidites, which is interpreted to reflect lobe migration (younging direction from left to right). (f) Base of the mass transport deposit shown at the top of the sedimentary log in (a) (above) and showing extensive soft-sediment deformation of thin turbidite sandstone beds. Interpreted as a relatively far-travelled slope-derived slump (younging direction towards the lower right). An inset shows the colour scheme and nomenclature of the facies and facies associations, plus symbols of the main primary sedimentary structures (mainly after Puasa 2018). (d) Puasa (2018) / Imperial College London. (e, f) Courtesy of Howard Johnson

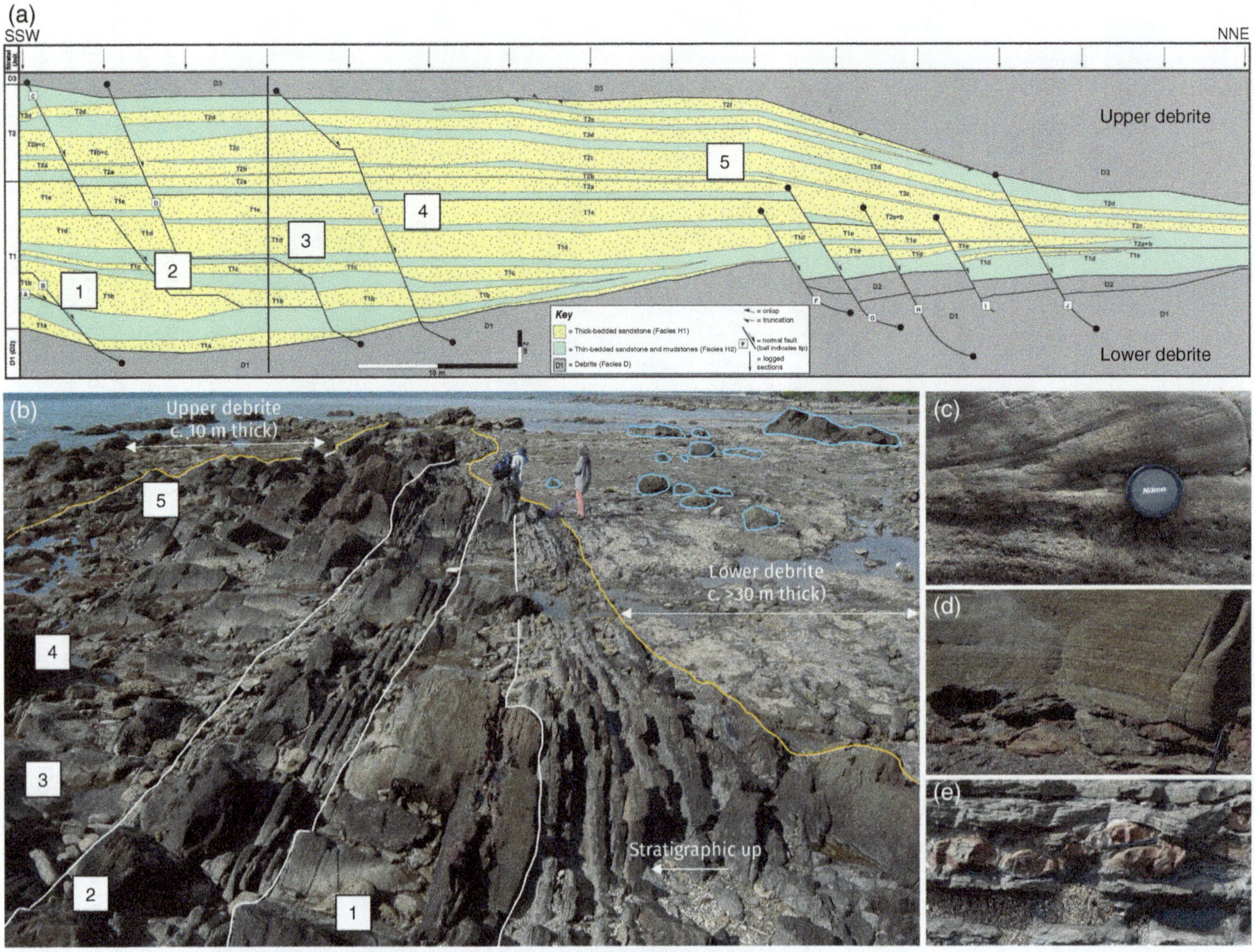

Figure 2.67 Slope to base-of-slope debrites with intercalated channelised sand-rich turbidites (structural dip c. 70° towards the WNW – away from viewer in (a) and to the left in (b). (a) Foreshore map illustrating the large-scale stratigraphic architecture, comprising: (i) lower debrite (right) with an irregular, slightly mounded top (right; grey shaded); (ii) sandstone-rich turbidite unit (yellow = sandstone; green = thin-bedded sandstones and mudstones), with a pronounced lenticular geometry, including thinning of all sandstone and mudstone layers towards NNE; and (iii) upper debrite unit with a scoured base, downcutting towards the NNE (from Jackson and Johnson 2009). (b) Photograph of (a), looking towards the NNE. Sandstone beds 1–5 are numbered in both (a) and (b). The inferred regional sediment transport direction was towards the WNW, approximately to the left and downwards into the foreshore. Hence, the geometry of the sandstone-rich turbidite unit reflects a combination of (i) drape onto a pre-existing debrite surface (lower debrite – pale blue lines highlight large blocks); (ii) channel erosion and pinch-out towards the channel margin (towards the NNE/away from viewer) plus thickening towards the channel axis (towards the SSW/towards the viewer); and (iii) erosion by the younger debrite bed (Upper debrite) towards the NNE (away from viewer). (c) A single, tabular cross-bed set in coarse- to very coarse-grained sandstone, overlying a coarse-grained (granule-rich) layer (below the lens cap). (d) High-density turbidite comprising a coarse-grained lag with large (5–15 cm) red-brown siltstone clasts and overlying parallel-laminated coarse-grained sandstones, with abundant mm-scale, organic-rich laminae (mangrove debris). (e) A muddy debrite with large, red-brown siltstone clasts and an irregular top surface draped by a parallel-laminated sandstone. Courtesy of Howard Johnson

The deep-water mud-dominated interval (containing Units 1a and 1b) passes gradationally upward into the c. 350–400 m-thick, sandstone-rich Unit 2 at the top of Stage III (Figure 2.65). This interval is characterised by multiple coarsening/sandier-upward units (i.e. parasequences and parasequence sets), interpreted as a prograding coastal-deltaic facies succession. Overall, Unit 2 is increasingly sand-rich upwards before it is truncated by the Deep Regional Unconformity (DRU) (Figure 2.65a).

The delta front displays evidence of mixed fluvial and wave/storm influence (i.e. storm-flood processes). This is supported by erosively based, waning flow, current- to combined flow-dominated sandstone event beds (including hummocky cross-stratification), unidirectional basinward-directed currents (towards the W and NW), abundance of finely disseminated mangrove debris, sparse benthic microfauna and low levels of bioturbation (BI = 0–1) (Figure 2.68). In addition, soft-sediment deformation is common at a wide range of scales (Figure 2.69), including tens of metres-thick slumping, debrite beds infilling erosional gullies, loading (e.g. ball and pillow structures) and repeated foundering and rotation of sand-filled

Figure 2.68 Examples of undeformed Stage III delta-front sandstones in the distal (a, b) and proximal (c, d) settings. (a) Planar-parallel to wavy lamination with dark laminae reflecting concentrations of mangrove-rich plant debris (Pen is 15 cm long). (b) Parallel to low-angle, asymmetric (unidirectional) cross-stratification infilling shallow scours and with abundant finely disseminated mangrove-derived plant debris (Pen is 15 cm long). (c) Parallel, low-angle and cross-stratified sandstone, with unidirectional foresets dipping to the SW (right to left; double white arrow); cross-bed sets wedge-out downdip into low-angle accretion surfaces (yellow surfaces); muddy debrite beds (Mdb, double green arrow) frequently underlie proximal delta-front (mouth bar) facies. Blue pen for scale (white circle) = 14 cm long. (d) Large (m-scale) tabular cross-bed set in the centre (double white arrows) dip to the W (towards the axis of the Labuan-Berakas Syncline c. 20 km away); dark layers comprise concentrations of disseminated mangrove debris; the blue line marks a flooding surface (blue pen is 14 cm long – white circle). *Source:* Courtesy of Howard Johnson.

gullies and gutters (Figure 2.69c). Hence, the substrate was probably soupground-like, unstable and inhospitable for infauna. The larger-scale deformation may reflect tectonic influence, since slump scars are widespread throughout this tectonically active basin-margin setting.

2.6.2.6 Shoreline Diagnosis

Interpreting the type of shoreline starts with a process interpretation stemming from the facies analysis. Rather than using a global facies model of prograding coastal-deltaic deposits, in this case, the large amount of logged stratigraphy has provided a modified classification scheme, based on NW Borneo-specific facies successions. Nine facies successions have been distinguished based on the type and vertical arrangement of facies and facies associations, especially in and around the Labuan-Berakas Syncline (Collins et al. 2018). Importantly, this includes input from local modern equivalents, which have much in common with their Neogene counterparts (e.g. Brunei Bay, Baram Delta and adjacent flanking shoreface systems; Figure 2.61b, but see also Figures 8.1d and 8.6d). One advantage of using this scheme is that it is often easier to infer relative, rather than absolute, proportions of different processes. Hence, the Stage III succession (Unit 2) is compared with three end-member out of a total of nine facies succession examples (Figure 2.70): (i) FS 1 = wave-dominated; (ii) FS 3 = tide-dominated; and (iii) FS 4 = fluvial-dominated (original FS numbering scheme of Collins et al. 2018 is retained).

Hence, the deltaic interval of Stage III (Unit 2 in Figure 2.65) is classified as fluvial dominated with secondary wave influence in proximal successions, and with increasing wave-/storm-influenced conditions in more distal settings, as manifested by the occurrence of subordinate hummocky-type structures (Figure 2.71). The coeval occurrence of river floods and offshore storms, whereby each

Figure 2.69 Examples of soft-sediment deformation structures in Stage III delta-front deposits. (a) Large-scale mass transport deposit composed of chaotically folded sandstone blocks in various stages of deformation consisting of delta-front sandstones (similar to those in b–d). These and other nearby slumps have been correlated with seismic-scale clinoforms and slump scars based on detailed mapping of the western flank of the Jerudong Fault (Back et al. 2008). (Kuala Lurah Quarry, near Limau Manis, Brunei). (b) Gully fill comprising a basal erosion surface with chaotic debrite bed (Db), which is overlain by a parallel laminated to low-angle stratified sandstone (Lens cap 7 cm diameter). (c) Asymmetric sandstone-filled gully showing foundering, loading and lateral migration of the gully axis (initially from left to right; successive positions of the gully axis defined by black triangles and black arrows), which results in progressive over-steepening of the internal stratification (up to vertical). Distal mouth bar (in CSUS-S2). Blue line defines horizontal. Scale bar 15 cm. (d) The distal delta front is characterised by abundant sand-filled gutter casts at a wide range of shapes and sizes. The steep to near-vertical, sometimes overlapping, margins are a combination of initial incision (two examples illustrated) and subsequent loading during repetitive sandstone infill (see concave-upward stratification below lens cap – 7 cm diameter) (b–d Ganggarak Quarry, Labuan). The scale and geometry of these gutter casts resemble those in the Mesoproterozoic Roper Group, discussed earlier – e.g. Figure 2.56b–e).

enhances the effectiveness of the other, is common in depositional systems characterised by short river lengths, nearby uplifted hinterlands, high fluvial gradients and stormy shelf settings. In distal facies, this is often manifested in small-scale sandstone-filled gutters and gullies, which are especially abundant in Stage III and in other deposits interpreted as storm-flood-dominated (e.g. see the Roper Group in Case Study-1).

Applying the same approach to the overlying Stage IVA shoreline deposits highlights further variability to that seen in Stage III (Figure 2.71). Tide-influenced conditions follow the regional transgression (post-DRU and pre-MFS in Figure 2.71), probably along an embayed shoreline. Subsequently, wave-influenced shelf-deltas form along a more linear shoreline (post-MFS in Figure 2.71).

One aspect of the process-based facies interpretation that is not captured in the W-F-T scheme of Figure 2.70 is the abundance and diversity of soft-sediment deformation, which carries additional environmental significance. The wide range of soft-sediment deformation structures described earlier is interpreted to reflect an unstable delta front, which triggered slope collapse, formation of slump scars and the emplacement of mass transport deposits. The latter are abundant in Stage III delta front, lower slope and basin floor settings, which coincided with initial basin-fill of a tectonically active mini-basin. Similar processes form a >100

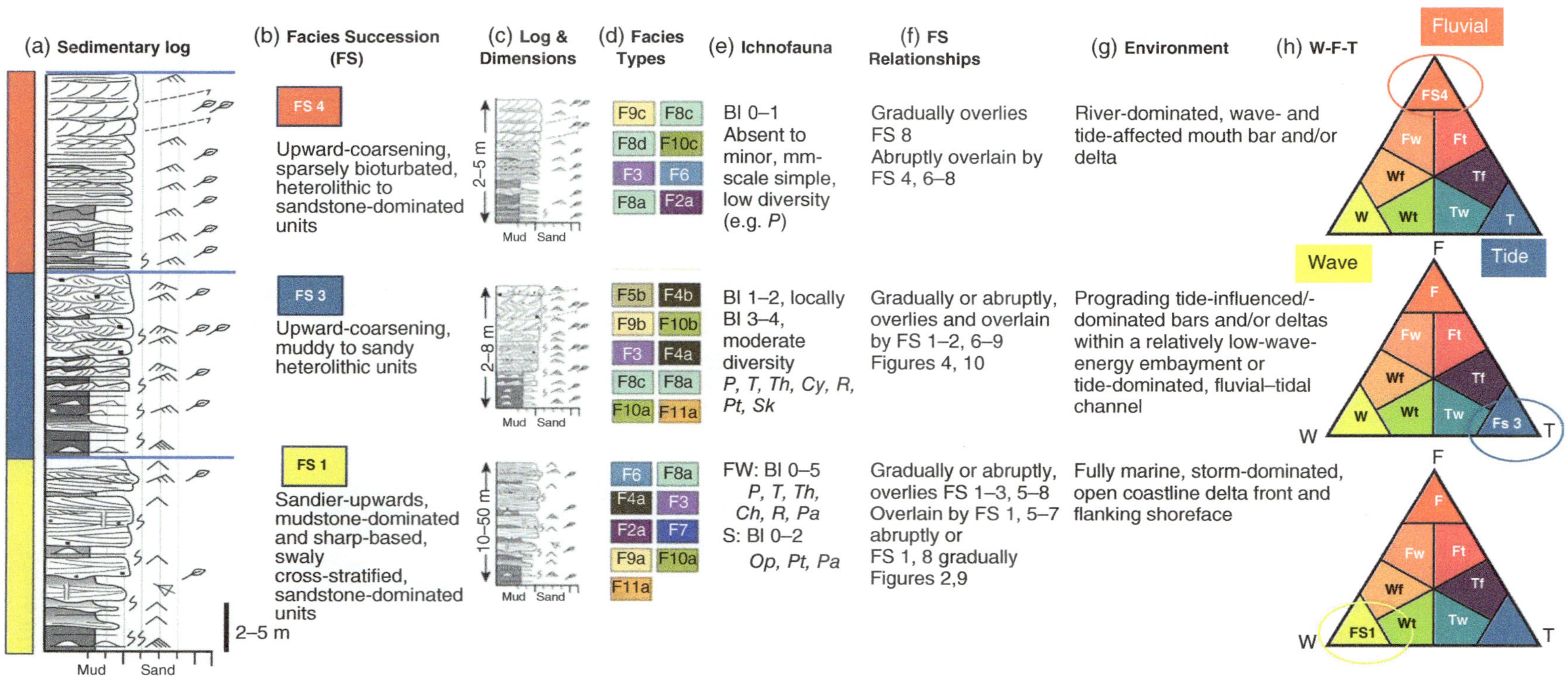

Figure 2.70 End-member progradational facies successions (FS) based on a synthesis of c. 3 km of logged outcrops through the Stage III to Stage IVA interval and lateral equivalents in the onshore Baram Delta Province (based on Collins et al. 2018, 2020; unpublished figure courtesy D.S. Collins). A summary of key sedimentological features and workflow, working from left to right (a to h) is used to distinguish facies successions in fluvial- (FS 4), tide- (FS 3) and wave-dominated (FS 1) settings in the Neogene of the Baram-Balabac Basin. The key components of the workflow are as follows: (a) simplified sedimentary logs (see legend in Figure 2.29); (b) facies succession summaries and terminology; (c) typical thickness range of each FS; (d) range of colour-coded facies types in each FS; (e) summary of ichnofauna types and biotrubation intensity (BI); (f) summary of vertical facies relationships; (g) environmental diagnosis; (h) summary of dominant and secondary processes are plotted on the Nyborg and Howell (2016), Ainsworth et al (2011) clastic coastal-deltaic classification scheme: dominant processes are in uppercase (fluvial = F; tide = T; and W = wave) and subordinate processes are in lower case (ftw).

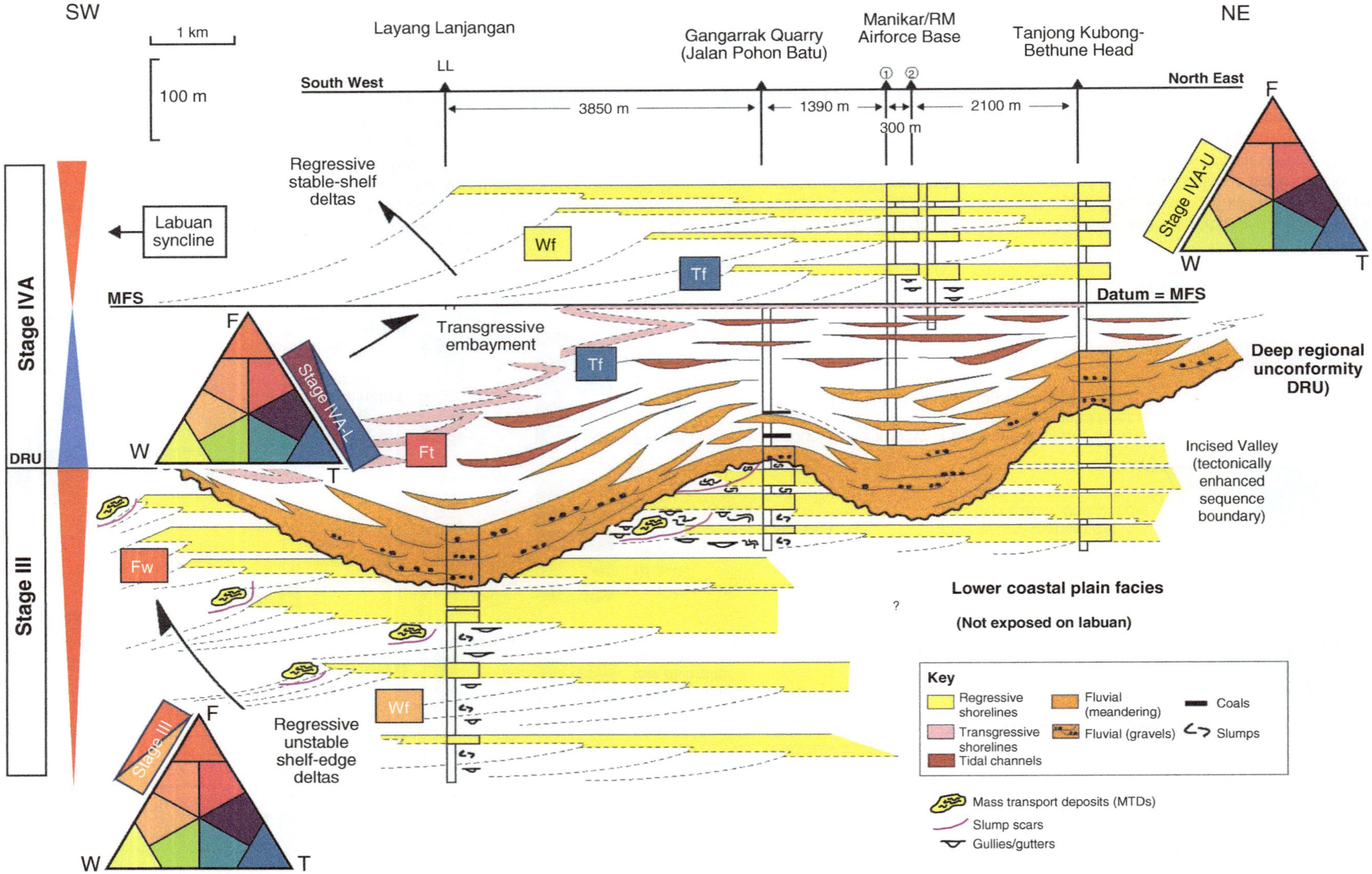

Figure 2.71 Synthesis of the coastal-shelf processes, depositional environments and sequence stratigraphic architecture in the Stage III to Stage IVA succession in the northern Labuan area (Malaysia). Outcrop descriptions are partly based on unpublished Imperial College London MSc theses in 2010 by D. Bloomer and E. Robinson. The basinward equivalents of the regional lowstand erosion surface (DRU), in the adjacent Labuan-Berakas Syncline, comprise westerly-dipping seismic foresets (see orange-coloured reflector interpretations in Figure 2.72). This shows the abrupt westward shift in facies as a result of DRU-related uplift.

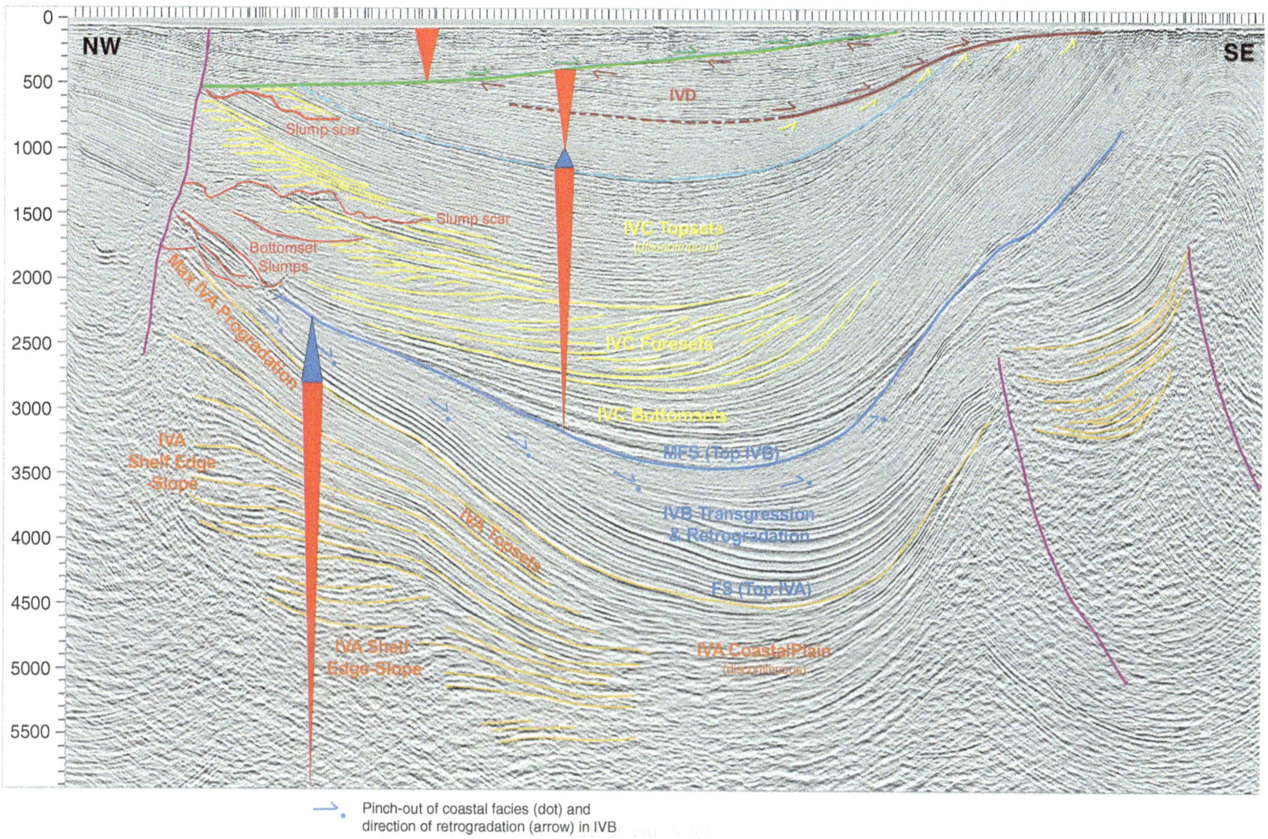

Figure 2.72 Sequence stratigraphic interpretation of the Mio-Pliocene infill of the Labuan-Berakas Syncline sub-basin based on van Vliet and Schwander (1987) and subsequent discussions with Arthur van Vliet. Note the prominent clinoform geometries of Stage IVA, which are the partial lateral equivalents of coastal plain deposits above the DRU on Labuan (Figure 2.65). The DRU must be close to the base of this seismic section. In this interpretation, the 'Stage IVA Coastal Plain' (see label in the basin axis) is the lateral equivalent of the outcropping incised valley that defines the DRU in north Labuan (Figure 2.71).

km-long, coalesced series of seismic-scale slump scars along fault-controlled shelf edges in the Miocene, which have been attributed to tectonically oversteepened slopes and linked to regional, tectonically induced unconformities (Levell and Kasumajaya 1985; Levell 1987). Similar features have also been identified in Stage IVA and younger shelf edges along other sub-basin-bounding faults (e.g. along the western edge of the seismic section in Figure 2.72 – see orange-coloured reflectors – lower left). The abundance of Stage III soft-sediment deformation provides early examples of similar instability along the Labuan-Berakas shelf edge. This is consistent with the location of the study area close to the basement-linked fault that initially defined the eastern margin of the Labuan-Berakas Syncline and which was later (Pliocene) reactivated as the Labuan Anticline (Figure 2.62b).

In addition to potential tectonic controls (e.g. slip along active faults), trigger mechanisms for soft sediment deformation could have had multiple environment-related causes, including the inferred high sedimentation rates, delta lobe switching, storm-wave loading and substrate instability caused by sediment density contrasts. Together, these are typical of high sediment supply storm-flood-dominated deltas (e.g. Collins et al. 2017, 2018).

2.6.2.7 Source-to-Sink System

The river system responsible for Stage III deposition was the palaeo-Padas River, which presently occupies a deep incised valley before terminating in Brunei Bay as a modern, tide-dominated bayhead delta (Figure 2.73b,c). During Stage III, this river flowed to the NW and discharged directly into the Labuan-Berakas Syncline. Braided fluvial deposits of the palaeo-Padas River are the outcrop manifestation on Labuan of the DRU, a tectonic event that caused widespread base-level fall and river incision (Figures 2.65 and 2.71). This resulted in a regional basinwards shift of facies belts during Stage IVA, when the deltaic system first reached the western margin of the Labuan-Berakas Syncline (Figures 2.64a and 2.72).

Although Stage III stratigraphy cannot be seismically imaged in the Labuan-Berakas Syncline due to its great burial depth, outcrop evidence highlights a large depositional

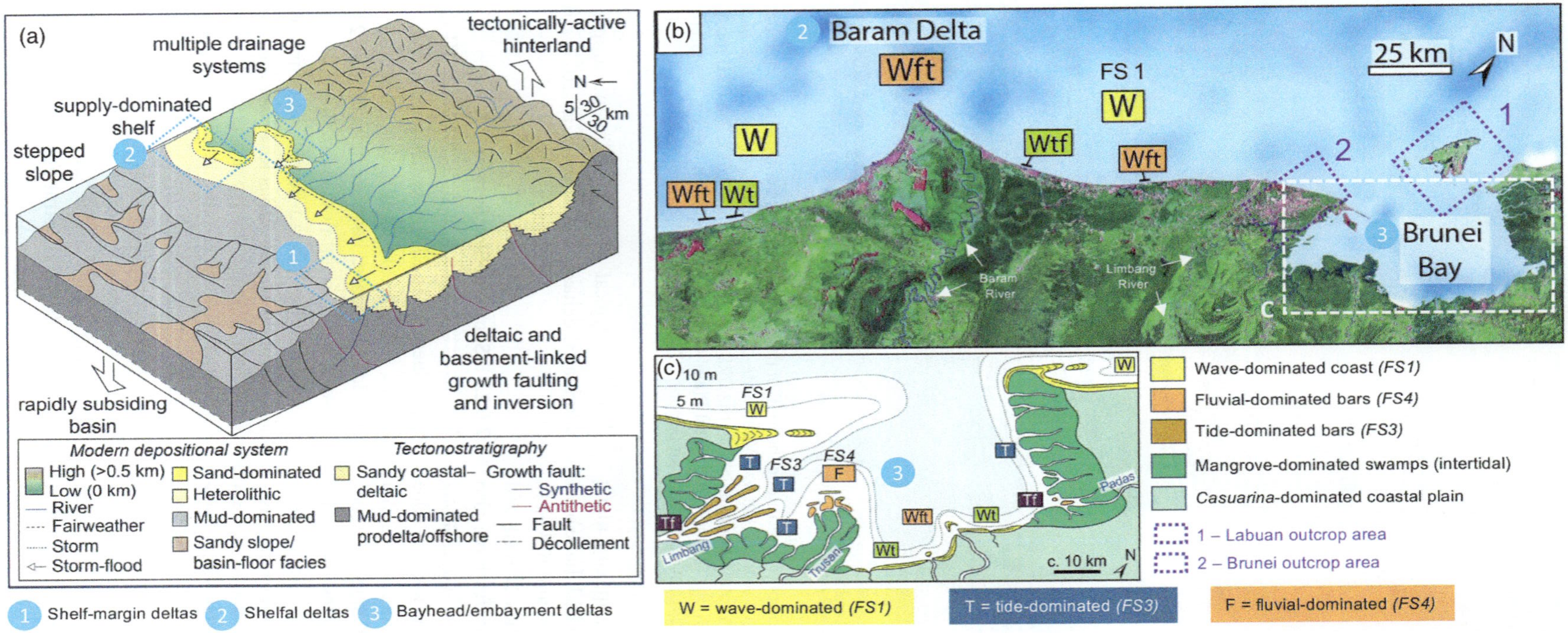

Figure 2.73 Integrated synthesis of modern and ancient (Neogene) depositional environments in NW Borneo, with emphasis on coastal-deltaic systems. (a) 3D depositional model of the Neogene depositional system. The modern length of the sediment routing system is around 300–500 km and its elevation (hinterland to basin floor) is c. 3–6 km. The model highlights unstable shelf-edge deltas controlled by underlying active fault systems (pale blue box/circle number 1), stable shelf deltas (pale blue box/circle number 2; see modern analogues in b and c) and bayhead/embayment deltas (pale blue box/circle number 3; see modern analogue in c). Repetitive tectonic uplift of the hinterland ensured sediment availability, while coeval basinward subsidence provided optimum accommodation space for the accumulation of c. 1 km-thick, prograding deltaic clastic wedges. From Collins et al. (2017) / John Wiley & Sons / CC BY 4.0. (b) Present-day, coastal–deltaic depositional system in the Baram Delta-Brunei Bay area, showing the Baram River shelf delta and the wave-dominated flanking strandplain. Further along strike (to the NE) is the coastal embayment of Brunei Bay. Several rivers discharge to the shoreline and display different morphologies: the Baram River (largest drainage basin area) feeds the wave-dominated, fluvial-to-tide-influenced Baram Delta (labelled Wft), and the Limbang River (second largest drainage basin area) discharges into Brunei Bay as a tide-dominated bayhead delta (see sketch map in c, labelled FS3). The white dashed box defines the area shown in c. From Collins et al. 2018. (c) Schematic map of Brunei Bay with interpreted process-dominated/-influenced depositional environments and geomorphological elements using the W-F-T letter classification scheme (Vakarelov and Ainsworth 2013). Simplified end-member facies successions are highlighted: wave- (FS1), tide- (FS3) and fluvial-dominated (FS4). From Collins et al. 2018.

system comprising regressive deltas with coeval basin floor fans and slope. Seismic evidence also shows similar inferred stratal geometries, seismic facies and depositional system throughout the continued infill of the Labuan-Berakas Syncline (e.g. Figure 2.64), especially during Stages IVA, IVB and IVC when deposition of around 8 km of sediment occurred in c. 7 Myr! The outcrops summarised here document the facies evidence during the initiation of this long-lived system, while seismic stratigraphy information documents the architecture and evolution of the subsequent large-scale basin-fill (Figure 2.72).

2.6.2.8 Sequence Stratigraphy

The sequence stratigraphy of Neogene outcrops on Labuan, and its lateral equivalents in the adjacent Labuan-Berakas Syncline, is summarised in Figures 2.65, 2.71 and 2.72. The succession overlying the DRU is represented by, first, an incised valley-fill and, second, by a large-scale transgressive to regressive cycle (Figure 2.65). The transgressive phase (Figure 2.61; Unit 3) comprises a fluvial- to tide-influenced coastal plain. Floodplain mudstones contain *in situ* mangrove-dominated coals and there is an upward increase in fluvio-tidal/estuarine channel sandstones, bayhead deltas and subtidal embayment mudstones with abundant drifted mangrove debris. The dominance of mangroves, both as *in situ* coals and as reworked material, supports a lower coastal plain with brackish water conditions.

The succeeding regressive cycle (Figure 2.65, Unit 4) represents a progradational stack of mainly wave-dominated shorelines, possibly representing flanking shoreface environments, but relatively distant from a direct fluvial sediment supply. The absence of soft-sediment deformation structures places this succession in a relatively stable, shelfal setting, possibly reflecting a temporary cessation of active basin-margin tectonics. Alternatively, it represents a flanking shoreface succession that is lateral to an axial river mouth system.

The fluvial interval that directly overlies the DRU contains a classical rising base level succession of braided to meandering river deposits, but it is thin (10–40 m thick; Figure 2.71). This begs the question: "where are the Stage IVA lowstand deposits"? The answer can be found in Figure 2.72, at the base of the seismic section (lower left/western side) in the form of prominent W-dipping seismic foresets and coeval aggradational topsets; flattened seismic sections on the flooding surface at the top of Stage IVA display a rising shoreline trajectory, typical of rising base level and high sediment supply conditions. This interval is around 30 km basinward of the Labuan outcrops (Figure 2.62b), suggesting that the axis of the Labuan-Berakas Syncline sub-basin had initially shifted westwards in response to uplift related to the DRU.

This cyclic pattern of interconnected tectonics and sedimentation continues to the present-day (Figure 2.73a) albeit distorted by the global impact of the Holocene transgression, which has led to the abandonment of the most recent phase of progradation (Late Pleistocene shelf-edge deltas). This transgression is being countered by today's rivers, which are building a wide spectrum of shelfal deltas along open coasts (e.g. Baram Delta) and bayhead deltas in embayed and/or drowned parts of the lower coastal plain (e.g. Brunei Bay) (Figure 2.73b,c). In the case of the NW Borneo margin, the modern and the ancient are so closely intertwined that each informs the other in equal amounts.

2.6.2.9 Lessons Learnt

1) Facies analysis of outcrops combined with nearby subsurface data of equivalent stratigraphy plus analogous modern environments at surface provide a powerful combination for understanding depositional systems (Figure 2.73). In addition, the modern analogue bridges the scale-gap between outcrops, cores, well logs and seismic datasets.
2) Rock-based facies analysis (using outcrops and/or cores, including facies-calibrated well logs (i.e. levels 2 and 3 in Table 2.1) provides the best means of capturing sedimentological data needed for the interpretation workflow of processes, facies, facies associations, architectural elements and depositional environments up to the scale of individual genetic units: this provides the building blocks of both large-scale sequence stratigraphic models and smaller reservoir-scale models.
3) Regional facies analysis, including seismic stratigraphic methods, can provide unrivalled insight into larger-scale facies successions and depositional systems (i.e. levels 3 and 4 in Table 2.1). For example, modern seismic data (e.g. Figure 2.72) can reveal significant levels of detail of stratigraphic architecture and depositional environments.
4) Integration of rock-based facies analysis, subsurface reservoir mapping, time-lapse satellite imagery and modern geomorphology enabled the dynamics of storm-flood deposition to be applied to the interpretation of shallow-marine event beds and storm-flood-dominated deltas in the Miocene of NW Borneo. This was enabled by re-evaluating a large and, in part, new outcrop database within a consistent facies framework.
5) Three distinctly different delta systems were recognised by combining process type with the broader tectono-stratigraphic setting (Figure 2.71):
   i) Fluvial-dominated deltas along an unstable tectonically active basin margin

   ii) Tide-dominated deltas in an embayed, transgressive coastline
   iii) Wave-dominated deltas along an open-coast and stable shelf
6) This shows the value of evaluating coastal-deltaic deposits relative to existing process-based classification schemes (Figure 2.70) as well as to larger-scale sequence stratigraphic models (Figure 2.71). Starting with an interpretation framework is good practice, both for posing questions to be directed at new outcrops and for distinguishing where and how new data may deviate from established models.
7) Finally, combining small-scale (item 2) and large-scale (item 3) facies analysis approaches (i.e. 'bottom-up' and 'top-down') provides one of the best means of interpreting ancient depositional environments.

2.6.3 Case Study-3: Quaternary, Sunda Shelf – Seismic Geomorphology

2.6.3.1 Aim

The aim of this case study is to demonstrate the combination of regional seismic facies analysis with modest borehole calibration in reconstructing an entire lowstand sediment routing system. The study area is part of the Sunda Shelf, SE Asia (Figure 2.74), which is the largest humid-tropical epicontinental sea in the world (2400 × 1200 km). Prior to the Holocene large parts of the Sunda Shelf were emergent due to a c. 120 m eustatic sea-level fall, corresponding to the Late Pleistocene Last Glacial Maximum (Figure 2.75). The Late Pleistocene stratigraphy represents one of the best-preserved examples of a lowstand systems tract, directly attributable to glacio-eustacy.

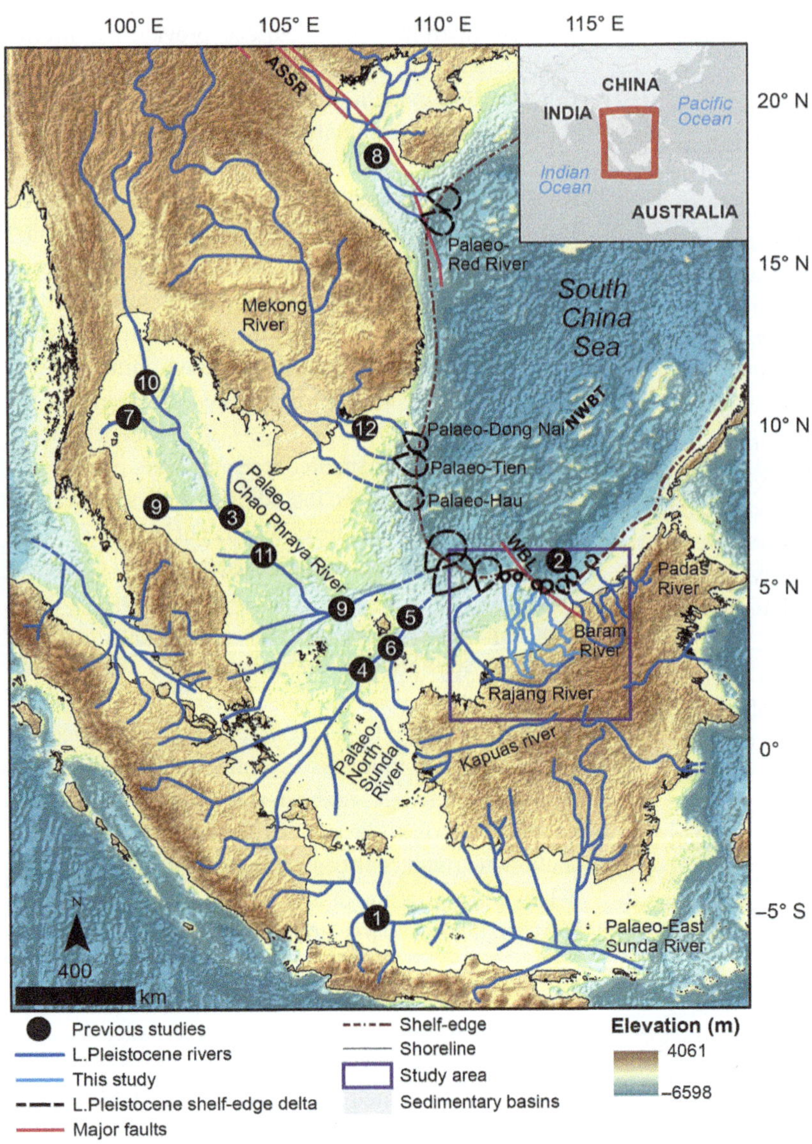

Figure 2.74 Digital elevation map of SE Asia, showing Late Pleistocene river networks and shelf-edge deltas, the majority of which flowed to the eastern margin of the Sunda Shelf and into the South China Sea; hinterland = brown, shelf areas = yellow/light blue; slope and deep-water = darker blue. Previous studies of incised valleys, including several seismic stratigraphic investigations, are numbered as follows: (1) Posamentier (2001); (2) Hiscott (2001); (3) Miall (2002); (4) Hanebuth et al. (2002); (5) Hanebuth et al. (2003); Wong et al. (2003); (6) Hanebuth and Stattegger (2004); (7) Samorn (2006); (8) Tanabe et al. (2006); Wetzel (2017); Wetzel et al. (2021); (9) Puchała et al. (2011); (10) Reijenstein et al. (2011); (11) Alqahtani et al. (2015, 2017); (12) Tjallingii et al. (2010); Thanh et al. (2021); Wetzel et al. (2021). The regional palaeo-river locations are compiled from Voris (2000), Lambiase and Cullen (2013), Thanh et al. (2021) and Wetzel et al. (2021). The Central Luconia Province study area is located in the centre of the purple box. The NWBT: NW Borneo Trough. *Source:* Mat Yusoff et al. (2024) / John Wiley & Sons / CC BY-NC 4.0.

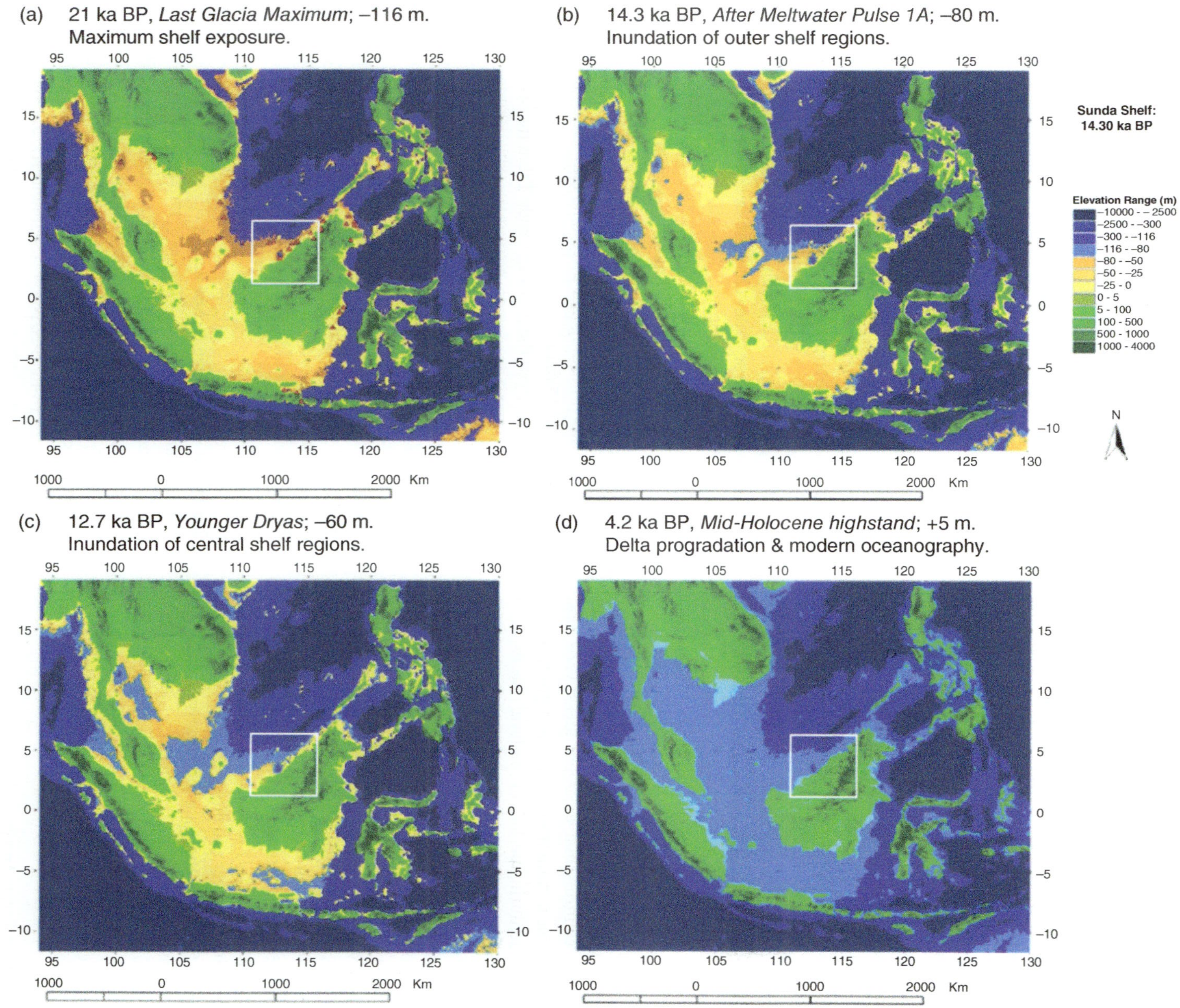

Figure 2.75 Reconstruction of palaeo-coastlines on the Sunda Shelf based on present-day bathymetric contours (from the National Geophysical Data Center – NGDC). Sathiamurthy and Voris (2006); modified from Hanebuth et al. (2011). (a) Last Glacial Maxima (21 ka BP) showing maximum shelf exposure. (b) Initial drowning of outer shelf areas during the Early Holocene (14.3 ka BP). (c) Continued drowning with exploitation of pre-Holocene basin axes (e.g. Malay/Gulf of Thailand Basin (12.7 ka BP)). (d) Mid-Holocene highstand of sea level, which is close to that of the present-day (4.2 ka BP). The Central Luconia study area is highlighted by the white box.

2.6.3.2 Data

This is a subsurface dataset, comprising 3D seismic data (c. 4000 km^2), high-frequency shallow 2D seismic data (pinger and boomer) and geotechnical boreholes 50–150 m deep. The 3D seismic data was used mainly to map the vertical and lateral distribution of seismic geomorphic features, while borehole data enabled integration of seismic facies and lithofacies.

2.6.3.3 Regional Geological Setting

The study area is located along the south-east margin of the shelf (Figures 2.74 and 2.75), where it merges with the Central Luconia Province of NW Borneo.

The Quaternary overlies a thick (0.5–2.0 km), clastic-dominated Neogene succession, comprising eight cycles (Cycles I–VIII), mostly regressive and bounded by marine flooding surfaces and/or their correlative unconformities (Figure 2.76).

2.6.3.4 Sunda Shelf: Quaternary Geological Setting

The Sunda Shelf is a broad, low-gradient continental shelf, located in an equatorial setting with a tropical-humid climate, high precipitation and high sediment discharge. The shelf is c. 600 km wide in its central part, narrowing to c. 300 km wide in the south-east (Figure 2.74).

Pollen data confirms that emergent areas were mostly covered by tropical rainforest, while mangrove swamps were widespread during periods of sea-level rise, especially in the Late Quaternary.

Sea level was around 120 m below the present-day level during the Last Glacial Maximum (LGM), resulting in shorelines located close to the present-day shelf-edge (Figure 2.75a). The broader and more stable parts of the Sunda shelf were dissected by long rivers (c. >1000 km) with large drainage networks (c. >1,000,000 km^2), such as those of the Chao Phraya, North Sunda and East Sunda rivers (Voris 2000; Alqahtani et al. 2015). In contrast, shelfal areas closer to tectonically active regions, such as the study area, tend to be narrower and dissected by the shorter rivers (Voris 2000). Several of the main modern rivers were active trunk rivers during the LGM and can be identified by subtle topographic variations across the submerged shelf, extending to the shelf-edge around the periphery of the deep-water South China Sea (Figures 2.74 and 2.75a).

2.6.3.5 Seismic Facies Analysis

A c. 500 km-long source-to-sink profile of the late Quaternary of this narrow shelf is differentiated into the following six segments (Figure 2.77): (i) hinterland; (ii) upstream/bypassed incised valleys; (iii) mid-stream incised valleys; (iv) down-stream incised valleys to shelf-edge deltas; (v) slope; and (vi) basin floor.

Segment I – Hinterland

The hinterland mainly comprises intensely folded and deeply weathered mud-rich sediments (the Rajang Range), covered in rainforest and subjected to an ever-wet, humid-tropical climate. Today there are approximately eleven major rivers with separate catchment areas, of which nine flow northwards into the study area, with approximately orthogonal trends to the regional NE-SW structural strike of the Rajang Range.

Segment II – Upstream/Bypassed Incised Valleys

This segment (Figure 2.77b) is a 40–80 km-wide zone immediately offshore from the present-day coastline (Figure 2.78). It comprises a 50–150 m thick, unconformably bounded Quaternary succession, including a series of prominent incised valleys, which can be traced updip into their modern river equivalents (e.g. the palaeo-Tatau drainage basin). These valleys have an overall N to NW orientation and cut across the Mio-Pliocene fold trends in the pre-Quaternary succession (Figure 2.78a). For example, three low-sinuosity incised valleys dissect the Temana reverse-fault-bounded anticline (Figure 2.78a), while dense networks of smaller dendritic channels drain the interfluve areas and flow in orthogonal directions into the adjacent valleys (Figure 2.78b,c).

Borehole data from the 2 to 4 km-wide Tatau incised valley (Figure 2.79a) show (i) a 20–30 m-thick valley with steep margins and filled with active (sand-rich) to inactive (mud-rich) facies, and (ii) mud-rich with peat layers and metre-scale fining-upward sand bodies of the dendritic channel-fills (Figure 2.79b).

Segment II is dominated by fluvial erosion, incision and sediment bypass, with limited preservation of floodplain deposits and accretionary fluvial features. Those sediments that are preserved could reflect rising base level towards the latest part of the LGM/earliest Holocene.

Segment III – Midstream Incised Valleys

This c. 50 km-wide segment (Figure 2.77b) coincides with a north to north-east thickening of the Quaternary/Cycle VIII progradational clastic wedge (Figure 2.76a). Incised valleys increase in width from <5 km (Segment II) to up to 10 km (Segment III), and the orientation of the valley axes changes from S-N/NNW (Segment II) to SW-NE (Segment III). The NE-trending palaeo-Tatau valley captures several smaller N-trending rivers in this segment (Figure 2.77b). Adjacent floodplain mudstones contain peat swamps dominated by freshwater tropical rainforest vegetation.

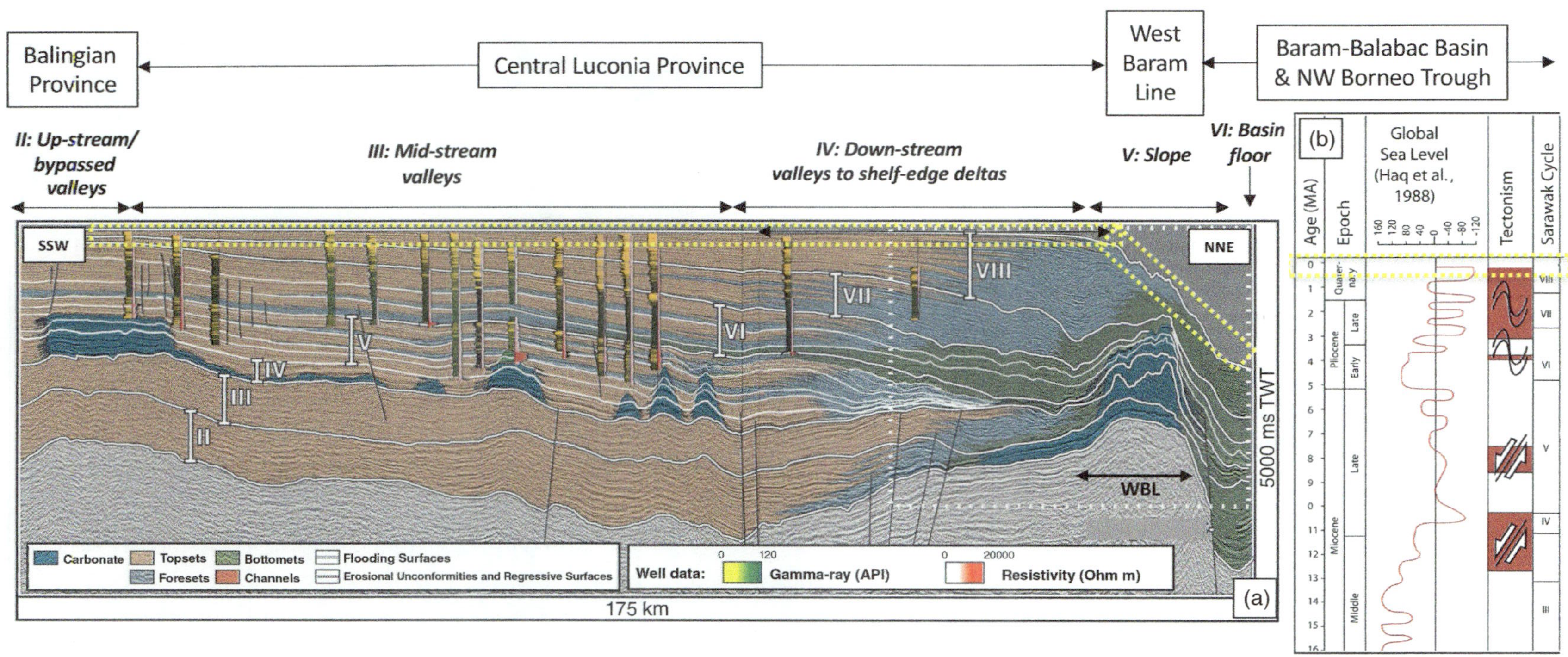

Figure 2.76 Regional geological setting of the SE Sunda Shelf (a) Interpreted 175 km long SSW-NNE, seismic section illustrating the large-scale stratigraphic architecture of the Neogene succession in the Central Luconia Province, mainly comprising: (i) pre-carbonate clastic interval (Cycles II/III = early Miocene), (ii) main carbonate interval (Cycle IV/V = middle-late Miocene) and (iii) younger clastic-dominated interval (Cycles VI, VII and VIII = Pliocene to Quaternary) (Koša et al. 2015). The approximate location of the late Quaternary depositional segments II to VI and geological provinces are shown along the top. The West Baram Line fault zone = WBL. This seismic line is located in Figure 2.77b. (b) Stratigraphic framework linking chronostratigraphy, biostratigraphy, global sea level, tectonics and cycle stratigraphy (Sarawak Cycles) (*Source:* Adapted from Koša et al. (2015)). This study is focused on the late Quaternary interval (dotted yellow box).

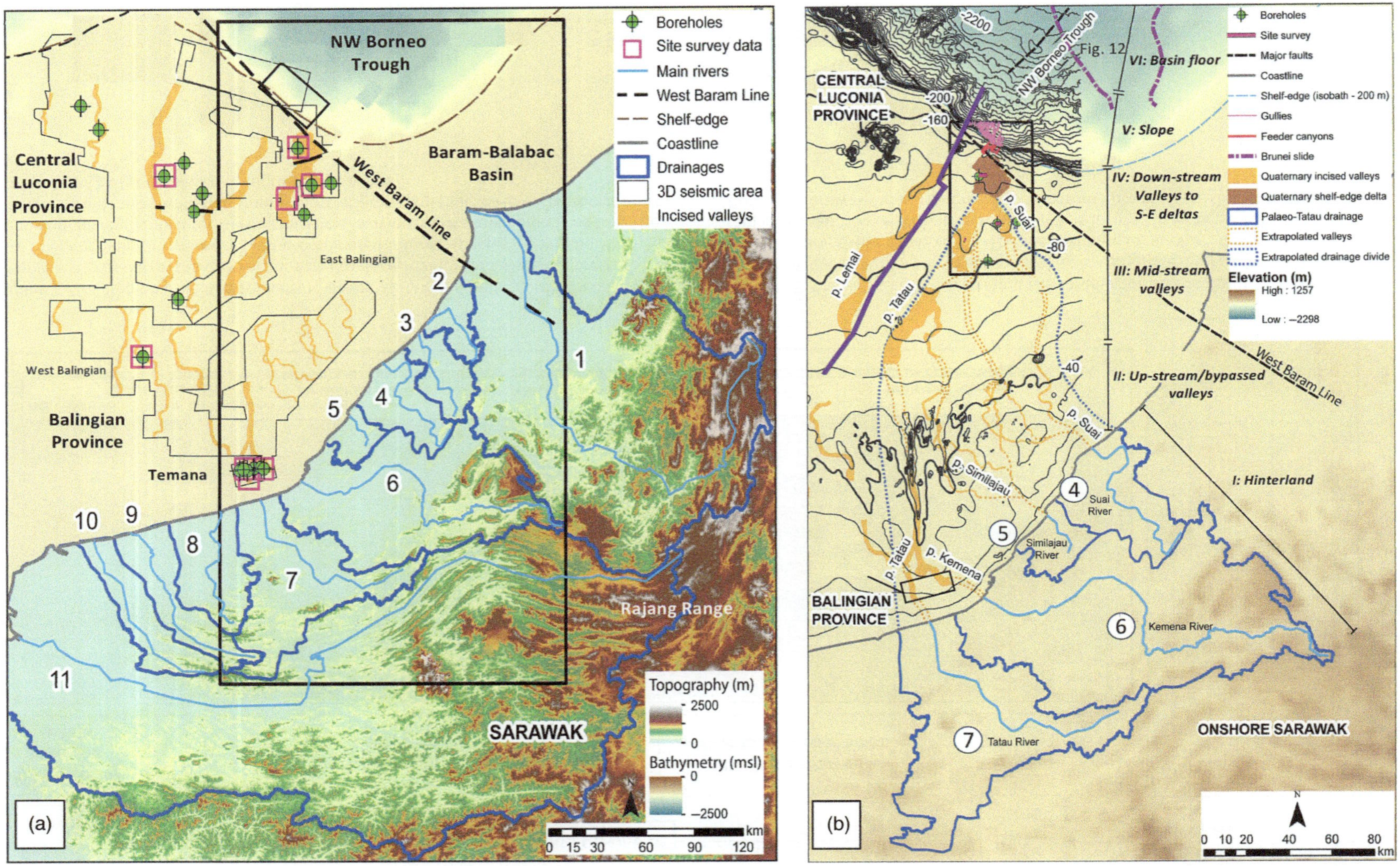

Figure 2.77 Late Quaternary incised valleys, gross depositional environments and geological setting. *Source:* Mat Yusoff et al. (2024) / John Wiley & Sons / CC BY-NC 4.0. (a) 30 m digital elevation model (DEM) highlighting modern rivers (light blue lines) and their drainage basin boundaries (darker blue lines), and late Pleistocene palaeo-valleys mapped from seismic data; numbered modern rivers are as follows: (1) Baram; (2) Sibuti; (3) Niah; (4) Suai; (5) Similajau; (6) Kemena; (7) Tatau; (8) Lemai; (9) Mukah; (10) Oya; (11) Rajang. The area of the rectangle is shown in b. Fine-lined black polygons are outlines of 3D seismic surveys used in this study. (b) Detailed reconstruction of the late Pleistocene paleo-Tatau drainage basin area based on 3D seismic interpretation and comprising four river systems (numbered 4–7); the inferred calibration with equivalent modern rivers is also shown (dashed lines correspond to 3D seismic data gaps). Note the downstream location of the confluence of the palaeo-Suai incised valley with the larger palaeo-Tatau system (orange-coloured valleys inside the black box). Immediately downstream of this confluence is the Tatau–Suai (T-S) shelf-edge delta (dark brown area). I–VI represent six source-to-sink segments discussed in the text. The Brunei submarine landslide is mapped from Gee et al. (2007). The seismic line in Figure 2.76 is in purple.

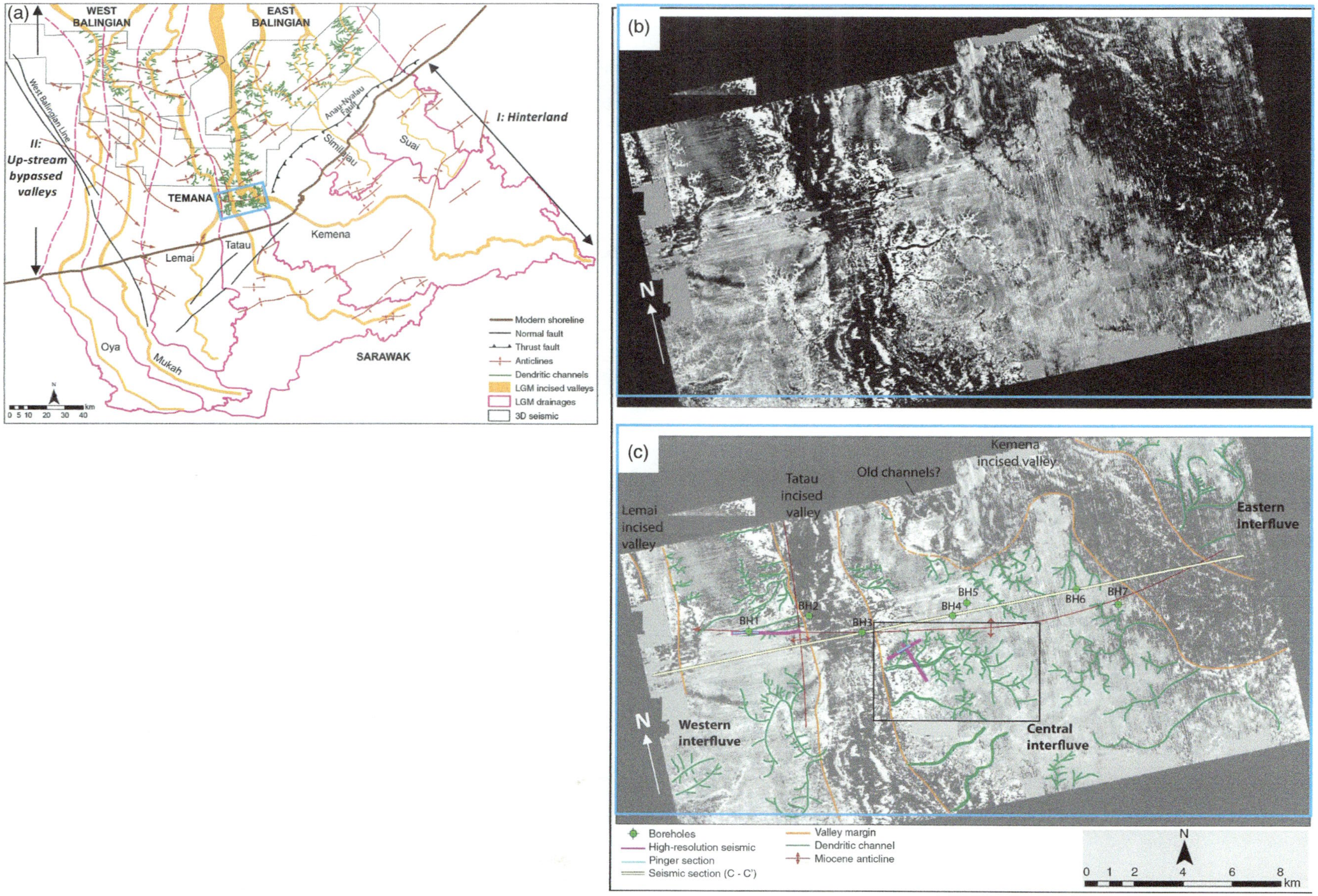

Figure 2.78 Segments I and II – upstream incised valleys and tributaries (*Source:* Mat Yusoff (2023) / Imperial College London). (a) Distribution of incised valleys (orange) and drainage basin areas (pink). (b) Uninterpreted 80 ms TWT seismic time-slice amplitude map of the Temana area. (c) Interpreted version of (c) showing the outline of three incised valley (dark grey tones/brown lines) and interfluve areas with dendritic channels (green) draining into the incised valleys in Segment II.

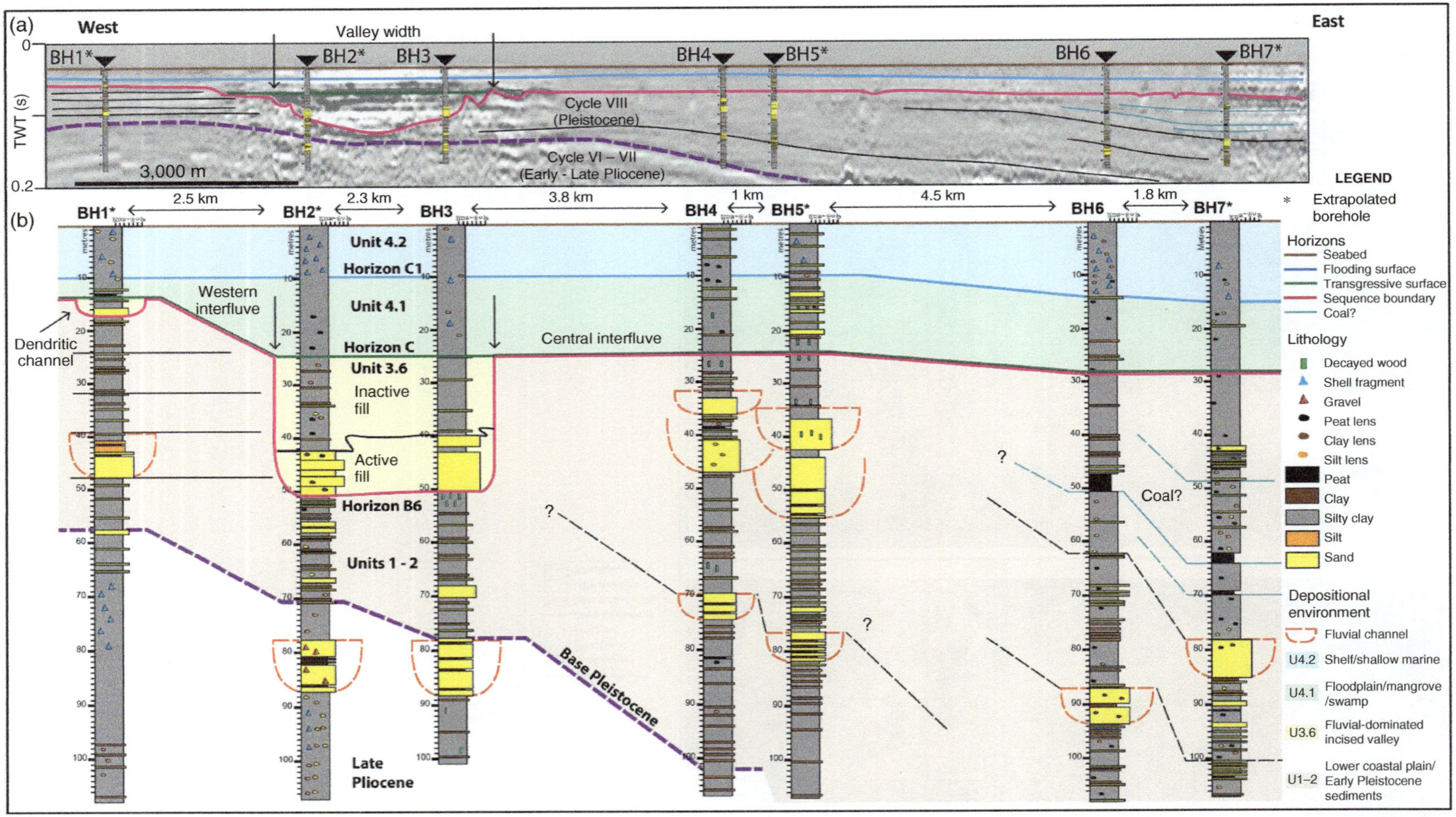

Figure 2.79 Seismic section and borehole correlation panel from Segment II – Up-stream/bypassed valleys (location approximately 15 km offshore and immediately north of the modern Tatau River mouth – see Figure 2.78 for details) (*Source:* Habibah Hanan Mat Yusoff (2023) / Imperial College London). (a) A 17 km-wide, E-W-oriented (depositional strike) interpreted seismic section. (b) A well correlation panel comprising seven, c. 100 m-deep geotechnical boreholes. The correlation panel is located along the crest of a reverse-fault-bounded and internally faulted anticline (the Temana Field in Figure 2.78). The stratigraphy comprises lower coastal plain (LCP) late Pliocene deposits (Cycles VI and VII) unconformably overlain by relatively thin Pleistocene LCP sediments (Cycle VIII). The latter (Units 1–2) thin southwards (c. 50–150 m thick) and include the 25 m-thick/c. 2.5–3 km-wide palaeo-Tatau incised valley (Unit 3.6; inferred LGM). The succession is capped by c. 10–30 m-thick Holocene lower coastal plain to shallow-marine mudstones (Units 4.1 and 4.2).

Segment III is fluvially dominated and reflects an increase in accommodation and a change in regional depositional dip (towards the NE). The original NW-SE-trending incised valleys are now oriented SW-NE, which is normal to the West Baram Line, suggesting renewed influence of this long-lived tectonic lineament on regional depositional trends.

Segment IV – Downstream Incised Valleys to Shelf-Edge Deltas

This segment is best illustrated by the palaeo-Tatau incised valley, which continues to widen downstream and captures several smaller, N-flowing rivers (Figure 2.77b), including the smaller Suai incised valley ('4' in Figure 2.77). Seismic amplitude variations on 3D seismic time-slice maps (Figure 2.80) and geoseismic cross sections (Figure 2.81) show that the main palaeo-Tatau valley-fill is up to 50–60 m thick (Figure 2.80c), it displays an overall upward fining facies trend and has a multistorey infill with a two-tier architecture (Figure 2.81b). The youngest part of the valley-fill is dominated by high sinuosity fluvial channels with point bars, which contribute to the downstream valley widening (Figure 2.77b). Tributaries, most notably the Suai incised valley, are narrower, thinner and dominated throughout by meandering rivers with a simpler internal stratigraphic architecture (Figures 2.80b and 2.81a).

Immediately downdip of the confluence of the Tatau and Suai valleys is a major downstream facies change into the Tatau-Suai shelf-edge delta (Figure 2.80b). This occupies a 20–30 km-wide area dominated by overall thinning of the Late Pleistocene interval (Figure 2.80c), while seismic facies characteristics display thinner, lower relief and more elongate to lobate geometries (Figure 2.80b,c). These are interpreted as coalesced sand-dominated mouth bars and fluvial–tidal distributary channels, which are flanked by peat-rich mudstones with mangrove vegetation in the tide-influenced lower delta plain.

Segment IV passes gradually down the palaeo-Tatau valley axis from a fluvial-dominated updip area (Figure 2.82a) to an increasingly tide-influenced downdip area, based on the abundance of mangrove material and valley widening, especially in the upper tier (Figure 2.81b). Variations in basal valley incision are the main control on thickness, which shows an overall downdip decrease (Figure 2.76c). This trend continues downdip into the Tatau-Suai delta, which is most conspicuously marked by a rapid reduction of meander belt planform geometries and an increase in elongate to lobate shapes. The downdip limit of the Tatau-Suai delta is abrupt and is defined by the West Baram Line fault zone, which is a long-lived (since the late Oligocene) tectonically controlled shelf-edge.

Segment V – Slope

The tectonically controlled slope coincides with the West Baram Line (Figures 2.76a and 2.83a). The slope is a NW-SE-trending zone that is c. 100 km-long, 20–40 km-wide and dominated by a variety of erosional features. These display a strong SW-NE orientation and range in size from closely spaced gullies (Figure 2.83b), through to larger channels with slumps and into canyons (km-wide and up to c. 1 km deep; Figure 2.83c). The West Baram Line slope merges to the east/south-east with the NW-dipping accretionary slope of the palaeo-Baram Delta (Figure 2.83a). This zone trends NE-SW and comprises an extensive gravity-driven fold-thrust-belt (Figure 2.84) with abundant mass transport complexes (MTCs), including the large Brunei Slide (Figures 2.83a,b 2.84a,b).

The Segment V slope represents a major basin-margin feature separating an extensive delta-fronted coastal plain from a deep-water basin floor (NW Borneo Trough; Figures 2.83a and 2.85). There were two major sediment supply systems. The first was directed to the north-west, with supply from the Baram Delta via a fold-thrust belt, which was lateral to the NW Borneo Trough (Figures 2.84c,d; context discussed in Section 2.6.2/Case Study-2). The second was via the West Baram Line wrench-fault zone, which supplied sediment to the north-east from the Central Luconia rivers described earlier, which were axial to the NW Borneo Trough (Figure 2.85; see also slope channel trends in Figures 2.80b and 2.83a).

Segment VI – Basin Floor

The termination of this source-to-sink profile is the deep-water NW Borneo Trough (up to c. 4 km deep). The floor of this trough is flat and receives sediment from dilute gravity flows, MTCs and hemipelagic suspension fall-out (Figure 2.85). Gravity flows from the Central Luconia Province depositional system, *via* the West Baram Line canyons, are transported to the NE, along the axis of the trough (Figure 2.85). Gravity flows from the fold-thrust belt of the Baram-Balabac depositional system are supplied laterally, along the SE edge of the basin, from NW-flowing river systems, including the Baram and Padas rivers, among others (Figures 2.83a and 2.85).

2.6.3.6 Source-to-Sink

High-resolution seismic facies analysis has been used to synthesise the main depositional elements along a coeval source-to-sink transect through the late Quaternary of the Sunda Shelf and into an adjacent deep-water basin (the NW Borneo Trough). An unexpected outcome of this work, and hence its value, was to identify how movement along an ancient wrench-fault zone (the West Baram Line) was reactivated in the Quaternary to a degree that it changed the orientation of rivers and their incised valleys. In so doing, a second sediment source was provided for the

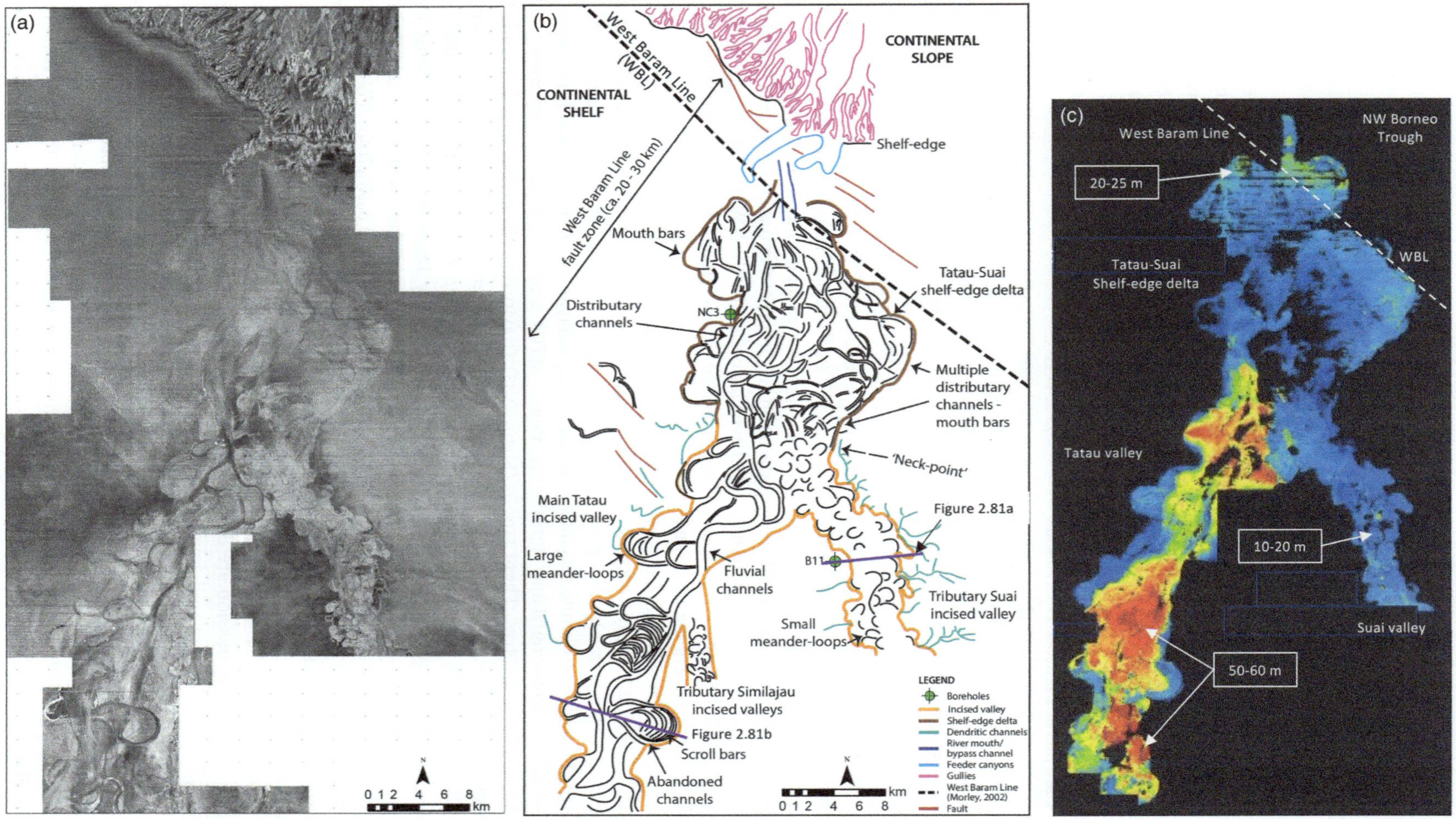

Figure 2.80 Segment IV – Mid- to down-stream valleys to shelf-edge deltas (*Source:* Mat Yusoff et al. (2024) / John Wiley & Sons / CC BY-NC 4.0). (a) Uninterpreted time-slice map at 30 ms TWT (c. 24 m depth beneath the seabed); the light grey and dark grey shades show maximum and minimum seismic amplitudes, respectively. (b) Line drawing summary of the interpretation of (a) showing: (i) Tatau incised valley comprises high sinuosity and narrow fluvial channels with clear scroll bars, (ii) the tributaries of the Similajau and Suai incised valleys comprise small-sized abandoned channels/meander-loops inside the valley margins and dendritic channels in the interfluve, (iii) Tatau–Suai shelf-edge delta is dominated by mouth bars and fluvial–tidal distributary channels, which terminated at the West Baram Line, which defines the shelf-edge, and (iv) the SW-NE-oriented gullies and channels (pink) dominate the upper-slope area. (c) Isopach map of downstream Tatau, Suai incised valleys and Tatau–Suai (T-S) shelf-edge delta generated from 3D seismic data using Paleoscan software. The Tatau incised valley fill is up to 60 m thick, compared to the 10–20 m-thick Suai valley fill and T-S shelf-edge delta (20–25 m thick).

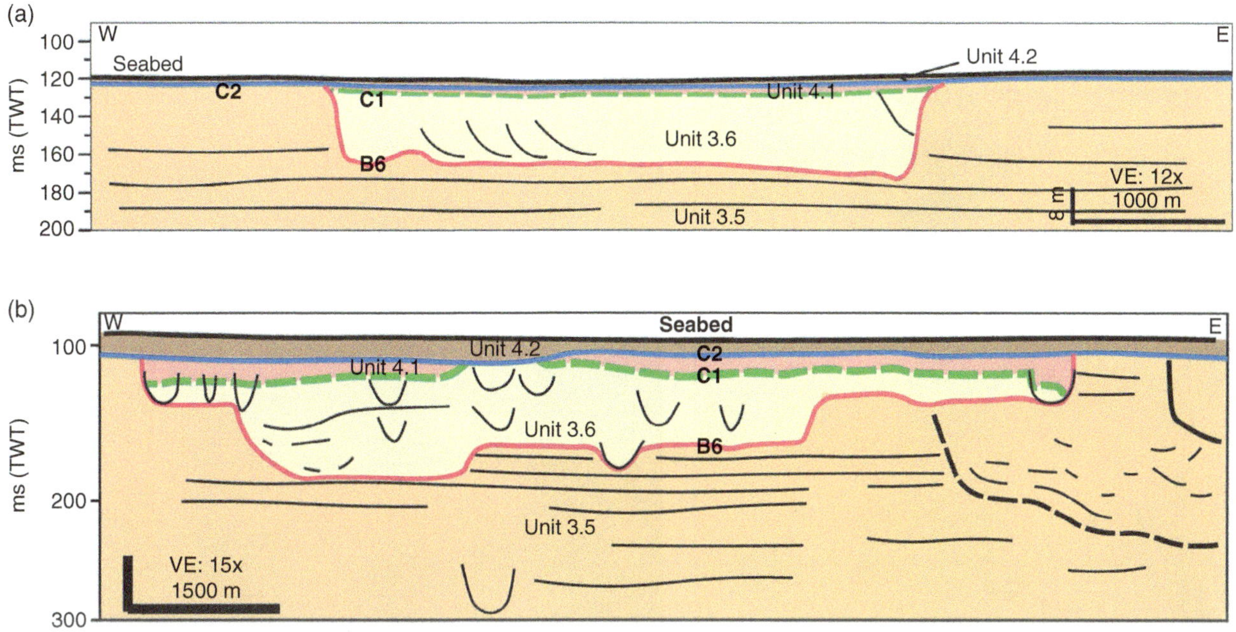

Figure 2.81 Segment IV – Down-stream valleys showing typical cross-sectional geometries of the incised valleys (Mat Yusoff et al. (2024) / John Wiley & Sons / CC BY-NC 4.0). (a) Geoseismic cross section through the c. 10–20 m-thick palaeo-Suai incised valley (Unit 3.6), comprising valley-confined, meanderbelt deposits with lateral accretion surfaces, point bars and clay plugs. (b) Geoseismic cross section through the deeper/thicker palaeo-Tatau incised valley (Unit 3.6), showing a two-tier architecture: (i) lower tier of valley-confined, multistorey channel sand bodies, of inferred braided fluvial origin, and (ii) upper tier dominated by meandering channels with lateral accretion, point bars and mud-filled abandoned channels, which extends laterally beyond the valley limits of the lower tier. Unit 4.1 comprises brackish water mudstones and Unit 4.2 shallow-marine mudstones (inferred transgressive Holocene mud layer). See Figure 2.80 for line locations.

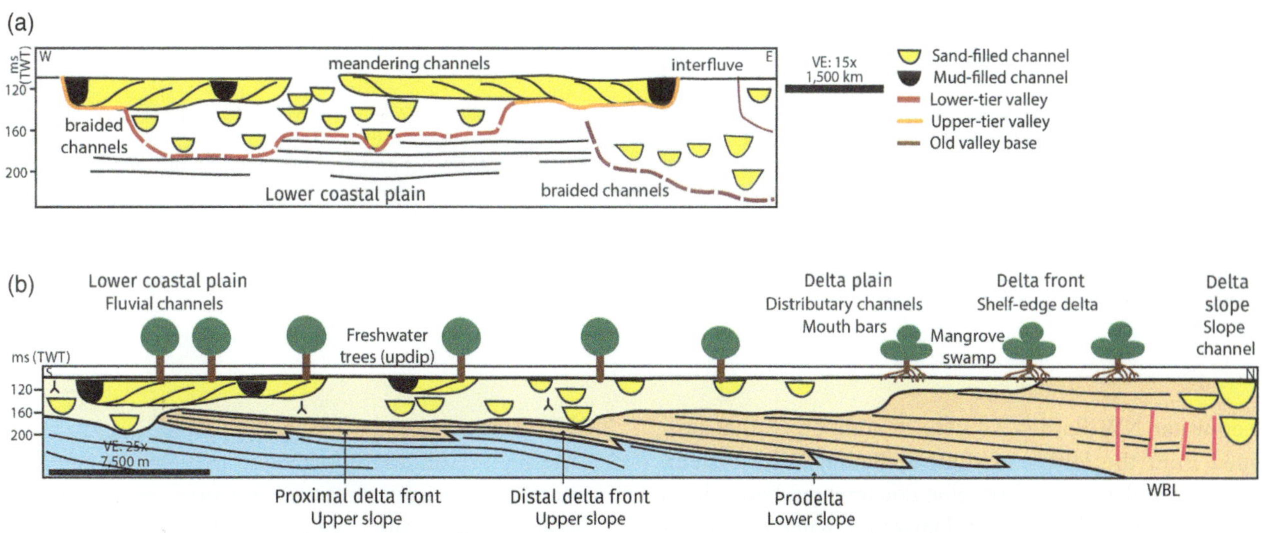

Figure 2.82 Segment III-IV – Mid- to down-stream valleys to shelf-edge deltas, showing simplified geoseismic interpretations of the lower reaches of the palaeo-Tatau incised valley system (*Source:* Mat Yusoff et al. (2024) / John Wiley & Sons / CC BY-NC 4.0). (a) Schematic depositional strike line showing the typical upstream incised valley fluvial architecture, comprising: (i) confined, narrow and, possibly, isolated sand-filled braided channels in the lower-tier of the valley (base defined by dashed red lines); and (ii) unconfined, wider, meandering channels with point bars and abandoned channel-fills (black) in the upper-tier of the valley (base defined by solid orange line). (b) Schematic depositional dip line in a downstream position showing an incised valley to the left (relatively landward) and its downstream termination and interfingering relationship with shelf-edge delta to upper-slope deposits (to the right).

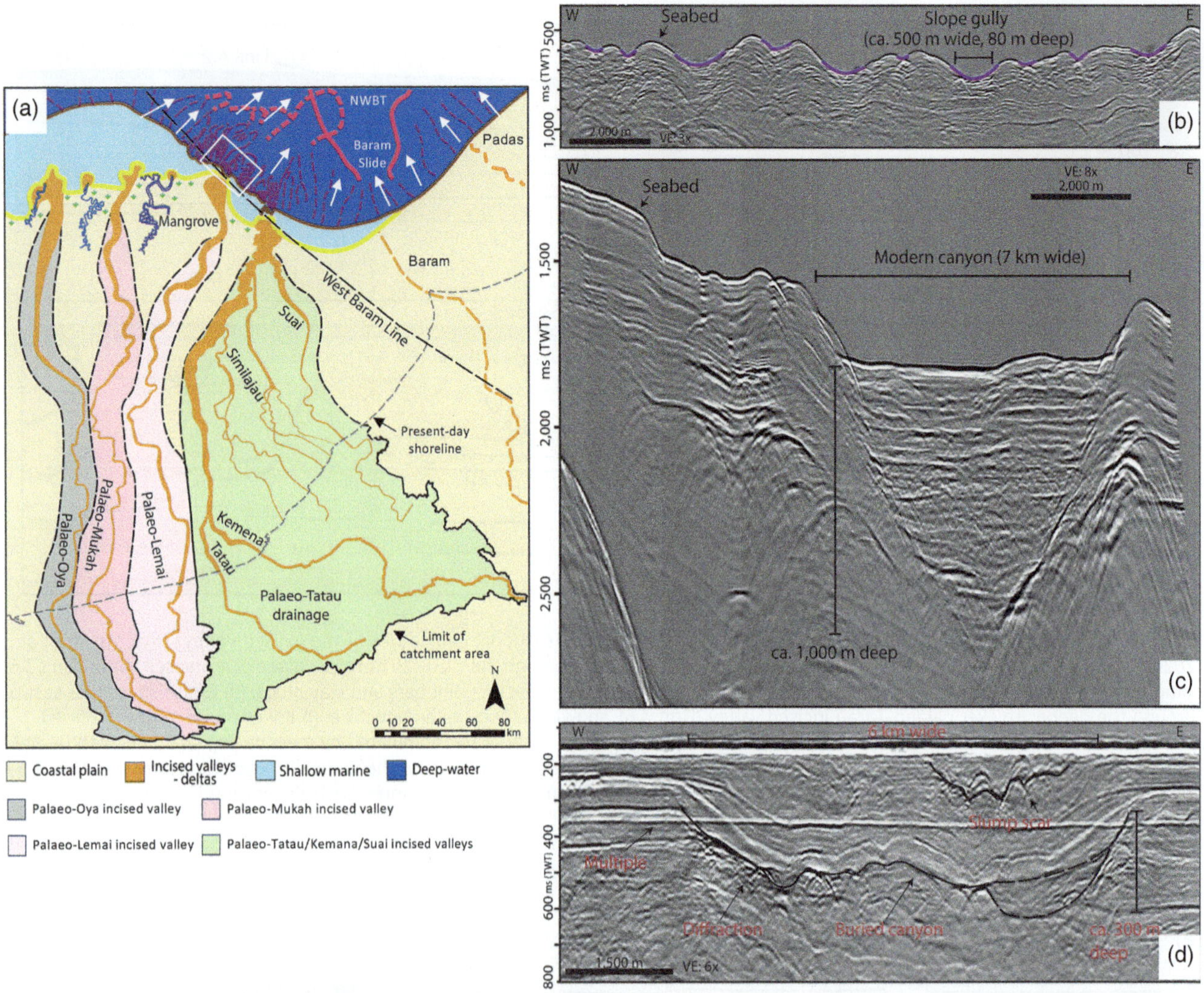

Figure 2.83 Segment V – Slope (West Baram Line). Seismic lines (b), (c) and (d) are all from within the white rectangle (*Source:* Habibah Hanan Mat Yusoff (2023) / Imperial College London). (a) Late Pleistocene (c. LGM) palaeogeography of the Central Luconia and Balingian area, and its extension into the SW Baram-Balabac Basin and NW Borneo Trough (NWBT). Schematic representation of slope gullies and canyons (dark red lines); Brunei Slide (Figure 2.84a, b) shown by the lighter red lines; white arrows show schematically the main directions of deep-water transport. White rectangle shows location of parts (b) to (d), this figure. (b) Strike-oriented (NW-SE) seismic section through erosional gullies (tens of metres relief) in the upper slope of the palaeo-Lamai shelf-edge delta (white box). (c) Canyon (c. 7 km wide and 1 km deep) incised into the slope of the palaeo-Lamai shelf-edge delta (white box). (d) Slumps and channels in upper-slope deposits palaeo-Lamai shelf-edge delta (white box).

deep-water NW Borneo Trough, in addition to that provided from the longer established Baram Delta system. Most of the detailed information documented here comes from around the Last Glacial Maximum, when the Sunda Shelf became the largest exposed humid-tropical intra-cratonic area in the world. It represents a major low-latitude lowstand systems tract. Today, following the Holocene transgression, it is the world's largest submerged humid-tropical intra-cratonic area!

2.6.3.7 Lessons Learnt

1) The high-resolution information provided by multiple merged 3D seismic datasets, supported by lithological and facies calibration with geotechnical borehole and other site survey data, provides an unusually rich sedimentological database, despite this not being a primary intention of the original data collection.
2) The 3D seismic amplitude data expressed in the form of time slices and/or flattened horizon surfaces provides

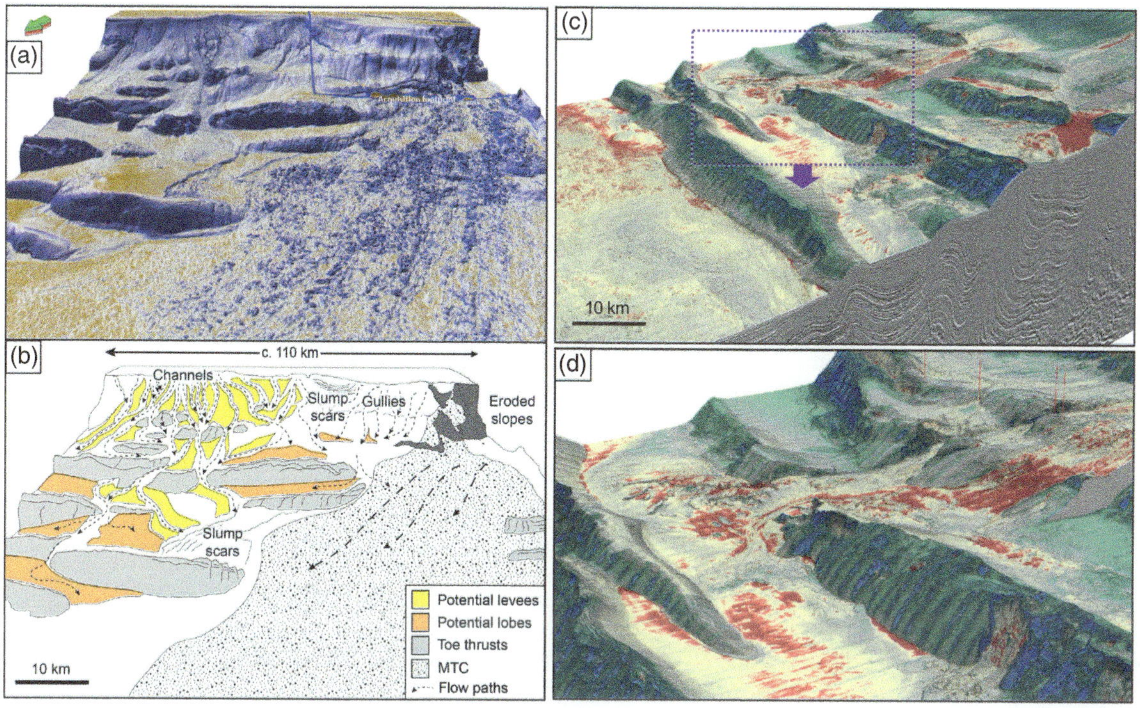

Figure 2.84 Segment V – Slope system with seismic amplitude examples from the fold-thrust belt in front of the Baram Delta part of the NW Borneo margin. (a) A 3D image of the present-day seabed bathymetry of deep-water Brunei with a dip attribute map overlay (dark blue and brown colours indicate steep and gentle seabed gradients, respectively); this view faces southwards, looking from the area of the NW Borneo Trough towards the slope, with the shelf edge in the distance. It shows a large mass transport complex (MTC) in the west (part of the giant 'Brunei Slide') and steeply dipping thrust faults and slump scars in the east (dark blue areas). *Source:* (from Puasa 2018) / Imperial College London. (b) An interpreted line drawing of (a), highlighting surface gullies (dashed lines), in the upper-slope areas, ponded areas with potential turbidite sand-rich lobe accumulations (orange) and possible channel-levees (yellow). The 'speckled' appearance of the amplitude map over the MTC captures the km-scale dimensions of displaced blocks. Note the E-W-oriented toe-thrusts and associated folds in the east and south-east. From Puasa 2018. (c) An RMS amplitude extraction from just below the seabed and draped onto the seabed horizon. This present-day system shows reservoir variability and ponding of turbidite sands in the structural lows. This resembles the depositional setting of the Neogene fold and thrust belt of NW Borneo. (d) Enlargement of part of C, showing details of potential sandstone reservoir channels and lobes, plus spectacular images slump scars along oversteepened parts of the fold-thrust belt, (c) and (d) are courtesy of CGG.

exceptional quality planform information, which often allows direct mapping and interpretation of a range of geomorphic features – an old and very familiar lesson. However, in this example, it enabled mapping and quantification of the size, shape, extent and lateral relationships of an entire sediment routing system.

3) The relatively short length of the sediment transport pathway (c. 500 km) is typical of tectonically active margins around the world and contrasts with the much longer distances encountered in passive margins (e.g. Congo river-fan system is c. 6000 km long and the Mississippi River-delta-fan system is 5000 km long).
4) The quality of the planform information collected from Quaternary sedimentary environments makes it particularly suitable for depositional and reservoir analogue purposes, especially in the underlying successions, provided that gross depositional environments and geological setting remained similar; in this case, it applies to a remarkably high level.
5) One of the main limitations of this type of data, especially for analogue purposes, is the limited vertical resolution of conventional seismic data (e.g. 10–15 m). Borehole data may partly address this problem, but usually only in very limited areas (i.e. the site survey area itself). The lack of other borehole information, especially biostratigraphy and conventional well log data, is a further limitation.
6) The late Pleistocene of the Sunda Shelf (c. 20–120 ka BP) comprises one main progradational clastic wedge (with three smaller-scale cycles) bounded by two major transgressions: (i) 110–130 ka BP and (ii) 0–20 ka BP (Holocene) (Hanebuth 2011). The final phase of progradation coincides with the LGM (19–21 ka BP), which caused a glacio-eustatic sea level fall of c. 120 m. This

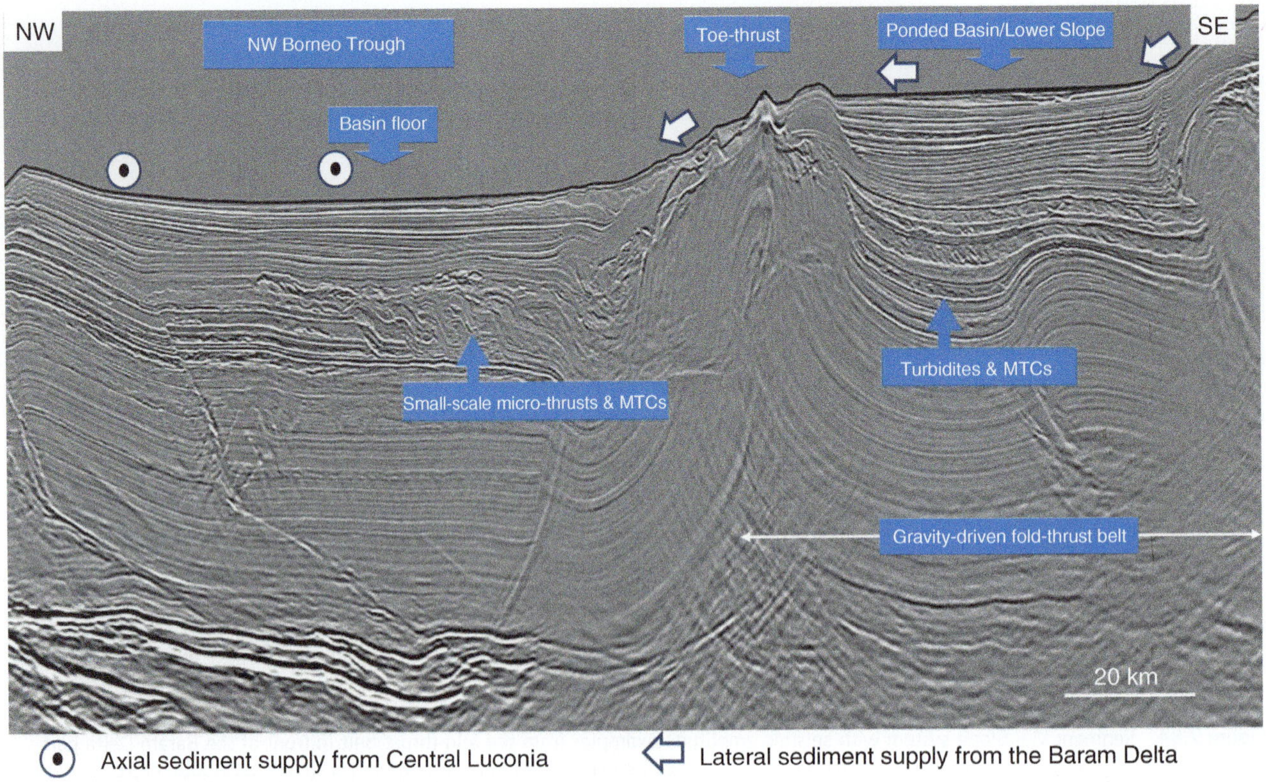

Figure 2.85 Segment VI – basin floor, comprising the NW Borneo Trough (NWBT) and its complex Miocene-Recent fold and thrust belt imaged in a 3D seismic line taken from a multisensor broadband seismic dataset from deep-water Sabah. Note examples of (i) soft sediment deformation (small-scale thrusts and MTCs) in front of the leading toe-thrust, and (ii) intercalated MTC (chaotic seismic facies) and turbidite sands (high-continuity seismic facies) in the adjacent ponded basin. *Source:* Photo courtesy of PETRONAS and TGS.

is widely considered to have been the primary cause of widespread emergence and fluvial incision throughout the Sunda Shelf during the late Pleistocene (Figures 2.82 and 2.83). Hence, it is concluded that the final phase of progradation (the Cycle VIII clastic wedge of the study area) was dominated by a glacio-eustatic sea-level fall, albeit superimposed on a longer-term cycle of uplift, rotation and denudation.

2.6.4 Case Study 4 – Carbonates: Carboniferous Carbonate Platform System (M-Factory, Sierra del Cuera, NW Spain): Progradation-Aggradation-Retrogradation Pattern

2.6.4.1 Aim

Use of facies to evaluate the sedimentation processes across a microbial boundstone-dominated carbonate platform (M-factory *sensu* Schlager 2003) and showing differences from the classical present-day tropical flat-topped carbonate platforms (T-factory platforms *sensu* Schlager 2003). The study demonstrates (i) continuous upper slope to deepened, down to 300–400 m, aphotic depth, platform margin sediment production along (ii) a high-relief platform with (iii) steepened platform flanks and (iv) planar clinoforms; (iv) early diagenetic marine cementation processes; and (v) limited sediment export towards the surrounding basin.

2.6.4.2 Data

The Pennsylvanian of the Sierra del Cuera has been studied extensively as this outcrop is an excellent analogue for various late Palaeozoic hydrocarbon reservoirs (e.g. Tengiz, Kashagan, Karachaganak fields, Pricaspian Basin and Kazakhstan). The subsurface occurrences also consist of M-factory flat-top platforms showing a high-relief, a deepened platform margin, and steep planar slopes. Studies of the Sierra del Cuera include analyses of the large-scale structures (Bahamonde et al. 1997, 2000 and Kenter et al. 2003, 2005), detailed sedimentological and microfacies analysis (Della Porta et al. 2002, 2003, 2004) and diagenetic studies (Immenhauser et al. 2002; Van der Kooij et al.

2009). In addition, two extensive summaries were published in two excursion guides by Della Porta et al. (2017, 2021) and Wright & Della Porta (2021).

2.6.4.3 Tectonic Setting

Bahamonde et al. (1997, 2015 and references therein) summarised the development of the Cantabrian foreland basin (NW Spain) in response to the continental collision of Gondwana and Laurentia during the Late Devonian-Carboniferous. The foreland basin originally showed a linear SW-NE-oriented geometry that acquired its present-day arch shape during the closure of the Ibero-American Arc (Stewart 1995; Weil et al. 2000; Gutiérrez-Alonso et al. 2008).

The Carboniferous sedimentary basin infill of the Cantabrian Zone includes (from bottom to top): (i) laterally continuous Mississippian-lowest Pennsylvanian series, approximately 300 m thick, containing condensed deposits (Baleas, Vegamian and Alba Formations; Tournaisian to lower Serpukhovian) and lime-mudstones (Barcaliente Formation; upper Serpukhovian-lower Bashkirian); (ii) siliciclastic-dominated foredeep successions, up to 6000 m thick, of Pennsylvanian age (Bashkirian–Moscovian); and (iii) Pennsylvanian high-relief carbonate platforms dominated by microbial boundstone (Bashkirian-lower Kasimovian Sierra del Cuera/Picos de Europa carbonate platform) that became established on the craton-ward margin of the basin. Subsequently, these deposits were superimposed by foredeep successions (Bahamonde et al. 1997, 2000, 2007; Della Porta et al. 2003, 2004; Kenter et al. 2003).

As deformation progressed a series of imbricated thrust-sheets developed that became tilted and compartmentalised. The Cantabrian Zone was originally subdivided into several tectono-stratigraphic provinces but based on detailed analysis of the Variscan tectonic structure a new subdivision was proposed with three main units (Alonso et al. 2009; Bahamonde et al. 2015): (i) Somiedo, (ii) Bodón-Ponga and (iii) Pisuerga-Carrión. The Bodón-Ponga Unit comprises the Bodón and Aramo nappes, the Central Asturian Coalfield, the Ponga Nappe and Picos de Europa Provinces (Figure 2.86).

2.6.4.4 Stratigraphy

Thick clastic wedges consisting of turbiditic and deltaic successions (Folds and Nappes, Central Coal Basin and Pisuerga-Carrión domains) characterised the subsiding proximal zones as extensive carbonate platforms developed in the more distal and shallow areas (Picos de Europa and Ponga Nappe domains; Colmenero et al. 1993; Bahamonde et al. 1997). The margins of these platforms are exposed in the extreme northeast of the Ponga Nappe Unit (Figure 2.86).

The Sierra del Cuera is a classical M-factory carbonate platform with a flat-top, deepened platform margin and steep planar slopes; the platform developed during the Carboniferous in the synorogenic foreland basin of Asturias (N-Spain) (Bahamonde et al. 1997, 2000, 2015). The outcrop is characterised by a zone with horizontal-bedded platform deposits that transfers to a margin with steep clinoforms (15° to 32°) with a relief of up to 500 m; succeeded by a sharp transition to a low-angle toe-of-slope, where beds with slope-derived material interfinger with basin sediments (Figure 2.87).

The Sierra del Cuera carbonate platform displayed five successive development stages (Bahamonde et al. 2000; Della Porta et al. 2003, 2004). The first Bashkirian stage started with the flooding of the pre-existing Serpukhovian platform and the development of a low-angle ramp with numerous microbial mud-mound accumulations, together with gravity deposits, for example, breccias and calciturbidites (Figure 2.88a). The successive influx of siliciclastic sediment during the early Moscovian stage buried part of the carbonate ramp and restricted carbonate sedimentation (Figure 2.88). The third Moscovian stage is marked by aggradation and progradation of the entire flat-top carbonate system. The fourth stage, latest Moscovian to earliest Kasimovian, is marked by the drowning of the entire platform that successively, fifth stage, was covered by marly calcareous ramp sediments.

2.6.4.5 Facies

Four main sedimentological domains characterise the outcrop (Figure 2.87): (i) flat-top platform deposits composed of thick algal boundstone, skeletal packstone to grainstone and peloidal micrite wackestone; these deposits have a stacked repetitive appearance; (ii) platform margin and upper-slope facies consisting of microbial boundstone; these deposits cover a bathymetric range from the platform margin down to approximately 300–350 m downslope; (iii) slope deposits comprising predominantly of margin and upper-slope-derived rudstones and breccias together with extensive early marine cements of botryoidal aragonite and radiaxial fibrous calcite; and (iv) a toe-of-slope to basin environment in which a cyclic alternation occurs with spiculitic siltstones, calci-mudstone, packstone to grainstone gravity flows and rudstone breccias (Bahamonde et al. 2000; Della Porta et al. 2002, 2003, 2004).

The diagenesis and stable isotope geochemistry studies of Immenhauser et al. (2002) discussed the stable isotope variations of the inner platform deposits as van der Kooij et al.

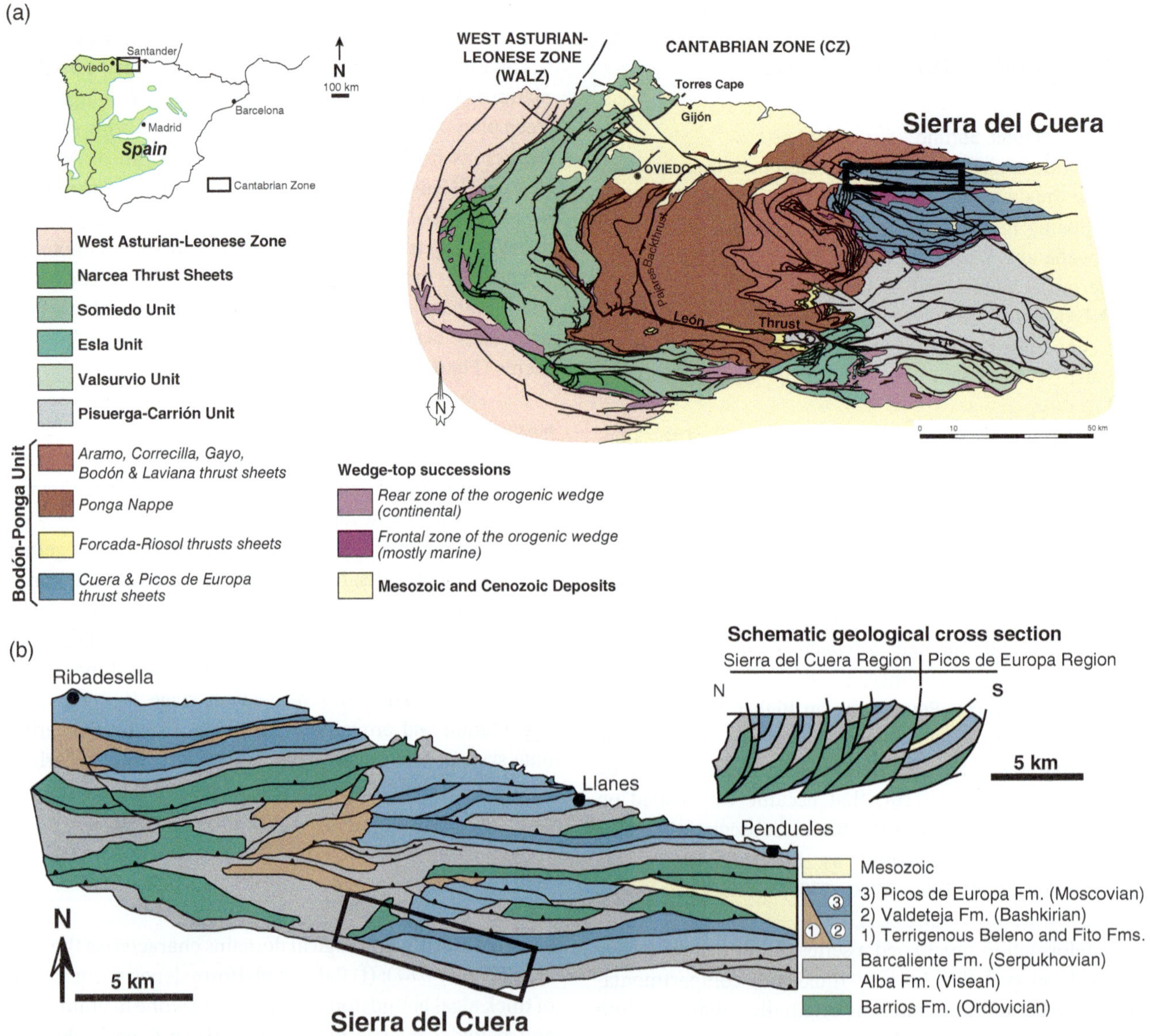

Figure 2.86 (a) Location map of the Cantabrian Mountains Zone (Asturias, NE Iberian Massif, Spain). Geological map with the Sierra del Cuera (SDC) and the Picos de Europa (south of Sierra del Cuera) carbonate platform outcrops and tectonostratigraphic units of the Cantabrian Zone (modified after Julivert 1971; Alonso et al. 2009; Bahamonde et al. 2015). The SDC is part of the Bodón-Ponga Unit (Alonso et al., 2009), formerly labelled as Ponga Nappe and Picos de Europa provinces (Julivert, 1971; Pérez-Estaún et al., 1988). (b) Geological map and cross section of the SDC Variscan thrust sheet (modified after Bahamonde et al. 1997, 2004 / CINECA; modified from Della Porta et al. 2017, 2021) / CINECA.

(2009, 2010) examined the diagenetic cementation and dissolution history of the slope sediments.

2.6.4.6 Depositional Model

The depositional model for the Sierra del Cuera (Figure 2.88) highlights the sedimentation patterns of an M-factory, slope shedding *sensu* Kenter et al. (2005), in which the upper slope to margin transect is the main sediment production area. In addition, the microbial boundstone dominance of the system leads to the development of a flat-top platform with gradual basinward deepening margins and planar clinoforms (e.g. Della Porta et al. 2003, 2004; Kenter et al. 2005). The facies distribution patterns, platform morphology and sediment export patterns of M-factory carbonate platforms were

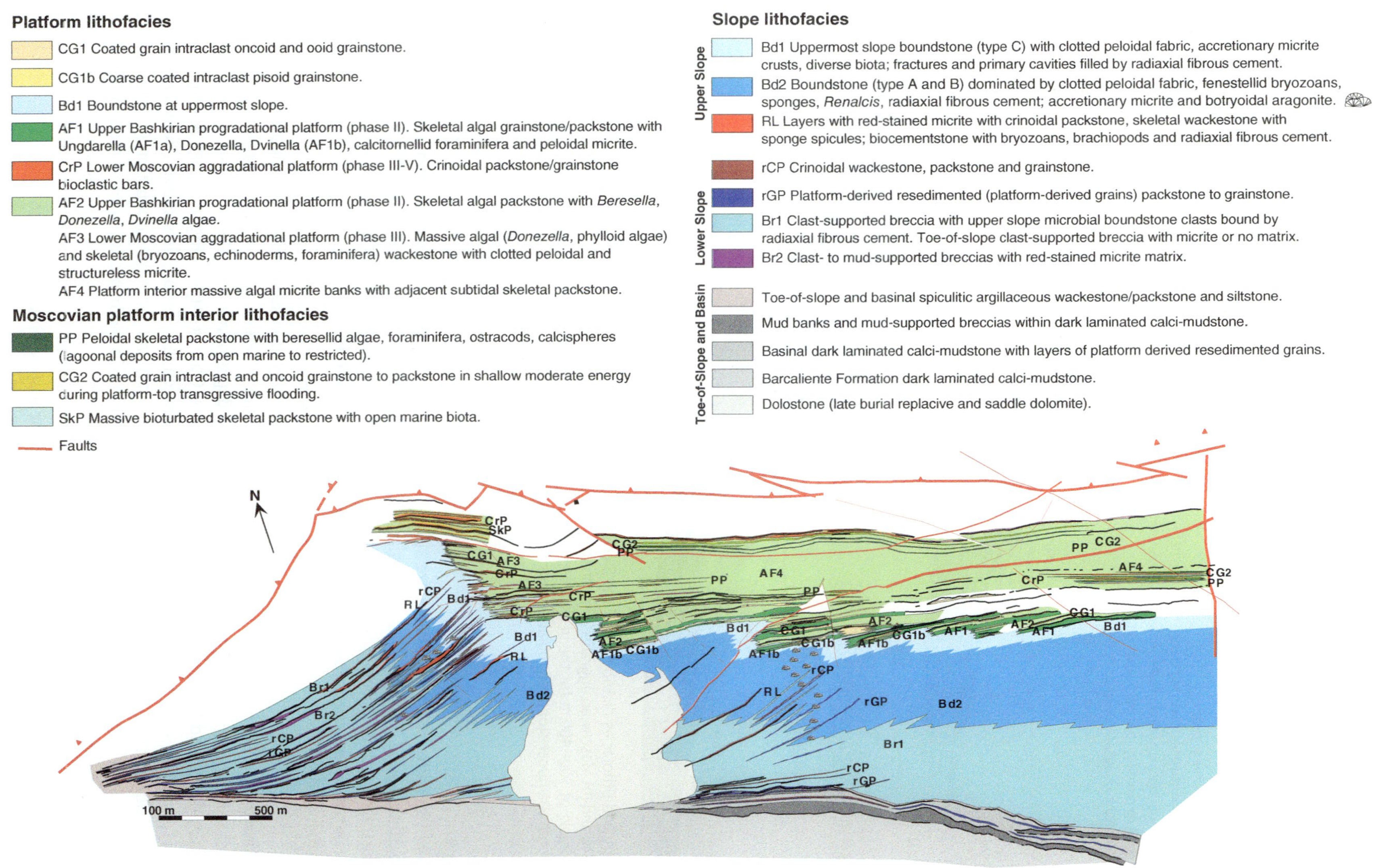

Figure 2.87 Detailed lithofacies mapping of the Sierra del Cuera platform (modified from Della Porta et al. 2003; in Della Porta et al. 2021). The platform succession clearly builds up through time but also shows alternating intervals with prograding and aggrading geometries. Facies variations in platform interior deposits, platform margin and slope deposits are marked by various colours and lithofacies abbreviations. Note the lack of a barrier at the platform edge, the gradual basinward deepening margins and planar clinoforms. The transition from the Barcaliente Fm at the base (grey colours) to the overlying toe-of-slope facies belt shows a clear downlap surface. The thickness of the entire platform-slope edifice differs significantly from the basin deposits marking the sediment production and depositional nature of an M-factory carbonate platform with limited sediment export.

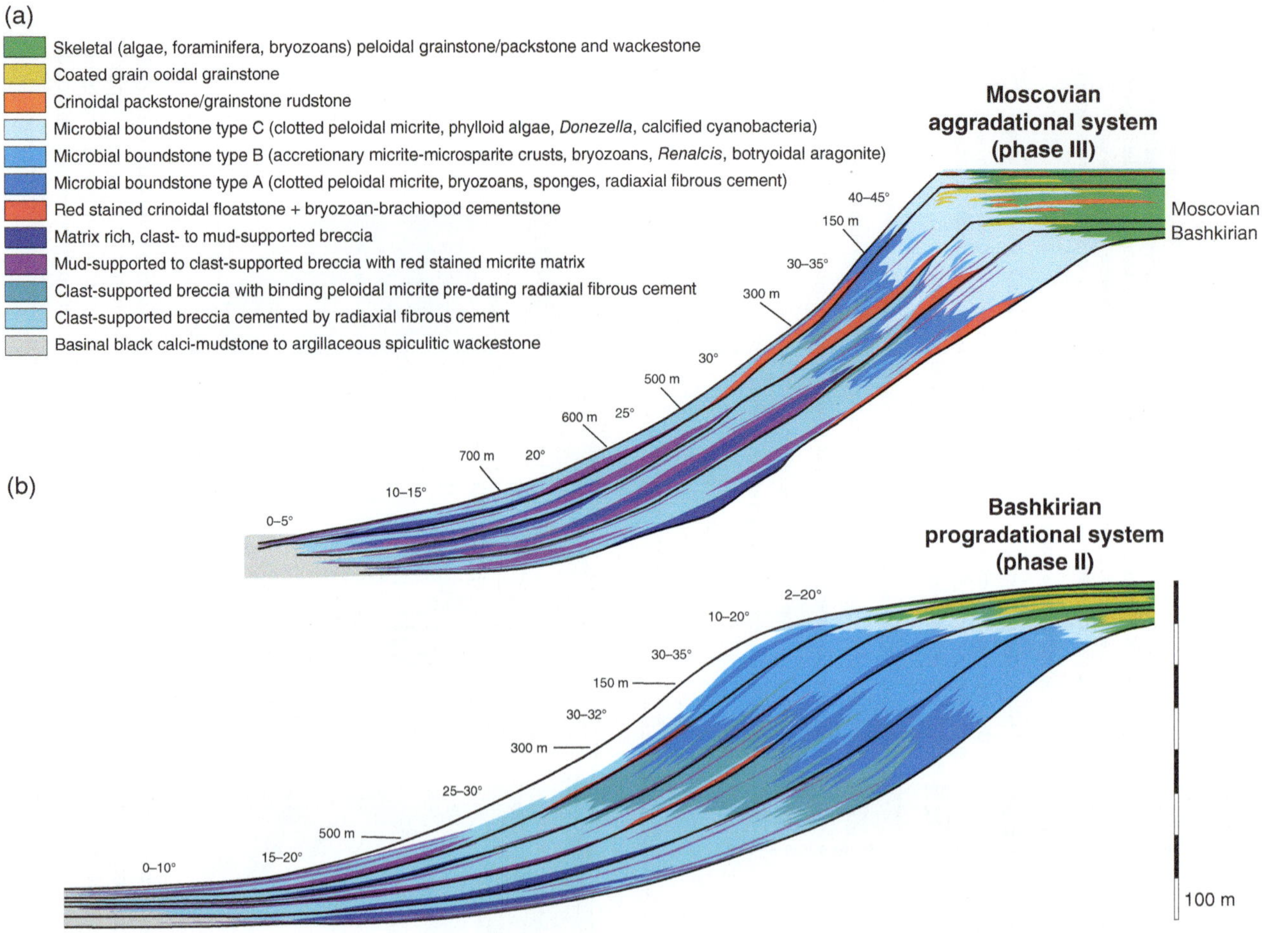

Figure 2.88 Geometry and lithofacies distribution of the Sierra del Cuera platform slope and margin during the (a) Bashkirian progradational phase II; and (b) Moscovian aggradational phase III. Note the palaeo-water depth (calculated considering 0 m at the time-equivalent platform break) of 300–400 m for the in situ microbial boundstone accumulation passing downslope to detrital breccias of reworked lithoclasts from the upper-slope microbial boundstone (modified after Della Porta et al. (2017, 2021) / CINECA; in Wright and Della Porta 2021).

defined at this carbonate platform. These patterns are also consistent with sedimentation patterns in Devonian (Canning Basin, Australia), Triassic (Dolomites, Italy) and Jurassic (Jbel Bou Dahar, Morocco) carbonate platforms (see Chapter 11 for more details).

2.6.4.7 Lessons Learned

The Sierra del Cuera case study highlights sedimentation patterns of a M-factory dominated flat-top carbonate platform. The position of the main sediment production area on the upper slope to platform margin together with the deepened platform margin and the planar clinoforms make the Sierra del Cuera outcrop a good analogue for various late Palaeozoic subsurface systems (see overviews by Della Porta et al. 2017, 2021, and Wright and Della Porta 2021).

2.6.5 Case Study 5 – Mixed System: Cenozoic Mixed System of the Gulf of Papua (Papua New Guinea): Interaction of Siliciclastic and Carbonate Sedimentation Patterns

2.6.5.1 Aim

This case study discusses a mixed siliciclastic and carbonate system with different styles of sediment production, transport and deposition within the Cenozoic sedimentary system in the Gulf of Papua (Figure 2.89). A comparison is made between the contrasts in sediment delivery (siliciclastic sediments) and sediment production (carbonates) and the variations in sediment dispersal of these different sediment populations, their variations in slope angles, as well as the differences in timing of sediment production and input for both sediment populations.

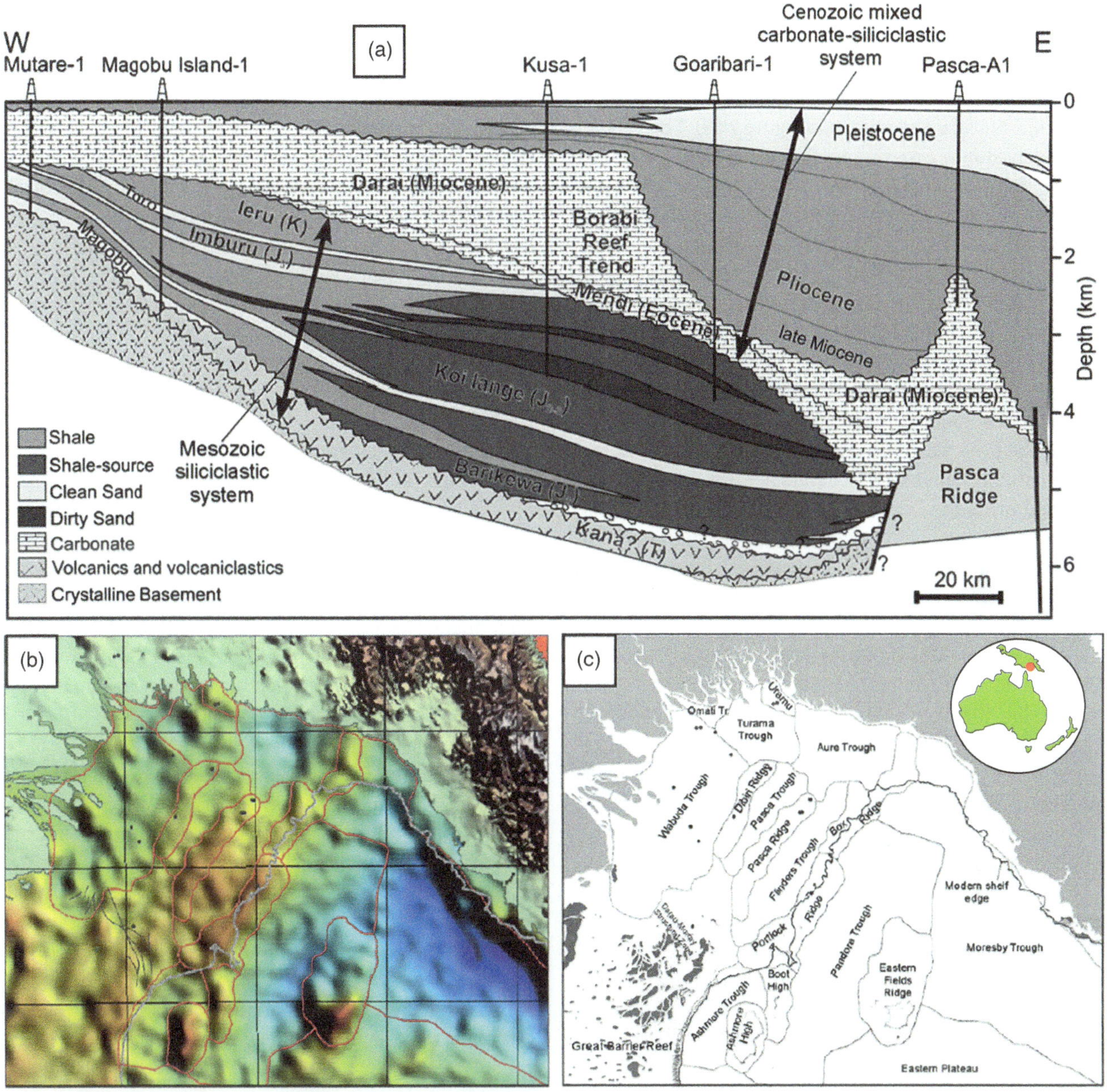

Figure 2.89 (a) Schematic geologic cross section of Papuan Basin (modified from Gordon et al. 2000); (b) gravity map of the study area with superimposed structural features, black dots are industrial wells; (c) main structural features in the Gulf of Papua. Tcherepanov et al. (2010) / with permission of Elsevier.

2.6.5.2 Data

The dataset studied by Tcherepanov et al. (2008a, 2008b, 2010) included a set of seismic grids, and gravity (Figure 2.89b), well and high-resolution multibeam bathymetry datasets. In addition, the results of Morgan (2005) study were incorporated. The seismic data included a series of surveys acquired over several decades that were re-processed by Fugro Multi Client Services Inc. In total, 13,763 km of pre-stack data and 9049 km of post-stack data were re-processed. In addition, borehole datasets from 20 wells drilled in the Gulf of Papua (courtesy Fugro Multi Client Services Inc.) were analysed; the borehole data included biostratigraphic data, strontium dating, core and cutting logs, and numerous petrophysical logs (Tcherepanov et al. 2008b).

2.6.5.3 Tectonic Setting

The basement of the Papuan Basin consists of Palaeozoic metamorphic rocks and Permian–Triassic granites that are overlain by Middle Triassic to Middle Jurassic volcanoclastic and siliciclastic sediments (Figure 2.89). The Papuan Basin evolved into a passive continental margin positioned on the north-eastern corner of the Australian Plate (Tcherepanov et al. 2008b). The Jurassic to Cretaceous time interval includes a several kilometres-thick marine siliciclastic sedimentary succession (Figure 2.89). Spreading in the Coral Sea (SW of the research area) in the Late Cretaceous-Paleocene resulted in the development of a mixed carbonate-siliciclastic depositional system that prevailed during the Cenozoic. Successive periods of regional thermal uplift exposed major parts of the basin. In the middle-late Eocene the area was re-flooded and a full-marine, carbonate-dominated environment developed. The arc-continent collision during the Oligocene, resulted in the uplift of the fold and thrust belt along the north-eastern margin of the Gulf of Papua and the development of a foreland basin (Hamilton 1979; Pigram et al. 1989; Hall 2002; Cloos et al. 2005; Jablonski et al. 2006). The carbonate sedimentary system was most extensive during the late Oligocene (Chattian) to the early middle Miocene (Langhian). Exposure during the Serravallian (late middle Miocene) and the successive re-flooding resulted in a backstepping carbonate sedimentary system and partial drowning of the northern central part of the system. Ultimately, the entire carbonate sedimentary system was buried by siliciclastics supplied from the north during the late Miocene to Pleistocene. This large siliciclastic influx resulted from renewed uplift during the Central Range Orogeny in the early late Pliocene. Continued carbonate deposition was restricted to the southern region, comprising several carbonate banks (northern part Great Barrier Reef) and atolls (Ashmore, Boot, Portlock and Eastern Fields Reef), which were beyond the influence of siliciclastic sediment supply. In recent times, an almost continuous carbonate barrier system developed along the narrow southern shelf edge of the Papuan Peninsula (Figure 2.89c).

2.6.5.4 Stratigraphy

The Late Cretaceous to Holocene evolution of the Gulf of Papua margin exhibits four phases: (i) the first phase (Late Cretaceous–Paleocene) was marked by tectonic processes in which rifting created grabens and uplifted structural blocks that in a later stage served as pedestals for carbonate factory development; (ii) the second phase and varying response to diagenetic processes (Eocene–middle Miocene) displays active neritic carbonate accumulation (Figure 2.90). Eustatic fluctuations controlled the development of the large-scale sedimentary geometries of the carbonate system; (iii) the third phase (late Miocene–early Pliocene) depicted the extensive demise of the carbonate platforms in the central part of the Gulf of Papua. This demise can originate from one or combination of several factors, for example, eustatic sea-level fluctuations, increased tectonic subsidence, uplift, a drastic influx of siliciclastics, or environmental and climatic changes; (iv) the fourth phase (late Pliocene–Holocene) was marked by the input of siliciclastic sediments causing the burial of drowned and/or active carbonate platforms; some platforms in the vicinity of the shelf edge have survived and remain active today (Figure 2.90).

2.6.5.5 Facies

The differences in depositional styles between siliciclastic sedimentary systems and carbonate systems are well displayed in the seismic profiles with clear morphologies (Figure 2.90). The morphology of carbonate systems stands out in the seismic profiles with parallel horizontal lagoon reflectors, a distinct rim at the platform margin and high-angle, steep exponential slopes with clear clinoforms. The siliciclastics show a totally different configuration as the siliciclastic submarine fan deposits display low-angle, prograding parallel reflectors sourced from one direction, together with semi-horizontal layered, ponded turbidite deposits (Figure 2.90). Hence, the seismic reflection patterns of both sedimentary systems vary considerably, which mainly reflects different source-to-sink systems and varying response to diagenetic processes (Figure 2.90).

The siliciclastic sedimentary system relied on the influx through a series of rivers delivering varying amounts of sediments linked to orogenic and erosional processes to the shelf that developed extensively through time. Sediments infilled existing topography and bypassed the early carbonate sedimentary systems. Ultimate a shelf-edge delta system formed that in the early late Pliocene prograded over 150 km in one million years (Figure 2.86; Wei et al. 2019). Other features observed were submarine fans, slope sediments and ponded turbidites. Sometimes drowned reefs acted as barriers for siliciclastic input.

In the mid-Pliocene, the carbonate sedimentary system displayed the growth of carbonate mounded coralgal edifices in response to eustatic sea-level transgressions flooding the prograding shelf edge. In the southern part of the study area, these mounded edifices are replaced by flat-topped positive high-relief features that are several km wide, which are interpreted as short-lived small carbonate platforms (Figure 2.91). Isolated carbonate platforms developed, some of which drowned or were fully or partially covered by siliciclastics, while others were not buried,

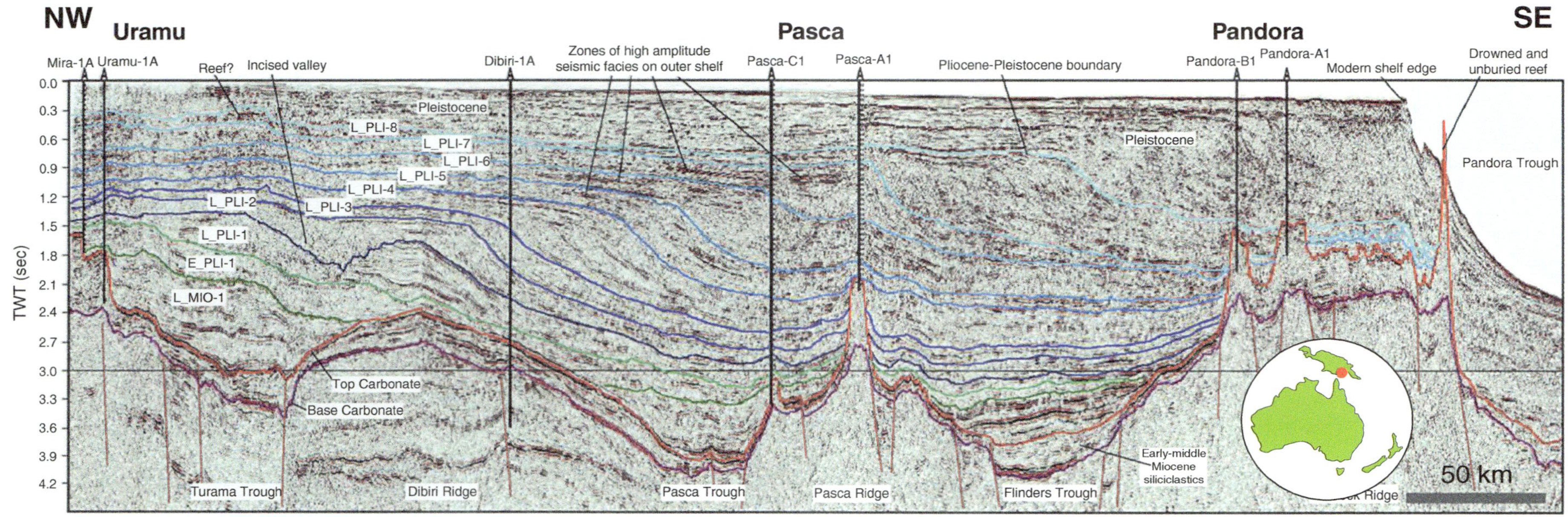

Figure 2.90 NW-SE orientated composite seismic section across the Gulf of Papua showing prograding siliciclastic sequences overlying a carbonate system morphology positioned at the base of the profile (red reflector). The carbonate system positioned furthest away from the entry point of the siliciclastics remained active the longest with the drowned and unburied reef in the Pandora Trough (right-side figure) as the latest carbonate sedimentation expression. The late Pliocene and Pleistocene prograding siliciclastic sequences infill the Pasca and Flinders Trough from the NW. Note high-amplitude zones close to shelf breaks in the late Pliocene and their absence in the Pleistocene. *Source:* Tcherepanov et al. (2010) / with permission of Elsevier.

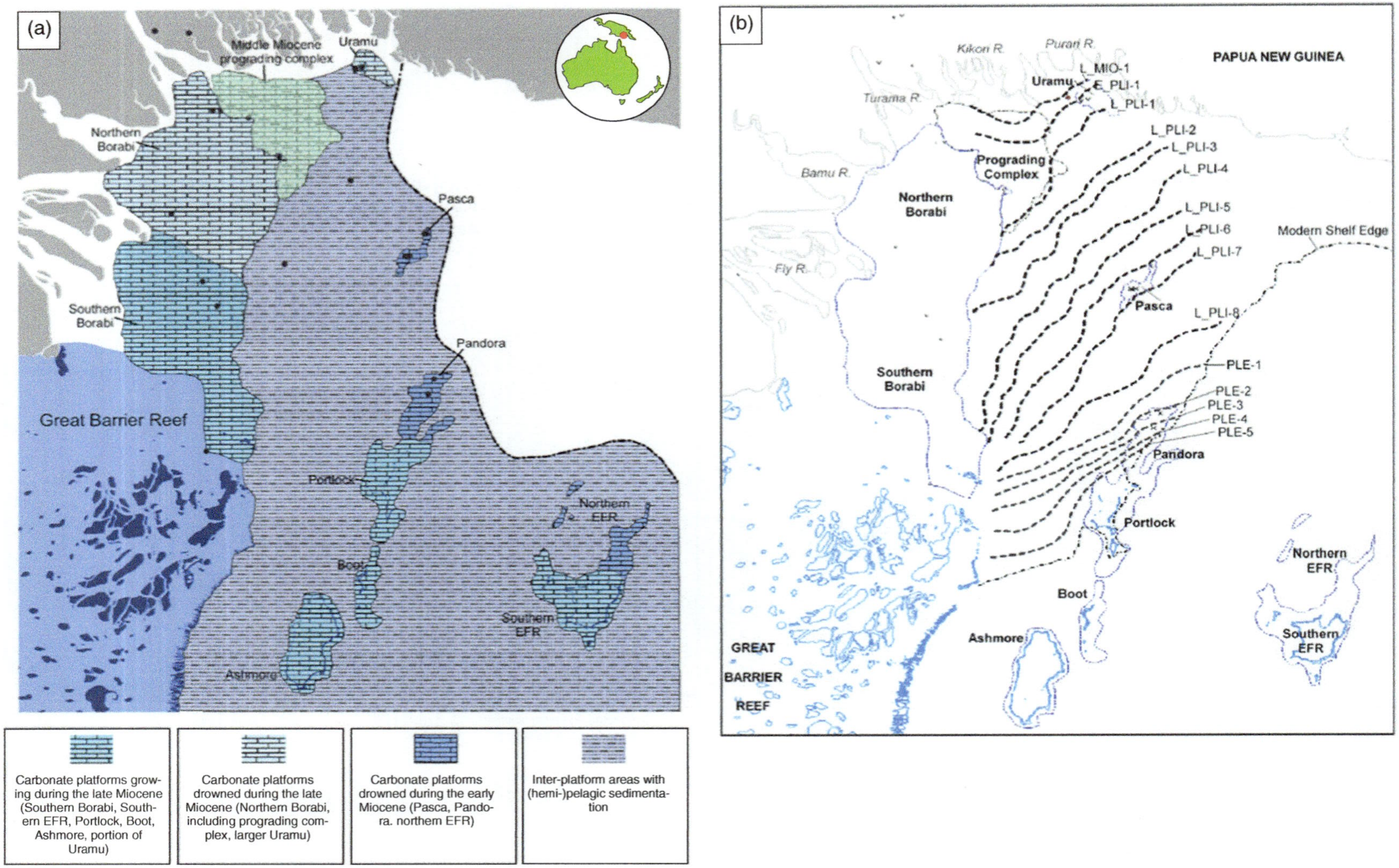

Figure 2.91 (a) Carbonate system configuration in the Gulf of Papua during the late Miocene before the major stage of siliciclastic influx; (b) siliciclastic shelf-edge progradation during the late Miocene–Pleistocene. Present-day river systems are indicated. Pandora is a drowned atoll, see Fig. 2.90. Active isolated carbonate platforms (atolls) positioned seaward of the shelf edge include Ashmore Reef, Boot Reef, Portlock Reef and Northern and Southern Eastern Fields Reef (EFR). *Source:* Tcherepanov et al. (2008a, 2008b, 2010).

remained active and developed into atolls over time (Webster et al. 2004a, 2004b).

2.6.5.6 Depositional Model

Based on seismic, gravity, multibeam bathymetry, well datasets and Landsat imagery, a detailed evaluation was made of the depositional processes in the mixed carbonate-siliciclastic sedimentary system of the Gulf of Papua (Papua New Guinea). The interplay of tectonic processes, eustatic sea-level variations together with the *in situ* carbonate production and siliciclastic sediment supply determined the depositional processes within the mixed sedimentary system (e.g., Harris et al. 1996; Jorry et al. 2008; Mallarino et al. 2021).

2.6.5.7 Lessons Learned

The Gulf of Papua seismic profiles show the interaction between tectonic processes, eustatic sea-level variations and the response of siliciclastic – and carbonate depositional systems to these processes. The interaction between contrasting sedimentary systems with fluctuations in siliciclastic input linked to orogenic processes as carbonate sediment production in the Gulf of Papua relies on areas within the photic zone on which carbonate producers can settle. The formation of horst and graben structures and their impact on directing siliciclastic input (graben infill) but also enabling carbonates to establish themselves on horst blocks; a pattern that at present is also well established in the Red Sea (Piller and Mansour 1990; Emmermann et al. 1999; Petrovic et al. 2023). In mixed systems, carbonates and siliciclastics can function side by side, but in the Gulf of Papua switches in response to the tectonic regime resulted in the increased input of siliciclastic sediments that almost destroyed the entire carbonate depositional system (Tcherepanov et al. 2008a, 2008b, 2010).

Further Reading

Anderton, R. (1985). Clastic facies models and facies analysis. In: *Sedimentology: Recent Developments and Applied Aspects*, Special Publication No. 18 (ed. P.J. Brenchley and B.P.J. Williams), 31–47. Geological Society.

Catuneanu, O. (2022). *Principles of Sequence Stratigraphy*, 2e, 494. Elsevier.

Collinson, J.D. and Mountney, N.P. (2019). *Sedimentary Structures*, 4e, 352 pp. Edinburgh: Dunedin Academic Press.

Flügel, E. (2010). *Microfacies of Carbonate Rocks – Analysis, Interpretation and Application*, 984. Heidelberg–Dordrecht–London–New York: Springer.

James, N.P. and Bone, Y. (2011). *Neritic Carbonate Sediments in a Temperate Realm – Southern Australia*, 254 pp. Dordrecht–Heidelberg–London–New York: Springer.

James, N.P. and Dalrymple, R.W. (ed.) (2010). *Facies Models 4*, GEOtext 6, 586. St. John's: Geological Association of Canada.

James, N.P. and Jones, B. (2016). *Origin of Carbonate Sedimentary Rocks*, 446. Chichester: Wiley.

Moore, C.H. (2001). *Carbonate Reservoirs: Porosity Evolution and Diagenesis in a Sequence Stratigraphic Framework*, 444. Amsterdam, London, New York, Oxford, Paris, Shannon, Tokyo: Elsevier.

Morse, J.W. and Mackenzie, F.T. (1990). *Geochemistry of Sedimentary Carbonates*, 707. Amsterdam: Elsevier, Developments in Sedimentology.

Posamentier, H.W. and Allen, G.P. (1999). Siliciclastic sequence stratigraphy – concepts and applications. SEPM concepts in sedimentology and paleontology no. 7, 210.

Posamentier, H.W. and Walker, R.G. (eds.) (2006). Facies models revisited. SEPM Special Publication, Volume 84, 527. https://doi.org/10.2110/pec.06.84

Schlager, W. (2005). *Carbonate Sedimentology and Sequence Stratigraphy*, 200. Tulsa, OK: SEPM (Society for Sedimentary Geology).

Scholle, P.A. and Ulmer-Scholle, D.S. (2003). *A Color Guide to the Petrography of Carbonate Rocks*, 474. Tulsa, OK: American Association of Petroleum Geologists.

Suter, J.R. and Clifton, H.E. (1999). The Shannon sandstone and isolated linear sand bodies: interpretations and realizations. In: *Isolated Shallow Marine Sand Bodies: Sequence Stratigraphic Analysis and Sedimentologic Interpretation*, Special Publication 64 (ed. K.M. Bergman and J.W. Snedden), 321–356. SEPM.

Tucker, M.E. and Jones, S.J. (2023). *Sediment Petrology*, 448. Hoboken, NJ: Wiley.

Tucker, M.E. and Wright, V.P. (2009). *Carbonate Sedimentology*, 2e, 496. Oxford: Wiley.

Van Wagoner, J.C., Mitchum, R., Campion, K. & Rahmanian, V. (1990) Siliciclastic sequence stratigraphy in well logs, cores, and outcrops: concepts for high-resolution correlation of time and facies. AAPG Methods in Exploration Series No. 7, 55.

Walker, R.G. (1992). Facies, facies models and modern stratigraphic concepts. In: *Facies models – response to sea level change* (ed. R. G. Walker and N.P. James), 1–14. St. John's: Geological Association of Canada.

Wright, V.P. and Della Porta, G. (ed.) (2021). *Field Guides to Exceptionally Exposed Carbonate Outcrops*, 792. Ghent: International Association of Sedimentologists.

References

Abbott, S.T. and Sweet, I.P. (2000). Tectonic control on third order sequences in a siliciclastic ramp style basin: an example from the Roper Superbasin (Mesoproterozoic), Northern Australia. *Australian Journal of Earth Sciences* 47: 637–657.

Ainsworth, R.B., Vakarelov, B.K., and Nanson, R.A. (2011). Dynamic spatial and temporal prediction of changes in depositional processes on clastic shorelines: toward improved subsurface uncertainty reduction and management. *American Association of Petroleum Geologists Bulletin* 95: 267–297.

Allen, J.R.L. (1970). *Physical Processes of Sedimentation*, 248. London: Unwin.

Allen, J.R.L. (1982). *Sedimentary Structures: Their Character and Physical Basis*, Developments in Sedimentology, vol. 30A & B., 612. Elsevier Science Ltd. ISBN: 978-0444419354.

Allen, J.R.L. (1985). *Principles of Physical Sedimentology*, 272. London: George Allen & Unwin.

Allen, P.A. and Allen, J.R. (2013). *Basin Analysis: Principles and Practice to Petroleum Play Assessment*, 3e, 640. Wiley.

Allen, P.A., Eriksson, P.G., Alkmim, F.F. et al. (2015). Classification of basins, with special reference to Proterozoic examples. In: *Precambrian Basins of India: Stratigraphic and Tectonic Context*, Memoirs, vol. 43 (ed. R. Mazumder and P.G. Eriksson), 5–28. London: Geological Society.

Allgöwer, A.M. and Lignum, J.S. (2019). Fine-scale variations in distributary mouth-bar morphology in the Middle Triassic Caley Member of the Bedout Sub-basin, Western Australia. In: *The Sedimentary Basins of Western Australia V: Proceedings of the Petroleum Exploration Society of Australia Symposium* (ed. M. Keep and S.J. Moss), 23. PESA: Perth, WA.

Alonso, J.L., Marcos, A., and Suárez, A. (2009). Palaeogeographic inversion resulting from large out of sequence breaching thrusts: the León Fault (Cantabrian Zone, NW Iberia). A new picture of the external Variscan Thrust Belt in the Ibero-Armorican Arc. *Geologica Acta* 7: 451–473.

Alqahtani, F.A., Jackson, C.A.-L., Johnson, H.D., and Som, M.R.B. (2017). Controls on the geometry and evolution of humid-tropical fluvial systems: insights from 3D seismic geomorphological analysis of the Malay Basin, Sunda Shelf, Southeast Asia. *Journal of Sedimentary Research* 87: 17–40.

Alqahtani, F.A., Johnson, H.D., Jackson, C.A. L., and Som, M.R.B. (2015). Nature, origin and evolution of a Late Pleistocene incised valley-fill, Sunda Shelf, Southeast Asia. *Sedimentology* 62: 1198–1232.

Andresen, N., Reijmer, J.J.G., and Droxler, A. W. (2003). Timing and distribution of calciturbidites around a deeply submerged carbonate platform in a seismically active setting (Pedro Bank, Northern Nicaragua Rise, Caribbean Sea). *International Journal of Earth Sciences - Geologische Rundschau* 92 (4): 573–592.

Anselmetti, F., Eberli, G.P., and Ding, Z.-D. (2000). From the Great Bahama Bank into the Straits of Florida: a margin architecture controlled by sea-level fluctuations and ocean currents. *Geological Society of America Bulletin* 112 (6): 829–844.

Archie, G.E. (1952). Classification of carbonate reservoir rocks and petrophysical considerations. *American Association of Petroleum Geologists Bulletin Bulletin* 36: 278–298.

Arnott, R.W. and Southard, J.B. (1990). Exploratory flow-duct experiments on combined-flow bed configurations, and some implications for interpreting storm-event stratification. *Journal of Sedimentary Research* 60: 211–219.

Asquith, D.O. (1970). Depositional topography and major marine environments, Late Cretaceous, Wyoming. *American Association of Petroleum Geologists Bulletin* 54: 1184–1224.

Baas, J.H., Best, J.L., and Peakall, J. (2016). Predicting bedforms and primary current stratification in cohesive mixtures of mud and sand. *Journal of the Geological Society* 173: 12–45.

Back, S., Strozyk, F., Kuklan, P. A. and Lambiase, J. J., 2008. Three-dimensional restoration of original sedimentary geometries in deformed basin fill, onshore Brunei Darussalam, NWBorneo. *Basin Research* 20: 99–117. https://doi.org/10.1111/j.1365-2117.2007.00343.x.

Bacon, M., Sim, R., and Redshaw, T. (2007). *3-D Seismic Interpretation*, 225. Cambridge University Press.

Bahamonde, J.R., Colmenero, J.R., and Vera, C. (1997). Growth and demise of Late Carboniferous carbonate platforms in the eastern Cantabrian Zone, Asturias, northwestern Spain. *Sedimentary Geology* 110: 99–122.

Bahamonde, J.R., Kenter, J.A.M., Della Porta, G. et al. (2004). Lithofacies and depositional processes on a high, steep-margined Carboniferous (Bashkirian-Moscovian) carbonate platform slope, Sierra del Cuera, NW Spain. *Sedimentary Geology* 166: 145–156.

Bahamonde, J.R., Vera, C., and Colmenero, J. R. (2000). A steep-fronted Carboniferous carbonate platform: clinoformal geometry and lithofacies (Picos de Europa, NW Spain). *Sedimentology* 47: 645–664.

Bahamonde, J.R., Merino-Tomé, O.A., and Heredia, N. (2007). A Pennsylvanian microbial boundstone-dominated carbonate shelf in a distal foreland margin (Picos de Europa Province, NW Spain). *Sedimentary Geology* 198: 167–193.

Bahamonde, J.R., Merino-Tomé, O., Della Porta, G., and Villa, E. (2015). Pennsylvanian carbonate platforms adjacent to deltaic systems in an active marine foreland basin (Escalada Fm., Cantabrian Zone, NW Spain). *Basin Research* 27: 208–229.

Balaguru, A. and Lukie, T. (2012). Tectono-stratigraphy and development of the Miocene delta systems on an active margin of Northwest Borneo, Malaysia. *Warta Geologi* 38: 127–129.

Bassant, P., Van Buchem, F., Strasser, A., and Lomando, A. (2004). A comparison of two early Miocene carbonate margins: the Zhujiang carbonate platform (subsurface, South China Sea) and the Pirin Platform (outcrop, Southern Turkey). In: *Integration of Outcrop and Modern Analogs in Reservoir Modeling, Volume AAPG Memoir 80* (ed. G.M. Grammer, P.

M. Harris, and G.P. Eberli), 153–170. Tulsa, OK: AAPG.

Bassant, P., van Buchem, F.S.P., Strasser, A., and Görür, N. (2005). The stratigraphic architecture and evolution of the Burdigalian carbonate–siliciclastic sedimentary systems of the Mut Basin, Turkey. *Sedimentary Geology* 173 (1–4): 187–232.

Bassant, P., Janson, X., Gurbuz, K., Eriş, K. and Van Buchem, F. (2021). Mut Basin, Turkey: diverse carbonate depositional styles and mixed systems. In: *Field Guides to Exceptionally Exposed Carbonate Outcrops* (ed. V.P. Wright and G. Della Porta), 645–694. Ghent: International Association of Sedimentologists.

Bathurst, R.G.C. (1971). *Carbonate Sediments and their Diagenesis*, Developments in Sedimentology 12, 620. Elsevier Scientific Publishing Co.

Bathurst, R.G.C. (1975). *Carbonate Sediments and their Diagenesis*, Developments in Sedimentology 12, 2e, 633. Elsevier Scientific Publishing Co.

Beard, D.C. and Weyl, P.K. (1973). Influence of Texture on Porosity and Permeability of Unconsolidated Sand. *American Association of Petroleum Geologists Bulletin* 57: 349–369.

Bergman, K.M. (1994). The Shannon sandstone in Hartzog Draw-Heldt Draw fields (Cretaceous, Wyoming, USA) reinterpreted as lowstand shoreface deposits. *American Association of Petroleum Geologists Bulletin* 64: 184–201.

Bergman, K.M. and Walker, R.G. (1987). The importance of sea-level fluctuations in the formation of linear conglomerate bodies; Carrot Creek Member of Cardium Formation, Cretaceous Western Interior Seaway, Alberta, *Canada. Journal of Sedimentary Petrology* 57 (4): 651–665.

Betzler, C., Reijmer, J.J.G., Bernet, K. et al. (1999). Sedimentary patterns and geometries of the Bahamian outer carbonate ramp (Miocene-Lower Pliocene, Great Bahama Bank). *Sedimentology* 46 (6): 1127–1144.

Betzler, C., Lindhorst, S., Eberli, G.P. et al. (2014). Periplatform drift: the combined result of contour current and off-bank transport along carbonate platforms. *Geology* 42 (10): 871–874.

Bhattacharya, J.P., Howell, C.D., MacEachern, J.A., and Walsh, J.P. (2020). Bioturbation, sedimentation rates, and preservation of flood events in deltas. *Palaeogeography, Palaeoclimatology, Palaeoecology* 560: 110049.

Billeaud, I., Caline, B., Livas, B. et al. (2014). The carbonate-evaporite lagoon of Al Dakhirah (Qatar): an example of a modern depositional model controlled by longshore transport. In: *Sedimentary Coastal Zones from High to Low Latitudes: Similarities and Differences*, Geological Society, London, Special Publications, 388 (ed. I.P. Martini and H.R. Wanless), 561–587. London: The Geological Society of London.

Bol, A.J. and van Hoorn, B. (1980). Structural styles in western Sabah offshore. *Bulletin of the Geological Society of Malaysia* 12: 1–16.

Bosence, D. (2005). A genetic classification of carbonate platforms based on their basinal and tectonic settings in the Cenozoic. *Sedimentary Geology* 175 (1–4): 49–72.

Boyd, R., Dalrymple, R., and Zaitlin, B.A. (1992). Classification of clastic coastal depositional environments. *Sedimentary Geology* 80: 139–150.

Bouma, A.H. (1962). *Sedimentology of Some Flysch Deposits: A Graphic Approach to Facies Interpretation*, 168. Elsevier.

Bryant, S., Cade, C., and Mellor, D. (1993). Permeability prediction from geologic models. *AAPG Bulletin* 77: 1338–1350.

Burchette, T.P. and Wright, V.P. (1992). Carbonate ramp depositional systems. *Sedimentary Geology* 79 (1): 3–57.

Burgess, P.M. and Steel, R. (2008). Stratigraphic forward modelling of basin-margin clinoform systems: implications for controls on topset and shelf width and timing of formation of shelf-edge deltas. In: *Recent Advances in Models of Siliciclastic Shallow-Marine Stratigraphy*, vol. 90 (ed. G.J. Hampson, R.J. Steel, P.M. Burgess, and R.W. Dalrymple), 35–45. SEPM Special Publications.

Burgess, P.M. and Prince, G.D. (2015). Non-unique stratal geometries: implications for sequence stratigraphic interpretations. *Basin Research* 27: 351–365.

Busson, J., Joseph, P., Mulder, T. et al. (2019). High-resolution stratigraphic forward modeling of a Quaternary carbonate margin: controls and dynamic of the progradation. *Sedimentary Geology* 379: 77–96.

Cade, C.A., Evans, I.J., and Bryant, S.L. (1994). Analysis of permeability controls: a new approach. *Clay Minerals* 29: 491–501.

Caracciolo, L. (2020). Sediment generation and sediment routing systems from a quantitative provenance analysis perspective: Review, application and future development. *Earth-Science Reviews* 209: 103226.

Catuneanu, O. (2022). *Principles of Sequence Stratigraphy*, 2e, 501. Amsterdam: Elsevier Science.

Christ, N., Immenhauser, A., Amour, F. et al. (2012). Characterization and interpretation of discontinuity surfaces in a Jurassic ramp setting (High Atlas, Morocco). *Sedimentology* 59: 249–290.

Christ, N., Immenhauser, A., Wood, R.A. et al. (2015). Petrography and environmental controls on the formation of Phanerozoic marine carbonate hardgrounds. *Earth-Science Reviews* 151: 176–226.

Cloos, M., Sapiie, B., Quarles van Ufford, A. et al. (2005). Collisional delamination in New Guinea: the geotectonics of subducting slab breakoff. GSA Special Paper 400, 51pp.

Coleman, J.M. and Prior, D.B. (1982). Deltaic environments of deposition. In: *Sandstone Depositional Environments*, vol. 31 (ed. P. A. Scholle and D.R. Spearing), 139–178. AAPG Memoir.

Collins, D.S., Johnson, H.D., Allison, P.A. et al. (2017). Coupled 'storm-flood' depositional model: application to the Miocene–Modern Baram Delta Province, north-west Borneo. *Sedimentology* 64: 1203–1235.

Collins, D.S., Johnson, H.D., Allison, P.A., and Damit, A.R. (2018). Mixed process, humid-tropical, shoreline–shelf deposition and preservation: middle Miocene–Modern Baram Delta Province, northwest Borneo. *Journal of Sedimentary Research* 88: 399–430.

Collins, D.S., Johnson, H.D., and Baldwin, C. T. (2020). Architecture and preservation in the fluvial to marine transition zone of a mixed-process humid-tropical delta:

Middle Miocene Lambir Formation, Baram Delta Province, north-west Borneo. *Sedimentology* 67: 1–46.

Collinson, J.D. and Mountney, N. (2019). *Sedimentary Structures*, 4th edition. Dunedin, 340 pp.

Colmenero, J.R., Agueda, J.A., Bahamonde, J.R. et al. (1993). Evolucion de la cuenca de Antepais Namuriense y Westfaliense de la zona Cantabrica, nordoeste de España. *Comptes Rendus XII ICC-P* 2: 175–190.

Colombera, L. and Mountney, N.P. (2020a). On the geological significance of clastic parasequences. *Earth-Science Reviews* 201: 103062.

Colombera, L. and Mountney, N.P. (2020b). Accommodation and sediment-supply controls on clastic parasequences: a meta-analysis. *Sedimentology* 67: 1667–1709.

Crombez, V., Kunzmann, M., Faiz, M. et al. (2023). Stratigraphic architecture of the 1420-1210 Ma Velkerri and Kyalla Formations (Beetaloo sub-basin, Australia). *AAPG Bulletin* 107: 1901–1932.

Cronin, B., Nemcok, M., Doran, H. (2023). Transform margin source–sink clastic deposystems. *Geological Society of London's Special Publications* 524 (1): 387. https://doi.org/10.1144/SP524-2021-196.

Cross, T.A. and Homewood, P.W. (1997). Amanz Gressly's role in founding modern stratigraphy. *GSA Bulletin* 109: 1617–1630.

Cullen, A.B. (2010). Transverse segmentation of the Baram-Balabac Basin, NW Borneo: refining the model of Borneo's tectonic evolution. *Petroleum Geoscience* 16: 3–29.

Dalrymple, R.W. (2010). Interpreting sedimentary successions: facies, facies analysis and facies models. In: *Facies Models 4*. GEOtext 6 (ed. N.P. James and R. W. Dalrymple), 3–18. Geological Association of Canada.

Darman, H. and Handoyo, K. (2013). "Deltaic" reservoir characteristics of giant fields of the Kutei and Baram Basins, Borneo. AAPG Search and Discovery Article #20191.

de Kruijf, M., Slootman, A., de Boer, R.A., and Reijmer, J.J.G. (2021). On the settling of marine carbonate grains: review and challenges. *Earth-Science Reviews* 217: 103532.

Della Porta, G., Bahamonde, J.R., Kenter, J. A.M., and Verwer, K. (2017). The Sierra del Cuera (Pennsylvanian microbial platform margin) in Asturias, north Spain. *AAPG Bulletin* 101: 543–551.

Della Porta, G., Kenter, J.A.M., and Bahamonde, J.R. (2004). Depositional facies and stratal geometry of an Upper Carboniferous prograding and aggrading high-relief carbonate platform (Cantabrian Mountains, N. Spain). *Sedimentology* 51: 267–295.

Della Porta, G., Kenter, J.A.M., Bahamonde, J.R. et al. (2003). Microbial boundstone dominated carbonate slope (Upper Carboniferous, N Spain): Microfacies, lithofacies distribution and stratal geometry. *Facies* 49: 175–208.

Della Porta, G., Kenter, J.A.M., Immenhauser, A., and Bahamonde, J.R. (2002). Lithofacies character and architecture across a Pennsylvanian inner-platform transect (Sierra de Cuera, Asturias, Spain). *Journal of Sedimentary Research* 72: 898–916.

Della Porta, G., Bahamonde, J.R., Kenter, J. A.M., and Verwer, K. (2017). The Sierra del Cuera (Pennsylvanian microbial platform margin) in Asturias, north Spain. *AAPG Bulletin* 101 (4): 543–551.

Della Porta, G., Bahamonde, J.R., Kenter, J.A. M., and Verwer, K. (2021). Pennsylvanian high-relief microbial boundstone-dominated carbonate platform margins (Sierra del Cuera, Asturias, Spain). In: *Field Guides to Exceptionally Exposed Carbonate Outcrops* (ed. V.P. Wright and G. Della Porta), 71–127. IAS (2021) International Association of Sedimentologists. ISBN 978-1-7398845-0-5.

Doust, H. and Sumner, H.S. (2007). Petroleum systems in rift basins - a collective approach in Southeast Asian basins. *Petroleum Geoscience* 13: 127–144.

Dravis, J. (1979). Rapid and widespread generation of recent oolitic hardgrounds on a high energy Bahamian platform, Eleuthera Bank. *Bahamas: Journal of Sedimentary Petrology* 49 (1): 195–208.

Dravis, J.J. (1996). Rapidity of freshwater calcite cementation—implications for carbonate diagenesis and sequence stratigraphy. *Sedimentary Geology* 107 (1): 1–10.

Droxler, A.W. and Schlager, W. (1985). Glacial versus interglacial sedimentation rates and turbidite frequency in the Bahamas. *Geology* 13: 799–802.

Duke, W.L. (1985). Hummocky cross-stratification, tropical hurricanes, and intense winter storms. *Sedimentology* 32: 167–194.

Duke, W.L., Arnott, R.W., and Cheel, R.J. (1985). Shelf sandstones and hummocky cross-stratification; new insights on a stormy debate. *Geology* 19: 625–628.

Dumas, S., Arnott, R.W.C., and Southard, J.B. (2005). Experiments on oscillatory-flow and combined-flow bed forms: implications for interpreting parts of the shallow-marine sedimentary record. *Journal of Sedimentary Research* 75: 501–513.

Dunham, R.J. (1962). Classification of carbonate rocks according to depositional texture. In: *Classification of Carbonate Rocks*, AAPG Memoir 1 (ed. W.E. Ham), 108–121, American Association of Petroleum Geologists Tulsa, Oklahoma.

Ebanks, W.J., Scheihing, M.H., Atkinson, C. D. et al. (1992). *Flow Units for Reservoir Characterization Development Geology Reference Manual*, Methods in Exploration Series, 282–285. American Association of Petroleum Geologists.

Eberli, G.P. (2000). The record of Neogene sea-level changes in the prograding carbonates along the Bahamas transect - Leg166 synthesis. In: *Proceedings of the Ocean Drilling Program, Scientific Results Leg 166*, vol. 166 (ed. P.K. Swart, G.P. Eberli, M.J. Malone, and J.F. Sarg), 167–177. College Station, TX: Ocean Drilling Program.

Eberli, G.P., Anselmetti, F.S., Kroon, D. et al. (2002). The chronostratigraphic significance of seismic reflections along the Bahamas Transect. *Marine Geology* 185 (1–2): 1–17.

Eberli, G.P., Swart, P.K., Malone, M.J. et al. (1997). *Proceedings Ocean Drilling Program, Initial Reports Leg 166*, 850. College Station, TX: Ocean Drilling Program.

Elliott, T. (1974). Interdistributary bay sequences and their genesis. *Sedimentology* 21: 611–622.

Elliott, T. (1975). The sedimentary history of a delta lobe from a Yoredale (Carboniferous) cyclothem. *Proceedings of* the Yorkshire Geological Society 40: 505–536.

Elliott, T. (1976). The morphology, magnitude and regime of a Carboniferous fluvial-distributary channel. *Journal of Sedimentary Research* 46: 70–76.

Emmermann, P.P., Reijmer, J.J.G., and Andresen, N. (1999). Sedimentation rates of Late Quaternary periplatform sediments based on aragonite/calcite ratios: Sudanese Red Sea vs. Pedro Bank (Caribbean). In: *On the Determination of Sediment Accumulation Rates*, GeoResearch Forum 5 (ed. P. Bruns and H. C. Hass), 67–86. Trans Tech Publications.

Enos, P. and Samankassou, E. (1998). Lofer cyclothems revisited (Late Triassic, Northern Alps, Austria). *Facies* 38: 207–228.

Enos, P. and Samankassou, E. (2021). Florida Bay: Modern analogue for Lofer cyclothems? *Sedimentology* 69: 254–281.

Epting, M. (1980). Sedimentology of Miocene carbonate buildups, Central Luconia, offshore Sarawak. *Bulletin of the Geological Society of Malaysia* 12: 17–30.

Fauquembergue, K., Mulder, T., Reijmer, J. et al. (2024). Sedimentology of modern Bahamian carbonate slopes: summary and update. *Geochemistry, Geophysics, Geosystems* 25: e2023GC011077.

Fischer, A.G. (1964). The Lofer Cyclothems of the Alpine Triassic. *Kansas Geological Survey Bulletin* 169: 107–149.

Flügel, E. (1982). *Microfacies Analysis of Limestones*, 375. Berlin: Springer Verlag.

Flügel, E. (2010). *Microfacies of Carbonate Rocks – Analysis, Interpretation and Application*, 984. Heidelberg-Dordrecht-London-New York: Springer Science & Business Media.

Folk, R.L. (1959). Practical petrographic classification of limestones. *American Association of Petroleum Geologists Bulletin* 43: 1–38.

Folk, R.L. (1974). *Petrology of Sedimentary Rocks*, 182. Austin, TX: Hemphill Publishing Company.

Fournier, F., Borgomano, J., and Montaggioni, L.F. (2005). Development patterns and controlling factors of Tertiary carbonate buildups: Insights from high-resolution 3D seismic and well data in the Malampaya gas field (Offshore Palawan, Philippines). *Sedimentary Geology* 175 (1–4): 189–215.

Fraser, S.I., Robinson, A.M., Johnson, H.D. et al. (2003). Upper Jurassic. In: *The Millennium Atlas: Petroleum Geology of the Central and Northern North Sea* (ed. D. Evans, C. Graham, A. Armour, and P. Bathurst), 157–189. Geological Society of London.

Gallois, R. (2019). The stratigraphy of the Permo-Triassic rocks of the Dorset and East Devon Coast World Heritage Site, U. K. *Proceedings of the Geologists Association* 130: 274–293.

Galloway, W.E. (1975). Process framework for describing the morphologic and stratigraphic evolution of deltaic depositional systems. In: *Deltas: Models for Exploration* (ed. M.L. Broussard), 87–98. Houston, TX: Houston Geological Society.

Garrison, J.R. Jr. and Van den Bergh, T.C.V. (2004). High-resolution depositional sequence stratigraphy of the upper Ferron sandstone last chance delta: an application of coal-zone stratigraphy. In: *Regional to Wellbore Analog for Fluvial-Deltaic Reservoir Modeling: The Ferron Sandstone of Utah*, AAPG Studies in Geology, vol. 50 (ed. T.C. Chidsey Jr., R.D. Adams, and T. H. Morris), 125–192. American Association of Petroleum Geologists, Tulsa, Oklahoma.

Gartrell, A., Torres, J, and Hoggmascall, N. (2011). A regional approach to understanding basin evolution and play systematics in Brunei – unearthing new opportunities in a mature basin. In *International Petroleum Technology Conference*, Bangkok, Thailand. IPTC 15171, 1–5 p.

Gartrell, A., Torres, J, and Hoggmascall, N. (2012). A regional approach to understanding basin evolution and play systematics in Brunei – unearthing new opportunities in a Mature Basin. AAPG Search and Discovery Article #10467.

Gawthorpe, R.L. and Leeder, M.R. (2000). Tectono-sedimentary evolution of active extensional basins. *Basin Research* 12: 195–218.

Gee, M.J.R., Uy, H.S., Warren, J. et al. (2007). The Brunei slide: A giant submarine landslide on the North West Borneo Margin revealed by 3D seismic data. *Marine Geology* 246: 9–23.

Ginsburg, R.N. (2001). *Subsurface Geology of a Prograding Carbonate Platform Margin, Great Bahama Bank: Results of the Bahamas Drilling Project, Society for Sedimentary Geology Special Publication*, 271. Tulsa, OK: SEPM (Society for Sedimentary Geology).

Ginsburg, R.N. and Lowenstam, H.A. (1958). The influence of marine bottom communities on the depositional environment of sediments. *The Journal of Geology* 66 (3): 310–318.

Goldhammer, R.K., Dunn, P.A., and Hardie, L.A. (1990). Depositional cycles, composite sea-level changes, cycle stacking patterns, and the hierarchy of stratigraphic forcing: examples from Alpine Triassic platform carbonates. *Geological Society of America Bulletin* 102 (5): 535–562.

Gordon, S.A., Huizinga, B.J. and Subtlette, V. (2000). Petroleum potential of the Southern Gulf of Papua. In: *Papua New Guinea's Petroleum Industry in the 21st Century* (ed. P.G. Buchanan, A.M. Grainge and R.C.N. Thornton). In: Proceedings of the Fourth PNG Petroleum Convention, 205–218. Port Moresby: PNG Chamber of Mines and Petroleum.

Gowland, S. and Riding, J.B. (1991). Stratigraphy, sedimentology and palaeontology of the Scarborough Formation (Middle Jurassic) at Hundale Point, North Yorkshire. *Proceedings of the Yorkshire Geological Society* 48 (4), 375–392.

Grammer, G.M., Ginsburg, R.N., and McNeill, D.F. (1991). Morphology and development of modern carbonate foreslopes, Tongue of the Ocean, Bahamas. In: *Proceedings Transactions of the 12th Caribbean Geological Conference*, 27–32. St. Croix, U.S. Virgin Islands: Miami Geological Society.

Grammer, G.M., Ginsburg, R.N., Swart, P.K. et al. (1993). Rapid growth rates of syndepositional marine aragonite cements in steep marginal slope deposits, Bahamas and Belize. *Journal of Sedimentary Petrology* 63 (5): 983–989.

Granjeon, D. and Gervais, V. (2023). Forward and inverse stratigraphic forward modelling in exploration and appraisal workflows: Insights from the Miocene carbonate platforms, Central Luconia Basin, South China. In: *Cenozoic Isolated Carbonate Platforms - Focus Southeast Asia*, vol. 114 (ed. E.C. Rankey and M.C. Pöppelreiter), 314–333. SEPM Society for Sedimentary Geology SEPM Special Publication.

Gressly, A. (1837). Observations géologiques sur les terrains des chaines jurassiques du Canton de Soleure, et des contrées limitrophes: Verhandlungen der allgemeinen Schweizerischen Gesellschaft für die Gesammten Naturwissenschaften, in ihrer einundzwanzigsten Jahresversammlung zu Solothurn, den 25, 26, 27 Juli 1836. *Gedruckt bei Joseph Tschan, Solothurn* 1837: 126–132.

Gressly, A. (1838). Observations géologiques sur le Jura soleurois: Nouveaux mémoires de la Société Helvetique des Sciences Naturelles. *Neuchâtel* 2: 349.

Grotzinger, J.P. and Milliken, R.E. (2012). The sedimentary rock record of Mars: distribution, origins, and global stratigraphy. SEPM Special Publication No. 102, p. 1–48.

Grundvåg, S.A., Jelby, M.E., Olaussen, S., and Śliwińska, K.K. (2021). The role of shelf morphology on storm-bed variability and stratigraphic architecture, Lower Cretaceous, Svalbard. *Sedimentology* 68: 196–237.

Gutiérrez-Alonso, G., Fernández-Suárez, J., Weil, A.B. et al. (2008). Self-subduction of the Pangaean global plate. *Nature Geoscience* 1: 549–553.

Hadler-Jacobsen, F., Groth, A., Hearn, R.E., and Liestøl, F.M. (2010). Controls on and expressions of submarine fan genesis within a high accommodation margin setting, Santos Basin, Brazil—A high-resolution seismic stratigraphic and geomorphic case study. In: *Seismic Imaging of Depositional and Geomorphic Systems: Gulf Coast Section Society for Sedimentary Geology Foundation Annual Bob F. Perkins Research Conference Proceedings*, vol. 30 (ed. L.J. Wood, T.T. Simo, and N.C. Rosen), 572–615. https://doi.org/10.5724/gcs.10.30.0572.

Hajek, E.A. and Straub, K.M. (2017). Autogenic sedimentation in clastic stratigraphy. *Annual Review of Earth and Planetary Sciences* 45: 681–709.

Halaj, E., Leszek Paja, L.K., and Papiernik, B. (2022). Simulation study of the Lower Cretaceous geothermal reservoir for aquifer thermal energy storage. *Environmental Geochemistry and Health* 44: 2253–2279.

Hall, R. (2002). Cenozoic geological and plate tectonic evolution of Southeast Asia and the southwest Pacific: computer-based reconstructions, model and animations. *Journal of Asian Earth Sciences* 20: 353–431.

Hamilton, W. (1979). Tectonic of the Indonesian Region, U.S. *Geological Survey Professional Paper* 1078: 218–270.

Hampson, G.J. (2000). Discontinuity surfaces, clinoforms, and facies architecture in a wave-dominated, shoreface- shelf parasequence. *Journal of Sedimentary Research* 70: 325–340.

Hampson, G.J., Burgess, P.M. and Howell, J. A. (2001). Shoreface tongue geometry constrains history of relative sea-level fall: examples from Late Cretaceous strata in the Book Cliffs area, *Utah. Terra Nova* 13 (3): 188–196.

Handford, C.R. and Loucks, R.G. (1993). Carbonate depositional sequences and system tracts – responses of carbonate platforms to relative sea-level changes. In: *Carbonate Sequence Stratigraphy: Recent Developments and Applications, Volume AAPG Memoir 57* (ed. R.G. Loucks and J.F. Sarg), 3–41. Tulsa, OK: American Association of Petroleum Geologists.

Hanebuth, T.J., Stattegger, K., and Saito, Y. (2002). The stratigraphic architecture of the central Sunda Shelf (SE Asia) recorded by shallow-seismic surveying. *Geo-Marine Letters* 22: 86–94.

Hanebuth, T.J.J. and Stattegger, K. (2004). Depositional sequences on a late Pleistocene-Holocene tropical siliciclastic shelf (Sunda Shelf, southeast Asia). *Journal of Asian Earth Sciences* 23: 113–126.

Hanebuth, T.J.J., Voris, H.K., Yokoyama, Y. et al. (2011). Formation and fate of sedimentary depocentres on Southeast Asia's Sunda Shelf over the past sea-level cycle and biogeographic implications. *Earth-Science Reviews* 104: 92–110.

Harms, J.C. and Fahnestock, R.K. (1965). Stratification, bed forms, and flow phenomena (with examples from the Rio Grande). *AAPG Bulletin* 12: 84–115.

Harris, P., Pattiaratchi, C., Keene, J. et al. (1996). Late Quaternary deltaic and carbonate sedimentation in the Gulf of Papua Foreland Basin: response to sea-level change. *Journal of Sedimentary Research* 66: 801–819.

Harris, P.M., Purkis, S.J., Ellis, J. et al. (2015). Mapping bathymetry and depositional facies on Great Bahama Bank. *Sedimentology* 62: 566–589.

Heinz, J. and Aigner, T. (2003). Hierarchical dynamic stratigraphy in various Quaternary gravel deposits, Rhine glacier area (SW Germany): implications for hydrostratigraphy. *International Journal of Earth Sciences* 92: 923–938.

Helland-Hansen, W. and Martinsen, O.J. (1996). Shoreline trajectories and sequences; description of variable depositional-dip scenarios. *Journal of Sedimentary Research* 66: 670–688.

Helland-Hansen, W. and Hampson, G.J. (2009). Trajectory analysis: concepts and applications. *Basin Research* 21: 454–483.

Helland-Hansen, W., Steel, R.J., and Sømme, T.O. (2012). Shelf genesis revisited. *Journal of Sedimentary Research* 82: 133–148.

Henstra, G.A., Grundvåg, S-A., Johannessen, E.P., Kristensen, T.B., Midtkandal, I., Nystuen, J.P. Rotevatn, A., Surlyk, F., Sæther, T. and Windelstad, J. (2016). Depositional processes and stratigraphic architecture within a coarse-grained rift-margin turbidite system: The Wollaston Forland Group, east Greenland. *Marine and Petroleum Geology* 76: 187–209.

Hillgärtner, H. (1998). Discontinuity surfaces on a shallow-marine carbonate platform (Berriasian, Valanginian, France and Switzerland). *Journal of Sedimentary Research* 68 (6): 1093–1108.

Hiscott, R. (2001). Depositional sequences controlled by high rates of sediment supply, sea-level variations, and growth faulting: the Quaternary Baram Delta of northwestern Borneo. *Marine Geology* 175 (1–4): 67–102.

Hornung, J. and Aigner, T. (2002). Reservoir architecture in a terminal alluvial plain: an outcrop analogue study (Upper Triassic, southern Germany) Part II: cyclicity, controls and models. *Journal of Petroleum Geology* 25: 151–178.

Hubbard, R.J. (1988). Age and significance of sequence boundaries on Jurassic and early Cretaceous rifted continental margins. *AAPG Bulletin* 72: 49–72.

Hubbard, R.J., Pape, J., and Roberts, D.G. (1985). Depositional sequence mapping as a technique to establish tectonic and stratigraphic framework and evaluate hydrocarbon potential on a passive continental margin. In: *Seismic Stratigraphy II*, vol. 39 (ed. O.R. Berg and D.G. Woolverton), 79–92. American Association of Petroleum Geologists Memoir.

Immenhauser, A., Kenter, J.A.M., Ganssen, G. et al. (2002). Origin and significance of isotope shifts in Pennsylvanian carbonates (Asturias, NW Spain). *Journal of Sedimentary Research* 72: 82–94.

Jablonski, D., Pono, S., and Larsen, O.A. (2006). Prospectivity of the deepwater Gulf of Papua and surrounds in Papua New Guinea (PNG)—A new look at a frontier region. *APPEA Journal* 46: 1–22.

Jackson, C.A.-L. and Johnson, H.D. (2009). Sustained turbidity currents and their interaction with debrite-related topography; Labuan Island, offshore NW Borneo, Malaysia. *Sedimentary Geology* 219: 77–96.

Jackson, M.J., Sweet, I.P., and Powell, T.G. (1988). Studies on petroleum geology and geochemistry, Middle Proterozoic McArthur Basin, northern Australia: petroleum potential. *The APEA Journal* 28: 283–302.

James, D.M.D. (1984). *The Geology and Hydrocarbon Resources of Negara Brunei Darussalam*, 164 pp. Muzium Brunei.

Javaux, E.J., Knoll, A.H., and Walter, M.R. (2001). Morphological and ecological complexity in early eukaryotic ecosystems. *Nature* 412: 66–69.

Jelby, M.E., Grundvåg, S.A., Helland-Hansen, W. et al. (2020). Tempestite facies variability and storm-depositional processes across a wide ramp: towards a polygenetic model for hummocky cross-stratification. *Sedimentology* 67: 742–781.

Johnson, H.D. and Levell, B.K. (1995). Sedimentology of a transgressive, estuarine sand complex: Lower Cretaceous Woburn Sands (Lower Greensand), southern England. In: *Sedimentary Facies Analysis*, vol. 22 (ed. A. G. Plint), 17–46. Special Publication of the International Association of Sedimentologists.

Johnson, H.D., Johns, D.R., Menpes, S. et al. (2024). Facies analysis of regressive-transgressive deltaic cycles from the mesoproterozoic Roper Group, Greater McArthur Basin, Australia. *Conference Proceedings: Central Australian Basins Symposium IV Volume 2*, 34 pp. PESA Petroleum Energy Society of Australia, Energy Geoscience. https://doi.org/10.36404/LIDU7850.

Johnson, K., Barnett, A., and Wright, V.P. (2010). An evaluation of existing carbonate pore system classifications and rock-typing approaches. AAPG Search and Discovery Article #50342.

Jorry, S.J., Droxler, A.W., Mallarino, G. et al. (2008). Bundled turbidite deposition in the central Pandora Trough (Gulf of Papua) since Last Glacial Maximum: linking sediment nature and accumulation to sea level fluctuations at millennial timescale. *Journal of Geophysical Research* 113 (F01S19): 1–15.

Jorry, S.J., Jouet, G., Edinger, E.N. et al. (2020). From platform top to adjacent deep sea: New source-to-sink insights into carbonate sediment production and transfer in the SW Indian Ocean (Glorieuses archipelago). *Marine Geology* 423: 106144.

Julivert, M. (1971). Décollement tectonics in the Hercynian cordillera of northwest Spain. *American Journal of Science* 270: 1–29.

Kendall, C.G. (2014). Sequence stratigraphy. In: *Encyclopedia of Marine Geosciences*. Dordrecht: Springer Science + Business Media https://doi.org/10.1007/978-94-007-6644-0_178-1.

Kenter, J.A.M., van Hoeflaken, F., Bahamonde, J.R. et al. (2003). Anatomy and lithofacies of an intact and seismic-scale Carboniferous carbonate platform (Asturias, NW Spain): Analogues of hydrocarbon reservoirs in the Pricaspian Basin (Kazakhstan). In: *Paleozoic Carbonates of the Commonwealth of Independent States (CIS): Subsurface Reservoirs and Outcrop Analogs*, SEPM Special Publication 74 (ed. W.G. Zempolich and H.E. Cook), 185–207. Tulsa, OK: SEPM (Society for Sedimentary Geology).

Kenter, J.A.M., Harris, P.M., and Della Porta, G. (2005). Steep microbial boundstone-dominated platform margins—examples and implications. *Sedimentary Geology* 178: 5–30.

Kleinhans, M.G. (2005). Phase diagrams of bed states in steady, unsteady, oscillatory and mixed flows. In: *SANDPIT, Sand Transport and Morphology of Offshore Mining Pits, EU 5th Framework Programme 1998–2002* (ed. L.C. van Rijn, R.L. Soulsby, P. Hoekstra, and A.G. Davies), Q1–Q15. Brussels: EU.

Koša, E. (2015). Sea-level changes, shoreline journeys, and the seismic stratigraphy of Central Luconia, Miocene-present, offshore Sarawak, NW Borneo. *Marine and Petroleum Geology* 59: 35–55.

Koša, E., Warrlich, G.M.D., and Loftus, G. (2015). Wings, mushrooms, and Christmas trees: the carbonate seismic geomorphology of Central Luconia, Miocene–present, offshore Sarawak, northwest Borneo. *AAPG Bulletin* 99: 2043–2075.

Krebs, W.N. (2011). Upper Tertiary chronosequence stratigraphy of offshore Sabah and Sarawak, NW Borneo, Malaysia. *Bulletin of the Geological Society of Malaysia* 57: 39–46. https://doi.org/10.7186/bgsm2011006.

Kuenen, P.H. and Migliorini, C.I. (1950). Turbidity currents as a cause of graded bedding. *Journal of Geology* 58: 91–127.

Kulpecz, A.A. (2008). Controls on facies architecture in the Middle Jurassic Ravenscar Group, Yorkshire Coast, UK. Doctoral dissertation, Imperial College London.

Lambiase, J.J. and Cullen, A.B. (2013). Sediment supply systems of the Champion "Delta" of NW Borneo: Implications for deepwater reservoir sandstones. *Journal of Asian Earth Sciences* 76: 356–371.

Laugié, M., Michel, J., Pohl, A. et al. (2019). Global distribution of modern shallow-water marine carbonate factories: a spatial model based on environmental parameters. *Scientific Reports* 9 (1): 16432.

Leckie, D.A. and Krystinik, L.F. (1989). Is there evidence for geostrophic currents preserved in the sedimentary record of inner to middle shelf deposits? *Journal of Sedimentary Petrology* 59: 862–870.

Leeder, M.R. and Gawthorpe, R.L. (1987). Sedimentary models for extensional tilt-block/half-graben basins. In: *Continental Extensional Tectonics, Geological Society*, Special Publication No. 28 (ed. M.P. Coward, J.F. Dewey, and P.L. Hancock), 139–152. John Wiley & Sons Ltd, European Association of Geoscientists & Engineers and International Association of Sedimentologists.

Legler, B., Johnson, H.D., Hampson, G.J. et al. (2013). Facies model of a fine-grained, tide-dominated delta: Lower Dir Abu Lifa Member (Eocene), Western Desert, Egypt. *Sedimentology* 60: 1313–1356.

Levell, B.K. (1987). The nature and significance of regional unconformities in the hydrocarbon-bearing Neogene sequences offshore. *Bulletin-Geological Society of Malaysia* 21: 55–90.

Levell, B.K., Johnson, H.D., Collins, D.S., and van Cappelle, M. (2019). Deposition and preservation of fluvio-tidal shallow-marine sandstones: a re-evaluation of the Neoproterozoic Jura Quartzite (western Scotland). *Sedimentology* 67: 173–206.

Levell, B.K. and Kasumajaya, A. (1985). Slumping at the late Miocene shelf-edge offshore West Sabah: a view of a turbidite basin margin. *The Bulletin of the Geological Society of Malaysia* 18: 1–29.

Lucia, F.J. (1999). *Carbonate Reservoir Characterization – An Integrated Approach*, 226. Springer.

Lucia, F.J. (2007). *Carbonate Reservoir Characterization – An Integrated Approach*, 2e, 336. Springer.

Lukasik, J.J., James, N.P., McGowran, B., and Bone, Y. (2000). An epeiric ramp: low energy, acool-water carbonate facies in a Tertiary inland sea, Murray Basin, South Australia. *Sedimentology* 47: 851–881.

Lunt, P. (2022). Field and well evidence for major unconformities in north Sarawak, compared to southwest Sabah, Malaysia. *The Bulletin of the Geological Society of Malaysia* 74: 69–83.

MacEachern, J.A. and Pemberton, S.G. (1994). Ichnological aspects of incised valley-fill systems from the Viking Formation of the Western Canada Sedimentary Basin, Alberta, Canada. In: *Incised Valley Systems: Origin and Sedimentary Sequences*, vol. 51 (ed. R.W. Dalrymple, R. Boyd, and B.A. Zaitlin), 129–157. SEPM Society for Sedimentary Geology.

Mallarino, G., Francis, J.M., Jorry, S.J. et al. (2021). Timescale dependent sedimentary record during the past 130 kyr from a tropical mixed siliciclastic–carbonate shelf edge and slope: Ashmore Trough (southern Gulf of Papua). *Sedimentology* 68: 2606–2648.

Mat Yusoff, H.H. (2023). *3D Seismic Geomorphology of the Late Quaternary Fluvial to Deltaic Depositional Systems within a Source-to-Sink Framework of Central Sarawak, Malaysia*, 277. Imperial College London https://doi.org/10.25560/105481.

Mat Yusoff, H.H., Johnson, H.D., Lonergan, L. et al. (2024). Seismic stratigraphy of Late Pleistocene incised valleys and adjacent environments, eastern Central Luconia Province, offshore north-west Borneo. *Sedimentology* 71: 319–354.

McKie, T. and Williams, B. (2009). Triassic palaeogeography and fluvial dispersal across the northwest European Basins. *Geological Journal* 44: 711–741.

McNeil, M.A., Nothdurft, L.D., Dyriw, N.J. et al. (2021). Morphotype differentiation in the Great Barrier Reef Halimeda bioherm carbonate factory: Internal architecture and surface geomorphometrics. *The Depositional Record* 7: 176–199.

McPhee, C., Reed, J., and Zubizarreta, I. (2015). Chapter 9, v. 64, 449–517. Developments in Petroleum Science. In: *Capillary Pressure. Core Analysis: A Best Practice Guide* (ed. C. McPhee, J. Reed and I. Zubizarreta). Elsevier.

Medici, G., West, L.J., and Mountney, N.P. (2016). Characterizing flow pathways in a sandstone aquifer: Tectonic vs sedimentary heterogeneities. *Journal of Contaminant Hydrology* 194: 36–58.

Miall, A.D. (2010). *The Geology of Stratigraphic Sequences*, 2e, 522. Berlin: Springer-Verlag.

Miall, A.D. (2016). *Stratigraphy – A Modern Synthesis*, 454. Springer.

Miall, A.D. (2002). Fluvial systems in the Malay time-slice analysis. *American Association of Petroleum Geologists Bulletin* 7: 1201–1216.

Middleton, G.V. (1973). Johannes Walther's law of correlation of facies. *BGSA* 84: 979–988.

Milliman, J.D. (1974). *Recent Sedimentary Carbonates: Part 1 Marine carbonates*, 375. Berlin-Heidelberg-New York: Springer Verlag.

Minken, J., Thompson, M., Woodward, J. et al. (2019). Tectonics of the Bedout Sub-basin and deposition of the Archer and Lower Keraudren Formations; influence of tectonics in a rapidly deposited succession. In: *The Sedimentary Basins of Western Australia V: Proceedings of the Petroleum Exploration Society of Australia Symposium* (ed. M. Keep and S.J. Moss), 26. SEPA, Perth, WA.

Mitchum, R.M.J., Vail, P.R., and Sangree, J.B. (1977). Stratigraphic interpretation of seismic reflection patterns in depositional sequences. In: *Seismic Stratigraphy: Applications to Hydrocarbon Exploration*, AAPG Memoir 26 (ed. C.E. Payton), 117–133. American Association of Petroleum Geologists.

Mitchum, R.M.J., Sangree, J.B., Vail, P.R., and Wornardt, W.W. (1991). Sequence stratigraphy in Late Cenozoic expanded sections, Gulf of Mexico. In: *Sequence Stratigraphy as an Exploration Tool: Concepts and Practices in the Gulf Coast*, vol. 11 (ed. J.M. Armentrout and B.F. Perkins), 163–197. SEPM Gulf Coast Section Publication https://doi.org/10.5724/gcs.90.11.

Mjøs, R. and Prestholm, E. (1993). The geometry and organization of fluviodeltaic channel sandstones in the Jurassic Saltwick Formation, Yorkshire, England. *Sedimentology* 40: 919–935.

Morgan, G.D. (2005). Sequence stratigraphy and structure of the Tertiary limestones in the Gulf of Papua, Papua New Guinea.

PhD thesis, The University of New South Wales, Sydney, Australia, 294 pp.

Morley, C.K., Crevello, P., and Ahmad, Z.H. (1998). Shale tectonics and deformation associated with active diapirism: the Jerudong Anticline, Brunei Darussalam. *Journal of Geological Society of London* 155: 475–490.

Morton, A., Knox, R., and Frei, D. (2016). Heavy mineral and zircon age constraints on provenance of the Sherwood Sandstone Group (Triassic) in the eastern Wessex Basin, UK. *Proceedings of the Geologists' Association* 127: 514–526.

Mulder, T., Joumes, M., Hanquiez, V. et al. (2017). Carbonate slope morphology revealing sediment transfer from bank-to-slope (Little Bahama Bank, Bahamas). *Marine and Petroleum Geology* 83: 26–34.

Munson, T.J. (2016). Sedimentary characterisation of the Wilton package, greater McArthur Basin, Northern Territory. Northern Territory Geological Survey, Record 2016-003, 151 p.

Munson, T.J. and Revie, D. (2018). Stratigraphic subdivision of the Velkerri Formation, Roper Group, McArthur Basin, Northern Territory. Northern Territory Geological Survey, Record 2018-006.

Nagtegaal, P.J.C. (1978). Sandstone-framework instability as a function of burial diagenesis. *Journal of the Geological Society* 135: 101–105.

Nichols, G. (2023). *Sedimentology and Stratigraphy*, 3e, 544. Wiley.

Nutman, A.P., Bennett, V.C., Friend, C.R.L., and Van Kranendonk, M. (2019). The Eoarchean legacy of Isua (Greenland) worth preserving for future generations. *Earth-Science Reviews* 198: 102923.

Nyberg, B. and Howell, J.A. (2016). Global distribution of modern shallow marine shorelines. Implications for exploration and reservoir analogue studies. *Marine and Petroleum Geology* 71: 83–104.

Orton, G.J. and Reading, H.G. (1993). Variability of deltaic processes in terms of sediment supply, with particular emphasis on grain size. *Sedimentology* 40: 475–512.

Patruno, S., Hampson, G.J., and Jackson, C. A. (2015). Quantitative characterisation of deltaic and subaqueous clinoforms. *Earth-Science Reviews* 142: 79–119.

Payton, G.E. (ed.) (1977). Seismic Stratigraphy: applications to hydrocarbon exploration. *AAPG Memoir* 26: 516.

Petrovic, A., Reijmer, J.J.G., Alsaihati, S.H.M. et al. (2023). Sediment dynamics and geomorphology of a submarine carbonate platform canyon system situated in an arid climate setting. *Sedimentology* 70: 2241–2271.

Pigram, C.J., Davies, P.J., Feary, D.A., and Symonds, P.A. (1989). Tectonic controls on carbonate platform evolution in southern Papua New Guinea: Passive margin to foreland basin. *Geology* 17: 199–202.

Piller, W.E. and Mansour, A.M. (1990). The northern Bay of Safaga (Red Sea, Egypt): an actuopaleontological approach. II: Sediment analysis and sedimentary facies. *Beiträge zur Paläontologie von Österreich* 16: 1–102.

Pittet, B. and Gorin, G.E. (1997). Distribution of sedimentary organic matter in a mixed carbonate-siliciclastic platform environment: Oxfordian of the Swiss Jura Mountains. *Sedimentology* 44 (5): 915–937.

Playton, T.E. and Kerans, C. (2015a). Late Devonian Carbonate Margins and Foreslopes of the Lennard Shelf, Canning Basin, Western Australia, Part A: Development During Backstepping and the Aggradation-To-Progradation Transition. *Journal of Sedimentary Research* 85 (11): 1334–1361.

Playton, T.E. and Kerans, C. (2015b). Late Devonian carbonate margins and foreslopes of the Lennard Shelf, Canning Basin, Western Australia, Part B: Development during progradation and across the Frasnian–Famennian biotic crisis. *Journal of Sedimentary Research* 85 (11): 1362–1392.

Plint, A.G. (1988). Sharp-based shoreface sequences and 'offshore bars' in the Cardium formation of Alberta: their relationship to relative changes in sea level. In: *Sea-Level Changes: An Integrated Approach*, SEPM, Special Publication 42 (ed. C.K. Wilgus, B.S. Hastings, C.G.S. C. Kendall, et al.), 357–370. Society of Economic Paleontologists and Mineralogists.

Pomar, L. and Hallock, P. (2008). Carbonate factories: a conundrum in sedimentary geology. *Earth-Science Reviews* 87: 134–169.

Posamentier, H.W. (2001). Lowstand alluvial bypass systems : Incised vs. unincised. *AAPG Bulletin* 10: 1771–1793.

Powell, J.H. (2010). Jurassic sedimentation in the Cleveland Basin: a review. *Proceedings of the Yorkshire Geological Society* 58: 21–72.

Powell, T.G., Jackson, M.J., Sweet, I.P. et al. (1987). Petroleum geology and geochemistry, middle Proterozoic McArthur Basin. Bureau of Mineral Resources, Australia, Record 1987/048.

Puasa, I. (2018). *Fine-Grained Deep-Water Turbidites, Baram Delta Province, Offshore Brunei Darussalam*, 278. Imperial College London https://doi.org/10.25560/82011.

Puchała, R.J., Porbski, S.J., Śliwiński, W.R., and August, C.J. (2011). Pleistocene to Holocene transition in the central basin of the Gulf of Thailand, based on geoacoustic survey and radiocarbon ages. *Marine Geology* 288: 103–111.

Purdy, E.G. (1963a). Recent calcium carbonate facies of the Great Bahama Bank. 1. Petrography and reaction groups. *Journal of Geology* 71 (3): 334–355.

Purdy, E.G. (1963b). Recent calcium carbonate facies of the Great Bahama Bank. 2. Sedimentary facies. *Journal of Geology* 71: 472–497.

Purser, B.H. (1973). *The Persian Gulf: Holocene Carbonate Sedimentation and Diagenesis in a Shallow Epicontinental Sea*, 471. New York, Heidelberg, Berlin: Springer-Verlag.

Purser, B.H. and Evans, G., 1973. Regional sedimentation along the Trucial coast, SE Persian Gulf. In: *The Persian Gulf* (ed. B.H. Purser), 211–231. Berlin: Springer.

Rankey, E.C. and Doolittle, D.F. (2012). Geomorphology of carbonate platformmarginal uppermost slopes: Insights from a Holocene analogue, Little Bahama Bank, Bahamas. *Sedimentology* 59: 2146–2171.

Rankey, E.C. and Reeder, S.L. (2010). Controls on platform-scale patterns of surface sediments, shallow Holocene platforms, Bahamas. *Sedimentology* 57: 1545–1565.

Rankey, E.C. and Pöppelreiter, M.C. (2023). *Cenozoic Isolated Carbonate Platforms - Focus Southeast Asia*, SEPM Special Publication 114, 348. SEPM Society for Sedimentary Geology.

Rankey, E.C., Schlaich, M., Mokhtar, S. et al. (2019). Seismic architecture of a Miocene isolated carbonate platform and associated off-platform strata (Central Luconia Province, offshore Malaysia). *Marine and Petroleum Geology* 102: 477–495.

Razin, P., Taati, F., and van Buchem, F.S.P. (2010). Sequence stratigraphy of Cenomanian–Turonian carbonate platform margins (Sarvak Formation) in the High Zagros, SW Iran: an outcrop reference model for the Arabian Plate. In: *Mesozoic and Cenozoic Carbonate Systems of the Mediterranean and the Middle East: Stratigraphic and Diagenetic Reference Models*, Volume Special Publications 329 (ed. F.S.P. Van Buchem, K.D. Gerdes, and M. Esteban), 187–218. London: The Geological Society of London.

Read, J.F. (1985). Carbonate platform facies models. *American Association of Petroleum Geologists Bulletin* 69 (1): 1–21.

Reading, H.G. (1978). Facies. In: *Sedimentary Environments and Facies*, 1e (ed. H.G. Reading), 4–14. Oxford: Blackwell.

Reading, H.G. (ed.) (1978a). *Sedimentary Environments and Facies*, 1e, 557. Oxford: Blackwell Scientific Publications.

Reading, H.G. (ed.) (1986b). *Sedimentary Environments and Facies*, 2e, 615. Oxford: Blackwell Scientific Publications.

Reading, H.G. (ed.) (1996). *Sedimentary Environments: Processes, Facies and Stratigraphy*, 3e, 688. Oxford: Blackwell Science Ltd.

Reading, H.G. (2001). Clastic facies models, a personal perspective. *Bulletin of the Geological Society of Denmark* 48: 101–115.

Reading, H.G. and Orton, G.J. (1991). Sediment calibre: a control on facies models with special reference to deep-sea depositional systems. In: *Controversies in Modern Geology* (ed. D.W. Müller, J.A. McKenzie, and H. Weissert), 85–111. London: Academic Press.

Reading, H.G. and Richards, M. (1994). Turbidite systems in deep-water basin margins classified by grain size and feeder system. *AAPG Bulletin* 78: 792–822.

Reijenstein, H.M., Posamentier, H.W., and Bhattacharya, J.P. (2011). Seismic geomorphobgy and high-resolution seismic stratigraphy of inner-shelf fluvial, estuarine, deltaic, and marine sequences, Gulf of Thailand. *AAPG Bulletin* 95: 1959–1990.

Reijmer, J.J.G. (2021). Marine carbonate factories: review and update. *Sedimentology* 68 (5): 1729–1796.

Reijmer, J.J.G., Sprenger, A., Ten Kate, W.G. H.Z. et al. (1994). Periodicities in the composition of Late Triassic calciturbidites (Eastern Alps, Austria). In: *Orbital Forcing and Cyclic Sequences*, Volume Special Publication No. 19 (ed. P.L. De Boer and D.G. Smith), 323–343. London: International Association of Sedimentologists – Blackwell.

Reijmer, J.J.G., Betzler, C., Kroon, D. et al. (2002). Bahamian carbonate platform development in response to sea-level changes and the closure of the Isthmus of Panama. *International Journal of Earth Sciences* 91: 482–489.

Reijmer, J.J.G., Swart, P.K., Bauch, T., Otto, R., Reuning, L., Roth, S. and Zechel, S. (2009). A re-evaluation of facies on Great Bahama Bank I: new facies maps of western Great Bahama Bank. In: Perspectives in Carbonate Geology: A Tribute to the Career of Robert Nathan Ginsburg (eds. P.K. Swart, G.P. Eberli, and J.A. McKenzie), Special Publication 41 International Association of Sedimentology, 29–46. *Oxford: Wiley-Blackwell-IAS.*

Reineck, H.E. and Singh, I.B. (1973). *Depositional Sedimentary Environments*, 454. Berlin, Heidelberg: Springer.

Reolid, J., Betzler, C., Eberli, G.P., and Grammer, G.M. (2017). The importance of microbial binding in Neogene-Quaternary steep slopes. *Journal of Sedimentary Research* 87 (5): 567–577.

Richards, M., Bowman, M., and Reading, H.G. (1998). Submarine fan systems I: characterization and stratigraphic prediction. *Marine and Petroleum Geology* 15: 689–717.

Ringrose, P. and Bentley, M. (2015). *Reservoir Model Design – A Practitioner's Guide*, 249. Springer.

Roberts, H.H. (1997). Dynamic changes of the Holocene Mississippi River delta plain: the delta cycle. *Journal of Coastal Research* 13: 605–627.

Røgen, B., Fabricius, I.L., Japsen, P. et al. (2005). Ultrasonic velocities of North Sea chalk samples: influence of porosity, fluid content and texture. *Geophysical Prospecting* 53: 481–496.

Saller, A.H., Dickson, J.A.D., Rasbury, E.T., and Ebato, T. (1999). Effects of long-term accommodation change on short-term cycles, Late Paleozoic platform limestones, West Texas. In: *Advances in Carbonate Sequence Stratigraphy: Application to Reservoirs, Outcrops, and Models*, Volume SEPM Special Publication v. 63 (ed. P.M. Harris, A. Saller, and T. Simo), 227–246. Tulsa: SEPM.

Samankassou, E. and Enos, P. (2019). Lateral facies variations in the Triassic Dachstein platform: A challenge for cyclostratigraphy. *The Depositional Record* 5: 469–485.

Sandall, S.T. (1996). *The geology and hydrocarbon resources of Negara Brunei Darussalam: Bandar Seri Begawan, Brunei Darussalam*, 243. Brunei Museum: Brunei Shell Petroleum Company.

Samorn, H. (2006). Fluvial reservoir architecture from near-surface 3D seismic data, Block B8/32, Gulf of Thailand. MS thesis, Colorado School of Mines, 156 p.

Sathiamurthy, E. and Voris, K.H. (2006). Maps of Holocene sea level transgression and submerged lakes on the Sunda Shelf. *Tropical Natural History* 2: 1–44.

Satterley, A.K. (1996a). Cyclic carbonate sedimentation in the Upper Triassic Dachstein Limestone, Austria: the role of patterns of sediment supply and tectonics in platform-reef-basin system. *Journal of Sedimentary Research* 66 (2): 307–323.

Satterley, A.K. (1996b). The interpretation of cyclic successions of the Middle and Upper Triassic of the Northern and Southern Alps. *Earth-Science Reviews* 40: 181–207.

Saxena, S. and Betzler, C. (2003). Genetic sequence stratigraphy of cool water slope carbonates (Pleistocene Eucla Shelf, southern Australia). *International Journal of Earth Sciences (Geologische Rundschau)* 92: 482–493.

Schieber, J., Bose, P.K., Eriksson, P.G. et al. (eds.) (2007). Atlas of microbial mat features preserved within the siliciclastic rock record. In: *Atlases in Geoscience 2*, 311. Elsevier Science .

Schlager, W. (2003). Benthic carbonate factories of the Phanerozoic. *International Journal of Earth Sciences* 92: 445–464.

Schlager, W. (2005). *Carbonate Sedimentology and Sequence Stratigraphy*, 200. Tulsa, OK: SEPM (Society for Sedimentary Geology), SEPM Concepts in Sedimentology and Paleontology.

Schlager, W., Reijmer, J. J., & Droxler, A. (1994). Highstand shedding of carbonate platforms. *Journal of Sedimentary Research*, 64(3b), 270-281.

Schwarzacher, W. (1947). Über die sedimentäre Rhythmik des Dachsteinkalkes bei Lofer. *Verhandlungen der Geologische Bundesanstalt* 10–12: 175–188.

Schwarzacher, W. (1954). Die Großryhtmik des Dachsteinkalkes von Lofer. *Tschermaks Mineralogische Und Petrographische Mitteilungen* 4 (1–4): 44–54.

Sech, R.P., Jackson, M.D., and Hampson, G.J. (2009). Three-dimensional modeling of a shoreface-shelf parasequence reservoir analog: Part 1. Surface-based modeling to capture high-resolution facies architecture. *AAPG Bulletin* 93: 1155–1181.

Sim, R. and Bacon, M. (2014). *Seismic Amplitude: An Interpreter's Handbook*, 271. Cambridge University Press.

Simons, D.B., Richardson, E.V., and Nordin, C.F. (1965). Sedimentary structures generated by flow in alluvial channels. In: *Primary Sedimentary Structures and Their Hydrodynamic Interpretation*, vol. 12 (ed. G.V. Middleton), 34–52. SEPM Special Publications.

Sixsmith, P.J., Hampson, G.J., Gupta, S. et al. (2008). Facies architecture of a net transgressive sandstone reservoir analog: The Cretaceous Hosta Tongue, New Mexico. *AAPG Bulletin* 92: 513–547.

Slatt, R.M. (2006). *Stratigraphic Reservoir Characterization for Petroleum Geologists, Geophysicists, and Engineers*, 478. Elsevier.

Slootman, A., de Kruijf, M., Glatz, G. et al. (2023). Shape-dependent settling velocity of skeletal carbonate grains: implications for calciturbidites. *Sedimentology* 70 (6): 1683–1722.

Snedden, J.W. and Bergman, K.M. (1999). Isolated shallow marine sand bodies: deposits for all interpretations. In: *Isolated Shallow Marine Sand Bodies: Sequence Stratigraphic Analysis and Sedimentologic Interpretation*. SEPM, Special Publication 64 (ed. K.M. Bergman and J.W. Snedden), 1–11. SEPM Society for Sedimentary Geology.

Snedden, J.W. and Dalrymple, R.W. (1999). Modern shelf sand ridges: from historical perspective to a unified hydrodynamic and evolutionary model. In: *Isolated Shallow Marine Sand Bodies: Sequence Stratigraphic Analysis and Sedimentologic Interpretation*. SEPM, Special Publication 64 (ed. K.M. Bergman and J.W. Snedden), 13–28. SEPM Society for Sedimentary Geology.

Sneider, R.M. (1990). Reservoir Description of Sandstones. In: *Sandstone Petroleum Reservoirs (Casebooks in Earth Sciences)* (ed. J.H. Barwis, J.G. McPherson, and J.R.J. Studlick), 1–3. New York, NY: Springer.

Sneider, R.M., Tinker, C.N., and Meckel, L.D. (1978). Deltaic environment reservoir types and their characteristics. *Journal of Petroleum Technology* 30: 1538–1546.

Sneider, R.M., Tinker, C.N. and Richardson, J.G. (1984). Reservoir geology of sandstones. Short Course, Soc. Petrol. Engineers, 59th Annual Technical Conference, Houston (September, 1984), variously paginated.

Soltanian, M.R. and Ritzi, R.W. (2014). A new method for analysis of variance of the hydraulic and reactive attributes of aquifers as linked to hierarchical and multiscaled sedimentary architecture. *Water Resources Research* 50: 9766–9776.

Southard, J.B. and Boguchwal, L.A. (1990). Bed configuration in steady unidirectional water flows; Part 2, Synthesis of flume data. *Journal of Sedimentary Research* 60: 658–679.

Sprague, A.R., Patterson, P.E., Hill, R.E. et al. (2002). The physical stratigraphy of fluvial strata: a hierarchical approach to the analysis of genetically related stratigraphic elements for improved reservoir prediction. (Abs) AAPG Annual Meeting, Official Program, p. A167.

Stewart, S.A. (1995). Paleomagnetic analysis of fold kinematics and implications for geological models of the Cantabrian/Asturian arc, north Spain. *Journal of Geophysical Research: Solid Earth* 100: 20079–20094.

Strasser, A., Pittet, B., Hillgärtner, H., and Pasquier, J.-B. (1999). Depositional sequences in shallow carbonatedominated sedimentary systems: concepts for a high-resolution analysis. *Sedimentary Geology* 128: 201–221.

Strick, R.J.P., Ashworth, P.J., Sambrook Smith, G.H. et al. (2019). Quantification of bedform dynamics and bedload sediment flux in sandy braided rivers from airborne and satellite imagery. *Earth Surface Processes and Landforms* 44: 953–972.

Strohmenger, C.J., Al-Mansoori, A., Al-Jeelani, O. et al. (2010). The sabkha sequence at Mussafah Channel (Abu Dhabi, United Arab Emirates): Facies stacking patterns, microbial-mediated dolomite and evaporite overprint. *GeoArabia* 15 (1): 49–90.

Strohmenger, C.J., Shebl, H., Al-Mansoori, A. et al. (2011). Facies stacking patterns in a modern arid environment: a case study of the Abu Dhabi sabkha in the vicinity of Al-Qanatir Island, United Arab Emirates. In: *Quaternary Carbonate and Evaporite Sedimentary Facies and their Ancient Analogues: A Tribute to Douglas James Shearman*, Volume Int. Assoc. Sedimentol. Spec. Publ. 43 (ed. C.G.S.C. Kendall and A.S. Alsharhan), 149–182. Ghent: International Association of Sedimentologists.

Sullivan, M.D., Van Wagoner, J.C., Jennette, D.C. et al. (1997). High resolution sequence stratigraphy and architecture of the Shannon Sandstone, Hartzog Draw Field, Wyoming: implications for reservoir management. In: *Shallow Marine and Nonmarine Reservoirs: Sequence Stratigraphy, Reservoir Architecture, and Production Characteristics. SEPM, Gulf Coast Section, 18th Annual Research Conference, Proceedings* (ed. K.W. Shanley and R.F. Perkins), 331–344.

Suter, J.R. and Clifton, H.E. (1999). The Shannon Sandstone and isolated linear sand bodies: interpretations and realizations. In: *Isolated Shallow Marine Sand Bodies: Sequence Stratigraphic Analysis and sedimentologic Interpretation*. SEPM, Special Publication 64 (ed. K.M. Bergman and J.W. Snedden), 321–356. Society of Economic Paleontologists and Mineralogists, Tulsa, Oklahoma.

Swart, P.K., Eberli, G.P., Malone, M.J., and Sarg, J.F. (2000). *Proceedings of the Ocean Drilling Program, Leg 166, Scientific Results*, 213. College Station, TX: Ocean Drilling Program.

Swart, P.K., Reijmer, J.J.G. and Otto, R. (2009). A re-evaluation of facies on Great Bahama Bank II: variations in the δ13C, δ18O and mineralogy of surface sediments. In: *Perspectives in Carbonate Geology: a Tribute to the Career of Robert Nathan Ginsburg* (eds. P.K. Swart, G.P. Eberli, J.A. McKenzie and I. Jarvis), 41, 47–59. International Association of Sedimentologists Special Publications. Chichester: Wiley-Blackwell.

Tan, D.N.K. and Lamy, J.M. (1990). Tectonic evolution of the NW Sabah continental margin since the Late Eocene. *The Bulletin of the Geological Society of Malaysia* 27: 241–260.

Tanabe, S., Saito, Y., Lan Vu, Q. et al. (2006). Holocene evolution of the Song Hong (Red River) delta system, northern Vietnam. *Sedimentary Geology* 187: 29–61.

Tcherepanov, E.N., Droxler, A.W., Lapointe, P. et al. (2008a). Neogene evolution of the mixed carbonate-siliciclastic system in the Gulf of Papua, Papua New Guinea. *Journal of Geophysical Research: Earth Surface* 113: F01S21.

Tcherepanov, E.N., Droxler, A.W., Lapointe, P., and Mohn, K. (2008b). Carbonate seismic stratigraphy of the Gulf of Papua mixed depositional system: Neogene stratigraphic signature and eustatic control. *Basin Research* 20: 185–209.

Tcherepanov, E.N., Droxler, A.W., Lapointe, P. et al. (2010). Siliciclastic influx and burial of the Cenozoic carbonate system in the Gulf of Papua. *Marine and Petroleum Geology* 27: 533–554.

Thompson, M., Woodward, J., Minken, J., and Ryan, R. (2019). Unravelling the lower Keraudren petroleum system in the Bedout Sub-basin: some early observations. In: *The Sedimentary Basins of Western Australia V: Proceedings of the Petroleum Exploration Society of Australia Symposium* (ed. M. Keep and S.J. Moss), 2019 20. Perth, WA.

Tinterri, R. (2011). Combined flow sedimentary structures and the genetic link between sigmoidal-and hummocky-cross stratification. *GeoActa* 10: 1–43.

Thanh, N.T., Cuong, D.H., Stattegger, K. et al. (2021). Depositional sequences of the Mekong river delta and adjacent shelf over the past 140 kyr, southern Vietnam. *Journal of Asian Earth Sciences* 206: 104634.

Tillman, R.W. and Martinsen, R.S. (1984). The Shannon shelf-ridge sandstone complex, Salt Creek Anticline area, Powder River Basin, Wyoming. In: *Siliciclastic Shelf Sedimentation*, SEPM Special Publication 34 (ed. R.W. Tillman and C.T. Siemers), 85–142.

Tillman, R.W. and Martinsen, R.S. (1987). Sedimentologic model and production characteristics of Hartzog Draw Field, Wyoming. In: *Reservoir Sedimentology*. SEPM Special Publication 40 (ed. R.W. Tillman and K.J. Weber), 15–112. Society of Economic Paleontologists and Mineralogists, Tulsa, Oklahoma.

Tjallingii, R., Stattegger, K., Wetzel, A., and Phach, P. (2010). Infilling and flooding of the Mekong River incised valley during deglacial sea-level rise. *Quaternary Science Reviews* 29: 1432–1444.

Tonks, L.H. (1922). Recent Notes on the Dogger Sandstone of the Yorkshire Coast. *Transactions. Leeds Geological Association* 29–33.

Torres, J., Gartrell, A. and Hoggmascall, N. (2011). Redefining a sequence stratigraphic framework for the Miocene to present in Brunei Darussalam: roles of local tectonics, eustacy and sediment supply. *International Petroleum Technology Conference*, Bangkok, Thailand. IPTC 15167, pp. 1–12.

Tucker, M.E., Gallagher, J., and Leng, M.J. (2009). Are beds in shelf carbonates millennial-scale cycles? An example from the mid-Carboniferous of northern England. *Sedimentary Geology* 214: 19–34.

Tucker, M.E. and Wright, V.P. (2009). *Carbonate Sedimentology*, 2e, 496. Oxford: Wiley.

Underhill, J.R. and Partington, M.A. (1993). Jurassic thermal doming and deflation in the North Sea: Implications of the sequence stratigraphic evidence. In: *Petroleum Geology of Northwest Europe: Proceedings of the 4th Conference: Geological Society*, vol. 4 (ed. J.R. Parker), 337–345. London: Petroleum Geology Conference Series.

Vail, P.R., Bubb, J.N., Hatlelid, W.G. et al. (1977). Seismic stratigraphy and global changes in sea level, parts 1-11. In: *Seismic Stratigraphy – Applications to Hydrocarbon Exploration*, AAPG, Memoir 26 (ed. C.E. Payont), 49–212. American Association of Petroleum Geologists, Tulsa, Oklahoma.

Vakarelov, B.K. and Ainsworth, R.B. (2013). A hierarchical approach to architectural classification in marginal-marine systems: bridging the gap between sedimentology and sequence stratigraphy. *AAPG Bulletin* 97: 1121–1161.

van Buchem, F.S.P., Allan, T.L., Laursen, G.V. et al. (2010). Regional stratigraphic architecture and reservoir types of the Oligo-Miocene deposits in the Dezful Embayment (Asmari and Pabdeh Formations) SW Iran. *Geological Society, London, Special Publications* 329: 219–263.

van den Berg, J.H., Boersma, J.R., and Van Gelder, A. (2007). Diagnostic sedimentary structures of the fluvial-tidal transition zone-Evidence from deposits of the Rhine and Meuse. *Netherlands Journal of Geosciences* 86: 287–306.

van den Berg, J.H. and Nio, S.D. (2010). *Sedimentary Structures and their Relation to Bedforms and Flow Conditions*, 138. EAGE Publications.

Van der Kooij, B., Immenhauser, A., Csoma, A. et al. (2009). Spatial geochemistry of a Carboniferous platform-margin-to-basin transect: Balancing environmental and diagenetic factors. *Sedimentary Geology* 219: 136–150.

Van der Kooij, B., Immenhauser, A., Steuber, T. et al. (2010). Controlling factors of volumetrically important marine carbonate cementation in deep slope settings. *Sedimentology* 57: 1491–1525.

van Vliet, A. and Schwander, M.M. (1987). Stratigraphic interpretation of a regional seismic section across the Labuan syncline and its flank structures, Sabah, North Borneo. In: *Atlas of Seismic Stratigraphy*, vol. 27 (ed. A.W. Bailey), 163–167. American Association of Petroleum Geologists, Studies in Geology.

Veeken, P.C.H. and van Moerkerken, B. (2013). *Seismic Stratigraphy and Depositional Facies Models*, 494. EAGE Publishing.

Voris, H.K. (2000). Maps of Pleistocene sea levels in Southeast Asia: Shorelines, river systems and time durations. *Journal of Biogeography* 27: 1153–1167.

Walker, R.G. (1978). Deep-water sandstone facies and ancient submarine fans: models for exploration for stratigraphic traps. *AAPG Bulletin* 62: 932–966.

Walker, R.G. (ed.) (1979). Facies models. Geosci. Can. Reprint Ser., 1, Geol. Soc. Can., Waterloo, Ontario.

Walker, R.G. (ed.) (1984a). Facies models, 2e. Geosci. Can. Reprint Ser., 1, Geol. Soc. Can., Waterloo, Ontario.

Walker, R.G. (1984b). General introduction: facies, facies sequences and facies models. In: *Facies Models Second Edition*, Geoscience Canada Reprint Series 1 (ed. R. G. Walker), 1–9. Geological Association of Canada, Waterloo.

Walker, R.G. (1984c). Turbidites and associated coarse clastic deposits. In: *Facies Models*, Geoscience Canada Reprint Series 1, 2e, 171–188. Geological Association of Canada, Waterloo.

Walker, R.G. and Bergman, K.M. (1993). Shannon Sandstone in Wyoming; a shelf-ridge complex reinterpreted as lowstand shoreface deposits. *Journal of Sedimentary Research* 63: 839–851.

Walker, R.G. and Plint, A.G. (1992). Wave- and storm-dominated shallow marine systems. In: *Facies, Facies Models and Modern Stratigraphic Concepts*. In: Walker, R.G. and James, N.P., eds., Facies Models – Response to Sea Level Change (ed. R.G. Walker), 219–238. St. John's: Geological Association of Canada.

Walker, R.G. and James, N.P. (1992). *Facies Models – Response to Sea Level Change*, 409. Geological Association of Canada.

Walther, J. (1894). *Einleitung in die Geologie als historische Wissenschaft*, vol. Band 3, 535–1055. Lithogenesis der Gegenwart: Jena, G. Fischer.

Weber, K.J. and Van Geuns, L.C. (1990). Framework for constructing clastic reservoir simulation models. *Journal of Petroleum Technology* 1990: 1248–1297.

Weber, K.J. and Van Geuns, L.C. (2005). *Petroleum Geology/Production*, 308–330. Elsevier https://doi.org/10.1016/b0-12-369396-9/00249-5.

Webster, J.M., Wallace, L., Silver, E. et al. (2004a). Drowned carbonate platforms in the Huon Gulf, Papua New Guinea. *Geochemistry, Geophysics, Geosystems* 5: Q11008.

Webster, J.M., Wallace, L., Silver, E. et al. (2004b). Coralgal composition of drowned carbonate platforms in the Huon Gulf, Papua New Guinea; implications for lowstand reef development and drowning. *Marine Geology* 204: 59–89.

Wei, E.A., Driscoll, N.W., and Slingerland, R. L. (2019). Oceanographic currents, differential subsidence, and physiography control three-dimensional clinothem growth in the Gulf of Papua, Papua New Guinea. *Marine Geology* 407: 164–180.

Weil, A.B., Van der Voo, R., van der Pluijm, B.A., and Parés, J.M. (2000). The formation of an orocline by multiphase deformation: a paleomagnetic investigation of the Cantabria–Asturias Arc (northern Spain). *Journal of Structural Geology* 22: 735–756.

Wellner, R., Beaubouef, R., Van Wagoner, J.C. et al. (2005). Jet-plume depositional bodies – the primary building blocks of Wax Lake Delta. *Transactions. Gulf Coast Association of Geological Societies* 55: 867–909.

Wells, J.T. and Coleman, J.M. (1987). Wetland loss and the subdelta life cycle. *Estuarine, Coastal and Shelf Science* 25: 111–125.

Wetzel, A., Feldens, A., Unverricht, D. et al. (2021). Late Pleistocene sea-level changes and the formation and fill of bent valleys incised into the shelf of the western South China Sea. *Journal of Asian Earth Sciences* 206: 104626.

Wilber, R.J., Milliman, J.D., and Halley, R.B. (1990). Accumulation of bank-top sediment on the western slope of Great Bahama Bank: rapid progradation of a carbonate megabank. *Geology* 18 (10): 970–974.

Williams, G.D. (1993). Tectonics and seismic sequence stratigraphy: an introduction. In: *Tectonics and Seismic Sequence Stratigraphy*, Geological Society Special Publication No. 71 (ed. G.D. Williams and A. Dobb), 1–13.

Williams, H.D., Burgess, P.M., Wright, V.P. et al. (2011). Investigating carbonate platform types: multiple controls and a continuum of geometries. *Journal of Sedimentary Research* 81 (1–2): 18–37.

Wilson, A., Côté, A., Richards, B. and Altmann, C. (2023). Borehole image features map depositional environment change in Mesoproterozoic deltas: insights from the Beetaloo Sub-Basin. CABS IV Special Publication. Volume 1, Exploring Australia's Resource Frontier: Beetaloo Sub-Basin. 10p. https://doi.org/10.36404/YTBN7230

Wilson, J.L. (1975). *Carbonate Facies in Geological History*, 471. Berlin: Springer Verlag.

Wilson, R.A.M. (1964). The Geology and Mineral Resources of the Labuan and Padas Valley area, Sabah, Malaysia. Geological Survey of the Borneo Region of Malaysia. Memoir 17: 150.

Wong, H.K., Lüdmann, T., Haft, C. et al. (2003). Quaternary sedimentation in the Molengraaff paleo-delta, northern Sunda Shelf (southern South China Sea). In: *Tropical Deltas of Southeast Asia—Sedimentology, Stratigraphy, and Petroleum Geology* (ed. F.H. Sidi, D. Nummedal, P. Imbert, et al.), 201–216. SEPM (Society for Sedimentary Geology).

Wong, R.H.F. (1997). Sequence stratigraphy of the upper Miocene stage IVE in the Labuan-Paisley Syncline, northwest Sabah Basin. *The Bulletin of the Geological Society of Malaysia* 41: 53–60.

Wright, V.P. and Della Porta, G. (ed.) (2021). *Field Guides to Exceptionally Exposed Carbonate Outcrops*, 792. Gent:

International Association of Sedimentologists, Ghent University.

Yamin Ali, M. (2014). An integrated analysis of the depositional control, sedimentology and diagenesis of cenozoic carbonates from the Sarawak Basin, East Malaysia. Imperial College London, PhD thesis, 467p. https://doi.org/10.25560/29605

Zakaria, A.A., Johnson, H.D., Jackson, C.A.-L., and Tongkul, F. (2013). Sedimentary facies analysis and depositional model of the Palaeogene West Crocker submarine fan system, NW Borneo. *Journal of Asian Earth Sciences* 76: 283–300.

Zampetti, V., Schlager, W., Konijnenburg, J.-H., and Everts, A.-J. (2004). Architecture and growth history of a Miocene carbonate platform from 3D seismic reflection data; Luconia province, offshore Sarawak, Malaysia. *Marine and Petroleum Geology* 21: 517–534.

Ziegler, P.A. (1982). *Geological Atlas of Western and Central Europe*, 256. Amsterdam: Elsevier.

Ziegler, P.A. (1988). Evolution of the Arctic-North Atlantic and the Western Tethys. *AAPG Memoir* 43: 198.

Ziegler, P.A. (1990). *Geological Atlas of Western and Central Europe*, 2e, 239. Amsterdam: Elsevier.

3

The Application of Biogenic Structures to Facies Analysis

James A. MacEachern[1], Murray K. Gingras[2], and Kerrie L. Bann[3]

[1] Department of Earth Sciences, Simon Fraser University, Burnaby, British Columbia, Canada
[2] Department of Earth and Atmospheric Sciences, University of Alberta, Edmonton, Alberta, Canada
[3] Ichnofacies Analysis Inc., Calgary, Alberta, Canada

Summary

Biogenic structures are widespread and abundant in most sedimentary facies and yet are markedly underutilised by many sedimentologists. This is unfortunate because as macroscopic palaeontologic elements, they can be identified and interpreted with the same level of ease as are primary physical structures and yield valuable insights into the physico-chemical conditions of the depositional environment. As such, the ability to describe and interpret animal-sediment interactions should be regarded as a crucial skill for any researcher undertaking sedimentary facies analysis. Accurate taxonomy is certainly important, but this chapter cannot hope to provide the reader with a complete grounding in the discipline. Developing such expertise requires training and experience. That said, valuable palaeoenvironmental insights can be achieved by assessing bioturbation intensity and its distribution, recognising trace-fossil characteristics indicative of an animal's behaviour(s), evaluating trace-fossil diversity, appraising ichnofabric characteristics and discerning recurrent trace-fossil suites that allow their assignment to the appropriate ichnofacies. This 'applied ichnology' approach is foundational to a comprehensive description and interpretation of sedimentary facies and facilitates the assessment of depositional rate, physical energy, continental versus paralic/marine conditions, subaerial versus subaqueous deposition and variations in salinity and oxygenation. Additionally, biogenic structures can also play a crucial role in the identification of and differentiation between autogenic and allogenic discontinuities, essential for the sequence stratigraphic assessment of facies successions and facies associations. Finally, this chapter showcases three case studies wherein ichnological datasets are integrated with physical sedimentology to yield high-resolution descriptions and depositional interpretations of the facies succession. So armed, the reader will be better equipped to gain a deeper appreciation of the remaining chapters of this volume.

3.1 Introduction

It has become increasingly clear that for most, if not all sedimentary environments, the integration of biogenic structures into an interpretation leads to a more holistic and detailed understanding of the physical, chemical and biological processes responsible for sedimentary facies. The renowned sedimentologist, James Howard (1975, p. 131) perhaps stated it best – '... traces are often given short shrift by sedimentologists. This situation is unfortunate, and indeed unfair, to the study of sediments, because the contained lebensspuren *are* sedimentary structures, (albeit biologically formed) and should receive attention equal to that devoted to structures developed by physical processes. Trace fossils often supply evidence of sedimentological conditions that is superior to information gained only by the study of physical structures. If this is not sufficient reason for sedimentologists to be concerned with the study of ichnology, perhaps they can be prodded into it by virtue of the fact that the nefarious beasts creating the biogenic structures have a nasty habit of destroying their beloved physical structures, and they should at least attempt to identify the enemy!'

Indeed, the late Harold G. Reading ended his third edition of this very same book with the statement 'In addition, there is a need for closer integration of physical sedimentary processes with biological aspects, particularly ichnofacies analysis and palaeoecology, within a high-resolution

Reading's Sedimentary Environments: Processes, Facies and Stratigraphy, Fourth Edition. Edited by Bruce Levell.

Companion website: www.wiley.com/go/levell4e

chronostratigraphic framework' (Reading 1996, p. 571). This chapter seeks to address this need by summarising the utility of biogenic structures, and best practices approaches to their integration with physical lithofacies analysis. This chapter outlines the range of biogenic structures that are routinely employed in palaeoecological assessments of sedimentary facies, but with a focus on applied ichnology.

3.1.1 Ichnology and Biogenic Structures

Biogenic structures can be regarded as all features arising from the interactions of life with geologic substrates, although the vast majority of these are associated with sedimentary materials (i.e. biogenic sedimentary structures). The most widely encountered biogenic structures in the sedimentary record are those generated by the activities of animals and properly fall into the discipline of ichnology, historically regarded as a rather eclectic science. Plant-generated structures such as rhizoliths are also produced by the activity of life and so are regarded as part of the trace-fossil assemblage. However, because the discipline of ichnology is governed by the International Commission on Zoological Nomenclature (ICZN), these botanically generated structures are not formally named as ichnotaxa. In this chapter, most of the focus is directed towards the use of ichnological data (e.g. trace fossils, bioturbation intensity, ichnological diversity and ichnofacies analysis) for the purposes of providing integrated characterisations of facies and refining facies interpretations.

Like other palaeontological disciplines, a fundamental tension persists between a focus and reliance upon *taxonomic* ichnology versus an emphasis on *applied ichnology*. Taxonomic ichnology is primarily focused on classifying trace fossils into ichnogenera and ichnospecies, based on their morphological characteristics. By contrast, applied ichnology is paramount for facies analysis, in that it addresses – among other things – issues surrounding the ecology/palaeoecology of sedimentary facies as expressed by the biogenic structures. That said, the ability to effectively employ ichnology to solve palaeoenvironmental problems is founded on reliable taxonomy. As such, while this chapter focuses on the use of ichnological datasets to better understand the sedimentary record, practitioners ultimately must develop the skills to accurately identify the various trace fossils. Would-be students of the discipline are directed to several outstanding volumes that cover aspects of taxonomy, including Frey (1975), Ekdale et al. (1984), Bromley (1990), Seilacher (2007), Buatois and Mángano (2011) and Knaust and Bromley (2012).

Trace-fossil taxonomy is governed by the ICZN, and individual trace fossils are named using the Linnean binomial nomenclature of ichnogenus and ichnospecies. This can cause some confusion on the part of non-specialists, who erroneously may begin to think of the trace fossils as the animals themselves. *Planolites beverlyensis*, for example, does not 'prefer' muddy substrates, nor does 'it' burrow into anything. The *trace-maker* may do these things, but the trace fossil is merely a record of that activity. Further, because the ICZN names the structures *as* fossils, modern traces technically are not assigned ichnogenus and ichnospecies names, regardless of their possessing a form identical to their fossil counterparts. It is not helpful to obsess about these philosophical nuances in a chapter such as this, and so the details of taxonomy and its challenges are downplayed here; we focus instead on the *application* of biogenic structures for the purposes of facies analysis.

In many cases, particularly subsurface studies using core datasets or most two-dimensional outcrop exposures, the three-dimensional characteristics of structures cannot be reliably determined, limiting the identification of many trace fossils to the ichnogenus level. Fortunately, few trace fossils have ichnospecies that reflect markedly different animal behaviours and therefore, they possess broadly similar palaeoecological implications. In most routine facies work, diagnosis to the ichnogenus level is sufficient for deriving their palaeoenvironmental significance.

Applied ichnology has a long tradition, founded originally on the work of Adolf (Dolf) Seilacher in the 1950s. Many of the original studies were published in German but were eventually synthesised in later manuscripts published in English. Seilacher's studies were augmented by a rich tradition of neoichnological work by Rudolf Richter and later by Hans-Erich Reineck, employing the North Sea and Waddenzee as natural laboratories (e.g. Richter 1928; Reineck 1967). These efforts led to research headed by Robert Frey and James Howard in North America along the coast of Georgia, USA (e.g. Howard and Frey 1973). These early neoichnological efforts not only provided a foundation for many of the ethological (behavioural) bases for trace-fossil interpretations, but greatly improved our understanding of infauna distribution and the distribution of their traces. This has been relied upon heavily for identifying the majority of shallow-marine and brackish-water settings.

Most work focused on animal-sediment relations, which are underpinned by substrate consistency/grain size and animal behaviours (e.g. feeding strategy, dwelling styles and locomotion, among others). Such interactions align positively with notions of palaeoecology, facies analysis and palaeoenvironmental interpretation. Given that siliciclastic intervals commonly lack abundant body fossils but contain numerous trace fossils, it was inevitable that applied ichnology would begin to have a significant impact on sedimentary facies analysis.

A daunting issue for many classical (particularly siliciclastic-oriented) sedimentologists can be the rapid introduction and use of numerous unfamiliar palaeontological terms and concepts. To that end, we provide some commonly employed terms and concepts we use in this chapter in the following:

Biogenic structures: Biogenic structures are features that show evidence of organism activity (animal or plant), other than the production of body parts (Frey and Wheatcroft 1989). They encompass: (i) biogenic sedimentary structures; (ii) bioerosion structures; and (iii) other evidence of organism activity. As stated above, the discipline of ichnology covers the majority of these features. Ichnology is the study of traces produced by organisms on or within a substrate, and includes all aspects of bioturbation, bioerosion and biodeposition (e.g. Bromley 1990; Buatois and Mángano 2011). Some biostratification structures (e.g. stromatolites, thrombolites, biogenically graded bedding and microbiologically induced sedimentary structures [MISS]) are not strictly regarded as organism traces by most ichnologists, as they do not reveal the functional anatomy of the organism itself (Frey and Pemberton 1985). Hence, one can say that while not all biogenic sedimentary structures are traces, all traces are biogenic structures.

Biogenic sedimentary structures encompass all features produced by the activity of life on or within unconsolidated substrates (e.g. bioturbation and biodeposition). By contrast, *bioerosion structures* record those features generated by the mechanical or biochemical boring of organisms into lithified, compacted or rigid substrates (e.g. hardgrounds, firmgrounds and woodgrounds, respectively). Such bioerosion structures may also be produced into isolated shells, bones and logs. The grouping regarded as '*other evidence of activity*' is necessarily vague in the attempt to be fully inclusive and includes such structures as spider webs, egg cases and wasp nests, which are unlikely to be encountered in most facies studies.

Trace fossil: A discrete biogenic structure produced by the activity of an organism and recording some aspect of the organism's functional anatomy. Structures possessing a recurring morphology by which it can be identified are assigned an ichnogenus and potentially ichnospecies name under the rules of the ICZN. Trace fossils produced by the activity of an organism but lacking distinctive morphological characteristics, or those that are poorly preserved, may be assigned to behavioural categories (e.g. escape, dwelling, locomotion) rather than given an ichnogenus or ichnospecies name. Such behavioural terms (e.g. fugichnia, domichnia and repichnia, respectively), however, are not italicised. Named trace fossils can still be assigned to such behaviours (e.g. *Skolithos linearis* is regarded as 'domichnia' or a dwelling structure). Note that many correspond to a combination of behaviours (e.g. dwelling *and* surface deposit feeding or locomotion *and* mobile deposit feeding). Some trace fossils (e.g. fugichnia) have no formal ichnogenus or ichnospecies names assigned.

Trace-fossil assemblage: This is a general term used to refer to collections of trace fossils preserved in a sedimentary facies without any assumptions regarding the *time of emplacement* (i.e. their contemporaneity) or recurrence in the stratigraphic record (Buatois and Mángano 2011). This is the most general grouping of trace fossils and imparts the least detail regarding the characterisation of a facies and its palaeoenvironmental interpretation.

Trace-fossil suite: This term refers to a grouping of trace fossils that represent animal activity within a sedimentary unit or facies during specific time frames. Often considered synonymous with an 'ichnocoenose' – the activity of a single biological community – here, we define a trace-fossil suite as a time-averaged record of colonisation and bioturbation within a bed or bedset of a *single sedimentary facies*, following MacEachern and Bann (2008). This concept includes both long-term colonisation periods (spanning years to centuries) and short-term episodes (from hours to months).

In settings with slow deposition, pervasive bioturbation reveals the progressive cross-cutting of mid- and shallow-tier structures by deep-tier structures. This reflects prolonged colonisation within a single trace-fossil suite in an aggrading environment. In contrast, short-term trace-fossil suites, which form following brief depositional events (e.g. tempestites, flood deposits and sediment-gravity flows), are interlayered with pre-existing ambient (pre-event) suites. These event and ambient suites, while part of the same facies, should be assessed separately. The event suite reflects the benthic community's response to the depositional event, whereas the pre-event suite provides insight into the overall palaeoenvironment.

Significant interruptions in deposition, marked by stratigraphic omission surfaces, can be identified by juxtaposed trace-fossil suites. In such cases, a younger suite (e.g. borings) cross-cuts an older suite (e.g. softground traces from the original deposition), signifying a boundary between genetically unrelated facies.

It is our assertion that trace-fossil suites offer the most valuable palaeoenvironmental data for facies analysis.

Ichnofabric: Ichnofabrics encompass all substrate texture and internal structure resulting from bioturbation and bioerosion across various scales. Ichnofabrics are linked to concepts of organism tiering position (i.e. how deeply the animal descended into the substrate) and the timing of burrowing activities (Ekdale et al. 2012). These are determined primarily through the observance of cross-cutting ichnofossils and by establishing 'constituent diagrams' (see Section 3.3.1).

Ichnofacies: The ichnofacies concept is a multidimensional framework underpinned by recurring, facies-controlled groupings of biogenic structures reflecting animal responses (ethology) to a range of palaeoenvironmental conditions. They serve as theoretical constructs at the hierarchical level of facies models and encompass animal-sediment interactions with the physical processes and chemical conditions that operated within the depositional environment (see Section 3.3.2).

Cryptic bioturbation or cryptobioturbation: Cryptic bioturbation is more or less synonymous with cryptobioturbation. These express situations where bioturbation is pervasive, such that individual trace fossils and sedimentary laminae cannot be identified. It is normally associated with high biomasses at the time of colonisation or with prolonged exposure to bioturbating infauna. The terms have also been used to describe the subtle repositioning of sediment grains (normally sand), such that laminae become blurred or fuzzy. This process is associated with infauna that are sufficiently small that they do not substantively displace sand grains when moving through the sediment (Pemberton et al. 2008).

Ichnology is ideally suited to the concepts of facies analysis. Just as primary sedimentary structures, erosion structures and soft-sediment deformation structures are the products of physical and/or chemical processes operating within the environment, so too are trace fossils and biodeformation structures the products of organism interactions with the substrate in response to the various physical, chemical and biological attributes of the environment. The description and characterisation of facies necessarily, then, require the inclusion of biogenic structures. As ethology (organism behaviour) lies at the heart of applied ichnology, a careful treatment of ichnological characteristics yields crucial insights into the processes that operated in the ancient environment and is vital for the palaeoenvironmental interpretation of the facies.

3.2 Applications of Ichnology to Facies Analysis

3.2.1 Identification of Trace-Fossil Assemblages and Suites

In facies analysis, the integration of ichnology (or of other biogenic structures) helps to highlight subtle differences between broadly similar depositional facies and helps to refine their palaeoenvironmental interpretation(s). Students new to the discipline often ask at what level of detail should trace-fossil observations be made. Like lithofacies analysis, the scale at which such observations are made is dictated by the scope of the study and the level of detail desired.

At the broadest scale is the identification of the *trace-fossil assemblage* and other biogenic features, coupled with an 'average' intensity of bioturbation. The assemblage consists of *all* identified biogenic structures within a sedimentary unit, regardless of their palaeoecological linkages. Trace-fossil assemblages tend to be useful only for the analysis of lithologically and ichnologically homogeneous facies. As facies become increasingly complex (e.g. composite bedsets, sporadically distributed burrowing, vertical and/or lateral changes in the types and/or proportions of ichnogenera), the use of broad-scale trace-fossil assemblages inevitably leads to less accurate characterisation of the facies and typically to less precise or even flawed depositional interpretations.

For composite bedsets, in particular (e.g. heterolithic facies), trace fossils need to be grouped in the context of the bed types with which they are associated. Specifically, this is not merely which trace fossils occur within a bed; but rather, *from which bed they subtend* (i.e. the colonisation surface). For example, in a heterolithic bedset, a mudstone bed might be cross-cut by *Diplocraterion* or *Ophiomorpha*, but such structures may record colonisation of an overlying sandstone bed. The facies analyst must investigate the character of the trace fossil and the nature of its infill and then discern the beds with which these traces are associated. In this way, the researcher identifies the coeval and genetically related groups of traces (the trace-fossil suites) making up the facies and will be able to resolve the combination of temporally and spatially varying physical, chemical and/or biological processes responsible for the observed facies characteristics.

It is valuable to assess the relative abundance of each ichnofossil type comprising the suite. There are various ways to approach this (e.g. quantitative, semi-quantitative and qualitative), but it is generally useful to have a sense of the dominant versus subordinate versus uncommon elements of the suite, particularly when assessing the prevailing organism responses to the sedimentary environment. More experienced workers may assign a quantitative or qualitative value to each element of the suite in order to derive a greater level of detail. As stated earlier in the chapter, however, correct ichnotaxonomy is necessary. Would-be students of the discipline must guard against assigning identifications for uncertain or questionable features. Identifications that are uncertain should be clearly annotated as such in the list of ichnogenera (we use a '?' in front of the name; e.g. ?*Rosselia*), so that an uncertain diagnosis does not carry undue weight in the ultimate characterisation of the facies and its interpretation. Practitioners should not become discouraged when a biogenic feature cannot be identified to a particular ichnogenus, especially in subsurface datasets. Accurate identifications require specific orientations that highlight diagnostic features. Additionally, some large and complex structures may only be transected in core through a small part of the overall feature, precluding their reliable identification. In general, many biogenic structures encountered in cored intervals cannot be reliably named.

3.2.2 Trace Fossils Represent Animal Behaviours

Dolf Seilacher's pioneering work in ichnology laid the foundation for understanding trace fossils as indicators of in situ animal behaviour (ethology) within sedimentary environments. His ethologic classification of trace fossils and the concept of ichnofacies, introduced in English in 1964, revolutionised the discipline of ichnology by emphasising the relationships between animal behaviours and their resulting biogenic structures (Seilacher 1964). Seilacher's ichnofacies highlighted the sensitivity of epifaunal and infaunal communities to various physico-chemical conditions in their habitats, providing insights that continue to shape the discipline of ichnology in the present day.

Many trace fossils encompass a spectrum of behaviours that may blend into one another (e.g. Müller 1962; Frey et al. 1987; Bromley 1996; Gingras et al. 2007; Plotnick 2012; and Vallon et al. 2016). These behaviours are categorised into distinct patterns that reveal the inferred animal strategies (Figure 3.1). The main categories encompass fugichnia (escape), cubichnia (resting), repichnia (locomotion), pascichnia (grazing), agrichnia (farming), fodinichnia (feeding) and domichnia (dwelling) and represent the highest level of trace-fossil classification, many of which were established by Seilacher (1953). Subtle transitions to additional behaviours have also been recognised, including navichnia (sediment swimming) for instratal movement through fluid sedimentary media, digestichnia (defecating) for faecal strings and faecal pellets, calichnia (nesting) for vertebrate and invertebrate colonial dwelling structures, praedichnia (predation) for indications of carnivory, and equilibrichnia (equilibrium-adjustment) for actively repositioned (mainly dwelling) structures owing to incremental but recurring sedimentation and/or substrate erosion. The arrows on Figure 3.1 showcase many of the most commonly associated behaviours. For example, resting structures may shift to locomotion, permanent dwelling to equilibration owing to rapid and repeated sedimentation at the bed, or deposit feeding associated with a permanent dwelling structure.

Correspondingly, many trace fossils encountered in sedimentary facies do not record a single ethology, and the practitioner must always consider the range of ethologies each may represent when attempting to characterise the palaeoenvironmental implications of the ichnological suite (Figure 3.2). To achieve the most highly refined interpretation, these ethological implications are assessed in the context of the prevailing physical characteristics of the sedimentary facies to which they belong. It is important to note that most of these behaviours are linked with the sedimentological characteristics of the depositional environment (e.g. high-energy and/or dynamic conditions or lower-energy and stable benthic habitats), and this integral relationship forms the foundation for the ichnofacies concept (see Section 3.3.2).

Despite the potential of the ichnofossil record to elucidate the evolution of behaviour, this has not been fully realised. Plotnick (2012) pointed to the disconnect between ichnology and modern behavioural biology, lamenting the lack of shared concepts, methods and literature between the two disciplines. Neoichnological work has strived to bridge this divide by more fully integrating the two. This has been particularly useful in the development of continental and coastal margin ichnofacies, owing to the relative ease of investigating the trace-makers responsible for observed biogenic structures. In permanently subaqueous settings, this is less easily achieved, with logistical costs rising astronomically as one investigates increasingly deep-water habitats. Hence, much of what is 'known' about marine benthic ethology, biogenic structures and their linkages to trace-makers is largely inferred (see Frey 1975; Ekdale et al. 1984; Bromley 1990).

3.2.3 Interpreting Burrow Distributions

Interpreting biogenic fabric can be a daunting task for researchers who lack familiarity with ichnology. The apparent chaos observed in highly burrowed sediment, coupled with often unfamiliar trace-fossil taxonomy, can instil a level of apprehension in those attempting to integrate bioturbation with their primary sedimentary fabrics. Despite these challenges, it is important to recognise that straightforward interpretations can be applied to most bioturbated media. Indeed, much of the valuable information latent in the ichnofossil suite is accessible even to non-specialists.

To aid in this interpretation process, a framework for understanding generalised bioturbated textures has been developed (see Figure 3.3). This chart provides a structured approach to deciphering some of the common complexities associated with bioturbated facies.

As ichnologists, our first-order categorisation of lithified sediment encompasses three main types of sediment fabric: burrowed, unburrowed, and chaotic through to apparently structureless (massive). Unburrowed strata often result from processes such as rather extreme sediment shifting, high depositional energies, elevated sedimentation rates or environments characterised by oxygen-deficient conditions.

The initial observations of burrowed media include two key parameters: (i) the distribution of trace fossils; and (ii) the size of the ichnofossils and diversity of the ichnological suite. The distribution of trace fossils is closely linked to the physical sedimentary processes present in the depositional

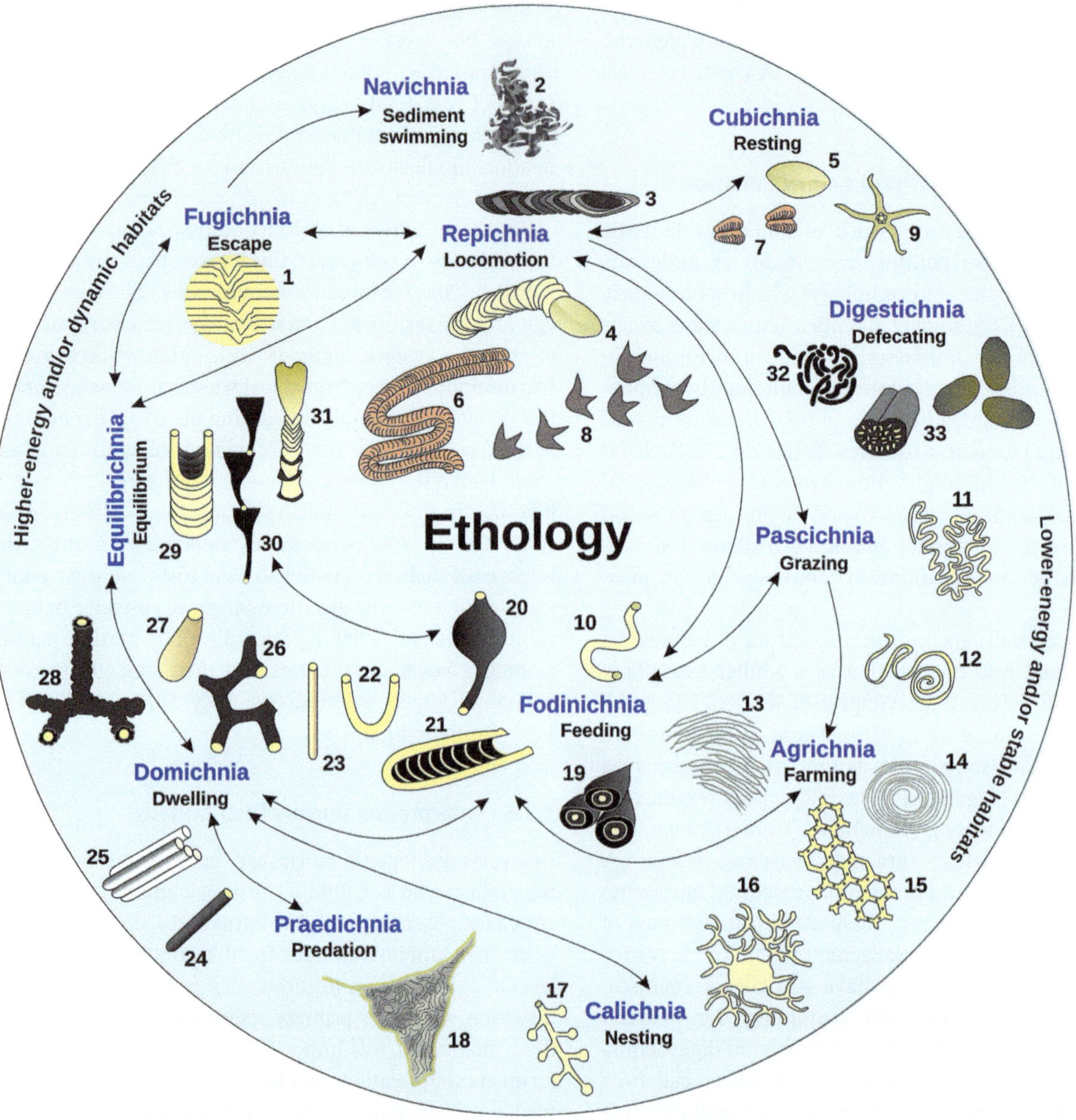

Figure 3.1 Ethological wheel showing the range of faunal behaviours. *Source:* Adapted from Müller 1962; Frey et al. 1987; Bromley 1996; and Gingras et al. 2007. Numerous behavioural styles are overlapping or interact, depending upon details of the depositional setting, and form the basis of the ichnofacies concept. Representative trace fossils and biogenic structures include: (1) fugichnia, (2) navichnia, (3) *Scolicia*, (4) *Protovirgularia*, (5) *Lockeia*, (6) *Cruziana*, (7) *Rusophycus*, (8) tetrapod trackways, (9) *Asteriacites*, (10) *Planolites*, (11) *Nereites* missouriensis, (12) *Spirophycus*, (13) *Urohelminthoida*, (14) *Spirorhaphe*, (15) *Paleodictyon*, (16) *Termitichnus*, (17) wasp nests, (18) *Piscichnus*, (19) *Asterosoma*, (20) *Rosselia*, (21) *Rhizocorallium*, (22) *Arenicolites*, (23) *Skolithos*, (24) *Palaeophycus*, (25) *Schaubcylindrichnus*, (26) *Thalassinoides*, (27) *Gastrochaenolites*, (28) *Ophiomorpha*, (29) retrusive-protrusive *Diplocraterion*, (30) re-equilibrated *Rosselia*, (31) *Siphonichnus*, (32) faecal strings and (33) faecal pellets.

system. Thoroughly or pervasively burrowed sediment, for example, suggests a relatively high infaunal biomass under conditions of low sediment accumulation. In contrast, moderately abundant to rare but evenly distributed trace fossils indicate higher sedimentation rates with a lower biomass, yet in overall constant depositional conditions.

The size of the traces and diversity of the suite in burrowed media offer insights into the chemical aspects of the depositional waters, although high-energy settings feature lowered diversities as well. In marine deposits, for instance, large trace fossils often indicate favourable conditions such as food availability, high dissolved oxygen contents and stable salinity levels. Conversely, diminutive trace fossils typically point to environments characterised by some combination of salinity and/or oxygen stress. The presence of diverse ichnogenera in burrowed media is typically

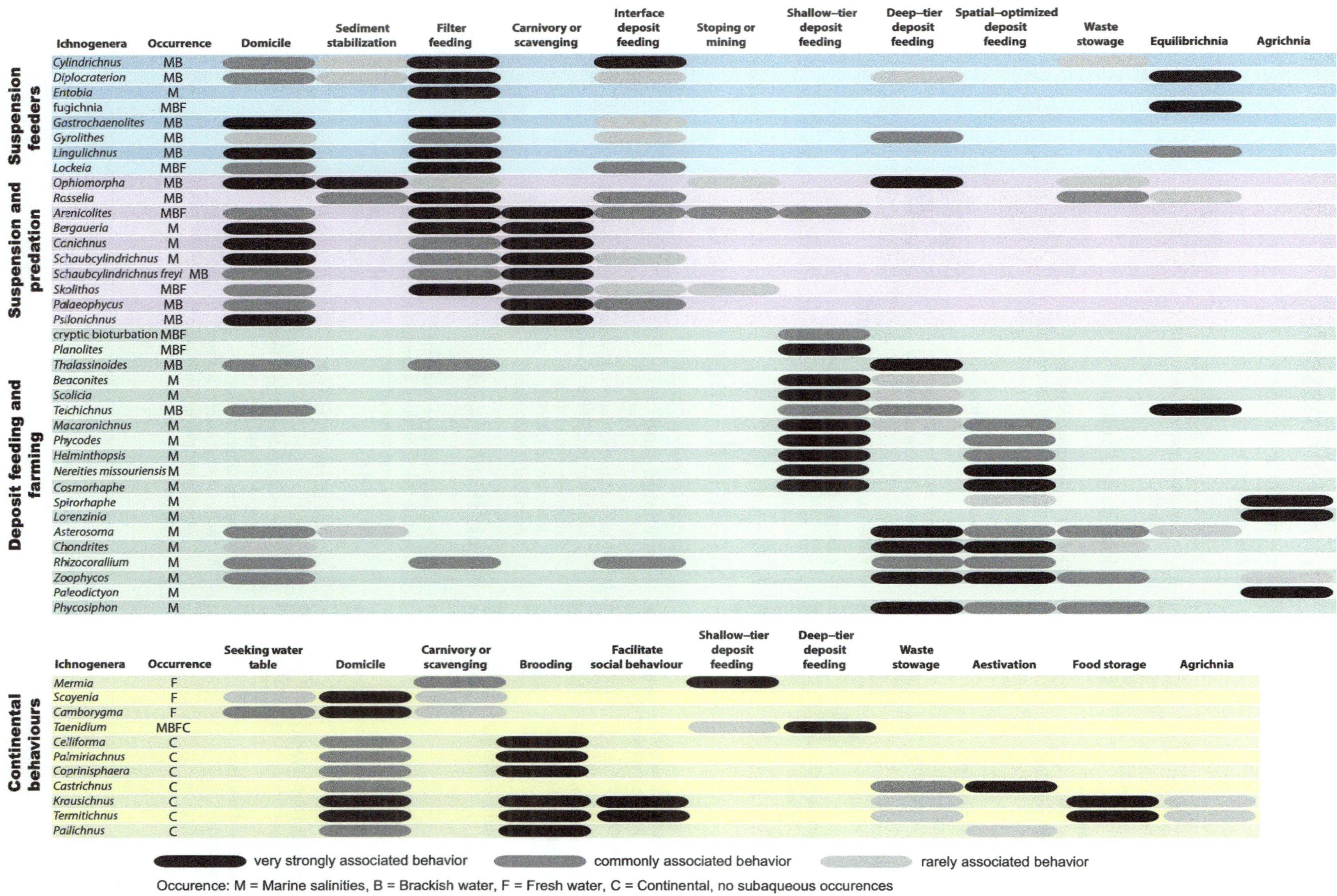

Figure 3.2 Ethological chart of some common ichnotaxa. The chart showcases the range of behaviours assigned to these structures, as well as some of the uncertainty associated with such assignments (see Gingras et al. 2007). Note that the 'Spatial-Optimised Deposit Feeding' also includes surface grazing (pascichnia). Ichnotaxa showing multiple occurrences (e.g. MBFC) and/or a broad range of behaviours tend to be regarded as facies crossing and commonly occur in multiple ichnofacies (see MacEachern et al. 2007). By contrast, those with limited occurrences (e.g. M) and showing a narrow range of behaviours (e.g. agrichnia) tend to be more specialised and environmentally restricted, and so typically occur in fewer ichnofacies. *Source:* Adapted from MacEachern et al. (2010).

Burrowed

BI 5–6, evenly distributed

Suggests some combination of low sedimentation rate and dependable food supply.

- large ichnofossils and high diversity (e.g. proximal offshore, temperate zone intertidal flat)

- small ichnofossils and low diversity (e.g. distal bay, low-salinity lagoon, temperate zone intertidal flat)

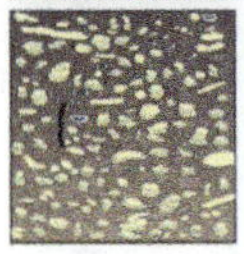

BI 3–4, evenly distributed

Somewhat elevated sedimentation rates and dependable food supply.

- large ichnofossils and high diversity (e.g. bay, distal prodelta)

- small ichnofossils and low diversity (e.g. central basin of a wave-dominated estuary)

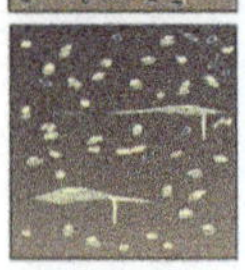

BI 0 and 1–4, sporadically distributed

Variability in sedimentation rates and depositional energies. Potential variance in physico-chemical parameters.

- large ichnofossils and high diversity (e.g. proximal prodelta, distal delta front)

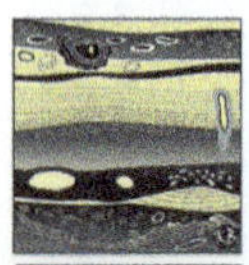

- small ichnofossils and low diversity (e.g. inner estuary)

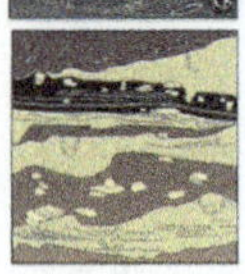

BI 0 and 4–6, lam–scram

Variability in sedimentation rates and depositional energies. Potential variance in physico-chemical parameters.

- large ichnofossils and high diversity (e.g. lower shoreface)

- small ichnofossils and low diversity (e.g. freshened bay)

Mottled/massive

unimodal grain size distribution

- lack of grain size variability explains absence of discernible fabric

load-casted bedding

- normally explained as rapid sedimentation onto a pore–water–rich sediment

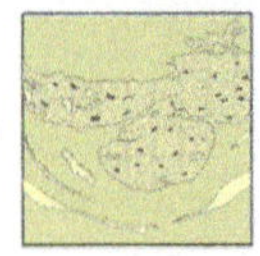

convolute bedding

- best explained as bedding deformed due to slumping

rare burrowed contacts and poorly defined burrows

- cryptic bioturbation

Stratified, unburrowed

planar to wavy parallel laminated sand

- sand deposited under high-energy conditions (e.g. UFR plane bed, storm-generated HCS), accumulated rapidly, and/or deposited in fresh water

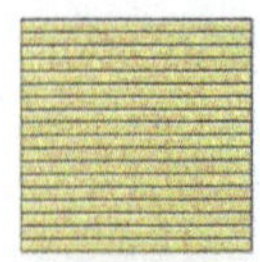

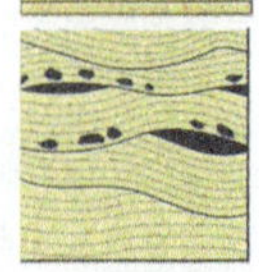

cross-stratified sand

- the sand was deposited under rapid flow conditions, accumulated rapidly, and/or was deposited in fresh water

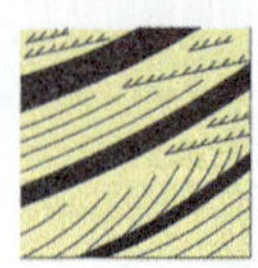

planar laminated mud

- can indicate deposition in low-oxygen settings, high rates of sedimentation, and/or accumulation in fresh water, where penetrative bioturbation is comparatively rare

Figure 3.3 Some simple interpretations that can be drawn from various sedimentary fabrics. Here, the fabrics are categorised into the basic types of sedimentary fabric: burrowed, mottled to massive-appearing and unburrowed and stratified. For burrowed strata, observations focus on bioturbation intensity and ichnofossil distribution and diversity, reflecting sedimentary processes and chemical aspects of the depositional environment. For mottled to massive-appearing sediment, key characteristics include the nature of the bedding contacts and the presence or absence of subtle bioturbation, which are crucial for reaching a viable interpretation. Finally, for stratified media, the absence of bioturbation must be explained as a result of extreme physico-chemical stress such as physical sedimentation conditions or very low oxygenation levels near the bed. Details of the lithology and primary sedimentary structures provide the basis for discerning the likely cause(s) of an environment lacking a benthic community. *Source:* Gingras et al. (2007) / GeoScienceWorld.

indicative of normal marine salinities and full oxygenation of the depositional bottom waters. This diversity is also commonly associated with environments rich in food resources, such as marine algae and plankton. Such environments support a range of feeding behaviours among infauna, contributing to the proliferation of trace-fossil forms.

Within the realm of burrowed media lies an impressive range of variability, however. These may be expressed through a continuum ranging from subtle to marked changes in bioturbation intensity, burrow distribution, trace-fossil diversity and shifts in burrow size within stacked beds and bedsets of the facies, reflecting the magnitude of temporal and spatial variations in the physico-chemical parameters characterising the depositional environment.

In contrast, sedimentary materials that appear structureless (massive) raise questions about their formation processes. Such appearances can be ascribed to various factors, including a lack of sufficient grain-size variation to define stratification, high sedimentation rates, mechanical mixing due to sediment-gravity flows or liquefaction, or extensive biogenic churning. Each of these processes leaves distinct signatures on the sedimentary record, which can be deciphered through careful analysis. Furthermore, the presence of cryptobioturbation (see Section 3.1.1), characterised by vague or even hidden burrow structures or bioturbated bedding contacts, may be particularly challenging to identify, yet is an important observation. In sandstone, cryptic bioturbation is often associated with the rapid and thorough reworking of sandy substrates by infauna and meiofauna (minute but not microscopic organisms) in order to exploit marine-derived planktonic resources (e.g. Pemberton et al. 2008). By contrast, cryptic bioturbation in fine-grained sediment results from successive reworking of slowly aggrading seafloors or intertidal flats, typically by small, shallow-tier or epifaunal animals such as small polychaetes, foraminifera, nematodes and copepods (e.g. Levin et al. 2003; Lacoste et al. 2018).

Overall, understanding the range of common biogenic fabrics is essential for deciphering palaeoecology and process sedimentology. The principles outlined earlier are expanded upon in the process ichnology section later.

3.2.4 Process Ichnology and Its Neoichnological Underpinning

Process ichnology emphasises that ichnofossils are biogenic sedimentary structures which, like primary physical sedimentary structures (e.g., ripple marks, cross-bedding), provide valuable insights into physical aspects of the sedimentary environment, including sedimentation rates, substrate consistency, water turbidity, and hydrodynamic conditions. However, unlike most primary sedimentary structures, ichnofossils uniquely offer interpretations of chemical and biological parameters of past environments, including sediment pore-water chemistry, oxygenation levels, salinity, and nutrient availability. This chemical insight arises from the biological activity of organisms interacting directly with the sediment substrate. Consequently, process ichnology represents an integrative ichnological approach aimed at interpreting both physical and chemical parameters of sedimentary environments. The accuracy and robustness of these interpretations rely significantly on neoichnological studies, which document modern lateral and vertical ichnological trends and the behavioural responses of infauna to observable environmental conditions and stressors.

Many observations that are tied to process ichnology are founded on inductive interpretations of commonly observed phenomena: a good example of this are escape traces (fugichnia), which are interpreted as animal responses to sedimentation events based on their tendency to occur as upwards propagating biodeformational structures that cross-cut sharp bedding contacts. The first 'Process Ichnology' paper was Gingras et al. (2011); however, it and subsequent papers on the topic simply unify many earlier studies that focused on using trace fossils to improve process sedimentological interpretations.

Ichnological datasets that have proven useful to process ichnology studies include:

1) trace-fossil distributions;
2) ethological diversity;
3) the significance of burrow linings;
4) trace-fossil sizes;
5) presence or absence of retrusive or protrusive spreiten;
6) post-depositional compaction of trace fossils;
7) intensity and spatial distribution of bioturbation.

Process ichnology data are preferably collected systematically in vertical associations so that trends in trace-fossil sizes, distribution and diversity can be identified in the context of observed sedimentological changes. From these data, higher-resolution estimates can be made for the determination of sedimentation rates, temporal variation in sedimentation rate, sediment consistency and aspects of the bottom- and interstitial-water chemistries. Gingras et al. (2011), Dashtgard and Gingras (2012) and Timmer et al. (2016) fully outlined the interpretive bases of trace-fossil datasets, the most important of which are presented later.

What sets process ichnology apart is the focus on using ichnological trends to contribute to an understanding of the evolving physico-chemical characteristics of sedimentary environments over time. Such trace-fossil interpretations are not the sole province of process ichnology, and ichnofacies and ichnofabric research constitute critical foundations for assessing process ichnology datasets.

An important characteristic of process ichnology is that the methodologies and interpretations employed are designed to be used by non-ichnologists in a manner akin to the interpretation of physical sedimentary structures. As such, a focus on readily assessed characteristics such as trace-fossil size, bioturbation intensity, ichnodiversity and characteristics of trace-fossil distributions is preferred (Figure 3.3). In this context, the documentation of ichnological characteristics at regular intervals is important. This is particularly appropriate for the description and interpretation of successions from drill core, especially for thick intervals.

3.2.5 Assessing Bioturbation Intensity and Its Spatial Variations

It is a common misconception that intensity of bioturbation and trace-fossil diversity are always positively correlated. This is not the case and so details of the trace-fossil suites or assemblages must be kept separate from assessments of bioturbation intensity. Many sedimentologists evaluate bioturbation intensity in a purely qualitative fashion, employing terms such as 'heavily bioturbated', 'weakly burrowed' and 'thoroughly bioturbated'. There are, however, two frameworks that are semi-quantitative, based on the percentage of primary fabric disrupted by biogenic activity: ichnofabric index (ii), based on the work of (Droser and Bottjer 1986) and Bioturbation Index (BI; Figure 3.4), originally defined as grades of bioturbation by Reineck (1963) and refined and popularised by Taylor and Goldring (1993). The authors of this chapter prefer BI values, although there are numerous proponents of the ichnofabric index. While there are subtle distinctions, they are not materially different and for most purposes, either would serve the needs of the practitioner.

The Bioturbation Index goes from a value of 0 to 6, based on grades of bioturbation deemed discernible to the naked eye. The framework was originally developed using box cores of modern bioturbated sediment. The grades do not follow a uniform pattern of values, nor are there widely accepted subdivisions within each grade, which some workers find problematic. Nevertheless, until a more quantitative and readily measurable framework is established, we have elected to employ the system here. In the framework, BI 0 equates to 0% bioturbated; BI 1 = >0% and ≤5%); BI 2 = >5% and ≤30%; BI 3 = >30% and ≤60%; BI 4 = >60% and ≤90%; BI 5 > 90% and ≤99%; and BI 6 > 99% – 100%. Most researchers employ visual inspection charts that show the midpoint of each grade range (Figure 3.4), but it is clear that a number of BI grades (BI 2–4) have rather large ranges and so might appear markedly different if at the upper end of BI 2 or the lower range of BI 4, for example.

Additionally, whether the BI value appears in sandy versus muddy facies also leads to markedly different expressions. In stratified sandstone units, for example, estimates of BI intensity are significantly easier, as the primary fabric is readily observed, particularly if well stratified. By contrast, it may be more difficult to discern BI values in pure mudstone units, where primary fabric can be exceedingly subtle and there is little in the way of lithologic contrast to highlight biogenic structures. With increasing silt and sand content within mudstone intervals, BI values are more accurately discerned, albeit most workers generally estimate the % of trace fossils and biogenic disruptions rather than the proportion of primary fabric biogenically reworked.

Low BI values generally record more physico-chemically stressful conditions operating in an environment, whereas elevated BI values are generally recorded from more stable and benign settings. The particular stress or stresses, however, may be highly variable and so not readily interpretable based only on BI. The integration of the trace-fossil suite and particularly the diversity and the proportion of specialised versus facies-crossing elements with the BI values is crucial for any meaningful interpretation of the environmental conditions. That said, stable environments rather than highly dynamic ones, settings with slow rates of deposition, and the presence of uniform salinity (in the marine realm), availability of water (in the continental realm) and prevalence of oxygen all favour facies displaying higher BI values. Increases in deposition rate, physical energy operating at the bed, reductions in oxygenation, reductions and fluctuations in salinity, as well as paucity of food among others will tend to lead towards lowered BI values. Which of these stresses or combination of stresses might be operating in the environment can be more readily discerned in the context of the ichnological suite as well as the physical sedimentology of the facies. An effective facies analyst must make use of all available data.

Newcomers to the discipline commonly ask at what scale one should record BI values. Should they be recorded at the laminaset scale, bed scale, bedset scale or facies scale? In the case of ichnofabric analysis, for example, a BI value is assigned to each ichnofabric encountered (see Section 3.3.1). In more facies-based studies, the level of detail runs the full range listed above. In our experience, greater detail is preferable to a more generalised analysis, though obviously there are limits, which again, are dictated by the level of detail being undertaken in a study. In most cases, it is better to give a range of BI for a unit. The more homogeneous the sedimentary unit happens to be, the more a single value or a narrow range of values is likely to be sufficient. However, as a unit becomes increasingly variable, the recording of a single value or even a range of values becomes less useful from the perspective of interpretation. If a facies were to vary from BI 1 to 4 (5–90% burrowed), one would want to know which components of the facies tend to be more thoroughly burrowed and which

Bioturbation Index (BI): A measure of bioturbation intensity				
%	BI	Characteristics	Mudstone facies	Sandstone facies
0	0	Bioturbation absent		
1–5	1	Sparse bioturbation, bedding distinct, few discrete traces		
5–30	2	Uncommon bioturbation, bedding distinct, low trace density		
30–60	3	Moderate bioturbation, bedding boundaries sharp, traces discrete, overlap rare		
60–90	4	Common bioturbation, bedding boundaries indistinct, high trace density with overlap common		
90–99	5	Abundant bioturbation, bedding completely disturbed (just visible)		
99–100	6	Complete bioturbation, total biogenic homogenization of sediment		

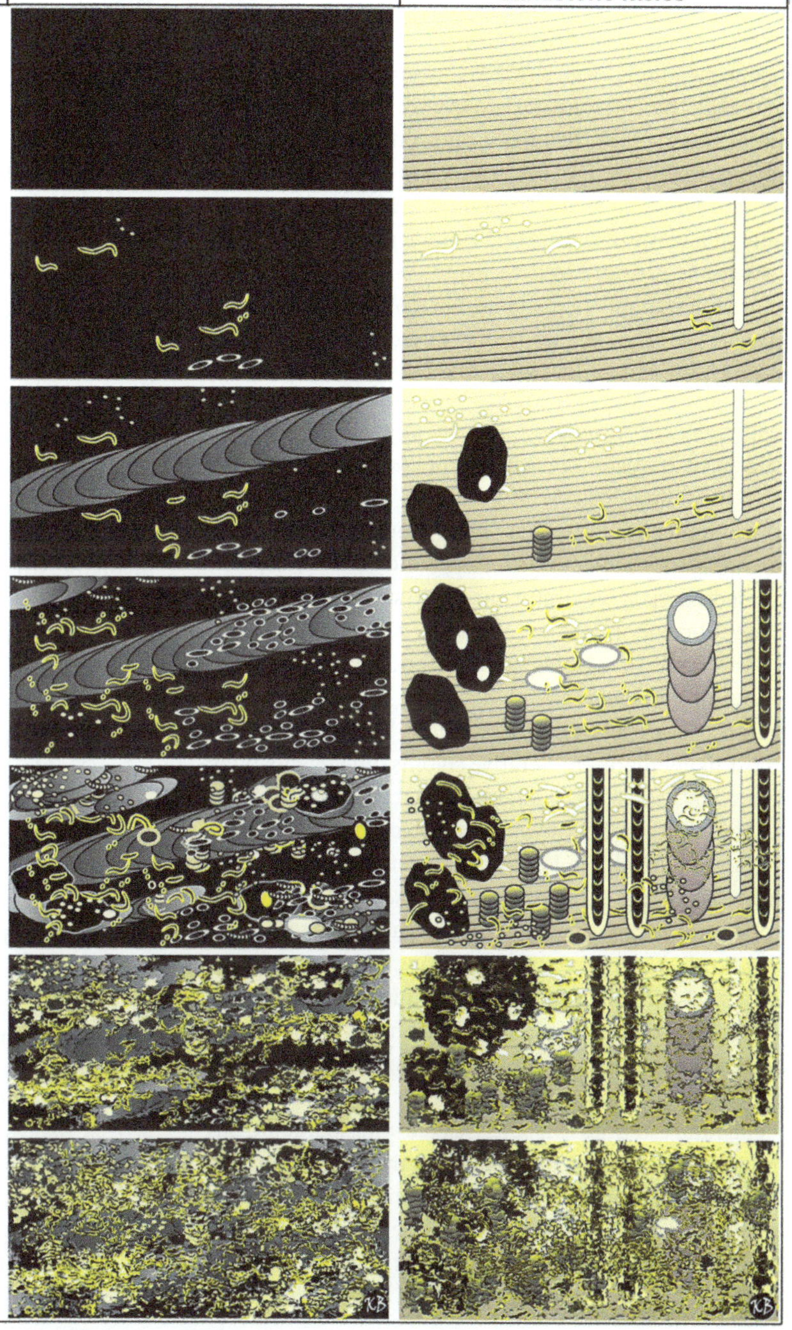

Figure 3.4 Bioturbation Index (BI) scale, used for assessing grades of bioturbation intensity. Diagram shows the range of primary fabric disrupted by biogenic structures (%), the BI index from 1 to 6, general characteristics of the BI value, and a visual expression of the types of fabrics encountered within mudstone facies dominated by horizontally oriented burrows, and sandstone facies dominated by both vertically and horizontally oriented burrows. *Source:* Adapted from Reineck (1963); Taylor and Goldring (1993) and Bann et al. (2004).

tend to be only sparsely bioturbated. What is the nature of this variability? Do the variations follow particular layer types (beds, laminaset or stratification types) or is the range of BI related to progressive or incremental changes to a succession of beds? When a facies consists of a heterolithic bedset, for example, it is generally more useful to provide a BI range for *each* lithologic unit. In this way, specific details of the suite as well as the BI values to which they are related are more likely to provide the necessary level of detail required to provide an effective means of interpreting each facies component and, in turn, a more accurate assignment of the depositional environment.

Trace fossils can be distributed: (i) homogeneously; (ii) regularly heterogeneously; or (iii) sporadically heterogeneously (Gingras et al. 2011). Homogeneous distributions imply that the trace fossils are evenly distributed at the scale of observation. Regular heterogeneous distributions recognise that trace fossils are concentrated along or within bedding planes, so that the bioturbated zones are spaced evenly in a vertical succession. Sporadically heterogeneous distributions are used to describe trace-fossil distributions that are not spatially predictable. Each of these distribution patterns has significance from a sedimentary process viewpoint.

Homogeneous distributions: If food is available and dissolved oxygen contents sufficient, continuous and slow sedimentation lead to bioturbated, evenly distributed trace fossils. Continuous, highly burrowed sediment can be associated with slope, shelf and offshore deposits. In sedimentary environments that suffer variations in hydraulic energy, sediment supply, salinity fluctuations or variations in bottom-water oxygen, homogeneous distributions are generally absent.

Regularly heterogeneous distributions: Such distributions most likely represent iterative colonisation events punctuated by depositional conditions that are less favourable to colonisation. This style of bioturbation is the result of regular variability in local physico-chemical parameters, such as settings influenced by tidal, seasonal/annual, or longer-term climatic (e.g. El Niño/La Niña) cycles.

Sporadically heterogeneous distributions: Persistent, spatiotemporal variability in physico-chemical conditions lead to sporadic heterogeneous distributions. BI is characteristically variable. Event beds associated with turbidites, tempestites, hyperpycnites and recurrent river flooding into deltas and estuaries can be associated with sporadic distributions. Trace fossils associated with these sedimentological event beds typically comprise ichnogenera made by an infaunal community that was abruptly buried; albeit more rarely, animals carried in the sediment flow also may colonise bed tops following emplacement of the event bed.

3.2.6 Identification of Oxygen Stress

In marine and paralic realms, reduced oxygen settings include the continental slope and distal shelf, thermohaline-stratified lagoons and, to a lesser degree, stratified bays and estuaries and abandoned tidal and distributary channels. All animals are aerobic, and so the very presence of bioturbation attests to at least intermittent oxygen availability at the sediment–water interface. Lowered dissolved oxygen in the bottom waters is clearly expressed in the ichnological assemblage, through aspects such as: (i) reduction in ichnogenera diversities; (ii) reduced trace-fossil abundances and lower BI values; (iii) decreasing burrow diameters; and (iv) reduced depth of burrow penetration into the substrate (e.g. Rhoads and Morse 1971; Bromley and Ekdale 1984; Diaz and Rosenberg 1995; Savrda 2007). These characteristics are all consistent with a physiological trend of diminution (size reduction of the animal and correspondingly, their traces), as the ability of the environment to meet the metabolic demands of the infaunal assemblage is progressively lowered.

Neoichnological studies of lowered oxygen settings are limited, but they do confirm that lowered diversity and progressive diminution follow lowered oxygen gradients. Rhoads and Morse (1971) synthesised observations from low-oxygen regions of the Black Sea, Gulf of California and adjacent to Southern California, and identified three biofacies associated with different concentrations of dissolved oxygen: (i) an azoic region ($\leq$0.1 ml l^{-1}); (ii) a low diversity, small, soft-bodied infauna (0.3–1.0 ml l^{-1}); and (iii) a diverse (calcareous) fauna (>1.0 ml l^{-1}).

There is strong evidence suggesting that even moderate oxygen stress can impact the burrowing assemblage. Studying a muddy shelf in the Gulf of Mexico, Dashtgard et al. (2015) showed that the diversity and density of burrows correlated with the dissolved oxygen concentration of the bottom water. The degree of bioturbation and trace diversities were substantially lowered in waters that contained dissolved oxygen levels below 4.3 ml l^{-1}.

The analysis of very small-scale bioturbation with petrographic techniques has been used to refine interpretations of organic-rich shale units. For example, Egenhoff and Fishman (2013) reported the presence of ichnofossils through nearly all depositional environments in the Devonian upper Bakken Shale Member, suggesting dysoxic conditions during its deposition. Likewise, Biddle et al. (2021) demonstrated that microscopic bioturbation could be observed in several levels of the Devonian Canol Formation and posited that intermittent oxygen windows allowed sporadic sediment colonisation. In both examples, burrows were sub-mm-scale, indicative of deposit feeding, and existed in low-diversity suites of one or two trace-fossil types.

Several papers have contended that some ichnogenera can be associated with reduced-oxygen settings, although some studies question the viability of assigning any particular ichnogenera with lowered oxygen (e.g. Wignall 1993; Martin 2004). The ichnofossil most commonly ascribed to low-oxygen settings is *Chondrites* (e.g. Bromley and Ekdale 1984). The linking of *Chondrites* to lowered oxygen settings is, in part, based on the fact that it is characteristically small in diameter and does commonly occur in low-diversity assemblages. Bromley and Ekdale (1984) suggested that '*Chondrites* typically represents the last trace fossil in a bioturbation sequence. This indicates that the burrow system was produced deep within the sediment in the anaerobic zone below the surficial oxidised zone that was characterised by freely circulating and oxidizing pore waters'. This study has been mischaracterised as evidence that *Chondrites* is a low-oxygen associated trace fossil. However, this disregards the fact that in fine-grained sediments, many modern polychaetes readily burrow into suboxic sediments simply by irrigating their burrows from oxygenated waters above the sediment–water interface (Zorn et al. 2006). It is worth noting that burrows that are actively backfilled and isolated from the sediment–water interface rely on dissolved oxygen in the pore waters. As such, some trace fossils like *Scolicia*, *Planolites* and *Taenidium* are likely to be excluded from low-oxygen assemblages.

Bromley and Ekdale (1984) suggested that suites dominated by *Chondrites* may be indicative of strongly reduced oxygenation (e.g. dysoxia) and suggested that because infaunal elements tend to become diminutive under depleted oxygen conditions, irrigation and aeration of the sediment substrate also declines, leading to a rise in the position of the redox potential discontinuity. This, in turn, reduces the vertical extent of burrows, resulting in shallow tier bioturbation. Anoxic conditions lead to no bioturbation whatsoever, and the accumulation of organic-rich, well-laminated sediments. Excluded are those forms of *Chondrites* with a preferred chemosymbiotic lifestyle as, for instance, described by Knaust (2009). Using late Palaeozoic deep-water facies, Ekdale and Mason (1988) suggested that certain types of trace fossils could be used to assess a relative degree of oxygen reduction. Dwelling burrows, such as *Thalassinoides*, *Rhizocorallium* and *Diplocraterion* evidence well-oxygenated conditions. Dysoxia leads to trace-fossil suites dominated by grazing traces, including low-diversity assemblages of *Gordia*, *Helminthopsis*, *Planolites* and *Phycosiphon*. Finally, dysaerobic bottom waters coupled with anoxic pore waters were accompanied by a switch from grazing to deep-tier deposit-feeding (e.g. *Chondrites* and *Zoophycos*).

Importantly, Martin (2004) pointed out that the majority of ancient dysaerobic ichnocoenoses have not been linked to independent evidence of palaeo-oxygenation (such as geochemical redox indicators). As such, the associated ichnofossil assemblages may be controlled by factors other than oxygenation, such as substrate consistency. Savrda (2007) stated that organism responses to oxygenation gradients may be more complex than previous models supposed. Some oxygen minimum zones, for example, show that: (i) bioturbation may be intense even at low levels of oxygenation (Levin et al. 2003); and (ii) burrow depths and burrow diameters may not correspond directly to benthic oxygenation (Smith et al. 2000). Increasingly, research efforts endeavour to integrate geochemical redox proxies with bioturbation trends and so our understanding of low oxygen trace-fossil assemblages continues to improve (e.g. Birchenough et al. 2012; Dunkel et al. 2022).

To summarise, the most compelling ichnological evidence for low oxygen levels in depositional environments can be observed through trends in the reduction of trace fossil sizes and decreased ichnodiversity, possibly in conjunction with the presence of characteristic trace fossils such as *Chondrites*, *Gordia* and *Helminthopsis*. This evidence is strengthened further when correlated with geochemical indicators for redox conditions, such as chemical redox indicators and the high preservation of organic matter.

3.2.7 Identification of Salinity Stress

Brackish-water-associated trace-fossil assemblages were first recognised by Pemberton et al. (1982). Later studies (see Beynon et al. 1988; Buatois et al. 2005; MacEachern and Gingras 2007) established the recognition criteria for brackish-water trace-fossil assemblages:

- Brackish-water settings typically show a reduction in the number and diversity of animal species. This results from the need to osmoregulate in lowered and fluctuating salinities. Not all animals can osmoregulate, and for those that can, there is a metabolic cost.
- Brackish-water suites primarily consist of marine organisms. Although some insects have radiated into hypersaline environments, marine organisms dominate the brackish realm. A major reason for this is that there are many species of marine animals that have become well adapted to living in brackish water: insects are outcompeted.
- Settings characterised by reduced and/or fluctuating salinity show a pronounced size reduction of fauna compared to their fully marine counterparts. This, again, relates to the metabolic cost of osmoregulation and selective pressures favouring smaller animal sizes. Additionally, salinity-stressed environments are characterised by elevated mortality rates in fauna, resulting in a higher proportion of biogenic structures produced by juvenile rather than adult individuals.
- Brackish-water benthic communities contain a predominance of infaunal rather than benthic organisms. It is thought that infaunalisation dampens the magnitude of salinity fluctuation experienced by the animal, because pore water salinities vary less markedly than do the waters above the sediment–water interface.
- Most fauna inhabiting brackish-water settings correspond to omnivores or trophic generalists. There is an abundant food supply brought in by tidal currents that de-emphasises the importance of specific feeding behaviours.

- Despite the reduced diversity of species inhabiting brackish-water settings, many marginal-marine environments support a large biomass. This, again, can be related to the presence of an abundant and regularly replenished food resource.
- The fauna of brackish-water environments predominantly employs an r-selected (i.e. opportunistic) strategy in population recruitment. Rapid colonisation styles are required in what can be a daily to seasonally dynamic sedimentary environment.

The scientific and geological bases for the brackish-water model are summarised in MacEachern and Gingras (2007). Notably, such assemblages have been recently recognised as sufficiently recurrent that they are now recognised as attributable to the newly erected *Teichichnus* Ichnofacies (Gingras et al. 2025). The details are summarised in the ichnofacies part of this chapter in Section 3.3.2.

Finally, it is important to recognise that ichnological trends towards diminution and lowered diversity can be caused by stresses other than those related to salinity. Salinity stress is discernible from lowered oxygen conditions (discussed in Section 3.2.6) in two ways: (i) some ichnogenera (e.g. *Arenicolites*, *Cylindrichnus*, *Gyrolithes*, *Skolithos* and *Teichichnus*) are common in salinity stressed environments, but not typically observed in oxygen-stressed locales; and (ii) salinity stress is normally associated with paralic settings and therefore may be accompanied by many of the facies elements associated with the continent-to-marine transition (e.g. subaerial exposure surfaces, evidence for waves and tides, an abundance of terrestrial organic detritus).

3.3 Ichnological Frameworks

Trace-fossil researchers have two main approaches to ichnological analysis: an ichnofabric approach and an ichnofacies approach. Each brings their own unique perspective into the utility of biogenic structures for describing and interpreting the sedimentary record.

3.3.1 The Ichnofabric Approach

The ichnofabric concept arose in the 1980s, primarily in an effort to describe bioturbation distributions and sedimentological context outside of the ichnofacies framework. The reasons for avoiding ichnofacies were 2-fold: (i) at that time, ichnofacies were seen as flawed, owing to a debated association with bathymetry; and (ii) the interpretive value of departures from archetypal ichnofacies was not widely understood, and so ichnofacies were seen as too confining to capture ichnological variation in many sedimentary environments. The latter point was put forth by Ekdale (1988) '... not all environmental situations are represented in the nine standard ichnofacies, so many trace-fossil associations in the real world cannot be easily fit into those categories'.

Through the 1980s, the concept of ichnofabrics was further developed. A definition of the term is provided by Bromley and Ekdale (1986): 'Ichnofabric is that aspect of a sediment's texture and internal structure that arises from bioturbation and bioerosion at all scales (Ekdale and Bromley 1983). A suite of organism traces (biogenic sedimentary structures) represents the preserved evidence of the activities (feeding, crawling, dwelling, etc.) of a particular benthic community that inhabited the sediment at any one time'. Later on, the definition of ichnofabric came to include the 'overall texture of a bioturbated sediment' (Taylor et al. 2003). It is worth noting that ichnofabrics and ichnofacies are defined on the basis of similar observations. The major criterion that differentiate the practices is that an ichnofacies is required to be recurrent through space and time and is treated as a framework to which trace-fossil suites are assigned.

The ichnofabric method focuses on the identification of trace fossils and the establishment of their cross-cutting relationships. Cross-cutting trace fossils are used to interpret depth (tier) relationships. An important tenet of the ichnofabric concept is that organisms that share depth tiers exploit common resources using similar behaviours. This underpins the ichnoguild concept, which is based on the trace-fossil morphologies, the (interpreted) nature of the food resource exploited and observed spatial usage (summarised in Bromley 1996). Ichnofabric analyses aim to provide high-resolution food-exploitation interpretations of ichnofossil assemblages, which in turn refines characterisations of prevailing sedimentary conditions, because the link between tiering and colonisation patterns reflects the nature of food distribution in, on and above the seafloor.

Beyond establishing depositional energies, very successful efforts have been made in discerning the link between resultant ichnofabric and oxygenation (e.g. Bromley and Ekdale 1984; Wignall 1991; Savrda 2007). Ichnofabric studies also address, to lesser degrees, sedimentation rate (Bromley 1996; Taylor et al. 2003), substrate consistency (Goldring 1995) and the impact of diagenetic processes (Taylor et al. 2003). Finally, in refining local (depositional) environmental interpretations, ichnofabrics have been used to enhance sequence stratigraphic interpretations (Malpas et al. 2005; Morris et al. 2006).

Ichnofabric data collection is meant to be on a bed-by-bed basis (Figure 3.5). Ideally, this is done by producing an ichnofabric constituent diagram for each bed (Figure 3.5 – lower left; Taylor and Goldring 1993). The constituent diagram

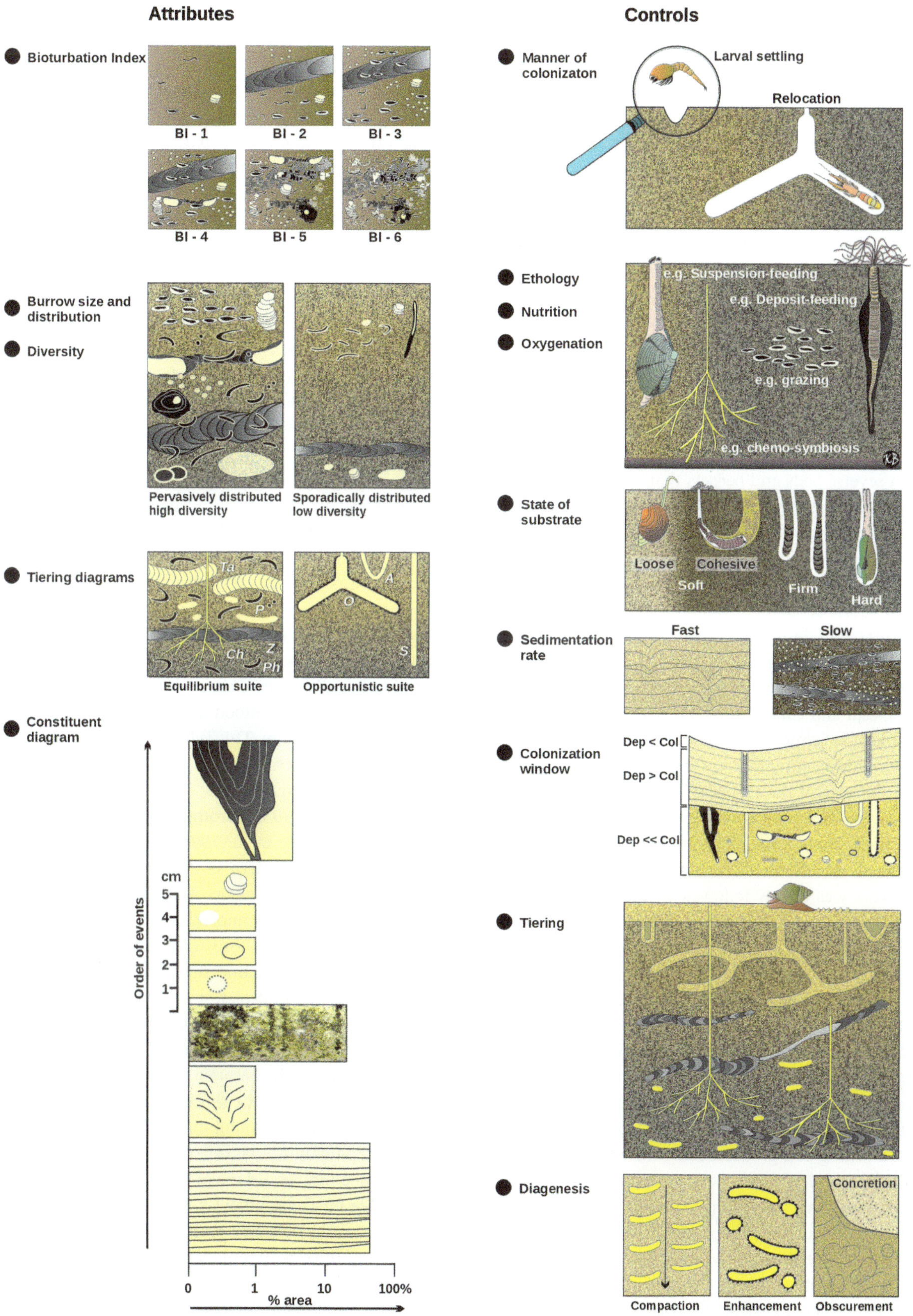

Figure 3.5 Framework for ichnofabric analysis, outlining ichnological attributes and animal-sediment controls. *Source:* Taylor et al. 2003 and McIlroy 2004. Aspects of bioturbation index, trace-fossil size, trace-fossil diversity and tiering relationships are crucial for construction of a constituent diagram. See text for an explanation of the construction of the constituent diagram and assignment of the resulting ichnofabric.

provides a visual means for describing and comparing ichnofabrics by: (i) assessing the remaining primary fabric; (ii) documenting the tiering; and (iii) recognising the type of bioturbation. The collected data are presented along the horizontal axis that reports the area % occupied by the sedimentary structures and burrow types. On the vertical axis, events are shown in ascending order, from initial sedimentation to subsequent modification by bioturbation. The order of burrowing is obtained from observed cross-cutting relationships. Relative thicknesses of the sedimentary units and the burrow diameters are also shown on the vertical axis. In short, constituent diagrams record many observations that may be useful for environmental interpretation.

An initial goal of ichnofabrics and ichnofabric constituent diagrams was to enable bed-by-bed sedimentological interpretations of sedimentary successions. In practice, this is seldom done, because sedimentary environmental interpretations using ichnofabrics, as with ichnofacies, are drawn from several beds and bedsets that form facies successions or facies associations. Additionally, it has been noted that ichnofabrics for many settings may be highly patchy in their distributions, making their mapping or correlation to adjacent units problematic (McIlroy 2007). Finally, ichnofabric constituent diagrams are time consuming to construct, and despite their criticality in defining an ichnofabric, very few published studies have been based on such data collection (notable exceptions include Callow et al. 2013; Desai and Chauhan 2021).

An example of an ichnofabric constituent diagram is seen in Figure 3.5. This diagram displays the order of events and the final percentage of the features. It can be ascertained that the bed consists of moderately bioturbated, laminated sandstone. First, wavy parallel laminated sandstone was deposited. The lamination makes up 60% of the final fabric of the bed. This was closely followed by the formation of fugichnia that were generated when organisms disrupted the primary lamination in their attempt to escape burial during the rapid accumulation of the sand. Fugichnia make up 1% of the final fabric of the bed. Organisms then reworked the substrate but did not leave any discrete structures. The result is a mottled burrow texture that makes up 30% of the final fabric. After this, a series of tubular burrows were introduced, notably *Macaronichnus* isp. (representing a deposit-feeding structure), *Palaeophycus tubularis* (representing the dwelling of a predaceous carnivore), *Planolites* (representing a mobile deposit-feeding structure) and *Teichichnus* (representing a dwelling structure of a deposit feeder). These four burrow types make up 4% of the final fabric of the bed. The final structure to be produced, and which cross-cuts all other features, is robust *Rosselia socialis* (a dwelling and feeding structure of a surface detritus-feeding organism). This makes up 5% of the final fabric of the bed. The naming of ichnofabrics is somewhat subjective, so different workers might choose different names for similar ichnofabrics, but in this instance, one might choose something like 'mottled burrow-*Rosselia socialis* ichnofabric'. The difficulty with this method of analysis arises when assigning the ichnofabric to a particular environmental interpretation. Although various parameters are shown in the figure, such as probable storm deposition (wavy-parallel lamination), episodic deposition (fugichnia) and return to fair-weather conditions (mottled burrow fabric and subsequent bioturbation), no definitive palaeoenvironmental interpretation can be made without taking into account the sedimentological and ichnological features of the adjacent beds. Upon doing this, the resulting suites can be assigned to particular ichnofacies that will more accurately pinpoint the likely depositional environment.

3.3.2 The Ichnofacies Approach

The ichnofacies concept is a multidimensional framework underpinned by recurring, facies-controlled (i.e. environmentally related) groupings of biogenic structures reflecting animal responses to the physical, chemical and biological characteristics of a depositional environment (see MacEachern et al. 2007, 2012a, and Buatois and Mángano 2011 for reviews). The ichnofacies concept has evolved significantly since Seilacher (1953) but remains one of the most misunderstood and misused concepts in the discipline of ichnology. This is largely owing to various workers having approached the concept in markedly different fashions (e.g. Seilacher 1967; Frey 1978; Byers 1982; Pemberton and Frey 1985; Bromley and Asgaard 1991; Goldring 1993; Lockley et al. 1994; McIlroy 2004; Hunt and Lucas 2007; MacEachern et al. 2007, 2012a; MacEachern and Bann 2020). Lockley et al. (1994) argued that there are three scales of ichnofacies (only two of which are mentioned here), with those at the finest scale broadly corresponding to palaeoichnocoenoses (i.e. trace-fossil suites) and those at the broadest scale reflecting the so-called 'Seilacherian ichnofacies' (*sensu* Bromley 1990).

The palaeoichnocoenose scale of usage has little application beyond a local study, because it can only characterise units within a particular rock succession. While it may be useful for identifying and mapping individual facies in a single stratigraphic unit, it lacks both geographic and temporal recurrence that would make such constructs universally applicable for facies analysis. Usage of ichnofacies at this scale mirrors the outdated practice of establishing stratigraphically confined sedimentary facies (e.g. Moore 1949). Today, the use of 'facies' is almost exclusively unconfined, permitting its application to rocks of any age and at any location on Earth. While the details of a particular trace-fossil suite are certainly crucial in the characterisation and interpretation of an individual facies, such suites are more broadly understood within the context of the archetypal ichnofacies – an approach we deem preferable to the erecting of a new ichnofacies for every stratigraphic unit in each geographic location.

It is at the broadest scale wherein ichnofacies display both temporal and spatial recurrence, are stratigraphically unconfined and therefore, we believe, have the greatest utility in facies analysis. We regard the archetypal Seilacherian ichnofacies to correspond to theoretical constructs at the hierarchical level approaching that of facies models, encompassing animal-sediment interactions in response to the physical processes and chemical conditions that operated within the depositional environment or subenvironment. Trace-fossil suites, by contrast, more closely track the details of discrete benthic communities that occupied and disrupted the beds or bedsets of an individual sedimentary facies. Hence, ichnotaxa making up the different suites may vary markedly from geographic location to location, and certainly throughout Earth history, but the ethological *implications* of these elements ultimately allow their assignment to an ichnofacies. Hence, no trace-fossil suite is of itself an ichnofacies; nor does it define one.

One of the earliest expressions of the original marine ichnofacies was in the context of changing animal responses as a function of variations in water depth. Although it seems clear that Seilacher (1967) intended that the distribution of these ichnofacies be regarded in the context of proximal–distal relationships rather than to water depth *per se*, some workers regarded the trace fossils within the ichnofacies to be palaeobathymeters (e.g. Weimer and Hoyt 1964; Farrow 1966; Pickett et al. 1971), leading some workers to imply a bathymetric control on ichnofacies distributions (Lockley et al. 1987; Ekdale 1988). It was soon apparent, however, that bathymetry itself has little direct influence on the distribution of trace-fossil suites (or ichnocoenoses) and therefore on their requisite ichnofacies. Frey et al. (1990) argued that the ichnofacies relationship to water depth is entirely passive, in that any relationship with bathymetry is at best indirect. Organisms respond to a number of environmental factors (e.g. substrate consistency, sediment grain size, energy conditions at the bed, food resource types and availability, water turbidity, water salinity, depositional rates and oxygenation), which *tend* to change with water depth. How the trace-makers responded to these changing conditions is expressed by the particular trace-fossil suites encountered. The recurring organism responses to the combination of temporally and spatially recurring physico-chemical factors within the environment form the modern basis of the Seilacherian ichnofacies (e.g. Frey and Pemberton 1987; Frey et al. 1990, MacEachern et al. 2007, 2012a). The continental ichnofacies, with the possible exception of suites attributable to the *Mermia* Ichnofacies, are independent of such proximal–distal trends.

Earlier expressions of the ichnofacies concept, which were dominated by marine ichnofacies, also appealed to a single over-arching control on organism behaviours and, therefore, on the resulting suites of each ichnofacies. This is now recognised to be overly simplistic and masks the true scope of the concept's validity. Furthermore, the initial approach focused primarily on the food-resource paradigm for differentiating marine ichnofacies, which mainly considered the role of physical energy operating at the sediment–water interface. This gave the false impression that food delivery and substrate consistency/grain size were the sole controls on benthic communities and bioturbation patterns. For example, high-energy and shifting non-cohesive substrates with food particles held in suspension, leading to the benthic community being dominated by permanent vertical dwelling for the purpose of filter feeding, is the basis of the *Skolithos* Ichnofacies. More recently developed marine ichnofacies (e.g. the *Rosselia* and *Phycosiphon* ichnofacies of MacEachern and Bann 2020 and the *Teichichnus* Ichnofacies of Gingras et al. 2025), however, have demonstrated the need to incorporate numerous other aspects of the sedimentary environment in order to explain their distinction from the original Seilacherian ichnofacies, which has highlighted a deeper insight – that the original marine ichnofacies (e.g. *Skolithos* Ichnofacies, *Cruziana* Ichnofacies and *Zoophycos* Ichnofacies) pre-supposed environmental settings that were also characterised by fully marine salinities, stable environmental conditions, uniform sedimentation rates and persistent levels of dissolved oxygen, among others.

It must be made clear that each ichnofacies is not diagnostic of any single sedimentary environment, but rather reflects the complex interactions of organisms in response to the spatial and temporal variations of the physical, chemical and biological factors that operated within the environment. The *Skolithos* Ichnofacies, for example, is not diagnostic of a wave-dominated shoreface. While suites from such an environment might be attributable to the *Skolithos* Ichnofacies, those of other shallow-marine, sand-dominated palaeoenvironments (e.g. sheltered sand spits, washover fans or barrier islands) may likewise be assigned to the ichnofacies. It is the faunal responses to a set of particular environmental conditions that define an ichnofacies. The more adept the practitioner is in assessing the 'process ichnology' of a trace-fossil suite (Gingras et al. 2011), the more accurate will be the palaeoenvironmental interpretation. Just as the assignment of a facies model to a sedimentary succession does not constitute the ultimate goal of facies analysis, but rather serves as the framework within which facies can be appraised (see Chapter 2), so too do Seilacherian ichnofacies only comprise an overarching framework within which stacked trace-fossil suites can be evaluated and interpreted. Like sedimentary facies models, the ichnofacies must be able to accommodate the breadth of animal-sediment responses to the full range of processes operating within the depositional environments. To achieve this, case studies from modern environments and ancient successions must be distilled (Figure 3.6; see Walker 1984; MacEachern et al. 2010) to showcase their commonalities. To reiterate, in order to be a viable ichnofacies, it must recur through geologic time and have widespread geographic distribution.

3.3.2.1 Assigning Trace-Fossil Suites to a Seilacherian Ichnofacies

It is a common misconception that the ichnogenus serving as the namesake of an ichnofacies must be present in the suite in order to assign it to the ichnofacies. Neither is it the case that the mere presence of the namesake ichnogenus requires the suite to be assigned to that ichnofacies. Many ichnogenera occur in more than one ichnofacies. Rather, the namesake ichnogenus is merely a placeholder that represents the most common ethology deemed characteristic of the ichnofacies. Any suite dominated by ichnogenera corresponding to the full range of organism behaviours diagnostic of an ichnofacies meets the requirement for its assignment to that ichnofacies. It is the *ethological implications* of the ichnotaxa present that ultimately allow assignment to an ichnofacies (see Figure 3.2).

The authors recommend that in any study the details of the trace-fossil suite be identified first. This includes taxonomic identifications as well as the relative abundance of each element. These, combined with bioturbation intensities and their spatial distributions, provide crucial information regarding the sedimentary environment. Nonetheless, the assignment of a trace-fossil suite to its appropriate ichnofacies is, at some level, somewhat subjective. Many ichnogenera are less environmentally restricted and therefore may occur in suites reflecting different Seilacherian ichnofacies. In many cases, the relative proportion of ichnogenera must be assessed before a trace-fossil suite can be assigned to an ichnofacies. The lower the diversity of the trace-fossil suite, the more challenging it is to assign to an ichnofacies, although there are exceptions (e.g. *Skolithos-Diplocraterion* pipe rock attributable to the *Skolithos* Ichnofacies). As a rule of thumb, however, suites that consist of only three or less ichnogenera may defy reliable assignment to an ichnofacies and should be assessed solely on the basis of the components present, rather than the broader palaeoecological implications hosted by an inferred ichnofacies. Additionally, the more facies-crossing (less environmentally restricted) an ichnogenus happens to be (e.g. *Palaeophycus*, *Planolites*, *Skolithos*, *Teichichnus* and *Thalassinoides*), the less influence it exerts on determining the most appropriate ichnofacies for the suite. A low-diversity suite consisting solely of *Planolites*, for example, defies any meaningful ichnofacies assignment. By contrast, ichnogenera such as *Asterosoma*, *Cosmorhaphe*, *Lorenzinia*, *Scolicia* or *Zoophycos* reflect more specialised ethological strategies that are more restricted environmentally and so impart a greater influence on the suite's ichnofacies assignment. Such ichnogenera accompanied by greater diversity of elements are more readily assigned to an archetypal ichnofacies.

Trace-fossil suites attributable to a given ichnofacies can vary markedly, depending on the particular ichnogenera present, the age of the stratigraphic interval, the climatic conditions of the setting, the type of sedimentary basin, as well as a host of other factors. In this way, trace-fossil suites reside at the facies level. Just as bedsets consisting of trough cross-stratified sandstone may occur in a range of sedimentary environments, so too may some trace-fossil suites be assigned to different ichnofacies, depending upon the stacked character of the suites and their recurrence in the sedimentary facies. The more dynamic a sedimentary environment happens to be, the more variable are the intercalated trace-fossil suites likely to be. This is particularly true for environments that experience temporal variations in physical energy, chemical conditions (e.g. salinity), deposition rates or sediment delivery styles. Stable low-energy settings such as a muddy shelf, for example, may consist of thick intervals expressed by a single trace-fossil suite (Chapter 9). By contrast, mixed river- and wave-influenced delta fronts (Chapter 8) are highly dynamic and may show the influences of river floods, hypopycnal plumes, hyperpycnal discharge, variations in deposition rate, fluctuations in salinity and the effects of periodic storms, in addition to periods of relatively stable shallow-marine conditions. As such, the resulting facies show marked juxtaposition of a number of discrete suites that nonetheless show recurring intercalations with one another. In these situations, it is the variability of the suites and their interbedded character that allows assignment to an ichnofacies. It takes some measure of experience to discern how to combine such suites within the same ichnofacies as opposed to interpreting stacked ichnofacies. However, this is likewise the case when evaluating hierarchies of heterolithic lithofacies (Section 2.8), the only difference being that sedimentologists have learned to recognise most of the recurring patterns that identify the bedsets that constitute a single facies.

The details of the trace-fossil suite are crucial for assessing the range of processes and environmental conditions that controlled the benthic community, in order to help pinpoint the most likely depositional environment. The arrangement of over- and underlying trace-fossil suites helps to constrain

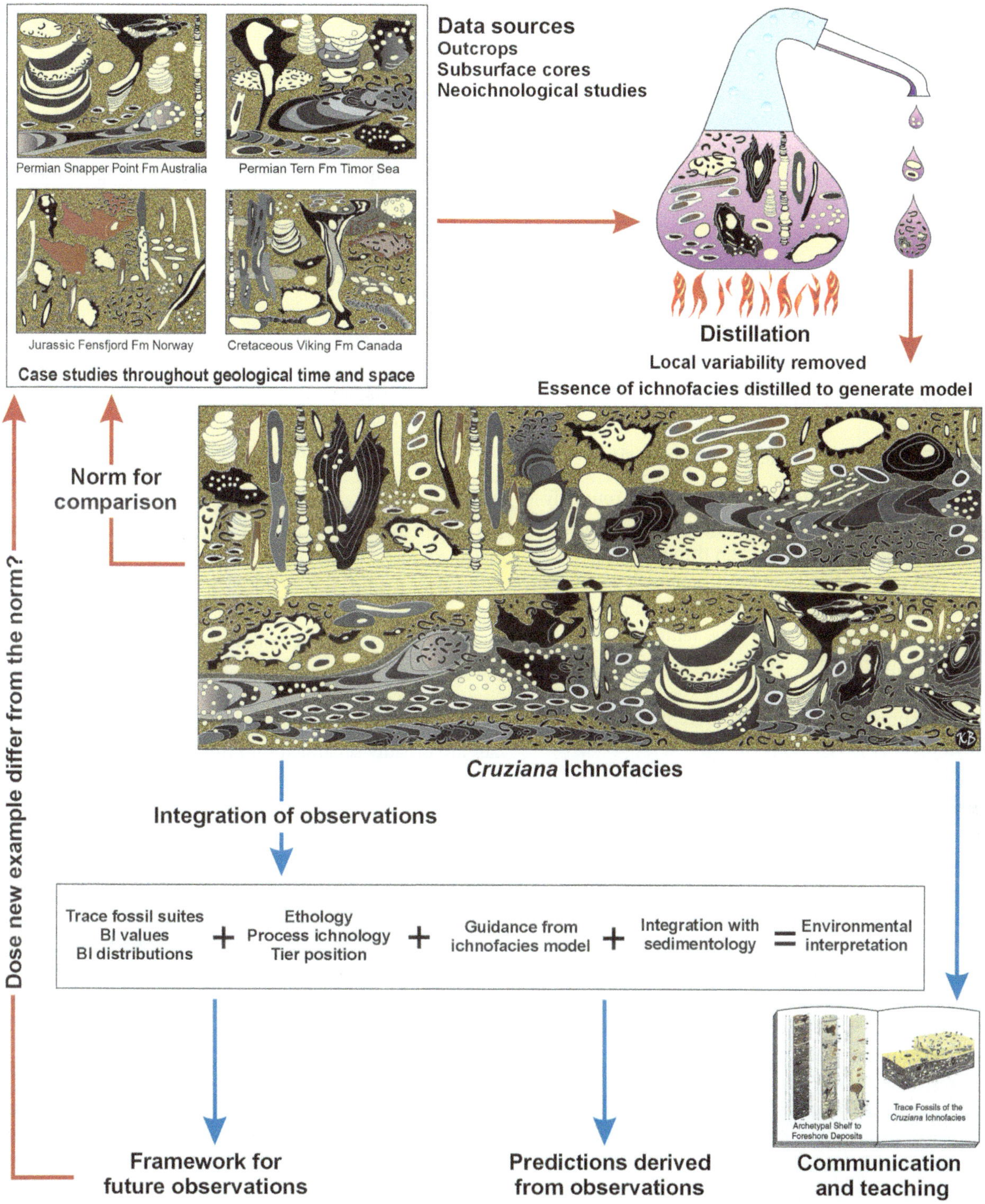

Figure 3.6 The construction and application of Seilacherian ichnofacies. The figure is divided into two linked loops (red and blue arrows). The construction of an ichnofacies follows the red arrows. Ichnological data are collected from a variety of sources (outcrop, core and neoichnological analyses). The greater the geographic distribution and the greater the range of geological ages of the examples, the better the quality of the resulting product. To construct the ichnofacies, local variability of the individual examples is removed and the 'essence' of the ichnofacies is distilled – it is what all suites attributable to the ichnofacies have in common. Once built, the ichnofacies model serves as a norm for comparison so that any new example can be evaluated to see whether it conforms to the ichnofacies and/or whether it contains features that depart from the archetype. The lower loop (blue arrows) outlines the applications of the ichnofacies. The model provides an integration of observations that form the facies interpretation equation (black box). From this equation, a framework for future observations permits the evaluation of new examples and allows predictions of trace-fossil suites and ichnofacies distributions in adjacent areas. Finally, the model facilitates teaching and communication. The ichnofacies concept provides a basis for understanding the ways in which organisms respond to their environment and forms a theoretical basis for evaluating observed trace-fossil suites. *Source:* Adapted from Walker (1984b) and MacEachern et al. (2010).

the range of possibilities. Discrete suites that can be used to differentiate storm-dominated delta-front, sheltered distal delta-front, distributary mouth bar or distributary channel subenvironments, for example, may all be attributable to the archetypal *Rosselia* Ichnofacies, because they all record animal-sediment responses to fluvial sediment influx and the accompanying physico-chemical stresses introduced into an otherwise sand-dominated, fully marine environment (e.g. Bann et al. 2008). This commonality, however, does not diminish the insights that each suite brings to our understanding of the setting's subenvironments and its facies. In our experience, accurate interpretation of the sedimentary environment is often only achieved by integrating the ichnological dataset with the physical sedimentological characteristics of the facies.

3.3.3 The Seilacherian (Archetypal) Ichnofacies

At present, there are 18 Seilacherian ichnofacies routinely employed for characterising depositional environments that extend from the continental realm through to the deep sea. Additional ichnofacies have been proposed, although most of these either reside at the level of trace-fossil suites or have not been demonstrated to have temporal and spatial recurrence. Interested readers are directed to McIlroy (2004), Buatois and Mángano (2011) and Knaust and Bromley (2012) for summaries. Likewise, there is considerable variation in the expression of the ichnofacies names by different workers. However, the authors here consider the archetypal ichnofacies to be proper names, and so we elect to express them by italicising the ichnogenus namesake in the title and by capitalising the word 'Ichnofacies'.

3.3.3.1 Continental Ichnofacies

Currently, there are seven widely accepted Seilacherian ichnofacies spanning a range of continental settings. Some of these are founded on relatively few ancient examples and can be regarded to be early in their development. Nonetheless, they constitute a solid basis for tackling the animal-sediment responses to the rigours of occupying terrestrial conditions and future work will certainly serve to refine them further (see Buatois and Mángano 2011 and Genise 2017 for summaries). Glacial settings are even less well characterised ichnologically, with examples mainly limited to descriptions and interpretations of trace-fossil assemblages and individual trace fossils (Netto et al. 2012), and ichnofacies expressed in the context of existing continental and marine archetypes. Given the complex range of physico-chemical stresses associated with climatic conditions, freshwater input, elevated deposition rates and subaqueous scour, one might expect to see discrete ichnofacies proposed for some glacial environments in the future.

The scope of animal-sediment interactions in the continental realm span the full range of feeding, locomotion, dwelling and nesting strategies. Although a number of ichnogenera have marine and marginal marine counterparts, the purposes of their constructions may not be equivalent (e.g. most continental organisms occupying vertical dwellings are not known to undertake filter-feeding strategies) and therefore care must be taken when attempting to provide ethological interpretations of the resulting continental suites (Figure 3.2). In order to properly employ continental trace fossils to interpret continental environments, the complex relationships between environmental processes (e.g. climate, physico-chemical stresses and biological interactions), substrate consistency and trace-fossil ethology must be understood (Hasiotis 2007). See also Chapter 5, Sections 5.4.4 and 5.11.

Most continental ichnofacies and the ichnological suites upon which they are founded are primarily controlled by the presence of water, as flowing entities, ponded bodies, the water table and/or soil moisture (Buatois and Mángano 1995a; Genise 2000; Hasiotis 2007; Dubois et al. 2012). In the marine realm, this is not a significant factor! Beyond the availability of moisture, issues of food resources, climate controls (and its seasonal variations) on temperature and precipitation, vegetation patterns and substrate consistency/durability all come into play (Genise et al. 2010; Buatois and Mángano 2011). Hence, continental ichnofacies are broadly separated into those that are permanently subaerial (*Celliforma*, *Coprinisphaera*, *Camborygma*, *Octopodichnus-Entradichnus* and *Termitichnus* ichnofacies), those intermittently inundated by water (*Scoyenia* Ichnofacies) and those permanently or predominantly subaqueous (*Mermia* Ichnofacies). Within persevering subaerial environments, climate and its overriding control on the position of the groundwater table, availability of soil moisture and substrate cohesion largely dictate the remaining ichnofacies groupings.

3.3.3.2 Permanently Subaerial Continental Settings

At present, five of the seven continental ichnofacies record animal-sediment relations in permanently subaerial settings. Of these, four are associated with palaeosols and one is associated with arid (eolian) looseground environments.

Termitichnus *Ichnofacies*

The *Termitichnus* Ichnofacies (Figure 3.7) was originally proposed by Smith et al. (1993) as a subdivision of the *Scoyenia* Ichnofacies (the only continental ichnofacies at the time), in order to accommodate trace-fossil suites associated with firm or compacted, permanently subaerial substrates. Buatois and Mángano (1995b) elevated this to an

Figure 3.7 Schematic block diagram of the *Termitichnus* Ichnofacies. The ichnofacies is characteristic of palaeosols of closed forests with elevated water tables and warm humid climates. It is commonly expressed by trace-fossil suites that include the ichnogenera such as *Daimonelix* (*Da*), *Edaphichnium* (*Ed*), *Fleaglellius* (*Fl*), *Krausichnus* (*Kr*), *Termitichnus* (*Tm*), *Vondrichnus* (*Vo*), as well as unnamed vertebrate dwelling burrows (vdb) and root traces (rt).

archetypal ichnofacies for trace-fossil suites associated with all palaeosols. This was later revised by Genise et al. (2000), who proposed restricting *Termitichnus* Ichnofacies to accommodating trace-fossil suites in palaeosols of closed forests with elevated water tables and vegetation reflecting warm, humid conditions, because the ichnofacies as defined by Smith et al. (1993) did not adequately express the range of variability encountered in palaeosols of different climatic regimes. *Coprinisphaera*, *Celliforma* and *Camborygma* ichnofacies were later erected to address these different palaeosol settings (see below).

The *Termitichnus* Ichnofacies is characterised by the dominance of chambers and chambered burrow systems (e.g. *Fleaglellius*, *Krausichnus*, *Termitichnus* and *Vondrichnus*) ostensibly assigned to termite nests, with subordinate J-, Y- and U-shaped vertical dwelling burrows related to other insects and arachnids (e.g. *Macanopsis*), pelleted deposit-feeding burrows (e.g. *Edaphichnium*) of probable earthworms, various vertebrate dwelling burrows including *Daimonelix* and root traces. Some occurrences of the ichnofacies may reflect moderate to incipiently developed palaeosols associated with fluvial channel belt successions.

Few ancient examples are known, with the best-known ones being late Palaeogene to Neogene in age (e.g. Genise and Bown 1994).

Coprinisphaera *Ichnofacies*

Genise et al. (2000) proposed the *Coprinisphaera* Ichnofacies (Figure 3.8) to accommodate trace-fossil suites and animal-sediment interactions of palaeosols that were distinctly different from those of the originally proposed *Termitichnus* Ichnofacies. Occurrence of the *Coprinisphaera* Ichnofacies arose concomitantly with the development of open habitats dominated by low vegetation (mostly grasses) during the Palaeogene. With the appearance and expansion of open grasslands came the diversification of large mammal herbivores and with them, the dung beetles that made use of their faeces (Genise et al. 2016). Such palaeosols form in different fluvial and eolian environments that range from semi-arid and cold steppes to humid and warm savannas.

The ichnofacies encompasses trace-fossil suites dominated by the breeding and pupation of multiple insect taxa, including dung beetle brood balls (e.g. *Coprinisphaera*, *Eatonichnus* and *Monesichnus*), bee cells and nests (e.g. *Celliforma*, *Ellipsoideichnus*, *Palmiraichnus*, *Rosellichnus* and *Uruguay*), moth pupation chambers (e.g. *Teisseirei*), cicada feeding chambers (e.g. *Feoichnus*), ant nests (e.g. *Attaichnus*), termite nests (e.g. *Coatonichnus*, *Microfavichus* and *Tacuruichnus*), traces of cleptoparasites and social parasites, various vertebrate dwelling burrows and root traces. Occurrences containing termite nests are regarded to record humid and warm settings. Examples of the ichnofacies are restricted to the Cenozoic and record the most recent middle Palaeogene to early Neogene innovation into palaeosol colonisation (Genise 2017).

Celliforma *Ichnofacies*

The *Celliforma* Ichnofacies (Figure 3.9; Genise 2000) is the earliest continental ichnofacies to be dominated by

Figure 3.8 Schematic block diagram of the *Coprinisphaera* Ichnofacies. The ichnofacies is characteristic of palaeosols associated with open habitats and grasslands. It is commonly expressed by trace-fossil suites that include the ichnogenera *Attaichnus* (*At*), *Celliforma* (*Ce*), *Coprinisphaera* (*Cop*), *Eatonichnus* (*Ea*), *Ellipsoideichnus* (*Ell*), *Monesichnus* (*Mo*), *Palmiraichnus* (*Pam*), *Rosellichnus* (*Ros*), *Tacuruichnus* (*Tac*), *Teisseirei* (*Tei*) and *Uruguay* (*Uru*), as well as unnamed vertebrate dwelling burrows (vdb) and root traces (rt).

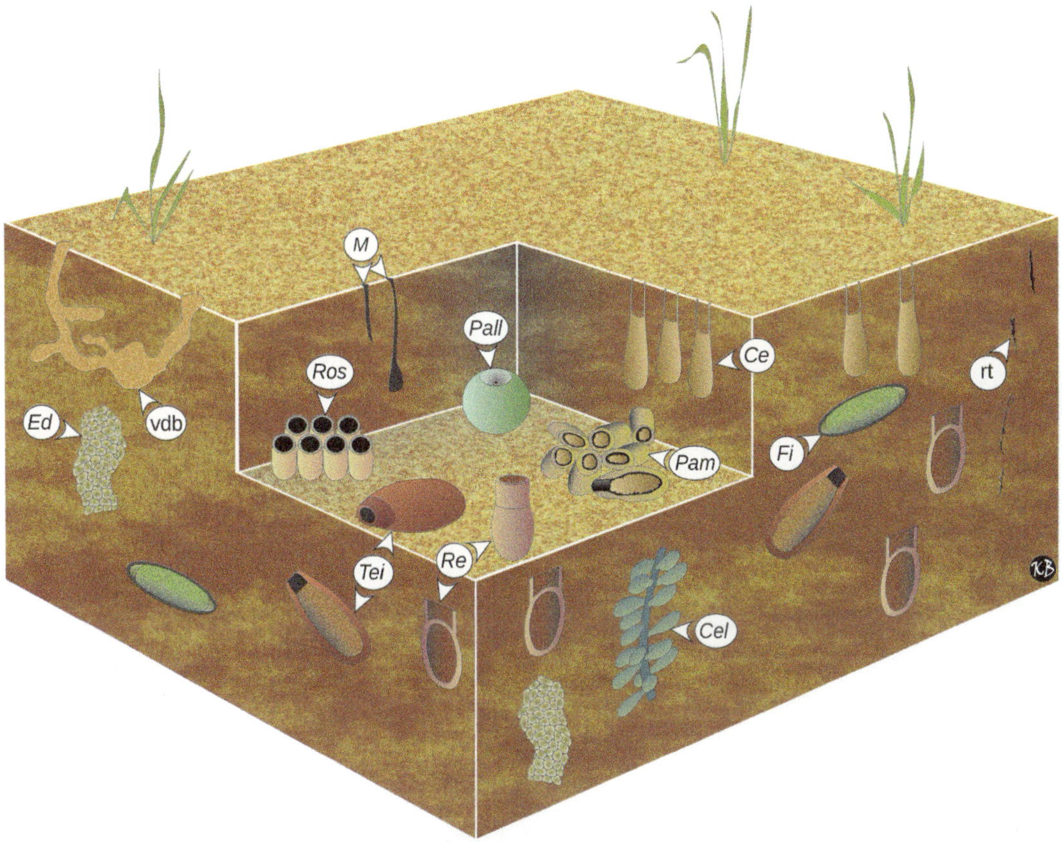

Figure 3.9 Schematic block diagram of the *Celliforma* Ichnofacies. The ichnofacies is characteristic of predominantly insect-generated burrows associated with arid or semi-arid, carbonate-rich, well-aerated palaeosols. It is commonly expressed by trace-fossil suites that include the ichnogenera *Celliforma* (*Ce*), *Edaphichnium* (*Ed*), *Fictovichnus* (*Fi*), *Macanopsis* (*M*), *Pallichnus* (*Pall*), *Palmiraichnus* (*Pam*), *Rebuffoichnus* (*Re*), *Rosellichnus* (*Ros*) and *Teisseirei* (*Tei*), as well as unnamed vertebrate dwelling burrows (vdb) and root traces (rt).

insect-generated trace fossils. It is well represented in carbonate-rich palaeosols reflecting arid or semi-arid conditions such as scrublands and woodlands, as well as some palustrine environments (e.g. Alonso-Zarza 2003). Some attributable suites occur in calcretes. All occurrences demonstrate the presence of large patches of bare, well-drained and aerated soils with sparse vegetation. Although its original designation was for calcareous-rich palaeosols, any palaeosol characterised by dry, well-aerated conditions could host trace-fossil suites attributable to the *Celliforma* Ichnofacies.

Common structures are those reflecting bee cells (e.g. *Cellicalichnus*, *Celliforma*, *Palmiraichnus* and *Rosellichnus*), wasp nests and cocoons (e.g. *Fictovichnus*), beetle and moth pupation chambers (e.g. *Pallichnus*, *Rebuffoichnus* and *Teisseirei*), as well as subordinate nests of social insects such as ants and termites (e.g. *Socialites*), vertebrate dwelling burrows, rhizoliths and rhizolith balls (e.g. possible ant fungus gardens; Genise et al. 2010). Secondary elements may include pelleted mobile deposit-feeding structures (e.g. *Edaphichnium*), vertical dwelling structures of predators (e.g. *Macanopsis*), vertebrate dwelling burrows and coprolites. Suites are generally of moderate diversity, with dominant ichnogenera occurring in high abundance. The ichnofacies is known from a number of Cenozoic intervals, though with only two examples published from Late Cretaceous intervals (e.g. Alonso-Zarza et al. 2011; Martin and Varricchio 2011). As duration times of subaerial exposure diminish, this association may give way to the *Scoyenia* Ichnofacies (Buatois and Mángano 2011).

Camborygma *Ichnofacies*

The *Camborygma* Ichnofacies (Figure 3.10; Genise et al. 2016; Buatois et al. 2020) constitutes one of the most recent additions to the continental ichnofacies framework and encompasses trace-fossil suites associated with palaeosols and pedogenically modified media characterised by fluctuating but generally elevated water tables. Such environments are closely associated with various fluvial systems, reflecting adjacent wetlands, mossy bogs, marshes and swamps, as well as peatlands from evergreen forests, scrubs

Figure 3.10 Schematic block diagram of the *Camborygma* Ichnofacies. The ichnofacies is characteristic of palaeosols and pedogenically modified media showing fluctuating but generally elevated water tables. It is commonly expressed by trace-fossil suites that include the ichnogenera *Camborygma* (*Ca*), *Castrichnus* (*Cas*), *Cellicalichnus* (*Cel*), *Dagnichnus* (*Dag*), *Eatonichnus* (*Ea*), *Edaphichnium* (*Ed*), *Loloichnus* (*Lol*), *Rebuffoichnus* (*Re*) and *Taenidium* (*Ta*), as well as unnamed vertebrate dwelling burrows (vdb) and root traces (rt).

and grasslands. These may be closely associated with ponds, floodplains, crevasse splays, levees and abandoned fluvial channels.

Biogenic structures are dominated by dwellings (e.g. *Camborygma* and *Loloichnus*) and breeding traces (e.g. *Dagnichnus*) formed by crayfishes as well as pelleted deposit-feeding structures (e.g. *Edaphichnium*) and aestivation chambers (e.g. *Castrichnus*) of earthworms. Subordinate elements are those generated by insects (including their larvae) such as bee cells (e.g. *Cellicalichnus*), beetle pupation chambers (e.g. *Rebuffoichnus*), dung beetle nests (e.g. *Eatonichnus*), and meniscate deposit feeding structures (e.g. *Taenidium*). Suites also commonly display the presence of various vertebrate dwelling structures and roots. Crayfishes require water to collect in their burrows in order for them to respirate and so they typically extend their structures below the water table. The trace-fossil suites attributable to the ichnofacies are generally of low diversity but with high abundances of individual ichnogenera. Examples of the *Camborygma* Ichnofacies are not known from strata older than the Mesozoic (Buatois et al. 2020).

Octopodichnus-Entradichnus *Ichnofacies*

The *Octopodichnus-Entradichnus* Ichnofacies (Figure 3.11) arises from two independent studies that proposed archetypal ichnofacies for eolian dune deposits: the *Octopodichnus* Ichnofacies (Hunt and Lucas 2007) and the *Entradichnus* Ichnofacies (Ekdale et al. 2007). The *Octopodichnus* Ichnofacies is primarily based on studies of the Permian Coconino Sandstone of Arizona and consists predominantly of traces generated by arthropod trackways assigned mainly to spiders and scorpions. The *Entradichnus* Ichnofacies is based on studies of the Jurassic Navajo Sandstone of Utah with some comparison to other eolianites and comprises vertical dwelling structures of arthropods as well as meniscate burrows. Buatois and Mángano (2009) chose to combine the two proposed ichnofacies into a single archetype, suggesting that the differences may spring from temporal changes in burrowing behaviours within eolian regimes as well as possible geographic variations. This was further refined by Krapovickas et al. (2016) who, in their review of numerous published studies of eolianites, elected to retain the *Octopodichnus-Entradichnus* Ichnofacies and proposed that trace-fossil suites record five major phases of animal colonisation of desert environments during the Phanerozoic.

The *Octopodichnus-Entradichnus* Ichnofacies encompasses trace-fossil suites that record faunal activities in mobile and temporary stabilised subaerial sandy substrates, subject to repeated erosion, deposition, marked climatic seasonality and prolonged periods of aridity. Trace-fossil distributions and variations in ichnological suites for any given sedimentary unit can be characterised by the environmental stresses related to the availability and abundance of water and its temporal fluctuations, as well as nutrient availability, character of the substrate, and the dominant organisms present in the palaeocommunity. The presence of water is also important in enhancing the preservation of biogenic structures.

Attributable suites may also contain trackways of arthropods (e.g. *Octopodichnus* and *Paleohelcura*) and tetrapods (e.g. *Brasilichnium* and *Chelichnus*), instratal meniscate locomotion structures of insect larvae (e.g. *Entradichnus*), vertical single-entrance dwellings (e.g. *Digitichnus* and *Skolithos*), U-shaped dwellings (e.g. *Arenicolites* and *Diplocraterion*) and

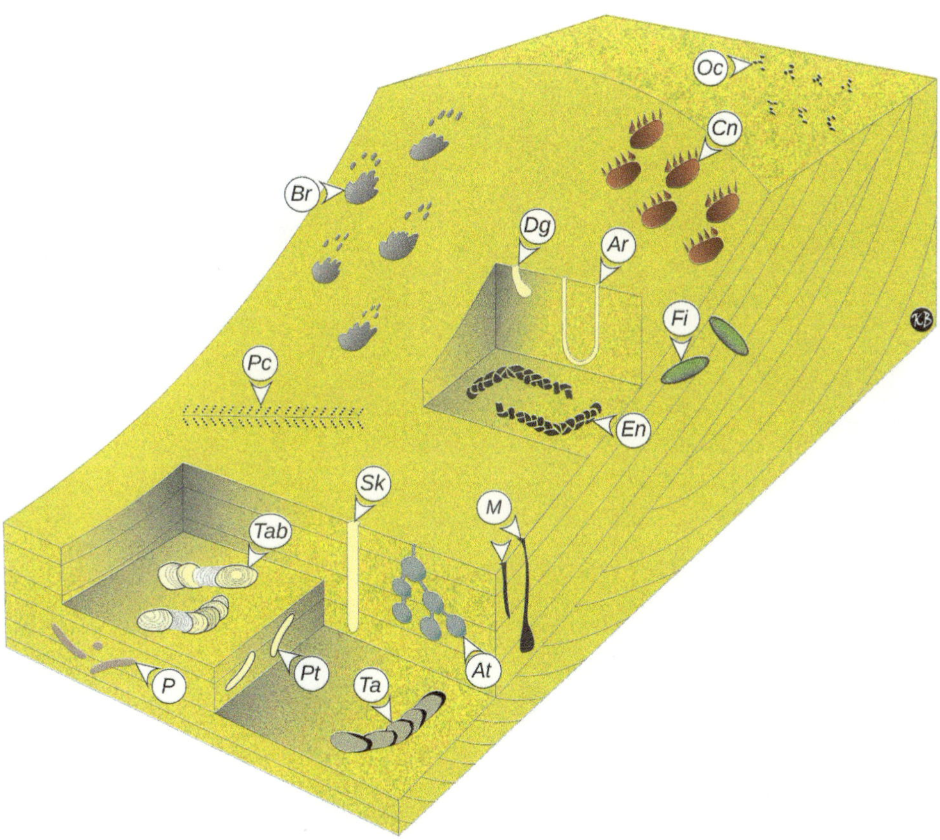

Figure 3.11 Schematic block diagram of the *Octopodichnus-Entradichnus* Ichnofacies. The ichnofacies is characteristic of eolian dune environments. It is commonly expressed by trace-fossil suites that include the ichnotaxa *Arenicolites* (*Ar*), *Attaichnus* (*At*), *Brasilichnium* (*Br*), *Chelichnus* (*Cn*), *Digitichnus* (*Dg*), *Entradichnus* (*En*), *Fictovichnus* (*Fi*), *Macanopsis* (*M*), *Octopodichnus* (*Oc*), *Palaeophycus tubularis* (*Pt*), *Paleoheicura* (*Pc*), *Planolites* (*P*), *Skolithos* (*Sk*), *Taenidium bowni* (*Tab*) and *Taenidium* isp. (*Ta*).

the horizontal tubes of predators and deposit feeders (e.g. *Palaeophycus*, *Planolites* and *Taenidium*). That said, with the paucity of examples, additional recurring ichnogenera might be expected to be encountered, not least *Macanopsis* (reflecting vertical dwelling structures of spiders and wasps), *Fictovichnus* (e.g. wasp burrows), *Attaichnus* (e.g. ant nests), and meniscate structures of cranefly larvae (*Entradichnus meniscus*; Seike et al. 2010), based on a study of modern eolian deposits (see Ahlbrandt et al. 1978). Wetter interdune areas may see the transition from the *Octopodichnus-Entradichnus* Ichnofacies to the *Scoyenia* Ichnofacies (Krapovickas et al. 2016). Trace-fossil suites associated with coastal dunes, by contrast, may show a basinward transition to suites attributable to the *Psilonichnus* Ichnofacies (see Section 3.3.3.5). With so few studies published to date, this ichnofacies can be considered to be in its infancy and revisions and refinements are sure to be expected.

3.3.3.3 Intermittently Subaerially Exposed Conditions

At present, only a single ichnofacies (*Scoyenia* Ichnofacies) encompasses trace-fossil suites associated with intermittently subaerially exposed settings. Such settings are affected by elevated water tables, looseground substrates and the periodic occurrence of flowing freshwater.

Scoyenia *Ichnofacies*

The *Scoyenia* Ichnofacies (Figure 3.12) was the first continental ichnofacies erected (Seilacher 1967), based on

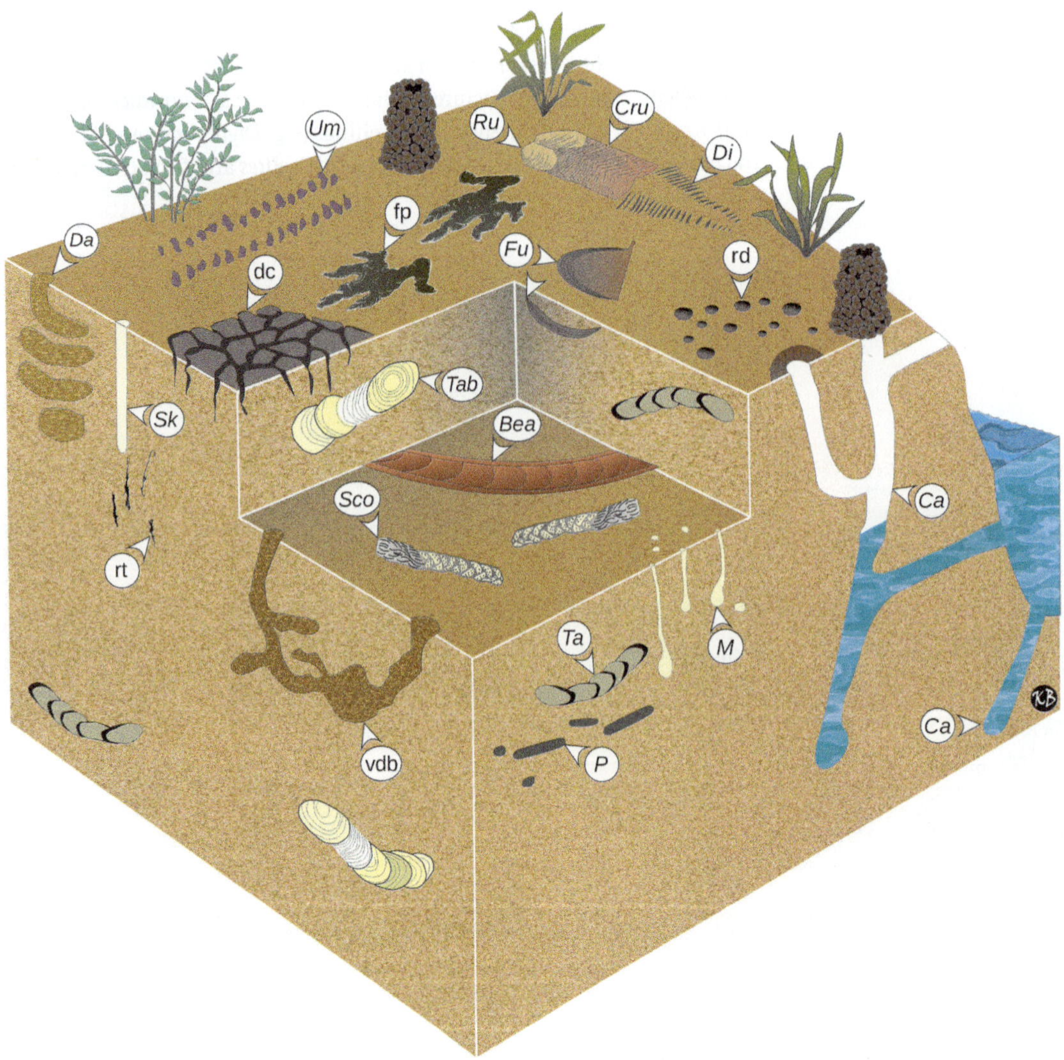

Figure 3.12 Schematic block diagram of the *Scoyenia* Ichnofacies. The ichnofacies is characteristic of unconsolidated looseground substrates intermittently exposed to flowing freshwater, such as those associated with fluvial environments. It is commonly expressed by trace-fossil suites that include the ichnotaxa *Beaconites* (*Bea*), *Camborygma* (*Ca*), *Cruziana* (*Cru*), *Daimonelix* (*Da*), *Diplichnites* (*Di*), *Fuersichnus* (*Fu*), *Macanopsis* (*M*), *Planolites* (*P*), *Rusophycus* (*Ru*), *Scoyenia* (*Sco*), *Skolithos* (*Sk*), *Taenidium bowni* (*Tab*), *Taenidium* isp. (*Ta*) and *Umfolozia* (*Um*), as well as vertebrate foot prints (fp), unnamed vertebrate dwelling burrows (vdb) and root traces (rt). In addition, facies may show desiccation cracks (dc) and rain-drop impressions (rd).

meniscate burrows, bilobed traces and arthropod trackways. It stood as the sole freshwater ichnofacies for nearly three decades, owing to the focus of most ichnologists on marine and marginal marine settings. As a result, later workers embraced the complexity of animal-sediment interactions in the continental realm, and the *Scoyenia* Ichnofacies was redefined to accommodate only trace-fossil suites associated with intermittent flowing water in unconsolidated substrates (Buatois and Mángano 1995a).

Characteristic environments include channel belt complexes, particularly floodplain positions adjacent to channels, crevasse splays, levee and overbank deposits and the flanks of flood basin ponds. Additionally, suites attributable to the ichnofacies also occur in association with some braidplains and ephemeral river systems. Finally, lake margins may also host the ichnofacies where they are intermittently exposed (see Chapter 6, Section 6.4.4)

Representative trace-fossil suites predominantly display horizontal meniscate structures generated by deposit-feeding trace-makers (e.g. *Ancorichnus*, *Beaconites*, *Scoyenia* and *Taenidium*), locomotion structures reflected by bilobed trails and trackways of both vertebrate and invertebrate organisms (e.g. *Cruziana*, *Diplichnites*, *Rusophycus* and *Umfolozia*), vertical dwelling structures (e.g. rare *Camborygma*, *Macanopsis* and *Skolithos*), horizontal and inclined deposit-feeding structures (e.g. *Fuersichnus* and *Planolites*), various vertebrate dwelling structures including *Daimonelix* and root structures (see Buatois and Mángano 2011 for summary). Trace-fossil suites tend to be of low to moderate diversity, although with ichnogenera locally occurring in high abundances. Suites typically record changes in substrate consistency and water saturation levels, which are common to environments formed in close proximity to active fluvial channel systems. Examples of the *Scoyenia* Ichnofacies remain the most common expression of continental trace-fossil suites in the rock record by far, with examples as old as the Ordovician (Johnson et al. 1994).

3.3.3.4 Subaqueous Continental Ichnofacies

At present, only a single ichnofacies (*Mermia* Ichnofacies) encompasses trace-fossil suites associated with permanently or predominantly subaqueous freshwater conditions.

Mermia *Ichnofacies*

Mermia Ichnofacies (Figure 3.13) was erected by Buatois and Mángano (1995a, 1995b) and records trace-fossil suites associated with permanently or predominantly subaqueous settings in the continental realm. Such systems are overwhelmingly freshwater. The ichnofacies is most representative of permanent lakes, although suites attributable to the ichnofacies have been identified from ephemeral ponds on alluvial plains, margins of hypersaline lakes and carbonate lakes, among others (e.g. Buatois and Mángano 2011 and Scott et al. 2012 and references therein). Trace-fossil suites from larger lacustrine systems tend to be more diverse than those of smaller lakes or fluvial ponded water bodies of fluvial systems (Buatois and Mángano 2007).

The ichnofacies encompasses trace-fossil suites that are dominated by horizontal and sub-horizontal deposit-feeding and grazing structures developed by mobile trace-makers at or near the sediment–water interface (e.g. *Circulichnus*, *Cochlichnus*, *Helminthoidichnites*, *Helminthopsis*, *Mermia*, *Planolites*, *Treptichnus* and *Vagorichnus*). Insect trackways, bilobate structures and other locomotion structures commonly form subordinate elements of the suites (e.g. *Diplichnites*, *Maculichna* and *Undichna*; Scott et al. 2012). Bivalve resting structures (e.g. *Lockeia*) are present locally, as well as rare feeding structures of fish (e.g. *Piscichnus*). Dwelling structures such as *Palaeophycus* and *Trichichnus* are generally uncommon, but can occur in sandier, shallow-water zones of the lacustrine system, where they may be accompanied by *Polykladichnus* or *Skolithos*. Suites tend to show high to moderate diversities, though these decline markedly in lacustrine systems prone to thermal stratification and lowered oxygenation, particularly in deeper-water positions. Ichnogenera show a general lack of specialised feeding or grazing patterns. Bioturbation intensities tend to be low to moderate, with higher intensities accompanied by higher trace-fossil diversities, consistent with better oxygenated parts of the lake (see Buatois and Mángano 2011 and Scott et al. 2012 for summaries).

Like the marine realm, lacustrine systems typically show a marked change in depositional conditions from proximal to distal positions. At present, however, no recurring suites of the *Mermia* Ichnofacies have been identified to distinguish between shallow and deep lacustrine environments (Buatois and Mángano 2011). Marked variations in the resulting biogenic patterns would be expected, based upon whether the lake is low oxygen, saline and/or stratified as compared to fresh and fully oxygenated. In low-oxygen and saline lakes, simple trails and diminutive burrows would extend to the edge of the shoreline. By contrast, in well-mixed and freshwater lakes, simple trails of *Mermia* would be restricted to relatively deeper water and full relief infaunal bioturbation can be expected in shallower lake deposits. Additionally, it has been suggested that a 'continental *Skolithos* Ichnofacies' may mark shallow, sandy lacustrine shorelines and lake edges (e.g. Scott et al. 2012), consisting of ichnogenera assigned to *Arenicolites*, *Polykladichnus*, *Psilonichnus* and *Skolithos*. The basis for the assignment of these vertical to inclined structures to the *Skolithos* Ichnofacies is related only to their morphology and orientation and, to our mind, only mimics the archetypal *Skolithos* Ichnofacies. The Seilacherian archetype was based on the

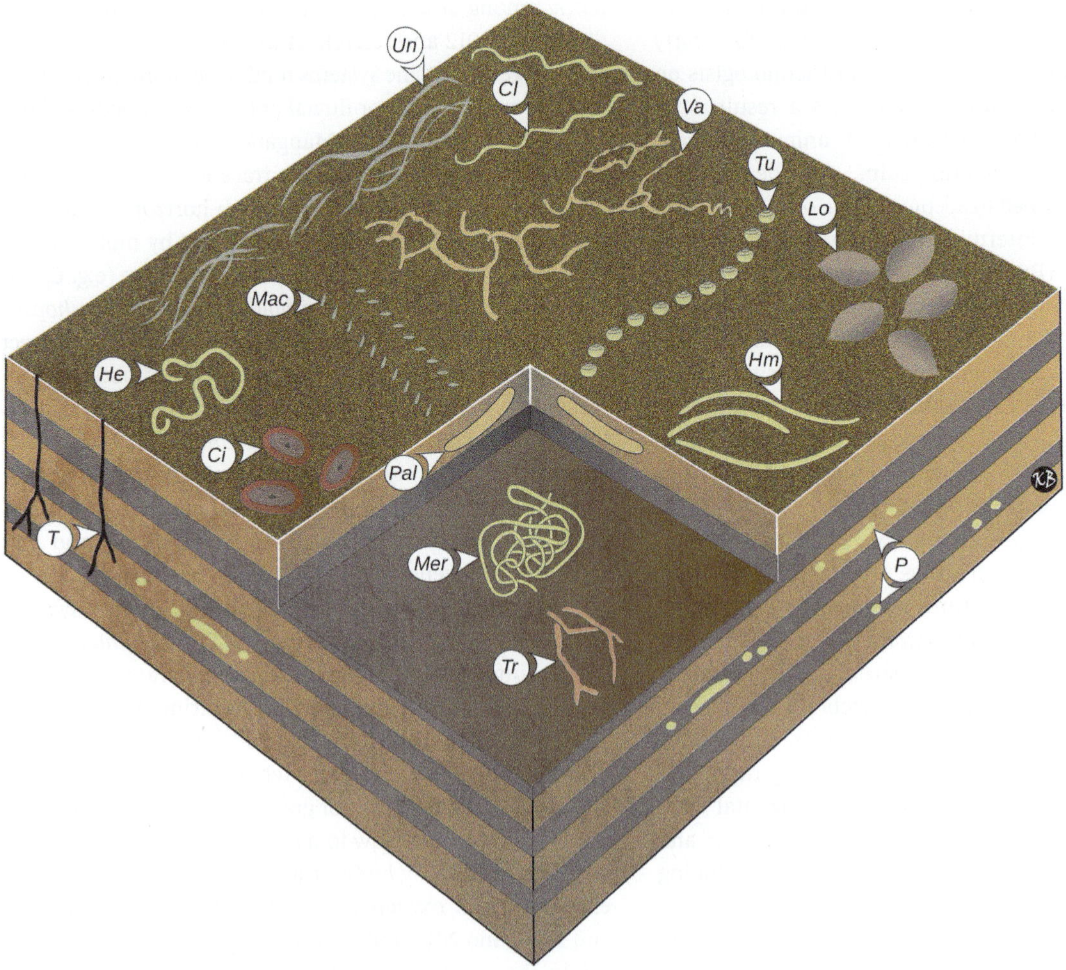

Figure 3.13 Schematic block diagram of the *Mermia* Ichnofacies. The ichnofacies is characteristic of permanently subaqueous freshwater conditions associated with lacustrine environments. It is commonly expressed by trace-fossil suites that include the ichnogenera *Circulichnus* (*Ci*), *Cochlichnus* (*Cl*), *Helminthoidichnites* (*Hm*), *Helminthopsis* (*He*), *Lockeia* (*Lo*), *Maculichna* (*Mac*), *Mermia* (*Mer*), *Planolites* (*P*), *Treptichnus* (*Tr*), *Trichichnus* (*T*), *Undichna* (*Un*) and *Vagorichnus* (*Va*).

predominance of more deeply penetrating, permanent dwelling structures for the primary purpose of suspension/filter feeding and presupposed high-energy, *fully marine* conditions (Seilacher 1967). The suites of the continental expression do not reflect these behavioural characteristics and instead, predominantly record dwellings for the purposes of passive carnivory and deposit feeding. While we generally discourage the usage of the 'continental *Skolithos* Ichnofacies' and would recommend, instead, the creation of a distinct lacustrine shoreline ichnofacies that captures the full range of animal-sediment responses in these settings, the clarification of such suites as discrete from their marine counterparts serves its purpose for now, pending future refinement of lacustrine trace-fossil associations.

Additionally, the impacts of lacustrine deltas on oxygenation, sedimentation rates, suspended sediment load and variations in physical energy can be expected to affect lacustrine suites as well and may ultimately give rise to a discrete ichnofacies for such depositional regimes.

The oldest suites of *Mermia* Ichnofacies occur in the Carboniferous. Palaeozoic suites are dominated by surface trails, which are accompanied by increased numbers of infaunal structures through the Mesozoic and Cenozoic (Buatois and Mángano 2011).

3.3.3.5 Marine and Marginal Marine Ichnofacies

Currently, there are eight accepted marine and marginal marine Seilacherian ichnofacies that accommodate a range of animal-sediment responses to saline and brackish-water softground environments. Three of these (*Phycosiphon*, *Rosselia* and *Teichichnus* ichnofacies) are in their early stages of development, but so far have demonstrated recurrence in both space and geologic time.

Psilonichnus ***Ichnofacies***

The *Psilonichnus* Ichnofacies (Figure 3.14) was erected by Frey and Pemberton (1987) to accommodate trace-fossil suites associated with backshore, largely subaerial environments periodically inundated by saline water during storms and exceptional tides. As such, representative ichnological suites record a combination of marine, marginal marine and non-marine conditions. Typical environments include the beach backshore, coastal dunes, washover fans and supratidal flats. Such environments are exposed to marked changes in energy conditions, sediment types and weather conditions, with maritime eolian processes alternating with marine influences during heightened tidal periods and storms. Salinities may be highly variable, owing to subaerial exposure and freshwater precipitation.

Relatively few marine organisms can occupy such environments, with amphibious crabs of the Family Ocypodidae being the most notable. These crabs are both scavengers and surficial deposit feeders, and they typically excavate permanent vertical to inclined J-, Y- or U-shaped dwellings (e.g. *Psilonichnus*). Other biogenic structures are generated by essentially fully terrestrial organisms and include: (i) the vertical shafts of insects and spiders (e.g. *Macanopsis* and *Skolithos*); (ii) the horizontal tunnels of insects and tetrapods (e.g. *Thalassinoides*); and (iii) the ephemeral tracks, trails and faecal pellets of bivalves, gastropods, insects, reptiles, birds and mammals (e.g. *Archaeonassa*, *Lockeia* and *Protovirgularia*). The stressful nature of the depositional regime effectively demands infauna to operate as opportunists. Most behaviours reflect the activity of scavengers, surficial deposit feeders, predators and herbivores.

The other major group of biogenic structures relates to plant-root penetrations. The types of plants able to exploit these substrates range from intertidal halophytes on the distal margins of some washover fans, to maritime or terrestrial grasses, weeds, vines, shrubs, bushes and trees on backshore dunes. Suites attributable to the *Psilonichnus* Ichnofacies are typically of low diversity and low abundance, and made up of dwellings, locomotion and resting structures as well as coprolites.

Figure 3.14 Schematic block diagram of the *Psilonichnus* Ichnofacies. The ichnofacies is characteristic of backshore, largely subaerial environments periodically inundated with saline water during storms and exceptional tides. It is commonly expressed by trace-fossil suites that include ichnogenera such as *Archaeonassa* (*Arc*), *Lockeia* (*Lo*), *Macanopsis* (*M*), *Planolites* (*P*), *Protovirgularia* (*Pr*), *Psilonichnus* (*Pi*), *Skolithos* (*Sk*) and *Thalassinoides* (*Th*), as well as root traces (rt) and a wide range of vertebrate footprints and trackways (vfp).

The *Psilonichnus* Ichnofacies is best known from modern settings (e.g. Frey and Pemberton 1987; Netto and Grangeiro 2009), with only a few clear-cut examples preserved in the rock record (e.g. Nesbitt and Campbell 2006; Curran 2007; Myint 2007). This is largely owing to the low preservation potential of such deposits in the coastal regime. Furthermore, the low diversity and low abundance of the suites also make assignment to the ichnofacies challenging, particularly in light of the other ichnofacies to which it may be intergradational. To the extent that ocypodid crab burrows may occur in the uppermost foreshore or the upper part of estuarine point bars, the *Psilonichnus* Ichnofacies may slightly overlap with the *Skolithos* Ichnofacies or with the *Teichichnus* Ichnofacies. By contrast, because of its potentially large numbers of terrestrial traces, suites attributable to the *Psilonichnus* Ichnofacies may be broadly intergradational with the continental *Scoyenia,* 'continental *Skolithos*', *Entradichnus-Octopodichnus* and *Coprinisphaera* ichnofacies.

Teichichnus *Ichnofacies*

The *Teichichnus* Ichnofacies has recently been proposed as the archetype for trace-fossil suites associated with environments characterised by persistent brackish-water conditions (Gingras et al. 2025; Figure 3.15). The ichnofacies has a long history of conception, beginning with numerous biological and neoichnological studies of modern estuaries (e.g. Schäfer 1962; Dörjes and Howard 1975; Gingras et al. 1999; Dashtgard 2011), later augmented by evaluation of ancient estuarine and fluvio-tidal deposits of the rock record (e.g. Pemberton et al. 1982; Beynon et al. 1988; Buatois et al. 2005; MacEachern and Gingras 2007). Later studies included open bay, restricted bay, central basins of wave-dominated estuaries, tidal bays, heterolithic successions associated with lateral accretion of point bars and vertical accretion of channel margins of fluvio-tidal channels (summarised in MacEachern and Gingras 2007). The common element in these settings is one of salinity reduction, although fluctuations in salinity are exceedingly common. Concomitant with salinity fluctuations (euryhaline) are variations in substrate coherence, deposition rate, tidal and wave energy and food delivery to the different environments. Additionally, a number of low-energy brackish-water environments also experience variations in oxygenation, imparting additional stress on benthic communities. Hence, markedly different trace-fossil suites develop in the resulting facies and show sporadically distributed bioturbation with a wide range of BI values at the bed and bedset scales. These are characteristics of pronounced physico-chemical stress within the sedimentary environment.

Brackish-water settings are characterised by dominant euryhaline-tolerant species, resulting from osmoregulation-related selective pressures that reduce faunal diversity, favour diminution of trace-makers and promote infaunal lifestyles (summarised in Gingras and MacEachern 2012). Organisms adopt r-selected (opportunistic) strategies, which are broadly characterised by rapid colonisation, high population densities and monospecific assemblages. Tides in these settings provide abundant food through the transport of organic matter, sedimentation during slack water periods and proximity to marine environments. This increased food availability favours trophic generalists employing various feeding behaviours to exploit both intrastratal and surficial food resources.

Accordingly, suites of the *Teichichnus* Ichnofacies are recognised by reduced trace-fossil diversities, smaller trace-fossil sizes, dominance of facies-crossing elements, sporadically distributed trace fossils, locally elevated though commonly variable bioturbation intensities and evidence of opportunistic colonisation. Trace-fossil suites are dominated by dwellings strongly associated with surface deposit feeding (e.g. *Arenicolites*, *Cylindrichnus*, *Siphonichnus*, *Skolithos*), intrastratal deposit feeding (e.g. *Planolites*, *Teichichnus*, *Thalassinoides*) and subordinate filter feeding (e.g. *Skolithos*, *Siphonichnus*). In terms of variations within the *Teichichnus* Ichnofacies, three commonly observed, recurring assemblages are reported:

1) *Restricted and sheltered settings (e.g. central basins of wave-dominated, barred estuaries):* Typically associated with siltstone, clayey siltstone or sandy siltstone facies with minor sand laminae, this variant is marked by moderate to intense bioturbation (BI 3–5) coupled with sporadic, heterogeneous distributions. The trace-fossil suites include diminutive *Planolites* and *Teichichnus*, locally accompanied by other ichnogenera such as *Arenicolites, Cylindrichnus*, *Gyrolithes* and *Skolithos*.
2) *Heterolithic sedimentation with moderate sedimentation rates (e.g. IHS in tidally influenced brackish-water channels)*: Comprising heterolithic wavy bedded to interbedded sandstone and mudstone layers, this variant displays variable bioturbation intensities (BI 0–4). Common ichnogenera include *Arenicolites, Cylindrichnus*, *Gyrolithes* and *Skolithos* with localised presence of *Planolites* and *Teichichnus*.
3) *Moderate salinities with episodic deposition (e.g. moderately brackish unbarred bays)*: Characterised by lenticular to wavy bedding, oscillation ripple lamination and low-angle undulatory parallel lamination indicative of tempestite deposition, this variant shows highly variable bioturbation intensities (BI 0–5). The trace-fossil suite is dominated by moderate to common numbers of *Planolites*, *Schaubcylindrichnus freyi, Teichichnus* and *Thalassinoides*, accompanied by other facies-crossing ichnogenera in various abundances.

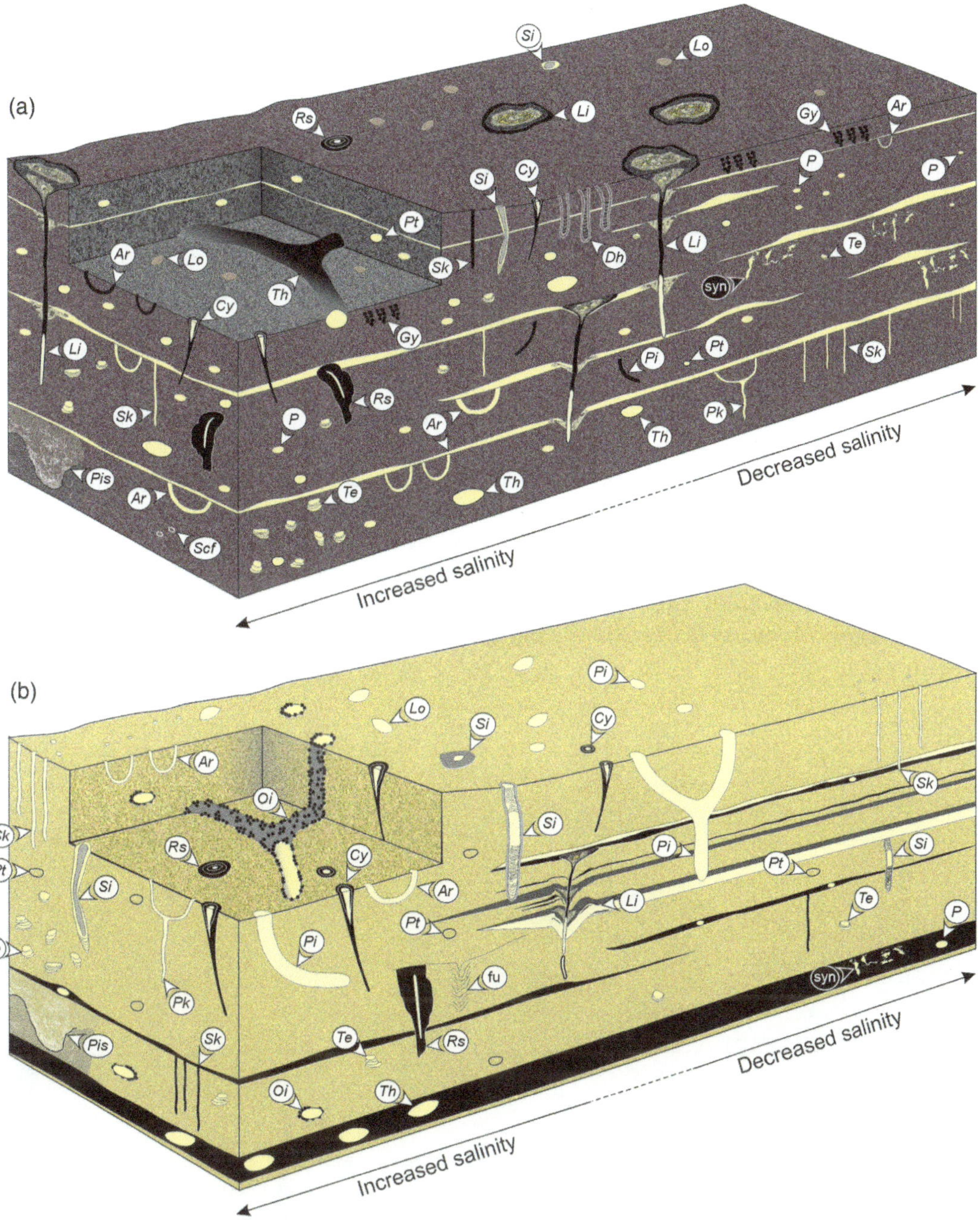

Figure 3.15 Schematic block diagrams of the *Teichichnus* Ichnofacies. The ichnofacies is characteristic of subaqueous and intertidal environments subject to fluctuating but generally reduced salinities (i.e. brackish-water conditions). Trace-fossil suites vary markedly depending upon the substrate consistency (a and b) and the degree of average salinity. The ichnofacies is generally expressed by suites of low diversity and trace fossils that are diminutive, commonly including ichnotaxa such as *Arenicolites* (*Ar*), *Cylindrichnus* (*Cy*), *Diplocraterion habichi* (*Dh*), fugichnia (fu), *Gyrolithes* (*Gy*), *Lingulichnus* (*Li*), *Lockeia* (*Lo*), *Ophiomorpha irregulaire* (*Oi*), *Palaeophycus tubularis* (*Pt*), *Piscichnus* (*Pis*), *Planolites* (*P*), *Polykladichnus* (*Pk*), *Psilonichnus* (*Pi*), *Rosselia socialis* (*Rs*), *Schaubcylindrichnus freyi* (*Shf*), *Siphonichnus* (*Si*), *Teichichnus* (*Te*) and *Thalassinoides* (*Th*). Note that syneresis cracks (syn) are increasingly common in mudstone units as salinity levels decrease. Bioturbation intensities are generally highly variable at the bed and bedset scales with diminished and sporadically distributed bioturbation reflecting fluctuating physico-chemical stress. Suites typically consist entirely of facies-crossing elements.

While suites of the *Teichichnus* Ichnofacies show evidence of pronounced physico-chemical stress, they lack intercalated beds containing fully marine ichnogenera because they record persistent salinity reduction, making such suites distinct from those of the *Rosselia* and *Phycosiphon* ichnofacies (see below).

Skolithos *Ichnofacies*

The *Skolithos* Ichnofacies (Figure 3.16) was erected by Seilacher (1967) and expresses trace-fossil suites indicative of relatively high levels of wave and/or current energy developed in clean, well-sorted, loose or shifting particulate substrates of marine environments. Abrupt changes in rates of deposition, erosion and physical reworking of sediments are characteristic of the typical settings. Such conditions commonly occur on shoreface and sheltered foreshores, but similar conditions may occur in a wide range of high-energy marine depositional environments (e.g. proximal wave-dominated delta fronts, sandy bars and spits, tidal inlets, sandy open bays, estuary mouth complexes and submarine fans).

Under such conditions, most trace-makers are suspension feeders and employ substrates mainly as anchoring media. Correspondingly, infaunal organisms typically construct deeply penetrating, more or less permanent domiciles as platforms for feeding or as refugia. Secondary behaviours include passive carnivory and surface detritus feeding, particularly as energy conditions at the seabed decline. Depth of burrowing in the intertidal zone is controlled, in part, by tidal range and height of the low-tide interstitial water column in the substrate. During low-tide, moist sediments at depth help to buffer organisms against desiccation and temperature shock, as well as providing respiratory water. In both intertidal and high-energy subtidal settings, deep burrowing is one means of escaping the instability of the ever-shifting substrate surface. Bioturbation intensities may be highly variable (BI 0–5), with much of the variation ostensibly related to the overall depositional energy and magnitude of erosional scour. Elevated BI values (BI 4–5) generally indicate slower rates of deposition and decreasing energy. Episodic deposition, particularly associated with variations in wave energy and recurrence of storms, commonly leads to both fugichnia and equilibrichnia. Tempestites generated under these conditions generally record opportunistic, top-down colonisation by trace-makers. As such settings are dominated by well-oxygenated waters, cryptic bioturbation by meiofauna (e.g. Pemberton et al. 2008) may also be locally abundant.

Suites attributable to the *Skolithos* Ichnofacies are characterised by: (i) predominantly vertical, cylindrical or U-shaped dwelling burrows; (ii) protrusive and retrusive

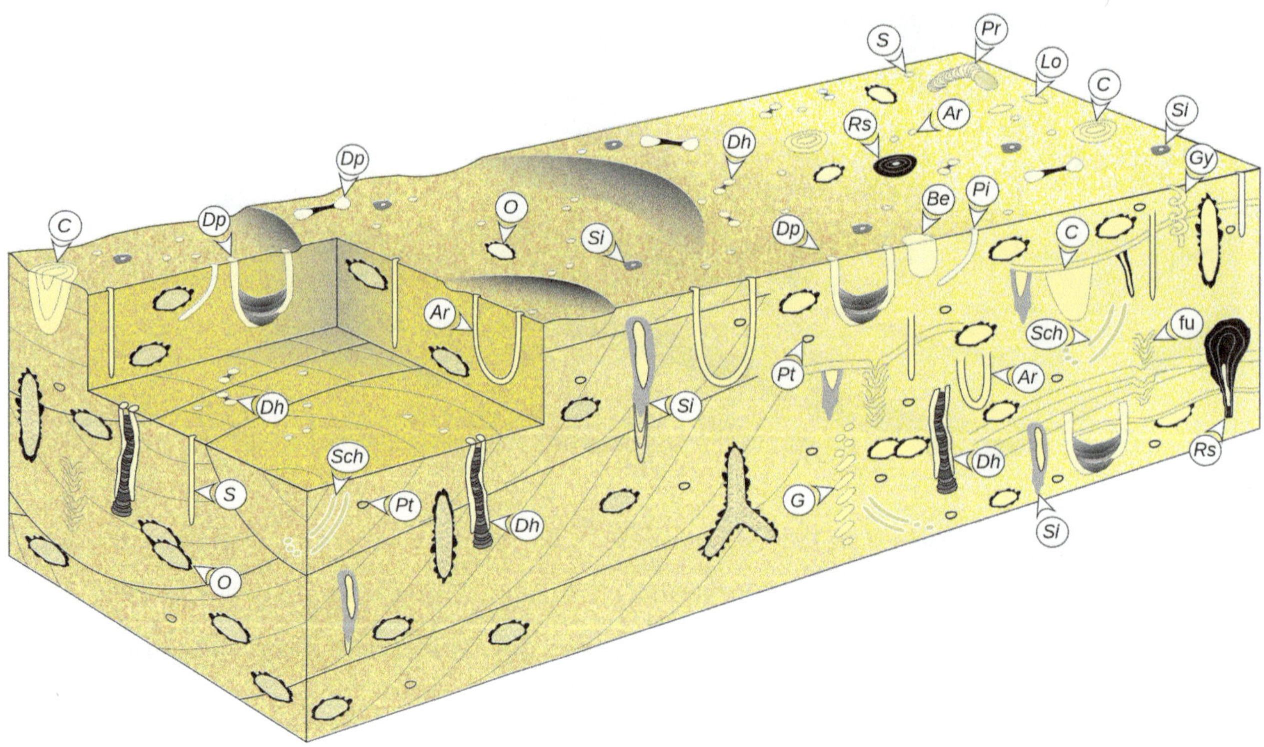

Figure 3.16 Schematic block diagram of the *Skolithos* Ichnofacies. The ichnofacies is characteristic of high-energy marine environments with clean, well-sorted, particulate (non-cohesive) substrates. The ichnofacies is generally expressed by trace-fossil suites that include ichnotaxa such as *Arenicolites* (*Ar*), *Bergaueria* (*Be*), *Conichnus* (*Co*), *Diplocraterion habichi* (*Dh*), *Diplocraterion parallelum* (*Dp*), fugichnia (fu), *Gyrolithes* (*Gy*), *Lockeia* (*Lo*), *Ophiomorpha* (*O*), *Palaeophycus tubularis* (*Pt*), *Protovirgularia* (*Pr*), *Rosselia socialis* (*Rs*), *Schaubcylindrichnus coronus* (*Sch*), *Siphonichnus* (*Si*) and *Skolithos* (*Sk*). Facies typically show a predominance of lined and unlined vertical to inclined dwelling burrows and branching dwelling burrows, with subordinate horizontal dwellings. BI values are variable but generally relatively low.

spreiten in some U-burrows, which record organism responses to substrate aggradation or degradation, respectively; (iii) minor horizontal structures; (iv) subordinate structures produced by mobile organisms; (v) low-diversity suites (although individual elements may be abundant); (vi) predominance of dwelling burrows constructed by suspension feeders or passive carnivores; and (vii) vertebrate traces in some low-energy intertidal settings. Typical ichnotaxa include permanent vertical dwelling structures of suspension feeders, carnivores and surface detritus feeders (*Arenicolites*, *Bergaueria*, *Conichnus*, *Diplocraterion*, *Gyrolithes saxonicus*, *Lingulichnus*, and *Skolithos*), branched vertical to horizontal dwelling structures of omnivores (*Ophiomorpha*), horizontal dwelling structures of predators (*Palaeophycus* and *Schaubcylindrichnus coronus*) and predation structures (*Piscichnus*). Possible deposit-feeding structures that can be associated with the *Skolithos* Ichnofacies include *Cylindrichnus*, *Macaronichnus*, *Rosselia*, *Siphonichnus* and *Taenidium*.

The *Skolithos* Ichnofacies ordinarily grades landward into supratidal or terrestrial zones and seaward into low-energy, subaqueous marine settings of the *Cruziana* Ichnofacies. The landward boundary tends to be more abrupt than that lying seaward. Moving basinward, increasing interstitial silt and clay within the sandy substrates in response to a general decline in depositional energy, as well as a decrease in sedimentation rates tend to increase the diversity of the trace-fossil suites. The resulting suites show an increase in the numbers of associated horizontal dwellings (e.g. *Palaeophycus*, *Rhizocorallium*, *Schaubcylindrichnus* and *Thalassinoides*) as well as mobile and sessile deposit-feeding structures (e.g. *Asterosoma*, *Chondrites*, *Macaronichnus*, *Phycosiphon*, *Rosselia* and *Taenidium*), generally with an accompanying increase in BI values. Finally, the *Skolithos* Ichnofacies may appear in slightly to substantially deeper-water deposits, wherever energy levels, food supplies and/or hydrographic and substrate characteristics are suitable (Crimes et al. 1981). Potential examples include submarine canyons, deep-sea fans and bathyal slopes swept by strong contour currents. As such, palaeobathymetric interpretations cannot be based solely on checklists of trace-fossil names. Rather, evaluation of associated physical sedimentary structures, stratigraphic position and other facies evidence is essential, even in normal beach-to-offshore successions.

Cruziana *Ichnofacies*

The *Cruziana* Ichnofacies (Figure 3.17) was erected by Seilacher (1967) and is characteristic of poorly sorted and unconsolidated cohesive (muddy) substrates in fully marine and permanently subtidal settings typified by uniform salinity. Such conditions generally range from low to moderate energy levels, reflecting settings near and below fair-weather (minimum) wave base but typically above storm wave base. The most common settings correspond to the offshore and extend to the very distal fringes of the lower shoreface. The ichnofacies may also occur in lower energy deeper waters. Some continental slope settings, for example, may host suites attributable to the *Cruziana* Ichnofacies (Hubbard et al. 2012). BI values are generally elevated (BI 4–6) and uniformly distributed, owing to generally slow and continuous rates of deposition, low-energy conditions and stable fully marine conditions. Episodic deposition (e.g. storm-generated event beds) typically leads to more variable and sporadically distributed bioturbation patterns. Physical sedimentary structures, where not modified or destroyed by bioturbation, most commonly include low-angle, undulatory, parallel and subparallel lamination (interpreted as hummocky cross-stratification (HCS) and oscillation ripple laminated sand, alternating with silty or sandy mudstone intervals.

As a result of reduced but not negligible energy levels, food supplies consist of both suspended and deposited components, with either fraction locally dominant or the two fully intermixed. Characteristic organisms, therefore, include suspension and deposit feeders (mobile and sessile), as well as mobile carnivores and scavengers. In response to lowered energy and cohesive substrates as well as an abundance of deposited food, trace-makers tend to construct their burrows horizontally rather than vertically, although scattered vertical or steeply inclined burrows persist in many suites. Trails of epibenthic and endobenthic foragers may also be common and reflect the abundance, diversity and accessibility of food. Surface grazing is also commonly associated but does not dominate the suites. Correspondingly, one of the hallmarks of the archetypal *Cruziana* Ichnofacies is high diversity, commonly with suites containing upwards of 10–12 ichnogenera manifested by large numbers of individual elements. Suites dominated by deposit-feeding structures but displaying low BI values and typified by low diversity are not attributable to the *Cruziana* Ichnofacies.

Suites attributable to the *Cruziana* Ichnofacies are recognised by: (i) a mixed association of horizontal, inclined and vertical structures, many of them corresponding to permanent to semi-permanent dwellings; (ii) structures constructed by mobile organisms; (iii) high diversity of ichnogenera; (iv) abundant numbers of traces; (v) predominance of deposit-feeding structures with subordinate grazing structures; and (vi) common overprinting of deep-tier over shallow-tier structures during continued burial. The dominant ichnogenera comprise both deposit-feeding structures and permanent dwellings of inferred deposit feeders. These encompass a highly diverse group that includes *Asterosoma*,

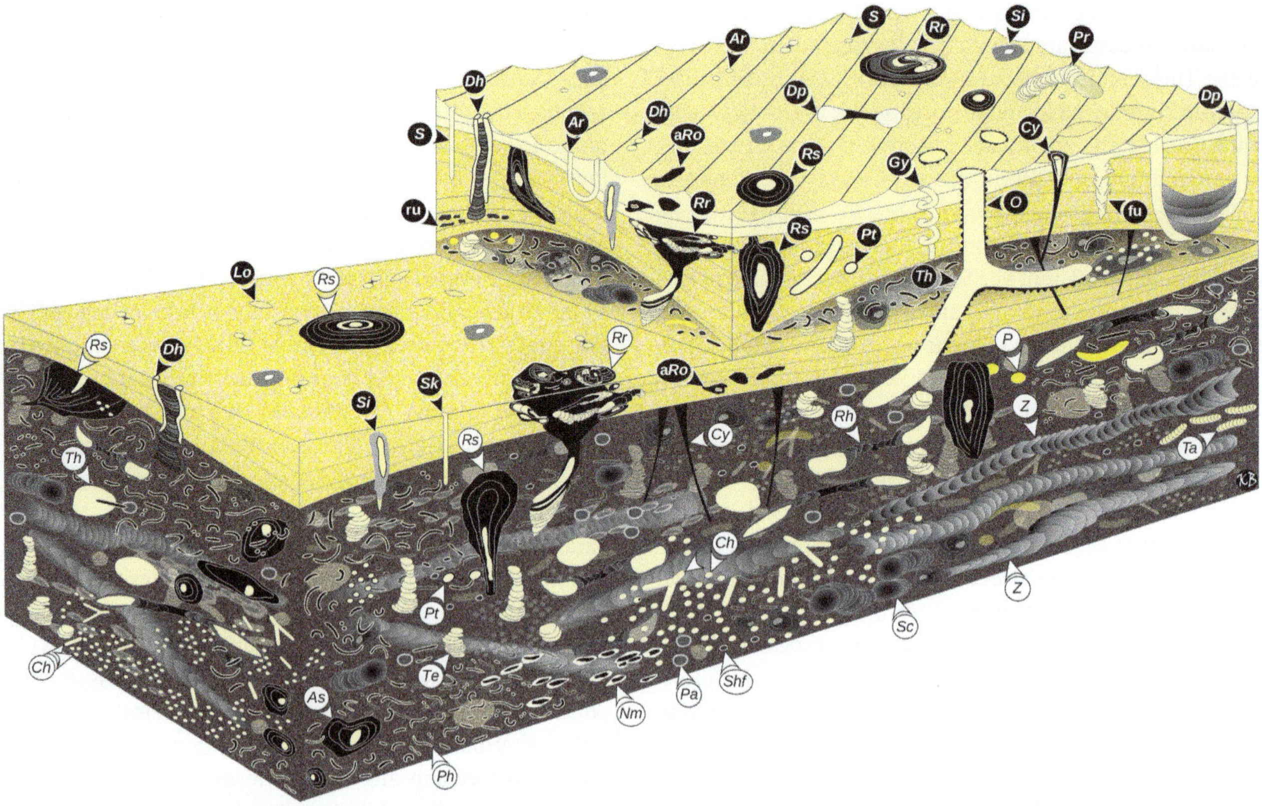

Figure 3.17 Schematic block diagram of the *Cruziana* Ichnofacies, showing the archetypal fair-weather suite (lower mudstone layer) and an opportunistic, post-storm suite (upper sandstone layer). The ichnofacies is characteristic of poorly sorted, unconsolidated cohesive substrates in fully marine subtidal settings. The ichnofacies is commonly expressed by trace-fossil suites that include ichnotaxa such as *Asterosoma* (*As*), *Chondrites* (*Ch*), *Gyrolithes* (*Gy*), *Lockeia* (*Lo*), *Nereites missouriensis* (*Nm*), *Palaeophycus heberti* (*Pa*), *Phycosiphon* (*Ph*), *Planolites* (*P*), *Protovirgularia* (*Pr*), *Rhizocorallium* (*Rh*), *Rosselia rotatus* (*Rr*), *Rosselia socialis* (*Rs*), *Schaubcylindrichnus freyi* (*Scf*), *Scolicia* (*Sc*), *Taenidium* (*Ta*), *Teichichnus* (*Te*), *Thalassinoides* (*Th*) and *Zoophycos* (*Z*). Tempestites commonly contain mudstone rip-up clasts (ru) and allochthonous fragments of *Rosselia* (a*Ro*) along their bases, and opportunistic, facies-crossing trace-fossil suites showing top-down colonisation patterns that include ichnotaxa such as *Arenicolites* (*Ar*), *Cylindrichnus* (*Cy*), *Diplocraterion habichi* (*Dh*), *Diplocraterion parallelum* (*Dp*), fugichnia (fu), *Ophiomorpha* (*O*), *Palaeophycus tubularis* (*Pt*), *Siphonichnus* (*Si*) and *Skolithos* (*Sk*). In most cases, fair-weather beds show pervasive and uniformly distributed bioturbation, with sporadically distributed burrowing indicative of the degree of storm influence on the setting.

Chondrites, *Cylindrichnus*, *Phoebichnus*, *Phycodes*, *Phycosiphon*, *Planolites*, *Rhizocorallium*, *Rosselia*, *Scolicia*, *Siphonichnus*, *Taenidium*, *Teichichnus*, *Thalassinoides* and *Zoophycos*, although this list is by no means exhaustive. Resting structures and trails associated with deposit feeding are also common and include ichnogenera such as *Cruziana*, *Gyrochorte*, *Lockeia*, *Rusophycus* and numerous others. Significant numbers of grazing structures such as *Gordia* and *Helminthopsis* may be common, with less common *Nereites missouriensis* present locally. Horizontal dwellings associated with passive carnivory such as some *Palaeophycus* and *Schaubcylindrichnus* may also be present, particularly in sandier substrates. Permanent dwelling structures of inferred suspension-feeding and omnivorous trace-makers are generally uncommon, although in some successions they comprise significant elements. The main ichnogenera include *Arenicolites*, *Diplocraterion*, *Ophiomorpha* and *Skolithos*.

In relatively shallow water, periodic scour by storm waves and renewed deposition following their cessation may lead to intercalation of tempestites within a succession of otherwise low-energy bioturbated mudstone. Event beds may show truncation of underlying burrows and the presence of escape structures. Tempestites commonly show opportunistic, post-storm top-down colonisation, consisting of facies-crossing dwelling structures such as *Diplocraterion*, *Ophiomorpha* and *Rosselia*. Eventually, the storm beds with their opportunistic suites are overlain by mudstone containing fair-weather suites of the *Cruziana* Ichnofacies. In the past, such recurring alternations were termed the 'mixed *Skolithos*-*Cruziana* Ichnofacies' but are better regarded as a composite suite rather than as an ichnofacies. Their recurring juxtaposition at the bed scale demonstrates that the suites record autogenic temporal and spatial changes in depositional process within the same sedimentary environment.

Proximal expressions of the *Cruziana* Ichnofacies: As depositional energies increase and substrates become sandier, suites of the *Cruziana* Ichnofacies increasingly incorporate permanent and semi-permanent vertical dwellings and progressively lose surface grazing structures. Such suites have been regarded as proximal expressions of the *Cruziana* Ichnofacies (see MacEachern et al. 2007 and MacEachern and Bann 2008 for summaries). These depositional conditions reflect an increase in suspended food particles and a more mobile substrate coupled with reduction in the amounts of deposited food, inhibiting surficial grazing. While deposit-feeding structures and horizontal dwellings of sessile deposit feeders continue to dominate the suites, vertical dwellings of suspension feeders and surface detritus feeders (e.g. *Diplocraterion*, *Rosselia* and *Skolithos*), branched dwellings of omnivores (e.g. *Ophiomorpha*) and horizontal dwellings of passive carnivores (e.g. *Palaeophycus* and *Schaubcylindrichnus*) become significant rather than subordinate components of the suite. Concomitantly, grazing structures such as *Helminthopsis* and *Nereites missouriensis* are largely absent. Such suites are transitional to more distal expressions of the *Skolithos* Ichnofacies.

Distal expressions of the *Cruziana* Ichnofacies: Similarly, as depositional energies decrease and substrates become muddier, suites of the *Cruziana* Ichnofacies increasingly incorporate surface grazing structures (e.g. *Gordia*, *Helminthopsis* and *Nereites missouriensis*) and progressively lose vertical and inclined dwelling structures (e.g. *Cylindrichnus*, *Diplocraterion*, *Rosselia* and *Skolithos*). Such suites have been regarded as distal expressions of the *Cruziana* Ichnofacies (see MacEachern et al. 2007 and MacEachern and Bann 2008 for summaries). Robust horizontal dwelling structures such as *Palaeophycus* and *Schaubcylindrichnus coronus* become uncommon, with more diminutive, thinly lined horizontal dwellings such as *Schaubcylindrichnus freyi* persisting. Deposit-feeding structures associated with lower-energy conditions (e.g. *Asterosoma*, *Chondrites*, *Phycosiphon* and *Zoophycos*) and deeper tier dwellings associated with muddy substrates (e.g. *Thalassinoides*) tend to dominate. Intrastratal, mobile deposit-feeding (e.g. *Planolites*, *Scolicia* and *Taenidium*) may be locally common. Such suites are generally accompanied by a decrease in diversity, reflecting diminishment in the range of ethological approaches to exploiting deposited food on cohesive substrates, but contain numerous ichnogenera that are positively correlated to fully marine conditions and tend to show elevated BI values reflecting slow, continuous deposition of mud. The presence of tempestites may increase the total number of ichnogenera within the trace-fossil association, but post-storm suites themselves remain of low diversity and stand in stark contrast to the ambient fair-weather suites. Suites recording distal expressions of the *Cruziana* Ichnofacies are transitional to those of the *Zoophycos* Ichnofacies.

Rosselia *Ichnofacies*

The *Rosselia* Ichnofacies (Figure 3.18; MacEachern and Bann 2020) expresses organism responses to the dynamic range of physico-chemical conditions associated with sandstone-dominated delta-front environments. The ichnofacies is best reflected by trace-fossil suites that occur in mixed river- and wave-influenced delta successions, as these highlight the interplay of normal marine conditions, shoaling waves, storm event cycles, sediment-gravity transport and episodic fluvial discharge into the basin. Within this setting, no single behavioural strategy is sufficient for characterising the resulting suites that typify the ichnofacies. Further, as the relative importance of fluvial, fair-weather wave, storm wave or tidal energy changes within a system, the resulting ichnological suites begin to show departures from this archetype (MacEachern and Bann 2023).

The *Rosselia* Ichnofacies is recognised by trace-fossil suites showing variable (BI 0–5; though typically BI 0–3) bioturbation intensities and sporadically distributed bioturbation recording high-energy, episodic deposition of sandstone-dominated heterolithic marine successions. Markedly variable BI values occur in both mudstone and sandstone beds, with BI values typically declining upwards in most delta-front successions, particularly as sandstone contents increase. Variable bioturbation intensities commonly record fluctuating deposition rates and event-style sediment emplacement.

Pre-event beds and bedsets display higher ichnological diversities, higher bioturbation intensities and trace-fossil suites positively correlated to normal marine conditions. These are juxtaposed with ichnological suites recording episodic, top-down opportunistic colonisation of event beds or with beds entirely devoid of bioturbation. The *Rosselia* Ichnofacies is recognised by bed-scale but not ubiquitous presence of trace-fossil suites containing ichnotaxa positively correlated to normal marine conditions, such as *Asterosoma*, *Chondrites* and *Phycosiphon* (in muddier intervals) and robust *Diplocraterion*, *Macaronichnus segregatis*, *Rhizocorallium* and *Scolicia* (in sandier successions). These ichnogenera, however, tend to constitute subordinate elements. As mudstone contents decline in response to increasing depositional energy, these specialised ichnotaxa become increasingly uncommon.

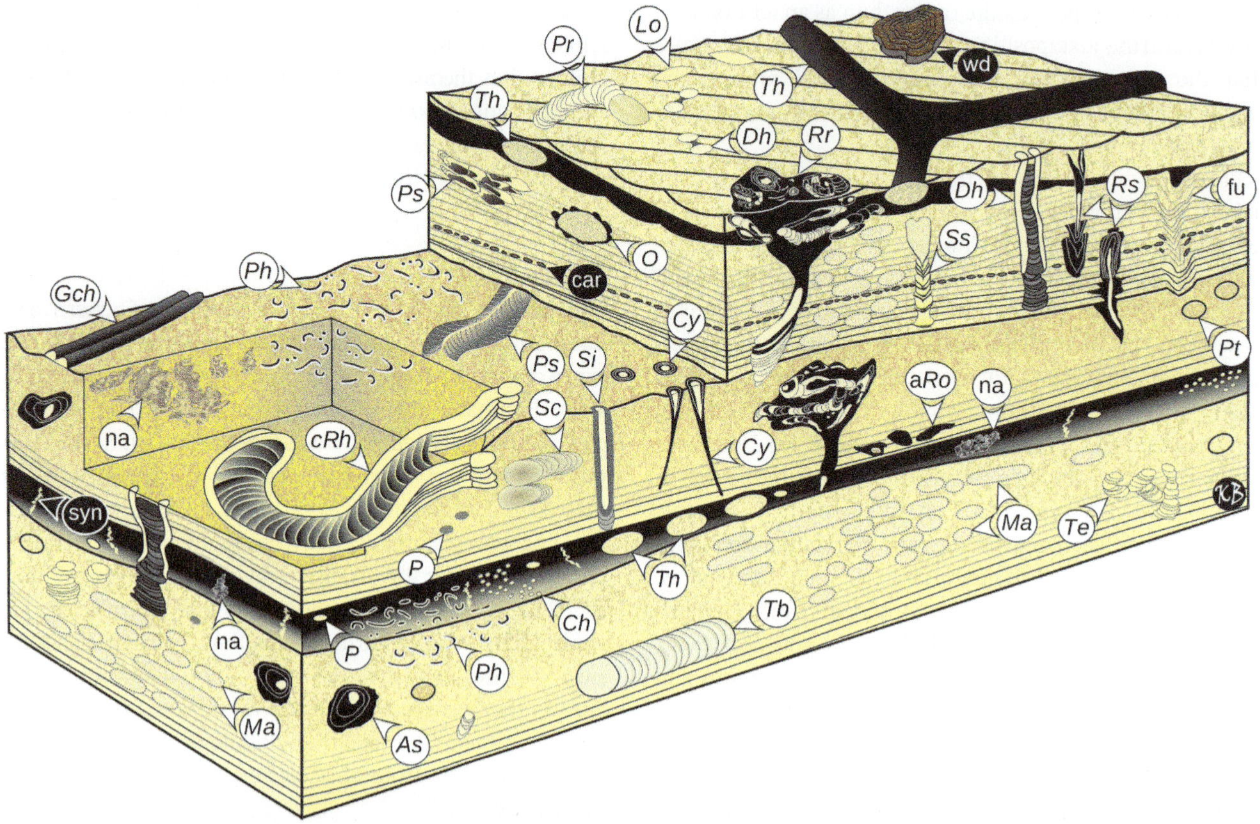

Figure 3.18 Schematic block diagram of the *Rosselia* Ichnofacies. The ichnofacies is characteristic of physico-chemically variable conditions associated with sandstone-dominated heterolithic bedsets deposited in marine delta-front environments. Such settings highlight the interplay of normal marine conditions, shoaling waves, storm events, sediment-gravity transport and episodic fluvial discharge into the basin. The ichnofacies is commonly expressed by complexly juxtaposed trace-fossil suites that include ichnotaxa such as allochthonous fragments of *Rosselia* (a*Ro*), *Asterosoma* (*As*), *Chondrites* (*Ch*), *Cylindrichnus* (*Cy*), *Diplocraterion habichi (Dh),* fugichnia (fu), *Gyrochorte* (*Gch*), *Lockeia* (*Lo*), *Macaronichnus* isp. (*Ma*), navichnia (na), *Ophiomorpha* (*O*), *Palaeophycus heberti* (*Pa*), *Palaeophycus tubularis* (*Pt*), *Phycosiphon* (*Ph*), *Planolites* (*P*), *Protovirgularia* (*Pr*), *Psammichnites* (*Ps*), complex *Rhizocorallium* (c*Rh*), *Rosselia rotatus* (*Rr*), *Rosselia socialis* (*Rs*), *Scolicia* (*Sc*), *Siphonichnus sersumdeorsum* (*Ss*), *Taenidium barretti* (*Tb*), *Teichichnus* (*Te*) and *Thalassinoides* (*Th*). Intercalated mudstone beds commonly contain syneresis cracks (syn) and sandstone beds generally contain abundant carbonaceous detritus (car) and wood fragments (wd). Facies typically show highly variable and sporadically distributed bioturbation at the bed and bedset scales, reflecting fluctuating physico-chemical stress in an otherwise marine environment.

The fully marine suites are juxtaposed at the bed and bedset scale with low-diversity trace-fossil suites dominated by facies-crossing elements such as *Planolites*, *Schaubcylindrichnus freyi*, *Teichichnus* and *Thalassinoides* and in mudstone layers and *Arenicolites*, *Cylindrichnus*, *Macaronichnus* isp., *Ophiomorpha* (particularly *O. irregulaire*), *Palaeophycus*, *Rosselia*, *Siphonichnus* and *Teichichnus* in sandstone beds. Such suites are characteristic of elevated physico-chemical stress. Event beds commonly display fugichnia (escape structures) and, if burrowed, show top-down colonisation. Although this colonisation pattern is common to all settings dominated by event-style deposition, event beds in these systems may be either sandstone or mudstone layers, which is atypical of many other shallow-marine environments. Rapidly emplaced sandstone event beds are commonly tempestites and hyperpycnites or other sediment-gravity flow deposits. Rapidly deposited mud is typically associated with fluid mud emplacement and/or its bedload transport. Mudstone layers of fluid mud origin may also show the presence of navichnia (sediment-swimming structures). Elevated deposition rates are reflected by the recurring presence of re-equilibration structures (e.g. bivalve readjustment, retrusive *Diplocraterion* and/or *Rosselia* dwelling tubes, complex, upwardly shifting *Rhizocorallium* and vertically stacked *Rosselia*). Cryptic bioturbation is common in some erosionally amalgamated tempestites.

Most ichnogenera representing domiciles are those constructed for the purposes of deposit feeding (surface detritus feeding or substratal probing) and carnivory (e.g. *Cylindrichnus*, *Palaeophycus*, *Rhizocorallium*, *Rosselia*, *Schaubcylindrichnus freyi*, *Siphonichnus*, *Teichichnus* and

Thalassinoides). Secondary ethologies are associated with interstratal and intrastratal mobile deposit feeding (e.g. *Macaronichnus* isp., *Planolites* and *Taenidium*). Locomotion and resting structures (e.g. *Gyrochorte*, *Lockeia* and *Psammichnites*) may be present locally. With increasing erosional amalgamation of sandstone beds, shallow-tier structures are preferentially removed. Prevalence of deposit-feeding behaviours and a paucity of ichnogenera positively correlated with suspension feeding strategies (e.g. *Bergaueria*, *Conichnus*, *Conostichus* and *Skolithos*) in otherwise sand-dominated (higher-energy) substrates may be an indicator of elevated water turbidity that inhibits effective filter feeding (Moslow and Pemberton 1988; MacEachern et al. 2005; MacEachern and Bann 2020), consistent with delta-front conditions. In settings of reduced physico-chemical stress, suites attributable to the *Rosselia* Ichnofacies may become intergradational with those associated with proximal expressions of the *Cruziana* Ichnofacies and distal expressions of the *Skolithos* Ichnofacies, particularly in wave-dominated deltas (MacEachern and Bann 2023).

Phycosiphon *Ichnofacies*

The *Phycosiphon* Ichnofacies (Figure 3.19; MacEachern and Bann 2020) was erected to express organism responses to the dynamic range of erosional and depositional conditions associated with mudstone-dominated heterolithic successions of marine prodeltaic environments. The ichnofacies is best reflected by trace-fossil suites that occur in prodeltas of mixed river- and wave-influenced deltas, as these display a balance between ambient (normal marine) conditions and

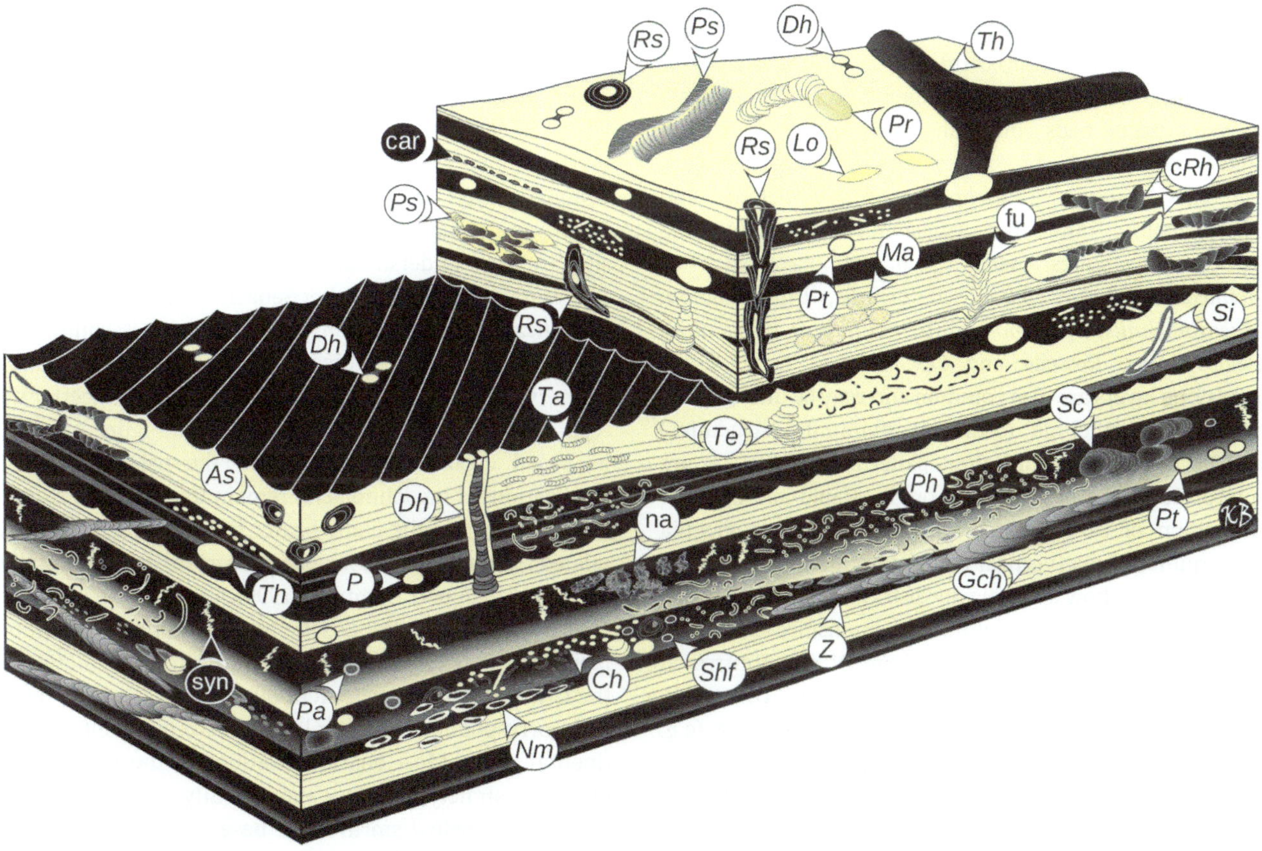

Figure 3.19 Schematic block diagram of the *Phycosiphon* Ichnofacies. The ichnofacies is characteristic of a dynamic range of erosional and depositional conditions associated with mudstone-dominated heterolithic bedsets of marine prodelta environments. Such settings highlight the interplay of normal marine conditions lying below fair-weather wave base and the physico-chemical stresses brought about by periodic fluvial discharge. The ichnofacies is commonly expressed by complexly juxtaposed trace-fossil suites that include ichnotaxa such as *Asterosoma* (*As*), *Chondrites* (*Ch*), *Diplocraterion habichi* (*Dh*), fugichnia (fu), *Gyrochorte* (*Gch*), *Lockeia* (*Lo*), *Macaronichnus* isp. (*Ma*), navichnia (na), *Nereites missouriensis* (*Nm*), *Palaeophycus heberti* (*Pa*), *Palaeophycus tubularis* (*Pt*), *Phycosiphon* (*Ph*), *Planolites* (*P*), *Protovirgularia* (*Pr*), *Psammichnites* (*Ps*), complex *Rhizocorallium* (cRh), *Rosselia socialis* (*Rs*), *Schaubcylindrichnus freyi* (*Scf*), *Scolicia* (*Sc*), *Siphonichnus* (*Si*), *Taenidium* (*Ta*), *Teichichnus* (*Te*), *Thalassinoides* (*Th*) and *Zoophycos* (*Z*). The relative importance of the different ichnogenera shifts as sandstone contents increase. Intercalated mudstone beds commonly contain syneresis cracks (syn) and sandstone beds generally contain abundant carbonaceous detritus (car). Facies typically show highly variable and sporadically distributed bioturbation at the bed and bedset scales, reflecting fluctuating physico-chemical stress in an otherwise low-energy marine environment.

the environmental stresses brought about by periodic fluvial discharge. The resulting juxtaposed ichnological suites are typified by pronounced changes in ethological characteristics, bioturbation intensities and faunal colonisation patterns, which record the complex interactions of fully marine basinal conditions with fluvial sediment influx in distal, low-energy marine environments that lie below fair-weather wave base. These variations occur at the bed and bedset scales and point to short-term but recurring alternations between stable marine conditions and periods of elevated physico-chemical stress. As the relative influence of fair-weather waves, storm waves, tidal energy and fluvial sediment influx changes, the resulting facies and trace-fossil suites depart from this archetype, albeit in predictable ways (MacEachern and Bann 2023).

The *Phycosiphon* Ichnofacies is recognised by sporadically distributed trace-fossil suites in heterolithic bedsets containing ichnotaxa positively correlated to fully marine conditions (e.g. *Asterosoma*, *Chondrites*, *Nereites missouriensis*, *Phycosiphon*, *Rhizocorallium*, *Scolicia* and *Zoophycos*). These beds are interstratified at the bedset scale with mudstone beds containing low-diversity trace-fossil suites dominated by facies-crossing elements (e.g. *Planolites*, *Schaubcylindrichnus freyi*, *Teichichnus* and *Thalassinoides*) and thin sandstone layers containing low-diversity suites made up of ichnotaxa such as *Arenicolites*, *Asterosoma*, *Cylindrichnus*, *Diplocraterion*, *Macaronichnus*, *Ophiomorpha* (particularly *O. irregulaire*), *Palaeophycus*, *Rosselia* and *Siphonichnus*. Dark, fissile mudstone layers are also common and may be unburrowed or display navichnia (sediment swimming structures) and/or top-down colonisation patterns. Such layers are interpreted to be of fluid mud origin. Rapid and episodic deposition of beds are typically expressed by fugichnia and equilibrichnia (vertical readjustment structures).

Most facies show sporadically distributed burrowing displaying highly variable bioturbation intensities (BI 0–5) at both the bed and bedset scales. Low BI or unburrowed intervals typically record rapid deposition, commonly accompanied by evidence of physico-chemical stress (e.g. reduced salinity, reduced O_2 and/or soupground conditions). Phytodetrital concentrations, synaeresis (subaqueous shrinkage) cracks, normal and inverse grading, soft-sediment deformation and sharp-based, carbonaceous, fissile mudstone layers commonly point to these concomitant stresses. Some sandy marine event beds, particularly tempestites, show the presence of cryptic bioturbation, associated with meiofaunal activity (e.g. Pemberton et al. 2008).

The dominant ethological groupings of suites attributable to the ichnofacies are mobile deposit-feeding (e.g. *Phycosiphon*, *Planolites*, *Macaronichnus* and *Scolicia*) and sessile deposit-feeding behaviours (e.g. *Asterosoma*, *Chondrites*, *Rosselia*, *Siphonichnus*, *Taenidium*, *Teichichnus*, *Thalassinoides* and *Zoophycos*), as well as domiciles used for deposit feeding or carnivory (e.g. *Palaeophycus*, *Rhizocorallium*, *Schaubcylindrichnus freyi* and *Thalassinoides*). Surface grazing (e.g. *Helminthopsis* and *Nereites missouriensis*) is typically a subordinate ethology, possibly owing to the generally elevated deposition rates in such settings. Locomotion structures (e.g. *Cruziana*, *Diplichnites*, *Gyrochorte* and *Psammichnites*) and resting structures (e.g. *Lockeia* and *Rusophycus*) may be common, although more readily identified from outcrop datasets. Ethologies are consistent with low-energy depositional conditions with accumulation of food resources largely from suspension settling. In settings of reduced physico-chemical stress, suites may be intergradational with those of the *Cruziana* Ichnofacies, particularly in weakly storm-influenced but wave-dominated delta settings (MacEachern and Bann 2023).

Zoophycos *Ichnofacies*

The *Zoophycos* Ichnofacies (Figure 3.20) was erected by Seilacher (1967) but has seen significant revision since its original inception (Seilacher 1978; Frey and Seilacher 1980; Bromley 1990). Of all recurring marine ichnofacies, the *Zoophycos* Ichnofacies remains the most debated and possibly the least understood. This arises largely from the fact that more so than in any other Seilacherian ichnofacies, the identification of the *Zoophycos* Ichnofacies is commonly dependent on the identification of the namesake ichnogenus within the suites. Given the ongoing uncertainty regarding the ethology of *Zoophycos* (see Figure 3.2) and its broad depositional occurrence, the palaeoecological implications of the ichnofacies remain unclear. As a result, its designation as the name-bearer for a supposedly depth-related ichnofacies has long been controversial (see Kotake 1991; Uchman and Demírcan 1999).

In popular depositional schemes, the *Zoophycos* Ichnofacies is typically portrayed as an intermediary between the *Cruziana* and *Nereites* ichnofacies, residing at a position corresponding more or less to the outer shelf to continental slope. Most characterisations of the ichnofacies place the archetype in low-energy, fully marine environments lying below storm wave base with silty to clayey substrates typical of many muddy shelves. However, the ichnofacies does occur in some sandy successions associated with turbidity currents and debris flows (Buatois and Mángano 1992) and so extends onto slope environments as well (Hubbard et al. 2012). As re-evaluated by Seilacher (1978) and Frey and Seilacher (1980), one of the major environmental controls suggested for the ichnofacies is lowered oxygen levels associated with organic debris accumulation in quiet-water settings: a scenario deemed

Figure 3.20 Schematic block diagram of the *Zoophycos* Ichnofacies. The ichnofacies is characteristic of low-energy, fully marine environments lying below storm wave base with silty to clayey substrates, typical of many muddy shelves. The ichnofacies is commonly expressed by trace-fossil suites that include ichnotaxa such as *Asterosoma* (*As*), *Chondrites* (*Ch*), *Nereites missouriensis* (*Nm*), *Phycosiphon* (*Ph*), *Planolites* (*P*), *Schaubcylindrichnus freyi* (*Scf*), *Scolicia* (*Sc*), *Spirophyton* (*Sp*), *Thalassinoides* (*Th*) and *Zoophycos* (*Z*). Unless settings are prone to significant reduction of dissolved oxygen near the bed, facies tend to be pervasively bioturbated.

common to muddy shelves and epeiric seaways. However, it has since been pointed out that like many deep-tier structures, it may merely indicate the ability of the *Zoophycos* trace-maker to extend its structure into oxygen-depleted substrates (see Buatois and Mángano 2011). The robust nature of the ichnogenus coupled with the elevated bioturbation intensities commonly associated with many of the suites argues against oxygen depletion of waters immediately above the substrate. Hence, the linkage of *Zoophycos* to opportunism and of the ichnofacies to physico-chemically stressed environments remains far from certain.

Suites attributable to the *Zoophycos* Ichnofacies are characterised by: (i) a dominance of simple to complex deposit-feeding structures; (ii) significant but generally subordinate proportions of shallow-tier grazing structures; (iii) lowered diversity, although with locally abundant individual traces; (iv) deep-tier structures generated by deposit feeders and bacterial farmers; and (v) horizontal to gently inclined, spreiten-bearing structures. Suites are dominated by *Zoophycos*, with common *Helminthopsis*, *Nereites missouriensis*, *Phycosiphon* and *Planolites*, and lesser *Asterosoma*, *Chondrites*, *Lophoctenium*, *Neonereites*, *Scolicia*, *Spirophyton*, *Taenidium* and *Thalassinoides*. In general, shelfal expressions of the *Zoophycos* Ichnofacies tend to consist of lower diversity suites than do their slope-related counterparts, although diversities may vary locally, and bioturbation intensities show marked variation in slope occurrences characterised by turbiditic or contouritic sedimentation (Hubbard et al. 2012).

Considering the above characteristics of the ichnofacies, together with the widespread distribution of individual specimens of *Zoophycos* in both shallow- and deep-water deposits (Frey et al. 1990; Olivero and Gaillard 1996; Uchman and Demírcan 1999), it is probable that *Zoophycos*-producing animals were simply broadly adapted in most ecologic respects. Some animals tolerated not only a considerable range of water depths but also numerous substrate types, variable food resources, and different energy and oxygen levels. Their traces, therefore, appear in the *Cruziana* through to *Nereites* ichnofacies (Crimes et al. 1981). Additionally, it is generally regarded that the ichnogenus shows a progressive shift from shallow-water positions to deeper waters throughout the Phanerozoic (Bottjer et al.

1988), making identification of the ichnofacies and its depositional assignment problematic for many Palaeozoic intervals (Buatois and Mángano 2011).

Nereites *Ichnofacies*

The *Nereites* Ichnofacies (Figure 3.21) was erected by Seilacher (1967) and more formally defined by Frey and Seilacher (1980) and Frey and Pemberton (1984). Suites attributable to the ichnofacies were originally identified from flysch deposits, but presumably this is more a reflection of a preservational bias, wherein ichnogenera at the sediment–water interface are cast on the soles of dilute, low-density turbidites. Two models have been proposed for preserving casts of these surficial and shallow-tier structures: the widely accepted model of subtle erosion of the uppermost millimetres of mud during emplacement of the turbidite, or the shock-wave model wherein the arrival of the turbidite draws unconsolidated mud at the surface back into suspension (Seilacher 2007). Highly dynamic settings prone to high-frequency emplacement of sediment gravity flows are less likely to accord sufficient environmental stability required for the development of suites diagnostic of the ichnofacies, particularly the grouping of surface traces known as the 'graphoglyptids' (see below). Studies of modern abyssal plains show the presence of graphoglyptids, which occur beyond the reach of most turbidity currents. Under such conditions, these shallow-tier structures are likely to generate mottled bioturbated fabrics that are cross-cut by mid-tier and deep-tier structures such as *Planolites*, *Teichichnus* and *Zoophycos* (Ekdale and Berger 1978). That being said, the palaeodepositional implications of the *Nereites* Ichnofacies are less equivocal than those of any other archetypal ichnofacies because while numerous trace fossils otherwise typical of shallow-water deposits may range *into* deep-sea deposits, the reverse is not ordinarily true. Suites of the *Nereites* Ichnofacies record animal-sediment interactions associated with base-of-slope turbidite systems at the fringes of terminal splays, crevasse splays and distal fan (lobe) margins. Suites also have been suggested to occur on channel levee

Figure 3.21 Schematic block diagram of the *Nereites* Ichnofacies, showing the sole of a turbidite that casts a series of graphoplyptids below and the return to ambient, non-turbiditic deposition characterised by more pervasive bioturbation and the predominance of deep-tier structures above. The ichnofacies is characteristic of deep-water fully marine and oxygenated environments favouring very slow rates of deposition and resource-limited conditions. The ichnofacies is commonly expressed by trace-fossil suites that include shallow-tier graphoglyptids and grazing structures such as *Cosmorhaphe* (*Cr*), *Gordia* (*Go*), *Lorenzinia* (*Lr*), *Nereites irregularis* (*Ni*), *Nereites missouriensis* (*Nm*), *Paleodictyon* (*Pd*), *Spirodesmos* (*Sd*), *Spirophycus* (*Sh*), *Spirorhaphe* (*Sr*) and *Urohelminthoida* (*Ur*), as well as deeper-tier structures such as *Chondrites* (*Ch*), *Phycosiphon* (*Ph*), *Planolites* (*P*), *Scolicia* (*Sc*), *Spirophyton* (*Sp*), *Thalassinoides* (*Th*) and *Zoophycos* (*Z*). Ambient deposition is dominated by pervasive and uniform bioturbation, with sporadically distributed burrowing marked by intercalation of sediment-gravity flow deposits.

successions, although this appears inconsistent with the required slow, continuous deposition and stable environmental conditions needed for the archetypal benthic community. It has also been suggested that a shallow-water expression of such suites may occur, but this has yet to be fully documented (see Buatois and Mángano 2011 for a summary).

Suites of the *Nereites* Ichnofacies are characterised by: (i) dominance of complex patterned structures referred to as graphoglyptids, which reflect organisms that farm bacteria and/or trap microorganisms; (ii) numerous sophisticated horizontal crawling/grazing traces and sinuous faecal castings; (iii) structures produced by deposit feeders, scavengers, or possibly harvesters (i.e. agrichnia); (iv) predominance of shallow-tier rather than deep-tier structures; and (v) high diversities but characterised by low abundances of individual ichnotaxa. Graphoglyptids are arguably the most diagnostic element of the suites attributable to the ichnofacies. Characteristic ichnogenera include *Acanthorhaphe*, *Cosmorhaphe*, *Desmograpton*, *Helminthorhaphe*, *Megagrapton*, *Paleodictyon*, *Paleomeandron*, *Spirorhaphe* and *Urohelminthoida*, among others. Grazing and deposit-feeding structures include *Gordia*, *Helminthopsis*, *Nereites irregularis* and *N. missouriensis*. Surface feeding traces include radially arranged structures (e.g. *Glockerichnus* and *Lorenzinia*), branched systems (e.g. *Polykampton*) and some simple meandering features (e.g. *Halopoa*). Mobile deposit-feeding structures include *Lophoctenium* and *Scolicia*.

Depth- and energy-related variables in a fully marine and oxygenated environment appear to be of paramount importance to the resulting trace-fossil suites. Depth and energy constraints favour very slow rates of deposition in a low-energy setting dominated by reduced rates of suspension settling of silt, clay and food resources in a stable environment. Animals exploiting these lower bathyal to abyssal environments have two principal concerns: (i) scarcity of food, relative to more abundant supplies in shallower settings; and (ii) periodic disruption by strong, down-canyon bottom currents or actual turbidity currents. Resource-restricted settings are typified by sophisticated feeding strategies, as expressed by graphoglyptids and grazing structures showing complex patterns. Episodic emplacement of event beds ultimately leads to juxtaposed suites, composed of two component parts: pre-turbidite and post-turbidite associations. In many regards, these mirror the pre- and post-storm suites of the shallow-marine realm. The pre-turbidite resident association is characteristic of quiet, normal conditions and is dominant wherever the substrate is free of the influence of turbidity currents. It tends to be overwhelmed or eliminated by pronounced erosion, turbulence or significant deposition, however, and replaced by the post-turbidite association following cessation of the sediment-gravity event. As the prevailing low-energy conditions are re-established, so too are the benthic communities responsible for the pre-turbidite trace-fossil suite. The pre-turbidite ichnogenera (e.g. *Cosmorhaphe*, *Lorenzinia*, *Nereites* and *Paleodictyon*) reflect a stable, equilibrium community, well adapted to quiet-water conditions and so correspond to the archetypal expression of the ichnofacies. By contrast, post-turbidite ichnogenera (e.g. *Diplocraterion*, *Ophiomorpha* and *Skolithos*) represent a more opportunistic, less stable community better adapted to colonisation of rapidly emplaced turbidites (Frey and Seilacher 1980).

The *Nereites* Ichnofacies has also been subdivided into the *Nereites* and *Paleodictyon* ichnosubfacies (Seilacher 1974) for distal fan turbidites and slightly more proximal turbidites containing casts of graphoglyptids, respectively. A third ichnosubfacies, termed the *Ophiomorpha rudis* subfacies, has been proposed for channel and proximal fan lobe deposits (Uchman 2009).

3.3.3.6 Substrate-Controlled Ichnofacies

Currently, three Seilacherian ichnofacies (*Glossifungites*, *Trypanites* and *Teredolites* ichnofacies) are recognised as indicators of marine to marginal marine colonisation (or recolonisation) of exhumed, compacted or cemented substrates. As such, the resulting ichnological suites post-date the original accumulation of the sedimentary media in which they occur and cross-cut any pre-existing softground trace-fossil suites. Such structures are regarded as 'palimpsest', in that they are superimposed structures created at different – in these cases, later – times. One must be cautious to ensure that the observed cross-cutting features are not just deep-tier structures overprinting earlier shallow-tier features, which is merely the result of ongoing deposition and burial of sediment or record deep-tier structures that extend into beds or bedsets that have been more deeply buried. Careful consideration of the details of the structures and the surfaces from which they subtend is required to discern the nature of the cross-cutting relationship, before assignment to the ichnofacies can be reliably made. Most suites attributable to the substrate-controlled ichnofacies commonly demarcate periods of stratigraphic omission and hence, may have important implications for the sequence stratigraphic evaluation of sedimentary succession (see Sections 3.4 and 4.4).

In many examples, these ichnofacies are associated with erosionally exhumed discontinuities in the rock record, although scenarios favouring autogenic development are also common. In the case of the *Glossifungites* Ichnofacies, some firmground suites may be intergradational with palimpsest softground suites, particularly where substrates pass from muddy to sandy, because while burial compaction may lead to firm or stiff mud, it may permit sandy units to retain their non-cohesive character. Likewise, suites attributable to the *Glossifungites* Ichnofacies may be intergradational with hardground suites of the *Trypanites* Ichnofacies, wherein differential compaction and/or cementation occur (e.g. Bromley 1975, 1990). The *Teredolites* Ichnofacies is somewhat more problematic within this grouping, as the consistency of the substrate may or may not fundamentally change prior to colonisation. Some woodground suites correspond to the colonisation of the xylic substrate prior to coalification, whereas others are associated with erosional discontinuities that have incised into coal seams (presumably xylic material coalified prior to exhumation; e.g. Bromley et al. 1984).

Glossifungites *Ichnofacies*

The *Glossifungites* Ichnofacies (Figure 3.22) was proposed by Seilacher (1967) and redefined by Frey and Seilacher (1980). It is characteristic of firm but unlithified substrates and predominantly occurs in dewatered and compacted mud deposits. Incipiently cemented sand intervals may also host firmground ichnogenera under some circumstances. In most siliciclastic settings, suites attributable to the *Glossifungites* Ichnofacies are the most widespread and pervasive expression of substrate-controlled suites. As currently defined, suites attributable to the ichnofacies record marine or marginal marine colonisation of firmground substrates, regardless of its origin. These suites commonly demarcate a number of erosional discontinuities of both sequence stratigraphic importance (e.g. transgressive surfaces of erosion, amalgamated sequence boundaries and flooding surfaces, regressive surfaces of erosion; Section 4.4) and autogenic derivation (e.g. cut-bank margins of tidal channels and periodically exposed intertidal flats).

Many ichnogenera occur in both softground and firmground substrates, requiring the worker to clarify the expression in any characterisation of the suites. The softground expression is presumed for most ichnogenera unless otherwise stated. Hence, communication of the firmground expressions is necessary in order to assign it to a suite of the *Glossifungites* ichnofacies. Suites of the *Glossifungites* Ichnofacies are characterised by: (i) sharp-walled and unlined dwelling structures that are predominantly vertical to sub-vertical, with or without spreite (e.g. firmground *Arenicolites*, *Diplocraterion*, *Gastrochaenolites* and *Skolithos*), reflecting suspension-feeding organisms and passive carnivores; (ii) common but subordinate horizontal and locally branched structures (*Glossifungites*, *Palaeophycus*, *Rhizocorallium*, *Spongeliomorpha* and *Thalassinoides*) of presumed deposit feeders, passive carnivores or

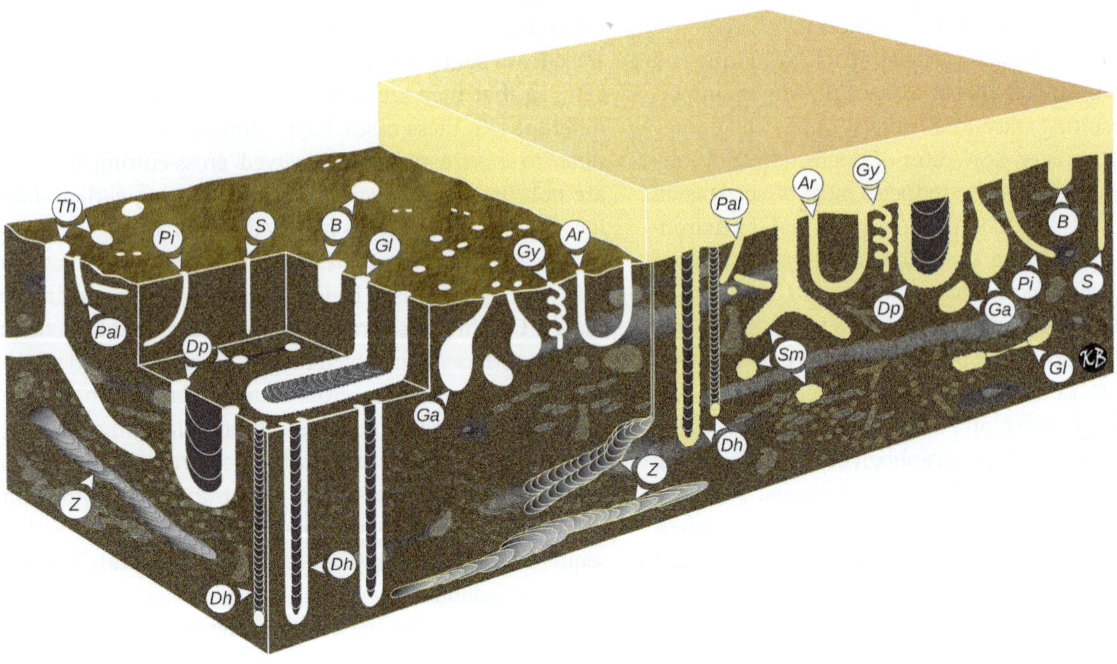

Figure 3.22 Schematic block diagram of the *Glossifungites* Ichnofacies. The ichnofacies is characteristic of the marine or marginal marine post-depositional colonisation of firm, compacted substrates. The ichnofacies is commonly expressed by trace-fossil suites that include firmground examples of ichnogenera such as *Arenicolites* (*Ar*), *Bergaueria* (*B*), *Chondrites* (*Ch*), *Diplocraterion habichi* (*Dh*), *Diplocraterion parallelum* (*Dp*), *Gastrochaenolites* (*Ga*), *Glossifungites saxicava* (*Gl*), *Gyrolithes* (*Gy*), *Palaeophycus* isp. (*Pal*), *Psilonichnus* (*Pi*), *Skolithos* (*Sk*), *Spongeliomorpha iberica* (*Sm*) and *Zoophycos* (*Z*). Firmground suites vary markedly in bioturbation intensity, cross-cut pre-existing trace-fossil suites and commonly demarcate stratigraphic omission surfaces.

omnivores; (iii) presence of scratch-marked walls on some ichnogenera; (iv) low-diversity suites with generally high abundances of individual ichnogenera; and (v) burrows typically showing passive infill that corresponds to detritus overlying an omission surface. The presence of robust vertical shafts within shaly intervals is anomalous, contrasting sharply with the predominantly horizontal and diminutive trace fossils typical of such muddy intervals. The presence of these vertical shafts commonly points to a firmground condition, because such structures are not capable of being maintained in a muddy substrate while it is soft. That said, workers need to ensure that the traces in question are not merely deep-tier elements extending down into compacted mud from a much higher level.

In siliciclastic settings, most firmground suites are associated with erosionally exhumed (dewatered and compacted) substrates and, hence, demarcate erosional discontinuities. The overwhelming majority of these omission suites originate in marine or marginal marine settings. Consequently, while a discontinuity may be generated in either subaerial or submarine settings, its colonisation occurs under marine-influenced conditions. The suites attributable to the *Glossifungites* Ichnofacies indicate that a depositional hiatus occurred between the erosional event and sedimentation of the overlying unit (Figure 3.23) during which benthic colonisation ensued. Significant depositional cover of the discontinuity generally precludes firmground colonisation. These aspects of the omission suite make it useful both in the recognition of the stratigraphic discontinuity and its genetic interpretation.

Nevertheless, several depositional systems are prone to autogenically produced erosional contacts that may be colonised by suites of the *Glossifungites* Ichnofacies. In submarine fans, this is due to the erosive capacity of turbidity currents, which need not result in deposition of a sedimentary unit. Eroded margins of laterally migrating tidal channels and tidal creeks also yield firmground suites of autogenic origin. In addition, long periods of depositional quiescence may also permit mudstone units to become somewhat 'firm' prior to colonisation (Sections 4.5 and 9.3.2). These suites, however, differ from those associated with more deeply exhumed allogenic firmgrounds. Autogenic *Glossifungites* Ichnofacies-demarcated surfaces tend to show evidence of burrow compaction, tend to be more deeply penetrating (reflecting minimal substrate stiffness) and possess ichnogenera less typical of firmground settings (e.g. *Chondrites*, *Palaeophycus*, and *Zoophycos*).

Trypanites *Ichnofacies*

The *Trypanites* Ichnofacies (Figure 3.24) was erected by Frey and Seilacher (1980) to encompass bioerosion of fully lithified marine substrates such as reefs, hardgrounds, rocky coasts, beach rock, unconformities and other omission surfaces. The ichnofacies has been extensively revised since, particularly with respect to its association with carbonate substrates (see Buatois and Mángano 2011 for a summary). Key here is that the ichnofacies corresponds to a discontinuity surface and not the mere presence of borings in allochthonous shells. Bromley and Asgaard (1993) erected the *Entobia* Ichnofacies and the *Gnathichnus* Ichnofacies to either serve as subsets of the *Trypanites* Ichnofacies, or as its replacement. The *Entobia* Ichnofacies is morphologically and ethologically identical to the archetypal *Trypanites* Ichnofacies (Buatois and Mángano 2011), in that it corresponds to long-term bioerosion of a

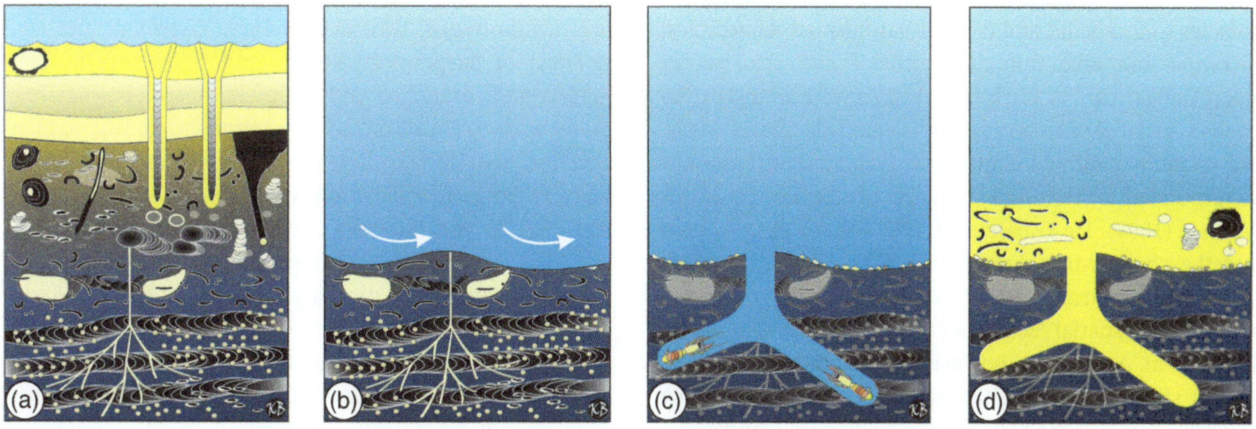

Figure 3.23 Schematic model showing development of stratigraphic omission reflected by the *Glossifungites* Ichnofacies. (a) Initial deposition of a sedimentary succession. (b) Base level change and associated erosional exhumation of buried substrate. (c) Recolonisation of the discontinuity with traces that cross-cut the original depositional fabric. (d) Resumption of deposition, with passive infill of the omission suite and its burial. Colonisation of the discontinuity occurs during the time gap separating substrate erosion and its subsequent burial. *Source:* MacEachern et al. (2007) / GeoScienceWorld.

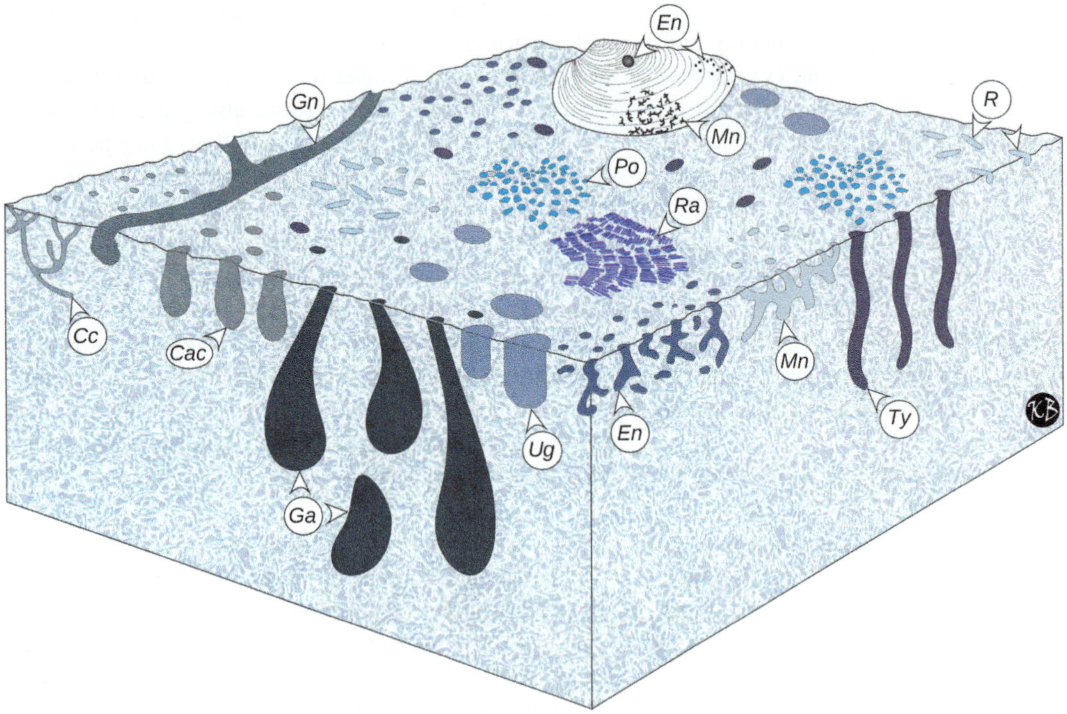

Figure 3.24 Schematic block diagram of the *Trypanites* Ichnofacies. The ichnofacies is characteristic of fully lithified marine substrates such as reefs, hardgrounds, rocky coasts, beach rock and discontinuity surfaces such as unconformities. The ichnofacies is commonly expressed by trace-fossil suites consisting of borings that include ichnotaxa such as *Caulostrepsis cretacea* (*Cac*), *Conchotrema* (*Cc*), *Entobia* (*En*), *Gastrochaenolites* (*Ga*), *Gnathichnus* (*Gn*), *Maeandropolydora* (*Mn*), *Podichnus* (*Po*), *Radulichnus* (*Ra*), *Rogerella* (*R*), *Trypanites* (*Ty*) and *Ubiglobites* (*Ug*). Suites dominated by surface borings and etchings fall into the *Gnathichnus* expression, whereas suites characterised by deep-tier borings constitute the archetypal *Trypanites* Ichnofacies. See the text for a discussion. Such suites of borings vary in bioturbation intensity, cross-cut pre-existing trace-fossil suites and commonly demarcate stratigraphic omission surfaces.

lithified surface (typically carbonate), with little or no contemporaneous sedimentation, allowing deep-tier dwelling structures to be excavated and maintained and superficial borings to be obliterated.

Buatois and Mángano (2011) regard the *Gnathichnus* Ichnofacies to be a valid and distinct hardground ichnofacies, encompassing predominantly surficial and shallow-tier bioerosion structures. Early occurrences were almost exclusively associated with borings on isolated clasts and shells, and so neither formed continuous mappable surfaces nor corresponded to 'facies' in the sense of Gressley (1838) or Walther (1894). De Gibert et al. (2007) have since shown that the shallow-tier expression is common in Mesozoic and Cenozoic strata and so demonstrates the necessary temporal and spatial recurrence required for an ichnofacies designation. At its heart, however, the *Gnathichnus* expression is still typified by shallow-tier borings that can be accommodated by the Seilacherian *Trypanites* Ichnofacies, in our opinion. That said, Buatois and Mángano (2011) have pointed out that the *Gnathichnus* expression points to much shorter term bioerosion interrupted by periods of rapid sedimentation compared to the archetypal *Trypanites* Ichnofacies and so may ultimately warrant its recognition as an archetypal ichnofacies.

Suites attributable to the *Trypanites* Ichnofacies are characterised by: (i) deep-tier, sharp-walled, unlined and passively filled cylindrical, vase- and tear-shaped borings (e.g. *Caulostrepsis*, *Gastrochaenolites*, *Trypanites* and *Ubiglobites*) of suspension feeders and passive carnivores; (ii) irregular branched and multi-chambered borings (e.g. *Conchotrema*, *Entobia* and *Meandropolydora*) of suspension feeders; (iii) shallow-tier irregular pits, attachment scars, raspings and gnawings (e.g. *Gnathichnus*, *Podichnus*, *Radulichnus* and *Rogerella*); (iv) shallow anastomose borings; (v) microborings of euendolithic microorganisms; (vi) low to moderate diversity; and (vii) high abundance of individual structures. The archetype, at least as defined here, is dominated by the deep-tier ichnogenera, commonly at the expense of the preservation of shallow-tier elements, whereas the *Gnathichnus* expression of the ichnofacies commonly lacks well-developed deep-tier ichnogenera and is dominated instead by the shallow-tier components. Deep-tier structures are generally produced by bivalves, polychaetes and sponges, whereas shallow-tier elements are generated by gastropods, chitons, irregular echinoids, brachiopods, limpets, barnacles and endolithic organisms such as cyanobacteria and heterotrophic bacteria. Suites attributable to the *Trypanites* Ichnofacies are commonly intergradational with those of the *Glossifungites* Ichnofacies and may cross cut former suites (Bromley 1975).

Hardgrounds may form on any lithified substrate, although they tend to be more common in carbonate successions (Section 11.3.2). The hardground may constitute a stratigraphic discontinuity, either as an erosional break or a hiatal surface reflecting depositional omission rather than erosional removal. In siliciclastic successions, most hardgrounds correspond to erosionally exhumed surfaces and may record a wide variety of discontinuities. By contrast, carbonate hardgrounds may reflect depositional omission or erosional exhumation. Carbonate settings are also prone to submarine cementation to form hardgrounds (commonly phosphatic) and may correspond to paraconformities or autogenic hiatal surfaces (e.g. Bromley 1975; Brett 1988; Wilson 2007). Drowning of carbonate regimes may also result in generation of transgressively related phosphatic hardgrounds (Coniglio and Dix 1992), although similar surfaces also can be generated by tectonic changes, changes in latitudinal position, changes in nutrient upwellings and numerous other factors. Extreme care should be exercised when assigning the sequence stratigraphic significance of trace-fossil suites corresponding to the *Trypanites* Ichnofacies.

Teredolites *Ichnofacies*

The *Teredolites* Ichnofacies (Figure 3.25) was erected by Bromley et al. (1984) and has been revised by Gingras et al. (2004). Suites of the ichnofacies encompass borings excavated into xylic (woody or coaly) substrates in marine or marginal marine conditions. Such substrates are referred to as woodgrounds, regardless of whether the borings were excavated prior to or following coalification (if at all) of the substrate. It is critical to identify whether the woodground borings are in situ (i.e. reflecting a substrate) or allochthonous (e.g. transported logs in coastal or marine environments). Only the assemblages associated with a substrate constitute the *Teredolites* Ichnofacies. Savrda et al. (1993) applied the term 'log-grounds' to high concentrations of allochthonous wood strewn across a depositional surface that were bored following their accumulation, to which the ichnofacies applies, in our opinion. These may form useful horizons for correlation, particularly where they mantle stratigraphic discontinuities such as transgressive surfaces of erosion.

The *Teredolites* Ichnofacies, as described from the rock record, is confined to marine and marginal marine settings. The namesake, *Teredolites*, reflects the structures of wood-boring bivalves, which in the present day only occur in brackish-water to fully marine environments. Suites attributable to the *Teredolites* Ichnofacies are characterised by: (i) dominance of clavate (club-shaped) borings; (ii) wall margins of borings ornamented with the texture of the

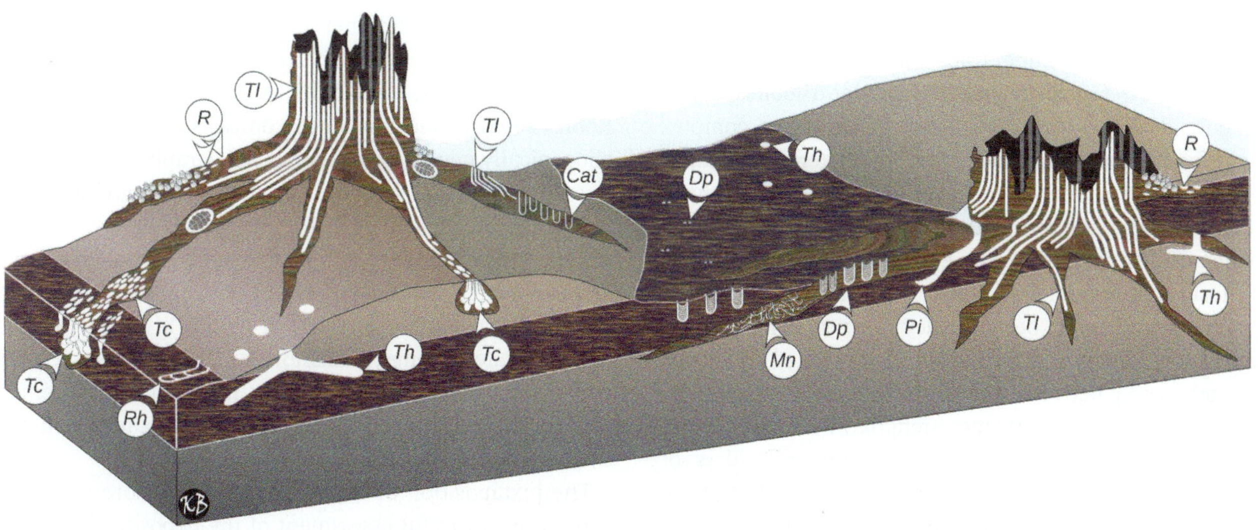

Figure 3.25 Schematic block diagram of the *Teredolites* Ichnofacies. The ichnofacies is characteristic of post-accumulation marine or marginal marine colonisation of xylic (woody or coaly) substrates. The ichnofacies is commonly expressed by trace-fossil suites that include borings and pseudo-borings such as the ichnogenera *Caulostrepsis taeniola* (*Cat*), *Diplocraterion parallelum* (*Dp*), *Maeandropolydora* (*Mn*), *Psilonichnus* (*Pi*), *Rhizocorallium* (*Rh*), *Rogerella* (*R*), *Teredolites clavatus* (*Tc*), *Teredolites longissimus* (*Tl*) and *Thalassinoides* (*Th*). Most rock record examples show predominantly *Teredolites* and *Thalassinoides* in most occurrences. Neoichnological studies, however, show a much wider range of borings as part of the suites. See the text for a discussion. Trace-fossil suites commonly record marine erosion along coal or lignite horizons and so commonly delineate transgressive surfaces of marine erosion.

woodground substrate; (iii) low-diversity and even monospecific suites; and (iv) high abundances of borings with only rare interpenetrations.

A striking feature of present-day woodground ichnocoenoses is the high diversity of boring and encrusting organisms comprising the faunal and floral community (Gingras et al. 2004). Although *Teredolites* is the common boring observed in xylic substrates in modern settings, structures akin to *Caulostrepsis*, *Meandropolydora* and *Rogerella* also are observed, as are the less common *Entobia* and rare *Psilonichnus* and *Thalassinoides*. Nevertheless, fossilised examples of such diverse suites have not yet been documented. Most occurrences of the ichnofacies in the ancient record are almost wholly dominated by *Teredolites*. Some rock record examples from the Jurassic, Cretaceous and Miocene include the ichnogenera *Diplocraterion* and *Thalassinoides* in suites attributable to the *Teredolites* Ichnofacies, although these are uncommon.

3.4 Ichnology and Sequence Stratigraphy

Unlike many body fossils, which enjoy widespread use in biozonation and biostratigraphic correlation, trace fossils have limited use in assigning stratigraphic ages, largely owing to the difficulty of assigning traces to specific trace-makers. Exceptions to this certainly exist, particularly in strata where fossil age control is otherwise absent, in situations where the trace fossils can be used to indicate the presence of individual trace-makers (e.g. some ichnospecies of *Cruziana* correlated to particular trilobites, or tetrapods 'identified' from their trackways), where complex trace fossils indicate a new behavioural pattern (e.g. *Oldhamia*) or serve as a first appearance datum (e.g. *Treptichnus pedum* marking the base of the Cambrian) (see Buatois and Mángano 2011 and Mángano et al. 2012 for reviews).

Most ichnogenera, however, have little or no biostratigraphic significance, because they are very long ranging and largely represent the activities of soft-bodied organisms that possess a patchy fossil record at best. In facies analysis, of course, this is their greatest strength, in that they represent similar animal-sediment interactions regardless of their age or geographic occurrence. Given their linkage to depositional environments and characteristics pointing to evidence of stratigraphic omission, trace fossils are well suited for providing insights into facies-driven stratigraphic frameworks such as sequence stratigraphy (see MacEachern et al. 2012b for a review).

Sequence stratigraphy (Section 4.4) can be regarded as a methodology that employs stratal stacking patterns and key bounding surfaces to erect a framework in which the depositional facies of sedimentary environments can be mapped and interpreted in the context of palaeogeography (Catuneanu 2006, 2022; Bhattacharya 2011). Stratal stacking patterns are a response to changes in the rate of the creation of accommodation and the rate of sediment supply, with the bounding surfaces corresponding to depositional breaks or changes in depositional architecture across which Walther's Law does not apply. Sequence stratigraphic methodology is employed at a number of scales, ranging from core, outcrop, geophysical well-logs, to seismic datasets. Ichnological data are largely restricted to core- and outcrop-scale observations and can be applied to sequence stratigraphic challenges at the same scale of observation as is physical sedimentology (MacEachern et al. 2012b). The strong facies control on the character and distribution of trace-fossil suites allows these data to be integrated with physical sedimentology in order to provide more precise depositional interpretations of the sedimentary successions with respect to their systems tracts within the sequence.

The ichnological identification and interpretation of key bounding surfaces has rested on the recognition of: (i) ichnologic omission suites; and (ii) the juxtaposition of softground trace-fossil suites that contravene Walther's Law (summarised in Pemberton et al. 2004 and MacEachern et al. 2010). Ichnological omission relies on the recognition of substrate-controlled ichnofacies (*Glossifungites*, *Trypanites* or *Teredolites* ichnofacies) and has a long history of application to sedimentary successions (e.g. Bromley 1975; Hayward 1976; Pemberton et al. 1980; Fürsich et al. 1981; Miller and Rehmer 1982; Pemberton and Frey 1985; MacEachern et al. 1992; MacEachern et al. 1999; Savrda et al. 2001; Pemberton et al. 2004; de Gibert and Robles 2005). In most instances, omission suites demarcate allogenically generated erosional stratigraphic breaks, and so commonly mark boundaries such as the regressive surface of marine erosion (RSME), the transgressive surface of marine erosion (TSE) and amalgamated or coplanar surfaces (e.g. subaerial unconformities overlain by transgressive surfaces; see MacEachern et al. 2012b for a review). In carbonate settings, omission suites may also record depositional hiatuses that may or may not be allogenic, and so their sequence stratigraphic significance can be challenging to ascertain.

The juxtaposition of trace-fossil suites is more nuanced and requires careful assessment of the depositional environments responsible for the underlying and overlying sedimentary facies. This requires evaluating vertically and laterally juxtaposed trace-fossil suites in order to discern those that contravene Walther's Law and is best achieved by integrating physical sedimentological facies analysis (e.g. MacEachern et al. 1992, Bhattacharya 1993; Taylor and Goldring 1993; MacEachern et al. 1999; McIlroy 2004, 2008; Pemberton et al. 2004; Rodríguez-Tovar et al. 2007; Fielding et al. 2008). Ongoing research continues to refine the role of trace-fossil analysis in the sequence stratigraphic evaluation of siliciclastic and carbonate successions.

3.5 Applying an Integrated Ichnological-Sedimentological Approach to Facies Successions

Ultimately, one must accurately describe the observed facies and facies successions with the aim of reaching a depositional interpretation. In our experience, biogenic data should not be collected nor interpreted in isolation from the physical sedimentological characteristics of the facies. Rather, it is essential that the two datasets be integrated with one another in order to yield a realistic depositional model. Three example case studies are presented (see Sections 3.5.1–3.5.3) that showcase the ways in which the two datasets can be merged to achieve a finely honed palaeoenvironmental interpretation within a sequence stratigraphic framework.

3.5.1 Case Study 1 – Lower Permian Pebbley Beach Fm, New South Wales, Australia

The Lower Permian (Artinskian to Sakmarian) Pebbley Beach Formation in the southernmost Sydney Basin of New South Wales, Australia, records deposition in shallow-marine to coastal environments at the close of the late Palaeozoic Gondwanan ice age. The formation crops out along numerous sea cliffs. Vertical facies transitions indicate significant basinward shifts of environment across complex and locally enigmatic erosion surfaces. The use of integrated ichnology and sedimentology permits the recognition of at least eight sequences. Stratigraphic sequences are thin (<10 m thick), condensed, incomplete and top-truncated, resulting from large amplitude (up to approximately 50 m) changes in relative sea level. Successions contain little if any record of lowstand systems tracts and, instead, are dominated by more substantial transgressive and highstand systems tracts that are erosionally truncated or entirely removed (Fielding et al. 2006).

The facies succession on the southern and northern ends of Pebbly Beach are described and expressed as an idealised core succession in this case study (Figure 3.26). In the example, the interval is complex and characterised by the juxtaposition of open marine and estuarine deposits in stacked, thin parasequences. Recognition of the ichnological signatures of the various facies and their related trace-fossil omission suites is crucial for the interpretation of the different marine and estuarine deposits owing to the paucity of preserved primary sedimentary structures and their general lithologic similarity.

The lowermost interval consists of sporadically bioturbated, well sorted, very fine- to fine-grained sandstone units with intercalated carbonaceous mudstone drapes and interbeds. Sandstone beds are erosionally based and contain wavy parallel lamination, HCS and oscillation ripples, as well as allochthonous (reworked) trace fossils and large petrified fossil trees. These beds are interpreted as tempestites (Bann et al. 2008). Massive beds devoid of bioturbation with common carbonaceous detritus are interpreted as hyperpycnites. Soft-sediment deformation features (e.g. convolute bedding, flame structures, ball and pillow structures and load structures) are common. The intervening mudstone drapes and interbeds locally show erosive bases, are structureless and contain abundant carbonaceous detritus and common syneresis cracks. Mud beds with sharp but non-erosional bases are interpreted as fluid mud deposits reflecting flocculation from buoyant (hypopycnal) plumes associated with river floods (MacEachern and Bann 2020). Erosionally based layers reflect bedload transport of fluid mud. Sandstone and mudstone beds show variable (BI 0–3) bioturbation intensities. Trace-fossil suites are relatively diverse (up to 16 ichnogenera and species), many positively correlated to normal marine conditions. Ethologies reflect mobile and sessile deposit feeding, surface detritus feeding and the dwellings of passive carnivores. Fugichnia and equilbrichnia are common elements of the suite. Vertical dwellings of suspension-feeding organisms are conspicuously absent. The fully marine suite is juxtaposed at the bed and bedset scales with low-diversity trace-fossil suites dominated by opportunistic, facies-crossing elements reflecting simple deposit-feeding behaviours. Tempestites contain ichnological suites that record episodic, top-down opportunistic colonisation. The juxtaposed suites are attributable to the *Rosselia* Ichnofacies. The facies records deposition in a dynamic, fully marine but intermittently physico-chemically stressed storm- and river-flood-dominated distal delta front.

The distal delta-front succession is truncated by an inclined erosion surface mantled by a single-clast-thick pebble lag, with carbonaceous mudstone drapes, concentrations of carbonaceous detritus, flattened coaly plant compressions and petrified fossil trees. The surface is overlain by interbedded and interlaminated, fine-grained sandstone and siltstone bedsets forming inclined heterolithic stratification (IHS) within a channelised body. Numerous individual channels <6 m deep and typically 100–250 m wide occur at this stratigraphic level. The erosion surface represents at least a 50 m base level fall (Fielding et al. 2006). As such, the area was subaerially exposed and

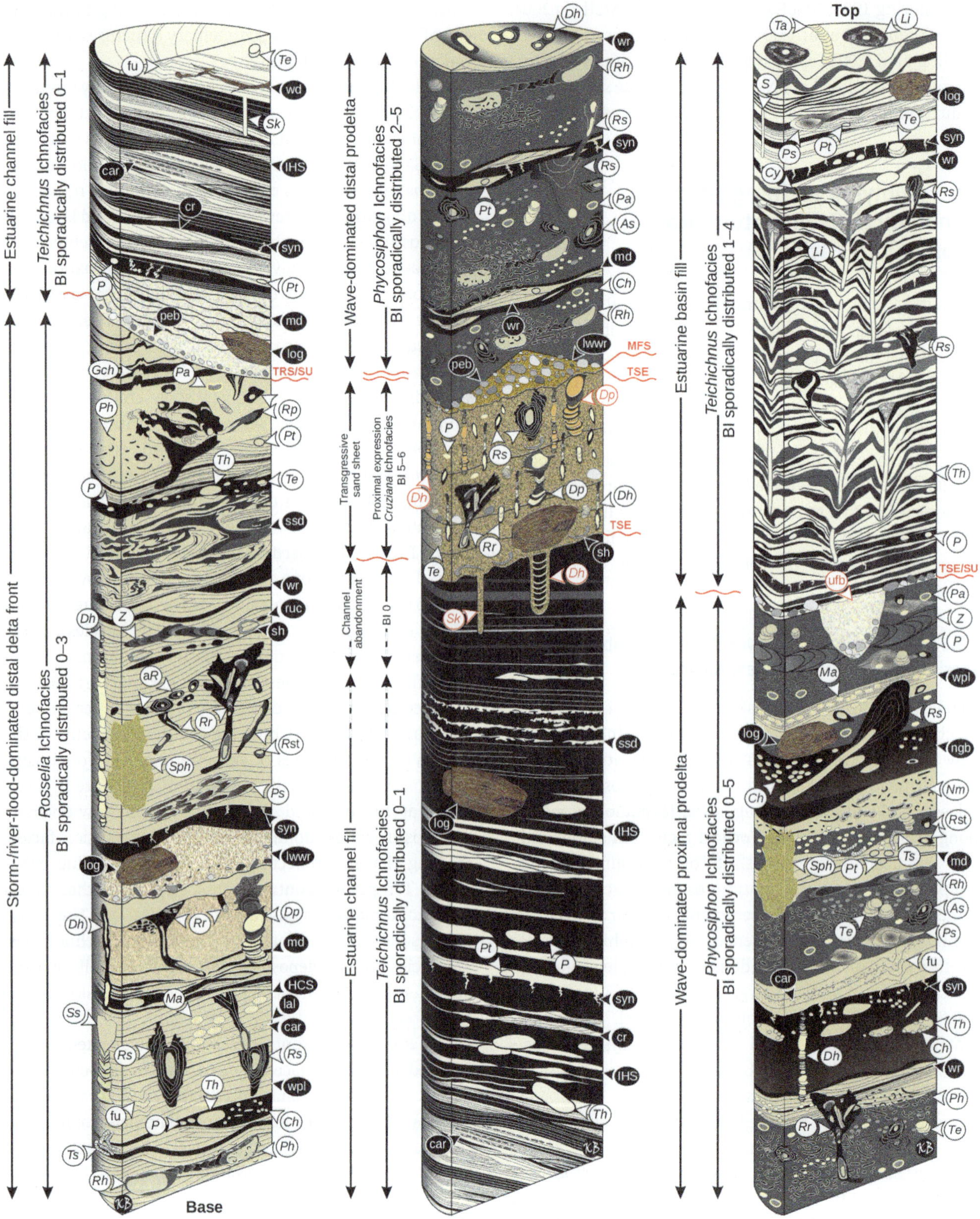

Figure 3.26 Case Study 1: Lower Permian Pebbley Beach Formation in the southernmost Sydney Basin, NSW, Australia. The schematic split core diagram shows the complex juxtaposition of transgressive and regressive shallow-marine to paralic cycles across a number of stratigraphic discontinuities. The base of the interval shows a regressive sandstone of the highstand systems tract (HST), erosionally truncated by an amalgamated tidal ravinement surface and subaerial unconformity (TRS/SU) andoverlain by an estuarine channel-fill complex of a transgressive systems tract. Following channel fill, regional base level rise led to a transgressive surface of erosion (TSE)

incised by channels to form a regional subaerial unconformity (SU). The subsequent accumulation of IHS records ensuing transgression, such that the basal surface represents an amalgamated tidal ravinement surface (TRS) and subaerial unconformity (TRS/SU; Chapter 8, Figure 8.29). The overlying IHS facies shows pinstripe laminated and lenticular, wavy and flaser bedded composite bedsets. Synaeresis cracks are abundant with localised small-scale soft-sediment deformation structures. The upper parts of the channel fills are mud dominated, locally with an uppermost unbioturbated, carbonaceous mudstone deposit. Bioturbation in the IHS is sparse (BI 0–1). The trace-fossil suite is of low diversity (5 ichnotaxa) and dominated by opportunistic facies-crossing elements that reflect head-up deposit-feeding and permanent dwellings for scavengers, passive carnivores and rare filter feeders. The trace-fossil suite records brackish-water conditions and is attributable to the *Teichichnus* Ichnofacies. The facies are interpreted to record the backfilling of sinuous estuarine channels during early phases of the transgressive systems tract (TST).

The estuarine channel deposits are erosionally overlain by a poorly sorted, muddy and pebble-bearing sandstone approximately 1 m thick. The erosion surface contains a firmground suite of vertical, unlined, fully marine trace fossils (e.g. *Diplocraterion habichi* and *Skolithos*) of the *Glossifungites* Ichnofacies and represents a transgressive surface of erosion (TSE). The overlying facies contains abundant shells and petrified fossilised logs and is pervasively and uniformly bioturbated (BI 5–6) with a moderately diverse (6 ichnotaxa) trace-fossil suite comprising the vertical domiciles of detritus-feeding and suspension-feeding organisms with lesser horizontal burrows of sessile deposit-feeders. The suite is representative of a proximal expression of the *Cruziana* Ichnofacies and is interpreted as a shallow-marine transgressive sand sheet deposited during the TST.

At the top of the unit is another erosion surface evidenced by truncated softground trace fossils, cross-cut by robust vertical trace fossils passively filled with the overlying gravel-bearing sediment, interpreted to represent a palimpsest softground omission suite. The surface also represents a TSE and is mantled by a transgressive pebble lag moulded into large-scale (wavelength > 1 m) oscillation ripples.

The transgressive lag is abruptly overlain by a sporadically but locally pervasively bioturbated (BI 0–5, mainly 3–5) mudstone-dominated heterolthic facies that records the return to normal regression and the end of transgression, corresponding to the maximum flooding surface (MFS). The heterolithic facies comprises sandstone beds with wavy parallel lamination, micro-HCS and oscillation ripples (distal tempestites) intercalated with structureless, unburrowed, carbonaceous mudstone drapes and interbeds (fluid mud from hypopycnal plumes associated with river floods). The facies contains a diverse (18 ichnotaxa) trace-fossil assemblage characterised by the juxtaposition of ichnological suites composed of the domiciles of fully marine detritus- and deposit-feeding organisms and passive carnivores and the structures of mobile and sessile deposit-feeders with trace-fossil suites reflecting opportunistic, facies-crossing structures of deposit feeders and passive carnivores. The suites are attributable to the *Phycosiphon* Ichnofacies. The facies was deposited below fair-weather wave base in a wave-dominated, distal prodelta passing upwards into a proximal prodelta during the HST.

Finally, the heterolithic prodeltaic deposits are truncated by a locally subtle erosion surface, particularly where both overlying and underlying facies are heterolithic. In some locations, the surface is marked by a gravel lag and an omission suite consisting of vertical, unlined, gravel-filled, cone-shaped firmground trace fossils of the *Glossifungites* Ichnofacies. The surface is complex and represents an amalgamated TSE/SU. The overlying heterolithic facies consists

Figure 3.26 demarcated by a firmground omission suite of the *Glossifungites* Ichnofacies and overlain by a transgressive sandstone sheet. This sandstone sheet was reworked during continued transgression and colonised with a palimpsest softground omission suite. A maximum flooding (MFS) marks the end of this transgressive systems tract (TST) and the onset of regression, expressed by prodeltaic deposits of a wave-dominated delta of the HST. These deltaic facies, however, were subaerially exposed and cross-cut by a subaerial unconformity during a major base level fall and a period of sediment bypass. Later transgression generated an amalgamated TSE/SU marked by a firmground omission suite of the *Glossifungites* Ichnofacies and overlain by an estuarine basin fill of the transgressive systems tract (TST). See text for a detailed description. Abbreviations for biogenic structures: allochthonous *Rosselia* (*aR*), *Asterosoma* (*As*), *Chondrites* (*Ch*), *Cylindrichnus* (*Cy*), *Diplocraterion habichi* (*Dh*), *Diplocraterion parallelum* (*Dp*), fugichnia (fu), *Gyrochorte* (*Gch*), *Lingulichnus* (*Li*), *Macaronichnus* isp. (*Ma*), *Nereites missouriensis* (*Nm*), *Palaeophycus heberti* (*Pa*), *Palaeophycus tubularis* (*Pt*), *Phycosiphon* (*Ph*), *Planolites* (*P*), *Psammichnites* (*Ps*), *Rhizocorallium* (*Rh*), *Rosselia* isp. (*Rp*), *Rosselia rotatus* (*Rr*), *Rosselia socialis* (*Rs*), *Rosselia* dwelling stalk (Rst), *Siphonichnus phiale* (*Sph*), *Siphonichnus sersumdeorsum* (*Ss*), *Skolithos* (*Sk*), *Taenidium synyphes* (*Ts*), *Teichichnus* (*Te*), *Thalassinoides* (*Th*), unidentified firmground burrow (ufb) and *Zoophycos* (*Z*). Trace fossils associated with omission suites and stratigraphic discontinuities are shown in red. Abbreviations for primary sedimentary features: carbonaceous detritus (car), current ripple (cr), hummocky cross-stratification (HCS), inclined heterolithic stratification (IHS), large wavelength wave ripple (lwwr), low-angle parallel lamination (lal), fossilised log (log), mudstone drape (md), normally graded beds (ngb), pebbles (peb), shells (sh), rip-up clast (ruc), soft-sediment deformation (ssd), syneresis crack (syn), wood fragments (wd), wavy parallel lamination (wpl) and wave ripple (wr). Abbreviations for stratigraphic surfaces: maximum flooding surface (MFS), transgressive surface of erosion (TSE), amalgamated tidal ravinement surface and subaerial unconformity (TRS/SU) and amalgamated transgressive surface of erosion and subaerial unconformity (TSE/SU).

of interbedded and interlaminated sandstone and mudstone. Wave-dominated combined flow ripples and micro-HCS are common, suggesting an increased incidence of currents augmenting oscillatory processes. Mudstone drapes and rhythmic sand-mud couplets, synaeresis cracks and soft-sediment deformation structures are locally common. Bioturbation intensities are highly variable (BI 0–4) and trace fossils are sporadically distributed. Locally, suites are moderately diverse (11 ichnotaxa) and composed of structures reflecting head-up deposit-feeding, and dwellings of scavengers, passive predators and minor filter-feeders. The suite is interpreted to record brackish-water conditions and is attributable to the *Teichichnus* Ichnofacies. The facies records deposition within an estuarine basin during the TST.

3.5.2 Case Study 2 – Lower Cretaceous Viking Formation, Alberta, Canada

The Lower Cretaceous (Upper Albian) Viking Formation is a hydrocarbon-producing siliciclastic succession that is widespread in the subsurface of Alberta, Canada, and has been intersected by hundreds of cored wells and many thousands of geophysical well logs. The bulk of the Viking Formation is represented by shallow-marine and paralic deposits within the systems tracts of four sequence stratigraphic sequences (e.g. Pattison and Walker 1994; Walker and Wiseman 1995; Dafoe et al. 2010; Schultz et al. 2022).

In the Kaybob Field of west-central Alberta, the formation is characterised by a sharp-based, coarsening-upward sandstone unit, erosionally overlying bioturbated mudstone and erosionally overlain by a granule conglomerate that passes into bioturbated mudstone. The interval records deposition within three systems tracts of two discrete sequences. Consider a cored interval to be described and interpreted, such as that shown in Figure 3.27.

The lowermost interval consists of two pervasively and uniformly bioturbated (BI 4–6) facies: a silty mudstone facies passing gradationally upwards into a sandy mudstone facies with rare, thin parallel laminated sandstone beds interpreted as tempestites. The trace-fossil suites of each facies are diverse and ichnogenera are abundant, reflecting fully marine and oxygenated conditions dominated by suspended sediment deposition below fair-weather wave base. Ichnogenera are dominated by deposit-feeding structures and subordinate grazing structures. The interval coarsens upward and shows a trace-fossil suite attributable to a distal expression of the *Cruziana* Ichnofacies giving way to a suite attributable to the archetypal *Cruziana* Ichnofacies. This mudstone-dominated interval is interpreted to record progressive shallowing from lower offshore to upper offshore conditions, and reflects the progradation of a distal parasequence during normal regression in the HST of Sequence 1.

This mudstone interval is erosionally overlain by a sandstone interval consisting of three stacked facies forming a coarsening-upward succession. The basal contact is demarcated by a firmground trace-fossil suite of the *Glossifungites* Ichnofacies. This stratigraphic discontinuity is sharply overlain by a pervasively and uniformly bioturbated (BI 4–5) muddy sandstone facies. The trace-fossil suite is of high diversity and consists of abundant ichnogenera reflecting a predominance of deposit-feeding structures with subordinate dwelling structures of suspension feeders and passive carnivores. The suite is consistent with fully marine salinities, oxygenated conditions and persistent energy at the bed, is attributable to the proximal expression of the *Cruziana* Ichnofacies and records deposition near to but generally above fair-weather wave base. Better-sorted sandstone layers containing erosionally based, oscillation ripples and HCS are intercalated and thicken upwards, which show lower bioturbation intensities (BI 0–3) and sporadically distributed burrowing. These sandstone beds display top-down colonisation with ichnogenera positively correlated to opportunistic trace-makers and are interpreted as tempestites. The facies is interpreted to record a fully marine, storm-influenced distal lower shoreface erosionally overlying the omission suite, demonstrating an abrupt basinward shift of facies associated with forced regression. The basal contact is interpreted as a RSME (see Section 3.4) and marks the base of the falling stage systems tract (FSST) of Sequence 1.

The muddy sandstone facies passes upwards into a laminated-to-burrowed (BI 0–5), moderately well-sorted sandstone facies. Burrowed sandstone and silty sandstone beds show BI 3–5 with a moderately diverse suite of trace fossils recording vertical and horizontal dwelling structures of suspension feeders and passive carnivores as well as burrows of mobile and sessile deposit-feeding and surface detritus-feeding organisms. This suite records fully marine, oxygenated and physically dynamic conditions lying above fair-weather wave base. Intercalated erosionally based, wavy parallel laminated and HCS-bearing sandstones show variably distributed and generally lower intensities of bioturbation (BI 0–3) with top-down colonisation patterns, consistent with the opportunistic colonisation of tempestites. The ichnological suite records a distal expression of the *Skolithos* Ichnofacies and is interpreted to record a moderately storm influenced proximal lower shoreface environment of the FSST (Chapter 4, Figure 4.31).

This facies is overlain by a facies characterised by erosionally amalgamated, well-sorted sandstone beds with HCS, wavy parallel lamination and rare wave ripples, interpreted as stacked tempestites. These beds show low

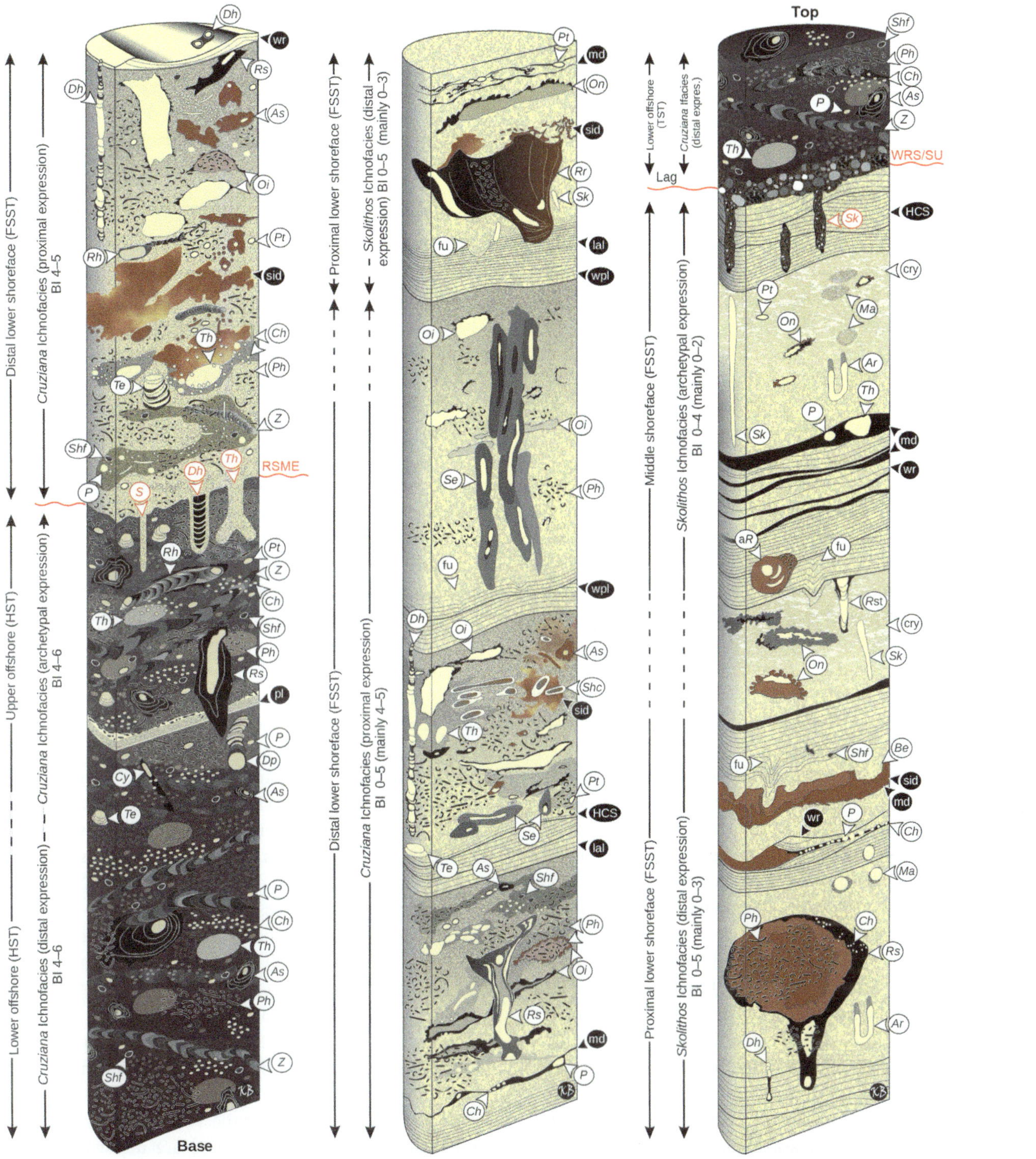

Figure 3.27 Case Study 2: Lower Cretaceous Viking Fm at Kaybob Field, Alberta, Canada. Schematic split core diagram showing bioturbated offshore mudstone facies of a highstand systems tract (HST), erosionally truncated by a regressive surface of marine erosion (RSME) and cross-cut by a trace-fossil omission suite (red) of the *Glossifungites* Ichnofacies. The overlying forced regressive lower-middle shoreface sandstone corresponds to the falling stage systems tract (FSST). The FSST is truncated by a co-planar surface that marks the lowstand subaerial unconformity (SU) erosionally reworked by the wave ravinement surface (WRS) at the base of the overlying transgressive systems tract (TST). The TST in this location is characterised by bioturbated offshore mudstone. See text for a detailed description. Abbreviations for biogenic structures: *allochthonous Rosselia* (*aR*), *Arenicolites* (*Ar*), *Asterosoma* (*As*), *Bergaueria* (*Be*), *Chondrites* (*Ch*), *cryptobioturbation* (*cry*), *Cylindrichnus* (*Cy*), *Diplocraterion habichi* (*Dh*), *Diplocraterion parallelum* (*Dp*), fugichnia (fu), *Macaronichnus* isp. (*Ma*), *Ophiomorpha irregulaire* (*Oi*), *Ophiomorpha nodosa* (*On*), *Palaeophycus tubularis* (*Pt*), *Phycosiphon* (*Ph*), *Planolites* (*P*), *Rhizocorallium* (*Rh*), *Rosselia rotatus* (*Rr*), *Rosselia socialis* (*Rs*), *Rosselia* dwelling stalk (*Rst*), *Schaubcylindrichnus coronus* (*Shc*), *Schaubcylindrichnus freyi* (*Shf*), *Siphonichnus eccaenis* (*Se*), *Skolithos* (*Sk*), *Teichichnus* (*Te*), *Thalassinoides* (*Th*) and *Zoophycos* (*Z*). Trace fossils associated with omission suites and stratigraphic discontinuities are shown in red. Abbreviations for primary sedimentary features: hummocky cross stratification (HCS), low-angle parallel lamination (lal), mudstone drape (md), planar parallel lamination (pl), siderite (sid), wavy parallel lamination (wpl) and wave ripple (wr). Abbreviations for stratigraphic surfaces: regressive surface of marine erosion (RSME), amalgamated wave ravinement surface and subaerial unconformity (WRS/SU).

bioturbation intensities (BI 0–2) and sporadically distributed burrowing, consistent with post-storm opportunistic colonisation. Tempestites locally show allochthonous fragments of burrows near their bases. The in situ ichnogenera predominantly record the dwelling structures of suspension feeders, passive carnivores and omnivores. Some cryptobioturbation is evident near the tops of some tempestites as well. Intercalated with these event beds are rare, bioturbated (BI 2–4) fair-weather sandstone beds containing trace-fossil suites dominated by permanent vertical and horizontal dwelling structures of suspension feeders, surface detritus feeders and passive carnivores. Subordinate intrastratal mobile deposit-feeding structures are also locally present. The fair-weather suite is consistent with the archetypal *Skolithos* Ichnofacies, reflecting relatively high-energy, fully marine and well-oxygenated conditions, punctuated by high-energy storm events, consistent with a storm-dominated middle shoreface environment. Interestingly, intercalated within this facies are dark, carbonaceous mudstone drapes and interbeds showing BI 0–1. Such layers are interpreted to record fluid mud delivered from buoyant (hypopycnal) plumes associated with an active delta lobe lying updrift of the shoreface. The three facies of the FSST record progressive shallowing and upward increase of energy, reflecting ongoing forced regression and shoreline progradation.

The sandstone interval of the FSST is erosionally overlain by a clast-supported granule to pebble lag. Sharp-walled, gravel-filled dwelling structures subtend from the contact and cross-cut the original softground suite of the underlying sandstone, constituting a palimpsest softground omission suite that marks the stratigraphic discontinuity. From regional correlations, it is clear that successive base level falls had shifted the shoreline well basinward this location and that the lowstand shoreline position ultimately resided some 100 km to the east of this location. As such, the Kaybob area was subaerially exposed during the onset of Sequence 2 and the succession truncated by a regional disconformity. This subaerial unconformity (SU), however, was erosionally modified during base level rise and ensuing transgression, during which marine colonisation of the discontinuity occurred. Wave ravinement (WRS) reworked lowstand sand and gravel into a transgressive lag, which mantled the unconformity to produce a composite surface (WRS/SU).

The transgressive lag passes upward rather abruptly into a pervasively bioturbated (BI 4–6), granule-bearing silty mudstone facies broadly similar to the lowermost facies of the underlying HST interval. The facies contains a diverse, fully marine trace-fossil suite attributable to a distal expression of the *Cruziana* Ichnofacies and is interpreted to reflect deposition well below fair-weather wave base. The silty mudstone facies is interpreted to reflect a lower offshore environment in the TST of Sequence 2.

3.5.3 Case Study 3 – Pleistocene Interval at Willapa Bay, Washington, USA

Pleistocene outcrops rim the eastern and northern margins of Willapa Bay, Washington in northwest United States. The sedimentology and the stratigraphy of the outcrops were well characterised by Clifton and others of the USGS throughout the 1970s and 1980s (Kvenvolden et al. 1979; Clifton 1983; Anima et al. 1989). Later work on the outcrops focused on integrating ichnology with the sedimentologic and stratigraphic characteristics to refine facies interpretations. Here, we present an idealised vertical succession from Pleistocene outcrops of Willapa Bay, Washington that displays a sedimentological record reflecting transgression in an estuary setting (Figure 3.28). Initiated at a subaerial unconformity (SU), the sequence includes lithic sandstone and gravel deposits transitioning into tidally influenced fluvio-estuarine mudstone with rare bioturbation, indicating low salinity and elevated sedimentation. The sequence progresses into estuarine tidal channels and intertidal flats with increasing bioturbation intensities and trace-fossil diversities, largely representing increasing salinity. Successive transitions are marked by TRSs. Firmground-associated trace fossils, such as *Psilonichnus* and *Thalassinoides*, indicate development of the *Glossifungites* Ichnofacies at some discontinuities. The uppermost deposits record energetic tidal inlet channels with sparse bioturbation – a result of ongoing transgression.

The lowermost part of the succession is demarcated by a subaerial unconformity (SU) and the upper bounding surface (which is not shown) is a flooding surface that gives way to lower shoreface deposits representing shoreline retreat at the open coast. It is important to state that no shoreface strata are actually preserved within the Pleistocene outcrops.

The sequence is initiated at the SU, which is overlain by lithic sandstone and gravel. These are unburrowed and contain abundant wood clasts. When examining the outcrops, distinguishing between fluvial and bayhead delta deposits is based on the lateral distribution of the gravel, but this is a challenging determination to make when analysing core. The gravel-bearing units are not always present in the outcrops and may be entirely removed by the ensuing TRS.

Above the TRS are intervals of mudstone-dominated inclined stratification, consisting of cm-scale parallel bedding with minor variations in silt and clay content.

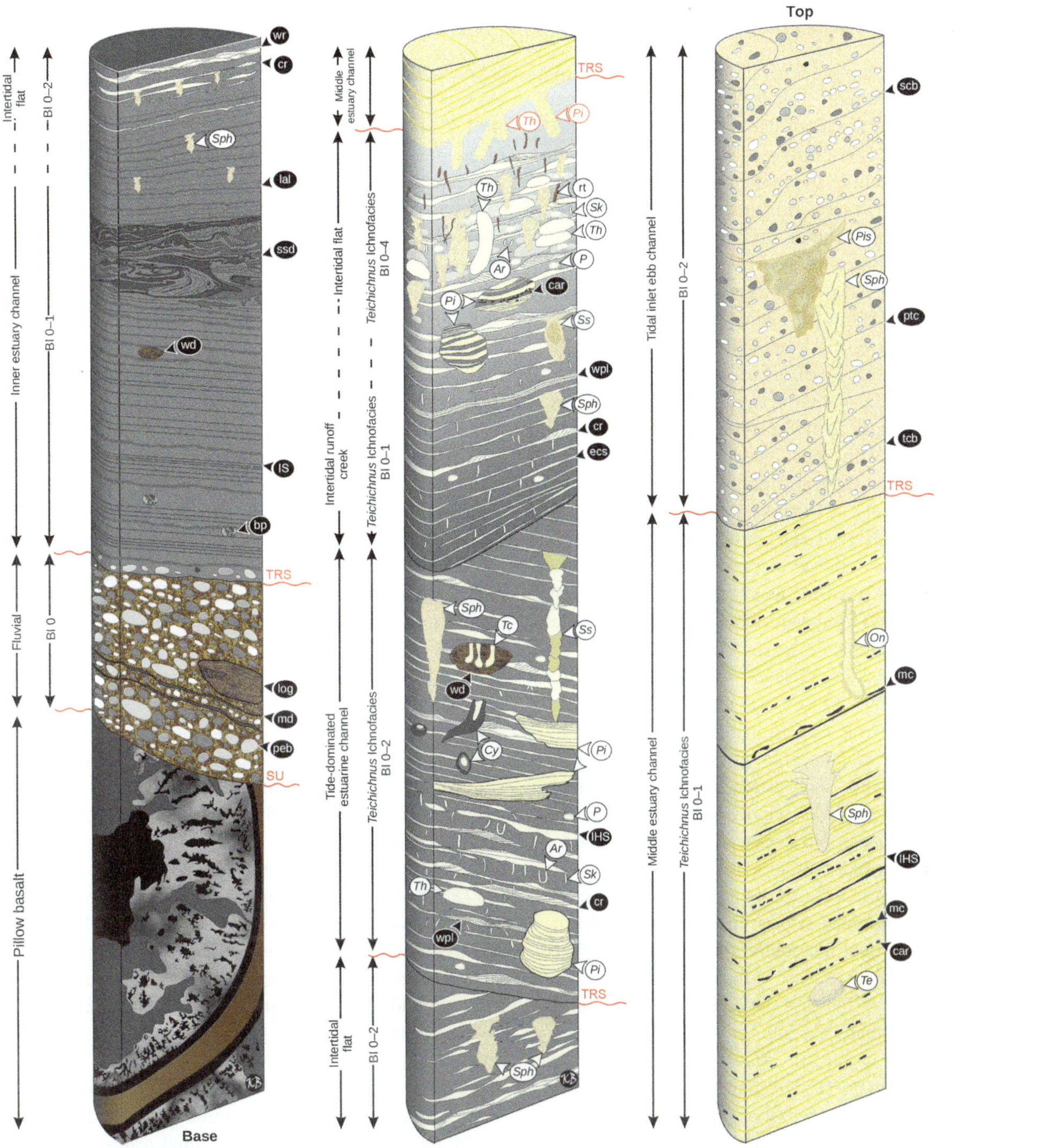

Figure 3.28 Case Study 3: Pleistocene strata exposed at Willapa Bay, Washington, USA. Schematic split core diagram shows a subtle stacking of inshore brackish-water paralic intervals, characterised by low-diversity suites of the *Teichichnus* Ichnofacies. The succession erosionally overlies pillow basalts across a regional subaerial unconformity (SU). A fluvial conglomeratic lag of the lowstand systems tract (LST) is erosionally truncated by a tidal ravinement surface (TRS) and overlain by mudstone-dominated inclined heterolithic stratification (IHS) reflecting an inner estuary channel fill of the transgressive systems tract. The interval shows erosionally amalgamated intertidal runoff creek deposits capped by tidal flats, which are themselves erosionally overlain by sandstone-dominated IHS deposited in middle estuary channels of the same transgressive systems tract. The erosion surface corresponds to a TRS and is marked by a palimpsest softground omission suite. The sandy IHS is ultimately erosionally truncated by a third TRS and overlain by planar tabular and trough cross-stratified pebbly sandstone deposits of the tidal inlet. The succession is interpreted to record progressive transgression of the inshore zone following the lowstand. See text for a detailed description. Abbreviations for biogenic structures: *Arenicolites* (*Ar*), *Cylindrichnus* (*Cy*), *Ophiomorpha nodosa* (*On*), *Piscichnus* (*Pis*), *Planolites* (*P*), *Psilonichnus* (*Pi*), root traces (rt), *Siphonichnus phiale* (*Sph*), *Siphonichnus sersumdeorsum* (*Ss*), *Skolithos* (*Sk*), *Teichichnus* (*Te*), *Teredolites clavatus* (*Tc*) and *Thalassinoides* (*Th*). Trace fossils associated with omission suites and stratigraphic discontinuities are shown in red. Abbreviations for primary sedimentary features: basalt pebble (bp), carbonaceous detritus (car), current ripple (cr), *eta* cross-stratification (ecs), inclined heterolithic stratification (IHS), inclined stratification (IS), low-angle parallel lamination (lal), fossil log (log), mudstone clast (mc), mudstone drape (md), pebbles (peb), sigmoidal cross-bedding (scb), soft-sediment deformation (ssd), planar tabular cross-bedding (ptc), trough cross-bedding (tcb), wave ripple (wr), wavy parallel lamination (wpl) and wood fragments (wd). Abbreviations for stratigraphic surfaces: subaerial unconformity (SU) and tidal ravinement surface (TRS).

Soft-sediment deformation is locally evident and unbored wood clasts are common. Pebbles are rare and confined to the lower part of the mudstone. Rare, medium-grained lithic sandstone beds are interbedded with the mudstone. Bioturbation is absent in the more inclined beds. Bedding gradationally – over several decimetres – becomes horizontal upwards. The horizontal stratification displays fine-grained sand pinstripes as well as rare lenticular and wavy bedding with low-relief current ripples. It is in these levels that very rare and diminutive *Siphonichnus* may be present. The inclined beds result from sedimentation in tidally influenced channels in the innermost estuary. The mud was derived from the adjoining river and reflects elevated sedimentation, owing to flocculation of the clay as the freshwater mixed with very low-salinity estuary waters. Likewise, the lithic sand and gravel are advected from the river, most likely at times of heightened fluvial discharge. Although elevated sedimentation rates were probably responsible for reduced bioturbation intensities, the general absence of trace fossils suggests that the waters were less than 1 ppt. The upwards transition into horizontal bedding signals shoaling of the channel and the establishment of the intertidal flat. The intertidal flat typically traps less flocculated mud and aggrades more slowly than the channel, thus allowing sand to winnow and concentrate from the mud. The *Siphonichnus*, most likely bivalve-generated, represent clams such as *Macoma balthica*, which are able to tolerate very low salinity levels and prefer the lowered sedimentation rates associated with microtidal and mesotidal intertidal flats.

The succession continues to the next (middle) column, wherein a tide-dominated estuary channel, intertidal run-off creek and intertidal flat continuum are depicted. This shows a tidal channel outboard of the mud-dominated fluvio-tidal channel (in this case the outer part of the inner estuary) that erosionally cut into the muddy unit across a TRS. The bedding within the estuary channel comprises cm-scale IHS that is primarily defined by a wavy and lenticular bedded composite bedset characterised by current ripple laminated, fine-grained sandstone interbedded with mudstone. Here, the sandstone is quartz arenite and was delivered to the channel via tidal currents, as opposed to being derived from the rivers. Unbored wood clasts and wood clasts bearing *Teredolites* occur locally. Bioturbation is conspicuous (BI 1–3), with diminutive *Arenicolites* and *Skolithos* abundant, *Cylindrichnus*, *Planolites* and *Psilonichnus* common, and localised *Thalassinoides* making up the trace-fossil suite. Higher in the beds, *Siphonichnus* are present. The primary assemblage of *Arenicolites*, *Planolites*, *Psilonichnus* and *Skolithos*, corresponds to the *Teichichnus* Ichnofacies and is consistent with colonisation in brackish-water conditions. The intertidal run-off creek cross-cuts the IHS of the estuary channel, signalling the establishment of an intertidal setting following the aggradation of the channel. Heterolithic beds become horizontal and vertically aggraded upwards (e.g. *eta* cross-stratification). Within the intertidal creek, sediment accumulation was more rapid than in the estuary channel, and so bioturbation is sparse (BI 1–2) and limited to diminutive *Arenicolites* or *Skolithos*. The transition into the intertidal flat is marked by horizontal bedding and higher intensities of bioturbation (BI 3–5). This trace-fossil suite is dominated by comparably robust *Psilonichnus*, *Siphonichnus* or *Thalassinoides*. Colonisation of the intertidal flat, though limited by lowered salinity, was encouraged by the lowered sediment accumulation rates and the rich food resources that were delivered to the sediment surface during each falling tide. Upwards, evidence for salt marsh may be present, normally expressed as illuviated mudstone and/or the presence of root traces.

The mud-dominated estuary channel to intertidal flat complex is truncated by a back-stepped middle estuary channel. The basal contact (TRS) is demarcated by firmground *Psilonichnus* and *Thalassinoides* of the *Glossifungites* Ichnofacies. This stratigraphic discontinuity is sharply overlain by inclined and stratified upper fine- to lower medium-grained sandstone in a laterally accreted bar of a middle estuary channel. The inclined stratification is defined by wavy parallel lamination, current ripple lamination and trough cross stratification that generally becomes of smaller-scale upwards. Bioturbation is sparse (BI 1). However, larger *Siphonichnus* and *Teichichnus*, as well as *Ophiomorpha* are observed locally. Their larger size and the presence of *Ophiomorpha*, in particular, point to increased salinities. The rare and sporadic distribution of these ichnofossils is attributed to the mobile/shifting sand associated with this larger and comparatively more energetic tidal channel.

The final stage of estuary backstepping imposed tidal inlet channel deposits over the middle estuary succession across another TRS. The inlet channel was much larger than those of the middle and inner estuary channels, and the tidal currents therein were correspondingly much more vigorous. As such, the discontinuity that separates the middle estuary and the inlet channels is uncolonised. Owing to persistent winnowing by tidal currents, strata of the inlet are pebbly compared to the sandstone units of the estuary. Bioturbation is normally absent (BI 0), although rare, very large, bivalve-associated *Siphonichnus* have been observed, cross-cut by *Piscichnus* (a vertebrate feeding trace), reflecting predation by large marine animals.

3.6 Conclusions

Biogenic structures are common elements of virtually every sedimentary facies encountered in present-day environments, throughout the Phanerozoic and into the Neoproterozoic. Unfortunately, they tend to be underutilised by sedimentologists, perhaps owing to the uneven training of ichnology at the undergraduate level. Most researchers of our generation came to the discipline during our graduate research. In recent years, this has begun to change, and the hope is that students will be introduced to the identification and interpretation of biogenic structures with the same rigour currently applied to physical sedimentary structures. On the whole, nothing is as sensitive to the environment and its physico-chemical conditions as the biota that occupy it. The ability to 'read' the record of animal-sediment interactions is a crucial skill for anyone hoping to achieve high-resolution palaeoenvironmental interpretations of sedimentary facies. While accurate taxonomy is important, it is the assessment of bioturbation intensity and its distribution, recognition of features indicative of the structure's ethology, evaluation of ichnological diversity with respect to both specialised and opportunistic (facies-crossing) elements, consideration of ichnofabric characteristics, and the ability to group trace fossils into recurring suites that can be assigned to the appropriate ichnofacies that are more foundational to a sound interpretation of depositional facies. This chapter has sought to outline the fundamental aspects of applied ichnology but does not attempt to encompass the full scope of the discipline. Instead, it should serve as a starting point for facies analysts seeking to integrate biogenic structures with physical sedimentology to enhance the description and interpretation of the sedimentary record.

Acknowledgements

This chapter benefited from the careful and positive reviews by Luis Buatois, Rodmar Ravnås, Bruce Ainsworth and editor Bruce Levell. This chapter is dedicated to the memory of our friend and mentor Dr. S. George Pemberton, who was a champion of trace-fossil applications to facies analysis and stratigraphy. George greatly admired H.G Reading and would have been proud of his contributions to this chapter, even if delivered indirectly through his training and research legacy.

References

Ahlbrandt, T.S., Andrews, S., and Gwynne, D.T. (1978). Bioturbation in eolian deposits. *Journal of Sedimentary Research* 48: 839–848. https://doi.org/10.1306/212F7586-2B24-11D7-8648000102C1865D.

Alonso-Zarza, A.M. (2003). Palaeoenvironmental significance of palustrine carbonates and calcretes in the geological record. *Earth-Science Reviews* 60: 261–298.

Alonso-Zarza, A.M., Genise, J.F., and Verde, M. (2011). Sedimentology, diagenesis and ichnology of cretaceous and palaeogene calcretes and palustrine carbonates from Uruguay. *Sedimentary Geology* 236: 45–61. https://doi.org/10.1016/j.sedgeo.2010.12.003.

Anima, R.J., Clifton, H.E., and Phillips, R. L. (1989). Comparison of modern and Pleistocene estuarine facies in Willapa Bay, Washington. In: *Modern and Ancient Examples of Clastic Tidal Deposits—A Core and Peel Workshop* (ed. G.E. Reinson), 1–19. Canadian Society of Petroleum Geologists.

Bann, K., Fielding, C.R., MacEachern, J. A., and Tye, S. (2004). Distinction between offshore marine and fine-grained coastal facies using integrated ichnology and sedimentology: the Permian Pebbley Beach Formation, southern Sydney Basin, Australia. In: *The Application of Ichnology to Palaeoenvironmental and Stratigraphic Analysis*, vol. 228 (ed. D. McIlroy), 179–212. The Geological Society of London, Special Publication.

Bann, K.L., Tye, S.C., MacEachern, J.A. et al. (2008). Ichnological and sedimentologic signatures of mixed wave- and storm-dominated deltaic deposits: examples from the early Permian Sydney Basin, Australia. In: *Recent Advances in Models of Siliciclastic Shallow-Marine Stratigraphy*, SEPM, Special Publication, vol. 90 (ed. G.J. Hampson, R.J. Steel, P.M. Burgess, and R.W. Dalrymple), 293–332. SEPM.

Beynon, B.M., Pemberton, S.G., Bell, D.D., and Logan, C.A. (1988). Environmental implications of ichnofossils from the lower cretaceous grand rapids formation, cold lake oil sands deposit. In: *Sequences, Stratigraphy, Sedimentology: Surface and Subsurface*, vol. 15 (ed. D.P. James and D.A. Leckie), 275–289. Canadian Society of Petroleum Geologists.

Bhattacharya, J.P. (1993). The expression and Interpretation of marine flooding surfaces and erosional surfaces in core; examples from the Upper Cretaceous Dunvegan Formation, Alberta Foreland Basin, Canada. In: *Sequence Stratigraphy and Facies Associations*, IAS Special Publication, vol. 18 (ed. H. W. Posamentier, C.P. Summerhayes, B.

U. Haq, and G.P. Allen), 125–160. 10.1002/9781444304015.ch8.

Bhattacharya, J.P. (2011). Practical problems in the application of the sequence stratigraphic method and key surfaces: integrating observations from ancient fluvial–deltaic wedges with quaternary and modelling studies. *Sedimentology* 58: 120–169. https://doi.org/10.1111/j.1365-3091.2010.01205.x.

Biddle, S.K., LaGrange, M.T., Harris, B.S. et al. (2021). A fine detail physico-chemical depositional model for Devonian organic-rich mudstones: a petrographic study of the Hare Indian and Canol Formations, Central Mackenzie Valley, Northwest Territories. *Sedimentary Geology* 414: 105838. https://doi.org/10.1016/j.sedgeo.2020.105838.

Birchenough, S.N., Parker, R.E., McManus, E., and Barry, J. (2012). Combining bioturbation and redox metrics: potential tools for assessing seabed function. *Ecological Indicators* 12 (1): 8–16.

Bottjer, D.J., Droser, M.L., and Jablonski, D. (1988). Palaeoenvironmental trends in the history of trace fossils. *Nature* 333: 252–255.

Brett, C.E. (1988). Paleoecology and evolution of marine hard substrate communities: an overview. *Palaios* 3 (4): 374–378.

Bromley, R.G. (1975). Trace fossils at omission surfaces. In: *The Study of Trace Fossils* (ed. R.W. Frey), 399–428. Berlin, Heidelberg: Springer Berlin Heidelberg https://doi.org/10.1007/978-3-642-65923-2_18.

Bromley, R.G. (1990). *Trace Fossils; Biology and Taphonomy*. London: Unwin Hyman.

Bromley, R.G. (1996). *Trace Fossils: Biology, Taphonomy and Applications*, 2e. London: Chapman & Hall https://doi.org/10.4324/9780203059890.

Bromley, R.G. and Asgaard, U. (1991). Ichnofacies: a mixture of taphofacies and biofacies. *Lethaia* 24: 153–163. https://doi.org/10.1111/j.1502-3931.1991.tb01463.x.

Bromley, R.G. and Asgaard, U. (1993). Two bioerosion ichnofacies produced by early and late burial associated with sea-level change. *Geologische Rundschau* 82: 276–280. https://doi.org/10.1007/BF00191833.

Bromley, R.G. and Ekdale, A.A. (1984). *Chondrites*: a trace fossil indicator of anoxia in sediments. *Science* 224: 872–874. https://doi.org/10.1126/science.224.4651.872.

Bromley, R.G. and Ekdale, A.A. (1986). Composite ichnofabrics and tiering of burrows. *Geological Magazine* 123: 59–65. https://doi.org/10.1017/S0016756800026534.

Bromley, R.G., Pemberton, S.G., and Rahmani, R.A. (1984). A cretaceous woodground: the *Teredolites* ichnofacies. *Journal of Paleontology* 58: 488–498.

Buatois, L.A. and Mángano, M.G. (1992). La oxigenación como factor de control en la distribución de asociaciones de trazas fósiles. Formación Kotich Point, Cretácico de Antártida. *Ameghiniana* 29: 69–84.

Buatois, L.A. and Mángano, M.G. (1995a). The paleoenvironmental and paleoecological significance of the lacustrine *Mermia* ichnofacies: an archetypical subaqueous nonmarine trace fossil assemblage. *Ichnos* 4: 151–161. https://doi.org/10.1080/10420949509380122.

Buatois, L.A. and Mángano, M.G. (1995b). Sedimentary dynamics and evolutionary history of a late carboniferous Gondwanic lake in north-western Argentina. *Sedimentology* 42: 415–436. https://doi.org/10.1111/j.1365-3091.1995.tb00382.x.

Buatois, L.A. and Mángano, M.G. (2007). Chapter 17: Invertebrate ichnology of continental freshwater environments. In: *Trace Fossils* (ed. W. Miller), 285–323. Amsterdam: Elsevier https://doi.org/10.1016/B978-044452949-7/50143-1.

Buatois, L.A. and Mángano, M.G. (2009). Applications of ichnology in lacustrine sequence stratigraphy: potential and limitations. *Palaeogeography, Palaeoclimatology, Palaeoecology* 272: 127–142. https://doi.org/10.1016/j.palaeo.2008.10.012.

Buatois, L.A. and Mángano, M.G. (2011). *Ichnology: Organism-Substrate Interactions in Space and Time*. New York: Cambridge University Press.

Buatois, L.A., Gingras, M., MacEachern, J.A. et al. (2005). Colonization of brackish-water systems through time: evidence from the trace-fossil record. *Palaios* 20: 321–347.

Buatois, L.A., Wetzel, A., and Mángano, M.G. (2020). Trace-fossil suites and composite ichnofabrics from meandering fluvial systems: the oligocene lower freshwater molasse of Switzerland. *Palaeogeography, Palaeoclimatology, Palaeoecology* 558: 109944. https://doi.org/10.1016/j.palaeo.2020.109944.

Byers, C.W. (1982). Geological significance of marine biogenic sedimentary structures. In: *Animal-Sediment Relations, Topics in Geobiology* (ed. P.L. McCall and M.J.S. Tevesz), 221–256. Boston: Springer https://doi.org/10.1007/978-1-4757-1317-6_5.

Callow, R.H.T., McIlroy, D., Kneller, B., and Dykstra, M. (2013). Integrated ichnological and sedimentological analysis of a late cretaceous submarine channel-levee system: the Rosario Formation, Baja California, Mexico. *Journal of Marine Petroleum Geology, Special Issue: Internal Architecture, Bedforms and Geometry of Turbidite Channels* 41: 277–294. https://doi.org/10.1016/j.marpetgeo.2012.02.001.

Catuneanu, O. (2006). *Principles of Sequence Stratigraphy*. Amsterdam: Elsevier.

Catuneanu, O. (2022). *Principles of Sequence Stratigraphy*, 2e. Cambridge: Elsevier.

Clifton, H.E. (1983). Discrimination between subtidal and intertidal facies in Pleistocene deposits, Willapa Bay, Washington. *Journal of Sedimentary Research* 53: 353–369.

Coniglio, M. and Dix, G.R. (1992). Carbonate slopes. In: *Facies Models: Response to Sea Level Change* (ed. R.G. Walker and N.P. James), 349–373. St. John's, Nfld.: Geological Association of Canada.

Crimes, T.P., Goldring, R., Homewood, P. et al. (1981). Trace fossil assemblages of deep-sea fan deposits, Gurnigel and Schlieren flysch (Cretaceous-Eocene),

Switzerland. *Eclogae Geologicae Helvetiae* 74: 953–995.

Curran, H.A. (2007). Ichnofacies, ichnocoenoses and ichnofabrics of quaternary shallow-marine to dunal tropical carbonates: a model and implications. In: *Trace Fossils: Concepts, Problems, Prospects* (ed. W. Miller III), 232–247. Amsterdam: Elsevier.

Dafoe, L.T., Gingras, M.K., and Pemberton, S.G. (2010). Wave-influenced deltaic sandstone bodies and offshore deposits in the Viking Formation, Hamilton Lake area, south-central Alberta, Canada. *Bulletin of Canadian Petroleum Geology* 58: 173–201.

Dashtgard, S.E. (2011). Neoichnology of the lower delta plain: Fraser River Delta, British Columbia, Canada: implications for the ichnology of deltas. *Palaeogeography, Palaeoclimatology, Palaeoecology* 307: 98–108.

Dashtgard, S.E. and Gingras, M.K. (2012). Marine invertebrate ichnology. In: *Trace Fossils as Indicators of Sedimentary Environments*, Developments in Sedimentology, vol. 64 (ed. D. Knaust and R.G. Bromley), 273–295. Elsevier.

Dashtgard, S.E., Snedden, J., and MacEachern, J.A. (2015). Under-bioturbated sediments on a mud-dominated shelf, evidence of hypoxia or a paucity of infauna? *Palaeogeography, Palaeoclimatology, Palaeoecology* 425: 128–138.

Desai, B.G. and Chauhan, S. (2021). Taphonomic pathways for the formation of bioturbated cycles in the early cretaceous wave-dominated deltaic environment: Ghuneri Member, Kachchh Basin, India. In: *Mesozoic Stratigraphy of India: A Multi-Proxy Approach, Society of Earth Scientists Series* (ed. S. Banerjee and S. Sarkar), 311–332. Cham: Springer International Publishing https://doi.org/10.1007/978-3-030-71370-6_11.

Diaz, R. and Rosenberg, R. (1995). Marine benthic hypoxia: a review of its ecological effects and the behavioural response of benthic macrofauna. *Oceanography and Marine Biology: An Annual Review* 33: 245–303.

Dörjes, J. and Howard, J.D. (1975). Fluvial-marine transition indicators in an estuarine environment, Ogeechee River-Ossabaw Sound. *Senckenbergiana Maritima* 7: 137–179.

Droser, M.L. and Bottjer, D.J. (1986). A semiquantitative field classification of ichnofabric. *Journal of Sedimentary Research* 56: 558–559.

Dubois, M.K., Goldstein, R.H., and Hasiotis, S.T. (2012). Climate-controlled aggradation and cyclicity of continental loessic siliciclastic sediments in Asselian–Sakmarian cyclothems, Permian, Hugoton embayment, USA. *Sedimentology* 59: 1782–1816. https://doi.org/10.1111/j.1365-3091.2012.01326.x.

Dunkel, C.A., Vázquez-Ortega, A., and Evans, J.E. (2022). Black shale–gray shale transitions in a Late Devonian shale succession, Central Appalachian Basin (Northern Ohio): sedimentary and geochemical evidence for terrestrial organic matter input driving anoxia events. *Palaeogeography, Palaeoclimatology, Palaeoecology* 608: 111271. https://doi.org/10.1016/j.palaeo.2022.111271.

Egenhoff, S.O. and Fishman, N.S. (2013). Traces in the dark—Sedimentary processes and facies gradients in the upper shale member of the Upper Devonian–Lower Mississippian Bakken Formation, Williston Basin, North Dakota, USA. *Journal of Sedimentary Research* 83: 803–824. https://doi.org/10.2110/jsr.2013.60.

Ekdale, A.A. (1988). Pitfalls of paleobathymetric interpretations based on trace fossil assemblages. *Palaios* 3: 464–472. https://doi.org/10.2307/3514720.

Ekdale, A.A. and Berger, W.H. (1978). Deep-sea ichnofacies: modern organism traces on and in pelagic carbonates of the western equatorial Pacific. *Palaeogeography, Palaeoclimatology, Palaeoecology* 23: 263–278. https://doi.org/10.1016/0031-0182(78)90096-2.

Ekdale, A.A. and Bromley, R.G. (1983). Trace fossils and ichnofabric in the Kjalby Gaard Marl, Upper Cretaceous, Denmark. *Bulletin of the Geological Society of Denmark* 31: 107–119.

Ekdale, A.A. and Mason, T.R. (1988). Characteristic trace-fossil associations in oxygen-poor sedimentary environments. *Geology* 16: 720–723. https://doi.org/10.1130/0091-7613(1988)016<0720:CTFAIO>2.3.CO;2.

Ekdale, A.A., Bromley, R.G., and Knaust, D. (2012). The ichnofabric concept. In: *Trace Fossils as Indicators of Sedimentary Environments*, Developments in Sedimentology, vol. 64 (ed. D. Knaust and R.G. Bromley), 139–155. Elsevier.

Ekdale, A.A., Bromley, R.G., and Loope, D.B. (2007). Chapter 35: Ichnofacies of an ancient erg: a climatically influenced trace fossil association in the Jurassic Navajo Sandstone, Southern Utah, USA. In: *Trace Fossils* (ed. W. Miller), 562–574. Amsterdam: Elsevier https://doi.org/10.1016/B978-044452949-7/50161-3.

Ekdale, A.A., Bromley, R.G., and Pemberton, S.G. (1984). Ichnology: the use of trace fossils in sedimentology and stratigraphy. SEPM (Society for Sedimentary Geology) Short Course Notes No. 15. https://doi.org/10.2110/scn.84.15

Farrow, G.E. (1966). Bathymetric zonation of Jurassic trace fossils from the coast of Yorkshire, England. *Palaeogeography, Palaeoclimatology, Palaeoecology* 2: 103–151. https://doi.org/10.1016/0031-0182(66)90011-3.

Fielding, C.R., Bann, K.L., MacEachern, J. A. et al. (2006). Cyclicity in the nearshore marine to coastal, Lower Permian, Pebbley Beach Formation, southern Sydney Basin, Australia: a record of relative sea-level fluctuations at the close of the Late Palaeozoic Gondwanan ice age. *Sedimentology* 53: 435–463. https://doi.org/10.1111/j.1365-3091.2006.00770.x.

Fielding, C.R., Frank, T.D., Birgenheier, L. P. et al. (2008). Stratigraphic imprint of the Late Palaeozoic ice age in eastern Australia: a record of alternating glacial and nonglacial climate regime. *Journal of the Geological Society of London* 165: 129–140. https://doi.org/10.1144/0016-76492007-036.

Frey, R.W. (ed.) (1975). *The Study of Trace Fossils: A Synthesis of Principles, Problems, and Procedures in Ichnology*. New York: Springer.

Frey, R.W. (1978). Behavioral and ecological implications of trace fossils. In: *Trace Fossil Concepts. SEPM (Society for Sedimentary Geology) Short Course Notes*, vol. 5 (ed. P.B. Basan), 43–56. 10.2110/scn.77.05.

Frey, R.W. and Seilacher, A. (1980). Uniformity in marine invertebrate ichnology. *Lethaia* 13: 183–207. https://doi.org/10.1111/j.1502-3931.1980.tb00632.x.

Frey, R.W. and Pemberton, S.G. (1984). Trace fossil facies models. In: *Facies Models*, Geoscience Canada Reprint Series (ed. R.G. Walker), 189–207. St. John's, Nfld.: Geological Association of Canada.

Frey, R.W. and Pemberton, S.G. (1987). The *Psilonichnus* ichnocoenose and its relationship to adjacent marine and nonmarine ichnocoenoses along the Georgia coast. *Bulletin of Canadian Petroleum Geology* 35: 333–357.

Frey, R.W., Howard, J.D., and Hong, J.S. (1987). Prevalent lebensspuren on a modern macrotidal flat, Inchon, Korea: ethological and environmental significance. *Palaios* 2: 571–593.

Frey, R.W. and Pemberton, S.G. (1985). Biogenic structures in outcrops and cores I: approaches to ichnology. *Bulletin of Canadian Society of Petroleum Geologists* 33: 72–115. https://doi.org/10.35767/gscpgbull.33.1.072.

Frey, R.W. and Wheatcroft, R.A. (1989). Organism-substrate relations and their impact on sedimentary petrology. *Journal of Geological Education* 37 (4): 261–279.

Frey, R.W., Pemberton, S.G., and Saunders, T.D.A. (1990). Ichnofacies and bathymetry: a passive relationship. *Journal of Paleontology* 64: 155–158.

Fürsich, F.T., Kennedy, W.J., and Palmer, T.J. (1981). Trace fossils at a regional discontinuity surface: the Austin/Taylor (Upper Cretaceous) contact in central Texas. *Journal of Paleontology* 55: 537–551.

Genise, J.F. (2017). *Ichnoentomology, Topics in Geobiology*. Cham: Springer International Publishing https://doi.org/10.1007/978-3-319-28210-7.

Genise, J.F. (2000). The ichnofamily Celliformidae for *Celliforma* and allied ichnogenera. *Ichnos* 7: 267–282. https://doi.org/10.1080/10420940009380166.

Genise, J.E. and Bown, T.M. (1994). New trace fossils of termites (Insecta: Isoptera) from the late Eocene-early Miocene of Egypt, and the reconstruction of ancient isopteran social behavior. *Ichnos* 3: 155–183. https://doi.org/10.1080/10420949409386386.

Genise, J.F., Mángano, M.G., Buatois, L.A. et al. (2000). Insect trace fossil associations in paleosols: the *Coprinisphaera* Ichnofacies. *Palaios* 15: 49–64. https://doi.org/10.1669/0883-1351(2000)015<0049:ITFAIP>2.0.CO;2.

Genise, J.F., Melchor, R.N., Bellosi, E.S., and Verde, M. (2010). Invertebrate and vertebrate trace fossils from continental carbonates. In: *Carbonates in Continental Settings: Facies, Environments, and Processes, Developments in Sedimentology*, vol. 61 (ed. A.M. Alonso-Zarza and L.H. Tanner), 319–369.

Genise, J.F., Bedatou, E., Bellosi, E.S. et al. (2016). The phanerozoic four revolutions and evolution of paleosol ichnofacies. In: *The Trace-Fossil Record of Major Evolutionary Events, Topics in Geobiology* (ed. M.G. Mángano and L.A. Buatois), 301–370. Dordrecht: Springer Netherlands https://doi.org/10.1007/978-94-017-9597-5_6.

de Gibert, J. and Robles, J.M. (2005). Firmground ichnofacies recording high-frequency marine flooding events (Langhian transgression, Vallè-Penedès Basin, Spain). *Geologica Acta* 3: 295–305.

de Gibert, J.M., Domènech, R., and Martinell, J. (2007). Bioerosion in shell beds from the Pliocene Roussillon Basin, France: implications for the (macro) bioerosion ichnofacies model. *Acta Palaeontologica Polonica* 52: 783–798.

Gingras, M.K. and MacEachern, J.A. (2012). Tidal ichnology of shallow-water clastic settings. In: *Principles of Tidal Sedimentology* (ed. R.A. Davis Jr. and R.W. Dalrymple), 57–77. Dordrecht: Springer Netherlands https://doi.org/10.1007/978-94-007-0123-6_4.

Gingras, M.K., Pemberton, S.G., Saunders, T., and Clifton, H.E. (1999). The ichnology of modern and Pleistocene brackish-water deposits at Willapa Bay, Washington: variability in estuarine settings. *Palaios* 14: 352–374.

Gingras, M.K., MacEachern, J.A., and Pickerill, R.K. (2004). Modern perspectives on the *Teredolites* Ichnofacies: observations from Willapa Bay, Washington. *Palaios* 10: 79–88.

Gingras, M.K., Bann, K.L., MacEachern, J.A. et al. (2007). A conceptual framework for the application of trace fossils. In: *Applied Ichnology*, SEPM Short Course Notes, vol. 52 (ed. J.A. MacEachern, K.L. Bann, M.K. Gingras, and S.G. Pemberton), 1–26. Tulsa: Society for Sedimentary Geology.

Gingras, M.K., MacEachern, J.A., and Dashtgard, S.E. (2011). Process ichnology and the elucidation of physico-chemical stress. *Sedimentary Geology* 237: 115–134.

Gingras, M.K., MacEachern, J.A., Dashtgard, S.E., and Bann, K.L. (2025). The *Teichichnus* Ichnofacies: its neoichnological basis and identification in the rock record. *Sedimentology* 72: 408–441.

Goldring, R. (1993). Ichnofacies and facies interpretation. *Palaios* 8: 403–405.

Goldring, R. (1995). Organisms and the substrate: response and effect. In: *Marine and Palaeoenvironmental Analysis from Fossils*, vol. 83 (ed. D.W.J. Bosence and P.A. Allison), 151–180. Geological Society of London Special Publication https://doi.org/10.1144/GSL.SP.1995.083.01.09.

Gressley, A. (1838). Observations géologiques sur le Jura Soleurois. *Neue Denkschriften der Allgemeinen Schweizerischen Gesellschaft für die Gesammten Naturwissenschaften* 2: 1–112.

Hasiotis, S.T. (2007). Continental ichnology: fundamental processes and controls on trace fossil distribution. In: *Trace Fossils* (ed. W. Miller), 268–284. Amsterdam: Elsevier https://doi.org/10.1016/B978-044452949-7/50142-X.

Hayward, B.W. (1976). Lower Miocene bathyal and submarine canyon ichnocoenoses from Northland, New Zealand. *Lethaia* 9: 149–162.

Howard, J.D. (1975). The sedimentological significance of trace fossils. In: *The Study of Trace Fossils* (ed. R.W. Frey), 131–146. New York: Springer-Verlag.

Howard, J.D. and Frey, R.W. (1973). Characteristic physical and biogenic sedimentary structures in Georgia estuaries. *AAPG Bulletin* 57: 1169–1184.

Hubbard, S.M., MacEachern, J.A., and Bann, K.L. (2012). Slopes. In: *Trace Fossils as Indicators of Sedimentary Environments*, Developments in Sedimentology, vol. 64 (ed. D. Knaust and R.G. Bromley), 607–642. Elsevier.

Hunt, A.P. and Lucas, S.G. (2007). Tetrapod ichnofacies: a new paradigm. *Ichnos* 14: 59–68. https://doi.org/10.1080/10420940601006826.

Johnson, E.W., Briggs, D.E.G., Suthren, R. J. et al. (1994). Non-marine arthropod traces from the subaerial Ordovician Borrowdale Volcanic Group, English Lake District. *Geological Magazine* 131: 395–406. https://doi.org/10.1017/S0016756800011146.

Knaust, D. (2009). Characterisation of a Campanian deep-sea fan system in the Norwegian Sea by means of ichnofabrics. *Marine and Petroleum Geology* 26: 1199–1211. https://doi.org/10.1016/j.marpetgeo.2008.09.009.

Knaust, D. and Bromley, R.G. (2012). *Trace Fossils as Indicators of Sedimentary Environments*, Developments in Sedimentology, vol. 64, 924. Elsevier.

Kotake, N. (1991). Non-selective surface deposit feeding by the *Zoophycos* producers. *Lethaia* 24: 379–385. https://doi.org/10.1111/j.1502-3931.1991.tb01489.x.

Krapovickas, V., Mángano, M.G., Buatois, L.A., and Marsicano, C.A. (2016). Integrated ichnofacies models for deserts: recurrent patterns and megatrends. *Earth-Science Reviews* 157: 61–85. https://doi.org/10.1016/j.earscirev.2016.03.006.

Kvenvolden, K.A., Blunt, D.J., and Clifton, H.E. (1979). Amino-acid racemization in quaternary shell deposits at Willapa Bay, Washington. *Geochimica et Cosmochimica Acta* 43: 1505–1520.

Lacoste, É., Piot, A., Archambault, P. et al. (2018). Bioturbation activity of three macrofaunal species and the presence of meiofauna affect the abundance and composition of benthic bacterial communities. *Marine Environmental Research* 136: 62–70. https://doi.org/10.1016/j.marenvres.2018.02.024.

Levin, L.A., Rathburn, A.E., Gutiérrez, D. et al. (2003). Bioturbation by symbiont-bearing annelids in near-anoxic sediments: implications for biofacies models and paleo-oxygen assessments. *Palaeogeography, Palaeoclimatology, Palaeoecology* 199: 129–140. https://doi.org/10.1016/S0031-0182(03)00500-5.

Lockley, M.G., Hunt, A.P., Meyer, C.A., and Donovan, S. (1994). Vertebrate tracks and ichnofacies concept: implications for paleoecology and palichnostratigraphy. In: *The Paleobiology of Trace Fossils* (ed. S. Donovan), 241–268. Baltimore: John Hopkins University Press.

Lockley, M.G., Rindsberg, A.K., and Zeiler, R.M. (1987). The paleoenvironmental significance of the nearshore *Curvolithus* ichnofacies. *Palaios* 2: 255–262.

MacEachern, J.A. and Gingras, M.K. (2007). Recognition of brackish-water trace-fossil suites in the Cretaceous Western Interior Seaway of Alberta, Canada. In: *Sediment–Organism Interactions: A Multifaceted Ichnology*, SEPM (Society for Sedimentary Geology) Special Publication, vol. 88 (ed. R.G. Bromley, L.A. Buatois, G. Mángano, et al.), 149–194. https://doi.org/10.2110/pec.07.88.0149.

MacEachern, J.A. and Bann, K.L. (2008). The role of ichnology in refining shallow marine facies models. In: *Recent Advances in Models of Siliciclastic Shallow-Marine Stratigraphy*, vol. 90 (ed. G. Hampson, R. Steel, P. Burgess, and R. Dalrymple), 73–116. SEPM Special Publication.

MacEachern, J.A. and Bann, K.L. (2020). The *Phycosiphon* Ichnofacies and the *Rosselia* Ichnofacies: two new ichnofacies for marine deltaic environments. *Journal of Sedimentary Research* 90: 855–886. https://doi.org/10.2110/jsr.2020.41.

MacEachern, J.A., Raychaudhuri, I., and Pemberton, S.G. (1992). Stratigraphic applications of the *Glossifungites* Ichnofacies: delineating discontinuities in the rock record. In: *Applications of ichnology to petroleum exploration, a core workshop: Society of Economic Paleontologists and Mineralogists*, Core Workshop, vol. 17 (ed. S.G. Pemberton), 169–198.

MacEachern, J.A. and Bann, K.L. (2023). Departures from the archetypal deltaic ichnofacies. In: *Ichnology in Shallow-Marine and Transitional Environments*, vol. 522 (ed. C. Cónsole-Gonella, S. de Valais, I. Díaz-Martínez, et al.), 175–213. Geological Society of London Special Publications https://doi.org/10.1144/SP522-2022-56.

MacEachern, J.A., Bann, K.L., Bhattacharya, J.P., and Howell, C.D. (2005). Ichnology of deltas: organism responses to the dynamic interplay of rivers, waves, storms and tides. In: *River Deltas—Concepts, Models, and Examples*, vol. 83 (ed. L. Giosan and J.P. Bhattacharya), 45–85. SEPM Special Publication.

MacEachern, J.A., Bann, K.L., Pemberton, S.G., and Gingras, M.K. (2007). The ichnofacies paradigm: high-resolution paleoenvironmental interpretation of the rock record. In: *Applied Ichnology, Short Course Notes*, vol. 52 (ed. J.A. MacEachern, K.L. Bann, M.K. Gingras, and S.G. Pemberton), 27–64. SEPM Society for Sedimentary Geology https://doi.org/10.2110/pec.07.52.0027.

MacEachern, J.A., Zaitlin, B.A., and Pemberton, S.G. (1999). A sharp-based sandstone of the Viking Formation, Joffre Field, Alberta, Canada: criteria for recognition of transgressively incised shoreface complexes. *Journal of Sedimentary Research* 69: 876–892.

MacEachern, J.A., Pemberton, S.G., Gingras, M.K. et al. (2010). Ichnology and facies models. In: *Facies Models 4* (ed. N.P. James and R.W. Dalrymple), 19–58. St. John's, Nfld.: Geological Association of Canada.

MacEachern, J.A., Bann, K.L., Gingras, M. K. et al. (2012a). The ichnofacies paradigm. In: *Trace Fossils as Indicators of Sedimentary Environments*, Developments in Sedimentology, vol. 64 (ed. D. Knaust and R.G. Bromley), 103–138. Elsevier 10.1016/ B978-0-444-53813-0.00004-6.

Malpas, J.A., Gawthorpe, R.L., Pollard, J. E., and Sharp, I.R. (2005). Ichnofabric analysis of the shallow marine Nukhul Formation (Miocene), Suez Rift, Egypt: implications for depositional processes and sequence stratigraphic evolution. *Palaeogeography, Palaeoclimatology, Palaeoecology* 215: 239–264. https://doi.org/10.1016/j.palaeo.2004.09.007.

MacEachern, J.A., Dashtgard, S.E., Knaust, D. et al. (2012b). Sequence stratigraphy. In: *Trace Fossils as Indicators of Sedimentary Environments, Developments in Sedimentology*, vol. 64 (ed. D. Knaust and R. Bromley), 157–194. Elsevier.

Mángano, M.G., Buatois, L.A., and McNaughton, R.B. (2012). Ichnostratigraphy. In: *Trace Fossils as Indicators of Sedimentary Environments*, Developments in Sedimentology, vol. 64 (ed. D. Knaust and R.G. Bromley), 195–212. Elsevier.

Martin, A.J. and Varricchio, D.J. (2011). Paleoecological utility of insect trace fossils in dinosaur nesting sites of the Two Medicine Formation (Campanian), Choteau, Montana. *Historical Biology* 23: 15–25. https://doi.org/10.1080/08912963.2010.505285.

Martin, K.D. (2004). A re-evaluation of the relationship between trace fossils and dysoxia. In: *The Application of Ichnology to Palaeoenvironmental and Stratigraphic Analysis*, vol. 228 (ed. D. McIlroy), 141–156. Geological Society of London Special Publication https://doi.org/10.1144/GSL.SP.2004.228.01.08.

McIlroy, D. (2008). Ichnological analysis: the common ground between ichnofacies workers and ichnofabric analysts. *Palaeogeography, Palaeoclimatology, Palaeoecology* 270: 332–338. https://doi.org/10.1016/j.palaeo.2008.07.016.

McIlroy, D. (2007). Lateral variability in shallow marine ichnofabrics: implications for the ichnofabric analysis method. *Journal of the Geological Society* 164: 359–369. https://doi.org/10.1144/0016-76492005-101.

McIlroy, D. (2004). Some ichnological concepts, methodologies, applications and frontiers. In: *The Application of Ichnology to Palaeoenvironmental and Stratigraphic Analysis*, vol. 228 (ed. D. McIlroy), 3–27. Geological Society of London Special Publication https://doi.org/10.1144/GSL.SP.2004.228.01.02.

Miller, M.F. and Rehmer, J. (1982). Using biogenic structures to interpret sharp lithologic boundaries: an example from the Lower Devonian of New York. *Journal of Sedimentary Research* 52: 887–895.

Moore, R.C. (1949). Meaning of facies. *Geological Society of America Memoir* 39: 1–34.

Morris, J.E., Hampson, G.J., and Johnson, H.D. (2006). A sequence stratigraphic model for an intensely bioturbated shallow-marine sandstone: the Bridport Sand Formation, Wessex Basin, UK. *Sedimentology* 53: 1229–1263. https://doi.org/10.1111/j.1365-3091.2006.00811.x.

Moslow, T.F. and Pemberton, S.G. (1988). An integrated approach to the sedimentological analysis of some Lower Cretaceous shoreface and delta front sandstone sequences. In: *Sequences, Stratigraphy, Sedimentology: Surface and Subsurface*, vol. 15 (ed. D.J. James and D.A. Leckie), 373–386. Canadian Society of Petroleum Geologists, Memoir.

Müller, A.H. (1962). Zur Ichnologie, Taxiologie und Ökologie fossiler Tiere, Teil 1. *Freiberger Forschungshefte, C* 151: 5–49.

Myint, M. (2007). The *Psilonichnus* Ichnofacies: an example from the Iwaki Formation, Shiramizu Group, Joban coal field, Japan. In: *Applied Ichnology*, SEPM Short Course Notes, vol. 52 (ed. J. A. MacEachern, K.L. Bann, M.K. Gingras, and S.G. Pemberton), 335–342. Tulsa: Society for Sedimentary Geology.

Nesbitt, E.A. and Campbell, K.A. (2006). The paleoenvironmental significance of *Psilonichnus*. *Palaios* 21: 187–196.

Netto, R.G. and Grangeiro, M.E. (2009). Neoichnology of the seaward side of Peixe Lagoon in Mostardas, southernmost Brazil: the *Psilonichnus* ichnocoenoses revisited. *Revista Brasileira de Paleontologia* 12: 211–224.

Netto, R.G., Benner, J., Buatois, L.A. et al. (2012). Glacial environments. In: *Trace Fossils as Indicators of Sedimentary Environments*, Developments in Sedimentology, vol. 64 (ed. D. Knaust and R.G. Bromley), 299–327. Elsevier.

Olivero, D. and Gaillard, C. (1996). Paleoecology of Jurassic *Zoophycos* from south-eastern France. *Ichnos* 4: 249–260. https://doi.org/10.1080/10420949609380135.

Pattison, S.A.J. and Walker, R.G. (1994). Incision and filling of a lowstand valley: late Albian Viking Formation at Crystal, Alberta, Canada. *Journal of Sedimentary Research* 64: 365–379.

Pemberton, S.G. and Frey, R.W. (1985). The *Glossifungites* ichnofacies: modern examples from the Georgia Coast, USA. In: *Biogenic Structures: Their Use in Interpreting Depositional Environments*, SEPM Special Publication, vol. 35 (ed. H. A. Curran), 237–259. Tulsa: Society for Sedimentary Geology.

Pemberton, S.G., Flach, P.D., and Mossop, G.D. (1982). Trace fossils from the Athabasca Oil Sands, Alberta, Canada. *Science* 217: 825–827.

Pemberton, S.G., Kobluk, D.R., Yeo, R.K., and Risk, M.J. (1980). The boring *Trypanites* at the Silurian-Devonian disconformity in southern Ontario. *Journal of Paleontology* 1258–1266.

Pemberton, S.G., MacEachern, J.A., and Saunders, T. (2004). Stratigraphic applications of substrate-specific ichnofacies: delineating discontinuities in the rock record. In: *The Application of Ichnology to Palaeoenvironmental and Stratigraphic Analysis*, vol. 228 (ed. D. McIlroy), 29–62. Geological Society of London Special Publication https://doi.org/10.1144/GSL.SP.2004.228.01.03.

Pemberton, S.G., MacEachern, J.A., Gingras, M.K., and Saunders, T.D.A. (2008). Biogenic chaos: cryptobioturbation and the work of sedimentologically friendly organisms. *Palaeogeography, Palaeoclimatology, Palaeoecology* 270: 273–279. https://doi.org/10.1016/j.palaeo.2008.01.036.

Pickett, T.E., Kraft, J.C., and Smith, K. (1971). Cretaceous burrows, Chesapeake and Delaware canal, Delaware. *The Journal of Paleontology* 45: 209–211.

Plotnick, R.E. (2012). Behavioral biology of trace fossils. *Paleobiology* 38: 459–473. https://doi.org/10.1666/11008.1.

Reading, H.G. (1996). *Sedimentary Environments: Processes, Facies and Stratigraphy*, 3e. Oxford: Blackwell Science.

Reineck, H.-E. (1967). Parameter von schichtung und bioturbation. *Geologische Rundschau* 56: 420–438. https://doi.org/10.1007/BF01848734.

Reineck, H.-E. (1963). Sedimentgefüge im Bereich der südlichen Nordsee. *Abhandlungen der Senckenbergischen Naturforschenden Gesellschaft* 505: 1–138.

Rhoads, D.C. and Morse, J.W. (1971). Evolutionary and ecologic significance of oxygen-deficient marine basins. *Lethaia* 4: 413–428. https://doi.org/10.1111/j.1502-3931.1971.tb01864.x.

Richter, R. (1928). Psychische reaktionen fossiler tiere. *Palaeobiologica* 1: 226–244.

Rodríguez-Tovar, F.J., Pérez-Valera, F., and Pérez-López, A. (2007). Ichnological analysis in high-resolution sequence stratigraphy: the *Glossifungites* ichnofacies in Triassic successions from the Betic Cordillera (southern Spain). *Sedimentary Geology* 198: 293–307.

Savrda, C.E. (2007). Trace fossils and marine benthic oxygenation. In: *Trace Fossils: Concepts, Problems, Prospects* (ed. W. Miller), 149–158. Elsevier.

Savrda, C.E., Ozalas, K., Demko, T.H. et al. (1993). Log-grounds and the ichnofossil *Teredolites* in transgressive deposits of the Clayton Formation (Lower Paleocene), Western Alabama. *Palaios* 8: 311. https://doi.org/10.2307/3515263.

Savrda, C.E., Browning, J.V., Krawinkel, H., and Hesselbo, S.P. (2001). Firmground ichnofabrics in deep-water sequence stratigraphy, tertiary clinoform-toe deposits, New Jersey slope. *Palaios* 16: 294–305.

Schäfer, W. (1962). *Sonderdruck aus: Aktuo-palaeontologie nach Studien in der Nordsee.* Frankfurt am Main: Kramer.

Scott, J.J., Buatois, L.A., and Mángano, M. G. (2012). Lacustrine environments. In: *Trace Fossils as Indicators of Sedimentary Environments*, Developments in Sedimentology, vol. 64 (ed. D. Knaust and R.G. Bromley), 379–417. Elsevier https://doi.org/10.1016/B978-0-444-53813-0.00013-7.

Seike, K., Hasegawa, H., and Ichinnorov, N. (2010). Preferred orientation of the trace fossil *Entradichnus meniscus* in eolian dune strata (Djadokhta Formation) at Tugrikiin Shiree, southern Mongolia and its paleoecological implications. *Island Arc* 19: 622–627. https://doi.org/10.1111/j.1440-1738.2010.00742.x.

Seilacher, A. (1953). Studien zur Palichnologie. II. Die fossilen Ruhespuren (Cubichnia). *Neues Jahrbuch für Geologie und Paläontologie (Abhandlungen)* 98: 87–124.

Seilacher, A. (1964). Biogenic sedimentary structures. In: *Approaches to Paleoecology* (ed. J. Imbrie and N. Newell), 296–316. New York: Wiley.

Seilacher, A. (1967). Bathymetry of trace fossils. *Marine Geology* 5: 413–428. https://doi.org/10.1016/0025-3227(67)90051-5.

Seilacher, A. (1974). Flysch trace fossils: evolution of behavioural diversity in the deep-sea. *Neues Jahrbuch für Geologie und Paläontologie (Monatsheft)* 4: 233–245.

Seilacher, A. (1978). Use of trace fossil assemblages for recognizing depositional environments. In: *Trace Fossil Concepts*, SEPM Short Course, vol. 5 (ed. P.B. Basan), 167–181. Tulsa: Society for Sedimentary Geology.

Seilacher, A. (2007). *Trace Fossil Analysis.* Berlin, Heidelberg: Springer https://doi.org/10.1007/978-3-540-47226-1.

Schultz, S.K., MacEachern, J.A., Catuneanu, O. et al. (2022). High-resolution sequence stratigraphic framework for the late Albian Viking Formation in central Alberta. *Marine and Petroleum Geology* 139: https://doi.org.10.1016/j.marpetgeo.2022.105627.

Smith, C.R., Levin, L.A., Hoover, D.J. et al. (2000). Variations in bioturbation across the oxygen minimum zone in the northwest Arabian Sea. *Deep Sea Research Part II: Topical Studies in Oceanography* 47: 227–257. https://doi.org/10.1016/S0967-0645(99)00108-3.

Smith, R.M.H., Mason, T.R., and Ward, J. D. (1993). Flash-flood sediments and ichnofacies of the Late Pleistocene Homeb Silts, Kuiseb River, Namibia. *Sedimentary Geology* 85: 579–599. https://doi.org/10.1016/0037-0738(93)90103-C.

Taylor, A., Goldring, R., and Gowland, S. (2003). Analysis and application of ichnofabrics. *Earth-Science Reviews* 60: 227–259. https://doi.org/10.1016/S0012-8252(02)00105-8.

Taylor, A.M. and Goldring, R. (1993). Description and analysis of bioturbation and ichnofabric. *Journal of the Geological Society* 150: 141–148. https://doi.org/10.1144/gsjgs.150.1.0141.

Timmer, E.R., Gingras, M.K., and Zonneveld, J.-P. (2016). Spatial and temporal significance of process ichnology data from silty-mudstone beds of inclined heterolithic stratification, Lower Cretaceous McMurray Formation, NE Alberta, Canada. *Palaios* 31: 533–548. https://doi.org/10.2110/palo.2015.089.

Uchman, A. (2009). The *Ophiomorpha rudis* ichnosubfacies of the *Nereites* ichnofacies: characteristics and constraints. *Palaeogeography, Palaeoclimatology, Palaeoecology* 276: 107–119.

Uchman, A. and Demírcan, H. (1999). A *Zoophycos* group trace fossil from Miocene flysch in Southern Turkey: evidence for a U-shaped causative burrow. *Ichnos* 6: 251–259. https://doi.org/10.1080/10420949909386456.

Vallon, L.H., Rindsburg, A.K., and Bromley, R.G. (2016). An updated classification of animal behaviour preserved in substrates. *Geodinamica Acta* 28: 5–20.

Walker, R.G. (1984a). *Facies Models*, Geoscience Canada Reprint Series 1, 2e. St. John's, Nfld.: Geological Association of Canada, 317 p. https://www.abebooks.com/magazines-periodicals/Facies-Models-second-edition-1984-Geological/20549848841/bd.

Walker, R.G. (1984b). General introduction: facies, facies sequences and facies models. In: *Facies Models*, Geoscience Canada Reprint Series 1, 2e (ed. R.G. Walker), 1–9. St. John's, Nfld.: Geological Association of Canada.

Walker, R.G. and Wiseman, T.R. (1995). Lowstand shorefaces, transgressive incised shorefaces, and forced regressions: examples from the Viking Formation, Joarcam field, Alberta. *Journal of Sedimentary Research* 65: 132–141.

Walther, J. (1894). *Einleitung in die geologie als historische wissenschaft: Beobachtungen über die bildung der gesteine und ihrer organischen einschlüsse*. G. Fischer.

Weimer, R.J. and Hoyt, J.H. (1964). Burrows of *Callianassa major* say: geologic indicators of littoral and shallow neritic environments. *Journal of Paleontology* 38: 761–767.

Wignall, P.B. (1993). Distinguishing between oxygen and substrate control in fossil benthic assemblages. *Journal of the Geological Society* 150: 193–196. https://doi.org/10.1144/gsjgs.150.1.0193.

Wignall, P.B. (1991). Dysaerobic trace fossils and ichnofabrics in the Upper Jurassic Kimmeridge Clay of Southern England. *Palaios* 6: 264–270. https://doi.org/10.2307/3514906.

Wilson, M.A. (2007). Macroborings and the evolution of marine bioerosion. In: *Trace Fossils* (ed. W. Miller), 356–367. Elsevier.

Zorn, M.E., Lalonde, S.V., Gingras, M.K. et al. (2006). Microscale oxygen distribution in various invertebrate burrow walls. *Geobiology* 4: 137–145. https://doi.org/10.1111/j.1472-4669.2006.00074.x.

4

The Stratigraphic Record: Preservation, Facies and Depositional Sequences

Bruce Levell[1] *and Frans van Buchem*[2]

[1] *Department of Earth Sciences, University of Oxford, Oxford, UK*
[2] *Physical Sciences and Engineering Division, King Abdullah University of Science and Technology, Thuwal, Saudi Arabia*

Summary

This chapter summarises how sedimentary facies and stratigraphic surfaces together generate stratigraphy on a variety of time scales. This analysis requires an assessment of (i) how much time is represented in a sedimentary layer (facies) and its bounding hiatuses (stratal surfaces), and (ii) how the stratification is related to the depositional and geological processes responsible for both the layer and the hiatuses. We address basic stratigraphic principles, such as base level, accommodation, sediment supply, sequence hierarchy and stacking patterns. We present three end-member ways of approaching the stratigraphic record as (i) a sequence of 'events', (ii) a consequence of the natural dynamics of sedimentary environments: 'autogenic' and (iii) formed through the action of external forcing: 'allogenic'. Allogenic processes are further treated under three headings: cyclostratigraphy driven by orbitally controlled climatic fluctuations (Milanković cycles), sequence stratigraphy driven by a combination of orbitally controlled and longer period sea level and subsidence variations that combine to affect relative sea level and sediment supply and tectonostratigraphy controlled by subsidence mechanisms and tectonic events. With this as background, the basics of sequence stratigraphic methodology are presented in more detail, since this is a key unifying concept in sedimentary geology enabling the interpretation of the stratigraphic record as a result of fundamental interactions of rates of accommodation space formation and sediment supply. In a final section, the recording of geological time in both sedimentary facies and stratal surfaces is explained through the action of a series of successive 'preservation filters'. Seven different time scales are distinguished, each characteristic of particular processes that dominate sedimentation, ranging from hours and days till hundreds of millions of years. Particular attention is paid to how deposits may get preferentially preserved, and the extent to which differential preservation may bias the record, with implications for the limits to interpretability.

4.1 Introduction

Facies analysis (Chapter 2) is a reductionist technique. It allows us to break down, or *dissociate*, any stratigraphic succession into individual interpretable elements, such as microfacies (fabric units), beds or indeed facies themselves, and is a main topic of this book. In this chapter, however, we are interested in how facies elements are genetically linked (or '*associated*') to make stratigraphy, for example, in facies associations, facies successions and sequences. To successfully accomplish this analysis, it is important to distinguish between two fundamental aspects of any sedimentary succession: (i) the sedimentary facies themselves and (ii) the nature and duration of the depositional breaks between them. In addition, all sedimentary environments and their associated stratigraphic successions are organised in hierarchies in time and space. This is a direct consequence of sedimentary processes that act over different time periods and geographical extents. These hierarchies also provide the structure for interpreting any succession of strata in terms of, for example, past environments, organic evolution, climate change, tectonic processes or predicting the geometry and lateral extent of rock units.

Stratigraphy is thus concerned with the temporal and spatial 'order' and 'relative position' of strata. Stratigraphy provides the organisational context within which sedimentary facies analysis (process–response studies) provides a dynamic process-based understanding of the rock units. The integration of sedimentology and stratigraphy occurs

Reading's Sedimentary Environments: Processes, Facies and Stratigraphy, Fourth Edition. Edited by Bruce Levell.

Companion website: www.wiley.com/go/levell4e

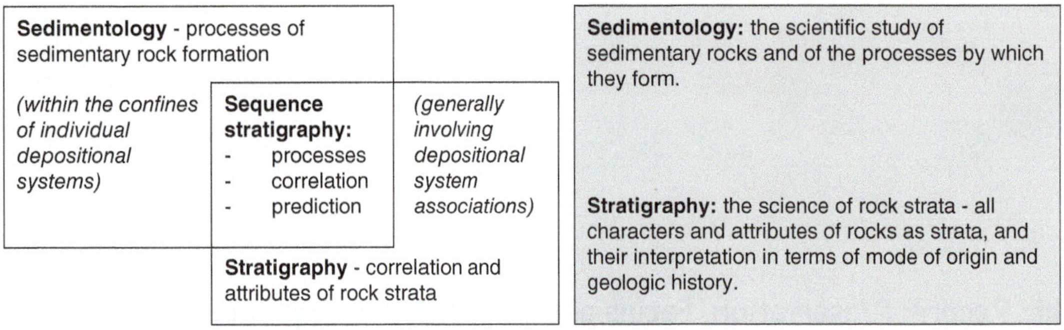

Figure 4.1 The complementary relationship of sedimentology and stratigraphy and the central, unifying, role of sequence stratigraphy are illustrated in this scheme (Catuneanu 2022 / with permission of Elsevier).

in the discipline of 'sequence stratigraphy', where sedimentary processes and genetic physical stratigraphic relationships are jointly interpreted within the hierarchy of temporal and spatial scales (Figure 4.1).

Here, we consider the depositional and preservational processes that construct a stratigraphic succession, with its associated stratigraphic gaps, through three lenses:

1) *Successions, sequences and stratigraphy – Sections 4.2 and 4.3*: this section considers (i) processes intrinsic to sedimentation in any given environment, such as the action of waves and tides or the effects of river floods (i.e. autogenic or 'autocyclic') and (ii) extrinsic controls, such as changes in sediment supply, tectonics, climate and sea-level change (i.e. allogenic or 'allocyclic').
2) *Sequence stratigraphic methodology – Section 4.4*: this concerns the analysis of physical stratigraphy in terms of the three-dimensional (3D) relationships between sedimentary deposits and key stratal surfaces and provides a systematic approach to interpreting these in the rock record.
3) *The preservation filter and the incompleteness of the sedimentary record – Section 4.5*: this more philosophical perspective considers in detail which sediments get preserved and why, which portions of geological time are actually recorded in a deposit, which are lost in bounding stratal surfaces and how the preserved record may therefore be biased.

Fuller treatments of modern stratigraphic thinking are given by Smith et al. (2015) and Miall (2016). The formalisms of stratigraphy are summarised and updated by the International Commission on Stratigraphy and in particular its various sub-commissions (https://stratigraphy.org/). A useful summary of modern stratigraphic procedure and techniques related to geological time is given by Smith et al. (2015) and Coe (2022). A good treatment of stratigraphy as the discipline supporting the erection of the Geological Time Scale is documented in Gradstein et al. (2020). The specific topic of sequence stratigraphy, with an emphasis on siliciclastics is synthesised by Catuneanu (2022).

4.2 Facies Successions and Sequences – Understanding Stratigraphic Architecture

The term 'stratigraphic architecture' refers to the vertical and lateral organisation of sedimentary strata and their time relationships. Understanding how different architectures arise involves the application of some fundamental concepts and methodologies that are applicable to the majority of the sedimentary systems and are briefly summarised here.

4.2.1 Sediment Routing Systems

Siliciclastic depositional environments[1] are all components of integrated sediment routing systems (Allen 2017), and the path from the source of sediment to its long-term resting place, or 'sink', is highly variable, ranging from tens to thousands of kilometres long and often covering complex geology (Figure 4.2). The siliciclastic system starts with weathering and erosion, the 'sediment-generation' process (Weltje 2012), which produces both grains and dissolved materials and passes through a 'transfer zone'; a chain of episodic transport, interim storage and erosion, frequently accompanied by yet more weathering (e.g. in soils). Eventually, the sediment reaches a given long-term depositional environment, or 'sink' where it accumulates (Caracciolo 2020).

Sediment in a routing system may be temporarily stored in transit in a variety of deposits, for example, floodplains, delta lobes or submarine slopes. It may then be released

1 A 'depositional environment' is broadly defined here as 'a geographic and/or palaeogeographic area with a distinctive set of sedimentary processes and geomorphic characteristics organised along a sediment routing system'.

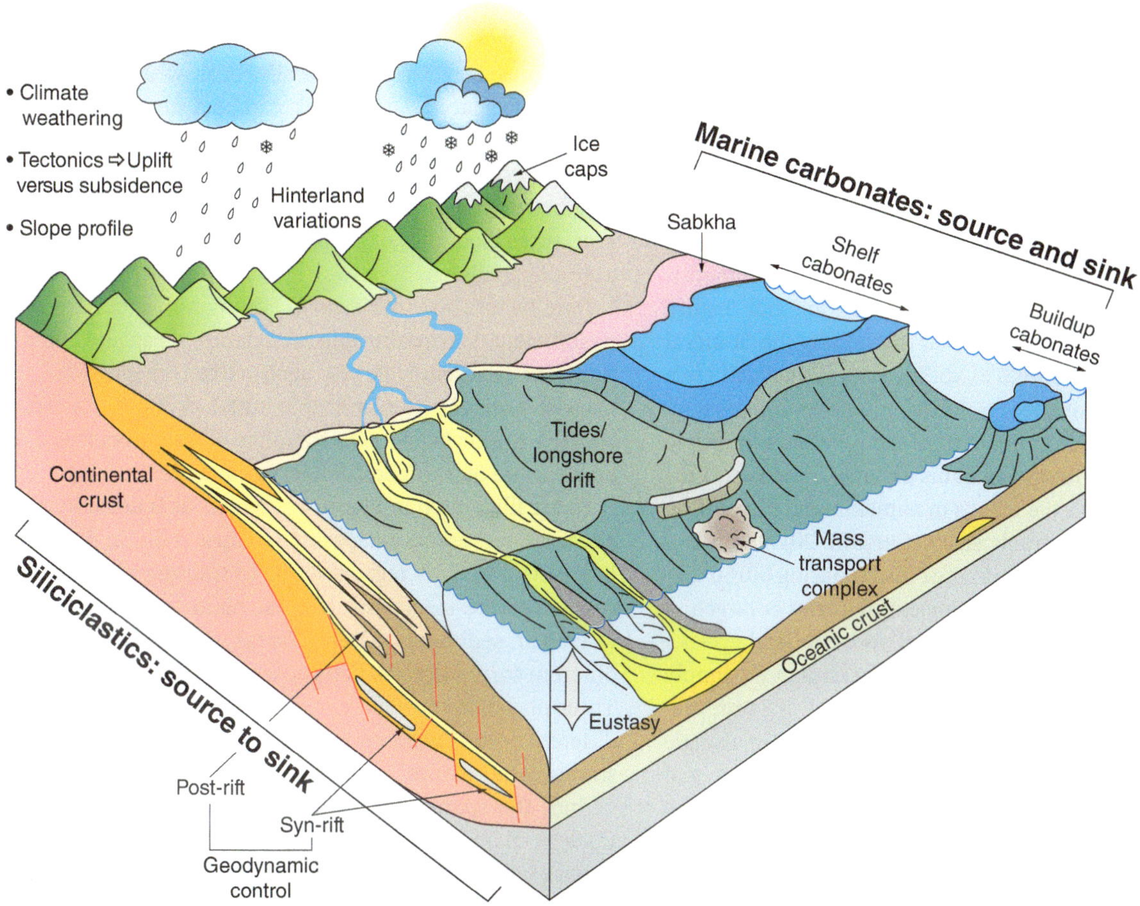

Figure 4.2 The concept of the siliciclastic and marine carbonate sediment routing systems. Weathering and erosion generate and supply sediment, which follows a route to both temporary and permanent sinks spread along transport paths. The routing system consists of connected segments within which sediment may be deposited, stored and re-eroded. Sediment is thus passed from one segment to another depending on both extrinsic (e.g. climate and uplift) and intrinsic (e.g. drainage basin changes and slope instability) controls. Sediment supply to any segment normally involves temporary storage in other segments (see Allen 2017). In contrast, carbonate sediments are mostly deposited in or close to their place of formation, with relatively little transportation. A carbonate depositional environment (right-hand side) is frequently both source and sink.

again to travel further by floods, coastal erosion or submarine slides, respectively. Re-erosion from temporary storage often occurs when some geomorphic 'threshold' is exceeded and there is a state change in a depositional environment. Examples include the change to or from a braided to meandering stream pattern as slope or discharge changes (Schumm 1973), when a submarine slope becomes oversteepened and therefore unstable (Prather et al. 2017), or a river channel aggrades, increasing the cross-valley slope such that it breaks out of the old channel sideways ('avulses') to form a new one (Törnqvist and Bridge 2002; Sinha et al. 2014). A consequence of storage and sharp thresholds is that the signal of changing sediment composition or volume due to external forcing by, say, climate or tectonic uplift, may become unrecognisable at the sink-end of the routing system. This may be due to the mixing of batches of sediment released from stores within a routing system under different conditions and with time lags. The input signal of climate change or tectonic uplift has been effectively 'shredded' by the combined effects of storage or 'buffering', sediment release due to threshold-exceedance or simply the random timing of high-energy transport events (Jerolmack and Paola 2010).

The geomorphic and other thresholds between 'states' of a depositional environment, the volumetric and timing effects of buffers (sediment storage) and the feedback between connected depositional environments are all subjects of active current research (Mississippi River system; Bentley Sr et al. 2016; Po River – Adriatic system; Amorosi et al. 2016; Cretaceous Western Interior Seaway; Lupin and Hampson 2020). Understanding such inter-relationships will become of increasing importance in modern systems responding to rapid external change. Man-made interventions such as the trapping of sediment and water by dams, lowering of coastal plains due to ground-water extraction, the embankment of rivers, dredging of downstream channels for navigation and man-made coastal defences all have complex knock-on effects that need to be placed in the context of the routing systems as a whole. (See for example Chadwick et al. (2020) for effects on river avulsion frequencies).

The interconnected processes and the sheer length of siliciclastic sediment routing systems can complicate the reconstruction of a sediment source area from the material found in the basinal sink. Selective extraction of minerals and grains means that sediment composition and texture may differ significantly from that implied by the 'mineral fertility' of the source area (its ability to source key minerals distinctive of provenance). Selective transport and erosion, chemical weathering and mixing of sediment from multiple source areas via tributary streams are also important. Furthermore, although the effects of abrasion and dissolution in transit are demonstrably small (Garzanti 2017), those of subsequent unstable mineral dissolution during burial diagenesis may be large. Mineralogical maturity is largely a complex function of multiple burial and erosional cycles.

The sediment routing system concept cannot be simply translated to carbonate systems. In carbonate systems, sediment generation and transport often occur within a single depositional environment (source *and* sink: James and Jones 2015). When carbonate sediment is transferred to adjacent environments, such as from a reef-front into deep water, transport distances and transport paths are relatively short and direct (often 100 s m to 10 s km). Carbonate production is best related to a series of carbonate-producing 'factories' based on the ecology of the primary carbonate producers (Chapter 11). There is admittedly still a trajectory from shallow water sources to deep-water sinks (Mulder et al. 2017), but the complex internal feedback between sediment production, trapping, binding and 'shedding' (off-platform transport) in each environment are primarily ecological and sensitive to small changes in the immediate depositional environment.

The deposition of evaporites (Chapter 10) is even more connected to the location of sediment formation and transportation is limited to occasional mass transport, with in situ precipitation in the sabkha environments or settling out of suspension in Salinas for example.

4.2.2 Sediments and Stratigraphic Breaks

A fundamental stratigraphic principle is the notion that time can be recorded in either a package of sediment or a stratigraphic surface. Obviously, this book is mostly focussed on the sediment packages and the depositional environments they represent, but when it comes to establishing the stratigraphic architecture of a system, understanding the significance of the *stratigraphic surfaces* is fundamental. The importance of establishing the time relationships between sediments and surfaces becomes particularly clear when correlating stratigraphic sections. Given that correlations typically deal with incomplete and widely spaced datasets this can be a surprisingly difficult task. There are, however, a number of powerful stratigraphic concepts and principles that facilitate this task, which are presented below. Before addressing those, however, three examples of correlations are presented to illustrate the challenge.

Two factors are decisive in stratigraphic correlation: (i) sufficient data quality and abundance, especially of indicators of time equivalence, such as can be provided by marker horizons (e.g. volcanic tuffs) or biostratigraphy and (ii) the level of understanding of the processes and architecture of the depositional system under investigation. Both are illustrated in Figure 4.3, which, as is usual in such exercises, represents a depositional dip section. In this case, it depicts a broadly regressive (R) to transgressive (T) succession of the siliciclastic coastal plain to coastal-shoreface/shelf sediments. (Depositional strike sections pose different correlation challenges, which are addressed later.) Three plausible correlations are proposed in Figure 4.3. The different interpretations largely arise from the choice of stratigraphic 'datum' horizon (or flattening horizon), which defines the spatial relationship between the vertical sections by making the assumption of tracking a depositional surface with that particular horizon. Correlation is always underpinned by the quality of the observations, including the facies interpretation, the identification of (erosional) surfaces and the availability of age dating. However, even the best descriptions may result in a range of non-unique interpretations, not least, as in this example, related to the erosion surfaces and their inferred lateral extent. Are these due to autogenic process and likely to be of limited lateral extent (e.g. avulsion of a fluvial channel) or of allogenic origin caused, for example, by changes in the base level and, therefore, more likely to have regional extent? Answers to these questions are in many cases provided by extending the observations in both time and space. In the case of Figure 4.4, which also represents a depositional dip-oriented section, the same facies successions can be interpreted as either a normal prograding deltaic succession (a) or a prograding delta incised by a later, and genetically disconnected, erosional valley (b). The implications for the potential presence of reservoir facies are very different: the first option is likely to result in greater facies complexity and poorer reservoir connectivity, whereas the second option could result in multistorey/multilateral sand bodies with higher overall reservoir connectivity and the potential to predict sediment throughput to a basinal turbidite fan.

Critical to both examples (Figures 4.3 and 4.4) and all correlation work is that the various correlation scenarios are

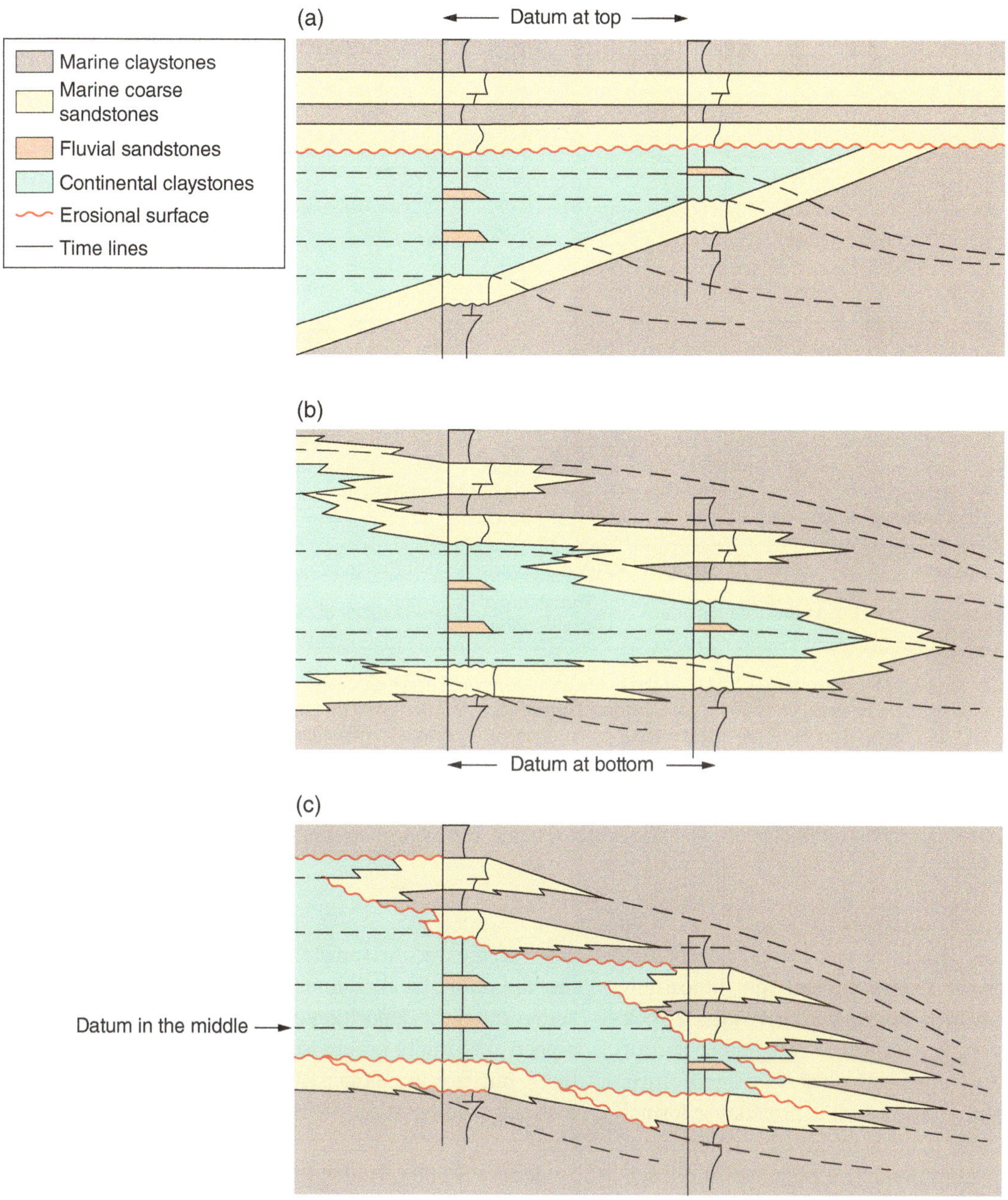

Figure 4.3 The depositional concept defines the correlation strategy. The same information can lead to very different stratigraphic correlations based on the choice of a depositional surface along which to align the section (Modified from Homewood et al. 2000). Ideally this 'datum' surface is constrained by relative or absolute ages of the facies units combined with water depth estimations. (a) Flattening on the uppermost sandstones. This interpretation is the simplest possible geometrically but disconnects the architecture of the uppermost sands from those beneath, requiring a disconformity. (b) Interprets the sandstones as part of a single prograding and retrograding coastal belt. (c) Adds genetic detail to the interpretation of (b) by suggesting that the coastal sand facies are progradational and that the transgressions are erosional, implying pulsed relative base-level changes, both falls and rises, as the causes of the progradation and retrogradation. Each of these interpretations predicts a different distribution and connectivity of the rock units away from the datapoints. The implied timelines (depositional surfaces at key moments) are shown. These, together with observations of erosion surfaces and some facies details, offer tests of the different interpretations and illustrate that correlation is an exercise in forming and testing multiple working hypotheses (Chamberlin 1890).

geologically feasible. This test is often addressed using analogous depositional systems (modern and/or ancient, surface or sub-surface) and/or numerical models of similar depositional settings (e.g. forward stratigraphic models and physical models). Hence, the experience of the practitioner and their understanding of the depositional system being investigated are other critical aspects. The term 'genetic stratigraphic correlation' is often used when

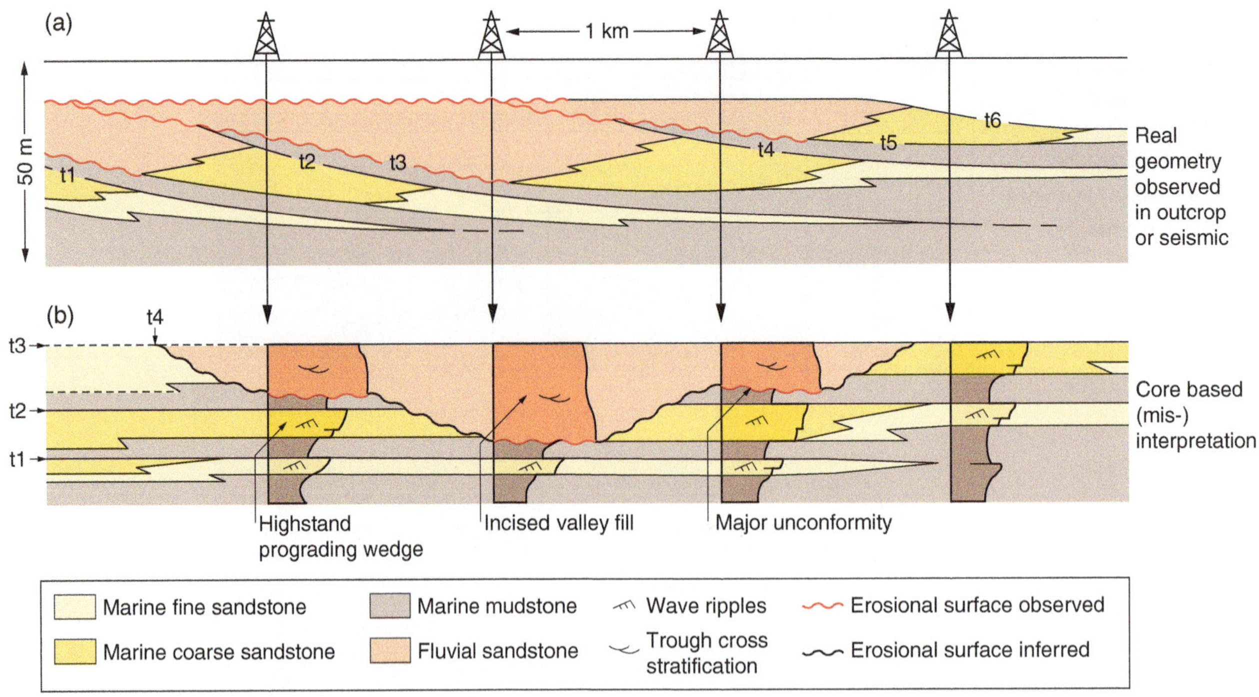

Figure 4.4 Two possible stratigraphic correlations based on the same lithological observations, but with differing interpretations of the erosional surfaces (Modified from Homewood et al. 2000). (a) Erosional surfaces are interpreted as generated by successive base-level falls during deltaic progradation. (b) Erosional surfaces are interpreted as connected, forming as a single major unconformity the base of a large, incised valley eroding and removing the upper parts of a prograding deltaic succession. Note that the two models have very different implications for the geometry of timelines representing depositional profiles (dipping in (a), and horizontal in (b)). The nature of the erosional surfaces also differs. Marine erosion by lowered wave base is implied in (a) whereas fluvial erosion is required in (b). These two implications of the correlation, together with more detailed facies observations, and more contextual information allow the differing working hypotheses to be tested.

physical correlation sections incorporate the results of sedimentary facies analysis. This contrasts with 'lithostratigraphic correlation', which simply connects similar rock types independently of their chronostratigraphic and/or depositional framework, often with no genetic significance.

Sequence stratigraphy (Section 4.3) provides a formal correlation methodology to develop robust and consistent interpretations, with high predictive potential and testability. Fundamental to this approach are the identification of (i) genetically related packages of strata, based on facies analysis, (ii) correlative stratigraphic surfaces based on notable facies juxtapositions and constrained by age dating and (iii) conceptual models of the depositional system.

4.2.3 Chronostratigraphic Diagrams and Age Dating

An important test of the understanding of time relationships between sediment packages and surfaces in correlations is the display of stratigraphic relationships against time in chronostratigraphic diagrams. This process forces decisions as to the relative timing of events, the nature of gaps in the record (erosional or non-depositional) and the defintion of time-lines along depositional profiles, across different depositional environments. These are all aspects that might otherwise remain implicit in cross sections but are fundamental to understanding the geometrical configuration of the strata. An added benefit of bracketing rock in time is that sedimentation rates can then be deduced. The introduction of absolute time in a stratigraphic succession allows the step from a description-focussed approach to a more quantitative one to be taken. This sets the spatial and temporal boundary conditions for the parameters of process dynamics, for example, subsidence, sea-level change and sediment supply variations. In this way, a predictive feedback loop between stratigraphy and sedimentology can be created.

Age dating of the rock succession is thus of prime importance, for both the reconstruction of the stratigraphic architecture and the quantitative approach to stratigraphy and sedimentology. An example of a chronostratigraphic diagram is given in Figure 4.5, and this topic will be further developed in Section 4.4. This book is not the place to provide an in-depth overview of different age dating techniques for the sedimentary record. For comprehensive

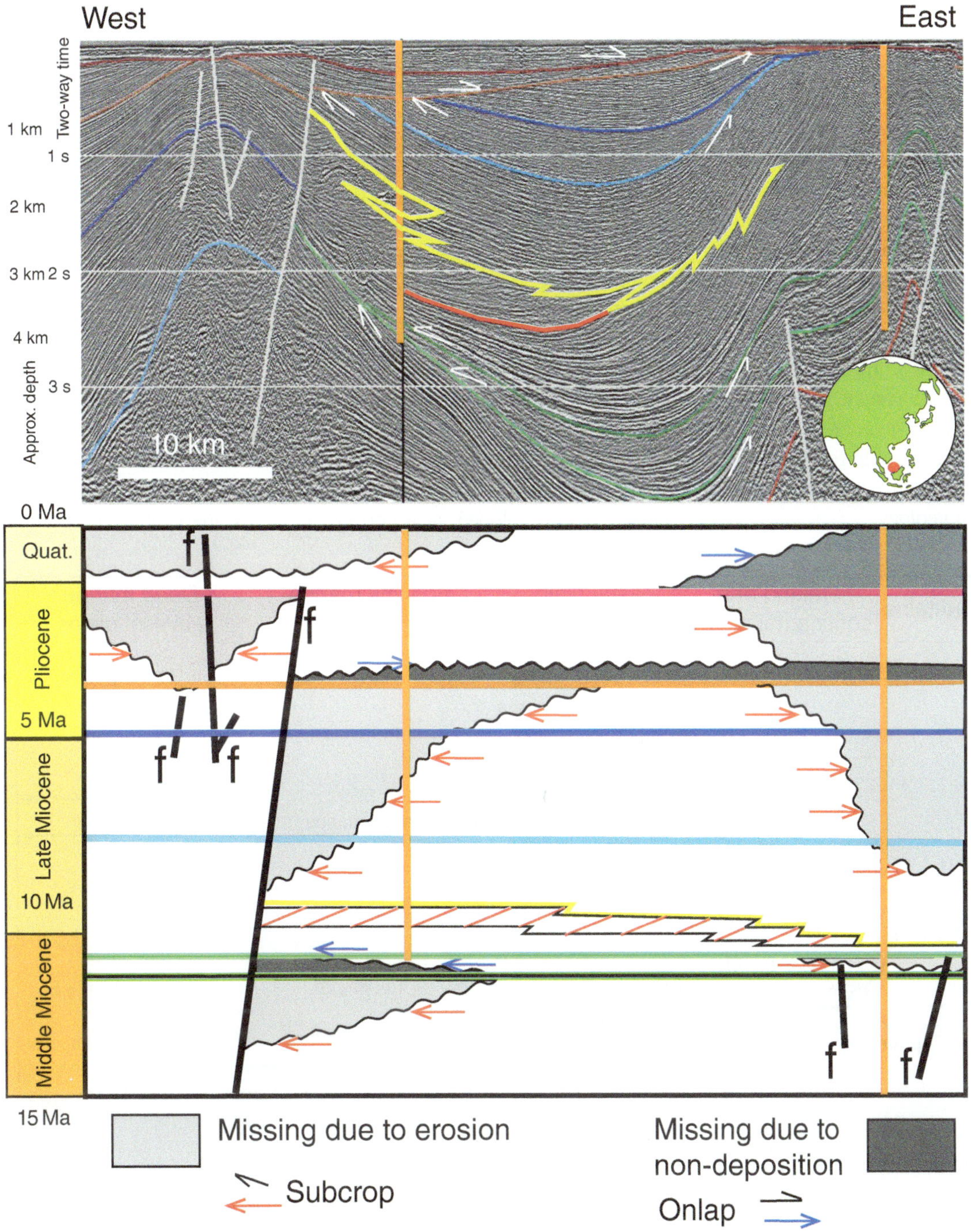

Figure 4.5 A seismic stratigraphic interpretation of a two-dimensional (2D) seismic line, and its translation into a chronostratigraphic or 'Wheeler' diagram (Wheeler 1964). The seismic section is displayed in two-way reflection time, which is here approximately converted to the depth scale on the left-hand side. Geometric observations of truncation and onlap allow the extents of individual rock units to be described and understood in terms of non-deposition, erosion and subsidence variations. These geometries can then be plotted against absolute rather than simply arbitrary time intervals if, as here, chronostratigraphic control is available. In this case, this is supplied by biostratigraphic data from two wells (orange vertical lines). Such diagrams are excellent for summarising breaks in sedimentation, how facies units cross timelines (diachroneity) and their causes, on all scales from outcrop to basin. This example is from the syn-depositional Labuan syncline situated on the shelf offshore Sabah, Malaysia, NW Borneo. The Wheeler diagram clearly shows the variable timings of uplift in the two flanks of the growth syncline. A westward-prograding delta (the delta topsets are shown in yellow and the delta slope 'clinoforms' as sloping red lines) crossed the growing syncline at around 10 Ma. This delta encountered continually increasing subsidence rates as it prograded westward towards the axis of the syncline. Subsidence constrained progradation until sufficient bottomset aggradation had occurred to allow the delta slope to prograde. Adapted from Levell and Kasumajaya (1985), Van Vliet and Schwander (1987) and Morison and Wong (2003).

overviews, the reader is referred to the detailed chapters in Gradstein et al. (2020), where the current status of biostratigraphy, stable isotope stratigraphy (C, O, Sr), magneto-stratigraphy, radioisotope geochronology and astrochronology are summarised and relevant literature is provided. Normally a combination of several of these dating methods is needed for robust and consistent age control.

Finally, the introduction of high-resolution timelines in sedimentary successions has also paved the way for the development of process-based stratigraphic forward modelling approaches (Section 4.4). This type of software is capable of forward modelling the 3D distribution of sediments in a basin in time-steps, controlled by physical, biological and chemical laws. And as such, it is one of the few options we have in sedimentary geology to test (falsify) our interpretations of the stratigraphic record and explore or even overcome any personal bias, much like sand-box experiments in structural geology.

4.2.4 Fundamental Stratigraphic Concepts

The following fundamental stratigraphic concepts, applicable to most sedimentary environments, are used in correlation and the reconstruction of stratigraphic architecture:

- Base level
- Accommodation space
- Stacking patterns as related to the varying ratio of the rates of Accommodation space (A) and Sediment supply (S)
- Temporal hierarchy of depositional sequences and processes

4.2.4.1 Base Level

The idea of base level is credited to Powell (1875) who led the Colorado River expeditions and was fascinated by the processes that formed in the Grand Canyon. It was further developed as a concept in landscape evolution by Davis

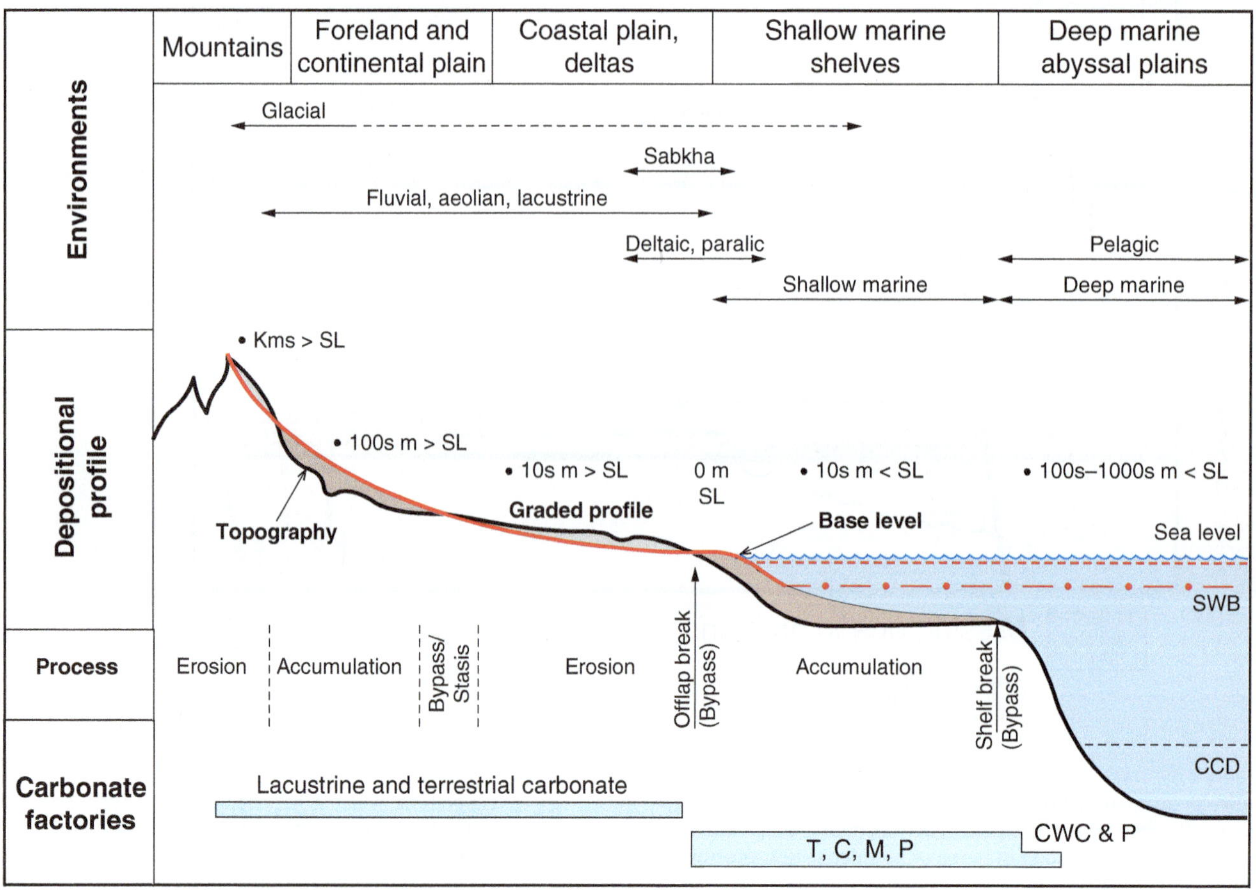

Figure 4.6 Base level in a depositional dip profile across the main depositional environments dealt with in this book. The red line is the graded profile of a river. Topography above this line is subject to erosion, and land below this line is subject to deposition. This graded profile is connected to the marine base level at the river mouth. The sea surface provides a level above which sediment cannot be deposited (depositional base level), whereas storm wave base (SWB) provides a level down to which it is subject to erosion (erosional base level). The relationship between topography/bathymetry along the profile and base level determines the sites of potential accumulation, stasis/bypass and erosion. These affect the throughput of sediment fluxes between linked depositional (storage) and erosional (release) settings in the routing system. Base levels vary through time as a function of tectonics, sea/lake level change and climate (e.g. fluvial and storm erosive power). For meaning of carbonate factory abbreviations see Chapter 11.

(1902). Base level (Figure 4.6), which is exemplified best by sea level or the graded profile of a river, is a surface below which sediment can be deposited and potentially preserved and above which it is subject to erosion. This duality was captured by Barrell (1917, p. 778): *'base level, local or regional, is that surface toward which the external forces strive, the surface at which neither erosion nor sedimentation takes place'.*

The base-level concept applies to both siliciclastic and carbonate deposition, although distinctions need to be made between the different carbonate factories (see Chapter 11). Physical base levels in cold-water carbonates are the same as for clastic sediments, with the erosional effects of shoaling surface waves and tidal channel erosion. Typical tropical carbonate build-ups may, however, thrive very near the sea surface under the continuous influence of surface waves due to the binding capacity of certain organisms in combination with early cementation. They have the ability to create what Pomar (2001) termed 'ecological accommodation space' above the erosional base level of the storm wave base. Early cemented, carbonate-rich aeolianites and beach-rock also resist erosion to a greater degree than equivalent siliciclastic sediments even when well above the erosional base level. In contrast, the ability of a carbonate system to fully fill accommodation space up to a depositional base level may be limited by the various ecological production/depth functions, which mostly decrease steeply with loss of light in the water column, making them sensitive to such environmental factors as clay influx.

Erosional Base Level

Erosional base level (Figure 4.7) can be regarded as a theoretical surface tangential to the deepest levels of erosion by sedimentary processes operating within a given depositional environment at a particular time. Any sediment above this level is subject to the possibility of erosion simply by autogenic processes. In practical considerations of preservation, we are concerned with the level down to which several metres or more can be removed. An example is the lateral migration of a river channel where the deepest line of the channel (the thalweg) defines the erosional base level of the river. In aeolian environments, the erosional process is wind deflation, and the erosional base level will be affected by wind speed and substrate erodibility. The latter may be a function of the cohesiveness of moist sand below the water table, or erodibility may be controlled by desert bed armours: non-erodible residual gravel lags (see Chapter 7). In shallow-marine settings, especially carbonates, early cementation may also be involved in controlling erodibility. Such erodibility constraints set an effective erosional limit, which may be above the theoretical erosional base level.

Depositional Base Level

Depositional base level (Figure 4.7) can be viewed as the highest topographic or bathymetric level up to which preservable deposition in a particular environment can normally take place. Despite the vagueness in this definition,

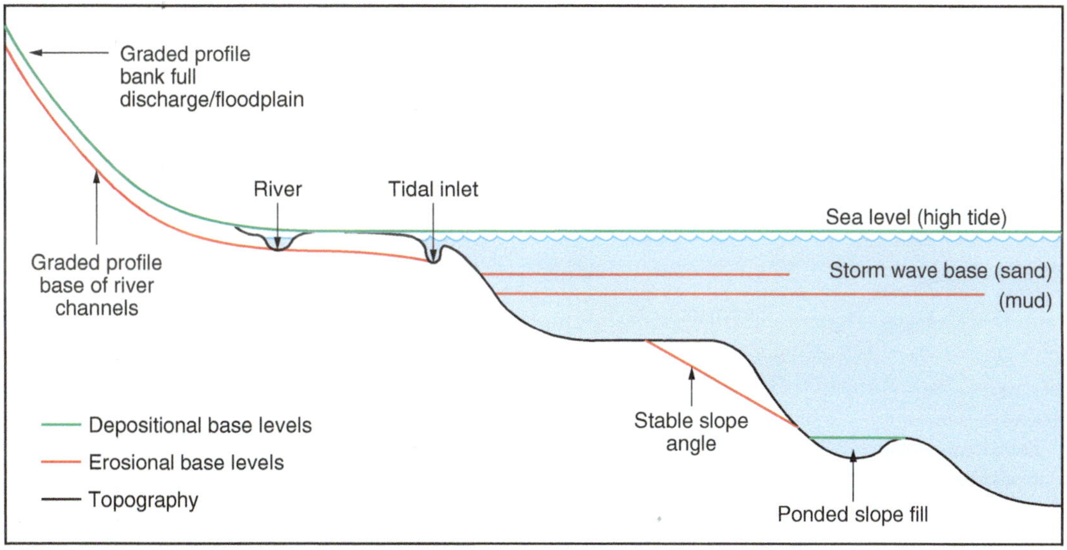

Figure 4.7 Base levels can be usefully separated into erosional and depositional base levels. Shallow-marine (coastal-shelf), coastal/delta plains and lacustrine/alluvial plain sediments do not accumulate significantly above sea/lake level, but they are liable to erosion down to wave base. Wave base, an erosional concept, varies with wave climate but is deeper for mud, which is more easily re-suspended than sand. Depth ranges are several metres to tens of metres. Sediments between the depositional base level of sea level and the erosional base level of wave base are temporarily stored and subject to potential erosion. Similarly, the erosional base level of a fluvial system is the base of the channel, while the depositional base level is on the flood plain (or bankfull discharge level). Although less commonly considered, stable slope angles also form local erosional base levels, and ponded mini-basin fills on slopes have local depositional base levels (Prather et al. 2017).

in coastal depositional environments, it would seem obvious that this depositional base level would be at or close to 'sea level'. In tidal settings, the depositional base level approximates to the level of the salt marsh/mangrove swamp, which is close to High Water Mean Spring Tide.

The concept of base level is most straightforward for coastal and shallow-marine realms and their onshore extension – the fluvial graded profile (Figures 4.6 and 4.7). A river will, dynamically, be 'striving' to respond to changing base levels. For example, as explained in Chapter 5, an erosional 'knick-point', with local gradient increase, moves upstream following a base-level fall. Aggradation of the coastal plain occurs during base-level rise due to the forced gradient decrease in the lower reaches of a river.

For deep-water environments, the ultimate base level is still sea level but more relevantly local base levels can also be defined by graded slopes. Slopes that are unstable due to slumping will erode to grade and those that are below grade can be 'healed' by deposition (Prather et al. 2017). Even in the pelagic realm, strong ocean-floor currents may set an effective base level on accumulation, albeit locally (Section 12.2).

The twin planes of erosional and depositional base levels define between them three categories of sediments. Those above the depositional base level are strong candidates for imminent erosion. Those between depositional and erosional base levels are likewise candidates for erosion and are not yet definitively preserved; they can be considered as potentially erodible 'stored' sediments or a 'buffer' volume of sediment available for re-distribution if conditions change. Sediments below the erosional base level are preserved accumulations, at least under the prevailing conditions, as they are still subject to the possibility of being eroded by processes of the longer-term preservation filter (Section 4.4).

4.2.4.2 Accommodation Space

Accommodation space is the volume available to be filled with sediment, between the base level above and the existing depositional or erosional surface (topography) below (Figure 4.8).

Although often represented in 2D, accommodation space is a 3D volume. The lower boundary is relatively straightforward: the topography of the previous depositional surface, as modified by any erosion, and the upper boundary is the lower of either a depositional base level or an erosional base level (Figure 4.7).

The main factors controlling accommodation space in the marine domain are eustatic sea-level change and tectonism. These increase and decrease accommodation space at variable and independent rates through time (Figure 4.8). An important distinction is made between

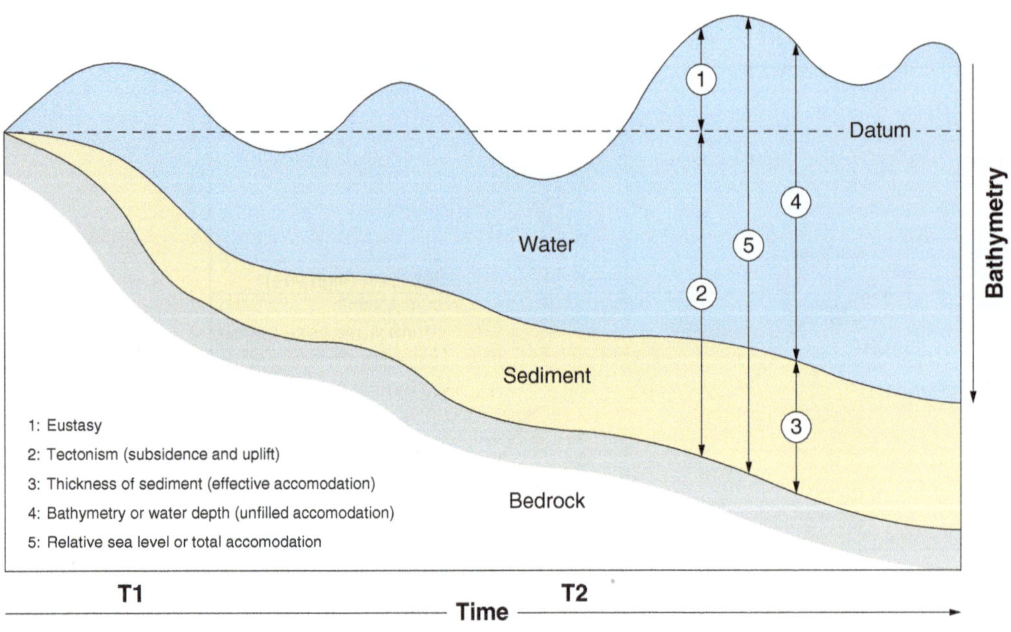

Figure 4.8 Accommodation space is the volume available at any given time to be filled by sediment. In marine environments, it increases as the sea (base)-level rises and as subsidence (aided by compaction) occurs. It decreases with sea-level fall, uplift, and sediment accumulation. The combination of subsidence and sea-level change (additive for sea-level rises and subtractive for falls) gives a 'relative sea level' (RSL) curve, which expresses the change in sea level relative to a previous depositional surface. The figure shows the history of sea level, subsidence, net sedimentation, accommodation space and relative sea-level variations over time at a single point (Modified from Homewood et al. 2000).

effective accommodation, which is filled by sediment in any given time increment, and unfilled accommodation space, representing the residual bathymetry or water depth.

Since base level is a dynamic concept, constantly shifting due to ongoing tectonic activity and sea-level change, the closer it approaches the depositional surface, the more gaps there can be in the record, due to lack of accommodation space. These time gaps represent erosion or times of temporary equilibria with the depositional surface at or close to the base level and marking sediment bypass through an environment or 'stasis' (Figure 4.6).

The impact of changing the base level on accommodation space in different depositional environments is illustrated in Figure 4.9. This figure shows the difference in relative (in-) completeness of the sedimentary record in three depositional environments along a single depositional profile from coastal marine to deep water as a function of changing sea level and subsidence.

The combination of subsidence and sea-level change (additive for sea-level rises and subtractive for falls) gives a 'relative sea level' (RSL) curve, which expresses the change in sea level relative to a previous depositional surface. Falls in RSL potentially generate substantial time gaps through erosion of the near-coastal sedimentary record. Erosional time gaps in the shallow-water setting are fewer and shorter. Due to the greater initial water depth, less time is spent close to base level during a given fall. In deep-water settings where the RSL fall is normally significantly less

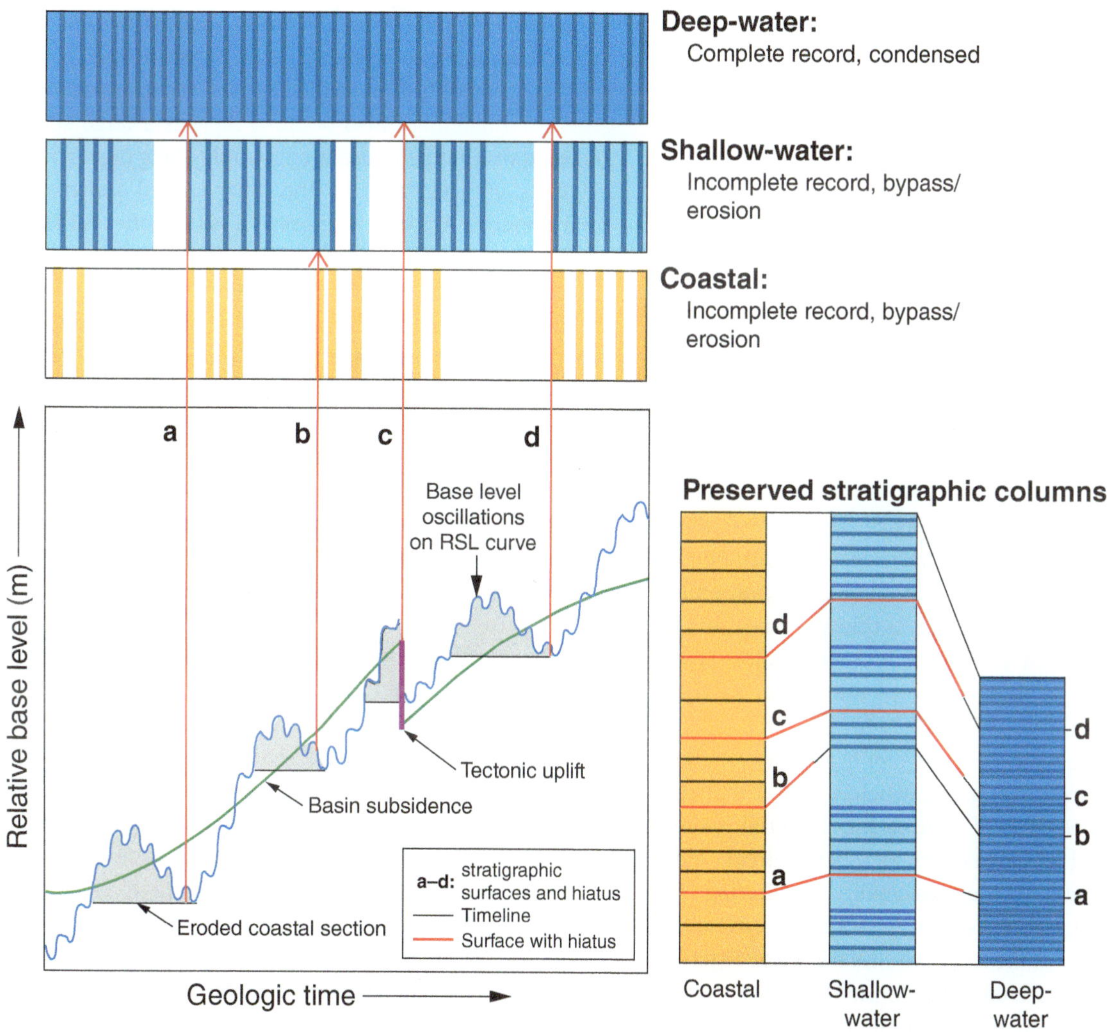

Figure 4.9 Stratigraphic continuity, as controlled by base level (Modified from Barrell 1917). The relative base-level curve comprises the addition of a basin subsidence curve (green) and a base-level/sea-level oscillation curve. In this case, there are two frequencies of base-level oscillation. An instantaneous tectonic uplift interrupts basin subsidence (magenta). Erosional removal of sediment during relative sea level (RSL) fall is assumed to be complete down to base level in coastal settings (grey shading under blue RSL curve). Erosion is less at these times in shallow water marine settings. Deposition in the relative base level rise periods is indicated in dark blue. Preserved sediments in time (top) and thickness (right) are shown in colour, with time gaps in white.

than the initial water depth, there may be no time gap at all. The full cyclicity of water depth change can therefore be recorded in deep-water settings (provided that depth indicators are encoded in the facies, chemistry or biology). The darker blue stripes indicate times of RSL rise with deeper-water deposition. In the shallow marine setting, only the longer and larger transgressions on overall rising-limbs of RSL are preserved from subsequent erosion, and these alternate with disconformities generated during RSL fall (red lines). The coastal record is the most incomplete due to its sensitivity to RSL change. Subsidence is held equal for all three locations. Consequently, even though less time is recorded by sediment in the coastal sequence than in the shallow marine sequence, they have the same thickness as shown in the preserved sedimentary sections illustrated on the right. The deeper marine section is shown as thinner due to decreased sediment supply distally: 'condensation'. A disruptive tectonic unconformity is shown in magenta. Its effect, in terms of missing time and section, is, in this example, mitigated because uplift happened to occur at a time of counteracting rapid sea-level rise.

4.2.4.3 Stacking Patterns and the Ratio of Rates of Change in Accommodation Space Versus Sediment Supply (A/S)

Stratigraphic architecture from the decimetre to kilometre scale is, to a large extent, the result of the changing ratio of rates of accommodation space creation or destruction (A) to sediment supply (S). A fast rate of space creation may well outpace a given rate of sediment supply even if its magnitude is small. The magnitude of space creation will affect the longevity of the induced changes (i.e. the duration of coastal retreat/transgression or coastal progradation/emergence). The occurrence of changes is thus governed, in the first place, by rates of change of accommodation space and sediment supply and, in the second place, by their magnitude.

Controlling factors on sediment supply are very different for different groups of sediment: siliciclastics, carbonates and evaporites. Supply of siliciclastic material (Allen 2017) is controlled by factors such as relief, bedrock, temperature and rainfall and their seasonality, weathering rate (including biological effects of vegetation and in soils), slope and the nature of the source to sink trajectory in terms of the transfer of material between environments downsystem. The relative supply of different grain-size materials, that is, gravel, sand and mud is also important, as are the transport mechanisms between environments.

Because of their biological nature, carbonate-producing organisms are very sensitive to changes in the environment, including temperature, oceanography (currents, nutrient supply), siliciclastic sediment supply (mud influx inhibits carbonate growth) and chemical composition of the water column. They have the capacity to exponentially increase production rates under favourable conditions or shut down under stress. Another modifier in the carbonate sedimentation rate is the different carbonate factories, which all have their particular response to environmental changes, particular grain-size spectrum and mineralogy (see Sections 11.4 and 11.5).

Although every case is unique, some generalisations can be made concerning the A/S ratio along a source-to-sink trajectory (Figure 4.10). Onshore, limited accommodation space is often the determining factor of stratigraphic architecture, with sediment supply in general being readily available and A/S often less than one (insufficient space) with much sediment bypassed towards the coast. In the transitional domain from coastal to shallow marine, it is more common to see a closer balance between A and S. This is in part not only because basinward accommodation is generally higher but also because there is a large spatial and temporal range in sediment supply along depositional strike among river mouths and other coastal and marine environments. A/S may be more or less than one and varies rapidly over space and time. This is the domain where sequence stratigraphic concepts were initially developed and are most applicable. Stratigraphic architecture in the deep marine domain is mostly controlled by variations in the sediment supply from the coast since accommodation space is readily available (Figure 4.10). Cyclostratigraphic signals also have an optimum window for preservation along the source-to-sink profile (Figure 4.10), for example, in positions where they are less likely to be truncated by shortage of accommodation space.

The standard stacking patterns of sedimentary strata, resulting from variations in A/S are retrogradation, aggradation and progradation (Figure 4.11). Note that these three geometries are not tied to RSL change alone. All three geometries are possible, as long as at least some accommodation space is created, due to sediment supply variations. For example, in an increasing sediment influx scenario, a retrogradational trend can change into an aggradation or even progradational trend, without any change in the rate of creation of accommodation space (Figure 4.11). Only in the case of negative accommodation space, the 'forced regression' (carbonates: Hunt and Tucker 1992; clastics: Plint and Nummedal 2000), can we be certain that progradation/regression is actually caused by an RSL change (a fall) rather than a sediment supply change.

The non-uniqueness of the stacking pattern is particularly clear in carbonate systems. Isolated carbonate platforms can display contemporaneous progradation, aggradation and retrogradation stacking patterns at

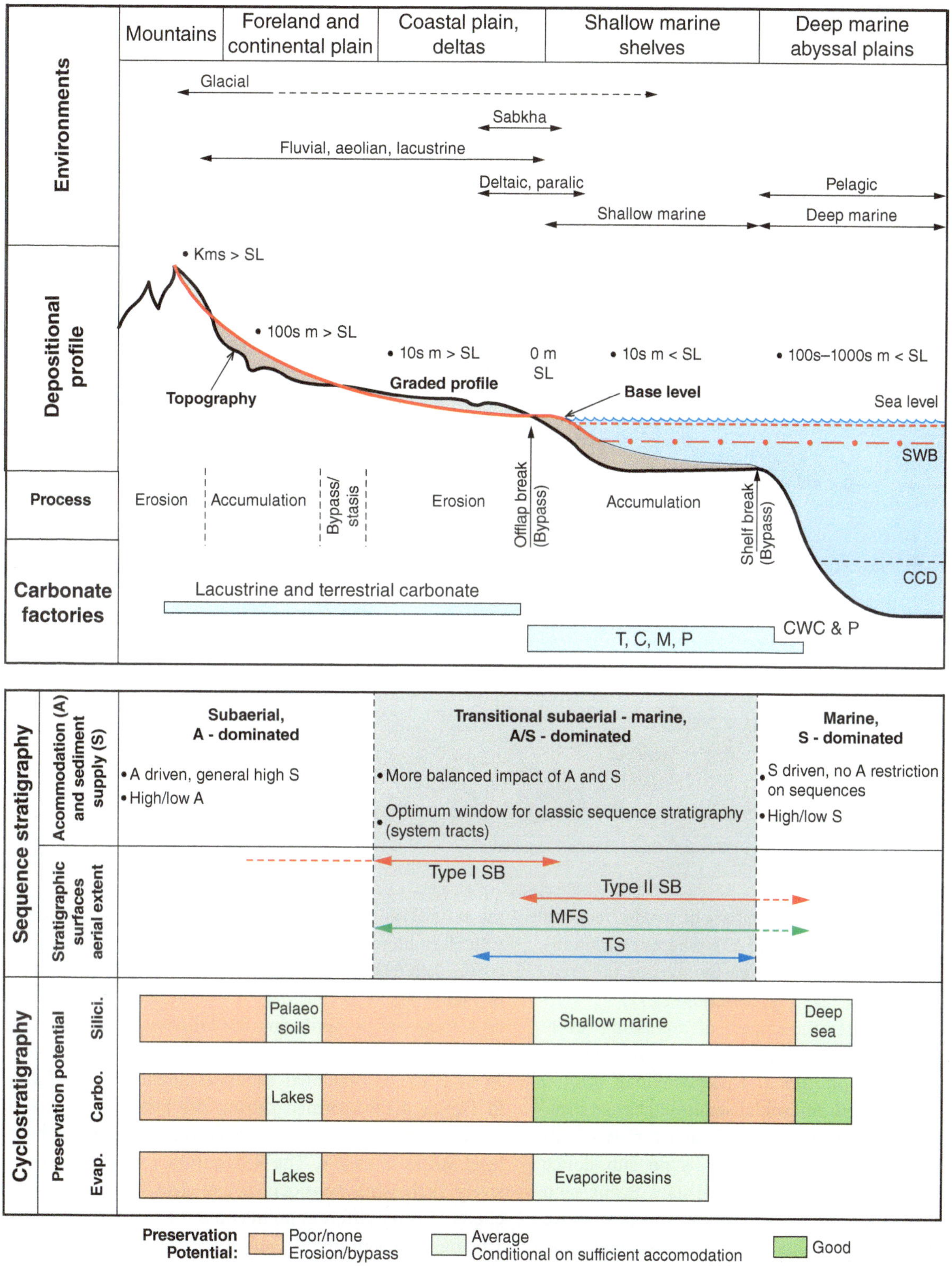

Figure 4.10 Theoretical source-to-sink depositional profile and indicative sediment fluxes in and out of storage along an idealised sediment routing system. An indication is given of the coverage of chapters in this book and the optimum zone of base-level-controlled environments for sequence stratigraphic analysis (grey-shaded area). The optimum zones/environments for the preservation of orbitally induced cyclicity are indicated in green. Sequence stratigraphic terminology is explained in Figures 4.31 and 4.35.

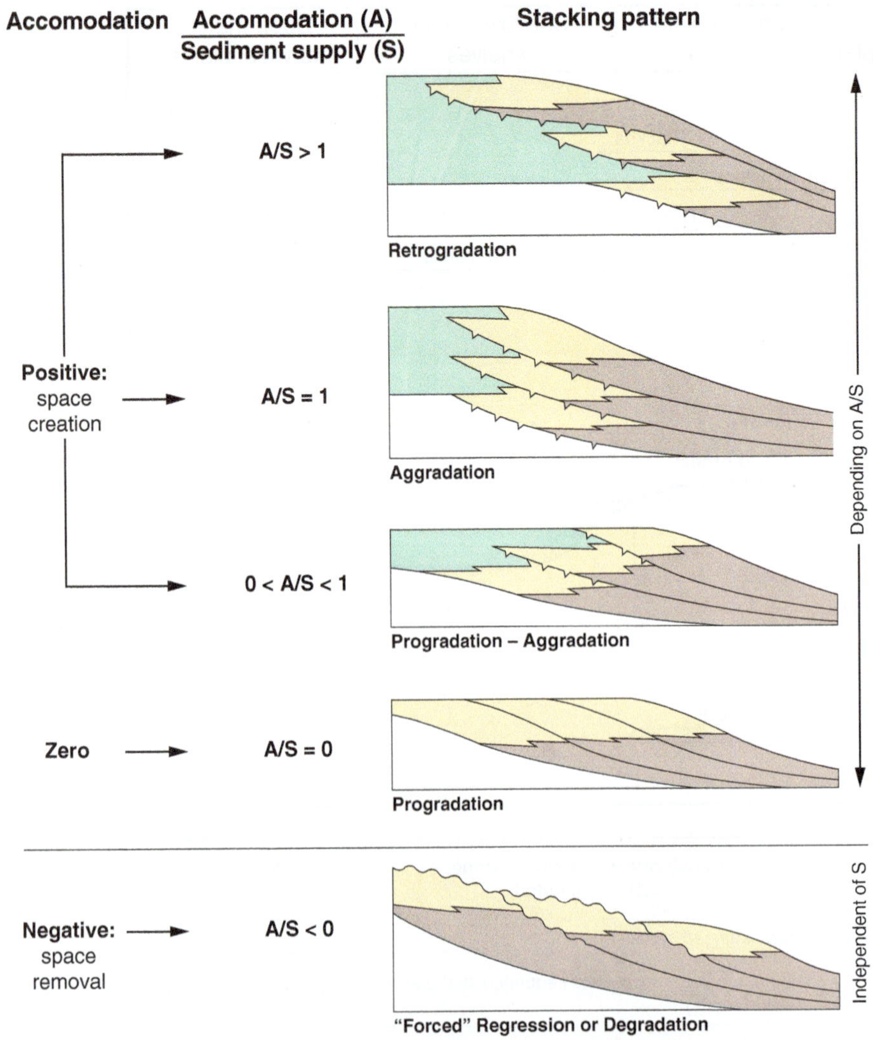

Figure 4.11 Stacking pattern geometries and the ratio of the rates of creation of Accommodation space to Sediment supply (A/S). Only the negative accommodation space case is uniquely indicative of a change in one of these two factors, namely a reduction in A. All the other sedimentary geometries are a product of the ratio of rates of accommodation space creation and sediment supply and therefore change between these geometries can occur through changes in either parameter or both at the same time (Modified from Homewood et al. 2000). Colour key is as for Figure 4.3. Individual progradational units separated by the thin black timelines are termed 'parasequences'. It is their mutual stacking pattern that allows the overall geometry of progradation, aggradation and retrogradation to be seen.

different sides as a result of the impact of wind-directions and/or sea currents (see Figure 11.67), whereas variations in carbonate production rate can easily outpace the creation of accommodation (see Figures 11.69–11.71).

Figure 4.12 illustrates the impact of increasing levels of accommodation space creation on stacking pattern and surface expression along a depositional profile. Figure 4.12a shows how sinusoidal sea-level change combines with steady subsidence to produce a variety of relative sea-level curves (1–6). If there is no subsidence (curve 1), the time spent in an erosional state during each base-level cycle (i.e. with the depositional surface above base level [red]), equals that in the depositional state when there is accommodation space available (white). The result may be a patchy or non-existent record (shown as an erosion surface in Figure 4.12b). As the rate of subsidence increases (curves 2 through 6), the duration and rate of RSL falls decrease (orange shading), until, in curve 6, there is only an RSL still-stand and no actual RSL fall (The rate of subsidence is large enough to equal that of the base-level fall). Also, the duration and amplitude of RSL rises increase as the subsidence rate increases from curves 1 to 6 (blue shading). The consequence, illustrated in Figure 4.12b, is that the erosion surfaces marked also by subsequent onlap during the fall/rise cycle (e.g. curve 3) are not formed in the high subsidence case of curve 6. In other words, erosional 'sequence boundaries' (red lines in Figure 4.12b) formed during the fastest rate of RSL fall are not created if subsidence rates counteract base-level fall as in case 6. The RSL curves 1–6, with progressively increasing subsidence, can also be visualised as characterising different points on a rotational basin flank: 1 being the basin margin and 6 the basin centre, as indicated in the two dip cross sections of Figure 4.12b. The progressive decrease of time spent in an erosional state towards the basin centre (6) is indicated by the upward tapering red triangles in Figure 4.12a. They are the time representation of basin margin lowstand erosion surfaces ('sequence boundaries') passing distally with higher subsidence rates into 'correlative conformities' towards

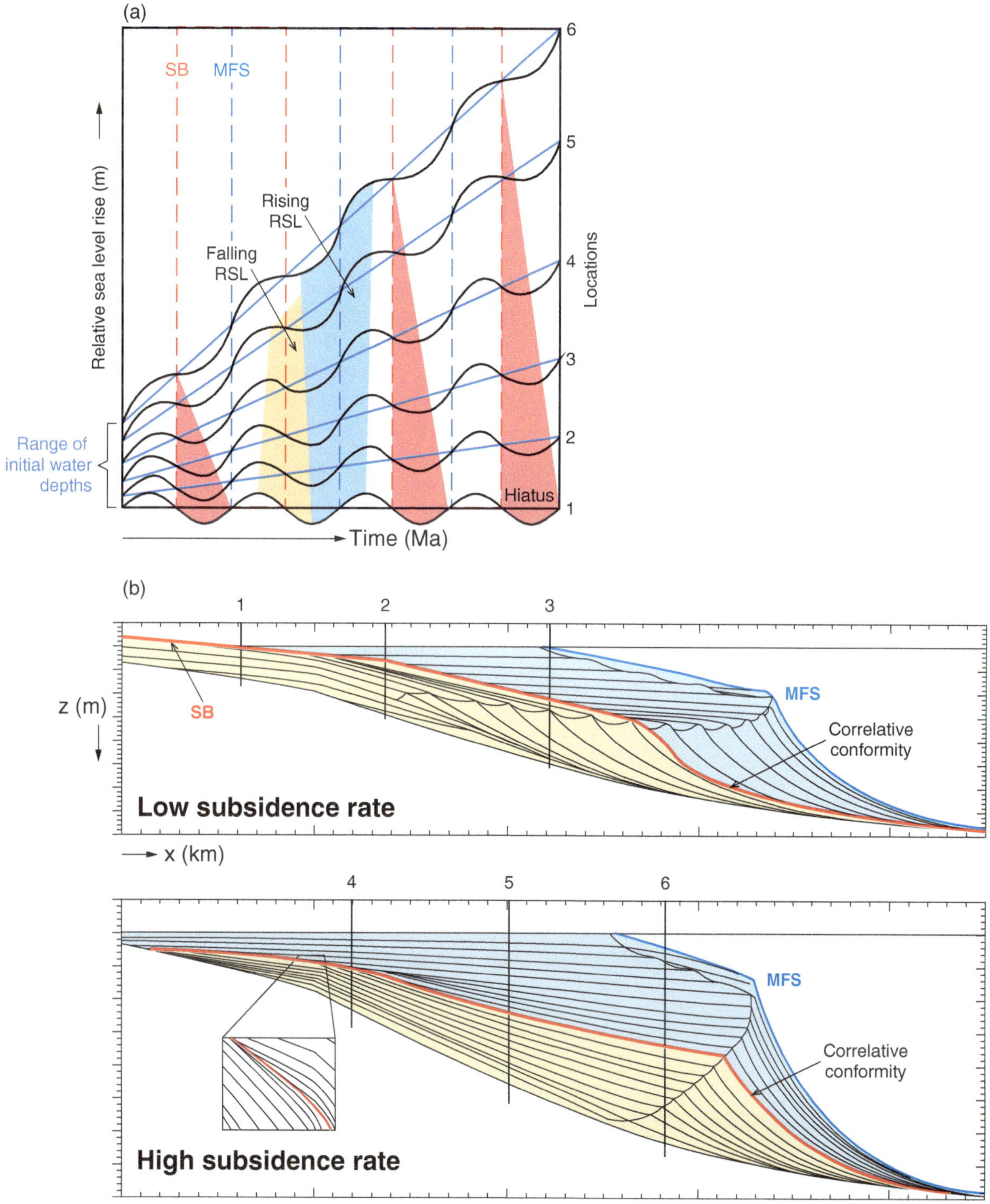

Figure 4.12 (a) Accommodation space as a control on depositional geometries: the fundamental background to sequence stratigraphy. (b) Modelled schematic depositional dip cross sections from basin margin to basin centre, showing the locations represented by RSL curves 1–6 in (a). Models from BASMOD (Lawrence et al. 1990). See text for explanation.

the basin centre. Note that base-level cycles are shown as symmetrical; this is for simplicity and is absolutely not a given, for example, many 100 kyr Quaternary sea-level cycles are markedly asymmetrical (with more rapid falls and slow rises).

The classic sequence stratigraphic model of Van Wagoner et al. (1990) and its subsequent modifications (e.g. Posamentier and Allen 1999; Catuneanu et al. 2009; Embry 2009; Catuneanu 2022) all reflect inferred spatial and temporal variations in A/S.

4.2.4.4 The Temporal Hierarchy of Depositional Sequences and Processes

As outlined in Section 2.2, beds with common characteristics are grouped into facies, which are then combined into facies associations. Sometimes these facies associations display patterns in vertical succession, commonly representing a depositional environment with autogenic cyclicity. Facies associations are further combined into facies successions and depositional sequences related to changes in external forcing (allogenic), such as climate, sediment supply, base level and subsidence. These in turn combine to form basin fills. Recognising this neat hierarchy is however complicated in real life because the stratigraphic record contains similar repeated successions and patterns at many different time and length scales (Figure 4.13). As Wolfgang Schlager pointed out delightfully: this is one reason why we put scales in our pictures (Schlager 2010)! Upward-fining units, upward-coarsening units and foresets are simple examples of features, which will have totally different interpretations depending on their position in a hierarchy of scales.

Scales can be described, as large, medium and small (Figure 4.13) but are more meaningful if related to a duration. The most commonly used hierarchy of sequences uses the terms *first order* (longest duration) to *fourth order* (shortest duration: Catuneanu 2022). This relative terminology has acquired loose time connotations (Figure 4.14). The significance of this is that it links the order to processes that act at different time scales: the main drivers of cyclicity define the duration of the orders.

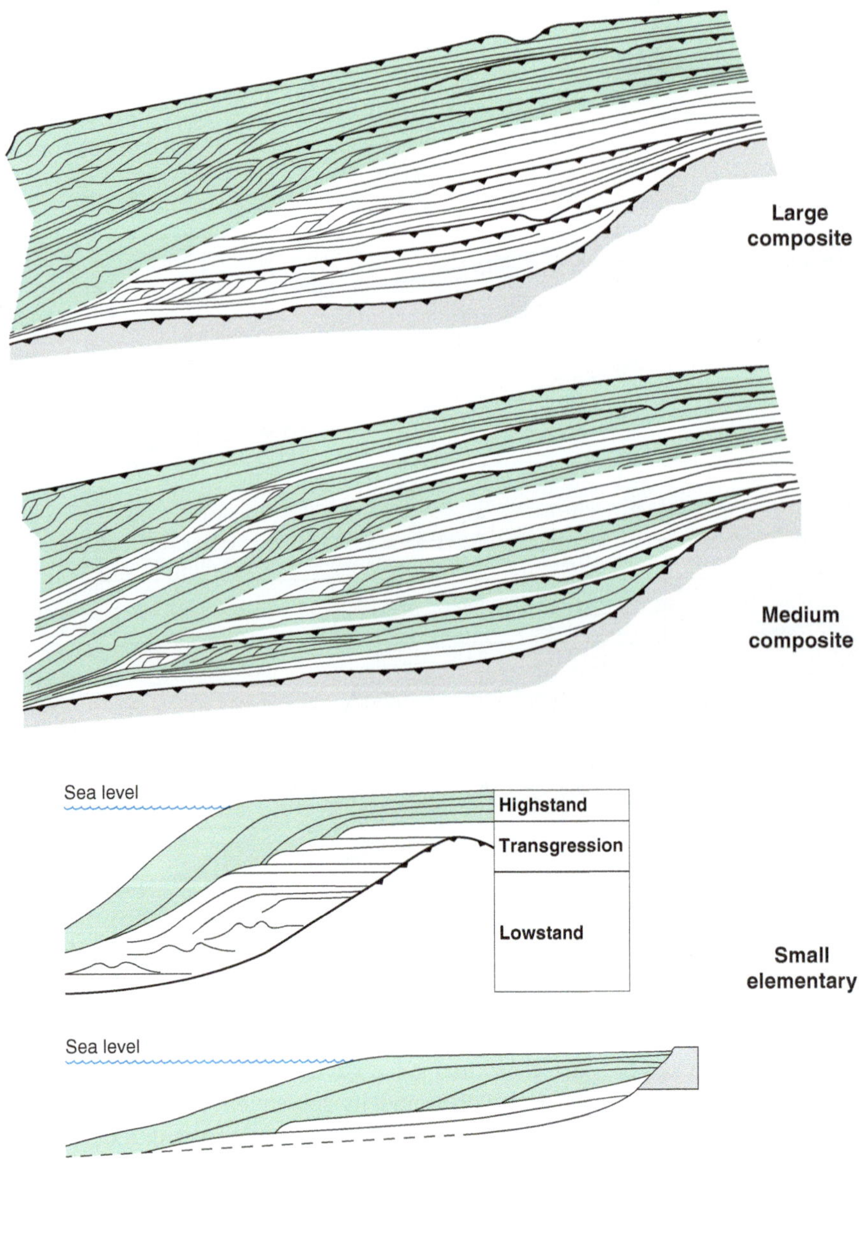

Figure 4.13 Illustration of the nested hierarchical nature of depositional sequences and their relative order, from small, elementary to medium and large composite depositional sequences (Modified from Duval et al. 1988). Highstand deposits are indicated in green.

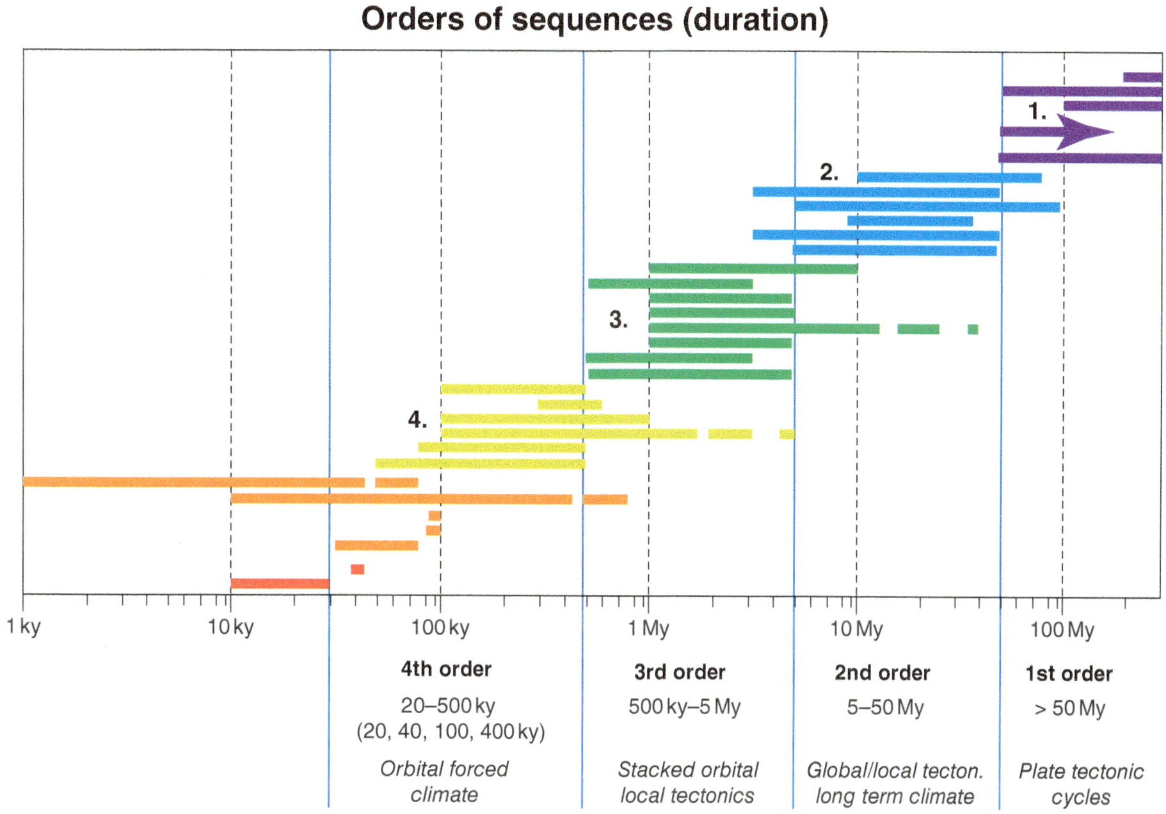

Figure 4.14 A consensus hierarchy of 'orders' of cyclic change in external forcings affecting sedimentation has emerged in the literature (e.g. due to sea level but also climate change). This figure (Modified from Schlager 2005) plots the published durations of different orders of sequences as seen by various authors with the oldest publication on top of each order. Generally, four orders are recognised.

The first- and second-order sequences (several tens of millions of years) are broadly controlled by plate tectonic processes, such as Mid-Ocean Ridge volume, mountain-building and supercontinent formation, which all change the volume of ocean basins and can be summarised in the Wilson cycle (Haq and Schutter 2008; van der Meer et al. 2022). On the other side of the spectrum, at the 4th order, are the orbitally forced climatic changes that act at regular cyclic beats of approximately 20, 40, 100 and 400 kyr and are clearly recognised to have influenced sedimentation in the Cenozoic and in some form for much of earlier Earth history (Hilgen et al. 2015). In between these is the third order with an arbitrarily defined range of 0.5–5 million years. Processes active at this time scale include stacked Milanković frequencies (800 kyr, 1.2 Myr, 2.4 Myr), other forms of climate-driven sea-level change and (often local) tectonic processes (Boulila et al. 2021). Classically, sequences are best recognised at the third-order level. It is over this time period that rates of sea-level change, subsidence and sediment supply overlap most closely, which all operate at millimetres per year or tens of metres per million years.

Although straightforward in concept, deducing these orders from the sedimentary record is not always easy. Notably at the shorter duration, 'fourth order' and higher levels, autogenic and allogenic processes interfere, necessitating a high-resolution age control of the succession and proper, 3D understanding of stratigraphic geometries. So, although a sea-level control on sequences might be expected to be stronger as the frequency and hence the rates of change increase, practically speaking over fourth order and shorter time periods purely autogenic, Milanković-band, and subsidence-related mechanisms are all acting together. This means that the relative importance of changes in accommodation space in causing unconformity-bounded units as opposed to episodic autogenic erosion and/or climate-related sediment-supply fluctuations cannot always be distinguished (Catuneanu and Zecchin 2013; Catuneanu 2019). These difficulties can manifest themselves in the use of the term 'parasequence' (Catuneanu and Zecchin 2020; Colombera and Mountney 2020), which is often applied to fourth-order sequences of uncertain allogenic or autogenic origin, but whose mutual stacking patterns are often used to help define third-order sequences (Figure 4.11).

4.3 The Stratigraphic Record – Different Perspectives

In the stratigraphic literature, there are broadly speaking three end-member ways in which the accumulation of sedimentary strata is interpreted (Figure 4.15). These are (i) 'accidental' or 'event' stratigraphy), which has a focus on individual, short-term, depositional events; (ii) autogenic (autocyclic) stratigraphy, which highlights the intrinsic and local nature of most sedimentation and preservation processes; and (iii) allogenic (allocyclic) stratigraphy, which stresses the importance of external, sub-regional or even global controlling factors on sedimentation. Allogenic stratigraphy can be further subdivided into 'cyclostratigraphy', 'sequence stratigraphy' and 'tectonostratigraphy'. Each of these approaches has a focus on a different main controlling process and consequently deals with a different duration (Figure 4.16).

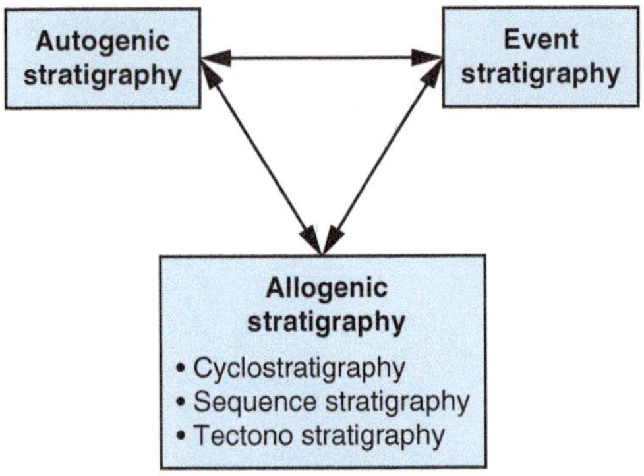

Figure 4.15 Three end-member perspectives on the accumulation of any given stratigraphic succession. The allogenic end-member is further subdivided into three. All illustrated perspectives are relevant to all successions. To understand the information recorded in the succession, it is the task of the sedimentologist to deconvolve the relative influence of the processes highlighted by these different concepts (see Figure 4.16).

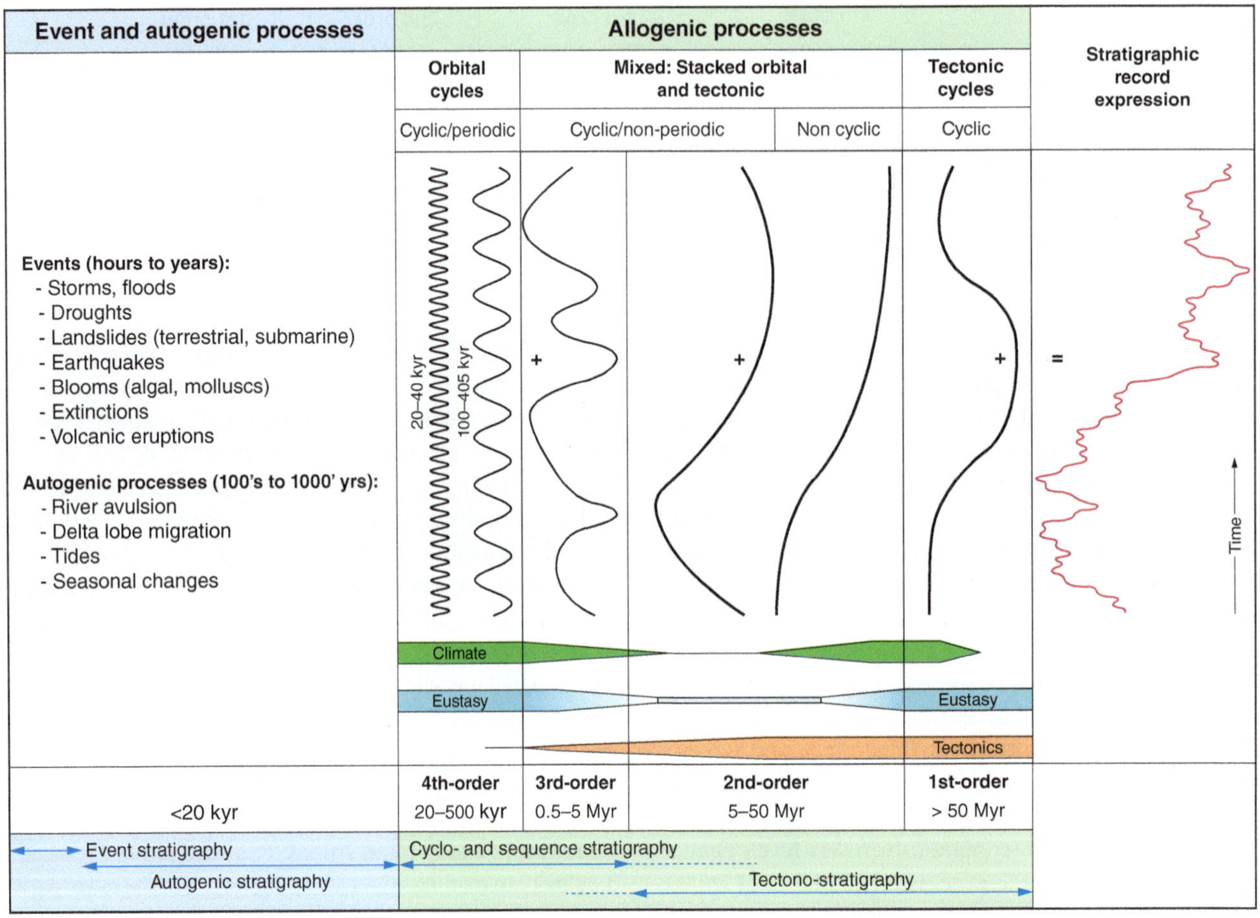

Figure 4.16 The stratigraphic record (schematically indicated by the red curve on the right side) consists of changes in properties (e.g. grain size, chemical signature, biota content and palaeobathymetry) on a variety of time frequencies. Interpretation of this record in terms of causation requires the deconvolution of this time series into the 'background noise' of 'random' events and autogenic processes and the high- and low-frequency (high and low order) changes in allogenic controls, such as climate sea level and sediment supply (Modified from Guillocheau et al. 2012).

These different types of 'stratigraphies' are by no means mutually exclusive, and none alone adequately explains all sediments, but together they perhaps can. Depending on geographical location and time interval, one set of processes may become the more dominant. There are healthy debates as to how any one of them may dominate a particular succession (see Smith et al. 2015). Sequence stratigraphy, because it provides a framework to combine these approaches, is the most commonly used methodology to describe and characterise stratigraphic architecture. Its basic premises are therefore treated in more detail in Section 4.3.3.

4.3.1 Accidental or Event Stratigraphy

'Accidental stratigraphy' (Miall 2015) refers to the products of distinct, recognisable events as time-correlative signals in an environment or a basin. Classic, if extreme, examples include meteorite impacts (Figure 4.17, Schmieder and Kring 2020), volcanic eruptions (Rougier et al. 2018), tsunamis and mega-slumps. Increasingly tighter age constraints based on higher accuracy absolute dating have established that many Earth system changes are both geologically speaking rapid and synchronous. These include stable isotope excursions in marine reservoirs (Cramer and Jarvis 2020; Grossman and Joachimski 2020; McArthur et al. 2020), sea-level changes recording glaciations (Röhling et al. 2014) and Heinrich events of ice-rafted materials in the deep sea (Hemming 2004). Longer-duration state-change events such as global extinctions, eruption of Large Igneous Provinces (Ernst et al. 2020), climate excursions (e.g. the Paleocene-Eocene Thermal Maximum: PETM; Dunkley Jones et al. 2018), disruptions in oceanic

Figure 4.17 A complex and dramatic event. The Cretaceous-Palaeogene boundary in an underground limestone mine at Geulhemmerberg, Maastricht, the Netherlands. At this shallow marine locality, the famous iridium and soot-rich clay layer is absent, likely due to erosion. Minor impact material is found in burrow fills. The site saw violent erosion of boulder-size pieces of a Maastrichtian firmground followed by chaotic deposition of locally cross-stratified coarse bioclastic calcarenites, succeeded by thin, fining-upward calcarenite interbeds. The lowest erosion surface, the event horizon, is arrowed. This event or series of events is attributed to storms (resulting from a highly disturbed atmosphere) that eroded the airfall impact deposits. (The upper grey claystone is not as thick as it appears, it forms the roof of the mine gallery from the dotted white line upward.) See Brinkhuis and Smit 1996; Smit and Brinkhuis 1996. Photo by Bruce Levell.

circulation (Keller and Barron 1983) and Oceanic Anoxic Events (Jenkyns 2010) also belong to this category. Such events can be global and provide an excellent means for correlation: often better than biostratigraphy with its inherent problems of variable fossil preservation, bio-provincialisation and multiple zonation schemes. (See the excellent review of global stratigraphic correlation principles by Smith et al. 2015). In general terms, interpretation of the stratigraphic record over the last 40 years has seen an ever-increasing realisation of the importance of exceptional events in driving Earth system change, in particular in geochemical cycles as revealed by anomalous excursions in stable isotope values and climate change driven by Large Igneous Provinces).

Some of these events are purely local and frequent, others global and relatively infrequent but the idea that stratigraphic successions are formed by stacking deposits from random events can be generalised to a view that a rock succession is the product of a time series of 'frozen accidents' (Miall 2015). This is in some ways an unattractive view of the stratigraphic record in that even if these events were neither accidents nor random, their individual causes might often be un-reconstructable. For example, take shallow marine storm deposits: Tian et al. (2019) document two cores from the East China Sea in water depths of 40 and 54 m, one records 15 typhoons or major storms over the last 140 years (5 of which can be named) and the other just 3. Suggested reasons for the difference are the grain size range available (coarser material, silt, which is not everywhere present makes a storm event recognisable), inshore topography (the presence of a 70 km wide strait between islands further inshore) and distance from the coast. Despite some relationship between storm magnitude and bed thickness in this example, it is unlikely that these causalities could ever be reconstructed from the stratigraphic record.

Spatial complexity can produce a similar effect of apparently random 'events', in a vertical succession. For example, in carbonate environments, the depositional surface is often a textural and compositional mosaic of facies related to local depositional conditions and the habitat preferences of marine organisms. Wilkinson and Drummond (2004) demonstrated that in the contrasting carbonate environments of Antigua and the Persian Gulf, modern maps of the surface facies-mosaic demonstrated an exponential distribution of the areas of the patches of individual facies. There might well be a relationship between the areal extent of a facies and its bed thickness, simply due to the likely persistence of a facies over time in a larger versus a smaller area. The argument is that the facies mosaic in map-view is in accord with the observed exponential distribution of bed-thicknesses of individual facies in ancient successions (e.g. Wilkinson et al. 1999).

There are many complications in recognising a stack of accidental stratigraphic events as being such. Paradoxically, repetitive patterns are produced by chaotic processes. Sets of non-linear processes can produce beautiful patterns of periodic 'sequential' behaviour (Gleick 1987; Hooke 2007; Bailey 2009). Patterns in a property, such as regularly increasing or decreasing bed-thickness, will arise simply by happenstance as well as deterministically due to, for example, increasing or decreasing proximality of a sediment source (Chen and Hiscott 1999; Prélat and Hodgson 2013). Scientists are trained to identify repeated patterns, such as the pervasive cyclicity seen in the sedimentary record. Hypotheses, perhaps subject to confirmation bias, are then made for a proximate, normally single, deterministic cause for the observed pattern. Current favourites are orbital-induced climate fluctuations and sea-level changes. In previous editions of this book, emphasis was placed on the intrinsic autogenic dynamics of sedimentary environments, which reflected the sedimentological thinking at the time (Reading 1996).

An explanation of patterns in strata as due to happenstance needs to be considered, even though it is intellectually unsatisfying (Bailey and Smith 2010). The record probably contains many spurious patterns of bedding that are simply complex functions of effectively random changes in currents, sediment availability and local topography. Such an 'accidental stratigraphy' should arguably always provide the null hypothesis against which alternative, deterministic, explanations are tested (Bailey 2009). But the formulation of null hypotheses and testing against them are not straightforward and will always be context dependent (Holbrook and Miall 2020).

Beerbower (1964) approached the problem of apparent randomness from a slightly different standpoint, that of a series of interacting signals from partially inter-related allogenic (external) controls:

> ....simultaneous operation of two or more allocyclic [allogenic] mechanisms will complicate the simple cyclic sequence. Some of the mechanisms, e.g., discharge variation induced by climatic change and slope variation resulting from glacial change in sea level, are at least partly interdependent. If these operate in phase because of their interdependence, they will invariably reinforce or cancel each other. On the other hand, mechanisms operating independently will vary in their interaction as they go in and out of phase. If their periods and intensities are irregular, the probability of exact repetition in a finite interval declines until the sequence of events and sediments approaches randomness.

Beerbower's view is essentially that of 'signal shredding' (Jerolmack and Paola 2010; Li et al. 2016). An associated suggestion is that system patterns (channel avulsion, channel and bar morphology and dimensions, etc.) change state at thresholds determined by sets of inter-related parameters and not due to single controls. Schumm (1973) introduced the idea of a 'morphodynamic turbulence' to describe this. Fortunately, there are specific statistical tests for assessing the significance of patterns and self-organisation of quantifiable properties such as composition (e.g. carbonate or organic content) and bed thickness. (See e.g. Schwarzacher (2005) for carbonate platform sequences and Chen and Hiscott (1999) for turbidite lobe successions.)

The accidental stratigraphy perspective raises challenging questions for sedimentologists. What are the limits of the interpretability of the record? What is the set of hypotheses that could all equally well explain a particular succession? And how can these multiple working hypotheses (Chamberlin 1890; Reading 1996) be tested, to avoid, in Chamberlin's words, 'the dangers of parental affection for a favourite theory'?

4.3.2 Autogenic Stratigraphy

Self-organisation within depositional systems to produce repeated, recognisably similar, depositional units or land-system elements such as delta lobes, fluvial channels and bars is normal and has been recognised for nearly 200 years as a potential cause of 'cyclic' sedimentation patterns.

Systematic lateral movement of depositional systems leads, by virtue of Walther's law (Section 2.1.2 and Section 4.2), to repeated vertical sedimentary successions of genetically related facies (Figure 4.18a). Typical examples of autogenic processes are progradation of tidal flats and banks, lateral migration of tidal and fluvial meanders and, classically, the progradation, abandonment, compaction and switching of delta lobes such as those of the Mississippi delta. Hajek and Straub (2017; their Table 1) present a set of links to time-lapse videos using satellite images of a variety of autogenic processes. Alternatively, successions may also evolve not by lateral migration but by simple vertical superposition of a different 'landscape system'. A classic example in the terrestrial realm would be climate changes, such as aridification and desertification or, as in Figure 4.18c, glaciation.

Autogenic changes may thus occur in combination with, or independent of, external forcing, but sea-level change can induce or alter an autogenic response. For example, it is likely that river avulsion rates increase with RSL rise (see the Rhine-Meuse example from Stouthamer and Berendsen 2001, Stouthamer and Berendsen 2007). In practice, there are few tests that we can perform to establish whether a recurrent pattern of facies in an ancient succession, or even a recent one, is autogenic or allogenic. The length scale over which correlation is possible is perhaps the best indicator. Autogenic successions would be expected to share a common stratigraphy only over the length scale of a single architectural element, such as a delta lobe but not further. If good time correlation can establish synchronicity of a control over a larger area an allogenic control is implied.

To further complicate matters, it is plausible that a short-term cyclicity (decades, centuries or a few millennia) of some external forcing was the trigger for apparently autogenic environmental shifts. For example, Nahon et al. (2019) illustrate an example of spit growth over the last 250 years and explore the alternatives of fully autogenic mechanisms and the influence of decadal climate oscillations combined with very low rates of relative sea-level rise as two plausible alternative explanations.

Vertical successions produced by lateral shifts in sub-environments are typically asymmetric (e.g. barrier island spit migration produces sharp-based, upward-fining inlet-fill successions). Deposition is followed by stasis whilst the locus of deposition, in this case the inlet, migrates elsewhere. The time scales of such autogenic change are to a degree predictable and often scale related (Jerolmack and Mohrig 2007). They are commonly related to self-organisation of a depositional system such that a pattern recurs (e.g. regular spacing of meander bends, aeolian dunes or tidal inlets on a barrier island coast). For example, six documented lobes of the Mississippi delta have each lasted 1–2000 years but the superimposed crevasse-splay lobes each lasted 150–200 years. Both processes formed asymmetric upward-coarsening successions, which were, respectively, 10–100 m and <10 m thick (Roberts 1997). In the Rhine-Meuse delta system, 86 river-channel avulsions occurred with inter-avulsion periods of 40–120 years and an average active life of each channel reach of 335 years (Stouthamer and Berendsen 2001, Stouthamer and Berendsen 2007; Pierik et al. 2017: Figure 4.19). However, autogenic processes are better described as leading to 'recurrent' rather than necessarily 'periodic' successions.

Over time, autogenic processes can, in principle, lead to stacking patterns of facies successions that are independent of external forcing. In the case of delta lobes, the gradient control on channel-switching triggers an offset stacking of the lobes, whereby channel avulsion tends to lead to deposition on the most compacted and subsided previous lobe, normally the oldest that is still available. As progradation occurs, the available accommodation space generally becomes less, and subsequent lobes will therefore be thinner with successively more proximal facies leading to systematic upward-thinning of successive successions in the

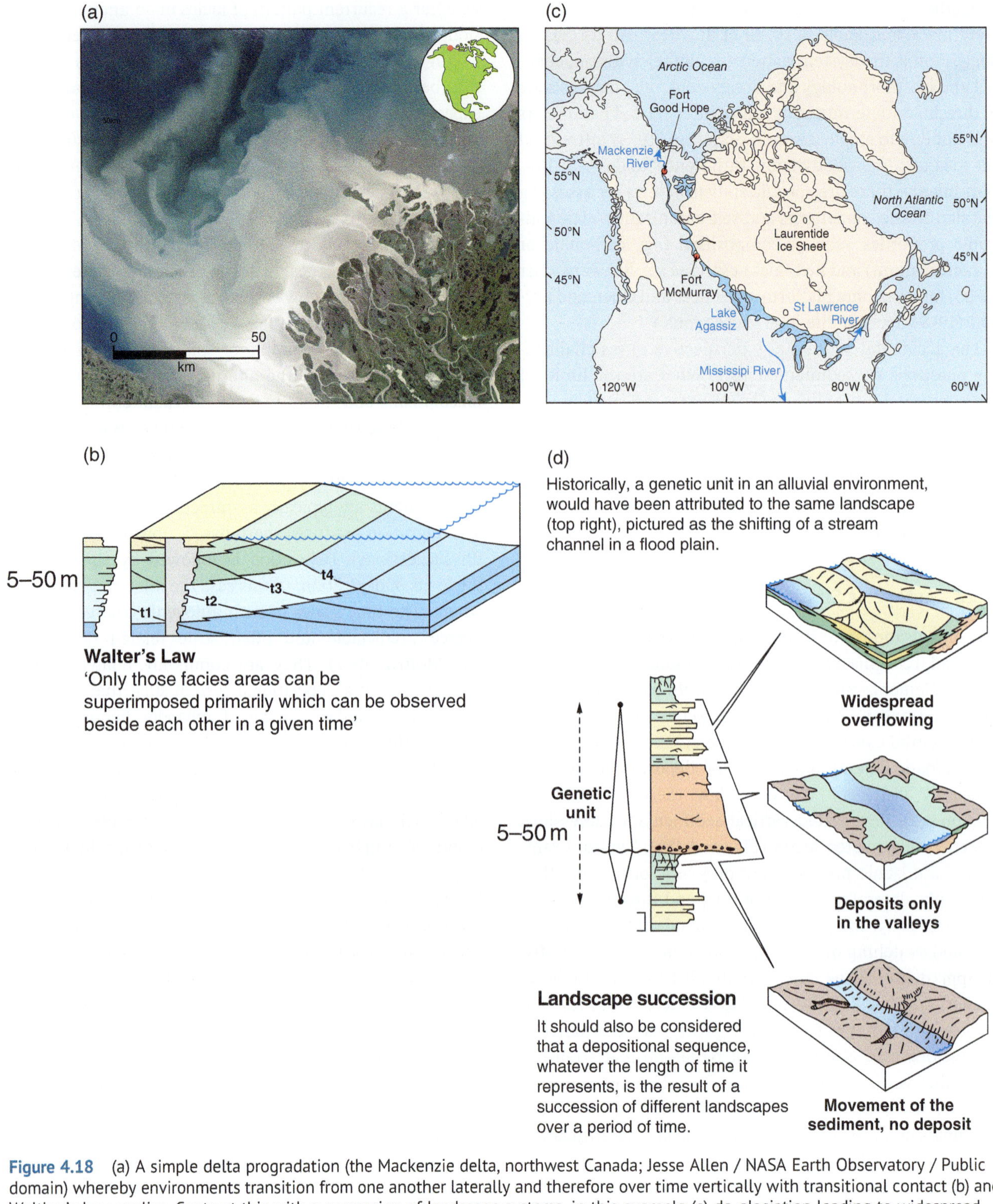

Figure 4.18 (a) A simple delta progradation (the Mackenzie delta, northwest Canada; Jesse Allen / NASA Earth Observatory / Public domain) whereby environments transition from one another laterally and therefore over time vertically with transitional contact (b) and Walther's law applies. Contrast this with a succession of landscape systems, in this example (c) de-glaciation leading to widespread lacustrine deposits abruptly overlie glacial sediments during down-wasting of the Laurentide ice sheet. These glacial lakes subsequently drained sub-glacially NE into the Hudson Bay area altering North Atlantic oceanic circulation patterns by suddenly adding substantial volumes of fresh water (Modified from Murton et al. 2010). The cartoon (d) illustrates a similar but slightly less dramatic change of landscapes and facies in a fluvial setting controlled by base-level changes (Modified from Homewood et al. 2000).

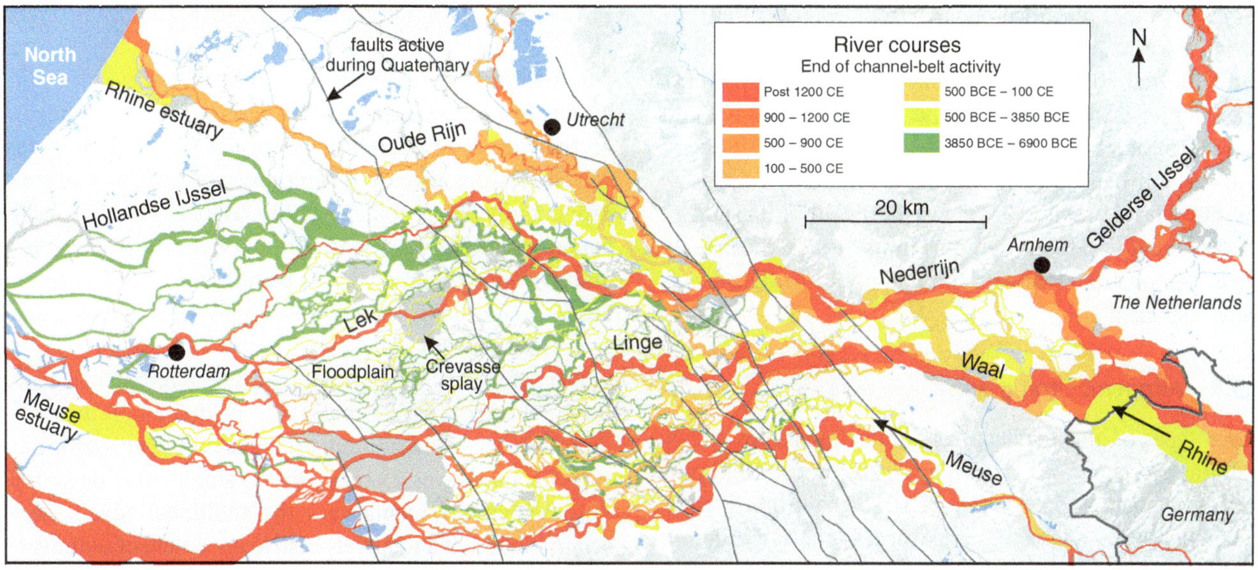

Figure 4.19 Evolution of channels in the Rhine–Meuse–Waal river systems of the Netherlands over the last 9000 years (Pierik et al. 2017 / with permission of Elsevier). Channels have frequently switched course and migrated within their meander belts, both classic examples of autogenic processes.

coastal plain: a 'progradational and aggradational stack of parasequences'. Somewhat differently, turbidite channel deposits tend to evolve into vertically aggradational stacks (Jobe et al. 2017). But the point is that autogenic processes can produce not only single characteristic successions but also hierarchies of repeated patterns.

Autogenic processes lead to stratigraphic successions of sediment, which are continuous on the time scale of the duration of a given architectural element, a channel or a lobe and have predictable hiatuses with durations of one to a few 'recurrences' between successive visitations of the same spot by the depositional system. The boundaries between these individual successions may be condensed sections, palaeosols, hardgrounds or erosion surfaces.

In this view of stratigraphy, gaps in the record are related to local cessation of sediment supply as the routing system (river, longshore drift and submarine channel) shifts elsewhere. Or they are related to erosion that is intrinsic to the depositional environment such as by laterally migrating shoals, river or tidal channels, or shoreface erosion by storms.

Autogenic effects are present in many sequences that have a dominant allogenic control. Both computer and flume modelling studies suggest that discrimination is not that straightforward (e.g. Hajek and Straub 2017). A 'compensation statistic', a measure of facies thickness variation across a stratigraphic section, may be used to assess the degree to which sedimentary units are responding to intrinsically created space (e.g. between lobes of sediment) rather than new accommodation space created by RSL rise (Trampush et al. 2017). The obvious way to distinguish allogenic from autogenic sequences is to demonstrate that separate environmental shifts are time-equivalent between successions, which come from separate depositional systems yet are affected by the same external factors. Outside the domain of short-lived isotope dating (Chapter 9), it is rare that time correlation is sufficiently accurate and precise to allow this. Modelling studies suggest that the combined presence of both autogenic and allogenic processes precludes the practical differentiation of either; certainly not from simple measurements such as cycle thickness (Hill et al. 2012).

Distinction may be possible when the external forcing has an unusually direct impact. For example, a palaeosol developed directly on shelf muds cannot be autogenic (Plint and Wadsworth 2003) and, more subtly, neither can regressive surfaces of marine erosion underlying shoreface sequences (Hampson 2016). Sediment provenance may also be useful in distinguishing purely local changes in sediment supply as a control from relative base-level change (Catuneanu and Zecchin 2013).

Autogenic processes can be overlooked in the search for evidence of external forcing. There are also reasonable arguments that some, and perhaps many, apparently autogenic changes in the recent past have, at a deeper level of causation, also been influenced by ongoing climate or sea-level change. The available detailed record of autogenic patterns is only the last 5 ka or so: a period of rapid sea-level and climate change, which has lasted just one-quarter of the shortest Milanković period! Distinguishing intrinsic versus extrinsic (and anthropogenic) controls over this time period even in recent sediments is not simple (e.g. Stefani and Vincenzi 2005; Amorosi et al. 2016 for the Po delta). The counter-factual ('What would have happened without these external forcings'?) remains speculative. However, until established otherwise, a parsimonious view is that the burden of proof, at least for shorter duration and/or smaller length scale stratigraphic units, lies on those seeking to demonstrate that extrinsic and not intrinsic controls were responsible.

4.3.3 Allogenic Stratigraphy

Three types of allogenic control are distinguished: cyclostratigraphy, sequence stratigraphy and tectonostratigraphy. Each one focuses on a different set of controlling parameters, typically addressing rock successions at a different time scale (Figure 4.16).

4.3.3.1 Cyclostratigraphy

Cyclostratigraphy is defined as '*the rhythmic variation in sedimentary deposits resulting from astronomically forced palaeoclimatic changes*' (Hinov 2013). Cyclostratigraphy developed out of the clear rhythmicity in climate signals seen, particularly in temperature-sensitive stable isotopes in the well-dated Plio-Pleistocene marine record. Subsequently, the same time periods of rhythmicity were detected encoded in the lithology, especially organic-richness, of pelagic sediments (see Chapter 13, Box 13.3 for the processes behind Milanković cyclicity). This spurred the search for other ways in which the primary insolation signal could be reflected in biology, chemistry or facies. The determination of specifically astronomical forcing is inextricably linked to the dominant astronomical periodicities of insolation variation, which requires (i) a sedimentary record that is continuous on the scale of the astronomical cycle over a statistically significant number of cycles, (ii) a means of transforming the stratigraphic record from thickness into time (an 'age model') and (iii) a demonstrable process link between the encoded signal in the rock and primary changes in solar insolation. A summary of the environments most suitable for the preservation of orbitally induced cyclicities is provided in Figure 4.10.

The most convincing examples of Milanković cyclicity are where the 'encoding process' can be established as well as the statistical recognition of the periodicity in the resulting sediment. A classic example is deep-sea pelagic sedimentation where (i) time gaps due to erosion and dissolution are often relatively small and (ii) the system oscillates between states of high and low phytoplankton productivity, which are insolation controlled (Hilgen et al. 2015). In these cases, the curve of expected solar insolation over time, a composite curve summing cyclic changes in all the orbital parameters, can be directly compared with the productivity and type of marine plankton. This is beautifully expressed in Figure 4.20 by the mixing of two lithological end-members: (i) black organic-rich mudstones that formed during times of very high organic productivity when organic matter was not oxidised in the water column or at the sea bed and hence was preserved; and (ii) light-coloured beds of plankton tests that accumulated when organic matter was largely re-cycled by oxidation or respiration.

The continuity requirement is quite severe, particularly for the shorter duration cycles (e.g. 25 kyr), and is most easily satisfied in environments unaffected by autogenic erosion or base-level change, for example, the deep sea. Deep-sea environments have the additional advantage of accumulation of a relatively restricted variety of sediments, direct relationships of sediment type, volume to sea surface productivity or temperature and excellent possibilities of simultaneously collecting stable isotope data often from tests of a single genus or even species. The establishment of a good record of Milanković cyclicity in the deep ocean demonstrates that astronomically forced climate changes must have also affected the rest of the planet. What is less clear is how they have been encoded (registered) in terrestrial and shallow marine sediments, where the sedimentary record is less complete. Nevertheless, Milanković cyclicity has been reported in a broad range of ancient stratigraphic successions back to the Precambrian and from a wide variety of depositional environments, including even alluvial settings where stratigraphic continuity is limited and autogenic processes pervasive (Hilgen et al. 2015).

Relating the observed stratigraphy of cyclical sedimentation to time requires some knowledge of the variations in accumulation rates. 'Rates need dates', and if dates are scarce and the periods of the cyclicity *in thickness* are in the same ratio as those of the orbital periods (i.e. 1:5:20) then that observation is sometimes taken to simultaneously demonstrate three conditions: continuous accumulation, a constant depositional rate and Milanković forcing. This simple use of 'beats' or ratios of rhythms in bed thickness or lithology as testimony to a Milanković 'control' is a hypothesis that needs further testing, as in the case of the proposed 10:1 ratio of cycles at Arabia Terra on Mars (Lewis et al. 2008), but it is not proof (Puetz et al. 2016).

Establishing an orbital control in alluvial sequences highlights some of the difficulties, including (i) understanding continuity on an appropriate time scale for the analysis, (ii) establishing the accumulation rates of different facies, including palaeosols, with which to convert thicknesses to time, (iii) understanding the encoding mechanisms (i.e. 'How did insolation change produce sedimentological

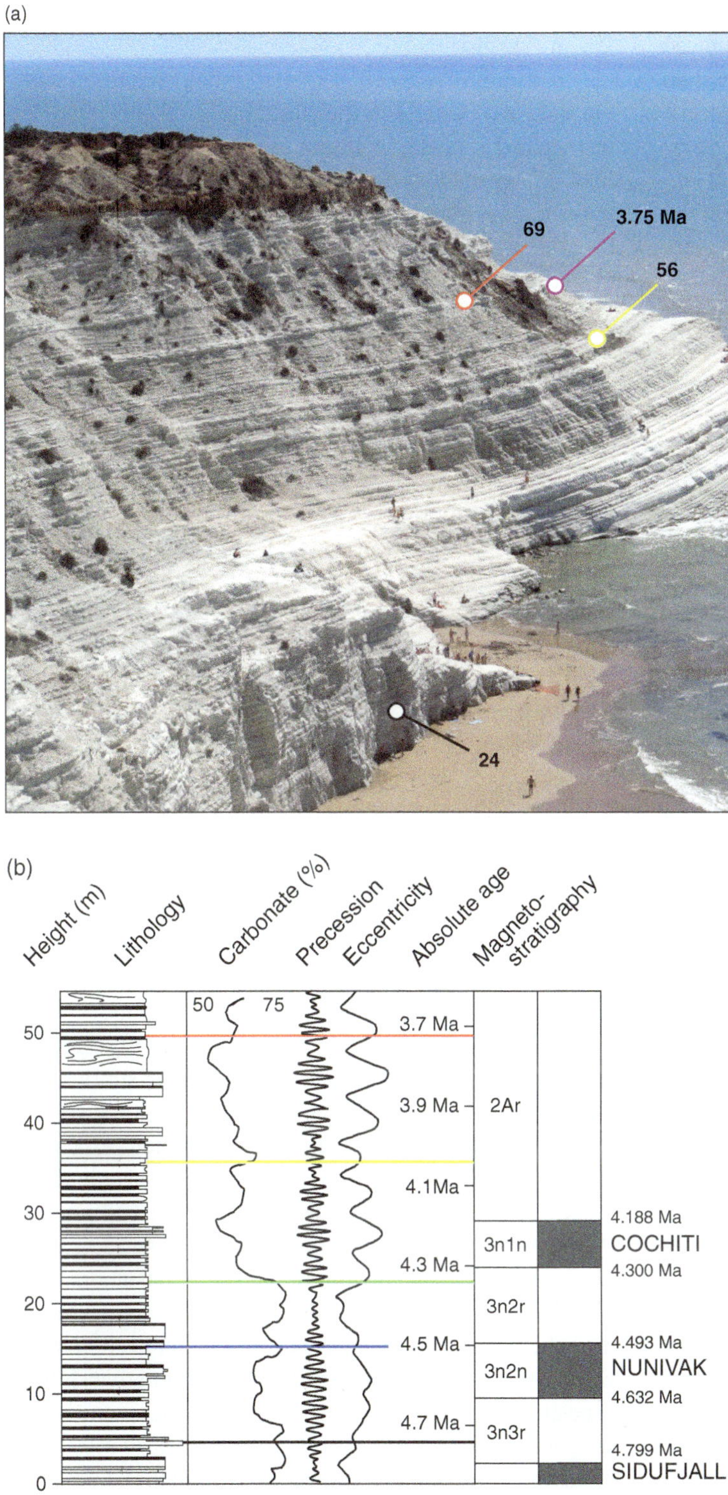

Figure 4.20 Milanković cyclicity in outcrop, Trubi Formation, Zanclian Stage Pliocene, Punta di Maiata, southern Sicily, Italy (Adapted from Langereis and Hilgen 1991; Hilgen 1991; and Hilgen et al. 2006). (a) Section of cyclically bedded limestones and marls with four facies: white (carbonate-rich), grey and beige (increasingly carbonate-poor) marls and occasional slumped marls. The entire section was deposited in deep water with abundant planktonic foraminifera. Numbers refer to the numbered precession-related cycles. The coloured arrows relate to the coloured horizons in (b). (b) Large-scale cyclicity of carbonate content is shown (the curve is an eight-point moving average) and is partly visible in the stepped profile of the cliff. This is related to the eccentricity cycle (400 kyr), whereas the individual cyclical alternations of grey/white and beige/white marls are regarded as the insolation-related encoding of precession cycles (20 kyr) in planktic carbonate accumulation on the sea floor. Correlation with calculated precession allows astronomical calibration of the time represented by the cycles. The sample resolution of the carbonate content is thought to be on the order of 5 kyr, which is a minimum estimate for the time scale of completeness of this record (Photo by Bruce Levell).

change'?) and (iv) testing against a null hypothesis of autogenic cyclicity. As an example of the sometimes-involved chain of reasoning, Milanković control was established for an alluvial sequence of Paleocene to Early Eocene palaeosols in the Bighorn Basin of Wyoming USA by Aziz et al. (2008). The cyclicity is based on patterns in rock colour (redness). Surprisingly, no hiatuses were found in the 70–80 m thick stacked palaeosol sections, which were therefore regarded as continuous. The independent absolute age transformation in this study was based on a log-linear correlation using a regression analysis of the tooth size of two fossil mammal species (R^2 of 0.26 and 0.57) versus absolute time that is tied to palaeomagnetic reversals found elsewhere in the same formation across 900 m of stratigraphy, which included the sampled interval. The accumulation rate was assumed to be the same over the 900 m as the 70 m. This analysis, encoding in colour as a proxy for a climate-controlled redox state, continuity, lack of gaps and steady rates, is what is required to construct the Milanković hypothesis. But fundamentally the argument is built on a power-spectral analysis of the periodicities of redness versus thickness, which shows the ratios 4:5:1.

Shallow marine clastic deposits are affected by a wide variety of processes (e.g. Woodroffe 2007) and would be expected also to be a difficult archive for extracting an orbital signal. A review of Quaternary shelf deposits by Lobo and Ridente (2014) suggests that the 100 ka cycle is well expressed by virtue of forced regressions related to rapid relative sea-level falls associated with glaciations, although the precise timing of progradation in each cycle is variable. The 20 kyr precession cycle is more variably expressed and often mixed with local autogenic processes.

Cyclical shallow water platform carbonates are a widely used archive, either based directly on facies themselves or indirectly, through attributes of the rock such as rock magnetic properties (Kodama and Hinov 2015). The 'encoding mechanisms' in such rocks, the explanations why combinations of ferromagnetic, ferrimagnetic and paramagnetic minerals are present in cyclically variable amounts ultimately related to insolation, are however 'complex'. Milanković rhythms have also been established in shallow marine mixed carbonate mudstone environments with the encoding mechanism being changes in storminess (Van Buchem et al. 1994; Weedon et al. 2018).

The distinction between autogenic and orbital controls depends on the spacing of the observed periodicities in time. Normally the ratio of periodicities from the power spectra is the trump card. Since ages are unknown at the bed scale (thicknesses stand as a proxy for elapsed time), this implies a knowledge of the accumulation rate and its variability in the studied section at the scale of the beds themselves. The quality of the independently determined thickness-to-time transform (the 'age model') is therefore key in this test.

The discrimination of an orbital signal from other possible signals contained in the stratigraphic record requires a test against a null hypothesis of what stratigraphy would look like in the absence of the orbital rhythms. This comparison process is referred to as extracting the 'signal' (Milanković rhythms) from the 'noise' (everything else, in particular, autogenic processes). Both signal and noise are expressed in terms of the attribute (normally a single attribute such as thickness) that is being tested for cyclicity. This attribute might also be lithology, chemical composition, colour or even, although rarely, facies. A range of possible stratigraphic models for the 'noise' is available. Gaussian white noise (a purely random stratigraphy with a normal distribution) is geologically less common but is used for longer time intervals. More common is red noise (a random pattern weighted towards lower frequencies, such as is seen in bed thickness statistics). Recurrent autogenic patterns or patterns derived from chaos are not modelled in the noise and should therefore, if present, appear in the power spectrum together with any orbital periodicity. This difficult subject is treated in depth by Weedon (2022).

4.3.3.2 Sequence Stratigraphy

Sequence stratigraphy is '*the study of genetically related facies within a framework of chronostratigraphically significant surfaces*' (Van Wagoner et al. 1990; see also Section 4.3), with the understanding that all chronostratigraphic surfaces (for example clinoforms, Figure 4.21) have at one time or another been depositional surfaces.

The origin of sequence stratigraphy as a discipline is closely related to the development of seismic stratigraphy, which grew in the mid-1970s from the realisation that for practical purposes (related in part to the limited vertical resolution of the then-available seismic data), seismic reflections could be treated as originating from bed surfaces that were essentially chronostratigraphic surfaces. This was a breakthrough in interpretation as reflections were therefore shown to reveal depositional geometries rather than the boundaries of larger-scale diachronous lithostratigraphic units. It became immediately apparent that shifts in depositional geometries were related to the creation and destruction of accommodation space as revealed in patterns of onlap (RSL rise) and truncation (RSL fall and erosion). Sea-level change was implicated, and the hunt was on for a global correlation of sea-level changes and a universal global sea-level curve.

The notion of cyclical change in accommodation space driven by sea-level change (Figure 4.12) led to the proposal that genetically significant depositional units could be separated by unconformities related to the reduction in

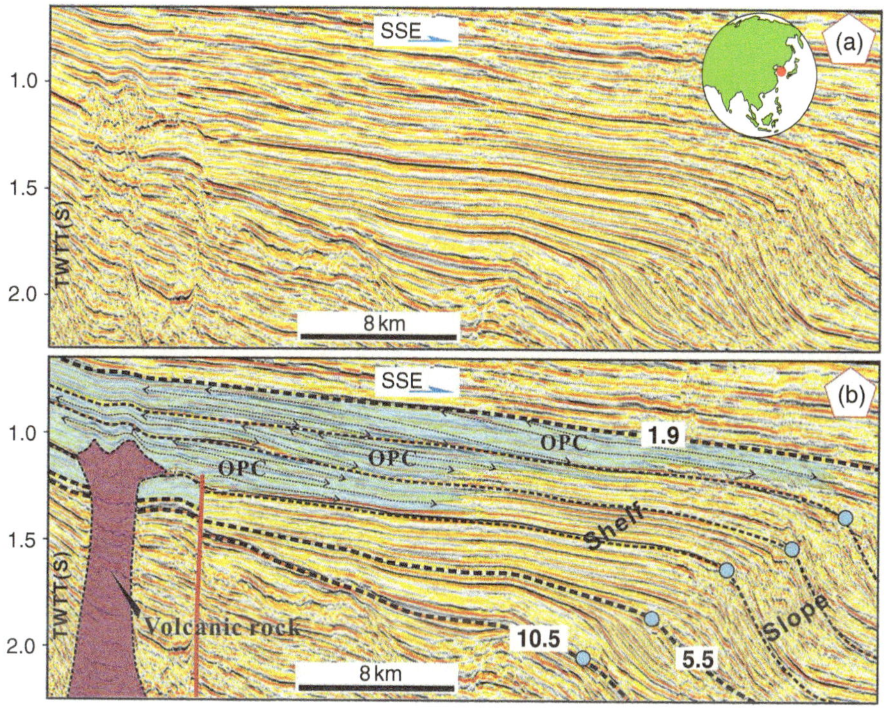

Figure 4.21 Seismic line through an approx. 3.5 km-thick siliciclastic progradational shelf margin succession (Mio-Pliocene, Qiangdongnan basin, southeastern China; Liu et al. 2019 / John Wiley & Sons). Shelf breaks are indicated with blue dots. Coastal progradation indicated by clinoforms (OPC = oblique progadational configuration) is shaded in blue. Repeated cycles of onlap are clearly evident above the black dashed horizons. The onlap indicates periodic increases in available accommodation space allowing the accumulation of shelfal sediments during periods of relative sea-level rise. It is not possible to tell from the figure alone if these RSL rises and increases in accommodation space were due to pulsed subsidence or pulsed sea-level change, hence we can speak only of periodic 'relative sea-level' change. Numbers are the ages of key horizons in Ma. The section is organised in depositional sequences (see Section 4.3).

accommodation space as sea level fell. The basic unit of sequence stratigraphy, the depositional sequence, was (and still is) therefore defined as '*a relatively conformable succession of genetically-related strata bounded at its top and base by unconformities or their correlative conformities*' (Mitchum et al. 1977). The initial concepts of sequence stratigraphy were defined for siliciclastic systems and were based on the assumption that the variation of A/S over time was controlled by significant, fast, changes in accommodation space due to eustasy and that sediment supply was broadly constant. Consequently, all stratigraphic stacking patterns would be globally correlatable. Acceptance of this assumption allowed the construction of the first approximately third-order global 'eustatic' sea-level curves.

The application of sequence stratigraphic methodology to outcrop sections (high-resolution sequence stratigraphy) highlighted the important role of sedimentary facies in determining the relative water depth of deposition, a key factor in the construction of models (e.g. Van Wagoner et al. 1990). Subsequently, the methodology was also applied to carbonate depositional systems (e.g. Sarg 1988; Schlager 2005; Tucker and Jones 2023) and to the terrestrial domain (e.g. Holbrook et al. 2006). The higher stratigraphic resolution obtained in the outcrop work demonstrated the fractal nature of the sedimentary record: 'depositional sequences' occur in a whole nested hierarchy of scales (Figure 4.13). But it also demonstrated that the stacking patterns of smaller-scale stratigraphic units might reveal larger-scale depositional patterns, such as progradation even when such large-scale (e.g. 100 m-thick) depositional geometries are not directly observed (e.g. inner carbonate platform cycle stacking; Figure 11.72). These sedimentologically focussed sequence stratigraphic studies led to a deeper appreciation of the role of A/S. First, it was emphasised that the rate of change of accommodation space (A) was determined by both eustasy and tectonics (i.e. both global and local impacts), and, second, that sediment flux (S) was a spatially and temporally variable component in both clastic and carbonate systems. Feedback between the location and timing of accommodation space formation and sediment flux was also recognised (e.g. due to gradient change in fluvial systems). Consequently, it is now accepted that stacking patterns and sequences are the result of the ratio between A/S and are thus by definition non-unique.

Global sea level has changed over time as a function of the changing volume and shape of the ocean basins and the changing storage of water outside the oceans (mostly as ice and a very minor part in aquifers related to continental flooding; see Davies et al. 2020). The problem of defining a globally applicable sea-level curve and quantifying its change over long periods of geological time has proven challenging. There are numerous issues, not least the chrono- and bio-stratigraphic demonstration of synchroneity, and the extraction of a quantifiable sea-level change from a facies record, which at best provides a view of changing bathymetry and coastal position. The most significant complicating factors include sediment supply, sediment dynamics, tectonics, compaction and isostasy. Despite these difficulties, however, an increasing number of highly detailed and age-dated case studies are providing support for the global occurrence of a number of short-term, high-magnitude sea-level fluctuations (e.g. Gale et al. 2002; Dromart et al. 2003; Snedden and Liu 2011; Maurer et al. 2013; Krencker et al. 2019; Bover-Arnal et al. 2022). These studies underline the potential significance of a eustatic sea-level curve and, at the same time, reveal the deficiencies of the classical often poorly documented 'global' curves (e.g. Ray et al. 2019; Davies et al. 2020; Simmons et al. 2020).

Sequence stratigraphy is now seen more broadly as a methodology for interpreting sedimentary facies and geometries based on changes in accommodation space and sediment supply rather than being linked exclusively to cycles of sea-level change (e.g. Neal and Abreu 2009; Burgess and Prince 2015).

The optimum window for the application of classic sequence stratigraphic methodology is the segment of a depositional profile, where rates of change in subsidence, sea level and sediment flux are similar, leading to the greatest sensitivity in outcome (progradation, aggradation, retrogradation and emergence) for a given small change in external forcing (Figure 4.10). Up-dip from this window, on a carbonate platform or alluvial plain, sediment is normally available, and accommodation space is the determining factor. In contrast, downdip from this window, in deeper water, there is plentiful space, and the limiting factor is sediment supply.

4.3.3.3 Tectonostratigraphy

Tectonostratigraphy (e.g. Hubbard 1988) is an approach to stratigraphic subdivision that seeks to understand large-scale stratigraphic units by viewing them in the context of their tectonic settings, specifically the temporal and spatial variations of subsidence and uplift (e.g. Figure 4.12). An example is distinguishing tectonostratigraphic units in rapidly deposited syn-rift deposits from later 'post-rift' or 'thermal sag' units deposited when faulting has diminished or ceased. The approach provides essential context for more detailed analyses and is a critical first step in regional studies; this is typically captured in a tectonostratigraphic chart summarising the stratigraphic distribution of facies, gross depositional environments, tectonics and sea-level fluctuations (Figures 4.22 and 4.23). Tectonostratigraphic units are particularly helpful in reconstructing and summarising deformed rocks in mountain belts or only loosely correlatable Precambrian sediments, for example, but they also provide the basis for excellent summary overview diagrams for any sedimentary basin.

Tectonic processes affect the source area of clastic sediments through their effect on relief, climate, weathering, slope angles, erosion rates and sediment supply (S) in terms of grain size range. Tectonic processes also disrupt drainage patterns and can cause or contribute to major shifts in provenance. In the sedimentary basins themselves, tectonic setting fundamentally shapes sedimentary sequences by influencing both accommodation space creation by changing the amplitude and distribution of uplifted and subsiding areas in both space and time.

Tectonic unconformities within sedimentary basins are a normal aspect of basin-forming processes, such as footwall uplift in faulted basins and forebulge migration in flexural basins (Allen and Allen 2013). In both cases, sediment supply patterns are simultaneously affected especially if uplifted areas become emergent. In rift basins, a relatively abrupt change in the pattern and rate of subsidence occurs when active faulting transitions to thermal subsidence. This is often accompanied by a widening of the area of active subsidence (White and McKenzie 1988). This creates accommodation space on the basin flanks and non-depositional hiatuses due to consequent sediment starvation in the basin centre. The Base Cretaceous Unconformity Complex of the North Sea is a classic example (Kyrkjebø et al. 2004). The so-called rift-drift transition on passive margins is similarly often a composite surface formed by coincident tectonic subsidence and uplift changes and related sediment supply reductions (Stoker et al. 2018).

Local stacks of unconformities over growing structures on a range of time scales are seen routinely on offshore seismic lines over syn-depositional structures related to deltaic growth faults or salt movement. Flexural isostasy related to shifting depositional loads would be expected to similarly be reflected in hiatuses, but its stratigraphic imprint is rarely documented. An exception is Plint et al.'s (2012) careful work in the Western Canadian (foreland) Basin where, based on exhaustive compilation of thousands of wells, they have been able to demonstrate a 'trampoline' tectonics, whereby c. 300 km depocentres change shape from elongated to rounded and shift on a time scale of 800 kyr or less. These changes occur both along strike and across dip and are interpreted as a response to the flexural loading patterns in the adjacent thrust pile of the Cordillera. On time scales of millions of years, by contrast, depocentres can be reasonably viewed as 'prismatic' (a wedge-shape that is continuous along-strike) in line with conventional models of foreland basin development.

Subsidence differentials also provide detailed controls on sedimentary geometries and facies belts at the scale of individual fault blocks, and the relay ramps between fault blocks, by controlling the local sediment routing systems (Leeder and Gawthorpe 1987). At even smaller scales,

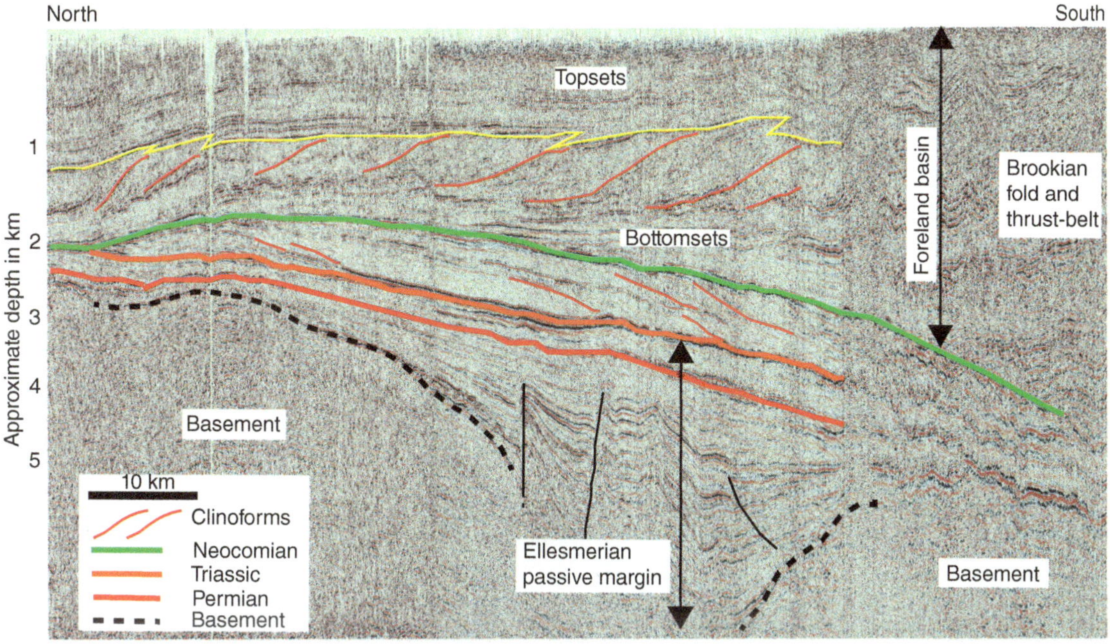

Figure 4.22 Four tectonostratigraphic units are shown each associated with a particular subsidence regime. The 'Ellesmerian' rift rests on the basement and represents Late Palaeozoic deposition with clear fault-control on sediment accumulation. It is followed by Permo-Triassic sediments of the Upper Ellesmerian deposited under relatively uniform (thermal sag) subsidence. Between the Triassic and the Early Cretaceous Neocomian, the 'Beaufortian' shows southward-prograding clinoforms and is clearly truncated to the north below the major Neocomian unconformity (green). The Early Cretaceous to Recent ('Brookian') section comprises a foreland basin sedimentary wedge related to over-thrusting of the Brookes Range fold-and-thrust belt, just visible on the south end of the line. The associated forebulge is expressed in the truncation and onlap visible to the north. Each tectonostratigraphic unit is controlled by a different subsidence regime, which sets the overall context for the stratigraphic architecture. For example, the northward-prograding seismic clinoforms of the Brookian represent a major increase in sediment input, overwhelming an initially deep-water foreland basin. NPRA line R8 United States Geological Survey (Miller et al. 2000 / U.S. Geological Survey).

individual fault slip events episodically generate accommodation space and therefore can preserve packages of sediment related to discrete quanta of subsidence. Each package is separated by relatively short inter-earthquake hiatuses when there was no subsidence. For example, co-seismic subsidence of tidal wetlands accompanied by evidence of tsunamis is well documented for the Cascadia margin. The active margin has a 10,000 year record of up to 19 large-magnitude subduction-related mega-thrust earthquakes with a recurrence interval of 230–480 years (Leonard et al. 2010; Walton et al. 2021). The coastal record of these events is one of subsidence due to arching at the leading edge of the overriding plate. Subsidence pulses have resulted in a thickness of c. 2–7 m of stacked tidal flat deposits, with many correlatable buried soils, each with a sharp upper surface and often overlain by liquified sand or muddy sand. The spacing of the buried soils, interpreted as the co-seismic subsidence, is from 25 cm to 1 m. Subsidence of this magnitude resulted in the sudden burial of salt-marsh peat horizons and/or deposition of tsunami deposits overlying salt-marsh peats (e.g. Kelsey et al. 2002; Graehl et al. 2015). A contemporary, correlatable, deep-water record consists of individual turbidites (Goldfinger 2011).

High-frequency tectonism, for example, co-seismic movements during earthquakes, is also recorded as event beds, such as landslides, debris flows turbidites or mass transport complexes. A spectacular 50 kyr long record of earthquakes is provided by the lacustrine seismites of the Dead Sea for example (Marco et al. 1996).

4.4 Sequence Stratigraphic Methodology

Sequence stratigraphy is here considered to be the most effective way to describe and analyse the stratigraphic architecture of a sedimentary succession and provides an important unifying concept for integrating the various branches of sedimentary geology (Figure 4.24).

This section provides a synthesis of the terminology and methodology used in sequence stratigraphic studies,

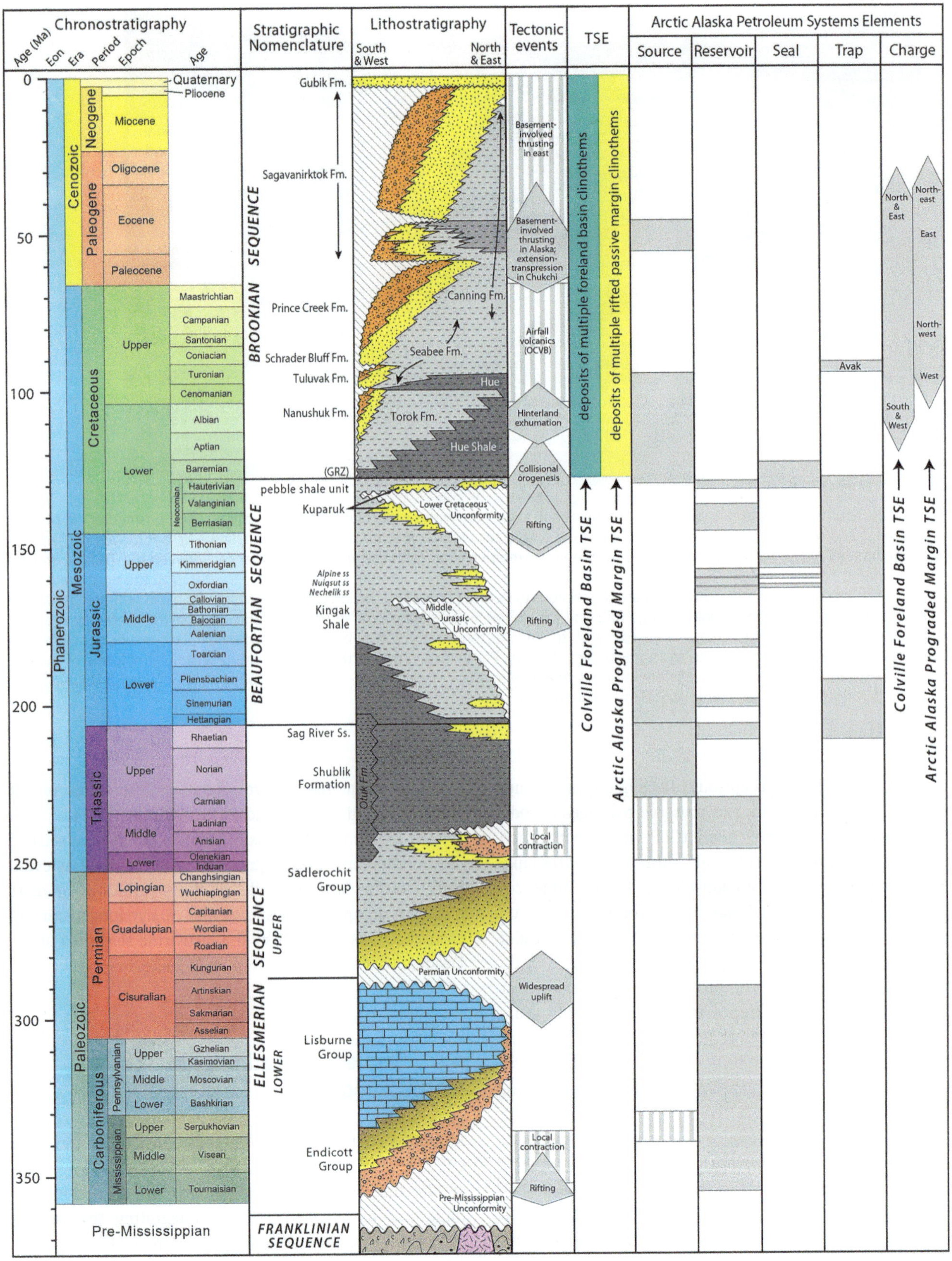

Figure 4.23 Summary of tectonostratigraphic diagram of North Alaska illustrating the tectonic sequences (TSE = tectono-sedimentary elements) illustrated in Figure 4.22, the timing of tectonic events and the timing of key elements of economic significance, in this case, the petroleum system against the chrono- and lithostratigraphy in a single diagram (Houseknecht 2024 / Geological Society of London / CC BY 4.0).

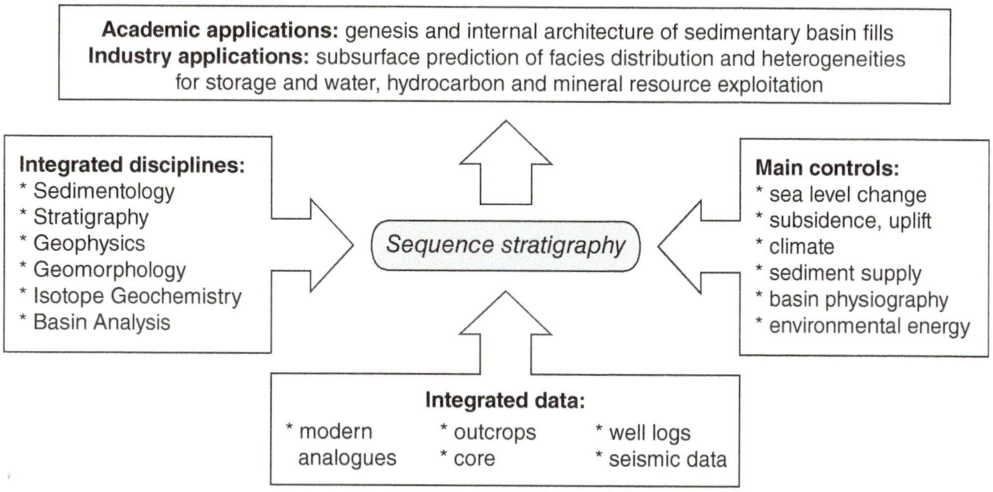

Figure 4.24 Sequence stratigraphy as a unifying concept (Catuneanu 2022 / with permission of Elsevier). Sequence stratigraphic analysis uses 'Integrated data' (the observations) and 'Integrated disciplines' (providing the process understanding and the concepts). The output of the analysis is an understanding, or a hypothesis, of the formation of stratigraphic architecture in terms of the relative importance of the 'Main controls'. Quantitative forward modelling of sedimentary architecture using these same controls (Section 4.4) allow these hypotheses to be tested for reasonableness.

followed by 3D illustrations of the variability of stratigraphic architecture using forward stratigraphic modelling software. More exhaustive reviews of sequence stratigraphy are provided elsewhere (e.g. Loucks and Sarg 1993; Van Wagoner et al. 1990; Emery and Myers 1996; Posamentier and Allen 1999; Embry 2009; Catuneanu 2019, 2022).

4.4.1 Terminology and Methodology

In addition to the general stratigraphic principles of base level, accommodation space, sediment flux, stacking pattern and hierarchy explained above, a methodology has been developed specifically for seismic and sequence stratigraphic analysis. This includes: (i) the formal characterisation of depositional geometries and stratal terminations (e.g. onlap and downlap), (ii) the characterisation of special stratigraphic surfaces (e.g. transgressive and flooding surfaces and sequence boundaries), (iii) the definition of system-specific conceptual models, notably the systems tracts and (iv) the representation of the sequence stratigraphic models in chronostratigraphic schemes. These are briefly summarised below.

4.4.1.1 Depositional Geometries, Clinoforms and Stratal Termination Patterns

The definition of a depositional sequence: '*a relatively conformable succession of genetically-related strata bounded at their top and base by unconformities or their correlative conformities*' (Mitchum et al. 1977) combines the sediment package with the delineating stratigraphic surfaces. It therefore introduces a focus on the stratal geometries and termination patterns, which require a descriptive terminology that is summarised here.

Most depositional surfaces are inclined, albeit at very low angles. Hence, they form 'depositional profiles' along which facies change as a function of water depth. Sedimentation via lateral movement of low-angle depositional surfaces leads to a vertical 1D facies succession, which can mirror the lateral facies succession (Walther's law; Figure 4.18). Surfaces with detectable depositional dips at large scale (tens of metres or more in height) are referred to as 'clinoforms' (or 'large-scale foresets'; Figure 4.25a). Clinoforms can be recognised on any data set, which presents pseudo-chronostratigraphic surfaces: correlations of biostratigraphic markers, volcanic tuffs or in an approximation at least, seismic reflections that track the reflective interfaces of bundles of strata. Clinoforms are particularly well expressed in clastic systems at delta fronts (Figures 4.5, 4.20 and 4.22) shorefaces, shelf-break slopes or in large submarine sediment drifts such as in contourites or mid-shelf mud belts. Strike-fed linear coasts generate elongate clinoforms. Lobate deltas generate aerially restricted clinoforms related to the mouth bar and/or the pro-delta platform that are crescentic or tongue-shaped in plan-view (see Chapter 8 for further details of coastal planform geometries). Clinoforms are also commonly observed in carbonate systems, along the margins of land-attached and land-detached platforms, intrashelf basins and various types and sizes of carbonate mounds.

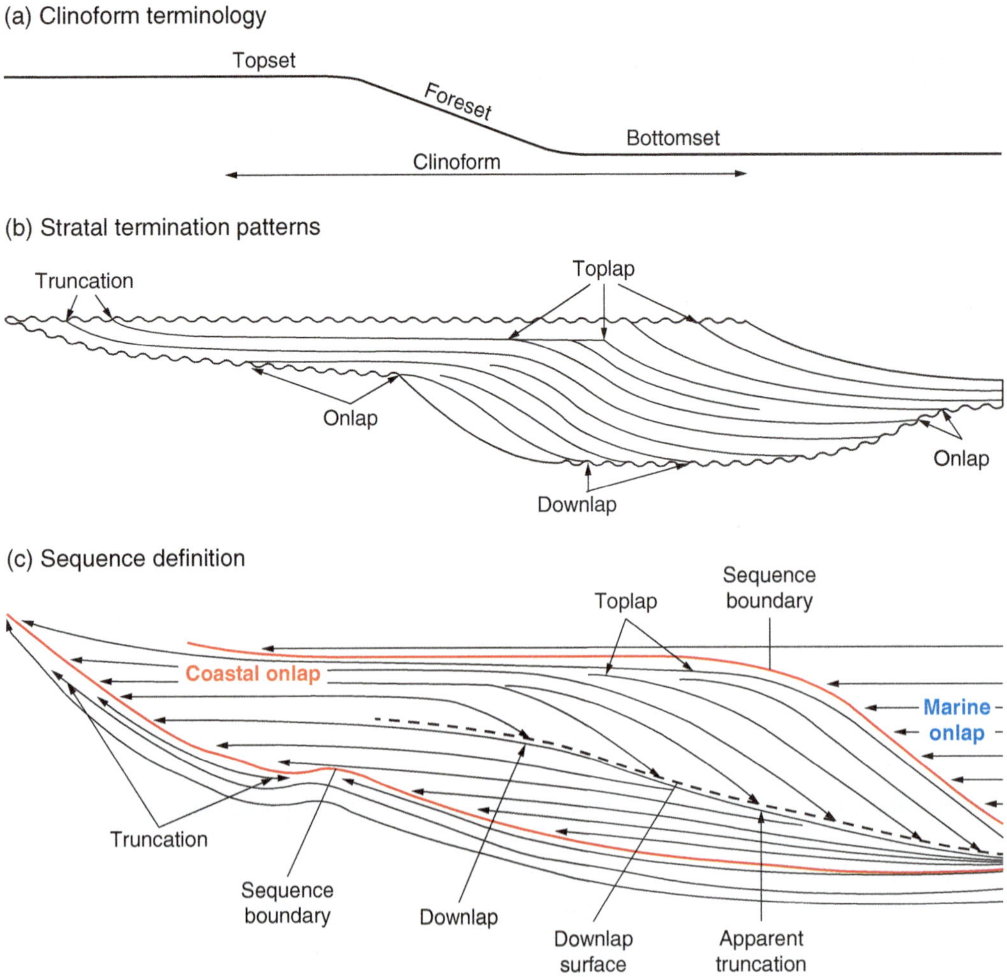

Figure 4.25 Stratal geometry terminology. (a) The clinoform depositional profile and naming for its different parts: topset, foreset and bottomset and (b) stratal terminations: onlap and truncation and for clinoforms: downlap and toplap. The rock interval between two clinoforms is sometimes called a clinothem. (c) Stratal termination patterns defining sequence boundaries (red) and the maximum flooding surface (dashed). 'Apparent truncation' is used where thinning due to condensation is suspected, as well as actual physical truncation. Adapted from Mitchum et al. (1977).

When available, stratal termination patterns are the basis of analyses of the geometrical relationship between different sediment packages. Four types are distinguished: (i) *onlap*, the horizontal termination of younger layers against dipping older layers, this may be, for example, marine onlap or in the marginal-marine domain (coastal onlap), (ii) *truncation*, termination of underlying strata beneath an erosion surface, (iii) *downlap*, the termination of more steeply dipping overlying strata (clinoforms) against lower strata or a surface with lower dip, (iv) *toplap*, termination of clinoform strata beneath an overlying lower dip surface (topset) often resulting from non-deposition or sedimentary bypass with perhaps only minor erosion (Figure 4.25b,c). These stratal termination patterns are one of the basic tools to identify and delineate depositional sequences as well as the systems tracts (see Section 4.4.1.2).

Clinoform sets are also of great use in tracking the trajectory of relative sea-level changes (Pellegrini et al. 2020). The clinoform roll-over point or 'offlap break' (i.e. where a depositional surface transitions from sub-horizontal to inclined) is a depth reference point, such as a shoreline, a shelf-edge break or the margin of a carbonate platform (Figure 4.26). The offlap break trajectories enable the recognition of prograding, aggrading, retrograding and degradation trends, which are fundamental for the definition of the sequences and their systems tracts (Figure 4.27).

4.4.1.2 Special Stratigraphic Surfaces

Three special stratigraphic surfaces are used to sub-divide depositional sequences: the sequence boundary (SB), the maximum flooding surface (MFS), and the transgressive surface (TS) or maximum regressive surface (MRS). All these surfaces can in principle be recognised as more-or-less

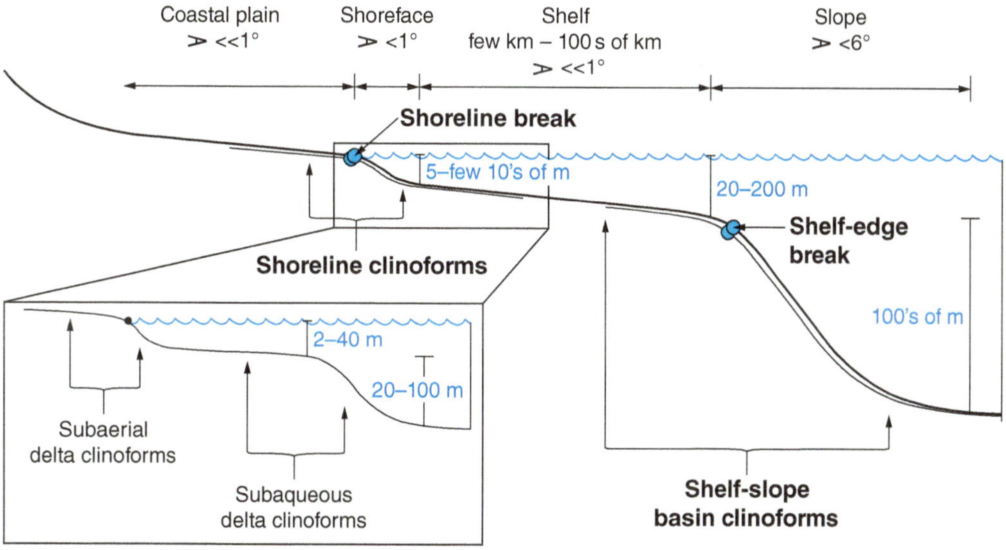

Figure 4.26 Schematic representation of shoreline and shelf-edge breaks (offlap break) and their geometries, dimensions and bathymetry (Adapted from Helland-Hansen and Hampson 2009).

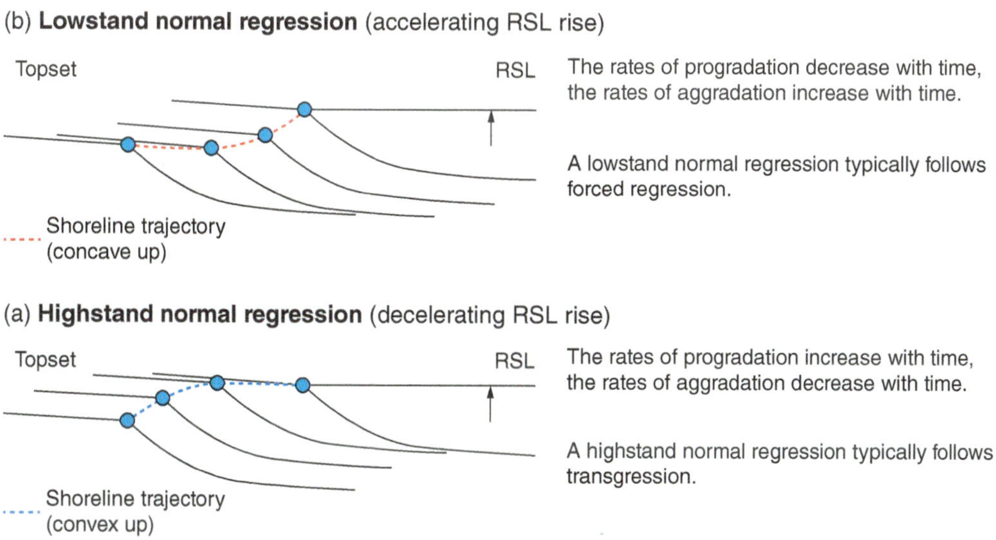

Figure 4.27 Tracking of offlap break is a good indicator of accommodation space evolution and helps define sequences and systems tracts. In this case, two extreme cases of 'normal' shoreline regression are illustrated (constant sediment supply): (a) accelerating relative sea-level (RSL) rise, which is the idealised stacking pattern of lowstand systems tracts and (b) decelerating RSL, which typifies highstand systems tracts (Adapted from Van Wagoner et al. 1990).

abrupt breaks in the succession of facies and by specific stratal termination patterns.

Sequence Boundary (SB)

The boundaries of depositional sequences mark, in theory, inflection points in RSL: the highest rates of change from high to low accommodation space. Sequence boundaries are complex, due, for example, to variable depths of erosion, locally composite in nature (i.e. formed by different physical processes), can be associated with considerable hiatuses and vary in expression along the depositional profile (Figure 4.28). Original definitions of the term emphasised sub-aerial exposure, fluvial incision and/or soil formation. Indeed, in the proximal, sub-aerial domain sequence boundary surfaces are normally erosive, associated with incised channel or larger valley systems, or define a surface of non-deposition and sediment bypassing. However, in the shallow-marine realm, the same surface can be entirely the result of marine processes (waves and tides). For example, in shelfal settings, base-level fall causes wave

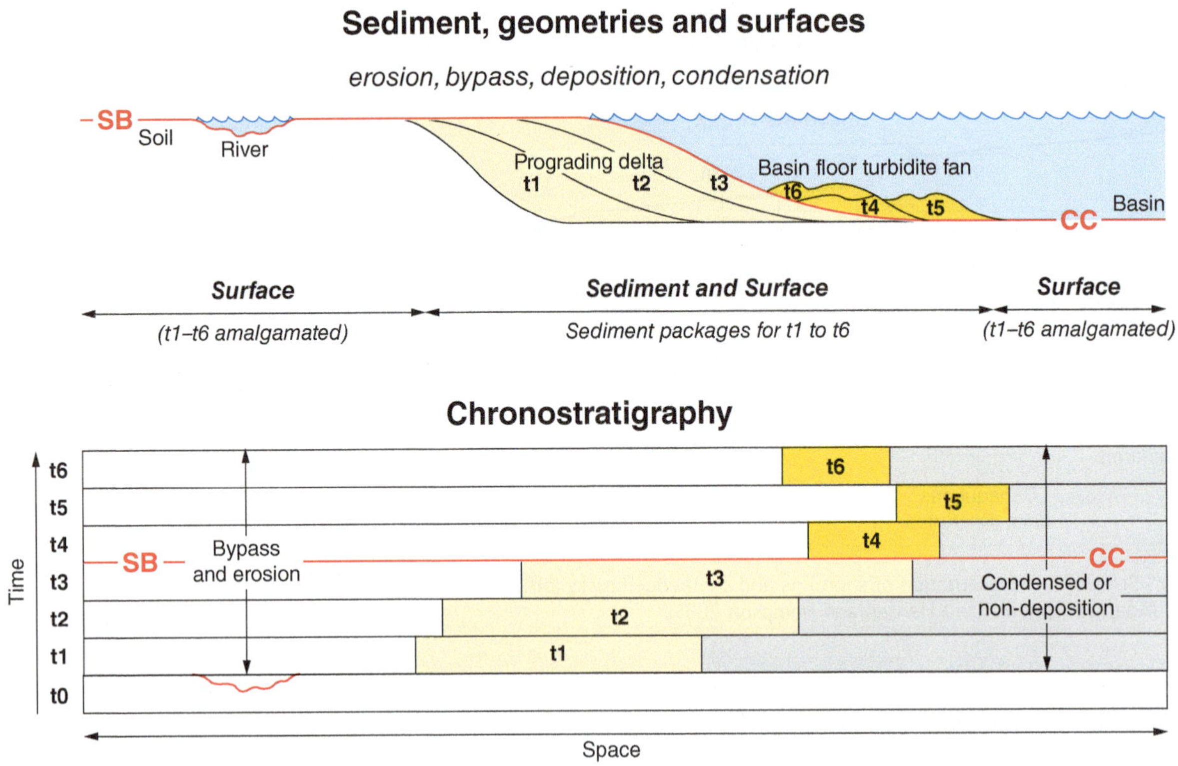

Figure 4.28 Time relationship between surfaces and sediment packages around a sequence boundary (SB) and its 'correlative conformity' (CC). See text for explanation.

base lowering, which, depending on water depth and wave climate, results in wave-related erosion during the formation of the new, lower, submarine base level. These submarine erosion surfaces have been termed regressive surfaces of marine erosion ('RSME') and are often laterally extensive with relatively planar geometries (Figure 4.32b, Walker and Plint 1992), in contrast to the deeply incised nature of their fluvial equivalents (e.g. Posamentier and Allen 1999).

The lateral equivalent of the sequence boundary extends into deeper-water areas, eventually extending below the erosional effects of base-level lowering; here, deposition is often continuous and the timeline equivalent to the onset of erosion at the sequence boundary is simply a conformable surface, known as the 'correlative conformity (CC)' (Figure 4.28). The time interval represented by the unconformity is represented basinward (at least in part) by sediment bypassed along the sequence boundary into deeper water. The amount of time represented by the sequence boundary surface varies along the depositional profile, for example (in Figure 4.28), (a) in the proximal subaerial domain time, intervals T1–T6 are amalgamated into the sequence boundary surface; (b) in the transitional domain (where both 'sediment & surface' are present), the sequence boundary surface is a conformable or slightly erosive surface on top of the prograding delta, representing a shorter hiatus (T4–T6); and (c) in the distal domain, there may be bypassed turbiditic sediments at the base of slope, while further basinwards the sequence boundary equates to the correlative conformity, which often comprises a thin, stratigraphically condensed interval of sediment or a surface of non-deposition.

The nature of erosion at a sequence boundary depends on the following factors:

- The amplitude of the relative sea-level fall: if the fall exceeds the shelf water depths and hence results in exposure of the shelf, there will be fluvial incision into the coast and shelf, perhaps with canyon formation at the shelf break. The LST takes the form of a lowstand wedge developed outboard of the previous shelf edge, and the sequence boundary is known as 'Type 1'. If not, and the sea level remains above the shelf break, it is 'Type 2' (Figure 4.29). In the Type 2 case, there may be less, or even no, sediment bypassing into deeper water, and there may be no LST present.
- The initial shape of the depositional profile: a ramp-like profile, whether in clastics or carbonates, will result in large lateral translations (regression) of the shoreline,

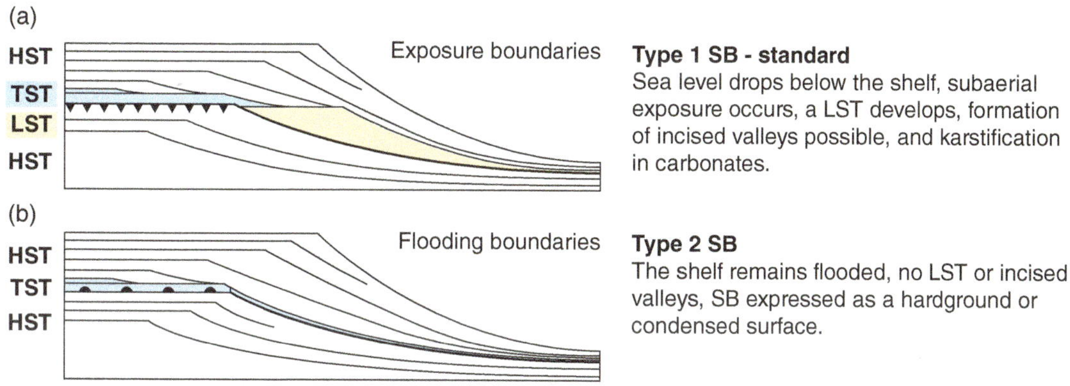

Figure 4.29 Sequence boundary types: (a) Type 1 sequence boundary, the sea level drops below the shelf, subaerial exposure occurs, and LST develops, formation of incised valleys possible and karstification in carbonates, and (b) Type 2 sequence boundary, the shelf remains flooded, no LST or incised valleys, sequence boundary expressed as a hardground or condensed surface (Adapted from Schlager 2005).

and potentially shallow-marine shelfal erosion for relatively modest relative sea-level falls.

- The duration of the relative sea-level low: a longer duration will, all things being equal (such as climate and erodibility), allow more complete lateral erosion of previously deposited sediments above the new, lower, base level.
- The erodibility of the material below the sequence boundary. For instance, the early cementation of carbonates may help them to survive physical erosion, but dissolution results in a karstified surface.

Maximum Flooding Surface (MFS)

A maximum flooding surface is defined as the surface that marks the time when the rate of relative sea-level rise is at its greatest. It corresponds theoretically to the depositional profile at the point in time when the coast reaches its most landward position and marks the turn-around point from transgression to regression. However, the position and behaviour of the coast are also controlled by other factors, especially rates of sediment supply, which are semi-independent of relative sea-level change, and not the same everywhere along the coast. Consequently, the maximum flooding surface will not be entirely synchronous, but this diachroneity is often within the chronostratigraphic error margin. Temporal change of sediment supply may represent a bigger problem since the maximum flooding surface is normally picked on the basis of evidence for the deepest water conditions in a particular facies succession. This works well if there is no fluctuation in the sediment supply, but not where there is a large increase causing progradation, and thus shallowing before the maximum rate of sea-level rise is reached. This situation is, in fact, a particular characteristic of carbonate systems (see Figure 11.71) but may also apply to some siliciclastic systems.

Maximum flooding surfaces are more recognisable than sequence boundaries in basinal settings, due to their often-condensed nature and a distinctive lithological response. For example, concentration of organic matter, mineralisation (phosphate, iron, silica) and macro- and/or microfossils, including faunal diversity and abundance peaks. This results in two important implications: (i) it makes them more easily dateable (Loutit et al. 1988), and (ii) they often coincide with pronounced and distinctive wireline log signatures in shelfal to basinal successions (e.g. gamma ray log 'spikes'), which aids well log correlation and calibration with seismic data (e.g. Partington et al. 1993). Extreme condensation may lead to a genuine depositional hiatus in the more distal parts. This is a consequence of high rates of accommodation space creation in onshore and inshore areas, which causes sediment trapping and reduced sediment supply down the routing system onto the shelf or beyond.

Transgressive Surface

The 'transgressive surface' or 'transgressive surface of erosion' is a marine erosion surface cut as RSL rises and the shoreline is forced to retreat landwards. The erosional processes involved are mainly shoreface erosion by shoaling waves, in addition to scouring by tidal currents in inshore environments (e.g. estuaries and tidal inlets) and storm erosion in offshore environments. It is thus a composite of multiple shallow-marine erosion surfaces, which are diachronous, younging in a shoreward direction. Identifying a single transgressive surface in multiple locations can therefore be difficult but is possible when confident biostratigraphic data are available.

In sequence stratigraphic terminology, the 'transgressive surface' has a specific meaning and generally equates to the 'maximum regressive surface' (Helland-Hansen and Martinsen 1996). This surface marks both the top of the regression associated with the lowstand and the regional onlap of the sedimentary system at the onset of the transgressive systems tract (see below). In classic sequence stratigraphy, this is the initial point of flooding of the continental shelf, and the erosion surface tracks the landward retreat of potentially multiple shoreface systems (Vail et al. 1991). In some situations, the shallow-marine erosion processes can also erode down to, and rework, the sequence boundary, resulting in a composite sequence boundary/transgressive surface (see Section 4.4.2 and Figures 4.32a,b). In shallow-marine successions, the maximum regressive surface also often marks the upper surface of the last major coarsening-upward progradational unit (Catuneanu 2002) and can thus form a good regional correlation marker.

4.4.2 Conceptual Models and Chronostratigraphic Schemes

Conceptual depositional models are one of the great contributions of sequence stratigraphy to sedimentary geology. They convey a concept in a schematic way that allows prediction and the posing of insightful questions about the stratigraphic architecture away from datapoints. This made such models especially popular in the petroleum industry, but they are equally important for any application concerned with the prediction of effective reservoir-seal pairs or reservoir geometry. These models are simplifications and should be regarded as guidelines, with the understanding that local variations in dimensions, 3D architecture and duration are likely to occur.

Among the unique sequence stratigraphic concepts are the 'systems tracts', the building blocks of the depositional sequences. A 'systems tract' is defined as *'a linkage of contemporaneous depositional systems, forming the subdivision of a (depositional) sequence'* (Brown Jr and Fisher 1977). Hence, systems tracts form an array of contemporaneous depositional environments, positioned along a depositional profile from proximal to distal, that characterise the time interval between the formation of the significant stratal surfaces described above (sequence boundary, transgressive surface and maximum flooding surface). Systems tracts may be defined by combinations of stratal termination patterns, stacking patterns, depositional surfaces and facies characteristics.

Depending on the dataset and the depositional system, systems tracts are recognised using different characteristics:

1) *Seismic interpretation* emphasises the observed depositional geometries analysed, using offlap break tracking and stratal termination patterns (Figure 4.30). Seismic facies characteristics, based on the frequency, continuity and amplitude of seismic reflectors that can be related broadly to depositional geometries (Xu and Haq 2022) are also used. The seismic geometries alone are generally insufficient for interpretation and need to be supplemented with information about the sedimentary system, ages and lithofacies. Offlap tracking, a powerful technique for understanding A/S is illustrated by the 'Accommodation succession sequences' of Figure 4.30.
2) *Outcrop-, core- and/or well log-based* studies focus on more detailed facies analysis (i.e. process- and environment-based interpretations based on an understanding of lithologies, primary sedimentary structures, bedding styles, ichnofabrics, biostratigraphic information, etc.). This analysis is combined with correlation and mapping of potential key stratal surfaces. See Figure 4.31 for a model for siliciclastic systems. The finer-scale sedimentological and stratigraphic information provides better insight into the physical dynamics, sediment flux (composition and rate) and other environment-sensitive aspects (e.g. climate and biochemical aspects, salinity and substrate conditions) of the sedimentary systems. These in turn may lead to a better appreciation of the relative contributions of A and S.
3) *In carbonate studies* both (a) and (b) apply, but there is the added complication of the impact of the different carbonate factories, and their specific sensitivities to sea-level and environmental changes (see Section 11.7). As an example, the depositional geometries and sea-level responses of a land-attached T-factory are shown in Figure 4.33. Note the larger variation of stacking patterns and depositional geometries than observed in marine siliciclastic systems is represented in Figure 4.30.

The four systems tracts referred to earlier (Figures 4.30–4.33) are discussed below in further detail, building on the sequence stratigraphic terminology and definitions of others (e.g. Van Wagoner et al. 1990; Vail et al. 1991; Hunt and Tucker 1992, 1995; Catuneanu 2022).

4.4.2.1 Lowstand Systems Tract (LST)

Defined as the deposits extending from the sequence boundary/ correlative conformity below, which marks the time of minimum accommodation space creation, to the transgressive surface above, marking the maximum and end of regression. The lowstand is typically recognised by a marine onlap contact against the lower sequence boundary (Figure 4.29). In the proximal domain of siliciclastic systems, sea-level falling below the delta top, causes increased erosion, fluvial incision, possibly shelf-edge deltas and bypass of sediment into deep water to feed submarine fans. The LST shelf break trajectories will be progradational with an increasing aggradational component, due to gradual accommodation space increase (Figures 4.30 and 4.31). The fall in base level can be efficient in mobilising large amounts of clastic sediment, which will create the lowstand deposits. In siliciclastic ramp systems,

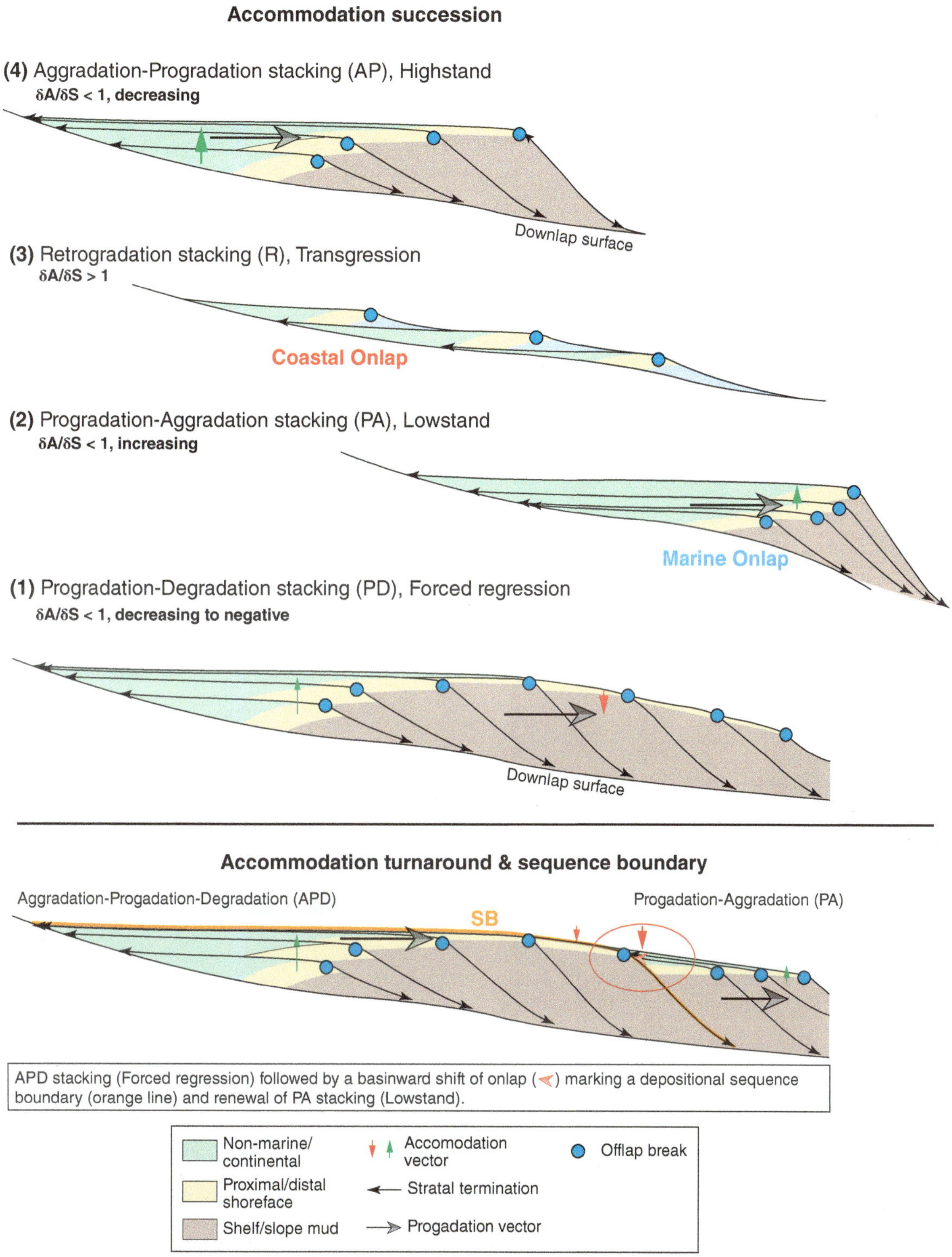

Figure 4.30 Parasequence stacking patterns and offlap break tracking are used to define the accommodation succession. Note the critical difference in accommodation space trend between a highstand (decreasing accommodation space) and a lowstand (increasing accommodation) systems tract. This defines the sequence boundary, and is also characterised as the accommodation turn around, see also Figure 4.31. Accommodation succession sequences are defined as 'regional sedimentary packages resulting from changes in rates of shelfal accommodation space creation and depositional fill in response to changes in sea level' (Modified from Neal and Abreu 2009).

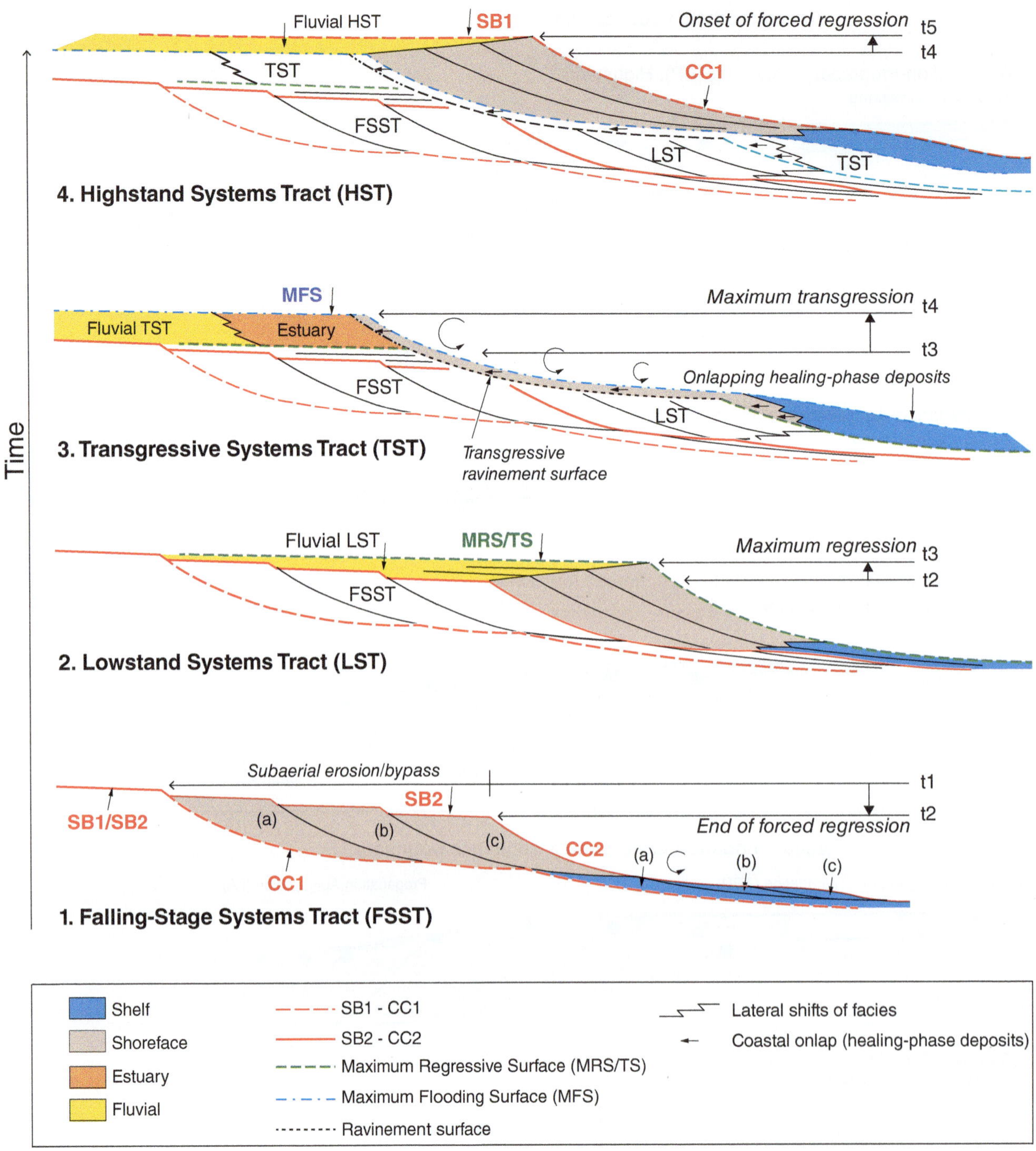

Figure 4.31 Schematic representation of the four systems tracts of a depositional sequence in a coastal marine siliciclastic system (Modified from Catuneanu, 2022). t1 to t5 are time steps and show relative sea level. In the FSST, (a), (b), and (c) show time equivalent, but spatially disconnected layers. In the TST, note the erosional impact of the transgressive ravinement surface, and the deposition of the onlapping healing phase deposits. CC = correlative conformity.

the lowstands will be expressed as downward shifts of facies belts (Figure 4.32), often being less dramatic in their expression than passive margin systems with basin floor fans.

In carbonate systems, sea-level lowstands can also be expressed as incised valleys of considerable dimensions cut into the platform top (e.g. Late Aptian, Qatar, up to 8 km wide and more than 60 m deep; Raven et al. 2010; Spain, 100 m deep; Bover-Arnal et al. 2014, Bover-Arnal et al. 2022). Lowstand deposits can be found as shelf margin wedges clinging to the margins of the exposed carbonate platforms and are limited in size due to the strongly reduced carbonate sediment production area (Figure 4.33).

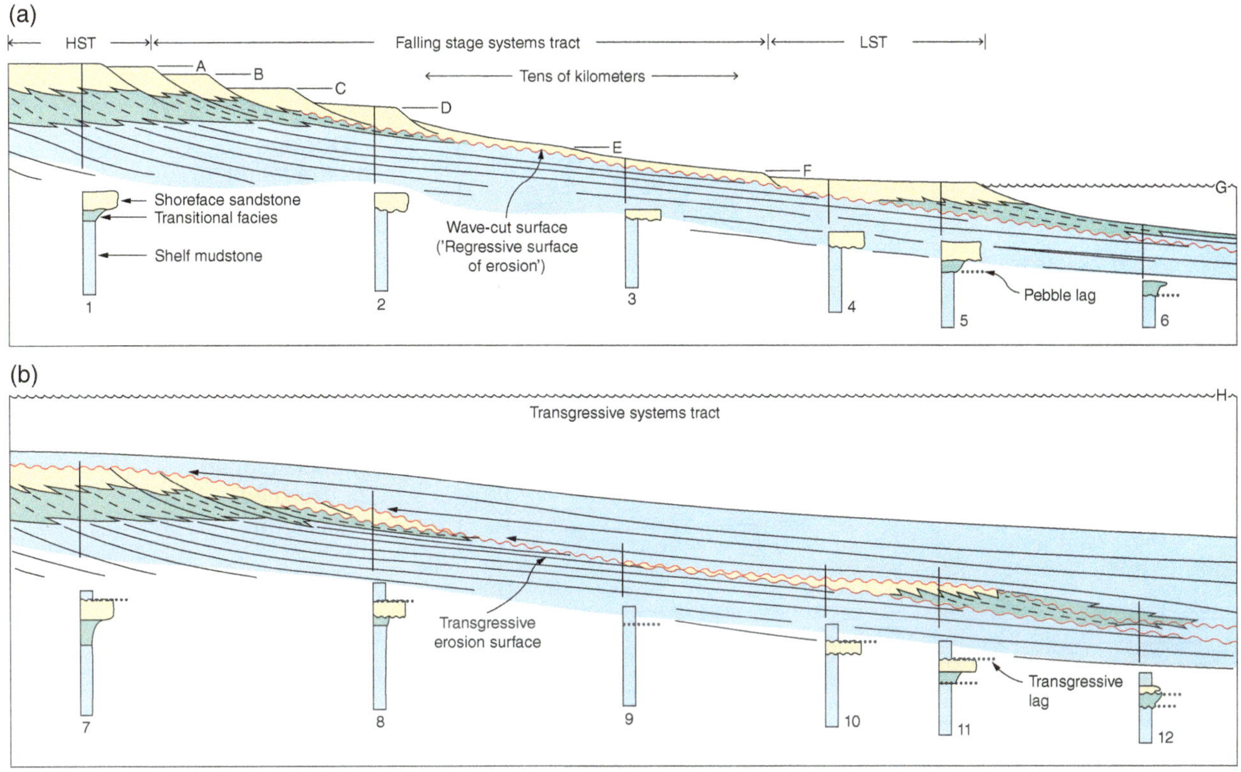

Figure 4.32 The impact of sea-level fall and rise in a siliciclastic ramp-type setting (Modified and reinterpreted from Walker and Plint, 1992, Plint 1988). (a) Timesteps A to F illustrate relative sea level fall (FSST), and G represents sea-level stabilization, and the development of a lowstand shoreface (LST). Note that during fall shore face sands (yellow) rest directly on shelf mudstones (column 3, regressive surface of erosion). (b) Sea level has risen to level H, the shoreline is driven landward and wave scour in the backstepping shore face has left a gentle undulating surface (the transgressive erosion or ravinement surface). The thin sand that was deposited during the fall in column 3, has been completely eroded during the transgression. As a result shoreface detached, coast parallel, lowstand sand bodies, are found encased in offshore marine mudstones.

On carbonate platform tops, lowstands are generally not present but represented by a composite surface of a combined sequence boundary and transgressive surface.

4.4.2.2 Transgressive Systems Tract (TST)

This extends from the transgressive surface up to the maximum flooding surface. In siliciclastic systems, periods of rapidly rising relative sea level are characterised by ascending transgressive shoreline trajectories and aggrading to retrogradational stacking patterns (Figures 4.30 and 4.31). Sediment is trapped onshore or in coastal zones due to the alluviation of flood plains associated with high rates of space creation. This leads to reduced offshore sediment supply, causing condensation and local non-deposition. The transgressive flooding may cause erosion by the retreating shoreface that removes part of the underlying (LST) sediments sometimes re-depositing these further offshore as 'onlapping-healing phase' shelf deposits (Figures 4.31 and 4.32).

In carbonate systems, transgression may have different effects depending on the dominant type of carbonate factory (Chapter 11). For example, T (tropical)-Factory-dominated carbonate platforms (Figure 4.33) may (a) keep up with the rise and aggrade over the full extent of the platform, with or without the formation of a rim, (b) keep up partially, forming a bioconstructed external rim and an internal depression (lagoon or the larger sized intra-shelf basin) due to local sediment starvation, (c) backstep on the shelf or (d) give up and drown completely, which would be the end of the carbonate sedimentation. The C (cold)-Factory carbonate system would behave more like a siliciclastic system (see Section 11.7).

4.4.2.3 Highstand Systems Tract (HST)

This extends from the maximum flooding surface to the sequence boundary above. Depending on the dataset and sedimentary system, a distinction can be made between an early HST, when accommodation space decreases but is larger than zero, and a late HST when accommodation space is destroyed. This latter case is also referred to as the 'Falling Stage Systems Tract' (Figures 4.31 and 4.32) and is a useful distinction to make when present (see Section 4.4.2.4). The early highstand is characterised by the turn-around from the retrograding ($A > S$) to the

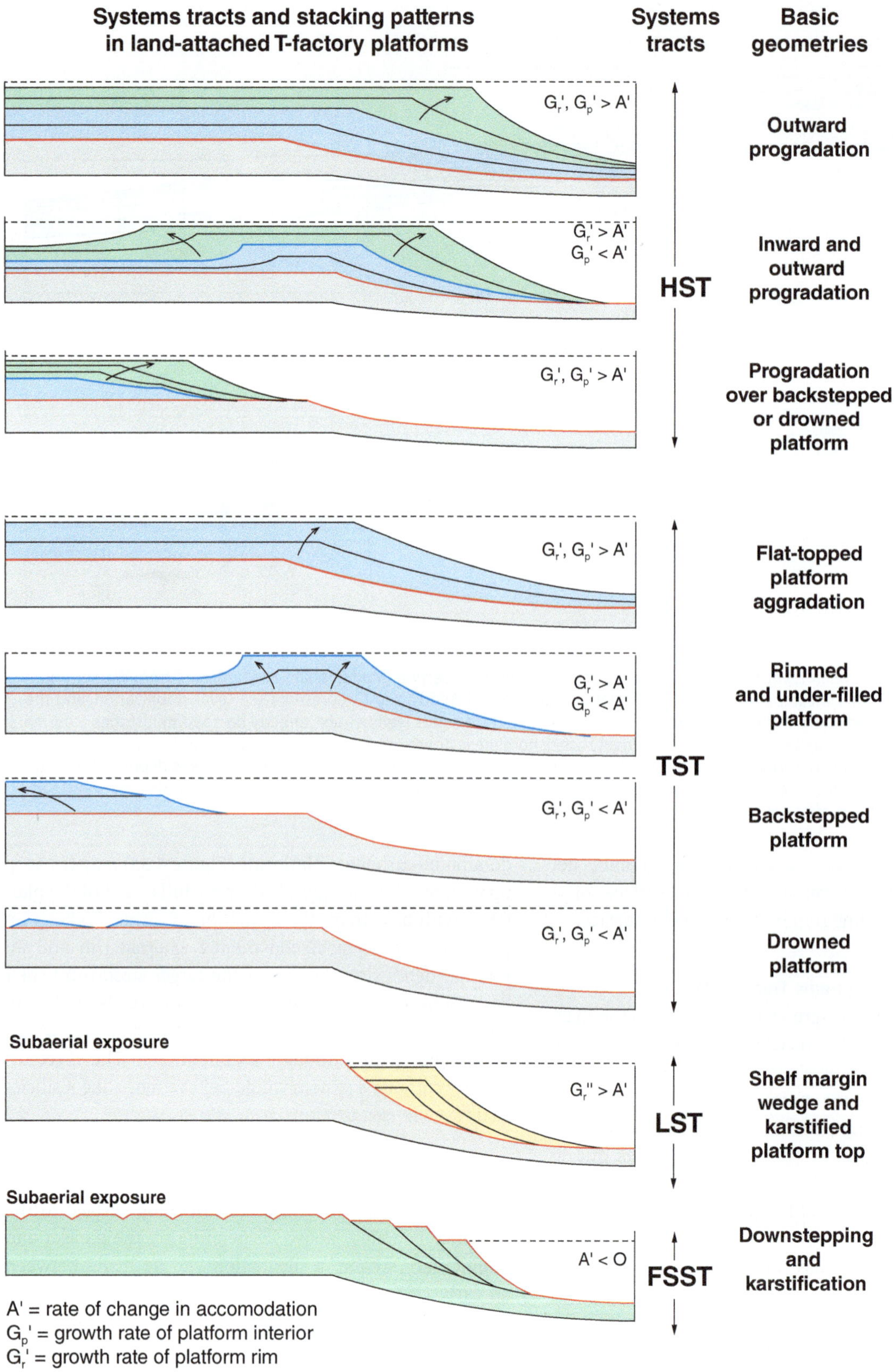

Figure 4.33 Stacking patterns, depositional geometries and systems tracts typical for land-attached T-factory (Tropical) carbonate systems (Modified from Schlager 2005, Hunt and Tucker 1992, 1995). Note that the colour scheme here represents not the lithology but the systems tracts – green: HST and FSST, yellow: LST and blue: TST.

aggrading/progradation stacking pattern (S > A). With gradually decreasing accommodation space and assuming a constant sediment supply, progradation will increase. In siliciclastic systems, it is important to take into account the potential for large spatial variation in S; for instance, the difference between a point-sourced delta, and a wave-dominated shoreline that shows a more even progradation of the shoreline over long distances (along strike).

In T-factory carbonate systems (Figure 4.33), highstands are characterised by the shedding of the surplus carbonate produced on the shallow platform top to the offshore domain and, if present, to the lagoons and intrashelf basins ('Highstand shedding').

4.4.2.4 Falling Stage Systems Tract (FSST)

This extends from the onset of base-level fall to the minimum sea level (Figures 4.30–4.32). The FSST was first identified by Hunt and Tucker (1992) in seismic scale outcrops of Mid-Cretaceous carbonates in S. France and has since been recognised in other carbonate case studies (e.g. Figure 11.69, Razin et al. 2010) and siliciclastic systems (Plint and Nummedal 2000). Depending on the definition of the chosen author, the sequence boundary can be placed below the FSST (SB1/CC1) or above the FSST (SB2/CC2; Figure 4.34).

The FSST can be well expressed in the siliciclastic ramp or shallow shelf setting (Figure 4.32) where the shoreline is forced to prograde when the relative sea level starts to fall below its previous level (forced regression), creating a sharp-based shoreface deposit (downward erosional step in contrast to the gradual facies change of the HST). This is separated from the underlying sequence by an erosion surface cut by waves at the shoreface. Such forced regressive shoreface sediments can locally be preserved from erosion at the next transgression if there is even a small amount of accommodation space created in the ensuing RSL rise. Such deposits are very important in epeiric sea and foreland basin settings (Plint and Nummedal 2000). Clinoforms may show toplap and possibly even downstepping depending upon the amount of relative sea-level fall. Enhanced erosion occurring in higher gradient fluvial systems may increase sediment supply at the same time.

On a historical note, depositional sequences have been defined using different special surfaces by different authors (Figure 4.34). The debate arose from practicality: the Exxon School led by P. Vail, worked mostly in the shelf and shelf margin settings, on passive margins with modest

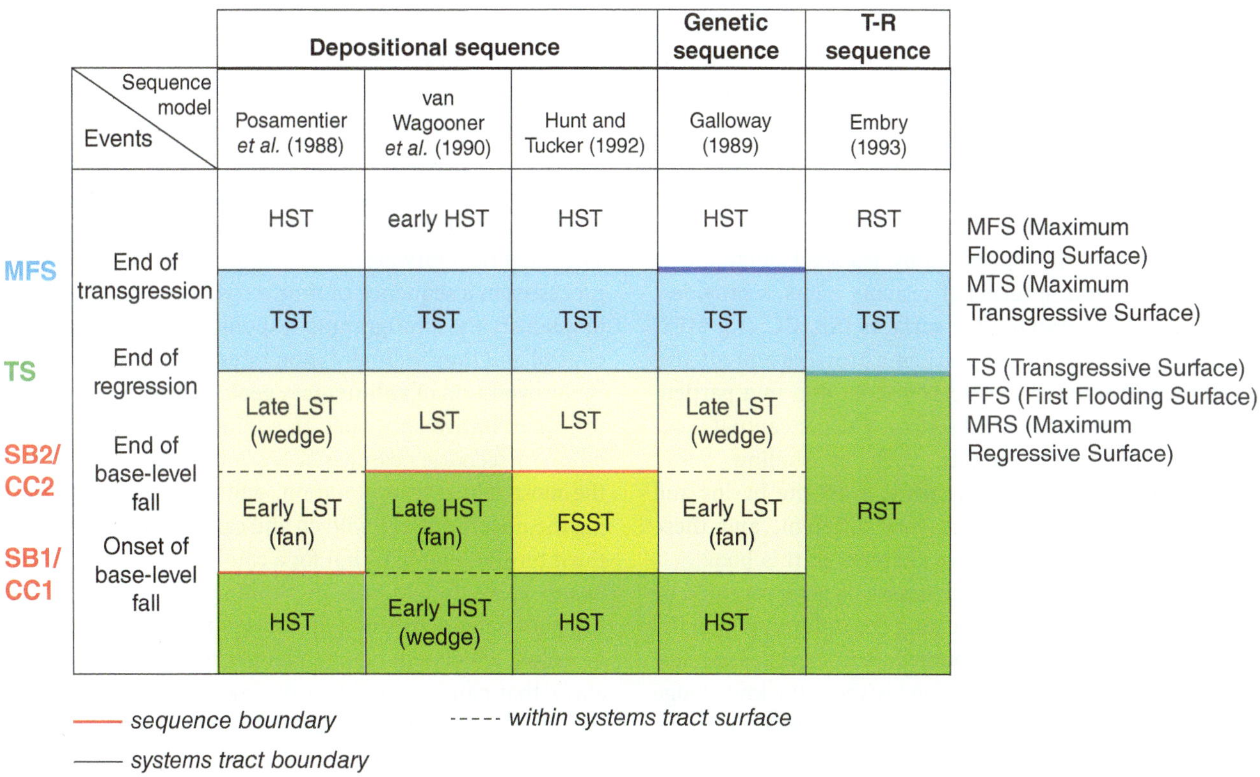

Figure 4.34 Diagram illustrating the position of special stratigraphic surfaces in relation to the systems tracts. Note that different authors choose to put their sequence boundary at different positions. The option favoured here is the one of Hunt and Tucker (1992). CC = correlative conformity.

subsidence rates where erosional SBs are well expressed. The 'Galloway' school worked mostly in the Gulf of Mexico where subsidence is exceptionally rapid, SBs are often difficult to spot but maximum flooding surfaces are clearly expressed. Hence, their 'genetic sequences' were defined as bounded by maximum flooding surfaces. Here, we prefer to follow the interpretation that places the sequence boundary on top of the FSST, an addition provided by Hunt and Tucker (1992) to the original sequence boundary-based model by Vail et al. (1991).

Sequence stratigraphy provides compelling explanations of the depositional geometries created by variations in accommodation space, both through time (due principally to base-level change) and in space (due principally to subsidence and sediment flux). It connects key surfaces of erosion or non-deposition at one place and time with correlative deposits elsewhere at the same time (e.g. Figure 4.28). Chronostratigraphic schemes (Figure 4.5) offer a powerful method to visualise these sediment–hiatus relationships and provide critical insights into the stratigraphic organisation of sedimentary systems. They also force the interpreter to introduce timelines across different depositional environments and by doing so, to take an important step from a descriptive to a quantitative approach in sedimentary geology, including sedimentation rates.

In Figure 4.35a, a siliciclastic shelf margin system is used to illustrate the comparison between a simple lithostratigraphic characterisation and a sequence stratigraphic interpretation of the same system. Through the introduction of timelines (Figure 4.35b), a detailed stepwise evolution of the sedimentary system is defined in three depositional sequences, which are here interpreted to have been primarily controlled by sea-level fluctuations. An interpretation in terms of systems tracts is provided, and the special stratigraphic surfaces (SB, CC, TS, MFS) are indicated. This idealised scheme provides a predictive model with regard to the facies distribution in a particular sedimentary and basinal setting. The conceptual models for other environments, such as shallow water carbonate systems, fluvial systems, evaporite basins and lacustrine settings, will look very different, and these are provided in the respective chapters of this book.

In Figure 4.35, note that the timescale is represented by 35 arbitrary time-slices, indicating the order in which the sediment packages were deposited. A first-pass, sequence stratigraphic model can be defined without the knowledge of absolute time, an approach applied in seismic stratigraphic interpretation before the availability of borehole-based information. Once age dating does become available, the chronostratigraphic scheme (Figure 4.35c) can be updated to absolute time: the true nature of the hiatuses becomes apparent, and sedimentation rates can be calculated.

4.4.2.5 Differences Between Clastic and Carbonate Sequence Stratigraphic Behaviour

In the foregoing, we have highlighted differences between carbonate and clastic sequence stratigraphic responses to changing relative sea levels. These differences are many and varied but centre on different feedbacks between accommodation space creation or destruction and sediment supply/production, these feedbacks also differ within carbonate systems as a function of the specific biological carbonate factory (see Chapter 11). Early cementation of carbonates, which affects erodibility can also be an important difference. In both systems, as explained earlier, the sequence stratigraphic response depends on the original depositional profile. This is also more varied among carbonate systems. Table 4.1 should therefore be treated as illustrative and incomplete rather than comprehensive and authoritative.

4.4.2.6 3D Forward Stratigraphic Modelling

The conventional illustration of sequence stratigraphic relationships is in 2D cross sections, such as those shown earlier. However, the 3D reality is much more complicated as captured in numerous high-resolution 3D seismic studies, which illustrate the significant variations of depositional geometries and geobodies in strike directions (see, for instance, Figures 8.42 and 9.38).

An additional way to explore the 3D complexity of depositional systems and the processes that control them is to use 3D numerical process-based Forward Stratigraphic Modelling (FSM) programmes (Burgess 2012). The numerical forward model reconstructs the geological processes in a sequence of time steps, using physical laws to quantify the stratigraphic response. The focus on time control and thus sedimentation rates is an important step in the evolution of sedimentary geology from a descriptive to a quantitative approach. In addition, the outcomes for different scenarios can assist the interpreter in choosing the most robust interpretation, while better understanding the uncertainties involved and can, in this way, reduce the interpretation bias (but for a cautionary note see Catuneanu 2020). In addition, by running multiple scenarios, the parameter sensitivity of stratigraphic expression can be tested, which will help to identify the diagnostic differences that can be sought in the rocks. Finally, the FSM models allow a fully characterised lithology and geometry-informed 3D volume to be constructed based on the application of physical, chemical and biological laws that can be used for self-consistent extrapolation and interpolation. This predictive aspect is used in industry,

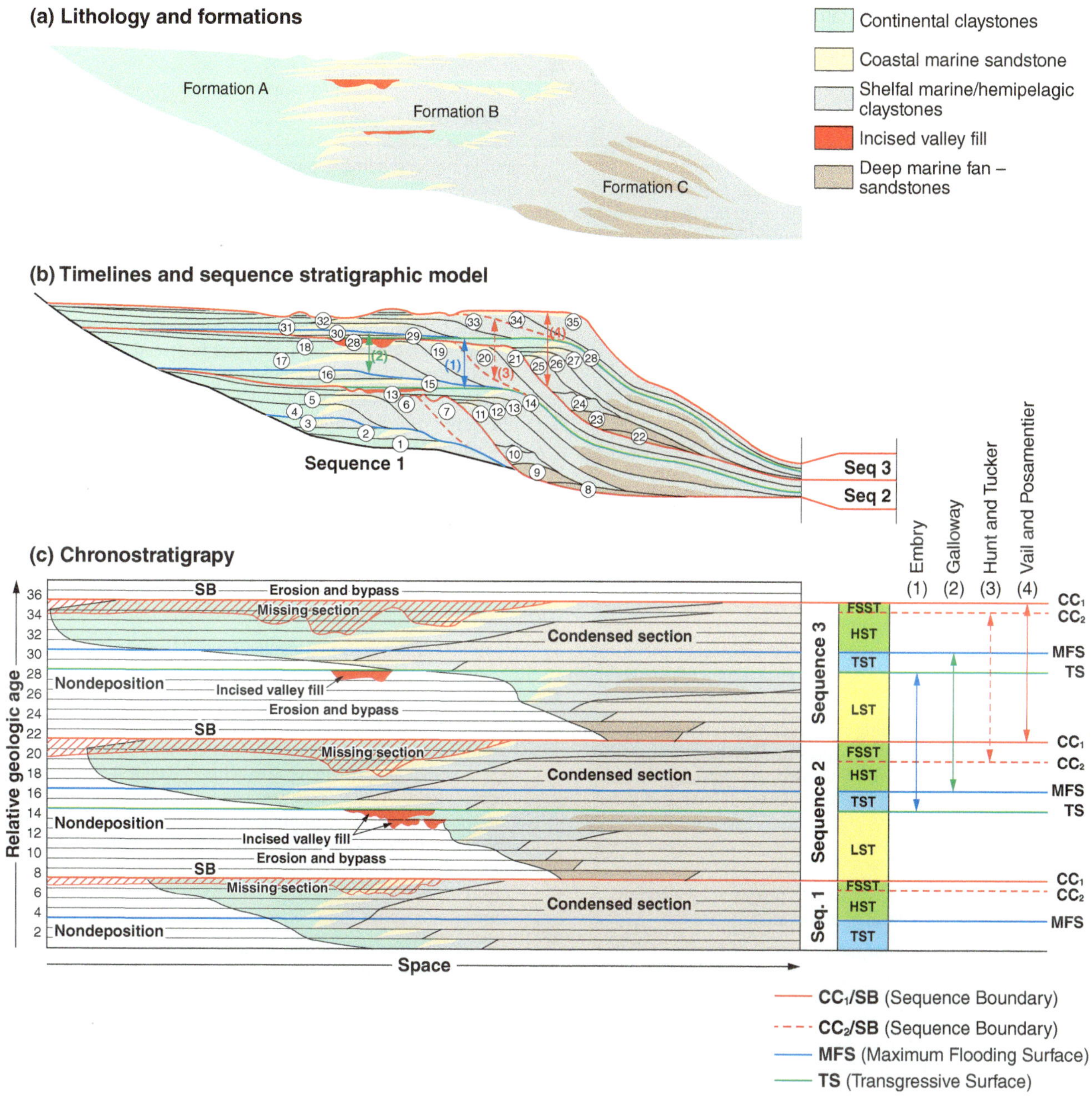

Figure 4.35 A siliciclastic shelf margin represented by (a) a lithostratigraphic interpretation with three formations. Note that this model has limited predictive power due to the lack of timelines. (b) A sequence stratigraphic interpretation, which subdivides this margin into 35 time slices organised in three depositional sequences. The logic of their distribution, a result of the interplay of A/S, makes this model predictive with respect to the facies distribution. (c) The chronostratigraphic representation of the sequence stratigraphic model illustrates the relationship among the stratigraphic surfaces, the hiatuses and the sediment packages (Modified from Abreu et al. 2017). CC = correlative conformity. This diagram basically encapsulates a sequence stratigraphic workflow: beginning with rock description (a), adding timelines and depositional geometries through biostratigraphic and seismic control (b) and explicitly interpreting age relationships using a chronostratigraphic chart (c).

traditionally in the hydrocarbon industry, but now also in the energy transition, where the identification of effective reservoir–seal pairs is an important step in all sub-surface storage projects.

Two main types of process-based FSM are distinguished (Figure 4.36):

1) Fluid-flow process models, which use physical equations to simulate the movement of water, sediment particles and geometrical elements (channels, bars) in 3D (e.g. Syvitski et al. 2010). These methods are applied, for instance, in coastal engineering, where study areas are relatively limited, and where high accuracy of the

Table 4.1 Some examples of differences between siliciclastic and carbonate sequence stratigraphic responses.

Sequence stratigraphic context	Siliciclastic system response	Carbonate system response
Highstand system tract	Aggrading to prograding geometries. Bypassing into deep water is reduced by inboard sediment trapping but the potential for bypassing increases over time as the rate of RSL rise slows.	Aggrading to prograding geometries. In some depositional profiles, this may be a period of increased sediment bypass into deeper water due to increased productivity as RSL rise enables continuing euphotic carbonate production. 'Highstand shedding' may be associated with carbonate turbidites at the base of the slope as a result. However, ramp settings by contrast may not see increased but rather reduced sediment production areas during RSL rise as large flat distal areas fall below the euphotic zone. Sensitivity to small water depth changes may lead to more clearly recorded cyclostratigraphic signals in platform top deposits with variations in 'ecological accommodation' space.
Maximum flooding surface	Condensation-Firmgrounds. Mud content is higher due to increased distality/depth.	Condensation-Hardgrounds. Clastic content (mud) may be higher due to lowered carbonate production and reduced dilution.
Transgressive system tract	Transgression is associated with increased alluviation and coastal sediment trapping reducing sediment transported offshore.	Transgression can increase surface area for carbonate production in shallow waters leading to the possibility of increased sediment export. Alternatively, drown platforms below the photic zone can lead to cessation of carbonate export. If both effects operate, minor differences in the pre-existing topography are emphasised (see Chapter 11 carbonate system depositional profiles).
Transgressive surface	Complete erosion of coastal sediments through shoreface erosion or 'ravinement' is possible to probable.	Cementation of, for example, aeolianites and beachrock may allow preservation of the coastal record.
Lowstand system tract	Prograding to aggrading geometries. Classically a period of bypassing of sediment into deep water with increased sediment supply is possible due to enhanced fluvial gradients.	Prograding to aggrading geometries. Sediment supply change depends on the depositional profile and the change in carbonate productive area with a given RSL fall. Rimmed platforms would typically seed reduce sediment production due to platform-top exposure for example. Some ramp systems might see increased sediment supply as flatter, previously sub-photic, domains become euphotic.
Falling stage system tract	System is driven to prograde by a regional reduction in accommodation space.	System is driven to prograde by a regional reduction in accommodation space.
Sequence boundary	Unconformity. Erosion. Pedogenesis, fluvial incision. Depositional system may become more concentrated at point sources with associated lobate accumulations (deltaic or submarine fans) due to local fluvial incision. Sharp juxtaposition of shallow water on deep water facies associated with those areas that remain depositional.	Unconformity. Erosion. Karstification. Terra Rossa. Incision by rivers is also possible. Sharp juxtaposition of shallow on deep facies is more prevalent and persistent along strike (linear systems).

transportation process is required for short time intervals. This type of modelling is difficult to condition on real well or seismic data.

2) Dynamic-slope or diffusion/advection process models, which simulate landscape (topography) evolution, and use mass conservation combined with empirical transport laws, that average over several processes to simulate the sediment transport in 3D (Granjeon 2014). The diffusion/advection algorithms transport and erode sediment as a function of the evolving (dynamic) slope. They do not simulate in detail the processes at the grain and sedimentary body scale but define belts where specific processes operate (e.g. the fluvial, coastal and shelf margin domains) and provide average geometries and facies content as an output. The simplicity of this method allows the user to fit the model outputs to real data without the need to specify multiple poorly constrained transport parameters and is applicable at the basin and prospect scale. It also covers longer time spans and is, thus, capable of modelling stratal patterns and stratigraphic architecture in 3D at the third-order level.

The choice of the most appropriate FSM application is closely linked to the scale of the study object

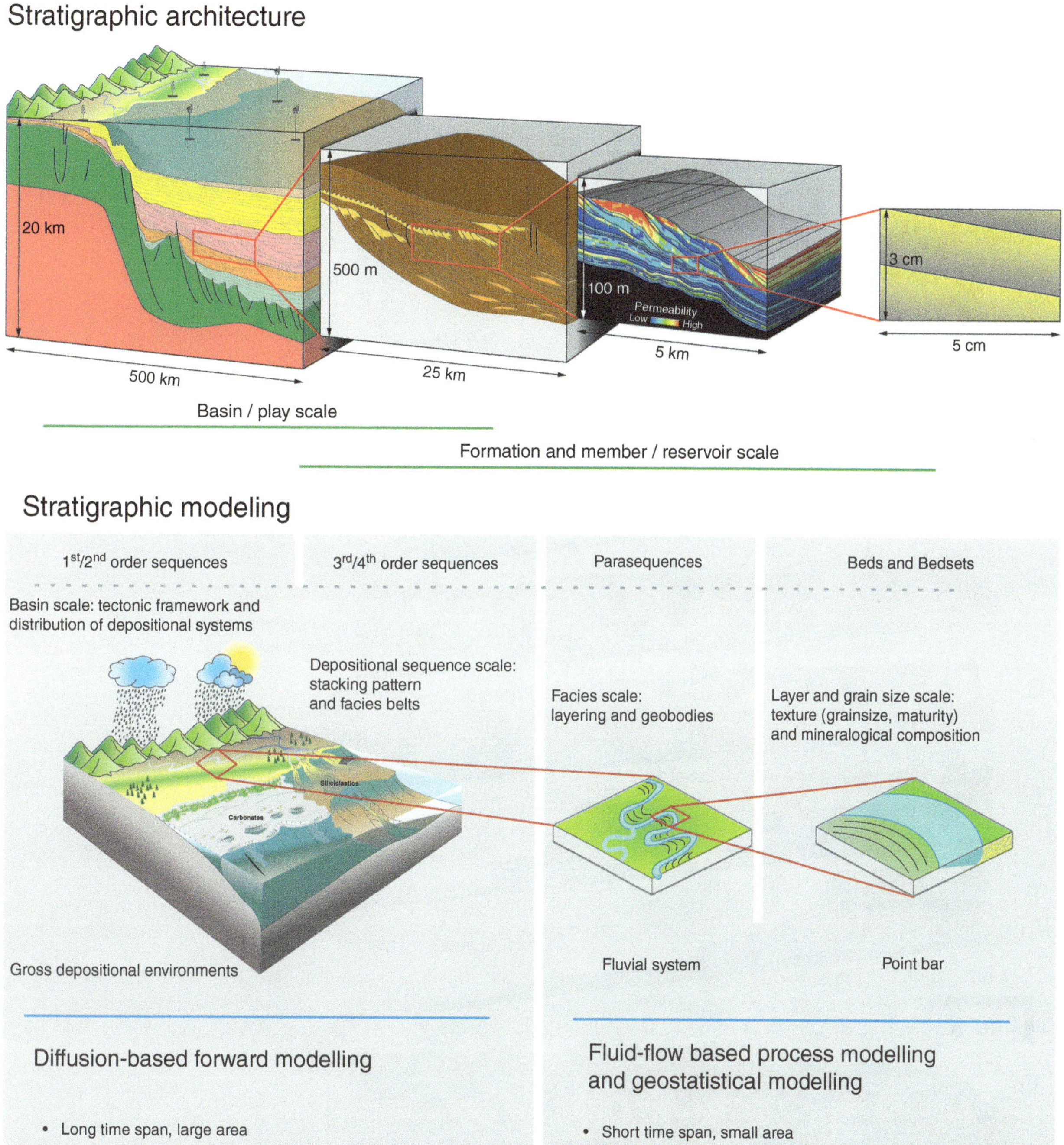

Figure 4.36 The hierarchical organisation of the stratigraphic record in different orders of sequences forms the basis for a nested stratigraphic forward modelling approach. Diffusion process modelling is most applicable at the second to fourth order sequence scale, whereas fluid-flow process modelling and geostatistical modelling are more applicable at the higher-order (<20 ka) scale. Combined, these models are capable of dealing with different scales of heterogeneities. The models at the larger scales can ideally provide boundary conditions for more detailed models at smaller scale. (Modified from van Buchem 2021).

(Figure 4.36). For the understanding of the stratigraphic organisation of, for instance, the avulsion patterns of a fluvial system, or for the progradation of a river-dominated delta, (Figure 4.37) the fluid-flow process models are best. For the modelling of a complete depositional system over a longer time range (20 kyr to millions of years), the diffusion/advection process modelling is more appropriate.

Fluid-Flow Process Modelling Example

A good example of a fluid-flow process-based approach to study 3D stratigraphic complexity is the study by Willis et al. (2021) of a river-dominated delta. This particular model addresses progradation at a stable sea level followed by an imposed linear sea-level rise (Figure 4.37). The model gives a good impression of the internal autogenic stratigraphic architecture at the parasequence (fourth-order) scale.

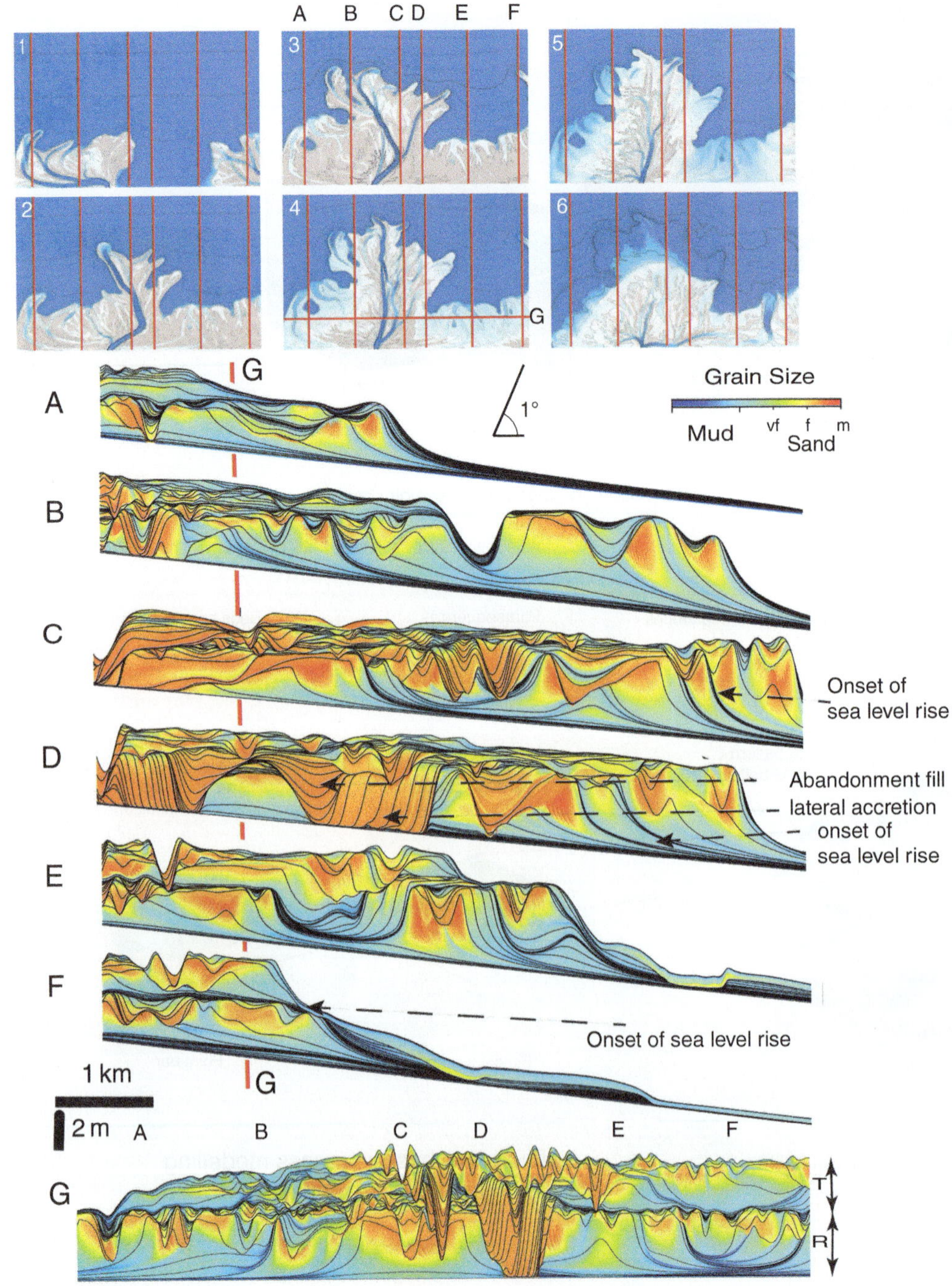

Figure 4.37 Fluid-flow process-based model of a river-dominated delta. The delta is allowed to prograde under conditions of constant sea level (up to maximum regression, map 3) followed by a rapid sea-level rise (to maximum transgression, map 6). Dip cross-sections A–F and the proximal strike cross-section G are located in maps 1–6. Colours represent grain size with red being medium sand and blue being mud. Black timelines represent every 100th modelled bed. Internal architecture of the sequences is dominated by autogenic behaviour, such as channel migration, point bar accretion, mouth bar progradation and infill of embayments. The allogenic control of linear sea-level rise after the time of map 3 is clearly visible as aggradation and transgression, especially on the delta margins (see strike cross-section G and dip sections A and F). Willis et al. (2021) / Society for Sedimentary Geology. Model parameters: Dimensions: 30 km by 10 km. Initial basin slope: 0.57°. Maximum basin water depth: 12.5 m. Magnitude of transgressive sea-level rise: 4 m. Run time: 65 simulated days.

Diffusion/Advection Process Modelling Example

Diffusion-process-based FSM has been used to illustrate some diagnostic differences in the stratal geometries generated by different controlling variables in a shallow marine siliciclastic delta system (Figures 4.38 and 4.39) and a shallow marine tropical, T-Factory, carbonate system (Figures 4.40 and 4.41). The software that has been used is DionisosFlow™ (Granjeon 2014).

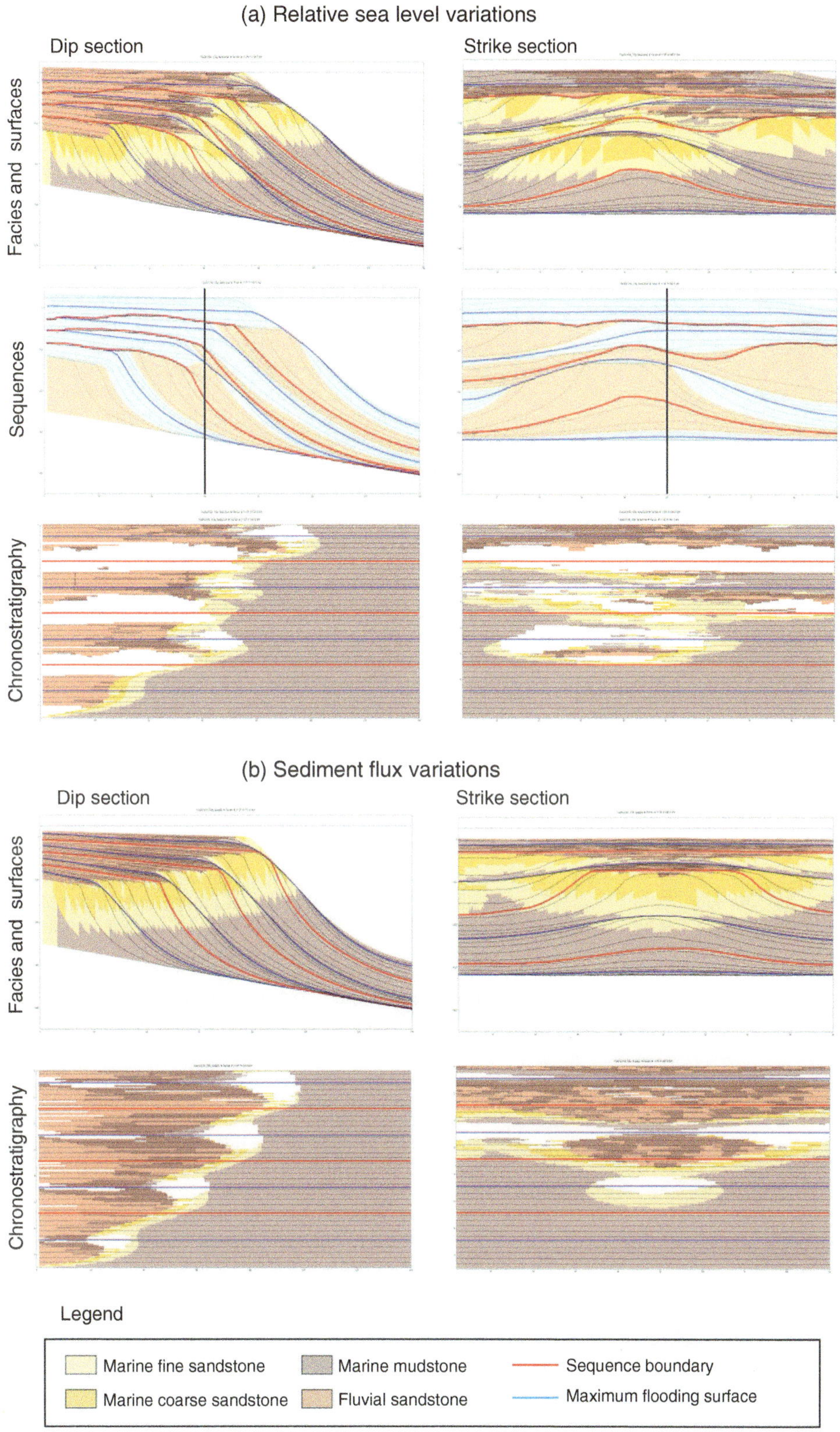

Figure 4.38 (a) 2D cross sections from 3D stratigraphic model of a point-sourced, fluvial-dominated delta under control of cyclical eustatic sea-level change (amplitude 25m). (Courtesy D. Granjeon, IFPEN). The black vertical line marks the intersection of dip and strike sections. The changes in accommodation space result in unconformities bounding depositional sequences (red) marked by both erosional truncation during relative sea-level fall and onlap during relative sea-level rise. Maximum flooding surfaces are in blue. The related gaps in the stratigraphic record affecting the shallower water areas, and the time relationship between these and the correlative conformities in the deeper-water sections are most clearly shown in the chronostratigraphic charts. The strike sections show clearly that the paucity of accommodation space in axial sections causes offset stacking as delta lobes are driven to low points at the falling stage, with the implication that sediment supply rates are laterally variable even though the overall sediment input rate is constant. (b) Same model parameters as in (a) but with a constant sea level, steady subsidence and cyclically varying sediment supply. Red lines represent times of maximum sediment input and blue lines minimum sediment input. The local hiatuses are related to sediment starvation and condensation and to reworking at the shelf edge rather than variations in accommodation space. The key stratigraphic difference with (a) is the lack of onlap and truncation. The strike section also shows that delta lobe switching is absent in this particular model because of the lack of a fluctuating accommodation space control (i.e. relative sea-level changes). This would however not be the case if subsidence rates were lower.

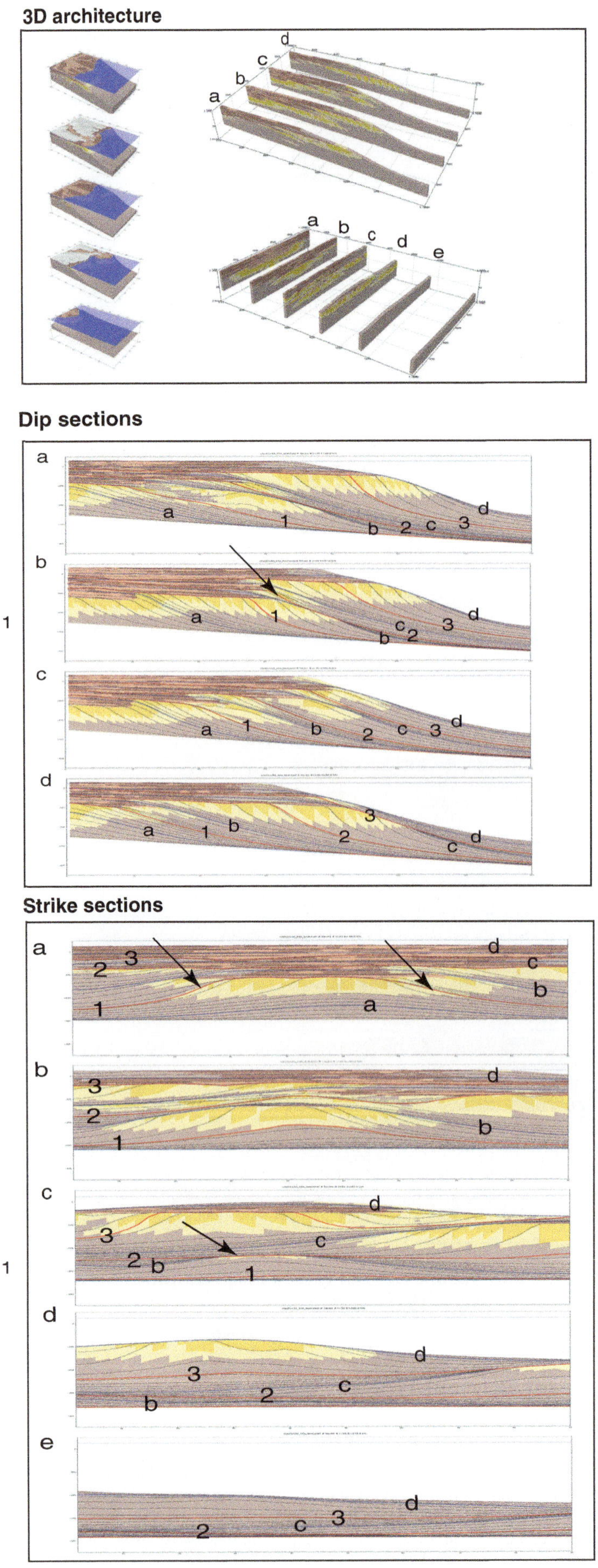

Figure 4.39 2D depositional dip and strike cross sections through a forward stratigraphic model of a siliciclastic delta system. (Courtesy D. Granjeon, IFPEN). Legend as in Figure 4.38. The model is based on sinusoidal sea-level change, a constant subsidence rate and a constant rate of sediment supply. Sequence boundaries (SBs) in red are numbered 1–3 and maximum flooding surfaces (MFSs) in blue are lettered a–d. Both are picked at maximum rates of sea-level change (falling and rising inflexion points, respectively). The delta shifts laterally to low points in the model as driven by changing accommodation space and therefore shows an 'offset stacking pattern'. The simple, and fundamental, offset pattern of lobe deposition has important complicating effects on the recognition of key stratal surfaces on both dip and strike sections. 3D diagrams and strike sections are important as these features are not visible on single dip sections. The set of dip sections shows clear anomalies with respect to classical patterns. For example, in dip sections (a and b), SB1 and SB3 have a more or less classical expression whereas SB2 (arrowed) is difficult to pick from either stratal geometries or facies due to a delta shift to the right during the lowstand. (See strike section (b)). In strike section (a), the SB1 occurs at the top of the deltaic sands whereas on the right hand side of the same delta lobe it occurs within the prograding sands (arrowed) again reflecting autogenic delta switching. The picking of the key surfaces based on facies change and stratigraphic geometries would be particularly difficult in the strike sections. For example, SB2 resembles a maximum flooding surface on strike section (c) (arrowed). In (b), SB2 (arrowed) is not obviously a key surface due to this delta switching. A note on key stratal surfaces: sequence boundaries (red) are in theory formed during the maximum rate of RSL reduction, which, at a slow subsidence rate is also the falling inflexion point of the sea-level curve. That is exactly how the red SBs are picked in this model. Empirically the hiatus at sequence boundary erosion surfaces will occur at roughly this time depending upon the erosional process, for example, knick-point migration up a river, lowering of wave base on the shelf, and so on. Likewise, the precise timing of the maximum flooding surface in facies terms at around the rising inflexion point depends on the spatial distribution of both A and S, and in a point source, S will clearly vary away from the river mouth. The position assigned to the maximum flooding surface in the rock record also depends upon the characteristics used: maximum landward position of the coast versus deepest water offshore sediment, or most condensed stratigraphic sequence for example.

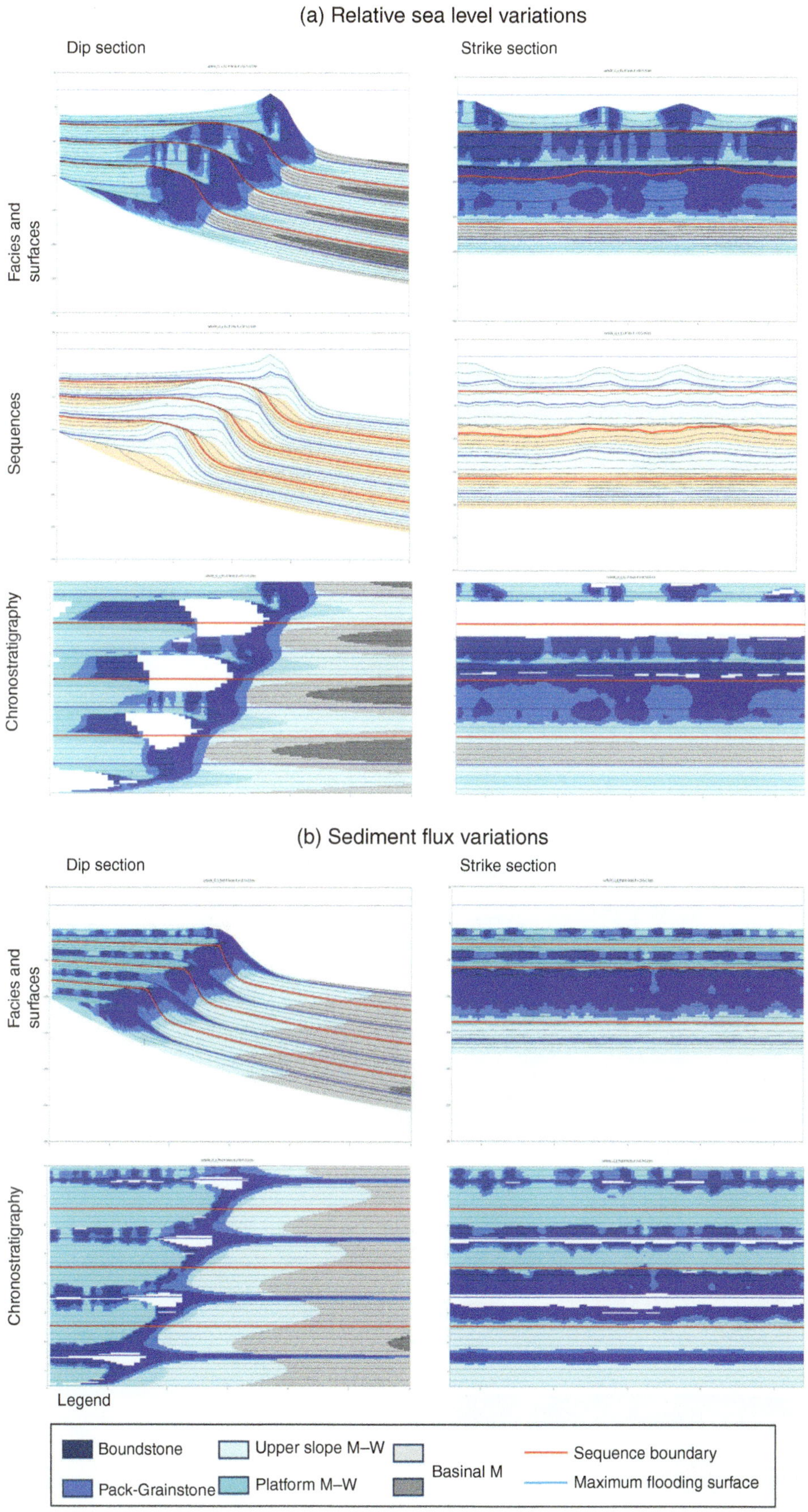

Figure 4.40 Forward stratigraphic model of a prograding, tropical, land-attached carbonate shelf system (Courtesy D. Granjeon, IFPEN). (a) 2D expression of a relative sea-level-driven sedimentation pattern, displaying facies, surfaces, sequences and a chronostratigraphic (Wheeler diagram) expression in both dip and strike direction. (b) 2D expression of a sediment flux variation driven sedimentation pattern, with constant subsidence and stable sea level, displaying facies, surfaces and a chronostratigraphic diagram in both dip and strike direction. Note the more pronounced hiatuses around the sequence boundary in (a) caused by the amplitude of the sea-level fluctuations (25 m) and the building of topography around the shelf margin, protecting a muddier lagoon in the back. This contrasts with the more continuous record in 9b, where the reduced topography shows a more equal distribution across the shelf of the boundstone-dominated reefal facies.

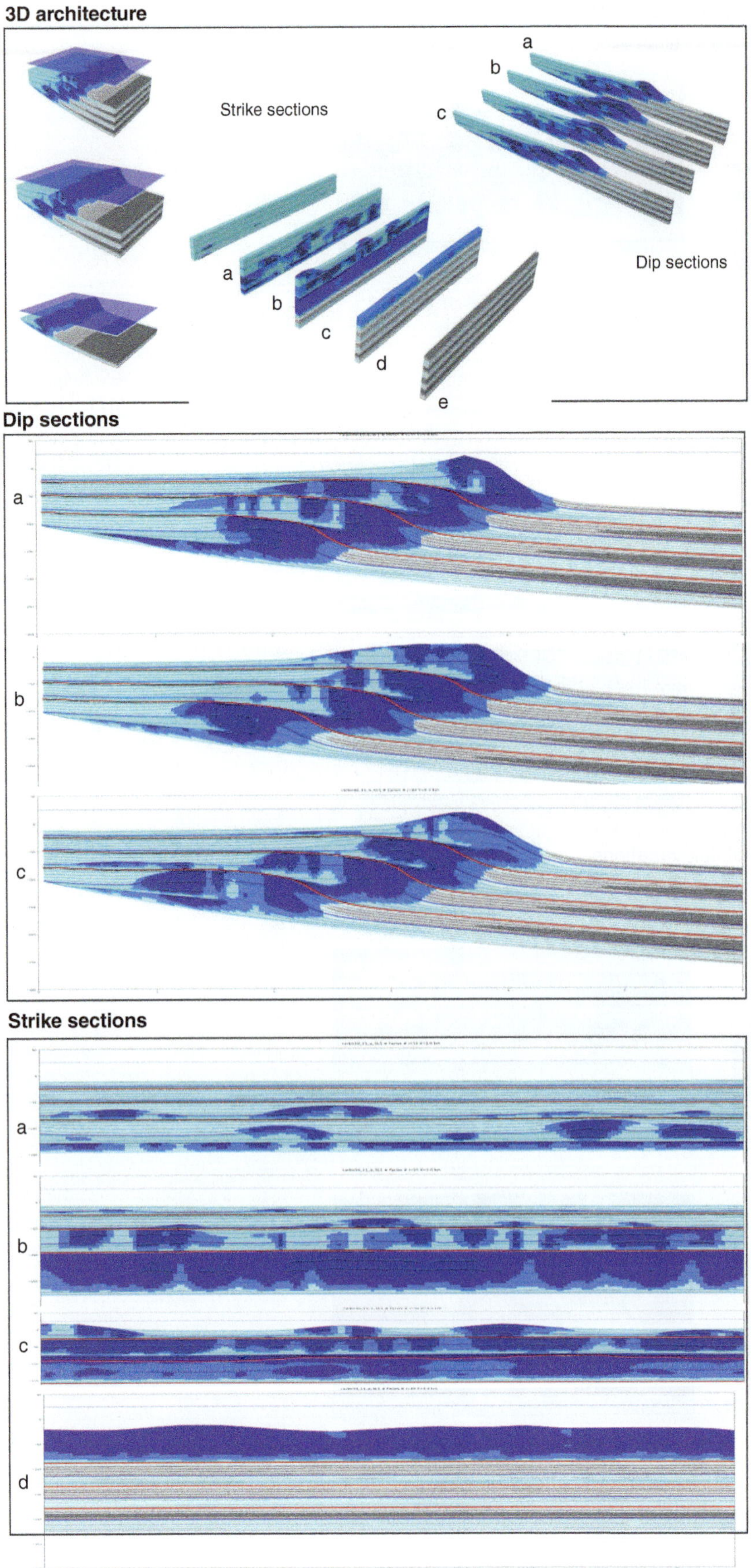

Figure 4.41 The 3D stratigraphic architecture in parallel dip and strike transects of a forward stratigraphic model of a prograding, tropical, land-attached carbonate shelf system (Courtesy D. Granjeon, IFPEN). Legend as in Figure 4.40. Note the similarity in architecture of the dip profiles, which explains the lateral continuity in the strike sections. The carbonates are mostly formed in situ in this shelf system, the reef builders along the margin consolidate early, whereas part of the loose and fine-grained sediment formed in the lagoon is transported offshore, through tidal currents and occasional storms, where they settle out of suspension or are further transported by long-shore currents. The depositional geometries of this carbonate system are thus in stark contrast with the point sourced, fluvial-dominated siliciclastic delta system illustrated in Figures 4.38 and 4.39.

Two scenarios have been tested, the impact of eustatic sea-level variations (25 m magnitude) and the impact of variations in the sediment supply/production. Compared to the fluid-flow process modelling, these case studies deal with a much longer time interval (400 kyr), which is organised in four sea-level cycles of 100 kyr. The mean sediment thickness is around 100 m. There is a difference in the parameters for the simulated domain length: (a) clastics: Lx = 150 km, Ly = 100 km and resolution = 2 km; and (b) carbonate: Lx = 7.5 km, Ly = 5 km and resolution = 100 m. The autogenic behaviour of fluid-flow process-based example shown in Figure 4.37, at parasequence scale, can be envisioned as embedded within this type of larger-scale diffusion process-based models.

The siliciclastic example very clearly shows the 3D stratigraphic complexity of the point-sourced, fluvial dominated delta system. It highlights the challenges for a high-resolution interpretation based only on 2D transects, and without proper time control, which are both common features of datasets in such depositional systems (Figure 4.38). Its overall architecture is in stark contrast with the carbonate example, a land-attached, tropical (T-factory) shelf system, that is characterised by in situ carbonate production, with a wave-dominated linear coastline. Sea-level variations remobilise sediment in siliciclastic systems, and create a much more heterogeneous architecture, with lateral facies changes and hiatuses than do variations in sediment supply (Figure 4.38). A similar pattern is observed in the carbonate case study (Figures 4.40 and 4.41).

4.5 The Sedimentary Record in Time – The Processes of Deposition and Preservation

In the preceding section, we highlighted the importance of considering the representation of time by both the sedimentary rock successions and the stratigraphic surfaces, and the spatial realtionships between these. Here, we focus on the array of processes operating over a wide range of time scales that control the incompleteness of the sedimentary record in any given succession (e.g. Miall 2016). We review depositional, and, in particular, preservation processes from the viewpoint of seven different timescales of accumulation (Figure 4.42) with an emphasis on siliciclastic examples. We also use this more philosophical approach to highlight some ongoing debates within stratigraphy.

Average net rates of accumulation of ancient sediments are on the order of millimetres per year. In contrast, modern sedimentation rates, although highly variable, may be tens of centimetres or even metres per year. Net accumulation is 2–3 orders of magnitude slower than deposition. An example is illustrative. The average duration of an ammonite bio-zone might be around 1 million years, and the rocks representing that zone might be 10–100 m thick. However, if the deposits accumulating in a year could each be 10 cm thick, then the accumulated sediment could be recording, on average, just one out of 1000–10,000 years. The bulk of geological time in any given vertical succession is represented by the bedding planes and not by the sediment. A representative column of preserved sedimentary rock with a thickness of 100 m to 1 km may thus contain beds that represent the order of just 10^{-3} to 10^{-5} of the geological time that elapsed during the accumulation of that entire sequence. This is highly simplified, but the idea that the geological tine recorded in sedimentary successions is more in the gaps (i.e. stratigraphic surfaces) than in the sediments themselves (the beds) is an old but sometimes forgotten concept. As Derek Ager (1993) wrote: 'Sedimentation in the past has often been very rapid indeed and very spasmodic. This may be called the *Phenomenon of the Catastrophic Nature of the Stratigraphical Record*'.

The accumulation of sediments needs to be considered across many orders of magnitude in time: from 10^{-3} years to 10^{9} years. Documentation of processes in modern depositional environments normally only stretches from 10^{3} to 10^{2} years, perhaps sometimes up to 10^{4} years. Our understanding of the effects on deposition of external factors, such as changes in sediment supply, climate and sea level (and their mutual interactions), is built on just a few hundred years of historical data. Predictions and interpretations are therefore based on extrapolations from these limited observations using conceptual models.

Data for net positive accumulation rates from a selection of siliciclastic sedimentary facies, uncorrected for compaction, were summarised by Sadler and Jerolmack (2015). The plotted data (Figure 4.42) clearly show that both thicknesses and sediment accumulation rates systematically *decrease* as the time period over which they are measured increases. Average accumulation rates measured over short time periods are high and those measured over long periods are low. Intuitively this is not surprising. We are used to recent deposits in modern settings or cores, which are comparable in thickness (e.g. 10 m) to units seen in outcrops which are the stratigraphic record of up to a million years. An obvious explanation is that the aggregate duration of gaps in the record becomes greater in sedimentary sequences, which represent longer periods of time. Preservation probability, or 'survivorship', decreases inexorably with time over most time scales.

This power law relationship means that preservation behaviour can be modelled as fractal: gaps of all sizes are present, with many shorter time breaks (Miall 2015).

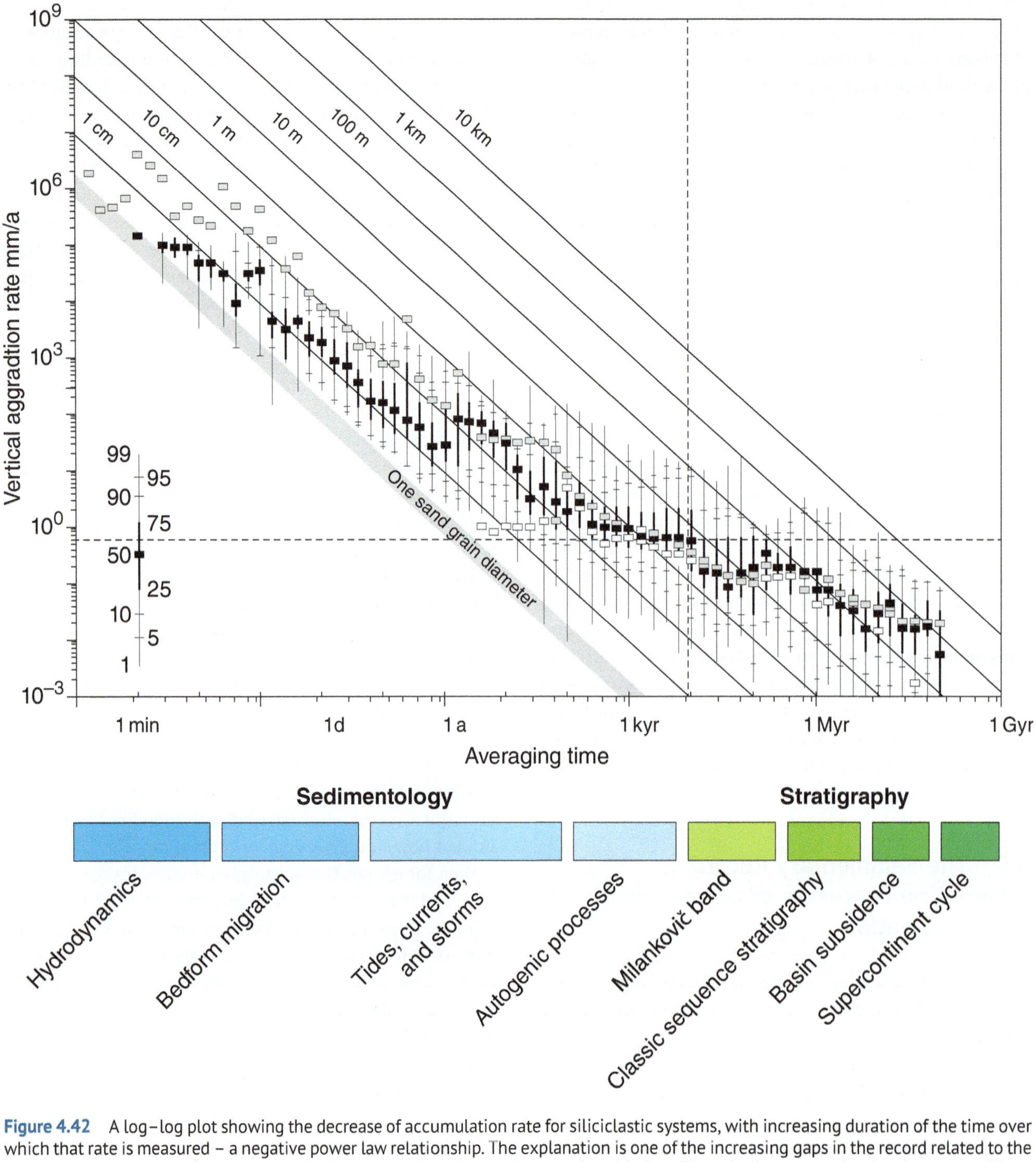

Figure 4.42 A log–log plot showing the decrease of accumulation rate for siliciclastic systems, with increasing duration of the time over which that rate is measured – a negative power law relationship. The explanation is one of the increasing gaps in the record related to the removal of sediment by a 'preservation filter' of progressively increasing severity with time. The spindles summarise data in groups of 5 per log cycle for the linear accumulation rates. The data are 14,206 determinations from floodplain accumulations (grey squares), 23,554 shelf accumulations (black squares), and 10,179 continental slope accumulations (open squares). Also shown are lines of equal thickness, which serve to illustrate the different combinations of rate and duration that can be responsible for a sedimentary section of a given thickness. Sadler and Jerolmack (2015) / Geological Society of London. The schematic durations of sedimentary and stratigraphic processes are shown following Miall (2016).

However, the relationship does not require fractal behaviour. It could also arise through periodic behaviours such as cyclic sedimentation on different time scales or by a purely random-walk process such as Bernoulli trials (coin-tossing).

In detail, the global data presented by Sadler and Jerolmack (2015) suggest two different power laws (Figure 4.42), a negative slope in log/log space of −1 up to time scales of around a day and a negative slope of −0.5 for longer times. A lesser negative slope indicates that thicknesses increase progressively more slowly with increasing time spans: the aggregate duration of the gaps is proportionately longer. The 'preservation filter' is increasingly effective in reducing the survivorship of deposits over time.

The rate of decay of sediment accumulation with elapsed time also varies between depositional settings. Deep-water sediments show a slower decrease in accumulation rate with increasing time (less missing record, less erosion) than, for example, fluvial sediments (Sadler and Strauss 1990). Fluvial sediments are characterised by the preservation of a few, more-or-less deeply erosive units deposited relatively quickly with large periods of unrecorded time: when the depositional surface is near or above base level (stasis). The key controlling variable in these two cases is the availability of accommodation space to allow preservation (Section 4.3).

Facies models as presented in this book typically derive from modern environments constructed from data on the left-hand side of Figure 4.42 with timescales of less than 10,000 years. These models are applied to deposits plotting on the right-hand side: sediments that have been preserved over tens to hundreds of millions of years. As Sadler and Strauss (1990, p. 471) warn: 'Consequently, when estimating completeness, we must be wary of the popular maxim that stratigraphical questions can be answered from experience in modern depositional environments'. This does not refer to process-based facies interpretations, which are reliably guided by 'experience in modern depositional environments'. Nonetheless it is a thought-provoking conclusion.

Tipper (2015) emphasises that many siliciclastic sedimentary systems (rivers are a good example) are, for much of the time, in *stasis*. Depositional and erosional processes such as currents, floods and storms are in action but, in a longer-term accumulation and preservation sense, nothing stratigraphically significant is happening. Sand bars are simply shuffling back and forth. Stasis is a consequence of the fact that, as parts of sediment routing systems, siliciclastic sedimentary systems can spend much of their time in quasi-equilibrium. Their sediment balance is neutral: transport-in broadly equals transport-out, and sediment is simply bypassed down the sediment routing system ('throughput systems' of Levell et al. 2020). These quasi-equilibrium conditions, for example, a meandering fluvial system or barrier island coast may shift abruptly to a new quasi-equilibrium state (e.g. braided streams or a strandplain, respectively) when intrinsic thresholds, such as channel gradient or amount of available sediment or key climatic variables, are exceeded. The related 'landscape system' approach to sedimentology is developed further in this book in Chapters 5 (fluvial), 7 (aeolian) and 14 (glacial). The notion of stasis is also applicable to carbonate sediments albeit some carbonates such as shallow water tropical carbonates are generally characterised by early cementation (see Chapter 11), making them liable to being preserved even when they do get out of equilibrium with changed conditions.

The Sadler plots also raise the question of which information may be missing from the record provided by those sediments that happen to be preserved? Do the sediments, the surfaces between beds, or perhaps the periods of unrecorded time, have any special significance, or are they simply a representative sampling of the range of conditions and processes at the depositional site?

Sadler and Jerolmack's data in Figure 4.42 describe 1D rather than 3D stratigraphic completeness: they are derived from one-dimensional vertical sedimentary sections. Clearly, during times of non-deposition or erosion at any given point, it is almost certain that sediment is being deposited somewhere else. In fact, forms of 'reciprocal sedimentation' whereby depositional sites alternate in time and space are ubiquitous in most sedimentary systems. Examples are currents seeking out the topographic lows between depositional lobes, with non-deposition on the highs or shifting of accumulation space between deep and shallow water in relation to relative sea-level change. 'Completeness' in terms of sampling geological time not only needs a specific timescale to be meaningful, it also requires a geography, as captured for example in two or even 3D 'Wheeler' diagrams (Figure 4.5; Wheeler 1964; Qayyum et al. 2015). Such diagrams plot stratigraphy against geological time rather than depth, thereby emphasising the gaps in the record and displaying their relationship to time recorded by sediments. The stratigraphic completeness of any single vertical section will obviously be less than that of a 3D stratigraphic unit, such as a formation or a basin-fill.

4.5.1 Milliseconds to Days – Hydrodynamic and Biological Processes

On the timescale of seconds or less, individual grains are selected for further transport or for deposition by virtue of their weight, size and shape. Such selection can be sufficiently precise to allow interpretations of a depositional process, or even an environment, based on texture (grain size analysis, sorting and roundness; e.g. Dunham 1962; Syvitski 1991; Pettijohn et al. 2012; Lokier and al Junaibi 2016). This classical approach is currently undergoing a modest resurgence of interest based on possibilities opened by automated digital analysis of sediment texture and grain texture and AI or machine-learning approaches.

The sediment that is selected for deposition is normally just a fraction of everything that is being transported. This fraction will depend largely on the transport mechanisms involved immediately prior to deposition (suspension, traction, saltation, etc.), the selection of specific grains by the bed itself (grain-fitting for sand or stickiness for organic-coated grains) and what happens after deposition (any re-erosion). The flow mechanism, velocity, turbulent structure and rates of acceleration or deceleration will all determine which fraction gets deposited and when this happens. If a flowing fluid (water, air or other gases such as in volcanic flows) decelerates rapidly to a standstill such as happens in some mass flows (Section 12.3.1), much of the entrained sediment will be deposited together and the resulting deposit will then be a representative sample of the transported load.

More commonly grains will settle out in an organised way according to settling velocity. Single grains settle when settling velocities exceed the upward component of the fluid velocity. Settling velocities depend on diameter, shape, the grain and fluid densities, and the fluid viscosity. Carbonate grains are somewhat special by virtue of their complex morphologies and in some cases densities (de Kruijf et al. 2021). For quartz spheres in water, settling velocities vary by four orders of magnitude (0.0001 to 1 $m\,s^{-1}$) for spheres that range from 0.01 to 10 mm in diameter (three orders of magnitude). Velocities to maintain a suspension are roughly 100 times greater in air (see Grotzinger et al. 2013, and Section 15.2.1, for a review of the terrestrial planets). The velocity, which maintains a suspension or prevents settling is, in fact, the time-averaged upward component of turbulence (which has a value close to that of the 'shear velocity' – the square root of the shear stress divided by the fluid density). The grains that settle together at any particular velocity will be those whose diameter, density and shape determine that they have approximately equivalent settling velocities. Complications to this simple picture exist in the real world settling of groups of grains and shearing flows (Amy et al. 2006) but, for example, dense grains, such as zircon, ilmenite or other heavy minerals, will have smaller diameters than the quartz with which they settle. Tabular grains, such as micas, will have relatively smaller masses. There are some complications with this picture due to the time intervals between the stronger upward flows in eddies. These 'waiting times' tend to be longer for the bigger eddies that suspend the coarser fraction. This means that the inertia of the coarser grains also plays a role in settling and can allow the larger grains to settle out even if they could in principle remain in suspension simply from average velocity considerations (Blatt et al. 1980). Concentrations of grains all settling together cause water to be forced upwards between clusters of settling particles, which can also prevent further settling: a partially 'fluidised' bed. Also, close to the bed, mutual collisions may cause shear dispersion of the grains further impeding final settling.

At the bed, some grains are selected for deposition from the shearing 'traction carpet' at the base of a flow by their shape and size. Those that can fit in spaces between grains already forming the bed are less likely to be subsequently removed, particularly in 'plane-bed' conditions. This property of grain selection can be seen for example in the ubiquitous 'York flagstones' (Late Carboniferous deltaic mouth-bar deposits) on streets in the United Kingdom where long-axis grain orientations result in a primary current lineation defining sets of stepped splitting-planes on the slabs. Grain selection by the bed allows for relationships between texture and process, which are exploited in facies interpretation (e.g. Collinson and Mountney 2019; also, for example, in aeolian dune cross-stratification, see Chapter 7).

The amount of sediment transported is a steeply increasing function of velocity. In fact, velocity raised to a power from three (the theoretical minimum) to six (empirical). Exceptionally high-velocity flow events with the competence (a measure of the maximum grain size that can be moved) and capacity (the volume that can be moved) to transport the coarser calibre grains would therefore be expected to dominate transport volumes, and products of these events will likely be over-represented in both erosion surfaces and deposits.

In contrast, the material forming the dissolved load of rivers, which comprises roughly 20% of the total annual river load (i.e. globally 3.8×10^9 out of 19×10^9 tonnes [Allen 2017]) may not be evidenced at all in direct deposits. Dissolved load is a particularly large fraction when it is derived from limestone terranes and tropical areas with intense chemical weathering.

The concept of alternating erosional surfaces and deposits in both time and space applies to all types of stratified sedimentary deposits even at the bedform scale (Figure 4.43: see also Section 7.3 for examples from aeolian cross-stratification, Paola and Borgman 1991, and Reesink et al. 2015).

Erosion and bioturbation both work to partly or completely remove and/or obscure the primary depositional record. Bioturbation can make any primary stratification un-recognisable and, therefore, often un-interpretable (Section 3.2.5). In that sense at least the depositional record is then not preserved.

Figure 4.43 Tidal cross-stratification from the modern Oosterschelde Estuary ('Bouwput de Schaar') in the SW Netherlands showing tidal bundles of mud drapes formed during slack water from diurnal cycles that themselves occur in spring/neap groups of week-scale duration. Bioturbation by echinoids is visible in the bottomsets. The section forms part of an abandoned tidal channel that was c. 17 m deep and infilled in the 17th or 18th century. (Photo by Bruce Levell.)

It might be expected that erosion occurs at a similar current velocity as that at which particles settle. This is however far from generally true, particularly for fine-grained cohesive sediments (Grabowski et al. 2011). Mud becomes denser and more cohesive as it de-waters under self-compaction, and, as it does so, it becomes more difficult to erode. Standard curves (Dade et al. 1992) suggest a 5–8-fold increase in threshold erosion velocities and a 100-fold decrease in erosion rates for compacted muddy sediments related to de-watering consolidation simply due to ageing at the depositional surface. De-watering consolidation may be estimated from trace fossils and ichnofabrics within the mudstone.

Some sediments serve to protect the units that happen to be beneath them. An example is the formation of pebble or granule pavements by wind deflation in desert environments. The wind-blown removal of sand from sandy gravel leads to a surface gravel armour, which becomes near-inerodible by the wind alone. River beds commonly display similar gravel armours, which may form following floods. In ancient fluvial and deltaic successions, it is common to see channel bases that have been eroded down to the level of a palaeosol or a coal seam but no further. Apparently, the peat or the soil was simply too cohesive and difficult to erode, and the channel consequently migrated and enlarged by lateral rather than vertical erosion. Other examples are hardgrounds formed by near-surface cementation, particularly in warm water carbonate settings (Section 11.1.1) or firmgrounds (Section 3.3.3.6). Biological factors such as burrowing and sediment cohesion or surface coatings due to biological materials such as extra-cellular polymeric substances (EPSs) also inhibit erosion.

Preservation of any particular bed from subsequent erosion or bioturbation can depend simply on chance, for example, that the next layer to be deposited was particularly thick: a thick succeeding layer protects those below from physical erosion simply by burial. More subtly, if a newly deposited bed remains in the zone of active bioturbation it can, in certain depositional settings, be rendered unrecognisable as a depositional unit within months. Only if the succeeding layer is thicker than the active zone of bioturbation will the deeper layer survive as a recognisable unit (Wheatcroft and Drake 2003).

4.5.2 Days to Centuries – 'Events'

Geology, as a science, is around 250 years old. Our experience in terms of scientific study and quantification of depositional processes and 'events' (floods, storms, slumps, earthquakes, etc.) is largely derived from this period. The Sadler plots suggest that over this time interval accumulation rate falls steadily with increase in elapsed time. The random walk process of preserving sediment that has been deposited reduces sediment accumulation from around 10 to 1 mm per year over a few hundred years. On average, 90% of the record disappears over a century. To understand preservation bias, such as the differential preservation of more energetic events or early cemented units, we have to understand the erosional processes of the 'preservation filter'. Deposition and erosion can occur under 'normal' or 'exceptional' conditions or any mix of the two. Figure 4.43, for example, shows a section of modern deposits preserved because they infill a relatively deep tidal channel and were therefore not (yet) re-eroded.

Transport capacity (the volume of material moved) is a steeply rising function of current velocity. It is therefore possible, or even likely, that exceptional erosion, for example, in a hurricane, is paired with exceptional amounts of (re-)deposition. Hence, it is possible *a priori* that recorded stratigraphy is dominated by high current velocity, high competence (ability to move large grains), high capacity and transport events. However, in general, we only know the thickness and texture of a bed in one or a few places and not its volume. Bed thickness in one locality may say as much, or more, about the bed's relative position along a sediment transport pathway (captured in terms like 'proximal–distal trends'), as about the total transport capacity or competence (and hence a possible measure of the 'magnitude' of the event). Moreover, preserved bed thickness, even if deemed to be a good proxy for the magnitude of a single event, may reflect the combined result of both deposition and erosion. Sylvester (2007) gives an illuminating summary of the statistical and interpretational complexities of bed thickness data from two turbidite successions. In general, thinner beds are more susceptible to modification, including complete removal by erosion. They lose their separate identities both through amalgamation with younger beds when intervening mudstones are eroded and through bioturbation. The preservation filter will, in general, bias the record towards the thicker beds particularly if accumulation rates are low enough and/or pre-depositional scour is high enough to remove the inter-storm/turbidite and minor-event deposits (Weedon et al. 2018).

Deducing process energy from bed thicknesses is non-trivial, even if there is no erosion (Figure 4.44). Patterns of bed thickness after modern storms at a given location can sometimes be related to storm magnitude and track (e.g. Tian et al. 2019) but other studies suggest that it is not the storm magnitude itself but specific coincidences of storms with currents that control the maximum bed thickness (Wiberg 2000). Both empirically and theoretically storm bed thickness is strongly controlled by water depth (Niedoroda et al. 1989). Therefore, to reconstruct relative storm magnitudes from thickness data, relative water depth and current patterns need to be independently determined.

Seeking a direct relationship between event magnitude and depositional thickness is further complicated by 'threshold effects', as illustrated when thick deposition is favoured not by transport power or proximality alone but when some sediment store happens to be full and available for re-transport. For example, Wheatcroft (1990) working on the Eel River shelf off northern California showed that rainstorms and associated river floods randomly encounter full or empty stores of sediment in the river flood basin, depending on the magnitude of preceding events. The volume or thickness of the resulting storm flood deposits on the shelf may therefore be unrelated to rainfall totals or flood magnitudes but be controlled simply by timing. Elsewhere, a severe offshore storm may occur during the dry season when the coast is starved of sediment. The first storms of autumn are supplied by the summer beach profile, whereas stronger winter storms attack already depleted beaches. Hence the maximum volume of transport may occur not during the most energetic events but during relatively large events that happen to be timed well to coincide with the highest availability of sediment. Timing of an earthquake capable of triggering submarine slope failure with respect to the existence of slopes that have, or have not, reached their limit of stability is another example (Lee 2009). In summary, state changes that relate rather to inheritance from previous events and/or to thresholds being exceeded may be of equal or greater importance than event magnitude in controlling deposition, erosion and preservation even at the bed scale. And these state changes may have little direct relation to specific event magnitudes.

A possible approach to understanding bias in the preservation of events might be to consider whether there is evidence for the same spectrum of event sizes in the ancient record as there is in the modern analogue. Analyses of natural event magnitudes and frequency or recurrence interval suggest that these follow particular distributions. Some aspects of natural systems are governed by normal distributions that occur symmetrically around some mean value. Magnitudes of many natural processes, such as storm intensities, are not so distributed. There are fewer more powerful storms, many more minor ones, and at some arbitrarily defined lower meteorological limit, an event is no longer termed a storm. For many sedimentological processes, it is the log of the physical event magnitude (e.g. wave height, and flood discharge), rather than the magnitude itself, that is normally or near-normally distributed (e.g. a Weibull distribution). Typically, log-normal distributions result from combinations of processes whose effects are multiplicative rather than additive.

Figure 4.44 Interpreting bed thickness trends. A turbidite unit comprising Bouma C–E divisions with two sets of five beds successively thinning upwards. Bedding plane surfaces of the three successive beds with climbing ripples in the upper set display identical ripple crest orientations (i.e. the same paleocurrent direction). It is tempting to relate these sets of beds to either a single event with pulsatory but declining flows or a series of very closely related events, whereby the sediment volumes delivered to this point successively decreased, accompanied by a reduction in current power. Three such units occur within 4 m of stratigraphy at this locality in the Namurian Bude Sandstone at High Mear Cliff, Bude, Cornwall, UK. Possible interpretations, under the assumption that thickness is related to volume, include decreasing volumes liberated by successive headwards retreating slumps or over-topping of a levee system by successively weaker pulses of a channelised turbidity current. Testing between these and other plausible hypotheses depends on context. (Photo by Bruce Levell.)

In principle, using such distributions, an extrapolation of event magnitude from historical to geological timescales should be possible. At some level, there will be maximum possible event sizes even over geological time, for example, in storm energy related to feasible terrestrial wind velocities or rainfall intensities. Frequently, the data do not exist to determine the magnitudes of the most extreme events close to these limits (Geist and Parsons 2014). Furthermore, the most extreme events with the greatest erosion and/or deposition are likely to be due to a coincidence of the timing of multiple events: a major wind-induced storm from a particular direction with a high tide (a storm surge) or a storm synchronous with a tsunami. This problem is very familiar to designers of offshore structures, who need design criteria for 'conjoint probabilities' of extreme events in, or even, beyond the tails of the probability distributions of the observed magnitudes of the separate events (Jonathan and Evans 2013). The question of what is exceptional and what is normal in forming the preserved depositional record over geological time is not simple.

As a practical example of the difficulties of deciphering what is exceptional and what is normal, consider the hundreds of metres thick, decimetre cross-bedded Precambrian Applecross Formation of the Torridon Group in the Northwest Highlands of Scotland. Paola et al. (2018) suggested that these sandstones represent normal sedimentation on the basis that there is nothing to differentiate the preserved sedimentary structures from those seen in sediments of modern sandy braided streams after a modest flood. This may well be the case, but how can we test that view? To a large extent in the clastic sedimentary record 'What is available to interpret is constrained by what was available to be deposited'. For example, the maximum velocities of the flows that deposited these sandstones cannot be determined from grain size unless all sizes were available to the flood. Decimetre-scale dunes are not particularly diagnostic, since they are stable over a large range of flow conditions (Collinson and Mountney 2019). Even if such indicators were present event magnitude assessment is problematic even for historic flood magnitudes (Baker 2008). Sambrook Smith et al. (2010), report that channel flow parameters of large floods are hydrologically rather similar to more normal floods: once a channel reaches its bank-full state, the flow becomes wider but maintains a similar depth and velocity and therefore similar bedforms. We might therefore get clues to flood magnitude by considering the geometry of the breaks in sedimentation and their lateral correlatability. But we can rarely trace such surfaces unambiguously. So, the question: 'Is that normal or exceptional?' often remains unanswered.

4.5.3 Centuries to Millennia – Autogenic Processes

Autogenic processes are those that cause change in a depositional environment simply due to their intrinsic behaviour. No external forcing is required, even though it may be active. Lateral migration of meandering channels and switching of delta lobes as rivers find shorter, steeper routes to the sea are classical examples (Figures 4.19, 4.37 and 4.45). Autogenic processes affect both depositional and erosional landscapes (Scheingross et al. 2020), and each results in shifts in both depocentres and areas of erosion. Facies models are used to conceptualise and summarise the dynamic autogenic behaviour of depositional environments (Section 2.4). Implicit in a good facies model or a depositional model of a specific case study is, therefore, a 'preservation model' that explains which deposits have the highest probability of being preserved in the resulting 3D stratigraphic architecture (Figure 4.37).

Figure 4.45 Autogenic successions. Coarsening-upward successions from mudstone to medium-grained sandstone. Interpreted as river-dominated deltaic mouth bars. Extensive slumping within the distal mouth bar facies (immediately above the three geologists – lower right). Middle Miocene, Gangarrak, Pohon Batu, Labuan, Sabah, Malaysia. (Photo by Howard Johnson.)

Autogenic preservation biases typically occur as a result of differential erosion either due to the intrinsic lateral migration of erosional sub-environments in space or due to the intrinsic likelihoods of episodic erosional events through time. A classic example is provided by the behaviour of barrier islands. Many barrier islands form on transgressive coasts (for example, on abandoned delta lobes), where their preservation potential is extremely low. Shoreface erosion by waves down to the storm wave base will generally remove the entire barrier-lagoon system. This process typically results in the preservation only of truncated infills of the deepest sub-environment, the tidal inlets, overlain by a ravinement surface and a lag of high energy shoreface sands overlain in turn by open-marine muds (Chapter 8). Even along stable or prograding coasts, the preserved record of the shoreface is strongly affected by the rate of lateral inlet migration. The deep tidal channels of the inlets, driven sideways by longshore drift and spit accretion on their up-drift sides, erode through any prograding coarsening-upwards shoreface sediments and replace them with fining-upward, inlet-fill, tidal-channel and bar deposits (Figure 8.9). This is typical. In all depositional environments, the higher preservation potential deposits are those from the lower topographic levels: deposits associated with the deepest erosional relief, most notably channels.

There are further implications of a preservation bias in environments with deep erosion. The biota, ichnofauna and preserved paleocurrent indicators, for example, will all tend to be those associated with stratigraphic units with the highest preservation potential. Hence, estuary tidal-channel fills formed under the strongest current conditions (e.g. Choi et al. 2013) will have high preservation potential. Although mutually evasive ebb and flood tidal channels are common in such settings, the strongest currents and deepest erosion commonly occur where the ebb tide is enhanced by fluvial currents, especially during major river-flood events. Fluvial currents may maintain seaward-directed flows, even during the flood tide cycle, preventing flood-tidal currents from entering the deeper fluvio-/ebb-dominated channels. This leads to channel bottoms with a predominance of (a) unidirectional, seaward-directed (ebb tide plus fluvial currents) bedforms (Fenster et al. 2001), (b) land-derived biota, such as plant fragments and (c) reduced ichnofauna diversity and abundance due to rapid rates of deposition and/or reduced salinities. As a salient detail the deepest erosion, by far, in such channel networks (with depths 2.5–7 times the channel average) is often associated with one very specific setting: that of the 'nodes' or channel confluences (Ferrarin et al. 2018). Allen

(1967) made a plea, that has gone largely unanswered, for systematic study of the differences in variance and direction of current indicators associated with architectural elements with different preservation potential (see also Rubin's work on aeolian paleocurrents: Section 7.3).

The shallowest parts of tidally influenced coastal systems (tidal flats and emergent tidal bars) are by far the most intensively studied part of tidal depositional systems, not least because of their relative ease of access during low tide. These areas typically record the day-to-day reversing tidal currents and intervening slack water periods. Sediments deposited in these areas show great variability in facies characteristics, which forms the basis of the classical tidal flat facies model (Section 8.3.2). The preservation potential of intertidal zone deposits is, particularly, in transgressive systems, low due to wave erosion in a similar manner to retreating barrier islands. Furthermore, the physical recognition criteria classically used to identify ancient tidal flats are derived from modern, high-latitude settings. In contrast, modern low-latitude tidal flats have much greater benthic species abundance and diversity (Dittmann 2002). This results in biogenic destruction of the physical features characteristic of high latitude settings due to deep and abundant burrowing macrofauna often with further intense modification by mangrove-related pedogenic activity. In such a case, the ability to identify an environment may be compromised due to geographic or climatic bias in the data set of diagnostic criteria.

In addition to autogenic change within a depositional environment, there can also be autogenic change between depositional environments. There is a general understanding both in geomorphology and sedimentology of the importance of thresholds (Schumm 1979) and 'state changes' in geomorphic systems. An oft-quoted example is that of slope or stream-power changes in fluvial systems causing the shift from braided to meandering. Another example is glacier surging (Section 14.5.6). Further examples are ecological community changes, such as between sea grass, oyster reefs or muddy bottoms in lagoons (McGlathery et al. 2013) or algal versus coral predominance in reefs (Woodroffe 2007). The wrong event at the wrong time may trigger a fundamental re-organisation in the depositional and erosional balance in a landscape system. This principle also applies to macroscopic carbonate systems, where passing environmental thresholds may cause an abrupt change from one mode of carbonate production to another (e.g. from the coral-dominated, shallow water tropical ecosystem (T-factory) to a microbial-dominated ecosystem (M-factory) by bringing in an excess of nutrients in the system. Threshold issues are not well documented even in contemporary geomorphology or oceanography (Phillips and Van Dyke 2016; Naylor et al. 2017).

4.5.4 Millennia to a Million Years – the Milanković Band

Typical subsidence rates are hundreds of metres per million years (i.e. mm per year). The time period 10,000–500,000 years typically translates to metres to tens of metres of normal accumulation. Repeated patterns of bed thickness, grain-size or sediment type on the 5–10 m scale are particularly common in the sedimentary record and were previously referred to as 'cycles' or 'sequences' (Schwarzacher 2000). (The word 'sequence' is nowadays conventionally reserved for use in its specific sequence stratigraphic meaning.) A variety of processes, such as subsidence, isostasy, climate change, glacial sea-level variations and autogenic processes, operate on this timescale.

'Milanković ' orbital cyclicity (see also Box 13.4) was introduced above. Here, we discuss some aspects of the recognition and preservation of Milanković cycles. Orbital variations operate on the Earth system through both the direct and indirect effects of insolation changes. Examples of direct effects are those of light on phytoplankton productivity or ocean temperature on carbonate productivity (De Boer and Smith 1994). Indirect effects include the glacial/interglacial cyclicity of sea level caused by melting of ice and climatic influences on weathering and sediment supply. Periodic change could of course equally be encoded as periodic gaps in the record as by periodic deposits, for example, through changes in erosive power related to climate (glacial advance, river floods and coastal storms, e.g.) or, again, climate-induced sea-level fluctuations inducing changes in accommodation space.

In some cases, the observed sedimentary changes occur only under extreme conditions when insolation cycles of different periods interfere constructively. For example, conjoint high seasonality from the precession cycle and low polar insolation from the eccentricity cycle induce glaciations. In most cases, the sedimentary record will be in part discontinuous at the relevant timescales, and hence the record of the driving cyclicity can be expected to be disrupted by gaps, which may themselves be periodic.

Invoking Milanković cyclicity as a causal mechanism follows from observation of a periodic relationship between different durations of sedimentary cyclicity. This requires transforming measured thicknesses to durations in time. This, in turn, requires knowledge of the sedimentation rate (if sedimentation was continuous), together with the location and duration of any time gaps (if sedimentation was discontinuous) and of variation in rates (if sedimentation was continuous but unsteady [Figure 4.46]). Most studies of Milanković cyclicity are based on periodic changes in *deposition* with less emphasis on periodicity of *preservation*, that is, changes in the degree of preservation or erosion. The

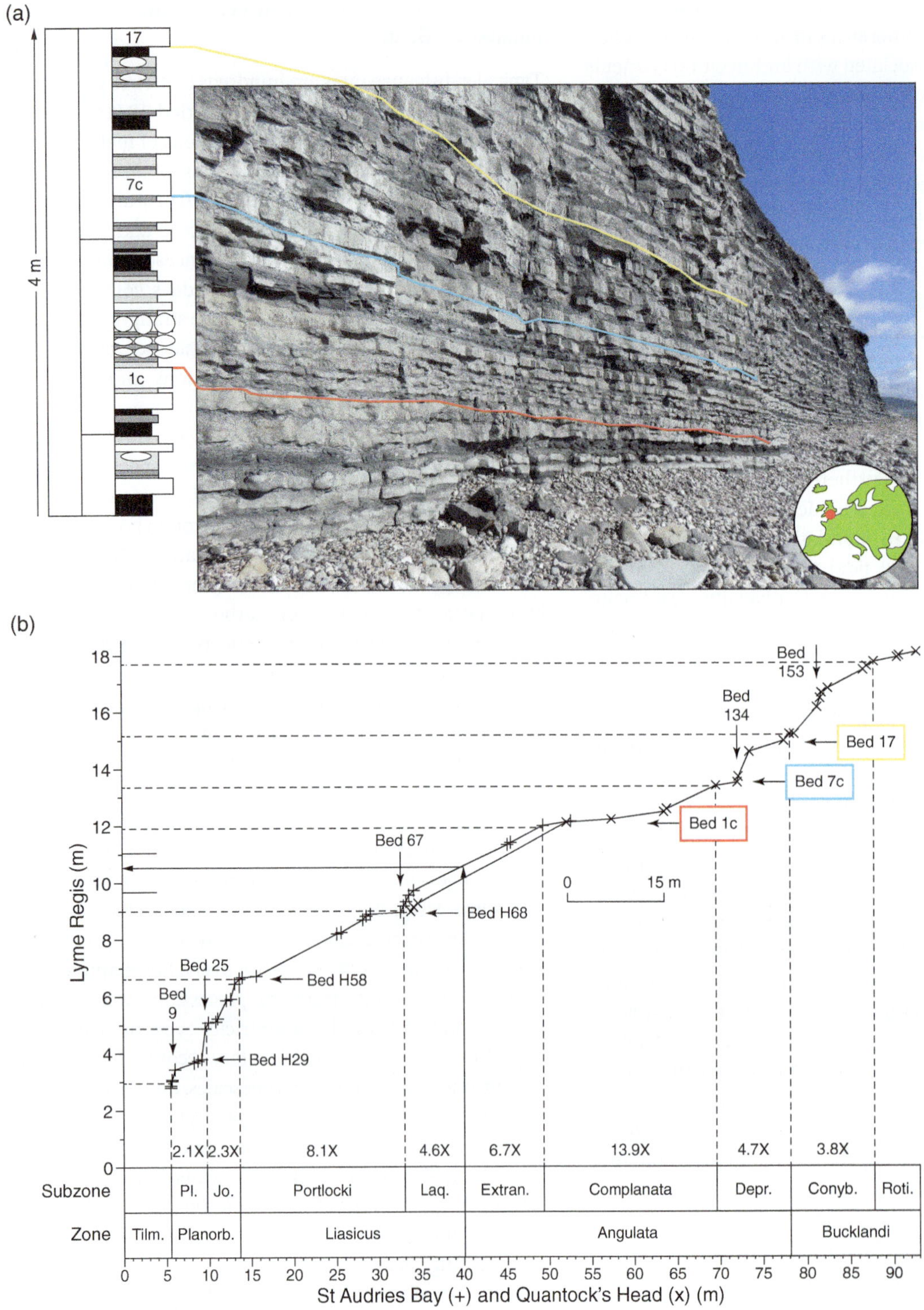

Figure 4.46 Milanković cyclicity at outcrop, Lyme Regis (Church Cliff section), Dorset, UK. (a) Rhythmically bedded, succession without obvious breaks consisting of four shallow marine facies: laminated limestones and concretionary limestones (white), marls, and organic-rich laminated mudstones (grey and black) from the Angulata ammonite zone of the Early Jurassic (Hettangian). (b) The three beds indicated are all candidates for disconformities based on comparison with the more complete section at St. Audries Bay 55 km to the north (Bristol Channel area). The plot shows the thicknesses of the same time intervals of rock in the two sections plotted against one another using detailed biostratigraphic correlation (a 'Shaw diagram'). Such diagrams highlight the relative thicknesses and importantly where a hiatus in one section is represented by the rock in another. Bed 1c in the Complanata subzone is equivalent to 15 m of section in St. Audries Bay. Beds 7c and 17 in the Depressa subzone may be associated with more modest depositional breaks. There are no obvious indications in the lithology, ichnofauna or palaeontology of breaks at these levels at Lyme Regis (photograph). The Angulata zone is likely around 700 kyr in duration. Note: these sections are tied to the absolute timescale but only indirectly via correlation with dated tuffs in Peru, which are in a different biogeographic province (Schaltegger et al. 2008). (Log and Shaw diagram are from Weedon et al. 2018) (Photo by Bruce Levell).

issue of completeness is clearly crucial. The problem is that the degree of incompleteness is difficult to quantify. Meyers and Sageman (2004) argue that gap presence and duration can be detected and estimated using frequency-domain analysis of the depth domain signal to detect missing cyclical segments, that is, breaks related to missing sections.

It is important to note that the concept of completeness is meaningless without an associated period of time. Few records are going to be complete at timescales of hours or even days. Perhaps Darwin, who struggled with the problem of incompleteness in demonstrating evolution, would have been content with a geological record that was complete at the timescale of the longevity of an animal species, such that every animal that existed could be preserved. For the modern climate modeller, a record complete at the level of an oceanographic 'decadal oscillation', an Earth-orbital period or the duration of a glaciation might be desired depending upon the question. For the determination of Milanković cyclicity of a given period (say precession), a record is complete if it preserves a recognisable sedimentary signal of each manifestation of that period. These and other vital statistical issues are clearly set out by Weedon (2022).

In sedimentary systems in which autogenic processes are important, the deconvolution of the impacts of Milanković forcing requires great care. Model studies can help in exploring natural system variability and give us clues as to whether different causes of cyclicity can be disentangled. Hill et al. (2009, 2012) provide an illustrative example of this analytical difficulty in interpreting marked cyclicity in superbly exposed peri-tidal carbonates.

Clear expressions of orbitally induced cyclicities are observed during 'Ice ages', periods in the geological record when the waxing and waning of large ice caps on both poles caused short-term (recently 20, 100 and 400 kyr), high-magnitude sea-level changes in the order of 50 to >150 m and resulted in successions of starkly contrasting depositional facies (e.g. the Late Carboniferous 'cyclothems'). These ice ages had varying durations and have been recorded throughout the geological record, with the best known in the Neoproterozoic (Cryogenian) Ordovician (e.g. Le Heron et al. 2018), Carboniferous-Permian (e.g. Rygel et al. 2008) and the Cenozoic, from the Oligocene onwards (Miller et al. 2020). There is also growing evidence for the presence of short-term, high-magnitude sea-level fluctuations in periods previously characterised as greenhouse that urges the reconsideration of the potential presence of, shorter duration, ice-caps (e.g. Dromart et al. 2003; Maurer et al. 2013; Krencker et al. 2019; Ray et al. 2019; Davies et al. 2020).

Milanković-modulated sea level and climate change have had a profound effect on not only our ice-age Earth but much of Earth history (see reviews for each Geological Period in Gradstein et al. 2020).

4.5.5 Million Years to Tens of Millions – Eustasy and Subsidence

The preservation of sediments in this time band depends on the creation and destruction of large volumes of accommodation space by subsidence or sea-level rise, or uplift or sea-level fall, respectively. The erosional processes that control preservation can be understood using the concept of base level.

The particular relevance of sea-level change as the cause of hiatuses is because rates of sea-level fall often exceed those of subsidence. High rates of sea-level fall have the potential to cause falls in RSL even in continuously subsiding sedimentary basins. For the same reason, glacio-eustasy is a particularly important mechanism as it is associated with high rates of global sea-level change, up to an order of magnitude faster than subsidence and with amplitudes of several tens of metres. Thermal expansion of sea water, due to temperature change, is a fast process that could account for global sea-level fluctuations in the order of several metres. Aquifer-eustasy, reflecting sea-level fluctuations driven by the flux of water into or out of continental aquifers, is unlikely to account for more than 5 m of variations in sea level (Davies et al. 2020).

Accommodation space changes over a few millions of years result in ca 10–200 m thick stratigraphic successions. Thick enough that they (i) are visible on seismic data, (ii) strongly affect sediments at the scale of mappable units: stratigraphic members and formations and (iii) are of great importance in economic geology. For these reasons, these are the sequences to which sequence stratigraphic methodologies have been most widely applied (see Section 4.3).

4.5.6 Tens to Hundreds of Millions of Years – Sedimentary Basin Band

Preservation on the timescale of tens to a hundred million years is normally a function of the persistence of basin subsidence. Basin analysis is not the subject of this book, and the interested reader is referred to Allen and Allen (2013). The stratigraphic expression of this time period is several 100 m–km thick successions. Formations and Groups with geometries that are best studied on seismic sections and through geological mapping.

Subsidence and uplift rates and therefore rates of relative sea-level rise and fall, and their pattern over time, vary greatly by basin type (Figure 4.47). Intra-cratonic basins subside particularly slowly over long periods whereas at the other extreme, strike-slip basins are sites of ten times faster subsidence but are very short lived. The preservation filter is particularly strong in slowly subsiding basins, where the slowly accumulating, thin, deposits are prone to removal during sea-level falls. Such slowly subsiding

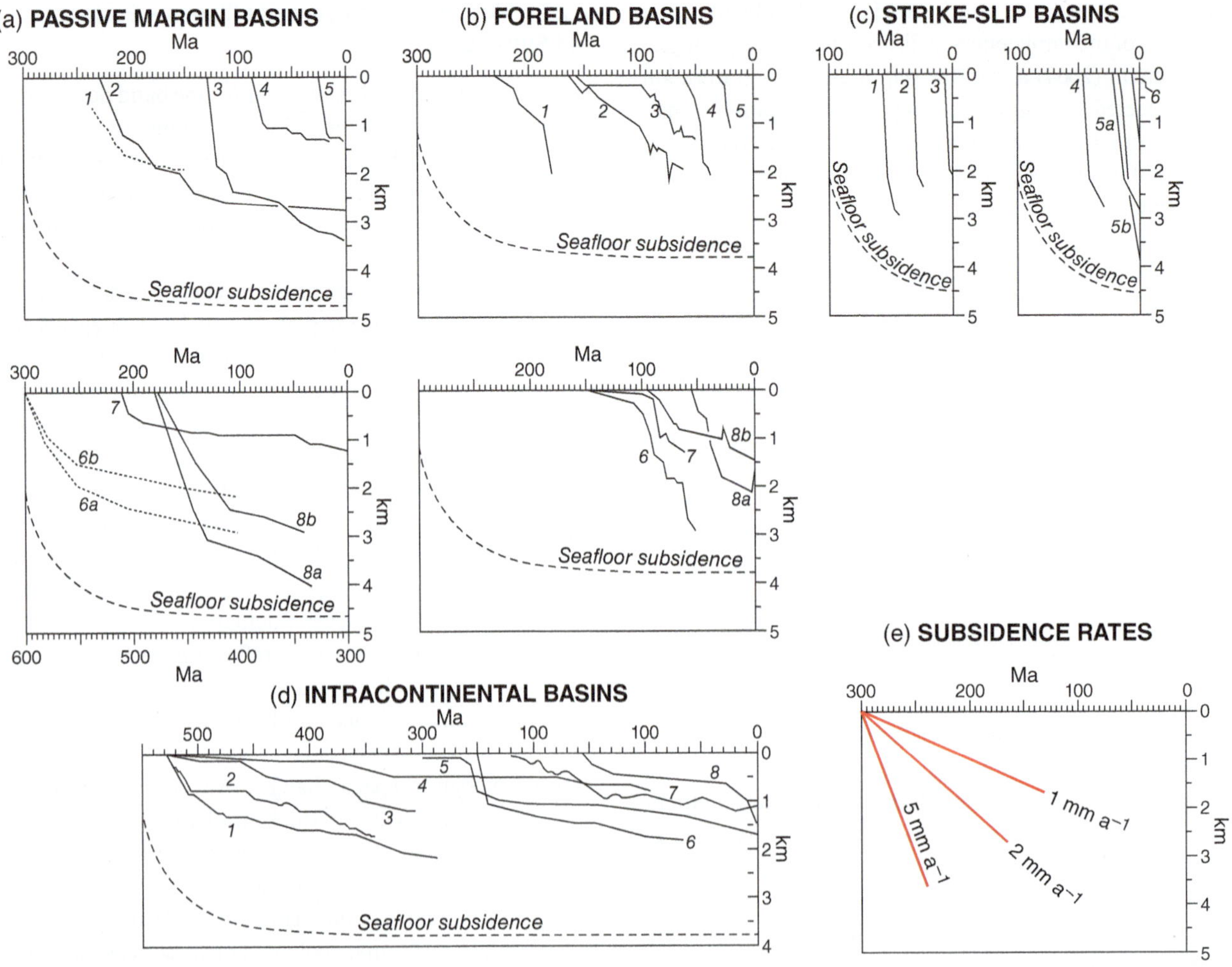

Figure 4.47 Subsidence curves in different basin types (a)–(e) (Xie and Heller 2009 / Geological Society of America). These curves have been made by calculating the subsidence that would have occurred if the sedimentary fill were replaced with water ('back-stripping'). This removes the local isostatic effect of the respective sedimentary loads. They, therefore, all show an estimate of tectonic driving-subsidence and reveal the differing subsidence behaviour of major types of sedimentary basins on a comparable basis. Numbers next to subsidence curves refer to the specific basins represented, consult the original reference for details).

areas with wide, flat, shelves (intra-cratonic basins and the distal parts of foreland basins) may therefore, for example, be prone to accumulating stacks of depositional sequences characterised by forced regressions (e.g. Plint and Nummedal 2000; Plint et al. 2011) where accommodation space during sea-level fall is only available on the former distal shelf and beyond (Section 4.3). In addition, sandstone deposition is often terminated by extensive shoreface retreat and erosional transgression during sea-level rises. This results in relatively thin (10–30 m), shallow-marine sandstones with sharp, erosive bases and tops, with 'blocky' well log responses, which are easily mistaken for fluvial channel deposits in the absence of core data. By contrast, in early, fault-controlled, rapidly subsiding phases of rift or strike-slip basins, high rates of relative sea-level rise lead to aggradational stacks of facies, with lateral progradation only occurring when very large sediment supply rates can balance or exceed the creation of accommodation space by subsidence; consequently, preservation potential is very high.

As well as erosion related to sea-level change, tectonic unconformities and disconformities are inherent to many basin formation processes. A good example is migrating forebulge unconformities in flexural foreland basins. Forebulge unconformities systematically remove the marginal portions of the foreland basin on its exterior side. The degree of preservation is related to the behaviour of the forebulge, which is a complex function of *inter alia* rheology, elastic thickness, thrust load and thrust migration rate (Allen and Allen 2013). In rift basins, a widening of the basin often occurs after the initial (syn-rift) phase of active faulting, when thermal subsidence and flexural loading become the dominant mechanisms for creating accommodation space: this change is manifested by the so-called 'steers-head' geometry (White and McKenzie 1988). The widened 'post-rift' basin is often associated with a retreat of active depositional systems to the margins of the basin, where new accommodation space is being created. Sediment starvation in the basin centre then leads to condensation and a disconformity or even angular unconformities above rotated fault blocks of the syn-rift phase.

Extensive unconformities are also related to tectonic processes independent of basin formation even in continental interiors where basins are relatively protected from plate boundary tectonics. Large-scale vertical uplifts of the crust related to mantle flow ('dynamic topography' or plumes) cause differential removal of sediments over large areas, for example, in Africa (Guillocheau et al. 2012).

Exceptional preservation is related to anomalously high global sea levels with a recurrence interval of roughly 100 Myr. or more that reduce continental freeboard substantially, as occurred in the Middle Ordovician and Late Cretaceous (e.g. Simmons et al. 2020). In the case of the United States, where long-period sequences were formalised by Larry Sloss (1949, 1963, 1988), these relatively long periods of slow rise coincided with episodes of gentle continent-wide warping and are separated by angular unconformities. The lowstand unconformities are thus tectonically enhanced (Burgess 2008).

4.5.7 Hundreds to Thousands of Million Years – Super-Continent Cycle/Planetary

The onset of modern plate tectonic processes around 3 Ga (Hawkesworth et al. 2019) brought with it modern sedimentary basin types such as rifts, passive margins and foreland fold and thrust belts. Even the older Archaean Greenstone belts dating back to 3.5 Ga, despite their unique igneous and sedimentary geochemistry, can plausibly be understood using the same sedimentary basin types (rifts, arc-related and marginal sea basins) that are used for modern basins (Eriksson et al. 1998). The enigmatic intracratonic basins seem to have largely been a feature of the Palaeozoic and are attributed by some (McKenzie et al. 2015) to the persistently convergent margins associated with Pangea assembly. In general, workers have managed to interpret even the most ancient sediments in terms of familiar basin types.

It is, however, the tectonic history of a basin over multiple plate tectonic cycles that is important for ultimate preservation. It would seem intuitive that the preserved volumes of sediment would decrease with time due, for example, to collision, mountain building and erosion through the plate tectonic cycle. And, indeed, Veizer and Mackenzie 2003 showed an exponential decline in sediment mass with age through the Phanerozoic. More recent analyses (Peters and Husson 2017) show that this model applies only to oceanic sediment and non-marine sediments on the continents. Such a progressive age-related decline in volume is not clear for shallow marine and coastal deposits on or adjacent to the continents. Preservation appears to have been contingent on the vagaries of the supercontinent cycle rather than any more systematic preservation filter (Figure 4.48).

Veizer and MacKenzie (2003) listed basin types whose 'oblivion ages' are less than about 300 million years. This list includes sediment accumulations directly on oceanic crust, as well as remnant ocean, back arc, trench slope, piggyback and transpressional basins. Oceanic deposits are an obvious example of preservation effects (Chapter 13). The record of oceanic pelagic sedimentation is largely younger than 100 Ma (W. Hay in Veizer and Mackenzie 2003). Older oceanic pelagic sediments, such as those of Precambrian age, are volumetrically small and tend to be found in fragments in suture zones. Mass-balance calculations suggest significant recycling of Palaeozoic inorganic pelagic carbonates into the mantle (Berner and Mackenzie 2011). (The oldest oceanic crustclaimed is a piece of Neo-Tethys in the eastern Mediterranean that dates to 340 Ma [Granot 2016; Seton et al. 2020].)

Basin types that are particularly favourable for long-term preservation include intra-cratonic and retro-foreland basins formed on cratons. These are not only themselves sites of slow subsidence for up to hundreds of millions of years but are formed in cratonic interiors and hence can be relatively protected from the hurly-burly of subsequent plate margin tectonics. Allen et al. (2015) attempted a classification of 25 selected basins from the Late Archaean and Proterozoic. This group includes 5 foreland basins, 5 cratonic basins and 15 rift-related basins of various types. Their conclusion is that preservation needs 'cratonisation'.

Insofar as facies associations relate to basin type (e.g. more alluvial fans in fault-bounded basins, shallow marine facies in cratonic basins), the tectonic bias in preservation potentially introduces both a lithologic and facies bias. There are no data on the relative volumetric records of specific facies relative to one another over geological time. Ronov et al. (1991a, 1991b) give data on broad sandstone petrography, suggesting, for example, that immature lithologies such as greywackes and arkoses increase in relative abundance with age.

Differential preservation of basins on stable cratons means that the older record is biased towards shallow epicontinental seas (epeiric seaways and epeiric embayments) rather than the shelf-slope systems characteristic of continental margin basins. Shallow sea conditions such as relatively warm water, and possibly increased evaporation rates, may therefore have been differentially preserved as well. A trend of an increasing proportion of carbonates to siliciclastics through the Phanerozoic is attributed to increasing degrees of preservation of shallow marine sediments in intra-cratonic and retro-foreland basins (Veizer and Ernst 1996). Enhancement of tidal amplitudes through propagation of the ocean tide over shallow water shelves is also to be expected. Low

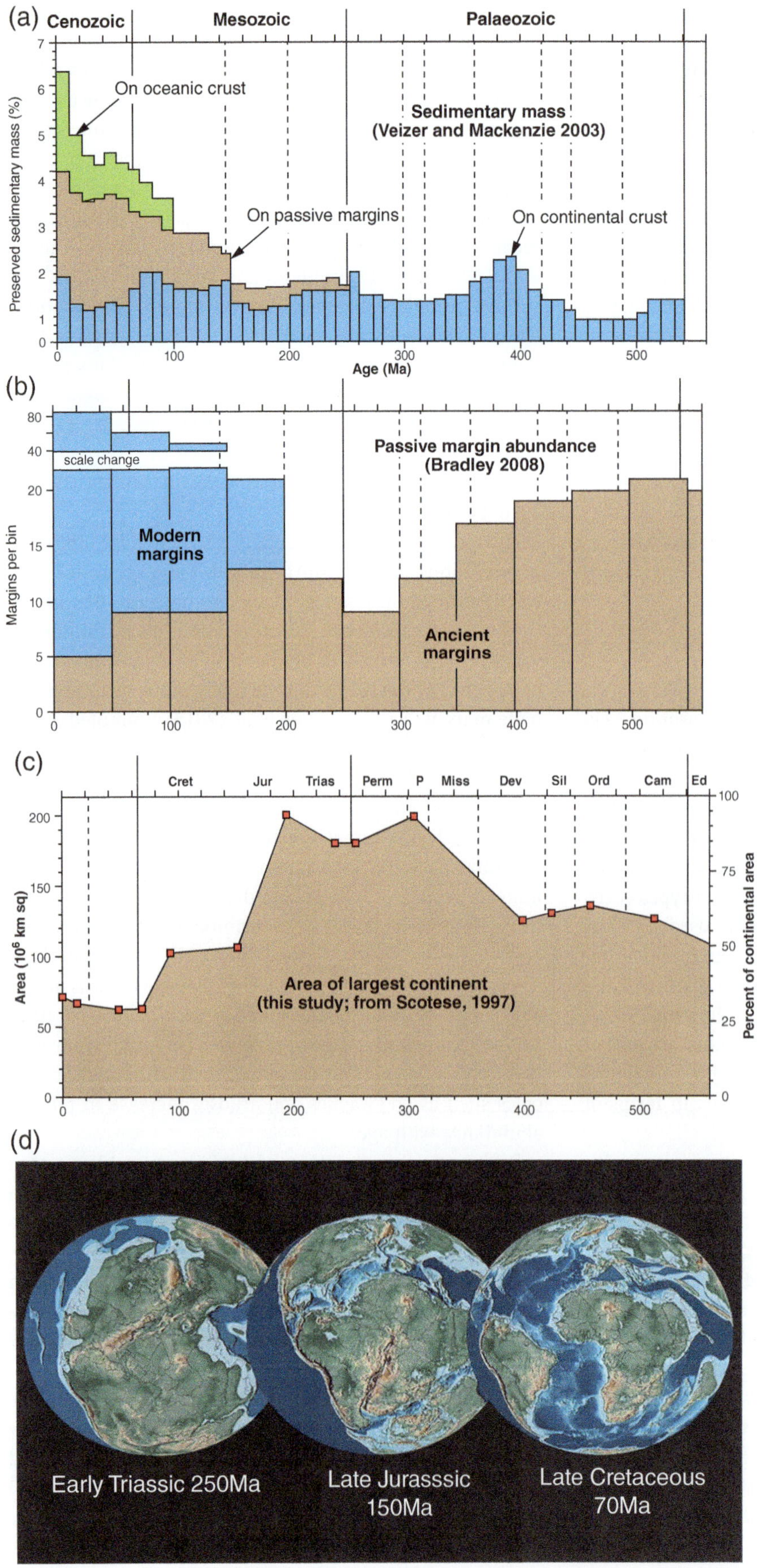

Figure 4.48 The age distribution of passive margins (a,b) is related to the supercontinent cycle (c) in particular the break-up of Pangea (d). The distribution over time of different basin types is contingent on plate movement history and leads to a bias in the record with cratonisation important for long-term preservation. Bradley (2008, 2011) / with permission of Elsevier. Part (c) courtesy of Chris Scotese.

depositional slope systems such as extensive fluvial braid plains and ergs may be over-represented with respect to higher slope systems characteristic of mountain belts (Bose et al. 2012), which have a larger chance of deformation, uplift and erosion.

The trend in increasing carbonate preservation with age, due to tectonic setting, is even more impressive than it looks at first sight since the Phanerozoic data suggest that carbonates are on average removed from the record some 1.5 times faster than siliciclastics, probably due to their more rapid weathering.

There are interpretational consequences of tectonic preservation bias. For example, the bias against oceanic sediment preservation constrains direct access to the conditions of the oceans themselves through time (Morse and Mackenzie 1990; Mackenzie and Morse 1992). This is to a degree mitigated by the preservation of at least parts of passive margin basins on the thinned continental crust (Bradley 2008). The majority of well-documented preserved passive margins is Neoproterozoic or younger with but a patchy record back to the Archaean (Bradley 2011). Nonetheless, tectonic preservation bias on these timescales constrains the reconstruction of conditions in the deep oceans. This contrasts, for example, with shallow seas which flooded large continental areas and whose deposits are well-preserved in intra-cratonic basins. Arguments therefore need to be presented as to how a potentially biased sampling of shallow epeiric seas is relevant to physical–chemical conditions in genuinely oceanic settings. An example of such an argument is that for Sr isotopes: the isotopic system, is well-mixed over relevant timescales irrespective of water depth and habitat (Prokoph et al. 2008).

The known sedimentary records on Earth date back to 3.7 Ga (the Isua super-crustals; Nutman et al. 2019) and the record on Mars (Chapter 15) to a similar date. The facies of the Isua Rocks are not distinctively different from those formed by modern depositional processes (e.g. cross-stratified sandstones, conglomerates and some debated stromatolitic forms). Nonetheless, secular changes in physical and chemical conditions over geological time have fundamentally affected sedimentary environments and facies. In Chapter 15, the perspectives of sedimentary preservation are broadened further to that of planetary bodies in our solar system.

4.6 Summary

In this chapter, we have attempted to set out in broad outline different lenses with which to look at the sedimentary record. We have discussed the difficulties of disentangling the intrinsic processes of deposition and erosion from externally forced deposition and erosion (autogenic versus allogenic processes), and the relationships of deposits at one place to non-deposition somewhere else at the same time and the possible biases introduced by differential preservation of sediments through time. It is our view that all these perspectives are helpful, and all need to be taken into account. We also believe that facies analysis and sequence stratigraphy in combination represents the best framework for the analysis of stratigraphic units, and it is these two approaches that are therefore highlighted in most chapters in the rest of this book (Table 4.2).

Table 4.2 Coverage of the general topics addressed in this chapter in the other chapters of this book.

Chapter	Facies	Events	Autogenic	Cyclostrat.	Allogenic Seq. Strat.	Tectonostrat.
2. Facies	xxx	x	xxx			
3. Biogenic sedimentary structures	xxx	x	xx		x	
5. Alluvial	xxx	x	xxx		xx	x
6. Lakes	xxx		xxx		xxx	x
7. Aeolian	xxx	xx	xxx		xxx	
8. Sandy shallow marine	xx	x	xxx		xxx	
9. Fine-grained shallow marine	xx	x	xx	x	xxx	xx
10. Evaporites	xxx	x	xx	x	xx	xx
11. Carbonates	xx		xxx	x	xxx	x
12. Deep-water clastics	xxx	xx	xxx	x	xx	x
13. Pelagic	xx	x		xx		
14. Glacial	x		xxx	x		
15. Extra-terrestrial	xx	x	xx			

Further Reading

Ager, D. (1993). *The Nature of the Stratigraphical Record*, 3e, 151. New York: Wiley ISBN: 0-471-93808-4.

Allen, P.A. (2017). *Sediment Routing Systems: The Fate of Sediment from Source to Sink*, 407. Cambridge University Press ISBN: 978-1-107-09199-3. https://doi.org/10.1017/9781316135754.

Allen, P.A. and Allen, J.R. (2013). *Basin Analysis: Principles and Application to Petroleum Play Assessment*, 619. Wiley ISBN: 978-0-470-67376-8.

Catuneanu, O. (2022). *Principles of Sequence Stratigraphy*, 2e, 494. Newnes: Elsevier.

Coe, A. (ed.) (2022). *Deciphering Earth's History: the Practice of Stratigraphy*, Geoscience in Practice, 349. Geological Society of London https://doi.org/10.1144/GIPI.

Gradstein, F.M., Ogg, J.G., Schmitz, M.D., and Ogg, G.M. (ed.) (2020). *Geologic Time Scale 2020*, 1357. Elsevier ISBN: 978-0-12-824362-6, Vol. 2: ISBN: 978-0-12-824363-3.

Hajek, E.A. and Straub, K.M. (2017). Autogenic sedimentation in clastic stratigraphy. *Annual Review of Earth and Planetary Sciences* 45: 681–709.

Miall, A.D. (2016). *Stratigraphy: A Modern Synthesis*, 454. Springer ISBN: 978-3-319-24302-3.

Paola, C., Ganti, V., Mohrig, D. et al. (2018). Time not our time: physical controls on the preservation and measurement of geologic time. *Annual Review of Earth and Planetary Sciences* 46: 409–438.

Schlager, W. (2005). *Carbonate sedimentology and sequence stratigraphy* (SEPM concepts in sedimentology and paleontology; no. 8). Society for Sedimentary Geology (SEPM). Tulsa, Oklahoma USA. 200. ISBN: 1-56576-116-2.

Smith, D.G., Bailey, R.J., Burgess, P.M., and Fraser, A.J. (ed.) (2015). Strata and time: probing the gaps in our understanding. *Geological Society, London, Special Publications* 404: 325.

References

Abreu, V., Feldman, H.R., Pedersen, K.H., and Neal, J.E. (2017). *Sequence Stratigraphy of Siliciclastic Systems: Atlas of Exercises – Solutions*, SEPM Concepts in Sedimentology and Paleontology #9, 2e.

Ager, D. (1993). *The Nature of the Stratigraphical Record*, 3e, vol. 42, 151. New York: Wiley ISBN: 0-471-93808-4.

Allen, J.R.L. (1967). Notes on some fundamentals of palaeocurrent analysis, with reference to preservation potential and sources of variance. *Sedimentology* 9 (1): 75–88.

Allen, P.A. (2017). *Sediment Routing Systems: The Fate of Sediment from Source to Sink*, 407. Cambridge University Press ISBN: 978-1-107-09199-3. https://doi.org/10.1017/9781316135754.

Allen, P.A. and Allen, J.R. (2013). *Basin Analysis: Principles and Application to Petroleum Play Assessment*, 619. Wiley ISBN: 978-0-470-67376-8.

Allen, P.A., Eriksson, P.G., Alkmim, F.F. et al. (2015). Precambrian basins of India: stratigraphic and tectonic context. In: *Precambrian Basins of India: Stratigraphic and Tectonic Context*, vol. 43 (ed. R. Mazumder and P.G. Eriksson), 5–28. The Geological Society of London.

Amorosi, A., Maselli, V., and Trincardi, F. (2016). Onshore to offshore anatomy of a late quaternary source-to-sink system (Po Plain–Adriatic Sea, Italy). *Earth-Science Reviews* 153: 212–237.

Amy, L.A., Talling, P.J., Edmonds, V.O. et al. (2006). An experimental investigation of sand–mud suspension settling behaviour: implications for bimodal mud contents of submarine flow deposits. *Sedimentology* 53: 1411–1434.

Aziz, H.A., Hilgen, F.J., van Luijk, G.M. et al. (2008). Astronomical climate control on paleosol stacking patterns in the upper Paleocene–lower Eocene Willwood Formation, Bighorn Basin, Wyoming. *Geology* 36 (7): 531–534.

Bailey, R.J. (2009). Cyclostratigraphic reasoning and orbital time calibration. *Terra Nova* 21 (5): 340–351.

Bailey, R.J. and Smith, D.G. (2010). Thematic set: scaling in stratigraphic data series: implications for practical stratigraphy. *First Break* 28 (1): 56–66.

Baker, V.R. (2008). Paleoflood hydrology: Origin, progress, prospects. *Geomorphology* 101 (1–2): 1–13.

Barrell, J. (1917). Rhythms and the measurements of geologic time. *Bulletin of the Geological Society of America* 28 (1): 745–904.

Beerbower, J.R. (1964). Cyclothems and cyclic depositional mechanisms in alluvial plain sedimentation. *Kansas Geological Survey Bulletin* 169 (1): 31–32.

Bentley, S.J. Sr., Blum, M.D., Maloney, J. et al. (2016). The Mississippi River source-to-sink system: perspectives on tectonic, climatic, and anthropogenic influences, Miocene to Anthropocene. *Earth-Science Reviews* 153: 139–174.

Berner, R.A. and Mackenzie, F.T. (2011). Burial and preservation of carbonate rocks over Phanerozoic time. *Aquatic Geochemistry* 17 (4): 727–733.

Blatt, H., Middleton, G.V., and Murray, R.C. (1980). *Origin of Sedimentary Rocks*, 2e, 782. Englewood Cliffs, NJ: Prentice-Hall Inc.

Bose, P.K., Eriksson, P.G., Sarkar, S. et al. (2012). Sedimentation patterns during the Precambrian: a unique record? *Marine and Petroleum Geology* 33 (1): 34–68.

Boulila, S., Haq, B.U., Hara, N. et al. (2021). Potential encoding of coupling between Milankovitch forcing and Earth's interior processes in the Phanerozoic eustatic sea-level record. *Earth-Science Reviews* 220: 103727.

Bover-Arnal, T., Salas, R., Guimera, J., and Moreno-Badenar, J.A. (2022). Eustasy in the Aptian world: a vision from the eastern margin of the Iberian Plate. *Global and Planetary Change* 214: 103849.

Bover-Arnal, T., Salas, R., Guimera, J., and Moreno-Bedmar, J.A. (2014). Deep incision in an Aptian carbonate succession indicates major sea-level fall in the Cretaceous. *Sedimentology* 61 (6): 1558–1593.

Bradley, D.C. (2008). Passive margins through earth history. *Earth-Science Reviews* 91 (1–4): 1–26.

Bradley, D.C. (2011). Secular trends in the geologic record and the supercontinent cycle. *Earth-Science Reviews* 108 (1–2): 16–33.

Brinkhuis, H. and Smit, J. (1996). The Geulhemmerberg Cretaceous/Tertiary boundary section (Maastrichtian type area, SE Netherlands); an introduction. *The Geulhemmerberg Cretaceous/ Tertiary boundary section (Maastrichtian type area, SE Netherlands). Geologie en Mijnbouw* 75: 101–106.

Brown, L.F. Jr. and Fisher, W.L. (1977). Seismic-stratigraphic interpretation of depositional systems: examples from Brazilian rift and pull-apart basins: Section 2. Application of seismic reflection configuration to stratigraphic interpretation. In: *Seismic Stratigraphic Applications to Hydrocarbon Exploration*, vol. 26 (ed. C.W. Payton), 53–62. American Association of Petroleum Geologists.

van Buchem, F. (2021). Forward stratigraphic modeling – closing the interpretation loop. In: *Exploration Handbook* (ed. R. Head and M. Simmons), 136–143. Halliburton.

Burgess, P.M. (2008). Phanerozoic evolution of the sedimentary cover of the North American craton. *Sedimentary Basins of the World* 5: 31–63.

Burgess, P.M. (2012). A brief review of developments in stratigraphic forward modelling, 2000–2009. In: *Principles of Geologic Analysis, Chapter 14* (ed. D.G. Roberts and A. Bally), 379–404. Elsevier B.V https://doi.org/10.1016/B978-0-444-53042-4.00014-5.

Burgess, P.M. and Prince, G.D. (2015). Non-unique stratal geometries: Implications for sequence stratigraphic interpretations. *Basin Research* 27 (3): 351–365.

Caracciolo, L. (2020). Sediment generation and sediment routing systems from a quantitative provenance analysis perspective: review, application and future development. *Earth-Science Reviews* 209: 103226.

Catuneanu, O. (2002). Sequence stratigraphy of clastic systems: concepts, merits, and pitfalls. *Journal of African Earth Sciences* 35 (1): 1–43.

Catuneanu, O. (2019). Model-independent sequence stratigraphy. *Earth-Science Reviews* 188: 312–388.

Catuneanu, O. (2020). Sequence stratigraphy in the context of the 'modeling revolution'. *Marine and Petroleum Geology* 116: 104309.

Catuneanu, O. (2022). *Principles of Sequence Stratigraphy*, 2e, 494. Newnes: Elsevier.

Catuneanu, O., Abreu, V., Bhattacharya, J. P. et al. (2009). Towards the standardization of sequence stratigraphy. *Earth-Science Reviews* 92 (1–2): 1–33.

Catuneanu, O. and Zecchin, M. (2013). High-resolution sequence stratigraphy of clastic shelves II: controls on sequence development. *Marine and Petroleum Geology* 39 (1): 26–38.

Catuneanu, O. and Zecchin, M. (2020). Parasequences: Allostratigraphic misfits in sequence stratigraphy. *Earth-Science Reviews* 208: 103289.

Chadwick, A.J., Lamb, M.P., and Ganti, V. (2020). Accelerated river avulsion frequency on lowland deltas due to sea-level rise. *Proceedings of the National Academy of Sciences* 117 (30): 17584–17590.

Chamberlin, T.C. (1890). The method of multiple working hypotheses. *Science* 15 (366): 92–96.

Chen, C. and Hiscott, R.N. (1999). Statistical analysis of turbidite cycles in submarine fan successions; tests for short-term persistence. *Journal of Sedimentary Research* 69 (2): 486–504.

Choi, K., Hong, C.M., Kim, M.H. et al. (2013). Morphologic evolution of macrotidal estuarine channels in Gomso Bay, west coast of Korea: Implications for the architectural development of inclined heterolithic stratification. *Marine Geology* 346: 343–354.

Coe, A. (ed.) (2022). *Deciphering Earth's History: the Practice of Stratigraphy*, Geoscience in Practice, 349. Geological Society of London https://doi.org/10.1144/GIPI.

Collinson, J.D. and Mountney, N.P. (2019). *Sedimentary Structures*, 340. Edinburgh, Scotland: Dunedin Academic Press.

Colombera, L. and Mountney, N.P. (2020). Accommodation and sediment-supply controls on clastic parasequences: a meta-analysis. *Sedimentology* 67 (4): 1667–1709.

Cramer, B.D. and Jarvis, I. (2020). Carbon isotope stratigraphy. In: *Geologic Time Scale 2020*, 309–343. Elsevier.

Dade, W.B., Nowell, A.R.M., and Jumars, P.A. (1992). Predicting erosion resistance of muds. *Marine Geology* 105 (1–4): 285–297.

Davies, A., Gréselle, B., Hunter, S.J. et al. (2020). Assessing the impact of aquifer-eustasy on short-term Cretaceous sea-level. *Cretaceous Research* 112: 104445.

Davis, W. (1902). Baselevel, grade and peneplain. *The Journal of Geology* 10 (1): 77–111.

De Boer, P.L. and Smith, D.G. (1994). Orbital forcing and cyclic sequences. In: *Orbital Forcing and Cyclic Sequences*, vol. 19, 1–14. Special Publications of the International Association of Sedimentologists.

Dittmann, S. (2002). Benthic fauna in tropical tidal flats–a comparative perspective. *Wetlands Ecology and Management* 10: 189–195.

Dromart, G., Garcia, J.P., Picard, S. et al. (2003). Ice age at the Middle–Late Jurassic transition? *Earth and Planetary Science Letters* 213 (3–4): 205–220.

Dunham, R.J. (1962). Classification of carbonate rocks according to depositional texture. In: *Classification of Carbonate Rocks*, vol. 1 (ed. W.E. Ham), 108–121. American Association of Petroleum Geologists.

Dunkley Jones, T., Manners, H.R., Hoggett, M. et al. (2018). Dynamics of sediment flux to a bathyal continental margin section through the Paleocene–Eocene Thermal Maximum. *Climate of the Past* 14 (7): 1035–1049.

Duval, B., Cramez, C., and Vail, P.R. (1988). Stratigraphic cycles and major marine source rocks. In: *Mesozoic and Cenozoic sequence stratigraphy of European Basins, Society for Sedimentary Geology*, Special Publication, vol. 60 (ed. P.C. De Graciansky, J. Hardenbol, T. Jacquin, and P.R. Vail), 43–51.

Embry, A.F. (2009). Practical sequence stratigraphy. *Canadian Society of Petroleum Geologists* 81: 79.

Emery, D. and Myers, K.J. (ed.) (1996). *Sequence Stratigraphy. v+ 297 pp.* Oxford: Blackwell Science.

Eriksson, P.G., Condie, K.C., Tirsgaard, H. et al. (1998). Precambrian clastic sedimentation systems. *Sedimentary Geology* 120 (1-4): 5–53.

Ernst, R.E., Bond, D.P.G., and Zhang, S.H. (2020). Influence of large igneous provinces. In: *Geologic Time Scale 2020* (ed. F.M. Gradstein, J.G. Ogg, M.D. Schmitz, and G.M. Ogg), 345–356. Elsevier 1357.

Fenster, M.S., FitzGerald, D.M., Kelley, J. T. et al. (2001). Net ebb sediment transport in a rock-bound, mesotidal estuary during spring-freshet conditions: Kennebec River estuary, Maine. *Geological Society of America Bulletin* 113 (12): 1522–1531.

Ferrarin, C., Madricardo, F., Rizzetto, F. et al. (2018). Geomorphology of scour holes at tidal channel confluences. *Journal of Geophysical Research: Earth Surface* 123 (6): 1386–1406.

Gale, A.S., Hardenbol, J., Hathway, B. et al. (2002). Global correlation of Cenomanian (Upper Cretaceous) sequences: Evidence for Milankovitch control on sea level. *Geology* 30 (4): 291–294.

Garzanti, E. (2017). The maturity myth in sedimentology and provenance analysis. *Journal of Sedimentary Research* 87 (4): 353–365.

Geist, E.L. and Parsons, T. (2014). Undersampling power-law size distributions: effect on the assessment of extreme natural hazards. *Natural Hazards* 72 (2): 565–595.

Gleick, J. (1987). *Chaos: Making a New Science*, 352. New York: Penguin Books.

Goldfinger, C. (2011). Submarine paleoseismology based on turbidite records. *Annual Review of Marine Science* 3: 35–66.

Grabowski, R.C., Droppo, I.G., and Wharton, G. (2011). Erodibility of cohesive sediment: The importance of sediment properties. *Earth-Science Reviews* 105 (3-4): 101–120.

Gradstein, F.M., Ogg, J.G., Schmitz, M.D., and Ogg, G.M. (ed.) (2020). *Geologic Time Scale 2020*, 1357. Elsevier ISBN: 978-0-12-824362-6, Vol. 2: ISBN: 978-0-12-824363-3.

Graehl, N.A., Kelsey, H.M., Witter, R.C. et al. (2015). Stratigraphic and microfossil evidence for a 4500-year history of Cascadia subduction zone earthquakes and tsunamis at Yaquina River estuary, Oregon, USA. *Bulletin* 127 (1-2): 211–226.

Granjeon, D. (2014). 3D forward modelling of the impact of sediment transport and base level cycles on continental margins and incised valleys. In: *From Depositional Systems to Sedimentary Successions on the Norwegian Continental Margin*, 453–472. IAS Special Publication.

Granot, R. (2016). Palaeozoic oceanic crust preserved beneath the eastern Mediterranean. *Nature Geoscience* 9 (9): 701–705.

Grossman, E.L. and Joachimski, M.M. (2020). Oxygen isotope stratigraphy. In: *Geologic Time Scale 2020*, 279–307. Elsevier.

Grotzinger, J.P., Hayes, A.G., Lamb, M.P., and McLennan, S.M. (2013). Sedimentary processes on Earth, Mars, Titan, and Venus. *Comparative Climatology of Terrestrial Planets* 1: 439–472.

Guillocheau, F., Rouby, D., Robin, C. et al. (2012). Quantification and causes of the terrigenous sediment budget at the scale of a continental margin: a new method applied to the Namibia–South Africa margin. *Basin Research* 24 (1): 3–30.

Hajek, E.A. and Straub, K.M. (2017). Autogenic sedimentation in clastic stratigraphy. *Annual Review of Earth and Planetary Sciences* 45: 681–709.

Hampson, G.J. (2016). Towards a sequence stratigraphic solution set for autogenic processes and allogenic controls: Upper Cretaceous strata, Book Cliffs, Utah, USA. *Journal of the Geological Society* 173 (5): 817–836.

Haq, B.U. and Schutter, S.R. (2008). A chronology of Paleozoic sea-level changes. *Science* 322 (5898): 64–68.

Hawkesworth, C., Cawood, P.A., and Dhuime, B. (2019). Rates of generation and growth of the continental crust. *Geoscience Frontiers* 10 (1): 165–173.

Helland-Hansen, W. and Hampson, G.J. (2009). Trajectory analysis: concepts and applications. *Basin Research* 21 (5): 454–483.

Helland-Hansen, W. and Martinsen, O.J. (1996). Shoreline trajectories and sequences; description of variable depositional-dip scenarios. *Journal of Sedimentary Research* 66 (4): 670–688.

Hemming, S.R. (2004). Heinrich events: massive late Pleistocene detritus layers of the North Atlantic and their global climate imprint. *Reviews of Geophysics* 42 (1): 1–43.

Hilgen, F.J. (1991). Extension of the astronomically calibrated (polarity) time scale to the Miocene/Pliocene boundary. *Earth and Planetary Science Letters* 107 (2): 349–368.

Hilgen, F., Brinkhuis, H., and Zachariasse, W.J. (2006). Unit stratotypes for global stages: the Neogene perspective. *Earth-Science Reviews* 74 (1–2): 113–125.

Hilgen, F.J., Hinnov, L.A., Aziz, H.A. et al. (2015). Stratigraphic continuity and fragmentary sedimentation: the success of cyclostratigraphy as part of integrated stratigraphy. *Geological Society, London, Special Publications* 404 (1): 157–197.

Hill, J., Tetzlaff, D., Curtis, A., and Wood, R. (2009). Modeling shallow marine carbonate depositional systems. *Computers & Geosciences* 35 (9): 1862–1874.

Hill, J., Wood, R., Curtis, A., and Tetzlaff, D.M. (2012). Preservation of forcing signals in shallow water carbonate sediments. *Sedimentary Geology* 275: 79–92.

Hinov, L.A. (2013). Cyclostratigraphy and its revolutionizing applications in the earth and planetary sciences. *Geological Society of America Bulletin* 125 (11–12): 1703–1734. https://doi.org/10.1130/B30934.1.

Holbrook, J.M. and Miall, A.D. (2020). Time in the Rock: A field guide to interpreting past events and processes from siliciclastic stratigraphy. *Earth-Science Reviews* 203: 103121.

Holbrook, J., Scott, R.W., and Oboh-Ikuenobe, F.E. (2006). Base-level buffers and buttresses: a model for upstream versus downstream control on fluvial geometry and architecture within sequences. *Journal of Sedimentary Research* 76 (1): 162–174.

Homewood, P.W., Mauriaud, P., and Lafont, F. (2000). *Best Practices in Sequence Stratigraphy for Explorationists and Reservoir Engineers*, 81. Editions Technip.

Hooke, J.M. (2007). Complexity, self-organisation and variation in behaviour in meandering rivers. *Geomorphology* 91 (3–4): 236–258.

Houseknecht, D.W. (2024). Arctic Alaska Basin, Hanna Trough and Beaufortian Rifted Margin Composite Tectono-Sedimentary Elements. *Geological Society, London, Memoirs* 57 (1): M57–M2018.

Hubbard, R.J. (1988). Age and significance of sequence boundaries on Jurassic and Early Cretaceous rifted continental margins. *AAPG Bulletin* 72 (1): 49–72.

Hunt, D. and Tucker, M.E. (1992). Stranded parasequences and the forced regressive wedge systems tract: deposition during base-level fall. *Sedimentary Geology* 81: 1–9.

Hunt, D. and Tucker, M.E. (1995). Stranded parasequences and the forced regressive wedge systems tract: deposition during base-level fall. - reply. *Sedimentary Geology* 95: 147–160.

James, N.P. and Jones, B. (2015). *Origin of Carbonate Sedimentary Rocks*. Wiley.

Jenkyns, H.C. (2010). Geochemistry of oceanic anoxic events. *Geochemistry, Geophysics, Geosystems* 11 (3).

Jerolmack, D.J. and Mohrig, D. (2007). Conditions for branching in depositional rivers. *Geology* 35 (5): 463–466.

Jerolmack, D.J. and Paola, C. (2010). Shredding of environmental signals by sediment transport. *Geophysical Research Letters* 37 (19): 1–5.

Jobe, Z.R., Sylvester, Z., Howes, N. et al. (2017). High-resolution, millennial-scale patterns of bed compensation on a sand-rich intraslope submarine fan, western Niger Delta slope. *Bulletin* 129 (1–2): 23–37.

Jonathan, P. and Evans, K. (2013). Statistical modelling of extreme ocean environments for marine design: a review. *Ocean Engineering* 62: 91–109.

Keller, G. and Barron, J.A. (1983). Paleoceanographic implications of Miocene deep-sea hiatuses. *Geological Society of America Bulletin* 94 (5): 590–613.

Kodama, K.P., and Hinnov, L.A. (2015). *Rock Magnetic Cyclostratigraphy*, 159. John Wiley & Sons, Ltd.

Kelsey, H.M., Witter, R.C., and Hemphill-Haley, E. (2002). Plate-boundary earthquakes and tsunamis of the past 5500 yr, Sixes River estuary, southern Oregon. *Geological Society of America Bulletin* 114 (3): 298–314.

Krencker, J.N., Lindstrom, S., and Bodin, S. (2019). A major sea-level drop briefly precedes the Toarcian oceanic anoxic event: implications for Early Jurassic climate and carbon cycle. *Scientific Reports, Nature Research* 9: 12518. https://doi.org/10.1038/s41598-019-48956-x.

de Kruijf, M., Slootman, A., de Boer, R.A., and Reijmer, J.J. (2021). On the settling of marine carbonate grains: review and challenges. *Earth-Science Reviews* 217: 103532.

Kyrkjebø, R., Gabrielsen, R.H., and Faleide, J.I. (2004). Unconformities related to the Jurassic–Cretaceous synrift–post-rift transition of the northern North Sea. *Journal of the Geological Society* 161 (1): 1–17.

Langereis, C.G. and Hilgen, F.J. (1991). The Rossello composite: a Mediterranean and global reference section for the Early to early Late Pliocene. *Earth and Planetary Science Letters* 104 (2–4): 211–225.

Lawrence, D.T., Doyle, M., and Aigner, T. (1990). Stratigraphic simulation of sedimentary basins: concepts and calibration. *AAPG Bulletin* 74 (3): 273–295.

Le Heron, D.P., Tofaif, S., and Melvin, J. (2018). The Early Palaeozoic glacial deposits of Gondwana: overview, chronology, and controversies. *Past Glacial Environments* 47–73.

Lee, H.J. (2009). Timing of occurrence of large submarine landslides on the Atlantic Ocean margin. *Marine Geology* 264 (1–2): 53–64.

Leeder, M.R. and Gawthorpe, R.L. (1987). Sedimentary models for extensional tilt-block/half-graben basins. *Geological Society, London, Special Publications* 28 (1): 139–152.

Leonard, L.J., Currie, C.A., Mazzotti, S., and Hyndman, R.D. (2010). Rupture area and displacement of past Cascadia great earthquakes from coastal coseismic subsidence. *Bulletin* 122 (11–12): 2079–2096.

Levell, B.K., Johnson, H.D., Collins, D.S., and Van Cappelle, M. (2020). Deposition and preservation of fluvio-tidal shallow-marine sandstones: A re-evaluation of the Neoproterozoic Jura Quartzite (western Scotland). *Sedimentology* 67 (1): 173–206.

Levell, B. and Kasumajaya, A. (1985). Slumping at the late Miocene shelf-edge offshore West Sabah: a view of a turbidite basin margin. *Bulletin of the Geological Society of Malaysia* 18: 1–29.

Lewis, K.W., Aharonson, O., Grotzinger, J.P. et al. (2008). Quasi-periodic bedding in the sedimentary rock record of Mars. *Science* 322 (5907): 1532–1535.

Li, Q., Yu, L., and Straub, K.M. (2016). Storage thresholds for relative sea-level signals in the stratigraphic record. *Geology* 44 (3): 179–182.

Liu, H., Meng, J., Zhang, Y., and Yang, L. (2019). Pliocene seismic stratigraphy and deep-water sedimentation in the Qiongdongnan Basin, South China Sea: Source-to-sink systems and hydrocarbon accumulation significance. *Geological Journal* 54 (1): 392–408.

Lobo, F.J. and Ridente, D. (2014). Stratigraphic architecture and spatio-temporal variability of high-frequency (Milankovitch) depositional cycles on modern continental margins: an overview. *Marine Geology* 352: 215–247.

Lokier, S.W. and al Junaibi, M. (2016). The petrographic description of carbonate facies: are we all speaking the same language? *Sedimentology* 63 (7): 1843–1885.

Loucks, R.G. and Sarg, J.F. eds., 1993. Carbonate sequence stratigraphy: recent developments and applications, *American Association of Petroleum Geologists Memoir* 57: 545

Loutit, T.S., Hardenbol, J., Vail, P.R., and Baum, G.R. (1988). Condensed sections: *The key to age-dating and correlation of continental margin* sequences. In: , SEPM Special Publication, vol. 42 (ed. C.K. Wilgus, B.S. Hastings, C.G.S.C. Kendall, et al.), 183–213.

Lupin, J.H. and Hampson, G.J. (2020). Sediment-routing controls on sandstone bulk petrographic composition and texture across an ancient shelf: Example from Cretaceous Western Interior Basin, Utah and Colorado, USA. *Journal of Sedimentary Research* 90 (10): 1389–1409.

Mackenzie, F.T. and Morse, J.W. (1992). Sedimentary carbonates through Phanerozoic time. *Geochimica et Cosmochimica Acta* 56 (8): 3281–3295.

Marco, S., Stein, M., Agnon, A., and Ron, H. (1996). Long-term earthquake clustering: A 50,000-year paleoseismic record in the Dead Sea Graben. *Journal of Geophysical Research: Solid Earth* 101 (B3): 6179–6191.

Maurer, F., Van Buchem, F.S., Eberli, G.P. et al. (2013). Late Aptian long-lived glacio-eustatic lowstand recorded on the Arabian Plate. *Terra Nova* 25 (2): 87–94.

McArthur, J.M., Howarth, R.J., Shields, G.A., and Zhou, Y. (2020). Strontium isotope stratigraphy. In: *Geologic Time Scale 2020*, 211–238. Elsevier.

McGlathery, K.J., Reidenbach, M.A., D'Odorico, P. et al. (2013). Nonlinear dynamics and alternative stable states in shallow coastal systems. *Oceanography* 26 (3): 220–231.

McKenzie, D., Daly, M.C., and Priestley, K. (2015). The lithospheric structure of Pangea. *Geology* 43 (9): 783–786.

van der Meer, D.G., Scotese, C.R., Mills, B. J. et al. (2022). Long-term Phanerozoic global mean sea level: insights from strontium isotope variations and estimates of continental glaciation. *Gondwana Research* 111: 103–121.

Meyers, S.R. and Sageman, B.B. (2004). Detection, quantification, and significance of hiatuses in pelagic and hemipelagic strata. *Earth and Planetary Science Letters* 224 (1–2): 55–72.

Miall, A.D. (2015). Updating uniformitarianism: stratigraphy as just a set of 'frozen accidents'. *Geological Society, London, Special Publications* 404 (1): 11–36.

Miall, A.D. (2016). *Stratigraphy: A Modern Synthesis*, 454. Springer ISBN: 978-3-319-24302-3.

Miller, J.J., Agena, W.F., Lee, M.W. et al. (2000). *Regional seismic lines reprocessed using post-stack processing techniques: National Petroleum Reserve, Alaska* (No. 2000-286). US Geological Survey.

Miller, K.G., Schmelz, W.J., Browning, J.V. et al. (2020). Ancient sea level. *Oceanography* 33 (2): 32–41.

Mitchum, R.M. Jr., Vail, P.R., and Thompson, S. III (1977). Seismic stratigraphy and global changes of sea level: Part 2. The depositional sequence as a basic unit for stratigraphic analysis: Section 2. Application of seismic reflection configuration to stratigraphic interpretation. In: *Seismic Stratigraphic Applications to Hydrocarbon Exploration*, vol. 26 (ed. C.W. Payton), 53–62. American Association of Petroleum Geologists.

Morison, K. and Wong, C.L. (2003). Sequence stratigraphic framework of Northwest Borneo. *Bulletin of the Geological Society of Malaysia* 47: 127–138.

Morse, J.W. and Mackenzie, F.T. (1990). *Geochemistry of Sedimentary Carbonates*, Development is Sedimentology, vol. 48, 689. Amsterdam: Elsevier.

Mulder, T., Joumes, M., Hanquiez, V. et al. (2017). Carbonate slope morphology revealing sediment transfer from bank-to-slope (Little Bahama Bank, Bahamas). *Marine and Petroleum Geology* 83: 26–34.

Murton, J., Bateman, M., Dallimore, S. et al. (2010). Identification of Younger Dryas outburst flood path from Lake Agassiz to the Arctic Ocean. *Nature* 464: 740–743. https://doi.org/10.1038/nature08954.

Nahon, A., Idier, D., Senechal, N. et al. (2019). Imprints of wave climate and mean sea level variations in the dynamics of a coastal spit over the last 250 years: Cap Ferret, SW France. *Earth Surface Processes and Landforms* 44 (11): 2112–2125.

Naylor, L.A., Spencer, T., Lane, S.N. et al. (2017). Stormy geomorphology: geomorphic contributions in an age of climate extremes. *Earth Surface Processes and Landforms* 42 (1): 166–190.

Neal, J. and Abreu, V. (2009). Sequence stratigraphy hierarchy and the accommodation succession method. *Geology* 37 (9): 779–782.

Niedoroda, A.W., Swift, D.J., and Thorne, J.A. (1989). Modeling shelf storm beds: Controls of bed thickness and bedding sequence. In: *Shelf Sandstone, Shelf Depositional Sequence and Petroleum Accumulation: A Symposium, SEPM, Tulsa, Oklahoma*, 15–39. SEPM.

Nutman, A.P., Bennett, V.C., Friend, C.R. et al. (2019). Cross-examining Earth's oldest stromatolites: Seeing through the effects of heterogeneous deformation, metamorphism and metasomatism affecting Isua (Greenland)~ 3700 Ma sedimentary rocks. *Precambrian Research* 331: 105347.

Paola, C. and Borgman, L. (1991). Reconstructing random topography from preserved stratification. *Sedimentology* 38 (4): 553–565.

Paola, C., Ganti, V., Mohrig, D. et al. (2018). Time not our time: physical controls on the preservation and measurement of geologic time. *Annual Review of Earth and Planetary Sciences* 46: 409–438.

Partington, M.A., Copestake, P., Mitchener, B.A., and Underhill, J.R. (1993). Biostratigraphic calibration of genetic stratigraphic sequences in the Jurassic–lowermost Cretaceous (Hettangian to Ryazanian) of the North Sea and adjacent areas. In: *Geological Society, London, Petroleum Geology Conference Series*, vol. 4, 371–386. The Geological Society of London.

Pellegrini, C., Patruno, S., Helland-Hansen, W. et al. (2020). Clinoforms and clinothems: Fundamental elements of basin infill. *Basin Research* 32: 187–205.

Peters, S.E. and Husson, J.M. (2017). Sediment cycling on continental and oceanic crust. *Geology* 45 (4): 323–326.

Pettijohn, F.J., Potter, P.E., and Siever, R. (2012). *Sand and Sandstone*. Springer Science & Business Media.

Phillips, J.D. and Van Dyke, C. (2016). Principles of geomorphic disturbance and recovery in response to storms. *Earth Surface Processes and Landforms* 41 (7): 971–979.

Pierik, H.J., Stouthamer, E., and Cohen, K.M. (2017). Natural levee evolution in the Rhine-Meuse delta, the Netherlands, during the first millennium CE. *Geomorphology* 295: 215–234.

Plint, A.G. and Nummedal, D. (2000). The falling stage systems tract: recognition and importance in sequence stratigraphic analysis. *Geological Society, London, Special Publications* 172 (1): 1–17.

Plint, A.G., Tyagi, A., McCausland, P.J. et al. (2012). Dynamic relationship between subsidence, sedimentation, and unconformities in mid-Cretaceous, shallow-marine strata of the Western Canada Foreland Basin: Links to Cordilleran tectonics. *Tectonics of Sedimentary Basins: Recent Advances* 480–507.

Plint, A.G. and Wadsworth, J.A. (2003). Sedimentology and palaeogeomorphology of four large valley systems incising delta plains, western Canada Foreland Basin: Implications for mid-Cretaceous sea-level changes. *Sedimentology* 50 (6): 1147–1186.

Pomar, L. (2001). Ecological control of sedimentary accommodation: evolution from a carbonate ramp to rimmed shelf, Upper Miocene, Balearic Islands. *Palaeogeography, Palaeoclimatology, Palaeoecology* 175 (1–4): 249–272.

Posamentier, H.W. and Allen, G.P. (1999). *Siliciclastic Sequence Stratigraphy – Concepts and Applications*, SEPM Concepts in Sedimentology and Paleontology Series no. 7, vii + 210 pp. Tulsa: SEPM (Society for Sedimentary Geology).

Powell, J.W. (1875). *Exploration of the Colorado River of the West and its Tributaries*, 291. Washington, DC: Report of the Smithsonoan Institution. Govt. Printing Office.

Prather, B.E., O'Byrne, C., Pirmez, C., and Sylvester, Z. (2017). Sediment partitioning, continental slopes and base-of-slope systems. *Basin Research* 29 (3): 394–416.

Prélat, A. and Hodgson, D.M. (2013). The full range of turbidite bed thickness patterns in submarine lobes: controls and implications. *Journal of the Geological Society* 170 (1): 209–214.

Prokoph, A., Shields, G.A., and Veizer, J. (2008). Compilation and time-series analysis of a marine carbonate δ18O, δ13C, 87Sr/86Sr and δ34S database through Earth history. *Earth-Science Reviews* 87 (3–4): 113–133.

Puetz, S.J., Prokoph, A., and Borchardt, G. (2016). Evaluating alternatives to the Milankovitch theory. *Journal of Statistical Planning and Inference* 170: 158–165.

Qayyum, F., Catuneanu, O., and de Groot, P. (2015). Historical developments in Wheeler diagrams and future directions. *Basin Research* 27 (3): 336–350.

Raven, M.J., van Buchem, F.S., Larsen, P.H. et al. (2010). Late Aptian incised valleys and siliciclastic infill at the top of the Shu'aiba Formation (Block 5, offshore Qatar). In: *Barremian-Aptian Stratigraphy and Hydrocarbon Habitat of the Eastern Arabian Plate*, vol. 2 (ed. F.S.P. van Buchem, M.I. Al-Husseini, F. Maurer, and H.J. Droste), 469–502. GeoArabia Special Publication no. 4.

Ray, D.C., van Buchem, F.S., Baines, G. et al. (2019). The magnitude and cause of short-term eustatic Cretaceous sea-level change: A synthesis. *Earth-Science Reviews* 197: 102901.

Razin, P., Taati, F., and Van Buchem, F.S.P. (2010). Sequence stratigraphy of Cenomanian–Turonian carbonate platform margins (Sarvak Formation) in the High Zagros, SW Iran: an outcrop reference model for the Arabian Plate. *Geological Society, London, Special Publications* 329 (1): 187–218.

Reading, H.G. (ed.) (1996). *Sedimentary Environments: Processes, Facies and Stratigraphy*, 3e. Wiley.

Reesink, A.J.H., Van den Berg, J.H., Parsons, D.R. et al. (2015). Extremes in dune preservation: Controls on the completeness of fluvial deposits. *Earth-Science Reviews* 150: 652–665.

Roberts, H.H. (1997). Dynamic changes of the Holocene Mississippi River delta plain: the delta cycle. *Journal of Coastal Research* 605–627.

Röhling, E.J., Foster, G.L., Grant, K.M. et al. (2014). Sea-level and deep-sea-temperature variability over the past 5.3 million years. *Nature* 508 (7497): 477–482.

Ronov, A.B., Yaroshevskiy, A.A., and Migdisov, A.A. (1991a). Chemical constitution of the Earth's crust and geochemical balance of the major elements. *International Geology Review* 33 (10): 941–1048.

Ronov, A.B., Yaroshevskiy, A.A., and Migdisov, A.A. (1991b). Chemical constitution of the Earth's crust and geochemical balance of the major elements (part II). *International Geology Review* 33 (11): 1049–1097.

Rougier, J., Sparks, R.S.J., Cashman, K.V., and Brown, S.K. (2018). The global magnitude–frequency relationship for large explosive volcanic eruptions. *Earth and Planetary Science Letters* 482: 621–629.

Rygel, M.C., Fielding, C.R., Frank, T.D., and Birgenheier, L.P. (2008). The magnitude of Late Paleozoic glacioeustatic fluctuations: a synthesis. *Journal of Sedimentary Research* 78 (8): 500–511.

Sadler, P.M. and Jerolmack, D.J. (2015). Scaling laws for aggradation, denudation and progradation rates: the case for time-scale invariance at sediment sources and sinks. *Geological Society, London, Special Publications* 404 (1): 69–88.

Sadler, P.M. and Strauss, D.J. (1990). Estimation of completeness of stratigraphical sections using empirical data and theoretical models. *Journal of the Geological Society* 147 (3): 471–485.

Sambrook Smith, G.H., Best, J.L., Ashworth, P.J. et al. (2010). Can we distinguish flood frequency and magnitude in the sedimentological record of rivers? *Geology* 38 (7): 579–582.

Sarg, J.F. (1988). Carbonate sequence stratigraphy. In: *Sea Level Changes—An Integrated Approach*, vol. 42 (ed. C.K. Wilgus et al.), 155–181. SEPM Special Publications.

Schaltegger, U., Guex, J., Bartolini, A. et al. (2008). Precise U–Pb age constraints for end-Triassic mass extinction, its correlation to volcanism and Hettangian post-extinction recovery. *Earth and Planetary Science Letters* 267 (1–2): 266–275.

Scheingross, J.S., Limaye, A.B., McCoy, S. W., and Whittaker, A.C. (2020). The shaping of erosional landscapes by internal dynamics. *Nature Reviews Earth and Environment* 1 (12): 661–676.

Schlager, W. (2005). *Carbonate sedimentology and sequence stratigraphy* (SEPM Concepts in Sedimentology and Paleontology; No. 8). Society for Sedimentary Geology (SEPM). Tulsa, Oklahoma USA. 200. ISBN: 1-56576-116-2.

Schlager, W. (2010). Ordered hierarchy versus scale invariance in sequence stratigraphy. *International Journal of Earth Sciences* 99 (1): 139–151.

Schmieder, M. and Kring, D.A. (2020). Earth's impact events through geologic time: a list of recommended ages for terrestrial impact structures and deposits. *Astrobiology* 20 (1): 91–141.

Schumm, S.A. (1973). Geomorphic thresholds and complex response of drainage systems. *Fluvial Geomorphology* 6: 69–85.

Schumm, S.A. (1979). Geomorphic thresholds: the concept and its applications. *Transactions of the Institute of British Geographers* 485–515.

Schwarzacher, W. (2000). Repetitions and cycles in stratigraphy. *Earth-Science Reviews* 50 (1–2): 51–75.

Schwarzacher, W. (2005). The stratification and cyclicity of the Dachstein Limestone in Lofer, Leogang and Steinernes Meer (Northern Calcareous Alps, Austria). *Sedimentary Geology* 181: 93–106.

Seton, M., Müller, R.D., Zahirovic, S. et al. (2020). A global data set of present-day oceanic crustal age and seafloor spreading parameters. *Geochemistry, Geophysics, Geosystems* 21 (10): e2020GC009214.

Simmons, M.D., Miller, K.G., Ray, D.C. et al. (2020). Phanerozoic eustasy. In: *Geologic Time Scale 2020*, 357–400. Elsevier.

Sinha, R., Sripriyanka, K., Jain, V., and Mukul, M. (2014). Avulsion threshold and planform dynamics of the Kosi River in north Bihar (India) and Nepal: A GIS framework. *Geomorphology* 216: 157–170.

Sloss, L.L. (1949). Integrated facies analysis. In: *Sedimentary facies in geologic history: Conference at meeting of the Geological Society of America held in New York, New York, November 11, 1948* (Vol. 39, p. 91). Geological Society of America.

Sloss, L.L. (1963). Sequences in the cratonic interior of North America. *Geological Society of America Bulletin* 74 (2): 93–114.

Sloss, L.L. (1988). Tectonic evolution of the craton in Phanerozoic time. *The Geology of North America* 2: 25–51.

Smit, J. and Brinkhuis, H. (1996). The Geulhemmerberg Cretaceous/Tertiary boundary section (Maastrichtian type area, SE Netherlands); summary of results and a scenario of events. *Geologie en Mijnbouw* 75: 283–293.

Smith, D.G., Bailey, R.J., Burgess, P.M., and Fraser, A.J. (ed.) (2015). Strata and time: probing the gaps in our understanding. *Geological Society, London, Special Publications* 404: 325.

Smith, A.G., Barry, T., Bown, P. et al. (2015). GSSPs, global stratigraphy and correlation. *Geological Society, London, Special Publications* 404 (1): 37–67.

Snedden, J.W. and Liu, C. (2011). Recommendations for a uniform chronostratigraphic designation system for Phanerozoic depositional sequences. *AAPG Bulletin* 95 (7): 1095–1122.

Stefani, M. and Vincenzi, S. (2005). The interplay of eustasy, climate and human activity in the late Quaternary depositional evolution and sedimentary architecture of the Po Delta system. *Marine Geology* 222: 19–48.

Stewart, A.D. (2002). The Later Proterozoic Torridonian rocks of Scotland: their sedimentology, geochemistry and origin. *Geological Society, London, Memoirs* 24: 130 pp.

Stoker, M.S., Holford, S.P., and Hillis, R.R. (2018). A rift-to-drift record of vertical crustal motions in the Faroe–Shetland Basin, NW European margin: establishing constraints on NE Atlantic evolution. *Journal of the Geological Society* 175 (2): 263–274.

Stouthamer, E. and Berendsen, H.J. (2001). Avulsion frequency, avulsion duration, and interavulsion period of Holocene channel belts in the Rhine-Meuse delta, the Netherlands. *Journal of Sedimentary Research* 71 (4): 589–598.

Stouthamer, E. and Berendsen, H.J. (2007). Avulsion: the relative roles of autogenic and allogenic processes. *Sedimentary Geology* 198 (3–4): 309–325.

Sylvester, Z. (2007). Turbidite bed thickness distributions: methods and pitfalls of analysis and modelling. *Sedimentology* 54 (4): 847–870.

Syvitski, J.P. (ed.) (1991). *Principles, Methods, and Application of Particle Size Analysis*, vol. 388. Cambridge: Cambridge University Press.

Syvitski, J.P., Slingerland, R.L., Burgess, P. et al. (2010). *Morphodynamic Models: An Overview*, 3–20. Boca Raton: Taylor & Francis.

Tian, Y., Fan, D., Zhang, X. et al. (2019). Event deposits of intense typhoons in the muddy wedge of the East China Sea over the past 150 years. *Marine Geology* 410: 109–121.

Tipper, J.C. (2015). The importance of doing nothing: stasis in sedimentation systems and its stratigraphic effects. *Geological Society, London, Special Publications* 404 (1): 105–122.

Törnqvist, T.E. and Bridge, J.S. (2002). Spatial variation of overbank aggradation rate and its influence on avulsion frequency. *Sedimentology* 49 (5): 891–905.

Trampush, S.M., Hajek, E.A., Straub, K. M., and Chamberlin, E.P. (2017). Identifying autogenic sedimentation in fluvial-deltaic stratigraphy: Evaluating the effect of outcrop-quality data on the compensation statistic. *Journal of Geophysical Research: Earth Surface* 122 (1): 91–113.

Tucker, M.E. and Jones, S.J. (2023). *Sedimentary Petrology*, 4e. Wiley.

Vail, P.R., Audemard, F., Bowman, S.A. et al. (1991). The stratigraphic signatures of Tectonics, Eustasy and Sedimentology - an overview. In: *Cycles and Events in Stratigraphy* (ed. G. Einsele, W. Ricken, and A. Seilacher), 617–659. Springer-Verlag.

Van Buchem, F.S.P., McCave, I.N., Weedon, G.P. et al. (1994). Orbitally induced small-scale cyclicity in a siliciclastic epicontinental setting (Lower Lias, Yorkshire, UK). In: *Orbital Forcing and Cyclic Sequences*, vol. 19, 345–366. International Association of Sedimentologists.

Van Vliet, A. and Schwander, M.M. (1987). Stratigraphic interpretation of a regional seismic section across the Labau syncline and its flanks structures, Sabah, north Borneo. In: *Atlas of Seismic Stratigraphy*, vol. 27-3 (ed. A.W. Bally), 163–167. American Association of Petroleum Geologists Studies in Geology.

Van Wagoner, J.C., Mitchum, R.M. Jr., Campion, K.M., and Rahmanian, V.D. (1990). *Siliciclastic Sequence Stratigraphy in Well Logs, Core, and Outcrops: Concepts for High-Resolution Correlation of Time and Facies*, American Association of Petroleum Geologists Methods in Exploration Series, vol. 7, 55 pp.

Veizer, J. and Ernst, R.E. (1996). Temporal pattern of sedimentation: Phanerozoic of North America. *Geochemistry International* 33: 64–76.

Veizer, J. and Mackenzie, F.T. (2003). Evolution of sedimentary rocks. In: *Treatise on Geochemistry: Sediments, Diagenesis and Sedimentary Rocks*, vol. 7, 369–407.

Walker, R.G. and Plint, A.G. (1992). Wave- and storm-dominated shallow marine systems. In: *Facies Models: Response to Sea Level Change* (ed. R.G. Walker and N.P. James), 219–238. Newfoundland: Geological Association of Canada.

Walton, M.A., Staisch, L.M., Dura, T. et al. (2021). Toward an integrative geological and geophysical view of Cascadia subduction zone earthquakes. *Annual Review of Earth and Planetary Sciences* 49: 367–398.

Weedon, G.P. (2022). Cyclostratigraphy: regular cycles detected and counted to measure geological time. In: *Deciphering Earth's History: The Practice of Stratigraphy* (ed. A. Coe), 161–179. Geological Society of London.

Weedon, G.P., Jenkyns, H.C., and Page, K.N. (2018). Combined sea-level and climate controls on limestone formation, hiatuses and ammonite preservation in the Blue Lias Formation, South Britain (uppermost Triassic–Lower Jurassic). *Geological Magazine* 155 (5): 1117–1149.

Weedon, G.P., Page, K.N., and Jenkyns, H. C. (2019). Cyclostratigraphy, stratigraphic gaps and the duration of the Hettangian Stage (Jurassic): insights from the Blue Lias Formation of southern Britain. *Geological Magazine* 156 (9): 1469–1509.

Weltje, G.J. (2012). Quantitative models of sediment generation and provenance: state of the art and future developments. *Sedimentary Geology* 280: 4–20.

Wheatcroft, R.A. (1990). Preservation potential of sedimentary event layers. *Geology* 18 (9): 843–845.

Wheatcroft, R.A. and Drake, D.E. (2003). Post-depositional alteration and preservation of sedimentary event layers on continental margins, I. The role of episodic sedimentation. *Marine Geology* 199 (1–2): 123–137.

Wheeler, H.E. (1964). Baselevel, lithosphere surface, and time-stratigraphy. *Geological Society of America Bulletin* 75 (7): 599–610.

White, N. and McKenzie, D. (1988). Formation of the" steer's head" geometry of sedimentary basins by differential stretching of the crust and mantle. *Geology* 16 (3): 250–253.

Wiberg, P. (2000). A perfect storm: formation and potential for preservation of storm beds on the continental shelf. *Oceanography-Washington DC-Oceanography Society* 13 (3): 93–99.

Wilkinson, B.H. and Drummond, C.N. (2004). Facies mosaics across the Persian Gulf and around Antigua—stochastic and deterministic products of shallow-water sediment accumulation. *Journal of Sedimentary Research* 74 (4): 513–526.

Wilkinson, B.H., Drummond, C.N., Diedrich, N.W., and Rotman, E.D. (1999). Poisson processes of carbonate accumulation on Paleozoic and Holocene platforms. *Journal of Sedimentary Research* 69: 338–350.

Willis, B.J., Sun, T., and Ainsworth, R.B. (2021). Contrasting facies patterns between river-dominated and symmetrical wave-dominated delta deposits. *Journal of Sedimentary Research* 91 (3): 262–295.

Woodroffe, C.D. (2007). The natural resilience of coastal systems: Primary concepts. In: *Managing Coastal Vulnerability* (ed. L. McFadden, E. Penning-Rowsell, and R.J. Nicholls), 45–60. Amsterdam: Elsevier.

Xie, X. and Heller, P.L. (2009). Plate tectonics and basin subsidence history. *Geological Society of America Bulletin* 121 (1–2): 55–64.

Xu, G. and Haq, B.U. (2022). Seismic facies analysis: past, present and future. *Earth-Science Reviews* 224: 103876.

5

Alluvial Sediments

Martin R. Gibling[1], Alessandro Ielpi[2], Massimiliano Ghinassi[3], and Luis A. Buatois[4]

[1] *Department of Earth and Environmental Sciences, Dalhousie University, Halifax, Canada*
[2] *Earth, Environmental and Geographic Sciences, University of British Columbia, Okanagan Campus, Kelowna, Canada*
[3] *Dipartimento di Geoscienze, Università degli Studi di Padova, Padova, Italy*
[4] *Department of Geological Sciences, University of Saskatchewan, Saskatoon, Canada*

Summary

The chapter covers the processes, facies and architecture of alluvial systems in mountain valleys, on fans where rivers enter depositional basins and on alluvial plains to the tidal limit in estuaries and deltas. A facies approach assists researchers in identifying the spectrum of fluvial styles and planforms in perennial to ephemeral rivers and in forested to relatively barren terrains. Also discussed briefly are colluvial processes that include gully erosion and sheet wash, meltwater flows and periglacial processes associated with permafrost. An alluvial facies approach aims at establishing *landscape systems*, which provide a bird's-eye view of former landscapes. Analysis commences with *facies* and *facies associations*, formed through a range of physical, chemical and biological processes. By applying palaeoflow and architectural data to the facies analysis, *geomorphic units* may be identified, especially a range of sediment bars. The units form the building blocks of alluvial deposits and assist in defining fluvial styles, placing river systems within the broader *basin architecture*. The level of interpretation that can be reached depends on the availability of data (scattered, intermediate or dense datasets). The chapter discusses upstream controls in mountainous areas, where relief and climate govern discharge and sediment supply, and fault movements divert rivers and trigger landslides and debris flows. Downstream controls include basin subsidence, the base level of lakes and oceans, and the landward extent of the backwater zone. Across river basins, short-lived weather events and longer climatic patterns govern flood frequency and magnitude, and autogenic controls (intrinsic to the fluvial system) include avulsion and channel confinement. Operating collectively, these factors cause channels and floodplains to aggrade and degrade, governing architecture and preservation potential. Alluvial facies and architecture are time-dependent because they are greatly affected by evolving ecosystems, with changes over time in vegetation and bioturbators that include large animals. A facies approach is applicable to economic alluvial resources (aquifers, hydrocarbon reservoirs, economic minerals and coalfields). The approach also has much to contribute to pressing social issues that include geohazard assessment and the Earth's carbon budget. Since the early Holocene, humans have profoundly altered river systems through agriculture, irrigation and river regulation with dams and other barriers. Consequently, many modern rivers are poor analogues for the ancient record, and anthropogenic effects must be considered in studying Holocene and modern sediments.

5.1 Introduction

Rivers have shaped Earth landscapes since the dawn of the hydrological cycle some 4.4 billion years ago. William Morris Davis wrote in 1909 that 'Geology is only an integration of a long series of past geographies; and we cannot properly appreciate those yesterdays without a good understanding of what is going on to-day'. His prescient statement underpins this chapter, which explores alluvial facies in a geomorphic context from mountain valleys to the tidal limit.

As knowledge of the rock record expanded through the 20th century, criteria were refined for identifying and interpreting river deposits, based on mapping and deployment of simple equipment in modern rivers (Figure 5.1). Further insight has arisen from dating Quaternary alluvium using radiocarbon, optically stimulated luminescence and cosmogenic-radionuclide dating techniques. Isotope geochronology of detrital minerals such as zircon and rutile is

Reading's Sedimentary Environments: Processes, Facies and Stratigraphy, Fourth Edition. Edited by Bruce Levell.

Companion website: www.wiley.com/go/levell4e

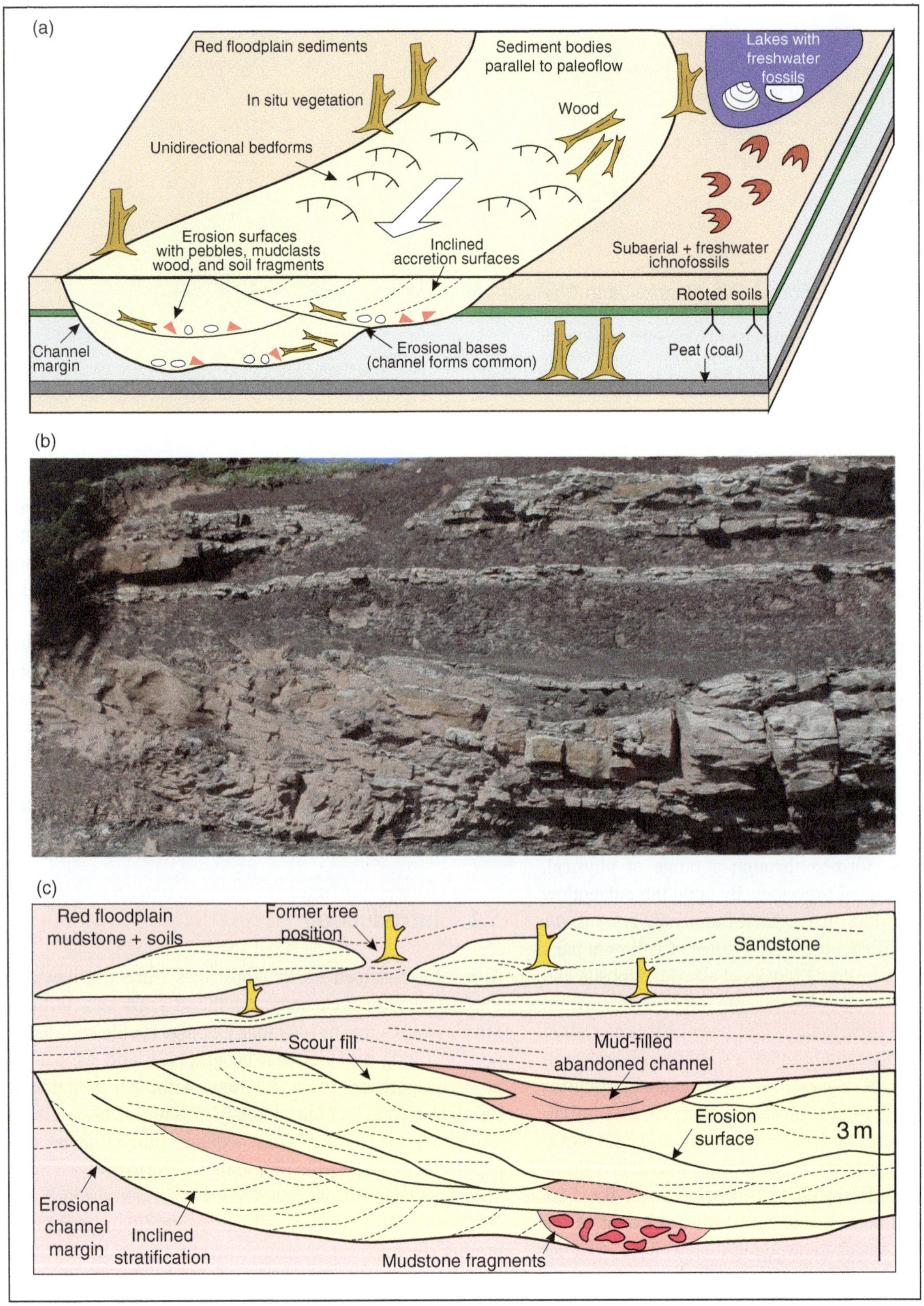

Figure 5.1 Recognising alluvial strata in the ancient record. (a) Diagnostic features. (b, c) Illustration of diagnostic features using uninterpreted and interpreted images from the Pennsylvanian Boss Point Formation, Joggins, Nova Scotia, Canada. The channel base is overlain by mud fragments from a collapsed bank, and a small sediment bar with inclined stratification is directed up the margin, showing that the channel was initially incised to its present depth (a self-confining channel). The channel aggraded vertically, and erosional surfaces and mud-filled hollows indicate fluctuating, probably seasonal flow. The overlying strata yield woody debris, tetrapod trackways, calcic palaeosols and desiccation cracks, and former tree positions are marked by mud-filled hollows. *Source:* From Martin Gibling (Chapter Contributor).

yielding insight into continent-scale sediment pathways. Space probes and surface rovers are documenting the ancient and modern river landscapes of other planets and moons such as Mars and Titan. Rivers witnessed the growth of hydraulic societies based on agriculture. In a world where human activity profoundly affects climate, aquifers and ecosystems, river science has never been more relevant.

Facies analysis commences with the recognition of recurring sediment types. In modern and Quaternary systems, this approach benefits from a known geomorphic setting and is aided by remotely sensed images, high-resolution dating and extant fauna with known environmental tolerances. Thin successions of strata may be analysed in three dimensions using cliffs, trenches, ground-penetrating radar (GPR), Light Detection and Ranging (LiDAR) and uncrewed aerial vehicles (UAVs or drones). In contrast, the study of ancient systems commonly lacks a geomorphic, chronological and environmental framework but benefits from extensive outcrop belts and arrays of cores and seismic profiles through considerable thicknesses of strata.

At the outset of any study, it is important to assess the quality and the spatial and temporal extent of the available dataset (Figure 5.2). Scattered datasets allow an assessment of facies and facies associations, whereas intermediate to dense datasets also allow an assessment of processes, facies architecture and geomorphic units. In outcrop-based studies, the relative prominence of vertical cliffs and bedding surfaces influences the observer's ability to identify many physical sedimentary structures, trace fossils and vegetation features (Shillito and Davies 2020, their Figure 12). Additionally, important discoveries are often made through repeated visits to outcrops that are prone to collapse (Shillito and Davies 2019a). Deformation, diagenesis and metamorphism also affect the appropriate level of interpretation. In view of data limitations and multiple possible processes, there may not always be a 'right answer'.

Harold Garner Reading (1924–2019) played a central role in establishing facies concepts, and his memorable field trips, lectures and writings introduced the present authors to these concepts. We have drawn closely on John Collinson's chapter in the first (1978) edition of Harold Reading's *Sedimentary Environments and Facies* (other formative texts are listed in Supplementary Table 5.S.1: 1), adding insights from nearly half a century of later research. The eight supplementary tables online provide fuller references on key topics (Supplementary Tables online). References cited in the supplementary material are incorporated in the reference section at the end of this chapter.

5.2 Terminology

5.2.1 Definitions

We use the adjective *fluvial* for processes and sediment related to river channels, derived from the Latin words *flumen* and *fluvius* (river or flood). These words denote the body of water flowing between two banks, or the course that the water follows. The broader term *alluvial* is derived from the Latin *alluvium* (from *ad luere*, to be washed against), which denotes deposits laid down by flowing water in

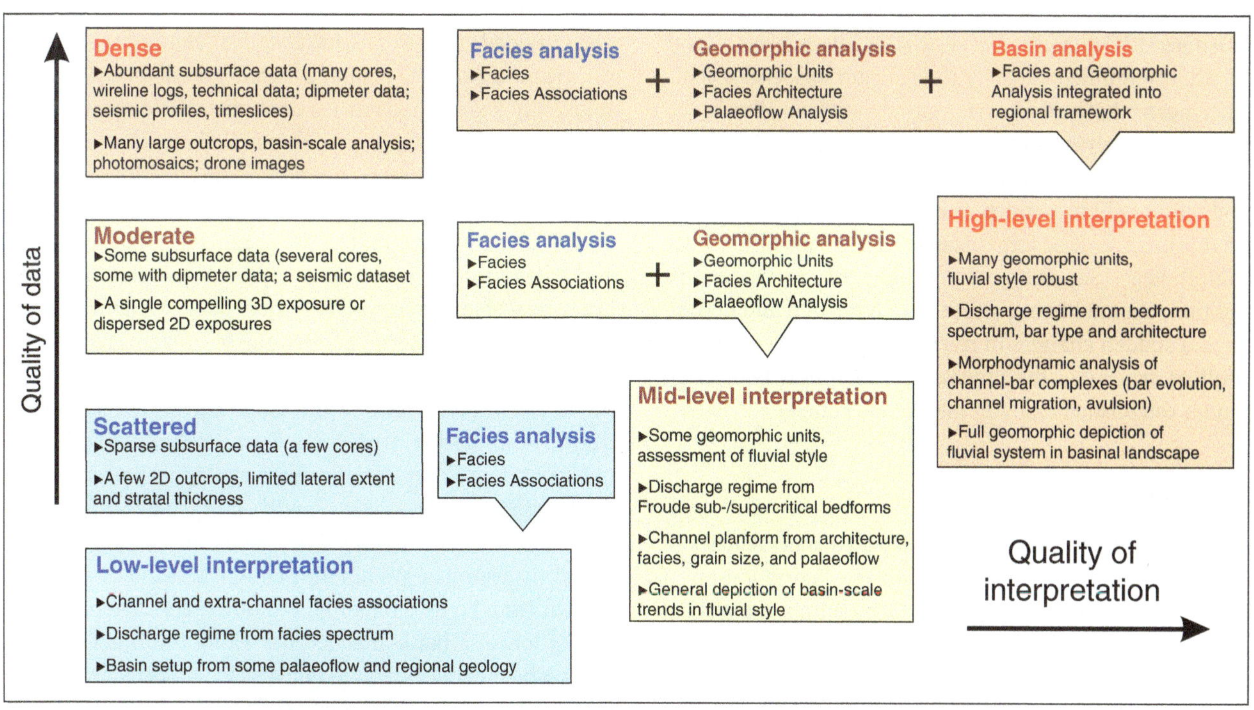

Figure 5.2 The level of interpretation appropriate for alluvial strata, depending on the quality of the available data (scattered, moderate or dense).

channels and floodplains. In this wider context, alluvium is associated with colluvial, periglacial, lacustrine and aeolian sediments. The *river corridor* is an ecosystem that includes three components with varied connectivity: the active channel, the floodplain and the hyporheic zone in pore space below and adjacent to the channel, where river and groundwater mix and interact. The fluxes of water, sediment, woody debris and biota interact within the river corridor, maintaining a patchy and heterogeneous distribution.

Alluvial channels are *self-formed*, generated within sediments that they have transported, deposited and are able to remobilise (Church 2006). *Confined channels* are unable to migrate freely due to banks of bedrock, older cohesive sediment, fibrous peat and resistant vegetation. *Competence* is the ability of a flow to mobilise sediment of a given size. *Bedload* corresponds with the coarser part of the river's load, moving by rolling, sliding or bouncing over the riverbed or through intermittent suspension. *Suspended load* corresponds with the finer load, and once entrained, may be transported for long distances. *Wash load* is a fraction of suspended load that is kept afloat by the thermally induced kinetic state of the flow, remaining suspended indefinitely. Bed material normally forms the channel bed and lower bars, whereas floods deposit suspended material on upper bars and overbank areas, influencing bank strength and enhancing sediment cohesion where fines are trapped interstitially in coarser bed sediment. *Dissolved and colloidal load*, including organic components, are also part of a river's load, and they include molecularly dissolved solutes produced by chemical weathering in the catchment. Rivers have a limited capacity to transport bedload, which is commonly <15% of the total load but <1% in large lowland river systems such as the Amazon (Constantine et al. 2014). Rivers have a large and usually unfilled capacity to transport suspended load, which means that sediment transport in lowland alluvial settings is effectively limited by the kinetic energy of the flow.

Bankfull stage refers to a condition at which discharge fills the channel right to its brim. *Formative stage* reflects the fact that most fluvial landscape change takes place at or above bankfull stage (Supplementary Table 5.S.1: 5). Under this condition, channel banks shift through erosion and deposition and sudden changes in river course take place through cut-off or *avulsion* – the process by which flow is diverted from an established channel into a new course.

5.2.2 Planforms

The earliest alluvial facies models depicted short reaches of meandering and braided rivers, with the later addition of models for anastomosing systems. They employed vertical successions of facies to reconstruct river planforms. Planform was considered the key to inferring geomorphic parameters such as gradient and discharge, and inferring the planform was often the endpoint of analysis. These models provided a simple, linear workflow to relate facies with depositional processes, and they proved useful for sparse subsurface datasets and basin-scale analysis. In reality, commonly recognised planforms are endmembers that geoscientists find convenient in converting the continuum of natural systems into a manageable set of archetypes (Miall 1985). Planforms change dynamically in space and over short periods (Carling et al. 2014; Holbrook and Allen 2021), and confinement results in planforms that the river might not otherwise adopt (Figure 5.3; Fryirs and Brierley 2018). Furthermore, the notion that planform can be inferred from a limited set of vertical facies successions has been discounted (Bridge 1985; Brierley 1989; Ethridge and Davidson 2011). Discharge variability is increasingly acknowledged as a major influence on the facies spectrum, independent of the planform (Fielding et al. 2018; see Section 5.9.6).

Nevertheless, channel planform remains an important attribute of rivers and the sediment bodies that they generate (Galeazzi et al. 2021; Lyster et al. 2022). Planform has traditionally been related to empirical parameters such as slope, discharge and bed grain size, but recent research has additionally highlighted the important effects of bank strength (Supplementary Table 5.S.1: 7; Ielpi and Lapôtre 2020; Candel et al. 2020a, 2020b; Lyster et al. 2022). We emphasise that the fluvial styles discussed below are selected from a continuum of planforms (Schumm 1981, 1985), with delta distributary systems discussed in Section 8.3. Fluvial style remains a topic of ongoing research, especially under strongly seasonal climates (Best and Fielding 2019) and with a range of vegetation.

Single channels (Figure 5.4a) may be classified into laterally stable, self-constraining, meandering, low-sinuosity and braided styles. *Laterally stable channels* and *self-constraining channels* display a high degree of *tortuosity*, with unusually sharp bends of variable size that display abrupt and irregular changes in channel direction (Schumm 1963). Tortuosity may be defined by the fraction of 'sharp' bends with $R_{curv}/w < 2.0$ (where R_{curv} is bend curvature and w is channel width; Candel et al. 2020a, 2020b). These two categories are sometimes termed 'fixed channels' and are typical of low-energy rivers where bank strength strongly influences channel form. Laterally stable systems have tortuous channels that barely migrate and rarely contain sediment bars. Self-constraining channels show a modest degree of migration, with a few bars attached to the banks.

Meandering channels have relatively high *sinuosity* (channel length divided by valley length), defined as >1.5 by Leopold and Wolman (1957). The bends are smooth and of relatively even size. Within the bends, a helicoidal flow component leads to erosion along the outer, concave bank and deposition of a bank-attached bar along the inner, convex bank. The meanders tend to increase in sinuosity and curvature until the flow cuts through them at the necks. Meandering channels maintain near-constant width through

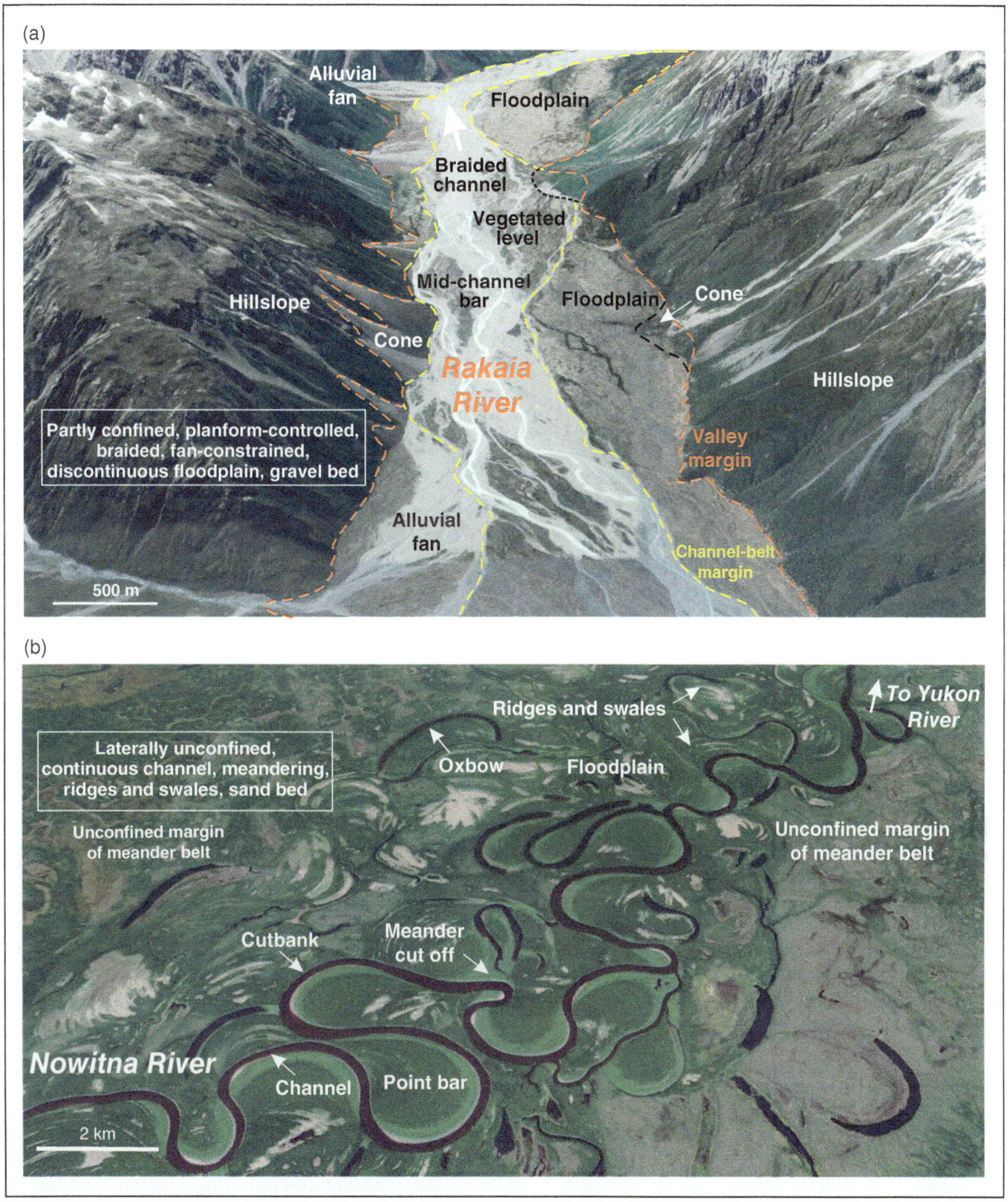

Figure 5.3 Classification of river style applied to the Rakaia River in New Zealand (a) and the Nowitna River in Alaska, USA (b). Classification reflects the degree of confinement of the channel belt; the continuity of the channel; the channel planform; distinctive geomorphic units (e.g. fan and floodplain); and bed-material texture. Facies associations in these settings are not closely linked to the channel planform. *Source:* Modified from Fryirs and Brierley (2018). Maps data: (a) Google, Maxar Technologies, CNES/Airbus; (b) CNES/Airbus, Landsat/Copernicus, Maxar Technologies.

concurrent erosion of the cutbanks (bank pull) and deposition of bank-attached bars (bar push) (Parker et al. 2011; Mason and Mohrig 2019). Based on a study of ~1000 meanders worldwide, meanders were found to migrate at rates in the order of 0.01–0.1 channel widths per year, with unvegetated meanders migrating as much as 10× faster than vegetated counterparts of similar size (Ielpi and Lapôtre 2020). Because tortuosity and sinuosity are different metrics, no simple parameter can currently be used to distinguish 'fixed' from meandering channels (Candel et al. 2020b).

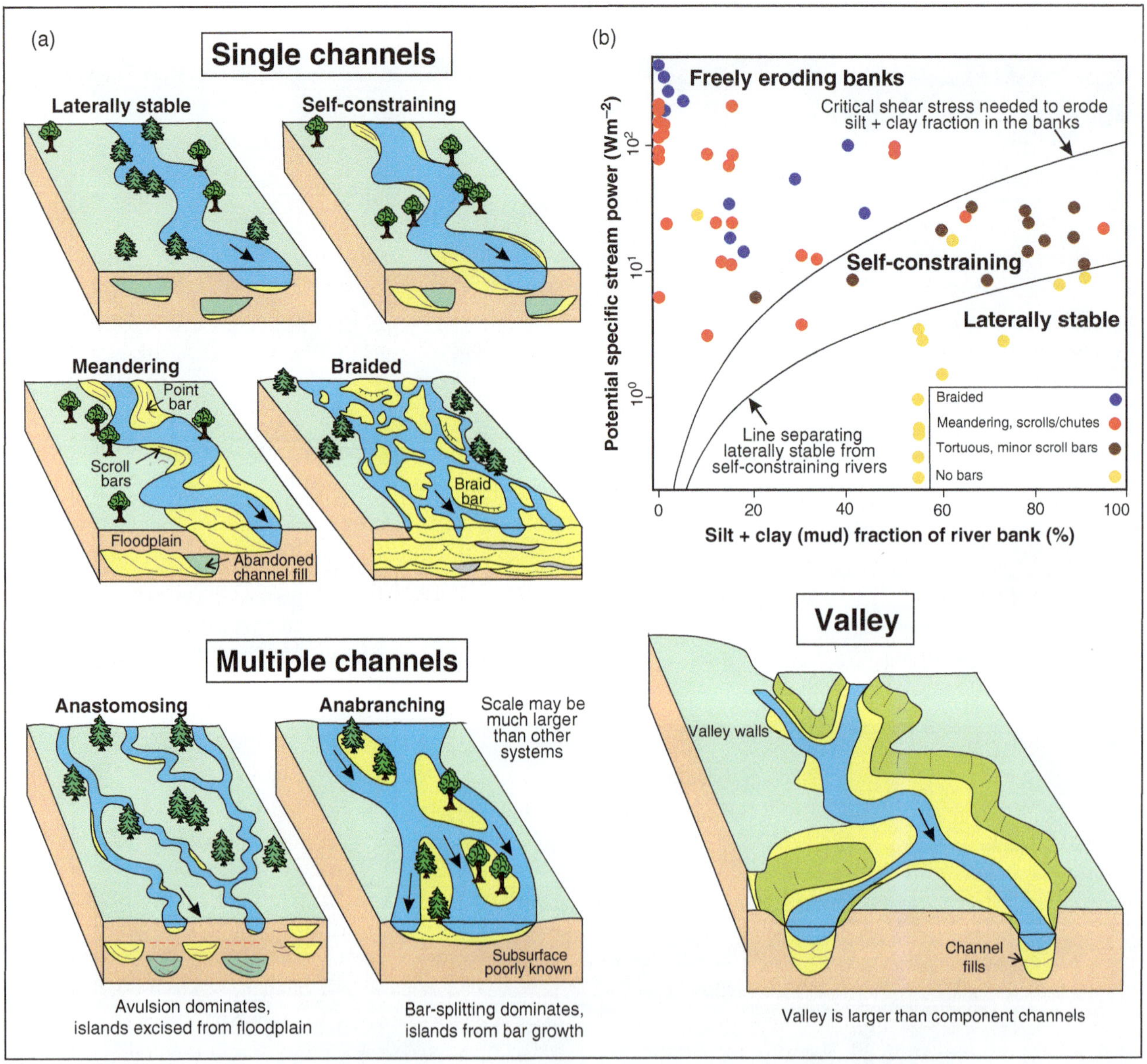

Figure 5.4 Channel planforms. (a) Planforms for single and multiple channels and valley definition. *Source:* Adapted from Carling et al. (2014). (b) Potential specific stream power plotted against the silt-clay fraction of the banks, for 70 modern channels. *Source:* Candel et al. 2020a / SAGE Publications / CC BY 4.0. The silt-plus-clay fraction is an average of multiple bank samples determined using sieving and pipette analysis. The cross-plot successfully predicted the planform for 87% of the channels.

Braided channels are single-thread at high stage when the bars are underwater, but they are multi-thread at low stage, when flow divides around poorly cohesive, mid-channel bars of unvegetated gravel and sand. The whole channel system may be sinuous on a larger scale (Rust 1978; Holbrook and Allen 2021). Braided channels may deposit sediment across a broad, relatively unconfined alluvial region termed a *braidplain*, as spectacularly observed in river systems of Iceland and New Zealand that are periodically swept by glacial meltwater. *Low-sinuosity channels* (<1.5) are transitional between braided and meandering, winding along relatively straight reaches between bank-attached bars, with some mid-channel bars. *Wandering gravelbed and sandbed systems* are transitional between braided and meandering systems (Church 2002, his Figure 4), shifting position like those of braided systems but containing larger and more stable vegetated bars. Wandering rivers are prominent in modern systems, especially in large rivers, but they have received relatively little attention in the ancient record (Galeazzi et al. 2021).

Multiple-channel systems (Figure 5.4a) may be classified by the process of channel formation. *Anastomosing channels* are discrete conduits that divide and rejoin within their floodplain deposits through successive *avulsions* (Makaske 2001), variously building alluvial ridges within floodbasins or incised into tough floodplain deposits. In contrast, *anabranching channel systems* divide and rejoin around long-lived vegetated islands, generated by bar growth and splitting, which are rarely inundated during floods (Carling et al. 2014). Anabranching is not closely related to avulsion, nor to the strength of alluvium, and the component channels may have varied planforms. Some braided systems contain vegetated islands that are inundated during floods (sometimes termed *island-braided* systems), but no clear boundary has been defined in the continuum between braided and anabranching systems.

Valleys (Figure 5.4a) are geomorphic features that are deeper than the channels that they contain. They range from deep conduits cut through mountain bedrock (Figure 5.3a) to shallow courses incised into the sediment of alluvial plains.

To a first approximation, the geometry of alluvial channels may be predicted from four parameters: bankfull discharge, slope, a friction factor and the stress required to entrain the most resistant materials (Dunne and Jerolmack 2020). Their planforms may be explored using a graph that plots potential specific stream power per unit width (Wm^{-2}, the product of discharge, slope, fluid density and gravity acceleration per unit of channel width) against the silt and clay (mud) fraction in the bank (Figure 5.4b; Candel et al. 2020a; see also Schumm 1963). The graph shows that meandering and braided planforms with sandy banks are present through a range of stream power. In contrast, self-constraining and laterally stable planforms are present where the banks have a higher mud fraction and where stream power is modest. Bed material is a prime determinant of channel morphology (Church 2006), but the nature of the bank sediment (muddy, sandy or gravelly), coupled with biological strengthening from rooted vegetation and biofilms, strongly influences river planform (Dunne and Jerolmack 2020).

5.3 Common Fluvial Styles Around the World

We start our conceptual voyage through river systems with brief 'snapshots' that build on the spectrum of alluvial styles discussed earlier (see Supplementary Table 5.S.2 for literature about the styles). What follows should not be considered a wholesale classification but rather a pragmatic account of fluvial styles with a range of processes and morphology. The examples are not neat constructs based on a single parameter, and they demonstrate varied sediment grade, climate and vegetation, in erosional settings (uplands), at basin margins, and on alluvial plains. Most of these fluvial styles may reasonably be inferred in the ancient record (see guidelines in Supplementary Table 5.S.2), although megafans and large rivers require a basin-scale analysis for identification. Most of the examples below exhibit perennial and moderately to strongly seasonal flow, but ephemeral rivers flow only for short periods.

5.3.1 Mountain Valleys, Megafans and Alluvial Fans

Mountain valleys contain channels that are braided in style or are partially confined by bedrock, terraces and cones of detritus with outsized landslide blocks. They typically experience large seasonal floods from precipitation and snowmelt. An example is the Kali Gandaki River in the High Himalaya of Nepal (Figure 5.5a), which occupies a valley with a relative relief of >5 km below the mountain tops in its watershed, among the deepest valleys on Earth. At times, late Quaternary rock avalanches blocked the Kali Gandaki and adjacent valleys with debris walls up to 300 m high and tens of kilometres long, and some are estimated to have emplaced as much as 23 km^3 of debris after beheading the highest peaks. During glacial periods, lobes of ice blocked the valleys locally, and meltwater laid down hundreds of metres of glacial debris in tributary valleys. As the supply of glacial sediment waned through the Holocene, the tributary rivers cut into the alluvium, generating terrace 'staircases' (Figure 5.5b,c) with progressively younger ages at lower levels.

The natural dams of debris and ice impounded the Himalayan rivers and formed lakes up to 200 m deep and >20 km long. The eventual collapse of the dams released catastrophic outburst floods that anomalously widened valleys in the mountains and on the plains below. In 2000, a lake impounded by a rock-avalanche dam on the Yarlung Tsangpo River released an outburst flood rated at 120,000 $m^3\ s^{-1}$, deeply eroding the bedrock, promoting secondary landslides and mobilising 5-m blocks into boulder bars kilometres long (Turzewski et al. 2019; Dong et al. 2024).

Megafans (also referred to as fluvial fans or distributive fluvial systems) are convex-up, river-generated landforms across which channels radiate from a valley exit point at a mountain front (Hartley et al. 2010a, 2010b; Weissmann et al. 2015). They typically have strong seasonal precipitation patterns and low gradients (generally about 0.1°–0.01°), and they are defined as extending for >30 km into an adjoining basin, in a continuum with smaller alluvial fans. Exceptionally large megafans are present east of the Andes in South America, where the Pilcomayo Megafan (the world's largest) has a length of 720 km and an area of 210,000 km^2. The Taquari Megafan of the Pantanal wetlands of Brazil has a length of 250 km and an area of 50,000 km^2.

Some of the best studied megafans are in the Himalayan Foreland Basin, where the Gandak, Kosi and Tista megafans extend southward to the axial Ganga River system, and

Figure 5.5 Kali Gandaki Valley, Nepal, and megafans in the Himalayan Foreland Basin. (a) Kali Gandaki Valley, Tatopani to Kagbeni. *Source:* Gibling (2021) / with permission of Dunedin Academic Press. (b) Kag Khola Valley. *Source:* Google, CNES/Airbus. (c) Terrace stratigraphy and radiocarbon dates in the Kagbeni area. *Source:* Adapted from Hurtado Jr. et al. (2001). (d) Megafans visible in Landsat 5 false-colour image from Fall 2019, using a combination of red, green and blue bands (4, 5 and 7). *Source:* Courtesy of Gary Weissmann. (e) Evolution of the Tista Megafan (at right in D) during the latest Pleistocene and Holocene. *Source:* Abrahami et al. (2018) / with permission of Elsevier.

megafans build northward from Peninsular India (Figure 5.5d). The Kosi Megafan is the best known, with a length of 150 km and an area of 10,300 km^2. The rivers pass downfan from gravelly braided systems, deeply incised in places, to sandy meandering systems, and they exhibit nodal avulsions that periodically generate new lobes, usually triggered by monsoon floods. Elsewhere on the megafans, smaller sinuous channels rise in the foothills or from monsoon-fed springs on the plains, and many are laterally stable or self-constrained by the tough surface mud. Lakes (dark patches in Figure 5.5d) occupy areas of higher water table below spring lines downfan. On the Tista Megafan (Figure 5.5e), major avulsions and river capture, possibly linked to earthquakes, generated three lobes during the past 10,000 years. Although difficult to establish, the subsurface record of these megafans is probably dominated by sandy sheets formed by major braided rivers during prolonged periods of intense monsoon rainfall, with intervals of narrow sediment bodies laid down by the smaller rivers.

In a different tectonic and climatic setting, the Okavango Megafan of Botswana extends for 150 km into a developing rift of the East African Rift System. The basin is endorheic (lacking connection to the ocean) and lies on the semi-arid fringe of the Kalahari Desert, losing most of its flow to infiltration and evapotranspiration. Fed by seasonal, subtropical flow from highlands in Angola, the megafan comprises meandering and anastomosing channels that are confined by vegetation in papyrus wetlands. Megafans are also prominent in a foreland basin south of the European Alps, where fluvioglacial activity during the Last Glacial Maximum generated a suite of megafans up to 70 km long and 3000 km^2 in area.

Alluvial fans are also radial and convex-up in shape but have radii of <30 km and gradients of about 1°–5°. Present in extensional and compressional geodynamic settings, they originate from modest watercourses that enter basins along breaks in slope at the foot of uplands, or where tributaries enter large valleys. They form isolated sediment bodies or coalesce laterally through avulsion into sloping blankets termed *bajadas* (Figure 5.6). In the active trans-extensional and endorheic basin of Death Valley (California), the alluvial fans of Anvil Spring Canyon and Warm Spring Canyon are up to 10 km in radius with areas of about 30 km^2, and their slopes decrease downfan from about 5° to 2.5° over an elevation change of some 250 m (Blair 1999a, 1999b). Coarse-grained sediment on the fans was laid down by several processes: stream flows in channels; sheetflows (locally transporting outsized boulders: Figure 5.6a,c); and debris flows (see definitions in Section 5.4.1). Channels divide and widen downfan, with a corresponding decrease in stream power and grain size as the discharge and slope of the individual channels decrease. Inactive tracts are sites of soil development and aeolian activity, and the lower parts of the fans interact with axial rivers, lakes and playas and aeolian dunes. In dryland areas, rivers may terminate as small fans and splays (Donselaar et al. 2013). The distributary pattern may reflect avulsion of major conduits, rather than the simultaneous flow of all the channels (North and Warwick 2007).

5.3.2 Braided and Meandering Rivers

Gravelly braided rivers occupy steep mountain valleys in the headwaters of river networks that dissect large orogens (Figure 5.3a), as well as on fans and alluvial plains. Their processes and facies overlap with those of fans, but their gradients are lower, typically <0.5°. Humid-climate examples in narrow valleys of the North American Cordillera include the Kicking Horse and Donjek rivers of Canada, up to 1 km wide, which featured in early facies models. Later GPR studies in the Sagavanirktok River of Alaska and the Kicking Horse River (Figure 5.7a,b) have integrated surface processes and sediments with the subsurface record. The rivers change their planforms rapidly during major floods that may be linked to glacial meltwater (Figure 5.7c). At the watershed scale, bed slope and grain size decrease downflow, and they commonly evolve into gravelly to sandy wandering rivers.

Sandy braided rivers share geomorphic similarities with their gravelly counterparts. They display a wide range of channel morphology and sinuosity, with sediment storage in bars along the banks and in the mid-channel zone. Large rivers such as the Brahmaputra–Jamuna system of Bangladesh have channel belts up to 20 km wide (Figure 5.7d), with sediment bars as much as 12 m high and several kilometres long. Within such enormous sediment accumulations, bar accretion has generated inclined bedsets up to 8 m thick and cross-sets up to 4 m thick from individual dunes (Best et al. 2003). In contrast, the South Saskatchewan River of Canada (Figure 5.7e) occupies a valley <1 km wide with braided channels 50–150 m wide and 3 m deep from the thalweg to the tops of the adjacent bars. These shallow channels reorganise their bed morphology during floods, leaving low-relief bar remnants with dune sets that are difficult to identify as barforms in the subsurface record (Parker et al. 2013).

Meandering rivers typically comprise a single, low-gradient channel with an adjoining floodplain, or are confined within a valley, as in the Pjakupur River of Russia, the Klarälven River of Norway and Sweden and the Beaver River of Canada (Figure 5.8a,b). Channel evolution is mainly controlled by the interplay between streamwise and helicoidal flows along the meanders (Figure 5.8a), with helicoidal flows becoming efficient at sinuosity >1.5. A range of meander-bend morphologies may be defined from scroll-bar patterns along the inner bank (Figure 5.8b; Ghinassi et al. 2016).

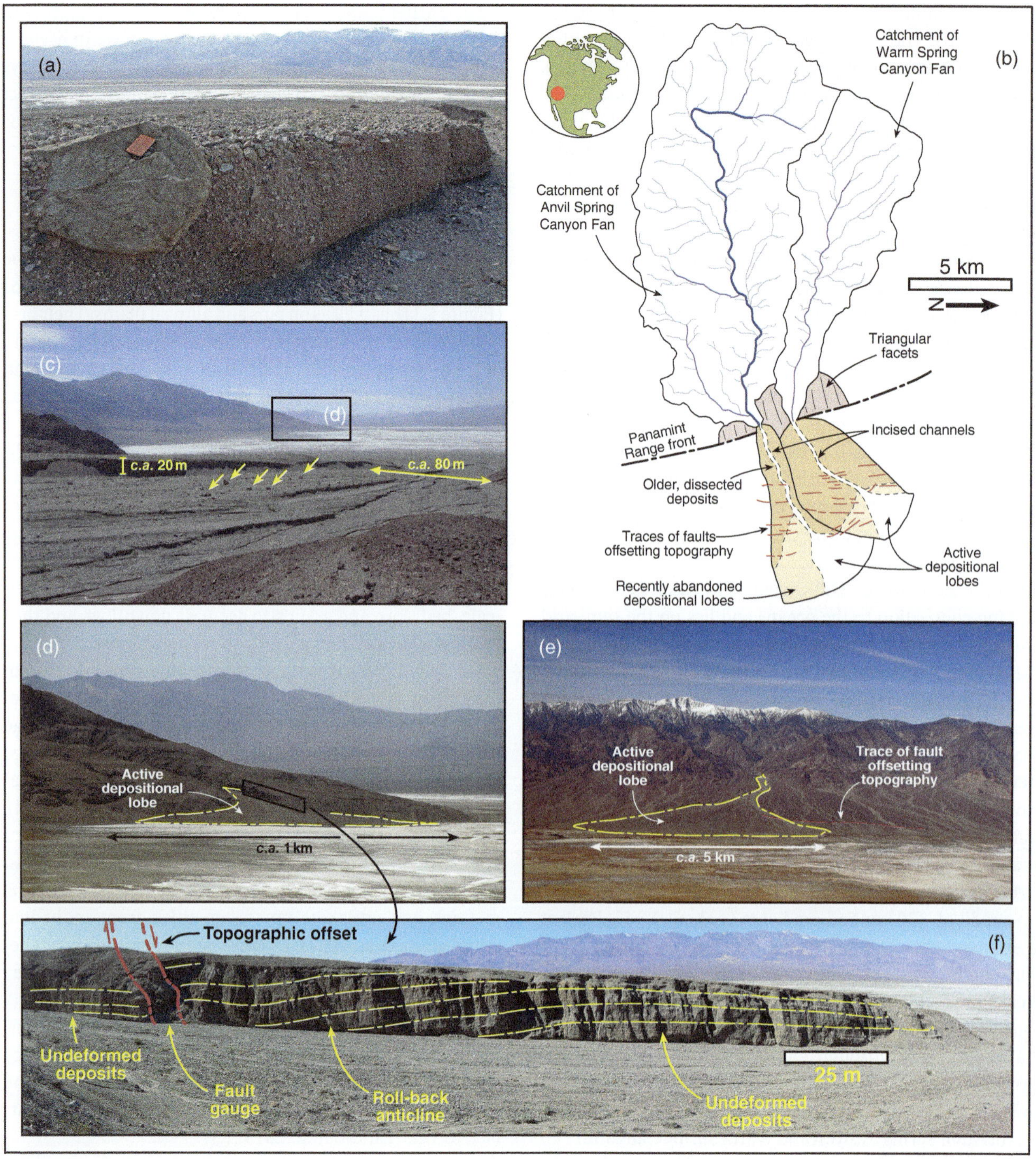

Figure 5.6 Arid alluvial fans in Death Valley, California, USA. (a) Cutbank 1.5 m high near Furnace Creek, showing gravel laid down by stream flows, with an outsized boulder. Notebook is 20 cm long. White surfaces in background are playa lakes. (b) Anvil Spring Canyon and Warm Spring Canyon alluvial fans. *Source:* Adapted from Blair (1999b): two coalescing alluvial fans where ongoing transtension along faults led to switches in depositional lobes. (c) Fan trunk channel at Copper Canyon, showing gravel sheets and outsized boulders (arrowed). (d) Mormon Point Alluvial Fan, located in (c). (e) Hanaupah Canyon Alluvial Fan with its catchment (including Telescope Peak) in the background. Note fault trace offsetting the surface of an abandoned depositional lobe, and the younger, active lobe to its hydrographic left. (f) Field panel located in (d), where older alluvial-fan strata are dissected and gently folded in a roll-back anticline related to a normal fault. *Source:* From Alessandro Ielpi (Chapter author).

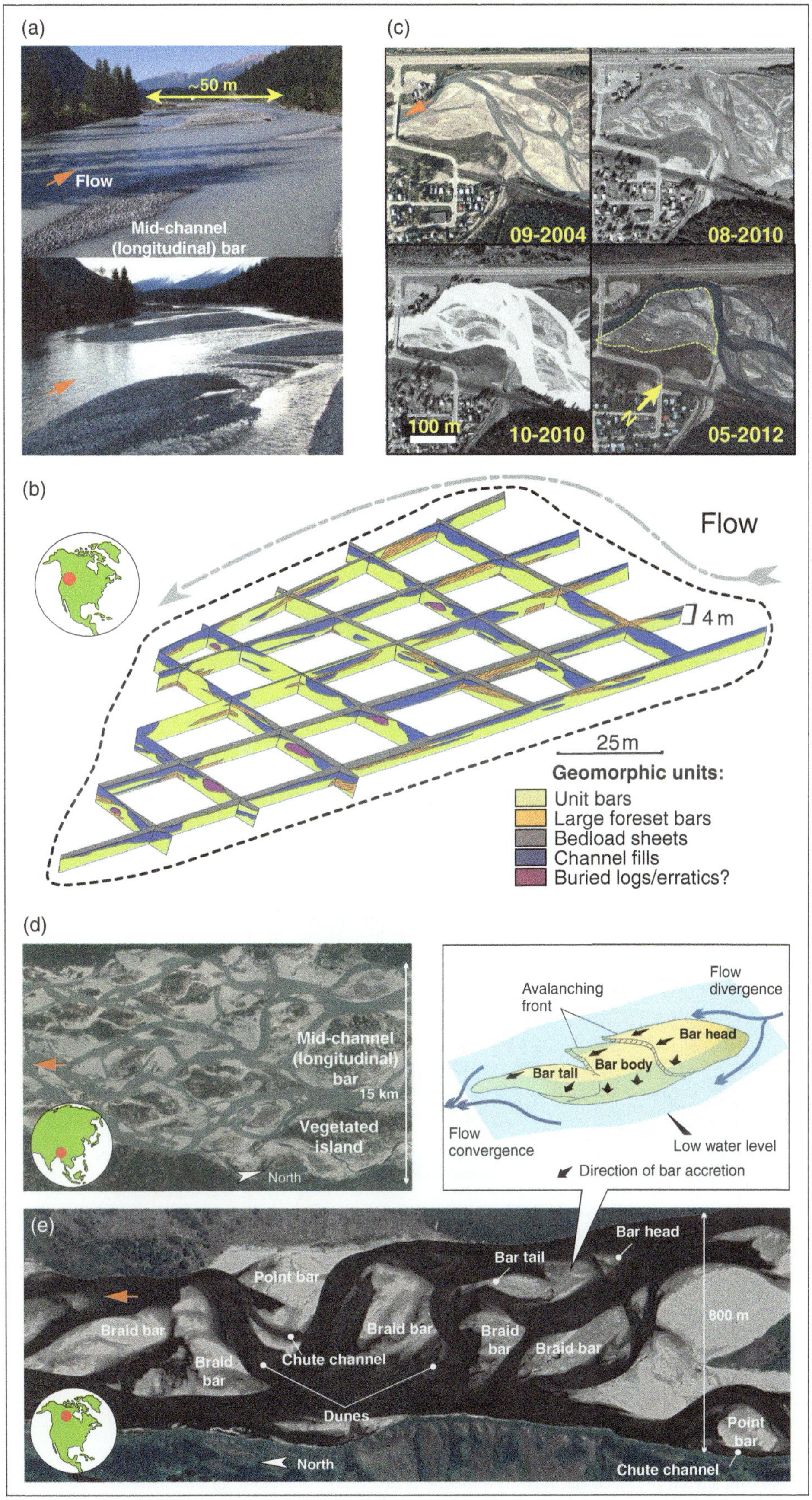

Figure 5.7 Gravelly and sandy braided rivers. (a) The gravelly, glacier-fed Kicking Horse River, Field, British Columbia, Canada, at high-flow (afternoon, August) and low-flow (morning, October). *Source:* From Alessandro Ielpi (Chapter author). (b) 3D grid of GPR profiles from the braid bar in (c). Geomorphic units are mainly unit bars and channel fills, with large foreset bars and bedload sheets. *Source:* Adapted from Cyples et al. 2020. (c) Planform changes from bar accretion, reworking, and flood stage in the Kicking Horse River. (d) Sandy braid bars of the Jamuna River, Bangladesh. (e) Sandy braid bars of the South Saskatchewan River, Canada, with bar planform model. *Source:* (d, e) Google, Maxar Technologies, CNES / Airbus.

Figure 5.8 Meandering, self-constraining, and ephemeral rivers. (a) Expansional bends of the freely meandering Pjakupur River, Russia, with line drawing of top point bar and sketch of helicoidal flow in a meander bend. (b) Translating or downstream-migrating meander bends of the valley-confined Beaver River, Canada, with line drawing above. (c) Tana River, Kenya, with floodwater in chute and crevasse channels and low-lying floodbasins. (d) Büyük Menderes River, Turkey, with crevasse splay spreading over a pointbar inside an oxbow lake formed by neck cutoff. (e) Self-constraining channel of Barwon River, Australia, with tortuous planform linked to tough bank material and vegetation. (f,g) Ephemeral channels in Buia plain, Danakil Depression, south of Massawa, Eritrea, during and after a rainstorm flood peak. *Source:* Maps data: (a) Google, Maxar Technologies, Landsat / Copernicus, CNES / Airbus; (b,c) Google, Maxar Technologies, CNES / Airbus; (e) Google, CNES / Airbus; (f,g) From Massimiliano Ghinassi (Chapter author).

Associated with the meanders are chute channels, levees, crevasse channels, crevasse splays and oxbow lakes (Figure 5.8c,d).

Examples of meandering rivers that have contributed to facies models include sand-mud systems of the Endrick River of Scotland, the Wabash River of Illinois and the Beatton and Peace rivers of Canada. Meandering rivers range in scale from Muddy Creek in Wyoming with a width of 5 m and a bankfull discharge of 1.2 $m^3 s^{-1}$ (studied by Dietrich et al. 1979 to understand sediment transport on bends), to the Mississippi River with a width of 1 km near Natchez and a discharge of c. 17,000 $m^3 s^{-1}$ (studied by Fisk 1944 to understand meander-belt development). Meandering channels traverse densely vegetated plains (Mamoré River in Bolivia, Sylvester et al. 2019) and unvegetated plains (Amargosa River in California, Ielpi 2019; Río Colorado in Bolivia, Li et al. 2015). Coarse-grained (coarse sand to pebble gravel) meandering rivers such as the Colorado River of Texas have point bars with prominent chute channels, generating large chute bars where the flow re-enters the main channel (McGowen and Garner 1970).

5.3.3 Self-Constraining, Laterally Stable and Anastomosing Rivers

Self-constraining and laterally stable rivers are present in low-gradient, low-energy settings. They exhibit a high degree of tortuosity where tough banks resist erosion. Such settings include inland and relatively arid basins with indurated mud and bank vegetation, such as the Barwon (Figure 5.8e) and Murrumbidgee rivers of Australia. They also include inland and coastal deltas with fibrous peat and vegetation, including some channel systems on the Okavango Delta of Botswana and the Rhine-Meuse Delta of western Europe.

Anastomosing rivers have laterally stable or self-confining channels with varied sinuosity. In its headwaters in western Canada, the Columbia River occupies a narrow mountain valley (Figure 5.9a) where, aided by log jams and beaver dams, abundant sediment from tributaries accumulates rapidly in narrow channels with levees that border peat-bearing wetlands (Figure 5.9b). Flowing within an alluvial ridge of its own making, the river periodically breaches the levees through crevasses. In the reach illustrated in Figure 5.9b, the river has avulsed two or three times per thousand years over the past 4000 years, with active channel lifetimes of 800–3000 years (Makaske et al. 2002, 2017). Anastomosing rivers are prominent in arid central Australia, where the laterally stable, sinuous channels are cut into tough floodplain mud and levees are small or absent (Knighton and Nanson 1993; Gibling et al. 1998). In this setting, the existence of multiple channels is an adaptation to transmit seasonal discharge across a low-gradient plain (Nanson and Huang 1999).

5.3.4 Ephemeral and Chains-of-Ponds Rivers

Ephemeral rivers of dryland regions display infrequent and short-lived floods with high discharge after occasional or seasonal rainfalls. With little vegetation to help stabilise their banks, they change their location and morphology rapidly during high-discharge events (Figure 5.8f), and they show strong downstream changes in grain size and channel morphology (Figure 5.8g). Ephemeral rivers usually form terminal systems in endorheic basins, interfingering with aeolian systems and ephemeral lakes, as in the Kobo Basin of Ethiopia (Billi 2007, 2008). Channel planform ranges from meandering to braided and bedload from gravel to sand, with high bedload transport rates (Laronne and Reid 1993; Hassan et al. 2009). The broad term 'sandbed river' may be a useful descriptor for such systems. Until recently, facies models were largely lacking for ephemeral rivers (Hasson et al. 2023), and their deposits have frequently been assigned to braided or low-sinuosity systems, even though distinctive braid bars are rare.

Ephemeral rivers typically decline in flow downstream. In the deserts of central Australia, evaporation may be an order of magnitude higher than precipitation (Tooth 2000), and the infiltration of water into the porous channel bed of Cooper Creek causes transmission losses as high as 75% along some reaches, with channels transitioning into areas of unconfined flow or 'floodouts' (Knighton and Nanson 1994; Tooth 2000).

Some river systems connect strings of standing-water bodies. In upland valleys of southeast Australia, *chain-of-ponds river systems* such as Mulwaree Ponds comprise pools up to 7 m deep that are linked by preferential flow paths through vegetated swamps (Williams and Fryirs 2020). The pools have existed for thousands of years during a period of declining fluvial activity, and they probably represent deep scours along a palaeoriver course. Rivers such as the Shannon of Ireland connect strings of large lakes.

5.3.5 Large Rivers

Large rivers are not simply scaled-up versions of small rivers, although they contain braided, meandering and anastomosing reaches (Ashworth and Lewin 2012; Lewin and Ashworth 2014a, 2014b; Latrubesse 2015). Large rivers such as the Amazon, Huang He (Yellow), Yangtze and

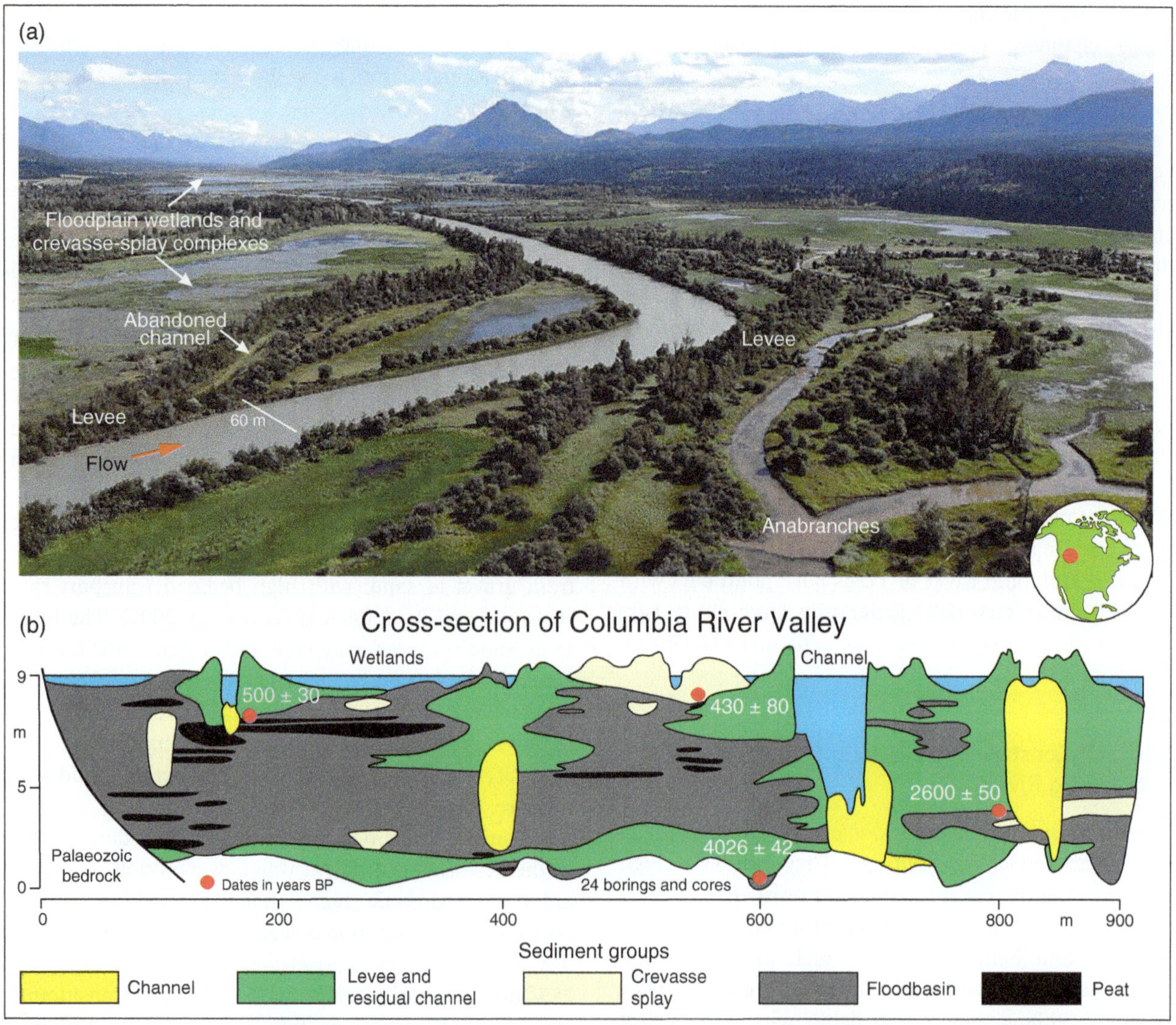

Figure 5.9 Anastomosing rivers. (a) Drone photograph of the Columbia River, Golden, British Columbia, Canada. *Source:* From Alessandro Ielpi (Chapter author). (b) Cross section of a nearby reach, showing multiple narrow channel bodies encased in floodbasin and levee deposits. *Source:* Makaske et al. (2017) / John wiley & Sons / CC BY 4.0.

Brahmaputra exceed the size of individual sedimentary basins, traversing a range of climatic belts and geological domains. Most large rivers have anabranches that divide and rejoin around alluvial islands that may have existed for millennia. Trees and log jams exert little influence on large anabranches, although cohesive muddy banks may limit their migration.

The world's largest river, the Amazon, became a connected drainage system from the Andes to the Atlantic as little as 11 million years ago (Figueiredo et al. 2009). The Amazon's discharge equals that of the next eight large-discharge rivers combined, with recent floods running to about 260,000 $m^3\ s^{-1}$ (Filizola et al. 2014). The river anabranches through 90% of the trunk system (Carling et al. 2014; Figure 5.10), with channels several kilometres wide between alluvial islands (Figure 5.10) and as deep as 100 m in places. Levees several kilometres wide impound extensive floodplains and lakes that vary in their hydraulic and biological connectivity with the trunk channel. Large tributaries from mountains and cratons supply variable quantities and grades of sediment, including large organic loads derived from the modern biomass with contributions from bedrock shales (Bouchez et al. 2010; Walcker et al. 2021). Many of these 'tributaries' are large enough to be considered separate alluvial systems, including the Tapajós (Figure 5.10), which terminates in a lake dammed by sediment laid down in the Amazon trunk channel.

Figure 5.10 Amazon River near Santarém, Brazil, with tributary junction of the Tapajós River. Annotations after Latrubesse (2015, Figure 6c) and information from Myrtle Shock. *Source:* Google Earth, Landsat/Copernicus.

5.4 Alluvial Processes

5.4.1 Sediment Release and Transport

In upland headwaters, sediment is mobilised mainly through gravity-driven processes, broadly termed *landslides* (see Keefer 1999 for a classification), which release large volumes of rock and regolith instantaneously, in geological terms, into mountain valleys. At slopes lower than about 25°, there is a linear relationship between erosion rate and slope, but at higher slopes, landslide activity is enhanced, and the relationship becomes exponential (Montgomery and Brandon 2002). Earthquakes have a profound effect on upland systems: the Assam 1950 earthquake in northern India is estimated to have released 47 km^3 of material (Keefer 1999). Megafloods generated by the breaching of impounded lakes and by extraordinary precipitation events act as major agents of erosion and deposition (Carling 1996; Hewitt et al. 2011). Elsewhere on the steep slopes of geologically young mountain ranges, deep-seated gravitational slope deformation straddles the boundary between landslides and shallow neotectonic deformation. Such deformation processes release large quantities of sediment to river valleys as they evolve over centuries to a few thousand years (Pánek and Klimeš 2016).

In areas of permafrost, slow downslope movement of the active layer generates *solifluction lobes* that transport regolith into depressions and valleys. Overland flow and gullying associated with enhanced surface runoff over permafrost, aided by aeolian activity, contribute to the formation of denudational landforms termed *cryopediments*, especially in areas of discontinuous permafrost and deep winter frost (Vandenberghe 2024). Such surfaces have slopes from a fraction of a degree to a few degrees, and they grade distally into alluvial terraces and valley floors. Although braided channels are often considered characteristic of permafrost zones, fluvial styles in such settings are diverse, ranging from single- to multi-channel forms (Vandenberghe 2024). *Rock glaciers* of ice-cemented debris are locally important features.

In upland valleys and on alluvial fans, mixtures of water and coarse, immature sediment are transported by a range of processes that yield distinctive deposits (Carling 1988; Brenna et al. 2020; Figure 5.11). *Debris flows* are mixtures of water, sediment, logs and other organic debris, moving largely as cohesive masses with a high sediment concentration and predominantly plastic behaviour (near-homogeneous internal shear). In the three other modes of flow shown in Figure 5.11, the bed may be largely static with some entrainment of sand (Phase 1), partially entrained and transported at a low rate (Phase 2), or fully

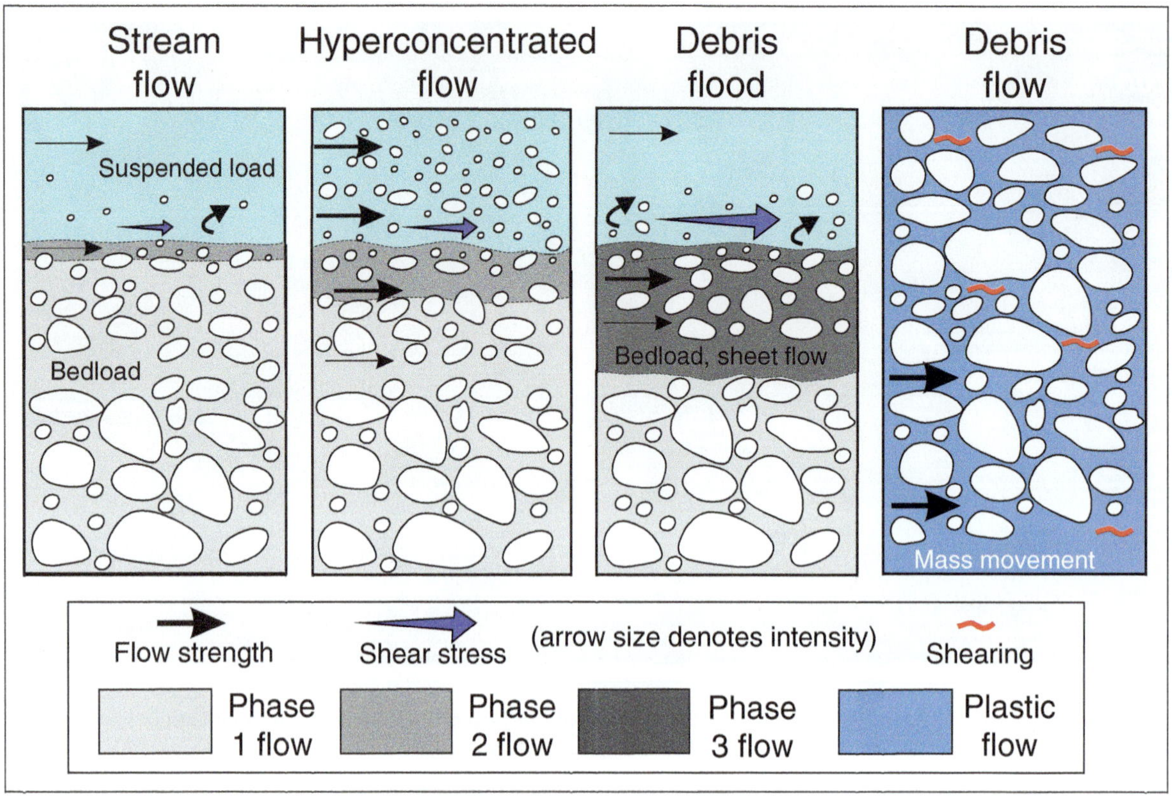

Figure 5.11 Flow modes in gravel systems. *Source:* Brenna et al. (2020, their Figure 1). Flow phases are described in Carling (1988).

mobile (Phase 3). *Stream flow* (also termed *water flow*) is the ordinary flow of rivers, where turbulence under viscous forces acts on grains transported as bedload and suspended load, variably at Phase 1–3. The suspended-sediment concentration is too low to influence the rheological properties of the flowing water. *Hyperconcentrated flows* are turbulent flows with intermediate plastic to viscous behaviour (cohesive to non-cohesive), highly charged with suspended sediment and typically at Phase 2–3. *Debris floods* are water-driven flows that mobilise sediment at Phase 3 to a considerable depth below the riverbed, once the coarser bed armour is removed. This results in high rates of bedload transport in a 'live bed' condition (Church and Jakob 2020). Debris floods may also result from dilution of debris flows (water injection), or from additional concentration in hyperconcentrated flows (sediment injection). Flows of several types commonly transport outsized boulders (Figure 5.6). On alluvial fans, *sieve deposits* form lobes where coarse sediment almost devoid of fines is transported downslope, triggered by rainfall that infiltrates into the permeable ground (Novak et al. 2023).

Examples from upland rivers illustrate deposits laid down by some of these processes (Table 5.1). Over three days in October 2018, Storm Vaia swept through northeastern Italy with high winds and >500 mm of precipitation, generating high-magnitude floods that laid down up to 2 m of gravel in the Piave River catchment (Brenna et al. 2020; Brenna et al. 2024; Figure 5.12a). Debris flows were prominent in most steep tributaries, depositing disorganised gravel with little preferred orientation (Figure 5.12c). In larger channels, considerable volumes of gravel were mobilised as debris floods under high shear stress, forming poorly stratified deposits. Clusters of steeply dipping, imbricated clasts were formed, and outsized boulders promoted local scouring (Figure 5.12b,d). Water flows in many reaches laid down stratified and imbricated gravel, especially as the floods waned (upper part of Figure 5.12d). In the Colorado Front Ranges over a week in September 2013, >450 mm of rain fell on saturated ground, triggering >1000 debris flows that mobilised surficial material and surged through upland channels (Coe et al. 2014). The deposits were poorly sorted and unstratified, with outsized boulders and minimal clast orientation (Figure 5.12g,h).

In lowland rivers, debris flows, hyperconcentrated flows and debris floods may occur in lowland rivers near basin margins, in active tributaries, and downflow from alluvial

Table 5.1 Sedimentary features formed in gravel under a range of processes.

	Debris flow	Debris flood	Hyperconcentrated flow	Stream flow
Sorting	Very poorly to poorly sorted. Outsized and angular clasts common, from slope debris	Poorly sorted	Moderately to well sorted	
Grading	Generally ungraded, or with reverse grading	Normal grading uncommon to common		
Clast preferred orientation	Absent to poor, a axes unoriented or parallel to flow	Absent to poor, a axes parallel to transverse to flow	Absent to poor, a axes rarely oriented	Prominent, a axes transverse to flow
Imbrication	Absent to poor, with steeper dips of imbricated clasts (ab planes) as flows transition to debris flows			Generally prominent, with shallow ab dips
Clast or matrix support	Matrix-support common	Clast-support with openwork		
Stratification	Unstratified, may show shear bands, slump folds and faults.	Poorly stratified	Unstratified to poorly stratified	Well stratified, with bedforms
Vegetation	Generally abundant	Present in modest amounts		

The presence or absence of fabric elements may depend on clast form. Modified from Brenna et al. (2020).

fans. However, sediment is mainly transferred by streamflow, generating a suite of *bedforms* in fine gravel, sand and coarse silt with a range of stratification styles and related sedimentary structures (Collinson and Mountney 2019; see photos in Section 5.6). The bedforms include ripples, dunes, hump-backed dunes and plane beds in subcritical to transcritical flows, and antidunes and cyclic steps in supercritical flows (Wang and Plink-Björklund 2020; Zellman et al. 2020; Walker and Holbrook 2023, their Figure 3; Supplementary Table 5.S.1: 2–3). Dunes are considered the primary building blocks of alluvium (Best and Fielding 2019). They tend to scale with shear stress and are sensitive to water depth, but the relationship between flow depth and bedform relief is non-linear (Bradley and Venditti 2017). Many dunes have low-angled leeside slopes (<10°), much less than the angle of repose that is commonly assumed (Cisneros et al. 2020). Groove and flute casts form where turbulent stream flows scour muddy substrates (Figure 5.13a).

Coarse gravel and mixed gravel-sand systems exhibit gravel clusters, transverse ribs, gravel dunes and bedload sheets that generate plane beds with varied grain size (Supplementary Table 5.S.1: 4). Gravel particles deposited under tractional conditions commonly show imbrication with their long (a) axes transverse to flow and their intermediate (b) axes dipping upflow, generally at shallow angles (c. 30°). Where the particles moved in saltation or suspension, their long axes may be parallel to flow and dip upstream. Imbricated bar deposits and channels with gravel dunes are illustrated from stream-flow deposits in the Tagliamento River of Italy (Figure 5.12e,f).

5.4.2 Variations in Flow Level

In alluvial channels, floods that exceed the bankfull level have been classically reported to occur about every 1.5 years, with a range from about 1–32 years (Williams 1978; Andrews 1980). Some channels inherited cross-sectional geometries that are out of equilibrium with the current discharge, for example, where post-glacial streams are 'underfit' in broad channels scoured by meltwater. Such channels may experience much longer bankfull recurrence intervals. Since the characteristic flood discharge governs the channel cross section at equilibrium, the cross section may shrink in tandem with decreased discharge downstream in strongly seasonal and ephemeral rivers. Over a 30 km stretch in the meandering ephemeral Río Colorado River of Bolivia, the channel decreases gradually in bankfull width from 70 to 6 m and in depth from 2 to 0.5 m, terminating in a belt of amalgamated crevasse splays and terminal splays (Donselaar et al. 2013). In the Mojave River of California, formative flows are transmitted from the humid headwaters to the arid terminus over short periods despite hydrologic loss (Ielpi et al. 2020).

Figure 5.12 Gravel deposits from 2018 Storm Vaia in the Piave River system of Italy (a–d), from normal flow conditions in the Tagliamento River, Italy (e, f), and from 2013 floods in the Colorado Front Ranges, USA (g, h). Photos (a–f) were taken in 2023, and photos (g, h) were taken a few days after the 2013 floods. The gravel deposits are attributed to facies Gms, Gcs and Gi. (a) Poorly sorted gravel in channel 30 m wide, fed by hillslope debris flows. (b) Imbricated pebble clusters with steep dips. (c) Debris-flow deposits in a steep tributary channel, overlying debris-flood deposits in the main channel. (d) Debris-flood deposits in main channel, with scour around outsized boulders, overlain by stratified gravel from the waning flow. (e) Gravel dunes with 30 cm amplitude in shallow channel. (f) Bar deposits at right of (e), with shallowly dipping imbrication and grain-size banding. (g, h) Debris-flow deposits choking upland creeks, with outsized clasts and little preferred orientation. *Source:* From Martin Gibling (Chapter author).

Figure 5.13 Processes associated with (a) high-strength flow, (b) low-stage flow, and (c) vegetation. From left to right: (a) Triassic, Sardinia, Italy; Joggins Formation, Pennsylvanian, Canada; Port Hood Formation, Pennsylvanian, Canada. (b) Great Basin, Nevada, USA; Husky Creek Formation, Proterozoic, Nunavut, Canada; Vega Formation, Jurassic, Spain. (c) Saskatchewan River, Saskatchewan, Canada; Chinle Formation, Triassic, Arizona, USA; Platte River, Nebraska, USA; Joggins Formation, Pennsylvanian, Canada. *Source:* From Martin Gibling, Massimiliano Ghinassi, and Alessandro Ielpi (Chapter authors). Facies codes in Table 5.4.

Unconfined flow may result where flow is released across the adjoining plain from confined channels (North and Davidson 2012), or from local precipitation in overbank areas. Overbank floodwater may return to channels down bar surfaces in the downstream part of meander bends, where gullying promotes headward erosion (Gay et al. 1998). Floods may sweep sand-sized mud aggregates eroded from soils into the channels, where they contribute to the bedload and generate unusually muddy channel fills (Rust and Nanson 1989).

The recurrence interval and duration of bankfull conditions also influence the groundwater system. During floods, water percolates through permeable bed, bank and floodplain sediments into the hyporheic zone, recharging the groundwater column and causing considerable transmission losses. At low stage, groundwater may flow back into the channel, maintaining the base flow, promoting the collapse of porous sandy banks in rivers such as the Brahmaputra, and contributing nutrients to the ecosystem (Supplementary Table 5.S.1: 6). A shallow groundwater lens below channel belts promotes the growth of in-channel and riparian (bank-line) vegetation.

Waning and prolonged low-stage flows generate a range of distinctive structures (Figure 5.13b). They include desiccation cracks, rain prints, rills formed where water drains from the floodplain or emerges from the bank sediments, and symmetrical ripples in standing water, commonly with planed-off crests where the bedforms are exposed. 'Swept ripples' may form where the upslope parts of ripple crests are stranded by falling water levels while the downslope parts continue to migrate, generating oblique crest-lines that do not represent the flow direction (Simon and Gibling 2017). As water levels fall, mud drapes plaster the banks and are indurated during the succeeding dry period. Bar slipfaces may be eroded during waning and low-flow periods, leaving reactivation surfaces once the next flow event covers the slipface with younger sediment.

5.4.3 Pedogenic, Colluvial and Aeolian Processes

Soil formed on floodplains and in abandoned channels may constitute a large proportion of alluvial successions (see Mack et al. 1993; Kraus 1999, and Sheldon and Tabor 2009 for soil and palaeosol description and classification, and photos later in this chapter, e.g. Figure 5.22). Carbonate-rich palaeosols and groundwater carbonate are locally common. Groundwater carbonate precipitates in seasonal climates at subsurface permeability contrasts such as channel and valley bases (Nash and Smith 2003). Thick, indurated regoliths (duricrusts) of carbonate, siliceous and ferruginous composition form in stable landscapes such as those of central Australia, locally confining and supplying debris to channels. Travertine and tufa form at hydrothermal and cold springs in river valleys, controlling knick-points and channel equilibrium profiles (Capezzuoli et al. 2014, see definitions in their Table 5.1).

Colluvial processes are common along the steep flanks of river valleys and along fault lines. They include rainwash, unconfined sheetflow, processes of mass movement and gullying of hillsides and floodplains (Supplementary Table 5.S.2; Nelson 1992). Although these processes are usually localised, gullied badlands cover thousands of square kilometres of young alluvium along Indian rivers such as the Chambal (Haigh 1984). Gully erosion may supply a disproportionate amount of sediment to river and coastal systems. For example, hillslope and alluvial gullies constitute as little as 0.1% of a catchment in Queensland, Australia, but supply 40% of the fine sediment exported from the river basin (Khan et al. 2023). Seismic shaking may cause surface rupture (Figure 5.14d), hillslope collapse and colluvial deposition, sediment liquefaction, convoluted laminae, and fields of small sand volcanoes, as in fault zones in the Mississippi valley (Obermeier et al. 1990). Dewatering more generally may also cause unstable sand to collapse.

Aeolian and fluvial processes interact where rivers cross aeolian dune fields and where sand blown out of channels forms source-bordering dunes (Figure 5.14b; Supplementary Table 5.S.1: 8). Windblown silt and fine sand are mobilised from glacial terrains, desert rivers, playa lakes and exposed shallow-marine tracts, covering interfluves with loess that blows or is washed into channels, modifying its grain-size distribution. Many of the world's large rivers, including the Rhine and Yellow, carry significant loess loads. Lava, volcanic ash and ignimbrites affect some river systems (Figure 5.14c). Lahars (debris flows from volcanic areas, some of them hot) may yield deposits 50 m thick, with runout distances that may exceed 100 km along river valleys. Many anthropogenic processes modify the dynamics of sediment transport and introduce markers in the form of *technofossils* that include stone tools, pottery, brick and plastic.

5.4.4 Biological Processes

A wide variety of animals and plants produce biogenic structures that yield important environmental information in alluvial settings (Figure 5.15; see Section 3.3.3). In contrast with most marine environments, trace-fossil suites are only locally abundant, and they tend to be monospecific or contain few ichnotaxa, with trace fossils restricted to discrete beds or bedding planes. However, palaeosols may display relatively high trace-fossil diversity (Bellosi et al. 2016;

Figure 5.14 Processes associated with (a) microbes, (b) aeolian activity, (c) volcanism, and (d) liquefaction and seismic action. From left to right: (a) Kicking Horse River, Field, Canada; Horton Bluff Formation, Mississippian, Canada. (b) Dandiero Basin, Eritrea (2 photos); Pliocene, Valdarno Basin, Italy; Cañizar Formation, Triassic, Spain; Halgaito Formation, Pennsylvanian / Permian, Utah, USA. (c) Pleistocene, Dandiero Basin, Eritrea. (d) Stoer Group, Proterozoic, Scotland, UK; Tynemouth Creek Formation, Pennsylvanian, Canada. *Source:* From all Chapter authors. Facies codes in Table 5.4, see Section 7.6.5 on aeolian-fluvial interactions.

Figure 5.15 Representative trace fossils in Pennsylvanian to Jurassic alluvial settings. (a) Tetrapod trackway *Ichniotherium sphaerodactylum*, lower Permian, Tambach Formation, Bromacker Quarry, Thuringian Forest, Germany. (b) Fish trail *Undichna* isp., middle Permian, Abrahamskraal Formation, Gansfontein palaeosurface, Karoo Basin, South Africa. (c) Therapsid burrow oriented oblique to stratification, Lower Triassic, Katberg Formation, New Lootsberg Pass, Karoo Basin, South Africa. (d) Claw-scratched ventral walls of a therapsid burrow, Middle Triassic, Burgersdorp Formation, Burgersdorp, Karoo Basin, South Africa. (e) Arthropleurid trackway *Diplichnites cuithensis*, Pennsylvanian, Tynemouth Creek Formation, New Brunswick, Canada. (f) *Tambia spiralis*, lower Permian, Tambach Formation, Bromacker Quarry, Thuringian Forest, Germany. (g) *Taenidium* isp., middle Permian, Alcotas Formation, Landete, Spain. (h) *Scoyenia gracilis* and raindrop imprints, Lower Jurassic, Kayenta Formation, Canyonlands National Park, Utah, USA. (i) *Treptichnus* isp., middle Permian, Abrahamskraal Formation, Gansfontein palaeosurface, Karoo Basin, South Africa. (j) Tetrapod trackway *Chirotherium barthii*, Middle Triassic, Solling Formation, replica of type surface from Winzer Quarry, *Chirotherium* Monument Hildburghausen, Thuringia, Germany. (k) Dinocephalian trackway, middle Permian, Abrahamskraal Formation, Gansfontein palaeosurface, Karoo Basin, South Africa. (a, e–g) = in abandoned channel deposits; (b, c, h–k) = in overbank deposits; (d) = in crevasse splay deposits. *Source:* From Luis Buatois (Chapter author).

Genise 2017). Discharge variability and the spacing of depositional episodes govern windows for colonisation, and the depth to the water table controls the type and preservation of biogenic structure (Supplementary Table 5.S.1: 13; D'Alessandro et al. 1987; Buatois and Mángano 2011). Many biological processes are focused along dynamic boundaries within fluvial systems: levees that separate channels from floodplains, and shorelines that separate alluvial ridges and islands from anabranches and wetlands (Tockner et al. 2003). In particular, riparian ecosystems connect the aquatic and terrestrial realms (Brown et al. 1979; Malanson 1993).

Active channels are generally unsuitable for colonisation by the benthos and for the preservation of biogenic structures (Figure 5.16). This reflects the fluctuating but overall high energy, coarse grain size, rapid changes in sedimentation and erosion and unstable banks and floors. In contrast, abandoned channels and overbank settings may be intensely bioturbated and yield well-preserved trace fossils that record episodes of submergence, desiccation, dewatering and vegetation growth (see Smith 1993; Savrda et al. 2000). Alluvial deposits essentially display three ichnofacies (*Skolithos*, *Scoyenia* and *Mermia*), illustrated in Figure 5.16 and discussed in Section 5.11.

Palaeosols associated with alluvial settings host additional ichnofacies (*Coprinisphaera*, *Celliforma*, *Termitichnus* and *Camborygma*) (Genise 2017), and the soil biotas and their corresponding biogenic structures reflect numerous factors. These include the vegetation type (e.g. grasslands, closed forests and scrub), the intensity of soil-forming processes such as laterisation, the position of the water table (e.g. well-drained, seasonally flooded and waterlogged soils), climatic factors, the brief or extended duration of subaerial exposure and macroevolutionary trends (see Section 5.16.1). These controls are particularly evident at the ichnofacies scale (Genise et al. 2000, 2016a;

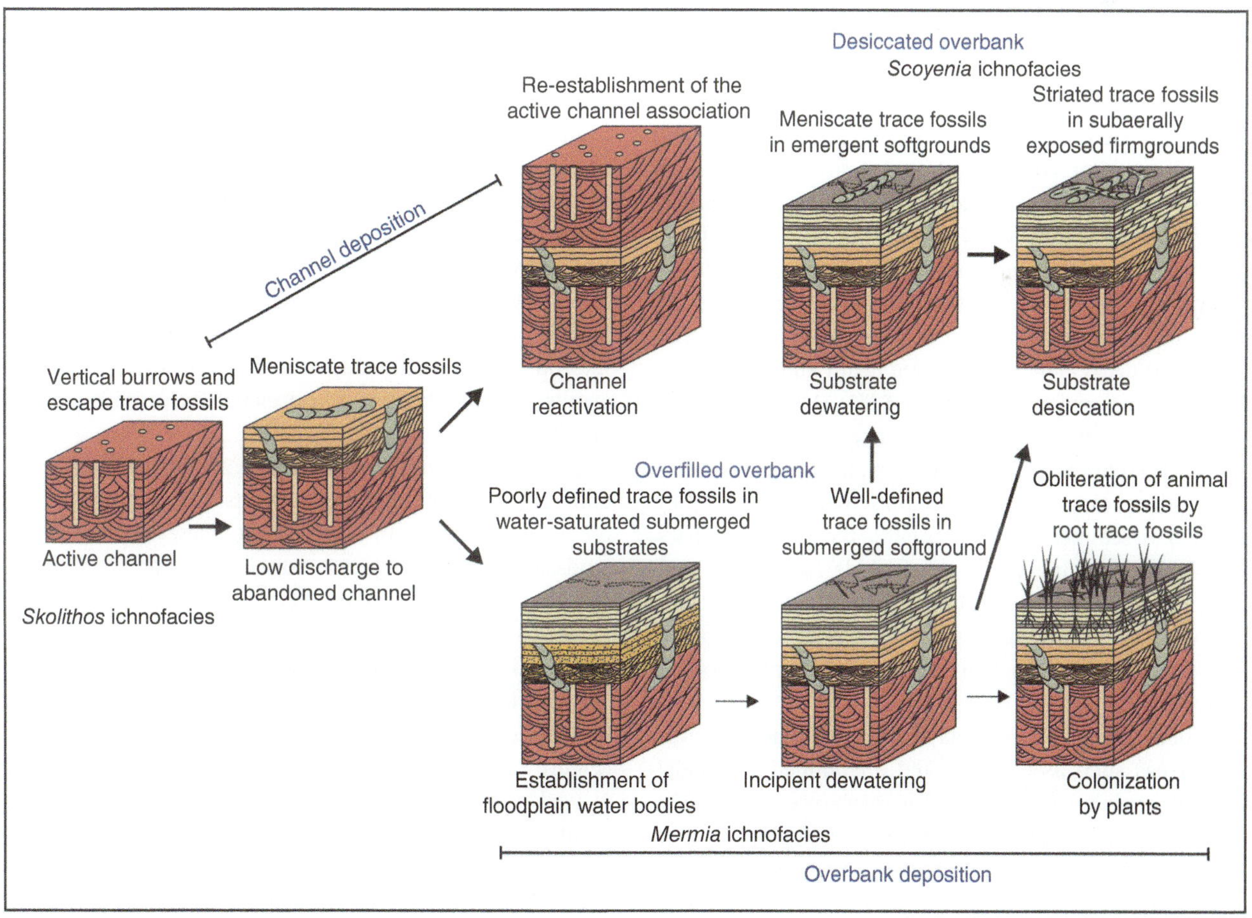

Figure 5.16 Taphonomic pathways for preservation of alluvial trace fossils. *Source:* Buatois and Mángano (2004) / Geological Society of London.

Genise 2017). The *Coprinisphaera* Ichnofacies, dominated by trace fossils of bees, wasps, ants and dung beetles, is present in soils formed under cold and dry to warm and humid conditions, where herbaceous communities encompass savannas, grasslands, prairies and steppes (Genise et al. 2000). The *Termitichnus* Ichnofacies, dominated by termite nests, is present in closed-forest soils under warm and humid conditions and a high water table (Genise 2017). The *Celliforma* Ichnofacies, dominated by halictid bee nests, is present in well-drained scrub and woodland soils, including calcretes and soils in paralic areas (Genise et al. 2000). The *Camborygma* Ichnofacies, dominated by crayfish and earthworm burrows, is associated with a range of vegetational settings that include forest, scrub and open herbaceous communities, especially in wetlands with a fluctuating but relatively high water table and a warm climate (Genise et al. 2016a). These ichnofacies are directly linked with the rise to dominance of specific insect groups (Figure 5.54).

Vegetation exerts a major influence on channel processes and bank strength (Supplementary Table 5.S.1: 9–10). Woody debris (Figure 5.13c), leaves, peat and charcoal are deposited in rivers and wetlands, and roots stabilise the upper parts of riverbanks and slump blocks, reducing bank erosion. Rhizoconcretions form where carbonate precipitates around roots, contributing to soil strength, and they are washed out into channels and gullies. *Vegetation-induced sedimentary structures* (VISS, Rygel et al. 2004) form where sediment is eroded and deposited around standing vegetation, yielding vegetation shadows, scours with centroclinal cross-stratification (dipping towards the tree trunk), and decay hollows from tree death and windthrow (Figure 5.13c; Supplementary Table 5.S.1: 11). Capable pioneering plants that resist disturbance, such as willow (*Salix* sp.) and horsetails (*Equisetum* sp., including calamitaleans in the geological record), colonise bars (Figure 5.13e) and build up sediment where they baffle the flow. Other bars form around 'key members' that include fallen trees and large logs, providing habitat for new trees in the 'floodplain large-wood cycle' (Collins et al. 2011). Log jams may block channels (Figure 5.13c) and promote overbank flooding and avulsion.

Microbial mats (Figure 5.14a) stabilise sediment in protected areas such as bar tops and abandoned channels. Although their organic material is rarely preserved, they leave *microbially induced sedimentary structures* (MISS, Noffke 2010), although other processes may produce some similar features (Davies et al. 2016). Other common organic components include the shells of freshwater organisms and fish and tetrapod bones. Many organisms act as ecosystem engineers (Supplementary Table 5.S.1: 12). For example, beaver dams cause ponding, and hippopotamuses trample through riverbanks and promote avulsion (as dinosaurs may have done).

5.5 Facies Analysis: Conceptual Approach

Our guiding concept to facies analysis is that of *Landscape Systems*, with a progressive increase in the scale of analysis (Figure 5.17) and the probability that modern fluvial features will be preserved somewhere in the geological record. This stepped, map-like approach is loosely hierarchical, with iterations between the steps. *Facies* (Chapter 2) are identified from key features in 1D to 2D outcrop sections or cores (Table 5.2). Each facies represents a distinctive and recurring sediment type generated as a small sediment patch by a single process or several possible processes. *Facies successions* are vertical stacks of facies that record progressive and relatively orderly change in hydrodynamic and other processes during deposition. *Facies associations* are groups of facies, not necessarily disposed in an orderly fashion, with a combination of processes that is diagnostic of a river sub-environment such as a channel reach or part of a floodplain. Facies successions typically form local components of broader facies associations. Following Walther's law, alluvial facies that are above or below each other in a facies succession or facies association may formerly have been laterally adjacent to each other on the alluvial surface. Exceptions include the deposits of debris flows and similar events, which may rest on unrelated facies with a stratigraphic break.

Geomorphic units are identified in 2D to 3D outcrops and data arrays, representing components that might have been observed obliquely in a bird's-eye view of the former landscape. Although facies associations and geomorphic units both comprise groups of facies, it is helpful to identify facies associations first because (i) scattered datasets allow only a low-level interpretation of facies groups, and (ii) identification of geomorphic units draws additionally on palaeoflow analysis to explore the *facies architecture* (Figure 5.2). The identification of geomorphic units may be viewed as refining the facies associations. A related approach identifies *architectural elements*, linked in part to the geomorphic setting. Finally, the broader *depositional system* may be established, aided by *comparison with modern and ancient systems*, after which a more extensive 3D study may delineate the *basin architecture*.

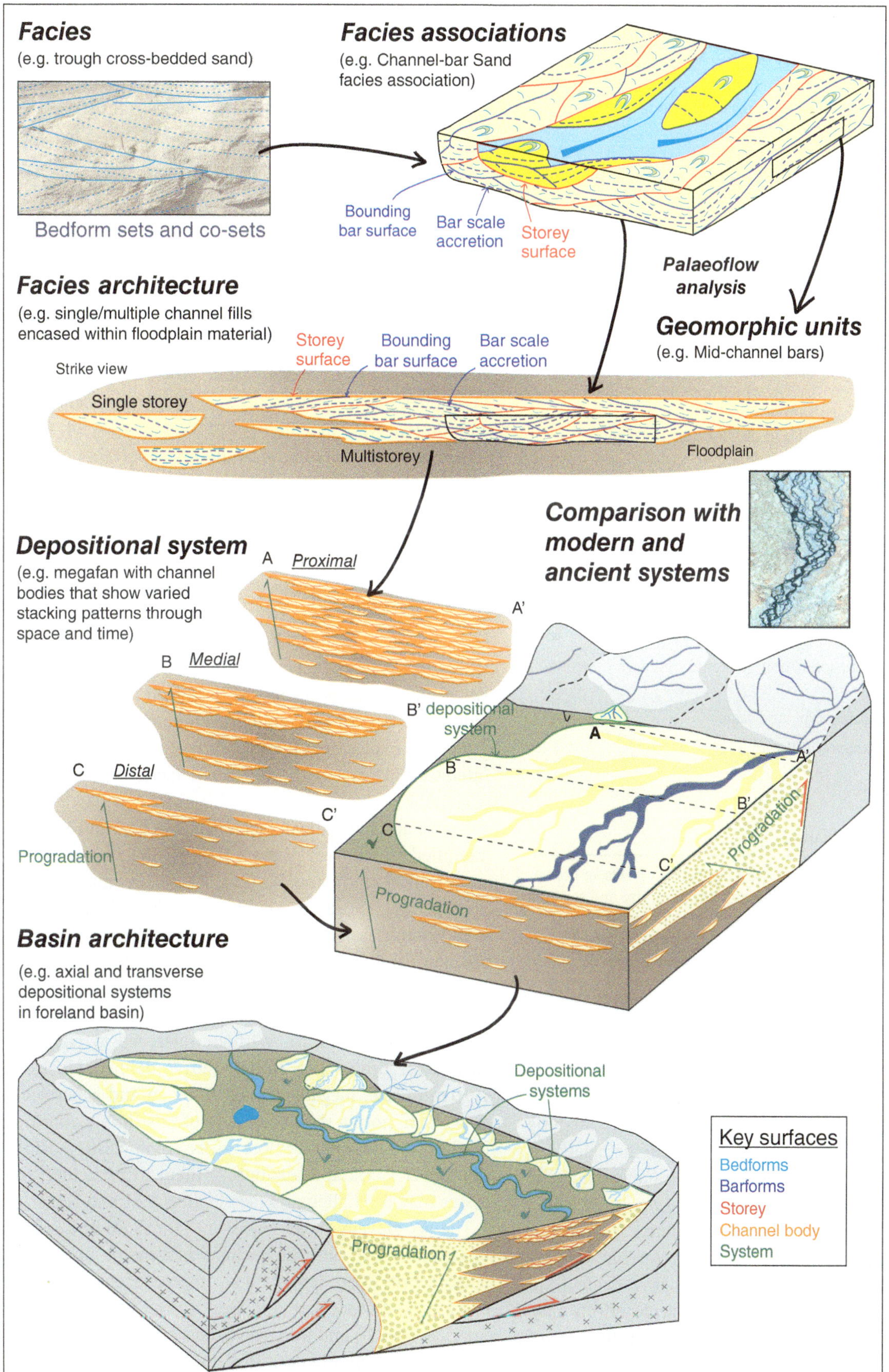

Figure 5.17 Landscape System for facies analysis, scaled up from facies to basin architecture. *Source:* Owen et al. (2017) / John Wiley & Sons.

Table 5.2 Descriptive features for clastic strata in alluvial sections.

Gravel and conglomerate
Gravel clast size: mean long axis, maximum long axis
Clast shape
Clast roundness
Clast sorting
Clast and matrix proportions
Clast types and percentages
Clast orientation: imbricated, not apparent
Clast grading: normal, inverse and not apparent
Stratification (may be structureless)
Sand and sandstone
Sand grain size
Grading: coarse-tail grading and distribution grading
Sand composition (may not be possible to determine)
Stratification: bed or lamina thickness and style (may be structureless)
Mud and mudstone
Proportions of silt and clay
Stratification and fissility (may be structureless)
All strata
Basal bed contact: erosional/non-erosional, relief on erosional surfaces
Sedimentary structures: type and scale
Colour: fresh, weathered and mottling
Body fossils
Ichnofossils
Concretions and cementation

The section can be recorded as a written log, on data sheets, or drawn in the field. See textbooks for classifications of grain size and sorting, grain form and stratification thickness and style. Collinson and Mountney (2019) illustrate many sedimentary structures. See Bridge (1993) for comments about the consistent use of appropriate terminology.

The level of detail required for each component of this approach depends on the purpose of the analysis (Table 5.3). For example, an interpretation of the palaeoenvironment requires a full analysis of facies and geomorphic units to constrain the range of processes, the geomorphic assembly and (where possible) the planform. In contrast, analysis of an aquifer or hydrocarbon reservoir requires an emphasis on petrographic detail to investigate fluid storage and flow. A study of anthropogenic effects may require high-resolution dating and the identification of technofossils.

5.6 Facies and Facies Successions

5.6.1 Facies Types

Table 5.4 lists 22 common facies, illustrated in Figures 5.18–5.22 (see Supplementary Table 5.S.6 for a fuller description). They are designed to be identified in the field, and they involve a first-order interpretation of depositional processes. The facies are largely *microscale* and *mesoscale* deposits (Bridge 1993) – informal terms that are commonly linked to the dimensions of small-scale (ripple) cross-sets and large-scale (dune) cross-sets, respectively. A few facies are *macroscale* deposits such as landslides and accumulations from large-volume flows. The facies are assigned codes (Miall 1978), but researchers should feel free to design their own facies names and codes for sediment types that are not included in Table 5.4. Bridge (1993) considered coding to be an imposition that required learning the facies language of 'acronymese'. Nevertheless, we have found facies codes to be useful shorthand, providing that the facies are described in detail and that the process of formation is evaluated (several processes may produce one facies, and one process may produce several facies). We specify grain size only generally because calibre varies widely in many facies and because sand and mud are often mixed in rapidly decelerating flows (Baas et al. 2016). We prefer the term 'facies' to 'lithofacies' because many sediment types contain biological components that provide key points of description and greatly aid the interpretation. Interstratified or closely associated facies may be linked (e.g. Sr/Sh).

Gravel and megagravel facies (Figures 5.19 and 5.20) are identified on the basis of grain size, sorting, degree of clast organisation and stratification. Clasts in mass-wasting deposits may exceed the largest size class of the Udden–Wentworth classification, in which boulders have an intermediate (b) axis of 25.6–409.6 cm. For larger clasts, Blair and McPherson (1999) extended the classification to define blocks (4.1–65.5 m), slabs (65.5 m to 1 km) and monoliths and megaliths (>1 km). Without reference to particular clast axes, Bruno and Ruban (2017) defined blocks as 1–10 m in size, megablocks as 10–100 m and superblocks as >100 m, noting the advantages of this simple approach in the field and for remote assessment. Distinguishing the gravel facies and attributing them to particular processes may be challenging (see Figures 5.11 and 5.12).

The six gravelly facies represent landslides and related processes (Gb), slow downslope movement of regolith (Gd), debris flow, debris flood, hyperconcentrated flows and stream flows (Gms and Gcs), gravel with prominent imbrication (Gi) and thin gravel lags on erosional surfaces (Gl). Matrix-supported gravel (Gms) may be prominent in

Table 5.3 Key features for varied purposes of alluvial analysis.

Purpose	Key features							
	Facies and facies associations	Geomorphic units	Architecture	Planform	Palaeoflow	Palaeobiology	Petrography and geochemistry	Chronology
Interpretation of palaeoenvironment	XXXX Processes	XXXX Fluvial style	XXXX Supports geomorphic units	XXX	XXXX Supports geomorphic units	XXXX	X	XX
Basin-scale architecture and tectonics	XXX	X	XXXX Supports depositional systems	X	XXXX Regional flow trends	X	X	XXX Biostratigraphy, physical dating
Provenance	XX		XX Supports depositional systems		XXXX Transport paths		XXXX Dense minerals, U-Pb dating of zircons; links to source areas	XXX Biostratigraphy, physical dating
Palaeoclimate	XXXX Climatic indicators	XXX Fluvial style	XXX	XX		XXXX Biomes	XXXX Indicator minerals, palaeosols, weathering	XXXX Biostratigraphy, physical dating, proxy records
Palaeobiology	XXXX Processes	XXX Fluvial style	XXX	XX		XXXX Biomes	XXX Palaeoclimatic features, ichnofossil identification and structure	XXX Biostratigraphy
Hydrocarbon reservoir and aquifer assessment	XXXX	XX	XXXX 3D sediment bodies, fluid compartments	X	XXXX 3D sediment bodies, fluid compartments	XX	XXXX Fabric, porosity, permeability, diagenesis, maturation, petrophysics	XXXX Biostratigraphy
Geohazards	XXXX Especially for upland valleys	XXX	XXX Supports facies analysis	X	X		XXX Clay analysis, grain size, geotechnical properties	XXXX High-resolution physical dating, proxy records
Human influence	XX	X	X	X	X	XXXX Palynology, domesticated organisms; palaeoclimate	XXXX Technofossils	XXXX High-resolution physical dating, proxy records

XXXX = very important; XXX = important; XX = valuable; X = minor; empty = not needed.

upland valleys and on alluvial fans, where the facies is commonly attributed to muddy debris flows. Imbrication is common in clast-supported gravel (Gcs) and, less commonly, in matrix-supported gravel (Gms), but gravels with an obvious imbricated fabric are worth identifying separately (Gi) because they provide information about bedload transport processes and palaeoflow direction. Similarly, gravel accumulations on erosional surfaces are worth

Table 5.4 Facies commonly present in alluvial and related successions.

Code	Facies	Description	Interpretation
Gb	**Boulder and block gravel/ megagravel**	Structureless to weakly stratified, clasts angular and unsorted, with outsized blocks	Debris from rock avalanches and falls, regolith, and glacial activity may evolve into sediment-gravity flows.
Gd	**Disorganised gravel**	Structureless to weakly stratified, clasts angular and poorly sorted, clast or matrix support	Talus from downslope movement of rock and soil, with solifluction deposits over frozen ground
Gms	**Matrix-supported gravel**	Structureless to moderately stratified, angular clasts supported by finer matrix. Fabric generally disorganised	Debris flows, local debris floods, hyperconcentrated flows and high-energy stream flows
Gcs	**Clast-supported gravel**	Weakly to well stratified, angular to rounded clasts, clast supported, with imbrication common	Debris floods, hyperconcentrated flows and stream flows
Gi	**Imbricated gravel**	Clast-supported and well stratified, imbrication prominent	Stream flows with mobilisation and adjustment of bedload clasts
Gl	**Gravel lag**	Thin layer on erosional surface, with local mud and carbonate clasts	Channel-base gravels, with some clasts eroded from banks and floodplains
Gt, St	**Trough cross-bedded gravel, sand**	Cosets of trough cross-beds, with sets >5 cm thick	Suites of 3D dunes, from Froude-subcritical stream flow
Gp, Sp	**Planar cross-bedded gravel, sand**	Planar cross-beds, as cosets or solitary sets >5 cm thick	Suites of 2D dunes, and solitary bars and scour fills, from Froude-subcritical stream flow
Gx, Sx	**Cross-bedded gravel, sand**	Cross-beds of indeterminate trough or planar form	2D and/or 3D dunes, from Froude-subcritical stream flow
Sr	**Cross-laminated sand**	Cosets of trough and planar cross-laminae, with sets <5 cm thick	2D and 3D ripples, from Froude-subcritical stream flow
Sh	**Plane-laminated sand**	Planar laminae with current lineation. Convex-up forms common	Plane beds and transitional antidunes, from Froude-transcritical stream flows
Sac	**Mound-and-scour sand**	Convex-up bedsets and scours, with current lineation	Transitional antidunes, antidunes, chutes and pools, and cyclic steps, from Froude-transcritical to -supercritical stream flow
Sm	**Massive sand**	Structureless to weakly stratified, commonly with cryptic convolute laminae	High-density currents, mass flow, dewatering and destruction of stratification by burrows and roots
Gs, Ss	**Scour-fill gravel, sand**	Sediment filling solitary or grouped scours	Filling of scours generated during floods or low-discharge periods
Fl	**Laminated mud**	Well laminated, with desiccation features, plants, roots, trace fossils and shells	Suspension deposits in shallow standing water
Fm	**Massive mud**	Structureless to weakly stratified, with soil features, roots, trace fossils and desiccation features	Suspension deposits, commonly exposed and altered by soil formation
Srt, Frt	**Rooted sand, mud**	Root traces predominant, preserved as carbon or drab mottles	Palaeosols, or strongly rooted sediment
Vc	**Coal (peat)**	Carbon-rich sediment with charcoal, leaves and wood	Histosols (peaty soils) or water-lain plant accumulations
Gv, Sv, Fv	**Organic-rich gravel, sand and mud**	Wood, leaves and charcoal prominent	Accumulation of sediment with organic detritus
Cn	**Nodular carbonate**	Type 1: Concretions, some around roots. Type 2: Carbonate lenses on channel bases, gullies and erosion surfaces	Type 1: Calcic soil horizons. Type 2: Carbonate precipitation at permeability contrasts
Cs	**Stratified carbonate**	Well stratified, with peloids, oncoids, shells and stromatolites. Local travertine and tufa	Precipitation through biological and physicochemical activity and evaporation in channels, springs and lakes
Ga, Sa, Fa	**Anthropogenic gravel, sand and mud**	Technofossils prominent, including brick, masonry, pottery and plastic	Human activity at the site or nearby, dating from the Palaeolithic onwards

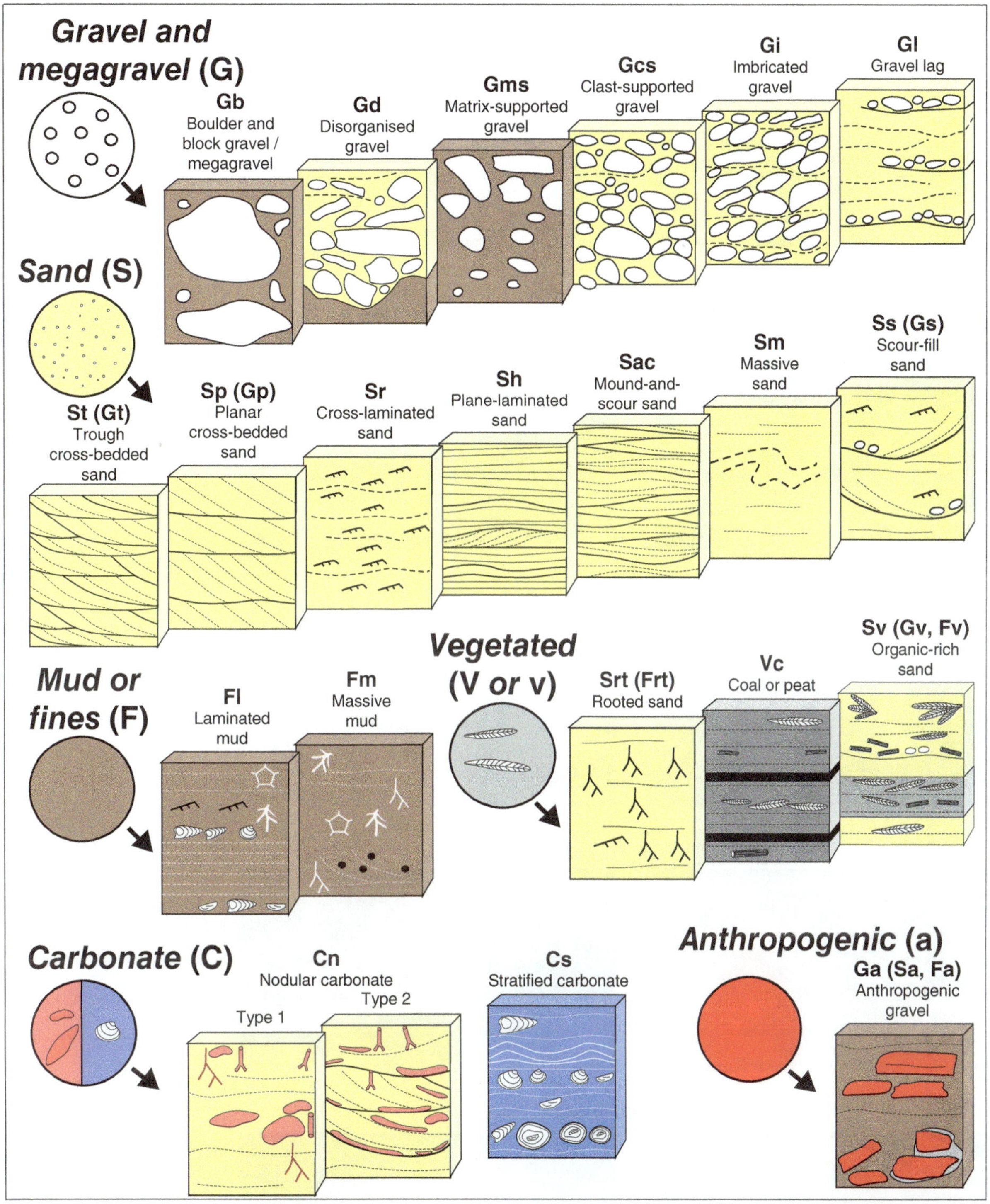

Figure 5.18 Facies in alluvial systems.

highlighting separately (Gl). Several sandy facies have less common gravelly equivalents (Gt, Gp, Gs and Gv). The term 'breccia' is sometimes applied to gravels with angular clasts, but the term is best restricted to coarse-grained deposits formed by tectonic, volcanic and diagenetic processes.

Eight facies are composed mainly of sand and coarse silt (Figures 5.20 and 5.21). Facies St, Sp and Sr represent

Figure 5.19 Gravel facies. From left to right: (a) Kali Gandaki River, Nepal. (b) Svalbard; Holocene, Salcombe, UK. (c) Tegnas Creek, Belluno, Italy; Ordovician, Arisaig, Canada. (d) Serengeti Plains, Tanzania; Pleistocene, Gateway, Colorado, USA. (e) Svalbard; Pleistocene, Dandiero Basin, Eritrea. *Source:* From Martin Gibling and Massimiliano Ghinassi (Chapter authors) and Clare Boston (b) at left. See Facies codes in Table 5.4.

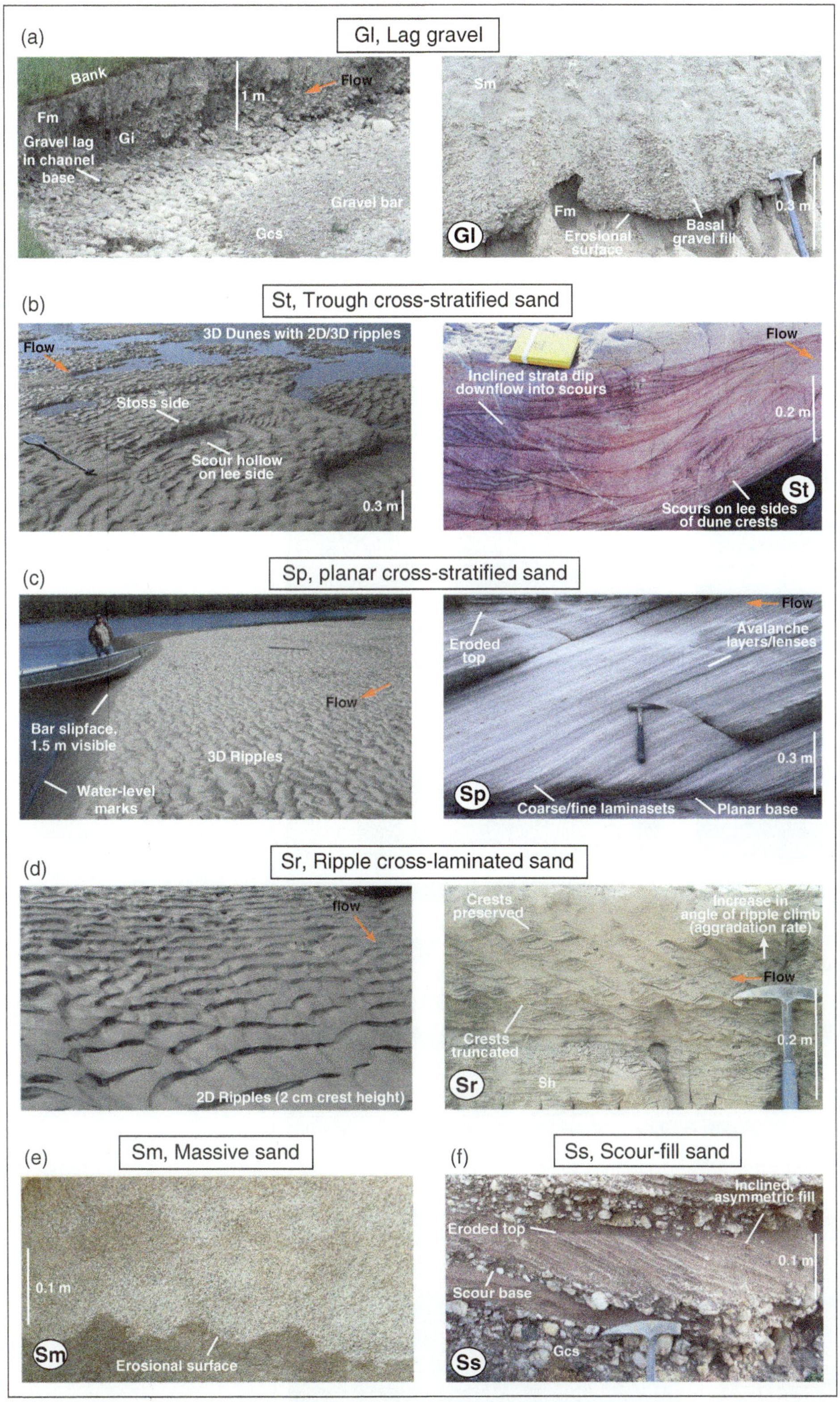

Figure 5.20 Gravel and sand facies. From left to right: (a) Orcia River, Tuscany, Italy; Pleistocene, Valdarno Basin, Italy. (b) Saskatchewan River, Saskatchewan, Canada; Burnside River Formation, Proterozoic, Nunavut, Canada. (c) Saskatchewan River, Saskatchewan, Canada; Hawkesbury Sandstone, Triassic, Australia. D) Kicking Horse River, Field, Canada; Atbara River, Sudan, showing a cross-laminated facies succession in a crevasse splay, with upward increase in the angle of ripple climb and the aggradation rate. (e) Pleistocene, Valdarno Basin, Italy. (f) Triassic, Sardinia, Italy. *Source:* From Martin Gibling, Massimiliano Ghinassi, and Alessandro Ielpi (Chapter authors). Facies codes in Table 5.4. (c) Left, after Gibling (2021), with permission from Dunedin Academic Press.

Figure 5.21 Sand and mud facies. From left to right: (a) Nelson Head Formation, Proterozoic, Northwest Territories, Canada (2 photos); Boss Point Formation, Pennsylvanian, Canada. (b) Stoer Group, Proterozoic, Scotland, UK; Entrada Sandstone, Jurassic, Utah; Green River Formation, Eocene, Utah (3 photos). (c) Clear Fork Formation, Permian, Texas; Trentishoe Formation, Devonian, UK. *Source:* From Martin Gibling and Alessandro Ielpi (Chapter authors), From Dario Ventra, and From Jianquao Wang and Piret Plink-Björklund. Facies codes in Table 5.4.

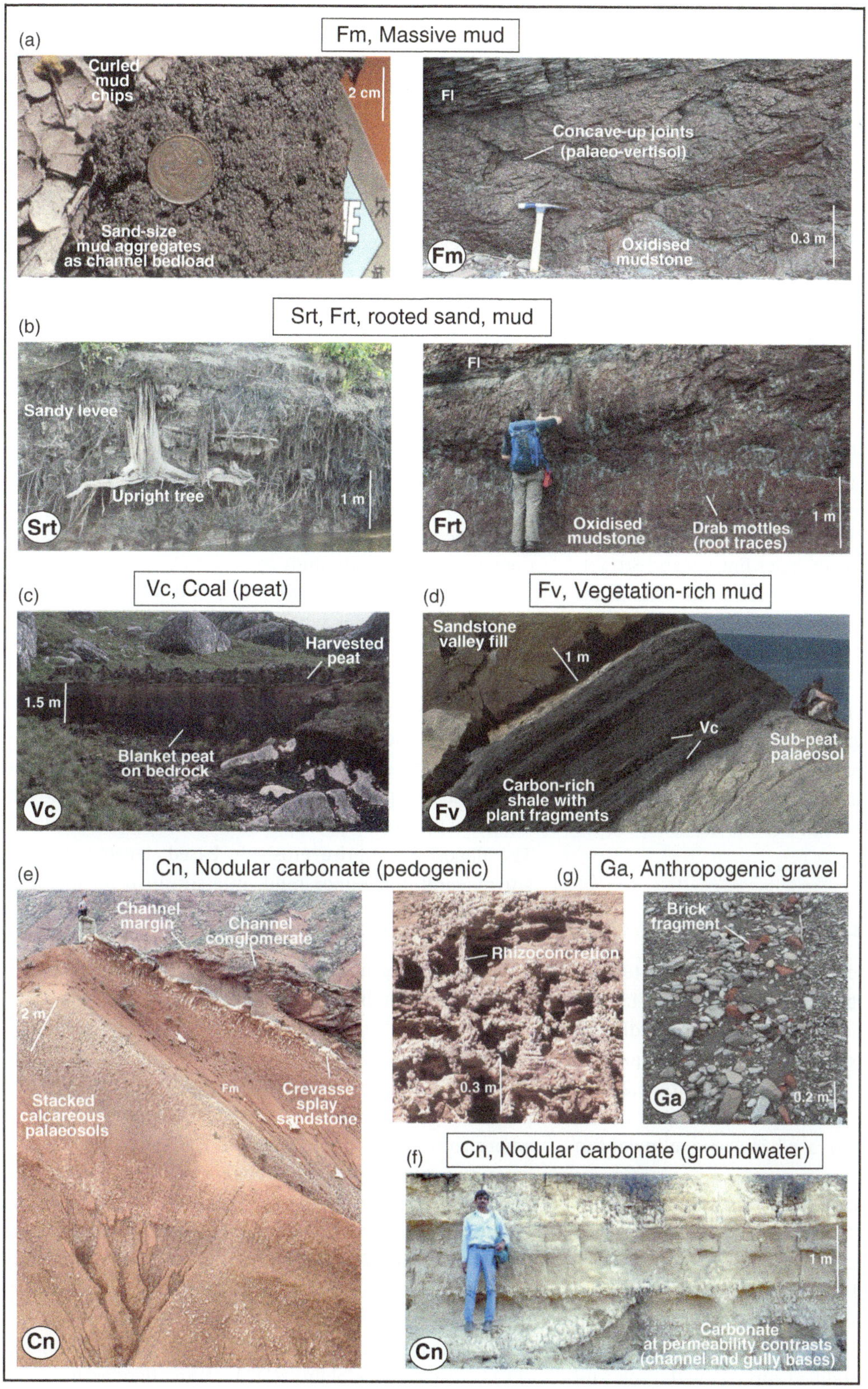

Figure 5.22 Mud, vegetation, carbonate, and anthropogenic facies. From left to right: (a) Cooper Creek, Queensland, Australia; Cheverie Formation, Mississippian, Canada. (b) Saskatchewan River, Saskatchewan, Canada; Cheverie Formation, Mississippian, Canada. (c) Healy Pass, Ireland. (d) Port Hood Formation, Pennsylvanian, Canada. (e) Eocene, Spain; Katapiri Formation, Quaternary, Queensland, Australia. (f) Ganga River plains, Quaternary, India. (g) Arno River, Italy. *Source:* From Martin Gibling and Massimiliano Ghinassi (Chapter authors). Facies codes in Table 5.4. See Section 6.5.2 for carbonates in lacustrine systems.

mesoscale and microscale bedforms (dunes and ripples) that form under Froude sub-critical flow conditions. Facies Sx may be a useful designation for cross-bedded facies that cannot be reliably assigned to St or Sp, especially in core and small or weathered outcrops. Facies Sh and Sac represent bedforms that formed under Froude transcritical to supercritical flow conditions. Facies Sm represents structureless sand formed by several processes, and Ss represents small scours filled with other facies, which should be specified.

Two mud facies (Figures 5.21 and 5.22) are distinguished by their laminated or massive nature (Fl and Fm). Vegetation is a prominent component of three facies suites (Figure 5.22): rooted sediment (Srt and Frt), peat or coal (Vc) and organic-sediment mixtures (Gv, Sv and Fv). Two carbonate facies are nodular carbonate (Cn) formed by soil-forming processes or groundwater cementation, and stratified carbonate (Cs) (Figure 5.22; lacustrine carbonate is described in Section 6.4.2). Palaeosol description and terminology may be appropriate for facies Fm, Cn and Srt/Frt, with combined codes where Cn contributes to other facies (e.g. St/Cn).

Anthropogenic sediment (Ga, Sa and Fa) is increasingly prominent in river deposits (Figure 5.22). Technofossils may be encountered at archaeological sites of late Pleistocene age onward, locally extending back to the Pliocene in the case of stone tools. It may be a judgement call whether, for example, to use St or Sa, depending on the prominence of the technofossils and the purpose of the study.

5.6.2 Facies Successions

Facies successions on a scale of centimetres to tens of metres represent upward changes in conditions, from a brief event to the formation of a substantial sediment body. Figure 5.23 shows examples from crevasse channels, crevasse splays (see also Sr image in Figure 5.20), a floodplain lake and a channel body. Box 5.1 illustrates a thicker facies succession in a Pleistocene alluvial-fan deposit.

A useful approach to facies successions is the concept of *fluvial aggradational cycles* (FACs), which may be grouped into FAC sets where facies trends are apparent (Atchley et al. 2004; Famubode and Bhattacharya 2016; Letourmy et al. 2021). These mappable units are on a scale of decimetres to a few metres, and they are bounded by erosional surfaces or pedogenically modified surfaces with roots and burrows, depending on the age of the strata. In many overbank settings (Figure 5.23e), successions commence with flooding surfaces that pass upward into laminated mud (Fl) or carbonaceous shale (Fv), laid down in floodplain lakes, and capped by rooted mud (Fm) or sandy crevasse splays and lacustrine shoreline deposits. Some channel sandstones capped by palaeosols may also be described as FACs.

5.7 Facies Associations

Facies associations are groups of genetically related facies that define a sub-environment, commonly a channel reach (Figures 5.24–5.27). Table 5.5 lists 13 facies associations (see Supplementary Table 5.S.7 for a fuller description). The associations draw on the first-order interpretation of individual facies to generate a second-order, process-based interpretation of facies groups. The associations overlap and are commonly interstratified, such that a set of strata may be divided into several associations or described as a single association with minor, contrasted facies sets. As with facies, researchers may prefer to define facies associations relevant to their study area.

Mass-flow association (MF) is found in valleys and on proximal alluvial fans where gravitational movements are prominent. Ancient examples of landslides are known from rift basins and basal valley fills cut into basement rocks, and they are associated in alluvial fans with the deposits of debris flows and related processes.

Channel-bar gravel (CBG), channel-bar sand (CBS) and channel-bar high-energy (CBH) associations collectively represent channel deposits on most alluvial fans and plains. The deposits comprise erosively based, lensoid units of gravel, sand and minor mud, interstratified on the scale of a few metres (Figures 5.25 and 5.26). CBS and CBH are distinguished by the relative proportions of low and high flow-strength facies, respectively. Despite interbedding and overlap between the two associations, it is worth attempting to distinguish them because they serve as a guide to flow dynamics, flood events and climatic parameters. Channel-bar fines (CBFs) are present on distal plains where the supply of coarse sediment is limited, or where rivers are supplied by loess. Pedogenic mud aggregates may be prominent components of these successions, and sedimentary structures may be difficult to identify due to compaction. Abandoned channels (ACs) may be large and prominent. Where preserved as remnants in channel belts, they may be included as minor components in channel associations CBS, CBG or CBH.

Three overbank associations are proximal floodplain (PF), distal floodplain (DF) and floodbasin (FB) (Figure 5.27). When working with dense datasets that include large outcrops, PF and DF may be readily distinguished (see Figure 5.25b), but the general term floodplain (FP) may be preferred where data are sparse or the

Figure 5.23 Facies successions. (a) Crevasse-channel fill, coarsening up from sand to gravel, Eocene, Boyabat Basin, Turkey. (b) Crevasse splay, coarsening-up to fining-up as flow waxed and waned; climbing ripples at the top show rapid aggradation, Atbara River, Sudan. (c) Hyperpycnites from sediment-charged flow in a floodplain lake, coarsening- to fining-up with interbedded Sh and Sr, Pleistocene, Dandiero Basin, Eritrea. (d) Channel fill fining up from gravel to sand, with mud fill after abandonment, Miocene, Sardinia, Italy. (e) Stacked fluvial aggradational cycles (FACs), Mississippian, Cheverie Formation, Canada. *Source:* From Massimiliano Ghinassi and Martin Gibling (Chapter authors). Facies codes in Table 5.4.

Box 5.1 Alluvial-Fan Lobes Advance and Retreat

Near Florence in Italy, sandy to gravelly alluvial fans of Pleistocene age advanced from upland exit points into a fault-bounded basin, where they coalesced into a bajada (Fidolini et al. 2013; Ghinassi and Ielpi 2016). During the 1500–1600s, both Leonardo da Vinci and Nicolas Steno gained inspiration from the splendid outcrops. The fans had radii of up to 7 km and a gradient of about 1.8° (2.4%). The strata can be mapped into belts 1–3 km wide of proximal, medial and distal deposits on the basis of downstream distance from the fan apex and the ratio of gravel to sand.

Through time, the rate of basin subsidence decreased and the rate of upland sediment supply increased. This caused fan lobes to prograde into the basin, generating a coarsening-up rhythm 30 m thick, with distal- and medial-fan deposits overlain by proximal-fan deposits. The medial fan deposits consist of sandy and gravelly braided-channel fills, and the proximal fan deposits consist of gravelly streamflow deposits. Later, reduced activity on the bounding faults caused retrogradation of the fans, generating a fining-up rhythm 15 m thick, with a transition to medial- and distal-fan deposits. Thus, the deposits can be considered large-scale facies successions. *Source:* Massimiliano Ghinassi (Chapter Contributor).

relationship to parent channels is uncertain. Floodbasin (FB) deposits are locally prominent in valleys, and some channel systems include chains of lakes and small ponds.

Colluvial (CV) (Figure 5.27) and periglacial (PG) associations are widespread in modern alluvial settings but have been identified more rarely in the geological record. The colluvial association includes gully fills and slope wash from rocky hillslopes, associated with MF, CBG and CBS and displaying a range of facies from gravel to mud (Nelson 1992; Figure 5.27e).

Before defining facies associations, it is again helpful to consider the data quality and the purpose of the study.

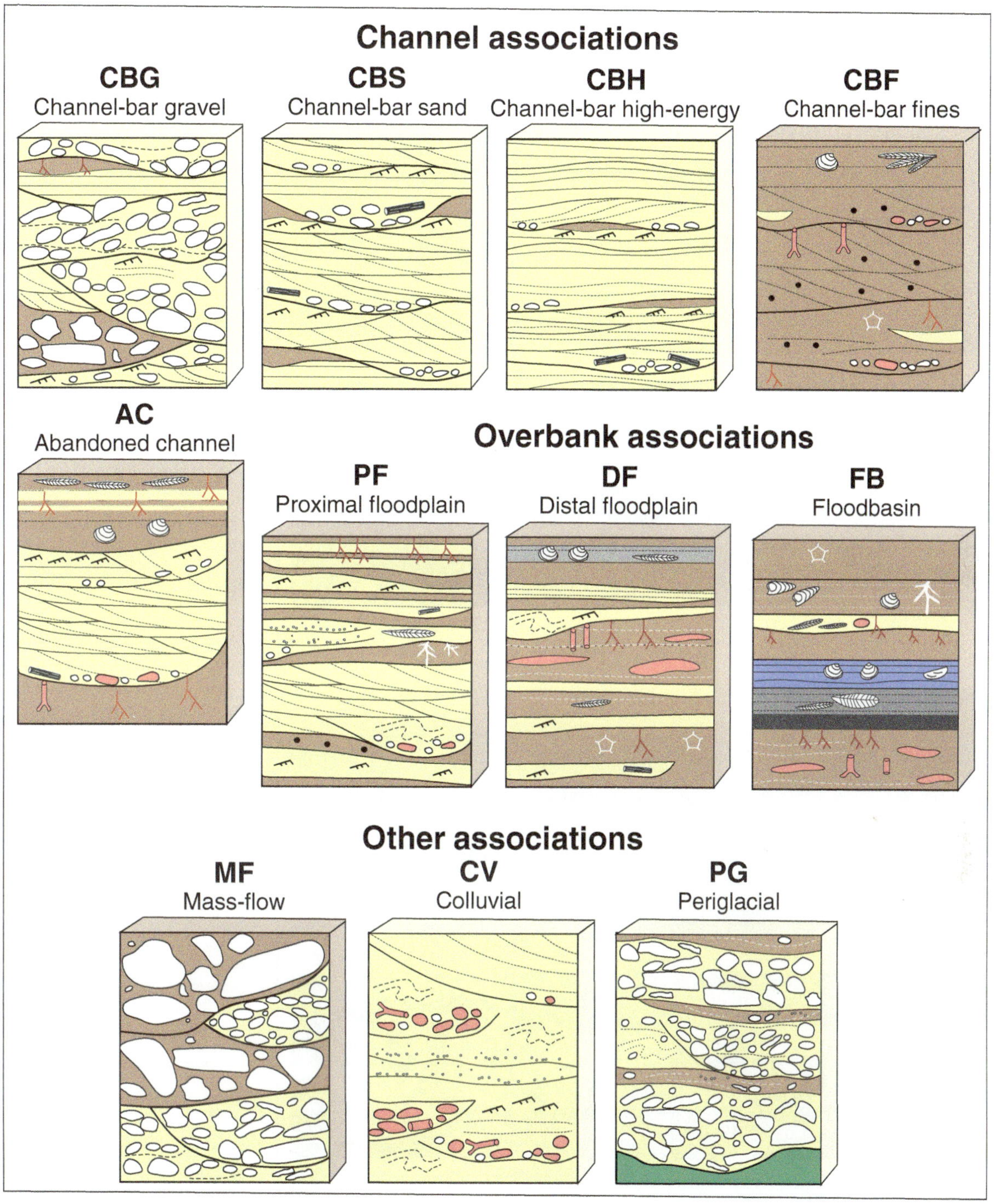

Figure 5.24 Facies associations in alluvial systems.

Facies associations provide a general but reliable interpretation where only scattered outcrops are available (Figure 5.2). A broad-based study might best be served with a small number of facies associations, whereas a more detailed study of reservoir permeability or palaeohydrology might benefit from identifying more associations at a finer scale. Table 5.6 provides examples of how facies associations (often termed 'facies groups') have been used for particular purposes. The example from the Eocene Green River Formation of Utah is a detailed process study of a fluvial system with high-discharge variability, whereas the example from the Quaternary of India is a

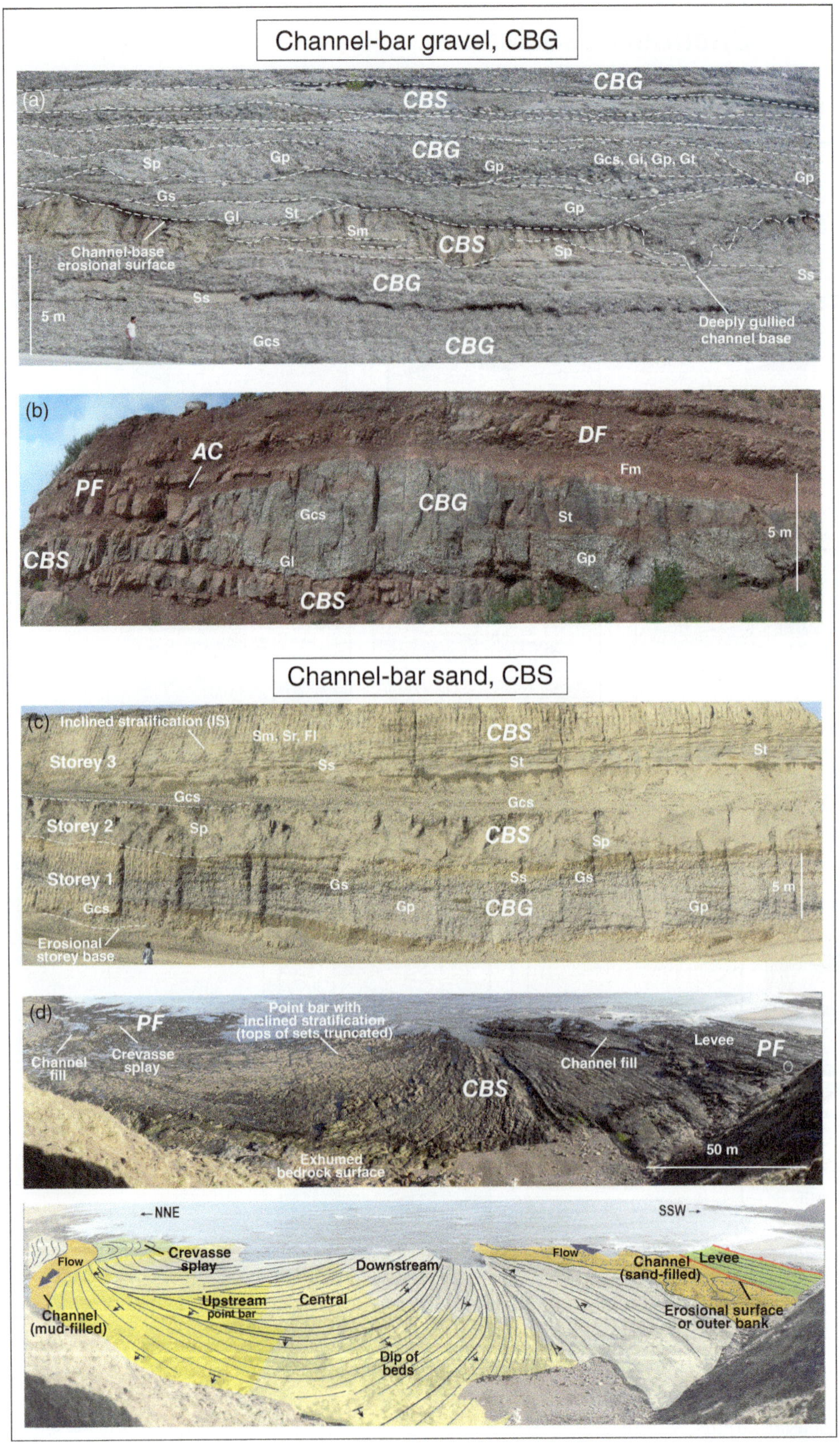

Figure 5.25 Facies associations of Channel-bar Gravel and Channel-bar Sand. Person circled in (d). (a) Eocene, Boyabat Basin, Turkey. (b) Sant Llorenç del Munt system, Eocene, Ebro Basin, Spain. (c) Pleistocene, Valdarno Basin, Italy. (d) Scalby Formation, Jurassic, UK. *Source:* From Massimiliano Ghinassi and Alessandro Ielpi (Chapter authors). (d) from Ielpi and Ghinassi (2014, Figure 4) / with permission from John Wiley & Sons. Codes in Tables 5.4 and 5.5.

Figure 5.26 Facies associations of Channel-bar Sand, Channel-bar High-energy, and Channel-bar Fines. (a) Proterozoic, Applecross Formation, UK, inclined strata and Sp sets denote mid-channel and bank-attached bars. Storeys are difficult to identify. (b) Pennsylvanian, Waddens Cove Formation, Canada, cut through resistant paleosols with slump blocks at the channel base. Storeys fill an initial incision and widen the channel tract. Person circled. (c) Cheverie Formation, Mississippian, Canada. (d) Clear Fork Formation, Permian, Texas, USA. *Sources:* From Alessandro Ielpi and Martin Gibling (Chapter authors). (d) Gibling (2021) / with permission of Dunedin Academic Press. Codes in Tables 5.4 and 5.5.

Figure 5.27 Facies associations of Abandoned Channel, Distal Floodplain, Floodbasin, and Colluvial. (a) Nelson Head Formation, Proterozoic, Northwest Territories, Canada. (b) Joggins Formation, Pennsylvanian, Canada, with lower fixed-channel body seen in cross-section and upper body in a section near-parallel to palaeoflow. (c) Cutler Group, Pennsylvanian to Permian, New Mexico, USA. (d) Brenta River terrace, Quaternary, Italy. (e) Quaternary, Ethiopia, unpublished dates on charcoal represent 2-sigma error ranges (see Section 5.17.1). *Sources:* From Alessandro Ielpi, Massimiliano Ghinassi and Martin Gibling (Chapter authors). Codes in Tables 5.4 and 5.5.

Table 5.5 Facies associations for alluvial and related successions.

Code	Facies association	Geomorphic setting
MF	**Mass-flow**	Hill slopes, upland valleys and proximal alluvial fans. Downslope movement as rock avalanches, rock falls, landslides and talus, with associated high-density flows
CBG	**Channel-bar gravel**	Gravelbed rivers in upland valleys, alluvial fans and proximal alluvial plains. Braided and wandering channels, with stream flows, debris floods, hyperconcentrated flows and local debris flows
CBS	**Channel-bar sand**	Sandbed and mixed-load rivers in upland valleys, alluvial fans and alluvial plains. A range of alluvial styles, including braided, meandering and anastomosing
CBH	**Channel-bar high-energy**	Sandbed rivers in upland valleys, alluvial fans and alluvial plains, commonly with strongly seasonal or ephemeral flow. Floods with Froude transcritical and supercritical flows prominent
CBF	**Channel-bar fines**	Fine-grained rivers on distal alluvial plains. Pedogenic mud aggregates common, and large silt loads where loess is available. Meandering, laterally stable and ephemeral styles
AC	**Abandoned channel**	Channel cut-off and avulsion in rivers with a range of styles, with suspension settling and coarser deposits from periodic flow
PF	**Proximal floodplain**	Overbank area with levees, crevasse splays and crevasse channels, adjacent to mainstem and tributary channels
DF	**Distal floodplain**	Overbank area with levee fringes, thin crevasse splays, floodplain fines and standing-water deposits, distal from mainstem and tributary channels
FP	**Floodplain**	Overbank area with alternate proximal and distal settings, or with setting unclear
FB	**Floodbasin**	Overbank area with wetlands (shallow lakes and swamps), crevasse splays and palaeosols
CV	**Colluvial**	Sand, silt and reworked gravel, on degrading hill slopes, channel banks, floodplains and terraces, with slope wash and gully erosion
PG	**Periglacial**	Lower hillslopes, valley bases and bedrock hollows, with solifluction from flow of active layer over frozen ground and upslope cryopediments

Table 5.6 Facies associations: examples linked to the purpose of the study.

Description	Interpretation	Purpose of study	Comparison with Chapter 5 approach
Sandstones and conglomerates with low-wavelength, low-angle lamination *Green River Formation, Eocene, Utah* (Wang and Plink-Björklund 2020). See Figure 5.21b, lower three photos.			
Nine facies, some attributed to plane beds, cyclic steps, antidunes and hydraulic jumps	Rivers with highly variable discharge and frequent supercritical flows	Focus on structures formed under trans- to supercritical flows	Corresponds to channel-bar high-energy facies association (CBH), with facies fully described (e.g. Sh and Sac). Facies and facies associations facilitate process analysis of bedforms.
Gully fills *Older Alluvium, Quaternary, India* (Gibling et al. 2005). See Figure 5.53.			
Mud, structureless sand and carbonate gravel. Erosion surfaces with groundwater carbonate cement, and deep incisions with inclined beds	Floodplain degradation and fills of deep gullies. Colluvial and streamflow sediment, and reworked carbonate	Analysis of colluvial, fluvial, aeolian and lake strata in cliffs, with physical dates. Compared with proxy monsoon records for palaeoclimatic trends	Corresponds to the Colluvial facies association (CV). Facies are briefly described (e.g. Gcs, Sm and Fm). Interpretations are generalised for palaeoclimatic analysis, rather than for process analysis.

more general account that was designed to link facies to palaeoclimate.

5.8 Geomorphic Units

The next step in the workflow is to identify *geomorphic units* within the facies associations (Supplementary Table 5.S.1: 14). Geomorphic units are mappable landforms generated by processes that yield distinctive geometry, sedimentary features and bounding surfaces. They are identified in modern settings by (i) their vertical position or formative flow stage relative to the channel bed, from low flow to the highest recorded flood level; (ii) their shape; (iii) the type of unit (e.g. bars and depressions); (iv) the vegetation cover (e.g. bare sediment, pioneering vegetation and forested); and (v) their bed-material texture and other sedimentological features. Geomorphic units are 3D sediment bodies, considered to be the 'building blocks' of alluvial successions, and they may be fitted together like pieces in a jigsaw puzzle (Wheaton et al. 2015; Fryirs and Brierley 2018). Comparable terms include *architectural elements* (Allen 1983a; Miall 1996, 2010), discussed further later; *morphostratigraphic units* (Brierley 1991); and *macroscale forms* (Bridge 1993). The identification of geomorphic units requires the integration of facies association data with architectural and palaeoflow information (Sections 5.9 and 5.10) to establish the 3D form and internal geometry of the sedimentary bodies. An important part of the analysis is the relationship between local flow direction inferred from sedimentary structures and accretion directions established from bedding surfaces.

Identification of geomorphic units in the geological record is challenging because dense datasets and relatively complete preservation are needed to determine their form. In some cases, exhumed bedrock surfaces provide exceptional 3D exposure of geomorphic units (Supplementary Table 5.S.1: 22). More commonly, however, interpretation is based on moderate and scattered 2D data from cliff faces and GPR lines, often with little information about how the strata are oriented with respect to palaeoflow (Willis and Tang 2010). The formative stage of flow must be inferred, and vegetation – a diagnostic feature for some geomorphic units – may be preserved only as woody debris and root traces. A particular problem in assessing geomorphic units is the fractal nature of much of the stratigraphic record (Holbrook and Miall 2020), where sediment bodies may be self-similar but have different scales. For example, channel-like forms range in scale from scours a few tens of centimetres deep to valleys hundreds of metres deep.

Where datasets are limited, researchers should avoid drawing definitive inferences, providing instead a detailed description and suggesting possibilities. However, where datasets are dense (Figure 5.2), it is worth attempting to identify geomorphic units, drawing on the wealth of information that is now available about fluvial processes and geomorphology. Fifty-three of the more common geomorphic units, from the large number listed by Fryirs and Brierley (2018), are defined in Table 5.7 (see Supplementary Table 5.S.8 for a fuller description). Many of the units are illustrated in Figures 5.28–5.37, and Supplementary Table 5.S.3 lists papers that describe the units in modern and ancient settings.

Table 5.7 Geomorphic units in alluvial and bedrock channels (modified from Wheaton et al. 2015 and Fryirs and Brierley 2018).

Code	Geomorphic unit	Setting
		Upland valleys
hs	**Hillslope**	Cut into bedrock or tough sediment at valley margin
dc	**Debris cone**	Dipping alluvial and colluvial cone, from hillslope mass wasting
tc	**Talus cone**	Steeply dipping cone at cliff base, from rock fall and slope wash
at	**Alluvial terrace**	Flat-lying alluvium banked against bedrock or tough sediment where valley incision has stranded sediment from the former channel
st	**Strath terrace**	Flat-lying alluvium resting on shelf of bedrock or tough sediment cut by lateral planation. Stranded by renewed valley incision
gy	**Gully**	Linear erosional depression cut into degrading sediment or palaeosol
	Erosional features on bedrock and tough sediment	
br, pp, sp	**Bedrock step or waterfall, plunge pool and step pool**	Change in slope and elevation of bedrock and tough sediment at an unconformity, forming distinct steps, large plunge pool below waterfall, or small pools
cc, rp	**Cascade, rapid**	Bedrock surface with a relatively steep gradient and small steps (cascade) or few or no steps (rapid)

Table 5.7 (Continued)

Code	Geomorphic unit	Setting
rn, gl	**Run, glide**	Gently dipping planar bedrock surface. Runs have steeper gradient and greater roughness than glides
ph	**Pothole**	Small indentation on bedrock surface, caused by abrasion
		Bank-attached setting
pb	**Point bar**	Bar along the convex side (inner bank) of a meander bend
ab	**Alternate bar**	Bar in low-sinuosity channel
cpb	**Counterpoint bar**	Sediment accumulation at a point-bar tail, thickening into the concave side of the next meander bend downflow
sb	**Scroll bar**	Flow-parallel mound atop a bank-attached bar, where unit bars are stranded as flow wanes or sediment builds up downflow from rooted or fallen vegetation
ab	**Accretionary bank**	Steep surface of bank or bench, with alternate layers of sand and mud from high- and low-stage flow
chc, chr, chb	**Chute channel, chute ramp, chute bar**	Chute channel on a bank-attached bar, with downflow chute ramp and local progradation into the main channel as a chute bar
tmb	**Tributary mouth bar**	Bar at junction of tributary and mainstem channel, locally with a confluence scour
fdb, fdp	**Forced bank-attached bar or pool**	Downflow from upright and fallen vegetation, large boulders and slump blocks along the banks
bc	**Bench**	Buildup below bankfull level in seasonal channel, with top broadly matching low-stage flow level
		Mid-channel setting
mcb	**Mid-channel bars**	Bars with varied form and orientation in active channels. Include longitudinal, transverse, diagonal, lobate and confluence bars, commonly amalgamated into compound bars. Bar types are difficult to identify in the ancient record.
ub	**Unit bar**	The building blocks of many larger bars, typically with dm-scale relief
gs, ss	**Gravel sheet, sand sheet**	Sediment sheets in braided and ephemeral rivers, where short-lived flow precludes formation of bars and larger bedforms
rf, pl	**Riffle, pool**	Shallow alluvial bed (riffle) from flow expansion, with pool downflow
bm, bcb, fmb, fmp	**Boulder mound, bedrock core bar, forced mid-channel bar or pool**	Sediment buildups and pools associated with obstacles that include outsized boulders, bedrock outcrops and rooted or fallen vegetation
cf	**Confluence scour**	Elongate hollow where major channels join or where tributaries join mainstem channels
cs	**Channel scour**	Elongate or equant hollow where flow converges around bars and at channel margins
btc, bth	**Bartop channel, bartop hollow**	Channels or hollows formed on bartops during falling stage
id	**Island**	Sediment built above bankfull level where mid-channel bars accrete, with baffling by rooted pioneer and climax plants and woody debris
sb	**Slump block**	Coherent sediment from bank failure, typically in the basal channel fill
		Floodplain and floodbasin
fp	**Floodplain**	Flat-lying strata laid down by overbank flows
lv	**Levee**	Streamwise, elevated mound above bankfull level at channel margin, where cross-margin flow expands, decelerates and loses competence
cvs	**Crevasse splay**	Lobe on floodplain where crevasse channel cuts through a levee
cvc	**Crevasse channel**	Incision through levee, funnelling channel sediment to the floodplain
fps	**Floodplain sand sheet**	

(Continued)

Table 5.7 (Continued)

Code	Geomorphic unit	Setting
		Sheet from unconfined flow in a poorly channelized setting, or from upland floods
abc	**Abandoned channel**	Channel reach disconnected briefly or permanently by avulsion and meander cut-off
fpc	**Floodplain channel**	Small channels that guide overbank flow
rd	**Ridge**	Former channel with levees and splays, incorporated into floodplain after avulsion
fpw, fpl, bsw	**Floodplain wetland, floodplain lake, backswamp**	Overbank area with high water table, forming vegetated wetlands, lakes and backswamps ponded by topography

Rock outcrops and large woody debris force flow separation and the generation of some geomorphic units. Channels of any scale may contain in-channel geomorphic units.

5.8.1 Upland Valleys and Erosional Bedrock Settings

Along upland valley margins, geomorphic units include *hillslopes* cut into bedrock and overlain by soil and thin sheets of slope wash. *Debris cones* are composed largely of colluvial and alluvial material, and *talus cones* are composed of debris from rock falls and slope wash, with short transport distances (Figures 5.5a and 5.19, Gb; see Vincent (2001) and Breda et al. (2007) for ancient examples). In Neogene talus deposits of northern Greece (Figure 5.29), the clasts comprise a single rock type (quartzite), and the inclined beds are thick and poorly stratified to unstratified. The clasts are angular and poorly sorted, with outsized fragments and minimal matrix in the coarse-grained, clast-supported layers. A few matrix-supported layers contain reworked soil.

Alluvial terraces form where the discharge of water and sediment fluctuates during glacial and other climatic rhythms (Lewin and Gibbard 2010). *Strath terraces* comprise shelves cut into bedrock or older alluvium and overlain by younger sediment (Figure 5.5b,c; see Bestland 1997 for an ancient example). Staircase terraces may form where rivers in upland areas incise through rising land or fluvio-glacial deposits (Figure 5.5b,c), or are affected by base-level lowering in coastal areas (Blum and Aslan 2006). Along late Cenozoic rivers, dating of terraces provides a crucial record of alluvial and loess accumulation, incision events and changes in discharge and palaeoclimate. Many terrace successions, such as those along the River Somme in France, yield important archives of hominin artefacts. *Gullies* with complex fills develop on semi-consolidated alluvium during degradational episodes in uplands and on alluvial plains (Figure 5.27e).

On valley floors, erosional geomorphic units are present on bedrock unconformities and on disconformities within semi-consolidated sediments. They include *bedrock steps* (waterfalls), *plunge pools*, *cascades*, *rapids* and *potholes* (Figure 5.30a–c). Figure 5.31 shows an example from the ancestral Orange River of South Africa, where Miocene gravels rest unconformably on Proterozoic bedrock, with diamonds eroded from kimberlite pipes concentrated in a range of geomorphic units (Jacob et al. 1999). Where particle size (D) approaches or exceeds the channel depth (h) (Church 1992), strong turbulence around bedrock projections and outsized clasts may force the generation of geomorphic units.

5.8.2 Bank-Attached and Mid-Channel Bars

Bars are among the most common geomorphic units in mid-channel settings and attached to the channel banks. Because many alluvial channels migrate laterally at rates that are orders of magnitude higher than their aggradation rates (Jerolmack and Mohrig 2007; Ielpi and Lapôtre 2020), the chance of preserving a bar remnant in the rock record is high, statistically speaking. Many bars scale broadly with channel size (Bridge and Lunt 2006). Their dimensions follow power–law relationships with channel width, and their elevation is comparable to the bankfull depth (Leopold and Maddock 1953; Greenberg et al. 2021).

Many bars, especially those in mid-channel settings, have discrete, curvilinear slipfaces where sediment transported across the bartop avalanches over the edge of the bar. This process generates single or stacked sets of inclined strata that are assigned to planar cross-bedded sand or gravel (Sp or Gp). Many large mid-channel bars grow by the accretion of smaller, lobate *unit bars*, which generate planar cross-sets from decimetres to metres thick and may also fill braid channels (Bridge and Lunt 2006). The structure and composition of unit bars may vary considerably in rivers with highly variable discharge (Herbert et al. 2020).

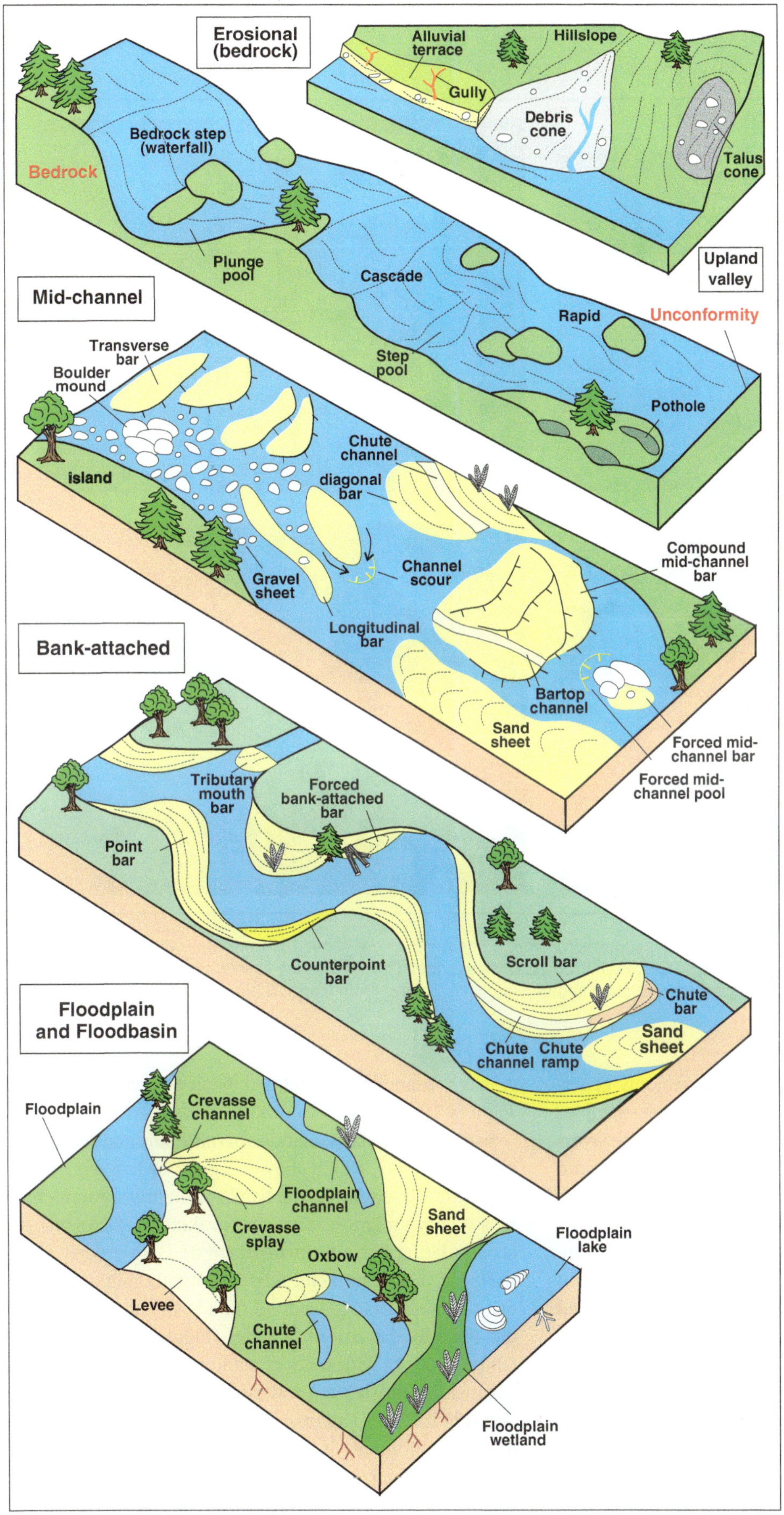

Figure 5.28 Sketch of geomorphic units in bedrock channels, alluvial channels and floodplains. These and other geomorphic units are listed in Wheaton et al. (2015) and Fryirs and Brierley (2018), and they are described in Table 5.7, with references in Supplementary Table 5.S.3.

Figure 5.29 Neogene talus deposits near Sidirokastro, northern Greece. The deposits are about 15 m thick (a), with layers of debris dipping away from a quartzite outcrop to the right. Geomorphic unit is tc (talus cone). The deposits are coarse, clast-supported gravels (b), with minor beds of finer, matrix-supported gravel with reworked red soil (c, d). *Source:* From Martin Gibling (Chapter author) (a–d).

In bank-attached bars, bar migration commonly generates dipping bedsets (6°–15° dips in Cretaceous examples in western Canada; Durkin et al. 2017) that comprise layers of cross-bedded, cross-laminated, plane-laminated and massive sand (St, Sp, Sr, Sh and Sm), with thin layers of massive and laminated mud (Fm and Fl). These bedsets are termed *inclined stratification* (IS), where the grain size is relatively uniform, and *inclined heterolithic stratification* (IHS), where layers of sand and mud are intercalated (Thomas et al. 1987; Figure 5.30g,i,j). Some inclined sets can be traced from the base to the top of a channel body, but in other cases inclined sets are only partially preserved, with their lower parts obscured by large bedforms and their upper parts eroded. Bank-attached bars in large fine-grained rivers such as the Paraná of South America contain inclined accretion surfaces with unusually low angles, mainly <3° but up to 8° (Sambrook Smith et al. 2015). Ancient examples of inclined bedsets are illustrated in Figures 5.25–5.27.

Bank-attached bars include *alternate bars* and *side bars* of sand and gravel, which show a regular or less regular pattern, respectively, along the banks of slightly sinuous channels. They migrate through avalanche faces at the accumulating bar tail downstream and are eroded at the bar head upstream. *Point bars* in laterally migrating sinuous channels yield gravelly to muddy IS and IHS sets that dip towards the parent channel (Figure 5.25d; Ielpi and Ghinassi 2014), with styles of planform growth that include expansion, translation (downstream migration) and rotation (Figure 5.32). *Expansion* increases the meander sinuosity and culminates in a meander neck cutoff, with the formation of an oxbow lake. *Translation* maintains the existing sinuosity of the meander bend as the point bar migrates, with downstream deposition and upstream

Figure 5.30 Geomorphic units in erosional and bank-attached settings. (a), (b) Iceland. (c) Serengeti Plains, Tanzania. (d) Great Basin, Nevada, USA. (e) Vermilion River, Canada. (f) Georgina River, Queensland, Australia. (g) Rangal Coal Measures, Permian, Queensland, Australia. (h) Saskatchewan River, Saskatchewan, Canada. (i) Horseshoe Canyon Formation, Cretaceous, Canada. (j) Sydney Mines Formation, Pennsylvanian, Canada. *Sources:* From Alessandro Ielpi, Massimiliano Ghinassi and Martin Gibling (Chapter authors). (f) Gibling (2021) / with permission of Dunedin Academic Press.

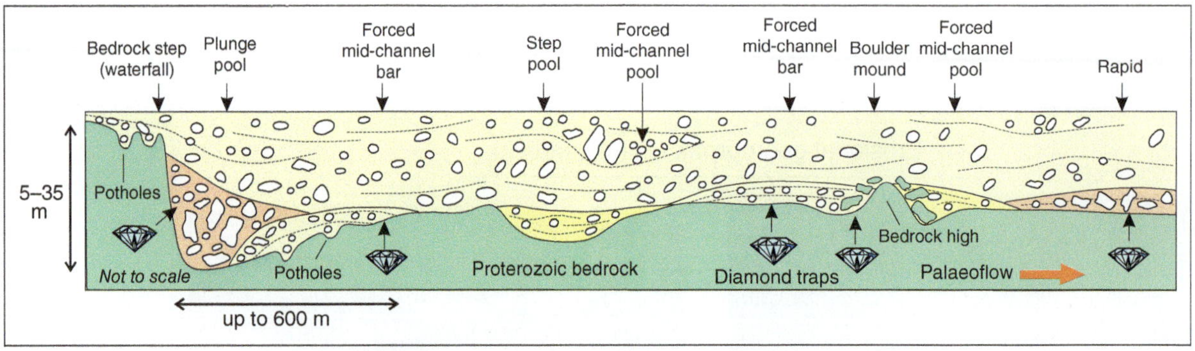

Figure 5.31 Miocene strata of the lower Orange River, South Africa, showing geomorphic units on the basal unconformity, associated with alluvial diamonds in open pits. Identification of geomorphic units is based on the interpretations of the original authors. *Source:* Adapted from Jacob et al. (1999).

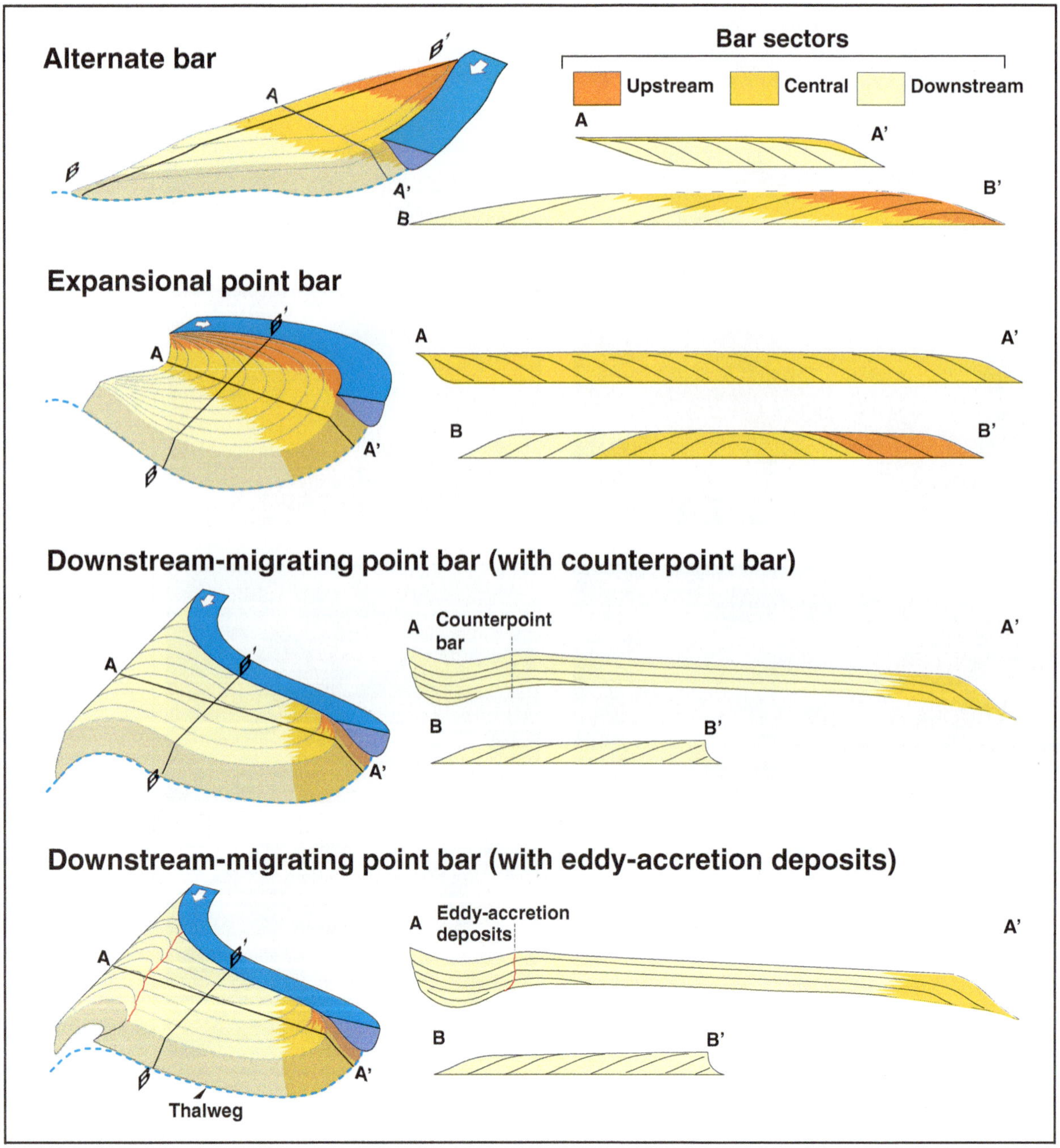

Figure 5.32 Planform morphology and cross-sectional views of bank-attached bar deposits (alternate bars and point bars). Bar head (upstream), body (central) and tail (downstream) are indicated, and cross sections show the spatial distribution of the deposits resulting from bar evolution. Counterpoint bars and eddy-accretion deposits are shown here with downstream-migrating point bars, but they are also associated with bars that show a different planform evolution. Source: Adapted from Ghinassi et al. (2016).

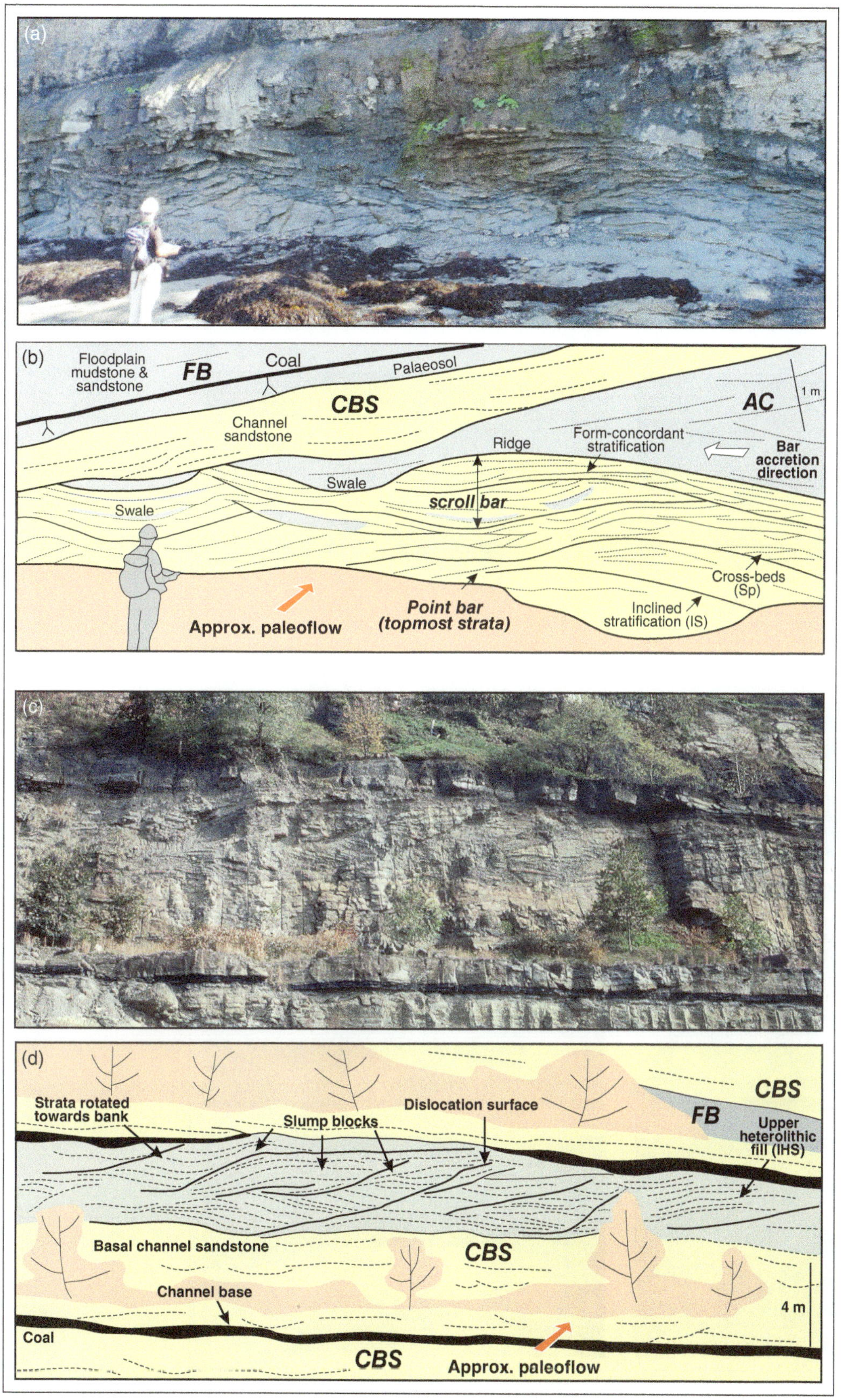

Figure 5.33 Scroll bar in the Sydney Mines Formation, Pennsylvanian, Canada (a, b), and slump blocks in the Kanawha Formation with Lower War Eagle Coal, Pennsylvanian, Mahan Exit in West Virginia, USA (c, d). *Source:* From Martin Gibling (Chapter author).

Figure 5.34 Geomorphic units in mid-channel settings. (a), (b) Tagliamento River, Italy. (c) Platte River, Nebraska, USA. (d) Cooper Creek, Queensland, Australia. (e), (f) Hawkesbury Sandstone, Triassic, New South Wales, Australia, people circled. (g) Boss Point Formation, Pennsylvanian, Canada. (h) Joggins Formation, Pennsylvanian, Canada. (i) Pleistocene, Dandiero Basin, Eritrea. (j) Lake Manyara area, Tanzania. *Sources:* From Roberto Pizzutti (a), Alessandro Fontana (b), Massimiliano Ghinassi and Martin Gibling (Chapter authors). (f) Gibling (2021) / with permission of Dunedin Academic Press.

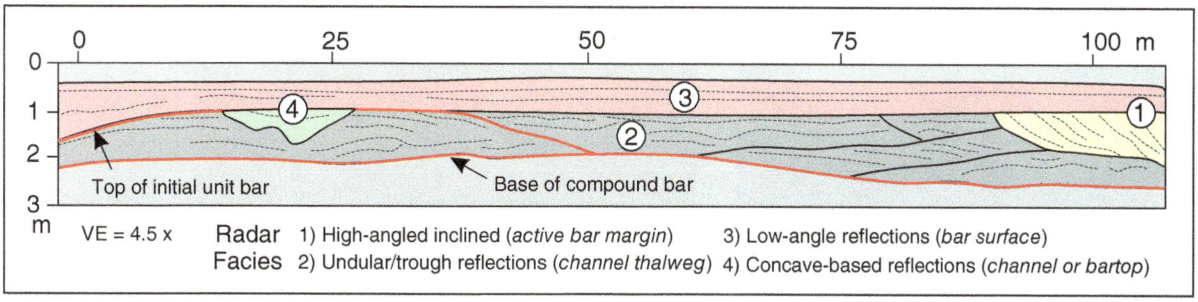

Figure 5.35 Ground penetrating radar (GPR) profile and radar facies for a compound bar in the South Saskatchewan River. The compound bar originated as a unit bar with a convex-up profile, cut by a bartop channel. The bar grew laterally and vertically through the addition of flat-lying channel thalweg and bar-surface deposits (principally dunes) and inclined bedsets generated at a bar-margin avalanche face. *Source:* Adapted from Parker et al. (2013).

erosion. *Rotation* can occur in combination with expansion or translation, causing the bend apex to approach or move away from the meander-belt axis. Flow may also bypass point bars along *chute channels*. Where chute channels re-enter the trunk channel, a *chute ramp* terminates in a *chute bar* with a high-angle (15–20°) slipface (McGowen and Garner 1970).

Counterpoint bars are finer grained deposits that extend from point-bar tails into the concave part of the next bend downriver. They have IHS sets that dip towards the channel. Where the bars transition to the next point bar downflow, a zone of reverse flow (upriver) may form an eddy in the concave bend, with sandy 'eddy-accretion deposits' filling a scour pool. Smith et al. (2011) gave eddy-accretion deposits separate status from counterpoint bars. The two types of deposit have mainly been identified in association with translational point bars, but they also accompany other modes of point-bar development.

Scroll bars form where unit bars migrate directly or obliquely up the surfaces of point bars and counterpoint bars during high-stage flow, generating slipfaces with inwardly directed sets of planar cross-strata. The bars are stranded at a high elevation as the flow wanes, contributing to the channelward accretion of the parent point or counterpoint bar. Scroll bars also form on point bars where sediment accretes downflow from vegetation during floods. The scroll bars generate metre-scale ridge-and-swale topography on the planar tops of the parent bars. Within the ridges, stratification surfaces are generally convex-upward in accord with the ridge crests (form-concordant), and the deposits consist of interbedded cross-sets, plane beds, ripple cross-lamination, scour fills and mud lenses (Figure 5.33a, b). The dipping stoss surfaces of the scroll bars extend downward into IS and IHS sets within the parent bars, and small channels may occupy the swales. Scroll bars are also present on some mid-channel bars.

Benches and *accretionary banks* (Figure 5.30f) are common in channels with seasonally variable discharge. The bench tops represent sustained flow at a level below bankfull. Inclined layers of sand form on the overlying accretionary banks at high-flow stage and are draped with mud as the water-level falls, generating IHS sets. Bank draping may be termed 'oblique accretion' because the drapes join the adjacent bar, bench or channel bed at an oblique, abrupt angle (Page et al. 2003).

Mid-channel bars have a range of planform styles that include longitudinal, transverse and diagonal forms. They are generated under high-stage formative flows (see Figure 5.12e,f, where gravel in the Tagliamento River was mobile at a considerable elevation), and they commonly amalgamate into compound bars (Figures 5.34a–d and 5.35). Bars in the gravelly Sagavanirktok River of Alaska are mainly compound bars, generated through the accretion of unit bars on their upstream, downstream and side areas, or by cutoffs of point bars (Bridge and Lunt 2006). Flow separates around the bar heads, where erosion predominates, and converges at the bar tails where avalanche fronts are common (Figure 5.7e). Lateral accretion takes place where sediment avalanches over streamwise bar margins. *Channel scours* are prominent where flow converges at bar tails, and they are commonly filled with cross-sets of facies Sp or Gp generated at avalanche fronts. Bartop facies include crudely stratified and imbricated gravel (Gi and Gcs) with thin sandy bedsets of various facies. The gravel may contain a muddy matrix laid down during waning flow (generating a closed-work fabric) or may lack fines (generating a porous, open-work fabric). Bar tops may be incised by small *bartop channels* scoured under shallow-water conditions. *Islands* form where mid-channel bars build to elevations at which permanent vegetation is established (Figure 5.34a,b), leading to small tracts of mature forest.

Figure 5.36 Geomorphic units in floodplain and floodbasin settings. (a) Mississippi River, Louisiana, USA. (b) Tuross River, New South Wales, Australia. (c, d) Baghmati River, Bihar, India. (e) Southern Thailand. (f, g) Channel Country, Queensland, Australia. (h) Saskatchewan River Delta, Saskatchewan, Canada. (i) Baraila Tal, Bihar, India. (j) Humboldt River, Nevada, USA. *Sources:* From Arthur Hanson (e), Alessandro Ielpi and Martin Gibling (Chapter authors). (a, e) Gibling (2021) / with permission of Dunedin Academic Press.

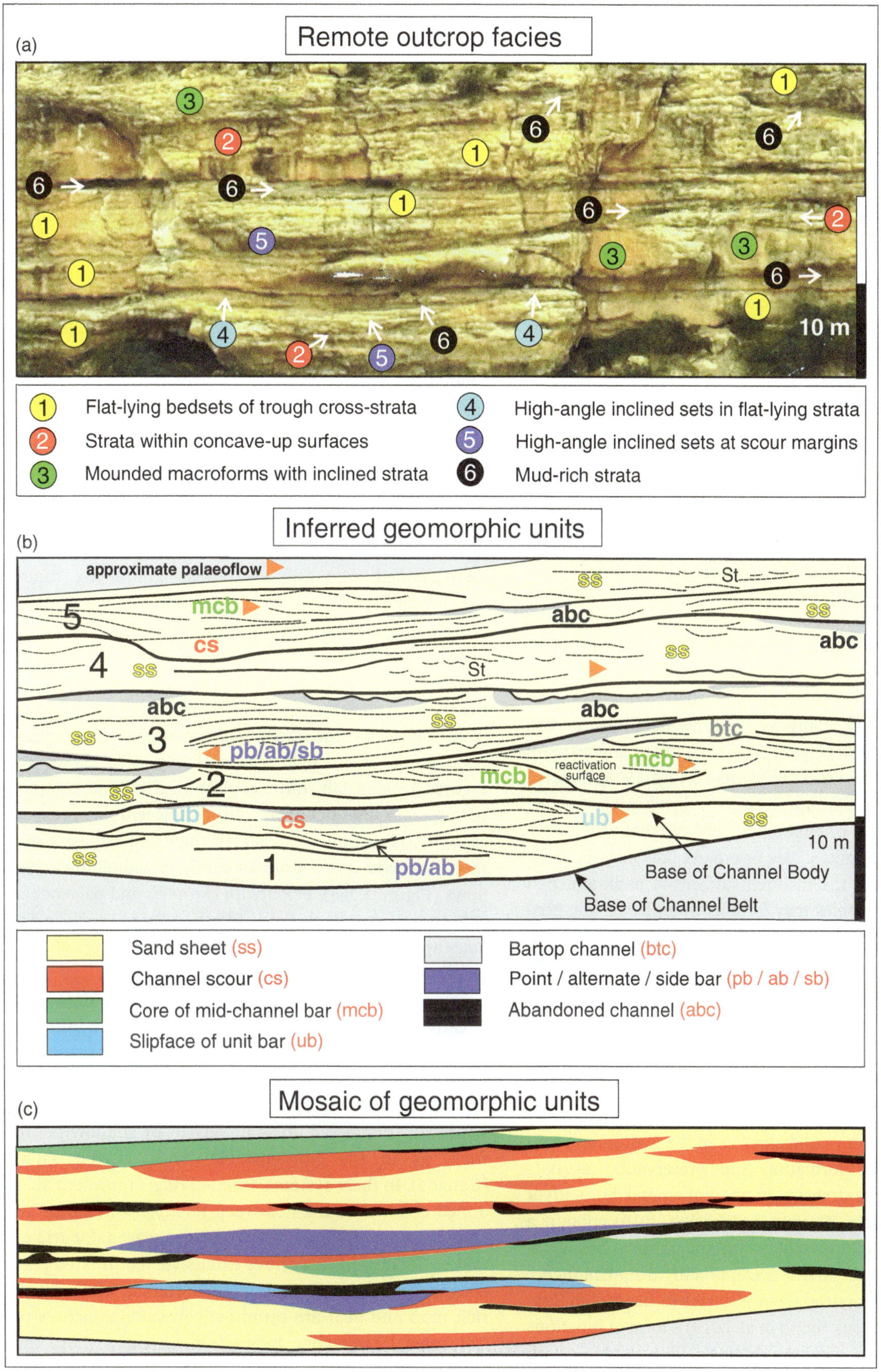

Figure 5.37 Remote outcrop facies (a) and inferred geomorphic units (b, c) for a braided-fluvial system, Triassic Katberg Formation, Bethulie, South Africa, from drone images of an inaccessible cliff (Gibling et al. 2023). Palaeoflow is to the right. Geomorphic unit map (c) shows that sand sheets and channel scours predominate (~70% of cliff area), with unit, mid-channel, and bank-attached bars (~20%) and abandoned channel fills (~10%). With permission from R. Jia, R. Gastaldo, and J. Neveling (a).

Although bar types may have distinctive planforms in modern rivers, few of these geometric forms yield diagnostic sedimentary features, and bar size exceeds that of many outcrops. Consequently, it is difficult to distinguish them in the geological record, especially in the deep subsurface, without exceptional exposure or dense datasets. Furthermore, both bars and channel fills contain stacked dune cross-sets, and flood reworking of bars in shallow channels hinders bar identification in subsurface deposits (Parker et al. 2013). In such cases, the term *channel-bar complexes* is a useful interpretation (e.g. Figures 5.26a and 5.27a). Bar types may be identified in exceptional exposures such as those from the Triassic Hawkesbury Sandstone of New South Wales (Australia), where large barforms with downstream-dipping surfaces are reasonably interpreted as transverse bars (Figure 5.34e,f; Rust and Jones 1987).

Where bedforms such as dunes are present on inclined bar surfaces in the geological record, palaeoflow directions have been used to distinguish mid-channel bars from bank-attached bars. This approach assumes that mid-channel bars have a prominent downstream flow component, whereas bank-attached bars have a prominent along-strike flow component (Miall 1994). However, quantitative investigations show that dunes migrate along the strike of inclined surfaces on both groups of bars (Ghinassi et al. 2016). Additionally, many mid-channel bars have curved slipfaces in planview (Figure 5.34c), with both downstream and across-stream dip components (Bridge 1993). Palaeoflow indicators on bar tops, formed as flow begins to wane, may not match the orientation of slipfaces that were active at higher stage (Figure 5.20c). Overbank flow may return to the channel down the inclined surfaces of bank-attached bars, and swept ripples may form as water levels fall, generating ripple-crest orientations that do not reflect the original flow direction. Thus, palaeoflow information must be carefully assessed in interpreting barforms.

5.8.3 Other Channel Units

Gravel sheets and *sand sheets* are common in many rivers, and they are especially prominent in shallow, seasonally flowing rivers where rapid changes in the hydrograph militate against the formation and preservation of bars (Figure 5.34j). Sand sheets in the ephemeral Luni River of northwest India (Carling and Leclair 2019) contain a high proportion of transcritical and supercritical bedforms, with many *channel scours*. Exceptionally deep *confluence scours* – up to 50 m deep on large rivers in the Indian Subcontinent (Sambrook Smith et al. 2019) – form at junctions between trunk and tributary channels, filled in places with *tributary mouth bars*. A confluence scour 30 m deep was documented at the junction of the Jamuna and Ganga rivers, five times the mean upstream channel depth (Best and Ashworth 1997).

Forced bars, *forced pools* and *boulder mounds* form in association with obstacles in the channel. These include standing trees, logs and woody accumulations, outsized boulders and bedrock outcrops (Figures 5.19d, 5.30d, and 5.33d, g–i). *Slump blocks* contain a range of facies (Figures 5.33c and 5.34d, g–i), and they are present where the banks are undercut or where an increase in pore-fluid pressure causes bank collapse. Slump blocks along the bank foot may protect the banks from erosion by the flow. In the Carboniferous example of Figure 5.33c, heterolithic point-bar deposits have failed along dislocation surfaces that flatten onto sandy beds below, with the slump-block strata rotated and dipping towards the bank.

5.8.4 Floodplains and Floodbasins

Levees form low-lying sediment wedges that border and are locally cut into by channel bodies. They are composed of interstratified mud and sand beds, commonly rooted, which were laid down by successive floods that overtopped the riverbank (Figure 5.36a,b). In small-scale examples, levee deposits may form distinctly convex-up sediment bodies. From low in the channel, sediment passes through the levees in *crevasse channels* (Figure 5.23a) that build *crevasse splays* onto the floodplain, sometimes with multiple flood phases in a single season (Figure 5.36c,d; see also Figures 5.20d, 5.23b, and 5.27b). Where sediment is transported onto the floodplain from adjacent uplands, for example, as a result of landslides, thick and extensive *sand sheets* are laid down from unconfined sediment-charged flows (Figure 5.36e). *Floodplain channels* and palaeochannels receive overbank flow (Figure 5.36f,g), as do *oxbow lakes* and *meander cutoffs* that were formerly part of trunk or tributary channels (Figure 5.36j).

Floodbasins are prominent in large river systems with negative floodplain topography, such as the Amazon, and on extensive inland deltas such as the Saskatchewan River Delta of Canada. They are also prominent in northern boreal river valleys where ice-stagnation complexes left locally depressed areas underlain by shallow permafrost, as in the Nordenskiold River of the Yukon (Canada). In these systems, isolated *ridges* represent abandoned channel courses between *floodplain wetlands* and *floodplain lakes* (Figure 5.36h). Under seasonally arid climates, floodbasins may have extensive, bioturbated playa fringes (Figure 5.36i). Floodplain wetlands with organic-rich mud and peat are prominent in valleys such as the Columbia River valley in western Canada, where they are banked against alluvial ridges (Figure 5.9). On a smaller scale, many alluvial tracts contain vernal pools unconnected to flow systems, where seasonal rainwater serves an important ecosystem function.

5.8.5 Geomorphic Units in the Geological Record

Geomorphic units have been identified with a high degree of confidence in the Jurassic Scalby Formation of the United Kingdom, where an extensive wave-cut platform and adjacent cliffs have exhumed a flat-lying meander belt (Figure 5.25d; Ielpi and Ghinassi 2014). Point bars, counter-point bars, scroll bars, levees, crevasse splays and crevasse channels are all present, as well as floodplain wetlands where peat (coal) accumulated. Channel width and depth and point-bar radius of curvature can all be measured accurately, and dinosaur footprints and in situ and transported vegetation provide clues for assembling the jigsaw puzzle of geomorphic units. Exhumed bedrock surfaces in other formations expose mid-channel bars (Cardenas et al. 2020), although examples are uncommon.

Based on GPR profiles in modern rivers, *radar facies* provide a means of identifying geomorphic units. Radar facies are mappable 3D units with reflection characteristics that differ from those of adjacent units (Huggenberger 1993), in the same manner as seismic facies but at a higher resolution. Unlike the facies discussed earlier, radar facies are sediment bodies that comprise assemblages of bedforms and sediment types, and good examples are given in Supplementary Table 5.S.1: 15. In a GPR profile through a compound bar in the braided South Saskatchewan River, Canada (Figure 5.35), Parker et al. (2013) identified four radar facies: (i) *high-angle inclined reflections*, attributed to migrating bar slipfaces; (ii) *undular and/or trough-shaped reflections*, attributed to sinuous-crested dunes and forming extensive sand sheets; (iii) *low-angle reflections* that dip upstream, downstream or laterally, attributed to low-amplitude dunes or unit bars that migrated onto bar surfaces; and (iv) *reflections of variable dip underlain by a concave reflection*, identified as the fills of channel scours, cross-bar channels or bartop hollows. The compound bar comprises a mounded unit bar cut by a bartop channel, with the later growth of sand sheets, a marginal slipface, and a capping veneer of dunes and small unit bars.

Figure 5.37 illustrates a Triassic channel-bar complex with five stacked channel bodies, exposed in a cliff in South Africa (Gibling et al. 2023). Palaeoflow is broadly to the right-hand side of the drone image. For such inaccessible outcrops, a useful approach is to identify *remote outcrop facies*, which correspond broadly with the radar facies identified in the South Saskatchewan River example and can be interpreted as geomorphic units (Figure 5.37b,c). The bulk of the Triassic outcrop (~70%) consists of flat-lying bedsets with trough cross-sets and discrete scour fills, which are interpreted as *sand sheets* (ss), *channel scours* (cs) and *bartop channels* (btc). Three remote outcrop facies with sets of inclined strata (~20%) are interpreted as the mounded cores of *mid-channel bars* (mcb), *unit bars* (ub) and bank-attached bars that may be *point, alternate or side bars* (pb/ab/sb). Lenses of mud-rich strata (~10%) are interpreted as the fills of *abandoned channels* (abc); their low thickness (average <1 m) suggests that the parent channels were largely filled at the time of abandonment.

Overall, the array of inferred geomorphic units in the South African example accords with a braided-fluvial system, but inferring geomorphic units involves some uncertainty (see also Adams and Bhattacharya (2005) and Chamberlin and Hajek (2019) for Cretaceous cliff examples). For example, (i) the planforms and types of mid-channel bars are unknown, and some may be part of larger compound bars, as in the Saskatchewan River example; (ii) discriminating point, alternate and side bars depends on planform exposures, which are not available in the outcrop belt; and (iii) some flat-lying strata interpreted as sand sheets may be longitudinal sections through channel scours or inclined bar strata.

5.8.6 Architectural Elements

Datasets may be scattered to moderate in quality, where they are restricted to small outcrops and cores. Identifying *architectural elements* (Miall 1996, 2010, 2014) provides a useful descriptive approach where datasets preclude a full architectural analysis. In such cases, identifying architectural elements provides a valuable low-resolution assessment of geomorphic units.

Miall (2010) recognised 14 elements to which upper-case codes were applied, based on bedform assemblages, accretionary surfaces and geometry (Table 5.8). Nine channel elements have descriptive names. They include three macroforms identified by accretion direction (lateral, downstream and upstream: LA, DA and UA), and four sheets identified by grain size and bedforms: gravel bars and bedforms (GB), sandy bedforms (SB), laminated sand sheets (LS) laid down under predominantly upper-regime flows and sediment-gravity flows (SG). Two other elements have scoured bases and concave-up form (channels and scour hollows, CH and HO). Difficulties in identifying the elements include (i) the problem of using accretion direction as a discriminator of macroform type, as discussed above; (ii) the challenge of defining channels and scour hollows within channel deposits, a fractal issue (Holbrook and Miall 2020); and (iii) the distinction of gravel bars and bedforms (GB) from sediment-gravity flows (SG), which requires a careful process analysis within the spectrum of flow types (Figures 5.11 and 5.12).

Five overbank elements have genetic names – levees crevasse splays, crevasse channels, floodplain fines and abandoned channels (LV, CR, CS, FF and CH(FF)). They are generally distinctive and correspond with some of the geomorphic units discussed above.

Table 5.8 Architectural elements in fluvial deposits.

Code	Name	Description
Channel architectural elements		
CH	**Channels**	Concave-up base, with internal concave-up surfaces common
GB	**Gravel bars and bedforms**	Generally tabular bodies, commonly interbedded with SB
SB	**Sandy bedforms**	Present in channel fills, crevasse splays and minor bars
UA	**Upstream-accretion macroform**	Rests on bar remnant or LA/DA deposit, with accretion surfaces dipping gently upstream
DA	**Downstream-accretion macroform**	Rests on flat or channelled base, with accretion surfaces dipping downstream; convex-up internal erosional surfaces
LA	**Lateral-accretion macroform**	Accretion surfaces oriented across channel, downlapping onto flat basal erosional surface
HO	**Scour hollows**	Scoop-shaped hollow with asymmetric fill
SG	**Sediment-gravity flows**	Lobe or sheet, typically interbedded with GB
LS	**Laminated sand sheet**	Sheet or blanket, composed mainly of upper-regime plane beds
Overbank architectural elements		
LV	**Levee**	Sediment wedge adjacent to channel, from overbank flooding
CR	**Crevasse channel**	Ribbon-shaped body, at break in channel margin
CS	**Crevasse splay**	Lens formed by progradation from crevasse channel onto floodplain
FF	**Floodplain fines**	Extensive sheet formed in floodplain ponds and swamps
CH (FF)	**Abandoned channel**	Ribbon-shaped body comparable in scale to active channel, formed by chute or neck cut-off

From Miall (1996, 2010, 2014).

5.9 Alluvial Architecture

Alluvial architecture describes the geometry of alluvial components and their assembly into 3D sediment bodies, which represent a composite and partially preserved history of river evolution (Potter 1967; Holbrook 2001; Gibling 2006; see Table 5.9 for some definitions). Only rarely are 'snapshots' available from unmodified architectural features 'frozen' at a particular moment in time; the exhumed outcrop of the Scalby Formation is an exceptional example (Figure 5.25d). However, the architectural approach described later enables researchers to work back from preserved alluvial deposits to the systems and events that formed them. Such an analysis also may establish, at least in outline, the transport direction of the ancient river system.

Alluvial architecture is best studied by tracing stratal surfaces at high resolution and with a high degree of precision. Tracing can be carried out using drafting software or geophysical packages, working with panoramic outcrop photos, 3D photogrammetric outcrop models (increasingly generated from drone images to avoid perspective problems), and high-resolution seismic, GPR and LiDAR profiles. Digital panoramic images may be annotated directly in the field, but under rough field conditions, it may work better to record information on serial photos mounted on cardboard, which can later be transferred to digital images. Where possible, outcrops should be visited later with tracings in hand, to discount shadows, joints, and cliff falls. Cores and petrophysical analysis may help in confirming subsurface observations for seismic and GPR profiles. The process of tracing yields considerable insight and should be considered an analytical method, not just a means of illustration. It also allows precise measurements of the thickness, and less commonly, the width of many features.

5.9.1 Channel Migration, Avulsion and Channel Belts

The architecture of sediment bodies depends not only on the geometry of the original channel but also on the processes involved in channel relocation, linked to deposition, erosion, rates of channel migration and avulsion events (Bridge 1993). Channels may migrate progressively across their floodplains as riverbanks retreat and bank-attached and mid-channel bars advance, generating a largely concordant set of bar accretion surfaces. Alternatively, channels may avulse as flow is diverted rapidly (*full avulsion*) or gradually (*partial avulsion*) from an established channel into a new course, leaving a fully or periodically unoccupied channel reach (Figure 5.38a; Slingerland and Smith 2004; Supplementary Table 5.S.1: 16).

Table 5.9 Architectural terms for alluvial systems.

Terminology for alluvial architecture	
Term	**Definition**
Channel body	3D body of unconsolidated or lithified channel sediment underlain by a prominent erosion surface. May be the fill of an individual channel or a composite deposit generated through migration and avulsion. Most channel bodies evolved through time and do not represent an instantaneous channel form filled over a short period.
Channel belt	Surface appearance or preserved deposits of a channel system with multiple components developed through time. Typically comprise multiple channel bodies. They include meandering- and braided-river systems with active, inactive and abandoned channels and bars.
Channel fill	General term for sediments within a channel body. The term need not imply the filling of a channel without change in its perimeter (contra Gibling 2006).
Storey	Prominent, erosionally based component of a channel belt, commonly the equivalent of a channel body. Although channel belts may contain many erosional surfaces, the term usually denotes a substantial channel deposit that rests on a prominent erosion surface. Storey tops are commonly truncated below an overlying storey but may be fully preserved where a channel avulsed. May be written as 'story' and 'stories'.
Single storey	Channel body composed of one storey
Multistorey	Channel body composed of more than one storey. Sometimes restricted to vertically stacked storeys, but more generally used for multiple storeys, however disposed
Multilateral	Channel body composed of more than one storey, where the storeys are disposed laterally. Multistorey may also be used for this situation.
Macroscale set, composite set	Alternative terms for storeys, focusing on bedset description (Bridge 1993)
Amalgamation	Linkage of two or more storeys across an erosional surface, typically with a sand-to-sand contact
Inclined stratification (IS), inclined heterolithic stratification (IHS)	Descriptive terms for inclined sets of strata, typically sandy (IS) or sandy and muddy (IHS) that represent surfaces of bar accretion, commonly bank-attached bars (Thomas et al. 1987). The sets comprise one or more facies and are generally 1–30 m thick. Distinct from angle-of-repose cross-strata (Sp, Gp)
Reactivation surface	Erosion surface cutting beds with depositional dips (bar surfaces), usually due to lowering of discharge. A resumption of sedimentation reactivates the bar, reforming the accretion surface.
Concentric fill	Infilling of a relatively narrow, single-storey channel (active or abandoned) by deposition on the floor and accretionary banks, progressively reducing the cross-sectional area.
Asymmetric fill	Infilling of a relatively narrow, single-storey channel by a bank-attached bar that accreted laterally more rapidly than the bank retreated.
Central body, wing	Main part of channel body, and a thin marginal part distinguished from central body where the basal scour shows an inflection point. Wings may represent channel widening at channel-body top, or a marginal levee-crevasse splay complex.
Bounding surfaces	Stratigraphic surfaces that bound bodies of strata, including channel bodies, storeys and geomorphic units
Sheets, ribbons	Fluvial bodies (usually channel bodies) with width:thickness ratio of >15:1 and <15:1, respectively
Ribbon sandstone tiers	Numerous ribbon channel bodies at the same stratigraphic level, commonly connected by thin sandy or silty sheets that form wings from the individual bodies (Kraus and Gwinn 1997). May indicate an anastomosing channel system, although it is not possible to show that flow took place simultaneously in all the channels.
Valley	Elongate topographic low that is larger than a single channel form. May be formed by fluvial activity, or rivers may occupy valleys generated by glacial or marine activity. Valleys are typically composite and, in coastal settings, are time-transgressive in both cross- and down-valley directions (Blum and Aslan 2006). Valley fill may correlate with mature palaeosols on the adjacent interfluve.
Composite valley-fill unconformity	Bounding surface that underlies a composite valley, formed over time. Thus, the valley did not have the form of the bounding surface at any point in time.
Clustering	Stratigraphic grouping of channel bodies in the same area through time

Source: Adapted from Gibling (2006), which provides source literature for some of the terms.

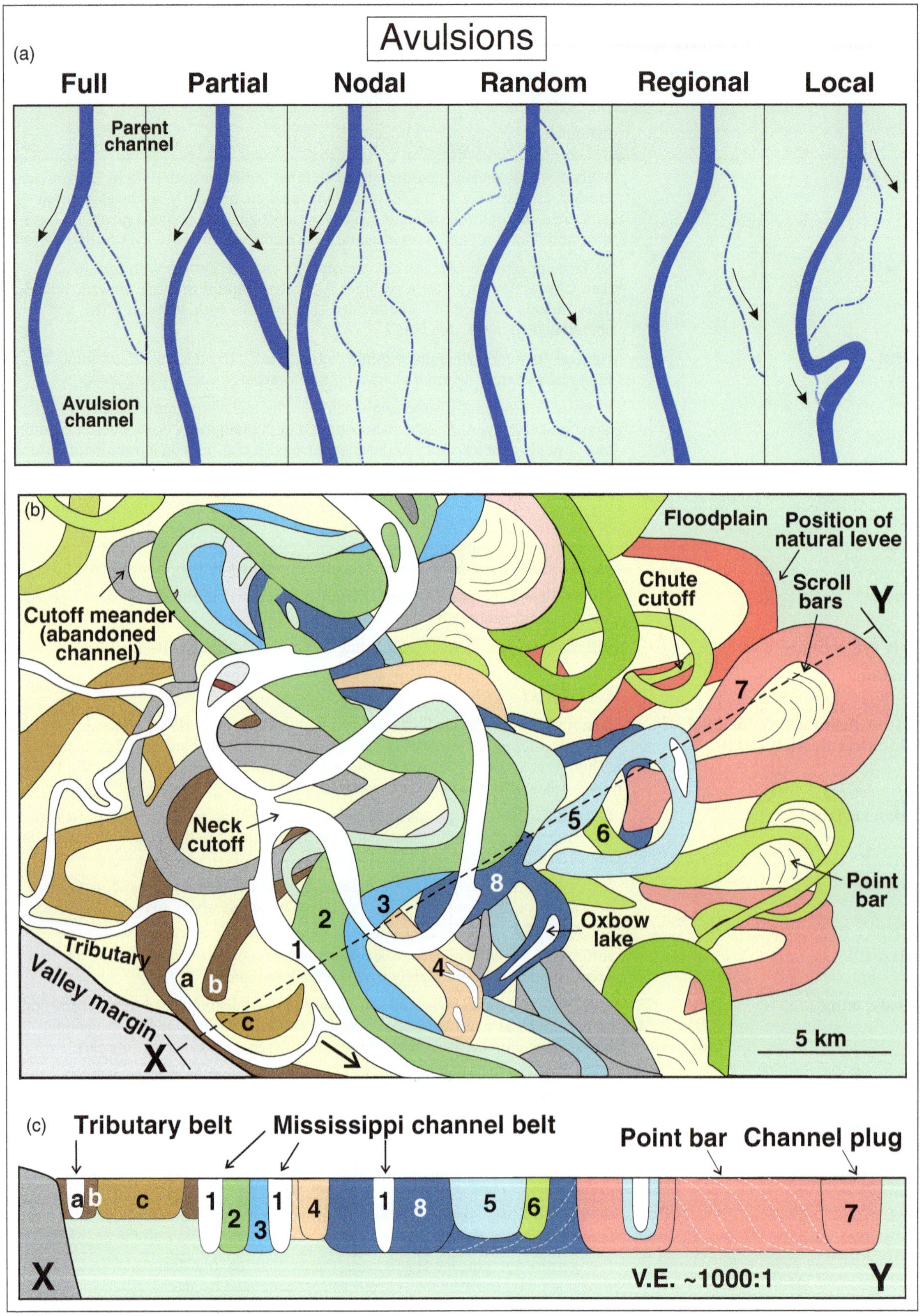

Figure 5.38 Avulsion dynamics. (a) Avulsion terminology. Source: Adapted from Slingerland and Smith (2004). (b) Map of the Mississippi River channel-belt near Natchez, Mississippi. Each colour represents the deposits of a meandering channel that was later transected by meander neck and chute cut-offs and by avulsion of longer channel reaches. Yellow background represents point-bar deposits, some of which can be linked to abandoned-channel fills. The ages of the courses are not well known, but the channel belt probably formed over several thousand years. Source: Adapted from Fisk (1944) in Gibling (2021), with permission from Dunedin Academic Press. (c) Hypothetical cross section along line X–Y in B), to illustrate the inferred architecture of the channel belt. The relative depth of incision of particular channel fills is judged from their width at surface. Inclined point-bar surfaces are likely to be preserved in the Mississippi system, which has a low discharge variance (Fielding et al. 2018).

Avulsions are thought to originate by two mechanisms (Gearon et al. 2024). *Alluvial-ridge superelevation* occurs when the channel bed has aggraded a threshold fraction of its depth above the distal floodplain, calculated by the alluvial-ridge height divided by channel depth. The second mechanism, *gradient advantage*, occurs when the local cross-channel slope is steeper than the down-channel slope, calculated by the ratio of the two slopes. In both cases, the channel switches to a new, steeper conduit that follows a path of least resistance. Superelevation values are highest in mountain-front fans where sediment aggrades rapidly, whereas gradient advantage is highest in floodplains and deltas where gradients are low. In general, upstream controls may be more important than downstream controls in governing the frequency of avulsions (Colombera and Mountney 2023).

Nodal avulsion occurs repeatedly at a fixed point, for example, at the head of a distributive fluvial system (megafans and alluvial fans), whereas *random avulsion* takes place anywhere along the river course. In a *regional avulsion*, the new channel affects the entire downstream course, whereas in a *local avulsion*, the new channel rejoins the parent channel downstream. Channel bodies may show *clustering* due to repeated local avulsions and the tendency for channels to occupy depressed areas generated by subsidence or differential compaction (Supplementary Table 5.S.1: 18).

Channel relocation may take place through *annexation* of an abandoned channel or an existing active channel (in a fashion akin to piracy). Relocation may also involve *incision* where the channel carves a new course across a floodplain, locally guided by gullies that propagate headward where overbank flow re-enters the channel. Avulsion by *progradation* generates an advancing avulsion belt where multiple channels gradually consolidate into a single master channel. *Retrogradational avulsion*, documented from vegetated sediment fans, is caused by channel blockage with sediment and wood, followed by a rapid upstream wave of channel filling and eventual breakout into a new course (Edmonds et al. 2022). Over a more local scale, a channel may relocate by *chute cutoff* through its own depositional banks, or by *neck cutoff* when two sinuous bends intersect and cut off an intervening meander.

The time between avulsions is typically hundreds to a few thousands of years, as in the Columbia River (Figure 5.9b) and lobes of the Tista Megafan (Figure 5.5e). However, avulsion periodicity may be as little as a few years to decades in hyperavulsive systems such as the Baghmati River of India (Sinha et al. 2005) and the Taquari Megafan of South America (Assine 2005).

Figure 5.37b is a redrawing of part of Harold Fisk's (1944) classic maps of the Mississippi River *channel belt*. The maps form an exemplary case study of how the interplay between migration and avulsion of meandering channels determines a river's stratigraphic record. Colours represent channel segments of similar age, developed over hundreds to a few thousand years. The channel belt is about 15 km wide with numerous short meander neck and chute cutoffs, as well as longer local avulsions of the channel and its tributaries. A hypothetical cross section (Figure 5.38c) shows a suite of channel fills generated by local and random avulsions that were variously full or partial. Within the abandoned meander loops, sets of inclined strata with scroll bars record systematic point-bar migration, which was restricted where the channel encountered abandoned channels filled with stiff mud plugs (Turnbull et al. 1966).

Braided channel belts evolve in a different manner. A study of the Missouri River showed that mid-channel bars have an average lifespan of about 8 years, with the majority migrating less than one bar-length downstream during their lifetime (Underwood 2022). About two-thirds of the bars were eroded and incorporated back into the bedload. Of the remaining bars, repeated accretion along particular areas of the bank generated *point assemblages* with multiple compound bars and lateral bars (Holbrook and Allen 2021). This docking process forced the river system into a sinuous planform, and the mid-channel bars were incorporated and preserved in the floodplain. Located in an upland valley, the Kicking Horse River provides an example of braided channel-belt development, during which a complex of bars, bedload sheets and channel fills built up through local avulsions and channel migration over tens of metres and in less than a decade (Cyples et al. 2020; Figure 5.7b).

5.9.2 Architectural Features

Channel deposits are composed of *storeys*, which are channel-bar complexes with erosional bases, preserved as near-complete successions or as truncated remnants. *Multistorey* channel bodies are composed of two or more storeys (Figure 5.39a), each of which consists of numerous geomorphic units and may comprise several facies associations. The term commonly denotes vertical stacking of storeys, as illustrated in the Columbia River where some channels were abandoned and later reoccupied (Figure 5.9b). Pennsylvanian multistorey channel bodies are illustrated in Figures 5.26b and 5.39b. The term *multilateral* is used where the storeys cut into each other at a similar stratigraphic level, generating a broad, composite channel body (Supplementary Table 5.S.1: 20). However, multistorey is widely used as a general term because some combination of vertical and lateral stacking is usual. Bridge (1993) suggested 'macroscale coset' or 'composite set' as descriptive terms in place of 'storey'. The term 'channel fill' is used here

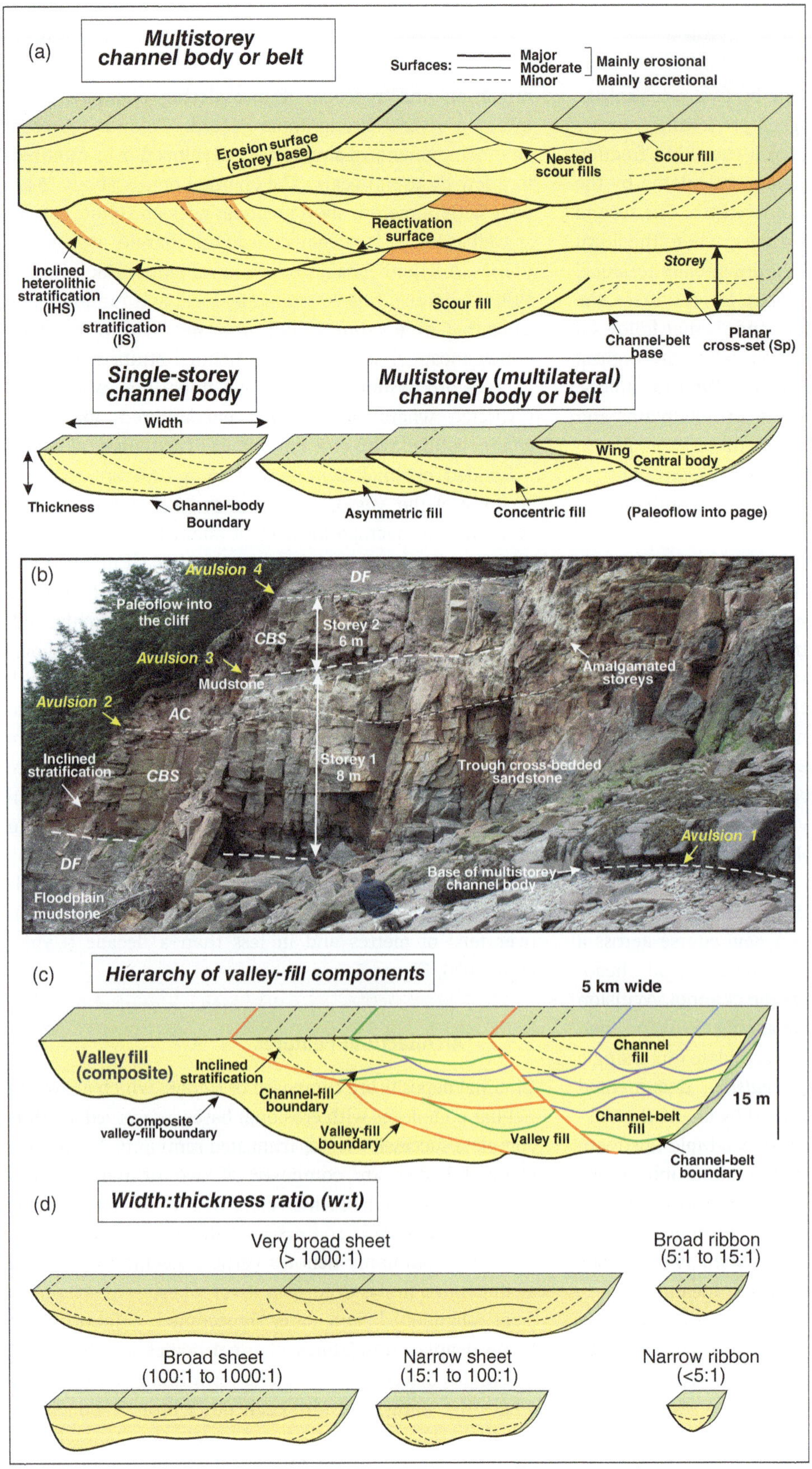

Figure 5.39 Alluvial architecture and terminology (a); channel body in the Pennsylvanian Boss Point Formation, Canada (b); large-scale composite channel body (c); and terms for channel-body geometry (d). Terminology from Potter (1967), Friend et al. (1979), Holbrook (2001) and Gibling (2006), and definitions in Table 5.9. In (b), a multistorey channel body (CBS) from a braided river lies between distal floodplain (DF) deposits. Two storeys of trough cross-bedded sandstone (St) and sandstone with wood fragments (Sv) are separated by an abandoned-channel fill (AC) but are amalgamated locally. Inclined stratification at lower left represents a mid-channel bar. Channel avulsion generated stacked storeys through reoccupation and abandonment. *Source:* From Martin Gibling (Chapter author) (b).

in a broad sense and does not imply the fill of a discrete channel form.

Multistorey architecture is common in alluvial-fan deposits, as illustrated in Permian examples from the United Kingdom, laid down under strongly seasonal and relatively arid conditions (Figure 5.40a). Attributed to the CBG Association, the stacked channel bodies have prominent erosional bases that cut down in places to form deeper hollows. The poorly sorted gravels (Gcs and Gms) contain minor erosional surfaces, and some channel fills are capped by sandstone layers (Sm). The fan architecture represents a mosaic of channels 1–2 m deep, which were swept by floods that laid down poorly organised successions of gravel and sand with outsized clasts and local imbrication. Although some inclined bedsets suggest the presence of bars, geomorphic units are difficult to identify in these channel-bar complexes. Analysis of clast parameters (lithology, size and roundness) and palaeoflow indicators (radial flow patterns) supports a fan origin.

Other channel bodies are *single storey* (Figure 5.39a), many representing 'fixed' channels (Figure 5.27b). Small channel bodies may display *asymmetric fills* generated by the migration of bank-attached bars, or *concentric fills* where sediment built up vertically and obliquely on accretionary banks, causing the channel to narrow and shallow through time. Some deposits consist of a *central body* and a *wing*. Wings develop where the channel eroded across its banks (see Figure 5.34j for an example from Tanzania), laying down a more extensive sheet of channel sediment, or where levees and crevasse splays advanced across the floodplain as the channel filled. Several small channel bodies may be present at the same stratigraphic level (*ribbon sandstone tiers*). This architecture suggests an anastomosing channel system, although it is rarely possible to demonstrate that the multiple channels flowed simultaneously (see Cuevas Martinez et al. 2010).

The development of a single-storey channel body from the Jurassic of India is illustrated in Figure 5.40b,c (Datta and Ghosh 2023). The initial incision was 4.5 m deep with a relatively steep cutbank (on the right-hand side) against semi-consolidated floodplain strata and palaeosols (note 1). A bank-attached bar with inclined heterolithic strata of sand and mud (IHS) prograded towards the cutbank. The bar was initially nearly the height of the incision (note 3) and was probably a laterally accreted point bar, with repeated incision and reactivation of the bar surface under seasonal changes in flow stage. The rate of sediment supply and bar migration outpaced the rate of cutbank retreat, and mud accumulated by vertical accretion in the thalweg, progressively raising the channel bed and the bar toe (notes 2 and 4). As the bar approached the cutbank, the channel narrowed, aided by draping of mud on the cutbank (note 5), and shallowed to little >1 m. The channel was finally abandoned, a shallow channel was cut on the bar top, and the river avulsed to a new location on the alluvial plain (notes 6–8). The channel body is a mosaic of geomorphic units that include a point bar (pb), accretionary bank (ab) and abandoned channel fill (abc). The minimal retreat of the cutbank suggests that the channel was self-constraining or laterally stable (a 'fixed' channel), with a combination of asymmetric and concentric filling.

Valleys are elongate topographic lows that host channel belts and are floored by prominent erosional surfaces (Figures 5.4, 5.5a and 5.9a; Supplementary Table 5.S.1: 21). They were usually generated by the occupying river, although 'underfit' rivers may occupy large valleys scoured by glaciers or meltwater floods, as in European and Siberian rivers associated with permafrost (Vandenberghe 2024). Strong and Paola (2008) distinguished two types of valley. *Topographic valleys* are representative of the surface topography at an instant of time, whereas *stratigraphic valleys* are defined by a stratigraphic surface that developed over time as a series of topographic valleys came into existence and were modified. The morphology of a stratigraphic valley never existed at any point in time and has resulted from cumulative geological processes. An example of an ancient stratigraphic valley is the Cretaceous Mesa Rica Sandstone of Colorado and New Mexico (Figure 5.39c; Holbrook 2001). The fluvial deposits comprise channel fills that combine to form channel-belts, which in turn form components of valley fills nested inside the composite valley. The base of the valley is a *composite valley-fill unconformity* (Figure 5.39c; Blum and Aslan 2006).

The geometry of channel bodies and valley fills can be represented in part by the ratio of width to thickness of the lithological unit, with categories of *sheets* and *ribbons* (Figure 5.39d; Potter 1967; Friend et al. 1979; Gibling 2006). The maximum or average thickness of the channel bodies can usually be determined from cores and outcrops (Figure 5.39b). However, the width of a channel body often exceeds the size of the outcrop or the core and wireline log array, leading to a minimum estimate of width and of the ratio of width to thickness. Width is measured normal to the average transport direction, determined from palaeoflow analysis. Only rarely (e.g. where channel bodies represent self-constraining channels) will the width and thickness of a channel body represent the width and depth of the parent channel, recorded at an instant in time (instantaneous dimensions). In such cases, the channel dimensions broadly reflect the discharge of water that must be transmitted on a frequent basis in the formative stage.

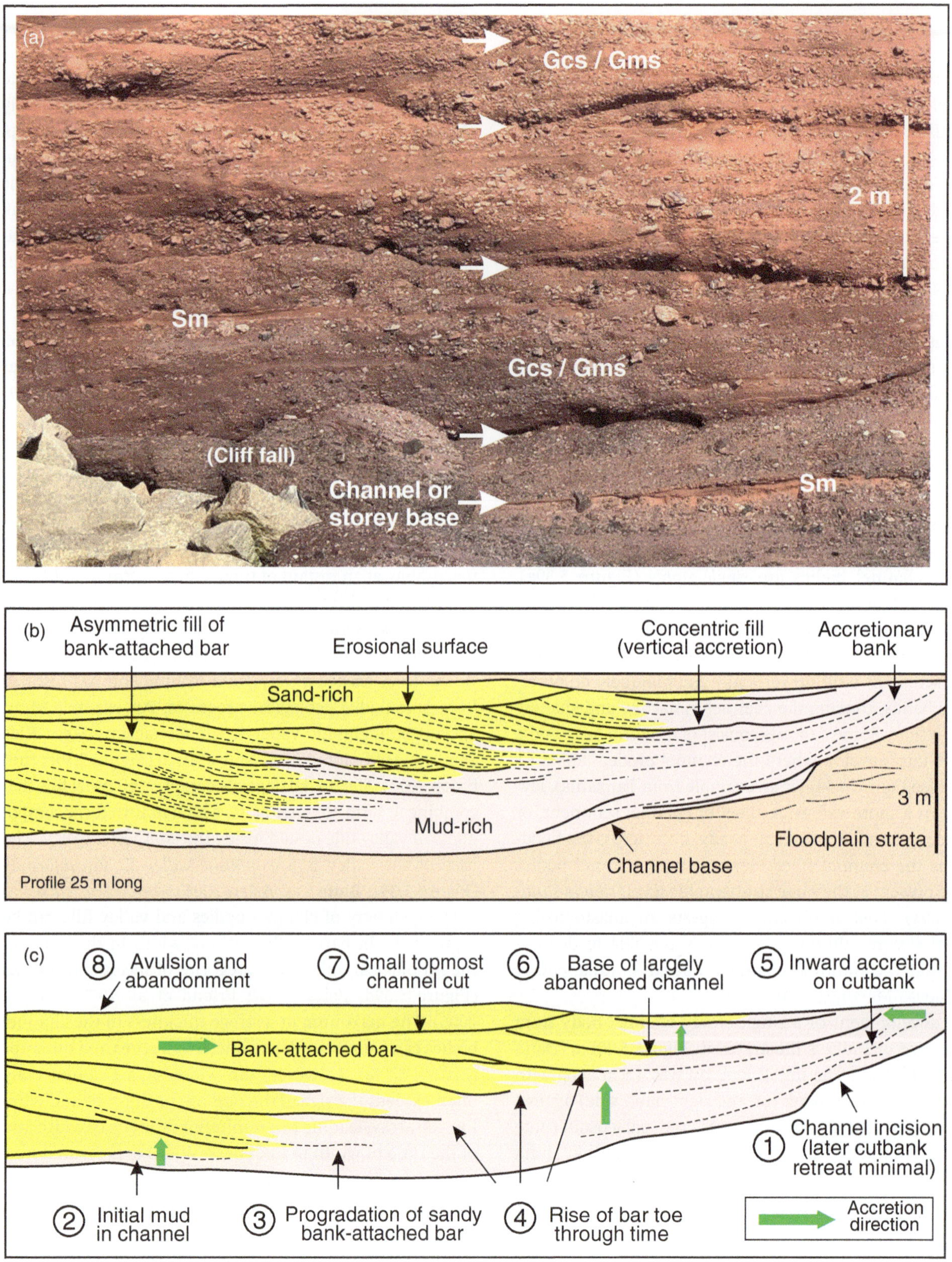

Figure 5.40 (a) Multistorey channel system in Permian alluvial-fan deposits of the Teignmouth Breccia, at Teignmouth, UK (Laming 1966), laid down as poorly organised gravelly deposits of the Channel-Bar Gravel Association, in a mosaic of small fan channels. Palaeoflow was broadly to the right of the photo, based on poorly developed imbrication. With permission from Martin Gibling. (b, c) Single-storey channel body in the Jurassic Dharmaram Formation, central India, shown in a cross section approximately normal to the palaeoflow (b) and with notes of channel-body development (c). The cutbank was largely stable (right-hand side), and migration of the sandy bank-attached bar progressively narrowed the channel, aided by vertical accretion of mud in the thalweg and minor inward accretion against the cutbank. Over time, the channel shallowed, accumulated a concentric fill and avulsed. The channel body was probably laid down in a self-constraining or laterally stable (fixed) channel. With permission for use of field photo (not shown) from Shantanu Datta and Parthasarathi Ghosh.

5.9.3 Bounding Surfaces

Bounding surfaces separate packages of strata and may be erosional or non-erosional (accretional). Miall (1996) used a hierarchical system of six bounding surfaces (Table 5.10), which is helpful conceptually. However, the system may be difficult to apply in practice (Bridge 1993), especially in core and where surfaces are discontinuous.

The approach used here is designed to document as objectively as possible the major architectural features and stacking patterns within the alluvial body, for practical use when tracing photomosaics and other profiles (Figure 5.39). Bounding surfaces may be distinguished as *major* or *moderate* in prominence (usually erosional) or as *minor* (usually depositional). They may be represented on a tracing using different line weights, styles and colours. The surfaces should be traced precisely at high resolution, not drawn as straight lines, and terminations and lateral changes in prominence are routine and should be recorded. The individual surfaces may then be interpreted using the categories of Table 5.10 and other terms (Figure 5.39a), although not necessarily using order numbers. Cross-set and coset boundaries are usually below the resolution of cliff tracings or are too detailed to include, but this level of detail should be collected for key areas of the outcrop.

Erosional surfaces include (i) the bases of channel bodies in contact with overbank and other strata; (ii) storey boundaries as through-going or extensive surfaces within channel bodies; (iii) minor channels or local scours that form less extensive or discontinuous surfaces; and (iv) inclined surfaces that represent the erosion and later reactivation of bar accretion surfaces, across which accretion direction may change considerably (Durkin et al. 2015). *Non-erosional surfaces* include (i) the flat-lying tops of bedsets that represent bars, channel sheets and other geomorphic units; (ii) inclined surfaces that represent bar increments; and (iii) the tops of channel bodies and storeys overlain conformably by overbank deposits and abandoned-channel fills. Interpretation depends on the orientation of the traverse with respect to the palaeoflow; for example, a storey boundary may appear concave-up in a transverse section normal to the palaeoflow but flat-lying in a longitudinal section parallel to the palaeoflow. Interpretation of the surfaces may involve uncertainty unless the exposure offers at least some degree of three-dimensionality.

5.9.4 Estimating Channel Parameters from Alluvial Architecture

The depth and width of the parent channel may be approximated from the preserved dimensions of storeys and inclined strata measured during field observations and from tracings. Six methods are illustrated in Figure 5.41.

The most complete measurements represent storeys with fully preserved accretion surfaces (Method 1), some of which pass upward into convex-up scroll bars. Abandoned, mud-filled channels are especially useful in providing the depth and width of the parent channel (Method 2), especially for oxbow lakes that were isolated abruptly during neck cutoffs. However, mud plugs may represent the final deposition in channels that were already largely filled with coarser sediment, and they may not reflect the dimensions of the active channels.

Table 5.10 Bounding surfaces in alluvial strata, modified from Miall (1996).

Bounding surfaces and orders	Comments
Ripple and dune cross-set boundaries (1)	Microscale (ripples) and mesoscale (dune) sets. Palaeoflow is similar across set boundaries.
Coset and bedset boundaries (2)	Microscale or mesoscale cosets of cross-strata and bedsets of other facies. Palaeoflow may diverge across coset boundaries.
Macroform (bar) increments (3)	Macroscale inclined stratum, formed by accretion on bar slipfaces. Include non-erosional increments; increments emplaced on erosional (reactivation) surfaces; and flat-lying increments within bars and other geomorphic units
Macroform boundaries (4)	Macroform sets of inclined strata within bars and on accretionary banks, commonly with convex-up top surfaces. Includes bounding surfaces of other types of geomorphic unit (e.g. sediment sheets)
Channel boundaries (5)	Group of macroscale sets, typically containing multiple geomorphic units
Channel-belt and valley boundaries, and boundaries associated with distributive fluvial systems (6)	Group of macroscale sets, typically containing multiple channel and floodplain bodies, as well as components from other types of depositional system (lacustrine, aeolian)

Orders 1–4 are associated with geomorphic units.

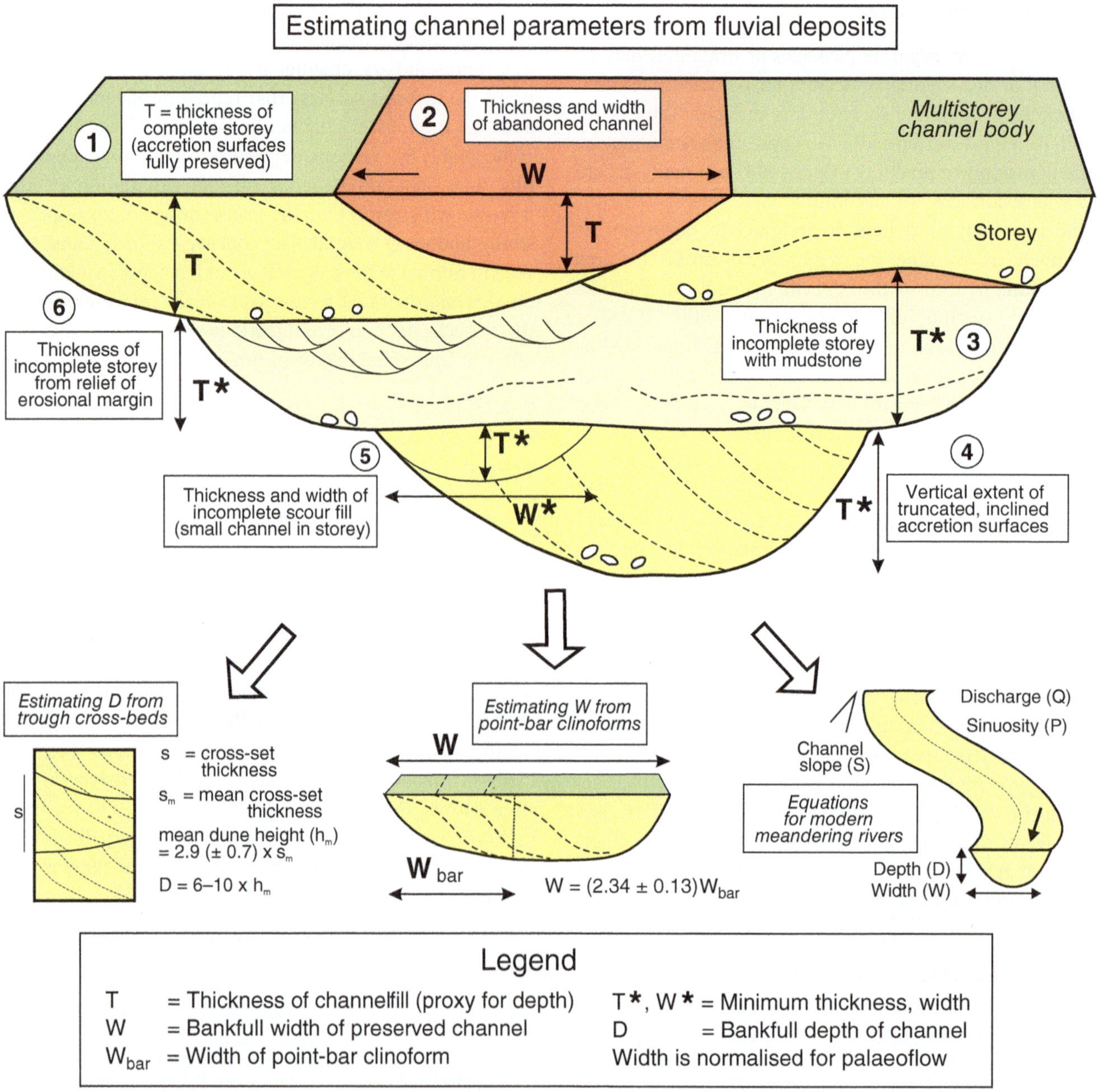

Figure 5.41 Estimation of channel parameters from fluvial deposits. A good approximation of the original channel depth and width may be obtained from storeys with fully preserved inclined surfaces and from abandoned-channel fills (1, 2). A minimum depth may be estimated from truncated storeys or from the relief of erosional margins (3–6), especially where the storey contains an upward succession of bedforms. See Leclair and Bridge (2001) for estimation of bankfull depth using grouped cross-bed thickness, Greenberg et al. (2021) for estimation of bankfull width using bar clinoform width, and Hampson et al. (2013) for equations that link channel parameters with discharge and other hydraulic features.

Methods 3–6 generate minimum width and depth measurements from incompletely preserved strata. Facies successions may be measured from an erosional base to the preserved top of an abandoned-channel fill (Method 3). Other measurements include the vertical extent of truncated accretion surfaces (Method 4) and the dimensions of truncated scour fills (Method 5), as well as the vertical relief on the margins of truncated storeys (Method 6). Channel bankfull width may be estimated using the width of point-bar clinoforms (IS and IHS surfaces) that have been fully preserved, when corrected for palaeoflow direction (lower centre of Figure 5.41; Greenberg et al. 2021).

For the Triassic strata of Figure 5.37, fully preserved channel and abandonment fills (Methods 1 and 2) were not identified. Truncated facies successions (Method 3) average 2.9 m in thickness (up to 5 m), and truncated accretion surfaces of bank-attached bars (Method 4) average 2 m in vertical extent, with some storey margins of similar height (Method 6). Overall, the Triassic channels were typically 2–3 m deep, and the mid-channel and bank-attached bars scale broadly with channel depth. The fills of prominent channel scours (Method 5) average 1.7 m in thickness, and they probably represent smaller anabranches or local scours. In the absence of palaeoflow data from the inaccessible cliff, true channel width could not be measured.

Flow depth may be estimated using a linear relationship with mean cross-set thickness (lower left of Figure 5.41). Cross-set thickness allows an estimate of original dune height which, as noted above, scales broadly with flow depth (Leclair and Bridge 2001). The relationships show a significant scatter of data and should be considered to yield a first-order estimate. Furthermore, some cross-sets may represent scour fills, especially in rivers with highly variable discharge, or may represent the avalanche front of unit bars rather than individual dune trains. Channel palaeodepth may be overestimated in such cases without careful field scrutiny.

Parameters that include width, depth, slope, sinuosity and discharge may be estimated using equations derived from modern rivers (Hampson et al. 2013; Long 2021; lower right of Figure 5.41). The equations use power–law relationships to connect the geometry of a channel body with the hydraulic geometry of the parent channel, provided that the planform can be inferred (Leopold and Maddock 1953, among many studies). As a starting point, the equations commonly use the *average bankfull channel depth* (the average depth of a river across its full width), from which parameters such as bankfull width, bankfull discharge, and drainage area may be estimated (Long 2021). In the ancient record, average bankfull channel depth may be determined in places, but more commonly, the thickness of channel bodies and facies successions provides an estimate of the *maximum channel depth*, from which average bankfull depth may be calculated using an equation provided by Long (2021). The results are subject to many sources of uncertainty related to the fluvial style, transport dynamics and climatic setting of the modern examples, especially where equations employ several calculated parameters that generate cumulative errors (Long 2021). Box 5.2 illustrates a quantitative approach to inferring channel planform in the ancient record, using the grain size of bed and bank sediment.

Seismic profiles and timeslices provide exceptional opportunities to determine channel planform and estimate channel width, depth and sinuosity, especially if supported by analysis of core and seismic facies (Supplementary Table 5.S.1: 24). In Quaternary examples from the Malay Basin, these parameters can be measured directly for low-sinuosity and meandering channels (Figure 5.42; Alqahtani et al. 2017). Tan et al. (2020) estimated the dimensions of mid-channel bars from timeslices in the Bohai Basin.

5.9.5 Compaction of Alluvial Sediment

Sand, mud and peat in channel and floodplain strata compact during burial, affecting estimation of the original channel depth. Compaction locally promotes a high rate of subsidence that is paced by floodplain and channel aggradation where sediment supply is sufficient, leading to the formation of multistorey channel bodies (Jerrett et al. 2023). The degree of compaction may be estimated from site-specific outcrop studies (Supplementary Table 5.S.1: 19) or from compaction curves for sandstone and shale in wells with normally pressured settings (Baldwin and Butler 1985). Based on mechanical settling, sand is commonly estimated to compact by 10%. Consequently, a decompaction factor of 1.1 is sometimes applied to sandstones (Ethridge and Schumm 1978), although the diagenetic history inferred from thin-section analysis may provide a better estimate (Long 2021). However, decompaction estimates may be problematic because (i) cementation of porous sand near the surface limits compaction during deeper burial; (ii) surface peat has variable density (van Asselen et al. 2010); (iii) sand and mud with different compaction ratios may be interbedded; and (iv) surface mud may harden during dry seasons, as in central Australia where mechanical excavators may be unable to dig trenches through the tough floodplain mud. In view of these issues, it may be unrealistic to use decompaction estimates to correct stratal thickness, unless a site-specific study is available.

5.9.6 Discharge Variance and the Facies Spectrum

Variation in river discharge profoundly influences the morphology, architecture and facies spectrum of channel bodies, which may show a poor correspondence with channel planform (Plink-Björklund 2015; Fielding et al. 2018; Hansford et al. 2020; Supplementary Table 5.S.1: 23). Most rivers show some variability in seasonal discharge due, for example, to variation in snowmelt or monsoon rainfall. However, long-term patterns of flow variation may be predictable (low discharge variance) or erratic (high discharge variance) as a result of weather patterns that include cyclonic storms and El Niño periods.

Where long discharge records are available from gauging stations, parameters of discharge variance may be calculated (Fielding et al. 2018). For example, the Amazon with contributions from many tributaries in the tropical belt has

Box 5.2 Estimating Channel Form

One approach for assessing the planform of channel deposits in the geological record (Lapôtre et al. 2019) uses field and lab analysis of bed and bank strata in well-exposed channel deposits. The worked example here is calibrated with data from the Amargosa River of Nevada and California (Ielpi 2019).

The grain size of the former riverbed is determined to be ~100 μm, with bar forms about 1 m high and current ripples with a wavelength of about 11 cm. Using the formulation of Lapôtre et al. (2017), the formation of ripples with this wavelength in grains 100 μm in size requires bed stresses (τ_b) of ~0.6 Pa. Assuming that bar height can be used as a 1:1 proxy for flow depth (h), bed slope under steady, uniform flow conditions can be estimated as $S = \tau_b/(\rho g h) \approx 6 \times 10^{-5}$.

The strata also contain mud drapes of fine-grained sediment with a grain size of ~10 μm. These muds are assumed to be representative of the sediments that provided strength to the riverbanks. The model predicts that the instantaneous channel had a width:depth ratio (W/h) of about 20. With an estimated flow depth of 1 m, the channel would have been about 20 m wide, with a shape that could have allowed relatively high sinuosity. With an increase of grain size to 50 μm in the finest intervals, the model predicts an equilibrium W/h > 200, suggesting deposition in a broader channel prone to braiding. *Source:* Alessandro Ielpi (Chapter Contributor).

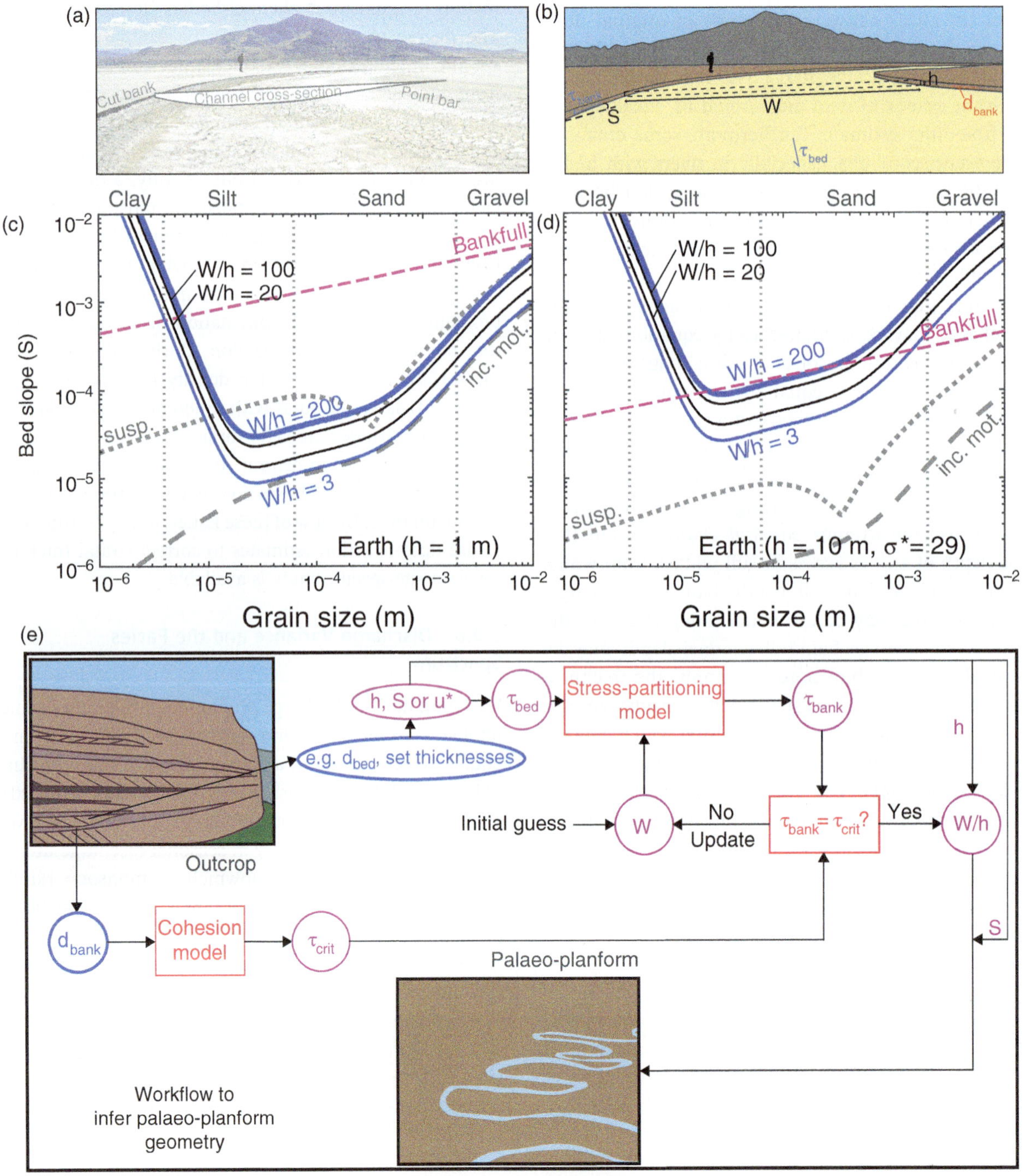

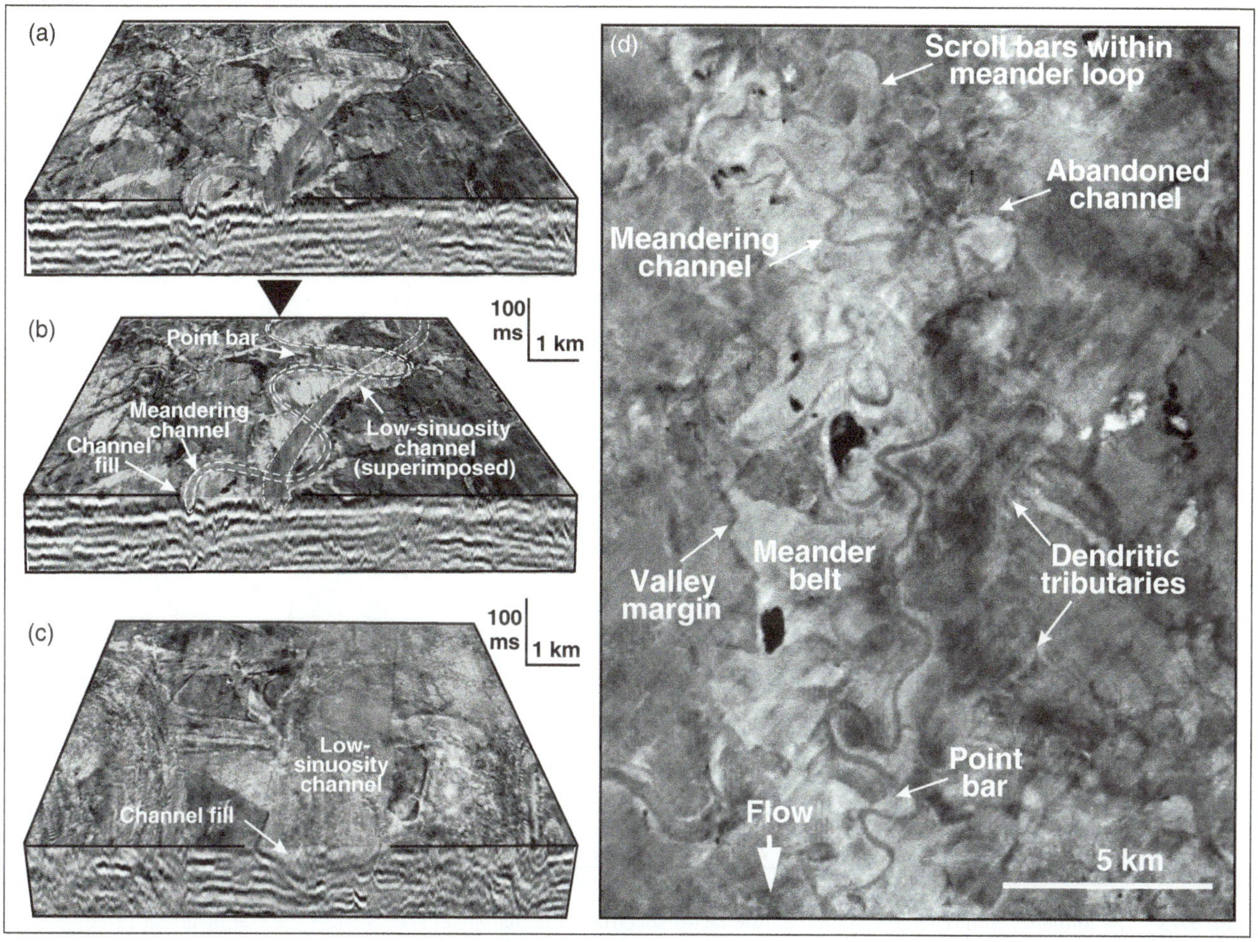

Figure 5.42 Seismic timeslices and profiles from Quaternary strata of the Malay Basin, SE Asia. (a, b) Large meandering channel up to 30 m deep and 600 m wide, with sinuosity <3.5, inclined stratification attributed to sandy point bars, traces of scroll bars and muddy topmost fills. The channel is partly confined with a valley, and a low-sinuosity channel is superimposed on the lower channel system. (c) Single-thread, single-storey channel filled with sand or gravel, up to 50 m deep and 3 km wide and with sinuosity <1.3. (d) Timeslice showing a meandering channel <20 m deep and 75–300 m wide. The meanderbelt is partly confined where meanders cut into the valley walls. Point bars, scroll bars and abandoned channels are visible. The bordering interfluve is crossed by tributaries with dendritic drainage patterns, indicating gully erosion. Modified from Alqahtani et al. (2017), by courtesy of F. Alqahtani.

exceptionally low variance, whereas the Hugh River of dryland Australia has an exceptionally high variance. Flood waves may rise and fall rapidly, causing disequilibrium in sediment deposition. In the Burdekin River in tropical Australia, flood waves can travel at 4.3 ms^{-1}, and a single flood in 1997 mobilised 6 m of riverbed sediment and redeposited an equivalent amount over an erosion surface (Alexander et al. 1999a; Fielding et al. 1999). Within the context of such variable systems, high-magnitude events may not be as formative as they are in other rivers.

Flood patterns have important consequences for river morphodynamics. A survey of meandering rivers across a range of climatic settings showed that short-lived, high-magnitude overbank floods tend to promote chute cutoff of meanders, whereas long-lasting, low-magnitude floods tend to promote neck cutoff (Maitan et al. 2024). As discharge variance increases, the effects of high-magnitude events become more evident in the preserved record (Figure 5.43). Macroforms such as sediment bars are fewer and show reduced relief, and their deposits become more difficult to identify. Erosional surfaces and lateral facies changes are prominent, along with Froude trans- and supercritical bedforms (facies Sh and Sac) and large woody debris. Upright in-channel vegetation may be present where water is accessible below dry channels (see Figure 5.1b,c).

Hansford et al. (2020) identified four river discharge patterns based on long-term flood magnitude and variability. Rivers with a relatively steady hydrology include rainforest rivers such as the Paraná of South America, which receives rain throughout the year. Rivers with a strongly seasonal hydrograph include snowmelt and monsoonal rivers, such as the Chao Phraya of Thailand with 4–5 months of strong monsoonal flow. Rivers with discharge controlled largely by single storms include some temperate and polar rivers,

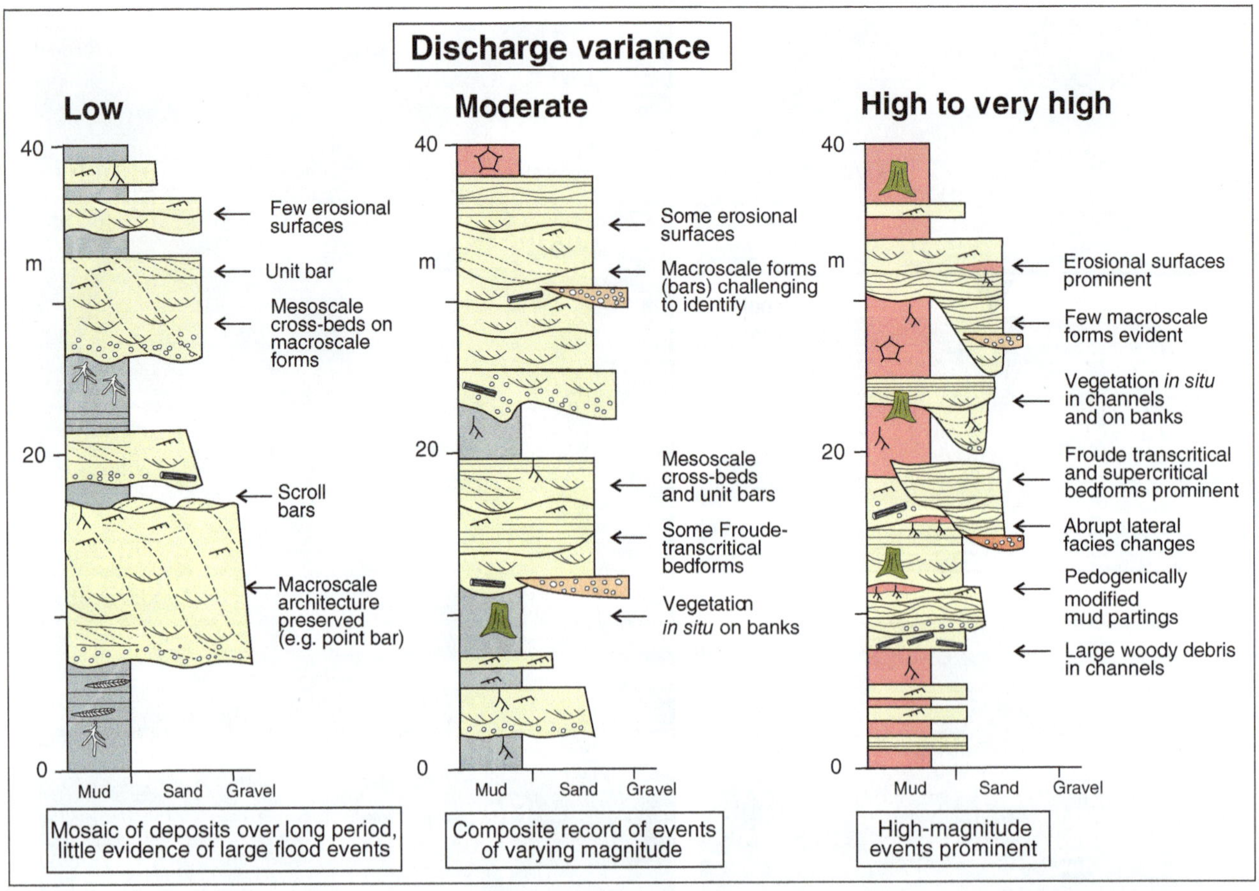

Figure 5.43 The influence of discharge variation on the facies spectrum. The numerical values show the coefficient of variance (CVQ_p) or the variance of peak discharge, calculated from the standard deviation of peak flood discharge over a prolonged period divided by the mean peak flood discharge. *Source:* Adapted from Fielding et al. (2018).

such as Little Beaver River of the United States. Finally, rivers with extremely erratic discharge, such as Smoky Hill River of the arid US Midwestern plains, may have flood discharge as much as 1500× the median discharge.

5.10 Palaeoflow Analysis

5.10.1 Palaeoflow Indicators

Palaeoflow indicators and accretion directions from large-scale inclined strata are crucial for establishing the original flow vectors of river systems. They may be compiled into regional datasets to assess sediment provenance. Palaeoflow analysis for a reach requires the measurement with a compass of linear indicators, many of which will approximate the local current direction. Trough and planar cross-beds (St and Sp), cross-lamination (Sr), current lineation on Sh and Sac surfaces and imbrication in gravel facies are the principal palaeocurrent indicators (Figure 5.44a–e), along with groove and flute casts on erosional surfaces. Less common indicators include obstacle marks, vegetation shadows and the orientation of wood fragments (variably parallel, transverse or oblique to flow). In the account below, an *azimuth* refers to a unique direction, whereas a *trend* refers to a linear feature with no unique direction unless corroborated from other information. Azimuths may be referred to as unipolar data and trends as bipolar data.

Cross-strata must be measured on bed surfaces that display the true 3D orientation of the cross-strata. This is to avoid collecting apparent dip directions on 2D faces which, in the case of trough cross-beds, may yield errors of as much as 90° (Figure 5.44a). Methods exist for integrating apparent dip directions on two faces, but in our experience, these methods are unreliable and to be avoided. As the photos in Figure 5.44a show, vertical sections through trough cross-beds are typically oblique to the flow direction, and palaeoflow is difficult or impossible to gauge accurately. In contrast, where stratification has been truncated at bed surfaces to generate a 'ridge-and-furrow' appearance, the orientation of trough axes and intersections between troughs can be measured as an azimuth to within a few

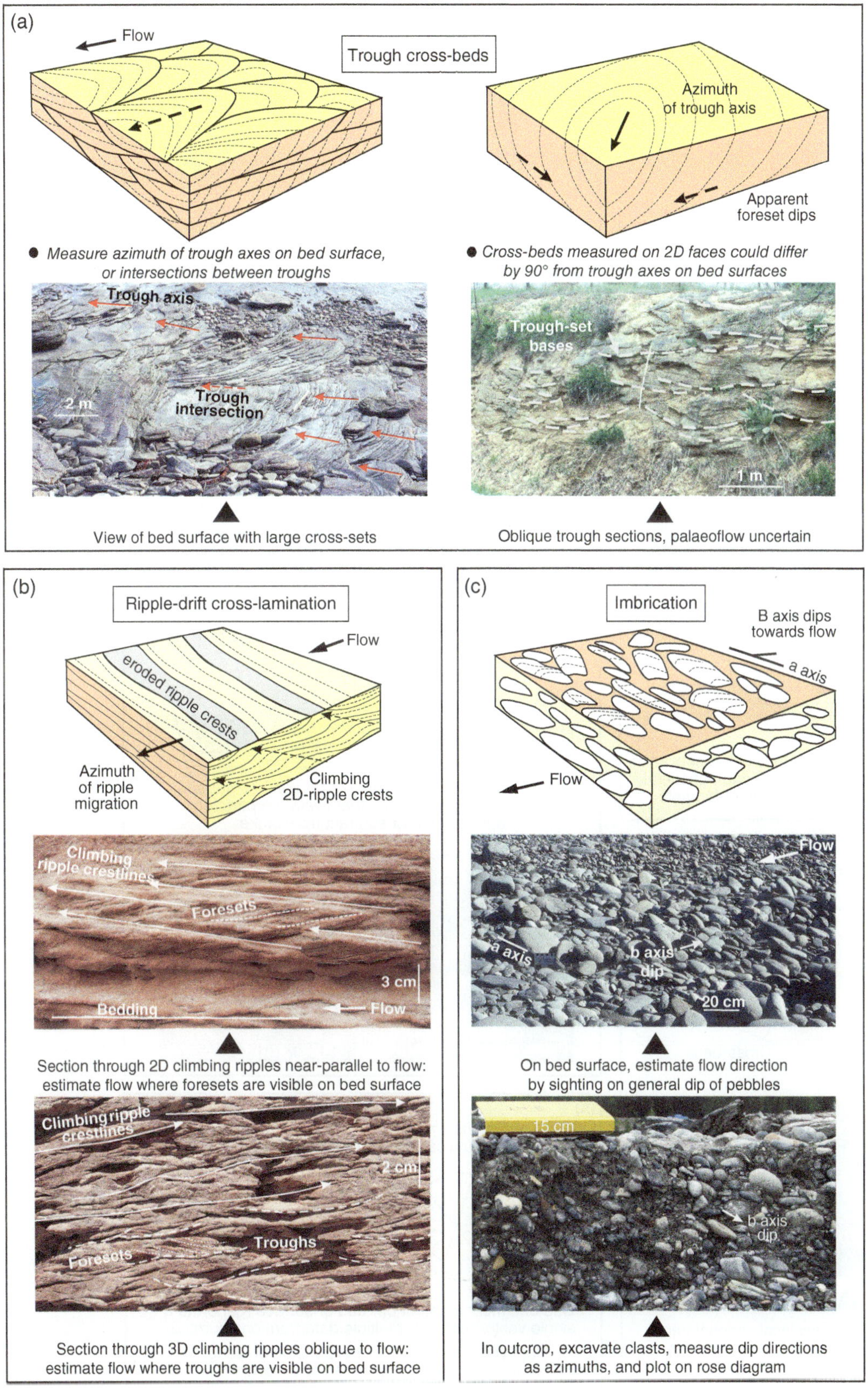

Figure 5.44 Palaeoflow measurement and analysis in alluvial strata. Trough cross-beds (a), ripple-drift cross-lamination (b), imbrication (c), planar cross-beds (d), primary current lineation (e), channel trends (f), measurement in dipping beds (g), and rose diagrams (h). *Sources:* Martin Gibling, Alessandro Ielpi and Massimiliano Ghinassi (Chapter authors).

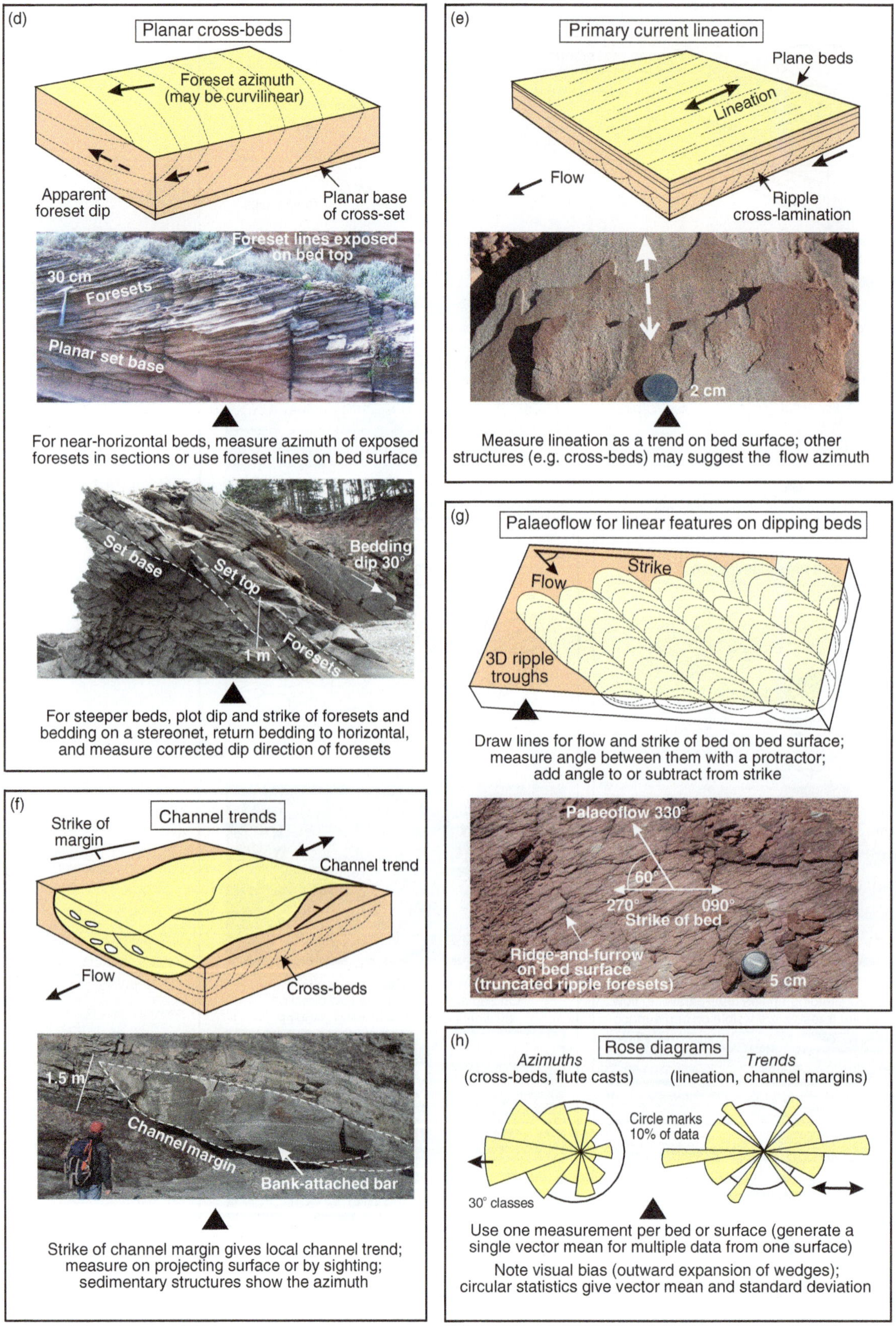

Figure 5.44 (Continued)

degrees. Similar reasoning applies to ripple cross-lamination laid down by 3D ripples. For planar cross-sets in near-horizontal beds (Figure 5.44d), the dip direction of the foresets may be measured as an azimuth where sufficient 3D exposure is available, or the traces of foresets truncated on bed surfaces may be measured, generating an azimuth for their dip direction. The orientation of cross-bed and erosional surfaces may also be measured approximately using dipmeters in wells with gently dipping strata.

Ripple-drift cross-lamination, formed by climbing ripples in aggrading sediment, presents a particular problem. As the ripples migrate downflow and the bed aggrades, the ripple crests track an angle of climb, generating apparent surfaces that dip upflow and may be mistaken for larger scale foresets (Figure 5.44b). For these structures, the best approach is to determine the apparent flow direction of the small ripple foresets in a vertical section, after which ripple foreset traces or troughs (2D and 3D ripples, respectively) on the bed surface can be used to measure the palaeoflow direction as an azimuth.

In gravel beds with imbrication of clast a–b planes, the dip direction (upflow) of clasts may be measured as a single azimuth by sighting on bed surfaces, first ensuring that numerous clasts show a consistent dip direction (Figure 5.44c). In high-energy flows that brought some clasts into suspension, the a-axis tends to be aligned with the flow and plunge upflow. Where gravels can be excavated, the dip direction of each imbricated clast, and/or the orientation and plunge of the a-axis, may be measured as azimuths and plotted on a rose diagram.

Current lineation can be measured as a trend on bed surfaces. The palaeoflow direction for these bipolar data may be assessed from associated unipolar data such as cross-bed orientations. When recording current-lineation trends (Figure 5.44e), measurements should be noted within a single hemisphere (180° range) to avoid later confusion in plotting.

Palaeoflow measurement in dipping beds poses a problem. Even at low angles but especially with dips of >30°, tilt substantially changes the orientation of inclined planar surfaces such as cross-beds and accretion surfaces. Two methods can be used to overcome this problem. One method is to plot the dip and strike of bedding and inclined beds on a stereonet and restore the bedding to horizontal; the correct dip direction of the foresets can then be determined (Figure 5.44d). The second method is to measure the structure as a linear feature in the field, as in the case of current lineation, groove casts, trough axes and ridge-and-furrow from dunes or ripples truncated on bed surfaces. The flow direction and the strike of the bedding are drawn on the dipping surface and the angle between them is measured with a protractor (Figure 5.44g). The latter method is direct and accurate to a few degrees, but it is mainly applicable to small-scale features.

Refined studies of architecture in suitable outcrops combine palaeoflow indicators with accretion directions determined from the strike and dip of inclined and inclined heterolithic sets (IS and IHS) and geomorphic bounding surfaces. Using the angular relationships between average palaeoflow from bedforms and the dip direction of accretion surfaces, downstream (0–60°), lateral (60–120°) and upstream (120–180°) accretion directions may be inferred (Miall 1994; Long 2006; Tamura et al. 2019).

Important directional evidence is provided also by the strike and dip of channel margins. These can often be measured using a compass on the projecting, inclined surface of the margin, or by sighting where the margin cannot be accessed or measured directly. The 3D orientation of channel bodies may be mapped using outcrop and subsurface data (Figure 5.44f). These orientations can be compared with palaeoflow and accretion directions from the channel fills.

As regards the number of measurements, the general rule is to collect as many measurements as possible, from all parts of a fluvial body. However, a few reliable measurements (e.g. trough axes on bed surfaces) outweigh many uncertain measurements or rough estimates from 2D sections: even a few unreliable measurements can seriously affect palaeoflow analysis. The rule of thumb is not to measure if in doubt. Many trough axes may be measured on a particular bed surface, but they represent a single flow system and should be averaged to provide a single azimuth for the surface.

Unipolar and bipolar forms (azimuths or trends) on the rose diagrams show a visual bias that needs to be evaluated because the area of the wedges expands outwards (Figure 5.44h). Freely distributed software programmes such as OpenStereo (http://igc.usp.br/openstereo/) have been developed by academics for statistical analysis of geometric data, generating vector means and standard deviations. Box 5.3 shows a useful approach to plotting palaeoflow data.

5.10.2 Ranking Palaeoflow Indicators

River channels yield a hierarchy of flow indicators at different scales. The importance of each type of indicator may be ranked by the proportion of the flow that it represents (Table 5.11; Allen 1966; Miall 1974). Ranks 1–6 (high to low) represent progressively smaller parts of the drainage system, lesser volumes of sediment transported and a

Box 5.3 Plotting Palaeoflow Data on Panoramas

Although palaeoflow results are commonly grouped in summary rose diagrams, individual palaeoflow measurements should first be plotted in their precise positions on stratigraphic panels, along with key surfaces, facies and fossils. As shown in the panel diagram, a useful convention is to plot each measurement as an azimuth or trend, with north upwards. Palaeoflow data can also be plotted in a similar manner alongside stratigraphic columns. This allows the researcher to assess the distribution and uniformity of the data, identifying changes in flow direction that may easily be masked in a rose diagram. Types of palaeoflow indicator should be plotted separately.

Rose diagrams can then be plotted for grouped data, for example, by storeys or channel bodies. In the Pennsylvanian example, inspection shows that both storeys of the channel body have a similar northward palaeoflow, slightly oblique to the trend of the cliff face. At three locations, the strike of inclined stratification (IS) is sub-parallel to the orientation of trough cross-beds and ripple cross-lamination, suggesting that the inclined stratification represents the accretion surfaces of bank-attached bars (point or alternate bars).

Concave-up scours are prominent in the cliff face, some cutting through mudstone into the underlying coal. Although scour orientation was not measured in this example, the near-symmetric form of the basal scour in the upper right suggests that the scour trends into the outcrop, in accord with other measurements.

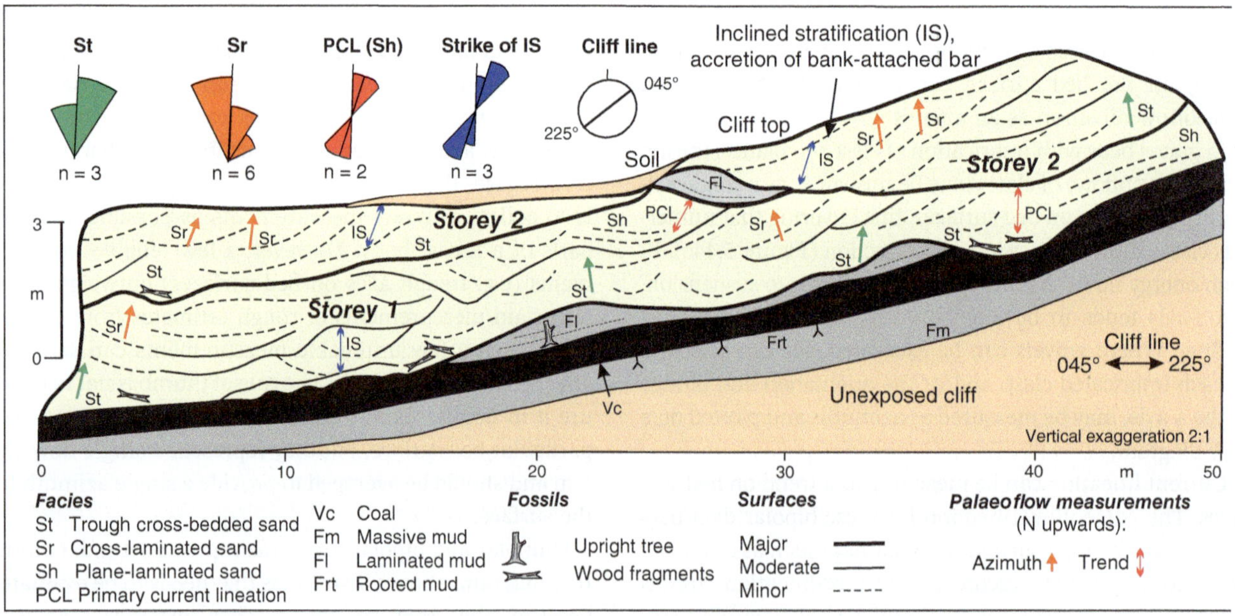

greater variance in palaeoflow direction. Among the most widely used indicators, dunes, imbricated gravel and current lineation in thick bedsets (rank 5) represent relatively large and formative discharge events. They commonly yield low palaeocurrent variance. In contrast, ripples (rank 6) represent low-discharge flows influenced by bed topography. They commonly yield high palaeocurrent variance, although ripple orientation may match the orientation of rank 5 data where low-stage flows followed channel axes. Cross-bed directions (rank 5) in crevasse channels and splays may be at right angles to cross-bed directions in the parent channel. Bar accretion sets (rank 4) represent large sediment volumes but may show a high directional variance depending on bar type and slipface orientation. More broadly, directional variance may be described by five categories at increasing scale (Almeida et al. 2023): bedforms, local flow shifts, bar forms and sinuosity in channels, distributive patterns in lobes and fans and basin-scale drainage patterns. Channel planforms may be inferred using palaeocurrent circular variance from large modern-river datasets (Galeazzi et al. 2021), although analysis for the ancient record is fraught with difficulties.

Table 5.11 Rank of palaeoflow indicators for fluvial systems, modified from Allen (1966) and Miall (1974).

Rank	Palaeoflow vector field	Examples of indicators
1	**Drainage system**	Numerous 3D channel bodies mapped over large outcrop belt or from seismic timeslices and subsurface data. Palaeotransport tracers (e.g. dated zircons, dense minerals and distinctive fossils) linked to particular source areas.
2	**River within the drainage system**	Several 3D channel bodies mapped over local area.
3	**Channel reach in the river**	Single mapped 3D channel body. Storeys within multistorey bodies
4	**Geomorphic units or macroforms in the reach**	Bar accretion sets (IS, IHS). 3D channel bodies from subsidiary channels
5	**Large-scale bedforms or mesoforms**	Dune cross-sets (St, Sp), imbricated gravel, current lineation in thick bedsets of Sh
6	**Small-scale bedforms or microforms**	Ripple cross-sets (Sr)

Indicators with higher numbers are likely to show a higher variance because they are generated from local flow fields that represent only a small proportion of the flow system.

5.11 Trace Fossils in Alluvial Systems

As noted above, trace fossils yield information about discharge variability, depositional episodes, time for colonisation and depth to the water table. These factors are discussed here with reference to fluvial styles in modern environments and the geological record (Figure 5.45a–f; Supplementary Table 5.S.1: 25–31; Table 5.S.4; see Buatois and Mángano 2011 and Melchor et al. 2012 for overviews).

Trace fossils are virtually absent in *alluvial fan* and *gravelly braided river* deposits. This scarcity reflects rapid sedimentation, high energy, coarse-grained sediments and episodes of reworking, which largely preclude colonisation by animals and preservation of trace fossils. Invertebrate and vertebrate trace fossils may be present in sheetflow deposits on distal fans. In *sandy braided rivers*, active channels are colonised by aquatic insects that produce simple vertical dwelling burrows, representing a continental expression of the *Skolithos* Ichnofacies. More abundant and diverse trace fossils characterise abandoned channels and overbank areas with longer colonisation windows. In this setting, arthropods produce horizontal to inclined meniscate backfilled burrows, horizontal and inclined to vertical simple burrows, horizontal bilobate trails and trackways, representing feeding, locomotion and dwelling behaviours in the *Scoyenia* Ichnofacies. Vertebrates are represented by footprints, including swim tracks.

Meandering river deposits with more stable channels and floodplains typically contain the most abundant and diverse fluvial trace-fossil suites. The *Scoyenia* and *Mermia* Ichnofacies are present in periodically desiccated and ponded areas, respectively. Point-bar, abandoned-channel and overbank deposits may display abundant horizontal to inclined meniscate burrows and vertebrate tracks, commonly overprinted by root traces. Low water tables favour meniscate burrows and arthropod and vertebrate trackways, whereas high water tables favour simple horizontal trails that record grazing behaviour. Where the water table is high, crayfish dwelling burrows have many branches and chambers, whereas those in areas of low or fluctuating water table have deep, simple burrows (Genise 2017).

For *anastomosing river* deposits, the few available studies document the *Scoyenia* Ichnofacies and suggest that climate is a major factor in determining the assemblage (see Krapovickas et al. 2009 and Valencia et al. 2024 for case studies from the Miocene of Argentina). Under arid to semiarid conditions, crabs and crayfish produce deep vertical dwelling burrows in abandoned channel fills, and overbank deposits are sparsely bioturbated, probably due to water scarcity. Extensive surfaces are covered with tetrapod and bird tracks. Under more humid conditions, overbank areas and abandoned channels contain abundant meniscate horizontal burrows.

Ephemeral river trace-fossil suites resemble those of sandy braided systems. Although generally unbioturbated, channel deposits locally contain simple vertical dwelling burrows and escape trace fossils of the *Skolithos* Ichnofacies. Abandoned-channel deposits host horizontal to inclined meniscate burrows of the *Scoyenia* Ichnofacies. Overbank areas show a similar range of trace fossils with

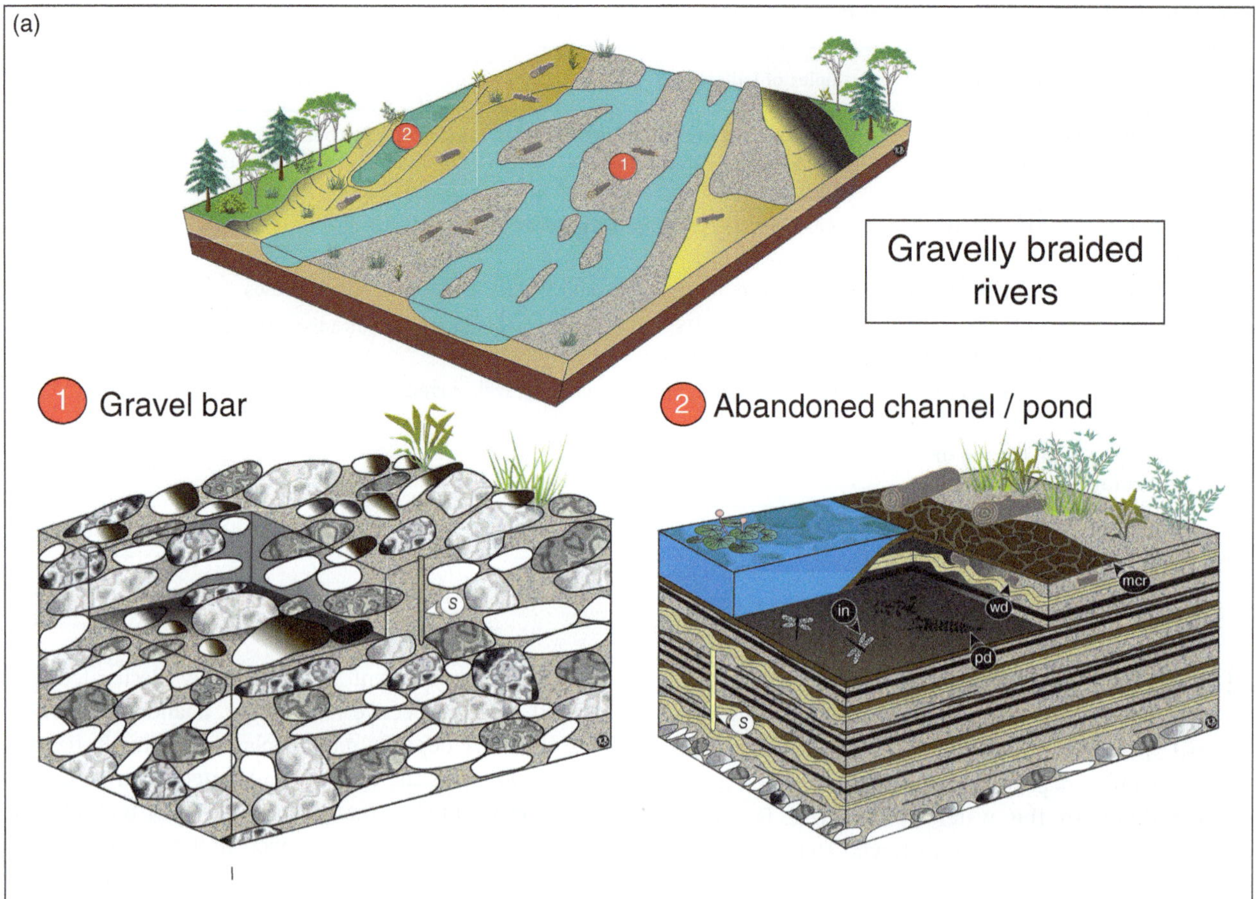

Ichnofossils

A	*Arenicolites*	Cru	*Cruziana*	P	*Palaeophycus*	S	*Skolithos*
Arc	*Archaeonassa*	Di	*Diplichnites*	Pc	*Paleohelcura*	Ta	*Taenidium*
Bea	*Beaconites*	fu	*fugichnia*	Pt	*Planolites*	Tb	*Treptichnus bifurcus*
Ca	*Camborygma*	He	*Helminthopsis*	Pro	*Protichnites*	Tp	*Treptichnus pollardi*
Cap	*Capayanichnus*	Hm	*Helminthoidichnites*	Ru	*Rusophycus*	Tr	*Treptichnus* (both species)
Cl	*Cochlichnus*	Lo	*Lockeia*	Sco	*Scoyenia*		

Other features

bf	Bone fragment	cr	Current ripples	m	Leaf mold	peb	Pebble
bi	Bivalve	crl	Climbing ripples	mc	Mudclast	rt	Root
br	Bedrock fragment	fi	Fish	mcr	Mudcracks	tt	Tetrapod track
bt	Bird track	ga	Gastropod	os	Ostracode	wd	Wood debris
co	Coprolite	in	Insect	pd	Plant debris	wf	Wood fragment

Figure 5.45 Trace-fossil assemblages in alluvial settings. Gravelly braided rivers (a), sandy braided rivers (b), meandering rivers (c, d), anastomosing rivers (e) and ephemeral rivers (f). The trace fossils represent modern and inferred ancient examples of each alluvial setting, and they are listed in Supplementary Table 5.S.4. Illustrations by Kerrie Bann.

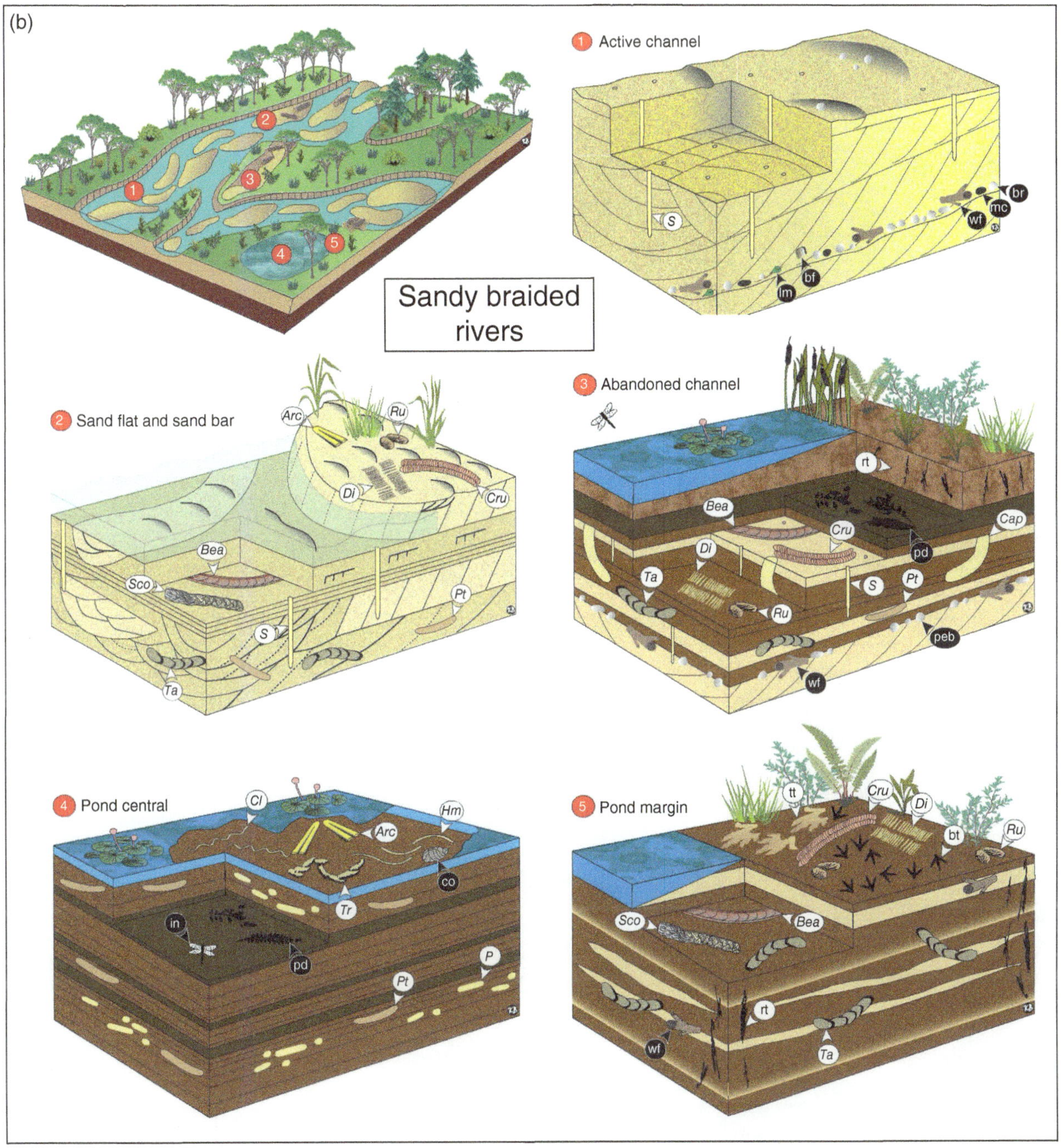

Figure 5.45 (Continued)

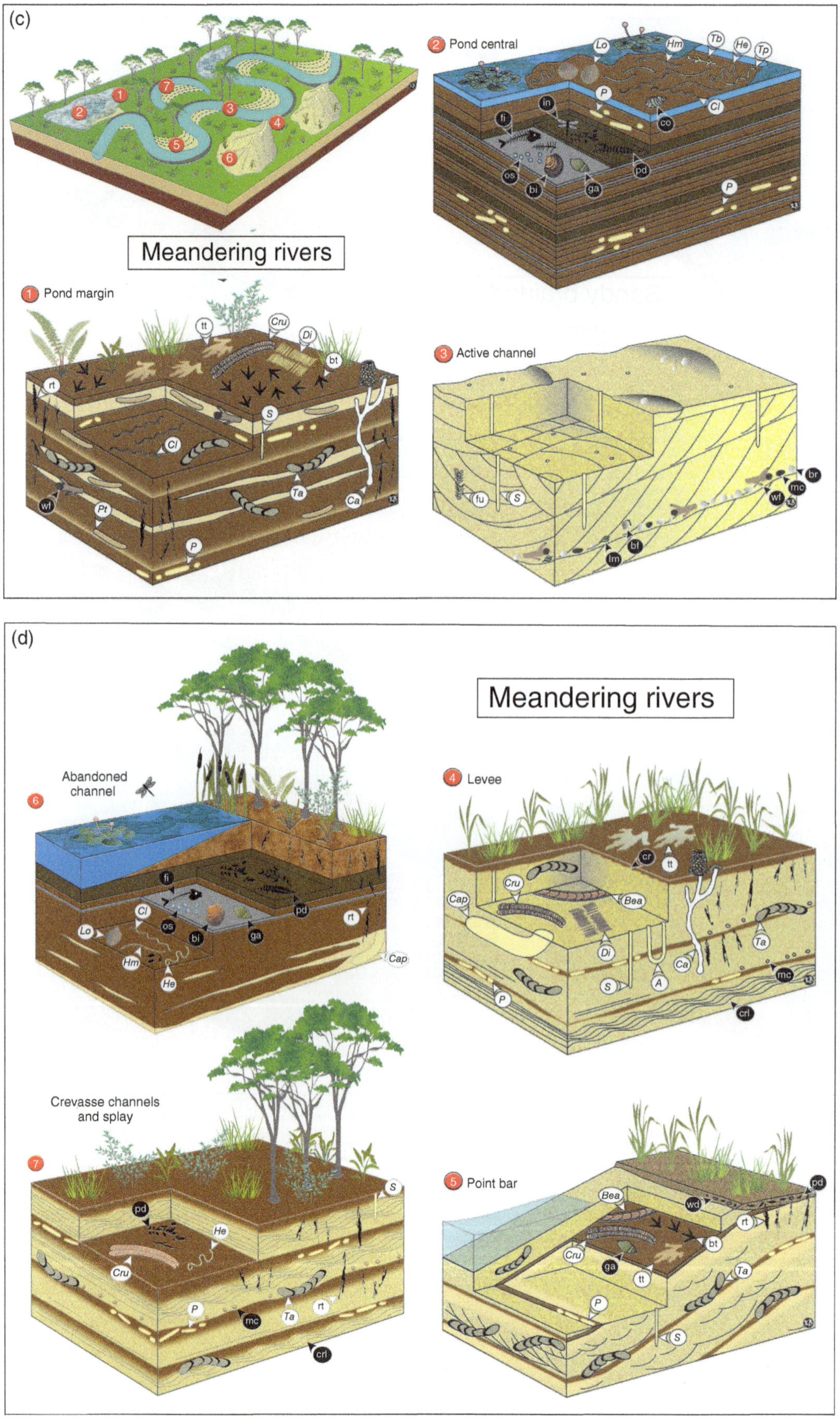

Figure 5.45 (Continued)

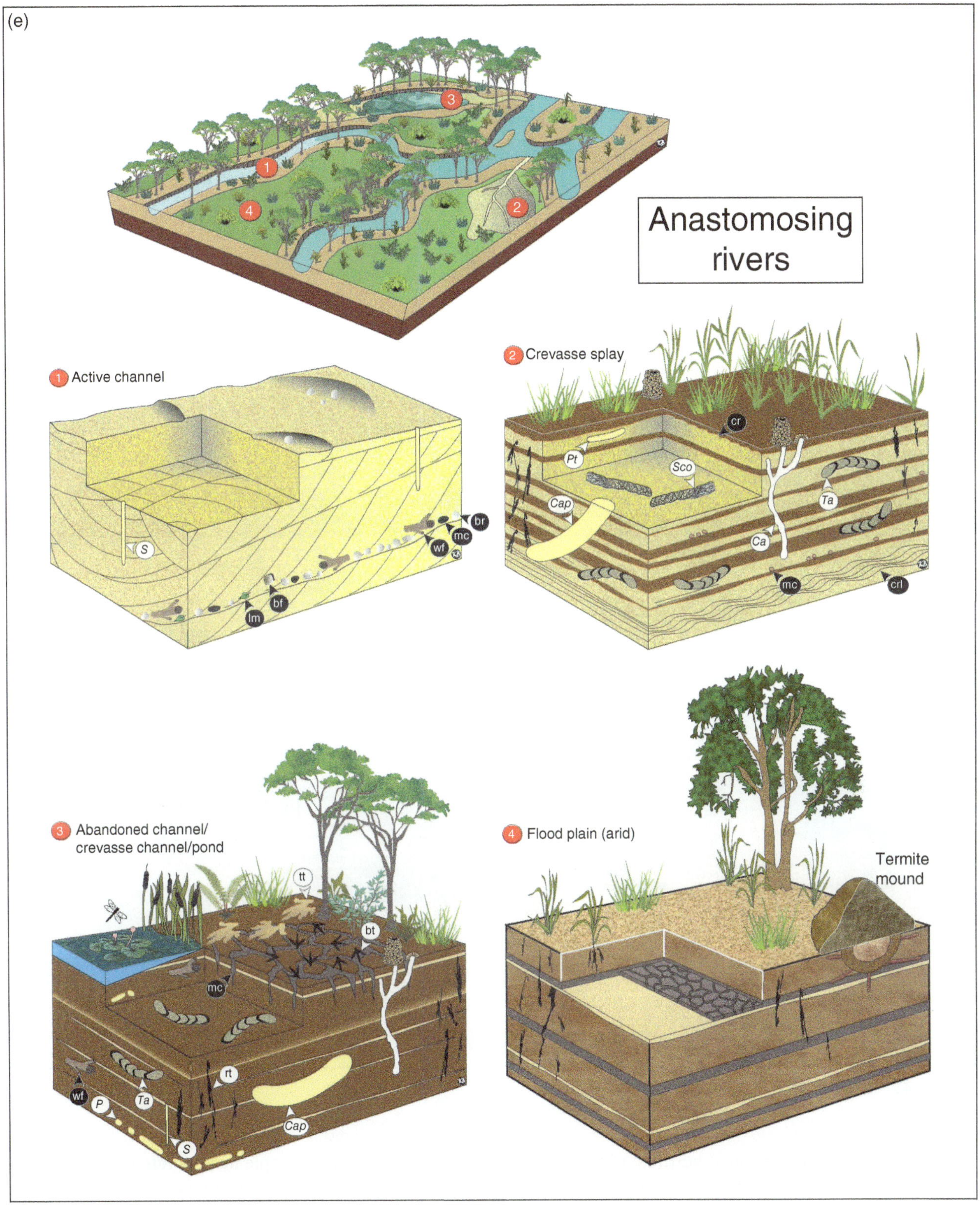

Figure 5.45 (Continued)

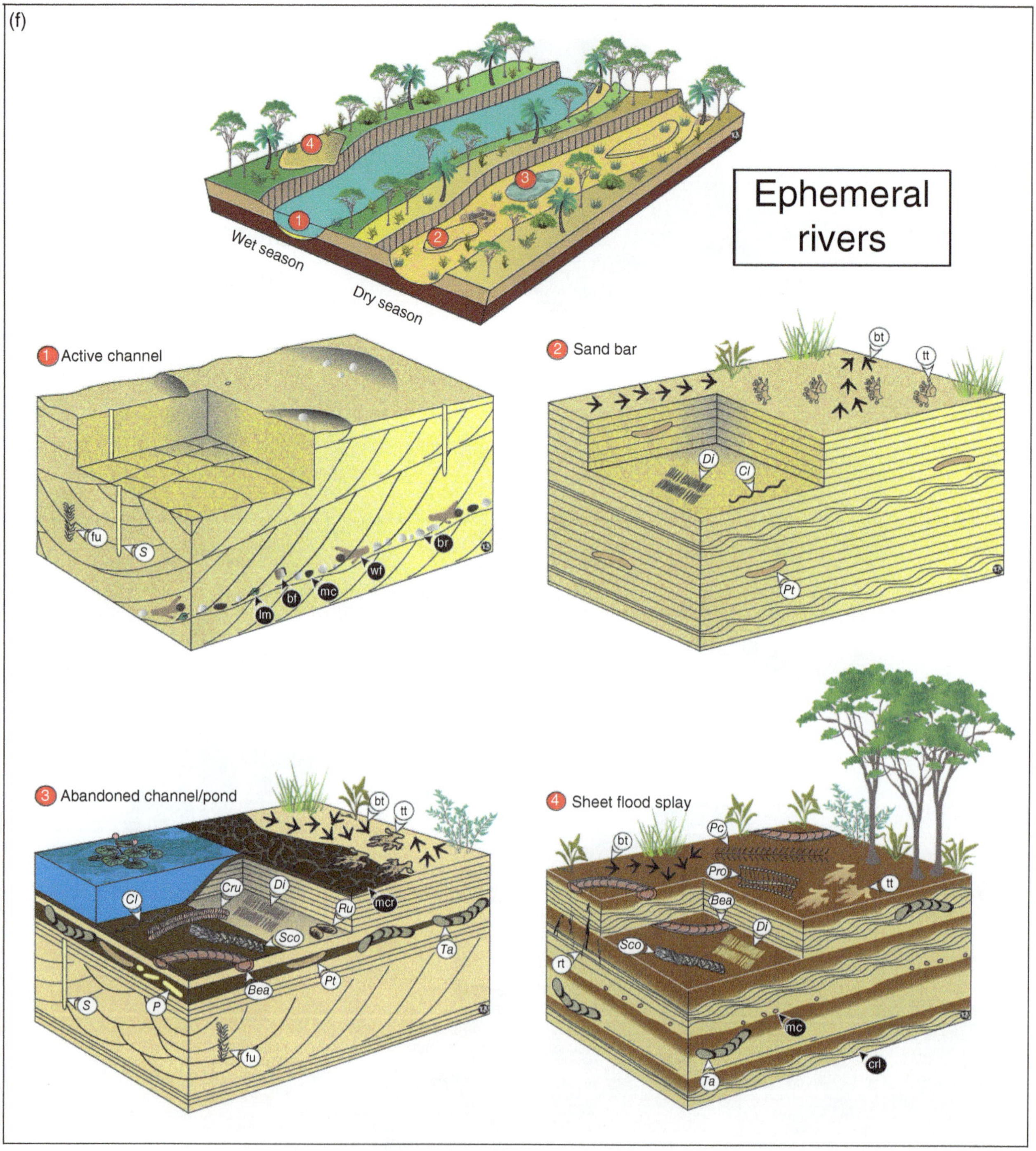

Figure 5.45 (Continued)

the addition of simple horizontal burrows, and sheetflow deposits contain abundant arthropod and vertebrate trackways, with some meniscate burrows, both attributed to the *Scoyenia* Ichnofacies. Many trace-forming organisms are resilient in arid environments that we would regard as extreme.

Overbank and standing water deposits commonly yield trace fossils of the *Mermia* Ichnofacies, indicative of soft substrates. In these settings, vermiform organisms and insect larvae produce simple invertebrate horizontal trails and burrows and branching burrow systems, recording grazing, feeding and dwelling ethologies. Bivalve resting and equilibrium traces are present, along with fish trails. Desiccated substrates are characterised by the *Scoyenia* Ichnofacies, with morphologies and burrow-wall characteristics, such as sharp scratch imprints or bioglyphs, which indicate firm substrates (Figure 5.16). Invertebrate trace fossils typically comprise striated meniscate horizontal burrows, branching burrow systems and bilobate trace fossils, as well as arthropod trackways and simple vertical burrows that may represent insect and arachnid nesting structures. Vertebrate trackways and burrows are also common, and some burrows contain skeletal remains that yield valuable insights into the behaviour and palaeoecology of alluvial biotas (Fernandez et al. 2013; Smith et al. 2021). This trace-fossil suite may cross-cut a suite emplaced in soft substrates before desiccation, including meniscate burrows without ornamented walls.

5.12 Basin Architecture

The final analytical step in Landscape System analysis focuses on depositional systems and basin architecture. Basins are areas of active subsidence in which sediment can be preserved over geological timescales of millions of years, and they underlie about 16% of the Earth's terrestrial surface, mainly in areas of arid climate (Nyberg and Howell 2015). In terms of tectonic setting, intracratonic basins constitute 49% of the total basin surface area, foreland basins 29%, and passive-margin, extensional, strike-slip and forearc basins <10% each (Nyberg and Howell 2015).

Weissmann et al. (2015) identified six large-scale fluvial *geomorphic elements* in inland basin fills (Figure 5.46): (i) megafans (broadly defined as >30 km long); (ii) fluvial fans and alluvial fans (<30 km long); (iii) incised megafans or alluvial fans, cut by an incised valley that is active under current climatic conditions; (iv) bajadas of coalesced fans; (v) axial rivers with tributaries; and (vi) interfan tributary rivers between megafans. These elements include all the fluvial styles described in Sections 5.2 and 5.3, and lakes and aeolian landforms may also be present. Avulsion histories play an important part in the analysis of geomorphic elements.

There is considerable debate about the relative importance of axial rivers and transverse, distributive fluvial systems (megafans and alluvial fans) in modern basin fills and in the geological record (Supplementary Table 5.S.1: 32). Large river systems such as the Amazon (Figure 5.10), Mississippi and their tributaries dominate long and wide alluvial tracts, commonly occupying subsiding, fault-bounded basins with a high potential for preservation of strata. However, rapidly aggrading megafans are prominent and volumetrically important in basins that receive large volumes of detritus from eroding mountain belts. For example, in the Himalayan Foreland Basin, the axial Ganga drainage forms a relatively narrow belt, constrained by megafans from the

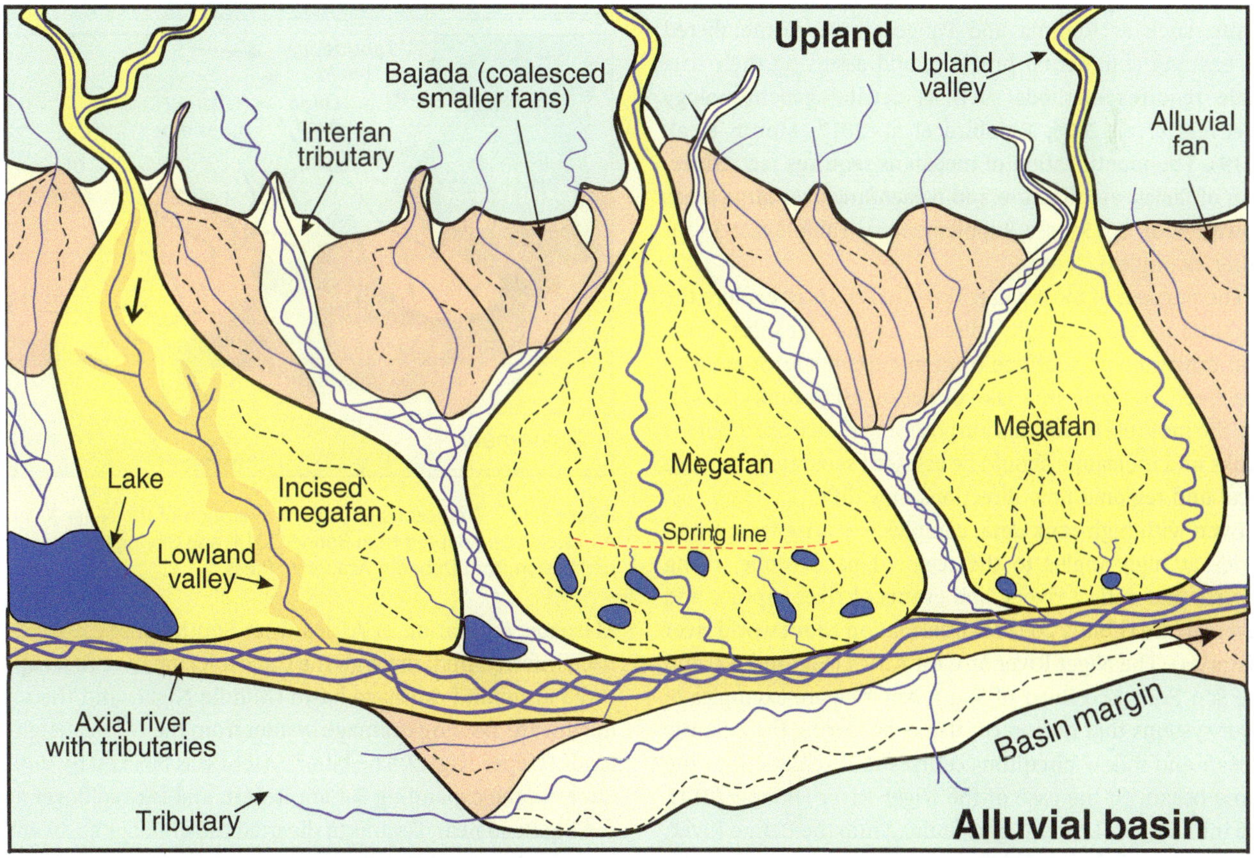

Figure 5.46 Basinal elements in a continental basin. *Source:* Adapted from Decelles and Cavazza (1999), Weissmann et al. (2015) and Gibling (2021) with permission from Dunedin Academic Press.

Himalaya to the north and the Indian Craton to the south (Figure 5.5d). Megafans do not necessarily require strong relief: the Taquari Megafan of South America borders modest uplands under a tropical climate with high precipitation. Some megafans were formerly more active during periods of greater monsoon intensity and strong fluvioglacial flow in Asia and the European Alps, or during earlier tectonic regimes in midwestern USA (Ogallala Apron: Galloway et al. 2011).

For the relatively low-lying continent of Australia, Nanson and Nanson (2025) recognised a suite of *Large Accretionary Fluvial Systems* (LAFS) that include both megafans and axial rivers. With areas >1000 km^2, the LAFS cover >6% of the continent, and they are variously unconfined or confined by bedrock and tough alluvium. The systems terminate in coastal deltas and endorheic fluvial and playa-lake settings, locally inferfingering with aeolian systems. The component channels are mainly braided and anabranching, locally meandering, and the LAFS have contributive and distributive patterns when traced downflow.

The scale of large rivers makes them difficult to identify in the ancient record. The Hawkesbury Sandstone of Australia (Figure 5.34e,f) was laid down by one of the largest river systems studied to date, but the scale of the preserved geomorphic units suggests deposition from a narrower and shallower river than the modern Brahmaputra (Miall and Jones 2003). Large rivers that crossed former supercontinents, such as Rodinia and Pangea, were dismembered by ensuing continental breakup, and assessing their true scale requires methods such as detrital geochronology (Veevers et al. 2006; Rainbird et al. 2017; Morón et al. 2019). The identification of megafans requires reconstruction of facies, architecture and palaeoflow over large areas (Owen et al. 2015a; see Supplementary Table 5.S.1: 33 for other examples).

The purpose of most palaeoflow analysis is to explore the regional palaeogeography, discerning the transport direction down the palaeoslope from mountain source areas to the receiving basin in a lake or ocean. High-rank datasets are the most useful for this purpose, and data from different ranks and indicators should be assessed separately because local and regional flow directions may differ.

Even with high-rank data, complexities arise from tributaries at high angles to the regional palaeoslope, rising thrusts that deflect rivers, and events of river capture and reversal (see Gibling 2021 for information about world river histories). The Niger River of West Africa (Figure 5.47) and the São Francisco River of South America are examples of river systems that rise on rift flanks bordering the Atlantic Ocean and follow circuitous courses to discharge into the same ocean. In the case of the Niger River (Bonne 2014), the initial river (Lower Niger) drained into the Benue River, which occupied a low-lying rift generated during the Mesozoic separation of Africa from South America. It has been inferred that, through the Cenozoic, the river first captured an inland drainage basin (Middle Niger) and then a northward-flowing drainage system from the Guinea Highlands (Upper Niger). The latter system was blocked by sand dunes in the expanding Sahara Desert, and lake spillover at the Big Bend near Timbuktu diverted the drainage eastward into the Niger. Although the river rises near the coast, its

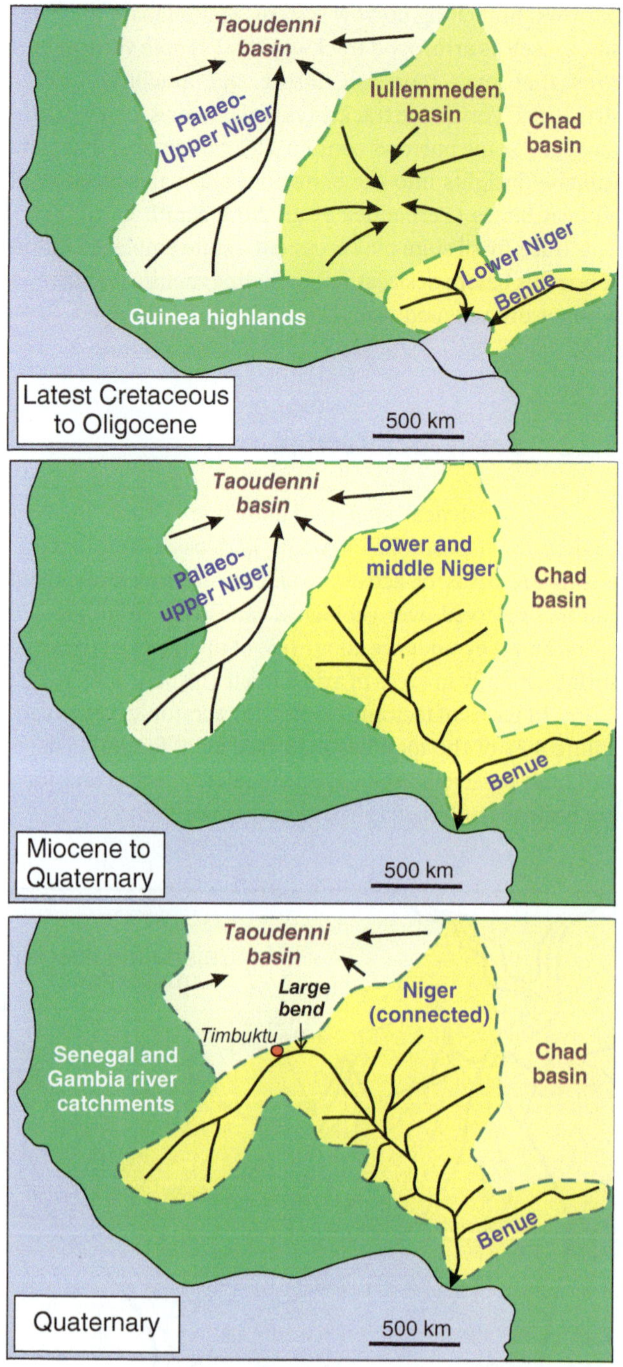

Figure 5.47 Cretaceous to present evolution of the Niger River, Africa. *Source:* Adapted from Bonne (2014) and Gibling (2021) with permission from Dunedin Academic Press.

inland course is >4000 km long and was finally connected only in the late Quaternary. Rivers in foreland basins, such as the Ucayali in the upper Amazon basin and the Ganga and Brahmaputra in the Himalayan Foreland Basin, follow the structural grain and axis of the basin, which may not represent the shortest path to the ocean. Blocked and diverted by glacial lobes during the Pleistocene, the Yukon River currently rises <100 km from the Pacific Ocean but drains north and west for >3000 km to the Bering Sea.

5.13 Controls on Alluvial Systems

5.13.1 Overview

Many factors influence alluvial systems over spatial scales from metres to basins and timescales from hours to tens of millions of years (Miall 2014; Figure 5.48). *Autogenic* factors are intrinsic to the river system with its self-organising mode, under which the system's internal dynamics produce an ordered and predictable response. *Allogenic* factors are imposed on the river system externally. As noted in Section 4.3.2, autogenic effects are always at play in alluvial systems and constitute a null hypothesis for controls, whereas positive evidence is needed to infer allogenic controls. Timescales (both durations and periodicities) are defined here as *short term* (operating over hours to a few decades), *medium term* (decades to about 10^5 years) and *long term* (more than about 10^5 years) (Supplementary Table 5.S.5).

At the outset, it is important to recognise that the alluvial record contains major imperfections that complicate the recognition of environmental signals (Jerolmack and Paola 2010). Such imperfections also complicate the distinction of autogenic from allogenic effects. Key problems include (i) buffering of the signal, during which the landscape responds in such a way as to damp down the magnitude of the effect; (ii) the incompleteness of the alluvial record due to erosional breaks and gaps in time (Miall et al. 2021); and (iii) the shredding of environmental signals due to stochastic processes of sediment transport, deposition and reworking, which vary in an unpredictable fashion to the point that the signal cannot be recognised (Straub et al. 2020). Sadler's (1981) classic work showed that stratigraphic completeness decreases exponentially with the length of the timescale under consideration (Section 4.5). Despite these difficulties, many bed surfaces can be considered 'true substrates', where ephemeral features such as raindrop impressions are remarkably well preserved, providing reliable environmental signals (Davies and Shillito 2021).

The incompleteness of the alluvial record may be partially rectified in regions where sufficient scattered archival niches, such as sites of prolonged flood accumulation, collectively provide exceptional records. Based on hundreds of physical dates from such isolated sites, alluvial strata can now be investigated using probability-based approaches that generate near-continuous records of river flooding. These can be matched against independent proxy records for sea level and climate. In some areas, historical data are available: for example, records of Nile flood levels date back to the 7th century CE (Macklin et al. 2012).

5.13.2 Autogenic Factors

Autogenic processes are the default alluvial dynamics and are always in operation. They are the 'builders of stratigraphy' – key operators in the 'stratigraphy machine' (Holbrook and Miall 2020; Miall et al. 2021). In through-going rivers such as the Rhine-Meuse and Mississippi, avulsion and cutoff typically operate on timescales of hundreds to a few thousand years (medium term), whereas in well-supplied fluvial fans and terminal systems, avulsion may operate on timescales of years to decades (short term). Autogenic factors also include the growth of geomorphic units such as bars, and the restriction of channel migration due to cohesive mud, tough palaeosols and groundwater cement. The Mississippi meander belt (Figure 5.38b,c) is an example of a channel body built by repeated avulsion and cutoff, with migration restricted in places by tough sediment. On alluvial fans, periods of aggradation spread sediment across the fan, whereas periods of incision in upfan areas release sediment to generate new aggradational lobes downfan (lateral versus vertical autogenic effects: Ventra and Nichols 2013).

Ecosystem effects are also intrinsic to alluvial systems. Plants and fungi bind substrates, and they adjust to forcing factors in concert with the fluvial landforms and animals with which they evolved. Microbes may also stabilise substrates and modulate sediment erosion and entrainment.

A reach's geological history is important (Church 1992), operating through features such as bedrock remnants and tough glacial deposits in the river tract (Figure 5.48). Channel dimensions may reflect earlier events such as the last major flood (Wolman and Miller 1960), and meltwater pulses and outburst floods anomalously enlarged many Eurasian valleys (Westaway and Bridgland 2010; Pickering et al. 2019). Scouring and bar growth influence how the reach responds to later events, affecting discharge and sediment input to the next reach downstream.

5.13.3 Allogenic Factors

5.13.3.1 Climate and Weather

Climate is a key driver of alluvial systems (Figure 5.48a). Climate encompasses daily weather patterns and events (including insolation, precipitation, temperature and air pressure) that are averaged regionally over at least 30 years, fluctuating over medium-term timescales. Influenced by the Earth's rotation, atmospheric circulation cells and ocean currents transfer heat from the tropics to the poles,

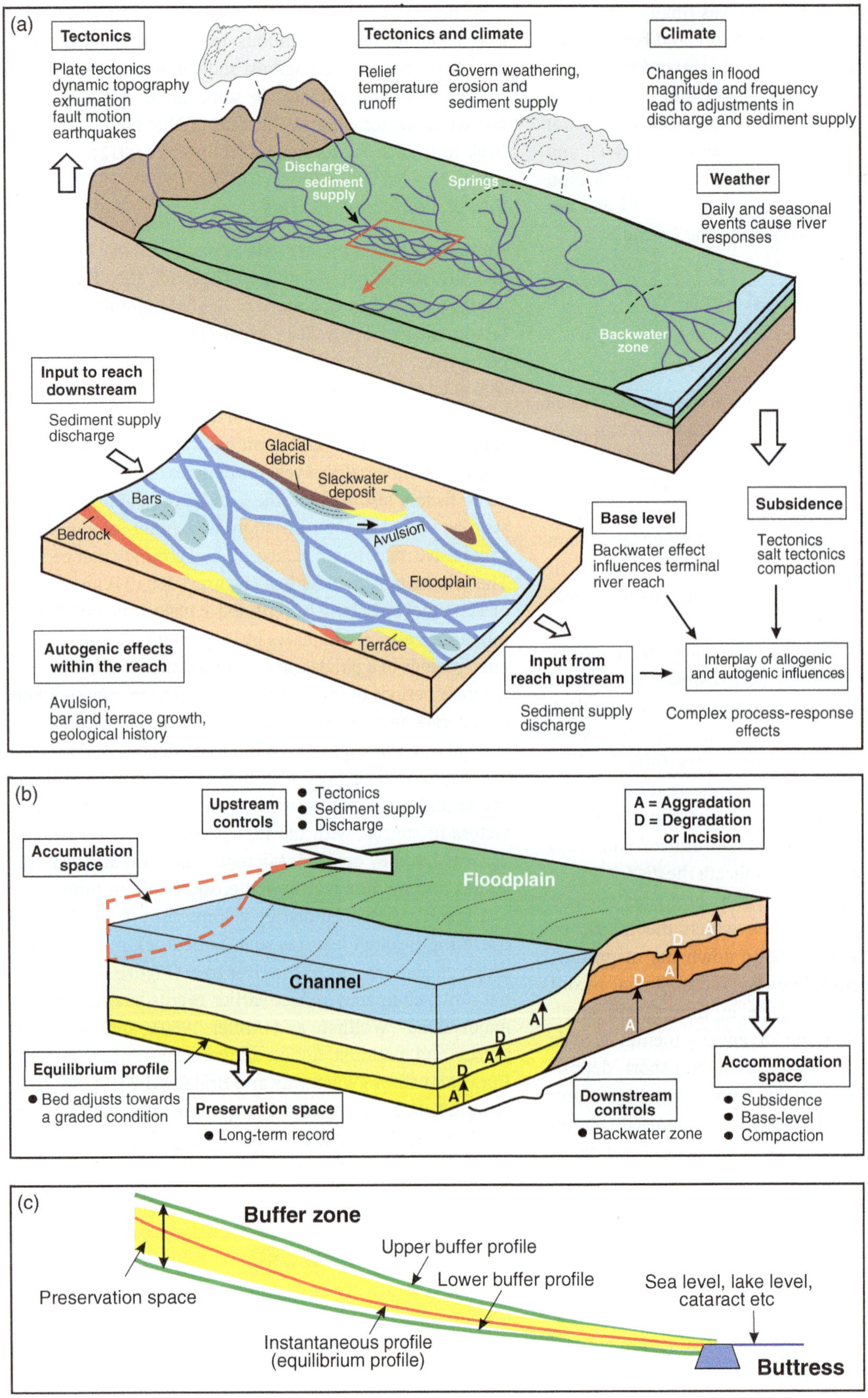

Figure 5.48 Controls on alluvial systems. (a) Major factors. (b) Alluvial response to upstream and downstream controls, from concepts in Blum and Törnqvist (2000). (c) Equilibrium profile graded to a downstream buttress, with a buffer zone upstream (preservation space) from the maximum depth of incision to the maximum height of accumulation, modified from Holbrook et al. (2006).

generating climate systems that include the trade winds, the North Atlantic Oscillation, El Niño Southern Oscillation, the Southwest Asian Monsoon and the East Asian Winter Monsoon. Climatic systems are closely linked to tectonically controlled elevation, and they operate in conjunction with glacial rhythms generated by orbital forcing, with sub-orbital shifts from changing glacial boundary conditions. Rivers respond to changes of air masses and storm tracks principally through changes in flood frequency and magnitude (Knox 2000; Macklin et al. 2012). These changes influence channel geometry and sediment calibre and storage, as well as generating alluvial terraces (Supplementary Table 5.S.1: 34), as in the glacial watershed of the Kali Gandaki River (Figure 5.5b,c). In the Mississippi Valley, braided-river regimes characterised much of the last glacial cycle as the river adjusted to glacially induced changes of discharge and sediment supply (Rittenour et al. 2007).

Weather events on a daily basis constitute climate patterns over longer timescales and generate recognisable features in the geological record, especially in response to discharge variation (Figure 5.43). It is helpful to distinguish weather events from longer-term climate patterns because individual weather events may profoundly affect rivers, with some individual storms altering a river's facies assemblage, cutoff style and geomorphic development (Hansford et al. 2020; Maitan et al. 2024). In Australia, a decade of large floods on the Manning River stripped an extensive floodplain area (Figure 5.49a; Nanson 1986). At Buckskin Gulch in Utah (Figure 5.49b), major floods laid down slackwater flood layers in a bedrock alcove due to several centuries of upstream river incision and sediment release (Harvey et al. 2011). Each layer represents a weather event, but the flood pattern represents climate change.

Figure 5.49 Examples of alluvial responses to weather, climate and tectonics. (a) Erosional surface at Charity Creek, Manning River system, New South Wales, Australia, where up to 4 m of floodplain deposits, dating within the past 400 years, were eroded down to the basal gravels by major floods associated with weather events from 1968 to 1978 (Nanson 1986). (b) Stacked flood layers at Buckskin Gulch, Paria River system, Utah, USA. Flood deposits built up in an alcove in the Navajo Sandstone, where slack water ponded at the mouth of a slot canyon. The stack was laid down in the past 350 years, linked to climate change (Site 1 of Ely 1992). (c) and (d) Yellow River in the Loess Plateau near Xining, China, with planation surfaces and gorge linked to tectonic effects. (c) Planation surface dated at 17–10 Ma (Miocene), with terraces younger than 6–10 Ma (Miocene) in the foreground along the Huang Shui, a tributary of the Yellow River (Wang et al. 2012). (d) Gorge cut through the Miocene planation surface, below Longyangxia Reservoir. *Source:* From Martin Gibling (Chapter author). (c) Gibling (2021) / with permission from Dunedin Academic Press.

5.13.3.2 Tectonic Activity and Related Processes

Tectonic activity causes crustal unroofing, leading to sediment production and basin subsidence, as well as imposing river gradients and a basinal framework. Subsidence rates are generally averaged over periods of thousands to millions of years based on the thickness of dated successions; longer chronostratigraphic successions are inherently less complete and thus yield apparently lower subsidence rates (the so-called Sadler Effect, after Sadler (1981); see Miall et al. (2021) for applications to fluvial systems). Faults may rupture the surface (Figure 5.14d), change gradients and divert rivers, and earthquakes liberate huge sediment volumes. Rivers also experience 'longitudinal tilting' and gradient change where they cross active fault zones, and 'lateral tilting' where fault zones draw the channel into a zone of subsidence (Holbrook and Schumm 1999; Bianchi et al. 2015). Furthermore, relief and temperature strongly influence sediment supply, thus linking tectonics and climate.

Other processes that influence topography include glacio-isostasy and crustal doming and flexure related to igneous dynamics. Salt withdrawal may cause river diversion where salt walls and diapirs rise and minibasins subside over geological time. Near-surface compaction of mud and peat enhances subsidence: for example, peat in the Saskatchewan River Delta compacted at 6 mm yr^{-1} over a century following loading by crevasse splays (van Asselen et al. 2010).

5.13.3.3 Base Level Change

Base level is a surface that separates domains subject to erosion and deposition. At the basin scale, base level corresponds broadly with the shoreline, although channels can erode below this level. Base level exerts a major control on erosion and deposition through the *backwater zone* at the terminal reach of a river. The length of the backwater zone (L_b) is defined as H/S_{ws}, where H is the mean channel depth and S_{ws} is the gradient of the water surface in the normal reach upstream of the backwater zone (Paola and Mohrig 1996). L_b corresponds to the length of the reach where the mean elevation of the channel bed is below mean sea level and where the gradients of the bed and the water surface diverge downstream. In practical terms, the backwater length describes how far upstream from the receiving basin a hydraulic response can propagate under prevailing conditions of normal, sub-critical flow in response to a change in stream power. The backwater length is up to 375 km for the Mississippi, and about 100 km for the Rhone and Texas Gulf Coast rivers (Blum and Törnqvist 2000; Fernandes et al. 2016). Where stream power declines at the backwater limit, nodal avulsion may take place (Supplementary Table 5.S.1: 17). In endorheic basins where flow terminates in ephemerally flooded playas, the boundary between erosional and depositional realms depends largely on discharge and local variations in slope.

5.13.3.4 Upstream and Downstream Controls

Upstream controls on a river system are linked primarily to climate and tectonics, which impose discharge and sediment supply on the river. *Discharge* is largely a function of precipitation or seasonal snowmelt from glacio-nival watersheds, but it is modulated by varied runoff rates linked to vegetation (a mediator of climate), transmission losses through porous substrates and exchange with groundwater. Most drainage originates in uplands, but tributaries on the Indian plains, for example, flow from springs fed by monsoon rains ('plains-fed rivers' of Sinha and Friend 1994). *Sediment supply* reflects catchment-wide processes of weathering and erosion, with contributions from slope failure and aeolian, volcanic and organic sources. Supply is mediated by rates of chemical weathering and reflects relationships between tectonic unroofing, topographic relief and hillslope erosion, as well as temperature and runoff (Hovius 1998; Montgomery and Brandon 2002). When sediment yield is normalised by catchment area, small and steeply sloping watersheds act as 'hot reactors' where sediment-charged rivers transport their load effectively to the receiving basin (Milliman and Syvitski 1992). In contrast, large, gently sloping watersheds show less effective transport, with prolonged storage on alluvial plains. *Downstream controls* operate largely through the backwater zone, in association with long-term basinal subsidence and compaction.

5.13.3.5 Centrality of Plate Tectonics

In the long term, plate tectonics plays a dominant role in organising land masses, generating upland source areas and governing transport routes to receiving basins. Crustal plates drift over a mantle with heterogeneous density and thermal properties, generating *dynamic topography* that affects the elevation and gradient of the land. Across southern Africa, a topographic anomaly of >1000 m (the Southern African Plateau) has been linked to the Cretaceous drift of Africa over a stationary region of mantle upwelling, resulting in tilting, denudation and drainage reorganisation of the Orange River (Braun et al. 2014). Along the Yellow River in China, periods of tectonic stability during the Miocene caused channels to migrate laterally and carve out *planation surfaces*, whereas later tectonic activity with ground uplift caused rivers to incise into the planation surfaces, generating gorges with terrace staircases (Figure 5.49c,d). Further back in time, the Neoproterozoic record of Laurentia, the ancestral core of North America, provides an example of continent-wide fluvial progradation and retrogradation related to pulses of crustal thickening and unroofing during the Grenvillian Orogeny (Rainbird et al. 2017).

In summary, long-term tectonic activity sets the framework for upstream controls, whereas medium- to long-term changes in base-level exert downstream controls. Climatic effects over a range of timescales, in conjunction with weather-related events, exert the predominant allogenic influence from uplands to the backwater zone. Rivers experience a complex interplay between allogenic and autogenic factors, and they are not simply passive responders to allogenic forcing because geological events in one reach may strongly influence the next reach downstream.

5.13.4 Inferring Controls in Deep Time

Two Carboniferous examples from Atlantic Canada (Figures 5.1 and 5.39b) illustrate how, under some circumstances, it may be possible to assess autogenic and allogenic effects. The examples are part of a 4 km succession in the fault-bounded Cumberland Basin, where tectonic activity and salt withdrawal (imaged in seismic profiles) caused rapid subsidence and generated stacked fluvial aggradational cycles. Alternation between facies associations of CBS and CBH is inferred to represent long-term, climatically forced changes in discharge variance. These tectonic and climatic controls are inferred from a basin record of several million years duration.

These controls, however, cannot be inferred for short successions nor for individual channel bodies, many of which probably had a lifespan of only hundreds to a few thousand years, based on circumstantial evidence from modern avulsion sites. In Figure 5.1, a laterally stable channel was established following avulsion into vegetated floodplain mud, and the accumulation space was filled through phases of incision and aggradation. In Figure 5.39b, a braided-fluvial channel-body consists of two storeys separated by an abandoned-channel fill; four avulsion events are recorded, with channels moving into and away from the area. Although some architectural features may reflect allogenic controls, possibly climate change within the lifespan of the channel body, no clear evidence supports this. Thus, the null hypothesis is also plausible, that avulsion coupled with weather-related floods is sufficient to explain the main architectural features of these relatively small examples.

Two other examples illustrate probable autogenic effects. In the Triassic Hawkesbury Sandstone of Australia, the channel body at the clifftop in Figure 5.34e displays superimposed storeys that show radical changes in palaeoflow direction of up to 135° (Rust and Jones 1987). The storeys probably reflect avulsion downstream from upland exits (see the Tista Fan parallel in Figure 5.5e). Although tectonic events may have triggered these avulsions, no firm evidence supports this hypothesis. For the Cretaceous Blackhawk Formation of Utah, Hampson et al. (2013) quantified channel-body parameters through a thick alluvial succession inland from the ancient coastline. They found little change in the parameters through base-level rhythms and concluded that autogenic factors were the predominant control on alluvial architecture.

5.14 River Profiles and Preservation

5.14.1 Graded Streams and the Equilibrium Profile

In the 1870s, John Wesley Powell, G. K. Gilbert and Clarence Dutton developed early fluvial concepts in the western USA (Blum and Törnqvist 2000). These included the *graded* condition for an alluvial channel that transmits the imposed water and sediment load without net change in the channel form (Church 2006). Such a channel adjusts its cross-sectional geometry and planform (principally width, depth, sinuosity and slope) in a dynamic equilibrium over a timescale of 10^2–10^3 years (Mackin 1948). The middle Yukon is an example of a graded river that flows through a largely unglaciated terrain (Froese et al. 2005). Along a reach of 1000 km, the gravel thickness varies only slightly (8–16 m), and the reach is situated too far upstream for sea-level influence and too far downstream for an oversupply of glaciofluvial sediment. The river has been in equilibrium since the Pliocene, acting as a 'conveyor-belt' (Blum and Törnqvist 2000) for sediment transport to the Bering Sea.

Depending on whether sediment supply exceeds or falls short of the maximum bedload transport rate, the channel adjusts towards an *equilibrium profile*. Under this condition, the bed maintains a near-steady form, neither aggrading nor degrading over a multiyear timescale that averages out brief responses to floods (Figure 5.48b). Floodplains are also affected by changes in the equilibrium profile, experiencing degradational episodes (disconformities) and aggradational episodes when overbank floods decrease or increase in frequency, respectively.

Changes in the longitudinal profile of a river can be envisaged using the concept of a buttress and a buffer zone (Figure 5.48c; Holbrook et al. 2006). The equilibrium profile is anchored downstream by a *buttress*, commonly sea level. Upstream, the equilibrium profile may increase or decrease in elevation depending on changes in discharge and sediment supply, generating a *buffer zone* that defines the preservation space. The lower limit of the buffer zone represents the maximum depth of incision, and the upper limit represents the maximum height of accumulation.

Accumulation space is the space that can potentially be filled under the current fluvial regime, extending from the channel base to the equilibrium floodplain height (Blum and Törnqvist 2000; Figure 5.48b). More broadly, *accommodation space* is the space potentially available for preservation of fluvial and other strata over a prolonged period due to changes in 'relative base-level'. Accommodation space encompasses the combined effects of subsidence, compaction and the space from the sediment surface to the water level of lakes and reservoirs (local base level) and the ocean (ultimate base level) (Jervey 1988). Patterns of coastal aggradation, retrogradation and progradation reflect the balance between accommodation space (A) and sediment supply (S), captured in the parameter A/S. For example, the Fraser River of Canada, fed by a large sediment supply from the Cordillera, prograded through the last glacial cycle, outpacing relative sea-level rise (low A/S: Williams and Roberts 1989). *Preservation space* lies below the lowest equilibrium profile and is the accumulation and accommodation space that was actually filled with sediment over a given prolonged period.

5.14.2 Sites of Enhanced Preservation Potential

The *preservation potential* of alluvial strata is an important factor in facies analysis (Figure 5.50). The term is defined as the likelihood that a sedimentary deposit will survive near-surface erosion, entering long-term preservation space during basin subsidence. *Preservation bias* is common because sediment has a higher preservation potential in some geomorphic settings than in others (see Section 4.5). Channel systems may change their equilibrium profiles rapidly (Figure 5.48b), greatly affecting the preservation potential of their sediments.

Regions undergoing long-term exhumation constitute about 84% of the Earth's terrestrial surface (Nyberg and Howell 2015). Sediment is prominent in these areas but has a low preservation potential (Figure 5.50a). Upland settings such as the Rakaia Valley of New Zealand (Figure 5.3a) contain alluvial deposits, and such deposits may be preserved as terraces (Figures 5.5c and 5.49c) and locally on unconformities, for example, in plunge pools cut into bedrock below waterfalls (Figure 5.31). Although intramontane basins may generate thick accumulations, the strata are commonly dismembered by faults, as along the Yellow River of China where Cenozoic basin fills were repeatedly eroded when faults connected separate drainage systems (Pan et al. 2011). Sediment may accumulate in valleys that act as sediment-transfer systems, such as the Kali Gandaki Valley in the Himalayan ranges of Nepal (Figure 5.5a–c), where 1400 m of alluvial, lacustrine and glacial strata accumulated over the past 10 million years, subdivided into formations by angular unconformities (Hurtado Jr. et al. 2001). Box 5.4 shows an example of an older Cenozoic valley fill.

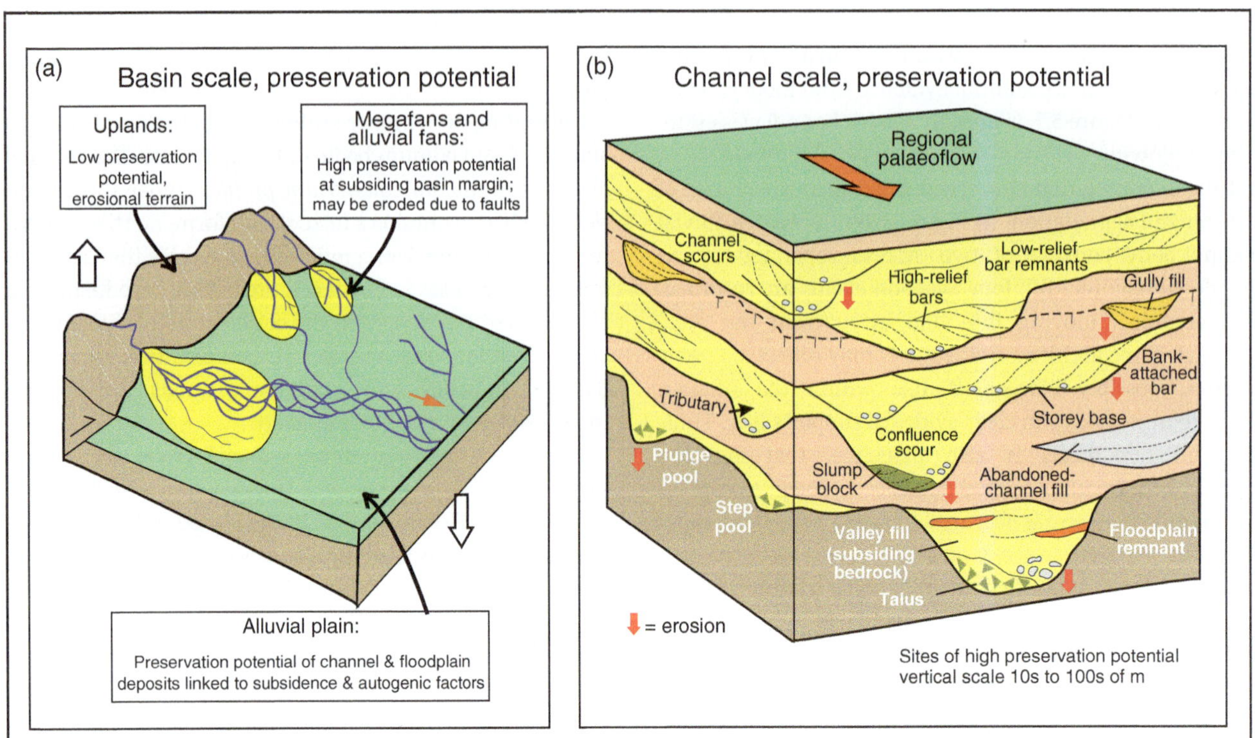

Figure 5.50 Preservation potential of alluvial strata. (a) Regional context from uplands to basins. (b) Local sites of enhanced preservation within river reaches.

Box 5.4 A Rare Sediment-Transfer Zone in the Rock Record

The Eocene to Oligocene Sis Formation and related rock units of Spain can be traced downflow through bedrock for 20 km. The formations constitute a valley fill >1400 m thick and up to 7.5 km wide (Vincent 2001), on a scale similar to the deposits of some Himalayan valleys. For at least 38 million years, the valley formed a sediment-transfer zone from the deforming thrust stacks of the Pyrenees to the adjoining foreland basin, where the valley probably supplied the shallow-marine Roda Sandstone (see Section 8.4.5).

The mainly gravelly fill of the valley was laid down by debris flows and stream flows in axial rivers. Alluvial fans built into the valley from both sides, and landslides emplaced bedrock masses 100 m long on the valley floor. Floodplain and lacustrine strata with coal seams are present locally. Valley-wide bounding surfaces form erosional disconformities along the valley axis and angular unconformities at the margins, marking events of drainage reorganisation linked to tectonic activity. Abrupt changes in clast composition across some surfaces denote river capture in the hinterland, with sudden changes in the sediment source. Tilting and deformation of the strata reflect tectonic activity during the active life of the valley system. *Source:* With permission from Stephen Vincent.

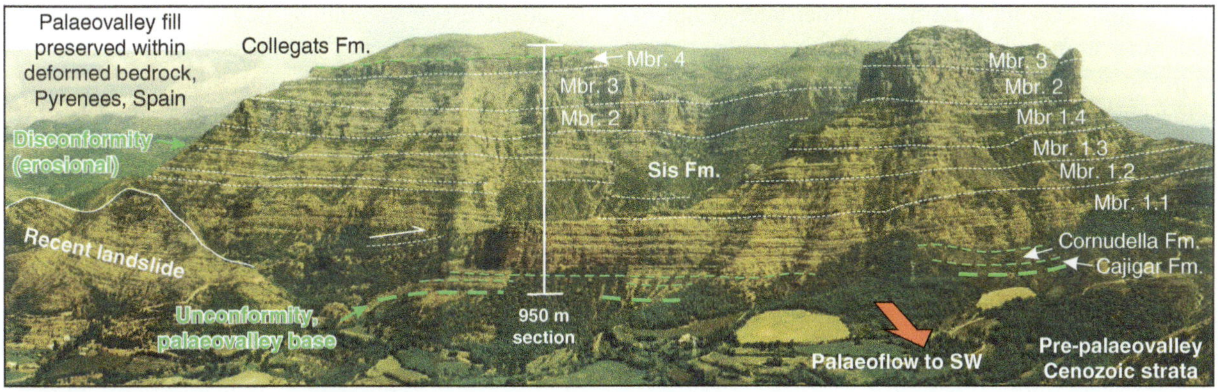

Megafan and alluvial fan deposits have a high preservation potential in rapidly subsiding depocentres such as the Himalayan Foreland Basin (Figure 5.5d) and the East African Rift System (Okavango Megafan). However, they may be uplifted and 'cannibalised' by basin-margin faults (Figure 5.6), which grow as 'blind thrusts' below alluvial plains.

On alluvial plains where sediment is aggrading, avulsion, subsidence and sediment supply influence the relative preservation potential of channel and floodplain deposits, as explored in numerical and analogue models (Supplementary Table 5.S.1: 35). On a local scale (Figure 5.50b), sites with a high preservation potential include valleys cut deeply into tough palaeosols (Figure 5.26b); deep confluence scours between trunk channels and tributaries; channel-belt and storey bases cut by avulsion or major floods; abandoned channels; and deep gullies with colluvial fills. In floodplain deposits, correlation of palaeosols and marker beds shows that depressions may be sites of preferential preservation, as in Miocene strata of Pakistan (Willis and Behrensmeyer 1994).

Bars have a variable preservation potential. In large rivers, high-relief braid bars are commonly well preserved (Best et al. 2003; Figure 5.34e,f), but bars in smaller rivers are commonly reworked and difficult to recognise, surviving as remnant cross-sets with accretionary surfaces trimmed off at the bar margins (Parker et al. 2013). Although bars are prominent at surface in the braided Platte River in Nebraska (Figure 5.34c), analysis of GPR profiles and cores shows that recognisable bars are few, whereas scour fills are widespread in the subsurface and have a high preservation potential (Horn et al., 2012a, 2012b). Similarly, for the Triassic braided-fluvial example of Figure 5.37, trough cross-bedded sand sheets and channel scour fills constitute the bulk of the cliff face, with only modest proportions of bar slipface deposits and mounded cores. In meandering rivers, bank-attached bars may be well preserved, with estimates of 35–85% of bar sediment preserved (Durkin et al. 2018), although capping scroll bars are usually eroded.

More broadly, the fluvial record shows a *strange ordinariness* (Paola et al. 2018), with fairly ordinary transport conditions commonly better represented than extreme events.

This paradox may reflect the presence of a morphodynamic hierarchy in many river systems, where higher order elements such as avulsion belts and bars provide local accommodation that enhances the preservation of lower-order elements such as dunes (Ganti et al. 2020). Along the Gulf of Mexico, avulsion may leave sinuous alluvial ridges that form topographic highs, bounding deeply incised tributary channel networks where ordinary alluvial strata may be preferentially preserved in the slowly subsiding coastal plain (Swartz et al. 2022).

5.15 Sequence Stratigraphy

5.15.1 Principles for Alluvial Systems

Sequence stratigraphy uses facies, key surfaces, stratal geometry and stacking patterns to understand sediment distribution from uplands to basins (Catuneanu 2006; Catuneanu et al. 2009). The discipline has shed new light on the alluvial controlling mechanisms discussed in Section 5.13, and it employs a rich terminology to identify critical surfaces and stratal packages (see Chapter 4 and Catuneanu (2006) for details).

In coastal settings, a sequence forms when accommodation space changes during a cycle of base-level fall and rise, during which eustasy interacts with local subsidence or uplift. A base-level fall causes the shoreline to shift seaward in a forced regression, triggering erosion and basinward accumulation (Falling Stage Systems Tract) and culminating in a Sequence Boundary. Early in base-level rise, the rate of sediment supply commonly outpaces the rate of rise, promoting a further seaward shift of the shoreline under aggradational conditions (Lowstand Systems Tract). As base-level rise outpaces the rate of sediment supply, sediment is deposited above an initial flooding surface as the shoreline shifts landward (Transgressive Systems Tract). As the rate of base-level rise decreases and is again outpaced by the rate of sediment supply, sediment is deposited above a Maximum Flooding Surface as the shoreline shifts seaward in a normal regression (Highstand Systems Tract). Although this terminology reflects base-level change, the systems tracts are closely related to sediment supply, compaction and autogenic river dynamics.

Rivers respond to coastal base-level change where they encounter the backwater zone. Models linked to subsidence rate and avulsion frequency (Wright and Marriott 1993) suggest that channel bodies tend to amalgamate and cluster when accommodation space is low (Falling Stage and Lowstand Systems Tracts) and are dispersed within coastal and floodplain strata when accommodation space is high (Transgressive and Highstand Systems Tracts). Data in the next section provide some support for these models. Low- and high-accommodation systems tracts have been defined from the degree of amalgamation of channel bodies (Martinsen et al. 1999), especially in inland areas beyond marine influence; the model is most effective where single-channel systems migrate and amalgamate rapidly.

The response of Quaternary alluvial systems to some forcing factors may be assessed by comparing fluvial facies and architecture with proxy records for sea level and climate, especially through marine-isotope stages (MIS) 1–6 of the last glacial cycle (from about 191 to 14 ka: Figure 5.51c). This timespan witnessed up to 120 m of sea-level change as ice sheets grew and melted. The Quaternary and Carboniferous case studies below (see Supplementary Table 5.S.1: 36–40 for references) represent Icehouse periods with high-frequency and high-amplitude eustatic changes (Rygel et al. 2008). The fluvial systems experienced correspondingly large contrasts in dynamics, backwater effects, and sediment transport and storage (Blum et al. 2013). In contrast, the Mesozoic represents a Greenhouse period with modest eustatic effects (Ray et al. 2019). In climatically sensitive lakes such as the Dead Sea, which experience rapid fluctuations in level, sediment transported by ephemeral dryland rivers may generate sequences on timescales as brief as centuries (Moran et al. 2023). Sequences may be fractal, exhibiting self-similar architectural patterns at radically different spatial scales. For example, sequences in the Dead Sea resemble the fill of the large Pelotas Basin of the South Atlantic (Moran et al. 2023, their Figure 14), and a full deltaic sequence formed in a few days in a roadside ditch in Canada (Posamentier et al. 1992).

5.15.2 Coastal Alluvial Examples

5.15.2.1 Po River Plain, Italy

The Po River of northern Italy laid down a thick succession in the rapidly subsiding Apennine foredeep (Amorosi et al. 2016a), feeding the offshore Adriatic mudbelt (see Sections 9.2.2 and 9.4.1). Proximal and distal profiles based on cores with age dates are shown in Figure 5.51a,b. The modern channel passes below mean sea level about 110 km from the river mouth. During periods of low discharge, rhythmic changes in river level from Adriatic tides can be observed up to 70 km inland, losing prominence as river discharge increases. Thus, the illustrated profiles may have lain within and landward of the backwater zone at earlier times, depending on discharge and the shoreline position at times of lowered sea level.

From the interglacial of MIS 5e to the Last Glacial Maximum of MIS 2, ice buildup and associated sea-level fall forced a regression on the Po plain, where channel deposits are assigned to the Falling Stage Systems Tract (FSST).

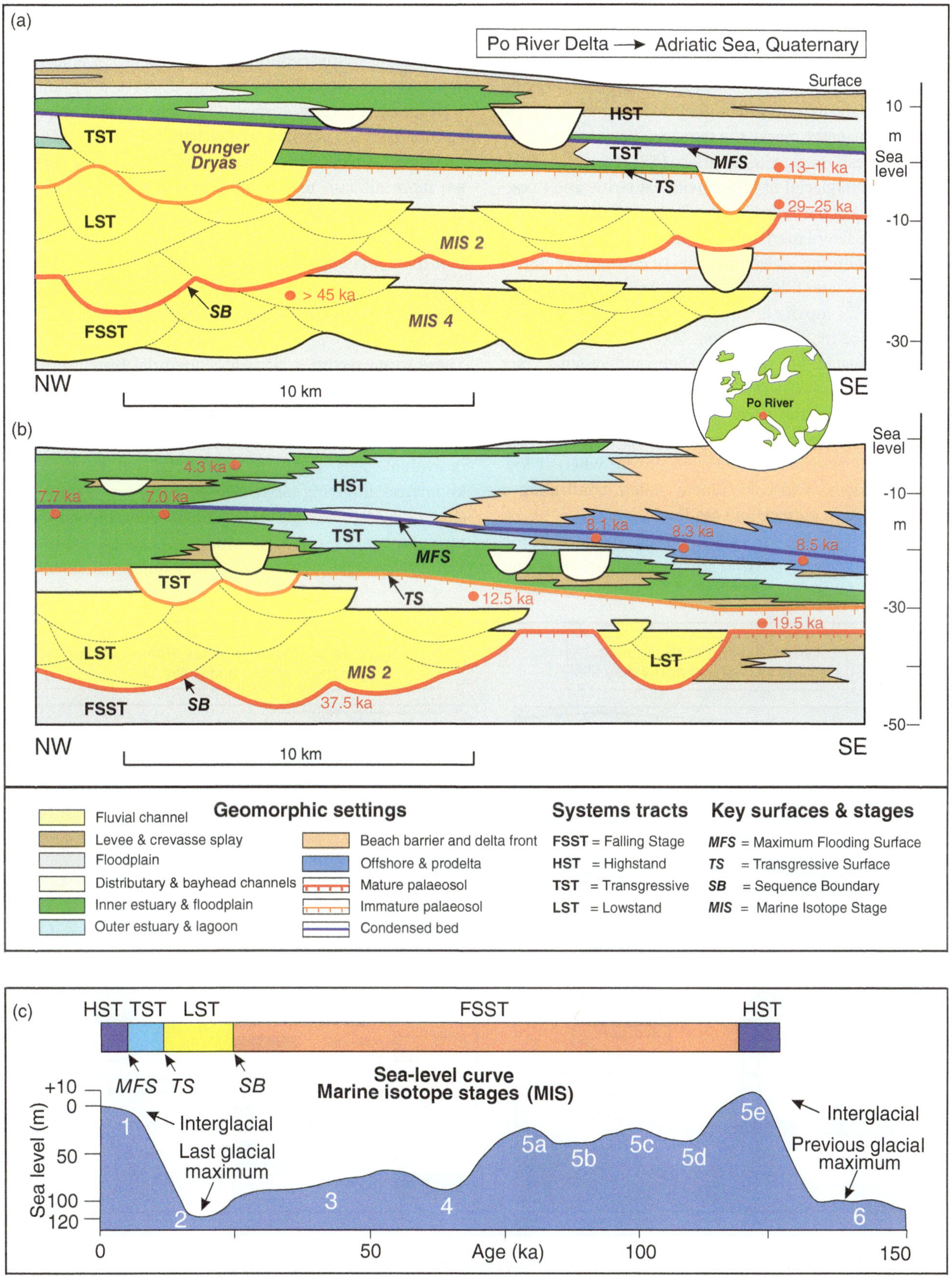

Figure 5.51 Sequence stratigraphy of coastal alluvial systems. (a)–(c) Po River plain and delta of Italy, showing the response of river systems to sea-level change during the last glacial cycle. Downriver proximal and distal profiles (a) and (b) are based on analysis of facies, pollen spectra and radiocarbon dates from cores, supplemented by subsurface methods to trace key surfaces and sediment bodies. *Source:* Adapted from Amorosi et al. 2016a). Sea-level curve in (c), based on dated oxygen-isotope proxy records from benthic foraminifera and sea-level data from coral-reef terraces. *Source:* Adapted from Waelbroeck et al. (2002).

The onset of the Last Glacial Maximum of MIS 2 is marked by a sequence boundary at the base of a multistorey valley fill 30 m thick and 20 km wide in the Lowstand Systems Tract (LST). Low accommodation during this period favoured amalgamation of channel bodies, and a mature interfluve palaeosol marks the sequence boundary.

A widespread transgressive surface (TS) records a transition to the interglacial of MIS 1, with estuarine and flood-basin deposits of the Transgressive Systems Tract (TST). An immature palaeosol marks the Younger Dryas, a short-lived cooling event during which the river cut down to generate a valley that was later filled. The Maximum Flooding Surface (MFS) lies at the top of a fossil-rich, condensed bed, a diachronous surface that formed about 9000–7000 years ago as the coast retreated. The overlying Highstand Systems Tract (HST) marks renewed progradation of the alluvial plain in a normal regression, during which the river switched position frequently to generate isolated channel bodies. Transgressive periods yield warm-climate pollen, whereas low sea-level periods yield cold-climate pollen, confirming a related change in climate and sea level.

In broad terms, the Po Valley deposits subsided at the rapid rate of 2.5 mm yr^{-1} over the past million years and more, generating unusually thick deposits (up to 130 m from MIS 5e to MIS 1: Campo et al. 2020). This phenomenon has allowed exceptionally high-resolution analysis of age and architecture. The system displays a prominent, widespread flooding surface (TS) and several sequence boundaries (SB) marked by the base of valley fills and a hierarchy of palaeosols, whereas other bounding surfaces are more difficult to identify and correlate. For practical applications to groundwater and engineering in this Quaternary setting, mapping lithological bodies is more important than applying complex terminology.

5.15.2.2 Pennsylvanian Strata of the North America Craton

These strata record a far-field, equatorial response to the Late Palaeozoic Ice Age in Gondwanaland. With long-term subsidence rates an order of magnitude lower than the Po Basin, the Illinois Basin contains thin sequences with individual beds that can be traced for hundreds of kilometres, implying external forcing on a continental scale (Figure 5.52a; Fielding et al. 2020). Although no high-resolution proxy record is available, spectral analysis of Pennsylvanian successions suggests a link to orbital forcing

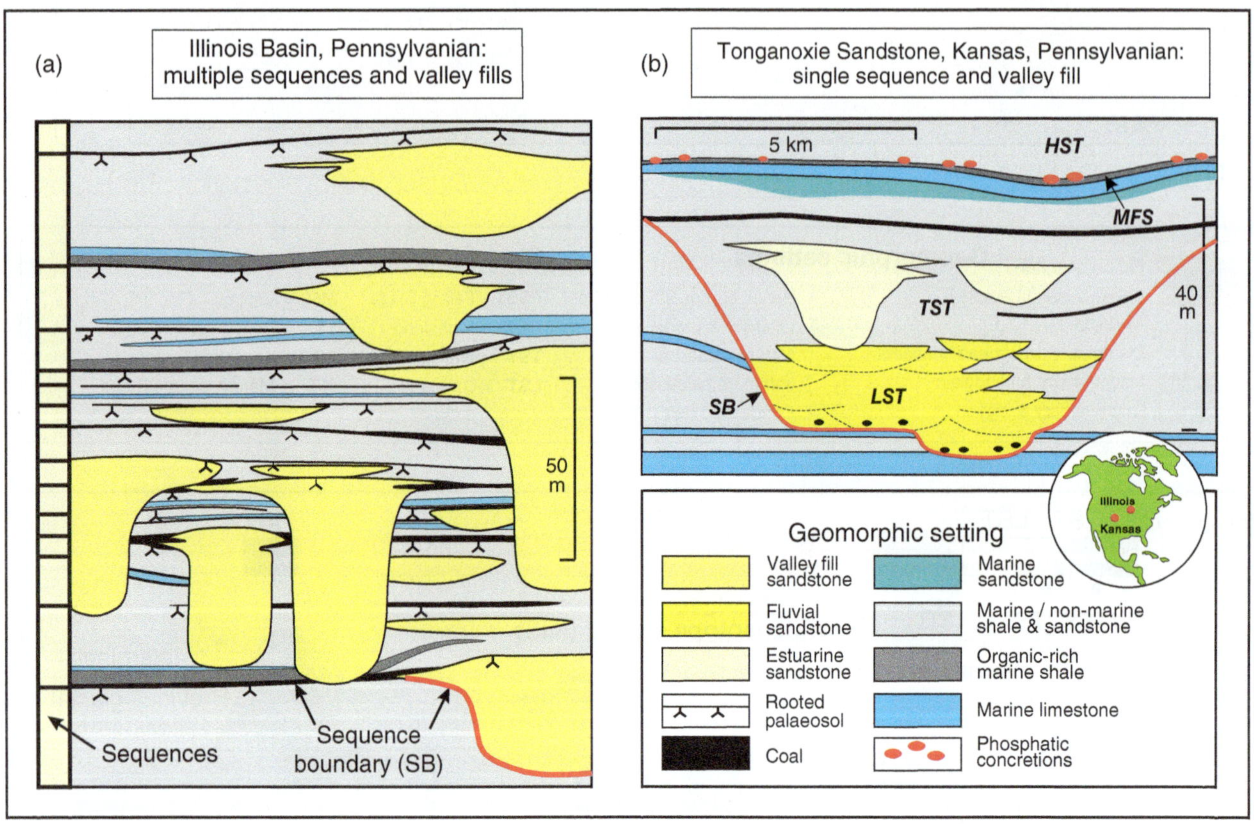

Figure 5.52 Pennsylvanian sequences from outcrops, coal mines and cores in the Illinois Basin (a). *Source:* Adapted from Fielding et al. (2020) and Kansas (b). *Source:* Adapted from Feldman et al. (1995). In these examples, valley fills pass upward from fluvial conglomerate and cross-bedded sandstone to estuarine sandstone and mudstone.

and glacial rhythms on timescales of 10^4–10^5 years (Le Cotonnec et al. 2020). Multistorey valley fills contain up to 70 m of fluvial and estuarine strata (LST and TST), and their bases mark sequence boundaries that can be traced laterally into interfluve palaeosols. Many are capped by extensive peats (now coals) that formed when rising sea-level ponded freshwater across the low-gradient craton. Succeeding organic-rich marine shales and bioclastic limestones mark the MFS. Xerophytic floras in the basal valley fills and vertic and calcic palaeosols at sequence boundaries indicate strongly seasonal climates during the LST, whereas TST and HST strata yield mixed dry forest and wetland taxa, suggesting a lower degree of seasonality.

The Pennsylvanian Tonganoxie Sandstone of Kansas is a multistorey valley fill, 40 m thick and 10 km wide, which can be traced for 240 km downslope and has been linked to glacio-eustatic forcing (Figure 5.52b; Feldman et al. 1995). Basal fluvial strata represent braided channels (LST), and overlying point-bar and tidal-flat deposits contain tidal rhythmites and agglutinated foraminifera (TST), with terrestrial and freshwater arthropod traces, grazing and feeding structures, fish traces and tetrapod trackways (Buatois et al. 1998b). Coals lie within and cap the valley fill. The MFS is marked by a dark shale with an open-marine fauna and phosphatic concretions, overlain by marine shales of the HST. Box 5.5 illustrates how alluvial facies and architecture evolved within Pennsylvanian sequences in Canada.

5.15.3 Inland Alluvial Examples

Climatic fluctuations linked to the Southwest Asian Monsoon, governed by insolation and glacial boundary conditions, have profoundly influenced Indian rivers. These effects are illustrated by cliff strata at Kalpi on the Yamuna River, which lies 1000 km inland from the Bay of Bengal. The river has a modern backwater zone 200 km long, and the Quaternary successions are too far upstream to have been affected by sea-level change.

In the cliff at Kalpi, channel, floodplain, floodbasin and colluvial facies associations span about 120,000 years based on luminescence dating (Figure 5.53a,b; Gibling et al. 2005). Four discontinuities define sequences of alternate aggradation and degradation. Two of the discontinuities are cemented by groundwater carbonate (facies Cn; Figure 5.53c), and a third comprises gully fills 20 m thick with inclined layers and slumps of colluvial material (Figure 5.53d). The fourth discontinuity marks the modern gullied surface and the presently incised Yamuna channel.

A proxy record from Arabian Sea cores captures changes in intensity of the Southwest Asian Monsoon, in general accord with the Marine Isotope Stages (Figure 5.53e). Despite uncertainties in correlating luminescence dates with the proxy record, inferences from Kalpi are supported from sites across the Indian subcontinent (Supplementary Table 5.S.1: 40). Fluvial facies characterise MIS 5 to 3, with discontinuities that probably correspond to periods of weakened monsoon and reduced discharge. The large gully fills date reliably to

Box 5.5 Linking Pennsylvanian Fluvial Styles to Sequence Stratigraphy

Stacked sequences in the Sydney Mines Formation are exposed along the Atlantic coast of Canada. Like the successions of Illinois and Kansas, they contain a standard set of systems tracts linked to the Late Palaeozoic Ice Age, but the exceptional cliff exposures reveal how changes in sea level and climate affected the fluvial systems (Batson and Gibling 2002).

Sequence boundaries are marked by thick calcic palaeosols and valley fills up to 30 m thick, representing landscapes that were exposed and dissected during sea-level lowering. The valleys are filled with braided-river sandstone (CBS) that accumulated in the LST as relative sea-level rose. Peat and poorly drained (hydromorphic) palaeosols formed in the TST as transgression drowned the interfluves, and economic coals are capped by organic-rich limestones with a restricted-marine fauna, marking the MFS. As the coastline readvanced, meandering channels generated point bars and scroll bars (Figure 5.33). An abrupt change is marked by red vertic palaeosols with pseudo-anticlines (formed by seasonal shrinkage and swelling of clays) and freshwater limestones, which denote dryland conditions. Associated channel bodies contain high-energy bedforms (facies Sh and Sac), rill casts and upright tree trunks, indicative of low-sinuosity rivers with high discharge variance (CBH). These strata are ascribed to the FSST, as lowered relative sea level forced a regression on the coastal plain below the next sequence boundary.

The calcic and vertic palaeosols and high-energy channel fills denote relatively arid conditions with strongly seasonal flow, whereas meandering channels, coals and hydromorphic palaeosols denote humid conditions with more sustained river flow. Thus, the sequences display climatic fluctuations in tandem with glacially driven changes in sea level, as in the Po Valley Quaternary example.

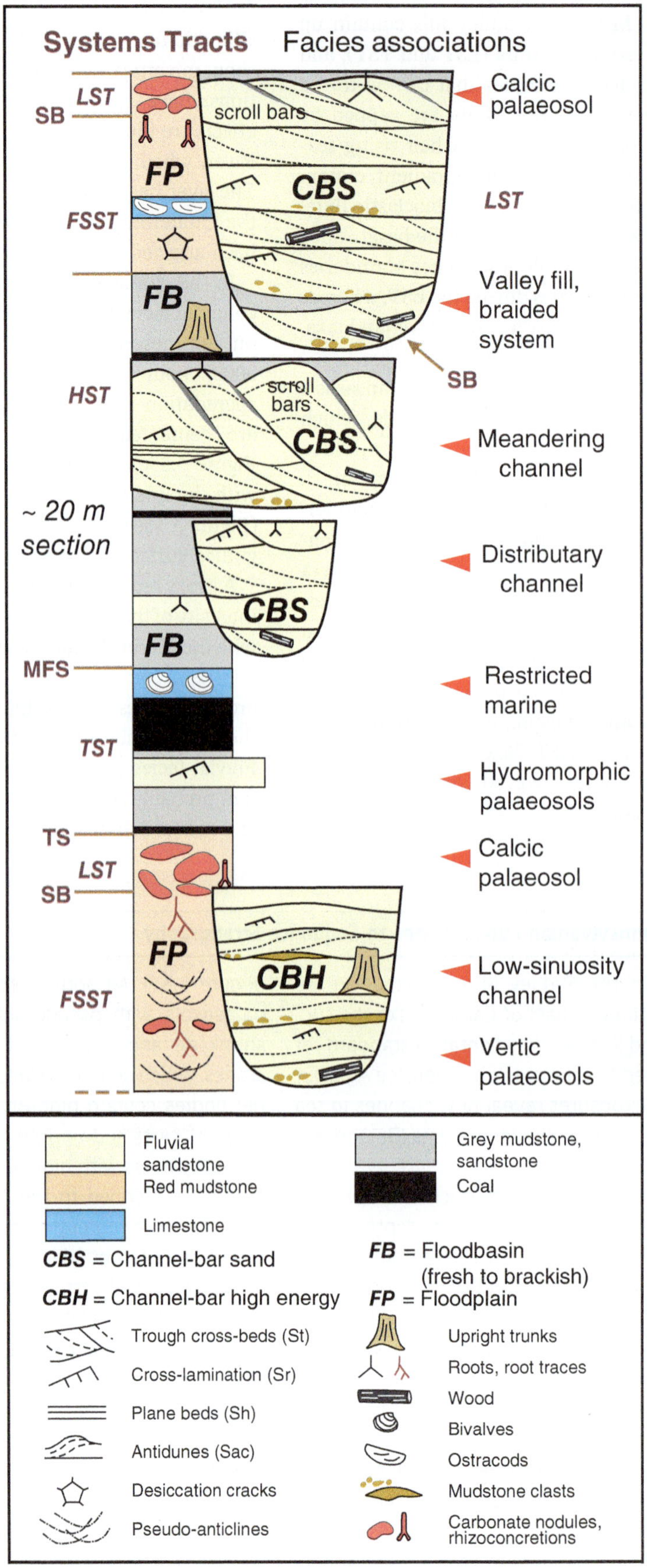

Systems Tracts
Facies associations
SB
LST
FSST
HST
~ 20 m section
MFS
TST
TS
LST
SB
FSST
FP
FB
scroll bars
CBS
CBH
Calcic palaeosol
LST
Valley fill, braided system
SB
Meandering channel
Distributary channel
Restricted marine
Hydromorphic palaeosols
Calcic palaeosol
Low-sinuosity channel
Vertic palaeosols
Fluvial sandstone
Red mudstone
Limestone
Grey mudstone, sandstone
Coal
CBS = Channel-bar sand
CBH = Channel-bar high energy
FB = Floodbasin (fresh to brackish)
FP = Floodplain
Trough cross-beds (St)
Cross-lamination (Sr)
Plane beds (Sh)
Antidunes (Sac)
Desiccation cracks
Pseudo-anticlines
Upright trunks
Roots, root traces
Wood
Bivalves
Ostracods
Mudstone clasts
Carbonate nodules, rhizoconcretions

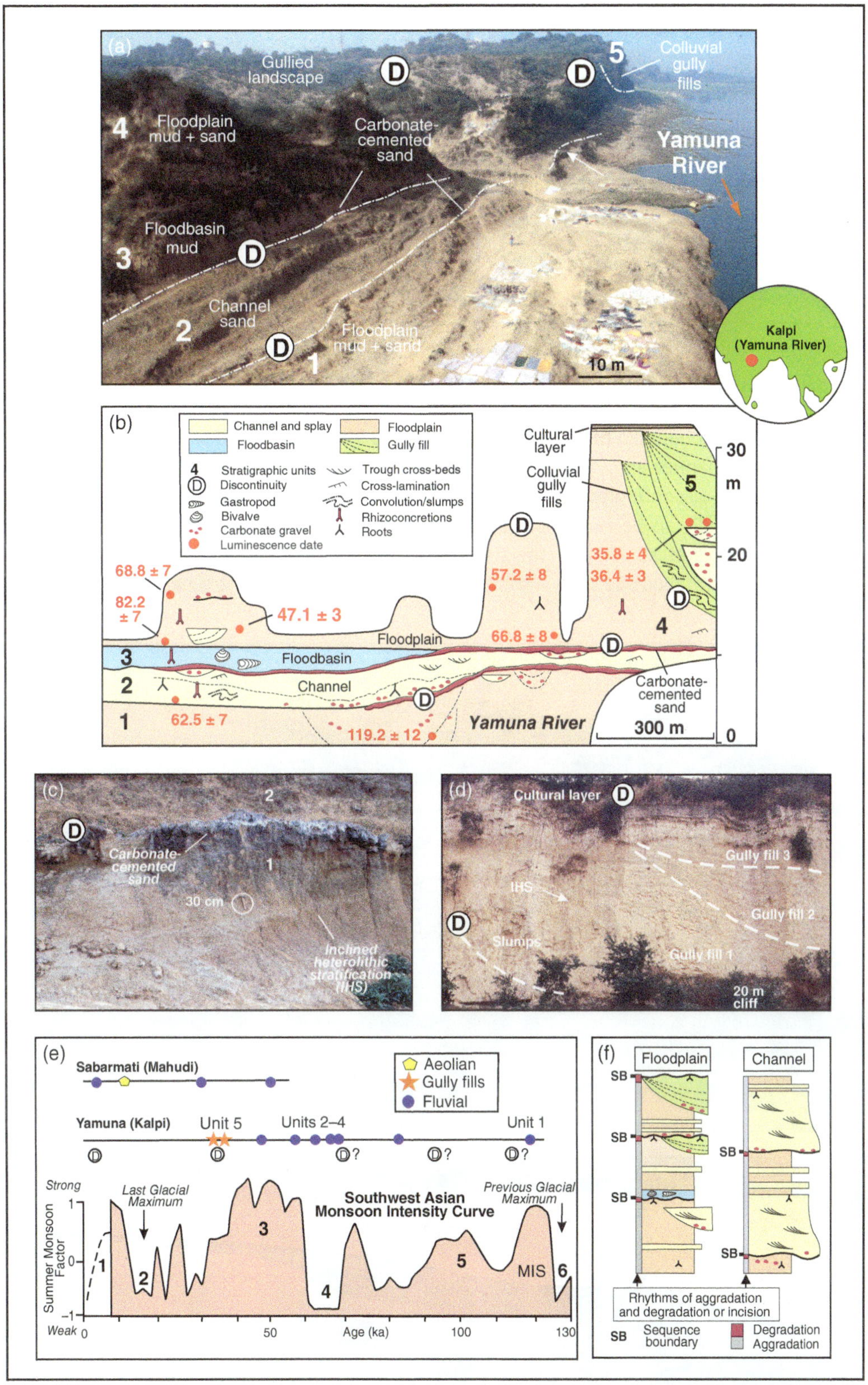

Figure 5.53 Sequence stratigraphy of inland alluvium in India. (a) (d) Sequences bounded by discontinuities in a cliff of the Yamuna River at Kalpi, Uttar Pradesh (Gibling et al. 2005). Chronology is from luminescence dates in thousands of years (ka) with large error bars. (e) Monsoon intensity from Clemens and Prell (2003), based on stacking of five dated proxy records from Arabian Sea cores. See Box 5.6 for Sabarmati data. (f) Sequences with rhythms of aggradation and degradation or incision northern India. MIS = Marine Isotope Stage (see Figure 5.51C). *Source:* From Martin Gibling (Chapter author).

about 36 ka, corresponding to a waning monsoon from MIS 3 to the Last Glacial Maximum of MIS 2.

Under the powerful but fluctuating summer monsoon, river equilibrium profiles adjusted to changing water and sediment discharge (see Figure 5.48b). These changes resulted in rhythms of aggradation and degradation or incision in channel and floodplain deposits, as summarised in Figure 5.53f (see Box 5.6 also). These rhythms fit the sequence-stratigraphic definition of sequences. A terminology that employs 'rhythms' is used because it is unfitting to identify systems tracts that are explicitly linked to sea-level change in such climatically generated sequences (Gibling et al. 2011). Some tectonic influence on sedimentation cannot be ruled out, but proxy tectonic records are difficult to apply, being limited to physical dating of a few faults and historical earthquake records.

Box 5.6 What Caused the Sabarmati River of Northwest India to Incise?

Mahudi on the Sabarmati River lies 150 km inland on the fringe of the Thar Desert, with a modern backwater zone of only about 20 km. The river has cut a spectacular 40-m cliff, with a fluvial lower section and an aeolian cap that includes vegetated dunes. The incised channel and the modern gullied surface mark a prominent discontinuity.

The timing of channel incision at Mahudi can be estimated using luminescence dates (Srivastava et al. 2001). Aeolian strata at the top of the cliff date to about 12 ka, and the oldest sediment trenched in bars in the channel dates to about 4.5 ka. Bracketed by these dates, the deep incision of the Sabarmati Valley dates to a period of intensified monsoon and increased river discharge after the Last Glacial Maximum (Figure 5.53e), when the river cut down rapidly through aeolian and fluvial sands. Later in the Holocene, the monsoon weakened, and sand bars built up in the channel when the river was unable to transport the available load. *Source:* Martin Gibling (Chapter Contributor).

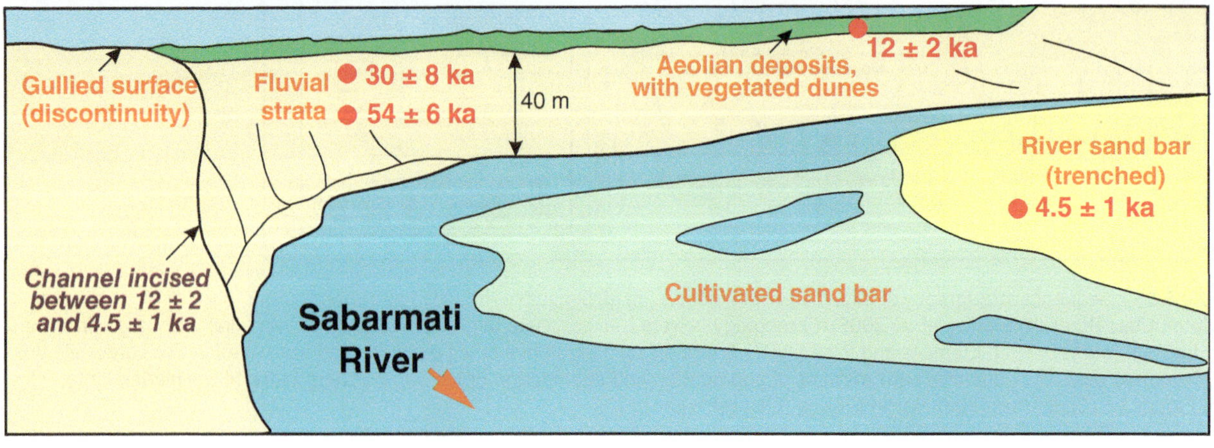

5.16 Geological Applications

5.16.1 Biotic Factors and Alluvial Landscapes Through Geological Time

Facies models are time-dependent because evolutionary innovations have contributed to the shaping of fluvial landscapes (Buatois and Mángano 2007; Buatois et al. 2022), where biological communities mediate weathering, soil formation and bank stability (Figure 5.54; Supplementary Table 5.S.1: 41–46). The oldest known fluvial strata form part of the Fig Tree and Moodies groups of South Africa and Eswatini, dating back to 3.28 billion years ago (Eriksson et al. 2006; Drabon and Lowe 2022). Microbial communities (cyanobacterial mats and crusts) have inhabited alluvial settings since at least the Neoarchean (Watanabe et al. 2000), although they apparently had little effect as ecosystem engineers (McMahon et al. 2017). Cambrian arthropods made temporary incursions into coastal terrestrial settings but did not establish alluvial communities (MacNaughton et al. 2002). Precambrian and early Palaeozoic landscapes had unstable banks with little mud (Ielpi and Ghinassi 2016), but Middle to Late Ordovician cryptospores indicate an incipient vegetation cover of non-vascular land plants that were cosmopolitan ecological generalists. No major innovations in animal–substrate interactions are apparent during this period.

From the Ordovician onwards, alluvial mudrock increased substantially as vegetation spread, profoundly affecting fluvial morphodynamics and bank strength (Davies and Gibling 2010a, 2010b; Gibling and Davies 2012; McMahon and Davies 2018). The Silurian-Devonian marks the establishment of vascular land plants, including forests with extensive root systems, charcoal from wildfires, vascular-plant peats and mycorrhizal communities, heralding an increase in terrestrial biomass and the rise in the Devonian of modern-style meandering rivers (Davies et al. 2021; Buatois et al. 2022). The increased mud content in alluvial and tidal systems suggests that land plants triggered a global increase in weathering and mud production. This profound effect was accompanied by enhanced mud retention through vegetation baffling, biochemical weathering within sand and gravel and flocculation aided by plant organics (McMahon and Davies 2018; Zeichner et al. 2021; Ielpi et al. 2022; Gurnell and Bertoldi 2022; McMahon et al. 2022). Trace-fossil diversity and disparity increased, with the appearance of the archetypal *Scoyenia* Ichnofacies (Buatois et al. 1998a; Minter et al. 2017) and deposit-feeding aquatic and terrestrial arthropods that included scorpions, trigonotarbid arachnids, arthropleurids (Figure 5.15e), kampecarid millipedes, centipedes and hexapods. The earliest tetrapod trackways are from the Devonian.

The Carboniferous marks the establishment of the main fluvial styles recognised today. Bioturbation was widespread in riparian habitats, and land plants diversified to include gymnosperms and deeper root systems that stabilised riverbanks. Sufficient large woody debris was present in channels to cause log jams that triggered avulsion (Gibling et al. 2010; Ielpi et al. 2014). Fixed channels became prominent, some representing anastomosing rivers, as did braided systems with vegetated islands (Davies and Gibling 2011, 2013). In overbank areas, trace-fossil diversity increased, and the *Mermia* Ichnofacies appeared in floodplain water bodies (Buatois et al. 1998a; Buatois and Mángano 2002; Minter et al. 2017). Annelids, arthropods, fish and molluscs diversified, along with new lineages of aquatic insects associated with new feeding modes that included herbivory. However, late Palaeozoic soils remained underexploited, with bioturbation largely limited to root traces and millipedes (Genise et al. 2016a).

Middle Permian channels were colonised by a deep infauna of the *Skolithos* Ichnofacies (Fitzgerald and Barrett 1986), and other trace fossils are well represented (Figure 5.15). The Capitanian and end-Permian mass extinctions adversely affected invertebrate bioturbation (Buatois et al. 2021), but burrowing may have helped Triassic vertebrates survive global warming and aridification, with abundant excavations produced by cynodonts, dicynodonts and procolophonoids (Figure 5.15c).

Cone-bearing, seed-producing gymnosperms and free-sporing lycopsids and ferns were the predominant Triassic vegetation. Aquatic insects increased in diversity, and crayfish and earthworms began to dominate soils, with increased depth and intensity of bioturbation in riparian habitats (Buatois et al. 1998a, 2016; Genise et al. 2016a). The Cretaceous radiation of angiosperms may have triggered significant changes in soil biotas, along with the Late Cretaceous advent of termites, ants, bees and cicadas (Genise et al. 2016a). Dung beetles, adapted to use large-mammal excrement, were established towards the Palaeogene–Neogene transition (Genise et al. 2000), and open-grass communities hosted mollisols – the characteristic fertile soils of grassland ecosystems (Genise et al. 2016a). By the Miocene, beavers were exhibiting woodcutting behaviour and dam building that would have transformed some fluvial landscapes (Rybczynski 2007).

Although modern rivers remain the basis for alluvial facies models, fluvial dynamics and styles would have been considerably different prior to the rise of terrestrial vegetation and other biological innovations. Thus, the age of a fluvial formation is important when inferring processes and architecture.

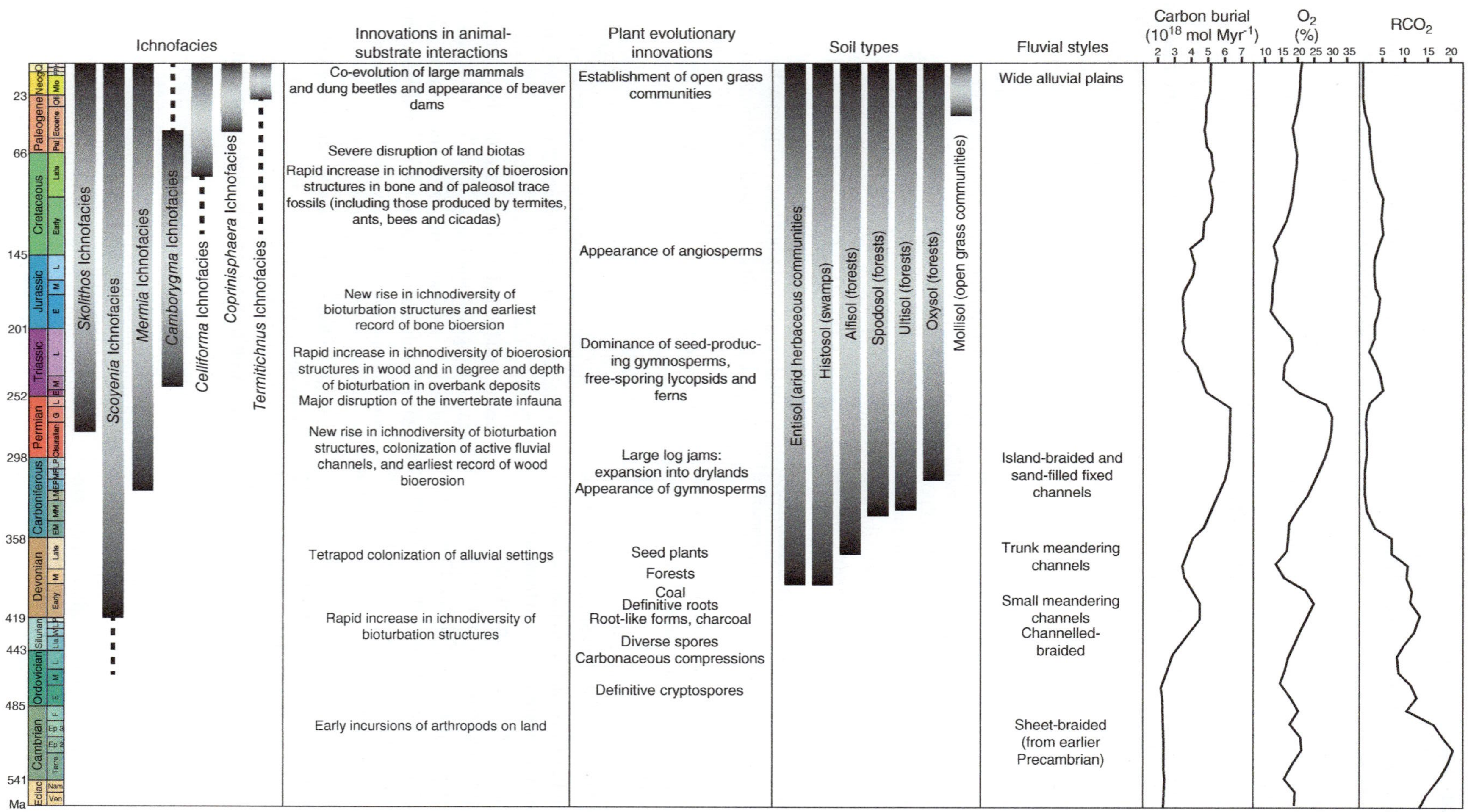

Figure 5.54 Biotic factors influencing fluvial landscapes through geologic time. Data from the following sources: ichnofacies (Buatois and Mángano 1993; Buatois et al. 1998a; Genise et al. 2016a; Minter et al. 2016a, 2016b, 2017); plant evolutionary innovations and fluvial styles (Davies and Gibling 2010a, 2011, 2013; Gibling and Davies 2012); soil types (Genise et al. 2016a); carbon burial (Berner 2003); and carbon dioxide and oxygen (Berner 2006).

5.16.2 Fluvial Biogeochemistry and the Earth's Carbon Budget

Rivers transport many organic components, and the rate at which these are passed oceanward, stored in floodplains or buried in wetlands and estuaries affects biogeochemical fluxes at the watershed scale. As such, rivers modulate some of Earth's fundamental cycles and the atmospheric CO_2 budget (Aufdenkampe et al. 2011; Supplementary Table 5.S.1: 47).

Fluvial processes affect carbon sinks and sources by modulating silicate weathering and by reworking substrates that contain photosynthetic organisms. Such processes not only consume atmospheric CO_2 but also release CO_2 by promoting oxidation of organic carbon. In a catchment, organic carbon comprises the dissolved fraction (DOC, typically finer than 0.7–0.22 μm) and the coarser particulate fraction (POC). Floods transfer POC as granular sediment to the floodplain, where it resides until it is oxidised or eroded by channel migration or undercutting. Two POC varieties are *biospheric POC*, produced by photosynthesis, and *petrogenic POC* (kerogen), eroded from organic-rich rocks. Kerogen is refractory and resists abrasion during transport, as do other forms of POC (Bouchez et al. 2010; Scheingross et al. 2019). The proportions of biospheric and petrogenic POC may be evaluated using $^{13}C/^{12}C$ systematics, but petrogenic POC has undergone considerable isotopic change during burial and is essentially ^{14}C-dead because its geological storage is typically far longer than the radionuclide's half-life. As such, water or sediment with a mixture of biospheric and petrogenic POC returns a ^{14}C age that bears little geological meaning, effectively reflecting their proportions (Torres et al. 2017).

Only a small fraction of the catchment's carbon flux reaches the ocean. Alluvial systems effectively consume CO_2 through burial of POC in abandoned channels, soils, floodbasins, lakes and peatlands, especially in major catchments such as the Amazon Basin (Aufdenkampe et al. 2011; Wójcicki 2022). Carbon release to the atmosphere is mainly through respiration and oxidation during residence in floodplains. However, processes such as meander migration, which erode bank-attached bars and floodplain sediments, demonstrably affect the quantities and residence timescales of organic carbon (Walcker et al. 2021; Ielpi et al. 2023; Barrera and Ielpi 2024). Rivers may mediate anthropogenic perturbations to the carbon cycle. For example, commercial timber harvesting, which focused historically on the riparian zone, promotes bank erosion and increases silicate weathering and organic carbon oxidation in soils, in addition to altering photosynthetic CO_2 uptake.

Many of the facies described in Table 5.4 provide insight into the carbon cycle. Gravels and sands commonly contain woody debris, and muddy and organic-rich facies additionally contain leaves, roots and peat. Freshwater carbonates (Cn, Cs) fix carbon through biotic and abiotic processes. As vegetation evolved through the Palaeozoic, a more intricate landscape developed with narrow multi-thread channels, abandoned channels and floodbasins. Such a diverse landscape would have affected groundwater biogeochemistry and promoted the stability of minerals such as calcite, siderite and pyrite during early diagenesis. These effects would in turn have influenced diagenesis during deeper burial.

5.16.3 Planetary Geology

Space probes to other parts of the Solar System reveal remarkable insights into alluvial processes (see Section 15.2.2). On Mars, formerly active alluvial systems date to 3.9 billion years ago (Balme et al. 2020) and include alluvial fans, meandering rivers with point bars and scroll bars, hillslope gullies, gigantic cataracts and canyons that dwarf the Grand Canyon on Earth. Some fluvial systems represent sustained, perennial flow with modest variation in peak discharge (Salese et al. 2020), whereas outburst flood channels carried flows estimated at a billion cubic metres a second (Baker 2001). On Titan, one of Saturn's moons, a hydrocarbon river system operates at low temperature (Burr et al. 2013). These fluvial systems and records provide a unique perspective on fluvial dynamics under conditions of gravity and fluid and atmospheric composition that differ radically from those on Earth. Evidence for meandering rivers on Mars implies that muddy sediments can constrain channels sufficiently to promote systematic migration, in the absence of vegetation (Matsubara et al. 2015; Lapôtre et al. 2019).

5.16.4 Aspects of Economic Exploration

Alluvial architecture, sediment texture and diagenesis are key factors for hydrocarbon reservoirs, aquifers, geothermal systems and CO_2 storage, as well as for the use of water and carbon dioxide to stimulate hydrocarbon production (Feldman 2024; Supplementary Table 5.S.1: 48–54). Box 5.7 illustrates some of these factors for a North Sea oilfield.

An important alluvial resource is the High Plains Aquifer of the midwestern USA, largely within the Miocene Ogallala Group. The strata cover nearly half a million square kilometres and supply 30% of irrigated US farmland. With an average saturated thickness of >60 m and zones of high transmissivity, the aquifer's drainable volume exceeds that of Lake Huron in the Great Lakes. Joeckel et al. (2014) described sandstone sheets laid down by low-sinuosity rivers, with channel and bar deposits of facies St and Sm that were rooted and burrowed. Floodplain palaeosols (Fm) and carbonate and mudstone lenses (Cs, Fl) in abandoned channels limit lateral and vertical flow, exerting a first-order control on groundwater yield and drawdown. Formerly recharged by precipitation, the aquifer is now overexploited, and the water table is dropping by >1 m per year in places. In another example of water supply, the exploitation of geothermal systems in alluvial formations depends on an efficient pumping strategy to overcome pump energy losses due to impermeable barriers (Willems et al. 2017). In these examples, architectural and palaeoflow analysis may aid in determining an efficient flow path that follows the trend of porous channel bodies.

Alluvial diamonds are mined from geomorphic units above bedrock in Miocene deposits of South Africa (Figure 5.31). In the Yukon goldfields of Canada, alluvial gold was similarly extracted from modern gravel bars and lags ('pay gravels'), traps where gold accumulated on an irregular bedrock surface, and older terraces above the level of the modern channel. The Archaean Witwatersrand goldfields of South Africa contain channel sandstones and conglomerates rich in detrital (placer) gold, which accumulated in scours and was concentrated by hydraulic sorting on plane beds, cross-bed foresets and the stoss surfaces of ripples. Other alluvial economic deposits include uranium minerals in sandstones, dimension (building) stone and aggregates from modern alluvium and glaciofluvial deposits.

Many alluvial successions contain economic coals (Figure 5.52), locally charged with coalbed methane. Coal extraction is aided by architectural analysis, facies maps and palaeoflow data. Channels commonly split the seams through avulsion or eroded the original peat to create 'washouts'. Porous channel bodies may flood mines with groundwater, connate brine and oil. In many underground mines, cemented channel bodies support shafts and tunnels (plural), but during retreat longwall mining, they may cantilever and crush the hydraulic supports that protect the coal face. Where sandstone bodies rest on the coal (e.g. Figure 5.22d), sparks from mining equipment risk methane ignition and explosions. On mine floors, scroll bars ('stone rolls') and palaeosols with swelling clays may generate unstable floors that affect the operation of equipment. In opencast mines, the strength of the highwall strata is an important factor in development.

5.16.4.1 The Challenges of Subsurface Correlation

Subsurface interpretations are challenging because of sparse information and the inherent architectural complexity of alluvial bodies (Bridge 2001; see Box 5.7), although seismic profiles and timeslices may inform the location and geometry of channel bodies. A variety of models may be used to assist correlation (see overviews in Bridge and Tye 2000; Tye 2013; Pyrcz et al. 2015).

Structure-imitating (object-based) models use the dimensions of analogues to condition reservoir models. The analogues include channels and sediment bodies in modern river plains (Tye 2004) and in outcrops and subsurface settings (Reynolds 1999; Gibling 2006; Keogh et al. 2007). For example, Gibling (2006) assessed the aspect ratio for channel bodies and valley fills of varied type, based largely on

Box 5.7 Architecture of a Triassic Fluvial Reservoir

The Skagerrak Formation occupies a rift system below the North Sea, where dryland fluvial deposits host oilfields sourced from Jurassic strata (McKie 2011). A lower unit comprises sandy terminal splays and desiccated playa lakes, and an upper unit comprises stacked channel belts up to 20 m thick and 3 km wide, laid down by low-sinuosity rivers (a, b). Borehole images show accretion surfaces generated by mid-channel and bank-attached bars. Correlation is difficult because channel-body width is similar to the well spacing.

Reservoir properties (C) are influenced by grain size, sorting, clay content and early cements in calcic palaeosols and groundwater carbonate. Amalgamated sand bodies in the upper unit have good lateral connectivity, but low-permeability barriers limit vertical connectivity and divide the reservoir into compartments with varied pressures and rates of pressure decline during production. Oil initially flowed from the highly permeable thalweg and lower bar deposits, but over time finer grained and less permeable upper bar and splay deposits released oil into more permeable pathways and maintained production. The unusually high porosity in these deep reservoirs (up to 35% at >3500 m below the sea floor) reflects the preservation of primary porosity where pore-fluid overpressures impeded compaction and early chlorite overgrowths impeded macroquartz cementation at depth.

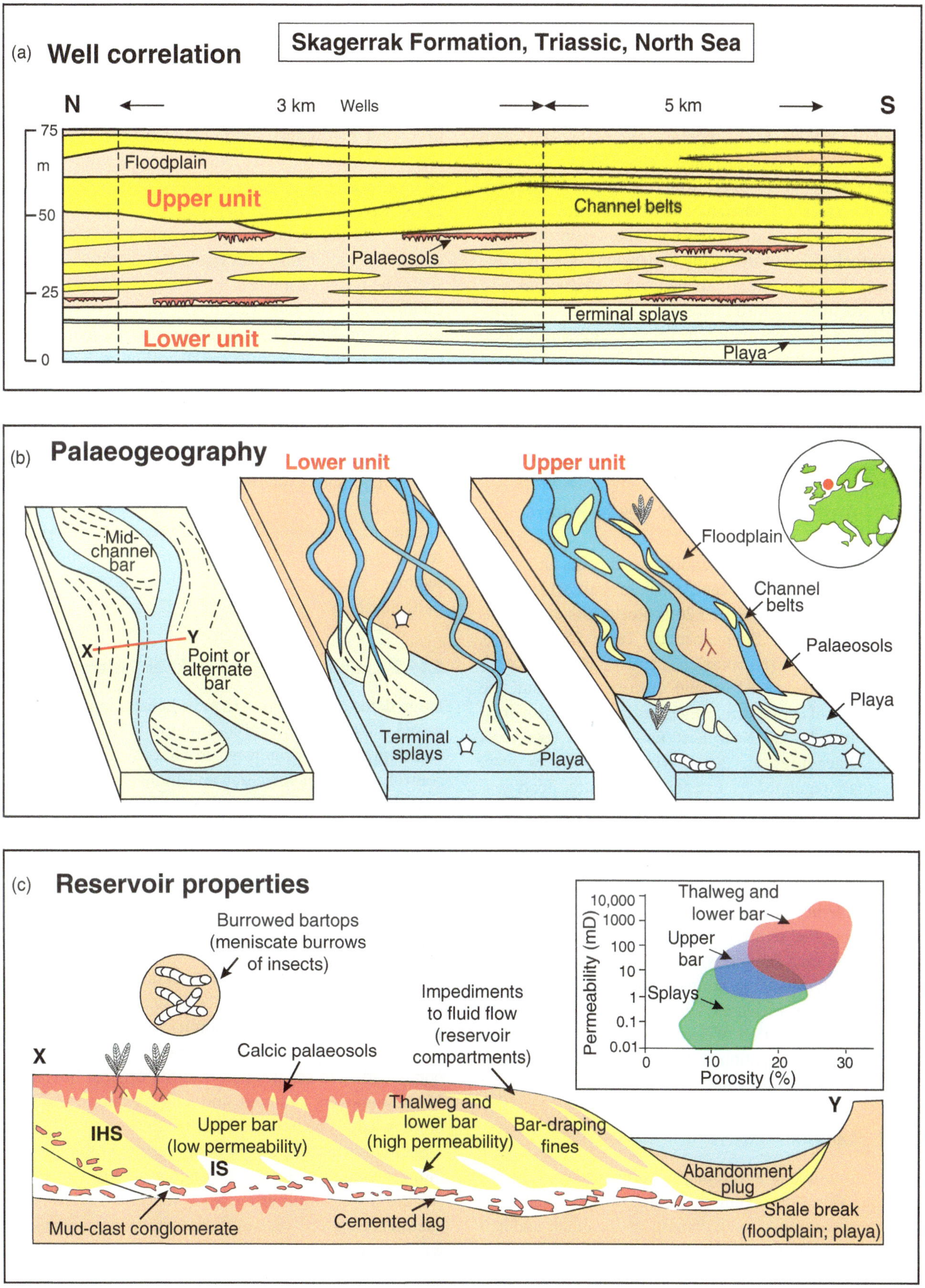
(a) Well correlation
Skagerrak Formation, Triassic, North Sea
N
3 km
Wells
5 km
S
75
m
50
25
0
Floodplain
Upper unit
Channel belts
Palaeosols
Terminal splays
Lower unit
Playa
(b) Palaeogeography
Lower unit
Upper unit
Mid-channel bar
X
Y
Point or alternate bar
Terminal splays
Playa
Floodplain
Channel belts
Palaeosols
Playa
(c) Reservoir properties
Burrowed bartops (meniscate burrows of insects)
Calcic palaeosols
Impediments to fluid flow (reservoir compartments)
X
IHS
Upper bar (low permeability)
IS
Thalweg and lower bar (high permeability)
Bar-draping fines
Y
Abandonment plug
Mud-clast conglomerate
Cemented lag
Shale break (floodplain; playa)
Thalweg and lower bar
Upper bar
Splays
Permeability (mD)
10,000
1000
100
10
1
0.1
0.01
0
10
20
30
Porosity (%)

outcrops that yield information about thickness and true width. Selecting an appropriate analogue requires an assessment of fluvial style and planform for the target reservoir, which may not be possible. *Process-based models* use mathematical descriptions to simulate fluvial processes of erosion, transport and accumulation, exploring system behaviour rather than reproducing particular cases (Bridge and Tye 2000; Karssenberg et al. 2001). *Rule-based models* (Pyrcz et al. 2015) draw on rules governing depositional processes to generate a time sequence (forward model), place objects in 3D space and constrain their fills.

Nevertheless, quantifying the dimensions of subsurface sediment bodies remains challenging. Bridge and Tye (2000) provided an instructive example from the Cretaceous Travis Peak Formation of Texas. Using object-based models, the channel bodies were initially estimated to be 4.8–9.6 km wide, but a process-based reassessment yielded a far narrower width range, between 0.4 and 1.7 km (see Colombera et al. 2014 for a review of these estimates).

5.17 Anthropogenic Changes to Rivers

Human activities have profoundly altered rivers and floodplains since the early Holocene (Macklin et al. 2012; Brown et al. 2013; Gibling 2018; Best 2019; Roebroeks et al. 2021; Syvitski et al. 2022; Supplementary Table 5.S.1: 55–60), especially since 1950 and possibly for >100,000 years. Landscape alteration centred around the use of fire and the development of agriculture, which has operated as an organised system in the Near East for the past 9800 years (Arranz-Otaegui et al. 2016). Thereafter, rivers were increasingly regulated for irrigated agriculture, flood control and navigation. Sedimentation rates increased as hillslopes collapsed through landslides and debris flows, weakened by deforestation and the planting of crops with ineffective root systems (Figure 5.55a). These processes generated *legacy sediments* that yield palynomorphs of cereal crops, the remains of domesticated animals, and technofossils such as pottery and brick. As anthropogenic influence intensified, a coupled interaction developed between human activities and riverine environments, including the formation of anthropogenic soils in South America through irrigation and mixing with kitchen waste (Schmidt et al. 2014; Goodbred et al. 2020).

5.17.1 Landscape Effects

An early example of anthropogenic influence comes from Ethiopia, where the Axum civilization was established in the first millennium CE with domesticated animals and irrigated farming (Butzer 1981; Darbyshire et al. 2003). The main settlement was largely abandoned by 800 CE, and later fluctuations in rainfall, vegetation and population contributed to the collapse of agricultural terraces, gully erosion and aggradation from soil wash in gullies and channels. Figure 5.27e illustrates gullying and aggradation in the Colluvial Facies Association, with dates of 1320–1670 CE from Unit 2. Although anthropogenic and climatic effects are intertwined, the gully fill of Unit 2 may match periods of depopulation and abandonment of farmland (Butzer 1981). Recent landscape degradation is demonstrated by Italian army cans of 1920s–30s vintage in Unit 4, after which the Unit 5 gully was cut to 5 m depth.

A more recent example is the Sacramento River system of California. In the river catchment, hydraulic mining that followed the 1849 gold rush released more than a billion cubic metres of debris into the river (Figure 5.55b; Gilbert 1917), raising flood levels and burying farmland in mining debris. River valleys were choked with some 20 m of sediment. After hydraulic mining was outlawed in 1884, channels cut down and terraces formed as sediment supply waned. Landforms that normally take centuries or millennia to form were generated within decades. Enhanced sedimentation from flash floods is a major problem worldwide, resulting from upland deforestation, bedrock mining and sediment extraction from river channels, as in parts of Bangladesh (Islam et al. 2022) and Thailand (Figure 5.36e).

5.17.2 Influence of Dams on River Systems

Large dams (defined as >15 m in height) number an estimated 58,000 worldwide and, along with many smaller dams, provide hydropower, irrigation, navigation, flood control and water for cities and industry. However, they have profoundly damaged river systems. An example is the Glen Canyon Dam on the Colorado River in Arizona (Figure 5.55c), which impounds Reservoir Powell, 300 km long. Although the reach below the dam appears pristine (Figure 5.55d), the river transmits a regulated flow of cold reservoir water that damages the ecosystem, which is now divided into small compartments by the many dams on the Colorado system. The reservoir loses large volumes of water through evaporation in the desert setting and by percolation into the surrounding, porous Navajo Sandstone. By 1840, in the northeastern USA, >65,000 dams with millponds had been constructed, and the original fluvial style of anastomosing channels and valley swamps passed out of knowledge (Walter and Merritts 2008).

Sediment trapping in reservoirs results in the release of 'hungry water', as in the Saskatchewan River of Canada (Figure 5.55e; Smith et al. 2016) where a formerly sandy channel below a power canal and turbines is now armoured with boulders. The river has incised for 80 km downstream,

Figure 5.55 Anthropogenic effects on rivers. (a) Landslide in the Himalayan Foothills, Sikha, Nepal, with house at upper left for scale. Although the landslide history is unclear, deforestation of the land for agriculture and terracing probably enhanced hillslope collapse in a monsoon setting and an area subject to large earthquakes. (b) Hydraulic mining in Sacramento River system, California, USA, in the 1880s, using water cannons supplied from dams to wash alluvium for gold. (c) Glen Canyon Dam and Reservoir Powell, Arizona, USA, designed as a 'cash register' hydropower dam to fund irrigation dams in the Colorado River Basin. The dam is one of many that divide the river and its ecosystems into compartments. The reservoir took 17 years to fill, during which time the Colorado River Delta (already depleted by water loss for irrigation) received virtually no water. (d) Horseshoe Bend downstream from the Glen Canyon Dam. The apparently pristine Colorado River is a regulated flow of cold reservoir water that damages the ecosystem. (e) 'Hungry water' re-entering the Saskatchewan River from the power canal and turbines below the E.B. Campbell Dam, Saskatchewan, Canada. Boils on the water surface denote strong turbulence. Formerly sandy, the riverbed is now armoured with coarse gravel. Note the flocks of pelicans and gulls feeding from the dead and fragmented fish emerging from the turbines. (f) Roaring River Fan, Colorado, USA, generated by collapse of the Lawn Lake Dam in 1982. Flows traveling at 15 m s^{-1} cut down 24 m in the upland valley, and sediment-gravity flows, unconfined sheet flows and stream flows generated an alluvial fan with 14 m of gravel and logs within a few hours. With permission from Martin Gibling. D by Carleton E. Watkins, USGS, public domain. *Source:* Gibling (2021) / with permission from Dunedin Academic Press.

contributing to reduced flood heights across the adjacent wetlands (Smith et al. 2016). Irrigation takeout at dams has resulted in many *anthropogenic seasonal rivers*. For example, the lower Yellow River in China experienced 226 no-flow days in 1997 (Xu 2004), whereas no-flow was previously recorded only 10× in four millennia. The collapse of dams such as the Lawn Lake Dam in Colorado in 1982 may cause substantial geomorphic change within a few hours (Figure 5.55f). Box 5.8 shows an example from the many smaller barriers that affect rivers.

A basin-scale example of anthropogenic river effects is the Mississippi system (Blum and Roberts 2009; Bentley et al. 2016; Edmonds et al. 2023). Large sediment volumes are trapped in reservoirs behind about 40,000 dams, especially behind large dams on the Missouri and Tennessee rivers. By the mid-1940s, most upland source areas were no longer delivering substantial sediment volumes to the lower Mississippi. Although the river still carries a substantial suspended load, much of the sediment originates from scouring of the riverbed by energetic, channelised water below the dams. Levees disconnect the sediment-bearing channel from the delta wetlands, starving the delta of vital sediment in the face of rising sea level, and hydrocarbon extraction accentuates compaction and enhances subsidence. The lower river system is adjusting its equilibrium profile, evolving to a supply-limited system. Many rivers, including the Ganga and Brahmaputra (Raff et al. 2023), face similar challenges in sustaining sediment supply.

Box 5.8 Barriers to River Flow

Humans have built structures on rivers since Neolithic times. On European rivers, more than one million instream barriers, mostly <2 m high, have been mapped by researchers who walked the river courses (Belletti et al. 2020). Such barriers break channel connectivity and affect sediment transport through upstream trapping and the downstream release of energetic, erosive water. They also divide the river into compartments that isolate aquatic organisms into small populations – a major ecological concern. Most barriers were built to control and divert flow (mills and navigation locks) or to stabilise channels and raise water levels (weirs and sluice gates). The photo shows a weir with a fall of <2 m and a lock system for navigation on the Thames at Goring, UK. Constructed in the 1500s to provide a head of water for a mill, this is one of the 45 locks on the river spaced about 8 km apart, many of which lack fish passages.

Source: Martin Gibling (Chapter Contributor).

5.17.3 Sand Mining and Fluvial Hazards

The mining of river sand for construction (often unregulated) greatly affects many rivers. Valued for its chlorine-free and angular nature, river sand is mined by subaqueous dredging or during dry periods. Sand extraction from rivers and coasts now greatly exceeds the global discharge of fluvial sediment. In the lower Mekong River, sand mining increases turbulence and scour, changes the channel geometry and destabilises riverbanks. Anthropogenic bedforms generated by extraction include linear scours and pockmarks up to 70 m wide and 10 m deep (Hackney et al. 2020).

Fluvial hazards have been greatly accentuated by anthropogenic effects. These include deforestation and changes in flood magnitude and frequency associated with climate change, as well as flash floods (Figure 5.8f,g) and flood waves from dam failures (Figure 5.55f). Destabilised hillslopes collapse through landslides, debris flows and bank slumps (Figures 5.19a, 5.33c,d, 5.36e, and 5.55a), especially in earthquake-prone areas.

The magnitude of these and many other anthropogenic changes on river sedimentology worldwide cannot be overestimated. Riverbeds coarsen where alluvium is eroded across farmland (Brooks et al. 2003) and fine where barriers impound water and cause deposition of fines, affecting fish and arthropod habitat (Lenders et al. 2016). Changes in discharge, flood patterns and sediment supply in regulated rivers affect the relative abundance of bedforms and geomorphic units (Erskine et al. 2017). Vegetation-rich facies (Gv, Sv, Fv and Vc) have been affected by deforestation, draining of swamps, loss of beaver dams and the widespread removal of woody 'snags' to improve navigation. And technofossils have become major sediment components, as in the case of microplastics (<5 mm in size), which are now abundant components in many European rivers (Kiss et al. 2022). Consequently, most modern rivers are problematic analogues for the ancient record, and any study of modern rivers and Holocene alluvium should evaluate possible anthropogenic changes.

5.17.4 River Restoration

Many modern rivers have largely lost their ecosystem function and are considered *zombie rivers*, strangled within artificial corridors (Brierley and Fryirs 2022). Restoration of the ecological integrity of a river corridor involves re-establishing the system's connectivity and *natural flow regime*, defined by the flow magnitude, frequency, duration, timing and rate of change (Poff et al. 1997). Much depends on the river's response to *disturbance* (episodic or continuous, moderate or extreme) and the *resilience* of the river corridor (the ability to recover to pre-existing conditions), which is often linked to the bank vegetation and woody debris. Such restoration has a foundation in facies analysis.

Regenerating the assemblages and patterns of geomorphic units is crucial for process-based river management (Brierley and Fryirs 2022). Although a restored river may not replicate exactly the former river corridor, it may be dynamically stable and ecologically functional, recognising the co-evolution of human and river systems in a *sociogeomorphic* understanding (Ashmore 2015). River sediment has a 'social life' (Parrinello and Kondolf 2021) with a context in human history.

5.18 Areas of Future Research

The facies approach to alluvial sedimentology is process based and requires a strong foundation in the physical and biological processes that generate facies. Consequently, our analysis is only as good as our understanding of the processes and facies that anchor the analytical pyramid. We draw on an assessment by Best and Fielding (2019) in suggesting some research recommendations.

Cross-beds generated by dune migration are ubiquitous in bars and channel deposits, but despite early studies (Sorby 1859), many aspects of dune scaling, geometry and amalgamation remain obscure. Many dunes have low-angle leesides (in places only a few degrees: Cisneros et al. 2020), and similar surfaces may have been misidentified in the alluvial record. Cross-beds may be preferentially preserved during flood recession and times of flow variability, when bedforms are out of equilibrium with the flow and adjusting to changing rates of sediment transport (Leary and Ganti 2020; Lyster et al. 2022). New approaches are quantifying bedload transport rates by integrating slope–velocity relationships, cross-bed thickness and bar heights and grain size (Mahon and McElroy 2018), with considerable application to the ancient record.

Fine-grained sediment influences bedforms, adds cohesion to bed and bank sediment and affects channel morphology. Bedform-existence diagrams may need to be modified because mixed sand and mud behaves differently from sediment with a limited grain-size range (Baas et al. 2016). Aided by natural organic chemicals, flocculation of river clay is probably important but is poorly understood. The amount of fine sediment preserved in riverbeds and estuaries may be a proxy for the amount supplied to ancient rivers (Brückner et al. 2021; Wysocki and Hajek 2021), of importance for understanding how the earliest vegetation affected weathering and landscapes. We also need to explore how river and bedform dynamics and sediment cohesion are affected by different types of vegetation, microbial mats and extracellular polymeric substances (EPS) secreted by microorganisms (van de Lageweg et al. 2018).

On a broader scale, more studies are needed of large rivers and of alluvial systems in seasonal and dryland settings. Linking surface and sub-surface records in modern rivers requires GPR and other techniques to assess the preservation potential of geomorphic units, for comparison with the ancient record. For fluvial systems in deep time, we need to know more about non-vegetated Precambrian and early Palaeozoic systems; how the rise of land plants affected biogeochemical fluxes; and how the evolution of angiosperm plants in the Mesozoic and grasses in the Cenozoic affected rivers. Did major extinctions reduce vegetation and terrestrial animals sufficiently to affect fluvial systems globally?

To counteract signal shredding, more alluvial archives need to be assembled to generate proxy tectonic, climatic and other environmental records. Finally, big-data analysis (e.g. Galeazzi et al. 2021) could usefully include more about river systems and, in conjunction with field analysis and flume, tank and numerical experiments, could improve predictive models for fluvial behaviour.

Further Reading

River dynamics

Ashworth, P.J. and Lewin, J. (2012). How do big rivers come to be different? *Earth-Science Reviews* 114: 84–107.

An evaluation of the world's large transcontinental rivers, and why they are so different from smaller rivers.

Blum, M.D. and Törnqvist, T.E. (2000). Fluvial responses to climate and sea-level change: a review and look forward. *Sedimentology* 47: 2–48.

A thoughtful account of how rivers respond to climate and base-level, with examples from the Gulf of Mexico and the Rhine-Meuse Delta.

Church, M. (2006). Bed material transport and the morphology of alluvial river channels. *Annual Review of Earth and Planetary Sciences* 34: 325–354.

An authoritative review of bed sediment transport in rivers, and how sediment calibre affects the form of alluvial channels.

Fryirs, K.A. and Brierley, G.J. (2018). What's in a name? A naming convention for geomorphic river types using the River Styles Framework. *PLoS One* 13 (e020): 1909.

The River Style Framework, with a consistent geomorphic approach to river description that can be applied to the ancient record.

Hansford, M.R., Plink-Björklund, P., and Jones, E.R. (2020). Global quantitative analyses of river discharge variability and hydrograph shape with respect to climate types. *Earth-Science Reviews* 200: 102977.

How variable river discharge correlates with global hydroclimate, recognising patterns of river behaviour that influence facies and architecture.

Weissmann, G.S., Hartley, A.J., Scuderi, L. A. et al. (2015). Fluvial geomorphic elements in modern sedimentary basins and their potential preservation in the rock record: a review. *Geomorphology* 250: 187–219.

How geomorphic elements that include megafans, smaller alluvial fans, axial and tributary rivers work together to fill sedimentary basins.

Anthropogenic effects on rivers

Best, J.L. (2019). Anthropogenic stresses on the world's big rivers. *Nature Geoscience* 12: 7–21.

An authoritative paper on the wide-ranging effects of humans on rivers, now pushed to the brink of ecosystem collapse over the past centuries.

Vegetation and rivers

Corenblit, D. and Steiger, J. (2009). Vegetation as a major conductor of geomorphic changes on the Earth surface: toward evolutionary geomorphology. *Earth Surface Processes and Landforms* 34: 891–896.

The rise of vegetation in the Palaeozoic, which profoundly affected rivers and generated a tight coupling between terrestrial biology and geomorphology.

Wohl, E. (2013). Floodplains and wood. *Earth-Science Reviews* 123: 194–212.

How wood interacts with fluvial systems, a vital process for channel and floodplain ecosystems.

Ichnofossils

Buatois, L.A. and Mángano, M.G. (2004). Ichnology of fluvio-lacustrine environments: animal-substrate interactions in freshwater ecosystems. In: *The Application of Ichnology to Palaeoenvironmental and Stratigraphic Analysis*, vol. 228 (ed. D. McIlroy), 311–333. Geological Society, London, Special Publication.

A comprehensive analysis of ichnofacies and their host sediments across a range of alluvial environments, with applications to sequence stratigraphy.

Genise, J.F. (2017). *Ichnoentomology: Insect Traces in Soils and Paleosols*, Topics in Geobiology, vol. 37. Springer.

Bioturbation in soils and palaeosols by insects (a major terrestrial group), including fascinating structures generated by bees, ants, termites, and dung beetles.

References

Abbe, T.B. and Montgomery, D.R. (2003). Patterns and processes of wood debris accumulation in the Queets river basin, Washington. *Geomorphology* 51: 81–107.

Abdala, F., Cisneros, J.C., and Smith, R.M. (2006). Faunal aggregation in the Early Triassic Karoo Basin: earliest evidence of shelter-sharing behavior among tetrapods? *Palaios* 21: 507–512.

Abernethy, B. and Rutherfurd, I.D. (1998). Where along a river's length will vegetation most effectively stabilise stream banks? *Geomorphology* 23: 55–75.

Abernethy, B. and Rutherfurd, I.D. (2000). The effect of riparian tree roots on the mass-stability of riverbanks. *Earth Surface Processes and Landforms* 25: 921–937.

Abrahami, R., Huyghe, P., van der Beek, P. et al. (2018). Late pleistocene – Holocene development of the Tista megafan (West Bengal, India): ^{10}Be cosmogenic and IRSL age constraints. *Quaternary Science Reviews* 185: 69–90.

Adams, M.M. and Bhattacharya, J.P. (2005). No change in fluvial style across a sequence boundary, Cretaceous Blackhawk and Castlegate Formations of central Utah, USA. *Journal of Sedimentary Research* 75: 1038–1051.

Alexander, J., Fielding, C.R., and Pocock, G.D. (1999a). Flood behaviour of the Burdekin River, tropical north Queensland, Australia. In: *Floodplains: Interdisciplinary Approaches*, vol. 163 (ed. J. Alexander and S.B. Marriott), 27–40. London: Geological Society, Special Publications.

Alexander, J., Fielding, C.R., and Jenkins, G. (1999b). Plant-material deposition in the tropical Burdekin River, Australia: implications for ancient fluvial sediments. *Palaeogeography, Palaeoclimatology, Palaeoecology* 153: 105–125.

Allen, J.R.L. (1966). On bed forms and palaeocurrents. *Sedimentology* 6: 153–190.

Allen, J.R.L. (1979). Studies in fluviatile sedimentation: an elementary geometrical model for the connectedness of avulsion-related channel sand bodies. *Sedimentary Geology* 24: 253–267.

Allen, J.R.L. (1983a). Studies in fluviatile sedimentation: bars, bar complexes and sandstone sheets (low-sinuosity braided streams) in the Brownstones (L. Devonian) Welsh Borders. *Sedimentary Geology* 33: 237–293.

Allen, J.R.L. (1983b). A simplified cascade model for transverse stone-ribs in gravelly systems. *Proceedings of the Royal Society A* 385: 253–266.

Allen, J.P., Fielding, C.R., Rygel, M.C., and Gibling, M.R. (2013). Deconvolving signals of tectonic and climatic controls from continental basins: an example from the late Paleozoic Cumberland Basin, Atlantic Canada. *Journal of Sedimentary Research* 83: 847–872.

Almeida, R.P., Galeazzi, C., Tavares Freitas, B. et al. (2023). Statistically robust multi-scale analysis and interpretation of fluvial paleocurrents. *12th International Conference on Fluvial Sedimentology*, Riva del Garda, Italy, Abstracts with Programme.

Alonso-Zarza, A.M., Genise, J.F., and Verde, M. (2014). Paleoenvironments and ichnotaxonomy of insect trace fossils in continental mudflat deposits of the Miocene Calatayud–Daroca Basin, Zaragoza, Spain. *Palaeogeography, Palaeoclimatology, Palaeoecology* 414: 342–351.

Alqahtani, F.A., Jackson, C.A.-L., Johnson, H.D., and Som, M.R.B. (2017). Controls on the geometry and evolution of humid-tropical fluvial systems: insights from 3D seismic geomorphological analysis of the Malay Basin, Sunda Shelf, Southeast Asia. *Journal of Sedimentary Research* 87: 17–40.

Amorosi, A. and Marchi, N. (1999). High-resolution sequence stratigraphy from piezocone tests: an example from the Late Quaternary deposits of the southeastern Po Plain. *Sedimentary Geology* 128: 67–81.

Amorosi, A., Colalongo, M.L., Fusco, F. et al. (1999). Glacio-eustatic control of continental-shallow marine cyclicity from Late Quaterary deposits of the southeastern Po Plain, Northern Italy. *Quaternary Research* 52: 1–13.

Amorosi, A., Pavesi, M., Ricci Lucchi, M. et al. (2008). Climatic signature of cyclic fluvial architecture from the Quaternary of the central Po Plain, Italy. *Sedimentary Geology* 209: 58–68.

Amorosi, A., Maselli, V., and Trincardi, F. (2016a). Onshore to offshore anatomy of a late quaternary source-to-sink system (Po Plain - Adriatic Sea, Italy). *Earth-Science Reviews* 153: 212–237.

Amorosi, A., Bruno, L., Cleveland, D.M. et al. (2016b). Paleosols and associated channel-belt sand bodies from a continuously subsiding late Quaternary system (Po Basin, Italy): new insights into continental sequence stratigraphy. *Geological Society of America Bulletin* 129: 449–463.

Amorosi, A., Bruno, L., Campo, B. et al. (2017). Global sea-level control on local parasequence architecture from the Holocene record of the Po Plain, Italy. *Marine and Petroleum Geology* 87: 99–111.

Andrews, E.D. (1980). Effective and bankfull discharges of streams in the Yampa River basin, Colorado and Wyoming. *Journal of Hydrology* 46: 311–330.

Arranz-Otaegui, A., Colledge, S., Zapata, L. et al. (2016). Regional diversity on the timing for the initial appearance of cereal cultivation and domestication in southwest Asia. *PNAS* 113: 14001–14006.

Ashley, G.M. and Symposium Chairperson (1990). Classification of large-scale subaqueous bedforms: a new look at an old problem. *Journal of Sedimentary Petrology* 60: 160–172.

Ashmore, P. (2015). Towards a sociogeomorphology of rivers. *Geomorphology* 251: 149–156.

Ashworth, P.J. and Lewin, J. (2012). How do big rivers come to be different? *Earth-Science Reviews* 114: 84–107.

van Asselen, S., Stouthamer, E., and Smith, N.D. (2010). Factors controlling peat compaction in alluvial floodplains: a case study in the cold-temperate Cumberland Marshes, Canada. *Journal of Sedimentary Research* 80: 155–166.

Assine, M.L. (2005). River avulsions on the Taquari megafan, Pantanal wetland, Brazil. *Geomorphology* 70: 357–371.

Atchley, S.C., Nordt, L.C., and Dworkin, S.I. (2004). Eustatic control on alluvial sequence stratigraphy: a possible example from the Cretaceous-Tertiary transition of the Tornillo Basin, Big Bend National Park, west Texas, USA. *Journal of Sedimentary Research* 74: 391–404.

Aufdenkampe, A.K., Mayorga, E., Raymond, P.A. et al. (2011). Riverine coupling of biogeochemical cycles between land, oceans, and atmosphere.

Frontiers in Ecology and the Environment 9: 53–60.

Baas, J.H., Best, J.L., and Peakall, J. (2016). Predicting bedforms and primary current stratification in cohesive mixtures of mud and sand. *Journal of the Geological Society* 173: 12–45.

Baker, V.R. (2001). Water and the martian landscape. *Nature* 412: 228–236.

Baldwin, B. and Butler, C.O. (1985). Compaction curves. *AAPG Bulletin* 69: 622–626.

Balla, A., Mohsen, A., Gönczy, S., and Kiss, T. (2022). Spatial variations in microfiber transport in a transnational river basin. *Applied Sciences* 12: https://doi.org/10.3390/app122110852.

Balme, M.R., Gupta, S., Davis, J.M. et al. (2020). Aram Dorsum: an extensive mid-Noachian age fluvial depositional system in Arabia Terra, Mars. *Journal of Geophysical Research: Planets* 125: e2019JE006244.

Barrera, M. and Ielpi, A. (2024). Floodplain organic-carbon dynamics modulated by meandering-channel migration: Vermilion River, Ontario, Canada. In: *Meandering Streamflows: Patterns and Processes Across Landscapes and Scales*, vol. 540 (ed. A. Finotello, P.R. Durkin, and Z. Sylvester), 145–162. Geological Society of London, Special Publication.

Batson, P.A. and Gibling, M.R. (2002). Architecture of channel bodies and paleovalley fills in high-frequency Carboniferous sequences, Sydney Basin, Atlantic Canada. *Bulletin of Canadian Petroleum Geology* 50: 138–157.

Belletti, B., Garcia de Leaniz, C., Jones, J. et al. (2020). More than one million barriers fragment Europe's rivers. *Nature* 588: 436–441.

Bellosi, E., Genise, J.F., González, M., and Verde, M. (2016). Paleogene laterites bearing the highest insect ichnodiversity in paleosols. *Geology* 44 (2): 119–122.

Benito, G., Macklin, M.G., Zielhofer, C. et al. (2015). Holocene flooding and climate change in the Mediterranean. *Catena* 130: 13–33.

Bentley, S.J. Sr., Blum, M.D., Maloney, J. et al. (2016). The Mississippi River source-to-sink system: perspectives on tectonic, climatic, and anthropogenic influences, Miocene to Anthropocene. *Earth-Science Reviews* 153: 139–174.

Berner, R.A. (2003). The long-term carbon cycle, fossil fuels and atmospheric composition. *Nature* 426: 323–326.

Berner, R.A. (2006). GEOCARBSULF: a combined model for Phanerozoic atmospheric O_2 and CO_2. *Geochimica et Cosmochimica Acta* 70: 5653–5664.

Best, J.L. (2019). Anthropogenic stresses on the world's big rivers. *Nature Geoscience* 12: 7–21.

Best, J.L. and Ashworth, P.J. (1997). Scour in large braided rivers and the recognition of sequence stratigraphic boundaries. *Nature* 387: 275–277.

Best, J.L. and Fielding, C.R. (2019). Describing fluvial deposits: linking processes to deposits and stratigraphy. In: *River to Reservoir: Geoscience to Engineering*, vol. 488 (ed. P.W.M. Corbett, A. Owen, A.J. Hartley, et al.), 151–166. Geological Society of London, Special Publications.

Best, J.L., Ashworth, P.J., Bristow, C.S., and Roden, J. (2003). Three-dimensional sedimentary architecture of a large, mid-channel sand braid bar, Jamuna River, Bangladesh. *Journal of Sedimentary Research* 73: 516–530.

Bestland, E.A. (1997). Alluvial terraces and paleosols as indicators of Early Oligocene climate change (John Day Formation, Oregon). *Journal of Sedimentary Research* 67: 840–855.

Bianchi, V., Ghinassi, M., Aldinucci, M. et al. (2015). Tectonically driven deposition and landscape evolution within upland incised valleys: Ambra Valley fill, Pliocene–Pleistocene, Tuscany, Italy. *Sedimentology* 62: 897–927.

Billi, P. (2007). Morphology and sediment dynamics of ephemeral stream terminal distributary systems in the Kobo Basin (northern Welo, Ethiopia). *Geomorphology* 85: 98–113.

Billi, P. (2008). Bedforms and sediment transport processes in the ephemeral streams of Kobo basin, Northern Ethiopia. *Catena* 75: 5–17.

Blair, T.C. (1999a). Sedimentary processes and facies of the waterlaid Anvil Spring Canyon alluvial fan, Death Valley, California. *Sedimentology* 46: 913–940.

Blair, T.C. (1999b). Sedimentology of the debris-flow-dominated Warm Spring Canyon alluvial fan, Death Valley, California. *Sedimentology* 46: 941–965.

Blair, T.C. and McPherson, J.G. (1999). Grain-size and textural classification of coarse sedimentary particles. *Journal of Sedimentary Research* 69: 6–19.

Blair, N.E., Leithold, E.L., Ford, S.T. et al. (2003). The persistence of memory: the fate of ancient sedimentary organic carbon in a modern sedimentary system. *Geochimica áået Cosmochimica Acta* 67: 63–73.

Blanchet, C.L., Osborne, A.H., Tjallingii, R. et al. (2021). Drivers of river reactivation in North Africa during the last glacial cycle. *Nature Geoscience* 14: 97–103.

Blom, A., Arkesteijn, L., Chavarrias, V., and Viparelli, E. (2017). The equilibrium alluvial river under variable flow and its channel-forming discharge. *Journal of Geophysical Research - Earth Surface* 122: 1924–1948.

Blum, M.D. and Aslan, A. (2006). Signatures of climate vs. sea-level change within incised valley-fill successions: quaternary examples from the Texas Gulf Coast. *Sedimentary Geology* 190: 177–211.

Blum, M.D. and Roberts, H.H. (2009). Drowning of the Mississippi Delta due to insufficient sediment supply and global sea-level rise. *Nature Geoscience* 2: 488–491.

Blum, M.D. and Törnqvist, T.E. (2000). Fluvial responses to climate and sea-level change: a review and look forward. *Sedimentology* 47: 2–48.

Blum, M.D., Martin, J., Milliken, K., and Garvin, M. (2013). Paleovalley systems: insights from Quaternary analogs and experiments. *Earth-Science Reviews* 116: 128–169.

Bonne, K.P.M. (2014). Reconstruction of the evolution of the Niger River and implications for sediment supply to the equatorial Atlantic margin of Africa during the Cretaceous and the Cenozoic. In: *Sediment Provenance Studies in Hydrocarbon Exploration and*

Production, vol. 386 (ed. R.A. Scott, H.R. Smyth, A.C. Morton, and N. Richardson), 327–349. London: Geological Society of London, Special Publication.

Bordy, E.M. and Paiva, F. (2021). Stratigraphic architecture of the Karoo river channels at the end-Capitanian. *Frontiers in Earth Science* 8: 521766. https://doi.org/10.3389/feart.2020.521766.

Bordy, E.M., Sciscio, L., Abdala, F. et al. (2017). First Lower Jurassic vertebrate burrow from southern Africa (upper Elliot Formation, Karoo Basin, South Africa). *Palaeogeography, Palaeoclimatology, Palaeoecology* 468: 362–372.

Bouchez, J., Beyssac, O., Galy, V. et al. (2010). Oxidation of petrogenic organic carbon in the Amazon floodplain as a source of atmospheric CO_2. *Geology* 38: 255–258.

Bradley, R.W. and Venditti, J.G. (2017). Reevaluating dune scaling relations. *Earth-Science Reviews* 165: 356–376.

Braun, J., Guillocheau, F., Robin, C. et al. (2014). Rapid erosion of the Southern African Plateau as it climbs over a mantle superswell. *Journal of Geophysical Research* 119: 6093–6112.

Breda, A., Mellere, D., and Massari, F. (2007). Facies and processes in a Gilbert-delta-filled incised valley (Pliocene of Ventimiglia, NW Italy). *Sedimentary Geology* 200: 31–55.

Brenna, A., Surian, N., Ghinassi, M., and Marchi, L. (2020). Sediment-water flows in mountain streams: recognition and classification based on field evidence. *Geomorphology* 371: 107413.

Brenna, A., Martini, I., Menapace, L. et al. (2024). Imbrication fabric as a diagnostic feature for the genetic classification of gravels deposited by fluid-gravity versus sediment-gravity subaerial flows. *Earth Surface Processes and Landforms* 49: 4088–4098.

Bridge, J.S. (1985). Paleochannel patterns inferred from alluvial deposits: a critical evaluation. *Journal of Sedimentary Petrology* 55: 579–589.

Bridge, J.S. (1993). Description and interpretation of fluvial deposits: a critical perspective. *Sedimentology* 40: 801–810.

Bridge, J.S. (2001). Characterization of fluvial hydrocarbon reservoirs and aquifers: problems and solutions. *Latin American Journal of Sedimentology and Basin Analysis* 8: 87–114.

Bridge, J.S. (2003). *Rivers and Floodplains*, 491. Oxford: Blackwell.

Bridge, J.S. (2006). Fluvial facies models: recent developments. In: *Facies Models Revisited*, Special Publication 84 (ed. H. W. Posamentier and R.G. Walker), 85–170. Tulsa, OK: Society for Sedimentary Geology (SEPM).

Bridge, J.S. and Lunt, I.A. (2006). Depositional models of braided rivers. In: *Braided rivers: Process, Deposits, Ecology and Management*, vol. 36 (ed. G. H.S. Smith, J.L. Best, C.S. Bristow, and G.E. Petts), 11–50. International Association of Sedimentologists, Special Publication.

Bridge, J.S. and Mackey, S.D. (1993). A revised alluvial stratigraphy model. In: *Alluvial Sedimentation*, vol. 17 (ed. M. Marzo and C. Puigdefábregas), 319–336. Oxford: Blackwell Scientific Publications.

Bridge, J.S. and Tye, R.S. (2000). Interpreting the dimensions of ancient fluvial channel bars, channels, and channel belts from wireline-logs and cores. *AAPG Bulletin* 84: 1205–1228.

Bridge, J., Collier, R., and Alexander, J. (1998). Large-scale structure of Calamus River deposits (Nebraska, USA) revealed using ground-penetrating radar. *Sedimentology* 45: 977–986.

Brierley, G.J. (1989). River planform facies models: the sedimentology of braided, wandering and meandering reaches of the Squamish River, British Columbia. *Sedimentary Geology* 61: 17–35.

Brierley, G.J. (1991). Floodplain sedimentology of the Squamish River, British Columbia: relevance of element analysis. *Sedimentology* 38: 735–750.

Brierley, G.J. and Fryirs, K.A. (2022). Truths of the Riverscape: moving beyond command-and-control to geomorphically informed nature-based river management. *Geoscience Letters* 9: https://doi.org/10.1186/s40562-022-00223-0.

Brooke, S., Chadwick, A.J., Silvestre, J. et al. (2022). Where rivers jump course. *Science* 376: 987–990.

Brookes, I.A. (2003). Palaeofluvial estimates from exhumed meander scrolls, Taref Formation (Turonian), Dakhla Region, Western Desert, Egypt. *Cretaceous Research* 24: 97–104.

Brooks, A.P., Brierley, G.J., and Millar, R. G. (2003). The long-term control of vegetation and woody debris on channel and flood-plain evolution: insights from a paired catchment study in southeastern Australia. *Geomorphology* 51: 7–29.

Brown, D.E., Lowe, C.H., and Pase, C.P. (1979). A digitized classification system for the biotic communities of North America, with community (Series) and association examples for the Southwest. *Journal of the Arizona-Nevada Academy of Science* 14: 1–16.

Brown, A., Toms, P., Carey, C., and Rhodes, E. (2013). Geomorphology of the Anthropocene: time-transgressive discontinuities of human-induced alluviation. *Anthropocene* 1: 3–13.

Brückner, M.Z.M., McMahon, W.J., and Kleinhans, M.G. (2021). Muddying the waters: modeling the effects of early land plants in Paleozoic estuaries. *Palaios* 36: 173–181.

Bruno, D.E. and Ruban, D.A. (2017). Something more than boulders: a geological comment on the nomenclature of megaclasts on extraterrestrial bodies. *Planetary and Space Science* 135: 37–42.

Buatois, L.A. and Mángano, M.G. (1993). Ecospace utilization, paleoenvironmental trends, and the evolution of early nonmarine biotas. *Geology* 21: 595–598.

Buatois, L.A. and Mángano, M.G. (2002). Trace fossils from Carboniferous floodplain deposits in western Argentina: implications for ichnofacies models of continental environments. *Palaeogeography, Palaeoclimatology, Palaeoecology* 183: 71–86.

Buatois, L.A. and Mángano, M.G. (2004). Animal-substrate interactions in freshwater environments: applications of ichnology in facies and sequence stratigraphic analysis of fluvio-lacustrine successions. In: *The Application of Ichnology to Palaeoenvironmental and Stratigraphic Analysis*, vol. 228 (ed. D. McIlroy), 311–333. Geological Society of London Special Publication.

Buatois, L.A. and Mángano, M.G. (2007). Invertebrate ichnology of continental freshwater environments. In: *Trace Fossils Concepts, Problems, Prospects* (ed. W.I. Miller), 285–323. Amsterdam: Elsevier.

Buatois, L.A. and Mángano, M.G. (2011). *Ichnology: Organism-Substrate Interactions in Space and Time*, 358. Cambridge: Cambridge University Press.

Buatois, L.A., Jalfin, G., and Acenolaza, F. G. (1997a). Permian nonmarine invertebrate trace fossils from southern Patagonia, Argentina: ichnologic signatures of substrate consolidation and colonization sequences. *Journal of Paleontology* 71: 324–336.

Buatois, L.A., Mángano, M.G., Maples, C. G., and Lanier, W.P. (1997b). The paradox of nonmarine ichnofaunas in tidal rhythmites: integrating sedimentologic and ichnologic data from the late Carboniferous of eastern Kansas, USA. *Palaios* 12: 467–481.

Buatois, L.A., Mángano, M.G., Genise, J. F., and Taylor, T.N. (1998a). The ichnologic record of the continental invertebrate invasion: evolutionary trends in environmental expansion, ecospace utilization, and behavioral complexity. *Palaios* 13: 217–240.

Buatois, L.A., Mángano, M.G., Maples, C. G., and Lanier, W.P. (1998b). Ichnology of an Upper Carboniferous fluvio-estuarine paleovalley: the Tonganoxie Sandstone, Buildex Quarry, eastern Kansas, USA. *Journal of Paleontology* 72: 152–180.

Buatois, L.A., Labandeira, C.C., Mángano, M.G. et al. (2016). The Mesozoic lacustrine revolution. In: *The Trace-Fossil Record of Major Evolutionary Changes, Vol. 2: Mesozoic and Cenozoic, Topics in Geobiology*, vol. 40 (ed. M.G. Mángano and L.A. Buatois), 179–263. Springer.

Buatois, L.A., Wetzel, A., and Mángano, M.G. (2020). Trace-fossil suites and composite ichnofabrics from meandering fluvial systems: the Oligocene Lower Freshwater Molasse of Switzerland. *Palaeogeography, Palaeoclimatology, Palaeoecology* 558: 109944.

Buatois, L.A., Borruel-Abadía, V., De la Horra, R. et al. (2021). Impact of Permian mass extinctions on continental invertebrate infauna. *Terra Nova* 33: 455–464.

Buatois, L.A., Davies, N.S., Gibling, M.R. et al. (2022). The invasion of the land in deep time: integrating Paleozoic records of paleobiology, ichnology, sedimentology, and geomorphology. *Integrative and Comparative Biology* 62: 297–331.

Buck, S.G. (1983). The Saaiplaas Quartzite Member: a braided system of gold- and uranium-bearing channel placers within the Proterozoic Witwatersrand Supergroup of South Africa. In: *Modern and Ancient Fluvial Systems* (ed. J.D. Collinson and J. Lewin), 549–562. International Association of Sedimentologists.

Burr, D.M., Perron, J.T., Lamb, M.P. et al. (2013). Fluvial features on Titan: insights from morphology and modeling. *Geological Society of America Bulletin* 125: 299–321.

Butler, D.R. (2006). Human-induced changes in animal populations and distributions, and the subsequent effects on fluvial systems. *Geomorphology* 79: 448–459.

Butler, D.R. (2018). Zoogeomorphology in the Anthropocene. *Geomorphology* 303: 146–154.

Butler, D.R. and Malanson, G.P. (2005). The geomorphic influences of beaver dams and failures of beaver dams. *Geomorphology* 71: 48–60.

Butzer, K.W. (1981). Rise and fall of Axum: a geo-archaeological interpretation. *American Antiquity* 46: 471–495.

Campo, B., Bruno, L., and Amorosi, A. (2020). Basin-scale stratigraphic correlation of late Pleistocene - Holocene (MIS 5e-MIS 1) strata across the rapidly subsiding Po Basin (northern Italy). *Quaternary Science Reviews* 237: 106300.

Candel, J.H.J., Kleinhans, M.G., Makaske, B., and Wallinga, J. (2020a). Predicting river channel pattern based on stream power, bed material and bank strength. *Progress in Physical Geography: Earth and Environment* 45: 253–278. https://doi.org/10.1177/0309133320948831.

Candel, J.H.J., Makaske, B., Kijm, N. et al. (2020b). Self-constraining of low-energy rivers explains low channel mobility and tortuous planforms. *The Depositional Record* 6: 648–669. https://doi.org/10.1002/dep2.112.

Cantil, L.F., Sánchez, M.V., Bellosi, E.S., and Genise, J.F. (2018). Beyond *Coprinisphaera*: fossil nests of dung beetles. *Lethaia* 51: 444–455.

Cantil, L.F., Bellosi, E.S., Sánchez, M.V. et al. (2020). The earliest burst of necrophagous dung beetles in South America revealed by the Cenozoic record of *Coprinisphaera*. *Lethaia* 53: 421–438.

Capezzuoli, E., Gandin, A., and Pedley, M. (2014). Decoding tufa and travertine (fresh water carbonates) in the sedimentary record: the state of the art. *Sedimentology* 61: 1–21.

Cardenas, B.T., Mohrig, D., Goudge, T.A. et al. (2020). The anatomy of exhumed river-channel belts: bedform to belt-scale river kinematics of the Ruby Ranch Member, Cretaceous Cedar Mountain Formation, Utah, USA. *Sedimentology* 67: 3655–3682.

Carling, P.A. (1988). The concept of dominant discharge applied to two gravel-bed streams in relation to channel stability thresholds. *Earth Surface Processes and Landforms* 13: 355–367.

Carling, P.A. (1996). Morphology, sedimentology and palaeohydraulic significance of large gravel dunes: Altai Mountains, Siberia. *Sedimentology* 43: 647–664.

Carling, P.A. (1999). Subaqueous gravel dunes. *Journal of Sedimentary Research* 69: 534–545.

Carling, P.A. and Breakspear, R.M.D. (2006). Placer formation in gravel-bedded rivers. *Ore Geology Reviews* 28: 377–401.

Carling, P.A. and Leclair, S.F. (2019). Alluvial stratification styles in a large, flash-flood influenced dryland river: the Luni River, Thar Desert, north-west India. *Sedimentology* 66: 102–128.

Carling, P.A., Jansen, J., and Meshkova, L. (2014). Multichannel rivers: their definition and classification. *Earth Surface Processes and Landforms* 39: 26–37.

Carter, D.C. (2003). 3-D seismic geomorphology: Insights into fluvial reservoir deposition and performance, Widuri field, Java Sea. *AAPG Bulletin* 87: 909–934.

Cartigny, M.J.B., Ventra, D., Postma, G., and Van Den Berg, J.H. (2014). Morphodynamics and sedimentary structures of bedforms under supercritical-flow conditions: New insights from flume experiments. *Sedimentology* 61: 712–748.

Catena, A.M., Hembree, D.I., Saylor, B.Z. et al. (2016). Paleoenvironmental analysis of the Neotropical fossil mammal site of Cerdas, Bolivia (middle Miocene) based on ichnofossils and paleopedology. *Palaeogeography, Palaeoclimatology, Palaeoecology* 459: 423–439.

Catena, A.M., Hembree, D.I., Saylor, B.Z. et al. (2017). Paleosol and ichnofossil evidence for significant Neotropical habitat variation during the late middle Miocene (Serravallian). *Palaeogeography, Palaeoclimatology, Palaeoecology* 487: 381–398.

Catuneanu, O. (2006). *Principles of Sequence Stratigraphy*, 375. Amsterdam: Elsevier.

Catuneanu, O., Abreu, V., Bhattacharya, J. P. et al. (2009). Towards the standardization of sequence stratigraphy. *Earth-Science Reviews* 92: 1–33.

Cendón, D.I., Larsen, J.R., Jones, B.G. et al. (2010). Freshwater recharge into a shallow saline groundwater system, Cooper Creek floodplain, Queensland, Australia. *Journal of Hydrology* 392: 150–163.

Chadwick, A.J., Lamb, M.P., Moodie, A.J. et al. (2019). Origin of a preferential avulsion node on lowland river deltas. *Geophysical Research Letters* 46: 4267–4277.

Chamberlain, C.K. (1975). Recent lebensspuren in nonmarine aquatic environments. In: *The Study of Trace Fossils. A Synthesis of Principles, Problems, and Procedures in Ichnology* (ed. R.W. Frey), 431–458. New York: Springer-Verlag.

Chamberlin, E.P. and Hajek, E.A. (2019). Using bar preservation to constrain reworking in channel-dominated fluvial stratigraphy. *Geology* 47: 531–534.

Chatanantavet, P., Lamb, M.P., and Nittrouer, J.A. (2012). Backwater controls of avulsion location on deltas. *Geophysical Research Letters* 39: https://doi.org/10.1029/2011GL050197.

Chesley, J.T. and Leier, A.L. (2018). Sandstone-body variability in the medial-distal part of an ancient distributive fluvial system, Salt Wash Member of the Morrison Formation, Utah, USA. *Journal of Sedimentary Research* 88: 568–582.

Church, M. (1992). Channel morphology and typology. In: *The River's Handbook: Hydrological and Ecological Principles* (ed. P. Calow and G.E. Petts), 126–143. Oxford: Blackwell.

Church, M. (2002). Geomorphic thresholds in riverine landscapes. *Freshwater Biology* 47: 541–557.

Church, M. (2006). Bed material transport and the morphology of alluvial river channels. *Annual Review of Earth and Planetary Sciences* 34: 325–354.

Church, M. and Jakob, M. (2020). What is a debris flood? *Water Resources Research* https://doi.org/10.1029/2020WR027144.

Cisneros, J., Best, J.L., van Dijk, T. et al. (2020). Dunes in the world's big rivers are characterized by low-angle lee-side slopes and a complex shape. *Nature Geoscience* 13: 156–162.

Clemens, S.C. and Prell, W.L. (2003). A 350,000 year summer-monsoon multi-proxy stack from the Owen Ridge, Northern Arabian Sea. *Marine Geology* 201: 35–51.

Coe, J.A., Kean, J.W., Godt, J.W. et al. (2014). New insights into debris-flow hazards from an extraordinary event in the Colorado Front Range. *GSA Today* 24: 4–10.

Coleman, J.M. (1969). Brahmaputra River: channel processes and sedimentation. *Sedimentary Geology* 3: 129–239.

Collins, B.D., Montgomery, D.R., Fetherston, K.L., and Abbe, T.B. (2011). The fluvial large-wood cycle hypothesis: a mechanism for the physical and biotic structuring of temperate forested alluvial valleys in the North Pacific coastal ecoregion. *Geomorphology* 139: 460–470.

Collinson, J.D. (1978). Alluvial sediments. In: *Sedimentary Environments and Facies* (ed. H.G. Reading), 15–60. Oxford: Blackwell Scientific Publications.

Collinson, J. and Mountney, N. (2019). *Sedimentary Structures*, 4e, 292. Edinburgh: Dunedin Academic Press.

Colombera, L. and Mountney, N.P. (2023). Downstream controls on coastal plain river avulsions: a global study. *Journal of Geophysical Research Earth Surface* 128: e2022JF006772. https://doi.org/10.1029/2022JF006772.

Colombera, L., Mountney, N.P., Felletti, F., and McCaffry, W.D. (2014). Models for guiding and ranking well-to-well correlations of channel bodies in fluvial reservoirs. *AAPG Bulletin* 98: 1943–1965.

Constantine, J.A., Dunne, T., Ahmed, J. et al. (2014). Sediment supply as a driver of river meandering and floodplain evolution in the Amazon Basin. *Nature Geoscience* 7: 899–903.

Corenblit, D. and Steiger, J. (2009). Vegetation as a major conductor of geomorphic changes on the Earth surface: toward evolutionary geomorphology. *Earth Surface Processes and Landforms* 34: 891–896.

Cosarinsky, M.I. (2021). A review of micromorphological studies of ant and termite's epigean nests located in neotropical soils of Argentina. *Journal of South American Earth Sciences* 110: 103380.

Craw, D. (2010). Delayed accumulation of placers during exhumation of orogenic gold in southern New Zealand. *Ore Geology Reviews* 37: 224–235.

Cuevas Martinez, J.L., Pérez, L.C., Marcuello, A. et al. (2010). Exhumed channel sandstone networks within fluvial fan deposits from the Oligo-Miocene Caspe Formation, South-east Ebro Basin (North-east Spain). *Sedimentology* 57: 162–189.

Cyples, N.N., Ielpi, A., and Dirszowsky, R. W. (2020). Planform and stratigraphic signature of proximal braided streams: remote-sensing and ground-penetrating-radar analysis of the Kicking Horse River, Canadian Rocky Mountains. *Journal of Sedimentary Research* 90: 131–149.

D'Alessandro, A., Ekdale, A.A., and Picard, M.D. (1987). Trace fossils in fluvial deposits of the Duchesne River Formation (Eocene), Uinta Basin, Utah. *Palaeogeography, Palaeoclimatology, Palaeoecology* 61: 285–301.

Darbyshire, I., Lamb, H., and Umer, M. (2003). Forest clearance and regrowth in northern Ethiopia during the last 3000 years. *The Holocene* 13: 537–546.

Darwin, C. (1881). *The Formation of Vegetated Mould Through the Action of Worms with Observation of Their Habitats*, 328. London: Murray.

Datta, S. and Ghosh, P. (2023). Depositional environments in a continental rift basin of India during the Triassic - Jurassic transition. *12th International Conference of Fluvial Sedimentology, Riva del Garda*, Italy, Abstract with Programme.

Davies, S.J. and Gibling, M.R. (2003). Architecture of coastal and alluvial deposits in an extensional basin: the Carboniferous Joggins Formation of eastern Canada. *Sedimentology* 50: 415–439.

Davies, N.S. and Gibling, M.R. (2010a). Cambrian to Devonian evolution of alluvial systems: the sedimentological impact of the earliest land plants. *Earth-Science Reviews* 98: 171–200.

Davies, N.S. and Gibling, M.R. (2010b). Paleozoic vegetation and the Siluro-Devonian rise of fluvial lateral accretion sets. *Geology* 38: 51–54.

Davies, N.S. and Gibling, M.R. (2011). Evolution of fixed-channel alluvial plains in response to Carboniferous vegetation. *Nature Geoscience* https://doi.org/10.1038/NGEO1237.

Davies, N.S. and Gibling, M.R. (2013). The sedimentary record of Carboniferous rivers: continuing influence of land plant evolution on alluvial processes and Palaeozoic ecosystems. *Earth-Science Reviews* 120: 40–79.

Davies, N.S. and Shillito, A.P. (2021). True substrates: the exceptional resolution and unexceptional preservation of deep time snapshots on bedding surfaces. *Sedimentology* 68: 3307–3356.

Davies, N.S., Liu, A.G., Gibling, M.R., and Miller, R.F. (2016). Resolving MISS conceptions and misconceptions: a geological approach to sedimentary surface textures generated by microbial and abiotic processes. *Earth-Science Reviews* 154: 210–246.

Davies, N.S., Berry, C.M., Marshall, J.E.A. et al. (2021). The Devonian landscape factory: plant-sediment interactions in the Old Red Sandstone of Svalbard and the rise of vegetation as a biogeomorphic agent. *Journal of the Geological Society* 178: jgs2020-225.

Davis, W.M. (1909). The lessons of the Colorado Canyon. *Bulletin of the American Geographical Society* 41: 345–354.

Davis, R.B., Minter, N.J., and Braddy, S.J. (2007). The neoichnology of terrestrial arthropods. *Palaeogeography, Palaeoclimatology, Palaeoecology* 255: 284–307.

DeCelles, P.G. and Cavazza, W. (1999). A comparison of fluvial megafans in the Cordilleran (Upper Cretaceous) and modern Himalayan foreland basin systems. *Bulletin of the Geological Society of America* 111: 1315–1334.

Dennehy, K.F., Litke, D.W., and McMahon, P.B. (2002). The high plains aquifer, USA: groundwater development and sustainability. In: *Sustainable Groundwater Development*, vol. 193 (ed. K.M. Hiscock, M.O. Rivett, and R.M. Davison), 99–119. Geological Society of London, Special Publications.

Devine, L.K. and Minter, N.J. (2022). Neoichnology of amphibious arthropods: effects of subaqueous and subaerial substrate conditions on trace morphology. *Palaios* 37: 585–605.

Dietrich, W.E., Smith, J.D., and Dunne, T. (1979). Flow and sediment transport in a sand bedded meander. *Journal of Geology* 87: 305–315.

Dong, X., Wang, X., Yang, L. et al. (2024). Quantitative assessment of the erosion and deposition effects of landslide-dam outburst flood. *Eastern Himalaya. Scientific Reports* 14: 7038. https://doi.org/10.1038/s41598-024-57894-2.

Donselaar, M.E., Cuevas Gozalo, M.C., and Moyano, S. (2013). Avulsion processes at the terminus of low-gradient semi-arid fluvial systems: lessons from the Rio Colorado, Altiplano endorheic basin, Bolivia. *Sedimentary Geology* 283: 1–14.

Drabon, N. and Lowe, D.R. (2022). Progressive accretion recorded in sedimentary rocks of the 3.28-3.23 Ga Fig Tree Group, Barberton Greenstone Belt. *GSA Bulletin* 134: 1258–1276.

Dunne, K.B.J. and Jerolmack, D.J. (2020). What sets river width? Science. *Advances* 6: eabc1505.

Durkin, P.R., Hubbard, S.M., Boyd, R.L., and Leckie, D.A. (2015). Stratigraphic expression of intra-point-bar erosion and rotation. *Journal of Sedimentary Research* 85: 1238–1257.

Durkin, P.R., Boyd, R.L., Hubbard, S.M. et al. (2017). Three-dimensional reconstruction of a meander-belt evolution, Cretaceous McMurray Formation, Alberta Foreland Basin, Canada. *Journal of Sedimentary Research* 87: 1075–1099.

Durkin, P.R., Hubbard, S.M., Holbrook, J., and Boyd, R. (2018). Evolution of fluvial meander-belt deposits and implications for the completeness of the stratigraphic record. *Bulletin of the Geological Society of America* 130: 721–739.

Eaton, B.C. and Giles, T.R. (2009). Assessing the effect of vegetation-related bank strength on channel morphology and stability in gravel-bed streams using numerical models. *Earth Surface Processes and Landforms* 34: 712–724.

Edmaier, K., Burlando, P., and Perona, P. (2011). Mechanisms of vegetation uprooting by flow in alluvial non-cohesive sediment. *Hydrology and Earth System Sciences* 15: 1615–1627.

Edmonds, D.A., Martin, H.K., Valenza, J. M. et al. (2022). Rivers in reverse: upstream-migrating dechannelization and flooding cause avulsions on fluvial fans. *Geology* 50: 37–41.

Edmonds, D.A., Toby, S.C., Siverd, C.G. et al. (2023). Land loss due to human-altered sediment budget in the Mississippi River Delta. *Nature Sustainability* https://doi.org/10.1038/s41893-023-01081-0.

Ely, L.L. (1992). *Large Floods in the Southwestern United States in Relation to Late-Holocene Climatic Variations*, 326. University of Arizona.

Eriksson, K., Simpson, E.L., and Mueller, W. (2006). An unusual fluvial to tidal transition in the Mesoarchean Moodies Group, South Africa: a response to high tidal range and active tectonics. *Sedimentary Geology* 190: 13–24.

Erskine, W., Turner, L., Rose, T. et al. (2017). Bedform maintenance and pool destratification by the new environmental flows on the Snowy River downstream of the Jindabyne Dam, New South Wales. *Journal and Proceedings of the Royal Society of New South Wales* 150: 152–171.

Ethridge, F.G. and Davidson, S.K. (2011). Interpretation of ancient fluvial channel deposits: review and recommendations. In: *From River to Rock Record: The Preservation of Fluvial Sediments and Their Subsequent Interpretation*, vol. 97 (ed. C.P. North, S. K. Davidson, and S. Leleu), 9–35. Tulsa, Oklahoma: SEPM Special Publication.

Ethridge, F.G. and Schumm, S.A. (1978). Reconstructing paleochannel morphologic and flow characteristics: methodology, limitations, and assessment. In: *Fluvial Sedimentology*, vol. 5 (ed. A.D. Miall), 703–721. Canadian Society of Petroleum Geologists Memoir.

Falcon-Lang, H.J., Nelson, W.J., Elrick, S. et al. (2009). Incised channel fills containing conifers indicate that seasonally dry vegetation dominated Pennsylvanian tropical lowlands. *Geology* 37: 923–926.

Falk, A.R., Hasiotis, S.T., Gong, E. et al. (2017). A new experimental setup for studying avian neoichnology and the effects of grain size and moisture content on tracks: trials using the domestic chicken (*Gallus gallus*). *Palaios* 32: 689–707.

Famubode, O.A. and Bhattacharya, J.P. (2016). Sequence stratigraphic analysis of the youngest nonmarine seuence in the Cretaceous Ferron Notom Delta, south central Utah, USA. *Journal of Sedimentary Research* 86: 168–198.

Feldman, H.R. (2024). Fluvial systems. In: *Handbook of Siliciclastic Facies* (ed. H. R. Feldman). SEPM Concepts in Paleontology and Sedimentology, 442 pp.

Feldman, H.R., Gibling, M.R., Archer, A. W. et al. (1995). Stratigraphic architecture of the Tonganoxie Paleovalley Fill (Lower Virgilian) in Northeastern Kansas. *AAPG Bulletin* 79: 1019–1043.

Fernandes, A.M., Törnqvist, T.E., Straub, K.M., and Mohrig, D. (2016). Connecting the backwater hydraulics of coastal rivers to fluvio-deltaic sedimentology and stratigraphy. *Geology* 44: 979–982.

Fernandez, V., Abdala, F., Carlson, K.J. et al. (2013). Synchrotron reveals Early Triassic odd couple: injured amphibian and aestivating therapsid share burrow. *PLoS One* 8: e64978.

Fidolini, F., Ghinassi, M., Aldinucci, M. et al. (2013). Fault-sourced alluvial fans and their interaction with axial fluvial drainage: an example from the Plio-Pleistocene Upper Valdarno Basin (Tuscany, Italy). *Sedimentary Geology* 289: 19–39.

Fielding, C.R. (2006). Upper flow regime sheets, lenses and scour fills: extending the range of architectural elements for fluvial sediment bodies. *Sedimentary Geology* 190: 227–240.

Fielding, C.R., Alexander, J., and McDonald, R. (1999). Sedimentary facies from ground-penetrating radar surveys of the modern, upper Burdekin River of north Queensland, Australia: consequences of extreme discharge fluctuations. In: *Fluvial Sedimentology VI*, vol. 28 (ed. N.D. Smith and J. Rogers), 347–362. International Association of Sedimentologists, Special Publication.

Fielding, C.R., Frank, T.D., Birgenheier, L. P. et al. (2008). Stratigraphic imprint of the Late Palaeozoic Ice Age in eastern Australia: a record of alternating glacial and nonglacial climate regime. *Journal of the Geological Society of London* 165: 129–140.

Fielding, C.R., Allen, J.P., Alexander, J., and Gibling, M.R. (2009). A facies model for fluvial systems in the seasonal tropics and subtropics. *Geology* 37: 623–626.

Fielding, C.R., Ashworth, P.J., Best, J.L. et al. (2012). Tributary, distributary and other fluvial patterns: what really represents the norm in the continental rock record? *Sedimentary Geology* 261: 15–32.

Fielding, C.R., Alexander, J., and Allen, J.P. (2018). The role of discharge variability in the formation and preservation of alluvial sediment bodies. *Sedimentary Geology* 365: 1–20.

Fielding, C.R., Nelson, W.J., and Elrick, S. D. (2020). Sequence stratigraphy of the late Desmoinesian to early Missourian (Pennsylvanian) succession of southern Illinois: insights into controls on stratal architecture in an icehouse period of Earth history. *Journal of Sedimentary Research* 90: 200–227.

Figueiredo, J., Hoorn, C., van der Ven, P., and Soares, E. (2009). Late Miocene onset of the Amazon River and the Amazon deep-sea fan: Evidence from the Foz do Amazonas Basin. *Geology* 37: 619–622.

Filizola, N., Latrubesse, E.M., Fraizy, P. et al. (2014). Was the 2009 flood the most hazardous or the largest ever recorded in the Amazon? *Geomorphology* 215: 99–105.

Fillmore, D.L., Lucas, S.G., and Simpson, E.L. (2010). Invertebrate trace fossils in semi-arid to arid braided-ephemeral-river deposits of the Mississippian middle member of the Mauch Chunk Formation, eastern Pennsylvania, USA. *Palaeogeography, Palaeoclimatology, Palaeoecology* 292: 222–244.

Fisk, H.N. (1944). *Geological Investigation of the Alluvial Valley of the Lower Mississippi River*, 78. Vicksburg: Mississippi River Commission.

Fitzgerald, P.G. and Barrett, P.J. (1986). *Skolithos* in a Permian braided river deposit, southern Victoria Land, Antarctica. *Palaeogeography,*

Palaeoclimatology, Palaeoecology 52: 237–247.

Francalanci, S., Lanzoni, S., Solari, L., and Papanicolaou, A.N. (2020). Equilibrium cross section of river channels with cohesive erodible banks. *Journal of Geophysical Research - Earth Surface* 125: https://doi.org/10.1029/2019JF005286.

Francalanci, S., Paris, E., and Solari, L. (2021). On the prediction of settling velocity for plastic particles of different shapes. *Environmental Pollution* 290: https://doi.org/10.1016/j.envpol.2021.118068.

Friend, P.F., Slater, M.J., and Williams, R. C. (1979). Vertical and lateral building of river sandstone bodies, Ebro Basin, Spain. *Journal of the Geological Society of London* 136: 39–46.

Froese, D.G., Smith, D.G., and Clement, D. T. (2005). Characterizing large river history with shallow geophysics: Middle Yukon River, Yukon Territory and Alaska. *Geomorphology* 67: 391–406.

Fryirs, K.A. and Brierley, G.J. (2018). What's in a name? A naming convention for geomorphic river types using the River Styles Framework. *PLoS One* 13: e020 1909.

Fuglewicz, R., Ptaszyliski, T., and Rdzanek, K. (1990). Lower Triassic footprints from the Swigtokrzyskie (Holy Cross) Mountains, Poland. *Acta Palaeontologica Polonica* 35: 109–164.

Fulton, I.M., Guion, P.D., and Jones, N.S. (1995). The application of sedimentology to deep-mined coal. In: *European Coal Geology*, vol. 82 (ed. M. K.G. Whateley and D.A. Spears), 17–43. Geological Society Publishing House.

Galeazzi, C.P., Almeida, R.P., and do Prado, A.H. (2021). Linking rivers to the rock record: channel patterns and paleocurrent circular variance. *Geology* 49: 1402–1407.

Galloway, W.E., Whiteaker, T.L., and Ganey-Curry, P. (2011). History of Cenozoic North American drainage basin evolution, sediment yield, and accumulation in the Gulf of Mexico basin. *Geosphere* 7: 938–973.

Galy, V., Beyssac, O., France-Lanord, C., and Eglinton, T. (2008). Recycling of graphite during Himalayan erosion: a geological stabilization of carbon in the crust. *Science* 322: 943–945.

Ganti, V., Hajek, E.A., Leary, K. et al. (2020). Morphodynamic hierarchy and the fabric of the sedimentary record. *Geophysical Research Letters* 47: e2020GL087921 https://doi.org/10.1029/2020GL087921.

Gao, L., Wang, X., Yi, S. et al. (2018). Episodic sedimentary evolution of an alluvial fan (Huangshui catchment, NE Tibetan Plateau). *Quaternary* 1: https://doi.org/10.3390/quat1020016.

Garwood, R.J., Oliver, H., and Spencer, A.R.T. (2020). An introduction to the Rhynie chert. *Geological Magazine* 157: 47–64.

Gay, G.R., Gay, H.H., Gay, W.H. et al. (1998). Evolution of cutoffs across meander necks in Powder River, Montana, USA. *Earth Surface Processes and Landforms* 23: 651–662.

Gearon, J.H., Martin, H.K., Delisle, C. et al. (2024). Rules of river avulsion change downstream. *Nature* 634: 91–95.

Genise, J.F. and Bown, T.M. (1994a). New Miocene scarabeid and hymenopterous nests and early miocene (santacrucian) paleoenvironments, patagonian Argentina. *Ichnos* 3: 107–117.

Genise, J.E. and Bown, T.M. (1994b). New trace fossils of termites (Insecta: Isoptera) from the late Eocene-early Miocene of Egypt, and the reconstruction of ancient isopteran social behavior. *Ichnos* 3 (3): 155–183.

Genise, J.F. (2017). Ichnoentomology: insect traces in soils and paleosols. In: *Topics in Geobiology*, vol. 37. Springer.

Genise, J.F., Mángano, M.G., Buatois, L.A. et al. (2000). Insect trace fossil associations in paleosols: the *Coprinisphaera* ichnofacies. *Palaios* 15: 33–48.

Genise, J.F., Bellosi, E.S., and Gonzalez, M.G. (2004). An approach to the description and interpretation of ichnofabrics in palaeosols. In: *The Application of Ichnology to Palaeoenvironmental and Stratigraphic Analysis*, vol. 228 (ed. D. McIlroy), 355–382. Geological Society Special Publication.

Genise, J.F., Melchor, R.N., Bellosi, E.S., and Verde, M. (2010). Invertebrate and vertebrate trace fossils from continental carbonates. In: *Carbonates in Continental Settings: Facies, Environments, and Processes*, Developments in Sedimentology, vol. 61 (ed. A.M. Alonso-Zarza and L. Tanner), 319–369. Elsevier.

Genise, J.F., Cantil, L.F., Dinghi, P.A. et al. (2013). The aestivation chamber of the giant earthworm *Glossoscolex bergi* (Glossoscolecidae) in the subtropical rainforest of Misiones (Argentina). *Ichnos* 20: 116–119.

Genise, J.F., Bedatou, E., Bellosi, E.S. et al. (2016a). The Phanerozoic four revolutions and evolution of paleosol ichnofacies. In: *The Trace-fossil Record of Major Evolutionary Changes, Volume 2: Mesozoic and Cenozoic, Topics in Geobiology*, vol. 40 (ed. M.G. Mángano and L.A. Buatois), 301–370. Springer.

Genise, J.F., Cantil, L.F., and Bellosi, E.S. (2016b). Lower Paleogene complex ant nests from Argentina: evidence for early polydomy in ants? *Palaios* 31: 549–563.

Genise, J.F., Bellosi, E.S., Cantil, L.F. et al. (2022). Middle Miocene Climate Transition as reflected by changes in ichnofacies and palaeosols from Patagonia, Argentina. *Palaeogeography, Palaeoclimatology, Palaeoecology* 594: 110932.

Ghinassi, M. and Ielpi, A. (2016). Morphodynamics and facies architecture of streamflow-dominated, sand-rich alluvial fans, Pleistocene Upper Valdarno Basin, Italy. In: *Geology and Geomorphology of Alluvial and Fluvial Fans: Terrestrial and Planetary Perspectives*, vol. 440 (ed. D. Ventra and L.E. Clarke), 175–200. Geological Society, London, Special Publication.

Ghinassi, M., Ielpi, A., Aldinucci, M., and Fustic, M. (2016). Downstream-migrating fluvial point bars in the rock record. *Sedimentary Geology* 334: 66–96.

de Gibert, J.M. and Sáez, A. (2009). Paleohydrological significance of trace fossil distribution in Oligocene fluvial-fan-to-lacustrine systems of the Ebro Basin, Spain. *Palaeogeography,*

Palaeoclimatology, Palaeoecology 272: 162–175.

Gibling, M.R. (2006). Width and thickness of fluvial channel bodies and valley fills in the geological record: a literature compilation and classification. *Journal of Sedimentary Research* 76: 731–770.

Gibling, M.R. (2018). River systems and the Anthropocene: a late Pleistocene and Holocene timeline for human influence. *Quaternary* 1: https://doi.org/10.3390/quat1030021.

Gibling, M.R. (2021). *River Planet*. Edinburgh: Dunedin Academic Press.

Gibling, M.R. and Bird, D.J. (1994). Late Carboniferous cyclothems and alluvial paleovalleys in the Sydney Basin, Nova Scotia. *Bulletin Geological Society of America* 106: 105–117.

Gibling, M.R. and Davies, N.S. (2012). Palaeozoic landscapes shaped by plant evolution. *Nature Geoscience* 5: 99–105.

Gibling, M.R. and Rust, B.R. (1993). Alluvial ridge-and-swale topography: a case study from the Morien Group of Atlantic Canada. In: *Alluvial Sedimentation*, Special Publication 17 (ed. M. Marzo and C. Puigdefabregas), 133–150. International Association of Sedimentologists.

Gibling, M.R. and Wightman, W.G. (1994). Paleovalleys and protozoan assemblages in a Late Carboniferous cyclothem, Sydney Basin, Nova Scotia. *Sedimentology* 41: 699–719.

Gibling, M.R., Nanson, G.G., and Maroulis, J.C. (1998). Anastomosing river sedimentation in the Channel Country of central Australia. *Sedimentology* 45: 595–619.

Gibling, M.R., Saunders, K.I., Tibert, N.E., and White, J.A. (2004). Sequence sets, high-accommodation events and the coal window in the Carboniferous Sydney Basin, Atlantic Canada. In: *Sequence stratigraphy, paleoclimate, and tectonics of coal-bearing strata*, vol. 51 (ed. J.C. Pashin and R.A. Gastaldo), 169–197. Tulsa, Oklahoma: AAPG, Studies in Geology.

Gibling, M.R., Tandon, S.K., Sinha, R., and Jain, M. (2005). Discontinuity-bounded alluvial sequences of the southern Gangetic Plains, India: aggradation and degradation in response to monsoonal strength. *Journal of Sedimentary Research* 75: 369–385.

Gibling, M.R., Bashforth, A.R., Falcon-Lang, H.J. et al. (2010). Log jams and flood sediment buildup caused channel abandonment and avulsion in the Pennsylvanian of Atlantic Canada. *Journal of Sedimentary Research* 80: 268–287.

Gibling, M.R., Fielding, C.R., and Sinha, R. (2011). Alluvial valleys and alluvial sequences: towards a geomorphic assessment. In: *From River to Rock Record: The Preservation of Fluvial Sediments and their Subsequent Interpretation* (ed. C.P. North, S. Davidson, and S. Leleu), 423–447. Tulsa, OK: SEPM Society for Sedimentary Geology.

Gibling, M.R., Jia, R., Gastaldo, R.A. et al. (2023). Braided-river architecture of the Triassic Swartberg member, Katberg Formation, South Africa: assessing age, fluvial style and paleoclimate after the end-Permian extinction. *Journal of Sedimentary Research* 93: 741–775.

Gilbert, G.K. (1917). *Hydraulic-Mining Debris in the Sierra Nevada*, Professional Paper, vol. 105, 154 pp. United States Geological Survey.

Glasspool, I.J., Edwards, D., and Axe, L. (2006). Charcoal in the Early Devonian: a wildfire-derived Konservat-Lagerstatte. *Review of Palaeobotany and Palynology* 142: 131–136.

Golombek, N.Y., Scheingross, J.S., Repasch, M. et al. (2021). Fluvial organic carbon composition regulated by seasonal variability in lowland river migration and water discharge. *Geophysical Research Letters* 48: e2021GL093416.

Goodbred, S.L.J. (2003). Response of the Ganges dispersal system to climate change: a source-to-sink view since the last interstade. *Sedimentary Geology* 162: 83–104.

Goodbred, S.L. Jr., Dillehay, T.D., Galvez Mora, C., and Sawakuchi, A.O. (2020). Transformation of maritime desert to an agricultural center: Holocene environmental change and landscape engineering in Chicama River valley, northern Peru coast. *Quaternary Science Reviews* 227: 106046.

Goodbred, S.L.J. and Kuehl, S.A. (2000). Enormous Ganges-Brahmaputra sediment discharge during strengthened early Holocene monsoon. *Geology* 28: 1083–1086.

Grant, N.T., Middleton, A.J., and Archer, S. (2014). Porosity trends in the Skagerrak Formation, Central Graben, United Kingdom Continental Shelf: the role of compaction and pore pressure history. *AAPG Bulletin* 98: 1111–1143.

Gray, E., Hartley, A.J., and Howell, J. (2020). The influence of stratigraphy and facies distribution on reservoir quality and production performance in the Triassic Skagerrak Formation of the UK and Norwegian Central North Sea. In: *Cross-Border Themes in Petroleum Geology I: The North Sea*, vol. 494 (ed. S. Patruno et al.). Geological Society, London, Special Publications https://doi.org/10.1144/SP494-2019-68.

Greb, S.F. and Chestnut, D.R.J. (1996). Lower and lower Middle Pennsylvanian fluvial to estuarine deposition, central Appalachian basin: effects of eustasy, tectonics, and climate. *Geological Society of America* 108: 303–317.

Greb, S.F., Andrews, W.M., Eble, C.F. et al. (2003). Desmoinesian coal beds of the Eastern Interior and surrounding basins: the largest tropical peat mires in Earth history. In: *Extreme depositional events: Mega end members in geologic time* (ed. M.A. Chan and A.W. Archer), 127–150. Boulder, CO: Geological Society of America, Special Paper 370.

Greenberg, E., Ganti, V., and Hajek, E.A. (2021). Quantifying bankfull flow width using preserved bar clinforms from fluvial strata. *Geology* 49.

Groenewald, G.H., Welman, J., and MacEachern, J.A. (2001). Vertebrate burrow complexes from the Early Triassic *Cynognathus* Zone (Driekoppen Formation, Beaufort Group) of the Karoo Basin, South Africa. *Palaios* 16: 148–160.

Gulliford, A.R., Flint, S.S., and Hodgson, D.M. (2014). Testing applicability of models of distributive fluvial systems or trunk rivers in ephemeral systems: reconstructing 3-D fluvial architecture in the Beaufort Group, South Africa. *Journal of Sedimentary Research* 84: 1147–1169.

Gurnell, A.M. and Bertoldi, W. (2022). The impact of plants on fine sediment storage within the active channels of gravel-bed rivers: a preliminary assessment. *Hydrological Processes* 36: e14637.

Hackney, C.R., Darby, S.E., Parsons, D.R. et al. (2020). River bank instability from unsustainable sand mining in the lower Mekong River. *Nature Sustainability* 3: 217–225.

Hackney, C.R., Vasilopoulos, G., Heng, S. et al. (2021). Sand mining far outpaces natural supply in a large alluvial river. *Earth Surface Dynamics* 9: 1323–1334.

Haigh, M.J. (1984). Ravine erosion and reclamation in India. *Geoforum* 15: 543–561.

Hajek, E.A., Heller, P.L., and Sheets, B.A. (2010). Significance of channel-belt clustering in alluvial basins. *Geology* 38: 535–538.

Hales, T.C., Ford, C.R., Hwang, T. et al. (2009). Topographic and ecologic controls on root reinforcement. *Journal of Geophysical Research* 114: F03013. https://doi.org/10.1029/2008JF001168.

Hampson, G.J., Jewell, T.O., Irfan, N. et al. (2013). Modest change in fluvial style with varying accommodation in regressive alluvial-to-coastal-plain wedge: Upper Cretaceous Blackhawk Formation, Wasatch Plateau, Central Utah, U.S.A. *Journal of Sedimentary Research* 83: 145–169.

Hansford, M.R., Plink-Björklund, P., and Jones, E.R. (2020). Global quantitative analyses of river discharge variability and hydrograph shape with respect to climate types. *Earth-Science Reviews* 200: 102977.

Hartley, A.J., Weissmann, G.S., Nichols, G.J., and Scuderi, L.A. (2010a). Fluvial form in modern continental sedimentary basins: distributive fluvial systems: reply. *Geology* https://doi.org/10.1130/G31588Y.1.

Hartley, A.J., Weissmann, G.S., Nichols, G.J., and Warwick, G.L. (2010b). Large distributive fluvial systems: characteristics, distribution, and controls on development. *Journal of Sedimentary Research* 80: 167–183.

Hartley, A.J., Owen, A., Swan, A. et al. (2015). Recognition and importance of amalgamated sandy meander belts in the continental rock record. *Geology* 43: 679–682.

Harvey, J.E., Pederson, J.L., and Rittenour, T.M. (2011). Exploring relations between arroyo cycles and canyon paleoflood records in Buckskin Wash, Utah: reconciling scientific paradigms. *Geological Society of America Bulletin* 123: 2266–2276.

Hasiotis, S.T. and Honey, J.G. (2000). Paleohydrologic and stratigraphic significance of crayfish burrows in continental deposits: examples from several Paleocene Laramide Basins in the Rocky Mountains. *Journal of Sedimentary Research* 70: 127–139.

Hassan, M.A., Marren, P.M., and Schwartz, U. (2009). Bar structure in an arid ephemeral stream. *Sedimentary Geology* 221: 57–70.

Hasson, M., Marvin, M.C., Gunn, A. et al. (2023). A depositional model for meandering rivers without land plants. *Sedimentology* https://doi.org/10.1111/sed.13121.

Hembree, D.I. (2009). Neoichnology of burrowing millipedes: linking modern burrow morphology, organism behavior, and sediment properties to interpret continental ichnofossils. *Palaios* 24: 425–439.

Hembree, D. (2016). Using experimental neoichnology and quantitative analyses to improve the interpretation of continental trace fossils. *Ichnos* 23: 262–297.

Hembree, D.I. and Bowen, J.J. (2017). Paleosols and ichnofossils of the Upper Pennsylvanian–Lower Permian Monongahela and Dunkard groups (Ohio, USA): a multi-proxy approach to unraveling complex variability in ancient terrestrial landscapes. *Palaios* 32: 295–320.

Hembree, D.I. and Hasiotis, S.T. (2006). The identification and interpretation of reptile ichnofossils in paleosols through modern studies. *Journal of Sedimentary Research* 76: 575–588.

Hembree, D.I. and Hasiotis, S.T. (2007). Paleosols and ichnofossils of the White River Formation of Colorado: Insight into soil ecosystems of the North American Midcontinent during the Eocene-Oligocene transition. *Palaios* 22: 123–142.

Hembree, D.I. and Hasiotis, S.T. (2008). Miocene vertebrate and invertebrate burrows defining compound paleosols in the Pawnee Creek Formation, Colorado, USA. *Palaeogeography, Palaeoclimatology, Palaeoecology* 270: 349–365.

Hembree, D.I., Martin, L.D., and Hasiotis, S.T. (2004). Amphibian burrows and ephemeral ponds of the Lower Permian Speiser Shale, Kansas: evidence for seasonality in the midcontinent. *Palaeogeography, Palaeoclimatology, Palaeoecology* 203: 127–152.

Herbert, C.M., Alexander, J., Amos, K.J., and Fielding, C.R. (2020). Unit bar architecture in a highly-variable fluvial discharge regime: examples from the Burdekin River, Australia. *Sedimentology* 67: 576–605.

Hewitt, K., Gosse, J., and Clague, J.J. (2011). Rock avalanches and the pace of late Quaternary development of river valleys in the Karakoram Himalaya. *Geological Society of America Bulletin* 123: 1836–1850.

Hickson, T.A., Sheets, B.A., Paola, C., and Kelberer, M. (2005). Experimental test of tectonic controls on three-dimensional alluvial facies architecture. *Journal of Sedimentary Research* 75: 710–722.

Hirst, J.P.P. (1991). Variations in alluvial architecture across the Oligo-Miocene Huesca fluvial system, Ebro Basin, Spain. In: *The Three-Dimensional Facies Architecture of Terrigenous Clastic Sediments and Its Implications for Hydrocarbon Discovery and Recovery* (ed. A.D. Miall and N. Tyler), 111–121. Concepts in Sedimentology and Paleontology, SEPM.

Hofmann, M.H., Wroblewski, A., and Boyd, R. (2011). Mechanisms controlling the clustering of fluvial channels and the compensational stacking of cluster belts. *Journal of Sedimentary Research* 81: 670–685.

Holbrook, J. (2001). Origin, genetic interrelationships, and stratigraphy over the continuum of fluvial channel-form bounding surfaces: an illustration from middle Cretaceous strata, southeastern Colorado. *Sedimentary Geology* 144: 179–222.

Holbrook, J.M. and Allen, S.D. (2021). The case of the braided river that meandered: bar assemblages as a mechanism for meandering along the pervasively braided Missouri River, USA. *Geological Society of America Bulletin* 133: 1505–1530.

Holbrook, J.M. and Miall, A.D. (2020). Time in the Rock: a field guide to interpreting past events and processes from siliciclastic stratigraphy. *Earth-Science Reviews* 203: 103121.

Holbrook, J. and Schumm, S.A. (1999). Geomorphic and sedimentary response of rivers to tectonic deformation: a brief review and critique of a tool for recognizing subtle epeirogenic deformation in modern and ancient settings. *Tectonophysics* 305: 287–306.

Holbrook, J., Scott, R.W., and Oboh-Ikuenobe, F.E. (2006). Base-level buffers and buttresses: a model for upstream versus downstream control of fluvial geometry and architecture within sequences. *Journal of Sedimentary Research* 76: 162–174.

Horn, J.D., Fielding, C.R., and Joeckel, R. M. (2012a). Revision of Platte River alluvial facies model through observations of extant channels and barforms, and subsurface alluvial valley fills. *Journal of Sedimentary Research* 82: 72–91.

Horn, J.D., Joeckel, R.M., and Fielding, C. R. (2012b). Progressive abandonment and planform changes of the central Platte River in Nebraska, central USA, over historical timeframes. *Geomorphology* 139–140: 372–383.

Horton, B.K. and DeCelles, P.G. (2001). Modern and ancient fluvial megafans in the foreland basin system of the central Andes, southern Bolivia: implications for drainage network evolution in fold-thrust belts. *Basin Research* 13: 43–63.

Hovius, N. (1998). Controls on sediment supply by large rivers. In: *Relative Role of Eustasy, Climate, and Tectonism in Continental Rocks* (ed. K.W. Shanley and P.J. McCabe), 3–16. Tulsa, Oklahoma: SEPM Special Publication 59.

Hsieh, S., Łaska, W., and Uchman, A. (2023). Intermittent and temporally variable bioturbation by some terrestrial invertebrates: implications for ichnology. *The Science of Nature* 110: 1–18.

Huggenberger, P. (1993). Radar-facies: recognition of facies patterns and heterogeneities within Pleistocene Rhine gravels, NE Switzerland. In: *Braided Rivers*, vol. 75 (ed. J.L. Best and C.S. Bristow), 163–176. Geological Society Special Publication.

Hurtado, J.M. Jr., Hodges, K.V., and Whipple, K.X. (2001). Neotectonics of the Thakkhola graben and implications for recent activity on the South Tibetan fault system in the central Nepal Himalaya. *Bulletin of the Geological Society of America* 113: 222–240.

Ielpi, A. (2019). Morphodynamics of meandering streams devoid of plant life: Amargosa River, Death Valley, California. *Geological Society of America Bulletin* 131: 782–802.

Ielpi, A. and Ghinassi, M. (2014). Planform architecture, stratigraphic signature and morphodynamics of an exhumed Jurassic meander plain (Scalby Formation, Yorkshire, UK). *Sedimentology* 61: 1923–1960.

Ielpi, A. and Ghinassi, M. (2016). A sedimentary model for early Palaeozoic fluvial fans, Alderney Sandstone Formation (Channel Islands, UK). *Sedimentary Geology* 342: 31–46.

Ielpi, A. and Lapôtre, M.G.A. (2020). A tenfold slowdown in river meander migration driven by plant life. *Nature Geoscience* 13: 82–86.

Ielpi, A. and Lapôtre, M.G.A. (2022). Linking sediment flux to river migration in arid landscapes through mass balance. *Journal of Sedimentary Research* 92: 695–703.

Ielpi, A., Gibling, M.R., Bashforth, A.R. et al. (2014). Role of vegetation in shaping Early Pennsylvanian braided rivers: architecture of the Boss Point Formation, Atlantic Canada. *Sedimentology* 61: 1659–1700.

Ielpi, A., Gibling, M.R., Bashforth, A.R., and Dennar, C.I. (2015). Impact of vegetation on Early Pennsylvanian fluvial channels: insight from the Joggins Formation of Atlantic Canada. *Journal of Sedimentary Research* 85: 999–1018.

Ielpi, A., Ventra, D., and Ghinassi, M. (2016). Deeply channelled Precambrian rivers: Remote sensing and outcrop evidence from the 1.2 Ga Stoer Group of NW Scotland. *Precambrian Research* 281: 291–311.

Ielpi, A., Lapôtre, M.G.A., Finotello, A., and Ghinassi, M. (2020). Planform-asymmetry and backwater effects on river-cutoff kinematics and clustering. *Earth Surface Processes and Landforms* 46: 357–370.

Ielpi, A., Lapôtre, M.G.A., Gibling, M.R., and Boyce, C.K. (2022). The impact of vegetation on meandering rivers. *Nature Reviews Earth and Environment* https://doi.org/10.1038/s43017-021-00249-6.

Ielpi, A., Viero, D.P., Lapôtre, M.G.A. et al. (2023). How is time distributed in a river meander belt? *Geophysical Research Letters* 50: e2022GL101285. https://doi.org/10.1029/2022GL101285.

Inman, D.L. and Jenkins, S.A. (1999). Climate change and the episodicity of sediment flux of small California rivers. *The Journal of Geology* 107: 251–270.

Islam, M.N., Haq, S.M.A., Ahmed, K.J., and Best, J.L. (2022). How do vulnerable people in Bangladesh experience environmental stress from sedimentation in the Haor Wetlands? An exploratory study. *Water Resources Research* 58: e2021WR030241. https://doi.org/10.1029/2021WR030241.

Jacob, R.J., Bluck, B.J., and Ward, J.D. (1999). Tertiary-age diamondiferous fluvial deposits of the lower Orange River valley, southwestern Africa. *Economic Geology* 94: 749–758.

Jerolmack, D.J. and Mohrig, D. (2007). Conditions for branching in depositional rivers. *Geology* 35: 463–466.

Jerolmack, D.J. and Paola, C. (2010). Shredding of environmental signals by sediment transport. *Geophysical*

Research Letters 37: https://doi.org/10.1029/2010GL044638.

Jerrett, R., Wooldridge, P., Owusu, K., and Duller, R. (2023). Floodplain peat growth limits lateral migration and forces aggradation of river channel. *12th International Conference on Fluvial Sedimentology*, Riva del Garda, Italy, Abstract with Programme.

Jervey, M.T. (1988). Quantitative geological modeling of siliciclastic rock sequences and their seismic expression. In: *Sea-Level Changes: An Integrated Approach,* (ed. C.K. Wilgus, B.S. Hastings, H. Posamentier, et al.), 47–69. Tulsa, Oklahoma: SEPM Special Publication 42.

Joeckel, R.M., Wooden, S.R.J., Korus, J.T., and Garbisch, J. (2014). Architecture, heterogeneity, and origin of late Miocene fluvial deposits hosting the most important aquifer in the Great Plains, USA. *Sedimentary Geology* 311: 75–95.

Johnson, M.F., Rice, S.P., and Reid, I. (2010). Topographic disturbance of subaqueous gravel substrates by signal crayfish (*Pacifastacus leniusculus*). *Geomorphology* 123: 269–278.

Jones, L.S. and Gustason, E.R. (2006). Dinosaurs as possible avulsion enablers in the Upper Jurassic Morrison Formation, east-central Utah. *Ichnos* 13: 31–41.

Jones, C.G., Lawton, J.H., and Shachak, M. (1994). Organisms as ecosystem engineers. *Oikos* 69: 373–386.

Jones, N.S., Guion, P.D., and Fulton, I.M. (1995). Sedimentology and its applications within the UK opencast coal mining industry. In: *European Coal Geology*, vol. 82 (ed. M.K.G. Whateley and D.A. Spears), 115–136. Geological Society, London, Special Publications.

Karssenberg, D., Törnqvist, T.E., and Bridge, J.S. (2001). Conditioning a process-based model of sedimentary architecture to well data. *Journal of Sedimentary Research* 71: 868–879.

Keefer, D.K. (1999). Earthquake-induced landslides and their effects on alluvial fans. *Journal of Sedimentary Research* 69: 84–104.

Kemp, G.P., Day, J.W., Rogers, J.D. et al. (2016). Enhancing mud supply from the Lower Missouri River to the Mississippi River Delta USA: Dam bypassing and coastal restoration. *Estuarine, Coastal and Shelf Science* 183: 304–313.

Kennedy, K.L., Gibling, M.R., Eble, C.F. et al. (2013). Lower Devonian coaly shales of northern New Brunswick, Canada: plant accumulations in the early stages of terrestrial colonization. *Journal of Sedimentary Research* 83: 1202–1215.

Keogh, K.J., Martinius, A.W., and Osland, R. (2007). The development of fluvial stochastic modelling in the Norwegian oil industry: a historical review, subsurface implementation and future directions. *Sedimentary Geology* 202: 249–268.

Khan, S., Bartley, R., Kinsey-Henderson, A., and Hawdon, A. (2023). Assessing Gully erosion and rehabilitation using multi temporal LiDAR DEMs: case study from the Great Barrier Reef catchments, Australia. International Soil and Water Conservation Research. https://doi.org/10.1016/j.iswcr.2023.06.005

Kiss, T., Gönczy, S., Nagy, T. et al. (2022). Deposition and mobilization of microplastics in a low-energy fluvial environment from a geomorphological perspective. *Applied Sciences* 12: https://doi.org/10.3390/app12094367.

Klausen, T.G., Ryseth, A.E., Helland-Hansen, W. et al. (2014). Spatial and temporal changes in geometries of fluvial channel bodies from the Triassic Snadd Formation of offshore Norway. *Journal of Sedimentary Research* 84: 565–585.

Knighton, A.D. and Nanson, G.C. (1993). Anastomosis and the continuum of channel pattern. *Earth Surface Processes and Landforms* 18: 613–625.

Knighton, A.D. and Nanson, G.C. (1994). Flow transmission along an arid zone anastomosing river, Cooper Creek, Australia. *Hydrological Processes* 8: 137–154.

Knox, J.C. (2000). Sensitivity of modern and Holocene floods to climate change. *Quaternary Science Reviews* 19: 439–457.

Kondolf, G.M. (1997). Hungry water: effects of dams and gravel mining on river channels. *Environmental Management* 21: 533–551.

Krapez, B. (1985). The Ventersdorp Contact placer: a gold-pyrite placer of stream and debris-flow origin from the Archaean Witwatersrand Basin of South Africa. *Sedimentology* 32: 223–234.

Krapf, C.B.E., Stollhofen, H., and Stanistreet, I.G. (2003). Contrasting styles of ephemeral river systems and their interaction with dunes of the Skeleton Coast erg (Namibia). *Quaternary International* 104: 41–52.

Krapovickas, V., Ciccioli, P.L., Mángano, M.G. et al. (2009). Paleobiology and paleoecology of an arid–semiarid Miocene South American ichnofauna in anastomosed fluvial deposits. *Palaeogeography, Palaeoclimatology, Palaeoecology* 284: 129–152.

Kraus, M.J. (1999). Paleosols in clastic sedimentary rocks: their geologic applications. *Earth-Science Reviews* 47: 41–70.

Kraus, M.J. and Gwinn, B. (1997). Facies and facies architecture of Paleogene floodplain deposits, Willwood Formation, Bighorn Basin, Wyoming, USA. *Sedimentary Geology* 114: 33–54.

Krummeck, W.D. and Bordy, E.M. (2018). *Reniformichnus katikatii* (new ichnogenus and ichnospecies): continental vertebrate burrows from the Lower Triassic, Main Karoo Basin, South Africa. *Ichnos* 25: 138–149.

Kumar, R. (1993). Coalescence megafan: multistorey sandstone complex of the late-orogenic (Mio-Pliocene) sub-Himalayan belt, Dehra Dun, India. *Sedimentary Geology* 85: 327–337.

Labandeira, C.C. (2005). Invasion of the continents: cyanobacterial crusts to tree-inhabiting arthropods. *Trends in Ecology & Evolution* 20: 253–262.

Labandeira, C.C. and Currano, E.D. (2013). The fossil record of plant-insect dynamics. *Annual Review of Earth and Planetary Sciences* 41: 287–311.

Labandeira, C.C. and Wappler, T. (2023). Arthropod and pathogen damage on fossil and modern plants: exploring the origins and evolution of herbivory on land. *Annual Review of Entomology* 68: 341–361.

van de Lageweg, W.I., McLelland, S.J., and Parsons, D.R. (2018). Quantifying biostabilisation effects of biofilm-secreted and extracted extracellular polymeric substances (EPSs) on sandy substrate. *Earth Surface Dynamics* 6: 203–215.

Laming, D.J.C. (1966). Imbrication, paleocurrents and other sedimentary features in the lower New Red Sandstone, Devonshire, England. *Journal of Sedimentary Petrology* 36: 940–959.

Lanier, W.P., Feldman, H.R., and Archer, A.W. (1993). Tidal sedimentation from a fluvial to estuarine transition, Douglas Group, Missourian - Virgilian, Kansas. *Journal of Sedimentary Petrology* 63: 860–873.

Lapôtre, M.G.A., Lamb, M.P., and McElroy, B. (2017). What sets the size of current ripples? *Geology* 45: 243–246.

Lapôtre, M.G.A., Ielpi, A., Lamb, M.P. et al. (2019). Model for the formation of single-thread rivers in barren landscapes and implications for pre-Silurian and martian fluvial deposits. *Journal of Geophysical Research* 124: 2757–2777.

Laronne, J.B. and Reid, I. (1993). Very high rates of bedload sediment transport by ephemeral desert rivers. *Nature* 366: 148–150.

Latrubesse, E.M. (2015). Large rivers, megafans and other Quaternary avulsive fluvial systems: a potential "who's who" in the geological record. *Earth-Science Reviews* 146: 1–30.

Lawton, T.F., Schellenbach, W.L., and Nugent, A.E. (2014). Late Cretaceous fluvial-megafan and axial-river systems in the southern Cordilleran foreland basin: Drip Tank Member of Straight Cliffs Formation and adjacent strata, southern Utah, USA. *Journal of Sedimentary Research* 84: 407–434.

Le Cotonnec, A., Ventra, D., Hudson, S., and Moscariello, A. (2020). Systematically variable geometry and architecture of incised-valley fills controlled by orbital forcing: a conceptual model from the Pennsylvanian Breathitt Group (Kentucky, USA). *Sedimentology* 67: 2149–2188.

Leary, K.C.P. and Ganti, V. (2020). Preserved fluvial cross strata record bedform disequilibrium dynamics. *Geophysical Research Letters* 47: e2019GL085910.

Leclair, S.F. and Bridge, J.S. (2001). Quantitative interpretation of sedimentary structures formed by river dunes. *Journal of Sedimentary Research* 71: 713–716.

Leeder, M.R., Mack, G.H., Peakall, J., and Salyards, S.L. (1996). First quantitative test of alluvial stratigraphic models: Southern Rio Grande rift, New Mexico. *Geology* 24: 87–90.

Lehner, B. et al. (2011). High-resolution mapping of the world's reservoirs and dams for sustainable river-flow management. *Frontiers in Ecology and the Environment* 9: 494–502.

Leier, A.L., DeCelles, P.G., and Pelletier, J. D. (2005). Mountain, monsoons, and megafans. *Geology* 33: 289–292.

Lenders, H.J.R., Chamuleau, T.P.M., Hendriks, A.J. et al. (2016). Historical rise of waterpower initiated the collapse of salmon stocks. *Scientific Reports* 6: 29269. https://doi.org/10.1038/srep29269.

Leopold, L.B. and Maddock, T. (1953). *The Hydraulic Geometry of Stream Channels and Some Physiographic Implications*, Professional Paper, 252. United States Geological Survey.

Leopold, L.B. and Wolman, M.G. (1957). *River Channel Patterns -- Braided, Meandering and Straight*, Professional Paper, vol. 282B, 39–85. United States Geological Survey.

Letourmy, Y., Driese, S.G., and Sims, J.R. (2021). Absence of evidence of climate-driven cycles in Carboniferous deposits of Joggins, Nova Scotia, Canada: influence of salt withdrawal tectonics on deposition and petrogenesis. *Journal of Sedimentary Research* 91: 167–185.

Lewin, J. and Ashworth, P.J. (2014a). Defining large river channel patterns: alluvial exchange and plurality. *Geomorphology* 215: 83–98.

Lewin, J. and Ashworth, P.J. (2014b). The negative relief of large river floodplains. *Earth-Science Reviews* 129: 1–23.

Lewin, J. and Gibbard, P.L. (2010). Quaternary river terraces in England: forms, sediments and processes. *Geomorphology* 120: 293–311.

Li, J., Bristow, C.S., Luthi, S.M., and Donselaar, M.E. (2015). Dryland anabranching river morphodynamics: Rio Capilla, Salar de Uyuni, Bolivia. *Geomorphology* 250: 282–297.

Liebault, F. and Piégay, H. (2002). Causes of 20th century channel narrowing in mountain and piedmont rivers of southeastern France. *Earth Surface Processes and Landforms* 27: 425–444.

Liebault, F., Clement, P., Piégay, H. et al. (2002). Contemporary channel changes in the Eygues basin, southern French Prealps: the relationship of subbasin variability to watershed characteristics. *Geomorphology* 45: 53–66.

Long, D.G.F. (2006). Architecture of pre-vegetation sandy-braided perennial and ephemeral river deposits in the Paleoproterozoic Athabasca Group, northern Saskatchewan, Canada as indicators of Precambrian fluvial style. *Sedimentary Geology* 190: 71–95.

Long, D.G.F. (2021). Trickling down the paleoslope: an empirical approach to paleohydrology. *Earth-Science Reviews* 220: 103740.

Lowe, D.G. and Arnott, R.W.C. (2016). Composition and architecture of braided and sheetflood-dominated ephemeral fluvial strata in the Cambrian-Ordovician Potsdam Group: A case example of the morphodynamics of early Phanerozoic fluvial systems and climate change. *Journal of Sedimentary Research* 86: 587–612.

Lowey, G.W. (2006). The origin and evolution of the Klondike goldfields, Yukon, Canada. *Ore Geology Reviews* 28: 431–450.

Lu, M., Ikejiri, T., and Lu, Y. (2021). A synthesis of the Devonian wildfire record: Implications for paleogeography, fossil flora, and paleoclimate. *Palaeogeography, Palaeoclimatology, Palaeoecology* 571: 110321.

Lunt, I.A. and Bridge, J.S. (2004). Evolution and deposits of a gravelly braid bar, Sagavanirktok River, Alaska. *Sedimentology* 51: 415–432.

Lyster, S.J., Whittaker, A.C., and Hajek, E. A. (2022). The problem of paleo-planforms. *Geology* 50: 822–826.

Mack, G.H., James, W.C., and Monger, H. C. (1993). Classification of paleosols. *Geological Society of America Bulletin* 105: 129–136.

Mackin, J.H. (1948). Concept of the graded river. *Geological Society of America Bulletin* 59: 463–512.

Macklin, M.G. and Lewin, J. (2020). The rivers of humankind. In: *Science, Faith and the Climate Crisis* (ed. S. Myers, S. Hemstock, and E. Hanna), 29–46. Bingley: Emerald Publishing Limited.

Macklin, M.G., Lewin, J., and Woodward, J.C. (2012). The fluvial record of climate change. *Philosophical Transactions of the Royal Society A* 370: 2143–2172.

MacNaughton, R.B. and Pickerill, R.K. (1995). Invertebrate ichnology of the nonmarine Lepreau Formation (Triassic), southern New Brunswick, eastern Canada. *Journal of Paleontology* 69: 160–171.

MacNaughton, R.B., Cole, J.M., Dalrymple, R.D. et al. (2002). First steps on land: Arthropod trackways in Cambrian-Ordovician eolian sandstone, southeastern Ontario, Canada. *Geology* 30: 391–394.

Mahon, R.C. and McElroy, B. (2018). Indirect estimation of bedload flux from modern sand-bed rivers and ancient fluvial strata. *Geology* 46: 579–582.

Maitan, R., Finotello, A., Tognin, D. et al. (2024). Hydrologically driven modulation of cutoff regime in meandering rivers. *Geology* 52: 336–340.

Makaske, B. (2001). Anastomosing rivers: a review of their classification, origin and sedimentary products. *Earth-Science Reviews* 53: 149–196.

Makaske, B., Smith, D.G., and Berendsen, H.J.A. (2002). Avulsions, channel evolution and floodplain sedimentation rates of the anastomosing upper Columbia River, British Columbia. *Sedimentology* 49: 1049–1071.

Makaske, B., Smith, D.G., Berendsen, H.J. A. et al. (2009). Hydraulic and sedimentary processes causing anastomosing morphology of the upper Columbia River, British Columbia, Canada. *Geomorphology* 111: 194–205.

Makaske, B., Lavooi, E., De Haas, T. et al. (2017). Upstream control of river anastomosis by sediment overloading, upper Columbia River, British Columbia, Canada. *Sedimentology* 64: 1488–1510.

Malanson, G.P. (1993). *Riparian Lanscapes*. Cambridge: Cambridge University Press.

Maples, C.G. and Archer, A.W. (1989). The potential of Paleozoic nonmarine trace fossils for paleoecological interpretations. *Palaeogeography, Palaeoclimatology, Palaeoecology* 73: 185–195.

Martel, A.T. and Gibling, M.R. (1993). Clastic dykes of the Devono-Carboniferous Horton Bluff Formation, Nova Scotia: storm-related structures in shallow lakes. *Sedimentary Geology* 87: 103–119.

Martinsen, O.J., Ryseth, A., Helland-Hansen, W. et al. (1999). Stratigraphic base level and fluvial architecture: Ericson Sandstone (Campanian), Rock Springs Uplift, SW Wyoming, USA. *Sedimentology* 46: 235–259.

Mason, J. and Mohrig, D. (2019). Differential bank migration and the maintenance of channel width in meandering river bends. *Geology* 47: 1136–1140.

Matsubara, Y., Howard, A.D., Burr, D.M. et al. (2015). River meandering on Earth and Mars: a comparative study of Aeolis Dorsa meanders, Mars and possible terrestrial analogs of the Usuktuk River, AK, and the Quinn River, NV. *Geomorphology* 240: 102–120.

McCarthy, T.S., Ellery, W.N., and Bloem, A. (1998). Some observations on the geomorphological impact of hippopotamus (*Hippopotamus amphibilus* L.) in the Okavango Delta, Botswana. *African Journal of Ecology* 36: 44–56.

McGowen, J.H. and Garner, L.E. (1970). Physiographic features and stratification types of coarse-grained point bars: modern and ancient examples. *Sedimentology* 14: 77–111.

McKie, T. (2011). Architecture and behavior of dryland fluvial reservoirs, Triassic Skagerrak Formation, central North Sea. In: *From River to Rock Record* (ed. S.K. Davidson, S. Leleu, and C.P. North), 189–214. Tulsa, Oklahoma: SEPM Special Publication 97.

McKie, T. (2014). Climatic and tectonic controls on Triassic dryland terminal fluvial system architecture, central North Sea. In: *From Depositional Systems to Sedimentary Successions on the Norwegian Continental Margin*, vol. 46 (ed. A. Martinius, R. Ravnas, J.A. Howell, et al.), 19–58. International Association of Sedimentologists, Special Publication.

McLoughlin, S., Mays, C., Vajda, V. et al. (2020). Dwelling in the dead zone—vertebrate burrows immediately succeeding the end-Permian extinction event in Australia. *Palaios* 35: 342–357.

McMahon, W.J. and Davies, N.S. (2018). Evolution of alluvial mudrock forced by early land plants. *Science* 359: 1022–1024.

McMahon, W.J., Davies, N.S., and Went, D.J. (2017). Negligible microbial matground influence on pre-vegetation river functioning: evidence from the Ediacaran-Lower Cambrian Series Rouge, France. *Precambrian Research* 292: 13–34.

McMahon, W.J., Davies, N.S., Kleinhans, M.G., and Mitchell, R.L. (2022). Paleozoic vegetation increased fine sediment in fluvial and tidal channels: Evidence from secular changes to the mudrock content of ancient point bars. *Geology* 51: 136–140.

Meade, R.A. and Moody, J.H. (2010). Causes for the decline of suspended-sediment discharge in the Mississippi River system, 1940–2007. *Hydrocarbon Processing* 24: 35–49.

Melchor, R.N., Bedatou, E., de Valais, S., and Genise, J.F. (2006). Lithofacies distribution of invertebrate and vertebrate trace-fossil assemblages in an Early Mesozoic ephemeral fluvio-lacustrine system from Argentina: implications for the *Scoyenia* ichnofacies. *Palaeogeography, Palaeoclimatology, Palaeoecology* 239: 253–285.

Melchor, R.N., Genise, J.F., Farina, J.L. et al. (2010). Large striated burrows from fluvial deposits of the Neogene Vinchina Formation, La Rioja, Argentina: a crab origin suggested by neoichnology and sedimentology.

Palaeogeography, Palaeoclimatology, Palaeoecology 291: 400–418.

Melchor, R.N., Genise, J.F., Buatois, L.A., and Umazano, A.M. (2012). Fluvial environments. In: *Trace Fossils as Indicators of Sedimentary Environments, Developments in Sedimentology*, vol. 64 (ed. D. Knaust and R.G. Bromley), 329–378. Amsterdam: Elsevier.

Metz, R. (2020). Trace fossils in fluvial deposits of the uppermost Stockton Formation (Late Triassic), Newark Basin, New Jersey. *Ichnos* 27: 142–151.

Miall, A.D. (1974). Paleocurrent analysis of alluvial sediments: a discussion of directional variance and vector magnitude. *Journal of Sedimentary Petrology* 44: 1174–1185.

Miall, A.D. (1978). Lithofacies types and vertical profile models in braided river deposits: a summary. In: *Fluvial Sedimentology, Memoir 5* (ed. A.D. Miall), 597–604. Canadian Society of Petroleum Geology.

Miall, A.D. (1985). Architectural-element analysis: a new method of facies analysis applied to fluvial deposits. *Earth-Science Reviews* 22: 261–308.

Miall, A.D. (1994). Reconstructing fluvial macroform architecture from two-dimensional outcrops: examples from the Castlegate sandstone, Book Cliffs, Utah. *Journal of Sedimentology Research* B64: 146–158.

Miall, A.D. (1996). *The Geology of Fluvial Deposits*, 582. New York: Springer-Verlag.

Miall, A.D. (2002). Architecture and sequence stratigraphy of Pleistocene fluvial systems in the Malay Basin, based on seismic time-slice analysis. *AAPG Bulletin* 86: 1201–1216.

Miall, A.D. (2010). Alluvial deposits. In: *Facies Models 4* (ed. N.P. James and R. W. Dalrymple), 105–137. St. John's, Nfld.: Geological Association of Canada.

Miall, A.D. (2014). *Fluvial Depositional Systems*, 316. Berlin: Springer International Publishing.

Miall, A.D. and Jones, B.G. (2003). Fluvial architecture of the Hawkesbury Sandstone (Triassic), near Sydney, Australia. *Journal of Sedimentary Research* 73: 531–545.

Miall, A.D., Holbrook, J.M., and Bhattacharya, J.P. (2021). The stratigraphy machine. *Journal of Sedimentary Research* 91: 595–610.

Millar, R.G. (2000). Influence of bank vegetation on alluvial channel patterns. *Water Resources Research* 36: 1109–1118.

Milliman, J.D. and Syvitski, J.P.M. (1992). Geomorphic/tectonic control of sediment discharge to the ocean: the importance of small mountainous rivers. *Journal of Geology* 100: 525–544.

Minter, N.J., Buatois, L.A., Mángano, M.G. et al. (2016a). The prelude to terrestrial invasion. In: *The Trace-Fossil Record of Major Evolutionary Changes, Vol. 1: Precambrian and Paleozoic, Topics in Geobiology*, vol. 39 (ed. M.G. Mángano and L.A. Buatois), 157–204. Springer.

Minter, N.J., Buatois, L.A., Mángano, M.G. et al. (2016b). The establishment of continental ecosystems. In: *The Trace-Fossil Record of Major Evolutionary Changes, Vol. 1: Precambrian and Paleozoic, Topics in Geobiology*, vol. 39 (ed. M.G. Mángano and L.A. Buatois), 205–324. Springer.

Minter, N.J., Buatois, L.A., Mángano, M.G. et al. (2017). Early bursts of diversification defined the faunal colonization of land. *Nature Ecology & Evolution* 1: https://doi.org/10.1038/s41559-017-0175.

Misra, P., Farooqui, A., Sinha, R. et al. (2020). Millennial-scale vegetation and climatic changes from an Early to Mid-Holocene lacustrine archive in Central Ganga Plains using multiple biotic proxies. *Quaternary Science Reviews* 243: 106474.

Mohrig, D., Heller, P.L., Paola, C., and Lyons, W.J. (2000). Interpreting avulsion processes from ancient alluvial sequences: Guadalope-Matarranya system (northern Spain) and Wasatch Formation (western Colorado). *GSA Bulletin* 112: 1787–1803.

Montgomery, D.R. and Brandon, M.T. (2002). Topographic controls on erosion rates in tectonically active mountain ranges. *Earth and Planetary Science Letters* 201: 481–489.

Montgomery, D.R. and Piégay, H. (2003). Wood in rivers: interactions with channel morphology and processes. *Geomorphology* 51: 1–5.

Moran, M.G., Holbrook, J., Lensky, N.G. et al. (2023). Century-scale sequences and density-flow deltas of the late Holocene and modern Dead Sea coast, Israel. *Sedimentology* 70: 1945–1980.

Morón, S., Cawood, P.A., Haines, P.W. et al. (2019). Long-lived transcontinental sediment transport pathways of East Gondwana. *Geology* 47: 513–516.

Mountney, N.P. and Russell, A.J. (2004). Sedimentology of cold-climate aeolian sandsheet deposits in the Askja region of northeast Iceland. *Sedimentary Geology* 166: 223–244.

Mumpy, A.J., Jol, H.M., Kean, W.F., and Isbell, J.L. (2007). Architecture and sedimentology of an active braid bar in the Wisconsin River based on 3-D ground penetrating radar. In: *Stratigraphic Analyses Using GPR*, vol. 432 (ed. G.S. Baker and H.M. Jol), 111–131. Geological Society of America, Special Paper.

Nadon, G.C. (1998). Magnitude and timing of peat-to-coal compaction. *Geology* 26: 727–730.

Nadon, G.C. and Issler, D.R. (1997). The compaction of floodplain sediments: timing, magnitude and implications. *Geoscience Canada* 24: 37–43.

Naito, K. and Parker, G. (2019). Can bankfull discharge and bankfull channel characteristics of an alluvial meandering river by cospecified from a flow duration curve? *Journal of Geophysical Research - Earth Surface* 124: 2381–2401.

Nakayama, K., Fielding, C.R., and Alexander, J. (2002). Variations in character and preservation potential of vegetation-induced obstacle marks in the variable discharge Burdekin River of north Queensland, Australia. *Sedimentary Geology* 149: 199–218.

Nanson, G.C. (1986). Episodes of vertical accretion and catastrophic stripping: a model of disequilibrium flood-plain development. *Geological Society of America Bulletin* 97: 1467–1475.

Nanson, G.C. and Huang, H.Q. (1999). Anabranching rivers: divided efficiency leading to fluvial diversity. In: *Varieties of Fluvial Form* (ed. A.J. Miller and A. Gupta), 219–248. Chichester: Wiley.

Nanson, G.C. and Nanson, R.A. (2025). A continental review of Australia's Cenozoic megafans and associated large accretionary fluvial systems. *Earth-Science Reviews* 2025: 105084.

Nanson, G.C., Rust, B.R., and Taylor, G. (1986). Coexistent mud braids and anastomosing channels in an arid-zone river: Cooper Creek, central Australia. *Geology* 14: 175–178.

Nascimento, D.L., Netto, R.G., and Indicatti, R.P. (2021). Neoichnology of mygalomorph spiders: Improving the recognition of spider burrows in the geological record. *Journal of South American Earth Sciences* 108: 103178.

Nash, D.J. and Smith, R.F. (2003). Properties and development of channel calcretes in a mountain catchment, Tabernas Basin, southeast Spain. *Geomorphology* 50: 227–250.

Nelson, A.R. (1992). Lithofacies analysis of colluvial sediments – an aid in interpreting the recent history of Quaternary normal faults in the Basin and Range Province, western United States. *Journal of Sedimentary Petrology* 62: 607–621.

Nichols, G.J. (1987). Structural controls on fluvial distributary systems-the Luna System, northern Spain. In: *Recent Developments in Fluvial Sedimentology*, Special Publication No. 39 (ed. F.G. Ethridge, R.M. Flores, and M.D. Harvey), 269–277. Society of Economic Paleontologists and Mineralogists.

Niedźwiedzki, G., Szrek, P., Narkiewicz, K. et al. (2010). Tetrapod trackways from the early Middle Devonian period from Poland. *Nature Geoscience* 463: https://doi.org/10.1038/nature08623.

Nittrouer, J.A., Mohrig, D., and Allison, M. (2011). Punctuated sand transport in the lowermost Mississippi River. *Journal of Geophysical Research - Earth Surface* 116: F04025. https://doi.org/10.1029/2011JF002026.

Noffke, N. (2010). *Geobiology: Microbial Mats in Sandy Deposits from the Archean Era to Today*, 194. Berlin: Springer-Verlag.

North, C.P. and Davidson, S.K. (2012). Unconfined alluvial flow processes: recognition and interpretation of their deposits, and the significance for palaeogeographic reconstruction. *Earth-Science Reviews* 111: 199–223.

North, C.P. and Warwick, G.L. (2007). Fluvial fans: myths, misconceptions, and the end of the terminal-fan model. *Journal of Sedimentary Research* 77: 693–701.

Novak, A., Vrabec, M., Popit, T. et al. (2023). Facies analysis, depositional activity, and internal structure of sieve deposits on an active alluvial fan. *Earth Surface Processes and Landforms* 48: 647–664.

Nyberg, B. and Howell, J.A. (2015). Is the present the key to the past? A global characterization of modern sedimentary basins. *Geology* 43: 643–646.

Obermeier, S.F., Jacobson, R.B., Smoot, J. P. et al. (1990). Earthquake-induced liquefaction features in the coastal setting of South Carolina and in the fluvial setting of the New Madrid seismic zone. *United States Survey Professional Paper* 1504: 44.

Owen, A., Nichols, G.J., Hartley, A.J., and Weissmann, G.S. (2015a). Vertical trends within the Salt Wash distributive fluvial system. *Basin Research* 29: 64–80.

Owen, A., Nichols, G.J., Hartley, A.J. et al. (2015b). Quantification of a distributive fluvial system: the Salt Wash DFS of the Morrison Formation, SW USA. *Journal of Sedimentary Research* 85: 544–561.

Owen, A., Ebinghaus, A., Hartley, A.J. et al. (2017). Multi-scale classification of fluvial architecture: an example from the Palaeocene–Eocene Bighorn Basin, Wyoming. *Sedimentology* 64: 1572–1596.

Page, K.J., Nanson, G.C., and Frazier, P.S. (2003). Floodplain formation and sediment stratigraphy resulting from oblique accretion on the Murrumbidgee River, Australia. *Journal of Sedimentary Research* 73: 5–14.

Pan, B., Hu, Z., Wang, J. et al. (2011). A magnetostratigraphic record of landscape development in the eastern Ordos Plateau, China: Transition from Late Miocene and Early Pliocene stacked sedimentation to Late Pliocene and Quaternary uplift and incision by the Yellow River. *Geomorphology* 125: 225–238.

Pánek, T. and Klimeš, J. (2016). Temporal behavior of deep-seated gravitational slope deformations: a review. *Earth-Science Reviews* 156: 14–38.

Paola, C. (2000). Quantitative models of sedimentary basin filling. *Sedimentology* 47: 121–178.

Paola, C. and Mohrig, D. (1996). Palaeohydraulics revisited: palaeoslope estimation in coarse-grained braided rivers. *Basin Research* 8: 243–254.

Paola, C., Ganti, V., Mohrig, D. et al. (2018). Time not our time: physical controls on the preservation and measurement of geologic time. *Annual Review of Earth and Planetary Sciences* 46: 409–438.

Parker, G., Shimizu, y., Wilkerson, G.B. et al. (2011). A new framework for modeling the migration of meandering rivers. *Earth Surface Processes and Landforms* 36: 70–86.

Parker, N.O., Sambrook Smith, G.H., Ashworth, P.J. et al. (2013). Quantification of the relation between surface morphodynamics and subsurface sedimentological product in sandy braided rivers. *Sedimentology* 60: 820–839.

Parrinello, G. and Kondolf, G.M. (2021). The social life of sediment. *Water History* 12: 1–13.

Pérez-Arlucea, M. and Smith, N.D. (1999). Depositional patterns following the 1870s avulsion of the Saskatchewan River (Cumberland Marshes, Saskatchewan, Canada). *Journal of Sedimentary Research* 69: 62–73.

Pickering, J.L., Diamond, M.S., Goodbred, S.L. et al. (2019). Impact of glacial-lake paleofloods on valley development since glacial termination II: a conundrum of hydrology and scale for the lowstand Brahmaputra-Jamuna paleovalley system. *Geological Society of America Bulletin* 131: 58–70.

Pienkowski, G. and Niedzwiedzki, G. (2009). Invertebrate trace fossil assemblages from the Lower Hettangian of Sołtyków, Holy Cross Mountains, Poland. *Volumina Jurassica* 6: 109–113.

Platt, B.F., Hasiotis, S.T., and Hirmas, D.R. (2012). Empirical determination of physical controls on megafaunal footprint formation through neoichnological experiments with elephants. *Ichnos* 27: 725–737.

Plink-Björklund, P. (2015). Morphodynamics of rivers strongly affected by monsoon precipitation: review of depositional style and forcing factors. *Sedimentary Geology* 323: 110–147.

Poff, N.L., Allan, J.D., Bain, M.B. et al. (1997). The natural flow regime. *Bioscience* 47: 769–784.

Pollard, J.E. and Hardy, P.G. (1991). Trace fossils from the Westphalian D of Writhlington Geological Nature Reserve, nr. Radstock, Avon. *Proceedings of the Geologists Association* 102: 169–178.

Pollen, N. (2007). Temporal and spatial variability in root reinforcement of streambanks: accounting for soil shear strength and moisture. *Catena* 69: 197–205.

Poole, G.C., O'Daniel, S.J., Jones, K.L. et al. (2008). Hydrologic spiralling: the role of multiple interactive flow paths in stream ecosystems. *River Research and Applications* 24: 1018–1031.

Posamentier, H.W. (2001). Lowstand alluvial bypass systems: incised vs. unincised. *AAPG Bulletin* 85: 1771–1793.

Posamentier, H.W., Allen, G.P., and James, D.P. (1992). High-resolution sequence stratigraphy; the East Coulee Delta, Alberta. *Journal of Sedimentary Research* 62: 310–317.

Posamentier, H.W., Paumard, V., and Lang, S.C. (2022). Principles of seismic stratigraphy and seismic geomorphology I: extracting geologic insights from seismic data. *Earth-Science Reviews* 228: 103963.

Potter, P.E. (1967). Sand bodies and sedimentary environments: a review. *AAPG Bulletin* 51: 337–365.

Pratt-Sitaula, B., Burbank, D.W., Heimsath, A., and Ojha, T. (2004). Landscape disequilibrium on 1000 - 10,000 year scales, Marsyandi River, Nepal, central Himalaya. *Geomorphology* 58: 223–241.

Puigdefabregas, C. (1973). Miocene point bar deposits in the Ebro Basin, northern Spain. *Sedimentology* 20: 133–144.

Pyrcz, M.J., Sech, R.P., Covault, J.A. et al. (2015). Stratigraphic rule-based reservoir modeling. *Bulletin of Canadian Petroleum Geology* 63: 287–303.

Raff, J.L., Goodbred, S.L. Jr., Pickering, J. L. et al. (2023). Sediment delivery to sustain the Ganges-Brahmaputra delta under climate change and anthropogenic impact. *Nature Communications* 14: 2429. https://doi.org/10.1038/s41467-023-38057-9.

Rainbird, R.H., Rayner, N.M., Hadlari, T. et al. (2017). Zircon provenance data record the lateral extent of pancontinental, early Neoproterozoic rivers and erosional unroofing history of the Grenville orogen. *Bulletin of the Geological Society of America* 129: 1408–1423.

Rajchl, M. and Ulicny, D. (2005). Depositional record of an avulsive fluvial system controlled by peat compaction (Neogene, Most Basin, Czech Republic). *Sedimentology* 52: 601–625.

Ray, D.C., van Buchem, F.S.P., Baines, G. et al. (2019). The magnitude and cause of short-term eustatic Cretaceous sea-level change: a synthesis. *Earth-Science Reviews* 197: 102901.

Reesink, A.J.H., Best, J.L., Freiburg, J.T. et al. (2020). Interpreting pre-vegetation landscape dynamics: the Cambrian Lower Mount Simon Sandstone, Illinois, USA. *Journal of Sedimentary Research* 90: 1614–1641.

Reid, A.J., Carlson, A.K., Creed, I.F. et al. (2019). Emerging threats and persistent conservation challenges for freshwater biodiversity. *Biological Reviews* 94: 849–873.

Reynolds, A.D. (1999). Dimensions of paralic sandstone bodies. *AAPG Bulletin* 83: 211–229.

Rittenour, T.M., Blum, M.D., and Goble, R.J. (2007). Fluvial evolution of the lower Mississippi River valley during the last 100 k.y. glacial cycle: response to glaciation and sea-level change. *Geological Society of America Bulletin* 119: 586–608.

Rodrigues, S., Breheret, J.-G., Macaire, J.-J. et al. (2006). Flow and sediment dynamics in the vegetated secondary channels of an anabranching river: the Loire River (France). *Sedimentary Geology* 186: 89–109.

Roebroeks, W., MacDonald, K., Scherjon, F. et al. (2021). Landscape modification by Last Interglacial Neanderthals. *Science Advances* 7: eabj5567.

Rossetti, D.F., Zani, H., Cohen, M.C.L., and Cremon, E.H. (2012). A Late Pleistocene–Holocene wetland megafan in the Brazilian Amazonia. *Sedimentary Geology* 282: 276–293.

Roy, N.G., Sinha, R., and Gibling, M.R. (2012). Aggradation, incision and interfluve flooding in the Ganga Valley over the past 100,000 years: testing the influence of monsoonal precipitation. *Palaeogeography, Palaeoclimatology, Palaeoecology* 356–357: 38–53.

Rubinstein, C.V. and Vajda, V. (2019). Baltica cradle of early land plants? Oldest record of trilete spores and diverse cryptospore assemblages; evidence from Ordovician successions of Sweden. *GFF* 141: 181–190.

Rust, B.R. (1978). Depositional models for braided alluvium. In: *Fluvial Sedimentology*, vol. 5 (ed. A.D. Miall), 605–625. Canadian Society of Petroleum Geologists Memoir.

Rust, B.R. and Jones, B.G. (1987). The Hawkesbury Sandstone south of Sydney, Australia: Tirassic analogue for the deposit of a large, braided river. *Journal of Sedimentary Petrology* 57: 222–233.

Rust, B.R. and Nanson, G.C. (1989). Bedload transport of mud as pedogenic aggregates in modern and ancient rivers. *Sedimentology* 36: 291–306.

Rybczynski, N. (2007). Castorid phylogenetics: implications for the evolution of swimming and tree-exploitation in beavers. *Journal of Mammalian Evolution* 14: 1–35.

Rygel, M.C. and Gibling, M.R. (2006). Natural geomorphic variability recorded in a high-accommodation setting: fluvial architecture of the

Pennsylvanian Joggins Formation of Atlantic Canada. *Journal of Sedimentary Petrology* 76: 1230–1251.

Rygel, M.C., Gibling, M.R., and Calder, J. H. (2004). Vegetation-induced sedimentary structures from fossil forests in the Pennsylvanian Joggins Formation, Nova Scotia. *Sedimentology* 51: 531–552.

Rygel, M.C., Calder, J.H., Gibling, M.R. et al. (2006). Tournaisian forested wetlands in the Horton Group of Atlantic Canada. In: *Wetlands Through Time* (ed. S.F. Greb and W.A. DiMichele), 103–126. Boulder, CO: Geological Society of America Special Publication 399.

Rygel, M.C., Fielding, C.R., Frank, T.D., and Birgenheier, L.P. (2008). The magnitude of Late Paleozoic glacioeustatic fluctuations: a synthesis. *Journal of Sedimentary Research* 78: 500–511.

Sadler, P.M. (1981). Sediment accumulation rates and the completeness of stratigraphic sections. *Journal of Geology* 89: 569–584.

Salese, F., McMahon, W.J., Balme, M.R. et al. (2020). Sustained fluvial deposition recorded in Mars' Noachian stratigraphic record. *Nature Communications* 11: 1–8.

Sambrook Smith, G.H., Ashworth, P.J., Best, J.L. et al. (2006). The sedimentology and alluvial architecture of the sandy braided South Saskatchewan River, Canada. *Sedimentology* 53: 413–434.

Sambrook Smith, G.H., Best, J.L., Ashworth, P.J. et al. (2010). Fluvial form in modern continental sedimentary basins: distributive fluvial systems: Comment. *Geology* https://doi.org/10.1130/G31507C.1.

Sambrook Smith, G.H., Best, J.L., Leroy, J. Z., and Orfeo, O. (2015). The alluvial architecture of a suspended sediment dominated meandering river: the Rio Bermejo, Argentina. *Sedimentology* 63: 1187–1208.

Sambrook Smith, G.H., Nicholas, A.P., Best, J.L. et al. (2019). The sedimentology of river confluences. *Sedimentology* 66: 391–407.

Sánchez, M.V., Laza, J.H., Bellosi, E.S., and Genise, J.F. (2010). Ichnostratigraphy of middle Cenozoic Coprinisphaera from central Patagonia: insights into the evolution of dung beetles, herbivores and grass-dominated habitats. *Palaeogeography, Palaeoclimatology, Palaeoecology* 297: 633–648.

Sánchez, M.V., Bellosi, E.S., Genise, J.F. et al. (2021). An assemblage of large-sized insect traces in paleosols from the middle Miocene of northern Patagonia related to the climatic optimum. *Journal of South American Earth Sciences* 109: 103249.

Santos, M.G.M., Hartley, A.J., Mountney, N.P. et al. (2019). Meandering rivers in modern desert basins: Implications for channel planform controls and prevegetation rivers. *Sedimentary Geology* 385: 1–14.

Sarkar, S. and Chaudhuri, A.K. (1992). Trace fossils in middle to Late Triassic fluvial redbeds, Pranhita-Godavari Valley, south India. *Ichnos* 2: 7–19.

Savrda, C.E., Blanton-Hooks, A.D., Collier, J.W. et al. (2000). *Taenidium* and associated ichnofossils in fluvial deposits, Cretaceous Tuscaloosa Formation, eastern Alabama, southeastern USA. *Ichnos* 7: 227–242.

Scheingross, J.S., Hovius, N., Dellinger, M. et al. (2019). Preservation of organic carbon during active fluvial transport and particle abrasion. *Geology* 47: 958–962.

Schmidt, M.J., Py-Daniel, A.R., Moraes, C. D.P. et al. (2014). Dark earths and the human built landscape in Amazonia: a widespread pattern of anthrosol formation. *Journal of Archaeological Science* 42: 152–165.

Schumm, S.A. (1960). The effect of sediment type on the shape and stratification of some modern fluvial deposits. *American Journal of Science* 258: 177–184.

Schumm, S.A. (1963). A tentative classification of alluvial river channels. *United States Geological Survey Circular* 477: 10 pp.

Schumm, S.A. (1981). Evolution and response to the fluvial system, sedimentologic implications. In: *Recent and Ancient Nonmarine Environments: Models for Exploration*, vol. 31 (ed. F.G. Ethridge and R.M. Flores), 19–29. SEPM Special Publication.

Schumm, S.A. (1985). Patterns of alluvial rivers. *Annual Review of Earth and Planetary Sciences* 13: 5–27.

Scott, A.C. and Stephens, R.S. (2015). British Pennsylvanian (Carboniferous) coal-bearing sequences: where is the time? In: *Strata and Time: Probing the Gaps in our Understanding*, vol. 404 (ed. D.G. Smith, R.J. Bailey, P.M. Burgess, and A.J. Fraser), 283–302. Geological Society, Special Publications.

Sedell, J.R., Reeves, G.H., Hauer, F.R. et al. (1990). Role of refugia in recovery from disturbances: modern fragmented and disconnected river systems. *Environmental Management* 14: 711–724.

Sedorko, D., Alessandretti, L., Warren, L. V. et al. (2020). Trace fossils from the Upper Cretaceous Capacete Formation, Sanfranciscana Basin, Central Brazil. *Annales Societatis Geologorum Poloniae* 90: 247–260.

Shanley, K.W. and McCabe, P.J. (1991). Predicting facies architecture through sequence stratigraphy – an example from the Kaiparowits Plateau, Utah. *Geology* 19: 742–745.

Shanley, K.W. and McCabe, P.J. (1994). Perspectives on the sequence stratigraphy of continent strata. *AAPG Bulletin* 78: 544–568.

Shear, W.A. and Kukalová-Peck, J. (1990). The ecology of Paleozoic terrestrial arthropods: the fossil evidence. *Canadian Journal of Zoology* 68: 1807–1834.

Sheldon, N.D. and Tabor, N.J. (2009). Quantitative paleoenvironmental and paleoclimatic reconstruction using paleosols. *Earth-Science Reviews* 95: 1–52.

Shillito, A.P. and Davies, N.S. (2019a). Dinosaur-landscape interactions at a diverse Early Cretaceous tracksite (Lee Ness Sandstone, Ashdown Formation, southern England). *Palaeogeography, Palaeoclimatology, Palaeoecology* 514: 593–612.

Shillito, A.P. and Davies, N.S. (2019b). Death near the shoreline, not life on land: Ordovician arthropod trackways in the Borrowdale Volcanic Group, UK. *Geology* 47: 55–58.

Shillito, A.P. and Davies, N.S. (2020). The Tumblagooda Sandstone revisited: exceptionally abundant trace fossils and geological outcrop provide a window onto Palaeozoic littoral habitats before invertebrate terrestrialization. *Geological Magazine* 157: 1939–1970.

Sidor, C.A., Miller, M.F., and Isbell, J.L. (2008). Tetrapod burrows from the Triassic of Antarctica. *Journal of Vertebrate Paleontology* 28: 277–284.

Simon, S.S.T. and Gibling, M.R. (2017). Fine-grained meandering systems of the Lower Permian Clear Fork Formation of north-central Texas, USA: Lateral and oblique accretion on an arid plain. *Sedimentology* 64: 714–746.

Sinha, R. and Friend, P. (1994). River systems and their sediment flux, Indo-Gangetic plains, Northern Bihar, India. *Sedimentology* 41: 825–845.

Sinha, R., Gibling, M.R., Jain, V., and Tandon, S.K. (2005). Sedimentology and avulsion patterns of the anabranching Baghmati River in the Himalayan foreland basin, India. In: *Fluvial Sedimentology VII* (ed. M.D. Blum and S.B. Marriott), 181–196. Amsterdam, Special Publication 35: International Association of Sedimentologists, Elsevier.

Sinha, R., Kettanah, Y., Gibling, M.R. et al. (2009). Craton-derived alluvium as a major sediment source in the Himalayan Foreland Basin of India. *Geological Society of America Bulletin* 121: 1596–1610.

Skelly, R.L., Bristow, C.S., and Ethridge, F. G. (2003). Architecture of channel-belt deposits in an aggrading shallow sandbed braided river: the lower Niobrara River, northeast Nebraska. *Sedimentary Geology* 158: 249–270.

Slingerland, R. and Smith, N.D. (2004). River avulsions and their deposits. *Annual Review of Earth and Planetary Sciences* 32: 257–285.

Smalley, I., O'Hara-Dhand, K., Wint, J. et al. (2009). Rivers and loess: the significance of long river transportation in the complex event-sequence approach to loess deposit formation. *Quaternary International* 198: 7–18.

Smith, D.G. (1976). Effect of vegetation on lateral migration of anastomosed channels of a glacier meltwater river. *Geological Society of America Bulletin* 87: 857–860.

Smith, D.G. (1987a). Meandering river point bar lithofacies models: modern and ancient examples compared. In: *Recent Developments in Fluvial Sedimentology*, vol. 39 (ed. F.G. Ethridge, R.M. Flores, and M.D. Harvey), 83–91. Society of Economic Paleontologists and Mineralogists, Special Publication.

Smith, R.M. (1987b). Morphology and depositional history of exhumed Permian point bars in the southwestern Karoo, South Africa. *Journal of Sedimentary Petrology* 57: 19–29.

Smith, R.M. (1993). Sedimentology and ichnology of floodplain paleosurfaces in the Beaufort Group (Late Permian), Karoo sequence, South Africa. *Palaios* 8: 339–357.

Smith, N.D. and Smith, D.G. (1984). William River: an outstanding example of channel widening and braiding caused by bed-load addition. *Geology* 12: 78–82.

Smith, D.G., Hubbard, S.M., Lavigne, J.R. et al. (2011). Stratigraphy of counter-point-bar and eddy-accretion deposits in low-energy meander belts of the Peace-Athabasca Delta, northeast Alberta, Canada. In: *From River to Rock Record: The Preservation of Fluvial Sediments and their Subsequent Interpretation*, vol. 97 (ed. C.P. North, S. K. Davidson, and S. Leleu), 143–152. Tulsa, Oklahoma: SEPM Special Publication.

Smith, N.D., Morozova, G.S., Perez-Arlucea, M., and Gibling, M.R. (2016). Dam-induced and natural channel changes in the Saskatchewan River below the E.B. Campbell Dam Canada. *Geomorphology* 269: 186–202.

Smith, R.M., Angielczyk, K.D., Benoit, J., and Fernandez, V. (2021). Neonate aggregation in the Permian dicynodont *Diictodon* (Therapsida, Anomodontia): Evidence for a reproductive function for burrows? *Palaeogeography, Palaeoclimatology, Palaeoecology* 569: 110311.

Sorby, H.C. (1859). On the structure and origin of the Millstone-Grit in South Yorkshire. *Proceedings of the Yorkshire Geological Society* 3: 669–675.

Southard, J.B. and Boguchwal, L.A. (1990). Bed configurations in steady unidirectional water flows. Part 2. Synthesis of flume data. *Journal of Sedimentary Research* 60: 658–679.

Spencer, C.J., Davies, N.S., Gernon, T.M. et al. (2022). Composition of continental crust altered by the emergence of land plants. *Nature Geoscience* 15: 735–740.

Srivastava, P., Juyal, N., Singhvi, A.K. et al. (2001). Luminescence chronology of river adjustment and incision of Quaternary sediments in the alluvial plain of the Sabarmati River, north Gujarat, India. *Geomorphology* 36: 217–229.

Stear, W.M. (1983). Morphological characteristics of ephemeral stream channel and overbank splay sandstone bodies in the Permian Lower Beaufort Group, Karoo Basin, South Africa. In: *Modern and Ancient Fluvial Systems*, vol. 6 (ed. J.D. Collinson and J. Lewin), 405–420. International Association of Sedimentologists Special Publication.

Steeb, N., Rickenmann, D., Badoux, A. et al. (2017). Large wood recruitment processes and transported volumes in Swiss mountain streams during the extreme flood of August 2005. *Geomorphology* 279: 112–127.

Steel, R.J., Maehle, S., Nilsen, H. et al. (1977). Coarsening-upward cycles in the alluvium of Hornelen Basin (Devonian) Norway: Sedimentary response to tectonic events. *Geological Society of America Bulletin* 88: 1124–1134.

Steemans, P., Le Herisse, A., Melvin, J. et al. (2009). Origin and radiation of the earliest vascular land plants. *Science* 324: 353.

Stein, W.E., Berry, C.M., Hernick, L.V., and Mannolini, F. (2012). Surprisingly complex community discovered in the mid-Devonian fossil forest at Gilboa. *Nature* 483: 78–81.

Stein, W.E., Berry, C.M., Morris, J.L. et al. (2020). Mid-Devonian Archaeopteris roots signal revolutionary change in earliest fossil forests. *Current Biology* 30: 1–11.

Straub, K.M., Duller, R.A., Foreman, B.Z., and Hajek, E.A. (2020). Buffered, incomplete, and shredded: the challenges of reading an imperfect stratigraphic record. *Journal of Geophysical Research: Earth Surface* 125: e2019JF005079. https://doi.org/10.1029/2019JF005079.

Stricker, S. and Jones, S.J. (2018). Enhanced porosity preservation by pore fluid overpressure and chlorite grain coatings in the Triassic Skagerrak, Central Graben, North Sea, UK. In: *Reservoir Quality of Clastic and Carbonate Rocks: Analysis, Modelling and Prediction*, vol. 435 (ed. P.J. Armitage et al.), 321–341. Geological Society, London, Special Publications.

Stricker, S., Jones, S.J., and Grant, N.T. (2016a). Importance of vertical effective stress for reservoir quality in the Skagerrak Formation, Central Graben, North Sea. *Marine and Petroleum Geology* 78: 895–909.

Stricker, S., Jones, S.J., Sathar, S. et al. (2016b). Exceptional reservoir quality in HPHT reservoir settings: examples from the Skagerrak Formation of the Heron Cluster, UK, North Sea. *Marine and Petroleum Geology* 77: 198–215.

Strong, N. and Paola, C. (2008). Valleys that never were: time surfaces versus stratigraphic surfaces. *Journal of Sedimentary Research* 78: 579–593.

Strullu-Derrien, C., Kenrick, P., and Knoll, A.H. (2019). The Rhynie chert. *Curent Biology* 29: R1218–R1223.

Suman, H., Arens, S.K.M., Jensen, K. et al. (2022). A dynamic local-scale vegetation model for lycopsids (LYCOm v1.0). *Geoscience Model Development* 15: 2325–2343.

Swan, A., Hartley, A.J., Owen, A., and Howell, J. (2018). Reconstruction of a sandy point-bar deposit: implications for fluvial facies analysis. In: *Fluvial Meanders and Their Sedimentary Products in the Rock Record*, International Association of Sedimentologists, Special Publication, vol. 48 (ed. M. Ghinassi, L. Colombera, N. Mountney, et al.), 445–474.

Swartz, J.M., Cardenas, B.T., Mohrig, D., and Passalacqua, P. (2022). Tributary channel networks formed by depositional processes. *Nature Geoscience* 15: 216–221.

Sylvester, Z., Durkin, P.R., and Covault, J. A. (2019). High curvatures drive river meandering. *Geology* 47: 263–266.

Syvitski, J.P.M. and Milliman, J.D. (2007). Geology, geography, and humans battle for dominance over the delivery of fluvial sediment to the coastal Ocean. *Journal of Geology* 115: 1–19.

Syvitski, J.P.M., Ángel, J.R., Saito, Y. et al. (2022). Earth's sediment cycle during the Anthropocene. *Nature Reviews Earth and Environment* 3: 179–196.

Tamura, L.N., Almeida, R.P., Galeazzi, C. P. et al. (2019). Upper-bar deposits in large Amazon rivers: Occurrence, morphology and internal structure. *Sedimentary Geology* 387: 1–17.

Tan, M., Zhu, X., Liu, Q. et al. (2020). Multiple fluvial styles in Late Miocene post-rift successions of the offshore Bohai Bay Basin (China): Evidence from a seismic geomorphological study. *Marine and Petroleum Geology* 113: 104173.

Tandon, S.K. and Gibling, M.R. (1997). Calcretes at sequence boundaries in Upper Carboniferous cyclothems of the Sydney Basin, Atlantic Canada. *Sedimentary Geology* 112: 43–67.

Tandon, S.K., Sareen, B.K., Rao, M.S., and Singhvi, A.K. (1997). Aggradation history and luminescence chronology of late Quaternary semi-arid sequences of the Sabarmati basin, Gujarat, Western India. *Palaeogeography, Palaeoclimatology, Palaeoecology* 128: 339–357.

Tandon, S.K., Gibling, M.R., Sinha, R. et al. (2006). Alluvial valleys of the Gangetic Plains, India: Timing and causes of incision. In: *Incised Valleys in Time and Space* (ed. R.W. Dalrymple, D. A. Leckie, and R.W. Tillman), 15–35. Tulsa, OK: SEPM Society for Sedimentary Geology.

Thomas, R.G., Smith, D.G., Wood, J.M. et al. (1987). Inclined heterolithic stratification - terminology, description, interpretation and significance. *Sedimentary Geology* 53: 123–179.

Tockner, K., Ward, J.V., Arscott, D.B. et al. (2003). The Tagliamento River: a model ecosystem of European importance. *Aquatic Sciences* 65: 239–253.

Tooth, S. (2000). Process, form and change in dryland rivers: a review of recent research. *Earth-Science Reviews* 51: 67–107.

Tooth, S. and Nanson, G.C. (2000). The role of vegetation in the formation of anabranching channels in an ephemeral river, Northern plains, arid central Australia. *Hydrological Processes* 14: 3099–3117.

Torres, M.A., Limaye, A.B., Ganti, V. et al. (2017). Model predictions of long-lived storage of organic carbon in river deposits. *Earth Surface Dynamics* 5: 711–730.

Trimble, S.W. (1997). Stream channel erosion and change resulting from riparian forests. *Geology* 25: 467–469.

Turnbull, W.J., Krinitzsky, E.L., and Weaver, F.S. (1966). Bank erosion in soils of the Lower Mississippi valley. *Journal Soil Mechanics and Foundations Division, Proceedings American Society of Civil Engineers* 92: 121–136.

Turzewski, M.D., Huntington, K.W., and LeVeque, R.J. (2019). The geomorphic impact of outburst floods: Integrating observations and numerical simulations of the 2000 Yigong flood, eastern Himalaya. *Journal of Geophysical Research: Earth Surface* 124: 1056–1079.

Tye, R.S. (2004). Geomorphology: an approach to determining subsurface reservoir dimensions. *AAPG Bulletin* 88: 1123–1147.

Tye, R.S. (2013). Quantitatively modeling alluvial strata for reservoir development with examples from Krasnoleninskoye field, Russia. *Journal of Coastal Research* 69: 129–152.

Uchman, A., Pika-Biolzi, M., and Hochuli, P.A. (2004). Oligocene trace fossils from temporary fluvial plain ponds: an example from the Freshwater Molasse of Switzerland. *Eclogae Geologicae Helvetiae* 97: 133–148.

Underwood, N.J. (2022). An investigation of growth and movement patterns of mid-channel bars in the Missouri River. Unpublished M.Sc. thesis, Texas Christian University, Fort Worth, Texas, 54 pp.

Valencia, G.L., Buatois, L.A., Mángano, M. G. et al. (2024). Evaluating trace fossils, fluvial architecture, and colonization patterns in channel and overbank deposits from the Miocene Vinchina Formation, western Argentina. *Palaios* 39: 363–383.

Vandenberghe, J. (2013). Grain size of fine-grained windblown sediment: a powerful proxy for process identification. *Earth-Science Reviews* 121: 18–30.

Vandenberghe, J. (2024). Fluvial morphology in different permafrost environments – A review. *Quaternary 2024* 7: 15. https://doi.org/10.3390/quat7010015.

Vandenberghe, J., Sun, Y., Wang, X. et al. (2018). Grain-size characterization of reworked fine-grained aeolian deposits. *Earth-Science Reviews* 177: 43–52.

Veevers, J.J., Belousova, E.A., Saeed, A. et al. (2006). Pan-Gondwanaland detrital zircons from Australia analysed for Hf-isotopes and trace elements reflect an ice-covered Antarctic provenance of 700-500 Ma age, TDM of 2.0-1.0 Ga, and alkaline affinity. *Earth-Science Reviews* 76: 135–174.

Venditti, J.G., Nelson, P.A., and Dietrich, W.E. (2008). The domain of bedload sheets. *Marine and River Dune Dynamics* 3: 315–321.

Ventra, D. and Nichols, G.J. (2013). Autogenic dynamics of alluvial fans in endorheic basins: outcrop examples and stratigraphic significance. *Sedimentology* 61: 767–791.

Vincent, S.J. (2001). The Sis palaeovalley: a record of proximal fluvial sedimentation and drainage basin development in response to Pyrrenean mountain building. *Sedimentology* 48: 1235–1276.

Visscher, H., Looy, C.V., Collinson, M.E. et al. (2004). Environmental mutagenesis during the end-Permian ecological crisis. *PNAS* 101: 12952–12956.

Waelbroeck, C., Labeyrie, L., Michel, E. et al. (2002). Sea-level and deep water temperature changes derived from benthic foraminifera isotopic records. *Quaternary Science Reviews* 21: 295–305.

Walcker, R., Corenblit, D., Julien, F. et al. (2021). Contribution of meandering rivers to natural carbon fluxes: Evidence from the Ucayali River, Peruvian Amazonia. *Science of the Total Environment* 776: https://doi.org/10.1016/j.scitotenv.2021.146056.

Walker, S.P. (2017). Profiles and paleohydrology of supercritical rivers, the Triassic Dockum Group of west Texas. Unpublished M.Sc. thesis, Texas Christian University, 82 pp.

Walker, S. and Holbrook, J. (2023). Structures, architecture, vertical profiles, palaeohydrology and taphonomy of an upper-flow-regime-dominated fluvial system, the Triassic Dockum Group of the Palo Duro Canyon, Texas. *Sedimentology* 70: 645–684.

Walter, R.C. and Merritts, D.J. (2008). Natural streams and the legacy of water-powered mills. *Science* 319: 299–304. https://doi.org/10.1126/science.1151716.

Wang, J. and Plink-Björklund, P. (2020). Variable-discharge-river macroforms in the Sunnyside Delta Interval of the Eocene Green River Formation, Uinta Basin, USA. *Sedimentology* 67: 1914–1950.

Wang, X., Lu, H., Vandenberghe, J. et al. (2012). Late Miocene uplift of the NE Tibetan Plateau inferred from basin filling, planation and fluvial terraces in the Huang Shui catchment. *Global and Planetary Change* 88-89: 10–19.

Watanabe, Y., Martini, J.E.J., and Ohmoto, H. (2000). Geochemical evidence for terrestrial ecosystems 2.6 billion years ago. *Nature* 408: 575–578.

Webb, A.A. and Erskine, W.D. (2003). Distribution, recruitment, and geomorphic significance of large woody debris in an alluvial forest stream: Tonghi Creek, southeastern Australia. *Geomorphology* 51: 109–126.

Weissmann, G.S., Hartley, A.J., Nichols, G.J. et al. (2010). Fluvial form in modern continental sedimentary basins: distributive fluvial systems. *Geology* 38: 39–42.

Weissmann, G.S., Hartley, A.J., Scuderi, L. A. et al. (2013). Prograding distributive fluvial systems – Geomorphic models and ancient examples. In: *New Frontiers in Paleopedology and Terrestrial Paleoclimatology* (ed. S.G. Driese and L. C. Nordt), 131–147. Tulsa, OK: SEPM Special Paper 104.

Weissmann, G.S., Hartley, A.J., Scuderi, L. A. et al. (2015). Fluvial geomorphic elements in modern sedimentary basins and their potential preservation in the rock record: a review. *Geomorphology* 250: 187–219.

Westaway, R. and Bridgland, D.R. (2010). Causes, consequences and chronology of large-magnitude palaeoflows in Middle and Late Pleistocene river systems of northwest Europe. *Earth Surface Processes and Landforms* 35: 1071–1094.

Wheaton, J.M., Fryirs, K.A., Brierley, G.J. et al. (2015). Geomorphic mapping and taxonomy of geomorphic landforms. *Geomorphology* 248: 273–295.

Whiting, P.J., Dietrich, W.E., Leopold, L.B. et al. (1988). Bedload sheets in heterogeneous sediment. *Geology* 16: 105–108.

Widera, M. (2015). Compaction of lignite: a review of methods and results. *Acta Geologica Polonica* 65: 367–378.

Willems, C.J.L., Nick, H.M., Donselaar, M. E. et al. (2017). On the connectivity anisotropy in fluvial Hot Sedimentary Aquifers and its influence on geothermal doublet performance. *Geothermics* 65: 222–233.

Williams, G.P. (1978). Bank-full discharge of rivers. *Water Resources Research* 14: 1141–1154.

Williams, R.T. and Fryirs, K.A. (2020). The morphology and geomorphic evolution of a large chain-of-ponds river system. *Earth Surface Processes and Landforms* 45: 1732–1748.

Williams, H.F.L. and Roberts, M.C. (1989). Holocene sea-level change and delta growth: Fraser River delta, British Columbia. *Canadian Journal of Earth Sciences* 26: 1657–1666.

Willis, B. (1993). Evolution of Miocene fluvial systems in the Himalayan foredeep through a two kilometer-thick succession in northern Pakistan. *Sedimentary Geology* 88: 77–121.

Willis, B.J. and Behrensmeyer, A.K. (1994). Architecture of Miocene overbank deposits in northern Pakistan. *Journal of Sedimentary Research* B64: 60–67.

Willis, B.J. and Tang, H. (2010). Three-dimensional connectivity of point-bar deposits. *Journal of Sedimentary Research* 80: 440–454.

Wilson, A., Flint, S., Payenberg, T. et al. (2014). Architectural styles and sedimentology of the fluvial Lower Beaufort Group, Karoo Basin, South Africa. *Journal of Sedimentary Research* 84: 326–348.

Wisshak, M., Volohonsky, E., Seilacher, A., and Freiwald, A. (2004). A trace fossil assemblage from fluvial Old Red deposits (Wood Bay Formation; Lower to Middle Devonian) of NW-Spitsbergen, Svalbard. *Lethaia* 37: 149–163.

Wohl, E. (2011). Threshold-induced complex behavior of wood in mountain streams. *Geology* 39: 587–590.

Wohl, E. (2013). Floodplains and wood. *Earth-Science Reviews* 123: 194–212.

Wójcicki, K.J. (2022). Current and paleo sources of organic material within fluvial features of the meandering Ruda River, Poland. *Catena* 219: 106636.

Wolman, M.G. and Miller, J.P. (1960). Magnitude and frequency of forces in geomorphic processes. *Journal of Geology* 68: 54–74.

Wooldridge, C.L. and Hickin, E.J. (2005). Radar architecture and evolution of channel bars in wandering gravel-bed rivers: Fraser and Squamish rivers, British Columbia, Canada. *Journal of Sedimentary Research* 75: 844–860.

Wright, V.P. and Marriott, S.B. (1993). The sequence stratigraphy of fluvial depositional systems: the role of floodplain sediment storage. *Sedimentary Geology* 86: 203–210.

Wysocki, N. and Hajek, E.A. (2021). Mud in sandy riverbed deposits as a proxy for ancient fine-sediment supply. *Geology* 49: 931–935.

Xu, J. (2004). A study of anthropogenic seasonal rivers in China. *Catena* 55: 17–32.

Zeichner, S.S., Nghiem, J., Lamb, M.P. et al. (2021). Early plant organics increased global terrestrial mud deposition through enhanced flocculation. *Science* 371: 526–529.

Zellman, K.L., Plink-Björklund, P., and Fricke, H.C. (2020). Testing hypotheses on signatures of precipitation variability in the river and floodplain deposits in the Paleogene San Juan Basin, New Mexico, USA. *Journal of Sedimentary Research* 90: 1770–1801.

Zhu, L., Chen, D., Hassan, M.A., and Venditti, J.G. (2022). The influence of riparian vegetation on the sinuosity and lateral stability of meandering channels. *Geophysical Research Letters* 49: e2021GL096346.

6

Lake Environments

Chenglin Gong[1], Shunli Li[2], Ronald J. Steel[3], Xinghe Yu[4], Luis A. Buatois[5], and Xiaomin Zhu[6]

[1] *College of Geosciences, China University of Petroleum (Beijing), Beijing, China*
[2] *School of Energy Resources, China University of Geosciences, Beijing, China*
[3] *Department of Geological Sciences, Jackson School of Geosciences, University of Texas, Austin, TX, USA*
[4] *School of Energy Resources, China University of Geosciences, Beijing, China*
[5] *Department of Geological Sciences, University of Saskatchewan, Saskatoon, Canada*
[6] *College of Geosciences, China University of Petroleum (Beijing), Beijing, China*

Summary

Present-day lakes contain less than 0.02% of the water in the hydrosphere and occupy less than 2% of the Earth's surface. Lake deposits are also volumetrically a minor component of the stratigraphic record, yet their geological significance is far greater than these numbers suggest. Lakes and lake deposits host economic resources of minerals (e.g. lithium) and hydrocarbons. They contain long and continuous continental climate records and are significant sources of food and biodiversity.

In this chapter, we summarise lacustrine environments, their processes, facies, ichnology and sequence stratigraphy in both modern and ancient settings. For this, we use the pre-eminent set of lacustrine environments in the geological record: Chinese lakes. Lake environments are highly diverse and have been classified according to the hydrology, tectonic setting and genesis. Depth, volume and area are the three key parameters of lake basins, whereas density, temperature and salinity represent the three controlling properties of lake waters. Lake processes are governed largely by lake and groundwater hydrology, flow mechanisms (i.e. surface waves, lake currents, seiches, sediment-gravity flows, hypo- and hyperpycnal flows), chemistry (e.g. inorganic precipitation) and biochemical processes (e.g. biogenic carbonate production). Siliciclastic lacustrine facies are divided into shallow-lake deposits (e.g. beach and nearshore-bar deposits and a variety of sub-aerial and subaqueous deltaic deposits) and deep-lacustrine deposits (e.g. sub-lacustrine canyons, channel and channel-levee systems, mass-transport complexes [MTCs] and different kinds of sub-lacustrine fans). Facies of carbonate and mixed carbonate–siliciclastic systems are similarly sub-divided into littoral (carbonate shoals, siliciclastic shoals and delta deposits), sublittoral (microbial carbonate or sandstone/mudstone) and profundal (sediment-gravity flow deposits and organic-rich zones). Lacustrine faunas include annelids, aquatic to semi-aquatic insects, crustaceans, molluscs, fish, mammals and birds. Lacustrine systems are markedly different from marine systems, as short-term variations in sediment and water discharge cause lake levels to change, as well as longer-term external force. These variations cause differences in sequence stratigraphic expression in different lake types (i.e. over-filled, balanced-fill and under-filled), tectonostratigraphic setting (e.g. syn-rift and post-rift) and sediment supply (i.e. siliciclastics, carbonates or mixed carbonate–siliciclastics). Sequence stratigraphy and source-to-sink studies provide two complementary schools of predictive stratigraphy for the identification of sand-prone (reservoir) and mud-prone (source rock/seal) portions of lacustrine successions. From a source-to-sink perspective, emphasis is placed on sediment flux control through changes in tectonic uplift, climate, catchment area and bedrock geology. Sequence stratigraphy additionally emphasises sediment partitioning and accommodation sinks, with controls through tectonic subsidence and lake-level change.

In summary, lakes are not just small oceans. The different and varied interactions and feedbacks of sediment and water supply, physical processes, chemistry and accommodation among lake types require different models to summarise and understand their facies and stratal records.

Reading's Sedimentary Environments: Processes, Facies and Stratigraphy, Fourth Edition. Edited by Bruce Levell.
 Published 2026 by John Wiley & Sons Ltd.
Companion website: www.wiley.com/go/levell4e

6.1 Introduction

6.1.1 What are Lake Environments?

A lake is 'an inland body of very slowly moving or standing water occupying a depression in the Earth's crust' (Kelts 1988). Lakes form where there is a supply of water to a topographic low on the land surface which, at least temporarily, exceeds the loss of water from that basin by outflow, seepage and evaporation. Lakes currently contain less than 0.02% of the water in the hydrosphere and occupy <2% of the Earth's surface (i.e. $250 \times 10^4\ km^2$). The world's largest lake is the Caspian Sea (374,000 km^2), the deepest is Lake Baikal in Russia (1642 m), and one of the larger shallow lakes, Lake Chad in central Africa, is only 7–8 m deep but up to 26,000 km^2 (Bridge and Demicco 2008).

A large-scale perspective of the tectonic origin of many lakes is provided by comparing: (i) the suite of rift-basin lakes that developed along the Mesozoic Atlantic extensional margins (Beglinger et al. 2012), (ii) the modern East African Rift-related lakes, (iii) the Neogene Laramide lakes that succeeded the foreland basin of the Western Interior seaway (Cather et al. 2019), and (iv) the long string of Miocene Paratethys lakes (Pannonian, Dacian, Black Sea, and Caspian lakes) that developed alongside Alpine uplifts during ocean closure (Sztanó et al. 2013; Fongngern et al. 2016, 2018). However, despite these varied large-scale tectonic controls, we agree with Talbot and Allen (1996) that from a facies perspective perhaps the two most outstanding aspects of lake sediments are (i) their dynamic response in process and lithology to a generally high frequency of lake-level changes resulting in abrupt and frequent facies changes and (ii) their sensitivity to climate change and therefore their value as continental climate archives.

6.1.2 Recent Sedimentological Research on Lakes

Since the publication of AAPG Memoir 50 on lacustrine sedimentation (Katz 1990), we have witnessed an expansion of research on lakes, through the efforts of individual researchers as well as workgroups in academia and industry (e.g. the 'Global Palaeo-environmental Archives in Lacustrine Systems-International Geological Correlation Program Project 324' (1994)). Recent examples of lake research are represented by the 'International Continental Scientific Drilling Program', for example, in the Cretaceous Songliao Basin (Feng et al. 2013; Wang et al. 2013) and the Dead Sea Deep Drilling Project (Goldstein et al. 2020; Lu et al. 2020, 2021). This work is summarised in several books (e.g. Gierlowski-Kordesch and Kelts 1994, 2000; Cohen 2003) and in several noteworthy reviews (e.g. Bohacs et al. 2003; Håkanson 2007; Renaut and Gierlowski-Kordesch 2010; Baganz et al. 2012; Scott et al. 2012; Smith and Carroll 2015). Also notable is the tribute to the work of Elizabeth Gierlowski-Kordesch that recently appeared in a new series of volumes on palaeolimnology (Rosen et al. 2021). Recent trends include (i) research on ancient lacustrine systems that integrate stratigraphy, sedimentology, biofacies (including ichnofacies) and organic geochemistry, with a focus on hydrocarbon exploration (e.g. Bohacs et al. 2003; Smith and Carroll 2015), (ii) the use of seismic data to obtain 3D perspectives on the development and evolution of lacustrine basins and their stratigraphies (e.g. Sztanó et al. 2013; Gong et al. 2019), (iii) research on physical, chemical and biological controls and their depositional responses in lacustrine settings (e.g. Rosen et al. 2021) and (iv) studies of modern and Quaternary lakes regarding sediment dispersal, organic matter production and preservation and temporal evolution of lake hydrology often with a focus on (palaeo) climate research (e.g. Vandekerkhove et al. 2020).

6.2 Environments

6.2.1 Morphology of Modern Lakes

Depth, volume and area are three simple but key parameters that affect many physical, chemical, and biological processes in lakes. Ranges in maximum depth, mean depth, volume and area of 69 lakes from Bohacs et al. (2003) show the following ranges: maximum depth: 1–1741 m (average 134 m), mean depth: 1–730 m (average 53 m), volume: 1–78,200 km^3 (average 1952 km^3) and area: 500–374,000 km^2 (average 10,585 km^2) (Figure 6.1a,b). The power-law relationships among volume, mean depth and maximum depth have correlation coefficients of $R^2 > 75\%$ (Figure 6.1c,d), suggesting that maximum and mean lake depths play a pivotal role in determining lake volume. These dimensions of modern lakes show no strong relationship with other physical parameters, such as climate, latitude, water chemistry, drainage basin area, altitude or the origin of the lakes (Bohacs et al. 2003).

6.2.2 Properties of Modern Lakes

Density, temperature and salinity are the three main properties of lake waters that affect sediment transport, water-column stability, biological activity, organic-matter production and preservation and hence facies and stratigraphy. These properties are detailed below.

6.2.2.1 The Density of Lake Waters

The average density of lake water is $1.025\ kg\ m^{-3}$ and is a function largely of temperature and, to a lesser extent, of salinity and sediment concentration. Freshwater is most dense at 4°C and is less dense at both higher and lower temperatures.

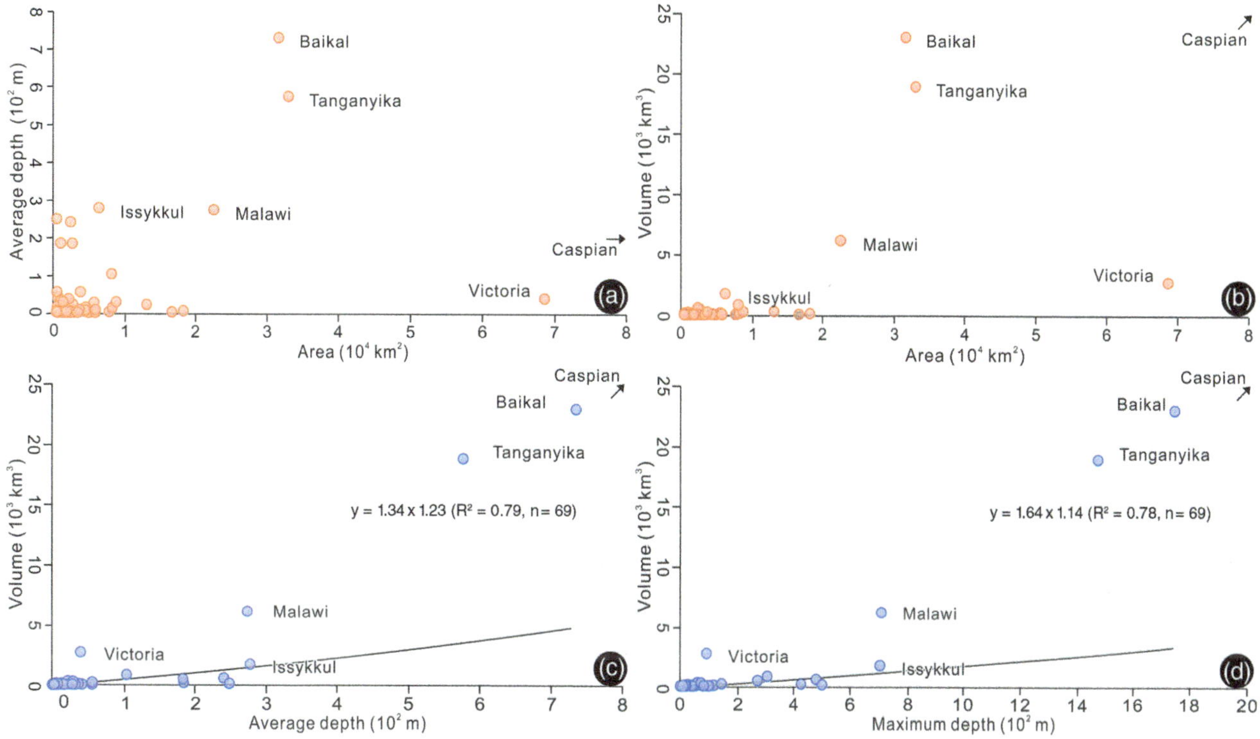

Figure 6.1 Relationships of surface area, average depth, maximum depths and volume for 69 modern lakes from Bohacs et al. (2003). There is a weak relationship of water depth to lake area (a and b) but a power-law relationship ($R^2 > 75\%$) among volume, mean depth and maximum depth (c and d). Maximum and mean lake depths play a pivotal role in determining lake volume. The largest lakes by area are not necessarily the lakes with the greatest depth. For example, Lake Baikal (Russia) has the greatest maximum water depth (1642 m) and mean water depth (730 m) but is seventh in area (31,500 km^2).

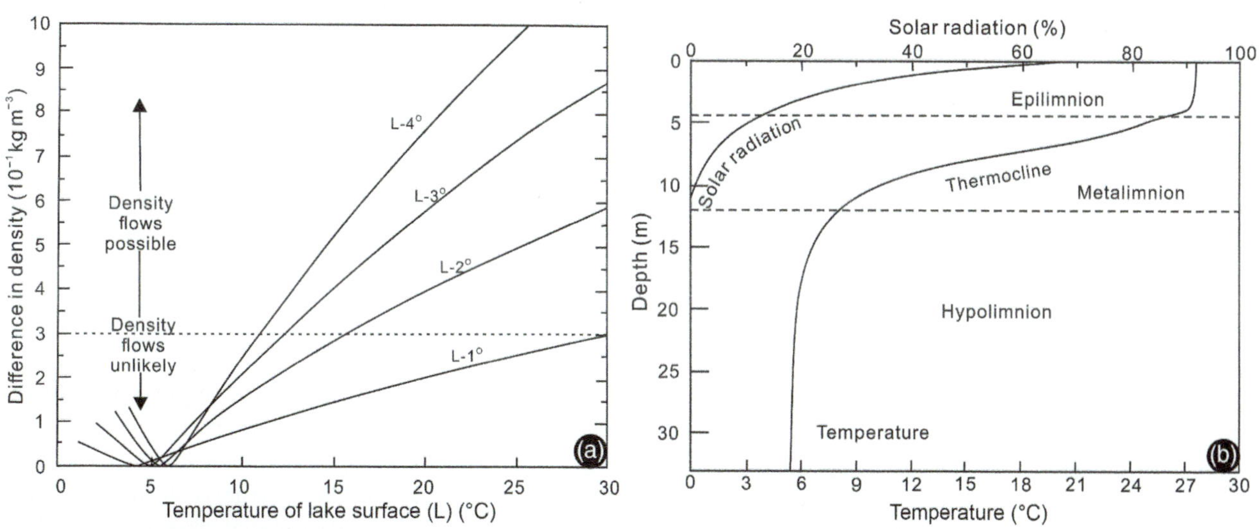

Figure 6.2 (a) Relationship between the difference in density and temperature of an inflow water that is 1, 2, 3 or 4 °C cooler than lake surface water (adapted from Axelsson 1967 and Talbot and Allen 1996). For a given temperature difference, the density difference is greatest at higher surface temperatures (L) with the possibility of direct density underflows of inflowing cool water and/or density stratification. Density differences of as little as 0.3 kg m^{-3} can impede mixing. (b) The temperature profile of Crooked Lake, Indiana, 18 July 1964 (Talbot and Allen 1996 / John Wiley & Sons).

Differences in water density play an important role in the physical, chemical and biological dynamics of lakes and their interactions (Figure 6.2a). The main hydrological differences between non-saline lakes and oceans are a consequence of freshwater having a maximum density at 4°C, significantly above its freezing point of 0°C (Figure 6.2a) whereas seawater does not. The density of fresh lake water decreases both above and below 4°C.

The rate of increase of density with falling temperature increases as water temperature decreases. Figure 6.2a shows the density differences of surface waters at a temperature '*L*' with waters that are from 1° to 4° cooler. Hot tropical lake waters, therefore, tend to become density-stratified more easily than temperate lakes. The density of lake water also increases with increasing concentrations of dissolved salts, but the salinity-induced density stratification is important only in certain highly saline lakes or lakes influenced by deep-water hydrothermal springs (e.g. Lake Kivu, DR Congo/Rwanda).

6.2.2.2 The Temperature of Lake Waters

The temperature of lake water is determined by the source of heat, which is mainly solar radiation, although geothermal heat flow is also significant in some deep rift lakes (e.g. Malawi, Tanganyika, Kivu and Baikal). Lake-water temperature has a major influence on biologically important factors, such as dissolved oxygen, mineral content, water clarity and water circulation.

The vertical temperature profile of a lake is a direct response to the penetration of solar radiation (Figure 6.2b). Limnologists refer to the upper, warmer, water mass as the '*epilimnion*', the transitional layer as the '*metalimnion*' and the colder, denser, bottom water as the '*hypolimnion*' (see Figure 6.2b). The upper and lower water masses are separated by a surface ('*thermocline*') across which the temperature gradient is greatest. The thermocline occurs in the metalimnion, a transition zone between distinct water zones that differ in their density, chemical composition and biochemical processes (i.e. it defines lake stratification).

A study of several lakes in Wisconsin, USA and central Canada shows a simple relationship between the depth of the summer thermocline and the maximum wind fetch (the distance over which the wind blows uninterrupted by topography). Seasonal changes in air and surface-water temperatures, and turbulence induced by winds break down any stratification in the upper layers, lower the thermocline and overturn the water column. Lakes in which water circulates completely to the bottom during winter cooling are termed '*holomictic*', whereas those that have only a partial circulation, leaving a permanently stagnant bottom layer, are termed '*meromictic*'. Lakes with persistent water-column stratification that overturn very seldomly are termed '*oligomictic*'; these are typically tropical lakes with very high surface water temperatures.

6.2.2.3 The Salinity of Lake Waters

The salinity of lake water refers to the amount (by weight) of salts dissolved in water. It is governed by the net contributions from drainage basin run-off, wind-blown material and aerosols, as well as the balance between evaporation and precipitation. Salinity is arguably the main controlling factor on lacustrine flora and fauna with diversity dramatically decreasing towards the hypersaline end-member. The salinity of lake water increases as the ratio of the volumes of evaporation to net inflow increases. Lake-water salinity is prone to changing rapidly and frequently, especially in closed hydrography basins (Renaut and Gierlowski-Kordesch 2010). This contrasts with marine waters whose salinity varies little and has been stable for more than one billion years. Lake-water salinity can also vary laterally, especially in shallow stratified lakes – fresher near river inputs and more saline in restricted bights. For example, Great Salt Lake, Utah, the largest hypersaline lake in North America also hosts the largest freshwater wetlands in the western USA (around the Bear River delta).

The most important sources of salts in all lake waters are (i) salts derived directly by ice, water, and wind transport or solution from bedrock and soil; and (ii) salts in dust, rain and snow falling into the lake. Lake water is classified as 'fresh' with a salinity of <0.05%, 'brackish' with a salinity of 0.05–3%, 'saline' with a salinity of 3–5% and 'brine' with a salinity of >5% up to a maximum of 26–28%. Both freshwater lakes and saline lakes occur from the Equator to the Poles. Ancient lakes can be classified roughly by salinity based on their sediment composition: lakes with mainly clastic deposits, such as silt and small amounts of carbonate, are equivalent to freshwater and brackish lakes. Lakes with mainly carbonate deposition are equivalent to saline lakes. Lakes containing mainly sulphates and chlorides are equivalent to brine-forming playas. Freshwater lakes occur wherever a surface or subsurface outflow maintains an open drainage, and where the annual recharge from the precipitation, runoff and groundwater is greater than, or balanced by, water loss from outflow and evaporation. Saline lakes, in contrast, form in closed or intermittently closed drainage systems, in which the rate of evaporation or groundwater loss exceeds the amount of annual recharge from precipitation, runoff and groundwater.

6.2.3 Diversity and Classification of Lakes

Lakes are highly diverse, occur in a variety of geological settings and can be categorised in many different ways, according to: (i) hydrology (i.e. open and closed lakes), (ii) tectonic setting (e.g. fault-bounded lakes, depression lakes and transitional lakes), (iii) genesis (i.e. glacial lakes, volcanic lakes, barrier lakes, wind lakes, karstic lakes and tectonic lakes) and (iv) salinity (i.e. freshwater lakes, brackish lakes, salt lakes and playa) (Talbot and Allen 1996; Yu et al. 2018).

6.2.3.1 Classification of Lakes According to Hydrology

Hydrologically, lakes are classified as either 'open' or 'closed'. Open lakes exist wherever a surface or subsurface outflow maintains open drainage: water thus flows through the system and annual inflow and the direct precipitation equal or exceed the sum of the annual outflow and evaporation. Closed lakes, in contrast, exist in closed or intermittently closed drainage systems, in which the annual inflow and the direct precipitation are not outpaced by the annual outflow and evaporation: water is thus trapped in the system. Importantly, lake basins are neither always hydrologically closed nor always hydrologically open. For example, Lake Malawi in the East African rift system is currently filled approximately to its spill point, with stable water levels fluctuating from 1 to $2\,\mathrm{m\,a^{-1}}$. However, high-resolution seismic data suggest fluctuations of as much as 100 m in the last few hundred years and that over the past several thousand years, lake levels dropped by as much as 300 m (Late Pleistocene to Early Holocene: Carroll and Bohacs 1999). Hydrologically open and closed conditions may thus be recorded at the time-scale of individual stratal parasequences (Carroll and Bohacs 1999; Bohacs et al. 2000). From an ecological perspective, hydrologically closed lakes are more stressful than open lakes, because their salinities can both fluctuate greatly and become very high (Gore 1989). As a result, closed lake deposits display both less intensity of bioturbation and less diversity of biogenic structures (Buatois and Mángano 1998).

6.2.3.2 Classification of Lakes According to Tectonic Setting

Lakes occur in a wide variety of tectonic settings from extensional to compressional. In active tectonic areas, lakes may develop in rift basins (e.g. Lake Malawi in East Africa), strike-slip to pull-apart basins (e.g. Dead Sea), foreland basins (e.g. Laguna Mar Chiquita in central Argentina) and intra-arc basins (e.g. Lake Titicaca, Bolivia). In rift lakes, hot springs can exert a strong control on the distribution of biogenic structures (Buatois et al. 2017). Lakes on cratons form in open topography and are normally large in area and shallow, for example, Lake Chad in Sahelian Africa, which is up to $26{,}000\,\mathrm{km^2}$ in area and Kati Thanda-Lake Eyre, an ephemeral cratonic lake in South Australia, which occupies an area up to $11{,}088\,\mathrm{km^2}$ (Olariu et al. 2021).

6.2.3.3 Classification of Lakes According to Genesis

On the basis of their genesis, lakes can be grouped into glacial lakes (Chapter 14; Section 14.5.9), volcanic lakes, barrier lakes, wind lakes, karstic lakes and the above-mentioned tectonic lakes. Many but not all glacial lakes are related to de-glaciation or direct glacier melting (e.g. Kashmir Great Lakes in Himalayan mountains). Lakes of glacial origin may be pro-glacial (e.g. Lake Malaspina, Alaska) or formed by ice-damming, barriers composed of moraine (e.g. Finger Lakes, New York State), ice scour, freeze-thaw or valley glaciation (e.g. Pitt Lake, British Columbia). These glacial lakes are generally small, with the notable exceptions of Great Bear, Great Slave, Athabasca, Winnipeg and the Laurentian Great Lakes (or indeed the Holocene pro-glacial Lake Agassiz [$260{,}000\,\mathrm{km^2}$] in Alberta/Saskatchewan, Figure 4.18). Glacial lakes are also well known from both Late Palaeozoic and Pleistocene lake deposits (Netto et al. 2012). They commonly host abundant ichnofauna, dominated by simple vermiform grazing trails, fish trails and arthropod trackways. Lakes of volcanic origin can be formed by lava damming (e.g. Lake Kiva) or crater explosion and collapse (e.g. Crater Lake, Oregon). Volcanic lakes are often small in area but can be very deep. Crater Lake, the deepest in the United States has a maximum depth of 594 m. Changbaishan Lake, the deepest in China, has an average depth of 204 m. Barrier lakes or landslide-dam lakes are caused by natural damming of a river by landslides, such as debris flows and rock avalanches, or volcanic eruptions. They may also be called 'quake lakes' if the damming landslides were caused by an earthquake (e.g. Tang's Mountain landslide – a dam lake formed by the 2008 Sichuan earthquake, China). In mountainous areas, some of these lakes were very large and long-lived (e.g. in the Tsangpo catchment in the Tibetan Plateau). For example, those in the Kali Gandaki and Marsyandi valleys in Nepal were tens of kilometres long and $\sim$200 m deep. Lakes may also be created by the action of wind (e.g. the original landscapes of Les Landes in southern France and the former Lake Faguibine in Mali) and can also develop as a result of the dissolution of underlying strata, permafrost melting or even meteorite impact (e.g. Lake Bosumtwi, Ghana).

6.3 Processes

Lake processes presented in this section include the hydrology, physical, chemical and biochemical processes operating in lake waters.

6.3.1 The Hydrology of Lake Waters

Whenever surface-layer (epilimnion) waters reach 4°C in a stratified lake, they become more dense than the underlying water. This results in density instability and causes the lake water column to overturn or mix (Figure 6.2b). The less-dense bottom waters move upwards to the surface, whereas the dense surface waters sink. Epilimnion temperatures drop during the autumn, and at some point, the lake may become isothermal, without density stratification (the water is 'isopycnic'). In this condition, the wind stress over the surface of a lake can cause the mixing of even deep lakes all the way to the bottom, renewing the oxygen in the deep-lake waters. This phenomenon is referred to as 'turnover' (Bridge and Demicco 2008). During winter, epilimnion temperatures drop further, and the epilimnion becomes less dense than the 4°C bottom waters. It floats on top of the deep-lake waters and ice may develop (Renaut and Gierlowski-Kordesch 2010). During spring, epilimnion temperatures rise, the lake warms up and again reaches an isothermal, non-density-stratified condition. The deep oxygen is once again renewed, resulting in another turnover. In most temperate lakes, this spring turnover is enough to keep the lake oxygenated all the way to the bottom. By contrast, most tropical lakes remain density-stratified throughout the year, with only rare occurrences of turnover. In extreme cases, a lack of turnover may result in the excessive accumulation of magmatic CO_2 in the lake bottom, which in turn may trigger a so-called *limnic eruption* with potentially catastrophic consequences. For example, a limnic eruption at Nyos Crater Lake killed 1746 people through carbon dioxide asphyxiation in Cameroon in 1984 (Kusakabe et al. 1989; Sigvaldason 1989; Kusakabe 2017). Density stratification can also develop because of differences in the composition of water masses. For example, most saline lakes are permanently density stratified (Gulati and Zadereev 2017).

Lake hydrology can be classified based on the timescale of water-column turnover. Most lakes are '*holomictic*'; that is, at least once each year there is a physical mixing of surface and deep waters. A holomictic lake with mixing occurring once per year is called a '*monomictic*' lake; a lake with mixing occurring twice a year (typically spring and autumn) is called a '*dimictic*' lake. In some so-called '*meromictic*' lakes, however, the layers of the lake water can remain unmixed for years, decades or centuries. The water column of meromictic saline lakes can usually be divided into three layers: the bottom layer known as the '*monimolimnion*', the middle layer referred to as the '*chemocline*' and the top layer called the '*mixolimnion*'. The water in the monimolimnion of the lake circulates little and is generally hypoxic (Section 9.2) and saltier than the rest of the lake.

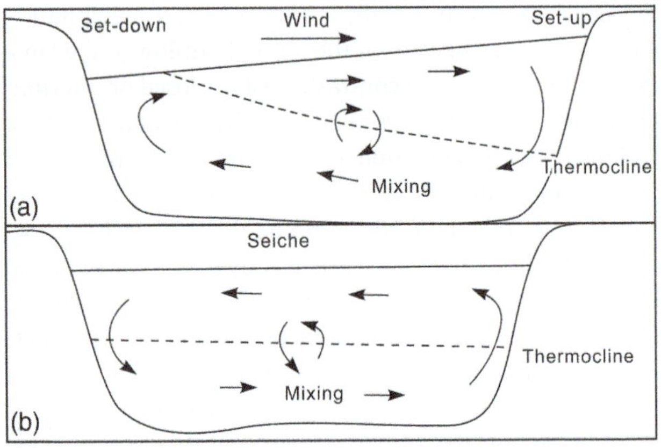

Figure 6.3 Schematic illustration of the development of lake water currents and seiches in an originally stratified lake created by a sustained wind from one direction (a) during and (b) after strong wind Bridge and Demicco (2008).

6.3.2 Physical Processes

The principal forces acting to move lake waters are those due to hydraulic gradients, wind stress (Figure 6.3) and displacement related to sedimentation processes, which all cause horizontal and vertical pressure gradients. Causes of lake water movement include surface waves, lake currents, seiches, sediment-gravity flows, hyperpycnal flows, Coriolis forces and lunar tides, as well as settling of detrital and organic particles in relatively immobile or stagnant water.

6.3.2.1 Surface Waves

Surface waves are formed by wind interacting with the lake surface. The dimensions of surface movement (e.g. wave heights, velocities of wind-driven currents) are a function of the wind strength, wind duration and the distance over which the wind blows unimpeded (i.e. 'fetch'). The energy of lake-surface waves is fed by the frictional drag of the air moving over the water, the direct force of the wind on the upwind faces of waves and pressure differences at the air–water interface. Losses of lake-wave energy occur mainly because of the turbulence in the water and, to a smaller extent, the effects of the viscosity. Lake waves have a height of several metres in some large lakes, with the greatest heights, of up to 7 m, reported in Lake Michigan in September 2011. Waves have a limited effect on the bottom in deep lakes, although they may strongly affect the shorelines. They give rise to rip currents that are, often simply called rips (or, misleadingly, a rip tide), a specific kind of return flow that can occur near beaches with breaking waves. Rip currents are notorious for their horizontal, offshore flow that sweep sediments and unsuspecting swimmers into deep water.

Although lake waves are normally much less strong and persistent than those on continental shelves, they similarly cause the drift of coarse-grained sediments along the shoreline and transport fine-grained sediments in suspension offshore. Their effectiveness varies according to the relative orientation of the shoreline and the maximum fetch (Bridge and Demicco 2008). Lake waves can mix lake surface layers or the entire isothermal water column with uniform density, which promotes the circulation of nutrients and helps to prevent thermal stratification (Figure 6.3).

6.3.2.2 Lake Currents

Lake currents have various origins, the most important being the wind-driven circulation. However, currents set up by river inflow, littoral warming or cooling and the hydrographic slope from river inflows to the outflowing spillways may also be significant. Wind stress, mostly exerted during storms, sets up complex water motion. In the nearshore zone, currents are strong (of the order of 0.30 m s^{-1} in the Great Lakes in the United States/Canada following storms) and are directed parallel to the coast, whereas deeper currents tend to be weaker and lack a preferred direction. The typical circulation pattern is one of the closed gyres lying on either side of the deepest axis of the lake basin. Wind-driven surface currents can be very effective in distributing suspended sediments, especially in lakes with minimal density contrast between inflows and lake waters. Inflow rivers themselves, together with geostrophic effects (Coriolis forces as discussed later), can provide a major driving force for water circulation and consequently for sedimentation patterns. This effect has been demonstrated even in relatively small Swiss lakes by Wright and Nydegger (1980). Wind-produced currents have been proposed to be responsible for the deposition of thick sediment drifts in Lake Cardiel of Argentina.

6.3.2.3 *Seiches* (from Talbot and Allen 1996)

Sustained wind from one direction forces the surface water of lakes to flow downwind, resulting in elevation of the water surface (*set-up*) and associated downwelling at the leeward side of the lake (Figure 6.3a). There is a complementary down-drawing of the water surface (*'set-down'*) and associated upwelling of cold water at the windward side of a given lake (Figure 6.3a). Such a lake water circulation may tilt the thermocline (Figure 6.3a) and trigger compensating motions in the water layers below. When the wind stress drops, the water flows back across the lake as an unopposed gradient current, initiating a periodic rocking motion. Such oscillations ('sloshing') of the water surface are known as a *'seiche'* (Figure 6.3b). Seiches can have amplitudes of >1 m and periods ranging from a few minutes to a few hours. The largest seiche observed in Europe was 1.87 m in amplitude in Lake Geneva in 1841. Seiches have been shown to be capable of significant sediment re-suspension when they break or surge over the bottom topography. Moreover, in stratified lakes, seiches can set up similar waves along the thermocline, resulting in so-called *'internal seiches'* (Renaut and Gierlowski-Kordesch 2010). Internal seiches have a period and amplitude much greater than an ordinary seiche and can have periods of the order of days to weeks, with amplitudes of up to 10–50 m. Oscillating waves have effects that reach water depths (wave base) that are about half of the wavelength of the waves. They may be particularly effective in causing local erosion, even at depths of several hundred metres, especially if they are focused on topographic features. Internal seiches can cause nutrient-rich deep waters to be 'eroded' and mixed upwards into the photic zone, triggering phytoplankton blooms, which may provide a significant biogenic component to local sediments (Talbot and Allen 1996).

6.3.2.4 Sediment-Gravity Flows

Mixtures of sediment and water brought in by a river or reworked from a lake delta may flow as dense, sediment-laden underflows along the lake bottom. Such flows can be episodic (sediment-gravity flows) or more persistent (hyperpycnal flows) (Figure 6.4). The recognition of persistent hyperpycnal underflows dates back to Forel (1892). Episodic gravity flows are normally ascribed to failures or mass-wasting processes triggered by earthquakes, catastrophic collapses or local slope instabilities (Schnellmann et al. 2002, 2005; Vandekerkhove et al. 2020; Cao et al. 2021). All of these mechanisms of failure can evolve downslope into various types of sediment-gravity flows and turbidity currents (Figure 6.4; Section 12.3). For example, lake sediments in Eklutna Lake in Alaska (USA) reveal the presence of spectacular failure-triggered turbidites caused by the 1964 Great Alaska earthquake (Vandekerkhove et al. 2020).

Sub-lacustrine sediment-gravity flows are normally divided into debris flows with plastic rheology ('C, D and E' in Figure 6.4) and turbidity currents with Newtonian rheology ('F' in Figure 6.4). Both debris flows and turbidity currents are normally accompanied by, and evolve from, two main kinds of mass transport (i.e. slides where there is little internal disturbance and slumps where the moving strata are deformed) ('A' and 'B' in Figure 6.4). Debris flows are further subdivided into muddy debris flows ('C' in Figure 6.4) and sandy debris flows ('D' in Figure 6.4), according to their mud contents. Turbidity currents are further subdivided into high-density turbidity currents ('E' in Figure 6.4) and low-density turbidity currents ('F' and 'G' in Figure 6.4), according to their density. Low-density turbidity currents are further subdivided into surge-like turbidity currents ('F' in Figure 6.4) and quasi-steady turbidity currents ('G' in Figure 6.4), according to the strength of sedimentary dynamic and flow duration (Mulder and Alexander 2001). Failure-triggered turbidites were preserved in Dongying and Huimin Sags of the Bohai Bay Basin, a Mesozoic–Cenozoic continental rift basin that covers an area of about 20×10^4 km^2 located in eastern China and were considered as depositional responses to different types of sediment-gravity flows, with the addition of quasi-steady hyperpycnal flows as discussed below (Figure 6.4) (Cao et al. 2021).

6.3.2.5 Hyperpycnal Flow

Forel (1892) postulated that the inflowing river water does not always mix vertically with lake waters and introduced the concepts of *hyperpycnal* (under-) flows, *homopycnal* flows and *hypopycnal* (over-) flows for the first time. Bates (1953) defined hypopycnal flows as buoyant plumes flowing at the water surface (i.e. the density of the river inflow including its sediment load is less than

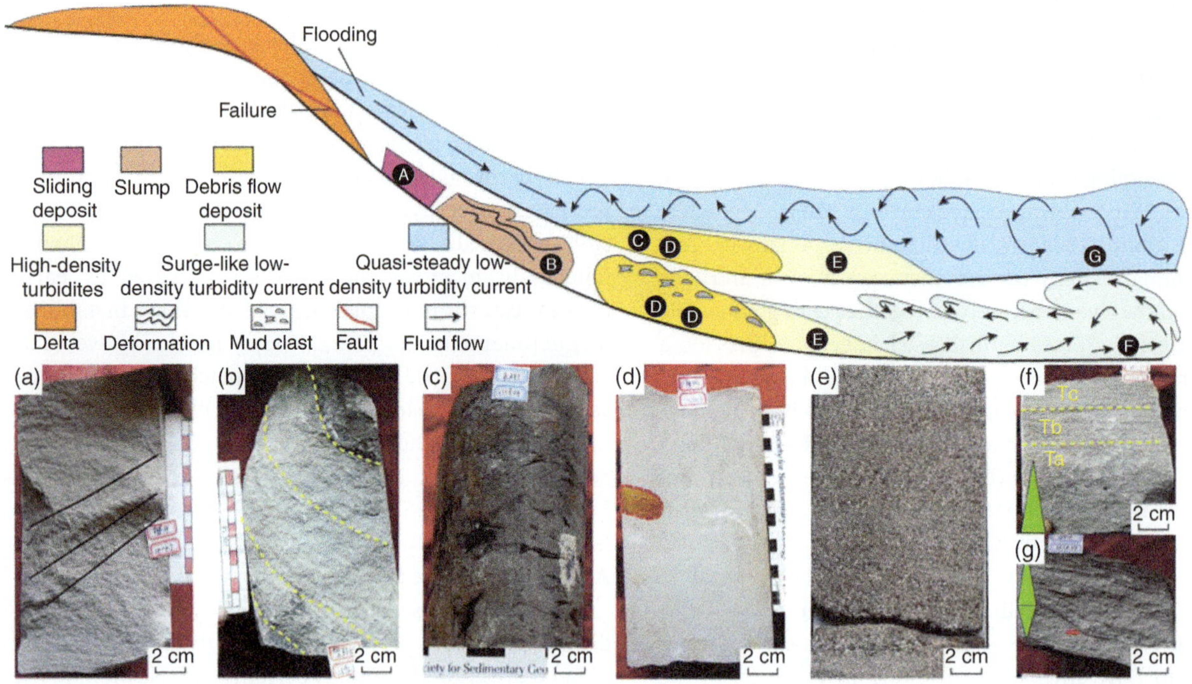

Figure 6.4 Schematic illustration of the development of sediment-gravity flows, and their resultant deposits in a lacustrine rift basin (Cao et al. 2021 / with permission from Elsevier). Two overlapping suites of sediments are represented, those generated by slope failure (A–F) and those that may be generated directly by river floods (C–G). Sediment-gravity-flow deposits in sub-lacustrine environments are typified by (a) Interpreted as slide blocks with incompletely modified beds (Well Fan 133; 2404.7 m); (b) interpreted as slumped blocks with internal folding and deformation (Well Fanxi 183; 2890.1 m); (c) interpreted as muddy debris-flow deposits with structureless mud clasts (Well Dongke 1; 3348.5 m); (d) a structureless sandy debris-flow deposit with floating mud clasts (Well Shang 548; 3240.2 m;); (e) high-density turbidites with internal plane-lamination (Well Shi 100; 3087.7 m); (f) turbidite with incomplete Bouma sequence created by low-density turbidity currents (Well Shi 129; 3438.7 m); and (g) turbidites with inverse-normal graded beds deposited by quasi-steady low-density turbidity currents (Well Niu 876; 3374.5 m). Core examples (a)–(c) and (e)–(g) are from the Dongying Sag of the Bohai Bay Basin, and core example (d) is from the Huiming Sag of the Bohai Bay Basin.

that of ambient lake waters) and homopycnal flows as flows of river effluent with a density in the same range as the density of the fluid in the receiving basin. A hyperpycnal flow is an underflow current of riverine materials (together with whatever is eroded by the flow from delta fronts or the basin floor) transported downslope due to its excess density compared to that of the receiving water body (Mulder et al. 2003). This definition of hyperpycnal flow excludes surge-type turbidity currents. In freshwater lakes, the suspended sediment concentration needed to produce a hyperpycnal underflow is quite low (1 kg m^{-3}) (Mulder et al. 2003; Bhattacharya 2006). Such flows are normally triggered during large or extreme floods, which bring significant fluxes of sediments towards the basin and represent a large proportion of the fine-grained event beds deposited in lake and marine basins (Mulder et al. 2003).

6.3.2.6 Lunar Tides

Although true tides (i.e. changes in water level caused by the gravitational forces of the sun and moon) occur in lakes, most bodies of lake water are too small for this effect to be significant. For example, tides are only 0.03 m even in Lake Baikal and Lake Superior. The largest Great Lakes tides are <5 cm in amplitude. These minor tidal variations are normally masked by the greater fluctuations in lake levels produced by wind and barometric pressure changes. In some lakes, the time period between the 'high' and 'low' of a seiche may be between four and seven hours. As this is very similar to the six-hour time period of the tides on the ocean, it is frequently mistaken for a tide.

6.3.2.7 Coriolis Forces

The Coriolis effect is an apparent deflection of the path of an object that moves within a rotating coordinate system. The horizontal deflection effect is zero at the equator and greatest at the poles. Coriolis effects may deflect the direction of water movement to be oblique to the direction of the wind or river inflow. In some very large and deep lakes, such deviation can reach a maximum at the surface of 45°, decreasing down from the surface and thus defining the so-called *Ekman spiral*. The balance of the pressure-gradient force with the Coriolis force frequently produces '*geostrophic flows*' at right angles to the pressure gradient, with low pressure on the left (Northern Hemisphere). Coriolis effects are significant only in very large lakes at higher latitudes, causing, for example, flow into a lake to move towards the right in the Northern Hemisphere and the left in the Southern Hemisphere.

6.3.3 Chemical Processes

6.3.3.1 Chemical Composition

Non-marine waters are dominated by four cations (calcium, magnesium, sodium and potassium) and three anions (bicarbonate, sulphate and chloride). Freshwater lakes contain Na^+ and Ca^{2+} as dominant cations along with highly variable concentrations of K^+ and Mg^{2+} and most anions. The supply and concentration of solutes depend mainly on the bedrock and sediment composition of the drainage basin, along with precipitation, weathering, groundwater flow, residence time, evaporation, dissolution–precipitation reactions and biological activity (Jones and Deocampo 2004). Saline lakes are generally dominated by either bicarbonate or sulphate and this is determined by the starting concentration of Ca^{2+} because calcite is the first mineral to form upon evaporation (Jones and Deocampo 2004). Sedimentary carbonate bedrock yields the most bicarbonate, but igneous and metamorphic terranes can also produce some bicarbonate ions through silicate mineral weathering by hydrolysis. Sulphate ions are mainly the products of sulphide mineral oxidation (e.g. pyrite) but are also released by the weathering of sulphate evaporites (gypsum and anhydrite) and are present in rainfall (from marine aerosols or pollution). Most chloride originates from marine aerosols and connate groundwaters but can also derive from halite dissolution and fluid inclusions within evaporites. The relative proportions of these and other ions are principally determined by the bedrock lithotypes of the catchment region, evaporative concentration, outgassing and changes in temperature and water-column stratification.

6.3.3.2 pH Conditions

The pH of lake waters influences the release of nutrients from incoming sediments, the bio-availability of such nutrients, the viability and vigour of organisms, syn- and post-depositional precipitation of authigenic minerals and the fate of toxic heavy metals. The optimum pH for most aquatic organisms ranges from 7 to 8.5. Open lakes usually have pH values ranging from 6 and 9 because most of them are strongly buffered by the CO_2–HCO_3–CO_3^{2-} system (Talbot and Allen 1996). Nearly all lake waters with pH values of less than 4 occur in volcanic regions that receive strong mineral acids, particularly sulphuric acids produced by the weathering of sulphide minerals. The most highly acidic lakes (pH < 3) occur in volcanic regions where H_2S and other gases dissolve in lake waters and are oxidised to produce sulphate ions and protons, (Renaut and Gierlowski-Kordesch 2010). Low pH values are also found in natural waters rich in dissolved organic matter, such as big pools.

6.3.3.3 The Relationship of Chemical Processes to Biological Activity

Chemical characteristics and processes of lake waters are closely inter-related with various sorts of biological activity. The dissolved oxygen content of a lake is of fundamental importance to organic productivity since oxygen is essential to the metabolism of all aerobic aquatic organisms. Oxygen concentrations are at the same time strongly dependent upon the rates of organic matter production and decomposition. Dissolved oxygen is supplied to the water body by chemical equilibration with the atmosphere and by photosynthesis and is consumed by aerobic respiration. Surface waters of low productivity (*oligotrophic*) lakes are nearly always close to saturation with respect to oxygen. However, in highly productive (*eutrophic*) lakes, the waters may be either supersaturated, with oxygen due to high rates of photosynthesis, or undersaturated due to the respiration and oxidation of organic matter.

Where and when the oxygen content of surface waters is sufficient to allow aerobic processes to function normally, the productivity of lake waters is largely determined by the availability of key nutrients, notably phosphorus, nitrogen and silica. Phosphorus and nitrogen have probably always been vital nutrients and complete consumption of one or the other may severely limit organic growth, even when conditions are otherwise favourable. Silica is important because diatoms are a major group of lacustrine phytoplankton. The nutrient budget of many large lakes depends largely upon the regeneration processes via overturning and circulation (e.g. ammonium regeneration), although nutrients may be renewed also by river inflow and rainfall.

6.3.4 Biochemical Processes

In some open lakes where the clastic sediment flux is low, biochemical processes predominate, leading to the deposition of largely biochemical sediments. Biochemical processes in lakes are performed by microbes, as well as macroflora and fauna, and lead in particular to the precipitation of carbonate minerals, which occurs in almost any climate or tectonic settings (Gierlowski-Kordesch 2010). The deposition of carbonate sediments depends heavily on the presence and abundance of carbonate rocks or Ca-bearing volcanic rocks in the catchment area or in the bedrock through which groundwater migrates (e.g. Deocampo and Renaut 2016). Conversely, the lack of influence of biological processes may prevent the precipitation of carbonate minerals. The main contributors of detrital calcite and aragonite to lake sediments are the shells of bivalves, gastropods and ostracods, carbonate precipitated by charophytes (freshwater green algae) and carbonate grains from drainage basins. In many lakes, pelagic calcite or aragonite 'whitings' form periodically by processes analogous to those on shallow-marine carbonate platforms (Sections 6.4.2 and 11.6) (Renaut and Gierlowski-Kordesch 2010). Crystal-nucleation in the lake water column can be triggered by evaporative concentration, seasonal warming or de-gassing of CO_2, all of which decrease $CaCO_3$ solubility. Most carbonate precipitation in lakes, however, is probably bio-induced by photosynthetic algae and bacteria that remove CO_2. Phosphatic material is supplied from fish and other vertebrate bones and coprolites.

In addition, vegetation is an important organic component in lacustrine settings. Many shallow lakes contain peat layers and accumulations of detrital woody material, which may also be transported to deeper water by gravity processes. Finer organic material (e.g. algal and microbial contributions and palynomorphs) forms much of the hydrocarbon-generating material ('oil shale') in lacustrine environments.

6.4 Facies

Based on their facies characteristics, lacustrine systems can be classified as dominantly siliciclastic, carbonate or mixed clastic-carbonate systems.

6.4.1 Facies of Siliciclastic Systems

Facies of siliciclastic lacustrine systems are illustrated in Figure 6.5 and are typified, for example, by those of the modern Qinghai Lake (an alkaline salt lake of about 4000 km^2 and an average depth of about 20 m: the largest lake in China) and are composed of facies developed in two main sub-environments, namely lake margins and lake-centre areas (Figures 6.5 and 6.6).

6.4.1.1 Lake Margin Facies

Sediment deposition along the margins of siliciclastic lacustrine basins is concentrated at and around river mouths and includes beach and nearshore-bar deposits, different kinds of subaerial and subaqueous deltaic deposits and other nearshore deposits (Figures 6.5 and 6.6). The deposits of these shore-zone and nearshore environments display an intimate

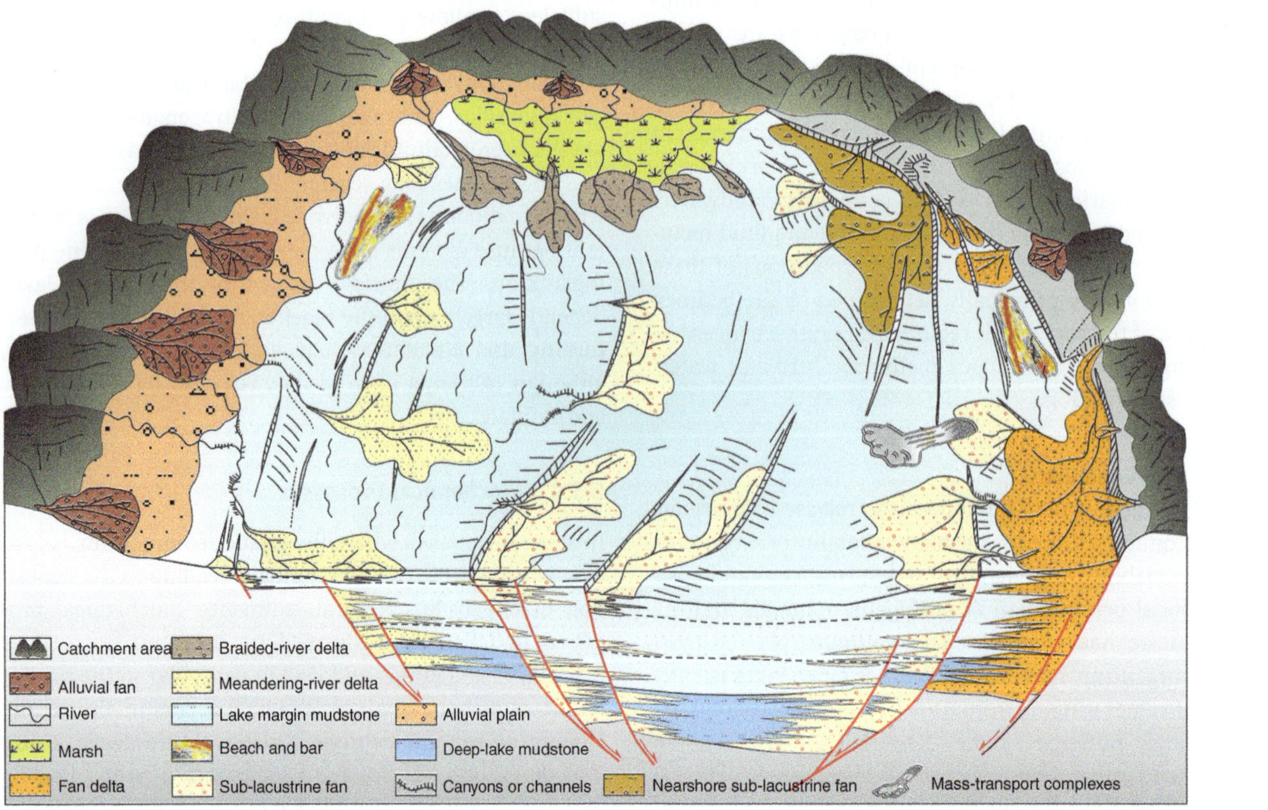

Figure 6.5 Schematic representation of main facies associations of siliciclastic systems developed in lakes. Lake margin facies are composed mainly of beach and other bar deposits, fan deltas, braided-river deltas and meandering-river deltas. Deep-lake facies consist mainly of nearshore sub-lacustrine fans, sub-lacustrine canyons or channels, sub-lacustrine base-of-slope fans and deep-lacustrine mass transport complexes.

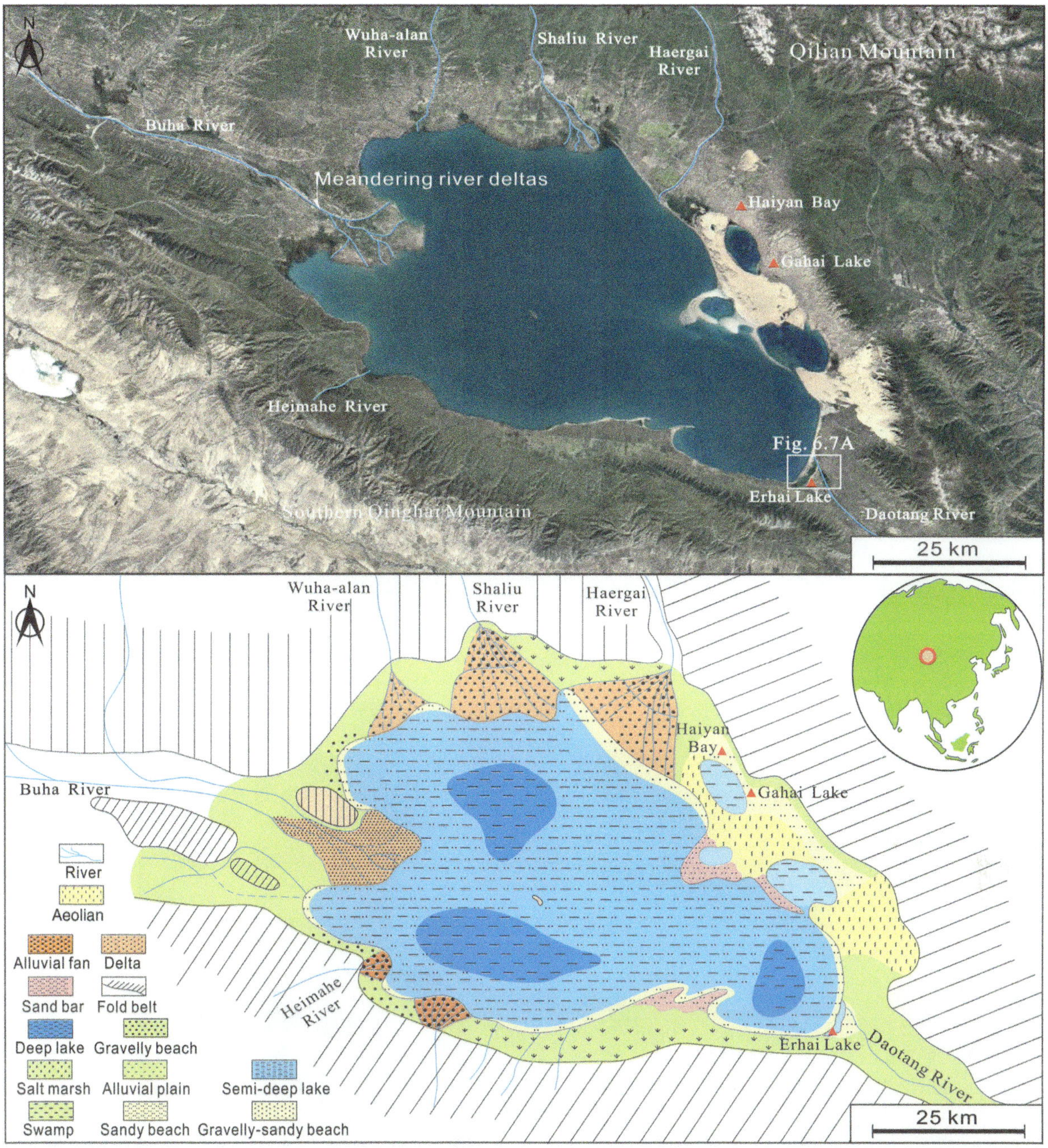

Figure 6.6 Satellite image and accompanying line drawing (Courtesy of Jing Wang) showing depositional environments and facies in modern Qinghai Lake, an endorheic (closed- basin) lake in Qinghai Province of northwestern China at an altitude of 3194 m on the Qinghai-Tibet plateau. The average depth is 21 m, and the maximum depth is 33 m. Note that the alluvial fans in the south, fed by steep, short-headed, streams are small, whereas the fans in the north are large. The strongest winds are from the NW. Wave-dominated environments are visible to the SE of the lake where the fetch is greatest. The linear course of the Buha River follows active faults. Also shown is the location of Figure 6.7a. Google Earth Image Landsat/Copernicus.

stratigraphic interfingering (decimetre- to metre-scale) with the adjacent, muddier, lake deposits offshore.

(1) Beach and Other Bar Deposits

The term *beach-bar deposits* is generic and used herein to describe beaches and associated nearshore bars developed along some gently sloping lake margins of shallow lacustrine basins (Figures 6.5 and 6.6) (e.g. Otvos 2020; Jiang et al. 2011; Schwartz 2012). Specifically, beaches refer to sandy or gravelly deposits formed by shoaling waves and at the same time reworked and drifted laterally by longshore currents. They form characteristically linear sandy lithosomes (2–5 m thick) and thin- to medium-bedded (bed thicknesses of 1–30 cm) layers. Related bars commonly form in the breaker zone, sometimes responding to the interaction between onshore flows and offshore-directed rip currents. Well-exposed outcrops of modern lake beaches and associated bars within coastal embayments or shorelines are shown in Figures 6.6 and 6.7 from the Erhai section of the Qinghai Lake. Such beaches and bars are commonly formed from detrital materials carried originally into the basin by river deltas or fan deltas (Houser and Greenwood 2005; Jiang et al. 2011;

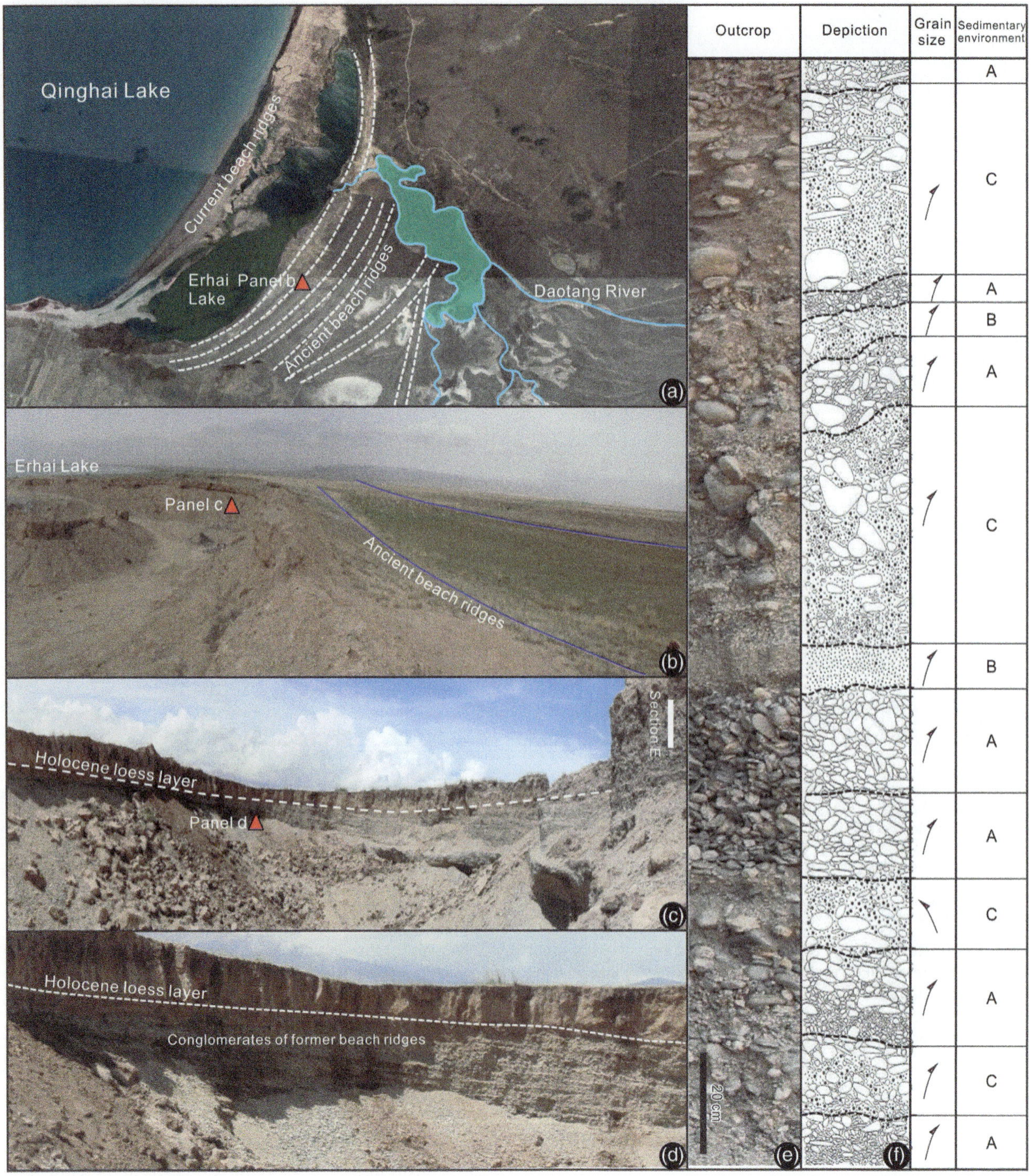

Figure 6.7 Sedimentary characteristics of beach-ridges in the Qinghai Lake, northwest China. (a) Google Earth Image © CNES/Airbus (see Figure 6.6 for location) showing plan view of the beach-ridges in the southeastern Qinghai Lake, (b–d) outcrop photos showing details of the gravelly deposits of ancient beaches. (e) and (f) are sedimentary logs showing grain-size change and sedimentary environment interpretations. Abbreviations: A = swash zone; B = backwash zone; and C = surf zone from breaking waves. Google Earth Image © 2023 Maxar Technologies.

Goulart and Calliari 2013). Beaches and associated bars represent an important kind of hydrocarbon reservoir in lacustrine basins, especially in eastern China and in the Uinta Basin, northern Utah, USA (Morgan and Bereskin 2003; Jiang et al. 2011).

The characteristic features of beaches and associated bars are (i) their formation of the upper part of upward-coarsening units (Figures 6.6 and 6.7), (ii) their high degree of textural maturity and (iii) their diagnostic sedimentary structures, such as flat and low-angle lamination and wave-generated ripple lamination (Tamura 2012; Song et al. 2018). Some of these shallow-lacustrine, high-energy deposits may locally display vertical burrows (Buatois and Mángano 1998). The sandy beach deposits can also contain some interbedded muddy layers. The sandy layers are generally composed of very low-angle cross-stratified or parallel-stratified, fine-grained sand and silt. The associated bars also show upward-coarsening and consist of very thick-bedded (>2 m) sandy layers with fewer intervening mud layers. These sandy layers show planar cross-strata or undulose stratification, and are composed of medium- to fine-grained sand (Jiang 2018). They also contain interbedded coquinas of low-diversity (but high abundance) shelly materials (e.g. gastropods) (Jiang 2018).

Beach and associated bar deposits can be composed of sand and/or gravel and may be direct-wave dominated, alongshore-current dominated or hybrid types. Their formation and geometry are controlled mainly by the coastal topography, waves and alongshore currents. Their stacking patterns are determined by these factors, as well as climate, lake-level variations and tectonic activity (Houser and Greenwood 2005; Jiang et al. 2011; Tamura 2012; Song et al. 2018).

(2) Fan Deltas

Nemec and Steel (1988) have defined a fan delta as a prism of sediment delivered by an alluvial fan and deposited mainly or entirely subaqueously at the interface between the active fan and a standing body of water (Figure 6.5). Fan deltas usually form on relatively steep lake margins with border faults and are often fed by ephemeral rivers (Figure 6.5). Fan deltas are interpreted to be created by a mix of hyperconcentrated stream flows and debris flows on the fans and fan-delta plains, which continue subaqueously with the addition of other sediment-gravity flows, including turbidity currents onto the subaqueous topset and delta front areas (Renaut and Gierlowski-Kordesch 2010). However, fan-delta front sediments are often reworked and sorted by wave action and alongshore currents, becoming mainly sand-dominated deposits.

Fan deltas typically display sigmoidally developed topset (fan-delta plain), foreset (delta front) and bottomset/prodelta strata, and their deposits show an overall upward-coarsening succession whose thickness reflects the approximate water depth in the area. Fan-delta plains, commonly occurring close to syn-depositional faults in rift basins, are characterised by conglomeratic and sandy distributary channels with cross-stratified, upward-fining infill and finer-grained inter-distributary areas showing low textural and compositional maturity. Fine-grained inter-distributary floodplain deposits may be preserved between the distributaries. In front of, and eventually lying below the delta-plain deposits, the foresets of the fan-delta front dominate the succession and consist of structureless or graded sandstone and conglomerate, usually in sheet-like beds but at times also channelled and interbedded with siltstone and mudstone. The prodelta facies are mainly mudstone and very fine to fine-grained sandstone (Tan and Plink-Björklund 2021). High hydrodynamic energy, strong erosion and coarse grain size are typically detrimental to the formation and/or preservation of biogenic structures. However, arthropod trackways and trails, insect nests and bird tracks are preserved locally in deposits formed in delta-plain pools, terminal deltaic lobes and abandoned channels (e.g. de Gibert and Sáez 2009).

Recent research on outcrops along the margin of the Permian Junggar Basin (Olariu et al. 2022) also illustrates some of the above points. Junggar lacustrine fan deltas developed in response to the early uplift of the Tian Shan mountain range. Stream-flood and debris-flow deposits in alluvial fan topsets change downslope to sheet-like pebbly sandstone, sandstone and interbedded parallel-laminated lacustrine siltstone and mudstone in the fan-delta foresets and then to organic-matter-rich laminated mudstone and graded thin sandstone in the bottomsets (Olariu et al. 2022). Cross-sections through inter-distributary embayments between the main fan-delta fairways also show downslope grain-size changes but in thinner packages. The development of fan deltas at these lake margins was controlled mainly by active fault movement, causing high relief and locally abundant sediment yield in the catchments of the basins. One of the remarkable features of this lake-margin stratigraphy, occurring where fan deltas protrude <10 km into the lake system, is the presence of repeated cyclicity between thin-bedded, very fine-grained sandstone hyperpycnites of the fan-delta toe and the parallel-laminated organic-matter-rich mudstone of the lake (Olariu et al. 2022). Cyclic deposits of this type stack in units, some 2–5 m thick, with more than 100 fine-grained lake cycles recognised in the muddy lacustrine Middle Permian Upper Lucaogou Formation along the southern margin of the Junggar Basin (Figure 6.8). Although many of these cycles are incomplete, the ideal facies succession starts with thin laminated dark organic-matter-rich mudstone at the base that is likely formed below the wave base in deeper parts of the lake and siltstone-sandstone or mudstone with subaerial exposure structures (mud cracks) at the top (Figure 6.8) (Olariu et al. 2022).

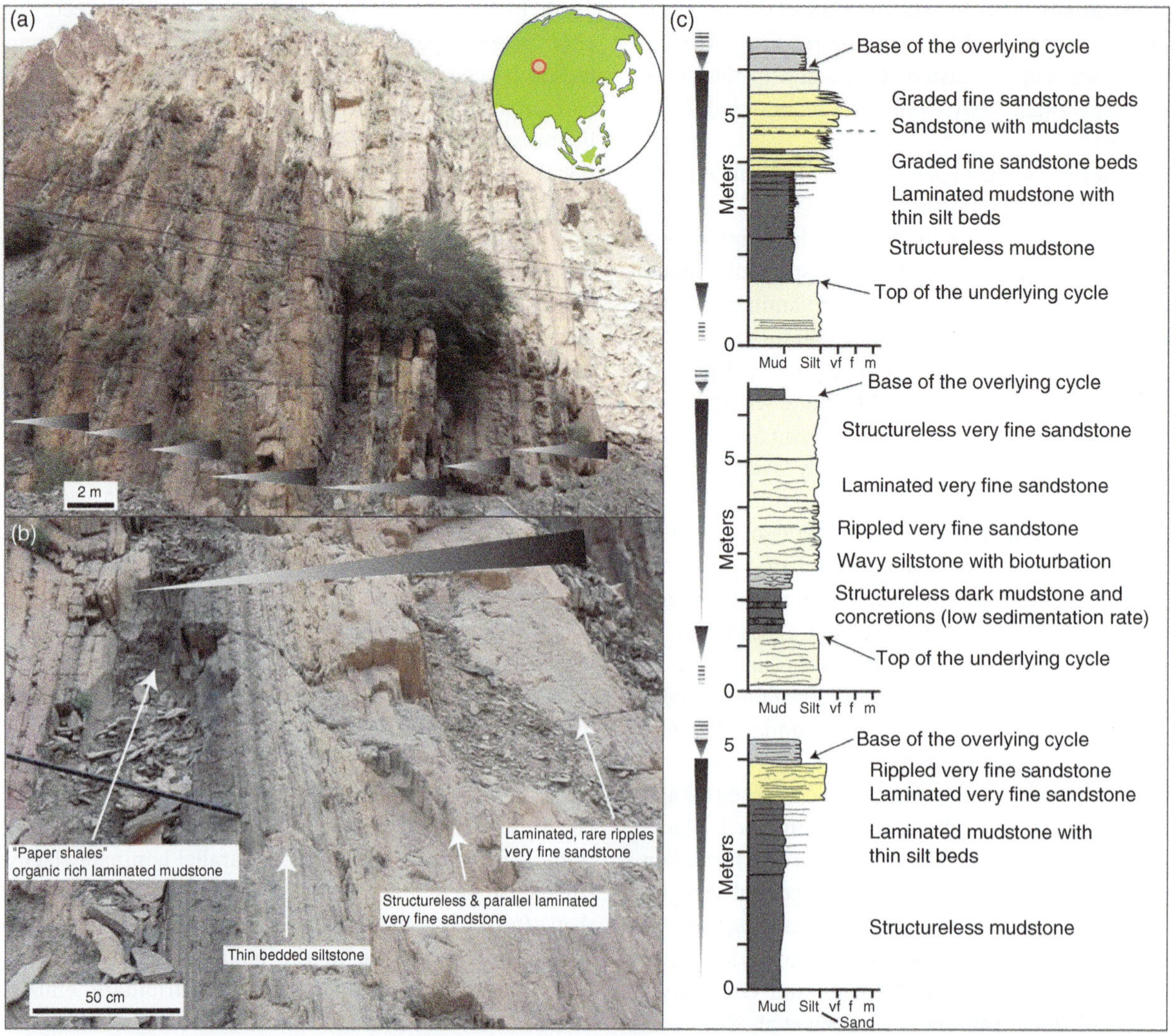

Figure 6.8 Spectacular vertically-dipping fan-toe to lake-mudstone cycles dominate the muddy lake-margin stratigraphy of the Permian Lucaogou Formation in the Junggar Basin (Olariu et al. 2022 / with permission from Elsevier). (a) Multiple thickening (and shallowing) upward cycles (a few metres thick), each showing relatively thin dark mudstones passing upwards to thicker light-coloured mudstone and very fine-grained sandstone, Jingjingzigou section. (b) A single mudstone-dominated cycle (<2 m thick) with organic-rich 'paper' shale passing up to laminated siltstone and very fine-grained sandstones. The beds are younger to the right in both (a) and (b). (c) Examples of lacustrine cycles (few metres-thick) from the Ya'an Road outcrop. These lacustrine cycles exhibit an overall upward-coarsening/shallowing pattern and typically start with thin laminated dark organic-matter-rich mudstone at the base and siltstone-sandstone or mudstone with subaerial exposure structures (mud cracks) at the top.

(3) Braided-River Deltas

Braided-river deltas (Figure 6.5) are distinguished from fan-deltas (which are fed by alluvial fans) and are defined as sand- and gravel-rich deltas formed directly by the progradation of a purely fluvial, braided alluvial plain, typically associated with high coarse sediment loads and widely variable water discharge (McPherson et al. 1987). Braided-river deltaic deposits are usually characterised by laterally extensive, coarse-grained and poorly sorted sand bodies that are closely associated with lacustrine muds (e.g. Min et al. 2020). As with the other types of lacustrine deltas, braided-river deltas have well-developed topsets, delta-front foresets and bottomsets and are commonly seen to develop on relatively gentle slopes of lake margins. Major delta-scale (200–300 m in thickness) clinoform sets of the Oligocene Bohai Bay braided-river deltas (Figure 6.9) are well-imaged seismically and show both flat (prograding) and rising (aggrading) topset-to-foreset roll-over trajectories. The lower foresets and bottomsets of Oligocene Bohai Bay comprise turbidite channel-lobe complexes and occasional slumps, and in the Triassic Ordos braided-river deltas, they comprise debris-flow deposits (see also Feng et al. (2020) for seismic examples of the latter).

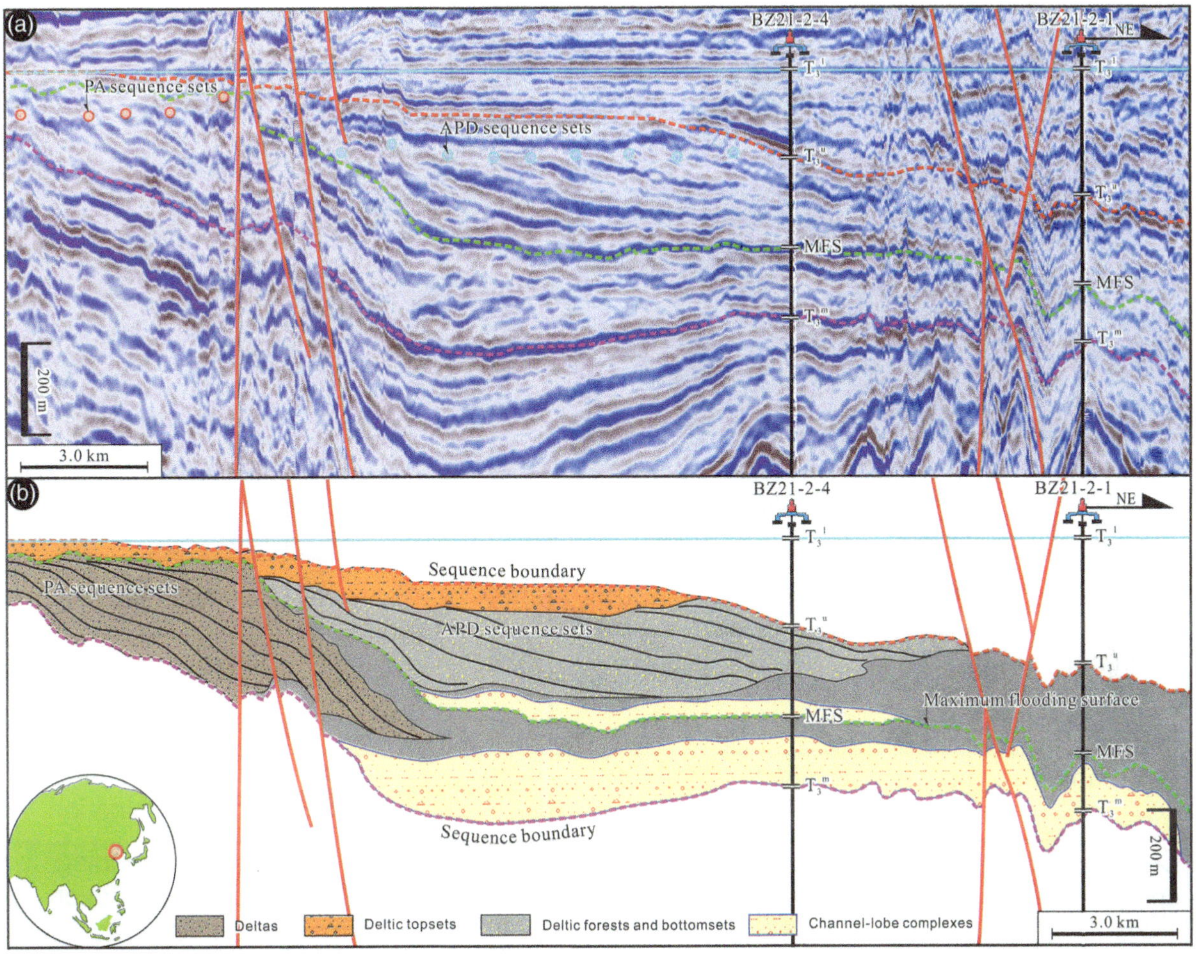

Figure 6.9 (a) Seismic section and interpreted geological cross-section (b) along depositional dip showing stratal terminations used to recognise and correlate the hierarchy of surfaces of a composite sequence in the Oligocene Dongying Formation of the Bohai Bay Basin, eastern China. The red dotted line shows onlap terminations above and truncation terminations below and is a sequence boundary. The purple dotted line has truncation terminations below and is also a sequence boundary. The green dotted line is a basal surface of multiple downlap terminations; a maximum flooding surface (MFS). The MFS separates the underlying progradational–aggradational sequence sets from the aggradational–progradational–degradational sequence sets above.

The braided-river delta plains are dominated by distributary channels with bedload pebbly sandstone and conglomerate, and some hyperconcentrated flows developed during strong flooding (Feng et al. 2020). Local channel-fill sandstone is cross-stratified or structureless and shows large-scale soft-sediment deformation structures. Braided-river delta fronts contain a variety of facies associations representing the terminal portions of distributary channels, distributary mouth bars and distal mouth bars with thin turbidites. The delta-front sediments are mainly pebble conglomerate, pebbly sandstone, siltstone and grey mudstone, with minor carbonaceous mudstone. The prodelta deposits consist of dark grey mudstone and siltstone with subhorizontal and wavy lamination. These deposits may contain abundant simple grazing trails (Buatois and Mángano 1995a), thin debris-flow deposits and turbidite beds.

(4) Meandering-River Deltas

Modern analogues of meandering-river deltas associated with intermittent streams are illustrated in Figure 6.6 (Buha meandering-river delta in the Qinghai Lake) and Figure 6.10a,b (Neales meandering-river delta in Kati Thanda-Lake Eyre, Australia). Modern examples associated with perennial streams include the Volga River delta, the northern Caspian Sea in Russia, the Bear River delta in Great Salt Lake, Utah, the Omo River delta, Lake Turkana, Kenya and Kaidu River delta in Bosten Hu, western China.

Lacustrine deltas with meandering-river distributaries also have well-developed topsets, foresets and bottomsets and are stratigraphically manifested as an overall coarsening-upward succession (Figure 6.11). Channel networks display a multitude of bifurcations, in which an increasing

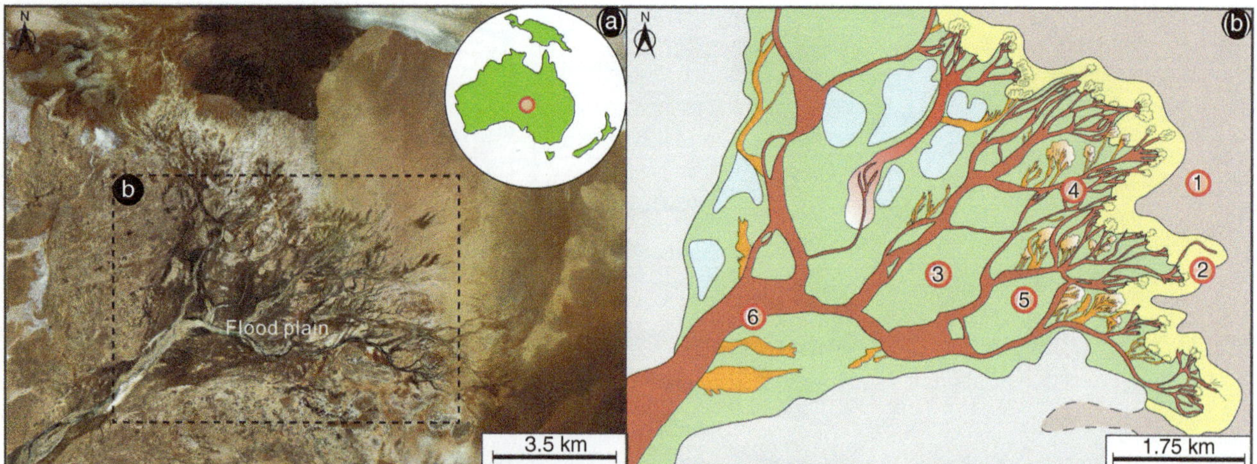

(c)

Depositional environment		Grain size	Length	Width	Thickness	Type log	Structures
①	Prodelta	Clay	>4000 m	>4000 m	cm		Flat laminae
②	Delta front	Silt and clay vf sand	>300 m	<500 m	<0.3 m		Trough x-bedding ripple x-lamination flat laminae
③	Overbank	Silt to vf sand	500 – 4000 m	<1500 m	<2 m		Hummocky x-strat. ripple x-lamination
④	Crevasse splay	Silt to m sand	300 – 4000 m	<50–250 m	0.5–1.5 m 2 cm		Trough x-bedding ripple x-lamination fining upward fining outward
⑤	Distributary channel	f-vc sand	NA	<50–400 m	<1–5 m		Trough x-bedding ripple x-lamination
⑥	Main channel pointbar	vc sand	<600 m	<300 m	<5 m		Trough x-bedding ripple x-lamination fining upward

Figure 6.10 (a) Google Earth image illustrating a modern meandering river delta (the Neales Delta), in periodically desiccated Kati Thanda-Lake Eyre, Australia, (b) interpretative map of the depositional elements of Neales river delta and (c) a tabulation of depositional element characteristics (a, b and c are all from Olariu et al. 2021 / John Wiley & Sons). Google Earth Image © 2023 Maxar Technologies.

number of channels become ever-narrower and shallower before passing into shallow standing water with almost no gradient on the delta fronts. This style of bifurcating channel network is well illustrated by the modern examples of the Neales delta in Kati Thanda-Lake Eyre (Figure 6.10). Lacustrine deltas of this type have a delta plain, delta front and prodelta. The delta plain contains subaerial distributary channels, subaerial levees and crevasse splays, channel point bars and swamps (e.g. Min et al. 2020). Distributary channels are generally incised into the underlying delta plain and are commonly filled with metres-thick, structureless or low-angle cross-bedded sandstone (e.g. Lang et al. 2004; Min et al. 2020; Olariu et al. 2021) (Figure 6.10c). Vertical burrows and escape trace fossils are locally present in channel-fill deposits, as well as root traces recording soil formation after channel abandonment (e.g. Melchor 2004). Inter-distributary and other swampy areas are generally dominated by dark or black, carbonaceous mudstone and thin coal seams (in post-Silurian strata). The delta-front portion of these

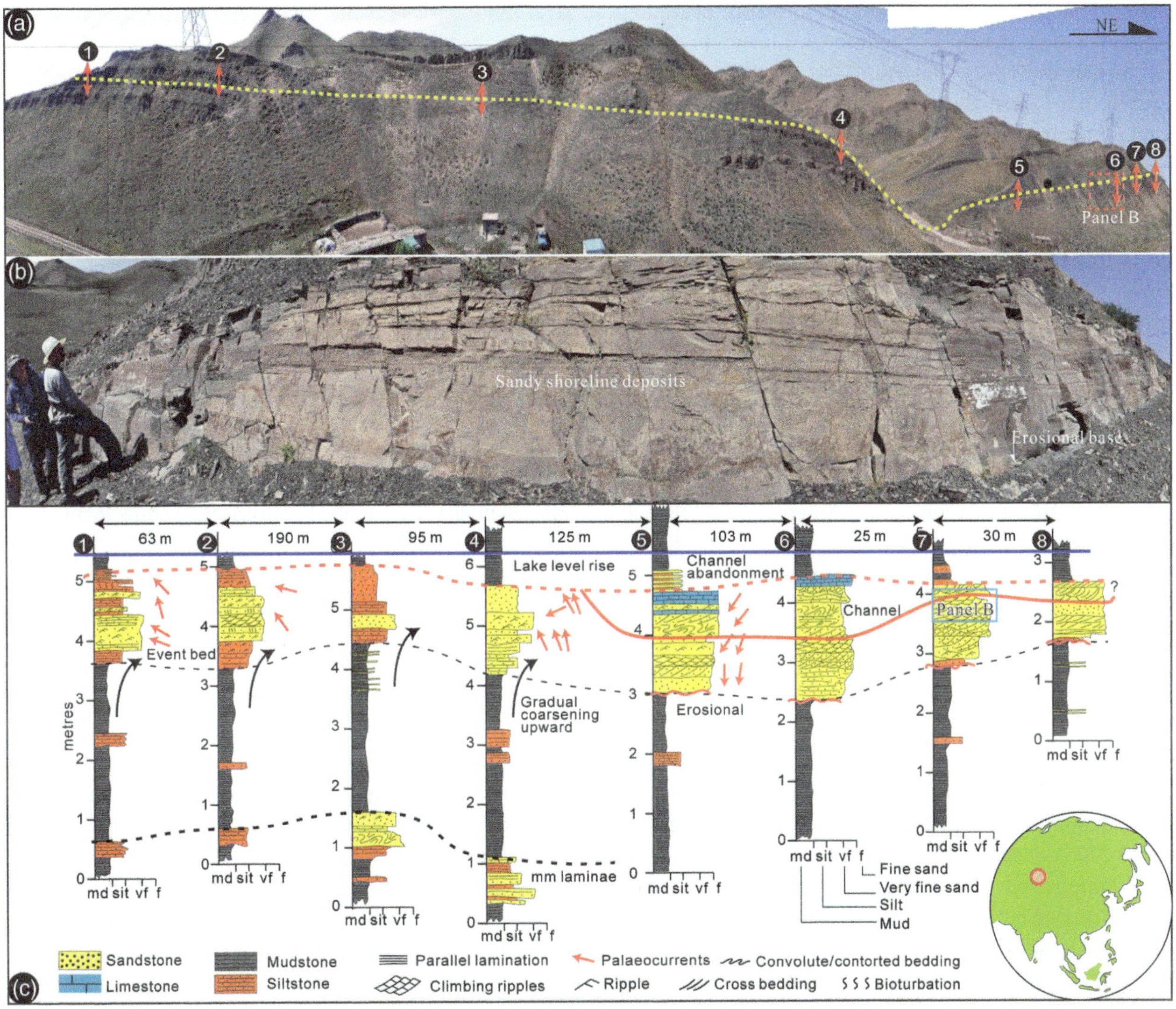

Figure 6.11 Meandering-river delta front deposits. (a) Photomosaic image (c. 1 km wide) showing an outcrop of sandy lacustrine shoreline deposits (from the mud-prone Early Permian Tashikula Formation in the southern Junggar Basin of northwestern China), showing locations of the measured sections in (c). The sandstone units seen in (a) are laterally continuous for kilometres without obvious pinch-outs. (b) Outcrop image showing a close-up view of an about 3 m thick outcrop of the sandy mouth-bar and shoreline deposits close to log '6' in (a). These sandy mouth-bar and shoreline deposits are commonly encased in mudstones, with thicknesses of metres to tens of metres. (c) Eight measured sections within part of the Tashikula Formation, correlated along depositional strike. The sandstone unit seen in this figure is sharp-based and erosional where thickest and sandiest but it locally exhibits an overall upward-coarsening transition from the underlying mudstones (see Olariu et al. 2021 / John Wiley & Sons for full details).

lacustrine deltas includes the deposits of subaqueous terminal distributary channels, distributary mouth bars, subaqueous levees, subaqueous crevasse splays and distal mouth bars. Subaqueous distributary channels do not incise deeply and comprise medium- to fine-grained sandstones with low-angle cross-stratification or parallel-lamination (e.g. Lang et al. 2004; Min et al. 2020) (Figure 6.10c). Well-exposed examples of subaqueous distributary channels cutting down into mouth bar deposits are shown in Figure 6.11, from the Early Permian Tashikula Formation in the southern Junggar Basin (Olariu et al. 2021). These mouth bar deposits show an abundance of current-ripple cross-lamination, together with channelised, cross-stratified sandstone which laterally continues for kilometres without obvious pinch-outs and displays an overall upward-coarsening pattern (Olariu et al. 2021) (Figure 6.11b,c). These cross-stratified sandstones are bounded at their bases by erosional surfaces and represent subaqueous distributaries, with incision depths ranging from decimetres to metres (Figure 6.11a,c).

Delta front deposits in freshwater lakes may contain relatively diverse trace fossil associations dominated by simple horizontal trails with the subordinate presence of feeding and locomotion structures produced by invertebrates, as well as fish trails (e.g. Melchor 2004). Vertical U-shaped burrows have been documented in mouth bar deposits (e.g. Mángano et al. 1994). Prodelta deposits of meandering river deltas are similar to those of the braided deltas, though in places finer-grained (Figure 6.10c) (e.g. Lang et al. 2004; Min et al. 2020). Simple grazing trails, fish trails and arthropod trackways occur in lacustrine prodelta deposits (e.g. Trewin et al. 2002).

(5) Delta–Lake Interactions

Delta–lake interactions are expressed stratigraphically in two ways. First, river plumes interact with the body of standing water when sediment-laden fresh water enters the lake. Such delta–lake interactions often gave rise to a subaqueous accumulation of sediments, and this is described and predicted by river-mouth jet theory (Wright 1977). Jet flows have high momentum and can travel long distances without dissipating. Because of the contrast in density between the river inflow and ambient lake waters, three types of sediment dispersal response may occur, as discussed above: hyperpycnal (under-), hypopycnal (over-) and homopycnal flows. Hyperpycnal underflows bypass the shoreline or mouth bar area to generate hyperpycnal (inertia-dominated) sediment-gravity flow accumulations (Bates 1953; Mulder et al. 2003; Bhattacharya 2006). Suspended sediment, boosted also by freshwater lofting from the initial hyperpycnal flow (Zavala 2020), is carried out into the receiving lake basin as a buoyant plume, resulting in hypopycnal (buoyancy- and friction-dominated) sediment dispersal. In some homopycnal settings, there is commonly a greater degree of mixing between the river and the standing body of water. Generally, hyperpycnal underflows are very common at the river mouth in a freshwater lacustrine basin. In shallow-lake areas, friction at the bed causes rapid flow deceleration and the development of a mouth bar that causes the associated distributary channel to bifurcate (Olariu and Bhattacharya 2006).

The second type of delta–lake interaction is the reworking and sometimes complete removal of delta-front sediments by lake waves. In this scenario, delta-front sediments can be transported alongshore from the river mouth. In some lakes under monsoon climates, the mouth bar of a delta can be extended for several kilometres in an alongshore direction, passing into laterally equivalent shoreface deposits. For example, in the Qinghai Lake, the Haergai and Daotang River deltas have been shaped as elongated mouth bars by the persistent northwesterly monsoons (Figure 6.6). The ability of lake waves to extend the river-mouth bar depends on the ratio of sediment deposition by river floods to the transport capacity of the longshore drift, and also the obliquity of wave approach (Bhattacharya and Giosan 2003).

(6) Other Nearshore Deposits

In addition to the above-mentioned lake margin facies, lacustrine basins also feature other nearshore facies, including wetlands, sandflats and mudflats. Swamps and marshes are the main types of wetlands that form along waterbodies containing rich hydric soils. Marshes are wetlands, continually or frequently flooded by nearby running water, that are dominated by emergent soft-stem vegetation and herbaceous plants. Swamps are wetlands consisting of saturated soils or standing water and are dominated by water-tolerant woody vegetation, such as shrubs, bushes and trees. Freshwater swamps form along large rivers or lakes. They are critically dependent upon rainwater and seasonal flooding to maintain natural water-level fluctuations.

Extensive sandflats or mudflats are particular characteristics of saline lakes (where they are referred to as playa or inland sabkhas (Section 7.3.14) and can include evaporite minerals accumulated in semi-arid to arid climates. They are very low gradient features, which merge landwards into alluvial fans or talus aprons along mountain fronts (e.g. intermontane basins of western USA, Iran and China) or, where the relief is low, into alluvial or desert plains (e.g. central Australia). The flats are dry for much of the time, being covered by water only during periods of flooding or during high lake levels. Sandflat and mudflat deposits are typically host to abundant trace fossils, mostly arthropod and tetrapod trackways (Gierlowski-Kordesch 1991; Buatois and Mángano 1998, 2004; Mancuso et al. 2020, 2022). Most of the clastic sediment associated with sandflats or mudflats is of fluvial origin, introduced by unconfined sheet floods, although windblown dust, trapped on the sticky hygroscopic surface may also represent a significant contribution. The northern margin of Kati Thanda-Lake Eyre is a vast playa occupying an area of about 5000 km^2, which has formed from sediments carried to the basin by exceptional floods originating mainly in subtropical Queensland.

6.4.1.2 Deep-Lake Facies

In lakes with sufficient synoptic relief (on the order of hundreds of metres), a full suite of deep-water environments can develop (Figures 6.5 and 6.12). Such lakes are typically tectonic in origin. Deepwater lakes contain a series of sub-environments that are usually linked downslope within clinothems that are much larger in scale than the 200–300 m

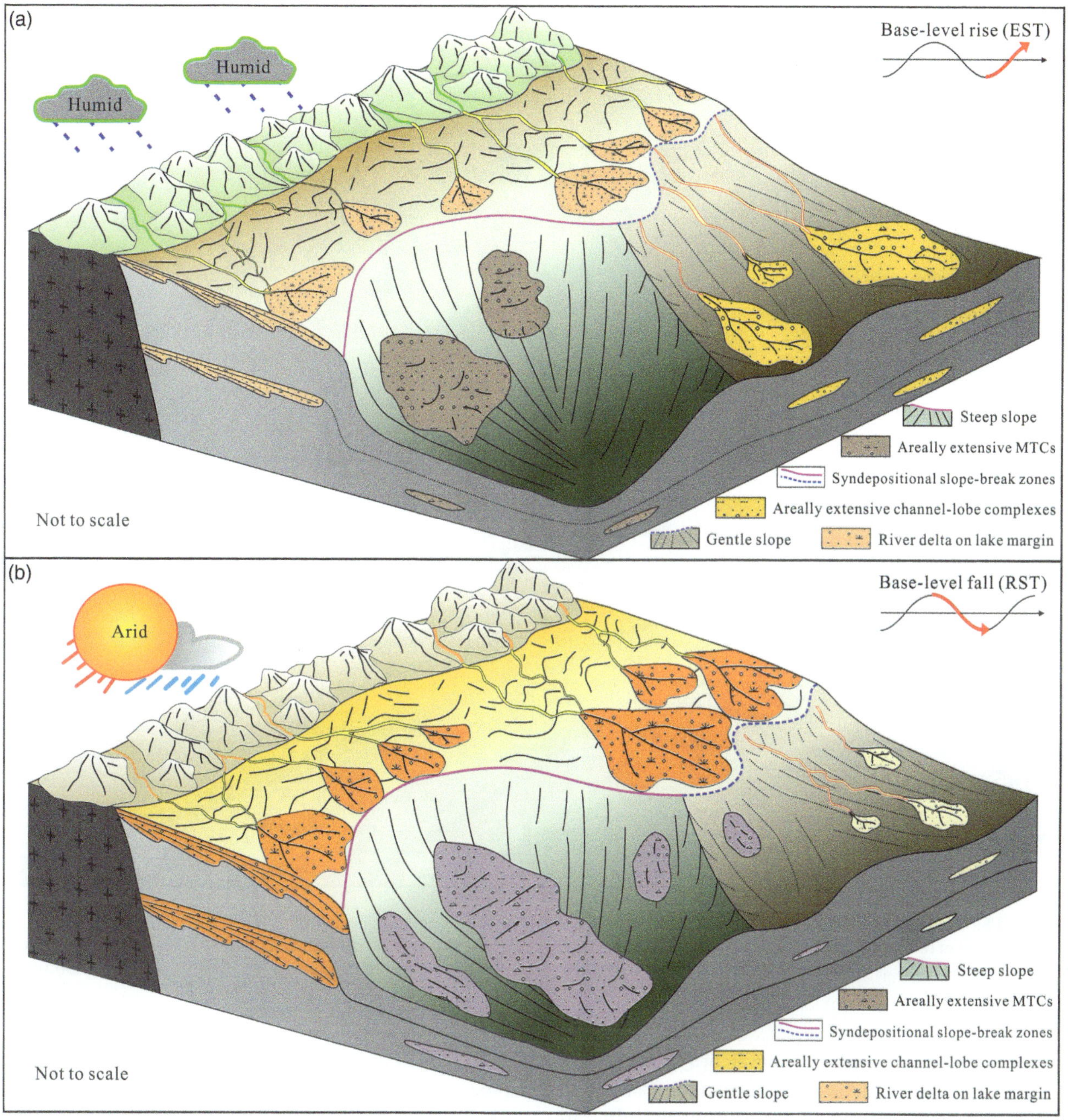

Figure 6.12 Schematic illustration of favourable conditions for the development of different types of sub-lacustrine fans (channelised and non-channelised fans) in (a) expanding (humid) systems tracts (EST) and (b) regressive (arid) systems tracts (RST). Base level and climate (precipitation) are linked as the lake basin can be closed.

thick delta-scale clinothems illustrated above (Figure 6.9). River channels and nearshore deltas or fans give way (usually at depths >150–200 m) to deepwater slope channels and channel-levee systems and eventually to mass transport complexes (MTCs), toe-of-slope fans and other deep-lake facies (Figures 6.5 and 6.12). The formation of these deep-lake facies (mostly turbidites) is related to sediment movement triggered by storms, tsunamis, floods and rapid sediment loading, as well as earthquakes.

(1) Nearshore Sub-lacustrine Fans

Nearshore sub-lacustrine fans (several hundreds of metres to a few kilometres in width and a few kilometres in length) commonly occur on steep slopes as fan-shaped sandstone

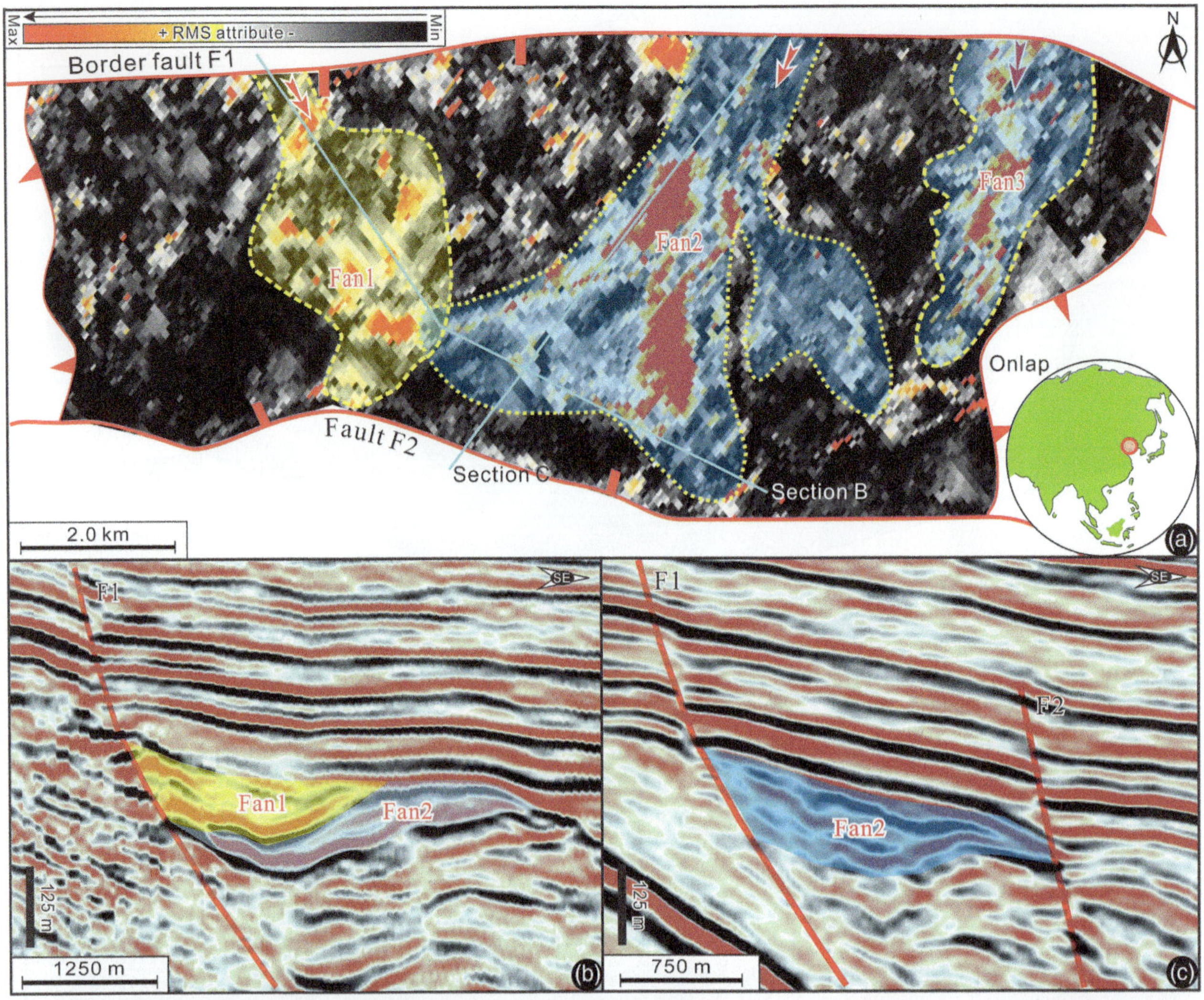

Figure 6.13 Nearshore fault-bounded, sub-lacustrine fans in the Paleogene Dongying Formation, Bohai Bay Basin, as imaged in RMS-attributes of reflection amplitude along a horizon (a) and seismic sections (b and c). These fault-bounded, sub-lacustrine fans are seen in RMS-attribute maps as lobate to tongue-shaped, high RMS accumulations and seismic sections as wedge-shaped, high-amplitude reflection packages.

and conglomerate accumulations developed from the footwalls of major border faults in rift basins (e.g. Wang et al. 2014; Zhang et al. 2019a). They have been interpreted as relatively deep-water, coarse-grained sediment gravity-flow lobes dispersed locally by landslides, debris flows and turbidity currents during seasonal floods. Well-imaged nearshore sub-lacustrine fans, related to major border faults, are seen on seismic RMS-attribute[1] maps as lobate to tongue-shaped, high RMS areas and seismic cross-sections as high-amplitude reflection packages in the Oligocene Dongying Formation of the Bohai Bay Basin (Figure 6.13). Nelson et al. (1999) also described late Quaternary non-channelised sub-lacustrine, coarse-grained aprons comprising thick sandstone and conglomerate units transitioning distally to fine-grained sandstones in the active rift basins of Lake Baikal, Russia. Nearshore sub-lacustrine fans have been subdivided into inner, middle and outer fan segments, with pebbly sandstone and sandstone fining down fan into dominant mudstone over a few kilometres (Zhang et al. 2019a).

Zhang et al. (2019a) recognised two types of nearshore sub-lacustrine fans on the northern steep slope of the Dongying Depression, Bohai Bay Basin, namely channelised and non-channelised sub-lacustrine fans. Channelised nearshore sub-lacustrine fans contain straight-to-weakly meandering channels that are fronted by a terminal splay and fed by larger-scale gullies with gentle gradients (Zhang et al. 2019a). In contrast, non-channelised nearshore sub-

1 Root-mean-squared (RMS) is a measure of an attribute, in this case normally seismic reflection amplitude, that is the square root of the average of the squares of a series of measurements.

lacustrine fans are lobate in plan view and are characterised by a lack of channels on the inner fan (Zhang et al. 2019a). These non-channelised nearshore sub-lacustrine fans are fed by small-scale gullies with steep gradients. Differences in types and distribution between these two types of nearshore sub-lacustrine fans were ascribed to the intensity of fault movement, climatic changes, slope topography and source supply (Zhang et al. 2019a).

(2) Sub-lacustrine Canyons or Channels and Channel-Levee Systems

A sub-lacustrine canyon or channel is defined herein as a V- or U-shaped erosional scour that has cut its way subaqueously out onto the deepwater lake floor. A channel-levee system is a similar channel that is flanked by wedge-shaped levees along each side (Sztanó et al. 2013; Fongngern et al. 2016, 2018). Sub-lacustrine channels commonly occur on the gentle slopes of deepwater lacustrine basins and can be highly sinuous or straight in plan view (Figure 6.14) (Feng et al. 2010a; Sztanó et al. 2013; Fongngern et al. 2016; Pan et al. 2017; Xian et al. 2018). Sub-lacustrine channels or channel-levee systems, as well as canyons, serve as the major conduits for the delivery of sediments and carbon from shallow lake margins down to lake floors (Sztanó et al. 2013; Fongngern et al. 2016, 2018; Dodd et al. 2019). They are formed by,

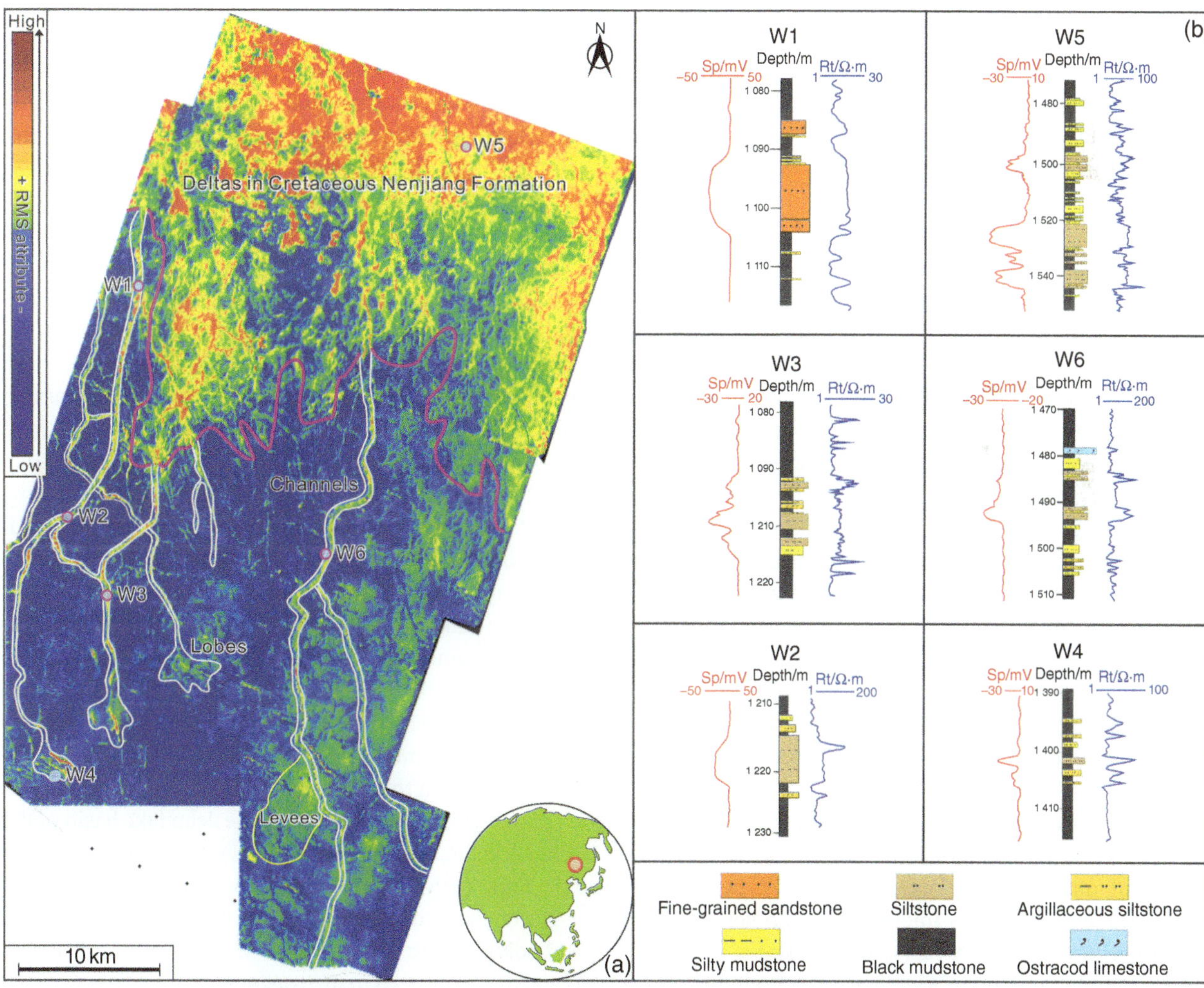

Figure 6.14 (a) RMS-attribute map of reflection amplitude along a horizon showing plan-view geomorphological expression of delta-to-fan systems in the Nen 1 Member of the Late Cretaceous Nenjiang Formation, Songliao Basin, NE China (from Pan et al. 2017). Sub-lacustrine fans are composed of channels expressed here as reddish-coloured, narrow and tortuous, high RMS ribbons. Subaqueous fan terminal lobes are expressed as bird-foot-shaped, high RMS accumulations. (b) Well-log patterns of prodelta (well W5), sub-lacustrine channels (wells W1, W2, W3 and W6), and terminal lobes (well W4). The channel fills are fine sandstones to siltstones from 10 to 4 m in thickness. See Pan et al. (2017) / with permission of Elsevier for details. Logs shown in panel (b) of this figure are spontaneous potential (SP) and resistivity (Rt).

and host, subaqueous density currents (Fongngern et al. 2018; Dodd et al. 2019), particularly those generated during flooding from the river (i.e. hyperpycnal flows) (Pan et al. 2017; Xian et al. 2018).

Slope to basin-floor channels have been seismically imaged in plan view as narrow, tortuous, high-RMS amplitude ribbons with tens of kilometres long sand fairways (Fongngern et al. 2016). In seismic sections, they are seen as wide (several tens to a few hundred metres), low relief (a few to several tens of metres) and U-shaped erosional features filled by high-amplitude continuous reflection packages. Sztanó et al. (2013) have shown that sub-lacustrine canyons that developed during periods of lacustrine clinothem aggradation fed meandering, bifurcated channels at the toe of slope region in Neogene Lake Pannon, Hungary. Deep-lake channels recognised in the Late Cretaceous Nenjiang Formation in the Songliao Basin of NE China have an average length of more than 80 km, a width of 100–900 m and are filled by fine-grained sandstone characterised by spontaneous potential (SP) well-log patterns with sharp bases and tops (see wells W1 and W2 in Figure 6.14). In another example, late Miocene-Pliocene deep-lacustrine channels with a depth of 25–50 m and width of 100–200 m were recognised in the western Dacian Basin of Romania and were found within 50–100 m deep, 0.8–4.0 km wide sub-lacustrine canyons (Fongngern et al. 2018).

In some ancient systems, the levees that flank sub-lacustrine channels are also seismically recognisable, due largely to large water depths of up to several hundred metres (e.g. in the seismic data of Sztanó et al. 2013 and Fongngern et al. 2016, 2018). Levees flanking both sides of sub-lacustrine channels show moderate to high RMS-attribute patches in seismic plan view. Modern-day sub-lacustrine channel-levee complexes dominated by low-density sediment-gravity flows were observed in Lake Mead (USA) connected to and following the drowned Colorado River (Twichell et al. 2005). They comprise fine-grained sandstone interbedded with silty and black mudstone (see well W6 in Figure 6.14).

Deepwater slopes in lacustrine basins, like marine slopes, are generally mud-prone, though the slope channels themselves stand out as slope-bypassing fairways, becoming filled eventually with muds interbedded with thin sands with many rip-up clasts. Extensive muddy slopes punctuated by sandy channels have been illustrated from the Miocene Dacian Basin, Romania (Fongngern et al. 2018), a ParaTethys lake basin sandwiched between the larger Pannonian and Black Sea lake basins. Sub-lacustrine channels containing hyperpycnal flows were recently recognised in the Triassic of southern Ordos Basin, central China by Xian et al. (2018), where they were interpreted to change character downdip from erosive, bedload-dominated hyperpycnal flows to mixed-load flows and, finally, to hypopycnal flows dominated by suspended load.

(3) Sub-lacustrine Channel-Lobe Complexes

There are two main types of deep-lacustrine fans: channelised sub-lacustrine fans (i.e. channel-lobe complexes) and non-channelised sub-lacustrine fans (e.g. MTCs) (Figures 6.5 and 6.12). Sub-lacustrine channel-lobe complexes or base-of-slope fans are fan-shaped or elongated sediment gravity-flow accumulations on the lake floor, in the vicinity of the base-of-slope (Figures 6.5 and 6.12). Sub-lacustrine base-of-slope fans are modest in size up to tens of square kilometres and are many times smaller than most submarine fans (Figures 6.5 and 6.12) (Feng et al. 2010a; Fongngern et al. 2016; Pan et al. 2017; Dodd et al. 2019). Seismically well-imaged sub-lacustrine channel-lobe complexes are illustrated in Figures 6.14a, 6.15 and 6.16. They were usually generated by long-distance, initially channelised turbidity currents that became unconfined, spreading from the sourcing channels (Pan et al. 2017). Outcrops of sub-lacustrine deepwater lobes have been recognised in the Middle Permian Lucaogou Formation in the southern Junggar Basin, Northwest China (Figure 6.17), the Early Cretaceous Gomo member, Candeias Formation in Brazil (Bohacs 2012) and the Early Cretaceous lacustrine rift basin of Lingshan Island, Eastern China (Li et al. 2020). These deep-lake lobes were formed predominantly by debris flows and both high- and low-density turbidity currents (Figure 6.17) (Li et al. 2020; Shan et al. 2021). The thin-bedded turbidites of sub-lacustrine fans or fan lobes may host abundant grazing trails and feeding burrows (e.g. Buatois et al. 1996).

Sub-lacustrine terminal lobes or base-of-slope lobes are seismically imaged in planform as lobate high RMS-attribute accumulations linked updip to high to moderate RMS-attribute ribbons with moderate to low sinuosities (see Figures 6.14a, 6.15 and 6.16 for some excellent examples). Other seismic examples of base-of-slope lobes or fan lobes are from the Late Miocene Lake Pannon, SE Hungary (Sztanó et al. 2013), the Miocene deep-lake basin of the Dacian foreland basin, Romania (Fongngern et al. 2016, 2018) and Early Cretaceous deep-lacustrine settings of the North Falkland Basin (Dodd et al. 2019). Sztanó et al. (2013) divided distal fan lobes occurring within Neogene Lake Pannon into two main categories, namely toe-of-slope lobes and slope-detached lobes fed by major leveed channels. Toe-of-slope lobes accumulated after just short-distance transport in front of progradational clinothems and were created by mud-poor turbidity currents.

Sztanó et al. (2013) recognised two main categories of distal fan lobes within Neogene Lake Pannon deposits, namely

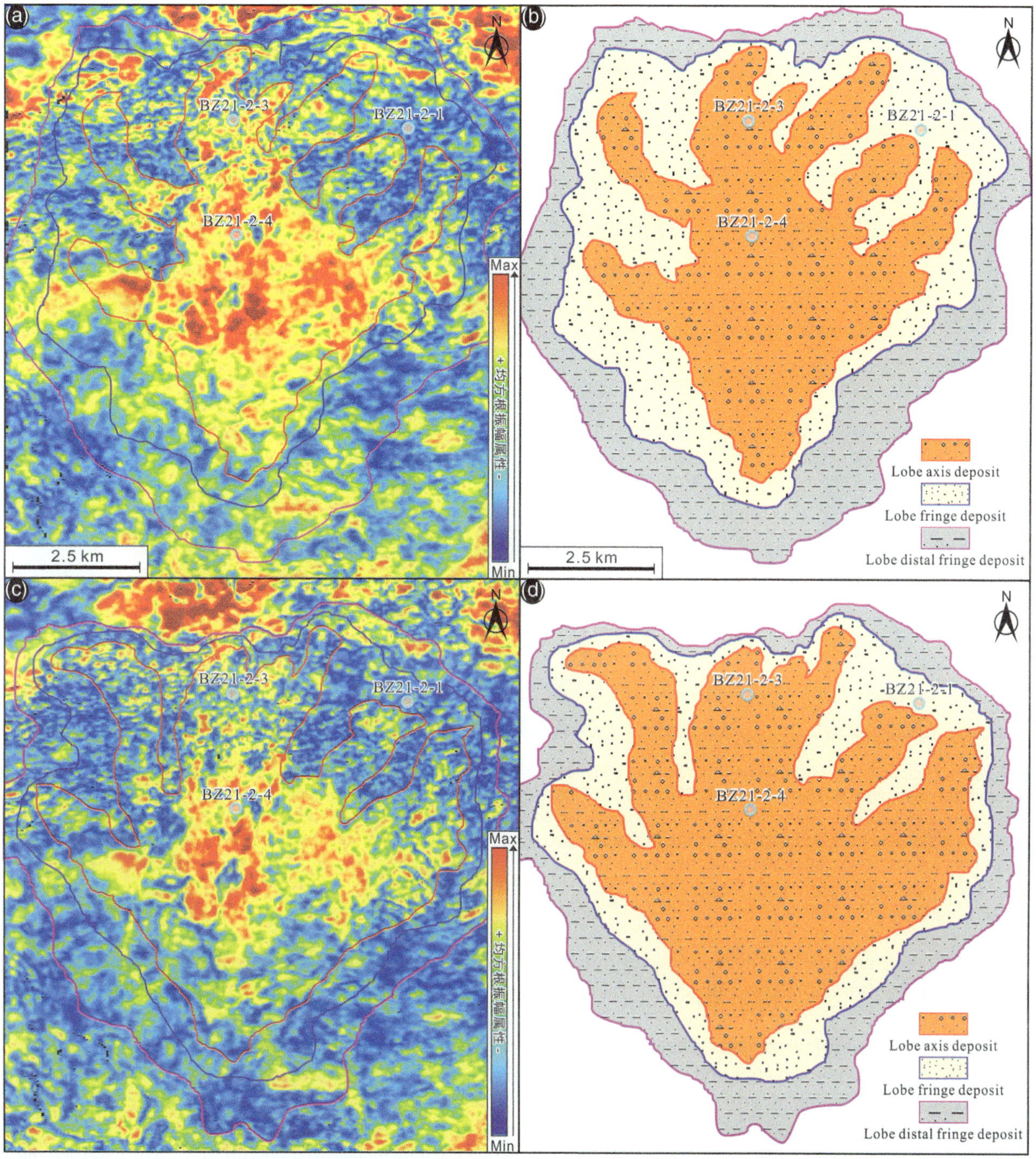

Figure 6.15 (a and c) RMS-attribute map and their corresponding interpretations (b and d) illustrating seismic geomorphology and associated sedimentologic interpretation of terminal lobes of extensive channel-lobe complexes flowing from the South. 'Hot' colours represent thicker sand: a wireline log is shown in Figure 6.20. These terminal lobes developed within the expanding (humid) systems tract (EST) (see Figure 6.9) of the Oligocene Dongying Formation, Bohai Bay Basin and are seen to develop in front of syn-depositional slope-break zones on a gentle slope.

toe-of-slope lobes and slope-detached lobes fed by major leveed channels. Toe-of-slope lobes accumulated at the toe-of-slope after just short-distance transport and were deposited by mud-poor, non-effective turbidity currents. Toe-of-slope lobes are seen to occur in front of progradational clinothems. Large slope-detached lobes, in contrast, accumulated at the terminations of leveed channels after long-distance transport (i.e. 30 km from the shelf edge)

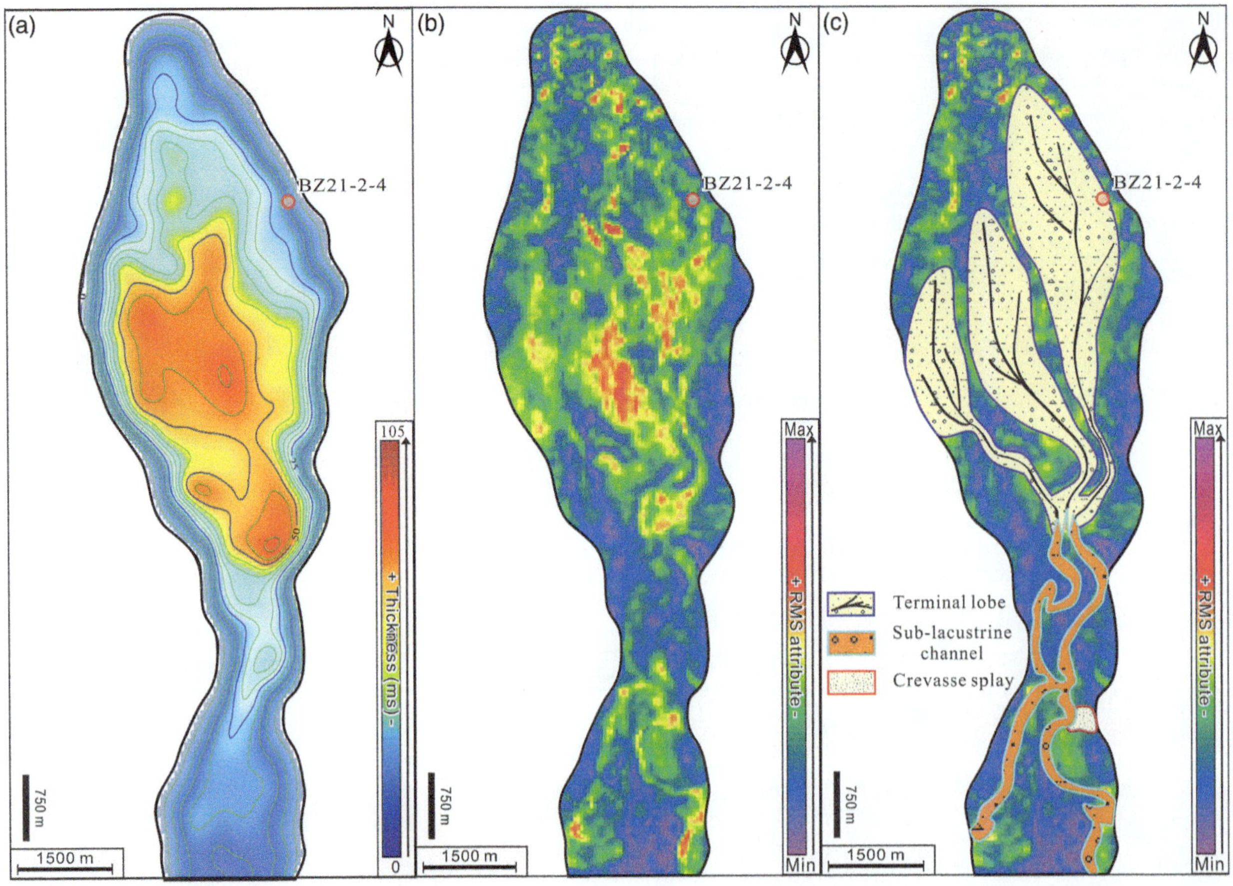

Figure 6.16 (a) Isopach map of localised channel-lobe complexes shown in (b). (b) and (c) RMS-attribute-extraction maps taken 10 ms above the maximum flooding surface (MFS) shown in Figure 6.9, illustrating plan-view geomorphological appearance and architectures of sub-lacustrine base-of-slope fans. The well-imaged areally localised channel-lobe complexes developed within the regressive (arid) systems tract (RST) of $SQd_2^{\,l}$ of the Oligocene Dongying Formation, Bohai Bay Basin and are seen to develop in front of syn-depositional slope-break zones on the gentle slope.

(Sztanó et al. 2013). They were built by mud-bearing turbidity currents flowing through canyons or leveed channels (Sztanó et al. 2013). Toe-of-slope lobes are seen to occur in front of the aggradational clinothems. These fan examples are traceable for tens of kilometres and are stacked into several tens to a few hundred metres thick successions. (See Figure 6.9 for a cross-sectional view of deep-lacustrine channel-lobe complexes.)

Oligocene deep-lacustrine terminal lobes of the Bohai Bay Basin contain lobe axis, lobe fringe, and distal lobe-fringe deposits, which exhibit dendritic high RMS-attribute patterns reflective of thicker sands in the lobe axes (Figure 6.15). These lobes were drilled by well BZ21-2-3, in which four main stages of lobe growth are recognised (Figure 6.18). Each of these four stages exhibits a fining-upward stacking trend that is composed of older lobe axis deposits, changing upwards into lobe fringe facies and/or distal lobe-fringe deposits (Figure 6.18). Lobe axis facies, the dominant component of the Oligocene Bohai Bay lobes, comprise poorly sorted, coarse-grained sub-rounded, pebbly sandstone and well-sorted structureless, fine- to coarse-grained, normally graded sandstone (Figure 6.18). The lobe fringe deposits are dark grey mudstones interbedded with thin beds of well-sorted, structureless, very fine to fine-grained sandstone or parallel-laminated siltstone. Distal lobe-fringe facies are thick (up to 20–30 m), blocky, dark grey mudstones.

Spectacular outcropping sub-lacustrine lobes of Cretaceous (Aptian) age in Lingshan Island off east China are composed of turbidites, debrites and hybrid event beds (Li et al. 2020). Similar interpretations were also reached by Shan et al. (2021) for outcrops of the Middle Permian Lucaogou Formation in the southern Junggar Basin, NW China, which documented deep-lake lobes made up of hybrid event beds, and deposits of both high- and low-density turbidity currents (Figure 6.17).

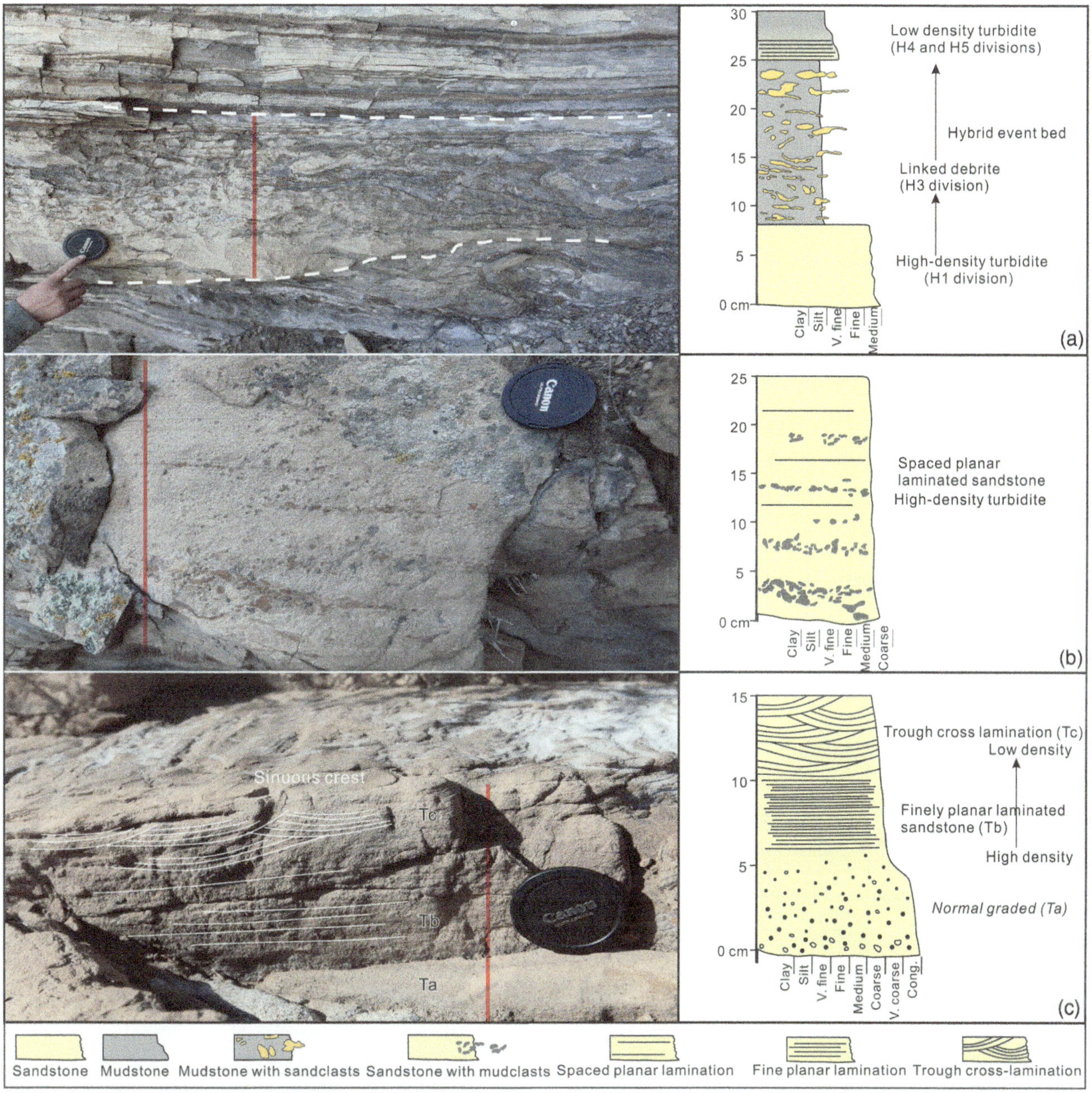

Figure 6.17 Field photographs (Courtesy of Dr. Xin Shan) and stratigraphic logs showing (a) hybrid event beds, (b) high-density turbidites and (c) low-density turbidites in the Middle Permian Lucaogou Formation in Southern Junggar Basin of NW China. Hybrid event beds in (a) contain three divisions: basal high-density turbidite (H1), linked debrite (H3) and mud cap (H5). High-density turbidites seen in (b) are expressed as spaced plane-laminated sandstones with an overall coarsening- and fining-upward pattern. Low-density turbidites seen in panel (c) are composed of normally graded and massive sandstone in the lower part, passing upwards to fine-scale plane-laminated sandstone, and finally to cross-laminated sandstone in the upper part. The red bars in the photographs indicate the locations of the sedimentary logs on the right.

(4) Sub-lacustrine Mass-Transport Complexes

As with their submarine counterparts, sub-lacustrine MTCs form fan-shaped or elongated sediment gravity-flow accumulations made up of blocks, slides, slumps and debris flows that originated from shelf-break or slope instability and collapse and were re-deposited on the lake floor (Figures 6.5 and 6.12). MTCs are seismically imaged as chaotic, discontinuous reflections with variable seismic reflection amplitudes and display irregular cross-sectional geometries (Figure 6.19). (See also Zhang et al. 2016; Pan et al. 2019 for some classic seismic examples of sub-lacustrine MTCs.) Seismically well-imaged deep-lacustrine

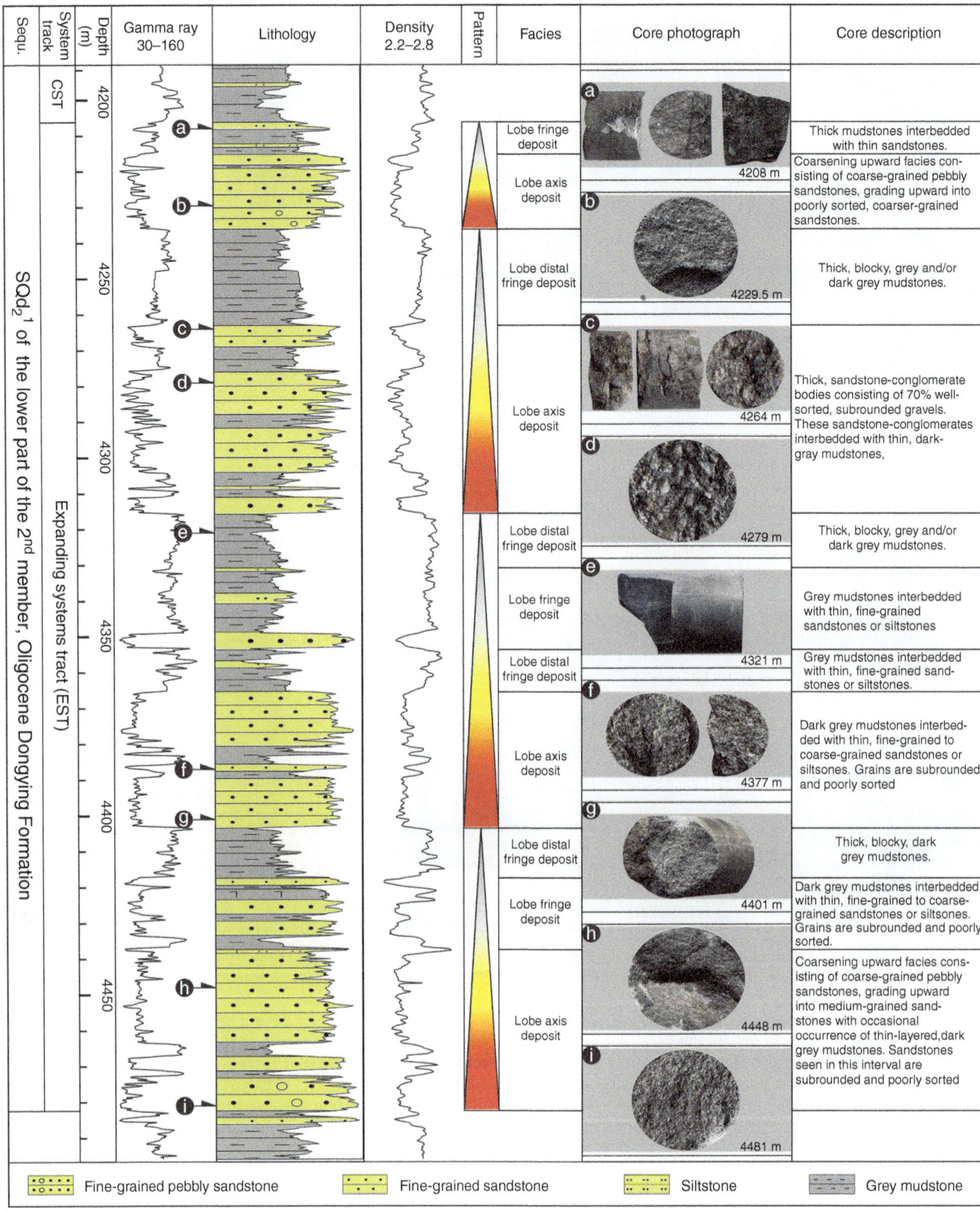

Figure 6.18 Lithological and sequence stratigraphic log of the BZ21-2-3 well (see Figure 6.15 for location), showing the development and sedimentologic interpretation of seismically well-imaged terminal lobes of areally extensive channel-lobe complexes shown in Figure 6.15.

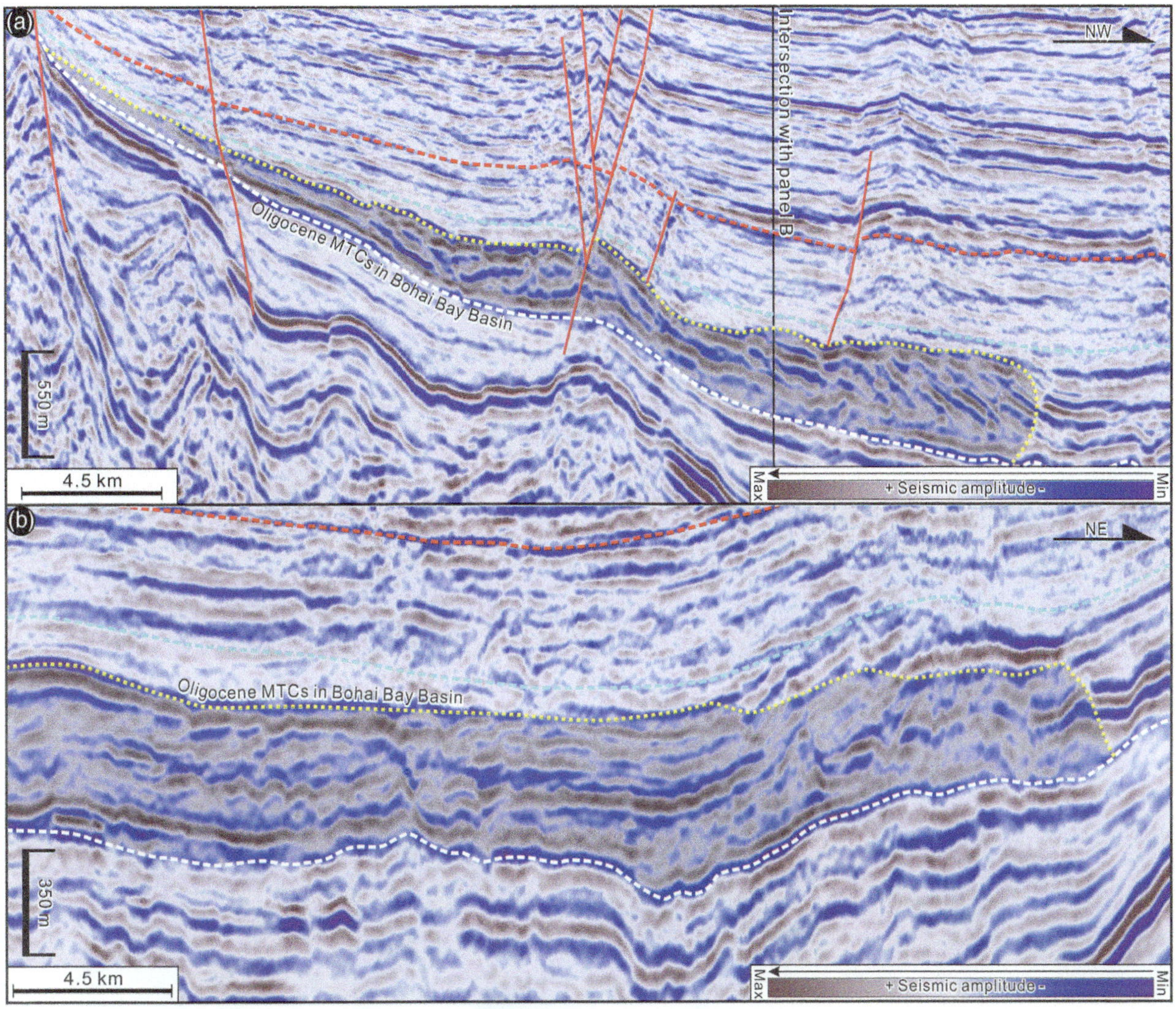

Figure 6.19 Seismic transects along (a) depositional dip and (b) strike showing cross-sectional appearance of mass-transport complexes (MTCs; shaded brown) developed in the Oligocene Bohai Bay Basin offshore China. These lacustrine MTCs are made up of chaotic, low-amplitude reflections with variable seismic reflection continuity and exhibit sheet-like cross-sectional geometry.

MTCs shown in Figures 6.19 and 6.20 are from the Oligocene Dongying Formation of the Bohai Bay Basin. They are imaged as chaotic, low-amplitude and semi-transparent seismic reflections with sheet-like cross-sectional geometries in both depositional dip and strike views (Figure 6.19a,b, respectively). They are bounded at their base by a laterally continuous, bed-parallel undeformed seismic reflection and thin in an upslope direction. These MTCs appears in RMS-attribute maps as fan-shaped or elongated moderate to low RMS-attribute accumulations (the muddy debris-flow deposits) with sporadic occurrence of irregular high RMS-attribute patches (the sandy debrites) (Figure 6.20).

Recently, the world's largest sub-lacustrine landslide deposits, the Qijia MTCs, were recognised in the Late Cretaceous Nengjiang Formation of the Songliao Basin by Pan et al. (2019). The MTC occupies an area of more than 300 km^2, has a thickness of 20–200 m and an estimated volume of 30 km^3. The main body of the Songliao MTC zone is composed of slumps, detached blocks, stacked blocks and drifted blocks. The Songliao Lake MTCs have been interpreted as triggered by a combination of rapid deposition and delta slope progradation (Pan et al. 2019) or earthquakes caused by volcanic activity (Zhang et al. 2016). Similar to the MTCs illustrated in Figure 6.19, the Qijia MTCs were mapped as discontinuous and internally chaotic seismic facies containing compressional structures in the downslope region, irregular top and basal surfaces and erosional grooves in the basal shear surfaces (Pan et al. 2019). Qijia MTCs are composed primarily of mudstone but with

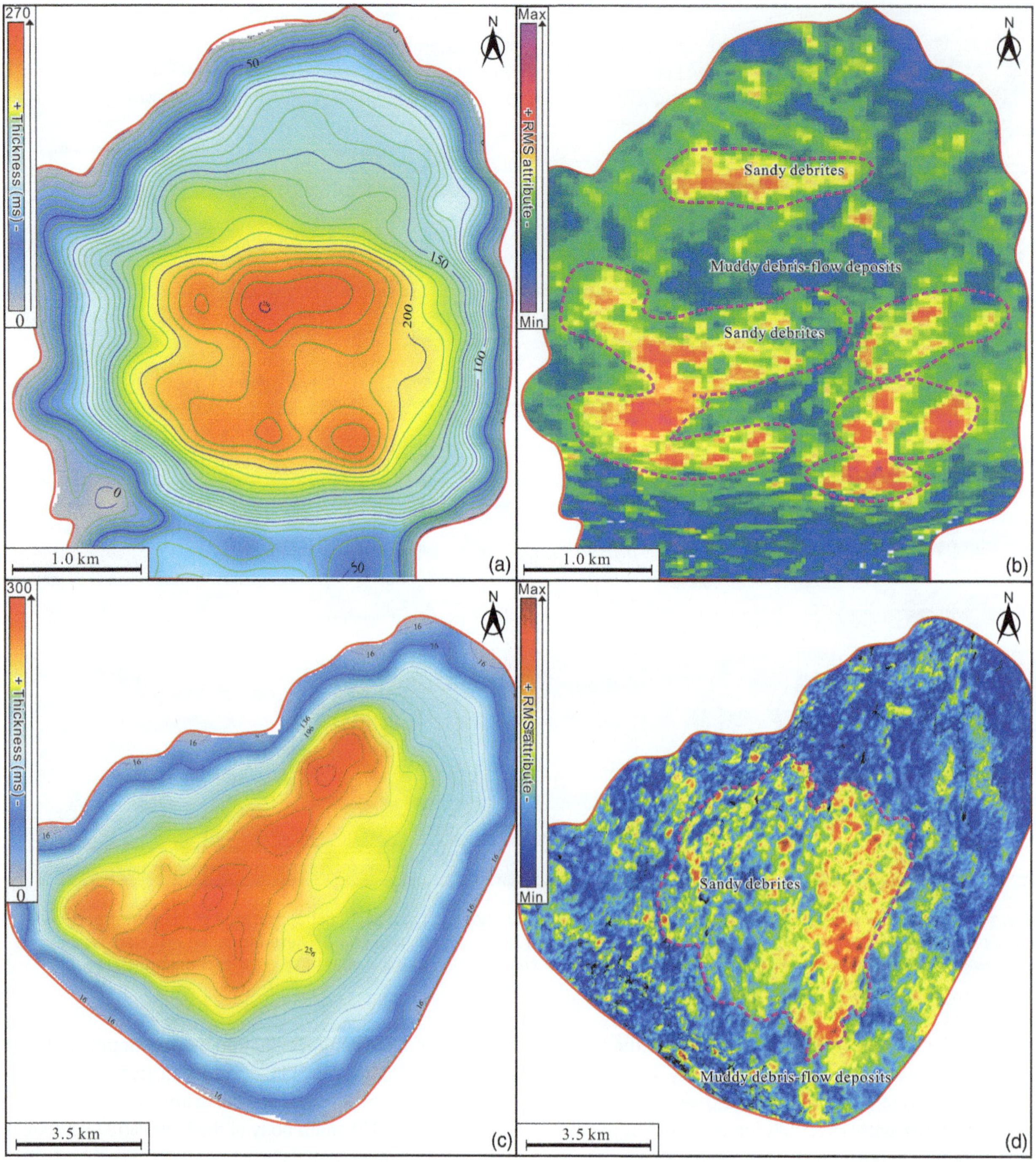

Figure 6.20 Isopach maps (a and c) and RMS-attribute-extraction maps (b and d) of areally extensive MTCs developed within SQd_2^l and SQd_2^u units, respectively, of the Oligocene Dongying Formation, Bohai Bay Basin. The MTCs of (a) and (b) are developed in front of syn-depositional slope-break zones on the gentle slope illustrated in Figure 6.a (expanding (humid) system tract – EST) and those of (c) and (d) in front of the slope in Figure 6.b (regressive (arid) system tract – RST).

sporadic occurrence of sandstone units (<16 m thick). These sparse sandy layers were blanketed by the surrounding MTCs and constituted stratigraphic traps that proved to be successful drilling targets for hydrocarbons.

(5) Other Deep-Lake Facies

In addition to the above-mentioned deep-lake facies, the central reaches of a lake are in some cases beyond the influence of rivers and sediment-gravity flows and can feature pelagic and hemipelagic muds and organic-matter-rich mudstone or 'oil shale'. Oil shale is an organic-matter-rich, fine-grained sedimentary rock containing kerogen, from which liquid hydrocarbons can be produced (Calvo et al. 1998). Oil Shales are often finely laminated so-called 'paper' shales. A very well-documented example of oil shale is the Mahogany Zone of the Eocene Green River Formation in the Uinta and Piceance Basins USA, which has up to 35% total organic carbon (TOC) (Renaut and Gierlowski-Kordesch 2010; Rosenberg et al. 2015). The Mahogany Zone is the richest of several oil-shale intervals. It can be recognised and correlated over 50,000 km^2 and was deposited during the cooling phase of the Eocene Climatic Optimum (Renaut and Gierlowski-Kordesch 2010; Rosenberg et al. 2015). It is well known for its thermally immature oil-shale resources and is one of the most prolific source rocks in the world (Johnson et al. 2010; Tänavsuu-Milkeviciene and Sarg 2012; Rosenberg et al. 2015).

Deep-lake sediments can be silt and clay transported by fluvial systems or littoral erosion. Aeolian grains, pollen and even tephra can be preserved in these deep-lake deposits, making valuable marker layers that can be used for stratigraphic correlation. In many lake systems, organic-matter-rich mudstone or oil shale is deposited in the most distal lake areas. The organic geochemistry of lacustrine oil shale normally features planktonic or benthic algal deposits.

Horizons of distal oil shale are themselves important markers for regional stratigraphic correlation. In lakes with waters with high uranium and thorium contents, oil shales are characterised by high gamma-ray and resistivity log motifs, as typified by the oil shale occurring immediately above the basal bounding surface of the Triassic Chang 7 Formation in the Ordos Basin (Feng et al. 2020). These deposits include skeletal and organic debris from planktonic, nektonic and benthic organisms, all predominantly algal. Trace fossils, typically grazing trails and feeding burrows, occur in these profundal deposits, representing periods when bottom waters were not oxygen depleted (Buatois and Mángano 1998, 2009).

6.4.2 Facies of Carbonate Lakes

In parts of lacustrine basins that are not overwhelmed by clastic material, chemical or biochemical deposits will often dominate. In low-salinity lakes, these are typically carbonate, siliceous or iron mineral accumulations, while more saline water bodies can produce a wide range of carbonate, evaporite and silicate minerals. Allochthonous detrital carbonates are commonly mixed with siliciclastic particles and preserved in the marginal systems, which are not further discussed in this section. Calcium carbonate precipitation is by far the most widespread chemical sedimentary process in lakes. Controls on carbonate precipitation whether evaporative biogenic or related to inflow water chemistry may be strongly seasonal (e.g. Lake Van, Turkey, McCormack et al. 2019).

6.4.2.1 Lacustrine Carbonates

Calcareous sediments are typically formed by one or more of four processes (Ahr 2008; Tänavsuu-Milkeviciene and Sarg 2012): (i) biogenically mediated, 'inorganic' precipitation – generally associated with plant photosynthesis or, less commonly, induced physically by evaporation, changes in temperature or mixing of water masses, (ii) biogenic production of calcareous shells, biogenic benthic encrustations (stromatolites and microbialites) or skeletal elements of living organisms, (iii) the input of allochthonous clastic carbonate particles derived from fluvial drainage basins and (iv) post-depositional/early-diagenetic precipitation from pore water. Note that lakes host no plankton with carbonate tests, in distinct contrast to marine systems. The only lacustrine plankton with hard tests are siliceous (i.e. diatoms in the Cenozoic).

Calcite is the commonest lacustrine carbonate mineral, but high-magnesian calcite and aragonite are also widespread. Moreover, normally rare carbonate minerals, such as vaterite ($CaCO_3$), monohydrocalcite ($CaCO_3{\cdot}H_2O$) and ikaite ($CaCO_3{\cdot}6H_2O$), are recorded from lakes. There is also circumstantial evidence that primary dolomite may form in some shallow, saline water bodies and that dolomite is widespread in oil shales and likely in many other lacustrine carbonates (Aziz et al. 2010; Bustillo et al. 2017), including biologically mediated dolomite in Qinghai Lake (Deng et al. 2010). Generally, carbonate deposits are seen to be most voluminous in lakes bordered by limestone bedrock (Renaut and Gierlowski-Kordesch 2010). Carbonates formed in marginal palustrine areas may contain abundant root traces and insect nests, typically the bee cell *Celliforma* and the coleopteran pupal chamber *Fictovichnus* (Genise et al. 2016; Genise 2017).

6.4.2.2 Biogenically Mediated ('Inorganic') Precipitation

The most important control on primary carbonate precipitation is exercised by the CO_2—HCO_3^-—CO_3^{2-} system. Removal of CO_2, accomplished most effectively by photosynthesis, raises pH and promotes carbonate precipitation. Removal of CO_2 by purely physical de-gassing into the atmosphere is less important and relatively slow (Ahr 2008). Carbonate precipitation can also be caused by the warming of lake waters, leading to carbonate supersaturation or the mixing of two waters of different ionic compositions (e.g. inflow and lake waters). These processes are also of lesser significance than photosynthesis.

Because of its association with photosynthesis, primary carbonate precipitation is restricted to shallow, marginal areas and the photic zone of deep-lake centres. An annual cycle of calcite precipitation has been described from Lake Zurich (Naeher et al. 2013) and the Greifensee in Switzerland (Müller et al. 2016) and other temperate-zone lakes. Late spring and summer diatom blooms abstract large quantities of dissolved CO_2, causing marked increases in pH and carbonate ion activity. This leads to supersaturation with respect to calcite. Photosynthetically induced precipitation in association with phyto- and picoplankton blooms is a major carbonate-producing process in the open waters of temperate, tropical and even Antarctic lakes (Lawrence and Hendy 1989). Simple chemical supersaturation is apparently not in itself sufficient to cause carbonate precipitation. Phytoplankton or similar substrates also appear to be required to provide nuclei around which crystal formation can commence, even in lakes that are highly supersaturated with respect to all the common carbonate minerals. Once precipitation begins, it may proceed very rapidly and the concentration of suspended carbonate crystals can cause surface waters to become cloudy: events known as 'whitings'.

Inorganic carbonate may precipitate where spring waters enter a lake or are ponded on an evaporative playa surface. Extensive carbonate precipitation can result from water-mass mixing or CO_2 outgassing, forming travertine or tufa mounds and chimneys. The nomenclature is confused but travertine is a form of terrestrial limestone, which is towards an abiotic, hydrothermal, end-member, and tufa is a biotic, cooler-water end-member (Capezzuoli et al. 2014). The most spectacular developments of this sort are found in some Pleistocene lakes in the western USA (e.g. Mono Lake), where massive, cathedral-like carbonate towers and pinnacles are tens of metres high. Although microbial textures (e.g. thrombolites, oncolites) can be recognised in these structures, discharge of calcium-rich spring waters along fractures into lakes rich in dissolved HCO_3^- was interpreted to be the primary cause of carbonate precipitation (Petryshyn et al. 2016).

Calcite, monohydrocalcite, Mg-calcite and aragonite may all be produced by direct precipitation. Which mineral precipitates is apparently determined by water chemistry, although the controlling factors are not known with certainty. In most temperate lakes, as well as in tropical Lakes Turkana and Malawi (East Africa), the carbonate is calcite. In Lakes Tanganyika (East Africa) and Van (Turkey), it is aragonite, and in Kivu (East Africa), aragonite and monohydrocalcite. The Mg/Ca ratio of the lake water may be significant: calcite precipitates when the ratio is low and Mg-calcite or aragonite at higher ratios (Choudens-Sanchez and Gonzalez 2009). Experimental evidence, however, indicates that other ions, notably sulphate and phosphate, can also influence carbonate mineralogy. In general, calcite is characteristic of low-salinity lakes, while Mg-calcite and aragonite are precipitated in brackish and saline water bodies. The minerals form silt-sized crystals that may have a variety of morphologies. They settle as a rain of discrete grains or in larger aggregates (pellets) of faecal origin. In some carbonate-precipitating lakes, much of the calcareous mud may be deposited in pellet form. Because of its association with phytoplankton blooms, 'inorganic' carbonate precipitation tends be episodic (despite permanent supersaturation) occurring in seasonal bursts. During the intervening periods, other sorts of sediment may accumulate. In these situations, the carbonates typically occur as discrete laminae, which alternate with other lithologies. Sediments consisting of carbonate interlaminated with clastic material, diatomaceous ooze, organic matter, or even evaporites are characteristic facies in the distal parts of a number of lakes.

Ooids are another form of inorganic carbonate. They are much less common than the carbonate silts described above. Calcitic ooids commonly have a cortex in which the crystals have a radial fabric or no preferred orientation at all. In common with marine ooids, lacustrine ooids are restricted to shallow, wave-agitated marginal areas. In Lake Tanganyika and the Great Salt Lake, littoral ooid shoals and beaches are widespread in areas starved of clastic sediments. During the Eocene, ooid-cemented sands or bioclastic sands were widespread in the Bohai Bay Basin, East China.

6.4.2.3 Biogenic Production-Metazoans Macrophytes and Microbes

Unlike the pelagic environment in the oceans, there are no common calcareous planktonic organisms in lakes. Some freshwater ostracods are free swimming, and one group of calcified planktonic green algae (charophytes) is known, but there are few records of these as major sediment producers (Külkylüoğlu 2005). The bulk of biogenic carbonate production is therefore due to benthic organisms and is thus restricted to shallow, mainly marginal areas. The major sediment producers are molluscs, ostracods and green algae of the charophyte group. Carbonates also precipitate extensively in association with benthic microbial communities (discussed in the following section).

Gastropods are the most widespread type of lacustrine molluscs. After death, gastropods tend to float because decomposition of the body tissue produces gas inside the shell. Shells may then drift ashore and accumulate in large numbers to form littoral shell beds. Sublittoral accumulations can also occur in areas where gastropods are abundant. These may form because of clastic sediment starvation or by winnowing of finer sediment, perhaps due to a slight fall in lake level. Some of the shallow platforms in Lake Tanganyika carry shell lags covering several hundred square kilometres (Cohen 1989).

Ostracods are the most common arthropods in the rock record, with more than 200 genera and 2000 species occurring in the continental realm, and they are widely reported from ancient lake strata (Cohen 1989; Martens et al. 2008; de Jesus e Sousa et al. 2019). Various species live in the water column or on the lake floor in proximal to distal areas. Species with calcareous tests can form millimetre to centimetre-thick carbonate layers in mudstone beds, as well as beds of wackestone, packstone and grainstone (e.g. Bohacs et al. 2000, 2007a).

Charophytes are a widespread group of aquatic algae, which are most common in freshwater lakes, although living specimens have also been recorded from saline waterbodies. At times in the past, they have been major carbonate sediment producers in lakes and are still of importance today. Some charophytes (principally species of *Chara* and *Nitella*) produce calcified female reproductive cells (oogonia) and may have carbonate-coated stems; up to 60% of the dry weight of the plant can be calcite. Other macrophytes may also become encrusted with calcite. This calcification process seems to be the result of a particular metabolism employed by some plants that inhabit alkaline waters. Many carbonate-producing lakes are characterised by relatively high pH (8.5–9.5) and, where the pH exceeds 8.7, the amount of dissolved CO_2 available for photosynthesis is very limited. A number of aquatic macrophytes and cyanobacteria have the ability to metabolise HCO_3^-, which in alkaline waters is present in much greater concentrations than CO_2. Under such conditions, these plants can out-compete species that are unable to utilise HCO_3^-. Calcification seems to be an integral part of bicarbonate assimilation, hence the widespread occurrence of calcite crusts on the aquatic macroflora of alkaline lakes. 'Meadows' of charophytes and plants, such as *Potamogeton*, can be a prolific source of calcitic sediment, yielding sands and muds with a gritty texture due to the presence of oogonia and calcified stem fragments. Similarly, abundant carbonate is produced in the lakes of the Florida Everglades (which have limestone bedrock) through the photosynthetic utilisation of HCO_3^- by cyanobacteria.

In hard-water lakes with little clastic input, extensive shallow-water carbonate deposits (commonly referred to as marl)[2] can be formed from the accumulation of biogenic and inorganic carbonates. In Littlefield Lake, Michigan, a flat, shore-attached platform is covered with oncoids and bordered by a swamp, which is accumulating peat. The platform passes lakeward into a 'marl bench slope', extensively colonised by *Chara*, and the site of algal mud accumulation. In this lake the slope facies grades into gastropod and ostracod-bearing mud in deeper water (10 m), but in other water bodies, there can be an abrupt change in gradient at the platform margin, causing large amounts of shallow-water carbonate material to be redeposited into deeper water by mass flows and turbidity currents. Marshes often form in shallows and on the deltas of inflowing rivers, such that carbonate sedimentation is associated with the accumulation of large quantities of organic matter that form peats. Similar conditions exist in the Florida Everglades, where there are some very extensive swamps, where seasonal and longer-term periods of low water cause extensive desiccation of the sediments (Pederson et al. 2018).

Microbial carbonates are produced by microbial growth and metabolism. They often form mounded or domal deposits but also occur as laminated mats (Walter and Allwood 2005). In the Uinta Basin, the Eocene Green River Formation shows typical microbialite and associated carbonates deposited in the sublittoral area, mainly comprised of stromatolites and thrombolites (Table 6.1; Figures 6.21 and 6.22) (Thompson et al. 1990; Sarg et al. 2013; Eljalafi and Sarg, 2021). The microbes involved encompass a range of microscopic organisms including bacteria, fungi, algae and protozoans. They can be present in almost any wet environment, are often prolific and are significant in the rock record. Microbial carbonates are formed by two major processes: grain-trapping and biologically mediated precipitation.

Microbialites typically develop on firm or lithified substrates. They may encrust boulders or cobbles along beaches and at the toes of alluvial fans or outcrops of basement rocks. They form lithified lacustrine sediments, such as beachrock, or lacustrine deposits that become indurated due to exposure and pedogenesis. Some very substantial microbialite structures are known from modern lakes. In Lake Tanganyika in the East African Rift, thrombolitic lithoherm mounds with diameters of up to 2 m have coalesced to form reef-like features at depths of 15–50 m, with plan-view areas of several hundred square kilometres (Cohen 1989). Studies of Quaternary lacustrine microbialites suggest that they may form over a depth range of 0–30 m. Spectacular examples are reported from Lake Van, where they are apparently forming at water depths down to 100 m (Kempe et al. 1991). These match in scale the famous tufa pinnacles of Lake Mono in the western USA and, like these, form where calcium-rich springs

2 Marl is a mixture of between one-third and two-thirds lime mud and siliclastic mud.

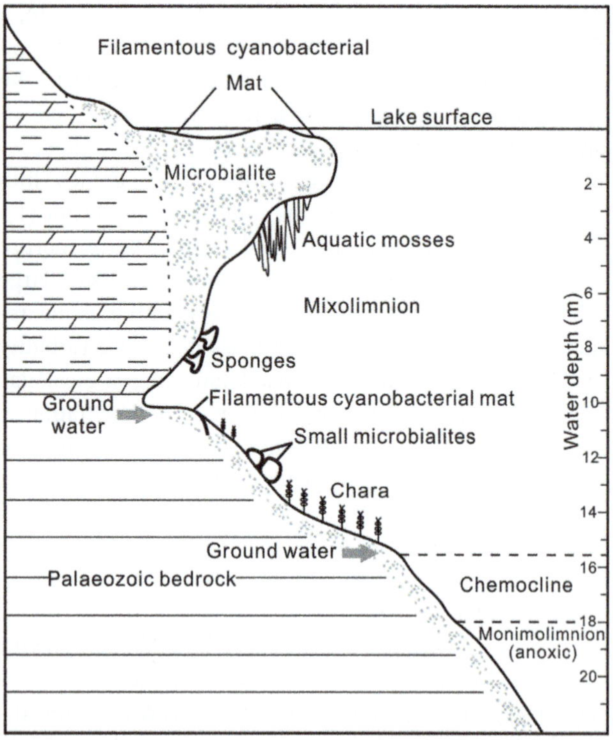

Figure 6.21 Profile of Green Lake New York, USA, at Deadman's Point illustrating local stratigraphy, points of suspected groundwater entry and the present-day position of the chemocline. The lake is stratified and anaerobic bacteria are the only life beneath the chemocline (Adapted from Thompson et al. 1990). See text for a detailed explanation.

emerge into alkaline lake water. However, cyanobacteria have played a much greater role in the development of the Lake Van structures, hence their designation as microbialites. Microbialites are also observed in ancient lake deposits, including the Miocene dolomite microbialites in the Madrid and Duero Basins (Sanz-Montero et al. 2009), Hot Spring limestone, Idaho (Bohacs et al. 2013), Eocene Green River Formation (Awramik and Buchheim 2015), Early Cretaceous microbialites of the Lagoa Feia Formation, offshore Brazil (a major hydrocarbon reservoir; Muniz and Bosence 2015) and the Mesoproterozoic stromatolites in the Middlebrun Bay Member of the Rossport Formation, Sibley Group, in Ontario, Canada (Czaja et al. 2019).

Where microbes coat grains, such as skeletal debris, pebbles, wood fragments or even reworked fragments of earlier microbialites, oncoids may form. These are concentrically laminated structures, commonly 1–50 mm in diameter. Their size and shape initially reflect the form of the nucleus, but the addition of successive growth laminae can build up a substantial layer of carbonate, resulting in oncoids as much as 0.3 m in diameter. Because concentric growth requires that the clasts are periodically rolled, oncoids are characteristic of areas of wave or current activity and mainly form in shallow, marginal parts of lakes.

One of the well-documented examples of modern lake carbonates is from Green Lake, New York State, USA, which has a water depth of 55 m at its deepest point. Green Lake is small (0.26 km^2) and is surrounded by steeply rising hills

Table 6.1 Facies summary of lacustrine margin carbonates recognised in the Green River Formation.

Facies Association	Lithofacies	Lithofacies	Subfacies		Dominant fabrics or grains	Depositional environment
Lacustrine microbial carbonates	G	Stromatolites	G1	Fine-grained stromatolite	Thinly to micro-laminated	Lower littoral to lower sublittoral
			G2	Columnar stromatolite	Columnar, thinly to micro-laminated	Lower sublittoral
			G3	Agglutinated stromatolite	Agglutinated	Upper littoral to upper sublittoral
			G4	Dendrolitic stromatolite	Dendrolitic, digitate, clotted	Middle sublittoral
			G5	Spheroidal/oncoidal stromatolite	Spheroidal, concentrically laminated	Upper to lower littoral
	H	Thrombolite	H1	Arborescent thrombolite	Coarse clotted	Lower littoral to upper sublittoral
			H2	Coarse-grained thrombolite	Coarse clotted	Middle sublittoral
			H3	Fine-grained thrombolite	Fine clotted	Lower sublittoral

See Figure 6.2 for idealised model of the lateral distribution of carbonate facies listed in this table.
Source: Adapted from Eljalafi and Sarg (2018). See Fig 6.22 for locations of facies associations G and H.

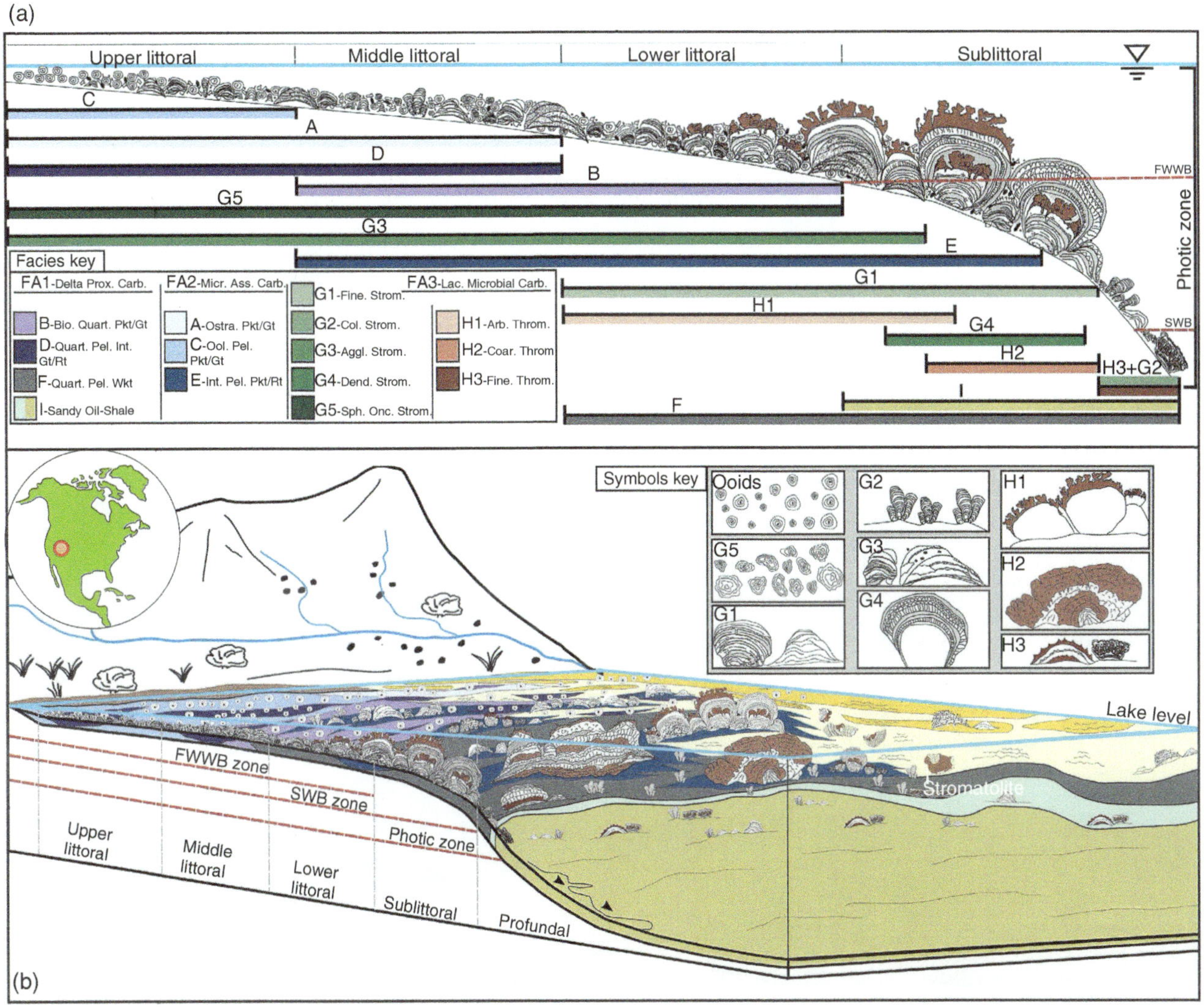

Figure 6.22 (a) 2D depositional profile along the depositional dip showing the transition from upper littoral to lower sublittoral environments in the Green River Formation, Uinta Basin United States (from Eljalafi and Sarg 2021). (b) 3D conceptual depositional model illustrating the lateral variability of the various carbonate deposits and their relationship with point-sourced siliciclastic deltaic sands of Green River Formation in the Uinta Basin (Eljalafi and Sarg 2021 / Society for Sedimentary Geology (SEPM)). See Table 6.1 for microbial abbreviations in legend. In (a): Quart = quartzose, Pkt = packstone, Gt = grainstone, Wkt = wackstone, Pel = peloidal, Int = intraclastic, Ostra = ostracod, Ooi = ooidal." In (b): FWWB = fair weather wave base, SWB = storm wave base.

to the northwest and southeast, with widespread occurrences of seasonally formed, laminated carbonate in the deeper part of the lake. The major recharge is comprised of groundwater, which is saline and contains high concentrations of Ca^{2+}, Mg^{2+} and SO_4^{2-} attributed to the dissolution of the surrounding dolomite, gypsum and halite deposits (Thompson et al. 1990). A model of two groundwater inputs was supported by the stratigraphic and sedimentological studies of Thompson et al. (1990), who suggested that a shallow seep enters the mixolimnion along the Syracuse–Vernon Formation boundary (~10 m depth), and a deeper more saline input enters along the contact between the green and red units of the Vernon Shale (~18 m) (Figure 6.21). Both intrinsic (microbial metabolisms) and extrinsic (temperature and chemical input) factors dictate the carbonate saturation index and ultimately the precipitation of carbonate minerals within the microbialite structure. Microbialite ledges (e.g. thrombolites) with thicknesses of 2–8 m are currently forming at the lake margin and extend up to 7 m into the lake from a shoreline outcrop of Palaeozoic rocks (Figure 6.23). Samples collected from the thrombolite platform at Green Lake (Figure 6.23a,b) show lithified clotted structure overlain by microbial mats made up of trapped and bound carbonate sediment, filamentous bacteria and precipitates (Figure 6.23c,d). The SEM micrographs show a little EPS (extra-cellular polymeric substances; ‘a’ in

Figure 6.23 Field photographs showing carbonate deposits in Green Lake, New York, USA. (a and b) Outcrop photos of thrombolite platform. (c and d) Outcrop photos showing a lithified clotted structure overlain by a microbial mat. (e to h) SEM Micrographs of samples present little extra-cellular polymeric substances (EPS) (a), carbonate grains (b) and zones of precipitation (c) (from Patterson 2014).

Figure 6.23e,f), carbonate grains ('b' in Figure 6.23e,g) and zones of precipitation ('c' in Figure 6.23e,h) (Patterson 2014). Illustrative ancient examples of well-documented lacustrine carbonates include the Bonneville palaeolake (the predecessor of Great Salt Lake: Hart et al. 2004) and Siling-Co Lake in the Qinghai-Tibetan Plateau (Morinaga et al. 2013), both of which have been studied as indicators for palaeoclimates and lake-level changes.

Benthic communities of microphytic algae and cyanobacteria are relatively common in lakes and can form substantial organo-sedimentary bodies generically called 'microbialites'. They may have internal structures that are stromatolitic (fine, planar or wavy lamination), oncolitic (concentrically laminated), thrombolitic (clotted texture) or cryptic (no obvious internal structure). Although some of these strongly resemble marine stromatolites and oncoids, a major difference is that in most lacustrine examples, the trapping and binding of discrete sedimentary particles seem to have played little part in the development of the structures. They are formed mainly by carbonate precipitation on and within the organically formed structure. It is still a matter of debate whether these are true biogenic deposits. Although microbialites may be produced by a variety of algae and cyanobacteria, none of the genera seems to be an obligate calcifier. Calcification apparently occurs mainly as a by-product of the organism's photosynthetic activity.

6.4.3 Facies of 'Inter-related' Carbonate–Siliciclastic Systems

Generally, carbonate depositional environments require conditions of low turbidity and clear water where or when siliciclastic sediment supply is minimal. Thus, the deposition of the two sediment types is often mutually exclusive.

However, the close association of clastic and carbonate sedimentary units is common in lacustrine stratigraphy (Tänavsuu-Milkeviciene and Sarg 2012; Du et al. 2020). Depositional inter-relations of clastics and carbonates occur at different temporal and spatial scales: (i) clastic material dispersed within a carbonate bed, (ii) clastic units interbedded with carbonate beds; and (iii) lateral transition between clastic and carbonate beds as time-equivalent facies changes. One of the primary reasons for the stratigraphic interbedding is rapid lake-level changes, causing extensive regressive and transgressive lateral lake shoreline movements accompanied by short-term increases and decreases of siliciclastic deposition at any one location (Olariu et al. 2021).

Inter-related clastic-carbonate deposits are widespread in the Bohai Bay Basin (Zhao et al. 2014). They are generally found in the First and Second Members of the Eocene-Oligocene Shahejie Formation and represent high-quality hydrocarbon reservoirs. The Eocene Green River Formation, deposited in one of a series of large Laramide Lake basins in eastern Utah, westernmost Colorado and southwestern Wyoming, contains typical clastic-carbonate association stratigraphic successions with abundant oil shale. The beginning of continuous lacustrine deposition in the Piceance Creek basin was marked by a distinctive ostracod, oolite and mollusc-rich bed, the Long Point Bed. The Douglas Creek Member consists mainly of basin margin limestone, sandstone and mudstone. The main part of the Green River Formation, the Parachute Creek Member, consists dominantly of thinly laminated dolomitic oil-shale deposits interbedded with sparse siltstone and sandstone beds and well-bedded and disseminated evaporite (halite, nahcolite ($NaHCO_3$) and dawsonite ($NaAlCO_3(OH)_2$)).

6.4.3.1 Inter-related Carbonates and Clastics in the Littoral and Sublittoral Zones

The littoral and sublittoral zones above storm wave base are generally defined as shallow-lake depositional environments (Renaut and Gierlowski-Kordesch 2010). The sedimentary facies in these zones in the Green River Formation consist mainly of seven facies:

1) Shoreline mudstone characterised by mudstone with intraclasts, microbial carbonate and laminated to homogeneous mudstone and siltstone, with bed thicknesses up to tens of centimetres.
2) Shoreline sandstone characterised by gradational or sharp-based fine- to medium-grained sandstone with low-angle cross-stratification, hummocky and swaley cross-stratification and wave-ripple cross-lamination.
3) Delta-front deposits consisting of sandstone beds that are laterally and vertically separated by mudstone and siltstone, in units up to several tens of metres thick, showing fining upwards (distributary channel fills: Figure 6.24a), coarsening upwards (river mouth bars: Figure 6.24b) or, in some cases, 'no-trend' facies successions.
4) Carbonate shoals, composed dominantly of structureless or plane-parallel, non-skeletal, coated calcitic (Suriamin 2010) 'grainstone and packstones' and coquina beds, which even when only a few centimetres to tens of centimetres thick, can be traced for hundreds of metres along the outcrop (Figure 6.24c).
5) Microbial carbonate comprising the millimetre- to centimetre-scale, thinly laminated limestone interbedded with boundstone displaying small, vertically oriented, shrub-like features and massive limestone displaying clotted fabrics (Figure. 6.24d).
6) Littoral to sublittoral siliciclastics composed of parallel-laminated or structureless mudstone and very fine-grained sandstone (Figure 6.24e), together with fine-grained sandstone lenses.
7) Littoral to sublittoral oil shale consisting of the laminated silt-rich oil shale and undulating, wavy-structured oil-shale deposits (Tänavsuu-Milkeviciene and Sarg 2012).

The various facies associations in the Uinta/Piceance Basins comprise mixed clastic-carbonate and oil-shale deposition in relatively shallow lake environments. In contrast, deposition along the basin margin was characterised by relatively high siliciclastic input but nonetheless with the formation of thick microbial carbonate. During the stages of cyclic sedimentation proposed by Tänavsuu-Milkeviciene and Sarg (2012), lake marginal deposits were characterised by prograding wave-dominated deltaic and shoreline sandstone, grading upwards into littoral to sublittoral siliciclastics and then to littoral to sublittoral oil shale (Stage 1: Fresh lake). Alternatively, the littoral to sublittoral siliciclastics may pass upwards into microbial carbonates, which are then also capped by silty, littoral to sublittoral oil shale (Stage 2: Transitional lake). Both stages emphasise transgression and landward movement of lake shorelines. This is logical in that the influx of siliciclastics was accompanied by water volumes that also raised the lake level and pushed the shoreline landwards.

6.4.3.2 Inter-related Clastics and Carbonates in the Profundal Zone

The littoral and sublittoral zones below the storm wave base are defined as deep lake or 'profundal' areas of deposition (Renaut and Gierlowski-Kordesch 2010). These zones in the same Piceance Creek basin example (Tänavsuu-Milkeviciene and Sarg 2012) consist mainly of four facies:

1) Laminated oil shale composed of light brown to dark brown, black or gray beds with spectacular millimetre-thick rhythmic lamination (Figure 6.24f).

Figure 6.24 Outcrop photos of the mixed clastic and carbonate successions in the Eocene Green River, Piceance Creek Basin. (a) Fine to medium-grained, cross-stratified sandstone interbedded with mudstone in a distributary channel of the delta fronts. (b) Distributary mouth bar showing internal sigmoid-shaped clinoforms. (c) Grainstone interbedded with calcareous mudstone of carbonate shoals, showing concave-up channel-fill at the base. (d) Microbial carbonate showing mounded stromatolite and thrombolite; see also detailed features in (c) and (d). (e) Littoral to sublittoral oil shale showing dark-gray mudstone interbedded with light-gray, laminated siltstone containing bioclasts. (f) Dark, laminated oil shale deposited in the profundal zone with total organic carbon (TOC) up to 35%.

2) Soft-sediment-deformed oil shale characterised by folds and overturned strata, as well as disrupted deposits, dolomitic kerogen-rich oil-shale deposits with angular, sub-angular angular, sub-angular and sub-rounded carbonate and siliciclastic clasts, as well as flat organic-matter clasts with their long axes parallel to bedding surfaces.
3) Siliciclastic turbidites consisting dominantly of well-sorted siltstone to sandstone forming centimetre- to metre-thick normally graded or ungraded depositional units.
4) Subaqueous evaporites composed of up to tens of centimetre-thick beds with scattered crystals of nahcolite and halite.

The oil-shale deposits change in time and space across the basin from being 'rich' to 'lean' and represent variable rates of organic-matter production, preservation and dilution. Although not thermally mature, the clay-rich or carbonate-rich (dolomitic marlstone) oil shale of the Green River Formation are estimated to be the largest oil-shale deposits in the world today, with an estimated total in-place volume of potential oil of about 1.5 trillion barrels (Johnson et al. 2010). Three main types of oil shale facies are well developed in the profundal zone, including laminated oil-shale, oil-shale breccia, and disturbed oil-shale (Feng 2011). Laminated oil-shale accumulated through the settling of suspended sediments, whereas oil-shale breccia is interpreted as mass-transport deposits (Feng 2011). Oil shale typically lacks trace fossils (e.g. Scott and Smith 2015), and none were found in laminated oil shale in the Green River Formation of the Piceance Creek Basin (Feng 2011).

6.4.4 Fossil Biotas and Ichnology

6.4.4.1 Ichnofacies Model and Trace-Fossil Distribution in Lacustrine Settings

A relatively rich benthos capable of producing biogenic structures is present in modern lakes. This fauna includes annelids (e.g. oligochaetes, leeches), aquatic to semi-aquatic insects (e.g. dipterans and coleopterans), crustaceans (e.g. ostracods) and molluscs (e.g. bivalves and gastropods) (Cohen 2003; White and Miller 2008). A few vertebrates may at times impact the substrate and produce traces although they are mostly nektonic in habit (i.e. fish, amphibians and aquatic reptiles).

Two archetypal ichnofacies are dominant in lake systems: *Scoyenia* and *Mermia* (Buatois and Mángano 1995b, 1998, 2004). The *Scoyenia* Ichnofacies (Figure 3.12) is dominated by meniscate burrows, simple horizontal and vertical burrows and vertebrate and arthropod trackways (Figure 6.25e,h). Simple trails may also be present (Figure 6.25d) but are not dominant. The *Scoyenia* Ichnofacies (Figure 3.12) indicates sediment periodically exposed to air or periodically inundated and is characteristic of lake margins. The *Mermia* Ichnofacies (Figure 3.13) is characterised by simple horizontal grazing trails (Figure 6.25i) and feeding burrows (Figure 6.25a), and to a lesser extent, fish trails (Figure 6.25c). This ichnofacies indicates aquatic freshwater conditions and is typical of the permanent subaqueous portions of a lacustrine basin. Both ichnofacies correspond to low-energy settings. More rarely, a continental expression of the *Skolithos* Ichnofacies, characterised by vertical simple or U-shaped burrows and escape trace fossils, may be present in high-energy portions of lakes affected by strong waves or near the mouth of distributary channels (Mángano et al. 1994; Buatois and Mángano 2011). Also, the *Celliforma* Ichnofacies (Figure 3.9) may occur in palustrine carbonate and lake-margin strata as indicated by insect nests, such as bee cells (Figure 6.25g) and coleopteran pupal chambers, together with widespread root traces (Bohacs et al. 2007b; Genise et al. 2010; Scott and Smith 2015; Genise 2017).

Trace-fossil distribution in lacustrine settings reflects the interplay of abiotic and biotic factors controlling the diversity of lake biotas (Cohen 2003). Abiotic factors include energy, light, oxygen, temperature, salinity, substrate and nutrients, whereas biotic factors include competition, grazing, predation and symbiosis. Lake hydrology is a major control and, as a result, significant changes are seen between trace fossil distribution in hydrologically open (Figure 6.26) and hydrologically closed (Figure 6.27) lakes (Buatois and Mángano 1998). Typically, the density and diversity of lacustrine benthos reach a maximum in the oxygenated sublittoral zone, which in most lakes is between 2 and 4 m deep, but may be up to 18 m deep in large oligotrophic lakes (Martin et al. 2005). Trace-fossil distribution is overprinted by taphonomic factors, with low-energy areas showing the highest preservation potential. Lithologic interfaces in fine-grained deposits tend to promote the preservation of delicate, very shallow-tier trace fossils (Buatois and Mángano 1998).

6.4.4.2 Biotic Factors Influencing Lake Evolution in Deep Time

Biotic factors have influenced lacustrine evolution in deep time (Figure 6.28). Although our knowledge of the earliest lacustrine systems on Earth is patchy and discrimination between lacustrine and marine sediments can be extremely difficult in the absence of clear biological criteria, there is very early evidence of microbial communities in lakes as revealed by the presence of stromatolites in the Neoarchean (Buick 1992; Lepot et al. 2008; Wilmeth et al. 2019). The fact that these stromatolites, interpreted as implying oxygenic photosynthesis, predate the Great Oxidation Event suggests that Archean lakes may have served as oxygen oases (Wilmeth et al. 2019). Stromatolites continued to be the dominant evidence of life in Proterozoic lacustrine ecosystems (e.g. Fedorchuk et al. 2016). Other evidence of life in Precambrian lakes is revealed by the occurrence of 'Microbially Induced Sedimentary Structures' or MISS (e.g. Callow et al. 2011) and terrestrial microfossils (Strother et al. 2011; Battison and Brasier 2012; Strother and Wellman 2021).

Overall, the evolution of lacustrine biotas during the early Palaeozoic may have been inhibited by the combination of the osmoregulation challenges of freshwater environments and limited nutrient delivery and high sediment loads because of a virtually nonexistent vegetation cover

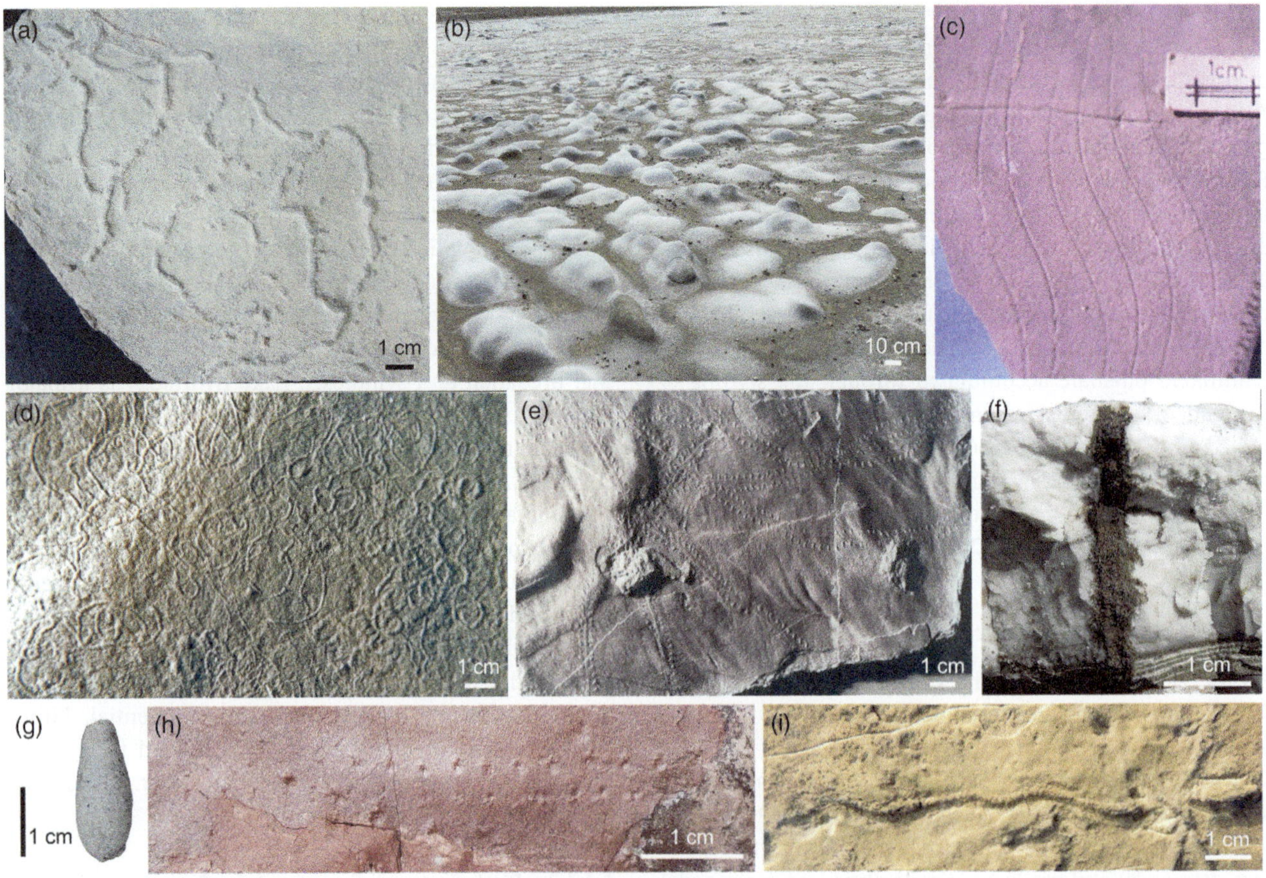

Figure 6.25 Representative trace fossils of lacustrine environments. (a) The burrow system *Vagorichnus anyao* preserved at the base of lacustrine turbidites, Lower Jurassic Anyao Formation, central China. (b) Fossilised flamingo nests in lake margin deposits, forming extensive mounded surfaces, Pleistocene High Magadi Beds, Lake Magadi, Kenya. (c) The fish trail *Undichna insolentia* (epirelief) in underflow current deposits, upper Carboniferous Agua Colorada Formation, western Argentina. (d) The looping trail *Mermia carickensis* in lake margin deposits, Devonian Old Red Sandstone, Dunure, Scotland. (e) The arthropod trackway *Siskemia elegans* in lake margin deposits, Devonian Old Red Sandstone, Dunure, Scotland. (f) Vertical burrow, probably produced by beetles, in magadiite of a hypersaline lake, Pleistocene High Magadi Beds, Lake Magadi, Kenya. (g) The bee cell *Celliforma* isp. in palustrine carbonates, Paleogene Queguay Formation, Rio Negro, Uruguay. (h) The arthropod trackway *Diplichnites* isp. in Playa Lake deposits, Middle Permian Hornburg Formation, Konberg Quarry, Germany. (i) The meandering trail *Helminthopsis hieroglyphica* preserved at the base of lacustrine turbidites, Early Jurassic Anyao Formation, central China.

(Figure 6.28) (Miller and Labandeira 2002; Cohen 2003; Buatois et al. 2022). Early Palaeozoic lakes were mostly shallow and rapidly filled with sediment (Cohen 2003). Stromatolites continued to be common in Cambrian lakes (e.g. White and Youngs 1980). Systematic evaluation of ichnologic data from continental strata reveals that lake environments were essentially barren of animal life during the Cambrian–Ordovician (Minter et al. 2016a, 2017). By the Silurian, there is sparse ichnologic evidence that suggests colonisation of ephemeral lakes as indicated by the presence of trackways and simple burrows, representing the onset of the *Scoyenia* Ichnofacies (e.g. Wright et al. 1995; Marriott and Wright 2004; Davies et al. 2006). The trace-fossil record is therefore consistent with the record of body fossils of freshwater vertebrates and arthropods, the latter including trigonotarbid arachnids, arthropleurids and centipedes (Minter et al. 2016b; Buatois et al. 2022). The earliest charophytes have been recorded from upper Silurian deposits, representing a source of food that may have fuelled colonisation by both vertebrates and invertebrates (García 1994). Mid-Palaeozoic lakes experienced an increase in nutrient availability, resulting from a rise in phosphate weathering and terrestrial humic matter and triggering an increased complexity in lacustrine food webs (Cohen 2003). This was accompanied by a rapid increase in the diversity and abundance of lacustrine vertebrate (including the earliest amphibians) and invertebrate biotas, as well as charophytes during the Devonian. A wide variety of fish and arthropods (e.g. large chelicerates) together with a lower diversity of molluscs (both bivalves and gastropods)

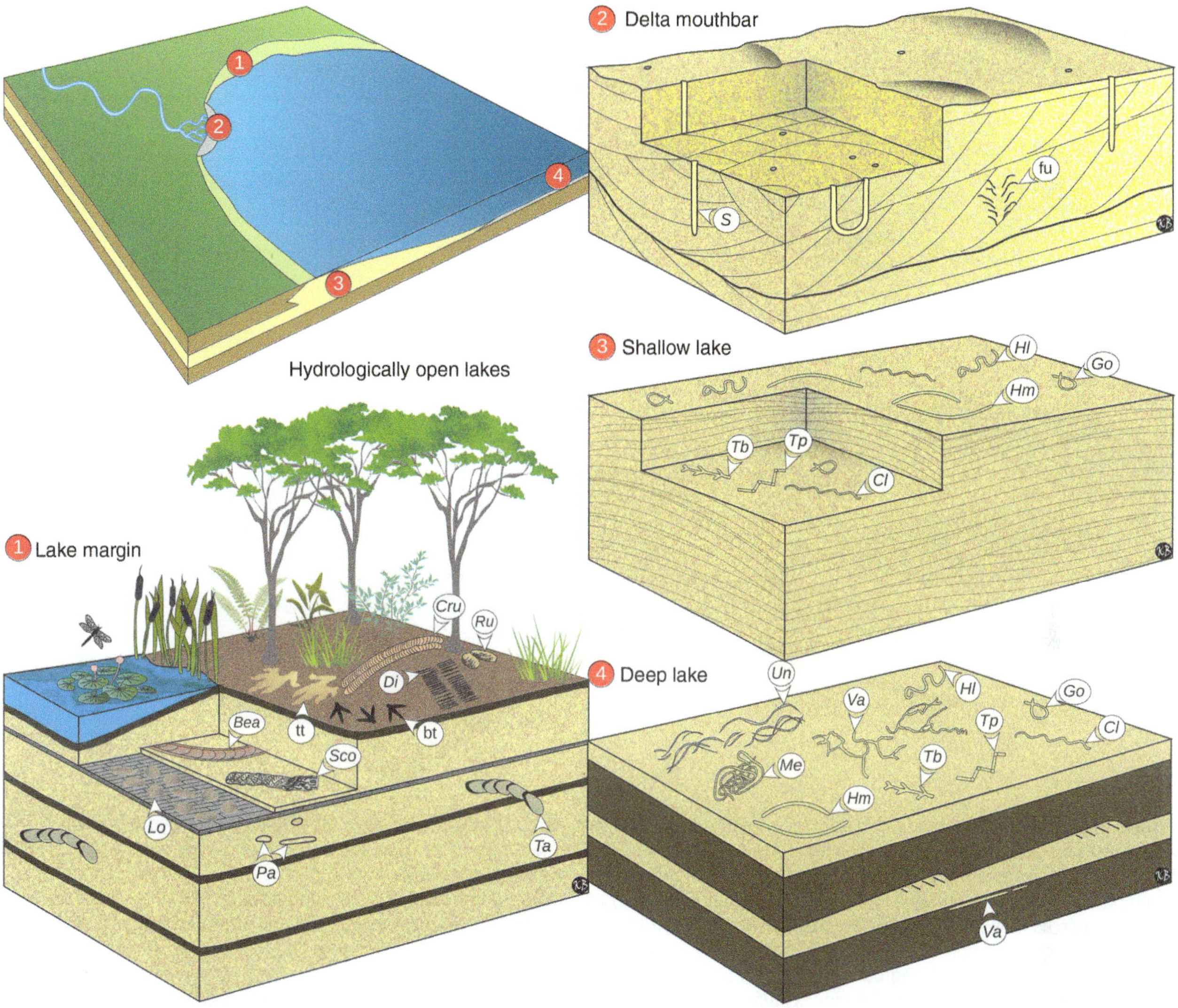

Figure 6.26 Schematic reconstruction of trace-fossil distribution in hydrologically open lakes. Such lakes typically contain more diverse trace-fossil assemblages than closed lakes. Low-energy lake-margin deposits (1) are characterised by meniscate trace fossils, such as *Beaconites* (*Bea*), *Taenidium* (*Ta*) and *Scoyenia* (*Sco*), bilobate trace fossils, such as *Cruziana* (*Cr*) and *Rusophycus* (*Ru*), arthropod trackways, such as *Diplichnites* (*Di*) and simple burrows, such as *Palaeophycus* (*Pa*). Limestone beds may contain high-density assemblages of the bivalve resting trace *Lockeia* (*Lo*). Tetrapod (Tt) and bird (Bt) tracks are also common in lake-margin deposits, but the overall composition of vertebrate ichnofaunas is highly variable according to geologic age. High-energy shoreline areas, such as deltaic mouth bars (2), may contain vertical burrows, including *Skolithos* (*Sk*), *Arenicolites* (*Ar*) and escape trace fossils (fu). Shallow lake deposits (3) tend to contain similar trace fossils to deep-lake deposits but may be less bioturbated and have less diverse assemblages, particularly if affected by wave action during both fair-weather and storm times. The grazing trails *Gordia* (*Go*), *Cochlichnus* (*Co*), *Helminthopsis* (*He*) and *Helminthoidichnites* (*Hl*) are dominant, typically associated with the feeding trace *Treptichnus*, most commonly represented by *T. pollardi* (*Tp*) and *T. bifurcus* (*Tb*). Deep-lake deposits (4) are also dominated by grazing trails, such as *Mermia* (*Me*), *Gordia* (*Go*), *Cochlichnus* (*Co*), *Helminthopsis* (*He*) and *Helminthoidichnites* (*Hl*), and feeding traces, such as *Vagorichnus* (*Va*) and *Treptichnus*, typically *T. pollardi* (*Tp*) and *T. bifurcus* (*Tb*). The fish trail *Undichna* (*Un*) is also common. Vertebrate and invertebrate trace fossils. Not to scale. Trace fossil block diagrams in this chapter are courtesy of Kerry Bann.

characterised Devonian lakes. Various feeding types were established, namely predation, suspension feeding and detritus/deposit-feeding, although evidence of herbivory is lacking (Cohen 2003). Ichnologic evidence suggests extended colonisation of lake margins by a diverse arthropod epifauna (Buatois and Mángano 1993; Buatois et al. 1998, 2022; Labandeira 2002; Minter et al. 2016b, 2017).

Late Palaeozoic lakes were characterised by an increase in diversity, notably crustaceans, molluscs and fish (Figure 6.28) (Gray 1988). Diversification was accompanied by a further increase in the complexity of food webs, albeit still with no evidence of herbivory (Labandeira 2002; Cohen 2003). Also, ichnologic evidence indicates colonisation of permanent subaqueous offshore and deep-water

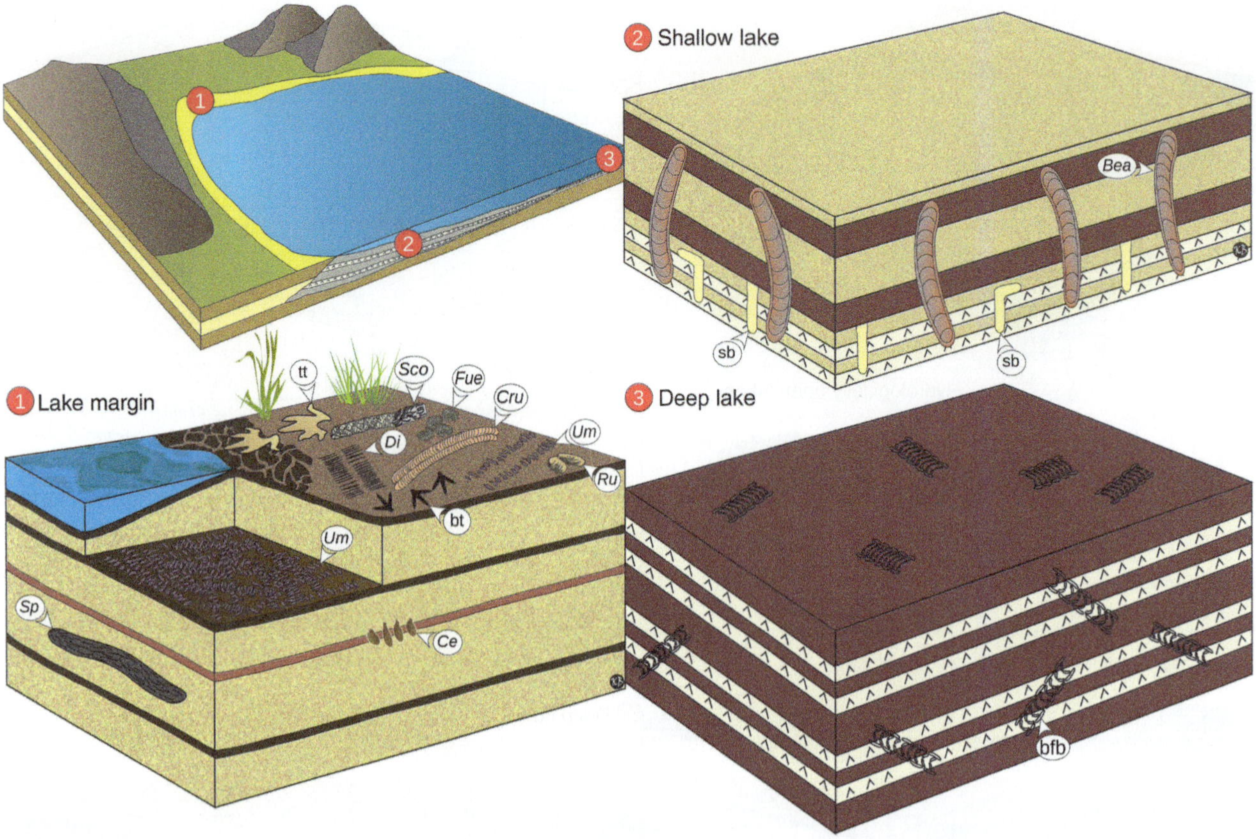

Figure 6.27 Schematic reconstruction of trace-fossil distribution in hydrologically closed lakes. Ichnodiversity in hydrologically close lakes is typically lower than in open lakes. Lake-margin deposits (1) may contain the meniscate and ornamented trace fossil *Scoyenia* (*Sco*), the banana-shaped burrow *Fuersichnus* (*Fue*), the bilobate trace fossils *Cruziana* (*Cr*) and *Rusophycus* (*Ru*), the ornamented branching burrow *Spongeliomorpha* (*Sp*), the bee cell *Celliforma* (*Ce*) and arthropod trackways, such as *Diplichnites* (*Di*) and *Umfolozia* (*Um*). Omission surfaces may be delineated by extremely high densities of arthropod trackways. Tetrapod (Tt) and bird (Bt) tracks are abundant, but the overall composition of vertebrate ichnofaunas is highly variable according to geologic age. Abundance of burrows having bioglyphs reflects dominant firmground conditions. Shallow lake deposits (2) are typically barren as a result of hypersaline conditions. Occasionally, the backfilled burrow *Beaconites filliformis* (*Bea*) and horizontal to vertical simple burrows (sb) may penetrate evaporite layers from overlying colonisation surfaces. Deep-lake deposits (3) are also sparsely bioturbated because of hypersalinity, although locally monospecific suites of backfilled burrows (bfb) have been recorded.

lacustrine areas, signalling the first occurrence of the *Mermia* Ichnofacies (Buatois and Mángano 1993; Buatois et al. 1998). Colonisation in distal lacustrine settings may have been favoured by increased delivery of food supply. However, progressive global ariditisation during the assembly of Pangea resulted in the abundance of hypersaline lakes, which were particularly stressful for animal colonisation (Buatois et al. 2022). Moreover, the Capitanian and end-Permian mass extinctions affected lacustrine biotas, impacting particularly the crustaceans and aquatic insects (Labandeira 1999). Arid conditions persisted in the Early Triassic and recovery from the end-Permian mass extinction in lacustrine environments may have taken up to 10 My, which is coincident with the so-called 'coal gap' (Zhao et al. 2020). This is consistent with the existence of a remarkably poor lacustrine trace-fossil record for the Early Triassic (Buatois et al. 2016).

Subsequent to the above decline in diversity, an event referred to as the Mesozoic Lacustrine Revolution was responsible for the diversification and increase in disparity of aquatic macrophytes, aquatic insects and teleost fish (Figure 6.28) (Cohen 2003). This was paralleled by a profound change at the ecosystem scale, with the rise to dominance of herbivory and intensification of predation pressures. In particular, the evolution of chironomids was significant with respect to burrowing intensity and food and nutrient redistribution (Cohen 2003; Buatois et al. 2016). This is consistent with an increase in the degree of infaunalisation, as indicated by the presence of deep-tier trace fossils in fully subaqueous lacustrine deposits by the

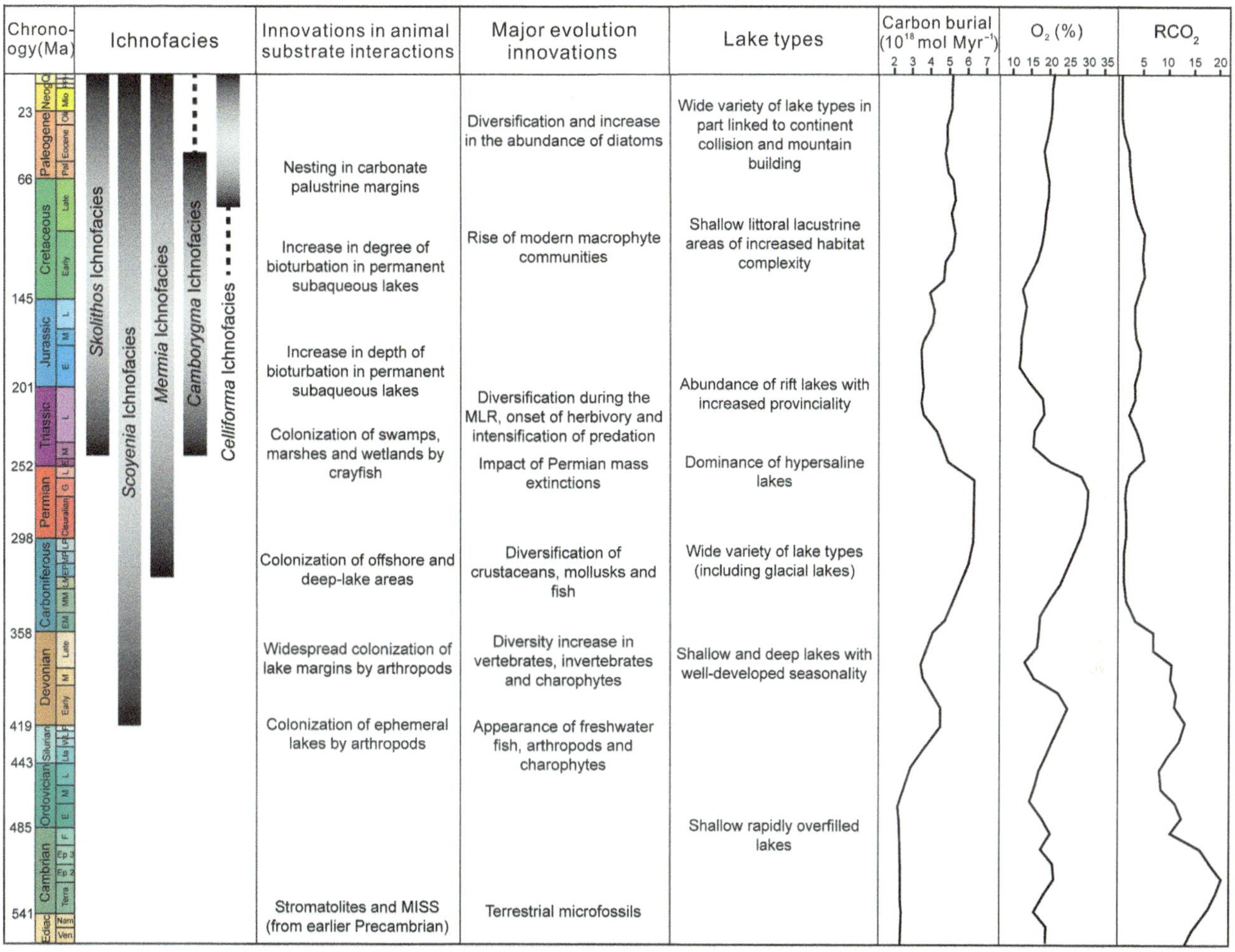

Figure 6.28 Biotic factors influencing lake evolution in deep time. Data from the following sources: ichnofacies and innovations in animal–substrate interactions (Buatois and Mángano 1993; Buatois et al. 1998, 2016; Genise et al. 2016; Minter et al. 2016a,b, 2017), major evolutionary events and lake types (Labandeira 2002; Cohen 2003), carbon burial and carbon dioxide and oxygen.

mid-Mesozoic (Buatois et al. 1996; Voigt and Hoppe 2010). Lake-margin sediments were in places affected by dinosaur trampling. Increased provinciality resulting from the break-up of Pangea promoted lacustrine endemism (Cohen 2003).

These trends were disrupted by the end-Cretaceous mass extinction to some degree, most likely due to acidification (Figure 6.28) (Cohen 2003). The basic lacustrine ecologic structure, however, seems to have been affected very little, and aquatic ecosystems were essentially buffered from extinction (Cohen 2003; Buatois et al. 2016). In fact, ichnologic evidence suggests the continuation of Mesozoic trends into the Cenozoic. Regardless, a significant Neogene innovation was the diversification and increase in abundance of diatoms in lacustrine basins (the only major lacustrine phytoplanktonic organism with hard parts), which impacted silica cycling (Cohen 2003). Trampling in lake margins continued but was produced by large mammals and nesting birds, most notably flamingos (Figure 6.25b).

6.5 Lacustrine Sequence Stratigraphy

6.5.1 Key Differences Between Lakes and Oceans

Numerous studies reveal that lakes should not be treated as just small oceans and that lakes differ from oceans in several significant ways (e.g. Bohacs et al. 2000, 2003; Sztanó et al. 2013; Gong et al. 2019; Zhang et al. 2019b), most notably:

1) *Frequent and rapid lake-level oscillations*: Lakes contain much smaller volumes of water than oceans and are thus much more sensitive to climatic fluctuations

(Bohacs et al. 2000; Zhang et al. 2019b). Lake levels, therefore, vary more frequently and rapidly than sea levels in marine basins (300 m change in 15,000 yr is not uncommon), thus causing lake shoreline regression and transgression to happen both rapidly and often (Gong et al. 2019; Zhang et al. 2019b; Olariu et al. 2021). For example, the shoreline of Lake Chad retreated up to 18 km in 9 months due to a 3 m lake level fall, and the lake area fluctuated 92% between 1966 and 1985. The general result of such abrupt lake-level oscillations in the ancient record is that shoreline and delta-front strata are commonly only thinly developed in lacustrine strata (there is no time for amplification of thickness by subsidence) and that progradation and retrogradation are laterally extensive relative to basin size, leading to Olariu et al. (2021) referring to 'telescoping' deltas in the Uinta and Junggar lake basins.

2) *Tight coupling between lake level and sediment discharge into the lake:* Wet and dry periods have an important control not only on water and sediment supply to the lake but also on the lake level; this does not generally happen in marine basins where sea level and sediment supply are not obviously linked (Carroll and Bohacs 1999; Zhang et al. 2019b). When lake-level falls, during a dry period, shorelines are forced to regress rapidly, and also sediment and water supply decrease. In contrast, lake-level rise during wet periods usually causes shoreline transgression, despite the increase in both sediment and water discharge (e.g. Lyons et al. 2011; Gong et al. 2019). Unlike in marine basins, the sediment supply to some closed lakes during wet periods is directly linked to the coeval lake-level rise; this coupling tends to cause lake-shoreline strata accumulated during overall transgressions to be relatively thick because peak sediment yield rates occur at the beginning of transgressions.
3) *Lakes are prone to receive hyperpycnal underflows:* Many lakes have waters with lower salinity than seawater. Sediment-laden, freshwater river discharge into freshwater to brackish lakes will therefore much more easily generate hyperpycnal flows during strong flood periods, even within the river-mouth reaches (Buatois and Mángano 1995a; Mulder et al. 2003; Zavala 2020). For this reason, sedimentation within many lacustrine basins tends to be dominated by hyperpycnal flows.

The above differences between oceans and lakes make it clear that lake basins, although they can be analysed with the same sequence stratigraphic methodology as marine basins, behave differently (e.g. Carroll and Bohacs 1999, 2001; Gong et al. 2019). This hypothesis has been highlighted by a comparison between the architectures of marine versus lacustrine aggradational to progradational clinothem pairs (*sensu* Gong et al. 2019). As shown in Figure 6.29, nine lacustrine clinothem pairs are recognised in the Late Miocene Lake Pannon (PCP_1 to PCP_9 in Figure 6.29a). In Lake Pannon, the aggradational (rising) topset-to-foreset rollover trajectory (through a height of tens of metres) is commonly linked downdip with efficient partitioning of sediment into the clinothem bottomsets, whereas the progradational (flat to slightly falling) topset-to-foreset rollover trajectory tends to link downdip to little or no partitioning of sediment into clinothem bottomsets. The opposite occurs in the marine aggradational to progradational clinothem pairs, where very thick bottomsets develop in front of progradational shelf-edge trajectories rather than aggradational shelf-edge trajectories (Sztanó et al. 2013; Gong et al. 2019). This marked contrast in architectural styles of marine versus lacustrine clinothem pairs, ultimately due to the conjoint base level and sediment supply changes in lakes, draws attention to the pitfalls of simply applying conventional marine sequence stratigraphic models to lacustrine basins (Sztanó et al. 2013; Gong et al. 2019).

6.5.2 Sequence Stratigraphy of Siliciclastic Lake Systems

From a sediment-supply perspective, lake basins can be classified into three types: over-filled, balanced-fill and under-filled categories. From a subsidence perspective, lake basins in rifts can be divided into two main regimes, namely, syn-rift and post-rift. The temporal and spatial expressions of sequence stratigraphy in siliciclastic lake basins are detailed below. (See also the section on aeolian sequence stratigraphy, Section 7.7, which also addresses climate linkages to sediment supply, accommodation, and preservation).

6.5.2.1 Sequence Stratigraphic Expression in Over-Filled Lake Basins

Over-filled lake basins, developed when the rate of sediment supply exceeds the rate of increase of potential accommodation space, are generally characterised by mixed fluvio-lacustrine facies associations developed in freshwater with open hydrology (Table 6.2; Figure 6.30) (Carroll and Bohacs 1999; Bohacs et al. 2000, 2003). Over-filled lakes with open hydrology imply that the basin is filled to the sill-level to produce a permanently filled lake basin. In permanently over-filled lake basins, stratal patterns are dominated by variations in sediment supply, and erosive sequence boundaries are non-existent or minimally developed. Systems tracts at the third-order timescale (several millions of years) superficially resemble classical (Figure 4.29). Type 2 marine siliciclastic sequences (i.e. shelf-margin, transgressive and highstand

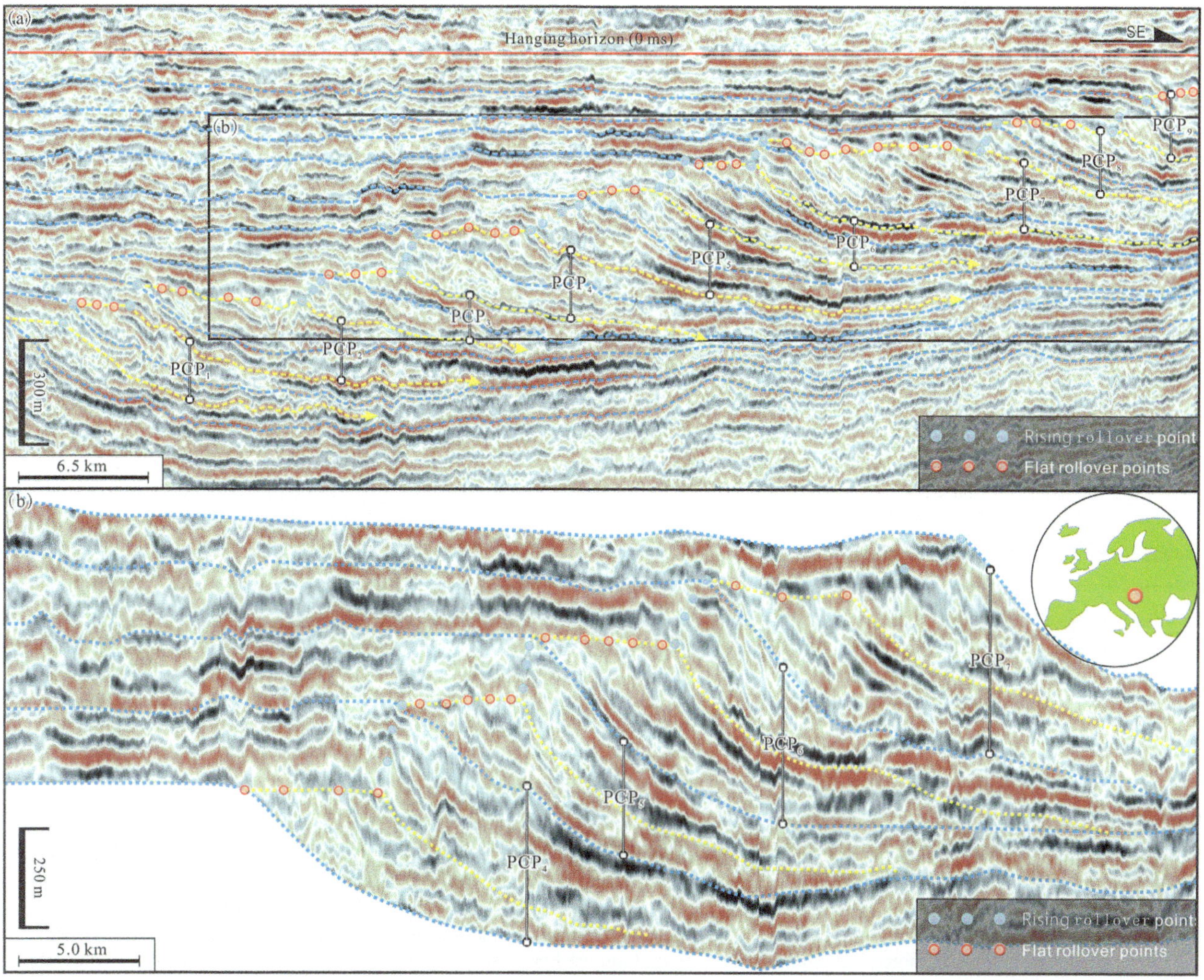

Figure 6.29 Depositional dip seismic transect (a) and associated enlarged seismic profile (b) from the Makó Trough in the Pannonian Basin, showing aggradational to progradational lacustrine clinothem pairs (PCP1–PCP9) with alternating series of flat (red dots) and rising (blue dots) shelf-edge trajectories. Note that rising rollover trajectories are fronted by aggradational bottomsets, but that their flat counterparts are not.

systems tracts) with a lack of basinally restricted lowstand deposits (Table 6.2; Figure 6.30) (Bohacs et al. 2000).

Depositional environments of over-filled lake basins are characterised by fluvio-lacustrine facies associations, well-defined paralic deposits (i.e. delta-plain sediments and coals and distributary channel fills) and deep-water mudstone or micrite with turbiditic successions near the basin centres (Table 6.2; Figure 6.30) (Renaut and Gierlowski-Kordesch 2010). In the absence of anoxia, organic matter preservation is generally low, and it is composed of terrestrial components with lacustrine planktonic material (Table 6.2).

Hyperpycnal flows and associated turbidity currents tend to oxygenate the deep levels of over-filled lakes, promoting the presence of rich benthos (Buatois and Mángano 1998). Also, these lakes are hydrologically open and fresh, and no stress due to hypersalinity occurs, further contributing to the establishment of lacustrine biotas. Land-derived organic matter constitutes the main source of nutrients and favours deposit-feeding strategies. From an ichnologic perspective, overfilled lakes are characterised by well-developed softground suites (Buatois and Mángano 2004, 2009; Scott and Smith 2015). Hyperpycnites typically host diverse trace-fossil suites. Vertical trends in ichnofacies succession may aid in delineation of lacustrine parasequences (Buatois and Mángano 2004, 2009). Permanent subaqueous lacustrine deposits containing the *Mermia* Ichnofacies are typically replaced laterally by lake-margin facies hosting the *Scoyenia* Ichnofacies. Overfilled lakes usually do not experience desiccation, and, therefore, firmgrounds are

Table 6.2 A summary of characteristics of different lake-basin types, including overfilled, balanced-fill and underfilled lake basins (modified from Bohacs et al. 2000; Renaut and Gierlowski-Kordesch 2010).

Sequence expression	Overfilled	Balanced-fill	Underfilled
Rationale	Very similar in geometry and development to Exxon (standard) marine siliciclastic sequences	Somewhat similar to or distinctly different from Exxon (standard) marine siliciclastic sequences	Distinctly different from Exxon (standard) marine siliciclastic sequences
Facies association	Mixed fluvial-lacustrine facies association	Fluctuating-profundal facies association	Evaporative facies association
Stratal stacking	Dominantly progradational parasequences, many poorly defined	Well- to poorly defined aggradational parasequences; flooding surfaces marked by lithological contrasts; thick highstand successions	Parasequences of progradational clastic sediments and vertically aggrading chemical sediments that form during desiccation cycles
A/S ratio	Sediment + water supply (S) > accommodation space (A)	Sediment + water supply < potential accommodation space	Sediment and water supply approximately balance potential accommodation space
Fluvial input	Maximum fluvial input	Variable fluvial input	Minimum fluvial input
Climate	All climatic zones; least common in arid climates	Most common in semi-arid climates and arid; can occur in other climatic zones	All climatic zones
Hydrology	Open drainage (surface and/or subsurface inflow and outflow)	Closed or partly closed drainage; groundwater recharge commonly important; frequent wet-dry cycles	Open to closed drainage; water inflow can periodically fill accommodation space but is not always balanced by outflow
Water column	Typically freshwater	Typically saline water	Dilute to saline water
Lake level	Permanently stable or may occasionally fall slightly below sill height	Typically unstable; seasonal and longer-term fluctuations frequent	Stable to unstable; climatically driven fluctuations are common
Lithologies	Sandstone, siltstone, mudstone, marlstone, carbonate, coal and coaly shale	Sandstone, siltstone, mudstone, marlstone, carbonates, evaporites and kerogenite	Sandstone, siltstone, mudstone, marlstone, carbonates, kerogenite, evaporites and coal
Organic matter	Mixed terrigenous and microbial; low to moderate TOC; mixed type I–III kerogen	Predominantly type I kerogen, with type I–III mixtures near flooding surfaces	Predominantly type I kerogen
Hydrocarbons	Oil and gas; waxy low-sulphur oils	Mainly oil (moderate to high low-sulphur oils)	Mainly oil (low-sulphur)
Ichnology trends	Abundant softground suites Profundal zone: *Mermia* Ichnofacies Sublittoral zone: Depauperate *Mermia* Ichnofacies Lake margin: *Scoyenia* Ichnofacies Delta-mouth sand bars: *Skolithos* Ichnofacies	Widespread firmground suites overprinted to softground suites Lowstand lake margin: *Scoyenia* Ichnofacies Transgressive and highstand shallow lake carbonates: depauperate *Mermia* Ichnofacies Paralic carbonates: *Celliforma* Ichnofacies Delta-mouth sand bars: *Skolithos* Ichnofacies	Widespread firmground suites Lowstand desiccated substrates: *Scoyenia* Ichnofacies Calcretes: *Celliforma* Ichnofacies Eolian dunes: *Octopodichnus-Entradichnus* Ichnofacies
Examples	Luman Tongue and upper Laney Mb., Green River F. (Eocene, Wyoming); Hongyanchi F. (Permian, Junggar Basin, China); Waterfall F. (Jurassic, Culpeper Basin, Virginia); Lake Baikal (Holocene, Siberia, Russia)	Wilkins Peak Mb., Green River F. (Eocene, Wyoming); Lucagou Formation F. (Permian, Junggar Basin, China); Passaic F. (Triassic, Newark Basin, New Jersey); Qingshankou 1 F. (Cretaceous, Songliao Basin, China); Lake Bogoria (Pleistocene-Holocene, Kenya)	Tipton and lower Laney Mbs., Green River F. (Eocene, Wyoming); Jingjingzigou F. (Permian, Junggar Basin, China); Shahejie 4 (Oligocene, Bohai Bay Basin); East Berlin F. (Jurassic, Hartford Basin, Connecticut); Lake Malawi (Pleistocene-Holocene, E. Africa Rift)

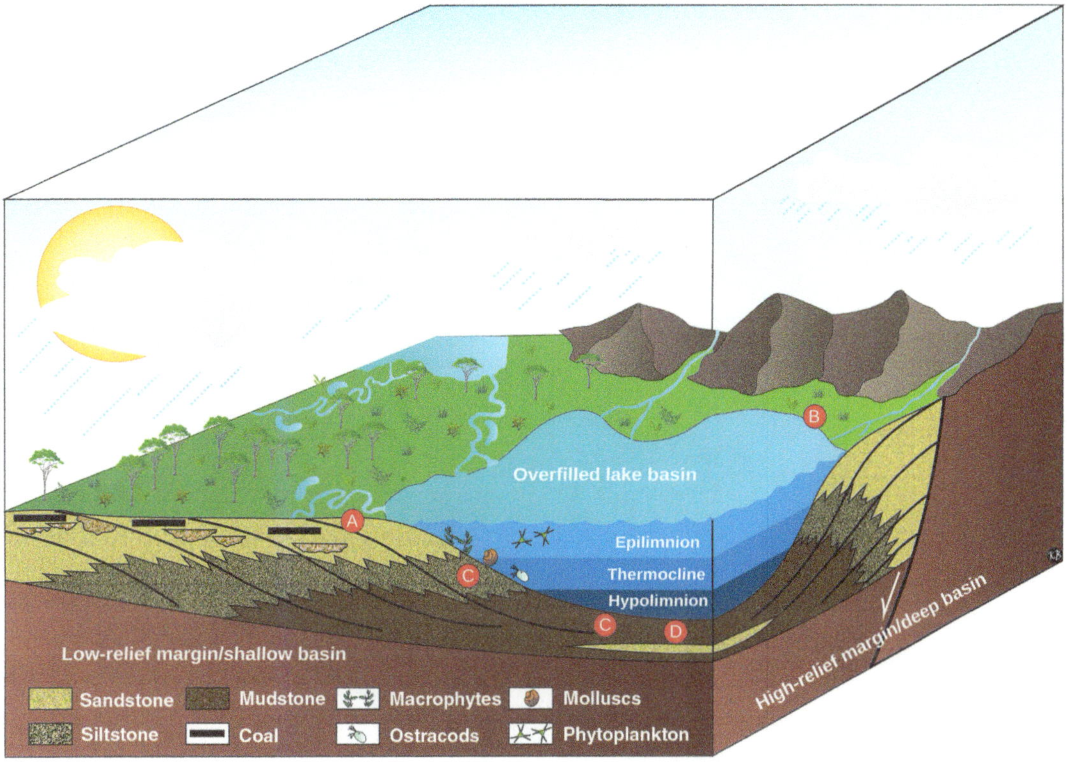

Figure 6.30 Schematic diagram illustrating key sedimentologic and ichnologic features of over-filled lake basins (see Table 6.2), including persistently open hydrology, freshwater lake chemistry and progradational shoreline architecture (from Bohacs et al. 2000; Renaut and Gierlowski-Kordesch 2010). Ichnofacies display a consistent distribution (Buatois and Mángano 2004, 2009). The continental equivalent of the *Skolithos* Ichnofacies may occur in high-energy shoreline and deltaic mouth bar deposits (A). The *Scoyenia* Ichnofacies is dominated by softground suites and illustrates low-energy lake-margin deposits (B). The depauperate *Mermia* Ichnofacies occurs as colonisation suites in both tempestites and turbidites (C). The archetypal *Mermia* Ichnofacies is present in underflow current deposits, thin-bedded turbidites and fine-grained suspension fallout deposits reflecting the activity of the resident benthos (D).

rare. As a result, only the pre-desiccation softground suite of the *Scoyenia* Ichnofacies occurs. Continuous wave action or proximity to deltaic distributaries may result in the latter being laterally replaced by the *Skolithos* Ichnofacies. Regarding vertebrate ichnology, of the three types of tetrapod-track taphonomic modes (TTTMs), high fidelity mode (HFM), moderate fidelity mode (MFM) and low fidelity mode (LFM), defined by Mancuso et al. (2022), the former two are dominant in the delta plain deposits.

6.5.2.2 Sequence Stratigraphic Expression in Balanced-Fill Lake Basins

Balanced-fill lake basins develop when the rate of sediment supply is about equal to the rate of increase of potential accommodation (Table 6.2; Figure 6.31) (Carroll and Bohacs 1999; Bohacs et al. 2000, 2003). In balanced-fill lake basins, water and sediment inputs are only periodically, sufficient to fill the lake to the sill level and even to generate surface outflows, but significant lake-level drops below the sill level also occur. Systems tracts and sequence boundaries at third-order timescales can be both superficially similar to and distinctly different from classical marine siliciclastic sequences (Table 6.2; Figure 6.31). Sequence boundaries tend to be marked by large basinward shifts of facies relative to the basin size but with limited incision and erosion. When the rivers dry up, the lake level falls. The erosive and load-carrying power of the river therefore decreases with such a lake-level fall but increases during a lake-level rise (Bohacs et al. 2000). This is in marked contrast with the explicit assumptions for classical marine siliciclastic sequences where sea level and sediment supply are only weakly linked if at all (Posamentier and Vail 1988). Lowstand systems tracts in shallow balanced-fill lake basins accumulate relatively thin aggradational parasequences that can be either carbonate or clastic-dominated, whereas deep balanced-fill lake basins can accumulate relatively aggradational lake-floor fan deposits (Table 6.2; Figure 6.31). Transgressive systems tracts tend to be relatively thick and generally dominate the wet-dry successions because of the conjoint lake-level rise and sediment accumulation. Lacustrine highstand systems tracts are generally dominated by fine-grained, organic-matter-rich lake

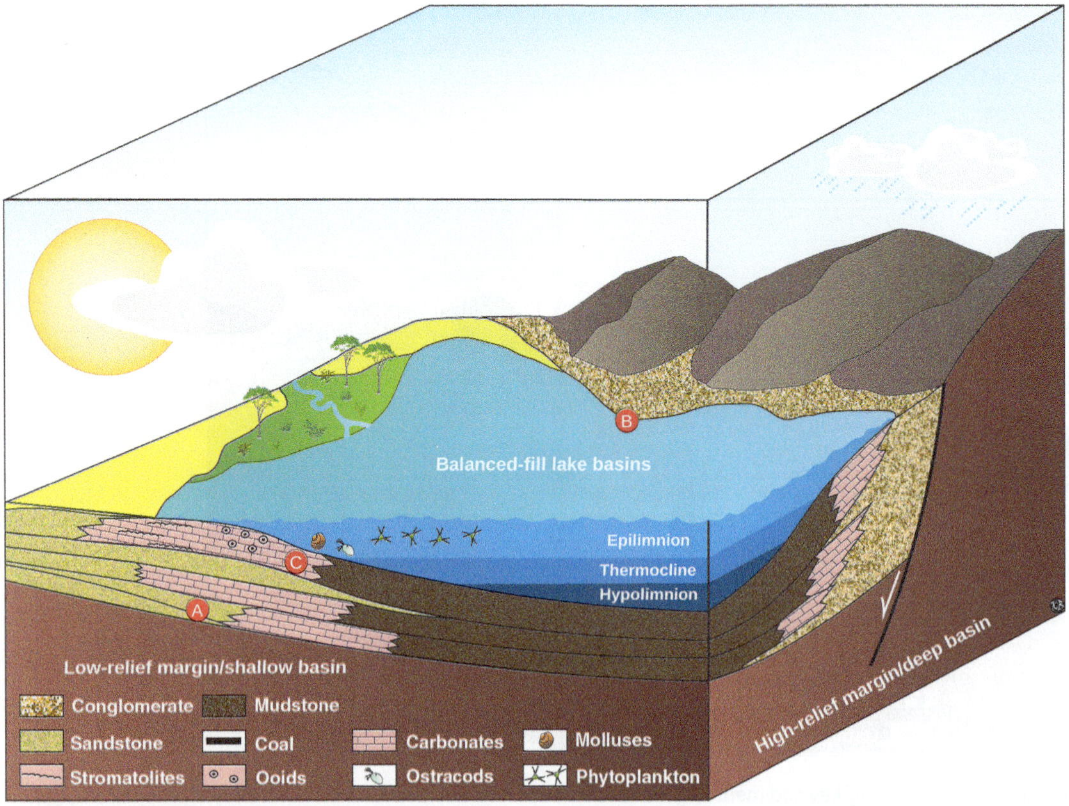

Figure 6.31 Schematic diagram showing key sedimentologic and ichnologic features and summary of balanced-fill lake basins (see Table 6.2), including intermittent open hydrology, fluctuating lake water chemistry, progradational and aggradational shoreline architecture, a close relation to fluctuating-profundal facies association and environmental distributions of ichnofacies (from Bohacs et al. 2000; Renaut and Gierlowski-Kordesch 2010). Balanced-fill lake basins develop when the rates of sediment and water supply and accommodation are roughly in balance over the time span of sequence development. Ichnofacies display a consistent distribution (Buatois and Mángano 2004, 2009). The *Scoyenia* Ichnofacies is represented by both softground and firmground suites and illustrates low-energy lake-margin deposits (A). The continental equivalent of the *Skolithos* Ichnofacies may occur in high-energy shoreline and deltaic mouth bar deposits (B). The depauperate *Mermia* Ichnofacies and the *Celliforma* Ichnofacies may be present in transgressive and highstand carbonates, the latter in palustrine settings (C).

deposits that become extensive when the lake level is high and vary laterally from carbonate to clastic-dominated (Table 6.2; Figure 6.31). They can occasionally exhibit some progradation, expressed as sigmoidal to oblique progradation on seismic data. The stacking of parasequences records a combination of progradation of clastic sediments and vertical aggradation of chemical sediments (mostly carbonates) (Table 6.2; Figure 6.31).

Depositional environments of balanced-fill lake basins are characterised by 'fluctuating profundal' facies associations, reflecting the differences between shallow and deep-lake environments (Table 6.2) (Renaut and Gierlowski-Kordesch 2010). The basins are characterised by wet-dry cycles, are primarily sediment-flux driven and are characterised by sediment supply being intimately linked to the lake level, since the same rivers supply both the water that increases the depth and the sediment to the lake in an intermittently open hydrologic system (Bohacs et al. 2000). In shallow balanced-fill lakes or during lowstands, both carbonate and siliciclastic rocks accumulate in a variable range of depositional environments within relatively thin aggradational parasequence sets. Deep balanced-fill lakes can also accumulate thicker aggradational parasequence sets associated with turbidite deposits in profundal areas and progradational stacking associated with littoral zones. Organic matter is dominated by microbial production with relatively high preservation, potentially because of the meromixis that commonly is induced by low salinity or fresh runoff during the highstands of relatively saline lakes (Table 6.2) (Renaut and Gierlowski-Kordesch 2010).

Balanced-fill lakes show a complex combination of stress factors that negatively impact the benthos (Buatois and Mángano 2004, 2009). During lowstands, lake hydrology is closed, resulting in an increase in salinity and during transgression and highstand, lake stratification results in dysoxic conditions (Bohacs et al. 2000). From an ichnologic standpoint, both situations are detrimental to the establishment of a burrowing biota. For example, trace-fossil suites in thin-bedded turbidites of balanced-fill lakes are typically less developed, if present at all, compared to overfilled lakes (Buatois and Mángano 2009). Trace fossils may occur locally in transgressive and highstand carbonates, but ichnodiversity is low. The ichnofauna, illustrating a depauperate *Mermia* Ichnofacies, is produced by epifaunal rather than infaunal animals. These characteristics suggest permanently anoxic interstitial waters (i.e. a redox discontinuity surface very close to the lake bottom), with brief periods of oxygenated bottom waters (e.g. Buatois et al. 2000). In contrast, the abundance of firmgrounds in desiccated lake margins promotes the presence of suites representing the *Scoyenia* Ichnofacies particularly during lowstands (Buatois and Mángano 2004, 2009; Scott and Smith 2015). In some instances, the desiccation suite is seen cross-cutting the pre-desiccation suite. The *Skolithos* Ichnofacies may occur in delta mouth-bar deposits formed during highstand, deltaic progradation (Mángano et al. 1994). Palustrine carbonates may contain the *Celliforma* Ichnofacies (Genise 2017). Regarding vertebrate ichnology, the three types of TTTMs are present in the lake margin deposits, but only HFM and MFM are present in delta plain deposits (Mancuso et al. 2022).

6.5.2.3 Sequence Stratigraphic Expression in Under-Filled Lake Basins

Under-filled lake basins develop when the rate of sediment supply is less than the rate of increase of potential accommodation and are particularly characterised by evaporitic or biochemical lithofacies associations because of the hydrologically closed, saline-to-hypersaline water setting (Table 6.2; Figure 6.32) (Carroll and Bohacs 1999; Bohacs et al. 2000, 2003). The stacking of parasequences in

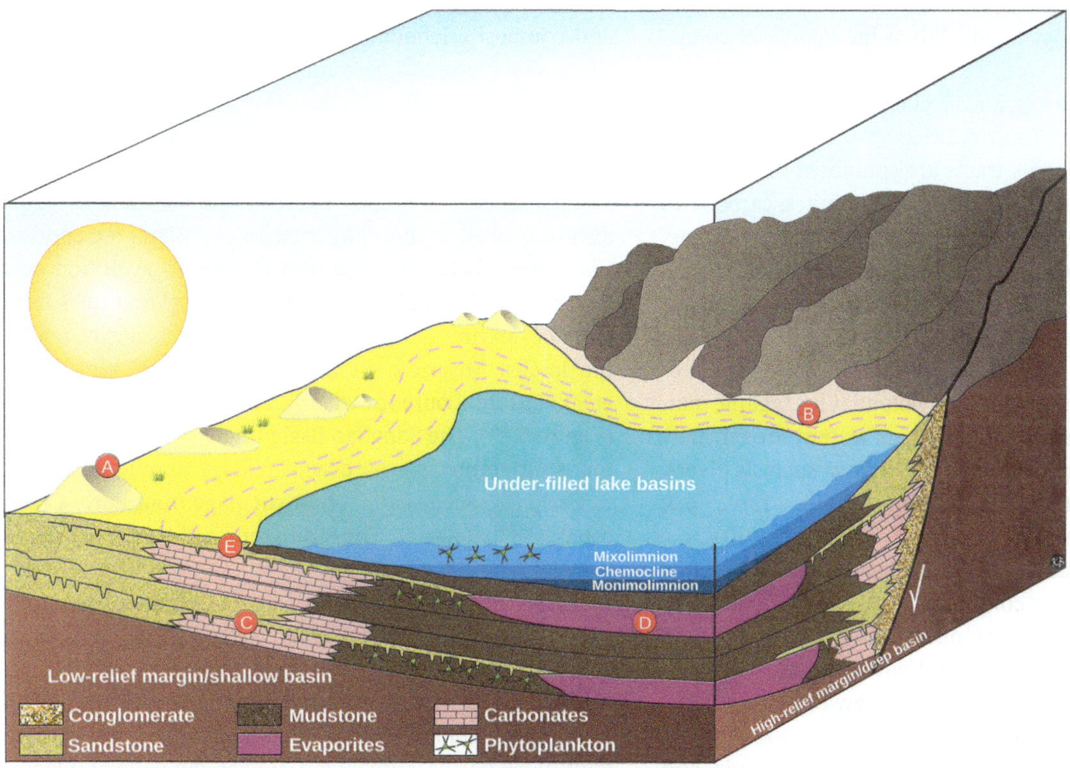

Figure 6.32 Schematic diagram illustrating key sedimentologic and ichnologic features and summary of underfilled lake basins (see Table 6.2), including persistently closed hydrology, characteristic chemical stratification, high solute content of lake waters and dominantly aggradational shoreline architecture, a close relation to evaporative facies association and environmental distributions of ichnofacies (modified from Bohacs et al. 2000 (reprinted by permission of the AAPG whose permission is required for further use); Renaut and Gierlowski-Kordesch 2010). Underfilled lake basins occur when rates of accommodation consistently outstrip available water and sediment supply. Ichnofacies display a consistent distribution (Buatois and Mángano 2004, 2009). The *Octopodichnus–Entradichnus* Ichnofacies illustrates associated aeolian dune deposits flanking the lake (A). Halophyte plant colonisation may characterise lake-margin deposits producing localised root structures (B). The depauperate *Mermia* Ichnofacies may rarely occur in shallow lake deposits but is generally absent due to hypersalinity. The *Celliforma* Ichnofacies may be present in associated calcretes (C). Central lake deposits are typically unbioturbated, although monospecific suites of burrows produced by salinity-tolerant insects may be present locally (D). The *Scoyenia* Ichnofacies is represented by firmground suites in lowstand desiccated substrates in lake-margin deposits (E).

under-filled lake basins is dominantly aggradational. Systems tracts and sequence boundaries of under-filled lake basins are again distinctly different from classical marine siliciclastic sequences. Sequence boundaries are subtly expressed in under-filled lake basins, whereas flooding surfaces are typified by distinct lithologic contrasts. As with balanced-fill lake basins, sediment supply is tightly linked to the lake level, decreasing drastically with lake-level fall but increasing during lake-level rise (Bohacs et al. 2000). Lake-lowstand deposition in under-filled lake basins is generally restricted to evaporites or other chemical/biogenic sediments formed in remnant pools in the highest subsidence areas of the basin. Early transgressive systems tracts tend to be thin, widespread clastic sheets (representing river floods), whereas late transgressive and highstand systems tracts can be thick and typically contain fluvial channels and shoreline carbonates.

Depositional environments of under-filled lake basins are characterised by evaporites and very saline playas and mudflats (shallow-and-under-filled) to perennial saline lakes (deep-and-under-filled). When the lake level is low, the systems tracts comprise evaporites, mudflat and sandflat deposits, palaeosols and aeolianites, with ubiquitous desiccation features (Table 6.2; Figure 6.32). During lake-level highstand, systems tracts are composed of perennial saline to hypersaline lake deposits of laminites, carbonates and evaporites, sublittoral organic-matter-rich mudstone, littoral bioherms or stromatolites and reworked beach deposits containing siliciclastics or carbonates (Renaut and Gierlowski-Kordesch 2010). Organic productivity can be high in this lake type, but desiccation, oxidation and erosion can reduce the preservation potential for organic matter. Fauna and flora intolerant of high salinity are typically absent except near sites of dilute inflow such as rivers, springs and groundwater seepage zones (Table 6.2; Figure 6.32) (Renaut and Gierlowski-Kordesch 2010).

In under-filled lakes, the hydrologically closed regime results in hypersaline conditions that are detrimental to the benthos (Buatois and Mángano 2004, 2009). Lowstand evaporite pools are very stressful environments that typically lack biogenic structures. Only a few organisms, such as aquatic chironomids and beetles (Figure 6.25f), show adaptations to these conditions (Rodriguez-Aranda and Calvo 1998; Uchman and Álvaro 2000; Sanz-Montero et al. 2013; Buatois et al. 2020). Hypersalinity in bottom waters persists during transgression and highstand, precluding the establishment of a subaqueous *Mermia* Ichnofacies. In rare instances, a depauperate expression of this ichnofacies may be present in thin, very shallow-water, deposits mantling flooding surfaces at the base of parasequences (Buatois and Mángano 2004, 2009). The *Mermia* Ichnofacies is abruptly replaced upwards by the *Scoyenia* Ichnofacies, as a result of the lake regression (Metz 1996, 2000). The *Scoyenia* Ichnofacies is widespread in lowstand desiccated substrates, being characterised by meniscate striated burrows, and arthropod and tetrapod trackways delineating trampled omission surfaces (Metz 1996; Farlow and Galton 2003; Szajna and Hartline 2003; Bohacs et al. 2007; Minter et al. 2007; Scott et al. 2007; Scott and Smith 2015). On the contrary, occurrences of lake-margin mudflat and sandflat deposits having trackways that do not display evidence of trampling indicate a short period of exposure (Mancuso et al. 2020). In the case of tetrapod trackways, although the three types of TTTMs are present in both lake-margin and central lake deposits, HFM is dominant in the former, and MFM and LFM are characteristics of the latter (Mancuso et al. 2022). High-energy ephemeral sandflat deposits may show elements of the *Skolithos* Ichnofacies (Mancuso et al. 2020). Terrestrial suites, typically representing the *Celliforma* Ichnofacies, may be present if soil formation occurs on top of lake-margin carbonates (Genise et al. 2010, 2016; Genise 2017). Palimpsest surfaces and composite ichnofabrics may be present, representing successive bioturbation events that reflect desiccation.

6.5.2.4 Tectonostratigraphic Evolution and Expression of Siliciclastic Lacustrine Rift Basins

The sequence stratigraphic expression of lake deposits must also be viewed in their larger tectonostratigraphic context. We review here the tectonostratigraphic evolution and expression of lake basins as represented by examples from the non-marine, intra-continental Songliao Basin, one of the largest Mesozoic continental basins in China (Feng et al. 2010b; Lourenço et al. 2021; Rapozo et al. 2021) (Figure 6.33). The Songliao Basin underwent two main tectonostratigraphic stages: a syn-rift stage from Late Jurassic to Early Cretaceous and a post-rift stage from Early Cretaceous to Quaternary. The stratigraphy of these two stages is shown in Figure 6.33.

The syn-rift stage was related to the initiation of subduction between the Izanagi and Pacific Plates (Feng et al. 2010b) and was stratigraphically manifested as fault-block basins characterised by a large number of isolated, NNE-trending fault blocks of 'basin-and-range' style. Within these fault-block basins, alluvial fan, fluvial, fan delta and lacustrine clastics are intercalated with volcanic and volcaniclastic rocks, with coal-bearing source rocks and sandstone, conglomerate and volcanic reservoirs. The post-rift unit was related to lithospheric cooling and extension, modulated by three main compressional events. This post-rift unit is typically 3000–4000 m thick (maximum thickness 6000 m), extends beyond the rift blocks and laps onto the basin margins to form a large relatively uniformly subsiding basin (Figure 6.33). The post-rift succession contains large, thick,

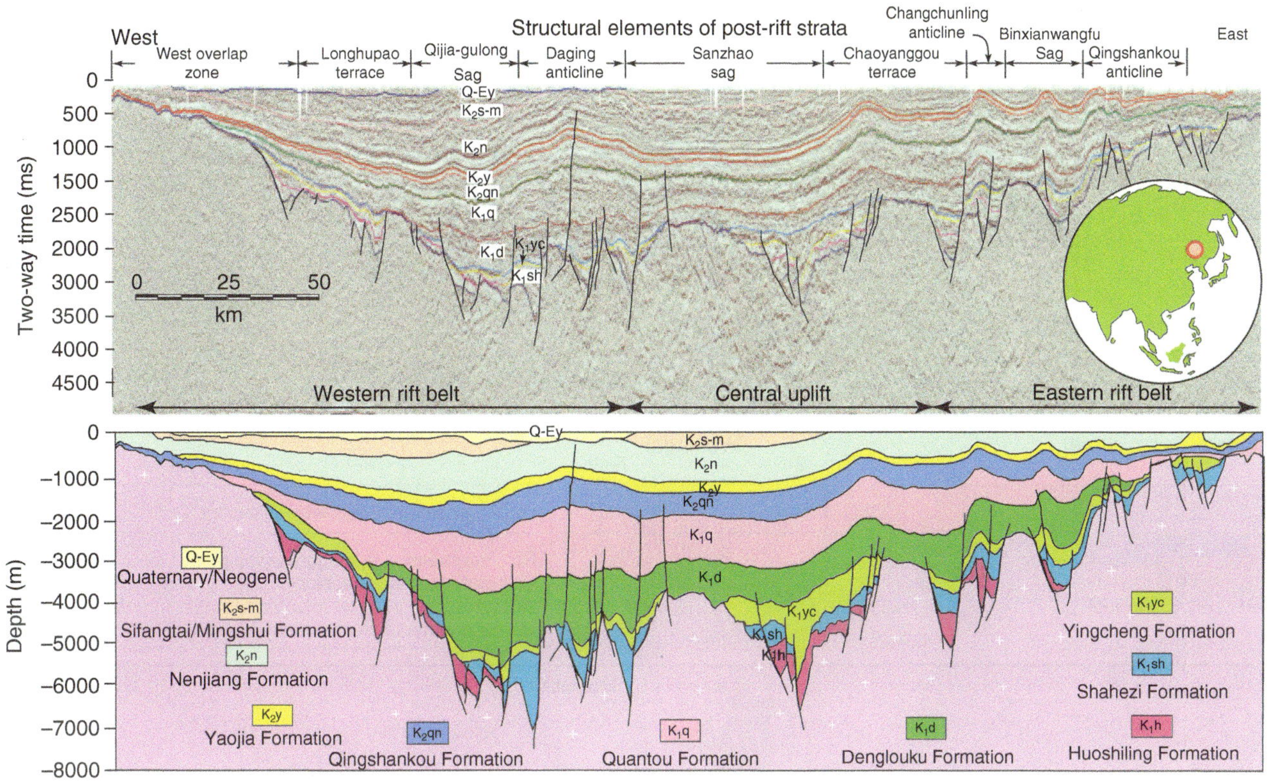

Figure 6.33 Regional 2D seismic line and interpretation illustrating the tectonostratigraphic units and sequence stratigraphic expression of the non-marine, intracontinental Songliao Basin of northeastern China (Feng et al. 2010b / John Wiley & Sons). The tectonostratigraphic history of the Songliao basin can be divided into two main units, including a syn-rift tectonostratigraphic unit consisting of Jurassic Huoshiling (J_3h), the Early Cretaceous Shahezi (K_1sh) and Yingcheng (K_1yc) formations and a post-rift tectonostratigraphic unit composed of Denglouku (K_1d), Quantou (K_1q), Qingshankou (K_2qn), Yaojia (K_2y), Nenjiang (K_2n), Sifangtai (K_2s) and Mingshui (K_2m) Formations.

organic-matter-rich lacustrine source rock, among the most prolific in the world (Feng et al. 2010b), together with fluvial and deltaic sandstone reservoirs. Because of the episodic compressional events that alternately caused subsidence and uplift, reservoirs are sandwiched between source rocks, providing direct fluid communication from source to reservoir (Feng et al. 2010b).

6.5.3 Sequence Stratigraphy of Carbonate or Mixed Carbonate–Siliciclastic Systems

The sequence stratigraphy of carbonate or mixed siliciclastic–carbonate lacustrine systems is still far from well understood, though there is the notable exception of the Eocene Green River Formation in the Piceance Basin of Colorado (Tänavsuu-Milkeviciene and Sarg 2012; Sarg et al. 2013; Tänavsuu-Milkeviciene et al. 2017; Eljalafi and Sarg 2018, 2021), the Uinta Basin of eastern Utah (Keighley et al. 2003) and the Bridger-Washakie Basins of Wyoming (Bohacs et al. 2007a).

The Early to Middle Eocene Green River Formation in the Piceance Basin is, therefore, taken as an example to illustrate the sequence stratigraphic expression of a lacustrine succession of mixed carbonates and siliciclastics. The lacustrine carbonates are interbedded with fluvial and lacustrine sandstones and oil shale of the Douglas Creek Member. Two facies associations are well developed within dominantly littoral to upper sublittoral environments: lacustrine margin carbonates consisting of six packstone to rudstone lithofacies (Figure 6.22) and lacustrine microbial carbonate composed of stromatolitic and thrombolitic lithofacies (Table 6.1; Figure 6.22) (Sarg et al. 2013; Eljalafi and Sarg 2018, 2021).

6.5.3.1 Sequence Stratigraphic Surfaces

Regionally extensive sequence boundary (SB), transgressive surface (TS) and (maximum) flooding surface ((M) FS) were recognised (Figure 6.34) (Tänavsuu-Milkeviciene et al. 2017; Eljalafi and Sarg 2018, 2021). Sequence boundaries, commonly best expressed along the basin margin, are marked by low gamma-ray peaks, a sharp basinward shift of facies association and changes in Fischer Assay logs (i.e. reduction in organic richness). They occur at the bases of sharp-based sandstone or sharp-based carbonate that have

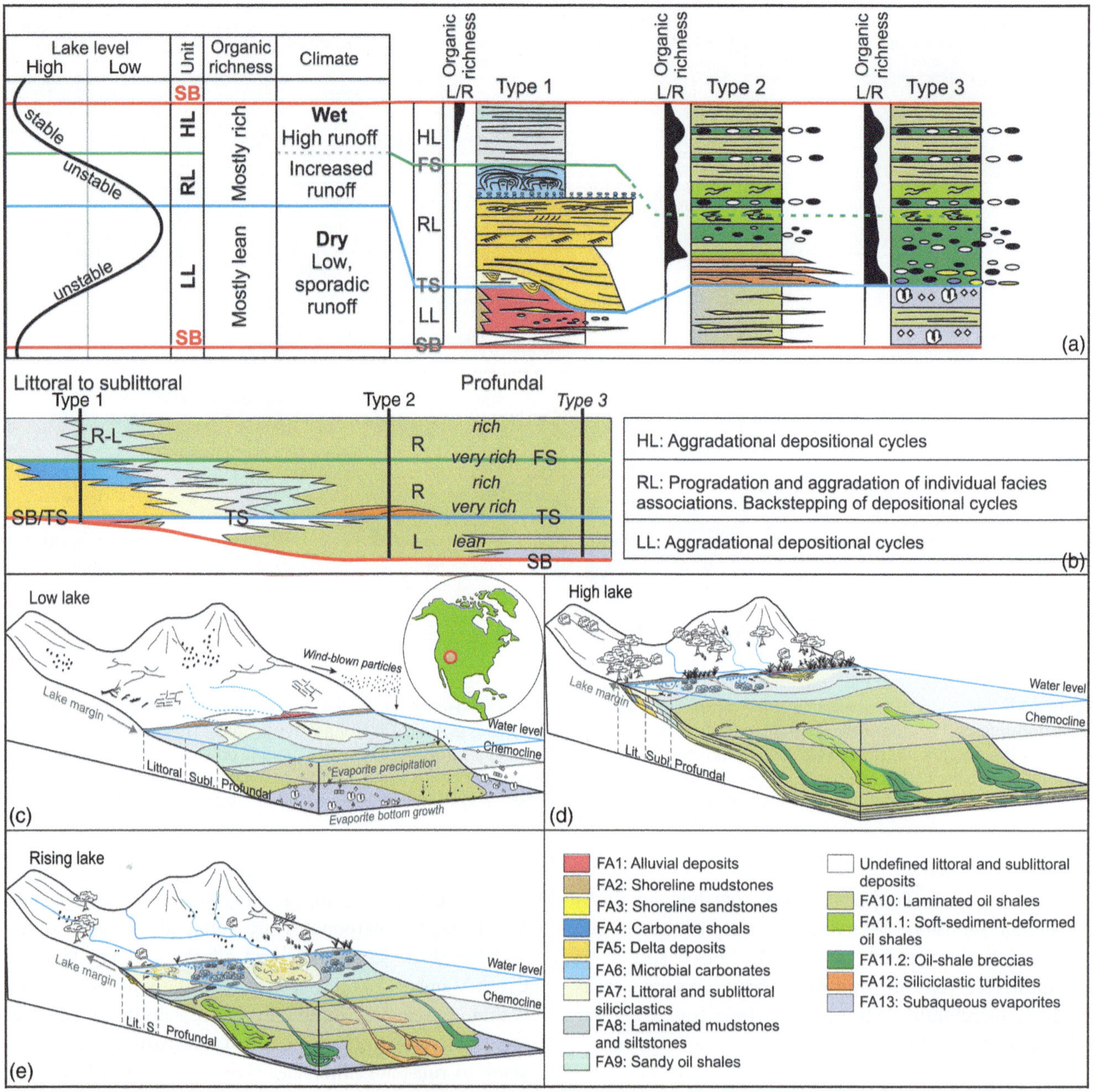

Figure 6.34 (a,b) Schematic portrayal of three idealised depositional sequences in Piceance Basin, W. Colorado with characteristic facies-association distribution, sequence boundary (SB), transgressive surface (TS) and main flooding surface (FS): type 1 characterised by fining- and deepening-upward littoral and sublittoral deposits, type 2 characterised by profundal deposits without evaporites and type 3 characterised by profundal deposits with evaporites (Tänavsuu-Milkeviciene et al. 2017 / Society for Sedimentary Geology (SEPM)). (c–e) Schematic portrayal of idealised palaeogeography for the periods of low lake level, rising lake level and high lake level (from Tänavsuu-Milkeviciene et al. 2017). Abbreviations: LL = low lake level; RL = rising lake level, HL = high lake level; SB = sequence boundary (SB); transgressive surfaces (TS); and main flooding surface (FS).

eroded into the underlying siltstone, mudstone and oil-shale deposits. Sequence boundaries are traced basinwards as correlative conformities below littoral and sublittoral siliciclastics or laminated siltstone and mudstone and are marked by sharp shifts from rich oil shale (below) to lean oil shale (above) in the profundal part of the basin. Transgressive surfaces commonly coincide with the sequence boundaries but in a few places are marked by the upward transition from alluvial deposits to the first significant deltaic deposits (Figure 6.34b) (Tänavsuu-Milkeviciene et al. 2017; Eljalafi and Sarg 2018, 2021). The same prominent transgressive surfaces also mark the turnaround from lean oil-shale deposits to rich oil-shale deposits in the profundal zone (Figure 6.34b). Flooding surfaces formed during the lake-level rise are marked by lag deposits along the basin margin (Figure 6.34b). Maximum flooding surfaces along the lake margin are clearly marked by the sudden appearance of laminated mudstone and siltstone or oil-shale deposits but are hard to recognise in the profundal zone (Figure 6.34b) (Tänavsuu-Milkeviciene et al. 2017; Eljalafi and Sarg 2018, 2021).

6.5.3.2 Sequence Stratigraphic Expression

The stratigraphic sequence in the Green River Formation of the Piceance Basin was divided into three parts by Tänavsuu-Milkeviciene et al. (2017) and Eljalafi and Sarg (2018, 2021) based on the distribution of facies associations and stacking patterns (Figure 6.34c–e). The lower interval is characterised by evaporite deposits in the basin centre, suggesting times of low lake level (Figure 6.34b,c). At the lake margin, the interval of low lake level is commonly marked by an erosional sequence boundary formed by erosion prior to the subsequent deposition of overlying lake-margin sandstone or carbonate (Figure 6.34a,b). The middle interval of each sequence in the profundal zone is marked by a rapid increase in rich oil-shale deposits and concurrent decrease of evaporite deposits and records the change from low lake level to rising water levels (Figure 6.34d). The next interval of further rising lake level is bounded at its base by the transgressive surface and top by a maximum flooding surface (MFS) (Figure 6.34a,b) and is marked by a change from lean oil-shale to rich oil-shale deposits (Figure 6.34b,d). The upper interval of a sequence is characterised by the widespread occurrence of mudstone and/or rich oil-shale deposits, suggesting times of high lake level (Figure 6.34b,e).

6.5.4 The Application of Predictive Stratigraphy to Understanding Lacustrine Systems

The prediction of sand-prone (reservoir) and mud-prone (source rock/seal) segments of stratigraphic sequences is of importance in the exploration and production of resources. Sequence stratigraphic and source-to-sink studies represent two complementary schools of predictive stratigraphy (Catuneanu et al. 2009; Martinsen et al. 2017).

6.5.4.1 Application of Sequence-Stratigraphic Principles

The sequence-stratigraphic approach involves the subdivision of thick stratal successions into repetitive stratal packages that reflect how sediment supply and accommodation interact. It is applicable to both lacustrine and marine systems (Lin et al. 2001; Catuneanu et al. 2009). Three types of lacustrine basin successions were recognised by Lin et al. (2001) and are discussed below:

1) Fluvial-shallow-lacustrine basin successions, which developed during the initial rifting stage, are composed mainly of lowstand alluvial-fan and highstand shallow-lacustrine deposits.
2) Shallow-lacustrine basin successions, which formed during the early and late periods of rifting, are dominated by shallow-water fan deltas, braid deltas and shallow-lacustrine deposits.
3) Deep-lacustrine basin successions, which developed in response to rapid tectonic subsidence during the maximum-rifting stage, are characterised by thick, deep-lacustrine mudstones, fan deltas, braid deltas and sub-lacustrine fan deposits.

A four-phase sequence stratigraphic approach is proposed for predicting the distribution of source rocks, reservoirs and seals in lacustrine settings (Mitchum and Vail 1977).

The first step is to utilise physical observations of facies, lithologic associations and stacking patterns in borehole data to predict and recognise stratigraphically important physical surfaces (e.g. bounding surfaces of fining-upward and coarsening-upward stacking trends). The second step is to use physical observations of stratal terminations in seismic data to predict and recognise stratigraphically important physical surfaces (e.g. truncation surfaces, toplap surfaces, onlap surfaces and downlap surfaces) (Bohacs 1998; Catuneanu et al. 2009, 2011; Catuneanu 2019). The third step is to utilise seismic-well ties to define sequence stratigraphic frameworks and correlate stratigraphically important physical surfaces (e.g. parasequence-set boundaries, maximum flooding surfaces and sequence boundaries). The fourth step is to use an understanding derived from the sequence framework to interpret depositional environments from log and seismic data and make predictions of environmental boundaries and the distribution of exploration play elements (e.g. source rocks, reservoirs and seals), regardless of which sequence stratigraphic model is employed.

This four-step approach is used here to predict the occurrence and development of sub-lacustrine fans in the Oligocene Bohai Bay Basin.

First, the Oligocene Dongying Formation contains four successions, each of which is composed of a lower fining-upward stacking trend as marked by the blue triangles in Figures 6.35 and 6.36 and an upper coarsening-upward stacking trend as marked by the red triangles in Figures 6.35 and 6.36. The lower fining-upward stacking trends are typified by fine-grained pebbly sandstones and fine-grained sandstone, fining upwards into mudstone interbedded with

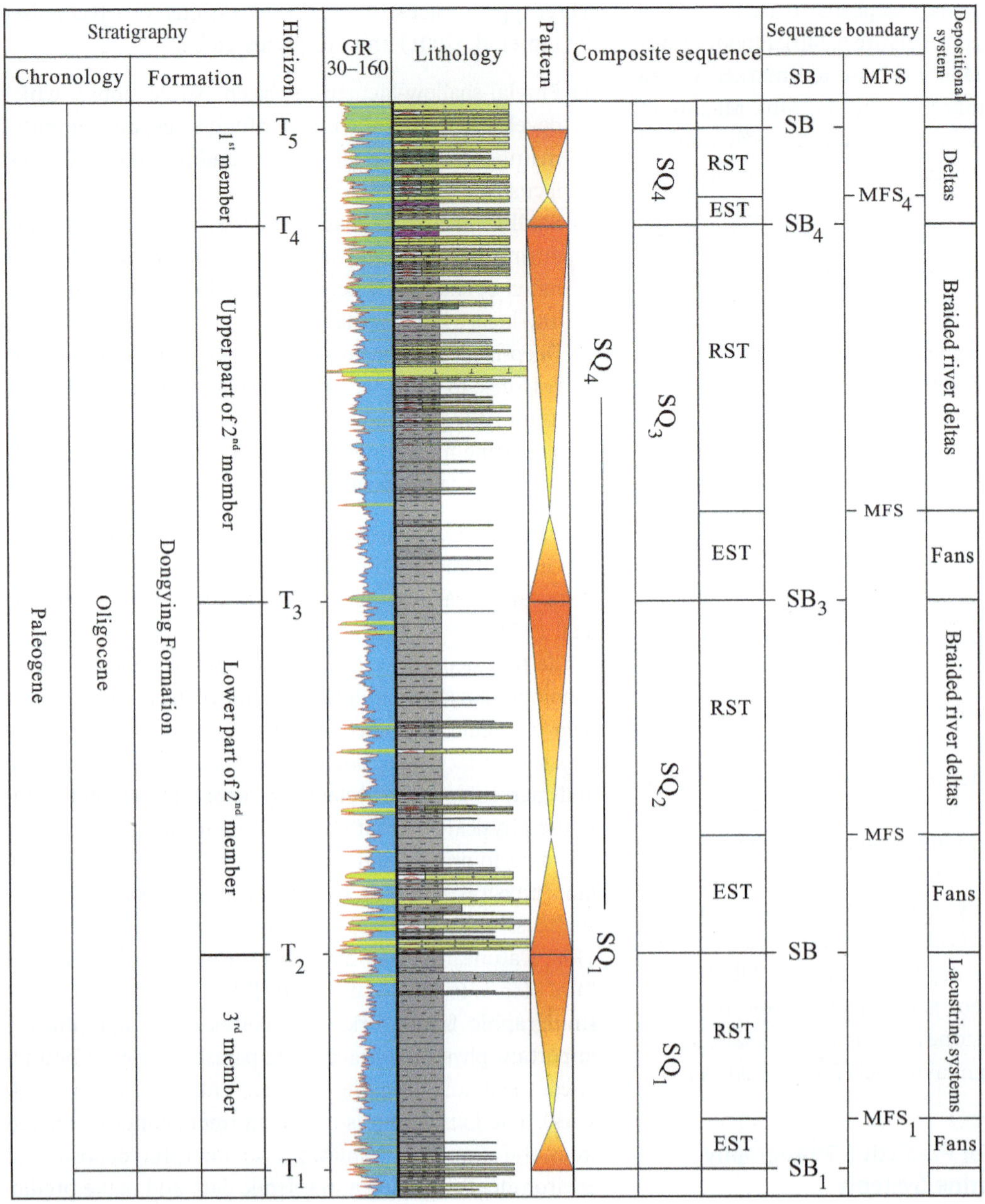

Figure 6.35 Lithological and sequence stratigraphic chart of well 19-6-16 showing lithologic associations and resultant vertical stacking patterns of Oligocene Dongying Formation in the Bozhong sub-basin. For key to lithologies refer to Figure 6.36. Regular and inverted triangles in this figure denote, respectively, fining-upward and coarsening-upward stacking trends. Abbreviations: SB = sequence boundary; MFS = maximum flooding surface.

siltstone and argillaceous siltstone or silty mudstone. The basal bounding surfaces of these four fining-upward stacking trends and upper bounding surfaces of coarsening-upward stacking trends define candidate sequence boundaries (SB_1–SB_5) (Figures 6.35 and 6.36). The upper coarsening-upward stacking trends are represented by mudstone interbedded with siltstone and argillaceous siltstone and silty mudstone, coarsening upwards into fine-grained pebbly sandstone and/or fine-grained sandstones The basal bounding surfaces of these four coarsening-upward stacking trends and the upper bounding surfaces of fining-upward stacking trends are candidate maximum flooding surfaces (MFS_1–MFS_4) (Figures 6.35 and 6.36).

Second, on seismic data, five unconformities (T_1–T_5) and the four main conformable surfaces of MFS_1–MFS_4 were recognised (Figure 6.37). Strata below these five unconformity surfaces terminate landwards by truncation and toplap terminations. Strata above them terminate landward

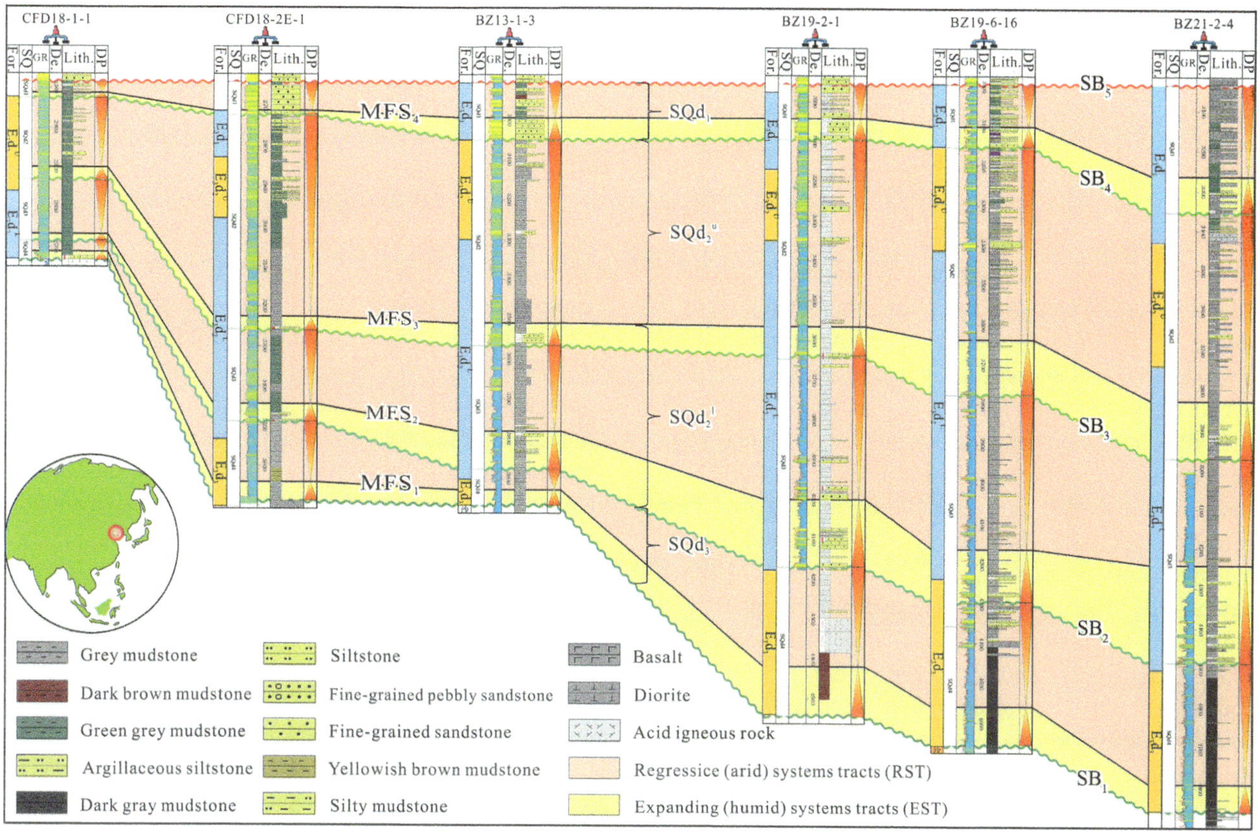

Figure 6.36 Well-to-well correlation profile illustrating lithologic associations and resultant vertical stacking patterns of the Oligocene Dongying Formation in the Bozhong sub-basin. Regular and inverted triangles in this figure denote, respectively, fining-upward and coarsening-upward stacking trends. Abbreviations: For. = Formation; SQ = sequence; GR = gamma ray; Lith. = lithology; DP = depositional patterns.

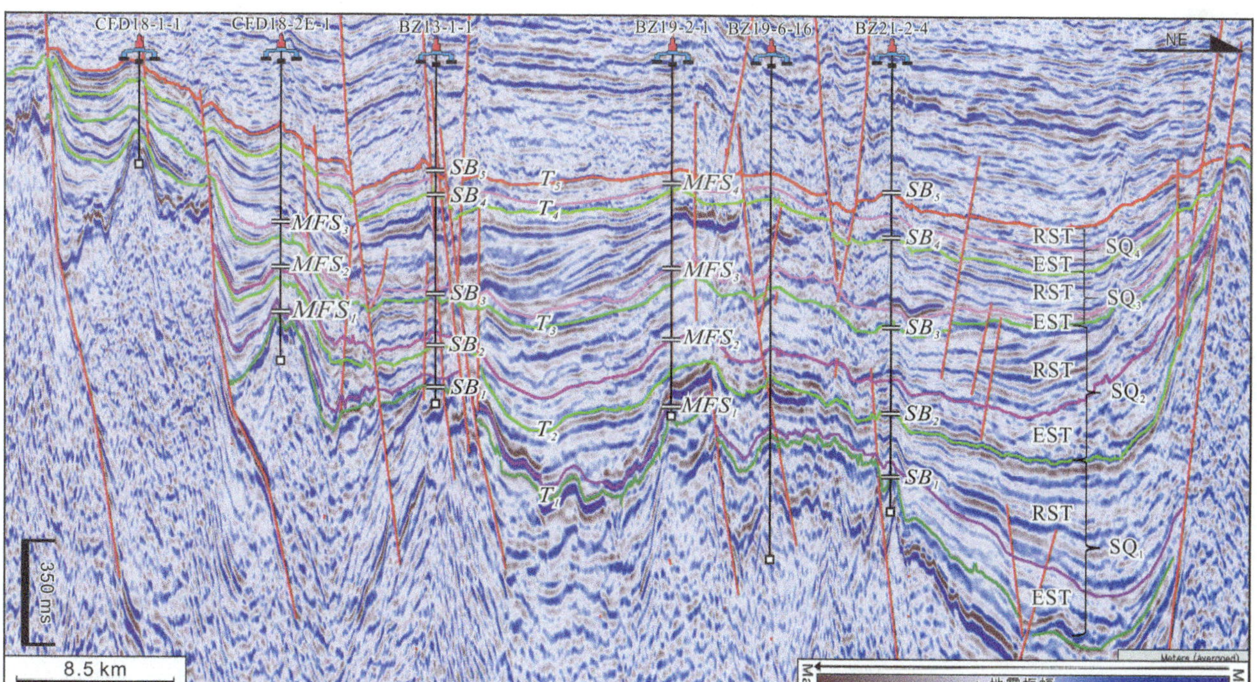

Figure 6.37 Seismic-well tie transect showing stratal terminations and associated physical surfaces of Oligocene Dongying Formation in the Bozhong sub-basin. Figure 6.36 shows lithologic associations and resultant vertical stacking patterns of the wells in this figure.

by onlap terminations but basinwards by downlap terminations (Figure 6.37). Truncation and toplap terminations are, respectively, predictors of prominent erosion and non-deposition (sedimentary by-passing) with perhaps only minor erosion, strongly suggesting that these five unconformity surfaces are sequence boundaries. Strata above the significant seismic reflection surfaces of MFS_1, MFS_2, MFS_3 and MFS_4 terminate basinwards by downlap terminations. Downlap terminations are common indicators of progradation, strongly suggesting that these seismic reflection surfaces are the MFSs.

Third, seismic-well ties shown in Figures 6.36 and 6.37 tie the two sets of surfaces recognised from logs and The Dongying Formation can therefore be divided into four main depositional sequences: SQ_1, SQ_2, SQ_3 and SQ_4 by sequence boundaries of SB_1, SB_2, SB_3, SB_4 and SB_5. These four main depositional sequences can be further subdivided into lower and upper parts by maximum flooding surfaces of MFS_1, MFS_2, MFS_3 and MFS_4. The lower parts of the four sequences display an overall fining-upward stacking trend, which is suggestive of retrogradational to aggradational parasequences and probably humid climactic conditions. They can, therefore, be best considered as expanding (humid) systems tracts (i.e. ESTs in Figures 6.35–6.37). The upper parts of the four sequences display an overall coarsening-upward stacking trend, which is suggestive of progradational parasequences and probably arid climactic conditions. They can, therefore, be best considered as regressive (arid) systems tracts (i.e. RSTs in Figures 6.35–6.37).

The fourth step is to use an understanding derived from the sequence framework to interpret depositional environments from log and seismic data and make predictions of environmental boundaries and the distribution of exploration play elements (e.g. source rocks, reservoirs and seals), regardless of which sequence stratigraphic model is employed. Specific to the Oligocene Bohai Bay case, the ESTs of Oligocene Dongying depositional sequences (i.e. SQ_2 and SQ_3) fostered areally extensive, channelised sub-lacustrine fans (i.e. channel-lobe complexes) seen on the gentle slope in front of syndepositional slope-break zones and areally extensive, non-channelised sub-lacustrine fans (i.e. MTCs) seen on the steep slope in front of syndepositional slope-break zones (Figure 6.12a). Areally extensive, channelised sub-lacustrine fans within an EST are represented by the seismically well-imaged substantial channel-lobe complexes of Figure 6.15, which are 130 km^2 in plan-view area and 0–380 m in thickness (averaging 320 m). Areally extensive, non-channelised sub-lacustrine fans within an EST are represented by the MTCs imaged in Figure 6.19, which are 36 km^2 in plan-view area and 0–450 in thickness (averaging 300 m). In contrast to the ESTs, the RSTs of the Dongying depositional sequences (i.e. SQ_2 and SQ_3) fostered only areally localised, channelised sub-lacustrine fans on the gentle slope in front of syndepositional slope-break zones, but larger areally extensive, non-channelised sub-lacustrine fans on the steep slope in front of syndepositional slope-break zones (i.e. MTCs) (Figure 6.19). Regionally localised, channelised sub-lacustrine fans within an RST are represented by the smaller channel-lobe complexes imaged in Figure 6.16, which are 23.88 km^2 in plan-view area and 0–100 m in thickness (averaging 40 m). Regionally extensive, non-channelised sub-lacustrine fans within an EST are represented by the well-imaged MTCs of Figure 6.20c,d, which are 180 km^2 in plan-view area and 0–450 m in thickness (averaging 342 m).

6.5.4.2 Application of the Source-to-Sink Approach

The source-to-sink (S2S) approach emphasises the semi-quantitative and integrated analysis of siliciclastic sediment transport and storage through linked sedimentary systems from their ultimate upstream sources to the ultimate downstream sediment sinks (Sømme et al. 2009; Martinsen et al. 2017; Gong et al. 2021). The 'linked sedimentary systems' are generally analysed using sequence stratigraphy. The source-to-sink approach explores when and how the clastic sediment budget is partitioned by volume, facies and grain size between different segments of the overall dispersal system (e.g. upland fans or alluvial plains, coastal and delta plains, shore zones, shelf, shelf edge, deep-water slope and basin floor) (Sømme et al. 2009; Martinsen et al. 2017; Gong et al. 2021). The above-mentioned key differences between lakes and oceans strongly suggest that S2S concepts and models for marine basins cannot be directly applied to lakes. New concepts and new architectural models of S2S systems need to be developed and used as the basis for making quantifiable predictions of reservoirs in lacustrine basins.

Some of the most interesting studies of lacustrine source-to-sink systems come from the Palaeogene Bohai Bay Basin by Xu (2013) and Xu et al. (2017a, 2017b, 2020) (Figure 6.38). As graphically illustrated in Figure 6.38a, the entire Palaeogene Bohai Bay source-to-sink system is composed of three main segments, namely source areas in the continental realm, depositional sinks in the deep-lake setting and with intermediate delivery conduits between the two (Xu 2013; Xu et al. 2017a, 2020). As demonstrated by Xu (2013) and Xu et al. (2017a, 2020), the delivery of terrigenous sediments from the continental realm into the Bohai Bay deep to form high-quality reservoirs (mainly fan deltas, braided river deltas, and sub-lacustrine fans) was achieved by three main types of sediment dispersal conduit (Xu et al. 2017a, 2020).

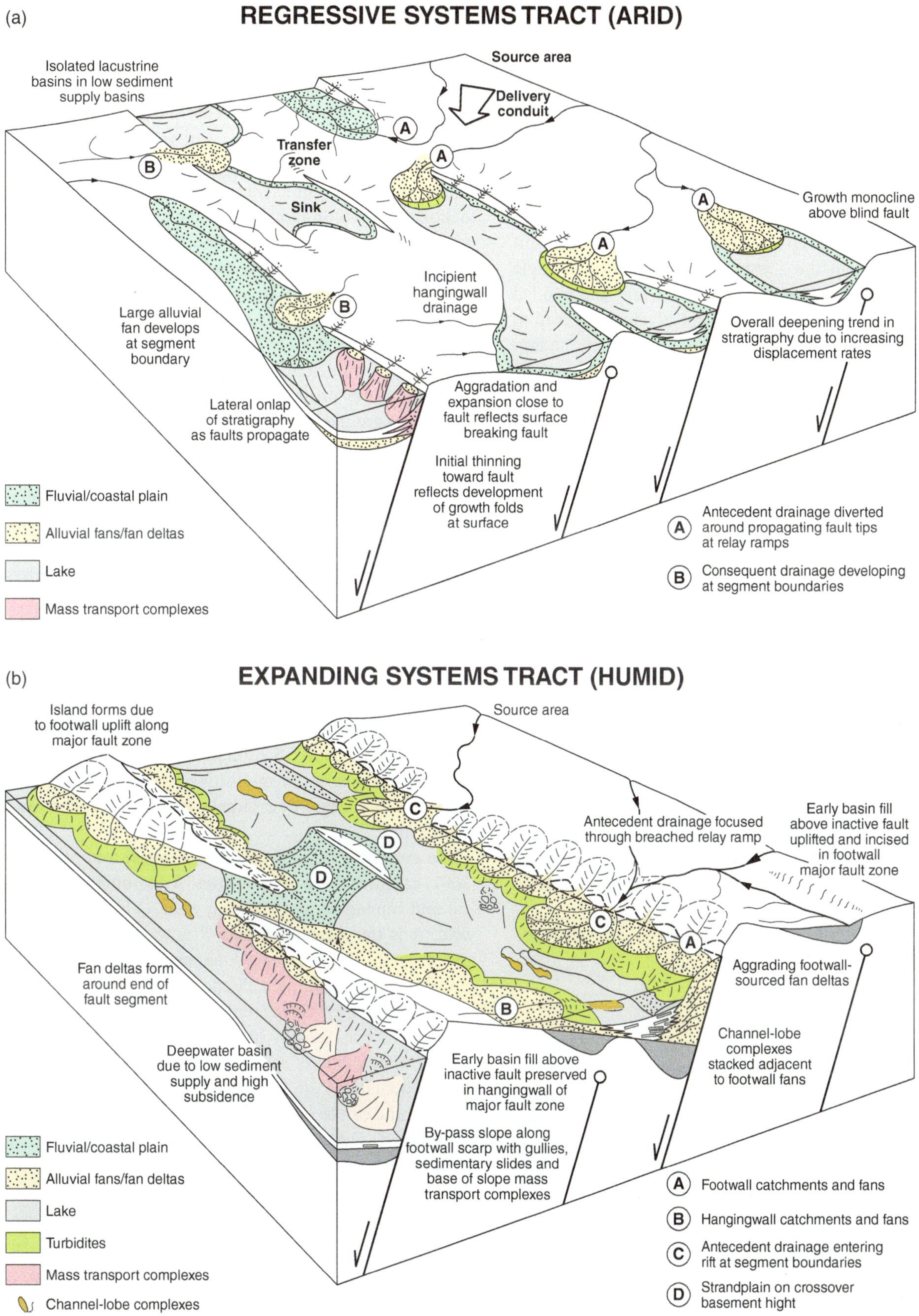

Figure 6.38 Conceptual models of a lake system developed in a continental rift basin based on Gawthorpe and Leeder (2000). The diagrams illustrate the variety of source-to-sink sedimentary systems that co-exist at any given time in such basins. Such systems change with lake level between (a) regressive and (b) expanding systems tracts. Note that sediment routing systems may be variably transverse or longitudinally fed, are determined locally by features such as fault relay ramps and may also change greatly as fault linkages evolve.

The first type occurs mainly in the zones of overlap and intersection of two individual boundary faults, which form relay ramps and rectangular corners, respectively. Fan deltas of 40.2 km^2 in area and 9.5 km^2 in area, respectively, were seen to develop in front of the rectangular corners formed by two boundary faults along the southern Shijiutuo uplift and in front of the relay ramps set up by overlapping normal faults in QHD 29-2 block. The second conduit type developed in the accommodation zone of a single fault (Figure 6.38b,c). Fan deltas of 3–16 km^2 and braided deltas of up to 110 km^2 in area, respectively, were seen to occur in front of the accommodation zones of a boundary fault. The third conduit type is valleys on either gentle or steep slopes. Valleys on gentle slopes normally fed large-scale deltas and coeval channel-lobe complexes and small-scale delta foreset slumps, whereas their counterparts on steeper slopes commonly favoured large-scale MTCs (Gawthorpe and Leeder 2000; Xu et al. 2017a, 2020).

In all these cases, delivery conduits between deltas and fans collectively formed the dispersal entry points for unconfined sediment-gravity flows, thus producing large-scale and high-quality reservoirs in the terminal depositional 'sinks' of the basin (Figure 6.38). The identification and understanding of the concept of three types of main delivery conduit in the Palaeogene Bohai lacustrine systems increased the success rate of reservoir prediction from 42% to 80%, resulting in significant hydrocarbon discoveries (Xu 2013; Xu et al. 2017a, 2017b, 2020). From an exploration standpoint, it should be possible in lacustrine basins to identify these sediment dispersal conduits worldwide even on coarse grids of 2D seismic data, which are reliable predictors of sub-lacustrine sand accumulations in deep-lake areas.

6.6 Controls

In this section, we explore the controls on lacustrine systems from a source-to-sink perspective. Specifically, we look at the controls on the sediment yield of the catchment source areas and the partitioning of sediments into the depositional sinks. An alternative approach that considers the interactions of allogenic and autogenic controls and depositional-system responses (Muto et al. 2007) in the stratal record is well documented in numerous studies dealing with lacustrine successions (e.g. Owen et al. 2019).

6.6.1 Controls on Sediment Yield

Tectonic uplift, climate (rainfall and temperature), catchment area and erodibility of bedrock geology are the major external environmental forcings controlling sediment yield (in tonnes m^{-2}) in source areas for a sediment-routing system to a lacustrine sink (e.g. Heins and Kairo 2007; Allen 2017). High terrain relief and extensive catchment size especially tend to produce large volumes of detritus that can be moved to the sinks (Zhu et al. 2017). Sustained deformation and uplift in tectonically active areas lead to exhumation, sediment production and landscape changes at $\geq 10^5$ yr timescales. Signals of tectonic uplift and deformation, therefore, require a correspondingly long record from which to decode the signal to either understand the tectonic history (Romans et al. 2016) or deconvolve it from other external environmental forcings.

Temperature, precipitation and their fluctuations are the second group of factors that influence sediment yield, impacting much of the S2S sediment-routing system in at least three ways (e.g. Bull 1991; Heins and Kairo 2007; Allen 2017). First, decreasing effective precipitation towards more arid climates decreases sediment yield and sediment flux through the system, whereas increasing precipitation towards a more humid climate generally causes increased sediment discharge in the rivers (e.g. Syvitski and Milliman 2007). Second, vegetation cover in the drainage area of the lake basin is also strongly affected by climate and sometimes increases sediment yield, presumably through soil formation. Very thick and widespread vegetation can eventually decrease sediment yield (Rogers and Schumm 1991; Hooke 2000). Third, the temperature is positively correlated with the sediment discharge, due largely to increased chemical weathering.

The empirical BQART model of Syvitski and Milliman (2007) captures these relationships: B accounts for geological and human factors affecting yield, Q is the sediment yield, A is the catchment area, R is the maximum relief and T is the basin-averaged temperature. The BQART model estimates sediment discharges to lakes or oceans, rather than simply rates of sediment production by erosion of the landscape. Sensitivity estimates using BQART show that the magnitude of sediment supply changes from climate fluctuations at the glacial–interglacial timescale (100 Ky) and can be on the order of 20–30% (Blum and Hattier-Womack 2009). This suggests that cooler climates likely have reduced total sediment supply, whereas warmer climates have likely increased sediment supply, for a specific S2S sediment-routing system.

6.6.2 Controls on Depositional Sinks

Tectonic subsidence, lake-level changes and pre-existing landscape morphology are the three main external forcings in the depositional sink of a lacustrine system. Tectonic subsidence affects the S2S system in three main ways. First, long-term tectonic subsidence produces and maintains the accommodation space, which greatly impacts the type of lacustrine basin (i.e. overfilled, balanced-fill and underfilled) (e.g. Bohacs et al. 2000, 2003). Second, hard- and soft-linkage transfer zones can exert fundamental controls on the routing and flow direction of rivers, location of syntectonic and post-tectonic clastic sediment entry points and erosion of catchment (provenance) areas (Moustafa and Khalil 2017). Third, the formation and development of sub-lacustrine fans (especially nearshore fan deltas) are commonly related directly to the activity of major border faults.

Lake levels are very sensitive to both climatic and seasonal and weather fluctuations because lakes contain small water masses (Bohacs et al. 2000; Gong et al. 2019; Zhang et al. 2019b). First, precipitation and temperature change strongly impact lake level through the evaporation/precipitation balance, the volume of surface run-off and its sediment load, the direct delivery of sedimentary particles by wind and rain and the influx of atmospheric solutes (e.g. chloride). Second, chemical sedimentation in lakes is strongly influenced by climate. Lakes in arid areas are dominated by the precipitation of gypsum, halite and other salts due to the significant evaporation. However, carbonate sedimentation generally occurs in relatively humid regions. Third, the temperature also influences lake sedimentation. In cold areas, the lake surface can be frozen, which decreases both sediment input and wave activity, only allowing fine-grained sediment fallout in the still lake water. In contrast, warm areas with more rainfall allow greater water and sediment discharge in river systems and into lakes.

Tectonic subsidence and climate affect lacustrine base levels, although subsidence normally operates at a much slower pace, often an order of magnitude slower (mm a^{-1}). Note that small base-level changes will lead to rapid transgression or regression over substantial distances in the case of gentle gradients from shoreline across the lake plain. Birgenheier et al. (2021) show an elegant example of overall transgressive lacustrine sequences, tens of metres thick, with an upward change from fluvial channels and lake-plain sediments to shoreline and sub-lacustrine carbonates as lake-level rises with sediment input (from the Eocene Green River, Uinta Basin). Importantly, Birgenheier et al. (2021) show that the periodic influx of sediments to the lake (expressed as amalgamated fluvial channels) corresponded with significant negative organic-$\delta^{13}C$ isotope excursions, suggesting warm hyperthermal climate spikes close to the well-known Paleocene-Eocene climate optimum. If correct, this illustrates that not only were the sequences recording alternating lake-level rise and fall,

Figure 6.39 NASA satellite images (NASA / Public Domain) showing a comparison of the Aral Sea in 1989 (a) and 2014 (b). The Aral Sea was formerly the fourth largest lake in the world (68,000 km^2) lying amidst the deserts of Central Asia but had largely dried up by the 2010s. The Aral Sea began shrinking in the 1960s after the rivers that fed it were diverted by Soviet irrigation projects.

but that the lake-level changes were ultimately driven by the warm and wet climate pulses.

6.6.3 Other Controls on Sediment-Routing Systems

We explore here some highlights of 'anthropogenic forcing', humans, as geological agents, have profoundly impacted the evolution of lakes (Wilkinson 2005; Price et al. 2011). The study of anthropogenic change has become an important approach to achieving sustainable development (Ruddiman 2013; Dirzo et al. 2014; Lewis and Maslin 2015). The combined effects of climate change and direct human activity (Chen et al. 2013) are illustrated by changes in the hydrological processes and the resulting ecological and environmental disasters, for example, the Aral Sea (Figure 6.39). The Aral Sea was an endorheic lake lying between Kazakhstan and Uzbekistan, which began shrinking in the 1960s and had largely dried up by the 2010s (Wæhler and Dietrichs 2017). Formerly the fourth largest lake in the world with an area of 68,000 km^2, the Aral Sea began shrinking after the rivers that fed it were diverted by irrigation projects (cf. Figure 6.39a,b). The water and sediment discharge decreased significantly resulting in an unnatural starved lake basin. By 1997, the sea had declined to 10% of its original size, splitting into four lakes: the North Aral Sea, the Eastern and Western basins of the once far larger South Aral Sea and the smaller Barsakelmes Lake. The biodiversity and sedimentation have also changed due to the decrease in water level and subsequent increase in salinity.

6.7 Geological Applications

The volume of sediments trapped in lakes is small compared with sediments in marine basins. Lacustrine deposits, therefore, are of only minor importance volumetrically in the overall stratigraphic record (Boggs 2005). Nevertheless, lakes and lake deposits are significant and important (e.g. Talbot and Allen 1996; Carroll and Bohacs 1999; Bohacs et al. 2000, 2003).

First, the economic importance of research on lakes and lake deposits far outweighs their abundance in the geological record (Talbot and Allen 1996). Lake deposits host significant economic resources, such as hydrocarbons, coal, uranium, iron and other mineral resources. Ancient lacustrine basins contain extensive evaporites that have provided sites for uranium fixation. Organic-matter-rich lake deposits (highly important source rocks worldwide) account for more than 20% of the current worldwide hydrocarbon production (e.g. Katz 1990; Carroll and Bohacs 1999, 2001; Bohacs et al. 2000, 2003). Lacustrine source rocks and reservoirs are a significant segment of current and future exploration opportunities, especially in areas of rapid market growth, such as in China, southeast Asia, Africa and South America (Bohacs et al. 2000, 2003). Moreover, the preservation of large amounts of limnic coals and lignites in some ancient lake deposits is still an important fuel source. A number of modern and ancient saline lake basins are important sources for a variety of evaporites, including nitrates and borates. In particular, lithium minerals, mostly extracted from brines, are becoming critical for the development of green technologies (Brooks 2020). The deposits of saline-alkaline lakes can be rich in diatomites and zeolites, both of which are today widely used in industrial processes.

Second, lacustrine systems are the single most important large integrated S2S systems in the continental realm and occur from Precambrian to Holocene (Zhu 2008; Jiang 2010; Yu et al. 2018). Lake deposits occur across areas of up to 30,000 km^2 in the Cretaceous and Permian systems of western China (Bohacs et al. 2003).

Third, lake deposits record a delicate sensitivity to palaeoclimate change otherwise unmatched in continental stratigraphy (Cohen 2003). They provide long and continuous continental climate records and are thus used widely to address palaeoclimate questions, leading to continental lake drilling projects around the world (e.g. Williams et al. 1997; Lyons et al. 2011, 2015). Some of these studies have provided valuable insights into the potential links between climate change and hominin evolution (e.g. Owen et al. 2018; Potts et al. 2020; Muiruri et al. 2021). A very recent drilling project in China (Continental Scientific Drilling Project of Cretaceous Songliao Basin) has retrieved a continuous terrestrial archive of Earth's most recent 'greenhouse' period, providing an unbroken record of climatic changes during the age of dinosaurs.

Fourth, lakes function as 'traps' for nutrients transported by surrounding rivers and are thus significant sources of fish and biodiversity. Plants and animals living in lakes are preserved as trace fossils. Lakes are important water resources for humans and play a significant role in maintaining the balance between regional ecosystems (Liu et al. 2020). However, it should be noted that there are hazards arising from lakes. These include natural outburst floods and water displacement due to landslides into lakes and reservoirs but also failures of artificial tailings ponds and dams.

Last but not least, many modern lakes respond to environmental changes in quite a dramatic fashion. The shrinking of the Aral Sea illustrates such a case. Examples of rapid lake expansion are known as well, as illustrated by the fast expansion of lakes located in the endorheic basin of the Qinghai-Tibet Plateau (Liu et al. 2021). An increase in the total lake area from 27,281.10 km^2 in 2000 to 33,974.85 km^2 in 2017 has been measured, implying a rate of 348.45 $km^2\ a^{-1}$. This remarkable expansion has been attributed to an increase in precipitation coupled with permafrost degradation.

In summary, lake systems, processes and their rock record are exceedingly variable and complex due to their relatively small volume, rapid response times and tight linkages among processes. Such systems can, however, be readily analysed and understood using standard approaches of facies analysis and sequence stratigraphy that integrate all relevant physical, chemical and biogenic aspects. Lake systems do require their own models to summarise and generalise the resulting observations and interpretations for use in prediction. Lakes are not just small oceans.

Further Reading

Baganz, O.W., Bartov, Y., Bohacs, K.M., and Nummedal, D. (2012). Lacustrine sandstone reservoirs and hydrocarbon systems. In: *Lacustrine Sandstone Reservoirs and Hydrocarbon Systems*, AAPG Memoir, vol. 95 (ed. O.W. Baganz, Y. Bartov, K. Bohacs, and D. Nummedal), 1–11. American Association of Petroleum Geologists.

Bohacs, K.M., Carroll, A.R., Neal, J.E., and Mankiewicz, P.J. (2000). Lake-basin type, source potential, and hydrocarbon character: an integrated sequence stratigraphic–geochemical framework. In: *Lake Basins through Space and Time*, American Association of Petroleum Geologists Studies in Geology, vol. 46 (ed. E.H. Gierlowski- Kordesch and K.R. Kelts), 3–34. American Association of Petroleum Geologists.

Bohacs, K.M. (2012). Relation of hydrocarbon reservoir potential to Lake-Basin type: an integrated approach to unraveling complex genetic relations among fluvial, lake-plain, lake margin, and lake center strata. In: *Lacustrine Sandstone Reservoirs and Hydrocarbon Systems*, AAPG Memoir, vol. 95 (ed. O.W. Baganz, Y. Bartov, K. Bohacs, and D. Nummedal), 13–56. American Association of Petroleum Geologists.

Buatois, L.A. and Mángano, M.G. (2004). Ichnology of fluvio-lacustrine environments: animal-substrate interactions in freshwater ecosystems. In: *The Application of Ichnology to Palaeoenvironmental and Stratigraphic Analysis*, Geological Society, London, Special Publication, vol. 228 (ed. D. McIlroy), 311–333. Geological Society London.

Carroll, A.R. and Bohacs, K.M. (1999). Stratigraphic classification of ancient lakes: balancing tectonic and climatic controls. *Geology* 27: 99–102.

Dodd, T.J.H., Mccarthy, D.J., and Richards, P.C. (2019). A depositional model for deep-lacustrine, partially confined, turbidite fans: early Cretaceous, North Falkland Basin. *Sedimentology* 66: 53–80.

Eljalafi, A. and Sarg, J.F. (2021). Depositional system and lake-stage control on microbialite morphology, Green River Formation, eastern Uinta Basin, Colorado and Utah, USA. *Journal of Sedimentary Research* 91 (6): 636–661.

Gawthorpe, R.L. and Leeder, M.R. (2000). Tectono-sedimentary evolution of active extensional basins. *Basin Research* 12: 195–218.

Gong, C., Sztano, O., Steel, R.J. et al. (2019). Critical differences in sediment delivery and partitioning between marine and lacustrine basins: a comparison of marine and lacustrine aggradational to progradational clinothem pairs. *Geological Society of America Bulletin* 131: 766–781.

Lyons, R.P., Scholz, C.A., Buoniconti, M.R., and Martin, M.R. (2011). Late Quaternary stratigraphic analysis of the Lake Malawi Rift, East Africa: an integration of drill-core and seismic-reflection data. *Palaeogeography, Palaeoclimatology, Palaeoecology* 303: 20–37.

Olariu, C., Zhou, C., Steel, R. et al. (2021). Controls on the stratal architecture of lacustrine delta successions in low-accommodation conditions. *Sedimentology* 68: 1941–1963.

Olariu, C. and Bhattacharya, J.P. (2006). Terminal distributary channels and delta front architecture of fluvial-dominated delta systems. *Journal of Sedimentary Research* 76: 212–233.

Pan, S., Liu, H., Zavala, C. et al. (2017). Sub-lacustrine hyperpycnal channel-fan system in a large depression basin: a case study of Nen 1 Member, Cretaceous Nenjiang Formation in the Songliao Basin, NE China. *Petroleum Exploration and Development* 44: 911–922.

Sarg, J.F., Tanavsuu-Milkeviciene, K., Humphrey, J.D., and Suriamin, H. (2013). Lithofacies, stable isotopic composition, and stratigraphic evolution of microbial and associated carbonates, Green River Formation (Eocene), Piceance Basin, Colorado. *AAPG Bulletin* 97: 1937–1966.

Tänavsuu-Milkeviciene, K., Sarg, J.F., and Bartov, Y. (2017). Depositional cycles and sequences in an organic-matter-rich lake basin: Eocene Green River Formation, Lake Uinta, Colorado and Utah, U.S.A. *Journal of Sedimentary Research* 87: 210–229.

Tan, C. and Plink-Björklund, P. (2021). Morphodynamics of supercritical flow in a linked river and delta system, Daihai Lake, Northern China. *Sedimentology* 68: 1606–1639.

Vandekerkhove, E., van Daele, M., Praet, N. et al. (2020). Flood-triggered versus earthquake-triggered turbidites: a sedimentological study in clastic lake sediments (Eklutna Lake, Alaska). *Sedimentology* 67: 364–389.

Wang, C., Feng, Z., Zhang, L. et al. (2013). Cretaceous paleogeography and paleoclimate and the setting of SKI borehole sites in Songliao Basin, northeast China. *Palaeogeography, Palaeoclimatology, Palaeoecology* 385: 17–30.

Xu, C., Du, X., Xu, W., and Zhao, M. (2017a). Source-to-sink system and its sedimentary records in the continental rift basins: an example from the Paleogene in the Bohai Sea Area, China. *Interpretation* 5: ST35–ST51.

References

Ahr, W.M. (2008). *Geology of Carbonate Reservoirs*, 1–12. New York: Wiley.

Allen, P.A. (2017). *Sediment Routing Systems: The Fate of Sediment from Source to Sink*, 196–206. Cambridge, UK: Cambridge University Press.

Awramik, S.M. and Buchheim, H.P. (2015). Giant stromatolites of the Eocene Green River Formation (Colorado, USA). *Geology* 43: 691–694.

Axelsson, V. (1967). The Laitaure Delta, a study of deltaic morphology and processes. *Geografiska Annaler Series A, Physical Geography* 49: 1–127.

Aziz, H.A., Sanz-Rubio, E., Calvo, J.P. et al. (2010). Palaeoenvironmental reconstruction of a middle Miocene alluvial fan to cyclic shallow lacustrine depositional system in the Calatayud Basin (NE Spain). *Sedimentology* 50: 211–236.

Baganz, O.W., Bartov, Y., Bohacs, K.M., and Nummedal, D. (2012). Lacustrine sandstone reservoirs and hydrocarbon systems. In: *Lacustrine Sandstone Reservoirs and Hydrocarbon Systems*, AAPG Memoir, vol. 95 (ed. O.W. Baganz, Y. Bartov, K. Bohacs, and D. Nummedal), 1–11. American Association of Petroleum Geologists.

Bates, C.C. (1953). Rational theory of delta formation. *Bulletin of American Association of Petroleum Geology* 37: 2119–2162.

Battison, L. and Brasier, M.D. (2012). Remarkably preserved prokaryote and eukaryote microfossils within 1 Ga-old lake phosphates of the Torridon Group, NW Scotland. *Precambrian Research* 196: 204–217.

Beglinger, S.E., Doust, H., and Cloetingh, S. (2012). Relating petroleum system and play development to basin evolution: West African South Atlantic basins. *Marine and Petroleum Geology* 30: 1–25.

Bhattacharya, J.P. (2006). Deltas. In: *Facies Models Revisited*, SEPM Special Publication, vol. 84 (ed. H.W. Posamentier and R.G. Walker), 237–292. SEPM Society for Sedimentary Geology.

Bhattacharya, J.P. and Giosan, L. (2003). Wave-influenced deltas: geomorphological implications for facies reconstruction. *Sedimentology* 50: 187–210.

Birgenheier, L.P., Vanden Berg, M.D., Plink-Björklund, P. et al. (2021). Climate impact on fluvial-lake system evolution, Eocene Green River Formation, Uinta Basin, Utah, USA. *Geological Society of America Bulletin* 132: 562–587.

Blum, M.D. and Hattier-Womack, J. (2009). Climatic change, sea-level change and fluvial sediment supply to deepwater depositional systems. In: *External Controls on Deep-Water Depositional Systems*, Society for Sedimentary Geology (SEPM) Special Publication, vol. 92 (ed. B. Kneller, O.J. Martinsen, and B. McCaffrey), 15–39. SEPM Society for Sedimentary Geology.

Boggs, S. Jr. (2005). *Principles of Sedimentology and Stratigraphy*, 4e, 228–234. NJ, USA: Pearson Prentice Hall.

Bohacs, K.M. (1998). Contrasting expressions of depositional sequences in mudrocks from marine to non-marine environs. In: *Mudstones and Shales*, Characteristics at the Basin Scale Schweizerbart'sche Verlagsbuchhandlung, vol. 1 (ed. J. Schieber, W. Zimmerle, and P. Sethi), 32–77. Stuttgart: Schweizerbart Science Publishers.

Bohacs, K.M. (2012). Relation of hydrocarbon reservoir potential to lake-basin type: n integrated approach to unraveling complex genetic relations among fluvial, lake-plain, lake margin, and lake center strata. In: *Lacustrine Sandstone Reservoirs and Hydrocarbon Systems*, AAPG Memoir, vol. 95 (ed. O.W. Baganz, Y. Bartov, K. Bohacs, and D. Nummedal), 13–56. American Association of Petroleum Geologists.

Bohacs, K.M., Carroll, A.R., Neal, J.E., and Mankiewicz, P.J. (2000). Lake-basin type, source potential, and hydrocarbon character: an integrated sequence stratigraphic–geochemical framework. In: *Lake Basins Through Space and Time*, American Association of Petroleum Geologists Studies in Geology, vol. 46 (ed. E.H. Gierlowski-, Kordesch and K.R. Kelts), 3–34. American Association of Petroleum Geologists.

Bohacs, K.M., Carroll, A.R., and Neal, J.E. (2003). Lessons from large lake systems-thresholds, nonlinearity, and strange attractors. In: *Extreme Depositional Environments: Mega end Members in Geologic Time*, GSA Special Paper, vol. 370 (ed. M.A. Chan and A.W. Archer), 75–90. Geological Society of America.

Bohacs, K.M., Grabowski, G.J. Jr., Carroll, A.R. et al. (2007a). Lithofacies architecture and sequence stratigraphy of the Green River Formation, greater Green River Basin, Wyoming and Colorado. *Mountain Geologist* 44: 39–60.

Bohacs, K.M., Hasiotis, S.T., and Demko, T.M. (2007b). Continental ichnofossils of the Green River and Wasatch Formation, Eocene, Wyoming: a preliminary survey, proposed relation to lake-basin type, and application to integrated paleo-environmental interpretation. *Mountain Geologist* 11: 79–108.

Bohacs, K.M., Lamb-Wozniak, K., Demko, T.M. et al. (2013). Vertical and lateral distribution of lacustrine carbonate lithofacies at the parasequence scale in the Miocene Hot Spring Limestone, Idaho, USA: an analog addressing reservoir presence and quality. *AAPG Bulletin* 97: 1967–1995.

Bridge, J. and Demicco, R. (2008). *Earth Surface Processes, Landforms and Sediment Deposits*, 462–472. Cambridge, UK: Cambridge University Press.

Brooks, K. (2020). Lithium minerals. *Geology Today* 36: 192–197.

Buatois, L.A. and Mángano, M.G. (1993). Ecospace utilization, paleoenvironmental trends and the evolution of early nonmarine biotas. *Geology* 21: 595–598.

Buatois, L.A. and Mángano, M.G. (1995a). Sedimentary dynamics and evolutionary history of a Late Carboniferous Gondwanic lake in north-western Argentina. *Sedimentology* 42: 415–436.

Buatois, L.A. and Mángano, M.G. (1995b). The paleoenvironmental and paleoecological significance of the lacustrine *Mermia* ichnofacies: an archetypical subaqueous nonmarine trace fossil assemblage. *Ichnos* 4: 151–161.

Buatois, L.A. and Mángano, M.G. (1998). Trace fossil analysis of lacustrine facies and basins. *Palaeogeography, Palaeoclimatology, Palaeoecology* 140: 367–382.

Buatois, L.A. and Mángano, M.G. (2004). Ichnology of fluvio-lacustrine environments: animal-substrate interactions in freshwater ecosystems. In: *The Application of Ichnology to Palaeoenvironmental and Stratigraphic Analysis*, Geological Society, London, Special Publication, vol. 228 (ed. D. McIlroy), 311–333. Geological Society of London.

Buatois, L.A. and Mángano, M.G. (2009). Applications of ichnology in lacustrine sequence stratigraphy: potential and limitations. *Palaeogeography, Palaeoclimatology, Palaeoecology* 272: 127–142.

Buatois, L.A. and Mángano, M.G. (2011). *Ichnology: Organism-Substrate Interactions in Space and Time*. Cambridge University Press, 358 p.

Buatois, L.A., Mángano, M.G., Wu, X., and Zhang, G. (1996). Trace fossils from Jurassic lacustrine turbidites of the Anyao Formation (Central China) and their environmental and evolutionary significance. *Ichnos* 4: 287–303.

Buatois, L.A., Mángano, M.G., Genise, J.F., and Taylor, T.N. (1998). The ichnologic record of the continental invertebrate invasion: evolutionary trends in environmental expansion, ecospace utilization, and behavioral complexity. *Palaios* 13: 217–240.

Buatois, L.A., Mángano, M.G., Fregenal-Martinez, M.A., and de Gibert, J.M. (2000). Short-term colonization trace-fossil assemblages in a carbonate lacustrine konservat-lagerstätte (Las Hoyas fossil site, Lower Cretaceous, Cuenca, central Spain). *Facies* 43: 145–156.

Buatois, L.A., Labandeira, C.C., Mángano, M.G. et al. (2016). The Mesozoic lacustrine revolution. In: *The Trace-Fossil Record of Major Evolutionary Changes, vol. 2: Mesozoic and Cenozoic. Topics in Geobiology*, vol. 40 (ed. M.G. Mángano and L.A. Buatois), 179–263. Springer.

Buatois, L.A., Renaut, R.W., Scott, J.J., and Owen, R.B. (2017). An unusual occurrence of the trace fossil *Vagorichnus* preserved in hydrothermal silica at Lake Baringo, Kenya Rift Valley: taphonomic and paleoenvironmental significance. *Palaeogeography, Palaeoclimatology, Palaeoecology* 485: 843–853.

Buatois, L.A., Renaut, R.W., Owen, R.B. et al. (2020). Animal bioturbation preserved in Pleistocene magadiite at Lake Magadi, Kenya Rift Valley, and its implications for the depositional environment of bedded magadiite. *Scientific Reports* 10: 6794.

Buatois, L.A., Davies, N.S., Gibling, M.R. et al. (2022). The invasion of the land in deep time: integrating Paleozoic records of paleobiology, ichnology, sedimentology, and geomorphology. *Integrative and Comparative Biology* 62: 297–331.

Buick, R. (1992). The antiquity of oxygenic photosynthesis: evidence from stromatolites in sulphate-deficient Archaean lakes. *Science* 255: 74–77.

Bull, W.B. (1991). *Geomorphic Responses to Climatic Change*, 326. London, UK: Oxford University Press.

Bustillo, M.A., Armenteros, I., and Huerta, P. (2017). Dolomitization, gypsum calcitization and silicification in carbonate–evaporite shallow lacustrine deposits. *Sedimentology* 64: 1147–1172.

Callow, R.H., Battison, L., and Brasier, M. D. (2011). Diverse microbially induced sedimentary structures from 1 Ga lakes of the Diabaig Formation, Torridon Group, northwest Scotland. *Sedimentary Geology* 239: 117–128.

Calvo, J.P., Rodriguez-Pascua, M., Martin-Velazquez, S. et al. (1998). Microdeformation of lacustrine laminite sequences from Late Miocene formations of SE Spain: an interpretation of loop bedding. *Sedimentology* 45: 279–292.

Cao, Y., Jin, J., Liu, H. et al. (2021). Deep-water gravity flow deposits in a lacustrine rift basin and their oil and gas geological significance in eastern China. *Petroleum Exploration and Development* 48: 286–298.

Capezzuoli, E., Gandin, A., and Pedley, M. (2014). Decoding tufa and travertine (fresh water carbonates) in the sedimentary record: the state of the art. *Sedimentology* 61 (1): 1–21.

Carroll, A.R. and Bohacs, K.M. (1999). Stratigraphic classification of ancient lakes: balancing tectonic and climatic controls. *Geology* 27: 99–102.

Carroll, A.R. and Bohacs, K.M. (2001). Lake-type controls on petroleum source rock potential in nonmarine basins. *AAPG Bulletin* 85: 1033–1053.

Cather, S.M., Heizler, M.T., and Williamson, T.E. (2019). Laramide fluvial evolution of the San Juan Basin, New Mexico and Colorado: paleocurrent and detrital-sanidine age constraints from the Paleocene Nacimiento and Animas formations. *Geosphere* 15: 1641–1664.

Catuneanu, O. (2019). Model-independent sequence stratigraphy. *Earth-Science Reviews* 188: 312–388.

Catuneanu, O., Abreu, V., Bhattacharya, J.P. et al. (2009). Towards the standardization of sequence stratigraphy. *Earth-Science Reviews* 92: 1–33.

Catuneanu, O., Galloway, W.E., Kendall, C.G.S.C. et al. (2011). Sequence stratigraphy: methodology and nomenclature. *Newsletters on Stratigraphy* 44: 173–245.

Chen, X., Bai, J., Li, X. et al. (2013). Changes in land use/land cover and ecosystem services in Central Asia during 1990–2009. *Current Opinion in Environmental Sustainability* 5: 116–127.

Choudens-Sanchez, V.D. and Gonzalez, L.A. (2009). Calcite and aragonite precipitation under controlled instantaneous supersaturation: elucidating the role of $CaCO_3$ saturation state and Mg/Ca ratio on calcium carbonate polymorphism.

Journal of Sedimentary Research 79: 363–376.

Cohen, A.S. (1989). The taphonomy of gastropod shell accumulations in large lakes: an example from Lake Tanganyika, Africa. *Paleobiology* 15: 26–44.

Cohen, A.S. (2003). *Paleolimnology: The History and Evolution of Lake Systems*, 500. New York: Oxford University Press.

Czaja, S., Bartley, S., and Fralick, P.H. (2019). Preservation of carbon isotopes in kerogen from thermally altered Mesoproterozoic lacustrine microbialites. *Canadian Journal of Earth Sciences* 56: 1017–1026.

Davies, N.S., Sansom, I.J., and Turner, P. (2006). Trace fossils and paleoenvironments of a late Silurian marginal-marine/alluvial system: the Ringerike Group (Lower Old Red Sandstone), Oslo Region, Norway. *PALAIOS* 21: 46–62.

Deng, S., Dong, H., Lv, G. et al. (2010). Microbial dolomite precipitation using sulfate reducing and halophilic bacteria: results from Qinghai Lake, Tibetan Plateau NW China. *Chemical Geology* 278 (3-4): 151–159.

Deocampo, D.M. and Renaut, R.W. (2016). Geochemistry of African soda lakes. In: *Soda Lakes of East Africa* (ed. M. Schagerl), 77–93. Springer International Publishing Switzerland.

Dirzo, R., Young, H.S., Galetti, M. et al. (2014). Defaunation in the Anthropocene. *Science* 345: 401–406.

Dodd, T.J.H., Mccarthy, D.J., and Richards, P.C. (2019). A depositional model for deep-lacustrine, partially confined, turbidite fans: early Cretaceous, North Falkland Basin. *Sedimentology* 66: 53–80.

Du, X., Liu, H., Huang, X. et al. (2020). A near-shore clastic-carbonate mixing mode in a continental rift basin (early Oligocene, eastern Shijiutuo Uplift, Bohai Bay Basin, China): sedimentology, reservoir characteristics and exploration practice. *Journal of Palaeogeography* 9: 54–71.

Eljalafi, A., and Sarg, J.F (2018). Lacustrine microbialite architectural and chemostratigraphic trends: green river formation, eastern Uinta Basin, Colorado and Utah. *AAPG Annual Convention and Exhibition, Salt Lake City, Utah*, May 20–23, 2018, Search and Discovery Article #51522 (2018). Website accessed January 2021.

Eljalafi, A. and Sarg, J.F. (2021). Depositional system and lake-stage control on microbialite morphology, Green River Formation, eastern Uinta Basin, Colorado and Utah, USA. *Journal of Sedimentary Research* 91 (6): 636–661.

Farlow, J.O. and Galton, P.M. (2003). Dinosaur trackways of Dinosaur State Park, Rocky Hill, Connecticut. In: *The Great Rift Valleys of Pangea in Eastern North America*, vol. 2 (ed. P.M. Letourneau and P.E. Olsen), 248–263. Columbia University Press.

Fedorchuk, N.D., Dornbos, S.Q., Corsetti, F.A. et al. (2016). Early non-marine life: evaluating the biogenicity of Mesoproterozoic fluvial-lacustrine stromatolites. *Precambrian Research* 275: 105–118.

Feng, J. (2011). Source rock characterization of the Green River oil shale, Piceance Creek Basin, Colorado. MSc thesis. Colorado School of Mines, Golden, CO, 1–23 pp.

Feng, Z., Zhang, S., Timothy, A.C. et al. (2010a). Lacustrine turbidite channels and fans in the Mesozoic Songliao Basin, China. *Basin Research* 22: 96–107.

Feng, Z., Jia, C., Xie, X. et al. (2010b). Tectonostratigraphic units and stratigraphic sequences of the nonmarine Songliao basin, northeast China. *Basin Research* 22: 79–95.

Feng, Z., Wang, C., Graham, S. et al. (2013). Continental scientific drilling project of Cretaceous Songliao Basin: scientific objectives and drilling technology. *Palaeogeography, Palaeoclimatology, Palaeoecology* 385: 6–16.

Feng, C., Sun, M., Liu, C. et al. (2020). Seismic sedimentary characteristics and filling structure of the Upper Triassic Yanchang Formation, Ordos Basin, China. *Interpretation* 8: 259–274.

Fongngern, R., Olariu, C., Steel, R.J., and Krézsek, C. (2016). Clinoform growth in a Miocene, Para-tethyan deep lake basin: thin topsets, irregular foresets and thick bottomsets. *Basin Research* 28: 770–795.

Fongngern, R., Olariu, C., Steel, R.J. et al. (2018). Subsurface and outcrop characteristics of fluvial-dominated deep-lacustrine clinoforms. *Sedimentology* 65: 1447–1481.

Forel, F.A. (1892). *Le Léman: Monographie Limnologique*, Géographie, Hydrographie, Géologie, Climatologie, Hydrologie, 543 pp., vol. 1. Lausanne: F. Rouge.

García, A. (1994). Charophyta: their use in paleolimnology. *Journal of Paleolimnology* 10: 43–52.

Gawthorpe, R.L. and Leeder, M.R. (2000). Tectono-sedimentary evolution of active extensional basins. *Basin Research* 12: 195–218.

Genise, J.F. (2017). *Ichnoentomology: Insect Traces in Soils and Paleosols. Topics in Geobiology*, vol. 37. Dordrecht: Springer.

Genise, J.F., Melchor, R.N., Bellosi, E.S., and Verde, M. (2010). Invertebrate and vertebrate trace fossils in carbonates. In: *Carbonates in Continental Settings. Developments in Sedimentology*, vol. 61 (ed. A.M. Alonso-Zarza and L. Tanner), 319–369. Amsterdam: Elsevier.

Genise, J.F., Bedatou, E., Bellosi, E.S. et al. (2016). The Phanerozoic four revolutions and evolution of paleosol ichnofacies. In: *The Trace-Fossil Record of Major Evolutionary Events Volume 2. Topics in Geobiology*, vol. 40 (ed. L.A. Buatois and M.G. Mángano), 301–370. Dordrecht: Springer.

de Gibert, J.M. and Sáez, A. (2009). Paleohydrological significance of trace fossil distribution in Oligocene fluvial-fan-to-lacustrine systems of the Ebro Basin, Spain. *Palaeogeography, Palaeoclimatology, Palaeoecology* 272: 162–175.

Gierlowski-Kordesch, E. (1991). Ichnology of an ephemeral lacustrine/alluvial plain system: Jurassic East Berlin Formation, Hartford Basin, USA. *Ichnos* 1: 221–232.

Gierlowski-Kordesch, E.H. (2010). Lacustrine carbonates. In: *Carbonates in Continental Settings: Facies, Environments and Processes:*

Developments in Sedimentology, vol. 61 (ed. A.M. Alonso-Zarza and L. Tanner), 1–101. Amsterdam: Elsevier.

Gierlowski-Kordesch, E. and Kelts, K. (1994). *Global Geological Record of Lake Basins*, vol. 1, 427. Cambridge University Press, Cambridge.

Gierlowski-Kordesch, E.H. and Kelts, K.R. (2000). *Lake Basins Through Space and Time*, American Association of Petroleum Geologists Studies in Geology, vol. 46. American Association of Petroleum Geologists, 648 p.

Goldstein, S.L., Kiro, Y., Torfstein, A. et al. (2020). Revised chronology of the ICDP Dead Sea deep drill core relates drier-wetter-drier climate cycles to insolation over the past 220 kyr. *Quaternary Science Reviews* 244: 106460.

Gong, C., Sztano, O., Steel, R.J. et al. (2019). Critical differences in sediment delivery and partitioning between marine and lacustrine basins: a comparison of marine and lacustrine aggradational to progradational clinothem pairs. *Geological Society of America Bulletin* 131: 766–781.

Gong, C., Li, D., Steel, R.J. et al. (2021). Delta-to-fan source-to-sink coupling as a fundamental control on the delivery of coarse clastics to deepwater: insights from stratigraphic forward modelling. *Basin Research* 33: 2960–2983.

Gore, P.J. (1989). Toward a model for open-and closed-basin deposition in ancient lacustrine sequences: the Newark Supergroup (Triassic-Jurassic), eastern North America. *Palaeogeography, Palaeoclimatology, Palaeoecology* 70: 29–51.

Goulart, E.S. and Calliari, L.J. (2013). Medium-term morphodynamic behavior of a multiple sand bar beach. *Journal of Coastal Research* 65: 1774–1779.

Gray, J. (1988). Evolution of the freshwater ecosystem: the fossil record. *Palaeogeography, Palaeoclimatology, Palaeoecology* 62: 1–214.

Gulati, R.D. and Zadereev, E.S. (2017). Conclusion: ecology of meromictic lakes. In: *Ecology of Meromictic Lakes, Ecological Studies*, vol. 228 (ed. R.D. Gulati, E.S. Zadereev, A.G. Degermendzhi, et al.), 379–398. Springer International Publishing AG.

Håkanson, L. (2007). Lake environments. In: *Environmental Sedimentology*, 3e (ed. C. Perry and K. Tylor), 109–143. Oxford: Blackwell.

Hart, W.S., Quade, J., Madsen, D.B. et al. (2004). The $^{87}Sr/^{86}Sr$ ratios of lacustrine carbonates and lake-level history of the Bonneville paleolake system. *Geological Society of America Bulletin* 116: 1107–1119.

Heins, W.A. and Kairo, S. (2007). Predicting sand character with integrated genetic analysis. In: *Sedimentary Provenance and Petrogenesis: Perspectives from Petrography and Geochemistry*, Geological Society of America, vol. 420 (ed. J. Arribas, S. Critelli, and M.J. Johnsson), 345–379. Geological Society of America Special Paper.

Hooke, R.L. (2000). Toward a uniform theory of clastic sediment yield in fluvial systems. *Geological Society of America Bulletin* 112: 1778–1786.

Houser, C. and Greenwood, B. (2005). Profile response of a lacustrine multiple barred nearshore to a sequence of storm events. *Geomorphology* 69: 118–137.

de Jesus e Sousa, A., de Souza Carvalho, I., and Pedrão Ferreira, E. (2019). Non-marine ostracod biostratigraphy of Cretaceous rift lake deposits (Sousa Basin, Brazil): paleogeographical implications and correlation with Gondwanic basins. *Journal of South American Earth Sciences* 96: 102345.

Jiang, Z. (2010). *Sedimentology*, 2e, 153–170. Beijing: Petroleum Industry Press (in Chinese).

Jiang, Z. (2018). *Sedimentary Dynamics of Windfield-Source-Basin System*, 121–202. Springer.

Jiang, Z., Liu, H., Zhang, S. et al. (2011). Sedimentary characteristics of large-scale lacustrine beach-bars and their formation in the Eocene Boxing Sag of Bohai Bay Basin, East China. *Sedimentology* 58: 1087–1112.

Johnson, R.C., Mercier, T.J., Brownfield, M.E., and Self, J.G. (2010). Assessment of in-place oil shale resources in the Eocene Green River Formation, Uinta Basin, Utah and Colorado, in U.S. Geological Survey Oil Shale Assessment Team, ed., Oil Shale Resources of the Uinta Basin, Utah and Colorado: U.S. Geological Survey Digital Data Series DDS-69-BB, 153 p.

Jones, B.F. and Deocampo, D.M. (2004). Geochemistry of Saline Lakes. In: *Treatise on Geochemistry*, vol. 5 (ed. H.D. Holland and K.K. Turekian), 393–424. Amsterdam: Elsevier.

Katz, B.J. (1990). Controls on distribution of Lacustrine source rocks through time and space. In: *Lacustrine Basin Exploration: Case Studies and Modern Analogs*, AAPG Memoir, vol. 50 (ed. B.J. Katz), 61–76. American Association of Petroleum Geologists.

Keighley, D.D., Flint, S.S., Howell, J., and Moscariello, A. (2003). Sequence stratigraphy in lacustrine basins: a model for part of the Green River Formation (Eocene), Southwest Uinta Basin, Utah, USA. *Journal of Sedimentary Research* 73: 987–1006.

Kelts, K. (1988). Environments of deposition of lacustrine petroleum source rocks: an introduction. In: *Lacustrine Petroleum Source Rocks*, Geological Society of London Special Publication, vol. 40 (ed. A.J. Fleet, K. Kelts, and M.R. Talbot), 3–26. Geological Society of London.

Kempe, S., Kazmierczak, J., Landmann, G. et al. (1991). Largest known microbialites discovered in Lake Van, Turkey. *Nature* 349: 605–608.

Külkylüoğlu, O. (2005). Ecology and phenology of freshwater ostracods in Lake Glky (Bolu, Turkey). *Aquatic Ecology* 39: 295–304.

Kusakabe, M. (2017). Lakes Nyos and Monoun gas disasters (Cameroon): limnic eruptions caused by excessive accumulation of magmatic CO_2 in crater lakes. *GEochemistry Monograph Series* 1: 1–50.

Kusakabe, M., Ohsumi, T., and Aramaki, S. (1989). The Lake Nyos gas disaster: chemical and isotopic evidence in waters and dissolved gases from three Cameroonian crater lakes, Nyos, Monoun and Wum. *Journal of Volcanology and Geothermal Research* 39: 167–185.

Labandeira, C.C. (1999). Insects and other arthropods. In: *Encyclopedia of Paleontology* (ed. R. Singer), 603–624. Chicago: Fitzroy Dearborn.

Labandeira, C.C. (2002). Paleobiology of predators, parasitoids, and parasites: death and accommodation in the fossil record of continental invertebrates. In: *The Fossil Record of Predation*, vol. 8 (ed. M. Kowalewski and P.H. Kelley), 211–249. Paleontological Society Papers.

Lang, S.C., Payenberg, T.H.D., Reilly, M.R.W. et al. (2004). Modern fluvial analogues for dryland sandy fluvial lacustrine deltas and terminal splay reservoirs. *APPEA Journal* 44: 329–356.

Lawrence, M.J.F. and Hendy, C.H. (1989). Carbonate deposition and Ross Sea ice advance, Fryxell basin, Taylor Valley, Antarctica. *New Zealand Journal of Geology and Geophysics* 32: 267–278.

Lepot, L., Benzerara, K., Brown, G.E., and Philippot, P. (2008). Microbially influenced formation of 2,724-million-year-old stromatolites. *Nature Geoscience* 1: 118–121.

Lewis, S.L. and Maslin, M.A. (2015). Defining the Anthropocene. *Nature* 519: 171–180.

Li, S., Zhu, X., Li, S. et al. (2020). Trigger mechanisms of gravity flow deposits in the Lower Cretaceous lacustrine rift basin of Lingshan Island, Eastern China. *Cretaceous Research* 107: 104269.

Lin, C., Kenneth, E., Li, S. et al. (2001). Sequence architecture, depositional systems, and controls on development of lacustrine basin fills in part of the Erlian basin, northeast China. *AAPG Bulletin* 85: 2017–2043.

Liu, W., Ma, L., and Abuduwaili, J. (2020). Anthropogenic influences on environmental changes of lake Bosten, the largest inland freshwater lake in China. *Sustainability* 12: 1–14.

Liu, W., Xie, C., Zhao, L. et al. (2021). Rapid expansion of lakes in the endorheic basin on the Qinghai-Tibet Plateau since 2000 and its potential drivers. *Catena* 197: 104942.

Lourenço, M.C.M., Jardim De Sá, E.F., Córdoba, V.C., and Pichel, L.M. (2021). Multi-scale tectono-stratigraphic analysis of pre- and syn-rift sequences in the Rio do Peixe Basin, NE Brazil. *Marine and Petroleum Geology* 130: 105127.

Lu, Y., Bookman, R., Waldmann, N., and Marco, S. (2020). A 45 kyr laminae record from the Dead Sea: implications for basin erosion and floods recurrence. *Quaternary Science Reviews* 229: 106143.

Lu, Y., Moernaut, J., Waldmann, N. et al. (2021). Orbital- and millennial-scale changes in lake levels facilitate earthquake-triggered mass failures in the Dead Sea Basin. *Geophysical Research Letters* 48: e2021GL093391.

Lyons, R.P., Scholz, C.A., Buoniconti, M.R., and Martin, M.R. (2011). Late Quaternary stratigraphic analysis of the Lake Malawi Rift, East Africa: an integration of drill-core and seismic-reflection data. *Palaeogeography, Palaeoclimatology, Palaeoecology* 303: 20–37.

Lyons, R.P., Scholz, C.A., Cohen, A.S. et al. (2015). Continuous 1.3-million-year record of East African hydroclimate, and implications for patterns of evolution and biodiversity. *Proceedings of the National Academy of Sciences* 112: 15568–15573.

Mancuso, A.C., Krapovickas, V., Benavente, C.A., and Marsicano, C.A. (2020). An integrative physical, mineralogical and ichnological approach to characterize underfilled lake-basins. *Sedimentology* 67: 3088–3118.

Mancuso, A.C., Marsicano, C.A., and Krapovickas, V. (2022). Tetrapod-track taphonomic modes (TTTM) applied to Triassic lacustrine systems from southern South America. *Journal of African Earth Sciences* 196: 104735.

Mángano, M.G., Buatois, L.A., Wu, X. et al. (1994). Sedimentary facies, depositional processes and climatic controls in a Triassic lake, Tanzhuang Formation, western Henan Province, China. *Journal of Paleolimnology* 11: 41–65.

Marriott, S.B. and Wright, V.P. (2004). Mudrock deposition in an ancient dryland system: Moor Cliffs Formation, Lower Old Red Sandstone, southwest Wales, UK. *Geological Journal* 39: 277–298.

Martens, K., Schon, I., Meisch, C., and Horne, D.J. (2008). Global diversity of ostracods (Ostracoda, Crustacea) in freshwater. *Hydrobiologia* 595: 185–193.

Martin, P., Boes, X., Goddeeris, B., and Fagel, N. (2005). A qualitative assessment of the influence of bioturbation in Lake Baikal sediments. *Global and Planetary Change* 46: 87–99.

Martinsen, O.J., Sømme, T.O., and Groth, A. (2017). Development of predictive stratigraphy: sequences, source-to-sink, and back to seismic. In: *Sequence Stratigraphy: The Future Defined*, vol. 36 (ed. B. Hart, N.C. Rosen, D. West, et al.), 7–8. SEPM Gulf Coast Section Publications.

McCormack, J., Nehrke, G., Jöns, N. et al. (2019). Refining the interpretation of lacustrine carbonate isotope records: implications of a mineralogy-specific Lake Van case study. *Chemical Geology* 513: 167–183.

McPherson, J.G., Shanmugam, G., and Mouola, R.J. (1987). Fan-deltas and braid deltas: varieties of coarse-grained deltas. *Geological Society of America Bulletin* 99: 331–340.

Melchor, R.N. (2004). Trace fossil distribution in lacustrine deltas: examples from the Triassic rift lakes of the Ischigualasto-Villa Unión basin, Argentina. In: *The Application of Ichnology to Palaeoenvironmental and Stratigraphic Analysis*, Geological Society, London, Special Publications, vol. 228 (ed. D. McIlroy), 335–354. Geological Society London.

Metz, R. (1996). Newark basin ichnology: the late Triassic Perkasie member of the Passaic formation, Sanatoga, Pennsylvania. *Northeastern Geology and Environmental Sciences* 18: 118–129.

Metz, R. (2000). Triassic trace fossils from lacustrine shoreline deposits of the Passaic Formation, Douglassville, Pennsylvania. *Ichnos* 7: 243–318.

Miller, M.F. and Labandeira, C.C. (2002). Slow crawl across the salinity divide: delayed colonization of freshwater ecosystems by invertebrates. *Geological Society of America Today* 12: 4–10.

Min, W., Deng, H., Xia, S., and Ye, D. (2020). The transition from meandering river delta to braided river delta: depositional characteristics and controlling factors from the second member of Shanxi Formation to the eighth member of Lower Shihezi Formation in Southeastern Ordos Basin. *Geological Journal* 55: 8122–8140.

Minter, N.J., Krainer, K., and Lucas, S.G. (2007). Paleoecology of an Early Permian playa lake trace fossil assemblage from Castle Peak, Texas, USA. *Palaeogeography, Paleoclimatology, Palaeoecology* 246: 390–423.

Minter, N.J., Buatois, L.A., Mángano, M.G. et al. (2016a). The prelude to terrestrial invasion. In: *The Trace-Fossil Record of Major Evolutionary Changes vol. 1: Precambrian and Paleozoic. Topics in Geobiology*, vol. 39 (ed. M.G. Mángano and L.A. Buatois), 157–204. Springer.

Minter, N.J., Buatois, L.A., Mángano, M.G. et al. (2016b). The establishment of continental ecosystems. In: *The Trace-Fossil Record of Major Evolutionary Changes, vol. 1: Precambrian and Paleozoic. Topics in Geobiology*, vol. 39 (ed. M.G. Mángano and L.A. Buatois), 205–324. Springer.

Minter, N.J., Buatois, L.A., Mángano, M.G. et al. (2017). Early bursts of diversification defined the faunal colonization of land. *Nature Ecology & Evolution* 1: 0175.

Mitchum R.M.Jr., and Vail, P.R., 1977. Seismic stratigraphy and global changes of sea-level, part 7: stratigraphic interpretation of seismic reflection patterns in depositional sequences. In: Payton, C.W. (Eds), *Seismic Stratigraphy - Applications to Hydrocarbon Exploration*: American Association of Petroleum Geologists American Association of Petroleum Geologists Memoir 26, pp. 135–144.

Morgan, C.D. and Bereskin, S.R. (2003). Characterization of petroleum reservoirs in the Eocene Green River Formation, Central Uinta Basin, Utah. *Mountain Geologist* 40: 111–127.

Morinaga, H., Itota, C., Isezaki, N. et al. (2013). Oxygen-18 and carbon-13 records for the last 14,000 years from Lacustrine carbonates of Siling-Co (Lake) in the Qinghai-Tibetan Plateau. *Geophysical Research Letters* 20: 2909–2912.

Moustafa, A.R. and Khalil, S.M. (2017). Control of extensional transfer zones on syntectonic and posttectonic sedimentation: implications for hydrocarbon exploration. *Journal of the Geological Society* 174: 318–335.

Muiruri, V.M., Owen, R.B., Lowenstein, T.K. et al. (2021). A million-year vegetation history and palaeoenvironmental record from the Lake Magadi Basin, Kenya Rift Valley. *Palaeogeography, Palaeoclimatology, Palaeoecology* 567: 110247.

Mulder, T. and Alexander, J. (2001). The physical character of subaqueous sedimentary density flows and their deposits. *Sedimentology* 48: 269–299.

Mulder, T., Syvitski, J.P.M., Migeon, S. et al. (2003). Marine hyperpycnal flows: initiation, behavior and related deposits. A review. *Marine and Petroleum Geology* 20: 861–882.

Müller, B., Meyer, J.S., and Gächter, R. (2016). Alkalinity regulation in calcium carbonate-buffered lakes. *Limnology and Oceanography* 61: 341–352.

Muniz, M.C. and Bosence, D.W.J. (2015). Pre-salt microbialites from the Campos Basin (offshore Brazil): image log facies, facies model and cyclicity in lacustrine carbonates. In: *Microbial Carbonates in Space and Time: Implications for Global Exploration and Production*, Geological Society, London, Special Publications, vol. 418 (ed. D.W.J. Bosence, K.A. Gibbons, D.P. Le Heron, et al.), 221–242. Geological Society London.

Muto, T., Steel, R.J., and Swenson, J.B. (2007). Autostratigraphy: a framework norm for genetic stratigraphy. *Journal of Sedimentary Research* 77: 2–12.

Naeher, S., Gilli, A., North, R.P. et al. (2013). Tracing bottom water oxygenation with sedimentary Mn/Fe ratios in Lake Zurich, Switzerland. *Chemical Geology* 352: 125–133.

Nelson, C.H., Karabanov, E.B., Colman, S.M., and Escutia, C. (1999). Tectonic and sediment supply control of deep rift lake turbidite systems: Lake Baikal, Russia. *Geology* 27: 163–166.

Nemec, W. and Steel, R.J. (1988). What is a fan delta and how do we recognize it? In: *Fan Deltas: Sedimentology and Tectonic Settings* (ed. W. Nemec and R.J. Steel), 3–13. Blackie and Son.

Netto, R.G., Benner, J., Buatois, L.A. et al. (2012). Glacial environments. In: *Trace Fossils as Indicators of Sedimentary Environments*, Developments in Sedimentology, vol. 64 (ed. D. Knaust and R.G. Bromley), 299–327. Amsterdam: Elsevier.

Olariu, C. and Bhattacharya, J.P. (2006). Terminal distributary channels and delta front architecture of fluvial-dominated delta systems. *Journal of Sedimentary Research* 76: 212–233.

Olariu, C., Zhou, C., Steel, R. et al. (2021). Controls on the stratal architecture of lacustrine delta successions in low-accommodation conditions. *Sedimentology* 68: 1941–1963.

Olariu, C., Zhang, Z., Zhou, C. et al. (2022). Conglomerate to mudstone lacustrine cycles revealed in Junggar Basin, Northwest China: Middle Permian Lucaogou and Jingjingzigou formations. *Marine and Petroleum Geology* 136: 105473.

Otvos, E.G. (2020). Beach ridges: definitions and significance. *Geomorphology* 32: 83–108.

Owen, R.B., Muiruri, V.M., Lowenstein, T.K. et al. (2018). Progressive aridification in East Africa over the last half million years and implications for human evolution. *Proceedings of the National Academy of Sciences* 115: 11174–11179.

Owen, R.B., Renaut, R.W., Muiruri, V.M. et al. (2019). Quaternary history of the Lake Magadi Basin, southern Kenya Rift: tectonic and climatic controls. *Palaeogeography, Palaeoclimatology, Palaeoecology* 518: 97–118.

Pan, S., Liu, H., Zavala, C. et al. (2017). Sub-lacustrine hyperpycnal channel-fan system in a large depression basin: a case study of Nen 1 Member, Cretaceous Nenjiang Formation in the Songliao Basin, NE China. *Petroleum Exploration and Development* 44: 911–922.

Pan, S., Liu, C., Li, X. et al. (2019). Giant sub-lacustrine landslide in the Cretaceous Songliao Basin, NE China. *Basin Research* 31: 1066–1082.

Patterson, M. (2014). Geomicrobial investigation of thrombolites in Green Lake, New York and Highborne Cay, Bahamas. Master thesis. University of Connecticut, pp. 54–75.

Pederson, C.L., Klaus, J.S., Swart, P.K., and McNeill, D.F. (2018). Deposition and early diagenesis of microbial mud in the Florida Everglades. *Sedimentology* 66: 1989–2010.

Petryshyn, V.A., Rivera, M.J., Agić, H. et al. (2016). Stromatolites in Walker Lake (Nevada, Great Basin, USA) record climate and lake-level changes ~35,000 years ago. *Palaeogeography, Palaeoclimatology, Palaeoecology* 451: 140–151.

Posamentier, H.W. and Vail, P.R. (1988). Eustatic controls on clastic deposition II: sequence and systems tracts models. In: *Sea-Level Changes: An Integrated Approach*, SEPM Special Publication, vol. 42 (ed. C.K. Wilgus, B.S. Hastings, H. Posamentier, et al.), 125–154. Society of Economic Paleontologists and Mineralogists.

Potts, R., Dommain, R., Moerman, J.W. et al. (2020). Increased ecological resource variability during a critical transition in hominin evolution. *Science Advances* 6 (43): eabc8975.

Price, S.J., Ford, J.R., Cooper, A.H., and Neal, C. (2011). Humans as major geological and geomorphological agents in the Anthropocene: The significance of artificial ground in Great Britain. *Philosophical Transactions of the Royal Society A: Mathematical, Physical and Engineering Sciences* 369: 1056–1084.

Rapozo, B.F., Córdoba, V.C., and Antunes, A.F. (2021). Tectono-stratigraphic evolution of a cretaceous intracontinental rift: example from Rio do Peixe Basin, north-eastern Brazil. *Marine and Petroleum Geology* 126: 104899.

Renaut, R.W. and Gierlowski-Kordesch, E.H. (2010). Lakes. In: *Facies Models 4* (ed. N.P. James and Dalrymple), 541–575. Newfoundland, Canada: Geological Association of Canada.

Rodriguez-Aranda, J.P. and Calvo, J.P. (1998). Trace fossils and rhizoliths as a tool for sedimentological and palaeoenvironmental analysis of ancient continental evaporite successions. *Palaeogeography, Palaeoclimatology, Palaeoecology* 140: 383–399.

Rogers, R.D. and Schumm, S.A. (1991). The effect of sparse vegetative cover on erosion and sediment yield. *Journal of Hydrology* 123: 19–24.

Romans, B.W., Castelltort, S., Covault, J.A. et al. (2016). Environmental signal propagation in sedimentary systems across timescales. *Earth-Science Reviews* 153: 7–29.

Rosen, M.R., Finkelstein, D.B., Boush, L. P., and Pla-Pueyo, S. (2021). Introduction to limnogeology: progress, challenges, and opportunities: a tribute to Elizabeth Gierlowski-Kordesch. In: *Limnogeology: Progress, Challenges and Lasticpportunities: A Tribute to Elizabeth Gierlowski-Kordesch* (ed. M.R. Rosen, D.B. Finkelstein, L.P. Boush, and S. Pla-Pueyo), 3–16. Switzerland: Springer Nature.

Rosenberg, M.J., Birgenheier, L.P., and Vanden Berg, M.D. (2015). Facies, stratigraphic architecture, and lake evolution of the Oil Shale bearing Green River Formation, Eastern Uinta Basin, Utah. In: *Stratigraphy and Paleolimnology of the Green River Formation, Western USA* (ed. M.E. Smith and A.R. Carroll), 211–249. London: Springer.

Ruddiman, W.F. (2013). The Anthropocene. *Annual Review of Earth and Planetary Sciences* 41: 45–68.

Sanz-Montero, M.E., Rodríguez-Aranda, J. P., and del Cura, M.A.G. (2009). Bioinduced precipitation of barite and celestite in dolomite microbialites: examples from Miocene lacustrine sequences in the Madrid and Duero Basins, Spain. *Sedimentary Geology* 222: 138–148.

Sanz-Montero, M. E., Calvo J.P., Angeles García del Cura M., Ornosa C., Outerelo R., Pablo Rodríguez-Aranda J., 2013. The rise of the diptera-microbial mat interactions during the Cenozoic: consequences for the sedimentary record of saline lakes. *Terra Nova*, v. 25, p. 465–471.

Sarg, J.F., Tanavsuu-Milkeviciene, K., Humphrey, J.D., and Suriamin, H. (2013). Lithofacies, stable isotopic composition, and stratigraphic evolution of microbial and associated carbonates, Green River Formation (Eocene), Piceance Basin, Colorado. *AAPG Bulletin* 97: 1937–1966.

Schnellmann, M., Anselmetti, F.S., Giardini, D. et al. (2002). Prehistoric earthquake history revealed by lacustrine slump deposits. *Geology* 30: 1131–1134.

Schnellmann, M., Anselmetti, F.S., Giardini, D., and Mckenzie, J.A. (2005). Mass movement-induced fold-and-thrust belt structures in unconsolidated sediments in Lake Lucerne (Switzerland). *Sedimentology* 52: 271–289.

Schwartz, R.K. (2012). Bedform, texture, and longshore bar development in response to combined storm wave and current dynamics in a nearshore helical flow system. *Journal of Coastal Research* 28: 1512–1535.

Scott, J.J. and Smith, M.E. (2015). Trace Fossils of the Eocene Green River Lake Basins, Wyoming, Utah, and Colorado. In: *Stratigraphy and Paleolimnology of the Green River Formation, Western USA*, Syntheses in Limnogeology, vol. 1 (ed. M.E. Smith and A.R. Carroll), 313–350. Springer.

Scott, J.J., Renaut, R.W., Owen, R.B., and Sarjeant, W.A.S. (2007). Biogenic activity, trace formation, and trace taphonomy in the marginal sediments of saline, alkaline Lake Bogoria, Kenya Rift Valley. In: *Sediment-Organism Interactions: A Multifaceted Ichnology*, Society for Sedimentary Geology Special Publication, vol. 88 (ed. R.G. Bromley, L.A. Buatois, M.G. Mángano, et al.), 309–330. Society for Sedimentary Geology.

Scott, J.J., Buatois, L.A., and Mángano, M. G. (2012). Lakes. In: *Trace Fossils as Indicators of Sedimentary Environments*, Developments in Sedimentology, vol. 64 (ed. D. Knaust and R.G. Bromley), 379–417. Amsterdam: Elsevier.

Shan, X., Yu, X., Jin, L. et al. (2021). Bed type and flow mechanism of deep water sub-lacustrine fan fringe facies: an example from the Middle Permian Lucaogou Formation in Southern Junggar Basin of NW China. *Petroleum Science* 18: 339–361.

Sigvaldason, G.E. (1989). International Conference on Lake Nyos Disaster, Yaounde, Cameroon 16–20 March 1987: Conclusions and Recommendations. *Journal of Volcanology and Geothermal Research* 39: 97–107.

Smith, M.E. and Carroll, A.R. (2015). Introduction to the Green River Formation. In: *Stratigraphy and Paleolimnology of the Green River Formation, Western USA* (ed. M.E. Smith and A.R. Carroll), 1–12. London: Springer.

Sømme, T.O., Helland-Hansen, W., Martinsen, O.J., and Thurmond, J. (2009). Relationships between morphological and sedimentological parameters in source-to-sink systems: a basis for predicting semi-quantitative characteristics in subsurface systems. *Basin Research* 21: 361–387.

Song, F., Su, N., Yang, S. et al. (2018). Sedimentary characteristics of thick layer lacustrine beach-bars in the Cenozoic Banqiao Sag of the Bohai Bay Basin, East China. *Journal of Asian Earth Sciences* 151: 73–89.

Strother, P.K. and Wellman, C.H. (2021). The Nonesuch Formation Lagerstätte: a rare window into freshwater life one billion years ago. *Journal of the Geological Society* v: 178.

Strother, P.K., Battison, L., Brasier, M.D., and Wellman, C.H. (2011). Earth's earliest non-marine eukaryotes. *Nature* 473: 505–509.

Suriamin, H. (2010). Facies, diagenesis, and geochemistry of the Eocene Green River Formation carbonates in the Piceance Creek basin, Colorado. MSc thesis. Colorado School of Mines, Golden, CO, 108 pp.

Syvitski, J.P.M. and Milliman, J.D. (2007). Geology, geography, and humans battle for dominance over the delivery of fluvial sediment to the coastal ocean. *The Journal of Geology* 115: 1–19.

Szajna, M.J. and Hartline, B.W. (2003). A new vertebrate footprint locality from the Late Triassic Passaic Formation near Birdsboro, Pennsylvania. In: *The Great Rift Valleys of Pangea in Eastern North America* (ed. P.M. LeTourneau and P.E. Olsen), 264–272. Columbia University Press.

Sztanó, O., Szafián, P., Magyar, I. et al. (2013). Aggradation and progradation controlled clinothems and deep-water sand delivery model in the Neogene Lake Pannon, Makó Trough, Pannonian Basin, SE Hungary. *Global and Planetary Change* 103: 149–167.

Talbot, M.R. and Allen, P.A. (1996). Lakes. In: *Sedimentary Environments: Processes, Facies and Stratigraphy*, 3e (ed. H.G. Reading), 83–124. Oxford: Blackwell.

Tamura, T. (2012). Beach ridges and prograded beach deposits as palaeoenvironment records. *Earth-Science Reviews* 114: 279–297.

Tan, C. and Plink-Björklund, P. (2021). Morphodynamics of supercritical flow in a linked river and delta system, Daihai Lake, Northern China. *Sedimentology* 68: 1606–1639.

Tänavsuu-Milkeviciene, K. and Sarg, J.F. (2012). Evolution of an organic-matter-rich Lake Basin – Stratigraphy, Climate and Tectonics: Piceance Creek Basin, Eocene Green River Formation. *Sedimentology* 59: 1735–1768.

Tänavsuu-Milkeviciene, K., Sarg, J.F., and Bartov, Y. (2017). Depositional cycles and sequences in an organic-matter-rich Lake Basin: Eocene Green River Formation, Lake Uinta, Colorado and Utah, USA. *Journal of Sedimentary Research* 87: 210–229.

Thompson, J.B., Ferris, F.G., and Smith, D.A. (1990). Geomicrobiology and sedimentology of the mixolimnion and chemocline in Fayetteville Green Lake. *New York: Palaios* 5: 52–75.

Trewin, N.H., Macdonald, D.I.M., and Thomas, C.G.C., 2002. Stratigraphy and sedimentology of the Permian of the Falkland Islands: lithostratigraphic and palaeoenvironmental links with South Africa: *Journal of the Geological Society*, v. 159, p. 5-19.

Twichell, D.C., Cross, V.A., Hanson, A.D. et al. (2005). Seismic architecture and lithofacies of turbidites in Lake Mean (Arizona and Nevada, USA), an analogue for topographically complex basins. *Journal of Sedimentary Research* 75: 134–148.

Uchman, A. and Álvaro, J.J. (2000). Non-marine invertebrate trace fossils from the Tertiary Calatayud-Teruel Basin, NE Spain. *Revista Espanola de Paleontologia* 15: 203–218.

Vandekerkhove, E., van Daele, M., Praet, N. et al. (2020). Flood-triggered versus earthquake-triggered turbidites: A sedimentological study in clastic lake sediments (Eklutna Lake, Alaska). *Sedimentology* 67: 364–389.

Voigt, S. and Hoppe, D. (2010). Mass occurrence of penetrative trace fossils in Triassic lake deposits (Kyrgyzstan, Central Asia). *Ichnos* 17: 1–11.

Wæhler, T.A. and Dietrichs, E.S. (2017). The vanishing Aral Sea: health consequences of an environmental disaster. *Tidsskrift for den Norske Lægeforening* 137 (18): https://doi.org/10.4045/tidsskr.17.0597.

Walter, M.R. and Allwood, A.C. (2005). Biosediments and biofilms. In: *Encyclopedia of Geology* (ed. R.C. Selley, L.R.M. Cocks, and L.R. Plimer), 279–294. Amsterdam: Elsevier.

Wang, C., Feng, Z., Zhang, L. et al. (2013). Cretaceous paleogeography and paleoclimate and the setting of SKI borehole sites in Songliao Basin, northeast China. *Palaeogeography, Palaeoclimatology, Palaeoecology* 385: 17–30.

Wang, Y., Cao, Y., Ma, B. et al. (2014). Mechanism of diagenetic trap formation in nearshore sub-lacustrine fans on steep rift lacustrine basin slopes. *Petroleum Science* 11: 481–494.

White, D.S. and Miller, M.F. (2008). Benthic invertebrate activity in lakes: linking present and historical bioturbation patterns. *Aquatic Biology* 2: 269–277.

White, A.H. and Youngs, B.C. (1980). Cambrian alkali playa-lacustrine sequence in the northeastern Officer Basin, South Australia. *Journal of Sedimentary Research* 50: 1279–1286.

Wilkinson, B.H. (2005). Humans as geologic agents: a deep-time perspective. *Geology* 33: 161–164.

Williams, D.F., Peck, J., Karabanov, E.B. et al. (1997). Lake Baikal record of continental climate response to orbital insolation during the past 5 million years. *Science* 278: 1114–1117.

Wilmeth, D.T., Corsetti, F.A., Beukes, N.J. et al. (2019). Neoarchean (2.7 Ga) lacustrine stromatolite deposits in the Hartbeesfontein Basin, Ventersdorp Supergroup, South Africa: implications for oxygen oases. *Precambrian Research* 320: 291–302.

Wright, L.D. (1977). Sediment transport and deposition at river mouths: a synthesis. *Geological Society of America Bulletin* 88: 857–868.

Wright, R.E. and Nydegger, P. (1980). Sedimentation of detrital particulate matter in Jakes: influence of currents produced by inflowing rivers. *Water Resources Research* 16: 597–601.

Wright, J.L., Quinn, L., Briggs, D.E.G., and Williams, S.H. (1995). A subaerial arthropod trackway from the Upper Silurian Clam Bank Formation of Newfoundland. *Canadian Journal of Earth Sciences* 32: 304–313.

Xian, B., Wang, J., Gong, C. et al. (2018). Classification and sedimentary characteristics of lacustrine hyperpycnal channels: Triassic outcrops in the south Ordos Basin, central China. *Sedimentary Geology* 368: 687–682.

Xu, C. (2013). Controlling sand principle of source-to-sink coupling in time and space in continental rift basins: basic idea, conceptual systems and controlling sand models. *China Offshore Oil and Gas* 25: 1–11. (in Chinese with English abstract).

Xu, C., Du, X., Xu, W., and Zhao, M. (2017a). Source-to-sink system and its sedimentary records in the continental rift basins: an example from the Paleogene in the Bohai Sea Area, China. *Interpretation* 5: ST35–ST51.

Xu, C., Du, X., Zhu, H. et al. (2017b). New advances of the "Source-to-Sink" system research in sedimentary basin. *Oil and Gas Geology* 38: 1–11. (in Chinese with English abstract).

Xu, C., Du, X., and Zhu, H. (2020). *The Role of Source-to-Sink Systems in Determining the Occurrence and Development of Lacustrine Sandstone Reservoirs in Continental Rift Basins: Theory and Applications*, 55–208. Beijing: Science Press (in Chinese).

Yu, X., Li, S., and Li, S. (2018). *Clastic Hydrocarbon Reservoir Sedimentology, Advances in Oil and Gas Exploration & Production*, 417–459. Springer.

Zavala, C. (2020). Hyperpycnal (over density) flows and deposits. *Journal of Palaeogeography* 9: 267–287.

Zhang, C., Wei, W., Zhang, S. et al. (2016). Architecture of lacustrine mass-transport complexes in the Mesozoic Songliao Basin, China. *Marine and Petroleum Geology* 78: 826–835.

Zhang, J., Olariu, C., Steel, R., and Kim, W. (2019b). Climatically controlled lacustrine clinoforms: theory and modelling results. *Basin Research* 32: 240–250.

Zhang, X., Zhu, X., Lue, Z. et al. (2019a). An early Eocene subaqueous fan system in the steep slope of lacustrine rift basins, Dongying Depression, Bohai Bay Basin, China: depositional character, evolution and geomorphology. *Journal of Asian Earth Sciences* 171: 28–45.

Zhao, X.Z., Li, Q., Jiang, Z.X. et al. (2014). Organic geochemistry and reservoir characterization of the organic matter-rich calcilutite in the Shulu Sag, Bohai Bay Basin, North China. *Marine and Petroleum Geology* 51: 239–255.

Zhao, X., Zheng, D., Xie, G. et al. (2020). Recovery of lacustrine ecosystems after the end-Permian mass extinction. *Geology* 48: 609–613.

Zhu, X. (2008). *Sedimentary Petrology*, 4e, 276–290. Beijing: Petroleum Industry Press (in Chinese).

Zhu, X., Li, S., Liu, Q. et al. (2017). Source to sink studies between the Shaleitian Uplift and surrounding sags: perspectives on the importance of hinterland relief and catchment area for sediment budget, western Bohai Bay Basin, China. *Interpretation* 5: ST65–ST84.

7

Aeolian Environments

Nigel P. Mountney[1], Grace I. E. Cosgrove[1], and Luca Colombera[2]

[1] School of Earth and Environment, University of Leeds, Leeds, UK
[2] Dipartimento di Scienze della Terra e dell'Ambiente, Università di Pavia, Pavia, Italy

Summary

This chapter describes the main types of aeolian sedimentary systems that are active today, including in the world's major hot- and cold-climate deserts and in both arid and humid coastal settings. It considers the main factors that influence where major aeolian sediment systems are developed on Earth today, and it discusses how aeolian sedimentary systems interact with neighbouring, coevally active non-aeolian environments, such as dryland rivers, saline lakes and marine shorelines. This chapter examines factors that have governed the construction, accumulation and preservation of aeolian sedimentary systems throughout Earth history. It considers how ancient aeolian environments are recognised and classified.

This chapter explains processes of aeolian erosion, transport and deposition. Sand deposits of aeolian origin are commonly mature in terms of both their composition and the textural properties of the sediment grains. Yet, many exceptions to this trend are documented. Aeolian transport leads to the development of ripples, dunes and larger-scale megadunes. Bedforms of distinctly differing scales may occur superimposed. Aeolian interdune flats and hollows developed between dunes may be dry, damp or wet, depending on their interaction with the water table or whether they act as pathways for fluvial flood events. In places, extensive aeolian sandsheets devoid of dune-scale bedforms may develop.

This chapter describes aeolian and related lithofacies in terms of formative processes. It explains how units of these facies occur in associations that indicate a common genetic origin. In many cases, facies associations occur as sedimentary bodies with distinctive geometric forms: these are architectural elements, which are themselves delineated by bounding surfaces. The study of these architectural elements and their bounding surfaces in ancient successions yields insight into the evolution of formative aeolian landforms and the mechanisms that controlled their construction, accumulation and preservation. In ancient aeolian successions, a variety of sedimentary structures can be used to reconstruct sediment transport direction, and for certain types of aeolian dunes, this may be related to palaeowind direction. The chapter demonstrates how facies models are used to reconstruct aeolian palaeoenvironments from evidence in the ancient rock record. Models for dry, wet, stabilised and mixed aeolian–fluvial aeolian systems are each presented.

Autogenic and allogenic controls that govern aeolian systems are discussed in the context of a sequence stratigraphic framework. Many ancient aeolian successions are arranged into sequences, each bounded by a so-called supersurface that represents a sequence boundary. Internally, aeolian sequences may be divided into systems tracts, for example, that record wetting-upward or drying-upward climatic trends. Static facies models record the operation of temporally and spatially invariable sets of controls on sedimentation; these models can be invoked to explain and predict relatively simple stratigraphic architectures. Dynamic facies models account for temporally and/or spatially variable controls on sedimentation; these models explain and predict the complex aeolian stratigraphic architectures that are common in many preserved aeolian successions.

From an applied standpoint, relatively porous and permeable aeolian successions may serve as important aquifers that might be exploited for drinking water, irrigation of farmland or the generation of geothermal energy. Aeolian successions may alternatively serve as reservoirs for the exploitation of hydrocarbons, repositories for carbon capture and underground storage, freshwater aquifers, geothermal reservoirs, or potential sites for safe hydrogen storage.

Reading's Sedimentary Environments: Processes, Facies and Stratigraphy, Fourth Edition. Edited by Bruce Levell.

Companion website: www.wiley.com/go/levell4e

7.1 Introduction

7.1.1 What Are Aeolian Environments?

The adjective 'aeolian' ('eolian' in US English) describes processes and products of erosion, entrainment, transportation and deposition by the wind. Aeolian processes operate in a variety of erosional and depositional settings, notably hot and cold climatic deserts, sandy coastlines, fluvial and glacial outwash plains, and volcanic regions. These processes also supply sediment to the deep ocean. Although aeolian processes operate under the influence of various climate regimes, they are currently most active and widespread in arid hot and cold deserts (Figure 7.1). In these settings, dry sandy substrates that lack significant vegetation cover are prone to sediment erosion, transport and deposition via the wind. The great deserts of the Sahara in North Africa and the Rub' al Khali in Arabia cover 9.2×10^6 and 6.5×10^5 km^2, respectively.

Aeolian environments include a suite of landforms and depositional settings: (i) dune bedforms and their larger counterparts called megadunes or draas, (ii) low-relief interdunes that form corridors or hollows between dunes, and (iii) extensive sheet-like bodies of dust and silt (loess) and sand (sandsheet). Dunes may occur singularly but more commonly occur in groups arranged into widespread dune fields called sands seas or ergs (Wilson 1973). A variety of related non-aeolian landforms may occur in close association with aeolian landforms, including alluvial fans, ephemeral, intermittent or perennial rivers, transient or permanent lakes, oases and salt lakes (chotts) in desert settings, salt flats (sabkha) and a range of coastal systems (Figure 7.2). Aeolian sediments – especially those that comprise bedforms in large dune fields – are commonly composed dominantly of quartz sand grains, which are relatively resilient to the attrition caused by inter-particle collisions during aeolian transport. Aeolian deposits composed of lithic fragments (e.g. basalt), carbonate and gypsum sand grains and even pumice are known. Active aeolian systems are also present on other planetary bodies in our solar system: Mars (e.g. Greeley et al. 2018), Venus (e.g. Kreslavsly and Bondarenko 2017) and Titan (e.g. Lucas et al. 2019). Ancient preserved aeolian successions are documented on Mars (e.g. Banham et al. 2018).

This chapter explains processes of aeolian erosion, transport and deposition. It describes aeolian and related lithofacies and explains how units of these facies occur within larger-scale architectural elements, which themselves

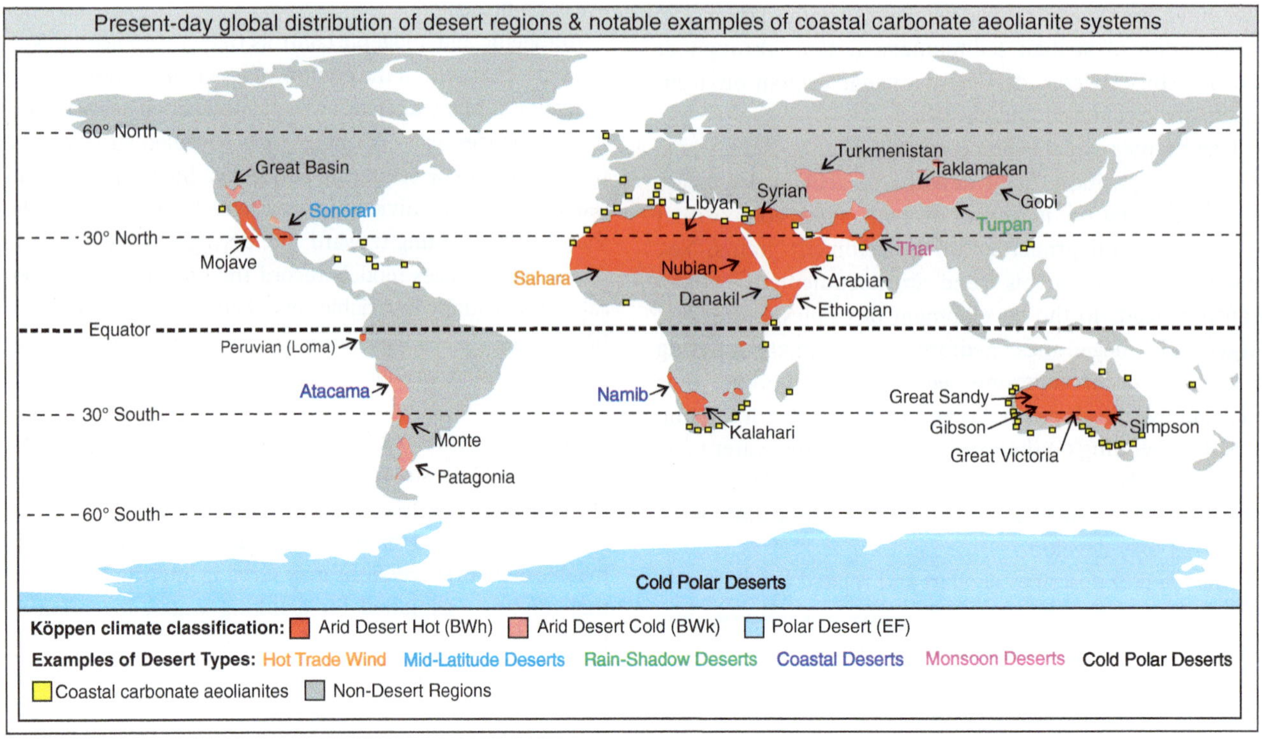

Figure 7.1 Global distribution of arid lands. Köppen climate classification of hot deserts, cold deserts and polar deserts. Notable examples of modern deserts are named and coloured according to desert type: hot trade wind deserts, mid-latitude deserts, rainshadow deserts, coastal deserts, monsoon deserts and cold polar deserts. Locations of notable examples of coastal carbonate systems and aeolianites are also indicated.

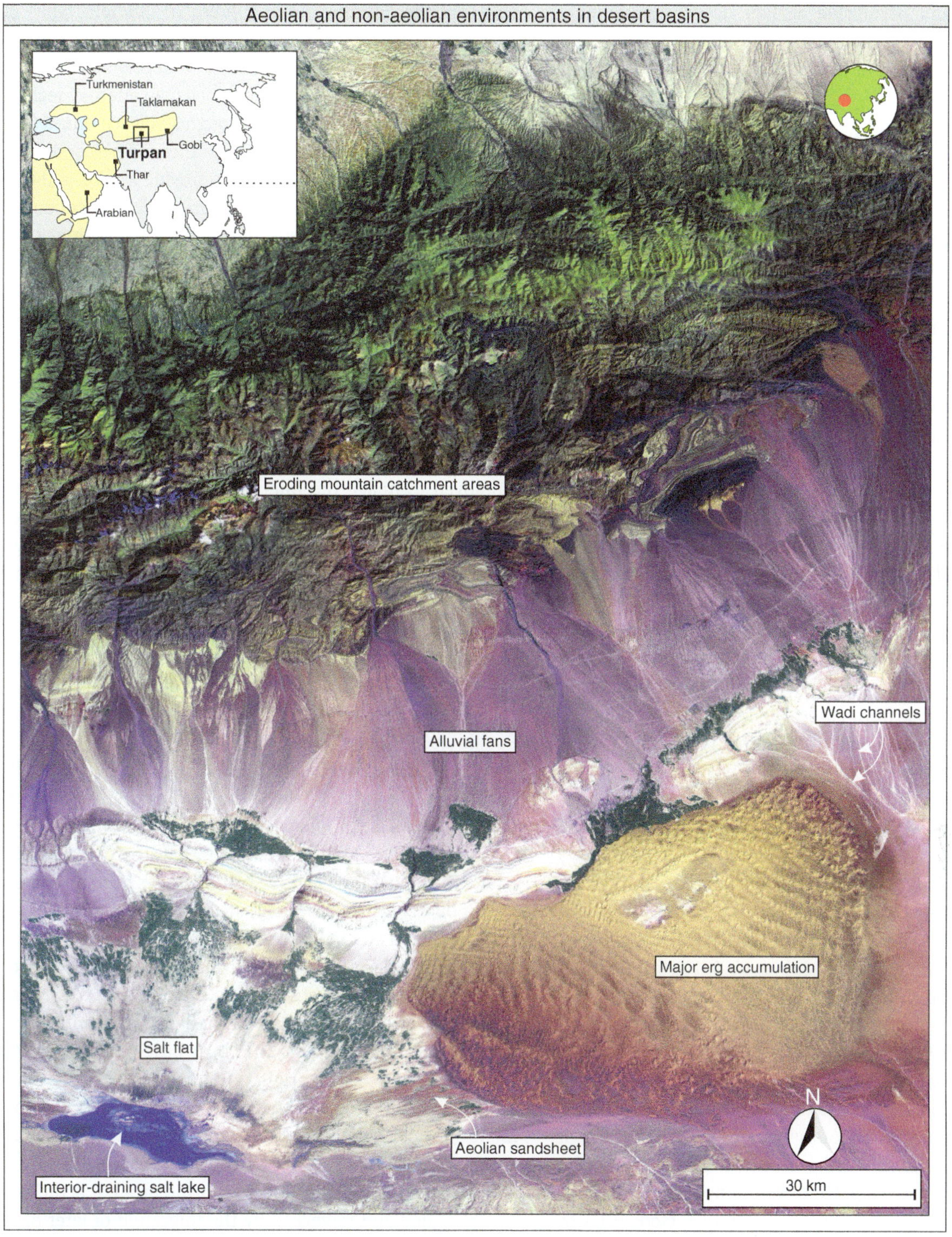

Figure 7.2 Examples of aeolian and neighbouring non-aeolian sedimentary environments in a desert basin. Satellite image of the Turpan Depression, Bogda, northern China. This arid system hosts a variety of eroding and depositing environments, including eroding mountain catchment areas, alluvial fans, wadi channel systems, an interior draining salt lake, salt flats (inland sabkha), aeolian sandsheets and a major aeolian dune field (or erg). Courtesy of NASA Earth Observatory collection.

record the construction, accumulation and preservation of deposits and the evolution of formative landforms. Autogenic and allogenic controls that govern aeolian systems are discussed in the context of a sequence stratigraphic framework. The chapter demonstrates how facies models and sequence stratigraphic models for aeolian sedimentary systems are used to reconstruct palaeoenvironments from evidence in the ancient rock record. The chapter considers how porous and permeable aeolian deposits may serve as important aquifers, reservoirs for hydrocarbons, and repositories for CO_2 and hydrogen.

7.1.2 Approaches to the Study of Aeolian Sedimentary Systems

Aeolian sedimentary systems and their accumulated deposits are studied either directly or indirectly (e.g. by remote sensing) via analyses in different natural settings: (i) modern environments in which aeolian processes are operating to generate a suite of aeolian landforms; (ii) recent and Quaternary landscapes in which relict landforms that are no longer fully active are present; (iii) ancient sedimentary successions exposed as outcrops; (iv) ancient sedimentary successions present in the subsurface and known from cored intervals and well-log records recovered from drilled wells, else imaged using seismic reflection or other geophysical methods; (v) extra-terrestrial environments (both modern and ancient [Section 15.2.3]). Results from physical and numerical modelling experiments provide additional insight.

Methodologies for the study of modern and recent aeolian sedimentary systems include: (i) direct observation and sampling, (ii) shallow coring, (iii) low-level aerial photography, (iv) remote sensing from space satellites, (v) studies of the internal structure of aeolian landforms (especially dunes) by examining natural exposures, else via trenching or by using Ground Penetrating Radar (GPR), (vi) the measurement of wind regimes and their relation to landform evolution, and (vii) process-based and geometric-based numerical modelling. The availability of high-resolution satellite and aerial-photograph imagery (e.g. McKee 1979), and more recently via resources such as Google Earth (e.g. Butler 2006; Yu and Gong 2012; Fisher et al. 2012), has enabled detailed regional-scale studies of geomorphological relationships within aeolian systems (e.g. Tooth 2007; Bullard et al. 2011; Al-Masrahy and Mountney 2013a). The global coverage of such data means that comprehensive and systematic analyses can now be undertaken.

The evolution and migratory patterns of dunes through time can be analysed using repeated surveys. Notably, time-series data are available from multiple generations of aerial photography or satellite imagery (e.g. Google Earth Engine), and such datasets reveal temporal changes (e.g. Mountney and Russell 2006; Ewing and Kocurek 2010). However, because large-scale aeolian bedforms undergo relatively slow rates of growth and migration, their temporal evolution and migratory behaviour is difficult to study directly. One solution to this has been the development of computer models that simulate temporal behaviour to predict how different bedform types and patterns of migration will generate, accumulate and preserve different types of cross-bedded sets (Rubin 1987a; Rubin and Carter 2006).

Analyses of the internal structure of deposits underlying modern aeolian bedforms have involved a variety of techniques. Trenches have been dug through stabilised dunes (e.g. McKee 1966) in orientations both parallel and perpendicular to the prevailing wind direction to provide a pseudo-three-dimensional (3D) view of sedimentary structures produced by aeolian bedforms. Where the upper parts of dunes are planed to a horizontal surface (in some cases by natural deflation), they may reveal their internal structure in plan view. Wind reversals associated with major storm events can expose the internal cores of bedforms in blowouts that leave exposed sections. Where river systems flow into dune fields, flash floods may undercut dune flanks causing them to collapse and leave exposed sections.

Geophysical techniques such as GPR have been widely employed to image the internal structure of modern and recently active dunes (e.g. Fu et al. 2019; Bristow et al. 2000). This method has the potential to provide a 3D view; the subsurface depth of imaging is typically 10–20 m and there is a trade-off between resolution and penetration depth. Recent and Quaternary deposits are dated using a variety of techniques, including radiometric dating methods, notably C^{14} (e.g. Stokes and Swinehart 1997) and optically stimulated luminescence (OSL) (e.g. Bristow and Pucillo 2006; Lomax et al. 2011).

In the ancient record, there are numerous well-documented examples where the preservation of aeolian bedform topography allows examination of both the original bedform morphology and its internal structure, notably where the topography of dune bedforms has been preserved in situ by a catastrophic event, such as rapid marine flooding (e.g. Benan and Kocurek 2000; Strömbäck et al. 2005) or inundation by lava flows (e.g. Jerram et al. 2000a,b; Michel et al. 2025 Mountney et al. 1999a,b; Scherer 2002). More commonly, ancient preserved aeolian dune and interdune deposits form sets of strata that bear little resemblance to the morphology of the original landforms responsible for their generation. These sets most commonly record the migration and climb of dunes and interdunes over one another whereby only the lowermost parts of the bedforms are preserved; the upper portions of the original bedforms are truncated and reworked by the migration of later bedforms. The study of ancient successions requires careful observation of stratal relationships to establish how aeolian dune, interdune, sandsheet and other related non-aeolian sub-environments interacted to generate an accumulated succession. This commonly involves the measurement of one-dimensional (1D) graphic sedimentary logs and, where exposure allows, the construction of two-dimensional (2D) stratigraphic panels (Box 7.1). Panels can be arranged in a

Box 7.1 Commonly employed techniques in the study of ancient preserved aeolian successions

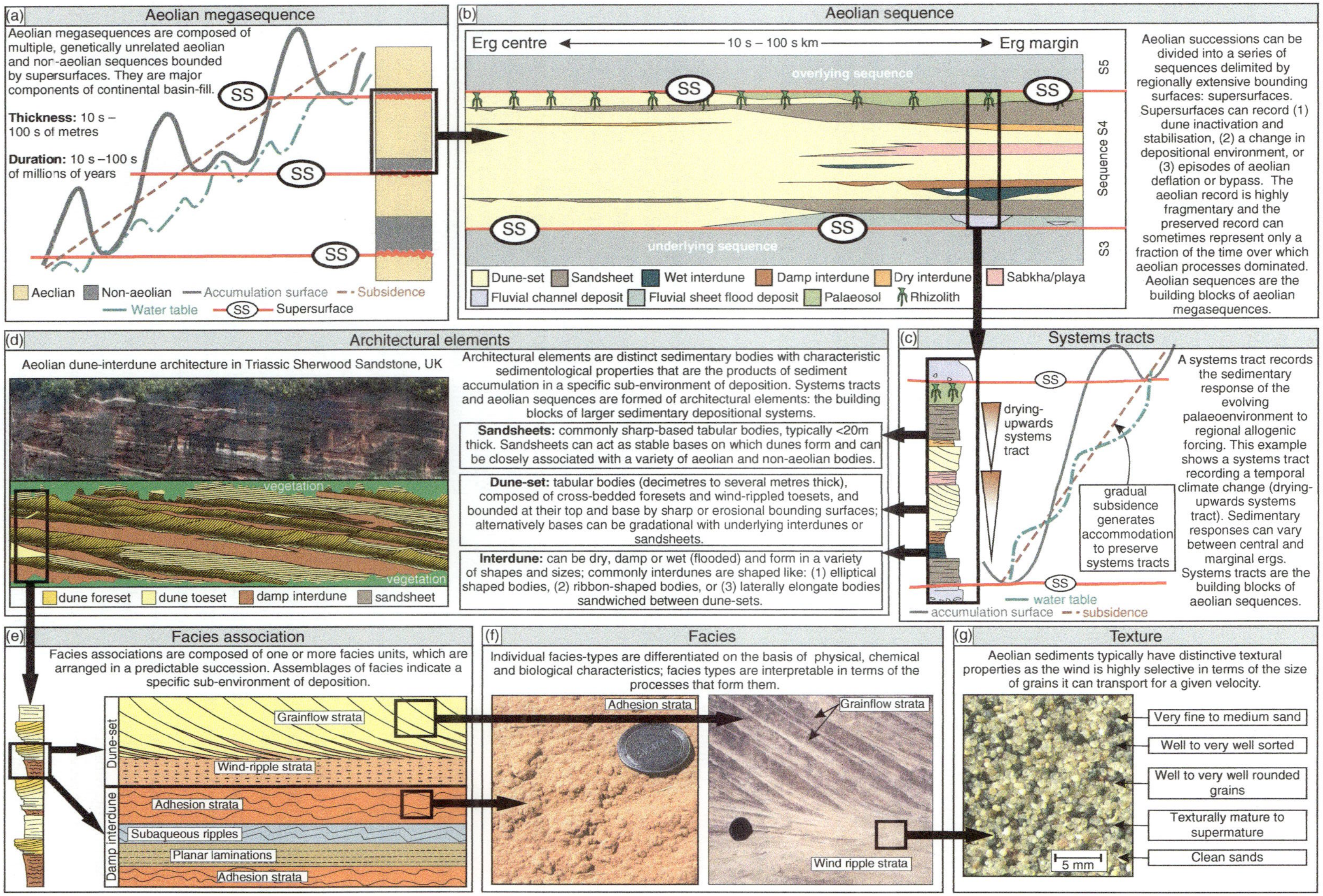

Box 7.1 Commonly employed techniques in the study of ancient preserved aeolian successions. Scales of aeolian observation from largest to smallest. (a) Aeolian megaseuence depicted on a time (x-axis) versus thickness/height (y-axis plot), with the resultant stratigraphic column at right; (b) Aeolian megaseuence depicted on a time (x-axis) versus thickness/height (y-axis plot), with the resultant stratigraphic column at right sequence showing variability between central- and marginal-erg settings. (c) Aeolian systems tract showing drying-upward depositional cycles in marginal-erg setting. (d) Aeolian architectural elements; image and panel shows dune and interdune architectural elements from the Triassic Sherwood Sandstone Group (UK). (e) Aeolian facies association showing a sedimentary log of part of the Sherwood Sandstone Group and the typical facies associations found in dune-set and damp interdune elements. (f) Aeolian facies showing adhesion strata (commonly found in damp interdune deposits) and grainflow strata (commonly found in dune foreset deposits). (g) Aeolian sand grain texture commonly found in grainflow strata.

variety of orientations to provide pseudo-3D insight into the accumulated stratigraphy (e.g. Kocurek et al. 1991b; Crabaugh and Kocurek 1993). The analysis of 3D architectural relationships determined from outcropping stratigraphy may be augmented by the collection of digital datasets using LiDAR and photogrammetry techniques (e.g. Pierce et al. 2016).

Subsurface aeolian successions are examined using numerous techniques: (i) recovering cores or core plugs from boreholes; (ii) geophysical methods with data recorded and logs, notably gamma-ray logs, density-neutron logs, sonic logs and resistivity logs; (iii) borehole imaging methods, such as formation micro-imaging (FMI), which is especially useful for analysis of the magnitude and azimuth of foreset dip in cross-bedded sets; (iv) seismic reflection methods; (v) GPR methods applied to shallow subsurface successions. Multiple combinations of these methods are routinely used for the analysis of subsurface aeolian successions (e.g. Cowan 1993; Silva and Fernandez 2004; Meadows 2006; Besly et al. 2018).

Extra-terrestrial aeolian systems are studied using mappers and imagers on orbiting spacecraft (e.g. Mars Reconnaissance Orbiter, MRO). Since 1997, Martian aeolian systems and successions have been examined using data acquired by multiple generations of ground-based rovers: NASA Opportunity, Spirit, Curiosity and Perseverance (Section 15.3.6).

See Section 7.S.1 in the Supplementary Information for a brief summary of the history of aeolian sedimentological research. See Section 7.S.2 in the Supplementary Information for a brief overview of particular challenges in the study of aeolian sedimentary systems.

7.2 Environments

7.2.1 Present-Day Aeolian Environments

7.2.1.1 Desert Aeolian Environments

Aeolian processes and resultant deposits are most widespread in the world's climatic desert regions. Approximately 30% of the present-day land surface is characterised by arid or semi-arid climatic conditions, and aeolian sand deposits cover 20% of these regions (i.e. 6% of the earth surface). In these regions, low amounts of precipitation result in a dry substrate with relatively sparse vegetation cover. This allows the wind to entrain and transport loose surface sediment. Most larger deserts are subdivided into a series of distinct geomorphic elements, defined by net sediment erosion, transport (bypass) or deposition. Geomorphic elements characterised by net aeolian sand deposition may develop as sand seas or 'ergs'. Typically, these areas are characterised by dune fields, though some are dominated by sandsheets or 'zibars' (low-relief mound-shaped bedforms). Desert sand seas range in size from a few square kilometres to 5.6×10^5 km^2 in the case of Rub' al Khali erg, Arabia (Wilson 1973).

Present-day aeolian-influenced hot trade-wind deserts, such as those of the Sahara (Figure 7.3a) and Kalahari of northern and southern Africa, the Great Sandy Desert of Australia and the Rub' al Khali of Arabia, are developed between latitudes of 15° and 30° where dry downwelling winds of the Hadley atmospheric cells circulate. Dry winds dissipate cloud cover, allowing more sunlight to heat the land, thereby limiting surface colonisation by vegetation. Mid-latitude deserts occur dominantly in continental interiors, between latitudes of 30° and 50°, poleward of the subtropical high-pressure zones and far from maritime effects of the world's oceans. Mid-latitude deserts experience a wide range of annual temperatures. Examples include the Sonoran Desert of south-western North America (Figure 7.3b) and the Gobi Desert of northern China and Mongolia. Rain-shadow deserts form in the lee (downwind 'shadow') of mountain ranges upon which moisture carried by the prevailing wind falls. For example, the Taklamakan Desert in the Turpan Depression (Figure 7.3c) is the rain-shadow desert of the Tian Shan Mountains, China. Coastal deserts develop on the western edges of continents at tropical latitudes, especially where cold ocean currents parallel the coast act to remove moisture such that drier air masses move inland. The Namib Desert (Figure 7.3d) of south-western Africa (influenced by the cold coastal Benguela current) and the Atacama Desert of Chile (influenced by the cold Humboldt current) are two examples; although these deserts are arid or hyperarid overall, fogs produced by upwelling cold currents commonly move inland, blanketing coastal areas. Fog blankets limit the solar radiation that reaches the desert surface, and as a result, coastal deserts can experience wide temperature ranges. Monsoon deserts develop in response to wind systems with pronounced seasonal reversals driven by temperature variations between continents and oceans; examples include the Rajasthan Desert of India and the Thar Desert of Pakistan (Figure 7.3e). Cold polar deserts are extensive (5×10^6 km^2); most are dominated by bedrock or gravel plains (Figure 7.3f; Wolfe 2022). Sand seas are not widespread, though extensive regions of snow dunes are developed in places. Where ground temperatures frequently fluctuate above and below zero Celsius, patterned ground can develop due to freeze–thaw action (Section 14.3.7).

Studies of spatial variations in landform type across modern desert sand seas are numerous. Spatial changes in dune type across the Great Sand Dunes of Colorado, USA, occur in a predictable sequence, over 12 km, from a zone of small,

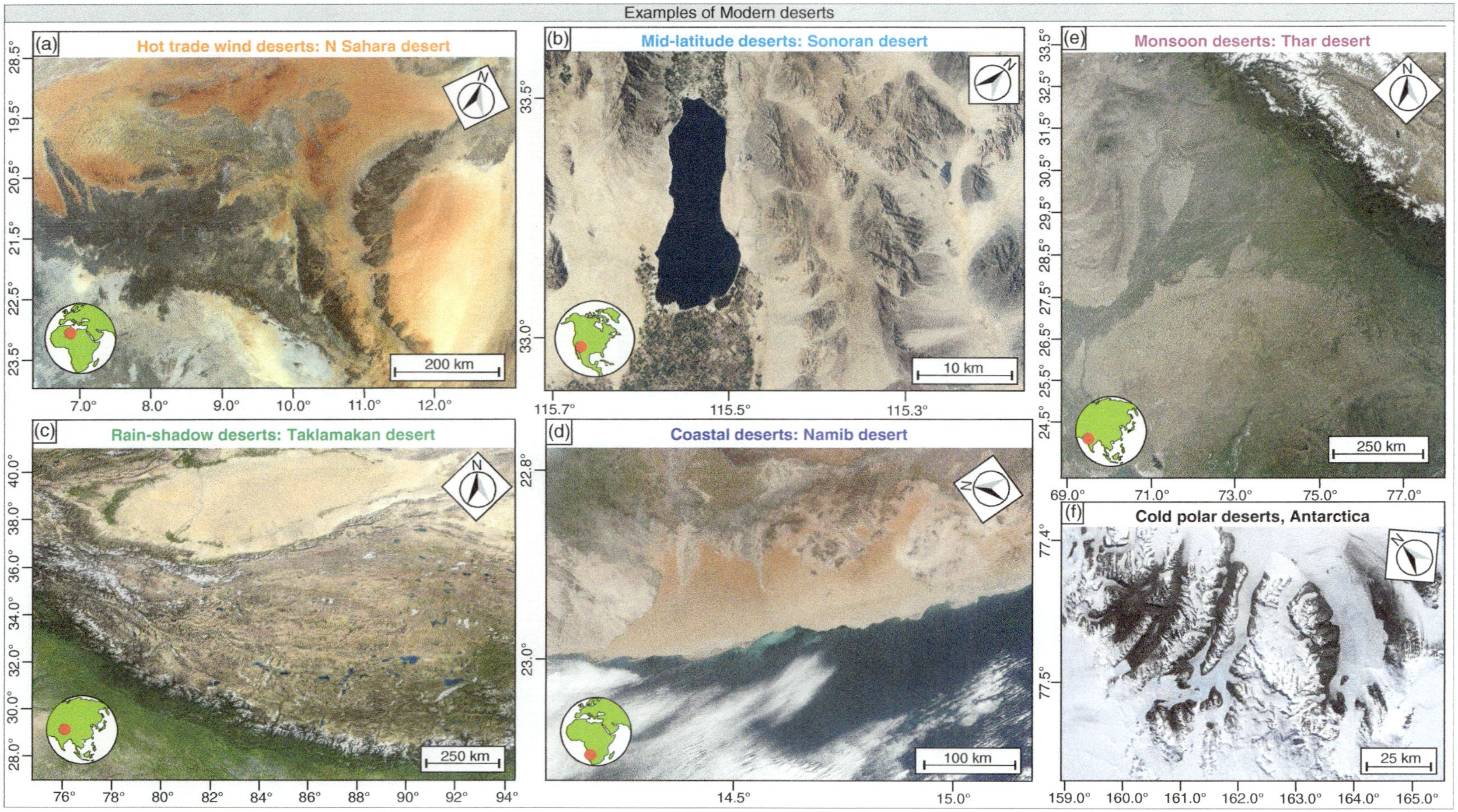

Figure 7.3 Examples of modern desert types. (a) Sahara Desert, Libya Algeria. True-colour NASA Terra Moderate Resolution Imaging Spectroradiometer (MODIS) image. (b) Sonoran Desert, southern California, USA. The Salton Sea – California's largest lake – is situated in the centre of the image and is bordered by blocky patches of farmland. Courtesy of the International Space Station (ISS) Crew Earth Observations Facility (NASA-JSC). (c) Taklamakan Desert and the surrounding region, western China. The Tibetan Plateau (centre) prevents rainfall from the southern South Asian monsoon passing the Himalayas (indicated in the image by the snow-covered mountains close to the bottom of the image), giving rise to an arid climate on the leeward (downwind) side to the north of the mountain range. Courtesy of the International Space Station (ISS) Crew Earth Observations Facility (NASA-JSC). (d) Namib Desert, Namibia. Note the cloud belt (fog) that develops due to moisture condensation in cold air over the Atlantic Ocean. (e) Thar Desert, northwest India. Parts d and e are true-colour NASA Terra Moderate Resolution Imaging Spectroradiometer (MODIS), courtesy of NASA Visible Earth. (f) McMurdo Dry Valleys, Antarctica; courtesy of NASA GSFC Oceans and Ice Branch and the Landsat 7 Science Team.

spatially isolated, partly cemented dunes, through a zone of undulating barchan, parabolic and transverse dunes with intervening interdunes, to a zone of large, actively accumulating transverse and star dunes separated by only minor interdune depressions at the downwind margin (Andrews 1981; Fryberger 1990a). Complex morphological arrangements of star dunes in the Gran Desierto sand sea, Mexico, accumulated under the influence of multidirectional wind regimes through the merging of smaller crescentic and reversing dunes (Lancaster et al. 1987). Remote sensing techniques (e.g. Breed and Grow 1979; Breed et al. 1979, 1987) provide an opportunity to appreciate the spatial variability of dunes and associated elements within aeolian systems. Dune fields of the Rub' al Khali, Arabia, are characterised by bedforms that vary systematically from their centres to their margins (Al-Masrahy and Mountney 2013b).

Most large aeolian dune and megadune bedforms are currently not in equilibrium with local wind regimes. The present-day humid (interglacial) climate means that many modern aeolian dune systems are currently partly or largely stabilised by surface crusts and/or vegetation and are only partly active. Relatively few thick aeolian dune successions are actively accumulating today; it is therefore difficult to directly relate modern dune systems to the widespread aeolian accumulations observed in the ancient record (Lancaster 1998). The slow response time of large aeolian dunes (and of sand seas composed of many such dunes) to changing environmental conditions may mean that states of disequilibrium are commonly recorded in ancient deposits.

7.2.1.2 Non-desert Aeolian Environments

Aeolian processes are also widespread in many humid settings where there exists a surfeit of sediment available for transport and the wind velocity is sufficiently great to overcome resistive forces induced by the presence of moisture on sediment surfaces or the stabilising influence of vegetation. Two notable settings are sandy coastlines (Pye 1983) and pro-glacial regions (McKenna-Neuman and Gilbert 1986).

Coastal dune fields may be composed of siliciclastic material, carbonates or mixed carbonate and siliciclastic material (Rodríguez-López et al. 2014). Siliciclastic coastal dune fields typically occur along humid, mid-high latitude coasts, whereas carbonate dune fields form preferentially along mid-low latitude coasts, especially those bordering productive shallow-marine platforms, which provide an ample supply of carbonate sediment.

Dune types along coasts vary; their size and morphology are dependent on vegetation cover, sand supply, wind regime and coastal environment. Sand blown inland from beaches typically forms vegetated embryonic foredunes and juvenile 'yellow' dunes initially. Older, mature, vegetated and commonly partly or fully stabilised 'grey' dunes develop further inland along progradational coasts. Post-depositional aeolian erosion of coastal dunes can lead to the formation of blowouts and parabolic dunes, which are especially common in so-called transgressive (inland retreating) dune fields (Hesp 1999). Along cliffed coasts, climbing dunes can form sand ramps (e.g. Clemmensen et al. 1997). 'Echo' dunes develop on the upwind sides of scarps or cliffs but may be separated from the scarp by a rolling eddy current of air that maintains a sand-free corridor. Many coastal dune fields can be classified as 'wet' because the groundwater table typically lies close to the surface. In many cases, the level of the groundwater table is linked to the sea level, especially in subsiding coastal basins (Kocurek et al. 2001; Mountney and Russell 2009). In other examples, dune dynamics are influenced by high precipitation rates and the formation of perched groundwater tables (Pedersen and Clemmensen 2005; Clemmensen et al. 2009). Dune fields develop both on retreating and prograding coasts: prograding systems typically develop successive rows of stabilised dune ridges (e.g. Bristow and Pucillo 2006; Madsen et al. 2007; Reimann et al. 2011); retreating systems typically experience phases of transgressive dune formation in the form of inland migrating parabolic dunes (Clemmensen et al. 2001; Pedersen and Clemmensen 2005).

Carbonate-prone aeolian systems are common in low- and mid-latitude settings, especially along tropical coastlines (e.g. Oman and the United Arab Emirates; Besler 1982) and along coastlines of the Western Mediterranean region (Clemmensen et al. 1997; Fornós et al. 2009; Andreucci et al. 2010a,b, 2014). Carbonate sand is prone to lithification due to partial dissolution and re-precipitation (Guern and Davaud 2005), especially in warm and humid settings. This is a mechanism of preservation that is rare in siliciclastic systems. Carbonate-cemented successions are commonly called 'aeolianites' (Brooke 2001). Similarly, gypsum dunes (which also tend to be readily cemented and lithified) may develop downwind and inland from evaporitic salt lakes (e.g. White Sands, New Mexico; McKee 1966) (Section 10.3.7).

Aeolian dune fields composed of lithic clasts of basalt and pumice have been recorded from active volcanic provinces (Mountney and Russell 2004) and from proglacial outwash plains and delta plains developed in humid climatic settings, for example, in southern Iceland (Mountney and Russell 2006, 2009). Here, high wind speeds enable significant aeolian sand transport, despite the overall humid climate setting and a generally high water-table level.

7.2.2 Ancient Aeolian Environments

7.2.2.1 Quaternary Aeolian Environments

Quaternary aeolian systems provide a link between modern and recently active aeolian systems and their ancient counterparts preserved in the rock record. Studies of episodes of construction and accumulation of Quaternary sand seas are supplemented by analyses of changes in associated neighbouring environments, such as lakes, rivers and coasts (e.g. Alsharan et al. 1998). Associated palaeoclimate indicators include organic plant remains, pollen spores and fossil biota. Collectively, this demonstrates how climatic conditions have changed considerably over vast areas of the Earth's desert and associated dryland regions in the recent geologic past. For example, during the last glacial maximum (LGM), approximately 18–20 ka, many of the world's desert regions were considerably more arid than they are today (e.g. Li et al. 2014). Presently, c. 10% of the land between 30° N and 30° S is covered by sand seas; during the LGM, sand seas in two vast belts occupied almost 50% of this land area.

The Nebraska Sand Hills, USA, represent the remnant of a stabilised late Quaternary dune field, covering 57×10^5 km^2, the largest sand sea in the Western Hemisphere (Ahlbrandt and Fryberger 1980; Muhs and Budahn 2019). Transverse dune ridges with up to 120 m relief and smaller, spatially isolated barchans are separated by low-lying interdune areas. These relict palaeo-dunes are now stabilised by vegetation; the region currently receives c. 500 mm of precipitation annually. A series of blowouts and fluvial drainage pathways expose the interior of many of the bedforms; analysis of foreset azimuths indicates uniform palaeodune migration to the southeast.

The coastal Akchar aeolian system of Mauritania represents the amalgamation of several separate ergs composed of complex linear megadunes, the accumulation and partial preservation of which have been controlled by eustatic and climatic variations during the Late Pleistocene and Holocene (Kocurek et al. 1991a). Dating by C^{14} methods (supported by the presence of archaeological artefacts) indicates that the aeolian accumulations are associated with three separate constructional erg-building phases, each partially preserved in a complex mosaic. Some of the sand used for the construction of more recent aeolian bedforms was derived from the partial cannibalisation of older deposits.

7.2.2.2 Ancient Aeolian Environments

The preserved ancient record of aeolian sedimentary systems extends from the Archean to the Quaternary, and example successions are recorded globally (Figure 7.4); see Rodríguez-López et al. (2014) for a review. The majority of identified ancient aeolian successions represent accumulations of inland dune fields, though numerous ancient coastal aeolian successions are also known. Throughout Earth history, the accumulation and preservation of aeolian successions have been controlled by numerous tectonic and environmental controls (Figure 7.5a,b): (i) supercontinent assembly, break-up and associated orogenic events and basin development (Figure 7.5c), (ii) repeated global climatic changes, including alternation of long-lived icehouse and greenhouse episodes and shorter-term, climate-driven glacio-eustatic changes (Figure 7.5c) and (iii) the presence or absence of terrestrial vegetation (Figure 7.5c). These controls have governed the palaeogeographical location of land masses, the climate that prevailed therein and thereby the type of aeolian system that developed and the nature of aeolian interactions with adjoining environments. Collectively, environmental forcing mechanisms dictated the timing and location of desert belts in which major aeolian systems were constructed, accumulated and preserved (Cosgrove et al. 2021a,b and 2022a,b; Figure 7.5d).

7.2.3 Dry, Wet and Stabilised Aeolian Systems

Both present-day aeolian systems and their preserved successions are commonly classified as dry, wet or stabilised (Kocurek and Havholm 1993). Dry aeolian systems develop where accumulation is controlled solely by aerodynamic configuration (Figure 7.6a). Downwind airflow deceleration causes sediment-laden winds to deposit some of their load. Dry aeolian systems are widely documented in both modern desert settings and the rock record. Wet aeolian systems develop where accumulation is controlled by progressive water-table rise concomitant with ongoing aeolian activity – notably, dune migration (Figure 7.6b). Wet systems are increasingly recognised in the rock record, though there remain fewer documented examples than of dry systems. Stabilised aeolian systems develop where physical, chemical or biogenic agents promote aeolian system construction and accumulation (Figure 7.6c). For example, where vegetation or surface cements act to restrict the transport capacity of the wind. Although widespread in present-day settings, stabilised systems are less commonly recognised in the rock record.

7.2.4 Aeolian System Interactions with Non-aeolian Environments

Aeolian sedimentary systems may interact with other types of sedimentary systems, especially close to the margins of aeolian dune fields. In particular, aeolian dunes, interdunes and sandsheets commonly interact with alluvial, fluvial,

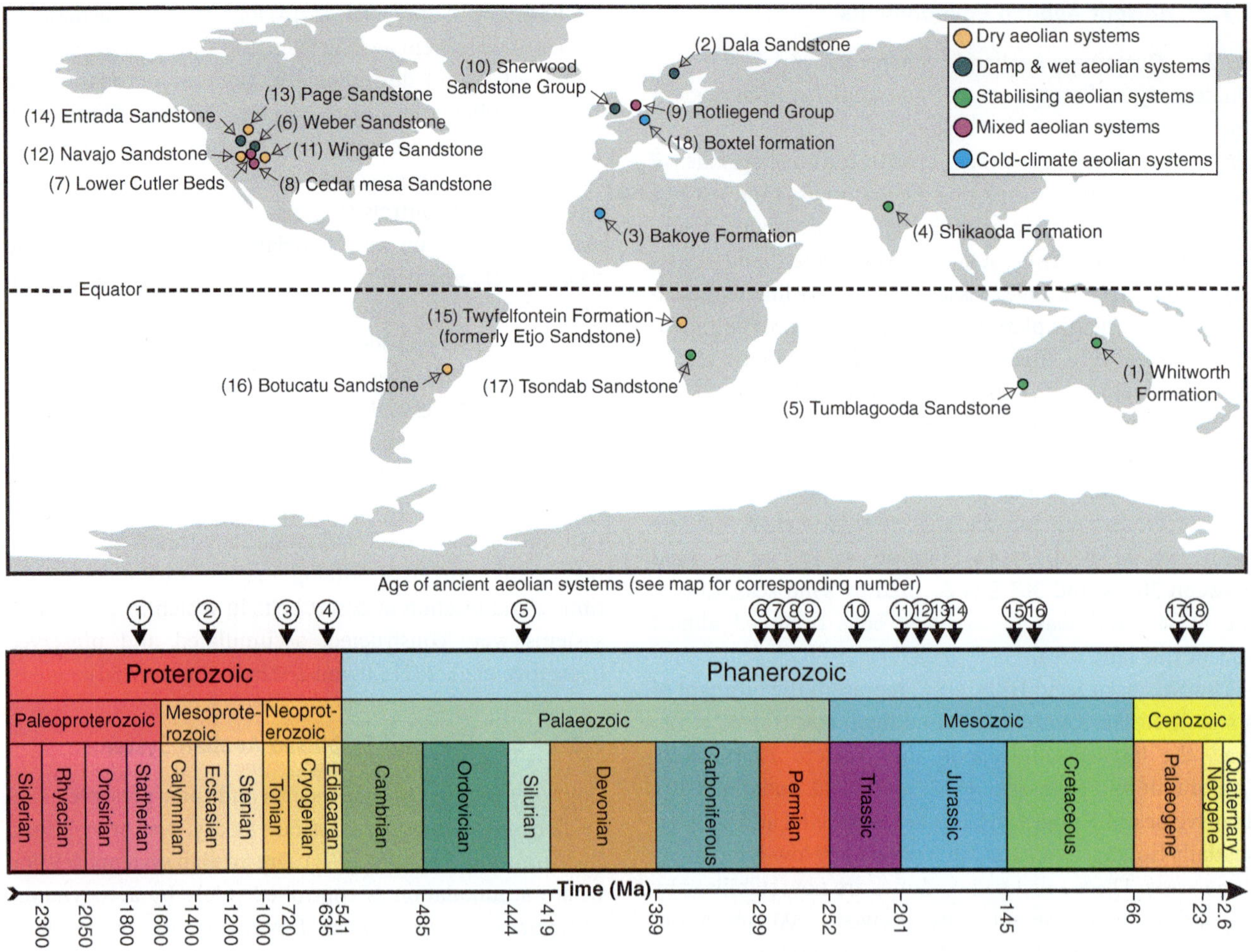

Figure 7.4 Selected examples of ancient aeolian successions. The numbers assigned to each example in the geological map are reported on the geological timescale to indicate the approximate age of the aeolian successions. The coloured spots indicate the aeolian system type. The term 'mixed' refers to desert systems that vary in type spatially and/or temporally and may include components of dry, damp, wet and stabilising systems. See Rodríguez-López et al. (2014) for a detailed review.

lacustrine and coastal landforms (Figure 7.7). Sediments of aeolian origin are commonly reworked by non-aeolian processes, and vice versa. Significant diversity in the arrangement and type of interaction of competing depositional sedimentary systems is recognised in modern desert dune fields and their marginal areas, and these give rise to complex yet predictable geomorphological patterns that commonly vary over space and time (Figure 7.8; e.g. Lancaster 1989; Cooke et al. 1993; Bullard and Livingstone 2002; Al-Masrahy and Mountney 2015). The record of these interactions is also recognised in the ancient sedimentary record (e.g. Langford and Chan 1988, 1989; Herries 1993; Chakraborty and Chaudhuri 1993; Mountney and Jagger 2004; Spalletti and Veiga 2007; Jordan and Mountney 2010, 2012; Spalletti et al. 2010), where spatial and temporal changes in the type of interaction between aeolian dune and associated desert sub-environments are known to have resulted in the preservation of complex arrangements of sedimentary deposits and stratigraphic architectures.

7.3 Processes and Landforms

7.3.1 Aeolian Sediment Entrainment

Aeolian sediment entrainment occurs when and where wind forces are able to overcome the effects of surface stabilisation. For grains of sand and finer grade, aerodynamic lift and drag promote entrainment, whereas particle weight, friction and cohesive inter-particle forces retard entrainment. Drag and lift are generated by fluid flow over and around particles; whereas weight, friction and cohesion are determined by particle size, density, shape, packing, moisture content, mineralogy and degree of

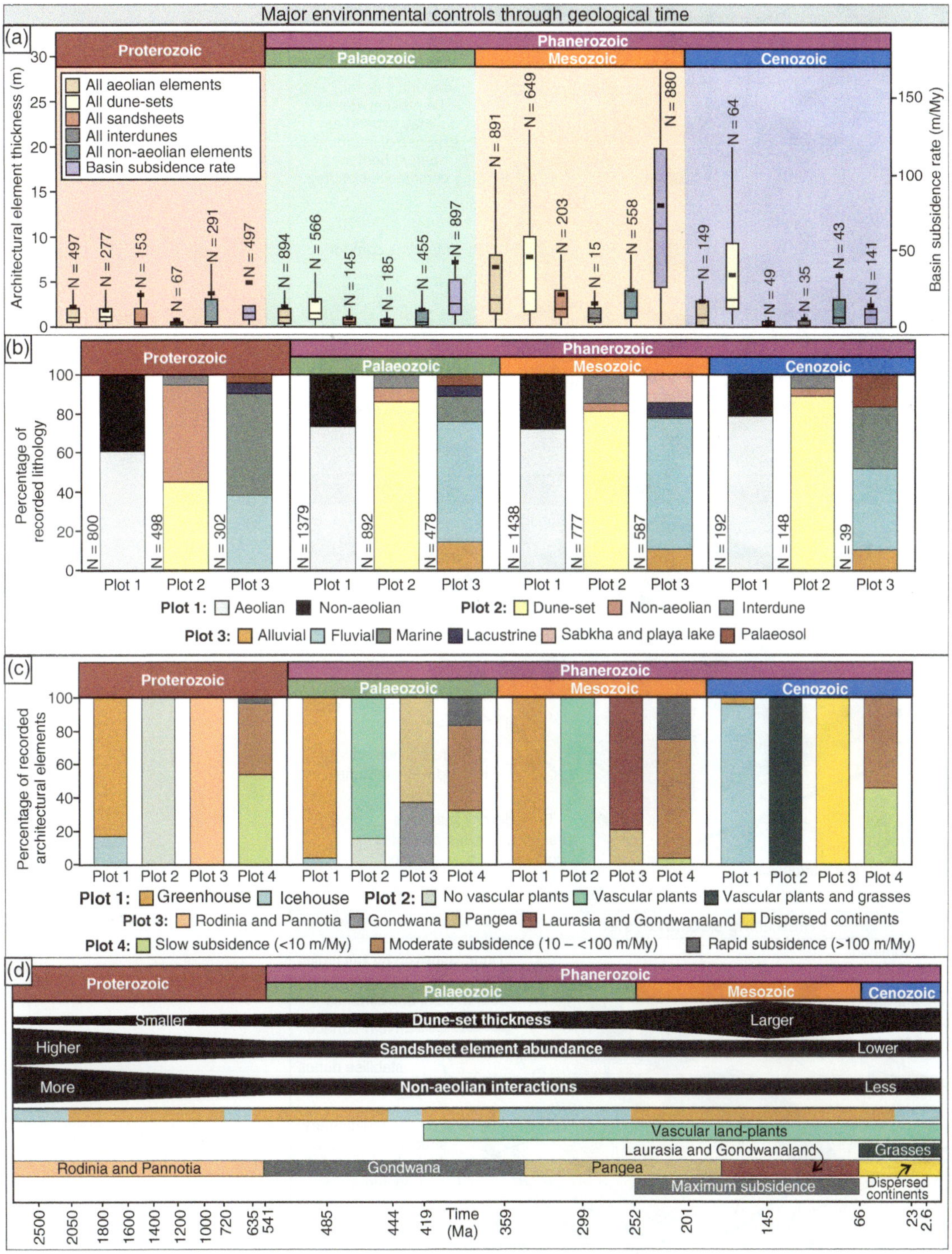

Figure 7.5 Major environmental controls that have operated through Earth history and their relationship to the accumulation and preservation of major aeolian systems. Variations through geological time. (a) Thicknesses of aeolian and non-aeolian architectural elements in relation to rates of basin subsidence (outliers are not presented). (b) Percentage of aeolian (and related) architectural element types in aeolian successions. (c) Percentage of recorded architectural elements in relation to a variety of environmental conditions. (d) Environmental forcings governing aeolian construction, accumulation and preservation through geological time. Adapted from Cosgrove et al. (2022a).

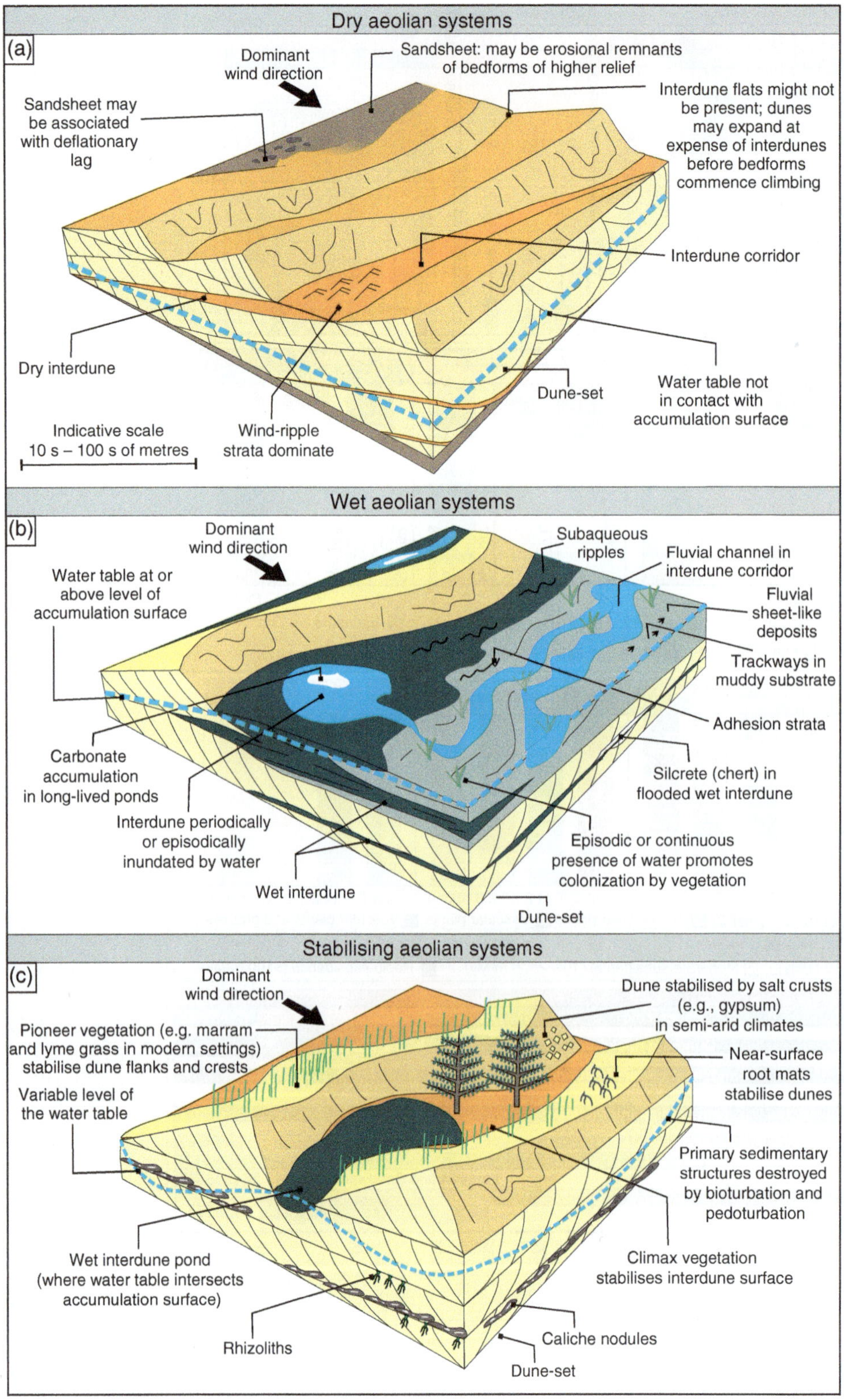

Figure 7.6 Characteristic sedimentary features of aeolian systems: (a) dry, (b) wet and (c) stabilising.

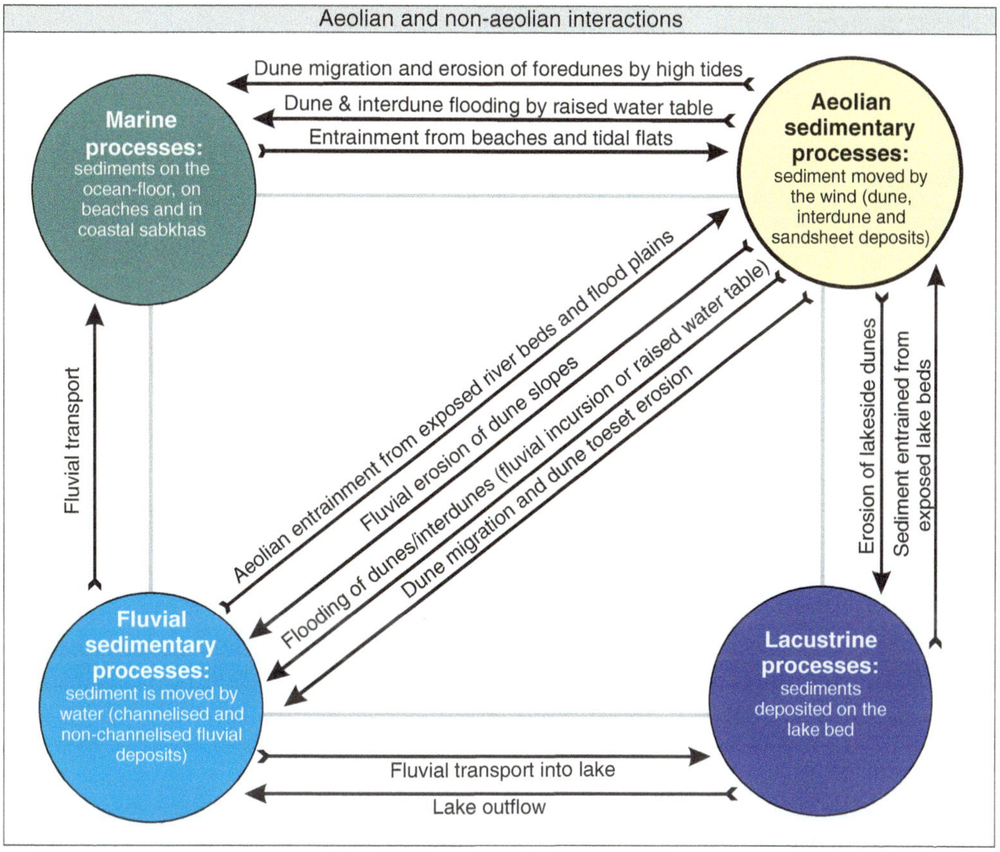

Figure 7.7 Key interactions between the principal environments of sediment storage for sand-sized material. Adapted from Bullard and Livingstone (2002).

inter-particle cementation (Lancaster 1995). Where a fluid flows over a solid granular surface, a boundary layer develops because of friction close to the interface; the time-averaged velocity profile increases from zero at the surface itself to approach the mean wind velocity some distance above the surface. For laminar flows, the rate of increase in velocity with height above the surface is linear, whereas for turbulent flows, typical for wind, flow mixing in the boundary layer results in a steeper (parabolic) velocity gradient just above the surface and, hence, greater shear stress close to the surface (Bagnold 1941). Thus, turbulence provides a mechanism by which initial grain entrainment into the airflow can occur at relatively modest wind velocities. Ballistic impact of landing grains can induce motion in other grains lying at rest on the bed. Sheltering of grains between adjacent neighbouring grains can inhibit motion.

7.3.2 Aeolian Sediment Transport

The construction of dunes and sandsheets requires the availability of abundant sand grains, which are mostly transported as bedload prior to deposition. By contrast, silt and clay fractions are mostly transported aloft in suspension (Figure 7.9). Under most wind conditions, gravel and coarser sediments are not moved and are left *in situ* as a deflation lag deposit: a 'reg' or 'serir'. Sand carried as bedload may be transported by one or more of the following processes: (i) rolling or creep due to drag directly on the bed; (ii) 'saltation', whereby particles are temporarily lifted into the airflow and then collide with other particles in the flow, thereby inducing a chain reaction of interparticle collisions; or (iii) 'reptation', whereby saltating particles that return to the bed impact and nudge other particles and cause them to hop short distances downwind (Figure 7.9; Figure 15.5). Of these processes, saltation is especially important in the aeolian transport of sand; wind impact ripples (see Section 7.3.5) owe their distinctive form to this process.

Observations in modern deserts suggest that a patch of sand must be at least 4–6 m long to develop into a dune bedform. The wind must be sufficiently retarded by the rugosity of the patch and its associated saltation cloud to induce further deposition to occur. Each patch, which will typically have aeolian ripples moving over its surface, grows

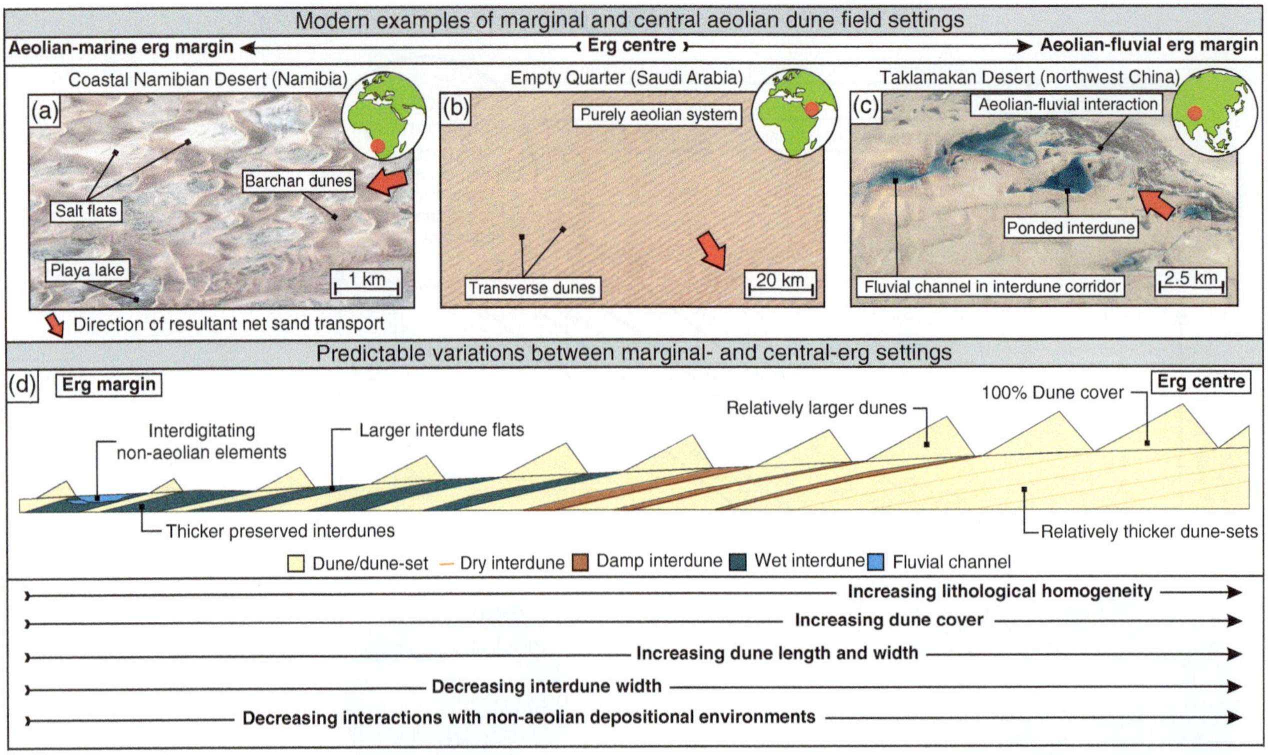

Figure 7.8 Modern examples of marginal and central aeolian dune field settings. (a) Barchan dunes isolated playa lakes (interdune ponds) in the coastal part of the Namib Desert (aeolian–marine marginal-erg). (b) Transverse dunes in the Empty and Quarter of Saudi Arabia (central-erg). (c) Barchanoid dunes with fluvial incursions along interdune corridors and interdune ponds in the Gurbantünggüt Desert, China (aeolian–fluvial marginal-erg). Courtesy of Google Earth. Non-aeolian interactions tend to increase towards erg-margins, whereas erg-centre settings tend to remain aeolian dominated. Red arrows in a–c indicate the dominant aeolian transport direction. (d) Predictable variations in aeolian architecture between marginal- and central-erg settings. Adapted from Mountney and Jagger (2004).

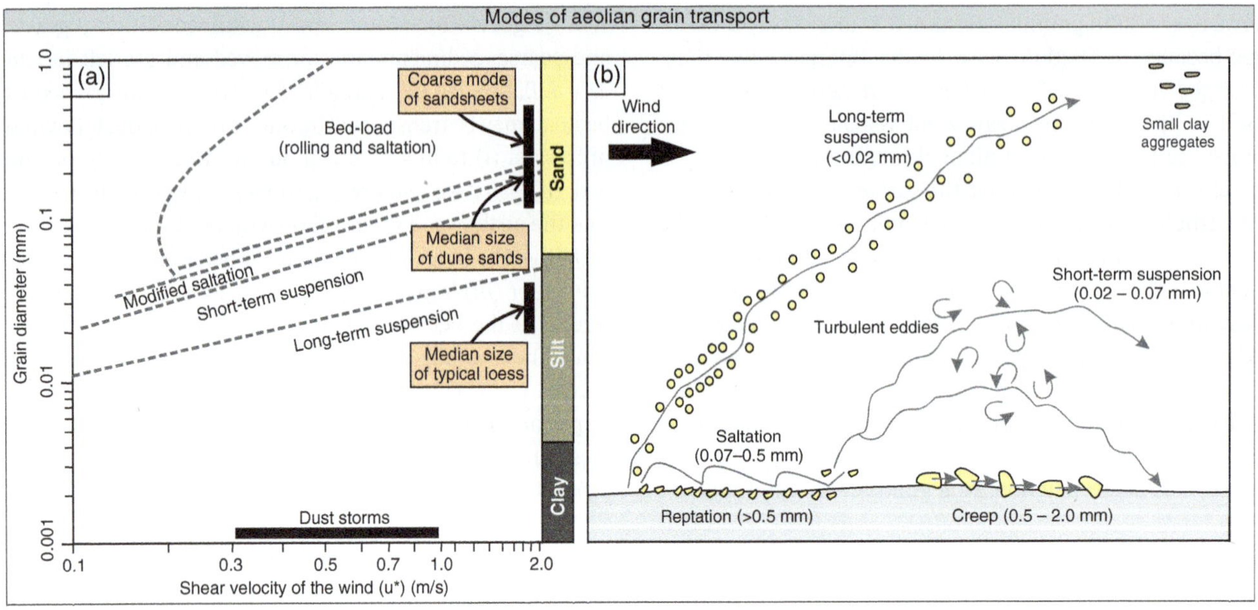

Figure 7.9 Modes of aeolian grain transport. (a) Quartz spheres of different diameters at different shear velocities of the wind (u*). Modified after Tsoar and Pye (1987). (b) Schematic representation of the main modes of aeolian sediment transport with typical grain sizes indicated. Adapted from Pye and Tsoar (1990).

to a critical height, dependent on grain size and wind strength, whereupon a dune form, potentially with a slipface, may ultimately develop.

Ripples, dunes and megadunes ('draas') that are migrating under the influence of the airflow comprise the sediment that is in transport. Although such bedforms may be large and slow-moving, over longer timescales (months, years or longer), they can move significant distances. Ripples, dunes and megadunes commonly develop into a 'train' of regularly spaced and repeated bedforms that record the response of a sand bed to the shearing action of the wind. A train of bedforms records a stage towards dynamic equilibrium in response to a varying but consistent fluctuation in the flow pattern (e.g. Jimenez et al. 1999). See also Section 15.2.1 for a comparative review of aeolian transport in the solar system.

7.3.3 Aeolian Sediment Composition and Texture

Aeolian sediments commonly exhibit distinctive grain size, shape and sorting characteristics that can be useful in their recognition (Pye 1982; Lancaster 1986). The wind is highly selective in terms of the size of grains that it can carry (itself related to grain density and therefore weight) for a given velocity. This is reflected in the texture of aeolian deposits, which are commonly exclusively composed of very fine to medium sand grains that are well or very well sorted and texturally mature to supermature; such aeolian sediments are described as 'clean sands' (Figure 7.10a–c).

During transport, inter-particle collisions result in high rates of grain abrasion; less resistant minerals, such as mica and some friable lithic fragments, readily break down to dust, whereas more resistant grains – especially quartz (including chert) – become highly rounded as angular grain edges are abraded. Quartz has no preferred cleavage angle; as such, quartz sand grains tend to (i) develop high sphericity, (ii) possess surfaces that are dull (frosted) as a result of repeated grain collisions and (iii) exhibit conchoidal fractures in the form of 'upturned plates' as a result of repeated inter-particle grain collisions (Vos et al. 2014). Aeolian grains with these properties commonly exhibit a 'millet seed' texture (Figure 7.10a,b). Certain resilient lithic clast types, notably basalt, can also form 'clean' aeolian sand. Although many aeolian sands are mono-mineralic, composition and texture ultimately depend on the nature of the source material, the availability of that material for aeolian transport, cumulative transport distance and wind gustiness. For example, aeolian sands composed of basalt grains are common in Iceland (Figure 7.10c). Rarely, where aeolian sand deposits have been derived from local sources (i.e. have undertaken only a short distance of transport), they may be texturally and/or compositionally immature (Figure 7.10d).

Although aeolian dunes composed of a minor but nevertheless significant fraction of friable, cleavable sand grains of feldspar, mica or silt-clay aggregates may develop in areas close to a sediment source, they are virtually absent in settings more distant from the source. However, deposits of dry, sand-size clay and silt aggregates commonly develop as parabolic dunes immediately downwind of dried-out lake beds and are called 'lunettes'. On wetting, such dunes may become solid masses but may retain their internal structure (e.g. cross-bedding).

Silt grains make up a significant fraction of some aeolian deposits. Silt is the dominant grain fraction in 'loess' deposited from a suspended load. In sand-dominated aeolian deposits, silt may still comprise a minor but significant proportion. Silt may originate from aeolian abrasion associated with ballistic impacts, glacial grinding prior to aeolian reworking, frost shattering, salt weathering, fluvial and colluvial comminution and aeolian weathering of silt-rich bedrock (e.g. Pye and Sperling 1993, Muhs and Bettis 2003, Bullard et al. 2004, Soreghan et al. 2008).

The movement of coarse sand and granules via creep typically occurs only sporadically during episodes of high winds. Notably, sandsheet deposits developed where creep is a dominant process may be characterised by poorly to moderately sorted sediments (Pye and Tsoar 1990).

Aeolian deflation involving the winnowing of sand-grade sediment may leave a surface dominated by a poorly sorted, relatively coarse-grained lag of granules and small pebbles – a reg or serir (McFadden et al. 1987). The concentration of larger clasts on deflationary surfaces can protect or 'armour' the deflating surface, thereby preventing further winnowing of the underlying sand. Regs or serirs may commonly develop wind-faceted cobbles and boulders (ventifacts) with distinctive upwind-facing scalloped and abraded surfaces that are useful palaeocurrent indicators.

Although sedimentary textures can aid the recognition of ancient aeolian deposits, they are most useful as recognition criteria when additionally associated with diagnostic sedimentary structures. The intimate association of many aeolian systems with fluvial, lacustrine and coastal environments means that sediments with aeolian textures are commonly reworked by non-aeolian processes. For example, the Kuiseb River of Namibia captures aeolian sand moving north from the Namib Sand Sea and transports it downstream before releasing it to the shallow-marine coastal realm (e.g. Goudie 1972; Ward 1983). As such, these fluvial, coastal and shallow-marine deposits all possess aeolian grain characteristics, which were developed earlier in the sediment transport history. Conversely, aeolian processes commonly locally rework sediments derived from nearby fluvial, coastal and shallow-marine environments. Locally

Figure 7.10 Examples of aeolian sediment texture. (a) Very well-sorted aeolian sand composed dominantly of quartz and lithic (basalt) grains. The well-rounded and highly spherical form of the grains, together with their dull (frosted) appearance, is typical for transport via aeolian saltation. Many quartz grains have surfaces marked by percussion impact 'pits'. Modern sand, Namib Desert. (b) 'Millet seed' texture of quartz sand grains giving rise to a supermature grain composition and texture that is typical of many aeolian deposits, especially in dune-field centre settings. (c) Compositionally and texturally supermature aeolian sand composed dominantly of basalt grains. Modern, Iceland. (d) Example of an aeolian sand deposit composed of detritus derived from a local source. Note the moderate sorting, the varied grain shapes and the presence of grains with angular surfaces. Not all aeolian deposits are supermature in terms of their grain composition and/or texture. Modern, Iceland.

reworked aeolian sediments can retain grain textures and clast types that are indicative of earlier modes of sedimentation where transport distances are short.

7.3.4 Aeolian Bedforms: Overview

Aeolian ripples, dunes and megadunes (draa) form a hierarchy of equilibrium bedforms and one or more of these types are present in the great majority of depositional aeolian environments. Bedform size provides a basis for initial bedform classification and description (Figure 7.11). Almost all dunes and megadunes have ripples present on many parts of their sloping surfaces, and ripples are also common on sandsheet surfaces and in interdunes. Three morphologically distinct types of small-scale ripple structure occur on wind-blown surfaces: impact ripples, megaripples and adhesion ripples. Each is also recognised in the vertical section, where they produce different types of aeolian ripple lamination or stratification (see Section 7.4).

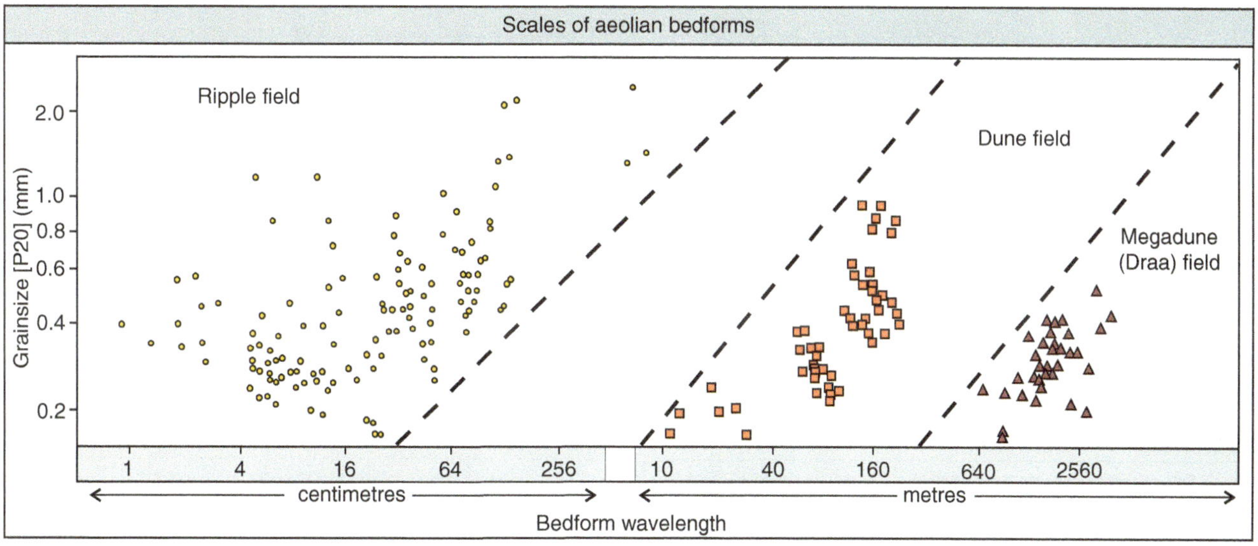

Figure 7.11 Three distinct scales of aeolian bedform that commonly occur on Earth: ripples, dunes and megadunes (draa). (a) Grain size (coarsest 20th percentile) versus wavelength for aeolian bedforms. Adapted from Wilson (1972). Note that bedforms developed at scales between the three fields depicted in the graph do occur on Earth but are less common.

7.3.5 Aeolian Ripples

7.3.5.1 Aeolian Impact Ripples

Impact ripples – also called 'ballistic ripples' – are the most widespread ripple type developed in aeolian environments (Hunter 1977a,b, 1981). They have low relief and form preferentially from the coarser-grained fraction of the sand in transport. They typically have wavelengths of 50–200 mm and heights of 5–10 mm, rarely up to 30 mm. They exhibit a high ripple-form index (ratio of wavelength to height), typically between 12 and 50. They may have straight or curved (in some cases sinuous), continuous crestlines, usually oriented transverse to the local wind direction. The coarsest sand grains tend to be concentrated upon or adjacent to the ripple crests. Impact ripples are slightly asymmetrical in profile, though with their lee faces inclined at relatively low gradients (<20°), below the angle of repose for loose dry sand.

Impact ripples develop mostly by saltation (longer hopping of sand grains downwind), supplemented by reptation (shorter nudges of sand grains). Sand grains act as high-momentum impacting particles under the influence of wind shear; they cause grains at rest on the bed to hop downwind (Figure 7.12). For a given wind velocity, saltating grains in motion are restricted to a narrow size range, and the distance that they jump downwind (their 'path length') is similar for most of the sediment in transport. For this reason, impact ripples begin to form with spacings that are determined by the saltation path length (Figure 7.12). Thus, ripple spacing is proportional to wind velocity and grain size (and mass). Minor surface perturbations in bed surface roughness act as the trigger required to initiate ripple development; sand grains that lie in a slightly elevated position are plucked from the bed into the airflow and induce the onset of saltation whereby a chain reaction of grain collisions and resultant motion is initiated (Figure 7.9). Impact ripples readily develop in just a few minutes under conditions of sustained grain saltation. Once initiated, the ripples themselves grow and steepen into bedforms because upwind-facing stoss slopes act as an impact zone that catches incoming saltating grains, whereas downwind facing lee slopes act as a shadow zone where grain impacts are minimal (Figure 7.12c). The ballistic impact of grains landing in the impact zone causes other grains to creep up the stoss slope towards the ripple crest until an impact causes them to launch into the airflow and hop downwind to the next ripple. Coarser grains tend to concentrate at ripple crests. Finer grains become preferentially trapped in ripple-trough shadow zones where the effects of wind shear are at a minimum. Air temperature, which influences the viscosity of the airflow, can exert a modest influence on the grain size of ripples: slightly coarser sand ripples are more common in cold settings.

Aeolian impact ripples may be differentiated from subaqueous ripples because the former (i) typically have high ripple form-index (wavelength-to-height ratio) of 25–40+ and (ii) are commonly characterised by inverse grading that results from the migration of coarser-grained ripple crests over finer-grained ripple troughs (Figure 7.12). More complex ripple patterns may develop from the merging of

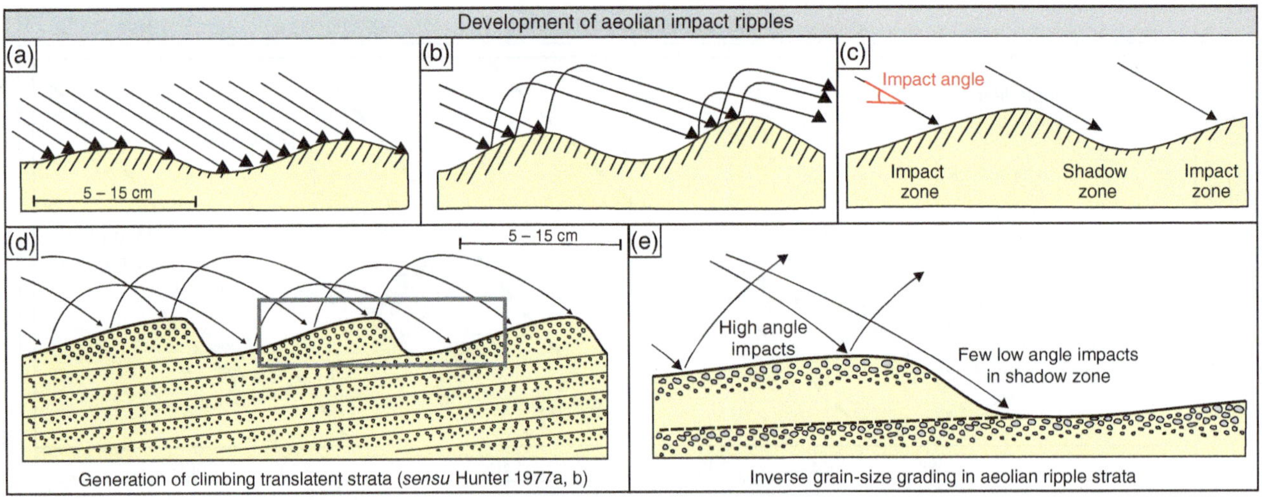

Figure 7.12 Model to explain the development of aeolian impact ripples. (a) Perturbations on a bed expressed as surface roughness influence how grains landing from saltation arrange themselves on the bed. This initiates the onset of ripple development. (b) Impact ripple spacing is determined by the saltation path length, which is a function of grain size, shape, density, mean wind velocity and gustiness close to the bed. (c) Aeolian impact ripples have upwind-facing stoss slopes subject to the impact of incoming grains and downwind-facing lee slopes where shadow zones develop. Adapted from Anderson (1987). (d) Wind ripples generated by the ballistic impact of incoming (i.e. landing) saltating grains. The ripple spacing relates in a general way to the saltation path length, which is the characteristic distance that individual grains hop downwind during saltation. Coarser grains preferentially accumulate in the impact zone on the upper part of the ripple stoss slopes and at ripple crests. Finer grains preferentially accumulate in the shadow zone on the lower parts of ripple lee slopes and ripple troughs. The migration of impact ripples thereby generates sub-parallel lamination. The grey box indicates the region highlighted in detail in Part e. (e) The impact angle of saltating sand grains differs between stoss and lee slopes. High-angle impacts on the stoss slope of the impact ripples promote the creep of coarser grains towards the ripple crest. Downwind-facing lee slopes form a shadow zone where relatively few low-angle impacts occur, thereby allowing the accumulation of finer grains in ripple troughs. As ripples migrate downwind, this sorting mechanism generates lamination with inverse grading.

smaller and larger ripples that undergo different migration rates. Larger ripple forms may perturbate the airflow such that air turbulence encourages the development of complex patterns of superimposed smaller bedforms; these are types of interference ripples.

7.3.5.2 Aeolian Megaripples

Where the supply of sediment for aeolian transport is restricted to coarse sand, gravel and small pebbles, and where the wind blows with sufficient intensity to move these grain sizes, aeolian megaripples (sometimes called granule ripples) may develop (Fryberger et al. 1992). Although these bedforms also develop due to the impact of saltating and reptating grains, as described for impact ripples, they typically have more sinuous crestlines. Wavelengths and heights are typically up to 5 and 0.35 m, respectively. The largest documented examples on Earth have wavelengths up to 43 m and heights up to 2.3 m (Milana 2009). A continuum of ripple sizes exists from impact ripples developed in fine sand (and rarely in coarse silt) to granule and pebble megaripples. Aeolian megaripples formed of pebble-grade sediments are rare, and most are composed of clasts of low-density lithology types, notably pumice (e.g. Mountney and Russell 2004; Milana 2009).

7.3.5.3 Adhesion Ripples

When dry sand is blown across a damp or wet sediment surface, some grains will stick (adhere) to that surface on impact. This process of grain adhesion results in the generation of a range of structures including adhesion ripples (Hunter 1973; Kocurek and Fielder 1982), adhesion warts (Olsen et al. 1989) and adhesion plane beds (Figure 7.13a–c). Adhesion ripples accrete in an upwind direction and are characterised by irregular bumpy or corrugated sediment surfaces (Figure 7.13d–h). Given that they require a damp surface to develop, their long-term preservation potential is low to moderate: if the sediments in which these structures are formed dry-out, they are prone to later aeolian reworking. Adhesion structures may be difficult to differentiate from microbially induced sedimentary structures (MISS) (Basilici et al. 2020).

7.3.6 Aeolian Dunes

Dune-scale aeolian bedforms are classified according to a number of separate criteria, including their size, morphology (shape, wavelength, height), spacing (frequency), orientation relative to net sand transport direction, type of migratory behaviour and form of any superimposed bedforms.

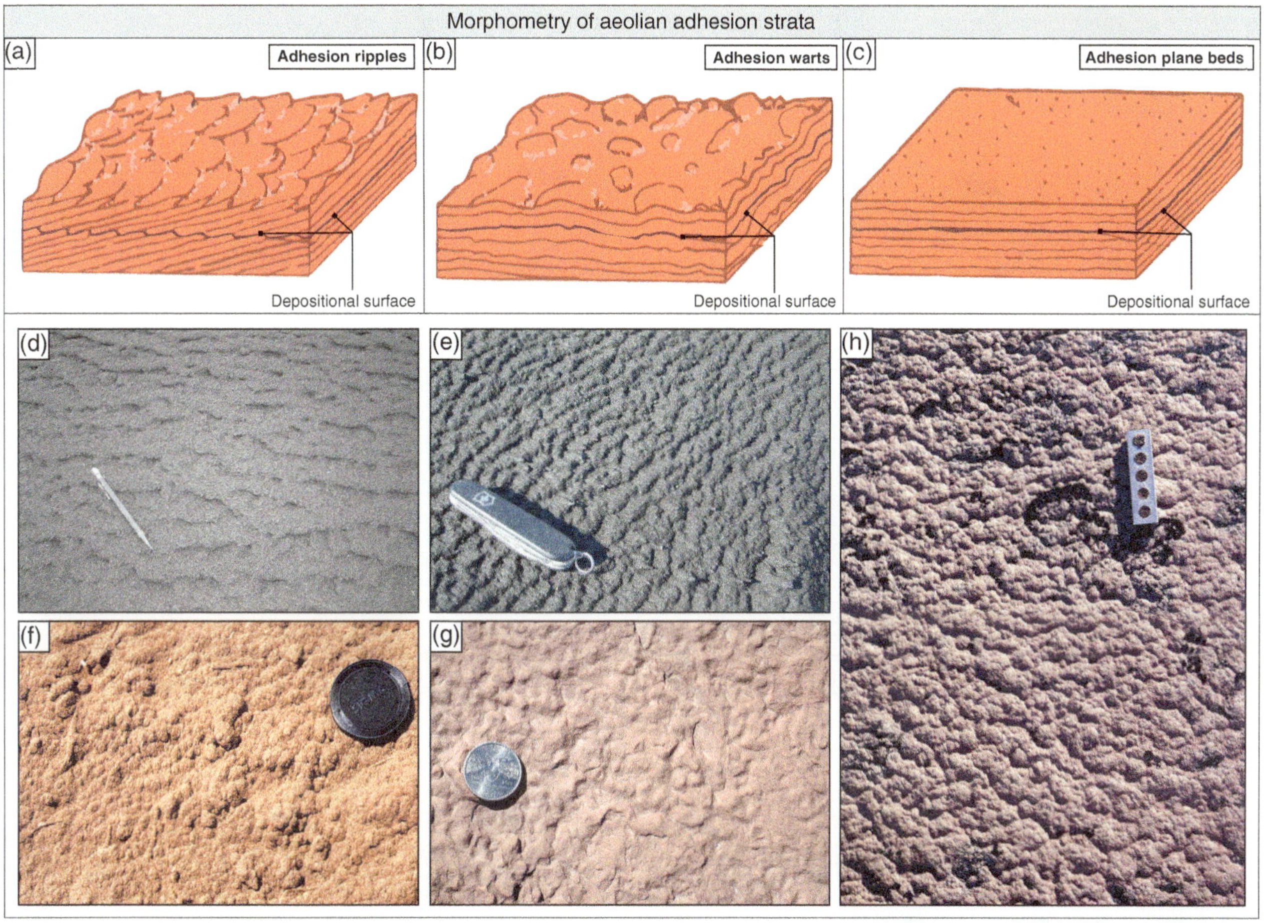

Figure 7.13 Examples of aeolian adhesion structures. Morphometry of (a) aeolian adhesion ripples, (b) aeolian adhesion warts and (c) aeolian plane beds. Modified after Kocurek and Fielder (1982). (d) Adhesion ripples developed on a damp sand surface. The steeper slopes face upwind. Wind direction was bottom-to-top in this photo. Modern, Iceland. (e) Adhesion ripples developed on a damp sand surface. The steeper slopes face upwind. Wind direction was right-to-left in this photo. Modern, Iceland. (f) Adhesion warts generated on a surface of damp sand over which dry sand has been blown. Indian Creek, Utah, USA. (g) Adhesion warts preserved on an upper bedding surface. Permian Cedar Mesa Sandstone, Utah, USA. (h) Upper bedding surface of sandstone showing an irregular small-scale morphology interpreted as wind adhesion warts. Independence Fjord Group, Proterozoic, North Greenland. Parts f–h after Collinson and Mountney (2019).

Aeolian dunes have wavelengths of c. 5–250 m and commonly develop as trains of regularly spaced bedforms (Figure 7.14). Dunes form topographic features that disrupt the primary airflow; as the flow moves up the windward (stoss) slope, it accelerates, thereby causing an increase in transport rate and promoting transport to the dune crest. As the flow moves over the crest, it passes to the lee-side depression, expands and decelerates, reducing transport capacity and promoting deposition on the lee slope. This is the mechanism by which aeolian dunes migrate downwind over time, though migration in directions oblique to the resultant wind is also common. Flow separation from the bedform surface occurs at or beyond the crest, leading to the development of a 'brink' – a distinct break in slope. Flow re-attachment occurs a distance of a few dune heights downwind. Thus, a separation cell exists in the dune lee within which turbulent secondary airflow can occur. Flow turbulence encourages the development of ripples and erosional scour hollows on the dune lee flank (especially in the lower, 'plinth', region). This can lead to complex migratory behaviour, whereby ripples and scour hollows migrate along the dune lee (else oblique to it). Downwind of the re-attachment point, renewed flow acceleration means that sediments present on an interdune surface in front of an advancing migrating dune may potentially be eroded, thereby providing a local sediment supply for the next dune downwind in the train.

Morphologically simple dunes are characterised by a single windward stoss-slope inclined at 8–16° and a lee-slope inclined at 20–36° (Figure 7.14). Lee slopes commonly face

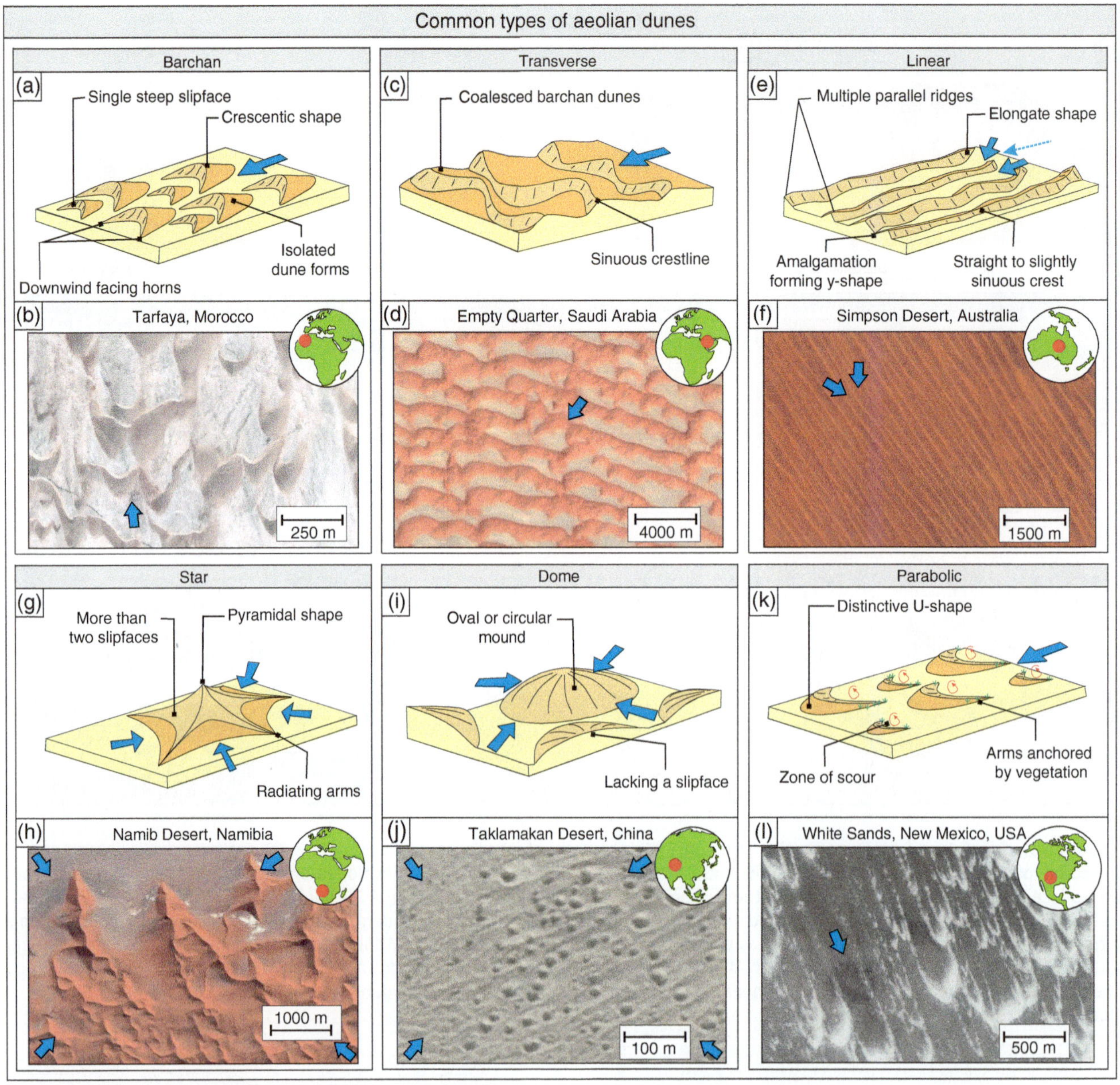

Figure 7.14 Common types of aeolian dunes. Morphometry, terminology for description and plan-view examples seen in aerial imagery. See main text for the explanation Courtesy of google earth.

downwind or oblique to the resultant wind direction. More elaborate morphological dune forms may possess several stoss and/or lee-slopes, and these may face various orientations. Lee-slopes commonly comprise a slipface inclined at or close to the angle of repose (rest) for loose, dry sand, down which sand episodically avalanches. However, not all dunes possess such slipfaces: some are described as 'slipface-less' (i.e. generated at or degraded to an angle less than repose for dry loose sand grains). Dunes exhibit a wide variety of morphological forms that reflect the combined effects of a number of controlling factors, including wind strength and directional variability on diurnal to seasonal (and longer) timescales, sediment supply and the availability of sediment for transport.

Dunes may be classified as mobile (actively migrating), active but anchored (for example, attached to a cliff, large boulder or a patch of vegetation) or stabilised (i.e. inactive relict landforms) (Figure 7.15). Mobile dunes are classified according to the number of lee faces they possess and the orientation of their crestlines relative to the dominant wind direction. Many common dune types are classified according to these criteria. Transverse dunes (including barchan dunes and sinuous-crested barchanoid dune ridges) possess a single, gently inclined, upwind-facing stoss slope, a steeper downwind-facing lee slope and a crestline oriented approximately perpendicular to the prevailing wind direction (Figure 7.14a–d). Longitudinally oriented linear dunes (sand ridges) have one or two lee faces and a crestline arranged approximately parallel to the prevailing wind (Figure 7.14e,f). Star (pyramid) dunes have three or more lee faces (Figure 7.14g,h). Although the assignment of dunes as purely transverse or longitudinal is useful for simple classification schemes, it is potentially misleading because the net sand transport direction is commonly oblique, resulting in the generation of oblique bedforms (Hunter et al. 1983).

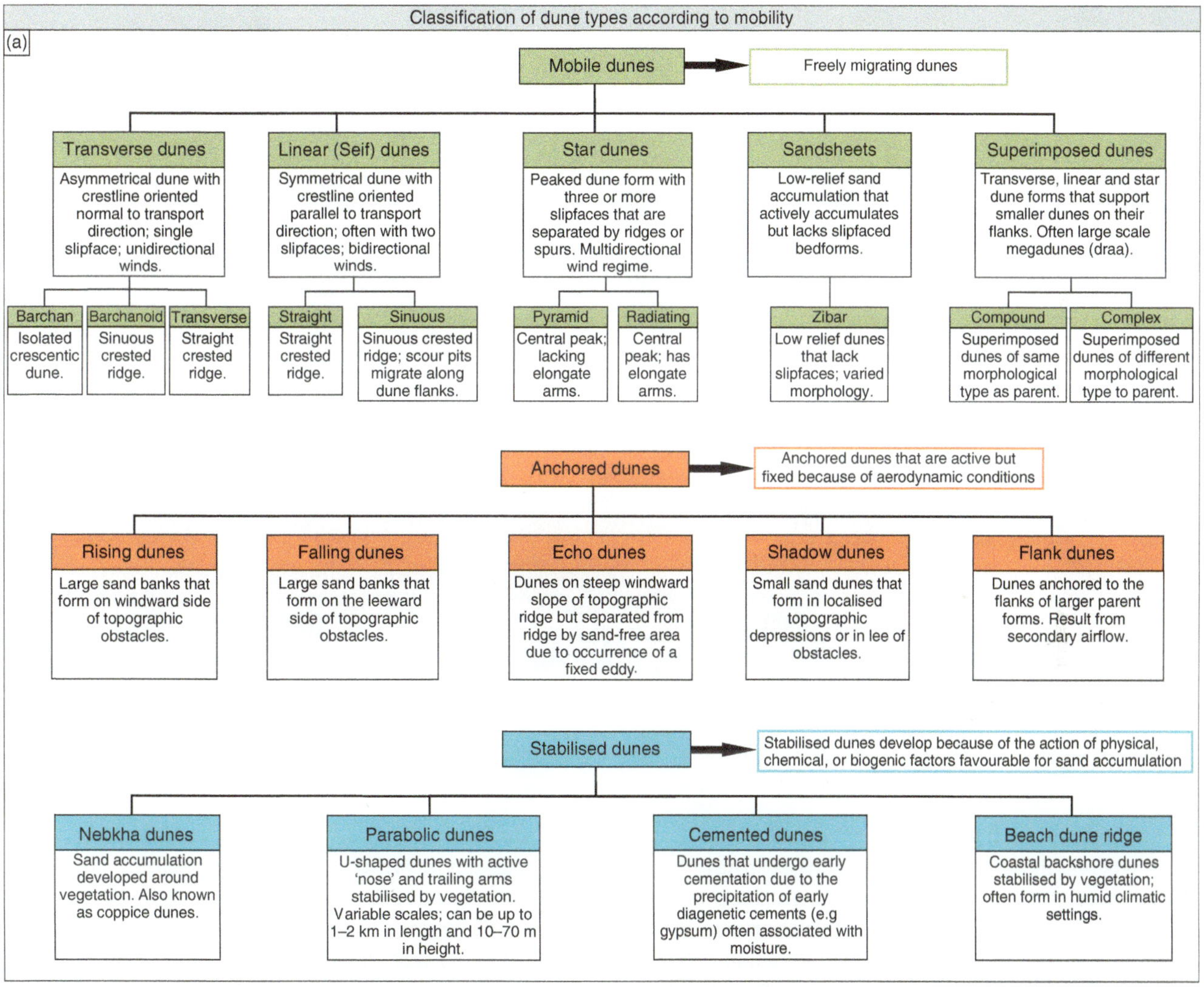

Figure 7.15 Aeolian dune types classified according to mobility. (a) Classification scheme. Adapted from Mountney (2006b). The scheme is flexible, and overlap can exist between the three groups. For example, nebkha and parabolic dunes could be classed as anchored by vegetation. Many dunes are neither transverse nor linear but are oriented oblique to the resultant sand transport direction. (b) Mobile isolated barchan dunes on Mars. Image taken by HiRISE Mars Reconnaissance Orbiter. Courtesy of NASA / JPL / University of Arizona / Public domain. (c) Examples of sinuous-crested linear dunes from the Namib Sand Sea. Courtesy of NASA's Earth Observatory (image ISS047-E-23405) / Public domain. (d) Example of mobile star dunes in the Rub' Al Khali Desert. Courtesy of Google Earth. (e) Examples of rising and falling anchored dunes from northwestern Saint Anthony, Idaho, USA. Courtesy of Google Earth. (f) Example of an anchored shadow dune from the Qinghai-Tibet Plateau. Modified after Xiao et al. (2015) / Springer Nature. (g) Example of an echo dune. Modified after Tsoar (2008) / Springer Nature. (h) Example of a stabilised nebkha dune in the Mojave National Preserve, San Bernardino County, California, USA; the dune is formed around a honey mesquite (*Prosopis glandulosa*). Courtesy of Chris M Morris / flickr / CC BY 2.0.
(i) Example of a parabolic dune on the shore of Poyang Lake, East China. Modified after Han et al. (2022). (j) Example of beach ridges and swales with stabilised and highly vegetated dunes developed on the ridges. Brodies Crossing, Makakatana strand plain. Modified after Botha et al. (2018) / Elsevier.

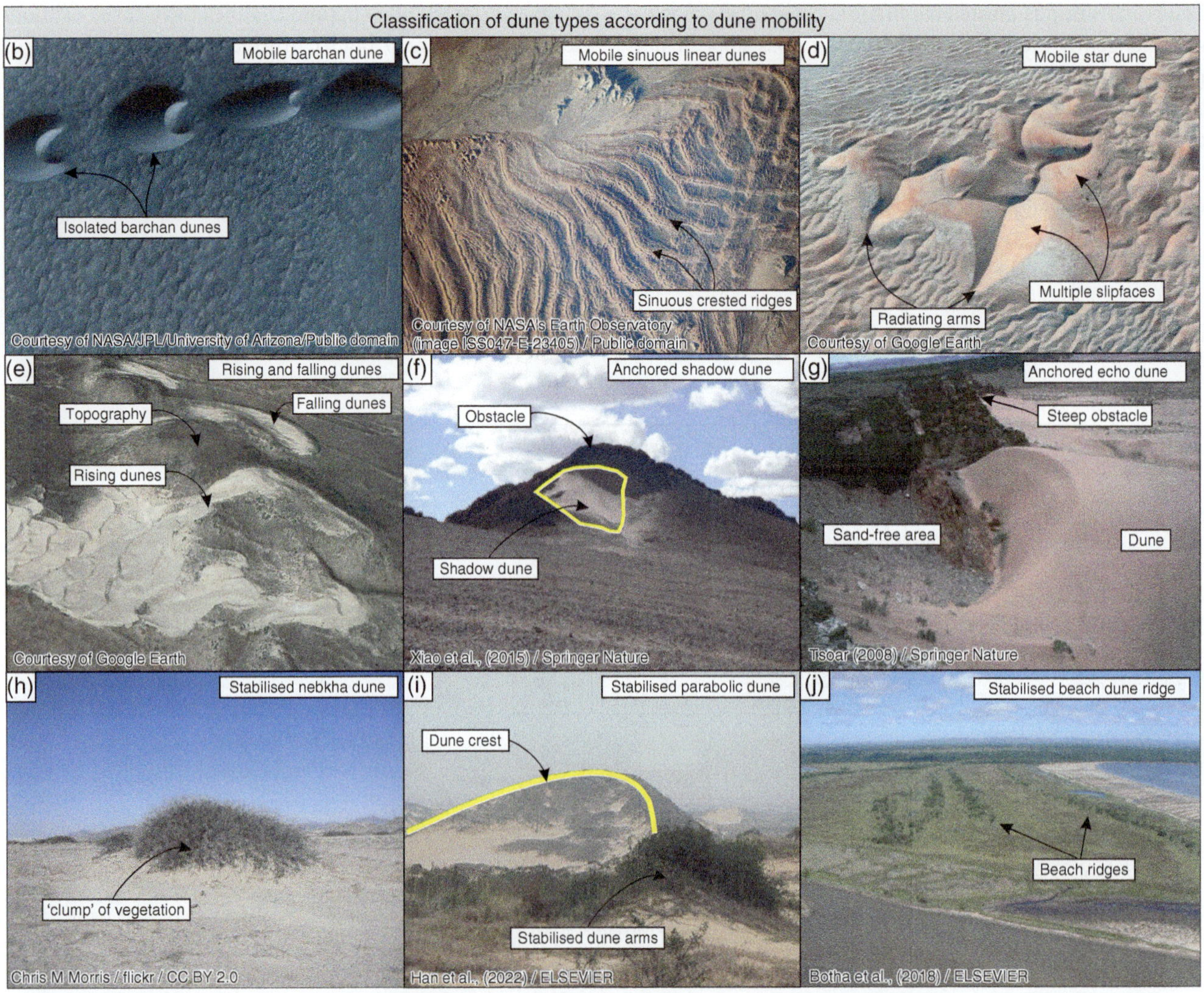

Figure 7.15 (Continued)

Dune bedforms that possess straight crestlines are classed as 2D, whereas those with curved, sinuous, cuspate or lobate crestlines are 3D (Rubin 1987a). Examples of 3D, transverse bedforms include spatially isolated, crescent-shaped barchan dunes that open downwind and are commonly surrounded by corridors of sand-free ground (Figure 7.14a). Transverse barchanoid dune ridges (Figure 7.14b–d) possess curved (in some cases sinuous) transverse crestlines; in plan view, the crestline curvature of successive bedforms may be either in phase or out of phase. In plan view, parabolic dunes form U-shapes that close downwind (i.e. have downwind convexity) (Figure 7.14k,l); this type is common in areas where vegetation acts to anchor or stabilise the outer dune limbs but is insufficient to halt the migration of the central part of the dune (Figure 7.15b). Seif dunes (from the Arabic for sword) are a type of 3D longitudinal (linear) bedform for which a curved (in some cases sinuous) crestline migrates downwind along the bedform crest in a motion similar to that of a snake. The spacing of seif dunes is typically approximately twice their mean width. The crests of many such dunes exhibit a regular plan-view curvature (in some cases referred to as a crestline sinuosity), with slipfaces developed on alternate flanks. Upwind ends of ridges are rounded, and along their length, Y-shaped bifurcations at angles of 30 to 50° may be present; these tend to open upwind. At their downwind ends, the ridges tend to be pointed. Most linear dunes (including seifs) are aligned close to parallel to the resultant vector of the effective winds. However, as they migrate, such types can creep ('crab') slowly sideways under the influence of complex wind patterns (Figure 7.16). Dome-shaped dunes have less steeply inclined upper flanks such that they present convex-up

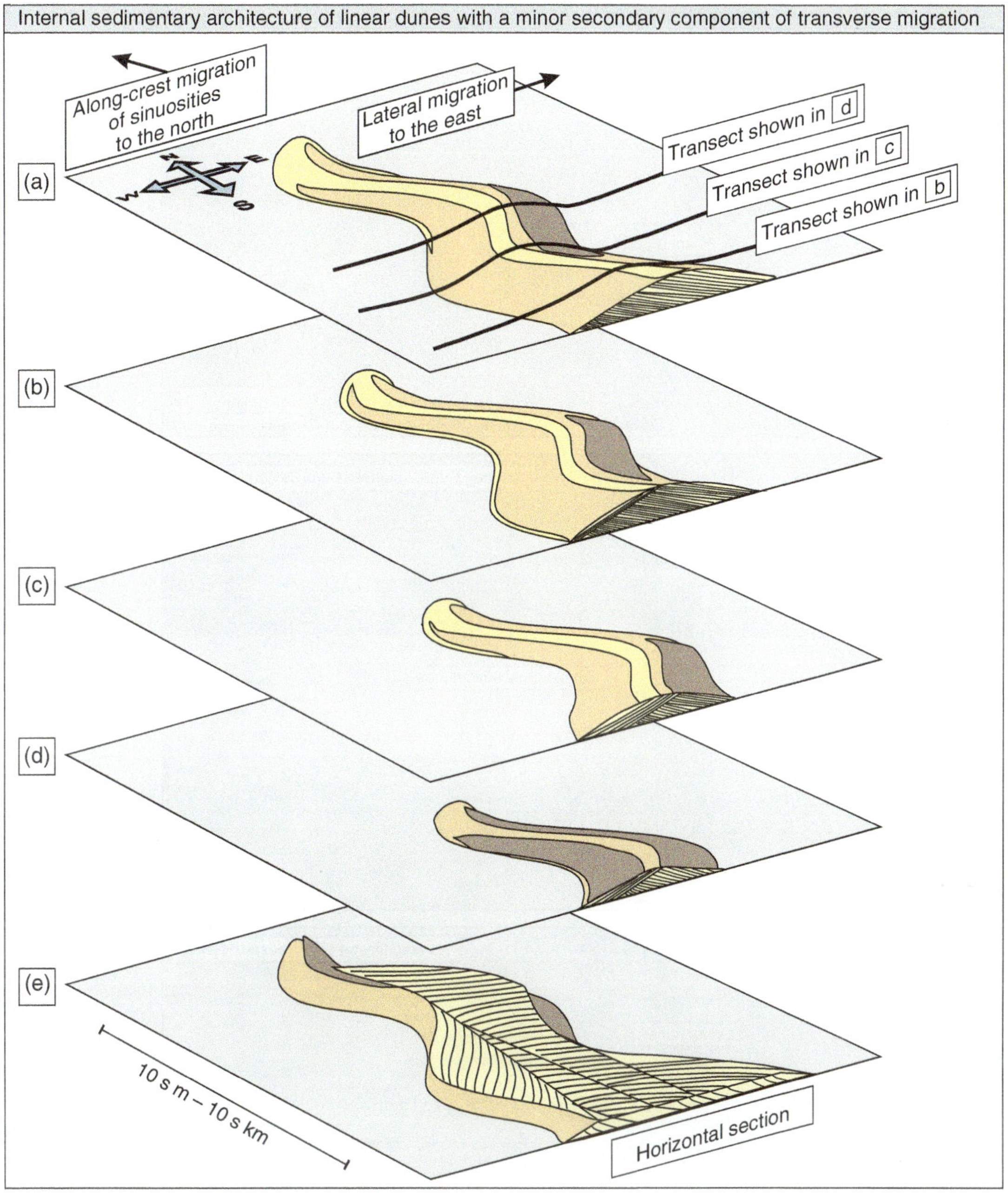

Figure 7.16 Longitudinal linear dunes (including seifs) experience a dominant direction of sand transport and migration that is close to parallel to the overall trend of the bedform crestline. However, many such bedforms additionally have a minor secondary component of transverse migration to the crestline trend. As such, bedforms have a resultant direction of migration that is a slightly oblique trend of the bedform crestlines. (a–e) Example depicting the pattern of compound cross-stratification deposited by a linear dune that is elongating to the north whilst simultaneously migrating laterally to the east. The component of eastward lateral migration results in the preferential accumulation of cross-bedding that dips approximately to the northeast. (a–d) Vertical sections through different parts of a single dune at an instant in time. (e) Horizontal section through the structure shown in Part a. Note the pattern of cross-stratification produced by along-crest migration of planform sinuosities. Lateral migration causes the preferential preservation of beds deposited on the eastern flank on the dune. Adapted from Rubin et al. (2008).

forms when viewed side-on (Figure 7.14i,j). They commonly develop as approximately circular or elliptical forms in plan view. 'Zibars' are low ridges without slipfaces, commonly of coarse-grained, hard-packed sand (Biswas 2005; Figure 7.17). They are usually aligned approximately transverse to the wind and commonly occur in the corridors between higher-relief bedforms, else as independent bedforms present on aeolian sandsheets.

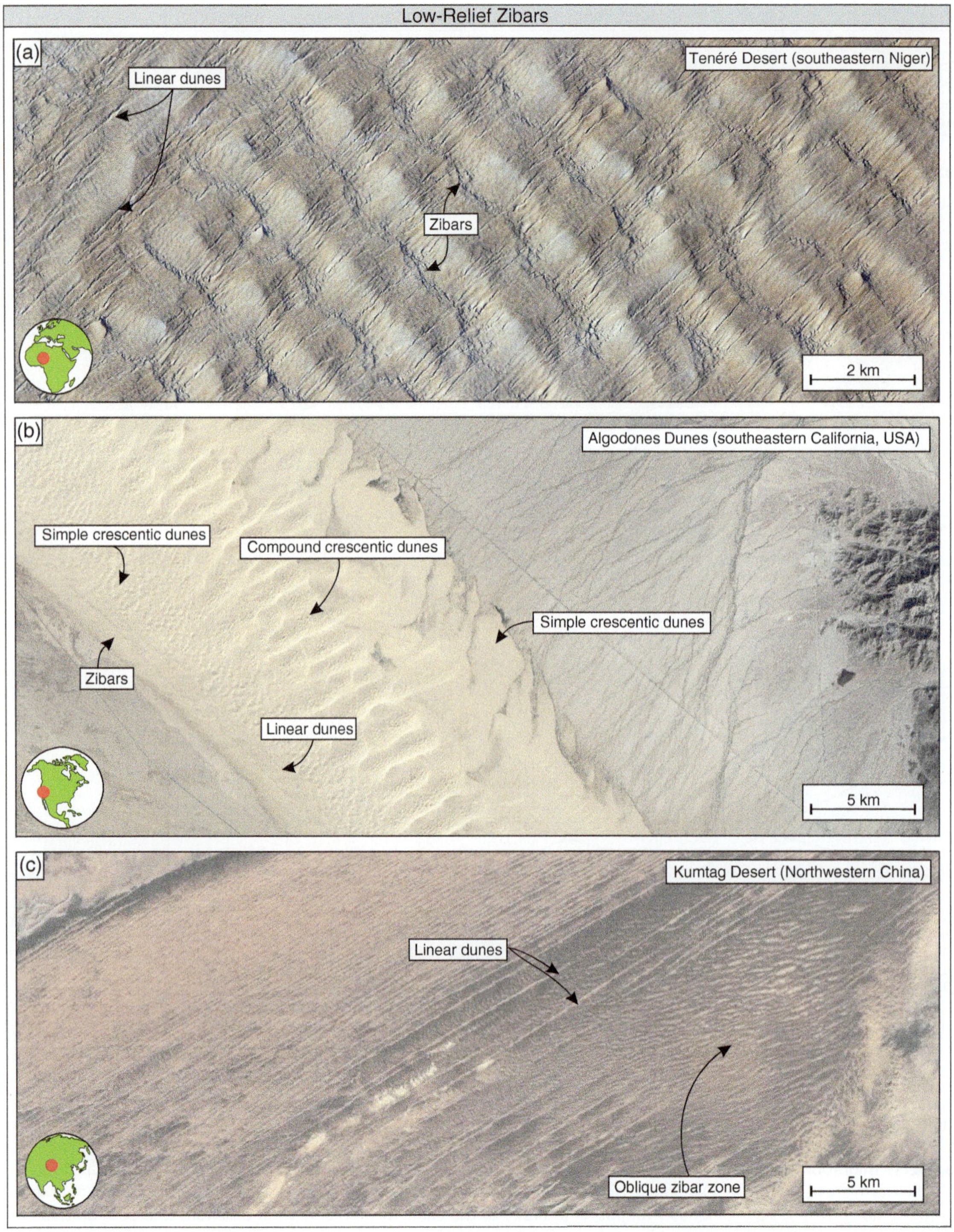

Figure 7.17 Examples of low-relief zibar bedforms. (a) Tenéré Desert, Niger. Photograph ISS022-E-5258 courtesy of the ISS Crew Earth Observations Experiment and Image Science and Analysis Laboratory (Johnson Space Centre). (b) Algodones Dunes, California, USA. courtesy of the NASA Earth Observatory (NASA EO-1 Team). (c) Kumtag Desert, China. Courtesy of Google Earth.

The style of migratory behaviour of mobile dunes can also be used for further classification. Dunes that migrate in a constant direction and at constant speed are *invariable*, whereas dunes that undergo temporal changes in migration direction, speed, asymmetry and/or steepness are *variable* (Rubin 1987b). A temporal change in the style of migratory dune behaviour is one potential explanation for the origin of geometrically complex bed sets delineated by erosional bounding surfaces in the stratigraphic record (see Section 7.5 and Box 7.2).

Box 7.2 Scheme for the classification of aeolian bedforms

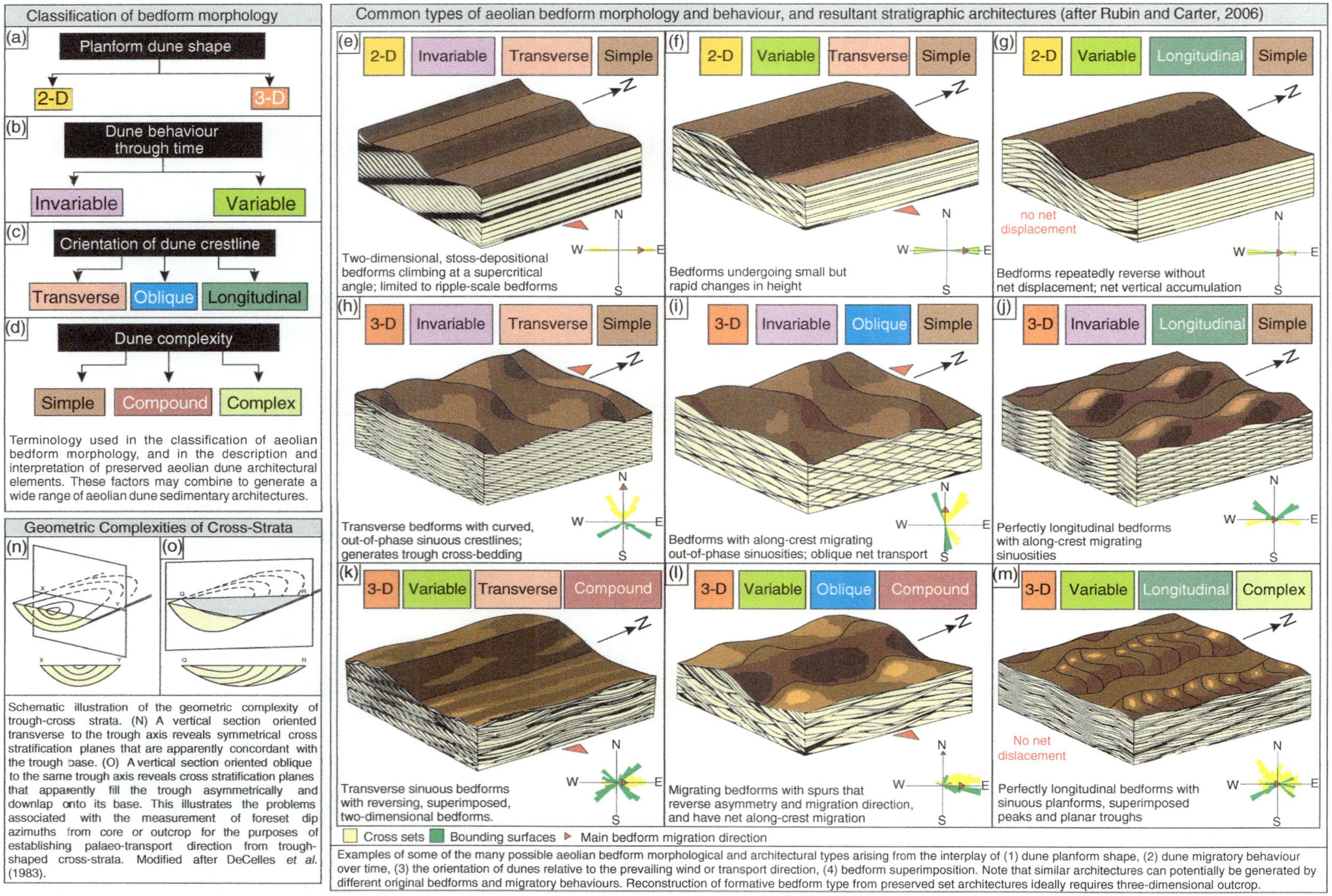

Box 7.2 Scheme for the classification of aeolian bedforms based on planform shape (a), nature of migratory bedform behaviour over time (b), orientation of bedform crestline relative to dominant wind or sand transport direction (c), and presence or absence of superimposed forms (d). Common types of aeolian bedform morphology and behaviour and resultant stratigraphic architectures are shown (e-m). Schematic illustrations of the geometric complexity of trough-cross strata are shown in parts (n) and (o). Models have been generated using the 'Bedforms' software of Rubin (1987b) and Rubin and Carter (2006).

Dunes are commonly arranged into a network ('aklé') in which there are linked transverse, longitudinal and oblique components (Figure 7.18). For example, sinuous, transverse ridges commonly display alternating linguoid and barchanoid sectors in plan view (i.e. convex and concave downwind, respectively). These curved crestline segments may occur either in or out of phase in relation to those developed on adjacent ridges. Elsewhere there may be straight-crested dunes (or megadunes) oriented transverse to the wind and, close to them, dome-shaped dunes with

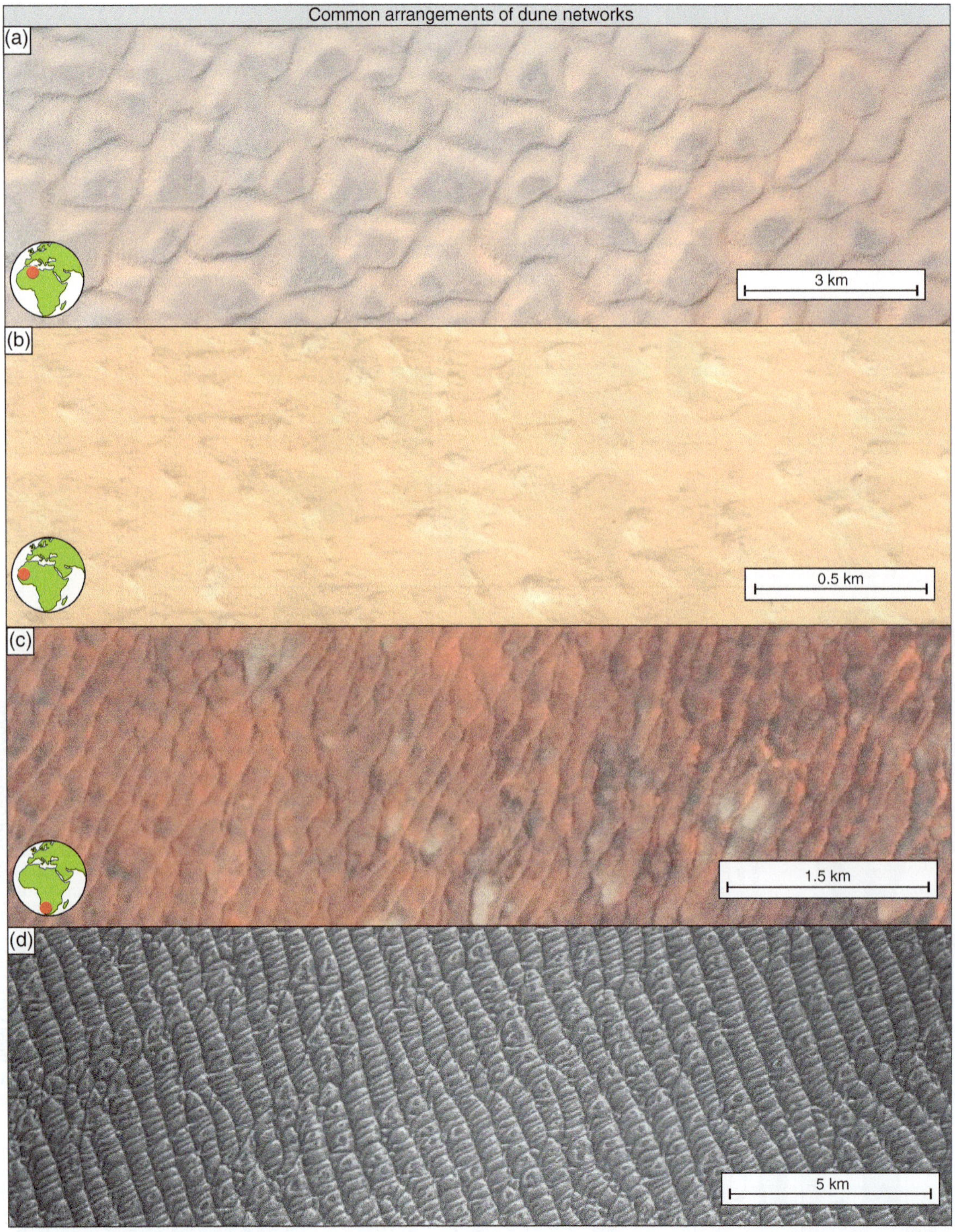

Figure 7.18 Examples of common arrangements of bedforms in dune networks. (a) Fish-scale dune network (aklé), Jebil National Park, Tunisia. (b) Fish-scale dune network (aklé), Akle Aoukar, Hodh Ech Chargui, Mauritania. (c) Dendritic (branched) pattern of linear dunes from the southwestern Kalahari Desert. a–c taken from Google Earth. (d) Network dunes on Mars formed from complex patterns of star dunes. CTX P13_006153_2608 (NASA/JPL/MSSS).

many minor slipfaces and rounded flanks inclined at low angles. Given the morphological complexity of many aeolian bedforms, it can be useful to plot the orientation of lee slopes as dip-azimuth data on a stereogram. This may enable subtle trends regarding bedform morphology and arrangement in relation to prevailing winds to be discerned.

Many aeolian dunes are comparable in size with aqueous dunes or sandwaves, and their preserved deposits may be difficult to discriminate (Charru et al. 2013). However, certain small-scale aeolian (and related) stratification types, notably distinctive wind-ripple strata and adhesion ripple strata, are useful diagnostic indicators (Section 7.4).

7.3.7 Aeolian Megadunes

Megadunes (draa) are bedforms at a larger scale than dunes, with wavelengths of 500–5000 m and heights in excess of 50 m. In some cases, megadune heights exceed 250 m, for example in parts of large sand seas of the Gobi, Rub' al Khali and Namib deserts (Figure 7.19) where aeolian sediment supply and transport rates are high. Aeolian megadunes are the tallest bedforms on Earth; there are no bedforms of aqueous origin of equivalent scale.

Megadunes are described using the same terminology as for dunes but may be additionally characterised by the presence of superimposed dune-scale bedforms on their flanks. 'Simple' megadunes lack superimposed dunes, whereas 'compound' forms possess superimposed dunes of the same morphological type, and 'complex' forms possess superimposed dunes of a different type (McKee 1979). Some megadunes may be slipface-less, though smaller superimposed dunes developed upon them may themselves have slipfaces. Common examples of complex megadunes include (i) linear (i.e. longitudinal) types that have smaller transverse ridges or spurs aligned across their slipface-less flanks; and (ii) star-draa and aklé (ridge complexes) with radiating arms, which support smaller, superimposed transverse barchanoid dunes that occur as fields of child bedforms across the flanks of the parent forms (Figure 7.19e). The oblique migration of superimposed dunes over larger, more slowly moving megadunes (Figure 7.19f) is common. The interaction of migrating megadunes and superimposed dunes is one potential explanation for the origin of geometrically complex compound cosets in the ancient record (see Section 7.5). The deposits of megadunes are interpreted from the ancient rock record (e.g. Mountney et al. 1999b).

7.3.8 Relation Between Bedform Type and Wind Direction and Variability

Four terms are commonly used to classify the energy and directional properties of the wind and to relate it to the form of particular dune types. The equivalent sand thickness (EST) is a measure of the size of an aeolian bedform: it is the thickness that the sand of which it is composed would be if it were spread evenly over its basal area. Drift potential (DP) is a measure of the total sand-moving capability of the wind, without regard to wind direction. Resultant drift potential (RDP) is a measure of the resultant or net sand-moving capability of the wind in the resultant drift direction (RDD) and can be determined by plotting sand-drift rose diagrams (Fryberger 1979). The ratio of RDP/DP (sometimes referred to as the unidirectionality index or UDI) provides an indicator of wind variability; values approaching unity (RDP/DP > 0.8) signify low wind variability (i.e. unidirectional winds) and low values (RDP/DP < 0.3) signify high wind variability. RDP/DP values plotted against EST yield a discriminant plot for different dune and megadune types (Figure 7.20), which suggests that bedform type is controlled by both wind regime and sand availability for aeolian transport (Wasson and Hyde 1983).

Barchan, longitudinal and seif bedforms are associated with relatively low sand availability, whereas transverse barchanoid ridges and star forms tend to develop where abundant sand is available. Barchan and transverse forms indicate a low diversity of effective wind directions, whereas star forms indicate a high diversity.

Transverse dunes tend to develop under conditions of unidirectional winds characterised by high RDP/DP values and are described as sand-transporting bedforms. By contrast, star dunes tend to develop in response to variable winds (low RDP/DP values), for example, winds that blow from different directions during different seasons and are sand-accumulating (storing) bedforms; they migrate only very slowly, if at all. The presence of sand stabilising agents, such as vegetation or chemical crusts on bedform slopes may complicate the relationship between wind regime and resultant morphology.

7.3.9 Aeolian Interdunes

Aeolian interdunes are depressions or low-relief areas present between adjoining dunes or megadunes. They may take the form of isolated depressions or occur as extensive 'interdune flats'. Interdunes may be sites of net sediment erosion (deflation), bypass or deposition, depending on the nature of the airflow that passes over them, the level of the water table relative to the interdune surface and the activity of non-aeolian processes within them.

Dry interdunes are topographic depressions (hollows) or flat areas between aeolian dunes where the surface exhibits no significant evidence of sedimentation that is influenced or controlled by moisture. Dry interdunes are dominated by wind ripples, though aeolian plane-beds may also occur (Figure 7.21a). Dry interdunes that are subject to episodic deflation may be characterised internally by granule and pebble deflation lags (Kocurek 1981). The absence of near-surface moisture means that the degree to which

Figure 7.19 Examples of aeolian megadunes (draa). (a) Megadunes and interdune lakes in the Badain Jaran Desert (Gobi Desert), China. Image from the NASA Earth Observatory (Landsat data from the US Geological Survey). (b) Megadunes in the Rub' al Khali Desert, Saudi Arabia. Courtesy of Google Earth. (c) and (d) Complex linear draa with numerous superimposed bedforms, mostly oriented oblique to the trend of the main crest line, northern Namib Sand Sea, Namibia. (e) Pyramid-shaped star draa with radiating linear ridges, some supporting superimposed transverse ridges, Central Namib Sand Sea, Namibia. (f) Oblique migration of superimposed dunes over larger, more slowly moving megadunes, eastern Namib Sand Sea, Namibia NASA Earth Observatory / Public domain.

dry interdunes are colonised by vegetation is typically minimal and animal trackways are not common. The geometry of present-day dry interdunes, such as those of the Namib Sand Sea (Lancaster and Teller 1988), is largely controlled by the spacing and plan-view shape (morphology) of adjoining dunes. Present-day dry interdunes range from spatially isolated hollows (Figure 7.21a) that are completely surrounded by dunes, to narrow but elongate corridors that

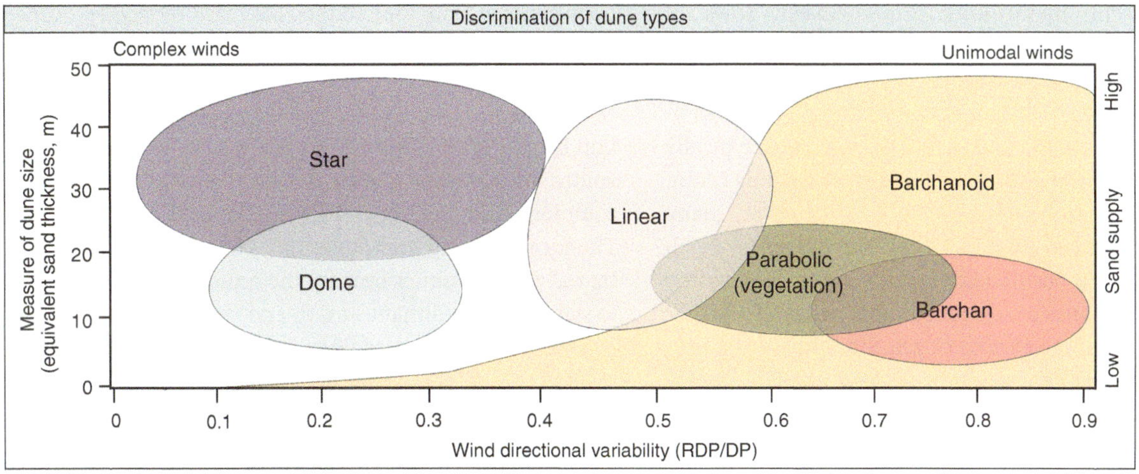

Figure 7.20 Discrimination of dune types in relation to a plot of equivalent sand thickness (EST) versus resultant drift potential (RDP)/ drift potential (DP). Adapted from Wasson and Hyde (1983).

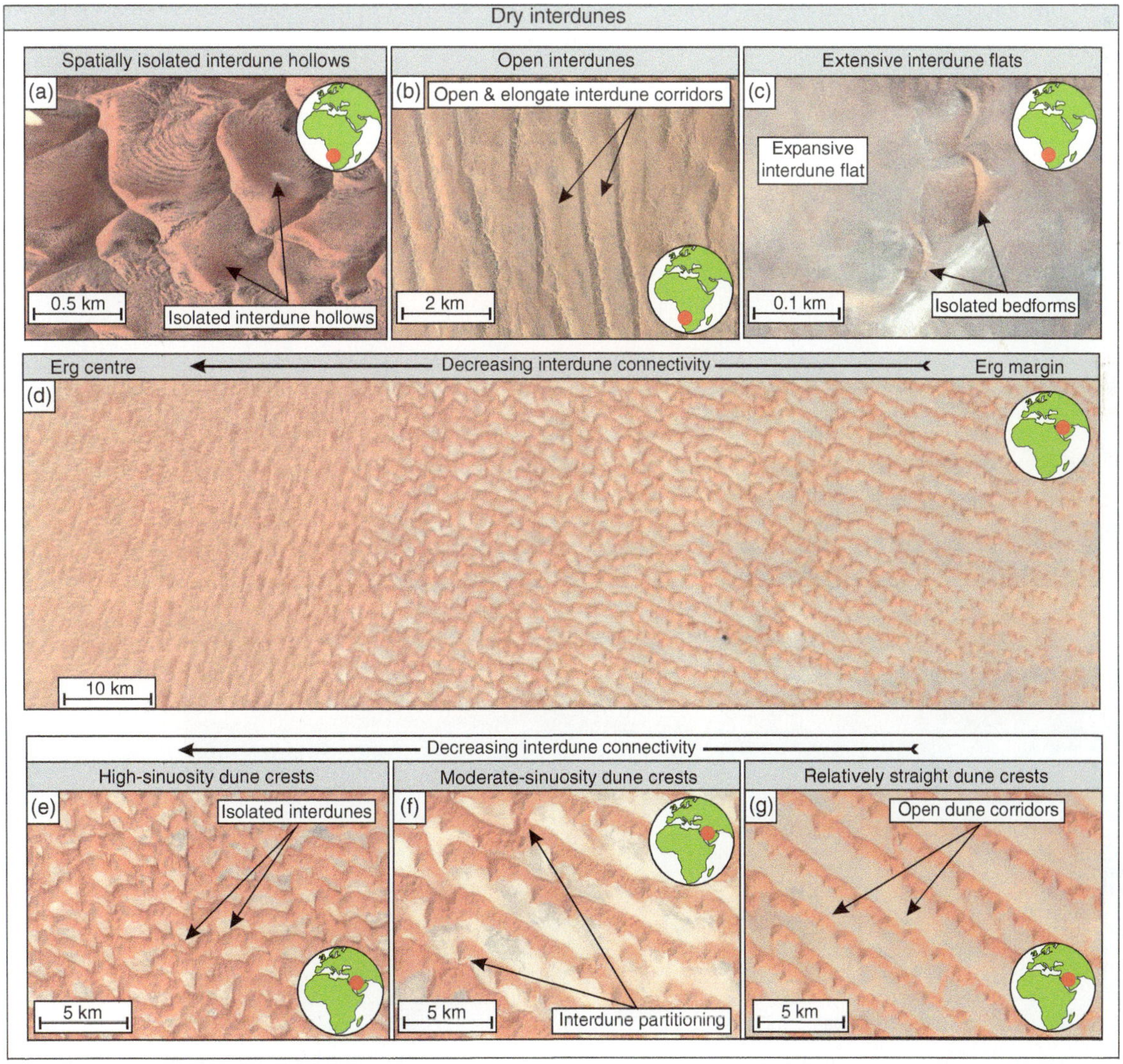

Figure 7.21 Examples of dry interdunes. (a–c) Namib Desert, Namibia. (d–g) Rub' Al Khali Desert, Saudi Arabia. (d) Transect from erg-centre to erg-margin showing changes in interdune connectivity. (e–g) Changes in interdune connectivity associated with changes in the amount of dune crestline sinuosity. Courtesy of Google Earth.

may extend for many kilometres between rows of transverse or linear dunes (Figure 7.21b), to extensive interdune flats completely surrounding isolated bedforms (Figures 7.21c). The width, length and degree of interconnection of adjacent dry interdune corridors typically decrease from a dune-field (erg) margin towards its centre as neighbouring dunes increase in size; however, many exceptions are known (Figure 7.21d). The degree to which dry interdunes are partitioned is partly influenced by the curvature of crestlines of adjacent dune forms, and whether they have protruding spurs or ridges. Straight-crested bedforms promote the generation of straight and continuous interdune corridors, whereas bedforms with highly curved or sinuous crestlines tend to partition interdune corridors into separate isolated hollows (Figure 7.21e–g).

Damp interdunes are topographic depressions (hollows) and flat areas between aeolian dunes where the accumulation surface is in contact with the capillary fringe of the water table such that sedimentation is influenced by the presence of moisture and loose sand may adhere to the damp surface upon contact. Damp interdunes are characterised by a range of adhesion structures and, in some cases, salt precipitation/dissolution structures (Ahlbrandt and Fryberger 1981; Kocurek 1981). Near-surface moisture encourages colonisation by plants and animals; a variety of root structures, rhizoliths, burrows and surface traces on a variety of scales are common (Ahlbrandt et al. 1978; Loope 1988; Hasiotis 2002). The form of damp interdunes may differ substantially from that of dry interdunes because the presence of moisture acts to stabilise the sediment surface and restricts the availability of sand within damp settings for aeolian transport. Documented examples of water table-controlled damp interdune elements in coastal settings include Padre Island, Texas (Hummel and Kocurek 1984; Kocurek et al. 1992), the Oregon Coast (Hunter et al. 1983; Figure 7.22a) and the Dhahran area of Saudi Arabia (Fryberger et al. 1983; Figure 7.22b). Examples from inland settings include parts of the Namib Desert (Lancaster and Teller 1988; Figure 7.22c) and White Sands, New Mexico (Simpson and Loope 1985; Fryberger et al. 1988).

Figure 7.22 Examples of water-table-controlled damp interdunes. (a) Oregon Coast, USA. (b) Northern Saudi Arabia Courtesy of Google Earth. (c) Ingólfshöfði, southern Iceland. (d) Skeiðarársandur, southern Iceland. Spade for scale is 1 m.

Wet interdunes are topographic depressions (hollows) and flat areas between aeolian dunes where the water table reaches or exceeds the level of the accumulation surface for protracted episodes such that the interdune is continuously or intermittently inundated by water (Figure 7.23a). Mud, silt and sand-grade sediment may be supplied to wet interdunes via aeolian or fluvial processes. Standing water in wet interdunes forms ponds or lakes, often termed 'oases'. Non-marine carbonate sediments or freshwater chert may accumulate in long-lived ponds (e.g. Driese 1985).

Figure 7.23 Examples of water-table-controlled wet interdunes. (a) Wet interdune pond, Lençóis Maranhenses, northeast Brazil. (b) Lençóis Maranhenses, northeast Brazil. (c) Badain Jaran Desert, Inner Mongolia, China. (d) Skeiðarársandur, southern Iceland. (e) Sossusvlei, Namib Desert, Namibia. Parts b and c from Google Earth.

Sedimentary structures associated with wet interdune settings include subaqueous current and wave ripples (sometimes with mud drapes), wavy laminae (Kocurek 1981), contorted bedding (Doe and Dott 1980), desiccation cracks, raindrop impressions, hard-pan crusts, mud flakes and mud curls (Fryberger 1990b). Chemical precipitates may accumulate in hypersaline wet interdune ponds, though they may be prone to later dissolution, in which case crinkly lamination or pseudomorphs (e.g. after halite or gypsum) may be evident. The continuous or episodic presence of water encourages plant colonisation and animal activity, and a wide variety of biogenic structures are common (Ahlbrandt et al. 1978; Gradziński and Uchman 1994; Carmona et al. 2018). The preservation potential of delicate features, such as burrows and animal trackways, may be enhanced where such features occur upon or within a damp, muddy substrate.

Interdunes may be flooded for a variety of reasons: ephemeral flash flooding from intra-erg rainfall events, fluvial inundation from beyond the erg margin, localised and temporary water-table elevation associated with intra- or extra-dune field rainfall events (Langford 1989; Langford and Chan 1989), seasonal variations in regional groundwater table and longer-term variations in water-table level due to changes in the balance between subsidence and regional climate variation (Kocurek and Havholm 1993). In some settings, flooding may occur where infiltration capacity is exceeded because of the presence of a low-permeability horizon at or beneath the interdune surface; this may promote the generation of a perched water table. Documented examples of permanently and episodically flooded wet interdunes in coastal settings include Lençóis Maranhenses, Brazil (Figure 7.23b), the tidally flooded back-barrier dune field of Guerrero Negro, Mexico (Fryberger 1990c), Padre Island, Texas (Hummel and Kocurek 1984; Kocurek et al. 1992) and the Jafurah area of Saudi Arabia (Fryberger et al. 1984, 1988; Fryberger 1990c). Examples from inland settings include the Duero Basin dune field, central Spain (García-Hidalgo et al. 2002), dune-dammed palaeovalleys in the Nebraska Sand Hills (Figure 7.23c; Loope et al. 1995) and parts of the Skeleton Coast dune field of northern Namibia (Stanistreet and Stollhofen 2002; Krapf et al. 2003; Alkathery 2016).

Interdunes may transition either spatially or temporally from dry to damp to wet types. Interdunes are 'open' where they form elongated corridors between dunes (Figure 7.24a); in some cases, open interdune flats may interconnect with neighbouring interdune flats. Open interdune corridors are susceptible to fluvial, lacustrine or marine incursion where they extend laterally to meet the boundaries of neighbouring sedimentary systems. Interdunes are 'closed' where they are entirely surrounded by topographically elevated aeolian dunes and are thereby isolated from neighbouring interdunes (Figure 7.24b).

7.3.10 Aeolian Sandsheets

Aeolian sandsheets are areas of sand accumulation that lack significant development of high-relief dune-scale or larger bedforms but are instead characterised by relatively flat areas of impact wind ripples, megaripples (Fryberger et al. 1992) or plane beds. In some cases, low-relief ridges and domes (zibars) with gently inclined sloping surfaces (dip $<10°$; relief no more than a few metres) are present (Nielson and Kocurek 1986). Sandsheet development is controlled by a variety of factors. Commonly there exists an overall positive net sediment budget, which is restricted by the limited supply and/or availability for the transport of sand (Breed et al. 1987; Khalaf 1989; Kocurek and Nielson 1986). The presence of vegetation in some sandsheets acts to restrict sand movement and may promote the construction of small 'nebkhas' (anchored embryonic dunes). In some aeolian sandsheets, the water table may lie close to the surface (Fryberger et al. 1988). Some sandsheets are the erosional remnants of what were previously higher relief bedforms; they are therefore the product of a negative sediment budget and widespread deflation (Nielson and Kocurek 1986; Pye and Tsoar 1990). Sandsheets armoured by granule and pebble deflation lags are indicative of ongoing deflation (Cosgrove et al. 2021b). Individual sandsheets vary in extent from localised sand patches ($<1\ km^2$) to major geomorphic features, such as the $>1 \times 10^5\ km^2$ Salima Sandsheet of the eastern Sahara (Breed et al. 1987; Maxwell and Haynes 1989). Other large sandsheets in hot desert settings are documented from the Ténéré Desert, Niger (Warren 1971) and parts of the Namib Desert (Lancaster 1985). Sandsheets also form important geomorphic elements in cold-climate deserts (Koster 1988; Wolfe 2022), such as those in Iceland (Mountney and Russell 2004, 2006, 2009).

7.3.11 Loess

Loess is a clastic sedimentary deposit, predominantly composed of silt-sized grains (dust). The accumulation of thick loess sheets requires a dust source, adequate wind energy to transport the dust, a suitable accumulation area where wind velocities are reduced to allow deposition from suspension, and conditions that are sustained for a protracted period of time (Pye 1995). Notable settings for loess generation and accumulation include dryland river valleys and proglacial or periglacial regions. Loess accumulates preferentially in areas of wind shadow where fine sediment fractions fall from suspension. Loess deposits are mostly homogeneous and may be loosely cemented soon after

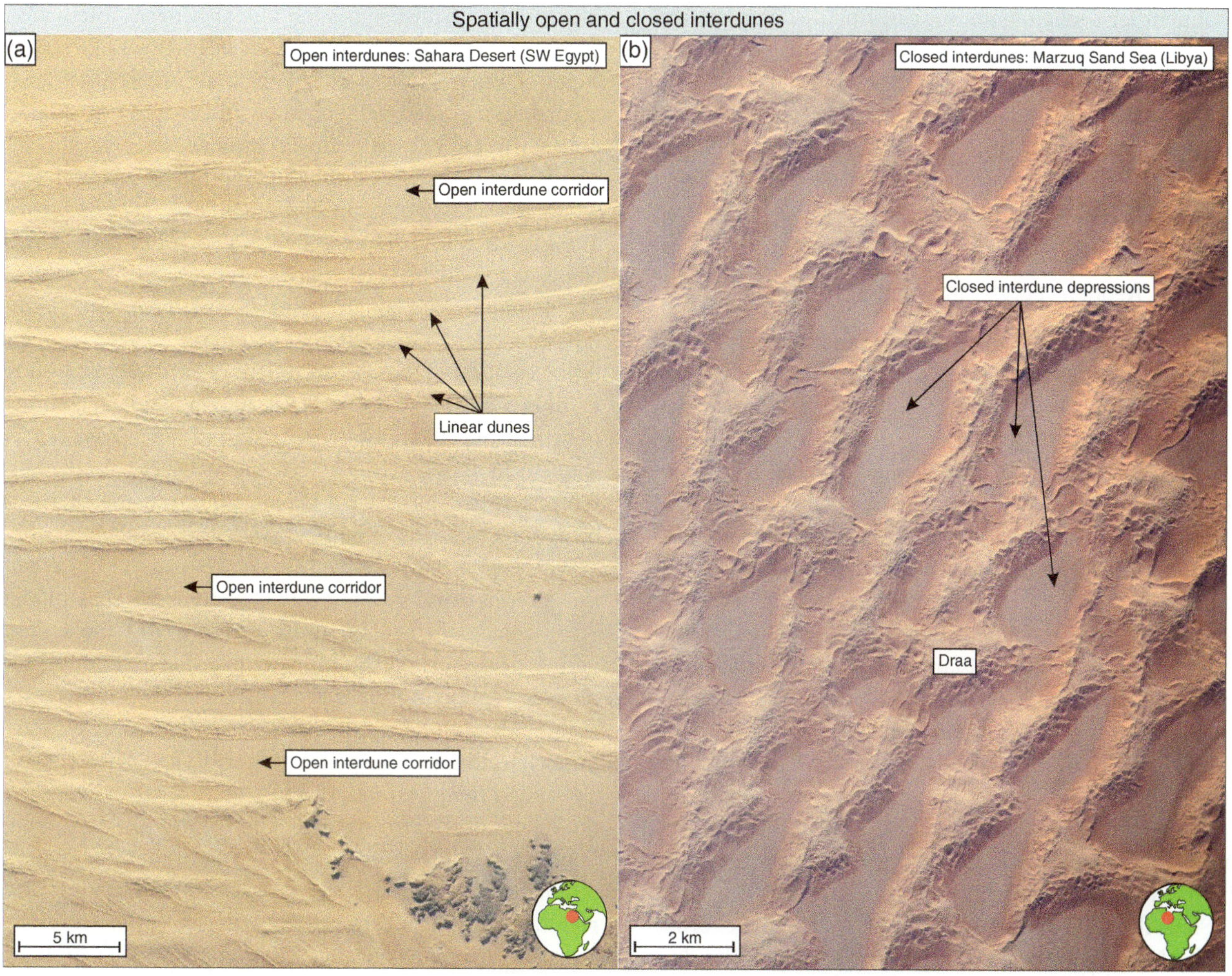

Figure 7.24 Spatially closed and open interdunes. (a) Open interdunes, Sahara Desert, SW Egypt. Courtesy of the NASA Earth Observatory (image ISS031-E-30783). (b) Closed interdunes, Marzuq Sand Sea, Libya. Courtesy of the NASA Earth Observatory (image ISS018-E-14770).

accumulation, for example, by calcium carbonate. The largest and best-known accumulation is on the Loess Plateau of northwest China, which covers an area of 6.4×10^5 km^2. Other extensive loess deposits occur across the Great Plains of North America, Central Europe and parts of Russia and Kazakhstan. Weakly consolidated loess accumulations, like those of the Loess Plateau, are prone to erosion, especially incision by the headward retreat of gullies during rainstorms. Loess accumulations many tens of metres thick are recorded near the Missouri River, Iowa, USA, and up to 300 m thick along parts of the Yellow River valley, China. Loess generally blankets older deposits, though accumulations may thicken into valleys and thin over rises. Loess deposits may form thick banks against cliff lines. Quaternary loess ridges are recorded aligned with palaeowinds that prevailed during the LGM: these are 'paha ridges' (North America) or 'greda ridges' (Europe).

Loess deposits make extremely fertile agricultural soil due to their mineral content: such deposits have a varied mineral content, drain water readily and are easily tilled by agricultural machinery. However, ploughed loess soils are prone to wind erosion, which in the past has led to the catastrophic loss of formally productive agricultural land, infamously the Dust Bowl of prairies in the Central United States in the 1930s, which contributed to long-term economic depression.

7.3.12 Aeolian–Alluvial Fan Interaction Processes

Alluvial-fan systems commonly form laterally extensive overlapping aprons ('bajadas') along mountain fronts that flank basins in which significant aeolian processes are active. Alluvial fans may contribute significant sources of sediment for aeolian landform construction, as is the case for the Mojave River, southeast California (Blair and McPherson 2009; Belnap et al. 2011) and the alluvial-fan systems that border parts of the Rub' Al-Khali sand sea, Saudi Arabia. Aeolian processes commonly rework alluvial sediments upon fan surfaces (e.g. Poplar Fan, China; Zhang et al. 2021).

7.3.13 Aeolian–Fluvial Interaction Processes

Permanent, intermittent and ephemeral fluvial systems (see also Sections 5.3.4 and 5.4.3) occur in many dryland regions (Powell 2009), and in many cases, they interact with aeolian systems, including in parts of Australia, India, Saudi Arabia and the southwestern United States (e.g. Schenk and Fryberger 1988; Tooth 2000, Glennie 1987, 2005; Nanson et al. 2002). Numerous studies have documented types of interaction between aeolian and fluvial systems in modern settings (e.g. Langford 1989; Trewin 1993; Stanistreet and Stollhofen 2002; Bullard and McTainsh 2003; Al-Masrahy and Mountney 2015). Several studies have explicitly focused on types of interaction between contemporaneously active aeolian and fluvial systems (e.g. Frostick and Reid 1987; Cooke et al. 1993; Tooth 2000; Bull and Kirkby 2002; Parsons and Abrahams 2009; Reid and Frostick 2011; Liu and Coulthard 2015; Santos et al. 2019). Many such fluvial systems exhibit complex and long-lived interactions with aeolian dunes.

Some fluvial systems generate significant supplies of sediment for aeolian-dune construction, as in the Kelso dune field, Mojave desert of California (Sharp 1966; Kocurek and Lancaster 1999). Other fluvial systems bound and limit the spatial extent of dune fields. They may remove significant volumes of aeolian-delivered sediment accumulated in dry riverbeds; seasonal floods act to flush aeolian sand from river channels and rework that material downstream. Examples include the Kuiseb River of Namibia that delimits the northern margin of the main Namib Sand Sea (Goudie 1972; Ward 1983) and the Hoanib, Ugab and Uniab Rivers that interact with aeolian dune fields of the Skeleton Coast, northern Namibia (Stanistreet and Stollhofen 2002; Krapf et al. 2003). Aeolian dunes are commonly present alongside dryland river courses; aeolian sand is derived from the local deflation of river sediment. Such systems are common in Australia and parts of India where dune deposits can locally include shell fragments of freshwater organisms (e.g. gastropods).

Aeolian and fluvial processes in desert settings rarely operate independently: they are usually dynamically linked and exhibit a range of sedimentary interactions. The diverse range of system interactions gives rise to considerable complexity in terms of geomorphology, sedimentology and preserved stratigraphy. Al-Masrahy and Mountney (2015) identify 10 distinct types of fluvial–aeolian interaction (Figure 7.25): (i) fluvial incursions aligned parallel to the trend of linear chains of aeolian dune forms; (ii) fluvial incursions oriented perpendicular to the trend of aeolian dunes; (iii) bifurcation of fluvial systems around the noses of aeolian dunes; (iv) through-going fluvial channel networks that cross entire aeolian dune fields; (v) flooding of dune fields due to regionally elevated water-table levels associated with fluvial floods; (vi) fluvial incursions emanating from a single point source into dune fields; (vii) incursions emanating from multiple sheet sources; (viii) cessation of the encroachment of entire aeolian dune fields by fluvial systems; (ix) termination of fluvial channel networks in aeolian dune fields; and (x) long-lived versus short-lived types of fluvial incursion. These interaction types form the basis for a classification scheme that can be applied to desert dune-field systems generally.

Where externally sourced fluvial systems cannot reach the interior parts of dry aeolian systems because of the great density of aeolian dunes present and the closed nature of associated interdune depressions, the opportunity for aeolian sediment reworking via fluvial processes is limited. Minor fluvial streams may, however, develop in such settings in response to localised surface run-off associated with rainfall events that occur within the dune field itself. Streams associated with intra-dune-field flooding are highly ephemeral. Reworking of aeolian sediment by ephemeral flows is of limited extent, and resultant deposits are composed solely of fluvially reworked aeolian sand (Svendsen et al. 2003; Strömbäck et al. 2005).

Where externally sourced fluvial systems are able to penetrate into the interior of aeolian dune systems, numerous morphological controls act to govern the distance and type of fluvial incursion: (i) morphological dune type, which defines the length and continuity of individual dune segments; (ii) the orientation of dunes relative to the direction of fluvial flooding; (iii) the form of interdune corridors that are present between dune segments, defined in terms of their width and length, and spatial changes in these parameters that dictate whether such features are classified as open or closed morphological elements; and (iv) the type and rate of aeolian dune and interdune migration relative to the frequency of fluvial flood events. The penetration distance of fluvial incursions into the margins of aeolian dune fields is greatest for regularly spaced trains of relatively straight-crested aeolian dunes for which bedforms are separated by broad interdune flats and where fluvial systems impact the dune-field margin at an angle whereby flood waters associated with high-magnitude events can pass relatively unhindered along open interdune corridors. Open interdune corridors play an important role where they occur adjacent to the path of fluvial systems passing into

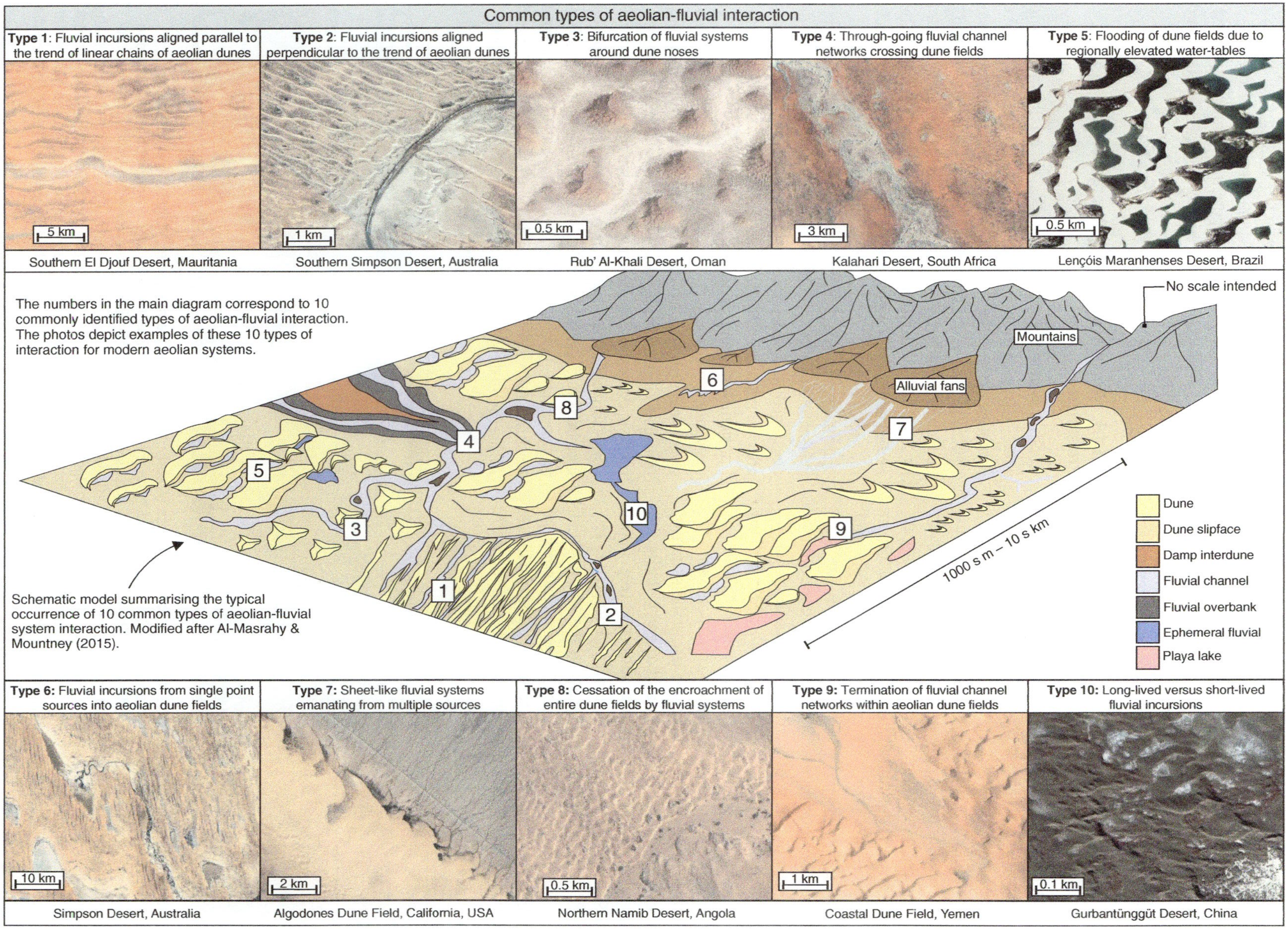

Figure 7.25 Ten common types of aeolian–fluvial interaction. Adapted from Al-Masrahy and Mountney (2015).

aeolian dune fields (e.g. Hoanib River in Skeleton Coast, north-western Namibia; Stanistreet and Stollhofen 2002): they act as catchments for excess water during flood events, thereby acting to buffer flood discharge. In cases where interdune corridors terminate in closed depressions, they may act to pond flood waters, the suspended-load deposits of which commonly form mudstone layers that are relatively resistant to erosion due to their cohesive nature (Loope et al. 1995; McKie et al. 2010; Höyng et al. 2014). Evaporitic salt deposits may also accumulate in such settings.

Increases in the frequency and/or magnitude of fluvial flood events within otherwise aeolian-dominated sedimentary systems – most obviously in response to climate change – impact continued aeolian system activity. Increased fluvial discharge and stream power promote erosion of earlier aeolian deposits. Fluvial reworking of aeolian sediment, its transport downstream and its ultimate re-deposition in areas where floods terminate influence the supply and availability of sediment of a calibre suitable for later renewed aeolian construction elsewhere in the overall system. Increased fluvial flood activity restricts the potential for aeolian dune migration (Pickup et al. 2002; Bullard and McTainsh 2003). The increased presence of water in otherwise dryland settings provides conditions suitable for vegetation colonisation, promotes the stabilisation of interdune flats, and dune flanks, and thereby limits the capability of the wind to erode such substrates (e.g. Levin et al. 2009). Similarly, the deposition of mud drapes via settling from suspension over wide areas in the aftermath of repeated flood events will also limit the availability of underlying sediment for aeolian transport (Purvis 1991). Frequent floods will recharge the groundwater beneath the aeolian dune field, thereby raising the water table, possibly to the level whereby formerly dry interdunes become damp or wet. An elevated water table tends to limit the availability of sediment for aeolian transport. However, it also increases the preservation potential of the aeolian bedforms that gradually subside beneath it (e.g. Mountney and Russell 2009).

7.3.14 Aeolian–Lacustrine Interaction Processes

Aeolian systems border many lakes, especially in dryland settings (see Section 6.5.2 'underfilled lake basins'). Lake shorelines can serve as a sediment source for dune construction. Evaporitic lakes may have a fringing salt flat of halite, gypsum, potash or other salts, in some cases referred to as sabkha (Section 3.15). Evaporitic crusts may be reworked by aeolian processes. Some small lakes and ponds may exist as transient features, in some cases flooding on a seasonal or decadal basis in response to fluvial floods that originate outside the dune field. The interdune ponds developed at Sossusvlei in the central Namib Sand Sea are one such example (Figure 7.26). In other cases, small transient lakes and ponds may flood due to a local rise in the water table, possibly in the aftermath of rainfall events. Over longer timescales (10^2–10^3 years), lake shorelines may transgress into adjoining aeolian dune fields during wet climatic episodes, especially along low-lying interdune corridors. Conversely, aeolian dunes may migrate over exposed lake beds and salt flats during dry climatic episodes. Present-day examples include Kati Thanda–Lake Eyre, Central Australia (Wopfner and Twidale 1988), Lake Chad, Central Africa (Schuster et al. 2009) and White Sands, New Mexico,

Figure 7.26 Examples of desert (playa) lakes. (a) Satellite image of part of the eastern Namib Desert, Namibia, with notable ephemeral interdune lakes (Sossusvlei and Deadvlei) highlighted. Note the major interdune corridor to the right of the image, along which fluvial streams pass during flood events. The cream colour is a calcrete pan that has been precipitated in low-lying interdune depressions that are subject to repeated flooding and evaporation of flood waters. Courtesy of Google Earth. The yellow star indicates the top of a dune from which the photo in b was taken. (b) The ephemeral interdune lake of Sossusvlei in the immediate aftermath of a flood event. (c) The ephemeral interdune lake of Deadvlei. This interdune no longer floods since a dune migrated to close off the neck of the interdune corridor. As a result the trees that once grew on the interdune flat have now died. Image modified after Diego Delso / Wikimedia Commons / CC BY-SA 4.0.

USA where gypsum deposited at the lake shoreline is blown inland to construct gypsum dunes that then undergo early cementation (McKee 1966). Ancient examples of aeolian–lacustrine interactions include the Jurassic Sergi Formation, Brazil (Scherer et al. 2007) and the Devonian Clair Group, Scotland (Nichols 2005).

7.3.15 Aeolian–Sabkha Processes

Sabkhas are low-relief regions where accumulation occurs wholly or partly as a result of evaporite precipitation, commonly of halite or gypsum, though in some cases as calcium carbonate (Smoot and Castens-Seidell 1994; Goodall et al. 2000; Alsharhan and Kendall 2003) (Section 10.3.5). Coastal and inland sabkhas commonly occur within and adjacent to desert aeolian environments (Figure 7.27), for example, inland salt flats that fringe playa-lake basins surrounded by aeolian dunes. Sabkha sedimentation involves interactions between chemical (precipitate) and aeolian processes that result in the generation of a variety of wavy and crinkly lamination types due to sediment disturbance by salt growth structures. Smaller-scale structures include salt concretions (Figure 7.28a) and gypsum desert rose crystals that grow at a water table interface in the shallow subsurface (Figure 7.28b); larger-scale salt structures include teepees (salt growth ridges) that develop as a polygonal pattern on evaporitic surfaces (Figure 7.28c). Sand may adhere to damp salt surfaces. Sandstone pseudomorphs after halite (hoppers) and gypsum (desert rose) are common (Figure 7.28d,e).

7.3.16 Aeolian–Marine Processes

Coastal aeolian systems may directly border and interact with marine shorelines. Coastal aeolian systems may be subject to construction or erosion. Beaches can act as an important source of sediment for dune construction directly inland. Longshore drift can provide a continuous supply of sediment for dune construction. Conversely, waves and tides can erode dunes that encroach onto a shoreline. Inland water tables can be influenced by the sea level such that coastal aeolian dune fields may have wet (i.e. flooded) interdunes, some with freshwater; others

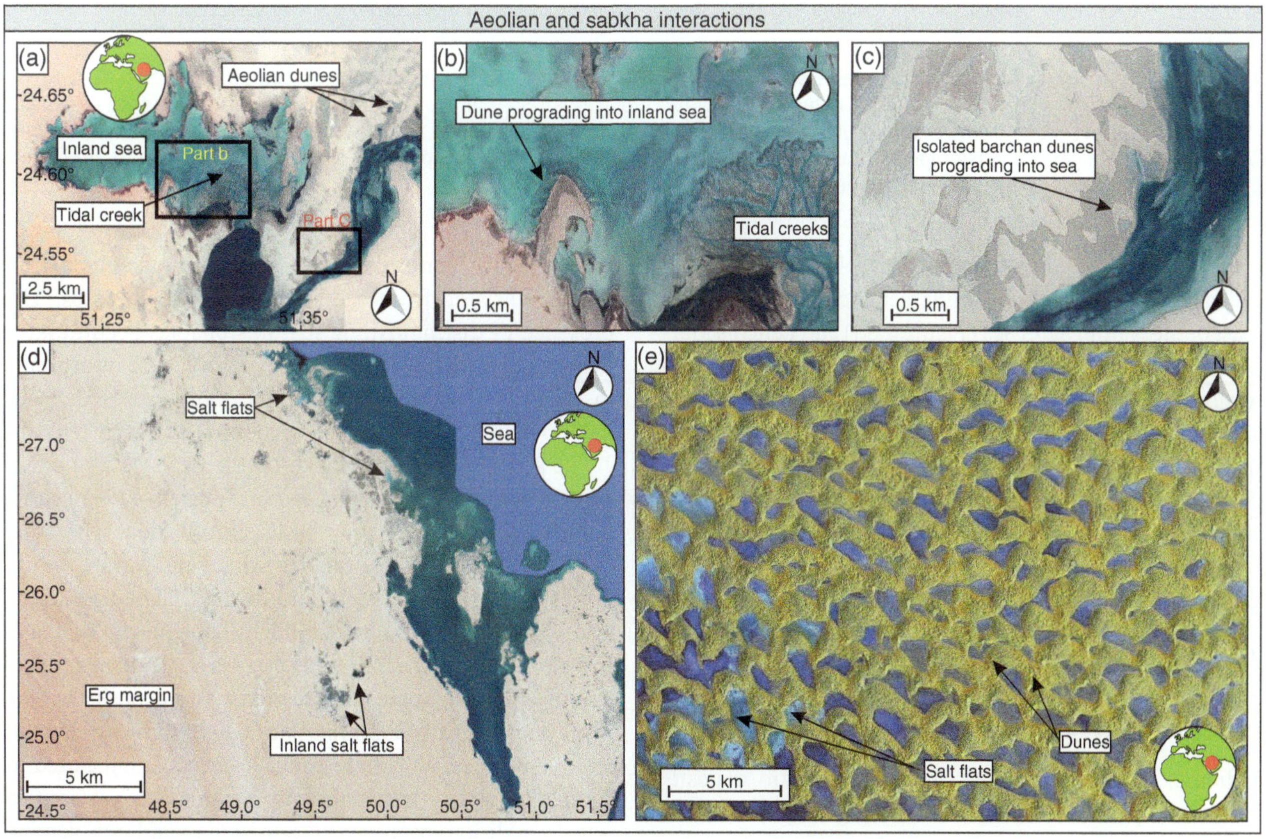

Figure 7.27 Examples of aeolian and sabkha interactions. (a) Eastern desert coast of Qatar showing interactions between a tidal lagoon along the Persian Gulf and adjoining aeolian dunes. (b) Interactions between tidal creeks and aeolian dunes that have prograded to form a peninsula into the inland sea. (c) Isolated barchan dune prograding towards the coastline of the inland sea. (d) Modern coastal sabkha, Bahrain Coast. (e) False colour image of aeolian dunes and salt flats in the Rub' Al Khali, Saudi Arabia. Image from NASA NSSDCA Photo Gallery. Images a–d courtesy of Google Earth.

Figure 7.28 Chemical sedimentary structures associated with aeolian environments. (a) Halite crystals forming extensive floating mats on the surface of a shallow playa lake. United Arab Emirates. Courtesy of Stephen Lokier. (b) A large desert rose, a type of gypsum crystal structure that grows in the shallow subsurface at the water-table interface in arid settings. (c) Polygonal salt growth structures forming pressure ridges due to isotropic expansion of a surface crust. Courtesy of Stephen Lokier. (d) Sandstone pseudomorphs after halite (cubic hopper crystals). (e) Sandstone pseudomorphs after gypsum.

with brackish water. Present-day examples where aeolian and marine systems directly interact include the western coast of the Namib Sand Sea (Figure 7.3d) and Lençóis Maranhenses on the north-eastern coast of Brazil. Ancient successions that record coastal aeolian systems commonly preserve a sedimentary record of the effects of sea-level rise and associated marine transgression. Other ancient successions record aeolian advance over former marine shorelines due to regression.

7.4 Facies

7.4.1 Introduction to Aeolian Facies Analysis

Although the grains that comprise aeolian deposits commonly possess distinctive mineralogical and textural properties, they can usually only be reliably used for palaeoenvironmental reconstruction when associated with diagnostic sedimentary structures. Collectively, the textural and structural characteristics of sedimentary deposits enable the recognition of distinctive facies. Aeolian facies comprise small-scale units of aeolian stratification arising from a distinct suite of processes. The presence of distinguishing features enables accumulated deposits of aeolian origin to be recognised and interpreted in terms of formative processes, in both recent deposits and ancient sedimentary successions (Hunter 1977a, b).

The stratified sets that make up the deposits of modern aeolian bedforms and sandsheets and ancient aeolian strata dominantly comprise four basic stratification types that occur in various configurations: wind-ripple strata, grainflow strata, grainfall strata and aeolian plane-bed strata. These stratal types are readily identifiable in modern aeolian settings (Figure 7.29) but can be difficult to reliably identify in some ancient successions. The deposits of aeolian dunes typically comprise packages of these fundamental stratal types, and these packages occur intercalated as complex mosaics that commonly make up cross-stratified sets. In addition, numerous other aeolian and non-aeolian but intimately related facies types occur: massive (i.e. internally structureless) deposits; deposits related to adhesion of aeolian grains to a damp surface; deposits arising from the migration of megaripples; deposits associated with

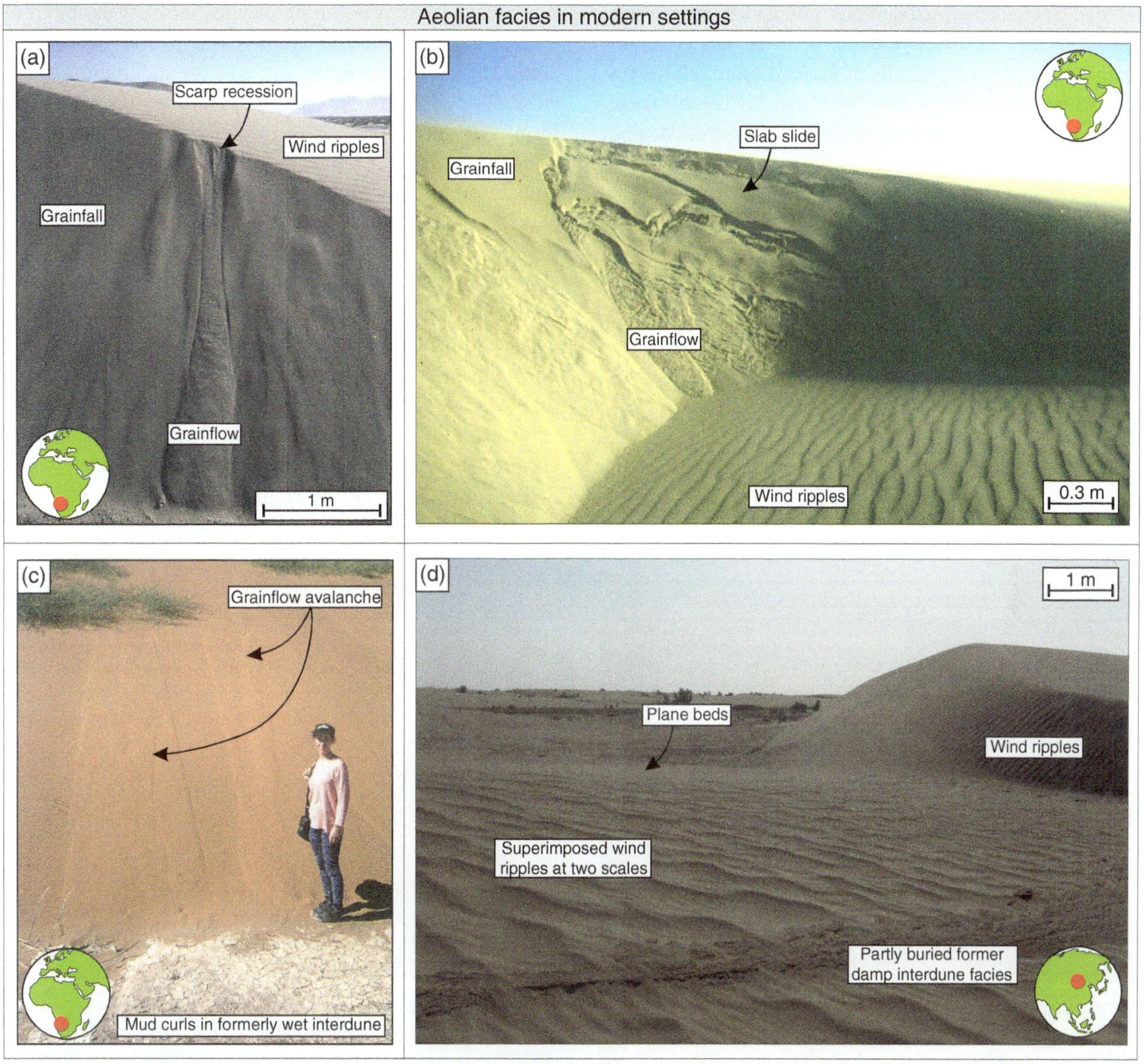

Figure 7.29 Aeolian facies and their occurrence. (a) Aeolian dune lee slope largely blanketed by grainfall deposits. Failure of the oversteepened lee has generated a tongue-shaped grainflow avalanche deposit. The point of failure receded back up-slope to the crest of the dune via a process called scarp recession. Note the wind ripples developed on the stoss slope of the dune. Huab Basin, northern Namibia. (b) Aeolian dune lee slope largely blanketed by grainfall deposits. Failure of the oversteepened lee has generated a slab slide. The slabs of sand have degenerated downslope into grainflow avalanche deposits. Wind ripples are developed on the dry interdune flat. Skeleton Coast, northern Namibia. (c) Grainflow avalanches down the lee slope of a dune. The dune is slowly advancing over a thin mud deposit with mud curls; this fine-grain sediment was originally deposited from suspension in a wet interdune pond. Sossusvlei region, Namib Desert, Namibia. (d) Wind ripples oriented across the lee slope of a dune that lacks an active slipface. Two scales of superimposed wind ripples are present in the dry interdune, as are aeolian plane beds in the distance. Gobi Desert, southern Mongolia.

soft-sediment deformation via physical means; deposits associated with chemical precipitation and the growth of nodules and concretions of various types; deposits associated with sediment disturbance of biogenic origin due to both animal bioturbation and the growth and preservation of plant-root structures; numerous closely related non-aeolian facies types that occur in close association with the abovementioned types. Many of these fundamental facies types occur collectively in facies associations that share a common genetic origin and are therefore valuable indicators of palaeoenvironment. Facies units commonly occur in predictable vertical or lateral successions that may be interpreted in terms of the temporal and spatial evolution of geomorphic elements within evolving aeolian systems. Such facies associations and facies successions may form the constituent parts of architectural elements, which are larger components of aeolian sedimentary systems with distinctive geometries; these architectural elements record the preserved expression of discrete aeolian sub-environments.

7.4.2 Aeolian Ripple Stratification

The tractional processes that generate impact wind ripples give rise to various types of wind-ripple stratification (Figure 7.30). *Rippleform laminae* (Hunter 1977a) occur where grain size differentiation enables the form of the ripples to be observed, either entirely or in part (Figure 7.31). Where ripple-scale bedforms climb at a supercritical angle (i.e. a climb trajectory that is steeper than the inclination of the stoss slope of the ripple), they may preserve lamination that records both the lee slope and the stoss slope of the rippleform. In this case, in-phase waviness may be apparent (Figure 7.31). Where the climb is at a critical angle (i.e. a climb trajectory that is the same as the inclination of the stoss slope of the ripple), the shape of the original rippleform is just preserved in full. Where the climb is at a subcritical angle (i.e. a climb trajectory that is less than the inclination of the stoss slope of the ripple), only the lower part of the original ripple and its foreset will be preserved (Figures 7.30a and 7.31). However, the uniformity of grain size that typifies many aeolian sands means that internal laminae cannot necessarily be distinguished, and *translatent stratification* results from ripple migration: the deposit arising from the passage of each ripple lacks internal structure (Figure 7.30b). Wind-ripple strata may exhibit a weak inverse grading, in part because the finest material tends to accumulate in sheltered ripple troughs, whereas coarser grains concentrate on the ripple crests (Figure 7.12), and

Figure 7.30 Wind-ripple stratification. (a) Aeolian truncated ripple-foreset cross-laminae. Jurassic Navajo Sandstone, Utah, USA. (b). Subcritically climbing translatent strata for which no ripple-foreset cross-laminae are evident. Permian Cedar Mesa Sandstone, Utah, USA. (c) Pinstripe lamination representing the accumulated deposits of aeolian ripples that climbed over one another at very low subcritical angles. Each stripe records the passage and accumulation of one ripple in a train, whereby slightly finer-grained fractions represent the passage of the ripple troughs and slightly coarser-grained fractions represent the passage of the ripple crests. Twyfelfontein Formation (formerly Etjo Sandstone), Namibia. (d) Packages of wind-ripple strata on the low-angle-inclined lower flanks of a cross-bedded aeolian dune set. Jurassic Navajo Sandstone, Utah, USA.

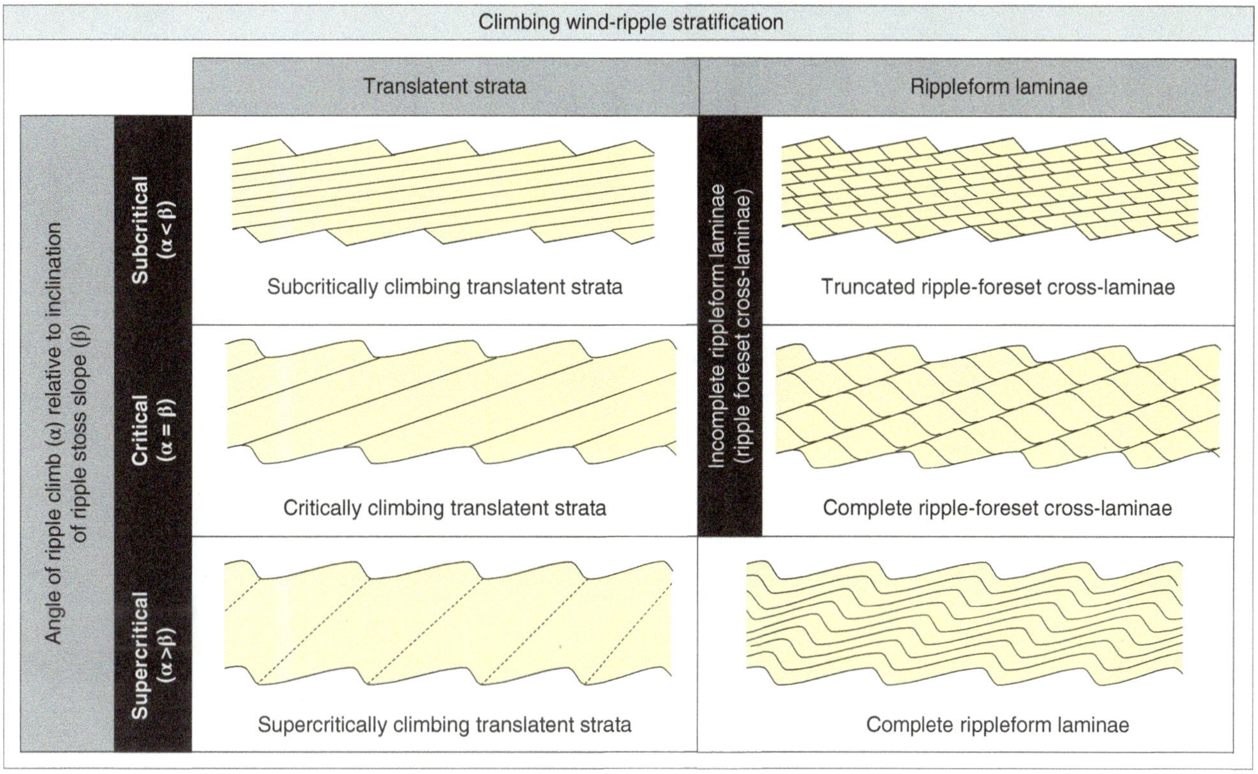

Figure 7.31 Classification of wind-ripple stratification according to the angle of ripple climb relative to the inclination of the stoss slopes of the bedforms and the presence or absence of cross lamination. Adapted from Hunter (1977a).

partly because finer grains tend to settle between coarser grains, resulting in a *pour-in* texture due to kinetic sieving. This means that the base of the ripple stratum is often characterised by a distinct thin layer of finer material. This feature is called pseudo-lamination (Hunter 1973) (Figure 7.30b). A characteristic *pinstripe lamination* (Fryberger and Schenk 1988) is preserved where a train of ripples migrating at a very low angle of climb accumulate thin but stacked traces of each ripple form, perhaps only one or two grains thick (Figure 7.30c). Contacts between laminae may be erosional or non-erosional and sharp or gradational.

Aeolian ripple strata form widespread deposits in sand-sheets, dry interdunes and on dune and megadune slopes inclined more gently than the angle of repose for dry, loose sand (Figure 7.30d). The strata develop on dune stoss slopes, at dune crests, on the convex-downwind projections of sinuous dunes and megadunes (e.g. the 'horns' of barchans and the 'noses' of parabolic dunes) and on sand drifts in the lee of obstacles. Packages of wind-ripple stratified deposits commonly occur as units 10 mm to several metres thick. Porosity in some non-compacted, wind-rippled sand may average around 39%, although this varies according to grain shape and grain-size distribution and is commonly less. Wind-ripple strata are commonly found interbedded with grainflow and grainfall strata, in many cases in rhythmic bundled packages.

7.4.3 Aeolian Grainflow Stratification

The gravitational collapse of dry, loose sand on aeolian dune and megadune lee slopes inclined at the angle of repose (typically 32°–36°) generates grainflow sand avalanches (Figures 7.29 and 7.32). Grainflows gather momentum as they pass down the lee slope and may run out onto lower-angle-inclined bedform plinths and flanks. Grainflow avalanches generate sediment units that are inclined at 18–36° within cross-bedded foreset deposits (the inclination is reduced by sediment compaction in ancient strata). Each grainflow unit is typically up to 10–50 mm thick but thins to a point of pinch-out both up- and down-slope and along strike. Individual grainflow units may be many metres long – several tens of metres in the case of large megadunes – and up to several metres wide. They form tongue- or cone-shaped structures, which display en-echelon, scalloped or tabular shapes in the deposits of large dunes and lenses in small dunes. The strata have sharp, flat or concave-upwards bases; some may be erosive. Most grainflow deposits are internally structureless (Figure 7.32a),

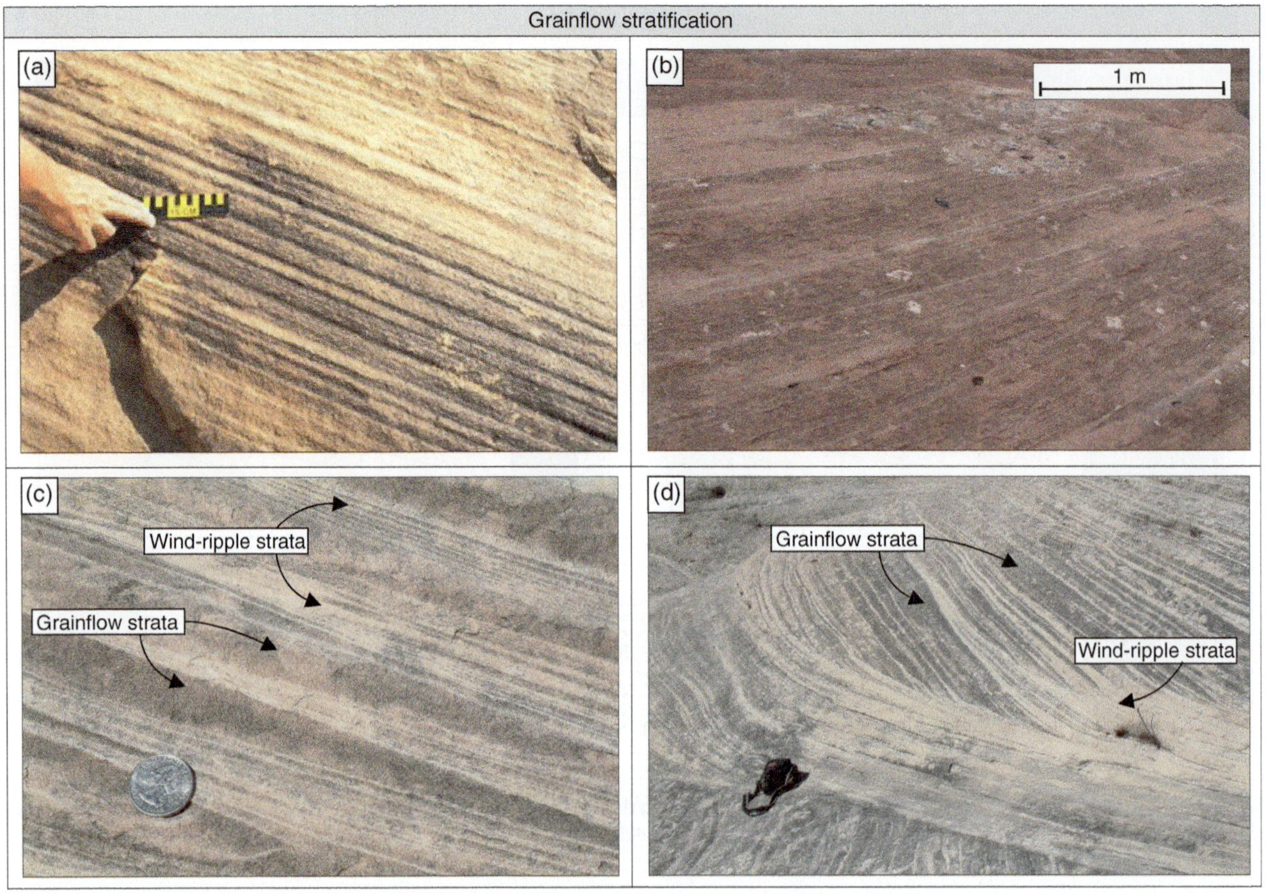

Figure 7.32 Aeolian grainflow stratification. (a) Thick package of multiple stacked aeolian grainflow avalanche deposits, Twyfelfontein Formation (formerly Etjo Sandstone), Namibia. (b) Grainflow strata with many individual stratum delineated by thin grainfall laminae. Jurassic Entrada Sandstone, Utah, USA. (c) Internally structureless grainflow deposits intercalated with wind-ripple translatent strata. (d) Wider view of the same intercalated package of grainflow and wind-ripple strata revealing their arrangement in the lower part of a cross-bedded aeolian dune set. (c and d) Permian Cedar Mesa Sandstone, Utah, USA.

though they may preserve inverse grading normal to the base due to dispersive pressures and kinetic sieving generated by colliding grains (Figure 7.32b). Grainflow deposits may also display longitudinal grading, with coarser and more rounded grains occurring further downslope. Notably, at the foreset toe, coarser grains and relatively dense heavy-mineral grains may occur at the top of the layer. In non-compacted dune sands, grain packing is usually very loose, and porosity is very high (up to 45%). Slump-degradation grainflows, where internal structure is destroyed as the flow travels downslope, are characterised by a chaotic wedge of loosely packed sediment up to a few metres wide that typically thickens downslope before eventually pinching out. Scarp-recession grainflows, where an initial point of failure generates a scarp that then retreats back up-slope towards the dune brink-line, form tongue-like bodies that rarely exceed 0.5 m in width but may extend almost the full height of a lee slope (Figure 7.29a). Where developed in very well-sorted sand, the boundaries between successive avalanches may not be evident, and only amalgamated (compound) grainflow units are evident. Elsewhere deposits of individual grainflows may be interbedded with grainfall or wind-ripple strata (Figure 7.32c,d).

7.4.4 Aeolian Grainfall Stratification

Aeolian grainfall stratification occurs where sediment falls from suspension as a result of decreasing wind velocity and reduced sediment carrying capacity. Gravity-driven grainfall occurs commonly when the wind carries clouds of saltating grains over a dune brink (Figure 7.33a). Here, a reduction in wind transport capacity occurs as a consequence of airflow separation within the lee-side eddy on the upper part of a dune lee slope. Grainfall may additionally occur on the flanks or apron of a dune where secondary cross-wind airflow is insufficient to form ripples (Figures 7.33b,c). Grainfall deposits are most commonly composed of very fine sand. Grainfall strata

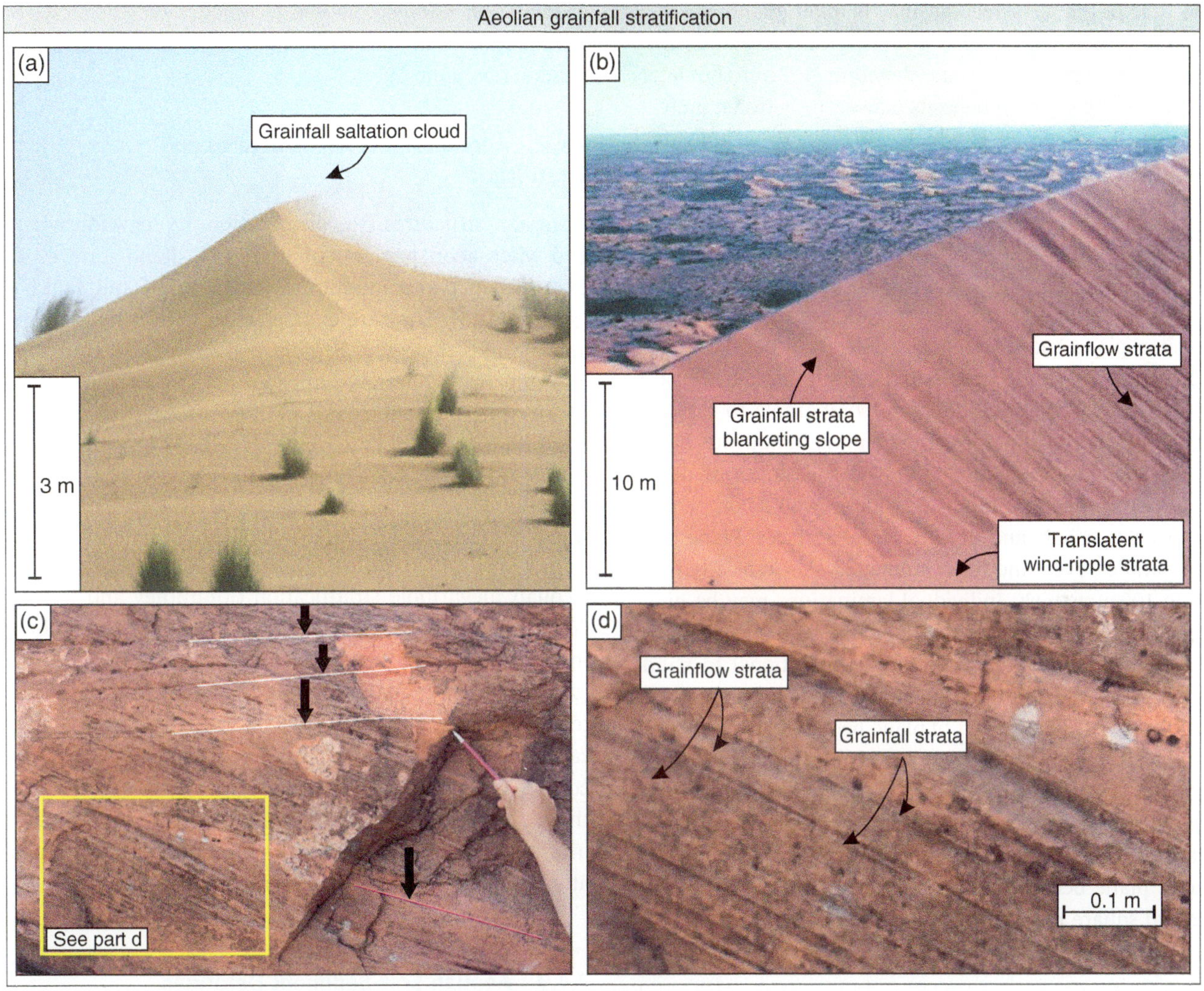

Figure 7.33 Grainfall and grainfall stratification. (a) Gravity-driven grainfall occurs commonly where the wind carries clouds of saltating grains over a dune brink. The airflow decelerates as it passes into the lee-side depression resulting in a reduction in sediment carrying capacity. Modern dune, Namib Desert, Namibia. (b) Grainfall may additionally occur on the flanks or apron of a dune where secondary cross-wind airflow is insufficient to form ripples. Arm of a star draa, northwest Gran Desierto, Mexico. Brookfield et al., (2010) / Geological Association of Canada. (c) Ancient example of grainfall stratification extending uninterrupted down to the base of the lee slope and blanketing grainflow stratification. Jurassic Page Sandstone, Arizona, USA. The white lines indicate set-bounding surfaces. The pink line indicates a reactivation surface, which records a period of lee-face erosion bounded by packages representing lee-face deposition. (d) Detail of individual units of grainflow stratification, each separated by thin, tabular units of grainfall stratification. Cardenas et al., (2019) / Society for Sedimentary Geology (SEPM).

drape and mantle pre-existing aeolian topography on dune and megadune surfaces, on aeolian sandsheets and on dry interdune surfaces. Grainfall deposits commonly form as foresets that dip at 20–36° on dune lee slopes. Deposits may possess indistinct but parallel internal lamination because of grain-size variations produced during fluctuations in wind energy. However, more commonly, grainfall deposits are internally structureless due to the uniformity in size of the suspended sediment load. Grading is not usually observed. Some units of grainfall strata exhibit a wedge-shaped geometry that is thickest just leeward of the dune brinkline and thins down the lee slope. Repeated grainfall deposition on the upper lee slope is an important mechanism by which the slope of a dune attains and exceeds the angle of repose, thereby inducing the reworking of grainfall strata by grainflow avalanches. Grainfall units in non-compacted dune sands are typically moderately packed and porosity may reach 40%. Grainfall strata have non-erosional contacts and individual units commonly blanket the upper parts of dune lee slopes for distances of tens of metres along slope, enabling them to be differentiated from individual avalanche strata. On small bedforms, grainfall strata commonly extend uninterrupted down to the base of the lee slope and may blanket grainflow strata (Figure 7.33d). On larger bedforms, grainfall wedges may be cut out by grainflow strata. Very thin (1–2 mm) but laterally extensive grainfall deposits may occur between individual grainflow deposits, in which case they enable individual grainflow event beds to be readily distinguished. Elsewhere, thin grainfall deposits may be eroded and truncated by later grainflows.

7.4.5 Aeolian Plane-Bed Stratification

Aeolian plane beds (sometimes called 'flat beds') develop when ripple-forming processes are inhibited during gale-force winds; such conditions are comparable to those of the lower part of the upper flow regime in aqueous flows. Plane-bedded lamination occurs on the slopes of dunes and megadunes that dip at 0°–15°, as well as on aeolian sandsheets, zibars and dry interdune surfaces. Plane beds are parallel, even and form laminated sets, typically from 10 to 100 mm thick. Individual laminations may be highlighted by slight variations of grain type and size. Contacts may be sharp but non-erosional or gradational to wind-ripple strata. Porosity in non-compacted sands is typically less than 30%. Thicker sets of plane-bed laminae occur only rarely and most commonly on upwind-facing stoss, gently sloping lee slopes of dunes, else on dry interdune surfaces. Plane-bed lamination seen in section appears similar to translatent ripple stratification. However, on bedding surfaces, plane beds have no appreciable surface relief over several square metres or more.

7.4.6 Aeolian Massive Sand and Sandstone

Some aeolian deposits are massive (i.e. structureless and homogeneous). Although rare, truly structureless aeolian sands can accumulate where aeolian sediment falls from suspension, typically where a location experiences a sustained reduction in wind velocity, perhaps on the lee-side of obstacles. Conditions of sediment fallout at a stable rate need to be sustained for a protracted time period to enable the accumulation of a substantial thickness (>0.1 m) of massive sand as a set. More commonly, what appear to be apparently structureless aeolian sands might have been generated originally via wind-ripple or grainflow processes, but the uniform grain size means that there is little opportunity for grain-size differentiation and segregation during the process of deposition; as such, stratification planes may not be evident (e.g. Campos-Soto et al. 2022). In yet other cases, massive aeolian sands may be a product of homogenisation due to post-depositional physical grain rearrangement, for example, via fluidisation associated with water escape (Kasse 2002). Finally, highly weathered rocks can appear to be massive in outcrop; yet examination using specialised techniques such as x-rays or examination under ultraviolet light might reveal faint stratification (Hamblin 1965).

7.4.7 Aeolian Adhesion Structures and Stratification

Adhesion structures (see also Section 3.5) develop where and when aeolian sediment sticks to a damp surface, for example, in damp interdunes (Section 3.9) and sabkhas (Section 3.15). They are characterised by low-relief ridges and mounds that grow by adhesion to their upwind edge and thereby undertake upwind growth (Kocurek and Fielder 1982). Capillary rise of moisture helps to trap further grains by maintaining a damp surface. Adhesion strata form crinkly and wavy laminae (Figure 7.13).

7.4.8 Aeolian Megaripple Stratification

Aeolian megaripple stratification (Mountney and Russell 2004) occurs as thin sets (< 0.4 m) of cross-strata, where inclined foresets dip at angles substantially less than repose (usually <15°). Inclined strata in the toeset may be asymptotic to the lower bounding surface. Set boundaries typically have a directly overlying deposit that is of slightly coarser grain size. Sets of megaripple cross-stratification that occur stacked on top of each other record the passage of a train of megaripples. However, diastem surfaces separating different sequences are common.

7.4.9 Aeolian Soft-Sediment Deformed Stratification

Post-depositional soft-sediment deformation structures of physical origin are common in aeolian strata (e.g. Strömbäck et al. 2005; Bryant et al. 2016; Chan et al. 2019). Such deformation structures take the form of intradune folding, indicative of near-surface liquefaction, or slumping of moderately cohesive, moist sands on a dune lee slope in response to wetting of the sediment surface (Doe and Dott 1980). Liquefaction results from elevated pore-water pressures as the wetting front infiltrates into highly porous dune sands (Figure 7.34a–d). Loosely packed grainflow laminae are particularly susceptible to liquefaction by the collapse of grain packing due to mechanical loading associated with the wetting event. Deformation due to surface collapse occurs where the upper dune lee slope fails due to a decrease in the angle of internal friction in response to wetting (Hunter et al. 1983; Loope et al. 2001), resulting in a range of brittle failure structures including slab slides (Figure 7.34e,f). Larger-scale deformation structures usually indicate liquefaction below the water table (McKee

Figure 7.34 Examples of soft-sediment deformation structures in aeolian sands and sandstones. (a) and (b) Small scale overturned bedding in wind-ripple- and grainflow-stratified aeolian dune sandstones. (c) Larger scale liquefaction structure affecting several dune sets. (a–c) Permian Cedar Mesa Sandstone, Utah, USA. (d) Sand fluidisation and injection resulting from mass-wasting by gravitational collapse and sliding of the slopes of water-saturated dunes, which loaded down-dip strata. Permian Hopeman Sandstone, Scotland. Hurst et al., (2008) / John Wiley & Sons. (e) and (f) Brittle failure due to slab slide: detachment of very weakly indurated sand slabs along a plane of weakness. Modern dune lee slope, Askja, Iceland. Black outline of person for scale.

et al. 1971), which may originate from the loading of saturated sand by an advancing or overriding dune (Collinson 1994; Horowitz 1982) or by elevated groundwater pore pressure due to increased head gradient, for example, driven by rainfall in basin-margin upland areas (Hurst and Glennie 2008). Fluidisation associated with water escape through sediments can lead to grain re-arrangement and homogenisation of deposits.

7.4.10 Related Non-aeolian Stratification Types

Non-aeolian processes and structures are common in damp and wet interdune settings (Ahlbrandt and Fryberger 1981; Kocurek 1981). Sedimentary structures of physical origin (Figure 7.35) include desiccation cracks and polygons, raindrop imprints, mud flakes and curls, parallel laminated muds, wave and current ripple forms (and stratification), cross-strata, wavy lamination of subaqueous origin and water-escape and other soft-sediment deformation structures (Ahlbrandt et al. 1978; Langford 1989; Langford and Chan 1989). Sedimentary structures of chemical origin (Figure 7.36) include evaporitic salt crusts, fenestral porosity, sandstone pseudomorphs of salt minerals especially halite (hopper crystals) and gypsum (desert rose) and evaporite precipitation structures such as 'teepees' (Kocurek 1981). Sedimentary structures of biogenic origin (Figure 7.37) include animal footprints, trackways, burrows and crawling traces (Ahlbrandt et al. 1978; Hasiotis 2002), plant-root structures and rhizoliths (Loope 1988), algal growth structures and MISS (Basilici et al. 2020, 2021). Palaeosols are also widespread in many desert systems and are collectively referred to as Aridisols. They commonly include calcrete, silcrete or gypcrete deposits.

7.4.11 Facies Associations

Routinely, multiple depositional units, each comprising a single facies, occur together as a facies association, the occurrence of which has a common genetic origin. These groups of genetically related facies are collectively indicative of the deposits of particular sub-environments within a larger aeolian system (Figure 7.38).

Figure 7.35 Examples of non-aeolian but related stratification types and sedimentary structures of physical origin. (a) Desiccation cracks on the surface of an episodically flooded (wet) interdune. Sossusvlei, Namib Desert, Namibia. (b) Aqueous climbing-ripple strata of ephemeral stream origin. Note the preserved form sets. Southern Namibia. (c) Mud curls and raindrop imprints. Southeast Spain. (d) Desiccation crack in mudstone in the upper part of a wet interdune element. The interudne element is overlain by an aeolian dune element, sand from which has filled the crack. Permian lower Cutler beds, Utah, USA.

Figure 7.36 Examples of non-aeolian but related stratification types and sedimentary structures of chemical origin. (a) Laminar calcrete bed forming the topmost deposit of an episodically flooded (wet) interdune. Deadvlei, Namib Desert, Namibia. (b) Sandstone pseudomorphs after gypsum in the form of desert roses. (c) Silcrete (chert) bed precipitated in an episodically flooded wet interdune element. (d) Palaeosol horizon with well-developed calcrete nodules within a damp interdune element. (b–d) Permian Cedar Mesa Sandstone, Utah, USA.

Cross-Stratified Aeolian Dune (or Megadune) Facies Association

Cross-stratified sets of strata of aeolian dune and megadune origin form a common facies association. Cross-stratified sets or cosets of strata occur in a variety of arrangements of stratal packages of fundamental aeolian facies: wind-ripple, grainflow and grainfall stratification, possibly also with aeolian plane-bedded stratification. These strata occur in inclined packages that comprise the foresets of cross-stratified dune and megadune sedimentary geobodies – composite stratal packages arising as a consequence of aeolian dune accumulation (Figure 7.38). The internal facies units can be arranged in many different ways, depending on the morphology of the original bedform and its migratory behaviour in response to the wind regime and sediment delivery over its surface (Section 5.2).

7.4.11.1 Aeolian Interdune Facies Association

This facies association may comprise a broad variety of facies types that singly or collectively are indicative of accumulation in low-relief depressions and flat areas present between aeolian dunes (Figure 7.38). The facies present reveal information about the nature of the substrate at the time of sediment accumulation. An association dominated by packages of aeolian wind-ripple strata and aeolian plane stratification is typical of a dry interdune setting. An association dominated by adhesion-ripple strata and wavy lamination, in some cases with bioturbation, rhizoliths or pedogenesis, is typical of a damp (i.e. water-table-controlled) interdune setting. An association dominated by aqueous-ripple strata and chemically precipitated nodules, in some cases with mudstone layers and desiccation cracks, and in other cases with carbonate facies, is typical of a wet (i.e. flooded) interdune setting.

7.4.11.2 Aeolian Sandsheet Facies Association

This facies association dominantly comprises facies units of coarse wind-ripple and aeolian plane-bed stratification that occur inclined at low angles (<10°) and commonly close to

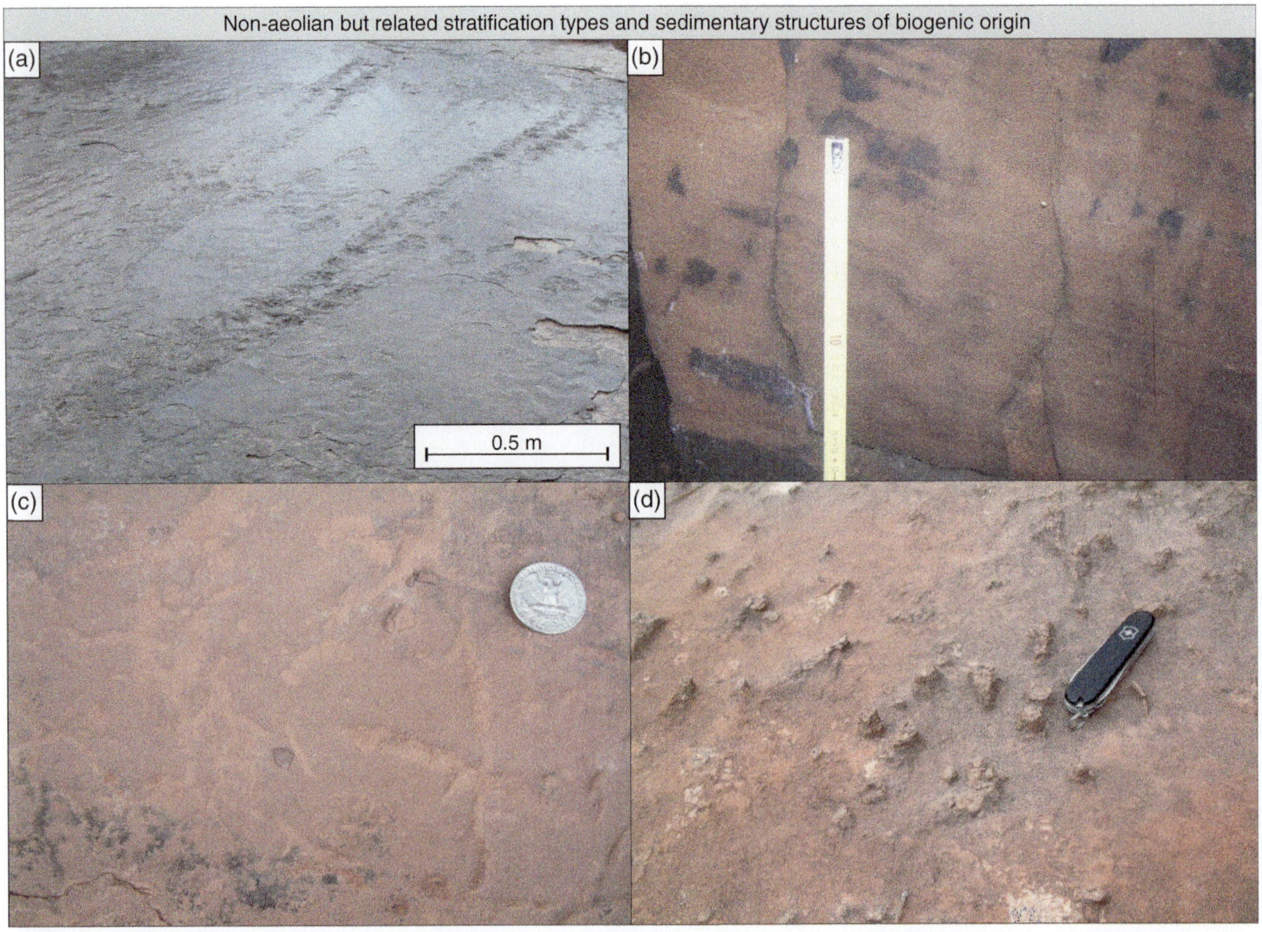

Figure 7.37 Examples of non-aeolian but related stratification types and sedimentary structures of biogenic origin. (a) Vertebrate (reptile) trackway on an aeolian dune-set bedding surface. Permian Cutler Group, Utah, USA. (b) Vertebrate indenter mark (footprint impression) in aeolian dune toeset deposits. Triassic Sherwood Sandstone Group, England. (c) Near-surface interstitial burrows in damp interdune strata. (d) Fossilized root structures (rhizoliths). (c and d) Permian Cedar Mesa Sandstone, Utah, USA.

palaeo-horizontal (Figure 7.38). This association represents the accumulated deposits of sand flats upon which high-relief dune-scale bedforms were not developed.

7.4.12 Facies Successions

A facies succession describes the occurrence of an association of facies units in a predictable order or arrangement – either vertically or laterally. Such successions provide insight into the mechanics by which aeolian sub-environments evolve over time and space. For example, a vertical facies succession through an aeolian dune facies association might reveal the gradual upward transition from a dune base, itself constructed upon an underlying interdune deposit, to a low-angle-inclined dune plinth dominated by wind-ripple strata and to a more steeply inclined dune foreset dominated by grainflow strata. A gradual trend of upward steepening indicates a dune lee slope with a broad plinth. An abrupt steepening indicates a dune with an active lee slope that extends close to the toe of the dune. As another example, facies units that comprise an interdune association might occur in a predictable vertical succession that follows Walther's law (Section 2.1.2; Section 4.2 and Figure 4.18). The vertical succession might be interpreted to record the migration of laterally adjacent interdune sub-environments that came to overlie one another as the system migrated: deposits of a dry interdune margin might be overlain by deposits of a damp interdune fringe, which might, in turn, be overlain by deposits of a wet (i.e. flooded) pond at the centre of a shallow interdune depression; this entire succession might be overlain by the cross-stratified deposits of a succeeding dune.

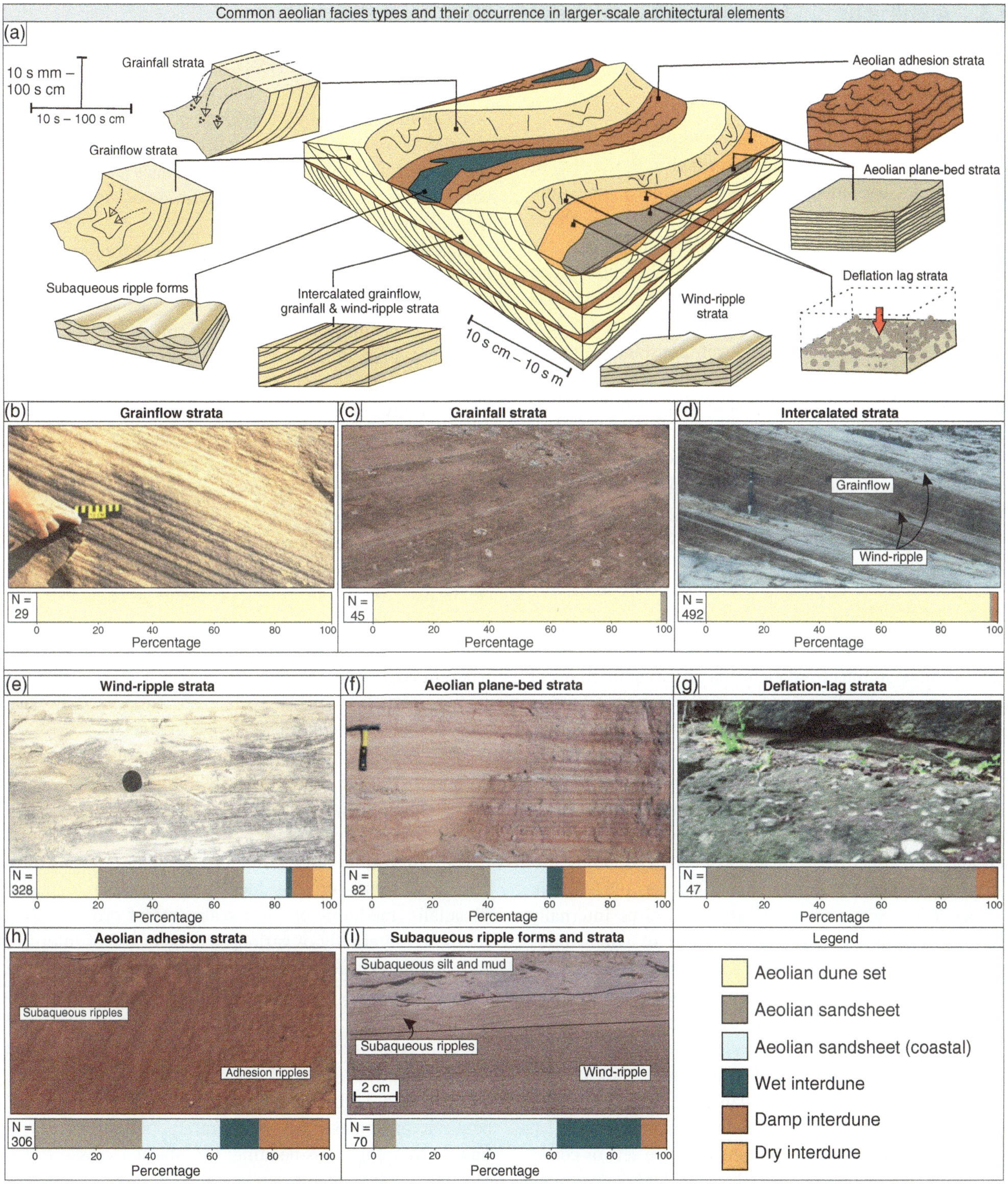

Figure 7.38 Examples of common types of aeolian facies and their typical distribution in larger-scale architectural elements. (a) Facies model. (b–i) Photo examples of the facies types illustrated in Part a. The bar chart beneath each image shows proportions indicating the frequency of occurrence of each facies in common architectural elements. Quantitative data from the Database of Aeolian Sedimentary Architecture (DASA). (b) Grainflow stratification. Jurassic Twyfelfontein Formation (formally Etjo Sandstone), northern Namibia. (c) Mixed grainflow-grainfall stratification. Permian Cedar Mesa Sandstone, Utah, USA. (d) Intercalated wind-ripple and grainflow stratification. Jurassic Guará Formation, Brazil. Adapted from Scherer and Lavina (2005). (e) Wind-ripple stratification; dune-plinth deposits. Permian Cedar Mesa Sandstone, Utah, USA. (f) Aeolian plane-bed lamination. Cretaceous Sanfranciscana Basin, Brazil. Mescolotti et al., (2019) / The Scientific Electronic Library Online (SciELO) / CC BY 4.0. (g) Deflation-lag strata. Lower Triassic, West German Basin, Germany. Bourquin et al., (2009) / John Wiley & Sons. (h) Bedding plane with symmetrical aqueous ripples on the left-hand side and aeolian adhesion ripples on the right-hand side, revealing emergence. Silurian Mereenie Sandstone, northern Territory, Australia. Shillito et al, (2021) / The Geological Society of London. (i) Subaqueous ripple forms and stratification overlying wind-ripple stratification. Recent deposits, Upper Tulum Valley, central-western Argentina. Dal' Bó et al., (2015) / The Scientific Electronic Library Online (SciELO) / CC BY 4.0.

7.5 Architectural Elements and Bounding Surfaces

7.5.1 Overview

Architectural elements – building blocks of larger sedimentary depositional systems – are described in terms of their internal facies assemblage (a facies association), the nature of any internal set-bounding surfaces, their external shape (e.g. tabular), their geometry (i.e. dimensions) and the nature of external bounding surfaces. In aeolian-dominated successions, the following types of architectural elements are commonly recognised: (i) various dune elements (e.g. Karpeta 1990), (ii) dry, damp and wet interdune elements, (iii) sandsheet elements, (iv) sabhka and playa lake elements and (v) other related non-aeolian elements. This section considers the range of architectural elements commonly present in aeolian sedimentary successions and the form of their internal and external bounding surfaces. Additionally, this section discusses factors that influence the development of these architectural elements and their bounding surfaces.

7.5.2 Aeolian Dune Elements

Aeolian dune architectural elements are accumulations of strata generated by the migration of hierarchies of aeolian bedforms of differing sizes and shapes, moving at varying rates and in varying directions relative to one another. In situations where the net aeolian sediment budget is positive, accumulation occurs such that bedforms may climb over one another at various angles, thereby accumulating sets and cosets of cross-strata and associated bounding surfaces (Brookfield 1977; Rubin and Hunter 1982; Rubin 1987a). These sets and cosets are themselves internally composed of smaller-scale facies units of wind-ripple, grainflow, grainfall and plane-bedded strata, the relative proportions and distributions of which are determined by how processes operated upon the formative bedforms. Thus, dune elements are larger-scale stratal units composed of a variety of arrangements of smaller-scale bodies.

In preserved successions, aeolian dune (and megadune) elements most commonly take the form of tabular (i.e. sheet-like) bodies composed of one or more sets or cosets of cross-strata (Kocurek 1981; Karpeta 1990; Mountney 2006b). Individual examples of such elements may be several metres to (rarely) decametres thick and may extend laterally for several hundred metres to several kilometres in directions both parallel and perpendicular to the mean azimuth of the cross-beds they contain (Figure 7.39).

Internally, aeolian dune elements comprise one or more types of cross-bedded sets, with each set bounded by a sharp or erosional bounding surface (Figure 7.40; also see Section 7.5.7). Where multiple sets are present in an element, they form a coset of strata. Externally, aeolian dune elements are almost always bounded at their top by sharp or erosional surfaces, and similar bounding surfaces may also define the bases of some dune elements. In other cases, the base of an aeolian dune element may be gradational with deposits of an underlying interdune or sandsheet element (Figure 7.41).

Stratal packages of aeolian wind-ripple, grainflow, grainfall and plane-bed stratification commonly occur together as the internal components of larger cross-stratified sets. Such cross-bedding develops as a consequence of the migration of aeolian dunes and megadunes, whereby depositional processes operating on bedform lee slopes progressively build inclined deposits as the bedform advances (Figure 7.41a–c). The stratification planes that form the cross-bedding provide a record of the former positions and shape of the bedform lee slope and the processes that operated on that slope.

Studies of modern cross-bedded aeolian sands demonstrate that a number of associated sedimentary features can be used as reliable indicators of large-scale aeolian bedform activity. Information from modern dunes is greatly augmented by studies of ancient aeolian sandstones, wherein large-scale sets of cross-bedded strata are the most striking feature of many aeolian successions. Individual cross-bedded sets commonly range in thickness from 0.1 to 10 m and may rarely be considerably thicker, up to 52 m in one example from the Twyfelfontein Formation (formerly the Etjo Sandstone) of Namibia (Mountney and Howell 2000).

The following are common sedimentary features associated with cross-bedded aeolian dune deposits: (i) small-, medium- or large-scale sets of cross-bedding composed internally of various arrangements of interbedded or intercalated packages of grainflow, grainfall and wind-ripple facies (rarely plane-bedded or massive facies) (Figure 7.41c–e); (ii) the occurrence of cross-bedding of multiple types in close association, including one or more of tabular-planar (Figure 7.40a), wedge-planar and trough types (Figure 7.40b), the relative proportions of which vary between examples; (iii) more rarely, units of convex-upward dipping cross-bedding; (iv) stacked cross-bedded sets that are separated by erosional bounding surfaces that are either near-horizontal or inclined at low to moderate angles in orientations that vary from sub-parallel to highly oblique relative to the foresets that they truncate; and (v) relatively thin units of flat- or low-angle-inclined wind-ripple stratification that occur in the lower parts of larger-scale cross-bedded sets and separate successive cross-bedded sets.

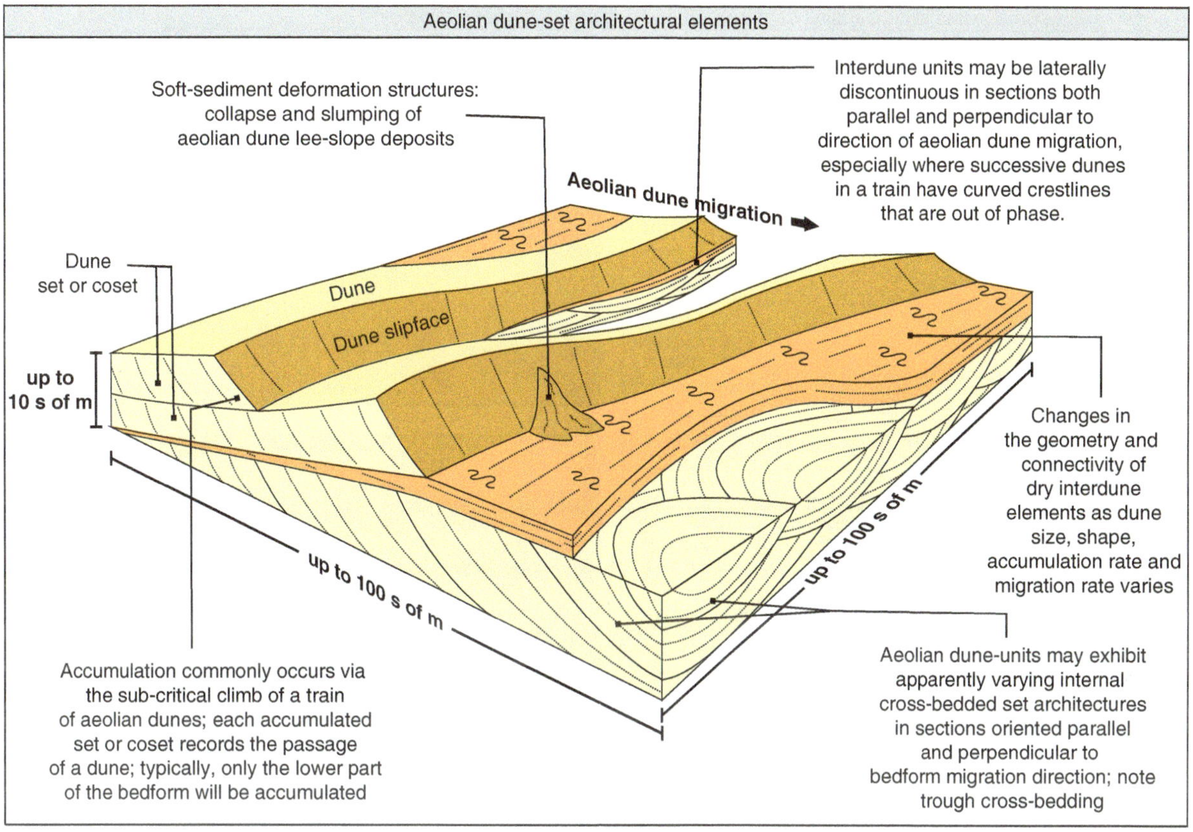

Figure 7.39 Summary model depicting the typical three-dimensional geometry and internal anatomy of aeolian dune architectural elements and associated dry interdune elements accumulated via bedform climbing. Adapted from Cain (2009) and Wakefield (2010).

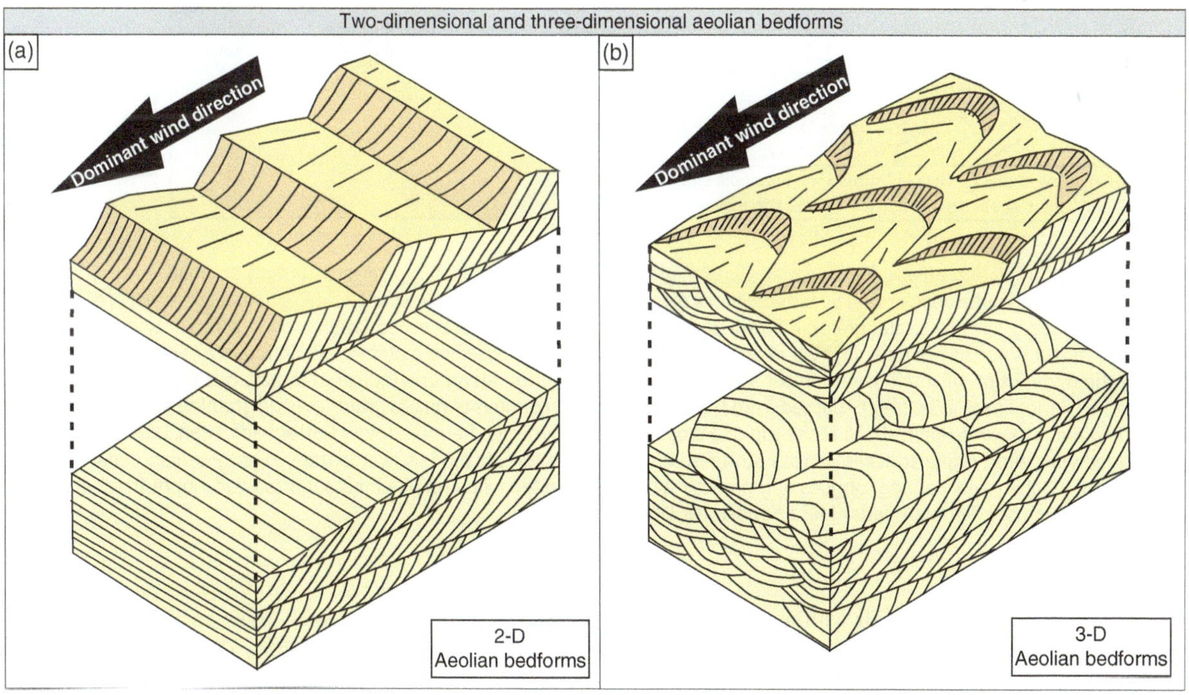

Figure 7.40 Classification of aeolian dune type according to variability in form. (a) Two-dimensional transverse bedforms produce sets of tabular cross-bedding as they migrate and climb over one another to generate an accumulation. (b) Three-dimensional transverse bedforms with crestline curvatures that are out of phase between successive bedforms in a train produce sets of trough cross-bedding as they migrate and climb over one another to generate an accumulation. Adapted from Allen (1970).

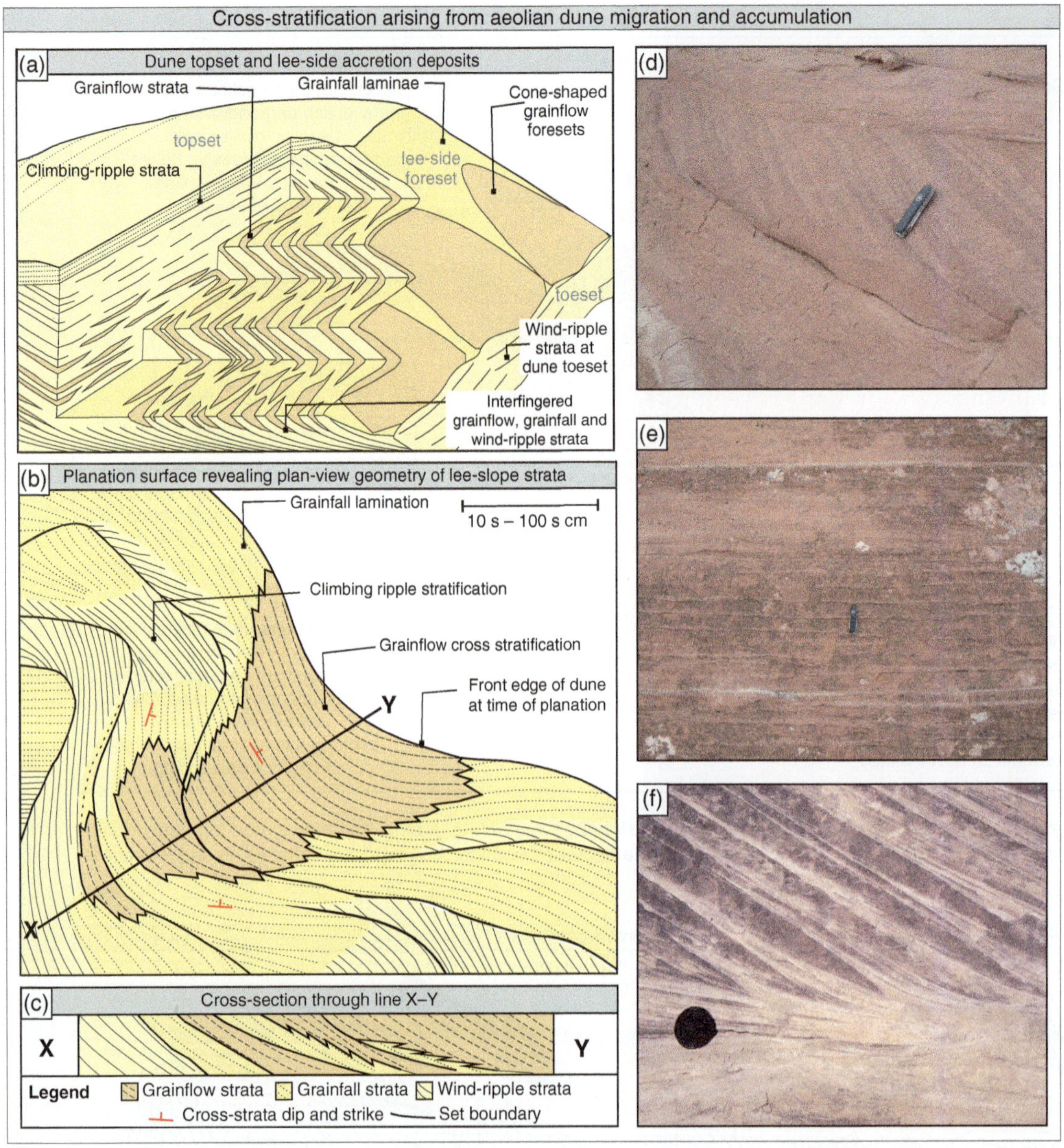

Figure 7.41 Cross-stratification arising from aeolian dune migration and accumulation. Typical distribution of small-scale lamination types in aeolian dune deposits. (a) Schematic block diagram showing the different small-scale structures of foresets of different type: simple cone-shape grainflow strata, grainfall strata and climbing-ripple strata. Plane-bed lamination is commonly developed on exposed dune crests but is not shown here. (b) Map of dune foreset cross-strata exposed on a planed-off sinuous transverse or barchanoid dune, showing the distribution of small-scale foreset structures. (c) Cross section along line indicated in part b. Simplified from an exposure on Padre Island, Texas, USA. (a–c) Adapted from Hunter (1977a). (d) Tongues of mixed grainflow and grainfall strata interbedded with and pinching out into wind-ripple strata at the toe of a cross-stratified dune set. Permian lower Cutler beds, Utah, USA. (e) Grainflow strata interbedded with wind-ripple strata in a cross-stratified dune set. The foreset dip is towards the viewer. Permian Cedar Mesa Sandstone, Utah, USA. (d) Sharply defined tongues of grainflow strata interbedded with and pinching out into wind-ripple strata at the toe of a cross-stratified dune set. Permian Cedar Mesa Sandstone, Utah, USA.

The orientation of inclined cross beds within aeolian bedforms, and the geometry and interrelationship of adjoining sets of cross-bedding, varies with bedform type. Transverse dunes and megadunes are composed internally of planar-tabular or broad ('open') trough cross-bedded sets with inclined foresets oriented with their strike within 15° of the trend of the bedform crestline (Figure 7.40). Curved- or sinuous-crested barchanoid ridges are composed internally of markedly 3D trough cross-bedded sets for which the internal architecture is characterised by foresets whose directions of maximum dip (dip-azimuths) vary radially around the trough-shaped features, in some cases by up to 90° within a single scallop-shaped trough (Figure 7.40b). Dome-shaped bedforms exhibit cross-bedded sets whose foresets are convex-up, typically dominated by wind-ripple strata. The internal structure of linear and star-shaped bedforms is complicated: it is commonly characterised by a mosaic of overlapping and intersecting sets, in some cases with foresets inclined in opposing directions (Figure 7.16).

Where bedforms migrate over one another – typically as part of a bedform train – they commonly climb at low angles such that the lowermost cross-bedded deposits of one bedform are accumulated as a second bedform overrides them. Cross-strata generated by the first bedform are partly truncated by the passage of the second. The level of truncation ('cut-off') generates a set of cross-strata bounded at its top by an erosional surface (Figure 7.40), a so-called 'bounding surface' (Brookfield 1977; Rubin and Hunter 1982; see Section 7.5.7).

The migration of simple, 2D, 'invariable' transverse bedforms (i.e. straight-crested transverse bedforms of fixed form that move in a uniform direction at a constant rate) generates cross-stratification characterised by constant foreset dip azimuths (Rubin and Carter 2006; Box 7.2). By contrast, the migration of simple, 3D invariable transverse bedforms, such as barchan dunes, parabolic dunes and curved-crested barchanoid ridges, generates cross-stratification characterised by a range of foreset dip azimuths that vary around a mean that coincides with the direction of bedform migration (Rubin and Carter 2006; Box 7.2). The range in the observed foreset dip azimuths increases as the amount of crestline curvature or sinuosity increases.

The predicted pattern of foreset dip azimuths for actively migrating linear dunes, especially seif dunes, is of two modes about 120° apart (Bagnold 1941; McKee 1966; Selley 1985). However, cross-bedded sets with such foreset arrangements are rare in the ancient record, chiefly because most linear dunes have minor but persistent components of transverse motion, causing them to shift laterally over prolonged periods, resulting in the preferential preservation of foresets dipping in the direction of net movement (Figure 7.16) and generating a more unimodal arrangement of dip azimuths (Rubin and Hunter 1985).

Migration of variable bedform types results in more complicated patterns of cross-strata whereby numerous erosional bounding surfaces may partition and subdivide cross-bedded sets and cosets (see Section 7.5.7; Box 7.2). Wind reversals may locally preserve foresets with azimuths oriented up to 180° from the mode, especially in the lower parts of the fills of larger scour troughs. Moreover, where deposition occurs from time-to-time on stoss slopes, rare, less steeply inclined laminae with azimuths at 180° from the mode can accumulate locally.

Where bedforms of different scales migrate over one another at different rates and commonly in different directions, even more complicated patterns of cross-stratification are generated (Rubin and Hunter 1983; Box 7.2). This is especially the case where two scales of bedform (parent and child) are both 3D in form and undertake variable migration.

Aeolian cross-bedding commonly occurs in a manner whereby the lower toesets are asymptotic to the bounding surface that defines the base of the set. This can be used to help establish a stratigraphic way up in potentially overturned ancient successions. However, other aeolian cross-bedded sets have planar foresets that meet the lower set-bounding surface with an abrupt angular relationship; such sets do not obviously provide an indication of way up. Moreover, rare convex-upward cross-bedding can occur (Thompson 1969; Campos-Soto et al. 2022).

Relating the morphology and migratory behaviour of modern bedforms to the architecturally complex bed-set and bounding-surface geometries that they generate is important for understanding the environmental significance of ancient aeolian strata, but doing so is far from straightforward. Although the external morphology of modern aeolian bedforms is readily apparent, the internal architecture of their bed-sets is difficult to predict and sparsely documented. Geometrical computer simulations (e.g. Rubin and Carter 2006) demonstrate that bedforms of similar external morphologies can generate radically different patterns of cross-bedding because they undertake different migratory behaviour through time. Furthermore, the amount of a bedform that is accumulated as a bed-set (i.e. not eroded) following the passage of multiple bedforms in a train is typically only a small fraction (usually <10%) and selectively the basal-most part of the entire bedform. As such, the reconstruction of bedform morphologies from bed-set architectures usually relies on the assumption that preserved sets (i.e. the sets that dominate in the ancient geologic record) adequately record the depositional processes that occurred on the upper (non-preserved) parts of the formative bedforms.

7.5.3 Dry Interdune Elements

In preserved successions, the geometry of dry interdune elements is determined by (i) the plan-view shape of the geomorphic interdune feature, (ii) the longevity of that interdune, (iii) the rate at which it migrated through space and (iv) the rate of aggradation (Mountney 2006a,b). The latter two aspects together define the angle (trajectory) at which the interdune element climbs through the stratigraphy

(Figure 7.39). In dry aeolian systems, dunes tend to grow to a size whereby they cover the accumulation surface and interdune flats are reduced to isolated depressions before climbing and accumulation commence. Resultant dry interdune architectural elements tend to be relatively thin and typically form only the basal-most metre or less of accumulated deposits above a bounding surface that defines the base of the interdune element (Figure 7.41f).

7.5.4 Damp Interdune Elements

In preserved successions, the geometry of damp interdune elements develops in a similar manner to that of dry interdunes (Section 7.5.3). However, because damp interdunes are associated with wet aeolian systems for which the relative water-table rise controls accumulation, dunes did not necessarily cover the surface prior to accumulation commencing. As such, damp interdune elements in ancient successions can potentially be thicker and more extensive than their counterparts in dry aeolian systems. Elliptical-shaped but isolated damp interdune hollows present between dunes with curved crestlines generate ribbon-shaped bodies as they translate through space over time (Figure 7.42). By contrast, laterally elongated interdune corridors that develop and migrate in front of straight-crested transverse dunes generate sheets. The Jurassic Entrada Sandstone, Utah, USA is one example where damp interdune elements are up to 4 m thick and are laterally

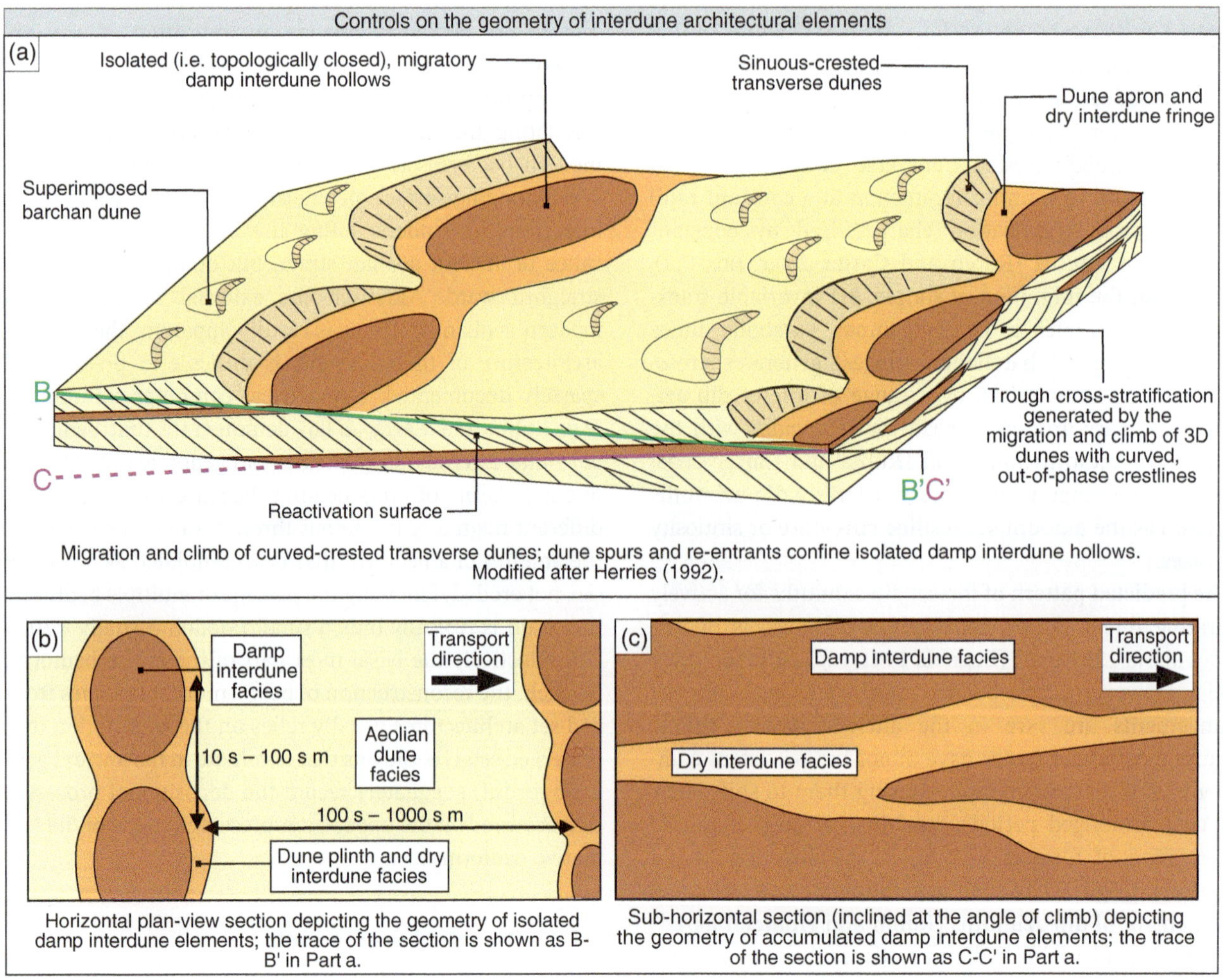

Figure 7.42 Relationship between the morphology of interdunes and the geometry of their accumulated architectural elements. (a) Elliptical-shaped but isolated (i.e. closed) damp interdune hollows present between dunes with curved crestlines. (b) Horizontal plan view section depicting the geometry of isolated damp interdune elements intersected on surface B–B′. The location of the surface is indicated in part a. (c) Sub-horizontal section (inclined at the angle of climb) depicting the geometry of accumulated damp interdune elements on surface C–C′. The location of the surface is indicated in part a. The interdunes generate ribbon-shaped architectural elements as they translate through space over time. In contrast to the geometries depicted in the figure, laterally elongated (i.e. open) interdune corridors developed in front of straight-crested transverse dunes tend to generate sheet-like architectural elements.

extensive for hundreds of metres to several kilometres (Kocurek 1981, Kocurek and Crabaugh 1993).

7.5.5 Wet (Flooded) Interdune Elements

In preserved successions, the geometry of damp interdune elements develops in a similar way as that of dry interdune elements (Figure 7.43; Section 7.5.3). However, where interdunes are wet due to flooding, aeolian system evolution may be disrupted by (i) limited availability of aeolian sediment for continued dune construction and (ii) dune migration stalling due to flooding in interdune ponds. In some cases – notably where interdune flooding is caused by fluvial incursion – wet interdune elements may be expressed as bodies that record the instantaneous shape of the interdune pond or lake at the time of the flood event, as is the case in part of the Permian Cedar Mesa Sandstone, Utah, USA (Mountney and Jagger 2004) and part of the Jurassic Navajo Sandstone, south-western, USA (Herries 1993). In other cases – notably where there is a sustained high water table but where dune migration is not curtailed – thick and laterally extensive wet interdune elements can accumulate to form laterally extensive and continuous elements. Interdune elements may host fluvial deposits where rivers were able to penetrate between dunes along interdune corridors that were open to the edge of a dune field.

7.5.6 Aeolian Sandsheet Elements

Sandsheet architectural elements rarely exceed 10–20 m thickness, either in modern deserts or in preserved ancient successions. However, although rare, exceptionally thick sandsheet elements (~50 m) have been documented in the Proterozoic Bandeirinha Formation (Simplicio and Basilici 2015). In the ancient stratigraphic record, preserved sandsheet elements commonly occur stratigraphically adjacent to aeolian dune elements, else underlie them and represent the accumulated records of precursor environments upon which later aeolian dune systems were constructed and accumulated (Fryberger et al. 1979; Khalaf et al. 1984), as is the case for the margins of several Australian dune fields (Brookfield 1970). Additionally, in many ancient aeolian systems, sandsheet elements are commonly closely associated with a variety of aeolian and non-aeolian bodies and typically occur: (i) as laterally extensive bodies, vertically and laterally intercalated with zibar (slipface-less proto-dune) elements (Nielson and Kocurek 1986), for example, in the São Sebastião Formation, Brazil (Ferronatto et al. 2019); (ii) as sharp-based tabular bodies with internal deflationary surfaces, for example in the Agrio Formation, Argentina (Veiga et al. 2002); (iii) as isolated sand bodies, partly eroded by channelised ephemeral fluvial elements, for example, the Guará Formation, Brazil (Scherer and Lavina 2005); (iv) in coastal settings, as thin sheets (< 1 m thick) with internal bounding surfaces associated with marine incursions, for example, the Shikaoda Formation, India (Chakraborty and Chakraborty 2001); or (v) as thin (<2 m) sheets cyclically interbedded with fine-grained lacustrine deposits, for example, in the Middle Buntsandstein, southern North Sea (Clemmensen 1991).

7.5.7 Bounding Surfaces

A variety of erosional or non-depositional but sharply defined bounding surfaces truncate and delimit aeolian architectural elements and subdivide them internally. Four distinct scales of bounding surface are differentiated in aeolian dune successions based on their shape, their lateral extent, their orientation relative to the strata (especially cross-strata) that they truncate and their relation to one another (Figure 7.44). In some cases, bounding surfaces delimit simple sets. However, in many cases, they define a nested hierarchy of stratal units whereby cosets of strata are internally composed of a mosaic of sets bounded by surfaces of different types (Figure 7.44). From lowest to highest in rank, these surfaces are reactivation surfaces, superposition (commonly 'superimposition') surfaces, interdune migration surfaces and supersurfaces (Figure 7.45).

Reactivation, *superimposition* and *interdune migration* bounding surfaces occur as a product of autogenic (i.e. intrinsic) bedform adjustment and migratory behaviour (Brookfield 1977). Erosive bounding surfaces are generated where bedforms (or parts thereof) scour into pre-existing deposits as they migrate; in some situations, surfaces are sharply defined but non-erosional. Regionally extensive *supersurfaces* most obviously result from allogenic (i.e. extrinsic) changes to controlling parameters, such as the rate of aeolian sediment supply, aeolian sediment availability and the sediment carrying capacity of the wind (Loope 1985; Langford and Chan 1988; Kocurek 1988, 1996; Fryberger 1993).

Reactivation surfaces divide cross-bedded aeolian dune elements into smaller stratal packages. They are characterised by planar, concave or scalloped-shaped erosion planes that typically dip at 10°–20°, in most cases at inclinations less than the cross-strata that they truncate. In sections perpendicular to aeolian transport, reactivation surfaces trend parallel or sub-parallel to the cross-strata; in some cases, they can be traced for 10 to >100 m along depositional strike (Figure 7.44). In sections parallel to aeolian transport, they may extend the full height of a set or may be restricted to its basal part, in which case they are commonly characterised by a sweeping (asymptotic) base (Figure 7.44). Although some reactivation surfaces occur apparently randomly distributed, others exhibit

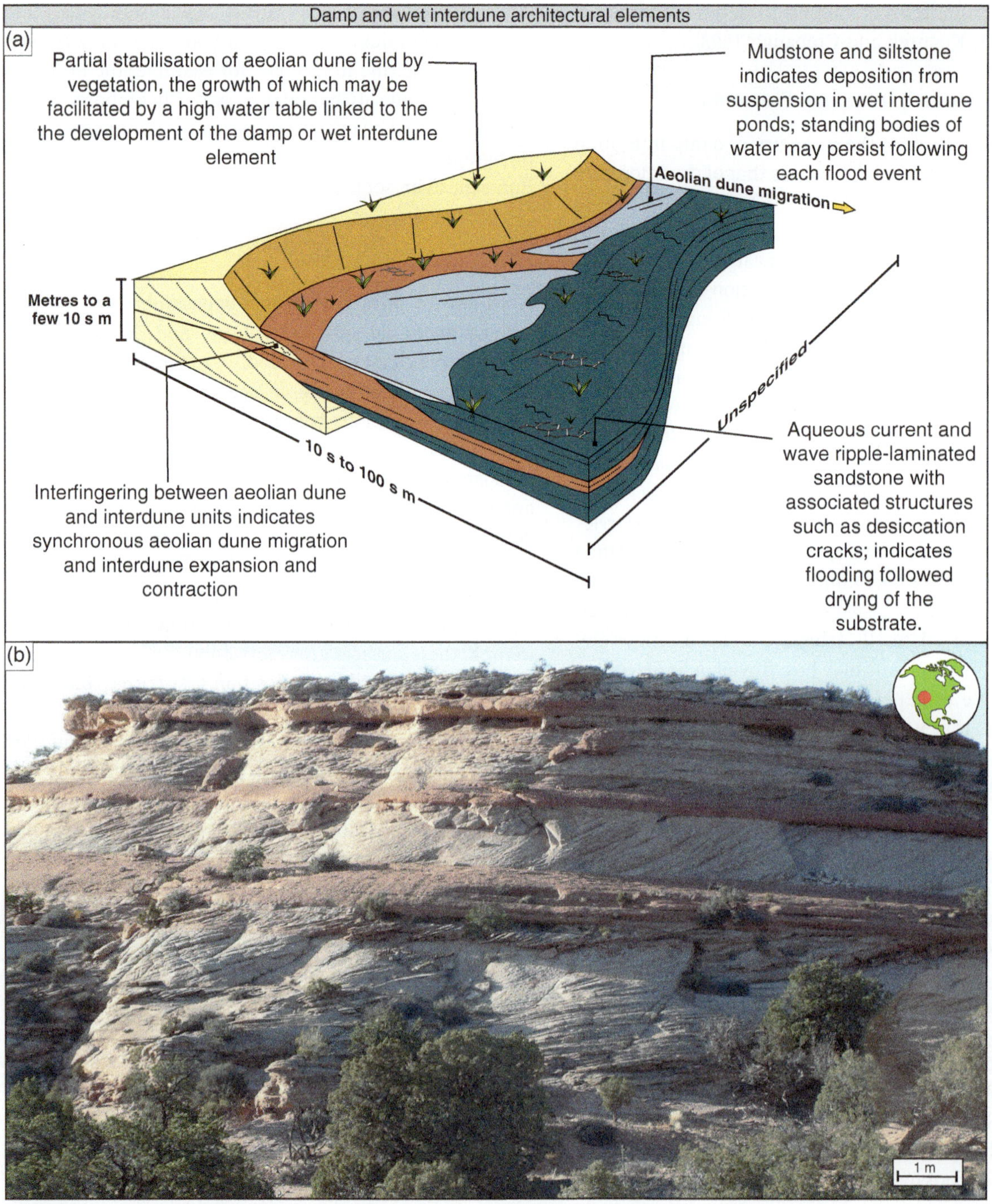

Figure 7.43 (a) Summary model depicting the typical three-dimensional geometry and internal anatomy of damp and wet interdune architectural elements. Adapted from Cain (2009). (b) Example of horizontally bedded wet interdune architectural elements (red) between cross-bedded aeolian dune elements (off-white). Jurassic Navajo Sandstone, southeast Utah, USA.

remarkably regular spacings, suggestive of a periodic formative mechanism. Overlying cross-strata exhibit either a concordant, downlap or onlap relationship (Figure 7.44). Reactivation surfaces result from episodic or periodic lee-slope erosion followed by renewed sedimentation associated with a change in bedform migration direction, migration speed, asymmetry or steepness (Rubin 1987b; Rubin and Carter 2006). These changes are common because airflow on dune lee slopes is commonly subject to turbulent modification and is rarely steady. In cases where the flow fluctuation is periodic such that regularly spaced reactivation surfaces are generated, the driver is

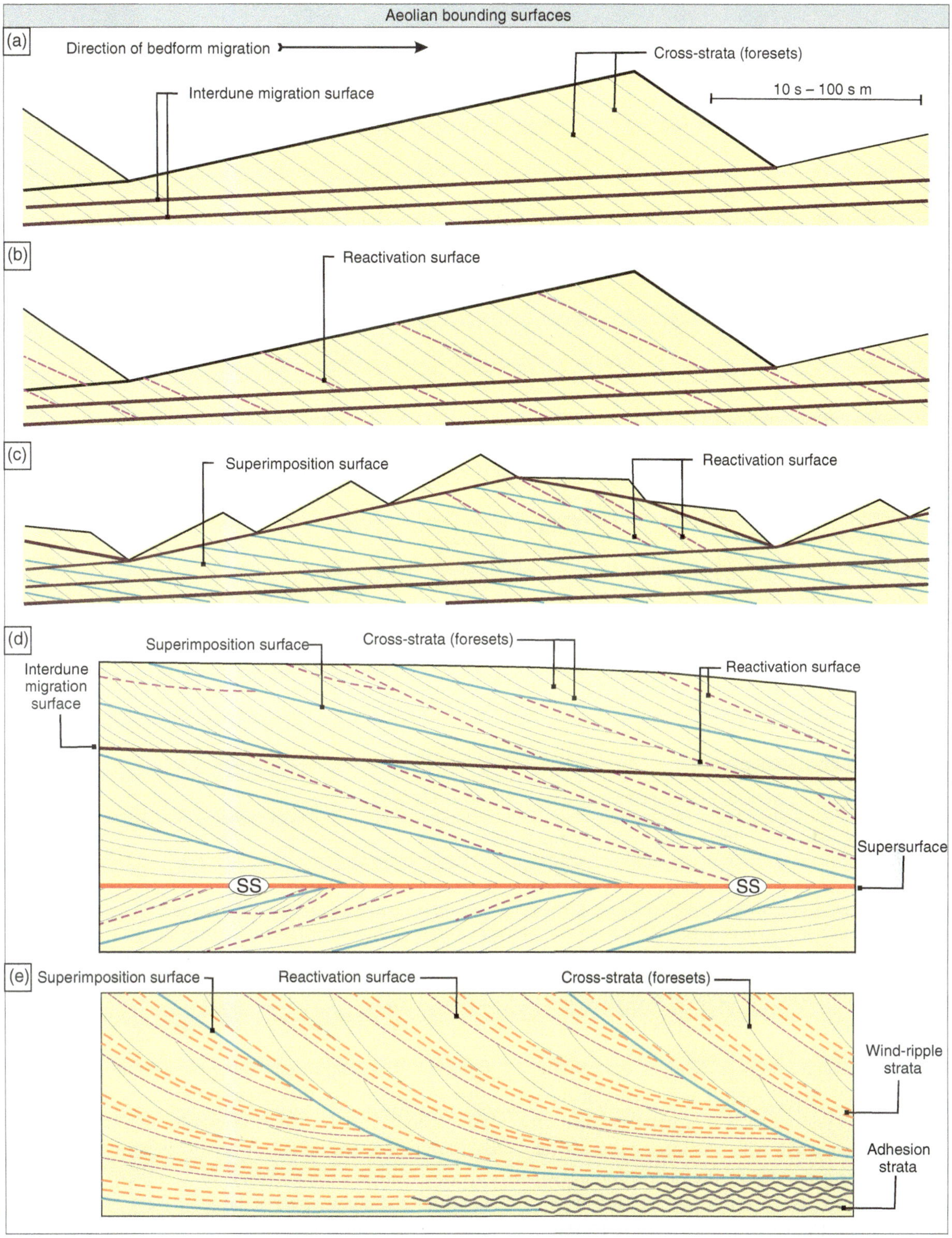

Figure 7.44 Definition diagrams explaining the hierarchical system used for the description of aeolian bounding surfaces and showing their relationships. (a) Interdune migration surfaces arise as a consequence of dune migration and climb. The lower parts of each bedform in a train are accumulated. The migrating interdune depressions each scour an interdune migration surface as they move. (b) Reactivation surfaces represent partial deflation of a bedform lee slope and arise in response to changes in bedform migration direction, steepness, speed and/or asymmetry. These changes may occur periodically, in which case, reactivation surfaces may have common spacing. (c) Superimposition surfaces are generated by the migration of superimposed bedforms and/or scour pits over a larger parent bedform that itself might be migrating, albeit more slowly. (d) Compound cross-bedded cosets of aeolian strata, depicting the hierarchy of bounding surface types described in parts a–c. Adapted from Brookfield (1977). In addition, this figure depicts a higher-order supersurface (SS) that truncates all lower-order bounding surfaces. (e) A coset of scalloped cross-strata with multiple scales of internal cyclicity and associated bounding surfaces. Two distinct scales of bounding surface are evident within the coset. Note how bounding surfaces at the base of the sets pass down the dip into corrugated surfaces. This relationship, which can potentially occur at a variety of scales, is indicative of aeolian dune migration that occurs synchronously with accumulation in damp, water-table-controlled interdunes. Based on observations from the Jurassic Entrada Sandstone, NE Utah, USA. Adapted from Crabaugh and Kocurek (1993).

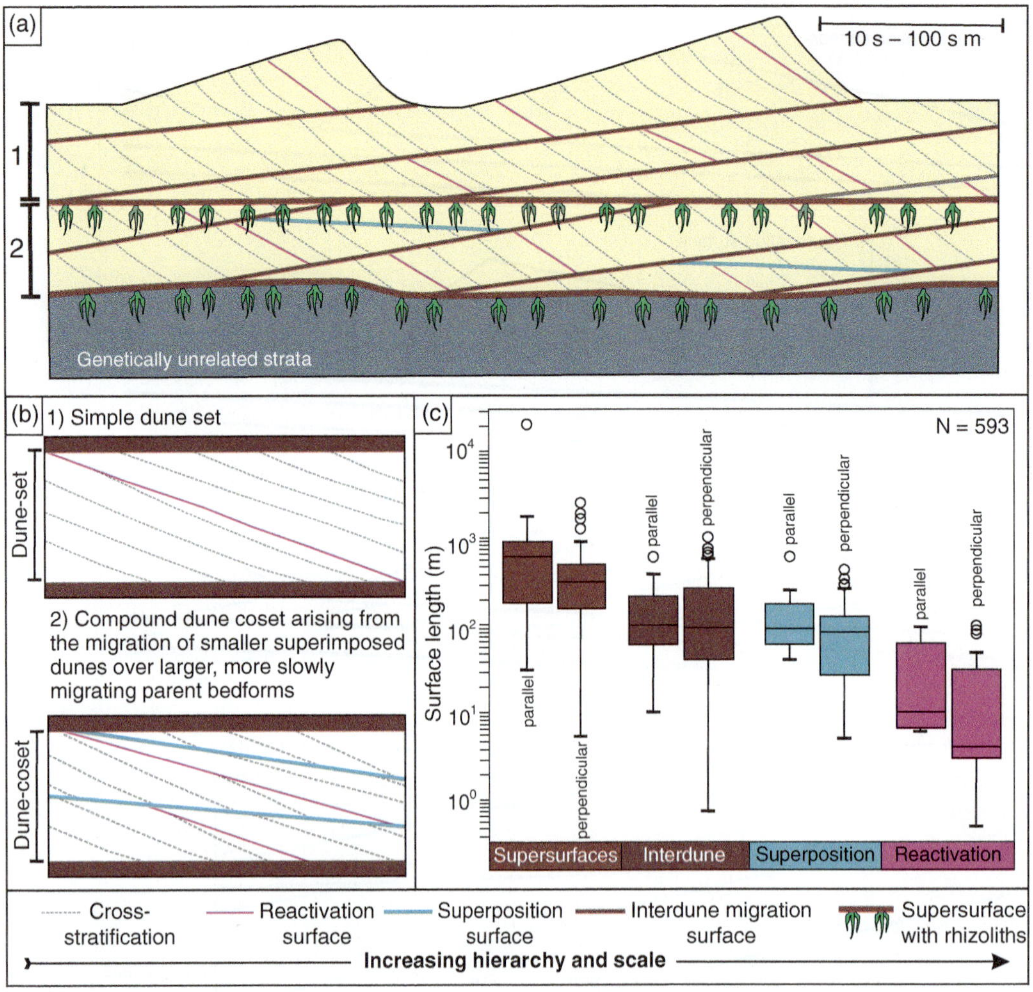

Figure 7.45 Summary of length-scales of aeolian bounding surfaces of different types. (a) Hierarchy of bounding surfaces generated by climbing dunes and interdunes; a higher-order supersurface truncates genetic sequences of aeolian accumulation. (b) Relationships between surfaces in simple aeolian dune sets and compound aeolian dune cosets. (c) Box and whisker plot showing recorded lengths of aeolian bounding-surface types. For each type, the left-hand and right-hand box plots show the recorded lengths of bounding surfaces in direction oriented parallel and perpendicular to the main direction of bedform migration, respectively. Data from the Database of Aeolian Sedimentary Architecture (DASA); see Cosgrove et al. (2021a).

commonly seasonal wind reversals (Hunter and Rubin 1983; Loope et al. 2001; Scherer and Lavina 2005; Bristow 2019). Nested reactivation surfaces that occur on two or more scales are associated with compound cosets of cross-bedding and are generated by the interaction of two or more influencing processes operating with different periodicities (Figure 7.44e; Crabaugh and Kocurek 1993; Jones et al. 2021).

Superimposition surfaces compartmentalise cosets of aeolian dune strata (Figure 7.44). They are characterised by planar to highly scallop-shaped erosion surfaces that dip in a wide range of orientations. In sections parallel to aeolian transport, superimposition surfaces may appear similar to reactivation surfaces and their identification can be problematic. However, in sections perpendicular to aeolian transport, superimposition surfaces are most commonly oriented oblique to the cross-strata that they truncate. Where both reactivation and superimposition surfaces are developed, the latter always truncates the former. Superimposition surfaces result from either migration of superimposed dunes over a larger parent bedform, else migration of scour troughs along the lee slope of a bedform that is itself migrating (Rubin and Hunter 1983; Rubin 1987b; Rubin and Carter 2006). The degree of curvature of these bounding surfaces gives some measure of the 3D shape of the bedforms or scour troughs responsible for carving them as they migrated. Although superimposed dunes and scour troughs can theoretically migrate directly up or down the lee slope of a parent bedform, oblique migration is more common because secondary airflow tends to direct these features obliquely across the lee slope of a larger host (parent) bedform.

Interdune migration surfaces are characterised by erosional surfaces inclined at low angles (typically <1° though more commonly just a few tenths of a degree) (Figure 7.44). These surfaces may extend in the direction of aeolian transport for distances of hundreds of metres to several kilometres (Figure 7.45); where the orientation and continuity of an outcrop permits, they might be observed to climb through the stratigraphy. Interdune migration surfaces bound simple sets or compound cosets of cross-strata. In sections parallel to aeolian transport, they may be planar- to slightly scalloped-shaped. In sections perpendicular to transport, they may be planar, concave, irregular or moderately to highly scalloped. Interdune migration surfaces truncate both superimposition and reactivation surfaces. Interdune migration surfaces result from the migration of bedforms separated by interdunes. The surfaces are carved by the passage of an erosive scour that defines the interdune trough between successive bedforms (Figure 7.44, Kocurek 1981). The depth to which the interdune trough scours as it migrates influences the extent to which deposits of the preceding bedform are eroded. Scallop- and trough-shaped interdune migration surfaces arise from the migration of curved or sinuous-crested bedforms. In many successions, dry, damp or wet interdune architectural elements may be present directly above interdune migration surfaces. In other successions, there may be no interdune element: the interdune migration surface may delimit two stacked aeolian dune elements, signifying the migration of an interdune hollow (but no extensive interdune flat) between successive dunes in a train.

Supersurfaces are either sharp but non-erosional or erosional surfaces that in most cases truncate all the above-mentioned types of aeolian bounding surface. Many supersurfaces have lateral extents of hundreds to many thousands of km^2; they may bound entire aeolian successions or significant parts thereof (Figures 7.44 and 7.45; Kocurek 1991; Havholm et al. 1993). Sedimentary features commonly associated with and developed upon supersurfaces include desiccation cracks and polygonal fractures, bioturbation, rhizoliths, halokinetic (salt) structures and palaeosols (Loope 1985; Talbot 1985; Mountney 2006b; Scherer and Lavina 2006; Basilici et al. 2009; Dal' Bo et al. 2010), all of which yield important palaeoenvironmental information regarding the nature of the accumulation surface at the time of supersurface formation. Supersurfaces may be flat and planar with no appreciable relief, else may exhibit considerable local relief, making their recognition and tracing problematic. In some settings, it is possible to correlate supersurfaces laterally to a point where they pass into deposits of adjacent non-aeolian palaeoenvironments or to relate them to basin-wide events, such as marine transgressions or regressions. Supersurfaces are generated when aeolian accumulation ceases and is replaced either by sediment bypass or deflation due to a switch from a positive to a neutral or negative sediment budget (Kocurek 1988; Langford and Chan 1988; Mountney 2012), else when the system is stabilised, for example, by vegetation growth; such surfaces are referred to as deflation, bypass and stabilisation supersurfaces, respectively.

Deflation supersurfaces develop in response to a lowering of the accumulation surface in response to a negative net sediment budget (Figure 7.46). Deflation occurs until one of the following conditions arises: (i) the net sediment supply becomes neutral or positive again; (ii) further deflation is prevented because the accumulation has been deflated (eroded) down to the water table or a lithified substrate (e.g. bedrock); (iii) where a mixture of grain sizes are present but deflation only erodes the finer grain fractions so that a coarse-grain deflation lag (reg, serir or desert pavement) develops as an armoured surface that prevents further deflation; (iv) a chemogenic or biogenic crust develops. The rate of deflation may be controlled by the rate of water-table fall. Where the water table acts to limit the extent of aeolian deflation, the resultant deflation supersurface is sometimes called a 'Stokes' Surface' (Stokes 1968).

Bypass supersurfaces develop where accumulation of the aeolian system ceases but no appreciable deflation (i.e. erosion) occurs. In some cases of bypass, the aeolian system may be replaced by a non-aeolian environment but for which there is neither accumulation nor erosion (Figure 7.46). In other cases of bypass, the aeolian system may persist but without accumulation; for example, aeolian dunes may continue to be present and may migrate but not accumulate. Flood surfaces (*sensu* Langford and Chan 1988) are a type of bypass supersurface developed where fluvial systems pass along interdune corridors between migrating but non-climbing aeolian dunes; fluvial deposits may be locally preserved in association with such surfaces.

Stabilisation supersurfaces develop where one or more physical, chemical or biogenic stabilising agent renders an aeolian landscape inactive. The preserved stratigraphic expression of stabilisation surfaces is commonly relict dune topography (Talbot 1985). Quaternary examples include parts of the Sahara (Kocurek et al. 1991a; Swezey et al. 1999; Swezey 2001), whereas part of the Permian Cedar Mesa Sandstone represents an example from the ancient geologic record (Langford et al. 2008).

Although most supersurfaces are considered to develop in response to allogenic controls, for example, in response to climate change, they might also potentially develop in response to the migration and climb at a very low angle of entire dune fields (Porter 1986). Through the application of Walther's law, two coexisting but spatially distinct

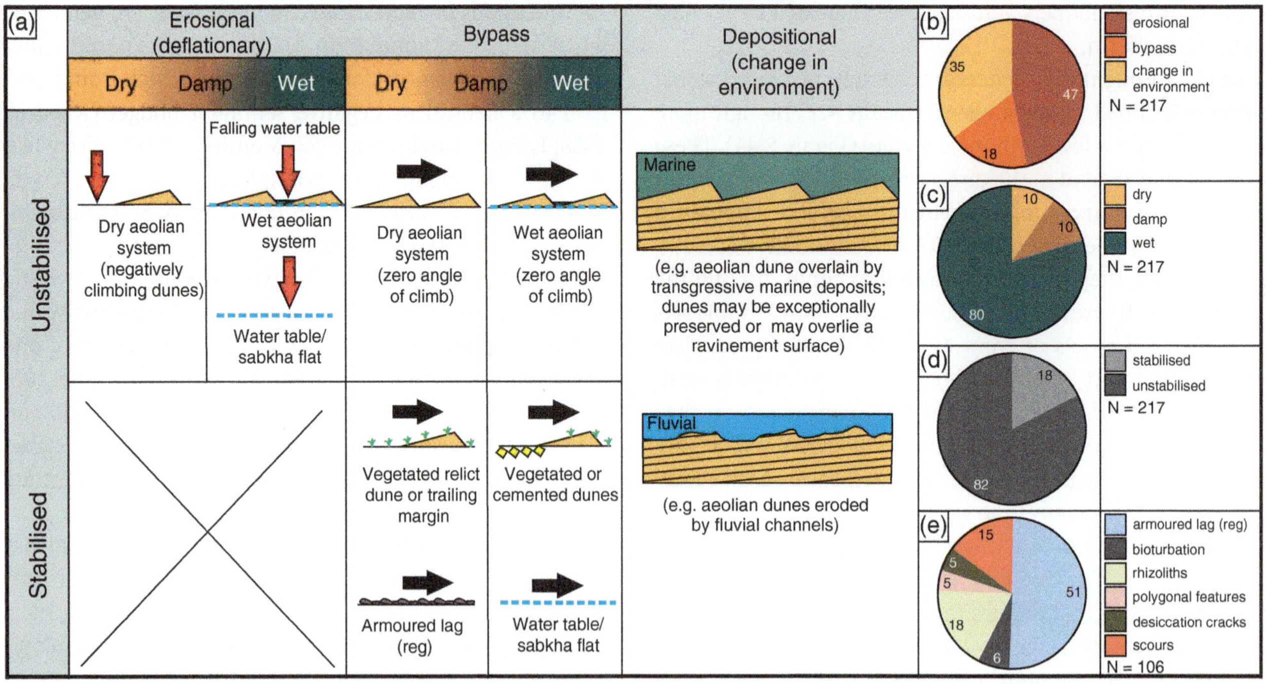

Figure 7.46 Classification of aeolian supersurfaces. (a) Classification based on the sediment budget, the relative wetness of the substrate (dry, damp, wet) and the nature of the substrate (stabilised or unstabilised). Red arrows indicate erosion of the substrate (deflation); black arrows indicate sediment bypassing. Adapted from Kocurek and Havholm (1993). (b–e) Pie charts classifying 217 supersurfaces based on data from 55 ancient aeolian case studies. (b) Classification based on sediment budget and nature of change of accumulation surface. (c) Classification based on relative wetness of the substrate. (d) Classification based on the nature of the substrate. (e) Dominant sedimentary structures associated with supersurfaces. Data from the Database of Aeolian Sedimentary Architecture (DASA); see Cosgrove et al. (2021a).

aeolian dune fields – else two spatially separated sub-environments of an erg – such as a fore-erg and a central-erg – may migrate over time such that their accumulated deposits may come to be stacked one upon the other. The surface delineating the two separate successions is a supersurface in that (i) it separates successions of two genetically unrelated aeolian systems and (ii) it is a surface of higher order than all other bounding surfaces contained in those successions.

7.5.8 Aeolian Depositional Elements: Larger-Scale Sediment Bodies

Groups of genetically related architectural elements may themselves occur together whereby they comprise a composite depositional element. A composite depositional element may be composed internally of a group of architectural elements of the same order, with each element juxtaposed adjacent to its neighbouring elements. For example, aeolian dune and interdune elements commonly occur together in a manner that records the migration and climb of a train of dunes, each separated by interdune flats (Figure 7.47). Alternatively, a depositional element might be composed of a hierarchically nested suite of elements and bounding surfaces. For example, a playa-lake element might be nested inside an interdune element; this might record the presence of a transient shallow lake in the centre of an interdune. Depositional elements might comprise sub-environments of entire aeolian erg systems: back-erg, central-erg, fore-erg and lateral-erg (Box 7.3).

7.6 Aeolian Facies Models

7.6.1 Overview

Facies models may be expressed in multiple ways: as idealised 1D graphic logs that depict successions of facies; as 2D panels that depict spatial relationships between adjoining elements; as facies entity–relationship diagrams; as 3D block models that relate the geometry of bodies of accumulated strata to the morphology and behaviour of the landforms responsible for their generation (e.g. Walker 1979). Distinct aeolian facies models can be devised for dry, wet and stabilised aeolian system types, as well as for aeolian systems that interact with other environments, such as fluvial systems. Bespoke, customised facies models may be constructed via selection of appropriate aspects from many case-study examples (Cosgrove et al. 2021a; Box 7.4).

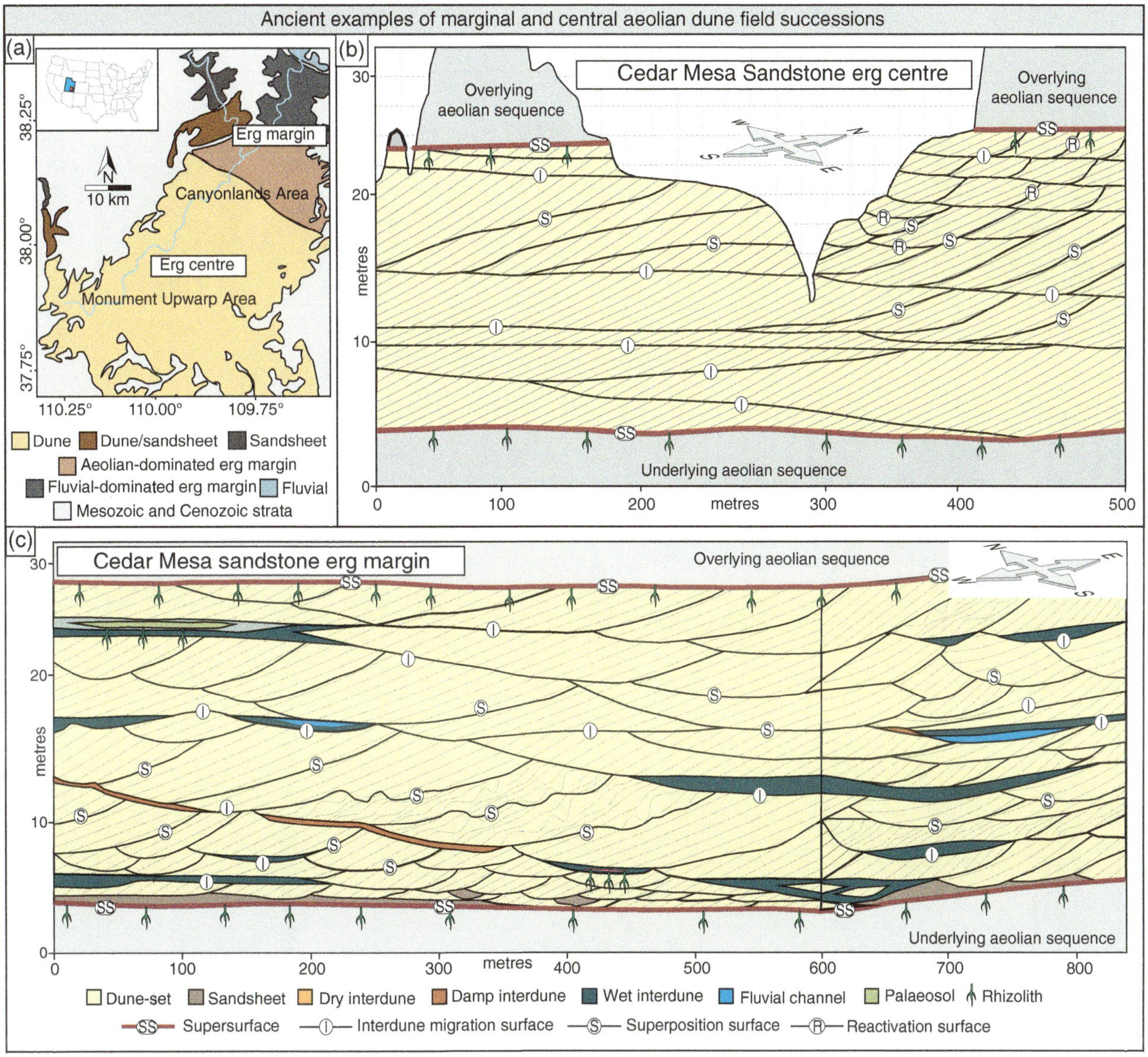

Figure 7.47 Preserved expression of aeolian stratigraphic architecture and variations between central and marginal aeolian dune field settings, Permian Cedar Mesa Sandstone, Utah, USA. (a) Map showing the distribution of central- and marginal-erg deposits in the Cedar Mesa Sandstone. The inset shows a map of the USA highlighting the state of Utah in blue and the location of the study area in red. (b) Architectural panel depicting aeolian (and related) architecture in the palaeo-erg-centre setting. (c) Architectural panel depicting aeolian architecture in the palaeo-erg-margin setting. Note the greater number of non-aeolian elements (including fluvial and palaeosol elements) and the presence of damp and wet interdunes. Adapted from Mountney and Jagger (2004).

7.6.2 Dry Aeolian System Facies Model

Accumulation in dry aeolian systems most commonly occurs once dunes have grown to a size such that they cover the accumulation surface and former interdune flats are reduced to isolated depressions between dunes or megadunes. The bedforms then climb with respect to the basal surface on which they are developed (which itself may be horizontal or inclined) to accumulate a succession (Figure 7.48a). The Permian Cedar Mesa Sandstone in the White Canyon region of Utah, USA, is a well-exposed example of a largely dry aeolian succession (Figure 7.49). The angle of climb of the system is determined by the ratio between the rate of rise of the accumulation surface and the rate of bedform migration. Relatively steep angles of climb enable larger proportions of the original bedforms to be preserved, though climb angles rarely exceed 1° and are usually only a few tenths of a degree; typically, no more than the basal 10–15% (commonly less – Romain and Mountney 2014) of a bedform is accumulated as a dune set (Figures 7.48a and 7.49). Thus, the facies units accumulated

Box 7.3 Preserved depositional elements representative of aeolian system sub-environments

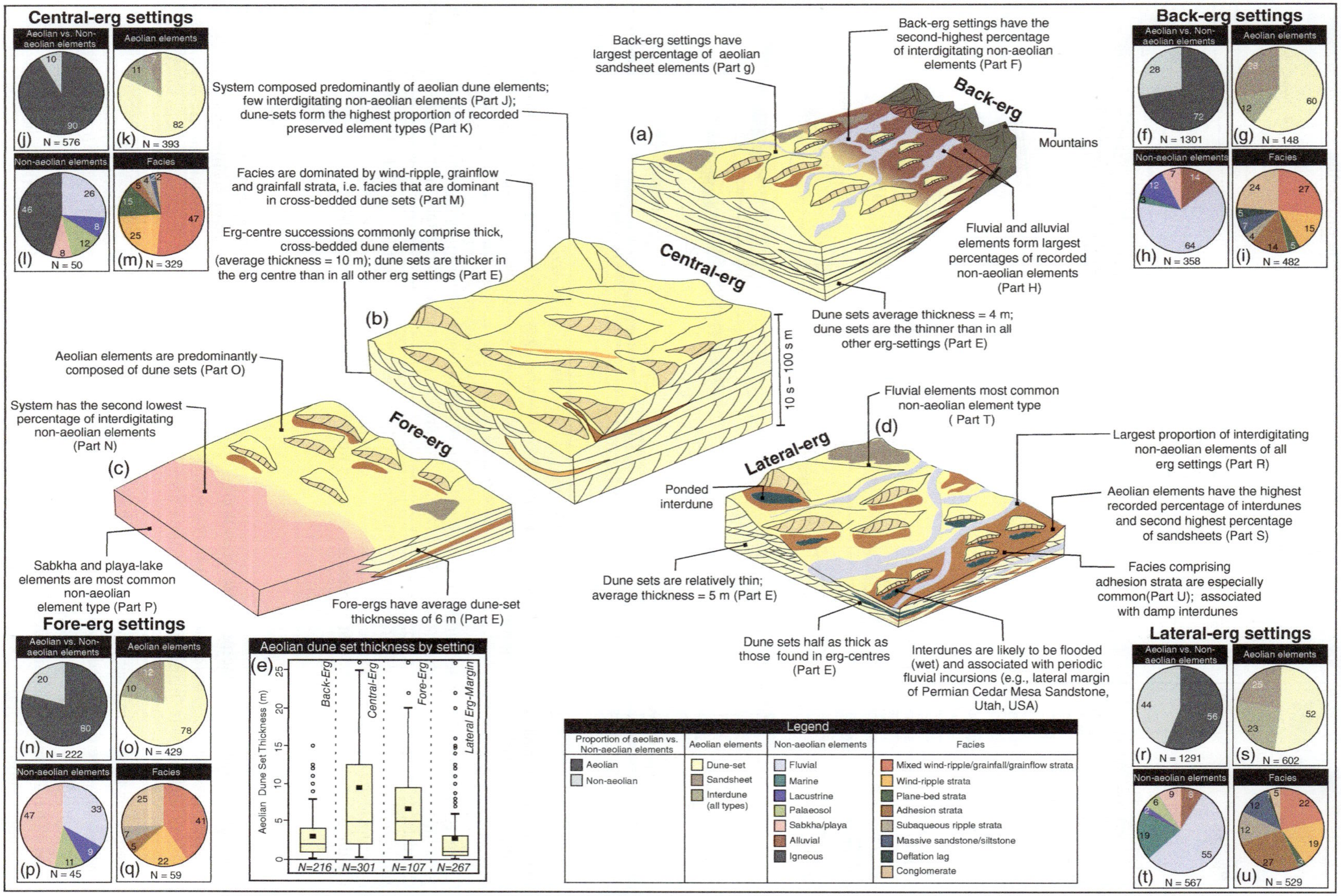

Box 7.3 Large-scale depositional elements representing (a) back-erg, (b) central-erg, (c) fore-erg and (d) lateral-erg sub-environments of aeolian dune-field systems. (e) Preserved aeolian dune-set thicknesses recorded in ancient successions for the four depositional-element types. (f)–(u) Pie charts depicting proportions of aeolian versus non-aeolian architectural elements, aeolian dune elements, non-aeolian element types and lithofacies units for preserved successions of the four depositional-element-element types. Information from the Database of Aeolian Sedimentary Architecture (DASA). See Cosgrove et al. (2021a) for details.

Box 7.4 Quantitative aeolian facies models

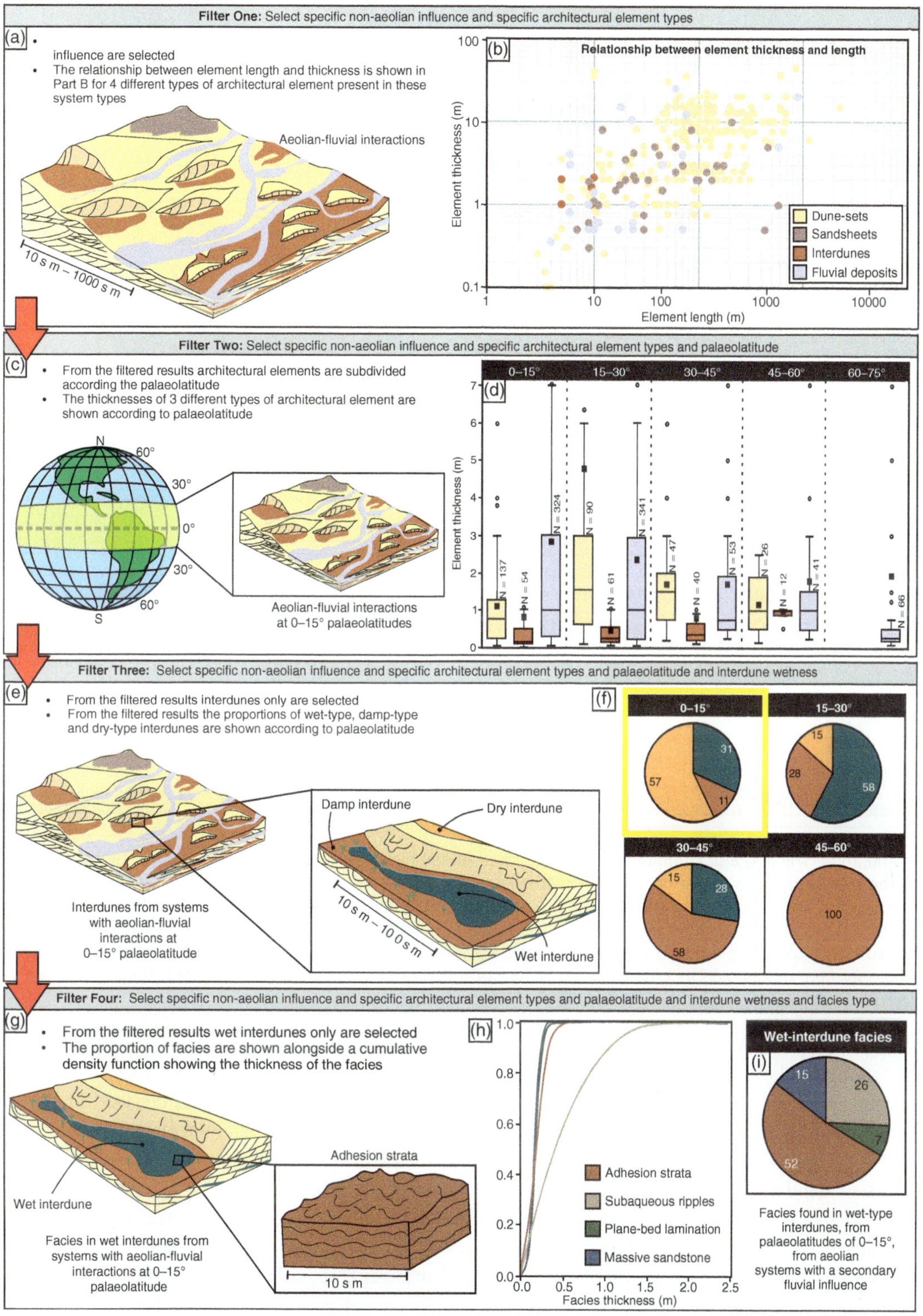

Box 7.4 Illustrated example of the application of a database-informed approach to build a quantitative aeolian facies model for a bespoke class of aeolian system. From a large database of aeolian successions, results are returned based on sets of specific queries. A series of filters can be applied to queries to build bespoke and specialised models. Information from the Database of Aeolian Sedimentary Architecture (DASA). See Cosgrove et al. (2021a) for details. (a) Conceptual diagram showing an aeolian system with fluvial interactions. (b) Scatter-plot showing the relationship between the length and thickness of various architectural elements, coloured by type. (c) Conceptual diagram showing how aeolian systems with associated fluvial interactions are filtered to only consider those form a specified palaeolatitude. (d) Box and whisker plot showing thicknesses of various architectural elements according to palaeolatitude. (e) Conceptual diagram showing how only interdune architectural elements are selected (from aeolian systems with fluvial interactions subdivided according to palaeolatitude). (f) Pie-charts showing the percentage of wet-type, damp-type, and dry-type interdune element by palaeolatitude. (g) Conceptual diagram showing how wet interdune elements from palaeolatitudes of 0–15° are selected to examine their constituent facies unit types (in this case adhesion strata); (h) Cumulative density plot showing the thickness of facies types found within wet interdune elements. (i) Pie chart showing the percentage of facies found within wet interdune elements.

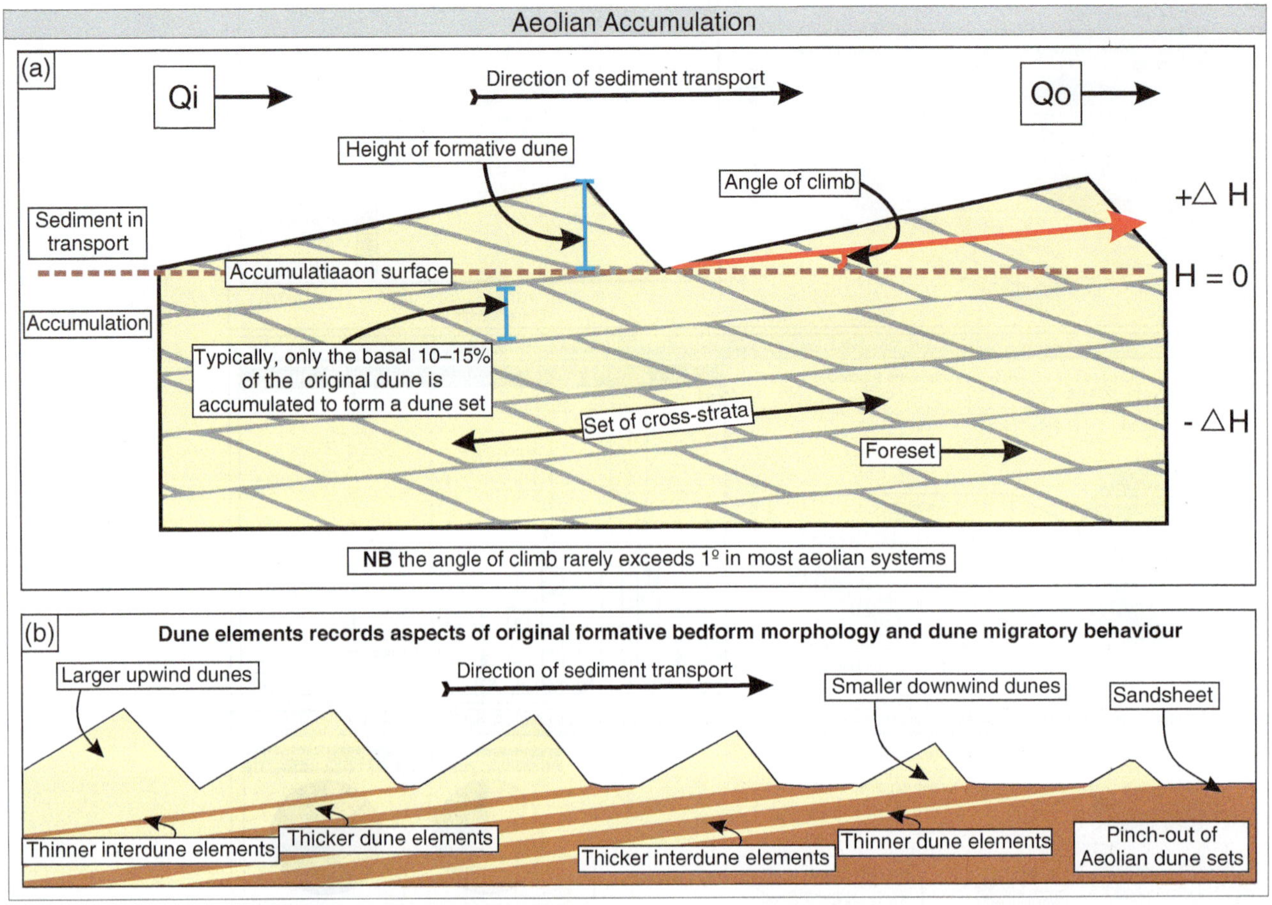

Figure 7.48 Aeolian accumulation. The angle of climb of the system is determined by the ratio between the rate of rise of the accumulation surface and the rate of bedform migration. Relatively steep angles of climb enable larger proportions of the original bedforms to be preserved, though climb angles rarely exceed 1 and are usually only a few tenths of a degree; typically, only the basal 10–15% of a climbing bedform is preserved as a dune set. (a) Subcritical climb of aeolian dunes to generate an accumulation. Q_i is the sediment flux in. Q_o is the sediment flux out. (b) Reduction of bedform size and associated increase in the interdune extent to the outer downwind limit of an aeolian fore-erg margin and its effects on the nature of the accumulated stratigraphic record. Note the constant angle of climb in this example.

via bedform climbing occur as aeolian cross-bedded dune architectural elements that record the processes that operated on the lower part of the dune lee slope at the time of accumulation; facies units of wind-ripple, grainflow and grainfall strata dominate. Reactivation and superimposition bounding surfaces may be present within climbing dune cosets. As dunes migrate, the interdune hollows in front of their toes scour erosional, low-angle-inclined interdune migration surfaces; these partition dune sets (or cosets). The migration and climb of a train of aeolian dunes generate an accumulation comprising multiple stacked aeolian dune sets or cosets, and each is a dune architectural element (Figure 7.49d,e). The 3D geometry of these dune elements records aspects of both the morphology of the original formative bedforms and the migratory behaviour of those bedforms. For example, sinuous-crested transverse dunes typically generate trough-shaped elements; a temporal and downwind decrease in the size of the bedforms may result in the accumulated sets thinning to a point of pinch-out down-system (Rubin and Hunter 1982; Figure 7.48b). Episodes of accumulation via the migration and climb of a train of dunes generate an aeolian sequence. Eventually, an accumulation episode will cease and be replaced by bypass, deflation or stabilisation, which will generate a supersurface (Figure 7.49a,b). Supersurfaces in dry aeolian systems most commonly develop as a consequence of changed aerodynamic conditions. For example, deflation occurs when the airflow passing over a dune field is not fully laden with sediment to its maximum potential carrying capacity. Repeated episodes of accumulation and supersurface

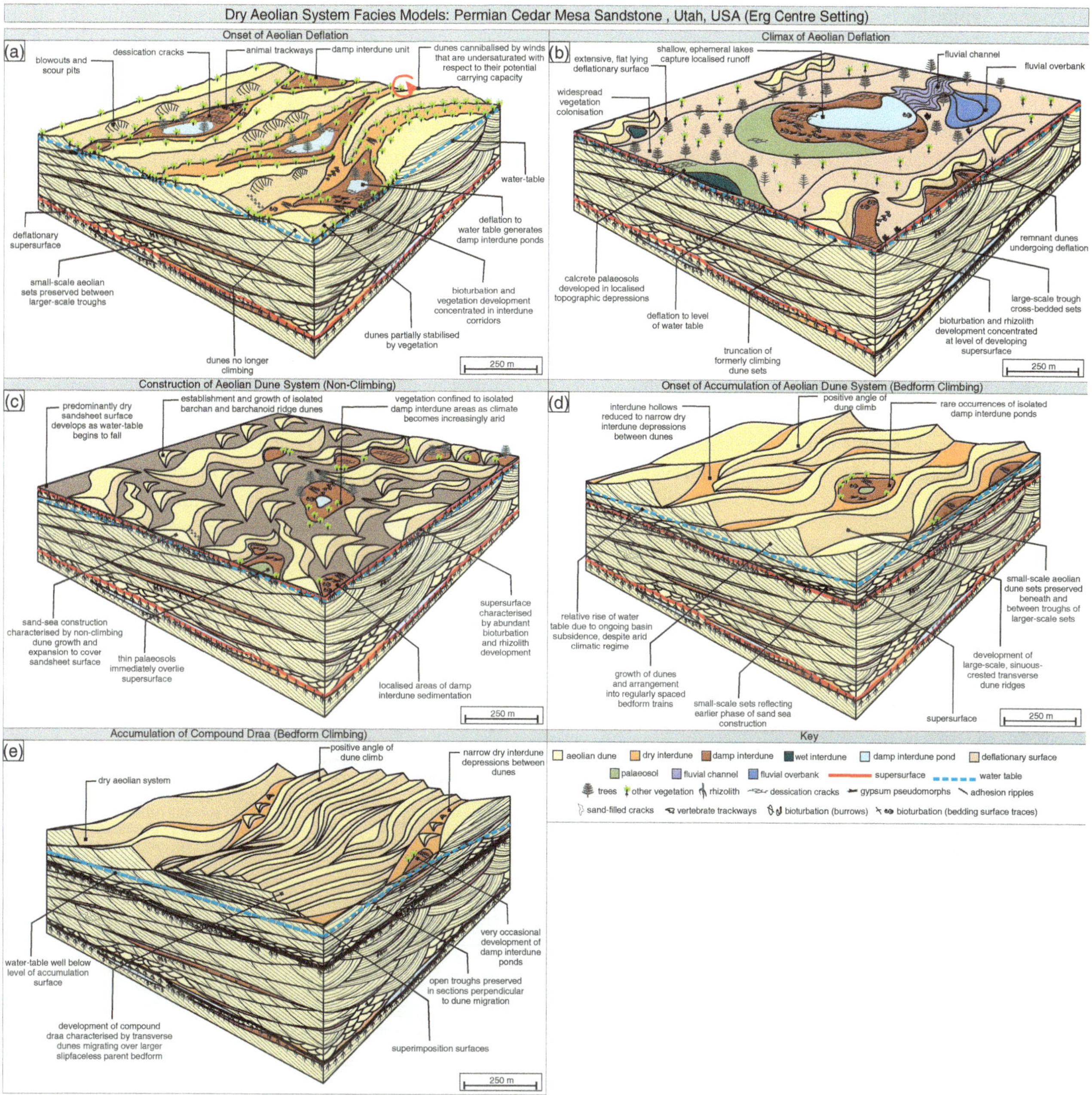

Figure 7.49 Summary facies models for a mostly dry aeolian system; an example from the erg-centre region of the Permian Cedar Mesa Sandstone, White Canyon region, southern Utah, USA. (a) Onset of regional aeolian dune-field deflation. (b) Climax of regional aeolian dune-field deflation; deflationary supersurface development. (c) Re-establishment of aeolian dune system; non-climbing dune construction and migration. (d) Onset of aeolian dune-field accumulation. (e) Climax of aeolian dune-field accumulation; development of compound mega-bedform (draa) with superimposed bedforms. Modified after Mountney (2006a).

generation can result in the accumulation of multiple vertically stacked or laterally offset (in some cases overlapping) aeolian sequences, each recording accumulation of a distinct aeolian dune field. Although many dry aeolian successions apparently accumulated via bedform climbing, some successions record alternative mechanisms of accumulation and preservation, for example, where relict dune forms are preserved in situ (Section 7.7.4).

The Jurassic Page Sandstone of Utah and northern Arizona is a dry aeolian system represented by separate supersurface-bounded sequences (Havholm and Kocurek 1994), each composed of one or more sets of climbing dune strata and an absence of significant interdune flat strata. Collectively, the aeolian sequences comprise a composite aeolian succession that interfingers with marine units of the neighbouring Carmel Formation, indicating a coastal setting subject to repeated transgression and regression (Blakey et al. 1996). In many cases, the supersurfaces are directly overlain by sabkha and subaqueous strata and are of great lateral extent. They are characterised by polygonal fractures and a corrugated, erosional relief. The aeolian-dominated sequences collectively record multiple episodes of accumulation; rather than being stacked vertically, they occupy laterally offset palaeo-depocentres. Dry aeolian strata accumulated under a plentiful sand supply during periods of marine regression; the exposed coast and shelf acted as sand source. Deflation to the water table and supersurface generation occurred because of reduced aeolian sand supply during the onset of episodes of marine transgression (Havholm and Kocurek 1994).

7.6.3 Wet Aeolian System Facies Model

Accumulation in wet aeolian systems is characterised by cross-stratified aeolian dune sets that are separated by interdune units exhibiting evidence of sedimentation at or close to the water table. Accumulation controlled by water-table rise is indicated by a climb through the stratigraphy of both dune elements and intervening interdune elements in the general direction of aeolian transport, as seen, for example, in parts of the Jurassic Entrada Sandstone, Utah, USA (Kocurek 1981; Kocurek and Day 2018) and the Triassic Helsby Sandstone Formation, UK (Mountney and Thompson 2002; Cosgrove et al. 2025). In wet aeolian systems, a progressive water-table rise that occurred synchronously with dune migration may be indicated by intertonguing of the toesets of advancing dune strata with strata accumulated in the adjacent interdune (Pulvertaft 1985). Changes in either the rate of water-table rise or the rate of dune migration cause the angle of climb to vary (Figure 7.50). The relative level of the water table is determined by both absolute water table changes, most obviously governed by the climatic regime, and subsidence, whereby an accumulation can slowly subside through a fixed (static) water table. The water table in the subsurface is not necessarily horizontal or planar: it may exhibit a subdued rise and fall, in some cases in phase with overlying

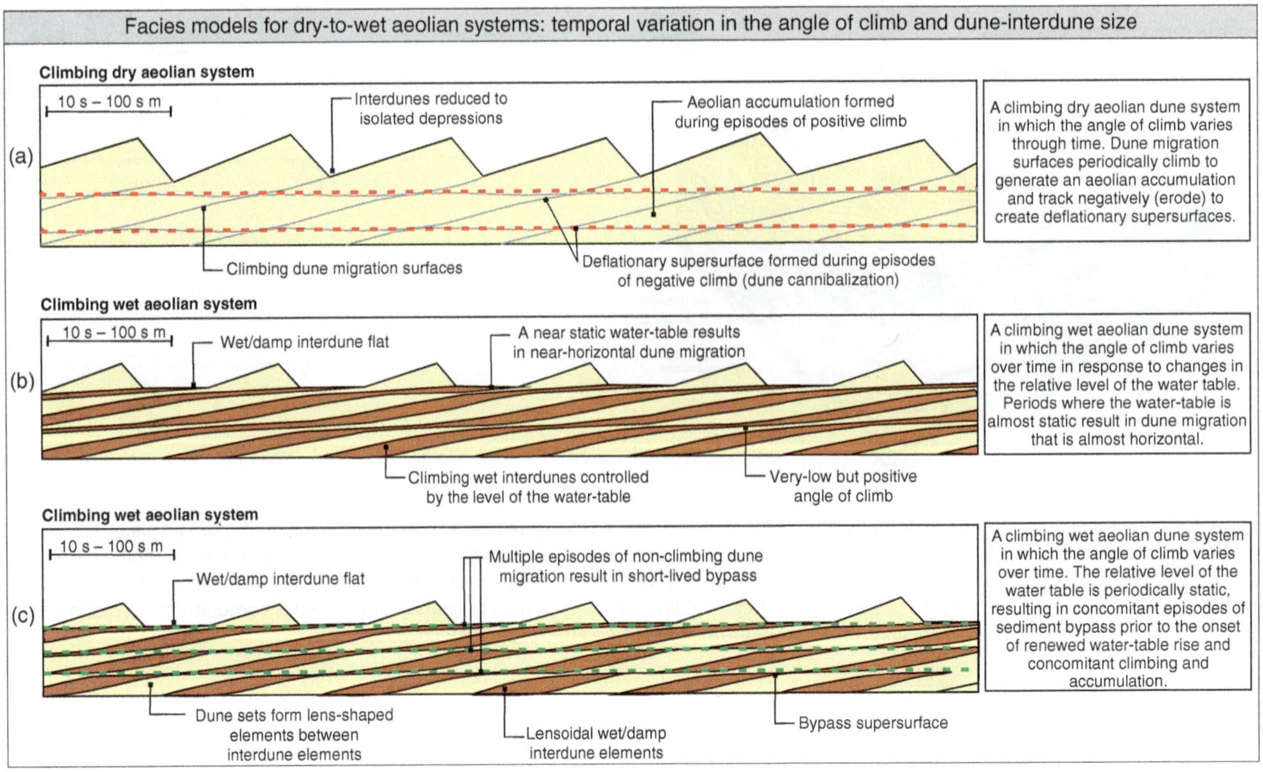

Figure 7.50 Summary facies models for dry-to-wet aeolian systems. Temporal variation in the angle of system climb due to changes in the ratio between the rate of change of the level of the accumulation surface and the rate of dune migration and associated changes in dune and interdune size. (a) Dry aeolian system subject to repeated episodes of positive climb (accumulation) and negative climb (deflation). (b) Water table-controlled wet aeolian system. Periods where the relative level of the water table is almost static result in dune migration with a positive but very low angle of climb. (c) Water table-controlled wet aeolian system. Periods where the relative level of the water table is static result in dune migration without climbing to generate a bypass supersurface. Adapted from Mountney (2012).

dune relief. Increases and decreases in the rate of relative water-table rise and associated changes in the availability of sediment suitable for dune construction cause interdunes to expand at the expense of dunes and vice versa (Figure 7.50b,c). Where climatic regimes fluctuate on a cyclical basis, interdune units preserved in wet aeolian successions record this change. As interdune units expand and contract through time, so conditions at the accumulation surface vary from relatively drier, through damp, to relatively wetter (and back) in response to subtle variations in the level of the water table relative to the level of the accumulation surface.

7.6.4 Stabilised Aeolian System Facies Model

Accumulation in stabilised aeolian systems is enabled by one or more physical, chemical or biogenic agents that act to enable accumulation. Vegetation that colonises dunes is a common stabilising agent in many present-day settings and is also considered to be responsible for enabling the accumulation of some ancient successions (Figure 7.51). Part of the Tertiary Tsondab Sandstone, which underlies much of the modern Namib Desert, is characterised by sets of cross-stratified dune strata that contain abundant plant root structures (Kocurek et al. 1999). Plants act to decrease the near-surface wind velocity, thereby promoting sediment accumulation and minimising opportunities for later aeolian winnowing; their roots act as a sand binding agent, thereby preventing deflation (Figure 7.51).

Prior to the evolution of vegetation on land and its widespread colonisation in the Silurian to Devonian, microbial (bacterial) mats may have acted as important biogenic agents that had potential to stabilise some aeolian substrates. Some Proterozoic aeolian successions are characterised by MISS that demonstrate the likely role of microbial and bacterial mats in enabling progressive accumulation in stabilised aeolian systems, including the Mesoproterozoic Mangabeira and Galho do Miguel formations,

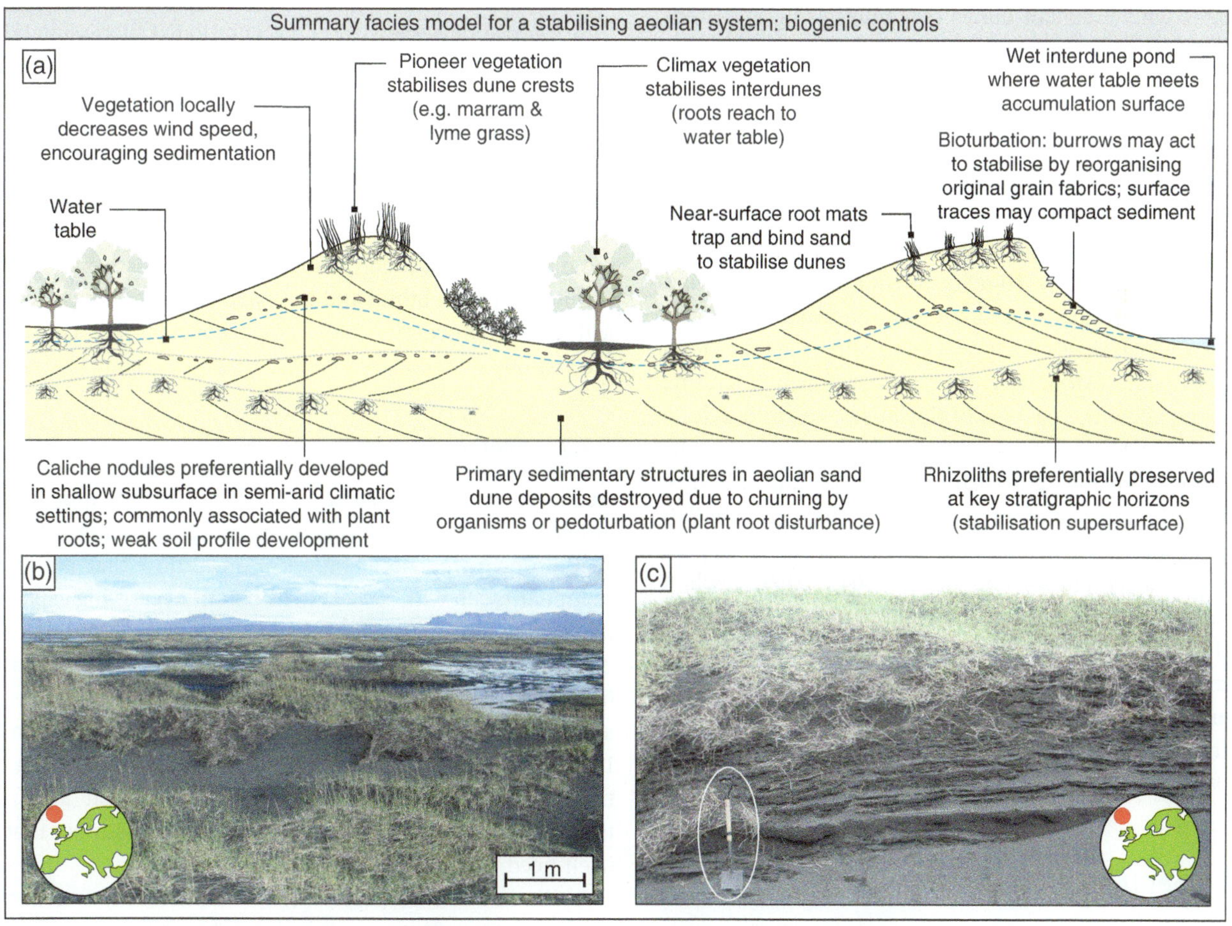

Figure 7.51 Summary facies model for a stabilising aeolian system. Biogenic stabilising agents in aeolian successions. (a) Schematic model illustrating types and locations of some common biogenic stabilising agents. (b) Example of root mats acting to stabilise aeolian dune sediment surfaces. (c) Aeolian accumulation enabled by root mats and networks that trap and bind sand. (b and c) Skeiðarársandur dune field, southern Iceland. Modified after Bristow and Mountney (2021) and Mountney and Russell (2009).

Brazil (Bállico et al. 2017a,b; Basilici et al. 2021; Mesquita et al. 2021) and the Neoproterozoic Venkatpur Sandstone Formation, Southern India (Basilici et al. 2020).

Physical and chemical factors are also important stabilising agents. The presence of coarse-grained lags, mud drapes and sheets (Figure 7.35a,c), a high water table, periodic flooding and surface or near-surface cementation by chemical precipitates (Figure 7.36) all encourage accumulation in warm-climate stabilised aeolian desert systems (e.g. Basilici and Dal' Bo' 2014). Permafrost may play an important role in enabling accumulation in cold-climate stabilised systems.

7.6.5 Aeolian–Fluvial System Interaction Facies Model

The relative dominance of aeolian versus fluvial activity is highly variable over a range of spatial and temporal scales; this gives rise to complex arrangements of aeolian and fluvial landforms and architectural elements. The frequency and persistence of fluvial flooding control the period of occupancy of interdune corridors by active fluvial systems. In cases where aeolian dunes continue to migrate either during a single flood or between successive floods, fluvially flooded interdunes may leave a record of laterally amalgamated flood deposits. In non-climbing (i.e. non-accumulating) aeolian systems, such behaviour favours the development of bypass supersurfaces (e.g. flood surfaces of Langford and Chan 1988). In aeolian systems that climb at low angles (i.e. where a modest component of vertical accumulation is coincident with on-going aeolian dune and interdune migration), thin intercalations of vertically stacked, sheet-like fluvial and aeolian elements tend to accumulate (Mountney 2012). Where open interdune corridors allow the passage of fluvial systems into a dune field, the scale and connectivity of fluvial flood deposits tend to gradually diminish with increasing distance towards the interior of an aeolian dune field (Figure 7.8) as open interdune corridors transition to spatially isolated depressions (Mountney and Jagger 2004). Where aeolian dunes act as natural dams, they encourage fluvially supplied floodwater to ponds creating temporary lakes, located either at the outer margin of a dune field or in its interior (Figure 7.26). Ponding of fluvial floodwater between or behind dunes permits the accumulation of clay and silt deposits and potentially of salts if the water salinity is high.

7.7 Sequence Stratigraphy

7.7.1 Overview

Where deposits of aeolian sedimentary systems accumulate they may form a stratigraphic record. Many accumulated aeolian sedimentary successions can be divided into a series of separate aeolian sequences, each delineated by a regionally extensive bounding surface (supersurface), indicative of regional and, in many cases, long-lived cessations of aeolian accumulation. Thus, many aeolian successions are internally characterised by multiple episodes of aeolian dune-field construction and accumulation, but these were apparently punctuated by episodes when the dune fields became inactive, were stabilised (e.g. by vegetation), were bypassed, else in some cases, were partly deflated. The aeolian stratigraphic record is therefore fragmentary: significant episodes of time may be represented by episodes of aeolian bypass or deflation (Rodríguez-López et al. 2014). For many aeolian successions, it is likely that the preserved stratigraphic record represents only a minor fraction of the total duration over which aeolian processes operated; indeed, aeolian system preservation into the long-term rock record should be considered the exception, rather than the norm. For this reason, care must be taken in interpreting overall palaeoenvironmental conditions from the aeolian stratigraphic record: the preserved deposits may not be representative of the conditions that prevailed over the entire time interval represented by an accumulated succession (cf Ager 1981). Interpretation of the origin, control upon and environmental significance of aeolian sequences and their bounding supersurfaces is undertaken through a sequence stratigraphic analysis. The preservation of desert aeolian systems in the rock record is a three-stage process that requires (i) the construction of an aeolian system (e.g. a dune field), (ii) its accumulation to form a succession and (iii) its long-term burial and incorporation into the geological record (Kocurek 1999). These concepts are considered below.

7.7.2 Aeolian System Construction

Aeolian dune-field construction involves the initiation and growth of aeolian bedforms, commonly in response to a spatial downwind deceleration of an airflow that is fully sediment-laden with respect to its potential carrying capacity. Airflow deceleration results in a reduction in sediment carrying capacity, thereby inducing the growth of aeolian bedforms. However, the construction of aeolian dunes alone does not necessarily generate an accumulation. Indeed, aeolian dunes undergoing construction and migration technically constitute sediment bodies that are in transport. These bedforms are constructed on a surface that itself need not necessarily aggrade over time.

Aeolian dune-field (sand-sea) construction is a function of sediment supply, the availability of that sediment for aeolian transport and the transport capacity of the wind (Kocurek and Lancaster 1999). Sediment supply is the volume of sediment of a suitable grain size for aeolian transport generated per unit time. This sediment may form either a contemporaneous or time-lagged source of material with which to construct an aeolian system and may be derived from a variety of pre-existing aeolian or non-aeolian sources (Kocurek 1999; Figure 7.52). Sediment availability is the susceptibility of surface grains to aeolian entrainment; it is influenced by stabilising factors such as the presence of vegetation, mud drapes, coarse-grained lags, an elevated water table or surface binding and cementing agents. The transport capacity of the wind refers to its sand-carrying capacity, which increases with wind power (i.e. velocity raised to the power ~3 and also dependent on mean speed and gustiness). Where sediment availability is limited, the airflow may be undersaturated with respect to its potential sediment load, and the wind will potentially erode by entraining available sediment as it passes over a surface. Conversely, an airflow that is fully

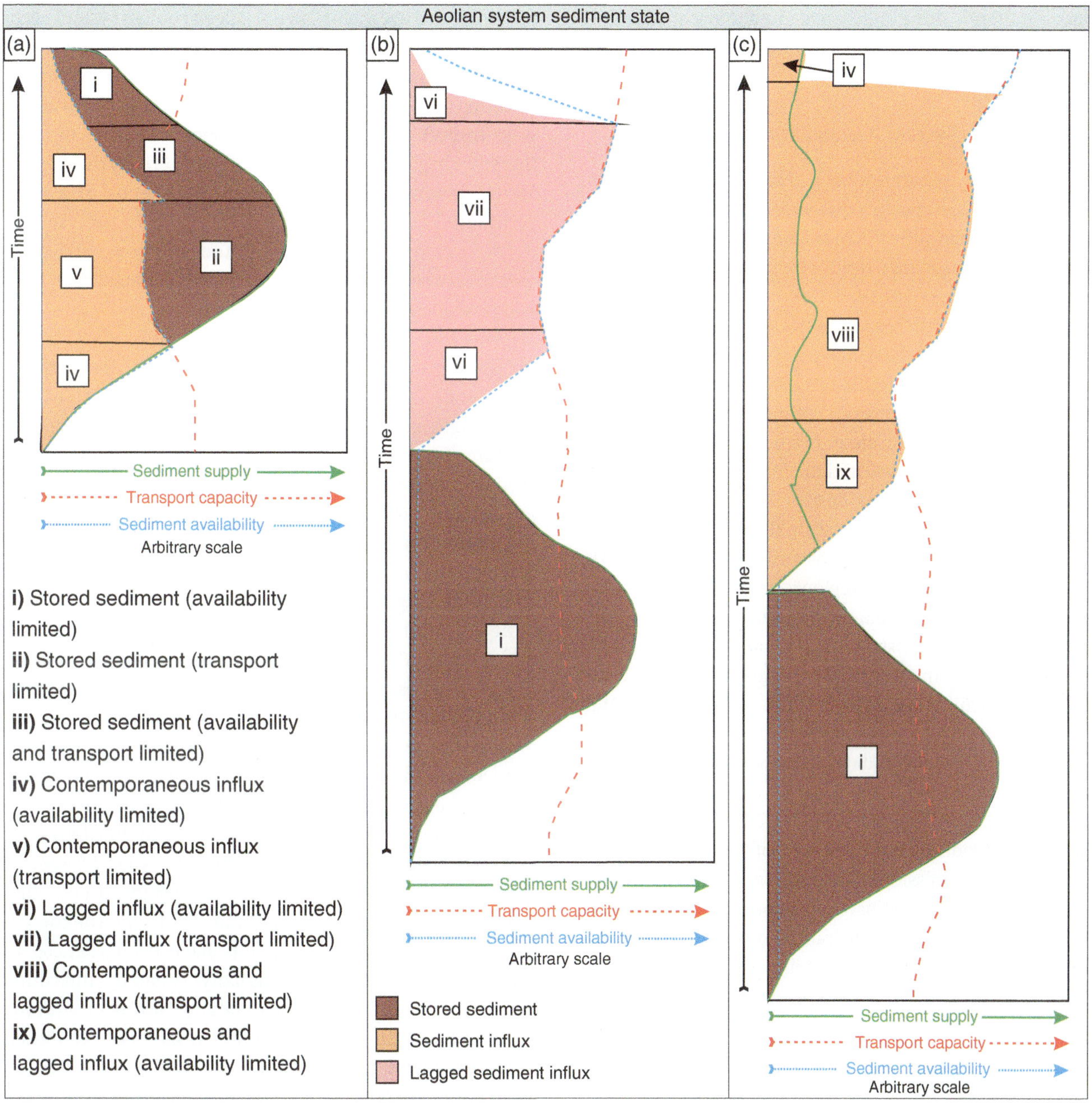

Figure 7.52 Definition diagram for classes of aeolian system sediment state determined by sediment supply, sediment availability and the transport capacity of the wind over time. (a) Aeolian construction contemporaneous with the generation of sediment supply. (b) Aeolian construction separated in time from an earlier period of sediment generation. (c) Aeolian construction sourced by both a contemporaneous sediment supply and erosion of previously stored sediment. Adapted from Kocurek and Lancaster (1999).

saturated with sediment and undergoes deceleration must drop some of its load, thereby encouraging aeolian construction and potential accumulation.

Aeolian sedimentary systems may be under construction at one location, whilst contemporaneously undergoing deflation elsewhere. For example, winds that are undersaturated with respect to their potential carrying capacity may deflate the upwind margin of a dune field, carry that material downwind, decelerate and then construct part of the downwind margin of the same dune field. As such, entire dune fields may translate over space and time (cf. Porter 1986). Similarly, dune fields may undergo construction, bypass or deflation at their lateral margins (Loope 1984; Langford and Chan 1988, 1989).

7.7.3 Aeolian System Accumulation

Accumulation to generate a body of strata occurs where an aeolian system has a positive net sediment budget such that upstream sediment influx exceeds downstream outflux and the accumulation surface rises over time. Accumulation of migrating bedforms (e.g. dunes) commonly occurs because of bedform climb with respect to the accumulation surface. The angle of climb is determined by the ratio between the rate of rise of the accumulation surface and the rate of downwind bedform migration. For most 'climbing' aeolian dune systems, the accumulation rate is small compared to the bedform migration rate and the resultant angle of climb is low: subcritical climbing ensues. As they migrate in a 'train', following bedforms truncate the upper parts of preceding bedforms leading to substantial sediment reworking. Where flat interdune regions are present, these too can migrate and accumulate in front of advancing dunes.

Alternative mechanisms for aeolian accumulation to bedform climbing include (i) the gradual accretion of sediment to the flanks of large linear or star-shaped dunes that are themselves mostly non-migratory (Argüello Scotti and Veiga 2019), (ii) the inundation of aeolian dune fields by lava flows associated with large volcanic eruptions (Jerram et al. 2000a,b; Michel et al. 2025; Scherer 2000) and (iii) the flooding and inundation of aeolian dune fields by marine transgression, in some cases leading to the reworking of the upper parts of flooded dunes by wave activity and the subsequent draping of marine sediments over aeolian sediments (Glennie and Buller 1983; Benan and Kocurek 2000; Strömbäck et al. 2005; Rodríguez-López et al. 2012b).

7.7.4 Aeolian Sequences and Supersurfaces as Sequence Boundaries

The continued construction and accumulation of an aeolian succession requires the following: (i) the maintenance of an upwind sediment supply; (ii) the continued availability of that supply for aeolian sediment transport to the site of deposition and accumulation; (iii) the maintenance of an airflow that is fully saturated with respect to its potential sediment carrying capacity; and (iv) the downwind deceleration of that airflow over the site of accumulation such that sedimentation remains ongoing (Kocurek 1999; Kocurek and Lancaster 1999). Given these conditions, an aeolian system can continue to accumulate and can potentially generate a thick succession of multiple elements as a *sequence*. However, these conditions cannot be maintained *ad infinitum* and eventually an episode of aeolian system accumulation must cease. A cessation in accumulation might occur in response to exhaustion of an upwind sand supply, resulting in a switch to an airflow that is undersaturated with respect to its potential transport capacity and will therefore begin to cannibalise (deflate or erode) the existing accumulation. Where cessation of an episode of aeolian accumulation is marked by the generation of a supersurface (by definition of regional extent), the bounding surface is a type of *sequence boundary* that caps and delineates an individual sequence of aeolian accumulation.

From this discussion, a field of accumulation, a line of bypass and a field of deflation can be defined (Mountney 2012; Figure 7.53). The ratio between dune wavelength and dune spacing (the latter being the sum of dune and interdune) defines the proportion of the accumulation surface that is covered with actively migrating aeolian bedforms (cf. Lancaster 1988) and is expressed as a dimensionless parameter from zero to one. The ratio between the rate of vertical accumulation and the rate of horizontal migration defines the angle of climb of the dune and interdune system: negative values indicate a fall in the accumulation surface over time (deflation); a value of zero equates to sediment bypass; positive values define systems that are 'climbing' and actively accumulating. By plotting these ratios graphically, the range of possible dune-interdune stratal configurations that can be generated by the operation of temporally and spatially invariable systems is defined (Figure 7.54). Within this definition diagram, fields representing particular types of aeolian systems can be identified. Dry aeolian systems occur when the ratio of dune size to interdune size is one (or thereabouts), such that interdune flats have been reduced to small, sporadic (typically spatially isolated) interdune depressions between dunes. Many lithostratigraphic formations of dominantly aeolian origin are composed internally of multiple aeolian sequences, each separated by a sequence boundary (a supersurface). The Permian Cedar Mesa Sandstone, Utah, USA, is one well-documented example (Loope 1985; Langford and Chan 1993; Mountney 2006b).

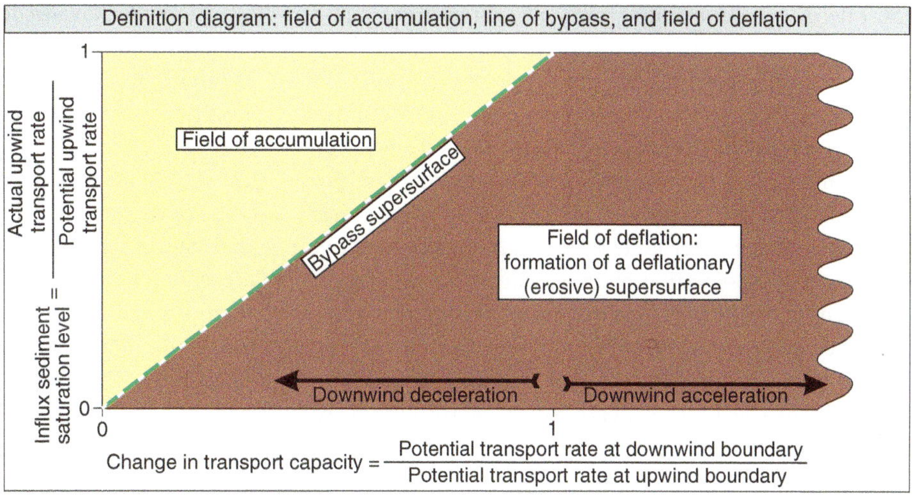

Figure 7.53 Field of accumulation, line of bypass and field of erosion (deflationary supersurface generation) defined by the sediment saturation level of the influx and the change in transport capacity downwind in the system. The slope of the line of bypass is +1. Adapted from Kocurek and Havholm (1993).

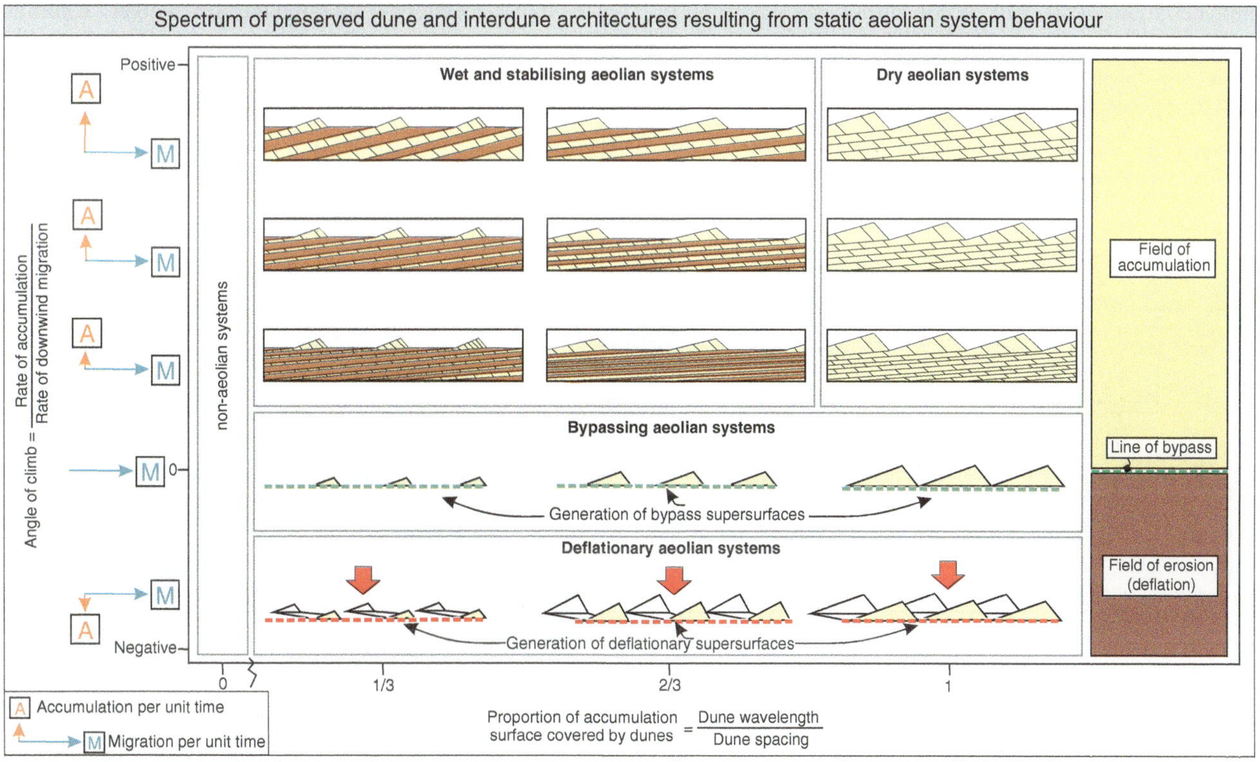

Figure 7.54 Spectrum of preserved dune and interdune architectures resulting from temporally and spatially invariable (i.e. static) aeolian system behaviour. The angle of bedform climb defines fields of accumulation and deflation, with bypass occurring when the angle of climb is zero. Within the field of accumulation, preserved sedimentary architecture is partly determined by the proportion of the accumulation surface covered by dunes. Accumulating dry aeolian systems require 100% dune cover. Accumulating wet or stabilising systems have less than 100% dune cover. The angle of climb is determined by the ratio of the vertical accumulation rate and bedform migration rate. The stratal configurations are scale-independent and can potentially occur in systems of any size. Adapted from Mountney (2012).

7.7.5 Aeolian System Preservation

The accumulation of aeolian sequences does not automatically ensure their long-term preservation. Indeed, evidence from the Quaternary record (e.g. Kocurek et al. 1991a; Kocurek 1999) argues that entire ergs may be constructed and accumulated to considerable thicknesses, only to be later deflated to leave no record. Long-term preservation requires that the body of strata is placed below a regional baseline, beneath which erosion does not occur. Thus, the rate of generation of accommodation and the rate at which aeolian accumulations fill that space is a fundamental control on preserved architectural style (Howell and Mountney 1997; Havholm and Kocurek 1994; Mountney and Howell 2000; Cosgrove et al. 2022b). The principal agent that generates long-term accommodation and promotes preservation is subsidence (Figure 7.55). This in turn can drive relative water-table rise and sea- or lake-level rise.

Long-term preservation of an aeolian succession can be achieved in many ways, but two mechanisms are widely identified. First, where an accumulation occurs within a subsiding basin in which accommodation is available to be filled, it can be placed beneath a so-called baseline of erosion – the level down to which deflation can proceed due to aerodynamic configuration (Kocurek and Havholm 1993). Second, where an accumulation is placed beneath the water table, thereby protecting the aeolian accumulation from later deflation. This can occur either via subsidence through a static water table or via a short-term (in geological terms) absolute rise in water-table level.

Additionally, aeolian successions can be preserved through a variety of exceptional mechanisms. Inundation of entire dune fields by marine incursion is known to have been the mechanism for preservation through a sudden change in environment (Eschner and Kocurek 1986). For example, part of the Permian Leman Sandstone of the Southern North Sea was preserved (with partly denuded relict dune topography) following rapid transgression by the Zechstein Sea, resulting in aeolian dune sediments being partly reworked and before being overlain by marine shale and carbonate deposits (Glennie and Buller 1983) and preserving dune topography (Strömbäck and Howell 2002). Marine flooding was also responsible for the preservation of part of the Permian White Rim Sandstone (Huntoon and Chan 1987; Chan 1989), part of the Jurassic Entrada Sandstone in New Mexico (Benan and Kocurek 2000) and part of the Jurassic Norphlet Sandstone known from the subsurface of the Gulf of Mexico (Godo 2017; Story 1998), for which relict dune topography is imaged using geophysical methods. Another mechanism for exceptional preservation of aeolian successions is via rapid burial of actively accumulating dune fields by lava flows, leading to preservation of largely intact relict dune forms, as occurred in the Twyfelfontein Formation (formerly the Etjo Sandstone), Namibia (Jerram et al. 2000a, b) and the Botucatu Formation, Brazil (Scherer 2002).

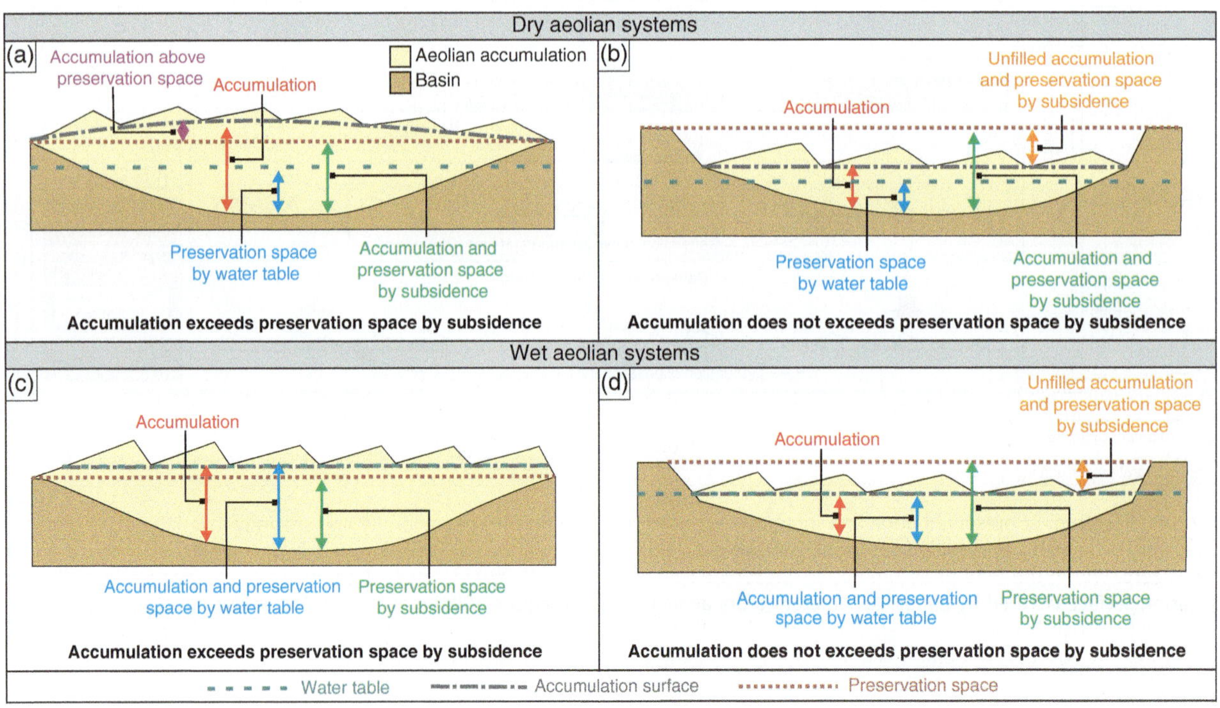

Figure 7.55 Components of accumulation and preservation space for dry and wet aeolian systems. (a) Dry aeolian system where the accumulation has built above the preservation space because a positive net sediment budget exists. Long-term preservation potential of that part of the accumulation above the preservation space line is low. (b) Dry aeolian system where the accumulation has not filled the available preservation space. (c) Wet aeolian system where water-table rise has enabled the accumulation to build above the preservation space due to subsidence. A fall in the water-table level will result in deflation. (d) Wet aeolian system where the water-table level is below the preservation space line, and some preservation space remains unfilled. Adapted from Kocurek and Havholm (1993).

7.7.6 Static Versus Dynamic Aeolian Facies Models

At a fundamental level, the geometry and architecture of dune and interdune elements that form accumulations through bedform climbing are determined by four variables (Mountney 2012): (i) dune extent in the direction of migration, (ii) interdune extent in the direction of migration, (iii) rate of bedform migration and (iv) rate of vertical accumulation – itself a response to a progressive relative rise in the water table for wet aeolian systems (Figure 7.56). Where these four variables are constant over both space and time, the system can be considered 'static' (Figure 7.56a): simple, invariable dune and interdune architectures are generated. Note that static aeolian systems should not be confused with stabilised aeolian systems (*sensu* Kocurek and Havholm 1993); the bedforms in static aeolian systems experience migration but in a spatially and temporally invariable manner. Changes in one or more of the four variables listed above over either space or time give rise to complexity in the resultant preserved stratigraphic architecture and the system can be considered dynamic. In these circumstances, spatially variable, temporally variable, both spatially and temporally variable system states and resultant dune and interdune architectures may arise (Figure 7.56b–d).

7.7.6.1 Static Aeolian Systems and Preserved Successions

In static aeolian systems, the arrangement of architectural elements is relatively simple (Figure 7.57). Many of the aeolian facies models developed in the 1970s and 1980s (e.g. Kocurek 1981) are static in that they do not account for the effects of variations in sediment supply and the transport capacity of the wind and, as a consequence, do not incorporate the effects of temporal changes in dune and interdune size, migration rate or accumulation rate (Figure 7.57). Nevertheless, static models remain useful because they define end-member configurations of possible stratal architectures.

7.7.6.2 Dynamic Aeolian Systems and Preserved Successions

Dynamic facies models incorporate the effects of spatial and/or temporal variations in one or more of the four above-mentioned controlling variables. Arrangements of architectural elements in preserved examples of dynamic system successions are more complicated than for static systems. Underlying controls that are known to change spatially and/or temporally include the rate of generation of an aeolian sediment supply, the availability of sediment for aeolian sediment transport and the carrying capacity of the wind, which together define the sediment state of a dune-interdune system and govern its state of construction, thereby defining dune and interdune size, and migration rate (Kocurek 1999; Kocurek and Lancaster 1999). The following act to determine the mechanism and rate by which aeolian dune and interdune accumulation occur (Kocurek and Havholm 1993): (i) the relative level of the water table and its rate of change, (ii) the level of the equilibrium height (below which net sedimentation takes place); and (iii) the distribution of stabilising factors such as vegetation. The rate of subsidence partly defines how accommodation is generated and the extent to which dune-interdune accumulations are preserved long-term. These variables are unlikely to all remain constant over large areas or over long time periods. Therefore, most aeolian systems should be

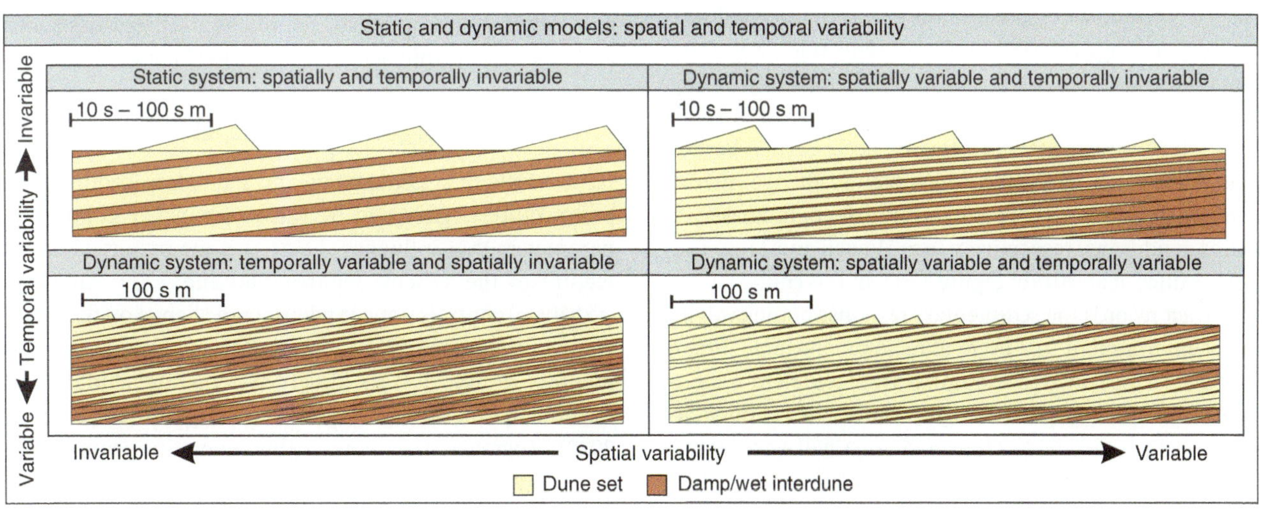

Figure 7.56 Examples of dune–interdune stratigraphic architectures arising from various combinations of temporal and spatial variability and invariability. Adapted from Mountney (2012).

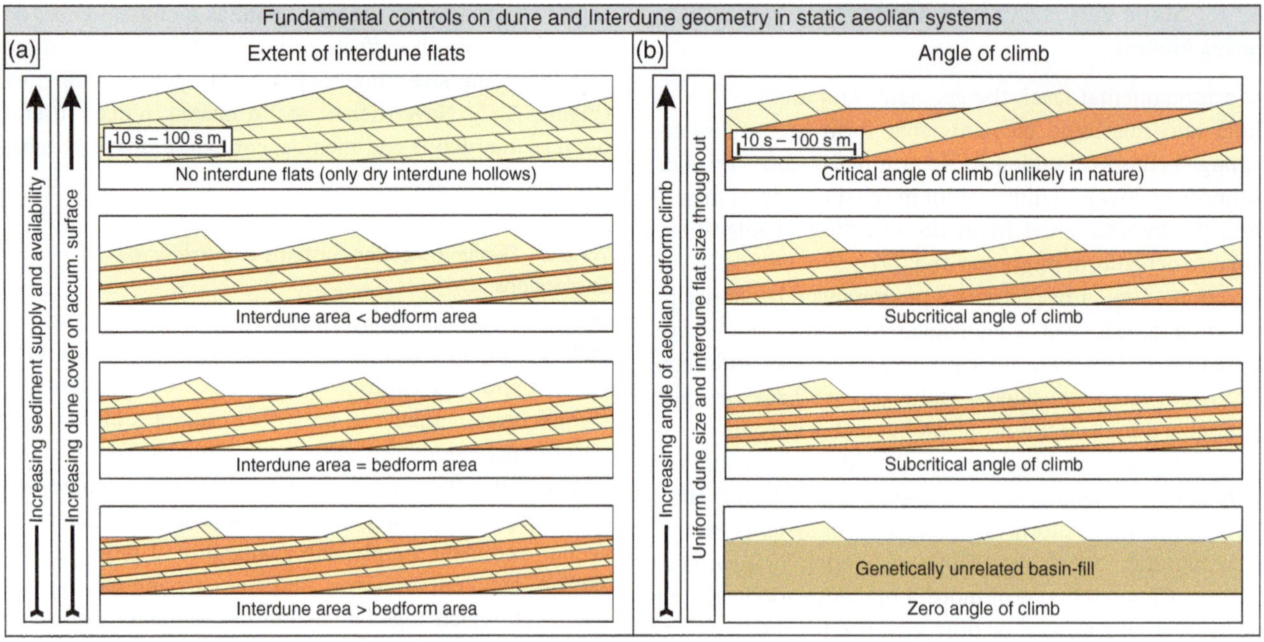

Figure 7.57 Fundamental controls on dune and interdune geometry in spatially and temporally invariable (i.e. static) aeolian systems. (a) Models depicting varying proportions of the accumulation surface covered by aeolian dunes and interdunes. The angle of climb is the same in each example. (b) Models depicting dunes and interdunes of a uniform size where the accumulated thickness of dune and interdune elements varies as a function of the angle of climb. Adapted from Kocurek and Havholm (1993).

considered dynamic, and the complexity inherent in documented examples of preserved stratal architectures of such systems demonstrates this (Figures 7.58 and 7.59; e.g. Mountney 2006a, 2012; Mountney and Thompson 2002).

7.7.7 Aeolian Systems Tracts and Sequences

Aeolian and related successions commonly record systematic vertical changes in the relative influence of particular environmental controlling conditions; such accumulations may be represented by bodies of strata accumulated during a particular stage referred to as a systems tract. For example, an aeolian succession may be subdivided into drying- and wetting-upward systems tracts (Figure 7.60; Box 7.5). For each of these systems tracts, the accumulated vertical facies succession might record a systemic change in environmental conditions to progressively drier or progressively wetter conditions, respectively (e.g. Howell and Mountney 1997; Bállico et al. 2017b; Figure 7.61). In this context, a systems tract records the sedimentary response of the evolving palaeoenvironment to a cycle of regional allogenic forcing (climate change in this example). One complete cycle of change defines a sequence. Significantly, the sedimentary response to region-wide climate change will vary from a dune-field centre setting, to its inner- and outer-dune-field margin, and to neighbouring non-aeolian settings. Yet, despite a varied sedimentary response in each of these locations, each can be interpreted in terms of a predictable response to overall allogenic climate control: a drying-upward systems tract records evidence for increasing aridity over time and a wetting-upward systems tract records evidence for increasing climatic humidity over time, albeit as different types of sedimentary succession in different locations (Figure 7.61). Together, these two systems tracts define a sequence. Importantly, if an allogenic driver of environmental change (e.g. climate) is of regional extent, the recognition of systems tracts and sequences that record the sedimentary response to such change can be potentially used as a chronostratigraphic correlation tool (Figure 7.61; Box 7.5). Gradual long-term basin subsidence might serve to generate the accommodation required to accumulate and preserve stacked systems tracts within aeolian sequences (Figure 7.55).

7.7.8 Aeolian Sequence Stratigraphic Models

At the scale of systems tracts and sequences, sedimentary architecture is governed by variations in key controlling parameters that influence the gross-scale preserved architecture of the system (Figures 7.60 and 7.61). Allogenic (extrinsic) controls impose widespread change on the behaviour of the system. Two major external controls on sedimentary architecture in desert systems are climate and tectonic basin setting (Figure 7.62; e.g. Abrantes et al. 2020). Together, they govern the generation, supply and availability of sediment for transport, and the subsidence of evolving basins in which aeolian systems are developed. As such, these parameters determine the equilibrium height (base level) to which an aeolian system may accumulate (Kocurek and Havholm 1993); at this equilibrium height, for a particular set of controlling conditions, neither

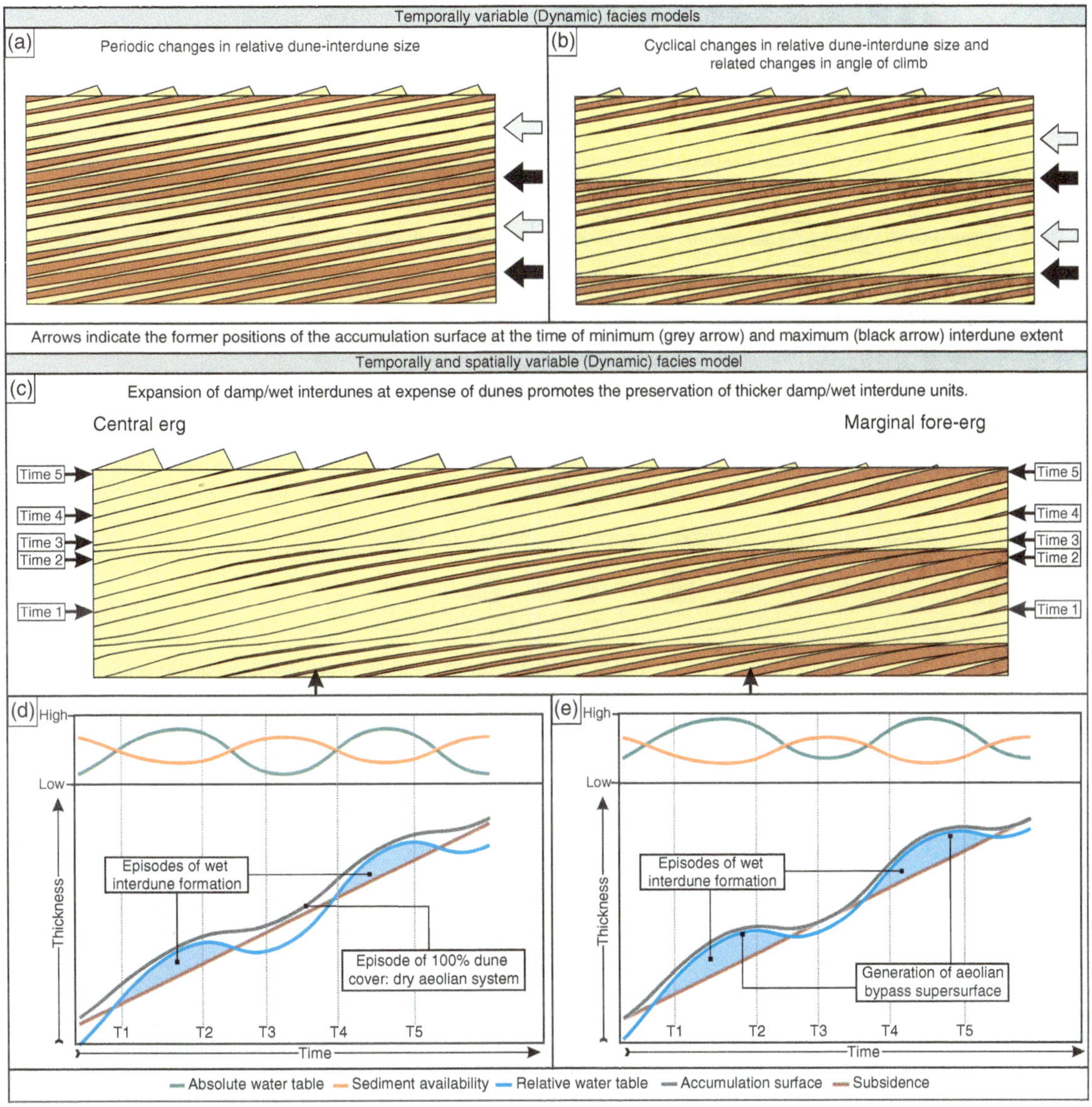

Figure 7.58 Dynamic facies models for aeolian dune and interdune architecture. (a) Cyclic temporal variation in dune and interdune size. For a fixed rate of dune migration, damp/wet interdune expansion and contraction occur during periods of accelerating and decelerating rates of relative water-table rise, respectively. (b) The combined effects of cyclic temporal variation in dune and interdune size and associated change in angle of climb of the system. The angle of climb steepens as the rate of relative water table rise increases and reduces to zero when the relative water-table level is static. (c–e) Model for dune and interdune architecture resulting from spatial and temporal variation in controlling parameters. (c) Model incorporating both temporal variations of relative dune and interdune size and angle of climb in response to cyclical changes in the rate of relative water table change and sediment availability. Additionally, the model incorporates the effects of a spatial decrease in dune coverage of the accumulation surface from a central erg setting to the outer limit of a fore-erg. (d) and (e) Barrel diagrams illustrate the possible configuration of controlling parameters responsible for generating the preserved succession in the erg centre and erg margin areas, respectively. In Parts a, b and c, the direction of bedform migration is from left to right. Adapted from Mountney (2012).

net accumulation nor net erosion will occur. Temporal changes in these parameters will have different effects on different parts of the desert system (Figure 7.61). For example, a switch to a more humid climate may result in playa-lake expansion due to a rise in the water table. At the same time, the elevation of the water table may restrict the availability of sediment for aeolian transport, causing dunes to cease accumulating and interdunes to expand or become flooded. Thus, various sub-environments within the desert system expand or contract and accumulate or become eroded over time as external controls vary. This is reflected in the preserved sedimentary architecture by large-scale shifts in the position of facies belts, which may in many cases be expressed vertically as drying-upward or wetting-upward systems tracts. Where environmental conditions lead to episodes of bypass (i.e. non-deposition), deflation (i.e. erosion) or stabilisation, supersurfaces will develop (Figures 7.60 and 7.61; Section 7 5.7).

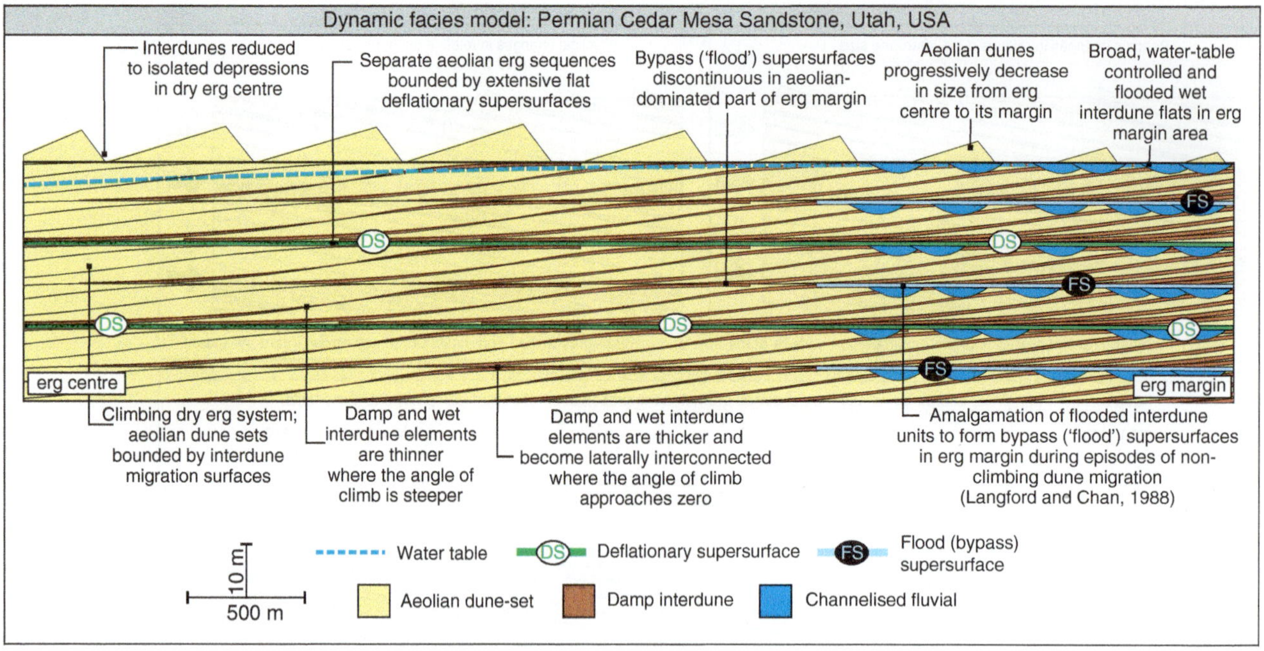

Figure 7.59 Example of a dynamic facies model to account for the origin of preserved aeolian successions. Example of preserved aeolian sedimentary architecture relating to the temporally and spatially variable aeolian system of the lateral palaeo-erg margin of the Permian Cedar Mesa Sandstone, Indian Creek Region, Utah, USA. Only major bounding surfaces are shown. Based on data from Langford and Chan (1988) and Mountney and Jagger (2004). Mountney (2012).

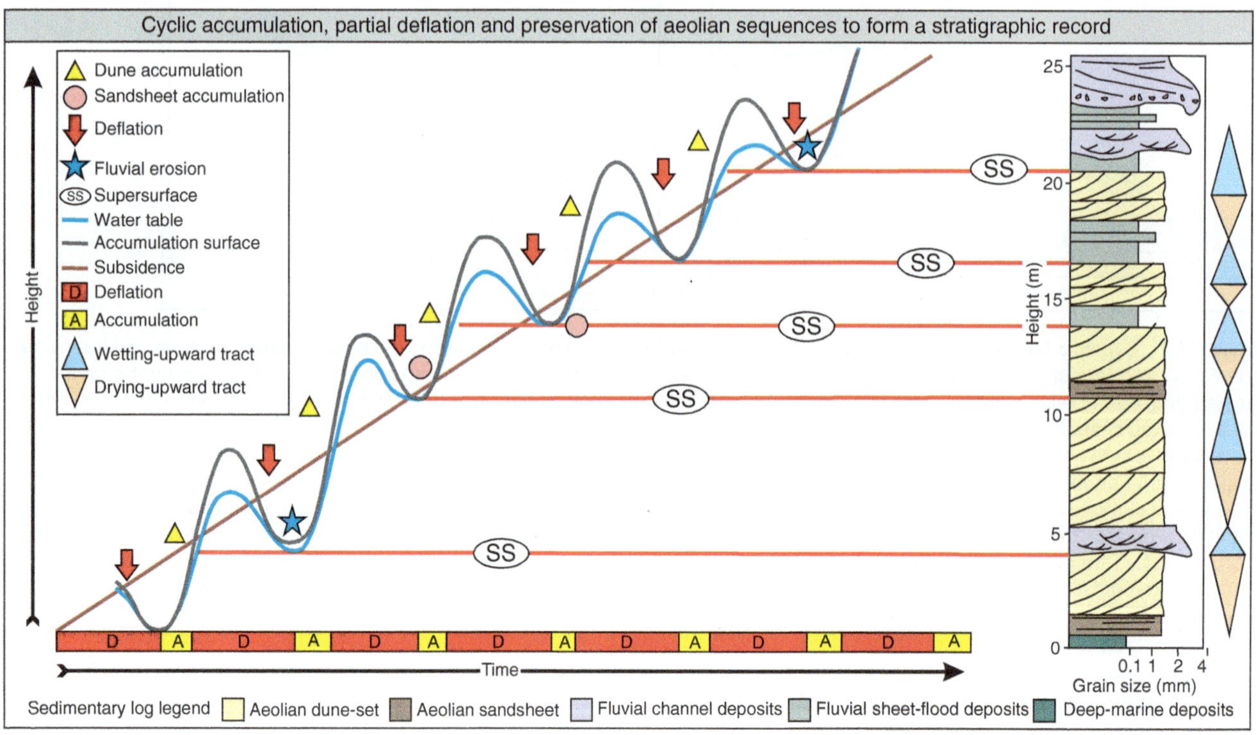

Figure 7.60 Example of the cyclic accumulation, partial deflation and preservation of aeolian sequences to form a stratigraphic record. Note the drying- and wetting-upward system tracts. Example from the Cretaceous (Hauterivian) Avilé Member of the Agrio Formation, Neuquén Basin, Argentina. The Avilé Member spans c. 0.5 million years (29.5–29 Ma). The high-frequency changes in the relative level of the water table have been attributed to Milankovitch cyclicity. However, due to a lack of biostratigraphic makers, the exact duration of the sequences remains undetermined. Adapted from Veiga et al. (2002). For other examples of wetting- and drying-upward system tracts within aeolian sequences, see Bállico et al. (2017b) and Scherer and Lavina (2005).

Box 7.5 Sequence-stratigraphic approach to correlation in aeolian and related successions

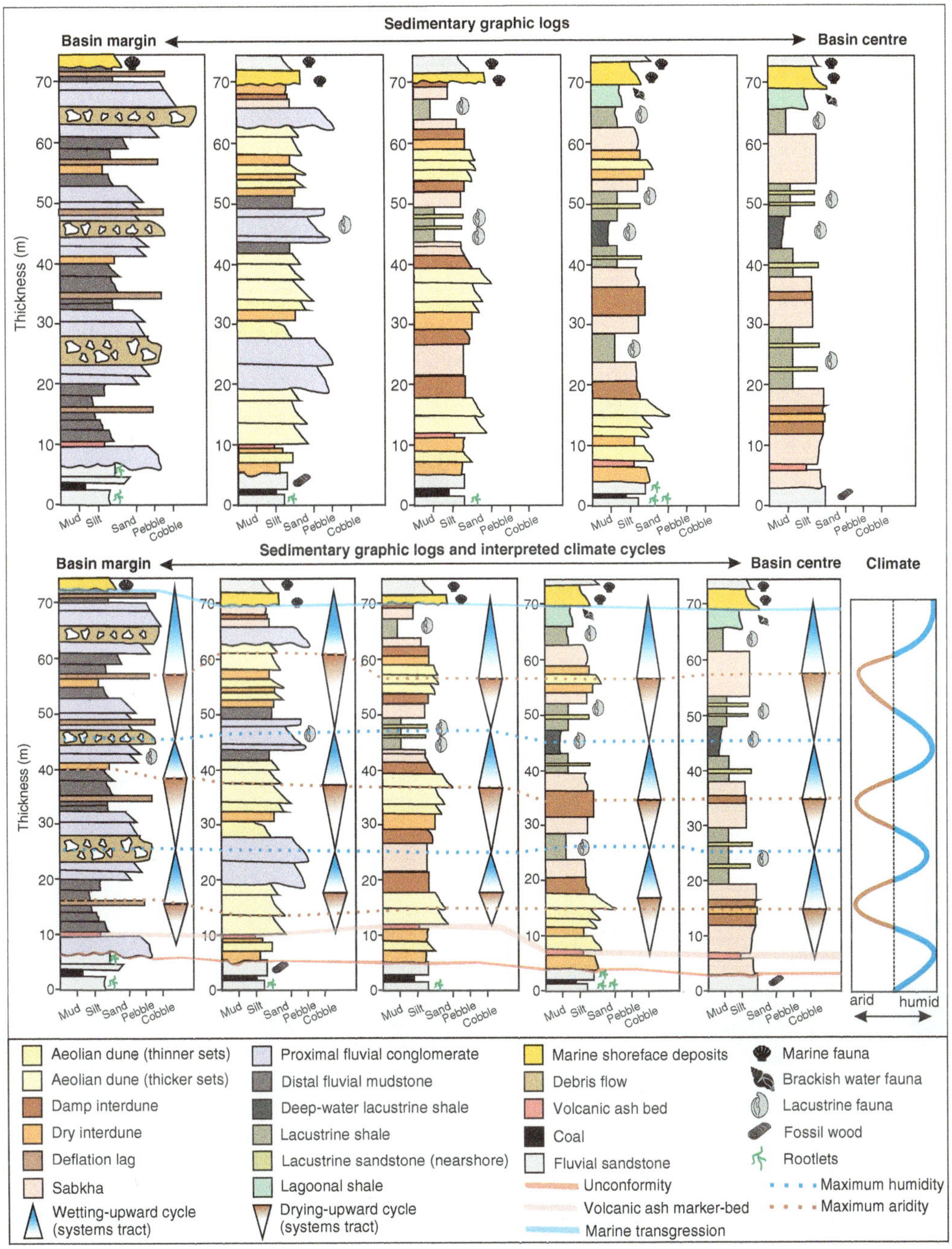

Box 7.5 Example exercise to demonstrate the application of a sequence-stratigraphic approach to the interpretation of a series of one-dimensional graphic sedimentary logs through a preserved succession. The five logs are located from the margin to the centre of a hypothetical sedimentary basin. Alluvial fans dominated at the basin margin; a dryland playa lake system occupied the basin centre; aeolian dune fields were developed between. In each log, the vertical succession of preserved facies records the preserved systematic response of each of the sedimentary environments to repeated cyclic changes in regional climate from dry to wet and back. The facies successions reveal drying- and wetting-upward cycles (systems tracts). Cycle boundaries represent points of maximum and minimum relative aridity. Assuming that climate change operated across the basin, cycle boundaries can be used tentatively as a correlation tool in the absence of biostratigraphic or other age information. Note the unconformity close to the base of the succession. Note the marine transgressive flooding surface close to the top of the succession. Note the single volcanic ash bed that might also be used as an event-bed marker for correlation.

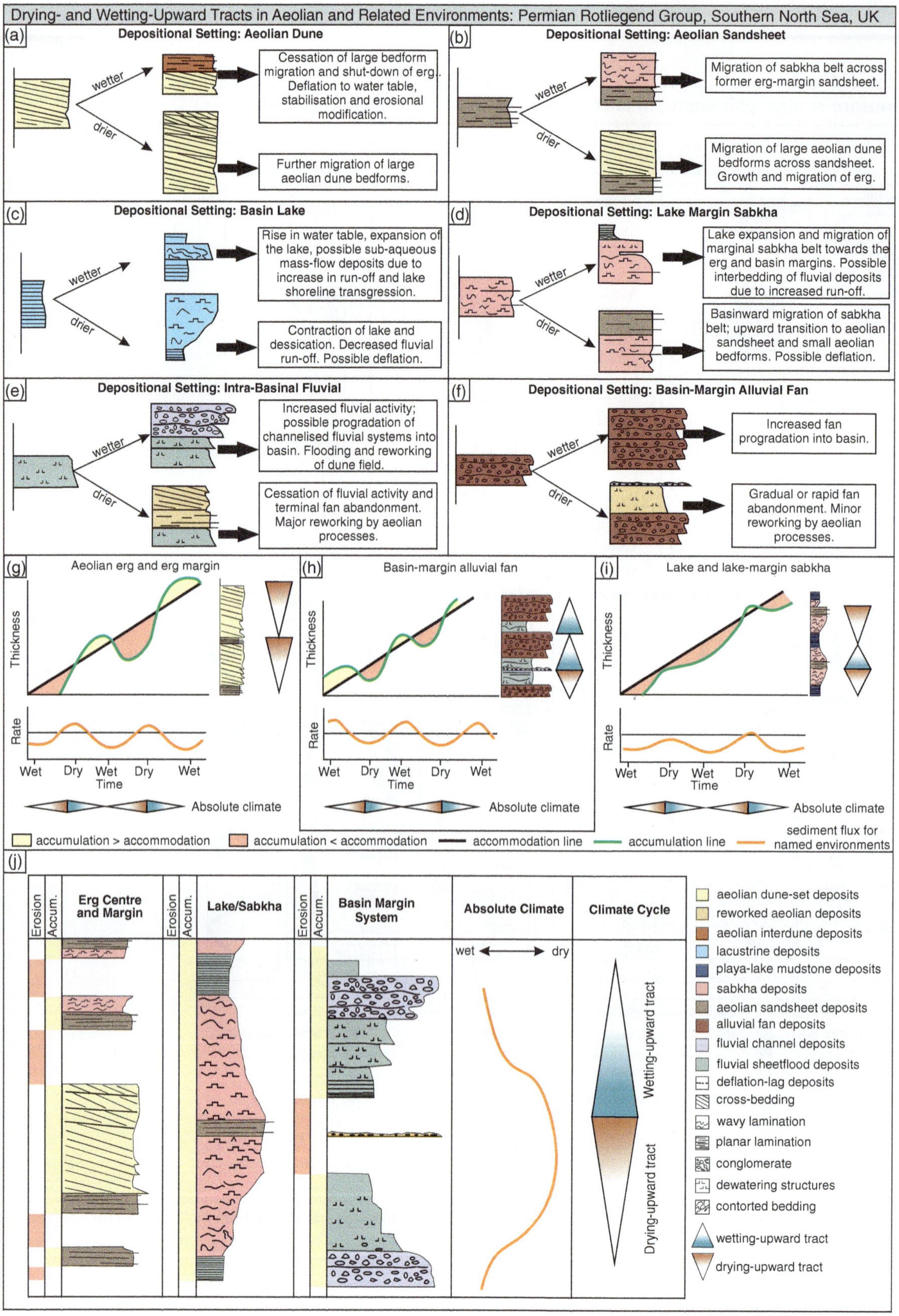

Figure 7.61 Drying- and wetting-upward tracts in aeolian and related dryland environments developed in response to cyclic changes in climate. (a–f) Predicted idealised process response of aeolian and related dryland environments to climate change resulting in increased (drier) or decreased (wetter) aridity. (g–i) Barrell diagrams illustrating episodes of accumulation and deflation in response to cyclic climate changes. Long-term preservation is enabled by the generation of accommodation via a steady rate of background basin subsidence. Note how different environments may preserve only drying-upward cycles or both drying- and wetting-ward cycles in response to the same overall climate change that operated region-wide across a basin. (j) Idealised sedimentary expression of one complete climate cycle for aeolian, lake/sabkha and basin-margin fluvial depositional environments. Note the differing timing of episodes of erosion and accumulation. The graphic logs are idealised and have no vertical scale implied, though preserved expressions of a climate cycle are typically metres to a few tens of metres thick. Example from the Permian Rotliegend Group, Southern North Sea, UK. Adapted from Howell and Mountney (1997).

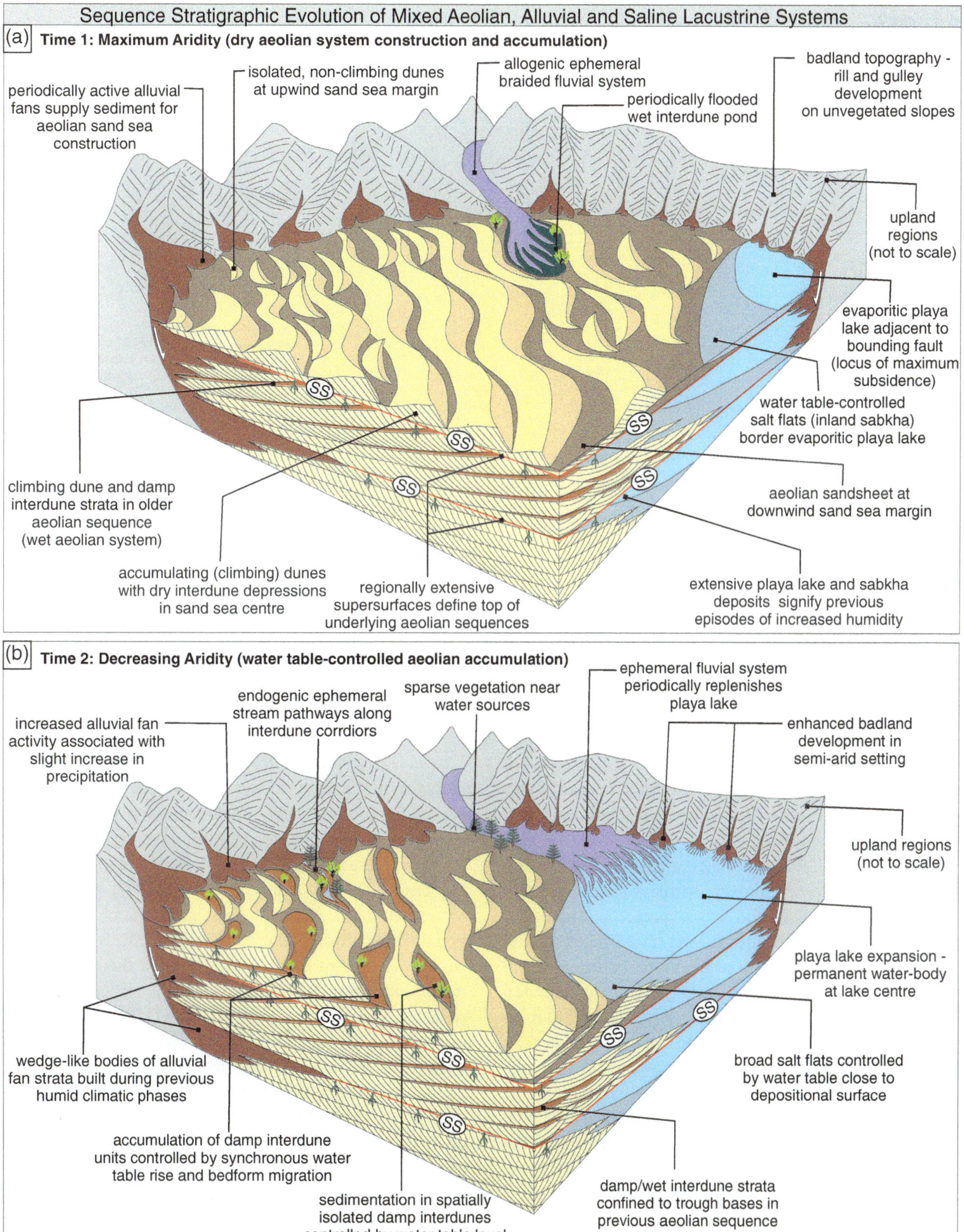

Figure 7.62 Sequence stratigraphic evolution of mixed aeolian, alluvial and saline lacustrine sedimentary systems. (a) Episode of maximum aridity; dry aeolian system construction and accumulation. (b) Episode of decreasing aridity; water table-controlled accumulation of a wet aeolian system. (c) Episode of minimum aridity; aeolian deflation and supersurface development; expansion of basin-margin alluvial fans and basin-centre lake system. Adapted from (2004) and Lancaster and Mountney (2021).

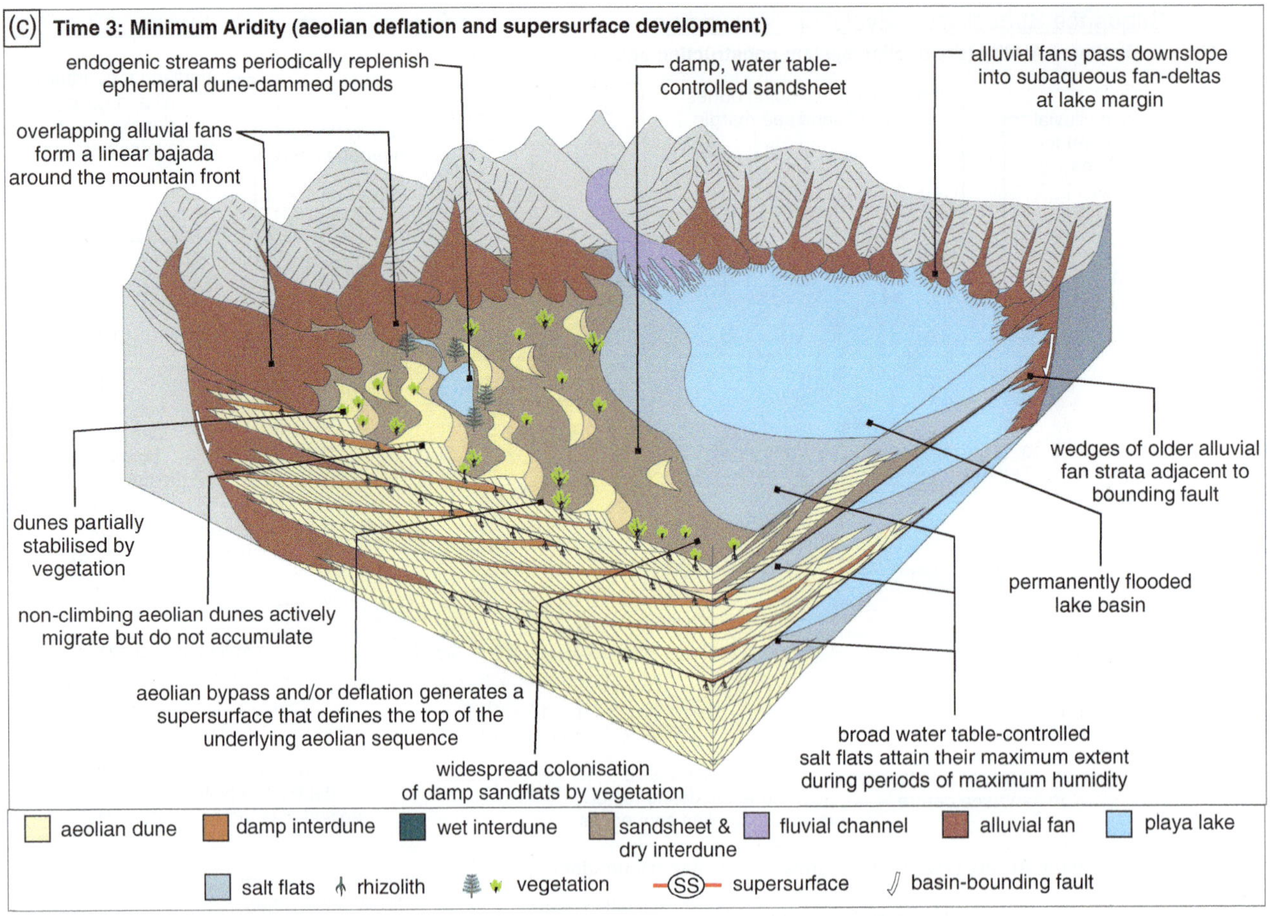

Figure 7.62 (Continued)

The approach outlined above enables a sequence stratigraphic framework to be applied as a tool for regional correlation between coevally active neighbouring sub-environments. In desert settings, playa lake, sabkha, dune-field margin, dune-field centre and alluvial fan settings may coexist in close proximity to each other. Resultant sedimentary successions and their sequence boundaries or their correlative conformities can be traced between sub-environments (Box 7.5). The sedimentary sequences themselves may comprise systems tracts that are characterised by predictable vertical facies successions. Horizons of maximum relative aridity and maximum relative humidity can potentially be identified and define the boundaries of systems tracts. Using this approach, sequence stratigraphic models have been developed and applied (Box 7.6). Such models serve two important roles. First, they explain the complicated origins of the aeolian (and related) stratigraphic record by accounting for autogenic system behaviour (e.g. aeolian dune and interdune bedform climbing) that is itself modified by long-term allogenic forcing (e.g. climatic cyclicity). Second, they enable chronostratigraphic correlation because region-wide and synchronous changes in climate and related environmental variables result in predictable changes in resultant accumulated stratigraphy between neighbouring depositional environments.

Parts of the Permian Rotliegend Group of NW Europe – including in the subsurface of the Southern North Sea – record a long-lived episode of sedimentation in a variety of aeolian, alluvial, sabkha and playa-lake depositional systems that were coevally active in a large and slowly subsiding basin (George and Berry 1993, 1997; Howell and Mountney 1997). Although these arid to semi-arid depositional systems were detached from the effects of sea level change, long-term cyclic fluctuations in the regional climatic regime resulted in variations in the vertical facies successions laid down (Figure 7.61a–f). The effects of increasing and decreasing aridity upon depositional processes are recorded as systems tracts. Climatic cycles, bounded at points of minimum and maximum aridity are defined (Figure 7.61g–i). These climatic cycles may be correlated between different depositional environments that developed coevally. This enables depositional systems to be considered chronostratigraphically (Figure 7.61j). Although the internal architectures of sedimentary cycles are controlled by the intrinsic processes that operate in the respective depositional sub-environments, overall, they each record relative climate changes that can be correlated across the basin using a sequence stratigraphic approach (Howell and Mountney 1997). The preserved thickness of climatic cycles identified in the stratigraphy is influenced

Box 7.6 A cycle of aeolian sequence development under greenhouse climate conditions

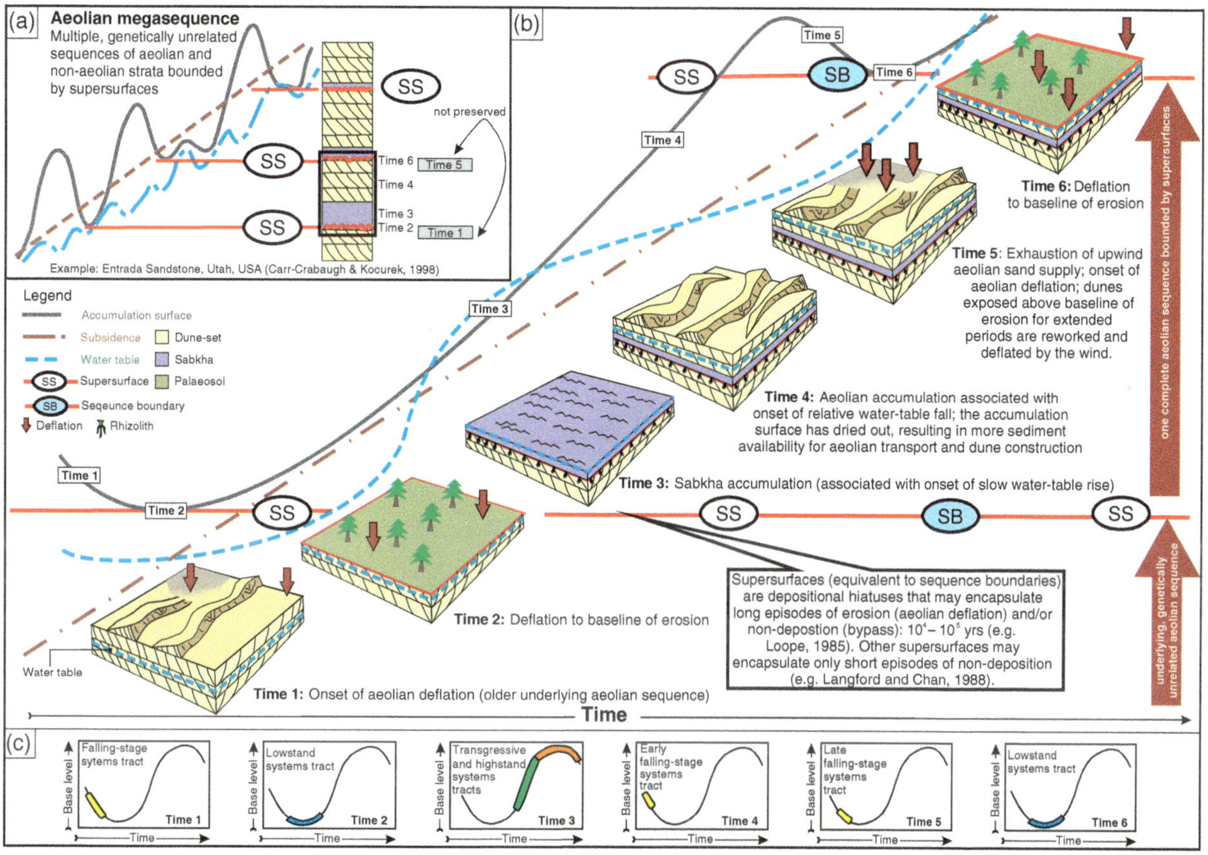

Box 7.6 Example of a cycle of aeolian sequence development under greenhouse climate conditions. (a) Aeolian megasequence comprising multiple genetically distinct sequences in the Jurassic Entrada Sandstone, Utah, USA. The black box overlying part of the illustrated sedimentary succession is shown in detail in Part (b); the notations of 'Time 1–6' also refer to Part (b). (b) Cycle of aeolian accumulation (accumulation of a sequence) and partial deflation (development of a deflationary supersurface) and associated changes in the water-table level and basin subsidence. (c) Common sequence stratigraphic terminology here applied to explain the cycle of aeolian accumulation and deflation.

in part not only by the rate of accommodation creation but also by the long-term preservation potential of deposits accumulated under differing climate regimes.

Sequence stratigraphic models have been proposed for many aeolian successions, including for coastal aeolian systems that are known to have responses to repeated changes in the relative sea level, such as the lower Cutler beds, Utah, USA (Jordan and Mountney 2010, 2012, Wakefield and Mountney 2013; Figure 7.63). See Section 7.S.3 in Supplementary Information for details.

7.8 Geological Applications

7.8.1 Subsurface Aeolian Reservoir Characterisation

In the subsurface, thick aeolian successions of porous, permeable, cross-bedded sandstone act as important reservoirs for oil, gas, and water and hosts for hydrothermal metalliferous deposits. Major hydrocarbon-producing provinces that exploit reserves hosted in aeolian reservoirs include the Pennsylvanian-Permian Weber Sandstone of Colorado and Wyoming (USA), the Permian Unayzah Formation of Saudi Arabia (Melvin et al. 2010), the Permian Rotliegend Group of the southern and central North Sea Heward 1991) and onshore Netherlands and Germany, the Triassic Sherwood Sandstone Group of the East Irish Sea Basin (Meadows 2006), the Jurassic Norphlet Sandstone onshore in Alabama and adjacent states and offshore in the Gulf of Mexico (USA) (Kugler and Mink 1999), and the Jurassic of the Neuquén Basin of Argentina (Veiga and Spalletti 2007).

Aeolian successions also form important groundwater aquifers, for example, deposits of the Sherwood Sandstone Group in the Cheshire Basin, UK (Bloomfield et al. 2006). The overall porous and permeable nature of aeolian groundwater reservoirs requires careful modelling of fluid-flow rates and pathways during subsurface water abstraction. Over-abstraction, for example, as a consequence of excessive farmland irrigation in dryland regions, can result in over depletion of potable water reserves in subsurface aeolian

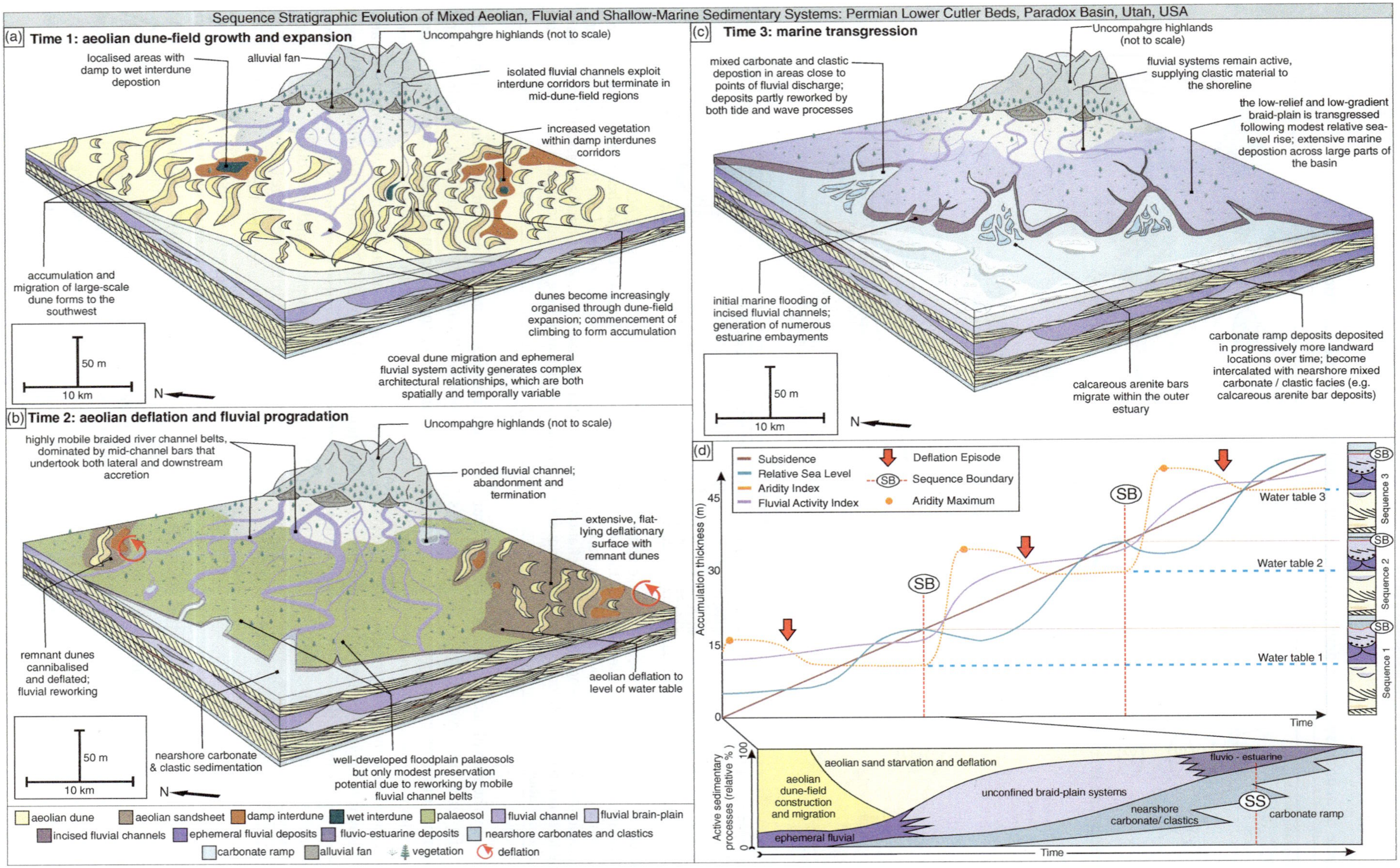

Figure 7.63 Sequence stratigraphic evolution of mixed aeolian, fluvial and shallow-marine sedimentary systems. Example from the Permian lower Cutler beds, Paradox Basin, Utah, USA. (a) Episode of aeolian dune-field construction and accumulation. (b) Episode of aeolian dune-field deflation and fluvial progradation. (c) Episode of marine transgression. (d) Sequence stratigraphic model to account for the generation of the preserved sedimentary succession. Adapted from Jordan and Mountney (2012).

reservoir successions and contamination of the groundwater due to recharge by saltwater influx.

Aeolian successions are well-suited for development as low-enthalpic geothermal reservoirs where warm water at depth is recovered to the surface via a producer well and is replaced by cold water via an injector well (e.g. Triassic Sherwood Sandstone Group, UK). Aeolian successions are also well-suited for underground storage of hydrogen and other volatile fluids that can be held in reserve for use at a later date. Moreover, some aeolian successions that form either depleted hydrocarbon reservoirs or deep saline groundwater aquifers are well-suited to long-term underground carbon sequestration as part of carbon capture and storage programmes.

To develop subsurface aeolian reservoir successions most effectively, the heterogeneous nature of the deposits must be evaluated (Yan et al. 2024; Figure 7.64). Characterisation of the lithological complexity of subsurface aeolian reservoirs requires the accurate determination of porosity and permeability distributions, which are themselves influenced by numerous geological parameters, including the lithology (Figures 7.64 and 7.65), geometry, dimensions and spatial distributions of aeolian and associated non-aeolian facies units and architectural elements. Outcropping analogous successions are commonly used to determine the likely range of dimensions of aeolian architectural elements, as well as their internal facies composition and arrangement. The synthesis of data from a large number of case studies permits the development of composite geological analogues that capture stratigraphic and sedimentological variability (Figure 7.66). This approach to geological characterisation of the subsurface can be applied in the fields of hydrocarbon exploration,

Figure 7.64 Core photograph representative of aeolian facies in the Permian Auk Formation, Central North Sea, UK. Black strata are oil-stained deposits of dune lee-slope grainflow origin; white strata are oil-bleached; red strata have not been in contact with oil due to lower permeability. Photographs of the curated reference set of cores held by Talisman Sinopec Energy UK Ltd.

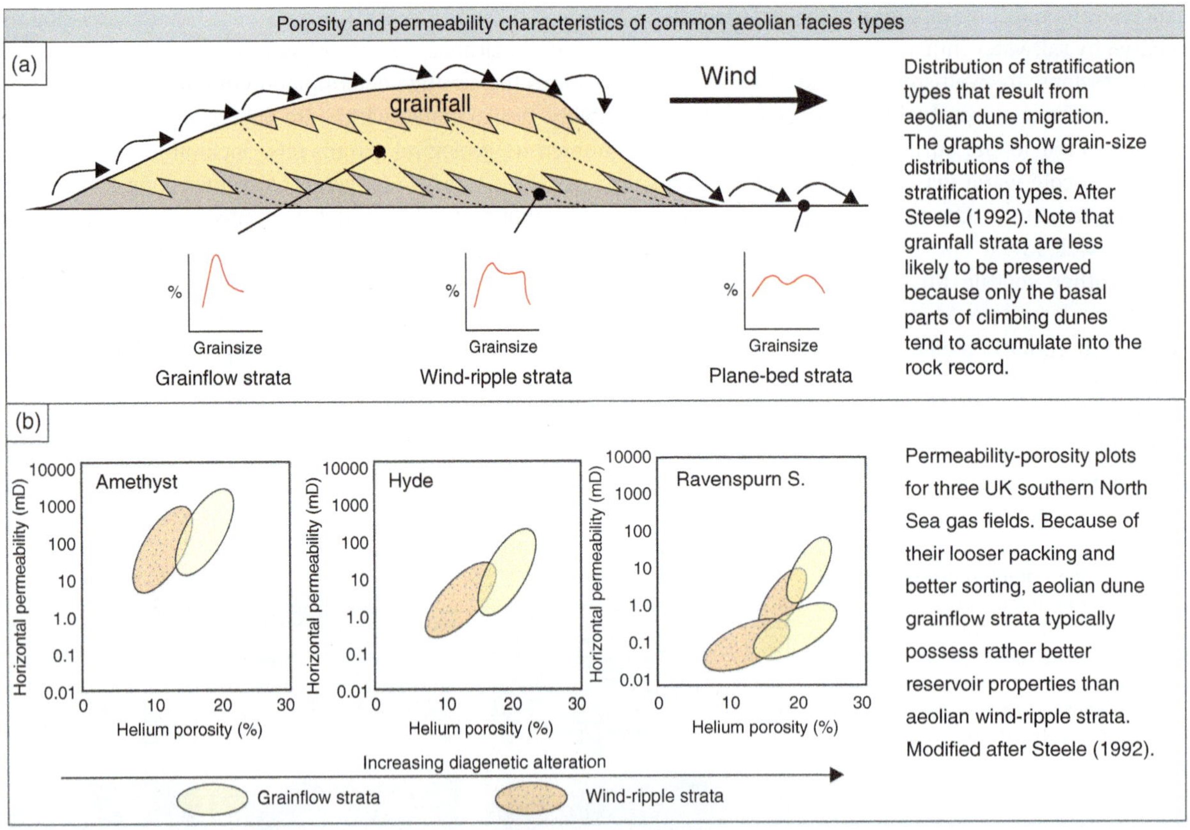

Figure 7.65 Typical porosity and permeability characteristics of common aeolian facies types. (a) Depositional settings for grainflow strata, dune-plinth wind-ripple strata and aeolian plane beds. (b) Examples of porosity-permeability cross-plots for three subsurface hydrocarbon fields from the southern North Sea, UK sector. Figure in part Courtesy of Richard Steele.

development and production, carbon capture and storage (CCS), deep geothermal reservoir development and groundwater aquifer management (e.g. Medici et al. 2016, 2019a,b).

To characterise subsurface successions, a database-informed approach to facies analysis using many case-study examples can be used in the following ways. First, to quantify the interdigitation and 3D stacking of relatively more porous and permeable aeolian dune elements (dominated by packages of grainflow strata) with non-dune elements (including interdune, fluvial and sabkha deposits) (Figure 7.66a,b; Nagtegaal 1979; Lindquist 1983; Krystinik 1990; Crabaugh and Kocurek 1993; Herries 1993; Shebi 1995). Second, to record the presence and nature of aeolian bounding surfaces at different scales, and their occurrence between and within specific depositional, architectural and facies elements (Figure 7.66c,d). Aeolian bounding surfaces can act as barriers to fluid migration, by exhibiting prominent grain-size contrasts (e.g. Shebi 1995; Ciftci et al. 2004). Bounding surfaces can also transmit percolating meteoric waters, resulting in the precipitation of carbonate and silicate cements (e.g. Chandler et al. 1989; North and Prosser 1993) and the acceleration of other diagenetic processes, such as the compaction and mechanical infiltration of clays and chlorite cementation (e.g. Bongiolo and Scherer 2010; De Ros and Scherer 2013). Third, to characterise, the statistically most likely configuration and orientation of aeolian facies units, such as those composed of packages of wind-ripple, grainfall and grainflow strata. The 3D arrangement of these aeolian facies elements can significantly impact horizontal and vertical permeability due to inter-facies variability in grain size, sorting and packing (Hunter 1977a, b; Chandler et al. 1989; Prosser and Maskall 1993; Howell and Mountney 2001; Pickup and Hern 2002; Besly et al. 2018).

Some aeolian reservoir successions have historically been considered to form relatively homogeneous bodies of sand due to their inherent high net-to-gross character (e.g. Glennie 1972). However, many aeolian successions have been shown to be lithologically highly heterogeneous at multiple scales (e.g. McCaleb 1979; Weber 1987; Fryberger et al. 1990d; Prosser and Maskall 1993; Taggart et al. 2010; Godo 2017); their successful development and exploitation require detailed characterisation of flow properties, pathways and behaviour. For example, the Permian Auk Formation of the Central North Sea is an oil-bearing aeolian reservoir succession for which development and production depend on accurate prediction of the 3D geometry and connectivity of relatively more permeable and porous dune architectural elements that are partitioned by less permeable and porous interdune elements that act as baffles and barriers to flow (Figure 7.67; Besly et al. 2018).

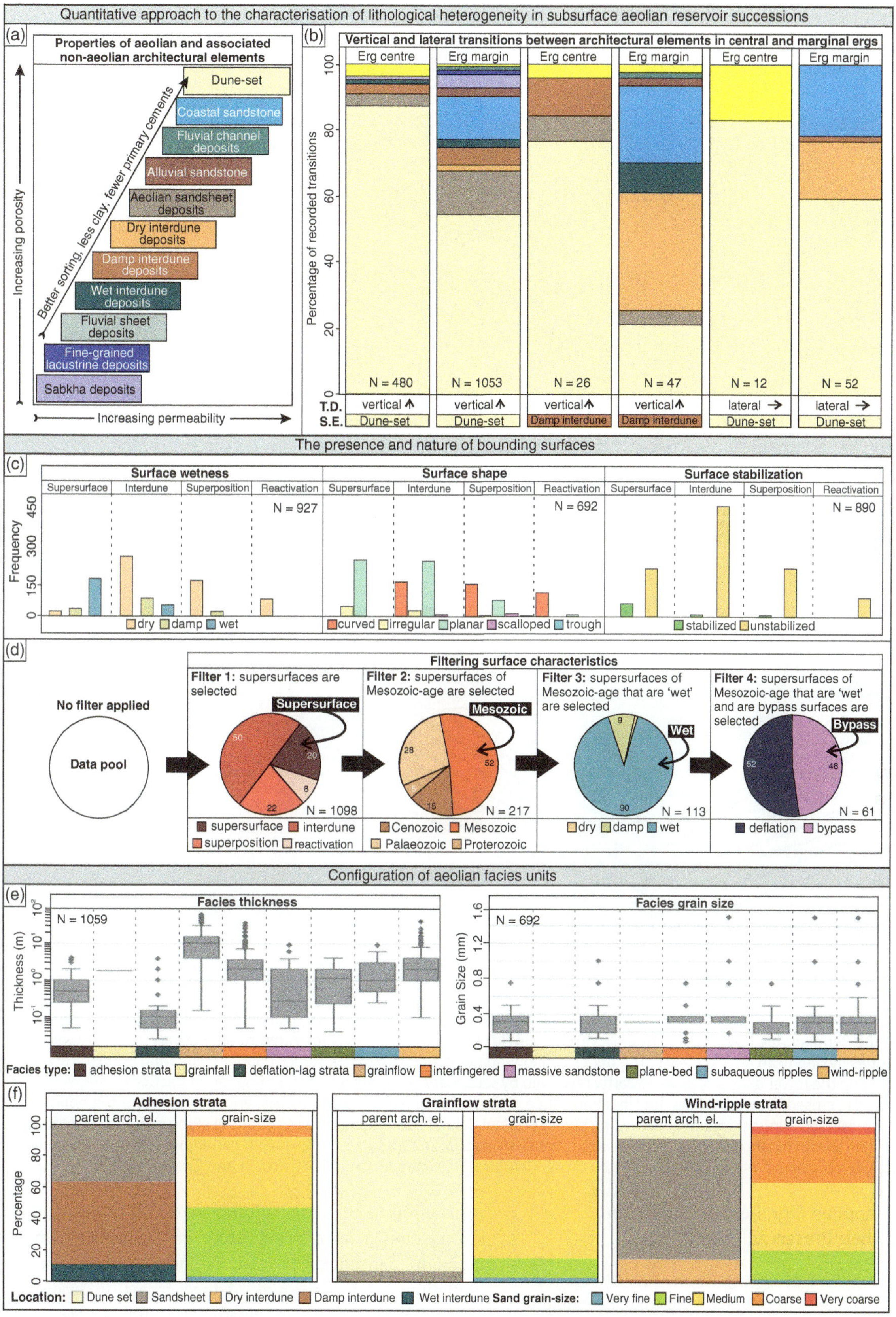

Figure 7.66 Example application of a quantitative approach to the characterisation of lithological heterogeneity at multiple scale in an aeolian succession. Data are from the Database of Aeolian Sedimentary Architecture (DASA). See Cosgrove et al. (2021a) for details. Also, see Box 7.4.

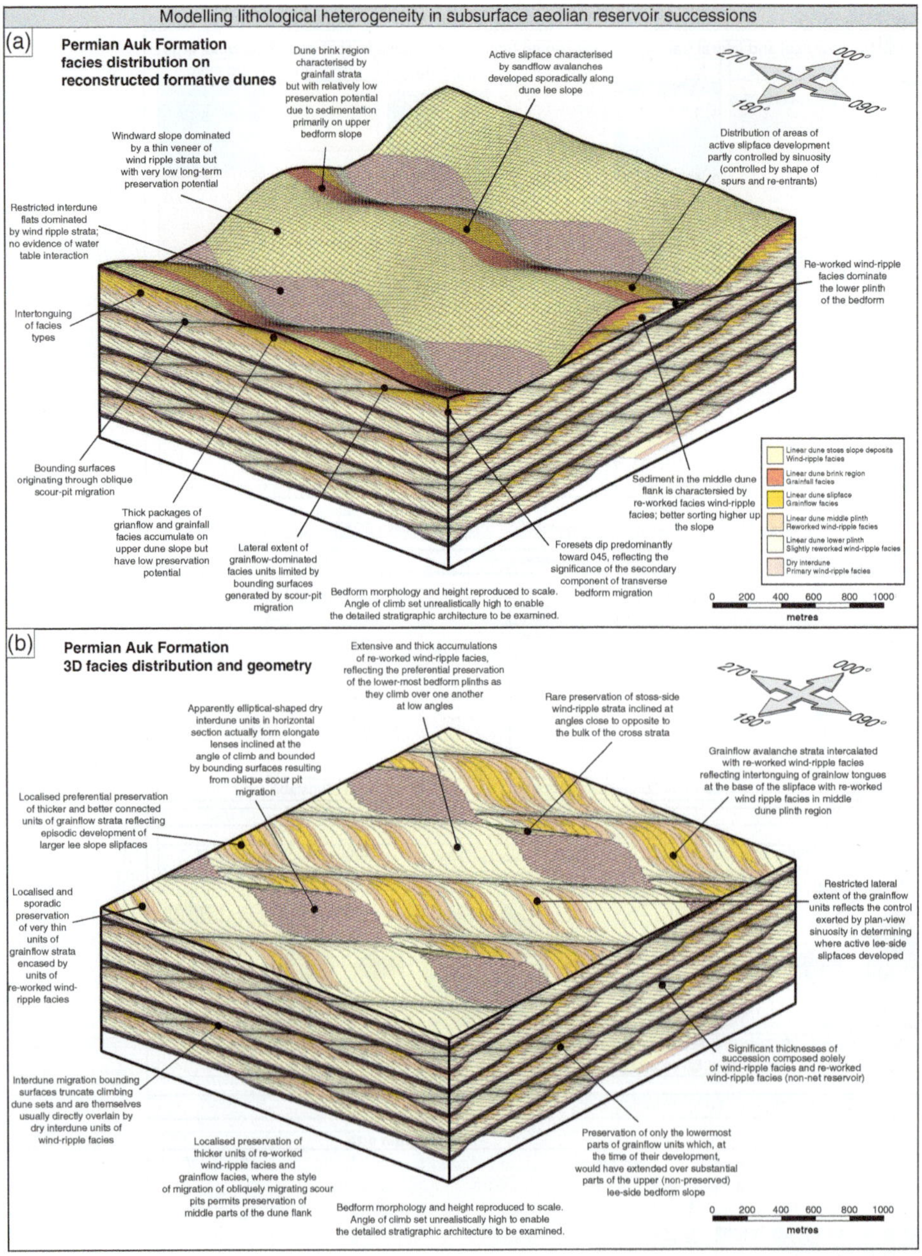

Figure 7.67 Facies models to explain the origins of lithological heterogeneity in a subsurface aeolian reservoir succession. The models relate to part of the Permian Auk Formation, central North Sea, UK. (a) The formative dunes migrated to an arbitrary 'north'. Along-crest-migrating sinuosities moved to an arbitrary 'east'. The dunes had low-angle-inclined lower lee slopes populated by wind ripples. Re-worked wind ripples higher up the lower dune slopes were partly reworked by secondary winds such that their finer grain fractions were winnowed. Note the elliptical (i.e. closed) interdune flats. (b) Horizontal slice revealing the resultant stratigraphic architecture. Note the oblique orientation of the preserved dune sets relative to the direction of original dune migration; this is a result of the along-crest movement of the sinuosities. Also, note the dominance of units of wind-ripple strata in the accumulated succession; units of grainflow strata are less common and are not well interconnected. Model grid constructed using 'Bedforms' software (Rubin 1987b; Rubin and Carter 2006).

7.8.2 Wider Applied Significance of Aeolian Systems and Their Preserved Successions

Aeolian sands are used as aggregates in construction, and pure quartz sand of aeolian origin is used to make valuable commodities like glass. Aeolian desert plains are known to host precious minerals, such as diamonds as placer deposits. Construction of roads and buildings on aeolian substrates requires careful planning and innovative engineering geology solutions. Dryland regions are currently responding to ongoing anthropogenically driven climate change and changes in land use. Desertification is a major problem in many dryland regions: aeolian erosion (deflation) of topsoil may reduce formally productive agricultural regions to dustbowls and desert dune encroachment over formerly productive agricultural land is reducing crop yields. Elsewhere aeolian deserts are being utilised as sites for solar power generation. See Section 7.S.1 in the Supplementary Information for further details.

Further Reading

Bristow, C.S., Bailey, S.D., and Lancaster, N. (2000). The sedimentary structure of linear sand dunes. *Nature* 406: 56–59.

Brookfield, M.E. (1977). The origin of bounding surfaces in ancient aeolian sandstones. *Sedimentology* 24: 303–332.

Crabaugh, M. and Kocurek, G. (1993). Entrada sandstone: an example of a wet aeolian system. In: *The Dynamics and Environmental Context of Aeolian Sedimentary Systems*, vol. 72 (ed. K. Pye), 103–126. Geological Society of London Special Publication.

Fryberger, S.G., Krystinik, L.F., and Schenk, C.J. (1990). Tidally flooded back-barrier dunefield, Guerrero Negro area, Baja California, Mexico. *Sedimentology* 37: 23–43.

Havholm, K.G. and Kocurek, G. (1994). Factors controlling aeolian sequence stratigraphy: clues from super bounding surface in the Middle Jurassic Page Sandstone. *Sedimentology* 41: 913–934.

Hunter, R.E. (1977a). Basic types of stratification in small eolian dunes. *Sedimentology* 24: 361–387.

Jordan, O.D. and Mountney, N.P. (2012). Sequence stratigraphic evolution and cyclicity of an ancient coastal desert system: the Pennsylvanian-Permian lower Cutler beds, Paradox Basin, Utah, USA. *Journal of Sedimentary Research* 82: 755–780.

Kocurek, G. (1996). Desert aeolian systems. In: *Sedimentary Environments: Processes, Facies and Stratigraphy*, 3e (ed. H.G. Reading), 125–153. Oxford: Blackwell Science.

Kocurek, G. and Havholm, K.G. (1993). Eolian sequence stratigraphy - a conceptual framework. In: *Siciliclastic Sequence Stratigraphy*, vol. 58 (ed. P. Weimer and H.W. Posamentier), 393–409. American Association of Petroleum Geologists Memoir.

Kocurek, G. and Lancaster, N. (1999). Aeolian system sediment state: theory and Mojave Desert Kelso dune field example. *Sedimentology* 46: 505–515.

Lancaster, N. (1988). Controls on eolian dune size and spacing. *Geology* 16: 972–975.

Langford, R. and Chan, M.A. (1988). Flood surfaces and deflation surfaces within the Cutler Formation and Cedar Mesa Sandstone (Permian), Southeastern Utah. *Geological Society of America Bulletin* 100: 1541–1549.

Loope, D.B. (1985). Episodic deposition and preservation of eolian sands: a late Paleozoic example from Southeastern Utah. *Geology* 13: 73–76.

McKee, E.D. (1979). A Study of global sand seas, geological survey professional paper, 1052, Washington, United States Government Printing Office, 421 p.

Mountney, N.P. (2006b). Eolian facies models. In: *Facies Models Revisited*, vol. 84 (ed. H.W. Posamentier and R.G. Walker), 19–83. Society for Economic Paleontologists and Mineralogists, Special Publication.

Mountney, N.P. (2012). A stratigraphic model to account for complexity in aeolian dune and interdune successions. *Sedimentology* 59: 964–989.

Rodríguez-López, J.P., Clemmensen, L., Lancaster, N. et al. (2014). Archean to Recent aeolian sand systems and their preserved successions: current understanding and future prospects. *Sedimentology* 61: 1487–1534.

Rubin, D.M. and Carter, C.L. (2006). Bedforms and cross-bedding in animation, SEPM. In: *Atlas*, vol. 2. DVD.

Rubin, D.M. and Hunter, R.E. (1982). Bedform climbing in theory and nature. *Sedimentology* 29: 121–138.

Scherer, C.M.S. (2000). Eolian dunes of the Botucatu Formation (Cretaceous) in southernmost Brazil: morphology and origin. *Sedimentary Geology* 137: 63–84.

References

Abrantes, F.R., Basilici, G., and Soares, M. V.T. (2020). Mesoproterozoic erg and sand sheet system: architecture and controlling factors (Galho do Miguel Formation, SE Brazil). *Precambrian Research* 338: 105592.

Ager (1981). *The Nature of the Stratigraphic Record*. New York: Wiley.

Ahlbrandt, T.S. and Fryberger, S.G., 1980, Eolian Deposits in the Nebraska Sand Hills, *United States Geological Survey Professional Paper*, 1120A-C.

Ahlbrandt, T.S. and Fryberger, S.G. (1981), Sedimentary features and significance of interdune deposits. In: Ethridge, F.G. and Flore, R.M., Eds., *Recent and Ancient Non-Marine Depositional Environments: Models for Exploration*, Society of Economic Paleontologists and Mineralogists Special Publication, 31, 293–314.

Ahlbrandt, T.S., Andrews, S., and Gwynne, D.T. (1978). Bioturbation in eolian deposits. *Journal of Sedimentary Petrology* Vol. 48: 839–848.

Alkathery, M., (2016). Modern and Ancient Fluvial-Aeolian Interactions in Dune-Field Margin Systems: Analysis and Implications. Unpublished M. Phil. Thesis, University of Leeds, UK, 196 pages.

Allen, J.R.L. (1970). A quantitative model of climbing ripples and their cross-laminated deposits. *Sedimentology* 14: 5–26. (Figure only).

Al-Masrahy, M.A. and Mountney, N.P. (2013a). Remote sensing of spatial variability in aeolian dune and interdune morphology in the Rub' Al-Khali, Saudi Arabia. *Aeolian Research* 11: 155–170.

Al-Masrahy, M.A. and Mountney, N.P. (2013b). Spatial variability in eolian dune and interdune morphology in the Rub' al-Khali dunefield, Saudi Arabia. *AAPG Search and Discovery*, 50830, 1–4.

Al-Masrahy, M.A. and Mountney, N.P. (2015). A classification scheme for fluvial–aeolian system interaction in desert-margin settings. *Aeolian Research* 17: 67–88.

Alsharhan, A. and Kendall, C.G.S.C. (2003). Holocene coastal carbonates and evaporites of the southern Arabian Gulf and their ancient analogues. *Earth Science Reviews* 61: 191–243.

Alsharhan, A.S., Glennie, K., Whittle, G.L. and Kendall, C.G. St. C. (1998), Quaternary Deserts and Climate Change. Proceedings of the International Conference on Quaternary Deserts and Climate Change, Balkema, Rotterdam, pp. 621.

Anderson, R.S. (1987). A theoretical model for aeolian impact ripples. *Sedimentology* 34: 943–956.

Anderson, R.S. (1988). The pattern of grainfall deposition in the lee of aeolian dunes. *Sedimentology* 35: 175–188.

Andreucci, S., Clemmensen, L.B., Murray, A.S., and Pascucci, V. (2010a). Middle to late Pleistocene coastal deposits of Alghero, northwest Sardinia (Italy): Chronology and evolution. *Quaternary International* 222: 3–16.

Andreucci, S., Clemmensen, L.B., and Pascucci, V. (2010b). Transgressive dune formation along a cliffed coast at 75 ka in Sardinia, Western Mediterranean: a record of sea-level fall and increased windiness. *Terra Nova* 22: 424–433.

Andreucci, S., Panzeri, L., Peter Martini, I. et al. (2014). Evolution and architecture of a West Mediterranean Upper Pleistocene to Holocene coastal apron-fan system. *Sedimentology* 61: 333–361.

Andrews, S. (1981). Sedimentology of Great Sand Dunes, Colorado, Society of Economic Paleontologists and Mineralogists Special Publication, 31, 279–291.

Argüello Scotti, A. and Veiga, G.D. (2019). Sedimentary architecture of an ancient linear megadune (Barremian, Neuquén Basin): Insights into the long-term development and evolution of aeolian linear bedforms. *Sedimentology* 66: 2191–2213.

Bagnold, R.A. (1941). *The Physics of Blown Sand and Desert Dunes*, 265. London: Methuen and Company.

Bállico, M.B., Scherer, C.M.S, Mountney, N.P., Souza, E.G., Chemale, F., Pisarevsky, S.A. and Reis, A.D. (2017a). The wind-pattern circulation as a palaeographic indicator: the case study of the 1.5-1.6 Ga Mangabeira Formation, São Francisco Craton, *Northeast Brazil. Precambrian Research*, 298, 1–15.

Bállico, M.B., Scherer, C.M.S, Mountney, N.P., Souza, E.G., Reis, A. D., Raja Gabaglia, G.P. and Magalhães, A.J.C. (2017b). Drying-upward cycles in a Mesoproterozoic aeolian erg-margin succession: Mangabeira Formation, Espinhaço Supergroup, Brazil. *Sedimentary Geology*, 349, 1–14.

Banham, S.G., Gupta, S., Rubin, D.M. et al. (2018). Ancient Martian aeolian processes and palaeomorphology reconstructed from the Stimson formation on the lower slope of Aeolis Mons, Gale crater, Mars. *Sedimentology* 65: 993–1042.

Basilici, G. and Dal' Bó, P.F.F. (2014). Influence of subaqueous processes on the construction and accumulation of an aeolian sand sheet, *Earth Surface Processes and Landforms*, 39, 1014–1029.

Basilici, G., Fuhr-Dal, B.P.F., and Ladeira, F.S.B. (2009). Climate-induced sediment-palaeosol cycles in a Late Cretaceous dry aeolian sand sheet; Marilia Formation (north-west Bauru Basin, Brazil). *Sedimentology* 56: 1876–1904.

Basilici, G., Soares, M.V.T., Mountney, N. P., and Colombera, L. (2020). Microbial influence on the accumulation of Neoproterozoic aeolian-dominated depositional systems: Venkatpur Sandstone Formation, Telangana State, southern India. *Precambrian Research* 347: 105854.

Basilici, G., Mesquita, A.F., Soares, A.V.T. et al. (2021). A Mesoproterozoic hybrid dry-wet aeolian system: Galho do Miguel Formation, SE Brazil. *Precambrian Research* 359: 106216.

Bateman, M.D., Holmes, P.J., Carr, A.S. et al. (2004). Aeolianite and barrier dune construction spanning the last two glacial-interglacial cycles from the southern Cape coast, South Africa. *Quaternary Science Reviews* 23: 1681–1698.

Bateman, M.D., Thomas, S.G., and Singhvi, A.K. (2003). Extending the aridity record of the Southwest Kalahari: current problems and future perspectives. *Quaternary International* 111: 37–49.

Becerril-Piña, R. and Mastachi-Loza, C.A., (2021), Desertification: Causes and countermeasures. In: *Life on Land*, 219–231. Cham, Springer International Publishing.

Belnap, J., Munson, S.M., and Field, J.P. (2011). Aeolian and fluvial processes in dryland regions: the need for integrated studies. *Ecohydrology* 4: 615–622.

Benan, C.A.A. and Kocurek, G. (2000). Catastrophic flooding of an aeolian dune field: Jurassic Entrada and Todilto Formations. *Ghost Ranch, New Mexico, USA, Sedimentology* 47: 1069–1080.

Besler, H. (1982). The north-eastern Rub' al Khali within the borders of the United Arab Emirates. *Zeitschrift für Geomorphologie* 26: 495–504.

Besly, B., Romain, H.G., and Mountney, N. P. (2018). Reconstruction of linear dunes from ancient aeolian successions using subsurface data: Permian Auk Formation, Central North Sea, UK. *Marine and Petroleum Geology* 91: 1–18.

Biswas, A. (2005). Coarse aeolianites: Sand sheets and zibar-interzibar facies from the Mesoproterozoic Cuddapah Basin, India. *Sedimentary Geology* 174: 149–160.

Blair, T.C. and McPherson, J.G. (2009). Processes and forms of alluvial fans. In: *Geomorphology of Desert Environments* 2e (ed. A.J. Parsons and A.D. Abrahams). Dordrecht: Springer Nature, 413–467.

Blakey, R.C. and Middleton, L.T. (1983). Permian shoreline eolian complex in central Arizona: dune changes in response to cyclic sea-level changes, In: Brookfield, M.E. and Ahlbrandt, T.S. Elsevier, Amsterdam, 38, 551–581.

Blakey, R.C., Havholm, K.G., and Jones, L. S. (1996). Stratigraphic analysis of eolian interactions with marine and fluvial deposits, Middle Jurassic Page

Sandstone and Carmel Formation, Colorado Plateau, USA. *Journal of Sedimentary Research* Vol. 66: 324–342.

Bloomfield, J.P., Moreau, M.F., and Newell, A.J. (2006)., Characterization of permeability distributions in six lithofacies from the Helsby and Wilmslow sandstone formations of the Cheshire Basin, UK. In: *Fluid Flow and Solute Movement in Sandstones: the Onshore UK Permo-Triassic Red Bed Sequence*, vol. 263 (ed. R.D. Barker and J. H. Tellam), 83–101. Geol. Soc. London Spec. Publ.

Bongiolo, D.E. and Scherer, C.M. (2010). Facies architecture and heterogeneity of the fluvial–aeolian reservoirs of the Sergi formation (Upper Jurassic), Recôncavo Basin, NE Brazil. *Marine and Petroleum Geology* 27: 1885–1897.

Botha, G.A., Porat, N., Haldorsen, S. et al. (2018). Beach ridge sets reflect the late Holocene evolution of the St Lucia estuarine lake system, South Africa. *Geomorphology* 318: 112–127. (Figure only).

Bourquin, S., Guillocheau, F., and Péron, S. (2009). Braided rivers within an arid alluvial plain (example from the Lower Triassic, western German Basin): recognition criteria and expression of stratigraphic cycles. *Sedimentology* 56: 2235–2264. (Figure only).

Breed, C.S. and Grow, T, 1979, Morphology and distribution of dunes in sand seas observed by remote sensing, In: McKee, E.D., Ed., *A study of global sand seas*, United States Geological Survey Professional Paper, 1052, 253–302.

Breed, C.S., Fryberger, S.G., Andrews, S., McCauley, C., Lennartz, F., Gebel, D. and Horstman, K. (1979). Regional studies of sand seas, using Landsat (ERTS) imagery, In: McKee, E.D., Ed., *A study of global sand seas*: United States Geological Survey Professional Paper, 1052, 305–397.

Breed, C.S., McCauley, J.F., and Davis, P. A. (1987). Sand sheets of the eastern Sahara and ripple blankets on Mars, in: frostick l.e. and reid i., eds., desert sediments ancient and modern, Geological Society of London Special Publication. 35: 337–359.

Bristow, C.S. (2019). Bounding surfaces in a barchan dune: annual cycles of deposition? Seasonality or erosion by superimposed bedforms, *Remote Sensing*, 11, 965. (Figure only)

Bristow, C.S. and Mountney, N.P. (2013). Aeolian stratigraphy. In: *Treatise on Geomorphology, Earth Systems and Environmental Sciences*, vol. 11 (ed. J. Shroder), 246–268. Figure only.

Bristow, C.S. and Mountney, N.P. (2021). Aeolian Landscapes, Aeolian Stratigraphy. In: *Treatise on Geomorphology* (ed. J. Shroder), 7982. San Diegeo, USA: Second Edition, Elsevier ISBN 9780128182345 (Figure only).

Bristow, C.S. and Pucillo, K. (2006). Quantifying rates of coastal progradation from sediment volume using GPR and OSL: the Holocene fill of Guichen Bay, south-east South Australia. *Sedimentology* 53: 769–788.

Bristow, C.S., Bailey, S.D., and Lancaster, N. (2000). The sedimentary structure of linear sand dunes. *Nature* 406: 56–59.

Bristow, C.S., Lancaster, N., and Duller, G. A.T. (2005). Combining ground penetrating radar (GPR) surveys and optical dating to determine dune migration in Namibia. *Journal of the Geological Society of London* 162: 315–321.

Bristow, C., Pugh, J., and Goodall, T. (1996). Internal structure of aeolian dunes in Abu Dhabi determined using ground-penetrating radar. *Sedimentology* 43: 995–1003.

Brooke, B. (2001). The distribution of carbonate eolianite. *Earth-Science Reviews* 55: 135–164.

Brookfield, M.E. (1970). Dune trends and wind regime in central Australia, *Zeitschrift für Geomorphologie Supplementband*, 10, 121–153.

Brookfield, M.E. (1977). The origin of bounding surfaces in ancient aeolian sandstones. *Sedimentology* 24: 303–332.

Brookfield, M.E. and Silvestro, S. (2010). Eolian systems, In: Dalrymple, R.W. and James, N.P., *Facies Models*, Fourth Edition, Geological Association of Canada, pp. 139–166.

Bryant, G., Cushman, R., Nick, K., and Miall, A. (2016). Paleohydrologic controls on soft-sediment deformation in the Navajo Sandstone. *Sedimentary Geology* 344: 205–221.

Bull, L.J. and Kirkby, M.J. (2002). Dryland rivers: hydrology and geomorphology of semi-arid channels. John Wiley & Sons.

Bullard, J.E. and Livingstone, I. (2002). Interactions between aeolian and fluvial systems in dryland environments. *Area* 34: 8–16.

Bullard, J.E. and McTainsh, G.H. (2003). Aeolian-fluvial interactions in dryland environments: examples, concepts and Australia case study. *Progress in Physical Geography* 27: 471–501.

Bullard, J.E., McTansih, G.H., and Pudmenzky, C. (2004). Aeolian abrasion and modes of fine particle production from natural red dune sands: An experimental study. *Sedimentology* 51: 1103–1125.

Bullard, J.E., White, K., and Livingstone, I. (2011). Morphometric analysis of aeolian bedforms in the Namib Sand Sea using ASTER data. *Earth Surface Processes and Landforms* 36: 1534–1549.

Butler, D. (2006). The web-wide world. *Nature* 439: 776–778.

Cain, S.A. (2009). *Sedimentology and stratigraphy of a terminal fluvial fan system: the Permian Organ Rock Formation, South East Utah*, Unpublished PhD thesis, vol. 56, 1774–1800. Keele University.

Cain, S.A. and Mountney, N.P. (2009). Spatial and temporal evolution of a terminal fluvial fan system: the Permian Organ Rock Formation, South East Utah, USA. *Sedimentology* 56: 1774–1800.

Cain, S.A. and Mountney, N.P. (2011). Downstream changes and associated fluvial-aeolian interactions in an ancient terminal fluvial fan system: the Permian Organ Rock Formation, SE Utah. In: *From River to Rock Record, SEPM Special Publication*, vol. 97 (ed. S. Davidson, S. Leleu, and C. North), 165–187.

Campos-Soto, S., Benito, M.I., Mountney, N.P. et al. (2022). Where humid and arid meet: Sedimentology of coastal siliciclastic successions deposited in apparently contrasting climates. *Sedimentology*, in press 69: 975–1027.

Cardenas, B.T., Kocurek, G., Swanson, T. and Hughes, C.M., (2019). Preservation of autogenic processes and allogenic forcings in set-scale aeolian architecture II: The scour-and-fill dominated Jurassic Page Sandstone, Arizona, U.S.A., Journal of Sedimentary Research, 89, 741-760. (Figure only)

Carmona, N.B., Ponce, J.J., and Wetzel, A. (2018). Biogenic structures of unioniform bivalves in wet-interdune deposits (late Miocene–early Pliocene, Argentina). *Palaios* 33: 431–440.

Carr-Crabaugh, M. and Kocurek, G. (1998). Continental sequence stratigraphy of a wet eolian system: A key to relative sea-level change, In: Shanley, K. and McCabe, P., Eds., *Relative Roles of Eustasy, Climate, and Tectonism in Continental Rocks*, Society of Economic Paleontologists and Mineralogists Special Publication, 59, 213–228.

Chakraborty, T. and Chakraborty, C. (2001). Eolian-aqueous interactions in the development of a Proterozoic sand sheet: Shikaoda Formation, Hosangabad, India. *Journal of Sedimentary Research* 71: 107–117.

Chakraborty, T. and Chaudhuri, A.K. (1993). Fluvial-aeolian interactions in a Proterozoic alluvial plain: example from the Mancheral Quartzite, Sullavai Group, Pranhita-Godavari Valley, India, In: Pye, E., Ed., *The dynamics and environmental context of aeolian sedimentary systems, Geological Society of London Special Publication*, 72, 127–141.

Chan, M.A. (1989). Erg margin of the Permian White Rim Sandstone, SE Utah. *Sedimentology* 36: 235–251.

Chan, M.A. and Archer, A.W. (1999). Spectral analysis of eolian foreset periodicities – implications for Jurassic decadal-scale paleoclimatic oscillators. *Palaeoclimates* 3: 239–255.

Chan, M.A. and Archer, A.W. (2000). Cyclic eolian stratification on the Jurassic Navajo Sandstone, Zion National Park: periodicities and implications for palaeoclimate. In: Sprinkel, D.A., Chidsey, T.C. and Anderson P.B., Eds. *Geology of Utah's Parks and Monuments, Utah Geological Association Publication* 28: 607–617.

Chan, M.A. and Kocurek, G. (1988). Complexities in eolian and marine interactions - processes and eustatic controls on erg development. *Sedimentary Geology* 56: 283–300.

Chan, M.A., Hasiotis, S.T., and Parrish, J. T. (2019). Enigmatic clastic pipe swarms and implications for fluidization dynamics in aeolian deposits. *Sedimentology* 66: 513–535.

Chandler, M.A., Kocurek, G., Goggin, D.J., and Lake, L.W. (1989). Effects of stratigraphic heterogeneity on permeability in eolian sandstone sequence, Page Sandstone, northern Arizona. *American Association of Petroleum Geologists Bulletin* 73: 658–668.

Charru, F., Andreotti, B., and Claudin, P. (2013). Sand Ripples and Dunes. *Annual Review of Fluid Mechanics* 45: 469–493.

Chrintz, T. and Clemmensen, L.B. (1993). Draa reconstruction, the Permian Yellow Sands, northeast England, In: Pye, K. and Lancaster, N., Eds., *Aeolian Sediments Ancient and Modern*, International Association of Sedimentologists Special Publication, 16, 51–161.

Ciftci, B. N., Aviantara, A.A., Hurley, N.F. and Kerr, D.R. (2004). Outcrop-based three-dimensional modeling of the Tensleep Sandstone at Alkali Creek, Bighorn Basin, Wyoming, In: Grammer, G.M., Harris, P.M. and Eberli, G.P., Eds., *Integration of outcrop and modern analogues in reservoir modelling*, American Association of Petroleum Geologists Memoir, 80, 235–259.

Clemmensen, L.B. (1987). Complex star dunes and associated aeolian bedforms, Hopeman Sandstone (Permo-Triassic), Moray Firth Basin, Scotland, in: Frostick, L.E. and Reid, I., desert sediments: ancient and modern, Geological Society of London special publication, 35, 213–231. (Figure only)

Clemmensen, L.B. (1988). Aeolian morphology preserved by lava cover, the Precambrian Mussartut Member, Eriksfjord Formation, South Greenland. *Bulletin of the Geological Survey of Denmark* 37: 105–116.

Clemmensen, L.B. (1989). Preservation of interdraa and plinth deposits by the lateral migration of large linear draas (Lower Permian Yellow Sands, northeast England). *Sedimentary Geology* 65: 139–151.

Clemmensen, L.B. (1991). Controls on aeolian sand sheet formation exemplified by the Lower Triassic of Heligoland, Acta Mechanica Supplementum, 2, 161–170.

Clemmensen, L.B. and Blakey, R.C. (1989). Erg deposits in the Lower Jurassic Wingate Sandstone, northeastern Arizona - oblique dune sedimentation. *Sedimentology* 36: 449–470.

Clemmensen, L.B., Fornós, J.J. and Rodríguez-Perea, A. (1997). Morphology and architecture of a Late Pleistocene cliff-front dune, *Mallorca, Western Mediterranean, Terra Nova*, 9, 251–254.

Clemmensen, L.B., Murray, A., Heinemeier, J., and De Jong, R. (2009). The evolution of Holocene coastal dunefields, Jutland, Denmark: A record of climate change over the past 5000 years. *Geomorphology* 105: 303–313.

Clemmensen, L.B., Olsen, H., and Blakey, R.C. (1989). Erg-margin deposits in the Lower Jurassic Moenave Formation and Wingate Sandstone, southern Utah. *Geological Society of America Bulletin* 101: 759–773.

Clemmensen, L.B., Øxnevad, I.E.I. and de Boer, P.L. (1994). Climatic controls on ancient desert sedimentation: some late Palaeozoic examples from NW Europe and the western interior of the USA, In: de Boer, P.L. and Smith, D.G., Eds., *Orbital Forcing and Cyclic Sequences*, International Association of Sedimentologists Special Publication, 19, 439–457.

Clemmensen, L.B., Pye, K., Murray, A., and Heinemeier, J. (2001). Sedimentology, stratigraphy and landscape evolution of a Holocene coastal dune system, Lodbjerg, NW Jutland, Denmark. *Sedimentology* 48: 3–27.

Colombera, L., Mountney, N.P., and McCaffrey, W.D. (2013). A quantitative approach to fluvial facies models: methods and example results. *Sedimentology* 60: 1526–1558.

Collinson, J.C. and Maltman, A. (1994). Sedimentary deformational structures. In: *The Geological Deformation of Sediments* (ed. A. Maltman), 95–125. London: Chapman and Hall.

Collinson, J.C. and Mountney, N.P., 2019. *Sedimentary Structures*. Fourth Edition, Dunedin Academic Press, 340 pages.

Cooke, R., Warren, A., and Goudie, A. (1993). *Desert Geomorphology*, 526. London: University College London Press.

Corbett, I. (1993). The modern and ancient pattern of sandflow through the southwestern Namib deflation basin, In: Pye, K. and Lancaster, N., Eds. *Aeolian Sediments Ancient and Modern, International Association of Sedimentologists Special Publication* 16: 45–60.

Cosgrove, G.I.E., Colombera, L., and Mountney, N.P. (2021a). A Database of Aeolian Sedimentary Architecture for the characterization of modern and ancient sedimentary systems. *Marine and Petroleum Geology* 127: 104983.

Cosgrove, G.I.E., Alkathery, M., Al-Masrahy, M., Colombera, L., Mountney, N.P., and Thompson, D.B. (2025). Quantitative sedimentological analysis of the aeolian-fluvial Triassic Sherwood Sandstone Group, Cheshire Basin, UK. *Sedimentology* 72: 1233–1274.

Cosgrove, G.I.E., Colombera, L., and Mountney, N.P. (2021b). Quantitative analysis of the sedimentary architecture of eolian successions developed under icehouse and greenhouse climatic conditions. *Geological Society of America, Bulletin* 133: 2625–2644.

Cosgrove, G.I.E., Colombera, L., and Mountney, N.P. (2022a). Eolian stratigraphic record of environmental change through geological time. *Geology* 50: 289–294.

Cosgrove, G.I.E., Colombera, L., and Mountney, N.P. (2022b). The role of subsidence and accommodation generation in controlling the nature of the aeolian stratigraphic record. *Journal of the Geological Society* 179: jgs2021-042.

Cosgrove, G.I.E., Colombera, L., Mountney, N.P. et al. (2023). Precambrian aeolian systems: A unique record? *Precambrian Research* 392: 107075.

Cowan, G. (1993). Identification and significance of aeolian deposits within the dominantly fluvial Sherwood Sandstone Group of the East Irish Sea Basin UK, In: North, C.P. and Prosser, D.J., Characterization of Fluvial and Aeolian Reservoirs, *Geological Society of London Special Publication*, 73, 231–245.

Crabaugh, M. and Kocurek, G. (1993). Entrada Sandstone: an example of a wet aeolian system. In: Pye, K., Ed., *The Dynamics and Environmental Context of Aeolian Sedimentary Systems, Geological Society of London Special Publication*, 72, 103–126.

Dal' Bó, P.F.F. and Basilici, G. (2015). Intermontane eolian sand sheet development, Upper Tulum Valley, central-western Argentina, Brazilian. *Journal of Geology* 45: 97–115. (Figure only).

Dal' Bó, P.F.F., Basilici, G., and Angélica, R.S. (2010). Factors of paleosol formation in a Late Cretaceous eolian sand sheet paleoenvironment. *Marília Formation, Southeastern Brazil, Palaeogeography, Palaeoclimatology and Palaeoecology* 292: 349–365.

Davies, N.S., Berry, C.M., Marshall, J.E.A., Wellman, C.H. and Lindemann, F-J. (2021). The Devonian landscape factory: plant-sediment interactions in the Old Red Sandstone of Svalbard and the rise of vegetation as a biogeomorphic agent, *The Journal of the Geological Society*, 178, jgs2020-225. (Figure only)

Day, M. and Zimbelman, J.R. (2021). Ripples, megaripples, and TARs, oh, my! Recommendations regarding Mars aeolian bedform terminology. *Icarus* 369: 114647.

DeCelles, P.G., Langford, R.P., and Schwartz, R.K. (1983). Two new methods of palaeocurrent determination from trough cross-stratification. *Journal of Sedimentary Petrology* 53: 629–642.

del Valle, L., Fornós, J.J., Pomar, F. et al. (2020). Aeolian-alluvial interactions at Formentera (Balearic Islands, Western Mediterranean): The late Pleistocence evolution of a costal system. *Quaternary International* 566-567: 271–283.

Doe, T.W. and Dott, R.H. Jr. (1980). Genetic significance of deformed cross bedding – with examples from the Navajo and Weber Sandstones of Utah. *Journal of Sedimentary Petrology* Vol. 50: 793–811.

Doorncamp, J.C., Brunsden, D., and Jones, D.K.C. (1980). *Geology, Geomorphology and Pedology of Bahrain*. Norwich: Geobooks.

dos Reis, A.D., dos Santos Scherer, C.M., do Amarante, F.B., Rossetti, M.D.M.M., Kifumbi, C., de Souza, E.G., Ferronatto, J.P.F. and Owen, A. (2019). Sedimentology of the proximal portion of a large-scale, Upper Jurassic fluvial-aeolian system in Paraná Basin, southwestern Gondwana, *Journal of South American Earth Sciences*, 95, 102248.

De Ros, L.F. and Scherer, C.M.S. (2013). Stratigraphic Controls on the Distribution of Diagenetic Processes, Quality and Heterogeneity of Fluvial-Aeolian Reservoirs from the Recôncavo Basin, Brazil. In: *Linking Diagenesis to Sequence Stratigraphy* (ed. S. Morad, J. M. Ketzer, and L.F. De Ros). Wiley.

Draut, A.E., 2012, Effects of river regulation on aeolian landscapes, Colorado River, southwestern USA, *Journal of Geophysical Research*, 117, F02022 (Figure only)

Driese, S.G. (1985). Interdune pond carbonates, Weber Sandstone (Pennsylvanian-Permian), northern Utah and Colorado. *Journal of Sedimentary Petrology* Vol. 55: 187–195.

Enge, H.D., Buckley, S.J., Rotevatn, A., and Howell, J.A. (2007). From outcrop to reservoir simulation model: workflow and procedures. *Geosphere* 3: 469–490.

El-Wahab, A., Al-Rashed, A.R., and Al-Dousari, A. (2018). Influences of physiographic factors, vegetation patterns and human impacts on aeolian landforms in arid environment. *Arid Ecosystems* 8: 97–110.

Ellwood, J.M., Evans, P.D., and Wilson, I. G. (1975). Small scale aeolian bedforms. *Journal of Sedimentary Petrology* Vol. 45: 554–561.

Eschner, T.B. and Kocurek, G. (1986). Marine destruction of eolian sand seas - origin of mass flows. *Journal of Sedimentary Petrology* Vol. 56: 401–411.

Eschner, T.B. and Kocurek, G. (1988). Origins of relief along contacts between eolian sandstones and overlying marine strata. *American Association of Petroleum Geologists Bulletin* 72: 932–943.

Evan, A.T., Flamant, C., Fielder, S., and Doherty, O. (2014). An analysis of aeolian dust in climate models. *Geophysical Research Letters* 41: 5996–6001.

Evans, G., Kendall, G. St C. and Skipwith, P. (1964). Origin of coastal flats, the sabkha of the Trucial Coast, *Persian Gulf, Nature*, 202, 759–761.

Ewing, R.C. and Kocurek, G. (2010). Aeolian dune interactions and dune-field pattern formation: White Sands Dune Field. *New Mexico, Sedimentology* 57: 1199–1219.

Ferronatto, J.P.F., Scherer, C.M.S., Souza, E.G., Reis, A.D., and Mello, R.G. (2019). Genetic units and facies architecture of a Lower Cretaceous fluvial-aeolian succession, São Sebastiäo Formation, Jatobá Basin, Brazil. *Journal of South American Earth Sciences* 89, 158–172.

Fisher, G.B., Amos, C.B., Bookhagen, B., Burbank, D.W. and Godard, V., 2012, Channel widths, landslides, faults, and beyond: The new world order of high-spatial resolution Google Earth imagery in the study of earth surface processes. Geological Society of America Special Papers, 492, 1–22.

Fornós J., Clemmensen, L.B., Gómez-Pujol.L. and Murray, A.S. (2009). Late Pleistocene carbonate aeolianites on Mallorca, Western Mediterranean: a luminescence chronology, *Quaternary Science Reviews*, 28, 2697–2709.

Frank, A. and Kocurek, G. (1996). Toward a model for airflow on the lee side of aeolian dunes. *Sedimentology* 43: 451–458.

Frostick, L.E. and Reid, I. (1987). Tectonic control of desert sediments in rift basins ancient and modern. *Geological Society, London, Special Publications* 35: 53–68.

Fryberger, S.G. (1978). Techniques for the evaluation of surface wind data in terms of eolian sand drift. *United States Geological Survey Open File Report* 78-405: 25.

Fryberger, S.G. (1979). Dune forms and wind regime, In: McKee, E.D., Ed. *A study of global sand seas, United States Geological Survey Professional Paper* 1052: 137–169.

Fryberger, S.G., Fryberger, S.G., Krystinik, L.F., and Schenk, C.J. (1990a). Great Sand Dunes depositional system – an overview. In: *Modern and Ancient Aeolian Deposits: Petroleum Exploration and Production* (ed. S.G. Fryberger, L.F. Krystinik, and C.J. Schenk), 1–9. Denver: Society of Economic Paleontologists and Mineralogists.

Fryberger, S.G., Fryberger, S.G., Krystinik, L.F., and Schenk, C.J. (1990b). Role of water in aeolian deposition. In: *Modern and Ancient Aeolian Deposits: Petroleum Exploration and Production* (ed. S.G. Fryberger, L.F. Krystinik, and C.J. Schenk), 5–11. Denver: Society of Economic Paleontologists and Mineralogists.

Fryberger, S.G., Fryberger, S.G., Krystinik, L.F., and Schenk, C.J. (1990c). Coastal eolian deposits of Oregon, USA, Guererro Negro, Mexico and Jafurah Sand Sea, Saudi Arabia. In: *Modern and Ancient Aeolian Deposits: Petroleum Exploration and Production* (ed. S.G. Fryberger, L.F. Krystinik, and C.J. Schenk), 11–15. Denver: Society of Economic Paleontologists and Mineralogists.

Fryberger, S.G., Fryberger, S.G., Krystinik, L.F., and Schenk, C.J. (1990d). Ancient eolian deposits of the Navajo and Weber Sandstones, northeast Utah and Colorado. In: *Modern and Ancient Aeolian Deposits: Petroleum Exploration and Production* (ed. S.G. Fryberger, L.F. Krystinik, and C.J. Schenk), 19–12. Denver: Society of Economic Paleontologists and Mineralogists.

Fryberger, S.G. (1993). A review of aeolian bounding surfaces, with examples from the Permian Minnelusa Formation, USA. In: *Characterization of Fluvial and Aeolian Reservoirs*, vol. 73 (ed. C.P. North and J.D. Prosser), 167–197. *Geological Society of London Special Publication.*

Fryberger, S.G. and Schenk, C. (1981). Wind sedimentation tunnel experiments on the origins of aeolian strata. *Sedimentology* 28: 805–821.

Fryberger, S.G. and Schenk, C.J. (1988). Pin stripe lamination - a distinctive feature of modern and ancient eolian sediments. *Sedimentary Geology* 55: 1–15.

Fryberger, S.G., Ahlbrandt, T.S., and Andrews, S. (1979). Origin, sedimentary features, and significance of low-angle eolian 'sand sheet' deposits, Great Sand Dunes National Monument and vicinity, Colorado. *Journal of Sedimentary Petrology* 49: 733–746.

Fryberger, S.G., Al-Sari, A.M., Clisham, T. J. et al. (1984). Wind sedimentation in the Jafurah sand sea. *Saudi Arabia, Sedimentology* 31: 413–431.

Fryberger, S.G., Al-Sari, A.M., and Clisham, T.J. (1983). Eolian dune, interdune, sand sheet, and siciliclastic sabkha sediments of an offshore prograding sand sea. *Dharan Area, Saudi Arabia, American Association of Petroleum Geologists Bulletin* 67: 280–312.

Fryberger, S.G., Hesp, P., and Hastings, K. (1992). Aeolian granule ripple deposits. *Namibia, Sedimentology* 39: 319–331.

Fryberger, S.G., Krystinik, L.F. and Schenk, C.J (1990). Tidally flooded back-barrier dunefield, Guerrero Negro area, Baja California, Mexico, *Sedimentology*, 37, 23–43.

Fryberger, S.G., Schenk, C.J., and Krystinik, L.F. (1988). Stokes surfaces and the effects of near-surface groundwater-table on aeolian deposition. *Sedimentology* 35: 21–41.

Fu, T., Wu, Y., Tan, L. et al. (2019). Imaging the structure and reconstructing the development of a barchan dune using ground-penetrating radar. *Geomorphology* 341: 192–202.

García-Hidalgo, J.F., Temiño, J., and Segura, M. (2002). Holocene eolian sediments on the southern border of the Duero Basin (Spain): origin and development of an eolian system in a temperate zone. *Journal of Sedimentary Research* 72: 30–39.

George, G.T. and Berry, J.K. (1993). A new lithostratigraphy and depositional model for the Upper Rotliegend of the UK sector of the Southern North Sea, In: North, C.P. and Prosser, D.J., Eds., Characterization of Fluvial and Aeolian Reservoirs, *Geological Society of London Special Publication*, 73, 291–319.

George, G.T. and Berry, J.K. (1997). Permian (Upper Rotliegend) synsedimentary tectonics, basin development and palaeogeography of the southern North Sea, In: Ziegler, K., Turner, P. and Daines, S.R., Eds., *Petroleum Geology of the Southern North Sea: Future Potential, Geological Society of London Special Publication*, 123, 31–61.

Glennie, K.W. (1970). *Desert sedimentary environments, Developments in Sedimentology, 14.* Amsterdam: Elsevier.

Glennie, K.W. (1972). Permain Rotliegendes of North West Europe interpreted in light of modern desert sedimentation studies. *American Association of Petroleum Geologists Bulletin* 56: 1048–1071.

Glennie, K.W. (1987). Desert sedimentary environments, present and past - a summary. *Sedimentary Geology* 50: 135–165.

Glennie, K.W. and Glennie, K.W. (1990). Lower Permian – Rotliegend. In: *Introduction to the Petroleum Geology of the North Sea* (ed. K.W. Glennie), 120–152. Oxford: Blackwell Science.

Glennie, K.W., (2005). *The Desert of Southeast Arabia: Desert Environments and Sediments*, Gulf Petrolink, Bahrain, 215p.

Glennie, K.W. and Buller, A.T. (1983). The Permian Weissliegend of NW Europe - the partial deformation of aeolian dune sands caused by the Zechstein Transgression. *Sedimentary Geology* 35: 43–81.

Godo, T. (2017). The Appomattox field: Norphlet Aeolian sand dune reservoirs in the deep-water Gulf of Mexico. In: *Giant Fields of the decade 2000–2010*, vol. 113 (ed. R.K. Merrill and C.A. Sternbach), 29–54. AAPG Memoir.

Goldhammer, R.K., Oswald, E.J. and Dunn, P.A. (1994). High-frequency, glacioeustatic cyclicity in the middle Pennsylvanian of the Paradox Basin: an evaluation of Milankovitch forcing, In: De Boer, P.L. and Smith, D.G., Eds., *Orbital Forcing and Cyclic Sequences*, International Association of Sedimentologists, Special Publication, 19, 243–283.

Goodall, T.M., North, C.P., and Glennie, K.W. (2000). Surface and subsurface sedimentary structures produced by salt crusts. *Sedimentology* 47: 99–118.

Goudie, A. (1972). Climate, weathering, crust formation, dunes, and fluvial features of the Central Namib Desert, near Gobabeb, South West Africa, Madoqua, 1972, 15–31.

Goudie, A.S. (2014). Desert dust and human health disorders. *Environment international* 63: 101–113.

Gradziński, R. and Uchman, A. (1994). Trace fossils from interdune deposits—an example from the Lower Triassic aeolian Tumlin Sandstone, central Poland. *Palaeogeography, Palaeoclimatology, Palaeoecology* 108: 121–138.

Greely, R. and Iverson, J.D. (2009). *Wind as a Geological Process on Earth.* Mars: Cambridge University Press.

Greeley, R., Lancaster, N., Lee, S. and Thomas, P. (2018). Martian aeolian processes, sediments, and features, *Mars*, 730–766, University of Arizona Press.

Grohmann, C.H., Garcia, P.B.G., Affonso, A.A. and Albuquerque, R.W., 2020, Dune migration and volume change from airborne LiDAR, terrestrial LiDAR and structure from motion-multi view stereo, *Computers and Geoscience*, 143, 104569 (Figure only)

Guern, P.L. and Davaud, E. (2005). Recognition of ancient carbonate wind deposits: lessons from a modern analogue. *Chrissi Island, Crete, Sedimentology* 52: 915–926.

Gunatilaka, A. and Mwango, S. (1987). Continental sabkha pans and associated nebkhas in southern Kuwait, Arabian Gulf, In: Frostick, L.E. and Reid, I., Eds., *Desert Sediments Ancient and Modern, Geological Society of London Special Publication*, 35, 187–203.

Hamblin, W.K. (1965). Internal structures of "homogeneous" sandstones, 175, *University of Kansas.*

Han, Z., Ren, Y., Li, X. et al. (2022). Formation of parabolic dunes on the shore of Poyang Lake in East China. *Geomorphology* 397: 108023. (Figure only).

Hargitai, H. (2015). Seif, In: Hargitai, H. and Kereszturi, A., *Encyclopedia of Planetary Landforms*, Springer, New York, NY. (Figure only)

Hasiotis, S.T. (2002). *Continental Trace Fossils*, Society of Economic Paleontologists and Mineralogists, Short Course Notes, 51, 132.

Havholm, K.G. and Kocurek, G. (1994). Factors controlling aeolian sequence stratigraphy: clues from super bounding surface in the Middle Jurassic Page Sandstone. *Sedimentology* 41: 913–934.

Havholm, K.G., Blakey, R.C., Capps, M., Jones, L.S., King, D.D. and Kocurek, G. (1993). Aeolian genetic stratigraphy: an example from the Middle Jurassic Page Sandstone, Colorado Plateau, In: Pye, K. and Lancaster, N., Eds., *Aeolian Sediments Ancient and Modern*, International Association of Sedimentologists Special Publication, 16, 87–107.

Hayward, R.K., Mullins, K.F., Fenton, L.K. et al. (2007). Mars global digital dune database and initial science results. *Journal of Geophysical Research: Planets* 112: E11.

Heckel, P.H., Dennison, J.M. and Ettensohn, F.R. (1994). Evaluation of evidence for glacio-eustatic control over marine Pennsylvanian cyclothems in North America and consideration of possible tectonic effects, In: Dennison, J.M. and Ettensohn, F.R., *Tectonic and eustatic controls on sedimentary cycles, Tulsa: Concepts in Sedimentology and Paleontology*, 4, 65–87, SEPM (Society for Sedimentary Geology).

Herries, R.D. (1992). Sedimentology of Continental Erg Margin Interactions, Unpublished Ph.D. Thesis, University of Aberdeen, pp. 205.

Herries, R.D. (1993). Contrasting styles of fluvial-aeolian interaction at a downwind erg margin: Jurassic

Kayenta-Navajo transition, northeastern Arizona, USA, In: North, C.P. and Prosser, J.D., Eds., *Characterization of Fluvial and Aeolian Reservoirs, Geological Society of London Special Publication*, 73, 199–218.

Herries, R.D. and Cowan, G. (1997). Challenging the 'sheetflood' myth: the role of water-table-controlled sabkha deposits in redefining the depositional model for the Ormskirk Sandstone Formation (Lower Triassic), East Irish Sea Basin, In: Meadows, N.S., Trueblood, S.P., Hardman, M. and Cowan, G., Eds., *Petroleum Geology of the Irish Sea and Adjacent Areas, Geological Society of London Special Publication*, 124, 253–276.

Hesp, P.A. (1999). The beach backshore and beyond. In: *Handbook of Beach and Shoreface Morphodynamics* (ed. A.D. Short), 145–169. John Wiley and Sons, Ltd.

Heward, A.P. (1991). Inside Auk – the anatomy of an eolian oil reservoir. In: *The Three-Dimensional Facies Architecture of Terrigenous Clastic Sediments and its Implications for Hydrocarbon Discovery and Recovery* (ed. A.D. Miall and N. Tyler), SEPM Concepts Sedimentol. Palaeontol., 3, 44–56.

Horowitz, D.H. (1982). Geometry and origin of large-scale deformation structures in some ancient wind-blown sand deposits. *Sedimentology* 29: 155–180.

Howell, J.A. (1992). Sedimentology of the Rotliegend Supergroup of the UK Southern North Sea. Unpublished PhD Thesis, University of Birmingham, 345 pp.

Howell, J.A. and Mountney. N.P. (1997). Climatic cyclicity and accommodation space in arid to semi-arid depositional systems: An example from the Rotliegend Group of the Southern North Sea. In: Ziegler, K., Turner, P. and Daines, S.R., Eds., *Petroleum Geology of the Southern North Sea: Future Potential, Geological Society of London Special Publication*, 123, 63–86.

Howell, J.A. and Mountney, N.P. (2001). Aeolian grain flow architecture: hard data for reservoir models and implications for red bed sequence stratigraphy. *Petroleum Geoscience* 7: 51–56.

Howell, J.A., Martinius, A.W., and Good, T.R. (2014). The application of outcrop analogues in geological modelling: a review, present status and future outlook. *Geological Society, London, Special Publications* 387: 1–25.

Höyng, D., D'Affonseca, F.M., Bayer, P. et al. (2014). High-resolution aquifer analog of fluvial–aeolian sediments of the Guarani aquifer system. *Environmental Earth Sciences* 71: 3081–3094.

Hummel, G. and Kocurek, G. (1984). Interdune areas of the Back-Island dune field, North Padre-Island, Texas. *Sedimentary Geology* 39: 1–26.

Hunter, R.E. (1973). Pseudo-crosslamination formed by climbing adhesion ripples. *Journal of Sedimentary Research* Vol. 43: 24–30.

Hunter, R.E. (1977a). Basic types of stratification in small eolian dunes. *Sedimentology* 24: 361–387.

Hunter, R.E. (1977b). Terminology of cross-stratified sedimentary layers and climbing-ripple structures. *Journal of Sedimentary Research* Vol. 47: 697–706.

Hunter, R.E. (1981). Stratification styles in eolian sandstones: Some Pennsylvanian to Jurassic examples from the western interior USA, In: *Recent and Ancient Non-Marine Depositional Environments, Models for Exploration*, Society of Economic Paleontologists and Mineralogists Special Publication, 31, 315–329.

Hunter, R.E. (1985). A kinematic model for the structure of lee-side deposits. *Sedimentology* 32: 409–422.

Hunter, R.E. and Richmond, B.M. (1988). Daily cycles in coastal dunes. *Sedimentary Geology* 55: 43–67.

Hunter, R.E. and Rubin, D.M. (1983). Interpreting cyclic cross-bedding, with an example from the Navajo Sandstone, In: Brookfield, M.E. and Ahlbrandt, T. S., Eds., *Eolian Sediments and Processes*, Developments in Sedimentology, 38, 429–454.

Hunter, R.E., Richmond, B.M., and Alpha, T.R. (1983). Storm-controlled oblique dunes of the Oregon Coast. *Geological Society of America Bulletin* 94: 1450–1465.

Huntington, E. (1907). Some characteristics of the glacial period in non-glaciated regions. *Geological Society of America Bulletin* 18: 351–388.

Huntoon, J.E. and Chan, M.A. (1987). Marine origin of paleotopographic relief on eolian White Rim Sandstone (Permian). *Elaterite Basin, Utah, American Association of Petroleum Geologists Bulletin* 71: 1035–1045.

Hurst, A. and Glennie, K.W. (2008). Mass-wasting of ancient aeolian dunes and sand fluidization during a period of global warming and inferred brief high precipitation: the Hopeman Sandstone (late Permian), Scotland. *Terra Nova* 20: 274–279.

Hwang, I.G. and Heller, P.L. (2002). Anatomy of a transgressive lag: Panther Tongue sandstone. *Star Point formation, central Utah, Sedimentology* 49: 977–999.

IPCC, 2019, Summary for Policymakers, In: Shukla, P.R., Skea, J., Calvo Buendia, E., Masson-Delmotte, V., Pörtner, H.- O., Roberts, D.C., Zhai, P., Slade, R., Connors, S., van Diemen, R., Ferrat, M., Haughey, E., Luz, S., Neogi, S., Pathak, M., Petzold, J., Portugal Pereira, J., Vyas, P., Huntley, E., Kissick, K., Belkacemi, M., Malley, J., Eds., Climate Change and Land: an IPCC special report on climate change, desertification, land degradation, sustainable land management, food security, and greenhouse gas fluxes in terrestrial ecosystems (Figure only).

Jackson, D.T., Costas, S., González-Villanueva, R., and Cooper, A. (2019). A global 'greening' of coastal dunes: An integrated consequence of climate change? *Global and Planetary Change* 182: 103026.

Jagger, A., 2003, Sedimentology and Stratigraphic Evolution of the Permian Cedar Mesa Sandstone, SE Utah, Unpublished PhD thesis, University of Keele, 391 p.

Jerram, D.A., Mountney, N.P. and Stollhofen, H. (1999). Facies architecture of the Etjo Sandstone Formation and its interaction with the basal Etendeka flood basalts of NW Namibia: implications for offshore analogues, In: Cameron, N., Bate, R. and Clure, V., Eds., Oil and Gas Habitats of the South Atlantic, Geological Society of London Special Publication, 153, 367–380.

Jerram, D.A., Mountney, N.P., Howell, J. A. et al. (2000a). Death of a sand sea: An active aeolian erg systematically buried by the Etendeka flood basalts of NW Namibia. *Journal of the Geological Society* 157: 513–516.

Jerram, D.A., Mountney, N., Howell, J., and Stollhofen, H. (2000b). The fossilized desert: recent developments in our understanding of the lower Cretaceous deposits in the Huab Basin, NW Namibia. *Communications of Geological Survey of Namibia* 12: 269–278.

Jimenez, J.A., Maia, L.P., Serra, J., and Morais, J. (1999). Aeolian dune migration along the Ceara coast, north-eastern Brazil. *Sedimentology* 46: 689–701.

Johnson, J., 1999, Nipissing Beach Trail, accessed 7th March 2022, http://individual.utoronto.ca/john_johnston/ipperwash/home.html (Figure only)

Jones, F.H., Scherer, C.M.S., and Kifumbi, C. (2021). Aeolian dunes morphodynamics and wind regime reconstruction in mid-latitudes of the Gondwana during Early Permian, Aracaré Formation, Sergipe-Alagoas Basin, Brazil. *Aeolian Research* 50: 100672.

Jordan, O.D. and Mountney, N.P. (2010). Styles of interaction between aeolian, fluvial and shallow marine environments in the Pennsylvanian-Permian Lower Cutler Beds, southeast Utah, USA. *Sedimentology* 57: 1357–1385.

Jordan, O.D. and Mountney, N.P. (2012). Sequence stratigraphic evolution and cyclicity of an ancient coastal desert system: the Pennsylvanian-Permian lower Cutler beds, Paradox Basin, Utah, USA. *Journal of Sedimentary Research* 82: 755–780.

Kamola, D.L. and Huntoon, J.E. (1994). Changes in rate of transgression across the Permian White Rim Sandstone, southern Utah. *Journal of Sedimentary Research* Vol. 64B: 202–210.

Karpeta, W.P. (1990). The morphology of Permian palaeodunes - a reinterpretation of the Bridgnorth Sandstone around Bridgnorth, England, in the light of modern dune studies. *Sedimentary Geology* 69: 59–75.

Kasse, C. (2002). Sandy aeolian deposits and environments and their relation to climate during the Last Glacial Maximum and Late glacial in northwest and central Europe. *Progress in Physical Geography* 26: 507–532.

Khalaf, F. (1989). Textural characteristics and genesis of the aeolian sediments in the Kuwaiti Desert. *Sedimentology* 36: 253–271.

Khalaf, F.I., Gharib, I.M., and Al Hashash, M.Z. (1984). Types and characteristics of recent surface deposits of Kuwait, Arabian Gulf. *Journal of Arid Environments* 7: 9–33.

Kocurek, G. (1981). Significance of interdune deposits and bounding surfaces in aeolian dune sands. *Sedimentology* 28: 753–780.

Kocurek, G. (1984). Origin of 1st-order bounding surfaces in aeolian sandstones – Reply. *Sedimentology* 31: 125–128.

Kocurek, G. (1988). First-order and super bounding surfaces in eolian sequences - bounding surfaces revisited. *Sedimentary Geology* 56: 193–206.

Kocurek, G. (1991). Interpretation of ancient eolian sand dunes. *Annual Review of Earth and Planetary Sciences* 19: 43–75.

Kocurek, G. (1996). Desert aeolian systems, In: Reading, H.G, Ed., *Sedimentary Environments: Processes, Facies and Stratigraphy*, third edition, Oxford, Blackwell Science, 125–153.

Kocurek, G. (1999). The aeolian rock record (Yes, Virginia, it exists but it really is rather special to create one). In: *Aeolian Environments Sediments and Landforms* (ed. A.S. Goudie, I. Livingstone, and S. Stokes), 239–259. Chichester: John Wiley and Sons Ltd.

Kocurek, G. and Crabaugh, M. (1993). Significance of thin sets of eolian cross-strata – discussion. *Journal of Sedimentary Petrology* 63: 1165–1169.

Kocurek, G. and Day, M. (2018). What is preserved in the aeolian rock record? A Jurassic Entrada Sandstone case study at the Utah-Arizona border. *Sedimentology* 65: 1301–1321.

Kocurek, G. and Dott, R.H. (1981). Distinctions and uses of stratification types in the interpretation of eolian sand. *Journal of Sedimentary Petrology* Vol. 51: 579–595.

Kocurek, G. and Fielder, G. (1982). Adhesion structures. *Journal of Sedimentary Petrology* Vol. 52: 1229–1241.

Kocurek, G. and Havholm, K.G., 1993, Eolian sequence stratigraphy - a conceptual framework, In: Weimer, P. and Posamentier, H.W., Eds., *Siciliclastic Sequence Stratigraphy*, American Association of Petroleum Geologists Memoir, 58, 393–409.

Kocurek, G. and Hunter, R.E. (1986). Origin of polygonal fractures in sand, uppermost Navajo and Page Sandstones. *Page, Arizona, Journal of Sedimentary Petrology* Vol. 56: 895–904.

Kocurek, G. and Nielson, J. (1986). Conditions favourable for the formation of warm-climate aeolian sand sheets. *Sedimentology* 33: 795–816.

Kocurek, G. and Lancaster, N. (1999). Aeolian system sediment state: theory and Mojave Desert Kelso dune field example. *Sedimentology* 46: 505–515.

Kocurek, G., Havholm, K.G., Deynoux, M., and Blakey, R.C. (1991a). Amalgamated accumulations resulting from climatic and eustatic changes, Akchar Erg, Mauritania. *Sedimentology* 38: 751–772.

Kocurek, G., Knight, J. and Havholm, K. (1991b). Outcrop and semi-regional three-dimensional architecture and reconstruction of a portion of the eolian Page Sandstone (Jurassic). In: Miall, A. D. and Tyler, N., Eds., *The Three-Dimensional Facies Architecture of Terrigenous Clastic Sediments and its Implications for Hydrocarbon Discovery and Recovery*, Society of Economic Paleontologists and Mineralogists

Concepts in Sedimentology and Paleontology, 3, 25–43.

Kocurek, G., Lancaster, N., Carr, M., and Frank, A. (1999). Tertiary Tsondab Sandstone Formation: preliminary bedform reconstruction and comparison to modern Namib Sand Sea dunes. *Journal of African Earth Sciences* 29: 629–642.

Kocurek, G., Robinson, N.I., and Sharp, J. M. Jr. (2001). The response of the water table in coastal aeolian systems to changes in sea level. *Sedimentary Geology* 139: 1–13.

Kocurek, G., Townsley, M., Yeh, E. et al. (1992). Dune and dune-field development on Padre-Island, Texas, with implications for interdune deposition and water-table-controlled accumulation. *Journal of Sedimentary Petrology* Vol. 62: 622–635.

Koster, E.A. (1988). Ancient and modern cold-climate aeolian sand deposition: a review. *Journal of Quaternary Science* 3: 69–83.

Krapf, C.B., Stollhofen, H., and Stanistreet, I.G. (2003). Contrasting styles of ephemeral river systems and their interaction with dunes of the Skeleton Coast erg (Namibia). *Quaternary International* 104: 41–52.

Kreslavsly, M.A. and Bondarenko, N.V. (2017). Aeolian sand transport and aeolian deposits on Venus: A review. *Aeolian Research* 26: 29–46.

Krystinik, L. F., 1990, Development geology in eolian reservoirs, In: Fryberger, S.G., Krystinik, L.F., and Schenk, C.J., Eds., *Modern and ancient eolian deposits: Petroleum Exploration and Production*, Rocky Mountain section, Society for Economic Paleontologists and Mineralogists, Denver, Colorado, 1–12.

Kugler, R.L. and Mink, R.M. (1999). Depositional and diagenetic history and petroleum geology of the Jurassic Norphlet Formation of the Alabama coastal waters area and adjacent federal waters area. *Marine Georesources and Geotechnology* 17: 215–232.

Lallier, F., Caumon, G., Borgomano, J. et al. (2012). Relevance of the stochastic stratigraphic well correlation approach for the study of complex carbonate settings: application to the Malampaya buildup (Offshore Palawan, Philippines), In: Garland, J., Neilsen, J.E., Laubach, S.E., and Whidden, K.J., Eds. *Advances in Carbonate Exploration and Reservoir Analysis, Geological Society of London Special Publication* 370: 265–275.

Lancaster, N. (1985). Winds and sand movements in the Namib sand sea. *Earth Surface Processes and Landforms* 10: 607–619.

Lancaster, N. (1986). Grain-size characteristics of linear dunes in the southwest Kalahari. *Journal of Sedimentary Petrology* Vol. 56: 395–400.

Lancaster, N. (1988). Controls on eolian dune size and spacing. *Geology* 16: 972–975.

Lancaster, N. (1989). *The Namib Sand Sea: Dune forms, processes, and sediments*, 200. Rotterdam.

Lancaster, N. (1995). *Geomorphology of Desert Dunes*, 290. New York: Routledge.

Lancaster, N., Alsharhan, A.S., Glennie, K. et al. (1998). Dune morphology, chronology and Quaternary climate change. In: *Quaternary Deserts and Climate Change, Proceedings of the International Conference on Quaternary Deserts and Climate Change* (ed. A.S. Alsharhan, K. Glennie, G.L. Whittle, and C.G.S.C. Kendall), 339–349. Rotterdam: Balkema.

Lancaster, N. and Baas, A. (1998). Influence of vegetation cover on sand transport by wind: field studies at Owens Lake, California. *Earth Surface Processes and Landforms*, The Journal of the British Geomorphological Group, 23, 69–82.

Lancaster, N. and Mountney, N.P. (2021). Aeolian Processes and Landforms. In: Alderton, D. and Elias, S. (eds), *Encyclopedia of Geology, Second Edition*, Elsevier, 809–829. https://doi.org/10.1016/B978-0-08-102908-4.00031

Lancaster, N. and Teller, J.T. (1988). Interdune deposits of the Namib Sand Sea. *Sedimentary Geology* 55: 91–107.

Lancaster, N., Greeley, R., and Christensen, P.R. (1987). Dunes of the Gran Desierto sand sea, Sonora, Mexico. *Earth Surface Processes and Landforms* 12: 277–288.

Langford, R.P. (1989). Fluvial-aeolian interactions: part 1, modern systems. *Sedimentology* 36: 1023–1035.

Langford, R. and Chan, M.A. (1988). Flood surfaces and deflation surfaces within the Cutler Formation and Cedar Mesa Sandstone (Permian), southeastern Utah. *Geological Society of America Bulletin* 100: 1541–1549.

Langford, R.P. and Chan, M.A. (1989). Fluvial-aeolian interactions: part 2, ancient systems. *Sedimentology* 36: 1037–1051.

Langford, R.P. and Chan, M.A. (1993). Downwind changes within an ancient dune sea, Permian Cedar Mesa Sandstone, southeast Utah, In: Pye, K. and Lancaster, N., Eds., *Aeolian sediments: ancient and modern*, International Association of Sedimentologists Special Publication, 16, 109–126.

Langford, R.P., Pearson, K.M., Duncan, K. A. et al. (2008). Eolian topography as a control on deposition incorporating lessons from modern dune seas: Permian Cedar Mesa Sandstone, SE Utah, U.S.A. *Journal of Sedimentary Research* 78: 410–422.

Lapotre, M.G.A., Ewing, R.C., Lamb, M.P. et al. (2016). Large wind ripples on Mars: A record of atmospheric evolution. *Science* 353: 55–58.

Levin, N., Tsoar, H., Herrmann, H.J. et al. (2009). Modelling the formation of residual dune ridges behind barchan dunes in North-east Brazil. *Sedimentology* 56: 1623–1641.

Li, Q., Wu, H., Yu, Y. et al. (2014). Reconstructed moisture evolution of the deserts in northern China since the Last Glacial Maximum and its implications for the East Asian Summer Monsoon. *Global and Planetary Change* 121: 101–112.

Li, G., Wu, C., Rodríguez-López, J.P. et al. (2018). Mid-Cretaceous aeolian desert systems in the Yunlong area of the Lanping Basin, China: Implications for palaeoatmosphere dynamics and paleoclimatic change in East Asia. *Sedimentary Geology* 364: 121–140.

Li, X-A., Wang, L., Yan, Y., Hong, B. and Li, L-C. (2019). Experimental study on the disintegration of loess in the Loess Plateau of China, *Bulletin of Engineering Geology and the Environment*, 78, 4907–4918. (Figure only)

Lindquist, S.J. (1983). Nugget Formation reservoir characteristics affecting production in the overthrust belt of southwestern Wyoming. *Journal of Petroleum Technology* 35: 1355–1365.

Liu, B. and Coulthard, T.J. (2015). Mapping the interactions between rivers and sand dunes: implications for fluvial and aeolian geomorphology. *Geomorphology* 231: 246–257.

Lomax, J., Hilgers, A., and Radtke, U. (2011). Palaeoenvironmental change recorded in the palaeodunefields of the western Murray Basin, South Australia: new data from single grain OSL-dating. *Quaternary Science Reviews* 30: 723–736.

Loope, D.B. (1981). Products and Processes of Ancient Arid Coastline: Lower Cutler Group (Permian). *Southeastern Utah, American Association of Petroleum Geologists Bulletin* 65: 950–951.

Loope, D.B. (1984). Eolian Origin of Upper Paleozoic Sandstones, Southeastern Utah. *Journal of Sedimentary Petrology* Vol. 54: 563–580.

Loope, D.B. (1985). Episodic deposition and preservation of eolian sands – a late Paleozoic example from southeastern Utah. *Geology* 13: 73–76.

Loope, D.B. (1988). Rhizoliths and ancient aeolianites. *Sedimentary Geology* 56: 301–314.

Loope, D.B., Eisenberg, L. and Waiss, E. (2004). Navajo sand sea of near-equatorial Pangea: tropical westerlies, slumps, and giant stromatolites, In: Nelson, P. and Erslev, E.A., Eds., *Field trips in the Southern Rocky Mountains, USA*, Geological Society of America, Field Guide, 5, 1–13.

Loope, D.B., Rowe, C.M., and Joeckel, R. M. (2001). Annual monsoon rains recorded by Jurassic dunes. *Nature* 412: 64–66.

Loope, D.B., Swinehart, J.B., and Mason, J. P. (1995). Dune-dammed paleovalleys of the Nebraska Sand Hills – intrinsic versus climatic controls on the accumulation of lake and marsh sediments. *Geological Society of America Bulletin* 107: 396–406.

Lucas, A., Rodriguez, S., Lemonnier, F. et al. (2019). Texture and composition of Titan's equatorial sand seas inferred from Cassini SAR data: Implications for aeolian transport and dune morphodynamics. *Journal of Geophysical Research: Planets* 124: 3140–3163.

Luo, W., Zhao, W., Liu, B., and Zhou, H. (2020). Climate change and agricultural trade in central Asia: evidence from Kazakhstan. *Ecosystem Health and Sustainability* 6: 1844550.

Madsen, A.T., Murray, A.S., and Andersen, T.J. (2007). Optical dating of dune ridges on Rømøs, a barrier island in the Wadden Sea. *Denmark, Journal of Coastal Research* 23: 1259–1269.

Maidwell, F.T. (1914). Some sections in the Lower Keuper of Runcorn Hill. *Proceedings of Liverpool Geological Society* 12: 40–52.

Mainguet, M. and Chemin, M.-C. (1983). Sand seas of the Sahara and Sahel: An explanation of their thickness and sand dune type by the sand budget principle, In: Brookfield, M.E. and Ahlbrandt, T.S., Eds. *Eolian Sediments and Processes, Developments in Sedimentology* 38: 353–363.

Marrero-Rodríguez, N., García-Romero., L., Sánchez-García, M.J., Hernández-Calvento, L. and Pérez-Chacón Espino, E., (2020). An historical ecological assessment of land-use evolution and observed landscape change in an arid aeolian sedimentary system, *Science of the Total Environment*, 716, 137087

Maxwell, T.A. and Haynes, C.V. (1989). Large-scale, low-amplitude bedforms (chevrons) in the Selima Sand Sheet. *Egypt, Science* 243: 1179–1182.

Maynard, J.R. and Leeder, M.R. (1992). On the periodicity and magnitude of Late Carboniferous glacio-eustatie sea-level changes. *Journal of the Geological Society* 149: 303–311.

McCaleb, J. A. (1979). The role of the geologist in exploitation: Vail Colorado, American Association of Petroleum Geologists Reservoir Fundamentals School.

McFadden, L.D., Wells, S.G., and Jercinovich, M.J. (1987). Influences of eolian and pedogenic processes on the origin and evolution of desert pavements. *Geology* 15: 504–508.

McKee, E.D. (1966). Structures of dunes at White Sands National Monument, New Mexico (and a comparison with structures of dunes from other selected areas). *Sedimentology* 7: 3–69.

McKee, E.D., 1979, A Study of Global Sand Seas, Geological Survey Professional Paper, 1052, Washington, United States Government Printing Office, 421 p.

McKee, E.D. and Moiola, R.J. (1975). Geometry and growth of the White Sands dune field, New Mexico. *Journal of Research of the United States Geological Survey* 3: 59–66.

McKee, E.D. and Tibbitts, J. (1964). Primary structures of a seif dune and associated deposits in Libya. *Journal of Sedimentary Research* Vol. 34: 5–17.

Mckee, E.D., Douglass, J.R., and Rittenhouse, S. (1971). Deformation of lee side laminae in eolian dunes. *Geological Society of America Bulletin* 82: 359–378.

McKenna Neuman, C. and Gilbert, R. (1986). Aeolian processes and landforms in glaciofluvial environments of southeastern Baffin Island, NWT, Canada. In: *Aeolian Geomorphology: Binghamton Geomorphology Symposium 17* (ed. W. G. Nickling). Routledge.

McKie, T., Jolley, S.J., and Kristensen, M. B. (2010). Stratigraphic and structural compartmentalization of dryland fluvial reservoirs: Triassic Heron Cluster. *Central North Sea, Geological Society, London, Special Publications* 347: 165–198.

Meadows, N.S. (2006). The correlation and sequence architecture of the Ormskirk Sandstone Formation in the Triassic Sherwood Sandstone Group of the East Irish Sea Basin. *NW England, Geological Journal* 41: 93–122.

Medici, G., West, L.J., and Mountney, N.P. (2016). Characterizing flow pathways in a sandstone aquifer: tectonic vs sedimentary heterogeneities. *Journal of Contaminant Hydrology* 194: 36–58.

Medici, G., West, L.J., and Mountney, N.P. (2018). Characterization of a fluvial aquifer at a range of depths and scales: the Triassic St Bees Sandstone Formation, Cumbria, UK. *Hydrogeology Journal* 26: 565–591.

Medici, G., West, L.J., Mountney, N.P., and Welch, M. (2019a). Review: Permeability of rock discontinuities and faults in the Triassic Sherwood Sandstone Group (UK): insights for management of fluvio-aeolian aquifers worldwide. *Hydrogeology Journal* 27: 2835–2855.

Medici, G., West, L.J., and Mountney, N.P. (2019b). Flow heterogeneities and permeability development in the Triassic UK Sherwood Sandstone Group; insights for reservoir quality. *Geological Journal* 54: 1361–1378.

Melchor, R.N. and Umazano, A.M. (2013). Continental trace fossils from the Miocene-Pliocene Río Negro Formation (Río Negro and Buenos Aires provinces), Argentina, In, Melchor, U., *SLIC 2013 - Post-symposiun Fieldtrip Guidebook, Chapter: Continental trace fossils from the Miocene-Pliocene Río Negro Formation (Río Negro and Buenos Aires provinces)*, Instituto de Ciencias de la Tierra y Ambientales de La Pampa, pp. 35–48. (Figure only)

Melvin, J., Sprague, R.A., and Heine, C.J. (2010). From bergs to ergs: The late Paleozoic Gondwanan glaciation and its aftermath in Saudi Arabia. In: *Late Paleozoic Glacial Events and Postglacial Transgressions in Gondwana*, vol. 468 (ed. O.R. López-Gamundí and L.A. Buatois), 37–80. Boulder: Special Paper of the Geological Society of America.

Mescolotti, P.C., Varejão, F.G., Warren, L. V., Ladeira, F.S.G., Giannini, P.C.F. and Assine, M.L., (2019). The sedimentary record of wet and dry eolian systems in the Cretaceous of Southeast Brazil: stratigraphic and paleogeographic significance, *Brazilian Journal of Geology*, 49, 1–20. (Figure only)

Mesquita, A.F., Basilici, G., Soares, M.V.T. et al. (2021). Hybrid dry-wet interdune deposition in Precambrian aeolian systems: Galho do Miguel Formation. *Southeast Brazil, Sedimentary Geology* 425 (106007): 106007–106021.

Michel, R.D.L., Scherer, C.M.S., dos Reis, A.D., Kifumbi, C., Cechetti, R.P., and Jones, F.H. (2025). Latest Triassic Active Aeolian Dune Field Preserved by CAMP-Related Lava Flows. *Terra Nova* 12754.

Middleton, L.T. and Blakey, R.C. (1983). Processes and controls on the intertonguing of the Kayenta and Navajo formations, northern Arizona: Eolian-fluvial interactions, In: Brookfield, M.E. and Ahlbrandt, T.S., Eds., *Eolian Sediments and Processes*, Developments in Sedimentology, 38, 613–634.

Milana, J.P. (2009). Largest wind ripples on Earth? *Geology* 37: 343–346.

Moghaddam, M.H.R., Sedighi, A., Fasihi, S. and Firozjaei,K., 2018, Effect of environmental policies in combating aeolian desertification over Sejzy Plain of Iran, *Aeolian Research*, 35, 19–28.

Mountney, N.P. (2004). The sedimentary signature of desert systems and their response to environmental change. *Geology Today* 20: 101–106.

Mountney, N.P. (2006a). Periodic accumulation and destruction of aeolian erg sequences in the Permian Cedar Mesa Sandstone, White Canyon, southern Utah. *Sedimentology* 53: 789–823.

Mountney, N.P. (2006b). Eolian Facies Models, In: *Facies Models Revisited, Society for Economic Paleontologists and Mineralogists, Special Publication* (ed. H.W. Posamentier and R.G. Walker), 84: 19–83.

Mountney, N.P. (2012). A stratigraphic model to account for complexity in aeolian dune and interdune successions. *Sedimentology* 59: 964–989.

Mountney, N. and Howell, J. (2000). Aeolian architecture, bedform climbing and preservation space in the Cretaceous Etjo Formation, NW Namibia. *Sedimentology* 47: 825–849.

Mountney, N.P. and Jagger, A. (2004). Stratigraphic evolution of a wet aeolian system: the Permian Cedar Mesa Sandstone, SE Utah. *Sedimentology* 51: 713–743.

Mountney, N.P. and Russell, A.J. (2004). Sedimentology of aeolian sandsheet deposits in the Askja region of northeast Iceland. *Sedimentary Geology* 166: 223–244.

Mountney, N.P. and Russell, A.J. (2006). Coastal aeolian dune-field development and response to periodic fluvial inundation, Sólheimasandur, southern Iceland. *Sedimentary Geology* 192: 167–181.

Mountney, N.P. and Russell, A.J. (2009). Aeolian dune field development in a water table-controlled system: Skeiðarársandur, southern Iceland. *Sedimentology* 56: 2107–2131.

Mountney, N.P. and Thompson, D.B. (2002). Stratigraphic evolution and preservation of aeolian dune and damp/wet interdune strata: an example from the Triassic Helsby Sandstone Formation, Cheshire Basin, UK. *Sedimentology* 49: 805–834.

Mountney, N.P., Howell, J.A., Flint, S.S., and Jerram, D.A. (1998). Aeolian and alluvial deposition within the Mesozoic Etjo Sandstone Formation, north west Namibia. *Journal of African Earth Sciences* 27: 175–192.

Mountney, N.P., Howell, J.A., Flint, S.S., and Jerram, D.A. (1999a). Climate, sediment supply and tectonics as controls on the deposition and preservation of the aeolian-fluvial Etjo Sandstone Formation. *Namibia, Journal of the Geological Society of London* 156: 771–777.

Mountney, N.P., Howell, J.A., Flint, S.S., and Jerram, D.A. (1999b). Relating eolian bounding-surface geometries to the bed forms that generated them: Etjo Formation, Cretaceous, Namibia. *Geology* 27: 159–162.

Muhs, D.R. and Bettis III, E.A. (2003). Quaternary loess-paleosol sequences as examples of climate-driven sedimentary extremes. In: *Extreme Depositional Environment: Mega End-Members in Geological Time*, vol. 370 (ed. M.A. Chan and A.W. Archer), 53–74. Special paper of the Geological Society of America.

Muhs, D.R. and Budahn, J.R. (2019). New geochemical evidence for the origin of North America's largest dune field, the Nebraska Sand Hills, central Great Plains, USA. *Geomorphology* 332: 188–212. (Figure only).

Nagtegaal, P.J.C. (1979). Relationship of facies and reservoir quality in Rotliegends desert sandstones, southern North Sea region. *Journal of Petroleum Geology* 2: 145–158.

Nanson, G., Tooth, S., and Knighton, A.D. (2002). A global perspective on dryland rivers: perceptions, misconceptions and distinctions. *Dryland Rivers: Hydrology and Geomorphology of Semi-arid Channels* 22: 17–54.

Newell, J.A. (2018). Rifts, rivers and climate recovery: a new model for the Triassic of England, *Proceedings of the Geologists Association*, 129, 352–371. (Figure only)

Nichols, G.J. (2005). Sedimentary evolution of the lower Clair Group, Devonian, west of Shetland; climate and sediment supply controls on fluvial, aeolian and lacustrine deposition, In: Doré, A.G., and Vining, B.A., Eds., Petroleum Geology of Northwest Europe, Proceedings of the Conference, published by the Geological Society of London, 6, 95–967.

Nickling, W.G., McKenna Neuman, C., and Lancaster, N. (2002). Grainfall processes in the lee of transverse dunes. *Silver Peak, Nevada, Sedimentology* 49: 191–209.

Nielson, J. and Kocurek, G. (1986). Climbing zibars of the Algodones. *Sedimentary Geology* 48: 1–15.

Noffke, N., 2021, Microbially induced sedimentary structures in clastic deposits: implication for the prospection for fossil life on Mars, *Astrobiology*, 866–892. doi:https://doi.org/10.1089/ast.2021.0011 (Figure only), 21

Noffke, N., and Awramik, S.M., 2013, Stromatolites and MISS—differences between relatives. *GSA Today*, 4–9. DOI: https://doi.org/10.1130/GSATG187A.1 (Figure only), 23

North, C.P. and Prosser, D.J. (1993). Characterization of fluvial and aeolian reservoirs: problems and approaches. *Geological Society of London, Special Publication* 73: 1–6.

Olsen, H., Due, P.H., and Clemmensen, L. B. (1989). Morphology and genesis of asymmetric adhesion warts - a new adhesion surface structure. *Sedimentary Geology* 61: 277–285.

Øxnevad, I.E.I. (1991). *Aeolian and mixed aeolian-subaqueous sedimentation in modern and ancient sub-tropical desert basins: examples from the Sahara and Permo-Triassic of NW Europe*, Unpublished Doctor of Science thesis, 161 p. University of Bergen, Norway.

Parsons, A.J. and Abrahams, A.D. (2009). Geomorphology of desert environments. In: *Geomorphology of desert environments*, 3–7. Dordrecht: Springer.

Paz, M., Ponce, J.J., Mángano, M.G. et al. (2021). The Vaca Muerta transgression (Upper Jurassic), Neuquén Basin, Argentina: Insights into the evolution and timing of aeolian–marine transitions. *Sedimentology* 68: 2732–2764.

Pedersen, K. and Clemmensen, L.B. (2005). Unveiling past aeolian landscapes: a ground-penetrating radar survey of a Holocene coastal dunefield system. *Thy, Denmark, Sedimentary Geology* 177: 57–86.

Pickup, G.E. and Hern, C.Y. (2002). The development of appropriate upscaling procedures. *Transport in Porous Media* 46: 119–138.

Pickup, G., Marks, A., Bourke, M. (2002). Paleoflood reconstruction on floodplains using geophysical survey data and hydraulic modelling, Water Science and Application 5, 47–60.

Pierce, C.S., Howell, J., Rieke, H. and Healy, D. (2016). Geomodeling of Aeolian Reservoir Analogues-A Virtual-outcrop Based Study of the Jurassic Page Sandstone Formation, In: Second Conference on Forward Modelling of Sedimentary Systems, European Association of Geoscientists & Engineers, April, pp. 483.

Porter, M.L. (1986). Sedimentary record of erg migration. *Geology* 14: 497–500.

Posamentier, H.W., Jervey, M.T., and Vail, P.R. (1988). Eustatic controls on clastic deposition I—conceptual framework. In: .

Powell, D.M., Parsons, A.J., and Abrahams, A.D. (2009). Dryland rivers: processes and forms. In: *Geomorphology of desert environments* (ed. A.J. Parsons and A.D. Abrahams), 333–373. Dordrecht: Springer.

Prosser, D.J. and Maskall, R. (1993). Permeability variation within eolian sandstone: a case study using core cut sub-parallel to slipface bedding, the Auk field, central North Sea, UK. *Characterization of fluvial and eolian reservoirs, Geological Society Special Publication* 73: 377–397.

Pulvertaft, T.C.R. (1985). Aeolian dune and wet interdune sedimentation in the middle Proterozoic Dala-sandstone, Sweden. *Sedimentary Geology* 44: 93–111.

Pursar, B.H. and Evans, G. (1973). Regional sedimentation along the Trucial Coast, Persian Gulf. In: *The Persian Gulf* (ed. B.H. Pursar), 211–231. Berlin: Springer.

Purvis, K. (1991). Stoss-side mud-drapes: deposits of interdune pond margins. *Sedimentology* 38: 153–156.

Pye, K. (1982). Negatively skewed aeolian sands from a humid tropical coastal dunefield, northern Australia. *Sedimentary Geology* 31: 249–266.

Pye, K. (1983). Coastal dunes. *Progress in Physical Geography* 7: 531–557.

Pye, K. (1995). The nature, origin and accumulation of loess. *Quaternary Science Reviews* 14: 653–667.

Pye, K and Sperling, C.H.B. (1983). Experimental investigation of silt formation by static breakage processes: the effect of temperature, moisture and salt on quartz dune sand and granitic regolith. *Sedimentology* 30, 49–62.

Pye, K. and Tsoar, H.. (1987). The mechanics and geological implications of dust transport and deposition in deserts, with particular reference to loess formation and dune sand diagenesis in the northern Negev, Israel, In: Frostick, L.E. and Reid, I., Eds., *Desert Sediments Ancient and Modern, Geological Society of London Special Publication*, 35, 139–156.

Pye, K. and Tsoar, H. (1990). *Aeolian sand and sand dunes*, 396. London: Unwin Hyman Limited.

Qian, G., Yang, Z., Luo, W. et al. (2020). Morphological and sedimentary characteristics of dome dunes in northeastern Qaidam Basin, China. *Geomorphology* 350: 106923. (Figure only).

Radebaugh, J., Lorenz, R.D., Lunine, J.I., Wall, S.D., Boubin, G., Reffet, E., Kirk,

R.L., Lopes, R.M., Stofan, E.R., Soderblom, L., Allison, M., Janssen, M., Paillou, P., Callahan, P., Spencer, C. and the Cassini Radar Team (2008). Dunes on Titan observed by Cassini Radar. *Icarus* 194: 690–703.

Rankey, E.C. (1997). Relations between relative changes in sea level and climate shifts: Pennsylvanian-Permian mixed carbonate-siliciclastic strata, western United States. *Geological Society of America Bulletin* 109: 1089–1100.

Ray, N. and Adams, J.M. (2001). A GIS-based vegetation map of the world at the last glacial maximum (25,000-15,000 BP). *Internet Archaeology* 11: https://doi.org/10.11141/ia.11.2.

Reading, B.G. (1978). *Sedimentary environments and facies*, Blackwell Scientific Publications, Oxford, London, Edinburgh, Victoria.

Reid, I. and Frostick, L.E. (2011). Channel form, flows and sediments of endogenous ephemeral rivers in deserts. In: *Arid zone geomorphology: Process, form and change in drylands*, 301–332.

Reimann, T., Tsukamoto, S., Harff, J. et al. (2011). Reconstruction of Holocene coastal foredune progradation using luminescence dating—An example from the Świna barrier (southern Baltic Sea, NW Poland). *Geomorphology* 132: 1–16.

Rodríguez-López, J.P. (2008). Sedimentología y evolución del sistema desértico arenoso (erg) desarrollado en el margen occidental del Tethys durante el Cretácico Medio. Cordillera Ibérica., Provincias de Teruel y Zaragoza, PhD Thesis, Universidad Complutense de Madrid_Consejo Superior de Invesrigaciones Científicas (CSIC), p. 500.

Rodríguez-López, J.P., Meléndez, N., de Boer, P.L., and Soria, A.R. (2010). The action of wind and water in a mid-Cretaceous subtropical erg-margin system close to the Variscan Iberian Massif, Spain. *Sedimentology* 57: 1315–1356 (Figure only).

Rodríguez-López, J.P., Liesa, C.L., Van Dam, J. et al. (2012a). Aeolian construction and alluvial dismantling of a fault-bounded intracontinental aeolian dune field (Teruel Basin, Spain); a continental perspective on Late Pliocene climate change and variability. *Sedimentology* 59: 1536–1567.

Rodríguez-López, J.P., Meléndez, N., de Boer, P.L., and Soria, A.R. (2012b). Controls on marine-erg margin cycle variability: aeolian-marine interaction in the Mid-Cretaceous Iberian Desert System. *Spain, Sedimentology* 59: 466–501.

Rodríguez-López, J.P., Meléndez, N., de Boer, P.L. et al. (2013). Spatial variability of multi-controlled aeolian supersurfaces in central-erg and marine-erg-margin systems. *Aeolian Research* 11: 141–154. (Figure only).

Rodríguez-López, J.P., Clemmensen, L., Lancaster, N., Mountney, N.P., and Veiga, G. (2014). Archean to Recent aeolian sand systems and their preserved successions: current understanding and future prospects. *Sedimentology* 61: 1487–1534.

Romain, H. and Mountney, N.P. (2014). Reconstruction of three-dimensional eolian dune architecture from one-dimensional core data through adoption of analog data from outcrop. *American Association of Petroleum Geologists Bulletin* 98: 1–22.

Ross, R. (2017). Climate Change through Earth History, In: Zabel, I.H.H., Duggan-Haas, D. and Ross, R., *The Teacher-Friendly Guide to Climate Change*, Paleontological Research Institution, Special Publication, 53, 53–90. (Figure only)

Rubin, D.M. (1987a). Cross-bedding, bedforms and palaeocurrents. Society of Economic Paleontologists and Mineralogists, Concepts in Sedimentology and Paleontology, 1, 187.

Rubin, D.M. (1987b). Formation of scalloped cross-bedding without unsteady flows. *Journal of Sedimentary Petrology* Vol. 57: 39–45.

Rubin, D.M. and Carter, C.L. (2006). Bedforms and Cross-Bedding in Animation, SEPM, Atlas 2, DVD.

Rubin, D.M. and Hunter, R.E. (1982). Bedform climbing in theory and nature. *Sedimentology* 29: 121–138.

Rubin, D.M and Hunter, R.E. (1983). Reconstructing bedform assemblages from compound crossbedding, In: Brookfield, M.E. and Ahlbrandt, T.S., Eds., *Eolian Sediments and Processes, Developments in Sedimentology*, 38, 407–427, Elsevier.

Rubin, D.M. and Hunter, R.E. (1984). Origin of 1st-order bounding surfaces in aeolian sandstones – Reply. *Sedimentology* 31: 128–132.

Rubin, D.M. and Hunter, R.E. (1985). Why deposits of longitudinal dunes are rarely recognized in the geologic record. *Sedimentology* 32: 147–157.

Rubin, D.M., Tsoar, H., and Blumberg, D. G. (2008). A second look at western Sinai seif dunes and their lateral migration. *Geomorphology* 93: 335–342. (Figure only).

Santos, M.G.M., Hartley, A.J., Mountney, N.P., Peakall, J., Owen, A., Merino, E. R., and Assine, M.L. (2019). Meandering rivers in modern desert basins: Implications for channel planform controls and prevegetation rivers. *Sedimentary Geology* 385: 1–14. doi: 10.1016/j.sedgeo.2019.03.011

Schenk, C.J., Fryberger, S.G., Krystinik, L. F., and Schenk, C.J. (1990). Eolian deposits of North Padre Island, Texas. In: *Modern and Ancient Aeolian Deposits: Petroleum Exploration and Production*, vol. 10 (ed. S.G. Fryberger, L.F. Krystinik, and C.J. Schenk), 1–7. Denver: Society of Economic Paleontologists and Mineralogists.

Schenk, C.J. and Fryberger, S.G. (1988). Early diagenesis of eolian dune and interdune sands at White Sands, New Mexico. *Sedimentary Geology* 55: 109–120.

Scherer, C.M.S. and Lavina, E.L.C. (2005). Sedimentary cycles and facies architecture of aeolian-fluvial strata of the Upper Jurassic Guará Formation, southern Brazil. *Sedimentology* 52: 1323–1341.

Scherer, C.M.S. and Lavina, E.L.C. (2006). Stratigraphic evolution of a fluvial-eolian succession: the example of the Upper Jurassic-Lower Cretaceous Guará and Botucatu formations, Paraná Basin, southernmost Brazil. *Gondwana Research* 9: 475–484.

Scherer, C.M.S. (2000). Eolian dunes of the Botucatu Formation (Cretaceous) in southernmost Brazil: morphology and origin. *Sedimentary Geology* 137: 63–84.

Scherer, C.M.S. (2002). Preservation of aeolian genetic units by lava flows in the Lower Cretaceous of the Paraná Basin, southern Brazil. *Sedimentology* 49: 97–116.

Scherer, C.M.S., Lavina, E.L.C., Dias Filho, D.C. et al. (2007). Stratigraphy and facies architecture of the fluvial-aeolian-lacustrine Sergi Formation (Upper Jurassic), Recôncavo Basin, Brazil. *Sedimentary Geology* 194: 169–193.

Schuster, M., Duringer, P., Ghienne, J-F., Roquin, C., Sepulchre, P., Moussa, A., Lebatard, A-E., Mackaye, H.T., Likius, A., Vignaud, P. and Brunet, M., 2009, Chad Basin, *Paleoenvironments of the Sahara since the Late Miocene*, 341, 603–611

Selley, R.C. (1985). *Elements of Petroleum Geology*. New York: W.H, Freeman and Company.

Shanley, K.W. and McCabe, P.J. (1994). Perspectives on the sequence stratigraphy of continental strata. *American Association of Petroleum Geologists Bulletin* 78: 544–568.

Sharp, R.P. (1963). Wind ripples. *Journal of Geology* 71: 617–636.

Sharp, R.P. (1966). Kelso dunes, Mojave Desert, California. *Geological Society of America Bulletin* 77: 1045–1074.

Shebi, M.A. (1995). The impact of reservoir heterogeneity on fluid flow in the Tensleep Sandstone of the Bighorn Basin: Resources of southwestern Wyoming, field conference guidebook. *Wyoming Geological Association* 343–359.

Shillito, A.P. and Davies, N.S., (2021). The Silurian inception of inland desert ecosytems: trace fossil evidence from the Mereenie Sandstone, Northern Territory, Australia, *Journal of the Geological Society*, 178, jgs2020-243. (Figure only)

Silva, B.A. and Fernandez, G.C.A. (2004). The Piramboia and Botucatu formations (Triassic-Jurassic) in the subsurface of eastern Entre Rios Province. *Revista de la Asociación Geológica Argentina* 59: 141–151.

Simplicio, F. and Basilici, G. (2015). Unusual thick eolian sand sheet sedimentary succession: Paleoproterozoic Bandeirinha formation, Minas Gerais. *Brazilian Journal of Geology* 45: 3–11.

Simpson, E.L. and Eriksson, K.A. (1993). Thin eolianites interbedded within a fluvial and marine succession - Early Proterozoic Whitworth Formation, Mount Isa inlier, Australia. *Sedimentary Geology* 87: 39–62.

Simpson, E.L. and Loope, D.B. (1985). Amalgamated interdune deposits, White-Sands, New-Mexico. *Journal of Sedimentary Petrology* Vol. 55: 361–365.

Smith, D. B. and Taylor, J. C. M., 1992, Permian, In: *Atlas of Palaeogeography and Lithofacies*, The Geological Society, London, Memoir (ed. J.C.W. Cope, J. Ingham, P.F. Rawson), 13, 87–96. (Figure only).

Smoot, J.P., Castens-Seidell, B., (1994). Sedimentary features produced by efflorescent crusts, Saline Valley and Death Valley, California. In: *Sedimentology and Geochemistry of Modern Ancient Saline Lakes* (ed. R. Renaut and W.M. Last). Special Publication of the Society of Economic Paleontologists and Mineralogisits, Tulsa, OK, USA, 50, 73–90.

Soreghan, G.S. and Dickinson, W.R. (1994). Generic types of stratigraphic cycles controlled by eustacy. *Geology* 22: 759–761.

Soreghan, G.S., Soreghan, M.J., and Hamilton, M.A. (2008). Origin and significance of loess in late Paleozoic western Pangaea: A record of tropical cold? *Palaeogeography Palaeoclimatology Palaeoecology* 268: 234–259.

Spalletti, L.A. and Veiga, G.D. (2007). Variability of continental depositional systems during lowstand sedimentation: an example from the Kimmeridgian of the Neuquén Basin. *Argentina, Latin American Journal of Sedimentology and Basin Analysis* 14: 85–104.

Spalletti, L.A., Limarino, C.O., and Colombo Piñol, F. (2010). Internal anatomy of an erg sequence from the aeolian - fluvial system of the De La Cuesta Formation (Paganzo Basin, northwestern Argentina). *Geologica Acta* 8: 431–447.

Stanistreet, I.G. and Stollhofen, H. (2002). Hoanib River flood deposits of Namib Desert interdunes as analogues for thin permeability barrier mudstone layers in aeolianite reservoirs. *Sedimentology* 49: 719–736.

Stollhofen, H., Bachmann, G.H., Barnasch, J., Bayer, U., Beutler, G., Matthias, F., Kästner, M., Legler, B., Radies, D., 2008, The evolution of a Playa to a Sabkha system (Permian, North German Basin), In: *Littke, R., Bayer, U., Gajewski, D., Nelskamp, S., Dynamics of complex intracontinental basins. The Central European Basin System*, Springer, Verlag, p. 181–210. (Figure only)

Stokes, S., Thomas, S.G., and Washington, R. (1997). Multiple episodes of aridity in southern Africa since the last interglacial period. *Nature* 388: 154–158.

Stokes, S. and Swinehart, J.B. (1997). Middle-and late-Holocene dune reactivation in the Nebraska Sand Hills. *USA, The Holocene* 7: 263–272.

Stokes, W.L. (1968). Multiple parallel-truncation bedding planes - a feature of wind deposited sandstone formations. *Journal of Sedimentary Petrology* Vol. 38: 510–515.

Story, C. (1998). Norphlet geology and 3-D geophysics: Fairway Field, Mobile Bay, Alabama. *Leading Edge* 17: 243.

Strömbäck & Howell (2002). Predicting distribution of remobilized aeolian facies using sub-surface data: The Weissliegend of the UK Southern North Sea. *Petroleum Geoscience* 8: 237–249.

Strömbäck, A., Howell, J.A., and Veiga, G. D. (2005). The transgression of an erg - sedimentation and reworking/soft-sediment deformation of aeolian facies: the Cretaceous Troncoso Member, Neuquen Basin, Argentina. *Geological Society of London Special Publication* 252: 163–183.

Sullivan, R., Kok, J.F., Katra, I., and Yizhaq, H. (2020). A broad continuum of Aeolian impact ripple morphologies on Mars is enabled by low wind dynamic pressures. *Journal of Geophysical Research Planets* 125: e2020JE006485.

Svendsen, J., Stollhofen, H., Krapf, C.B., and Stanistreet, I.G. (2003). Mass and hyperconcentrated flow deposits record dune damming and catastrophic breakthrough of ephemeral rivers, Skeleton Coast Erg, Namibia. *Sedimentary Geology* 160: 7–31.

Sweet, M.L. (1992). Lee-face air-flow, surface processes, and stratification types - their significance for refining the use of eolian cross-strata as paleocurrent indicators. *Geological Society of America Bulletin* 104: 1528–1538.

Sweet, M.L. and Kocurek, G. (1990). An empirical-model of aeolian dune lee-face air-flow. *Sedimentology* 37: 1023–1038.

Swezey, C. (2001). Eolian sediment responses to late Quaternary climate changes: temporal and spatial patterns in the Sahara. *Palaeogeography Palaeoclimatology Palaeoecology* 167: 119–155.

Swezey, C., Lancaster, N., Kocuerk, G. et al. (1999). Response of aeolian systems to Holocene climatic and hydrologic changes on the northern margin of the Sahara: a high-resolution record from the Chott Rharsa basin. *Tunisia, The Holocene* 9: 141–147.

Taggart, S., Hampson, G.J., and Jackson, M.D. (2010). High-resolution stratigraphic architecture and lithological heterogeneity within marginal aeolian reservoir analogues. *Sedimentology* 57: 1246–1279.

Talbot, M.R. (1985). Major bounding surfaces in aeolian sandstones - a climatic model. *Sedimentology* 32: 257–265.

Tang, L., Jones, S., and Gluyas, J. (2017). Facies Architecture of the Fluvial-Aeolian Buchan Formation (Upper Devonian) and Its Implications on Field Exploration: A Case Study from Ardmore Field, Central North Sea, UK. *International Journal of Geosciences* 08: 902–924 (Figure only).

Terrell, F.M. (1972). Lateral facies and paleoecology of Permian Elephant Canyon Formation, Grand County, Utah. *Brigham Young University Geological Studies* 19: 3–44.

Thompson, D.B. (1969). Dome shaped aeolian dunes in the Frodsham Member of the so-called 'Keuper' Sandstone Formation [Scythian - ?Anisan: Triassic] at Frodsham, Cheshire (England). *Sedimentary Geology* 3: 263–289.

Thompson, D.B. (1970a). Sedimentation of the Triassic (Scythian) red pebbly sandstones in the Cheshire Basin and its margins. *Geological Journal* 7: 183–216.

Thompson, D.B. (1970b). The stratigraphy of the so-called Keuper Sandstone Formation (Scythian-?Anisian) in the Permo-Triassic Cheshire Basin. *Journal of the Geological Society London* 126: 151–181.

Tooth, S. (2000). Process, form and change in dryland rivers: a review of recent research. *Earth-Science Reviews* 51: 67–107.

Tooth, S. (2007). Arid geomorphology: investigating past, present and future changes. *Progress in Physical Geography* 31: 319–335.

Trewin, N.H. (1993). Controls on fluvial deposition in mixed fluvial and aeolian facies within the Tumblagooda Sandstone (Late Silurian) of Western-Australia. *Sedimentary Geology* 85: 387–400.

Tsoar, H. (2008), Types of aeolian sand dunes and their formation, In: Balmforth, N.J. and Provenzale, A., *Geomorphological Fluid Mechanics, Lecture notes from the Gran Combin summer school, Saint Oyen, Aosta, Italy, June 19–30, 2000*, pp. 403–429. (Figure only)

Tsoar, H. and Pye, K. (1987). Dust transport and the question of desert loess formation. *Sedimentology* 34: 139–153.

Veiga, G.D. and Spalletti, L.A. (2007). The Upper Jurassic (Kimmeridgian) fluvial/aeolian systems of the southern Neuquén Basin, Argentina. *Gondwana Research* 11: 286–302.

Veiga, G.D., Spalletti, L.A., and Flint, S. (2002). Aeolian/fluvial interactions and high-resolution sequence stratigraphy of a non-marine lowstand wedge: the Avile Member of the Agrio Formation (Lower Cretaceous), central Neuquen Basin, Argentina. *Sedimentology* 49: 1001–1019.

Vos, K., Vandenberghe, N., and Elsen, J. (2014). Surface textural analysis of quartz grains by scanning electron microscopy (SEM): From sample preparation to environmental interpretation. *Earth Science Reviews* 128: 93–104.

Venus, J.H., Mountney, N.P., and McCaffrey, W.D. (2015). Synsedimentary salt diapirism as a control on fluvial system evolution: an example from the Permian Cutler Group, SE Utah, U.S.A. *Basin Research* 27: 152–182.

Wakefield, O.J.W. (2010). Aeolian, Fluvial and Shallow Marine Sedimentary System Interactions in the Permian Cutler Group, Southeast Utah, USA. Unpublished Ph.D. Thesis, Keele University, UK (figure only).

Wakefield, O.J.W. and Mountney, N.P. (2013). Stratigraphic architecture of back-filled incised-valley systems: Pennsylvanian-Permian lower Cutler beds, Utah, USA. *Sedimentary Geology* 298: 1–16.

Walker, R. (1979). *Facies models*. Geological Association of Canada, Geoscience Canada.

Ward, J.D. (1983). A reappraisal of the Cenozoic stratigraphy in the Kuiseb valley of the central Namib desert. In: *Late Cainozoic palaeoclimates of the southern hemisphere* (ed. J.C. Vogel), 455–463. Rotterdam: Balkema.

Warren, A. (1971). The dunes of the Ténéré Desert. *Geographical Journal* 137: 458–461.

Wasson, R.J. and Hyde, R. (1983). Factors determining desert dune type. *Nature* 304: 337–339.

Weber, K. J., (1987). Computation of initial well productivities in eolian sandstone on the basis of a geological model, Leman gas field, UK, In: Tillman, R.W. and Weber, K.J., Eds., *Reservoir sedimentology*, Society for Economic Paleontologists and Mineralogists Special Publication, 40, 333–354.

Werner, B.T. and Kocurek, G. (1999). Bedform spacing from defect dynamics. *Geology* 27: 727–730.

Wilson, I.G. (1971). Desert sandflow basins and a model for the development of ergs. *Geographical Journal* 137: 180–199.

Wilson, I.G. (1972). Aeolian bedforms – their development and origins. *Sedimentology* 19: 173–210.
Wilson, I.G. (1973). Ergs. *Sedimentary Geology* 10: 77–106.
Wolfe, S.A. (2022). Cold-Climate Aeolian Environments. In: *Treatise on Geomorphology* 2e (ed. J.F. Shroder), 195–219. New York: Elsevier.
Wopfner, H. and Twidale, C.R. (1988). Formation and age of desert dunes in the Lake Eyre depocentres in central Australia. *Geologische Rundschau* 77: 815–834.
Xiao, J., Qu, J., Yao, Z. et al. (2015). Morphology and formation mechanisms of the sand shadow dunes on the Qinghai-Tibet Plateau. *Journal of Arid Land* 7: 10–26. (Figure only).
Xu, H., Liu, Y., Kuang, H. and Peng, N., (2019). Late Jurassic fluvial–eolian deposits from the Tianchihe Formation, Ningwu–Jingle Basin, Shanxi Province, China. *Journal of Asian Earth Sciences*, 174, 245–262.
Yan, N., Colombera, L., Cosgrove, G.I.E., and Mountney, N.P. (2024). A 3D forward stratigraphic model of aeolian dune evolution for prediction of lithofacies heterogeneity. *Computers and Geosciences* 187: 105595.
Yu, L. and Gong, P. (2012). Google Earth as a virtual globe tool for Earth science applications at the global scale: progress and perspectives. *International Journal of Remote Sensing* 33: 3966–3986.
Yu, X., Wang, C., Bertolini, G., Liu, C. and Wang, J., (2021). Damp- to dry aeolian systems: sedimentology, climate forcing, and aeolian accumulation in the late Cretaceous Liyou Basin, South China, *Sedimentary Geology*, 426, 106030 (Figure only).
Zhang, X-X., Claiborn, C., Lei, J-Q., Vaughan, J., Wu, S-X., Li, S-X., Liu, L-Y., Wang, Y-D., Huang, S-Y. and Zhou, J., (2020). Aeolian dust in Central Asia: Spatial distribution and temporal variability, *Atmospheric Environment*, 238, 117734
Zhang, Y., Colombera, L., Mountney, N.P. et al. (2021). Sedimentation of open-framework gravels in alluvial-fan settings: Quaternary Poplar Fan, northwest China. *Marine and Petroleum Geology* 134: 105376.

Supplementary Information

7.S.1 History of Aeolian Sedimentological Research

Early interpretations suggesting an aeolian origin for ancient sedimentary successions comprising large-scale sets of cross-bedded sandstone include those made for the "Lower Keuper" – now part of the Triassic Sherwood Sandstone Group – of the Cheshire Basin, England (Maidwell 1914). However, Bagnold (1941) was first to investigate in detail the physical processes responsible for aeolian sediment entrainment, transport, and deposition, and to relate this to different aeolian desert dune types and their deposits in his classic book "The Physics of Blown Sand and Desert Dunes". McKee and Tibbitts (1964) and McKee (1966) documented in detail the internal facies arrangement and stratification of modern dunes by excavating trenches. This work revealed how variable patterns of dune construction and migration generated complex cross-stratified sets and cosets. Thompson (1969) used this new understanding to demonstrate an approach for the reconstruction of the original morphology of large dome-shaped aeolian dunes from cross-stratified deposits preserved in the same Triassic "Lower Keuper" succession that Maidwell had contemplated more than 50 years earlier. Stokes (1968) recognised that many aeolian accumulations underwent episodic deflation (erosion) down to the level of the water table, and that laterally extensive bounding surfaces (later to become known as 'Stokes' surfaces) represented the culmination of regional deflation events.

Although these works recognised and described the 3D architectural complexity in sets of aeolian strata, it was a further decade before unifying models were developed. Throughout the 1970s, there existed a growing need for predictive aeolian facies models because a number of significant hydrocarbon provinces exploited reserves in subsurface aeolian reservoirs. For example, the Permian Rotliegend Group of the Southern North Sea was a major source of natural gas (Glennie 1972) and a better understanding of sedimentary reservoir architecture was crucial for the successful exploitation of these reserves. At White Sands, New Mexico, McKee and Moiola (1975) demonstrated that large aeolian bedforms climbed downwind over one another at very low angles as they migrated, such that the basal parts of aeolian dunes accumulated as stratal sets as they were overridden by succeeding dunes. Brookfield (1977) introduced a model to explain and account for the origin and geometric interrelationship of compound sets (cosets) of strata in terms of a hierarchy of bounding surfaces that originated in response to interdune migration in front of large bedforms, the migration of smaller superimposed child bedforms over larger parent bedforms, and the episodic reactivation of dune lee slopes in response to changes in bedform morphology and/or behaviour. Hunter (1977) proposed criteria for the identification of small-scale aeolian sedimentary structures and related these to formative aeolian processes, thereby enabling many hitherto ambiguous successions to be ascribed unequivocally to an aeolian origin. Throughout the 1970s, studies of the sedimentary features of modern sand seas were conducted (e.g., Glennie 1970). Studies by Wilson (1971, 1972, 1973) on sand-sea dynamics and response to external controls began to consider aeolian environments holistically. Aeolian research in the 1970s culminated in the publication of 'A Study of Global Sand Seas' (McKee, 1979), which remains an invaluable reference. Technological advances in the late 1970s provided a new view of aeolian systems through the acquisition of remotely sensed satellite data (e.g., Breed & Grow, 1979; Breed et al. 1979), and via wind monitoring instruments that enabled complicated bedform morphological types to be related to patterns of airflow (e.g., Fryberger 1979). At the time of publication of the first edition of "Sedimentary Environments and Facies" (Reading 1978) and the first edition of "Facies Models" (Walker, 1979), numerous simple 3D aeolian facies models had been proposed. Although based on observations and understanding from a relatively limited number of modern and outcrop studies, these models provided a basis for interpretation, a norm for comparison, and a tool for prediction.

Pioneering research in the early 1980s led to the development of more sophisticated facies models that were tested and refined via the collection of detailed stratigraphic datasets from outcrop (e.g., Kocurek 1981; Loope 1981). Studies of modern aeolian systems highlighted that the relatively small proportion of facies within interdune regions were often key to environmental interpretation (Ahlbrandt and Fryberger 1981; Hummel and Kocurek 1984). Rubin and Hunter (1982) outlined a model to explain the mechanics of aeolian accumulation during dune migration through the process of *bedform climbing*. Considerable research focused on an assessment of the environmental significance of aeolian bounding surfaces, and in particular, whether major surfaces of great lateral extent were the product of regional deflation of entire aeolian successions (or large parts thereof) to the water table, as originally suggested by Stokes (1968) – so-called *supersurfaces*. Kocurek (1984) and Rubin and Hunter (1984) proposed that surfaces arising from both climbing bedform migration and regional deflation could both occur. Talbot (1985) and Loope (1985) demonstrated how many regionally extensive supersurfaces were likely to be a product of climatic and related environmental change. Kocurek (1988) proposed that supersurfaces of regional extent originated through a variety of mechanisms that usually resulted in the termination of erg accumulation; such surfaces effectively bounded separate *erg sequences* in the ancient record. Langford and Chan (1988, 1989) demonstrated how repeated fluvial incursions into the margins of aeolian dune fields could result in curtailment of bedform climbing and the generation of extensive "flood surfaces". The development of 3D geometric stratigraphic modelling techniques (e.g., Rubin and Hunter 1983; Rubin 1987a), which enabled simulation of the generation of complex bed-set geometries, provided a means to explore how relatively common types of bedform morphology and migratory behaviour could generate the many complicated bed-set architectures observed in outcrops through autogenic mechanisms. Notably, these new techniques helped in the recognition that accumulations of linear (longitudinal) dunes were apparently under-represented in the rock record because such bedforms usually have an additional small component of transverse motion (i.e., parallel to their crestlines), and it is this component that is preferentially accumulated as sets of strata (Rubin and Hunter 1985). By the end of the 1980s, 3D aeolian facies models had been developed for specific types of aeolian systems, including those characterised by transverse, oblique and longitudinal dune-scale bedforms (*sensu* Hunter et al. 1983), as well as megadunes (draa) over which superimposed dunes migrated.

In the 1990s, a conceptual framework allowed for the development of dynamic facies models that accounted for spatial and/or temporal variability in controlling parameters. Kocurek and Havholm (1993) discussed how sediment flux within an aeolian system controls whether a sediment body undergoes accumulation, bypass or deflation. Additionally, this work showed how aeolian systems could be classified as dry, wet, or stabilising depending on the nature of the agents that controlled accumulation. Fryberger (1993) provided a valuable overview of aeolian bounding-surface types and summarised the terminology used in their description and interpretation. Kocurek and Lancaster (1999) argued that the construction of aeolian systems is dependent on the generation of a sediment supply, the availability of that supply for aeolian transport, and the ability of the wind to transport that sediment and build the system. George and Berry (1997) and Howell and Mountney (1997) demonstrated that the rate of creation of accommodation played a major role in determining the extent to which aeolian accumulations may be preserved. Crabaugh and Kocurek (1993), Havholm and Kocurek (1994) and Clemmensen et al. (1994) each demonstrated that the accumulation and preservation of separate erg sequences in the ancient record is controlled by allogenic, cyclical and repeating changes in climate, alongside additional environmental factors, such as ground-water level and rates of sediment supply. The culmination of this research was presented in Kocurek (1999), who outlined the conditions required for (i) aeolian system construction, (ii) the accumulation of a body of strata, and (iii) preservation of that body of strata by long-term placement beneath a regional baseline of erosion.

Throughout the 1990s, researchers documented the response of aeolian systems to Quaternary climate change. Alsharhan et al. (1998) and papers therein explored how aeolian systems respond to external forces such as changes in precipitation, in the level of the ground-water table, in wind velocity and, in the case of coastal ergs, changes in sea level. One important conclusion from this work has been the recognition that most modern aeolian dune systems are currently still responding to late Quaternary climate change, and are therefore out of equilibrium with the current Holocene climatic regime (Lancaster 1998).

The application of geophysical techniques, such as ground penetrating radar (GPR), has enabled the detailed internal architecture of modern bedforms to be established in three dimensions, and is helping to reveal the mechanisms by which modern dunes generate sets of strata (Bristow et al. 1996, 2000). Dating techniques, for example using radiometric methods like C^{14} and optically stimulated luminescence (OSL), are helping to determine the migration histories of large dunes, under both present-day wind regimes and during previous windier glacial episodes (Bristow et al. 2005; Pedersen and Clemmensen 2005; Bristow and Pucillo 2006; Lomax et al. 2011).

Since the mid-2000s, aeolian research has benefited from the extensive documentation of hitherto little-studied ancient successions, especially from South America and Asia. In particular, these studies have helped to constrain (i) patterns of atmospheric circulation (e.g., Hasegawa et al., 2010), (ii) regional palaeogeographic reconstructions (e.g., Bállico et al. 2017a), and (iii) the impact of palaeoclimate change (e.g., Li et al. 2018).

Aeolian sedimentological models are now applied to addressing some of society's most pressing issues: (i) improved understanding of the contribution of aeolian processes to atmospheric dust emissions, which has significant implications for meteorological conditions, climate change, biogeochemical cycles, and human health (e.g., Evan et al. 2014; Goudie 2014; Zang et al. 2020); (ii) forecasting and monitoring the expansion of desert aeolian environments (desertification) and associated degradation and of formerly fertile agricultural land, including the loss of fertile soil and biodiversity (e.g., Becerril-Piña and Mastachi-Loza 2021; Moghaddam et al. 2018; Luo et al. 2020); (iii) the response of coastal aeolian systems to sea-level rise, which has important implications for rates of coastal erosion, and surface stability and ecology (e.g., del Valle, et al. 2020; Jackson et al. 2019); and (iv) improved understanding of anthropogenic drivers of aeolian landform change and how to mitigate them (e.g., El-Wahab et al., 2018; Marrero-Rodríguez et al., 2020).

Fundamental and applied research in aeolian sedimentary geology now benefits from a long history of study, for which sedimentary processes and products from many examples have been recorded as detailed accounts. One fundamental challenge is how to make the best use of all of this information. Cosgrove et al. (2021a) demonstrated the application of a relational Database Aeolian Sedimentary Architecture (DASA), which stores data on the sedimentological, stratigraphic and geomorphological aspects of many aeolian systems, including data extracted from the published literature. Databases of this type record the architecture and spatio-temporal evolution of a broad range of modern and recently active aeolian systems, and of their preserved deposits in ancient successions. This approach allows direct quantitative comparisons between multiple case studies that describe aeolian systems and their preserved sedimentary architectures. Database-informed investigative techniques are applied to elucidate the influences and controls that govern aeolian system construction, accumulation and preservation in the geological record (Cosgrove et al. 2021b–d, 2022).

7.S.2 Challenges in the Study of Aeolian Sedimentary Systems

Recognition of ancient aeolian deposits can sometimes be straightforward. For example, very large-scale sets of cross bedding exposed across much of the southwestern United States have long been ascribed an aeolian dune origin (Huntington 1907, referenced in McKee, 1979). Yet, in many cases, recognition is problematic: it can be difficult to differentiate deposits of aeolian origin from those of sandy fluvial and shallow-marine origin (e.g., Baars 1979; Elwood 1975). Building detailed facies models that are able to relate the preserved architecture of aeolian successions to the processes and landforms responsible for generating the sedimentary environment is difficult for six main reasons (Rodríguez-López et al. 2014). First, aeolian dune environments are commonly dominated by a relatively small number of fundamental lithofacies types, but there exist a great many ways in which these facies types can be arranged geometrically as the constituent parts of architectural elements that are generated via the migration of aeolian dunes and interdunes over one another to leave a resultant accumulation. Different bedform types migrating at different rates, in different directions, and under the influence of different controls (e.g., climate, tectonics), lead to highly variable lithofacies arrangements. Second, although associated non-aeolian facies (e.g., water-lain deposits of alluvial, fluvial, interdune pond, lake or coastal lagoon origin) make up only a relatively small proportion of most aeolian successions, their presence is critical for determining the mechanisms that controlled accumulation of the overall sedimentary environment. Third, sediment bypass and/or erosion is an inherent part of aeolian bedform migration and almost all preserved successions are highly fragmentary. In many preserved aeolian successions, aeolian accumulations are bounded by supersurfaces, which are paraconformities (diastems) representing long-lived depositional hiatuses. For many aeolian systems, the amount of time represented by supersurfaces is considerably greater than that represented by the aeolian accumulations themselves (cf., Ager 1981, Loope 1985). Fourth, certain selected parts of aeolian landforms have a considerably greater chance of long-term preservation, whereas other parts are only very rarely preserved into the long-term rock record. For example, the lowermost toesets of dunes are commonly accumulated where dunes migrate and climb over one another at low angles. By contrast, the crests of dunes are typically only preserved in exceptional circumstances, for example, where dunes are passively flooded by marine transgression or inundated by lava flows. Geologists must attempt to reconstruct palaeoenvironments from a preserved sedimentary record that is inherently biased. Fifth, most aeolian successions are characterised by marked lateral variations in bed-set thickness and, in many cases, limited lateral continuity. As such, traditional 1D sedimentary graphic logs are not particularly useful for detailed interpretation of formative dune type, and such logs are of limited use in lithostratigraphic correlation. Instead, interpretations of environments of deposition are best supported where outcrops can be examined in three dimensions, such that the distribution of facies within architectural elements (geobodies) can be established, and the three-dimensional geometry and topology of those elements can be determined. Sixth, although the external forms of modern aeolian bedforms can be documented from aerial photographs and satellite images, the organisation of the internal structure of underlying deposits is less well known. Conversely, in the ancient rock record, bed-sets of presumed aeolian dune (and megadune) origin are characterised by a variety of types of facies associations, but the morphology of the bedforms responsible for their generation cannot usually be determined directly as it is typically not preserved. Thus, it is difficult to relate the external morphology and migratory behaviour of aeolian bedforms to their deposits, and *vice versa*.

The remainder of this chapter describes common types of aeolian sedimentary systems and deposits, explains methods and techniques used in their analysis, and discusses how and why aeolian systems and their preserved successions are of great societal and economic value.

7.S.3 Integrated Case-study Example: Pennsylvanian to Permian Successions of the Paradox Basin, Utah, USA

Although coastal aeolian, fluvial, nearshore and shallow marine sub-environments are each known to respond differently to variations in climate, eustasy and tectonics, typically they each do so in a predictable manner (Kocurek and Havholm 1993; Posamentier et al. 1988; Shanley and McCabe 1994). In many basin settings where aeolian sedimentary systems develop, these three fundamental allogenic controls on sediment system evolution share an intimate and interdependent relationship with each other, such that the individual effects of each cannot necessarily be discerned, and instead a combined and interlinked depositional system response is recorded in the preserved stratigraphy (Blakey and Middleton 1983; Chan and Kocurek 1988).

The effects of climate, tectonics and eustasy are a focus of study for developing models with which to better account for the regional stratigraphic evolution of coevally active aeolian, fluvial and shallow marine systems. Complex feedback loops are known to operate between each control (Kocurek 1999), and these play a significant role in determining resultant preserved architecture. Further, the relative influence of, and interplay between, these control mechanisms is likely to vary significantly between the various sub-environments present in a system (Blakey et al. 1996).

Interdependent climatic and eustatic controls on sedimentation are widely documented in a variety of palaeoenvironmental settings, and such controls are known to operate at various intensities and over various timescales such that they enable the generation and preservation of regularly repeating cycles in the stratal record (e.g., Goldhammer et al. 1994; Heckel et al. 1994; Soreghan and Dickinson, 1994). Understanding the time significance of these cycles can provide a method in correlating sediment packages of shallow marine origin with coeval packages of non-marine origin that accumulated in more landward settings (e.g., Blakey et al. 1996).

Although many stratigraphic models for coastal systems stress the importance of eustasy as a primary control on accumulation, in the case of the Permian lower Cutler beds of the Paradox Basin, Utah, USA, it is unlikely to have been the only control. The lower Cutler beds are a ~200 m-thick succession of aeolian, fluvial, coastal and shallow-marine origin (Jordan and Mountney 2010, 2012; Wakefield and Mountney 2013).

The effects of eustasy decrease with increasing distance inland from the shoreline, and given that a substantial part of the depositional system lay significantly landward (>50 km) of the shoreline (Loope 1984), it may not have been a dominant influence. However, its effect may still be expressed, for example, as fluvial downcutting due to a lowering of base-level in response to relative sea-level fall (cf. Shanley and McCabe 1994). Thus, the role of climate and tectonics was potentially dominant on more landward parts of the aeolian and alluvial plain (cf. Kocurek 1999), whereas the effects of relative sea-level change became increasingly dominant in coastal, shoreline and more basinward areas. Each of these fundamental controls is considered to have interacted collectively to determine the style of accumulation in the lower Cutler beds succession.

Depositional cycles (sequences) preserved in the lower Cutler beds occur in a repeating, predictable pattern, indicative of systematic changes in the factors controlling sedimentation through time. A sequence-stratigraphic framework is used to illustrate how cyclic changes in controlling parameters were responsible for generating the preserved sequence architecture. Fundamental to this approach is the recognition that climate and eustasy were interdependent and operated in tandem to control sequence development during accumulation of the lower Cutler beds. The model presented in Figure 7.18, and discussed below, accounts for the accumulation of a single sequence deposited during a cycle of climate and relative sea-level change, and where there is little evidence for substantial and significant relative sea-level fall at the end of the episode of sequence accumulation.

7.S.3.1 Arid Climatic Phase (Lowstand)

Aeolian systems are interpreted to have been at their most extensive across the basin during episodes of relative sea-level lowstand. Sea-level lowstand exposed former shoreline deposits and these acted as a supply for aeolian transport by prevailing winds blowing to the southeast (Loope 1984). During lowstand events, dune fields were constructed across the low-relief coastal plain and eventually accumulated through the migration and climb of a series of well-organised dune bedforms and associated interdunes. The predominantly dry nature of the interdunes and the great extent of the dune fields indicates that episodes of lowstand occurred synchronously with climatic aridity. This supports the idea that lowstand events were likely driven by glacio-eustasy, and that lowstands in relative sea-level occurred during the Pennsylvanian and Permian due to periodic high-latitude icecap growth associated with periods of climatic aridity in low latitude regions (e.g., Blakey 2009; Heckel et al. 1994; Maynard and Leeder 1992). In addition to climatic aridity, it has also been suggested that lowstands were associated with increased wind speeds because of compression of the monsoonal climate cell (Loope et al., 2004), which would have also promoted dune-field construction and accumulation.

Despite the overall arid climate, dryland fluvial systems apparently remained active at these times, though on a limited scale, with only episodic incursions into the margins of dune fields by ephemeral streams (Jordan and Mountney 2010). Where these fluvial systems cut through the dune fields, they exploited low-relief interdune corridors (cf. Cain and Mountney 2011; Langford and Chan 1988; Mountney and Jagger 2004). Rarely, more significant flood events might have reached the coastline and were probably associated with localised and temporary elevation of the water table, resulting in the colonization of interdune corridors by vegetation, as demonstrated by the presence of small rhizoliths and mottling in some sinterdune deposits. Fluvial systems were likely to have been activated by precipitation in the Uncompahgre highlands that bounded the Paradox Basin to the northeast, rather than by direct precipitation on the basin plain (Cain and Mountney 2009).

7.S.3.2 Semi-Arid to Sub-Humid Climatic Phase (Early Transgression)

As relative sea level began to rise, it served to inhibit the development of the dune-field systems in two principal ways. Firstly, transgression progressively shut down the aeolian sediment supply as the low-lying coastal plain was progressively flooded. Thus, winds blowing to the southeast along the coastline (Loope, 1984) would have been undersaturated with respect to their potential sand transport capacity (*sensu* Kocurek and Lancaster 1999), resulting in the onset of dune-field deflation, and eventually culminating in the generation of a deflation surface (Loope 1985) as the dunes were gradually cannibalised and the sand was transported to locations progressively further downwind (cf. Mountney 2006a). Secondly, the rise in sea level would have induced a rise in the inland water table (cf. Kocurek et al. 2001; Mountney 2006b), albeit with a lag time between the two events. The elevated water table would have acted to stabilise sediment at the damp surface and increase the amount of vegetation covering both the dune-flank and interdune regions, thereby further reducing the availability of sediment for aeolian transport (cf. Kocurek and Lancaster 1999; Lancaster and Baas 1998).

Episodes of relative sea-level rise were associated with a transition to a more humid climate, as demonstrated by a transition to predominantly fluvial sedimentation in the upper parts of most sequences (Figure 7.18b), and an increase in the incidence of root mottling in fine-grained sediments. Increased fluvial activity could also have been at least partly responsible for the cannibalization of aeolian dune deposits (Jordan & Mountney, 2010). As larger fluvial systems became established an extensive braid-plain comprising unconfined channel networks and vegetated overbank regions developed over the deflated former aeolian surface. At positions where fluvial channels reached the coastline, partial marine back-flooding of low-lying channels occurred during the early stages of sea-level rise (Jordan and Mountney 2010; Terrell 1972).

7.S.3.3 Humid Climatic Phase (Late Transgression and Highstand)

The low-lying and low-relief nature of the continental basin led to extensive flooding of the coastal plain in response to only modest rises in relative sea level. Marine flooding would have initially inundated the lowest-lying parts of the shoreline system, primarily in the deepest-cut channels of fluvial systems that reached the shoreline (Figure 7.18c). As the transgression continued, marine waters eventually flooded surrounding interfluve regions. Within small incised valleys, the interplay between combined fluvial and marine processes resulted in the development of localised estuarine systems with a variety of shallow marine strata infilling the fluvially-cut channels (Jordan and Mountney 2010; Rankey 1997; Terrell 1972; Wakefield and Mountney 2013). In more landward parts of the basin, fluvial systems remained active during transgression and acted as the principal agents responsible for supplying coarse-grained clastic detritus to the shallow marine realm, where it was mixed with carbonate sediment to form shallow marine strata (sandy calcarenite facies). The transgression would have displaced the coastline several tens of kilometres landward in response to only modest relative sea-level rise and, as a result, many of the sand bars of estuarine origin were likely reworked by shoreface processes as the transgression progressed (cf. Hwang and Heller 2002). As water depth gradually increased, the accumulation of micritic, bioclastic wackestone facies became dominant in more distal settings, as recorded in the successions of the Canyonlands and Grabens districts.

The lower Cutler beds are a cyclic succession that records repeated oscillations in relative sea level and climatic conditions as a result of glacio-eustatic variability. During lowstand times, arid climatic conditions prevailed and aeolian processes and landforms were dominant. Episodes of transgression were characterised by more humid episodes in which fluvial systems became dominant in landward parts of the basin before being subject to transgression and the onset of accumulation of carbonate and mixed clastic-carbonate sediments across large parts of a low-relief coastal plain. At least 12 separate cycles (sequences) recording these oscillating conditions are recognised in the uppermost 80 m of the succession, with the final transgression forming a maximum flooding surface and representing the end of the lower Cutler beds sedimentation. The sequences are stacked and arranged into 3 sets. The lowermost sequence set is progradational and records the overall basinward advance of the shoreline with each successive sequence in the set signifying an increasing non-marine dominance, and a reduction in the thickness and likely landward extent of thin marine units. The upper two sequence sets both exhibit an upward decrease in mean preserved sequence thickness, and both record an upward decrease in the proportion of non-marine deposits in each successive sequence. Within these upper two sets, each successive sequence is characterised by flooding surfaces that extend successively further landwards (a retrogradational pattern), which resulted in larger areas of the basin being transgressed with each successive flood event, and culminating in a maximum flooding surface at the top of the lower Cutler beds.

8

Sand-Rich Coasts and Shallow Clastic Seas

Daniel S. Collins[1] *and Gary J. Hampson*[2]

[1] *Shell International Ltd., London, UK*
[2] *Department of Earth Science and Engineering, Imperial College, London, UK*

Summary

This chapter outlines the processes, facies, depositional architectures and stratigraphic organisation of sand-rich coastal and siliciclastic shallow-marine environments. The environments addressed extend from the landward limit of marine influence in coastal deposits (up-dip) to the marine shelf edge (down-dip). We describe the depositional environments of shorelines such as deltas, strandplains, tidal flats, estuaries and barrier islands, and sand-rich components of their neighbouring shelves and ramps. We use a simple classification scheme for these depositional environments to emphasise the interplay of fluvial, wave and tidal processes in controlling both coastal morphology and sediment dynamics. Sediment is commonly supplied by rivers, wind and glaciers to the shoreline, where it is then reworked and redistributed. Sediment redistribution includes partitioning of sandy and muddy grain size fractions into different coastal and shallow-marine environments. Sand is transported and deposited by fair-weather and storm waves, tides, river-mouth currents and oceanographic currents. We describe systematically those aspects of fluvial, wave and tidal processes that generate diagnostic characteristics in the resulting deposits.

Representative examples of facies, facies associations and facies successions are first documented using vertical successions and a set of standardised facies associations. Further case studies introduce 2D and 3D architectural elements and larger-scale stratigraphic patterns from accumulated successions to provide a practitioners' guide to the description and interpretation of marginal-to-shallow-marine sands and sandstones. The chapter builds on much previous work documenting end-member depositional systems that are dominated by a single process, by explicitly addressing depositional systems in which multiple processes operate together. Case studies are specifically selected to characterise the products and morphologies of these mixed-process depositional systems.

The chapter also explores the broader controls on the deposition, preservation and stratigraphic record of sand-rich coastal and shallow-marine deposits and their related strata. These include the differentiation between intrinsic (autogenic) behaviour of depositional systems and external (allogenic) factors such as sea level, sediment supply, climate, physiographic configuration of shorelines and seas, and tectonic basin evolution. We emphasise this wide range of controls, as a vital addition to the focus on relative sea-level changes that are implicit in conventional sequence-stratigraphic models. Furthermore, we highlight the role of sand-rich shorelines, shelves and ramps in sediment-routing systems that encompass other depositional environments from source to sink, notably alluvial, mud-rich coastal and deepwater environments.

The structured description and interpretation of marginal-to-shallow-marine sands and sandstones, as presented here, provide a toolkit with many applications. These include reconstructing the configuration and evolution of marine sedimentary basins and deciphering the environmental histories recorded in basin-fill strata. The same methods can be used to understand past and predict future coastal change, which will profoundly affect large populations living in vulnerable coastal regions. Interpretation of marginal-to-shallow-marine strata in the subsurface is also fundamental for efficient use of energy resources (e.g. hydrocarbons and geothermal fluids), sustainable management of groundwater and of the shorelines themselves and effective storage of CO_2 and hydrogen.

Reading's Sedimentary Environments: Processes, Facies and Stratigraphy, Fourth Edition. Edited by Bruce Levell.
 Published 2026 by John Wiley & Sons Ltd.
Companion website: www.wiley.com/go/levell4e

8.1 Introduction

Coasts form a broad zone that extends from the landward limit of marine processes to the seaward limit of shoreline processes. Sand-rich coasts and shallow seas occur at the transition between land and ocean. This transition is critical in linked Earth system processes, for example, the carbon and hydrological cycles and the sediment routing system. Sand-rich coastal deposits provide a record of this interface through Earth history and host significant natural resources. This chapter provides a practitioners' guide to sedimentological processes, facies analysis and characterisation of stratigraphy. The chapter links directly to not only those on neighbouring segments of the sediment routing system: alluvial sediments (Chapter 5), fine-grained coastal and marine sediments (Chapter 9) and deep-water detrital sediments (Chapter 12) but also those reviewing aeolian (Chapter 7) and glacial sediments (Chapter 14). Many coasts and shallow seas contain components made of carbonate sediments (Chapter 11), as well as those made of siliciclastic sand and mud, and therefore need to be considered using approaches described in Chapters 8, 9 and 11. The chapter also builds on techniques and concepts in the chapters on physical facies analysis (Chapter 2), integration of biogenic sedimentary structures (Chapter 3) and the stratigraphic record and sequence stratigraphy (Chapter 4).

Coastal and shallow-marine sediments provide the classical set of facies and sequence stratigraphic models for siliciclastic sedimentology. The study of these sediments has involved evolving paradigms (e.g. from facies analysis and 'cyclothems' of autogenic origin to 'sequences' formed by relative sea-level change and other external or allogenic controls). In this fourth edition, we emphasise the need to consider interactions of multiple interacting sedimentary processes and include four key developments.

1) *Re-assessing allogenic and autogenic effects*: A critical mass of case studies applying sequence-stratigraphic methods to high-resolution outcrop and subsurface datasets, especially 3D seismic data, has resulted in testing of early sequence stratigraphic methodologies. These show clearly the need for explicit consideration of multiple allogenic (external) controls and autogenic (internal) behaviours of depositional systems operating at multiple spatio-temporal scales. Good overviews of sequence stratigraphy that include treatments of shoreline-shelf strata, largely with an allogenic focus, include Reynolds (1996), Coe et al. (2003) and Catuneanu (2022).
2) *Sediment routing systems* (Allen 2017) link erosional sediment sources to depositional sediment sinks (Chapter 4). This emphasises sediment supply to, storage on and transport across the shoreline-shelf segment and its subdivisions (e.g. across the 'littoral energy fence' and sediment bypass beyond the shelf edge and into deep water). It also informs consideration of sediment provenance and grain-size partitioning from shoreline to shelf edge.
3) *Mixed-influence process regimes* are a new focus building on extensive previous work on end-member process regimes (wave-, tide- or fluvial-dominated). Excellent textbook chapters on wave-, tide- and fluvial-dominated shorelines and shelves with predominantly end-member process regimes include Bhattacharya (2010), Boyd (2010), Dalrymple (2010) and Plint (2010).
4) *New applications* of sedimentology will be more diverse, driven by the transition to low-carbon energy, geological storage of waste, exploration for critical minerals including placer and sedimentary ore deposits, and geoscientific support for sustainable development (e.g. groundwater, pollution control and mitigation, shoreline management and analogue studies of past responses to climate and sea-level change). These are replacing applications to hydrocarbon exploration and production.

A shoreline is simply the boundary between land and water, at the edge of a lake, ocean or other water body. A shelf is any broad, relatively shallow (typically <200 m) submarine platform that lies seaward of the shoreline. The seaward limit of the shelf (i.e. shelf edge) is sometimes defined by a pronounced break in slope and an associated abrupt increase in water depth. Where such a shelf-slope break is absent, the submarine platform seaward of the shoreline is referred to as a 'ramp'. Shelves and ramps dip gently seaward, commonly with a slope of ≤0.01°, which is similar to the gradient of the coastal plain that lies landward of the shoreline. Indeed, many shelves and ramps are drowned coastal plains. We currently live in a global highstand of sea level, and the present-day distribution of marine shelves and ramps reflects this condition. A range of depositional environments typically co-exists along a shoreline or shelf (Figure 8.1). Consequently, shoreline-shelf systems are morphologically complex and illustrate considerable, fractal-like variability in plan view across spatial scales, from thousands to a few kilometres (Figure 8.1).

Over geological time scales, shoreline-shelf morphology and preserved stratigraphy are controlled by a dynamic equilibrium between the rate of creation (or removal) of accommodation space and the rate and character of sediment supply – the 'A/S' ratio (Figure 8.2; Section 4.2.4) (Swift and Thorne 1991; Muto and Steel 1997). Variables influencing this ratio include the rate of relative sea-level change, the rate and character of sediment input, and rates of sediment transport (Figure 8.3). Over shorter time scales, hydrodynamic processes, principally wave (fair-weather and storms), tide and fluvial (river) processes, act to supply and rework sediment (Figure 8.3). Sediment transport to the coast, particularly of sand, may be decoupled from sediment transport to the shelf by virtue of the 'littoral energy fence': the mean landward-directed bottom stress formed

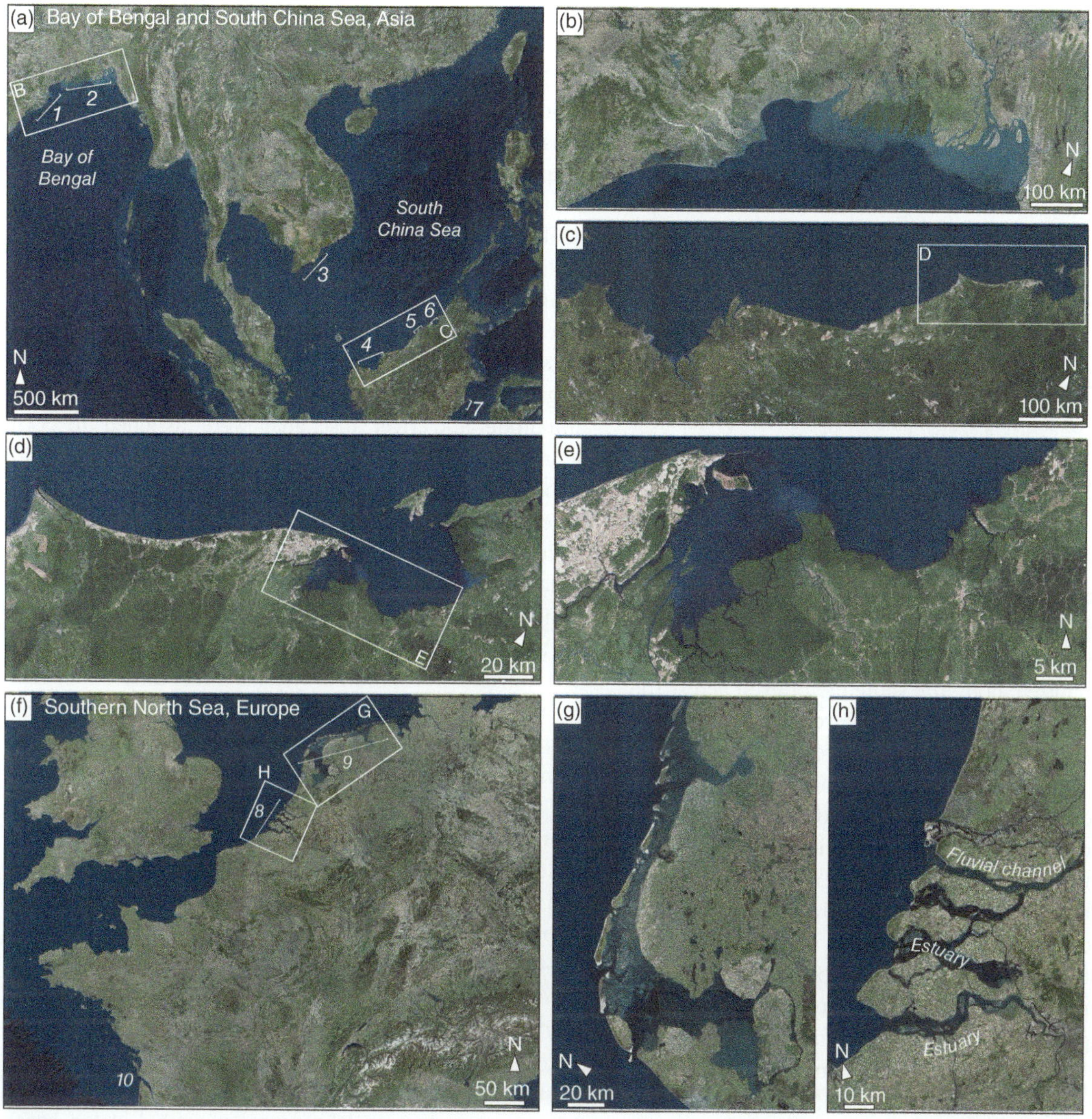

Figure 8.1 Satellite imagery illustrating variability and complexity in shoreline morphology across spatial scales, with examples from 'supply-dominated regimes' (Swift and Thorne 1991) in the Bay of Bengal and South China Sea, Asia (a–e) and the 'accommodation-dominated regime' of the Southern North Sea, Europe (f–h). Shoreline morphology and depositional elements vary between transgressive and regressive shorelines (Figure 8.2) and with the relative balance of wave, tide and fluvial processes (Figure 8.3). Modern shorelines only represent a 'snap-shot' in the range of forms and processes that may be preserved but are a useful proxy for facies distribution and geometry within ancient systems. Shoreline systems referred to in this chapter (Figure 8.1a,f) include the following: 1. Mahanadi–Brahmani delta; 2. Ganges–Brahmaputra delta; 3. Mekong delta; 4. Rajang delta; 5. Baram delta; 6. Brunei Bay; 7. Mahakam delta; 8. Rhine delta, which contains several smaller estuaries; 9. western Wadden Sea and Friesian Islands; 10. Gironde estuary. Google / Maxar Technologies.

by shoaling and breaking waves, which tends to retain sand close to the shoreline (Allen 1970). In general, 'supply dominated shoreline-shelf margins' (Swift and Thorne 1991; e.g. Figure 8.1a–e) are characterised by high sediment supply and are generally characterised by regressive (seaward-advancing) shorelines (Figure 8.2b) in which sediment is transported to delta fronts, shorefaces and shelves by river-mouth bypassing across the 'littoral energy fence'. Under such supply-dominated conditions, the prevailing hydrodynamic process regime may transport sediment along and across, and sculpt the morphology of, a linked shoreline-shelf system (Figure 8.3). However, shelves offshore of regressive coasts may also be sediment starved where sediment is retained near the shoreline.

'Accommodation-dominated shoreline-shelf margins' (Swift and Thorne 1991) are characterised by high rates of sea-level rise and/or sediment dispersal, and typically form transgressive (landward-retreating) shorelines (Figure 8.1f–h) characterised by landward reworking and transport of sediment (Figures 8.2 and 8.3). This results

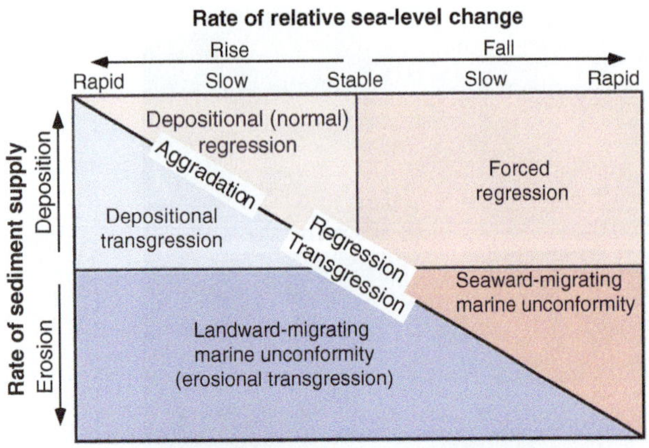

Figure 8.2 Plot showing the conditions required for development of regressive and transgressive shorelines (adapted from Curray 1964; Boyd et al. 1992) and reflecting the interplay between the rates of relative sea-level change and sediment supply.

in shoreface erosion and sediment-starved shelves that may leave unburied an older, relict morphology that is not in equilibrium with the prevailing hydrodynamic process regime. However, sediment in transgressive systems may also be transported along the shoreline or even across the shelf depending on the prevailing hydrodynamic processes (e.g. by tides, Figure 8.3). Changes in relative sea level can also impact sediment transport to shorelines and shelves through their effect on backwater zone hydraulics and coastal-plain avulsion (e.g. Chatanantavet et al. 2012). The backwater zone (Section 5.13.3) refers to the downstream reach of a river in which the mean elevation of the river channel bed is below mean sea level (Paola and Mohrig 1996); it describes the reach in which a hydraulic response, due to a perturbation in stream power at the shoreline, propagates upstream under conditions of normal (i.e. sub-critical) river flow (e.g. Chatanantavet et al. 2012; Lamb et al. 2012; Fernandes et al. 2016; Ganti et al. 2016).

Deposits accumulated landward of the shoreline are marginal marine (or paralic), whereas shallow-marine deposits accumulate seaward of the shoreline. Given sufficient accommodation, repeated regressive-transgressive transits of the shoreline both construct and maintain the platform morphology of the shelf and result in interfingering of shallow-marine and marginal-marine deposits.

8.2 Environments

8.2.1 Shoreline Types

Shoreline morphology and resultant facies and stratigraphic architecture form a continuum that reflects the interaction of sediment supplied principally by rivers and reworking by basinal wave and tidal processes, combined with longer-term changes in accommodation versus sediment supply (e.g. Coleman and Wright 1975; Galloway 1975; Hayes 1979; Boyd et al. 1992; Ainsworth et al. 2011; Vakarelov and Ainsworth 2012; Dashtgard et al. 2021; Collins et al. 2021a; Zuchuat et al. 2023). Process classifications of modern shorelines are generally qualitative and subjective, based on considering shoreline morphology and the occurrence of 'characteristic' depositional elements (Figure 8.4; Box 8.1). However, statistical analysis illustrates that wave height, tidal range and fluvial discharge variations conform with geomorphologically based classification of shorelines (Harris et al. 2002; Short 2006). Nyberg and Howell (2016) recently performed the first, systematic, semi-automated classification of global shoreline process regime (Figure 8.5). Available measurements of wave height, tidal amplitude and river discharge were assigned to 5 km shoreline segments. These data were supplemented with algorithms accounting for shoreline roughness (e.g. highly rugose shorelines are more fluvial dominated). The automated classification had an 85% success rate compared to a manual classification (e.g. Vakarelov and Ainsworth 2012).

Figure 8.4 shows a series of generalised models illustrating the morphology and depositional elements of regressive and transgressive shoreline systems under different process regimes. Shoreline types in these process-based classification schemes and models are referred to by well-established names (Figure 8.4).

Fluvial-dominated (F) regressive systems are typically characterised by progradational, highly lobate deltas that scale in size with river discharge and that contain active and abandoned lobes fed by distributary channels (e.g. Mississippi and Wax Lake Deltas, USA; Coleman and Gagliano 1964; Roberts 1997; Wellner et al. 2005). Increasing tidal influence (Ft) results in more funnel-shaped and straighter distributary channels (Gugliotta and Saito 2019) and, between distributaries, more sinuous tidal channels and potentially tidal flats. Increasing wave influence (Fw) is marked by a less rugose, more cuspate lobe geometry with isolated beach ridges potentially developing into localised, connected beach ridge sets (strandplains). Mixed-process deltas are characterised by combinations of morphological features, the predominance and geometry of which depend on the relative strength of the influencing processes. For example, the mixed wave- and tide-influenced Mekong delta, Vietnam, shows relatively straight, slightly funnel-shaped distributary channels, due to tidal influence, as well as interdistributary beach ridges, due to wave influence (Figure 8.7).

Wave-dominated (W) regressive systems are characterised by curvilinear shorelines with land-attached strandplains containing well-developed beach ridge sets (e.g. Nayarit strandplain, Pacific coast of Mexico; Curray et al. 1969). Increasing fluvial influence (Wf) results in minor shoreline protrusion around the active distributary, which may be deflected in the dominant wave-driven longshore direction, giving rise to delta asymmetry (Bhattacharya and Giosan 2003). Increasing tidal range (Wt) widens the foreshore,

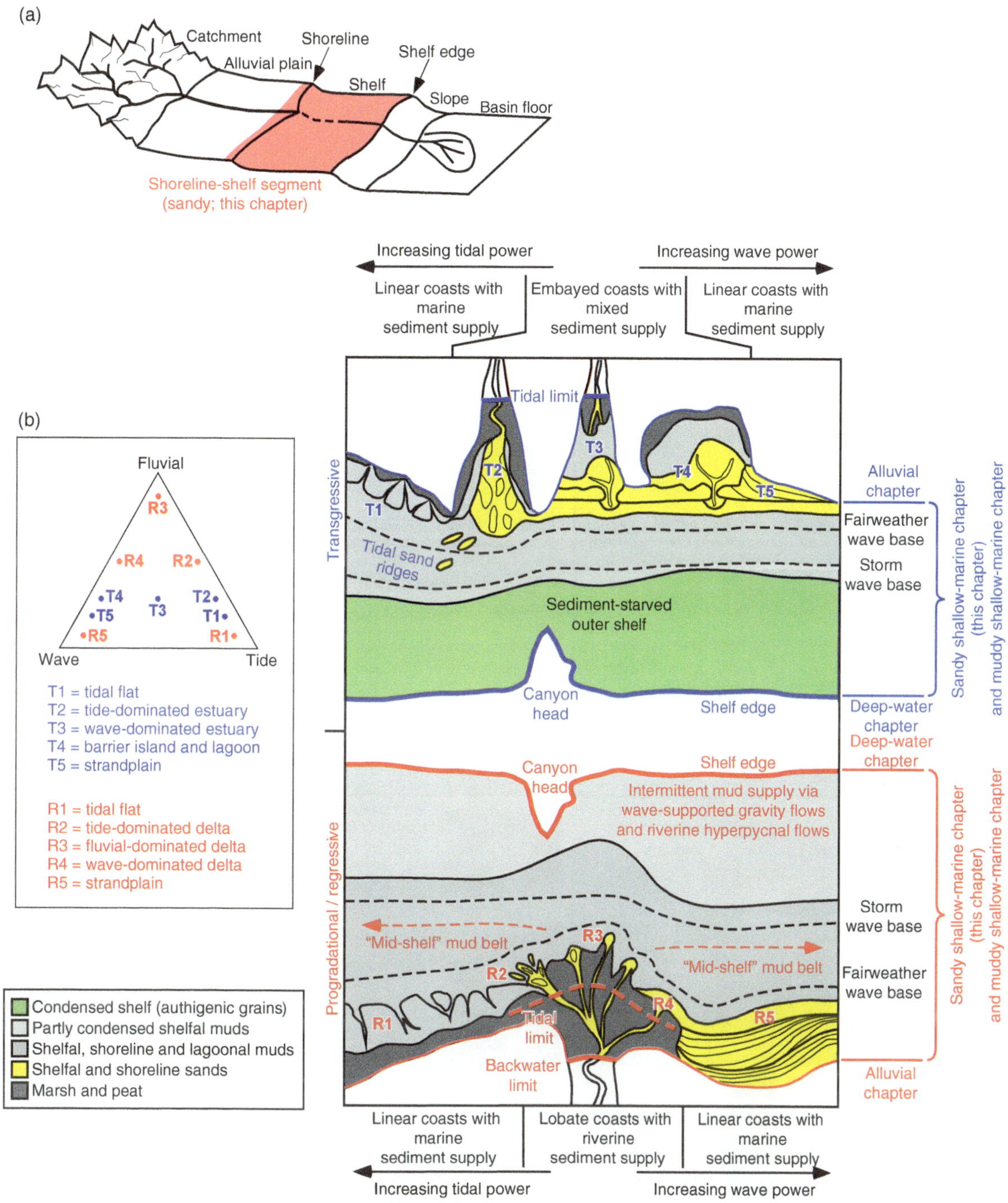

Figure 8.3 Context of sedimentary environments and their deposits addressed in this and related chapters. (a) Conceptual sketch of a sediment routing system from source to sink, highlighting the shoreline-shelf segment. Note that glacial and aeolian sediment-transport processes are not represented. (b) Plan views of idealised depositional environments along transgressive (upper, blue) and regressive (lower, red) shorelines and shelves (Adapted from Boyd et al. 1992). These environments are plotted on a ternary diagram showing the relative influence of wave, tide and fluvial (river) processes. Sand-rich shorelines and shelves addressed in this chapter are shown in yellow, and their mud-rich equivalents (Chapter 9) are shown in grey. The landward boundary between coastal environments (this chapter) and alluvial environments (Chapter 5) is taken at the tidal limit in non-deltaic coastal plains and the backwater limit in deltaic coastal plains. The seaward boundary between shoreline-shelf environments (this chapter) and deep-water environments (Chapter 12) is taken at the shelf edge.

which is exposed at low tide (Dashtgard et al. 2012; Vakarelov et al. 2012) and results in closer spacing of tidal inlets that intersect the shoreline and may link to tidal flats and lagoons landward of the foreshore. Waterside Beach in the Bay of Fundy, Canada, is an example of a tidally modulated, wave-dominated shoreface where tidal creeks and salt marshes have formed in macrotidal conditions (tidal range >4 m; Figure 8.8; e.g. Dashtgard et al. 2006).

Tide-dominated (T) regressive and transgressive systems are typically characterised by digitate shorelines with tidal

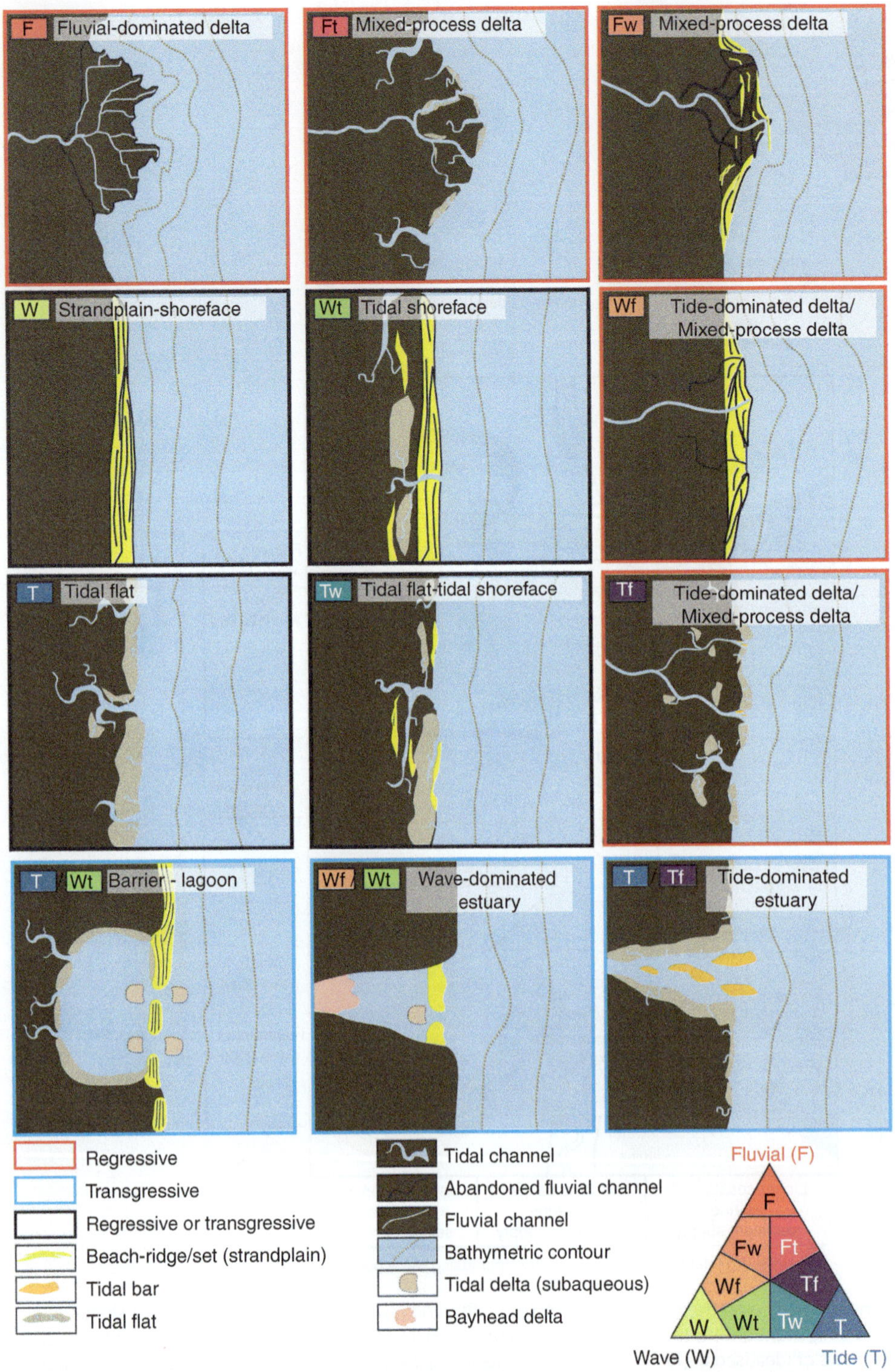

Figure 8.4 Generalised models from modern systems (Ainsworth et al. 2011 / The American Association of Petroleum Geologists) illustrating the variation in plan view shoreline morphology and distribution of depositional elements with changes in process regime. An overview of their classification is given in Box 8.1. The models represent points in a continuum of morphological and depositional style that is also influenced by sediment grain size. Sub-regions of several models could be classified in different process categories, as illustrated in modern systems (Figures 8.5 and 8.6).

channels, unvegetated tidal flats and wetland vegetation tolerant of salinity variations. Tidal inlets may intersect shorelines, notably in mixed tide-wave-influenced (Tw or Wt) transgressive barrier-lagoon systems (e.g. East Friesian Islands, Figure 8.9). Tidal bars in river mouths (e.g. in the transgressive Gironde estuary; Figure 8.10) and tidal ridges in delta front and shelf areas may also occur. Increasing wave influence (Tw) is expressed by less rugose shorelines that typically include isolated sandy beach ridges. Fluvial influence (Tf) results in wide, funnel-shaped and relatively

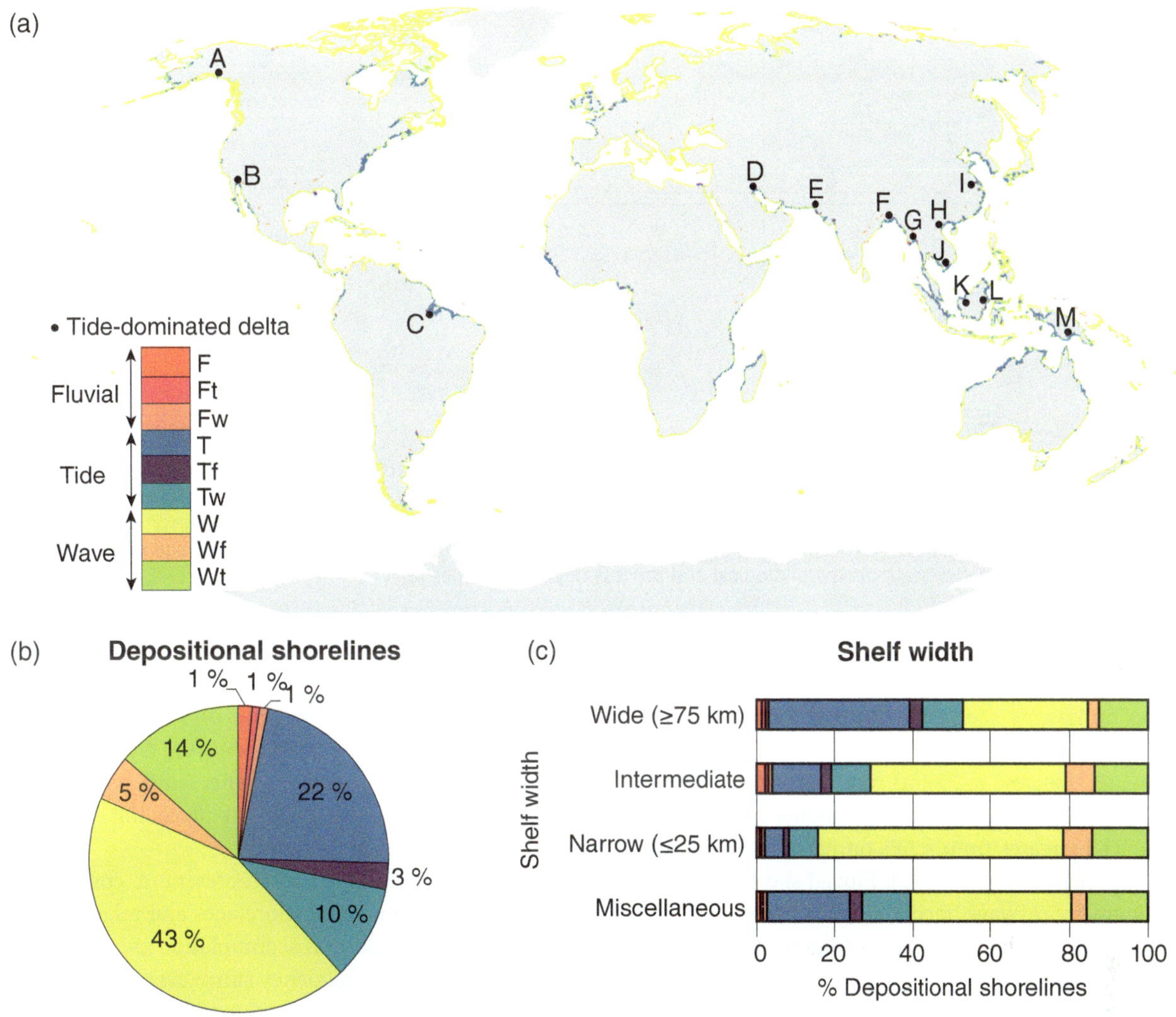

Figure 8.5 Quantification of modern global shoreline process regimes (Adapted from Nyberg and Howell 2016), using the two-tier ternary diagram of Figure 8.4. (a) Process classification of modern shorelines (see Figure 8.4) showing the locations of tide-dominated deltas (Goodbred and Saito 2012): (A) Copper, (B) Colorado, (C) Amazon, (D) Shatt-al-Arab, (E) Indus, (F) Ganges–Brahmaputra, (G) Irrawaddy, (H) Red River, (I) Yangtze, (J) Mekong, (K) Rajang, (L) Mahakam and (M) Fly. (b) Proportion of each process category along depositional shorelines. (c) Relationship between binned shelf width and proportion of the process categories along depositional shorelines. The 'miscellaneous' class is for shallow seas and seaways (e.g. Baltic Sea, Red Sea).

straight distributary channel(s) (Gugliotta and Saito 2019) and potentially shoreline-perpendicular tidal bars within and offshore of the distributary channels.

Modern shoreline morphologies and depositional elements are dynamic and only represent a 'snap-shot' of the forms and processes that may be preserved within a given depositional system (e.g. Howell et al. 2008). Furthermore, modern shorelines developed during a global highstand of sea level, which does not represent large periods of the geological past. For example, most modern deltas are fed by rivers that debouch particularly large volumes of sediment, while many smaller-load rivers are currently characterised by estuaries. A markedly different distribution of environments existed during the last global sea-level lowstand some 20 ka ago, when the sea level was approximately 120 m lower. Importantly, the preserved deposits of a depositional system may derive from a much smaller area than the system as whole (e.g. preferential preservation of channel systems with erosional bases). Process classification based on the facies of an ancient depositional system may therefore be biased by the few preferentially preserved sub-environments.

Shoreline morphology and depositional elements have been qualitatively and, to some extent, quantitatively linked to process regime in both regional (100–1000s km) and local (1–10s km) assessments of modern shoreline systems using the Nyberg and Howell (2016) and Ainsworth et al. (2011) process classification schemes. Examples include the Mitchell delta in Australia (Nanson et al. 2013), Brunei Bay (Collins et al. 2018a), Mekong delta (Figure 8.7; Gugliotta and Saito 2019) and over 400 coastal systems in Asia (Vakarelov and Ainsworth 2012). According to Nyberg and Howell (2016), 28% of global shorelines are depositional, of which 62% are wave dominated, 35% tide dominated and 3% river dominated (Figure 8.5A,B). Globally, shelf width is a major control: 90%

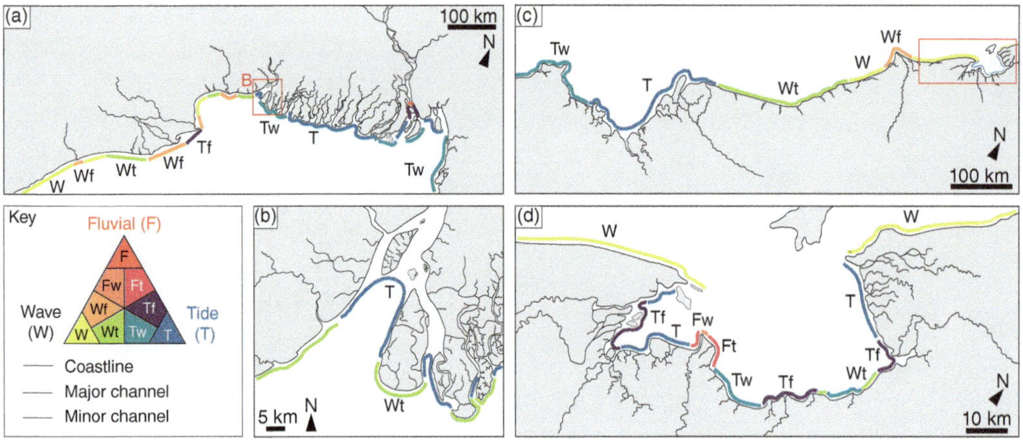

Figure 8.6 Modern shoreline examples showing changes in plan view geomorphology linked to process regime across spatial scales. (a) Northeastern India and Bangladesh, including the Mahanadi–Brahmani (wave-dominated) delta and Ganges–Brahmaputra (tide-dominated) delta (see also Figure 9.3), which includes both wave- and tide-dominated shoreline features at smaller scale (b) (cf. Figure 8.4). (c) Northwest Borneo, including the Rajang delta and embayment (tide-dominated), Baram delta (wave-dominated) and mixed-process Brunei Bay, which exhibits smaller-scale geomorphological and process regimes changes (d).

of shorelines on narrow shelves (≤25 km) are wave dominated and <5% tide dominated, whereas over 30% of shorelines on wide shelves (>75 km) are tide dominated (Figure 8.5C). Considering just the depositional shorelines, tide dominance increases from <20% on narrow shelves to >50% on wide shelves (Figure 8.5C). Fluvial shorelines are correspondingly more wave modified on narrow shelves and more tide modified on wide shelves (Figure 8.5B,C).

8.2.2 Shelf Types

Shelf morphology and width are related to both structural evolution and sediment accumulation (e.g. Helland-Hansen et al. 2012). In this section, we consider sediment dispersal and reworking on parts of the shelf that are distal from the coeval shoreline. Shoreline transits across the shelf and associated stratigraphic patterns that reflect construction of depositional shelves are considered in Section 8.5.

Sediment dispersal, morphology and facies architecture on subaqueous shelves are related to the interaction of multiple processes, notably fair-weather and storm waves, tidal currents and oceanic currents (e.g. Johnson and Baldwin 1996) (Figure B8.1.1d in Box 8.1). Most modern shelves are wave dominated (80%), with a subordinate proportion dominated by tides (17%) or oceanic currents (3%) (Swift et al. 1986). Note that these proportions of dominant processes on shelves differ from those of the shorelines that fringe the same shelves (Section 8.2.1; Nyberg and Howell 2016). Shelves associated with regressive shorelines tend to be characterised by high sediment supply ('supply-dominated regime' of Swift and Thorne 1991), such that shoreline-supplied sediment may accumulate in coeval shelf deposits (Figure 8.3). In contrast, shelves associated with transgressive shorelines are sediment starved ('accommodation-dominated regime' of Swift and Thorne 1991). On such sediment-starved shelves, the prevailing processes commonly rework older, relict sediment accumulations that were deposited during the preceding shoreline regression into 'palimpsest' deposits.

Shelves dominated by waves represent a contiguous extension of wave-dominated shorefaces and related shorelines. Sediment dispersal and accumulation are controlled by high-magnitude, low-frequency storm events, which are separated by fair-weather periods. Sand is typically trapped close to the shoreline, within the 'littoral energy fence', by onshore transport driven by shoaling fair-weather waves (e.g. Swift and Thorne 1991). Offshore sand transport occurs when the littoral energy fence is overcome during major storms by storm-generated downwelling currents ('relaxation' currents, previously thought to represent 'storm-surge ebb' currents). These are subsequently deflected sub-parallel to the shoreline by Coriolis acceleration, to form 'geostrophic' currents (Swift et al. 1986). Sediment transport efficiency is high under such conditions, due to a combination of vigorous wave orbital motion that keeps fine sand, silt and clay in suspension and currents. The currents may form solely in response to the wind-driven circulation that generates the storm waves (e.g. Swift et al. 1986), or they may contain a contribution from dense, sediment-laden hyperpycnal flows caused by high, storm-driven fluvial discharge and/or re-suspension of shelf sediments (e.g. Traykovski et al. 2000). The resulting current- and wave-enhanced sediment gravity flows are particularly important in transporting mud across shelves (e.g. Traykovski et al. 2000; Macquaker et al. 2010) and can account for the formation of broad mud belts on the mid or outer parts of subaqueous shelves (e.g. Plint et al. 2012; Sections

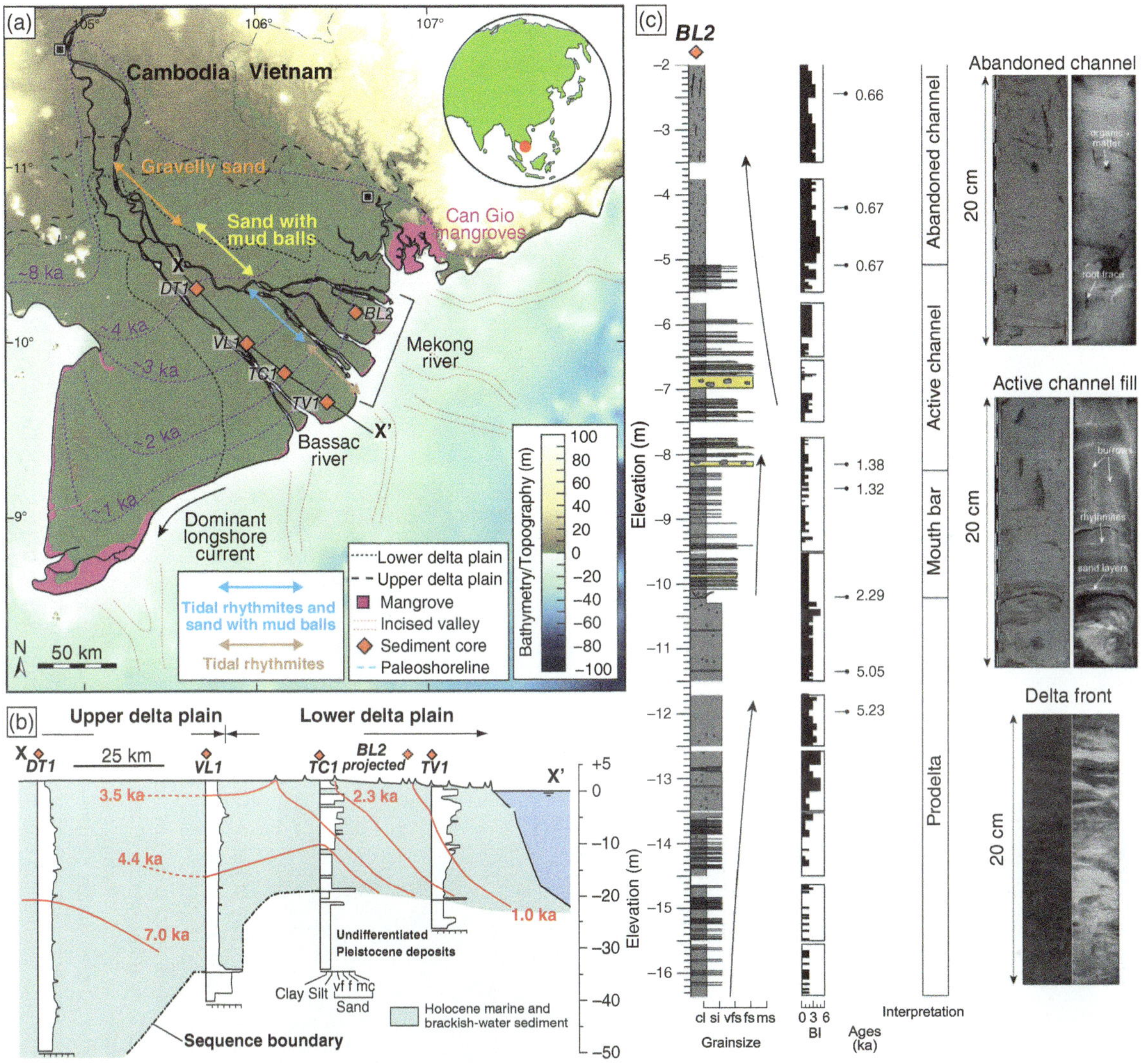

Figure 8.7 Example of a modern mixed-process delta: Mekong delta, Vietnam (labelled '3' in Figure 8.1a). (a) Map of the Mekong delta with palaeoshorelines (dashed purple) (adapted from Zoccarato et al. 2018 with data from Ta et al. 2002 and Tamura et al. 2012) and dominant facies type measured along the Mekong river (coloured arrows; from Gugliotta et al. 2017). (b) Shore-perpendicular transect through the upper and lower delta plain adjacent to the Bassac river (labelled X–X′ in Figure 8.7a) with isochronous lines (Tamura et al. 2012 / Geological Society of America, modified after Ta et al. 2005). (c) Sedimentary log of core BL2 with radiocarbon ages illustrating sedimentological and stratigraphic evolution over the last ~5000–6000 years (labelled BL2 in Figure 8.7d) (Gugliotta et al. 2021 / Springer Nature / CC BY 4.0). Selected photographs (left) and X-radiographs (right) from bottom to top are representative of delta fronts/mouth bars (from Can Gio; Collins et al. 2021b / with permission of Elsevier), active- and abandoned-channel fills (from BL2; Gugliotta et al. 2021 / Springer Nature / CC BY 4.0).

9.1.2, 9.2.2, and 9.2.3). Offshore and along-shelf sand transport during major storms may result in laterally extensive, shoreface-attached sand beds on the shelf (e.g. Morton 1981; Grundvåg et al. 2021), and relict sands on the shelf may also be reworked and sculpted by storm waves (e.g. Swift and Field 1981).

Tide-dominated shelves are swept by strong tidal currents and are associated with large transverse (sand waves) and longitudinal bedforms (sand ridges, sand ribbons) where sufficient sand is available (e.g. Stride 1974; Belderson et al. 1982). On supply-dominated sandy shelves, of which there are few modern examples, sand is supplied by deltas and then reworked from the delta front by tidal currents (e.g. Cretaceous Bilá Hora, Jizera, Teplice and Březno formations in Bohemian Basin, Czech Republic, Uličný 2001; Eocene Baronia Sandstone, Ager Basin, Spain, Olariu et al. 2012a; Eocene Esdolomada Member, Tremp-Graus Basin, Spain, Olariu et al. 2012b). On accommodation-dominated shelves, of which there are currently many due to recent sea-level rise, sand is reworked from relict accumulations, including incised valleys, over-stepped barrier islands, and shelf-edge deltas (e.g. Snedden and Dalrymple 1999). Tidal reworking is most pronounced in settings where tidal currents are geographically or bathymetrically constrained, such as straits (e.g. Messina Strait, Figure 8.11). Here, tidal current velocities are increased

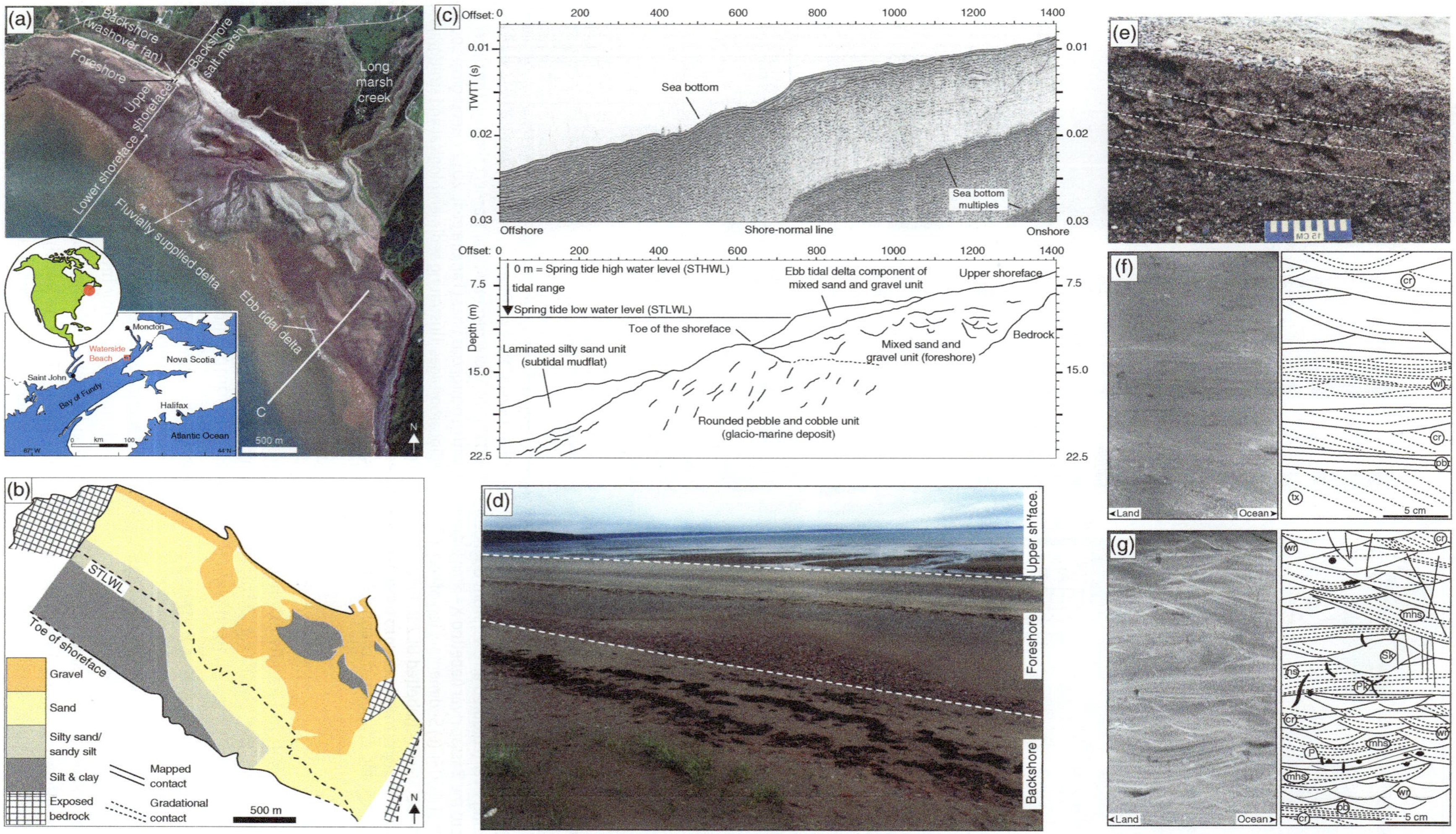

Figure 8.8 Example of a modern tidally-modulated shoreface: Waterside Beach, Bay of Fundy, Canada. (a) Satellite image taken at low tide (Dashtgard et al. 2021 / The Geological Association of Canada). (b) Simplified sediment distribution map of the intertidal and subtidal components (Dashtgard et al. 2009 / Society for Sedimentary Geology). Sediment calibre is determined by both river feeder systems and local, reworked, very coarse glacial deposits. (c) Shore-normal shallow seismic-reflection profile and line drawing with x30 vertical exaggeration showing the modern shoreface above a post-glacial transgressive succession (located in Figure 8.8a; Dashtgard et al. 2007 / with permission of Elsevier). (d) Photo of 100–200 m wide, 2°–4° dipping, backshore–foreshore–upper shoreface during falling tide (Dashtgard et al. 2012 / with permission of Elsevier). (e) Photograph of pebbly sand and gravel layers that dip seaward (to right) in foreshore deposit (Dashtgard et al. 2012 / with permission of Elsevier). (f, g) Box core X-ray images and line drawing interpretations of: (f) upper-shoreface (middle intertidal) deposit with plane-bedded sand (pb), shore-parallel current ripples (cr), wavy-parallel laminae (wl), and offshore-directed trough cross-beds (tx); and (g) lower-shoreface (lower intertidal and subtidal) deposit displaying wave ripples (wr), offshore-directed current ripples (cr), plane beds (pb), possible hummocky cross-stratification (hs) and micro-hummocky cross-stratification (mhs) in sand with intervening mud flasers and cross-cutting trace fossils (*Skolithos*, Sk; *Polykladichnus*, Pk; *Planoites*, P) (Dashtgard et al. 2009 / Society for Sedimentary Geology).

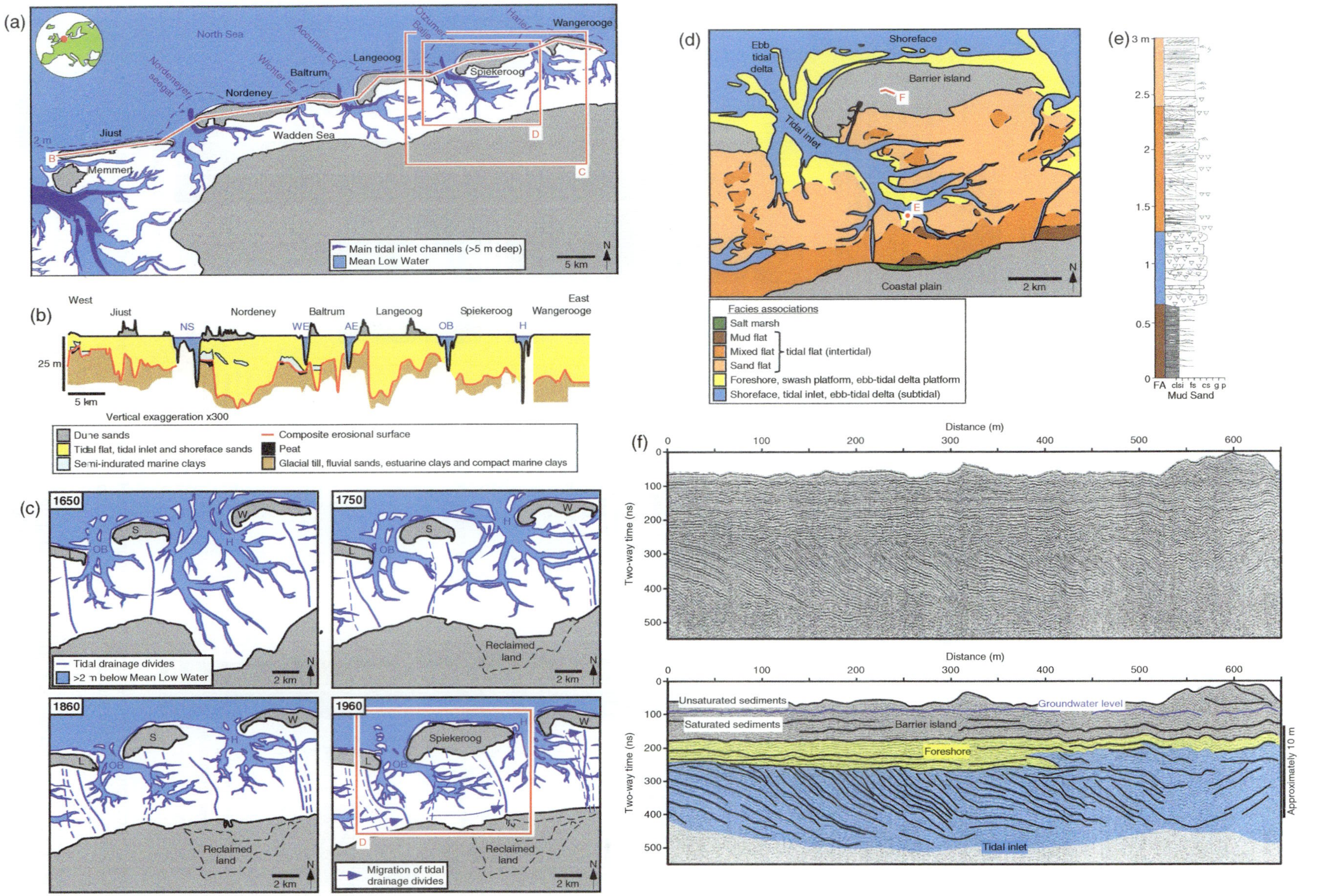

Figure 8.9 (a) Map of the East Friesian barrier islands and associated tidal inlets and tidal flats of the eastern Wadden Sea (labelled '9' in Figure 8.1f–g) and (b) cross section through Quaternary deposits of the East Friesian Islands (adapted from Fitzgerald and Penland 1987). (c) Historical changes in the morphology of tidal drainage systems adjacent to and landward of Spiekeroog barrier island (adapted from Fitzgerald and Penland 1987). Lateral migration of the tidal inlets explains the composite erosion surface underlying the tidal flat, tidal inlet and shoreface sands (Figure 8.9B). (d) Map of surficial sediment distribution of Spiekeroog barrier island and back-barrier tidal flats (adapted from Eitner 1996). (e) Facies succession from a core through tidal flat deposits (adapted from Chang et al. 2006; Flemming 2012), with facies associations similar to FA5 in Table 8.1. (f) Uninterpreted (upper) and interpreted (lower) ground-penetrating radar (GPR) profile through the central part of the Spiekeroog barrier island and an underlying, laterally accreting tidal inlet channel (adapted from Tronicke et al. 1999). The core (Figure 8.9e) records local regression of back-barrier tidal flats, while the GPR (Ground Penetrating Radar) profile (Figure 8.9f) records local regression and lateral migration of the barrier island complex.

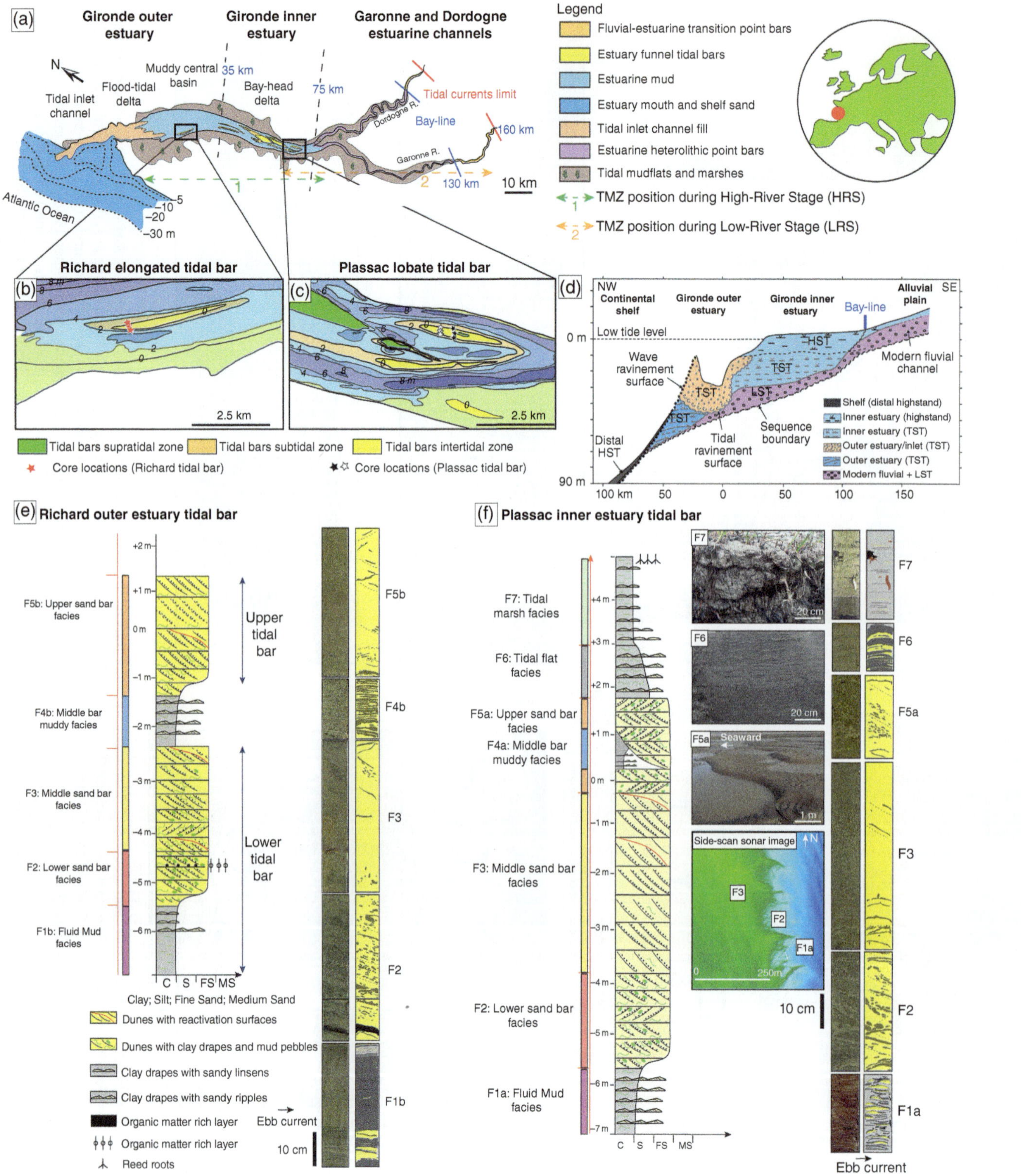

Figure 8.10 (a) Map and depositional environments of the wave-dominated Gironde estuary, France (labelled '10' in Figure 8.1f) and detailed bathymetric maps of (b) the elongated 'Richard' tidal bar and (c) the lobate 'Plassac' tidal bar (Virolle et al. 2020 / with permission of Elsevier). (d) Cross section illustrating Quaternary fluvio-estuarine deposits and stratigraphy along the axis of the Gironde incised valley fill (adapted from Allen and Posamentier 1993; Lericolais et al. 2001; Virolle et al. 2020). Synthetic sedimentary logs and facies successions through (e) the 'Richard' tidal bar and (f) the 'Plassac' tidal bar (Virolle et al. 2020 / with permission of Elsevier). To the right of the logs are photos and images of the Plassac bar surface (Figure 8.10f only), providing context for the facies, and photographs and interpretive line drawings of representative cored intervals through the facies (Figure 8.10e, f). TMZ, Turbidity maximum zone; TST, transgressive systems tract; LST, lowstand systems tract.

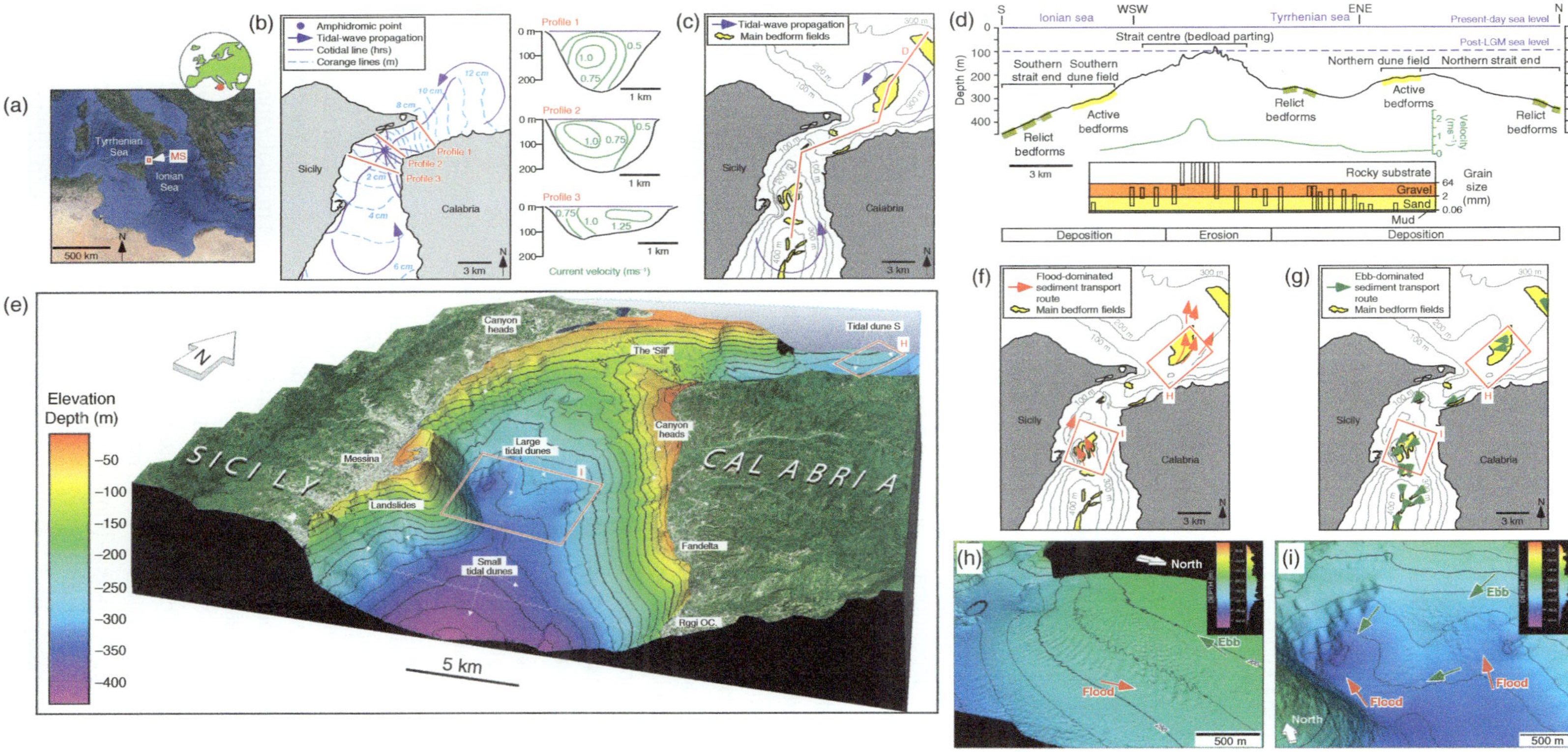

Figure 8.11 (a) Satellite image locating the modern Messina Strait (labelled 'MS') in the central Mediterranean Sea. Google / Maxar Technologies. (b) Map of the M_2 tidal wave in the Messina Strait (adapted from Androsov et al. 2002), showing the amphidromic point around which the tidal wave propagates. Cotidal lines indicate where the tidal wave has the same phase as it rotates around the amphidromic point. Tidal range is zero at each amphidromic point and increases outward, as indicated by corange lines. Three bathymetric profiles (right) show the horizontal velocity of the M_2 tidal component during one single flood or ebb tidal phase (adapted from Mosetti 1992 in Longhitano 2018a). (c) Simplified map and (d) along-axis bathymetric profile of the seabed in the Messina Strait, showing the position of the main bedform fields (adapted from Longhitano 2018a), sediment grain size distribution and tidal current velocity trend (adapted fromNesteroff and Rawson 1987). Relict bedforms may have been active during relative sea-level lowstand at the Last Glacial Maximum (Longhitano 2018a / with permission of Elsevier). (e) Block model showing the main bathymetric features of the modern Messina Strait, including the main bedform fields and landslides, canyons and fan deltas that supply sediment from the strait margins (Longhitano 2018b). (f, g) Sediment transport routes in the strait reconstructed from bedform morphology during (f) the flood stage and (g) the ebb stage (adapted from Longhitano 2018a). (h, i) Multibeam bathymetric views of the (h) northern and (i) southern dune fields (Figure 8.11d–g) showing interference and superimposition of two dune populations with different orientations, attributed to the dominant flood and subordinate ebb tidal currents (Longhitano 2018a / with permission of Elsevier). Flood and ebb tidal currents are oriented approximately perpendicular to each other.

by funnelling and shoaling of the tidal wave, such that extensive fields of transverse sandy bedforms may be developed if sufficient sand is available (e.g. Rossi et al. 2017b; Longhitano 2018a).

Oceanic current-dominated shelves are swept by strong currents from the deep ocean that impinge on the outer shelf. Strong current reworking results in sand bypass and development of a gravel lag (e.g. Agulhas current, offshore eastern South Africa; Flemming 1980), whereas weaker reworking results in the development of a sand sheet and bedforms (e.g. Kuroshio current, offshore southern Japan; Nishida et al. 2020).

For both supply- and accommodation-dominated shelves, the volume of sediment that is stored on the shelf versus that transferred beyond the shelf edge is strongly controlled by shelf width, shelf-edge depth and the amplitude of sea-level change. For very wide shelves, those in excess of 400 km, only high-amplitude sea-level falls in supply-dominated systems can deliver sediments to deeper water (Zhang et al. 2019a).

Box 8.1 Process-Based Classification of Shorelines and Shelves

There is no single classification of clastic shorelines. Here we briefly review the most commonly used classification schemes and explain their history because of their use in older literature. Several classifications exist that variously reflect changes in: (i) relative sea level; (ii) river mouth processes and marine reworking by tides, fair-weather waves and storm processes; (iii) volume and grain size of sediment supply; (iv) tectonic setting; and (v) gravitational processes.

An overarching classification recognises net-depositional versus net-erosional shorelines on medium-to-long timescales (hundreds to tens of thousands of years) (Curray 1964) (Figure 8.2). Boyd et al. (1992) subdivided shorelines into regressive (or progradational) and transgressive (or retrogradational) and related the associated changes in sediment body geometry and distribution to variations in the balance between rates of sediment supply and accommodation creation (Figure 8.3). Further subdivision is based on recognising the interplay between fluvial sediment input against tide and wave reworking, plotted as a ternary process diagram. These two subdivisions, transgressive versus regressive and the ternary process classification, permit a schematic dynamic model of the expected evolution in depositional shoreline types during relative sea-level change (Figure B8.1.1a; Dalrymple et al. 1992). The central ternary plot is a combination of both transgressive (estuaries and barrier-lagoons) and regressive (deltas and strandplain-shorefaces) shorelines that is linked to process regime in a generic sense (Dashtgard et al. 2021). During regression, estuaries tend to be replaced by deltas and strandplains, whereas transgression leads to development of estuaries, barrier-lagoon systems and embayments.

The classification of shoreline systems on a ternary process diagram was first formulated for modern deltas by Galloway (1975), building upon previous models (Fisher et al. 1969) and statistical analyses (Coleman and Wright 1975). These seminal studies recognised general differences in delta front stratigraphy, areal facies distribution and delta plain morphology between end-member river-, wave- and tide-dominated deltas. Subsequent studies have improved our understanding of variability within the spectrum of mixed-energy systems (e.g. Zuchuat et al. 2023). In classifying modern and ancient deltaic shorelines, the process regime of the delta front (if preserved) takes precedence as the process regime of the delta plain may be different (Elliot 1986). More recently, the ternary process classification has been formalised into a three-tier classification (Ainsworth et al. 2011), in which primary ('dominant'), secondary ('influencing') and tertiary ('affecting') processes are interpreted. There is also a simpler two-tier classification in which only the primary and secondary processes are interpreted (Figure B8.1.1b) (Nyberg and Howell 2016; Collins et al. 2021a). Both ternary classifications can be applied qualitatively, and possibly quantitatively, across the spectrum of transgressive and regressive shoreline types. However, due to ambiguity in process interpretations of sedimentary structures (Box 8.2) and common limitations in the completeness of ancient data, the two-tier ternary process classification is easier to adopt in studies of ancient systems.

The morphology of non-deltaic shorelines is also strongly linked to the relative effectiveness of wave and tidal processes. Different types of strandplain-shorefaces and tidal flats may be discriminated based on tidal range and wave height; for example, such shorelines with <2 m tidal range (microtidal) tend to be wave dominated, those with 2–4 m tidal range (mesotidal) tend to be mixed wave–tide, and those with >4 m tidal range (macrotidal) tend to be tide dominated (Hayes 1975, 1979). More recently, conceptual models have been developed that differentiate the relative impact of storm and fair-weather waves on preserved strata along non-fluviatile shorelines (Figure B8.1.1c) (Yang et al. 2005; Dashtgard et al. 2009, 2012, 2021). Further offshore, a few clastic shelves are also impacted by oceanic currents, in addition to tidal currents and oscillatory waves, justifying a four-component diagram (Figure B8.1.1d) (Johnson and Baldwin 1996).

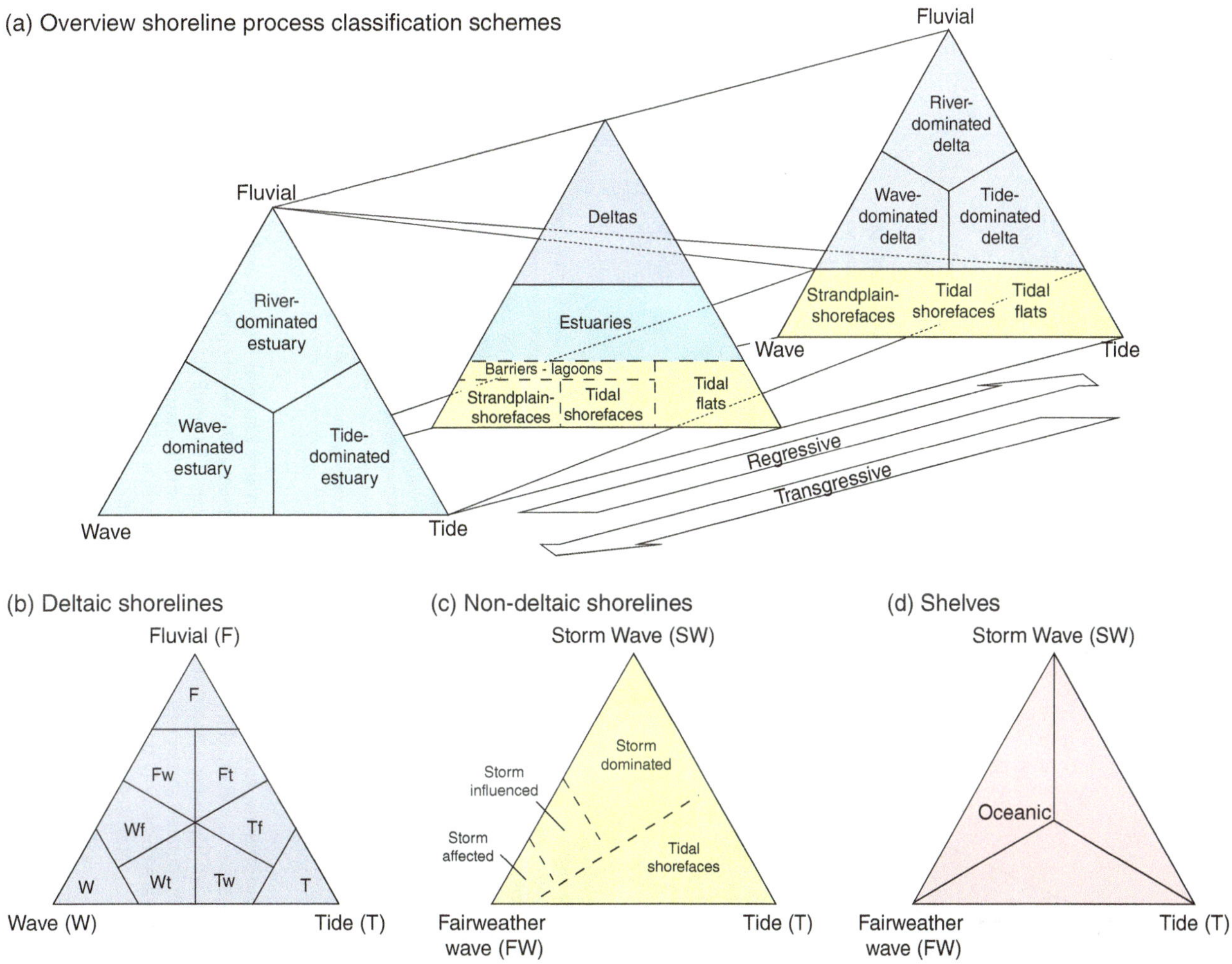

Figure B8.1.1 Ternary classifications relating to hydrodynamic processes of shorelines and shelves, adapted from (a) Dalrymple et al. (1992) and Boyd et al. (1992), (b) Nyberg and Howell (2016) and Collins et al. (2021a), (c) Yang et al. (2005) and Dashtgard et al. (2009, 2012) and (d) Johnson and Baldwin (1996).

As understanding of process interactions increases, additional process components may be added to classification diagrams, including glacial and aeolian processes that commonly act on high-latitude and arid coastlines (see Section 7.3.16). Many aspects of the classification schemes are also relevant to evaporite- and carbonate-dominated coastlines in which sand is the dominant grain size and physical processes dominate over biological and chemical processes (see Sections 10.2.4 and 11.2.3). Waves and tides are the dominant physical processes acting on such coastlines, and fluvial processes that introduce freshwater and detrital sediment are largely absent.

Grain size controls the effectiveness of reworking processes and the consequent shoreline morphology and depositional elements. Process-based classifications have therefore been extended to include variations related to grain size for deltaic shorelines (Figure B8.1.2a; Reading and Orton 1991; Orton and Reading 1993) and shelves (Figure B8.1.2b; Johnson and Baldwin 1996). This turns the classic triangular diagram into a ternary prism. Whatever the diagram, systems are habitually plotted as points, but each system invariably contains sub-environments that are subject to different wave, tide and/or fluvial conditions. As a result, the process regime of many systems cannot be defined from a single vertical section (Section 8.4). Furthermore, as shoreline systems undergo regression or transgression, the process regime may evolve in relation to the basin physiography (geometry and bathymetry), sediment load and/or climate changes (Elliot 1986). Ancient shoreline and shelfal systems plotted as points represent the 'average' process regime over time. Plotting systems as polygons is more representative of the spatial and temporal process variability within an individual system (Figure B8.1.2b).

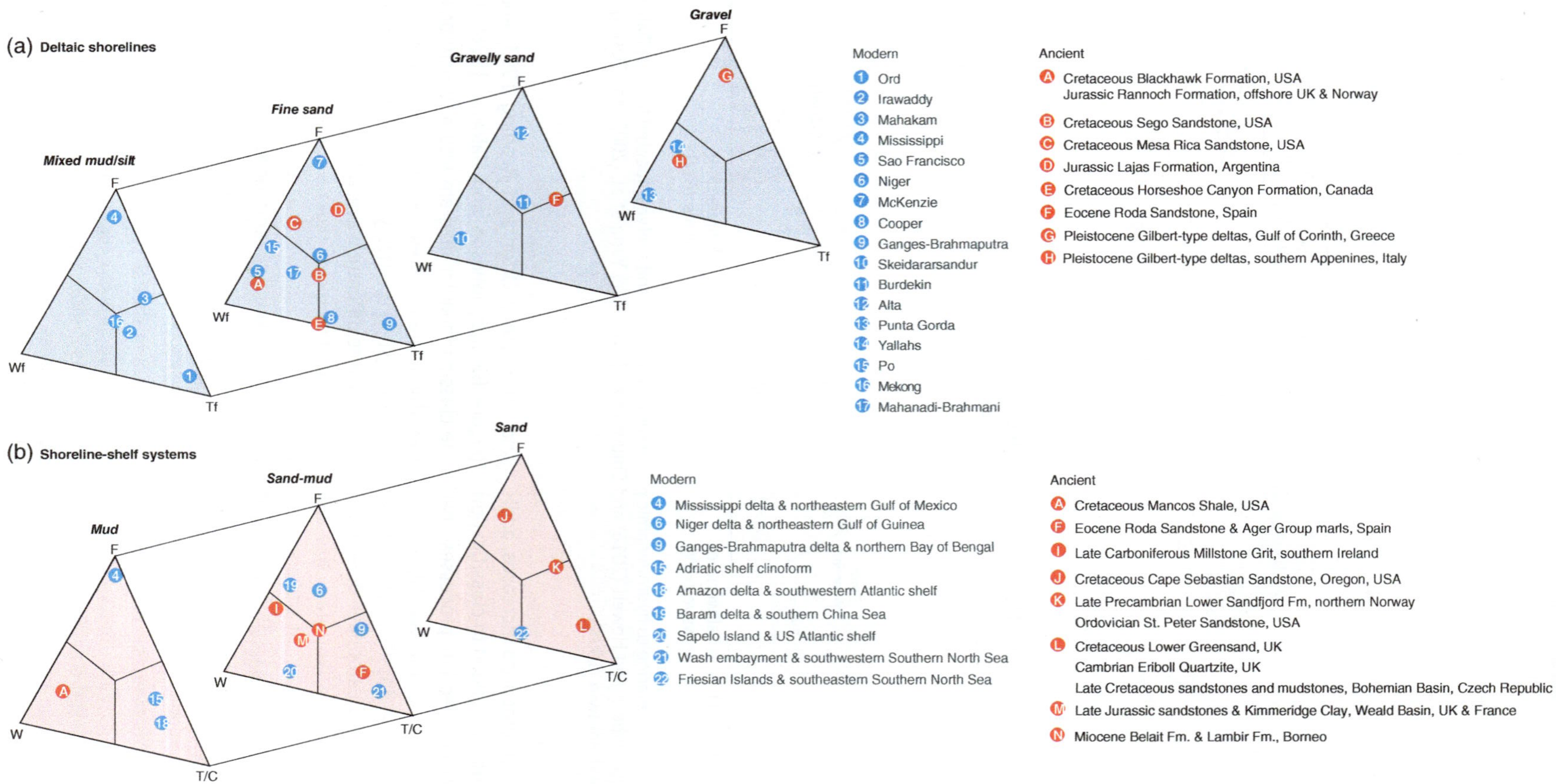

Figure B8.1.2 Ternary classifications of hydrodynamic processes on shorelines and shelves, emphasising variations in grain size between modern and ancient systems, for (a) deltaic shorelines (adapted from Orton and Reading 1993) and (b) shoreline-shelf systems (adapted from Johnson and Baldwin 1996). For clarity, these examples are shown as points, which represent the centroids of larger regions that encompass process variability in each example. Deltaic shorelines (Figure B8.1.2a) occupy the upper part of the progradational panel in Figure B8.1.1a and do not include strongly wave- and tide-dominated shorelines. Shoreline-shelf systems (Figure B8.1.2b) combine classification schemes for shorelines (Figure B8.1.1a) and shelves (Figure B8.1.1d), the deposits of which are commonly transitional and are not readily differentiated in many ancient strata. Selected deltaic shorelines and shoreline-shelf systems, including those that illustrate this chapter, are placed in both ternary classifications.

8.3 Processes

Next, we review the main physical processes that operate in sand-rich marginal- and shallow-marine environments (Figure 8.3) with reference to idealised shoreline-shelf transects that provide illustrative examples of where wave, tidal or fluvial processes are individually dominant (Sections 8.3.1–8.3.3) or act in combination (Section 8.3.4). Aeolian (Chapter 7) and evaporative (Chapter 10) processes may be important in arid and semi-arid climatic settings, and glacial processes (Chapter 14) in cold settings. Oceanic processes occur on shelves and in shallow seas characterised by density-driven circulation of different water masses. These processes are however not considered here but are described in Chapters 9, 12 and 13. More details of the bedforms and physical sedimentary structures described in this section are available in Collinson and Mountney (2019).

8.3.1 Physical Processes I: Waves and Wind

Surface waves are generated by wind blowing over the water surface. They increase in amplitude and wavelength with wind speed and the duration and fetch length over which the wind ruffles the surface. In deep water, waves move water particles in nearly circular orbits that decrease progressively in diameter with depth. In shallow water, the circular orbital motion is distorted as it impinges on the seabed, to produce a back-and-forth motion. The water depth at which the wave-induced motion moves sediment of a particular grain size (e.g. sand) on the seabed is termed wave base (or effective wave base) and occurs where water depth is approximately half the dominant surface wavelength. Wave base is thus shallower during normal, fair-weather conditions characterised by short wavelength surface waves, and deeper during storm events, when surface waves are of longer wavelength (Figure 8.12). Fair-weather and storm wave base are commonly portrayed in interpretations of ancient sediments as two distinct and temporally invariant water depths, but in reality, there is a continuous distribution of wave heights and wavelengths between fair-weather and storm waves (Figure 9.14; Storms 2003; Peters and Loss 2012). Thus, fair-weather wave base and storm wave base are best conceptualised as water depths at which there is, respectively, a specific high probability and a specific low probability of wave-induced current velocity exceeding a threshold erosion velocity for sand or mud at the seabed (Peters and Loss 2012). In addition to water depth, wave base is therefore also dependent on the grain size of sediment on the seabed and is deeper for mud than for sand, because mud can normally be more easily resuspended (Figure 8.12; see also Section 9.3.3). Irrespective of grain size, the depths of both storm and fair-weather wave base reflect wave climate, fetch length, wind strength and persistence, shelf width and palaeogeographic setting; for example, wave base is deeper in open oceans than in sheltered bays (e.g. Peters and Loss 2012) and wave energy may be attenuated by propagation across wide shelves (e.g. Ainsworth et al. 2011). The seaward-sloping area in front of the shoreline where shoaling waves interact with sediment on the seabed (i.e. above wave base) is termed the 'shoreface' (e.g. Figure 8.8c). In Figure 8.12, wave processes are illustrated along a shoreline-shelf transect through a regressive strandplain fronted by a wave-dominated shoreface that faces the open ocean (also labelled 'R5' in Figure 8.3).

At and above fair-weather wave base, in the oscillatory wave zone, the back-and-forth motion of sand generated by the wave orbitals is symmetrical. Straight-crested, symmetrical ripples are generated on the seabed (Clifton 1976). The orbital motion becomes asymmetrical in the shallower water of the shoaling wave zone, where wave motions consist of a shorter, stronger landward-directed stroke and a longer, weaker seaward-directed stroke. Symmetrical ripples thus pass landward into asymmetrical ripples and dunes (Clifton 1976). Frictional drag of the seabed increases the asymmetry of wave motions in progressively shallower water, causing the surface waves to oversteepen and break in the surf zone (Clifton 1976). Landward movement of water in the shoaling wave and surf zones generates unidirectional currents. Shoaling fair-weather waves therefore generate net-landward movement of sand and gravel at the sea bed on the upper part of a shoreface (Figure 8.12b). In the same water depths, mud is carried as suspended sediment and transported alongshore and offshore.

Where wave crests are nearly parallel to the shoreline (i.e. wave approach is nearly shoreline-perpendicular), circulatory cells are generated in which wave-induced landward-directed currents are deflected into longshore currents that pass laterally into strong rip currents, flowing seaward through narrow channels (Figure 8.13; Komar 1976). Where wave approach is oblique to the shoreline, continuous longshore currents are generated (Figure 8.13; Komar 1976). In combination with the shoaling waves, these unidirectional currents generate longshore bars aligned obliquely along the shoreline (Figures 8.12b and 8.13) and locally dissected by shoreline-perpendicular rip channels. The currents transport sand in ripples and dunes onshore over the crest of the bars, alongshore in troughs between the bars, and offshore in rip channels (Davidson-Arnott and Greenwood 1976;

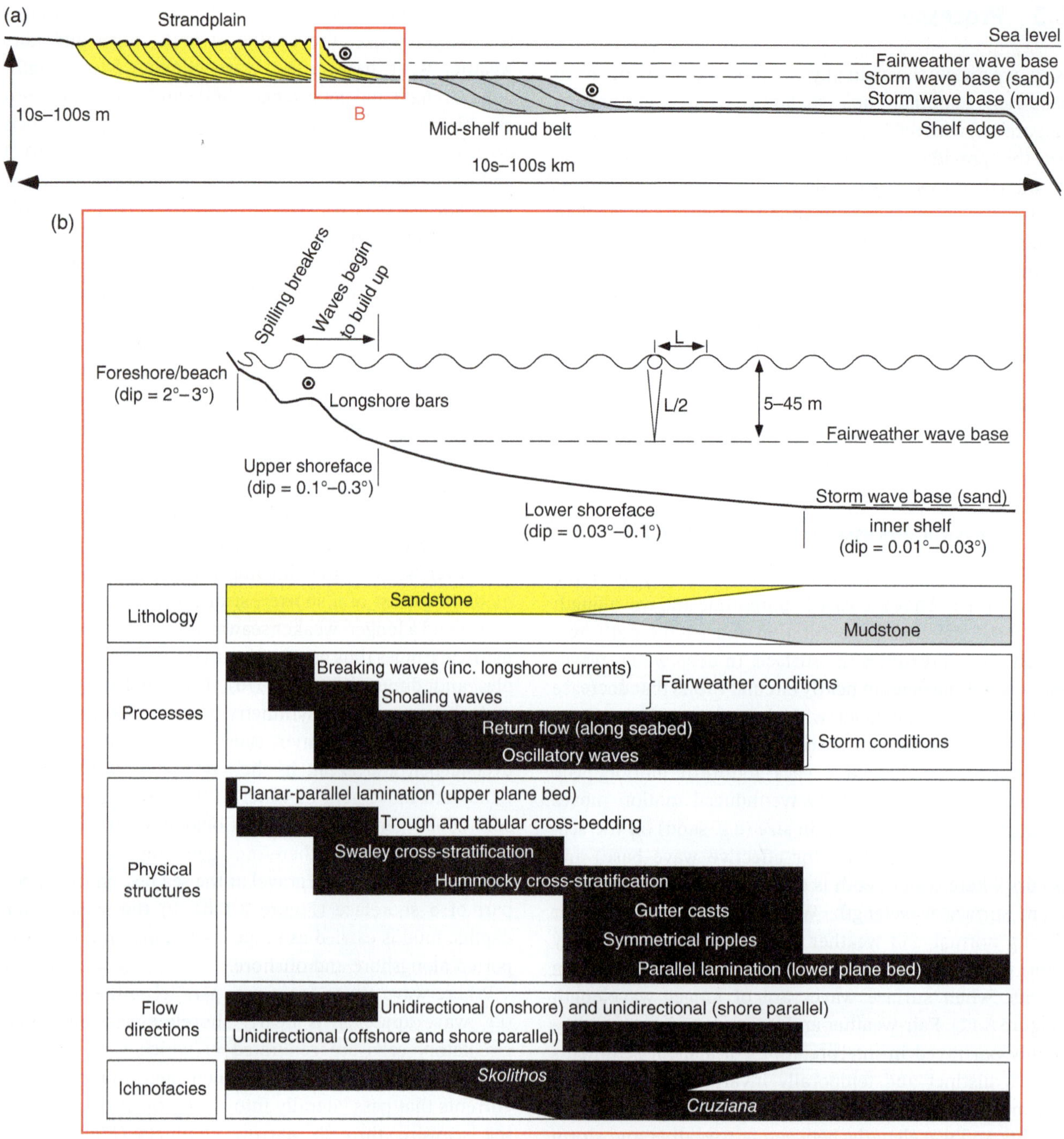

Figure 8.12 Idealised shoreline-shelf transect through a regressive, wave-dominated strandplain, shoreface and shelf: (a) vertically exaggerated shoreline-shelf transect from landward (tidal limit) to seaward (shelf edge); and (b) detail of vertically exaggerated transect through shoreface fronting the strandplain (labelled 'R5' in Figure 8.3), illustrating wave processes and products.

Hunter et al. 1979). At the foreshore, each breaking wave produces a high-velocity, landward-directed swash flow followed by seaward-directed backwash in the 'swash zone'. Swash-backwash flow is sufficiently strong to transport pebbles and to generate upper plane bed conditions in sand (Figure 8.12b).

Fair-weather wave processes result in a concave-upward shoreface profile in which steeper slopes characterise shallower water, where progressively higher current velocities are generated (Figure 8.12b). The development of longshore bars depends on the degree to which wave energy is reflected from, or dissipated at, the upper shoreface and

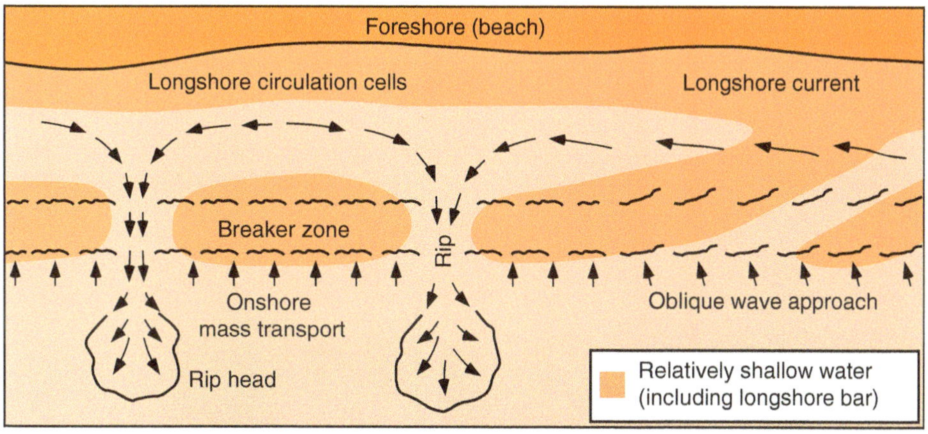

Figure 8.13 Schematic plan view showing water circulation generated by fair-weather waves on the upper shoreface and foreshore (Figure 8.12c) (Komar 1976 / GeoScienceWorld). As waves cross the shoreface, they transport water onshore. Shoreline-perpendicular wave approach (left side of figure) sets up circulation cells that transport water and sand parallel to the shoreline in a trough(s) developed behind a bar(s), and then seaward in narrow rip-current systems that erode shallow channels that dissect the bar(s). Rip currents dissipate seaward of the breaker zone. Oblique wave approach (right side of figure) sets up a unidirectional longshore current that sculpts elongate shoreline-oblique bars and is capable of transporting huge volumes of sand along the coast.

foreshore (or beach) (Wright et al. 1979). 'Dissipative' shorelines in which shoaling wave energy is absorbed by viscous drag in the substrate (e.g. shallow slopes in fine sand) are associated with one or more longshore bars, and their upper-shoreface deposits are characterised by cross-stratification with onshore, longshore and, locally, offshore-directed palaeocurrents. In contrast, steeper coarser or harder 'reflective' shorelines have smooth, steep shoreface profiles that lack longshore bars, and their upper-shoreface deposits are characterised by planar–parallel lamination (Isla et al. 2020).

During storm events, strong winds push water onshore as a 'storm surge' (or 'coastal set-up'), which erodes sand from the upper shoreface and foreshore. This sand is transported either onshore (e.g. to wash-over fans in lagoons, landward of the foreshore) or offshore, onto the lower part of the shoreface and contiguous shelf. The storm surge generates a hydraulic pressure gradient which drives offshore-directed unidirectional currents (relaxation flow) along the seabed. These offshore-directed bottom currents may be augmented locally by high fluvial discharge at river mouths ('storm floods' of Collins et al. 2017; Jelby et al. 2020) and may also be deflected to a shoreline-parallel orientation by Coriolis acceleration on the shelf (Swift et al. 1986). The bottom currents transport sand to the lower shoreface and inner shelf, where they may erode distinctive steep-sided 'gutter casts' and scours into any underlying muds (Figure 8.12b). Storms also generate long-wavelength surface waves, which impinge on the seabed above storm wave base as powerful oscillatory and combined flows (i.e. with both oscillatory and unidirectional components, Figure 8.14; Allen 1985; Swift et al. 1986). Such flows are interpreted to generate graded sand beds that contain hummocky cross-stratification capped by symmetrical ripples (Figure 8.14; Duke 1985; Dumas et al. 2005). On the distal part of the lower shoreface and inner shelf, hummocky cross-stratified sand beds are overlain by bioturbated muds that record quiescent fair-weather conditions (Figures 8.12b and 8.15a). On the proximal part of the lower shoreface, fair-weather mud deposits are removed by later storm-wave erosion, resulting in successions of amalgamated, hummocky and swaley cross-stratified sand beds (Figures 8.12b and 8.15a). During storm events, mud is suspended over large parts of the shelf, above the storm wave base for mud. This suspended mud can move obliquely downslope over the low gradients of the shelf, thus bypassing the inner shelf and accumulating on the mid-to-outer shelf (Traykovski et al. 2000). Downslope transport by muddy, wave-supported gravity flows is capable of generating a mid-shelf mud belt that is genetically linked to, but spatially distinct, and perhaps separated by a bypass surface, from, a coeval sand-rich shoreface (Figure 8.12a; Sections 9.1.2, 9.2 and 9.4.4).

Figure 8.15A presents a facies model and schematic vertical facies succession developed from the fair-weather and storm wave processes described earlier (after Clifton 1976; Davidson-Arnott and Greenwood 1976; Hunter et al. 1979; Hampson and Storms 2003; Dashtgard et al. 2021;). The distributions of morphological elements (e.g. foreshore and shoreline-oblique bars) are modified on shorefaces with a high tidal range (Figure 8.15b; after Dashtgard et al.

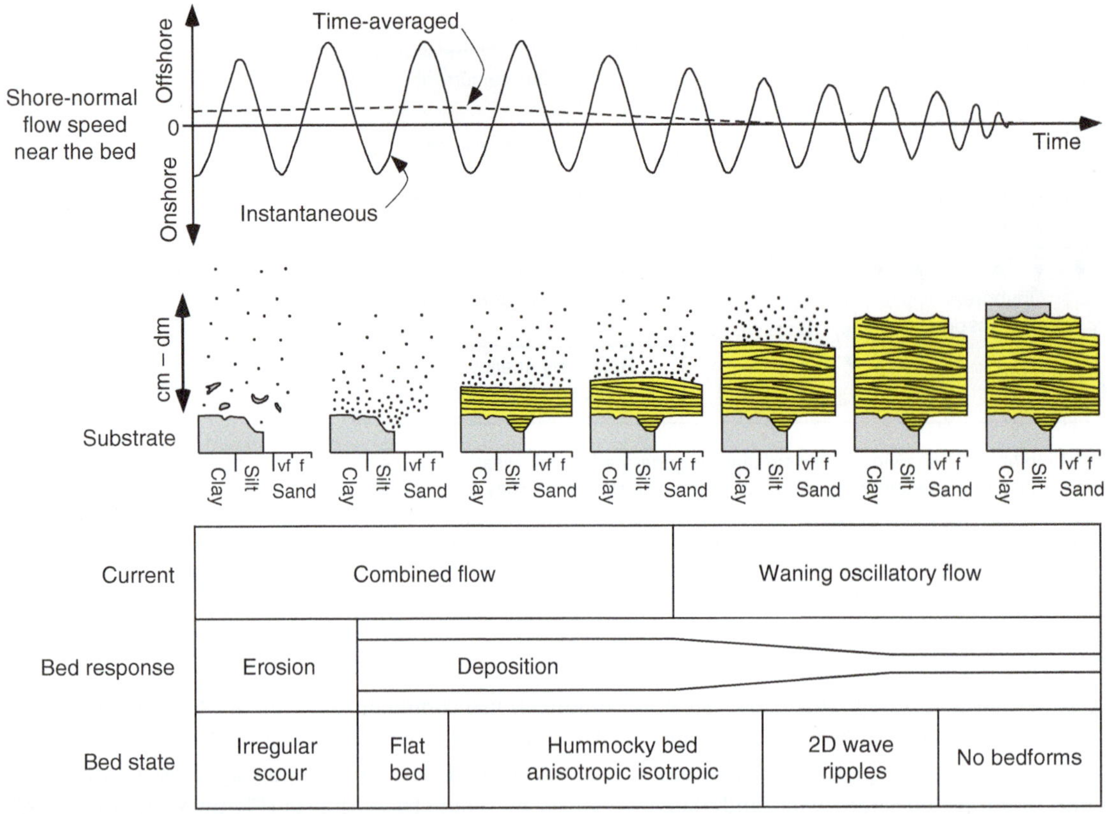

Figure 8.14 Physical process model for an idealised event bed developed in fine sand as a result of storm-generated combined-flow (adapted from Cheel 1991; Duke et al. 1991; Plint 2010).

2012, 2021; Angus et al. 2020), such as in modern Waterside Beach, Bay of Fundy (Figure 8.8). Shoaling-wave, breaking-wave and surf zones shift laterally as water depth changes during a tidal cycle, resulting in the possibility of repetitive interfingering of foreshore, upper shoreface and lower shoreface facies-association belts (indicated by double-headed blue arrow in Figure 8.15b). Similarly, tidal variations in water depth modulate the wavelength and amplitude of hummocky cross-stratification on open-coast tidal flats (Figure 8.15c; after Yang et al. 2005; Dashtgard et al. 2021). Thus, the interaction of wave and tidal processes may be diagnosed by the character and/or repetitive interfingering of depth-dependent sedimentary structures and facies (Figure 8.15a–c).

Barrier-island systems, such as the modern East Friesian Islands, northern Germany, are fronted by shorefaces (cf. Figure 8.15a), and barrier islands are separated by tidal inlets that allow exchange of water between the sea and a back-barrier lagoon (Figures 8.9a–d and 8.15d; after Hayes 1979). Tidal inlets migrate laterally, due to wave-generated longshore drift (Figure 8.13) that accretes sediment on the updrift margin of the inlets (Figures 8.9c and 8.15d; after Hoyt and Henry 1967).

8.3.2 Physical Processes II: Tides

Tides are generated by the gravitational attraction of the Moon and Sun on water in Earth's oceans and lakes. Tidal waves currently have a range of cyclical periodicities, including semi-diurnal (12.42 hours; approximately half-daily), diurnal (approximately daily) and neap-spring (13.66 days) (Figure 8.16a). These periods are readily explained using equilibrium tidal theory, in which two oceanic bulges are generated on the side of Earth beneath the Moon and on the opposite side of the Earth. Semi-diurnal and diurnal tidal periods arise from the daily rotation of Earth on its tilted axis, during which the two oceanic bulges can be simplified as staying fixed relative to the Moon above an eastward-rotating Earth (Figure 8.16b,c). The neap-spring tidal period arises from the positions of the Moon and Sun relative to Earth as the Moon orbits Earth during a lunar month (Figure 8.16d). Dynamic tidal theory accounts for the complicating effects of ocean bathymetry and geography on tides. In plan view, tidal waves entering a semi-enclosed sea from the ocean rotate around a series of amphidromic points (or tidal nodes) at which the tidal wave has zero amplitude (Figures 8.11b and 8.17). The amphidromic points have locations that reflect the bathymetry and

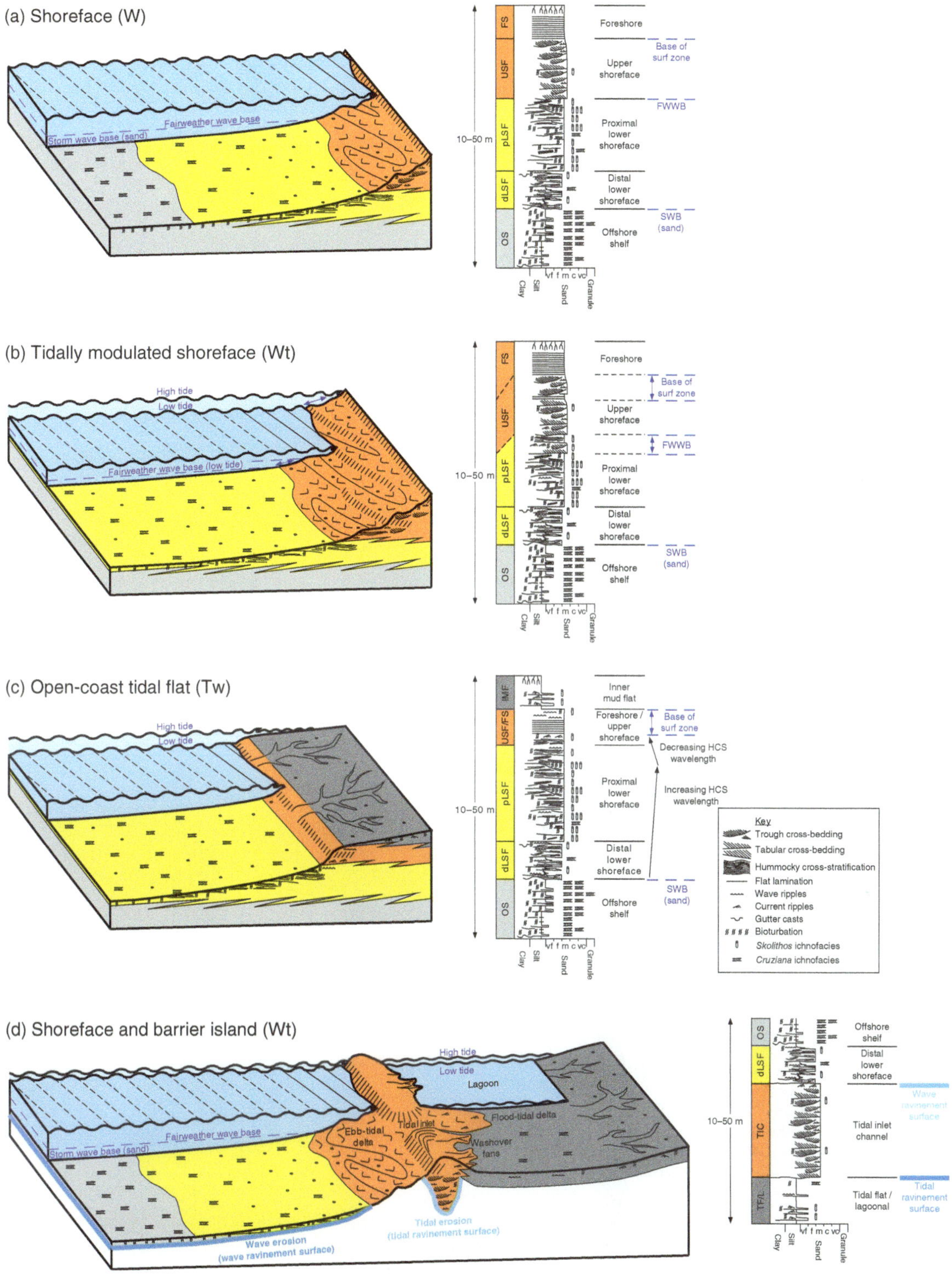

Figure 8.15 Facies models and schematic facies successions for wave-dominated shorefaces and related environments with variable tidal influence. (a) Wave-dominated shoreface with little tidal influence (e.g. modern Sapelo Island, Georgia, USA, and Cretaceous Blackhawk Formation, Utah, USA), characterised by oblique wave approach and development of elongate shoreline-oblique bars (Figure 8.13) (adapted from Hampson and Storms 2003; Dashtgard et al. 2021). (b) Wave-dominated, tidally modulated shoreface (e.g. modern Waterside Beach, Bay of Fundy, Figure 8.8, and Jurassic Grès de Châtillon, Boulonnais, France), characterised by facies belts that shift laterally as water depth changes during a tidal cycle (adapted from Dashtgard et al. 2012, 2021; Angus et al. 2020). In the resulting facies succession, morphological elements (e.g. foreshore and shoreline-oblique bars) and related deposits repetitively interfinger vertically, such that facies associations have poorly defined boundaries. (c) Tide-dominated, wave-influenced open-coast tidal flat (e.g. modern Baeksu tidal flat, South Korea), characterised by modulation of the wavelength and amplitude of hummocky cross-stratification as water depth changes during tidal cycles that coincide with a winter storm(s) (adapted from Yang et al. 2005; Dashtgard et al. 2021). (d) Wave-dominated shoreface and barrier island system developed under microtidal or mesotidal conditions (e.g. modern East Friesian Islands, Figure 8.9), characterised by lateral migration of tidal inlets driven by wave-generated longshore drift (Figure 8.13) (adapted from Hoyt and Henry 1967; Hayes 1979). Facies models and schematic facies successions are for regressive shorelines in Figure 8.15a–c, and for transgressive shorelines in Figure 8.15d.

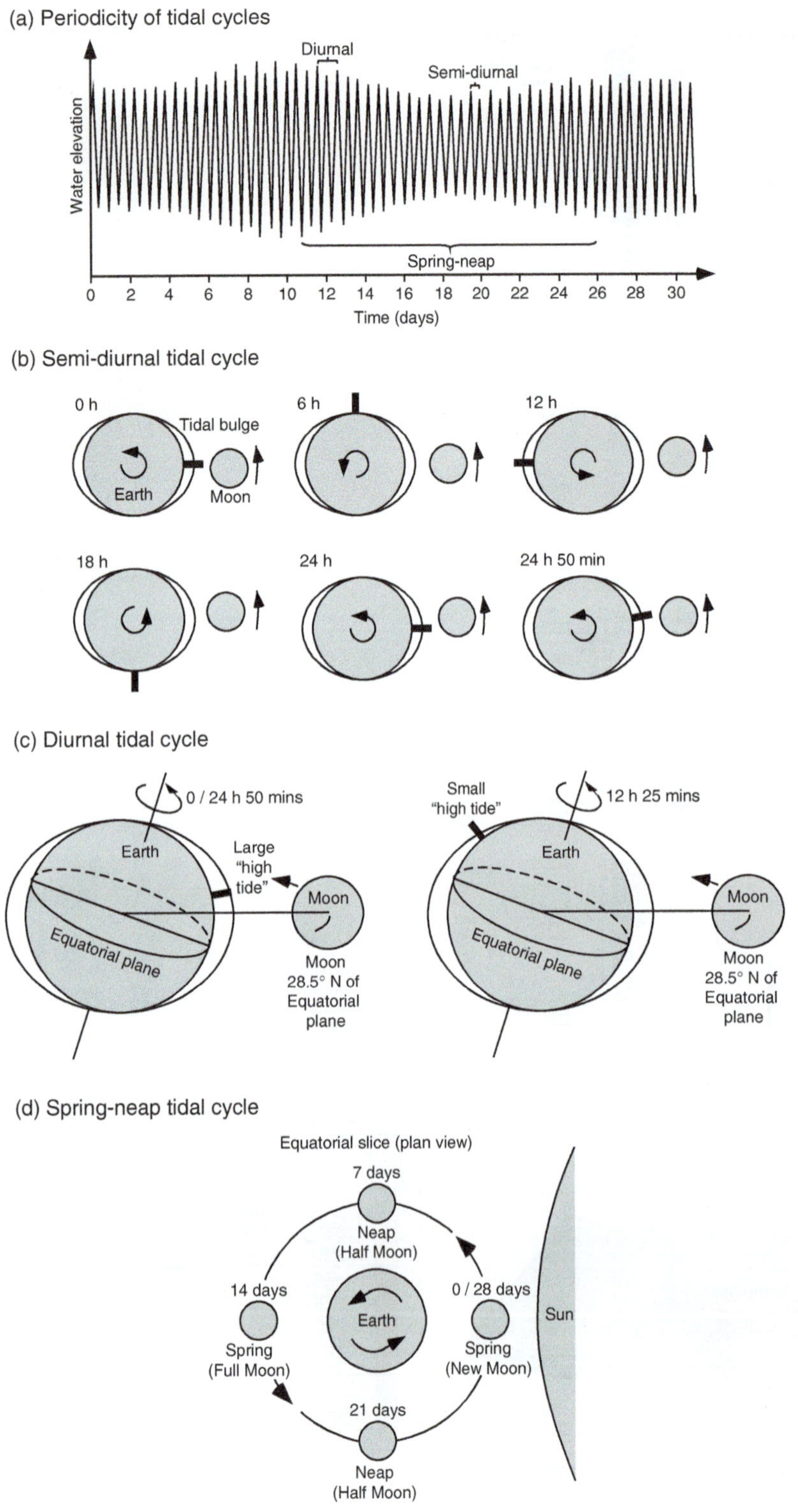

Figure 8.16 (a) Schematic tidal variations in water level at a coastal tide gauging station (adapted from Dalrymple 2010), indicating semi-diurnal, diurnal and spring-neap tidal cycles. (b–d) Schematic diagrams illustrating the origin of tidal constituents using equilibrium tidal theory (modified from Wells 2008). (b: Principal lunar semi-diurnal M_2 tidal constituent) Water forms two tidal bulges on opposite sides of the Earth, due to the gravitational attraction of the Moon combined with the centrifugal force of the rotating Earth–Moon system. As the Earth rotates on its axis, any given point on the Earth's surface (e.g. black rectangle) will pass under two tidal bulges. The Moon rotates about the Earth in the same direction as the Earth rotates about its axis, such that it takes the point 24 hours and 50 minutes to return to its original position relative to the tidal bulges (giving an M_2 period of 12.42 hours). (c: Principal lunar diurnal O_1 tidal constituent) Due to the tilt of the Earth's axis, any given point on the Earth's surface (e.g. black rectangle) will pass through different chords of the two tidal bulges during each rotation of the Earth on its axis. Consequently, the two 'high tides' in each lunar day are not equal to each other (diurnal inequality), giving an O_1 period of 12.42 hours. (d: Synodic spring-neap tidal cycle caused by M_2 and S_2 tidal constituents) Synodic spring tides occur when the Sun and the Moon are above the same line of the Earth's longitude (conjunction) or are 180° of longitude apart (opposition). Neap tides occur when the overhead positions of the Sun and the Moon are separated by 90° of the Earth's longitude.

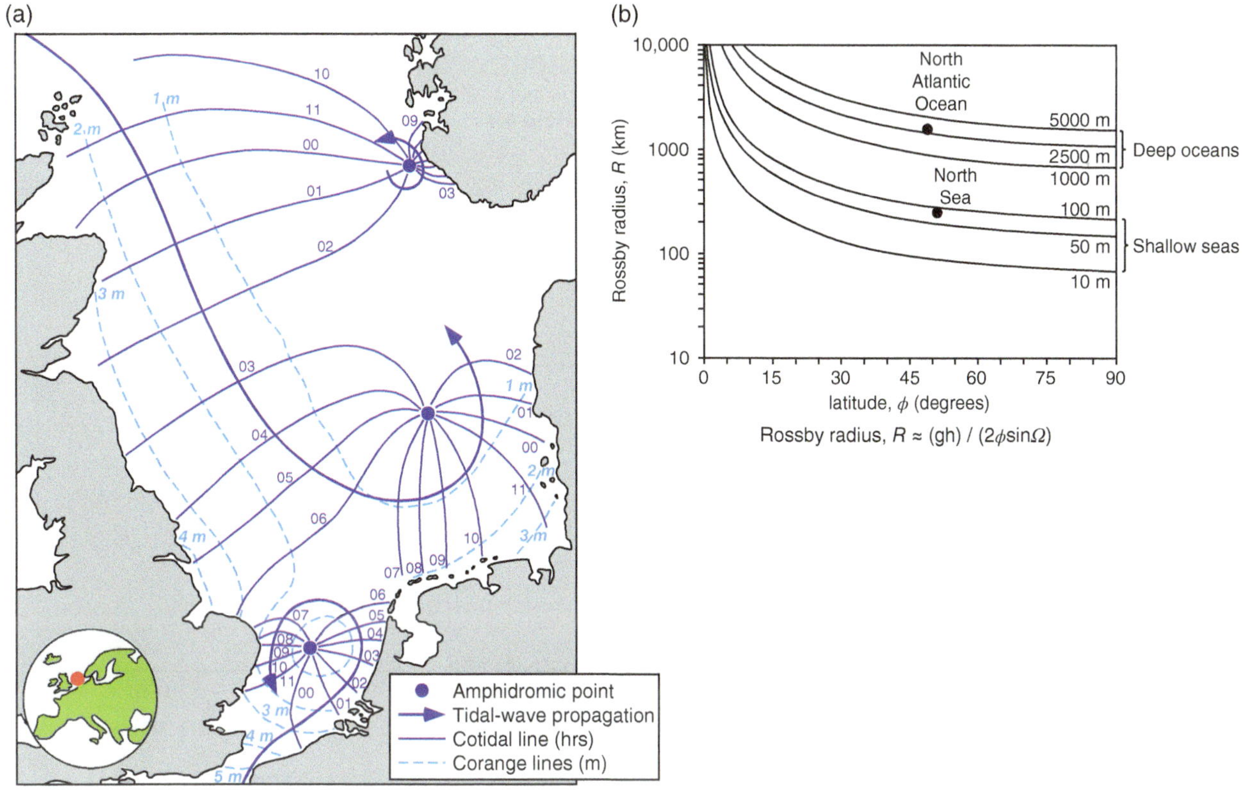

Figure 8.17 (a) Amphidromic tidal system in the North Sea (modified from Houbolt 1968). Cotidal lines show the times of high water in lunar hours, and indicate the direction of tidal-wave propagation around each amphidromic point. Tidal range is zero at each amphidromic point and increases outward, as indicated by corange lines. The highest tidal ranges occur in coastal embayments. (b) Plot of Rossby Radius, R, for differing latitudes and water depths (mofified from Wells 2008). Rotation of the tidal wave around an amphidromic point(s) occurs where the width of a water body exceeds $2R$. ϕ is the latitude, Ω is Earth's rotation rate (7.27×10^5 rad s^{-1}).

coastal geography of the sea and the deflection of the tidal wave by Coriolis acceleration (Figure 8.17b): counterclockwise in the northern hemisphere (e.g. Figure 8.17a) and clockwise in the southern hemisphere. As tidal waves rotate around amphidromic points, which typically occur in relatively deep water, their amplitudes increase towards shallower bathymetry (e.g. Figure 8.17a). Amplitudes and current velocities also increase in geographical constrictions (e.g. straits and embayments), where the volume of the tidal wave is forced through a smaller cross-sectional area. The amplitude of tidal waves is greatest at coastlines, where tidal range is further affected by coastline rugosity and near-shore bathymetry (e.g. Figure 8.17a). Tidal current ellipses are rotary in open-ocean settings becoming rectilinear (normally shoreline-perpendicular) in coastal settings, such that tidal currents alternate from onshore-directed (flood tide) to offshore-directed (ebb tide) within each tidal cycle. Tidal range can be amplified by resonance, when the period of a tidal wave is similar to the natural period of a water body in a particular geometrical configuration of the coastline (Figure 8.18; e.g. straits, gulfs, seas or even oceans, such as the modern Atlantic). Tidal processes are illustrated below along a shoreline-shelf transect through a transgressive, tide-dominated tidal flat and shelf, containing tidal sand ridges, that faces the open ocean (Figure 8.19, labelled 'T1' in Figure 8.3).

In coastal environments, the activity of tides is characterised by (i) evidence of periodic (i.e. regularly repeating) fluctuations in current velocities and (ii) two opposed strongest current directions (i.e. flood and ebb tides). Periodicity in current velocities can be expressed as rhythmical alternations of sand and mud deposition, with sand being transported during periods of high current velocity (e.g. as the flood tide rises or the ebb tide falls) and mud, including clay floccules, being deposited from suspension or from highly concentrated, cohesive suspensions of mud (i.e. fluid mud) during slack water periods when current velocity is negligible, especially within the turbidity maximum zone (TMZ) (Figure 8.20; see also Section 9.3.1). Mud aggregates can also be transported by bedload transport, commonly following erosional reworking of previously deposited mud layers. Rhythmicity may occur across a range of scales

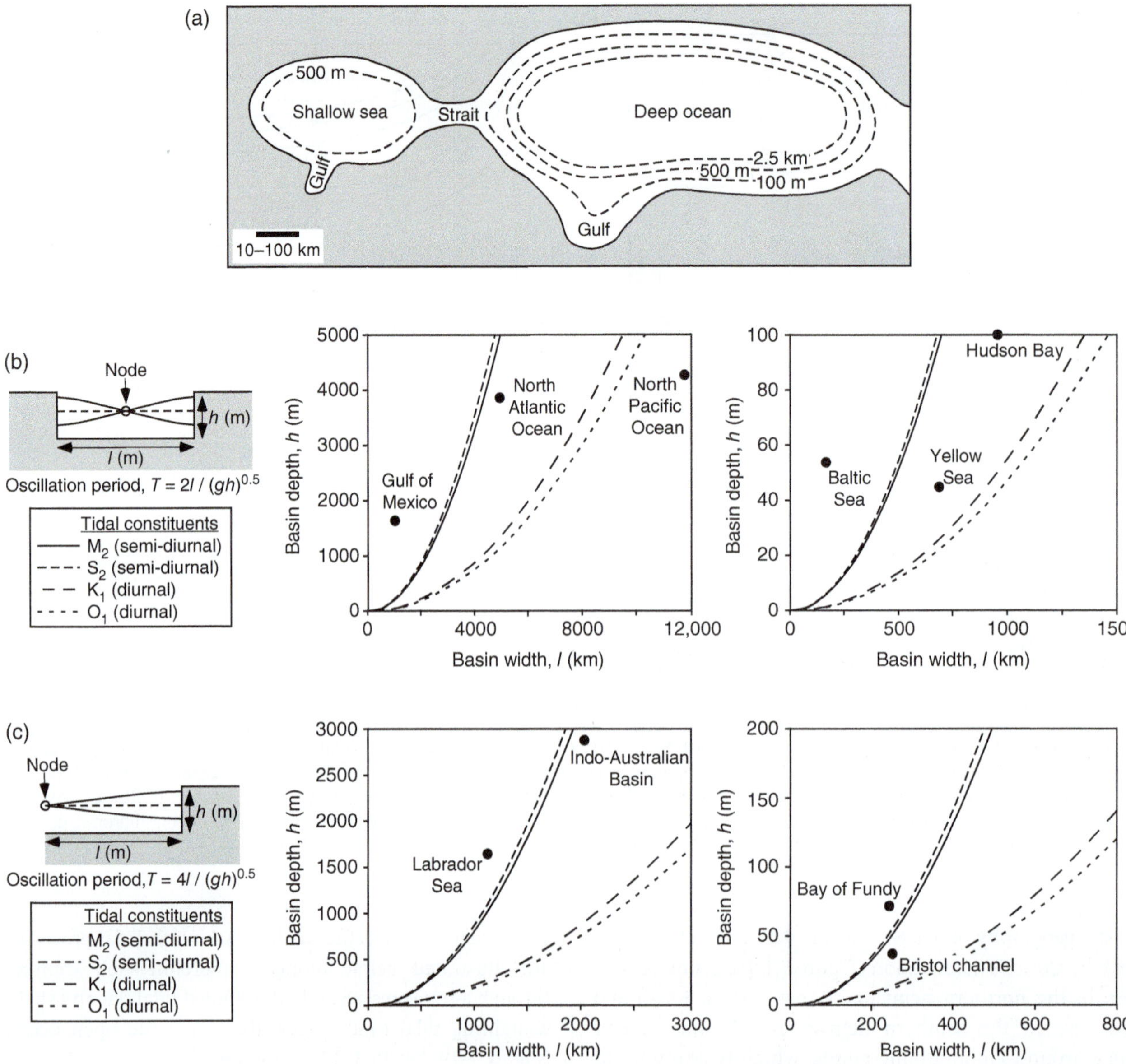

Figure 8.18 (a) Schematic plan view of water bodies through which tidal waves propagate, showing their geometrical configuration. Oceans and seas are 'enclosed' water bodies, but oceans are more areally extensive than seas. Gulfs (or embayments, bays) are partly enclosed by land and connected to an ocean or sea. Straits are narrow passageways connecting two oceans and/or seas. (b, c) Schematic cross sections (left) and plots (centre, right) of basin width and depth at which the forcing period of a tidal wave corresponds to the natural oscillation period of the water body in the basin: (b) half-wave oscillator geometries for oceans and seas; and (c) quarter-wave oscillator geometries for gulfs (modified from Wells 2008). Plots show oscillator geometries for high-amplitude semi-diurnal (principal lunar M_2, principal solar S_2) and diurnal (lunar-solar K_1, principal lunar O_1) tidal constituents. g is Earth's gravitational acceleration (9.81 m s^{-2}).

in fully preserved examples, from a sand-mud couplet (representing half of a tidal cycle; Figure 8.20a) to paired mud drapes and sand-mud couplets (representing a full tidal cycle; Figure 8.20b) to a series of sand-mud couplets exhibiting a periodic change in sand-lamina thickness and mud-drape spacing within a cross-set (representing a spring-neap tidal cycle; Figure 8.20c) (Visser 1980). More commonly, not all scales of rhythmicity are represented (e.g. Jaeger and Nittrouer 1995), such that it may not be possible to assign rhythmites to a tidal cycle of particular periodicity. Formation and preservation of tidal rhythmites are enhanced when non-tidal currents are absent, and abundant sand and mud are available (e.g. due to high riverine sediment supply; Gugliotta et al. 2022). For example,

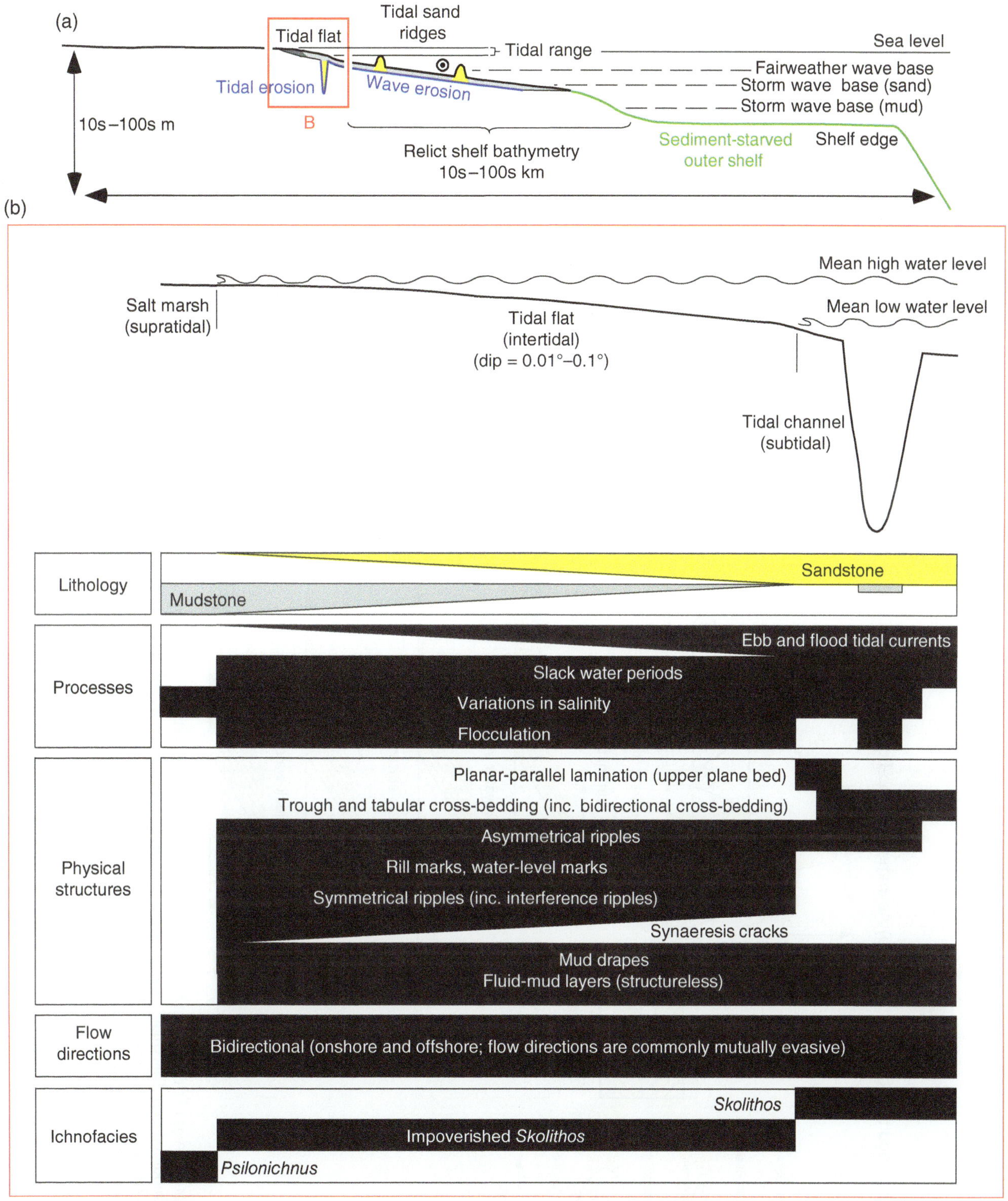

Figure 8.19 Idealised shoreline-shelf transect through a transgressive, tide-dominated tidal flat and shelf, which contains tidal sand ridges, developed under a temperate climate: (a) vertically exaggerated shoreline-shelf transect from landward (tidal limit) to seaward (shelf edge); and (b) detail of vertically exaggerated transect through tidal flat (labelled 'T1' in Figure 8.3), illustrating tidal processes and products.

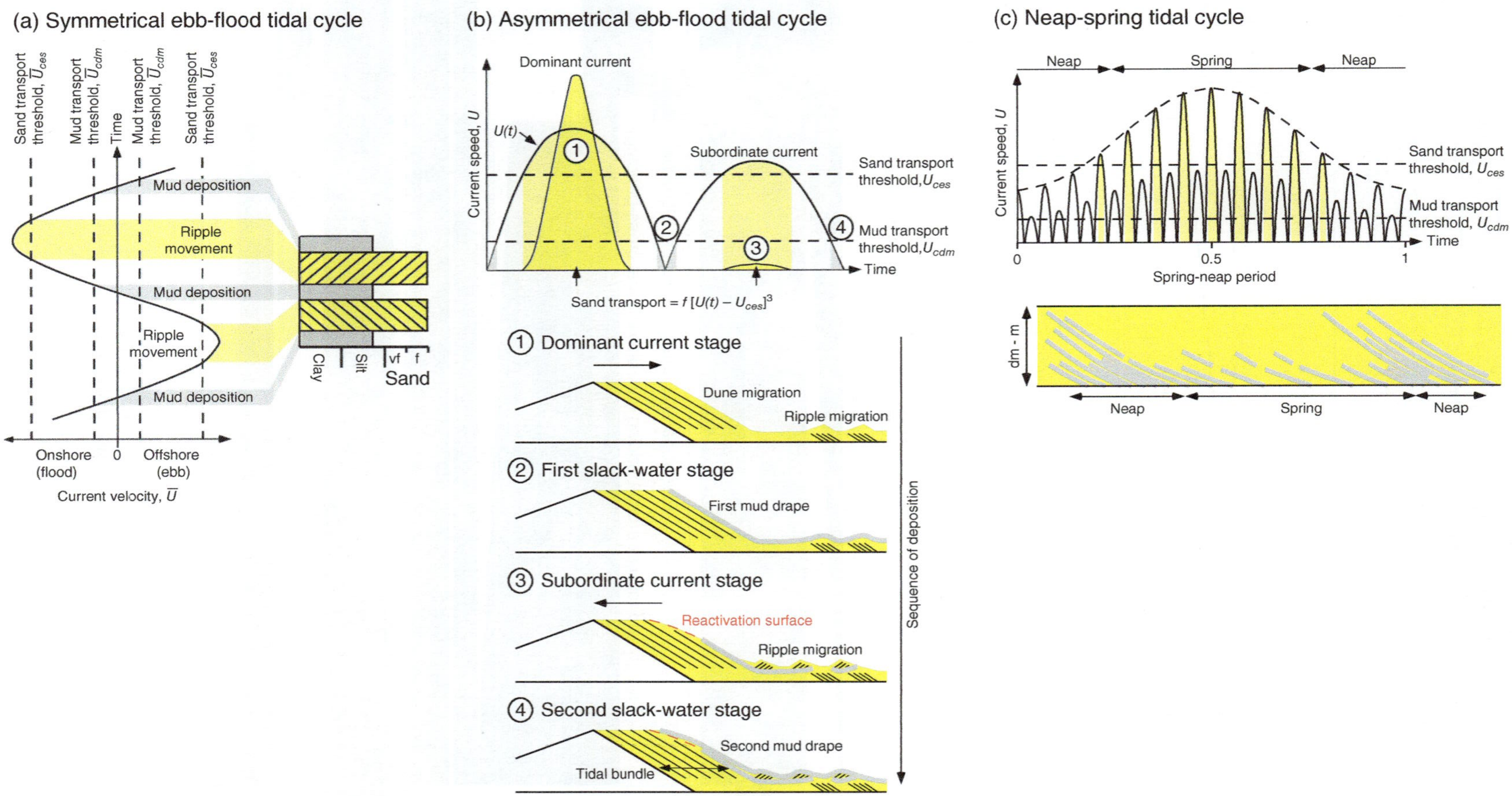

Figure 8.20 Schematic curves of tidal current velocity or speed, sediment transport and corresponding tidal rhythmites: (a) a single ebb-flood tidal cycle with symmetrical ebb and flood tidal currents (modified from Dalrymple et al. 1991); (b) a strongly asymmetrical ebb-flood tidal cycle, with dominant and subordinate tidal current directions, resulting in deposition of an ebb-flood tidal couplet characterised by a double mud drape (adapted from Visser 1980; Allen 1982); and (c) a spring-neap tidal cycle that contains strongly asymmetrical ebb-flood tidal cycles, which is characterised by thick sand-dominated spring tidal bundles and thin, mud-dominated neap tidal bundles (adapted from Allen 1982).

differences in the proportions of available sand and mud may be expressed in the formation of flaser, wavy and lenticular bedding, in addition to variations in current velocity (Reineck and Wunderlich 1968) or flow deceleration and expansion (Baas et al. 2016). Two opposed current directions may be expressed as cross-strata in which the palaeocurrent direction alternates by 180° in successive cross-sets (i.e. 'herringbone cross-stratification'). More commonly, one palaeocurrent direction is more abundant and/or is represented by larger cross-sets, whereas the opposed palaeocurrent direction is less abundant and/or is represented by smaller cross-sets; sand transport capacity varies with the cube of current velocity above the sand transport threshold, hence minor differences in velocity between dominant and subordinate tidal currents are exaggerated in their deposits (e.g. Figure 8.20b) (Visser 1980). This difference in the abundance and scale of cross-sets preserving the two current directions reflects asymmetry in the velocity of opposed tidal currents, such that one current is dominant and the other is subordinate, at one place (e.g. Figure 8.20b). Equally commonly, the two opposed tidal flows may be spatially separated in three dimensions, in mutually evasive ebb- and flood-tidal channels, or affected by local topography, such that the two current directions differ by much less than 180° (e.g. ebb currents diverted around flood-tide-generated bedforms). Circumstantial supporting evidence for tides may be provided by evidence of fluctuating, brackish or hypersaline salinities, such as subaqueous shrinkage cracks and/or impoverished trace-fossil and body-fossil assemblages (e.g. of low diversity or small size, even though abundance may be high).

Tidal flats are widespread along tidal shorelines and may even form the shoreline itself on those coasts that have a large tidal range (>1 m, but normally mesotidal (>2 m) and macrotidal (>4 m); Hayes 1975, 1979) and some protection from waves (Figures 8.19 and 8.21). The seaward-lying, subtidal part of a tidal flat is submerged throughout a tidal cycle and is thus subjected to high-velocity tidal currents (Figures 8.19b and 8.21). Ebb and flood tidal currents reach higher parts of the flats via flow through channels (Figures 8.9 and 8.21). These channels are generally highly sinuous and oriented perpendicular to the coastline but may be deflected by local topography so that they are oriented sub-parallel to the coastline (e.g. behind barrier islands; Figures 8.9a,c,d). Migration of these channels by cut-bank erosion and point-bar accretion removes large swathes of tidal flat deposits, replacing them with channel fills characterised by heterolithic lateral-accretion sets from the tidal point-bars and migration of sand dunes along the channel thalwegs (Figure 8.21). Sand dunes and ripples can

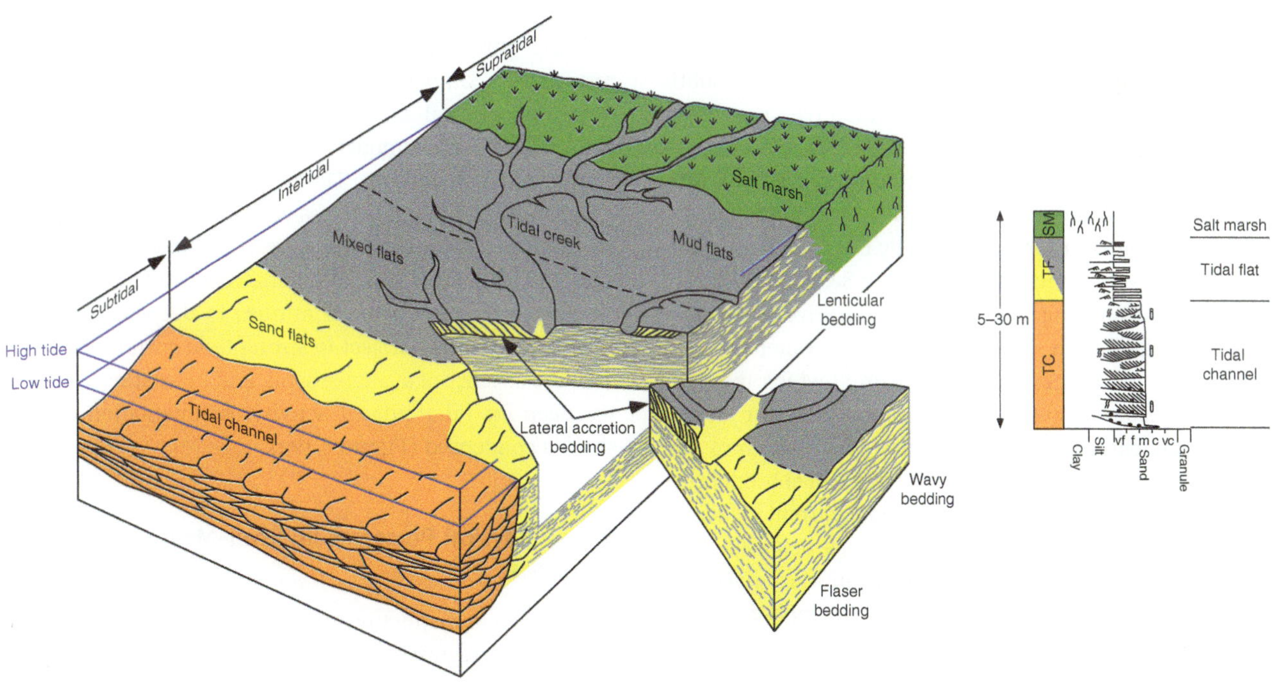

Figure 8.21 Facies model and schematic facies succession for a tidal flat and adjacent tidal channel (modified from Dalrymple 2010). The fining-upward facies succession represents regression of the tidal flat as a result of tidal channel migration. See Figure 8.15 for key to facies succession.

also form and migrate outside of the channels on the sand flats, with mud deposited as drapes during slack-water periods in discontinuous flasers (Van Straaten 1954; Reineck and Singh 1973) or as the remnants of ephemeral fluid mud sheets that accumulate during seasonal periods of low wave energy and mud consolidation, due to intense solar activity and biological stabilisation (Yang et al. 2005; Baek et al. 2018). Subtidal sand flats fronting the open ocean may be exposed to high wave energy and thus contain sedimentary structures generated by fair-weather and storm waves in water depths that vary over a tidal cycle (e.g. tidally modulated hummocky cross-stratification in Figure 8.15c; Yang et al. 2005). Sand content decreases landward in the intertidal zone (Figures 8.9d, 8.19b and 8.21), as the duration of submergence and exposure to tidal currents decreases and due to settling lag (Section 9.3.1; Figure 9.12). This leads to an overall upward fining succession of similar thickness to the tidal range as the flats prograde. The higher intertidal mixed and mud flats are characterised, respectively, by wavy and lenticular bedding (Van Straaten 1954; Reineck and Singh 1973). The wavelength and crestline shape of sandy, asymmetrical ripples changes with mud content (Baas et al. 2019). Cryptic evidence of short-lived subaerial exposure, such as rill marks, water-level marks and flat-topped ripples, may also be present. Intertidal mixed and mud flats are dissected by tributary networks of narrow, steep-sided meandering tidal gullies and creeks (Figures 8.7D and 8.21), which allow water to drain from the flats as the ebb tide falls. Tidal creeks also migrate laterally and drain seaward into subtidal channels. In temperate climates, tidal flats pass landward into supratidal salt marshes (Figures 8.19b and 8.21), characterised by root-penetrated muds and salt-tolerant (halophyte) flora, whereas in arid climates, the supratidal domain is characterised by displacive precipitation of evaporite minerals (Chapter 10; Section 10.3.5).

Tidal effects propagate further inland in rivers and estuaries. The 'tidal limit' is defined not by salt-water penetration but by the farthest point upstream at which the river flow is affected by tidal fluctuations in water depth. Flood-tidal currents that enter such channels from the ocean are subject to less bottom friction than ebb-tidal currents, which results in higher flood-tidal current velocities and net-onshore transport of sediment, particularly of re-suspended mud, in the absence of significant river discharge (e.g. Nowacki et al. 2015). During high-discharge river floods, these conditions are reversed and net-offshore sediment transport occurs (e.g. Gugliotta and Saito 2019). The fluvial-to-marine transition zone in a fluvio-tidal channel system is characterised by a landward-to-seaward increase and then decrease in sinuosity, landward-to-seaward deepening and then shallowing, and seaward-widening of the river channel (e.g. distributary channels in Mekong Delta, Figure 8.7a,d; Dalrymple and Choi 2007; Gugliotta and Saito 2019). The peak in channel sinuosity and depth occurs at the location of bedload convergence between the inner, fluvially dominated part of the channel (marked at normal river stage by seaward transport of sediment) and the outer, tidally dominated part of the channel (marked at normal river stage by landward transport of sediment) (Dalrymple and Choi 2007).

On open shelves and seaways, tidal currents may sculpt a range of bedforms (Belderson et al. 1982), including fields of tidal sand ridges and dunes (Figures 8.19a and 8.22), such as in the modern Straits of Messina (Figure 8.11). Active tidal sand ridges are very large asymmetrical bedforms up to 40 m high, several kilometres wide, tens of kilometres in length, and with linear or gently curved crestlines (Figure 8.22a,b; Houbolt 1968; Stride et al. 1982). Sand ridges are oriented obliquely sub-parallel (<20°) to mutually evasive flood and ebb tidal bottom currents, thus generating unequal secondary currents across the ridge (Figure 8.22a), such that sediment transport is dominated by flood and ebb tidal currents on opposite faces of the ridge (Figure 8.22b). The steeply dipping side of the ridge is depositional and faces obliquely in the direction of highest tidal current velocity and net transport of sand (Figure 8.22a). The gently dipping side of the ridge is erosional, due to ridge migration. Simple and compound tidal dunes (Figure 8.22b) migrate down-current over the ridges, resulting in compound cross-stratification internally. The formation of tidal sand ridges requires large volumes of sand, which in most modern shelf seas have been reworked from older shelf deposits. Therefore, tidal sand ridges commonly overlie erosion surfaces with discontinuous gravelly lags that develop during transgression (Snedden and Dalrymple 1999). Ridges will become inactive (moribund) if water depth on the shelf continues to deepen, such that they are no longer subject to strong tidal currents.

In the absence of available sand, and in locations that are too deep for tidal current activity, sedimentation on the shelf is condensed (Figure 8.19a).

8.3.3 Physical Processes III: River-Mouth Processes

Deltas form as distinct protruberances where rivers supply sediment to regressive shorelines (labelled 'R2', 'R3' and 'R4' in Figure 8.3). In this section, river-mouth processes are illustrated along a shoreline-shelf transect through a regressive, fluvial-dominated delta developed on the inner shelf (Figure 8.23, labelled 'R3' in Figure 8.3).

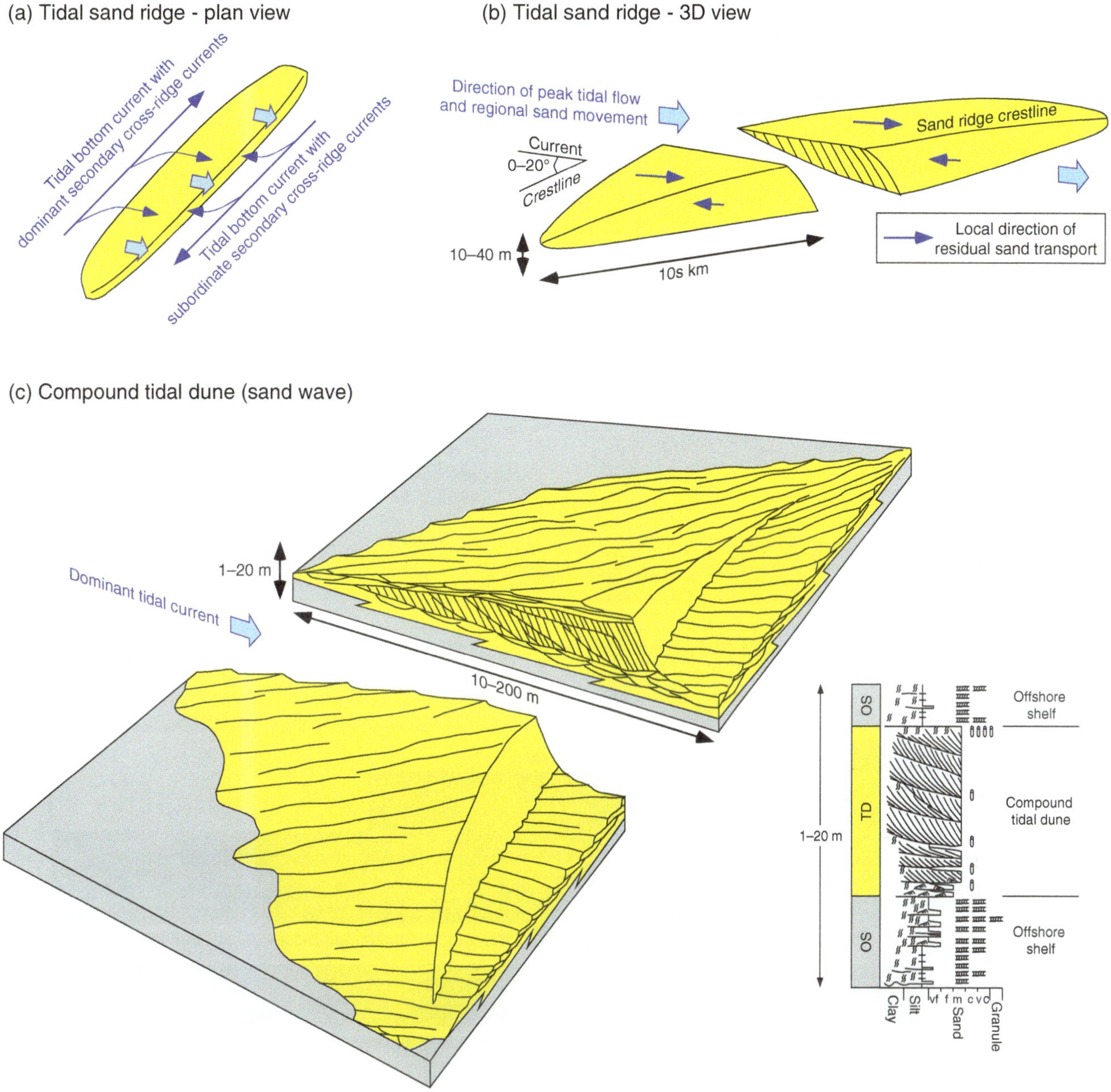

Figure 8.22 (a, b) Model of sediment transport around an active tidal sand ridge (adapted from Stride et al. 1982; Dalrymple 2010). Mutually evasive tidal currents flow obliquely sub-parallel to the ridge, generating unequal flood and ebb tidal currents that flow obliquely up and over the ridge. Residual sand transport is thus in opposite directions on either face of the ridge; net sand transport and ridge movement are in the direction of peak tidal flow. (c) Facies model and schematic facies succession for a compound tidal dune (sand wave) (adapted from Houbolt 1968; Berné et al. 1991; Olariu et al. 2012a). See Figure 8.15 for key to facies succession.

A distributary river channel network and related delta plain form the most landward part of a delta. In general, the distributary channel network forms by new channel formation (avulsion) in its landward part and by channel bifurcation in its downstream part (Figure 8.24b; Jerolmack and Swenson 2007). The apex of a deltaic distributary network is ideally defined by an upstream avulsion node of its trunk channel (Figures 8.3 and 8.24b). The location of this upstream avulsion node is influenced by several interacting factors, including pre-existing topography (Hartley et al. 2017), delta progradation (Moodie et al. 2019), river-flood variability (Chadwick et al. 2019) and basinal hydrodynamic processes in the backwater reach of the river (Jerolmack and Swenson 2007; Ganti et al.

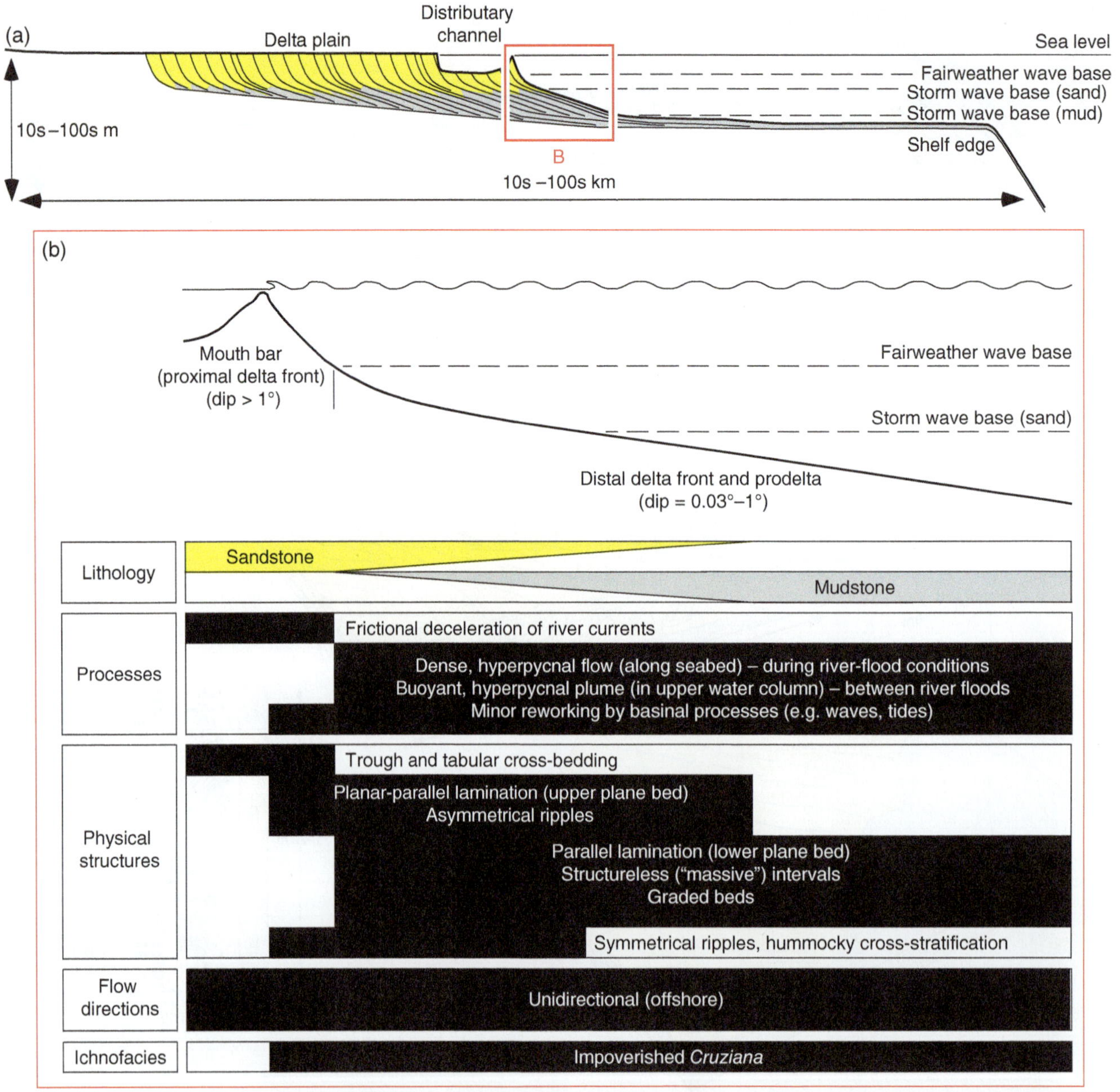

Figure 8.23 Idealised shoreline-shelf transect through a regressive, fluvial-dominated delta and shelf: (a) vertically exaggerated shoreline-shelf transect from landward (backwater limit) to seaward (shelf edge); and (b) detail of vertically exaggerated transect through mouth bar and delta front (labelled 'R3' in Figure 8.3), illustrating fluvial processes and products.

2016). The delta plain is the floodplain and its extensions laterally and to the coast. The lower delta plain extends from the shoreline to the limit of tidal influence (e.g. brackish water) outside of the main river channel (Figure 8.3), and the upper delta plain lies landward of direct marine influence. In fluvially dominated systems, the main distributary channels tend to be deep, narrow, straight and of limited lateral mobility (Fernandes et al. 2016), such that the resulting channelised sediment bodies typically have a width:depth ratio of 5–30 (Gibling 2006). Downstream channel bifurcation is pronounced in fluvial-dominated deltas, and distributaries become progressively narrower and shallower with successive downstream branching (Olariu and Bhattacharya 2006; Figure 8.24c), although this trend may be modified by scour due to dynamic drawdown of the water surface during river floods in the

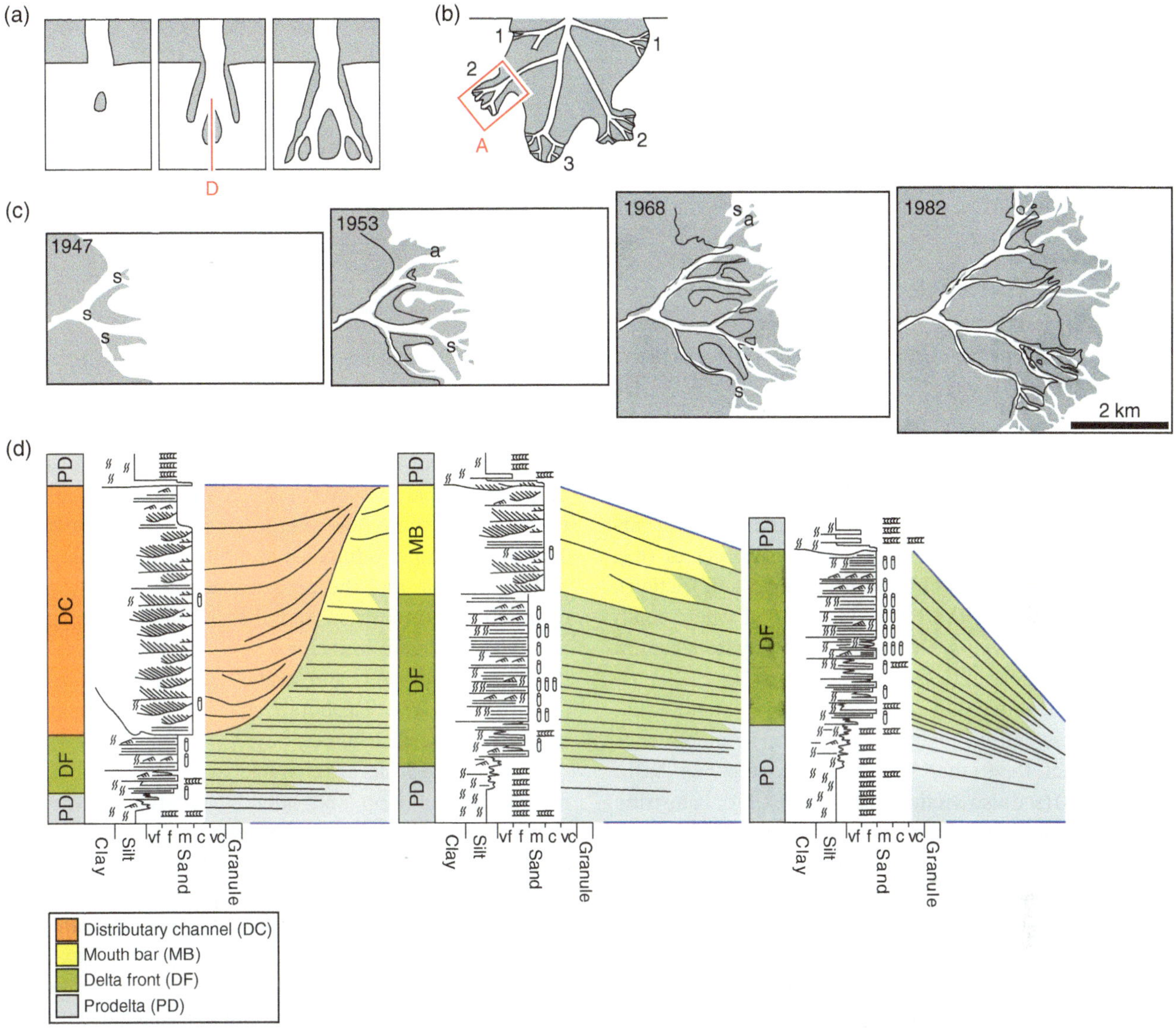

Figure 8.24 Delta morphology and growth due to interacting fluvial processes. (a, b) Idealised distributary channel network growth in a prograding fluvial-dominated delta lobe by: (a) splitting of distributary channels around mouth bars, resulting in a progressive basinward decrease in channel width and length (modified from Jerolmack 2009); and (b) repeated nodal avulsion at the apex of the delta, creating new trunk distributaries with branching tips and associated delta lobes (numbered from oldest, 1 to youngest, 3) (modified from Jerolmack 2009). (c) Serial maps illustrating the growth of the lacustrine, fluvial-dominated Mossy Delta, Saskatchewan, Canada, in which distributary channels bifurcate as a result of local channel avulsion (labelled a) and splitting around mouth bars (labelled s) (modified from Edmonds and Slingerland 2007). (d) Schematic depositional-dip-oriented cross section and vertical facies successions through a distributary channel, mouth bar and associated delta front and prodelta deposits (modified from Wellner et al. 2005); note that the mouth bar and delta front deposits contain clinoforms that dip in all directions away from the central axis of the mouth bar. See Figure 8.15 for key to facies successions.

backwater reach of a river (Lamb et al. 2012). Distributary channels at all scales are generally filled with cross-bedded and cross-laminated sands that record the seaward migration of dunes and ripples. Seaward-decreasing mean flow velocity, for example in the backwater reach of a distributary channel, results in a seaward decrease in sand content and a corresponding seaward increase in mud content of its fill (e.g. Martin et al. 2018). Mud is preferentially partitioned into the delta plain, which also contains thin, cross-stratified, sheet sands that record sheet flood and crevasse splay deposition during river floods. Lower delta plain deposits contain trace fossil, faunal and floral assemblages that record brackish water and/or episodic ingress of fully marine water.

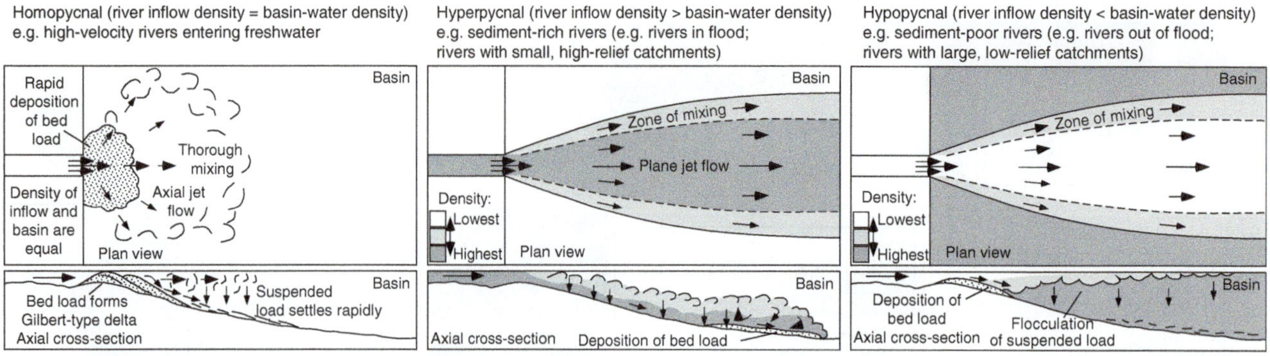

Figure 8.25 Different modes of interaction between sediment-laden river water and basin water, determined by the relative density of the two water bodies (adapted from Bates 1953; Wright 1977).

Processes at the river mouth are governed by: (i) the discharge of water, suspended sediment and bedload, and their temporal variations; (ii) the related interactions of variable river outflow with variable basinal processes (e.g. expressed in the effects of buoyancy, inertia and friction); (iii) position within the basin, in terms of water depth and proximity to the shelf edge; and (iv) hydrodynamics of the receiving basin (e.g. waves and tides).

River outflow may have the same density (homopycnal flow), be less dense (hypopycnal flow) or be more dense (hyperpycnal flow) than water in the receiving basin (Bates 1953) (see also Sections 6.3.2 and 6.4.1 for lacustrine deltas). Homopycnal flow is characterised by thorough mixing of river outflow and basinal water at the river mouth, expressed as the deposition of bedload in coarse-grained, steep-fronted deltas (i.e. Gilbert-type deltas, as first described by Gilbert 1885) due to rapid flow deceleration (Figure 8.25). Hypopycnal flow is characterised by the development of a buoyant plume of suspended mud, which is transported away from the river mouth on top of the basinal water (Figure 8.25). Hyperpycnal flow is expressed as a sediment-laden plume that bypasses sediment past the river mouth and transports it into deeper water under gravity and the inertia of the dense flow (Figure 8.25) (Mulder and Syvitski 1995). Homopycnal flow is rare but occurs most commonly in lacustrine deltas. Many deltas exhibit both hyperpycnal and hypopycnal flow conditions at different times, reflecting changes in river outflow due to seasonal precipitation changes and storm-induced floods. Mouth bars (Figure 8.24a) initiate where confined flow exits the mouth of a distributary channel to become an unconfined jet, with the result that velocity decreases due to flow expansion and frictional deceleration (Bates 1953; Wright 1977; Wellner et al. 2005). Bedload is deposited as a bar directly downstream of the channel mouth (e.g. Wellner et al. 2005; Edmonds and Slingerland 2007) irrespective of homopycnal, hypopycnal and hyperpycnal flow conditions. The shape of the mouth bar, however, reflects dispersion of the river plume due to these flow conditions (e.g. Reynolds 2022). The river channel bifurcates around the mouth bar as a delta advances, and the repeated action of this process of mouth bar deposition and channel bifurcation generates a downstream-branching network of channels that distributes sediment and water over the delta front (Figure 8.24a). The distributary channels may be separated by shallow interdistributary bays.

Mouth bar deposits are characterised by cross-strata that record the initiation and later aggradation of the bar (Figure 8.24d). The inner part of the bar, where flow is constricted by the margins of the distributary channel, is characterised by stacked, landward-accreting cross-beds that record seaward migration of dunes over the upstream face of the bar (e.g. in the modern Atchafalaya and Wax Lake Deltas; Van Heerden and Roberts 1988; Wellner et al. 2005). The outer part of the bar and associated delta front, where flow expands beyond the confines of the channel mouth and thus decelerates, is characterised by smaller cross-beds and current-ripple cross-lamination. Planar–parallel lamination may form at the bar crest during river flood stages, when upper plane bed conditions are developed, while the bar crest may be subaerially exposed at low flow stage. As the mouth bar grows, it aggrades and accretes laterally, downstream and/or upstream (Van Heerden and Roberts 1988; Wellner et al. 2005; Olariu and Bhattacharya 2006). The resulting shallowing of water depth forces flow from the distributary channel around, rather than over, the mouth bar; this in turn causes the distributary channel to bifurcate downstream (Figure 8.24a; Van Heerden and Roberts 1988; Wellner et al. 2005; Olariu and Bhattacharya 2006).

The mouth bar constitutes the most proximal part of the subaqueous delta (i.e. the delta front). In fluvial-dominated deltas, the delta front is steeply dipping (up to angle-of-slope-stability, both in coarse-grained Gilbert-type deltas and fine-grained delta fronts dominated by slumping of unconsolidated mud), because basinal processes do not significantly redistribute the river-supplied sediment. Seaward of the mouth bar, delta front deposition is characterised by short-term (commonly hourly to seasonal) changes in sediment flux, flow velocity, salinity and temperature that record, for example, the alternation between hyperpycnal and hypopycnal flow conditions, and corresponding changes in freshwater influx. During each episode of hyperpycnal flow conditions, a gravity-flow bed is deposited that records either decreasing (waning) or increasing-to-decreasing (waxing-to-waning) flow velocity, which may reflect the hydrograph of its formative river flood (Mulder et al. 2003). If the river flood occurs during a storm, a current- or wave-enhanced sediment gravity-flow bed (Section 8.3.1) may be deposited. A typical waning-flow gravity flow bed may comprise, from base to top, some of the following divisions, depending on its location on a proximal-to-distal profile: structureless sand (Ta), planar–parallel-laminated sand (Tb), current-ripple cross-laminated sand (Tc), and parallel-laminated mud (Td). Alternation of these intervals within a bed indicates more complex temporal variations in flow velocity (e.g. waxing-to-waning). Gravity-flow beds decrease in thickness, erosional amalgamation and maximum grain size from the proximal to distal delta front (Figure 8.24d; Wellner et al. 2005; Enge et al. 2010; Olariu et al. 2010), and suspension fall-out of mud from the buoyant plume occurs over the delta front during intervening hypopycnal flow conditions. The distal delta front grades seaward into the prodelta, where thick mud deposits accumulate under alternating hyperpycnal and hypopycnal flow conditions. Waves and tides may rework beds deposited on the mouth bar, delta front and prodelta. For example, storm waves may keep river-supplied mud in suspension such that it can be transported seaward and/or parallel to the shoreline for large distances over the low-gradient shelf as a wave-supported gravity flow, to form a mid-shelf mud belt (Figures 8.3 and 8.25b,c; see also Section 9.4.4). Bioturbation intensity and diversity tend to be low in delta front and prodelta deposits, despite plentiful nutrients, because of the physico-chemical stress induced by high net-sedimentation rates and variable salinity, although bed-scale variation in bioturbation may be pronounced.

A fluvial-dominated delta advances seaward to build out a sediment body that has an overall lobate geometry in plan view (e.g. 'R3' in Figure 8.3) and a digitate shoreline in which distributary channels separated by interdistributary bays extend seawards as finger-like protrusions. Delta lobes build out rapidly, due to high, but localised, sediment supply via a distributary channel (e.g. Van Heerden and Roberts 1988; Figure 8.24c). As a delta lobe builds out, its distributary channel becomes longer and also elevated above the delta plain. Such channel extension and elevation drives avulsion, in which the channel switches to a new position that provides a shorter, lower-lying route to the shoreline (e.g. to the head of an interdistributary bay). Avulsion occurs at multiple scales, from avulsion of a minor, downstream distributary channel that affects only a small part of the delta (e.g. Figure 8.24c) to avulsion of a trunk distributary channel that drives abandonment of an entire delta lobe (e.g. Figure 8.24b). Distributary channel avulsion and associated lobe switching are autogenic processes that are inherent to river-dominated deltas, often over historical time scales (e.g. within the Mississippi Delta; Coleman and Gagliano 1964) (Figure 8.24a–c), but they are suppressed in wave- and tide-dominated deltas in which sediment redistribution away from the distributary channel mouths limits the rapid lengthening of distributary channels (Figures 8.26 and 8.27). Waves rework sand from mouth bars and redistribute it along the shoreline in beach ridges that form flanking strandplains and barrier islands (e.g. Rodriguez et al. 2000), thus limiting distributary channel bifurcation (Figure 8.26a). The resulting delta morphology reflects the interaction between fluvial discharge and wave-generated longshore drift (Bhattacharya and Giosan 2003; Figure 8.26a). With increasing tidal influence, distributary channels exhibit seaward-increasing width (Figure 8.26b), increased but relatively low sinuosity (Figure 8.26b), and either seaward-decreasing (Gugliotta and Saito 2019; Figure 8.26b) or seaward-increasing depth (Dalrymple and Choi 2007) (e.g. distributary channels in Mekong Delta, Figure 8.7a,d).

The position of a delta within the basin is important, because basinal hydrodynamics may differ between shallow and deep water (e.g. due to the dissipation or amplification of tides and waves), and because the shelf edge may be gravitationally unstable. Such instability generates extensional growth faults and large-scale failures that are characteristic of shelf-edge deltas, but that are rarer in deltas developed on the inner and mid-shelf.

8.3.4 Physical Processes IV: Mixed Processes

The idealised shoreline-shelf transects described earlier (Figures 8.12, 8.19 and 8.23) belie the fact that mixed process regimes are common, even in shorelines that are generally considered to be 'end-members' dominated by a single physical process (e.g. Zuchuat et al. 2023). For example, the subtidal portions of many tidal flats are influenced by waves (Figure 8.15c; Yang et al. 2005), storm waves are important in redistributing mud supplied by fluvial-dominated deltas across and along the shelf

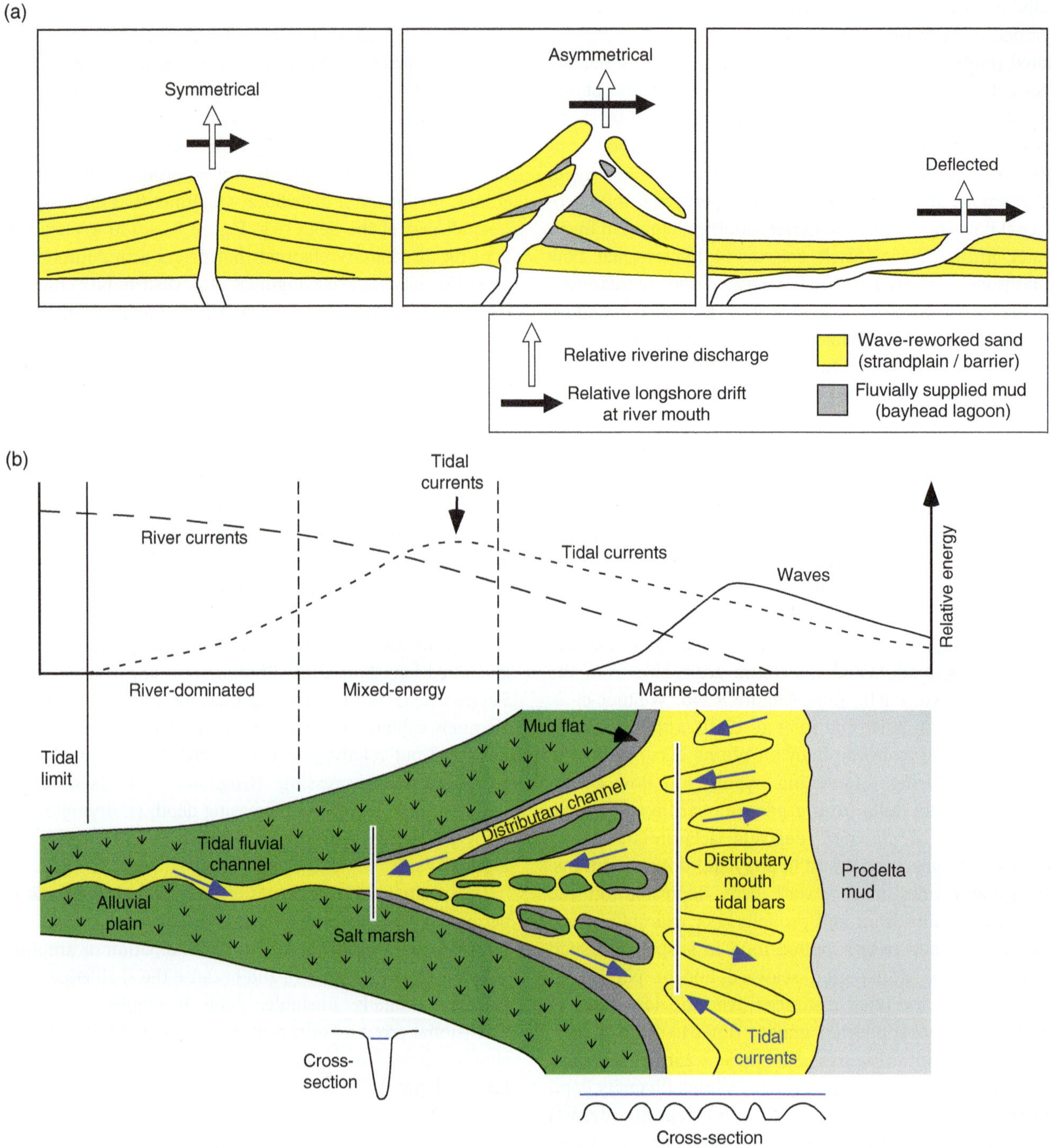

Figure 8.26 Delta morphology and growth due to the interaction of fluvial, wave and tidal processes: (a) sketch maps of sandy, wave-influenced deltas, in which shoreline morphology and the degree of plan-view delta asymmetry are controlled by the relative influence of river discharge and wave-generated longshore drift (modified from Bhattacharya and Giosan 2003); (b) sketch map of a tide-dominated delta and accompanying plot of longitudinal variation in the relative influence of physical processes: river currents, tidal currents and waves (after Dalrymple and Choi 2007 / with permission of Elsevier).

(Figure 8.27a–d; Jaramillo et al. 2009) and tides modify sediment transport dynamics near the mouths of fluvial-deltaic distributary channels (Figure 8.27a–d; Gugliotta et al. 2017). Below, spatial variations in wave, tide and river-mouth processes in an overall mixed process regime are illustrated along a shoreline-shelf transect through a transgressive, wave-dominated estuary and associated shelf (Figure 8.28, labelled 'T3' in Figure 8.3).

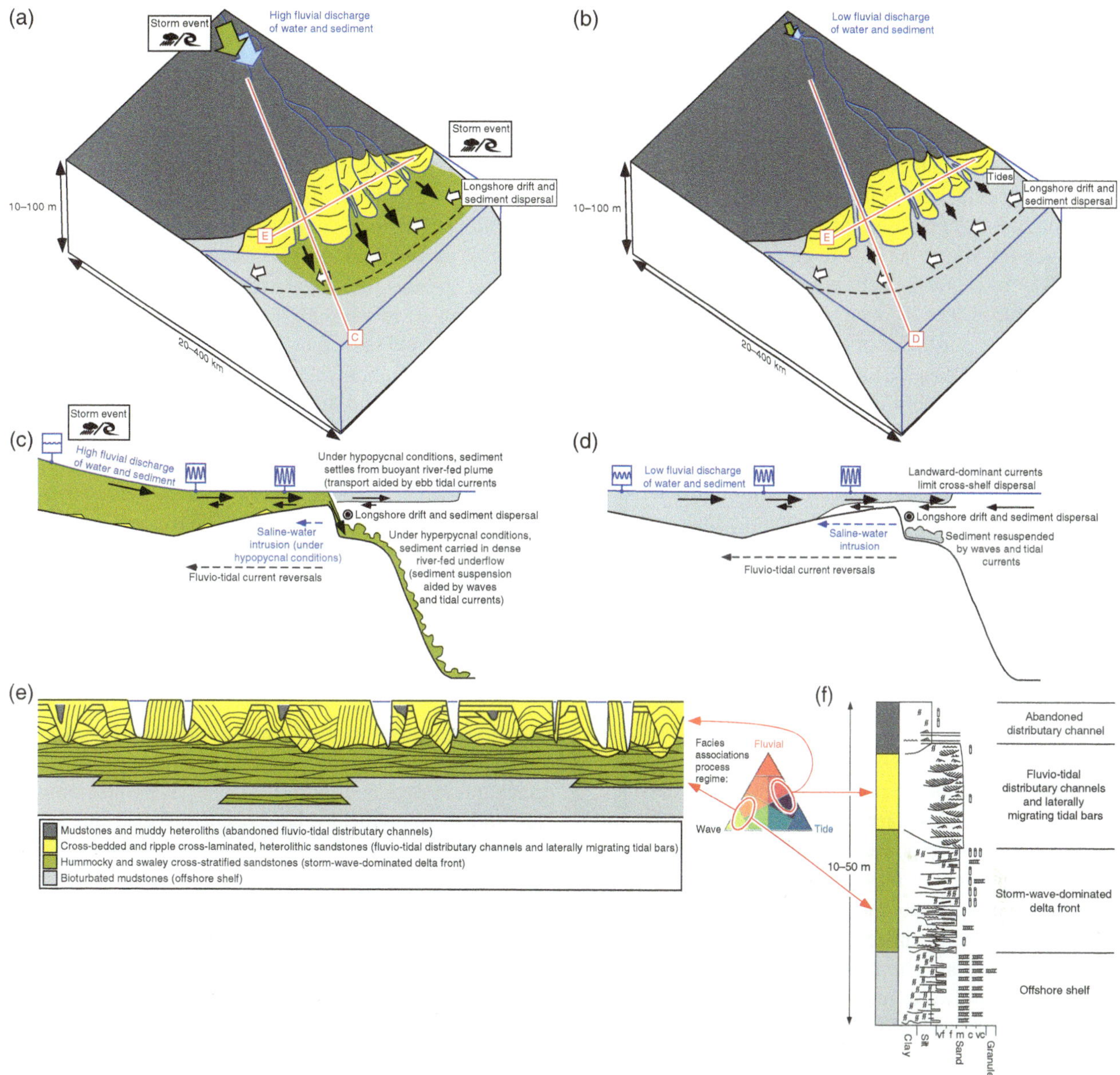

Figure 8.27 Conceptual summary of physical processes, facies model and schematic facies successions of a delta with mixed fluvial, tide and wave influence (e.g. modern Mekong Delta, Vietnam, Figure 8.7; Cretaceous Sego Sandstone, Utah and Colorado, USA, Figure 8.40). (a, b) Three-dimensional block diagrams and (c, d) shoreline-perpendicular cross sections through trunk distributary, mouth bar and adjacent shelf, illustrating differences in sediment transport between periods of high fluvial discharge during coastal storms (a, c) and intervening periods of low fluvial discharge (b, d) (adapted from Ta et al. 2005; Collins et al. 2017; Eidam et al. 2017; Gugliotta and Saito 2019). (e) Shoreline-parallel cross section of facies architecture and (f) schematic vertical facies succession through regressive, mixed-influence deltaic succession supplied with abundant sand (adapted from Willis and Gabel 2001, 2003; Legler et al. 2014; Van Cappelle et al. 2016).

A corresponding 3D facies model for a wave-dominated estuary is shown in Figure 8.29 (after Boyd 2010), comparable in many aspects to the modern Gironde Estuary (Figure 8.10). As described later, the constituent sub-environments of the estuary tend to be dominated by a single process, but the estuary as a whole records a mixed process regime.

The most landward part of the wave-dominated estuary is dominated by fluvial processes, exemplified by a channel fill and associated point bars containing cross-strata that indicate dominant seaward-directed (fluvial and ebb-tidal) flow (Figures 8.28b and 8.29). Tidal influence is expressed by deposition of rhythmical mud drapes, cross-stratification indicating subordinate landward-directed (flood-tidal) flow, and bioturbation of impoverished but locally variable intensity and diversity (cf. Dalrymple and Choi 2007; La Croix and Dashtgard 2015). Wave influence is absent. The tidally influenced fluvial channel passes seaward into a bayhead delta that builds into a lagoon in the central part of the estuary (i.e. estuary central basin) (Figures 8.28b and 8.29). The bayhead delta is fluvially dominated, indicated by a mouth bar composed of seaward-directed cross-stratification (cf. Figure 8.25), with mud drapes and bioturbation style respectively reflecting tidal influence and brackish salinities. The lagoon has low, but mixed, energy. Mud is abundant, due to flocculation in brackish water, and is kept in turbulent suspension in the water column by the activity of tides and subordinate waves (defining the turbidity maximum zone). Tidal flats (cf. Figures 8.19b and 8.21) are developed along the lateral margins of the lagoon. The lagoon passes seaward into a barrier island complex that is fronted by a wave-dominated shoreface (cf. Figures 8.12b and 8.15d). Sand is supplied to the barrier island either directly by aeolian processes or by wave processes during storms and then reworked by wind to form aeolian dunes during fair-weather periods (Figures 8.28b and 8.29). Storm waves may breach the aeolian dunes to deposit wash-over fans that extend from the landward side of the barrier island into the lagoon. One or more tidal inlets cut through the barrier island, allowing exchange of water between the open ocean and lagoon (Figures 8.9 and 8.15d). Tidal inlets cut deeply into the shoreface (Figures 8.28b and 8.29) and migrate laterally in response to accretion on one of their margins of sand supplied by wave-generated longshore currents (Figures 8.9c, 8.28b and 8.29). Flood and ebb tides flow at high velocity through the inlets and build the so-called tidal deltas where tidal currents decelerate on the respective landward and seaward sides of each inlet. Flood-tidal deltas are characterised by cross-strata indicating landward-directed (flood-tidal) flow and mud drapes deposited during slack water tidal periods. Tidal inlets are dominated by cross-strata that record both seaward-directed (ebb-tidal) and landward-directed (flood-tidal) flows and contain only sparse mud drapes (Moslow and Tye 1985). Ebb-tidal deltas are largely reworked by wave action on the shoreface, although residual sand accumulations may track their transit across the shelf during transgression (e.g. 'shoal retreat massifs'; Swift 1975b). Barrier island complexes at the mouths of wave-dominated estuaries are poorly preserved during transgression, due to: (i) deep, localised tidal erosion at the base of laterally migrating and retreating tidal inlets; and (ii) shallow, widespread wave erosion at wave base (Swift 1968) (Figures 8.28 and 8.29). Estuaries also tend to trap sediment from fluvial and marine sources during transgression, such that the coeval distal shelf is sediment-starved (Figure 8.28a).

8.3.5 Biological Processes

Biological processes are expressed in bioturbation and in assemblages of body fossils and bioclastic grains. Bioturbation intensity and diversity reflect the *in situ* faunal response to physical process regime and to physico-chemical stresses (e.g. local sedimentation rate, salinity, oxygenation) (Section 3.2). This response is exemplified by the archetypal ichnofacies (e.g. *Psilonichnus*, *Skolithos* and *Cruziana* ichnofacies, illustrated respectively in Figures 3.12, 3.14 and 3.15, with ichnofacies distributions shown in Figures 8.12, 8.19 and 8.28) and recurring departures from these archetypes (e.g. impoverished *Skolithos* and *Cruziana* ichnofacies, with distributions shown in Figures 8.19, 8.25 and 8.28) that characterise shallow-marine and marginal-marine environments (e.g. MacEachern and Bann 2008; Gingras et al. 2011). Bioturbation also overprints sedimentary structures (Box 8.2) and thus controls the degree to which the record of physical processes is preserved, although tiering relationships within a burrow assemblage may result in ichnofabrics that indirectly reflect physical bedding patterns (e.g. Taylor et al. 2003). Many sand-rich deltaic and estuarine environments are characterised by diminutive or impoverished ichnofacies and by sporadically heterogeneous distributions of bioturbation, due to the instability and impersistence of physico-chemical conditions in response to interacting wave, tide and/or riverine processes (MacEachern and Bann 2008). Associated physico-chemical stresses in these environments include temporal variations in salinity, water turbidity and deposition rate. Mixed wave–tide shorelines may also exhibit reduced bioturbation relative to their wave- and tide-dominated counterparts, because the former undergo more pronounced variations in water depth, deposition rate and erosion (Dashtgard et al. 2009, 2012, 2021).

Body fossils and bioclastic grains provide a partial record of the skeletal fauna that lived in, or was transported into, shallow-marine and marginal-marine environments. The types and diversity of body fossils can be used to interpret the palaeo-ecology of these environments, and the abundance of bioclastic material tends to decrease as local sedimentation rate increases (e.g. Kidwell et al. 1986). The preservation and taphonomy of body fossils and bioclasts reflect the energy with which physical and chemical processes operated in the depositional environment, and in the environments through which the body fossils and bioclasts had been transported (e.g. Brett and Baird 1986;

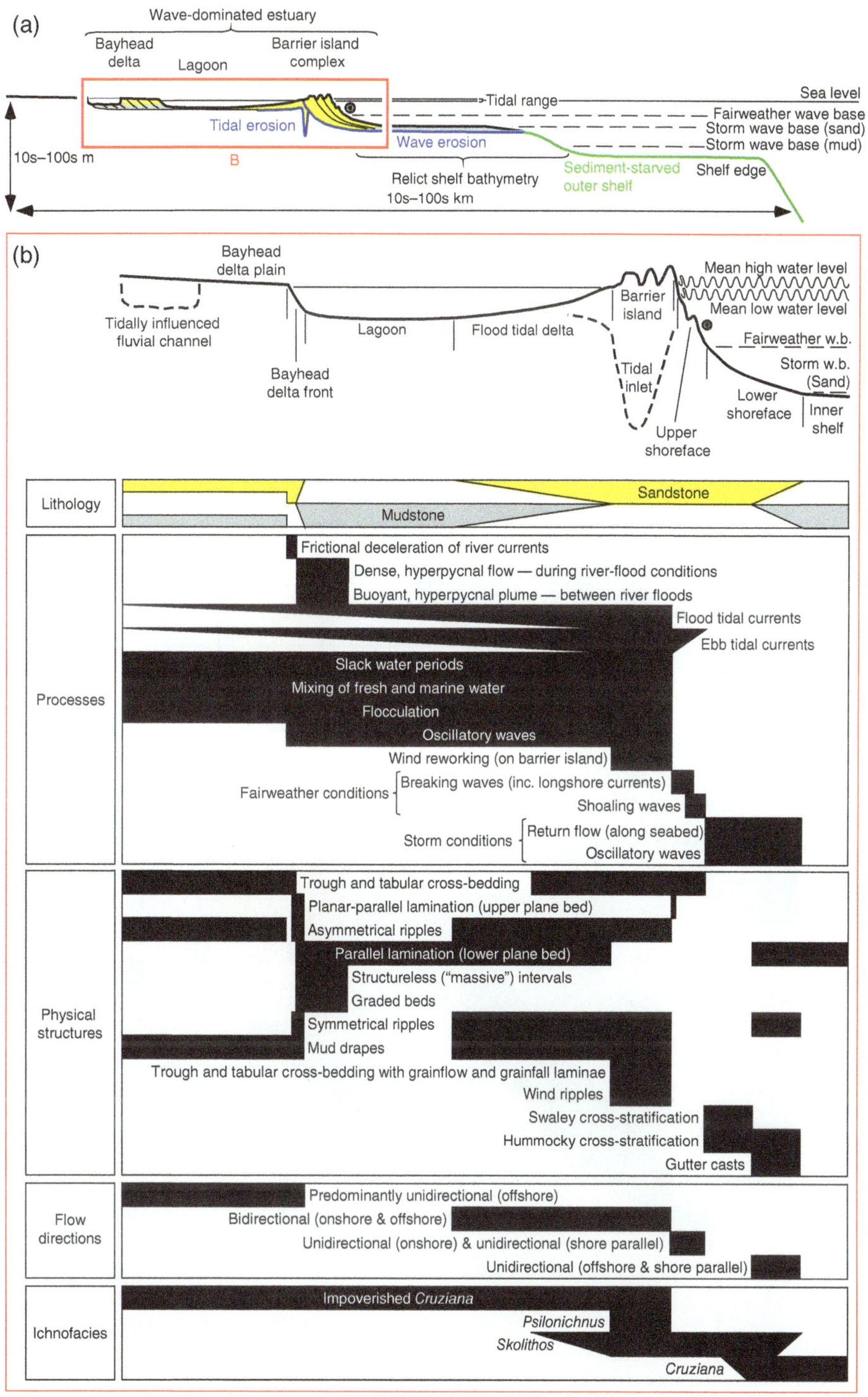

Figure 8.28 Idealised shoreline-shelf transect through a transgressive, wave-dominated estuary and shelf: (a) vertically exaggerated shoreline-shelf transect from landward (tidal limit) to seaward (shelf edge); and (b) detail of vertically exaggerated transect through estuary (labelled 'T3' in Figure 8.3), illustrating mixed process regime (wave, tide and fluvial) and products.

Meldahl and Flessa 1990; Chiarella and Longhitano 2012; Schwarz et al. 2018). Segregation of bioclastic and siliciclastic grain populations in shallow-marine sandstones implies differential transport of these grain types by currents that vary periodically in strength (e.g. tides, storm events), whereas a single combined population of these grain types reflects *in situ* production of carbonate grains and/or thorough mixing by bioturbation (Chiarella and Longhitano 2012; Schwarz et al. 2018).

Detrital carbonaceous grains occur in many shallow-marine and marginal-marine sandstones, due to the introduction and transport of plant material reworked from the

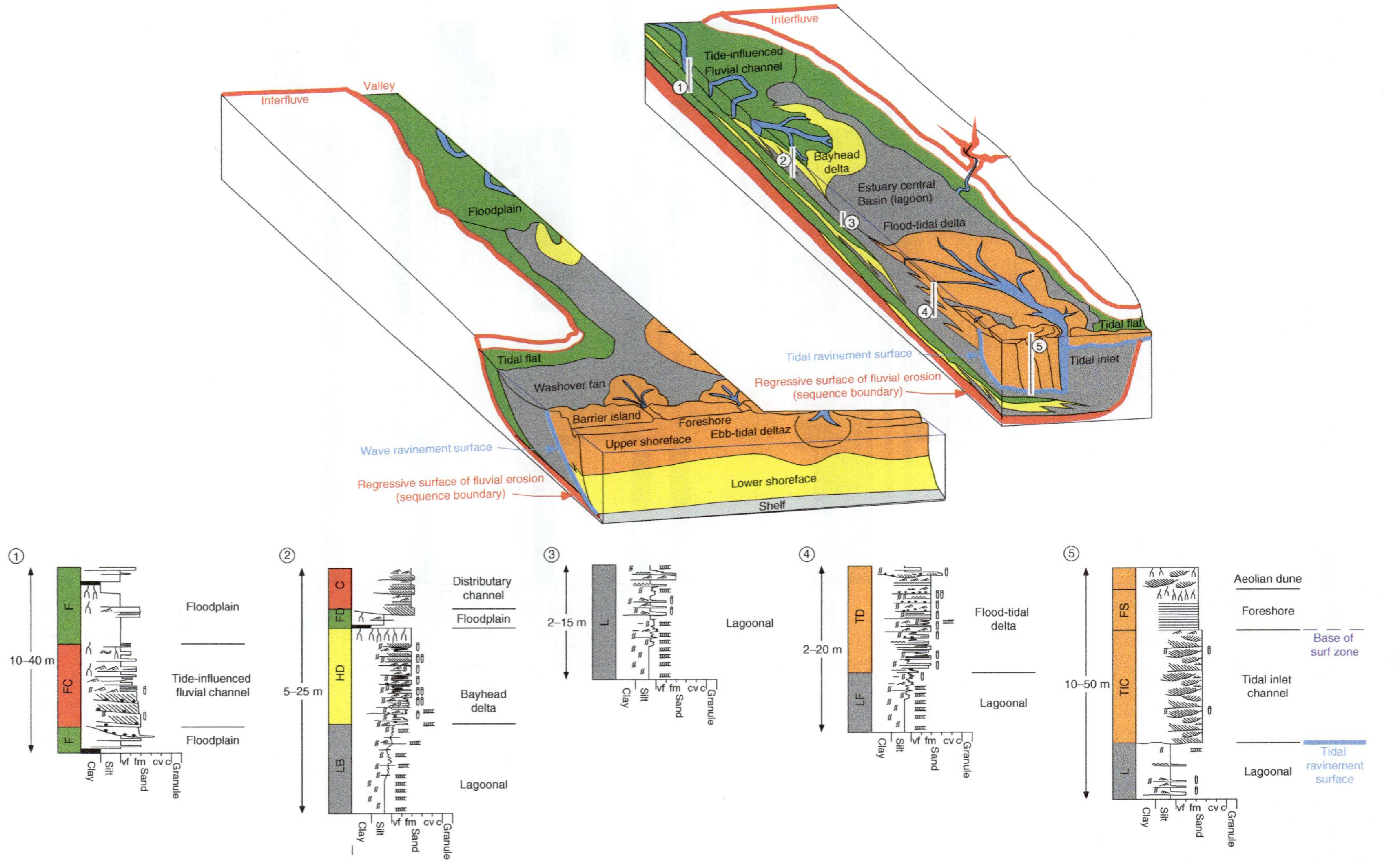

Figure 8.29 Facies model and schematic facies successions for a wave-dominated estuary, developed in a shore-perpendicular incised valley (modified from Boyd 2010). Facies successions are labelled from 1 (proximal) to 5 (distal) along the longitudinal axis of the estuary. See Figure 8.15 for key to facies successions. The development of bounding surfaces (regressive surface of fluvial erosion, wave ravinement surface, tidal ravinement surface) is illustrated in Figure 8.48. Estuaries are typically several hundred metres up to ten kilometres wide, and several kilometres to several tens of kilometres in length.

Box 8.2 Sedimentary Structures as Indicators of Process

Table B8.2.1 Inventory of physical sedimentary structures, flow indicators and ichnofacies in sand-rich coasts and clastic seas (e.g. Figures 8.12, 8.19, 8.23 and 8.28). The relative abundance of these features developed under wave, tide and fluvial (river) process regimes is shown (cf. adapted from Rossi et al. 2017a) and indicates the degree to which a particular sedimentary structure, flow indicator or ichnofacies can be used to diagnose a formative process. Grain size constraints on the formation of sedimentary structures (e.g. adapted from Yoshida et al. 2007; Baas et al. 2016), the direction of flow indicators and formative physico-chemical conditions of ichnofacies are indicated.

		Wave	Tide	Fluvial	Grain-size range
Physical structures	Planar-parallel lamination (upper plane bed)	Common	Common	Common	Sand and gravel
	Trough and tabular cross-bedding	Common	Common	Common	Sand and gravel
	Symmetrical ripple cross-lamination	Common	Absent	Absent	Sand and gravel
	Asymmetrical ripple cross-lamination	Rare	Common	Common	Sand; dimensions may reflect clay content
	Parallel lamination (lower plane bed)	Common	Common	Common	Mud and sand
	Swaley cross-stratification	Common	Absent	Absent	Very fine to fine sand
	Hummocky cross-stratification	Common	Absent	Absent	Very fine to fine sand
	Gutter casts	Common	Absent	Rare	Sand and gravel; require mud-rich substrate
	Flaser, wavy and lenticular bedding	Rare	Common	Rare	Mud and sand; may reflect availability of mud and sand
	Mud drapes	Rare	Common	Rare	Mud
	Rhythmic and "bundled" mud and sand laminae	Rare	Common	Rare	Mud and sand; may reflect availability of mud and sand
	Rill marks, water-level marks	Absent	Rare	Rare	Sand
	Synaeresis cracks	Absent	Rare	Absent	Clay
	Graded beds	Rare	Rare	Common	Mud, sand and gravel
	Inversely graded beds	Absent	Absent	Rare	Mud, sand and gravel
	Structureless sandstone beds	Rare	Rare	Common	Sand
	Structureless mudstone beds	Rare	Rare	Rare	Mud
	Soft-sediment deformation	Rare	Rare	Common	Mud, sand and gravel

		Wave	Tide	Fluvial	Flow direction
Flow indicators	Unidirectional cross-strata	Common	Common	Common	Onshore-, offshore- or alongshore-directed
	Bidirectional cross-strata	Rare	Common	Rare	Onshore- and offshore-directed (but may be deflected)

(Continued)

Table B8.2.1 (Continued)

		Wave	Tide	Fluvial	Physico-chemical conditions
Ichnofacies	*Psilonichnus*	Common	Common	Rare	Fluctuating energy and salinity, intermittent exposure
	Skolithos	Common	Rare	Rare	High energy, marine salinity
	Cruziana	Common	Rare	Rare	Low energy (with episodic high energy), marine salinity
	Impoverished *Skolithos*	Absent	Common	Rare	High energy, brackish or fluctuating salinity
	Impoverished *Cruziana*	Absent	Common	Common	Low energy, brackish or fluctuating salinity

Most physical sedimentary structures, flow indicators and ichnofacies are not unique indicators of a wave, tide or fluvial process. Instead, combinations of these features are required to diagnose a process regime (e.g. Figures 8.12, 8.19, 8.25 and 8.28). Consequently, many interpretations of process regime are inherently uncertain. Where data are available, interpretation of process regime should be based on: (i) the types and relative abundance of sedimentary structures; (ii) spatial patterns of flow direction(s); (iii) ichnological and palaeontological evidence of physico-chemical conditions (e.g. salinity, oxygenation and water turbidity); and (iv) contextual information from facies successions, architectural elements and stratigraphic bounding surfaces.

coastal plain by rivers (e.g. Bustin 1988) or transgressive marine erosion. Much reworked plant material takes the form of sand-sized carbonaceous flakes, which are hydraulically equivalent to clay and quartz silt grains due to their low density and platy habit (Leithold and Hope 1999), and which settle from suspension to drape sandy bedforms. Coarser material commonly comprises wood fragments, which are relatively durable. Sand-sized wood fragments behave hydraulically like quartz fine sand grains (Leithold and Hope 1999), but larger fragments, including waterlogged branches and tree trunks, also occur (e.g. MacDonald and Jefferson 1985). Finer carbonaceous grains are commonly adsorbed on to clays and dispersed by waves, tides and ocean currents into mud-rich environments (Bustin 1988).

8.3.6 Chemical Processes

Chemical processes are expressed by the *in situ* inorganic precipitation of authigenic minerals at or close to the sediment surface. Such precipitation requires specific physico-chemical conditions. For example, glauconite, an iron potassium phyllosilicate mineral, is a characteristic component of some shallow-marine sandstones ('greensands') and was precipitated under reducing, low-oxygen conditions and slow sedimentation rates (Odin and Matter 1981; Stonecipher 1999; Chafetz and Reid 2000). Glauconitic sandstones may therefore record hiatuses in sedimentation. Careful petrographic analysis is required to demonstrate that authigenic minerals precipitated during early diagenesis, such that they record conditions in the depositional environment, rather than during late, burial-related diagenesis. Environment-related diagenesis may include dissolution of aragonite and calcite bioclasts and associated re-precipitation of calcite cement (e.g. Bjørkum and Walderhaug 1990); thus biological and chemical processes may be intimately linked. Along sandy, arid and semi-arid sabkha coastlines, precipitation of evaporite minerals during environment-related diagenesis may generate displacive nodules and enterolithic folds of gypsum and anhydrite in the supratidal zone, and surficial halite crusts. These mixed siliciclastic-evaporite coastal deposits are treated in Section 10.3.5.

8.4 Facies, Facies Associations, Facies Successions and Facies Models

Using the facies analysis approach outlined in Chapter 2, we here present eight case studies of vertical facies successions representative of sand-rich marginal- and shallow-marine environments, with an emphasis on mixed -process regime systems. Subsequent sections (8.4.10–8.4.12) address 2D and 3D architecture. This approach allows us to: (i) highlight the diagnostic links to modern process outlined in Section 8.3; (ii) discuss interpretational ambiguities; and (iii) complement previous work on end-member, wave-, tide- or fluvial-dominated depositional systems (e.g. Bhattacharya 2010; Boyd 2010; Dalrymple 2010; Plint 2010).

Our eight facies successions are described using nine sandstone-dominated facies associations (FA1–9) and three mudstone-dominated facies associations (FA10–12) (Table 8.1). These 12 facies associations, and variations on them, are widely documented in shallow-marine strata, and are stacked both vertically and laterally in the case studies (Sections 8.4.1–8.4.8). Two aspects of the generalised facies associations are worthy of note. First, the grain size characteristics of the facies association reflect not simply the competence of depositional currents but the calibre of the available sediment at the site of deposition. These together determine the sedimentary structures that are generated (Box 8.2). Thus, the pairs FA1 and FA2, and also FA6 and FA7, have different grain size characteristics and therefore different sedimentary structures and facies, but similar process regimes. Second, the facies and architectural characteristics of mudstone-dominated facies associations, and their vertical and lateral stratigraphic relationships with interstratified sandstone-dominated facies associations, are particularly important for providing the palaeo-environmental context of persistent low-energy conditions in which to interpret sand transport and deposition. A wider variety of mudstone-dominated, shallow-marine facies associations is documented in Chapter 9. We emphasise that the 12 generalised facies associations chosen, and summarised in Table 8.1, are illustrative rather than complete.

8.4.1 Prograding Wave-Dominated Shorefaces (W) – Facies Succession 1 (Composed of FA1, 2 and 4)

The first example of a facies succession comprises a coarsening-upward succession dominated by wave-generated structures (Figure 8.30a) and developed as part of a sandy, wave-dominated deltaic and strandplain shoreline (example A in Figure B8.1.2a, Box 8.1, classified as a 'strandplain shoreface'

Table 8.1 Summary description and interpretation of generalised sandstone-dominated facies associations (FA1–9) and commonly interstratified mudstone-dominated facies associations (FA10–12).

	Description	Process interpretation	Environmental interpretation	Process regime	
FA1	(i) Sparsely bioturbated, cross-stratified, medium-grained sandstone overlain by (ii) root-penetrated, planar-parallel-laminated, medium-grained sandstone (Figure 8.30a,f–g). *Skolithos* ichnofacies.	Sand deposition by dunes migrating in response to unidirectional currents, overlain by sand deposition under upper plane bed conditions at shoreline (e.g. due to breaking waves).	Wave-dominated upper shoreface and foreshore (W) or proximal wave-dominated delta front (Wf)	W	Wf
FA2	(i) Sparsely bioturbated, low-angle cross-stratified, fine-grained sandstone interstratified with (ii) swaley and hummocky cross-stratified, fine-grained sandstone (Figures 8.30a,e, 8.33a, d–f,h). Bioturbation is variable. *Skolithos* ichnofacies.	Episodic sand deposition under strong oscillatory currents, combined oscillatory-and-unidirectional currents, and/or unidirectional currents with dune development suppressed by high suspended sediment concentration; infaunal colonisation during periods between sand deposition.	Wave-dominated proximal lower shoreface (W) or proximal wave-dominated delta front (Wf)	W	Wf
FA3	(i) Sparsely bioturbated, cross-stratified, medium-grained sandstone interstratified with (ii) symmetrically rippled surfaces draped by weakly laminated to structureless mudstone (Figure 8.32d–f). *Skolithos* ichnofacies.	Sand deposition by dunes migrating in response to unidirectional currents, punctuated by periods of wave reworking and rapid mud deposition under shallower water conditions.	Tidally modulated, wave-dominated upper shoreface and foreshore or tidal inlet (Wt)	Wt	

(*Continued*)

Table 8.1 (Continued)

	Description	Process interpretation	Environmental interpretation	Process regime		
FA4	(i) Hummocky and swaley cross-stratified, fine-grained sandstone interstratified with (ii) variably bioturbated mudstone with sandstone lenses (Figures 8.30a–d, 8.32a–c, 8.33a–c,e–g). *Cruziana* ichnofacies.	Episodic sand deposition under strong oscillatory currents and/or combined oscillatory-and-unidirectional currents, alternating with periods of mud deposition.	Wave-(storm-)dominated distal lower shoreface (W) or distal wave-(storm-)dominated delta front (Wf)	W	Wf	
FA5	(i) Cross-stratified, medium-grained sandstone with mud-draped foresets, grading upwards into (ii) wavy-bedded and (iii) root-penetrated, lenticular-bedded, fine-grained sandstone and mudstone (Figure 8.7e). *Skolithos* ichnofacies passing upwards into *Cruziana* ichnofacies.	sand transport and deposition by ripples and dunes in response to bidirectional and/or unidirectional currents of fluctuating velocity, with mud deposition during periods of low flow velocity; vertical succession records upward decrease in net flow velocity in response to upward shallowing, culminating in colonisation by plants.	intertidal flats, tidal creeks and tidal channels (T)	T		
FA6	(i) Mudstone passing upwards into (ii) lenticular and (iii) wavy-bedded, fine-grained sandstone and mudstone, and (iv) cross-stratified, medium-grained sandstone with mud-draped foresets (Figure 8.33e–h). Bioturbation is variable (Figures 8.36 and 8.37a–d). *Skolithos* ichnofacies.	Sand transport and deposition down gentle depositional slope by ripples and dunes in response to bidirectional and/or unidirectional currents of fluctuating velocity, with mud deposition during periods of low flow velocity; vertical succession records upward increase in net flow velocity in response to upward shallowing.	Mixed-influence delta front, with combination of tide, river and wave influence (Ft, Fw)	Ft	Fw	
FA7	(i) Intensely bioturbated silty sandstone passing upwards into (ii) cross-stratified, medium-grained sandstone with mud-draped foresets interstratified with (iii) steeply dipping, structureless and graded beds of medium-to-coarse-grained sandstone (Figure 8.35). *Cruziana* ichnofacies passing upwards into *Skolithos* ichnofacies.	Sand transport and deposition down steep depositional slope by gravity flows, with local reworking by dunes migrating in response to unidirectional currents of fluctuating velocity; slow sand and silt deposition at base of depositional slope; vertical succession records upward shallowing.	Mixed-influence delta front, with both tide and river influence (Ft); dominated by gravity-flow beds due to coarse grain size (Gilbert-type delta)	Ft		
FA8	(i) Cross-stratified sandstone, grading upwards into (ii) current-ripple cross-laminated sandstone and (iii) root-penetrated mudstone (Figure 8.37a,b). Bioturbation is absent or sparse. *Skolithos* ichnofacies.	Sand transport and deposition by ripples and dunes in response to unidirectional currents; vertical succession records upward decrease in flow velocity in response to upward shallowing, culminating in mud deposition and colonisation by plants.	Fluvial-dominated deltaic distributary channel (F) with variable marine influence (Ft, Fw)	F	Ft	Fw
FA9	(i) Cross-stratified sandstone with bi-directional palaeocurrents and mud-draped foresets, grading upwards into (ii) flaser-, wavy and lenticular-bedded sandstone and mudstone (Figure 8.38). Bioturbation is sparse.	Sand transport and deposition by ripples and dunes in response to bidirectional currents of fluctuating velocity, with mud deposition during periods of low flow velocity; vertical succession records upward decrease in net flow velocity in response to upward shallowing.	Tidal channel (T) with variable fluvial influence (Tf)	T	Tf	

Table 8.1 (Continued)

	Description	Process interpretation	Environmental interpretation	Process regime
FA10	(i) Intensely bioturbated mudstone containing diverse trace-fossil assemblage, with rare (ii) sparsely bioturbated mudstone intervals and sandstone lenses (Figure 8.38). *Cruziana* ichnofacies.	Mud deposition under fully marine conditions, with episodes of rapid influx of mud.	Wave-(storm-)dominated offshore and offshore transition (W, Wf)	W Wf
FA11	(i) Sparsely bioturbated mudstone containing depauperate trace-fossil assemblage, with (ii) soft-sediment deformed intervals (Figure 8.37a,b).	Rapid deposition of mud under physico-chemically stressed conditions (e.g. low or fluctuating salinity), at (or near to) base of depositional slope.	Fluvial-dominated offshore and prodelta (F)	F
FA12	(i) Organic-rich, unbioturbated mudstone, with (ii) sparsely bioturbated mudstone containing depauperate trace-fossil assemblage and rare sandstone lenses.	Rapid deposition of mud under physico-chemically stressed conditions (e.g. poorly oxygenated), with high rate of organic matter accumulation and preservation.	Tide-dominated offshore (T) and lagoon (Tw, Tf)	T Tw Tf

In some cases, interpretations are non-unique, and the same sub-environment can contain different facies. Contextual information from facies successions and architectural elements that contain the facies associations is essential to support or refute a particular interpretation.

in Figure B8.1.1a, Box 8.1 and a 'storm-dominated shoreface' in Figure B8.1.1c, Box 8.1). The example succession is taken from the Late Cretaceous (Campanian) Blackhawk Formation, Utah, USA (Hampson et al. 2011). The succession represents an archetypal end-member of a sandy, wave-dominated shoreline (process regime: W). Cored modern analogues of similar facies successions with associated process observations and measurements include Ventura Beach, California, USA; Sapelo Island, Georgia, USA; and Kouchibouguac Bay, New Brunswick, Canada (Howard and Reineck 1981; Davidson-Arnott and Greenwood 1976).

The Blackhawk Formation was deposited on the western margin of the wide (c. 1500 km), shallow (c. 200 m) intracratonic Western Interior Seaway developed under a subtropical, humid climate (e.g. Hampson et al. 2008; Hampson 2010). Blackhawk Formation shorelines were oriented approximately north–south. They were fed with abundant quartz-rich sand by 150–450 km long rivers draining the youthful Sevier Orogenic Belt to the west and by southward-directed longshore drift (e.g. Hampson 2010; Hampson et al. 2014). The shorelines fronted a wide basin, with a large potential fetch length for easterly winds. Numerical modelling studies indicate that water circulation and associated sediment transport in the seaway were storm driven (e.g. Slingerland and Keen 1999).

The lower part of the succession comprises hummocky cross-stratified, fine-grained sandstones (Figure 8.30c) interbedded with sparsely bioturbated mudstones (Figure 8.30b) in a single facies association (Figure 8.30d; FA4 in Figure 8.30a, Table 8.1). Some hummocky-cross-stratified sandstones occur within steep-sided erosional scours (e.g. gutter or pot casts) (Figure 8.30d; Collins et al. 2017; Onyenanu et al. 2018). The trace-fossil assemblage is the proximal expression of the *Cruziana* ichnofacies (Figure 8.31). Hummocky cross-stratified sandstone facies (Figure 8.30c) record sand deposition by strong oscillatory or combined-flow currents during storm events (e.g. Duke 1985), implying formation under a wave-dominated process regime. The formation of hummocky cross-stratification has only rarely been observed in modern environments (Li and King 2007; Arora et al. 2018), and at smaller scales than many of their ancient counterparts, although similar structures have been generated in flume tank experiments (e.g. Dumas et al. 2005). Consequently, the precise origin of this structure is still debated (Yang et al. 2006; Quin 2011; Jelby et al. 2020). Variations in hummocky and/or swaley cross-stratification style within sandstone beds are attributed to unsteadiness in the formative oscillatory currents, downwelling flows and/or fluvial-discharge conditions (e.g. Dumas et al. 2005; Jelby et al. 2020). Intervening sparsely bioturbated mudstone and lenticular sandstone facies (Figure 8.30b) record rapid and episodic deposition of predominantly mud from suspension during the waning phase of storm events and during fair-weather periods (Onyenanu et al. 2018), thus implying deposition on the distal part of the lower shoreface or inner shelf, at a water depth that lay between fair-weather and storm

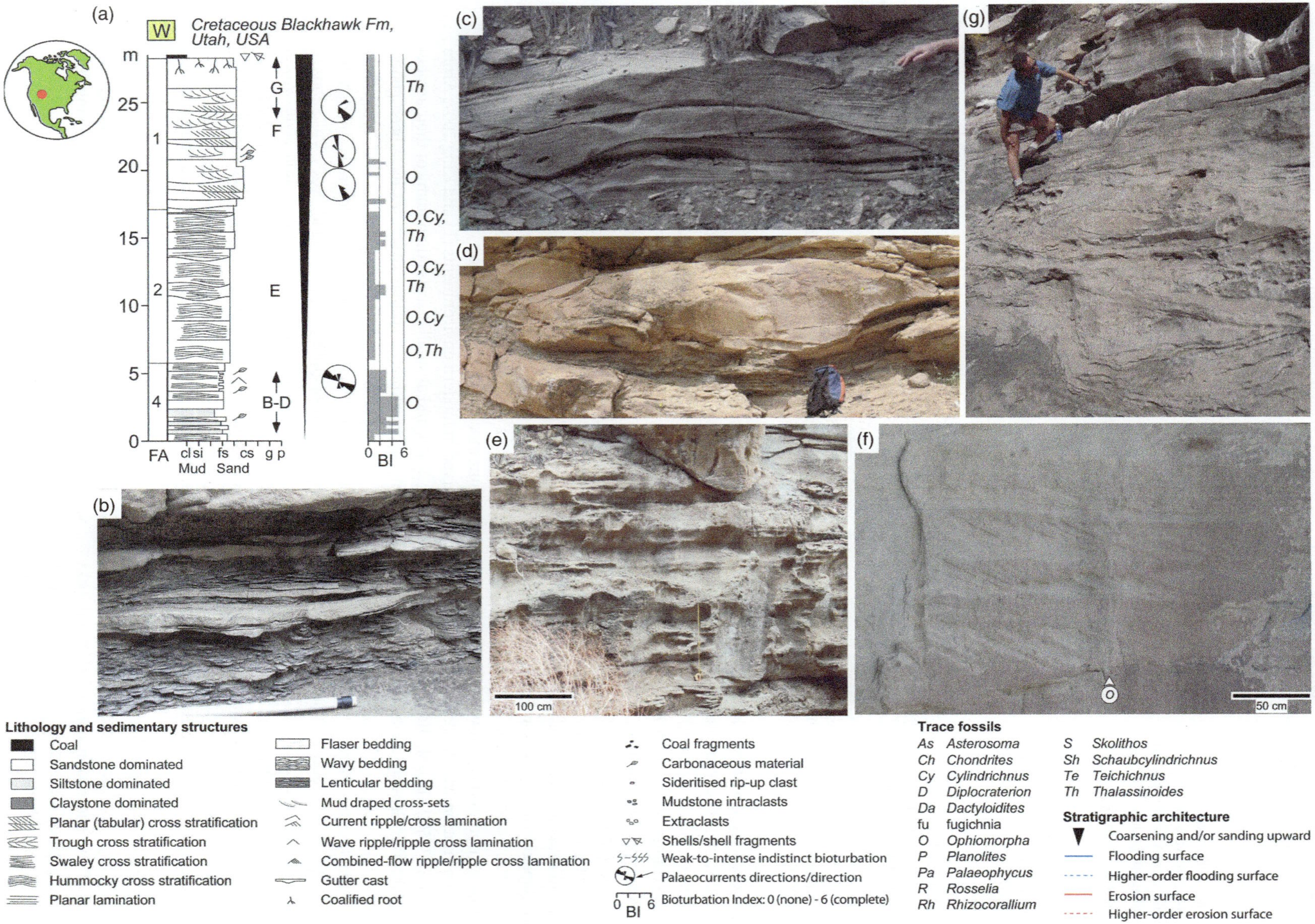

Figure 8.30 (a) Facies succession recording progradation of a wave-dominated shoreface (process regime: W) and (b–g) photographs of constituent facies and facies associations (Cretaceous Blackhawk Formation, Utah, USA). (b) Sparsely bioturbated mudstones and lenticular sandstones, and (c) hummocky-cross-stratified sandstones that are interbedded to form a facies association (FA4 in Figure 8.30a); (d) some of the hummocky-cross-stratified sandstones occur as discontinuous bodies with steep-sided erosional scours at their bases. (e) Amalgamated hummocky-cross-stratified sandstones that lack interbedded mudstones (FA2 in Figure 8.30a); sandstone bed tops are marked locally by bioturbated intervals. (f) Cross-bedded sandstone with sparse vertical burrows, that (**g**) is overlain by planar-parallel-laminated sandstone to form a facies association (FA1 in Figure 8.30a). Parts a, d and f modified from Hampson et al. (2011), Onyenanu et al. (2018) / John Wiley & Sons and Hampson et al. (2008) / John Wiley & Sons, respectively. Photograph credits parts b, c, e and g: Gary Hampson.

wave base (cf. Figures 8.12b and 8.15a; e.g. Kamola and Van Wagoner 1995). The scarcity of bioturbation in the mudstones may record a combination of high sedimentation rate and episodic freshwater input from rivers in flood during storms (Collins et al. 2017). Steep-sided erosional scours at the bases of some hummocky cross-stratified sandstone beds (Figure 8.30d) indicate pronounced erosion, due to high velocity, offshore-directed unidirectional currents, during sand transport (Eide et al. 2015; Onyenanu et al. 2018). The erosional scours were probably generated by storm-driven downwelling flow along the seabed, potentially supplemented by localised, high fluvial discharge from river mouths at the shoreline (e.g. Collins et al. 2017).

The central part of the succession comprises amalgamated beds of hummocky and/or swaley cross-stratified, fine-grained sandstones that lack any interbedded mudstones (Figure 8.30e; FA2 in Figure 8.30a, Table 8.1). The trace-fossil assemblage constitutes the distal expression of the *Skolithos* ichnofacies (Figures 3.14, 8.31; MacEachern and Bann 2008). This facies association is interpreted to record episodic sand deposition by strong oscillatory and/or combined currents during storms, similar to the underlying facies association. The absence of mudstone is attributed to removal by erosion at the base of overlying sandstone beds, which is attributed to enhanced storm scour in the shallower water depths of the proximal part of the lower shoreface (cf. Figures 8.12b and 8.15a; e.g. Kamola and Van Wagoner 1995).

The upper part of the succession comprises sparsely bioturbated, well-sorted, medium-grained, cross-bedded sandstones with shoreline-parallel palaeocurrents (Figure 8.30f) that are sharply overlain by root-penetrated, well-sorted, lower medium-grained, planar-parallel-laminated sandstones that together define a third facies association (Figure 8.30g; FA1 in Figure 8.30a, Table 8.1). This facies association has a low-relief (up to 10s cm) erosional contact at its base and an archetypal expression of the *Skolithos* ichnofacies (Figures 3.14 and 8.31). The upward transition from cross-stratified sandstone facies to root-penetrated, planar-parallel-laminated sandstone facies records an increase in flow velocity and a decrease in water depth. In this case, palaeocurrents in the cross-stratified sandstone facies are oriented sub-parallel to the regional palaeoshoreline trend, implying that they record dune migration driven by unidirectional, longshore currents. The cross-stratified sandstones are interpreted to record the action of shoaling and breaking waves that approached the shoreline obliquely (cf. Figures 8.12b, 8.13 and 8.15a; e.g. Howard and Frey 1984; Kamola and Van Wagoner 1995). Overlying root-penetrated, planar-parallel-laminated sandstone facies are interpreted to record upper plane-bed flow due to swash and backwash of breaking waves on the foreshore (cf. Figures 8.12b and 8.15a), with subsequent overprinting by roots (e.g. Howard and Frey 1984; Kamola and Van Wagoner 1995). Thus, this facies association is interpreted to record the upper shoreface and foreshore of a wave-dominated shoreline.

The overall facies succession (Figure 8.30a) records upward shallowing, and thus progradation, of a wave-dominated shoreface (cf. Figure 8.15a).

8.4.2 Prograding and Retreating Wave-Dominated, Tide-Influenced Shorefaces (Wt) – Facies Succession 2 (Composed of FA3, 4 and 10)

Figure 8.32 illustrates two facies successions, a lower coarsening-upward succession overlain by an upper fining-upward succession, that are dominated by wave-generated structures but also record tidal influence (example M in Figure B8.1.2a, Box 8.1, classified as a 'tidal shoreface' in Figure B8.1.1a,c, Box 8.1). The successions were developed at the margin of a mixed sand-mud shelf and are taken from the Late Jurassic (Kimmeridgian-Tithonian) Grès de la Crèche and Grès de Châtillon, Boulonnais, northern France (Angus et al. 2020). The differences between these facies successions (process regime: Wt) and that of the archetypal wave-dominated shoreface (process regime: W) shown in Figure 8.30 are attributed to the influence of tides.

The Grès de la Crèche and Grès de Châtillon were deposited on the northeastern margin of the Weald Basin, one of a series of linked basins that formed the Laurasian Seaway, with the Tethys Ocean at its eastern end (e.g. Wignall et al. 1996; Proust et al. 2001; Angus et al. 2020). The Weald Basin lay at a palaeolatitude of 30°–35° N during deposition of the Grès de la Crèche and Grès de Châtillon, when it was subject to a warm, humid and possibly monsoonal climate (e.g. Wignall and Ruffell 1990; Hesselbo et al. 2009). Sand was transported only a short distance (<100 km) from the London–Brabant Massif source to the Grès de la Crèche and Grès de Châtillon shorelines, which were locally oriented approximately north–south (Wignall et al. 1996; Proust et al. 2001; Angus et al. 2020). Waves generated by prevailing westerly and southwesterly winds in the Weald Basin had a fetch length of 200–600 km. The tidal wave from the Tethys Ocean propagated through a 220-km wide strait at the southeastern entrance (Angus et al. 2020). Semi-diurnal tides were amplified by resonance in the shallow (18–33 m), wide (150–200 km), open embayment of the Weald Basin (Angus et al. 2020).

The lower facies succession (3–18 m) is superficially similar to that in Figure 8.30a, and it contains facies associations representing offshore mudstones and lower shoreface deposits in its lower part (FA10 and FA4 in Figure 8.32a,

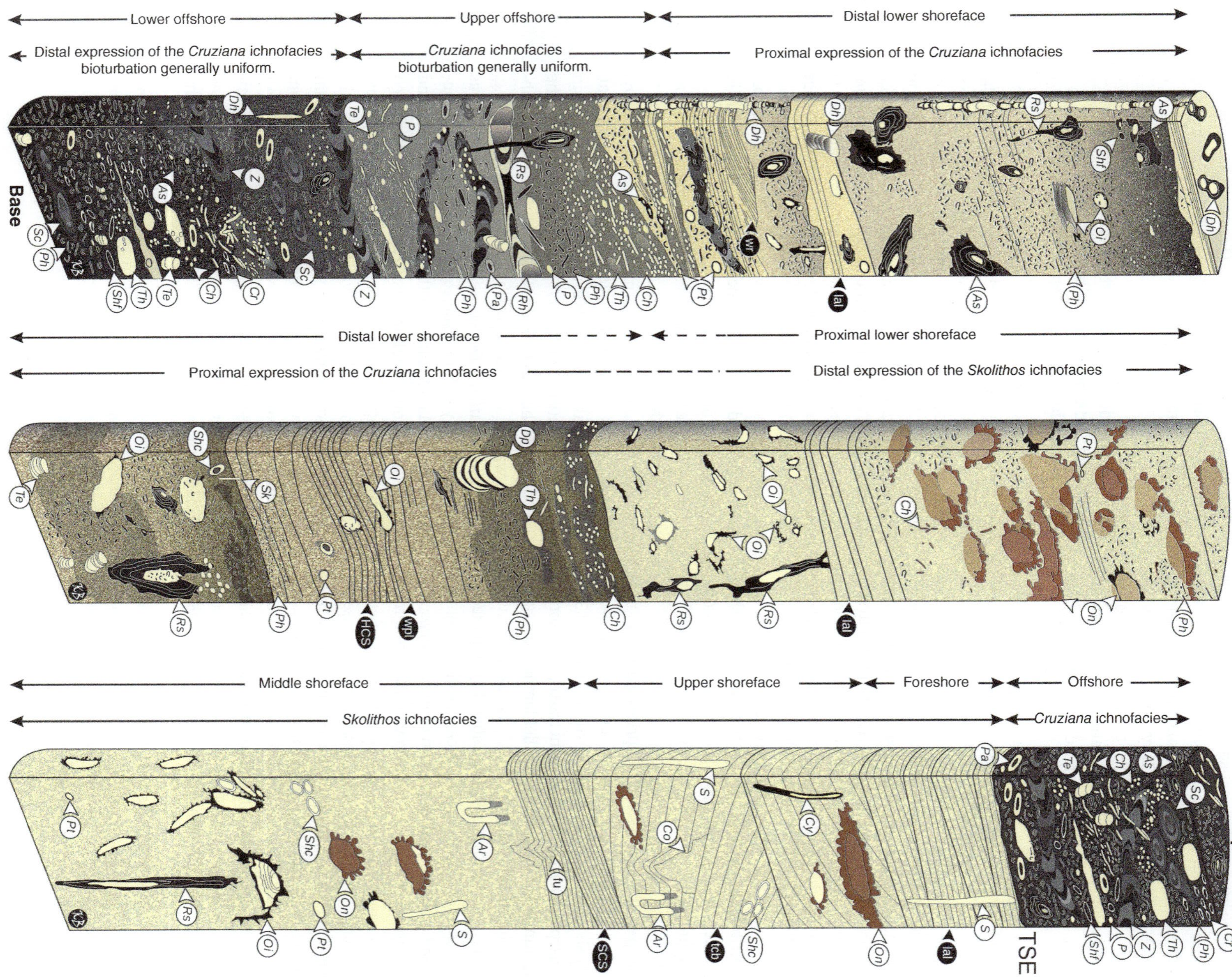

Figure 8.31 Schematic split-core diagrams of representative sections of a progradational, wave-dominated shoreface facies succession from the Cretaceous Blackhawk Formation, Utah, USA (Figure 8.30a). The succession comprises, from bottom left to top right: offshore; distal lower shoreface (FA4 in Figure 8.30a); proximal lower shoreface (FA2 in Figure 8.30a); upper shoreface and foreshore (FA1 in Figure 8.30a). Trace-fossil codes are as follows: *Arenicolites* (*Ar*), *Asterosoma* (*As*), *Conichnus* (*Co*), *Chondrites* (*Ch*), *Cosmorhaphe* (*Cr*), *Diplocraterion habichi* (*Dh*), *Diplocraterion parallelum* (*Dp*), fugichnia (fu), *Ophiomorpha irregulaire* (*Oi*), *Ophiomorpha nodosa* (*On*), *Planolites* (*P*), *Palaeophycus heberti* (*Pa*), *Palaeophycus tubularis* (*Pt*), *Phycosiphon* (*Ph*), *Rhizocorallium* (*Rh*), *Rosselia socialis* (*Rs*), *Skolithos* (*S*), *Scolicia* (*Sc*), *Schaubcylindrichnus coronus* (*Shc*), *Schaubcylindrichnus formosus* (*Shf*), *Teichichnus* (*Te*), *Thalassinoides* (*Th*) and *Zoophycos* (*Z*). HCS – hummocky cross-stratification; lal – low-angle lamination; SCS – swaley cross-stratification; tcb – trough cross-stratification; wpl – wavy parallel lamination. Figure drafted by Kerrie Bann.

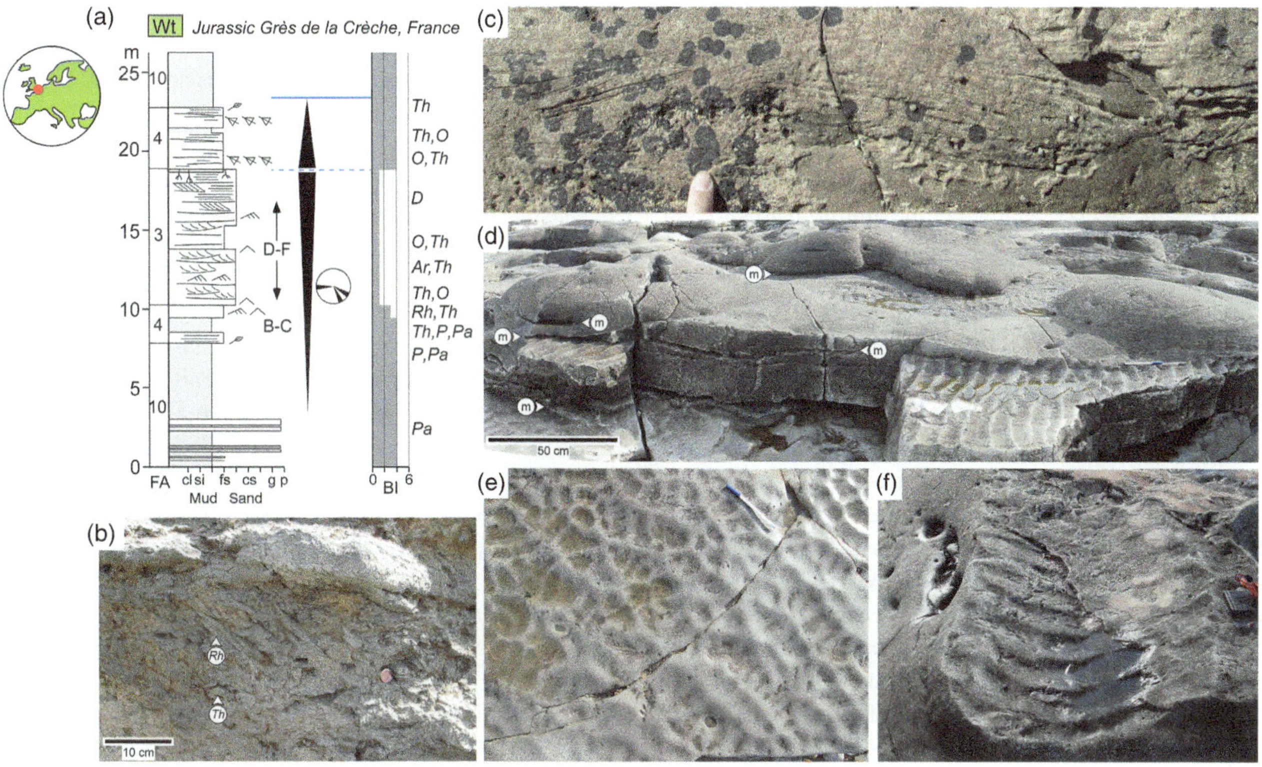

Figure 8.32 (a) Facies successions recording progradation of a wave-dominated, tidally modulated shoreface (process regime: Wt) and (b–f) photographs of constituent facies and facies associations (Jurassic Grès de la Crèche and Grès de Châtillon, Boulonnais, France). See Figure 8.30a for legend. (b) Intensely bioturbated sandstones and (c) hummocky-cross-stratified sandstones that are interbedded to form a facies association (FA4 in Figure 8.32a). (d–f) Sparsely bioturbated, cross-stratified sandstone facies interbedded with discontinuous drapes of sparsely bioturbated, weakly laminated mudstone facies (labelled 'm' in part d) (FA3 in Figure 8.32a). Mudstone drapes occur directly above symmetrically rippled surfaces that cap cross-bedded sandstones; (e) symmetrical interference and ladder ripples on bedding plane, and (f) symmetrical ripples overprinting trough-shaped erosion surfaces (Grès de Châtillon). Parts b, d–f after Angus et al. (2020). Photograph credit part c: Gary Hampson.

Table 8.1). These are overlain by FA3, which comprises sparsely bioturbated, cross-stratified, medium-grained sandstones with intercalations of planar-parallel-laminated, medium-grained sandstones (Figure 8.32a and Table 8.1). The cross-stratified sandstones contain symmetrically rippled surfaces draped by sparsely bioturbated, weakly laminated to structureless mudstones (Figure 8.32b–d). FA3 is interpreted to record the migration of dunes in response to unidirectional currents, with intermittent reworking by wave ripples followed by mud deposition during periods of low flow velocity (Wignall et al. 1996; Angus et al. 2020). Some of the symmetrically rippled surfaces exhibit interference and ladder ripples (Figure 8.32b–d), formed due to refraction and reflection in shallow water (pp. 355–372 in Reineck and Singh 1973). Repeated interbedding of sandstone cross-sets, wave-rippled surfaces and weakly laminated to structureless mudstones implies that fluctuations in flow velocity and water depth were persistent throughout deposition (Angus et al. 2020). Such fluctuations in flow velocity and water depth are tidally modulated, as represented in the facies model developed by Dashtgard et al. (2012) (Figure 8.15b). Palaeocurrents in the cross-bedded sandstone facies are oriented sub-parallel, onshore and offshore relative to the regional palaeoshoreline trend, while symmetrical ripple crests are aligned sub-parallel to the palaeoshoreline trend (Wignall et al. 1996; Angus et al. 2020). The top of the facies association is penetrated by rootlets. This first facies succession (Figure 8.32a) records upward shallowing, and thus progradation, of a wave-dominated, tidally modulated shoreface (cf. Figure 8.15b).

The upper facies succession is marked by a planar erosion surface at its base (18 m). The facies succession comprises, from base to top: proximal lower shoreface, distal lower shoreface and offshore deposits (FA4 and FA10 in Figure 8.32a, Table 8.1). This second facies succession records upward deepening, and therefore transgression, of the wave-dominated, tidally modulated shoreface. The erosion surface at 18 m separating the regressive facies succession from the overlying transgressive facies succession represents a significant break in deposition, such that Walther's law must be applied separately to the facies successions below and above the erosion surface. The surface is interpreted to have been cut by waves during transgression (i.e. a wave ravinement surface, *sensu* Swift 1975a), following maximum regression of the shoreface.

8.4.3 Prograding Wave-(Storm-)Dominated, Fluvial-Influenced Deltas (Wf) – Facies Succession 3 (Composed of FA2, 4 and 10)

This example (Figures 8.33a and 8.34) comprises a sandier-upward succession dominated by wave-generated structures and recording evidence of fluvial influence (example N in Figure B8.1.2a, Box 8.1, classified as a 'wave-dominated delta' in Figure B8.1.1a, Box 8.1 and a 'wave-dominated, fluvial-influenced delta' in Figure B8.1.1b, Box 8.1). The succession was developed on a mixed sand-mud shelf and is taken from the Middle Miocene (Serravallian) Belait Formation, northwest Borneo (Collins et al. 2017). The more heterolithic character of this facies succession (process regime: Wf) relative to that of the archetypal wave-dominated shoreface (process regime: W) in Figure 8.30 is attributed to the finer grain size and, possibly, increased proportion of mud supplied by rivers to the Belait Formation shelf (Collins et al. 2017).

The Belait Formation was deposited on a rapidly subsiding, narrow (<100 km wide) shelf in the Baram-Balabac Basin, adjacent to the Northwest Borneo Trough at the southern margin of the South China Sea (e.g. Cullen 2010; Collins et al. 2017). The Belait Formation and laterally equivalent Miri and Lambir formations record deposition in a similar 4°–5° N latitude, tectonic setting and humid-tropical monsoonal climate as the modern Baram Delta and its flanking shorelines (Figure 8.1a,c, d) (e.g. Lambiase and Cullen 2013; Collins et al. 2017). Sediment is transported only a short distance (10s–100s km) from an active, uplifting hinterland subjected to deep humid-tropical weathering in multiple, closely spaced river catchments (Collins et al. 2017). As a result, modern river runoff is dominated by suspended silt and clay (Milliman and Meade 1983). Transport of sandy bedload to the shoreline occurs during intense high-discharge events driven by monsoon-influenced storms, which may also cause contemporaneous shoreline-bypass of fluvially supplied sediment further offshore (Collins et al. 2017). Waves along the modern northwest Borneo shoreline are generated by the stronger northerly winds (e.g. Collins et al. 2017), giving rise to sand transport by fair-weather-wave-generated longshore drift, which is reflected in the cuspate geometry of the modern Baram delta (Figure 8.1d; e.g. Sandal 1996).

The lower facies association (FA10) is dominated by variably laminated-to-massive mudstones with millimetre-scale silty laminae, and subordinate centimetre-scale, sharp-based siltstone to very-fine grained sandstone layers displaying planar lamination and swaley cross-stratification. Bioturbation in mudstones is highly variable, ranging from impoverished to relatively diverse, which includes *Chondrites*, *Palaeophycus*, *Phycosiphon*, *Planolites* and *Thalassinoides*, and corresponds to a distal *Cruziana* ichnofacies (Figure 3.15; MacEachern and Bann 2008). Bioturbation styles in this facies association alternate between those of FA10 and FA11 in Table 8.1, and the facies association could have formed under a range of depositional energies, but is most likely to represent energetic conditions: a storm-dominated prodelta or delta-influenced offshore shelf below wave base (Bhattacharya and MacEachern 2009; Plint et al. 2012). The overlying, heterolithic facies association is similar to FA4 in Table 8.1 and comprises muddy heterolithics interbedded with sharp- and erosionally based, very fine-grained sandstone beds and bedsets (decimetre- to metre-thick) displaying swaley cross-stratification that is often slightly asymmetrical (FA4 in Figure 8.33b,d, Table 8.1). Gutter casts are notably abundant, several decimetres to metres thick and have long axes oriented approximately perpendicular to the palaeoshoreline (Figure 8.33c; Collins et al. 2017). Predominantly swaley-cross-stratified beds often slightly fine upwards and have increased bioturbation at bed tops. Together these characteristics are consistent with deposition by episodic, high energy, waning oscillatory-dominated combined flows (e.g. Arnott and Southard 1990). Given the geological setting, these combined flows were likely linked to a combination of storm and river flood-related flows ('storm-floods') (Collins et al. 2017). Bioturbation in sandstones is dominated by *Ophiomorpha* with associated minor ichnotaxa, which constitutes an impoverished *Skolithos* ichnofacies (Buatois et al. 2008; MacEachern and Bann 2008). This ichnofacies is atypical of fully marine shoreface to offshore successions (MacEachern and Bann 2008) and likely records high but fluctuating sedimentation rates and/or physico-chemical stress, such as reduced and/or fluctuating salinity (Section 3.2.7), consistent with combined storm and river influence. The uppermost facies association comprises predominantly swaley-cross-stratified sandstone in metre-scale, erosionally amalgamated bedsets, equivalent to FA2 in Table 8.1. Overall, the succession is interpreted to record deposition in a wave-(storm-)dominated, fluvial-influenced delta (Collins et al. 2017). The basal muddier facies association (FA10/11) is absent in the overlying facies succession in Figure 8.33a.

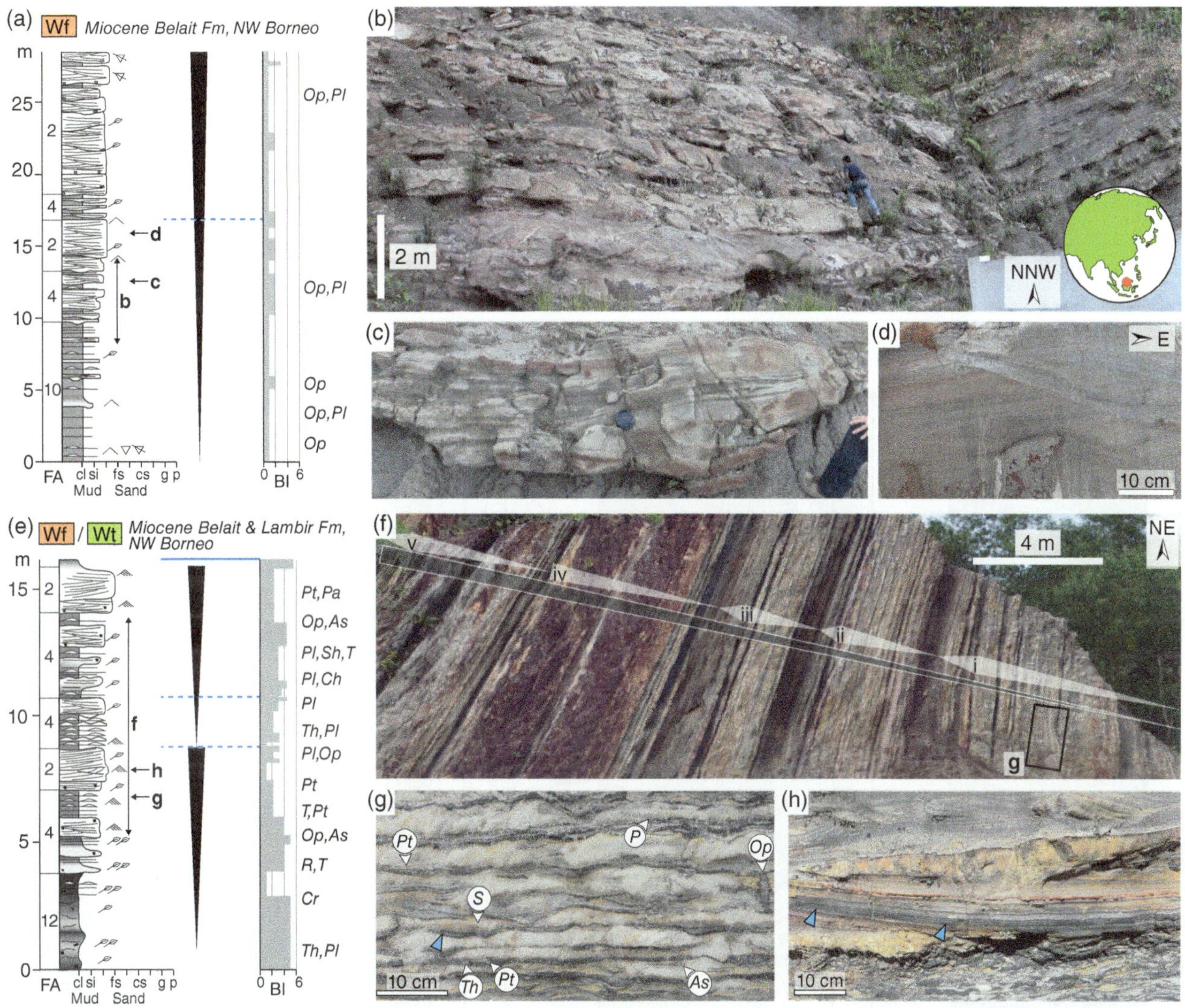

Figure 8.33 (a) Facies successions recording progradation of a wave-(storm-)dominated, fluvially influenced delta front (process regime: Wf) (Miocene Belait Formation, northwest Borneo). See Figure 8.30a for legend. Black upward coarsening symbol illustrates facies successions. (b–d) Representative photographs of constituent facies and facies associations in Figure 8.33a, including erosionally based swaley-cross-stratified sandstone beds with gutter casts. (e) Facies successions recording repeated progradation of a wave-dominated, tide-influenced delta front (process regime: Wf/Wt) (Miocene Lambir Formation, northwest Borneo). (f–h) Representative photographs of constituent facies and facies associations in Figure 8.33e illustrating repeated sandier-upward sets at facies association scale (white, numbered i-v) within the overall facies succession (black) (f), mud-draped (blue triangle) flaser-to-wavy bedded heterolithics (g), and complex intercalation of variably bioturbated muddy and sandy heterolithics and cross-stratified sandstone (h). Collins et al. (2017) / John Wiley & Sons / CC BY 4.0 and Collins et al. (2018a) / Society for Sedimentary Geology.

8.4.4 Prograding Wave-Dominated, Tide-Influenced Deltas (Wf, Wt) – Facies Succession 4 (Composed of FA2, 4 and 12)

Figure 8.33e shows two overall sandier-upward facies successions, each similar to that in Figure 8.33a (e.g. Figure 8.34) but with notable differences that reflect increased tidal influence (example N in Figure B8.1.2a, Box 8.1, classified as a 'wave-dominated delta' in Figure B8.1.1a, Box 8.1 and a 'wave-dominated, tide-influenced delta' in Figure B8.1.1b, Box 8.1). The successions were developed on the same mixed sand-mud shelf as the succession shown in Figure 8.33a and are from the Middle Miocene (Serravallian) Lambir Formation, northwest Borneo (Collins et al. 2020).

The geological context of the Lambir Formation is very similar to that of the Belait Formation (Section 8.4.3), sharing the same palaeolatitude, tectonic setting, climate and sediment-routing controls. Tidal influence is noted locally even

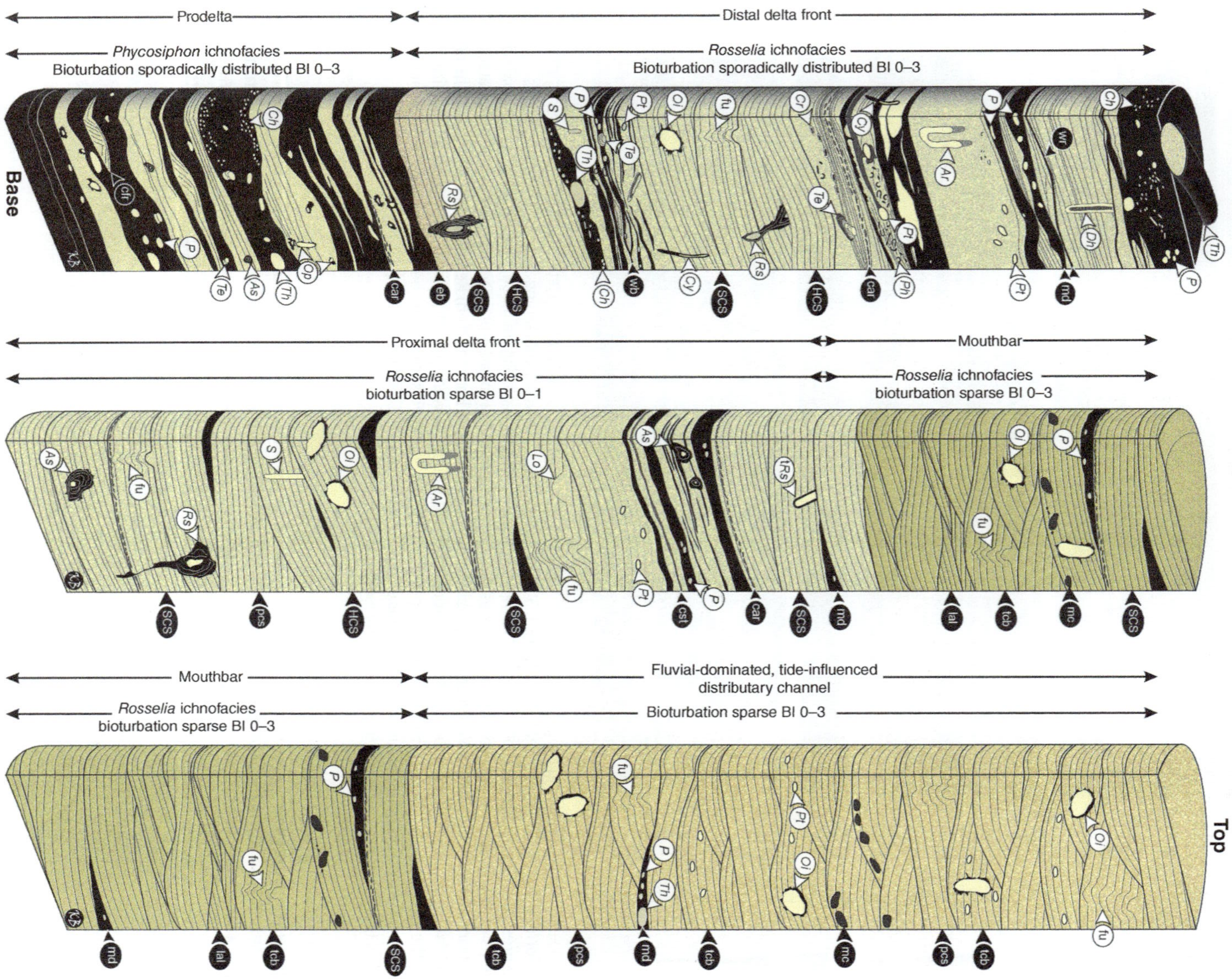

Figure 8.34 Schematic split-core diagrams of representative sections of a progradational, wave-dominated, fluvial- and tide-influenced delta front (process regime: Wf) from the Miocene Lambir Formation, northwest Borneo (cf. Figure 8.33). The succession comprises, from bottom left to top right: prodelta (FA10 in Figure 8.33a); distal delta front (FA4 in Figure 8.33e); proximal delta front to mouth bar (FA2 in Figure 8.33e); and fluvial-dominated, tide-influenced distributary channel (FA8 in Table 8.1). Trace-fossil codes are as follows: *Arenicolites* (*Ar*), *Asterosoma* (*As*), *Chondrites* (*Ch*), *Cosmorhaphe* (*Cr*), *Cylindrichnus* (*Cy*), *Diplocraterion habichi* (*Dh*), fugichnia (fu), *Locheia* (*Lo*), *Ophiomorpha irregulaire* (*Oi*), *Ophiomorpha* (*Op*), *Planolites* (*P*), *Palaeophycus tubularis* (*Pt*), *Phycosiphon* (*Ph*), *Rosselia socialis* (*Rs*), *Skolithos* (*S*), *Teichichnus* (*Te*), *Thalassinoides* (*Th*) and truncated *Rosselia socialis* (t*Rs*). car – carbonaceous debris; cfr – combined-flow ripple; cst – carbonaceous stringer; eb – erosional base; HCS – hummocky cross-stratification; lal – low-angle lamination; mc – mud clasts; md – mud drape; pcs – planar cross-stratification; SCS – swaley cross-stratification; tcb – trough cross-bedding; wb – wavy bedding; wr – wave ripple. Figure drafted by Kerrie Bann.

in the wave-dominated modern Baram delta (Figure 8.1d; e.g. Lambiase et al. 2002), but is more pronounced in coastal embayments that are sheltered from waves (e.g. modern Limbang River and Trusan River deltas in inner Brunei Bay, Figure 8.1e; Lambiase et al. 2003; Collins et al. 2018a).

Three key differences are noted between the facies successions shown in Figures 8.33e and 8.34 (process regime: Wf/Wt) and that of the wave-dominated, fluvial-influenced delta in Figure 8.33a (process regime: Wt). First, in the lowermost facies succession (0–8 m in Figure 8.33e), the basal mudstone facies association comprises massive mudstones with pervasive organic material (FA12 in Table 8.1), which contain abundant intertidal palynomorphs (in this case mangroves), consistent with subtidal–intertidal deposition adjacent to a low wave- and fluvial-energy lower coastal plain. Second, in both sandier-upward successions, FA4 (Table 8.1) is notably heterolithic with predominantly centimetre-thick, sharp-based siltstone–sandstone interbeds displaying combined-flow ripple cross-lamination and mud and/or carbonaceous drapes lacking stratigraphic organisation, pairing or rhythmicity (Figure 8.33g). Third, bioturbation in sandstone layers within FA4 and overlying FA2 is highly variable (Figure 8.33h), generally constituting a slightly impoverished *Cruziana* ichnofacies, though examples in the same facies succession may be pervasively bioturbated mostly by *Thalassinoides*. Though the overall facies and ichnofacies characteristics are suggestive of deposition in a wave-dominated delta, similar to the succession in Figure 8.33a, the subtle facies, ichnological and palynological differences, plus less preserved evidence for erosion, suggest delta progradation into a more restricted, tidally influenced sub-environment, perhaps an embayment or lagoon (Collins et al. 2018a). Brunei Bay is a direct modern analogue (Figure 8.1e), though the common occurrence of swaley-cross-stratification suggests greater wave exposure in the Lambir Formation, perhaps more like the northern lateral part of the Mekong delta (Figure 8.7).

8.4.5 Prograding Gilbert-Type, Fluvial-Dominated, Tide-Influenced Deltas (F, Ft) – Facies Succession 5 (Composed of FA7 and 10)

This example (Figure 8.35a) consists of two coarsening-upward facies successions, each characterised by steeply dipping, gravelly sandstones that were developed as part of a Gilbert-type, fluvial-dominated delta front which locally contains evidence for tidal influence (example F in Figure B8.1.2a, Box 8.1, classified as a 'river-dominated delta' in Figure B8.1.1a, Box 8.1 and a 'river-dominated, tide-influenced delta' in Figure B8.1.1b, Box 8.1). The successions are taken from the Early Eocene (Ypresian) Roda Sandstone, Spain, which was developed on a mixed sand-mud shelf (e.g. Martinius 2017). The successions illustrate aspects of an archetypal Gilbert-type delta (process regime: F), including the impact of coarse grain size on facies character and stratal geometry, and modification of the delta front by tides (process regime: Ft) (e.g. Joseph et al. 1993; López-Blanco 1996; Tinterri 2007; Leren et al. 2010).

The Roda Sandstone was deposited in a piggy-back basin on the northern margin of the South Central Pyrenean thrust sheet, in a relatively narrow (<60 km), elongate (c. 200 km long), west-east-oriented marine gulf that was open to the Atlantic Ocean at its western end, off the western lateral ramp of the thrust sheet (e.g. Martinius 2017). Basin configuration favoured resonant amplification of the M_2 (principal lunar) tidal constituent (Martinius 2017). Sediment supply to the Roda Sandstone was strongly focussed (e.g. Tinterri 2007), most likely via braided rivers routed through a long-lived, tectonically confined palaeovalley (e.g. Martinius 2017; Box 5.4). The Roda Sandstone Member was supplied with arkosic, gravelly sand that was transported just 20–60 km from an exhumed granitic pluton in the uplifting mountain chain (López-Blanco et al. 2003). Deposition occurred under stable, subtropical-to-tropical, arid-to-semiarid climatic conditions during the Early Eocene Climatic Optimum, at a latitude of 35° N (e.g. Martinius 2017). Evidence for subtropical-to-tropical conditions is provided by the presence of isolated coral reefs, large benthic foraminifers, calcareous green algae and mangrove-dwelling molluscs (e.g. Martinius and Molenaar 1991), which provided abundant bioclastic material that was reworked and incorporated into the Roda Sandstone.

The basal part of each succession consists of intensely to completely bioturbated siltstones with laminae of bioclastic, silty, very fine-grained sandstone (Figure 8.35a; FA10 in Table 8.1). These siltstones are sharply overlain by, and laterally interfinger with, commonly to completely bioturbated sandstone beds and sparse siltstone-drape interbeds with horizontal to shallow palaeo-seaward dips (Figure 8.35a,b; FA7 in Table 8.1). The sub-horizontal sandstone beds are gradationally overlain by steeply (10°–30°) palaeo-seaward-dipping, structureless sandstone beds (i.e. clinoforms) with sparsely to commonly bioturbated bed tops (Figure 8.35a,b; FA7 in Table 8.1). Locally, these sub-horizontal and steeply dipping sandstones contain lenticular sets and cosets of cross-stratified sandstone (FA7 in Table 8.1). The sets of cross-stratification contain mud-draped foresets, with mud drapes exhibiting rhythmical variations in thickness and spacing, and reactivation surfaces (Figure 8.35c). Palaeocurrents measured in the cross-beds are unidirectional, and generally oriented parallel and obliquely sub-parallel to clinoform strike; however, some are oriented up and down the clinoforms (i.e. palaeo-landward and palaeo-seaward, respectively).

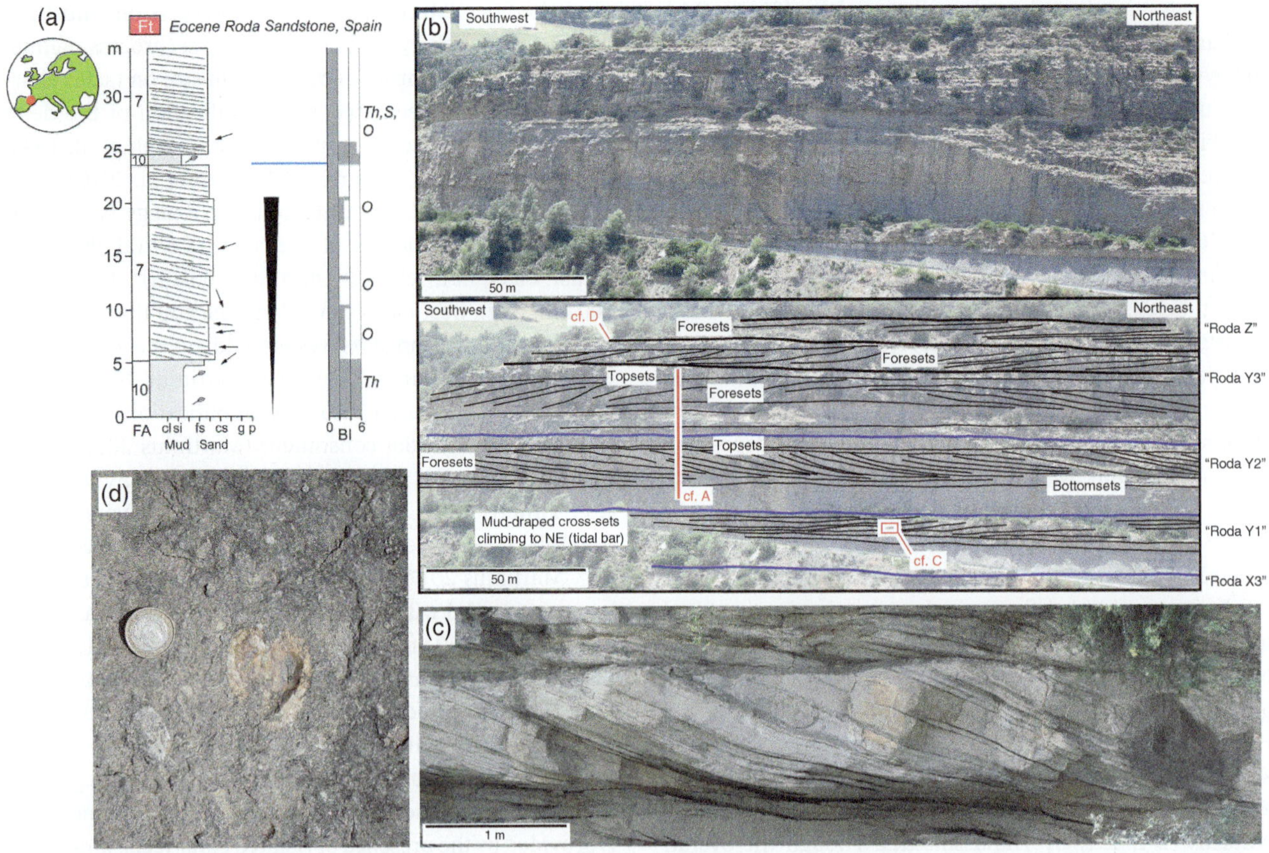

Figure 8.35 (a) Two vertically stacked facies successions recording progradation of a Gilbert-type, fluvial-dominated, tide-influenced delta front (process regime: Ft) and (b) photograph showing stratal geometries in vertically stacked, Gilbert-type delta-front successions (cf. Figure 8.35a) in 'Roda Y2, Y3 and Z' units, which are oriented oblique to depositional strike, and climbing dunes defining compound tidal dune in 'Roda Y1' unit. See Figure 8.30a for legend. Terminology of Roda Sandstone units modified from Leren et al. (2010). (c, d) Photographs of constituent facies and facies associations: (c) rhythmically spaced mud drapes along foresets of a set of cross-stratification, interpreted as a tidal dune deposit (from 'Roda Y1' unit); and (d) reworked and abraded bioclasts on top of cemented hardground (from top of 'Roda Y3' unit). Part (a) after Leren et al. (2010). Photograph credits parts b, c and d: Gary Hampson.

The steeply dipping sandstone beds are capped by intensely to completely bioturbated bioclastic sandstones (Figure 8.35d) that contain both reworked and *in situ* fossils of bivalves, including oysters, and corals, and common calcite-cemented nodular horizons. Trace-fossil assemblages are similar throughout FA7 (*Thalassinoides*, *Ophiomorpha* and *Skolithos*) and belong to the *Skolithos* ichnofacies (Figure 3.14). The detailed 2D and 3D architecture of sandstone bodies in the Roda Sandstone is presented in Section 8.4.11.

Each facies succession is interpreted to record regression of a sand-prone, Gilbert-type, fluvially dominated delta front (FA7 in Table 8.1) above offshore siltstones (FA10 in Table 8.1) (Figure 8.35a; e.g. Leren et al. 2010). Based on their dip and geometrical configuration, delta-front sandstone bodies are interpreted to represent, from base to top: sub-horizontal bottomsets, steeply dipping foresets and sub-horizontal topsets ('Roda Y2, Y3 and Z' units in Figure 8.35b; López-Blanco 1996; Tinterri 2007; Leren et al. 2010). Structureless delta-front sandstone beds are interpreted as the deposits of high-density turbidity currents and possibly cohesionless debris flows, while the *Skolithos* ichnofacies is consistent with bioturbation of a sandy substrate in a high-energy setting. Cross-beds with mud-draped foresets (Figure 8.35c) record the migration of dunes in response to unidirectional currents that fluctuated in velocity. Rhythmical variations in mud-drape thickness and spacing (Figure 8.35c) are attributed to periodic fluctuations in tidal current strength and define spring-neap tidal cycles based on their rhythmicity, while reactivation surfaces record periodic erosion by a subordinate current that flowed in a different direction to the dominant current (Yang and Nio 1989). Lenticular cosets of stacked, mud-draped cross-stratification define tidal bars and compound dunes ('Roda Y1' unit in Figure 8.35b; cf. Olariu et al. 2012b). The high bioclastic content and widespread occurrence of calcite cement in the topset beds are interpreted to record low sedimentation rates and hardground formation (Figure 8.35d), attributed to delta-lobe abandonment (Martinius and Molenaar 1991).

8.4.6 Prograding Fluvial-Dominated, Tide-Influenced Deltas (Ft) – Facies Succession 6 (Composed of FA6)

Figure 8.36a also illustrates a coarsening-upward, heterolithic facies succession that represents a fluvial-dominated delta front with tidal influence (example D in Figure B8.1.2a, Box 8.1, classified as a 'river-dominated delta' in Figure B8.1.1a, Box 8.1 and a 'river-dominated, tide-influenced delta' in Figure B8.1.1b, Box 8.1). The succession is taken from the Middle Jurassic (Bajocian–Bathonian) Lajas Formation, Argentina, which was developed on a mixed sand-mud shelf (Gugliotta et al. 2016). In contrast to the Roda Sandstone successions (Figure 8.35), which share a similar process regime (Section 8.4.5), the Lajas Formation succession is dominated by fine- to medium-grained sandstone. Thus, the differences between river-dominated, tide-influenced successions (process regime: Ft) in the Roda Sandstone (Figure 8.35) and Lajas Formation (Figure 8.36) can be attributed largely to differences in the grain size of their sediment supply. The modern Wax Lake Delta contains mouth bars of similar scale and lateral stacking to those documented in the Lajas Formation (Kurcinka et al. 2018), with cored successions of similar facies successions (cf. Figure 8.24d) and associated process measurements, although with less pronounced tidal influence (Wellner et al. 2005).

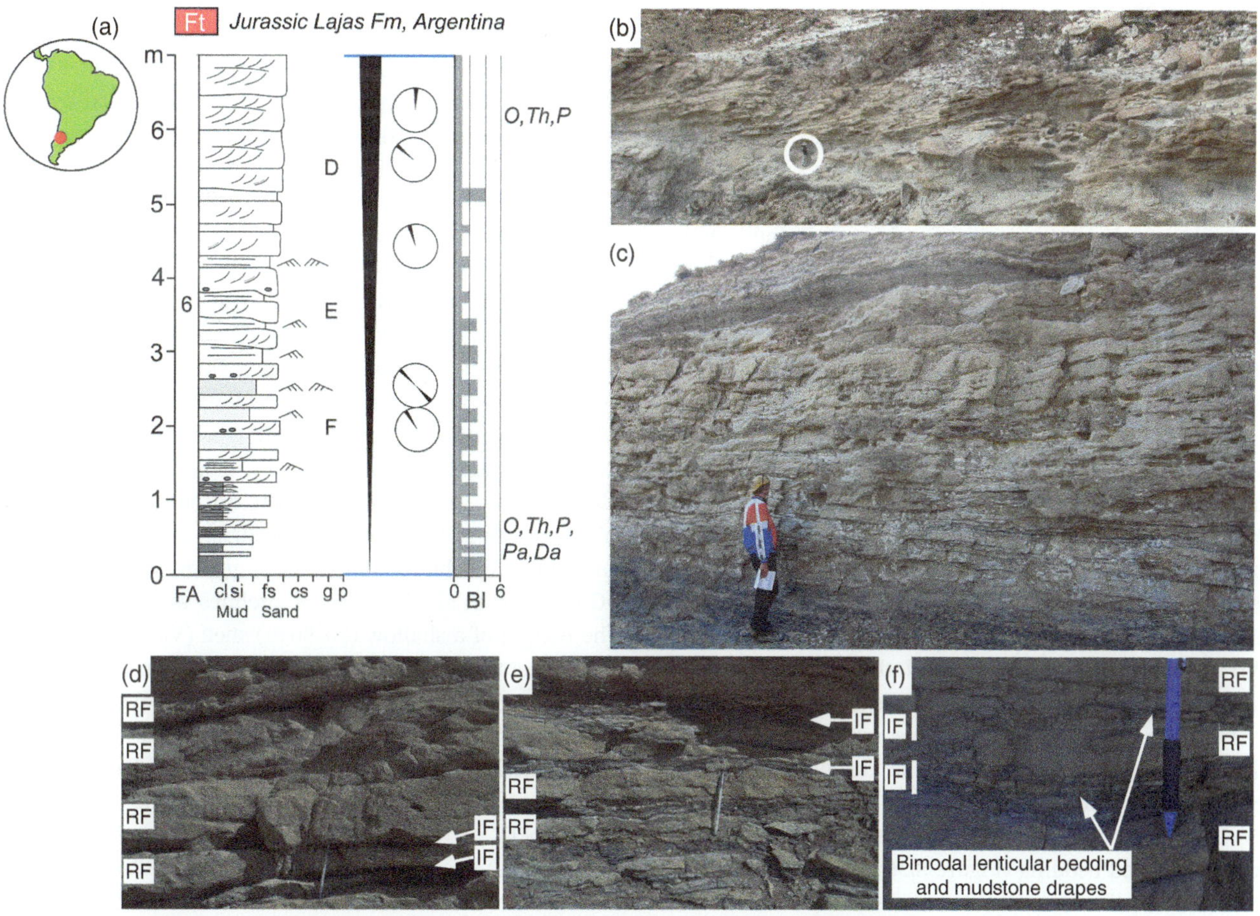

Figure 8.36 (a) Facies succession recording progradation of a fluvial-dominated, tide-influenced delta front (mouth bar) (process regime: Ft) (Jurassic Lajas Formation, Neuquén Basin, Argentina), and (b–f) photographs of constituent facies and facies associations. See Figure 8.30a for legend. (b) Coarsening-upward river-dominated mouth bars (process regime: F), characterised by their relatively steeply dipping clinoforms; (c) coarsening-upward tide-influenced mouth bars (process regime: Ft) with more gently dipping clinoforms; (d) amalgamated river-flood sandstone beds (labelled 'RF') and poorly preserved inter-flood mudstones (labelled 'IF') in the upper part of the tide-influenced mouth-bar unit; (e) river-flood sandstone beds ('RF') and inter-flood mudstones ('IF') in the medial part of the tide-influenced mouth-bar unit; and (f) tidal deposits in inter-flood deposits ('IF') of the tide-influenced mouth-bar unit, showing rhythmically spaced mudstone drapes and bi-directional current ripples. Parts a, d and e modified from Gugliotta et al. (2016); parts b and c Kurcinka et al. (2018) / American Association of Petroleum Geologists; part f Gugliotta et al. (2015) The Geological Society of London.

The Lajas Formation was deposited on a shelf at the eastern margin of a 120–240 km wide, embayed seaway developed in the Neuquen Basin and connected via several straits to the proto-Pacific Ocean further west (Franzese et al. 2003; McIlroy et al. 2005). The shorelines of the Lajas Formation were supplied with sand by rivers draining the North Patagonian Massif, >100 km to the southeast (e.g. Pujols et al. 2018; Estebenet et al. 2021). Deposition occurred at a palaeolatitude similar to the present day (approximately 39° S) under a warm, mainly arid climate of variable humidity (e.g. Martinez et al. 2002; Stukins et al. 2013). Tides are widely interpreted to have influenced deposition of the Lajas Formation, although the distribution and strength of tidal influence (or dominance) are debated (e.g. McIlroy et al. 2005; Gugliotta et al. 2016; Rossi and Steel 2016). Tides were potentially amplified in embayments along the Lajas Formation shoreline (McIlroy et al. 2005).

The facies succession consists of a coarsening-upward succession several metres thick (Figure 8.36a–c, FA6 in Table 8.1) containing decimetre-scale interbedding of sandstone beds that thicken upwards. Fine- to medium-grained sandstone beds typically have erosional bases, can contain mudstone clasts, show internal unidirectional, trough and planar-tabular cross-stratification and ripple cross-lamination oriented palaeo-seaward (in this case, towards the north and northwest) and have no or minor, low diversity bioturbation (BI 0–1). Carbonaceous and/or mudstone drapes are uncommon. Gradationally overlying very fine- to fine-grained sandstones may display millimetre-scale mudstone or carbonaceous and/or mica drapes that can show rhythmical spacing and be associated with bi-directional ripples. Bioturbation is more abundant but variable (BI 2–3 to BI 5–6), and more diverse compared with coarser-grained beds. Master bedding surfaces are inclined steeply (7°–16°) palaeo-seaward, parallel to cross-stratification, consistent with forward accretion (Gugliotta et al. 2015; Kurcinka et al. 2018). Overall, the facies, ichnofacies, inclined master bedding surfaces and coarsening-upward character suggest that these units represent mouth-bar deposits (Gugliotta et al. 2016). The alternating finer and coarser sandstone beds are interpreted to reflect fluctuations in river discharge forming intercalated river-flood and inter-flood bedding (Figure 8.36d–f), which, together with the uni-directional, seaward-oriented palaeocurrents, indicate river dominance. Rare carbonaceous and/or mud drapes and bioturbation may indicate tidal modulation of river flow, perhaps in the inner part of the fluvial-marine transition zone (e.g. Dalrymple and Choi 2007; Martinius and Gowland 2011). Finer-grained beds with rhythmically distributed drapes, bi-directional ripples and more abundant and diverse bioturbation are interpreted as interflood deposits formed during low river stage under transient dominance of tidal processes and marine conditions (Dalrymple et al. 2015; Gugliotta et al. 2016). Similar sandstone-dominated to heterolithic facies associations have been recognised and interpreted as river-flood and inter-flood deposits within the fluvial-marine transition zone of coastal rivers (e.g. Collins et al. 2020; Van Yperen et al. 2020).

8.4.7 Prograding Fluvial-Dominated, Wave-Influenced Deltas (Fw) – Facies Succession 7 (Composed of FA6, 8 and 11)

This succession (Figure 8.37a) comprises a coarsening-upward succession formed by current- and wave-generated structures and represents a fluvial-dominated delta front with wave influence (process regime: Fw) (example C in Figure B8.1.2a, Box 8.1, classified as a 'river-dominated delta' in Figure B8.1.1a, Box 8.1 and a 'river-dominated, wave-influenced delta' in Figure B8.1.1b, Box 8.1). The succession is taken from the Late Cretaceous (Cenomanian) Mesa Rica Sandstone, New Mexico, USA (Van Yperen et al. 2019) and was developed along a deltaic shoreline dominated by fine sand with a mixed sand-mud shelf. The facies successions in the Mesa Rica Sandstone (Figure 8.37) and the previous example, the Lajas Formation (Figure 8.36) are similar, but differ in the role of waves or tides in influencing the fluvial-dominated delta front.

The Mesa Rica Sandstone was deposited on the northwestern margin of the Tucumcari Basin, which constituted a flooded, intracontinental extension of the Gulf of Mexico during the Cenomanian (Van Yperen et al. 2019). Mesa Rica shorelines were oriented approximately southwest–northeast, with sand supplied principally by long (>800 km) rivers draining the Sevier Orogenic Belt to the northwest (Van Yperen et al. 2021). Deposition occurred at a palaeolatitude of approximately 35° N under a warm, humid climate (Chumakov et al. 1995). The shorelines were developed at the margin of a shallow (20–90 m) shelf (Van Yperen et al. 2019) that was contiguous with Tethys Ocean to the south (Blakey at https://deeptimemaps.com/). Waves reaching the Mesa Rica shorelines had a large potential fetch length for southerly winds, although they would have been frictionally damped as they travelled across the Tucumcari Basin shelf.

The facies succession shown in Figure 8.37a consists of three facies associations (FA11, 6 and 8) from base to top. The first consists of sparsely bioturbated, laminated mudstones containing a diverse bivalve fauna (FA11 in Table 8.1). Together these imply rapid sedimentation under conditions that were fully marine for part of the time (e.g. Gingras et al. 2011). FA11 passes gradationally upwards into FA6: intensely bioturbated very fine- to fine-grained sandstones and rarer, thin (<15 cm), normally graded sandstone beds, recording episodic sand transport by gravity flows (Table 8.1). A trace-fossil assemblage (*Thalassinoides*,

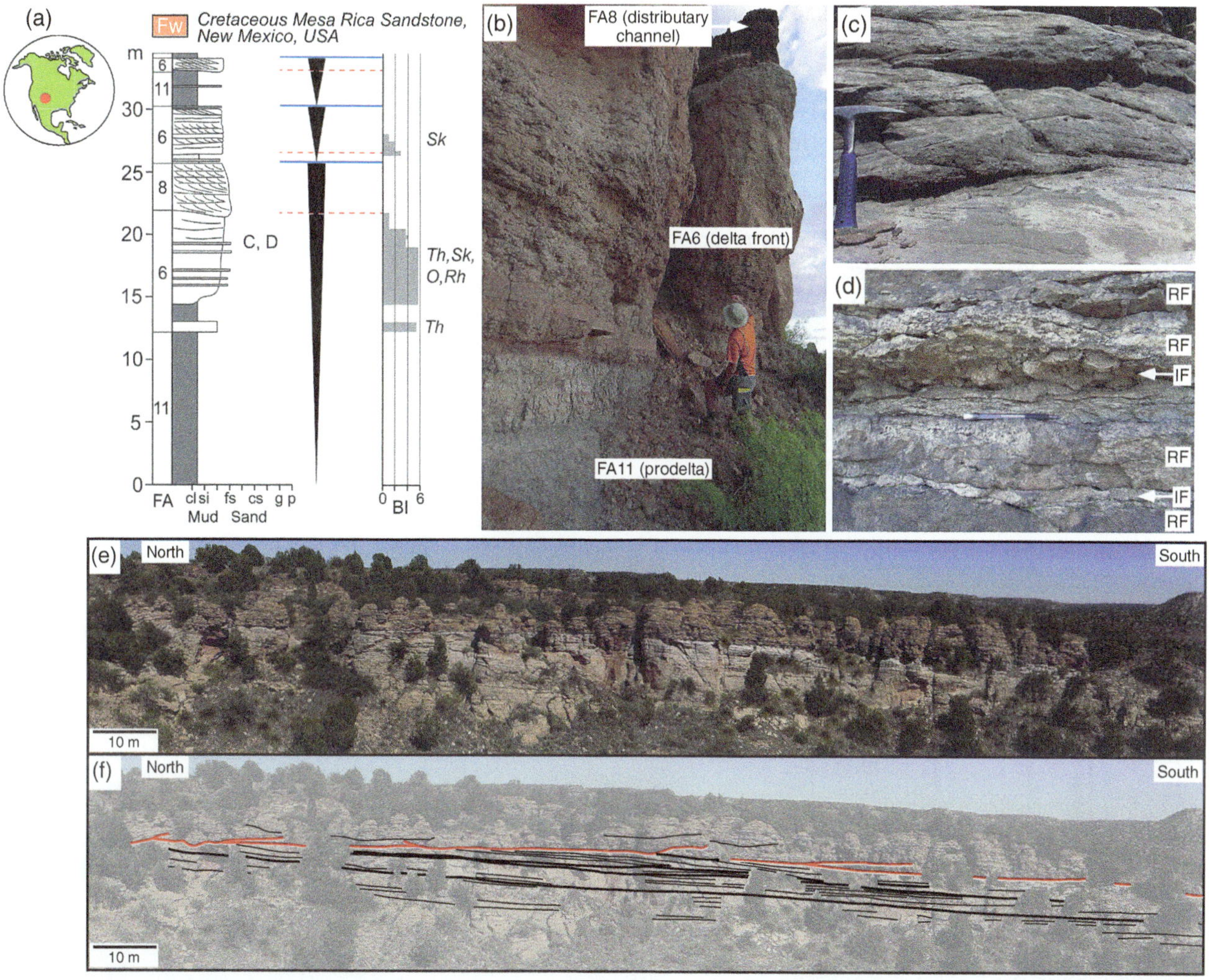

Figure 8.37 (a) Facies succession recording progradation of a fluvial-dominated, wave-influenced delta front to delta top (process regime: Fw) (Cretaceous Mesa Rica Sandstone, Dakota Group, New Mexico, USA), and (b–f) photographs of constituent facies and facies associations. See Figure 8.30a for legend. (b) Coarsening-upward succession consisting of prodelta (FA11), delta front (FA6) and distributary channels (FA8) (Figure 8.37a); (c) Trough cross-stratified pebbly sandstone in axial mouth-bar deposits (FA6); (d) thin- to thick-bedded (5–20 cm), fine-grained parallel-laminated and cross-stratified sandstone, and interbeds of bioturbated and asymmetrical ripple-laminated sandstone in mouth-bar fringe deposits (FA6); and (e) photopanel and (f) line-drawing of delta-front clinothems truncated at the erosional bases of laterally amalgamated distributary channels. Parts b, e and f van Yperen et al. (2019) / Society for Sedimentary Geology.; parts c and d van Yperen et al. (2020) / John Wiley & Sons / CC BY 4.0.

Ophiomorpha, *Cylindrichnus*, *Palaeophycus*, *Rhizocorallium* and *Skolithos*), constituting the proximal *Cruziana* ichnofacies (Figure 3.15; MacEachern and Bann 2008), suggests deposition under generally unstressed, fully marine conditions. Very fine- to medium-grained sandstones in the middle part of FA6 contain trough and tabular cross-stratification, which have generally palaeo-seaward-directed palaeocurrents, and planar-parallel-laminated beds (Figure 8.37c) with absent or sparse bioturbation (*Ophiomorpha*, *Skolithos*) constituting a restricted *Skolithos* ichnofacies (MacEachern and Bann 2008). These sandstones are intercalated laterally over several tens of metres with more heterolithic facies, with intercalations interpreted to record river discharge variations (Figure 8.37d). These bed-scale lithological intercalations are similar to those observed in the Lajas Formation (Section 8.4.6; Figure 8.36d–f), but the lower (<10%) mudstone content of the Mesa Rica Sandstone appears to have limited their preservation in off-axis areas (Van Yperen et al. 2019). FA6 records migration of 2D and 3D dunes alternating with upper plane bed conditions, in response to unidirectional currents, under high-energy conditions, high sedimentation rates, and reduced salinity and/or increased water turbidity (e.g. Gingras et al. 2011). Bedding in the middle part of FA6 exhibits moderate palaeo-seaward dips (2°–8°) that define upward-steepening clinoforms (Figure 8.37e,f). Clinoforms are erosionally truncated by FA8, which is composed of channelised, cross-bedded, fine- to medium-grained sandstone bodies with a simple cut-and-fill architecture (FA8 in Table 8.1; Figure 8.37e,f). Bioturbation is absent. FA8 is interpreted as the deposits of fluvial channels with a low-sinuosity planform geometry.

The overall facies succession is interpreted as a fluvially dominated, wave-influenced prodelta (FA11), delta front (FA6) and distributary channel (FA8). This interpretation is based partly on the distribution, dip and geometrical configuration of clinoforms, as discussed in Section 8.4.10. Variations in the degree of environmental stress recorded by trace-fossil assemblages are consistent with alternation between fully marine (potentially wave-dominated) and restricted (fluvial-dominated) conditions. The scarcity of mudstone and relatively high degree of sandstone sorting in the delta front are also consistent with wave influence, although the preserved primary structures are current-generated and imply fluvial dominance.

8.4.8 Prograding Mixed-Influence Delta Front Truncated by Tide-Dominated Fluvial Distributary Channel (Fw, Tf) – Facies Succession 8 (Composed of FA9 and 10/11)

Figure 8.38a shows a mudstone-dominated, coarsening-upward succession which is truncated by an erosionally based, channelised, fining-upward succession dominated by current-generated structures that contains evidence for tidal influence (example E in Figure B8.1.2a, Box 8.1). The succession is taken from the Late Cretaceous (Campanian) Horseshoe Canyon Formation, Alberta, Canada (Ainsworth 1994; Ainsworth et al. 2015), which comprises mixed-process deposits with evidence of wave, fluvial and tidal processes (Ainsworth et al. 2015). The modern

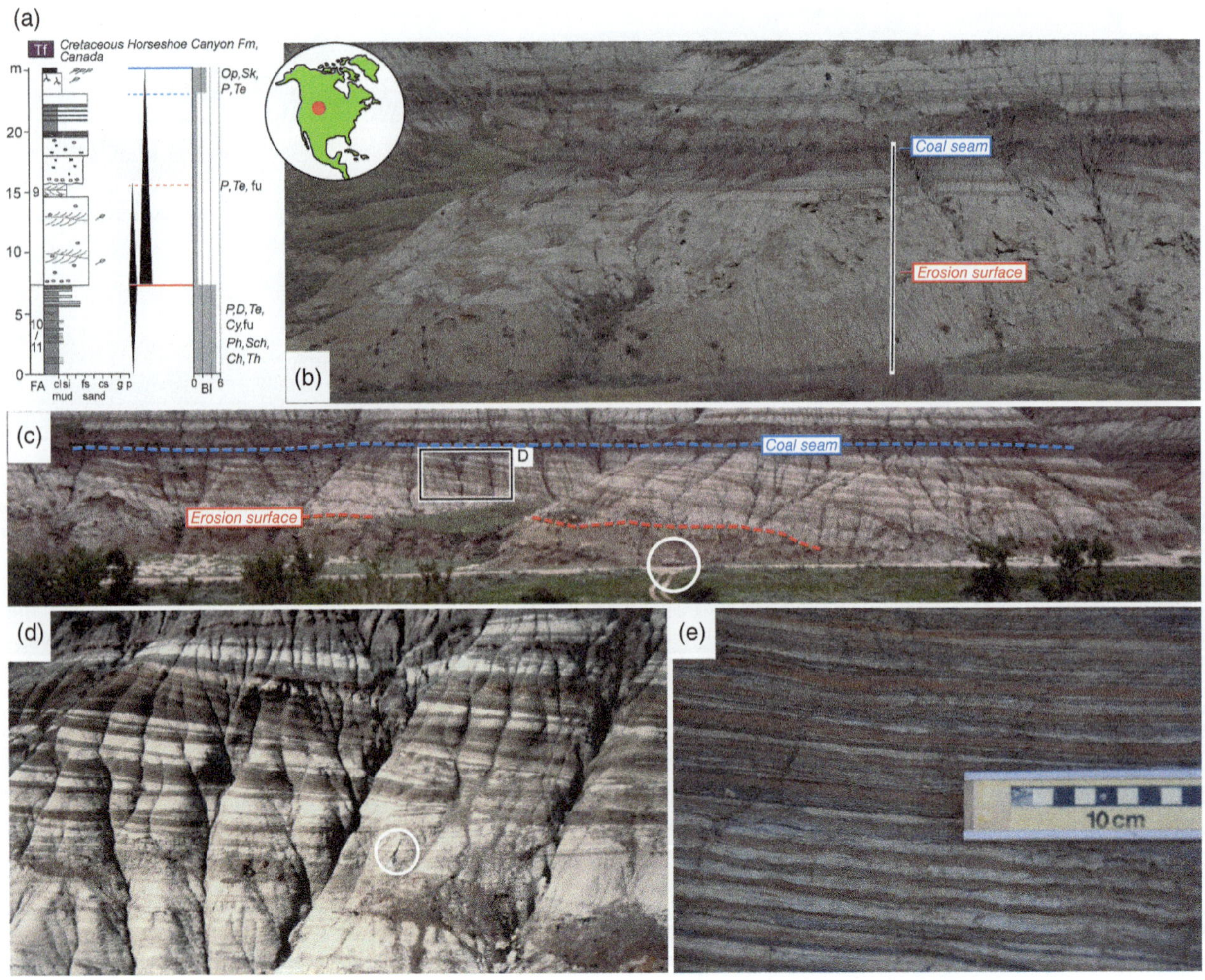

Figure 8.38 (a) Facies succession recording progradation of a mixed-influence delta front (FA10/11; process regime: Fw) erosionally overlain by a tide-dominated, fluvial-influenced distributary channel (FA9; process regime: Tf; cf. facies succession C in Figure 8.39) (Cretaceous Horseshoe Canyon Formation, Canada). See Figure 8.30a for legend. (b) Photograph locating the facies successions; (c, d) photographs illustrating inclined heterolithic stratification that record lateral accretion of the distributary channel margin (FA9); (e) alternating sandy and muddy beds attributed to daily tidal velocity fluctuations in the distributary channel (FA9). Car and shovel (1 m) indicated by white circles for scale in parts c and d, respectively. Parts a, c, d and e modified from Ainsworth et al. (2015); part b after Ainsworth et al. (2016) / Society for Sedimentary Geology.

mixed-influence Mitchell River Delta, Queensland, Australia (Nanson et al. 2013) is a partial analogue for the Horseshoe Canyon Formation (Ainsworth et al. 2015, 2016).

The Horseshoe Canyon Formation was deposited on the western margin of the wide (c. 1500 km), shallow (c. 200 m) intracratonic Western Interior Seaway at a palaeolatitude of approximately 60° N (e.g. Smith et al. 1981). Deposition occurred under a warm, humid climate (e.g. Quinney et al. 2013). Sand, with a relatively high proportion of volcanic fragments, was supplied by short (200–400 km long) rivers draining the southern Canadian Cordillera (Mack and Jerzykiewicz 1989). Horseshoe Canyon shorelines lay at the head of a broad (c. 800 km wide, 800 km long), southeast-facing embayment (e.g. Williams and Stelck 1975), which may have been conducive to tidal amplification (Ainsworth et al. 2015). Water circulation and associated sediment transport in the seaway itself were storm driven (e.g. Slingerland and Keen 1999).

The succession illustrated in Figure 8.38a consists of a lower mudstone-dominated facies association (FA10/11) sharply overlain by a fining-upward, sandstone-dominated facies association (FA9) topped by a coal. The lower facies association has a sandier-upward profile dominated by variably, often intensely, bioturbated mudstones with carbonaceous material. The sedimentary characteristics of interbedded siltstones and very-fine grained sandstones are wide ranging and include symmetrical ripples, hummocky cross-stratification and trough cross-stratification with mud drapes. Structureless mudstones occur locally, consistent with rapid fluid-mud deposition under high sediment concentrations, potentially during post-storm river floods (e.g. Bhattacharya and MacEachern 2009). Overall, this facies association is typical of a mixed-process delta front (process regime: Fw) (Ainsworth et al. 2015). These characteristics are a combination of facies associations 10 and 11 (Table 8.1). The overlying fining-upward succession (FA9) overlies a concave-up erosional surface and predominantly comprises trough cross-bedded to structureless, medium- to fine-grained sandstone. Inclined heterolithic strata are prominent, implying lateral accretion at the margin of the channel that cut the concave-up erosional surface (Figure 8.38c,d). Tidal influence is suggested by subordinate herringbone cross-stratification, paired mud and carbonaceous drapes (Figure 8.38e), and both onshore- and offshore-directed palaeocurrents. Fluvial influence is suggested by basal lags, abundant carbonaceous material, and a predominance of offshore-directed palaeocurrents (at correlatable exposures) (Ainsworth 1994; Ainsworth et al. 2015). FA9 is subdivided by minor erosional surfaces. Interbedded mudstone and sandstone facies together with *in-situ* rootlets towards the top of this association could represent tidal flat deposits. Overall, the fining-upward FA9 is interpreted as a tide-dominated fluvial distributary channel (process regime: Tf) (FA9 in Table 8.1) (Ainsworth et al. 2015).

8.4.9 Stacking of Facies Successions (1D)

The examples shown in Figures 8.30–8.38 illustrate some of the variation in single vertical facies successions that result from depositional process regime and the grain size of sediment supply. However, the process regime varies both spatially and temporally. Temporal variation is recorded by stacked facies successions. At their base and top, facies successions are bounded by surfaces marked by some or all of: (i) erosional truncation, downlap or onlap; (ii) abrupt change in process regime; and/or (iii) abrupt change in water depth, thereby breaking Walther's law (Section 4.3.2). Bounding surfaces vary in their geometry, lateral extent, and in the vertical facies juxtapositions across them. Therefore, bounding surfaces in 1D successions must be mapped or correlated in 2D and 3D, where data allow, to interpret architectural elements (Section 8.4.10) and, at larger scales, sequence stratigraphic units (Sections 8.5.3–8.5.5).

Figure 8.39 shows a vertical transect through several stacked facies successions from the Cretaceous Bearpaw and Horseshoe Canyon formations, Canada (example F in Figure B8.1.2a, Box 8.1), which comprises mixed-influence shallow-marine and marginal-marine strata (Section 8.4.8; Ainsworth et al. 2015). Most bounding surfaces in this example record maximum regression at the top of upward-shallowing facies successions (in red, sequence boundaries *sensu* Van Wagoner et al. 1990; Section 8.5.5) or maximum transgression at the top of upward-deepening facies successions (in blue, flooding surfaces *sensu* Van Wagoner et al. 1990; Section 8.5.5). Interpretations of bounding surfaces in this vertical transect are supported by regional mapping and correlation (see Section 8.4.12; Figure 8.42). Sequence boundaries are marked by either erosion (e.g. at base of facies succession C) or palaeosol development (e.g. capping facies successions D, F, G and H). Flooding surfaces coincide with underlying sequence boundaries where transgressive deposits are absent (e.g. capping facies successions F, G and H); here, flooding surfaces are marked by abrupt deepening. Facies association boundaries within facies successions are either gradational (e.g. in facies succession B) or locally erosional, bounding a channelised architectural element (e.g. in facies succession E), but do not correspond to abrupt shallowing or deepening when mapped regionally (Figure 8.42).

(a)

(b)

Figure 8.39 Vertically stacked facies successions and bounding surfaces (Cretaceous Bearpaw and Horseshoe Canyon formations, Canada; Ainsworth et al. (2015) / Society for Sedimentary Geology with interpreted process regimes of nine facies successions (A-I) and constituent facies associations (numbered) (cf. Table 8.1). Facies successions are: (A) a transgressive shoreline-to-offshore succession; (B) a regressive, mixed-influence shoreline succession; (C) a transgressive fluvio-tidal channel-fill complex (see Figure 8.38c–e); (D) a regressive tidal flat succession; (E) a transgressive tidal flat and channel-fill succession; (F) a regressive, wave-dominated shoreline succession; (G) a regressive tidal flat succession; (H) a regressive, wave-dominated shoreline succession; and (I) another regressive, wave-dominated shoreline succession. Bounding surfaces are identified from regional mapping and correlation (Figure 8.42).

8.4.10 Architectural Elements (2D–3D)

Where data afford sufficient spatial coverage and resolution, stratal termination relationships allow recognition of a hierarchy of architectural elements. The hierarchical arrangement of architectural elements is illustrated using a depositional-strike-oriented cross section (Figure 8.40) through the lower part of the Late Cretaceous (Campanian) Sego Sandstone, Utah, USA (after Willis and Gabel 2001), which crops out in continuous cliff-face exposures that allow stratal relationships to be traced with confidence. The high quality and large scale of the exposures allow a hierarchy of architectural elements to be identified and related directly to facies, facies associations and facies successions. The lower Sego Sandstone is interpreted to have been deposited by a mixed, tide- and wave-influenced delta (example B in Figure B8.1.2a, Box 1; Legler et al. 2014) over a period of 1.4 Myr (Willis and Gabel 2001).

At the lowest hierarchical level, architectural elements include inclined beds of cross-stratified sandstone (Figure 8.40a), which are stacked into sets bounded by channelised erosion surfaces overlain by lags of mud clasts, woody debris and shells (Figure 8.40b); these bedsets are interpreted to record downstream accretion and oblique lateral accretion of tidal bars or compound dunes within channels (Willis and Gabel 2001). Channel-fill architectural elements (Figure 8.40b) are laterally and vertically stacked into channel-fill complexes bounded by composite channelised erosion surfaces lined by discontinuous lags (Figure 8.40c), interpreted as the deposits of fluvio-tidal distributary or estuarine channel belts (Willis and Gabel 2001, 2003; Legler et al. 2014). Channel-belt architectural elements (Figure 8.40c) are stacked laterally, and erosionally overlie hummocky cross-stratified sandstones interbedded with bioturbated mudstones, to form coarsening-upward successions bounded below by laterally continuous, low-relief erosion surfaces lined by shell lags and overlain by bioturbated mudstones (Figure 8.40d). These coarsening-upward successions (Figure 8.40d) are interpreted to record regression of the storm-wave-dominated delta front and overlying fluvio-tidal distributary or estuarine channel belts of the delta plain, while their basal bounding surfaces record deepening and associated wave erosion (ravinement) during transgression (Willis and Gabel 2001, 2003; Legler et al. 2014). The successions (Figure 8.40d) thus record single episodes of regression and subsequent transgression and correspond to parasequences (Section 8.5.3) bounded by flooding surfaces (Section 8.5.5). Three such parasequences are stacked vertically in the lower Sego Sandstone, with the internal facies-association composition and thickness of successive parasequences implying long-term progradational to retrogradational stacking (Figure 8.40e; after Willis and Gabel 2003). The set of three stacked parasequences (Figure 8.40e) is bounded by two maximum flooding surfaces (Section 8.5.5).

It is rare that sufficient data and time are available for analysis at the level of detail described above. Instead, architectural analysis is often in part a prediction based on an existing hierarchical model or scheme of stratal geometrical relationships and associated facies, facies associations and facies successions. The interpreter uses stratal relationships, facies analysis and supporting data (e.g. palaeocurrent, ichnological, (micro)palaeontological, petrological and/or chronological analyses) to infer: (i) architectural elements that represent time-equivalent and laterally adjacent depositional environments, with bounding surfaces that record processes inherent to these environments (i.e. autogenic behaviours); and (ii) architectural elements that are genetically unrelated, with bounding surfaces that record external forcing (i.e. allogenic controls) and may have sequence stratigraphic significance (Section 8.5). Architectural elements that are autogenic exhibit spatial distributions that reflect their genetic linkage (i.e. the parent–child or sibling relationships of Vakarelov and Ainsworth 2013). High-resolution palaeogeographic maps that honour all available data, with reference to analogous modern depositional systems, allow effective tests of the interpreted genetic linkage between architectural elements.

8.4.11 Lobate Architectural Elements in Gilbert-Type, Fluvial-Dominated, Tide-Influenced Delta (F, Ft)

Figure 8.41 summarises architectural elements in the Eocene Roda Sandstone, Spain (after Yang and Nio 1989; Joseph et al. 1993; López-Blanco et al. 2003; Leren et al. 2010; Martinius 2017) (example F in Figure B8.1.2a, Box 8.1). The Roda Sandstone consists of at least three units comparable to parasequences (Section 8.5.3; 'Roda W', 'Roda X' and 'Roda Y' of Leren et al. 2010) and was deposited over a period of 0.7 Myr (López-Blanco et al. 2003). A parasequence typically comprises one sandier-upward succession or, more rarely, several such successions vertically stacked (Figure 8.35a,b), and records progradation of a Gilbert-type delta, as presented in Section 8.4.5. The average dip of delta-front foresets is towards the southwest, but lateral variations in foreset dip define broad arcs that are interpreted as delta lobes (e.g. 'lobe X1' and 'lobe X2' in Figure 8.41a) capped by laterally continuous, carbonate-cemented, intensely bioturbated bioclastic hardgrounds. Foresets are not truncated by channelised erosion surfaces at their tops (Leren et al. 2010), implying that delta lobes were fed by shallow, ephemeral terminal distributary channels, as in modern Gilbert-type deltas (e.g. Postma 1990), that were reworked by waves and tides

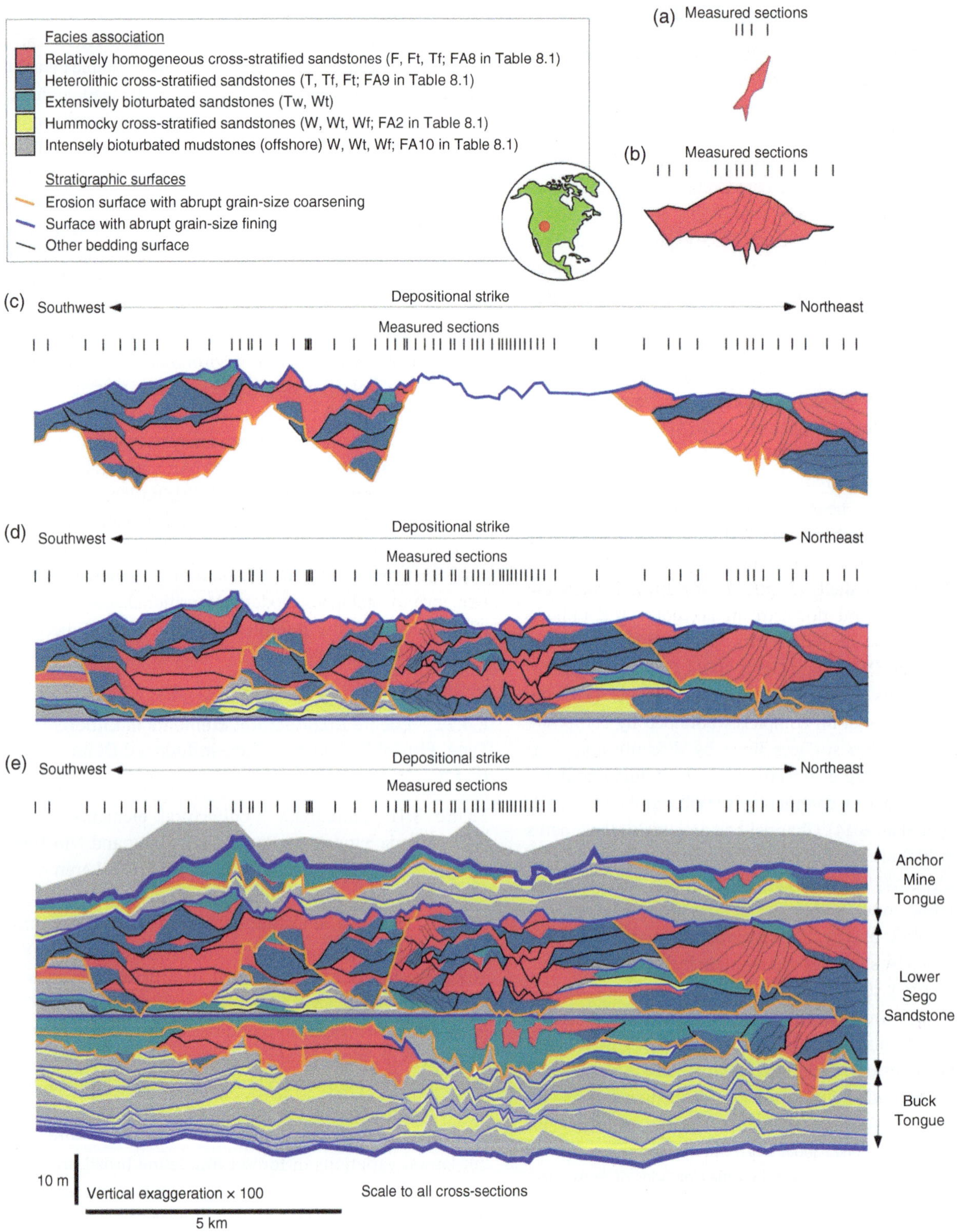

Figure 8.40 Cross sections oriented along depositional strike showing stratal geometries, architectural elements and facies associations in the lower part of the Sego Sandstone exposed in the Westwater Canyon area, Book Cliffs, Utah (modified from Willis and Gabel 2001). (a) Bedset architectural element, recording lateral accretion of beds at the margin of a channel. (b) Channelised architectural element, which contains multiple lateral-accretion bedsets (Figure 8.40a) bounded by inclined bedding surfaces that downlap the basal erosion surface. (c) Channel-complex architectural element, which comprises multiple vertically and laterally stacked channelised architectural elements (Figure 8.40b) and is bounded at its base by a composite erosion surface. (d) A parasequence, constituting an architectural element at a high hierarchical level; it comprises stacked channel-complex architectural elements and tabular delta-front architectural elements (Figure 8.40c) and is bounded at its base and top by flooding surfaces. (e) Multiple parasequences (Figure 8.40d) stacked within a parasequence set that contains the lower Sego Sandstone and bounded at its base and top by maximum flooding surfaces. The cross section(s) and measured sections were collected from continuous cliff-face exposures.

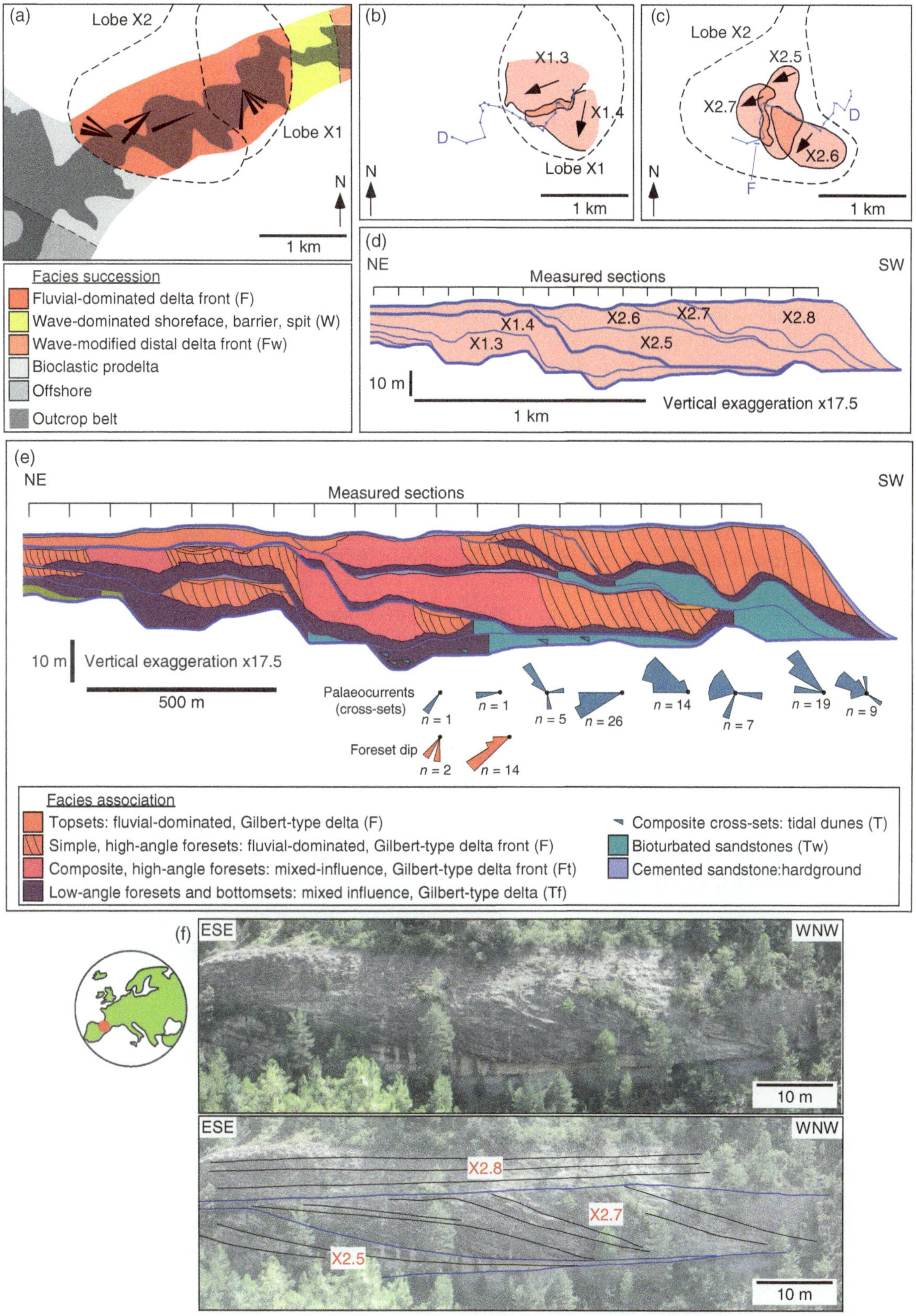

Figure 8.41 Architectural elements in the Eocene Roda Sandstone, northern Spain (see also Figure 8.35). (a) Palaeogeographic map with palaeocurrents of the Roda X parasequence at maximum regression. (b, c) The parasequence is composed of several Gilbert-type delta lobes, two of which are labelled ('lobe X1' and 'lobe X2'), which in turn comprise sub-lobes with arrows showing mean palaeoflow directions. Cross section along the outcrop belt showing (d) the two lobes and five sub-lobes, and (e) their constituent facies associations and related palaeocurrent data. (f) Photograph showing stratal geometries that define sub-lobes, including clinoform sets and clinoforms that bound smaller-scale architectural elements. Part a after Leren et al. (2010); parts b–e modified from Joseph et al. (1993), López-Blanco et al. (2003), Leren et al. (2010) and Martinius (2017); part f after Martinius (2017).

during delta-lobe abandonment, transgression and hardground development. The delta lobes are relatively large architectural elements in the Roda Sandstone, up to several tens of metres in thickness and several square kilometres in area ('lobes X1 and X2' in Figure 8.41a–d). Internally, each interpreted delta lobe contains smaller-scale lobate bodies ('sub-lobes X1.1–X1.4' and 'sub-lobes X2.5–X2.8' in Figure 8.41b–e) that are laterally offset from each other. These smaller-scale lobate architectural elements, or sub-lobes, are defined by onlap, downlap and truncation of topset-foreset-bottomset surfaces (Figure 8.41f) and locally by laterally continuous hardgrounds (Figure 8.41e). The lobes and sub-lobes are interpreted to record palaeogeographic re-organisation and compensational stacking of the delta system and its components during progradation, respectively in response to switching of trunk and terminal distributary channels. Portions of most sub-lobes are characterised by westward-directed to northwestward-directed (i.e. seaward-directed) cross-sets superimposed on their steeply dipping foresets (i.e. compound cross-strata; Figure 8.41e), which record reworking by tidal currents (process regime: Ft, Tf). The most extensive tidally reworked cross-sets contain bundled mud-draped foresets, in which mud-draped spacing records spring-neap tidal cyclicity (e.g. Figure 8.35c; Yang and Nio 1989), and occur in delta-front bottomsets as discrete lenses that are interpreted as tidal compound dunes and bars (Figure 8.41e; process regime: T).

8.4.12 Channelised and Lobate-to-Tabular Architectural Elements in Regressive, Mixed-Process Marginal Marine Strata (Wt, Tf, Ft)

Figure 8.42 summarises architectural elements in the mixed-process deposits of the Campanian Bearpaw and Horseshoe Canyon formations, Canada (Ainsworth et al. 2015 and references therein; example F in Figure B8.1.2a, Box 8.1). The integration of extensive outcrop-based observations (e.g. Figures 8.38 and 8.39) with core and wireline log data enabled interpretation of architectural elements and their bounding surfaces in cross sections (Figure 8.42a,b) and maps (Figure 8.42c). Architectural elements have either a laterally continuous, lobate-to-tabular geometry or a laterally discontinuous, channelised geometry (Figure 8.42).

Lobate-to-tabular elements include regressive shoreline deposits (within facies successions B, D, F, G, H and I in Figures 8.39 and 8.42) and transgressive shoreline deposits (within facies successions A, D, F, G, H and I in Figures 8.39 and 8.42). These elements each comprise either a single facies association, indicating a spatio-temporally uniform process regime during their deposition (Wt in facies succession A, and Tfw in facies successions D and G; Figure 8.42a), or two facies associations that grade into each other laterally, implying spatio-temporal variation in process regime during shoreline migration (Wtf and Ft in facies succession B, Tfw and Ft in facies succession E, Wtf and Wt in facies succession F, Tfw and Wt in facies succession H, and Ftw and Wt in facies succession I; Figure 8.42a and c). Locally, lobate-to-tabular elements are cut into by channelised elements (Figure 8.42). Regressive lobate-to-tabular elements are bounded at their bases by flooding surfaces, marked by a change from upward-deepening to upward-shallowing facies successions (Section 8.4.9), and at their tops by channelised erosion surfaces and/or palaeosols (Figure 8.42a,b). Transgressive lobate-to-tabular elements are bounded at their bases by palaeosols and at their tops by flooding surfaces (Figure 8.42a,b).

Channelised architectural elements are either transgressive (corresponding to facies successions C, J, L, M in Figures 8.39 and 8.42) or regressive (corresponding to facies successions K, N in Figure 8.42) and consist of a single facies association (Figure 8.42a). In map view, channelised elements vary in their width, sinuosity and orientation (e.g. Figure 8.42c), while palaeocurrent distributions in these elements reflect the process regime of their constituent facies associations (e.g. multi-directional palaeocurrents in tide-dominated facies succession C, and unidirectional palaeocurrents in fluvial-dominated facies succession B; Figure 8.42c). Channelised architectural elements are bounded at their bases by erosion surfaces, and at their tops by palaeosols and/or flooding surfaces (Figure 8.42a,b).

8.5 Stratigraphic Patterns and Organisation ('Descriptive Sequence Stratigraphy')

In this section, we focus on larger-scale stratigraphic patterns and organisation that are typical of sand-rich marginal- and shallow-marine strata. We highlight linkages to mud-rich shallow-marine strata (Chapter 9), alluvial strata (Chapter 5) and deep-water strata (Chapter 12). Our approach is descriptive because different controls are commonly interpreted for similar stratigraphic architectures, suggesting that stratigraphic architectures are non-unique (e.g. Burgess and Prince 2015). We follow this description with consideration of the genesis of stratigraphic architectures, and how they may be interpreted from available data, in Section 8.6. For further background in sequence stratigraphic concepts, see the dedicated section in Chapter 4, Section 4.4.

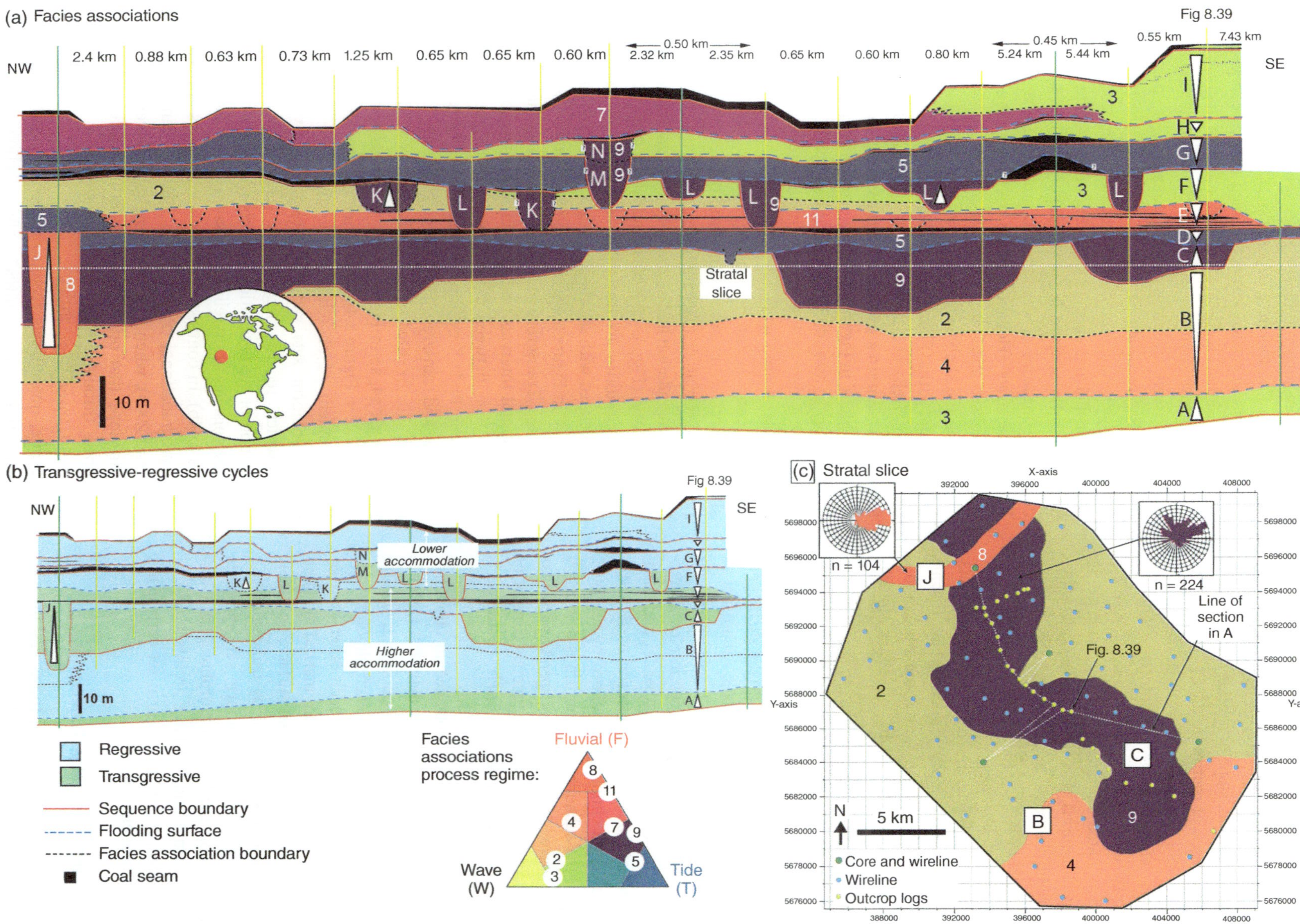

Figure 8.42 Three-dimensional interpretation of architectural elements in the Late Cretaceous Bearpaw and Horseshoe Canyon formations, Canada (modified from Ainsworth et al. 2015). Outcrop cross section showing (a) constituent facies associations (numbered) and interpreted process regime, and (b) facies successions (A–N) with intervening bounding surfaces (see also Figure 8.39). (c) Stratal slice showing fluvial channelised element with easterly directed palaeocurrents (red; facies succession J) incising wave- to fluvial-dominated (facies succession B) and fluvio-tidal channelised elements (facies succession C).

8.5.1 Clinoforms, Clinothems and Clinoform Sets

Clinoforms are a ubiquitous component of shallow-marine strata (Section 4.4). They are sloping depositional surfaces (Rich 1951) that in shallow-water strata represent a delta front, shoreface, shelf-break slope or similar morphological feature over several tens of metres in length (Patruno and Helland-Hansen 2018). The sediment volume bounded by two clinoforms is referred to as a clinothem. Clinoform geometry reflects the morphology of the dipping depositional surface. Wave-dominated shorefaces are characterised by seaward-dipping clinoforms that have a linear, shoreline-parallel geometry in plan view (e.g. Steventon et al. 2020), whereas more fluvial-influenced deltas contain lobate clinoforms in plan view that dip seaward and also laterally, from the lobe axis to its margins (e.g. in the Po Delta; Figure 8.43, after Correggiari et al. 2005). In many modern deltas that are subject to significant reworking by waves and tides, sand is retained near the shoreline in a sandy 'subaerial clinoform' (e.g. Po Delta; Figure 8.43, after Correggiari et al. 2005; example 15 in Figure B8.1.2a, Box 8.1) while fluvially supplied mud is bypassed across the inner shelf by wave, tide and/or current action to accumulate in a muddy 'subaqueous clinoform' or mid-shelf mud belt (e.g. Adriatic shelf clinoform; Figure 8.43, after Cattaneo et al. 2003; Amorosi et al. 2016; example 15 in Figure B8.1.2b, Box 8.1; see also Sections 9.2.2; 9.4.1). Mixed-process deltas may therefore be characterised by compound clinoforms with subaerial and subaqueous clinoform components (e.g. Swenson et al. 2005). The top of the subaerial clinoform is defined by sea level, whereas the top of the subaqueous clinoform is defined by the depth at which waves and/or tides resuspend a particular grain size. Sand-rich subaerial and subaqueous clinoforms tend to be steeper than their mud-rich counterparts (Patruno et al. 2015a).

Clinoforms and clinothems are arranged in clinoform sets, which record progradation of the delta front, shoreface, shelf-break slope or similar morphological feature (Figures 8.41e,f, 8.43). Clinoforms in each clinoform set are generally oriented sub-parallel to each other in cross section and in plan view. Their bases are marked by onlap or downlap onto an older depositional surface, and their tops either by aggradational toplap or by truncation at a younger erosional surface. Each clinoform set represents a distinct period of progradation of a morphologically distinct depositional feature and is a large architectural element. The clinothems within a clinoform set correspond to smaller architectural elements.

8.5.2 Shoreline and Shelf-Edge Trajectories

The transition between sub-horizontal topset and inclined foreset ('rollover point') can be traced between successive clinoforms to define the trajectory of this depositional break-in-slope. In subaerial deltas, the geometrical break-in-slope is close to the shoreline, which coincides with the transition from subaerial coastal plain deposits to subaqueous delta front or shoreface deposits in a facies succession. Transits of the shoreline across the shelf define regression or transgression (Figure 8.44a), and multiple regressive and transgressive shoreline transits result in the deposition of stacked regressive-transgressive tongues (or parasequences) (Figure 8.44b; Section 8.5.3). At a larger scale, the shelf edge is represented by a similar break-in-slope that separates shelfal and slope deposits (Figures 8.44c and 8.45). The trajectories of these breaks-in-slope provide a simple descriptive measure of stratigraphic architecture in shallow-marine strata (Helland-Hansen and Martinsen 1996; Helland-Hansen and Hampson 2009).

Shoreline and shelf-edge trajectories may exhibit a range of angles and paths (Figure 8.44; Helland-Hansen and Martinsen 1996; Helland-Hansen and Hampson 2009; see also Section 4.4). Within a single clinoform set (e.g. Roda Sandstone 'sub-lobes' in Figure 8.41e), trajectories are regressive but they may be ascending (i.e. upward-stepping; case a3 in Figure 8.44a), descending (i.e. downward-stepping or offlapping; case a5 in Figure 8.44a), or they may alternate between ascending and descending. Such trajectories can also be expressed in terms of the ratio between progradation and aggradation (ascending regressive trajectory) or between progradation and degradation (descending regressive trajectory) (Figure 4.30; Neal and Abreu 2009). Transgressive trajectories are invariably ascending and can be either associated with sediment accumulation (retrogradation; case a1 in Figure 8.44a) or non-deposition (case a2 in Figure 8.44a). Transgressive trajectories tend to characterise thin stratal intervals between vertically stacked clinoform sets (e.g. between clinoform sets in Figure 8.44b and between Roda Sandstone 'sub-lobes' in Figure 8.41e). Regressive and transgressive shoreline trajectories are both commonly documented, respectively, within a clinoform set (cases a3, a4, a5 in Figure 8.44a) and at its boundaries (cases a1, a2 in Figure 8.44a), and their occurrence reflects the movement of shorelines back and forth across a shelf. Clinoform sets can be stacked in a variety of ways, which record progradation (i.e. net-regression), aggradation or retrogradation (i.e. net-transgression) over multiple episodes of shoreline regression and transgression (Figure 8.44b). In contrast, shelf-edge trajectories are generally progradational (Figure 8.44c; Helland-Hansen and Hampson 2009), when reflecting long-term accretion of a break-in-slope that originated as a structural feature or developed through differential deposition on a ramp (Helland-Hansen et al. 2012).

Accurate, quantitative reconstruction of shoreline and shelf-edge trajectories requires careful selection of palaeo-horizontal stratigraphic datum surfaces, sediment

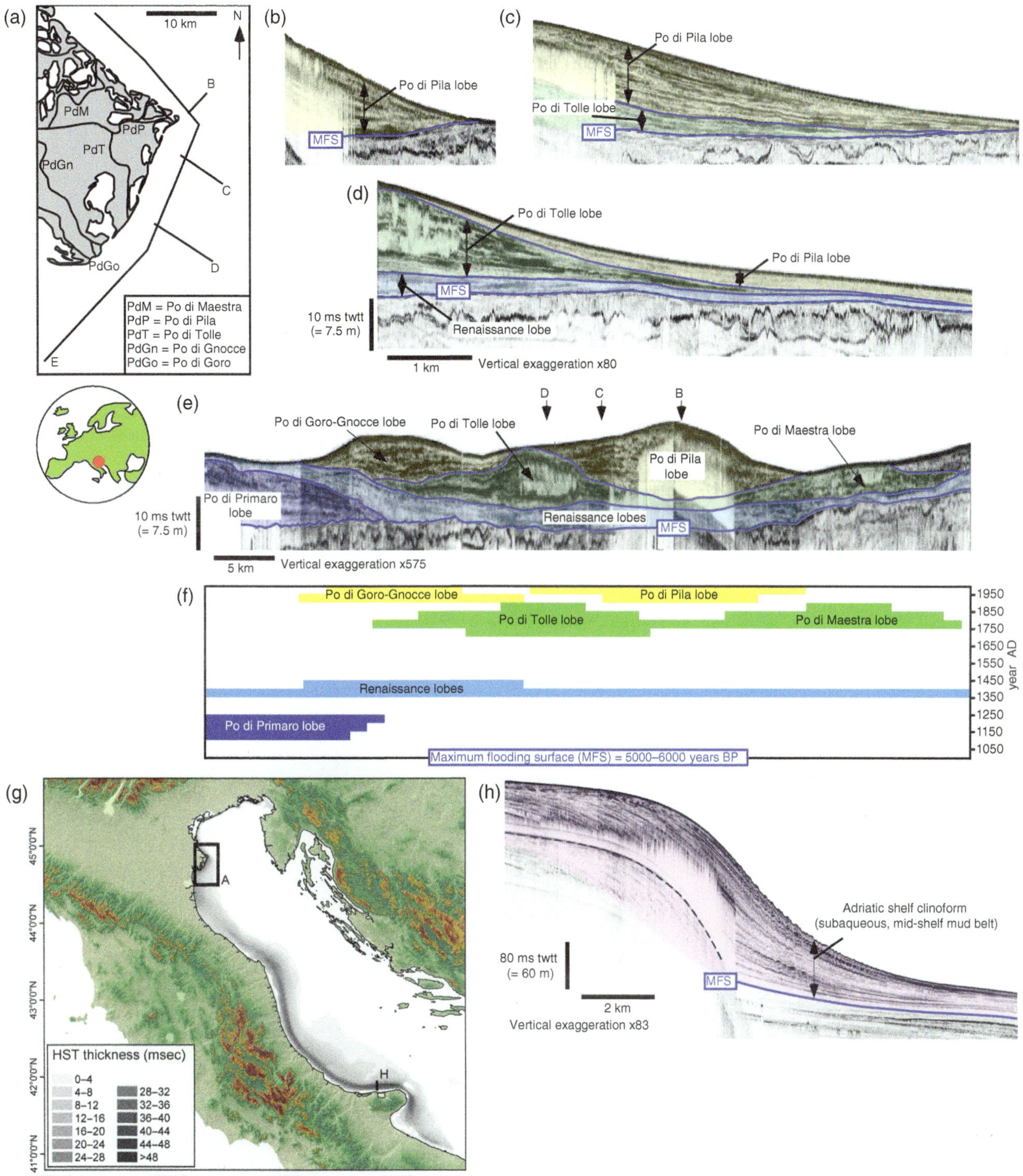

Figure 8.43 Clinoforms and clinoform sets in the Holocene-to-modern Po Delta and contiguous Adriatic shelf clinoform. (a) Map of the modern Po Delta, showing the location of chirp-sonar lines that are oriented (b–d) perpendicular and (e) parallel to the shoreline (modified from Correggiari et al. 2005). The river-dominated, wave-modified Po Delta is characterised by laterally offset stacking of subaerial clinoform sets, each recording progradation of a distinct delta lobe, as represented in (f) a chronostratigraphic interpretation of the shoreline-parallel chirp-sonar line. (g) Map of the modern Adriatic shelf clinoform and Po Delta (adapted from Cattaneo et al. 2003; Amorosi et al. 2016), showing the thickness of late Holocene strata (above MFS labelled in Figure 8.43b–f, and assigned to a highstand systems tract, HST) and location of (h) a chirp-sonar line on the mid-shelf that is oriented perpendicular to the shoreline (modified from Amorosi et al. 2016). The Adriatic shelf clinoform set is subaqueous, supplied by shoreline-parallel dispersal by oceanographic bottom currents of sediment largely from the multiple small subaerial deltas.

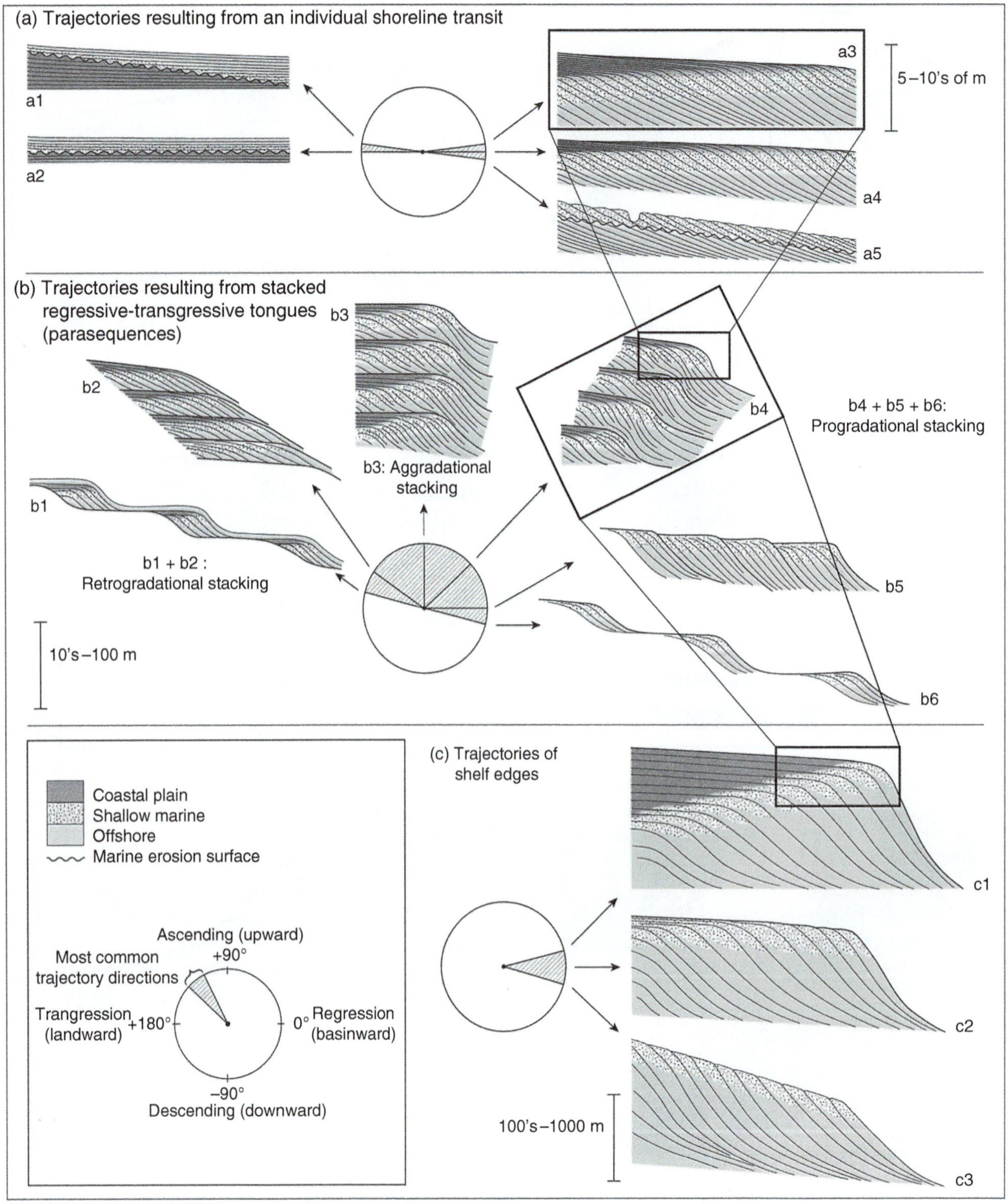

Figure 8.44 Conceptual sketches illustrating scales, hierarchy and common trajectory directions and angles of migrating shoreline and shelf-edge clinoforms (adapted from Helland-Hansen and Martinsen 1996; Helland-Hansen and Hampson 2009): (a) single transgressive or regressive shoreline transit across the shelf; (b) stacked regressive-transgressive shoreline tongues; and (c) shelf-edge migration. One example of how these three hierarchical scales are linked is illustrated. Note that the sketches only show simple trajectories in two dimensions, whereas more complex trajectories occur in forward stratigraphic models (e.g. Figures 4.37, 4.38 and 4.39) and real examples (e.g. Figure 8.45i, j). Clinoform trajectories allow interpretation of the relationship between accommodation and sediment supply (e.g. Figures 4.27 and 4.30).

decompaction and, for some applications, backstripping (i.e. sequential removal of younger strata to restore the depositional geometry of older stratigraphic surfaces) (e.g. Helland-Hansen and Hampson 2009; Prince and Burgess 2013). In such cases, it is possible to reconstruct detailed histories of shoreline and shelf-edge migration in ancient strata (Helland-Hansen and Hampson 2009), and even migration rates if high-resolution age data can be integrated (e.g. Patruno et al. 2015b). Qualitative interpretation of shoreline and shelf-edge trajectories (e.g. Figure 8.45) may only be able to distinguish regression and transgression with confidence, particularly in clinoform sets that are top-truncated by erosion during regression or subsequent transgression (e.g. Roda Sandstone 'sub-lobes' in Figure 8.41e). Such erosion may remove the shoreline or shelf-edge break-in-slope from some clinoforms and modifies the thickness of preserved topset and foreset strata. Shoreline and shelf-edge trajectories can vary along strike, due to lateral changes in the balance of accommodation and sediment supply (e.g. successively decreasing sediment supply in dip-oriented cross-sections in Figure 8.45f–h; Steventon et al. 2020), as is also reflected in stratal stacking patterns (Madof et al. 2016).

The interaction of waves, tides and fluvial processes may vary systematically as the shoreline transits across the shelf, with such variations recorded in the facies successions deposited during regression and transgression. On the shelf-slope margins of large ocean basins, shorelines prograding from the inner to outer shelf are generally exposed to more energetic wave conditions, as frictional attenuation of wave energy reduces with decreasing shelf width, and/or oceanic currents (e.g. Yoshida et al. 2007; Ainsworth et al. 2008). The resulting regressive shoreline deposits and clinothems become progressively more wave-dominated or current-dominated from the inner to the outer shelf in a clinoform set (e.g. Patruno et al. 2015c). In contrast, the corresponding shorelines developed during transgression from the outer to inner shelf are subject to progressively less wave influence and more tidal and fluvial processes locally (e.g. Yoshida et al. 2007; Ainsworth et al. 2008). Tides may be amplified in transgressive shoreline embayments, including drowned palaeovalleys that were incised by rivers during the preceding regression (e.g. Allen and Posamentier 1993; Zaitlin et al. 1994). There are many exceptions to the generalised model described above, because any change in process regime at the shoreline during regression or transgression is strongly dependent on the size, physiography and configuration of the shelf/ramp and adjacent basin. For example, in the intracratonic Western Interior Seaway of North America, shoreline regression is associated with narrowing of the seaway and correspondingly reduced wave fetch, combined with local funnelling, shoaling and resonance of tides (e.g. Yoshida et al. 2007). Wave-dominated shorelines (e.g. Blackhawk Formation; Figure 8.30) typify early regression, when the seaway was wide, but tide-dominated and fluvial-dominated shorelines were developed locally during late regression (Yoshida et al. 2007); the influence of shoreline trajectory and other controls on shoreline character are explored further in this case study in Section 8.6.1 (see also Figure 8.51).

Along shelf-slope margins, shoreline process regime is also partly related to shelf-edge trajectory and slope gradient. Wave-dominated shorelines are developed preferentially at shelf edges with steeply ascending regressive trajectories and are associated with gentle slope gradients (Paumard et al. 2020). Both wave-dominated and fluvial-dominated shorelines occur at shelf edges with gently ascending to falling regressive trajectories (Paumard et al. 2020), and fluvial-dominated (deltaic) shorelines at shelf edges are associated with steep slope gradients and sediment bypass to deep water (Dixon et al. 2012; Paumard et al. 2020) (Box 8.3).

8.5.3 Parasequences and High-Frequency Sequences (Regressive–Transgressive Tongues)

The smallest stratigraphic unit considered in most sequence stratigraphic models of shallow-marine strata is the parasequence (*sensu* Van Wagoner et al. 1990) or high-frequency sequence (*sensu* Zecchin and Catuneanu 2013), which records a single shoreline regression (cases a3, a4 or a5 in Figure 8.44a) and subsequent transgression (cases a1 or a2 in Figure 8.44a). Internally, a regressive–transgressive sandstone tongue consists of one or more architectural elements (including clinoform sets), which in combination exhibit a regressive shoreline trajectory, and overlying strata that record a transgressive shoreline trajectory. In plan view, wave-dominated shorelines generate one or more curvilinear clinoform sets that correspond to beach-ridge sets, whereas more morphologically complex shorelines generate multiple clinoform sets that are stacked laterally – for example, fluvial-dominated deltas may generate multiple lobate clinoform sets that record delta-lobe switching during a single period of overall shoreline regression (e.g. in the Po Delta; Figure 8.43a–e, after Correggiari et al. 2005). In cross section, an internal erosion surface may be developed that reflects a descending-regressive shoreline trajectory during part of the overall shoreline regression (as acknowledged in the term 'high-frequency sequence'). To avoid potentially confusing terminology in this and following sections, we use the established and widely used term 'parasequence' to refer to a stratigraphic unit that records one episode of shoreline advance and subsequent retreat. In outcrop, well-log and

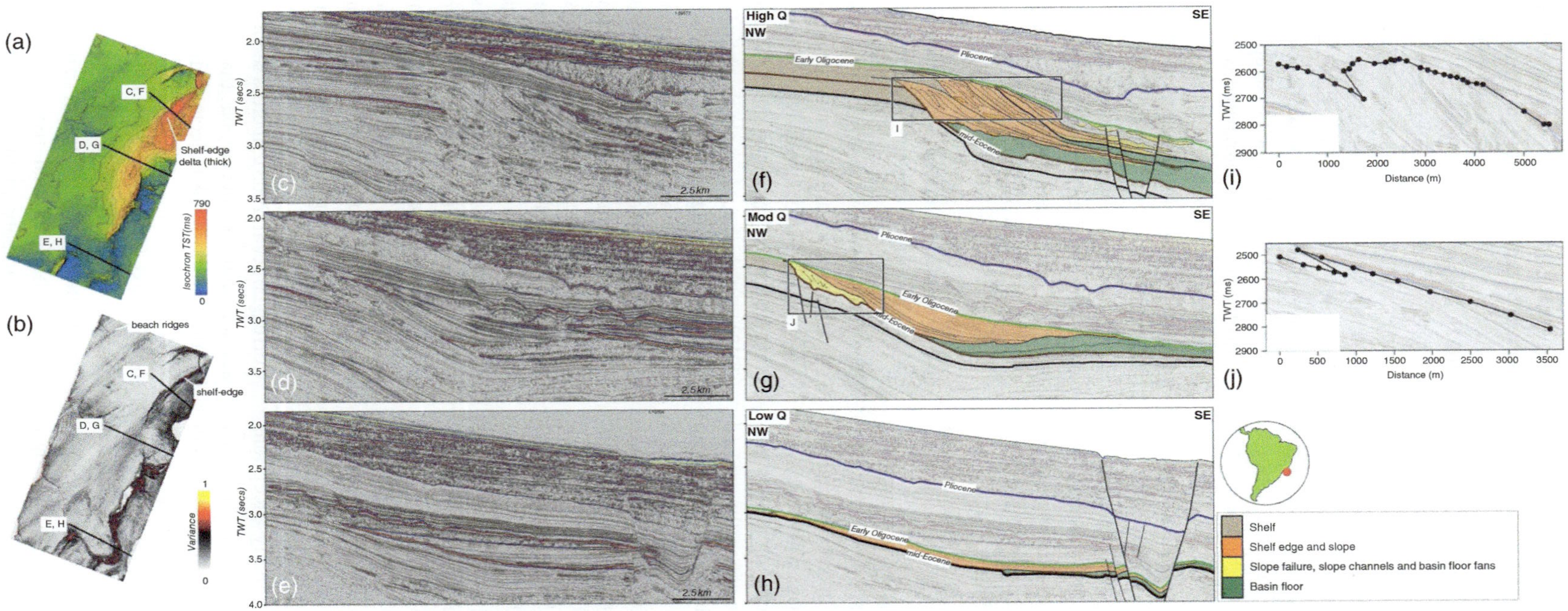

Figure 8.45 Illustration of shoreline and shelf-edge trajectories in 3D seismic data, Mid-to-Late Eocene strata, Santos Basin, offshore Brazil (Steventon et al. 2020 / Society for Sedimentary Geology). (a) Time thickness map (isochron, TST = true stratigraphic thickness) of Maastrichtian–Paleocene to early Oligocene strata (between black and green reflectors in Figure 8.45f–h) and (b) Maastrichtian–Paleocene at the base of the studied stratigraphic interval (black reflector in Figure 8.45f–h), showing the location of three seismic lines (Figure 8.45c–h). (c–e) Uninterpreted and (f–h) interpreted seismic lines oriented along depositional dip, showing along-strike variations in stratal thickness and architecture of shelf, shelf-edge, slope and basin-floor strata. (i, j) Successive shelf-edge positions and shelf-edge trajectories constructed from northern and central interpreted seismic lines oriented along depositional dip (Figure 8.45f–g). Note that the reconstructed shelf-edge trajectory is indicative and has not been corrected for time-depth conversion, decompaction or choice of palaeo-horizontal stratigraphic datum. High, medium and low Q refers to amount of sediment supply.

core data, a parasequence consists of an upward-shallowing, regressive facies succession that is overlain by an upward-deepening, transgressive facies succession or simply a transgressive surface of erosion in many cases, capped by a flooding surface (*sensu* Van Wagoner et al. 1990). Depending on the types and density of available data and the prevailing depositional process regime and resulting stratal geometries, parasequences are difficult to define consistently (Colombera and Mountney 2020a).

8.5.4 Parasequence Stacking Patterns and Parasequence Sets (Systems Tracts, Sequences)

Parasequences can be stacked in a variety of geometrical configurations (Figure 8.44b). In outcrop, well-log and core data, these stacking patterns are ideally expressed as a series of vertically stacked facies successions, each of which has the upward-shallowing to upward-deepening motif of a parasequence. A progradational stacking pattern, or net-ascending regressive shoreline trajectory (case b4 in Figure 8.44b; 'progradational parasequence set' *sensu* Van Wagoner et al. 1990), is marked by a more proximal assemblage of facies in each successive parasequence. For example, this progressively more proximal facies composition is expressed as an upward increase in sandstone content in successive parasequences in wave-dominated shoreface-shelf strata (Figure 8.46a). A retrogradational stacking pattern, or net-ascending transgressive shoreline trajectory (case b2 in Figure 8.44b; 'retrogradational parasequence set' *sensu* Van Wagoner et al. 1990), is marked by a more distal facies composition in each successive parasequence. In wave-dominated shoreface-shelf strata, this progressively more distal facies composition is expressed as an upward decrease in sandstone content in successive parasequences (Figure 8.46b). Although sequence stratigraphic models typically focus on parasequence stacking patterns in a depositional-dip orientation, they may also vary along depositional strike. Such depositional-strike-oriented

Figure 8.46 Parasequence stacking patterns in wave-dominated shoreface-shelf strata. (a) Progradational stacking pattern, from offshore shelf (Mancos Shale) through interpreted fluvial-dominated deltaic (Panther Tongue), wave-dominated shoreface (Storrs Tongue, Spring Canyon Member, Aberdeen Member) and coastal plain deposits (non-marine Blackhawk Formation). (b) Retrogradational stacking pattern, from interpreted coastal plain (Menefee Formation) to interbedded wave-dominated shoreface, tide-dominated channel fill and coastal plain deposits (Cliff House Sandstone). Individual parasequences are shown to the immediate right of the outcrop photos; parasequence stacking patterns are shown in black triangles. Photograph credits: Gary Hampson.

variations reflect non-linear (e.g. lobate) plan-view shoreline geometries and/or changes of the shoreline orientation (e.g. Madof et al. 2016; Zecchin and Catuneanu 2020).

Parasequence stacking patterns commonly define a recurring pattern, in which a series of progradationally stacked parasequences is overlain by a series of retrogradationally stacked parasequences (i.e. a 'genetic stratigraphic sequence' *sensu* Galloway 1989; see also Section 4.4). Such a unit is bounded top and bottom by surfaces of maximum shoreline transgression and condensed sedimentation (i.e. maximum flooding surfaces *sensu* Van Wagoner et al. 1990) (Figure 8.47). Intervals of consistent parasequence stacking pattern are one important way of recognising systems tracts (*sensu* Van Wagoner et al. 1990) (Figure 8.47). In widely applied sequence stratigraphic models, the progradational stacking of parasequences in the lower part of the unit records an initial ascending regressive trajectory (highstand systems tract), then a descending regressive trajectory (falling stage systems tract) and then a second ascending regressive trajectory (lowstand systems tract) (e.g. Coe et al. 2003; Catuneanu 2022; see also Section 4.4) (Figure 8.47a). On a basin margin with a shelf-slope morphology, the progradational stacking of parasequences may result in the shoreline reaching the shelf edge (e.g. Catuneanu 2022) (Figure 8.47b). The lower, progradational part of the unit is overlain by retrogradationally stacked parasequences in the upper part of the unit (transgressive systems tract; Figure 8.47). Parasequences that are stacked with a descending regressive trajectory (e.g. falling stage systems tract) may be capped by a widespread unconformity cut by fluvial erosion and marked by valleys separated by exposed interfluve surfaces (i.e. sequence boundary *sensu* Van Wagoner et al. 1990). These widespread erosional unconformities can be used to define the boundaries of stratal units characterised in shallow-marine strata by a series of retrogradationally stacked parasequences overlain by a series of progradationally stacked parasequences (i.e. a sequence *sensu* Van Wagoner et al. 1990).

Where the shoreline and associated rivers reach the shelf edge (Box 8.3), a shelf edge delta is established, and sediment may be bypassed beyond the shelf edge and into deep-water deposits on the slope and basin floor (Chapter 12). The depositional-dip-oriented trajectory of the shelf edge can also be tracked (Figure 8.44c) and records the stacking pattern of successive sequences (e.g. Figure 8.45).

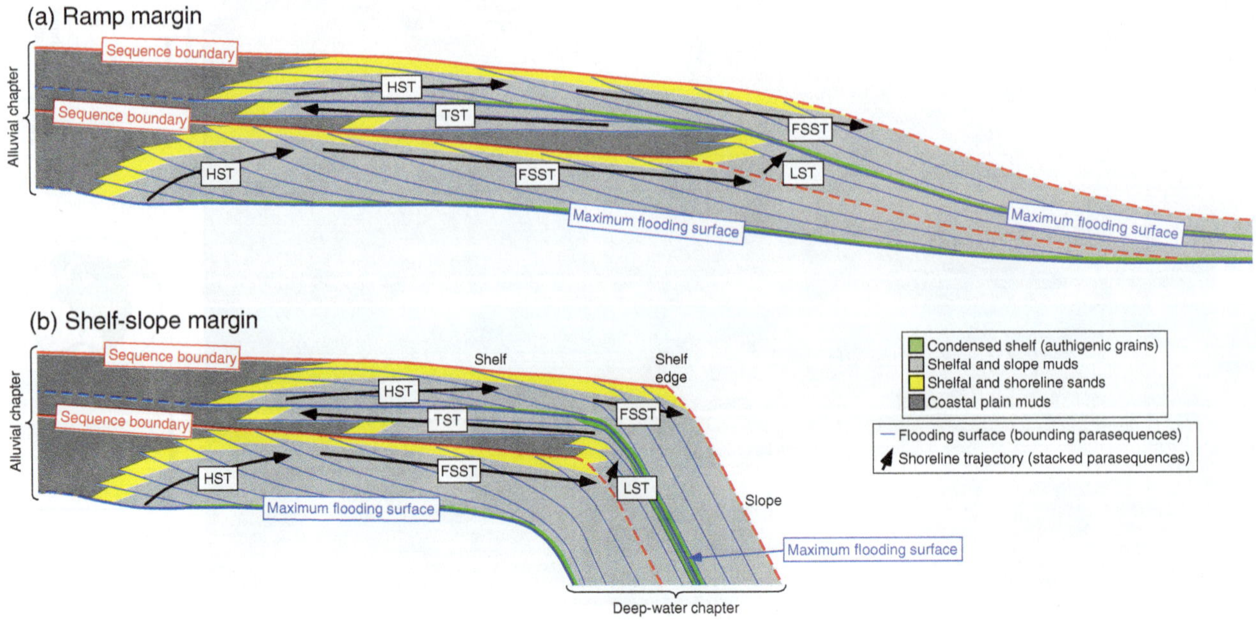

Figure 8.47 Depositional-dip oriented cross sections showing idealised stratal patterns and distributions of shallow-marine and related deposits in sequences developed in basins with (a) a ramp margin and (b) a shelf-slope margin (adapted from Van Wagoner et al. 1990; Plint and Nummedal 2000; Catuneanu 2022). Vertical exaggeration is approximately ×100; typical modern shelf/ramp and slope dips are 0.02° and 2°, respectively (e.g. Patruno and Helland-Hansen 2018). Stratal patterns in alluvial environments (Chapter 5) and deep-water environments (Chapter 12) occur in locations that lie up-dip and down-dip of the cross sections. Stratal patterns in mud-rich shallow-marine strata (shaded pale grey) are examined in more detail in Chapter 9. Systems tracts and their related shoreline trajectories, based on parasequence stacking patterns, are shown. HST = highstand systems tract, FSST = falling stage systems tract, LST = lowstand systems tract, TST = transgressive systems tract.

8.5.5 Character, Geometry and Extent of Bounding Surfaces (Flooding Surfaces, Maximum Flooding Surfaces, Sequence Boundaries)

Several types of stratigraphic surface are developed during deposition of parasequences and parasequence sets (e.g. Figure 8.47) and can be related to specific shoreline and shelf-edge trajectories (e.g. Figure 8.44).

Regressive shoreline trajectories may be associated with composite erosion surfaces at the base of fluvial, fluvio-deltaic and fluvio-estuarine deposits, which cut into underlying nearshore and shallow-marine (e.g. shoreface, delta front) deposits. These surfaces are cut by fluvial erosion as the shoreline advanced basinward (Figure 8.48a–e). In arid settings, fluvial erosion may be supplemented or replaced by aeolian deflation (e.g. Jordan and Mountney 2012; see also Sections 7.3.2, 7.5.7). The composite character of fluvial erosion surfaces is expressed in their widespread extent and, in some instances, deep relief relative to the dimensions of fluvial and estuarine palaeochannels that overlie them. Deep erosional relief at the surfaces may define valley morphologies (i.e. incised valleys *sensu* Van Wagoner et al. 1990). Physical experiments and modern analogues demonstrate that rivers initially cut downwards and then migrate laterally, with deposition (e.g. of river terraces) occurring as well as erosion during valley formation (Strong and Paola 2008; Blum et al. 2013). In vertical successions, these high-relief erosion surfaces may locally place vertically stacked (i.e. multistorey) fluvial and fluvio-estuarine palaeochannel-fill deposits directly above relatively distal shallow-marine deposits, which indicates an apparently abrupt shallowing of water depth in the preserved strata because proximal shallow-marine deposits are absent (e.g. Van Wagoner et al. 1990). There may also be a change in process regime across a regressive surface of fluvial erosion, in particular an upward increase in tidal influence that reflects tidal amplification due to shoaling and constriction of tidal currents in a palaeovalley or estuary. For example, high-relief erosion surfaces that place fluvial-dominated and tide-dominated sandstones above wave-dominated sandstones in the Sego Sandstone (Figure 8.40) have been interpreted as incised valley fills (Van Wagoner 1991). An alternative interpretation of deeply scoured, tidally influenced distributary channels is more consistent with depositional-dip-oriented stratigraphic architecture and facies relationships (Willis and Gabel 2003; Legler et al. 2014) and with the scaling relationships of modern channels and valleys (e.g. Blum et al. 2013). Incised valley fills typically comprise a lower fluvial component, which is assigned to the falling stage and/or lowstand systems tracts, and an upper estuarine component assigned to the transgressive systems tract (Section 5.6.2; Zaitlin et al. 1994) (cf. Figure 8.29). This interpretation is supported by studies of numerous retreating, Holocene-to-modern estuaries that infill Late Pleistocene fluvial valleys cut during relative sea-level fall and lowstand associated with the Last Glacial Maximum (e.g. the modern mixed-energy Gironde Estuary, Figure 8.10; and the Early Holocene wave-dominated estuary capping and overstepping Late Pleistocene fluvial deposits of the Po incised valley fill, Figure 5.45, Amorosi et al. 2016). A composite erosion surface with a stratigraphic-valley morphology (cf. Holbrook and Bhattacharya 2012; Van Yperen et al. 2021) may be associated laterally with tributary dendritic drainage networks (e.g. Figure 5.37e) and a well-developed palaeosol that developed in interfluve areas which did not undergo fluvial erosion (Figure 8.48d,e). Regressive surfaces of fluvial erosion constitute important stratigraphic surfaces (i.e. sequence boundaries *sensu* Van Wagoner et al. 1990) that commonly bound groups of shallow-marine parasequences arranged with a retrogradational–aggradational–progradational stacking pattern (Figure 8.47).

Composite erosion surfaces at the base of nearshore deposits may also be associated with regressive shoreline trajectories, particularly at the base of wave-dominated shoreface deposits (e.g. 'sharp-based shoreface' of Plint 1988). These erosional surfaces were cut by waves, tides or other marine currents as the shoreline advanced basinward (Figure 8.48a–e). In vertical successions, these erosion surfaces place proximal shallow-marine deposits directly above distal shallow-marine deposits, with the absence of medial deposits indicating an apparently abrupt shallowing of water depth in the preserved strata (e.g. Plint 1988). Recognition that this vertical succession lacks medial deposits requires a robust, detailed facies model. Regressive surfaces of marine erosion have also been interpreted as sequence boundaries (e.g. Plint 1988), although this interpretation is not straightforward (Bhattacharya 2011) (see also Section 9.6.9). In landward locations, river incision results in development of a regressive surface of fluvial erosion (sequence boundary *sensu* Van Wagoner et al. 1990) and sediment bypass to more basinward locations, where coeval shallow-marine strata are deposited with a descending regressive shoreline trajectory (i.e. falling stage systems tract; Figures 8.47 and 8.48e,f) (e.g. Plint and Nummedal 2000). If the descending regressive trajectory was steeper than the seaward dip of the shoreface or shelf at wave base, then a regressive surface of marine erosion is developed (e.g. Plint and Nummedal 2000; Storms and Swift 2003; Figure 8.48e). It is important to note that fluvial incision, river terrace formation and sediment bypass (i.e. stratigraphic-valley formation) in landward locations is correlative to forced regressive, shallow-marine deposition in basinward locations.

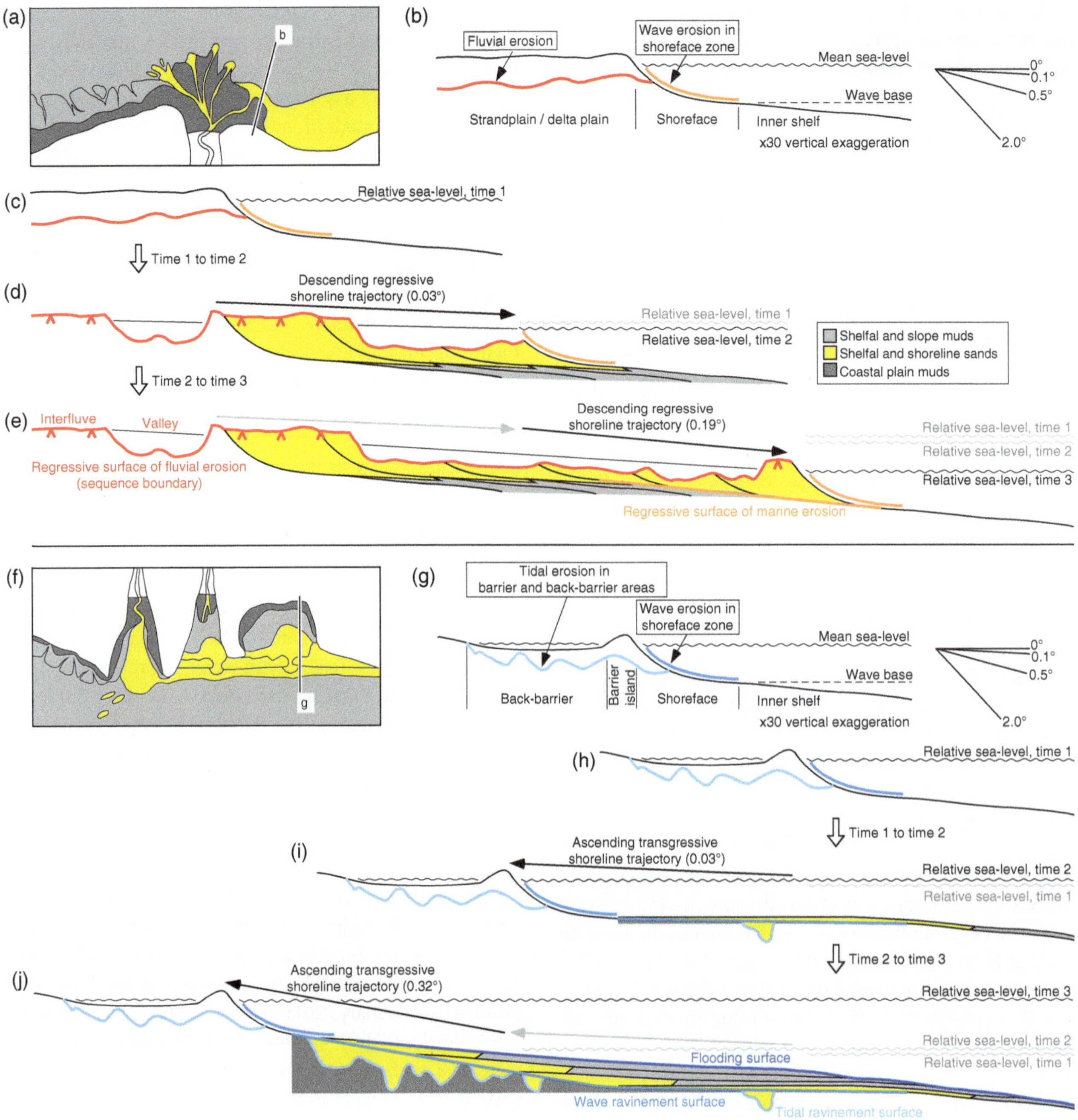

Figure 8.48 Conceptual sketches illustrating the development of stratigraphic surfaces in the (a–e) regressive and (f–j) transgressive components of an idealised shallow-marine parasequence. (a) Map and (b) depositional-dip oriented cross section through a regressive wave-dominated delta and contiguous strandplain and shoreface, showing geomorphological elements and the locations of fluvial and wave erosion. (c–e) Development of surface geomorphology, deposited strata and stratigraphic surfaces during conditions of (d) relatively gentle and (e) relatively steep descending regressive shoreline trajectory. The regressive surface of fluvial erosion has locally variable erosional relief associated with a valley and passes laterally into palaeosols developed at valley interfluves. The development of a regressive surface of marine erosion requires that the shoreline trajectory is steeper than the geomorphological dip of the shoreface or shelf at wave base. (f) Map and (g) depositional-dip oriented cross section through a transgressive wave-dominated barrier island and contiguous shoreface, showing geomorphological elements and the locations of tidal and wave erosion. (h–j) Sequence of cross sections showing the development of surface geomorphology, deposited strata and stratigraphic surfaces during conditions of (i) relatively gentle and (j) relatively steep ascending transgressive shoreline trajectory. The tidal ravinement surface is characterised by deep but localised erosional relief, whereas the wave ravinement surface has an extensive, planar geometry.

Transgressive shoreline trajectories may result in the development of erosion surfaces (i.e. ravinement surfaces *sensu* Swift 1968) cut by waves, tides and/or other marine currents. For example, the retreat of a wave-dominated barrier island and shoreface will generate a relatively deep, high-relief, localised erosion surface at the base of tidal inlets (i.e. tidal ravinement surface) and a relatively shallow, planar erosion surface at wave base (i.e. wave ravinement surface) (Figures 8.29 and 8.48g–j). Sediment released by wave erosion of the shoreface is transported landward to the back-barrier lagoon, into washover fans, for example, and seaward to the shelf (Figure 8.48i,j). Once on the shelf, this sediment can be reworked by tides and waves into compound dunes and bars. Continued retreat of the barrier island and shoreface, and the associated increase in water depth, will result in condensed sedimentation below wave base (i.e. flooding surface *sensu* Van Wagoner et al. 1990). The resulting vertical succession consists of back-barrier lagoonal deposits overlain across a tidal ravinement surface by tidal inlet deposits, which are in turn overlain across a wave ravinement surface by shoreface deposits that pass upwards into offshore mudstones containing a flooding surface (Figure 8.48j). The preserved thickness of the vertical succession and its components, and thus the degree to which the stratigraphic surfaces are distinct or amalgamated, depends on the upward climb of the transgressive shoreline trajectory (e.g. Cattaneo and Steel 2003; Sixsmith et al. 2008). A strongly ascending transgressive shoreline trajectory is characterised by a thick, well-preserved succession in which the tidal ravinement surface, wave ravinement surface and flooding surface are distinct from each other (Figure 8.48j). A weakly ascending or horizontal transgressive shoreline trajectory is characterised by a thin, poorly preserved succession in which the tidal ravinement surface, wave ravinement surface and flooding surface are amalgamated to form a composite surface (Figure 8.48i). In addition to their context in thin, net-transgressive facies successions, wave and tidal ravinement surfaces may be marked by firmground ichnofacies developed on transgressively exhumed substrates (MacEachern et al. 1992). Flooding surfaces are marked by condensed faunas, which may include biostratigraphic markers, and the potential occurrence of authigenic glauconite, ooidal ironstones and phosphate in locations subject to extended physical reworking during condensed sedimentation (e.g. Macquaker et al. 1996; Stonecipher 1999).

A hierarchy of stratigraphic surfaces is recognised, in the same way that a hierarchy of stratigraphic units is recognised. For example, individual parasequences are bounded by flooding surfaces, whereas groups of shallow-marine parasequences arranged with a progradational–aggradational–retrogradational stacking pattern are bounded by maximum flooding surfaces (Figure 8.47). Maximum flooding surfaces record a larger landward shoreline dislocation, have a larger areal extent and record a longer period of condensed sedimentation than flooding surfaces (e.g. Van Wagoner et al. 1990). Different types of stratigraphic surface are also more readily identified in different geographical regions of a basin. For example, regressive surfaces of fluvial erosion (i.e. sequence boundaries *sensu* Van Wagoner et al. 1990) are more clearly expressed in up-dip regions where coastal plain deposits interfinger with shoreline and nearshore deposits (Figure 8.47). Flooding surfaces and maximum flooding surfaces are more clearly expressed in regions where shoreline and nearshore deposits interfinger with offshore deposits. The condensed deposits that characterise flooding surfaces and maximum flooding surfaces are also the principal source of biostratigraphic data that allow regional correlation within and between basins. Different workers have used different types of stratigraphic surface to recognise and define the repeated stratigraphic patterns portrayed in Figure 8.47: regressive surfaces of fluvial erosion (Van Wagoner et al. 1990), maximum flooding surfaces (Galloway 1989) and maximum regressive surfaces (marking the boundary between a lowstand systems tract, LST and the overlying transgressive systems tract, TST in Figure 8.47; Embry 1995). These approaches are not contradictory and can potentially be used in combination (Section 4.4.1).

8.5.6 Representation of Time in Stratigraphic Architectures

It has long been recognised that the stratigraphic record is incomplete and biased, with gaps in deposition present at all scales of observation (Sadler 1981). Sequence stratigraphy provides a tool to characterise stratigraphic gaps and related chronostratigraphic relationships at the spatial and temporal scales represented by stacked parasequences and related bounding surfaces (Sections 8.5.4 and 8.5.5). In tectonically simple basin margins, regressive surfaces of fluvial erosion (i.e. sequence boundaries *sensu* Van Wagoner et al. 1990) are characterised by diachronous erosion, contemporaneous deposition and localised sediment storage (e.g. in fluvial terraces), and basinward bypass and deposition of sediment (e.g. Strong and Paola 2008; Holbrook and Bhattacharya 2012; Blum et al. 2013; Van Yperen et al. 2021) (Figure 8.49). Maximum flooding surfaces are characterised by condensed sedimentation or a depositional hiatus in basinal settings (Figure 8.49). The locus of sediment accumulation shifts basinwards during regression and landwards during transgression (Figure 8.49), although sediment erosion and deposition are in balance over long time scales (e.g. Sadler and Jerolmack 2015; Section 4.5).

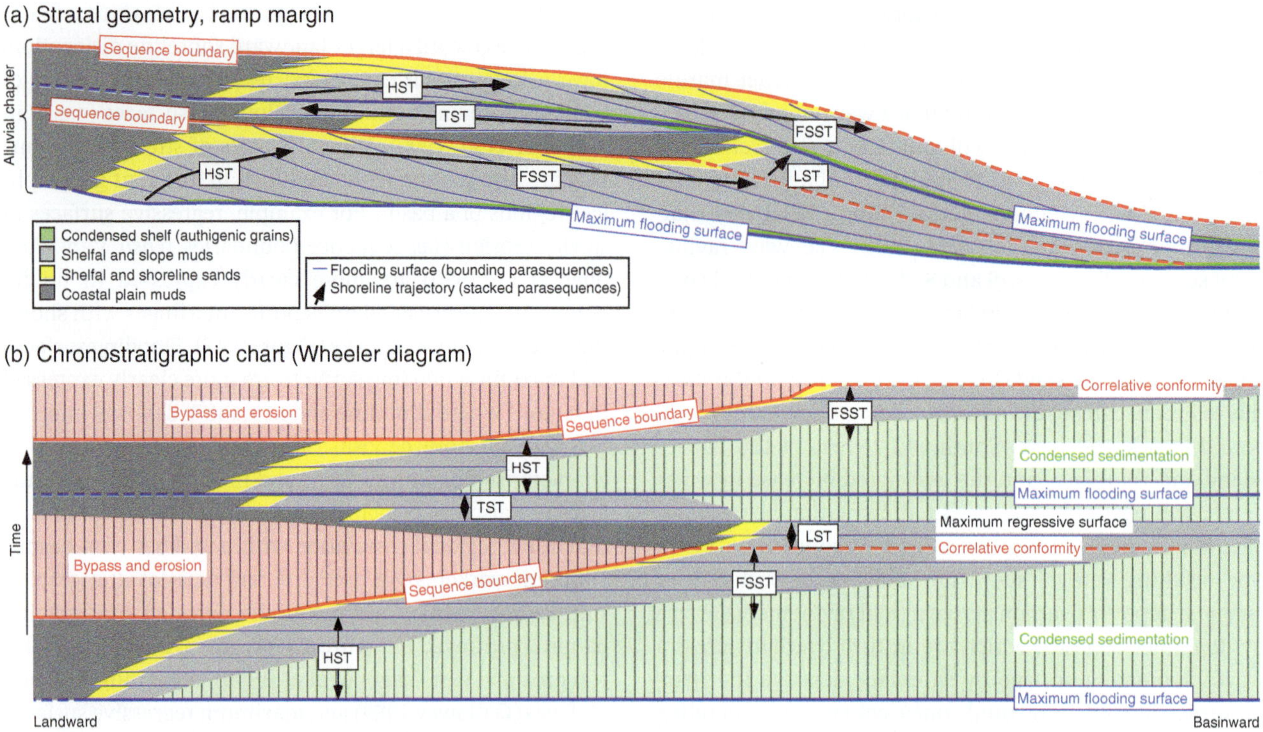

Figure 8.49 (a) Depositional-dip-oriented cross section showing idealised stratal patterns and distributions of shallow-marine and related deposits in sequences developed in a basin with a ramp margin (Figure 8.47a), and (b) a chronostratigraphic chart (Wheeler diagram) constructed from Figure 8.49a and showing time relationships between stratal units and stratigraphic surfaces (adapted from Van Wagoner et al. 1990; Plint and Nummedal 2000; Catuneanu 2022). An equivalent chronostratigraphic chart for a basin with a shelf-slope margin is presented in Figure 4.35.

A similar pattern is observed within individual parasequences. Regressive shallow-marine facies successions in the lower part of each parasequence are more completely preserved than their transgressive counterparts. Overall sediment bypass occurs in rivers that supply sediment to the advancing shoreline, and shoreline-shelf sands are relatively well preserved in thick facies successions (e.g. Figure 8.48c–e). Breaks in sedimentation within regressive facies successions may result from autogenic switching of river-mouth location (e.g. delta lobe switching; Figure 8.43a–f) and marine erosion (e.g. regressive surfaces of marine erosion; Figure 8.48e). In contrast, transgressive deposits in the upper part of each parasequence are characterised by relatively thin and incomplete facies successions that are punctuated by erosion surfaces (e.g. wave and tidal ravinement surfaces; Figure 8.48h–j). Spatial and temporal patterns of deposition in both regressive and transgressive settings are complicated by differential subsidence and topography across actively growing structures (e.g. half-graben rift basins; Gawthorpe and Leeder 2000; Gawthorpe et al. 2003) and mobile substrates (e.g. salt-withdrawal depocentres; Mannie et al. 2014), and by alongshore sediment transport.

Box 8.3 Sediment Storage and Bypass at the Shelf Edge

In sediment routing systems, the shelf edge represents a critical interface for sediment transport to slope and deepwater environments. The positions and trajectories of a shelf edge can be described in many datasets (Section 8.5.2), particularly those containing 3D seismic data (e.g. Figure 8.45). Within this stratigraphic-architectural context, predicting how, where and when sand is delivered to deeper water depends on the interplay of several parameters.

Early sequence stratigraphic models emphasised the role of reduced accommodation across shelves, primarily due to falling sea levels, in promoting fluvial incision, formation of 'shelf-edge deltas' and enhanced sediment transfer beyond the shelf edge (e.g. Vail et al. 1977; Van Wagoner et al. 1990). Accommodation on continental shelves is also linked to tectonic regime and the interaction of sedimentation with delta-related gravity tectonics. In many shelf-edge deltas, such as the Pleistocene Baram Delta (NW Borneo) and Niger Delta (Figure B8.3.1), sediment loading associated with high sedimentation rates initiates gravity-driven extensional faulting in outer-shelf

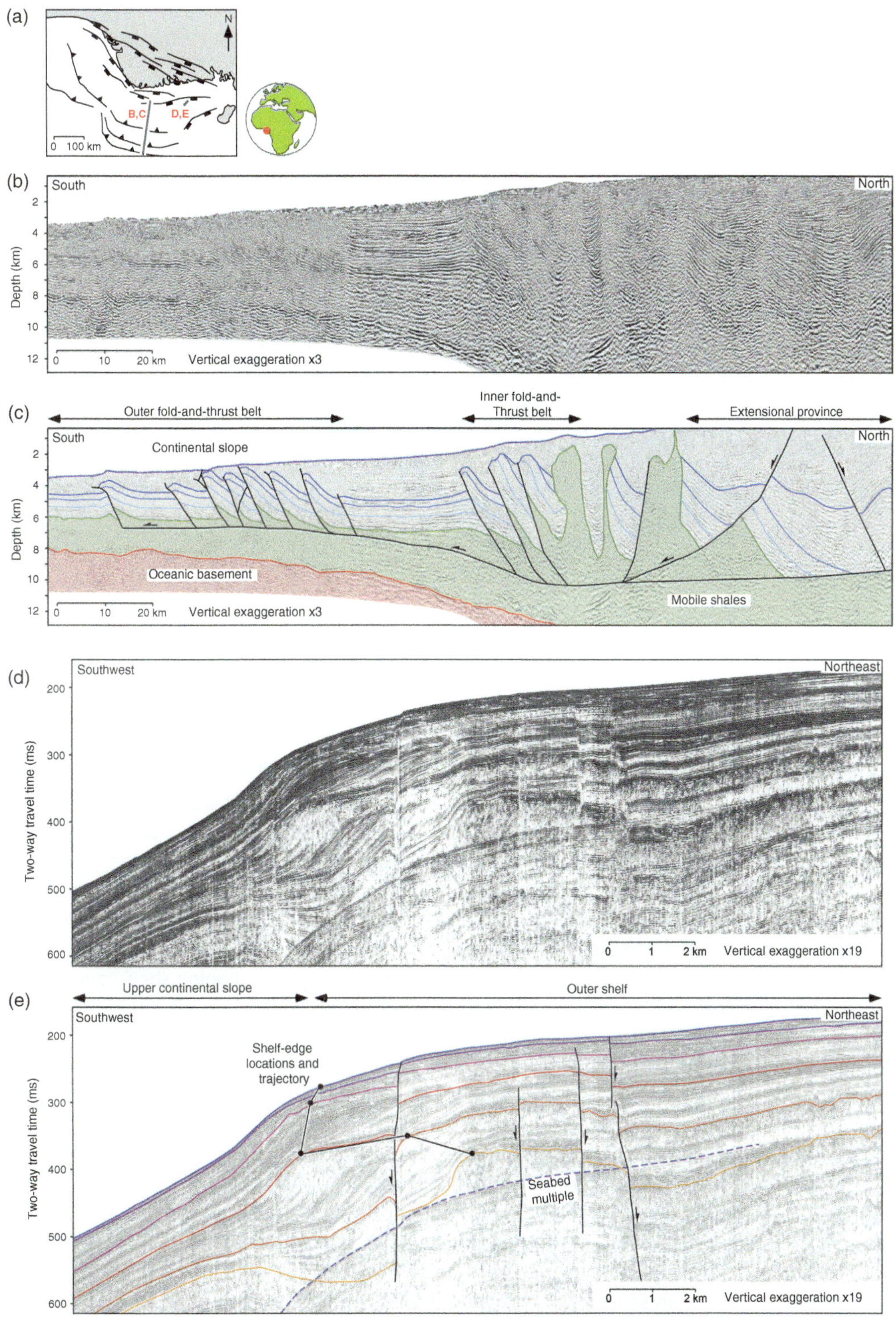

Figure B8.3.1 Stratigraphic architecture and regional structural-stratigraphic context of the Niger Delta and related shelf edge, offshore Nigeria. (a) Map showing the Niger Delta continental margin, major normal and thrust faults (adapted from Onuoha 1999; Corredor et al. 2005), and seismic lines (b–e). (b) Uninterpreted and (c) interpreted regional seismic profile (poststack migrated and depth converted) showing the link between the extensional province on the shelf and compression on the slope (Corredor et al. 2005 / American Association of petroleum Geologists). (d) Uninterpreted and (e) interpreted, very high-resolution sparker seismic profile across the outer shelf and upper slope in the extensional province, showing a net-regressive shelf-edge trajectory generated by progradational-to-aggradational stacking of successive deltaic clinoform sets (Riboulot et al. 2012/ with permission of Elsevier). Strata thicken across, and are offset by, extensional growth faults.

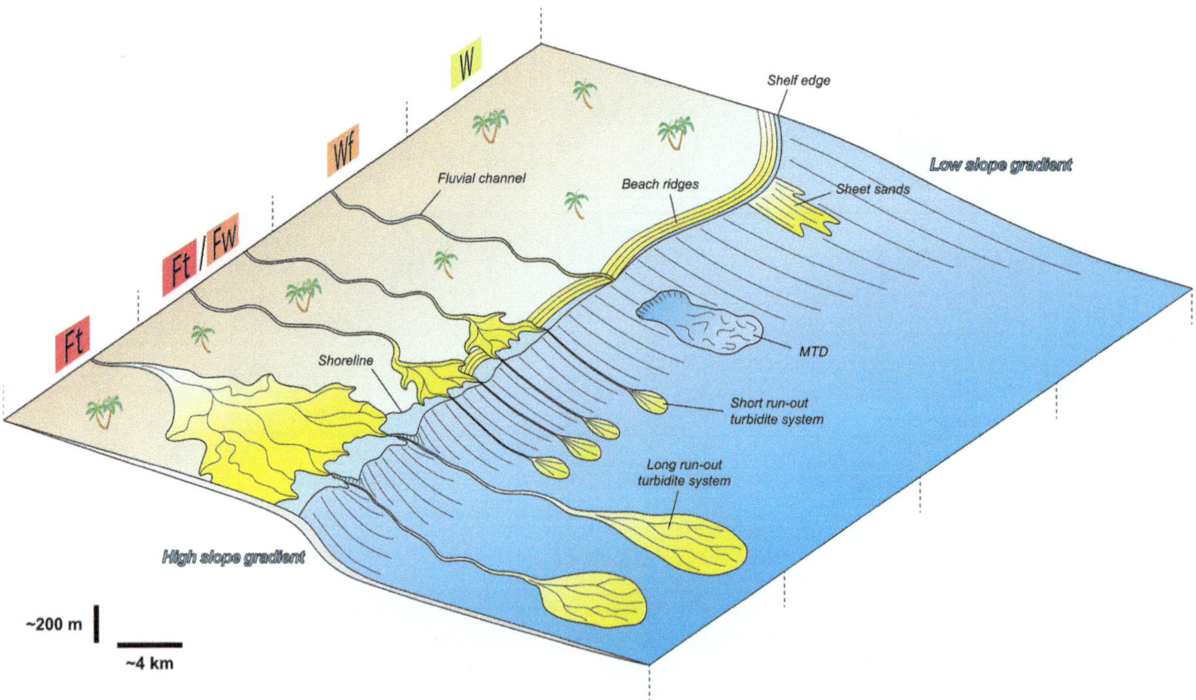

Figure B8.3.2 Simplified conceptual model showing the link between coastal-deltaic process regime, shelf-margin geometry and deep-water sediment transfer and system characteristics (Paumard et al. 2020 / Society for Sedimentary Geology). W/w, wave-dominated/wave-influenced; F/f, fluvial-dominated/fluvial-influenced; T/t, tide-dominated, tide-influenced.

areas, with the development of faults with downthrown sides towards the basin ('down-to-basin' or 'regional' faults) and/or faults towards the shoreline ('counter-regional' faults) that are often linked with mobile mud movement and basal detachments (Figure B8.3.1). Extension up-dip causes compression down-dip, which in the Niger Delta has formed fold-and-thrust belts on the outer shelf-to-upper slope (inner fold-and-thrust belt) and on the lower slope (outer fold-and-thrust belt) (Corredor et al. 2005; Figure B8.3.1). Increased subsidence and accommodation in the hanging walls of deltaic growth faults act to trap sediment on continental shelves (Figure B8.3.1), generating highly aggradational stacking of parasequences, reducing the rate of shoreline progradation (e.g. Olariu and Olariu 2015) and decreasing the potential supply of sediment to the shelf edge and thereby to deep water. For sand delivered to the outer shelf and across the shelf margin in these settings, seabed topography related to contemporaneous fold-and-thrust belts is an important control on sediment fairways and ponding on the slope (see Chapter 12).

As well as being a response to reduced accommodation, shelf-edge deltas have also been shown to develop under conditions dominated by sediment supply and on narrow shelves across a wide range of accommodation conditions (e.g. Burgess and Hovius 1998; Steel et al. 2003). Coastal processes also regulate the rate of shoreline transit across shelves, shelf-edge delta formation and sand delivery to deep water (Figure B8.3.2); fluvial-dominated coastlines are typically associated with more steeply dipping shelf-edge slopes and more long-lived basinwards sand supply via fluvial point sources whereas wave-dominated shorelines are linked to more gently dipping shelf-edge slopes with limited basinwards sand supply via line sources (e.g. Dixon et al. 2012; Gong et al. 2015; Paumard et al. 2020). Furthermore, shoreline-linked submarine canyons are important conduits for efficient sediment transfer to deep water and are most active when fluvial discharge is high, the continental shelf is narrower and steeper, and the magnitude of relative sea-level change (rate of accommodation change) is higher (e.g. Soutter et al. 2021).

Deciphering the controls on sediment partitioning from shelf to basin floor therefore requires a fully integrated and quantitative analysis of shelf-edge geometries and trajectories (Section 8.5.2), shelf-margin architectures (i.e. clinothems; Section 8.5.1) and shoreline-shelf process regime (Section 8.3), ideally leveraging 1D, 2D and (if available) 3D datasets, comparison to analogous surface and subsurface examples, and numerical models.

8.6 Controls on Stratigraphic Organisation

The commonly recurring patterns and organisation of shallow-marine sandstones described above are widely interpreted using sequence stratigraphic models that emphasise the role of relative sea level as a control (e.g. Van Wagoner et al. 1990). However, these recurring patterns are non-unique, in that similar patterns can be generated by multiple controls (e.g. Burgess and Prince 2015; Figure 8.50), which may not be distinguishable given the sparse sampling of many geological datasets. It is therefore important to separate description from interpretation (as attempted here, in descriptive Section 8.5 and interpretive Section 8.6). It is also important to follow a consistent, repeatable methodology in analysing stratigraphic patterns (Catuneanu et al. 2009), to use descriptive nomenclature (e.g. Helland-Hansen and Hampson 2009; Neal and Abreu 2009), and to consider multiple working hypotheses while interpreting stratigraphic patterns. Physical and numerical models have proven to be useful tools in testing the viability and implications of working hypotheses and concepts (e.g. Burgess et al. 2006; Paola et al. 2009; Zuchuat et al. 2022), although any such models are limited by their representation of the processes of interest. There remains a need to more closely integrate modelling studies into the interpretation of stratigraphic patterns in real-world datasets (e.g. Muto et al. 2016).

8.6.1 Multiple Controls

The most straightforward interpretation of stratigraphic patterns is that they are autogenic, generated by the internal dynamics and thresholds of depositional systems (Figure 8.50; Muto et al. 2007; Section 4.3). Switching of delta lobes (Figures 8.41, 8.43b–f and 8.50) is one example of an autogenically generated stratigraphic pattern, driven by the development of topographic gradients and backwater dynamics that drive avulsion of deltaic distributary channels on the delta plain (Section 8.3.3; e.g. Stouthamer et al. 2011; Ganti et al. 2016). Delta-lobe switching is reduced in frequency by the action of waves and tides, which transport sediment away from distributary mouths and thus suppress aggradation of the distributary-channel bed (Swenson 2005; Lentsch et al. 2018). At larger spatial and temporal scales, pulses of sediment storage and release by rivers on the coastal plain can drive shoreline migration (Figure 8.50; Kim et al. 2006). At still larger scales, typically related to parasequence stacking (Section 8.5.4), the lengthening of the coastal plain during overall shoreline progradation leads to a progressive increase in sediment retention here. Assuming constant rates of sediment supply and accommodation creation, increased sediment retention on the coastal plain reduces sediment supply to the shoreline, which slows progradation and can lead to retrogradation (i.e. shoreline autoretreat *sensu* Muto and Steel 1992) (Figure 8.50).

The generation, release and transport of clastic sediment in the up-dip parts of sediment routing systems determines

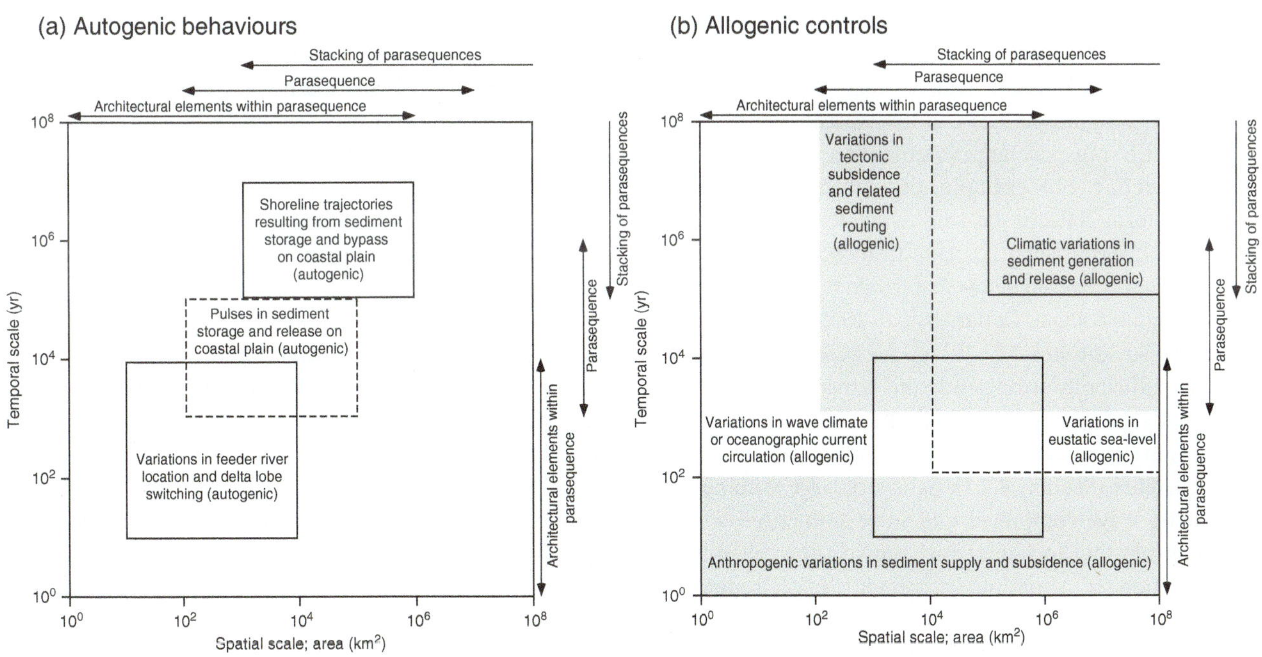

Figure 8.50 Summary of (a) potential autogenic and (b) potential allogenic mechanisms to generate stratigraphic patterns in shallow-marine strata (see text for supporting references). Note the overlap in spatial and temporal scales for different mechanisms.

the volume, grain size and rate of sediment supplied to coastal and shallow-marine environments (Figure 8.50; Section 4.2.1). The controls on sediment supply therefore include tectonic uplift, bedrock composition and climate of sediment source areas, which control the upstream release of sediment (e.g. Armitage et al. 2011), and the tortuosity and length of sediment routing systems that may buffer upstream signals in sediment supply during transport (e.g. Jerolmack and Paola 2010). For example, periods of falling and low water discharge due to climatic aridity may be characterised by river incision and sediment bypass on the coastal plain, the development of regressive surfaces of fluvial erosion (i.e. sequence boundaries *sensu* Van Wagoner et al. 1990) and shoreline progradation (e.g. Zhang et al. 2019b).

Basinal controls on accommodation include eustatic sea level, tectonic subsidence, compactional subsidence and isostatic loading and unloading by sediment, water and ice (Figure 8.50). These controls are often considered in combination as relative changes in sea level. Sediment dispersal in the basin, for example, due to antecedent topography, basin configuration and oceanographic setting, also influences stratigraphic patterns in shallow-marine strata. Tectonically driven variations in subsidence thus control sediment routing and local sediment supply, in addition to accommodation (e.g. in active extensional basins; Ravnås and Steel 1998; Gawthorpe and Leeder 2000; Figure 8.50). For example, narrow, structurally controlled basins can be associated with straits in which tidal current velocities are elevated due to funnelling and shoaling effects. If the tidal currents are sufficiently strong, they may rework and export sediment from tidal straits into neighbouring unconfined regions (e.g. Messina Strait, Figure 8.11; Longhitano 2018a, 2018b). Oceanographic currents can also transport large volumes of sediment parallel to the shoreline. For example, the subaqueous Adriatic shelf clinoform reflects the southward transport of mud by bottom-hugging shelf currents from the Po Delta and other, smaller subaerial deltas that fringe the Adriatic coast of Italy (Figure 8.43g,h; Cattaneo et al. 2003; Amorosi et al. 2016; see also Section 9.2.2). Where the oceanographic currents are sufficiently strong and sand is available for transport, then sand-rich shoreline-parallel subaqueous deltas can develop (Patruno et al. 2015a, c). Wave-generated longshore currents (Section 8.2.1) transport large volumes of sand along wave-dominated and wave-influenced shorelines, contributing to lateral variations in the magnitude and direction of shoreline movement (e.g. Rodriguez et al. 2001) and supplying sand to canyon heads along margins with a shelf-slope break, thus promoting sand transport and bypass to deep water (Covault et al. 2007). At larger scales (100–1000s km) in semi-enclosed marine basins, the configuration of landmasses and basin bathymetry influences key controls, such as the degree of connection to adjacent oceans and wind fetch, on tidal, wave and oceanographic processes and resultant sediment transport, deposition and preservation (e.g. Collins et al. 2021a; Zuchuat et al. 2022).

The range of natural controls that may influence shallow-marine stratigraphic architecture (Figure 8.50) can be illustrated using the widely exposed, well-dated and extensively studied strata of the Cretaceous Western Interior Seaway of North America (Figure 8.51). Figure 8.51b shows a palaeogeographic reconstruction for a timeslice representing c. 0.5 Myr of the middle Campanian (c. 79 Ma) along a portion of the western margin of the seaway (Van Cappelle et al. 2018 and references therein). Sediment was supplied from the west via multiple transverse drainages from a high-elevation plateau ('Nevadaplano' of DeCelles and Coogan 2006) fronted by the thin-skinned Sevier fold and thrust belt, and via axial drainages from the Mogollon Highlands to the southwest. Erosion and transport of sediment from these source areas occurred under a uniformly warm, humid, subtropical climate. High transverse sediment supply routed through the Sevier fold and thrust belt generated progradational stacking of parasequences in the northern and central parts of the palaeogeographic reconstruction (Figures 8.46a and 8.51b; Krystinik and DeJarnett 1995). In contrast, low axial sediment supply from the Mogollon Highlands is expressed in retrogradational parasequence stacking in the southern part of the palaeogeographic reconstruction (Figures 8.46b and 8.51b; Krystinik and DeJarnett 1995). A large, progradational tide-dominated delta in the northern part of the reconstructed palaeoshoreline (comprising the Haystack Mountains Formation, Morapos Sandstone, Berry Gulch Sandstone, Wise Gulch Sandstone; Figure 8.51b) was deflected towards the south via a combination of antecedent topography and wave-generated longshore currents (Figure 8.51c,d; Mellere and Steel 2000; Longhitano and Steel 2017). Strong tidal currents resulted from amplification by local shoaling and funnelling (Longhitano and Steel 2017) and potentially by resonance in the broad embayed bathymetry of the regional palaeoshoreline (Figure 8.51b; Dean et al. 2019). Progradational wave-dominated deltas and strandplains in the central part of the reconstructed palaeoshoreline (Blackhawk Formation, Castlegate Sandstone; Figures 8.46a and 8.51b) were supplied by sediment from multiple small transverse fluvial drainages (Pettit et al. 2019) and by southward-directed, wave-generated longshore drift (e.g. Hampson et al. 2014). Retrogradational wave-dominated strandplains and barrier islands in the southern part of the reconstructed palaeoshoreline (Cliffhouse Sandstone) were fed principally by southward-directed longshore drift (Figures 8.46b and 8.51b).

Increasingly, anthropogenic influence is also important in controlling water and sediment discharge to deltas (Syvitski and Saito 2007) and compactional subsidence of delta plains due to extraction of groundwater and

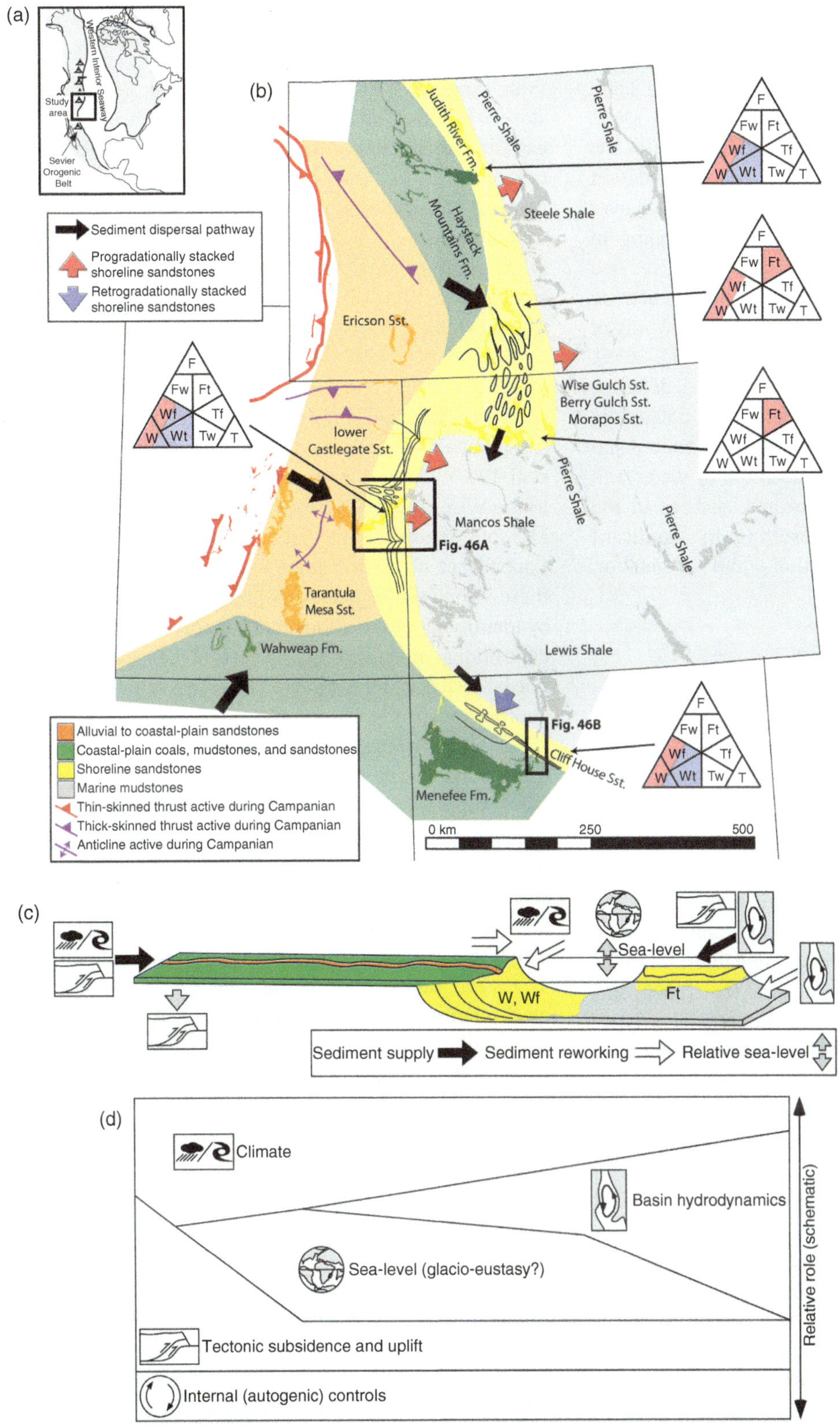

Figure 8.51 (a) Palaeogeographic map of the late Cretaceous Western Interior Seaway of North America (adapted from Kauffman and Caldwell 1993), locating the study area. (b) Mid-Campanian (c. 79 Ma) palaeogeographic map of the study area (modified from Van Cappelle et al. 2018 and references therein) showing interpretations of gross depositional environments (in bold colours over outcrop belts), palaeoshorelines and ternary process regimes (red – regressive; blue – transgressive) based on detailed published outcrop studies (see Van Cappelle et al. 2018 for details). The regional palaeoshoreline was marked by: (i) a large, progradational tide-dominated delta in southern Wyoming and northwestern Colorado; (ii) progradational wave-dominated deltas and strandplains in northeastern Utah (Figure 8.46a); and (iii) retrogradational wave-dominated strandplains and barrier islands fed by southward-directed longshore drift in northwestern New Mexico (Figure 8.46b). (c) Conceptual sketch summarising the impact of external (allogenic) controls on coastal and shallow-marine depositional systems along the Campanian palaeoshoreline and coeval strata (Figure 8.51b) (Hampson 2010 / John Wiley & Sons). Symbols illustrating the various controls are explained in Figure 8.51d. (d) Conceptual sketch illustrating the relative role of the main controls on stratigraphic architecture in coastal and shallow-marine strata along the Campanian palaeoshoreline (Figure 8.51d) (Hampson 2010 / John Wiley & Sons).

hydrocarbon resources (Syvitski et al. 2009) (Figure 8.50). For example, the Renaissance lobes of the Po Delta (Figure 8.43d–f) were abandoned due to diversion of the Po River course by Venetian engineers, and progradation of the modern Po Delta has slowed and then reversed over the last 50 years due to upstream sediment trapping in dams and mining of sand and gravel from the river bed (Simeoni and Corbau 2009). The effects of human activity are even more pronounced in the erosion and drowning of mega-deltas in Southeast Asia (e.g. Anthony et al. 2015). Changes in rainfall patterns linked to overgrazing, deforestation and variations in ocean temperatures, potentially related to anthropogenic climate change, have caused significant variability and an overall decrease in water and sediment discharge to shoreline systems, such as in the Sine-Saloum Delta of Senegal, resulting in significant changes in coastal sedimentation and vegetation (Descroix et al. 2020).

8.6.2 Interpretation Strategies

The occurrence of multiple controls (Figure 8.50) and complex interactions between these controls greatly complicates interpretation of stratigraphic patterns in shallow-marine strata. Simple sequence stratigraphic models may not be sufficient to account for and predict all of these patterns. How, then, should practitioners approach the interpretation of shallow-marine stratigraphic architecture? As a starting point, any interpretation of stratigraphic patterns must build on facies analysis and associated interpretations of depositional process regime. Catuneanu et al. (2009) outline the strengths and limitations of different data types (Section 2 of Catuneanu et al. (2009)) and the need to use a model-independent methodology for sequence stratigraphic interpretation (Section 3.1 of Catuneanu et al. (2009)). Practitioners should be aware, however, that sequence stratigraphic methodologies were developed with cycles in relative sea level assumed to the dominant control on stratigraphic patterns (e.g. Van Wagoner et al. 1990; Sections 3.2–3.5 in Catuneanu et al. 2009). Methodologies that are not anchored on this assumption include the use of shoreline and shelf-edge trajectories (e.g. Helland-Hansen and Martinsen 1996; Helland-Hansen and Hampson 2009) and comparable stacking patterns (e.g. Neal and Abreu 2009; Neal et al. 2016).

Sequence stratigraphic interpretation of shallow-marine sandstones may be ambiguous, because palaeo-water depth is insufficiently constrained by facies analysis and/or palaeogeographic context is poorly described by available data. For example, multiple shallow-marine sandstones in the Cretaceous Western Interior Seaway of North America appear to be encased in offshore mudstones (e.g. Shannon Sandstone; Suter and Clifton 1999). Facies analysis of these sandstones indicates deposition in shallow water by storm waves and/or tidal currents, but evidence of subaerial exposure such as rootlets is absent. The sandstones also appear to lie in locations up to 200 km seaward of the regional palaeoshoreline, based on exposed outcrop belts that are oriented west-east, perpendicular to the regional, north–south-oriented palaeoshoreline trend. These sandstones have variously been interpreted as offshore shelf ridges (Tillman and Martinsen 1984), 'detached' shoreline sandstones deposited during relative sea-level lowstands (Walker and Bergman 1993) and bars in the distal portion of tide-influenced deltas that were deflected southward, sub-parallel to the regional palaeoshoreline (Sullivan et al. 1997). This last interpretation is similar to that applied to the Wise Gulch Sandstone, Berry Gulch Sandstone and Morapos Sandstone in Figure 8.51b (Hampson 2010; Van Cappelle et al. 2018). Suter and Clifton (1999) appraise these various interpretations for the Shannon Sandstone by identifying observations that support or contradict each depositional model, and also aspects of the models that cannot be tested using available data. This careful, objective approach warrants wider use.

For a description of stratigraphic patterns generated using a model-independent methodology, three aspects of the stratigraphic patterns can be used to interpret potential controls. (i) The lateral extent of a stratigraphic pattern indicates the spatial scale of the underlying control (Figure 8.50). Evaluating this lateral extent requires robust correlation or mapping at local, regional or global scales using the available data. (ii) The timing and duration of a stratigraphic pattern indicates the synchronicity or diachroneity of the underlying control. A robust chronostratigraphic framework that is independent of the stratigraphic pattern is required to evaluate timing and duration (e.g. Miall 1992). (iii) Apparent organisation in a stratigraphic pattern implies that the underlying control is ordered, and potentially cyclical. However, robust statistical analysis of the stratigraphic pattern is required to demonstrate either order or cyclicity (e.g. Vaughan et al. 2011; Burgess 2016). It should also be noted that the duration and periodicity of a pattern is not necessarily diagnostic of a particular control, which limits the confidence and resolution of interpretation (Li et al. 2016).

It is not possible to distinguish the effects of potential controls in most, and arguably all, geological datasets. To illustrate this point, it is instructive to interpret stratigraphic patterns that are generated from physical modelling experiments in which the boundary conditions are known (e.g. the exercise presented in Martin et al. 2010); even in these cases, there are multiple viable interpretations of the observed stratigraphic patterns. Accordingly, practitioners should adopt interpretation strategies that acknowledge the complex nature of the stratigraphic record and attempt to mitigate uncertainty arising from limited data. These strategies may involve: the use of descriptive methods and nomenclature; explicitly stating the assumptions that underpin a particular stratigraphic interpretation; discussion of alternative interpretations; and use of numerical and physical modelling tools that allow concepts and hypotheses to be tested and related uncertainty to be constrained (e.g. Figures 4.37 and 4.39).

8.7 Applications

8.7.1 Reconstruction of Basin Configuration and Evolution

Understanding and reconstructing basin configuration and evolution can both inform and be informed by the interpretation of process regime of preserved shallow-marine strata, as exemplified by several of the studies summarised in Section 8.4. The spatial and temporal distribution of these strata constrain the position of the palaeoshoreline and nearshore deposits, and thus the extent, geometry and nearshore palaeobathymetry of shallow seas (e.g. Figure 8.51). Improved understanding of ancient palaeogeography is fundamental for understanding and quantifying the controls on shoreline type and strata distribution, selecting appropriate modern analogues (see Nyberg and Howell 2016), and for input into numerical models of ancient processes to understand the controls on, and make predictions of, ancient facies distributions (e.g. Collins et al. 2018b).

8.7.2 Understanding the Response of Shorelines to External Forcing

Marginal- and shallow-marine strata preserve a record of palaeoshoreline migration, which can be used to constrain histories of external forcing. Palaeoshoreline migration is reconstructed using a combination of stratal geometries, typically imaged in seismic data, and high-resolution chronostratigraphic data (Section 8.5) (e.g. Miller et al. 2013). External forcing can be simplified to the interplay of relative sea level, which combines controls on eustasy and subsidence, and sediment supply, which encompasses controls on sediment routing, storage and dispersal (cf. 'A/S' ratio of Swift and Thorne 1991; Muto and Steel 1997). Palaeoshoreline migration reconstructions can thus be used to infer local relative sea-level history, from the vertical component of palaeoshoreline migration, and local relative sediment supply history, from the 3D sediment volumes or 2D cross-sectional areas deposited during palaeoshoreline migration (e.g. Hampson et al. 2014).

Much effort has been expended in attempts to isolate the eustatic component of relative sea-level curves generated from palaeoshoreline migration, using a combination of techniques: decompaction; backstripping, to remove the effects of sediment and water loading; and calibration against oxygen isotope curves that provide a proxy for ice cap volume (e.g. Haq et al. 1988; Miller et al. 2005). However, it remains challenging to disentangle eustasy from the multiple controls on palaeoshoreline migration (Section 8.6). All eustatic sea-level curves have been the subject of debate and criticism.

It is perhaps more instructive to consider the effects of changes in relative sea level under specific climatic states. Icehouse periods are characterised by extensive polar ice sheets that wax and wane to generate high-frequency, high-magnitude, cyclical changes in eustasy. Numerical models suggest that such sea-level changes give rise to large regressive-transgressive transits of the shoreline across the shelf, with shelf-edge deltas and bypass to deep water occurring during many relative sea-level falls (Sømme et al. 2009; Sweet and Blum 2016). Fluvial incision of the shelf may limit delta avulsion during regression and favour development of point sources for sediment bypass beyond the shelf edge (Harris et al. 2020). In contrast, greenhouse periods lack continental glaciers and are characterised by low-magnitude, cyclical changes in eustasy (Ray et al. 2019). These sea-level changes give rise to small regressive–transgressive transits of the shoreline across the shelf, with shelf-edge deltas and bypass to deep water being established during various sea-level stands and over the duration of multiple relative sea-level cycles (Sømme et al. 2009; Sweet and Blum 2016). Delta avulsion during regression favours development of shelf-edge deltaic shorelines as line sources for sediment bypass to deep water (Harris et al. 2020). These experimental results have been corroborated by real-world stratigraphic architectures, for example, from late Cretaceous (Carvajal and Steel 2006) and Paleocene-to-Eocene (Zhang et al. 2016) strata developed under greenhouse conditions, although the role of sediment supply warrants further investigation.

8.7.3 Prediction of Future Coastal Change

Our understanding of how shorelines migrate in response to changes in relative sea level and sediment supply can also be applied to predicting future coastal change. Rising relative sea level along many shorelines makes them more vulnerable to flooding due to storm surges and high tides (e.g. Karim and Mimura 2008; Hache et al. 2021). Sediment discharge has been greatly modified in many rivers as a result of deforestation, dam construction and sand mining, and many delta plains are undergoing compactional subsidence due to extraction of groundwater and hydrocarbons (Syvitski et al. 2009; Dunn et al. 2019; Nienhuis et al. 2020). Deforestation has increased fluvial sediment supply and caused progradation of many deltas, whereas fluvial sediment retention and extraction has contributed to the retreat of many others (Nienhuis et al. 2020). Deltaic process regime leads to redistribution of marine sediment by waves and tides, and marine sediment supply may be sufficient to drive local shoreline progradation (e.g. in the Changjiang Delta; Leonardi et al. 2021, and part of the Ayeyarwady Delta; Anthony et al. 2019). Shoreline process regime is also linked to changes in the potential deposition of plastic and other pollutants in coastal sediments (Harris et al. 2021). In combination with rising global sea level, anthropogenic modifications to sediment supply and subsidence are increasing the vulnerability of many densely populated coastal regions to flooding and shoreline retreat. Our predictions of ongoing and future coastal change are underpinned by understanding of shallow-marine hydrological and sedimentological processes, in combination with remotely sensed monitoring data, analysis of global datasets (e.g. Caldwell et al. 2019; Nienhuis et al. 2020) and hydrodynamic modelling of local sediment transport and deposition. Such predictions can guide engineering projects to manage coastal sediment budgets, including managed shoreline retreat, sand nourishment of wave-dominated coasts, partial damming of tidal inlets and estuaries (e.g. Voordelta coastline of the southwest Netherlands; Elias et al. 2017), and controlled levee breaches and sediment diversions in fluvial-dominated deltas (e.g. Mid-Barataria Sediment Diversion in the Mississippi Delta; Cornwall 2021).

8.7.4 Characterisation of Subsurface Reservoirs and Storage Units

Shallow-marine sandstones host significant groundwater and hydrocarbon resources in the subsurface and are increasingly being considered for subsurface storage of CO_2 and hydrogen. The quality and behaviour of shallow-marine sandstones as reservoirs and storage units depend on their rock properties, in particular porosity and permeability, on their heterogeneity, which determines the distribution and spatial organisation of rock properties, and on the properties of the fluids that they host. Sedimentological heterogeneity occurs across multiple, hierarchically arranged, spatial scales (e.g. Nordahl and Ringrose 2008; Martinius 2017), which can be readily related to the spatial scales considered in the methods of facies analysis (Section 8.4) and stratigraphic analysis (Section 8.5), in order of descending scale: sequence stratigraphic units, architectural elements, facies association and facies (e.g. Figure 8.52a–c). Sedimentological heterogeneity also extends to smaller spatial scales within facies, related to sedimentary structures, sediment textures and grain size and shape (e.g. Figure 8.52d,e) and reflects diagenetic modification as deposited sediment is transformed into rock (e.g. carbonate-cemented concretions in Figure 8.52b,c). Identifying the types and distribution of heterogeneities across this range of spatial scales is challenging in many subsurface datasets, and data-rich modern and ancient analogues are often required to complement and extend subsurface data (e.g. Howell et al. 2008, 2014).

Numerical simulations of fluid flow illustrate the impact of sedimentological heterogeneity on hydrocarbon drainage and recovery in shallow-marine sandstone reservoirs (e.g. Larue and Legarre 2004; Howell et al. 2008; Sech et al. 2009; Graham et al. 2015), and numerical simulation results are borne out by real-world reservoir monitoring and performance studies (e.g. Cross et al. 2015; Reynolds 2017). Groundwater aquifers have typically been characterised and monitored in less detail than hydrocarbon reservoirs, but multi-scale sedimentological heterogeneity has a similar effect on controlling shallow-marine sandstone aquifer properties and connectivity, and groundwater flow (e.g. in the Brussels Sands aquifer; Huysmans et al. 2008; Possemiers et al. 2012). The increasing use and vulnerability of groundwater resources in urban and agricultural settings (e.g. for aquifer thermal energy storage, as well as for potable freshwater) will require more detailed future characterisation of heterogeneity in shallow-marine sandstone aquifers. The intrusion of saline water into Quaternary examples of such freshwater aquifers now threatens the water supply of densely populated coastal regions (e.g. Wang and Jiao 2012; Mabrouk et al. 2013; Tran et al. 2022).

Sedimentological heterogeneity is also important for geological storage of fluids, even though the goal is not to extract fluids from the subsurface (as in hydrocarbon reservoirs and groundwater aquifers). Storage of waste products such as CO_2 is intended to be permanent, whereas temporary storage is required for energy resources such as hydrogen and compressed air. Low-enthalpy geothermal energy can also be recovered and stored by extraction and injection of groundwater from and into shallow sedimentary aquifers. Any storage unit requires an overlying impermeable seal that retains its integrity during fluid injection and throughout storage duration. In shallow-marine strata, the distribution of such seals is typically related to parasequence stacking (e.g. Figure 8.52a). Movement of injected fluid away from the wellbore, and thus dissipation of fluid pressure that builds up during injection, requires that the storage unit is sufficiently permeable. The storage capacity of the unit requires porosity and may be influenced by heterogeneity that disperses and traps the plume of injected fluid. Such heterogeneity includes sandstone pinchouts, and impermeable facies such as cemented concretions and mudstones that form barriers and baffles at a range of scales (e.g. Figure 8.52b–d). The injected fluid may also react geochemically with mineral grains, causing dissolution or precipitation. The shallow-marine sandstone aquifers of the Johansen and Cook formations constitute the CO_2 storage unit in the Northern Lights project, offshore Norway (Figure 8.53a; Sundal et al. 2013, 2016). In addition to trapping in a structural closure beneath

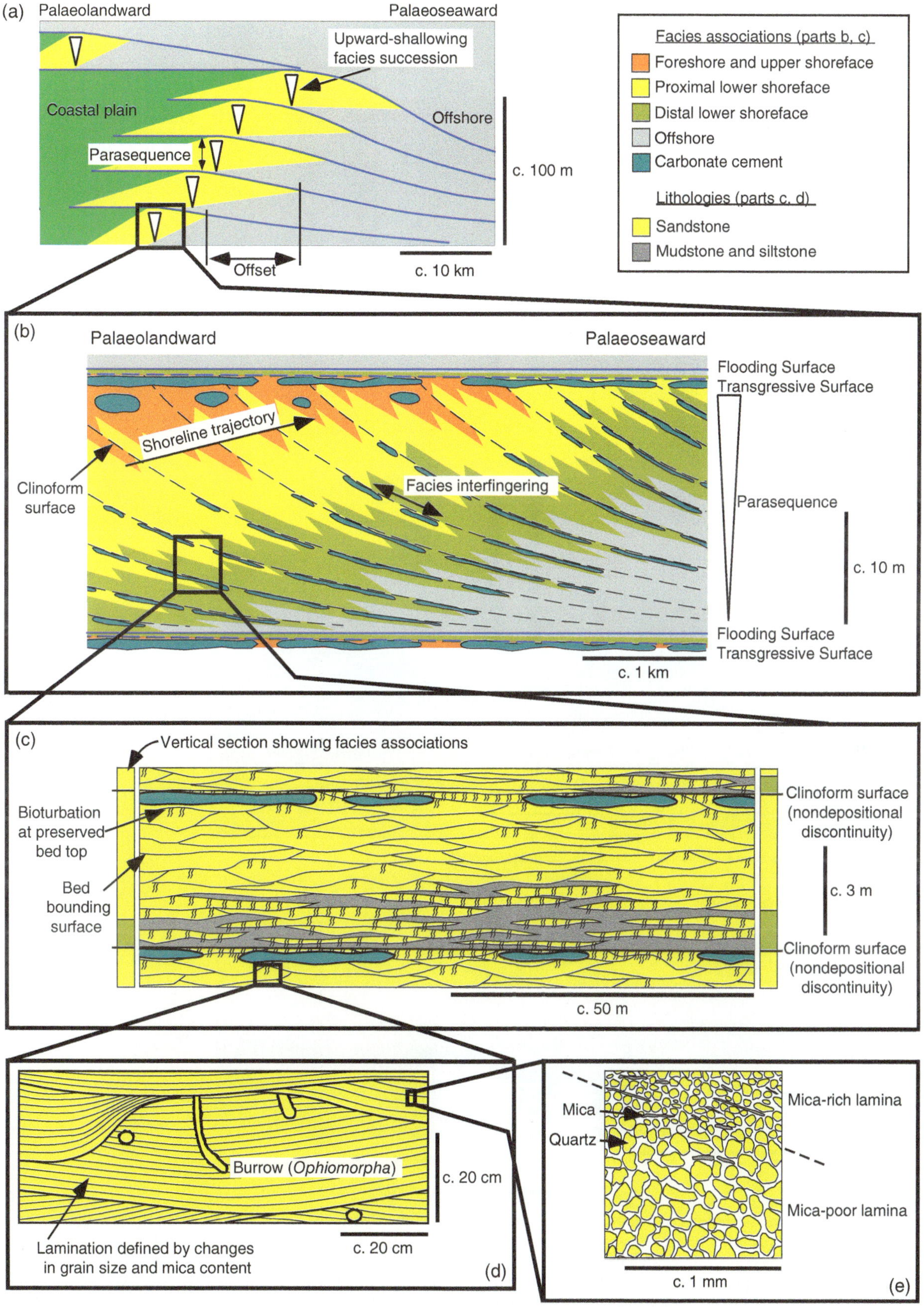

Figure 8.52 Generic hierarchy of heterogeneities within wave-dominated deltaic and shoreface-shelf sandstone reservoirs across a range of spatial scales (Sech et al. 2009 / The American Association of Petroleum Geologists): (a) stacked parasequences, (b) a single parasequence, (c) an architectural element bounded by clinoform discontinuity surfaces, (d) hummocky cross-stratification and (e) microscopic lamination. This figure is oriented along depositional dip and does not show along-strike variations in facies distributions and stratigraphic architecture.

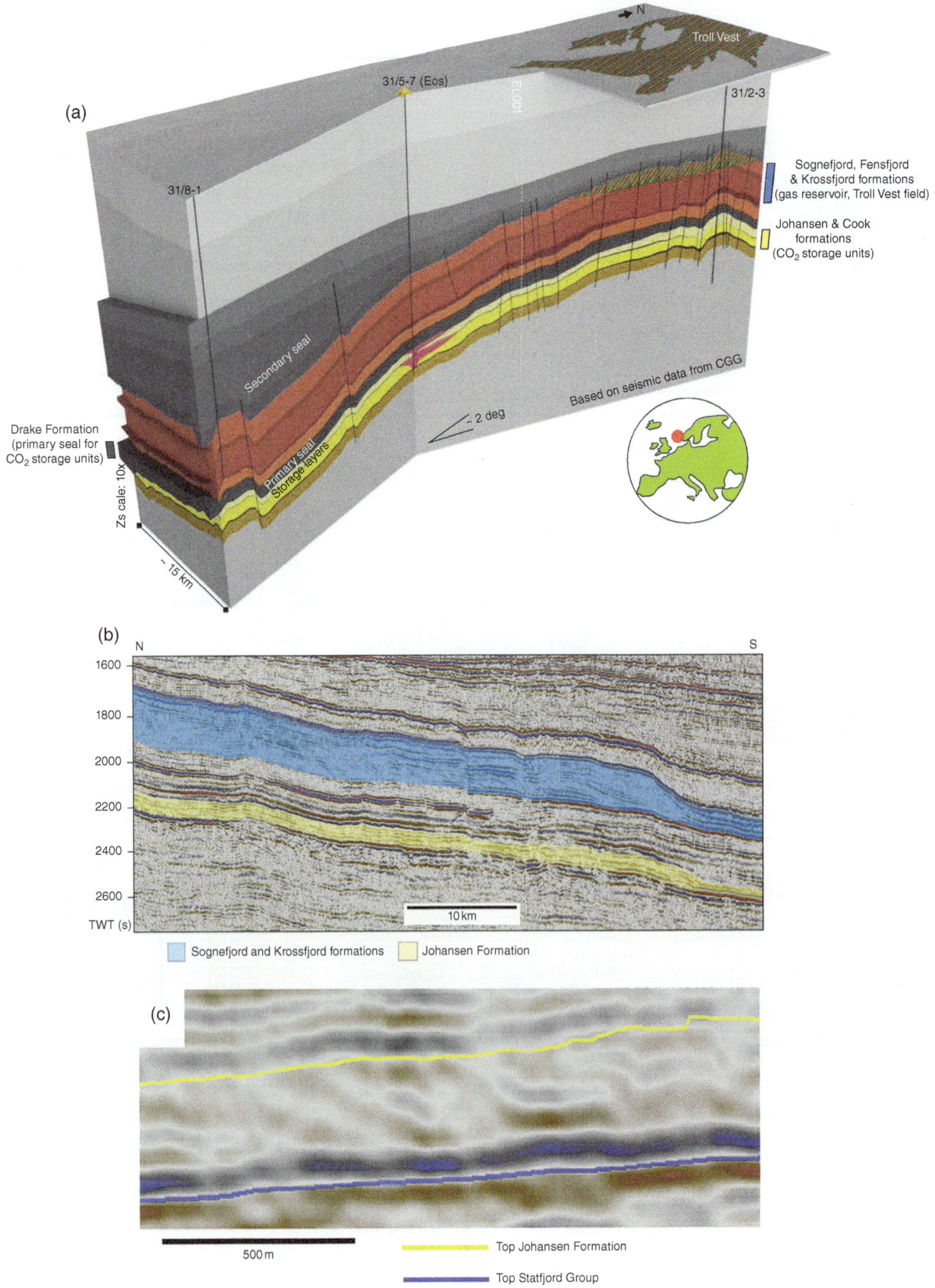

Figure 8.53 (a) Sketch of the subsurface configuration of the Johansen Formation CO_2 storage unit and overburden strata in the Northern Lights project, offshore Norway (after https://www.equinor.com/en/news/20201019-sharing-data-northern-lights.html). Predicted CO_2 plume extent after 37.5 Mt injection is illustrated in magenta. (b) Interpreted regional 2D seismic line showing the southward (down-dip) pinchout of the storage unit (shaded yellow), which (c) is locally expressed as a single clinoform set (Sundal et al. 2016 / with permission of Elsevier). (d, e) Core photos of facies in the Johansen Formation from the recently drilled Eos exploration well (31/5-7, Figure 8.53a): (d) cross-bedded sandstone with foresets that are rhythmically draped by carbonaceous material and mud (2738 m measured depth), and (e) patchily bioturbated, wavy- and lenticular-bedded sandstone and mudstone (2750 m measured depth) (Equinor ASA / https://www.equinor.com/en/news/20201019-sharing-data-northern-lights.html / Last Accessed on 11 July 2024). (f, g) Depositional models for the Johansen Formation: (f) early progradation as a wave-dominated delta, and (g) late progradation as a deflected wave-dominated delta fronted by a spit that lies seaward of a sheltered lagoon (Sundal et al. 2016 / with permission of Elsevier).

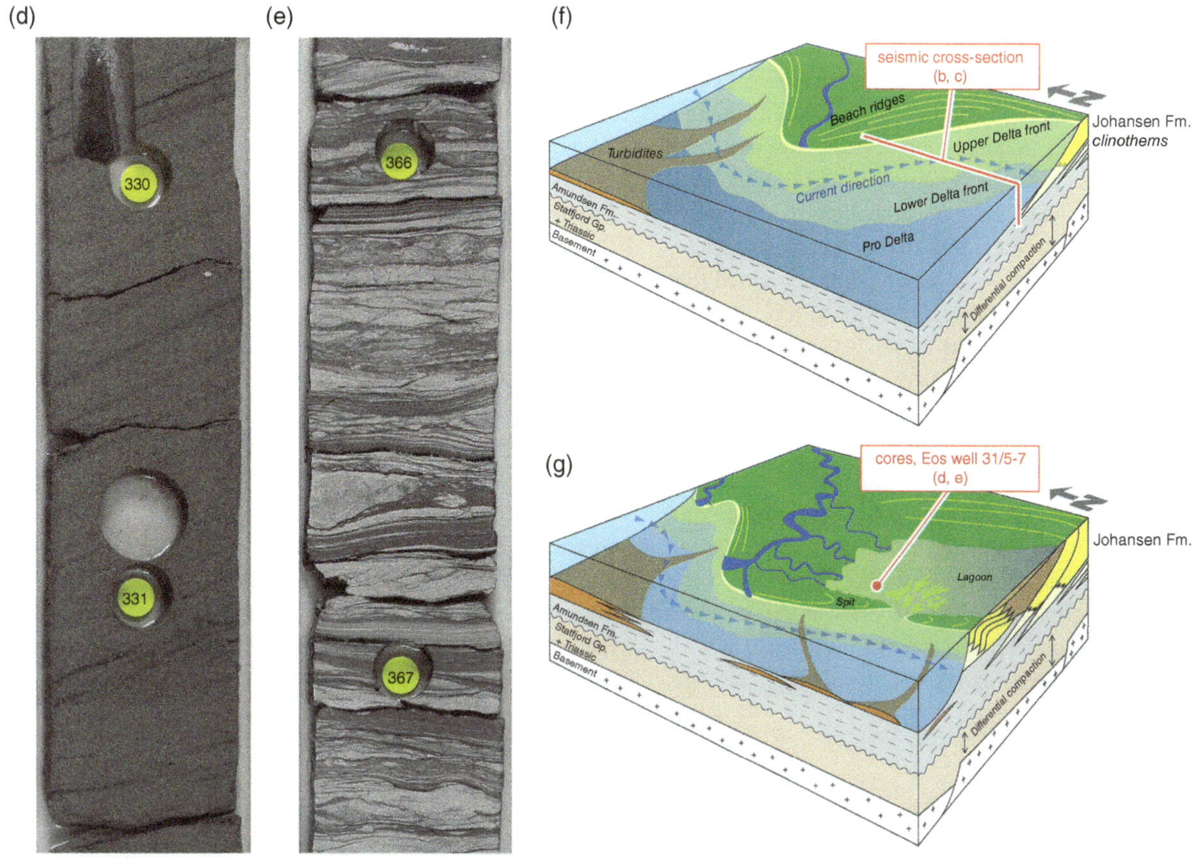

Figure 8.53 (Continued)

the regional mudstone seal of the Drake Formation (Figure 8.53a), it is envisaged that much CO_2 will be stored as a result of capillary trapping in the pore space and dissolution in the pore waters of the sandstone aquifers. These three trapping mechanisms will be enhanced by sedimentological heterogeneity in the Johansen Formation, including lateral pinchouts of parasequences (Figure 8.53b), facies interfingering, carbonate-cemented concretions and/or other heterogeneity along clinoforms in sandstone tongues (Figure 8.53c), and mudstones barriers and baffles within some sandstone facies (Figure 8.53d,e). Precipitation of carbonate minerals provides a fourth mechanism for trapping CO_2 in the Johansen Formation (Sundal and Hellevang 2019).

8.8 Future Research Directions

Further development of facies models, including characterisation and interpretation of architectural-element distributions and stratigraphic patterns, requires closer integration of observational data and numerical models. Sophisticated, physics-based hydrodynamic models can generate emergent morphodynamic behaviours and synthetic stratigraphic volumes that are directly comparable to observations of modern shoreline-shelf systems and their ancient counterparts (e.g. Van der Vegt et al. 2020; Willis et al. 2021); further iteration between observational data and modelling results is needed to test models and inform interpretation of real-world data. There is a need for continuing focus on mixed-process depositional environments, where complex interactions between different processes may occur in three dimensions, and on the transition zones between depositional environments (e.g. fluvial-marine transition zone, littoral energy fence, shelf edge), which are critical for understanding sediment routing and fluxes. Geomatic data for regional and global monitoring (e.g. Nyberg and Howell 2016; Caldwell et al. 2019; Nienhuis et al. 2020) will continue to become more widely accessible and will grow in volume. Integration of large-scale geomatic datasets will be increasingly important for the study of modern shorelines and shelves, and careful use of data science techniques will be fruitful in exploring the resulting data volumes.

Facies-based analysis of sand-rich shoreline-shelf systems will be important for a range of applications to ancient shallow-marine strata (e.g. for resource extraction and storage in the subsurface) and modern shorelines (e.g. monitoring and mitigating pollution). Again, data science and machine learning techniques have the potential to complement and extend traditional facies analysis of shallow-marine sandstones in the subsurface, including characterisation of uncertainty and integration with engineering data, provided that facies interpretations are formulated in a reproducible, quantitative form (cf. Ainsworth et al. 2018; Li and Plink-Björklund 2019; Colombera and Mountney 2020b). Sedimentological understanding will be fundamental to the engineering of natural shorelines, which will be required to protect vulnerable coastal regions and their populations from changes in sea level, climate and riverine water and sediment discharge (e.g. Syvitski et al. 2009; Dunn et al. 2019). Monitoring and modelling studies will need to be translated into sustainable engineering projects and management. This translation will require close integration with other disciplines (e.g. geomatics, geomorphology, biology, civil engineering and the social sciences) and effective communication and co-creation with all stakeholders, not least the local communities directly affected by coastal change.

Acknowledgements

We thank Bruce Ainsworth, Alessandro Amorosi, Kerrie Bann, Benjamin Brigaud, Antonio Cattaneo, Annamaria Correggiari, Shahin Dashtgard, Murray Gingras, Marcello Gugliotta, Chris Jackson, Howard Johnson, Colleen Kurcinka, Bruce Levell, Sergio Longhitano, James MacEachern, Björn Nyberg, Miquel Poyatos More, Michael Steventon and Anna Van Yperen for sharing and discussing material for figures; Luca Colombera, Marcello Gugliotta, Howard Johnson, Bruce Levell, Nigel Mountney, Piret Plink-Bjorklund, Maxime Virolle and Valentin Zuchuat for their insightful and critical reviews; and Howard Johnson and Bruce Levell for their encouragement and support.

Further Reading

Bergman, K.M. and Snedden, J.W. (ed.) (1999). *Isolated Shallow Marine Sand Bodies: Sequence Stratigraphic Analysis and Sedimentologic Interpretation*, SEPM, Special Publication, vol. 64. SEPM Society for Sedimentary Geology, 362 p.

Caldwell, R.L., Edmonds, D.A., Baumgardner, S. et al. (2019). A global delta dataset and the environmental variables that predict delta formation on marine coastlines. *Earth Surface Dynamics* 7: 773–787. Related open access dataset in supplemental data to paper: https://doi.org/10.5194/esurf-7-773-2019-supplement.

Catuneanu, O. (2022). *Principles of Sequence Stratigraphy*, 2e, 494p. Elsevier.

Clifton, H.E. (2006). A re-examination of facies models for clastic shorelines. In: *Facies Models Revisited*, SEPM, Special Publication, vol. 84 (ed. H.W. Posamentier and R.G. Walker), 293–337. SEPM Society for Sedimentary Geology.

Davis, R.A. Jr. and Dalrymple, R.W. (ed.) (2012). *Principles of Tidal Sedimentology*, 621p. Dordrecht: Springer.

Hajek, E.A. and Straub, K.M. (2017). Autogenic sedimentation in clastic stratigraphy. *Annual Review of Earth and Planetary Sciences* 45: 681–709.

Nienhuis, J.H., Ashton, A.D., Edmonds, D. A. et al. (2020). Global-scale human impact on delta morphology has led to net land area gain. *Nature* 577: 514–518. Related open access dataset: https://github.com/jhnienhuis/GlobalDeltaChange.

Nyberg, B. and Howell, J.A. (2016). Global distribution of modern shallow marine shorelines. Implications for exploration and reservoir analogue studies. *Marine and Petroleum Geology* 71: 83–104. Related open access dataset: https://dataverse.harvard.edu/dataset.xhtml?persistentId=doi:10.7910/DVN/4FUYGD.

References

Ainsworth, R.B. (1994). Marginal marine sedimentology and high resolution sequence analysis; Bearpaw–Horseshoe Canyon transition, Drumheller, Alberta. *Bulletin of Canadian Petroleum Geology* 42: 26–54.

Ainsworth, R.B., Flint, S., and Howell, J. (2008). Predicting coastal depositional style: influence of basin morphology and accommodation regime within a sequence stratigraphic framework. In: *Recent Advances in Models of Siliciclastic Shallow-Marine Stratigraphy*, SEPM, Special Publication, vol. 90 (ed. G.J. Hampson, R.J. Steel, P.M. Burgess, and R.W. Dalrymple), 237–263. SEPM Society for Sedimentary Geology.

Ainsworth, R.B., Vakarelov, B.K., and Nanson, R.A. (2011). Dynamic spatial and temporal prediction of changes in depositional processes on clastic shorelines: toward improved subsurface uncertainty reduction and management. *AAPG Bulletin* 95: 267–297.

Ainsworth, R.B., Vakarelov, B.K., Lee, C. et al. (2015). Architecture and evolution of a regressive, tide-influenced marginal marine succession, Drumheller, Alberta, Canada. *Journal of Sedimentary Research* 85: 596–625.

Ainsworth, R.B., Vakarelov, B.K., MacEachern, J.A. et al. (2016). Process-driven architectural variability in mouth-bar deposits: a case study from a mixed-process mouth-bar complex, Drumheller, Alberta, Canada. *Journal of Sedimentary Research* 86: 512–541.

Ainsworth, R.B., McArthur, J.B., Lang, S.C., and Vonk, A.J. (2018). Quantitative sequence stratigraphy. *AAPG Bulletin* 102: 1913–1939.

Allen, J.R.L. (1970). *Physical Processes of Sedimentation*, 248. London: Unwin.

Allen, J.R.L. (1982). Mud drapes in sand -wave deposits: a physical model with application to the Folkestone Beds (early Cretaceous, southeast England). *Proceedings of the Royal Society of London, Series A* 306: 291–345.

Allen, P.A. (1985). Hummocky cross-stratification is not produced purely under progressive gravity waves. *Nature* 313: 562–564.

Allen, P.A. (2017). *Sediment Routing Systems: The Fate of Sediment from Source to Sink*, 407p. Cambridge: Cambridge University Press.

Allen, G.P. and Posamentier, H.W. (1993). Sequence stratigraphy and facies model of an incised valley fill; the Gironde Estuary, France. *Journal of Sedimentary Research* 63: 378–391.

Amorosi, A., Maselli, V., and Trincardi, F. (2016). Onshore to offshore anatomy of a late Quaternary source-to-sink system (Po Plain–Adriatic Sea, Italy). *Earth-Science Reviews* 153: 212–237.

Androsov, A.A., Kagan, B.A., Romanenkov, D.A., and Voltzinger, N.E. (2002). Numerical modelling of barotropic tidal dynamics in the strait of Messina. *Advances in Water Resources* 25: 401–415.

Angus, L., Hampson, G.J., Palci, F., and Fraser, A.J. (2020). Characteristics and context of high-energy, tidally modulated, barred shoreface deposits: Kimmeridgian–Tithonian sandstones, Weald Basin, southern UK and northern France. *Journal of Sedimentary Research* 90: 313–335.

Anthony, E.J., Brunier, G., Besset, M. et al. (2015). Linking rapid erosion of the Mekong River delta to human activities. *Scientific Reports* 5: 1–12.

Anthony, E.J., Besset, M., Dussouillez, P. et al. (2019). Overview of the Monsoon-influenced Ayeyarwady River delta, and delta shoreline mobility in response to changing fluvial sediment supply. *Marine Geology* 417: 106038.

Armitage, J.J., Duller, R.A., Whittaker, A.C., and Allen, P.A. (2011). Transformation of tectonic and climatic signals from source to sedimentary archive. *Nature Geoscience* 4: 231–235.

Arnott, R.W. and Southard, J.B. (1990). Exploratory flow-duct experiments on combined-flow bed configurations, and some implications for interpreting storm-event stratification. *Journal of Sedimentary Research* 60: 211–219.

Arora, K., Goff, J.A., Wood, L.J. et al. (2018). Analysis of hummocky bedforms offshore Fire Island, New York, generated by superstorm Sandy. *Continental Shelf Research* 163: 12–22.

Baas, J.H., Best, J.L., and Peakall, J. (2016). Predicting bedforms and primary current stratification in cohesive mixtures of mud and sand. *Journal of the Geological Society* 173: 12–45.

Baas, J.H., Baker, M.L., Malarkey, J. et al. (2019). Integrating field and laboratory approaches for ripple development in mixed sand–clay–EPS. *Sedimentology* 66: 2749–2768.

Baek, Y.S., Chun, S.S., Chang, T.S., and Kim, J.K. (2018). The role of fluid mud in the formation of extensive mud sheets during summer on the Duuri macrotidal flat, west coast of Korea. *Geosciences Journal* 22: 19–32.

Bates, C.C. (1953). Rational theory of delta formation. *AAPG Bulletin* 37: 2119–2162.

Belderson, R.H., Johnson, M.A., and Kenyon, N.H. (1982). Bedforms. In: *Offshore Tidal Sands: Processes and Deposits* (ed. A.H. Stride), 27–57. London: Chapman & Hall Ltd.

Berné, S., Durrand, J., and Weber, O. (1991). Architecture of modern subtidal dunes (sand waves), Bay of Bourgneuf, France. In: *The Three-Dimensional Facies Architecture of Terrigenous Clastic Sediments, and Its Implications for Hydrocarbon Discovery and Recovery*, SEPM, Concepts and Models, vol. 3 (ed. A.D. Miall and N. Tyler), 244–261. SEPM Society for Sedimentary Geology.

Bhattacharya, J.P. (2010). Deltas. In: *Facies Models 4*, Geological Association of Canada, GEOtext, vol. 6 (ed. N.P. James and R.W. Dalrymple), 233–264. Geological Association of Canada.

Bhattacharya, J.P. (2011). Practical problems in the application of the sequence stratigraphic method and key surfaces: integrating observations from ancient fluvial–deltaic wedges with Quaternary and modelling studies. *Sedimentology* 58: 120–169.

Bhattacharya, J.P. and Giosan, L. (2003). Wave-influenced deltas: geomorphological implications for facies reconstruction. *Sedimentology* 50: 187–210.

Bhattacharya, J.P. and MacEachern, J.A. (2009). Hyperpycnal rivers and prodeltaic shelves in the Cretaceous seaway of North America. *Journal of Sedimentary Research* 79: 184–209.

Bjørkum, P.A. and Walderhaug, O. (1990). Geometrical arrangement of calcite cementation within shallow marine sandstones. *Earth-Science Reviews* 29: 145–161.

Blum, M., Martin, J., Milliken, K., and Garvin, M. (2013). Palaeovalley systems: insights from Quaternary analogs and experiments. *Earth-Science Reviews* 116: 128–169.

Boyd, R. (2010). Transgressive wave-dominated coasts. In: *Facies Models 4*, Geological Association of Canada, GEOtext, vol. 6 (ed. N.P. James and R. W. Dalrymple), 265–294. Geological Association of Canada.

Boyd, R., Dalrymple, R., and Zaitlin, B.A. (1992). Classification of clastic coastal depositional environments. *Sedimentary Geology* 80: 139–150.

Brett, C.E. and Baird, G.C. (1986). Comparative taphonomy: a key to paleoenvironmental interpretation based on fossil preservation. *Palaios* 1: 207–227.

Buatois, L.A., Santiago, N., Parra, K., and Steel, R. (2008). Animal–substrate interactions in an early Miocene wave-dominated tropical delta: delineating environmental stresses and depositional dynamics (Tacata Field, eastern Venezuela). *Journal of Sedimentary Research* 78: 458–479.

Burgess, P.M. (2016). Identifying ordered strata: evidence, methods, and meaning. *Journal of Sedimentary Research* 86: 148–167.

Burgess, P.M. and Hovius, N. (1998). Rates of delta progradation during highstands: consequences for timing of deposition in deep-marine systems. *Journal of the Geological Society* 155: 217–222.

Burgess, P.M. and Prince, G.D. (2015). Non-unique stratal geometries: implications for sequence stratigraphic interpretations. *Basin Research* 27: 351–365.

Burgess, P.M., Lammers, H., van Oosterhout, C., and Granjeon, D. (2006). Multivariate sequence stratigraphy: tackling complexity and uncertainty with stratigraphic forward modeling, multiple scenarios, and conditional frequency maps. *AAPG Bulletin* 90: 1883–1901.

Bustin, R.M. (1988). Sedimentology and characteristics of dispersed organic matter in Tertiary Niger Delta: origin of source rocks in a deltaic environment. *AAPG Bulletin* 72: 277–298.

Caldwell, R.L., Edmonds, D.A., Baumgardner, S. et al. (2019). A global delta dataset and the environmental variables that predict delta formation on marine coastlines. *Earth Surface Dynamics* 7: 773–787.

Carvajal, C.R. and Steel, R.J. (2006). Thick turbidite successions from supply-dominated shelves during sea level highstand. *Geology* 34: 665–668.

Cattaneo, A. and Steel, R.J. (2003). Transgressive deposits: a review of their variability. *Earth-Science Reviews* 62: 187–228.

Cattaneo, A., Correggiari, A., Langone, L., and Trincardi, F. (2003). The late-Holocene Gargano subaqueous delta, Adriatic shelf: sediment pathways and supply fluctuations. *Marine Geology* 193: 61–91.

Catuneanu, O. (2022). *Principles of Sequence Stratigraphy*, 2e. Elsevier, 494 p.

Catuneanu, O., Abreu, V., Bhattacharya, J.P. et al. (2009). Towards the standardization of sequence stratigraphy. *Earth-Science Reviews* 92: 1–33.

Chadwick, A.J., Lamb, M.P., Moodie, A.J. et al. (2019). Origin of a preferential avulsion node on lowland river deltas. *Geophysical Research Letters* 46: 4267–4277.

Chafetz, H.S. and Reid, A. (2000). Syndepositional shallow-water precipitation of glauconitic minerals. *Sedimentary Geology* 136: 29–42.

Chang, T.S., Flemming, B.W., Tilch, E. et al. (2006). Late Holocene stratigraphic evolution of a back-barrier tidal basin in the East Frisian Wadden Sea, southern North Sea: transgressive deposition and its preservation potential. *Facies* 52: 329–340.

Chatanantavet, P., Lamb, M.P., and Nittrouer, J.A. (2012). Backwater controls of avulsion location on deltas. *Geophysical Research Letters* 39: https://doi.org/10.1029/2011GL050197.

Cheel, R.J. (1991). Grain fabric in hummocky cross-stratified storm beds: genetic implications. *Journal of Sedimentary Petrology* 61: 102–110.

Chiarella, D. and Longhitano, S.G. (2012). Distinguishing depositional environments in shallow-water mixed, bio-siliciclastic deposits on the basis of the degree of heterolithic segregation (Gelasian, southern Italy). *Journal of Sedimentary Research* 82: 969–990.

Chumakov, N.M., Zharkov, M.A., Herman, A.B. et al. (1995). Climatic belts of the mid-Cretaceous time. *Stratigraphy and Geological Correlation* 3: 42–63.

Clifton, H.E. (1976). Wave-formed sedimentary structures – a conceptual model. In: *Beach and Nearshore Processes*, SEPM, Special Publication, vol. 24 (ed. R.A. Davis and R.L. Ethington), 126–148. SEPM Society for Sedimentary Geology.

Coe, A.L., Bosence, D.W.M., Church, K.D. et al. (2003). *The Sedimentary Record of Sea level Change*. Cambridge: Cambridge University Press, 288 p.

Coleman, J.M. and Gagliano, S.M. (1964). Cyclic sedimentation in the Mississippi River deltaic plain. *Gulf Coast Association of Geological Societies Transactions* 14: 67–80.

Coleman, J.M. and Wright, L.D. (1975). Modern river deltas: variability of processes and sand bodies. In: *Deltas: Models for Exploration* (ed. M.L. Broussard), 99–149. Houston Geological Society.

Collins, D.S., Johnson, H.D., Allison, P.A. et al. (2017). Coupled 'storm-flood' depositional model: application to the Miocene–Modern Baram Delta Province, north-west Borneo. *Sedimentology* 64: 1203–1235.

Collins, D.S., Johnson, H.D., Allison, P.A., and Damit, A.R. (2018a). Mixed process, humid-tropical, shoreline–shelf deposition and preservation: middle Miocene–Modern Baram Delta Province, northwest Borneo. *Journal of Sedimentary Research* 88: 399–430.

Collins, D.S., Avdis, A., Allison, P.A. et al. (2018b). Controls on tidal sedimentation and preservation: insights from numerical tidal modelling in the Late Oligocene–Miocene South China Sea, Southeast Asia. *Sedimentology* 65: 2468–2505.

Collins, D.S., Johnson, H.D., and Baldwin, C.T. (2020). Architecture and preservation in the fluvial to marine transition zone of a mixed-process humid-tropical delta: Middle Miocene Lambir Formation, Baram Delta Province, north-west Borneo. *Sedimentology* 67: 1–46.

Collins, D.S., Avdis, A., Wells, M.R. et al. (2021a). Prediction of shoreline–shelf depositional process regime guided by palaeotidal modelling. *Earth-Science Reviews* 223: 103827.

Collins, D.S., Nguyen, V.L., Ta, T.K.O. et al. (2021b). Sedimentary evolution of a delta-margin mangrove in Can Gio, northeastern Mekong River delta, Vietnam. *Marine Geology* 433: 106417.

Collinson, J.D. and Mountney, N.P. (2019). *Sedimentary Structures*, 4e. Edinburgh: Dunedin Academic Press, 352 p.

Colombera, L. and Mountney, N.P. (2020a). On the geological significance of clastic parasequences. *Earth-Science Reviews* 201: 103062.

Colombera, L. and Mountney, N.P. (2020b). Accommodation and sediment-supply controls on clastic parasequences: a meta-analysis. *Sedimentology* 67: 1667–1709.

Cornwall, W. (2021). Unleashing big muddy. *Science* 372: 334–337.

Corredor, F., Shaw, J.H., and Bilotti, F. (2005). Structural styles in the deep-water fold and thrust belts of the Niger Delta. *AAPG Bulletin* 89: 753–780.

Correggiari, A., Cattaneo, A., and Trincardi, F. (2005). The modern Po Delta system: lobe switching and asymmetric prodelta growth. *Marine Geology* 222: 49–74.

Covault, J.A., Normark, W.R., Romans, B.W., and Graham, S.A. (2007). Highstand fans in the California borderland: the overlooked deep-water depositional systems. *Geology* 35: 783–786.

Cross, N.E., Williams, Z.K., Jamankulov, A. et al. (2015). The dynamic behavior of shallow marine reservoirs: insights from the Pliocene of offshore North Trinidad. *AAPG Bulletin* 99: 555–583.

Cullen, A.B. (2010). Transverse segmentation of the Baram-Balabac Basin, NW Borneo: refining the model of Borneo's tectonic evolution. *Petroleum Geoscience* 16: 3–29.

Curray, J.R. (1964). Transgression and regression. In: *Papers in Marine Geology* (ed. R.L. Miller), 175–203. New York: Macmillan.

Curray, J.R., Emmel, F.J., and Crampton, P.J.S. (1969). Holocene history of a strand plain, lagoonal coast, Nayarit, Mexico. In: *Coastal Lagoons – A Symposium* (ed. A.A. Castanares and F.B. Phleger), 63–100. Mexico: Universidad Nacional Autónoma.

Dalrymple, R.W. (2010). Tidal depositional systems. In: *Facies Models 4*, Geological Association of Canada, GEOtext, vol. 6 (ed. N.P. James and R.W. Dalrymple), 201–232. Geological Association of Canada.

Dalrymple, R.W. and Choi, K. (2007). Morphologic and facies trends through the fluvial–marine transition in tide-dominated depositional systems: a schematic framework for environmental and sequence-stratigraphic interpretation. *Earth-Science Reviews* 81: 135–174.

Dalrymple, R.W., Makino, Y., and Zaitlin, B.A. (1991). Temporal and spatial patterns of rhythmite deposition on mudflats in the macrotidal Cobequid Bay-Salmon River estuary, Bay of Fundy. In: *Clastic Tidal Sedimentology*, Canadian Society of Petroleum Geologists, vol. 16 (ed. D.G. Smith, G.E. Reinson, B.A. Zaitlin, and R.A. Rahmani), 137–160. Canadian Society of Petroleum Geologists, Memoir.

Dalrymple, R.W., Zaitlin, B.A., and Boyd, R. (1992). Estuarine facies models; conceptual basis and stratigraphic implications. *Journal of Sedimentary Research* 62: 1130–1146.

Dalrymple, R.W., Kurcinka, C.E., Jablonski, B.V.J. et al. (2015). Deciphering the relative importance of fluvial and tidal processes in the fluvial–marine transition. In: *Fluvial-Tidal Sedimentology*, Developments in Sedimentology, vol. 68 (ed. P.J. Ashworth, J.L. Best, and D.R. Parsons), 3–45. Amsterdam: Elsevier.

Dashtgard, S.E., Gingras, M.K., and Butler, K.E. (2006). Sedimentology and stratigraphy of a transgressive, muddy gravel beach: Waterside Beach, Bay of Fundy, Canada. *Sedimentology* 53: 279–296.

Dashtgard, S.E., White, R.O., Butler, K.E., and Gingras, M.K. (2007). Effects of relative sea level change on the depositional character of an embayed beach, Bay of Fundy, Canada. *Marine Geology* 239: 143–161.

Dashtgard, S.E., Gingras, M.K., and MacEachern, J.A. (2009). Tidally modulated shorefaces. *Journal of Sedimentary Research* 79: 793–807.

Dashtgard, S.E., MacEachern, J.A., Frey, S.E., and Gingras, M.K. (2012). Tidal effects on the shoreface: towards a conceptual framework. *Sedimentary Geology* 279: 42–61.

Dashtgard, S.E., Vaucher, R., Yang, B., and Dalrymple, R.W. (2021). Wave-dominated to tide-dominated coastal systems: a unifying model for tidal shorefaces and refinement of the coastal-environments classification scheme. *Geoscience Canada* 48: 5–22.

Davidson-Arnott, R.G.D. and Greenwood, B. (1976). Facies relations on a barred coast, Kouchibouguac Bay, New Brunswick, Canada. In: *Beach and Nearshore Processes*, SEPM, Special Publication, vol. 24 (ed. R.A. Davis and R.L. Ethington), 149–168. SEPM Society for Sedimentary Geology.

Dean, C.D., Collins, D.S., van Cappelle, M. et al. (2019). Regional-scale paleobathymetry controlled location, but not magnitude, of tidal dynamics in the Late Cretaceous Western Interior Seaway, USA. *Geology* 47: 1083–1087.

DeCelles, P.G. and Coogan, J.C. (2006). Regional structure and kinematic history of the Sevier fold-and-thrust belt, central Utah. *Geological Society of America Bulletin* 118: 841–864.

Descroix, L., Sané, Y., Thior, M. et al. (2020). Inverse estuaries in West Africa: evidence of the rainfall recovery? *Water* 12: 647.

Dixon, J.F., Steel, R.J., and Olariu, C. (2012). Shelf-edge delta regime as a predictor of deep-water deposition. *Journal of Sedimentary Research* 82: 681–687.

Duke, W.L. (1985). Hummocky cross-stratification, tropical hurricanes, and intense winter storms. *Sedimentology* 32: 167–194.

Duke, W.L., Arnott, R.W., and Cheel, R.J. (1991). Shelf sandstones and hummocky cross-stratification; new insights on a stormy debate. *Geology* 19: 625–628.

Dumas, S., Arnott, R.W.C., and Southard, J.B. (2005). Experiments on oscillatory-flow and combined-flow bed forms: implications for interpreting parts of the shallow-marine sedimentary record. *Journal of Sedimentary Research* 75: 501–513.

Dunn, F.E., Darby, S.E., Nicholls, R.J. et al. (2019). Projections of declining fluvial sediment delivery to major deltas worldwide in response to climate change and anthropogenic stress. *Environmental Research Letters* 14: 084034.

Edmonds, D.A. and Slingerland, R.L. (2007). Mechanics of river mouth bar formation: implications for the morphodynamics of delta distributary networks. *Journal of Geophysical Research: Earth Surface* 112: F2.

Eidam, E.F., Nittrouer, C.A., Ogston, A.S. et al. (2017). Dynamic controls on shallow clinoform geometry: Mekong Delta, Vietnam. *Continental Shelf Research* 147: 165–181.

Eide, C.H., Howell, J.A., and Buckley, S.J. (2015). Sedimentology and reservoir properties of tabular and erosive offshore transition deposits in wave-dominated, shallow-marine strata: Book Cliffs, USA. *Petroleum Geoscience* 21: 55–73.

Eitner, V. (1996). Morphological and sedimentological development of a tidal inlet and its catchment area (Otzumer Balje, southern North Sea). *Journal of Coastal Research* 12: 271–293.

Elias, E.P., van der Spek, A.J., and Lazar, M. (2017). The 'Voordelta', the contiguous ebb-tidal deltas in the SW Netherlands: large-scale morphological changes and sediment budget 1965–2013; impacts of large-scale engineering. *Netherlands Journal of Geosciences* 96: 233–259.

Elliot, T. (1986). Deltas. In: *Sedimentary Environments and Facies*, 2e (ed. H.G. Reading), 113–154. Oxford: Blackwell.

Embry, A.F. (1995). Sequence boundaries and sequence hierarchies: problems and proposals. In: *Sequence Stratigraphy on the Northwest European Margin*, Norwegian Petroleum Society, Special Publication, vol. 5 (ed. R.J. Steel, V.L. Felt, E.P. Johannessen, and C. Mathieu), 1–11. Amsterdam: Elsevier.

Enge, H.D., Howell, J.A., and Buckley, S.J. (2010). The geometry and internal architecture of stream mouth bars in the Panther Tongue and the Ferron Sandstone Members, Utah, USA. *Journal of Sedimentary Research* 80: 1018–1031.

Estebenet, M.C.G., Naipauer, M., Pazos, P.J., and Valencia, V.A. (2021). U–Pb detrital zircon ages in the Lajas Formation at Portada Covunco: maximum depositional age and provenance implications for the Neuquén Basin, Argentina. *Journal of South American Earth Sciences* 110: 103325.

Fernandes, A.M., Törnqvist, T.E., Straub, K.M., and Mohrig, D. (2016). Connecting the backwater hydraulics of coastal rivers to fluvio-deltaic sedimentology and stratigraphy. *Geology* 44: 979–982.

Fisher, W.L., Brown, L.F. Jr., Scott, A.J., and McGowen, J.H. (1969). *Delta Systems in the Exploration for Oil and Gas*. Bureau of Economic Geology, University of Texas at Austin, 212 p.

Fitzgerald, D.M. and Penland, S. (1987). Backbarrier dynamics of the east Friesian Islands. *Journal of Sedimentary Research* 57: 746–754.

Flemming, B.W. (1980). Sand transport and bedform patterns on the continental shelf between Durban and Port Elizabeth (southeast African continental margin). *Sedimentary Geology* 26: 179–205.

Flemming, B.W. (2012). Siliciclastic back-barrier tidal flats. In: *Principles of Tidal Sedimentology* (ed. R.A. Davis Jr. and R.W. Dalrymple), 231–267. Dordrecht: Springer.

Franzese, J., Spalletti, L., Pérez, I.G., and Macdonald, D. (2003). Tectonic and paleoenvironmental evolution of Mesozoic sedimentary basins along the Andean foothills of Argentina (32°–54° S). *Journal of South American Earth Sciences* 16: 81–90.

Galloway, W.E. (1975). Process framework for describing the morphologic and stratigraphic evolution of deltaic depositional systems. In: *Deltas: Models for Exploration* (ed. M.L. Broussard), 87–98. Houston, TX: Houston Geological Society.

Galloway, W.E. (1989). Genetic stratigraphic sequences in basin analysis I: architecture and genesis of flooding-surface bounded depositional units. *AAPG bulletin* 73: 125–142.

Ganti, V., Chadwick, A.J., Hassenruck-Gudipati, H.J., and Lamb, M.P. (2016). Avulsion cycles and their stratigraphic signature on an experimental backwater-controlled delta. *Journal of Geophysical Research: Earth Surface* 121: 1651–1675.

Gawthorpe, R.L. and Leeder, M.R. (2000). Tectono-sedimentary evolution of active extensional basins. *Basin Research* 12: 195–218.

Gawthorpe, R.L., Hardy, S., and Ritchie, B. (2003). Numerical modelling of depositional sequences in half-graben rift basins. *Sedimentology* 50: 169–185.

Gibling, M.R. (2006). Width and thickness of fluvial channel bodies and valley fills in the geological record: a literature compilation and classification. *Journal of Sedimentary Research* 76: 731–770.

Gilbert, G.K. (1885). The topographic features of lake shores. *Annual Report of the US Geological Survey* 5: 75–123.

Gingras, M.K., MacEachern, J.A., and Dashtgard, S.E. (2011). Process ichnology and the elucidation of physico-chemical stress. *Sedimentary Geology* 237: 115–134.

Gong, C., Wang, Y., Pyles, D.R. et al. (2015). Shelf-edge trajectories and stratal stacking patterns: their sequence-stratigraphic significance and relation to styles of deep-water sedimentation and amount of deep-water sandstone. *AAPG Bulletin* 99: 1211–1243.

Goodbred, S.L. and Saito, Y. (2012). Tide-dominated deltas. In: *Principles of Tidal Sedimentology* (ed. R.A. Davis Jr. and R.W. Dalrymple), 129–149. Dordrecht: Springer.

Graham, G.H., Jackson, M.D., and Hampson, G.J. (2015). Three-dimensional modeling of clinoforms in shallow-marine reservoirs: Part 2. Impact on fluid flow and hydrocarbon recovery in fluvial-dominated deltaic reservoirs. *AAPG Bulletin* 99: 1049–1080.

Grundvåg, S.A., Jelby, M.E., Olaussen, S., and Śliwińska, K.K. (2021). The role of shelf morphology on storm-bed variability and stratigraphic architecture, Lower Cretaceous, Svalbard. *Sedimentology* 68: 196–237.

Gugliotta, M. and Saito, Y. (2019). Matching trends in channel width, sinuosity, and depth along the fluvial to marine transition zone of tide-dominated river deltas: the need for a revision of depositional and hydraulic models. *Earth-Science Reviews* 191: 93–113.

Gugliotta, M., Flint, S.S., Hodgson, D.M., and Veiga, G.D. (2015). Stratigraphic record of river-dominated crevasse subdeltas with tidal influence (Lajas Formation, Argentina). *Journal of Sedimentary Research* 85: 265–284.

Gugliotta, M., Kurcinka, C.E., Dalrymple, R.W. et al. (2016). Decoupling seasonal fluctuations in fluvial discharge from the tidal signature in ancient deltaic deposits: an example from the Neuquén Basin, Argentina. *Journal of the Geological Society* 173: 94–107.

Gugliotta, M., Saito, Y., Nguyen, V.L. et al. (2017). Process regime, salinity, morphological, and sedimentary trends along the fluvial to marine transition zone of the mixed-energy Mekong River delta, Vietnam. *Continental Shelf Research* 147: 7–26.

Gugliotta, M., Saito, Y., Ta, T.K.O. et al. (2021). Abandonment and rapid infilling of a tide-dominated distributary channel at 0.7 ka in the Mekong River Delta. *Scientific Reports* 11: 1–11.

Gugliotta, M., Saito, Y., Ta, T.K.O. et al. (2022). Late Holocene stratigraphic evolution and sedimentary facies of an active to abandoned tide-dominated distributary channel and its mouth bar. *Sedimentology* 69: 1151–1178.

Hache, I., Karius, V., and von Eynatten, H. (2021). Storm surge induced sediment accumulation on marsh islands in the southeastern North Sea: implications for coastal protection. *Estuarine, Coastal and Shelf Science* 263: 107629.

Hampson, G.J. (2010). Sediment dispersal and quantitative stratigraphic architecture across an ancient shelf. *Sedimentology* 57: 96–141.

Hampson, G.J. and Storms, J.E.A. (2003). Geomorphological and sequence stratigraphic variability in wave-dominated, shoreface-shelf parasequences. *Sedimentology* 50: 667–701.

Hampson, G.J., Rodriguez, A.B., Storms, J.E.A. et al. (2008). Geomorphology and high-resolution stratigraphy of progradational wave-dominated shoreline deposits: Impact on reservoir-scale facies architecture. In: *Recent Advances in Models of Siliciclastic Shallow-Marine Stratigraphy*, SEPM, Special Publication, vol. 90 (ed. G.J. Hampson, R.J. Steel, P.M. Burgess, and R.W. Dalrymple), 117–142. SEPM Society for Sedimentary Geology.

Hampson, G.J., Gani, M.R., Sharman, K.E. et al. (2011). Along-strike and down-dip variations in shallow-marine sequence stratigraphic architecture: Upper Cretaceous Star Point Sandstone, Wasatch Plateau, Central Utah, USA. *Journal of Sedimentary Research* 81: 159–184.

Hampson, G.J., Duller, R.A., Petter, A.L. et al. (2014). Mass-balance constraints on stratigraphic interpretation of linked alluvial–coastal–shelfal deposits from source to sink: example from Cretaceous Western Interior Basin, Utah and Colorado, USA. *Journal of Sedimentary Research* 84: 935–960.

Haq, B.U., Hardenbol, J., and Vail, P.R. (1988). Mesozoic and Cenozoic chronostratigraphy and cycles of sea level change. In: *Sea level Changes – An Integrated Approach*, SEPM, Special Publication, vol. 42 (ed. C.K. Wilgus, B. S. Hastings, C.G.S.C. Kendall, et al.), 125–154. SEPM Society for Sedimentary Geology.

Harris, P.T., Heap, A.D., Bryce, S.M. et al. (2002). Classification of Australian clastic coastal depositional environments based upon a quantitative analysis of wave, tidal, and river power. *Journal of Sedimentary Research* 72: 858–870.

Harris, A.D., Covault, J.A., Baumgardner, S. et al. (2020). Numerical modeling of icehouse and greenhouse sea level changes on a continental margin: sea level modulation of deltaic avulsion processes. *Marine and Petroleum Geology* 111: 807–814.

Harris, P.T., Westerveld, L., Nyberg, B. et al. (2021). Exposure of coastal environments to river-sourced plastic pollution. *Science of the Total Environment* 769: 145222.

Hartley, A.J., Weissmann, G.S., and Scuderi, L. (2017). Controls on the apex location of large deltas. *Journal of the Geological Society* 174: 10–13.

Hayes, M.O. (1975). Morphology of sand accumulation in estuaries: an introduction to the symposium. In: *Estuarine Research*, Geology and Engineering, vol. 2 (ed. L.E. Cronin), 3–22. London: Academic Press.

Hayes, M.O. (1979). Barrier island morphology as a function of tidal and wave regime. In: *Barrier Islands – from the Gulf of St. Lawrence to the Gulf of Mexico* (ed. S.P. Leatherman), 1–27. New York: Academic Press.

Helland-Hansen, W. and Hampson, G.J. (2009). Trajectory analysis: concepts and applications. *Basin Research* 21: 454–483.

Helland-Hansen, W. and Martinsen, O.J. (1996). Shoreline trajectories and sequences; description of variable depositional-dip scenarios. *Journal of Sedimentary Research* 66: 670–688.

Helland-Hansen, W., Steel, R.J., and Sømme, T.O. (2012). Shelf genesis revisited. *Journal of Sedimentary Research* 82: 133–148.

Hesselbo, S.P., Deconinck, J.F., Huggett, J.M., and Morgans-Bell, H.S. (2009). Late Jurassic palaeoclimatic change from clay mineralogy and gamma-ray spectrometry of the Kimmeridge Clay, Dorset, UK. *Journal of the Geological Society* 166: 1123–1133.

Holbrook, J.M. and Bhattacharya, J.P. (2012). Reappraisal of the sequence boundary in time and space: case and considerations for an SU (subaerial unconformity) that is not a sediment bypass surface, a time barrier, or an unconformity. *Earth-Science Reviews* 113: 271–302.

Houbolt, J.J.H.C. (1968). Recent sediments in the southern bight of the North Sea. *Geologie en Mijnbouw* 47: 245–273.

Howard, J.D. and Frey, R.W. (1984). Characteristic trace fossils in nearshore to offshore sequences, Upper

Cretaceous of east-central Utah. *Canadian Journal of Earth Sciences* 21: 200–219.

Howard, J.D. and Reineck, H.E. (1981). Depositional facies of high-energy beach-to-offshore sequence: comparison with low-energy sequence. *AAPG Bulletin* 65: 807–830.

Howell, J.A., Skorstad, A., MacDonald, A. et al. (2008). Sedimentological parameterization of shallow-marine reservoirs. *Petroleum Geoscience* 14: 17–34.

Howell, J.A., Martinius, A.W., and Good, T.R. (2014). The application of outcrop analogues in geological modelling: a review, present status and future outlook. In: *Sediment-body Geometry and Heterogeneity: Analogue Studies for Modelling the Subsurface*, Geological Society of London, Special Publication, vol. 387 (ed. J.A. Howell, A.W. Martinius, and T.R. Good), 1–25. Geological Society of London.

Hoyt, J.H. and Henry, V.J. (1967). Influence of island migration on barrier-island sedimentation. *Geological Society of America Bulletin* 78: 77–86.

Hunter, R.E., Clifton, H.E., and Phillips, R.L. (1979). Depositional processes, sedimentary structures, and predicted vertical sequences in barred nearshore systems, southern Oregon coast. *Journal of Sedimentary Research* 49: 711–726.

Huysmans, M., Peeters, L., Moermans, G., and Dassargues, A. (2008). Relating small-scale sedimentary structures and permeability in a cross-bedded aquifer. *Journal of Hydrology* 361: 41–51.

Isla, M.F., Schwarz, E., and Veiga, G.D. (2020). Record of a nonbarred clastic shoreline. *Geology* 48: 338–342.

Jaeger, J.M. and Nittrouer, C.A. (1995). Tidal controls on the formation of fine-scale sedimentary strata near the Amazon River mouth. *Marine Geology* 125: 259–281.

Jaramillo, S., Sheremet, A., Allison, M.A. et al. (2009). Wave-mud interactions over the muddy Atchafalaya subaqueous clinoform, Louisiana, United States: wave-supported sediment transport. *Journal of Geophysical Research: Oceans* 114: C04002.

Jelby, M.E., Grundvåg, S.A., Helland-Hansen, W. et al. (2020). Tempestite facies variability and storm-depositional processes across a wide ramp: towards a polygenetic model for hummocky cross-stratification. *Sedimentology* 67: 742–781.

Jerolmack, D.J. (2009). Conceptual framework for assessing the response of delta channel networks to Holocene sea level rise. *Quaternary Science Reviews* 28: 1786–1800.

Jerolmack, D.J. and Paola, C. (2010). Shredding of environmental signals by sediment transport. *Geophysical Research Letters* 37: L19401.

Jerolmack, D.J. and Swenson, J.B. (2007). Scaling relationships and evolution of distributary networks on wave-influenced deltas. *Geophysical Research Letters* 34: L23402.

Johnson, H.D. and Baldwin, C.T. (1996). Shallow siliciclastic seas. In: *Sedimentary Environments and Facies*, 3rde (ed. H.G. Reading), 232–280. Oxford: Blackwell.

Jordan, O.D. and Mountney, N.P. (2012). Sequence stratigraphic evolution and cyclicity of an ancient coastal desert system: the Pennsylvanian–Permian lower Cutler beds, Paradox Basin, Utah, USA. *Journal of Sedimentary Research* 82: 755–780.

Joseph, P., Hu, L.Y., Dubrule, O. et al. (1993). The Roda Deltaic Complex (Spain): from sedimentology to reservoir stochastic modelling. In: *Subsurface Reservoir Characterization from Outcrop Observations* (ed. R. Eschard and B. Doligez), 97–109. Paris: Editions Technip.

Kamola, D.L. and Van Wagoner, J.C. (1995). Stratigraphy and facies architecture of parasequences with examples from the Spring Canyon Member, Blackhawk Formation, Utah. In: *Sequence Stratigraphy of Foreland Basin Deposits: Outcrop and Subsurface Examples from the Cretaceous of North America*, AAPG, Memoir, vol. 64 (ed. J.C. Van Wagoner and G.T. Bertram), 27–54. American Association of Petroleum Geologists.

Karim, M.F. and Mimura, N. (2008). Impacts of climate change and sea level rise on cyclonic storm surge floods in Bangladesh. *Global Environmental Change* 18: 490–500.

Kauffman, E.G. and Caldwell, W.G.E. (1993). The Western Interior Basin in space and time. In: *Evolution of the Western Interior Basin*, Geological Association of Canada, Special Paper, vol. 39 (ed. W.G.E. Caldwell and E.G. Kauffman), 1–30. Geological Association of Canada.

Kidwell, S.M., Fuersich, F.T., and Aigner, T. (1986). Conceptual framework for the analysis and classification of fossil concentrations. *Palaios* 1: 228–238.

Kim, W., Paola, C., Swenson, J.B., and Voller, V.R. (2006). Shoreline response to autogenic processes of sediment storage and release in the fluvial system. *Journal of Geophysical Research: Earth Surface* 111: F01003.

Komar, P.D. (1976). *Beach Processes and Sedimentation*, 429p. NJ: Prentice Hall.

Krystinik, L.F. and DeJarnett, B.B. (1995). Lateral variability of sequence stratigraphic framework in the Campanian and Lower Maastrichtian of the Western Interior Seaway. In: *Sequence Stratigraphy of Foreland Basin Deposits: Outcrop and Subsurface Examples from the Cretaceous of North America*, AAPG, Memoir, vol. 64 (ed. J.C. Van Wagoner and G.T. Bertram), 11–26. American Association of Petroleum Geologists.

Kurcinka, C., Dalrymple, R.W., and Gugliotta, M. (2018). Facies and architecture of river-dominated to tide-influenced mouth bars in the lower Lajas Formation (Jurassic), Argentina. *AAPG Bulletin* 102: 885–912.

La Croix, A.D. and Dashtgard, S.E. (2015). A synthesis of depositional trends in intertidal and upper subtidal sediments across the tidal–fluvial transition in the Fraser River, Canada. *Journal of Sedimentary Research* 85: 683–698.

Lamb, M.P., Nittrouer, J.A., Mohrig, D., and Shaw, J. (2012). Backwater and river plume controls on scour upstream of river mouths: implications for fluvio-deltaic morphodynamics. *Journal of*

Geophysical Research: Earth Surface 117 (F1): F01002.

Lambiase, J.J. and Cullen, A.B. (2013). Sediment supply systems of the Champion 'Delta' of NW Borneo: implications for deepwater reservoir sandstones. *Journal of Asian Earth Sciences* 76: 356–371.

Lambiase, J.J., Rahim, A.A.b.A., and Peng, C.Y. (2002). Facies distribution and sedimentary processes on the modern Baram Delta: implications for the reservoir sandstones of NW Borneo. *Marine and Petroleum Geology* 19: 69–78.

Lambiase, J.J., Damit, A.R., Simmons, M. D. et al. (2003). A depositional model and the stratigraphic development of modern and ancient tide-dominated deltas in NW Borneo. In: *Tropical Deltas of Southeast Asia – Sedimentology, Stratigraphy, and Petroleum Geology*, SEPM, Special Publication, vol. 76 (ed. F.H. Sidi, D. Nummedal, P. Imbert, et al.), 109–123. SEPM Society for Sedimentary Geology.

Larue, D.K. and Legarre, H. (2004). Flow units, connectivity, and reservoir characterization in a wave-dominated deltaic reservoir: Meren reservoir, Nigeria. *AAPG Bulletin* 88: 303–324.

Legler, B., Hampson, G.J., Jackson, C.A. et al. (2014). Facies relationships and stratigraphic architecture of distal, mixed tide-and wave-influenced deltaic deposits: lower Sego sandstone, western Colorado, USA. *Journal of Sedimentary Research* 84: 605–625.

Leithold, E.L. and Hope, R.S. (1999). Deposition and modification of a flood layer on the northern California shelf: lessons from and about the fate of terrestrial particulate organic carbon. *Marine Geology* 154: 183–195.

Lentsch, N., Finotello, A., and Paola, C. (2018). Reduction of deltaic channel mobility by tidal action under rising relative sea level. *Geology* 46: 599–602.

Leonardi, N., Mei, X., Carnacina, I., and Dai, Z. (2021). Marine sediment sustains the accretion of a mixed fluvial-tidal delta. *Marine Geology* 438: 106520.

Leren, B.L., Howell, J., Enge, H., and Martinius, A.W. (2010). Controls on stratigraphic architecture in contemporaneous delta systems from the Eocene Roda Sandstone, Tremp-Graus Basin, northern Spain. *Sedimentary Geology* 229: 9–40.

Lericolais, G., Berné, S., and Féniès, H. (2001). Seaward pinching out and internal stratigraphy of the Gironde incised valley on the shelf (Bay of Biscay). *Marine Geology* 175: 183–197.

Li, M.Z. and King, E.L. (2007). Multibeam bathymetric investigations of the morphology of sand ridges and associated bedforms and their relation to storm processes, Sable Island Bank, Scotian Shelf. *Marine Geology* 243: 200–228.

Li, H. and Plink-Björklund, P. (2019). Applying information theory and Bayesian inference to paleoenvironmental interpretation. *Geophysical Research Letters* 46: 14477–14485.

Li, Q., Yu, L., and Straub, K.M. (2016). Storage thresholds for relative sea level signals in the stratigraphic record. *Geology* 44: 179–182.

Longhitano, S.G. (2018a). Between Scylla and Charybdis (part 1): the sedimentary dynamics of the modern Messina Strait (central Mediterranean) as analogue to interpret the past. *Earth-Science Reviews* 185: 259–287.

Longhitano, S.G. (2018b). Between Scylla and Charybdis (part 2): the sedimentary dynamics of the ancient, Early Pleistocene Messina Strait (central Mediterranean) based on its modern analogue. *Earth-Science Reviews* 179: 248–286.

Longhitano, S.G. and Steel, R.J. (2017). Deflection of the progradational axis and asymmetry in tidal seaway and strait deltas: insights from two outcrop case studies. In: *Sedimentology of Paralic Reservoirs: Recent Advances*, Geological Society of London, Special Publication, vol. 444 (ed. G.J. Hampson, A.D. Reynolds, B. Kostic, and M.R. Wells), 141–172. Geological Society of London.

López-Blanco, M. (1996). Estratigrafía secuencial de sistemas deltaicos en cuencas de antepaís: ejemplos de Sant Llorenç del Munt, Montserrat y Roda (Paleógeno: cuenca de antepaís surpirenaica). PhD thesis. Universidad de Barcelona, Barcelona, Spain, 238 p.

López-Blanco, M., Marzo, M., and Munoz, J.A. (2003). Low-amplitude, synsedimentary folding of a deltaic complex: Roda Sandstone (lower Eocene), South-Pyrenean Foreland Basin. *Basin Research* 15: 73–95.

Mabrouk, M.B., Jonoski, A., Solomatine, D., and Uhlenbrook, S. (2013). A review of seawater intrusion in the Nile Delta groundwater system–the basis for assessing impacts due to climate changes and water resources development. *Hydrology and Earth System Sciences Discussions* 10: 10873–10911.

MacDonald, D.I.M. and Jefferson, T.H. (1985). Orientation studies of waterlogged wood; a paleocurrent indicator? *Journal of Sedimentary Research* 55: 235–239.

MacEachern, J.A. and Bann, K.L. (2008). The role of ichnology in refining shallow marine facies models. In: *Recent Advances in Models of Siliciclastic Shallow-Marine Stratigraphy*, SEPM, Special Publication, vol. 90 (ed. G.J. Hampson, R.J. Steel, P.M. Burgess, and R.W. Dalrymple), 73–116. SEPM Society for Sedimentary Geology.

MacEachern, J.A., Raychaudhuri, I., and Pemberton, S.G. (1992). Stratigraphic applications of the *Glossifungites* ichnofacies: delineating discontinuities in the rock record. In: *Applications of Ichnology to Petroleum Exploration*, SEPM, Core Workshop, vol. 17 (ed. S.G. Pemberton), 169–198. SEPM Society for Sedimentary Geology.

Mack, G.H. and Jerzykiewicz, T. (1989). Provenance of post-Wapiabi sandstones and its implications for Campanian to Paleocene tectonic history of the southern Canadian Cordillera. *Canadian Journal of Earth Sciences* 26: 665–676.

Macquaker, J.H., Taylor, K.G., Young, T.P., and Curtis, C.D. (1996). Sedimentological and geochemical controls on ooidal ironstone and 'bone-bed' formation and some comments on their sequence-stratigraphical significance. In: *Sequence Stratigraphy in British Geology*, Geological Society of London, Special Publication, vol. 103 (ed. S.P. Hesselbo and D.N. Parkinson), 97–107. Geological Society of London.

Macquaker, J.H., Bentley, S.J., and Bohacs, K.M. (2010). Wave-enhanced sediment-gravity flows and mud dispersal across continental shelves: reappraising sediment transport processes operating in ancient mudstone successions. *Geology* 38: 947–950.

Madof, A.S., Harris, A.D., and Connell, S.D. (2016). Nearshore along-strike variability: is the concept of the systems tract unhinged? *Geology* 44: 315–318.

Mannie, A.S., Jackson, C.A.-L., and Hampson, G.J. (2014). Shallow-marine reservoir development in extensional diapir-collapse minibasins: an integrated subsurface case study from the Upper Jurassic of the Cod Terrace, Norwegian North Sea. *AAPG Bulletin* 98: 2019–2055.

Martin, J.M., Abreu, V., Neal, J.E., and Pratson, L. (2010). Exercise 13.1 – Application of sequence-stratigraphic methods to an experimental basin fill. In: *Sequence Stratigraphy of Siliciclastic Systems – the ExxonMobil Methodology*, SEPM, Concepts in Sedimentology and Paleontology, vol. 9 (ed. V. Abreu, J.E. Neal, K.M. Bohacs, and J.L. Kalbas), 199–208. SEPM Society for Sedimentary Geology.

Martin, J., Fernandes, A.M., Pickering, J. et al. (2018). The stratigraphically preserved signature of persistent backwater dynamics in a large paleodelta system: The Mungaroo Formation, North West Shelf, Australia. *Journal of Sedimentary Research* 88: 850–872.

Martinez, M.A., Quattrocchio, M.E., and Zavala, C.A. (2002). Análisis palinofacial de la Formación Lajas (Jurásico Medio), Cuenca Neuquina, Argentina: significado paleoambiental y paleoclimático. *Revista Española de Micropaleontologia* 34: 81–104.

Martinius, A.W. (2017). Multiscale Gilbert-type delta lobe architecture and heterogeneities: the case of the Roda Sandstone Member. *AAPG Bulletin* 101: 453–463.

Martinius, A.W. and Gowland, S. (2011). Tide-influenced fluvial bedforms and tidal bore deposits (late Jurassic Lourinhã Formation, Lusitanian Basin, Western Portugal). *Sedimentology* 58: 285–324.

Martinius, A.W. and Molenaar, N. (1991). A coral-mollusc (Goniaraea–Crassatella) dominated hardground community in a siliciclastic-carbonate sandstone (the Lower Eocene Roda Formation, southern Pyrenees, Spain). *Palaios* 6: 142–155.

McIlroy, D., Flint, S., Howell, J.A., and Timms, N. (2005). Sedimentology of the tide-dominated Jurassic Lajas Formation, Neuquén Basin, Argentina. In: *The Neuquén Basin, Argentina: A Case Study in Sequence Stratigraphy and Basin Dynamics*, Geological Society of London, Special Publication, vol. 252 (ed. G.D. Veiga, L.A. Spalletti, J.A. Howell, and E. Schwarz), 83–107. Geological Society of London.

Meldahl, K.H. and Flessa, K.W. (1990). Taphonomic pathways and comparative biofacies and taphofacies in a recent intertidal/shallow shelf environment. *Lethaia* 23: 43–60.

Mellere, D. and Steel, R. (2000). Style contrast between forced regressive and lowstand/transgressive wedges in the Campanian of south-central Wyoming (Hatfield Member of the Haystack Mountains Formation). In: *Sedimentary Responses to Forded Regressions*, Geological Society of London, Special Publication, vol. 172 (ed. D. Hunt and R. L. Gawthorpe), 141–162. Geological Society of London.

Miall, A.D. (1992). Exxon global cycle chart: an event for every occasion? *Geology* 20: 787–790.

Miller, K.G., Kominz, M.A., Browning, J. V. et al. (2005). The Phanerozoic record of global sea level change. *Science* 310: 1293–1298.

Miller, K.G., Mountain, G.S., Browning, J. V. et al. (2013). Testing sequence stratigraphic models by drilling Miocene foresets on the New Jersey shallow shelf. *Geosphere* 9: 1236–1256.

Milliman, J.D. and Meade, R.H. (1983). World-wide delivery of river sediment to the oceans. *Journal of Geology* 91: 1–21.

Moodie, A.J., Nittrouer, J.A., Ma, H. et al. (2019). Modeling deltaic lobe-building cycles and channel avulsions for the Yellow River delta, China. *Journal of Geophysical Research: Earth Surface* 124: 2438–2462.

Morton, R.A. (1981). Formation of storm deposits by wind-forced currents in the Gulf of Mexico and the North Sea. In: *Holocene Marine Sedimentation in the North Sea Basin*, International Association of Sedimentologists, Special Publication, vol. 5 (ed. S.-D. Nio, R.T.E. Shüttenhelm, and T.C.E. Van Weering), 385–396. Wiley.

Mosetti, F. (1992). Considerazioni sulla idrodinamica nello Stretto di Messina. *Rendiconti Lincei* 3: 269–281.

Moslow, T.F. and Tye, R.S. (1985). Recognition and characterization of Holocene tidal inlet sequences. *Marine Geology* 63: 129–151.

Mulder, T. and Syvitski, J.P. (1995). Turbidity currents generated at river mouths during exceptional discharges to the world oceans. *Journal of Geology* 103: 285–299.

Mulder, T., Syvitski, J.P., Migeon, S. et al. (2003). Marine hyperpycnal flows: initiation, behavior and related deposits. A review. *Marine and Petroleum Geology* 20: 861–882.

Muto, T. and Steel, R.J. (1992). Retreat of the front in a prograding delta. *Geology* 20: 967–970.

Muto, T. and Steel, R.J. (1997). Principles of regression and transgression; the nature of the interplay between accommodation and sediment supply. *Journal of Sedimentary Research* 67: 994–1000.

Muto, T., Steel, R.J., and Swenson, J.B. (2007). Autostratigraphy: a framework norm for genetic stratigraphy. *Journal of Sedimentary Research* 77: 2–12.

Muto, T., Steel, R.J., and Burgess, P.M. (2016). Contributions to sequence stratigraphy from analogue and numerical experiments. *Journal of the Geological Society* 173: 837–844.

Nanson, R.A., Vakarelov, B.K., Ainsworth, R.B. et al. (2013). Evolution of a Holocene, mixed-process, forced regressive shoreline: the Mitchell River delta, Queensland, Australia. *Marine Geology* 339: 22–43.

Neal, J. and Abreu, V. (2009). Sequence stratigraphy hierarchy and the accommodation succession method. *Geology* 37: 779–782.

Neal, J.E., Abreu, V., Bohacs, K.M. et al. (2016). Accommodation succession (δA/δS) sequence stratigraphy: observational method, utility and insights into sequence boundary formation. *Journal of the Geological Society* 173: 803–816.

Nesteroff, W.D. and Rawson, M. (1987). Dynamic of modern sediments in the Strait of Messina (Sicily) and South-West of Calabria. *Documents et travaux de l'Institut géologique Albert de Lapparent* 11: 211–223.

Nienhuis, J.H., Ashton, A.D., Edmonds, D.A. et al. (2020). Global-scale human impact on delta morphology has led to net land area gain. *Nature* 577: 514–518.

Nishida, N., Ajioka, T., Ikehara, K. et al. (2020). Postglacial stratigraphic evolution of a current-influenced sandy shelf: offshore Kujukuri strandplain, central Japan. *Sedimentology* 67: 559–575.

Nordahl, K. and Ringrose, P.S. (2008). Identifying the representative elementary volume for permeability in heterolithic deposits using numerical rock models. *Mathematical Geosciences* 40: 753–771.

Nowacki, D.J., Ogston, A.S., Nittrouer, C.A. et al. (2015). Sediment dynamics in the lower Mekong River: transition from tidal river to estuary. *Journal of Geophysical Research: Oceans* 120: 6363–6383.

Nyberg, B. and Howell, J.A. (2016). Global distribution of modern shallow marine shorelines. Implications for exploration and reservoir analogue studies. *Marine and Petroleum Geology* 71: 83–104.

Odin, G.S. and Matter, A. (1981). De glauconiarum origine. *Sedimentology* 28: 611–641.

Olariu, C. and Bhattacharya, J.P. (2006). Terminal distributary channels and delta front architecture of river-dominated delta systems. *Journal of Sedimentary Research* 76: 212–233.

Olariu, M.I. and Olariu, C. (2015). Ubiquity of wave-dominated deltas in outer-shelf growth-faulted compartments. *Journal of Sedimentary Research* 85: 768–779.

Olariu, C., Steel, R.J., and Petter, A.L. (2010). Delta-front hyperpycnal bed geometry and implications for reservoir modeling: Cretaceous Panther Tongue delta, Book Cliffs, Utah. *AAPG Bulletin* 94: 819–845.

Olariu, C., Steel, R.J., Dalrymple, R.W., and Gingras, M.K. (2012a). Tidal dunes versus tidal bars: the sedimentological and architectural characteristics of compound dunes in a tidal seaway, the lower Baronia Sandstone (Lower Eocene), Ager Basin, Spain. *Sedimentary Geology* 279: 134–155.

Olariu, M.I., Olariu, C., Steel, R.J. et al. (2012b). Anatomy of a laterally migrating tidal bar in front of a delta system: Esdolomada Member, Roda Formation, Tremp-Graus Basin, Spain. *Sedimentology* 59: 356–378.

Onuoha, K.M. (1999). Structural features of Nigeria's coastal margin: an assessment base of age data from wells. *Journal of African Earth Sciences* 29: 485–499.

Onyenanu, G.S.I., Jacquemyn, C.E., Graham, G.H. et al. (2018). Geometry, distribution and fill of erosional scours in a heterolithic, distal lower shoreface sandstone reservoir analogue: Grassy Member, Blackhawk Formation, Book Cliffs, Utah, USA. *Sedimentology* 65: 1731–1760.

Orton, G.J. and Reading, H.G. (1993). Variability of deltaic processes in terms of sediment supply, with particular emphasis on grain size. *Sedimentology* 40: 475–512.

Paola, C. and Mohrig, D. (1996). Palaeohydraulics revisited: palaeoslope estimation in coarse-grained braided rivers. *Basin Research* 8: 243–254.

Paola, C., Straub, K., Mohrig, D., and Reinhardt, L. (2009). The 'unreasonable effectiveness' of stratigraphic and geomorphic experiments. *Earth-Science Reviews* 97: 1–43.

Patruno, S. and Helland-Hansen, W. (2018). Clinoforms and clinoform systems: review and dynamic classification scheme for shorelines, subaqueous deltas, shelf edges and continental margins. *Earth-Science Reviews* 185: 202–233.

Patruno, S., Hampson, G.J., and Jackson, C.A. (2015a). Quantitative characterisation of deltaic and subaqueous clinoforms. *Earth-Science Reviews* 142: 79–119.

Patruno, S., Hampson, G.J., Jackson, C.A. L., and Whipp, P.S. (2015b). Quantitative progradation dynamics and stratigraphic architecture of ancient shallow-marine clinoform sets: a new method and its application to the Upper Jurassic Sognefjord Formation, Troll Field, offshore Norway. *Basin Research* 27: 412–452.

Patruno, S., Hampson, G.J., Jackson, C.A. L., and Dreyer, T. (2015c). Clinoform geometry, geomorphology, facies character and stratigraphic architecture of a sand-rich subaqueous delta: Jurassic Sognefjord Formation, offshore Norway. *Sedimentology* 62: 350–388.

Paumard, V., Bourget, J., Payenberg, T. et al. (2020). Controls on deep-water sand delivery beyond the shelf edge: accommodation, sediment supply, and deltaic process regime. *Journal of Sedimentary Research* 90: 104–130.

Peters, S.E. and Loss, D.P. (2012). Storm and fair-weather wave base: a relevant distinction? *Geology* 40: 511–514.

Pettit, B.S., Blum, M., Pecha, M. et al. (2019). Detrital-zircon U–Pb paleodrainage reconstruction and geochronology of the Campanian Blackhawk–Castlegate succession, Wasatch Plateau and Book Cliffs, Utah, USA. *Journal of Sedimentary Research* 89: 273–292.

Plint, A.G. 1988. Sharp-based shoreface sequences and 'offshore bars' in the Cardium formation of Alberta: their relationship to relative changes in sea level, *in* Wilgus, C.K., Hastings, B.S., Kendall, C.G.S.C., Posamentier, H.W., Ross, C.A. & Van Wagoner, J.C., eds., *Sea Level Changes – An Integrated Approach.*Society of Economic Paleontologists and Mineralogists SEPM, Special Publication 42, p. 357–370.

Plint, A.G. (2010). Wave- and storm-dominated shoreline and shallow-marine systems. In: *Facies Models 4,*

Geological Association of Canada, GEOtext, vol. 6 (ed. N.P. James and R.W. Dalrymple), 167–200. Geological Association of Canada.

Plint, A.G. and Nummedal, D. (2000). The falling stage systems tract: recognition and importance in sequence stratigraphic analysis. In: *Sedimentary Responses to Forded Regressions*, Geological Society of London, Special Publication, vol. 172 (ed. D. Hunt and R.L. Gawthorpe), 1–17. Geological Society of London.

Plint, A.G., MacQuaker, J.H., and Varban, B.L. (2012). Bedload transport of mud across a wide, storm-influenced ramp: Cenomanian–Turonian Kaskapau Formation, Western Canada Foreland Basin. *Journal of Sedimentary Research* 82: 801–822.

Possemiers, M., Huysmans, M., Peeters, L. et al. (2012). Relationship between sedimentary features and permeability at different scales in the Brussels Sands. *Geologica Belgica* 15: 156–164.

Postma, G. (1990). Depositional architecture and facies of river and fan deltas: a synthesis. In: *Coarse-Grained Deltas*, International Association of Sedimentologists, Special Publication, vol. 10 (ed. A. Colella and D.B. Prior), 13–27. Wiley.

Prince, G.D. and Burgess, P.M. (2013). Numerical modeling of falling-stage topset aggradation: implications for distinguishing between forced and unforced regressions in the geological record. *Journal of Sedimentary Research* 83: 767–781.

Proust, J.N., Mahieux, G., and Tessier, B. (2001). Field and seismic images of sharp-based shoreface deposits: Implications for sequence stratigraphic analysis. *Journal of Sedimentary Research* 71: 944–957.

Pujols, E.J., López, J.L., Stockli, D.F. et al. (2018). New insights into the stratigraphic and structural evolution of the middle Jurassic South Neuquén Basin from detrital zircon (U-Th)/(He-Pb) and apatite (U-Th)/He ages. *Basin Research* 30: 1280–1297.

Quin, J.G. (2011). Is most hummocky cross-stratification formed by large-scale ripples? *Sedimentology* 58: 1414–1433.

Quinney, A., Therrien, F., Zelenitsky, D.K., and Eberth, D.A. (2013). Palaeoenvironmental and palaeoclimatic reconstruction of the Upper Cretaceous (late Campanian–early Maastrichtian) Horseshoe Canyon Formation, Alberta, Canada. *Palaeogeography, Palaeoclimatology, Palaeoecology* 371: 26–44.

Ravnås, R. and Steel, R.J. (1998). Architecture of marine rift-basin successions. *AAPG Bulletin* 82: 110–146.

Ray, D.C., van Buchem, F.S., Baines, G. et al. (2019). The magnitude and cause of short-term eustatic Cretaceous sea level change: a synthesis. *Earth-Science Reviews* 197: 102901.

Reading, H.G. and Orton, G.J. (1991). Sediment calibre: a control on facies models with special reference to deep-sea depositional systems. In: *Controversies in Modern Geology* (ed. D.W. Müller, J.A. McKenzie, and H. Weissert), 85–111. London: Academic Press.

Reineck, H.E. and Singh, I.B. (1973). *Depositional Sedimentary Environments*. Berlin, Heidelberg: Springer, 454 p.

Reineck, H.E. and Wunderlich, F. (1968). Classification and origin of flaser and lenticular bedding. *Sedimentology* 11: 99–104.

Reynolds, A.D. (1996). Paralic successions. In: *Sequence Stratigraphy* (ed. D. Emery and K. Myers), 134–177. Oxford: Blackwell.

Reynolds, A.D. (2017). Paralic reservoirs. In: *Sedimentology of Paralic Reservoirs: Recent Advances*, Geological Society of London, Special Publication, vol. 444 (ed. G.J. Hampson, A.D. Reynolds, B. Kostic, and M.R. Wells), 7–34. Geological Society of London.

Reynolds, A.D. (2022). Variability of fluvially-dominated, fine-grained, shallow-water deltas. *Sedimentology* 69: 2779–2813.

Riboulot, V., Cattaneo, A., Berné, S. et al. (2012). Geometry and chronology of late Quaternary depositional sequences in the Eastern Niger Submarine Delta. *Marine Geology* 319: 1–20.

Rich, J.L. (1951). Three critical environments of deposition, and criteria for recognition of rocks deposited in each of them. *Geological Society of America Bulletin* 62: 1–20.

Roberts, H.H. (1997). Dynamic changes of the Holocene Mississippi River delta plain: the delta cycle. *Journal of Coastal Research* 13: 605–627.

Rodriguez, A.B., Hamilton, M.D., and Anderson, J.B. (2000). Facies and evolution of the modern Brazos Delta, Texas: wave versus flood influence. *Journal of Sedimentary Research* 70: 283–295.

Rodriguez, A.B., Fassell, M.L., and Anderson, J.B. (2001). Variations in shoreface progradation and ravinement along the Texas coast, Gulf of Mexico. *Sedimentology* 48: 837–853.

Rossi, V.M. and Steel, R.J. (2016). The role of tidal, wave and river currents in the evolution of mixed-energy deltas: example from the Lajas Formation (Argentina). *Sedimentology* 63: 824–864.

Rossi, V.M., Perillo, M.M., Steel, R.J., and Olariu, C. (2017a). Quantifying mixed-process variability in shallow-marine depositional systems: what are sedimentary structures really telling us? *Journal of Sedimentary Research* 87: 1060–1074.

Rossi, V.M., Longhitano, S.G., Mellere, D. et al. (2017b). Interplay of tidal and fluvial processes in an early Pleistocene, delta-fed, strait margin (Calabria, Southern Italy). *Marine and Petroleum Geology* 87: 14–30.

Sadler, P.M. (1981). Sediment accumulation rates and the completeness of stratigraphic sections. *The Journal of Geology* 89: 569–584.

Sadler, P.M. and Jerolmack, D.J. (2015). Scaling laws for aggradation, denudation and progradation rates: the case for time-scale invariance at sediment sources and sinks. In: *Strata and Time: Probing the Gaps in Our Understanding*, Geological Society of London, Special Publication, vol. 404 (ed. D.G. Smith, R.J. Bailey, P.M. Burgess, and A.J. Fraser), 69–88. Geological Society of London.

Sandal, S.T. (1996). *The Geology and Hydrocarbon Resources of Negara Brunei Darussalam*. Brunei Museum, Syabas,

Bandar Seri Begawan: Brunei Shell Petroleum Company, 243 p.

Schwarz, E., Veiga, G.D., Álvarez Trentini, G. et al. (2018). Expanding the spectrum of shallow-marine, mixed carbonate–siliciclastic systems: processes, facies distribution and depositional controls of a siliciclastic-dominated example. *Sedimentology* 65: 1558–1589.

Sech, R.P., Jackson, M.D., and Hampson, G.J. (2009). Three-dimensional modeling of a shoreface-shelf parasequence reservoir analog: Part 1. Surface-based modeling to capture high-resolution facies architecture. *AAPG Bulletin* 93: 1155–1181.

Short, A.D. (2006). Australian beach systems – nature and distribution. *Journal of Coastal Research* 22: 11–27.

Simeoni, U. and Corbau, C. (2009). A review of the Delta Po evolution (Italy) related to climatic changes and human impacts. *Geomorphology* 107: 64–71.

Sixsmith, P.J., Hampson, G.J., Gupta, S. et al. (2008). Facies architecture of a net transgressive sandstone reservoir analog: The Cretaceous Hosta Tongue, New Mexico. *AAPG Bulletin* 92: 513–547.

Slingerland, R. and Keen, T.R. (1999). Sediment transport in the Western Interior Seaway of North America: predictions from a climate-ocean-sediment model. In: *Isolated Shallow Marine Sand Bodies: Sequence Stratigraphic Analysis and Sedimentologic Interpretation*, SEPM, Special Publication, vol. 64 (ed. K.M. Bergman and J.W. Snedden), 179–190. SEPM Society for Sedimentary Geology.

Smith, A.G., Hurley, A.M., and Briden, J.C. (1981). *Phanerozoic Paleocontinental World Maps*, 106. Cambridge: Cambridge University Press.

Snedden, J.W. and Dalrymple, R.W. (1999). Modern shelf sand ridges: from historical perspective to a unified hydrodynamic and evolutionary model. In: *Isolated Shallow Marine Sand Bodies: Sequence Stratigraphic Analysis and Sedimentologic Interpretation*, SEPM, Special Publication, vol. 64 (ed. K.M. Bergman and J.W. Snedden), 13–28. SEPM Society for Sedimentary Geology.

Sømme, T.O., Helland-Hansen, W., and Granjeon, D. (2009). Impact of eustatic amplitude variations on shelf morphology, sediment dispersal, and sequence stratigraphic interpretation: icehouse versus greenhouse systems. *Geology* 37: 587–590.

Soutter, E., Kane, I., Hodgson, D.M., and Flint, S.S. (2021). Controls on submarine canyon connection to the shoreline: a numerical modelling approach. Preprint submitted to Geosphere. www.eartharxiv.com, https://doi.org/10.31223/X5P33S.

Steel, R., Porebski, S.J., Plink-Björklund, P., Mellere, D., and Schellpeper, M. (2003). Shelf-edge delta types and their sequence-stratigraphic relationships. Roberts, H.R., Rosen, N.C., Fillon, R.F., and Anderson, J.B., eds., *Shelf Margin Deltas and Linked Down Slope Petroleum Systems: Global Significance and Future Exploration Potential. SEPM, Gulf Coast Section, 23rd Bob F. Perkins Research Conference, Proceedings*, pp. 205–230.

Steventon, M.J., Jackson, C.A.-L., Hodgson, D.M., and Johnson, H.D. (2020). Lateral variability of shelf-edge and basin-floor deposits, Santos Basin, offshore Brazil. *Journal of Sedimentary Research* 90: 1198–1221.

Stonecipher, S.A. (1999). Genetic characteristics of glauconite and siderite: implications for the origin of ambiguous isolated marine sandbodies. In: *Isolated Shallow Marine Sand Bodies: Sequence Stratigraphic Analysis and Sedimentologic Interpretation*, SEPM, Special Publication, vol. 64 (ed. K.M. Bergman and J.W. Snedden), 191–204. SEPM Society for Sedimentary Geology.

Storms, J.E.A. (2003). Event-based stratigraphic simulation of wave-dominated shallow-marine environments. *Marine Geology* 199: 83–100.

Storms, J.E.A. and Swift, D.J.P. (2003). Shallow-marine sequences as the building blocks of stratigraphy: insights from numerical modelling. *Basin Research* 15: 287–303.

Stouthamer, E., Cohen, K.M., and Gouw, M.J. (2011). Avulsion and its implications for fluvial-deltaic architecture: insights from the Holocene Rhine-Meuse delta. In: *From River to Rock Record: The Preservation of Fluvial Sediments and Their Subsequent Interpretation*, SEPM, Special Publication, vol. 97 (ed. S.K. Davidson, S. Leleu, and C.P. North), 215–232. SEPM Society for Sedimentary Geology.

Stride, A.H. (1974). Indications of long term, tidal control of net sand loss or gain by European coasts. *Estuarine and Coastal Marine Science* 2: 27–36.

Stride, A.H., Belderson, R.H., Kenyon, N.H., and Johnson, M.A. (1982). Offshore tidal deposits: sand sheet and sand bank facies. In: *Offshore Tidal Sands: Processes and Deposits* (ed. A.H. Stride), 95–125. London: Chapman & Hall Ltd.

Strong, N. and Paola, C. (2008). Valleys that never were: time surfaces versus stratigraphic surfaces. *Journal of Sedimentary Research* 78: 579–593.

Stukins, S., Jolley, D.W., McIlroy, D., and Hartley, A.J. (2013). Middle Jurassic vegetation dynamics from allochthonous palynological assemblages: an example from a marginal marine depositional setting; Lajas Formation, Neuquén Basin, Argentina. *Palaeogeography, Palaeoclimatology, Palaeoecology* 392: 117–127.

Sullivan, M.D., Van Wagoner, J.C., Jennette, D.C. et al. (1997). High resolution sequence stratigraphy and architecture of the Shannon Sandstone, Hartzog Draw Field, Wyoming: implications for reservoir management. In: *Shallow Marine and Nonmarine Reservoirs: Sequence Stratigraphy, Reservoir Architecture, and Production Characteristics. SEPM, Gulf Coast Section, 18th Annual Research Conference, Proceedings* (ed. K.W. Shanley and R.F. Perkins), 331–344.

Sundal, A. and Hellevang, H. (2019). Using reservoir geology and petrographic observations to improve CO_2 mineralization estimates: examples from the Johansen formation, North Sea, Norway. *Minerals* 9: 671.

Sundal, A., Nystuen, J.P., Dypvik, H. et al. (2013). Effects of geological heterogeneity on CO_2 distribution and migration - a case study from the Johansen Formation, Norway. *Energy Procedia* 37: 5046–5054.

Sundal, A., Nystuen, J.P., Rørvik, K.L. et al. (2016). The Lower Jurassic Johansen Formation, northern North Sea–Depositional model and reservoir characterization for CO_2 storage. *Marine and Petroleum Geology* 77: 1376–1401.

Suter, J.R. and Clifton, H.E. (1999). The Shannon Sandstone and isolated linear sand bodies: interpretations and realizations. In: *Isolated Shallow Marine Sand Bodies: Sequence Stratigraphic Analysis and Sedimentologic Interpretation*, SEPM, Special Publication, vol. 64 (ed. K.M. Bergman and J.W. Snedden), 321–356. SEPM Society for Sedimentary Geology.

Sweet, M.L. and Blum, M.D. (2016). Connections between fluvial to shallow marine environments and submarine canyons: implications for sediment transfer to deep water. *Journal of Sedimentary Research* 86: 1147–1162.

Swenson, J.B. (2005). Relative importance of fluvial input and wave energy in controlling the timescale for distributary-channel avulsion. *Geophysical Research Letters* 32: L23404.

Swenson, J.B., Paola, C., Pratson, L. et al. (2005). Fluvial and marine controls on combined subaerial and subaqueous delta progradation: morphodynamic modeling of compound-clinoform development. *Journal of Geophysical Research: Earth Surface* 110: F02013.

Swift, D.J.P. (1968). Coastal erosion and transgressive stratigraphy. *Journal of Geology* 76: 444–456.

Swift, D.J.P. (1975a). Barrier-island genesis: evidence from the central Atlantic shelf, eastern USA. *Sedimentary Geology* 14: 1–43.

Swift, D.J. (1975b). Tidal sand ridges and shoal-retreat massifs. *Marine Geology* 18: 105–133.

Swift, D.J.P. and Field, M.E. (1981). Evolution of a classic sand ridge field: Maryland sector, North American inner shelf. *Sedimentology* 28: 461–482.

Swift, D.J.P. and Thorne, J.A. (1991). Sedimentation on continental margins, I: a general model for shelf sedimentation. In: *Shelf Sand and Sandstone Bodies: Geometry, Facies and Sequence Stratigraphy*, International Association of Sedimentologists, Special Publication, vol. 14 (ed. D.J.P. Swift), 3–31. Wiley.

Swift, D.J.P., Thorne, J.A., and Oertel, G.F. (1986). Fluid processes and sea-floor response on a modern storm-dominated shelf: middle Atlantic shelf of North America. Part II: response of the shelf floor. In: *Shelf Sands and Sandstones*, Canadian Society of Petroleum Geologists, Memoir, vol. 11 (ed. R.J. Knight and J.R. McLean), 191–211. Wiley.

Syvitski, J.P. and Saito, Y. (2007). Morphodynamics of deltas under the influence of humans. *Global and Planetary Change* 57: 261–282.

Syvitski, J.P., Kettner, A.J., Overeem, I. et al. (2009). Sinking deltas due to human activities. *Nature Geoscience* 2: 681–686.

Ta, T.K.O., Nguyen, V.L., Tateishi, M. et al. (2002). Holocene delta evolution and sediment discharge of the Mekong River, southern Vietnam. *Quaternary Science Reviews* 21: 1807–1819.

Ta, T.K.O., Nguyen, V.L., Tateishi, M. et al. (2005). Holocene delta evolution and depositional models of the Mekong River Delta, southern Vietnam. In: *River Deltas – Concepts, Models, and Examples*, SEPM, Special Publication, vol. 83 (ed. L. Giosan and J.P. Bhattacharya), 453–466. SEPM Society for Sedimentary Geology.

Tamura, T., Saito, Y., Nguyen, V.L. et al. (2012). Origin and evolution of interdistributary delta plains; insights from Mekong River delta. *Geology* 40: 303–306.

Taylor, A., Goldring, R., and Gowland, S. (2003). Analysis and application of ichnofabrics. *Earth-Science Reviews* 60: 227–259.

Tillman, R.W. and Martinsen, R.S. (1984). The Shannon shelf-ridge sandstone complex, Salt Creek anticline area, Powder River basin, Wyoming. In: *Siliciclastic Shelf Sediments*, SEPM, Special Publication, vol. 34 (ed. R.W. Tillman and C.T. Siemers), 85–142. SEPM Society for Sedimentary Geology.

Tinterri, R. (2007). The lower Eocene Roda Sandstone (south-central Pyrenees): an example of a flood-dominated river-delta system in a tectonically controlled basin. *Rivista Italiana di Paleontologia e Stratigrafia* 113: 223–255.

Tran, D.A., Tsujimura, M., Pham, H.V. et al. (2022). Intensified salinity intrusion in coastal aquifers due to groundwater overextraction: a case study in the Mekong Delta, Vietnam. *Environmental Science and Pollution Research* 29: 8996–9010.

Traykovski, P., Geyer, W.R., Irish, J.D., and Lynch, J.F. (2000). The role of wave-induced density-driven fluid mud flows for cross-shelf transport on the Eel River continental shelf. *Continental Shelf Research* 20: 2113–2140.

Tronicke, J., Blindow, N., Gross, R., and Lange, M.A. (1999). Joint application of surface electrical resistivity-and GPR-measurements for groundwater exploration on the island of Spiekeroog—northern Germany. *Journal of Hydrology* 223: 44–53.

Uličný, D. (2001). Depositional systems and sequence stratigraphy of coarse-grained deltas in a shallow-marine, strike-slip setting: the Bohemian Cretaceous Basin, Czech Republic. *Sedimentology* 48: 599–628.

Vail, P.R., Bubb, J.N., Hatlelid, W.G. et al. (1977). Seismic stratigraphy and global changes in sea level, parts 1-11. In: *Seismic Stratigraphy – Applications to Hydrocarbon Exploration*, AAPG, Memoir, vol. 26 (ed. C.E. Payont), 49–212. American Association of Petroleum Geologists.

Vakarelov, B.K. and Ainsworth, R.B. (2012). Predicting marginal marine reservoir architecture: examples from Asian shoreline systems. *International Petroleum Technology Conference*, paper 14472.

Vakarelov, B.K. and Ainsworth, R.B. (2013). A hierarchical approach to architectural classification in marginal-marine systems: bridging the gap between sedimentology and sequence stratigraphy. *AAPG Bulletin* 97: 1121–1161.

Vakarelov, B.K., Ainsworth, R.B., and MacEachern, J.A. (2012). Recognition of wave-dominated, tide-influenced shoreline systems in the rock record: variations from a microtidal shoreline model. *Sedimentary Geology* 279: 23–41.

Van Cappelle, M., Stukins, S., Hampson, G.J., and Johnson, H.D. (2016). Fluvial to tidal transition in mixed, tide-influenced and wave-influenced deltaic deposits: Cretaceous lower Sego Sandstone, Utah, USA. *Sedimentology* 63: 1333–1361.

Van Cappelle, M., Hampson, G.J., and Johnson, H.D. (2018). Spatial and temporal evolution of coastal depositional systems and regional depositional process regimes: Campanian Western Interior Seaway, USA. *Journal of Sedimentary Research* 88: 873–897.

Van der Vegt, H., Storms, J.E., Walstra, D. J.R. et al. (2020). Grain size fractionation by process-driven sorting in sandy to muddy deltas. *The Depositional Record* 6: 217–235.

Van Heerden, I.L. and Roberts, H.H. (1988). Facies development of Atchafalaya Delta, Louisiana: a modern bayhead delta. *AAPG Bulletin* 72: 439–453.

Van Straaten, L.M.J.U. (1954). Composition and structure of recent marine sediments in the Netherlands. *Leidse Geologische Mededelingen* 19: 1–108.

Van Wagoner, J.C. (1991). High-frequency sequence stratigraphy and facies architecture of the Sego Sandstone in the Book Cliffs of western Colorado and eastern Utah. In: *Sequence Stratigraphy Applications to Shelf Sandstone Reservoirs: Outcrop to Subsurface Examples*, AAPG, Field Guidebook, Day 1 (ed. J.C. Van Wagoner, D. Nummedal, C.R. Jones, et al.). American Association of Petroleum Geologists, unpaginated.

Van Wagoner, J.C., Mitchum, R.M., Campion, K.M., and Rahmanian, V.D. (1990). Siliciclastic sequence stratigraphy in well logs, cores, and outcrops. *AAPG, Methods in Exploration* 7: 55 p.

Van Yperen, A.E., Holbrook, J.M., Poyatos-Moré, M., and Midtkandal, I. (2019). Coalesced delta-front sheet-like sandstone bodies from highly avulsive distributary channels: the low-accommodation Mesa Rica Sandstone (Dakota Group, New Mexico, USA). *Journal of Sedimentary Research* 89: 654–678.

Van Yperen, A.E., Poyatos-Moré, M., Holbrook, J.M., and Midtkandal, I. (2020). Internal mouth-bar variability and preservation of subordinate coastal processes in low-accommodation proximal deltaic settings (Cretaceous Dakota Group, New Mexico, USA). *Depositional Record* 6: 431–458.

Van Yperen, A.E., Holbrook, J.M., Poyatos-Moré, M. et al. (2021). Low-accommodation and backwater effects on sequence stratigraphic surfaces and depositional architecture of fluvio-deltaic settings (Cretaceous Mesa Rica Sandstone, Dakota Group, USA). *Basin Research* 33: 513–543.

Vaughan, S., Bailey, R.J., and Smith, D.G. (2011). Detecting cycles in stratigraphic data: spectral analysis in the presence of red noise. *Paleoceanography* 26: PA4211.

Virolle, M., Féniès, H., Brigaud, B. et al. (2020). Facies associations, detrital clay grain coats and mineralogical characterization of the Gironde estuary tidal bars: a modern analogue for deeply buried estuarine sandstone reservoirs. *Marine and Petroleum Geology* 114: 104225.

Visser, M.J. (1980). Neap-spring cycles reflected in Holocene subtidal large-scale bedform deposits: a preliminary note. *Geology* 8: 543–546.

Walker, R.G. and Bergman, K.M. (1993). Shannon Sandstone in Wyoming; a shelf-ridge complex reinterpreted as lowstand shoreface deposits. *Journal of Sedimentary Research* 63: 839–851.

Wang, Y. and Jiao, J.J. (2012). Origin of groundwater salinity and hydrogeochemical processes in the confined Quaternary aquifer of the Pearl River Delta, China. *Journal of Hydrology* 438: 112–124.

Wellner, R., Beaubouef, R., Van Wagoner, J.C. et al. (2005). Jet-plume depositional bodies – the primary building blocks of Wax Lake Delta. *Transactions of the Gulf Coast Association of Geological Societies* 55: 867–909.

Wells, M.R. (2008). Tidal modelling of modern and ancient seas and oceans. PhD thesis. Imperial College London.

Wignall, P.B. and Ruffell, A.H. (1990). The influence of a sudden climatic change on marine deposition in the Kimmeridgian of northwest Europe. *Journal of the Geological Society* 147: 365–371.

Wignall, P.B., Sutcliffe, O.E., Clemson, J., and Young, E. (1996). Unusual shoreface sedimentology in the Upper Jurassic of the Boulonnais, northern France. *Journal of Sedimentary Research* 66: 577–586.

Williams, G.D. and Stelck, C.R. (1975). Speculations on the Cretaceous Palaeogeography of North America. In: *The Cretaceous System in the Western Interior of North America*, Geological Association of Canada, Special Paper, vol. 13 (ed. W.G.E. Caldwell), 1–20. Geological Association of Canada.

Willis, B.J. and Gabel, S.L. (2001). Sharp-based, tide-dominated deltas of the Sego Sandstone, Book Cliffs, Utah, USA. *Sedimentology* 48: 479–506.

Willis, B.J. and Gabel, S.L. (2003). Formation of deep incisions into tide-dominated river deltas: implications for the stratigraphy of the Sego Sandstone, Book Cliffs, Utah, USA. *Journal of Sedimentary Research* 73: 246–263.

Willis, B.J., Sun, T., and Ainsworth, R.B. (2021). Contrasting facies patterns between river-dominated and symmetrical wave-dominated delta deposits. *Journal of Sedimentary Research* 91: 262–295.

Wright, L.D. (1977). Sediment transport and deposition at river mouths: a synthesis. *Geological Society of America Bulletin* 88: 857–868.

Wright, L.D., Chappell, J., Thom, B.G. et al. (1979). Morphodynamics of reflective and dissipative beach and inshore systems, Southeastern Australia. *Marine Geology* 32: 105–140.

Yang, C.S. and Nio, S.D. (1989). An ebb-tide delta depositional model – a comparison between the modern

Eastern Scheldt tidal basin (southwest Netherlands) and the Lower Eocene Roda Sandstone in the southern Pyrenees (Spain). *Sedimentary Geology* 64: 175–196.

Yang, B.C., Dalrymple, R.W., and Chun, S. S. (2005). Sedimentation on a wave-dominated, open-coast tidal flat, south-western Korea: summer tidal flat–winter shoreface. *Sedimentology* 52: 235–252.

Yang, B., Dalrymple, R.W., and Chun, S. (2006). The significance of hummocky cross-stratification (HCS) wavelengths: evidence from an open-coast tidal flat, South Korea. *Journal of Sedimentary Research* 76: 2–8.

Yoshida, S., Steel, R.J., and Dalrymple, R. W. (2007). Changes in depositional processes – an ingredient in a new generation of sequence-stratigraphic models. *Journal of Sedimentary Research* 77: 447–460.

Zaitlin, B.A., Dalrymple, R.W., and Boyd, R. (1994). The stratigraphic organization of incised-valley systems associated with relative sea level change. In: *Incised-Valley Systems: Origin and Sedimentary Sequences*, SEPM, Special Publication, vol. 51 (ed. R.W. Dalrymple, R. Boyd, and B.A. Zaitlin), 45–62. SEPM Society for Sedimentary Geology.

Zecchin, M. and Catuneanu, O. (2013). High-resolution sequence stratigraphy of clastic shelves I: units and bounding surfaces. *Marine and Petroleum Geology* 39: 1–25.

Zecchin, M. and Catuneanu, O. (2020). High-resolution sequence stratigraphy of clastic shelves VII: 3D variability of stacking patterns. *Marine and Petroleum Geology* 121: 104582.

Zhang, J., Steel, R., and Ambrose, W. (2016). Greenhouse shoreline migration: Wilcox deltas. *AAPG Bulletin* 100: 1803–1831.

Zhang, J., Kim, W., Olariu, C., and Steel, R. (2019a). Accommodation-versus supply-dominated systems for sediment partitioning to deep water. *Geology* 47: 419–422.

Zhang, J., Burgess, P.M., Granjeon, D., and Steel, R. (2019b). Can sediment supply variations create sequences? Insights from stratigraphic forward modelling. *Basin Research* 31: 274–289.

Zoccarato, C., Minderhoud, P.S., and Teatini, P. (2018). The role of sedimentation and natural compaction in a prograding delta: insights from the mega Mekong delta. *Vietnam. Scientific Reports* 8: 1–12.

Zuchuat, V., Steel, E., Mulligan, R.P. et al. (2022). Tidal dynamics in palaeo-seas in response to changes in physiography, tidal forcing and bed shear stress. *Sedimentology* 69: 1861–1890.

Zuchuat, V., Gugliotta, M., Poyatos-Moré, M. et al. (2023). Mixed depositional processes in coastal to shelf environments: towards acknowledging their complexity. *The Depositional Record* 9: 206–212.

9

Fine-Grained Siliciclastic Coastal and Shallow-Marine Sediments

Bruce Levell[1] *and Guy Plint*[2]

[1] *Department of Earth Sciences, University of Oxford, Oxford, UK*
[2] *Department of Earth Sciences, The University of Western Ontario, London, Ontario, Canada*

Summary

In comparison to their sand-rich counterparts, siliciclastic mud-dominated marine systems have received much less attention, and marine mudstone facies models are less well-developed. Based on fine grain-size alone, many ancient mudstones were assumed to record passive, low-energy deposition by vertical settling in relatively deep water. Recent renewed interest, in part spurred by resource extraction, has highlighted the complexity of stratification in mudstones, on the thin-section scale. Modern techniques have made it possible to image and chemically analyse the morphology, composition and provenance of the multiplicity of grain types that form mudstones. Clay particles tend to cohere, through both physical and biological processes, so mud is often deposited in aggregate grains and is resistant to fluid scour. Erosion tears up the mud in chunks that are re-deposited as sand- to pebble-size intraclasts that, in highly compacted mudstones are very difficult to recognise! Thanks to extensive modern oceanographic studies, it is now clear that many shelf mudstones record active deposition from vigorous currents involving lateral transport over tens to hundreds of kilometres, in water sometimes only metres to a few tens of metres deep.

Holocene shelf mud bodies are broadly divided into delta-related and marine current dominated. They are accumulating rapidly during highstand in an icehouse world dominated by large and rapid eustatic sea-level changes. It is far from clear what of these deposits will survive the next eustatic fall, and their applicability as direct analogues to past greenhouse settings is debatable. With this caveat we present four modern muddy shelf systems illustrating the dispersal and accumulation of mud. They all post-date the Holocene transgression, being no more than about 11,000 years old. They are: the Amazon–Orinoco system off South America, the Po-Adriatic system of Italy, the Hwang He and Chiangjiang systems of East China and the Fly River in the Gulf of Papua. All are characterised by vigorous along-shelf transport of riverine mud, driven primarily by prevailing winds, supplemented by tides and thermohaline currents. Clays flocculate as fresh, turbid river water mixes with seawater. As the mud flocs slowly settle, they are transported both landward and along the shelf, accumulating to form a prism that accretes laterally as low-angle (c. 1°) clinoforms that downlap onto Holocene transgressive surfaces. Re-suspension of mud by storm waves can generate fluid mud so dense that, under the influence of gravity, it flows downslope as a thin, bottom-hugging hyperpycnal flow or 'current-wave-enhanced sediment-gravity flow' (CWESGF). In all four examples, mud bodies are a few tens of metres thick and accumulate to within c. 10–20 m of sea level, the upper limit being controlled by ambient wave energy.

Shelf mudstone facies depend on the transport processes, which are strongly linked to river floods and to storms or seasonal winds at sea. Where the river outflow floats (hypopycnal flow), planar-laminated or current-rippled silts and clays may alternate, reflecting fluctuations in the flood strength. Where the river outflow sinks as a hyperpycnal flow, the deposit may show an initial upward-coarsening from mud to very fine sand, recording the waxing flood, followed by fining as the flood wanes. The coarser fraction may show climbing current or wave ripples reflecting rapid aggradation, and abundant plant material belies a direct fluvial link. Thick (cm-scale) structureless and non-bioturbated mud beds may also indicate deposition from fluid mud underflows. In the absence of river floods, storm waves are capable of generating and maintaining fluid mud

Reading's Sedimentary Environments: Processes, Facies and Stratigraphy, Fourth Edition. Edited by Bruce Levell.

Companion website: www.wiley.com/go/levell4e

that will flow downslope as a CWESGF. As these flows drop below the zone of wave agitation, mud settles out, forming distinctive, erosive-based, normally graded beds characterised by current or combined-flow rippled coarse silt to very fine sand, parallel-laminated silt and clay, capped by a structureless mud. Large storms can also deposit sharp-based, normally graded beds of very fine sand to mud as a result of combined flows in which storm waves and geostrophic currents combine to transport sediment obliquely offshore.

In contrast to the Holocene examples, we present case studies of the stratigraphy and sedimentology of Cretaceous shelf mudstones deposited under greenhouse conditions. Sediment accumulation was strongly controlled by tectonic subsidence or uplift, modulated by high-frequency, low-amplitude (5–20+ m) eustatic changes in the c. 20–400 ka Milanković band. In these examples, sea-floor topography was commonly extremely subtle, with shallow water persisting far offshore. This reflected a situation where supply > accommodation, and efficient lateral transport of mud, primarily by wave-driven currents, was able to fill available accommodation. Prominent clinothems, typically with slopes of approx. 1°, tended to develop under conditions when accommodation was greater than supply, allowing deeper water to develop offshore, thus favouring gravity-driven processes of transport.

Because of generally low sea-floor slopes, these Cretaceous depositional sequences typically comprise very thin transgressive systems tracts, linked to rapid and extensive shoreline transgression and up-dip sediment storage in aggrading coastal plains. Most sediment accumulated during highstand, falling stage and lowstand times when the deposits typically form a continuum in which systems tracts shingle laterally. Shorelines migrated hundreds of kilometres across these muddy ramps, yet the transgressive passage of the shoreline may be marked by nothing more than a few millimetres of extra- or intra-basinal pebbles or granules, or sometimes no coarse lag at all. In other instances, rooted horizons or thin coals sandwiched within 'monotonous' marine mudstones betray cryptic lowstand and emergence events. Such evidence of repeated breaks in sedimentation stands as a warning to the view that shelf mudstone successions preserve a relatively complete record of climatic or oceanographic history.

9.1 Introduction

> *'Possibly many may think that the deposition and consolidation of fine-grained mud must be a very simple matter, and the results of little interest. However, when carefully studied experimentally, it is soon found to be so complex a question, and the results dependent on so many variable conditions, that one might feel inclined to abandon the inquiry, were it not that so much of the history of our rocks appears to be written in this language.'*
>
> *Henry Clifton Sorby, 1908.*

Mudstones form about 60% of all sedimentary rocks, but they can look distinctly unappealing. In the field, mudstones are often poorly exposed and seemingly lack lithological variety and sedimentary structures. Surely the depositional environment must have been as uneventful and boring as the rocks?

Perhaps not. Over the last three decades, 'shallow-marine' mudstones in particular and mudstones in general have received considerable attention. Modern investigations of their sedimentology and stratigraphy show that the depositional history was anything but monotonous. Instead of recording a 'steady, quiet vertical rain' of fine sediment, it is now clear that much mudstone deposition was highly energetic – and sediment transport was mainly lateral, rather than vertical. Sea-bed instruments show us how the topography of a muddy seabed degrades and aggrades during a storm or river flood, and X-ray images of cores from modern shelf mud illustrate details of erosional and depositional events, including the activity of dense and curiously gel-like 'fluid muds'. High-resolution shallow seismic profiles image complex clinoform architectures and discontinuities within Pleistocene–Holocene shelf mud deposits; features that track their responses to changes in climate and sea level. Similarly, detailed subsurface investigation of the stratigraphy of ancient mudstone units on a basin-scale, shows that many have complex internal architecture.

Understanding mudstone *sedimentology* requires bringing samples back to the lab for closer examination, but simply collecting coherent blocks of crumbly mudstone can be extremely difficult! Creative coring and stabilising techniques such as resin wax impregnation may be necessary. Stabilised samples are cut for thin sections and polished or ion-milled blocks are prepared for scanning electron microscopy (SEM). The SEM reveals an incredible heterogeneity of mud particles at the micrometre scale, including aggregate grains, reworked mud intraclasts of local provenance, far-travelled 'lithic' mudstone grains and various biogenic particles as well as fine detrital quartz, other mineral grains, organic matter and a wide variety of diagenetic minerals. Thin sections continue the story at the millimetre to centimetre scale, revealing erosion surfaces, lags and scours, subtle grain-size grading, completely structureless beds and even cross-lamination and minute burrows. At this scale, mudstones show astonishing lateral and vertical heterogeneity. What appears to be 'lamination' in the field is in fact a complex succession of thin beds, each recording a distinct sedimentary event. This is not 'quiet water sedimentation'!

The *stratigraphy* of mudstones is best deciphered using wireline logs from boreholes, and from large outcrop exposures, aided by biostratigraphic correlation. Combinations of well logs reveal subtle variations in porosity and geochemistry, emphasising depositional cyclicity. Ancient shelf mudstones are typified by subtle vertical coarsening and fining facies successions bounded by flooding surfaces. These commonly track shoreline proximity, changes in sediment accommodation and organic productivity or provenance changes. They are commonly proxies for relative sea-level change and may be expressions of depositional sequences, usually developed at the 10^4–10^5 year timescale. Core or outcrop observation may reveal extra-basinal lags on mud-on-mud flooding surfaces that provide cryptic evidence for subaerial emergence – despite the absence of nearshore facies, palaeosols or fluvial channels! Any student of ancient shallow-marine mudstone must keep in mind the distinct possibility that sometimes, their 'offshore mudstone' had once been dry land, and that a 'monotonous' mudstone succession cannot be assumed to provide a continuous record of deposition.

9.1.1 What Are 'Fine-Grained Siliciclastic Coastal and Shallow-Marine Environments'?

In this chapter, we address coastal and shelf environments dominated by deposition of *siliciclastic* mud. Modern examples, all fundamentally related to mud-rich rivers, include: muddy coasts and offshore bars (e.g. NE Brazil, Surinam and Guyana), chenier coasts (e.g. Louisiana, USA), mud-rich deltas (e.g. Ganges–Brahmaputra, Bangladesh; Mekong, Vietnam), mud deposits on the continental shelves, including mid-shelf mud blankets (e.g. Yellow Sea and East China Sea, China; Texas, USA; Adriatic, Italy) and outer-shelf mud deposits.

Chapter 11 addresses carbonate deposition including carbonate mud. Biological productivity in surface waters is also addressed in Chapter 13 (Pelagic sedimentation). The division between siliciclastic and biogenic mud deposition is entirely artificial and mixtures are the rule. These deposits also share many processes in common (see also Section 8.3; e.g. river, wave and tide-dominated coasts: Galloway 1975; Boyd et al. 1992; Ainsworth et al. 2011). Ichnofacies in muddy substrates are addressed in Section 3.3.2.

Reasons for highlighting mud-dominated systems separately include:

1) Understanding of modern mud-dominated systems has, as yet, resulted in few well-defined facies models.
2) There are few examples of ancient successions that are interpreted to have formed in the same settings as modern shallow-marine mud depocentres (e.g. mid-shelf mud blankets).
3) New research on mudstone deposition, in particular petrography, arose from the recent exploitation of shale-gas and shale-oil. This work is now informing seal-characterisation for containment: Carbon capture and storage (CCS) and nuclear waste (Gautschi 2017; Fisher et al. 2023).
4) Mudstones are commonly treated as archives of proxies for oceanic or atmospheric geochemistry and climate. Depositional process, stratigraphic continuity and completeness all need to be understood before any succession, even a mudstone, can be used as such an archive.
5) The simple equivalence of mud accumulation with low energy depositional systems is wrong. Sediment supply availability as a control on lithology needs to be considered as well as 'energy level'.

9.1.1.1 Mudstone Classification

Mud and mudstone are loosely defined terms referring to mixtures of silt and clay-size material comprising both detrital and biogenic material, including silica, clay minerals and carbonate. 'Mudstone' is a generalised grain-size term for sedimentary rocks composed of more than 50% silt and clay-size material, irrespective of composition (Figure 9.1; Lazar et al. 2015).

'Silt' and 'clay' are specific grain-size terms referring respectively to grains with diameters between 63 and 4 microns, and < 4 microns. However, both terms also have compositional connotations: silt often being composed largely of quartz, feldspar and mica, whereas clay is composed predominantly of clay minerals. This association of grain-size and composition has led to complexity in classifications of fine-grained rocks. Mudstone grains (which are commonly aggregates) range in diameter from less than 0.1 microns to greater than 100 microns. Clay-mineral fractions range from 10% to 100% (Aplin et al. 1999).

Chemically, mud comprises silicate or carbonate minerals in the silt fraction and a variety of clay minerals. Silt-sized particles also include mudstone 'intraclasts' or faecal pellets which may be carbonate (e.g. pelagic faecal pellets) or siliciclastic (e.g. benthic faecal pellets). Mudstone can also include a significant proportion, in some cases > 10%, of organic material, chiefly the remains of planktonic algae, but also including terrestrial organic matter such as leaf debris; such mudstones can constitute hydrocarbon source-rocks and 'oil shales'.

Genetically, components of mud can be: (1) detrital (allochthonous), transported in the water column in suspension or through the air (dust, pollen and volcanic ash), (2) biogenic, particles constructed by organisms that lived in the water column or sediment (e.g. algae, diatoms and radiolarians) and (3) authigenic minerals (such as silica, carbonates or pyrite precipitated post-burial, from pore waters in the sediment).

Several classification schemes are chemically based: distinguishing on the basis of five components, for example:

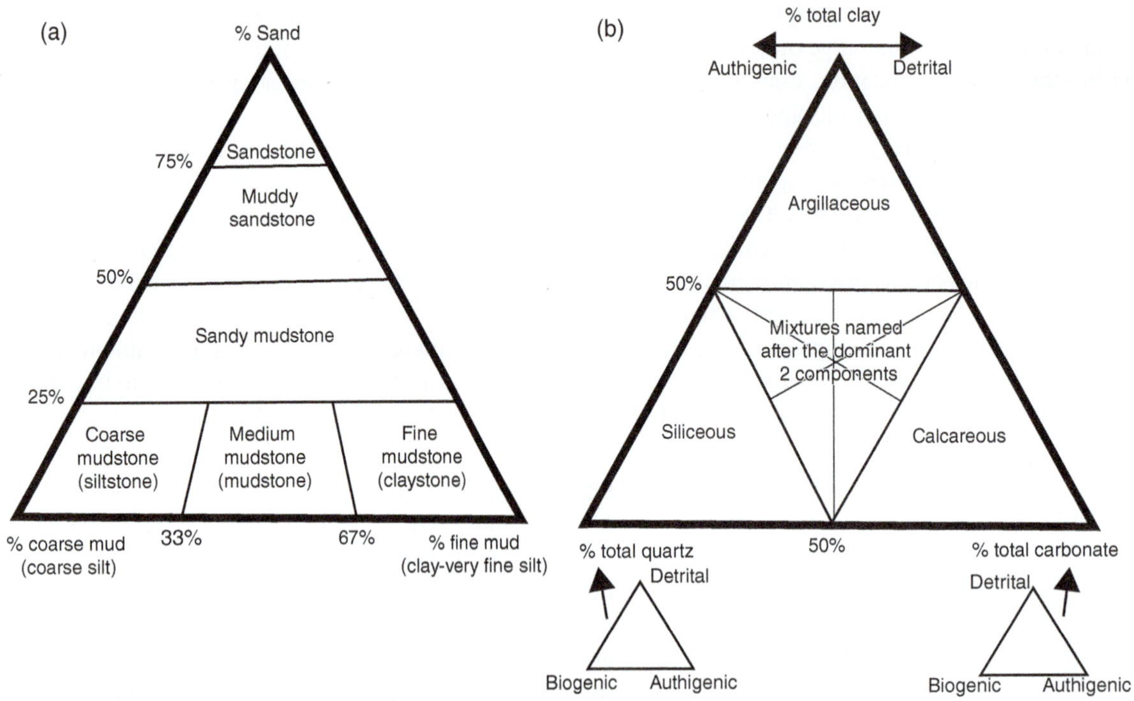

Figure 9.1 Mudstone classification (Lazar et al. 2015 / Society for Sedimentary Geology (SEPM)). (a) This ternary diagram provides a terminology for mudstones based on texture: relative proportions of sand, silt and clay-size grains. (b) This ternary diagram provides a terminology for fine-grained sedimentary rocks based on the proportions of argillaceous, siliceous and calcareous components.

clay, carbonate, silica, organic matter and the non-sheet-silicate minerals (primarily quartz and feldspar). Such classifications give rise to common terms such as 'marl' (mudstones with between 66% and 33% calcium carbonate), organic mudstone and siliceous mudstone.

From a facies and process perspective, important characteristics of mudstones are: grain-size, bedding structures, composition and grain origin. The comprehensive classification scheme of Figure 9.1 is based on texture, bedding type and chemical composition (Lazar et al. 2015).

9.1.2 Modern Shallow-Marine Mudstone Depositional Environments

Like all shelf deposits, marine muds need to be placed into a framework of accommodation space creation and sediment supply. Muds can only accumulate when there is sufficient mud supply and below 'base levels' set by the strength of marine currents and waves.

Walsh and Nittrouer (2009) grouped modern fine-grained coastal and shelf deposits ('mud depocentres') into five settings. These can be broadly grouped into point-source related (deltas and estuaries (1–3)) and marine current dominated (4–5):

1) Proximal-accumulation-dominated (PAD) systems are fine-grained deltas where mud is deposited largely in 'clinothems' (see Figure 4.25) lying directly off river mouths, such as the Po, Mississippi and Hwang He.
2) Estuarine-accumulation-dominated (EAD) systems are those where the fine-grained sediment is trapped within the estuary, such as Delaware Bay, USA.
3) Subaqueous-delta-clinoform (SDC) systems are sets of muddy clinothems detached from river mouths and supplied either along the coast (Mississippi (Figure 9.2), Amazon (Brazil), Po (Italy)) or across the shelf (Fly River, Papua New Guinea).
4) Marine-dispersal dominated (MDD) systems are shelf mud deposits detached from river mouths that may or may not have clinoforms. Examples are the mid-shelf mud blankets offshore Texas and those related to the Eel (California) River of the United States, as well as the mud blankets of the Yellow Sea (China).
5) Mud is also bypassed directly or indirectly across the shelf edge in 'canyon-captured' (CC) systems, typical of narrow shelves. Spectacular examples are the Ganges–Brahmaputra (Figure 9.3), where a third of the fluvial sediment is bypassed, and the Congo where perhaps two-thirds is bypassed (see Section 12.2.6; Figure 12.9).

The occurrence of these five systems is determined not only by the energy regime of the shelf (i.e. fluvial, wave and tidal currents and their relative and absolute strengths) but also by the sediment input rate and the availability of accommodation space due to the pre-existing physiography. Examples include drowned river valleys for EAD systems, narrow shelves for canyon-captured (CC) systems or shelf hollows

Figure 9.2 The Gulf of Mexico, USA, illustrating transport and re-suspension of mud on the continental shelf. Mississippi flood-derived detrital fine sediment plumes are dark brown, fading to green as the sediment settles or disperses, and may be accompanied by turbidity generated by surface phytoplankton. The sediment-laden water transitions to clearer water near the edge of the continental shelf. The preceding month was the wettest October in 115 years. Coast-wise advection of mud has led to deposition of up to 50 m of shelf mud in just the last 3500 years in the Texas mud belt (Weight et al. 2011). MODIS image from November 10, 2000. NASA / Wikimedia Commons / Public Domain.

Figure 9.3 Sediment outflow from the macrotidal Ganges–Brahmaputra–Meghna deltas after the monsoon-driven flood period in the northern Bay of Bengal (India and Bangladesh). Suspended sediment plumes are rectilinear, in-line with tidal currents flowing in and out of delta-front tidal channels. Coastwise advection is, however, visible towards the mouth of the Mahanadi River in the west. The mangroves of the Sundarbans are clearly visible in dark green. SeaWiFS image from November 25, 1999, NASA / https://visibleearth.nasa.gov/images/53502/northern-bay-of-bengal?size=largeS1999329061635.png / last accessed 02 January 2024

for marine dispersal-dominated (MDD) mud patches. As Walsh and Nittrouer carefully point out, the relative proportion of these systems at any given time is almost certainly related to the state and rate of relative sea-level rise or fall. Shelf-edge (PAD) deltas and canyon capture (CC) systems are probably more dominant at times of lower sea level, with narrow shelves, whereas MDD, SDC and EAD systems are more dominant when sea level is high (like now), shelves wider, and coasts are at highstand or transgressive.

Hanebuth et al. (2015) alternatively subdivided muddy systems by morphology into eight types: prodeltas (PAD above), 'subaqueous deltas' (SDC above: delta lobes disconnected from the river mouth), mud patches, belts, blankets, wedges, topographic mud entrapment sites and contourite drifts (broadly speaking, all grouped as MDD systems by Walsh and Nittrouer). A combination of the schemes of Walsh and Nittrouer and Hanebuth et al. is shown in Figure 9.4.

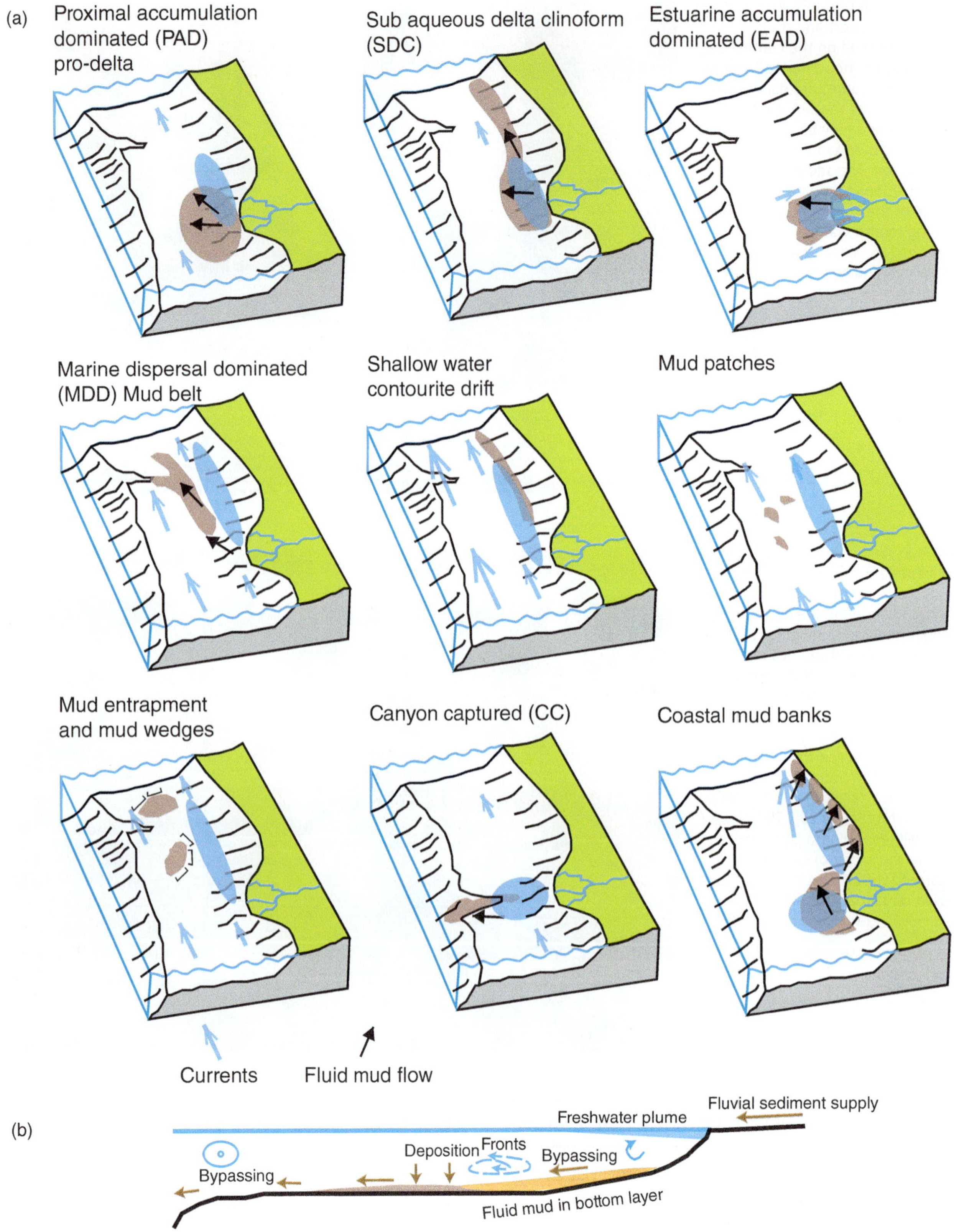

Figure 9.4 (a) Nine genetic types of shelf mud accumulation (brown) based on studies of modern accumulation sites. The 100-yr sediment accumulation rates vary from cm to mm per year. Suspended sediment plumes are shown in blue. Note that the Subaqueous Delta Clinoforms (SDC) include muddy clinothems that are physically detached from their supplying deltas and/or may extend down-drift for hundreds of kilometres from those deltas. (b) Summary of processes that can move mud across and along a shelf. Based on Walsh and Nittrouer (2009) and Hanebuth et al. (2015). Walsh and Nittrouer give the occurrences of studied examples in the world today, mostly from larger river systems, as: 16 EAD, 39 PAD, 9 CC, 32 MDD and 8 SDC.

9.1.3 Why Are Some Marine Depositional Environments Fine-Grained?

Muddy coasts and shelves are a direct response to fluvial input. Nyberg and Howell (2016) estimate that 26% of the world's *depositional* coastlines have significant fluvial sediment supply. The volumetric importance of mud is illustrated by comparing the ratios of fluvial bedload to suspended load. Typically for larger rivers these are in the proportion of 1:10 or less (Allen 2017). Such ratios imply that even 'coarse-grained' so-called 'bedload-dominated' rivers supply plentiful suspended sediment to the ocean (Syvitski and Milliman 2007).

Coasts with a large tidal range are also often mud-rich. In most cases, these are fluvio-tidal systems, which are muddy due to the large suspended load of rivers. However, the accumulation of mud along the coast can also be due to onshore transport of fine-grained sediment from the open sea as well as the result of grain-size sorting over long distances where transport is dominated by strong tidal currents.

Shoaling waves, in general, cause winnowing and offshore transport of the fine-grained material, particularly during storms, and can leave sand and gravel material, if present, concentrated on the coast as beaches (the 'littoral energy fence', Section 8.1). Mud-rich shelves are therefore commonly developed offshore of sandy coasts.

Each year, 19 billion tonnes of river-borne sediment is supplied to the oceans, about 66% from South East Asia alone (Figure 9.5). This percentage reflects the importance of high relief, high rainfall, tropical weathering and large drainage basins in supplying mud to basins. Substantial volumes of suspended sediment transported by rivers are stored in floodplains. Large drainage basins, such as the Congo, store proportionally more sediment (Baudin et al. 2020). About half of the sediment load of the Amazon is sequestered in its middle reaches (Allen 2017). In the United States, it has been estimated that as little as 10% of eroded material currently reaches the ocean (Holeman 1980). The sediment stored in river floodplains is partly released on human timescales in response to flood cycles (Wheatcroft and Drake 2003). However, in larger river systems, sediment in storage may only be released on longer timescales, in response to changes in base level, or to longer-term (Milanković band) changes in rainfall and river discharge (Goodbred 2003; Gibling *et al.* 2005, 2011). Overall, there is a relatively poor understanding of how variations in sediment supply to the ocean, on timescales of thousands to millions of years, are related to storage, buffering and threshold exceedance in drainage basins.

Monsoonal climates result in intense, seasonal peaks in the delivery of suspended sediment to the coast. The development of muddy coasts is favoured by intense pulses of supply which can overwhelm the ability of waves to winnow and redistribute mud to offshore areas. In combination with overall higher supply, it is therefore no surprise that the most extensive muddy coastal systems develop in tropical climates.

Human interaction with rivers and the coast has significantly altered fine-grained sediment routing systems, as is well illustrated by the effect of dams on large Chinese rivers (Section 9.2.3). However, Rodriguez et al. (2020) argue, on the basis of measured accumulation rates in coastal depocentres, that the amount of suspended sediment nourishing the coast is partially compensated because the dammed rivers then have diminished sediment loads, which gives them increased capacity to erode sediment in floodplain storage downstream of the dam.

9.1.4 Clay Minerals, Weathering and Provenance

Determining why a particular fluvial sediment supply is mudstone-dominated requires consideration of general factors such as relief, climate and bedrock composition in the source area as well as the weathering regime.

Clay has one of three origins (Millot 1970). *Autochthonous* clay minerals are precipitated from solution or via the crystallisation of amorphous materials, such as glassy volcanic ash. *Allochthonous* clay minerals are formed elsewhere and transported to the site of deposition. *Diagenetic or authigenic* clay minerals develop within the sediment subsequent to deposition.

Sea-bed maps of clay mineralogy show kaolinite to be abundant between the tropics, whereas chlorite and illite are prevalent at high latitudes. This distribution is interpreted broadly as a response to climate control on weathering and soil processes.

During weathering, there is progressive leaching of mobile cations out of the phyllosilicate lattice (Chamley 1989). Kaolinite is mainly formed in soils by complete removal of mobile cations as a result of intense leaching under a humid tropical climate. Illite, by contrast, is produced when hydrolysis is limited by a cold or arid climate. Chlorite is mainly a detrital residue from the weathering of low-grade metamorphic and mafic source rocks in alkaline environments. Like illite, chlorite characterises relatively dry and/or cold climates. Smectite forms during early weathering of unstable Fe-, Mg- and Ca-rich minerals in igneous or metamorphic rocks in temperate, warm to cool climatic zones characterised by moderate hydrolysis. Chlorite is unstable in warm, wet conditions.

Thiry (2000) examined the inter-relationships between source lithology, pedogenesis, climate, erosion and temporary sediment storage, concluding that detrital clays should not be uncritically interpreted as indicators of contemporaneous palaeoclimate. Allochthonous clay minerals may be related more to palaeo- than contemporary climates. For example, Argon dating ($^{40}Ar/^{39}Ar$) of minerals in soil profiles demonstrates that some modern weathering profiles in Brazil date back to the Paleocene (Vasconcelos and Carmo 2018)

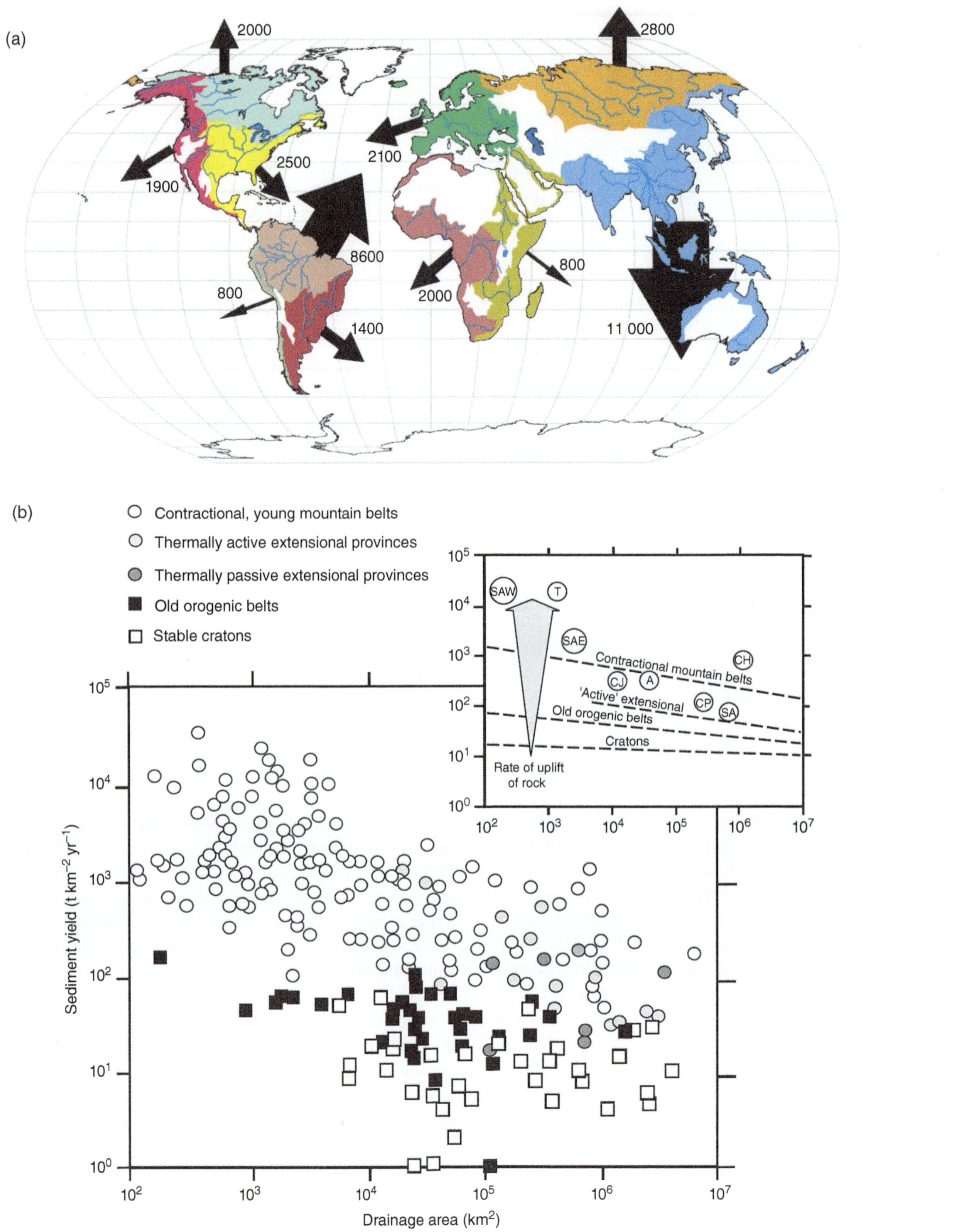

Figure 9.5 (a) Annual discharge of fluvial suspended sediment into the ocean. Sediment discharge is in millions of tonnes per year. The arrows are to scale. The global discharge is 19 billion tonnes per year. White areas are endo-rheic (i.e. that have no river outflow to the sea). Milliman and Farnsworth (2011) / Cambridge University Press. (b) Sediment yield versus drainage area by basin type. The figure shows the importance of relief, via increased weathering rates, stream power and increased landslide frequency in increasing sediment supply per unit area. SAW, Southern Alps New Zealand western flank; SAE, Southern Alps New Zealand eastern flank; T, Taiwan; CJ, Central Japan; A, Alps; CP, Colorado Plateau; SA, Southern Africa; CH, Central Himalaya. (Originally from Hovius 1998, modified by Allen 2017.)

and, worldwide, many are of Miocene age (Thiry 2000). On the order of one million years is required to form thick monomineralic soils, even under past climates with a higher CO_2 level. Release of this clay-rich material to the ocean by eventual fluvial erosion is therefore not necessarily a record of the climate at the time of deposition. Kaolinitic and bauxitic weathering of Late-Cretaceous carbonates across Europe is reflected in the appearance of those minerals in the Palaeogene succession in the North Sea Basin and in the Eocene–Oligocene strata in the Aquitaine Basin, tens of millions of years *after* they were formed (Thiry 2000).

Major elements in detrital minerals, as summarised in the Chemical Index of Alteration (CIA; Nesbitt and Young 1982), can record the extent of chemical weathering in the hinterland. This index is calculated using the molecular proportions of mobile alkali and alkaline earth metals relative to aluminium (aluminium is regarded as immobile: an assumption that breaks down in organic-rich muds (Macquaker et al. 2014)).

$$CIA = 100 * Al_2O_3/ (Al_2O_3 + CaO^* + Na_2O + K_2O)$$

(CaO^* is CaO associated with the silicate fraction (i.e. excluding carbonates).

The CIA assesses the breakdown of feldspars towards kaolinite, and the index increases with chemical weathering. The relative proportion of chemical weathering is, however, impacted by time, through the rate of erosion, as well as climate. In addition, high-relief areas will favour mechanical erosion irrespective of climate.

Heavy minerals, trace elements, such as Rare Earth Elements (REE), and isotopic methods using Sr, Sm, Os and Nd can all give important direct information on provenance, which in combination with the CIA and clay mineralogy can sometimes allow reconstruction of the entangled climatic and provenance signals preserved in fine-grained sediments (e.g. Clift et al. 2014; Garzanti et al. 2014).

9.1.5 Approaches to Sedimentological Research on Muddy Systems

In the field, weathering effects and poor exposure of mudstones commonly make it difficult to observe the details of texture, stratification and structure that are necessary to diagnose depositional processes and setting. Direct palaeocurrent information is scarce and very difficult to determine. It is sometimes unclear which features of the rock are depositional and which are post-depositional (e.g. colour, oxidation state and carbonate content). Consequently, facies analysis of fine-grained rocks, based on hand specimen observations, has developed slowly. For example, studies of cores from modern depositional environments, outside the deep sea, have often focussed on details of coarser-grained sediments as these were perceived to be more readily diagnostic of depositional conditions.

Mudstones, often with marine fossils, were long interpreted in terms of low-energy deposition in deeper water. Shepard (1932) showed, however, that modern continental shelves differ markedly from a theoretical concept of textural grading. He pointed out that the mid-shelf mud blankets of the South American and East China shelves were flanked to seaward by coarser-grained sediment that he correctly hypothesised to be relict and related to previous lowstands of sea level.

Early analytical studies of mudstones focussed on determining composition and mineralogy, largely using X-ray diffraction. This led to an understanding of the processes of near-surface alteration and of burial diagenesis. It also opened up provenance studies using clay mineralogy to distinguish the effects of alteration in soil profiles and the imprint of the parent rock. This approach was summarised by Weaver and Pollard (1989).

At the same time, the organic content of mudstones was extensively analysed by hydrocarbon industry scientists, largely focused on organic-rich shales as hydrocarbon source rocks. Characterisation of the organic matter allowed molecular matching to likely organic precursors (by identifying key bio-marker molecules) and their habitats, as well as determination of depth of burial and temperature exposure via thermal alteration of organic material. Hydrocarbon industry scientists also studied mudstones as fluid seals (e.g. Cartwright 1994).

Pioneering facies and process work on fine-grained sediments on the relatively easily accessible tidal flats of the Dutch Wadden Zee was undertaken by Van Straaten and Kuenen (1958) and Postma (1961). Important connections between oceanographers and sedimentologists were made by workers such as Nick McCave (McCave 1972, 1976), who drew attention to key questions regarding the erosion, deposition and transport of mud, pointing to the importance of flocculation, and the mid-shelf mud belts of modern shelves. Deployment of oceanographic equipment allowed the establishment of depositional models based on quantitative data from instruments, including shallow coring, measurement of suspended sediment concentration and current velocity. Complementary laboratory techniques involved the use of re-circulating benthic flumes to study the details of sediment transport, the use of short-lived isotopes (^{210}Pb (half-life 22.3 years); ^{137}Cs (30.2 years), ^{234}Th (1.9 years) and ^{10}Be,^{7}Be (half-lives of days)) to calculate accumulation rates and high-frequency shallow seismic surveys to establish stratal geometry. Such field- and lab-based studies are commonly complemented by sophisticated and increasingly reliable mathematical models of ocean circulation, including riverine and tidal inputs, and thermohaline and wind-driven currents. This fully integrated approach characterised the work of Charles Nittrouer and colleagues in the 1980s and 1990s: for

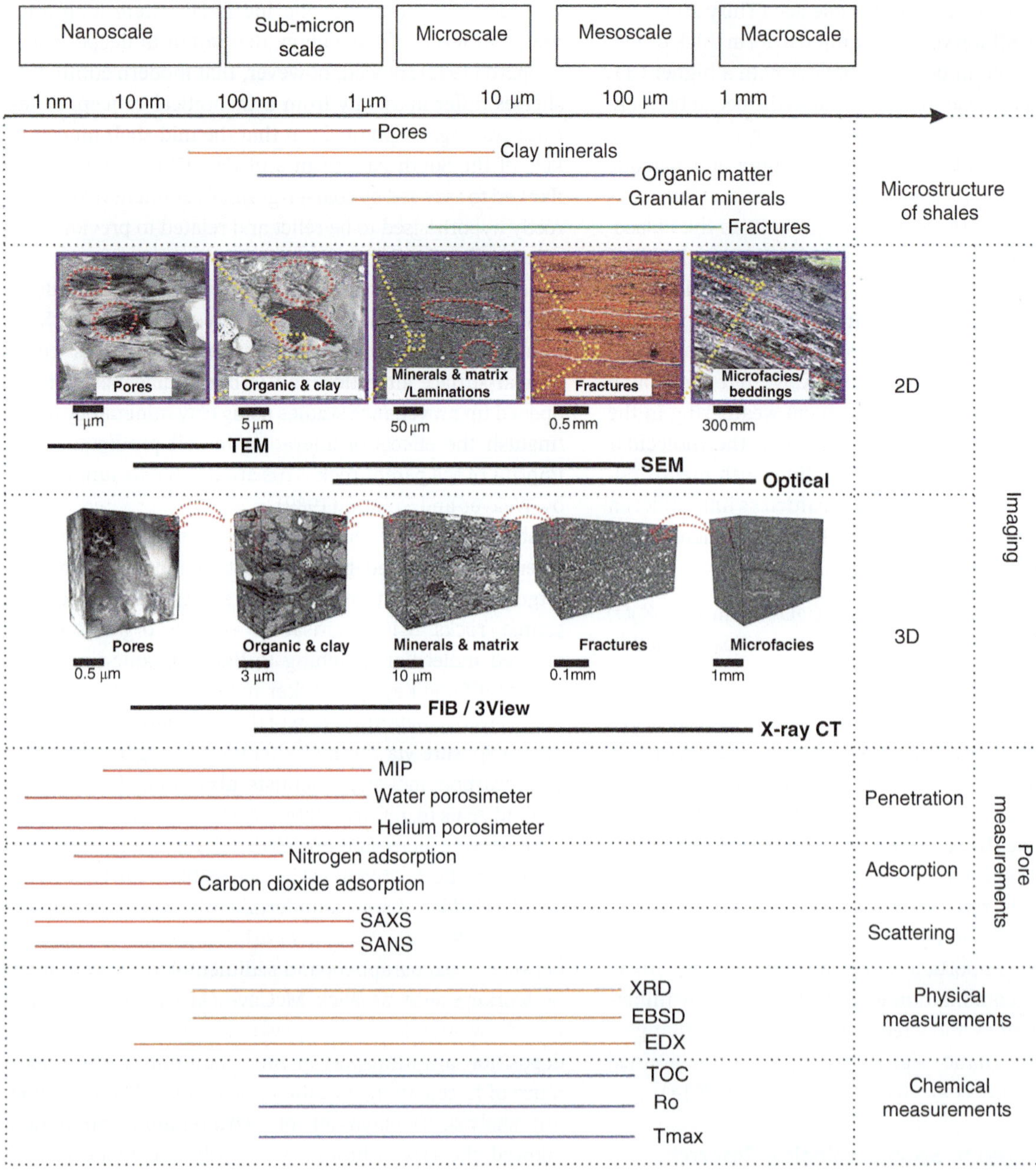

Figure 9.6 Analytical techniques applied to mudstones (Ma et al., 2017 / Geological Society of London / Public Domain). Note: great care has to be taken in such analyses to understand the effects of weathering on surface samples and of the sample preparation processes. Acronyms are as follows: TEM, transmission electron microscopy; SEM, scanning electron microscopy; FIB/3, view block face imaging; X-ray CT, X-ray computed tomography; MIP, membrane interface probe; SAXS/SANS, small-angle scattering techniques; XRD, X=ray diffraction; EBSD, electron backscatter diffraction; EDX, energy dispersive X-ray analysis; TOC, total organic carbon; Ro, reflectance measurement of thermal maturity; Tmax is the temperature of maximum pyrolysis yield.

example, the documentation of river flood deposits on the Eel River Shelf off California (Wheatcroft et al. 1996).

Recently, the exploitation of shale-gas and oil resources triggered a new wave of mudstone research. Thin-section analyses of mudstones, championed for decades by Joe Macquaker and Jurgen Schieber, are now mainstream and are supplemented by XRF (X-ray fluorescence and EDX (energy dispersive X-ray) or ICP-MS (inductively coupled plasma mass spectrometry) elemental mapping and ion microprobe isotope analysis, to give high-resolution geochemical data bearing on authigenic mineralogy and redox states. This has led to the recognition of a wide variety of mudstone fabrics and mudstone sedimentary structures. This was accompanied by flume research (Schieber and Southard 2009) into the hydrodynamic stability fields of these fabrics and structures in mudstones, siltstones and muddy sandstones.

A host of modern analytical techniques is now being applied to questions of mudstone composition and provenance (Dowey et al. 2020). Ion-beam milling is used for viewing and serially sectioning fabrics (Buckman et al. 2020). Quantitative mineralogical techniques now provide high throughput mineral phase maps of fine-grained rocks using EDX spectroscopy. These techniques together with traditional scanning or transmission electron imaging (showing texture), back-scattered electron imaging (revealing elemental composition) and X-ray computed tomography are now routinely used to produce sets of images that allow an understanding of fabric, porosity, permeability and composition in these exceptionally heterogeneous rocks (Figure 9.6; Ma et al. 2017). A remaining challenge is one of observational upscaling: microfacies studies need somehow to be generalised to larger rock units (Aplin and Macquaker 2011).

Another recent focus is on the role of the settling of aggregate grains ('marine snow') to the sea floor, as an element of the carbon cycle. The so-called 'ballast effect' operates through two processes: (1) desorption, or integration of carbon in mineral structures, which both extracts particulate or molecular carbon from seawater and protects it from oxidation (oxidation is also termed 're-mineralisation') by converting it into calcareous tests and (2) increasing the settling velocity of particles by adding mineral ballast, thereby reducing oxidation due to respiration as the particles sink through the water column. These processes add significantly to the effects of the biological 'carbon pump' (e.g. Legge et al. 2020).

9.2 Late Holocene Muddy Depositional Systems

Modern shelf sediment studies are perforce based on sediments deposited during the particular conditions of a eustatic sea-level rise of approximately 100–120 m amplitude which started after the last glacial maximum about 16,500 years ago. The last 5000 years is, in sequence stratigraphic terms, a highstand setting. Modern marine muds may therefore not provide direct analogues for ancient mudstone successions deposited in wide epeiric seas (Schieber 2016), during sea-level lowstands, or during greenhouse conditions with slower rates of sea-level change. Despite this, much can be learned about processes from Late Holocene mud-rich shelves and coasts.

9.2.1 The Amazon to Orinoco Mud-Transport System

The 1500 km long coast from the mouth of the Amazon ('Proximal Accumulation Dominated' with 'Subaqueous Clinoforms') to that of the Orinoco River on the passive continental margin of northeast South America is the longest continuous stretch of muddy coast ('Coastal Mud Belt') in the world (Figure 9.7). The Amazon River is conservatively estimated to annually supply about 1250 million tonnes of sediment to this coast, almost all of which is mud. The shelf down-drift (NW) from the mouth of the Amazon ('Marine Dispersal Dominated') is characterised by some 20–30 discrete mud banks, 10–50 km long, and up to 30 km wide, oriented obliquely to the coast (Eisma et al. 1991). These mud banks migrate alongshore at about 1 km per year and also shoreward. As the banks weld to the shore, they are colonised and stabilised by *Avicenna* mangroves on their landward sides. The volume of mud in each bank is about the same as that delivered by the Amazon in one year.

The Amazon River discharges onto a macrotidal shelf with a 6–10 m tidal range. The freshwater plume from the river is deflected to the northwest by along-shelf currents, driven largely by wind, augmented by tides. The peak discharge of suspended sediment occurs in January to April, which coincides with the strongest NE trade winds that drive the 'South Equatorial Current' towards the northwest. This current is so strong that for the first few hundred kilometres, mud is unable to settle. Only when the tidal range has diminished to 3 m, about 480 km NW from the mouth of the Amazon, do mud banks start to accrete (Orseau et al. 2020). Longshore transport is so effective that even as far NW as the Orinoco delta, ~ 50% of the sediment is sourced from the Amazon.

As the out-flowing Amazon freshwater floats above salt water, it sets up an estuarine type of circulation on the open shelf. (Although unconstrained by the margins of an estuary!) The river discharge is so great that, at times of peak discharge, plumes of turbid, fresh-to-brackish water extend some 200 km seaward. Suspended clay flocculates as fresh and salt water mix, and mud settles to the sea floor on the top of a muddy subaqueous delta platform. Near the seabed, suspended mud reaches densities of 10, up to 330 kg m^{-3}, forming a 'fluid mud'. Fluid mud is defined as a high-concentration (greater than 10 kg m^{-3}), colloidal suspension of particles ($< 63\ \mu m$) with a high percentage of clay particles ($< 2\ \mu m$). At such high concentrations, the settling of mud flocs is hindered by the simple difficulty of displacing water upwards as they fall ('hindered settling'). The fluid mud layer, which can be up to about 7 m thick, is periodically re-suspended by waves and tides, and eventually accumulates offshore from the river mouth on the prograding delta-front clinoforms extending to a depth of about 70 m, or along-shelf on the mud banks. The topset–foreset break of the subaqueous delta lies at a depth of 30–40 m, the depth being controlled by the ambient wave energy (Nittrouer et al. 1995).

The delta itself contains about 65% of the Amazon-derived sediment, with an additional 30% contained in the coastal mud-bank complex to the NW. The remaining

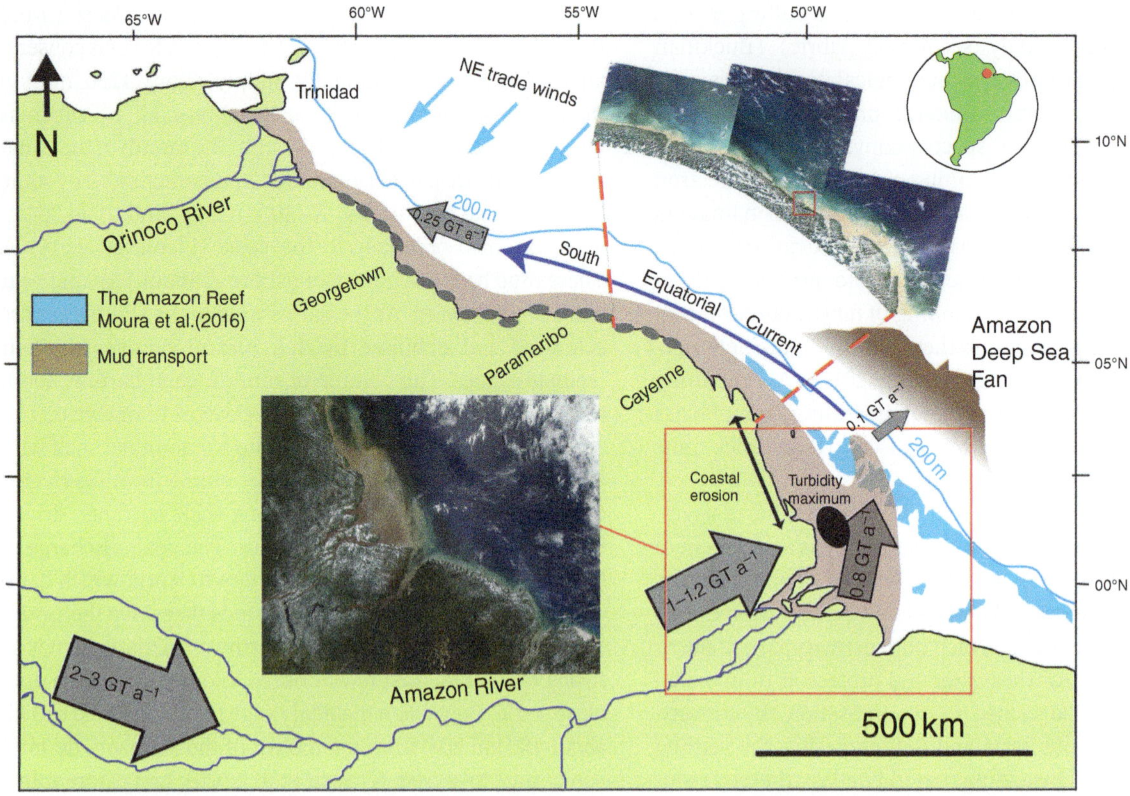

Figure 9.7 The Amazon–Orinoco system. Clay mineralogy suggests that up to 50% of the mud off the Orinoco may be Amazon derived (Van Andel 1967; Eisma et al. (1978). Coastal mud bank locations (grey ovals) are from Eisma et al. (1991). There are considerable uncertainties on the sediment fluxes (grey arrows). The Amazon River apparently sequesters some 1–2 billion tonnes per year in the lower Amazon floodplain (Allen 2017). Eisma et al. (1991) attribute 0.1 GTa^{-1} to transport in large mud banks through their migration and 0.15 GTa^{-1} to transport in suspension. The natural-colour image of the Amazon River mouth (large red box) was made by the Moderate Resolution Imaging Spectroradiometer (MODIS) on NASA's Terra satellite on July 29, 2020. This was outside the peak period of sediment delivery (January to April) but clearly shows the northward drift of fine suspended sediment on the subaqueous delta. Moura et al. (2016) document the extensive Amazon Reef system which grows partly under the freshwater and sediment plume. The Landsat 8 image of the NE coast is from Abascal-Zorrilla et al. (2018) (NASA / https://eoimages.gsfc.nasa.gov/images/imagerecords/147000/147327/amazondelta_tmo_2020211_lrg.jpg / last accessed 02 January 2024)

~ 5% of the Amazon mud travels directly off the shelf edge in the near-surface freshwater plume and is deposited over the Amazon deep-sea fan. The highest modern sedimentation rates (up to 100 mm per year) are found in a depocentre 200 km to the northwest of the river mouth. Much of the outer shelf is sediment starved, and in this region, Moura et al. (2016) documented lithified carbonate sediment with up to 30 m of relief. Significant carbonate sedimentation occurred during Pleistocene glacio-eustatic lowstands, and modern biogenic carbonate production, primarily by red algae and sponges, is favoured both by a permanent, near-bottom wedge of ocean water and by the along-shelf deflection of the muddy Amazon plume which otherwise would inhibit these organisms (Section 11.4.3).

9.2.2 The Po-Adriatic Mud-Transport System

The Adriatic Sea is 200 km wide and 800 km long and the western portion of its central and northern portions are a Pliocene to Quaternary foreland basin to the Apennine mountains, the frontal thrusts being onshore to the southwest (Figure 9.8). The Po River annually delivers to the Adriatic Sea 12–15 million tonnes of suspended sediment, together with ca 0.6 million tonnes of bedload. Smaller, but steep and sediment-rich rivers that drain the Apennines annually supply a further 32 million tonnes of suspended sediment: twice as much as the Po.

The large-scale water circulation of the micro-tidal Adriatic Sea is largely thermohaline. Three counterclockwise gyres are driven by the freshwater inputs and the strong northeasterly 'Bora' winds in the winter. These lead to a southeastward flow along the entire western shore for a distance of some 800 km. Storm waves can reach up to 12 m in height, the dominant storm wind being the 'Scirocco' from the southeast.

Suspended sediment from the Alps is deposited in: (1) the Po delta plain, (2) the subaqueous or prodelta clinothems, (3) in lagoonal and coastal deposits and (4) in a shelf mud

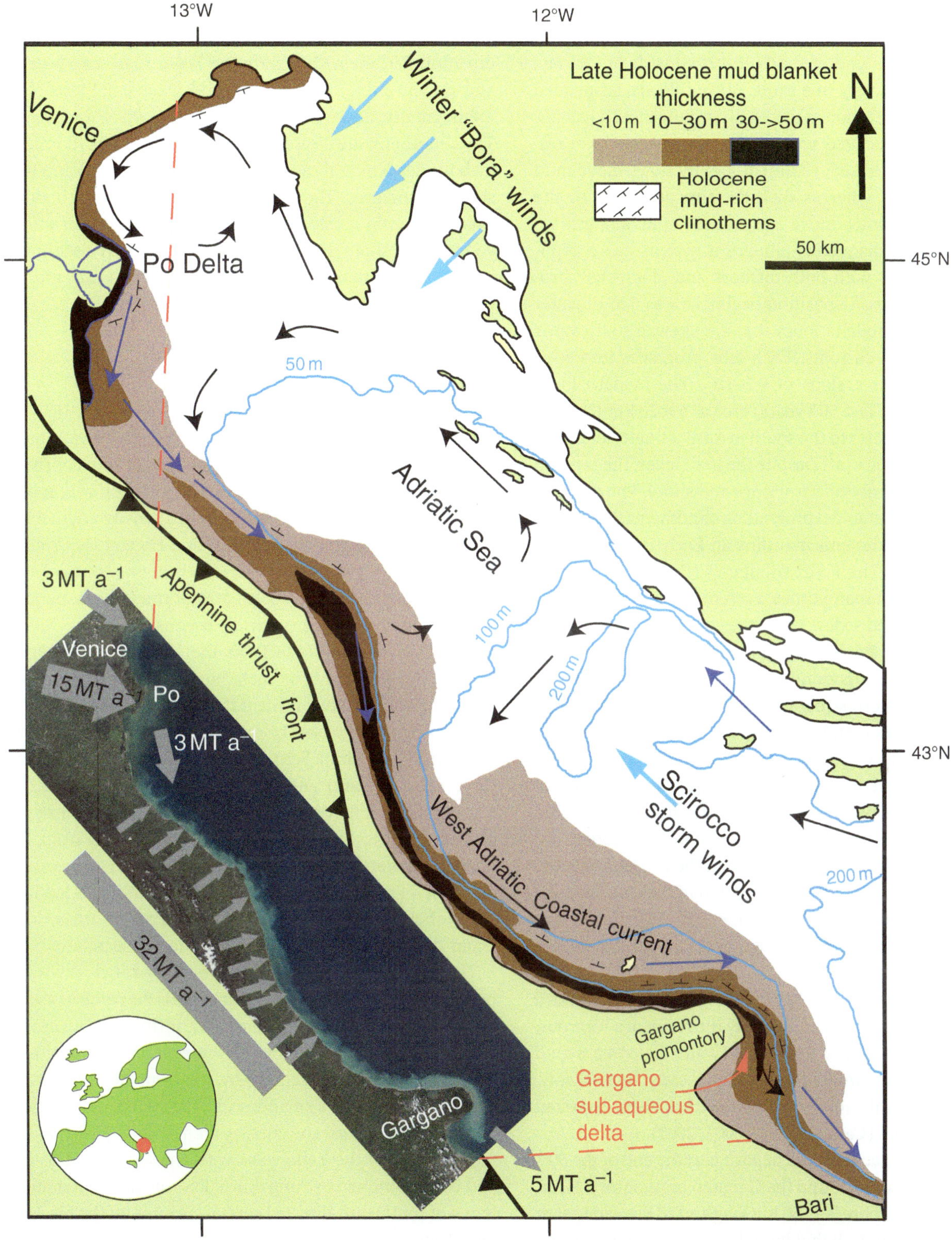

Figure 9.8 Suspended sediment routing system and mud belt of the Western Adriatic Sea between the Po Delta and the Gargano Promontory. Sediment delivery from small, steep drainage basins of the Apennine Mountains along the coast north of Gargano is more than twice that of the Po River. The accumulation rate in the entire mud belt system is about 42 Million tonnes per year. Sediment yield figures from Cattaneo et al. (2007), sediment isopach map of the Late Holocene mud deposit from Cattaneo et al. (2003). Clinoform dips are seaward and orthogonal to the overall transport direction, with sediment dispersal driven by a combination of wind-driven longshore and offshore downwelling currents. The true-colour satellite image is from the Moderate Resolution Imaging Spectroradiometer (MODIS) on NASA's Terra satellite, acquired on April 8, 2021, during the spring flood of the Po River. It shows the coast-hugging belt of high suspended sediment which is derived both from terrestrial run-off and possibly phytoplankton production due to coastal upwelling https://eoimages.gsfc.nasa.gov/images/imagerecords/43000/43478/Italy_AMO_2010096_lrg.jpg

belt to the southeast. The last is principally fed by the smaller muddy rivers draining the Apennines.

The suspended load of the Po River consists of 7% clay, 70% silt and 23% very fine sand. During floods, suspended sediment concentration at the Po River delta mouth reaches a maximum associated with mixing fronts just 2 km offshore. Wave-generated, sediment-gravity flows move mud directly offshore, whereas shelf currents generated by the northeasterly winds move suspended mud alongshore. The resulting subaqueous clinoform system has complicated geometries with both offshore and along-shelf prograding geobodies. Deposition begins in 4 to 6 m of water and occurs on prodelta lobes that are skewed southwards by the coastwise currents. Each lobe progrades into 30 m of water and has a slope of 0.5–1.5°. The modern lobes (deposited in the last 500 years) are ca. 5 × 20 km but coalesce in deeper water to the south to form a single delta platform of about 1000 km^2. Despite the low slopes, the modern prodelta displays abundant slumps and slides. The prodelta is composed of sigmoidal prograding clinothems that downlap onto relict sediments mantling an Early Holocene ravinement surface. The Po distributaries (there are currently five) have shifted many times, even in the last 500 years.

The fine-grained sediment on the continental shelf to the southeast of the Po is predominantly supplied by muddy Apennine rivers. Importantly these smaller drainage basins experience rainstorms that coincide with storm winds and waves. The waves rapidly disperse the sediment alongshore, preventing the development of subaerial deltas. This contrasts with the Po system that is extensive enough that it experiences rainstorms in the hinterland while the sea off the delta is calm, allowing deposition and progradation of prodelta lobes.

The volume of sediment deposited in the Po Delta over the last 6000 years is about the same as that of the coast-parallel shelf mud blanket to the SE. The latter is 50 km wide and 500 km long and lies mainly in 20–50 m of water. Seismic images reveal sigmoidal clinothems that dip offshore (NE) at 0.5°, with roll-over points in 20–30 m of water. The abrupt seaward downlap of the clinothems is mainly the result of a vigorous, downwelling along-shelf flow of cold water that sweeps mud to the SE, limiting progradation of the toesets. Remarkably, the 35 m thick mid-shelf mud blanket has been deposited in just 6000 years (Cattaneo et al. 2003, 2007).

The most spectacular mud deposit is at the end of the 500 km long mud patch around the Gargano promontory. Here, in the mid-shelf, there is a 150 km long and 20 km wide patch of riverine mud and silt that is totally disconnected from any coastal sources. This 'Gargano mud patch' (Cattaneo et al. 2003), accounts for 25 of the 180 km^3 of the western Adriatic Holocene mud blanket. The accumulation is up to 50 m thick, 8 m of this mud and silt being attributed to just the last 500 years. The rates of accumulation are the highest in the entire western Adriatic system: 10 mm/year. Successive mud additions have accreted seaward of one another with clinoform dips (1° max.) at right angles to the gross transport direction. The cross-shelf transport is due to Ekman veering and downwelling, and current and wave-enhanced sediment gravity flows (CWESGF; Sections 9.3.4 and 9.4.4). The clinothems are discrete depositional packages, bounded by both depositional (accumulation rate changes) and erosional boundaries. They demonstrate that the mud blanket was not continuously deposited and is also diachronous: younging offshore. Inboard of the mud patch, the sediment is silty sand whereas updrift it is bioclastic sand and, in places, there is bare bedrock.

9.2.3 The East China Sea Mud-Transport Systems

The large rivers of eastern China, which are exceptionally rich in fine-grained sediment, drain areas of actively rising mountains, and also areas blanketed by thick accumulations of Quaternary loess. Before being heavily engineered by man, the Changjiang (Yangtze) and Hwang He (Yellow River), annually supplied some 1600 million tonnes of suspended sediment to the ocean: 10% of the total suspended sediment delivery to the world's oceans by rivers. Collectively these shallow seas form the broadest continental shelf in the world and, as such, have been used as an analogue for ancient epicontinental seas (Figure 9.9).

Water depth increases from the northwest (Bohai Sea) to southeast (East China Sea). The Bohai Sea is surrounded on three sides by land and connected with the Yellow Sea through the Bohai Strait which has an average water depth of 18 m. In the Yellow Sea, the water deepens from the north and northwest to the southeast, with an average depth of 44 m. The East China Sea is broadly divided into inner shelf, outer shelf and Okinawa Trough areas, with water depth increasing from < 60 m to > 200 m.

The water circulation pattern is broadly anticlockwise in both Bohai Bay and the Yellow Sea. This is the result of a variety of interacting factors, including the intrusion of an arm of the warm Kuroshio thermohaline current on the east side of the Yellow Sea, the input of freshwater from the big rivers on the west side and, at a smaller scale, the system of local tidal currents. Tidal range, and current strength is amplified by the funnel-shape of the Yellow Sea with a tidal range of 3–9 m and correspondingly strong diurnal currents. The net effect is one of southward-directed coastal currents off the big river mouths to the west, and northward-directed currents off the northern Korean peninsula to the east.

In the East China Sea, the northward moving, warm water Kuroshio current dominates the outer shelf, but the coastal currents are return-flows directed southwestwards. As a result of these coastal currents, fine-grained sediment is moved consistently southward along the

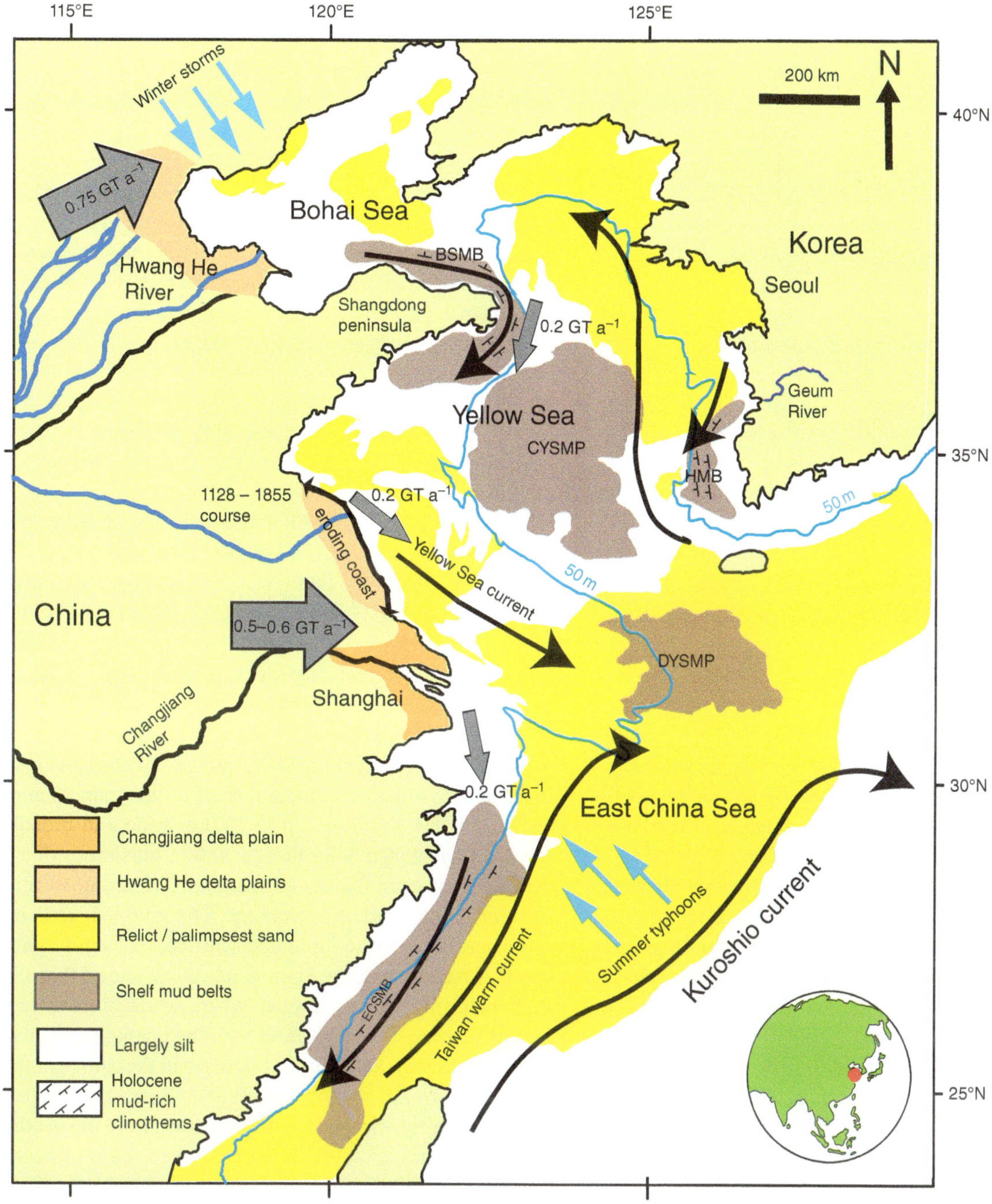

Figure 9.9 East China Seas sediment routing system for fine-grained marine sediments. BSMB: Bohai Sea Mud Belt, CYSMP: Central Yellow Sea Mud Patch, HMB: Hueksan Mud Belt, DYSMP: Distal Yellow Sea Mud Patch, ECSMB: East China Sea Mud Belt. The Holocene to Recent mud belts rest on relict sands (yellow) which have been locally reworked by strong tidal currents into linear sand ridges. The Mud Belts show clinoforms of various ages withoffshore dips (indicated; see Figure 9.10 for stratal architecture). Arrows indicate the main wind-driven and thermohaline currents (Yuan et al. 2008 with permission of Elsevier). This extensive suite of shallow seas has been cited as a modern model for ancient epicontinental seaways. Sediment fluxes are 'pre-industrial estimates'. Both the Changjiang (Yangtze) and Hwange He (Yellow River) have seen significant reductions in sediment flux over the last few decades due to irrigation and damming.

western sides of the Bohai, Yellow and East China seas in an enormous mud transport system. All these southerly flows of suspended sediment are strengthened in the winter by strong northerly winds and the associated waves. Overall, the wave climate of these shelf seas is moderate, with waves up to 4 m in the Yellow and Bohai seas, although the area is also subject to typhoons. Thus, overall, these are relatively high-energy shelf settings.

The various north- and south-flowing coast-parallel currents (Figure 9.9) limit the seaward advection of muddy sediment. For example, in the East China Sea, the seaward limit of the mud belt appears to be controlled by northeast-flowing warm water of the Taiwan Warm Current and the Kuroshio current. Upwelling beneath these warm currents and downwelling beneath the cold coastal current, particularly during winter, causes a vertical circulation cell which may trap mud on the inner continental shelf.

Annual sediment input from the Hwang He River, before civil engineering works, was 750 million tonnes. The river carried the astonishing load of 25 kg of sediment per cubic metre of water, which was 70 times that of the natural Mississippi and 100 times that of the Amazon! The high concentration of mud causes the river water to sink beneath the seawater as a hyperpycnal flow, resulting in mud being deposited as clinothems in the nearshore subaqueous delta. However cold winter winds from the north generate strong currents and waves that re-suspend this prodelta mud, forming fluid mud that is then transported along and across the shelf.

The Hwang He has experienced 20 channel avulsions that caused five significant shifts in the river mouth over the last 2500 years (Xue 1993). Natural avulsions have shifted the river mouth along the coast by as much as 300 km, whereas in 1128, during the Song Dynasty, a 750 km shift of the river mouth was engineered specifically to hamper an invading army! These changes in river mouth location have left a suite of abandoned delta lobes which have subsequently been reworked by waves and tides, the muddy sediment being transported southward along the coast (Zhou et al. 2014).

The very high sediment load causes the Hwange He Delta to prograde extremely rapidly. The post-1855 delta grew 50 km seaward with the subaqueous delta extending a farther 15 km. Because the coast is mesotidal, the delta is fringed by extensive tidal mud flats that range from 3 to 15 km wide. The modern delta is about 14 m thick and is fronted by a gently inclined clinoform where mud accumulates at the rate of about 4–9 mm per year. Distributary channels are very unstable, and individual delta lobes are active for about ten years.

Prior to engineering works, suspended sediment discharge from the Changjiang River was 500 million tonnes per year. Both the Changjiang and the Hwang He have experienced a reduction of 20% in suspended sediment discharge over the last 50 years, and the Changjiang has seen a further drop of at least 80% since the Three Gorges Dam was built. The scale of these anthropogenic changes (30,000 dams on the Changjiang and its tributaries) means that measurements of present sediment and water transport are not representative of those that generated the observed sediment bodies.

The Changjiang Delta progrades into 50 m of water, with a mean tidal range of 2.7 m. Sandy sediment is mainly confined to estuarine channels and bars in less than 10 m of water, whereas silt and clay dominate the subaqueous delta between 10 and 50 m. From July to September, sediment discharge is at a maximum, and this coincides with prevailing winds that drive the turbid plume to the northeast, when sediment accumulates at 50 mm per month. However, much of this sediment is re-suspended by more persistent shelf currents and transported back to the south, where the long-term accumulation rate of the subaqueous delta is 50 mm per year (Wright and Nittrouer 1995). Seismic images show that the Holocene subaqueous delta is essentially aggradational or may show extremely low-angle accretion surfaces, rather than clear clinoforms. Estimates suggest that the subaerial part of the Holocene delta has trapped about half the sediment, with another 20% in the nearshore subaqueous delta and the remaining 30% escaping to the offshore mud belt.

Three main Holocene mud belts are found along the coast in this complex of tidal shelf seas (Figure 9.9).

The Bohai Sea Mud Belt (Holocene, <7ka) has a maximum thickness of 15–40 m and lies in 20–60 m of water. The mud deposit has an aggradational geometry but this changes to a sigmoidal clinoform geometry in areas where cross-shelf flows are stronger.

The East China Sea or Changjiang Coastal Mud Belt (Liu et al. 2007, 2010; Liu et al. 2018; Xu et al. 2012) runs along the Zhejiang-Fujian continental shelf in less than 90 m of water. It is 800 km long, up to 100 km wide and broadly speaking has a sigmoidal profile, with a maximum thickness of 40 m near the 20–30 m isobath, thinning to zero between the 60 and 90 m isobaths. The mud belt rests on a transgressive surface mantled with coarser sandy sediment. A recent study by Shan et al. (2023) shows that the mud belt migrated landward when sea-level rose and moved seaward when sea-level fell. The opening and closing of the Taiwan Strait caused changes in the source of the mud, with an alternation between sediment contributed from the Changjiang delta in the northeast, from the nearby Minjiang River, and even from Taiwan to the south-east. The mud belt comprises both shallower- and deeper-water facies. The former predominantly comprising fluid-mud beds generated by waves and tidal currents. The deeper-water facies contain shell-rich and bioturbated muds, which were probably deposited by the coastwise Chinese coastal current and Taiwan Warm current. Accumulation was rapid between 8000 and 500 years ago, which may be linked to a wetter climate, whereas an additional rate increase, starting 200 years ago, may be anthropogenic.

The Heuksan Mud Belt (Figure 9.10) is located off South Korea in a region with tides of 2–8 m (Lee et al. 2015). The mud belt is 20–50 km wide and over 230 km long and lies in 20–100 m of water. The sediment has a mean thickness of

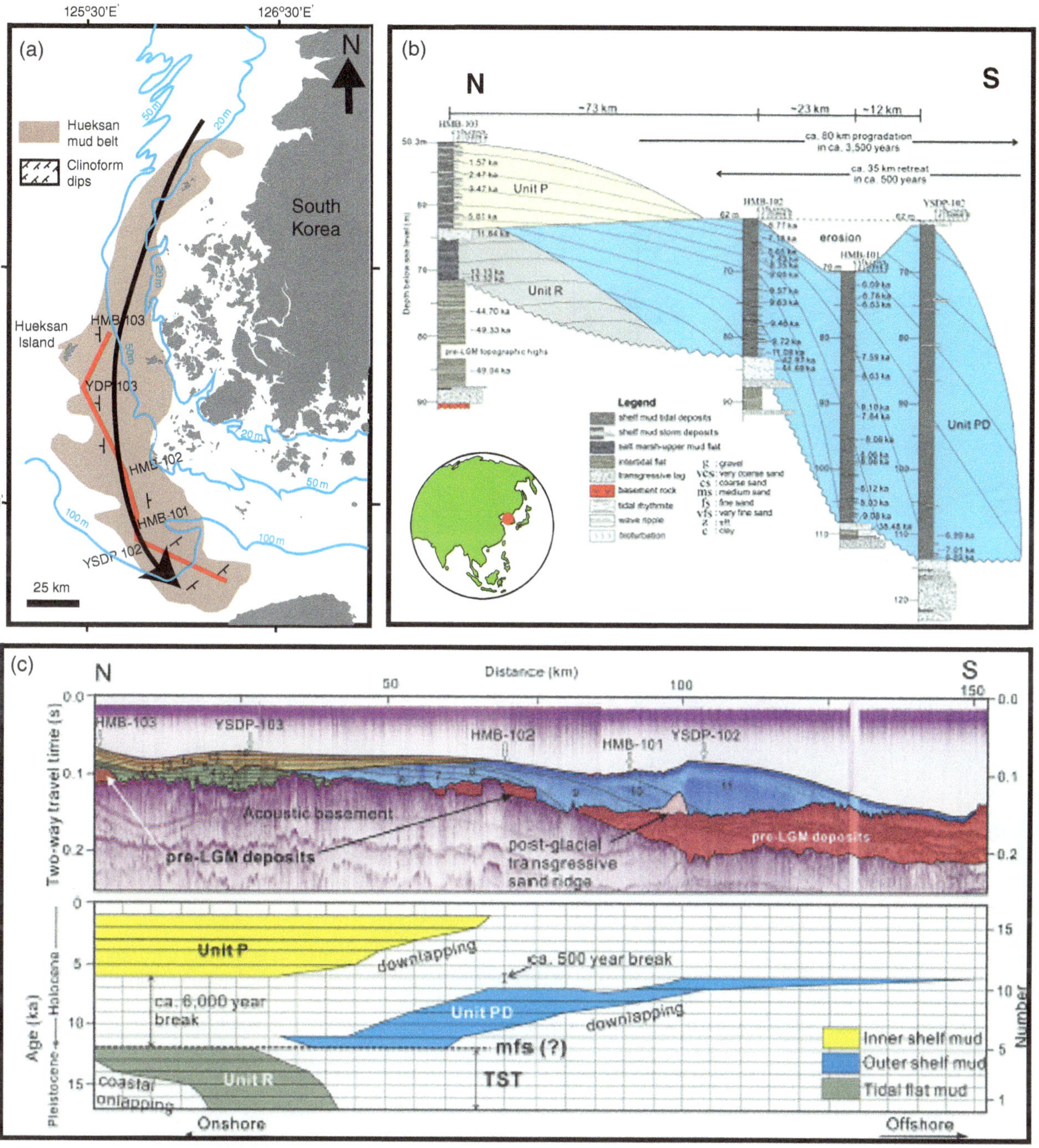

Figure 9.10 The Hueksan Mud Belt is supplied primarily by Korean rivers and the mud is driven southward by the Korean Coastal Current. The sediment body is up to 50 m thick and consists of a series of stacked clinoform units (see also Lee et al. 2015) resting on relict sands that were reworked into tidal sand ridges (as reflected in the 50 m isobath). The active mud belt (Unit P) extends as far south as the corehole HMB 102 and has seaward (west)-dipping clinoforms. Older Holocene units (PD and R) have southerly and southeasterly dipping clinoforms. Adapted from Chang and Ha (2015).

10 m, a maximum of 50 m. The clay mineralogy suggests derivation from Korean rivers. However, the mud belt begins some 100 km south of the Guem River mouth and is disconnected from it. Three mud units are recognised: (1) a lower, retrograding, transgressive tidal flat system (unit R), (2) a rapidly deposited shelf mud belt formed from 11 to 6 Ka b.p., which extends 80 km onto the shelf (unit PD) and (3) an upper shelf mud unit (P) that back-stepped some 35 km landwards. The abrupt break between units 2 (PD) and 3(P) is associated with an abrupt reduction in

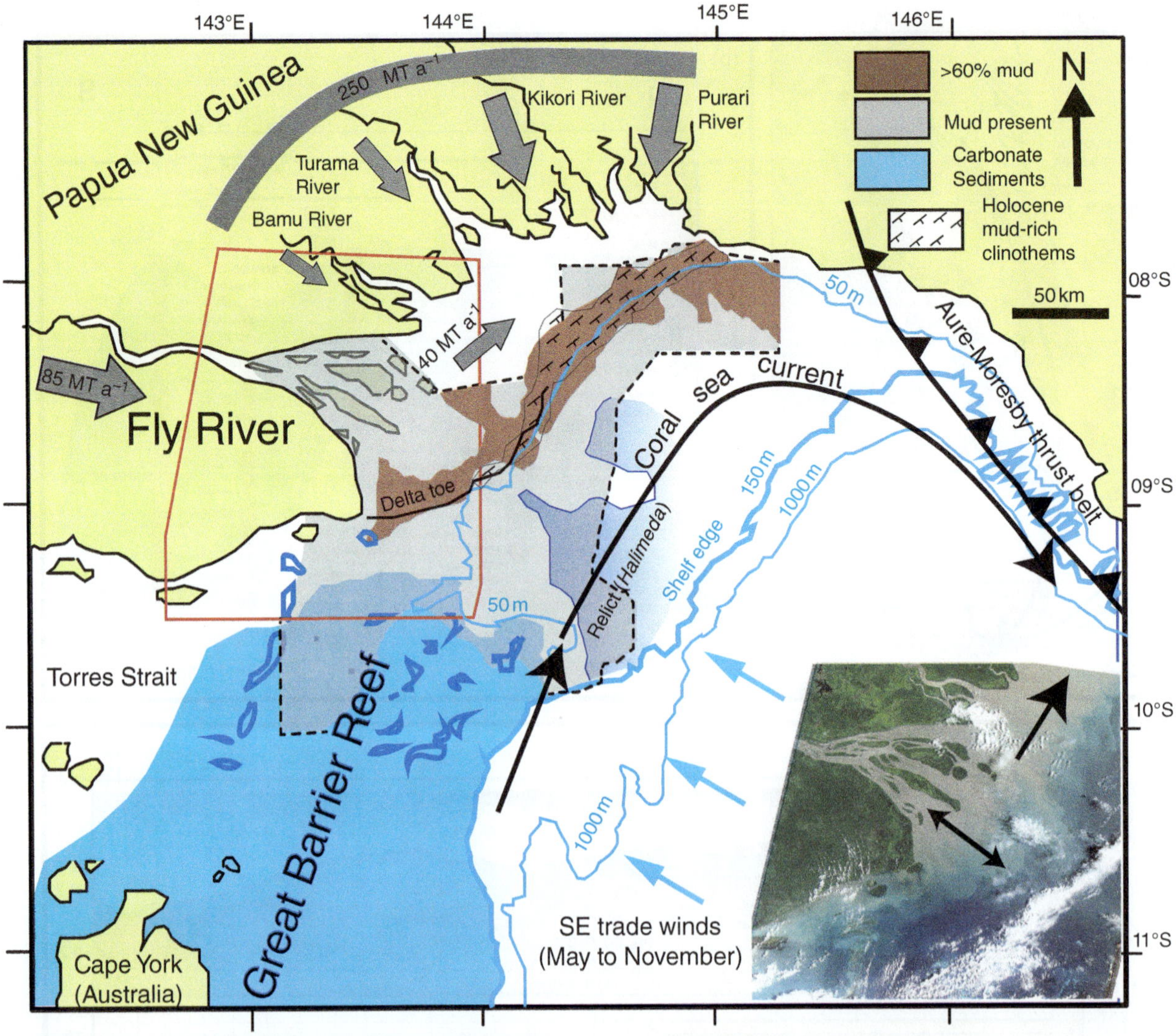

Figure 9.11 The Gulf of Papua is part of a foreland basin with steep drainage basins and rapid tectonic uplift of unconsolidated sediments in a tropical climate. The mud belt (brown) is composed of sets of Holocene to Recent, seaward-dipping, mud-rich, clinothems (Slingerland et al. 2008; Wei et al. 2019). The landward side of the mud belt is more sand-rich due to a combination of higher sand supply and winnowing by tidal currents. To seaward, the mud becomes mixed with carbonate sediments. The muddy plume emerging from the Fly system is deflected northward by an along-shelf current driven by the SE trade winds. This northward drift is illustrated by the Landsat 5 image taken on 11th Feb 2009 during the flood season. (https://eatlas.org.au/media/522.inset located by the red box). The double-headed arrow indicates the reversing tidal currents in the river mouths and the arrow to the NE indicates how the plume is swept away from the Great Barrier Reef by the Coral Sea Current, resulting in a relatively abrupt lateral transition between detrital muds and biogenic carbonates. Sediment fluxes from Milliman (1995) and Harris et al. (2004); Bathymetry from Septama et al. (2016). Sea bottom lithology from Harris et al. (1996 and 2004). See also Section 2.5.

accumulation rate from between 20 and 50 to 2 mm per year. Sedimentation subsequently resumed in unit P at up to 17 mm per year, building seaward-prograding clinothems in 40–75 m of water. These clinothems have prograded 80 km basinward in only 4.5 k yr.

The Central Yellow Sea Mud Deposit and the East China Sea Distal Mud Patch (Wang et al. 2014) occur in about 70 m of water, where they rest on relict sands and gravels of the Holocene transgression. Much of the sediment in these outer shelf mud deposits was supplied from the Hwang He and Changjiang rivers, with later addition of mud as a result of the erosion of the abandoned Hwang He delta lobe that had been active between 1128 and 1855 (Lim et al. 2007; Qiao et al. 2017). Both mud patches contain up to 75% biogenic particles. The location of the mud patches is probably controlled by large-scale gyres that are associated with both up- and down-welling currents that result in high nutrient flux and organic productivity.

9.2.4 The Fly River Mud Transport System of Papua New Guinea

The Fly River drains the tropical highlands of Papua New Guinea, from which it flows south and then east, debouching into the warm clear waters of the Gulf of Papua/Coral Sea (Figure 9.11). The Fly River and delta occupy an Oligocene to Recent foreland basin with southwestward-verging thrusts in the Papua Mountains that are formed by the convergence of Australia with the Papuan block (Ott and Mann 2015). The western Gulf of Papua is a 150 km wide, low-gradient continental shelf that narrows rapidly in the north. The shelf edge lies at about 160 m and is an extension of the Great Barrier Reef off eastern Australia (Harris et al. 1993, 1996). The shelf in the western Gulf of Papua comprises two physiographic steps and flats (Slingerland et al. 2008). The outer flat extends from 60 to 100 km from the coast beneath 80 to 120 m of water and is mostly covered by mud mixed with relict carbonates. This region is the eroded top surface of an approximately 50 ka old series of clinothems. To landward, in 25–80 m of water, lies a gently inclined (0.18°) muddy slope that marks the leading edge of prograding, Holocene to Recent mud-rich clinothems. The inner flat, 60 km wide, is smooth and rises gently from 25 m to the mangrove-fringed coast. The Fly River subaqueous delta is part of the clinothem complex, but it does not form an identifiable bathymetric lobe because the muddy plume from the Fly is strongly deflected to the north by the wind-driven longshore current.

The Fly River had a pre-industrial (before mining and deforestation) sediment discharge of 85 million tonnes per year, of which 90% was suspended sediment and 10% bedload. The Fly discharges into a meso-tidal sea with a tidal range of 3.5 m, amplified to 5 m at the apex of the delta. About half of the suspended sediment delivered by the Fly accumulates in deltaic environments and half is transported northeastward by the prevailing circulation in the Gulf of Papua. A small amount (6%) of this material is trapped by coastal mangroves.

Sediment initially deposited on the inner part of the Fly River delta is reworked by waves, tides and storm-generated geostrophic flows, all of which help to re-suspend the mud and transport it, via the distributary channels, to the subaqueous prodelta area where it accumulates on the clinoform in 18–60 m of water. Long-term upwelling and wind-driven currents tend to drive the mud obliquely across the clinoform towards the northeast. The upwelling of deep water moving northwest during the monsoon season helps keep the suspended sediment on the inner shelf.

The sedimentation rate varies from 10 to 20 mm per year on the delta front, rising to 40 mm per year on the prodelta. The clinoforms have a roll-over point at a depth of 25–40 m, and the toes can extend to 100 m water depth.

The northern Great Barrier Reef extends up to 50 km south of the Fly River. The interaction between the carbonate and clastic systems dates back to the Late Miocene but clastic input significantly increased in the Late Pliocene. The Fly River delivers 100 times more suspended sediment to the back of the barrier reef than any other river entering the back-reef 'lagoon'. Terrigenous mud content decreases rapidly southwards into the carbonate domain, with 60–90% calcium carbonate in surficial sediment 20 km south of the delta. The carbonate is derived from benthic foraminifera, bryozoans, molluscs, echinoderms, calcareous green algae including *Halimeda*, and occasional corals. The relatively sharp contact between carbonates to the south and the muddy prodelta to the north is due to the predominance of northward-directed sediment and freshwater advection during the SE trade-wind season.

These four well-studied modern examples illustrate the following points, which contrast starkly with the traditional concepts of mudstone deposition, namely: 'deep, quiet-water and continuous'.

1) The thickest and most rapidly deposited accumulations of mud associated with modern rivers are located in the subaqueous deltas which seem to trap between 50 and 65% of the river load. These mud deltas develop offshore where accommodation for mud is available below wave base, and they are commonly separated from sediment input points by bypass surfaces of erosion and non-deposition.
2) The relative timing of the season of peak fluvial sediment discharge and the highest energy level of the ocean is crucial in determining how much sediment accumulates near the river mouth, and how much is re-suspended and dispersed across and along the shelf. Sediment delivery to the shelf is maximised if river discharge and ocean energy levels peak simultaneously. Sediment is then transported offshore by current and wave-enhanced sediment gravity flows
3) Due simply to the huge volume of suspended sediment supplied, thick deposits of mud and silt are presently accumulating in inner shelf and even coastal settings on high energy shelves subject to open-ocean waves and meso- to macro-tidal conditions.
4) Mud transport occurs over several hundred kilometres along strike from river mouths. This is largely attributable to wind-driven shelf currents, with a small cross-shelf component due to downwelling, with Ekman veering at higher latitudes and current-wave enhanced sediment-gravity flows. These accumulations are both

hundreds of kilometres distant, and physically detached from, their riverine sources. There is thus considerable bypass and re-suspension, with mud accumulating in bathymetric lows where accommodation is available.

5) Recent (i.e. over the last 5 Ka) accumulation rates, measured over 100-year periods, are high (order of 10 mm per year). Civil engineering works have fundamentally altered many depositional systems and sedimentation rates. It is therefore important to use caution when applying modern sedimentation rates to the interpretation of sediment bodies that accumulated prior to human interference.
6) Accumulation of tens of metres of muddy sediment is occurring, and has occurred, on a timescale of 500–5000 years: Such bodies have not yet recorded a single half cycle of even the shortest Milanković frequency.
7) Accumulation of mud has taken place during the overall late Holocene transgression, in what is part of a transgressive- to highstand systems tract in sequence stratigraphic terminology. Nonetheless, most accumulations are characterised by prograding clinothems.
8) Most shelf mud deposits are diachronous because they accumulate laterally as low-angle clinothems. Clinoforms prograde perpendicular to the long-distance shelf sediment transport path.
9) Large-scale stratigraphic surfaces within the depositional units are commonly related to changes in riverine sediment supply. Supply variation may result from climate change or shifts in the position of distributaries. The latter may trigger cannibalisation of older mud deposits that are no longer nourished.

9.3 Processes

The oceanographic processes that affect mud transport in shallow-marine systems are the same as those that affect other clastic materials (see also Chapters 8, 10, 11 and 13). However, additional processes come into play with fine-grained materials, as discussed later.

9.3.1 Flocculation and Settling

When freshwater loaded with clay and silt mixes with salt water, the suspended clay particles combine in loose agglomerations or 'flocs' (McCave 1976, 1984). The generally negatively charged clay particles flocculate when they enter the higher ionic strength fluid – salt water. However, 'salt flocculation' is just one of five agglomerating processes that can operate in estuarine settings (Eisma 1986). The others are: gluing particles together with organic matter films and mucus, particle collision during settling, particle collision through Brownian motion (probably minor) and particle collision through turbulence. Organic matter in various forms is strongly implicated in the formation of larger flocs. A concise review of these processes is given by Grabowski et al. (2011).

All of these processes operate more effectively where fresh and salt water mix, as in estuarine channels, and collectively result in a region with a high concentration of suspended particulate material, termed a 'turbidity maximum'. High turbidity mixing zones may form in the inner estuary in areas that experience vigorous salt-water intrusion due to strong tidal currents, such as the Fly River at low stage. Vigorous mixing can also occur where freshwater protrudes far into the open sea, such as off the mouth of the Amazon. The position of the turbidity maximum will also move depending on river stage, tidal conditions and wind direction and strength.

An important consequence of the slow settling of flocs is that they travel long distances laterally as they sink through a flowing body of water. This effect can, over time, lead to a persistent landward mud transport over tidal flats, and shoreward fining in channels (Figure 9.12). Thus, a mud particle is carried landward on a flood tide even after it starts to settle as it sinks through the water column (the 'settling lag'). It can therefore reach areas of lower ebb velocities. Also, because of bed cohesion (amplified by microbial biofilms), the particle is not eroded until the ebb current has reached a higher velocity than that from which it settled (the 'scour lag'), and therefore the particle has less time to travel seaward. With each tidal cycle therefore, a sort of 'ratchet' process operates, moving mud particles progressively farther down the current deceleration path (normally landward), where they accumulate on tidal flats.

9.3.2 Erosion

For non-cohesive particles like sand and gravel, deposition occurs once current velocity decreases below the 'settling velocity', and re-entrainment occurs, at about the same velocity, as the flow accelerates. This behaviour is not generally true for silt and clay, which, for many reasons, can resist erosion, once they are deposited, as discussed below.

'Fluid stripping' is the direct action of the drag and lift forces exerted by a moving fluid on a non-cohesive sediment bed (Allen 1982). There is a near-linear relationship between bed shear-stress (proportional to current velocity squared) and the log of grain-size for non-cohesive sediments. In a much-published diagram, Hjulström (1934) sketched a relationship for cohesive sediment that showed that very fine sand (about 100 microns) is eroded at the lowest shear stress. In finer-grained sediment, a higher flow velocity is needed to erode the particles. This is mainly because of cohesive effects between the grains, although the fabric of the sediment also controls erodibility (Einsele et al. 1974; Young and Southard 1978).

'Corrasion' is erosion due to the impact of particles with the bed; it is analogous to sandblasting. Saltating granular

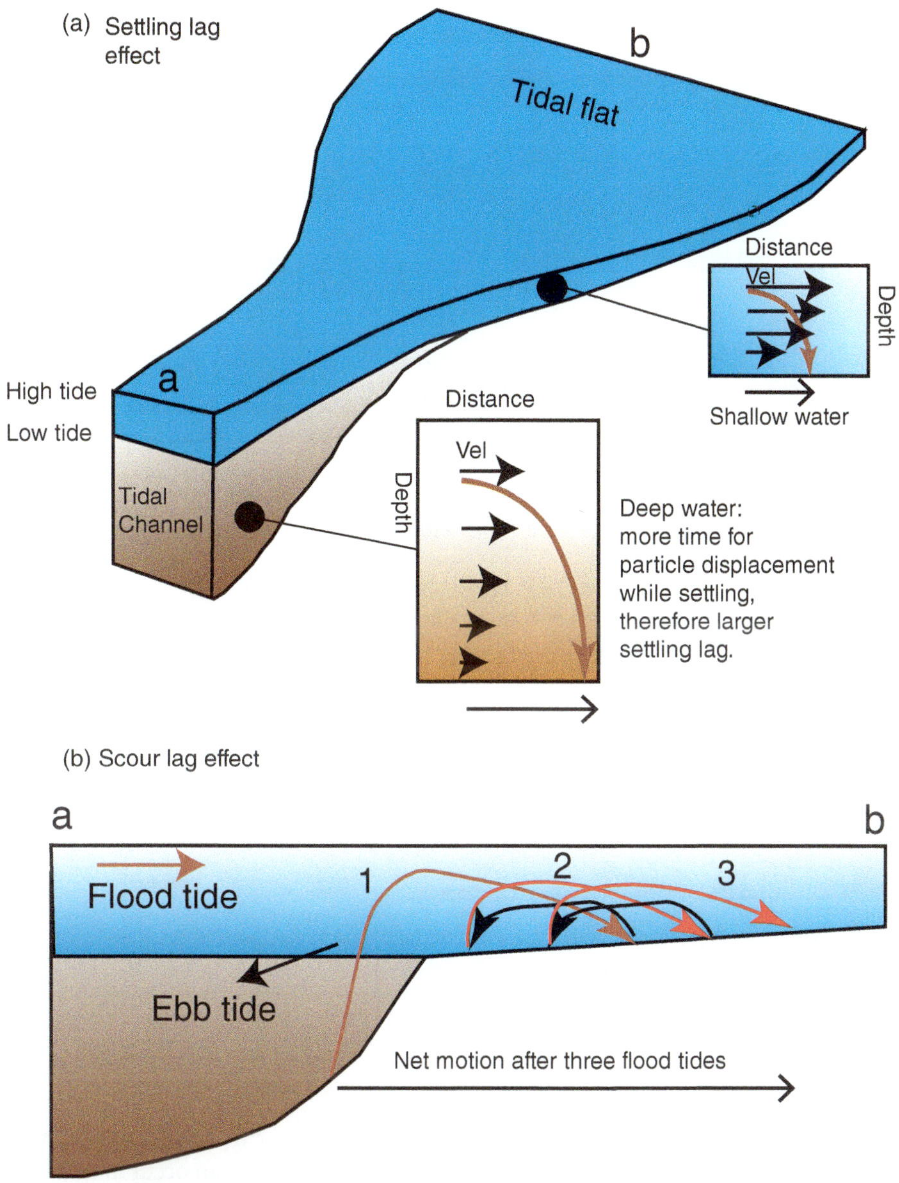

Figure 9.12 Net landward transport of fine-grained sediment in a tidal flat system. The diagram illustrates the water volume in a tidal channel relative to the surrounding tidal flat. Settling lag is the net distance travelled by a particle as it falls through the water: it increases with water depth or lower settling velocities. Scour lag can be due to differences between threshold erosion velocity (higher) and settling velocity (lower) for a particle of a given size, and/or the successive movement of particles into areas of shallower flow with greater bottom friction and slower velocity on each tidal cycle. Normally scour lag is the more important process controlling landward migration of particles. Other factors influencing the motion of particles include tidal current asymmetry and sediment concentration. Illustration based on Van Straaten and Kuenen (1958). The effect of these lags on different grain-size classes was investigated by Bartholdy (2000).

material will both lower the threshold of erosion and increase the erosion rate. Thompson and Amos (2002, 2004) showed that saltating shell material (the common cockle, *Cerastoderma*) increased the erosion rate of cohesive mud by a factor of 20.

On non-cohesive beds, such as sand, fluid stripping or corrasion removes individual grains, but cohesive muddy sediments tend to be eroded in flakes or blocks. Even soupy mud consisting of 70% water is eroded in chunks! Many of the eroded chunks disaggregate as they are carried in the flow (Schieber et al. 2010). Nonetheless, thin sections reveal that many 'fine-grained rocks' are in fact 'conglomeratic' consisting of reworked chips of mud (Figure 9.13) These intraclastic fabrics, which can be easily disguised by compaction, show that some ancient mudstones were deposited at much higher current velocities than would have been expected (Schieber et al. 2010).

Dade et al. (1992) showed that to erode a de-watered and consolidated mud required a 5- to 8-fold increase in the flow velocity compared to a non-consolidated mud. This is because

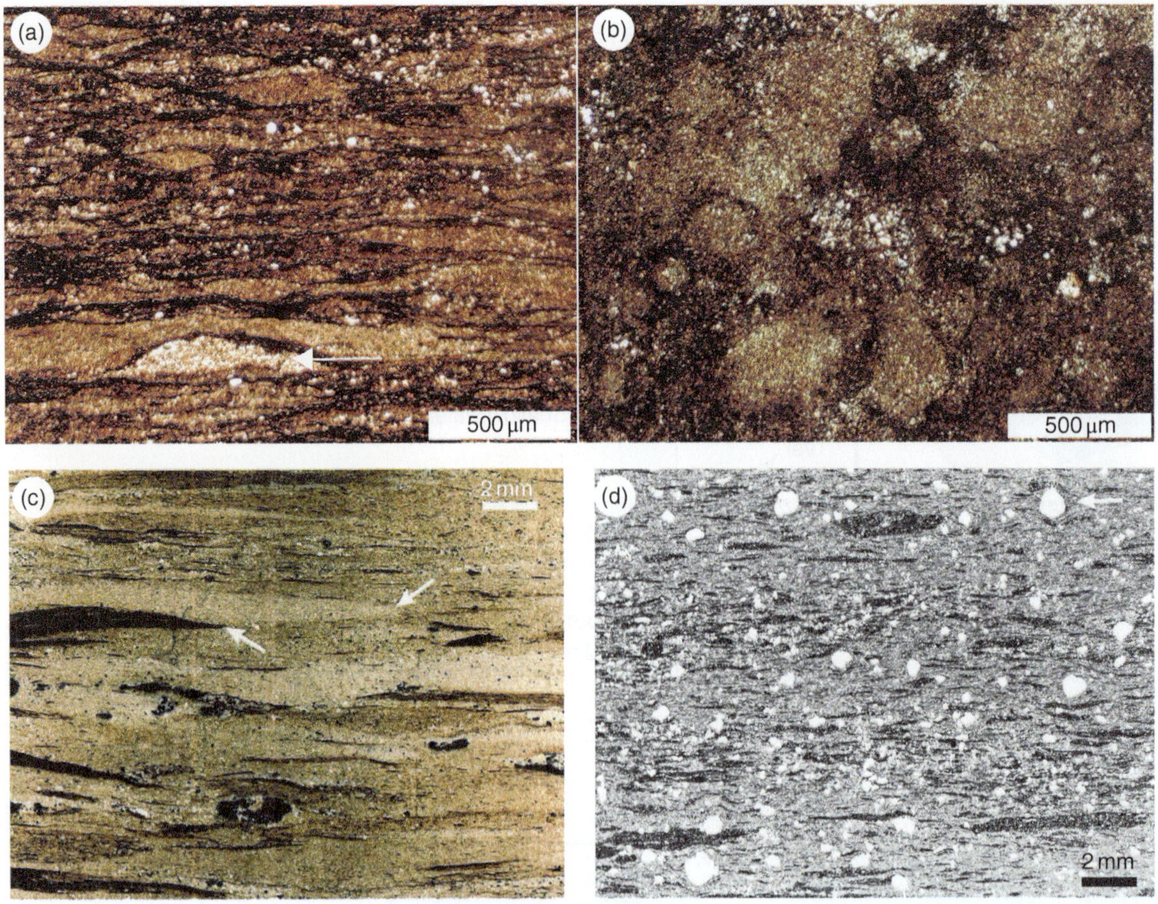

Figure 9.13 (a) Photomicrograph of lenticular laminated Proterozoic shale shows compressed mudstone rip-up clasts. View is parallel to bedding. (b) Photomicrograph perpendicular to bedding (same sample as in (a). Several mudstone clasts with irregular outlines are clearly visible (arrows). The sample is poorly sorted. (c) A bed of compacted mudstone clasts. Dark clasts are oil-shale fragments. Note compaction, bending of clasts, and tapered edges of clasts (arrows). (d) Shale with fragments of microbial mats (dark-coloured deformed particles and quartz grains (clear) in a matrix of grey mudstone. Arrow, upper right, points out differential compaction around quartz grain. From Schieber et al. (2010) / Society for Sedimentary Geology (SEPM).

compaction and burial strengthen the electrostatic attractions between particles (Kominz et al. 2011). A compacted mud will also erode 100 times more slowly than its unconsolidated equivalent. Biological activity affects erodibility in a number of important ways. Burrowers and surface grazers increase erodibility by loosening grain packing (churning), increasing water content, or creating surface irregularities. The formation of biofilms composed of microorganisms such as diatoms and bacteria, or the presence of the ubiquitous sticky extracellular polymeric substances (EPS: carbohydrates and proteins) that are secreted by a wide variety of organisms, can make sediment ten times more resistant to erosion.

It is clear that grain-size, cohesiveness, depth-dependent consolidation and biological stabilisation or de-stabilisation are all important factors in the stability of mud substrates, but subtle details of sediment fabric, layering, mineral and water geochemistry and temperature are additional factors that influence erodibility. Grabowski et al. (2011) concluded that it was not possible to make meaningful quantitative predictions of differential erodibility of muddy sediments. Nonetheless, it is clear from the deep sea that current erosion of muddy sediments can occur at all water depths down to several kilometres (Section 12.3.3) and erosion and re-suspension on continental shelves is a pervasive process in mud belt formation.

9.3.3 The Concept of Wave Base and Fine-Grained Sediment

Both erosion and transport in shallow-marine environments are controlled by the ability of wind-generated waves to produce turbulence, and current velocities sufficient to keep material in suspension, or to re-suspend it once deposited. A broadly agreed concept is that the water column can be differentiated into depth zones, delineated by fair-weather wave base and storm wave base (Section 8.3.1). Studies of modern oceans show, however, that there are no clear depth thresholds on wave or current action. Measurements from wave buoys in the Gulf of Mexico, Caribbean and Western Atlantic (Peters and Loss 2012) show

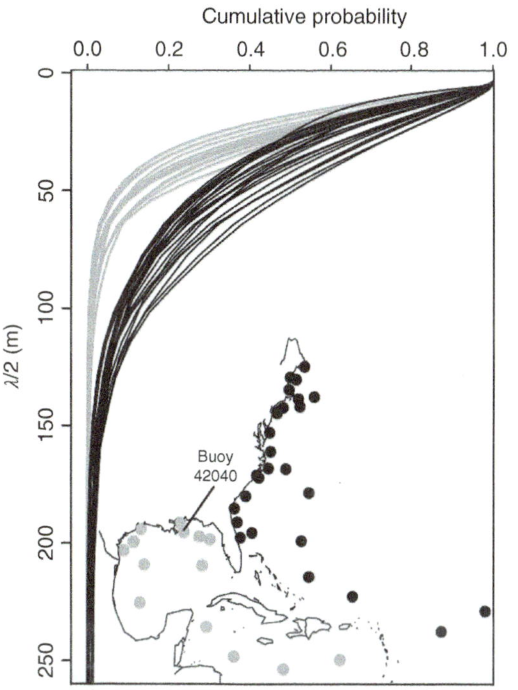

Figure 9.14 The concept of discrete wave bases for fair weather and storms and for the movement of sand and mud is useful in facies interpretation. However, data from wave buoys suggest that there is no discrete bimodality between fair weather and storm waves, rather a continuum. The probability that any given wave will encounter the sediment–water interface decreases as a continuous, smoothly varying function of water depth. Data are from >2.3 million spectral density estimates for the surface ocean collected from 32 buoys in the Caribbean, the Gulf of Mexico, and the western Atlantic. After Peters and Loss (2012) / GeoScienceWorld.

simply a continuously increasing probability that waves will reach the bottom with decreasing water depth, with no hint of storm/fair weather bimodality (Figure 9.14). However, the lower shoreface, the zone where wave base impinges on the seabed, is difficult to study (Anthony and Aasgaard 2020) especially in extreme weather, and marked breaks in slope and/or grain-size at the depth of the lower shoreface do suggest that the winnowing effect of wave base is a very real phenomenon (Wallace et al. 2010; Kinsela et al. 2022). The threshold of sediment motion is grain-size dependent, so the concept of wave base needs to be linked to a grain-size, as illustrated by Plint (2014), who showed that a shallower 'sand wave base' and deeper 'mud wave base' could be recognised in a vertical succession of Cretaceous prodelta to delta front sediments. Of course, waves do not operate independently of currents and *vice versa* and so 'combined flow' is the rule. Snedden et al. (1988) document the re-suspension of shelf mud by combined flows at depths of up to 90 m.

In addition to surface waves, internal waves of various sorts can be generated. Tidal waves may interact with bottom topography to set up waves in the pycnocline that are capable of re-suspending and transporting mud (see also Section 13.4.3). For example, on the Portuguese shelf, these internal waves can re-suspend shelf sediment at a depth of 70 m, generating a 10–20 m thick turbid bottom layer, even in the summer (Quaresma et al. 2007). This process may be an important mechanism by which shelf mud is moved onto the continental slope, and beyond.

9.3.4 Transport

Fine-grained marine sediments are deposited, re-suspended and re-transported many times before reaching their final site of accumulation. Four stages can be distinguished: (1) supply, often via river-mouth suspended-sediment plumes; (2) initial deposition close to the river mouth; (3) erosion, re-suspension and further transport, often repeated; and (4) long-term net accumulation. As has been emphasised in Section 9.2 seasonal variations in both strength and direction of currents and waves are of great importance in all these processes. Sediment transport in the wind-driven surface layer is relatively straight forward and easy to measure and observe from space, but once sediment settles below a certain depth its transport path will vary with changing oceanographic conditions (Section 9.2.3).

Satellite images show suspended sediment emerging from river mouths as pale brown plumes. These can float as 'hypopycnal' plumes, or, sink to various levels in the water. Each level being related to a separate thermohaline density contrast or 'pycnocline' which will arrest the sinking of riverine waters with appropriate density (controlled mainly by temperature, salinity and suspended sediment concentration; Figure 9.15). If the river water has a particularly high suspended sediment concentration, it will be denser than seawater and hence will flow along the seabed as 'hyperpycnal' flow or 'underflow'.

Suspended sediment in a floating plume will be transported tens of kilometres offshore if the river water is warm and the suspended sediment concentration is low, such as typifies seasonal floods with a less peaked discharge regime. Transport offshore will be further enhanced if winds and tides are weak. Under these conditions, mud settles gradually, over a long distance, helping to maintain a low shelf gradient. On the other hand, deposition of sediment close to the river mouth will be favoured by a river that carries a high suspended sediment concentration but with a relatively low discharge, when winds blow onshore, and tidal mixing is vigorous. Under these conditions, sediment tends to be deposited inshore, building a more steeply inclined inner shelf clinoform (Wright and Nittrouer 1995). The rapid fall-out of sediment results in the construction of classical clinothems that thin and downlap distally because they simply run out of sediment. Even though clinoforms usually have extremely low slopes (< 1°) the soupy mud is still very unstable, leading to ubiquitous sliding and slumping.

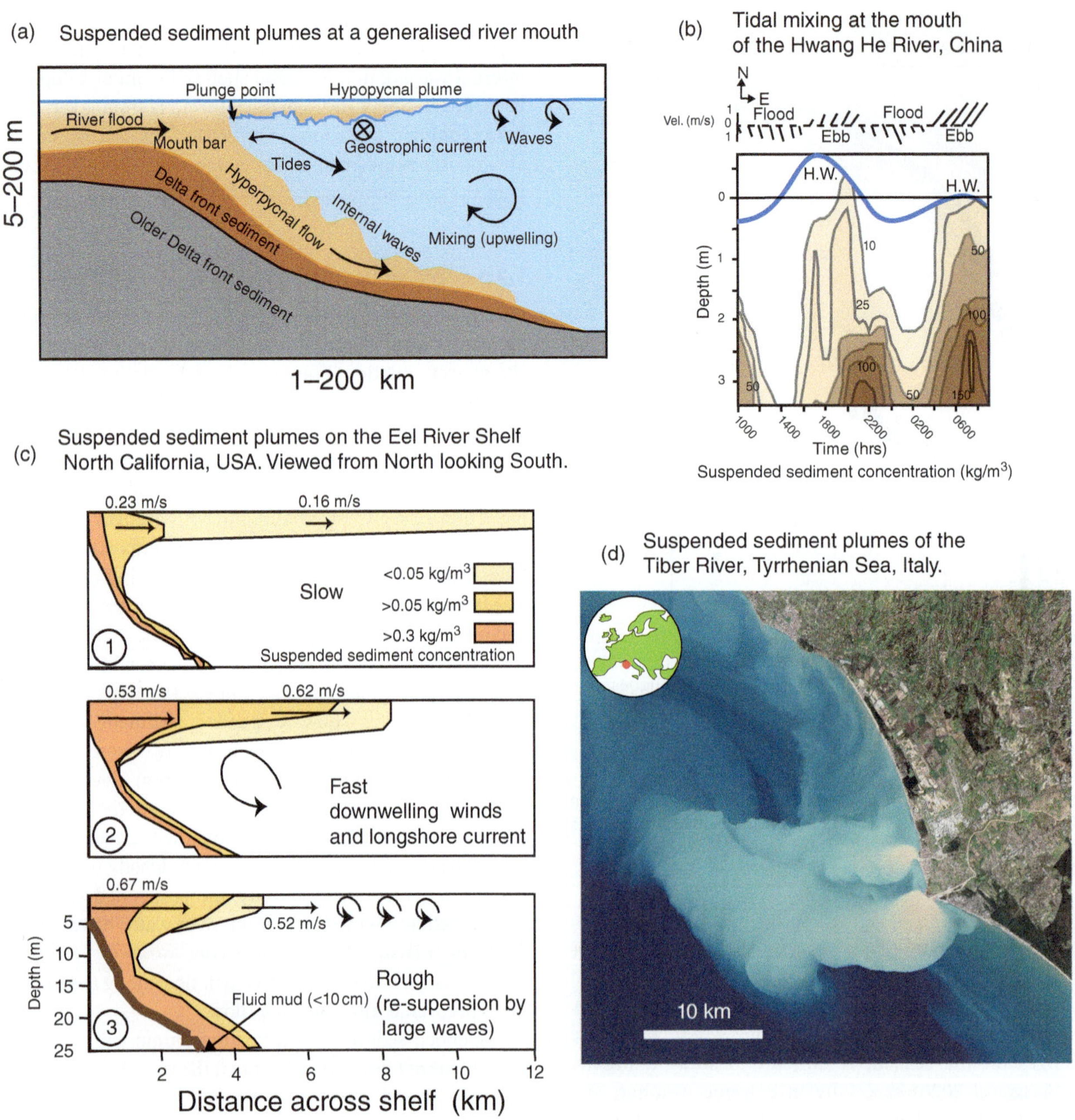

Figure 9.15 (a) Processes affecting sediment transport at a river mouth. Hyperpycnal flow in seawater requires an excess density of about 40 kg per cubic metre or more. The effects of tidal mixing, geostrophic flow, wind and wave mixing, and the formation of hyperpycnal flows are explored in (b) and (c). (b) Illustration of tidal effects on fine-grained sediment transport at a river mouth. Suspended sediment concentrations through two tidal cycles from a station in 3.5 m of water, 5 km off the mouth of the Hwang He River, China on 18–19 September 1995 when the river was at high stage (Adapted from Wang et al. 2010). The dark brown colours (approximately > 100 kg per cubic metre suspended sediment) chart the thickness of the hyperpycnal density layer through the tidal cycles. The density difference, driving hyperpycnal flow, also depends on the fresh/salt water mixing. The sediment flux from river to sea is predominantly driven by the bottom water motion (which has by far the highest sediment concentration) and by the ebb-plus-river currents flowing seaward. The rapid drop in concentration and hence in hyperpycnal flow at the end of the tidal cycle is associated with rapid dumping of sediment close to the river mouth. (c) Suspended sediment concentrations measured on the Eel River Shelf off northern California January 15, 1998, from Geyer et al. (2000) and Hill et al. (2007). The sections illustrate three different marine conditions on the shelf and their effects on fine-grained sediment transport. (1) northerly wind caused offshore flow (Ekman effect) and a wide low concentration plume; (2) strong southerly wind caused downwelling near the coast which had the effect of narrowing and accelerating the plume; (3) strong wind and rough sea caused rapid alongshore flow and re-suspension of sediment deposited from the plume. Suspended sediment concentration was sufficient to cause fluid mud to flow downslope across the seabed. (d) True-colour image of the Tiber River mouth on 5 February 2019, three days after heavy rainfall in Rome, Italy. The sediment plume stretches 28 km from the coast, carried northwest by wind-driven currents. Copernicus Sentinel-2B satellite. https://www.esa.int/ESA_Multimedia/Images/2019/02/Sediment_plume_at_sea

Thermohaline and wind-driven shelf-currents clearly transport muddy sediment broadly coast-parallel for hundreds of kilometres. It has been more difficult to establish how mud is transported *seaward* across the shelf. Aigner and Reineck (1982) and Swift et al. (1986) demonstrated the importance of storm winds that piled water up along the shoreline, resulting in a downwelling return flow that, under the influence of Coriolis deflection, flowed obliquely seaward across the shelf. These longer-term, unidirectional 'geostrophic flows' combine with high-frequency oscillatory currents generated by storm waves to produce 'combined flows'. The high bed shear stress exerted by oscillatory wave motion is very effective in re-suspending both mud and very fine sand, whereas the geostrophic component causes net unidirectional migration of particles obliquely across the shelf. As such flows decelerate, they deposit sharp-based, graded beds of sand, commonly with planar lamination, hummocky cross-stratification, or combined-flow ripples, all draped by mud.

Some studies of shallow-marine rocks in the 1970s and 1980s presented palaeocurrent evidence that seemed to indicate that turbidity currents had transported sand and even gravel across ancient shelves (e.g. Hamblin and Walker 1979; Wright and Walker 1981; Leckie and Walker 1982; Leckie and Krystinik 1989). Studies of modern shelves (Swift et al. 1986; Snedden and Swift 1991) showed, however, that offshore transport of sand was due to combined flows; the gradient of shelves was simply too low for auto-suspension of sand to be achieved. Duke (1990) re-interpreted these palaeocurrent data from ancient shelf successions and showed how they could be explained in terms of combined flows, rather than turbidity currents.

Although the notion of shelf sand transport by turbidity currents had been largely dispelled, it was realised, about 20 years ago, that significant, gravity-driven transport of *mud* did occur, in the form of fluid mud that could flow down slopes as low as 0.5 m per km. Sternberg and colleagues (Wright and Friedrichs 2006) showed that these hyperpycnal flows of fluid mud did not require auto-suspension for sustenance, like a true turbidity current, but could be initiated by the influx of sediment-laden river water into the nearshore region. Seaward, the mud could be maintained in suspension by the turbulent energy provided by waves, or in some cases, tidal currents. The result was a relatively thin (few decimetres to metres), bottom-hugging layer of turbid water that could transport mud far out across the shelf. Similarly, it was observed that storms *alone* could re-suspend muddy sediment on the shelf, producing a relatively dense, turbid bottom layer that would, under the influence of gravity, flow down-slope, typically seaward across the shelf. Such flows are the indigestibly named, 'current wave-enhanced sediment gravity flows' (CWESGF; Wright and Friedrichs 2006; Walsh and Nittrouer 2009; Macquaker et al. 2010; Plint 2014; Bhattacharya et al. 2020).

Direct or indirect evidence for wave- or current-supported sediment gravity flows has now been reported off the Amazon, Atchafalaya, Eel, Fly, Ganges–Brahmaputra, Klamath, Mad, Mississippi, Po, Rhone, Waiapu, Waipaoa, Yangtze and Hwang He Rivers, among others.

A key to these flows is a dynamic interplay between two factors: wave-generated turbulence, and the density and viscosity of the fluid mud. Wave-generated turbulence can disperse the bottom-hugging fluid mud and reduce the density stratification that drives flow down-slope, but, at the same time, wave energy is able to re-suspend more mud, thus increasing the density contrast that drives the flow. Alternatively, if wave energy decreases, the concentration of sediment in the lower part of the flow increases as the particles start to settle, thus increasing viscosity, and damping turbulence, allowing some sediment to be deposited. The removal of some suspended sediment then reduces viscosity and density, which allows greater turbulence, thereby allowing more sediment to be suspended. Over time, a dynamic equilibrium is established. The sedimentary result of this interaction between processes of transport and deposition is a distinctive tripartite, normally graded bed (Macquaker et al. 2010; Section 9.4.2).

Fine-grained sediment is also delivered to the ocean by the wind, as volcanic ash, and as autochthonous biogenic carbonate and silica. The load of aeolian dust transported in the atmosphere is about 2 Gt per year (cf. 19 Gt sediment per year by rivers), and of this perhaps 10% falls in the ocean. Recognition of an aeolian component is easier in pelagic sediments where discrete dust-enriched horizons can be recognised (Meijer et al. 2020). There may have been an increased contribution from aeolian dust before land surfaces were colonised by plants (Dalrymple et al. 1985).

9.3.5 Bedforms in Fine-Grained Sediments

A discussion of bedforms and their products in the rock record is beyond the scope of this book (see Collinson and Mountney (2019)). However, it is important to appreciate that the addition of even small amounts of suspended mud to flowing water transporting sand grains changes bedform stability fields significantly and alters the depositional products of a given strength current, wave or combined-flow event. Baas et al. (2015) showed, in flume experiments, that a flow carrying even a small proportion of suspended sediment will generate bedforms that differ from their clear-water counterparts, especially if cohesive mud is also present within the bed. Particularly noteworthy is that alternating sand and mud deposition occurs as the result of a continuously decelerating flow. The experiments of Baas et al. (2015) showed (1) alternating laminae or layers of sand and mud; (2) low-amplitude bed-waves, large current ripples and bed scours with intra-scour composite bedforms; (3) low-angle cross-lamination and long lenses and streaks of sand

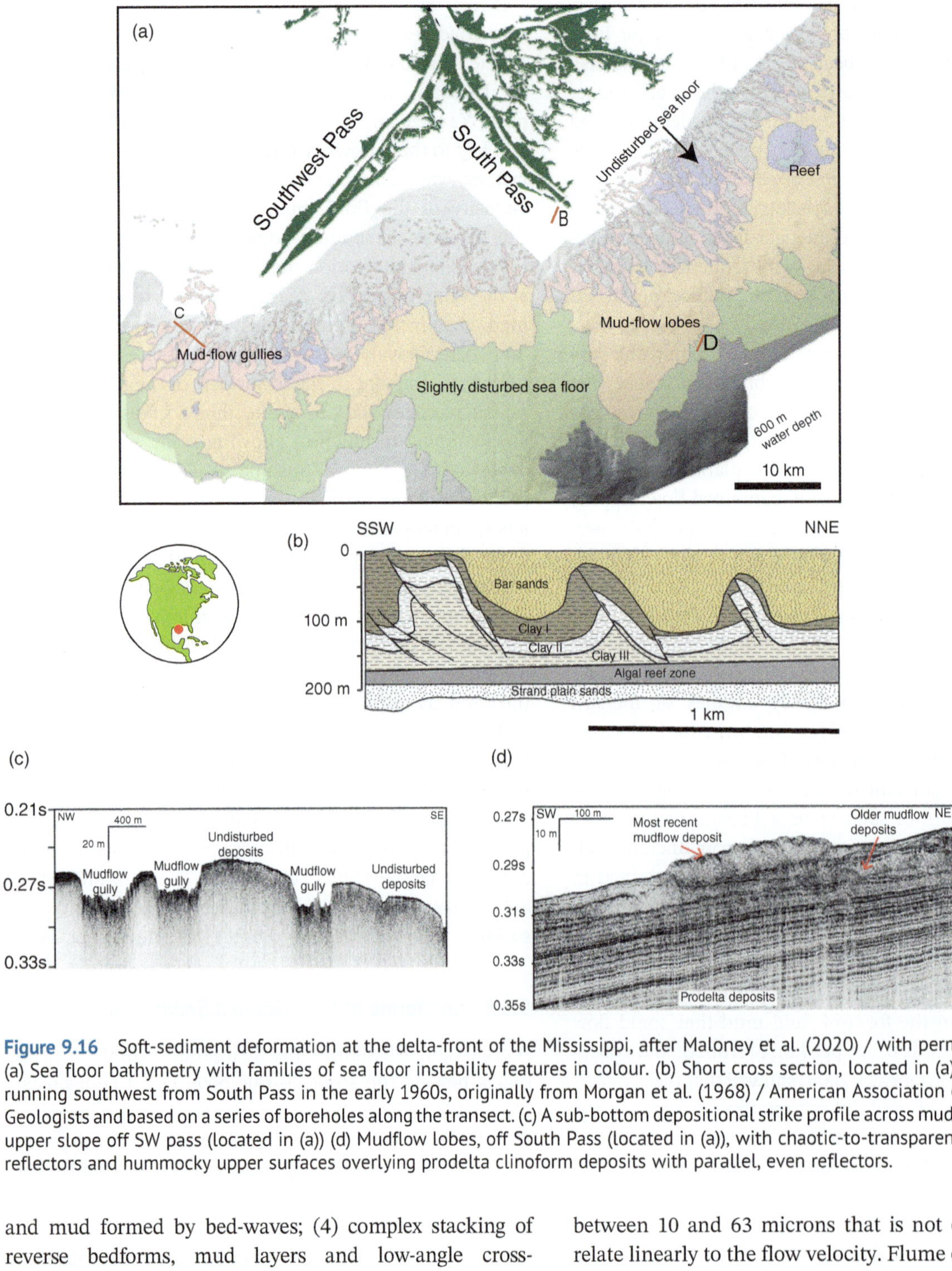

Figure 9.16 Soft-sediment deformation at the delta-front of the Mississippi, after Maloney et al. (2020) / with permission of Elsevier. (a) Sea floor bathymetry with families of sea floor instability features in colour. (b) Short cross section, located in (a) along a sand spit running southwest from South Pass in the early 1960s, originally from Morgan et al. (1968) / American Association of Petroleum Geologists and based on a series of boreholes along the transect. (c) A sub-bottom depositional strike profile across mud flow gullies on the upper slope off SW pass (located in (a)) (d) Mudflow lobes, off South Pass (located in (a)), with chaotic-to-transparent seismic reflectors and hummocky upper surfaces overlying prodelta clinoform deposits with parallel, even reflectors.

and mud formed by bed-waves; (4) complex stacking of reverse bedforms, mud layers and low-angle cross-lamination on the upstream face of bed scours; (5) planar bedding comprising stacked mud–sand couplets.

Schieber and Southard (2009) have demonstrated the formation of clay floccule ripples in flume experiments, with deposition of floccules occurring at current velocities normally associated with sand. Similar structures (albeit severely compacted) have been observed in the rock record (Schieber 2016).

In an attempt to reconstruct current velocities from sediments, McCave and Hall (2006) introduced the hypothesis of 'sortable silt': that the amount of sortable silt (i.e. silt between 10 and 63 microns that is not cohesive) should relate linearly to the flow velocity. Flume experiments support this hypothesis (McCave 2020).

9.3.6 Soft-Sediment Deformation

The high water content of fine-grained coastal and marine sediments makes them prone to slumping, sliding, loading and fluid injection (Figure 9.16). They may also retain and contain substantial overpressures. The mechanical properties of fine-grained sediments depend to a large extent on water content. Porosity, and therefore water content of freshly deposited mud is often 70–80%, pure silt is around 55% and fine sand around 45%. Half the surface porosity of mud is lost at a burial depth of between 500 and 1000 m (Kominz et al. 2011).

In addition to a high water or gas content, conditions that favour soft-sediment deformation include a near-critical condition for liquefaction/fluidisation or shear failure, a driving force such as a slope or unstable density gradient, and a trigger, such as an earthquake, or wave loading, able to exceed a particular failure threshold (Moretti et al. 2016). Small (outcrop) scale soft-sediment deformation structures vary widely (Collinson and Mountney 2019). Unfortunately, it is rarely possible to relate a deformation structure to a specific trigger mechanism.

Morgan et al. (1968) documented dramatic 'mud-lumps', which are low-relief islands of mud that develop in the Mississippi Delta (Figure 9.16). The prodelta mud rises diapirically in response to the load imposed by somewhat denser silty and sandy sediment deposited in distributary channels and mouth bars (Coleman et al. 1998; Maloney et al. 2020). The formation of mud diapirs has been observed to coincide with rapid sedimentation and loading that accompanies river floods.

9.3.7 Large-Scale Gas and Water Escape Structures

The escape of biogenic methane produced in shallowly buried organic-rich, fine-grained sediments, and thermal methane generated from deeply buried hydrocarbon source rocks, can produce impressive arrays of vertical fluid-escape pipes, primarily imaged in seismic profiles. The escaping methane and water can construct craters, pockmarks or mud volcanoes on the sea floor (Cartwright and Santamarina 2015).

Biogenic methane produced by archaea can be a feedstock for methanotrophs and is often associated with the production of isolated, bacterially mediated carbonate bodies around sea-floor seeps. Modern cold-water corals, such as *Lophelia* commonly grow close to hydrocarbon (predominantly methane) seeps (Hovland and Risk 2003). These corals have a symbiotic relationship with chemautotrophic bacteria which oxidise reduced sulphur and fix carbon (Vohsen et al. 2020). In the rock record, seep-associated carbonates can form isolated carbonate bodies clustered around mud volcanoes or craters (Allison et al. 2008). These carbonates are characterised by extremely light carbon isotopes (i.e. enriched in ^{12}C) due to isotopic fractionation by the methanogenic archaea.

9.3.8 Polygonal Faulting

Reflection seismic sections in thick mudstone units commonly show arrays of normal faults which, in plan view, form polygonal networks. The polygons are normally a few hundred metres across. These networks develop in fine-grained sediments ranging from chalk to pure siliciclastic muds (Cartwright 1994; Cartwright and Dewhurst 1998; King et al. 2022). The faults occur in tiers typically a few hundred metres thick. Outcrop examples include: the Oligocene Boom clay of Belgium (Verschuren 2019), and the Paleocene Khoman chalk in Egypt (Tewksbury et al. 2014).

Verschuren (2019) gave the following list of characteristics by which to identify a polygonal fault system at outcrop: (1) coeval, normal, dip-slip faults, (2) that are intraformational (syn-depositional) in age, (3) with random strikes, (4) directly correlatable to a genetically associated polygonal fault system and (5) lacking a demonstrable influence of either far-field tectonic stress or other non-intraformational deformation drivers, such as salt pillowing.

9.3.9 Biological Processes

The character of most Phanerozoic marine deposits is governed in part, by biological activity in the surface mixed layer (SML: see Chapter 3 for an extensive discussion of bioturbation processes). This layer, which may be up to 10 cm thick in muds, is under the influence of the surface and can vary in depth from zero (under extreme anoxic conditions) to over one metre when inhabited by deep-burrowing organisms (such as the shrimp *Callianassa*). Within the SML, burrowing organisms move material, and either increase or decrease erodibility. In a representative study, Hines and Comtois (1985) report animals burrowing to about 30 cm depth in both sand and mud, but 75% of organisms inhabited the upper 5 cm, and only 5% lived below 20 cm.

The thickness of the SML divided by the accumulation rate gives the residence time for particles within it. Residence time is important because it controls the physical, biological and chemical processes that affect particles before they are buried, potentially to become a preserved deposit. Physical processes include selective sorting of grains, based on size, shape, mineralogy and organic content. Particles moved upward are more likely to be eroded and transported away whereas downward movement improves preservability. Associated chemical processes are discussed later.

Bioturbation destroys the sedimentary record of depositional events by first destroying original layering and then, if continued, eliminates any residual depositional textural variations, leaving a totally homogenised sediment. The concept of 'biodiffusivity', by which sedimentary particles are dispersed by biological activity, can be quantified from the depth distribution of short-lived radioactive isotopes. Biodiffusivity varies by five orders of magnitude across various shelf sediments. However, there is a general relationship between bioturbation intensity, organic carbon delivery and sediment accumulation rate, such that high bioturbation rates and high sediment accumulation rates tend to correlate (Wheatcroft and Drake 2003). The obliteration of a recognisable sedimentary layer by bioturbation is related to the balance of these two competing rates, as well as to the thickness of the mixed layer. This balance is semi-quantified in the Bioturbation Index (B.I.; Taylor and Goldring 1993), that ranges from a value of 0 (no bioturbation) to 6 (total homogenisation) (see Figure 3.4).

Bentley et al. (2006) investigated how primary stratification of storm beds might escape destruction by bioturbation. The primary bedding fabric could only be preserved when sedimentation took place in pulses that deposited several centimetres at a time, thus overwhelming the infaunal macrofauna that were unable to fully rework these thicker beds. As a general rule, some recognisable stratification could be preserved if the average sedimentation rate exceeded about 1.5 cm/year. A comparable conclusion was reached in a study of the Po Delta region (Palinkas and Nittrouer 2007). This pulsed burial mechanism provides an explanation of well-stratified sediments that does not have to appeal to anoxic bottom water.

The thickness of the SML has varied over geological time. The so-called 'Agronomic Revolution' occurred largely in the Early Cambrian (Mángano and Buatois 2017), when marine sediment surfaces that formerly had been either bare or covered by microbial mats, were replaced by burrowed surfaces. This biotic revolution profoundly altered the nature of the SML, with significant geochemical implications. The evolution of burrowing infauna may have resulted in a decrease in the preservation of thin beds over time (e.g. from the Ordovician to the Devonian; Larson and Rhoads 1983). Such trends are difficult to substantiate because of changes in depositional setting as well as the secular trend of evolution.

In addition to direct effects on deposited sediment, organisms can also cause aggregation of fine particles into 'marine snow' within the water column. Marine snow is common in regions of high primary productivity and is largely due to the activity of filter-feeding organisms that extract suspended particles from the water, subsequently excreting indigestible waste as mucous-bound pellets that settle rapidly to the bottom (e.g. Kranck and Milligan 1980).

9.3.10 Biochemical Processes: Syn- to Post-depositional Redox Changes

The subject of near-surface redox changes and their biochemical significance is fascinating and deserves more attention than we are able to give it here. The reader is referred to the excellent review by Pufahl (2010).

The terminology used to describe redox conditions within sediments can be confusing. The following terms are applied to environments, facies or oxygenation levels: oxic (8.0-2.0 ml^{-1} O_2), dysoxic (2.0-0.2 ml^{-1}), suboxic (0.2-0.0 ml^{-1}) and anoxic (0.0 ml^{-1}). 'Euxinic' refers to anoxic conditions with free H_2S in the water column. The terms are similarly used to refer to specific microbial metabolisms, and the associated terminal electron acceptor coupled to the oxidation of organic carbon (e.g. Froelich et al. 1979; Canfield and Thamdrup 2009). Thus, oxic refers to aerobic respiration (where cells use O_2), suboxic refers to using nitrate (NO_3^-), whereas anoxic refers to using Mn (IV), Fe (III) or sulphate ($SO_4^{\ 2-}$) (Figure 9.17).

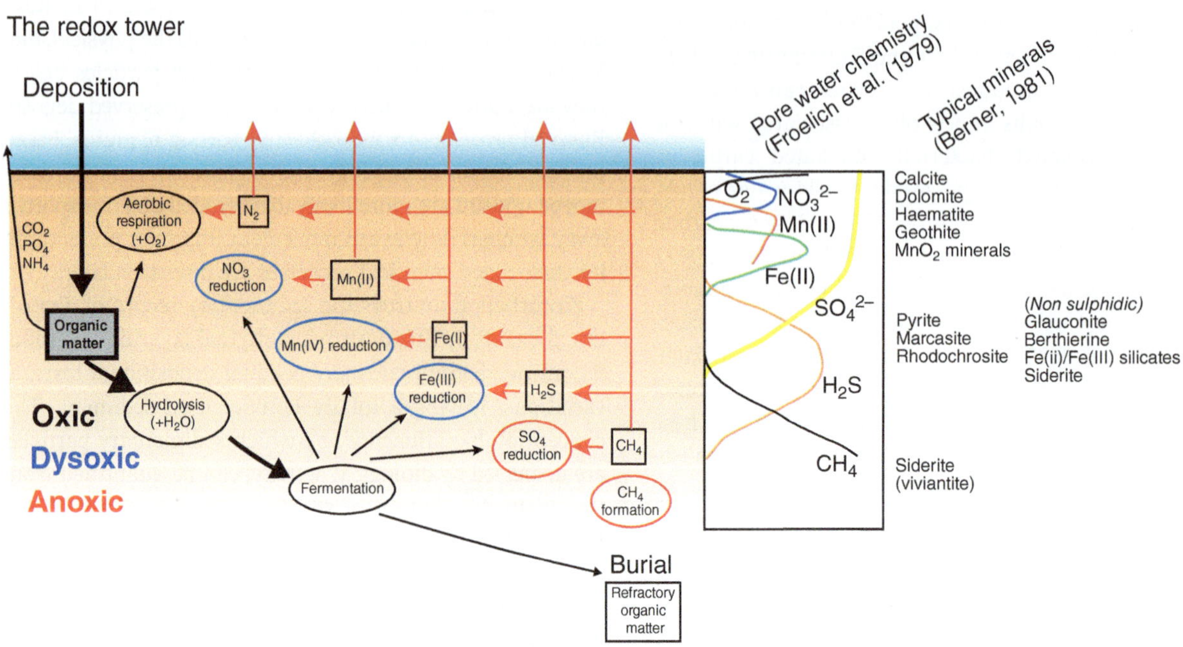

Figure 9.17 Typical sequence of bacterially mediated reduction processes during the early burial of fine-grained sediment and associated diagenetic mineral products. The red arrows reflect the fate of energy-rich bacterial substrates released during anaerobic mineralisation (Middelburg 2018 / Biogeosciences / CC BY 4.0.)

A substantial volume of low-density organic matter is inevitably deposited with fine-grained mineral matter with which it has the same settling velocity (hydraulic-equivalence). Organic matter is also produced within deposited sediments by fauna and flora inhabiting the SML. The presence of dispersed organic matter in sediment with heterogeneous, texturally controlled, oxygen exchange with surface water, and consumption of oxidants by bacterial communities, leads to variable redox states within the SML (Curtis 1980; Berner 1981, 1985; Curtis and Coleman 1986; Middelburg 2018). Field observation of chemical gradients within the SML led to the development of a concept known as the microbial redox 'tower'. This concept, has, in general terms been validated by genetic analysis of the bacterial communities with depth. It envisages a predictable succession (in space or time), of bacterially mediated redox reactions, based on their available free energy release ('thermodynamic sorting'). Each reaction is mediated by an ecological 'guild' of prokaryotes that consume electron acceptors. Aerobic bacteria first consume oxygen until it is depleted. Then, in the suboxic zone, de-nitrifying bacteria convert nitrate to nitrogen. Reduction of manganese and iron oxides follows in the anoxic zone, after which sulphate is reduced: (this is normally volumetrically the most important reaction), and finally, reduction of carbon dioxide produces methane. Fermentation is usually placed between denitrification and sulphate reduction (Orcutt et al. 2011). Such a redox tower predicts a sequence of early post-depositional reactions as the redox tower migrates upwards through the accumulating sediment column. Organic carbon is consumed during this process, being converted to CO_2 (a process referred to as 'mineralisation' because the CO_2 often ends up in diagenetic carbonates). Organic matter consumption is normally limited by the availability of the terminal electron acceptors (e.g. oxygen, nitrate and Fe (III)). Because these electron acceptors are depleted within the sediment faster than they can be supplied, it is possible for organic matter to escape oxidation and be preserved. The supply of electron acceptors, which provide the bacterial 'substrates', depends on the degree of exchange of pore water with the ocean, and on the rate and degree of sediment mixing by infauna. Modern studies using sea bottom observatories emphasise the dynamic nature of the boundaries between these zones in terms of their pore water chemistry (Ahmerkamp et al. 2017).

Diagenetic minerals produced by the redox tower are many and varied. Among the most common are microbial carbonate (calcite, ankerite or low-iron dolomite) which can be related theoretically to most phases of the redox tower. Micritic diagenetic carbonate is normally distinctively isotopically light (rich in C^{12} and O^{16}) due to biological isotopic fractionation effects. It may be found in dispersed form within the sediment or in concretionary carbonates. Pyrite, derived from reactions which successively add sulphide and elemental sulphur from sulphate reduction, to reduced iron (Fe (II)), is near ubiquitous.

The strong redox contrast between the surface and sediment lying as little as a few millimetres to a few decimetres below the sediment–water interface, provides a set of steep geochemical gradients that can act as ionic 'traps'. Concentrations, for example of Fe, Mn, or U, vary greatly because of different solubilities of Fe^{2+} and Fe^{3+}, Mn^{2+} and Mn^{4+}, U^{4+} and U^{6+}. These differing solubilities lead to diffusion and subsequent precipitation of oxidised iron and manganese. The processes that couple the C, O, Mn and Fe cycles during sediment diagenesis are, however, not fully understood in detail but include the formation of soluble Fe and Mn organic ligand complexes of variable redox states (e.g. Oldham et al. 2019).

The sulphate concentration in the pore-water declines rapidly with depth through the sulphate-reducing zone, causing some downward diffusion of H_2S. H_2S is an acid which can leach iron to form FeS and then pyrite (FeS_2). If the sediments have sufficient silica, chamosite may also be formed. Glauconite formation is complicated, but it contains both divalent (Fe^{2+}) and trivalent (Fe^{3+}) iron and oscillation of the redox boundary is probably required for genesis. Manganese does not form sulphides as stable as those of iron and has a greater tendency to be precipitated in the oxidised zone. However, iron minerals normally predominate as Fe is simply more abundant. Iron can also precipitate in the reduced state in a basic (high pH) environment. In carbonate-bearing mudstones with a relatively low sulphate/sulphide ratio, as commonly occurs around river mouths where freshwater dilutes saline pore water, siderite (Fe_2CO_3) can precipitate in the form of nodules. Siderite can however also be formed below the sulphate-reducing zone if there is excess iron, for example, by reaction with unstable aragonite from shelly material.

The formation of early diagenetic minerals has been shown to be closely linked to sedimentation rate. Extensive carbonate (calcite, dolomite or ankerite) -cemented horizons in ancient shelf mudstones are interpreted to record very slow sedimentation when anaerobic bacterial reduction of organic matter raised bicarbonate concentration, and pore-water alkalinity to the point where carbonate cements could precipitate (Irwin et al. 1977; Scotchman 1989; Macquaker and Gawthorpe 1993; Macquaker 1994). Sedimentation rate is also closely linked to relative sea-level changes and the development of depositional sequences. For example, Macquaker and Taylor (1996) and Taylor and Macquaker (2000) showed that ooidal ironstones

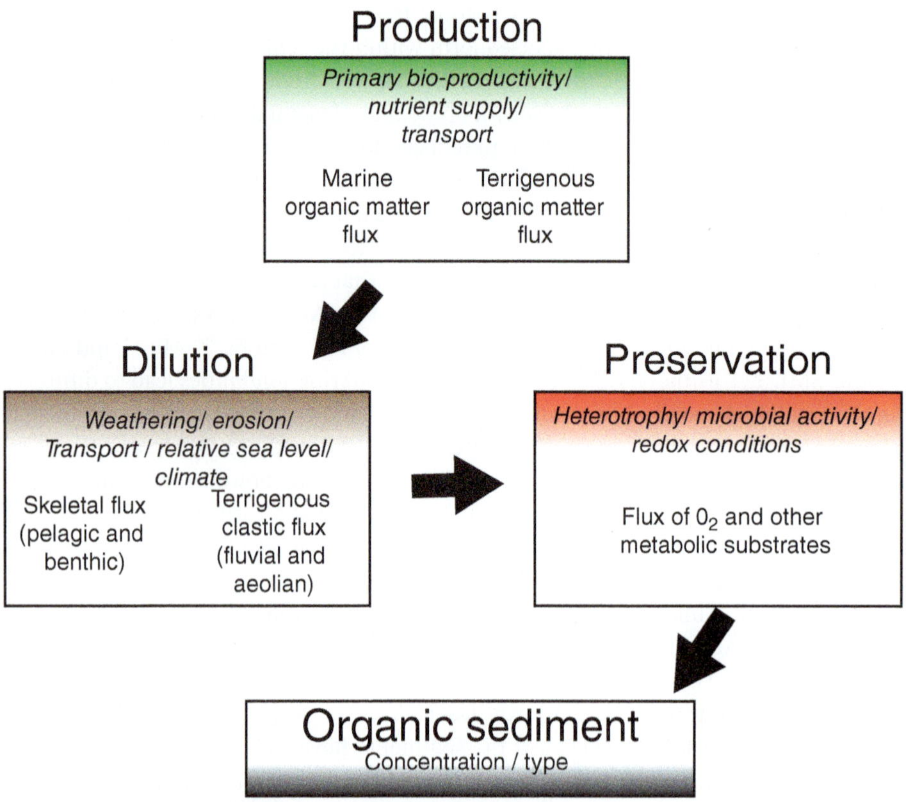

Figure 9.18 The concentration and type of organic matter found in sediments is a function of a balance of fluxes (rates of supply). The total organic carbon content (TOC %) of sediment is a function of the production flux both within the water column and the sediment itself, the dilution flux of inorganic sediment (whether siliciclastic, biogenic carbonate and silica or diagenetic) and the preservation of this organic matter from digestion, mineralisation and oxidation. Sageman et al. (2003) / with permission of Elsevier.

characterised by early diagenetic berthierine and siderite formed during the forced regressive (falling stage) and/or lowstand systems tract times when sedimentation rate was low, due to bypass, and the physical reworking rate was high. Under these sub-oxic conditions, organic matter oxidation took place, with abundant iron oxides but little free sulphide, favouring precipitation of berthierine and siderite. In contrast, nodular phosphate (apatite) cements, sometimes with associated bone beds, form at the maximum flooding surface between transgressive and highstand systems tracts, and also at major flooding surfaces, both settings where the detrital sedimentation rate is very low, and physical and/or biological reworking are protracted. These conditions favour remobilisation and precipitation of adsorbed and biogenic phosphate in the sub-oxic zone of the SML. In contrast, 'normal' relatively rapid mudstone sedimentation is dominated by anaerobic sulphate reduction with the formation of abundant sulphide and precipitation of pyrite as an early diagenetic cement (Figure 9.18).

The redox gradients discussed earlier can be thoroughly disrupted by bioturbation which can introduce oxygenated water deep into the SML (Aller 1994).

The geochemical evolution of ocean water through geological time has strongly affected diagenesis in the SML. The changes in oxygen saturation of seawater have had a marked effect on the interlinked geochemical cycles of, for example, carbon, sulphur iron and phosphorus. The manner in which these changes are reflected in the suite of diagenetic minerals formed at or just beneath the sea floor is currently the subject of intense research.

9.4 Modern Mudstone Facies

For modern environments, the hydraulic conditions are often very well documented, but relating these conditions to resultant sedimentary facies, through widely distributed, high-quality cores, is often not possible. Conversely, the stratigraphic record provides an abundance of rock to study, but it is commonly difficult to attribute lithologies and structures to specific depositional processes or environments. This section links processes with the resulting sedimentary facies using both modern and ancient observations.

9.4.1 Hypopycnal Deposits

A well-documented example of hypopycnal river flood deposits is provided by the Po River flood of October 2000. This flood deposited sediment on the prodelta in 6–29 m of water, with sediment extending 30 km along the coast, with a maximum thickness of 36 cm (Wheatcroft et al. 2006). The deposit comprised multiple, thick (1–10 cm) beds of silt and mud within the single 'flood bed' (Figure 9.19). These interbeds had sharp upper and

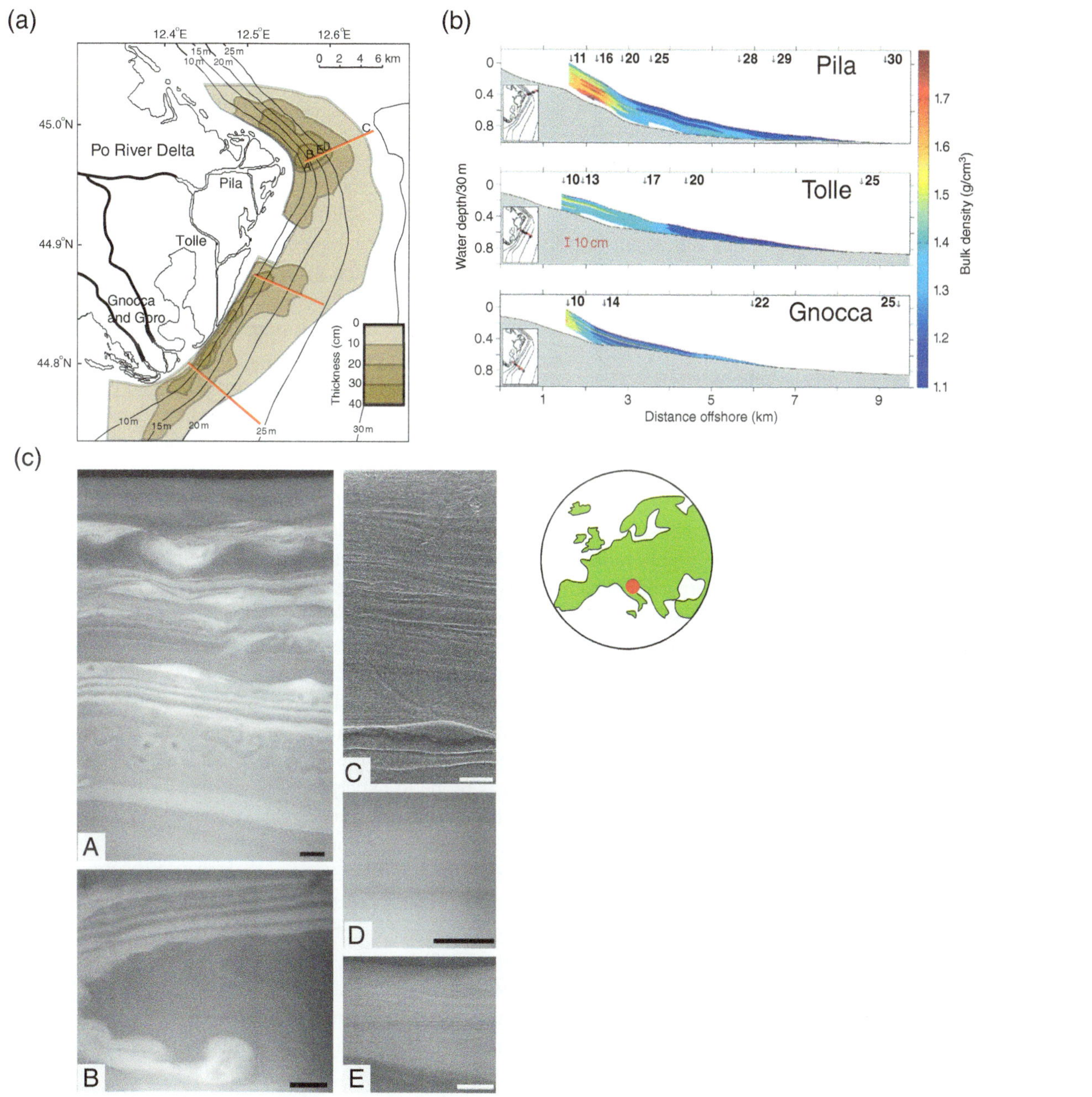

Figure 9.19 Flood deposits of the Po Delta, after Wheatcroft et al. (2006) / with permission of Elsevier. (a) Isopach map of autumn 2000 Po River flood deposit based on sampling in December 2000 and October 2001. (b) Shore-perpendicular depictions of the flood deposit thickness and bulk density of the Pila, Tolle and Gnocca distributary mouths. Sampling depths (in metres) are at the top of each panel and the *y*-axis is scaled at 1/30th the water depth (e.g., 0.4 =12 m). (c) Representative stratal types and physical sedimentary structures observed in X-radiographs along the northern transect in: (A) Soft-sediment deformation structures, lenticular bedding and lamination, (B) soft-sediment deformation structure, (C) enhanced X-radiograph showing lenticular bedding and lamination, (D) diffuse bedding in the lower portion (420 cm depth) of the flood deposit and (E) ripple cross-lamination in the correlatable coarse-silt layer (June 2001). The scale bars are 2 cm.

lower contacts and displayed plane-lamination, ripple cross-lamination, lenticular bedding and soft-sediment deformation structures, as well as dramatic changes in grain-size that implied rapid fluctuations in energy during deposition. A single, correlatable, occasionally cross-laminated and lenticular-bedded, coarse-silt layer occurred in the middle of the flood deposit. Bioturbation was scattered and increased in density when sampled during subsequent years (Wheatcroft et al. 2006). The complexity of the deposit reflects the long duration of sediment input from this big river (Palinkas 2009). The distribution of the deposit, directly off the river mouths and slightly to the south, strongly suggests that most deposition occurred directly from the surface plume, without further cross-shelf transport. The rapid vertical settling beneath the Po outflow plume might be in part due to the fact that significant floc formation occurs within the Po River itself, perhaps due to high organic content.

9.4.2 Hyperpycnal Flows and Their Deposits

A hyperpycnal flow comprises a turbulent mixture of sediment and freshwater that is denser than the surrounding water, and hence, upon leaving a river mouth, sinks and flows along the bottom of the water body (Mulder and Syvitski 1995). Hyperpycnal flows have proved frustratingly difficult to both observe and quantify. The highly turbid, dense river flows necessary to initiate hyperpycnal behaviour may only occur for a few hours, and deploying seabed instruments immediately in front of a flood-stage river mouth is difficult (Warrick and Milliman 2003).

It is necessary for freshwater to carry a suspended sediment load of 36–43 kg m^{-3} before it is dense enough to sink in seawater. A survey of 600 rivers revealed that only 15 had an *average* suspended sediment load of over 10 kg m^{-3} and hence were sufficiently 'dirty' that they had a high chance of exceeding the threshold for hyperpycnal behaviour during annual flood events (Milliman and Kao 2005). Such high suspended sediment concentrations are most common in small and medium-sized mountainous (tectonically active) drainage basins.

Hyperpycnal flows have been observed in aerial and satellite images that show dramatic plunging of turbid plumes below clear water. A particularly dramatic image by M.T. Gumundsson and F. Pálsson (in Mulder et al. 2003) shows a sediment plume that resulted from volcanic melting beneath the glacier Vatnajökull in southern Iceland. The resulting catastrophic meltwater flow (jökulhlaup) had a calculated density of 200 kg m^{-3} and sank almost immediately upon entering the Atlantic Ocean. In another example, Warrick and Milliman (2003) observed both a high suspended sediment concentration and strong offshore bottom flow one kilometre off the mouth of the Santa Clara River (California) within hours of a major flood, when there was no associated high wave energy. Direct sedimentary evidence of hyperpycnal flows on the California shelf is, however, elusive. Flows either transit the shelf without leaving an imprint, or the deposits do not survive post-depositional reworking by bottom currents, storms and bioturbation. Another example of hyperpycnal flow on the Eel River shelf in northern California appears to be recorded only by an erosion surface (Sommerfield and Nittrouer 1999).

Based on the sparse oceanographic information, and reasonable inferences about process, a facies model has been established for the deposits of hyperpycnal flows (Figure 9.20). A typical 'hyperpycnite' is a mainly fine-grained deposit characterised by an upward-coarsening and then upward-fining bed that reflects the increasing and then decreasing velocity of the fluvial, flood-related discharge (Mulder and Alexander 2001). Zavala (2020) lists diagnostic characteristics of 'sustained hyperpycnal flows'. These include evidence for sustained, pulsating flow rather than a single event, abundant fresh plant detritus indicating river flood input and a lack of evidence for hemipelagic fallout, which would indicate that the flow had lofted above the bottom. Sedimentary features considered indicative of hyperpycnites include: (1) composite beds with gradual and recurring changes in sediment grain-size and sedimentary structures, including climbing current or wave ripples reflecting high rates of bed aggradation; (2) a mixture of extra-basinal and intra-basinal components, especially including macerated plant material; (3) internal and discontinuous erosional surfaces. Such fine-grained beds may be associated with a host of related mass flow deposits depending upon the river load and depositional slope. Areas prone to hyperpycnal flows may have a particularly high sediment accumulation rate, which will suppress benthic infauna and bioturbation intensity. Such conditions were inferred from Cretaceous prodeltaic sediments that included thick, structureless 'fluid mud' beds, that were largely devoid of bioturbation (Bhattacharya and MacEachern 2009).

Muddy hyperpycnal flows are part of the spectrum of density flows and are further discussed in Section 6.3.2 (density underflows in lakes) and 14.5.9 (glaciolacustrine settings).

9.4.3 Flood Deposits from Small Rivers

Flood deposits from small rivers are illustrated by the Eel River Shelf off California (Hill et al. 2007). The Eel River drainage basin is steep and floored by easily erodible Franciscan mélange. Short, steep rivers have very high

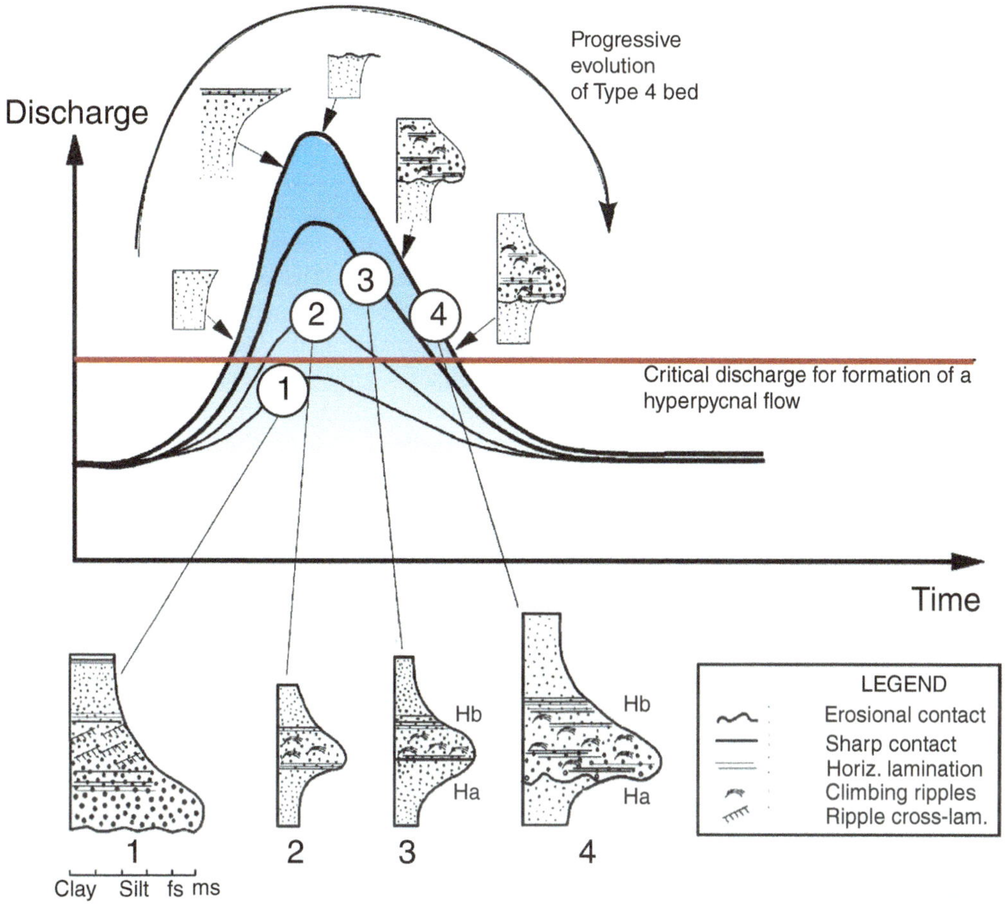

Figure 9.20 Model summarising potential behaviour of hyperpycnal flows of varied strength, and the variety of the resultant bed types (1–4) that may be deposited. 'Ha' and 'Hb' refer to subdivisions of a 'hyperpycnite' 'H'. From Mulder et al. (2003) / with permission of Elsevier.

suspended sediment concentrations, and they deliver this load in brief flood pulses lasting just hours or days, commonly at the same time as storm winds generate large waves on the adjacent shelf. Net sediment delivery in an event depends not only on rainfall amount and intensity but also on the history of previous floods. Sediment comes both from new erosion in the headwaters and from floodplain storage. The latter acts as a 'buffer' from which sediment is released when some threshold level of storage is exceeded. Hence, there need not be any direct relationship between sediment delivery (or bed thickness) and individual flood magnitude. Furthermore, actual deposition on the shelf during a flood varies with river discharge, sediment load, currents, distance from the river mouth and cross-shelf position.

Fine-grained deposits on the Eel shelf deposited from river floods in 1995 and 1997 reached 5–8 cm thick at around 50–110 m depth, 15–20 km from the river mouth; the deposit extended 50 km along the shelf (Wheatcroft et al. 1996; Sommerfield and Nittrouer 1999; Wheatcroft and Borgeld 2000). The shelf sediment layers represent about 25% of total river sediment discharged in these events. The remaining sediment is either deposited interbedded with sand in the inshore belt (shallower than 55 m), or advected even farther oceanwards to a shelf break at 100–130 m.

The 1995 and 1997 flood deposits were delivered by 3 of the 5 largest floods recorded during the 87-year discharge records of this area. Almost 25% of the cumulative sediment discharge to the mid-shelf during this 87-year period occurred during these 5 largest floods, which resulted in 40 cm of net deposition in the last 50 years. Flood beds are the dominant event beds in box cores that sample the upper ~50-cm of the seabed, seaward of the sand-silt transition (Sommerfield and Nittrouer 1999).

Mud appears to reach the mid-shelf in a two-step process. Sediment-laden water from river floods enters the sea in turbid hypopycnal plumes. These plumes are deflected northward by the Coriolis effect and southerly winds. However, the sediment plumes do not reach the mid-shelf but remain trapped against the coast within 5–10 km of the shore. Sediment deposited from the plume in this inshore band can be re-suspended in <15 m of water by storm waves, which form fluid-mud suspensions, up to 2 m thick and containing up to 100 kg m^{-3}. This fluid mud then flows seaward under the influence of gravity while waves keep sediment in suspension (i.e. a current-wave-enhanced sediment gravity flow).

Box cores show that the flood-related event beds on the Eel shelf are 1–8 cm thick, non-graded and have undulatory erosional bases (Bentley and Nittrouer 2003). Some beds are stratified and composed of 1–3 cm thick silt/clay layers with minor grain-size contrast, but with sharp contacts. A representative bed was composed of 2% sand, 60–70% silt and 30–40% clay with a uniform composition across the shelf: there was no seaward or along-shelf fining. Within each flood/storm deposit, the massive mud beds are interpreted to have been deposited from pulsatory fluid-mud flows whereas the more poorly sorted layers were deposited directly from suspension.

As discussed in Section 9.2.6, the potential for recognition and preservation of individual beds depends on the competing rates of accumulation and bioturbation, with the best bed preservation being due to fortuitous rapid burial by a subsequent, thick, flood deposit.

9.4.4 Fluid-Mud and Current-Wave-Enhanced Sediment Gravity Flow Deposits

The transport mechanisms and predicted stratigraphy resulting from muddy density underflows have been studied on the Atchafalaya shelf west of the Mississippi Delta (Jaramillo et al. 2009; Denommee et al. 2018). Fine-grained sediments deposited near the mouth of the Atchafalaya River are re-suspended by the combined effects of liquefaction induced by cyclic wave-loading of the seabed, and entrainment by strong currents. These suspended sediments are then transported westwards across the shelf by combined flows generated mainly by winter storms; hurricanes play a minor role. As a cold front approaches the coast, wave energy increases, re-suspending sediment (Rotondo and Bentley 2003; Jaramillo et al. 2009). As the cold front passes, the water column becomes stratified, and fluid mud develops near the seabed. Strong, wind-driven along- and on-shore currents persist for about two days, flowing along the shelf at up to 2 cm s^{-1}. Fluid mud, generated near the mouth of the Atchafalaya River persists for 9–12 hours, during which it flows across the shelf as a CWESGF at a velocity of up to 5 cm s^{-1} (Jaramillo et al. 2009). One flow of fluid mud, observed crossing the delta-front region, was shown in box cores to have caused erosion of 2–4 cm, followed by deposition of a 2 cm thick bed of unsorted, un-laminated silt and mud. The accumulation rate on the Atchafalaya shelf mud belt is up to 20 mm per year, compared with 10 mm per year on the outer shelf (Neill and Allison 2005). Given these rapid sedimentation rates, the foreset region of the subaqueous delta is subject to frequent soft-sediment slides.

Cores from the topsets of the mid-shelf clinothems off the Atchafalaya delta consist of cm- to dm-thick upward-fining event beds with eroded tops. The lower parts of beds contain continuous, wavy, cross-laminated silt layers overlain by normally graded clay-rich muds (Figure. 9.21). Some thick event layers contain a grain-size break but no discernible sedimentary structures. Bioturbation is moderate to intense. Some composite bedsets 30–40 cm thick do show the tripartite division suggested by Macquaker et al. (2010), involving a basal erosion surface overlain by a laminaset of silt and very fine-grained sand (unit A) with ripple laminae, abruptly overlain by a wavy parallel laminaset of silt- and clay-rich laminae (unit B) draped by structureless silt and clay (unit C). Such beds were also found on the clinothem foresets where they were accompanied by simple, normally graded non-laminated event beds. Although these deposits are reasonably interpreted as CWESGF deposits, they were not actually observed to be deposited during a storm.

9.4.5 Storm Deposits (Without Direct Input from River Floods)

Normally graded beds of sand, silt and clay, up to about 30 cm thick, can be deposited on shallow shelves as a result of combined flows generated during the waning stage of severe storms. The sediment in such beds may in part be freshly transported from the nearshore region, and in part reworked from the shelf (Morton 1981, 1988; Aigner and Reineck 1982; Snedden and Nummedal 1991). Overall, the sense of net sediment transport is shore-parallel to obliquely offshore, although at times, currents may transport sediment landward into coastal embayments (Bentley et al. 2006).

Hurricane Carla, one of the most powerful recorded hurricanes in the Gulf of Mexico, crossed the Texas shelf in September 1961 (Hayes 1967; Morton 1981). Snedden and Nummedal (1991) studied 23 box cores arranged in transects oriented both along and across the shelf. These cores revealed a bed of sand distributed along at least 200 km of the shelf to the SW of the hurricane track. This storm-deposited sand bed was thickest (7 cm) and coarsest (very fine sand) on the inner shelf and thinned and fined seaward to coarse silt, pinching out about 50 km from shore in about 75 m of water. The base of the bed was sharp and slightly erosive, and it typically showed an upward succession from plane parallel lamination or massive sand to gently inclined lamination with a gradational, bioturbated upper transition

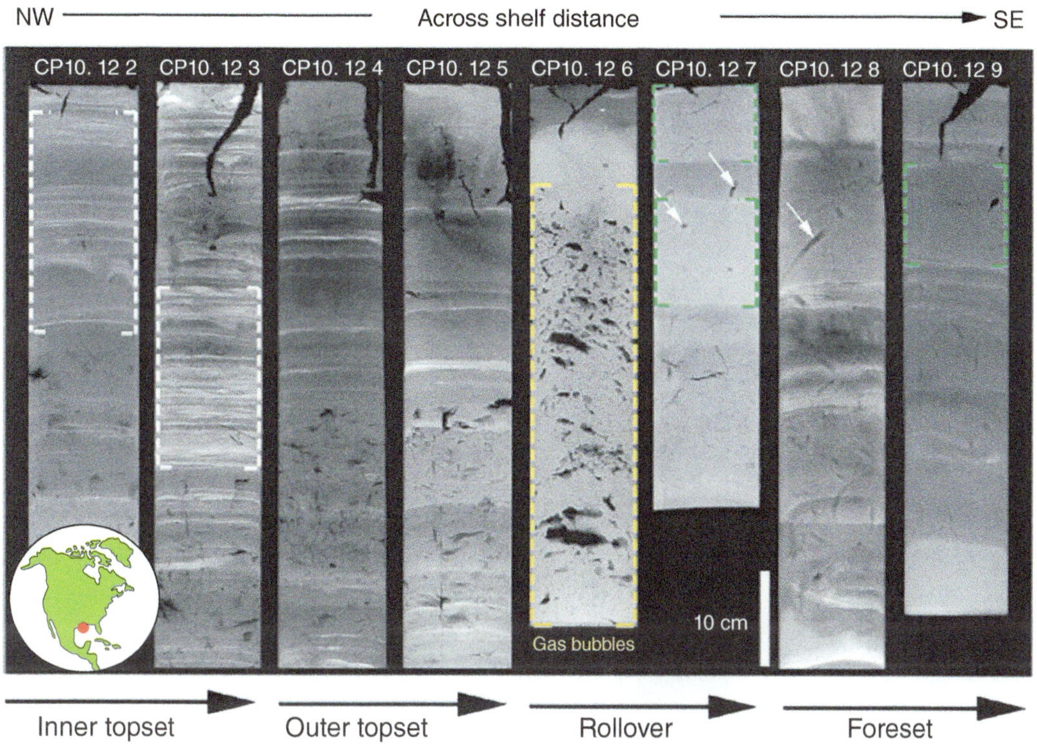

Figure 9.21 Current and wave-enhanced sediment gravity flows (CWESGF) in X-radiographs of Late Holocene mud (dark) and silt (light) from the SW Louisiana continental shelf. Cores were taken along the clinothem depositional profile approximately 1 km apart in water depths ranging from 1 to 9 m, increasing to the right. Inner and outer topsets contain higher amounts of laminated CWESGF candidates (white dashed brackets), whereas rollover and foreset cores show a higher proportion of thick, non-graded event beds (green brackets). Crustacean burrows (white arrows) indicate longer colonisation times in rollover and foreset muds. The clinothem rollover core contained abundant gas bubbles (yellow brackets). After Denommee et al. (2018) / Springer Nature.

to overlying mud. Although bottom currents were not measured during Hurricane Carla, comparison with subsequent storms, and model calculations, indicate that the Carla storm bed was generated by an along-shelf and obliquely offshore-directed combined flow with near-bottom current velocities that approached 3 m/sec.

Multiple storm deposits were studied in a 1.78 m long core taken in the updrift portion of the East China Sea mid-shelf mud belt (Section 9.2.3). It contains 15 layers, from 0.5 to 5.5 cm thick, each consisting of normally graded, coarse- to fine silt. The coarser fractions are better sorted, reflecting reworking by a bottom current whereas the finer components and inter-storm beds are more poorly sorted, reflecting settling from suspension. Five of these event beds could be attributed, via ^{137}Cs and ^{210}Pb dating (and sedimentation rate arguments), to typhoons and supertyphoons of known strength and landfall (Figure 9.22). Deposition of all these storm beds was associated with substantial storm surges of around 2 m or greater, with the thickest bed (5.5 cm) being associated with the largest surge (5.5 m). A second 1.78 m long core in the same mid-shelf mud-belt (54 m water depth, 100 km to the southwest), recorded only three different events, two of which could be linked to named typhoons. In these two cores, the event layers account, respectively, for 17% and 8% of the total sampled interval. During the period recorded by the 7 named event beds, 36 typhoons actually made landfall in the adjacent Chinese province of Zhejiang, implying that the sedimentary record provides a very incomplete record of storms! The coarse silt and small amount of sand reflect re-suspension of coarser inshore sediment, which was then transported offshore by the down-welling, wind-driven geostrophic flow. The storms did not appear to increase the transport capacity of the long-term, coastwise sediment transport system that is primarily responsible for building the mud belt itself (Tian et al. 2019).

Modern shelf studies show that storms capable of producing graded beds of sand and silt occur on timescales of years to a few decades. However, observation of the rock record, where the time represented by the succession is known, shows that the average interval between interpreted 'storm beds' is much longer. It is presently unknown what proportion of storm beds survive re-erosion, amalgamation or obliteration due to bioturbation, and so end up being preserved (e.g. Brenchley et al. 1979).

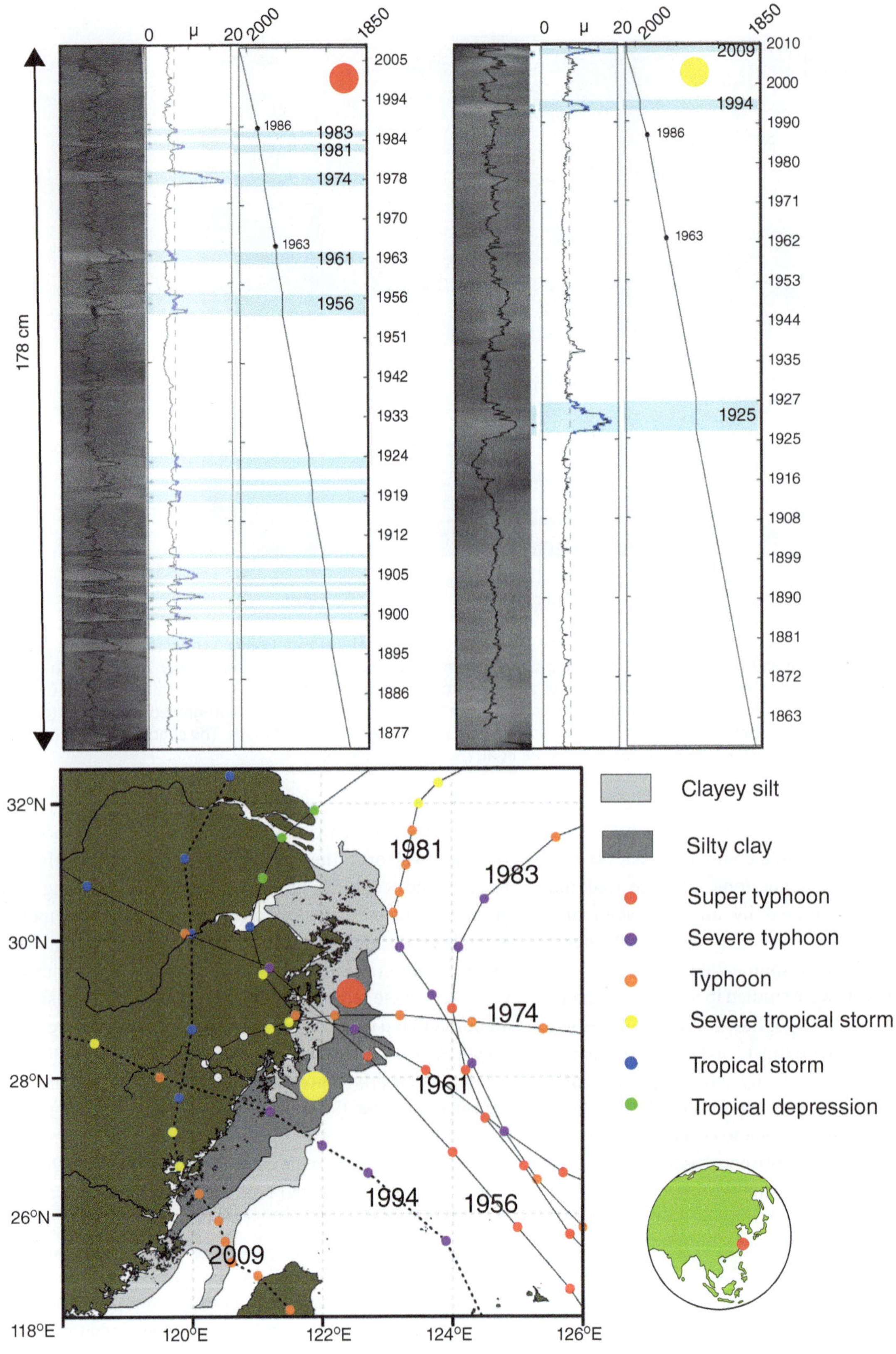

Figure 9.22 Typhoon deposits. Two 178 cm long cores from the East China Sea mud belt (Figure 9.9). Locations are shown with red and yellow dots. The X-radiographs show graded bedding with the coarser grain-sizes in light grey (see mean grain-size scale in microns). The accumulation rates have been calculated using ^{210}Pb dating that allows the responsible storms to be determined, and their tracks and changing intensities (shown by colour-coded dots) are shown. The preserved events have the common characteristics of high wind speed, long duration and strong storm surge: maximum wind speed at the centre of the typhoon >35 m^{-s}, >12-hour duration and >2 m in storm surge. During the period when the seven identified storm beds formed, 36 storms actually made landfall in this area. After Tian et al. (2019).

9.5 Late Holocene Facies-Associations

9.5.1 Coastal Mud Banks

Accreting muddy open coasts are characterised by convex-up cross-sectional profiles and a typical sequence of environments from landward to seaward. These comprise: (1) swamps composed of salt grasses (temperate climates), mangroves or nipa palms (tropics) and sometimes sandy or shelly 'chenier' (beach) ridges; (2) broad intertidal mudflats; and (3) shallow offshore muddy deposits.

Coastal mud banks typify the Amazon to Orinoco fine-grained sediment transport system, particularly along the coast of French Guiana and Surinam (Rine and Ginsburg (1985); Section 9.2.1), where mud (60–90% being high concentration fluid mud) is advected alongshore. The hydrodynamics of the mud banks are complex, with interaction of shore-parallel oceanic and seasonal wind currents and shore-normal tidal currents and jets of freshwater from rivers forming 'hydraulic groynes' (Orseau et al. 2020). The result is a suite of tens-of-kilometre-scale tabular mud banks (10–50 km long, 10–30 km wide and up to 5 m thick), which migrate both northwestwards along the coast, and shoreward (Figure 9.23). The long axes of the mud banks are oriented on average 24° counterclockwise relative to the coast, that is, ENE–WSW. The subtidal mud banks behave as bedforms. Their eastern, trailing, edges are relatively consolidated (mud density greater than 1350 kg m^{-3}) but nonetheless are eroded by waves which liquify the mud. The western, leading, edges are depositional and characterised by fluid mud (density less than 1250 kg m^{-3}) which dissipates up to 96% of the wave energy. Eventually, the inner parts of the mud banks accrete to the shore as they become colonised by stabilising mangroves (Rine and Ginsburg 1985; Eisma et al. 1991; Plaziat and Augustinus 2004; Allison and Lee 2004; Anthony et al. 2010).

The along- and obliquely onshore migration of mud banks leads to an alternation of shoreline progradation and mudbank stabilisation. Coastal erosion takes place when the migrating mud bank leaves the inter-bank portions of the coast unprotected from ocean waves. This alternation occurs on a roughly 30-year timescale (Orseau et al. 2020).

The vertical sedimentary sequence resulting from mud bank migration was described by Rine and Ginsburg (1985). The mud banks rest, in more than 10 m of water, on a transition zone characterised by planar and lenticular-laminated, bioturbated, thin-bedded, (1–4.5 cm) sand, silt and mud with vertical burrows. Overlying mudbank facies are 70–90% clay with less than 1% sand and show a limited range of structures: faint lamination or massive with macrofaunal burrows. Shells are extremely rare (Allison and Lee 2004). Very fine millimetre to submillimetre rhythmic dark laminae in laminasets of 12–13 layers related to tides were attributed by Debenay et al. (2007) to the binding effects of EPS, probably produced by diatoms. Biogenic traces are rare within the soupy bank muds, except in the intertidal zone, where mud accumulates during spring high tides, and roots from mangroves and

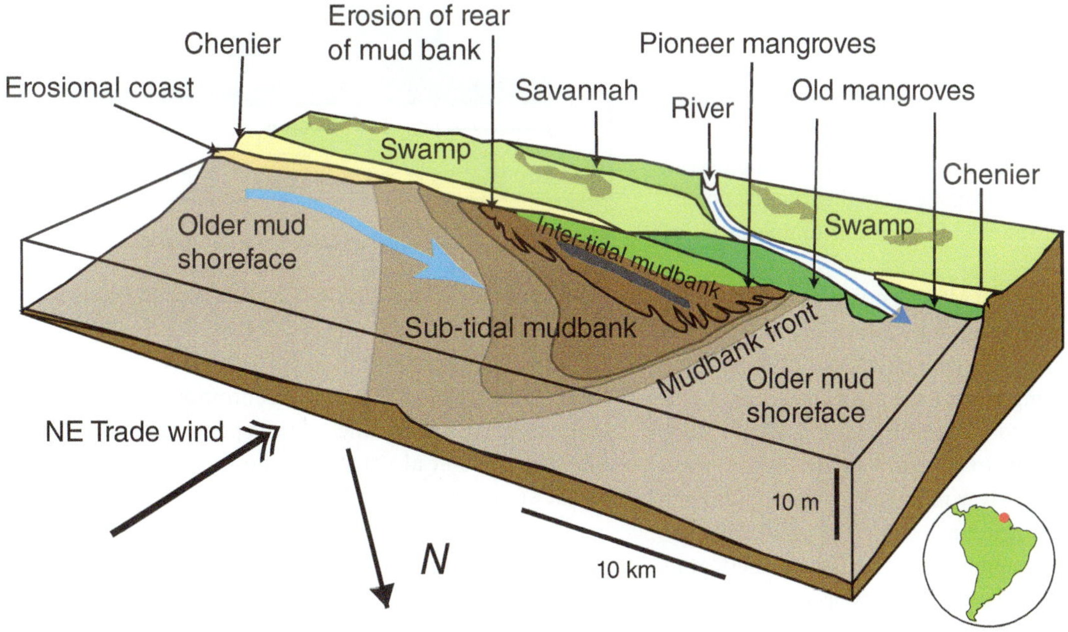

Figure 9.23 Mud banks on the north coast of South America (Plaziat and Augustinus 2004 / with permission of Elsevier.)

burrows from crabs are abundant. The mudbanks are covered with an array of tidal channels and gullies (Froidefond et al. 1988). Large-scale internal features include westward (landward) -dipping seismic reflectors within the banks. These appear to mark annual migration events (seasonal wind variations) which have generated consolidation differences.

By contrast, the sediments of the interbank areas are mud and silt with discontinuity features such as scour and fill structures and cross-lamination, and much more abundant bioturbation. These features suggest more frequent reworking by waves and tides, and the formation of firm surfaces as a result of erosion.

9.5.2 Chenier Plains and Mud Flats

Chenier coasts record the interruption of mudflat accretion by more energetic events.

On parts of the coast not protected by mud banks, wave and current winnowing of very fine sand supplied by the Amazon, and of medium-to-coarse-grained, fluvial sand from the Guiana Shield, leads to the formation of well-sorted, sometimes shell-rich, sand bodies at the strandline (Augustinus et al. 1989; Anthony et al. 2014). These sand banks are up to 2 m thick, alternate with the mud banks, and, as the coast accretes, form linear, erosively based sand bodies or 'cheniers' of the same length as the interbank areas (Figure 9.24). They are driven landwards by high tides and storms when high-angle cross-stratification develops on the landward wash-over slope. The predominant structure in both coarse- and fine-grained cheniers is, however, planar, very gently seaward-dipping swash lamination, with intercalated thin beds of coarse shell hash.

On many other coasts, cheniers are formed by the temporal alternation of conditions favourable for coastal mud accretion. These conditions are exemplified by the 200 km long and 30 km wide coastal plain of western Louisiana where 'cheniers' derive their name from the colonisation by oak trees (chêne, in French) of the better-drained soils of the sandy ridges.

Mudflat progradation in western Louisiana occurs when sediment supply is greater on the western half of the Mississippi River delta, whereas the cheniers are formed by transgressive winnowing when most of the major river discharge occurs on the eastern edge (Hijma et al. 2017). Modern coastal accretion, up to 70 m a^{-1} in the updrift area close to the Atchafalaya River mouth, is contemporaneous with coastal erosion 50 km downdrift (Huh et al. 2001)

Mud is added to the chenier plain by an along-shelf 'mud stream' driven by the wind and river flow, as well as by mud reworked from the shallow (< 10 m) part of the subaqueous delta as a result of liquefaction caused by winter storm waves. The high sediment concentration close to shore dampens wave energy and allows open-coast mud deposition (Jaramillo et al. 2009), despite the high wave energy (Draut et al. 2005).

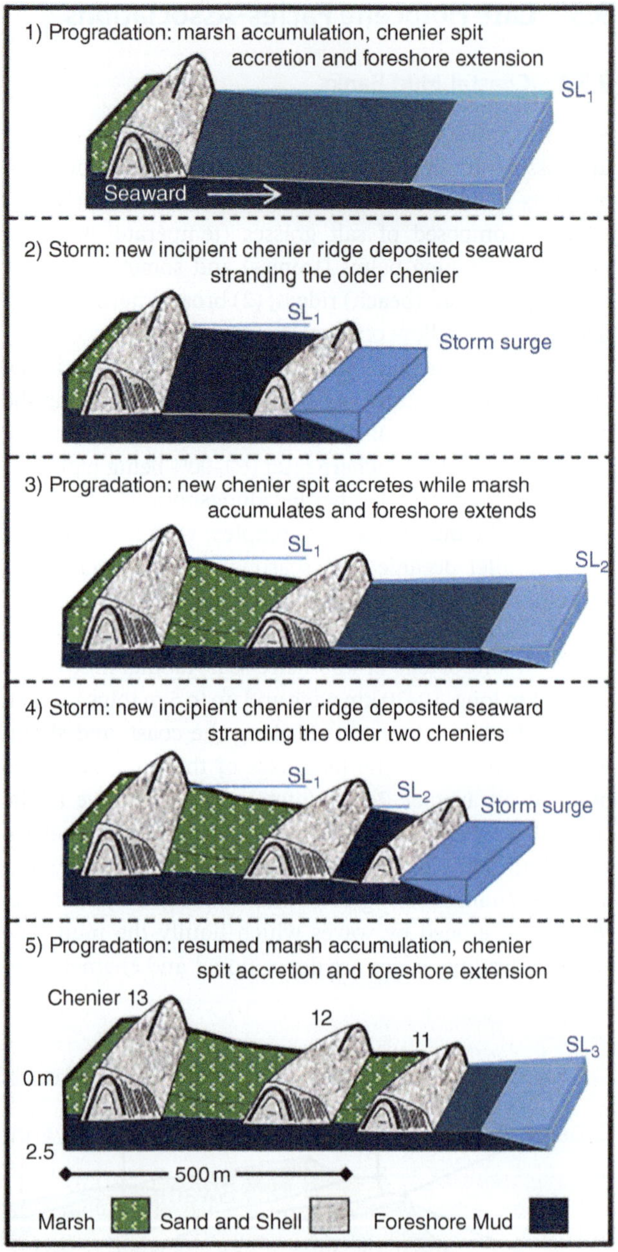

Figure 9.24 A five-stage model of chenier formation by means of alternating mud flat progradation and wave reworking (Dougherty and Dickson 2012 / with permission of Elsevier.)

9.5.3 Muddy Tidal Flats

Muddy open coast tidal flats have convex-up profiles with extremely low slopes, except where they grade into estuarine channels. Tidal flats typically fine inwards and upwards from fine sand deposited near low water, to mud at the high tide level (Van Straaten 1954; Evans 1965; Reineck 1967). Bedding styles, even in muddy flats, are dominated by wavy inter-bedded sand and mud with abundant bioturbation (Reineck and Wunderlich 1968; Fan 2012). Internal depositional breaks are common, with an irregular and random distribution of preserved depositional events such as daily, fortnightly or seasonal tidal increments of sedimentation (Fan 2012). The vertical accretion of temperate salt marsh is promoted by organic production and stabilisation by vegetation (Nyman et al. 2006). The preserved sequence depends on the spacing and migration rates of tidal channels and gullies which remove, by lateral erosion, the flat deposits. McClennen and Housley (2006) and Donnici et al. (2017) calculated that in principle, channel spacing (500 m to 1 km) and migration rates (10–20 m per century) were sufficient to completely rework mud flat deposits in the Venice lagoon down to a depth of 3 m in 500 years. These studies used shallow seismic to document substantial swaths of dipping, channel point bar reflectors demonstrating that extensive reworking by channels had indeed taken place.

9.5.4 Mangrove Coasts

Many tropical mud-rich coasts are fringed with mangroves which play an important role in sediment trapping, coastal aggradation and protection in response to storms and relative sea-level rise. Walsh and Nittrouer (2004) established a facies model for mangrove mud banks based on vibracores from the Fly River delta and the Kikori-Purari region of the northwestern Gulf of Papua. The composite vertical section is produced when mangrove-stabilised banks prograde into tidal channels. The high tidal muds (90% silt and clay) include more sand if the site is exposed to ocean waves. The sediment contains roots, pneumatophores (air tube-roots), abundant bioturbation by crustaceans, organic matter and some laminated sediments. The intertidal sandy muds have many mud laminae (>2 mm), probably deposited from fluid muds. Channel sediments are dominated by thin-bedded (>5 cm) sands, but also contain laminae of sand and mud. Organic matter is present in all facies and commonly forms discrete layers.

9.5.5 Prodelta Deposits of Large River Deltas

Progradational fine-grained deltas are characterised by depositional profiles with a large subaqueous delta platform, which is often physically separated from related distal shelf mud blankets. Progradation of the delta platform produces a coarsening-upward succession.

The facies association along such a depositional profile is summarised from examples in Section 9.2, namely the macro-tidal Amazon River prodelta and Amazon shelf (Kuehl et al. 1986; Jaeger and Nittrouer 1995); the microtidal Po delta (Wheatcroft et al. 2006); the Hwang He of China (Wright et al. 1990; Zhou et al. 2016; Wei et al. 2018); and the Fly River of Papua (Dalrymple et al. 2003; Harris et al. 2004).

General characteristics are:

1) Away from the river mouth, mean grain-size decreases from very fine sand to fine silt with a larger clay component in the distal mud facies.
2) Bioturbation intensity increases offshore.
3) Preservation of primary structures is greater in water less than 30 m deep.
4) Primary sedimentary structures are largely composed of plane-lamination with subordinate ripples and load structures.
5) There is more preservation of interlaminated sand and mud in tidal settings.
6) Massive beds of fluid mud, commonly up to 100 mm thick, may be found in various positions, including proximally in the river mouth or on the clinoforms of the subaqueous delta (Cacchione et al. 1995; Sternberg et al. 1996).
7) Organic-rich laminae and distributed organic matter are ubiquitous in the form of leaves, wood, etc. Decay of this material can lead to significant methanogenesis in prodelta muds.
8) The examples in Section 9.2 show lobate clinoform patterns in prodelta deposits, whereas the clinoform patterns from mud belts are linear.

The Atchafalaya River in Louisiana, a distributary of the Mississippi River, provides a good modern example of a muddy prodelta facies association. It has, over the last 100 years, constructed a subaqueous delta in water shallower than 25 m (Figure 9.25 see also the time lapse of the subaerial delta at: https://www.pbslearningmedia.org/resource/buac35-sci-deltas/deltas-atchafalaya-bay-atchafalaya-bay-deltas/).

Onshore are river-dominated sandy bay-head deltas. The subaqueous clinoform is muddy (70–100% finer than 63 micron) and extends approximately 25 km seaward of a shell reef across the mouth of Atchafalaya Bay. The clinothem is 3 m thick and has a maximum slope of about 1:550, and a narrow bottomset region only 500 m wide. The uppermost layer of the delta, within Atchafalaya Bay, is a 'yoghurt-like', under-consolidated mud, 2 m thick. The topset surface steepens from 1:2500 in the east to 1:1600 in the west and along-shore sediment dispersal by shelf currents is revealed by the progressive winnowing

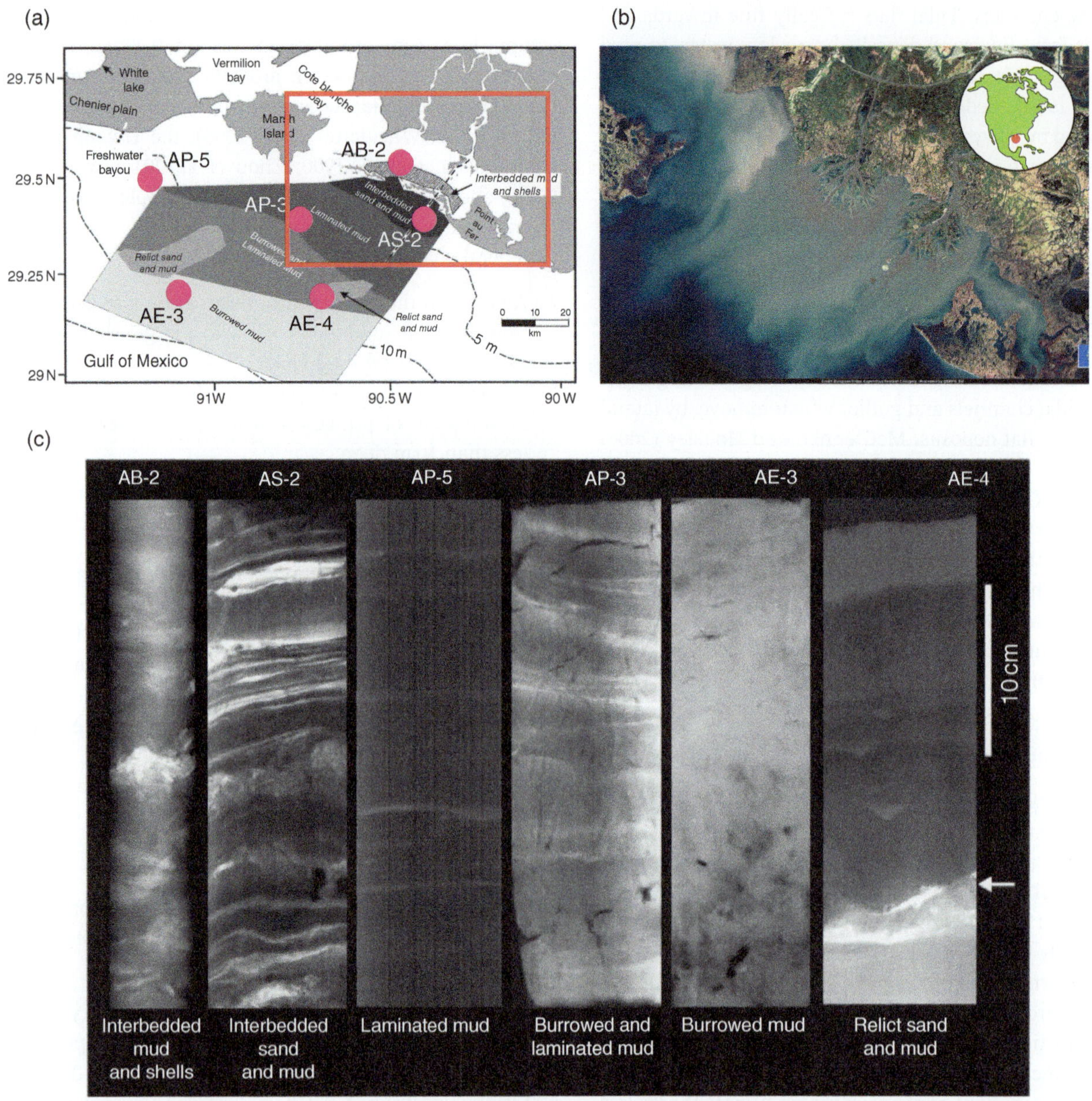

Figure 9.25 The subaqueous delta of the Atchafalaya River (Louisiana, USA). (a) The facies pattern is strongly related to sediment accumulation rates with decreased rates associated with increasing homogenisation through bioturbation. Laminated mud is deposited in depths of 5–10 m. Preferential coast-normal sorting along the dispersal system controls whether sand layers are present in the cores: silt/clay ratios also decrease to the west along the shelf transport path. (b) Satellite image taken 5 January 2021, located in the red box of (a), Credit: European Union, Copernicus Sentinel-2 imagery. https://www.copernicus.eu/en/media/image-day-gallery/atchafalaya-river-delta-louisiana-usa (c) Representative X-radiograph negatives for each facies mapped in (a). Bright layers in the X-radiographs are coarser sediments. The relict sand and mud example shows 220 mm of ephemeral surface mud (laminated mud) above an erosional interface (arrow) with underlying relict muddy sand. After Neill and Allison (2005).

of sand and coarse silt towards the west. Seven sedimentary facies are observed in the subaqueous delta, which are differentiated on the basis of the degree of biological destruction of physical stratification, which is inversely related to sediment accumulation rate, and by the proximity of an area to the riverine sand source. There is also a significant increase in shell content in Atchafalaya Bay relative to shelf facies (Neill and Allison 2005).

9.6 Stratal Architecture and Facies – Ancient Examples

9.6.1 Significant Surfaces

The first task in the facies analyses of ancient mudstone successions is to establish a correlation framework, especially one that can establish age relationships with other facies, such as coeval coastal deposits. Volcanic tuffs which, although usually rare, provide true chronostratigraphic surfaces (timelines) on which to establish correlation. Marine flooding surfaces are much more common, provide practical correlation surfaces recognisable at outcrop and in wireline logs and can be considered to approximate chronostratigraphic surfaces. The disadvantage of marine flooding surfaces is that they become unrecognisable as marine strata are correlated up-dip into non-marine deposits. To further complicate matters, shallow-marine mudstone successions may include subtle bypass surfaces that separate coeval up-dip and down-dip stratal packages, making correlation difficult.

Some hiatal surfaces in mudstone successions can lack coarse-grained lags or distinctive fossil assemblages and are marked only by an abrupt change in ichnofacies. Such 'omission' surfaces are discussed in Chapter 3.4: 'Substrate-controlled ichnofacies'.

Correlation of a succession of chronostratigraphic surfaces enables recognition of depositional geometries. In some ancient mudstone units, detailed correlation using closely spaced well logs shows that individual 'genetic' stratal packages can meet one another at very low angles through onlap, toplap or downlap, as is also shown by seismic profiles across modern muddy subaqueous deltas and mud belts. In other ancient mudstone units, equally detailed cross sections show that the succession accumulated essentially vertically, as a stack of near sheet-like bodies, and no systematic lap-out relationships, nor bathymetric breaks can be demonstrated. A variety of stratal stacking patterns and their associated facies drawn from the Cretaceous Western Interior Seaway are discussed below. The emphasis is on Canadian case studies that have utilised a publicly accessible database of wireline logs and cores chosen from hundreds of thousands of wells, coupled with extensive outcrop studies. These data present an unparalleled opportunity to reconstruct three-dimensional stratal geometry, facies distributions and palaeogeography.

9.6.2 Late Cretaceous Muddy Clinothems, Wyoming

Prograding muddy clinothems were described by Asquith (1970) and Winn et al. (1987) from Santonian to Early Maastrichtian strata in Wyoming. The Bighorn, Powder River, Washakie and Red Desert basins are separated by fault-bounded, basement-cored uplifts that were generated as a result of crustal contraction linked to flat subduction during the Laramide (Campanian to Eocene) stage of orogeny (Lawton 2019). Mudstone units grade updip into coeval deltaic and coastal plain facies, broadly ascribed to the Mesaverde Group. Using closely spaced resistivity logs, Asquith (1970) and Winn et al. (1987) showed that distinctive markers in the logs (some attributable to concentrations of volcanic ash) could be traced through the mudstone units, where they defined sets of clinothems, inclined typically at 0.5–1°, exceptionally at 2°, and up to about 270 m thick. Several distinct, clinothem units, arranged in both progradational and aggradational stacking patterns, were recognised by Asquith (1970; Figure 9.26).

Winn et al. (1987) showed that sandy deltaic facies predominate on the updip parts of clinothems (mostly to the left of what is shown in Figure 9.26), comprising heterolithic facies that include evidence of wave action. Sediments on the lower parts of clinothems include non-bioturbated to weakly bioturbated, normally graded mudstone beds which may be interstratified with sharp-based, graded sandstones showing classical features of turbidites, including loaded bases and fluid-escape structures. Plant debris and mudstone intraclasts are common in many beds. Winn et al. (1987) combined facies distributions and stratal geometry to infer that storm wave base in their studied area lay at 100–150 m water depth. The unbioturbated intervals were interpreted to have been deposited beneath water containing $< 0.1\,\mathrm{mg\,l^{-1}}$ dissolved oxygen. It was estimated that oxygenated water may have extended to a depth of 150–200 m, whereas the maximum depth of the basin was estimated at about 230 m.

More recently, Pyles and Slatt (2007) described submarine channels on the clinothem slope, and submarine fan sandstones at the base of slope. Soyinka and Slatt (2008) used slab samples and thin sections of mudstone to document four facies, comprising various associations of laminated and graded mudstone, siltstone and very fine sandstone. The facies were organised in both upward-fining and upward-coarsening lamina-sets on a vertical scale of millimetres to tens of millimetres. Subtle micro-erosion surfaces separate some lamina-sets. This association of features led Soyinka and Slatt (2008) to interpret the succession as largely the product of waxing and waning hyperpycnal flows that varied in strength in response to the varying volume and density of river floodwaters. True turbidites and hemipelagic deposits were also inferred to be present but composed only a minor proportion of these rocks.

The development of deeper water and the tall clinothems in the Late Cretaceous of Wyoming was mainly due to rapid tectonic subsidence that was associated with the development of basement-cored uplifts, resulting in the generation of accommodation faster than it could be filled (Winn et al. 1987; Devlin et al. 1993). Periods of eustatic rise augmented this accommodation, leading to the formation of widespread, fine-grained 'black shales', whereas eustatic fall and a lower subsidence rate promoted the progradation of deltas across the top of the muddy clinothems.

The Late Cretaceous delta-front sediments in Wyoming provide abundant evidence of storms which would be

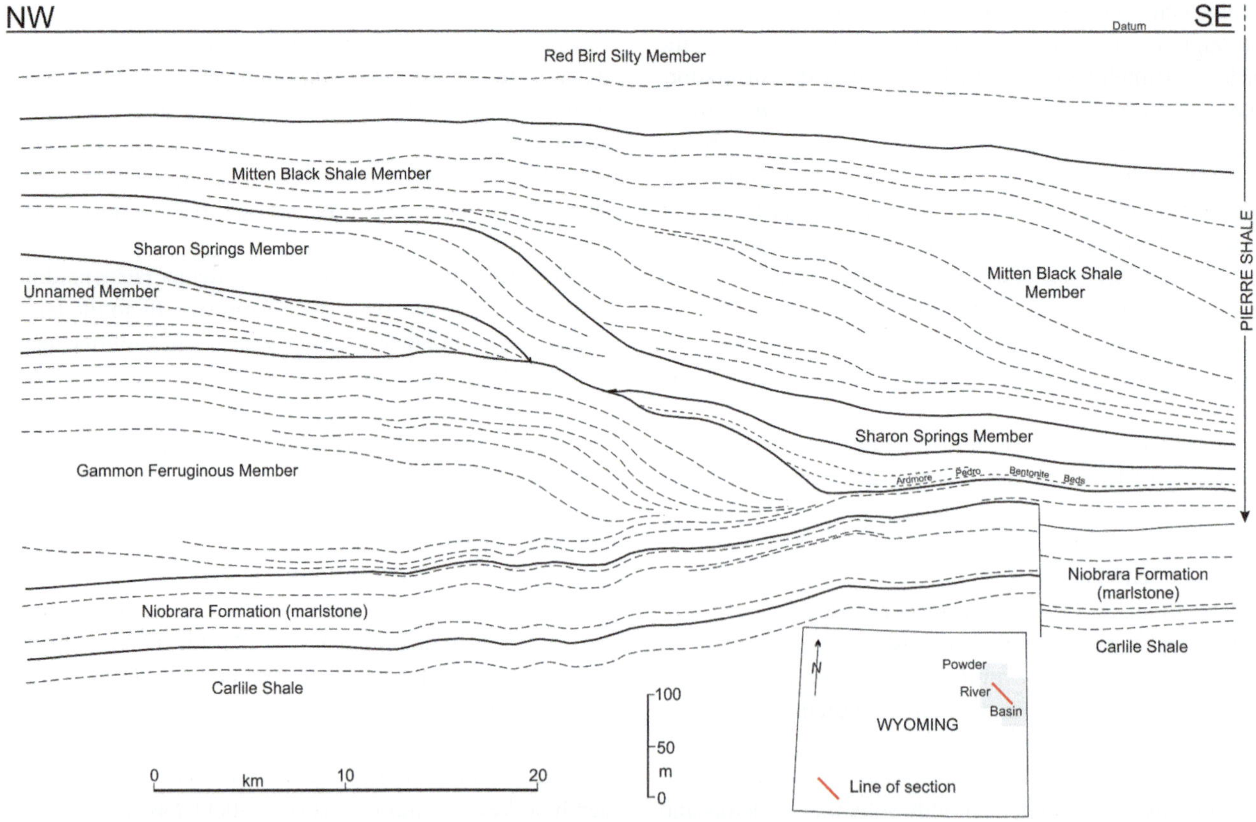

Figure 9.26 Cross section of Pierre Shale and Niobrara Formation in the Powder River Basin, Wyoming. The various members (except the sandy, Red Bird Silty Member), are mudstone-dominated and form a series of prograding and downlapping clinothem units (slope c. 1°). Clinothems have been interpreted to have developed when accommodation rate exceeded sediment supply, allowing deeper water basins to develop. The basins were filled laterally by deltaic systems. In shallower water, waves and storms were able to transport sediment, but much of the sediment transported into deep water is interpreted to have been carried by turbidity currents and by muddy hyperpycnal flows issuing from the mouths of rivers when in flood. Adapted from Asquith (1970).

expected to have set up strong along-shelf flows that would have affected the dispersal of mud. Studies of the Wyoming mud clinothems nevertheless focus strongly on sediment transport directly *down* the clinoform slope. No investigation appears to have considered the possibility that mud was transported parallel to slope (as occurs, e.g., in the mud clinothems in the Adriatic and South China seas), although Winn et al. (1987) did conclude that contour currents did not appear to have been an important transport process. Assessing the possibility of strike-transport of mud is difficult, but perhaps could be addressed through very detailed palaeocurrent observation of silt and sand ripples at outcrop.

9.6.3 Dunvegan Formation Clinothems, NW Alberta

In the Canadian sector of the Western Interior Foreland Basin, the Late Cretaceous Dunvegan Formation provides an example of unusually tall and steep prodelta clinothems. In the Early Cenomanian, obduction of an outboard terrane on the continental margin, resulted in rapid flexural subsidence and development of a relatively deep-water, under-filled condition in the foredeep in NE British Columbia and adjacent Alberta (Plint et al. 2009, 2012a). Into this trough, the Middle Cenomanian Dunvegan delta system prograded, in an axial, NW to SE direction (Bhattacharya and Walker 1991; Plint 2000; Plint et al. 2009). The problem of chronostratigraphic correlation within the diverse deltaic facies was solved by using an allostratigraphic approach. An allostratigraphic unit is '...defined and identified on the basis of its bounding discontinuities' (North American Commission on Stratigraphic Nomenclature 2005, p. 1578). By correlating regional marine flooding surfaces and their correlative updip subaerial unconformities, Bhattacharya and Walker (1991), Plint (2000, 2002) and Hay and Plint (2020) showed how the Dunvegan delta system comprised a succession of offlapping transgressive–regressive sequences that were deposited during the Middle Cenomanian, followed by major transgressive back-step and drowning of the system at about the Middle to Late Cenomanian boundary.

Allostratigraphic correlation showed that the muddy prodelta clinothems, up to about 100 m high and inclined at about 1°, downlapped relatively abruptly onto a highly radioactive, phosphatic condensed section (the 'Fish Scales Upper' log marker of Bhattacharya and Walker 1991). It was, however, not clear why the Dunvegan clinothems should downlap in such an abrupt manner. This question was addressed by Plint (2014), who correlated clinothems of Dunvegan allomember G from subsurface to outcrop, where oriented blocks of prodelta mudstone were collected. Large thin sections were cut in planes parallel, oblique and perpendicular to the regional clinothem slope. Mudstone microfacies were organised in thin event beds a few millimetres thick that were typically composed of an erosive-based layer of rippled silt, overlain by interlaminated silt and mud, then by dispersed silt grains floating in mud, and, locally, by structureless clay. About 75% of event beds showed an upward-fining successions whereas about 25% coarsened upward. Siltstone ripples were symmetrical in cross section, but internal cross-lamination showed a strong down-slope (SE-directed) preferred orientation. These observations were interpreted as evidence that silt transport took place from turbulent flow under the influence of storm waves, but with a superimposed downslope component of flow. The interlaminated silt and mud microfacies were interpreted to have been deposited from a dense laminar flow of fluid mud within which shear led to grain segregation. The overlying silt dispersed in mud represented 'gelling' of the fluid mud as flow ceased (Figure 9.27).

The Dunvegan microfacies are comparable to the tripartite divisions of CWESGF described in Section 9.3.4. The combination of palaeoflow and microfacies data from the Dunvegan rocks suggested that fluid mud travelled down the prodelta slope, primarily in wave-enhanced sediment gravity flows that were both initiated and then supported by storm wave action. As the flow descended below effective storm wave base, estimated at about 70 m water depth, the suspended mud was deposited. The basinward pinch-out of the clinothems could therefore be interpreted to approximate the depth of effective storm wave base for muddy sediment (Figure 9.27).

Although mud dispersal on the deeper part of the prodelta was controlled primarily by *gravity*, palaeoflow data from combined-flow ripples in very fine-grained sandstone beds in the upper, shallower part of the clinothems indicated that sand was transported approximately shore-parallel, under the influence of *combined flows*.

9.6.4 Ramp Style Deposition in a Semi-Enclosed Basin – The Middle Albian Harmon and Cadotte Alloformations

In the Middle Albian, the Polar Ocean flooded the Western Canada foredeep to form an elongated embayment extending southward in Alberta to about 54°N present latitude (Figure 9.28a).

This transgression is represented by marine mudstone of the Harmon alloformation in Alberta, whereas the subsequent regression is represented by strandplain sandstone of the Cadotte alloformation (Figure 9.28b). Allostratigraphic analysis showed that the Harmon and Cadotte alloformations were each composed of several allomembers bounded by regionally mappable marine flooding surfaces (Buckley and Plint 2013; Henderson et al. 2014; Buckley et al. 2016).

The Harmon alloformation is a mudstone-dominated wedge that, over 400 km, thins from > 150 m in the western foredeep to < 5 m on the forebulge in the east. The five constituent allomembers are also wedges that thin systematically eastward. In the west, allomembers comprise several stacked parasequences, typically 2–10 m thick and relatively sand- and silt-rich. Eastward, parasequences thin, become finer-grained and eventually lose their identity. No clinoform geometry, downlap surface or rollover point can be recognised. Ubiquitous wave- and combined-flow ripples in very fine sand and silt show that the sea floor in the area shown in Figure 9.28a lay above storm wave base. The combination of geometric and lithological characteristics indicate that deposition took place on an extremely low-gradient ramp. Gamma-ray log-based facies mapping around the southern periphery of the Harmon embayment suggests that the coastline consisted of elongate to lobate 'muddy sand' deltas that prograded into water only a few metres deep (Henderson et al. 2014). A core located about 50 km north of the Harmon shoreline (Figures 9.28a, 9.29) shows a succession of millimetre to centimetre-scale, erosive-based beds of siltstone grading up into silt-bearing and commonly structureless claystone. Bioturbation index is very low (BI 0-1). Many siltstone beds have a subtly lenticular, symmetrical profile and low-angle cross-lamination that dips in different directions in adjacent beds. The facies characteristics suggest that the sea floor was eroded by storm waves that formed millimetre-scale scours as well as centimetre-scale gutter casts.

Storm-generated combined flows are interpreted to have suspended and transported very fine sand and mud along and across the shelf. As each storm waned, coarser sediment settled, forming very low-amplitude combined-flow ripples, overlain by progressively finer silt and clay. Some of the dark clay-rich layers (Figures 9.29b–d) are structureless and probably represent the settling or gelling of fluid mud (cf. Bhattacharya and MacEachern 2009).

Palaeoflow data, mainly from outcrop on the western margin of the basin, include gutter casts that show that storm waves approached from the north, down the axis of the seaway (Figure 9.28a). Ripples indicate that geostrophic and combined flows transported fine sand both seaward (NE) and along-shelf (SE). The degree of bioturbation in the Harmon mudstones is extremely variable: BI 0–6 (Figure 9.30). Bioturbation tends to be more intense in the

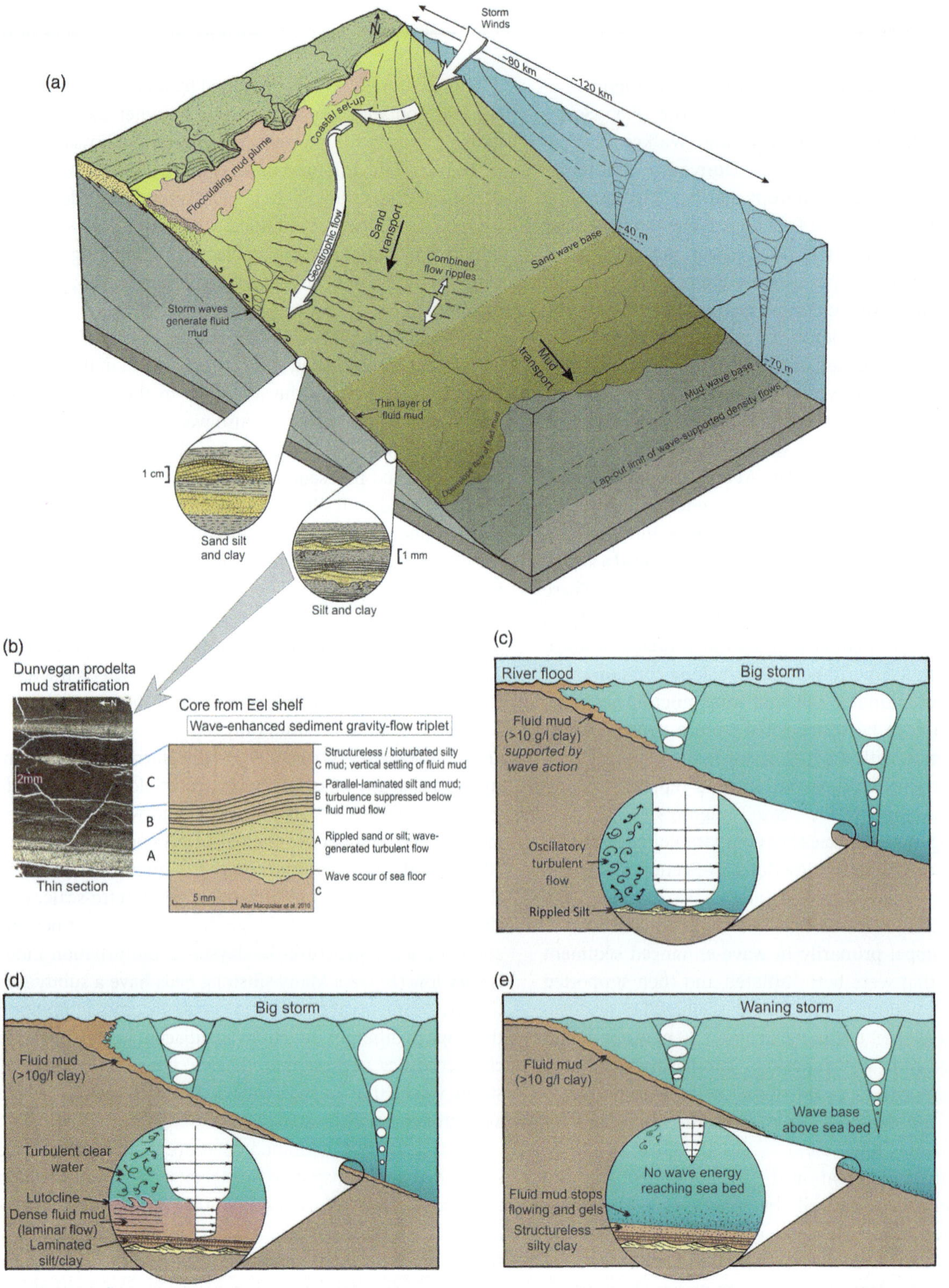

Figure 9.27 (a) Block diagram (Plint 2014 / John Wiley & Sons) summarises the spatial and bathymetric distribution of processes operating over the Dunvegan prodelta. Storm-generated combined flows transported fine sand shore-oblique in the nearshore region. Fluid mud, possibly generated by river floods, and also suspended by storm waves, flowed down the prodelta slope as a CWESGF. Mud remained suspended until it dropped below storm wave base, below which it was deposited as the flow slowed and gelled. (b). Thin section of Dunvegan prodelta mudstone showing tripartite division of event beds. These divisions can be matched with similar structures and textures observed in a core that sampled sediment inferred to have been deposited by a CWESGF on the modern Eel River shelf (Adapted from Macquaker et al. 2010). (c) River flood, combined with rising storm wave energy generates fluid mud in nearshore area of prodelta, while storm waves produce oscillation ripples in silt on the deeper part of the prodelta. (d) Downslope-flowing fluid mud reaches the deeper part of the prodelta where it causes ripples to migrate preferentially downslope, before suppressing turbulence at the bed, leading to the deposition of parallel-laminated silt and clay from laminar flow of mud below the lutocline. (e) Storm energy wanes and wave base lifts off the bottom, removing the energy source that kept mud in suspension. Fluid mud loses suspended material, decelerates and eventually gels, depositing structureless mud with suspended silt grains. Panels c-d-e inspired by Mackay and Dalrymple (2011) / Society for Sedimentary Geology (SEPM).

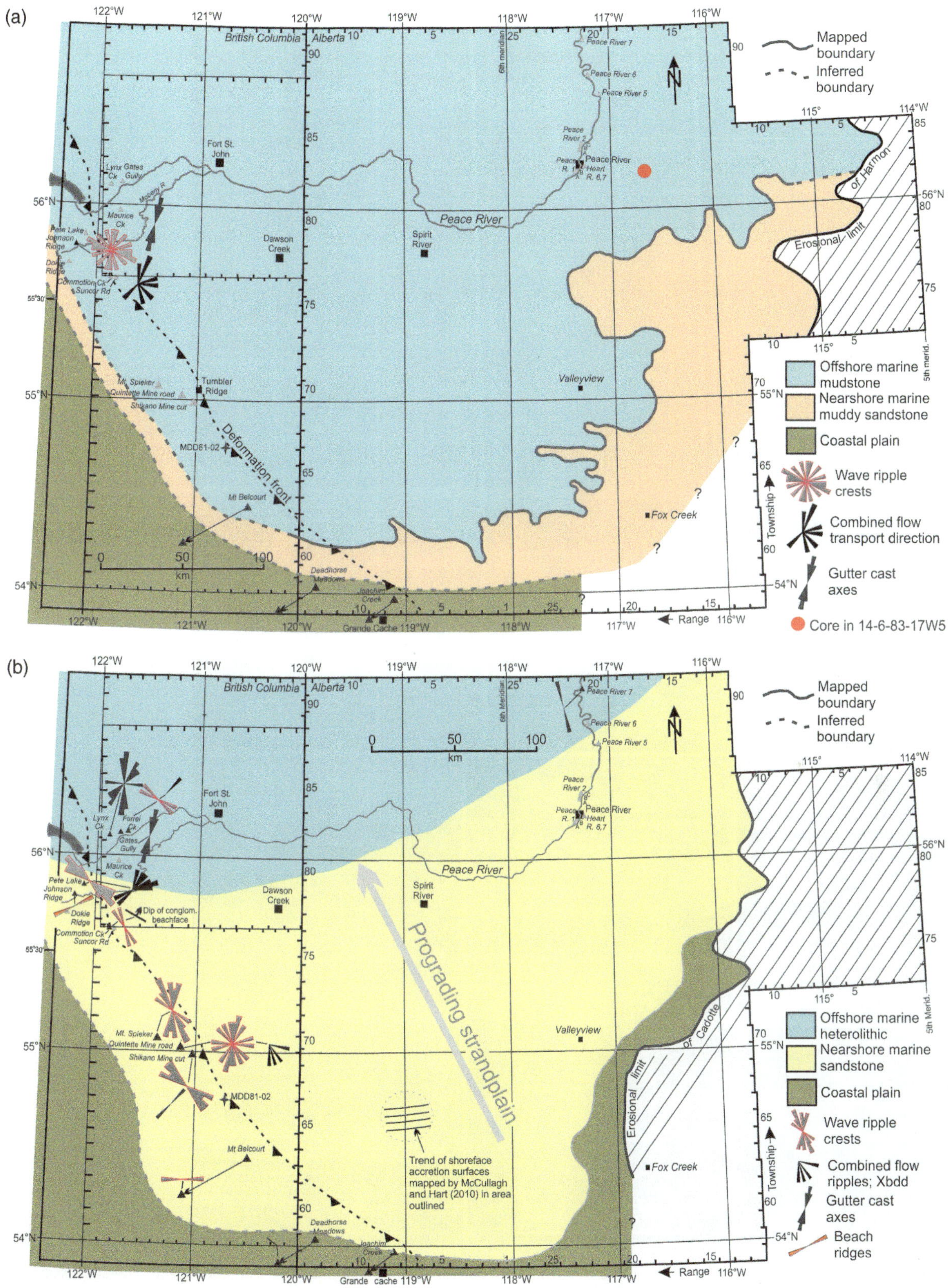

Figure 9.28 (a) Palaeogeography of the Middle Albian transgression that deposited marine mudstone of the Harmon alloformation. The distribution of interpreted depositional environments, palaeocurrent data and the location of the core shown in Figure 9.29 are shown. (b) Palaeogeographic extent of the regressive strandplain sandstone that forms the Cadotte alloformation, together with various palaeocurrent data. Figures adapted from Buckley et al. (2016).

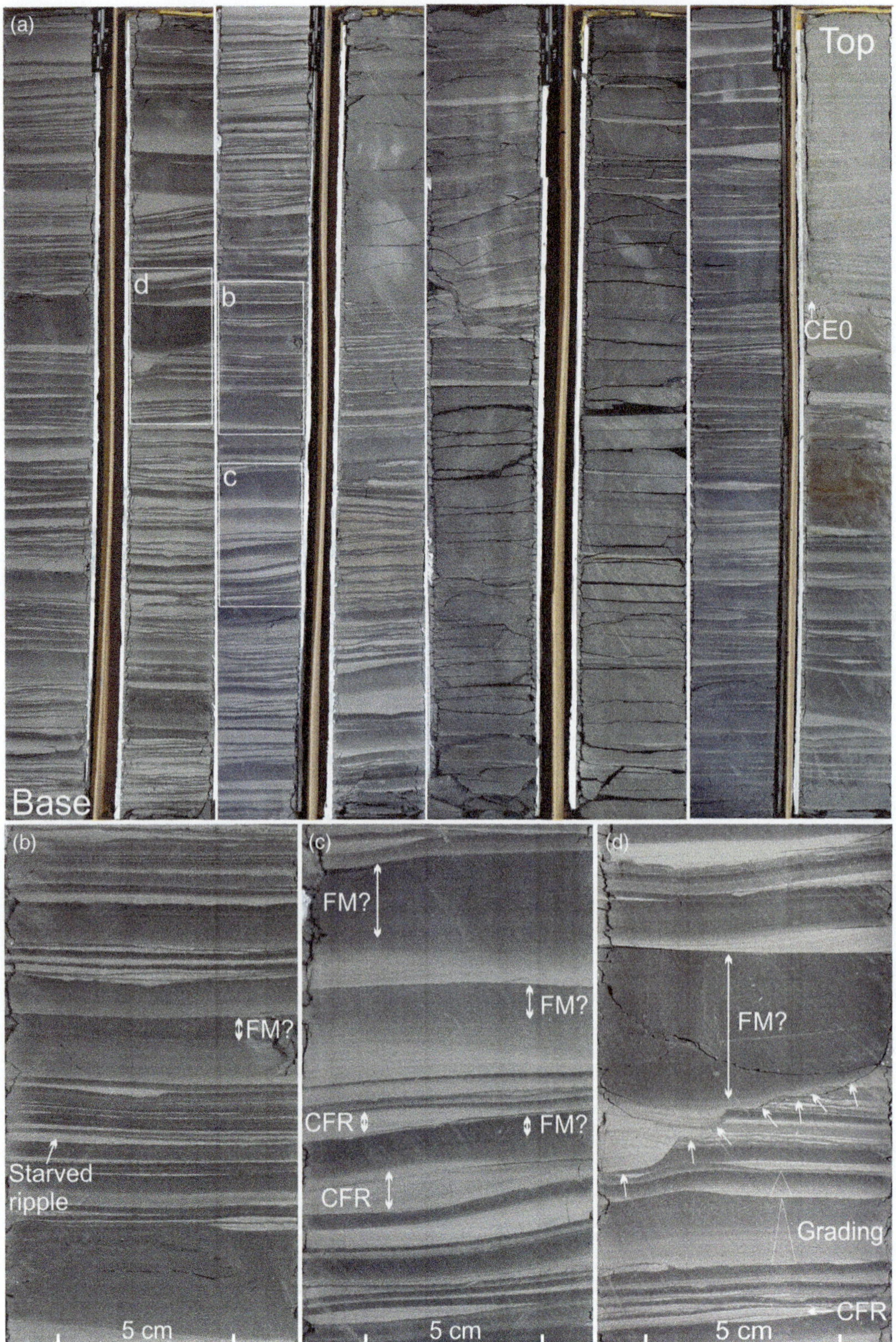

Figure 9.29 (a) Core from well 14-6-83-17W5 (389 to 395.2 m; each core sleeve = 75 cm), located in Figure 9.28a. The succession consists of millimetre to centimetre-scale, erosive-based, graded beds of coarse to medium-grained siltstone grading up into silt-bearing claystone. In the uppermost sleeve, discrete wave-rippled lenses of very fine-grained, cherty glauconitic sandstone are interstratified with silt and clay. Identical glauconitic sandstone, lacking mud, appears abruptly at surface CE0, which marks the base of the Cadotte alloformation. (b) Thin beds of graded siltstone, some of which display subtly erosive bases, massive fluid mud (FM), and very low-amplitude wave ripples. (c) Detail of millimetre to centimetre-scale graded beds of siltstone and claystone with well-developed form-discordant combined-flow wave ripple cross-lamination (CFR). (d) Detail of centimetre-scale gutter cast with steep, stepped walls that imply that the underlying sediment had significant cohesion. Note the absence of macroscopic bioturbation. Buckley et al. (2016) / John Wiley & Sons.

Figure 9.30 Rubbly weathering, intensely bioturbated (BI = 6) sandy siltstone facies near the top of the Harmon alloformation. Subtle coarsening and fining trends are indicated, with siderite nodules concentrated near the finest-grained portion of each succession, where they might represent minor maximum flooding surfaces.

western foredeep. Correlation shows, however, that the degree of bioturbation also varies laterally *within* the same allostratigraphic units, and therefore was probably controlled by spatially variable factors, such as proximity to distributary mouths, sedimentation rate and water turbidity (Buckley et al. 2016).

The regional stratal geometry shows that the Harmon sea floor lacked significant topography, except in the immediate vicinity of deltas. This suggests that the rates of sediment supply and accommodation were closely balanced, verging on oversupplied. Muddy sediment would have accumulated up to a depth determined by ambient wave energy, after which, surplus mud would have been repeatedly re-suspended and advected until it reached an area of higher accommodation. There is no evidence, in the form of tripartite beds, for the operation of CWESGF (Macquaker et al. 2010), or of upward-coarsening-then fining beds, typical of river flood-generated hyperpycnal flows (Figure 9.20). It may be that the slope on the Harmon sea-floor was so low that the effect of gravity on suspensions of sediment was outweighed by Coriolis deflection of storm-generated flows.

The Cadotte alloformation represents a wave-dominated strandplain sandstone that prograded rapidly across the top of the Harmon mudstone (Figures 9.28b, 9.29). Coeval coastal plain facies are essentially absent, suggesting deposition during relative sea-level fall (Buckley et al. 2016).

9.6.5 High-Accommodation, Mudstone-Dominated Systems

The foredeep in northeastern British Columbia provides two excellent examples of high-accommodation, mudstone-dominated systems that developed when the rate of flexural subsidence was high. Rapid subsidence rotated the basin floor down towards the orogen, causing updip alluvial systems to become aggradational as rivers strove to maintain grade towards the sea. The high accommodation rate also prevented sandy nearshore systems from prograding far into the basin and hence, sandy facies stacked vertically, close to the orogen; only mud and very fine sand were able to escape seawards to fill the more distal foredeep. The facies distribution observed in these rocks was accurately predicted by numerical models (Jordan and Flemings 1991) that overturned previous ideas about the relationship between 'coarse clastic tongues' and tectonic activity. The numerical models showed that coarse tongues prograded far into the foredeep when the rate of thickening in the orogen, and the rate of subsidence, was *low*. Conversely, thick marine mudstone successions did not record 'tectonic quiescence' but in fact indicated rapid tectonic loading, and resultant isostatic subsidence of the basin!

9.6.6 High-Accommodation Late Albian and Early Cenomanian Systems

A pulse of rapid subsidence during the Late Albian and Early Cenomanian accommodated a wedge of marine mudstone-dominated sediment up to 800 m thick in the most proximal foredeep in the Peace River region of NE British Columbia. The stratal architecture and chronostratigraphic relationships within these lithologically varied and highly diachronous rock units were established by allostratigraphic correlation, using wireline logs correlated to the principal outcrop sections (Angiel and Plint 2022). Spectral gamma-ray data from outcrop facilitated correlation to well logs. Marine flooding 'landward shift' surfaces were the principal allostratigraphic bounding surfaces, but in addition, a number of abruptly upward-shallowing or 'basinward shift' surfaces also formed mappable boundaries. Basinward shift surfaces record abrupt accommodation loss, and in some cases, subaerial exposure.

The geometry and lithology of the Upper Albian and Lower Cenomanian strata shown in Figure 9.31a represent only the proximal part of the foredeep (Angiel and Plint 2022); the eastward extension of the stratigraphy, 300

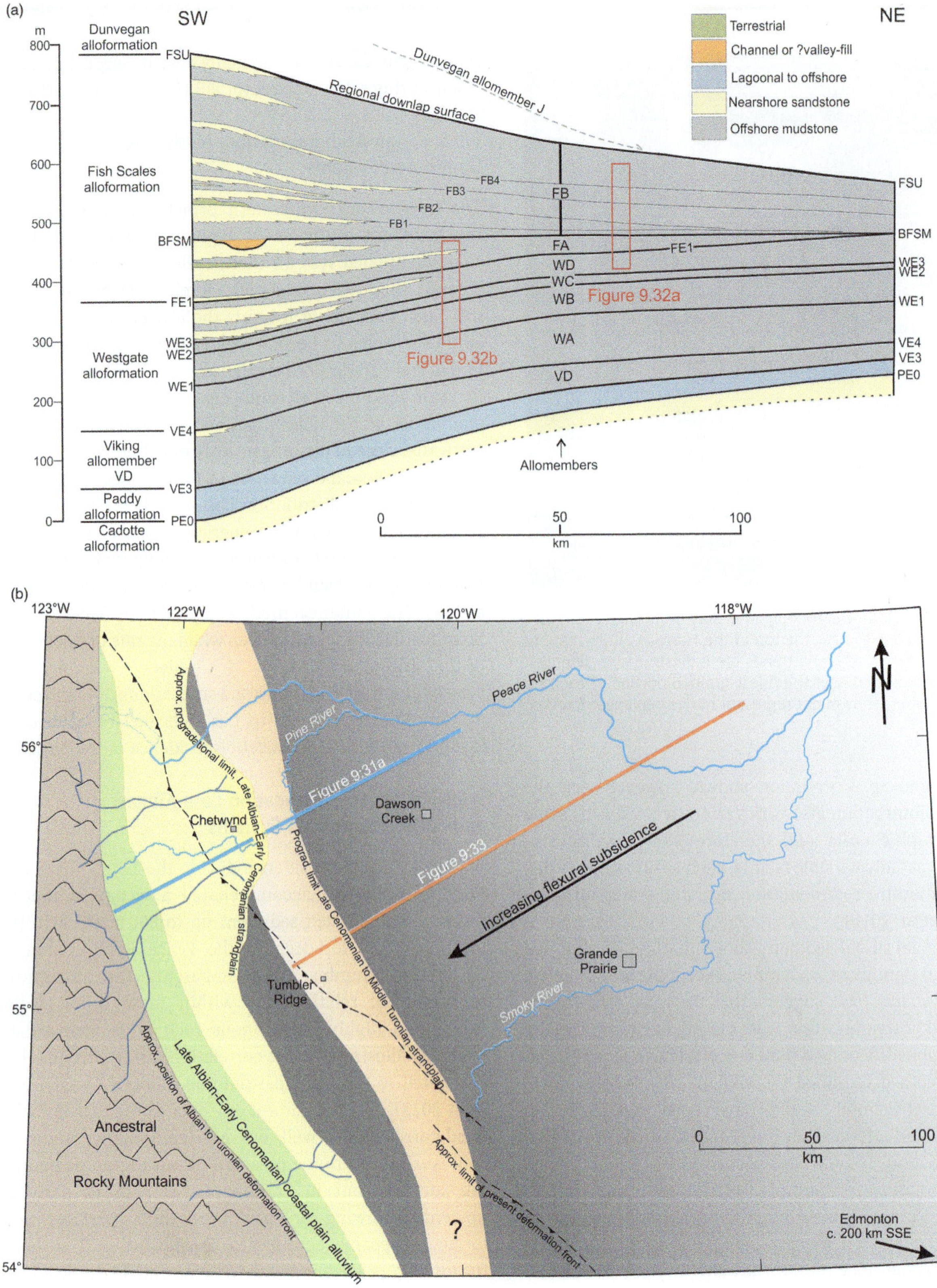

Figure 9.31 (a) Dip cross section, to scale, of Upper Albian and Lower Cenomanian rocks in the proximal foredeep of NE British Columbia, westward of 120° W (located in (b)). Datum is the subaerial unconformity surface BFSM. Deltaic sandstones are confined to the western, rapidly subsiding part of the foredeep. A variety of coarse to fine mudstone facies occupy most of the foredeep, generally becoming finer-grained and more organic-rich towards the NE. There is neither evidence of clinothems nor a downlap surface; all lap-out patterns involve onlap towards the forebulge. Red rectangles represent the positions of two outcrop photographs in Figure 9.32 (Angiel and Plint 2022 / The Canadian Society of Petroleum Geologists). (b) Palaeogeographic map of NE Alberta and adjacent British Columbia showing the location of the cross sections in Figures 9.31a and 9.33. Yellow shading indicates the generalised extent of delta-front sandstones of Late Albian to Early Cenomanian age (from Angiel 2013). Orange shading indicates the distribution of shoreface sandstones of Late Cenomanian to Middle Turonian age, shown in cross section in Figure 9.33 (from Varban and Plint 2008a).

km into Alberta, is illustrated in Roca et al. (2008). Allomembers VD, WA and WB are almost entirely mudstone and are interpreted to record the most rapid relative sea-level rise – which pushed the shoreline to the west of data control. The bulk of the rock consists of coarse to fine mudstone. In the west, the succession is thinly bedded (cm-scale) and very heterolithic, with a high proportion of siltstone and very fine-grained sandstone that forms millimetre to centimetre-scale beds. Sand- and siltstone beds are invariably sharp-based, commonly with wave- or combined-flow ripples, some being isolated in mud. Sand beds may have sharp tops, or may grade up into mudstone. Reverse- to normal grading was not observed.

Subtle siltier- and sandier-upward packages, each a few metres thick and bounded by flooding surfaces (i.e. parasequences), are ubiquitous (Figure 9.32). Towards the east, fine mudstone, and even claystone predominate, and some intervals have a high organic content, including both carbonaceous and phosphatic fish debris. Bioturbation is extremely variable, with BI 4–6 in more nearshore areas in the west, grading seaward to BI 0-1. However, bioturbation index also varies stratigraphically, with some mudstone units, particularly in allomembers VD, WA and FB lacking macroscopic bioturbation through many tens of metres of section. Benthic macrofauna, mainly inoceramid bivalves, are correspondingly scarce through much of the succession, although ammonites are locally common.

Stratal surfaces show no evidence of downlap or a clinoform morphology (Figure 9.31a). Surface FE1, which is a 'basinward shift' surface (expressed as an abrupt, upward-shallowing facies change), toplaps against BFSM as accommodation is lost towards the forebulge. Surface BFSM is an erosion surface mantled with extra-basinal pebbles and marks a major sea-level lowstand that led to subaerial exposure, and, in places, fluvial incision (Figures 9.31a; 9.32a). As parasequences within allomember FB thin and lap out towards the forebulge, they become increasingly radioactive due to an increasing content of pelagic organic matter and phosphatic fish debris, presumably in response to progressive trapping of clastics in the foredeep to the west.

Overall, it appears that the rate of sediment supply matched or exceeded the rate of subsidence throughout the Late Albian and Early Cenomanian such that the sea floor was maintained as a shallow, extremely low-gradient ramp. Storm-driven combined- and geostrophic flows appear to have been largely responsible for the re-distribution of mud across the foredeep. There is no evidence that the water depth ever exceeded storm wave base, even in the region of most rapid subsidence.

9.6.7 High-Accommodation Late Cenomanian to Late Turonian Systems

The Late Cenomanian to Middle Turonian Kaskapau alloformation is a mudstone-dominated wedge that occupies the foredeep in NE British Columbia and adjacent Alberta. The wedge-shaped geometry of the Kaskapau abruptly changes to a sheet-like geometry in the overlying Cardium alloformation (Figure 9.33). The Cardium is characterised by thin (< 10 m) but highly progradational (up to 180 km) strandplain sandstones, and the entire alloformation is partitioned by numerous, closely spaced unconformities that record high-frequency eustatic cycles (Hart and Plint 1993; Plint et al. 2022). The change from the Kaskapau 'wedge' to the Cardium 'sheet' is attributed to a marked decrease in the local rate of flexural subsidence that allowed the muddy basin floor to aggrade to a depth of about 10–20 m, at which point, modest eustatic fall was able to drive major coastal regression (Plint et al. 2022; Figure 9.33).

Here, we focus on the muddy Kaskapau wedge, within which outcrop and well log data allow broad facies belts to be mapped (Varban and Plint 2008a, 2008b). Clean, fine to medium-grained sandstone with swaley cross-stratification (SCS), planar lamination and localised conglomerate represent wave-dominated shoreface and distributary channel environments that extend up to 20 km eastward from the preserved basin margin. Within this belt, shoreface sandstones are commonly sharp-based, about 8 m thick, and separated by little or no mudstone. The stacked sandstones represent high-frequency oscillations of the shoreline (Figure 9.34). The clean sandstones grade seaward into a belt 10–30 km wide, dominated by decimetre-scale beds of very fine- to fine-grained sandstone with hummocky cross-stratification (HCS), planar lamination, gutter casts, wave and current ripples; interbedded mudstones are usually highly bioturbated. This facies represents deposition primarily from storms in an inner shelf position.

Heterolithic, cm-scale interbeds of very fine sandstone and mudstone form the next seaward belt, about 70 km wide. Sandstone beds are typified by parallel lamination, current and wave ripples. Ichnofauna include *Chondrites, Palaeophycus, Planolites, Teichichnus* and occasional *Thalassinoides*. Decimetre-scale siderite concretions are randomly scattered through the mudstone. The facies is interpreted to have been deposited between fair-weather and storm wave base, with sediment transport driven primarily by storm-generated combined flows on an open shelf.

From 100 to 200 km from the basin margin, the facies are laminated to bioturbated silty mudstones interstratified with

Figure 9.32 (a) Aerial view of parts of the Westgate and Fish Scales alloformations exposed in Pine River, British Columbia. Some of the main bounding surfaces are labelled, including the conglomerate-veneered BFSM surface (Base Fish Scale Marker). Courtesy of Piotr Angiel. (b) Aerial overview of parts of the Westgate and Fish Scales alloformations exposed in the Halfway River, British Columbia, showing, in allomember FA, the most progradational deltaic sandstone. Some major bounding surfaces are labelled; see Figure 9.31a for stratigraphic location of the photographs, and Angiel and Plint (2022) / Canadian Society of Petroleum Geologists for further details.

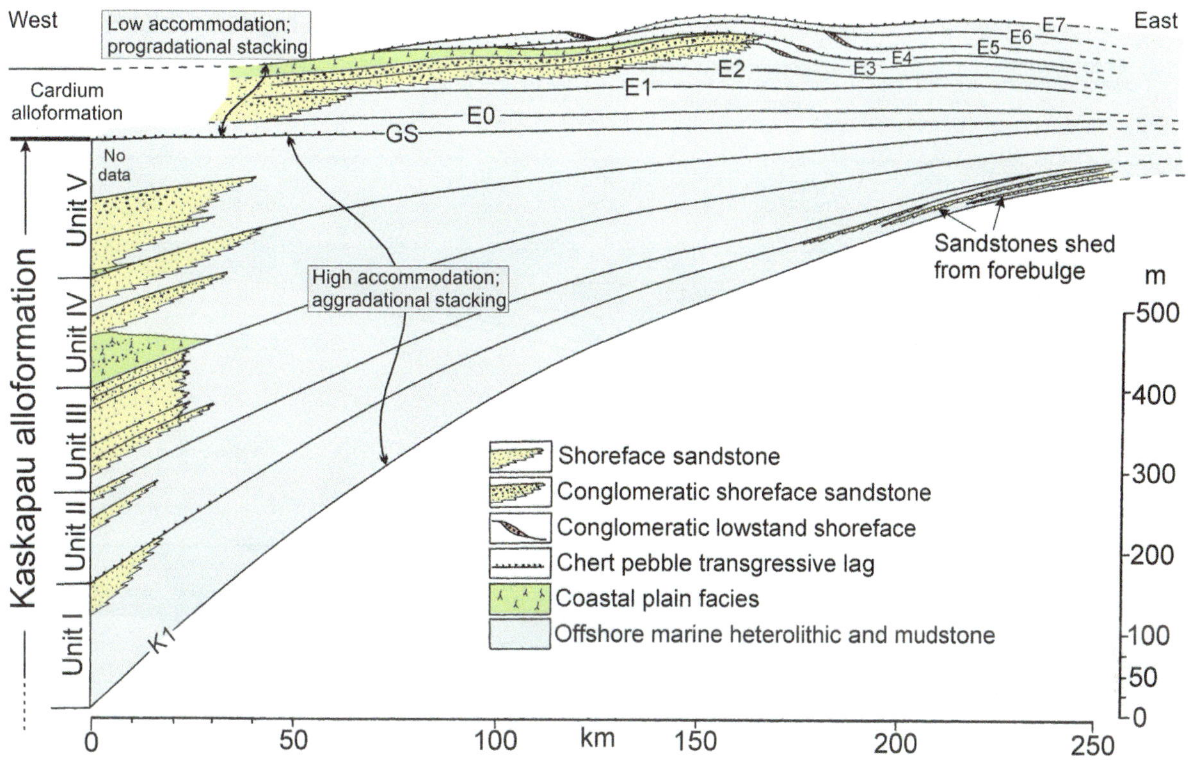

Figure 9.33 Scale dip cross section (located in Figure 9.31b), across the Cenomanian-Turonian foredeep in NE British Columbia and adjacent Alberta, contrasting the prominent wedge shape of the mudstone-dominated Kaskapau alloformation and the tabular shape of the overlying Cardium alloformation. The Cenomanian-Turonian boundary is at the junction of units I and II and the base of the Upper Turonian is at surface E1. During Kaskapau deposition, high accommodation inhibited coastal progradation, causing shoreface sandstones to stack vertically, while mud escaped seaward, filling the remainder of the basin. Subsidence rate diminished markedly above surface GS, allowing strandplain sandstones of the Cardium alloformation to prograde far across the foredeep. The five units of the Kaskapau alloformation are composed of 28 allomembers, none of which show evidence of downlap or clinoform stratal architecture. This implies that the rate of sediment supply exceeded that of accommodation, and that efficient, storm-driven redistribution of sediment maintained a near-planar ramp bathymetry. Varban and Plint 2008b / John Wiley & Sons.

siltstone and up to 20% very fine-grained sandstone in millimetre-scale laminae. Current, wave and combined-flow ripples, and parallel lamination characterise the sandstone laminae, the bases of which may show small-scale scours. The degree of bioturbation ranges from low to moderate; trace fossils include *Chondrites*, *Palaeophycus* and *Planolites*. The silt and sandstone laminae are suggestive of deposition from combined flows during storms, and the mudstones might also have been transported by geostrophic flows, as in many modern mid-shelf mud belts.

More than 200 km seaward, strata thin markedly onto the forebulge, and the sediments are bedded on a millimetre scale, weakly to unbioturbated, fine to coarse mudstone and siltstone with minor very fine sandstone. Sandstones and siltstones are typified by very small combined-flow ripples, with millimetre-scale scours on bed bases. Laminae rich in faecal pellets composed largely of coccoliths give a 'speckled' appearance to the mudstone. A distinctive mudstone unit that spans the Cenomanian-Turonian boundary was deposited during Oceanic Anoxic Event 2 and has from 6 to 11% TOC: this mudstone forms a major petroleum source-rock in this basin (Wallace-Dudley and Leckie 1995). Inoceramid bivalves are present in the mudstone and indicate some degree of oxygenation of the bottom water (Van Helmond et al. 2016), and ammonites and abundant fish debris indicate well-oxygenated and productive surface waters. Traced along-strike to warmer waters in the south, this organic-rich facies grades into limestone-marl rhythms, typical of units such as the Bridge Creek Limestone in Colorado (Sageman et al. 1997).

Thin section study of mudstone microfacies in this distal facies (i.e. east end of Figure 9.34; Plint et al. 2012b) shows that much of the clay mineral matter is packaged in both organo-minerallic aggregates (Figure 9.35d), as well as fine sand-sized intraclastic aggregates that are interstratified

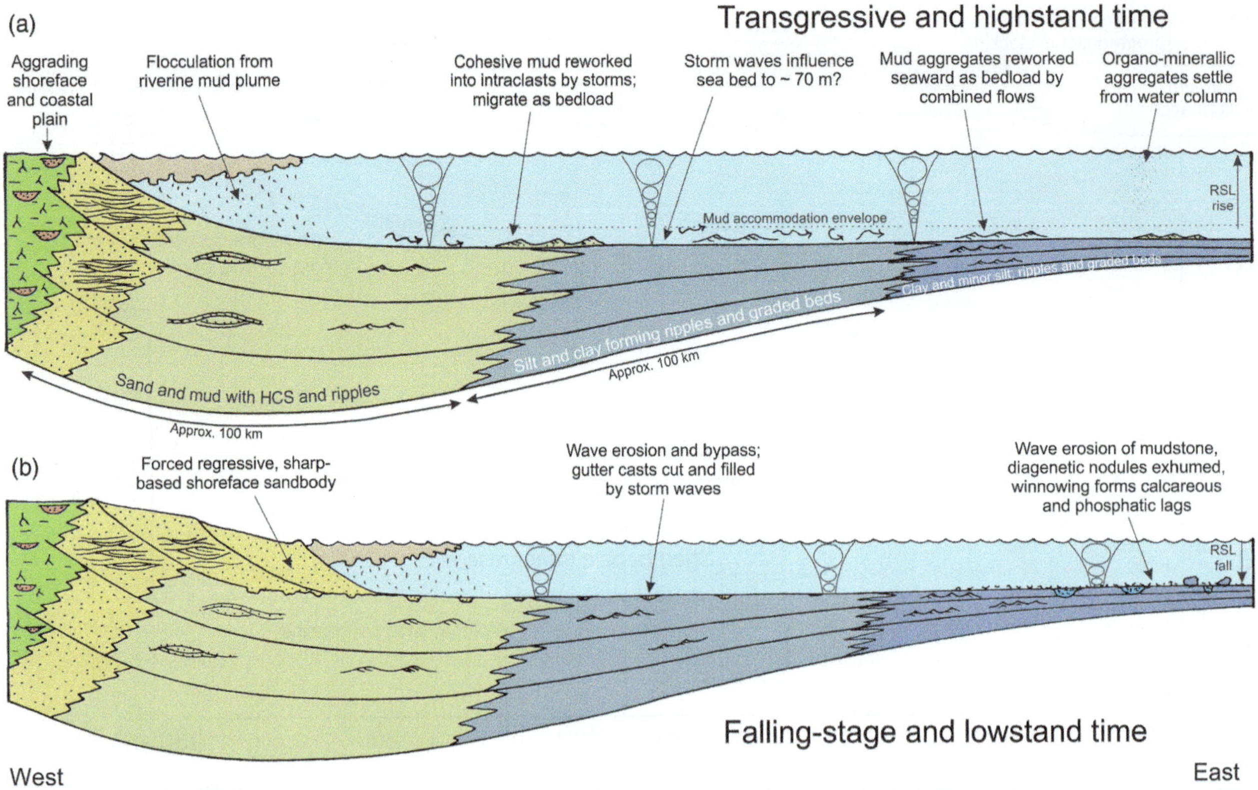

Figure 9.34 Schematic summary of the interpreted physiography, lithofacies and sequence-stratigraphic relationships across the Kaskapau ramp in relation to sea level. (a) During transgression and highstand, the depositional systems were largely aggradational; riverine clays flocculated and settled within a few tens of kilometres of shore. Semi-consolidated clays were repeatedly reworked by storm waves and advected along and across shelf by combined flows. (b) During falling-stage and lowstand time, relative sea-level fall caused sharp-based, forced regressive shoreface sandstones to prograde a short distance seaward; farther offshore, wave reworking formed gutter casts at the top of upward-shoaling successions. In the forebulge region, shallowing lead to more frequent wave scour of the clay-rich mudstone on the sea floor; wave-cut gutter casts were filled with winnowed bioclastic material and early diagenetic nodules were exhumed on erosion surfaces that constitute distal sequence boundaries. (Plint et al. 2012b / Society for Sedimentary Geology (SEPM).)

with silt-size quartz- and chert laminae that form both planar laminated and very low-angle, compacted ripple cross-laminated fabrics (Figure 9.35a,b). The intraclastic aggregates were inferred to have been reworked from shallowly buried but cohesive mud and then transported laterally as bedload during storms (Figure 9.35c). Although the sediment is dominated by mudstone, large HCS and gutter casts are present locally (Figure 9.36a). The sand-size grains that form these structures are entirely biogenic (comminuted bivalve shell debris, fish scales, teeth and bones; Figure 9.36b). These sedimentary structures show that storm waves were able to scour the sea floor, probably during sea-level lowstand, and form HCS at a distance of over 200 km from shore. The scarcity of siliciclastic sand in the region is therefore attributable to trapping in the rapidly subsiding foredeep, rather than lack of transport energy. Although benthic bivalves indicate an oxygenated seafloor, it is postulated that organic-rich mudstone was able to accumulate because, following storms, suspended material settled rapidly. The high organic content of the freshly deposited mud led to the development of sulphidic pore waters only a few millimetres below the sea floor, thus retarding bacterial degradation and promoting preservation of organic matter. Erosive-based, millimetre-scale storm beds seen in thin section show that the succession is suffused with micro-diastems that indicate that the sea floor was repeatedly reworked by storms in water only a few tens of metres deep (Figure 9.35a, b; 9.36c). There is no evidence for CWESGF in this environment, perhaps because the sea floor gradient was too low and/or because of a lack of dispersed clay particles to form a dense, bottom-hugging fluid mud.

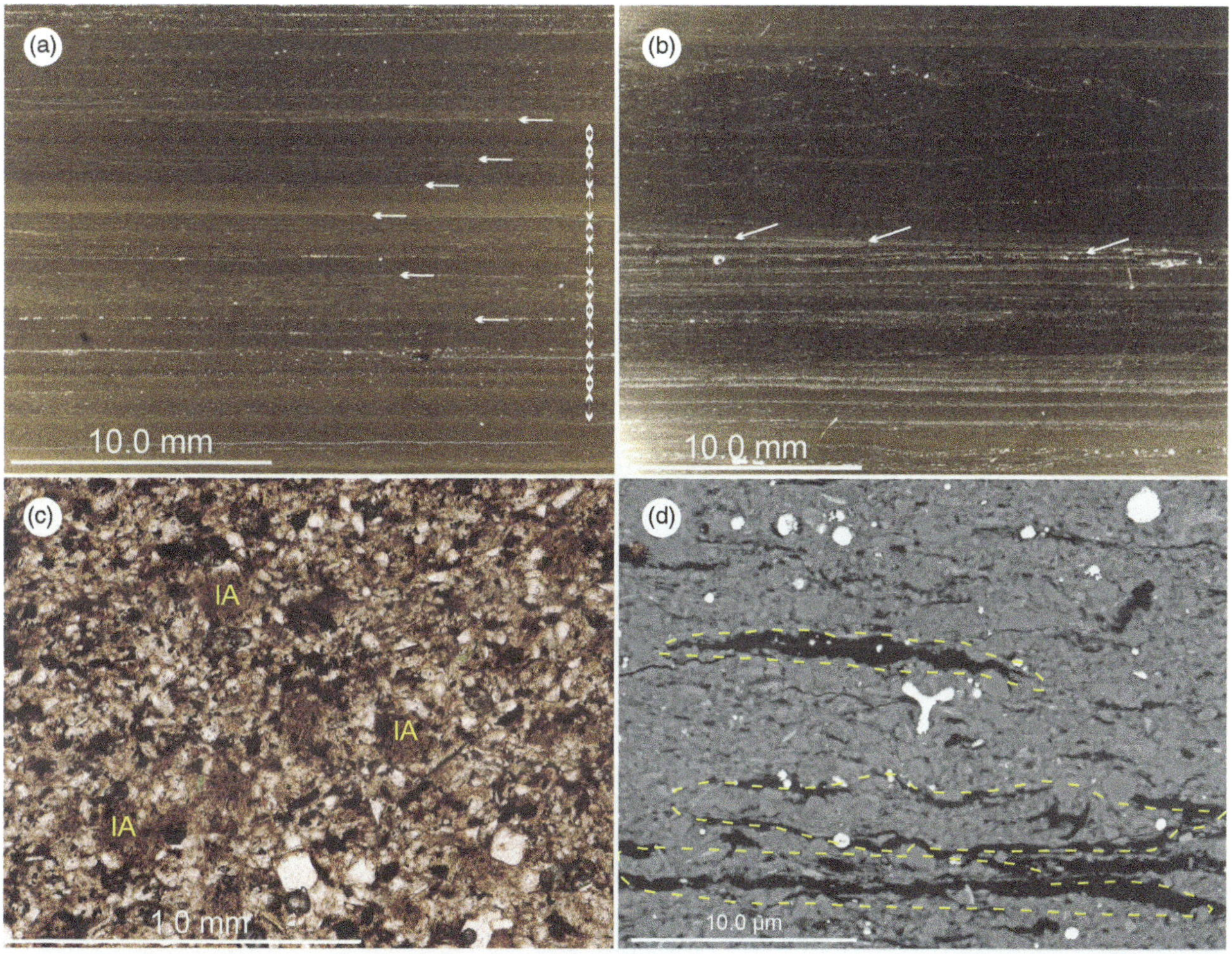

Figure 9.35 Microfacies details of Kaskapau mudstones from the forebulge region. (a) Thin section scan showing a succession of thin, normally graded beds (indicated by double-headed arrows), composed of fine mudstone beds. Individual beds are typically sharp-based and defined by discontinuous silt lags, indicated by arrows. The lamina-sets towards the base of individual beds contain abundant organo-mineralic aggregates, whereas the tops of the beds are typically homogenised. (b) Thin section scan showing a succession of thin-bedded, planar parallel-laminated and ripple cross-laminated fine to medium mudstone beds. Arrows indicate the termination of ripple cross-laminae. (c) Micrograph showing very fine sand and coarse silt-size framework grains that consist of intraclastic aggregates (IA) in addition to quartz, chert and woody fragments (opaque grains), in a matrix dominated by clay minerals. (d) Backscattered electron image, illustrating organo-mineralic aggregates, outlined by broken lines; organic matter appears black. (Plint et al. 2012b / Society for Sedimentary Geology (SEPM).)

9.6.8 A Moderate Accommodation, Coniacian Ramp, Alberta and Montana

In the very earliest Coniacian, an eustatic transgression flooded lowstand strandplain systems across the Western Canada Foreland basin. This transgressive event can be correlated, on the basis of multiple proxies, to Europe (Plint et al. 2022). A new pulse of flexural subsidence also commenced in the early Coniacian, although this resulted in no more than about 150 m of subsidence in the proximal foredeep. The lower part of the Lower Coniacian is assigned to the Cardium alloformation, up to the major sequence boundary 'E7' (Plint et al. 2022). The Coniacian strata above the E7 surface, up to the base of the Santonian are assigned to the lithostratigraphic Muskiki and Marshybank formations.

In NE British Columbia and NW Alberta, the Lower and Middle Coniacian are mudstone dominated with an essentially sheet-like geometry (Plint 1990; Plint and Norris 1991). Upper Coniacian strata include alluvial and shoreface deposits that to the south of about 54°N pass laterally into a more offshore, mudstone-dominated succession. This southern area was the subject of an integrated allostratigraphic, biostratigraphic and chemo-stratigraphic study that extended 750 km along the foredeep in western Alberta and NW Montana (Plint et al. 2017; Walaszczyk et al. 2017; Landman et al. 2017). The Coniacian strata are unusual in the Cretaceous of western Canada in being richly fossiliferous and therefore provide an excellent opportunity to compare biostratigraphic and allostratigraphic correlation methods.

Figure 9.36 Distal Kaskapau alloformation mudstone at Erin Lodge, Alberta. (a) Field photo showing gutter cast filled with phosphatic calcareous sandstone overlain by phosphatic sandstone moulded into a hummock (outlined). This bioclastic sandstone bed is laterally discontinuous on a scale of a few metres and is enclosed in fine mudstone. Metal boxes used to take samples of unconsolidated mudstone are visible. Scale bar = 20 cm. Photo courtesy of Piotr Angiel. (b) Micrograph (plane polarised light) of phosphatic sand shown in (a); detrital particles are phosphatic or calcareous; no sand-size quartz or chert is present. Bioclasts are rimmed by displacive gypsum which is the product of the surficial weathering of pyrite. (c) Micrograph showing succession of thin, normally graded beds composed of wavy laminated, silt-rich laminae interstratified with fine mudstone. Some silt laminae have scoured bases. Silt-size framework grains are composed of quartz, calcareous detritus and dark brown, intraclastic aggregates (IA). Plint et al. (2012b) / Society for Sedimentary Geology (SEPM).

Plint et al. (2017) recognised 24 allomembers in the Coniacian succession in the proximal foredeep of central and southern Alberta. Each allomember comprises one or more upward-shoaling parasequences (Figure 9.37a). Allomember boundaries were picked in wireline logs at surfaces that were consistently traceable over hundreds of kilometres (Figure 9.38a). However, the ease with which flooding could be recognised commonly changed along the strike of the foredeep in response to spatial changes in the lithology below and above the surface.

Observed at outcrop, four types of flooding surface were recognised:

1) Surfaces with no coarse-grained lag are sharp and planar and separate coarser-grained sediments below from finer-grained above. They are interpreted as simple

Figure 9.37 (a) A portion of the Lower Coniacian part of the Muskiki Member illustrating typical sandier-upward parasequences (arrows) bounded by flooding surfaces. The regionally mappable flooding surface CS3 is indicated. (b) Overview of surface CS1 showing, at right, a coarse-grained wave ripple composed of granules and small pebbles of chert and quartz that rests sharply on regional erosion surface CS1. Traced laterally over only 3 m, the coarse lag pinches out and surface CS1 is manifest simply as a sharp mud-on-mud contact; scale bar = 20 cm. (c) Detail of the coarse-grained wave ripple. Plint et al. (2017) / American Museum of Natural History.

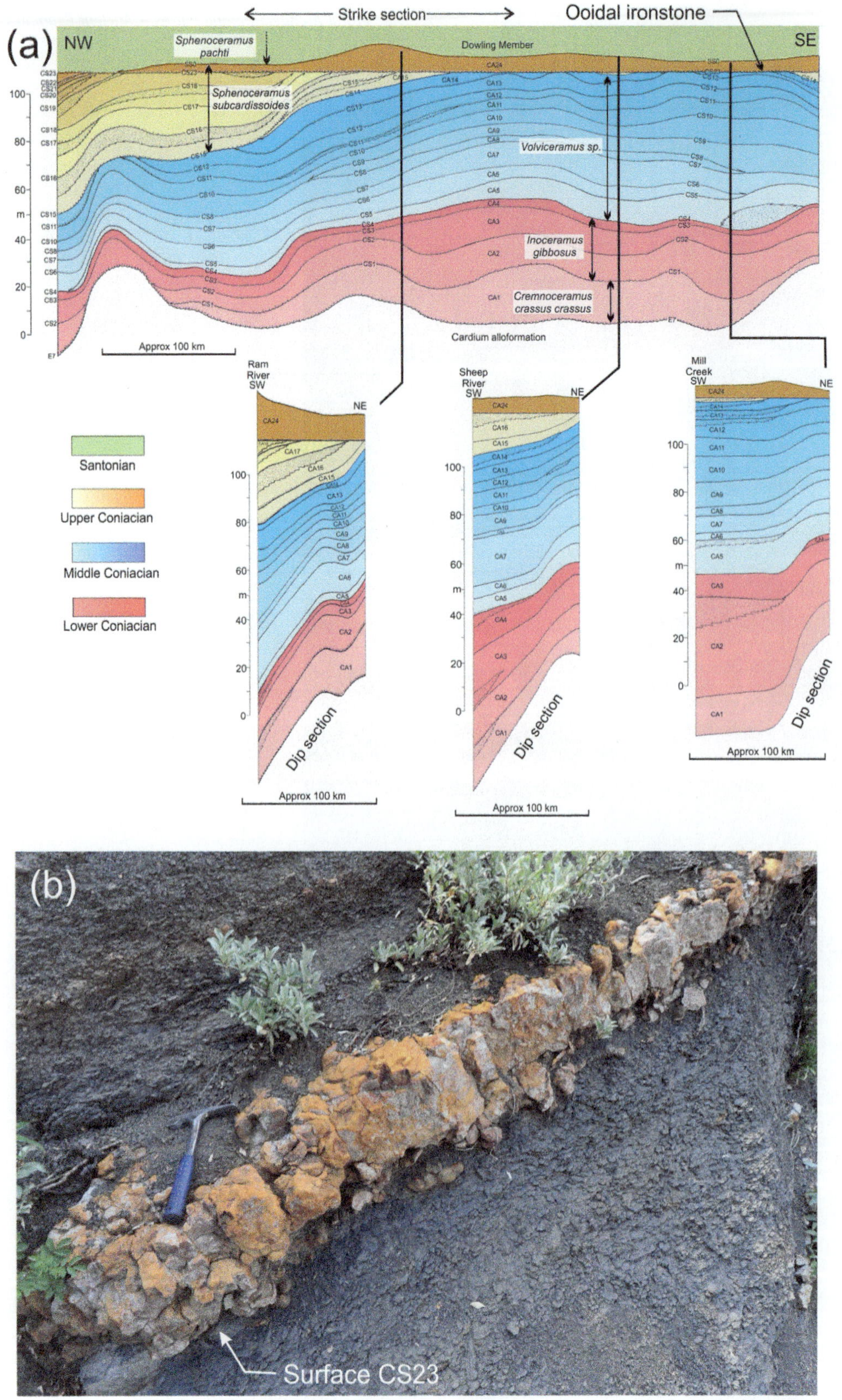

Figure 9.38 (a) Summary representation, to scale, of stratal geometry of Coniacian strata viewed in a ~ 750 km long strike section along the proximal foredeep in SW Alberta and NW Montana. Three shorter, dip-oriented sections show strata thickening towards the orogen in the SW. Lower Coniacian strata fill a saucer-shaped depocenter that thins to both NW and SE, whereas Middle Coniacian strata are thickest in the south and show minor erosional truncation towards the NW. Upper Coniacian strata fill a depocenter in the NW and are progressively truncated on surface CS23, such that they are largely absent over most of southern Alberta and northern Montana. The boundaries of molluscan biozones consistently follow specific flooding surfaces throughout the study area. (b) Siderite-cemented ooidal ironstone mantling erosion surface CS23 at Mill Creek in southern Alberta. The basal surface of the ironstone represents almost all of the Upper Coniacian – about two million years (Plint et al. 2017 / American Museum of Natural History).

flooding surfaces that record a relative sea-level rise and shoreline transgression, accompanied by a decrease in the volume and calibre of sediment supplied to the shelf. There is no evidence for a preceding relative sea-level fall.

2) Surfaces with lags of intra-basinal clasts have rounded pebbles of siderite or phosphatised materials, which were early diagenetic nodules, reworked from the substrate. The clasts indicate erosion and winnowing, to a depth of at least several decimetres, probably in response to relative sea-level fall
3) Surfaces with mixed intra- and extra-basinal clasts are mantled with locally derived siderite and phosphatic intraclasts, mixed with siliceous extra-basinal pebbles. Pebble lags are commonly laterally discontinuous on a scale of metres (Figure 9.37b,c). The extra-basinal clasts are absent from the underlying sediments and so could not have been concentrated by winnowing. Because of the asymmetry of wave-generated currents in shallow water, pebbles in the shoreface are not transported seaward across the inner shelf for more than a few hundred metres to a few kilometres (Clifton 2006). It must therefore be concluded that the presence of extra-basinal pebbles indicates that the surface was formerly subaerially exposed, at which time, rivers carried gravel to the lowstand shoreline (cf. Plint 1988).
4) Surface with ooidal ironstone and intra- and extra-basinal clasts is exemplified by surface CS23 (Figures 9.38a,b) that marks the regressive maximum of the Coniacian succession, just below the Coniacian-Santonian boundary. This surface has a *Glossifungites* ichnofacies (Chapter 3.4) and is overlain by 0.2–1.2 m of intensely bioturbated, structureless ooidal ironstone that also contains rounded extra-basinal pebbles. Regional correlation shows that, in the south, this ooidal ironstone overlies a major bevelling unconformity (Figures 9.38a,b) at which almost all Upper Coniacian strata are absent. The presence of extra-basinal chert pebbles dispersed through the ooidal ironstone indicates subaerial emergence at sea-level lowstand, whereas the ooidal ironstone developed under shallow-water conditions during early transgression when the supply of clastic sediment was at its lowest (cf. Donaldson et al. 1999; Plint et al. 2017).

In the study area of Plint et al. (2017), the facies of the Coniacian rocks broadly resemble the Kaskapau rocks described above. The sediments are strongly heterolithic, with decimetre-bedded sandstones with HCS and wave ripples in the nearshore region, becoming more thinly bedded offshore, eventually grading into millimetre-stratified mudstones and siltstone far offshore. Bioturbation is highly variable, from BI 0–6, but overall, becomes more intense upwards, perhaps related to overall shallowing and/or improved marine circulation. Sediment transport was driven primarily by storm-generated geostrophic and combined flows operating over a low-gradient ramp. Subsurface correlation of the Coniacian mudstones showed that subsidence varied markedly along the length of the foredeep, with flexural 'moats', typically 300–400 km wide, that were inferred to have developed adjacent to actively thickening thrust wedges in the orogen. Areas outside the 'moats' underwent little subsidence, or experienced subtle uplift, condensation or erosion (Figure 9.38a). Santonian mudstone-dominated ramp deposits, c. 200–300 m thick, that immediately overlie the Coniacian rocks of Figure 9.38a, show equally dramatic strike variation in the locus of flexural subsidence (Hu and Plint 2009; Plint et al. 2012a).

9.6.9 Forced Regression in Muddy Ramp Settings

In the more shore-proximal, northwestern part of the Coniacian study area documented by Plint (1990, 1991), flooding surfaces bounding mudstone-dominated offshore successions could be correlated landward, where they proved to be correlative with *two* erosion surfaces that lay below and above thin (<10 m) shoreface sandstone bodies. The erosion surface beneath the shoreface sandbody was marked by gutter casts but *lacked* an extra-basinal lag and was termed a 'regressive surface of marine erosion'. The upper, transgressive erosion surface capped the shoreface, and was mantled by an extra-basinal lag. To seaward, these two surfaces merge into a single, composite surface, locally incised by isolated gutter casts (Plint 1991). Thus, the apparently simple 'flooding surfaces' seen in the offshore area cryptically embody an initial phase of erosion or bypass due to relative sea-level fall, which was accompanied on the inner ramp by progradation of a sharp-based shoreface. Subsequent sea-level rise resulted in transgressive erosion and winnowing of a pebble lag from the shoreface sandbody, whereas farther offshore, mud passively mantled the flooding surface.

The recognition of sharp-based 'forced regressive' shoreface sandstone bodies in the Upper Coniacian rocks of western Canada (Plint 1991), as well as other examples from Turonian (Plint 1988) and Cenomanian rocks (Plint 1996; Plint and Kreitner 2019), showed that relative sea-level fall (probably due to eustasy) had played an important role in driving coastal progradation in these shallow ramp settings: These ideas were embodied in the concept of the falling stage systems tract (Plint and Nummedal 2000). Examples of falling stage progradation are well documented in seismic data from the Pleistocene and Holocene of the Sunda shelf (Miall 2002; Hanebuth et al. 2009; Alqahtani et al. 2015).

9.6.10 Chronostratigraphic Significance of Bounding Surfaces

A key conclusion to emerge from the integration of allostratigraphy and biostratigraphy in Coniacian strata was that the molluscan faunal changes that mark the boundaries of biozones, consistently occur at the *same* allostratigraphic surfaces, whether correlated within Western Canada (Plint et al. 2017; Figure 9.37a), or between Canada and Bohemia (Plint et al. 2022). This shows that, for practical purposes, marine flooding surfaces can have the same chronostratigraphic significance as palaeontological First Occurrence datums.

9.6.11 Muddy Chenier Coast, Alderson Member, (Upper Cretaceous) SW Saskatchewan

The Late Santonian to Early Campanian Lea Park Formation represents a major deltaic complex that prograded hundreds of kilometres from SW to NE across the western portion of the Cretaceous Interior Seaway in southern Alberta and adjacent Saskatchewan. Towards the NE, sandy deltaic strata grade seawards into heterolithic and mudstone-dominated facies assigned to the c. 180 m thick Alderson Member, of latest Santonian to Early Campanian age (Hovikoski et al. 2007, 2008). Although previously interpreted, because of its predominantly mudstone lithology, as an offshore deposit, the Alderson Member has been re-interpreted as a nearshore to coastal plain deposit (Hovikoski et al. 2008; Lemiski et al. 2011). Along-shore prograding subaqueous deltas are recognised on the basis of a subtle sandier-upward succession that progresses from highly bioturbated mudstone with a low-diversity, diminutive trace-fossil assemblage, into heterolithic very fine-grained, wave-rippled and laminated sandstone with load- and gutter-casts, interstratified on a millimetre to centimetre scale, with variably bioturbated mudstone and a somewhat more diverse trace-fossil assemblage (Figure 9.39). Beds of massive mudstone are common and suggest episodic emplacement of fluid mud, mainly as a result of re-suspension by storm waves. However, hyperpycnite beds attributable to river floods were not recognised.

Along-strike from the subaqueous delta facies lies a muddy shoreface succession in which bioturbated mudstone, rich in terrestrial plant debris, grades up through heterolithic facies interpreted to represent a muddy shoreface or tidal flat and is capped by homogeneous rooted mudstone indicative of an emergent coastal plain. This succession shows an upward *decrease* in wave action, reflecting the dissipation of wave energy in fluid mud that was concentrated in the nearshore zone (Figure 9.39). Neither sand-rich shoreface nor chenier facies were recognised by Hovikoski et al. (2008), who compared this ancient mud-dominated coast to the modern Louisiana, East China Sea and Brazil–Guyana coasts (Section 9.2).

9.6.12 Micro-facies Study, Turonian Muddy Ramp, Utah

An outcrop-based, microfacies study of the Turonian Tununk Shale Member of the Mancos Formation was undertaken by Li and Schieber (2018), who collected over 500 samples from three along-strike sections through a c. 220 m thick succession of Turonian mudstones exposed in the Henry Mountains in south-central Utah (Figure 9.40a). Samples were impregnated with epoxy resin, from which polished slabs and thin sections were prepared for microfacies analysis. In the field, lithology, grain-size, sedimentary structures, bioclastic content and bioturbation were noted. Critically, the dip direction of ripple foresets and the strike of wave ripple crests were also measured wherever they had not been disrupted by bioturbation. Previous regional mapping (Figure 9.40b,c) shows that in the Early Turonian, the contemporaneous shoreline was c. 110 km from the study area whereas by the Middle Turonian, the shore had prograded to within c. 70 km, decreasing to 10–20 km by the Late Turonian.

Four main facies were recognised:

1) Carbonate-bearing, silty and sandy mudstone (CSSM) facies has millimetre-scale interstratified mudstone and sharp-based beds of silt and sand, some showing starved ripple lamination (Figure 9.41a). Carbonate particles include inoceramid debris, foraminifera and coccolith-filled faecal pellets. Although BI is high (3–5), trace-fossil diversity is low, dominated by mud-swimming navichnia and diminutive *Chondrites*, *Phycosiphon* and *Planolites*. These characteristics suggest a relatively deep offshore setting, subject to occasional storm wave influence but with a dominantly soupy substrate and low sedimentation rate.
2) The silt-bearing, calcareous mudstone (SCM) facies overlies CSSM and has more common and laterally continuous calcareous, silt, or sand laminae from < 1 to a few mm thick (Figure 9.41b). Coarser grains are dominantly calcareous and include foraminifera and faecal pellets full of coccoliths. Sedimentary structures include wave and combined flow ripples, parallel-lamination and rare, normally graded layers. Bioturbation ranges from BI 1-2 to 3-4, with navichnia and *Phycosiphon* dominant in the lower part, whereas *Chondrites* and *Planolites* dominate above. Palaeocurrent indicators indicate along-shelf, storm-driven flows to the SSW

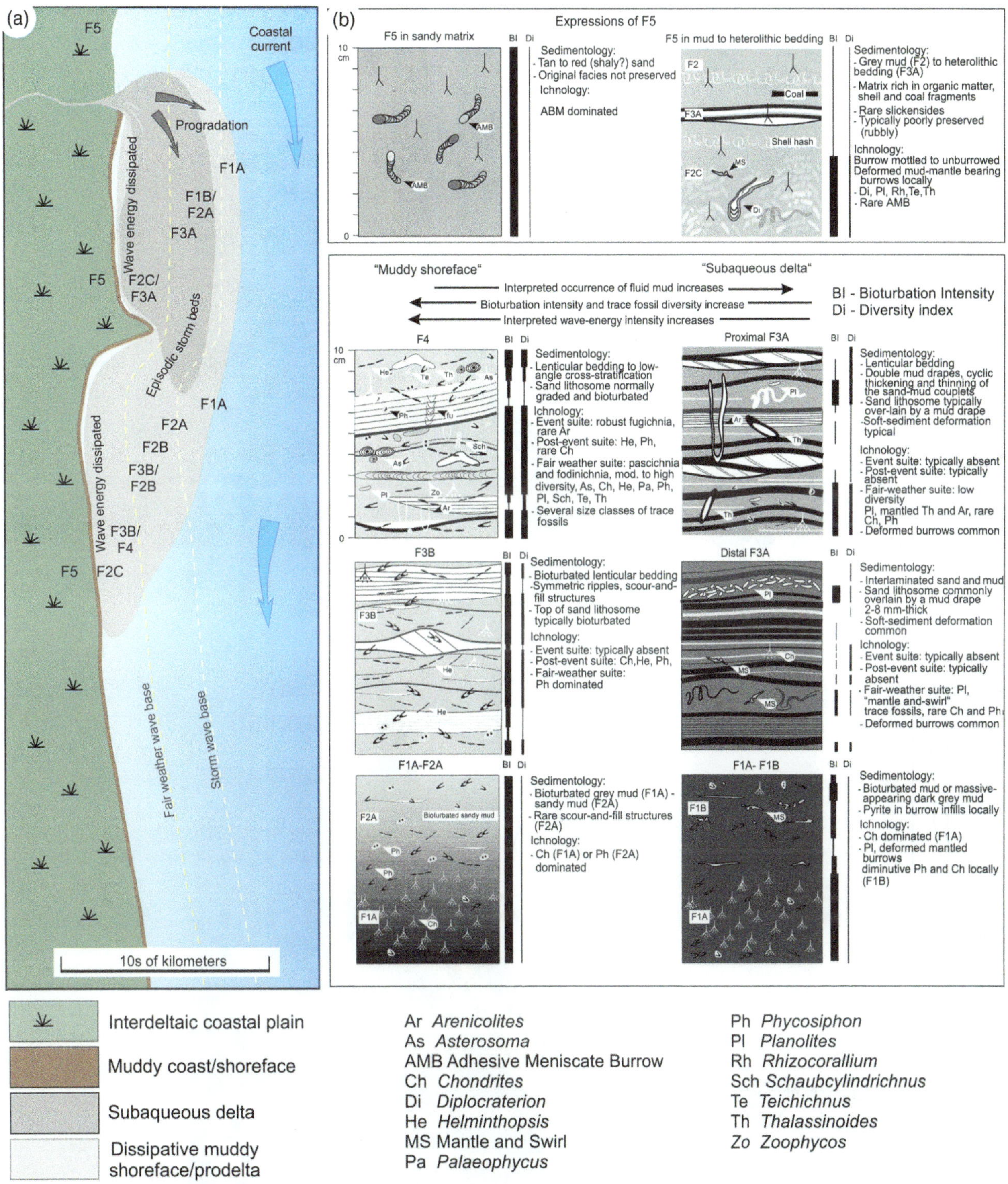

Ar *Arenicolites*
As *Asterosoma*
AMB Adhesive Meniscate Burrow
Ch *Chondrites*
Di *Diplocraterion*
He *Helminthopsis*
MS Mantle and Swirl
Pa *Palaeophycus*
Ph *Phycosiphon*
Pl *Planolites*
Rh *Rhizocorallium*
Sch *Schaubcylindrichnus*
Te *Teichichnus*
Th *Thalassinoides*
Zo *Zoophycos*

Figure 9.39 (a) Hypothetical palaeogeographic model for the Alderson Member showing a subaqueous delta developed down-drift from the mouth of a mud-dominated river. Waves and tides re-suspended and transported mud across the delta, constructing a mud-dominated shoreface and chenier plain on the down-drift side. Subaqueous fluid mud grades directly up into mudflat and terrestrial deposits. Adapted from Hovikoski et al. (2008) and Lemiski et al. (2011). (b) Summary of facies and ichnological suites in the various facies that span the offshore to nearshore environments represented in (a) (Hovikoski et al. 2008 / Society for Sedimentary Geology (SEPM)).

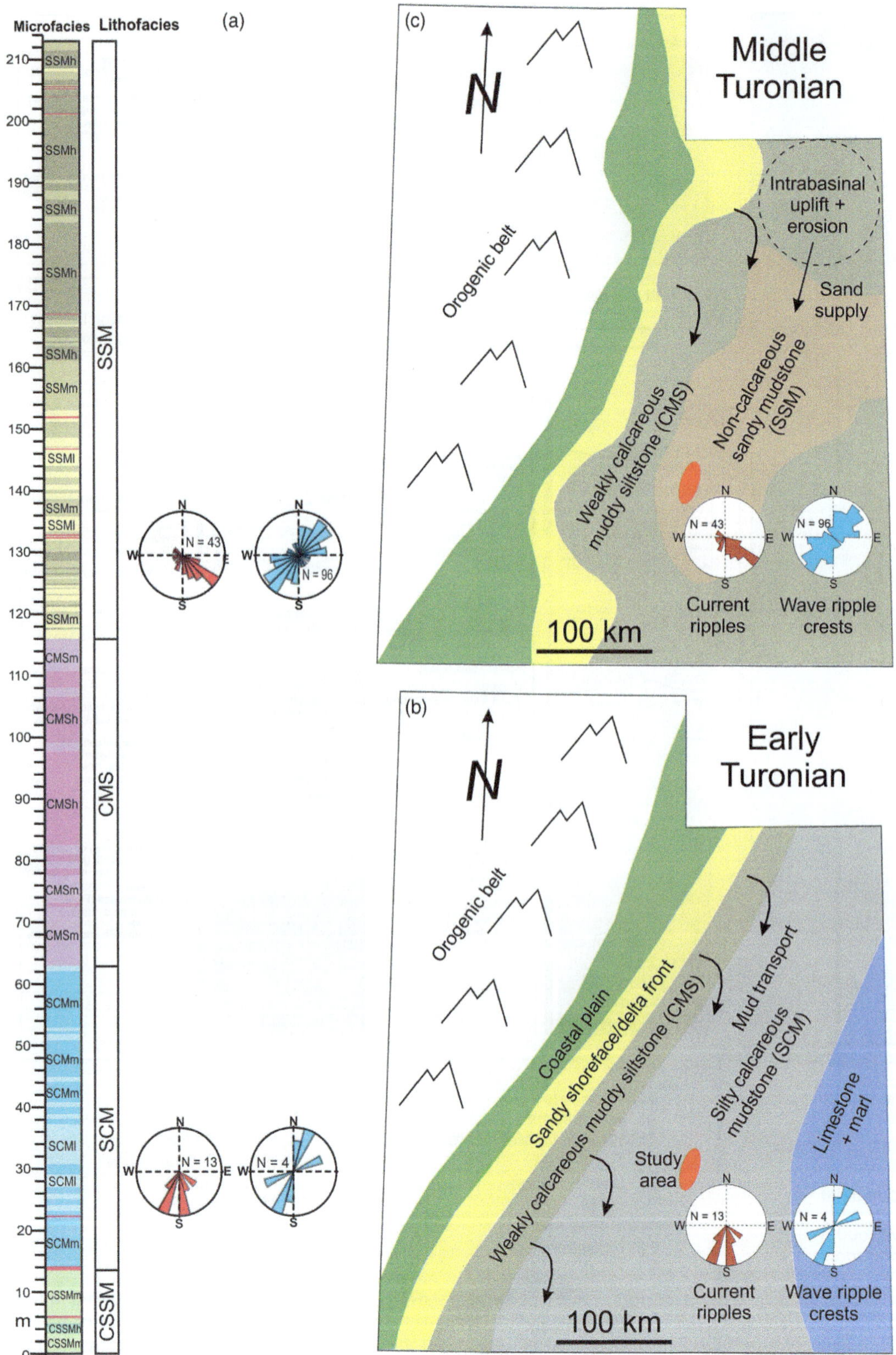

Figure 9.40 Stratigraphic and palaeogeographic setting of Turonian shelf mudstones of the Tununk Member of the Mancos Shale in Utah, USA. (a) Simplified stratigraphic section showing the vertical distribution of the four lithofacies illustrated in detail in Figure 9.41, plus summary palaeocurrent data for the SCM and SSM facies. (b) Outline map of Utah showing distribution of mudstone facies belts in relation to the palaeo-shoreline in the Early Turonian. Palaeocurrent data indicate waves approaching from the ESE and storm-driven, geostrophic currents flowing to the SSW, along the shelf. (c) Distribution of facies in the Middle Turonian when the shore was c. 70 km distant. Intrabasinal uplift in the NE led to erosion and shedding of sand to the SW, forming the SSM facies. Storm waves approaching from the SE drove downwelling geostrophic flows directed across the shelf to the south and southeast. Figures adapted from Li and Schieber (2018).

(Figure 9.40b). The preponderance of carbonate grains suggests deposition during peak transgression and minimal clastic flux. Laminae composed of foraminifera and/or quartz silt indicate bedload transport as a result of storm wave- and current winnowing and reworking in water estimated to have been about 70 m deep.

3) The carbonate-bearing, silty mudstone to muddy siltstone facies (CMS) is characterised by common silt laminae and lamina sets from a few mm to 20 mm thick (Figure 9.41c). Carbonate particles are subordinate and disappear up-section. Where preserved, wave ripple lamination is common, but structure is mostly lost to pervasive bioturbation. Bioturbation is higher relative to the underlying SCM facies, ranging from BI 3 to 6. Traces are dominantly feeding structures and include *Chondrites, Phycosiphon, Planolites, Schaubcylindrichnus, Thalassinoides* and *Zoophycos*. The increased proportion of detrital silt and sand relative to facies SCM indicates shallowing and increasing proximity of the shore, with sediment transport primarily by storm-driven geostrophic flows. The abundance of bioturbation indicates an abundant food supply and adequate oxygenation.

4) The non-calcareous, silty and sandy mudstone (SSM) facies comprises sharp-based silt and sand beds from a few millimetres to 14 cm thick. Structures include wave- and combined-flow ripples, starved ripples, HCS, normal and rare inversely graded beds and soft-sediment deformation; carbonate particles are rare (Figure 9.41d). Bioturbation ranges from BI 0-2 to 5-6 and includes the

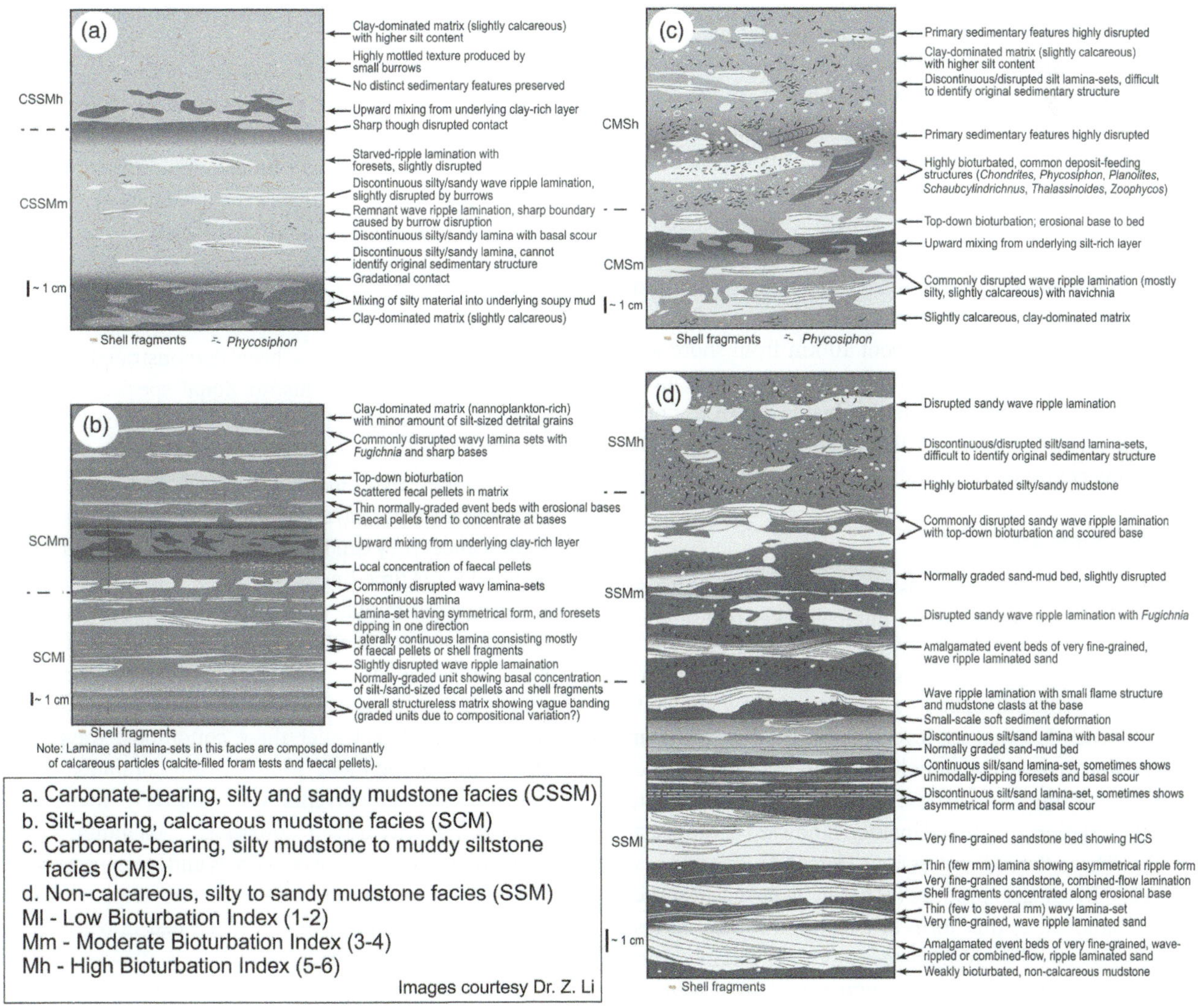

Figure 9.41 Graphical summaries of the four microfacies recognised in the Tununk Shale Member in the Henry Mountains, Utah. Adapted from Li and Schieber (2018). See Figure 9.40b,c, for palaeogeographic settings, and text for details.

feeding structures *Chondrites, Phycosiphon, Planolites, Schaubcylindrichnus, Teichichnus* and *Thalassinoides*, with dwelling structures *Ophiomorpha* and *Skolithos* being much less common. Ripples indicate dominant transport to the SE with a minor component to the SSW (Figure. 9.40c). The SSM facies is interpreted to represent relatively shallow inner shelf, grading to lower shoreface environment strongly affected by storms.

Li and Schieber (2018) emphasise the importance of along-shelf transport of mud, silt and sand under the dominant influence of storm-generated combined- and geostrophic flows. They propose deposition of the Tununk Shale in an elongate, shore-parallel 'Adriatic-style' (Section 9.2.2) prism of mud that downlapped and accreted to seaward (SE). However, no stratigraphic evidence was presented for such a stratal architecture, and it is possible that the Tununk Shale formed a more sheet-like muddy ramp, lacking clinothems, comparable to that mapped in the coeval Kaskapau Formation in the British Columbia foredeep (Varban and Plint 2005, 2008b).

Li and Schieber (2018) point out that the Tununk shales lack evidence for downslope-directed density-driven flows. In contrast, the stratigraphically overlying sediments of the river-dominated Ferron delta system contain common Bouma sequences and inverse-to-normally graded beds interpreted to have been deposited from surge-type turbidity currents and river flood-generated hyperpycnal flows; these flows died out within about 10 km from shore and did not affect the open shelf. These hyperpycnites are interstratified with graded, parallel-laminated to HCS storm deposits formed under oscillatory-dominant and combined-flow conditions (Li et al. 2015).

9.6.13 Summary

The study of Cretaceous mudstone-dominated successions in Western Canada has led to a number of key realisations:

1) Water depth in mud-rich, ramp settings in Western Canada was sufficiently shallow, and depositional gradients sufficiently low that sea-level fall of as little as 10 to 20 m was able to drive the shoreline tens to hundreds of kilometres offshore and expose shallow-marine deposits to soil formation and fluvial incision landward of the coastline. These 'forced' regressions (forced by relative sea-level fall) can also produce (a) basinally detached shoreface sandbodies that form at maximum lowstand, rest erosively on offshore mudstones and are variably truncated by the subsequent transgression. (b) Extensive wave-cut erosion surfaces cut at lowstand in offshore marine mudstone and (c) planar wave ravinement surfaces cut by landward translation of the shoreface during marine transgression; these may be accompanied by tidal ravinement surfaces associated with tidal inlets, or estuarine channels. The oscillation of the shoreline by hundreds of kilometres seaward and landward may be marked by no more than a subtle erosion surface, sometimes with a scattering of extra-basinal pebbles, that separates marine mudstone below from marine mudstone above.
2) Both modern and ancient studies suggest that wave-driven re-suspension in shallow water is a key factor in determining the geometry of mud-rich units. In modern seas, initial mud transport from rivers is dominantly along-shelf. Detached, offshore-prograding mud-rich foresets are constructed by re-suspension events that produce dense fluid mud that flows downslope. Ancient mudstones show pervasive evidence of erosion and re-deposition in the form of both erosional physical sedimentary structures and aggregate grains and clasts. The importance of multiple re-suspension events resulting from combined geostrophic and wave-induced flows helps explain why ancient mudstone units are so laterally extensive and also reveals the efficiency of sediment redistribution that ensured that all available accommodation was quickly filled: deep water never had a chance to develop except under conditions of exceptionally high subsidence rate.
3) In some instances, it has been demonstrated that the first occurrence of molluscan Zonal species coincides precisely with marine flooding surfaces that can be independently correlated through thousands of well logs. This implies that flooding surfaces and Zonal First Occurrence data have equivalent chronostratigraphic significance: they can be treated, for practical purposes, as proxies for time lines.
4) Relatively tall muddy clinothems, inferred to represent deposition in water c 100 to 150 m deep, appear to have been constructed in settings where subsidence rate exceeded sedimentation rate, allowing relatively deep water to develop. Under these conditions, wave- and storm-driven sediment transport by combined flows and geostrophic currents was relatively ineffective. Instead, gravity-driven processes, including river flood-generated hyperpycnal flows and storm-generated CWESGF transported sediment into deeper water, depositing beds that thinned and fined downslope. If the sediment pile accumulated to shallower depths, storm-driven transport became more efficient, gradually flattening and then eliminating the clinothems.

Further Reading

Aplin, A.C. and Macquaker, J.H.S. (2011). Mudstone diversity: origin and implications for source, seal, and reservoir properties in petroleum systems. *American Association of Petroleum Geologists, Bulletin.* 95: 2031–2059.

Baas, J.H., Best, J.L., and Peakall, J. (2015). Predicting bedforms and primary current stratification in cohesive mixtures of mud and sand. *Journal of the Geological Society* 173 (1): 12–45.

Cattaneo, A., Trincardi, F., Asioli, A., and Correggiari, A. (2007). The Western Adriatic shelf clinoform: energy-limited bottomset. *Continental Shelf Research* 27 (3-4): 506–525.

Grabowski, R.C., Droppo, I.G., and Wharton, G. (2011). Erodibility of cohesive sediment: the importance of sediment properties. *Earth-Science Reviews* 105 (3-4): 101–120.

Hanebuth, T.J., Lantzsch, H., and Nizou, J. (2015). Mud depocenters on continental shelves—appearance, initiation times, and growth dynamics. *Geo-Marine Letters* 35 (6): 487–503.

Hill, P.S., Fox, J.M., Crockett, J.S. et al. (2007). Sediment delivery to the seabed on continental margins. In: *Continental Margin Sedimentation: From Sediment Transport to Sequence Stratigraphy. International Association of Sedimentologists*, Special Publication *37*, (ed. C.A. Nittrouer et al.), 49–99. Blackwell Science, Oxford.

Lazar, O.R., Bohacs, K.M., Macquaker, J. H. et al. (2015). Capturing key attributes of fine-grained sedimentary rocks in outcrops, cores, and thin sections: nomenclature and description guidelines. *Journal of Sedimentary Research* 85 (3): 230–246.

Li, Z. and Schieber, J. (2018). Detailed facies analysis of the Upper Cretaceous Tununk Shale Member, Henry Mountains region, Utah: implications for mudstone depositional models in epicontinental seas. *Sedimentary Geology* 364: 141–159.

Mackay, D.A. and Dalrymple, R.W. (2011). Dynamic mud deposition in a tidal environment: the record of fluid-mud deposition in the Cretaceous Bluesky Formation, Alberta, Canada. *Journal of Sedimentary Research* 81: 901–920.

Mulder, T., Syvitski, J.P.M., Migeon, S. et al. (2003). Marine hyperpycnal flows: initiation, behaviour and related deposits. *A review. Marine and Petroleum Geology* 20: 861–882.

Neill, C.F. and Allison, M.A. (2005). Subaqueous deltaic formation on the Atchafalaya shelf, Louisiana. *Marine Geology* 214: 411–430.

Plint, A.G. (2014). Mud dispersal across a Cretaceous prodelta: storm-generated, wave-enhanced sediment gravity flows inferred from mudstone microtexture and microfacies. *Sedimentology* 61 (3): 609–647.

Plint, A.G., Macquaker, J.H., and Varban, B.L. (2012). Bedload transport of mud across a wide, storm-influenced ramp: Cenomanian-Turonian Kaskapau Formation, Western Canada Foreland Basin. *Journal of Sedimentary Research* 82 (11): 801–822.

Plint, A.G., Hooper, E.A., Grifi, M.D. et al. (2017). Integrated, high-resolution allostratigraphic, biostratigraphic and carbon-isotope correlation of Coniacian strata (Upper Cretaceous), Western Alberta and northern Montana. *Bulletin of the American Museum of Natural History* 414: 9–52.

Pufahl, P.K. (2010). Bioelemental sediments. In: *Facies Models 4* (ed. N.P. James and R.W. Dalrymple). Geological Association of Canada St. John's, NL, Canada, 586 p. ISBN-13: 978-1-897095-50-8; ISSN: 1208-2260. p 477–504.

Sageman, B.B., Murphy, A.E., Werne, J.P. et al. (2003). A tale of shales: the relative roles of production, decomposition, and dilution in the accumulation of organic-rich strata, Middle–Upper Devonian, Appalachian Basin. *Chemical Geology* 195 (1-4): 229–273.

Schieber, J. (2016). Mud re-distribution in epicontinental basins – exploring likely processes. *Marine and Petroleum Geology* 71: 119–133.

Schieber, J. and Southard, J.B. (2009). Bedload transport of mud by floccule ripples – direct observation of ripple migration processes and their implications. *Geology* 37: 483–486.

Slingerland, R., Driscoll, N.W., Milliman, J.D., Miller, S.R. and Johnstone, E.A., 2008. Anatomy and growth of a Holocene clinothem in the Gulf of Papua. *Journal of Geophysical Research: Earth Surface*, 113(F1).nn

References

Abascal-Zorrilla, N., Vantrepotte, V., Gensac, E. et al. (2018). The advantages of Landsat 8-OLI-derived suspended particulate matter maps for monitoring the subtidal extension of Amazonian coastal mud banks (French Guiana). *Remote Sensing* 10 (11): 1733.

Ahmerkamp, S., Winter, C., Krämer, K. et al. (2017). Regulation of benthic oxygen fluxes in permeable sediments of the coastal ocean. *Limnology and Oceanography* 62 (5): 1935–1954.

Aigner, T. and Reineck, H.-E. (1982). Proximality trends in modern storm sands from the Helgoland Bight (North Sea) and their implications for basin analysis. *Senckenbergiana Maritima* 14: 183–215.

Ainsworth, R.B., Vakarelov, B.K., and Nanson, R.A. (2011). Dynamic spatial and temporal prediction of changes in depositional processes on clastic shorelines: toward improved subsurface uncertainty reduction and management. *American Association of Petroleum Geologists, Bulletin* 95 (2): 267–297.

Allen, J.R.L. (1982). *Sedimentary Structures, Their Character and Physical Basis Volume 1*, 593. New York: Elsevier.

Allen, P.A. (2017). *Sediment Routing Systems: The Fate of Sediment from Source to Sink*, 407. Cambridge University Press.

Aller, R.C. (1994). Bioturbation and remineralization of sedimentary organic matter: effects of redox oscillation. *Chemical Geology* 114 (3-4): 331–345.

Allison, M.A. and Lee, M.T. (2004). Sediment exchange between Amazon mudbanks and shore-fringing mangroves in French Guiana. *Marine Geology* 208 (2-4): 169–190.

Allison, P.A., Hesselbo, S.P., and Brett, C. E. (2008). Methane seeps on an Early Jurassic dysoxic seafloor. *Palaeogeography, Palaeoclimatology, Palaeoecology* 270 (3-4): 230–238.

Alqahtani, F.A., Johnson, H.D., Jackson, C.A.L., and Som, M.R.B. (2015). Nature, origin and evolution of a Late Pleistocene incised valley-fill, Sunda Shelf, Southeast Asia. *Sedimentology* 62 (4): 1198–1232.

Angiel, P.A. (2013). Allostratigraphy, sedimentology and paleogeography of the Cretaceous Upper Fort St John Group (Upper Albian-Lower Cenomanian) in north-eastern British Columbia. Unpub. PhD thesis, University of Western Ontario, London, Canada, 402 pp.

Angiel, P.A. and Plint, A.G. (2022). An allostratigraphic framework for the Late Albian to Early Cenomanian upper Fort St. John Group across the foredeep of the Western Canada Foreland Basin, NE British Columbia. *Bulletin of Canadian Energy Geoscience* 69 (2): 77–140.

Anthony, E.J. and Aagaard, T. (2020). The lower shoreface: morphodynamics and sediment connectivity with the upper shoreface and beach. *Earth-Science Reviews* 210: 103334.

Anthony, E.J., Gardel, A., Gratiot, N. et al. (2010). The Amazon-influenced muddy coast of South America: a review of mud-bank–shoreline interactions. *Earth-Science Reviews* 103 (3-4): 99–121.

Anthony, E.J., Gardel, A., and Gratiot, N. (2014). Fluvial sediment supply, mud banks, cheniers and the morphodynamics of the coast of South America between the Amazon and Orinoco river mouths. *Geological Society, London, Special Publication* 388 (1): 533–560.

Aplin, A.C., Fleet, A.J., and Macquaker, J. H.S. (1999). Muds and mudstones: physical and fluid-flow properties. *Geological Society, London, Special Publication* 158 (1): 1–8.

Aplin, A.C. and Macquaker, J.H.S. (2011). Mudstone diversity: origin and implications for source, seal, and reservoir properties in petroleum systems. *American Association of Petroleum Geologists, Bulletin* 95: 2031–2059.

Asquith, D.O. (1970). Depositional topography and major marine environments, Late Cretaceous, Wyoming. *American Association of Petroleum Geologists, Bulletin* 54: 1184–1224.

Augustinus, P.G.E.F., Hazelhoff, L., and Kroon, A. (1989). The chenier coast of Suriname: modern and geological development. *Marine Geology* 90 (4): 269–281.

Baas, J.H., Best, J.L., and Peakall, J. (2015). Predicting bedforms and primary current stratification in cohesive mixtures of mud and sand. *Journal of the Geological Society* 173 (1): 12–45.

Bartholdy, J. (2000). Processes controlling import of fine-grained sediment to tidal areas: a simulation model. *Geological Society, London, Special Publication* 175 (1): 13–29.

Baudin, F., Rabouille, C., and Dennielou, B. (2020). Routing of terrestrial organic matter from the Congo River to the ultimate sink in the abyss: a mass balance approach. (André Dumont medallist lecture 2017). *Geologica Belgica* 23 (1-2): 41–52.

Bentley, S.J. and Nittrouer, C.A. (2003). Emplacement, modification, and preservation of event strata on a flood-dominated continental shelf: Eel shelf, Northern California. *Continental Shelf Research* 23 (16): 1465–1493.

Bentley, S.J., Sheremet, A., and Jaeger, J.M. (2006). Event sedimentation, bioturbation, and preserved sedimentary fabric: field and model comparisons in three contrasting marine settings. *Continental Shelf Research* 26 (17-18): 2108–2124.

Berner, R.A. (1981). A new geochemical classification of sedimentary environments. *Journal of Sedimentary Petrology* 51: 359–365.

Berner, R.A. (1985). Sulphate reduction, organic matter decomposition and pyrite formation. *Philosophical Transactions of the Royal Society of London* A315: 25–38.

Bhattacharya, J.P., Howell, C.D., MacEachern, J.A., and Walsh, J.P. (2020). Bioturbation, sedimentation rates, and preservation of flood events in deltas. *Palaeogeography, Palaeoclimatology, Palaeoecology* 560: 110049.

Bhattacharya, J.P. and MacEachern, J.A. (2009). Hyperpycnal rivers and prodeltaic shelves in the Cretaceous Seaway of North America. *Journal of Sedimentary Research* 79: 184–209.

Bhattacharya, J. and Walker, R.G. (1991). River-and wave-dominated depositional systems of the Upper Cretaceous Dunvegan Formation, north-western Alberta. *Bulletin of Canadian Petroleum Geology* 39 (2): 165–191.

Boyd, R., Dalrymple, R., and Zaitlin, B.A. (1992). Classification of clastic coastal depositional environments. *Sedimentary Geology* 80 (3-4): 139–150.

Brenchley, P.J., Newall, G., and Stanistreet, I.G. (1979). A storm surge origin for sandstone beds in an epicontinental platform sequence, Ordovician, Norway. *Sedimentary Geology* 22 (3-4): 185–217.

Buckley, R.A. and Plint, A.G. (2013). Allostratigraphy of the Peace River Formation (Albian) in north-western Alberta and adjacent British Columbia. *Bulletin of Canadian Petroleum Geology* 61: 295–330.

Buckley, R.A., Plint, A.G., Henderson, O.A. et al. (2016). Ramp sedimentation across a middle Albian, Arctic embayment: influence of subsidence, eustasy and sediment supply on stratal architecture and facies distribution, Lower Cretaceous, Western Canada Foreland Basin. *Sedimentology* 63: 699–742.

Buckman, J., Mahoney, C., Bankole, S. et al. (2020). Workflow model for the digitization of mudrocks. *Geological Society, London, Special Publication* 484 (1): 165–187.

Cacchione, D.A., Drake, D.E., Kayen, R.W. et al. (1995). Measurements in the bottom boundary layer on the Amazon subaqueous delta. *Marine Geology* 125 (3-4): 235–257.

Canfield, D.E. and Thamdrup, B.O. (2009). Towards a consistent classification scheme for geochemical environments, or, why we wish the term 'suboxic' would go away. *Geobiology* 7 (4): 385–392.

Cartwright, J.A. (1994). Episodic basin-wide fluid expulsion from geopressured shale sequences in the North Sea Basin. *Geology* 22: 447–450.

Cartwright, J.A. and Dewhurst, D.N. (1998). Layer-bound compaction faults in fine-grained sediments. *Geological Society of America, Bulletin* 110: 1242–1257.

Cartwright, J. and Santamarina, C. (2015). Seismic characteristics of fluid escape pipes in sedimentary basins: implications for pipe genesis. *Marine and Petroleum Geology* 65: 126–140.

Cattaneo, A., Correggiari, A., Langone, L., and Trincardi, F. (2003). The late-Holocene Gargano subaqueous delta, Adriatic shelf: sediment pathways and supply fluctuations. *Marine Geology* 193 (1-2): 61–91.

Cattaneo, A., Trincardi, F., Asioli, A., and Correggiari, A. (2007). The Western Adriatic shelf clinoform: energy-limited bottomset. *Continental Shelf Research* 27 (3-4): 506–525.

Chamley, H. (1989). Clay formation through weathering. In: *Clay Sedimentology*, 21–50. Berlin, Heidelberg: Springer.

Chang, T.S. and Ha, H.J. (2015). The Heuksan mud belt on the tide-dominated shelf of Korea: a supply-driven depositional system? *Geo-Marine Letters* 35 (6): 447–460.

Clift, P.D., Wan, S., and Blusztajn, J. (2014). Reconstructing chemical weathering, physical erosion and monsoon intensity since 25 Ma in the northern South China Sea: a review of competing proxies. *Earth-Science Reviews* 130: 86–102.

Clifton, H.E. (2006). A re-examination of facies models for clastic shorelines. In: *Facies Models Revisited*, vol. 84 (ed. H. W. Posamentier and R.G. Walker), 293–337. SEPM Special Publication (on CD).

Coleman, J.M., Roberts, H.H., and Stone, G.W. (1998). Mississippi River delta: an overview. *Journal of Coastal Research* 14 (3): 698–716.

Collinson, J.C. and Mountney, N.P. (2019). *Sedimentary Structures*, 340. Edinburgh, Scotland: Dunedin Academic Press.

Curtis, C.D. (1980). Diagenetic alteration in black shales. *Journal of the Geological Society of London* 137: 189–194.

Curtis, C.D. and Coleman, M.L. (1986). Controls on the precipitation of early diagenetic calcite, dolomite and siderite concretions in complex depositional sequences. In: *Roles of Organic Matter in Sediment Diagenesis*, vol. 38 (ed. D.L. Gautier), 23–33. SEPM Special Publication.

Dade, W.B., Nowell, A.R.M., and Jumars, P.A. (1992). Predicting erosion resistance of muds. *Marine Geology* 105 (1-4): 285–297.

Dalrymple, R.W., Narbonne, G.M., and Smith, L. (1985). Eolian action and the distribution of Cambrian shales in North America. *Geology* 13 (9): 607–610.

Dalrymple, R.W., Baker, E.K., Harris, P.T., and Hughes, M.G. (2003). Sedimentology and stratigraphy of a tide-dominated, foreland-basin delta (Fly River, Papua New Guinea). In: *Tropical Deltas of Southeast Asia-Sedimentology, Stratigraphy, and Petroleum Geology, SEPM Special Publication 76* (ed. F.H. Sidi, D. Nummedal, P. Imbert, et al.), 147–173. Oklahoma: Tulsa.

Debenay, J.P., Jouanneau, J.M., Sylvestre, F. et al. (2007). Biological origin of rhythmites in muddy sediments of French Guiana. *Journal of Coastal Research* 23 (6): 1431–1442.

Denommee, K.C., Bentley, S.J., and Harazim, D. (2018). Mechanisms of muddy clinothem progradation on the Southwest Louisiana Chenier Plain inner shelf. *Geo-Marine Letters* 38 (3): 273–285.

Devlin, W.J., Rudolph, K.W., Shaw, C.A., and Ehman, K.D. (1993). The effect of tectonic and eustatic cycles on accommodation and sequence-stratigraphic framework in the Upper Cretaceous foreland basin of southwestern Wyoming. In: *Sequence Stratigraphy and Facies Associations*, vol. 18 (ed. H.W. Posamentier, C.P. Summerhayes, B.U. Haq, and G.P. Allen), 501–520. International Association of Sedimentologists, Special Publication.

Donaldson, W.S., Plint, A.G., and Longstaffe, F.J. (1999). Tectonic and eustatic control on deposition and preservation of Upper Cretaceous ooidal ironstone and associated facies: Peace River Arch area, NW Alberta, Canada. *Sedimentology* 46 (6): 1159–1182.

Donnici, S., Madricardo, F., and Serandrei-Barbero, R. (2017). Sedimentation rate and lateral migration of tidal channels in the Lagoon of Venice (Northern Italy). *Estuarine, Coastal and Shelf Science* 198: 354–366.

Dougherty, A.J. and Dickson, M.E. (2012). Sea level and storm control on the evolution of a chenier plain, Firth of Thames, New Zealand. *Marine Geology* 307: 58–72.

Dowey, P.J., Osborne, M., and Volk, H. (2020). Application of analytical techniques to petroleum systems. *Geological Society, London, Special Publication* 484 (1): 1–7.

Draut, A.E., Kineke, G.C., Huh, O.K. et al. (2005). Coastal mudflat accretion under energetic conditions, Louisiana chenier-plain coast, USA. *Marine Geology* 214 (1-3): 27–47.

Duke, W.L. (1990). Geostrophic circulation or shallow marine turbidity currents? The dilemma of paleoflow patterns in storm-influenced prograding shoreline systems. *Journal of Sedimentary Petrology* 60: 870–883.

Einsele, G., Overbeck, R., Schwarz, H.U., and Unsöld, G. (1974). Mass physical properties, sliding and erodibility of experimentally deposited and differently consolidated clayey muds (approach, equipment, and first results). *Sedimentology* 21 (3): 339–372.

Eisma, D. (1986). Flocculation and de-flocculation of suspended matter in estuaries. *Netherlands Journal of Sea Research* 20 (2-3): 183–199.

Eisma, D., Augustinus, P.G.E.F., and Alexander, C. (1991). Recent and subrecent changes in the dispersal of Amazon mud. *Netherlands Journal of Sea Research* 28 (3): 181–192.

Eisma, D., Van der Gaast, S.J., Martin, J. M., and Thomas, A.J. (1978). Suspended matter and bottom deposits of the Orinoco delta: Turbidity, mineralogy and elementary composition. *Netherlands Journal of Sea Research* 12 (2): 224–251.

Evans, G. (1965). Intertidal flat sediments and their environments of deposition in the Wash. *Journal of the Geological Society* 121: 209–245.

Fan, D. (2012). Open-coast tidal flats. In: *Principles of Tidal Sedimentology* (ed. R.A. Davis Jr. and R.W. Dalrymple), 187–229. London: Springer.

Fisher, Q., Kaminskaite, I., and del Pino Sanchez, A. (2023). Shale barrier performance in petroleum systems: implications for CO_2 storage and nuclear waste disposal. *Geoenergy* 1 (1): 2023–2006.

Froelich, P., Klinkhammer, G.P., Bender, M.L. et al. (1979). Early oxidation of organic matter in pelagic sediments of the eastern equatorial Atlantic: suboxic diagenesis. *Geochimica et Cosmochimica Acta* 43 (7): 1075–1090.

Froidefond, J.M., Pujos, M., and Andre, X. (1988). Migration of mud banks and changing coastline in French Guiana. *Marine Geology* 84 (1-2): 19–30.

Galloway, W.E. (1975, 1975). Process framework for describing the morphologic and stratigraphic evolution of deltaic depositional systems. In: *Deltas-Models for Exploration* (ed. M.L. Broussard), 87–98. Houston, Texas: Houston Geological Society.

Garzanti, E., Padoan, M., Setti, M. et al. (2014). Provenance versus weathering control on the composition of tropical river mud (southern Africa). *Chemical Geology* 366: 61–74.

Gautschi, A. (2017). Safety-relevant hydrogeological properties of the claystone barrier of a Swiss radioactive waste repository: an evaluation using multiple lines of evidence. *Grundwasser* 22 (3): 221–233.

Goodbred, S.L. (2003). Response of the Ganges dispersal system to climate change: a source-to-sink view since the last interstade. *Sedimentary Geology* 162: 83–104.

Gibling, M.R., Tandon, S.K., Sinha, R., and Jain, M. (2005). Discontinuity-bounded alluvial sequences of the southern Gangetic Plains, India: aggradation and degradation in response to monsoonal strength. *Journal of Sedimentary Research* 75: 369–385.

Gibling, M.R., Fielding, C.R., and Sinha, R. (2011). Alluvial valleys and alluvial sequences: towards a geomorphic assessment. In: *From River to Rock Record: The Preservation of Fluvial Sediments and Their Subsequent Interpretation*, vol. 97 (ed. S.K. Davidson, S. Lelu, and C.P. North), 423–447. *SEPM Special Publication.*

Geyer, W.R., Hill, P., Milligan, T., and Traykovski, P. (2000). The structure of the Eel River plume during floods. *Continental Shelf Research* 20 (16): 2067–2093.

Grabowski, R.C., Droppo, I.G., and Wharton, G. (2011). Erodibility of cohesive sediment: The importance of sediment properties. *Earth-Science Reviews* 105 (3-4): 101–120.

Hamblin, A.P. and Walker, R.G. (1979). Storm-dominated shallow marine deposits: the Fernie-Kootenay (Jurassic) transition, southern Rocky Mountains. *Canadian Journal of Earth Sciences* 16: 1673–1690.

Hanebuth, T.J., Stattegger, K., and Bojanowski, A. (2009). Termination of the Last Glacial Maximum sea-level lowstand: The Sunda-Shelf data revisited. *Global and Planetary Change* 66 (1-2): 76–84.

Hanebuth, T.J., Lantzsch, H., and Nizou, J. (2015). Mud depocenters on continental shelves—appearance, initiation times, and growth dynamics. *Geo-Marine Letters* 35 (6): 487–503.

Harris, P.T., Baker, E.K., Cole, A.R., and Short, S.A. (1993). A preliminary study of sedimentation in the tidally dominated Fly River Delta, Gulf of Papua. *Continental Shelf Research* 13 (4): 441–472.

Harris, P.T., Hughes, M.G., Baker, E.K. et al. (2004). Sediment transport in distributary channels and its export to the prodeltaic environment in a tidally dominated delta: Fly River, Papua New Guinea. *Continental Shelf Research* 24 (19): 2431–2454.

Harris, P.T., Pattiaratchi, C.B., Keene, J.B. et al. (1996). Late Quaternary deltaic and carbonate sedimentation in the Gulf of Papua foreland basin; response to sea-level change. *Journal of Sedimentary Research* 66 (4): 801–819.

Hart, B.S. and Plint, A.G. (1993). Tectonic influence on deposition and erosion in a ramp setting: Upper Cretaceous Cardium Formation, Alberta foreland basin. *American Association of Petroleum Geologists Bulletin* 77: 2092–2107.

Hay, M.J. and Plint, A.G. (2020). High-frequency sequences within a retrogradational deltaic succession: Upper Cenomanian Dunvegan Formation, Western Canada Foreland Basin. *The Depositional Record* 6 (3): 524–551.

Hayes, M.O. (1967). Hurricanes as geological agents: Case studies of Hurricanes Carla, 1961, and Cindy, 1963. Bureau of Economic Geology. University of Texas, Report of Investigations No. 61. 58p.

Henderson, O.A., Vannelli, K.M., and Plint, A.G. (2014). Allostratigraphy and paleogeography of the Harmon and Cadotte members of the Peace River Formation (middle Albian), west-central Alberta and adjacent British Columbia. *Bulletin of Canadian Petroleum Geology* 62: 177–212.

Hijma, M.P., Shen, Z., Törnqvist, T.E., and Mauz, B. (2017). Late Holocene evolution of a coupled, mud-dominated delta plain–chenier plain system, coastal Louisiana, USA. *Earth Surface Dynamics* 5 (4): 689–710.

Hill, P.S., Fox, J.M., Crockett, J.S. et al. (2007). Sediment delivery to the seabed on continental margins. In: *Continental Margin Sedimentation: From Sediment Transport to Sequence Stratigraphy. International Association of Sedimentologists*, Special Publication *37*, (ed. C.A. Nittrouer et al.), 49–99. Blackwell Science, Oxford.

Hines, A.H. and Comtois, K.L. (1985). Vertical distribution of infauna in sediments of a sub-estuary of central Chesapeake Bay. *Estuaries* 8 (3): 296–304.

Hjulstrom, F. (1934). Studies of the morphological activity of rivers as illustrated by the River Fyris. *Bulletin of the Geological Institution of the University of Upsala* 25: 221–537.

Holeman, J.N. (1980). Erosion rates in the U.S. estimated by the Soil Conservation Service's inventory (abs.). *EOS (American Geophysical Union Transactions)* 61: 954.

Hovikoski, J., Lemiski, R.T., Gingras, M.K., Pemberton, S.G. and MacEachern, J.A. (2007). Preliminary notes on the depositional setting of the Late Cretaceous Alderson Member (Lea Park Formation). In: Summary of Investigations 2007. Vol. 1. Saskatchewan Geological Survey, Saskatchewan Industry Resources Misc. Report 2007-4.1 CD-ROM, Paper A-10. 11 pp.

Hovikoski, J., Lemiski, R.T., Gingras, M.K. et al. (2008). Ichnology and sedimentology of a mud-dominated deltaic coast: Upper Cretaceous Alderson Member (Lea Park Fm.) Western Canada. *Journal of Sedimentary Research* 78: 803–824.

Hovius, N. (1998). Controls on sediment supply by large rivers. In: *Relative Role of Eustasy, Climate, and Tectonism in Continental Rocks*, vol. 59 (ed. K.W. Shanley and P.J. McCabe), 3–16. SEPM Special Publication.

Hovland, M. and Risk, M. (2003). Do Norwegian deep-water coral reefs rely on seeping fluids? *Marine Geology* 198 (1-2): 83–96.

Hu, Y.G. and Plint, A.G. (2009). An allostratigraphic correlation of a mudstone-dominated syn-tectonic wedge: the Puskwaskau Formation (Santonian-Campanian) in outcrop and subsurface, Western Canada Foreland Basin. *Bulletin of Canadian Petroleum Geology* 57: 1–33.

Huh, O.K., Walker, N.D., and Moeller, C. (2001). Sedimentation along the eastern chenier plain coast: down drift impact of a delta complex shift. *Journal of Coastal Research* 17 (1): 72–81.

Irwin, H., Curtis, C., and Coleman, M.L. (1977). Isotopic evidence for source of diagenetic carbonates formed during burial of organic-rich sediments. *Nature* 269: 209–213.

Jaeger, J.M. and Nittrouer, C.A. (1995). Tidal controls on the formation of fine-scale sedimentary strata near the Amazon River mouth. *Marine Geology* 125 (3-4): 259–281.

Jaramillo, S., Sheremet, A., Allison, M.A. et al. (2009). Wave-mud interactions over the muddy Atchafalaya subaqueous clinoform, Louisiana, United States: wave-supported sediment transport. *Journal of Geophysical Research: Oceans* 114 (C4): C04002. https://doi.org/10.1029/2008JC004821.

Jordan, T.E. and Flemings, P.B. (1991). Large scale stratigraphic architecture, eustatic variation and unsteady tectonism: a theoretical evaluation. *Journal of Geophysical Research* 96: 6681–6699.

King, J.J., Roberts, D.T., Cartwright, J.A., and Levell, B.K. (2022). Numerical modelling of the growth of polygonal fault systems. *Journal of Structural Geology* 162: 104679.

Kinsela, M.A., Hanslow, D.J., Carvalho, R. C. et al. (2022). Mapping the shoreface of coastal sediment compartments to improve shoreline change forecasts in New South Wales, Australia. *Estuaries and Coasts* 45 (4): 1143–1169.

Kominz, M.A., Patterson, K., and Odette, D. (2011). Lithology dependence of porosity in slope and deep marine sediments. *Journal of Sedimentary Research* 81 (10): 730–742.

Kranck, K. and Milligan, T. (1980). Microflocs: production of marine snow in the laboratory. *Marine Ecology – Progress Series* 3: 19–24.

Kuehl, S.A., Nittrouer, C.A., and DeMaster, D.J. (1986). Distribution of sedimentary structures in the Amazon subaqueous delta. *Continental Shelf Research* 6: 311–336.

Landman, N.H., Plint, A.G., and Walaszczyk, I. (2017). Scaphitid ammonites from the Upper Cretaceous (Coniacian-Santonian) Western Canada Foreland Basin. *Bulletin of the American Museum of Natural History* 414: 104–172.

Larson, D.W. and Rhoads, D.C. (1983). The evolution of infaunal communities and sedimentary fabrics. In: *Biotic Interactions in Recent and Fossil Benthic Communities* (ed. M.J.S. Tevesz and P.L. McCall), 627–648. Boston, MA.: Springer.

Lawton, T.F. (2019). Laramide sedimentary basins and sediment-dispersal systems. In: *Sedimentary Basins of the United States and Canada*, 2nd Ed. (ed. A.D. Miall), 529–557. Elsevier.

Lazar, O.R., Bohacs, K.M., Macquaker, J. H. et al. (2015). Capturing key attributes of fine-grained sedimentary rocks in outcrops, cores, and thin sections: nomenclature and description guidelines. *Journal of Sedimentary Research* 85 (3): 230–246.

Leckie, D.A. and Walker, R.G. (1982). Storm- and tide-dominated shorelines in Cretaceous Moosebar-Lower Gates interval – outcrop equivalents of Deep Basin gas trap in Western Canada. *American Association of Petroleum Geologists, Bulletin* 66: 138–157.

Leckie, D.A. and Krystinik, L.F. (1989). Is there evidence for geostrophic currents preserved in the sedimentary record of inner to middle-shelf deposits? *Journal of Sedimentary Petrology* 59: 862–870.

Lee, G.S., Yoo, D.G., Bae, S.H. et al. (2015). Seismic stratigraphy of the Heuksan mud belt in the south-eastern Yellow Sea, Korea. *Geo-Marine Letters* 35 (6): 433–446.

Legge, O., Johnson, M., Hicks, N. et al. (2020). Carbon on the northwest European shelf: contemporary budget and future influences. *Frontiers in Marine Science* 7: 143.

Lemiski, R.T., Hovikoski, J., Pemberton, S. G., and Gingras, M.K. (2011). Sedimentological and reservoir characteristics of the low-permeability gas-charged Alderson Member (Hatton gas field, southwest Saskatchewan): Implications for reservoir development. *Bulletin of Canadian Petroleum Geology* 59: 27–53.

Li, Z. and Schieber, J. (2018). Detailed facies analysis of the Upper Cretaceous Tununk Shale Member, Henry Mountains region, Utah: implications for mudstone depositional models in epicontinental seas. *Sedimentary Geology* 364: 141–159.

Li, Z., Bhattacharya, J., and Schieber, J. (2015). Evaluating along-strike variation using thin-bedded facies analysis, Upper Cretaceous Ferron Notom Delta, Utah. *Sedimentology* 62: 2060–2089.

Lim, D.I., Choi, J.Y., Jung, H.S., Rho, K.C., and Ahn, K.S. (2007). Recent sediment accumulation and origin of shelf mud deposits in the Yellow and East China Seas. *Progress in Oceanography* 73 (2): 145–159.

Liu, J.P., Saito, Y., Kong, X. et al. (2010). Sedimentary record of environmental evolution off the Yangtze River estuary, East China Sea, during the last~ 13,000 years, with special reference to the influence of the Yellow River on the Yangtze River delta during the last 600 years. *Quaternary Science Reviews* 29 (17-18): 2424–2438.

Liu, J.P., Xu, K.H., Li, A.E.A. et al. (2007). Flux and fate of Yangtze River sediment delivered to the East China Sea. *Geomorphology* 85 (3-4): 208–224.

Liu, X., Li, A., Dong, J. et al. (2018). Provenance discrimination of sediments in the Zhejiang-Fujian mud belt, East China Sea: implications for the development of the mud depocenter. *Journal of Asian Earth Sciences* 151: 1–15.

Ma, L., Fauchille, A.L., Dowey, P.J. et al. (2017). Correlative multi-scale imaging of shales: a review and future perspectives. *Geological Society, London, Special Publications* 454 (1): 175–199.

Mackay, D.A. and Dalrymple, R.W. (2011). Dynamic mud deposition in a tidal environment: the record of fluid-mud deposition in the Cretaceous Bluesky Formation, Alberta, Canada. *Journal of Sedimentary Research* 81: 901–920.

Macquaker, J.H.S. (1994). A lithofacies study of the Peterborough Member, Oxford Clay Formation (Jurassic), UK: an example of sediment bypass in a mudstone succession. *Journal of the Geological Society* 151: 161–172.

Macquaker, J.H.S. and Gawthorpe, R.L. (1993). Mudstone lithofacies of the Kimmeridge Clay Formation, Wessex Basin, southern England: implications for the origin and controls of the distribution of mudstones. *Journal of Sedimentary Petrology* 63: 1129–1143.

Macquaker, J.H.S. and Taylor, K.G. (1996). A sequence-stratigraphic interpretation of a mudstone-dominated succession: the Lower Jurassic Cleveland Ironstone Formation, UK. *Journal of the Geological Society* 153: 759–770.

Macquaker, J.H.S., Bentley, S.J., and Bohacs, K.M. (2010). Wave-enhanced sediment-gravity flows and mud dispersal across continental shelves: reappraising sediment transport processes operating in ancient mudstone successions. *Geology* 38: 947–950.

Macquaker, J.H.S., Taylor, K.G., Keller, M.A., and Polya, D. (2014). Compositional controls on early diagenetic pathways in fine-grained sedimentary rocks: implications for predicting unconventional reservoir attributes of mudstones. *American Association of Petroleum Geologists, Bulletin* 98: 587–603.

Maloney, J.M., Bentley, S.J., Xu, K. et al. (2020). Mass wasting on the Mississippi River subaqueous delta. *Earth-Science Reviews* 200: 103001.

Mángano, M.G. and Buatois, L.A. (2017). The Cambrian revolutions: trace-fossil record, timing, links and geobiological impact. *Earth-Science Reviews* 173: 96–108.

McCave, I.N. (1972). Transport and escape of fine-grained sediment from shelf areas. In: *Shelf Sediment Transport: Process and Pattern* (ed. D.J.P. Swift, J.P. Duane, and O.H. Pilkey), 225–248. Geological Society of America.

McCave, I.N. (1976). Sea-floor deposition, erosion, and transportation. In: *The Benthic Boundary Layer*, 247–260. Boston, MA: Springer.

McCave, I.N. (1984). Size spectra and aggregation of suspended particles in the deep ocean. *Deep-Sea Research* 31: 329–352.

McCave, I.N. (2020). Direct measures of the vigour of ocean circulation via particle grain size. *Oceanography* 33 (2): 53–55.

McCave, I.N. and Hall, I.R., 2006. Size sorting in marine muds: processes, pitfalls, and prospects for paleoflow-speed proxies. *Geochemistry, Geophysics, Geosystems*, 7(10), doi: https://doi.org/10.1029/2006GC001284.

McClennen, C.E. and Housley, R.A. (2006). Late-Holocene channel meander migration and mudflat accumulation rates, lagoon of Venice, Italy. *Journal of Coastal Research* 22: 930–945.

Meijer, N., Dupont-Nivet, G., Licht, A. et al. (2020). Identifying eolian dust in the geological record. *Earth-Science Reviews* 103410.

Miall, A.D. (2002). Architecture and sequence stratigraphy of Pleistocene fluvial systems in the Malay Basin, based on seismic time-slice analysis. *American Association of Petroleum Geologists, Bulletin* 86 (7): 1201–1216.

Middelburg, J.J. (2018). Reviews and syntheses: to the bottom of carbon processing at the seafloor. *Biogeosciences* 15 (2): 413–427.

Milliman, J.D. (1995). Sediment discharge to the ocean from small mountainous rivers: the New Guinea example. *Geo-Marine Letters* 15 (3-4): 127–133.

Milliman, J.D. and Farnsworth, K.L. (2011). *River discharge to the coastal ocean: a global synthesis*, 384. Cambridge University Press.

Milliman, J.D. and Kao, S.J. (2005). Hyperpycnal discharge of fluvial sediment to the ocean: impact of super-typhoon Herb (1996) on Taiwanese rivers. *The Journal of Geology* 113 (5): 503–516.

Millot, G. (1970). *Geology of Clays*, 429. Berlin: Springer-Verlag.

Moretti, M., Alfaro, P., and Owen, G. (2016). The environmental significance of soft-sediment deformation structures: key signatures for sedimentary and tectonic processes. *Sedimentary Geology* 344: 1–4.

Morgan, J.P., Coleman, J.M., and Gagliano, S.M. (1968). Mudlumps: diapiric structures in Mississippi delta sediments. In: *Diapirism and diapirs. American Association Petroleum Geologists Memoir* 8: 145–161.

Morton, R.A., 1981. Formation of storm deposits by wind-forced currents in the Gulf of Mexico and the North Sea. In: S.-D. Nio, R. T. E. Shüttenhelm and, Tj. C. E. van Weering (Eds.), *Holocene Marine Sedimentation in the North Sea Basin*. International Association of Sedimentologists, Special Publication 5, Blackwell Science, Oxford, pp. 385–396

Morton, R.A. (1988). Nearshore responses to great storms. In: *Sedimentologic Consequences of Convulsive Geologic Events: Geological Society of America*, Special Paper, vol. 229 (ed. H.E. Clifton), 7–22.

Moura, R.L., Amado-Filho, G.M., Moraes, F.C. et al. (2016). An extensive reef system at the Amazon River mouth. *Science Advances* 2 (4): e1501252.

Mulder, T. and Alexander, J. (2001). The physical character of subaqueous sedimentary density flows and their deposits. *Sedimentology* 48 (2): 269–299.

Mulder, T. and Syvitski, J.P. (1995). Turbidity currents generated at river mouths during exceptional discharges to the world oceans. *The Journal of Geology* 103 (3): 285–299.

Mulder, T., Syvitski, J.P.M., Migeon, S. et al. (2003). Marine hyperpycnal flows: initiation, behaviour and related deposits. A review. *Marine and Petroleum Geology* 20: 861–882.

Neill, C.F. and Allison, M.A. (2005). Subaqueous deltaic formation on the Atchafalaya shelf, Louisiana. *Marine Geology* 214: 411–430.

Nesbitt, H.W. and Young, G.M. (1982). Early Proterozoic climates and plate motions inferred from major element chemistry of lutites. *Nature* 299 (5885): 715–717.

Nittrouer, C.A., Kuehl, S.A., Sternberg, R.W. et al. (1995). An introduction to the geological significance of sediment transport and accumulation on the Amazon continental shelf. *Marine Geology* 125 (3-4): 177–192.

North American Commission on Stratigraphic Nomenclature (2005). North American stratigraphic code. *American Association of Petroleum Geologists, Bulletin* 89 (11): 1547–1591.

Nyberg, B. and Howell, J.A. (2016). Global distribution of modern shallow marine shorelines. Implications for exploration and reservoir analogue studies. *Marine and Petroleum Geology* 71: 83–104.

Nyman, J.A., Walters, R.J., Delaune, R.D., and Patrick, W.H. Jr. (2006). Marsh vertical accretion via vegetative growth. *Estuarine, Coastal and Shelf Science* 69 (3-4): 370–380.

Oldham, V.E., Siebecker, M.G., Jones, M.R. et al. (2019). The speciation and mobility of Mn and Fe in estuarine sediments. *Aquatic Geochemistry* 25: 3–26.

Orcutt, B.N., Bach, W., Becker, K. et al. (2011). Colonization of subsurface microbial observatories deployed in young ocean crust. *The ISME Journal* 5 (4): 692–703.

Orseau, S., Zorrilla, N.A., Huybrechts, N. et al. (2020). Decadal-scale morphological evolution of a muddy open coast. *Marine Geology* 420: 106048.

Ott, B. and Mann, P. (2015). Late Miocene to Recent formation of the Aure-Moresby fold-thrust belt and foreland basin as a consequence of Woodlark microplate rotation, Papua New Guinea. *Geochemistry, Geophysics, Geosystems* 16 (6): 1988–2004.

Palinkas, C.M. (2009). The timing of floods and storms as a controlling mechanism for shelf deposit morphology. *Journal of Coastal Research* 25 (5): 1122–1129.

Palinkas, C.M. and Nittrouer, C.A. (2007). Modern sediment accumulation on the Po shelf, Adriatic Sea. *Continental Shelf Research* 27 (3-4): 489–505.

Peters, S.E. and Loss, D.P. (2012). Storm and fair-weather wave base: a relevant distinction? *Geology* 40 (6): 511–514.

Plaziat, J.C. and Augustinus, P.G. (2004). Evolution of progradation/erosion along the French Guiana mangrove coast: a comparison of mapped shorelines since the 18th century with Holocene data. *Marine Geology* 208 (2-4): 127–143.

Plint, A.G., 1988. Sharp-based shoreface sequences and "offshore bars" in the Cardium Formation: Their relationship to relative changes in sea level. In: C.K. Wilgus, B.S. Hastings, H.W. Posamentier, J.C. van Wagoner, C.A. Ross and C.G. St.C Kendall(eds.), *Sea Level Changes: An Integrated Approach*. Society of Economic Paleontologists and Mineralogists, Special Publication 42, pp. 357–370.

Plint, A.G. (1990). An allostratigraphic correlation of the Muskiki and Marshybank formations (Coniacian - Santonian) in the Foothills and subsurface of the Alberta Basin. *Bulletin of Canadian Petroleum Geology* 38: 288–306.

Plint, A.G. (1991). High-frequency relative sea level oscillations in Upper Cretaceous shelf clastics of the Alberta Foreland Basin: Evidence for a Milankovitch- scale glacio-eustatic control? In: *Sedimentation, Tectonics and Eustacy. International Association of Sedimentologists, Special Publication 12* (ed. D.I.M. Macdonald), 409–428. Oxford: Blackwell Science.

Plint, A.G. (1996). Marine and nonmarine systems tracts in fourth order sequences in the Early-Middle-Cenomanian, Dunvegan Alloformation, northeastern British Columbia, Canada *High Resolution Sequence Stratigraphy: Innovations and Applications* (ed. J. Howell and J.D. Aitken). Geological Society, London, Special Publication.

Plint, A.G. (2000). Sequence stratigraphy and paleogeography of a Cenomanian deltaic complex: the Dunvegan and lower Kaskapau formations in subsurface and outcrop, Alberta and British Columbia, Canada. *Bulletin of Canadian Petroleum Geology* 48 (1): 43–79.

Plint, A.G. (2002). Paleo-valley systems in the Upper Cretaceous Dunvegan Formation, Alberta and British Columbia. *Bulletin of Canadian Petroleum Geology* 50: 277–296.

Plint, A.G. (2014). Mud dispersal across a Cretaceous prodelta: storm-generated, wave-enhanced sediment gravity flows inferred from mudstone microtexture and microfacies. *Sedimentology* 61 (3): 609–647.

Plint, A.G. and Kreitner, M.A. (2019). High-frequency sequences, paleogeography, and syn-depositional tectonism on a shallow clastic ramp: Doe Creek and Pouce Coupe members of the Late Cenomanian Kaskapau Formation, Western Canada Foreland Basin. *Bulletin of Canadian Petroleum Geology* 67: 71–116.

Plint, A.G. and Norris, B. (1991). Anatomy of a ramp margin sequence: facies successions, paleogeography and sediment dispersal patterns in the Muskiki and Marshybank formations, Alberta Foreland Basin. *Bulletin of Canadian Petroleum Geology* 39: 18–42.

Plint, A.G., Macquaker, J.H., and Varban, B.L. (2012b). Bedload transport of mud across a wide, storm-influenced ramp: Cenomanian–Turonian Kaskapau Formation, Western Canada Foreland Basin. *Journal of Sedimentary Research* 82 (11): 801–822.

Plint, A.G. and Nummedal, D., 2000. The falling stage systems tract: recognition and importance in sequence stratigraphic analysis. In: D. Hunt and R.L. Gawthorpe (Eds.), *Sedimentary Responses to Forced Regressions*. Geological Society, London, Special Publication, 172(1), pp. 1–17.

Plint, A.G., Hooper, E.A., Grifi, M.D. et al. (2017). Integrated, high-resolution allostratigraphic, biostratigraphic and carbon-isotope correlation of Coniacian strata (Upper Cretaceous), Western Alberta and northern Montana. *Bulletin of the American Museum of Natural History* 414: 9–52.

Plint, A.G., Tyagi, A., Hay, M.J. et al. (2009). Clinoforms, paleobathymetry, and mud dispersal across the Western Canada Cretaceous foreland basin: evidence from the Cenomanian Dunvegan formation and contiguous strata. *Journal of Sedimentary Research* 79: 144–161.

Plint, A.G., Tyagi, A., McCausland, P.J.A. et al. (2012a). Dynamic relationship between subsidence, sedimentation, and unconformities in mid- Cretaceous, shallow-marine strata of the Western Canada Foreland Basin: links to Cordilleran Tectonics. In: *Recent Advances in the Tectonics of Sedimentary Basins* (ed. C. Busby and A. Azor), 480–507. Oxford: Wiley-Blackwell Publishing Ltd.

Plint, A.G., Uličný, D., Čech, S. et al. (2022). Trans-Atlantic correlation of Late Cretaceous high-frequency sea-level cycles. *Earth and Planetary Science Letters* 578: 117323.

Postma, H. (1961). Transport and accumulation of suspended material in the Dutch Wadden Sea. *Netherlands Journal of Sea Research* 1: 148–190.

Pufahl, P.K. (2010). Bioelemental sediments. In: *Facies Models*, Fourth Ed. (ed. N.P. James and R.W. Dalrymple), 477–503. Geological Association of Canada.

Pyles, D.R. and Slatt, R.M. (2007). Applications to understanding shelf edge to base-of-slope changes in stratigraphic architecture of prograding basin margins: Stratigraphy of the Lewis Shale, Wyoming, USA. In: *Atlas of deep-water outcrops: AAPG Studies in Geology*, vol. 56, CD-ROM (ed. T.H. Nilsen, R.D. Shew, G.S. Steffens, and J. R.J. Studlick), 19. https://doi.org/10.1306/12401031S5562471.

Qiao, S., Shi, X., Wang, G., Zhou, L., Hu, B., Hu, L., Yang, G., Liu, Y., Yao, Z. and Liu, S. (2017). Sediment accumulation and budget in the Bohai Sea, Yellow Sea and East China Sea. *Marine Geology* 390: 270–281.

Quaresma, L.S., Vitorino, J., Oliveira, A., and da Silva, J. (2007). Evidence of sediment resuspension by nonlinear internal waves on the western Portuguese mid-shelf. *Marine Geology* 246 (2-4): 123–143.

Reineck, H.E. (1967). Layered sediments of tidal flats, beaches, and shelf bottoms of the North Sea. In: *Estuaries*, vol. 83 (ed. G.H. Lauff), 191–206. American Association for the Advancement of Science.

Reineck, H.E. and Wunderlich, F. (1968). Classification and origin of flaser and lenticular bedding. *Sedimentology* 11: 99–104.

Rine, J.M. and Ginsburg, R.N., 1985. Depositional facies of a mud shoreface in Suriname, South America – a mud analogue to sandy, shallow-marine deposits. *Journal of Sedimentary Petrology*, 55, pp. 633–652.

Roca, X., Rylaarsdam, J.R., Zhang, H. et al. (2008). An allostratigraphic correlation of Lower Colorado Group (Albian) and equivalent strata in Alberta and British Columbia, and Cenomanian rocks of the Upper Colorado Group in southern Alberta. *Bulletin of Canadian Petroleum Geology* 56: 259–299.

Rodriguez, A.B., McKee, B.A., Miller, C.B. et al. (2020). Coastal sedimentation across North America doubled in the 20th century despite river dams. *Nature Communications* 11 (1): 1–9.

Rotondo, K.A. and Bentley, S.J. (2003). Deposition and resuspension of fluid muds on the western Louisiana inner continental shelf. *Transactions of the Gulf Coast Association of Geological Societies* 53: 721–731.

Sageman, B.B., Rich, J., Arthur, M.A. et al. (1997). Evidence for Milankovitch periodicities in Cenomanian-Turonian lithologic and geochemical cycles, Western Interior, U.S.A. *Journal of Sedimentary Research* 67: 286–302.

Sageman, B.B., Murphy, A.E., Werne, J.P. et al. (2003). A tale of shales: the relative roles of production, decomposition, and dilution in the accumulation of organic-rich strata, Middle–Upper Devonian, Appalachian Basin. *Chemical Geology* 195 (1-4): 229–273.

Schieber, J. (2016). Mud re-distribution in epicontinental basins – exploring likely processes. *Marine and Petroleum Geology* 71: 119–133.

Schieber, J. and Southard, J.B. (2009). Bedload transport of mud by floccule ripples – direct observation of ripple migration processes and their implications. *Geology* 37: 483–486.

Schieber, J., Southard, J.B., and Schimmelmann, A. (2010). Lenticular shale fabrics resulting from intermittent erosion of water-rich muds – interpreting the rock record in the light of recent flume experiments. *Journal of Sedimentary Research* 80: 119–128.

Scotchman, I.C. (1989). Diagenesis of the Kimmeridge Clay Formation, onshore UK. *Journal of the Geological Society* 146: 285–303.

Septama, E., Bentley, S.J., and Droxler, A.W. (2016). Conduits, timing and processes of sediment delivery across a high-relief continental margin: continental shelf to basin in Late Quaternary, Gulf of Papua. *Marine and Petroleum Geology* 72: 447–462.

Shan, X., Dalrymple, R.W., Shi, X. et al. (2023). Changjiang coastal mud-belt deposits in Taiwan Strait: controls on its distribution and facies. *Sedimentology* 70 (4): 1131–1163.

Shepard, F.P. (1932). Sediments of the continental shelves. *Bulletin of the Geological Society of America* 43 (4): 1017–1040.

Slingerland, R., Driscoll, N.W., Milliman, J.D. et al. (2008). Anatomy and growth of a Holocene clinothem in the Gulf of Papua. *Journal of Geophysical Research: Earth Surface* 113 (F1).

Snedden, J.W. and Nummedal, D. (1991). Origin and geometry of storm-deposited sand beds in modern sediments of the Texas continental shelf. In: *Shelf Sands and Sandstone Bodies. International Association of Sedimentologists, Special Publication 14* (ed. D.J.P. Swift, G.F. Oertel, R.W. Tillman, and J.A. Thorne), 283–308. Oxford: Blackwell Science.

Snedden, J.W. and Swift, D.J.P. (1991). Discussion of: Is there evidence for geostrophic currents preserved in the sedimentary record of inner to middle-shelf deposits? *Journal of Sedimentary Petrology* 61: 148–151.

Snedden, J.W., Nummedal, D., and Amos, A.F. (1988). Storm-and fair-weather combined flow on the central Texas continental shelf. *Journal of Sedimentary Research* 58 (4): 580–595.

Sommerfield, C.K. and Nittrouer, C.A. (1999). Modern accumulation rates and a sediment budget for the Eel shelf: a flood-dominated depositional environment. *Marine Geology* 154 (1-4): 227–241.

Soyinka, O.A. and Slatt, R.M. (2008). Identification and micro-stratigraphy of hyperpycnites and turbidites in Cretaceous Lewis Shale, Wyoming. *Sedimentology* 55: 1117–1133.

Sternberg, R.W., Cacchione, D.A., Paulso, B. et al. (1996). Observations of sediment transport on the Amazon subaqueous delta. *Continental Shelf Research* 16 (5-6): 697–715.

Swift, D.J.P., Han, G., and Vincent, C.E. (1986). Fluid processes and sea-floor response on a modern storm-dominated shelf: Middle Atlantic Shelf of North America. Part 1: the storm-current regime. In: *Shelf sands and sandstones.* Canadian Society of Petroleum Geologists, Memoir, vol. 11 (ed. R.J. Knight and J.R. Mclean), 99–119.

Syvitski, J.P. and Milliman, J.D. (2007). Geology, geography, and humans battle for dominance over the delivery of fluvial sediment to the coastal ocean. *Journal of Geology* 115 (1): 1–19.

Taylor, A.M. and Goldring, R. (1993). Description and analysis of bioturbation and ichnofabric. *Journal of the Geological Society* 150: 141–148.

Taylor, K.G. and Macquaker, J.H.S. (2000). Spatial and temporal distribution of authigenic minerals in continental shelf sediments: implications for sequence stratigraphic analysis. In: *Marine Authigenesis: From global to microbial*, vol. 66 (ed. C.R. Glenn, L. Prévôt-Lucas, and J. Lucas), 309–323. SEPM Special Publication.

Tewksbury, B.J., Hogan, J.P., Kattenhorn, S.A. et al. (2014). Polygonal faults in chalk: insights from extensive exposures of the Khoman Formation, Western Desert, Egypt. *Geology* 42 (6): 479–482.

Thiry, M. (2000). Palaeoclimatic interpretation of clay minerals in marine deposits: an outlook from the continental origin. *Earth-Science Reviews* 49 (1-4): 201–221.

Thompson, C.E.L. and Amos, C.L. (2002). The impact of mobile disarticulated shells of *Cerastoderma edulis* on the abrasion of a cohesive substrate. *Estuaries* 25 (2): 204–214.

Thompson, C.E. and Amos, C.L. (2004). Effect of sand movement on a cohesive substrate. *Journal of Hydraulic Engineering* 130 (11): 1123–1125.

Tian, Y., Fan, D., Zhang, X. et al. (2019). Event deposits of intense typhoons in the muddy wedge of the East China Sea over the past 150 years. *Marine Geology* 410: 109–121.

Van Andel, T.H. (1967). The Orinoco delta. *Journal of Sedimentary Petrology* 37 (2): 297–310.

Van Helmond, N.A.G.M., Sluijs, A., Papadomanolaki, N.M. et al. (2016). Equatorward migration of dinoflagellates during a cold spell within the Late Cretaceous supergreenhouse. *Biogeosciences* 13: 2859–2872.

Van Straaten, L.M.J.U. (1954). Composition and structure of marine sediments in the Netherlands. *Leidsche geologische mededeelingen* 19: 1–110.

Van Straaten, L. and Kuenen, P.H. (1958). Tidal action as a cause of clay accumulation. *Journal of Sedimentary Research* 28 (4): 406–413.

Vasconcelos, P.M. and Carmo, I.D.O. (2018). Calibrating denudation chronology through 40Ar/39Ar weathering geochronology. *Earth-Science Reviews* 179: 411–435.

Varban, B.L. and Plint, A.G. (2005). Allostratigraphy of the Kaskapau Formation (Cenomanian-Turonian) in subsurface and outcrop: NE British Columbia and NW Alberta, Western Canada Foreland Basin. *Bulletin of Canadian Petroleum Geology* 53: 357–389.

Varban, B.L. and Plint, A.G. (2008a). Palaeoenvironments, palaeogeography, and physiography of a large, shallow, muddy ramp: Late Cenomanian-Turonian Kaskapau Formation. *Western Canada Foreland Basin. Sedimentology* 55: 201–233.

Varban, B.L. and Plint, A.G. (2008b). Sequence stacking patterns in the Western Canada foredeep: Influence of tectonics, sediment loading and eustasy on deposition of the Upper Cretaceous Kaskapau and Cardium formations. *Sedimentology* 55: 395–421.

Verschuren, M. (2019). Outcrop evidence of polygonal faulting in Ypresian marine clays (Southern North Sea Basin) leads to a new synthesis. *Marine Geology* 413: 85–98.

Vohsen, S.A., Anderson, K.E., Gade, A.M. et al. (2020). Deep-sea corals provide new insight into the ecology, evolution, and the role of plastids in widespread apicomplexan symbionts of anthozoans. *Microbiome* 8 (1): 1–15.

Walaszczyk, I., Plint, A.G. and Landman, N.H., 2017. Inoceramid bivalves from the Coniacian and basal Santonian (Upper Cretaceous) of the Western Canada Foreland Basin. *Bulletin of the American Museum of Natural History*, 414, pp. 53–103.

Wallace, D.J., Anderson, J.B., and Fernández, R.A. (2010). Transgressive ravinement versus depth of closure: a geological perspective from the upper Texas coast. *Journal of Coastal Research* 26 (6): 1057–1067.

Wallace-Dudley, K.E. and Leckie, D.A. (1995). Sedimentology and source-rock potential of the lower Kaskapau Formation (Cenomanian-lowermost Turonian), north-western Alberta. *Geological Survey of Canada, Bulletin* 490: 60.

Walsh, J.P. and Nittrouer, C.A. (2004). Mangrove-bank sedimentation in a mesotidal environment with large sediment supply, Gulf of Papua. *Marine Geology* 208 (2-4): 225–248.

Walsh, J.P. and Nittrouer, C.A. (2009). Understanding fine-grained river-sediment dispersal on continental margins. *Marine Geology* 263 (1-4): 34–45.

Wang, H., Bi, N., Wang, Y. et al. (2010). Tide-modulated hyperpycnal flows off the Huanghe (Yellow River) mouth, China. *Earth Surface Processes and Landforms* 35 (11): 1315–1329.

Wang, Z., Yang, S., Wang, Q., Zhang, Z., Zhang, X., Lan, X., Li, R. and Huang, L. (2014). Late Quaternary stratigraphic evolution on the outer shelf of the East China Sea. *Continental Shelf Research* 90: 5–16.

Warrick, J.A. and Milliman, J.D. (2003). Hyperpycnal sediment discharge from semiarid southern California rivers: implications for coastal sediment budgets. *Geology* 31 (9): 781–784.

Weaver, C.E. and Pollard, L.D. (1989). *Clays, Muds, and Shales*. Developments in Sedimentology 44, 819 p. Elsevier, Amsterdam.

Wei, E.A., Driscoll, N.W., and Slingerland, R.L. (2019). Oceanographic currents, differential subsidence, and physiography control three-dimensional clinothem growth in the Gulf of Papua, Papua New Guinea. *Marine Geology* 407: 164–180.

Wei, G., Shohao, L., Jie, L. et al. (2018). The sedimentary facies and dynamic environment of the Diaokou lobe in the modern Huanghe River Delta of China. *Acta Oceanologica Sinica* 37 (11): 40–52.

Weight, R., Anderson, J.B., and Fernandez, R. (2011). Rapid mud accumulation on the central Texas shelf linked to climate change and sea-level rise. *Journal of Sedimentary Research* 81: 743–764. https://doi.org/10.2110/jsr.2011.57.

Wheatcroft, R.A. and Borgeld, J.C. (2000). Oceanic flood deposits on the northern California shelf: large-scale distribution and small-scale physical properties. *Continental Shelf Research* 20 (16): 2163–2190.

Wheatcroft, R.A. and Drake, D.E. (2003). Post-depositional alteration and preservation of sedimentary event layers on continental margins, I. The role of episodic sedimentation. *Marine Geology* 199 (1-2): 123–137.

Wheatcroft, R.A., Borgeld, J.C., Born, R.S. et al. (1996). The anatomy of an oceanic flood deposit. *Oceanography* 9 (3): 158–162.

Wheatcroft, R.A., Stevens, A.W., Hunt, L.M., and Milligan, T.G. (2006). The large-scale distribution and internal geometry of the fall 2000 Po River flood deposit: evidence from digital X-radiography. *Continental Shelf Research* 26 (4): 499–516.

Winn, R.D., Bishop, M.G., and Gardner, P.S. (1987). Shallow-water and sub storm-base deposition of Lewis Shale in Cretaceous Western Interior Seaway, south-central Wyoming. *American Association of Petroleum Geologists, Bulletin* 71: 859–881.

Wright, L.D. and Friedrichs, C.T. (2006). Gravity-driven sediment transport on continental shelves: a status report. *Continental Shelf Research* 26 (17-18): 2092–2107.

Wright, L.D. and Nittrouer, C.A. (1995). Dispersal of river sediments in coastal seas: six contrasting cases. *Estuaries* 18 (3): 494–508.

Wright, L.D., Wiseman, W.J. Jr., Yang, Z.S. et al. (1990). Processes of marine dispersal and deposition of suspended silts off the modern mouth of the Huanghe (Yellow River). *Continental Shelf Research* 10 (1): 1–40.

Wright, M.E. and Walker, R.G. (1981). Cardium Formation (U. Cretaceous) at Seebe, Alberta – storm-transported sandstones and conglomerates in shallow-marine depositional environments below fair-weather wave base. *Canadian Journal of Earth Sciences* 18: 795–809.

Xu, K., Li, A., Liu, J.P. et al. (2012). Provenance, structure, and formation of the mud wedge along inner continental shelf of the East China Sea: a synthesis of the Yangtze dispersal system. *Marine Geology* 291: 176–191.

Xue, C. (1993). Historical changes in the Yellow River delta, China. *Marine Geology* 113 (3-4): 321–330.

Young, R.N. and Southard, J.B. (1978). Erosion of fine-grained marine sediments: sea-floor and laboratory experiments. *Geological Society of America, Bulletin* 89 (5): 663–672.

Yuan, D., Zhu, J., Li, C., and Hu, D. (2008). Cross-shelf circulation in the Yellow and East China Seas indicated by MODIS satellite observations. *Journal of Marine Systems* 70 (1-2): 134–149.

Zavala, C. (2020). Hyperpycnal (over density) flows and deposits. *Journal of Palaeogeography* 9 (1): 1–21.

Zhou, L., Liu, J., Saito, Y. et al. (2014). Coastal erosion as a major sediment supplier to continental shelves: example from the abandoned Old Huanghe (Yellow River) delta. *Continental Shelf Research* 82: 43–59.

Zhou, L., Liu, J., Saito, Y. et al. (2016). Modern sediment characteristics and accumulation rates from the delta front to prodelta of the Yellow River (Huanghe). *Geo-Marine Letters* 36 (4): 247–258.

10

Marine Evaporite Systems

Stephen W. Lokier[1] and Marcello Natalicchio[2]

[1] School of Science, University of Derby, Derby, UK
[2] Department of Earth Sciences, University of Turin, Turin, Italy

Summary

In this chapter, we examine the environments, processes, facies, architectures, preservation and identification of marine evaporite sequences. Marine evaporites encompass a range of minerals that have precipitated from seawater in environments where the rate of evaporation equals or exceeds water replenishment. Dominant among these minerals are halite, gypsum and anhydrite. Evaporite systems develop at the coastline or in isolated marine basins, occurring in an array of depositional settings and in relation to a range of other sedimentary facies ranging from deep water pelagic, through shallow marine siliciclastics and carbonates to aeolian sediments. For many earth scientists, evaporite facies are something exotic and rarely encountered. The reasons for this are two-fold. First, although common throughout the stratigraphic record, evaporite sequences are under-represented at Earth's surface where exposure to undersaturated meteoric waters results in dissolution and removal. Second, access to the relatively few sites of present-day evaporite formation remains, for many, challenging. For these reasons, and in contrast with many other areas of sedimentology, early evaporite research primarily focused on experimental and theoretical approaches to elucidate the modes of evaporite formation, deposition, and preservation. Today, the most significant challenge to studying ancient evaporite sequences is the scarcity of present-day depositional settings that can be applied as analogues. This is particularly the case for the 'salt giants', thick basin-wide evaporite systems for which we have no contemporary equivalent. This chapter explores the controls on marine evaporite formation, considering the role of autogenic and allogenic processes on the development of evaporite facies geometries and stratigraphic architectures. To this end, we utilise the coastal evaporites of the southern Persian (Arabian) Gulf and the Late Miocene (Messinian) evaporites of the Mediterranean as case studies to provide representative examples of facies, facies associations and depositional architectures in a range of evaporite settings. Evaporite minerals have long been economically significant, not least in their role as the most effective seal to hydrocarbon plays. During the energy transition evaporites will play an important role in carbon sequestration, hydrogen storage and geothermal energy storage and extraction. Evaporite minerals are fundamental to the chemical industry as well as being critical to agriculture, food production and construction. Current research is also addressing evaporites on Mars, and other bodies in the solar system and assessing their potential to preserve evidence of life. Evaporite research has an exciting future. Long may they precipitate!

10.1 Introduction

'Evaporite' is a generic term encompassing a range of water-soluble minerals that have been concentrated and precipitated through the evaporation of water (Warren 2006). The dissolved salts precipitate under conditions of supersaturation because the rate of evaporation exceeds the rate of aqueous recharge – a situation best met where a body of water is isolated under arid climatic conditions. Evaporite deposits are broadly categorised as either marine or non-marine based on the relative role of seawater as the supplier of solutes. Non-marine evaporites (briefly discussed in Section 6.4.3) precipitate minerals from fluids that are dominantly continental in origin. Marine evaporites, the subject of this chapter, receive the vast majority of their solutes from seawater, though continentally

Reading's Sedimentary Environments: Processes, Facies and Stratigraphy, Fourth Edition. Edited by Bruce Levell.

Companion website: www.wiley.com/go/levell4e

derived groundwater and riverine discharge may also have some influence. The majority of marine evaporite deposits have formed either in arid coastal settings or within marine basins that have become, to a greater or lesser extent, isolated from the open ocean.

10.1.1 A Brief History of Evaporite Research

The saltiness of seawater has been the subject of discussion and debate since ancient times with Empedocles, Aristotle and Pliny all weighing in on the discourse. During the Renaissance, there was a shift from a philosophical to an experimental approach and, by the late 17th century, a range of analytical techniques was being deployed to precipitate, separate, identify and quantify the products of the evaporation of 'mineral waters'. It was the polymath Robert Boyle who first turned analytical attention to seawater (Boyle 1673). Although it was almost a hundred years before Antoine Lavoisier published the first analysis of the precipitates from seawater (Lavoisier 1772), followed a few years later by more quantitative analysis by Torbern Bergman (1784). Geochemical research made significant advances during the mid-19th century, with evaporation experiments by Giulio Usiglio (1849) and Jacobus van't Hoff (1912) laying much of the groundwork to establish the controls and boundary conditions for evaporite precipitation as well as the precipitation order and stability fields of many evaporite minerals in closed systems.

While experimental research continued apace, the investigation of ancient evaporite deposits was hampered by a lack of contemporary analogues with which to compare the array of evaporite deposits encountered in the sedimentary record. This situation was somewhat relieved during the late 1950s–early 1960s with the start of intensive research activity focusing on the Holocene to Recent evaporites of the extensive coastal salt flats at the southern shore of the Persian Gulf (e.g. Curtis et al. 1963; Shearman 1963; Evans et al. 1964; Kinsman 1964; Butler et al. 1965b). Then, during the 1970s, a significant leap forward in our understanding of ancient basinal evaporite systems was provided through the coring of the Mediterranean seafloor by the Glomar Challenger expedition revealing up to 2 km thick sequences of Late Miocene basinal evaporite deposits (Hsü et al. 1973b).

10.1.2 Approaches to the Study of Marine Evaporite Systems

Marine evaporite deposits are common throughout the stratigraphic record (Figure 10.1). However, the inherent tendency of evaporites to suffer rapid dissolution in the presence of rainfall or groundwater limits their preservation at outcrop to arid areas. Even in these settings, the exposed deposits will be subject to cycles of hydration and dehydration, are liable to be transient, and are unlikely to reflect the complete and original suite of minerals or structures (Lokier 2012; Court et al. 2017). Subsurface datasets, cores and seismic imaging have therefore been particularly important in developing our understanding of ancient evaporite systems. Cores provide evidence of mineralogies, textures and facies evolution while access to mines and other subsurface excavations has significantly contributed to our understanding of larger-scale fabrics, sedimentary structures and facies geometries. Remote data acquisition, for example, recent advances in seismic data acquisition and processing techniques, has been particularly useful in understanding large-scale facies geometries and the basin-scale context of evaporite deposits (e.g. Hudec et al. 2019; Haq et al. 2020). The main approaches and methodologies used in the study and elucidation of evaporite successions are summarised in Supplementary Table 10.S.1.

Perhaps the greatest challenge to the study of ancient marine evaporite deposits is the paucity of appropriate recent analogues (Warren 2021). Today, marine evaporite precipitation and deposition is limited to a small number of coastal settings. These fail to reflect the scale or diversity of environments represented within the geological record. In addition to the extensive coastal salt flats at the southern shore of the Persian Gulf, a second analogue was added in the 1980s in the form of Lake MacLeod in Western Australia (Handford et al. 1984; Logan 1987). These two Holocene to Recent systems are the largest areas of present-day coastal marine evaporite formation. However, when compared to the 'salt giants' of the Phanerozoic and Precambrian (Figure 10.1), these Recent systems are bordering on the insignificant with an aerial extent 3 orders of magnitude smaller and a stratigraphic thickness 1–2 orders of magnitude thinner than many of the evaporite systems in the stratigraphic record (Warren 2010).

10.2 Environments of Marine Evaporite Formation

Setting aside the ancient salt giants which include multiple depositional settings, marine evaporites may be subdivided into those that have formed along arid coastlines and those that have precipitated in restricted marine basins. Both these broad formational settings are composite: containing a diverse and complex array of interacting sub-environments controlled by a range of underlying climatic, geographical, bathymetric and oceanographic variables. In some instances, both coastal and basinal systems occur in tandem. However, regardless of the complexity of the system, in all cases the major source of solutes will be seawater.

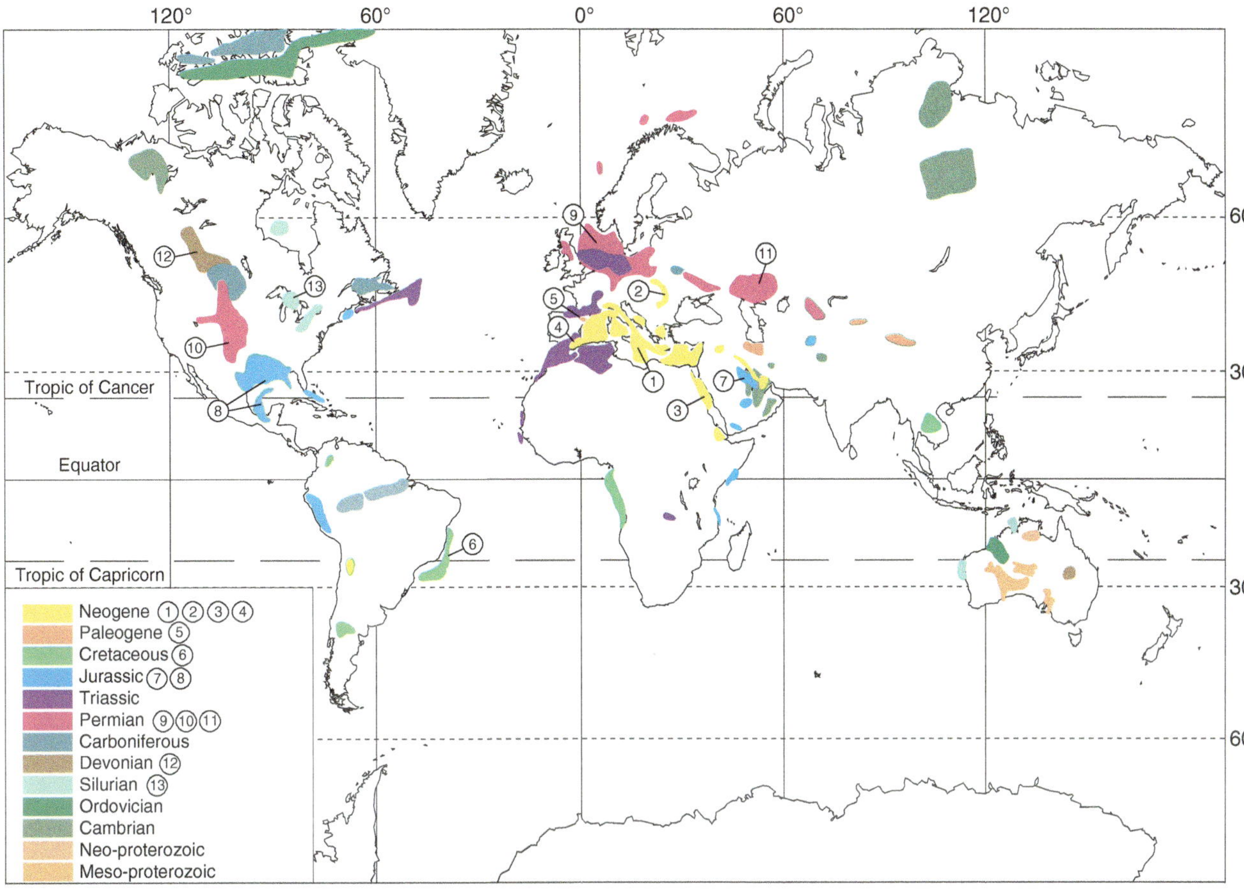

Figure 10.1 Major evaporite basins within the geological record. Data compiled from Jackson and Talbot (1991), Taberner et al. (2000), Tucker (2001) and Warren (2010). 1: Late Miocene Mediterranean Basin; 2: Middle Miocene Carpathian Foredeep Basin; 3: Neogene Red Sea Basin; 4: Late Miocene Granada Basin; 5: Eocene Catalan Basin; 6: Cretaceous Sergipe and Santos Basins; 7: Late Jurassic Arab Formation; 8: Jurassic Gulf of Mexico Basin ; 9: Permian Zechstein Basin; 10: Late Permian Delaware Basin; 11: Permian Pricaspian Basin; 12: Devonian Williston Basin; 13: Silurian Michigan Basin.

10.2.1 Drivers of Evaporite Precipitation

The precipitation of marine evaporite minerals occurs where dry undersaturated air allows the evaporation of significant volumes of seawater, producing concentrated brines under conditions of restricted dilution by fresh or marine water recharge. Achieving such significant evaporation (47% for the onset of calcium carbonate precipitation; Briggs (1958)) requires an arid climate and a degree of isolation or restriction from the global ocean or other sources of recharge. Evaporation rates will be highest in relatively shallow water bodies with large surface area-to-volume ratios.

10.2.1.1 Climatic Controls

Arid climates are the product of a combination of latitudinal (first order) and local-to-regional (second order) processes (Matthews and Perlmutter 1994). Today, the latitudinal distribution of Earth's arid zones is controlled by lower-atmosphere Hadley circulation (Hadley 1735). Dry air descending at the interface of the Hadley and Ferrel cells produces the arid sub-tropical high-pressure belts at latitudes of approximately 30° north and south of the equator. The circulation cells, and associated high-pressure zones, shift latitudinally with seasonal changes in global temperature – shifting north in the northern hemisphere summer and south in the southern summer. Arid conditions also result from more local to regional scale (second order) processes. Where air masses pass over topographic highs, the increase in altitude cools the air, increasing the relative humidity and resulting in the dumping of moisture on the windward side of the mountain range. On reaching the lee of the mountains, the descending air will warm with a resultant decrease in relative humidity and the potential for local arid conditions, e.g. the saline lakes in the rain shadow of the Andean cordillera (Alvarez et al.

2022). Similarly, cold upwelling ocean currents may cause a localised temperature inversion at the adjacent shoreline. Cold air with low relative humidity is held close to the ground, again resulting in the onset of arid conditions – e.g. the coastal evaporite systems of Namibia (Eckardt and Spiro 1999). In areas of extreme aridity, the rate of evaporation can be significant. For example, evaporation rates at the southern shore of the Persian Gulf and at Lake MacLeod average 2.75 and 2.6 m a^{-1} respectively (Warren 1991; Bottomley 1996).

The process of evaporation is accompanied by important feedback. With an increase in the salinity of the brine there is a decrease in the equilibrium water vapour pressure, this reduces the relative humidity that is required for evaporation to continue (Figure 10.2). In order to maintain evaporation, the air must become increasingly undersaturated as the salinity of the brine increases. However, where the overlying air is insufficiently undersaturated, the brine will absorb moisture directly from the atmosphere, thereby reducing salinity until a state of equilibrium is briefly achieved (Figure 10.3) (Kinsman 1976). In coastal settings, this process will inhibit the accumulation of deliquescent minerals, including halite, preferentially favouring the preservation of gypsum and, potentially, anhydrite deposits. The very low relative humidities (<35%) required to accumulate halite and the potassium salts (Schreiber and Tabakh 2000) are only likely where evaporation takes place in an isolated basin that is dominated by a continental climate.

10.2.1.2 Geographic Controls

In addition to climate, marine evaporite precipitation and preservation is also strongly controlled by the degree of connectivity and water interchange between the area of evaporite formation and the global ocean. If a thick sequence of evaporite strata is to be precipitated, then the basin or shoreline must be largely isolated from the global ocean but still retain a sufficient degree of connectivity to allow an overall net inflow of solute-bearing seawater to the basin. Isolation or restriction of the basin may be structural

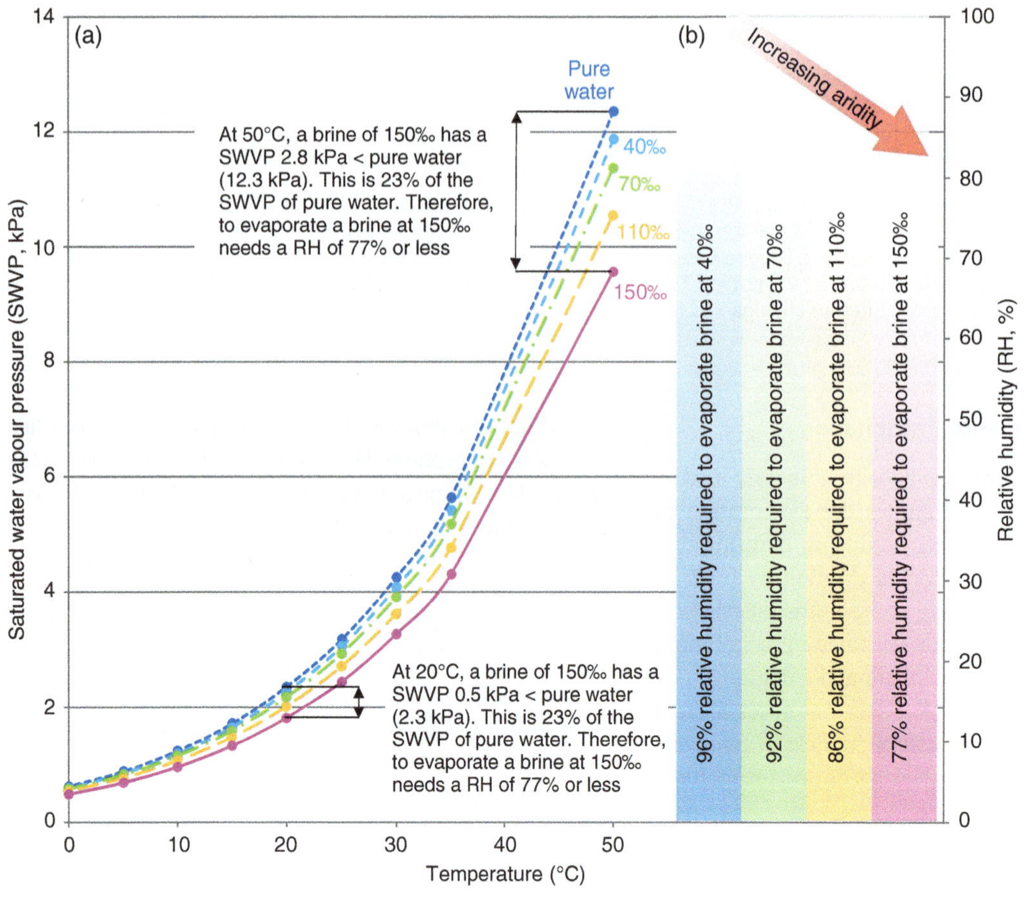

Figure 10.2 (a) Saturated water vapour pressure (SWVP) curves for pure seawater and four brines of increasing salinity. (b) The relative humidity (RH) required to evaporate brines of different concentrations is (i) lower than for evaporating pure water and (ii) is independent of temperature as explained for cases at 20 and 50°C in (a). Modified from Kinsman (1976), data from Arons and Kientzler (1954) and Lide (2004).

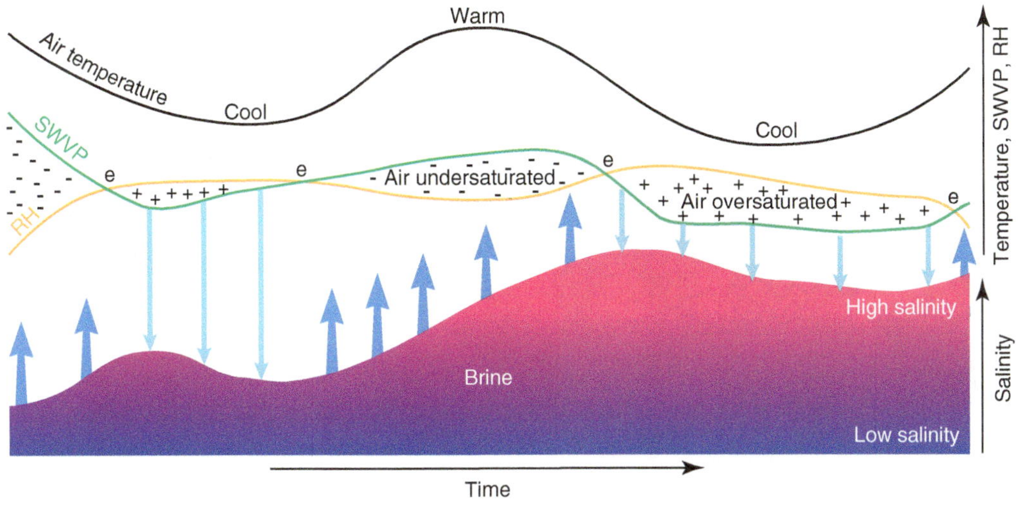

Figure 10.3 Cartoon showing the complex interactions between relative humidity (RH) and the saturated water vapour pressure (SWVP) of an evolving brine. RH rises as temperature falls. SWVP is a function of both temperature and brine salinity (Figure 10.2). Dark blue arrows indicate periods of evaporation of water from the brine to the atmosphere, pale blue arrows indicate periods when the brine absorbs moisture from the atmosphere. e = moments of equilibrium.

(e.g. due to tectonics or salt diapirism), geomorphological (e.g. a large embayment), a result of sea-level fluctuation or related to the development of a new landform such as a spit, bar or barrier island.

The connection between the evaporite basin and the open marine system may be through single or multiple surface channels or via the subsurface flow of seawater and brines through a permeable barrier (i.e. seepage; Figure 10.4). In the case of a channel connection, the exchange of waters is controlled by the morphology of the channel, particularly its cross-sectional area at the narrowest point, channel depth and channel length. For basinal waters to reach a state of saturation sufficient to initiate evaporite precipitation, the combined cross-sectional area of any channels must be at least orders of magnitude smaller than the evaporite basin's surface area (Lucia 1972). The rate of flow through permeable barriers is regulated by the permeability of the barrier, the cross-sectional area of the permeable zone, the width of the barrier through which the water must pass and the hydraulic head. For both channels and permeable barriers, the relationship between inflow and outflow will also be influenced by the regional tidal regime, storm surges and the morphologies of the evaporite and adjacent open marine basins.

Despite the infinite variety of evaporite basin morphologies and their degree and modes of isolation, there are two broad sub-divisions. (i) Basins that receive a net influx of seawater with no significant outflow of brines to the world's oceans. (ii) Evaporite basins that do see a significant outflow of brines to the ocean. In the case of these latter basins, the difference between the rates of seawater inflow and brine outflow, as a proportion of basin volume, will determine the residence time during which basinal waters are subject to evaporation. The surface area, volume and depth of the basin and the relative timings and duration of inflow and outflow all have a significant control on the degree of evaporation, the formation and retention of brines, and the precipitation and preservation of evaporite minerals. In addition to receiving water and solutes from the global oceans, in many cases evaporite basins will also have an influx of water from rivers and groundwater springs (Figure 10.4a,b). Where this influx is significant it is likely that there will be an influence on the chemistry of the precipitates. Aquicludes are important elements governing the evaporite mineral assemblage, particularly in inflow-dominated basins. These impermeable barriers (Figure 10.4c) prevent the seepage of brines from a basin and, conversely, limit the dilution of brine that would occur in the event of excessive influxes of fresh groundwater (Logan 1987).

As with any geological system, the geographical parameters of an evaporite basin are in constant flux. Relative sea levels will rise and fall – potentially isolating and reconnecting the basin and modifying the hydraulic head; channel morphology and barrier permeability will evolve – controlling the inflow of seawater and outflow of brine; tectonic and isostatic processes will influence the size and volume of the basin – controlling the rate of evaporation and residence time.

10.2.2 Present-Day Marine Evaporite Systems

Today, sites of evaporite mineral precipitation and accumulation are limited to localised instances of limited

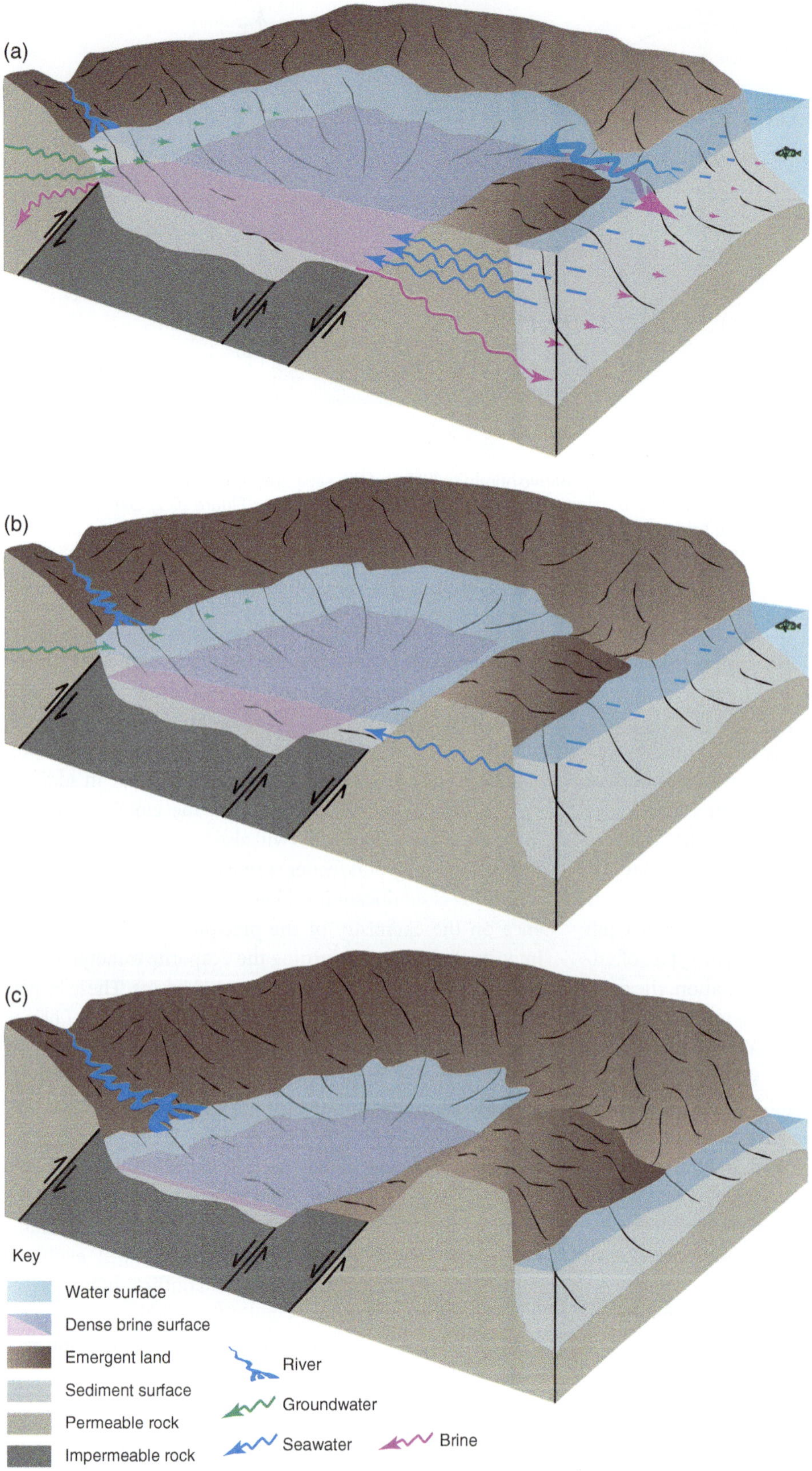

Figure 10.4 Cartoon showing the different modes of connection between a basin, the ocean and freshwater supply. (a) During a period of high sea level, the basin is connected to the ocean through both a channel and a permeable barrier. Freshwater is also supplied via a river and groundwater. Conversely, dense brines are able to flow out of the restricted basin via the same routes. (b) Following a small fall in sea level, the channel has now been exposed and connection between the basin and the ocean has been restricted to inflow through the permeable barrier. (c) With a further fall in sea level, there is no hydraulic head to drive water through the permeable barrier and the basin is now isolated from the marine system with water only being supplied from rivers and brine outflow no longer possible.

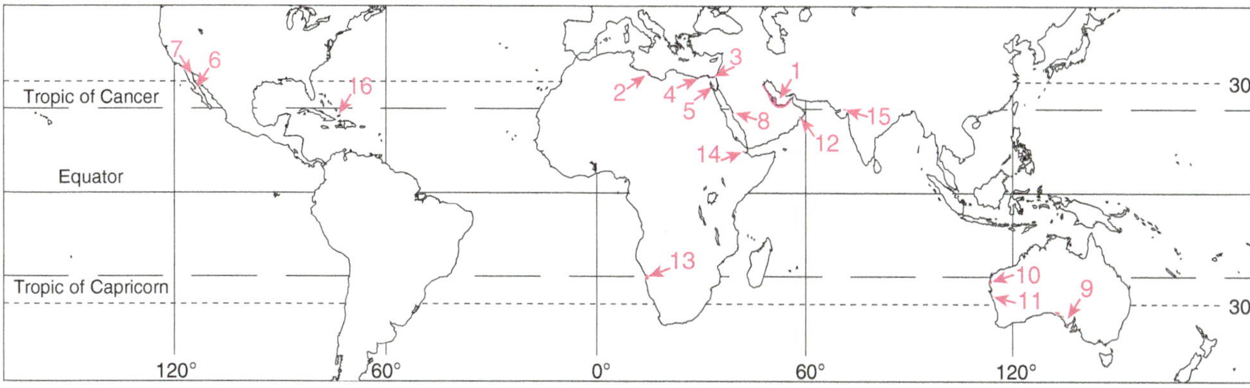

Figure 10.5 Sites of present-day marine evaporite formation. See Table 10.1 for brief descriptions of localities and key references. Note that the vast majority of marine evaporite formation is within the subtropics (15–30° N and S).

geographical extent. All of the present-day sites of marine evaporite precipitation are located at coastlines, dominantly within the subtropical zones (15–30 °N and S) (Figures 10.5, 10.6; Table 10.1), – there are no modern equivalents to the salt giants of the stratigraphic record (Figure 10.1). The familiar geological axiom that the 'present is the key to the past' cannot be applied directly in the case of salt giants. We may, instead, adopt the less eloquent claim that 'deciphering the Neogene guides us in deciphering the more distant past'. In many cases, as with aeolian sediments (Section 7.6), it is the relationships between evaporite bodies and the adjacent facies that are key to the interpretation of the nature of the evaporite system.

10.2.3 Temporal Distribution of Marine Evaporite Systems

Considering that marine evaporites form only a tiny proportion of present-day sedimentary systems, we may ask why the stratigraphic record of evaporite deposits (Hay et al. 2006) is so rich, particularly given a half-mass preservation time of evaporites of only ~100 Ma (Garrels and Mackenzie 1971). As discussed, the precipitation and preservation of evaporite minerals require the simultaneous presence of both an arid climate and an isolated or severely restricted basin. While atmospheric circulation, in the form of Hadley cells, still ensures the presence of subtropical highs offering an arid climate, today we lack the wide shelf basins and epeiric seas that were crucial to the formation of thick, laterally extensive marine evaporite successions (e.g. Silurian, Michigan Basin; Devonian, Williston Basin; Permian, Pricaspian Basin (Baillie 1955; Volozh et al. 2003; Harrison III et al. 2018)). This was not the case for large intervals of the geological past when tectonics and eustatic sea level interacted to produce suitable basins within arid climatic regions (Figure 10.6). In these basins, thick marine evaporite sequences were commonly formed, e.g. the Neogene evaporites of the Red Sea. Evaporite deposition was particularly enhanced with the formation of new oceanic basins during the break-up of Pangea (Lowell and Genik 1972; Rona 1982). Although insignificant compared to the stratigraphic record, the latitudinal distribution of marine evaporites within the subtropical zones has been stable since at least the Permian.

10.2.4 Coastal Marine Evaporites

At arid shorelines, evaporite minerals precipitate in a range of subaerial and subaqueous environments. These coastal evaporite systems can be morphologically complex, with different evaporite minerals precipitating simultaneously in adjacent sub-environments to form a complex mosaic of adjacent evaporite and non-evaporite facies.

The Bedouin Arabic term 'sabkha' traditionally describes a topographically flat area where a salt crust overlies a clay, silt or sand substrate (Figure 10.7) (Goodall and Al-Belushi 1998). In the context of coastal evaporites, the term 'sabkha system' is typically more broadly employed to describe both the low-lying coastal salt-flats of the upper intertidal to supratidal zone (the sabkha proper) as well as the rest of the intertidal zone and, on occasion, any adjacent lagoons.

Salinas are coastal ponds that are largely isolated from the marine system. Marine fluids are supplied to the salina either during tidal or storm surge flooding or via the inflow of marine groundwater through a permeable barrier. Salinas may lie within, or adjacent to, a sabkha system or may develop in isolation from tidal flats.

10.2.5 Basin-Scale Marine Evaporites

The stratigraphic record abounds with evaporite deposits that are hundreds of metres to kilometres thick and extend over entire sedimentary basins (Figure 10.1). These basin-scale evaporite deposits, 'salt giants', record the evaporite infilling of tectonically and hydrostatically isolated marine basins.

Table 10.1 Selected examples of present-day coastal evaporite formational settings.

Location	Brief description	Mode of connection	Minerals	Representative reference	Index to Figure 10.5
Arabian Shoreline, Persian Gulf	Supratidal salt flats locally associated with lagoons and barrier islands	Storm surge flooding. Increasing continental brine influence landward.	Gypsum, anhydrite, halite,	Evans (1969)	1
Mediterranean Coast, Libya	Sabkhas in isolated basins that were formerly lagoons that became totally isolated via sand barrier development.	Brines mostly of continental origin. Rare marine recharge of the seaward sabkha by storm flooding.	Gypsum, halite,	Rouse and Sherif (1980)	2
Mediterranean Coast, Egypt	Sabkhas developed shoreward of lagoons isolated by a sand barrier	Water supplied to lagoon through channels in sand barrier. Occasional flooding of sabkha by lagoon water.	Gypsum, halite,	Levy (1977)	3
Mediterranean Coast, Egypt	Shoreline parallel depressions hosting sabkha behind oolitic sand coastal ridges. Hosted in mixed carbonate siliciclastic.	Seepage through beach ridges	Gypsum, Halite (as ephemeral crust)	West et al. (1979)	4
Sinai, Egypt	Salinas at steep coastline.	Seepage through beach ridge and bedrock barriers	Gypsum, anhydrite, halite,	Gavish (1974, 1980)	5
Laguna Ojo de Liebre, Baja California, Mexico	Sabkhas in tidal flats and shallow basins	High tide flooding from lagoon	Gypsum, halite	Phleger (1969)	6
Ometepec, Baja California, Mexico	Sabkha and salina in a siliciclastic setting at the termination of an elongate gulf.	Storm surge flooding	Gypsum, anhydrite, halite	Lowenstein and Hardie (1985)	7
Red Sea Coast, Saudi Arabia	Innermost lagoon tidal flat sabkha	Network of channels	Gypsum, halite	Bahafzullah (1993)	8
South Australia	Coastal salinas developed in a Quaternary beach dune system	Marine derived groundwater passing through calcareous beach dunes.	Gypsum, halite	von der Borch et al. (1977); Warren (1982)	9
Lake MacLeod, Western Australia	Graben-hosted lake isolated from the Indian Ocean by calcareous dunes	Seawater flowing through cracks and caverns in calcareous barrier dunes	Gypsum, halite	Logan (1987)	10
Hutt Lagoon and Leeman Lagoon, Western Australia	Basins in Holocene dunes	Seawater flow through Pleistocene dunes	Gypsum, halite	Arakel (1980)	11
Barr Al Hikman, Oman coastline	Sabkhas developed on tidal flats adjacent to lagoons	Channels into lagoons. Marine flooding of sabkha during storms and high tides	Gypsum, halite	Mettraux (2010)	12
Namibia coast	Sabkhas developed on mud flats	Storm surge and high tide flooding	Gypsum, halite	Eckardt (1999)	13
Lake Asal, Djibouti	Rift basin lake lying 155 m below sea level. Separated from the Indian Ocean by 10 km wide volcanic ridge	Seawater passes through fractures in volcanic barrier.	Gypsum, halite	Gasse (1989)	14
Ranns of Kutch, Western India	Supratidal flat sabkha hosted in a shallow inlet.	Seasonal marine flooding by storm surges up inlet	Gypsum, halite	Glennie and Evans (1976)	15
West Caicos, Turks and Caicos	Salina between landward Pleistocene and seaward Holocene oolitic dunes.	Seawater flow through karstified underlying Pleistocene carbonates.	Gypsum, halite	Perkins et al. (1984)	16

Arakel (1980), Bahafzullah et al. (1993), Eckardt and Spiro (1999), Evans et al. (1969), Gasse and Fontes (1989), Gavish (1974), Glennie and Evans (1976), Levy (1977), Logan (1987), Phleger (1969), Rouse and Sherif (1980), Warren (1982), and West et al. (1979).

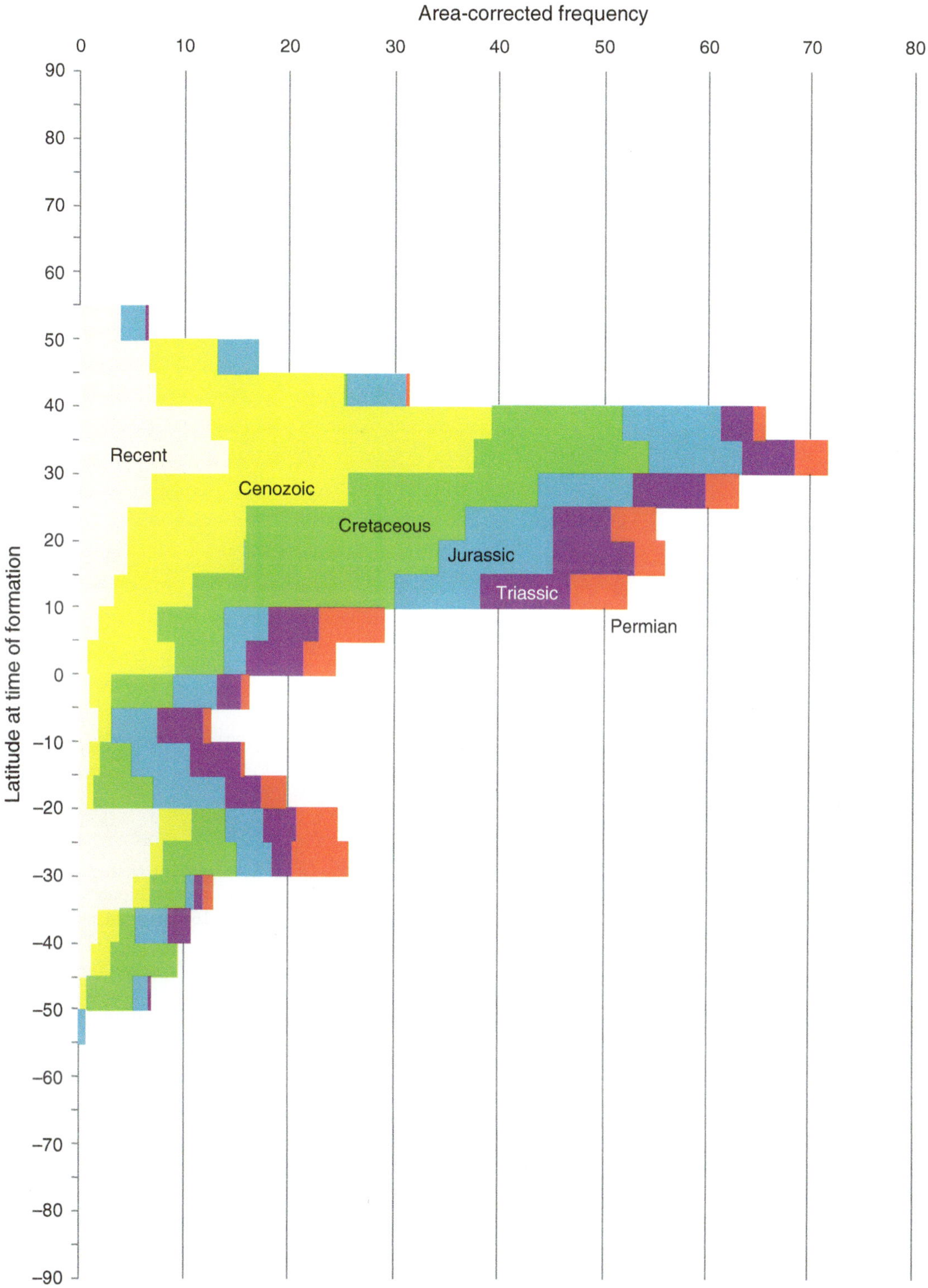

Figure 10.6 The aerial frequency of Permian to recent evaporites with palaeo-latitude (data from Ziegler et al. (2003)). Data counts are restricted to one site within each 5° latitude and longitude 'quadrangle' and were normalised to compensate for change in quadrangle area with latitude.

Figure 10.7 Photograph showing the flat, featureless nature of the supratidal zone of the Abu Dhabi Sabkha in the United Arab Emirates. The polygonal features with raised margins are halite polygons that have formed over a two-week period through lateral displacement during surface halite precipitation (see Lokier (2012) for a discussion of this process; person is 160 cm high). Photograph credit: Stephen Lokier.

Many salt giants are rimmed by carbonate shelves, sometimes with associated sabkhas, surrounding a central basinal evaporite sequence of repeated cycles of evaporite and non-evaporitic minerals recording recurrent episodes of sea level drop and flooding.

10.3 Processes of Marine Evaporite Formation

Evaporite minerals can nucleate at the water surface, within the water column, at the sediment surface or within the sediments themselves. In supratidal settings, precipitation occurs both in the sediment column as well as at the sediment surface. Before we consider the precipitation of evaporites in specific settings, we shall first discuss some generalities that may be applied to the precipitation of evaporites in a range of marine-related environments.

10.3.1 Evaporite Precipitation from Present-Day Seawater

The partial or complete evaporation of seawater can produce a diverse range of evaporite minerals (Table 10.2). The final mineral suite in any specific case will depend on the complex interplay of a range of variables including the original seawater chemistry, the supply of solutes, the evolution of the brine and the degree and rate of evaporation. In all cases, volumetrically the most abundant primary precipitates will be halite, calcite and aragonite, gypsum/anhydrite, and, in very low amounts, magnesium and potassium salts (e.g. kainite and carnallite).

Table 10.2 Marine evaporite minerals.

Mineral	Class	Formulae	Mohs scale hardness
Aragonite	Carbonate	$CaCO_3$	3.5–4
Calcite	Carbonate	$CaCO_3$	3
Dolomite	Carbonate	$CaMg(CO_3)_2$	3.5–4
Carnallite	Chloride	$KMgCl_3{\cdot}6(H_2O)$	2.5
Bischofite	Chloride	$MgCl_2{\cdot}6(H_2O)$	1.5-2
Halite	Chloride	NaCl	2-2.5
Kainite	Chloride	$MgSO_4{\cdot}KCl{\cdot}3(H_2O)$	2.5-3
Sylvite	Chloride	KCl	2.5
Tachyhydrite	Chloride	$CaMg_2Cl_6{\cdot}12(H_2O)$	2
Anhydrite	Sulphate	$CaSO_4$	3.5
Bassanite	Sulphate	$2CaSO_4{\cdot}(H_2O)$	2.5
Blödite	Sulphate	$Na_2Mg(SO_4)_2{\cdot}4(H_2O)$	2.5-3
Epsomite	Sulphate	$MgSO_4{\cdot}7(H_2O)$	2
Glauberite	Sulphate	$Na_2Ca(SO_4)_2$	2.5-3
Gypsum	Sulphate	$CaSO_4{\cdot}2(H_2O)$	2
Hexahydrite	Sulphate	$MgSO_4{\cdot}6(H_2O)$	2-2.5
Kieserite	Sulphate	$MgSO_4{\cdot}H_2O$	3.5
Langbeinite	Sulphate	$K_2Mg_2(SO_4)_3$	3.5-4
Leonite	Sulphate	$K_2Mg(SO_4)_2{\cdot}4(H_2O)$	2.5-3
Löweite	Sulphate	$Na_{12}Mg_7(SO_4)_{13}{\cdot}15(H_2O)$	2.5-3
Pentahydrite	Sulphate	$MgSO_4{\cdot}5(H_2O)$	2.5
Polyhalite	Sulphate	$K_2Ca_2Mg(SO_4)_4{\cdot}2(H_2O)$	3.5
Starkeyite	Sulphate	$MgSO_4{\cdot}4(H_2O)$	2-3
Syngenite	Sulphate	$K_2Ca(SO_4)_2{\cdot}H_2O$	2.5

Sodium and chloride ions are, by far, the dominant ions in present-day seawater, contributing more than 85% of the solute mass (Riley 1975). These are followed by sulphate, magnesium, calcium, potassium and bicarbonate that, together, contribute a further 14.1% (Table 10.3). The contributions of these seven ions, accounting for more than 99.7% of the ionic mass of present-day seawater, have remained relatively stable throughout the Quaternary with only small variations in the relative proportions of the minor components – Mg^+, Ca^{2-}, $SO^{2-}{}_4$ and K^+ (Lowenstein et al. 2003).

With the progressive evaporation of seawater in a closed basin system, the order of precipitation and mineralogy of the evaporite minerals is controlled by the ionic composition of the parent solute (Hardie and Eugster 1970). Evaporite minerals are precipitated in the order of increasing solubility, a sequence that is well established for modern seawater where chloride is the dominant anion.

Table 10.3 The major anions and cations in recent seawater with a salinity of 35‰ (Riley 1975).

Composition	Concentration in seawater at 35‰ (g kg^{-1})	Weight (%)
Cl^-	19.344	55.02
Na^+	10.773	30.64
$SO^{2-}{}_4$	2.712	7.71
Mg^{2+}	1.294	3.68
Ca^{2+}	0.412	1.17
K^+	0.399	1.13
HCO_3^-	0.142	0.4
Br^-	0.067	0.19
S^{r2+}	0.008	0.02
B	0.004	0.01
F^-	0.001	0.003

In a closed system, at the present day, a seawater volume reduction of over 46% (Briggs 1958) is required to precipitate the first of the evaporite minerals, calcium carbonate, in the form of the mineral aragonite ($CaCO_3$), at a salinity of 68‰ (Figure 10.8). During this process, the partitioning of (bi)carbonate into the aragonite leaves excess calcium in the brine (Figure 10.9). If evaporation should continue to concentrate the brine to a salinity of 157‰ (a volume loss of 81%), calcium sulphate will be the next mineral to precipitate. Typically, precipitation will be in the form of the hydrous phase, gypsum ($CaSO_4{\cdot}2H_2O$) though direct precipitation of the anhydrous form, anhydrite, ($CaSO_4$) is also possible. On reaching a salinity of 291‰ (90% volume loss) halite (NaCl), volumetrically the most significant mineral, will precipitate. The precipitation of halite depletes the brine in sodium ions, leaving a relative enrichment in magnesium, chloride and sulphate. Under continued evaporation, the excess sulphate combines with the concentrated magnesium to precipitate epsomite ($MgSO_4{\cdot}7H_2O$), the first of the magnesium sulphate salts, at a salinity of 350‰ and kieserite ($MgSO_4{\cdot}H_2O$; Figures 10.8 and 10.9). Continued evaporation beyond this point will see a theoretical increase of magnesium (Weldeghebriel et al. 2022) until the precipitation of potassium-magnesium chloride bitterns, for example, carnallite ($KCl{\cdot}MgCl_2{\cdot}6(H_2O)$).

It should be noted that, following the precipitation of halite, interactions between the brine, earlier metastable precipitates, rocks, sediments and other waters, including meteoric and continental waters, will introduce a significant level of complexity to the evolution of the evaporite sequence and the terminal mineral assemblage.

The evaporation of large volumes of seawater is required to precipitate volumetrically significant evaporite bodies. The complete evaporation of a 10 km deep column of seawater will precipitate just 143.46 m of evaporite minerals of which 132.47 m (92.3%) will be halite along with 6.81 m (4.7%) of potassium-magnesium chloride, 3.79 m (2.6%) of anhydrite and just 39 cm (0.3%) of calcite (Figure 10.10a) (Garrett 1996). However, complete evaporation of the water body is not assured – a basin can reach hydrologic equilibrium at any point – consequently, we may not observe the full spectrum of evaporite minerals. Equilibrium occurs where the amount of water removed by evaporation is balanced by the inflow of solute. If equilibrium is attained

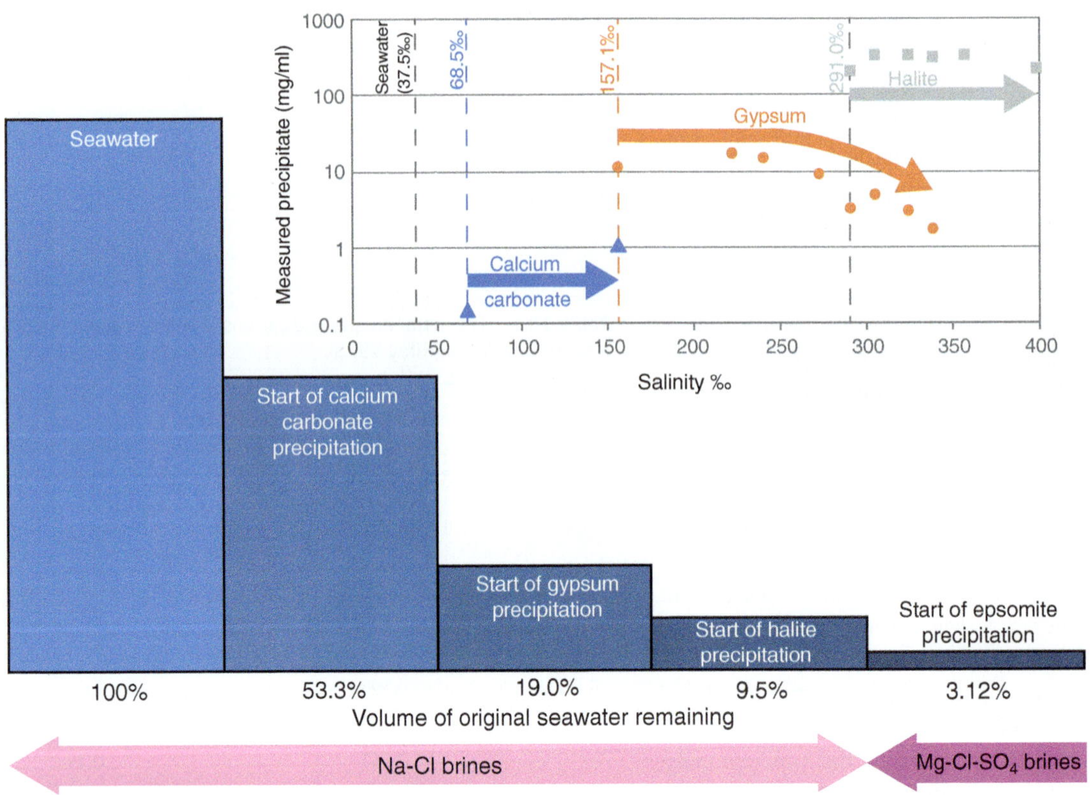

Figure 10.8 The order of evaporite minerals precipitation from seawater in a closed system together with the salinity at the commencement of each precipitation phase. Note the comparatively small volume of calcium carbonate precipitated when compared to halite (data compiled from Briggs (1958)). Note that there are overlaps in the precipitation of mineral phases with the quantity of measured precipitates initially rising with increasing salinity and then declining as the supply of ions becomes exhausted.

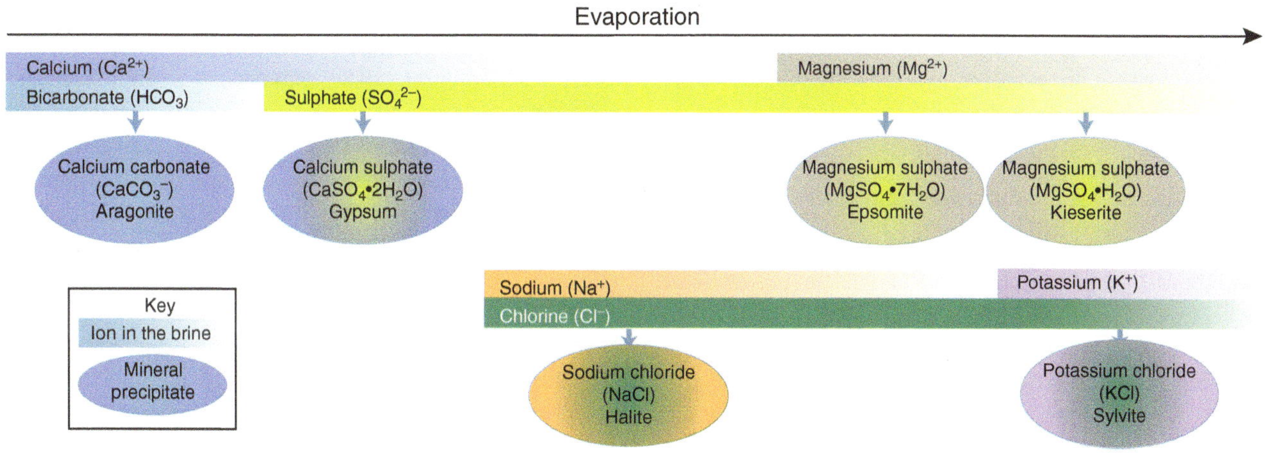

Figure 10.9 The partitioning of major ions into mineral phases during the evaporation of Quaternary seawater. Note that formulae are inclusive of the water of hydration.

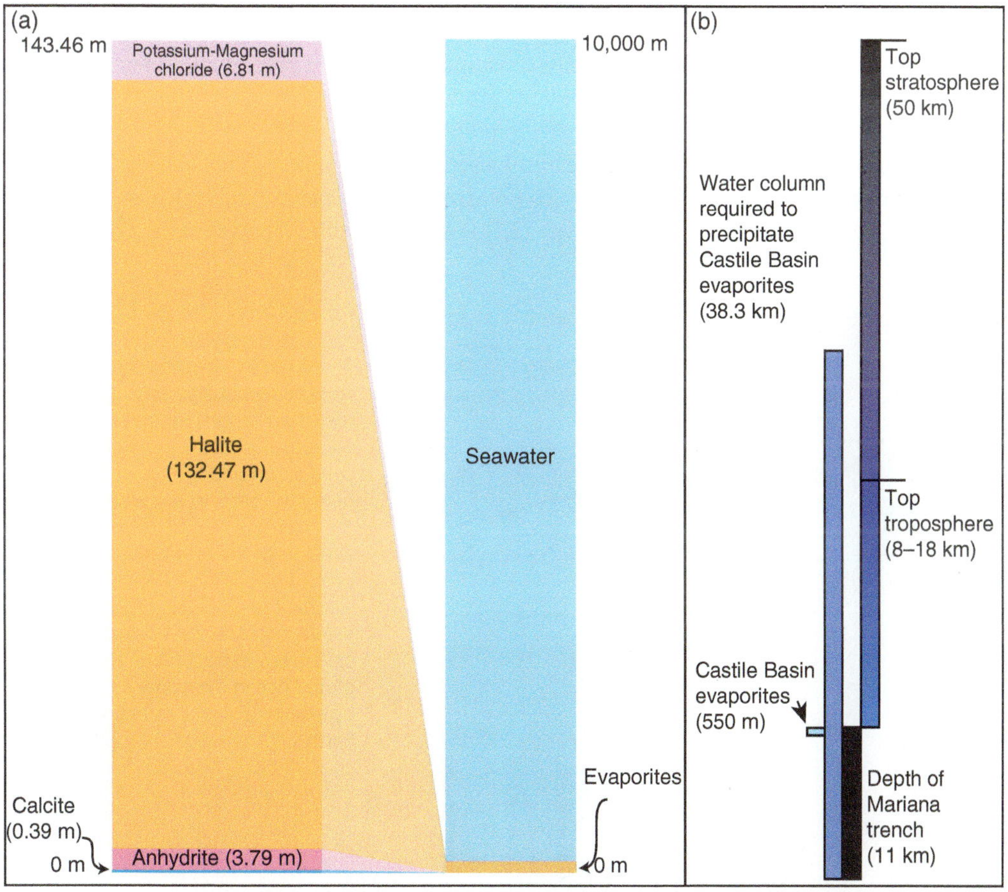

Figure 10.10 (a) The total evaporation of a 10,000 m column of seawater will precipitate 143.46 m of evaporites (data from Garrett (1996)). (b) The 'stratospherically high' water column that would be required to precipitate the Castile Formation in a closed evaporite basin.

within the salinity field of calcium sulphate precipitation, then the evaporite sequence will be dominated by gypsum and/or anhydrite, for example, as observed in the Late Permian Castile Formation (Kirkland et al. 2000).

10.3.2 Evaporite Precipitation from Ancient Seawaters

Our knowledge of the chemistry of ancient seawaters is largely founded on the study of small inclusions of fluids that became trapped within crystals during the rapid growth of evaporite minerals. Primary fluid inclusions, particularly those hosted in halite, offer an insight into both the ionic composition and temperature of the brines from which the evaporites precipitated (Roberts and Spencer 1995).

Fluid inclusion analysis reveals that the relative proportions of the ionic components of seawater have fluctuated over geological time – a theory first mooted by Leonardo da Vinci (1478–1518). These changes in ionic concentrations have occurred in phase with variations in the rate of seafloor spreading, volcanism and global sea level change, as well as the effect of salt sequestration in evaporite deposits themselves (Hallam et al. 1993; Lowenstein et al. 2001). Although fluctuating in their concentrations, sodium and chloride ions remained dominant throughout the Phanerozoic – thus ensuring halite as the major evaporite mineral. This was particularly the case prior to the Mesoproterozoic, and maybe even as late as the latest Ediacaran, when low sulphate concentrations inhibited gypsum precipitation, further enhancing the dominance of halite (Grotzinger and Kasting 1993; Blättler et al. 2020). Studies of ancient seawater salinities continue – while there is general consensus that the salinity of Palaeozoic seawater was similar to that of the modern oceans (Johnson and Goldstein 1993; Lowenstein et al. 2014), some studies (Hay et al. 2006) hypothesise Palaeozoic salinities up to 40% more saline than the present-day (Figure 10.11). Following this second hypothesis, elevated salinities, along with depressed sulphate concentrations (Figure 10.12), continued to favour halite precipitation at the expense of gypsum. Salinity has declined since the Early Permian, with the greatest declines corresponding to times of new ocean basin formation (Hay et al. 2006).

The relative ratios of magnesium and calcium ions in seawater exhibit a strong control on evaporative carbonate precipitation. 'Calcite seas' (Section 11.3.1) are characterised by elevated calcium (Mg^{2+}:Ca^{2+} ratio <2) along with depleted magnesium and sulphate (Figure 10.12). Calcium carbonate precipitates as calcite with the excess calcium reacting with sulphate to precipitate gypsum (Lowenstein et al. 2001). The remaining sulphate in the brine is continuously consumed by gypsum precipitation. Such sulphate depletion produces calcium chloride brines, inhibiting the precipitation of magnesium sulphate salts (e.g. epsomite and kieserite) and producing late-stage chloride salts rich in potassium, magnesium and calcium, including sylvite (KCl) and carnallite (Lowenstein et al. 2001;

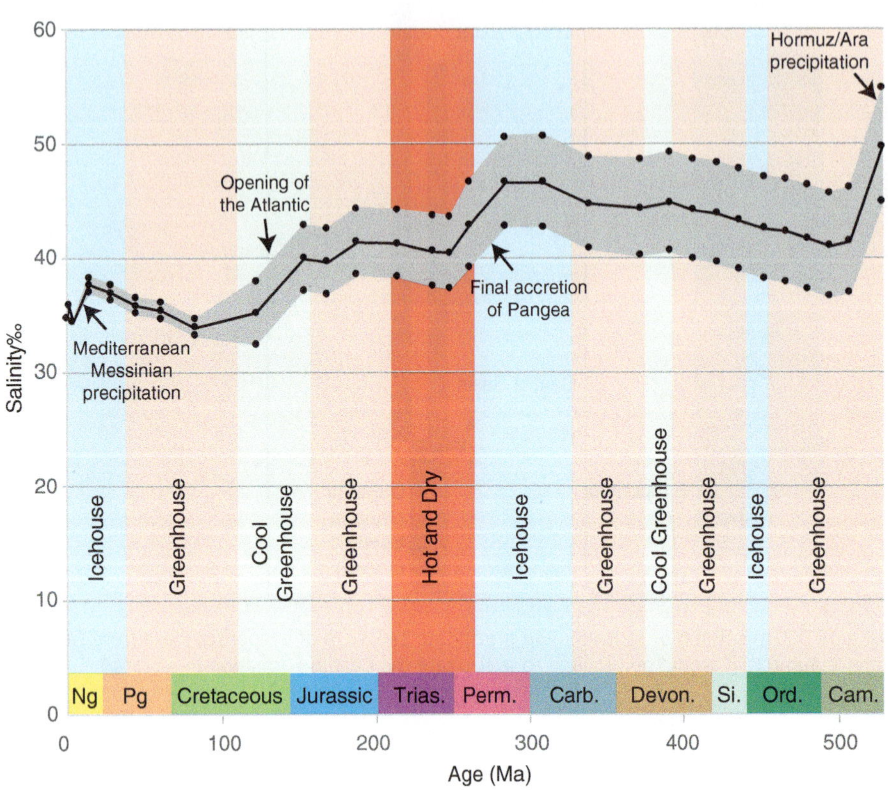

Figure 10.11 The decline in calculated seawater salinity over the Phanerozoic. The data is based on the two models of Hay et al. (2006) with the upper range being the case where water is not lost to the mantle during subduction and the lower range limit being the case where water is lost to the mantle. The black line records the average of these two models. The reconstructions are derived utilising secular temperature trends and mass balance calculations based on calculated extraction of salt from the global ocean over time as evidenced by evaporite deposits. Major phases in Earth's climate from Hay (2013).

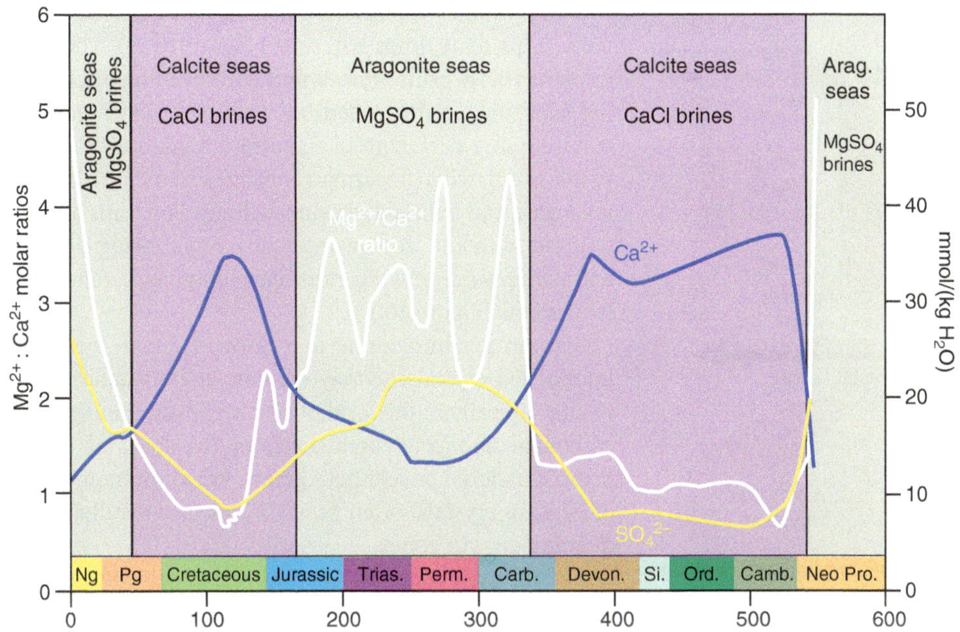

Figure 10.12 Secular variations in the Ca^{2+}, SO_4^{2-} and Mg^{2+}:Ca^{2+} ratios of seawater during the Phanerozoic as calculated from measurements of fluid inclusions in halite. Also plotted are the intervals dominated by $CaCO_3$ seas and $MgSO_4$ seas together with the primary non-skeletal carbonate precipitates (after Lowenstein et al. (2003); Lowenstein et al. (2001)).

Lowenstein et al. 2003). Calcite seas are typically associated with greenhouse climatic conditions (Figure 10.11). During these intervals, elevated sea levels flooded the continents creating epeiric seas and continental marine basins ideal for the precipitation and preservation of evaporites (Lowenstein et al. 2003). During periods with limited calcium in the oceans (Mg^{2+}:Ca^{2+} ratio >2), along with elevated magnesium and sulphate concentrations, calcium carbonate is precipitated as aragonite (Figure 10.12; Table 11.1). In these 'aragonite seas', elevated magnesium ion concentrations, together with sulphate-rich brines, promotes the precipitation of magnesium sulphate.

10.3.3 Modes of Evaporite Mineral Formation

The most abundant Phanerozoic marine evaporites are calcium sulphate, in the form of the minerals gypsum and anhydrite, and sodium chloride as halite (Warren 2010). While marine evaporite sequences may contain several other evaporite minerals (Table 10.2), these typically form only a small proportion of the deposits so, here, we shall concentrate on the precipitation modes of the two most common phases – gypsum and halite. Both of these minerals can be reworked to produce a range of sedimentary bedforms. Sediment grains can be in the form of primary crystals, particularly where crystals are forming at the air–water interface and settling to the sediment surface, or as eroded and reworked 'evapoclastic' material.

10.3.3.1 Marine Gypsum

Marine evaporite gypsum is typically deposited as selenite crystals (from Greek "selene", moon). Selenite has a diverse range of crystal habits that, in turn, can produce a wide variety of growth fabrics. These crystal forms and fabrics have inspired a plethora of descriptive and interpretative terminology. Here, we employ the most commonly used descriptive terminology.

Precipitation of gypsum may occur at the air–water interface, at the sediment surface or interstitially within the sediment column. Gypsum crystals nucleating at the surface of a brine will adopt an acicular prismatic habit. Elongate crystals will continue to grow, supported by surface tension, until their weight or a disturbance, such as a wave, causes them to founder and sink to the basin floor.

At the sediment surface, initial precipitation will be in the form of prismatic crystals, up to a few mm long. With the development of twins, growth will produce distinctive vertically orientated 'swallow-tail' (also termed 'arrowhead') selenite (Figure 10.13). These twinned gypsum crystals may occur singly but are more often found as dense accumulations forming palisades or conical clusters. Twinned selenite crystals may span the entire bed thickness, with crystals reaching up to 7 m in extreme cases (Schreiber 1978). In addition to twinning, gypsum crystals will also exhibit varying degrees of deformation at crystal faces. The degree of curvature is controlled by the presence of impurities (particularly organic matter, clay, magnesium and sodium), and the temperature and pH of the brine. Where the gypsum

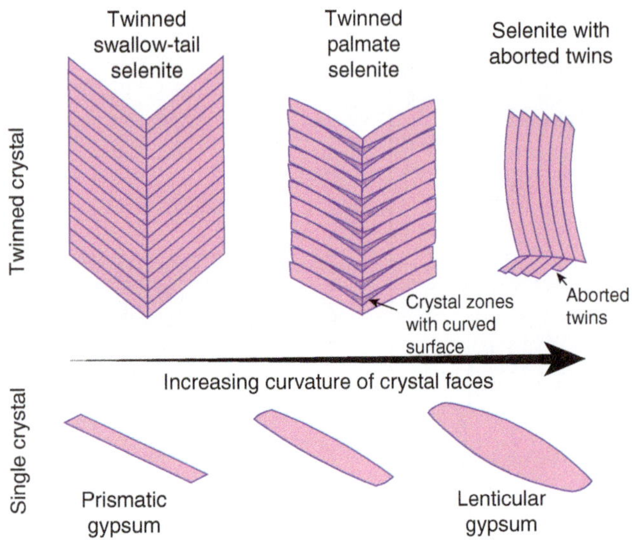

Figure 10.13 Variations in gypsum crystal morphology as a result of interactions between crystal twinning and the degree of curvature of the crystal face (modified from Schreiber (1978); Shearman and Orti-Cabo (1978)). Curvature of the crystal faces results from fluctuations in the temperature and pH of the brine or where impurities (such as organic matter, clay, magnesium and sodium) are present.

twins in swallow-tail selenite develop with curved crystal faces, a 'palmate' form will arise (Figure 10.13). The rapid growth of the gypsum crystals may entrap sediments, including biogenic particles, settling from the water column, resulting in a dark core or re-entrant angle in the case of swallow-tail twins (Shearman and Orti-Cabo 1978). In some instances, the twinned gypsum will develop with a prone attitude with the downward-facing twins being abortive and the upward-growing twins developing as parallel clusters or fans (Figure 10.13).

Gypsum accumulations may form a dense mosaic of interlocked crystals or may contain intercrystalline sediments, typically in the form of clastic gypsum, fine-grained carbonate and clay. Inclusions, in the form of brines, organic material or sediment grains, are a common feature of gypsum crystals, often occurring as face-parallel bands defining growth zones.

If an influx of water dilutes the brine to below the saturation state for gypsum, a period of dissolution will ensue. Dissolution will be focused at the upper surface of the evaporite deposit, effectively dissolving the gypsum crystals to produce a planar surface (Figure 10.14). Sediment inclusions will be liberated, and, along with any detrital material

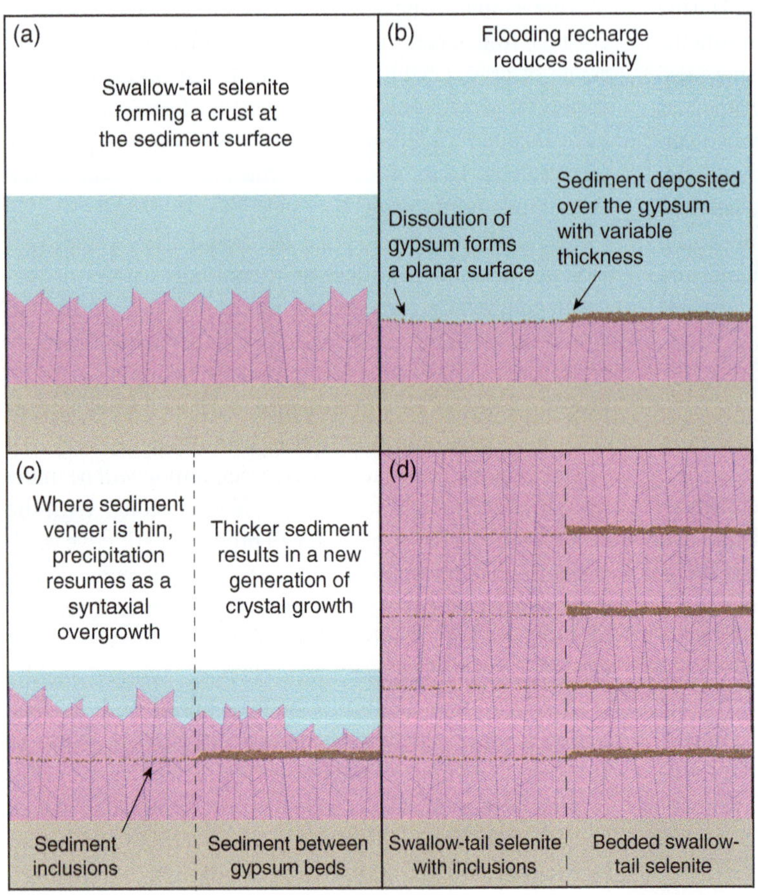

Figure 10.14 Effects of dissolution, sediment influx and recharge on gypsum development. (a) Gypsum precipitation at the basin floor forms a thick crust of densely packed swallow-tail selenite. (b) Recharge flooding of the basin results in dissolution at the upper surface of the gypsum to eventually produce a planar profile. Sediment may be deposited over this dissolution surface. (c) Where the sediment veneer is thin, crystal growth may resume as syntaxial overgrowths with sediment incorporated into crystals as bedding-parallel inclusions. Where the sediment is too thick, renewed crystal growth in optical continuity is precluded and new gypsum crystals will grow from the sediment surface. Repeated phases of dissolution and precipitation result either in thick beds of elongate selenite with sediment inclusions recording phases of dissolution or of thinner beds of gypsum separated by interbeds of non-evaporite sediments.

accompanying or subsequent to the flooding event, will produce a sediment veneer. Evaporation of the refreshed water body will result in a new phase of gypsum precipitation. Where the sediment veneer is thin or discontinuous, new precipitation may occur as syntaxial overgrowths in optical continuity with the truncated crystals with the veneer being incorporated as inclusions (Figure 10.14). Thicker sediment veneers will isolate the earlier evaporites from the new generation of crystal growth resulting in each successive gypsum crust developing in crystallographic isolation from its predecessor.

Gypsum is a common precipitate in the shallow subsurface of the sediment column with the presence of impurities producing a lenticular habit (Figure 10.13). Precipitation is concentrated at the top of the phreatic zone and into the overlying capillary fringe with crystal growth displacing, and often incorporating, the host sediment. Common inclusions include detrital carbonate, bioclasts, siliciclastic material and microbial mats, often preserving a relict sedimentary fabric. These poikilotopic textures provide a valuable insight into the environment of precipitation.

10.3.3.2 Marine Halite

Halite is precipitated in both subaerial and subaqueous settings. Under ideal conditions, halite precipitates as cubic crystals displaying uniform outward growth of cube faces. Where the brine is supersaturated, the cubic form is abandoned with preferential growth at crystal corners and, to a lesser extent, crystal edges, outpacing the growth of crystal faces to form hopper-shaped halite. Inclusion-defined growth bands are common in halite, these are formed by the incorporation of fluid, sediment and organic material during crystal growth.

At the brine surface, halite precipitates as cubic, platy and hopper crystals with inverted (hollow-sided) pyramidal hopper forms dominating. These crystals are supported by surface tension and, in the case of floating concave-upward hopper crystals, the displacement of brine. Under low-energy conditions, crystals coalesce to form rafts that may further amalgamate to cover the brine surface forming a halite crust over the aqueous body (Figure 10.15). Individual crystals and rafts will sink where either the buoyancy is exceeded, or the surface tension is disturbed – typically by waves. Foundering crystals and rafts settle through the water column, accumulating at the basin floor, along with bottom-nucleated crystals, as an unconsolidated halite sediment (Figure 10.16a). Where not subjected to remobilisation, this basin floor crystal 'cumulate' will be cemented and stabilised by syntaxial overgrowth (Figure 10.16b). Continuing precipitation results in competition for space with preferential precipitation at crystal corners, at the expense of slower-growing crystal faces, producing a chevron growth texture (Figures 10.16c,d and 10.17).

In the event of basin recharge reducing the salinity of the brine, precipitation will cease and the halite deposits will be subject to top-downward dissolution to produce a planar surface with sub-vertical microkarst pipes (Figure 10.16e). Sediments and organic matter may accumulate at this surface. With renewed evaporation, elevated salinities may result in the initial precipitation of a thin layer of fine gypsum crystals prior to the brine reaching halite saturation and a return to halite precipitation. Karstic voids are infilled by clear, inclusion-free, halite cement (Shearman 1970) (Figure 10.16e). Repeated cycles of precipitation and dissolution will produce a bedded or laminated halite deposit with surfaces defined by thin layers of gypsum or detrital sediment (Figure 10.16f) (e.g. Hovorka 1987).

Within the subaqueous and subaerial sediment column, halite is precipitated within pores and as displacive interstitial crystals. Where the rate of precipitation is rapid, host sediment can be incorporated into the crystals (Gornitz and Schreiber 1981).

In the subaerial realm, evaporation from the sediment surface produces thin ephemeral efflorescent crusts of acicular halite crystals. These crusts have a very low preservation potential, being subject to dissolution during rainfall or even episodes of high humidity (Lokier 2012). Buried subaerial halite is typically removed by groundwater in the shallow subsurface before entering the true burial realm.

10.3.4 Forming Thick Evaporite Sequences

Where a basin becomes fully isolated from the marine system, the complete evaporation of the water body produces an evaporite deposit of a thickness that is proportional to the height of the water column from which it was precipitated. To put this into context, if the Persian Gulf were to become isolated and the entire body of water was evaporated, then just 0.5 m of evaporites would be precipitated. From this observation, it is clearly unrealistic to interpret the thick evaporite sequences of the geological record as representing closed evaporite systems with short fluid residence times. For example, the precipitation of the 450–550 m thick evaporites of the Late Permian Castile Formation, would have required the complete evaporation of a water column between 31 and 38 km deep, a water column that would extend from the bottom of today's Mariana Trench into the stratosphere! Such stratospheric water depths are beyond unreasonable (Figure 10.10b).

A simple resolution is provided where a continuous or punctuated connection to the marine system permits the

Figure 10.15 Photograph of halite rafts amalgamating to form a surface 'crust' over a shallow (5 cm) brine pond in the Abu Dhabi sabkha. Photograph credit: Stephen Lokier.

continuous or batch replenishment of seawater into the basin. To maintain evaporite precipitation, the average rate of seawater influx, in relation to evaporation, must be sufficient to supply solutes but not so rapid as to dilute the basinal brines to the point where evaporation is halted, or dissolution occurs. Where the rate of inflow is continuous and maintained at relatively high levels then only the low solubility salts will be precipitated. Conversely, during episodes where seawater supply is outpaced by evaporation, major sea-level falls and, possibly, the complete desiccation of the basin will terminate subaqueous evaporite formation with a succession of high solubility minerals capping the evaporite sequence (Figure 10.18a). Brine outflow (a process termed reflux), either to the marine system or into the groundwater, is another important control on the evolution of the evaporite sequence. Where reflux rates are high, in relation to the rate of evaporation, this will reduce the residence time of the brines, effectively removing them from the basin before high-solubility minerals are precipitated (Figure 10.18c). The relative proportions of inflow, evaporation and brine outflow are controlled by many internal and external processes and are unlikely to be constant over the lifetime of an evaporite basin. These variations will result in an evolution in the mineralogy of the precipitated minerals and may cause the dissolution of previously precipitated minerals (Figure 10.18b). In interpreting evaporite sequences, care should be taken not to confuse the products of these processes with effects resulting from variations in climate. In the example of the Castile Formation, both recharge and reflux were via a permeable barrier that permitted the supply of the >9 million km^3 of water (Kirkland et al. 2000) that evaporated to precipitate the >10,000 km^3 of evaporite minerals within the basin (Kirkland 2003).

The rate of evaporite deposition, being the product of a variety of internal and external factors, can be highly variable. However, the nexus of conditions required to

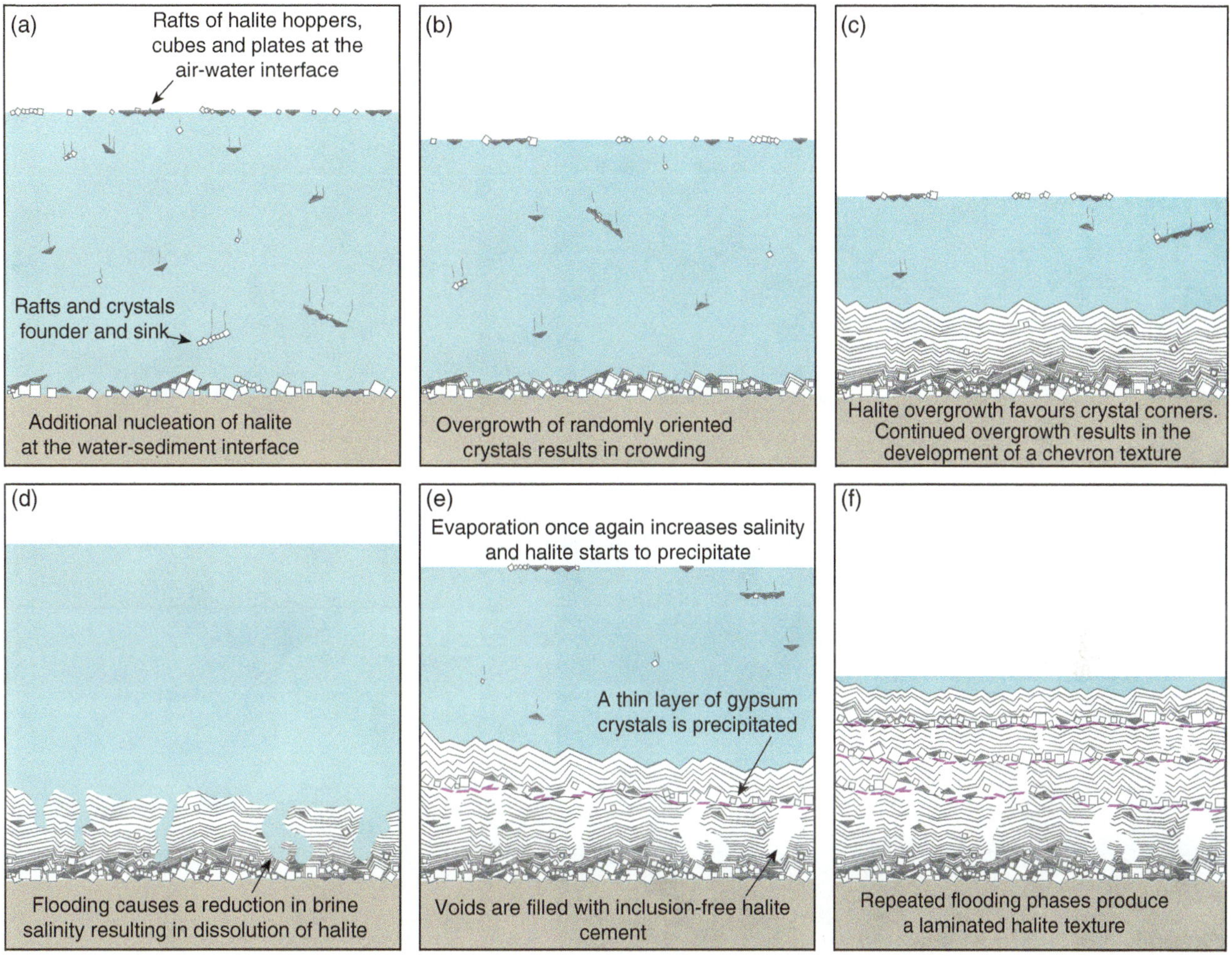

Figure 10.16 Precipitation, dissolution and preservation of subaqueous halite deposits. (a) Halite precipitated at the brine surface coalesces to form rafts that founder and sink to the basin floor. (b) Precipitation of halite at the sediment surface, along with the development of syntaxial overgrowths, results in crowding. (c) With limited space, crystal corner growth outcompetes crystal face growth. Continued overgrowth focused at crystal corners produces a chevron texture. (d) Basin recharge lowers salinity below halite saturation causing dissolution at the evaporite surface. Subvertical microkarst pipes develop from the planar corrosive surface. (e) As salinity once again increases, a brief period of gypsum precipitation is followed by a return to halite precipitation. Microkarst voids are filled with a clear, inclusion-free, halite cement. (f) Repeated phases of dissolution and precipitation produce a laminated or bedded halite – depending on the relative frequency, duration and intensity of precipitation and dissolution (modified from Handford (1991); Shearman (1970)).

precipitate a thick evaporite sequence ensures that most evaporite sequences are deposited over a relatively short time. The 450–550 m sequence of the Late Permian Castile Formation was deposited in just 175,000 years (Anderson 1982; Kirkland 2003) with an average net rate of 2.9 mm a^{-1} while the up to 2 km thick Messinian sequence from the Mediterranean was deposited in only 640 kyr (Krijgsman et al. 1999), a net rate of $\sim$3.1 mm a^{-1}. The precise dating of any rapidly deposited sedimentary sequence can be challenging – a difficulty further compounded by the limitations of a biostratigraphic approach in these stressed depositional environments. The inherently soluble character of evaporite minerals means that most precipitation rates and accumulation volumes are likely to be significantly underestimated. This inference is supported by two lines of evidence. First, the precipitation rates measured in recent evaporites settings of up to 100 mm a^{-1} (Schreiber and Hsü 1980) are significantly higher than those calculated from the sedimentary record of ancient systems. Second, the complete evaporation of a body of present-day seawater would yield evaporites with a halite-to-anhydrite ratio close to 35:1 (a ratio of 30:1 during the Permian). Yet, for many ancient systems the ratio is less than 1:1 this can be accounted for by incomplete evaporation, with brines refluxing from the basin, or via post-depositional dissolution of the more soluble phases.

Figure 10.17 Plan view photograph of recently precipitated halite in a desiccated very shallow sabkha pond showing cubic crystals at the pond floor, chevron growth from the pond margin (lower left of image) and foundered rafts. Also note the development of incipient efflorescent halite precipitation on the adjacent sediment surface. Photograph credit: Stephen Lokier.

10.3.5 Processes Specific to the Supratidal System

Situated at the interface between the marine and terrestrial realms, coastal sabkhas and salinas are subject to a complex interplay of terrestrial and marine environmental controls. Marine fluids are the primary source of solutes. Seawater is transported in the subsurface, as groundwater, and subaerially – flooding the coast during spring tides or through storm surges (Figure 10.19a).

10.3.5.1 Processes in Coastal Sabkhas

Coastal sabkha evaporites develop within a wide variety of host sediments. Mineralogically, the host sediment may be crudely considered as either carbonate or siliciclastic-dominated. Siliciclastic sediments will typically have a terrestrial origin, having been eroded from pre-existing units and transported to the sabkha setting – either directly, by rivers or wind, or along-shore. Carbonate sediments may also be erosional products but are more likely to be primary sediments that have been generated in a carbonate factory relatively close to the site of deposition. The composition and characteristics of the host sediment are unlikely to be uniform throughout the aerial extent of the sabkha. Primary carbonates are more commonly observed at the seaward margin of the sabkha with siliciclastic sediments becoming increasingly dominant landwards. Where carbonate sediments are present, these may undergo dissolution. Aeolian processes also provide a considerable influx of both fine-grained siliciclastic and carbonate sediment to the sabkha.

The grain size and sorting of the host sediment is another important attribute when considering sabkha development. Well-sorted coarser-grained sediments will have a higher permeability than their finer-grained poorly sorted counterparts. However, where pore sizes are too large, the capillary fringe will be suppressed thereby inhibiting

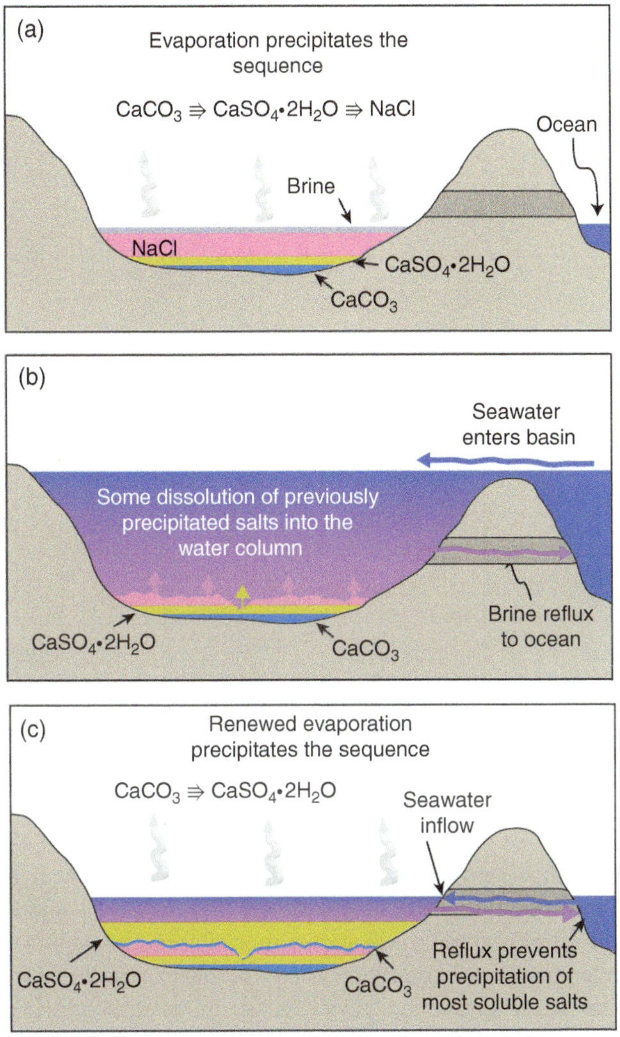

Figure 10.18 The effects of punctuated seawater supply and brine reflux in an evaporite basin. (a) During intervals where the basin is completely isolated from the ocean, brines will concentrate to the point of precipitating high-solubility evaporite minerals. (b) The renewed supply of seawater to the basin, for instance during an episode of relative sea-level rise, reduces the salinity of the brine resulting in the dissolution of the more soluble evaporite minerals. This process will create a subaqueous karstic topography. (c) Where there is both seawater inflow and brine reflux, the concentration of brines stays low such that only lower solubility evaporites will be precipitated. A balance of influx and reflux is needed for this situation to persist.

the transport of solutes to higher levels within the vadose zone of the sediment column.

Marine fluids dominate coastal sabkhas. However, depending on the local hydrology, the landward margin of the sabkha may also be influenced by continental groundwaters (Figure 10.19a) with the potential to precipitate evaporite minerals with a non-marine geochemical (e.g. isotope, or trace elements such as boron) signature.

Evaporite minerals are precipitated both at the sabkha surface and within the sediment column of the shallow subsurface. Subsurface evaporites are precipitated throughout the vadose zone, particularly in the capillary fringe immediately overlying the groundwater table. The interchange of fluids between the sabkha brines and the marine reservoir prevents the brines from reaching a salinity at which they can precipitate minerals of higher solubility than calcium sulphate – typically in the form of gypsum but also, more rarely, as anhydrite (Curtis et al. 1963). Crystal growth is primarily displacive but may also incorporate host-sediment, and even whole bioclasts, into the crystal to form a poikilotopic texture. The precipitation of these evaporite minerals may progressively cement the sediment, occluding pores, slowing the fluid flow and limiting the ion supply. At the sabkha surface, halite dominates, forming a millimetre to centimetre thick evaporite veneer (Figure 10.7). While ubiquitous over the surface of much of the sabkha, this halite crust is highly transient being rapidly dissolved by rainfall, overnight dew, or when inundated by storm surges (Lokier 2012). Coastal sabkha primary halite is highly unlikely to be retained into the sedimentary record (Court et al. 2017).

The presence of an adjacent body of warm shallow water will typically ensure that the relative humidity at the coast is too high to precipitate minerals more soluble than halite. Where seasonal reductions in humidity permit the precipitation of deliquescent minerals, these will be ephemeral – being removed during periods of higher humidities.

As with all shoreline sedimentary systems, coastal sabkhas are highly susceptible to sea-level fluctuation. Even a relatively small sea-level fall can isolate the sabkha from the marine system, effectively shutting down evaporite precipitation, or, where continental groundwater is available, resulting in the potential dissolution of marine evaporites and a shift to a more continental suite of evaporite minerals (see Section 6.4.3). Conversely, a rise in sea level will flood the sabkha, dissolving previously precipitated evaporites and shifting the zone of marine evaporite precipitation landward.

Progradation is also a significant control on coastal sabkha evolution. The limited amount of accommodation in these flat-lying, near base-level settings results in considerable potential for coastline progradation with rates exceeding 1 m a^{-1} being calculated from recent sabkha settings (Lokier and Steuber 2008). As the shoreline progrades, the landward margin of the sabkha will be increasingly isolated from marine-derived fluids, this will cause responses akin to those associated with sea-level fall.

During either sea-level fall or simple progradation, the landward portion of the sabkha may experience a fall in the level of the water table. Where the sediments are

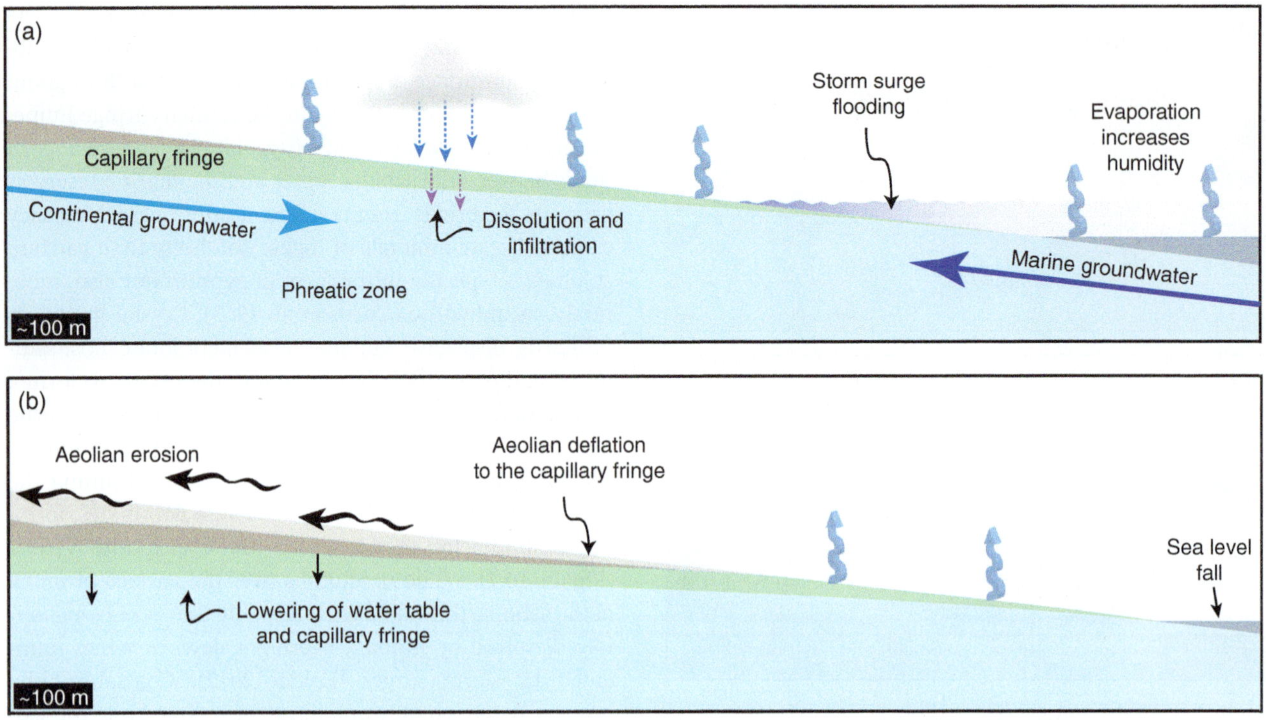

Figure 10.19 (a) Sources of solutes and transportation paths at the coastline. (b) Effects of sea-level fall on groundwater level and aeolian deflation in the landward sabkha.

uncemented, they will be susceptible to aeolian erosion and deflation down to the level of the capillary fringe at which point halite precipitation will form a protective veneer preventing further downward erosion (Figure 10.19b). Any gypsum exposed at the sabkha surface is liable to be desiccated, fragmenting to microcrystalline anhydrite with an increased susceptibility to aeolian erosion (Figure 10.20).

10.3.5.2 Processes in Coastal Salinas

As with sabkhas, coastal salinas are dominated by fluids of marine origin and form in association with a wide range of host sediments. Seawater is typically supplied to the salina by its flow through the porous barrier separating the basin from the marine system. This flow may be supplemented by flooding of the basin during episodic storm surges or periodic spring tides. Reflux will be minimal to absent. Salinas exist in a constant state of flux with a range of internal (e.g. storm surges, aeolian deflation, compaction) and external processes (basin filling, sea-level fluctuations, tectonic evolution and climatic changes), transitioning a salina along the continuum between perennial marine lake and desiccated brine pan.

Where the influx of fluids is dispersed over a wide area, then the distribution of evaporite minerals will be primarily controlled by water depth and seasonal climatic variations. Flooding recharge results in the partial dissolution of earlier basin-floor evaporites followed by the precipitation of a veneer of carbonate sediment. As evaporation increases the salinity, gypsum will start to precipitate with large selenite gypsum twins developing at the sediment surface in deeper water areas and finer bladed to acicular gypsum precipitating in shallower waters (Figure 10.21). The shallowest water areas are characterised by laminae comprising *in situ* or reworked gypsum crystals alternating with microbially precipitated carbonate sediment. The margin of the salina will be colonised by microbial communities forming mats (Figure 10.21), thrombolites and stromatolites. Landward of the microbial communities, the subaerial surface may have a thin crust of ephemeral efflorescent halite. Where brine reaches the saturation point with respect to halite, precipitation is concentrated at the brine surface with crystals foundering to the basin floor to produce halite cumulates overlying the earlier gypsum precipitates.

In cases where recharge is more geographically focused, evaporite mineralogy is typically zoned in relation to the distance from the point of fluid ingress. Carbonate will be the primary precipitate close to the source of marine fluids. With increasing distance from this source, the salinity of the brine will increase, thus precipitating gypsum. If the salina has a significant aerial extent, then the saturation point with respect to halite may be reached in more distal areas.

Figure 10.20 Photograph of a gypsum crystal that has been exposed at the surface in the intertidal zone of the Abu Dhabi sabkha. Anhydritisation has progressed along cleavage planes to fragment the crystal, increasing the surface area and enhancing weathering. Note the development of a pustular surface texture in the surrounding sediments, this results from efflorescent halite precipitation displacing grains at the sediment surface. Photograph credit: Stephen Lokier.

If a salina completely desiccates, evaporite precipitation will continue with displacive growth of lenticular gypsum in the shallow subsurface and the precipitation of a thin efflorescent halite crust at the surface. In ancient sedimentary sequences, distinguishing a desiccated salina from a sabkha can be challenging. Coastal salinas may also be episodically reconnected to the marine system, experiencing fully marine conditions with consequent deposition of marine facies (e.g. Perkins et al. 1994).

As with lacustrine settings (Section 6.4.3), salinas are highly responsive to a range of environmental controls, for example water balance, but extricating these records from the internal record of salina evolution can be challenging.

10.3.5.3 The Abu Dhabi Sabkha – A Present-Day Supratidal Evaporite System

Lying at the southern shore of the Persian (Arabian) Gulf, the Abu Dhabi sabkha is widely employed as the modern analogue from which to investigate the processes, facies, geometries and stratigraphic associations that are diagnostic of coastal evaporite settings in the geological record.

Following the Second World War, a flurry of hydrocarbon exploration brought increasing numbers of geologists to the shores of the Persian Gulf. During the 1950s and 1960s, considerable research was focused on the modern sedimentary environments of the Arabian coastline. Although sabkhas were found along much of this coast, logistical challenges, both in terms of transport links to the region as well as access to field sites, saw the sabkha proximal to Abu Dhabi Island becoming the target of many early evaporite studies. While subsequent studies have moved progressively further afield in an attempt to avoid anthropogenic disruption, the so-called Abu Dhabi sabkha remains the most studied and employed analogue for our understanding and interpretation of ancient sabkha systems. Much of the early work in the Abu Dhabi sabkha was driven by a research group based at the Imperial College of London. A key outcome of this group's research was the recognition of the importance of anhydrite and gypsum as primary components of the sedimentary system (Curtis et al. 1963; Shearman 1963; Evans et al. 1964; Kinsman 1965; Butler et al. 1965a).

To understand the context, uses and limitations of the Abu Dhabi sabkha as an analogue for ancient coastal evaporite formation, we must first appreciate its wider

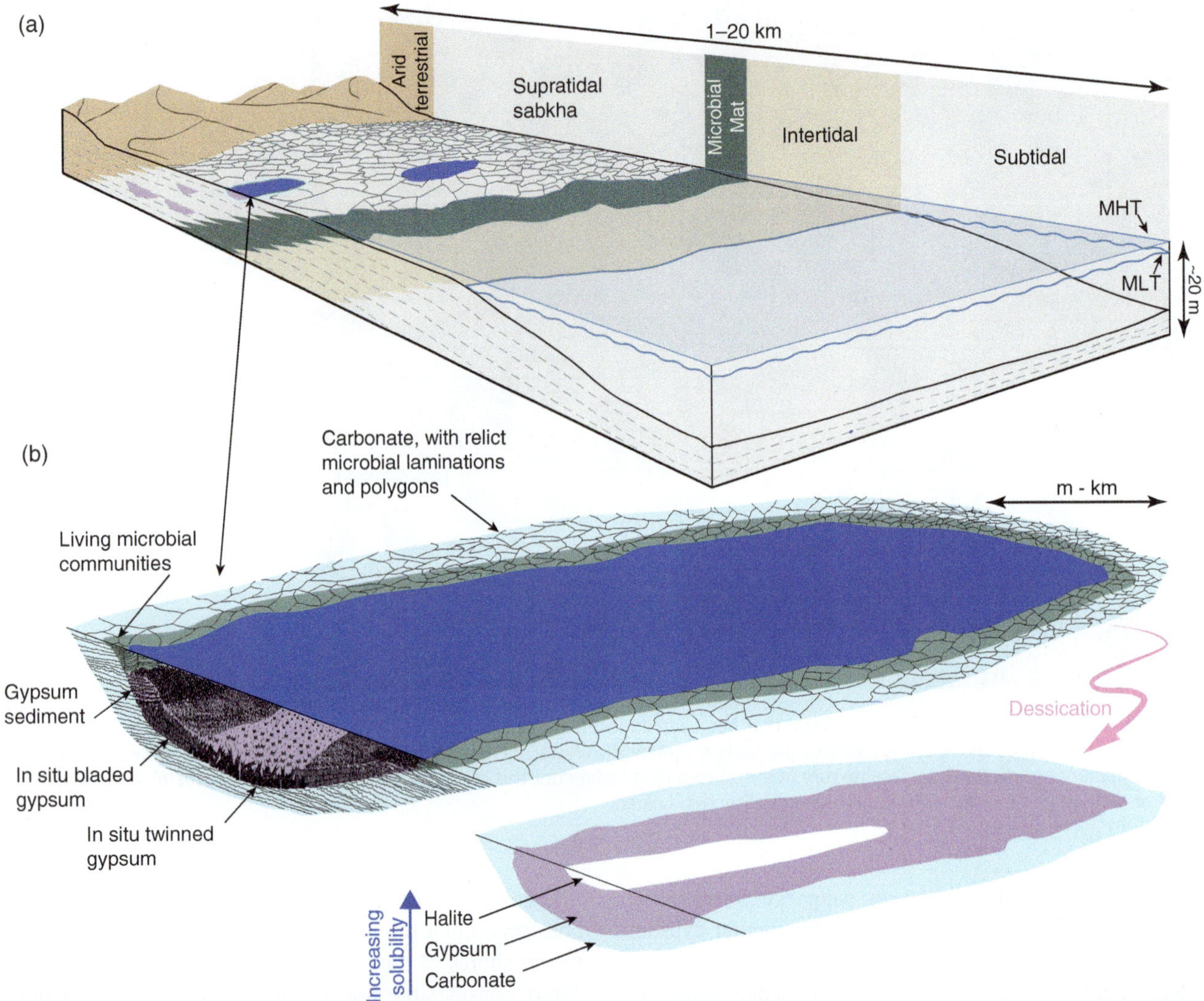

Figure 10.21 Facies architectures typically observed from sabkhas and salinas. (a) Sabkha facies form a series of parallel belts aligned parallel to the coastline (note the extreme vertical exaggeration). The vertical succession of observed facies (figure left) represents the lateral relationships of facies at the time of deposition (a 'Walther's law' succession. (b) Coastal salinas can form a sub-environment within sabkhas, as depicted here, or may develop in isolation. Salina facies show a succession in time through desiccation rather than progradation of adjacent facies belts as in (a). Where forming in sabkhas, following desiccation the salina facies are liable to overprinting by interstitial sabkha facies. MHT = mean high tide; MLT = mean low tide.

geographical and climatic setting within the context of the Persian Gulf. The Gulf is a foreland basin with a shallow epeiric sea flooding 227,000 km^2 of continental crust at the northern margin of the Arabian Plate. The northern shores of the Gulf are bounded by the coastal plains and Zagros Mountains of Iran while the low-relief shoreline of the Arabian Peninsula lies to the south. The western termination of the seaway is dominated by the prograding Shatt al-Arab delta, a site of considerable siliciclastic influx. In the east of this epicontinental sea, the narrow (46 km wide) Strait of Hormuz restricts water interchange with the Indian Ocean. The shallow (average 35 m) seafloor gently dips northwards from the Arabian coast towards deeper waters offshore of Iran, with water depths rarely exceeding 100 m. At the southernmost shore of the basin, an extremely low-angle ramp geometry dips north from a flat-lying supratidal coastal plain, through a laterally extensive intertidal complex into a shallow water (<10 m) subtidal environment. Numerous peninsulas and offshore islands locally protect the shore from open-marine conditions, creating a network of interconnected lagoons and channels. This is the setting of the Abu Dhabi sabkha.

Seawater flows into the Gulf at the sea surface through the Strait of Hormuz before heading west and north along the coast of Iran where high temperatures and dry offshore winds drive evaporation and increase salinity (Figure 10.22). On

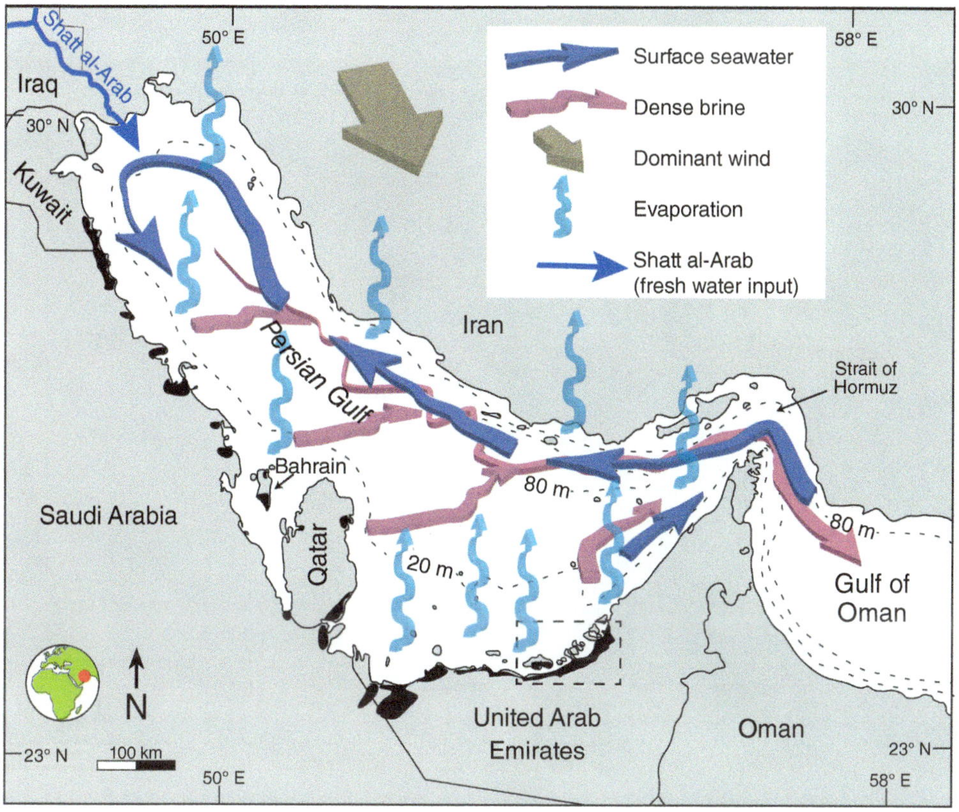

Figure 10.22 Map of the Persian (Arabian) Gulf region showing the large-scale circulation of seawater (average residence time is 3.5–8 years), the sabkhas of the Arabian shoreline (black) and the location of the Abu Dhabi sabkha satellite image in Figure 10.23 (dashed box).

reaching the western end of the Gulf, lower salinity is somewhat restored from the input of freshwater from the Tigris, Euphrates and Karun rivers (Shatt al-Arabi delta). Circulation now turns to the south and east, flowing along the southern shore of the basin where increasing evaporation again elevates the salinity. Dense saline water sinks, flowing at the sea floor towards the Strait of Hormuz and exiting into the Gulf of Oman below the fresher inflowing oceanic waters. Within the context of this complex circulation, bottom water flushing times range between 3.5 and 8 years (Sadrinasab and Kämpf 2004).

Sea-surface temperatures in open marine waters of the Gulf range between 20 and 35°C while in southern shore lagoons and tidal ponds water temperatures of up to 47°C are recorded (Ge et al. 2020; Paul et al. 2021). Salinities of 45 $g\,l^{-1}$ are typical along the open marine coast of Abu Dhabi (Lokier and Steuber 2009), in more restricted lagoons and tidal ponds salinities of up to 98 $g\,l^{-1}$ are recorded (Ge et al. 2020). Tides are microtidal, typically in thc rangc of 1–2 m except when enhanced during storm surges. The limited fetch of the Gulf restricts wave development.

Positioned below the descending northern Hadley cell, the region experiences an extremely arid climate. In Abu Dhabi, mean annual rainfall of just 7.2 cm is concentrated between January and March (Raafat 2007) and may be highly localised with some areas not receiving rain for more than a year. Mean annual evaporation of 275 cm (Bottomley 1996) exceeds rainfall by several orders of magnitude. Air temperatures in the Abu Dhabi sabkha range between 50°C in the summer falling to 7°C on winter nights, diurnal ranges are between 2 and 26°C (Lokier 2013). The adjacent warm shallow waters of the Gulf ensure that humidity is often high, particularly during summer months. Winds are dominated by the onshore northwesterly Shamal that can reach gale-force, particularly during July and August, driving storm surges many kilometres inland onto the low-relief coastal plain.

At the Arabian shoreline, the laterally extensive intertidal zone is characterised by tidal channels and locally developed tidal ponds before passing onshore, through a broad belt of microbial mats, into the supratidal sabkha (Figure 10.23). Landward of the microbial mats, the surface of the sabkha is covered by a thin ephemeral veneer of

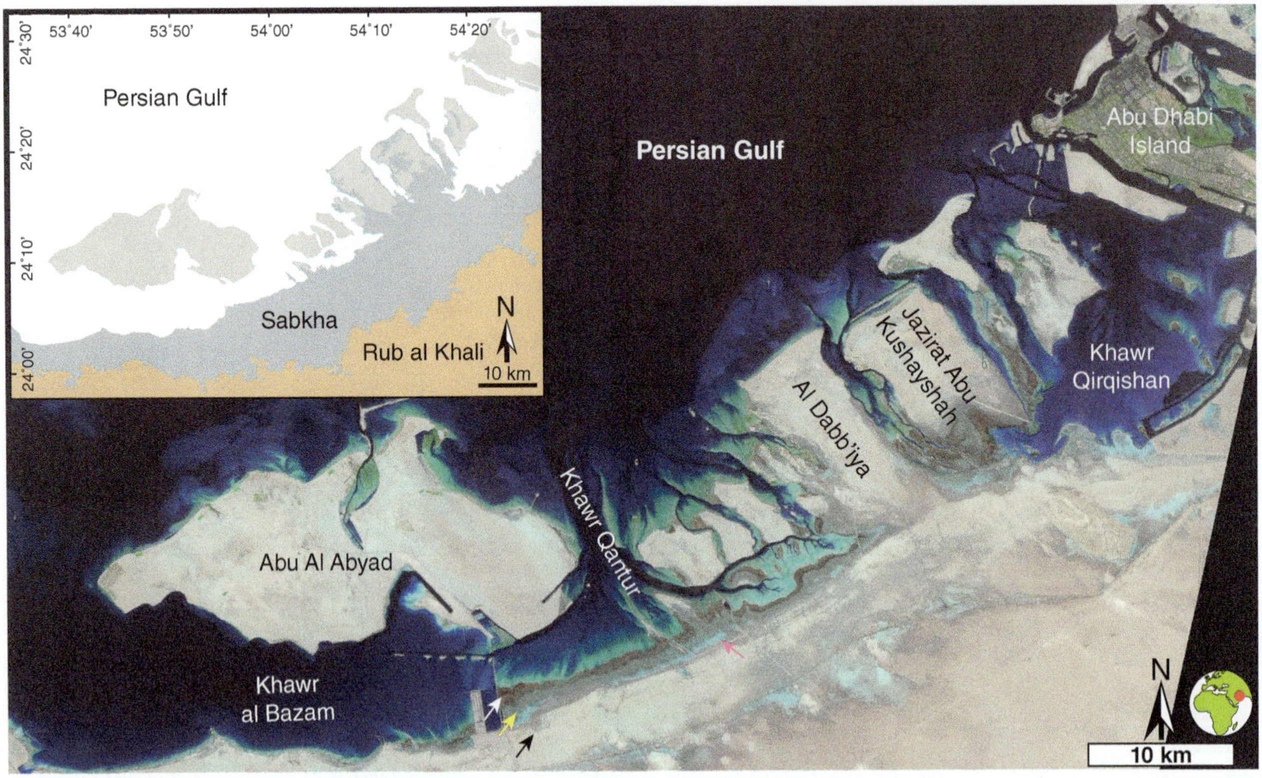

Figure 10.23 Satellite image showing the complexity of the Abu Dhabi coastline with lagoons restricted by barrier island and peninsulas. The microbial mat belt (white arrow), storm surge zone (yellow arrow) and fully supratidal zone (black arrow) are clearly visible, as is freshly precipitated halite (pink arrow). Landsat ETM image acquired 3rd July 2002. See Figure 10.22 for the regional context. The inset panel shows a sketch map highlighting the seaward and landward extent of the Abu Dhabi sabkha.

halite (Figure 10.7) that is underlain by Holocene sediments and evaporites. Continuing landward, the halite crust conceals progressively older sabkha deposits before passing over the palaeo-shoreline into a zone where the halite is covering Pleistocene aeolian dunes that have been deflated to the level of the groundwater capillary zone (Evans et al. 1964). The landward limit of the sabkha system is bounded by a transition to the aeolian dunes of the Rub al Khali. This supratidal sabkha flat forms a shoreline-parallel coastal plain which is up to 33 km wide at its widest point and covers an area of 1900 km^2. The featurelessness of the sabkha topography is exceptional, with an average slope angle of less than 0.05° in the supratidal setting 'steepening' to 0.5° in the intertidal zone (Court et al. 2017) (Figure 10.7).

At the time of the last glacial maximum, between 26.5 and 19 ka (Clark et al. 2009), global sea level was between 120 and 130 m lower than today (Fleming et al. 1998; Peltier and Fairbanks 2006; Solomon et al. 2007; Hanebuth et al. 2009) with the exposed Gulf Basin floor hosting an aeolian dune system that formed a northeast extension of the present-day Rub al Khali sand sea. Post-glacial transgression saw marine waters enter the Gulf by 12.5 ka (Lambeck 1996) and progressively flood the basin – reaching a peak of c. 1 m above present-day sea level at c. 5 ka ago before declining to slightly below present level by c. 1.4 ka (Evans et al. 1969; Kinsman 1969; Patterson and Kinsman 1977; Lokier et al. 2015; Paul and Lokier 2017). With flooding terminating the supply of fresh aeolian sediments, the newly established Arabian coastline was subjected to aeolian erosion with a consequent deflation of the coastal land surface to the level of the groundwater capillary fringe.

Coastal evaporite precipitation commenced during the post c. 5 ka forced regression (Evans et al. 1969; Kinsman 1969; Lokier et al. 2015) and continues today. Solutes are primarily supplied to the sabkha via subsurface flow of seawater. In the supratidal zone, high evaporation rates concentrate solutes in the uppermost phreatic zone and overlying capillary fringe to precipitate calcium sulphate in the form of gypsum. The displacive, interstitial, growth of these minerals within the sediment column results in upwards displacement that elevates the sediment surface into the vadose zone where gypsum dehydrates to anhydrite (Figure 10.24). This uplifted surface will be subjected to erosion either in the form of aeolian deflation, in the fully supratidal zone, or by waves and currents in the storm

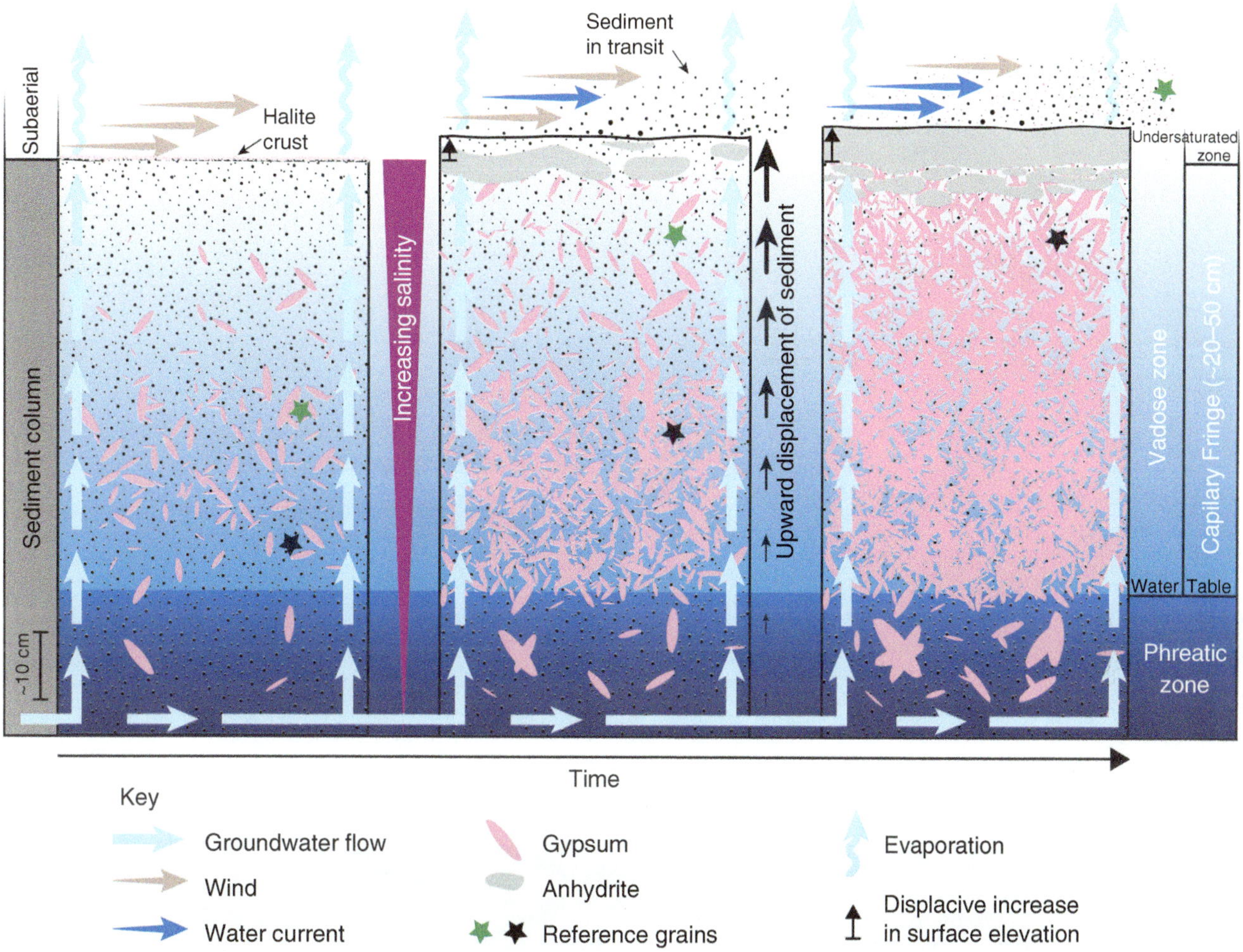

Figure 10.24 Evaporite precipitation in the subsurface of the Abu Dhabi sabkha. (a) Marine-derived groundwater is drawn upward through the capillary fringe and evaporates from the sediment surface. Displacive gypsum precipitation is concentrated above the phreatic zone and halite is precipitated at the sabkha surface. (b) Continued growth of interstitial gypsum displaces sediment upwards (note upward transport of starred grains). Gypsum that is transported into the vadose zone is progressively dehydrated to anhydrite. Where rainfall or flooding dissolves the halite crust, primary host sediments and uplifted evaporite are eroded by water or wind. (c) Over time, the 'evaporite conveyor' transports most of the host sediment to the surface where it is eroded, leaving a sediment column dominated by evaporite minerals.

surge zone. Over time, this 'sediment conveyor' will transport most of the host sediment from the capillary zone to the sediment surface leaving a capillary zone that is dominated by an internally thickened unit of displacive evaporites (Figure 10.24).

Evaporation from the sabkha surface precipitates an efflorescent halite crust that, with continued evaporation, expands laterally to produce polygons with uplifted tepee margins (Figures 10.7 and 10.25a; Figure 7.28c). This halite crust both protects the sabkha surface from aeolian deflation and traps aeolian sediment through its hygroscopic nature (Lokier 2012). The ephemeral halite crust is removed during rare rainfall or storm-surge flooding (Figure 10.25b).

The seaward margin of the sabkha is delineated by a relatively narrow (200–800 m wide) belt of microbial mat that is clearly visible in satellite images (Figure 10.23). The sediment-binding microbial communities must be regularly flooded by seawater to avoid desiccation; however excessive flooding would permit predation by grazing organisms. Under these constraints, the microbial mats are restricted to a narrow environmental range with a landward limit defined by the highest high tides. The mats display a range of surface textures (Figure 10.25c) and are ubiquitously anoxic at a depth of a few millimetres to centimetres. The microbial mat belt locally hosts intertidal ponds containing thrombolites and stromatolites (Figure 10.25d) that

Figure 10.25 Surface environments and facies of the Abu Dhabi sabkha. (a) Displacive lateral growth of halite produces uplifted tepee structures at polygon margins. (b) Rainfall dissolves the ephemeral halite crust of the supratidal zone, exposing anhydrite that will be susceptible to erosion. The polygonal arrangement of the exposed anhydrite mimics former halite polygon margins. (c) Microbial mats forming polygonal textures in the uppermost intertidal zone. (d) Tidal ponds hosted within the microbial mat zone contain thrombolites and stromatolites. (e) View seaward in the intertidal zone at low tide with isolated mangroves (*Avicennia marina*). Note how ebb-flow is concentrated into the tidal creek. (f) Hardgrounds lie close to or at the surface of much of the intertidal zone and extend into the shallow subtidal setting. In this image, the thin veneer (5–10 cm) of sediment has been removed during a recent storm. Lateral growth of the aragonite-cemented hardground produces low-relief margins (arrowed) defining large-scale polygons. In images with a perspective view, the centre of the scale bar indicates the scale at that distance. Photograph credits: Stephen Lokier.

have a much higher preservation potential due to early cementation (Paul et al. 2021).

Seaward of the microbial mats, a series of lagoons and tidal channels is characterised by pervasively bioturbated carbonate sediments (Figure 10.25e) and the presence of extensive surface and near-surface hardgrounds (Lokier and Steuber 2009; Paul and Lokier 2017; Ge et al. 2020).

The constraint of the landward margin of the microbial mats to the uppermost intertidal zone defines a datum that has been used to calculate an average progradation rate of $0.75\,\text{m}\,\text{a}^{-1}$ for the sabkha complex during the last 1.5 ka (Lokier and Steuber 2008). More recently, the Abu Dhabi sabkha has entered a phase of marine transgression with facies retrograding at rates of between 10 and $29\,\text{m}\,\text{a}^{-1}$ (Lokier et al. 2018).

Subsurface sedimentary facies distribution in the sabkha is complex; however, a general primary depositional, and diagenetic, sequence can be deduced (Figure 10.26). The lower limit of the section is marked by a well-cemented bioclastic grainstone containing abundant Cerithiidae gastropods and Peneroplidae foraminifera. This carbonate-cemented horizon shallows seaward to crop out on the sea bed as the present-day hardground (Figure 10.25f). The cemented horizon is succeeded by carbonate sediments akin to those observed overlying the hardground in the present-day lagoon and intertidal zone. These carbonate sediments are typically overlain by laminated microbial mats that are, in turn, superseded by a horizon of sand-to-gravel gypsum laths. This gypsiferous sediment locally grades up-section into banded or nodular anhydrite before terminating at a surface halite crust. There is a general trend of increasing anoxia down-section. This sedimentary sequence provides a perfect example of the application of Walther's Law (Walther 1894) with each diachronous facies belt younging in a seaward direction (Figure 10.21a).

Figure 10.26 A *typical* sabkha evaporite sequence as observed in the present-day Abu Dhabi sabkha. The sequence commences with a hardground (not seen) overlain by intertidal facies passing up-section into microbial and evaporite-dominated facies. The lowermost microbial mat in this progradational sequence has been dated (^{14}C) to 1273–1006 cal BP.

10.3.6 Processes Specific to the Basinal System

In the absence of modern analogues, our knowledge of the formation of basinal evaporite systems derives from the interpretation of evaporite sequences deposited within ancient salt giants. Covering areas of hundreds to thousands of square kilometres, these semi-enclosed seas would have possessed all of the geographic, climatic and hydrological complexity of modern-day marine basins of comparable size – such as the Mediterranean, Baltic, or Yellow Seas. Within the basin, seafloor topography produces numerous sub-basins separated by barriers or sills resulting in variable degrees of intra-basin connectivity. These marginal (shallow) and more distal (deep) sub-basins each have their own hydrologic narrative dependent on the degree of connectivity, amount of precipitation and riverine discharge, climate and rate of evaporation.

Basinal systems are dominated by marine water sourced from the open ocean, typically entering the basin through a relatively narrow shallow strait (Figure 10.27). As with other evaporite systems, sea-level has a strong control on the interchange of waters between the basin and the ocean (Figures 10.4 and 10.18). Episodes of relative sea-level fall may completely isolate marginal from distal sub-basins as well as the whole evaporite basin from the marine system. Conversely, periods of high relative sea level will allow the supply of seawater and, potentially, the outflow of basinal brines, for example as seen with the termination of Red Sea evaporite deposition as the Indian Ocean flooded through the Bab al-Mandab strait during the early Pliocene (Smith and Santamarina 2022) or the flooding of the Atlantic Ocean into the Mediterranean Sea via the Strait of Gibraltar at the Miocene-Pliocene boundary (see Section 10.6.4).

Given the scale and complexity of basinal evaporite systems, neither the onset nor termination of evaporite formation will necessarily be synchronous across the entire basin. Elevated marginal sub-basins will be the first areas to become at least partially isolated from the larger-scale basin system. Within these shallow sub-basins, evaporite formation will commence with the precipitation of carbonates. With continued evaporation, dense brines will sink to, and concentrate at, the basin floor where the calcium sulphate-saturated brines will precipitate coarse bottom-grown gypsum. Depending on local climatic, hydrological and topographical conditions, sub-basins may become periodically (seasonally) or permanently internally stratified by pycnoclines that may not necessarily coincide with any thermoclines and chemoclines (Figure 10.27; Section 6.2.2). Such complex internal stratification is a common feature of evaporitic basinal systems and modern hypersaline lakes (e.g. the Dead Sea; see Box 10.1), strongly controlling the deposition and distribution of evaporite minerals (Bąbel and Bogucki 2007). Where a seasonal pycnocline develops in marginal sub-basins, this will physically isolate the dense basin-floor brine from the atmosphere, hence limiting further concentration – a process that may cause a periodic interruption in bottom crystal growth.

Surprisingly, many ancient evaporite basins lack sulphate deposits at the basin depocentre. Such lack of gypsum may be explained where density-driven stratification facilitates substantial bacterial sulphate reduction in anoxic bottom waters. The resulting sulphate depletion within these waters prevents the precipitation of gypsum deposits below the chemocline (Bąbel and Bogucki 2007; de Lange and Krijgsman 2010; Dela Pierre et al. 2012; García-Veigas et al. 2018).

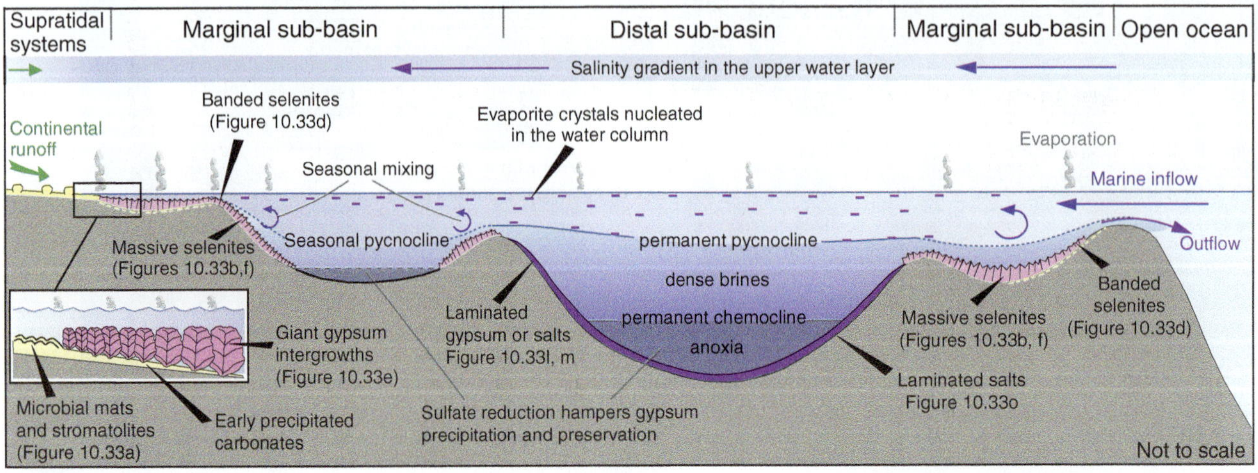

Figure 10.27 The main hydrologic controls on evaporite precipitation in the different portions of a marine basinal system. Note that the evaporite precipitation in marginal and distal sub-basins may be physically disconnected and diachronous. (Inset figure adapted from Ortí (2011)).

Box 10.1 The Dead Sea: An Analogue for Basinal Evaporite Systems?

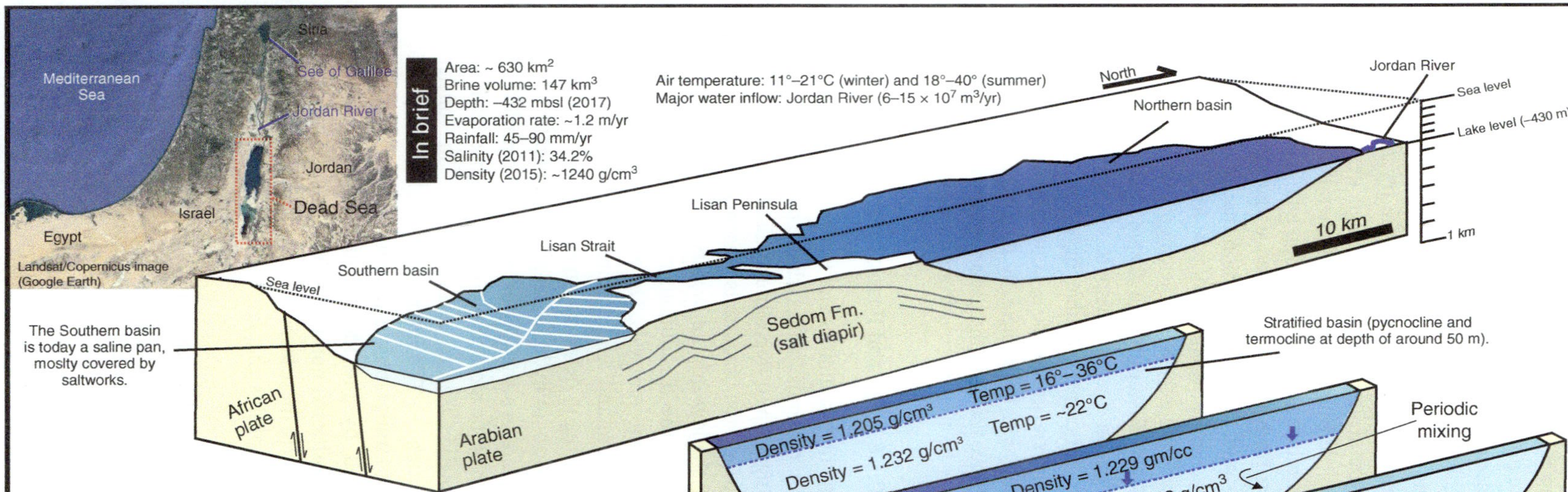

In brief

Area: ~ 630 km^2
Brine volume: 147 km^3
Depth: –432 mbsl (2017)
Evaporation rate: ~1.2 m/yr
Rainfall: 45–90 mm/yr
Salinity (2011): 34.2%
Density (2015): ~1240 g/cm^3

Air temperature: 11°–21°C (winter) and 18°–40° (summer)
Major water inflow: Jordan River (6–15 × 10^7 m^3/yr)

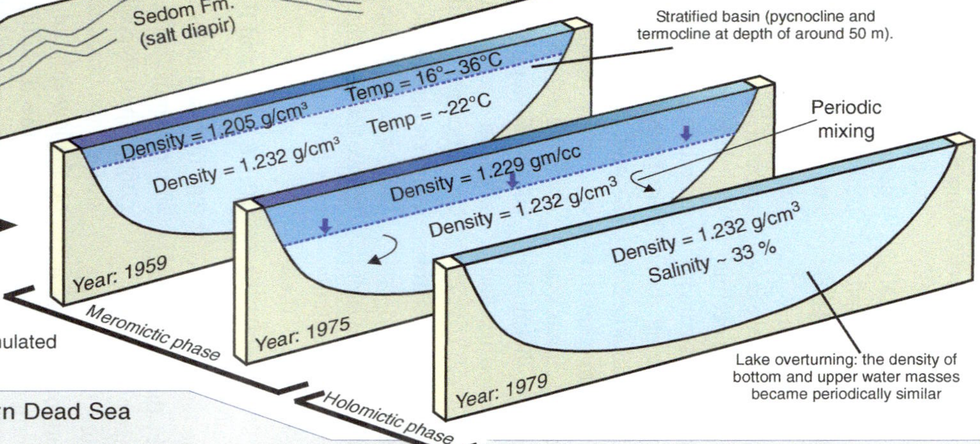

The recent stratification history of the Dead Sea

(from Warren 2016)

The Dead Sea was stratified (meromictic phase) until 1979 at which time an intensifying negative water budget caused density equalization of the upper and lower water layers – with consequent lake overturning (holomictic phase). Since 1979, huge volumes of evaporites have accumulated on the lake floor.

Seasonal control on evaporite precipitation in the northern Dead Sea

(from Sirota et al. 2017)

During the winter (a), cooling of the lake promotes mixing of the water mass and results in increase of the degree of halite saturation and precipitation of halite crystals. During the summer (b), a thermocline forms and separates warm upper water mass (epilimnion) from a deeper and cooler one (hypolimnion). In the epilimnion, halite saturation is depressed by the increasing temperature with consequent salt dissolution at the basin margins. The dissolved salt (DS) is delivered to the hypolimnion through diffusion flux promoting growth of coarse-grained halite at the basin floor. This model, explaining the thinning and disappearance of halite units toward basin margins, has implications for the interpretation of salt giants, such as the Mediterranean subbasins where thick Messinian salt deposits are only present at the depocenters.

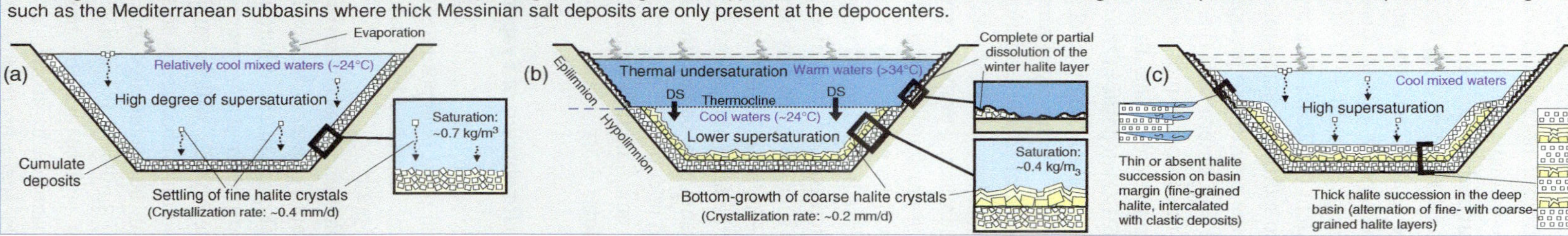

Throughout the basin, evaporite minerals are precipitated in suspension at the air–water interface, at the pycnocline (Raup 1982) and within the uppermost zone of the saturated layer immediately below the pycnocline. These crystals eventually sink to the basin floor where, especially in permanently stratified deeper water settings, precipitation of bottom-grown evaporites is limited or absent. These evaporite grains have the potential to form basin-wide cumulate deposits akin to those observed in shallow evaporite settings and salinas. If seawater inflow should decline to the extent that basin-floor brines become saturated with respect to sodium, magnesium, potassium or chlorine, then there is the potential to precipitate significant thicknesses of halite and bittern salts (i.e. potassium and magnesium salts) in the deeper part of the basin during the climax evaporitic phase (e.g. the thick carnallite and bischofite ($MgCl_2 \cdot 6H_2O$) deposits of the Cretaceous Sergipe and Santos Basins, Brazil (Szatmari et al. 2021)).

Changes in the supply of continentally derived water are also important to the evolution of evaporite basinal systems. An increase in freshwater supply will dilute basinal brines, hampering evaporite precipitation and, potentially, dissolving previously deposited evaporite minerals – these effects will be particularly acute in marginal sub-basins. An increase in continental runoff is typically accompanied by an increase in sediment supply to the basin. At basin margins such an event may be observed by the interdigitation of evaporites and terrestrially derived sediments – often in the form of localised deltas. In more distal settings, these events may be recorded in both bottom-grown selenites and in cumulate evaporites in the form of thin layers of clay and organic material that has settled from suspension.

10.3.7 Reworking of Evaporite Sediments

Evaporite sediments may be directly precipitated *in situ*, at their final point of deposition, or may be mobilised, transported and deposited in a manner akin to any other sedimentary grain.

Precipitation can occur on an exposed subaerial surface, in any depth of water (see Box 10.2) or within the sediment column (Figure 10.28). Where evaporite minerals are precipitated at the air–water interface or within the water column, the crystals will be subject to transportation prior to deposition. Subaqueous surface-nucleated crystals and crusts can be fragmented and transported by waves and currents. Similarly, evaporites precipitated at a subaerial sediment surface may be mobilised by aeolian erosion and transported as aeolian grains. Minerals precipitated within the sediment column, for instance in a sabkha setting, are typically less susceptible to early post-precipitation reworking. However, a fall in base level may result in exposure and aeolian deflation of the sediments, again mobilising evaporites as aeolian grains. All of these processes have the potential to produce evaporite clastic deposits with sedimentary structures analogous to those produced by any other sedimentary grains under similar transportation and depositional parameters. However, there are three caveats when considering the transportation of evaporite mineral grains. First, the high density of the brines and low density of evaporite minerals results in thresholds of mobilisation, transportation and deposition that are different from those associated with normal salinity seawater. Second, all of the common marine evaporite minerals are mechanically weak with a low hardness (typically <4) and well-developed cleavage (Table 10.2) making them highly susceptible to fragmentation (Figure 10.20) and abrasion during transport. Thirdly, all evaporite minerals are inherently soluble in water, therefore wherever the evaporite grains are in contact with undersaturated waters they will be subject to dissolution. These three factors reduce the time and distance over which evaporite minerals may be transported from their site of primary formation.

10.3.8 Syn and Post-depositional Processes

The high solubility and low mechanical strength of evaporite minerals makes them particularly susceptible to a range of post-depositional diagenetic processes. In the subsurface, the ubiquitous presence of pore fluids, in the form of shallow ground waters or deeper formational brines, drives the dissolution, mobilisation and precipitation of evaporite minerals. On entering the deeper burial realm, pressure and temperature increasingly also come to bear, resulting in deformation and mineral phase changes – particularly to hydrous evaporite minerals.

Here, we shall focus on the recognition and interpretation of the most common and diagnostically significant diagenetic features encountered in ancient evaporite sequences. The interpretation of the depositional context and stratigraphic evolution of evaporite systems is greatly aided by the identification of genetically related, co-depositional non-evaporite facies. A discussion of the evolution of these facies within the burial realm lies beyond the scope of this chapter.

10.3.8.1 Syndepositional and Early Post-depositional Processes

The term *diagenesis* is poorly defined being variously employed to sediments from the time of their genesis through to the equally poorly resolved transition to metamorphism. When considering evaporite systems, and the inherently unstable minerals therein, the boundaries

Box 10.2 How Deep Is Deep

The challenges of elucidating water depth from evaporite facies

The depth of deposition is one of the most difficult points to asses in deciphering ancient evaporite sequences. One could ask, "which are the features that can be used to reconstruct palaeodepth in basinal systems?" Here, we provide a list of sedimentological and palaeobiological features that are commonly accepted to discriminate among shallow- and deep-water evaporitic settings.

Shallow-water

Sedimentological features suggesting a shallow-water origin (depth: from 0 to few metres)

- Evidence of desiccation (e.g. mudcracks; shrinkage cracks), or very shallow water bird footprints (Figure.a).
- Presence of dissolution - reprecipitation surfaces (Figure b).
- Bedded halite: seasonal cyclical deposition in salt pans (Figure c).
- Presence of bottom current-related structures (cross-bedding and ripples; Figure d).
- Stromatolite-bearing evaporites (presence of phototrophic cyanobacteria; Figure e).

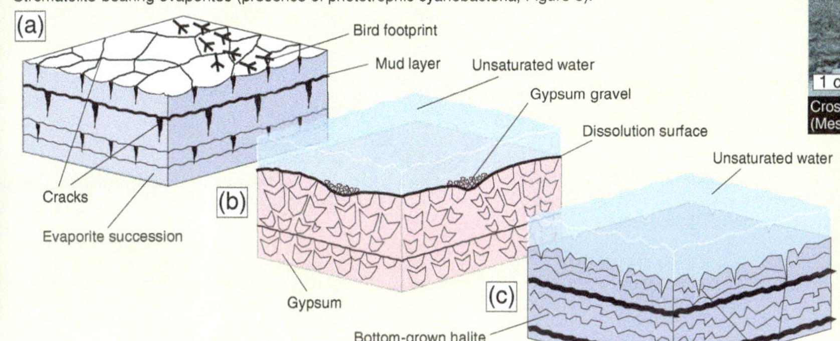

Sketches representing sedimentological evidence for desiccation (a) and episodic dissolution (b, c).

Cross lamination in fine grained clastic gypsum deposits (Messinian, Piedmont, Italy)

Wrinkled laminae of microbial carbonate selenitic gypsum (Badenian, Poland).

Deep-water

Sedimentological features suggesting a deep-water origin (depth: few metres to hundreds of metres)

- Laminated evaporites (cumulates): laminae continuous over large distances (several kilometers) or across the entire basin. The Permian Castile Formation (New Mexico, USA) is one of the most remarkable examples.
- Clastic evaporites (turbidites) deposited through fully subaqueous processes interbedded with deep water sediments.
- Absence of dissolution-reprecipitation surfaces (see above).

Well-defined alternation of continuous clastic fine-grained gypsum beds (white) and organic-rich shales (dark) (Messinian, Vena del Gesso, Apennines, Italy).

Cumulitic gypsum: alternation of fine-grained gypsum laminae and terrigenous, dolomite-rich, laminae (Messinian, Piedmont, Italy).

between depositional and diagenetic processes are particularly indistinct. During our earlier discussions on the genesis of evaporite minerals, facies and sequences, we have already touched on numerous examples of what may be considered syndepositional diagenetic processes (e.g. the phase change from gypsum to anhydrite observed in sabkha settings). We shall not dwell on these processes here, instead, we concentrate on processes that have not been previously discussed.

Under most scenarios, syndepositional or early diagenetic dissolution removes all evidence of the presence of former evaporite minerals. In situations where the morphology of the former crystal is retained as a mould, typically within fine-grained sediment (Figure 10.29a), it is possible to infer the primary, pre-dissolution, mineralogy. Such moulds are commonly infilled by a secondary mineral to form a pseudomorph, further enhancing their potential for preservation and for the recognition of the primary

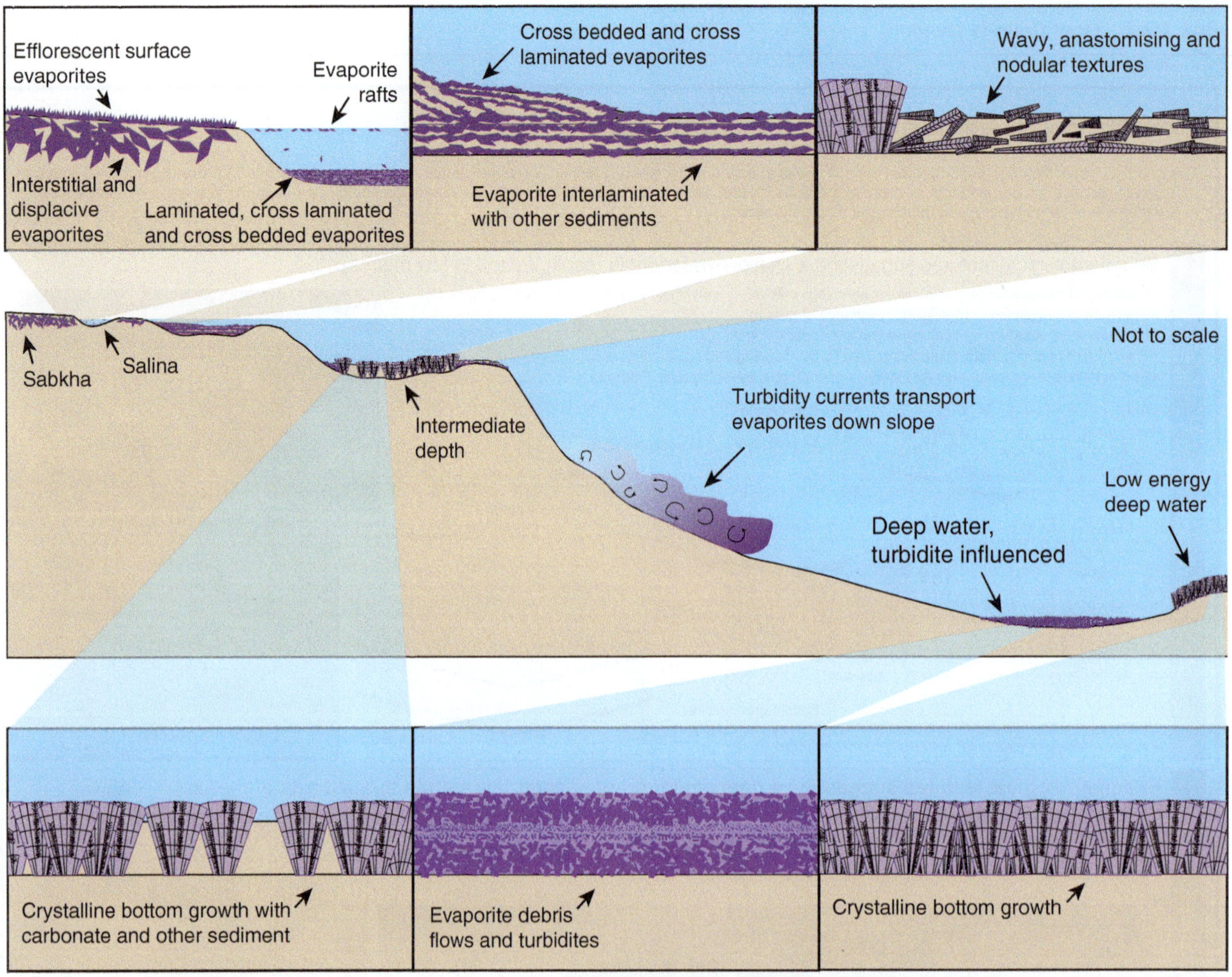

Figure 10.28 Marine evaporite facies in their depositional context (modified from interpretations of Schreiber et al. 1976).

evaporite phase (Figure 10.30). While high solubility minerals, such as halite and potassium and magnesium salts, are easily and rapidly dissolved and lost, less soluble evaporites, such as gypsum, will be subject to partial dissolution to produce crystals with a degraded or corroded appearance (Figure 10.29b).

Carbonate replacement of sulphate minerals is a common microbially driven diagenetic process that affects evaporite successions soon after their formation and during burial (Henchiri and Slim-S'Himi 2006). In evaporitic settings, bacteria can reduce the sulphate from the gypsum and employ organic matter, or other hydrocarbons, dispersed within the muddy sediments as a source of carbon (Aloisi et al. 2013). This process results in pervasive dissolution of sulphate minerals and the formation of carbonates (mostly dolomite; (Sanz-Montero et al. 2006)) and native sulphur (Rouwendaal et al. 2023) – minerals that are commonly observed in salt dome cap rocks (Caesar et al. 2019) and within evaporite successions (Rouchy et al. 1998).

Silicification, forming chert nodules and authigenic quartz, is also a common burial diagenetic process in evaporite sequences (Henchiri and Slim-S'Himi 2006). Silica may be derived either from the non-evaporitic sediments (i.e. biogenic silica or clays) or be supplied via circulating diagenetic fluids. However, the driving mechanisms for the silicification of evaporite minerals and their timing during diagenesis remain poorly understood (Henchiri and Slim-S'Himi 2006).

In subaerial settings, such as sabkhas and fully desiccated basins, high temperatures and arid conditions may result in the dehydration of hydrous evaporite minerals at the surface and in the shallowest subsurface above the capillary fringe. In marine-precipitated evaporites, the dominant mineral involved in this reversible process will be gypsum with euhedral lenticular crystals being replaced by masses of lath-shaped anhydrite microcrystals to form nodules and layers. In some cases, bassanite ($CaSO_4{\cdot}0.5H_2O$), calcium sulphate hemihydrate, is observed as the intermediate form

Figure 10.29 Diagenesis. (a) Cubic halite moulds within a peloidal carbonate (Messinian, Sicily); Photograph credit: Marcello Natalicchio. (b) Large lenticular gypsum crystals from the Abu Dhabi sabkha showing dissolution at crystal edges due to exposure to seawater. Note the preservation of primary cross-stratification retained from the host sediments through the retention of poikilotopic sediment inclusions. Photograph credit: Stephen Lokier. (c) Polished core showing cyclical succession (yellow dashed lines) of halite beds (Ha) intercalated with anhydrite layers (An) and mud-rich halite intervals (Mu); the thin anhydrite layer is folded and exhibits a typical enterolithic texture (arrows); the halite crystals are completely re-crystallised and there is no evidence of cubic crystals or chevron structures (Messinian, Sicily) (d) Thin section photograph of an area of the sample in Figure 10.29c (indicated by a dashed box) showing a sutured mosaic of re-crystallised halite crystals. (e) Thin section of a cumulitic halite partially recrystallised but still preserving an original hopper (Messinian, Sicily). (f) Nodular replacive anhydrite growth (arrows) within a large twinned selenite crystal (San Miguel de Salinas, Messinian evaporites, Spain). (g) Stratabound elongated anhydrite nodules (arrows) possibly representing former selenite beds intercalated with fine-grained siliciclastic sediments (Late Miocene evaporites, Granada Basin, Spain). Photograph credits (Figure 10.29c–g) Marcello Natalicchio.

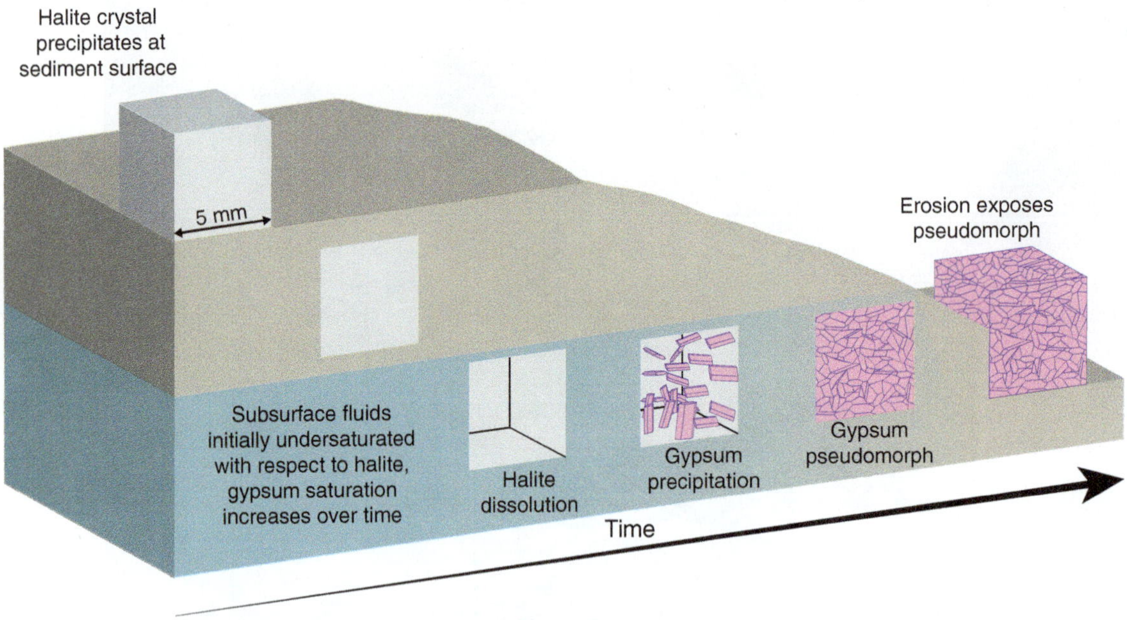

Figure 10.30 The process of formation of an evaporite pseudomorph, in this case – gypsum pseudomorphing after halite.

during the dehydration of gypsum to anhydrite (Butler 1970). Bassanite has also been described in association with ancient evaporites as a product of anhydrite hydration to gypsum (e.g. Middle Miocene evaporites, Poland; (Kasprzyk 2003)). Periods of dehydration and hydration will be controlled by prevailing climatic conditions with dehydration taking place during hot arid periods and rehydration following periods of rainfall or the onset of more humid conditions. This process may be seasonal.

In an open system, allowing free movement of water, the phase transition from gypsum to anhydrite results in a volume reduction of up to 38.5%, with the retrograde hydration of anhydrite to gypsum increasing mineral volume by up to 62.6% (Zanbak and Arthur 1986). These volume changes significantly disrupt the overlying and adjacent portion of the sediment column, deforming primary sedimentary structures and producing diagnostic nodular and enterolithic (intestine-like) textures. The volume increase associated with rehydration is particularly important as this will displace sediment upwards placing it beyond the capillary fringe and consequently promoting erosion (Figure 10.24).

10.3.8.2 Processes in the Deeper Burial Realm

In the subsurface, exposure to undersaturated ground water and formation fluids will result in the dissolution and removal of evaporite minerals. The effect of this process on the overlying or intercalated units will depend on their mechanical properties, the thickness of the evaporite units and the rate and distribution of evaporite removal. Where evaporite units are thin or dissolution is homogenous and relatively slow, then the overlying sediments, regardless of their rheology, will tend to uniformly subside with little, if any, evidence of disturbance (e.g. Anderson et al. 1978). In cases where dissolution is temporally or spatially heterogenous, differential foundering of the overlying units results in deformation and, potentially, unit-wide brecciation. Solution cavities may locally develop and propagate in an evaporite unit, the eventual failure of these voids will produce localised breccias in the collapsed overlying units to form a brecciated pipe (chimney) as observed in the Delaware Basin (Anderson et al. 1978). Where dissolution of evaporites occurs at relatively shallow depths, then surface subsidence will create accommodation for post-evaporite sediments (e.g. Smith 1980).

Establishing the former presence of subsurface evaporites and the mechanism of their removal is challenging. The strongest evidence for evaporite dissolution is provided by the occurrence of breccias in overlying insoluble units, yet unequivocally attributing these to dissolution, rather than other mechanisms, can be problematic. The presence of a solutional lag of insoluble residue at the base of, or as a matrix within, the breccias supports their interpretation as solution or evaporite collapse breccias. Recognition of pseudomorphs after evaporite minerals and the presence of evaporite cements within the brecciated units also support their interpretation as solution-related collapse breccias.

With increasing depth, compaction and pressure solution will cause halite units to recrystalise to form a dense sutured mosaic texture with impurities defining crystal boundaries (Figures 10.29c–e). Distinguishing a deep burial origin for such mosaic textures can, again, be challenging; however, where crystals exhibit stress-related flattening, a deep burial origin may be inferred.

During burial, gypsum will dehydrate to anhydrite with primary gypsum crystals being replaced by massive anhydrite units displaying a range of nodular (Figure 10.29f,g) and chicken-wire textures (Hardie and Lowenstein 2004). The process of anhydritisation is controlled by the salinity of pore fluids, the local geothermal gradient, and the lithostatic and fluid pressures. The phase change may be initiated at a depth of only a few metres and is typically complete by 1 km (Murray 1964; Hardie 1967) though, where geothermal gradients are low, gypsum may theoretically persist to depths in excess of 4 km (Jowett et al. 1993). The conversion of gypsum to anhydrite liberates large volumes of water of crystallisation that, while oversaturated in terms of calcium sulphate, is undersaturated with respect to other evaporite minerals and carbonates. These fluids have the potential to promote significant dissolution and reprecipitation in adjacent units. In open systems, where the liberated water is able to escape, a volume reduction of up to 38.5% is likely to result in the deformation of overlying and intercalated units. Where the system is closed, and the released water of crystallisation is unable to escape, an increase in the volume of up to 9.9% (Zanbak and Arthur 1986) can also drive deformation.

The burial-related loss of primary evaporite fabrics further complicates the interpretation of marine evaporite systems. However, some depositional fabrics may be retained, particularly those associated with inclusions, microbial activity and sedimentary structures. While useful, such structures should be treated with caution as there are often multiple potential modes of origin (Figure 10.31) – as always, an assemblage of genetically related features provides the best evidence on which to undertake a palaeoenvironmental interpretation.

10.3.8.3 Uplift and Exposure

Uplift and exhumation place evaporites in the meteoric realm. On exposure to undersaturated meteoric waters, high-solubility minerals, such as halite, will be susceptible to dissolution and hydration. Ongoing dissolution will produce an evaporite karst with a range of features akin to those observed in carbonate landscapes (see Section 11.3.2). In the case of complete dissolution and removal, the only evidence of the former presence of the evaporite unit may be a solutional lag or disruption of adjacent units.

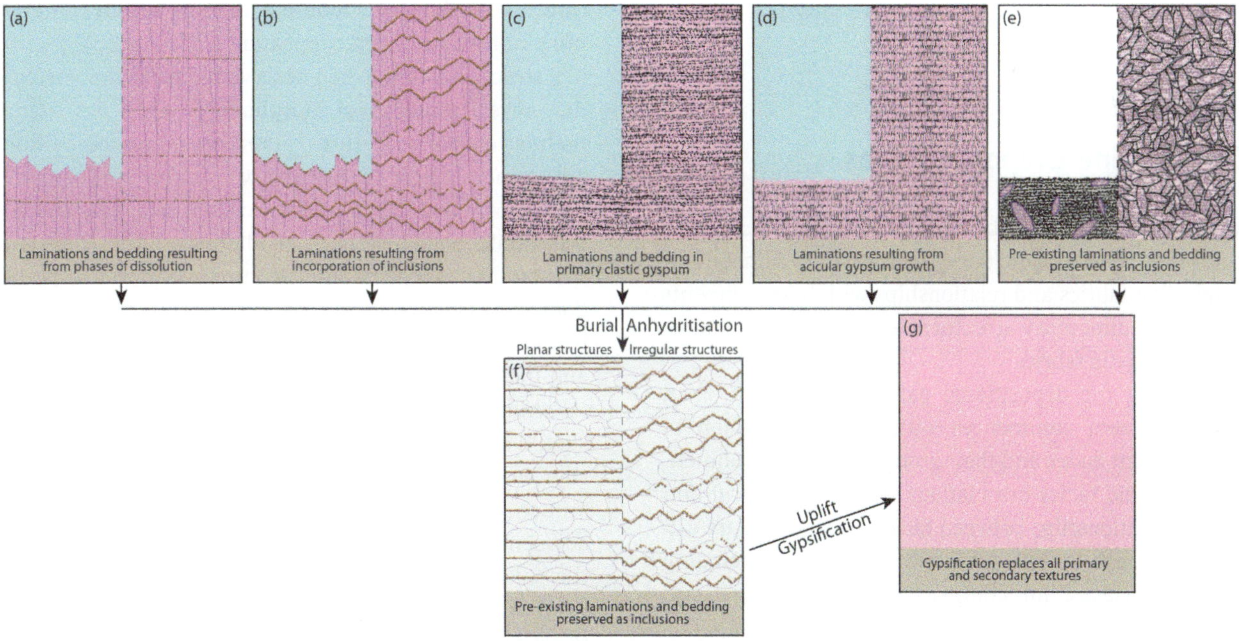

Figure 10.31 Laminations and bedding in primary evaporites may form in many ways including: during phases of dissolution (a), through the incorporation of sediment as inclusions (b), during deposition of evapoclastic grains (c), as primary growth layers (d) and through the incorporation of primary sedimentary structures during interstitial mineral growth (e). While burial anhydritisation may retain lamination or bedding, these relict fabrics are disconnected from their related formational evaporite fabrics, thus making interpretation highly problematic (f). The process of gypsification, during uplift, will typically completely remove all evidence of primary depositional fabrics (g). Note: In (a–e) the left panel demonstrates the mode of structure formation while the right panel shows the resultant sequence.

Anhydrite will undergo retrograde alteration, hydrating to gypsum to produce a variety of crystalline fabrics dominated by porphyroblastic and alabastrine textures. Coarsely crystalline gypsum porphyroblasts initially develop within the finer crystalline anhydrite texture through a solution-precipitation process that incorporates anhydrite relics within the gypsum crystal. Ongoing gypsification, via both solution-precipitation and direct hydration, will progressively replace the finely crystalline anhydrite groundmass to produce an alabastrine texture (Holliday 1970). As the evaporite undergoes further recrystallisation, initially anhedral crystals become progressively more subhedral to euhedral and textures become progressively coarser.

The process of gypsification may be expected to see a volume increase of up to 62.6% (Zanbak and Arthur 1986). However, the open system flush of meteoric fluids through the evaporite will remove excess sulphates thereby limiting the degree of volume change. On passing into adjacent facies, these solutes are liable to reprecipitate as vein-filling gypsum, often in the form of satin spar. Excess strontium released during gypsification may form celestite ($SrSO_4$) while the release of boron precipitates probertite, $NaCaB_5O_7(OH)_4{\cdot}3H_2O$ (Holliday 1970).

The complete replacement of primary and diagenetic fabrics during these processes further challenges our ability to interpret the depositional environment of the primary evaporite body (Figure 10.31). We must again rely on relationships with associated facies that have been less affected by diagenetic processes.

10.4 Marine Evaporite Facies

Evaporite facies are described and characterised based on a broad range of factors including their primary mineralogy, precipitation fabrics and relationships with host sediments as well as sedimentary structures and syndepositional to burial diagenetic fabrics.

Elucidating the depositional palaeoenvironment of an evaporite system commences with the identification of the component facies and their grouping into environmentally diagnostic facies associations. The geographical and temporal relationships between facies associations are utilised to establish the overall facies architecture, including bounding surfaces, for comparison with existing facies models.

10.4.1 Evaporite Facies

Coastal and basinal marine evaporite deposits are formed in an extensive array of geographic and environmental settings (Section 10.2). In some instances, the resultant facies are highly diagnostic of their environment of deposition. Yet, in many cases, facies are more generic and, if treated in isolation, less enlightening. Frequently, it is the associated non-evaporitic facies which are key to interpreting the depositional environment. The most common diagnostic marine evaporite facies are listed in Supplementary Table 10.S.2 along with their principal identifying characteristics. To discuss each of these facies in detail would require a volume in its own right (Warren 2016). However, it is worth commenting on some of the more important diagnostic features that may be observed when identifying the more common facies, particularly where the interpretation of these features is diagnostic or may be ambiguous or a source of confusion.

Evaporite mineralogy, textures and relationships: Establishing the age relationships between different evaporite mineral phases is essential to determining the mode of primary precipitation as well as the subsequent paragenetic history. In addition to documenting the evaporite mineralogy, it is important to also identify and record evaporite distribution and any textural relationships between different evaporite mineral phases. Evaporites may be distributed as simple isolated crystals, complex growth forms, layers, nodules or massive bodies (Figure 10.32). As in describing any texture, care should be taken to consider the distribution of evaporites in three dimensions. For example, in cross-section, enterolithic anhydrite has the appearance of folds and diapir-like structures, that, when observed in three dimensions may reveal a polygonal morphology. When describing nodular evaporite textures, attention should be paid to any mineralogical or textural variations between the nodule core and rim. Similarly, crystal cleavages should be examined for evidence of alteration. The transient nature of evaporite minerals commonly produces pseudomorphs (Figure 10.32b, Section 10.3.8), these should be described both in terms of the primary crystal habit and the secondary mineral phase.

Relationship to non-evaporite sediments: Observations should not be restricted to facies that are purely or primarily evaporitic in composition. In many evaporite sequences, it is the associated non-evaporitic host or adjacent facies that provides the key to fully interpreting the palaeoenvironment of deposition (Figures 10.32d,e, m). For this reason, it is important to fully describe the facies preceding, superseding and, where possible, laterally adjacent to the unit of interest. Particular attention should be paid to recording any fossils and sedimentary structures, including bioturbation, as well as the texture and mineralogy of the associated sediments. Where evaporites have been precipitated within the sediment

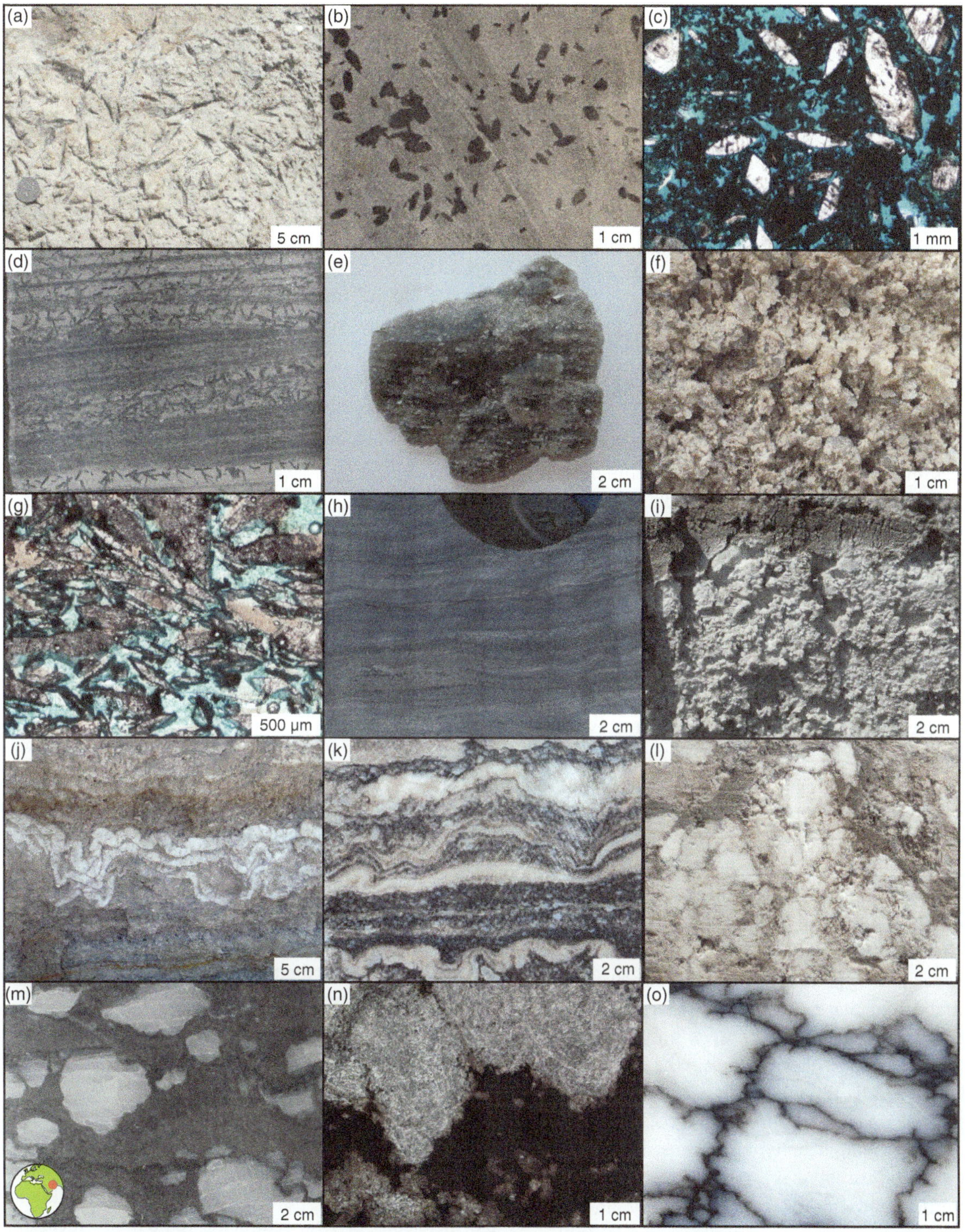

Figure 10.32 Representative images of evaporite facies associated with sabkha settings. (a) Field photograph of isolated gypsum crystals in a fine to medium carbonate sand matrix (Abu Dhabi sabkha, Recent). (b) Anhydrite pseudomorphs of isolated gypsum crystals in a peloidal packstone (Arab Formation, Jurassic). (c) Thin section of isolated gypsum crystals floating in a peloid-dominated carbonate sand. The sediment has been stabilised by blue resin (Abu Dhabi sabkha). (d) Isolated gypsum crystals in microbial-laminite (Arab Formation). (e) Large gypsum crystal preserving relict laminations as inclusions of sediment and microbial mat (Abu Dhabi sabkha). (f) Poorly sorted gypsum gravel from the shallow subsurface of the Abu Dhabi sabkha. The brown colour is derived from the host sediment. (g) Thin section image of the gypsum gravel as seen in Figure 10.32f. The sediment has been stabilised by blue resin (Recent). (h) Laminated anhydrite in core from the Arab Formation. (i) Massive anhydrite from the shallowest subsurface of the supratidal zone of the Abu Dhabi sabkha. The thin halite veneer is seen at the top of the image. (j) Enterolithic anhydrite in the shallow subsurface of the Abu Dhabi sabkha. (k) Bands of enterolithic anhydrite in core from the Arab Formation. (l) Nodular to chicken wire anhydrite structures in the shallow subsurface of the Abu Dhabi sabkha (Recent). (m) Nodular anhydrite hosted in peloidal packstone in core from the Arab Formation. (n) Thin section of nodular anhydrite in the Arab Formation. Note how the acicular anhydrite crystals give the nodules a fibrous texture. (o) Chicken-wire anhydrite in core from the Arab Formation. In this instance, nodule boundaries are defined by dolomite. Photograph credit: Stephen Lokier.

column, then any retained host sediments should be fully described. Relationships between primary or diagenetic features within the host sediment and the distribution of evaporite minerals should be documented. In extreme cases, host sediments may only be retained as poikilotopic inclusions within evaporite minerals (Figure 10.32e) or as fringes defining nodular chicken-wire textures (Figure 10.32o). In cases where evaporite minerals have grown at a former sediment surface then the nature of the surface should be recorded in detail.

10.4.2 Facies Associations

As is the case with many sedimentary systems, individual evaporite facies may be formed in a variety of different depositional settings. The depositional context is refined where facies are grouped within facies associations that are genetically constrained to specific formational settings. Here, we shall separately focus on the facies associations that are typically observed in coastal and basinal evaporite systems.

10.4.2.1 Coastal Evaporites – Sabkhas and Salinas

Coastal sabkha facies can be broadly subdivided into evaporite and non-evaporite facies with the latter typically forming the host sediments for the evaporites. Within sabkha settings, only the evaporite minerals that have been precipitated within the sediment column are likely to be preserved in the geological record. As previously discussed, subaerially precipitated efflorescent evaporites are ephemeral – rarely surviving beyond the next pluvial event. Similarly, any buried deliquescent minerals are rapidly dissolved and removed on contact with undersaturated groundwaters. Of the subsurface precipitates, gypsum is, by far, the volumetrically dominant evaporite mineral within sabkhas. While anhydrite may also be observed within the shallowest subsurface (Figure 10.32i), this is typically evanescent, being hydrated to gypsum on shallow burial.

Sabkha facies associations are typically constructed and described in the context of their position of deposition within the intertidal and supratidal zones of the coastal setting. While the precipitation of evaporite minerals only takes place in the supratidal setting (Section 10.3.5), the sediments that host this precipitation may have formed in a range of settings. By studying the textural and spatial relationships between the evaporite minerals, their host sediments and any proximal evaporite-free sediments, the broader depositional context and sequence stratigraphic framework can be defined.

The following provides a brief summary of the more common facies associations that may be identified in relation to a coastal evaporite system. This list is by no means exhaustive with a complete 'suite' of coastal evaporite facies associations likely to be encountered only rarely.

Subtidal facies association: peloidal packstones (in carbonate settings) or mudstones and fine sandstones (siliciclastic sediments) with common ostracods, molluscs, foraminifera and other marine skeletal material. Species may be halotolerant or halophilic (Fiorini and Lokier 2020). Typically, pervasively bioturbated (*Thalassinoides* and *Planolites*) – to the extent that bioturbation may be difficult to distinguish (Figure 10.32b). Sediments may be organic-rich and may provide evidence of seagrass rootlets (Pederson et al. 2021). In carbonate settings – laterally extensive hardgrounds may be observed (Ge et al. 2023).

Mid to lower intertidal facies association: peloidal and skeletal packstones (in carbonate settings) or mudstones and fine to coarse sandstones (siliciclastic sediments) with common gastropods, bivalves and other marine skeletal material. Sediments may be massive, laminated or cross-laminated. Common bioturbation and evidence of rootlets. Sediments are often organic-rich. Desiccation cracks may be observed in finer-grained facies. Intraclast conglomerates may be locally developed (Ge et al. 2020).

Upper intertidal facies association: laminated sediments with a variety of fabrics indicating the presence of microbial mats (Figure 10.32d; (Court et al. 2017)) or stromatolites (Paul et al. 2021). Sediments are typically organic-rich and may contain a diverse range of marine skeletal material, often with evidence of abrasion – halotolerant or halophilic species dominate. While desiccation cracks may be locally developed, bioturbation is absent.

Sabkha facies (supratidal) association: evaporite facies dominated by gypsum (anhydrite in the burial realm) precipitated as individual lenticular crystals (Figure 10.32a–c) resulting in massive (Figure 10.32f,g) or nodular (Figure 10.32i,l–o) textures within, and overprinting, the precursor subtidal to supratidal facies (Section 10.3.5; (Lokier and Steuber 2008)). Diagnostic features include poikilotopic inclusions of host sediments (potentially including organic matter and skeletal remains of halotolerant or halophilic organisms) within evaporite minerals (Figure 10.32e; (Court et al. 2017). Enterolithic textures may be observed (Figure 10.32j–k) but these can be difficult to distinguish from post-depositional evaporite emplacement.

Salina facies (supratidal) association: massive or laminated evaporites (Figure 10.32h), dominated by gypsum, sometimes occurring with interbeds of intertidal, sabkha or terrestrial facies (Section 10.3.9). This facies may be distinguished from the sabkha facies association by the lack of interstitial displacive gypsum growth within a host sediment. However, care should be taken as salina facies are particularly susceptible to overprinting by sabkha facies following complete desiccation.

The thickness of a single coastal evaporite parasequence (Figure 10.26) will be governed by the accommodation available at the time of deposition. For example, in the limited accommodation of the present-day Abu Dhabi sabkha, a complete sequence is typically on the order of 0.5–1.2 m (Lokier and Steuber 2008) while in the Late Jurassic Arab Formation of the UAE parasequences may reach up to 10 m in thickness (Grötsch et al. 2003).

10.4.2.2 Marine Basin Evaporites

Basinal evaporite facies are also commonly associated with non-evaporitic sediments, these sediments occur either as a host for the precipitation of evaporites within the sedimentary column, as sediments entrapped by the growth of evaporite minerals, or as beds or laminae intercalated within evaporite deposits (Section 10.3.3). The latter feature gives rise to cyclical evaporite-sediment alternations at the micro and macro-scale that are typical of basinal evaporite sequences (see Box 10.2).

Basinal evaporites can precipitate at the sediment–water interface as bottom-grown evaporites, or at either the air–water interface or within the water column. In the latter two cases, the crystals will settle to the basin floor to produce massive or laminated cumulate deposits. The significantly higher accommodation of basinal systems, when compared to the limited space in coastal settings, allows for the deposition and preservation of hundreds to thousands of metres thick evaporite sequences. Thick cumulate evaporite deposits will develop especially at the basin depocenter where sediments are less affected by erosion or dissolution during episodes of sea-level fall, and are isolated from continental freshwater input – particularly where the water column is permanently stratified (Bąbel and Bogucki 2007). Erosion of shallow water and basin margin evaporites will generate evapoclastic sediment that will be transported down-slope via debris flows, turbidites and slope failure. These allochthonous evaporite sediments may be admixed with autochthonous sediments and will form a range of sedimentary deposits dependent on their mode of emplacement.

The lack of modern analogues can make the interpretation of the bathymetric context one of the most difficult points to assess when deciphering ancient evaporite sequences (Hovorka et al. 2007; see Box 10.2). There are two main reasons behind this uncertainty: (i) the apparent paucity of fossil remains entrapped within evaporite minerals (see Box 10.4) and (ii) the often-ambiguous interpretation of many evaporite facies because of the absence of modern analogues. Many evaporite features interpreted as offering "conclusive evidence" of a shallow-water origin can be questioned if not accompanied by a thorough analysis of the evaporite succession trend (Hardie and Lowenstein 2004). One common example of ambiguous interpretation is represented by nodular or enterolithic anhydrite facies which can be confused as distinctive of subaerial sabkha environments or result from the diagenetic transformation of gypsum during burial. However, it is possible with, for instance, seismic data or, in the Messinian case, using present bathymetry (see Section 10.6.3) to categorise these basinal facies associations based on their relative position in relation to the basin margin (Figures 10.27 and 10.33, Supplementary Table 10.S.2).

Shallow-basin facies association: all of the previously described coastal evaporite facies can potentially be found in the shallowest parts of marine evaporite basins. Particularly diagnostic of this association is the presence of the stromatolitic facies comprising interlaminated carbonate and sulphate deposits formed by microbial mats with an undulate morphology (Figure 10.33a; (Peryt 1996)). Silt-to-sand sized, gypsum grains, in the form of 'gypsiltites' and 'gypsarenites', contain a range of sedimentary structures typically associated with shallow marine settings such as ripples and cross-stratification (Figure 10.34a). Sand-sized gypsum ooids (gypsolites), with a concentric internal structure, are rare features in modern saltern ponds (e.g. Guerrero Negro, Mexico (Vogel et al. 2009)) and have also been described from ancient marginal evaporite basins (Babel and Kasprzyk 1990). The concentric layers in these ooids reflect variations in the degree of agitation of supersaturated brines (Babel and Kasprzyk 1990). Analogous modern shoreline deposits characterise several present-day saline and restricted basins (e.g. the Dead Sea (Sirota et al. 2021) and Lake Assal (Castanier et al. 1992)) where halite ooids (halolites) are commonly associated with polygonal cracks indicative of repeated subaerial exposure. Fossil remains of hypersaline crustaceans (e.g. *Artemia salina*) may be observed while bird footprints can be a distinctive feature in intertidal deposits (Hardie and Eugster 1971; Manzi et al. 2016).

Where shallow-water carbonate factories are present within the basin, a range of distinctively shallow water carbonate sediments (e.g. oolitic grainstones, bioclastic rudstones) may be observed in association with the evaporitic facies.

Figure 10.33 Representative images of evaporite facies associated with basinal settings. (a) Gypsified microbial mat with stromatolitic morphology (Badenian Evaporites, Poland, Middle Miocene). (b) Giant twinned selenite crystals, massive selenite facies (San Miguel de Salinas, Spain; Messinian). (c) Bedded selenite facies (Sorbas Basin, Spain; Messinian). (d) Banded selenite facies (Caltanissetta Basin, Sicily; Messinian,). (e) A gypsum bed entirely made up of vertically oriented giant gypsum intergrowths (Badenian Evaporites, Poland, Middle Miocene). (f) Columnar twinned selenite crystals of the massive selenite facies (Piedmont Basin, Italy; Messinian). (g) Iso-oriented, curved single gypsum crystals, sabre gypsum (Sorbas Basin, Spain; Messinian). (h) Gypsum supercones laterally fringing into marls (Sorbas Basin, Spain; Messinian). (i) Outcrop view of a conical structure composed of the branching selenite facies (Piedmont Basin, Italy; Messinian). (j) Nucleation cone at the base of a gypsum bed (Sorbas Basin, Spain; Messinian). (k) Bedded gypsum facies forming a large dome at the top of the bed (Caltanissetta Basin, Sicily; Messinian). (l) Laminar gypsum facies composed of alternation of white/pink gypsum laminae and brown terrigenous laminae (Piedmont Basin, Italy; Messinian). (m) Bedded halite deposits formed by the cyclical alternation of halite beds and terrigenous layers (Realmonte mine, Sicily; Messinian). (n) Salt core showing a detail of the cycles of Figure 10.33m. Note the transition between the upper halite layer and the lower part with anhydrite layering. (o) Alternations of yellow potash salts, with halite and dark clay-rich halite layers. Photograph credits: Marcello Natalicchio.

Figure 10.34 Outcrop views of clastic gypsum facies. (a) Vertical succession of coarse-grained clastic gypsum (gypsrudite; GR) and massive (MG) and laminated (LG) sand-sized clastic gypsum (gypsarenite); the arrow indicates parallel to cross lamination (Messinian, Cyprus). (b) Matrix-supported breccia with sand-sized gypsum matrix (Messinian, Cyprus). (c) Clast-supported gypsrudite made up of gypsum clasts (gypsrudite) – note the swallow-tail gypsum crystals (arrows) (Messinian, Piedmont, Italy). (d) Chaotic deposit comprising metre- to hundreds of metre-scale gypsum blocks within a muddy matrix (Messinian, Piedmont, Italy). Photograph credits: Marcello Natalicchio.

Marginal-basin facies association: predominance of sulphate deposits dominated by massive (Figure 10.33b) and bedded selenite (Figure 10.33c) often interbedded with organic-rich deposits. Internally, gypsum units may be banded or bedded at a cm- to dm-scale (Figure 10.33d). The large variety of observed gypsum growth forms may be mostly related to internal basin stratification, to the development and fluctuation of a seasonal pycnocline (see Section 10.3.6), and/or to specific hydrological regimes. These facies include giant twinned selenite crystals (Figure 10.33b), gypsum intergrowths (Figure 10.33e; (Bąbel et al. 2010)), columnar gypsum (Figure 10.33f), curved (sabre) gypsum crystals (Figure 10.33g; (Peryt 1996; Bąbel 2002)), gypsum supercones (Figure 10.33h; (Dronkert 1976) and branching selenite (Figure 10.33i; (Lugli et al. 2010; Natalicchio et al. 2021)). Decimetre-scale nucleation cones are common at the base of gypsum beds (Figure 10.33j) and, locally, metre-scale domes at bed tops (Figure 10.33k). Laminar gypsum deposits are also a typical facies of basinal evaporites (Figure 10.33l; (Kirkland 2003; Natalicchio et al. 2021)).

More distal basin margin facies are dominated by laminated dolomicritic and organic-rich sediments potentially containing biomolecules of chemotrophic organisms (Natalicchio et al. 2017; Sabino et al. 2021; Pellegrino et al. 2023). Evapoclastic sediments showing variable fine to very coarse-grained textures are also common (Figure 10.34b–d). These deposits are the result of gravity-driven processes operating at a range of scales – including slumping, sliding and turbiditic resedimentation.

Distal-basin facies association: the previously described non-evaporitic organic-rich and dolomicritic facies can potentially be found in the deepest parts of marine evaporite basins. However, the more distal sectors of basin-wide evaporite systems are commonly represented by a thick succession of chloride, potassium and magnesium salts typically dominated by laminar (Figure 10.33m–o) and chevron halite. These evaporites frequently exhibit a bedded aspect as a result of the alternation with thin layers of polyhalite, anhydrite (after primary gypsum) and organic-rich mud (Figure 10.33m,n). Of secondary abundance are intercalated horizons of primary laminar gypsum nucleated in the water column during phases of seawater recharge. Redeposited clastic evaporite facies, mostly gypsiltites and gypsarenites deriving from turbiditic flows, can also be volumetrically significant.

10.4.3 Facies Successions

To gain an insight into the complete history of the evaporite basin, from initiation to demise, the evaporite facies must be studied within the context of their relationships with stratigraphically adjacent non-evaporite units. Evaporite successions can be broadly subdivided into those with a clear internal architecture and those that are internally heterogeneous.

Where the evaporite units are interbedded with non-evaporitic facies, or where they contain cyclic layers of evaporite minerals ordered by solubility, it is reasonable to infer that the supply of solutes was punctuated, with evaporite formation being controlled by episodic or periodic influx of fluids followed by intervals of evaporation. During marine recharge, normal marine salinity may be achieved within basin surface waters – allowing the development, or return, of carbonate factories. Following isolation, evaporation increases the salinity, eventually ensuring the demise of any carbonate factories and the initiation of evaporite precipitation. As evaporation continues, the surface elevation of the water will fall (evaporitive drawdown) and the sequence of marine evaporite minerals (Usiglio 1849) will be precipitated. Cyclical seawater recharge to the basin will repeat this process to produce the 'classical' repeated packages of carbonate, anhydrite, halite and sylvite – often referred to as *Usiglio Cycles* as observed in the Santos Basin, Brazil (Rodriguez et al. 2018) or Zechstein of northern Europe (Van Gent et al. 2011).

Where the evaporites are internally homogenous, it is reasonable to assume that they either represent a single formational event or, in the case of thick evaporite sequences, the continuous, steady-state, supply of solutes, for example, the Louann Salt of the Gulf of Mexico (Hudec and Jackson 2011) or reflux, as hypothesised for the thick anhydrite sequences of the Castile Formation (King 1947; Kirkland et al. 2000).

10.4.3.1 Coastal Sabkha Successions

The *typical* coastal sabkha evaporite succession will comprise a simple sequence of subtidal to intertidal facies associations grading up-section into gypsum gravel that is, in turn, succeeded by nodular, enterolithic or massive anhydrite (Figure 10.26). When treated in isolation, these evaporite fabrics are not conclusively diagnostic of a specific evaporite setting. Instead, it is their relationship with the host sediments that is critical to establishing a coastal sabkha origin. The presence of microbial textures, mud cracks or rootlets is also consistent with a continental or coastal sabkha setting. An interpretation as a marine coastal sabkha is best established through the identification of marine bioclasts within the host sediments. However, care should be taken to confirm that these sediments, and their skeletal grains, are contemporaneous with sabkha formation and do not represent an earlier deposit that has been overprinted by continental sabkha evaporites (e.g. due to a relative sea-level fall).

The thickness of the evaporite-dominated assemblage is primarily controlled by the height of the capillary fringe which is, in turn, influenced by several factors including the grain size and degree of sorting of the host sediments, the evaporation rate and the local tidal regime. With time, the precipitation of evaporites will occlude porosity, reduce permeability and thereby limit capillary action. Once all of the interstitial accommodation (pore space) has been filled, only displacive precipitation is possible, this will elevate the ground surface by a critical few centimetres to tens of centimetres, potentially cutting off the uppermost part of the sequence from the capillary regime. Where the water table is relatively stable, the zone of evaporite precipitation will be of limited thickness, typically on the order of less than 1 m (Lokier and Steuber 2008). In cases where the level of the water table rises, there is the potential for undersaturated groundwater brine to dissolve previously precipitated evaporites. Equally, a fall in the level of the water table will potentially move the capillary fringe below the zone of evaporation resulting in cessation of evaporite precipitation.

10.4.3.2 Basinal Successions

As we have seen, within marine evaporite basins, there are significant differences between the evaporite successions developed at the basin margin and in more distal, basin centre, areas. While different in terms of facies and unit thickness, different areas of the basin may share a common broadly synchronous cyclical pattern of alternating evaporite and non-evaporite facies recording the history of basin filling.

Particularly fine examples of basin margin evaporite sequences can be found in the Silurian Michigan Basin (Briggs et al. 1980; Pope et al. 2000), along the margin of the Eocene Catalan Potash Basin (the Odena Gypsum lower member, Spain (Rosell and Pueyo 1997)), in the Badenian (Middle Miocene) Carpathian Foredeep Basin (the Lower autochthonous unit in Poland and Ukraine (Bąbel et al. 2011) and in the Messinian Mediterranean marginal basins where up to 16 evaporite-shale cycles are observed (the Primary Lower Gypsum unit (Lugli et al. 2010); Figure 10.35a,b; also see Section 10.6.3).

Considering such examples, a typical marginal evaporite succession starts with a lower unit made up of cyclical alternations of metre to decametre thick gypsum beds (often diagenetically altered to anhydrite in the burial realm) and thin

Figure 10.35 Vertical successions of basinal evaporite deposits. (All examples from the Messinian). (a) Outcrop view of a planar gypsum bed lying on top of hemipelagic non-evaporitic deposits (Sorbas Basin, Spain). (b) The 'Vena del Gesso' ridge formed of up to 16 lithological cycles (Northern Apennines, Italy). (c) A vertical gypsum crystal facies assemblage in a single lithological cycle. Note the nucleation cones at the base of the bed "(black arrows)" and the supercones at the top ("yellow dotted lines" Sorbas Basin, Spain). Photograph credits: Marcello Natalicchio.

layers of non-evaporite shallow marine sediments (typically mudstones or marls) (Figure 10.35a-c). These sediments will normally record a shallowing-upward trend reflecting sea-level fall in concert with increasing brine saturation. The total thickness of these lower-unit marginal sulphate-dominated successions often exceeds 100 m.

Within the lower marginal unit, the base of a typical evaporite cycle can be characterised by a cm-thick stromatolitic limestone (Lugli et al. 2010; Ortí 2011) lying above a non-evaporitic organic-rich shale or marl. This carbonate unit is typically succeeded by vertically oriented twinned selenite crystals or gypsum intergrowths forming a massive selenite facies. The preferential orientation and large dimension of the gypsum crystals suggests competitive subaqueous growth (Ortí 2011) at the basin floor within a brine that was isolated below a relatively stable pycnocline lying at a depth of metres to tens of metres (Figure 10.36a,b) (Bąbel 2004). An up-section transition to banded (bedded) gypsum facies reflects fluctuations in the depth of the pycnocline resulting in episodes of dissolution followed by syntaxial growth with renewed precipitation (Figure 10.36c,d). In some marginal basin sequences, the bedded selenite facies are intercalated with gypsum microbialite-selenite layers reflecting the periodical blooming of cyanobacterial communities inhabiting the sediment surface as a mat in shallow oxygenated waters (Bąbel 2005). Part of the evaporite cycle can be characterised by units containing curved selenite crystals (sabre gypsum; Figure 10.33g), gypsum supercones (Figure 10.33h) or, specifically for the Messinian basin, branching selenite facies (Figure 10.33i). These characteristic facies are indicative of specific hydrological regimes within the evaporite basin. Curved gypsum is interpreted as forming where basin floor brine currents have preferentially supplied ions to the crystals thereby favouring their growth in an upstream direction (Bąbel 2002) (Figure 10.36e,f). A similar influence by basin floor currents has been tentatively proposed for the asymmetrical structure of the branching selenite facies (Lugli et al. 2010). These coarse crystalline selenite facies are locally intercalated with laminar cumulate gypsum formed at the oxycline and subsequently deposited on the seafloor, lacking any sedimentary structure indicative of tractive bottom currents (Figures 10.33l and 10.36g,h) (Natalicchio et al. 2021). Such a relationship suggests a syngenetic origin of the branching selenite facies immediately below or at the seafloor (Natalicchio et al. 2021). The evaporite cycle terminates with a mud or marl interval that indicates the onset of conditions unfavourable to evaporite precipitation.

A common feature in many ancient evaporite basins is the presence of an unconformable upper unit dominated by clastic gypsum facies. These units are interpreted as having been formed as sea level lowering exposed and eroded previously deposited marginal gypsum units, with the resultant evapoclastic deposits being transported down-slope and redeposited in the subaqueous environment by gravity currents (Figure 10.34).

Within the more distal regions of marine evaporite basins, the evaporitic sequence is characterised by thick evaporite units dominated by chloride minerals along with minor gypsum cumulate deposits. These distal evaporite deposits may reach thicknesses on the order of hundreds to thousands of metres, resulting in lower units becoming increasingly susceptible to deformation (halokinesis; Figure 10.37a). As previously discussed (see Section 10.3.1), the presence of significant thicknesses of halite, along with potassium- and magnesium-rich salts requires significant volumes of seawater to be supplied to the basin, either as episodic flooding events or as continuous, but limited, inflow.

The base of a typical distal evaporite sequence such as in the Upper Eocene Catalan Potash Basin (Taberner et al. 2000), the Miocene salt succession of the Granada basin (Spain; (García-Veigas et al. 2013)) and the Caltanissetta basin (Sicily; (Garcia-Veigas et al. 1995; Lugli et al. 1999) can be characterised by a unit of fine-grained deposits (shales and marls) intercalated with anhydrite layers representing the diagenetic product of primary gypsum deposits as evidenced by the localised preservation of pseudomorphs after gypsum crystals (Garcia-Veigas et al. 1995). The overlying sequence is typically characterised by a shallowing-upward trend of evaporite deposition that may record repeated desiccation events. The lower unit of each sequence mostly comprises massive to poorly bedded halite formed via the settling of platy halite crystals in a deep, stratified, basin (Figure 10.36g,h). These cumulate deposits are overlain by an upper unit of internally stratified (cm- to dm-scale) halite beds intercalated with fine-grained siliciclastic layers. The upper unit records the bottom growth of halite hoppers in shallow water hypersaline evaporite pans experiencing an influx of terrigenous sediment as a result of short-term climate fluctuation at the seasonal scale (Figures 10.33m and 10.36i,j; (Manzi et al. 2012)). The lower and upper halite units are sometimes separated by a laterally discontinuous unit of potassium and magnesium salts that record the point at which all of the available sodium had been consumed in the precipitation of halite with a consequent switch to the precipitation of 'bittern salts (Figure 10.37b). Depending on the hydrologic conditions within the basin, the terminal stage of salt deposition may be represented by the onset of continental conditions or by the deposition of a cap unit of primary gypsum deposits representing a new phase of seawater influx to the evaporite basin.

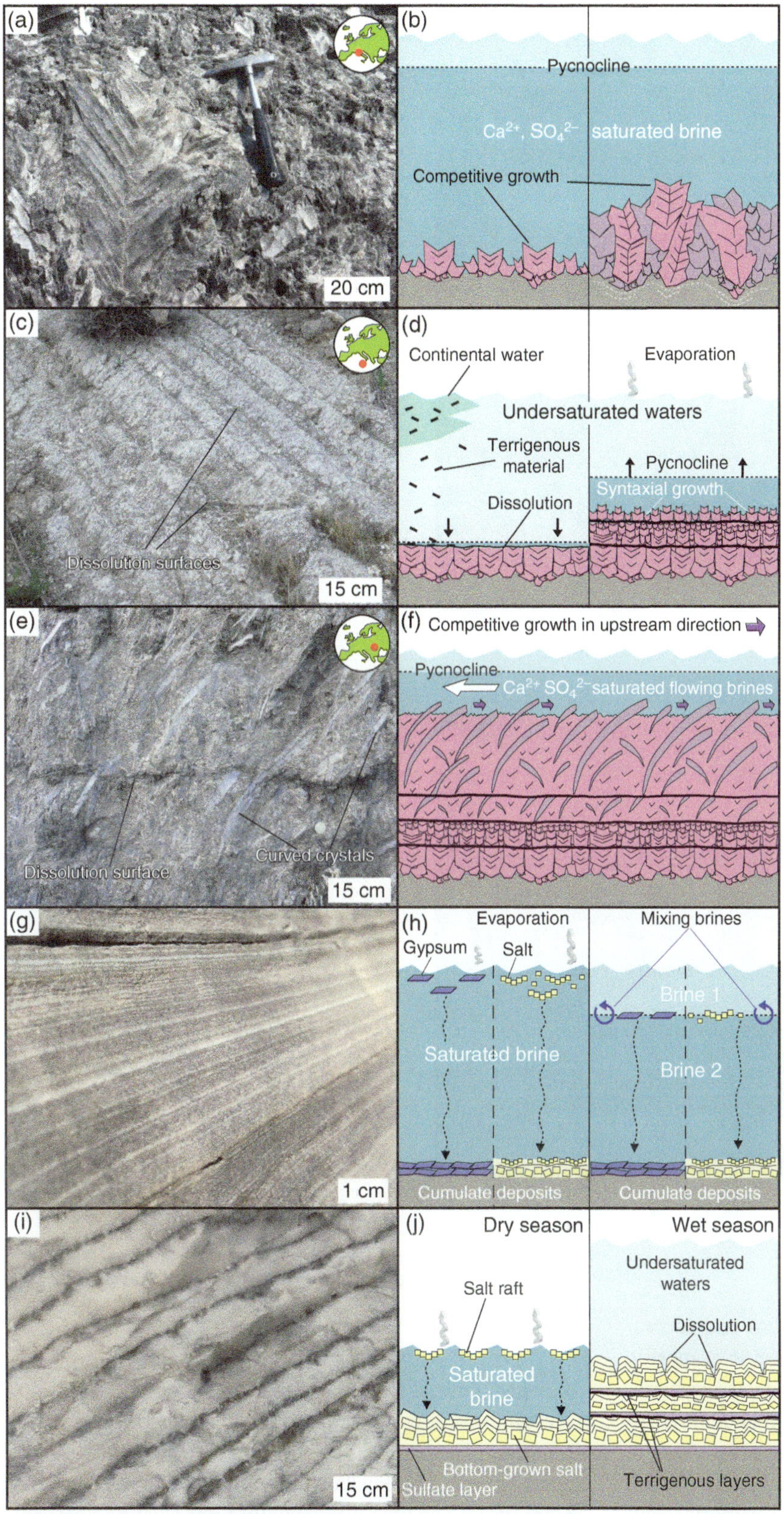

Figure 10.36 Modes of growth of different evaporite facies in marine basinal settings. (a,b) Massive selenite developed below a stable pycnocline (Messinian, Piedmont, Italy). (c,d) Banded selenite recording precipitation and dissolution in relation to a fluctuating pycnocline (Messinian, Sicily, Italy). (e,f) Curved gypsum formed under prevailing basin floor brine currents (modified from (Bąbel 2002); Middle Miocene Badenian, Poland). (g) Laminar cumulate gypsum (Messinian, Piedmont, Italy). (h) Sketches showing the formation of evaporite crystals at the air–brine interface (left) and at the interface of two mixing brines (right). These processes can be considered for the nucleation of both gypsum and halite-producing cumulate deposits at the basin floor, i.e. laminar gypsum and massive laminated salts, respectively. (i,j) Salt cyclicity within a bedded halite unit showing the repetition of halite beds, layers of sulphate minerals (mostly anhydrite) and terrigenous sediments (Messinian, Realmonte mine, Sicily, Italy). (j) The preferential bottom growth of crystals (hoppers and chevron textures) during dry periods (left) and phases of dissolution and deposition of the thin terrigenous layers during wet periods (right). Photograph credits: Marcello Natalicchio.

Figure 10.37 Basinal evaporite successions (Realmonte mine, Sicily; Late Miocene). (a) Folded layered halite. (b) Alternation of kainite rich intervals (yellow colour) and halite-rich intervals (white-grey).

10.4.4 Facies Architecture

Evaporite facies and, even, facies associations and successions are rarely uniquely diagnostic of a specific evaporitic setting (Table 10.2). By studying the overall facies architecture, that is, the two- and three-dimensional geometries of facies distribution, of the evaporite system, we can have the best opportunity to elucidate the depositional context of the evaporite deposits. Broadly speaking, coastal evaporites will be deposited parallel to the coastline, passing laterally seaward into marine sediments. Conversely, basin evaporites will display a broadly concentric distribution throughout the basin, typically with the most soluble salts located at the basin centre. Obviously, such relationships can only be established when considering the depositional architecture at the basin scale.

10.4.4.1 Coastal Marine Facies Architecture

Coastal evaporites are dominated by sulphate minerals, in the form of gypsum and anhydrite, precipitated in close relationship with siliciclastic or carbonate facies. Evaporite minerals may be precipitated within pre-existing host sediments, in the case of sabkhas, or within shallow ponds of varying degrees of connectivity to the marine realm – salinas. A dominance by sabkhas, salinas, or a combination of these evaporite systems is dependent primarily on the topography of the coastline, particularly the slope angle, along with the local tidal regime. The three-dimensional facies architecture of these coastal evaporite systems is also strongly controlled by their sequence stratigraphic context.

The study of coastal evaporites requires the analysis of the full suite of facies within the depositional system. Not all of these facies will contain evaporite minerals, nevertheless, their characterisation and the establishment of the architectural relationships between facies is essential to the accurate interpretation of the environment of deposition.

Sabkha facies belts: as the supply of fluids and, hence, solutes, is from the marine realm, it is no surprise that the broad-scale facies architecture is that of a series of shoreline-parallel facies belts. These facies transition landward from shallow marine, often restricted, subtidal facies, through an intertidal zone to a fully supratidal terrestrial setting (Figure 10.21a). The nature of the subtidal and intertidal facies will be dependent on a range of factors including the mineralogy of the sediments, prevailing energy regimes and the coastal morphology. Stressed conditions in the uppermost intertidal zone may facilitate the development of extensive microbial communities as mats, thrombolites and stromatolites (Figure 10.25c, d) – providing an important stratigraphic datum marking mean high tide (Lokier and Steuber 2008). Sulphate mineral precipitation within the shallow subsurface of the supratidal zone overprints pre-existing depositional fabrics to produce a range of nodular, enterolithic, chicken wire and massive evaporite textures (Figures 10.26 and 10.32). Further landward, with increasing distance from the marine realm, evaporite mineral precipitation declines with a transition to terrestrial.

Coastal salina facies geometries: salina facies development is controlled by the hydrological regime of the salina, specifically, the source of the water and the frequency and duration of flooding. At one extreme of this continuum lie dry, desiccated, salinas, these are rarely flooded (tens of years) and bear a close affinity to sabkhas. At the other extreme are salinas that are constantly flooded, being refreshed with every Spring tide, thus maintaining subaqueous conditions at the basin floor. Salina facies are controlled by both the topography of the basin and the source of the fluids. In cases where abundant fluids are supplied to the salina, either as surface floods or from numerous inflows, facies geometries are largely controlled by the morphology of the basin with facies concentrically contoured in relation to depth. In cases where lower volumes of brines enter the salina at a point source, facies are typically arranged concentrically in relation to this source. The 'ideal' salina zonation will comprise halite at the salina centre passing outward to gypsum and marginal carbonates that may contain evidence of microbial communities. The vertical succession at the centre of such a salina will pass from carbonates to large in situ gypsum crystals followed by laminated gypsum overlain by halite and potentially capped by a gypsum aeolianite (Figure 10.21b). These facies geometries record basin filling, producing an overall shallowing-upward evaporite sequence. However, closer inspection may reveal phases of both precipitation and dissolution – recording periods of desiccation and flooding, respectively. While shallow water evaporite textures are relatively easy to identify, placing such textures within the context of a salina can be challenging. As potash salts have not been identified from present-day coastal salinas, their presence within a salina sequence would hint at a continental origin.

10.4.4.2 Basinal Facies Architecture

Following the isolation of a marine basin, evaporitive drawdown will initially result in a period of non-deposition prior to the precipitation of the first evaporites, in the form of carbonate and sulphate minerals. As evaporation progresses, continued drawdown sees brines recede towards the deeper water basin centre, successively precipitating a layered sequence of progressively more soluble evaporite minerals (gypsum and halite) concluding with a cap of potash salts at the basin centre. The result of this model of basin desiccation will be a sequence of monomineralic evaporites precipitated from a water body of shrinking aerial extent, with a broadly concentric distribution of evaporite minerals, often referred to as a bullseye evaporite distribution.

While concentric bullseye geometries are observed in numerous ancient evaporite deposits, many basins are more geographically complex. The inherited internal physiographic architecture of the basin exerts a significant control on the hydrological regime and, consequently, the development of the vertical and lateral facies architecture of the evaporite deposits. Marginal sub-basins are preferentially characterised by the deposition of carbonate and sulphate evaporites during the early evaporative phases whereas late-phase chlorides dominate in more distal basins. The considerable thickness of halite in basinal evaporite deposits makes them inherently susceptible to deformation through halokinesis, akin to the thick Jurassic salt sequence of the Gulf of Mexico (Hudec et al. 2013). This, together with the effects of burial anhydritisation of primary gypsum facies, (see Section 10.3.8) can make the reconstruction of the basin-wide architecture particularly challenging (Kasprzyk 2003).

Despite the infinite variety of potential marine evaporite basin bathymetries and architectures, it is always useful to consider the marginal and distal deposits separately.

Marginal sub-basin facies geometries: the depositional geometry of the marginal evaporite units is intrinsically controlled by the sub-basin topography. Where the basin margin is wide and laterally extensive, this will facilitate the deposition of tabular mixed carbonate-evaporite units over a wide area. Narrow basin margins are more likely to accommodate an elongate, shoreline parallel, lens of mixed carbonate-evaporites thinning both landward and basinward (Ortí). This elongate lens geometry results from the combination of limited space for evaporite accumulation towards the coast and a basinward thinning of the non-evaporitic components (e.g. carbonates, organic-rich shales). Internal stratification of the water column is one of the primary parameters controlling the development of the evaporite facies architecture within marginal sub-basins. The development and stability of water column stratification is, in turn, driven by the fluctuating hydrologic interaction between evaporation, the supply of continental freshwater and the influx of seawater as well as prevailing climatic regimes. In basin margin areas, an input of continental waters may depress the pycnocline thereby promoting the partial dissolution of more proximal gypsum deposits (i.e. banded selenite facies; Figures 10.36c, d) and a basinward shift of the microbial carbonate factory. In deeper areas of marginal sub-basins, a chemocline between oxygenated and anoxic waters may develop. Below the chemocline, a high rate of microbial sulphate reduction limits gypsum precipitation but, conversely may promote carbonate production to form laminated peloidal wackestones and dolomitic organic-rich shales (Pellegrino et al. 2023). An idealised margin-to-basin profile commences with near-shore carbonates comprising microbialites (Bąbel 2005) and other carbonate facies sourced from coeval carbonate factories at the basin margin (Roveri et al. 2020). These carbonate units interdigitate, downslope, with banded and massive selenite facies precipitated below the pycnocline (Figure 10.38a). More distal areas of the marginal sub-basin will be characterised by carbonates and organic-rich shales deposited below the chemocline. Lenticular gypsum crystals may locally grow displacively within these fine-grained sediments. The landward transition of this profile is rarely preserved, typically being removed during subsequent subaerial erosion. As the limited accommodation of the shallow basin margin is filled, progradation will shift the intersection lines of both the pycnocline and chemocline interfaces against the sediment surface basinward. This will produce a diachronous shallowing-upward sequence characterised by basal dolomitised organic-rich shales overlain by deep water carbonates, massive and banded gypsum and, ultimately, capped by shallow water microbial carbonates.

Distal sub-basin facies geometries: the inherent solubility of the evaporite minerals precipitated in distal sub-basins typically precludes their occurrence in outcrop. For this reason, these evaporite sequences are best studied in mines and cores and from the integration of borehole and seismic data.

The geometry of evaporite deposits in distal sub-basins is strongly dependent on the basin's hydrological regimes, which, in turn, are intimately related to the palaeotopography of the basin and, in particular, the balance between seawater inflow and brine outflow. The considerable thickness of distal sub-basin evaporite successions required a considerable input of solutes to the basin, either through a narrow strait or by marine groundwater seepage (with a hydraulic head aided, for instance, by evaporitic draw-down within the basin). In addition, the deficit of salt observed in ancient basinal systems (i.e. the volume ratio of halite to anhydrite; Section 10.3.6) implies that a loss of NaCl brine, through outflow or reflux via permeable sediments, occurred during evaporation of the basin. Here, we shall consider the evaporite facies geometries formed during the filling of distal evaporite sub-basins.

A common feature of distal sub-basin sequences is a basal evaporite unit composed of carbonate and anhydrite, with a thickness of tens to hundreds of metres. The anhydrite derives from the diagenetic transformation of former gypsum deposits during burial (Section 10.3.15). These could represent either laminated cumulates of gypsum crystals that nucleated in the water column, as observed in the Permian Castile Formation, USA (Kirkland 2003), fine to coarse-grained evapoclastic turbidite deposits, as possibly developed in the deep Mediterranean Messinian basins (Roveri et al. 2014a), or bedded selenite facies deposited in marine lagoons, as interpreted for the Tortonian evaporites of the Granada Basin (García-Veigas et al. 2013). Regardless of the mode of formation, the deposition of the basal anhydrite unit is interpreted to be coeval with the selenite facies deposited within marginal sub-basins.

As the basin becomes increasingly restricted, the brines will eventually become supersaturated with respect to halite. During the initial stages of halite precipitation, small crystals, precipitated at the air–water interface or at the pycnocline, settle to the basin floor producing a laterally extensive aggrading sequence of thick, structureless, cumulate deposits onlapping the basin margin (Figure 10.38b; e.g. Caltanissetta Basin, Sicily; (Garcia-Veigas et al. 1995)). In the Permian Castile Formation, USA, these basal halite units are represented by a unit comprising seasonally varved anhydrite intercalated with bedded halite that is laterally continuous for tens of kilometres (Kirkland 2003).

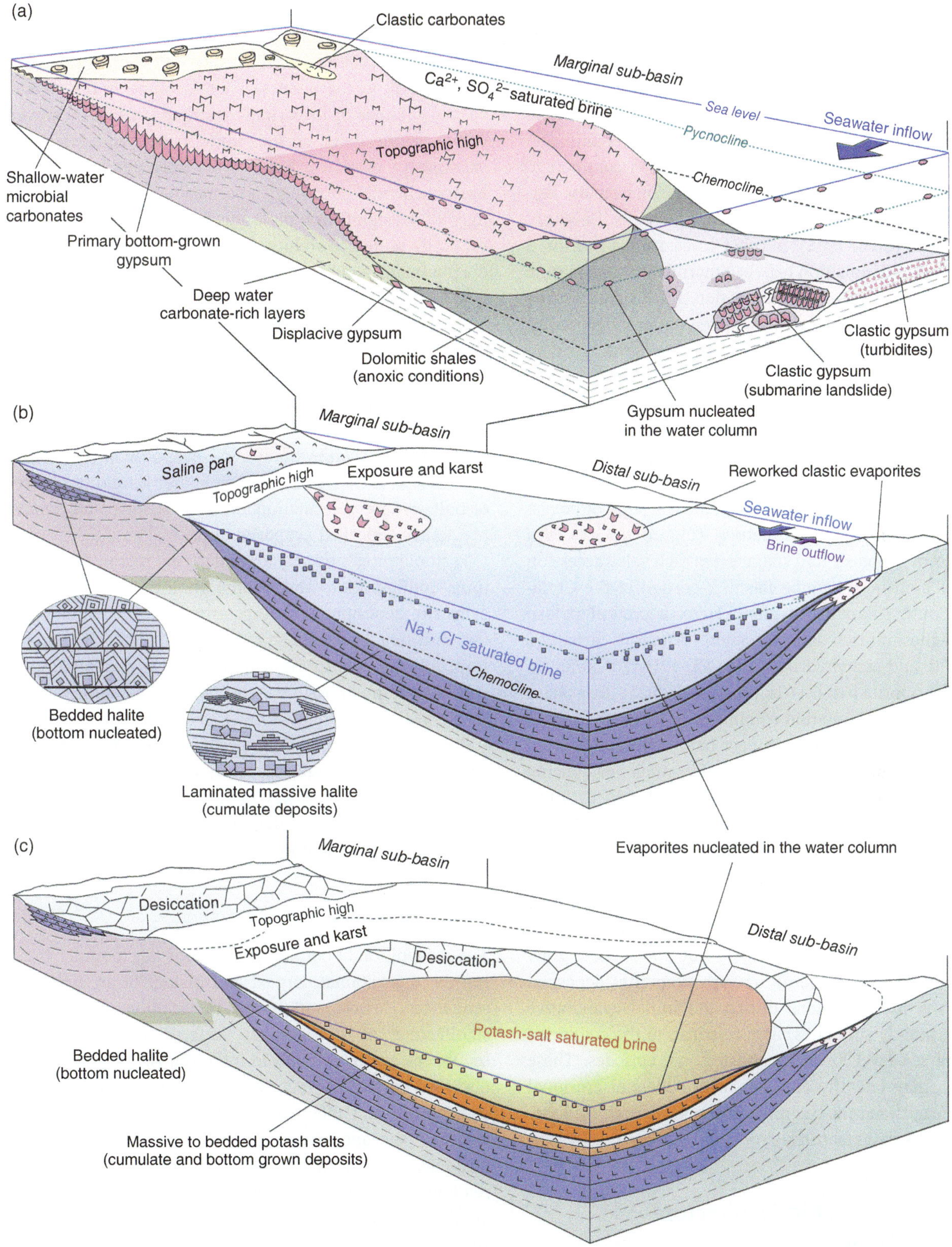

Figure 10.38 Simplified sketch of the facies architecture of basinal evaporite systems. (a) Idealised margin-to-basin profile of a marginal sub-basin showing the transition from shallow water carbonates (microbialite and clastic), bottom-grown gypsum, deep water carbonates to deeper dolomitic shales. Evapoclastic deposits are also represented in distal settings. (b) First phase of cumulitic evaporite (mainly halite) deposition in a deep distal sub-basin. Contemporaneous with the deposition of bottom-grown bedded halite in marginal sub-basins. (c) Phase of major isolation of the distal sub-basin resulting in increasing salinity of the brine with consequent deposition of potash salts. The marginal areas of the basin experience episodes of subaerial exposure and erosion.

In shallower, marginal, areas of the distal sub-basins, seasonal climatic fluctuations and continental water influx may result in the periodic mixing of water masses to form bedded halite consisting of alternating bottom-grown halite strata intercalated with thin, mud-rich layers. The transition between a persistently deep-water, lower halite unit and an upper shallower-water halite unit is common to many ancient basins, such as at the transition from the Permian Castile to the Salado Formation in the Delaware Basin (Lowenstein 1988). In other cases, such a transition may reflect the transition between marine and non-marine conditions during halite precipitation. In this context, the upper halite unit exhibits a completely different geochemical signature (e.g. bromide content and $\delta^{34}S$) and its deposition is affected by the recycling of previously deposited evaporites by non-marine waters (Cendón et al. 2004; García-Veigas et al. 2013). During shallowing, locally depressed areas of the distal sub-basin may form saline pans filled with brines enriched in Mg^+, K^+, Cl^- and other minor residual solutes that precipitate as a suite of potash salts with lenticular geometries (Figure 10.38c). An example of a potash and magnesium-rich salt succession is Brazil's offshore Sergipe Basin, a portion of the Cretaceous South Atlantic salt giant. Here, a cyclical potash and magnesium-rich salt succession (primarily carnallite, bischofite and tachyhydrite ($CaMg_2Cl_6{\cdot}12H_2O$)) exhibits a cyclical evaporite pattern (Szatmari et al. 2021). This phase of bittern salt precipitation commonly occurs at the transition between the lower and upper halite unit and is often associated with intrasalt unconformities evidencing episodes of exposure and subaerial dissolution (Warren 2010). Intermittent phases of marine re-flooding will recharge the basin with new solutes that will potentially precipitate as shallow water bottom-grown halite.

The facies architecture of distal sub-basins is complicated by the presence of clastic evaporite (evapoclastic) deposits that interdigitate with the primary salt units. These gravity-driven deposits primarily rework evaporites deposited across the basin margins during the initial evaporative phases (mainly sulphates) as a response to the combined effect of sea level lowering and tectonic episodes. These deposits may laterally correlate with subaerial erosional unconformities at the basin margin, as evidenced by dissolution and karstic features.

10.5 Stratigraphic Architecture of Evaporite Sequences

The fundamental role of seawater as a supply of solutes underlies a complex association between sea level and evaporite stratigraphic architecture. Both coastal and basinal marine evaporite systems have the potential to produce stratigraphic architectures that accurately record relative sea-level change, particularly where there is sustained connectivity to the open sea, for instance in the case of coastal sabkhas or in partially restricted basins. However, where connectivity to the world ocean is limited, for example in evaporite basins with seawater influx via a narrow channel or through a permeable barrier, then there may be a significant disconnect between the resultant evaporite sequence architecture and the record of sea level during its formation.

In addition to sea level, the prevailing climate at the location of deposition also has a considerable control on evaporite precipitation and preservation. Both long-term, basinal, and short-term, Milanković, oscillations control phases of evaporite precipitation and dissolution. During periods of increased aridity, enhanced evaporation concentrates brines, thereby increasing evaporite precipitation. Conversely, increased rainfall or higher humidity reduces or halts evaporite formation and may dissolve and remove previously deposited evaporite units.

Tectonic processes play an underlying role in the formation, configuration and geomorphological evolution of evaporite depositional settings as well as the evolution of the resultant evaporite sequence. Evaporites are known from all sedimentary basin types. Those precipitating in subsiding ramp or basin systems have the potential to build thick stratigraphic sequences.

Recognising the potential disconnect between evaporite stratigraphy and eustatic sea level, we will discuss evaporite sequences in terms of key bounding surfaces and stratigraphic trends and consider some of the various scenarios under which such features may develop.

10.5.1 Bounding Surfaces

The recognition of bounding services is key to understanding the depositional context and history of an evaporate sequence. However, if treated in isolation, the relevant significance of stratigraphic surfaces within evaporite sequences can be difficult to interpret. As an example, an observed karstified evaporite surface may record a period of exposure and meteoric dissolution, either during sea-level fall or evaporitive drawdown. Alternatively, this surface may result from subaqueous dissolution where rising sea levels refreshed basin brines and reduced salinity. Yet another scenario may see a deepening of the pycnocline, subaqueously exposing the basin floor evaporites to undersaturated brines with consequent dissolution. It is only when such surfaces are interpreted in the context of their relationship to the stratigraphic sequence that their significance can be elucidated.

10.5.1.1 Subaerial Erosion Surfaces

Subaerial exposure may occur as a result of relative sea-level fall, desiccation of a basin or lowering of local base level. During exposure, evaporites are subjected to a range of solutional and mechanical erosional processes, the nature and degree of which is primarily controlled by the prevailing climatic conditions – primarily temperature and humidity.

Under wet conditions, undersaturated surface waters and high humidity will rapidly dissolve exposed and near-surface evaporite minerals, transporting solutes into water courses or the subsurface. Solutes that are liberated during dissolution may re-precipitate either in the shallow subsurface or as ephemeral efflorescent surface crusts. During the process of solutional deflation, a transient karstic landscape will evolve. Evaporite minerals are rarely exposed at the surface as they are typically mantled by a layer of sediment comprising insoluble residue, derived from inclusions and intercrystalline sediment liberated from the dissolved evaporite, admixed with autochthonous subaerial sediment. In situations where the entire evaporite sequence has been removed through subaerial dissolution, it is the recognition of this admixture of an allochthonous solutional lag and autochthonous subaerial sediments that provides evidence for the former presence of evaporites and the process of subaerial solution. Care should be taken not to confuse subaerial dissolution with subaqueous erosion.

Where arid conditions prevail, the structural weakness and low hardness of subaerially exposed evaporite minerals makes them particularly susceptible to aeolian erosion, transportation and abrasion. Eroded evaporites may be concentrated to form dunes (e.g. Von Der Borch et al. 1977), however, their mechanical weakness restricts the duration and distance of transportation, limiting occurrence to areas proximal to their source. In sabkha systems, lowering of the water table under arid conditions may result in aeolian deflation of the surface to the capillary fringe producing a planar low-relief surface with a significant (10s of km) lateral extent. Aeolian bounding surfaces are difficult to recognise in evaporite sequences as they are likely to be overprinted and obscured by dissolution during subsequent transgression or on entering the burial realm.

10.5.1.2 Subaqueous Erosion Surfaces

Subaqueous erosion may be mechanical, solutional, or a combination of these processes.

Evaporites precipitated above the wave base are susceptible to erosion and transportation, either down-slope (gravity-driven) or down-current. During transportation and subsequent deposition, these evapoclastic sediments are subject to the complete range of subaqueous sedimentary processes, consequently producing sedimentary structures akin to those seen in their carbonate and siliciclastic counterparts (Figure 10.34). At the site of erosion, this process is evidenced by toppled crystals and the presence of evapoclastic sediments as a matrix between crystals in growth position.

Subaqueous evaporite solution occurs where the saturation state of the brine is reduced, typically following a flooding recharge event, or through the lowering of a pycnocline. The undersaturated brines will corrode crystals and cements producing a planar profile potentially punctuated by short sub-vertical dissolution pipes (Figure 10.16d). A sediment veneer will develop over the dissolution surface with the thickness of this layer largely determining the mode of succeeding precipitation (Section 10.3.3).

10.5.1.3 Flooding Surfaces

Flooding surfaces are invariably recorded by dissolution. In subaqueous evaporite systems, the flooding event will dilute the basinal brines – resulting in the dissolution of evaporite minerals (Section 10.3.3). Distinguishing between dissolution resulting from a true 'flooding' event and that from episodic recharge or pycnocline lowering can be challenging.

The flooding of subaerially exposed evaporites may result in the dissolution of the entire evaporite sequence (Section 10.6.1). During the flooding event, an increase in energy at the shoreline will enhance erosion, reworking insoluble sediments and driving dynamic back-stepping of facies belts (Lokier et al. 2018). Transgressive deposits are likely to be thin, particularly in carbonate systems where there will be a time lag of hundreds to thousands of years before the carbonate factory is re-established (Tipper 1997; Kemp and Sadler 2014).

Determining if the flooding event affected a previously subaqueous or subaerial setting can be challenging, particularly when considering shallow water and coastal evaporites.

10.5.1.4 Cessation Surfaces

Cessation surfaces represent a period of time (hiatus) where there is an absence of, or very little, sediment deposition. For a cessation surface to occur during evaporite formation requires a period of time during which the precipitated minerals are in a state of chemical equilibrium with relation to their associated brines, neither precipitating nor being dissolved. Such a situation is extremely unlikely to exist for any extended period of time in any natural open evaporite system.

An evaporite cessation event in a shallow water evaporite system may be recorded by the deposition of non-evaporite sediments that mantle the evaporite precipitates without evidence of preceding dissolution. A return to evaporite precipitation will see a new generation of evaporite precipitation at this surface. In basins with low or zero sediment influx, precipitation will likely recommence in optical continuity with previous crystal growth with the pause in precipitation potentially recorded by sediment or brine inclusions within the precipitates. Distinguishing these cessation inclusions from other inclusions would be challenging. In sabkha systems, a cessation event is unlikely to be recorded in any form that may be recognised in the stratigraphic record.

10.5.2 Shallowing-Upward Stratigraphic Sequences

A shallowing-upward sequence may be recognised as an up-section shallowing of facies, decrease in bed thicknesses or, a transition to progressively more soluble evaporite mineralogies. These sequences may develop under a variety of depositional scenarios and in a range of geographical settings (Table 10.4). Relative sea-level fall and accommodation filling will both produce shallowing-upward sequences. In the case of a fall in relative sea level, the resultant sequence will typically show a downstepping progradational (forced regression) geometry with the most proximal areas being subject to exposure with consequent erosion and, potentially, a transition to terrestrial deposition. Evaporite and non-evaporite facies will be reworked and transported basinward. Where the shallowing-upward sequence is the result of the filling of accommodation, then progradation of facies belts is again likely with the vertical component of the facies trajectory being a significant indicator of the underlying relative sea-level context. Once accommodation is completely filled during a period of highstand, lowstand or stillstand, then the upper surface of the sequence will represent a period of non-deposition. In the case of shallowing as a result of desiccation of the basin (evaporitive drawdown), this event will be recorded in the up-section precipitation of progressively more soluble minerals. Biotic assemblages will record increasingly stressful conditions (i.e. communities exhibiting low diversity and high abundance of halophilic and halotolerant species). Where complete desiccation ensues, then the evaporite sequence will be susceptible to subaerial and meteoric processes and may be succeeded by the deposition of terrestrial facies.

10.5.3 Deepening-Upward Stratigraphic Sequences

Deepening-upward sequences may be considered as either gradual – typically reflecting rising relative sea level, or geologically instantaneous – as a result of the breaching or failure of a barrier (Table 10.4). Both of these scenarios will result in the creation of new accommodation and a simultaneous change in brine salinity. At the shoreline, transgression will result in the flooding of adjacent terrestrial systems with the dissolution of evaporite minerals and a transition to subaqueous deposition. In shallow marine settings, the creation of new accommodation will drive an initial up-section increase in bed thickness, however, where sea-level rise outstrips sedimentation then a transition to progressively deeper facies will be observed. Decreased salinity and the disruption of any previously established stratification of the water column will result in the dissolution of evaporite minerals at the basin floor – creating a subaqueous dissolution surface. Once stratification becomes re-established, evaporite precipitation below the pycnocline will resume. A reduction in salinity will see a transition in the biotic assemblage from halophilic organisms to an assemblage reflecting more normal marine conditions. In the case of a catastrophic, near-instantaneous, flooding event, the stratigraphy will record a sudden deepening of the environment, for example, shallow near-shore facies directly overlain by deep water distal sediments. Shorelines will dramatically 'leap' landward with a significant lateral discontinuity between facies belts (e.g. the Silurian of the Michigan Basin (Rine et al. 2020)). During this process, carbonate platforms will be drowned (Schlager 1981). Reduced salinity and disturbance of the pycnocline will cause basin-wide dissolution of evaporites (Schröder et al. 2005) and a dramatic transition to a more open-marine biota.

10.5.4 Thick Evaporite Sequences

A thick, repetitive sequence of evaporites is indicative of ongoing precipitation in a setting with significant accommodation (Table 10.4), as previously discussed for the Messinian of the Mediterranean (Roveri et al. 2014a). Such conditions are best met in a deep-water basin, potentially with a stable stratified water column. Evaporite precipitation may be punctuated by episodic influxes of clastic deposits sourced from more proximal areas – these events potentially record sea-level fluctuations that are not directly observed in the deeper water facies. Where the water column is stratified, pycnocline displacement may expose basin floor evaporites to undersaturated brines – resulting in dissolution.

Table 10.4 Sequence trends.

Sequence trend	Brief characterisation	Possible causes	Diagnostic features	Geographic context	Sequence stratigraphic context
Shallowing-upward sequence	Evaporites and related sediments record an up-section decrease in water depth and, potentially, a transition from subaqueous to subaerial deposition (regression).	Fall in relative sea level	In most proximal areas the shallowing sequence is potentially succeeded by an exposure surface with erosion or transition to terrestrial deposition. Eroded evaporites and other sediments are redeposited in more distal basinal settings. Proximal forced regression - downstepping geometry.	Sabkha salina basin margin	FSST
		Filling accommodation	Significant progradation of facies belts. Where in a HST no vertical component in progradation. Where accommodation totally filled upper surface represents period of non-deposition. No exposure unless followed by sea-level fall.		HST, HS, LST, LS, SS, TST
		Evaporitive drawdown	Up-section change in mineralogy to progressively more soluble minerals. Biotic assemblages become progressively more halophilic. Where complete desiccation occurs, the sequence may transition to terrestrial facies with evidence of subaerial exposure and the influence of meteoric fluids.	Salina basin margin basin centre	None
Deepening-upward sequence	The depositional sequence records a general up-section increase in water depth with proximal terrestrial areas being flooded and transitioning to marine deposition (transgression).	Rise in relative sea level	Proximal terrestrial areas flooded. Shallow water settings initial up-section increase in bed thickness before transition to deeper water facies. Carbonate platforms aggrade. Distal areas thinner beds. Subaqueous dissolution of previous evaporite minerals. Restricted halophilic biotic assemblage progressively replaced by more diverse open marine assemblage.	Sabkha salina basin margin basin centre	TST, MFS, HST
		Instantaneous basin recharge	Instantaneous flooding (terrestrial) and deepening (basin). Carbonate platform drowning. Death assemblages. Sediment starvation to distal areas - cessation surface. Dissolution of previously deposited evaporite minerals. Sudden switch from halophilic to marine biotic assemblage.		Barrier breachingBarrier failure
Thick evaporite sequence	Thick repeated units of bottom grown or cumulate evaporites building up a continuous succession.	Significant accommodation, stable pycnocline	Deposition of evaporites either as bottom growth precipitates or as cumulates. Possibly intercalated with non-evaporite (pelagic) and redeposited facies. Dissolution surfaces may record a fluctuating pycnocline.	Basin centre	Any

10.6 Evaporite Sequence Stratigraphic Models

Defining controls on the development of evaporite sequences is challenging and is best undertaken at a regional scale incorporating observations and evidence from genetically related non-evaporitic units. Within this context, we shall now consider the stratigraphic sequences formed as a response to sea-level change in both coastal and basinal marine evaporite systems. In each case, recognising the infinite variety of potential stratigraphic architectures, we shall initially focus on the development of idealised sequences and highlight situations that may result in misinterpretation or confusion. For each system, we shall utilise the process descriptions of two model evaporite systems – the Abu Dhabi sabkha, together with its adjacent marine setting, and the Mediterranean basin-wide deposits of the Messinian salt giant – to examine evaporite facies successions in more detail.

10.6.1 Coastal Marine Evaporites

Coastal marine evaporite systems are particularly sensitive to sea-level fluctuations. The limited accommodation that is typical of shallow water coastal evaporite systems will both limit the vertical thickness of the depositional sequence (Figure 10.39a) and, simultaneously, promote progradation thus producing thin evaporite units of significant lateral extent as seen today in the Abu Dhabi sabkha (Lokier and Steuber 2008) and in the Jurassic Arab Formation (Grötsch et al. 2003).

During a period of relative sea-level rise, transgression will drive retrogradation of facies belts with sabkha and shoreline carbonate precipitation back-stepping landward (Figure 10.39b,c). Inundation of previously deposited subaerial sabkha and salina evaporites by undersaturated seawater will result in evaporite dissolution. However, in isolated basins, such as flooded salinas, the local development of a pycnocline may slow or even halt dissolution. With relative sea-level fall, shorelines and their associated evaporites will be abandoned. The termination of solute supply and emplacement of evaporites into the meteoric realm will result in the dissolution and karstification of evaporite deposits – a fate that will be shared, at a slower rate, by carbonate lithologies (Figure 10.39d–f). All of the exposed facies will also be subject to aeolian and other erosional processes. Periods with relatively stable sea level, that is, highstands, lowstands and stillstands, will see significant progradation of the coastline and the associated development of a broad coastal sabkha system (Figure 10.39c,f). A succession of transgressive and regressive events will repeat this cyclic pattern of evaporite precipitation and dissolution – a sequence that is unlikely to produce a significant recognisable stratigraphic record of evaporite precipitation (Figure 10.39g).

In order to produce and preserve a thick coastal evaporite sequence, it is necessary to consider a model where accommodation is significant and overall evaporite precipitation is greater than evaporite dissolution. Such a scenario is best met where a partially restricted continental margin is undergoing prolonged active subsidence whilst maintaining a degree of isolation from the open marine system (Figure 10.40). Such isolation may be provided either by a barrier or a wide shallow continental shelf. Ongoing subsidence will generate the accommodation required for the precipitation of a thick sequence of shallow marine and coastal evaporites. Subsidence will also reduce the likelihood of exposure and consequent subaerial erosion of the evaporite minerals. Similarly, the presence of the barrier will prevent an incursion of undersaturated seawater that would dissolve evaporite minerals on the sea floor. Barriers may be maintained by the aggradation of carbonate facies with carbonate precipitation and accumulation in warm shelf seas also helping to assure a degree of restriction at the coastline (Figure 10.40b); Tucker (1991)). Restricted subsiding coastlines are potentially able to accumulate thick evaporite sequences with evaporite accretion occurring throughout all phases of the sea-level cycle.

Another scenario for the accumulation of a thick evaporite sequence is where a coastal topographic low maintains a basin floor below eustatic sea level and is isolated from the marine system by a permeable barrier. Under this model, as long as sea level remains above the level of the brine in the salina, the hydraulic head will be maintained, and the salina continues to be supplied with fluids during all relative sea-level phases (Figure 10.41). The salina will continue to precipitate evaporite deposits that become progressively thicker and onlap salina margins to build an aggrading evaporite sequence regardless of the sea-level phase in the adjacent ocean basin. Such a relationship is observed in the Sedom Formation deposited during the Pliocene-Early Pleistocene when, prior to becoming a continental basin, the Dead Sea region still received a reduced inflow of seawater from the Mediterranean through the Yezreel Valley (García-Veigas et al. 2009). A similar relationship has also been inferred for the thick evaporite deposits of the Permian Castile Formation when ocean water entered the Delaware Basin through seepage via porous reefal carbonate at the basin margin (Kirkland et al. 2000). For this reason, salina sequences should be treated with caution when interpreting relative sea level.

In both of these scenarios, it is conceivable that the barrier may be breached during rapid sea-level rise.

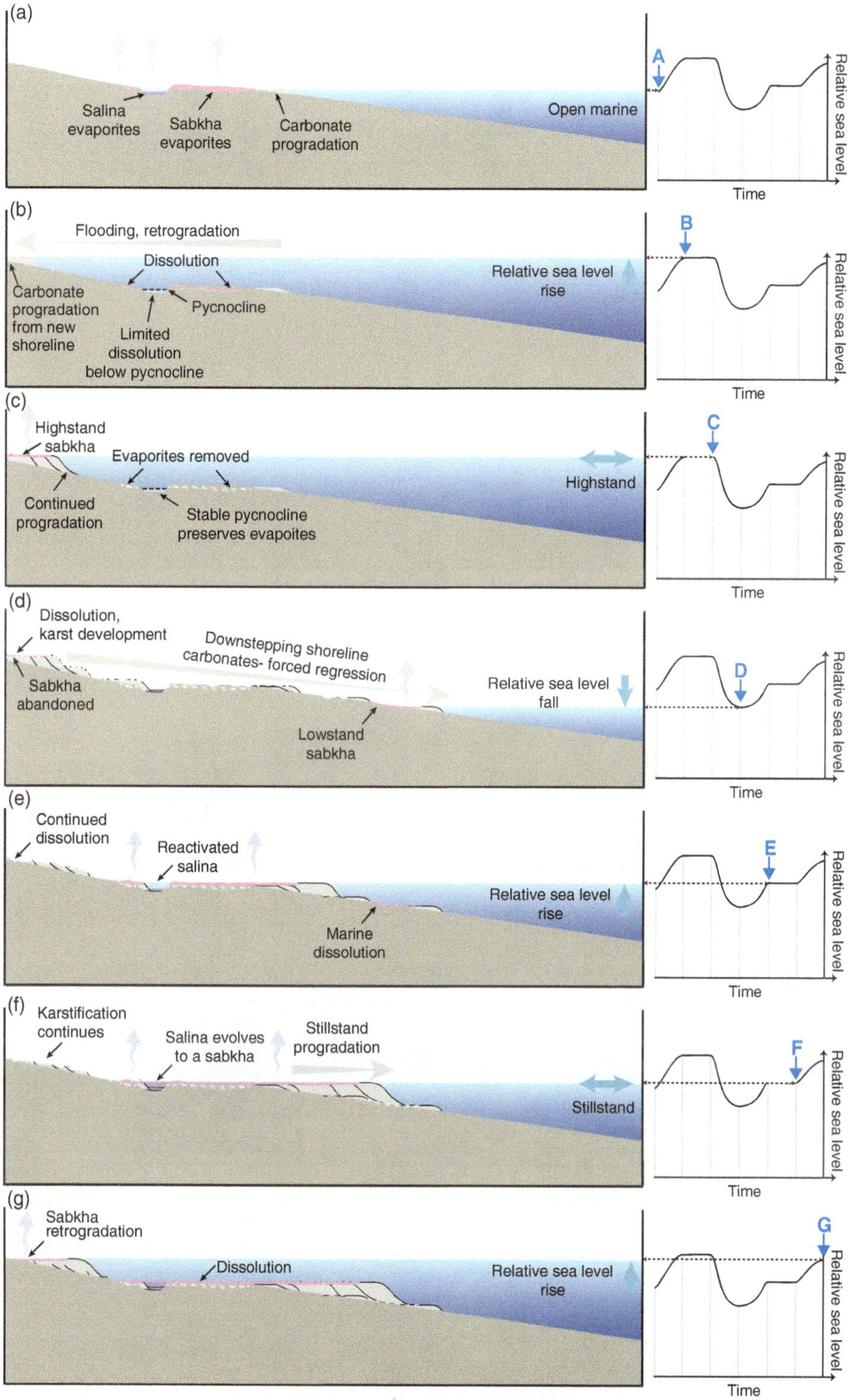

Figure 10.39 Sabkha and salina evolution with changes in relative sea level. (a) Initial state with sabkha and saline evaporites developing in the supratidal zone and carbonate progradation at the shoreline. (b) Relative sea-level rise results in transgression and the backstepping of facies belts. Previously precipitated evaporites are subject to dissolution by undersaturated seawater. (c) Formation of a highstand sabkha in the supratidal zone. Where a stable pycnocline exists, e.g. in an isolated basin, then submerged evaporites may be retained. (d) A fall in relative sea level exposes facies. Evaporites and associated carbonates will be subject to meteoric karstification. A lowstand sabkha will develop in the supratidal zone of the shoreline. (e) Renewed relative sea-level rise will flood the lowstand sabkha and dissolve evaporite minerals. Facies will again backstep landward with sabkhas and salinas reactivating. (f) Any periods of stillstand will promote progradation of the coastal sabkha system. Salinas will likely become infilled and terminate. (g) With the recommencement of relative sea-level rise, facies once again retreat landward with transgressed evaporites again susceptible to dissolution where fluids are undersaturated.

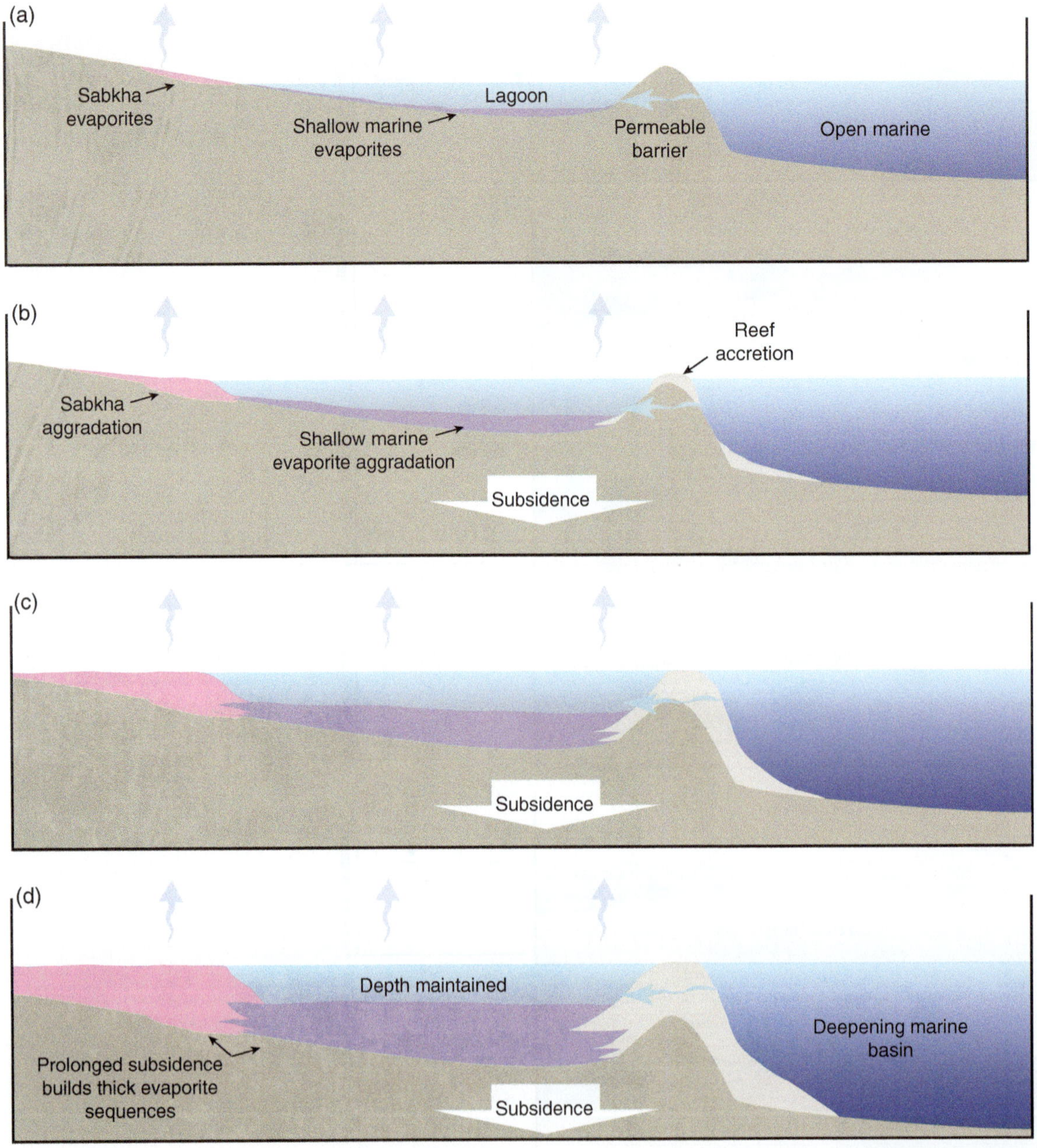

Figure 10.40 Development of a coastal evaporite stratigraphic sequence in a subsiding basin under sea-level stillstand. (a) Seawater supply to the coastline is restricted to flow through a permeable barrier. Elevated salinities drive carbonate precipitation in the lagoon (shallow marine) and at the shoreline (sabkha). (b) The development of an aggrading carbonate platform on shallow water topographic highs maintains basin restriction during subsidence. (c,d) As long as barrier accretion maintains restriction, evaporites will continue to be precipitated during ongoing subsidence to build a thick evaporite sequence.

However, even should such a scenario arise, the continued presence of the barrier will likely restrict near-shore circulation thereby promoting the establishment of a pycnocline that inhibits dissolution. Where dissolution does occur, then top-down dissolution of the evaporite sequence creates a subaqueous karstic topography and an associated dissolution lag that, together with pelagic sediment, will likely inhibit fluid flow and, again, limit the degree of dissolution.

It should be emphasised that these are just two examples of scenarios in which coastal marine evaporite sequences may be accumulated and preserved – there are, of course, many other situations under which coastal evaporite sequences may be formed.

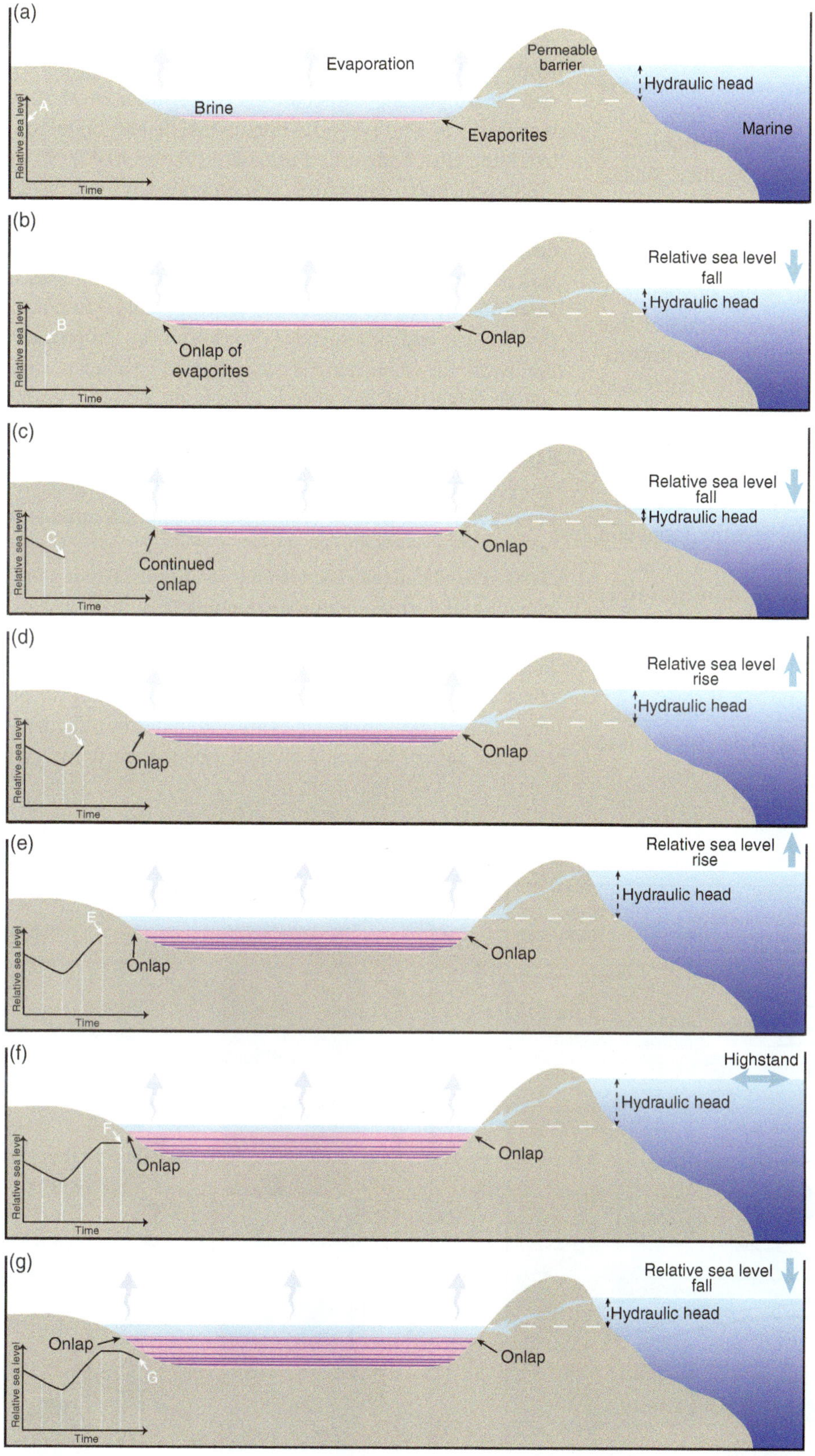

Figure 10.41 Example of continued onlapping evaporite deposition in an evaporite basin during various stages of a relative sea-level cycle. (a) Hydraulic head maintains inflow of seawater to the salina through a permeable barrier. (b,c) With a fall in relative sea level, hydraulic head is maintained as long as the floor of the salina remains lower than the adjacent sea surface. (d,e) With an increase in relative sea level, increasing hydraulic head ensures continued seawater supply to the salina. (f) Seawater will continue to be supplied during stillstand. (g) Again, as long as the basin floor remains below the relative sea level, the hydraulic head will continue to drive inflow.

10.6.2 The Jurassic Arab Formation – A Coastal Evaporite Case Study

During the Mesozoic, the Arabian Plate was flooded by the shallow tropical seas of the Neo-Tethys Ocean. Here, carbonate sedimentation was punctuated by episodic influxes of siliciclastic sediments, derived from the Arabian Shield, and periodic desiccation with associated evaporite formation (Sharland et al. 2001). Relative sea-level highstand during the Late Jurassic produced an upward-shallowing sequence under prevailing subtidal to peritidal conditions (Figure 10.42). At this time, the southern margin of this shallow tropical sea was characterised by a northward-dipping carbonate ramp geometry where inner ramp ooid shoals (Figure 10.43) isolated the coastline from open marine conditions (Hollis et al. 2017). This restriction from the marine system, along with arid conditions associated with a latitudinal position in the southern tropical zone, resulted in elevated evaporation rates and the formation of coastal sabkhas (Grötsch et al. 2003).

The Arab Formation comprises a wide range of facies that can be grouped into seven genetically linked facies associations (Supplementary Table 10.S.3). By applying the Abu Dhabi sabkha (see Section 10.3.5) as a present-day analogue, it is possible to interpret these Jurassic facies associations in terms of their overall depositional context (see Box 10.3).

At a broad scale, the upward-shallowing sequence records a transition from carbonate-dominated facies (the Arab D Member), through progressively more evaporite-rich interdigitating carbonate-evaporite units (Arab C-A members) to the evaporite-dominated deposition of the overlying Hith Anhydrite Formation (Figure 10.42).

Sub-tidal outer ramp facies are dominated by bioturbated mudstones and mud-rich peloidal packstones with locally developed hardgrounds. Bioclasts are common and are typical of normal marine conditions. The mid-ramp sees a transition to coarser grained facies with peloidal to ooidal packstones and interbedded wackestones. Progressing landward, the inner ramp is characterised by a facies association comprising cross-bedded and laminated ooid grainstones and packstones deposited in, or adjacent to, a shoal complex (Figure 10.43). Landward of the shoal complex, a protected, more restricted, environment was dominated by pervasively bioturbated carbonate mudstones, mud-rich packstones and locally developed microbial buildups. Bioclasts are indicative of hypersaline conditions. Passing into the intertidal zone, bioturbated mudstones and wackestones (both with abundant, low diversity, halophilic and halotolerant species) and microbially laminated facies are observed. Desiccation cracks and rootlets are locally developed. Supratidal, sabkha facies are dominated by anhydrite. Nodular anhydrite occurs as both isolated, clustered and

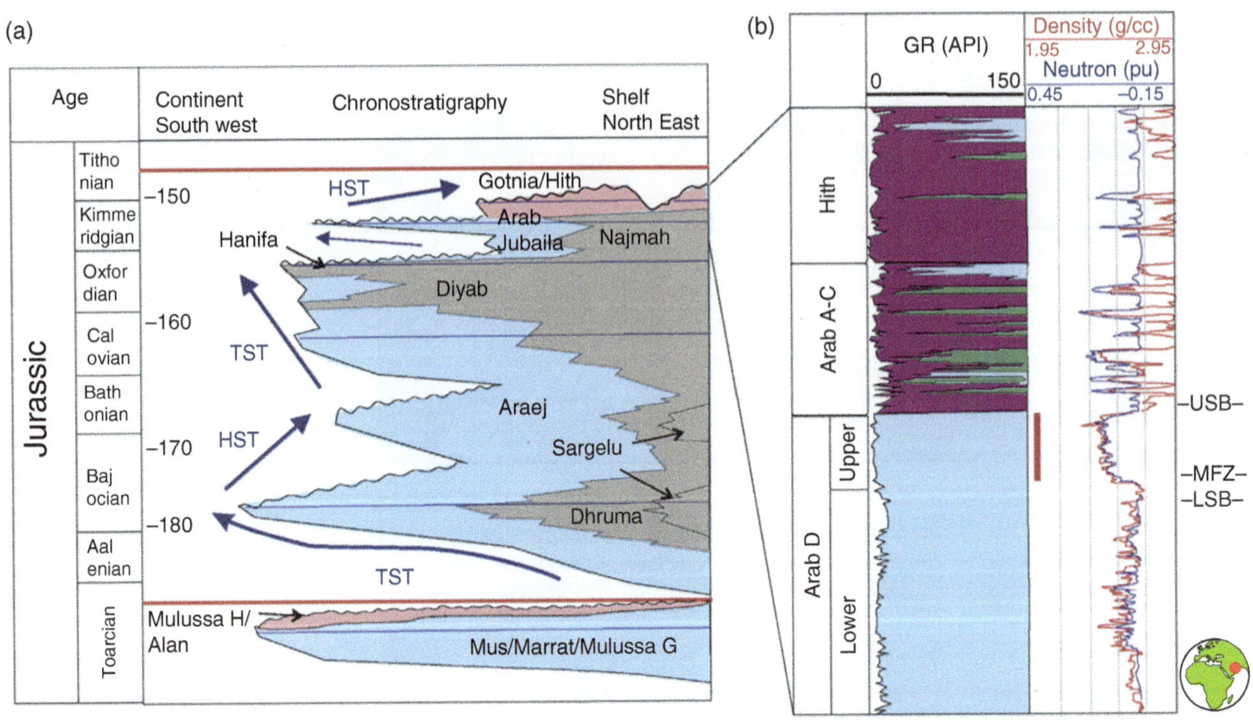

Figure 10.42 (a) The regional stratigraphy for the Arab and Hith formations. (b) Composite stratigraphy, along with representative Density and Neutron logs, for the Arab and Hith formations in the region of Abu Dhabi, UAE. LSB – lower sequence boundary, MFZ – maximum flooding zone, USB – upper sequence boundary (Hollis et al. 2017).

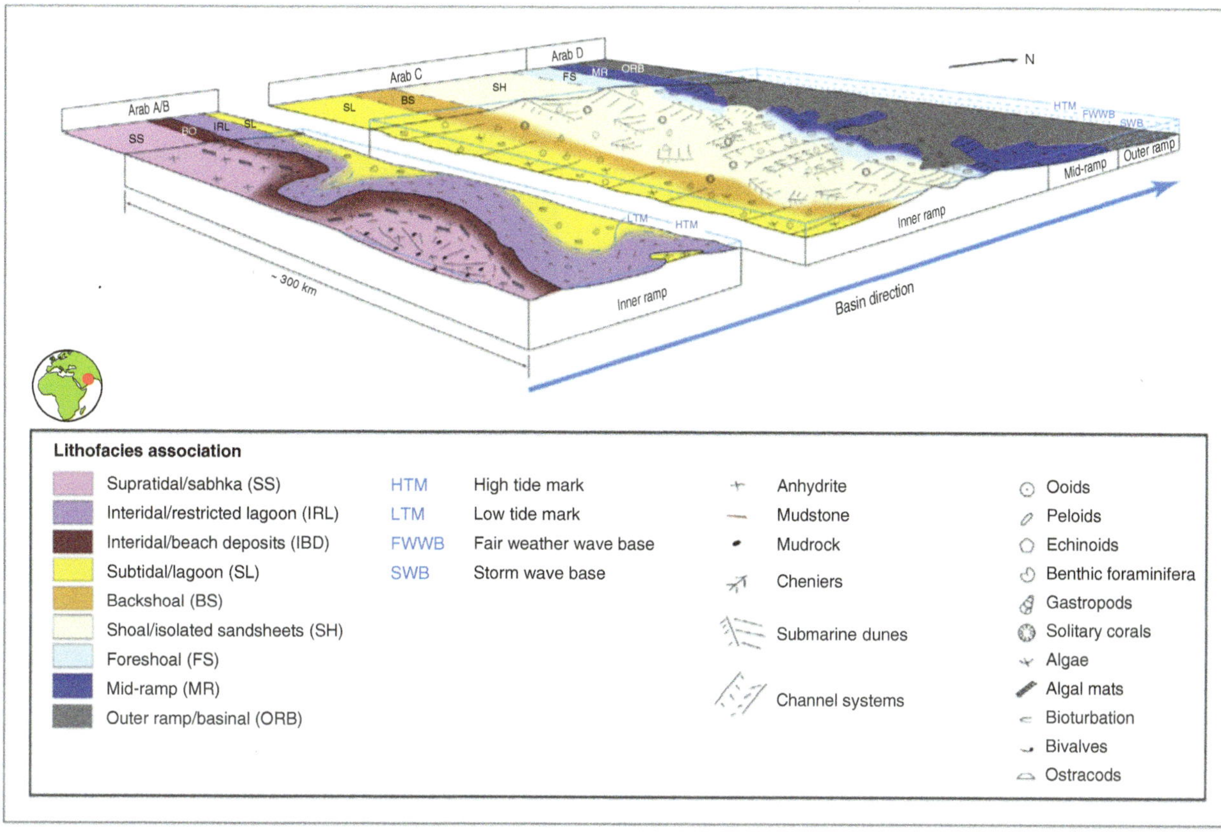

Figure 10.43 Conceptual palaeogeography of the southern shoreline of the Neo-Tethys Ocean during deposition of the Late Jurassic Arab Formation (Hollis et al. 2017).

fully amalgamated (i.e. chicken-wire) forms. Enterolithic anhydrite may record primary textures or be post-depositional in origin. Lenticular anhydrite, pseudomorphing after gypsum, is common at various ranges of scales and densities. Salina facies are dominated by laminated anhydrite that may be locally convolute indicating syndepositional remobilisation or post-depositional re-crystallisation.

Where preserved, the stacking patterns of peritidal facies can be useful particularly for elucidating cyclicity within a depositional system. This is particularly the case for the Arab A-C Members of the Arab Formation where the facies associations exhibit a distinctive cyclic stacking pattern with parasequences recording fourth- to fifth-order cyclicity (Figure 10.44; (Sharland et al. 2001; Grötsch et al. 2003)).

The base of each parasequence records transgressive flooding during relative sea-level rise. The flooding surface is immediately overlain by facies recording shallow subtidal conditions. As the newly created accommodation is filled, an upward shallowing profile ensues with the subtidal facies association being superseded by intertidal facies before climaxing with supratidal evaporite-dominated facies. Each upward-shallowing parasequence records the creation and infilling of accommodation and associated shoreline progradation before the next flooding event.

10.6.3 Basinal Marine Evaporites

The evolution of marine evaporite basin brines is controlled by the relationships between the supply of seawater to the basin, the rate of evaporation and the reflux of brines. In concert, these factors will govern both the salinity of the brines and the level of the brine surface within the basin. The decoupling of the basinal brine level from adjacent sea level allows basinal marine evaporites the potential to be deposited at any point within the sea-level cycle. A further degree of complexity is added where the basin is internally compartmentalised, producing sub-basins, or where several adjacent evaporite basins develop contemporaneously – scenarios that are envisaged for the Messinian of the Mediterranean. The stratigraphic sequences that develop in each of these sub-basins will potentially record very different sea-level histories, such is the case for the Permian Zechstein of Germany (Becker and Bechstädt 2006). The supply of seawater to the sub-basin will be controlled by a range of factors including the permeability of barrier

Box 10.3 Arab Formation interpretation

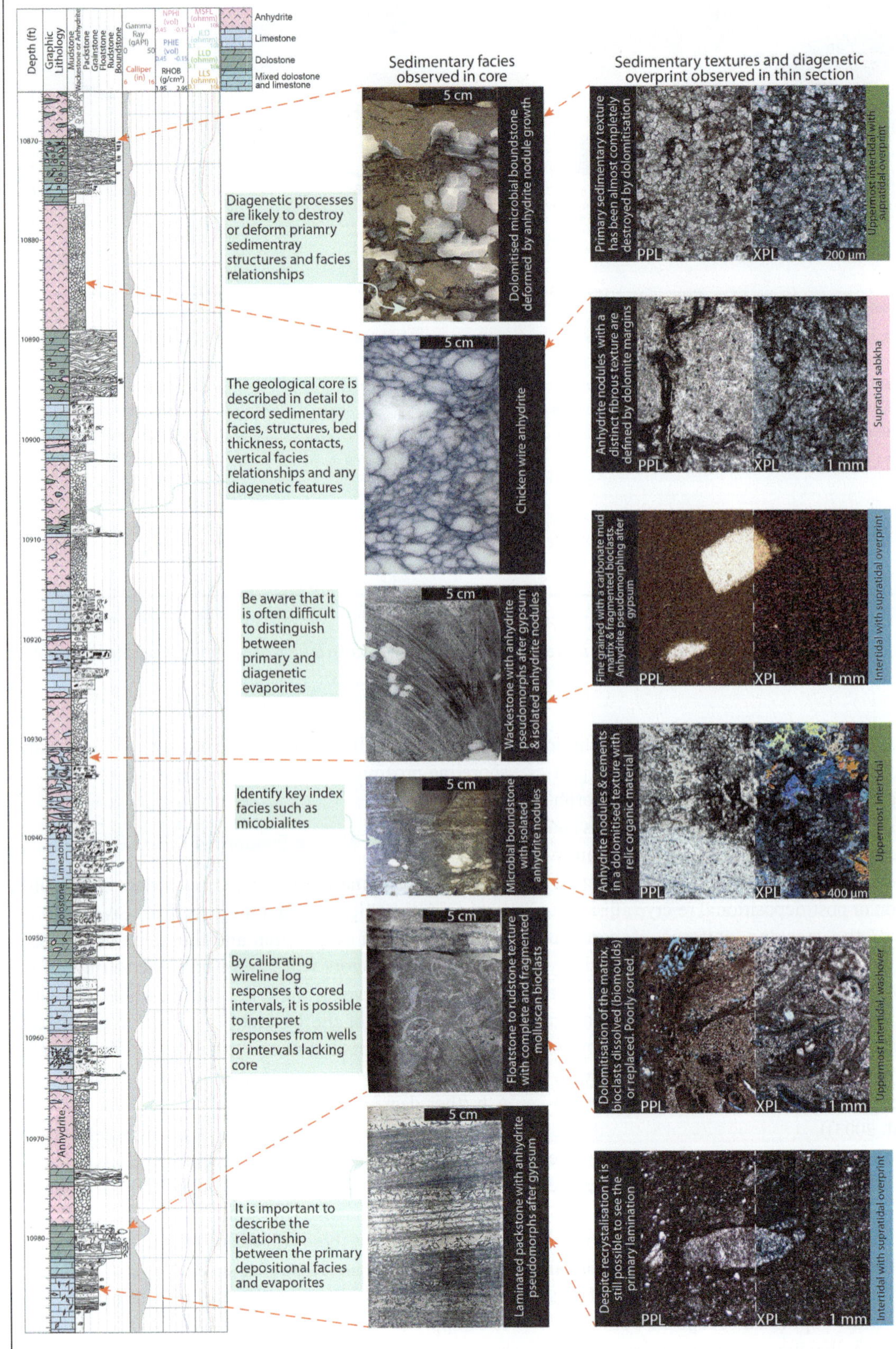

A case study exemplifying how core, wireline logs and microfacies can be applied to characterise the palaeoenvironment and sea level history of a coastal sabkha succession

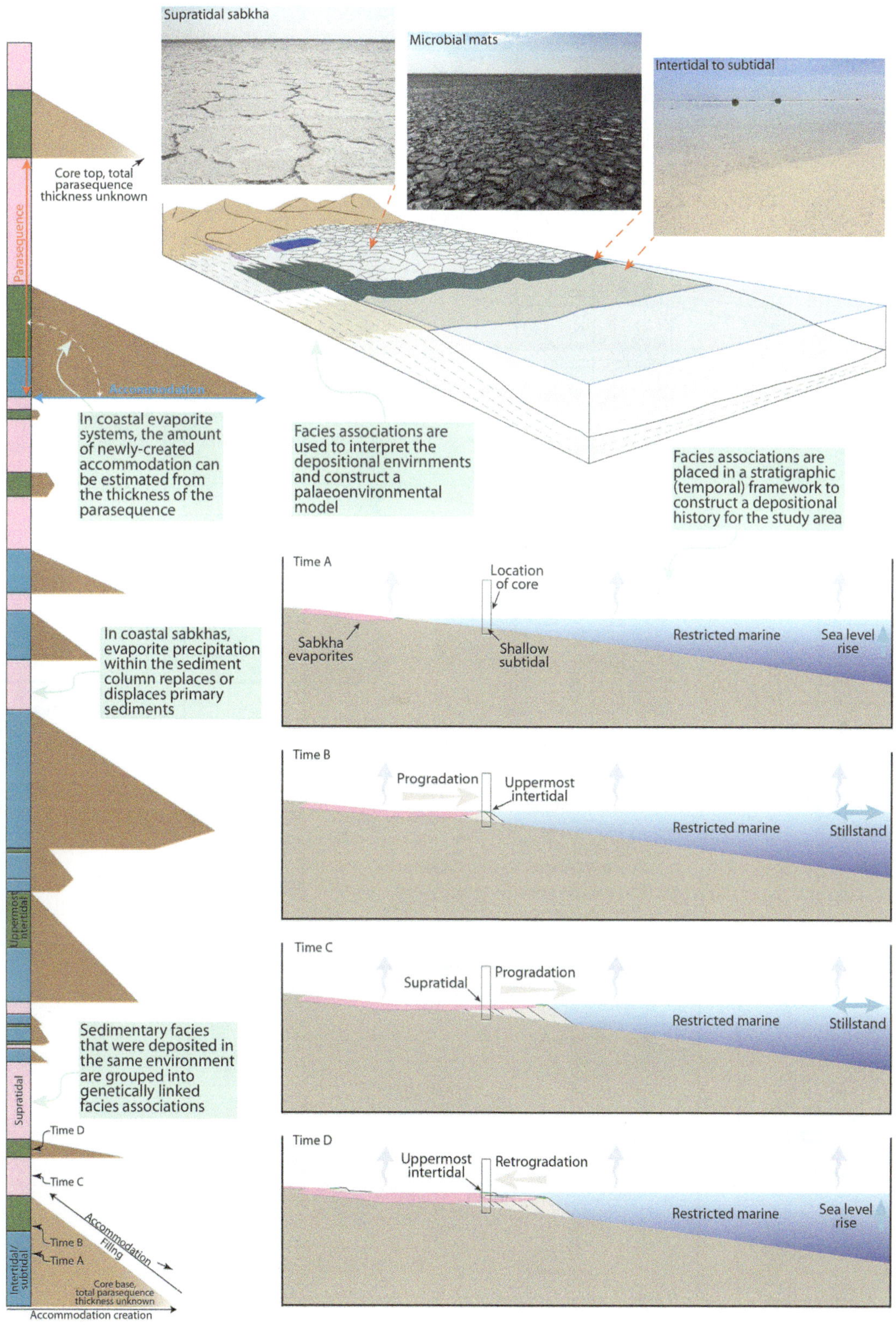

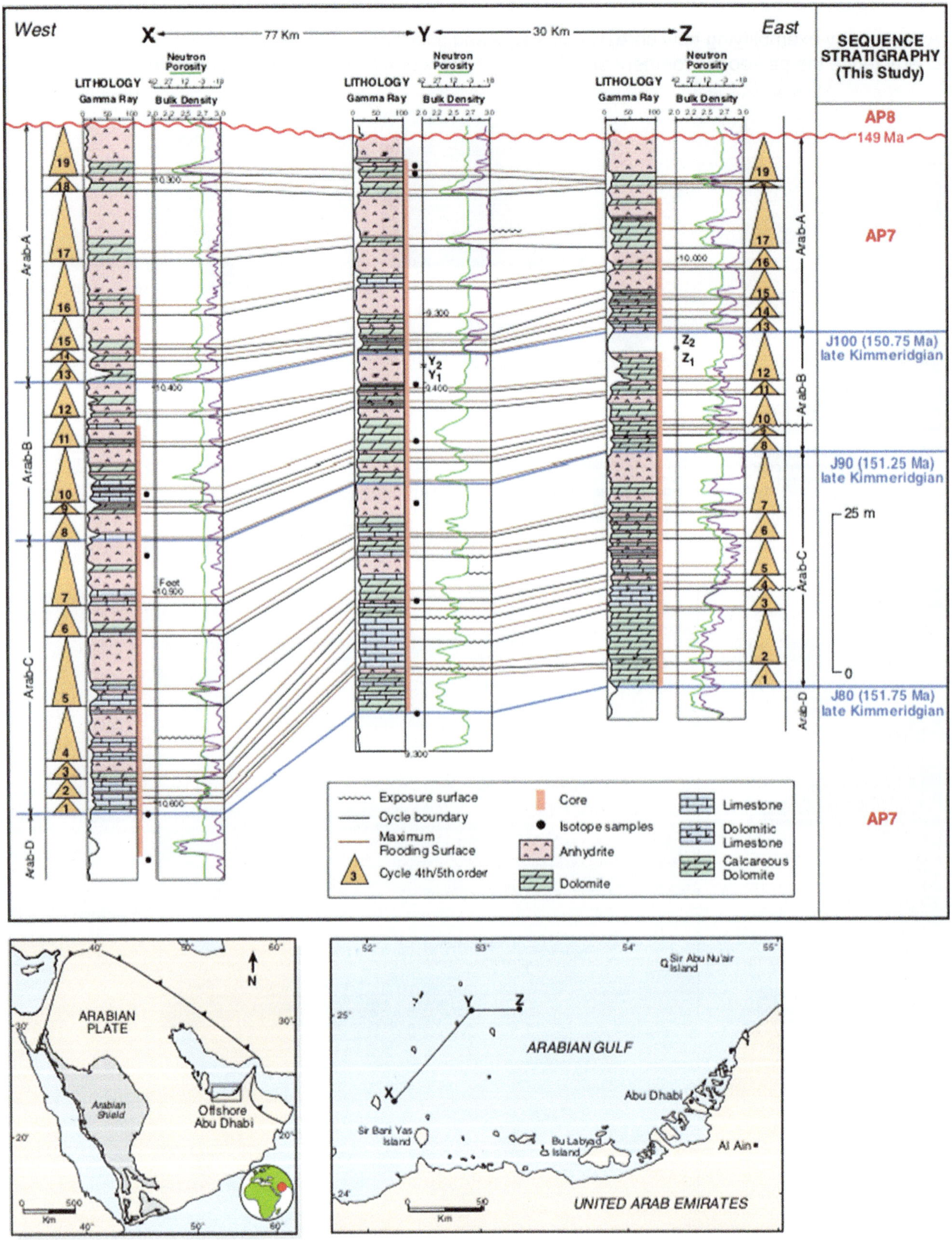

Figure 10.44 Cyclicity as interpreted from the Arab A-C members of the Late Jurassic Arab Formation in the offshore of Abu Dhabi, UAE (Sharland et al. 2001).

units, the hydraulic head, barrier topography – particularly the lowest elevation, and the distance between the basin and the open marine system. The large quantity of water required to produce a thick basinal evaporite sequence will typically prohibit the brines from reaching the point of halite saturation. Consequently, the evaporite fill will be dominated by gypsum, more soluble minerals are typically only precipitated on terminal desiccation. In those cases where the balance between seawater inflow and evaporation allows the brines to reach and maintain halite saturation, then thick halite sequences will be precipitated.

To demonstrate how seawater supply controls evaporite sequence formation, we shall consider the deposition of evaporite sequences in two hypothetical basins with similar geomorphologies but with different modes of connection to the global ocean system (Figure 10.45). These basins are identical in terms of morphology and climate, the only difference between them being the occurrence of a permeable horizon within the barrier complex of the basin on the left of our model (Figure 10.45). At the commencement of our example, the sea level is rising yet both of the basin systems are isolated from the ocean with sulphate precipitation taking place in deep sub-basin areas (Figure 10.45a). As sea level continues to rise, seawater will start to flow through the permeable barrier into the left basin (Figure 10.45b). At the point of inflow, there may be localised dissolution of previously deposited evaporites, but as long as the hydraulic head is maintained, the supply of fresh solutes will ensure that evaporite precipitation will continue in the deep sub-basin. The continued isolation of the right (impermeable barrier) basin system results in complete desiccation and the precipitation of a terminal halite and bittern salt sequence – note that this exposure is taking place during a period of relative sea-level rise. In the left basinal system, the deep sub-basin will continue to deepen without any perceptible change, eventually spilling into the topographically higher shallow sub-basin just before barrier overtopping (Figure 10.45c). In the right basinal system, the flooding of the deep sub-basin will be geologically instantaneous, however, this basin will be filled prior to the flooding of the topographically higher shallow sub-basin. This basin-scale flooding event will initially see the dissolution of previously precipitated evaporite units with the more soluble salts of the right basin being particularly susceptible to removal. Following flooding, the development of pycnoclines enhances renewed sulphate precipitation. Carbonate platforms, developed on shallow-water topographic highs, will interdigitate with the evaporite facies.

With a dramatic fall in sea level, both of the basins are once again isolated from the open marine system (Figure 10.45d). Shallow proximal sub-basins desiccate, precipitating a terminal salt phase and exposing carbonates to subaerial erosion. The permeable barrier unit will drain brines from the left deep sub-basin, although evaporite precipitation will continue during this process. In the isolated right basin, gypsum precipitation will continue uninterrupted. Renewed sea-level rise allows seawater to once again flow through the permeable barrier and reflood the left basin – gypsum precipitation continues but in doing so partially restricts the permeable zone (Figure 10.45e). Simultaneously, the right deep sub-basin reaches the point of complete desiccation, again precipitating a terminal salt sequence. Proximal sub-basins continue to experience dissolution. During subsequent stillstand, shallowing gypsum precipitation increasingly blocks the permeable horizon in the left basin barrier, restricting both the inflow of seawater and the outflow of brine (Figure 10.45f). The left deep sub-basin joins the shallow basins in becoming an area of subaerial erosion. As sea level once again starts to rise, occlusion of the permeable horizon isolates the left basin system with terminal desiccation precipitating halite and bittern salts. Subaerial erosion continues in all of the other basins (Figure 10.45g).

This example clearly demonstrates that, depending on the degree of connectivity to the open ocean system, contemporaneous evaporite sub-basins may produce significantly different stratigraphic records even though deposited under the same sea-level regime. Four cores or outcrops, each representing a different sub-basin from our hypothetical example, would exhibit very different stratigraphic sequences that, when taken in isolation, may be interpreted to define very different sea-level records (Figure 10.45). The correlation of these cores would be challenging, particularly in light of the difficulties in establishing a chronostratigraphic framework from an evaporite sequence.

As demonstrated, in our hypothetical scenario, the nature of the connection between the evaporite basin and the global ocean is key in defining specific switching points (e.g. flooding or isolation) in evaporite basin evolution. Even significant changes in relative sea level will have little to no effect on the evaporite basin unless such change either establishes or breaks the connection between the basin and the ocean. Flooding of the basin may occur at any point during rising sea level with the 'flooding event' rapidly filling the basin to close to global sea level over a geologically instantaneous interval (Figures 10.45 and 10.46). The stratigraphic profile observed in the basin is therefore disconnected from the regional or global sea-level cycle.

10.6.4 The Messinian Mediterranean Evaporites – A Basinal Case Study

In this section, we employ the well-preserved and comprehensively studied Late Miocene (Messinian) evaporites of

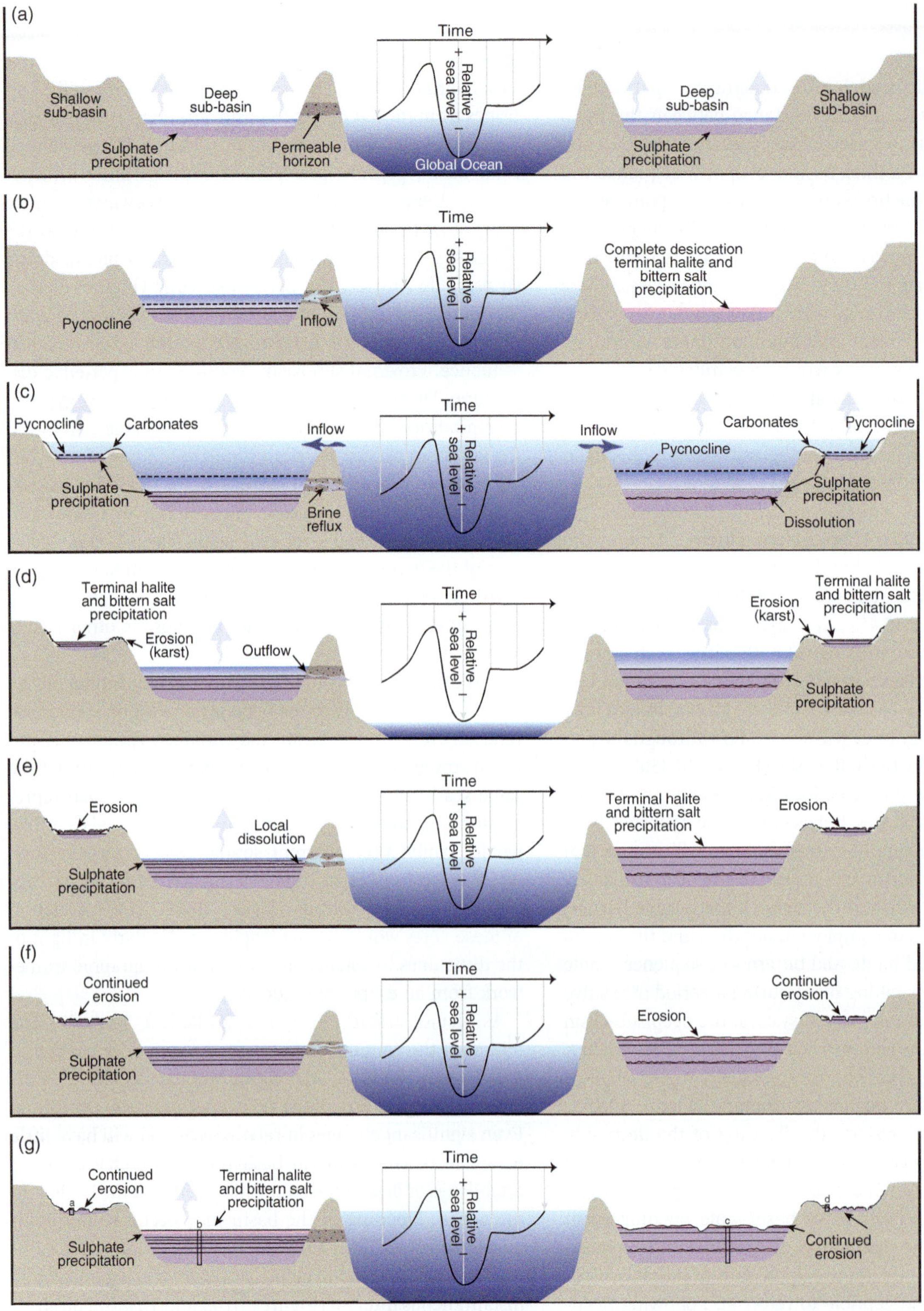

Figure 10.45 Examples of the deposition of different evaporite sequences in two morphologically identical basinal systems during different phases of a sea-level cycle. Note that the basin on the left has a permeable barrier horizon which is absent from the barrier of the basin on the right. (a,b) During sea-level rise, the left basins are progressively flooded as the hydraulic head, associated with rising sea level, ensures the supply of seawater to the basins via the permeable horizon. During this interval the basins to the right of the model remain isolated and desiccate. (c) The left shallow sub-basin is flooded via flow through the permeable horizon driven by increasing hydrostatic head. Overtopping of the barriers shortly prior to the highstand results in geologically instantaneous flooding of the right deep sub-basin followed by the shallow sub-basin. (d) Relative sea-level fall isolates the basins with shallow sub-basins desiccating. The left deep sub-basin experiences some brine outflow via the permeable barrier – staving off the precipitation of the more soluble salts. In the right deep sub-basin evaporation precipitates sulphate minerals. (e,f) Sea-level rise and stillstand resupply solutes to the left deep sub-basin allowing evaporite precipitation to continue. The right deep sub-basin remains isolated, desiccates and sees the onset of erosion. Erosion continues in both shallow sub-basins. (g) With the occlusion of the permeable barrier, the left deep sub-basin becomes isolated and desiccates. Erosion continues all of the other sub-basins. See the main text for a detailed discussion of the evolution of the basins. The boxes in panel (g) show the location of the 'core' profiles in Figure 10.46.

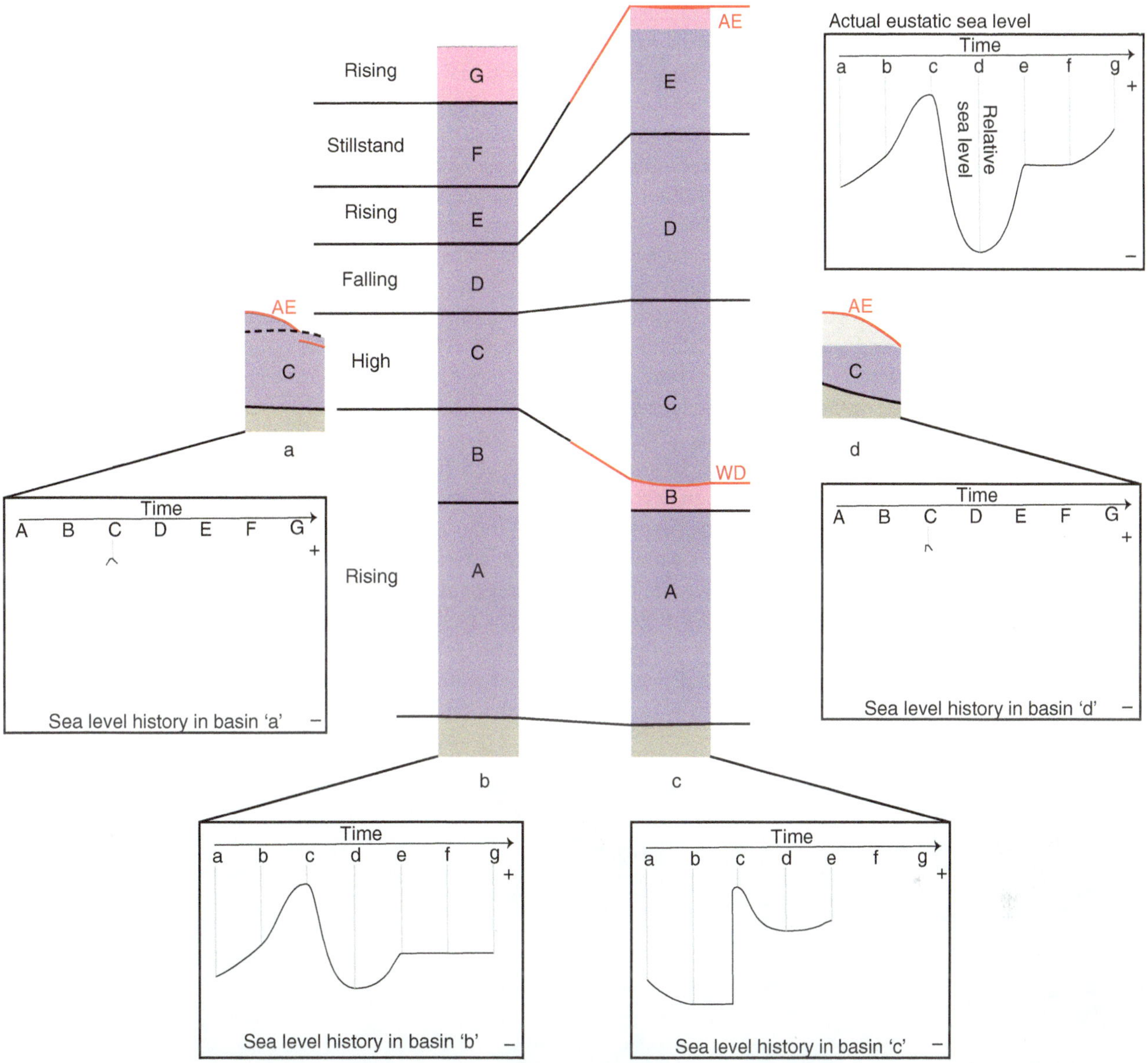

Figure 10.46 The final stratigraphic sequences that would be produced from the four sub-basins discussed in the hypothetical model shown in Figure 10.45 (see Figure 10.45g for the 'core' locations). The letters on the column indicate the time interval (panels in Figure 10.45) within which each unit was deposited. Assuming that the timing and depth of evaporite precipitation could be ascertained, the resultant sea-level history curves derived from the two deep sub-basin columns are very different both from each other and from the actual eustatic sea-level record, with the more isolated basin (core c) exhibiting the largest deviation from eustatic sea level. AE – subaerial erosion, WD – subaqueous dissolution.

the Mediterranean Basin as a case study to describe, interpret and understand the stratigraphic complexity of basinal evaporite systems.

The Late Miocene evaporitic outcrops of the Italian peninsula and islands were first described in detail during the latter half of the 19th century with the studies of Sebastiano Mottura (1871) and Luigi Baldacci (1886) being of particular significance. Extensive onshore studies in the mid-20th century resulted in the proposition of significant hydrological change during the Late Miocene, with a basin characterised by extensive evaporation and the formation of a series of isolated shallow marine basins around the Mediterranean region (Ogniben 1957). This model was further developed as the concept of the 'Messinian salinity crisis' (Selli 1960) a model that was supported when the Deep Sea Drilling Project (Hsü 1972; Hsü et al. 1973a) cored the upper layers of the offshore Mediterranean salt giant to reveal 'chicken-wire anhydrite' textures akin to those reported from recent sabkha settings. These basin-wide shallow water evaporites were inferred to have precipitated during a major desiccation event following the closure of the Strait of Gibraltar and restriction of seawater exchange between the Atlantic Ocean and the Mediterranean Sea (Hsü et al. 1973b). While the Mediterranean desiccation hypothesis attracted significant interest, some researchers remained sceptical, proposing an alternative formational model involving a deep water, non-desiccated basin origin for the Messinian evaporites (Schmalz 1969; Selli 1973). After the pioneering studies of the 1970s, controversies ensued until the early 21st century when the Mediterranean Science Commission workshop (CIESM 2008) integrated stratigraphic, geophysical and geochemical datasets to establish stratigraphic consensus. For the first time, an attempt was made to correlate onshore evaporites deposited in more marginal, peripheral, basins with the offshore, deep basin, record. This new model for the Messinian salinity crisis established three main stages of formation as characterised and distinguished by different evaporitic deposits.

Despite its relatively young age, establishing a comprehensive understanding of the architecture, environments and formational processes of this Late Miocene salt giant is still challenging (Krijgsman et al. 2024). To this end, we combine geographical and chemo-physical information from the present-day Mediterranean (Figure 10.47) with the available onshore and offshore datasets of the Messinian evaporites (Figure 10.48) in order to elucidate their depositional context.

The Mediterranean Sea is an enclosed basin (Figure 10.47) connected with the Atlantic Ocean only via the 13 km wide, c. 300 m deep Strait of Gibraltar. The complex geomorphological structure of the Mediterranean Basin is primarily a product of late Mesozoic to Cenozoic convergence between the African and European plates during the Alpine orogeny (Amblàs et al. 2004). The Mediterranean can be broadly subdivided into two main sub-basins, the Western Basin and the Eastern Basin, with areas of 0.85 and 1.65 million km^2 respectively (Coll et al. 2010). These basins are separated by the 145 km wide Strait of Sicily, a 400 m deep submarine plateau extending between the island of Sicily and the African continent. Within this context, a series of smaller sub-basins are separated by peninsulas and islands (Figure 10.47). The average water depth of the Mediterranean Sea is approximately 1500 m with the deepest point, in the Ionian Sea, at 5267 m.

In the modern Mediterranean, evaporation exceeds combined precipitation and river input, resulting in an elevated surface water salinity averaging 38‰. The main influx is Atlantic surface water entering at the Strait of Gibraltar

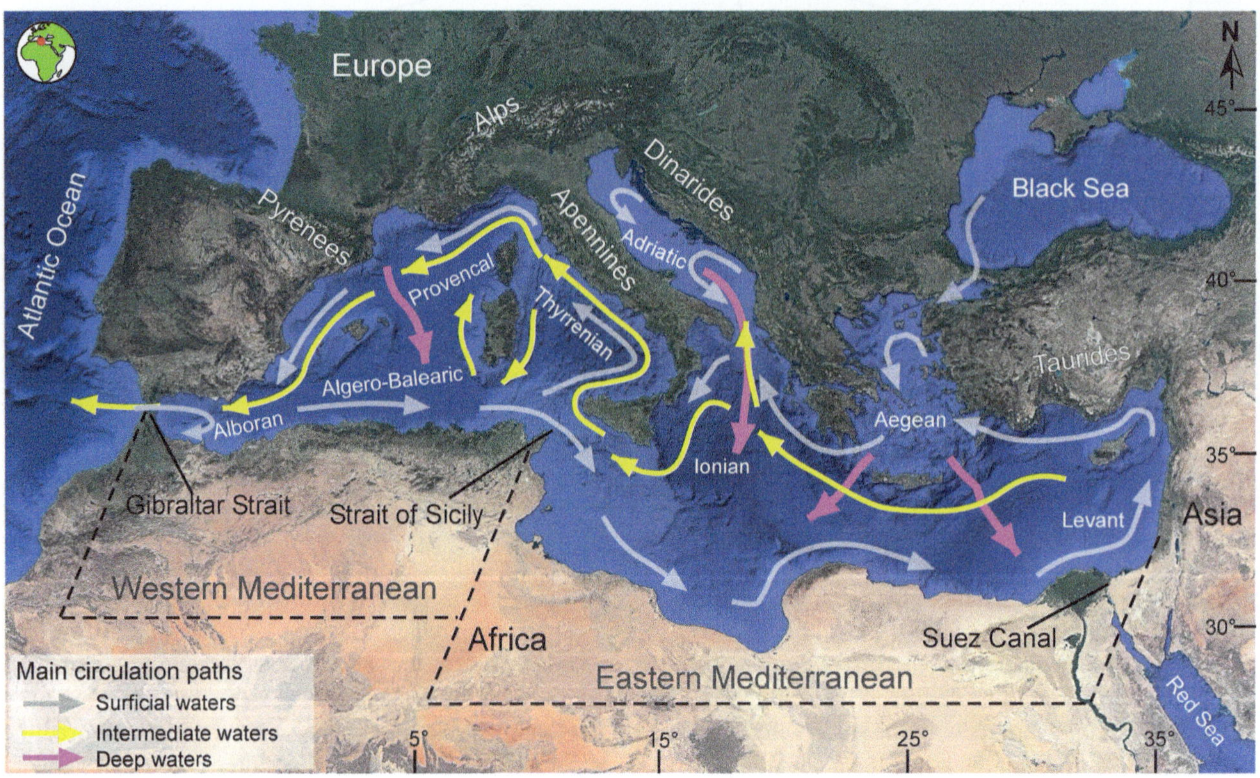

Figure 10.47 Satellite image of the Mediterranean Sea showing the main sub-basins and water circulation paths (modified from Bergamasco and Malanotte-Rizzoli (2010); Millot and Taupier-Letage (2005)). Google Earth Landsat/Copernicus image acquired 14th December 2015.

with an average temperature of 15°C and salinity of c. 36‰ (Bergamasco and Malanotte-Rizzoli 2010). The Atlantic waters circulate eastward along the southern shore, forming two major counterclockwise surface circulation cells, in the Western and Eastern basins, as well as several minor gyres within the sub-basins (Robinson et al. 2001). Within the Eastern Basin, the sinking of denser, higher salinity, water generates the Levantine Intermediate Current that travels westward, at a depth of 200 and 500 m, and contributes to the outflow of Mediterranean waters to the Atlantic (Figure 10.47) (Robinson et al. 2001). Dense water formation also occurs in the Gulf of Lyon, in the Southern Adriatic and the Aegean Sea (Bergamasco and Malanotte-Rizzoli 2010). The average residence time of Mediterranean seawater is between 50 and 100 years (Millot and Taupier-Letage 2005).

The Mediterranean Sea, positioned north of the junction of the northern Hadley and Ferrel cells, has a semi-arid climate characterised by mild wet winters and hot dry summers (Giorgi and Lionello 2008). Continental high-pressure systems dominate, generating storms over the entire region. Within the western Mediterranean, the North Atlantic Oscillation dominates, resulting in the north-westerly Mistral and northerly Tramontane. The Eastern Basin is dominated by the dry northerly Bora and Etesian winds. While to the south, the hot dry Sirocco, Ghibli and Khamsin southerlies blow from northern Africa. Precipitation over the Mediterranean is controlled strongly by the African and Asian monsoons. These complex wind systems play an important role in controlling the down-welling and upwelling of water masses throughout the Mediterranean (Bakun and Agostini 2001).

During the Late Miocene a succession of Mediterranean-Atlantic marine gateways, located in the region of the present-day Strait of Gibraltar (Figure 10.48b), became restricted and eventually closed (Flecker et al. 2015; Krijgsman et al. 2018). With progressive isolation from the Atlantic, there was a significant shift in deep-water circulation and ventilation throughout the Mediterranean Basin (Kouwenhoven and van der Zwaan 2006) resulting

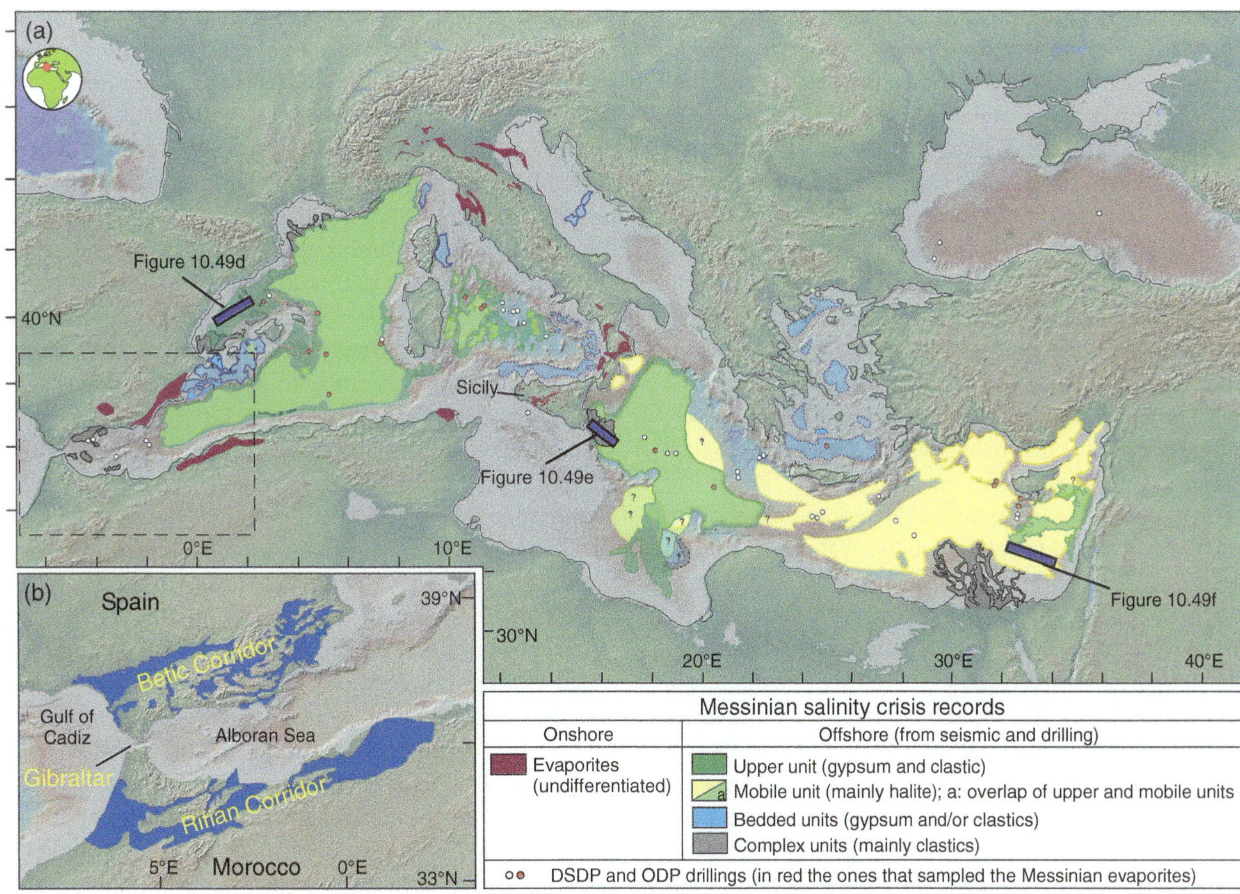

Figure 10.48 (a) Onshore and offshore distribution of the Messinian evaporites in the Mediterranean Basin (modified from Lofi (2018)). Blue rectangles indicate the location of the seismic line sketches shown in Figure 10.49. (b) Magnification of the region of the Strait of Gibraltar (dashed inset in (a) showing the main Late Miocene gateways (blue areas)); modified from Flecker et al. (2015).

in the development of stratified water columns in numerous sub-basins (García-Veigas et al. 2018; Sabino et al. 2021). These events coincided with a stable climatic period within a global cooling trend culminating during the Late Miocene (Herbert et al. 2016). The restricted Mediterranean basins were highly sensitive to short-term climate oscillations, resulting in the deposition of the cyclical evaporite deposits (mostly gypsum and halite) of the Messinian salinity crisis (Figures 10.35 and 10.37) (Krijgsman et al. 2024).

The distribution, thickness and internal architectures of the Messinian evaporite deposits vary greatly between the Western and Eastern Mediterranean and between shallow and deep regions (Figures 10.48 and 10.49a–e). Onshore exposures, representing shallow (<200 m) and intermediate (200 m–1 km) depth basins, contain an array of primary and redeposited evaporite units transitioning laterally into carbonate and organic-rich shale facies (Figure 10.49f). Offshore, km-thick bodies of Messinian evaporites (Figures 10.48 and 10.49) are buried beneath Pliocene to Quaternary terrigenous and marine sediments.

In marginal sub-basins, evaporite precipitation commenced, at 5.97 Ma (Roveri et al. 2014a), with the conformable deposition of the Primary Lower Gypsum unit (Lugli et al. 2010) over pre-evaporite shelf and slope deposits (Figure 10.35a,b). This gypsum unit comprises up to 16 precession-driven cycles cumulatively reaching several hundred metres in thickness. Each lithological cycle consists of a shale–gypsum couplet deposited during the more humid and more arid period of the precessional cycle, respectively (Natalicchio et al. 2019). Similarities between lithological cycles of the Primary Lower Gypsum unit described from marginal sub-basins across the Mediterranean are consistent basin-wide. This gypsum unit exhibits an up-section decrease in bed thickness recording a shallowing-upward trend. The typical sequence commences with two, relatively thin, gypsum beds comprising very large selenite crystals (Figure 10.36a). These are succeeded by three, thicker (up to 30 m), beds displaying an internal facies succession from basal massive selenite facies (Figure 10.33b) to upper banded selenite facies (Figure 10.33d), recording a shallowing-upward fluctuation of the pycnocline (Figure 10.36c,d). From the 6th cycle (5.84 Ma), facies within subsequent cycles became tripartite with gypsum supercones (e.g. in the Sorbas Basin, Spain; Figure 10.33h) and branching selenite facies (Figure 10.33i) superseding the massive and banded selenite (Natalicchio et al. 2021). The facies change at the 6th gypsum bed is employed as a synchronous correlatable marker horizon indicative of basin-wide hydrologic change – possibly due to the establishment of bottom brine currents, concomitant with an increased input of Atlantic water to the Mediterranean (Lugli et al. 2010). The Primary Lower Gypsum unit is excellently exposed in the Cordillera Betica (southern Spain; Figure 10.35a,c), the Italian Apennines (Figure 10.35b) and the Aegean peninsula and island basins.

In marginal (e.g. Piedmont Basin; (Dela Pierre et al. 2011)) and intermediate (e.g. Balearic Promontory; (Ochoa et al. 2015)) sub-basins, gypsum units typically thin basinward, laterally transitioning to dm-thick microbial carbonate and organic-rich shales deposited under sub-oxic–anoxic–euxinic conditions (Figure 10.38a) ((Dela Pierre et al. 2012; Natalicchio et al. 2017). Isotopic (García-Veigas et al. 2018; Aloisi et al. 2022; Guibourdenche et al. 2022) and geomicrobiological observations (Natalicchio et al. 2022) support gypsum precipitation from brines affected by microbial redox processes in a stratified water column.

The depth of deposition of the evaporite units is one of the most debated topics of the Messinian salinity crisis. For the Primary Lower Gypsum selenite facies, a proposed depth of <200 m was based on the interpretation of selenite-hosted filamentous, 'spaghetti-like' (Vai and Ricci Lucchi 1977) structures (Figure 10.50a,b; see Box 10.4) as fossilised benthic cyanobacteria from the photic zone (Panieri et al. 2010). The re-assignment of these filaments to large sulphur-oxidising bacteria (Dela Pierre et al. 2015), living at a wide range of water depths in modern settings (Bailey et al. 2009), indicates that the gypsum potentially formed at a greater range of depths than previously suggested. The presence of a well-preserved diverse diatom assemblage, as inclusions within bottom-grown gypsum crystals (Figure 10.50c) from several Mediterranean sub-basins, is consistent with gypsum growth in a relatively deep stratified setting with a deep chlorophyll maximum positioned in the sub-photic zone (Pellegrino et al. 2021). Further evidence for relatively deep evaporite precipitation is provided by the interpretation of coeval bedded gypsum units from the Balearic Basin as having precipitated offshore of the continental shelf (Ochoa et al. 2015). While none of these lines of evidence provide a conclusive water depth range for the deposition of these units, they all support an interpretation of precipitation in a subaqueous setting during the early phases of the Messinian salinity crisis.

The top of the Primary Lower Gypsum unit is truncated by a sharp erosional unconformity – the 'Messinian Erosional Surface' (Figure 10.49f). The unconformity is overlain by discontinuous lenses of a diverse range of evapoclastic facies ranging from fine-to-coarse-grained graded turbidite deposits to mountain-sized gypsum olistostromes embedded in a muddy matrix (Figure 10.34) (Roveri et al. 2014a). The sediments of this 'Resedimented Lower Gypsum' unit were derived from the erosion and down-slope transport of the marginal Primary Lower Gypsum deposits and reached thicknesses of up to several hundred metres at basin depocenters. These evapoclastic

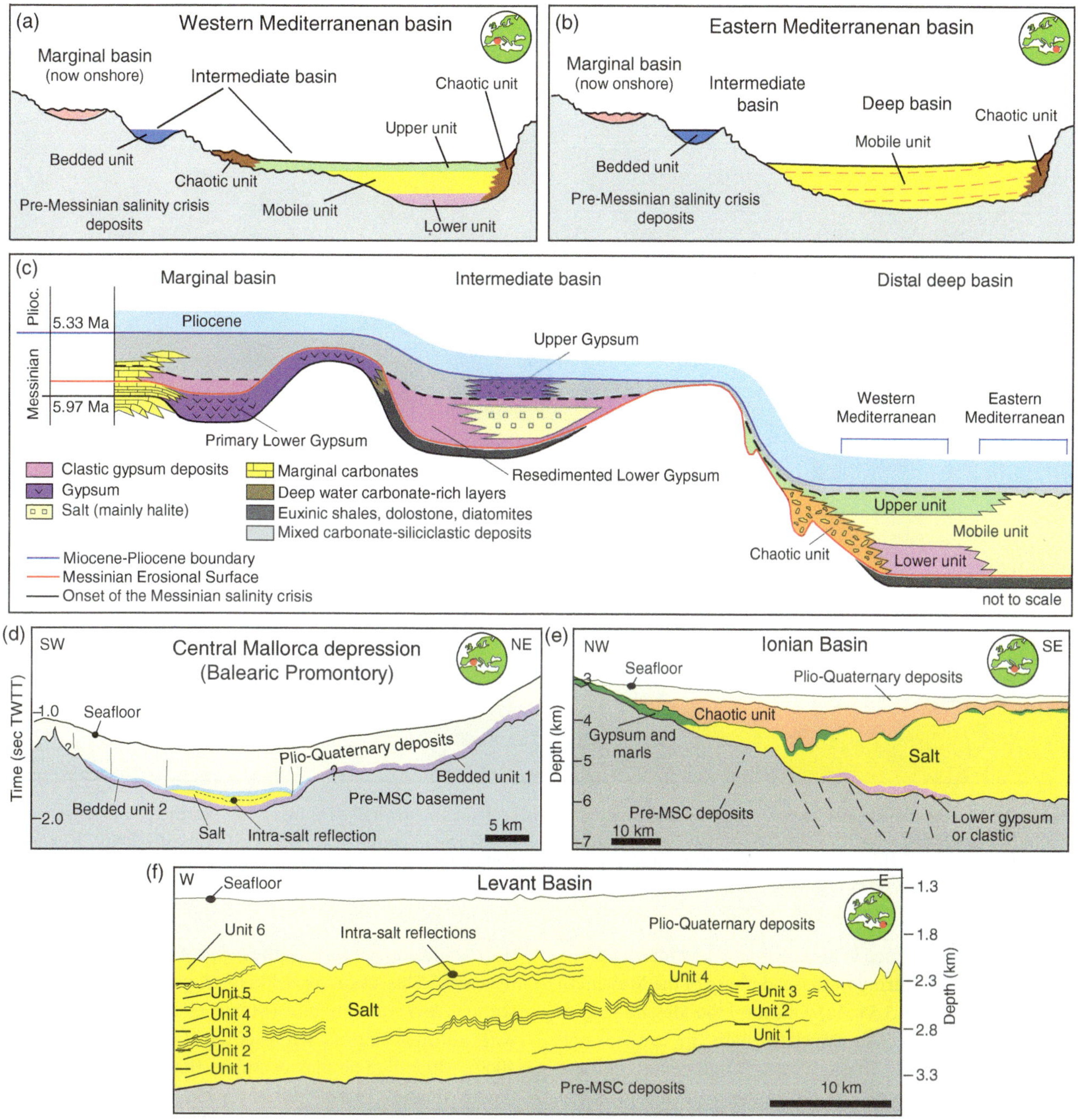

Figure 10.49 (a–c) Simplified line drawings of seismic lines from both the Western and the Eastern Mediterranean (see position in Figure 10.48). The 'Messinian trilogy' from the Central Mallorca depression (Balearic Promontory) comprising a basal bedded gypsum unit (unit 1), a middle salt unit (pinching out laterally) and a topmost discontinuous upper bedded gypsum unit (unit 2; modified from Raad et al. 2021). (b) Large-scale chaotic deposits resting on top of the Messinian salt and below the marine Plio-Pleistocene deposits (modified from Micallef et al. 2018). (c) Salt unit in the Levant Basin subdivided in up to six internal sub-units (modified from Gvirtzman et al. 2013). (d,e) Schematic cross-sections showing the vertical and lateral differentiation of the Messinian evaporite units between the marginal shallow and distal deep basins in the Western (D) and Eastern (E) Mediterranean. The deep basins are reconstructed from seismic and well-log data; modified from Lofi (2018). (f) Detail of the lateral transitions between the marginal and the intermediate basins primarily exposed in onshore basins (modified from Roveri et al. (2016)).

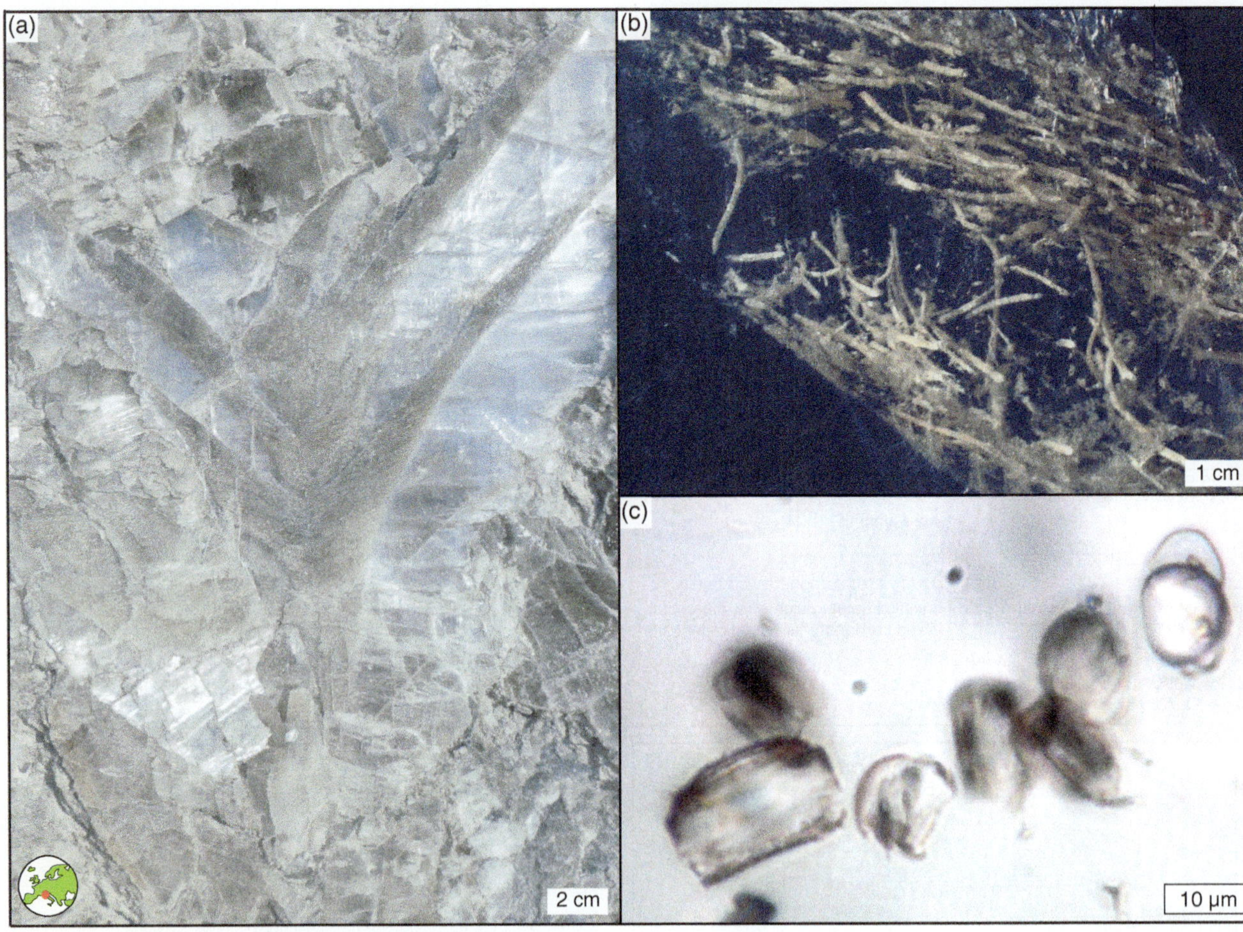

Figure 10.50 Messinian selenite crystal and its palaeobiological content. (a) swallow-tail selenite crystal (Northern Apennines). (b) Photomicrograph showing the 'spaghetti-like' structures interpreted as the remains of sulphide-oxidising bacteria (Dela Pierre et al. 2015). Modified from Natalicchio et al. (2022). (c) Photomicrograph showing nano-sized planktic diatoms preserved in Messinian gypsum. Modified from Pellegrino et al. (2021).

sediments accumulated in both marginal and intermediate-depth sub-basins during the second phase of the Messinian salinity crisis (5.60–5.55 Ma). Offshore, the Messinian Erosional Surface has been associated with the canyon incisions of the Nile and Rhone rivers. The discovery of these large canyons favoured the hypothesis of a significant sea-level drop during the second phase of the Messinian salinity crisis (e.g. Lofi et al. 2005). This interpretation of a significant relative sea-level fall is supported by the observed down-slope transitions of the Resedimented Lower Gypsum to >1 km thick halite deposits at numerous sites (e.g. Sicily, Calabria, Cyprus; Figure 10.49b,c). However, the origin of the unconformity, as either a subaerial or subaqueous erosion event, the amplitude of the inferred sea-level drop (Gvirtzman et al. 2022) and the possible deep-water deposition of halite (Sirota et al. 2018) leave the 'desiccated basin model' open to debate.

The precipitation of halite and potash salts in intermediate-depth sub-basins marks a period of major restriction of the Mediterranean Basin – the *acme* of the Messinian salinity crisis. This event was likely triggered by a period of stronger evaporation rate (Manzi et al. 2012) coincident with multiple steps of sea-level fall (Lofi et al. 2005; Ryan 2009). Modelling studies postulate that a narrow gateway (approximately 1 km wide) and a shallow sill depth (<10 m) were sufficient to block the Mediterranean outflow and increase the salinity of the Mediterranean brines (Meijer 2006; Topper and Meijer 2015). In the Sicilian onshore sub-basins (e.g. the Caltanissetta Basin), the salt succession, recording an overall shallowing-upward trend (Garcia-Veigas et al. 1995), commences with a massive cumulitic halite unit, likely deposited in a relatively deep stratified basin, that passes up-section to a lenticular body of fine-to-coarse grained potash salts dominated by kainite and carnallite with minor bischofite (Figure 10.37). At the Realmonte mine in Sicily, a proposed exposure surface at the top of the potash salt unit (Lugli et al. 1999), demonstrates that intermediate-depth sub-basins, locally

Box 10.4 Palaeobiology: A Key to Interpreting Ancient Evaporites?

Evaporite deposits have received scarce interest from palaeontologists. This is probably related to the fact that evaporitic basins, characterized by a fully hypersaline water mass, could not be tolerated by the majority of eukaryotic organisms. On the contrary, prokaryotes (bacteria and archaea) are able to developsurvival strategies even in extreme hypersaline environments. This is particularly well known in modern gypsum deposits in which a peculiar internal colour banding is the result of endoevaporitic col-onization of gypsum by different microbial communities (Figure a; Oren et al. 2009; Jahnke et al. 2014). "Salt-loving" (halophilic) archaea and bacteria have developed strategies allowing them to survive for millions of years within fluid inclusions in halite (Vreeland et al. 2007). In such a narrow micro-environment, these heterotrophic microbial communities survive by feeding on eukaryotes (i.e. the alga *Dunaliella*) entrapped within the inclusions (Lowenstein et al. 2011).

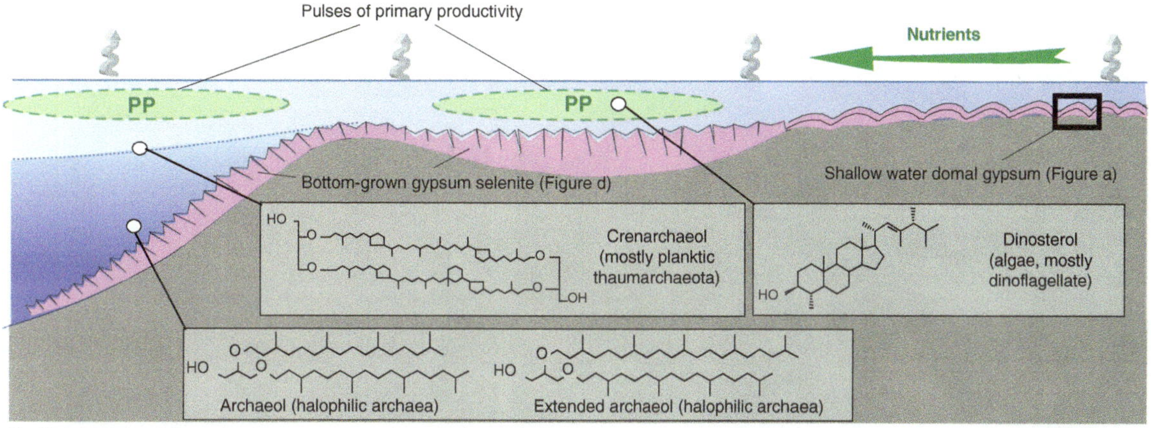

Simplified sketch of a shallow to deep transect in an evaporite basin showing distribution of the different gypsum lithofacies of Figs a and d. The green arrow indicates the continental input of nutrients (either fluvial or aeolian) promoting pulses of increased primary productivity (PP) in the upper water mass. The fate of the primary producers is to be trapped within gypsum growing at the basin floor. The chemical structures of some abundant archaeal and eukaryal molecular fossils found in Messinian gypsum also indicates evidence of the putative life position of their source organisms (Natalicchio et al., 2022).

Recent studies have shown that gypsum is an excellent mineral for the preservation of microfossils and organic matter (i.e. Schopf et al. 2012; Natalicchio et al. 2022; Pellegrino et al. 2021). In fact, the rapid growth of gypsum at the seafloor can entrap both the benthic life and organisms settling from the water column. This observation, together with the view of a deep water context for evaporite deposition characterized by a stratified water mass - allowing the expansion of eukaryotic organisms in the upper water layer - has increased palaeontologists interest in life preserved in gypsum.
The Messinian gypsum (Figure d) is an valuable example for unravelling the complextiy of life during the formation of a salt giant. The most common feature is an intricate network of filamentous microfossils (Figures e,f,g), also described as "spaghetti-like" structure (Vai and Ricci Lucchi 1977) and formerly assigned to cyanobacteria living in the photic zone, that is, at a maximum depth of 200 m (Panieri et al. 2010). Later on, Schopf et al. (2012) and Dela Pierre et al. (2015) suggested that the filaments may rather represent the remains of colorless sulfide-oxidizing bacteria such as *Beggiatoa* or Thioploca living at the seafloor; such interpretation has evidenced the importance of an active sulfur cycle in Messinian brines (Natalicchio et al. 2022). Interestingly, microfossils of planktonic microrganisms have also been found abundant in Messinian gypsum, including a diverse diatom assemblage (Figure 52; Pellegrino et al. 2021), often observed associated with clay-rich aggregates interpreted as marine snow floccules (Figures h,i; Dela Pierre et al. 2015).

Molecular fossils

The analysis of molecular fossils (lipid biomarkers) is a well-established approach to reconstruct the composition of microbial communities preserved in modern evaporites. However, very few studies report the molecular assemblage of ancient evaporite deposits (e.g. Christeleit et al. 2015; Isaji et al. 2019; Vasiliev et al. 2017). A recent investigation of the lipid inventory of the Messinian gypsum of the Mediterranenan area, evidenced the predominance of lipid sourced by a diversified archaeal community, including planktic halophilic archaea, planktic thaumarchaea, and a poorly constrained community of benthic archaea (Natalicchio et al. 2022). Such assemblages accompained by molecules sourced by algae (e.g. dinosterol) support the idea that part of the Messinian gypsum formed in a stratified basin typified by a normal marine to diluted upper water column and more saline deeper waters.

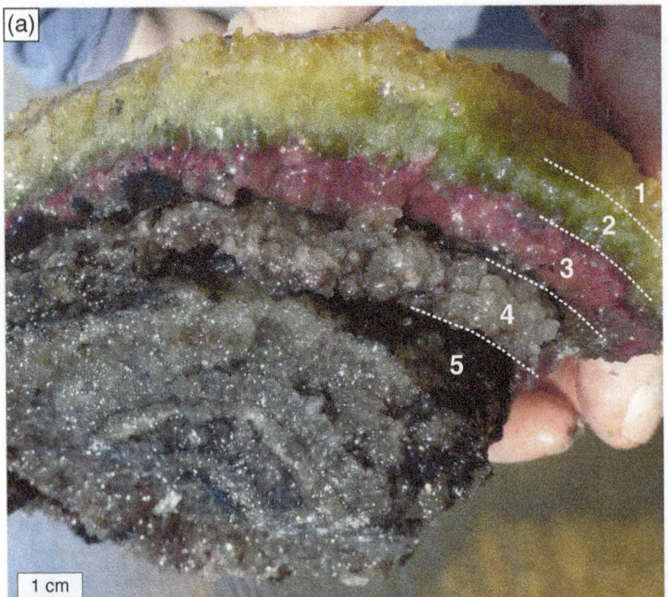

Modern gypsum dome showing colour banding related to the presence of different endolithic microbial communities (gypsum saltwork, Sicily, Italy); 1) yellow, unicellular cyanobacteria, 2) green, filamentous cyanobacteria, 3) pink, photosynthetic purple bacteria, 4) brown, green filamentous anoxygenic phototrophs, 5) black, sulphate-reducing bacteria (different microbial communities reconstructed by com-parison with Oren et al. (2009). Similar domal gypsum geometries are observed in modern sabkhas (Aref et al. 2014).

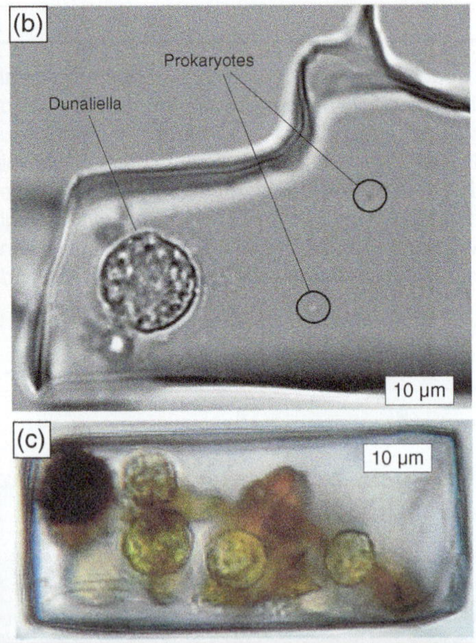

Primary fluid inclusions in Pleistocene halite from Saline Valley (b) and Death Valley (c) (California, USA; images courtesy of Tim Lowenstein) entrapping dunaliella cells and small prokaryotes (circled in Figure b). The orange and green colour of Dunaliella cells in Figure c suggest preservation of chlorophyll and ß-carotene.

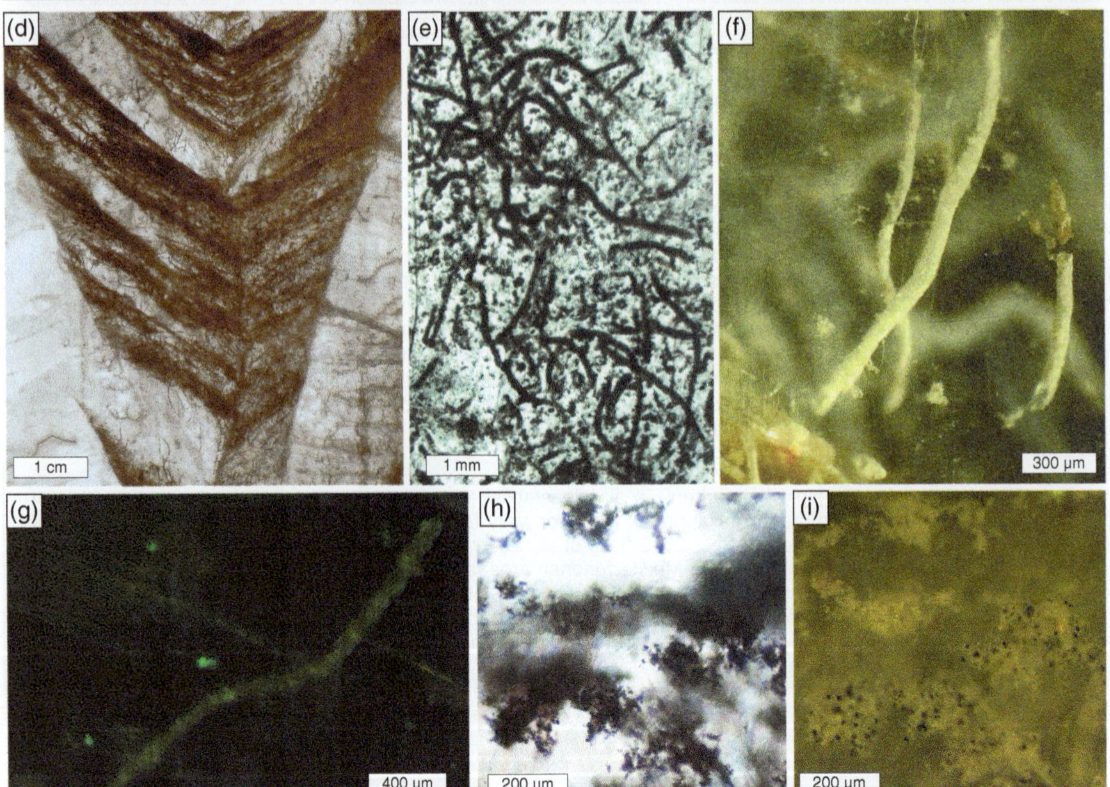

Palaeobiological content of the Messinian gypsum. d) Twinned selenite crystals; the darker colour of the core evidences the abundance of the filamentous (spaghetti-like) fossils. Optical (Figures e, f, h) and fluorescence (Figures g, i) microscope photomicrographs of filaments and marine snow aggregates trapped within Messinian gypsum. The black dots in Figure i are iron sulfide grains.

experienced rapid filling until their desiccation during the second phase of the crisis. The potash unit is overlain by a thick halite unit composed of extremely cyclical cm- to dm-thick halite-shale alternations (Figure 10.33m and 10.36i). These cyclical facies are interpreted as the product of annual or pluri-annual evaporation cycles deposited in shallow water, unstratified brine (Manzi et al. 2012).

The uppermost evaporite deposits, deposited during the final phase of the salinity crisis (also known as 'Lago-Mare' phase; 5.55–5.33 Ma), were only locally developed in shallow water sub-basins (e.g. Sicily and Cyprus). These Upper Gypsum deposits comprise up to seven cyclic gypsum-marl couplets, each of which is up to 10 m thick, and laterally grade into shallow water carbonate and siliciclastic deposits (e.g. in the northern Apennines and the Betic Cordillera). Both the gypsum and non-evaporite facies contain abundant brackish water fauna (ostracods, molluscs and dinocysts) characterised by a very low $^{87}Sr/^{86}Sr$ ratio in comparison with the global ocean. This indicates significant episodic freshwater dilution of the Mediterranean water mass, possibly sourced from the north-western freshwater Paratethys Basin (Andreetto et al. 2021), alternated with periods of intense evaporation.

The offshore deep basinal evaporites are the most poorly understood of the Messinian deposits (Figure 10.49). Our limited knowledge of the distribution and internal architecture of these units is largely derived from seismic imaging (Lofi 2018) supplemented by a few tens of metres of core, drilled from the topmost part of the salt giants during the 1970s and 1990s scientific drilling programs (Hsü et al. 1973c; Lugli et al. 2015) and some recent cuttings recovered during hydrocarbon exploration in the Levantine Basin of the Eastern Mediterranean (Meilijson et al. 2019). The deep basin evaporites exhibit substantially different geometries, thicknesses and seismic facies between the Western and the Eastern Mediterranean basins (Figures 10.48 and 10.49; (Lofi 2018)). In the Western Mediterranean, the offshore evaporites form what is termed the *Messinian trilogy* (Montadert et al. 1970) composed of two gypsiferous units, the Lower Unit and the Upper Unit, separated by the halite-dominated Mobile Unit. In the Eastern Mediterranean, only the Mobile Unit has been observed.

The onshore evaporites in marginal and intermediate-depth sub-basins and the offshore record of the deep sub-basins are physically disconnected with no relationship being exposed at the present-day Mediterranean margins. A broad-stroke, tentative correlation may be made between the offshore Messinian trilogy and the tripartite evaporite sequence of more marginal sub-basins (i.e. the Primary Lower Gypsum – salt unit – Upper Gypsum sequence of the Sicilian Basin). In the absence of definitive data, such correlations are highly speculative. Advancing our understanding of these deep water evaporite systems will require a targeted, large-scale scientific drilling expedition to the Mediterranean Basin (Camerlenghi and Aloisi 2020).

10.6.5 Where Does This Leave Us?

Given the above discussions, we may be forgiven for concluding that any attempt to place an evaporite sequence within a sequence stratigraphic framework is hopeless. Certainly, attempting to develop a 'one-size-fits-all' model is certainly pointless.

Coastal marine systems are highly responsive to sea-level change. Yet, the typically limited amount of accommodation and any antecedent topography are likely to complicate the primary sequence stratigraphic architecture in these systems. The stratigraphic record is further complicated by the particular susceptibility of these sequences to emergence dissolution during episodes of relative sea-level fall. By their very nature, many basinal and salina evaporite sequences have a significant disconnect from the global ocean system with sea level, in many cases, not being the dominant control on the stratigraphic architectural development.

However, as long as we remain aware of the limitations and pitfalls involved in the interpretation of these evaporite sequences, it *is* possible to establish detailed basin-specific sequence stratigraphic frameworks for evaporite depositional systems (e.g. Tucker 1991; Schröder et al. 2005; Becker and Bechstädt 2006).

10.7 Applications

Evaporite minerals and evaporite sequences are important to a diverse range of economic, scientific and societal applications. While a detailed discussion of all of the varied uses and benefits of evaporites lies beyond the scope of this chapter, we shall briefly consider some of their most significant applications. Many of these applications are, of course, also relevant to continental evaporites deposited in terrestrial settings.

10.7.1 Economic

The construction industry is a huge consumer of evaporite minerals. Gypsum and, to a lesser extent, anhydrite, are the main components of plasterboard and other construction materials – a typical new-build home in the USA contains over 7 Mt of gypsum (USGS 2018). Gypsum is also a critical component of cement and concrete where its presence slows the curing rate, allowing transportation and working.

The occurrence of economic accumulations of hydrocarbons in the earth's subsurface requires the presence of a low-permeability barrier to block their migration and escape at the surface. During burial, halite and anhydrite maintain the low permeability, high entry pressure and significant ductility that makes evaporite units some of the best seals for hydrocarbon plays. As well as forming the caprock for many of the world's most significant hydrocarbon accumulations, evaporites also contribute to hydrocarbon systems through the process of halokinesis, in particular, diapirism – producing a range of subsurface structures conducive to the accumulation and trapping of oil and gas (Sarg 2001; Schröder et al. 2005).

The vast majority of halite production is used for producing chemicals, particularly chlorine and caustic soda. These chemicals are, in turn, employed in the manufacture of a wide range of products including polyvinyl chloride (PVC), paints and detergents, as well as being required in the production of aluminium, paper, glass and ceramics and as a vital component of water treatment. Potassium chloride (sylvite) is also a significant feedstock to the chemical industry.

The agricultural sector is a major consumer of evaporite minerals. Potassium chloride is very important in the production of potash, over 97% of which is used in agriculture (Garrett 1996), primarily as fertiliser. Potassium chloride is also employed as an animal feed nutritional supplement. Gypsum, and to a lesser extent anhydrite, is widely used as a soil improver, as plant nutrition and a carrier medium for various agricultural dressings. Gypsum is also utilised to accelerate the reclamation of land that has experienced contamination through saltwater flooding. During turf production, halite is applied to suppress the growth of annual meadow grass, in concentrated doses halite will also kill unwanted plants – historically this was taken to an extreme where 'salting the earth' was employed to make the ground infertile.

Evaporite minerals and their products have broad medical applications as well as a substantial role in the cosmetics industry. Perhaps the most obvious medical applications of evaporites involve gypsum, in the form of plaster of Paris, as plaster casts to firmly hold healing fractures or its use to make dental impressions. Potassium chloride is employed to treat low blood potassium as well as being important in maintaining the electrolyte balance of dialysis fluids. Gypsum is a significant component of some toothpastes and is also present in a range of healthcare and beauty products including drugs, creams and shampoos. Potassium chloride and chemicals derived from halite are important for the production of soaps.

Halite is, perhaps, the best known and most widely employed 'geological' additive to foods where it is used as a flavour enhancer and preservative and to cure meat and fish. The food industry widely utilises potassium chloride as a nutrient supplement, stabiliser and thickener as well as for pH control. Potassium chloride is present across a range of foods including baby formulas, cereals, frozen desserts, meats, sports drinks, soups, sauces and snacks. Potassium chloride is also mixed with halite to produce low-sodium salt. As a dietary source of calcium, gypsum is added to a variety of foods including flour, bread, canned vegetables, ice creams and cheese. Gypsum is important to tofu production, is widely used in brewing beer and is very important to the process of wine making.

Evaporites have many other applications in industry and day-to-day life. The manufacturing sector utilises potassium chloride in metal and glass production, in electroplating and in producing pigments and dyes. Evaporites serve as fillers in a range of products, gypsum and anhydrite are used to manufacture paper with anhydrite also being used as a filler in plastics. The second largest use of halite is to manage ice, either as pure halite or, more commonly, in a mixture of sand and grit. Potassium chloride also has applications in de-icing products. Anhydrite is added as a drying agent to paints and varnishes. This list is far from exhaustive: evaporites infiltrate our daily lives.

10.7.2 Scientific

Evaporite sequences provide a high-resolution archive for deep-time palaeoenvironmental interpretation – particularly for the interpretation of palaeoclimates. The prevalence of fluid inclusions in evaporite minerals, particularly in halite, provides an important archive of ancient ocean water chemistry and also contributes to palaeoclimatic studies.

At the basin scale, innovations in the acquisition and processing of seismic data are allowing the imaging of evaporite sequences with increasing clarity. The illumination of both the pre-evaporite stratigraphy and the base of the evaporite sequence promises the potential to elucidate pre-evaporite palaeoenvironments and the landscapes in which evaporite precipitation was initiated. The three-dimensional imaging of internal layering and structures within the evaporite sequences will advance our understanding of evaporite basin evolution over time (e.g. in the Zechstein (Van Gent et al. 2011)). These factors build towards increasingly accurate models of evaporite deposition while simultaneously advancing our understanding of the role of tectonic processes in evaporite genesis.

Evaporite sequences and, in particular, fluid inclusions within salt minerals, provide an excellent repository for organic compounds and, potentially, even viable microorganisms that may be preserved for thousands to millions of years (Vreeland et al. 2007; Schubert et al. 2009; Lowenstein et al. 2011). These microorganisms (primarily algae, bacteria and halophilic archaea) were living in the brines from which the evaporites were formed, becoming entrapped during rapid mineral growth. Evaporite rocks provide an ideal natural laboratory from which to study these extremophiles and assess the tolerance of life to hypersaline conditions.

In light of these recent studies, the discovery of abundant chloride and sulphate minerals (including gypsum) on Mars (e.g. Clark et al. 2005; Osterloo et al. 2008) opens challenging new perspectives for the search for evidence of extra-terrestrial life, making evaporite deposits one of the primary targets for current and future extra-planetary missions (Benison 2019).

10.7.3 Societal

Evaporites have an important role to play during the transition from fossil fuels to alternative forms of energy generation. The sealing characteristics that have made evaporite units so important to hydrocarbon systems also make them effective storage solutions for the waste products of industry and energy generation alike. Depleted hydrocarbon reservoirs and engineered salt caverns provide an ideal storage solution for carbon capture and sequestration (CCS) as well as a site for hydrogen storage (Flesch et al. 2018; Stephenson et al. 2019; Krevor et al. 2023; Miocic et al. 2023).

The dry conditions and self-healing nature of massive halite units has also made them the preferred sites for the construction of deep geological repositories for the disposal of radioactive waste and numerous other hazardous materials.

Evaporite units are a significant target for geothermal energy extraction, both in the form of district heating and for electricity generation. In areas with low geothermal gradients, the cost of drilling boreholes forms a significant component of the development costs of a hydrothermal project. However, where halite is present, its high thermal conductivity increases the up-section transfer of heat, resulting in higher temperatures at relatively shallow depths (Raymond et al. 2022) and, consequently, reduced drilling costs. Of course, developing these resources is not without its challenges, geothermal fluids must be maintained in equilibrium with subsurface mineralogies if formation dissolution, precipitation clogging of production facilities and thermal short-circuiting are to be avoided (Moore and Holländer 2021). The presence of subsurface anhydrite can be particularly challenging as disruption of the geohydrology may initiate hydration and swelling, damaging both geothermal infrastructure and surface buildings (Sass and Burbaum 2010).

Evaporites in the shallow subsurface are susceptible to dissolution, particularly where there is significant subsurface water flow. The localised removal of these soluble units results in the collapse of the overlying land surface to produce a topography characterised by sinkholes and karst (Serridge Colin and Cooper Anthony 2023). Such collapses are a particular hazard in urban areas where recent anthropogenic activity has modified the natural subsurface hydrology, for instance, changing the level of the water table or introducing undersaturated fluids through increased watering of gardens. Under the context of the ongoing climate emergency, we can anticipate that areas of the world experiencing increased rainfall will likely be subject to intensified dissolution with the potential for greater instances of sinkhole collapse.

Further Reading

Bąbel, M. and Bogucki, A. (2007). The Badenian evaporite basin of the northern Carpathian Foredeep as a model of a meromictic selenite basin. *Geological Society, London, Special Publications* 285: 219–246.

Evans, G., Schmidt, V., Bush, P., and Nelson, H. (1969). Stratigraphy and geologic history of the sabkha, Abu Dhabi, Persian Gulf. *Sedimentology* 12: 145–159.

Kirkland, D.W., Denison, R.E., and Dean, W.E. (2000). Parent brine of the Castile Evaporites (Upper Permian), Texas and New Mexico. *Journal of Sedimentary Research* 70: 749–761.

Krijgsman, W., Rohling, E.J., Palcu, D.V. et al. (2024). Causes and consequences of the Messinian salinity crisis. *Nature Reviews Earth and Environment* 5: 335–350.

Hardie, L.A. and Lowenstein, T.K. (2004). Did the Mediterranean Sea dry out during the Miocene? A reassessment of the evaporite evidence from DSDP Legs 13 and 42A cores. *Journal of Sedimentary Research* 74: 453–461.

Logan, B.W. (1987). *The MacLeod Evaporite Basin, Western Australia: Holocene Environments, Sediments and Geological Evolution*. American Association of Petroleum Geologists.

Lowenstein, T.K., Timofeeff, M.N., Brennan, S.T. et al. (2001). Oscillations in Phanerozoic Seawater chemistry: evidence from fluid inclusions. *Science* 294: 1086.

Lugli, S., Manzi, V., Roveri, M., and Schreiber, B.C. (2010). The Primary Lower Gypsum in the Mediterranean: a new facies interpretation for the first stage of the Messinian salinity crisis. *Palaeogeography, Palaeoclimatology, Palaeoecology* 297: 83–99.

Ortí, F. (2011). Selenite facies in marine evaporites: a review. In: *Quaternary Carbonate and Evaporite Sedimentary Facies and their Ancient Analogues: A Tribute to Douglas James Shearman* (ed. C.G.S.C. Kendall, A.S. Alsharhan, I. Jarvis, and T. Stevens), 431–463. Chichester, West Sussex, UK: Wiley.

Purser, B.H. (1973). *The Persian Gulf - Holocene Carbonate Sedimentation and Diagenesis in a Shallow Epicontinental Sea*, 471. Berlin: Springer-Verlag.

Roveri, M., Flecker, R., Krijgsman, W. et al. (2014). The Messinian salinity crisis: past and future of a great challenge for marine sciences. *Marine Geology* 352: 25–58.

Schreiber, B.C. and Tabakh, M.E. (2000). Deposition and early alteration of evaporites. *Sedimentology* 47: 215–238.

Warren, J.K. (2006). *Evaporites: Sediments, Resources and Hydrocarbons*, 1036. Springer.

Warren, J.K. (2010). Evaporites through time: tectonic, climatic and eustatic controls in marine and nonmarine deposits. *Earth-Science Reviews* 98: 217–268.

Warren, J.K. (2016). *Evaporites: A geologic*. Springer.

References

Aloisi, G., Baudrand, M., Lécuyer, C. et al. (2013). Biomarker and isotope evidence for microbially-mediated carbonate formation from gypsum and petroleum hydrocarbons. *Chemical Geology* 347: 199–207.

Aloisi, G., Guibourdenche, L., Natalicchio, M., Caruso, A., Haffert, L., El Kilany, A. and Dela Pierre, F. (2022) The geochemical riddle of "low-salinity gypsum" deposits. *Geochimica et Cosmochimica Acta*, 327, 247–275.

Alvarez, M.D.P., Carol, E., Eymard, I. et al. (2022). Hydrochemistry, isotope studies and salt formation in saline lakes of arid regions: Extra-Andean Patagonia. *Argentina. Science of The Total Environment* 816: 151529.

Amblàs, D., Canals, M., Lastras, G. et al. (2004). Imaging the Seascapes of the Mediterranean. *Oceanography* 17: 144–155.

Anderson, R.Y. (1982). A long geoclimatic record from the Permian. *Journal of Geophysical Research: Oceans* 87: 7285–7294.

Anderson, R.Y., Kietzke, K.K. and Rhodes, D.J. 1978. Development of dissolution breccias, northern Delaware Basin, New Mexico and Texas, New Mexico Bureau Mines & Mineral Resources,.

Andreetto, F., Aloisi, G., Raad, F. et al. (2021). Freshening of the Mediterranean Salt Giant: controversies and certainties around the terminal (Upper Gypsum and Lago-Mare) phases of the Messinian Salinity Crisis. *Earth-Science Reviews* 216: 103577.

Arakel, A.V. (1980). Genesis and diagenesis of Holocene evaporitic sediments in Hutt and Leeman lagoons, Western Australia. *Journal of Sedimentary Research* 50: 1305–1326.

Arons, A.B. and Kientzler, C.F. (1954). Vapor pressure of sea-salt solutions. *Eos, Transactions American Geophysical Union* 35: 722–728.

Bąbel, M. (2002). Brine palaeocurrent analysis based on oriented selenite crystals in the Nida Gypsum deposits (Badenian, southern Poland). *Geological Quarterly* 46: 435–448.

Bąbel, M. (2004). Models for evaporite, selenite and gypsum microbialite deposition in ancient saline basins. *Acta Geologica Polonica* 54: 219–249.

Bąbel, M. (2005). Selenite-gypsum microbialite facies and sedimentary evolution of the Badenian evaporite basin of the northern Carpathian Foredeep. *Acta Geologica Polonica* 55: 187–210.

Bąbel, M. and Bogucki, A. (2007). The Badenian evaporite basin of the northern Carpathian Foredeep as a model of a meromictic selenite basin. *Geological Society, London, Special Publications* 285: 219–246.

Babel, M. and Kasprzyk, A. (1990). Gypsum ooids from the Middle Miocene (Badenian) evaporites of southern Poland. *Acta Geologica Polonica* 40: 215–239.

Bąbel, M., Olszewska-Nejbert, D., and Bogucki, A. (2011). Gypsum Microbialite Domes Shaped by Brine Currents from the Badenian Evaporites of Western Ukraine. In: *Advances in Stromatolite Geobiology* (ed. J. Reitner, N.-V. Quéric, and G. Arp), 297–320. Berlin Heidelberg, Berlin, Heidelberg: Springer.

Bąbel, M., Olszewska-Nejbert, D., and Nejbert, K. (2010). The largest giant gypsum inter growths from the Badenian (Middle Miocene) evaporites of the Carpathian Foredeep. *Geological Quarterly* 54: 477–486.

Bahafzullah, A., Fayed, L., Kazi, A., and Al-Saify, M. (1993). Classification and distribution of the red sea coastal sabkhas near Jeddah-Saudi Arabia. *Carbonates and Evaporites* 8: 23–38.

Bailey, J.V., Orphan, V.J., Joye, S.B., and Corsetti, F.A. (2009). Chemotrophic microbial mats and their potential for preservation in the Rock Record. *Astrobiology* 9: 843–859.

Baillie, A.D. (1955). Devonian System of Williston Basin1. *AAPG Bulletin* 39: 575–629.

Bakun, A. and Agostini, V.N. (2001). Seasonal patterns of wind-induced upwelling/downwelling in the Mediterranean Sea. *Scientia Marina* 65: 243–257.

Baldacci, L. (1886). *Descrizione geologica dell'isola di Sicilia*. Roma: Tipografia Nazionale.

Becker, F. and Bechstädt, T. (2006). Sequence stratigraphy of a carbonate-evaporite succession (Zechstein 1, Hessian Basin, Germany). *Sedimentology* 53: 1083–1120.

Benison, K.C. (2019). How to Search for Life in Martian Chemical Sediments and Their Fluid and Solid Inclusions Using Petrographic and Spectroscopic Methods. *Frontiers in Environmental. Science* 7.

Bergamasco, A. and Malanotte-Rizzoli, P. (2010). The circulation of the Mediterranean Sea: a historical review of experimental investigations. *Advances in Oceanography and Limnology* 1: 11–28.

Bergman, T. (1784). *Physical and chemical essays*. London: J. Murray.

Blättler, C.L., Bergmann, K.D., Kah, L.C. et al. (2020). Constraints on Meso- to Neoproterozoic seawater from ancient evaporite deposits. *Earth and Planetary Science Letters* 532: 115951.

Bottomley, N. (1996). Recent climate of Abu Dhabi. In: *Desert ecology of Abu Dhabi. A review and recent studies* (ed. P. E. Osborne), 36–49. Newbury: Pisces.

Boyle, R. (1673). *Tracts Consisting of Observations about the Saltness of the Sea: An Account of a Statical Hygroscope and Its Uses: Together with an Appendix about the Force of the Air's Moisture: a Fragment about the Natural and Preternatural State of Bodies. By the Honourable Robert Boyle*. In: *To All which is Premis'd a Sceptical Dialogue about the Positive Or Privative Nature of Cold: with Some Experiments of Mr. Boyl's Referr'd to in that Discourse. By a Member of the Royal Society*. E. Flesher.

Briggs, L.I. (1958). Evaporite facies. *Journal of Sedimentary Research* 28: 46–56.

Briggs, L.I., Gill, D., Briggs, D.Z., and Elmore, R.D. (1980). Transition from open marine to evaporite deposition in the Silurian Michigan Basin. In: *A. Nissenbaum), Developments in Sedimentology* (ed. H. Brines and E. Environments), 253–270. New York: Elsevier.

Butler, G.P. (1970). *Holocene gypsum and anhydrite of the Abu Dhabi sabkha, Trucial Coast: an alternative explanation of origin*, 120–152. In: *Third Symposium on Salt*.

Butler, G.P., Kendall, C.G.S.C., Kinsman, D.J.J. et al. (1965a). Recent anhydrite from the Trucial Coast of the Arabian Gulf. *Geological Society of London Circular* 120: 3.

Butler, G.P., Kendall, C.G.S.C., Kinsman, D.J.J. et al. (1965b). *Recent evaporite deposits along the Trucial Coast of the Arabian Gulf*, 91–94. *Geological Society of London Circular*.

Caesar, K.H., Kyle, J.R., Lyons, T.W. et al. (2019). Carbonate formation in salt dome cap rocks by microbial anaerobic oxidation of methane. *Nature Communications* 10: 808.

Camerlenghi, A. and Aloisi, V. (2020). Uncovering the Mediterranean Salt Giant (MEDSALT) - Scientific Networking as Incubator of Cross-disciplinary Research in Earth Sciences. *European Review* 28: 40–61.

Castanier, S., Perthuisot, J.-P., Rouchy, J.-M. et al. (1992). Halite ooids in Lake Asal, Djibouti: Biocrystalline build-UPS. *Geobios* 25: 811–821.

Cendón, D.I., Peryt, T.M., Ayora, C. et al. (2004). The importance of recycling processes in the Middle Miocene Badenian evaporite basin (Carpathian foredeep): Palaeoenvironmental implications. *Palaeogeography, Palaeoclimatology, Palaeoecology* 212: 141–158.

CIESM (2008) The Messinian salinity crisis from mega-deposits to microbiology. In: *A consensus report, in 33ème CIESM Workshop Monographs*, 33 (Ed F. Briand), pp. 168. CIESM, Monaco.

Clark, B.C., Morris, R.V., McLennan, S.M. et al. (2005). Chemistry and mineralogy of outcrops at Meridiani Planum. *Earth and Planetary Science Letters* 240: 73–94.

Clark, P.U., Dyke, A.S., Shakun, J.D. et al. (2009). The Last Glacial Maximum. *Science* 325: 710–714.

Coll, M., Piroddi, C., Steenbeek, J., Kaschner, K., Ben Rais Lasram, F., Aguzzi, J., Ballesteros, E., Bianchi, C.N., Corbera, J., Dailianis, T., Danovaro, R., Estrada, M., Froglia, C., Galil, B.S., Gasol, J.M., Gertwagen, R., Gil, J., Guilhaumon, F., Kesner-Reyes, K., Kitsos, M.-S., Koukouras, A., Lampadariou, N., Laxamana, E., López-Fé de la Cuadra, C.M., Lotze, H.K., Martin, D., Mouillot, D., Oro, D., Raicevich, S., Rius-Barile, J., Saiz-Salinas, J.I., San Vicente, C., Somot, S., Templado, J., Turon, X., Vafidis, D., Villanueva, R. and Voultsiadou, E. (2010) The Biodiversity of the Mediterranean Sea: Estimates, Patterns, and Threats. *PLOS ONE*, 5, e11842.

Court, W.M., Paul, A., and Lokier, S.W. (2017). The preservation potential of environmentally diagnostic sedimentary structures from a coastal sabkha. *Marine Geology* 386: 1–18.

Curtis, R., Evans, G., Kinsman, D.J.J., and Shearman, D.J. (1963). Association of dolomite and anhydrite in the recent sediments of the Persian Gulf. *Nature* 197: 679–680.

da Vinci, L. (1478-1518) The Notebook (The Codex Arundel), pp. 283.

de Lange, G.J. and Krijgsman, W. (2010). Messinian salinity crisis: A novel unifying shallow gypsum/deep dolomite formation mechanism. *Marine Geology* 275: 273–277.

Dela Pierre, F., Bernardi, E., Cavagna, S. et al. (2011). The record of the Messinian salinity crisis in the Tertiary Piedmont Basin (NW Italy): The Alba section revisited. *Palaeogeography, Palaeoclimatology, Palaeoecology* 310: 238–255.

Dela Pierre, F., Clari, P., Bernardi, E. et al. (2012). Messinian carbonate-rich beds of the Tertiary Piedmont Basin (NW Italy): Microbially-mediated products straddling the onset of the salinity crisis. *Palaeogeography, Palaeoclimatology, Palaeoecology* 344-345: 78–93.

Dela Pierre, F., Natalicchio, M., Ferrando, S. et al. (2015). Are the large filamentous microfossils preserved in Messinian gypsum colorless sulfide-oxidizing bacteria? *Geology* 43: 855–858.

Dronkert, H. (1976). Late Miocene evaporites in the Sorbas Basin and adjoining areas. *Mem. Soc. Geol. It.* 16: 341–361.

Eckardt, F.D. and Spiro, B. (1999). The origin of sulphur in gypsum and dissolved sulphate in the Central Namib Desert, Namibia. *Sedimentary Geology* 123: 255–273.

Evans, G., Kendall, C.G.S.C., and Skipwith, P. (1964). Origin of the coastal flats, the sabkha, of the Trucial Coast of the Persian Gulf. *Nature* 202: 759–761.

Evans, G., Schmidt, V., Bush, P., and Nelson, H. (1969). Stratigraphy and geologic history of the sabkha, Abu Dhabi, Persian Gulf. *Sedimentology* 12: 145–159.

Fiorini, F. and Lokier, S.W. (2020). Abnormal test growth in Larger Benthic Foraminifera from hypersaline coastal ponds of the United Arab Emirates. *Micropaleontology* 66: 151–156.

Flecker, R., Krijgsman, W., Capella, W., de Castro Martíns, C., Dmitrieva, E., Mayser, J.P., Marzocchi, A., Modestou, S., Ochoa, D., Simon, D., Tulbure, M., van den Berg, B., van der Schee, M., de Lange, G., Ellam, R., Govers, R., Gutjahr, M., Hilgen, F., Kouwenhoven, T., Lofi, J., Meijer, P., Sierro, F.J., Bachiri, N., Barhoun, N., Alami, A.C., Chacon, B., Flores, J.A., Gregory, J., Howard, J., Lunt, D., Ochoa, M., Pancost, R., Vincent, S. and Yousfi, M.Z. (2015) Evolution of the Late Miocene Mediterranean–Atlantic gateways and their impact on regional and global environmental change. *Earth-Science Reviews*, 150, 365–392.

Fleming, K., Johnston, P., Zwartz, D. et al. (1998). Refining the eustatic sea-level curve since the Last Glacial Maximum using far- and intermediate-field sites. *Earth and Planetary Science Letters* 163: 327–342.

Flesch, S., Pudlo, D., Albrecht, D. et al. (2018). Hydrogen underground storage—Petrographic and petrophysical variations in reservoir sandstones from laboratory experiments under simulated reservoir conditions. *International Journal of Hydrogen Energy* 43: 20822–20835.

García-Veigas, J., Cendón, D.I., Gibert, L. et al. (2018). Geochemical indicators in Western Mediterranean Messinian evaporites: Implications for the salinity crisis. *Marine Geology* 403: 197–214.

García-Veigas, J., Cendón, D.I., Rosell, L. et al. (2013). Salt deposition and brine evolution in the Granada Basin (Late Tortonian, SE Spain). *Palaeogeography, Palaeoclimatology, Palaeoecology* 369: 452–465.

Garcia-Veigas, J., Orti, F., Rosell, L. et al. (1995). The Messinian salt of the Mediterranean; geochemical study of the salt from the central Sicily Basin and comparison with the Lorca Basin, Spain. *Bulletin de la Société Géologique de France* 166: 699–710.

García-Veigas, J., Rosell, L., Zak, I. et al. (2009). Evidence of potash salt formation in the Pliocene Sedom Lagoon (Dead Sea Rift, Israel). *Chemical Geology* 265: 499–511.

Garrels, R. and Mackenzie, F.T. (1971). *Evolution of Sedimentary Rocks*. Norton and Company, New York: W. W.

Garrett, D.E. (1996). *Potash*, 734. Dordrecht: Springer.

Gasse, F. and Fontes, J.-C. (1989). Palaeoenvironments and palaeohydrology of a tropical closed lake (Lake Asal, Djibouti) since 10,000 yr B.P. *Palaeogeography, Palaeoclimatology, Palaeoecology* 69: 67–102.

Gavish, E. (1974). Geochemistry and mineralogy of a recent sabkha along the coast of Sinai, Gulf of Suez. *Sedimentology* 21: 397–414.

Ge, Y., Pederson, C.L., Lokier, S.W. et al. (2023). Radiaxial fibrous calcite forms via early marine-diagenetic alteration of micritic magnesium calcite. *Sedimentology* 70: 434–450.

Ge, Y., Pederson, C.L., Lokier, S.W. et al. (2020). Late Holocene to Recent aragonite-cemented transgressive lag deposits in the Abu Dhabi lagoon and intertidal sabkha. *Sedimentology* 67: 2426–2454.

Giorgi, F. and Lionello, P. (2008). Climate change projections for the Mediterranean region. *Global and Planetary Change* 63: 90–104.

Glennie, K.W. and Evans, G. (1976). A reconnaissance of the Recent sediments of the Ranns of Kutch, India. *Sedimentology* 23: 625–647.

Goodall, T.M. and Al-Belushi, J.D. (1998). A glossary of Arabic desert terminology used in southeastern Arabia. *In: Quaternary deserts and climate change* 611–619.

Gornitz, V.M. and Schreiber, B.C. (1981). Displacive halite hoppers from the Dead Sea; some implications for ancient evaporite deposits. *Journal of Sedimentary Research* 51: 787–794.

Grötsch, J., Suwaina, O., Ajlani, G. et al. (2003). The Arab Formation in central Abu Dhabi: 3-D reservoir architecture and static and dynamic modeling. *GeoArabia* 8: 47–86.

Grotzinger, J.P. and Kasting, J.F. (1993). New Constraints on Precambrian Ocean Composition. *The Journal of Geology* 101: 235–243.

Guibourdenche, L., Cartigny, P., Dela Pierre, F. et al. (2022). Cryptic sulfur cycling during the formation of giant gypsum deposits. *Earth and Planetary Science Letters* 593: 117676.

Gvirtzman, Z., Heida, H., Garcia-Castellanos, D. et al. (2022). Limited Mediterranean sea-level drop during the Messinian salinity crisis inferred from the buried Nile canyon. *Communications Earth & Environment* 3: 216.

Hadley, G. (1735). VI. Concerning the cause of the general trade-winds. *Philosophical Transactions of the Royal Society of London* 39: 58–62.

Hallam, A., Crame, J.A., Mancenido, M.O. et al. (1993). Jurassic Climates as Inferred from the Sedimentary and Fossil Record [and Discussion]. *Philosophical Transactions: Biological Sciences* 341: 287–296.

Handford, C.R. (1991) Marginal marine halite: sabkhas and salinas. In: *Developments in Sedimentology* (Ed J.L. Melvin), 50, pp. 1–66. Elesvier, Amsterdam.

Handford, C.R., Kendall, A.C., Prezbindowski, D.R. et al. (1984). *Salina margin tepees, pisoliths, and aragonite cements*, 523–527. Lake MacLeod, Western Australia: Their significance in interpreting ancient analogs. *Geology*.

Hanebuth, T.J.J., Stattegger, K., and Bojanowski, A. (2009). Termination of the Last Glacial Maximum sea-level lowstand: The Sunda-Shelf data revisited. *Global and Planetary Change* 66: 76–84.

Haq, B., Gorini, C., Baur, J. et al. (2020). Deep Mediterranean's Messinian evaporite giant: How much salt? *Global and Planetary Change* 184: 103052.

Hardie, L.A. (1967). The gypsum-anhydrite equilibrium at one atmosphere pressure. *The American Mineralogist* 52: 171–200.

Hardie, L.A. and Eugster, H.P. (1970). The evolution of closed-basin brines. *Mineralogical Society America Special Paper* 3: 273–290.

Hardie, L.A. and Eugster, H.P. (1971). The depositional environment of marine evaporites: A case for shallow, clastic accumulation. In: *Sedimentology, 16*, 187–220. Ltd: John Wiley & Sons.

Hardie, L.A. and Lowenstein, T.K. (2004). Did the Mediterranean Sea Dry Out During the Miocene? A Reassessment of the Evaporite Evidence from DSDP Legs 13 and 42A Cores. *Journal of Sedimentary Research* 74: 453–461.

Harrison, I.I.I. W.B., Voice, P.J., Grammer, G.M., Harrison, W.B., III and Barnes, D.A. (2018) Evaporite facies of the Michigan Basin. In: *Paleozoic Stratigraphy and Resources of the Michigan Basin*, 531, pp. 0. *Geological Society of America.*.

Hay, W.W. (2013). Documenting Past Climate Change. In: *Experimenting on a Small Planet: A Scholarly Entertainment* (ed. W.W. Hay), 140–179. Berlin Heidelberg, Berlin, Heidelberg: Springer.

Hay, W.W., Migdisov, A., Balukhovsky, A. N. et al. (2006). Evaporites and the salinity of the ocean during the Phanerozoic: Implications for climate, ocean circulation and life. *Palaeogeography, Palaeoclimatology, Palaeoecology* 240: 3–46.

Henchiri, M. and Slim-S'Himi, N. (2006). Silicification of sulphate evaporites and their carbonate replacements in Eocene marine sediments, Tunisia: two diagenetic trends. *Sedimentology* 53: 1135–1159.

Herbert, T.D., Lawrence, K.T., Tzanova, A. et al. (2016). Late Miocene global cooling and the rise of modern ecosystems. *Nature Geoscience* 9: 843–847.

Holliday, D.W. (1970). The petrology of secondary gypsum rocks; a review. *Journal of Sedimentary Petrology* 40: 734–744.

Hollis, C., Lawrence, D.A., Deville de Perière, M., and Al Darmaki, F. (2017). Controls on porosity preservation within a Jurassic oolitic reservoir complex, UAE. *Marine and Petroleum Geology* 88: 888–906.

Hovorka, S. (1987). Depositional environments of marine-dominated bedded halite, Permian San Andres Formation, Texas. *Sedimentology* 34: 1029–1054.

Hsü, K.J. (1972). Origin of saline giants: A critical review after the discovery of the Mediterranean Evaporite. *Earth-Science Reviews* 8: 371–396.

Hsü, K.J., Cita, M.B. and Ryan, W.B.F. (1973a) The origin of the Mediterranean evaporites. In: *Initial Reports of the Deep Sea Drilling Project 13, Part 2*. (Eds W.B. F. Ryan, K.J. Hsü and M.B. Cita), pp. 1203–1231. U.S. Government Printing Office, Washington D.C.

Hsü, K.J., Ryan, W.B.F., and Cita, M.B. (1973b). Late Miocene Desiccation of the Mediterranean. *Nature* 242: 240–244.

Hsü, K.J., Ryan, W.B.F. and Schreiber, B. C. (1973c) Petrography of a halite sample from hole 134—Balearic Abyssal Plain. In: *Initial Reports of the Deep Sea Drilling Project 13* (Eds *W.B.F. Ryan, K.J. Hsü, M.B. Cita, P. Dumitrica, J.M. Lort, W. Maync, W.D. Nesteroff, G. Pautot, H. Stradner, F.C. Wezel and A.G. Kaneps), pp. 708–711*. U.S. Government Printing Office.

Hudec, M.R., Dooley, T.P., Peel, F.J., and Soto, J.I. (2019). Controls on the evolution of passive-margin salt basins: Structure and evolution of the Salina del Bravo region, northeastern Mexico. *GSA Bulletin* 132: 997–1012.

Hudec, M.R. and Jackson, M.P.A. (2011). *The Salt Mine A Digital Atlas of Salt Tectonics*. American Association of Petroleum Geologists.

Hudec, M.R., Norton, I.O., Jackson, M.P. A., and Peel, F.J. (2013). Jurassic evolution of the Gulf of Mexico salt basin. *AAPG Bulletin* 97: 1683–1710.

Jackson, M.P.A. and Talbot, C.J. (1991). *A Glossary of Salt Tectonics*. Bureau of Economic Geology: The University of Texas at Austin.

Johnson, W.J. and Goldstein, R.H. (1993). Cambrian sea water preserved as inclusions in marine low-magnesium calcite cement. *Nature* 362: 335–337.

Jowett, E.C., Cathles, L.M. III, and Davis, B.W. (1993). Predicting Depths of Gypsum Dehydration in Evaporitic Sedimentary Basins 1. *AAPG Bulletin* 77: 402–413.

Kasprzyk, A. (2003). Sedimentological and diagenetic patterns of anhydrite deposits in the Badenian evaporite basin of the Carpathian Foredeep, southern Poland. *Sedimentary Geology* 158: 167–194.

Kemp, D.B. and Sadler, P.M. (2014). Climatic and eustatic signals in a global compilation of shallow marine carbonate accumulation rates. *Sedimentology* 61: 1286–1297.

King, R.H. (1947). Sedimentation in Permian Castile Sea. *AAPG Bulletin* 31: 470–477.

Kinsman, D.J.J. (1964) The recent carbonate sediments near Halat el Bahrani, Trucial Coast, Persian Gulf. In: *Deltaic and shallow marine deposits: Proceedings of the 6th International Sedimentological Congress The Netherlands and Belgium - 1963* (Ed L.M.J.U. Van Straaten), *Developments in Sedimentology*, pp. 185-192. Elsevier: Amsterdam, Amsterdam.

Kinsman, D.J.J. (1965). Dolomitization and evaporite development, including anhydrite, in lagoonal sediments, Persian Gulf. *Geological Society of America Special Paper* 82: 108–109.

Kinsman, D.J.J. (1969). Modes of formation, sedimentary assocuations, and diagnostic features of shallow-water and spratidal evaporites. *AAPG Bulletin* 53: 830–840.

Kinsman, D.J.J. (1976). Evaporites: Relative humidity control of primary mineral facies. *Journal of Sedimentary Petrology* 46: 273–279.

Kirkland, D.W. (2003). An explanation for the varves of the Castile evaporites (Upper Permian), Texas and New Mexico, USA. *Sedimentology* 50: 899–920.

Kirkland, D.W., Denison, R.E., and Dean, W.E. (2000). Parent Brine of the Castile Evaporites (Upper Permian), Texas and New Mexico. *Journal of Sedimentary Research* 70: 749–761.

Kouwenhoven, T.J. and van der Zwaan, G.J. (2006). A reconstruction of late Miocene Mediterranean circulation patterns using benthic foraminifera. *Palaeogeography, Palaeoclimatology, Palaeoecology* 238: 373–385.

Krevor, S., de Coninck, H., Gasda, S.E. et al. (2023). Subsurface carbon dioxide and hydrogen storage for a sustainable energy future. *Nature Reviews Earth & Environment.*.

Krijgsman, W., Capella, W., Simon, D. et al. (2018). The Gibraltar Corridor: Watergate of the Messinian Salinity Crisis. *Marine Geology* 403: 238–246.

Krijgsman, W., Hilgen, F.J., Raffi, I. et al. (1999). Chronology, causes and progression of the Messinian salinity crisis. *Nature* 400: 652–655.

Krijgsman, W., Rohling, E.J., Palcu, D.V. et al. (2024). Causes and consequences of the Messinian salinity crisis. *Nature Reviews Earth & Environment* 5: 335–350.

Lambeck, K. (1996). Shoreline reconstructions for the Persian Gulf since the last glacial maximum. *Earth and Planetary Science Letters* 142: 43–57.

Lavoisier, A. (1772). Mémoire sur l'usage de l'esprit-de-vin dans l'analyse des eaux minerales. *Memoires de l'Academie Royale des Sciences (Paris)* 555–563.

Levy, Y. (1977). Description and mode of formation of the supratidal evaporite facies in northern Sinai coastal plain. *Journal of Sedimentary Research* 47: 463–474.

Lide, D.R. (2004). *CRC Handbook of Chemistry and Physics*. CRC Press.

Lofi, J. (2018) *Seismic Atlas of theMessinian Salinity Crisis Markers in the Mediterranean Sea Volume 2*. Commission for the Geological Map of the World and Mémoires de la Société Géologie de France.

Lofi, J., Gorini, C., Berné, S. et al. (2005). Erosional processes and paleo-environmental changes in the Western Gulf of Lions (SW France) during the Messinian Salinity Crisis. *Marine Geology* 217: 1–30.

Logan, B.W. (1987). *The MacLeod Evaporite Basin, Western Australia: Holocene Environments*. American Association of Petroleum Geologists: Sediments and Geological Evolution.

Lokier, S. and Steuber, T. (2008). Quantification of carbonate-ramp sedimentation and progradation rates for the late Holocene Abu Dhabi shoreline. *Journal of Sedimentary Research* 78: 423–431.

Lokier, S. and Steuber, T. (2009a). Large-scale intertidal polygonal features of the Abu Dhabi coastline. *Sedimentology* 56: 609–621.

Lokier, S.W. (2012). Development and evolution of subaerial halite crust morphologies in a coastal sabkha setting. *Journal of Arid Environments* 79: 32–47.

Lokier, S.W. (2013). Coastal Sabkha Preservation in the Arabian Gulf. *Geoheritage* 5: 11–22.

Lokier, S.W., Bateman, M.D., Larkin, N.R. et al. (2015). Late Quaternary sea-level changes of the Persian Gulf. *Quaternary Research* 84: 69–81.

Lokier, S.W., Court, W.M., Onuma, T., and Paul, A. (2018). Implications of sea-level rise in a modern carbonate ramp setting. *Geomorphology* 304: 64–73.

Lokier, S.W. and Steuber, T. (2009b). Large-scale intertidal polygonal features of the Abu Dhabi coastline. *Sedimentology* 56: 609–621.

Lowell, J. and Genik, G.J. (1972). Sea-Floor Spreading and Structural Evolution of Southern Red Sea. *AAPG Bulletin* 56: 247–259.

Lowenstein, T.K. (1988) Origin of depositional cycles in a Permian "saline giant": The Salado (McNutt zone) evaporites of New Mexico and Texas. *Bulletin of the Geological Society of America*, 100, 592–608.

Lowenstein, T.K., Hardie, L.A., Timofeeff, M.N., and Demicco, R.V. (2003). Secular variation in seawater chemistry and the origin of calcium chloride basinal brines. *Geology* 31: 857–860.

Lowenstein, T.K., Kendall, B. and Anbar, A.D. (2014) The Geologic History of Seawater. In: *Treatise on Geochemistry* (Eds H. H.D. and T. K.K.) 2nd edn, 8, pp. 569–622. Elsevier, Oxford.

Lowenstein, T.K., Schubert, B.A., and Timofeeff, M.N. (2011). Microbial communities in fluid inclusions and long-term survival in halite. *GSA Today* 21: 4–9.

Lowenstein, T.K., Timofeeff, M.N., Brennan, S.T. et al. (2001). Oscillations in Phanerozoic Seawater Chemistry: Evidence from Fluid Inclusions. *Science* 294: 1086.

Lucia, F.J. (1972) Recognition of Evaporite—Carbonate Shoreline Sedimentation. In: *Recognition of Ancient Sedimentary Environments* (Eds J.K. Rigby and W.K. Hamblin), 16, pp. 0. SEPM Society for Sedimentary Geology.

Lugli, S., Manzi, V., Roveri, M., and Schreiber, B.C. (2010). The Primary Lower Gypsum in the Mediterranean: A new facies interpretation for the first stage of the Messinian salinity crisis. *Palaeogeography, Palaeoclimatology, Palaeoecology* 297: 83–99.

Lugli, S., Manzi, V., Roveri, M., and Schreiber, B.C. (2015). The deep record of the Messinian salinity crisis: Evidence of a non-desiccated Mediterranean Sea. *Palaeogeography, Palaeoclimatology, Palaeoecology* 433: 201–218.

Lugli, S., Schreiber, B.C., and Triberti, B. (1999). Giant polygons in the Realmonte Mine (Agrigento, Silily): Evidence for the desiccation of a Messinian Halite Basin. *Journal of Sedimentary Research* 69: 764–771.

Manzi, V., Gennari, R., Lugli, S., Roveri, M., Scafetta, N. and Schreiber, B.C. (2012) High-Frequency Cyclicity In the Mediterranean Messinian Evaporites: Evidence For Solar–Lunar Climate Forcing. *Journal of Sedimentary Research*, 82, 991–1005.

Manzi, V., Lugli, S., Roveri, M. et al. (2016). The Messinian salinity crisis in Cyprus: A further step towards a new stratigraphic framework for Eastern Mediterranean. *Basin Research* 28: 207–236.

Matthews, M.D. and Perlmutter, M.A. (1994). Global Cyclostratigraphy: An

Application to the Eocene Green River Basin. In: *Orbital Forcing and Cyclic Sequences* (Eds P.L. de Boer and D.G. Smith). *Special Publication of the International Association of Sedimentologists* 19: 459–481.

Meijer, P.T. (2006). A box model of the blocked-outflow scenario for the Messinian Salinity Crisis. *Earth and Planetary Science Letters* 248: 486–494.

Meilijson, A., Hilgen, F., Sepúlveda, J. et al. (2019). Chronology with a pinch of salt: Integrated stratigraphy of Messinian evaporites in the deep Eastern Mediterranean reveals long-lasting halite deposition during Atlantic connectivity. *Earth-Science Reviews* 194: 374–398.

Millot, C. and Taupier-Letage, I. (2005). Circulation in the Mediterranean Sea. In: *The Mediterranean Sea* (ed. A. Saliot), 29–66. Berlin Heidelberg, Berlin, Heidelberg: Springer.

Miocic, J., Heinemann, N., Edlmann, K. et al. (2023). Underground hydrogen storage: a review. *Geological Society, London, Special Publications* 528: SP528-2022-88.

Montadert, L., Sancho, J., Fail, J.P. et al. (1970). De l'âge tertiaire de la série salifère responsable des structures diapiriques en Méditerranée Occidentale (Nord-Est des Baléares). *In: C.R. Acad. Sci.* 271: 812–815.

Moore, K.R. and Holländer, H.M. (2021). Evaluation of NaCl and MgCl2 heat exchange fluids in a deep binary geothermal system in a sedimentary halite formation. *Geothermal Energy* 9: 8.

Mottura, S. (1871). Sulla formazione terziaria nella zona solfifera della Sicilia. *Memorie del Regio Romitato Reologico* 2: 50–140.

Murray, R.C. (1964). Origin and diagenesis of gypsum and anhydrite. *Journal of Sedimentary Petrology* 34: 512–523.

Natalicchio, M., Birgel, D., Dela Pierre, F. et al. (2022). Messinian bottom-grown selenitic gypsum: An archive of microbial life. *Geobiology* 20: 3–21.

Natalicchio, M., Birgel, D., Peckmann, J. et al. (2017). An archaeal biomarker record of paleoenvironmental change across the onset of the Messinian salinity crisis in the absence of evaporites (Piedmont Basin, Italy). *Organic Geochemistry* 113: 242–253.

Natalicchio, M., Dela Pierre, F., Birgel, D. et al. (2019). Paleoenvironmental change in a precession-paced succession across the onset of the Messinian salinity crisis: Insight from element geochemistry and molecular fossils. *Palaeogeography, Palaeoclimatology, Palaeoecology* 518: 45–61.

Natalicchio, M., Pellegrino, L., Clari, P. et al. (2021). Gypsum lithofacies and stratigraphic architecture of a Messinian marginal basin (Piedmont Basin, NW Italy). *Sedimentary Geology* 106009.

Ochoa, D., Sierro, F.J., Lofi, J. et al. (2015). Synchronous onset of the Messinian evaporite precipitation: First Mediterranean offshore evidence. *Earth and Planetary Science Letters* 427: 112–124.

Ogniben, L. (1957) *Petrografia della Serie Solfifera Siciliana e considerazioni geologiche relative.*

Ortí, F. (2011) Selenite facies in marine evaporites: a review. In: *Quaternary carbonate and evaporite sedimentary facies and their ancient analogues: A Tribute to Douglas James Shearman* (Eds C.G.S.C. Kendall and A.S. Alsharhan), *IAS Special Publication*, 43 pp. 431–463. John Wiley & Sons Ltd., Chichester, West Sussex, UK.

Osterloo, M.M., Hamilton, V.E., Bandfield, J.L. et al. (2008). Chloride-bearing materials in the southern highlands of mars. *Science* 319: 1651–1654.

Panieri, G., Lugli, S., Manzi, V. et al. (2010). Ribosomal RNA gene fragments from fossilized cyanobacteria identified in primary gypsum from the late Miocene, Italy. *Geobiology* 8: 101–111.

Patterson, R.J. and Kinsman, D.J.J. (1977). Marine and continental groundwater sources in a Persian Gulf coastal sabkha. In: *Reefs and Related Carbonates - Ecology and Sedimentology, AAPG Studies in Geology. SG4*, 381–397. AAPG.

Paul, A. and Lokier, S.W. (2017). Holocene marine hardground formation in the Arabian Gulf: Shoreline stabilisation, sea level and early diagenesis in the coastal sabkha of Abu Dhabi. *Sedimentary Geology* 352: 1–13.

Paul, A., Lokier, S.W., Sherry, A. et al. (2021). Erosion-initiated stromatolite and thrombolite formation in a present-day coastal sabkha setting. *Sedimentology* 68: 382–401.

Pederson, C.L., Ge, Y., Lokier, S.W., Swart, P.K., Vonhof, H., Strauss, H., Schurr, S., Fiorini, F., Riechelmann, S., Licha, T. and Immenhauser, A. (2021) Seawater chemistry of a modern subtropical 'epeiric' sea: Spatial variability and effects of organic decomposition. *Geochimica et Cosmochimica Acta*, 314, 159–177.

Pellegrino, L., Natalicchio, M., Abe, K., Jordan, R.W., Favero Longo, S.E., Ferrando, S., Carnevale, G. and Dela Pierre, F. (2021) Tiny, glassy, and rapidly trapped: The nano-sized planktic diatoms in Messinian (late Miocene) gypsum. *Geology*, 1.

Pellegrino, L., Natalicchio, M., Birgel, D. et al. (2023). From biogenic silica and organic matter to authigenic clays and dolomite: Insights from Messinian (upper Miocene) sediments of the Northern Mediterranean. *Sedimentology* 70: 505–537.

Peltier, W.R. and Fairbanks, R.G. (2006). Global glacial ice volume and Last Glacial Maximum duration from an extended Barbados sea level record. *Quaternary Science Reviews* 25: 3322–3337.

Perkins, R.D., Dwyer, G.S., Rosoff, D.B. et al. (1994). *Salina Sedimentation and Diagenesis: West Caicos Island*, 35–54. Dolomites: British West Indies. In.

Peryt, T.M. (1996). Sedimentology of Badenian (middle Miocene) gypsum in eastern Galicia, Podolia and Bukovina (West Ukraine). *Sedimentology* 43: 571–588.

Phleger, F.B. (1969). A Modern Evaporite Deposit in Mexicol. *AAPG Bulletin* 53: 824–829.

Pope, M.C., Grotzinger, J.P., and Schreiber, B.C. (2000). Evaporitic

Subtidal Stromatolites Produced by in situ Precipitation: Textures, Facies Associations, and Temporal Significance. *Journal of Sedimentary Research* 70: 1139–1151.

Raafat, H. (2007) Climate. In: *Physical Geography Sector Paper* (Ed A. Kumar), pp. 72–89. Environment Agency Abu Dhabi.

Raup, O.B. (1982). Gypsum precipitation by mixing seawater brines. *American Association of Petroleum Geologists Bulletin* 66: 363–367.

Raymond, J., Langevin, H., Comeau, F.-A., and Malo, M. (2022). Temperature dependence of rock salt thermal conductivity: Implications for geothermal exploration. *Renewable Energy* 184: 26–35.

Riley, J.P. (1975). Analytical chemistry of sea water. In: *J.P. Riley and G. Skirrow)*, 2nde (ed. C. Oceanography), 193–514. London: Academic Press.

Rine, M.J., McLaughlin, P.I., Bancroft, A. M. et al. (2020). Linked Silurian carbon cycle perturbations, bursts of pinnacle reef growth, extreme sea-level oscillations, and evaporite deposition (Michigan Basin, USA). *Palaeogeography, Palaeoclimatology, Palaeoecology* 554: 109806.

Roberts, S.M. and Spencer, R.J. (1995). Paleotemperatures preserved in fluid inclusions in halite. *Geochimica et Cosmochimica Acta* 59: 3929–3942.

Robinson, A.R., Leslie, W.G., Theocharis, A., and Lascaratos, A. (2001). *Mediterranean Sea Circulation. Ocean currents* 1: 1–19.

Rodriguez, C.R., Jackson, C.A.L., Rotevatn, A. et al. (2018). Dual tectonic-climatic controls on salt giant deposition in the Santos Basin, offshore Brazil. *Geosphere* 14: 215–242.

Rona, P.A. (1982). *Evaporites at passive margins*, 116–132. In: *Dynamics of Passive Margins.*

Rosell, L. and Pueyo, J.J. (1997) Second marine evaporitic phase in the South Pyrenean foredeep: the Priabonian Potash Basin. In: *Sedimentary Deposition in Rift and Foreland Basins in France and Spain (Paleogene and Lower Neogene)* (Eds G. Busson and B. C. Schreiber), pp. 358–387. Columbia University Press.

Rouchy, J.M., Taberner, C., Blanc-Valleron, M.M. et al. (1998). Sedimentary and diagenetic markers of the restriction in a marine basin: the Lorca Basin (SE Spain) during the Messinian. *Sedimentary Geology* 121: 23–55.

Rouse, J.E. and Sherif, N. (1980). Major evaporite deposition from groundwater remobilized salts. *Nature* 285: 470–472.

Rouwendaal, S.E., Birgel, D., Grossi, V. et al. (2023). Two modes of gypsum replacement by carbonate and native sulfur in the Lorca Basin, SE Spain. *Frontiers. Earth Science* 11.

Roveri, M., Flecker, R., Krijgsman, W. et al. (2014a). The Messinian Salinity Crisis: Past and future of a great challenge for marine sciences. *Marine Geology* 352: 25–58.

Roveri, M., Gennari, R., Lugli, S. et al. (2016). The Messinian salinity crisis: open problems and possible implications for Mediterranean petroleum systems. *Petroleum Geoscience* 22: 283.

Roveri, M., Lugli, S., Manzi, V. et al. (2020). Stratigraphic relationships between shallow-water carbonates and primary gypsum: insights from the Messinian succession of the Sorbas Basin (Betic Cordillera, Southern Spain). *Sedimentary Geology* 404: 105678.

Roveri, M., Manzi, V., Bergamasco, A. et al. (2014b). Dense shelf water cascading and messinian canyons: A new scenario for the mediterranean salinity crisis. *American Journal of Science* 314: 751–784.

Ryan, W.B.F. (2009). Decoding the Mediterranean salinity crisis. *Sedimentology* 56: 95–136.

Sabino, M., Dela Pierre, F., Natalicchio, M. et al. (2021). The response of water column and sedimentary environments to the advent of the Messinian salinity crisis: insights from an onshore deep-water section (Govone, NW Italy). *Geological Magazine* 158: 825–841.

Sadrinasab, M. and Kämpf, J. (2004). Three-dimensional flushing times of the Persian Gulf. *Geophysical Research Letters* 31.

Sanz-Montero, M.E., Rodríguez-Aranda, J. P., and Calvo, J.P. (2006). Mediation of Endoevaporitic Microbial Communities in Early Replacement of Gypsum by Dolomite: A Case Study from Miocene Lake Deposits of the Madrid Basin, Spain. *Journal of Sedimentary Research* 76: 1257–1266.

Sarg, J.F. (2001). The sequence stratigraphy, sedimentology, and economic importance of evaporite - carbonate transitions: a review. *Sedimentary Geology* 140: 9–42.

Sass, I. and Burbaum, U. (2010). Damage to the historic town of Staufen (Germany) caused by geothermal drillings through anhydrite-bearing formations. *Acta Carsologica* 39.

Schlager, W. (1981). The paradox of drowned reefs and carbonate platforms. *GSA Bulletin* 92: 197–211.

Schmalz, R.F. (1969). Deep-Water Evaporite Deposition: A Genetic Model1. *AAPG Bulletin* 53: 798–823.

Schreiber, B.C. (1978) Environments of subaqueous gypsum deposition. In: *Marine Evaporites, SEPM Short Course Notes*, 4, pp. 43-73. Society for Exploration and Petroleum Mineralogists.

Schreiber, B.C., Friedman, G.M., Decima, A., and Schreiber, E. (1976). Depositional environments of Upper Miocene (Messinian) evaporite deposits of the Sicilian Basin∗. *Sedimentology* 23: 729–760.

Schreiber, B.C. and Hsü, K.J. (1980) Evaporites. In: *Developments in Petroleum Geology* (Ed G.D. Hobson), 2, pp. 87–138. Barking, England.

Schreiber, B.C. and Tabakh, M.E. (2000). Deposition and early alteration of evaporites. *Sedimentology* 47: 215–238.

Schröder, S., Grotzinger, J.P., Amthor, J. E., and Matter, A. (2005). Carbonate deposition and hydrocarbon reservoir development at the Precambrian–Cambrian boundary: The Ara Group in South Oman. *Sedimentary Geology* 180: 1–28.

Schubert, B.A., Lowenstein, T.K., Timofeeff, M.N., and Parker, M.A.

(2009). How do prokaryotes survive in fluid inclusions in halite for 30 k.y.? *Geology* 37: 1059–1062.

Selli, R. (1960) Il Messiniano Mayer-Eymar 1867 : proposta di un neostratotipo. *Giornale di geologia.*, 28, 1–33.

Selli, R. (1973) An outline of the Italian Messinian. In: *Messinian Events in the Mediterranean* (Ed *C.W. Drooger),* Geodynamics Scientfic Reports 7, *pp. 150-171.* Kon. Ned. Akad. Wetnesch.

Serridge Colin, J. and Cooper Anthony, H. (2023). Natural and anthropogenic halite karst subsidence in north Cheshire, UK; comparison of Rostherne Mere, Melchett Mere, Tatton Mere and their surroundings. *Quarterly Journal of Engineering Geology and Hydrogeology* 56: qjegh2022-081.

Sharland, P.R., Archer, R., Casey, D.M., Davies, R.B., Hall, S.H., Heward, A.P., Horbury, A.D. and Simmons, M.D. (2001) *Arabian Plate Sequence Stratigraphy.*

Shearman, D.J. (1963). *Recent anhydrite, gypsum, dolomite, and halite from the coastal flats of the Arabian shore of the Persian Gulf*, 63–64. In: *Proceedings of the Geological Society of London.*

Shearman, D.J. (1970). Recent halite rock, Baja California, Mexico. *Transactions of the Institution of Mining and Metallurgy* B79: 155–162.

Shearman, D.J. and Orti-Cabo, F. (1978). Upper Miocene Gypsum: San Miguel de Salinas, southeast Spain. *Memorie della Societa Geologica Italia* 16: 327–333.

Sirota, I., Enzel, Y., and Lensky, N.G. (2017). Temperature seasonality control on modern halite layers in the Dead Sea: In situ observations. *Bulletin* 129 (9-10): 1181–1194.

Sirota, I., Enzel, Y., and Lensky, N.G. (2018). Halite focusing and amplification of salt layer thickness: From the Dead Sea to deep hypersaline basins. *Geology* 46: 851–854.

Sirota, I., Enzel, Y., Mor, Z. et al. (2021). Sedimentology and stratigraphy of a modern halite sequence formed under Dead Sea level fall. *Sedimentology* 68: 1069–1090.

Smith, A.R. (1980). Sulfur deposits in Ochoan rocks of the Gypsum Plain, southeast New Mexico and west Texas. In: *New Mexico Geological Society 31 st Annual Fall Field Conference Guidebook* (ed. P.W. Dickerson, J.M. Hoffer, and J. F. Callender), 277–283.

Smith, J.E. and Santamarina, J.C. (2022). Red sea evaporites: Formation, creep and dissolution. *Earth-Science Reviews* 232: 104115.

Solomon, S., Qin, D., Manning, M. et al. (2007). The Physical Science Basis. In: *Contribution of Working Group I to the Fourth Assessment Report of the Intergovernmental Panel on Climate.* Change, Cambridge.

Stephenson, M.H., Ringrose, P., Geiger, S. et al. (2019). Geoscience and decarbonization: current status and future directions. *Petroleum Geoscience* 25: 501–508.

Szatmari, P., Moré de Lima, C., Fontaneta, G. et al. (2021). Petrography, geochemistry and origin of South Atlantic evaporites: The Brazilian side. *Marine and Petroleum Geology* 127: 104805.

Taberner, C., Cendón, D.I., Pueyo, J.J., and Ayora, C. (2000). The use of environmental markers to distinguish marine vs. continental deposition and to quantify the significance of recycling in evaporite basins. *Sedimentary Geology* 137: 213–240.

Tipper, J.C. (1997). Modeling carbonate platform sedimentation—Lag comes naturally. *Geology* 25: 495–498.

Topper, R.P.M. and Meijer, P.T. (2015). Changes in Mediterranean circulation and water characteristics due to restriction of the Atlantic connection: A high-resolution ocean model. *Climate of the Past* 11: 233–251.

Tucker, M.E. (1991). Sequence stratigraphy of carbonate-evaporite basins: models and application to the Upper Permian (Zechstein) of northeast England and adjoining North Sea. *Journal of the Geological Society* 148: 1019–1036.

Tucker, M.E. (2001). *Sedimentary Petrology: an introduction to the origin of sedimentary rocks.* Oxford: Blackwell Science Ltd.

USGS (2018) *Minerals Yearbook, volume I, Metals and Minerals*, Reston, VA.

Usiglio, G. (1849). Analyse de l'eau de la Méditerranée sur les cotes de France. *Annalen Chemie* 27 (92–107): 172–191.

Vai, G.B. and Ricci Lucchi, F. (1977). Algal crusts, autochthonous and clastic gypsum in a cannibalistic evaporite basin: a case history from the Messinian of Northern Apennines. *Sedimentology* 24: 211–244.

Van Gent, H., Urai, J.L., and de Keijzer, M. (2011). The internal geometry of salt structures – A first look using 3D seismic data from the Zechstein of the Netherlands. *Journal of Structural Geology* 33: 292–311.

van't Hoff, J.H. (1912). *Untersuchungen uber die Bildungsverhaltnisse der ozeanischen Salzablagerungen, insbesondere das Stassfurter Salzlagers.* Leipzig.

Vogel, M.B., Des Marais, D.J., Turk, K.A. et al. (2009). The Role of Biofilms in the Sedimentology of Actively Forming Gypsum Deposits at Guerrero Negro, Mexico. *Astrobiology* 9: 875–893.

Volozh, Y., Talbot, C., and Ismail-Zadeh, A. (2003). Salt structures and hydrocarbons in the Pricaspian basin. *AAPG Bulletin* 87: 313–334.

Von Der Borch, C.C., Bolton, B., and Warren, J.K. (1977). Environmental setting and microstructure of subfossil lithified stromatolites associated with evaporites, Marion Lake, South Australia. *Sedimentology* 24: 693–708.

Vreeland, R.H., Jones, J., Monson, A. et al. (2007). Isolation of live cretaceous (121-112 million years old) halophilic Archaea from primary salt crystals. *Geomicrobiology Journal* 24: 275–282.

Walther, J. (1894). *Lithogenesis der Gegenwart*, 535–1055. Jena: Verlag Gustaf Fischer.

Warren, J.K. (1982). The hydrological significance of Holocene tepees, stromatolites, and boxwork limestones in coastal salinas in South Australia. *Journal of Sedimentary Petrology* 52: 1171–1201.

Warren, J.K. (1991). Sulfate dominated sea-marginal and platform evaporative settings: sabkhas and salinas, mudflats and salterns. In: *J.L. Melvin), Developments in Sedimentology* (ed. P. Evaporites and M. Resources), 69–187. Amsterdam: Elsevier.

Warren, J.K. (2006). *Evaporites: Sediments*, 1036. Resources and Hydrocarbons: Springer.

Warren, J.K. (2010). Evaporites through time: Tectonic, climatic and eustatic controls in marine and nonmarine deposits. *Earth-Science Reviews* 98: 217–268.

Warren, J.K. (2016). *Evaporites: A geological compendium*, 1814. Springer.

Warren, J.K. (2021). Evaporite Deposits. In: *Encyclopedia of Geology*, Seconde (ed. D. Alderton and S.A. Elias), 945–977. Oxford: Academic Press.

Weldeghebriel, M.F., Lowenstein, T.K., García-Veigas, J., and Cendón, D.I. (2022). [Ca2+] and [SO2−4] in Phanerozoic and terminal Proterozoic seawater from fluid inclusions in halite: The significance of Ca-SO4 crossover points. *Earth and Planetary Science Letters* 594: 117712.

West, I.M., Ali, Y.A., and Hilmy, M.E. (1979). Primary gypsum nodules in a modern sabkha on the Mediterranean coast of Egypt. *Geology* 7: 354–358.

Zanbak, C. and Arthur, R.C. (1986). Geochemical and Engineering Aspects of Anhydrite/Gypsum Phase Transitions. *Environmental and Engineering Geoscience, xxiii* 419–433.

Ziegler, A., Eshel, G., Rees, P.M. et al. (2003). Tracing the tropics across land and sea: Permian to present. *Lethaia* 36: 227–254.

Supplementary Table 10.S.1 Approaches and methodologies commonly employed in the study and elucidation of evaporite successions.

Technique	Particularly useful for	Limitations	Facies architecture	Stratigraphic evolution over time	Sedimentary structures	Sedimentary facies and textures	Mineralogy	Biotic relationships	Fluid inclusion studies	Early diagenesis	Burial diagenesis	Defining meteoric overprint	Primary water chemistry	Organic geochemistry	Surface	Shallow sub-surface	Deep sub surface	Remote sensing	Accessibility	Cost	Spatial scale
Outcrop (natural and man-made)	Gaining an understanding of large scale two and three dimensional facies geometries and the relationship with adjacent facies.	Minerals subject to dissolution or dehydration. Unlikely to represent full mineral suite at time of precipitation. Largely limited to arid settings.	✓	✓	✓	✓	✓	✓	✓	✓	✓	✓	✓		✓				Easy	Low	Various
Excavation pits and trenches	Examining high-resolution facies geometries and sedimentary structures.	Minerals subject to dissolution or dehydration. Unlikely to represent full mineral suite at time of precipitation. Largely limited to arid settings.	✓	✓	✓	✓	✓	✓	✓	✓	✓	✓	✓	✓		✓			Moderate	Low	Moderate
Mines	Lower degree of weathering than surface and shallow sub-surface observations. Potential to study primary mineralogies and fabrics.	Mineralogies affected by sub-surface fluids. Effects of pressure and temperature on mineral stability.	✓	✓	✓	✓	✓	✓	✓	✓	✓		✓	✓		✓	✓		Moderate	Low	Various
Recent analogues	Linking depositional facies, facies geometries and sedimentary structures to environmental	Lack of recent analogues for Saline Giants and Basinal Evaporites. Differences between recent and	✓	✓	✓	✓	✓	✓	✓	✓	✓		✓	✓	✓	✓			Easy	Low	Various

(Continued)

Supplementary Table 10.S.1 (Continued)

Technique	Particularly useful for	Limitations	Facies architecture	Stratigraphic evolution over time	Sedimentary structures	Sedimentary facies and textures	Mineralogy	Biotic relationships	Fluid inclusion studies	Early diagenesis	Burial diagenesis	Defining meteoric overprint	Primary water chemistry	Organic geochemistry	Surface	Shallow sub-surface	Deep sub surface	Remote sensing	Accessibility	Cost	Spatial scale
	controls. Studying early diagenetic processes.	ancient ocean chemistry.																			
Core	Studying hard-to reach sub surface formations, avoiding weathering effects, studying primary fabrics.	Effects of pressure and temperature on mineral stability. Evaporites rarely targeted in coring unless the subject of the investigation.	✓	✓	✓	✓	✓	✓	✓		✓		✓	✓		✓	✓		Difficult	High	Small
Well logs	Bulk facies analysis. Observing stratigraphic evolution. Standard tool in analysis of subsurface.	Potential for misinterpreting facies associations, relatively low resolution.		✓	✓	✓	✓												Difficult	High	Small
Seismic	Understanding large (basin) scale facies geometries.	Challenges in imaging within and below evaporite facies.	✓																Difficult	High	Large
Remote sensing	Large scale overview. When used in recent environments then high resolution evolution of facies architecture over time. Potential use in extraterrestrial settings.	Best used in tandem with ground-studies in order to calibrate results.	✓		✓													✓	Easy	Moderate	Large

Supplementary Table 10.S.2 Common evaporite facies and their distinguishing features.

Facies	Distinguishing features	Typical evaporite habits	Mode of genesis	Notes	Genesis					Environment				Common modes of Identification						Figures
					Subaerial		Subaqueous			Primary										
					Sediment surface	Sediment column	Sediment surface	Water column	Sediment column	Sabkha	Shallow marine	Marine basin	Diagenetic feature	Outcrop	Core	Hand specimen	Thin section	BHI logs	Density logs	
Gypsum crystals isolated in host sediment matrix	Randomly distributed single, twinned and aggregate gypsum crystals forming a discontinuous (isolated) fabric within the host sediment.	Lenticular gypsum – following burial often pseudomorphed by anhydrite	Growth of typically discoidal gypsum within previously-deposited host sediments. Crystal growth is initiated in interstitial voids and is largely displacive yet may also incorporate host grains to form a poikilotopic texture that has the potential to preserve primary sedimentary fabrics.	May locally overprint and obscure the primary sedimentary fabric. Conversely, large crystals potentially incorporate and preserve primary fabrics.		✓			✓	✓	✓	✓		✓	✓	✓	✓	✓		Figure 10.32a–e
Gypsum sand or gravel	The sediment is supported by randomly oriented sand to gravel grade gypsum crystals typically of a similar size – forming an equigranular texture. Some host sediment may be retained as an	Lenticular gypsum – during burial will dehydrate to anhydrite with the loss of the primary texture to form a chicken wire fabric.	Gypsum displacively growing within the shallow subsurface capillary fringe.	Destroys primary fabrics and potentially elevates the sediment surface. Referred to as gypsum mush in the older literature.		✓				✓				✓	✓	✓	✓	✓	✓	Figure 10.32f,g

(*Continued*)

Supplementary Table 10.S.2 (Continued)

Facies	Distinguishing features	Typical evaporite habits	Mode of genesis	Notes	Genesis: Subaerial		Genesis: Subaqueous			Environment: Primary				Common modes of Identification						Figures
					Sediment surface	Sediment column	Sediment surface	Water column	Sediment column	Sabkha	Shallow marine	Marine basin	Diagenetic feature	Outcrop	Core	Hand specimen	Thin section	BHI logs	Density logs	
	intercrystalline matrix.																			
Laminated gypsum/anhydrite	Crystal-supported laminated texture commonly intercalated with non-evaporite sediment. Where evaporite-dominated, laminations defined by changes in crystal size or detrital sediment content. Sand-grade gypsum crystals. Laminae surfaces may be irregular where coarser crystal grades are present.	Lenticular gypsum – burial anhydritisation may disrupt or completely destroy the laminated structure.	Gypsum crystals nucleate in the water column and then settle to the floor of the water body. Lack of deformation hints at early, pre burial, anhydritisation.	Where significant detrital material is present, the primary lamination may be preserved during burial anhydritisation.				✓		✓	✓	✓		✓	✓	✓	✓	✓	✓	Figures 10.32h and 10.32l
Enterolithic anhydrite	Single or multiple complexly folded layers and diapir-like structures of finely crystalline anhydrite.	Aggregate of acicular anhydrite often found with associated relic gypsum.	Either as a result of the displacive, sometimes seasonal, cyclical conversion of calcium sulphate between	Appearing as folds or diapir-like structures in cross section, these features are often polygonal when considered in three dimensions. In older literature sometimes described		✓				✓		✓	✓	✓	✓	✓				Figure 10.32j,k

	Typically within a host sediment.		gypsum and anhydrite or formed during anhydritisation in the burial realm.	as having a clotted/cream cheese-like texture.															
Nodular anhydrite	Irregularly shaped anhydrite nodules within a host sediment matrix. Nodules may be locally in contact but are typically 'floating' in the matrix. Nodules often flattened in the vertical plane.	Aggregate of acicular anhydrite. Where occurring in the Recent, then often found with associated relic gypsum.	Similar to enterolithic anhydrite, formed in the very shallow subsurface through the dehydration of precursor gypsum or during burial-related anhydritisation. Also formed as a direct primary precipitate during deeper burial diagenesis. Sediment may also be retained as inclusions within nodules.	Nodule formation will typically deform primary sedimentary structures but careful examination of the host sediment may elucidate diagnostic features. Nodules are often confined to layers within the sediment column.	✓			✓	✓	✓	✓	✓	✓	✓	✓		✓		Figure 10.26l–n
Chicken wire anhydrite	A mosaic of irregular anhydrite nodules separated by thin seams host sediment, organic material or diagenetic minerals. Individual nodules are typically flattened in the vertical plane.	Aggregate of acicular anhydrite.	Displacive and replacive growth of anhydrite during burial until only a small fraction of the host sediment is retained at the boundaries of adjacent nodules. Sediment may also be retained as inclusions within nodules.	Most of the host sediment, along with any primary sedimentary structures is totally removed. In some cases, variations in nodule size may preserve a relict of primary bedding.	✓				✓		✓	✓	✓	✓	✓	✓	✓	✓	Figure 10.26l,o
Massive anhydrite	A largely featureless mass of anhydrite lacking	Aggregate of acicular anhydrite.	Displacive and replacive growth of anhydrite during burial largely without	Small pockets of highly deformed host sediment may be locally retained. Sediment grains may	✓				✓		✓	✓	✓	✓	✓	✓	✓	✓	Figure 10.26i

(*Continued*)

Facies	Distinguishing features	Typical evaporite habits	Mode of genesis	Notes	Genesis					Environment				Common modes of Identification						Figures
					Subaerial		Subaqueous			Primary										
					Sediment surface	Sediment column	Sediment surface	Water column	Sediment column	Sabkha	Shallow marine	Marine basin	Diagenetic feature	Outcrop	Core	Hand specimen	Thin section	BHI logs	Density logs	
	recognisable textures or structures.		retention of the host sediment.	also be incorporated as inclusions within crystals.																
Carbonate and sulphate stromatolite	Laminated microbial carbonates, stromatolites, thrombolites, gypsified microbial mats.	Layering defined by alternation of micritic laminae, clotted fabric, fine grained gypsum crystals or palisades of small vertically oriented selenite, organic-rich layers and/or clay intercalation.	Poikilotopic and bottom-growth of gypsum during more evaporative phases, microbial-driven carbonate precipitation.	Modality of growth strongly dependant from salinity, terrigenous input and hydrodynamic regime.			✓			✓	✓	✓		✓	✓	✓	✓	✓		Figure 10.33a
Massive selenite	Massive beds of selenite gypsum from decimetre up to tens of meters thick; cone-like (nucleation cones) morphologies composed of clusters of gypsum crystals are common load structures at the bed base.	Aggregate of cm to m in size swallow-tail twinned crystals. The core of the crystals exhibits internal lamination.	Gypsum crystals nucleate at the basin floor and grow syntaxial in the water column. Crystals were submerged in brines saturated with calcium sulphate; unstratified basin or high position of the pycnocline.	Rapid growth of the gypsum traps sediment grains, fossils and fluids providing an important source of paleoenvironmental information.			✓				✓	✓		✓	✓	✓	✓	✓	✓	Figure 10.33b,e, f,j
Giant selenite twins and intergrowths	Large vertically oriented, twinned selenite crystals (up to 7 m high) or gypsum intergrowth (up to 3.5 m high)	Swallow tail twins with internal lamination. Gypsum intergrowth are untwined crystals joined in	Syntaxial bottom-growth. The large size of the crystals indicates a fluid with a lower state of supersaturation	The modality of gypsum intergrowth was described in detail by Babel (1987) with the surface of contact between the different crystals			✓				✓	✓		✓	✓	✓	✓			Figure 10.33b,e

	often forming palisades and associated with the massive gypsum facies.	pairs without crystallographic symmetry.	than deposits with numerous smaller crystals. Forming in an unstratified basin or indicating a high position of the pycnocline.	being described as a compositional surface.														
Banded selenite (bedded selenite)	Stratified selenite facies with a marked banding that may include thin intercalations of carbonate, clays, fine-grained gypsum or dissolution surfaces.	Small sized, vertically oriented shallow tail selenites (mm to few cm in size). The crystals at the top of the strata are truncated.	Syntaxial bottom-growth of the crystals with periodic fluctuation of the pycnocline. The small crystal size indicates high supersaturation.		✓				✓	✓		✓	✓	✓	✓	✓	✓	Figure 10.33c,d,k
Curved selenite (sabre gypsum)	Curved and inclined crystals. Crystals in a single bed are concordantly oriented.	Curved single crystals (few cm to m in size)	Upward crystal growth is simultaneous with lateral curving. Indicative of the presence of dense brine currents at the basin floor.	This facies is often associated with bedded selenites.	✓				✓	✓		✓	✓	✓	✓			Figure 10.33g
Branching selenite (super cones)	Large conical structures composed of selenite crystals. One side of the cone is generally more elongated. In one single bed, the elongated sides of the cones are concordantly oriented.	Small selenite crystals grouped as aggregates that develop to form m-scale cones.	Growing upward and simultaneously curving laterally. Indicative of a depressed periodically fluctuating pycnocline and the presence of dense brine currents at the basin floor.	This facies generally develops at the top of the gypsum cycle. Especially associated to the Mediterranean Messinian evaporites.	✓				✓	✓		✓	✓	✓	✓			Figure 10.33h,i

(Continued)

Facies	Distinguishing features	Typical evaporite habits	Mode of genesis	Notes	Genesis					Environment				Common modes of Identification						Figures
					Subaerial		Subaqueous			Primary										
					Sediment surface	Sediment column	Sediment surface	Water column	Sediment column	Sabkha	Shallow marine	Marine basin	Diagenetic feature	Outcrop	Core	Hand specimen	Thin section	BHI logs	Density logs	
Bedded halite (and other potash salts)	cm- to dm-thick salt layers intercalated with mm- to cm thick sulphate layers and fine grained, typically muddy, layers.	Well sorted mm-scale platy mm to cm scale hoppers of halite. Typically with well-developed fluid inclusion zonation.	Halite plates are formed at the air-water interface before sinking to the basin floor. Hopper crystals represent continued crystal growth at the basin floor. Intercalated non-evaporitic beds represent periods of non-precipitation or episodes of dissolution.	Dissolution features such as dissolution pits and truncation of the crystals are common.			✓	✓			✓			✓	✓	✓	✓	✓	✓	Figure 10.27m,n
Massive halite (and other potash salts)	Internally massive cm- to m-thick salt layers.	Well sorted mm-sized platy halite crystals with well-developed fluid inclusion zonation.	Nucleation of salt crystals at the air-water interface before settling to the basin floor.	Absence of dissolution feature and truncation surfaces indicates formation in a deep stratified basin.				✓			✓	✓		✓	✓	✓	✓	✓	✓	Figure 10.33o
Clastic gypsum	A wide range of redeposited evapoclastic facies including gypsarenites and large-scale chaotic deposits.	A wide array of fabrics akin to other clastic sediments.	Evapoclastic sediments reworked and transported by a range of subaqueous processes including waves, currents, debris flows, turbidity currents and submarine mass-wasting events.		✓		✓				✓	✓		✓	✓	✓	✓	✓	✓	Figure 10.34a–c

Supplementary Table 10.S.3 Examples of some of the more common evaporite stratigraphic sequences along with their diagnostic features and, where applicable, sequence stratigraphic context.

Lithofacies Association	Lithofacies
Outer ramp	Skeletal wackestone
	Skeletal packstone
	Massive mudstone
	Bioturbated mudstone
Mid ramp	Peloidal, ooidal and skeletal wackestones
	Peloidal, ooidal and skeletal packstones
	Massive peloidal skeletal packstone
Inner ramp	Ooid grainstone
	Ooid packstone
Lagoon	Bioturbated peloidal to skeletal packstone
	Bioturbated peloidal to skeletal wackestone
	Bioturbated peloidal to skeletal mudstone
	Microbial bindstone
Intertidal	Bioturbated peloidal to skeletal wackestone
	Bioturbated peloidal to skeletal mudstone
	Microbial bindstone
Sabkha	Nodular anhydrite
	Entrolithic anhydrite
	Lenticular disseminated anhydrite
Salina	Laminated anhydrite

11

Marine Carbonates

Frans van Buchem[1], John J.G. Reijmer[2], and Emmanuelle Vennin[3]

[1] Physical Sciences and Engineering Division, King Abdullah University of Science and Technology, Thuwal, Saudi Arabia
[2] Department of Earth Sciences, Faculty of Science, Vrije Universiteit Amsterdam, Amsterdam, The Netherlands
[3] Unité Mixte de Recherche Biogéosciences, University of Burgundy, Dijon, France

Summary

In this chapter, we present marine carbonate depositional environments by addressing (i) the geometrical nomenclature of land-attached and land-detached carbonate platforms and smaller geobodies, (ii) the definition of 'carbonate factories', (iii) facies models along end-member depositional profiles and (iv) a brief historical overview of the stratigraphic architecture of marine carbonate systems through time. This approach aims to capture the huge diversity of carbonate deposits in time and space in a systematic way, and to integrate different scales of observations and processes in consistent, predictive depositional and stratigraphic models.

This chapter organisation has been inspired by several important developments in the field of carbonate sedimentology. First, the proliferation of outcrop studies and three-dimensional (3D) seismic surveys over the past three decades has provided important new insights into the three-dimensional architecture of carbonate systems. Second, the definition of different carbonate factories, that each follow their own rules of carbonate deposition, has helped to group and focus observations and processes into consistent models. And third, the improvement of the age dating of marine carbonate deposits through high-resolution biostratigraphy, isotope stratigraphy and cyclostratigraphy allows for a higher time-resolution and thus a more quantitative approach to the depositional processes. This combined progress in understanding of the geometries (large and small), the carbonate factories and the age dating has opened the way to the application of stratigraphic forward modelling of which examples are provided in Chapter 4.

11.1 Introduction

11.1.1 Marine Carbonate Sediments and Environments

This chapter covers marine carbonate sediments, those formed in the shallow water marine domain and adjacent slopes, as well as carbonate mounds on the basin floor. Pelagic carbonates are treated in more detail in Chapter 13. Marine carbonates are predominantly of biological origin, deposited mostly *in situ* (in contrast with most siliciclastic sediments), and occur in all climatic belts from the warm tropics to the cold waters of high latitudes. Their biological origin, variability of facies expression and particular chemical compositions make them uniquely different from other sedimentary rocks, such as siliciclastics, evaporites and volcanic sediments.

Marine carbonates represent approximately 20% of the rock record (Morse and Mackenzie 1990). Spectacular modern examples of shallow-marine carbonate systems include the vast platforms of the Bahamas, the Maldives, the Saya de Malha Bank and the Australian Great Barrier Reef, and the smaller scale reef build-ups in the Red Sea, the Caribbean, the Indian Ocean and Pacific. All these are tropical examples, but modern carbonates also occur in cooler climates, where they consist of different organisms, forming different, often smaller, types of accumulations. Examples include shelly beaches in the temperate climates, or the aeolianites along the south and southwest Australian coastlines.

Carbonate-producing organisms are very sensitive to environmental changes, which makes them excellent recorders of changes in climate, sea level, oceanography and atmospheric conditions. Their responses to these

Reading's Sedimentary Environments: Processes, Facies and Stratigraphy, Fourth Edition. Edited by Bruce Levell.

Companion website: www.wiley.com/go/levell4e

changes are recorded in different ways. The long-term Phanerozoic biological evolution of carbonate producers saw their morphology and feeding systems continuously adapting to changing physio-chemical and ecological conditions. This has resulted in specific flora-faunal assemblages that thrive in particular water depth ranges, under varying water temperature, trophic, and salinity conditions. These assemblages are called 'carbonate factories', each characterised by specific and predictable architectures of sediment accumulation, and facies. Throughout the Phanerozoic, carbonate producers have gone through phases of global proliferation, and of near-global extinction. On shorter time scales, cyclic changes in depositional conditions, such as sea level, led to migration or replacement of biological assemblages, which are recorded as repetitive facies successions: one of the cornerstones of carbonate sequence stratigraphy.

Diagenetic processes (involving dissolution, re-crystallisation, compaction, and cementation), are another response of shallow-marine carbonate systems to environmental change and occur during deposition, subsequent burial, and/or re-exposure to atmospheric conditions. The chemically unstable nature of carbonates allows diagenesis, especially during early deposition and subaerial exposure, to profoundly change the texture and composition of the rock and its associated mechanical and petrophysical characteristics. A good example of early diagenesis is the formation of beachrock: the near syn-sedimentary cementation of carbonate sediments in the beach environment. Dissolution features are also common and occur on all scales. For instance, carbonate rocks host most of the largest subterranean cave systems, which result from dissolution by under-saturated waters (rainwater or hydrothermal fluids) often with further modification by subterranean rivers.

Carbonates not only record the changes in physical conditions during deposition but also, notwithstanding their chemical instability, many carbonate rocks have preserved aspects of their original chemical composition at the time of deposition, including the stable isotope ratios of many different elements (i.e. C, O, Ca, Mg, S, B, Sr and others). This is particularly valuable, since these ratios relate to the chemical composition of sea water at a given time, which was in turn in equilibrium with the atmosphere. Disturbances of that equilibrium, or of the isotopic composition of the major ocean or atmosphere reservoirs, can therefore be deduced from the carbonate sediments. These archives record changing global biogeochemical cycles and climatic events (such as greenhouse gas build-up and global warming and cooling).

The dating of carbonate sediments relies primarily on the evolution of the carbonate-producing organisms. Other methods include calibration with the aggregated global seawater isotopic records (carbon and strontium), and, more recently, U–Pb dating by laser ablation inductively coupled plasma mass spectrometry or LA–ICP–MS and, for younger rocks (<500 ka) on U–Th dating.

11.1.2 History of Carbonate Research

Historical overviews of the development of carbonate sedimentology can be found in Folk (1974a), Milliman (1974a), Wilson (1975), Tucker and Wright (1990), Schlager (2005), Reijmer (2021) and James and Pufahl (2026). Here we want to emphasise that one of the most characteristic aspects of modern carbonate sedimentology is the adoption of an integrated approach: combining information from different scales of observation, seismic to the pore scale and different disciplines, including palaeontology, ecology, geochemistry, geobiology, sequence stratigraphy and cyclostratigraphy.

The development of this integrated approach was facilitated by several factors. First, from the 1960s onward, oil companies needed three-dimensional (3D) stratigraphic models to predict facies heterogeneities in carbonate reservoirs at different scales. Subsurface data (seismic, geophysical logs and core) and seismic-scale outcrop analogues were instrumental in demonstrating the complexity of the 3D architecture of the carbonate constructions (e.g. Payton 1977; Vail and Mitchum 1977; Bally 1989; Sarg 1988; Simo et al. 1993; Zeng and Kerans 2003; Eberli et al. 2004a,b; Verwer et al. 2014). By combining the insights in the stratal geometries with palaeontology, palaeo-ecology and sedimentology, a very dynamic picture of facies models and different types of carbonate factories emerged (e.g. Wilson 1975; James 1983; Flügel 1984, 2010; Schlager 2003; Reijmer 2021). Second, the comparative sedimentological approach, studying the modern-day environment to understand the rock record, helped appreciate facies variations in space, and understand the influence of physical, biological and chemical factors on carbonate depositional systems (e.g. Ginsburg 1956, 1957, Ginsburg et al. 1958; Folk 1959, 1962, 1974a, b; Dunham 1962; Purdy 1963a,b; Flügel 1982b, 2004, 2010; Schlager 2005). Third, the advancement of geochemistry added insights into the early and burial diagenetic processes when tracing their diagenetic history, as well as the evolution of the global carbon cycle and different climatic proxies. And lastly, the increasingly quantitative approach of carbonate stratigraphy and related processes allows the addition of forward stratigraphic modelling to the interpretation loop: constraining stratigraphic models by physical, biological and chemical laws (Whitaker et al. 1997; Warrlich et al. 2002, 2005; Burgess and Wright 2003; Borgomano et al. 2008, 2020; Masiero et al. 2020; see also Chapter 4).

11.1.3 The Modern and the Ancient – Variations in Space and Time

Carbonate sedimentology follows the usual sedimentological approach, combining information from modern depositional environments with information from the rock record. Modern environments allow investigation of depositional processes (biological, chemical and physical) and study of the variation of resultant facies in 3D. It is self-evident that the modern-day carbonate-producing organisms are largely different from those organisms that were active in the geological past. What is comparable though, are the behaviours, and morphological and ecological adaptations to specific depositional conditions. There is, therefore, much about present-day carbonate-producing organisms and the environments in which they thrive, which is indeed applicable to the past.

On the other hand, the rock record helps to construct facies models through providing insight into the dynamics and evolution of depositional systems through space and time. High-resolution 3D seismic surveys, and the ever-growing number of seismic-scale outcrop studies have significantly improved understanding of the spatial variations of carbonate deposits and their stratigraphic architecture. The dual notion of facies variation in space and stratigraphic evolution through time is a powerful framework that guides much research in carbonate sedimentology.

11.1.4 Scope and Set-up

Much has been published about carbonate sedimentology and carbonate sequence stratigraphy. Non-exhaustive highly recommended sedimentological reference works include Bathurst (1975), Horowitz and Potter (1971), Milliman (1974a), Wilson (1975), Flügel (1982b, 2010), Scholle et al. (1983), Tucker and Wright (1990), Walker and James (1992), Wright and Burchette (1998), Scholle and Ulmer-Scholle (2003), James and Bone (2011) and James and Jones (2016). More stratigraphically focussed publications include Sarg (1988), Simo et al. (1993), Emery and Myers (1996), Homewood and Eberli (2000) and Schlager (2005). Publications specifically dealing with the diagenetic aspects and petrophysical characteristics of carbonate rocks include Moore (1989, 2001), Morse and Mackenzie (1990), Loucks and Sarg (1993), Anselmetti and Eberli (1993), Melim et al. (2002), Mackenzie (2005), Lucia (2007) and Ahr (2008). The IAS/SEPM field guides, a compilation of world-class outcrops that have been thoroughly studied, are valuable sources of information for carbonate systems (Wright and Della Porta 2021).

The purpose of this chapter is to provide a concise summary of (i) the physical, biological and chemical principles and processes of carbonate production and accumulation, (ii) detail the nature and terminology of carbonate platforms and facies models and (iii) provide a brief overview of the variation in facies models and stratigraphic expressions through time. We will emphasise some of the specific differences in approach when compared to siliciclastic systems. An extensive set of references has been provided in the text and in dedicated literature tables, to facilitate entry to the large body of literature that deals with carbonate systems.

Dedicated chapters deal with pelagic carbonates (Chapter 13), carbonates deposited in evaporitic environments (Chapter 10) and terrestrial/lacustrine carbonates (Chapters 5 and 6 (Sections 6.3.3 and 6.4.2)). Carbonate examples of facies analysis and sedimentological and sequence-stratigraphic studies are given in Chapters 2 and 4, respectively.

11.2 Descriptive Terminology and Classification Systems

For the purpose of comparison and classification of carbonate rocks, a consistent descriptive terminology is important. Wilson (1975) highlighted two essential points for such a terminology: *'(1) to separate expressions that describe the outer shape of the carbonate structures and geobodies from those describing their internal architecture and composition, and (2) to consider the scale, and clearly distinguish between the names used for smaller, discrete limestone geobodies and the ones used for the large regional configurations of carbonates'*.

Following these recommendations, a distinction is made between a nomenclature to describe (i) the *carbonate platforms* (edifices), which developed over a given time interval, in terms of their size and overall morphology, and (ii) the *depositional profile*, which existed at a given moment in time, and influenced the distribution and nature of the sedimentary facies and depositional environments in space. Typically, during the construction of a carbonate platform, be it land-attached or land-detached (Figure 11.1), the depositional profile evolves, for instance, from flat ramp profiles to steep-margin-bound platform profiles (e.g. Read 1985; Handford and Loucks 1993; Schlager 2005). Recognising these depositional profiles is important since it allows prediction of facies distribution within the overall carbonate edifices and can provide insights into the evolution of the factors controlling deposition.

The terminology we use below is designed to facilitate the study of carbonate systems through the systematic integration of scales of observation. We start with km or larger-scale understanding of the depositional geometrical context provided by 2D/3D seismic surveys, regional log correlations or, several-km scale, quality, outcrops. This

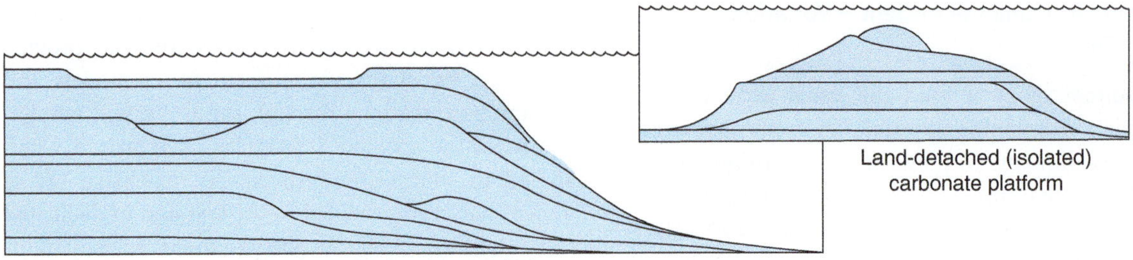

Figure 11.1 Schematic representation of the stratigraphic complexity of land-attached and land-detached carbonate platforms. Note that the internal architecture and facies distribution are the result of the changing depositional profile through time. The horizontal scale can vary from hundreds of meters to hundreds of kilometres, and the vertical scale from tens of meters to hundreds of meters.

is subsequently infilled with detail: sedimentary facies and depositional environments along the depositional profiles, using the macro- and microscopic information provided by outcrops, core and cutting material.

11.2.1 Carbonate Platforms and Geobodies

Insight into the overall morphology and size of large carbonate platforms has increased significantly through outcrop studies and ever-larger 3D seismic surveys and well-log datasets (e.g. Eberli et al. 2004a, b; Lukasik and Simo 2008; van Buchem et al. 2010a; Verwer et al. 2014; Hendry et al. 2021). The understanding of large-scale morphology also highlighted the role that tectonic setting plays in shaping these structures: by controlling the areas available for carbonate growth at any particular time, the accommodation space and the morphology of the sea floor (e.g. Wilson 2000, 2011; Wilson et al. 2000; Bosence 2005; Dorobek 2008). Four major tectonic settings for sedimentary basins are distinguished: (i) tectonically passive intra-cratonic or foreland basin settings (slow flexural subsidence), (ii) marine extensional or rift to passive margin settings, (iii) early active margin compressional settings and (iv) late active margin settings (Figure 11.2).

The proposed terminology for a high-level description of the carbonate system makes the distinction between 'land-attached' or 'land-detached' and adds the tectonic setting. *Land-attached* carbonate platforms vary in areal extent from hundreds to ten thousands of square kilometres, and in thickness from tens to thousands of metres, whereas the *land-detached isolated* carbonate platforms have a typical areal extent of one to thousands of square kilometres, and thicknesses varying from tens to thousands of metres (Figure 11.3). Land-attached platforms are typical for slow flexural subsidence settings, such as intra-cratonic basins and proximal portions of passive margins where they can be extensive, but they also occur along the faster-subsiding (interior) margins of foreland basins when they can keep up with subsidence. The often smaller, land-detached, isolated, carbonate platforms, on the contrary, are more typical in areas of high subsidence and faulting, where local highs or irregular antecedent topography provide multiple initiation points for growth of carbonate-producing organisms and accumulation of sediment (Figure 11.2).

Whereas the overall morphology of carbonate structures is mostly controlled by the tectonic context, the shapes and compositions of smaller-scale (tens of metres to kilometres) carbonate geobodies are governed by climate, sea-level fluctuations, ocean chemistry and biological evolution, and are often linked to the nature of the carbonate factory (see Section 11.4). Descriptive terms for carbonate geobodies have been used inconsistently and have also evolved through time (e.g. Lowenstam 1950; Chevalier 1961; Dunham 1970; Wilson 1975; Read et al. 1986; Riding 1989; Copper and Brunton 1991; Pomar 2001; Flügel and Kiessling 2002; James and Jones 2016). Here we provide definitions drawing from the literature cited above. We distinguish between geobodies using their outer shape, size and internal composition, and make the distinction between those that build a positive relief and those that are flat bedded (Figure 11.4). It should be pointed out here that geobodies with a positive relief have attracted relatively more attention, due to their rich faunal content and relevance for the modelling of carbonate systems. Despite their relief, however, much of the sediment accumulates in relatively flat strata in the platform interior, characterised by gradual lateral facies changes between the grainy and muddy fractions.

Reef: A massive or layered, laterally restricted carbonate bio-construction that forms *in situ* possesses wave and current-resistant strength, may have bathymetric relief, is built by the intergrowth of a community of sessile benthic organisms (such as corals, bivalves, serpulids, calcareous algae or microbes) and might contain internal cavities and syn-sedimentary cement. Reef structures are important features of many carbonate shelves and land-detached platforms and display a large variation in forms and dimensions (Figure 11.4). The following distinctions can be made:

Fringing and barrier reefs: these are reefs, as above, located at the offlap-break position of shelves and land-detached

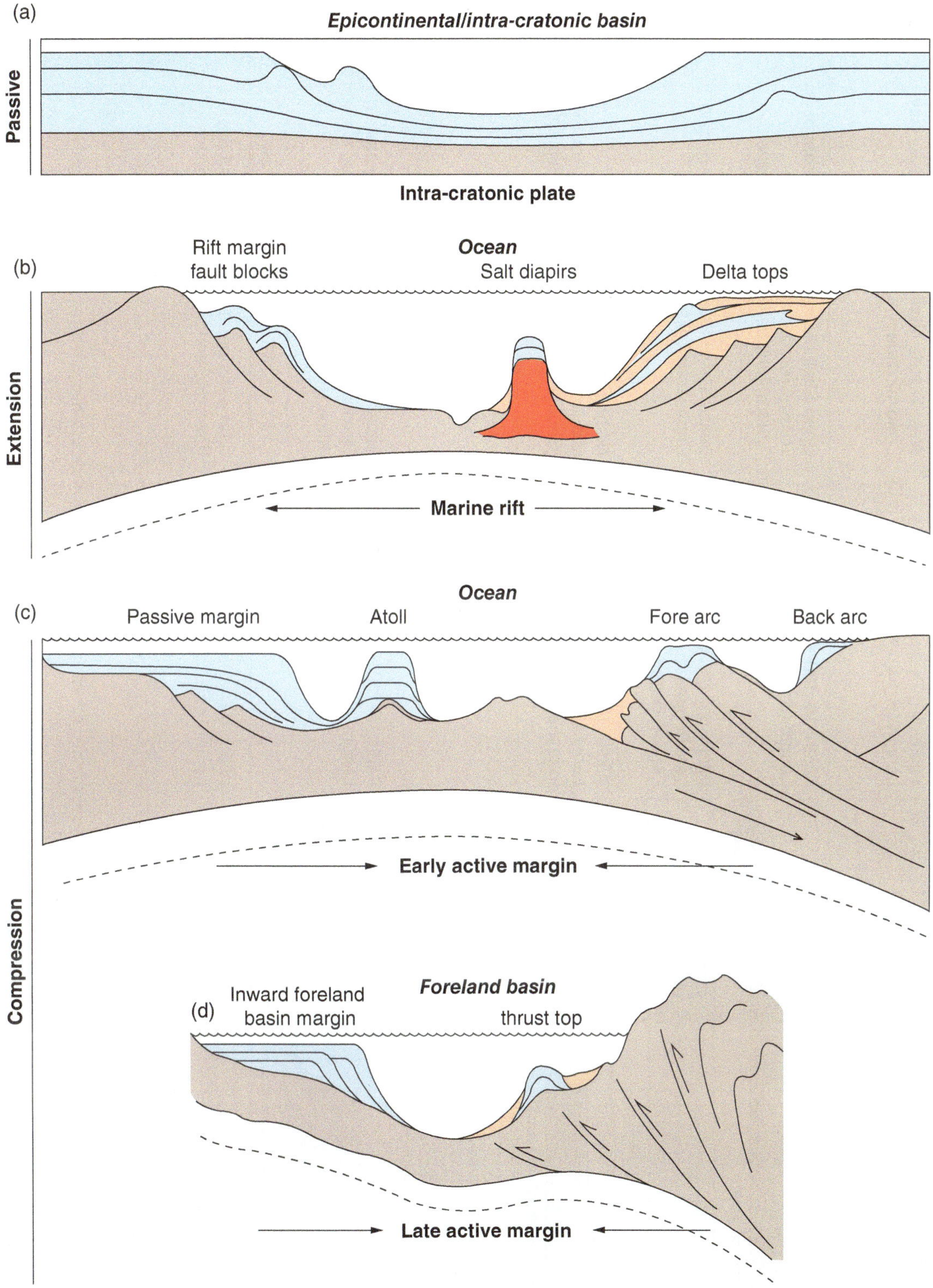

Figure 11.2 Schematic representation of four main types of geodynamic settings and the distribution of marine carbonate deposits. *Source:* Adapted from Bosence (2005), James and Jones (2016). Understanding of the geodynamic setting provides insights in key controlling parameters such as accommodation (space available for sedimentation), clastic supply and oceanography. Horizontal scale: hundreds to thousands of kilometres; vertical scale: multi-kilometre.

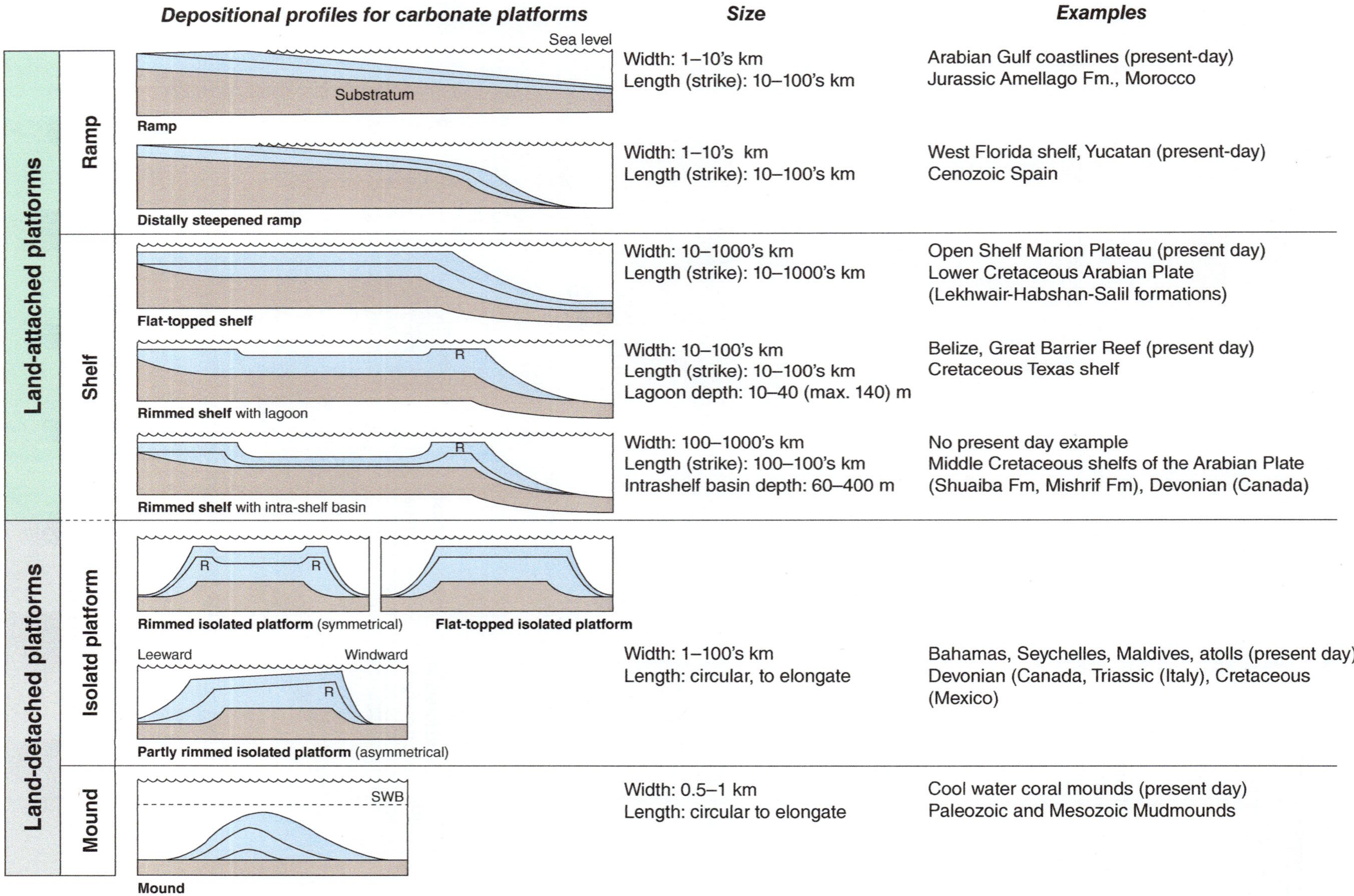

Figure 11.3 Main types of depositional profiles of carbonate systems. Note the distinction between land-attached and land-detached platforms. In land-attached carbonate systems, land-derived siliciclastic input can influence the sedimentation. In land-detached platforms and mounds, wind direction and ocean currents can have an important impact on the facies distribution. Note the difference between the shallower lagoon and the deeper intrashelf basin in the underfilled shelf. The Open Shelf falls in terms of depositional profiles in the category of flat-topped shelves without a rim, it has a limited width and is within reach of fair-weather wave base. R: rim, either bioconstructed or bioclastic. SWB: storm wave base.

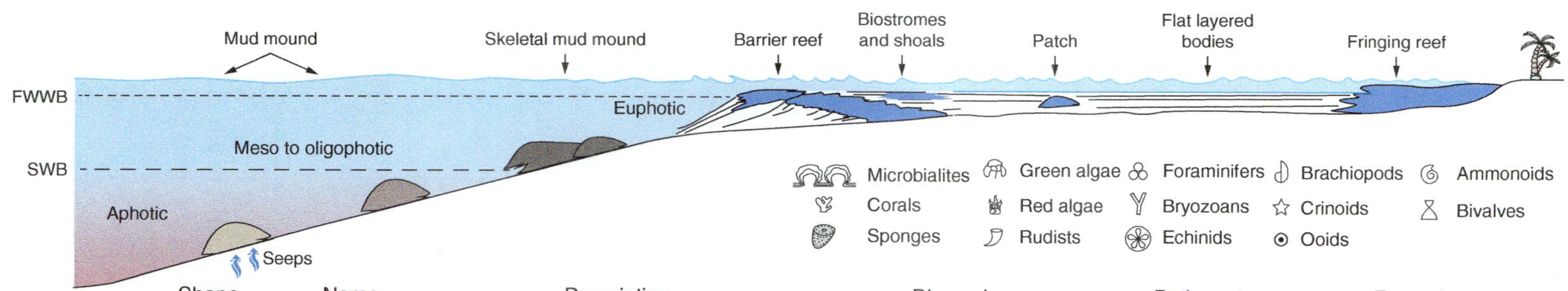

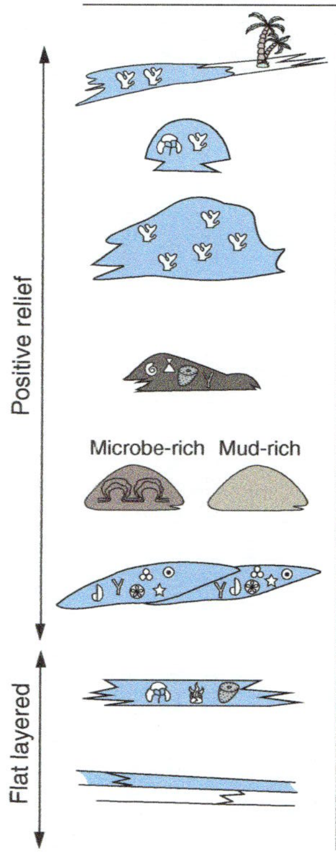

Shape	Name	Description	Dimensions	Bathymetry	Examples
	Fringing reef	Reef growing near shore along the slopes of older emergent area, without intermediate lagoonal belt	Width: 10' s m Length (strike): km's Thickness: m to 10's m	0–25 m	Fringing reef review (Kennedy and Woodroffe 2002)
	Patch 'reef' (bioherm)	Isolated reef generally occurring in lagoonal facies	Diameter: 10's m Thickness: 10's m	m's–25 m	Barret and Webster 2017
	Barrier reef	Reef-rimmed shelf margin with inner lagoon belt bordering the shoreline	Width: 10–100's m Length (strike): upto 1000's km Thickness: upto 100's m	0–50 m	Modern Great Barrier Reef, Australia (Mathews et al. 2007); Paleozoic reef barrier (Copper 1994); Phanerozoic review in Kiessling et al. 2002.
	Skeletal mud mound	Laterally restricted carbonate bioconstruction that formed in situ, possessed bathymetric relief, lacking large skeletal frame-builders	Diameter: 100's m – km's Thickness: 10 to 100's m	m's–150 m	Carboniferous Waulsortian mounds (Dorobek and Bachtel 2001; Wu and Chafetz 2002); Ireland (Devuyst and Lees 2001); Early-mid Cretaceous, Spain (Garcia-Mondéjar and Fernandez-Mendiola 1995, Neuweiler et al. 1999, 2002); modern carbonate mounds (Roberts et al. 2006)
Microbe-rich Mud-rich	Mud mound	Laterally restricted carbonate bioconstruction that formed in situ, possessed bathymetric relief mainly composed of mud and microbes	Diameter: 100's m–km's Thickness: 10 to 100's m	10's–1000 m	
	Shoals	Natural submerged ridge, bank, or bar that consists of, or is covered by, grainy unconsolidated material and rises from the bed of a body of water to near the surface; grainstone -packstone	Variable shape: 10–100's m Thickness: m to 10's m	0–10 m	Mississippian and Holocene ooid shoals (Handford 1988)
	Biostrome (lithosome)	Laterally extensive, bedded, blanket-like carbonate bioconstruction composed of skeletal remains, microbial components, in situ organisms and synsedimentary cement; boundstone. Specifically, when composed of rudists, the term lithosome is a synonyme.	Variable shape: 10–100's m Thickness: 10 m	m's to 40 m	Miocene coralline algal biostrome (Bosence and Pedley 1982)
	Flat layered bodies	Platform interior (lagoonal) lateral facies change (muddy and grainy), creating a flat-bedded layering. These bodies form volumetrically often the most important part of the platforms.	Variable shape: 10–100's m Thickness: 10–100's m	m's–40 m	Most interiors of carbonate platforms

Figure 11.4 Overview of terminology for common positive relief and flat layered carbonate geobodies along a carbonate shelf depositional profile. See text for further explanations.

platforms. A barrier reef separates the ocean from a protected inboard lagoon, a fringing reef has no inboard lagoon. These reefs can extend for hundreds of kilometres in strike direction or delineate the outer rim of carbonate platforms such as atolls.

Patch reef or bioherm: positive-relief carbonate bio-constructions created by *in situ* carbonate production and construction of an organic, and often cemented, framework, with lateral and vertical dimensions typically at the metre to decametre scale.

Skeletal mud-mound: positive feature of a mix of skeletal and muddy carbonate, created by sediment trapping and *in situ* production (Dorobek and Bachtel 2001; Wu and Chafetz 2002; Devuyst and Lees 2001; Neuweiler et al. 1999, 2023; Roberts et al. 2006). Sizes are in the metres to decametres range.

Lime mud-mound: a positive feature with a lime mud dominated matrix with additional binding organisms and bioclasts. The origin is a combination of hydrodynamic processes, induced trapping, and *in situ* production (Toomey and Finks 1969; Wilson 1975; Bourque 2001). Sizes are in the metres to decametres range.

Shoal/bar: generally flat to lenticular geobodies, with metre-scale elevation, consisting of a grainy carbonate facies, such as carbonate sands or skeletal debris, deposited in shallow water under the influence of waves and currents, locally with cross-bedding and or clinoforms, with lateral dimensions at the metres to decametres scale (e.g. Handford 1988). This is a very broad descriptive term, so we add, where possible, a process-descriptor, such as tidal bar, barrier island shoal, beach-strand shoal and lagoonal bioclastic-shoal.

Biostrome: a flat, layered, blanket-like carbonate bio-construction, with minimal or no 'synoptic' relief (i.e. relative to its immediate surroundings), composed of skeletal remains, microbial carbonate, calci-microbes, *in situ* organisms and syn-sedimentary cement, elongate or rounded in shape with a lateral extension of metres to hectometres, and thicknesses of decimetres to metres.

Flat-layered bodies: tabular bedding at the decimetre to metre scale encountered in the platform interior, characterised by lateral variations of the grainy and muddy sediment fractions that are not organically bound. These bodies stand out from the ones listed above, in that they are volumetrically often the most important part of the platform interiors. They can be tens to thousands of metres thick, and their spatial extent can cover the best part of tectonic plates (e.g. Palaeozoic platforms of North America and Mesozoic platforms of the Arabian Plate).

11.2.2 Depositional Profiles

Carbonate platforms, both land-attached and land-detached, can display different depositional profiles, which represent the geomorphology at a given moment in time as defined by the bathymetry and confinement, and being in part a function of the hydrodynamic regime. For marine systems, three morphological end-members profiles are distinguished: the ramp, the shelf and the mound (Figure 11.3). These depositional profiles form the basis of the carbonate facies models described in Section 11.5. It should be pointed out that carbonate platforms are normally the end result of a succession of different depositional profiles through time (Figure 11.1).

The *ramp depositional profile* (Figure 11.3) is defined as a surface inclined at 0.1°–1.5° or less sloping from the coast towards the open sea (Ahr 1973; Evans et al. 1973; Read 1985; Burchette and Wright 1992). A distinction can be made between distally steepened ramps (Figure 11.3) and those with local patch reefs or mud mounds in their distal parts (Read 1982, 1985). High hydrodynamic energy conditions occur close to the shoreline, where waves break. The width of the facies belts is generally large (hundreds of metres to kilometres) but may vary depending on palaeoslope and the hydrodynamic regime: the degree of wave, storm and tidal activity.

The *shelf depositional profile* (Figure 11.3) is characterised by a flat horizontal surface (platform top) delineated by a marked break in slope, ranging from a few degrees to up to 90° locally into deeper water, and a flatter toe-of-slope (Ginsburg and James 1974; Wilson 1975; Read 1985; Kenter 1990; Schlager 2005). Typical of the shelf profile is the margin, where wave energy is focussed. In most cases, the margin is marked by a rim of more or less continuous carbonate bio-constructions (reef assemblages) and/or bioclastic, grain-supported carbonate sand shoals (ooids, peloids and oncoids). Depending on the composition of the marginal facies, and the effectiveness of early cementation, different slope morphologies can be distinguished.

The flat part of the shelf may either be filled with sediments (close to sea level) or be under-filled creating a deeper water environment behind a barrier (Figure 11.3). The under-filled areas are *lagoons*, which can be mud- or grain-dominated and are characterised by a relatively lower-energy hydrodynamic regime below fair-weather wave base. Water depths in the lagoons can vary from several metres to up to 140 m, which is normally proportional to the aerial extent of the under-filled area. This can vary from several square kilometres to thousands of square kilometres (Sections 11.6 and 11.7).

Note that the shelf depositional profile may occur in the land-attached, as well as in the land-detached situation (Figure 11.3). The distinction between these two is vital, since the facies types and distribution will be controlled by different environmental conditions, such as siliciclastic influx in the land-attached setting, and wind-directions and sea currents in the land-detached situation.

The *mound depositional profile* (Figure 11.3) consists of an elevated structure with a rounded, hill-like top. This profile is characteristic of a special type of land-detached, isolated mound that forms in deeper-water settings, less sensitive to the influence of waves and storms. The deep-water mound varies in plan view from circular to arcuate or is irregular. Mounds may display single or multiple peaks with different heights, from several metres to over 300 m, and variable lengths, up to several hundreds of metres.

Transitions through time between these three depositional-profile end-members are common within any given carbonate setting (e.g. Read 1985; Eberli and Ginsburg 1989; Pomar 2001; Schlager 2005; Phelps et al. 2008; Williams et al. 2011). Ramp profiles may develop into shelf profiles, and vice versa, whereas deeper-slope, mound profiles may develop into flat-topped isolated platform profiles when they grow up into storm wave base (Figure 11.5). Situations where two profiles occur at the same time in one carbonate structure have also been observed, for instance, in the present-day Bahamas where the windward side shows a shelf depositional profile with a well-developed rim and the leeward side is characterised by a distally steepened ramp profile (e.g. Bosence 2005). A similar development is observed in the Maldives (Aubert and Droxler 1996; Belopolsky and Droxler 2003; Betzler et al. 2013a; Lüdmann et al. 2013).

11.2.3 Carbonate Texture Classification

The purpose of a carbonate texture classification system is to provide an objective, simple and clear set of rules to

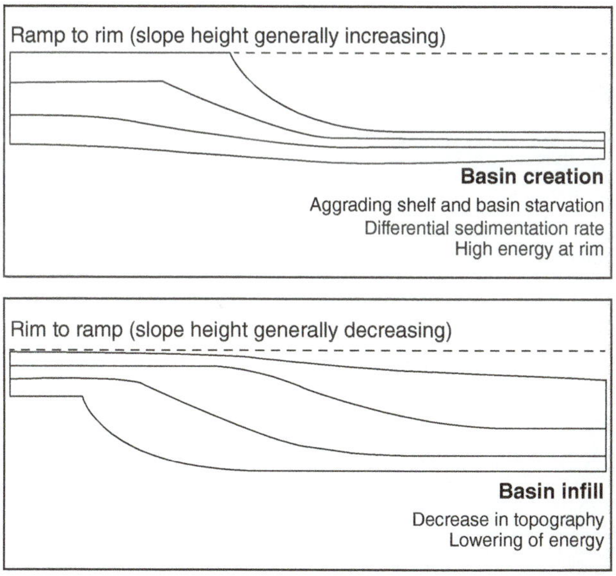

Figure 11.5 Illustration of two typical depositional profile trends encountered in carbonate systems (*Source:* Schlager 2005 / Society for Sedimentary Geology). In the upper part, topographic relief is created through differential accumulation rates (see Section 11.7), and in the lower part topography decreases through gradual infill.

describe the sedimentary rock with a hand lens in the field and under the optical microscope in the laboratory. The use of a descriptive rather than an interpretative, genetic, classification is preferable to avoid rushed interpretations in the observation stage.

The classification system that best fulfils these criteria, and which is used most, is the Dunham classification (Dunham 1962) as modified by Embry and Klovan (1971), Wright (1992) and Lokier and Al Junaibi (2016). Dunham (1962) based his classification on the depositional texture, defined as the ratio between matrix and grains, regardless of the size of these grains because, unlike in siliciclastic systems where the grain size is directly related to the hydraulic energy in the environment and the distance of transport, in carbonates larger grains can occur in low-energy environments and mud can be deposited *in situ*. Thus, Dunham interpreted the amount of mud within the sediment as a proxy for the hydraulic energy during deposition. Embry and Klovan (1971) added a notion of grain size with the introduction of floatstone and rudstone, defined as having more than 10% grains larger than 2 mm, whereby the floatstone is matrix-supported (< 2 mm) and the rudstone is grain-supported (> 2 mm), and the terms mudstone, wackestone, packstone and grainstone are reserved for a grain size smaller than 2 mm (Figure 11.6). In addition, they subdivided the boundstones, components organically bound during deposition, into bafflestone, bindstone and framestone (Figure 11.6). The last category deals with the chemical precipitation, crystalline carbonates for rocks that retain no or too little of the original texture, and cementstones for syndepositional cements that formed in voids and are non-replacive (Wright 1992).

A common modifier, which provides useful extra information, is the addition of the nature of the dominant grains, such as 'oolitic grainstones' or 'rudist floatstones'. Some precautions must be taken when considering mudstones. These are not only composed of micrite and may contain 90% of particles less than 63 μm of different origins which, if unrecognised, can result in incorrect interpretation. Also, some packstones may result from early or late diagenetic processes (compaction and selective dissolution) and not be depositional (Lokier and Al Junaibi 2016). In addition, in a two-dimensional view, such as in a thin section, grains may appear to float even though they are part of a grain-supported fabric in 3D. The 3D aspect has to be kept in mind in all carbonate petrographic analyses.

The majority of the more common skeleton-building biominerals are represented by carbonate phases. An overview of the chemical composition of skeletal material of carbonate-producing organisms is presented in Table 11.1. Other minerals used for skeleton-building include phosphates and silica.

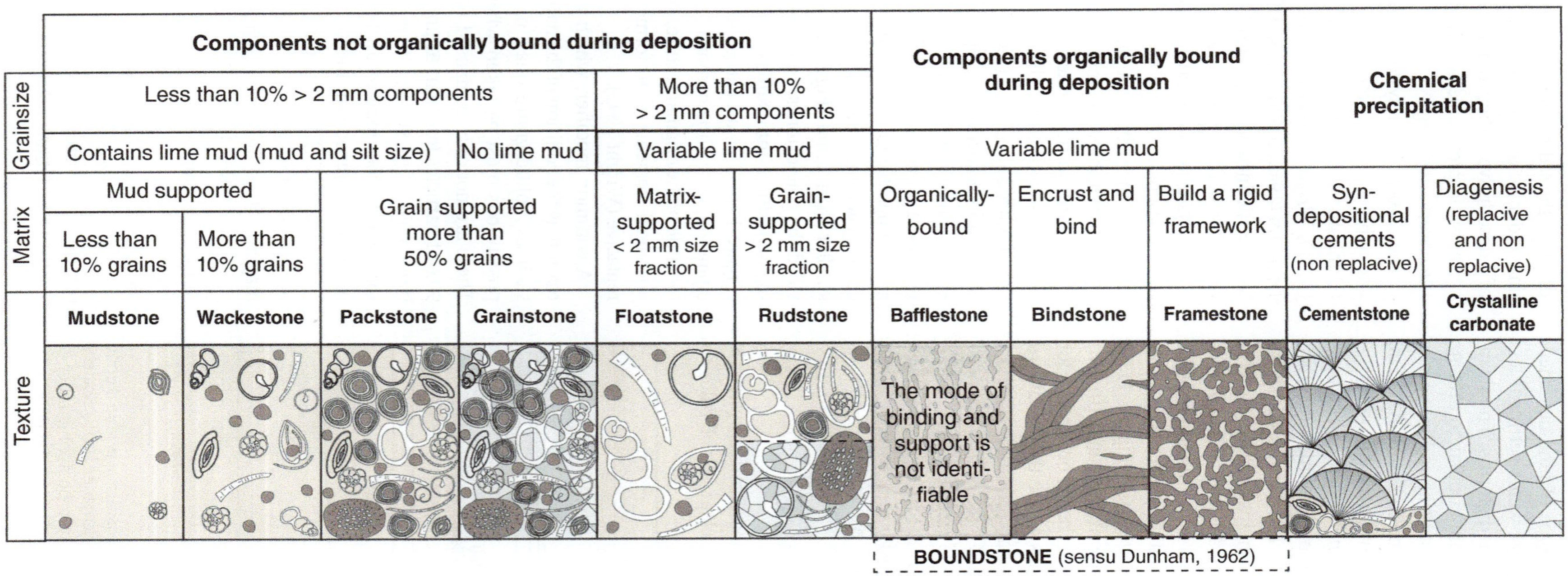

Figure 11.6 Carbonate texture classification scheme based on Dunham (1962), Embry and Klovan (1971), Wright (1992) and Lokier and Al Junaiba (2016). See text for further explanation. Note that the grain size cut-off value for floatstones and rudstones is here defined as >10% fraction of elements of >2 mm.

Table 11.1 Literature compilation of the most common carbonate-producing organisms and their primary chemical composition through time (based on Leadbeater and Riding 1986; Lowenstam and Weiner 1989; Mann et al. 1989; Carter 1990; Van de Poel and Schlager 1994; Stanley and Hardie 1998; Kosnik et al. 2008; Flügel 2010). Note that the early and burial diagenetic processes can significantly change this primary composition over time.

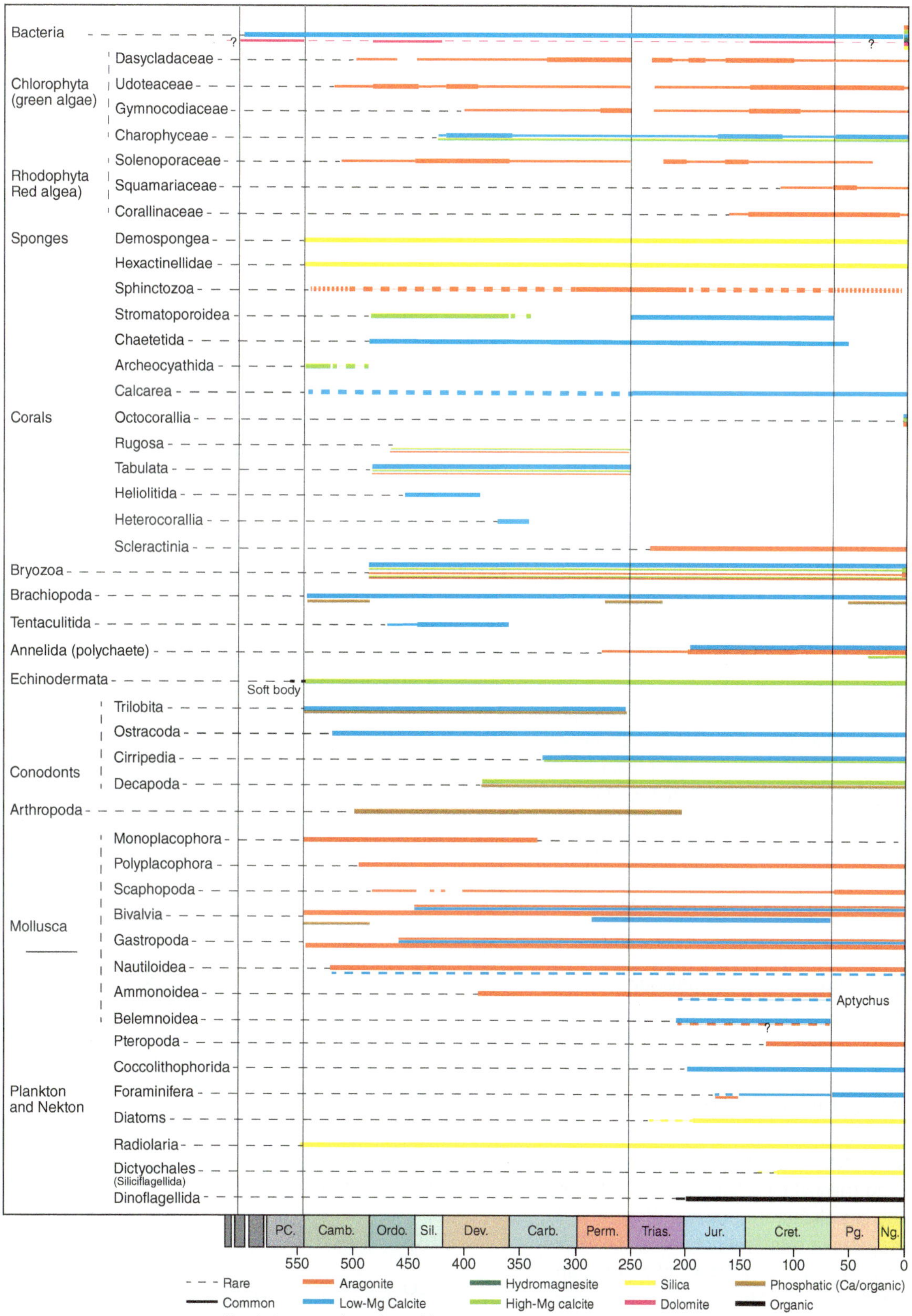

11.3 Chemical and Mineralogical Aspects

Carbonate sediments are prone to chemical and physical alterations both shortly after deposition and during shallow to deep burial. This alteration has an effect on the sedimentology (e.g. early cementation makes them less prone to reworking), petrophysics (porosity and permeability) and rock mechanical properties (elasticity and brittleness) of carbonates.

Carbonate rock properties are the result of the combination of (i) primary sedimentary facies, (ii) chemical overprint and (iii) mechanical deformation. Together, these alterations determine the ultimate rock type properties, which control the reservoir and/or sealing capacities. An essential part of a full rock description is, therefore, the porosity classification (an updated scheme is presented below), as well as a rock type definition for petrophysical characterisation.

By means of their chemical instability, carbonates can reveal events of stratigraphic importance, such as subaerial exposure by karstification or phases of little or no deposition through submarine hardground formation, or fluid migration events by different phases of cement formation or dissolution. In addition, despite diagenesis, there are many examples of carbonates which were buried as effectively closed systems and retained their initial geochemical composition, notably of the stable isotope values of Sr, Ca, Mg, C, O, B and S, which are valuable for the reconstruction of the depositional and climatic history.

Here we provide a brief overview of the key principles of the chemical processes and the mineralogical characteristics of carbonates. An in-depth discussion of this topic is beyond the scope of this chapter. Interested readers are referred to the large body of available literature (e.g. Moore 1989; Morse and Mackenzie 1990; Tucker and Wright 1990; Scholle and Ulmer-Scholle 2003; Mackenzie 2005; Moore and Wade 2013; James and Jones 2016; Nader 2017).

11.3.1 Mineralogy and Primary Composition

There are several calcium carbonate minerals: aragonite, low-magnesium calcite (LMC), high-magnesium calcite (HMC) and dolomite, among which the common component is the $CaCO_3$, and the differentiator the amount of Mg (Chave 1962; Tucker and Wright 1990). These minerals are organised in two different crystal structures, orthorhombic for aragonite and trigonal for calcite and dolomite. This explains their different chemical properties (Figure 11.7). The amount of magnesium (Mg^{2+}) contained in the calcite lattice can vary from 0 Mol% to 50 Mol% of $MgCO_3$, and the definition of LMC, HMC, VHMC (very high-magnesium calcite) and dolomite are based on arbitrary % cut-offs (Figure 11.7). During precipitation and diagenesis, Mg^{2+} can enter the structure of calcite by replacing calcium (Ca^{2+}) or leave the structure of HMC during early diagenesis, producing proto-dolomite and LMC (Melim et al., 2002).

The Mg^{2+}/Ca^{2+} ratio in a solution determines whether calcite or aragonite is formed (Berner 1975; Sandberg 1983; Morse and Mackenzie 1990; Veizer 1992). At a certain value of the Mg^{2+}/Ca^{2+} ratio in solution (around 10), aragonite precipitates faster than calcite (i.e. becomes kinetically favoured over calcite; Berner 1974). Mg^{2+} affects precipitation of calcite in two ways: first, although an increase in Mg^{2+} favours HMC precipitation, this mineral is more soluble than aragonite and hence aragonite forms. And second, Mg^{2+} inhibits the normal crystal growth habit of calcite and favours the aragonite growth form. This process favouring specific carbonate types acts at different scales. At the micro-environment scale, carbonate-secreting organisms synthesise organic molecules, in particular proteins which promote the precipitation of aragonite compared to calcite. That this process can go both ways is testified by the variation of the Mg^{2+} content of carbonate-producing organisms in a global dataset, which ranges from <1 to > 27 mole% $MgCO^3$ both among and within taxa (Van de Poel and Schlager 1994; Lebrato et al. 2016). In Table 11.1, a synthesis is provided of primary carbonate grains and their mineralogy through time. At the scale of the oceans, long-term changes in seawater chemistry affecting the Mg^{2+}/Ca^{2+} ratio have been observed, which favoured the precipitation of either calcite or aragonite at particular geological times ('calcite and aragonite seas'; Sandberg 1983; Hardie 1996, 2003; Stanley and Hardie 1998). For instance, times of active plate motion, high ocean spreading rates and marine volcanism are characterised by ocean water with a low Mg^{2+}/Ca^{2+} ratio because Mg^{2+} is stripped from seawater during serpentinisation at mid-ocean ridges. Whereas, during periods of slower ocean spreading, the Mg^{2+}/Ca^{2+} ratio is higher (Stanley and Hardie 1998; Stanley 2006).

The chemical reactions involved in the precipitation of carbonates require energy to drive the reaction (thermodynamics) and an excess or deficit of dissolved components ($Ca^{2+}{}_{(aq)}$, $Mg^{2+}{}_{(aq)}$, and $CO_3{}^{2-}{}_{(aq)}$; aq = aqueous).

Carbonate precipitation occurs as the result of a chain of chemical reactions (1–4) which involve carbonic acid (H_2CO_3) and carbon dioxide (CO_2):

$$CO_{2(g)} \leftrightarrow CO_{2(aq)} \tag{11.1}$$

$$CO_{2(aq)} + H_2O \leftrightarrow H_2CO_{3(aq)} \tag{11.2}$$

$$H_2CO_{3(aq)} \leftrightarrow HCO_3{}^-{}_{(aq)} + H^+{}_{(aq)} \tag{11.3}$$

$$HCO_3^-{}_{(aq)} \leftrightarrow CO_3{}^{2-}{}_{(aq)}\ H^+{}_{(aq)} \tag{11.4}$$

Chemical characteristics	Aragonite	Low-Mg calcite (LMC)	High-Mg calcite (HMC to VHMC)	Dolomite
Mineralogy	$CaCo_3$	$CaCo_3$	$CaCo_3$	$CaMg(Co_3)_2$
Cristalline system (and class)	Orthorhombic	Trigonal (Hexagonal scalenohedral	Trigonal (Hexagonal)	Trigonal (Rhombohedral)
Carbonate Calcium Magnesium Ca^{2+} = 0.98Å Mg^{2+} = 0.66Å Sr^{2+} = 1.12Å	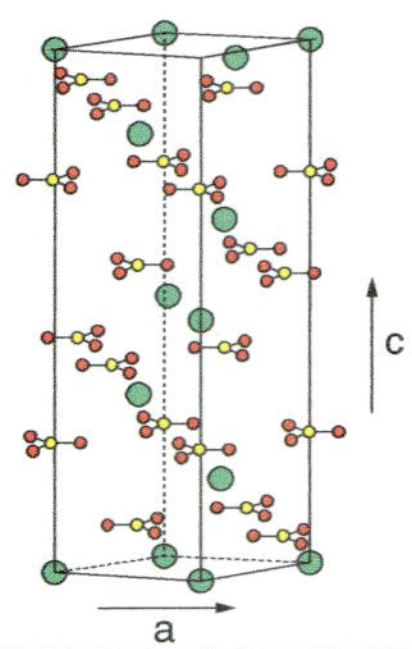	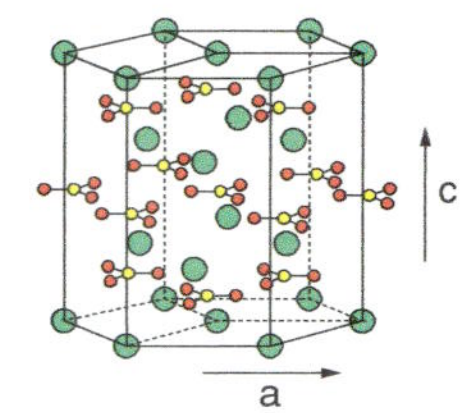		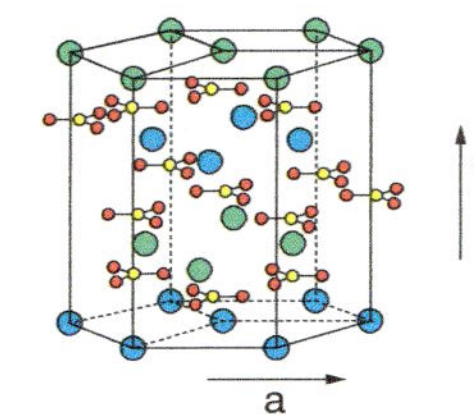 Stoichiometric dolomite
Trace elements	Sr, Ba, Pb, K	Mg, Fe, Mn, Zn, Cu	Fe, Mn, Zn, Cu	Fe, Mn, Zn, Cu
Mg:Ca	~2 for warm surface water	below ~2	~2 for warm surface water	~10:1
Mol% $MgCO_3$	Very low	<4	<4 a <18 (IMC=4–12 Mol% and HMC = 12–18 Mol%; VHMC = 38–50 Mol%)	40–50
Density	2.94	2.72	2.72	2.86
Hardness	3.5–4	3	3	3.5–4
Stability	Metastable, alters to calcite under aqueous conditions	Stable	Metastable	Stable
Occurrence	Shallow sea-water	Shallow cool water, deep sea-water and meteoric water	Shallow sea-water, VHMC (related to microbe and organic matter	Evaporite, early burial diagenesis
Mineral X-ray diffraction patterns	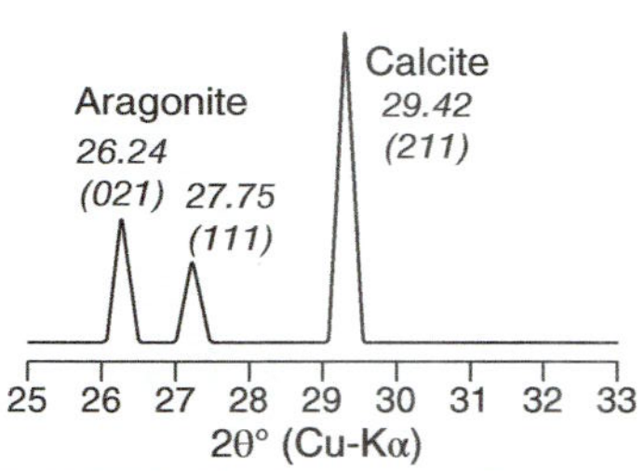	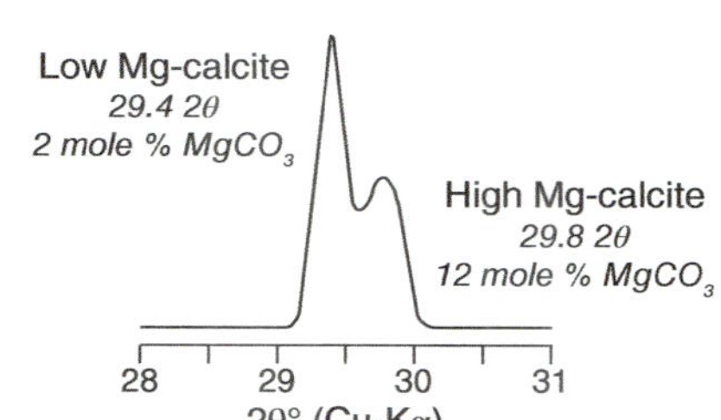		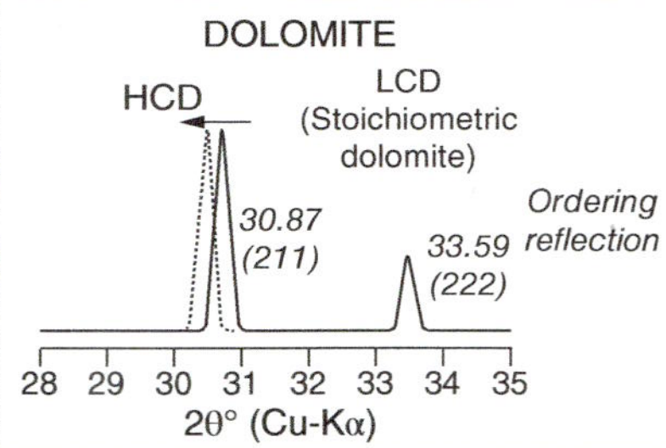

Figure 11.7 Mineralogical and geochemical characteristics of the main carbonate minerals aragonite, calcite and dolomite (*Source:* Adapted from Dickson 1965; Reeder 1983; Kacmarek et al. 2017). The calcites are further subdivided depending on the amount of Mg incorporated in the crystal lattice, as low-, high- and very high magnesium calcite (LMC, HMC and VHMC) (Kacmarek et al. 2017).

Equations (11.2) and (11.3) can be combined to give Equation (11.5) (Morse and Mackenzie 1990):

$$CO_{2(aq)} + H_2O \leftrightarrow HCO_3^-{}_{(aq)} + H^+{}_{(aq)} \tag{11.5}$$

To precipitate a mineral, a solution must be supersaturated with respect to the mineral phase (Eq. (11.6)). It means that the product of the activities of the chemical species, ions, constituting the mineral phase exceeds the solubility of the mineral phase.

$$K_s = \left(aCa^{2+}\right) \cdot \left(a\, CO_3{}^{2-}\right) / \left(a\, CaCO_3 \text{ solid}\right) \tag{11.6}$$

As the activity of the solid calcite is 1, the solubility constant (K_s) becomes

$$\boldsymbol{K_s} = \left(aCa^{2+}\right) \cdot \left(a\, CO_3{}^{2-}\right) \tag{11.7}$$

Knowledge of the carbonic acid equilibria (above) is essential for understanding ocean acidification (Lebrato et al. 2016), because the cascade of equilibria acts to buffer the system against large changes in pH. The saturation of a solution with calcite will increase with a local increase in the concentration of Ca^{2+} and/or the concentration of $CO_3{}^{2-}$. The concentration of $CO_3{}^{2-}$ can increase either following an increase in that of bicarbonate ions ($HCO_3{}^-$) at constant pH, or following an increase in pH at constant bicarbonate ion concentration. The pH of modern sea water is 7.5–8.4, which means that most of the carbonate is in the form of $HCO_3{}^-$. Several biological processes can increase the pH but are most effective locally. For example, photosynthesis in aquatic organisms (Eq. (11.8)), whether cyanobacteria or unicellular algae, generally tends to decrease the concentration of CO_2 around cells and favours the precipitation of calcite (Eq. (11.9)).

$$6CO_{2(g\ ou\ aq)} + 6H_2O_{(g\ ou\ aq)} \rightleftharpoons C_6H_{12}O_{6(s)} + 6O_{2(g)} \tag{11.8}$$

$$CaCO_3 + CO_2 + H_2O \underset{\text{Precipitation}}{\overset{\text{Dissolution}}{\rightleftarrows}} 2HCO_3^- + Ca^{2+} \tag{11.9}$$

The classic reversible carbonate reaction (Eq. (11.9)) shows that solid calcium carbonate, when in reaction with water and carbon dioxide, can either dissolve or precipitate. The main control on the reaction is the removal or addition of CO_2, inducing precipitation and dissolution, respectively. CO_2 solubility depends on temperature (increasing temperature favours the loss of CO_2, and thus carbonate precipitation), pressure (increasing pressure produces water with higher CO_2, and thus promotes dissolution) and biology (photosynthesis removes CO_2, and thus favours precipitation) whereas water percolating through soil absorbs CO_2 and implies dissolution.

The distribution of aragonite and calcite depends on the temperature and ionic composition of the water. Stable LMC dominates in non-marine and deep-sea carbonates; aragonite and HMC in shallow tropical settings; HMC in shallow temperate-warm non-tropical settings and HMC associated with LMC in cool-temperate non-tropical settings (Tucker and Wright 1990; James and Jones 2016). Dolomites occur in specific environmental settings like sabkhas and marine shallow water (McKenzie et al. 1980; Gunatilaka 1991; Bontognali et al. 2010; Petrash et al. 2017; Shalev et al., 2021).

11.3.2 Diagenesis

During diagenesis the chemical reactions summarised above may move either way, causing precipitation or dissolution of carbonates and thus transforming the primary mineralogical composition, texture, porosity and permeability. These transformations result from fluid–rock interactions and depend on the mineralogical composition of the host rock and the composition of the fluids (marine and meteoric waters, deep subsurface brines). In addition, these transformations may also be catalysed by microbially driven intervention (Petrash et al. 2017). Because both aragonite and HMC are metastable and HMC loses Mg^{2+} with time, most carbonate rocks will eventually evolve into LMC (Bischoff 1969; Budd 1988; Ries et al. 2009). The original mineral composition controls the 'diagenetic potential' and determines the time interval needed to reach an equilibrium level of lithification.

Alteration is an 'ongoing process' throughout the history of a sediment; it is, therefore, difficult to define a meaningful boundary between 'early' and 'late' diagenesis; a distinction introduced by Dickson (1984) and Bathurst (1983). We prefer to apply the following terminology following Choquette and Pray (1970) and Melim et al. (2002) (Figure 11.8):

1) *Early diagenesis*: taking place during or shortly after sedimentation within the stratigraphic unit, which, although not a closed system, provides the environmental context (P, T, mineral and fluid composition).
2) *Burial diagenesis:* taking place after burial of the sediment, when the chemical composition of the porewaters and surrounding rock in the burial domain will be altered by *in situ* processes as well as the impact of vertically or laterally migrating fluids.
3) *Exhumation diagenesis*: concerns rocks that after burial are exhumed through tectonic uplift or prolonged denudation and are successively influenced by subaerial exposure-induced diagenesis.

Despite the diversity of factors implicated in any particular diagenetic succession, these three conceptual end members help to organise and classify diagenetic processes in space and time. These processes include dissolution, micritisation, cementation, compaction, neomorphism, dolomitisation and the replacement of carbonate grains and matrix by carbonate and non-carbonate mineralogies. They will occur in different orders depending on local conditions. The common

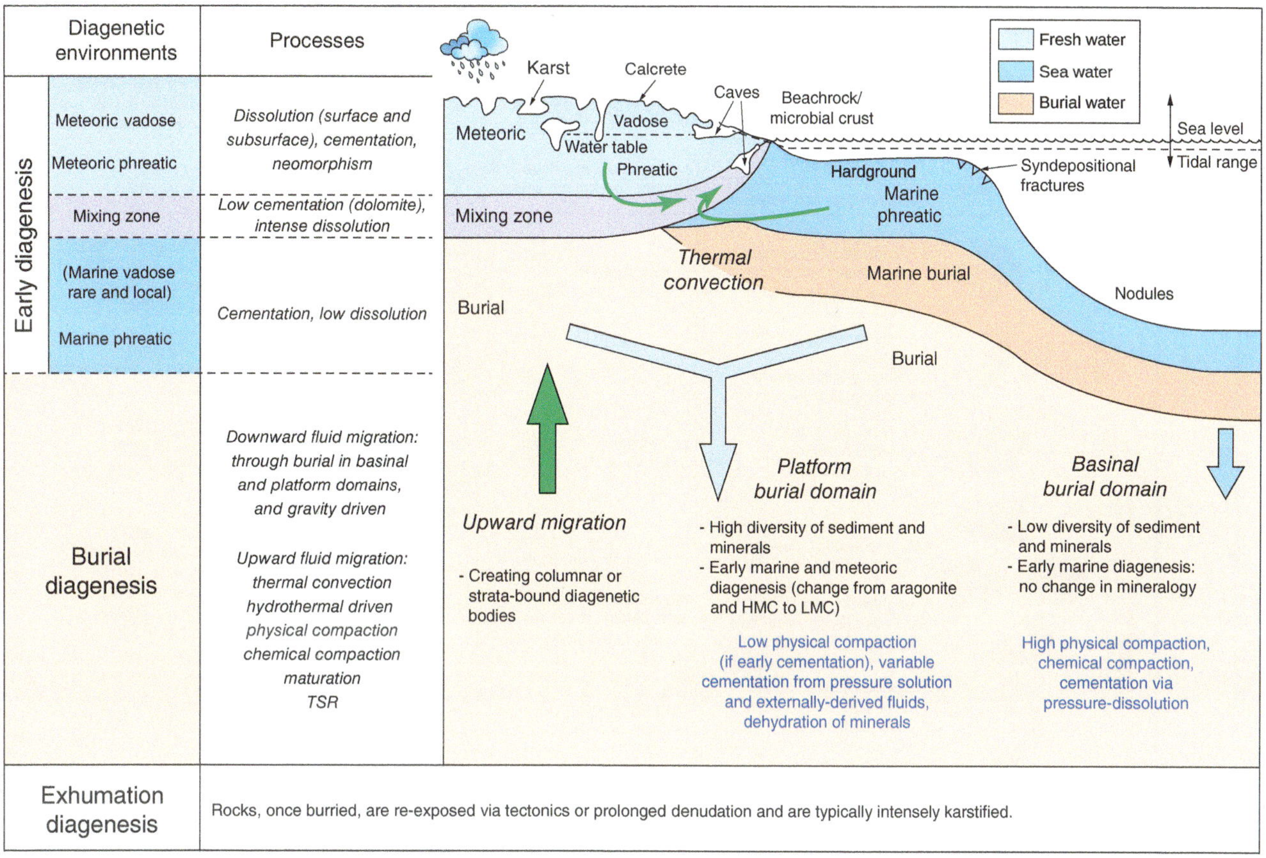

Figure 11.8 Summary diagram of main diagenetic environments and processes along a platform to basin depositional profile (based on Kohout 1967; Choquette and Pray 1970; Moore 1989; Moore and Wade 2013). These processes can take place at timescales varying from months to millions of years.

cement types (shape, crystal form, size, growth relation, etc.) and fabrics (i.e. symmetrical around grains, asymmetrical at grain contact, cement crusts and fans, cement mosaics in pores and syntaxial overgrowth cement) for the main diagenetic environments are summarised in Figure 11.9. Note that few fabrics are truly diagnostic for a certain environment, many have now been observed to occur in different environments (e.g. Melim et al. 2002).

Among the diagenetically affected carbonate rocks, dolostones (those with more than 75% of dolomite) are the most abundant. Given the kinetic barriers that inhibit low-temperature dolomite nucleation and precipitation, massive dolostones are usually attributed to slow burial or hydrothermal dolomitisation of calcite or aragonite precursors by Mg-rich fluids at temperatures of >100°C (Gregg et al. 2015). In contrast, some studies have asserted that dolomites can also form at low temperatures in marine porewaters during early diagenesis, that is, the sabkha model in evaporative settings (Warren 2000), and the organogenic dolomite model through microbial mediation (Vasconcelos et al. 1995). Overviews of dolomite formation are provided in Machel and Mountjoy (1986), Machel (2004), Gregg et al. (2015) and James and Jones (2016).

Compaction occurs both early, through grain rearrangement and water expulsion in muddy sediment, and late, through carbonate dissolution at grain contacts and formation of stylolites. For in-depth treatment of these processes and illustrations see, for example, Tucker and Wright (1990), Flügel (2004, 2010), and James and Jones (2016).

11.3.2.1 Early Diagenesis

Diagenesis can start within months of deposition, as in the formation of beach rocks in the intertidal zone (Dickson 1985; Shinn 2009), precipitation in microbialites (Riding 1991a, b; Dupraz et al. 2009; Pace et al. 2018; Diaz and Eberli 2021; Medina Ferrer et al. 2021), formation of hardgrounds on shallow platforms (Dravis 1979, 1996) and on the platform slope (Grammer et al. 1999). Early diagenetic alterations occur in relatively well-defined depth zones, from top down (Figure 11.8): (i) the *meteoric vadose zone*, which is that part of the sediment column that is above the groundwater table (level at which the hydrostatic pressure is equal to the atmospheric pressure), where both air and freshwater (rain, meltwater and runoff) fill the pore space. In this setting, meteoric water is held to the grains by capillary forces; (ii) the *meteoric phreatic zone*, which

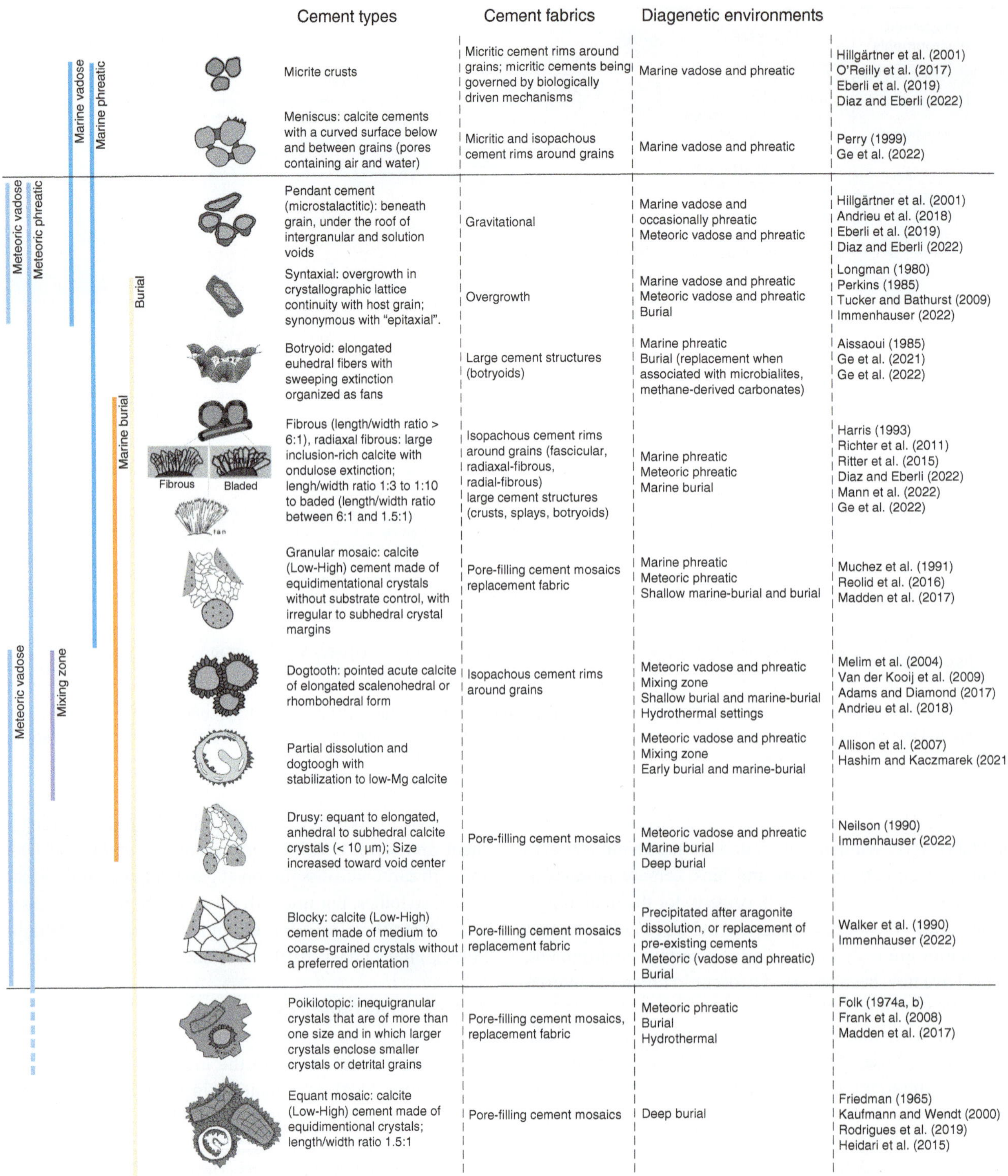

Figure 11.9 Illustration of typical cement types and their occurrence in the main diagenetic environments (*Source:* Adapted from Boggs 2001; James and Jones 2016; Moore and Wade 2013). Note that cement types are rarely diagnostic for one diagenetic environment.

is below the groundwater table where all pore space is constantly filled with freshwater; (iii) the *mixing zone*, where pore space is continuously filled with a mix of freshwater and marine waters; (iv) the *marine vadose zone*, where pore space is not continuously filled with marine waters. This is a rare case and limited to for instance tidal flats, where during low tide the marine water will recede; (v) the *marine phreatic zone*, where marine water permanently fills all the pore space. These environments may also involve various types of complex rock–fluid–microbial interaction (Pagel et al. 2014). The relative position of these zones will change with time and in space, depending on freshwater supply and sea-level fluctuations (Melim et al. 2002). For instance, because all marine carbonates are first formed in the marine domain, they will first experience diagenesis in the marine phreatic domain. Subsequently, depending on their position along the depositional profile, and depending on the magnitude of relative sea-level fall, they may become exposed to the meteoric domain, a process that may be repeated many times in environments sensitive to sea-level fluctuations such as the carbonate platform top. These events will create a diagenetic succession ('cement stratigraphy') locally captured in the cementation and dissolution histories (Purser and Loreau 1973; Longman 1980; Moore 1989). The interpretation of the diagenetic successions can, however, be complicated by the fact that many cements and fabrics thought to be indicative of phreatic meteoric diagenesis have now been proven to also occur in the marine-burial environment (Figure 11.9; Melim et al. 2002).

Meteoric Phreatic and Vadose Environments and Cements

The interaction between carbonate sediments and fresh meteoric water results in complex diagenetic patterns (e.g. Longman 1980, 1981; Esteban and Klappa 1983; James and Choquette 1988; Tucker and Wright 1990; Swart 2015; James and Jones 2016; Immenhauser 2022). Meteoric diagenesis induces changes in mineralogy, chemistry (e.g. Mg and Sr lost from HMC and aragonite), and stable isotope composition (Allan and Matthews 1982; Goldstein 1991).

The phreatic zone often exhibits a lens-shape geometry (Figure 11.8). Here cementation is enhanced by supersaturation of LMC caused by the dissolution of aragonite and the dissolution-reprecipitation and replacement of HMC. The different solubilities depend on Mg^{2+} content. Cements that form during meteoric diagenesis are generally clear (limpid) crystals that have few inclusions because the water contains comparatively few cations, other than Ca^{2+} and perhaps Mg^{2+}. Dissolution occurs during percolation of undersaturated meteoric water, acidified by atmospheric CO_2 and the degradation of organic matter. Meltwater, rainwater and runoff affect the meteoric vadose zone either in marine carbonate substrates or by percolation through soils and calcretes (Goudie 1996; Esteban and Klappa 1983; Wright and Tucker 1991). The end member of the dissolution stage is karstification (Trudgill 1985; Ford and Williams 1989; Radulović 2013). Typical cement types in the meteoric phreatic zone are dog-tooth, blocky, fibrous, microcrystalline and peloidal microcrystalline. The main fabrics, illustrated in Figure 11.9, are circumgranular, isopachous, gravitational, equant, granular and blocky mosaics and syntaxial overgrowth. The percolation process in the vadose zone induces irregular and asymmetrical crystals, mainly composed of pendant microstalactitic and meniscus LMC cements (Figure 11.9; Land 1970; Hillgärtner et al. 2001).

Climate has a strong influence on the effectiveness of vadose and phreatic meteoric diagenesis. In arid settings, alteration and cementation will be limited and mainly occur in the phreatic zone. In semi-arid settings, alteration in the vadose zone can be rapid, and will be slow in the phreatic zone. In wet climates, dissolution is very efficient, creating large caves and minimal mineral precipitation (James and Jones 2016). Sea-level fluctuations and tectonics can amplify the effectiveness of meteoric diagenesis through prolonging subaerial exposure. These dissolution and precipitation processes also occur during Exhumation diagenesis (see below; Choquette and Pray 1970). A typical expression of meteoric diagenesis is palaeokarst, which is often associated with regional stratigraphic and tectonic unconformities and may lead to enhanced porosity (Esteban and Klappa 1983; Esteban 1991; Loucks 1999; Klimchouk 2017). Water movement in the vadose zone can occur by conduit flow through fractures, and pervasive water movements can transport soil and other sediments into the karst network producing *Terra Rosa* type siliciclastic deposits (Benedetti et al. 1994).

Marine Phreatic Environments and Cements

The marine phreatic zone extends along the depositional profile from the shallow-water domain of carbonate platforms and ramps, where the seawater is oversaturated with respect to the $CaCO_3$, to deeper water (slope and basin) where the sea floor can intersect the lysocline (the depth at which the rate of dissolution of aragonite and calcite begins to increase dramatically), and down to the calcite compensation depth (CCD). A characteristic feature of the marine phreatic zone is the early cementation of the seafloor creating submarine hardgrounds (Melim et al. 2002; Christ et al. 2012). Modern shallow seawater is generally oversaturated with respect to both aragonite and Mg-calcite, and cementation between the grains is favoured by the oversaturation and seawater circulation through the pore space by tidal currents, waves or ocean currents. Precipitation is enhanced in shallow-water settings and is far more intense in warm waters (James and Jones 2016), although some considerable volumes of early marine cements have also been described in deep-water environments (Schlager and James 1978; Van der Kooij et al. 2010). In platform settings cementation is especially favoured at windward margins, where supersaturated upwelling waters leach reefs and oolitic sand shoals, thus providing the carbonate ions for early diagenetic cement growth (James and Ginsburg 1979; Marshall and Davies 1981). Inner ramps usually are also sites of intense cementation because of the high rate of water flow and evaporation processes. Dissolution is also observed just below the seafloor in both warm tropical and cool temperate environments, and even more in zones with low sedimentation rates when microbes reduce the amount of organic matter and enhance the production of acids (Coleman and

Raiswell 1993). In cool-water systems, this process can potentially remove all depositional or skeletal aragonite from the near-surface sediment (Jones et al. 2005). The marine phreatic domain can, however, also be stagnant (e.g. uncemented lagoon sediments and patch reefs) where there is little water movement on the sediment surface or in the pore fluids, creating reducing conditions, promoting micritisation of grains (induced by microbes, algae and fungi) and inhibiting cementation (Heckel 1983). The typical cements for the marine phreatic environment are fibrous, acicular, bladed, botryoidal, radiaxial, dogtooth and microcrystalline (Figure 11.9). The main fabrics are typically circumgranular, isopachous, crusts and botryoidal.

A marine vadose zone occurs only where seawater evaporation is efficient. Cementation here is frequently associated with microbial deposits and induces beach-rock formation (Strasser et al. 1989). There are cases in inter- and supratidal beach deposits where there is a minimum of freshwater influx. Cements may also develop on humid and arid tidal flats as cemented crusts frequently related to microbially induced cementation (Jahnert and Collins 2012; Bouton et al. 2016). Beach rocks are not only common in tropical, subtropical but also in temperate zones (Meyers 1987; El Moursi and Montaggioni 1994; Neumeier 1998; Beier 1985; Gischler 2008). Cements comprise microcrystalline aragonite and Mg-Calcite. In this diagenetic environment the cements are mainly gravitational such as dripstone or pendant and meniscus, and frequently associated with microbes (Figure 11.9). The cement types are frequently bladed relicts, incipient growth of dogtooth and radiaxal fibrous.

Mixing Zone

The boundary between the marine and the freshwater phreatic zones is the transitional mixing zone (Figure 11.8), where water is brackish (Back et al. 1984; Smart et al. 1988; Machel et al. 1990; Zhao et al. 2020). As a result of the non-linearity of the activity coefficients with additional species, the mixture of marine and meteoric waters that individually may have been previously saturated with respect to carbonate minerals becomes undersaturated, thus promoting dissolution. Although the mixing zone has been suggested to promote dolomitisation, there is little evidence of this in modern mixing zones (Melim et al. 2004). The thickness of this zone and the associated dissolution processes may vary with changes in permeability and flow rate. The petrographic fabrics typical for this diagenetic environment frequently mimic the overlying meteoric zone and the underlying marine-burial zone (Figure 11.9). Near the top of the zone, mouldic porosity is associated with blocky spar and deeper in the mixing zone, dogtooth cements are observed (Melim et al. 2002).

11.3.2.2 Burial Diagenesis

In the burial domain, which lies between the zone of early, 'near' surface, diagenesis above and the zone of low-grade metamorphism below, another set of diagenetic changes takes place (Scholle and Halley 1985; Immenhauser 2022). As mentioned above the boundary between early and burial diagenetic alterations is diffuse. However, most limestones have been affected by burial diagenetic processes, including even deep-water carbonates that were scarcely affected by early diagenesis.

The composition of the carbonate rocks that reach the burial domain is controlled by the primary characteristics of the sediment, including (i) the carbonate mineralogy (e.g. sediment altered during early diagenesis and cemented with LMC will resist physical compaction), (ii) the purity of the carbonate (e.g. presence of clay or organic matter facilitate compaction and stylolitisation), (iii) the grain size (e.g. coarse sediment is more resistant to compaction), (iv) early cementation, which increases the bearing strength of sediments and (v) porosity/permeability characteristics (e.g. fluids flow in larger volume and higher rate in coarse sediments). The inherited early diagenetic history will differ between the shallow neritic platform domain, where the pore fluids will evolve from a varying marine-freshwater composition to saline or hypersaline brines during burial, and the basin domain, where the sediments containing marine pore waters pass directly into the burial domain with a lesser early diagenetic overprint (Figure 11.8). Interestingly, Melim et al. (2002) were able to distinguish a transitional marine-burial diagenetic domain, which is enriched in marine pore fluids but mimics many aspects of diagenesis in meteoric pore fluids, most notably by producing a mainly low-Mg calcite limestone with blocky spar, neomorphism, microspar and mouldic porosity (Figure 11.9).

In the deeper burial domain factors such as pore pressure (lithostatic, hydrostatic and tectonic), temperature and thermochemical reactions become important. In a closed system induced by the presence of shales, evaporites or early lithification, an increase in pore-fluid pressure can inhibit burial diagenesis (e.g. Wanless 1983; Neveux et al. 2014; Smit et al. 2022). The increase in temperature during burial may induce calcite precipitation, whereas release of water from hydrated minerals and transformation of organic matter may lead to an increase in organic acids and thus carbonate dissolution. The pore-fluid chemistry may evolve through changes in the ionic concentration of the more soluble minerals inducing either dissolution or precipitation (Nogues et al. 2013). In the deep burial domain, thermochemical sulphate reduction (TSR), a redox reaction between sulphates and organic matter generating CO_2, H_2S, organic sulphur compounds, diamondoids and solid bitumen, is the dominant mechanism of hydrogen sulphide production (Machel et al. 1995). TSR takes place as soon as a geologic environment with the right 'ingredients' (e.g. anhydrite and hydrocarbon) is buried to depths that correspond to temperatures of about 100–140°C. Sulphate reduction also occurs through microbial processes (BSR or bacterial sulphate reduction) but then preferentially in a low-temperature diagenetic environment, in all phreatic environments. The common cement types for the burial domain are blocky, poikilotopic, dogtooth, radiaxial calcite and coarse dolomite and anhydrite (Figure 11.9). The main fabrics are drusy, equant and granular mosaics and syntaxial overgrowth. Dissolution can also occur under deep burial conditions as a result of mixing or cooling of brines (Esteban and Taberner 2003). When the deep-burial domain grades into that of metamorphism, the gradual recrystallisation of carbonates form meta-carbonates and subsequently marbles at temperatures of 200–250°C (Immenhauser 2022).

Fluid flow plays an important role in the early diagenetic and burial diagenetic domains and is driven by different processes during the overall history of a carbonate system. Starting with the syn-sedimentary and early diagenetic phase, fluid flow can be controlled by (i) evaporation in a closed lagoon, causing heavy seepage brines to penetrate the surrounding carbonates, (ii) tidal currents, generating a daily steady flow of marine, Mg-rich water, (iii) relative sea-level fluctuations, acting at different amplitudes varying from the metre to 100-m scale, and occurring at the 10 kyr to million years scale, (iv) thermal convection ('Kohout convection', Kohout 1967), generated by the temperature contrast of seawater and platform interior groundwaters warmed by geothermal heat flux (Figure 11.8). In the burial domain upward fluid migration can be driven by chemical and mechanical compaction, or compressive tectonic events and by hydrothermal fluids that actively move upward along faults or fractures creating columnar or stratabound diagenetic geobodies (e.g. Davies and Smith 2006; Smit et al. 2022). Another basin scale process of fluid flow is the gravity or hydrodynamic head-driven downward flow into the burial domain followed by upward flow after warming and chemical alteration at depth (Garven and Freeze 1984).

11.3.2.3 Exhumation Diagenesis

Exhumation of carbonate successions happens at major, mostly structurally controlled, stratigraphic unconformities. Exposure of these surfaces is long lasting, which in humid climates can lead to intense karstification with thick (bauxite) soil development and extensive vegetation cover (Loucks 1999; Palmer 1995), whereas under arid conditions diagenetic alterations are minor. The semi-arid climate zone takes an intermediate position with calcareous soils or calcretes/caliches (Esteban and Klappa 1983; James and Choquette 1988; Wright and Tucker 1991; Moore 2001; Alonso-Zarza and Tanner 2006) and minor karst systems (Legrand and Stringfield 1973; Morse and Mackenzie 1990).

11.3.3 Petrophysical Characteristics

The key petrophysical properties of carbonate rocks are lithology (limestone, dolomite, shale, etc.), porosity (macro- and micro-porosity), pore type (e.g. framework, inter- and intra-granular and microporosity), mineralogy (aragonite, HMC, LMC and dolomite), texture (grain supported versus matrix supported), grain-size distribution, permeability and water saturation (e.g. Anselmetti and Eberli 1993, 2001). Together these properties determine the reservoir properties as well as the elastic (acoustic) properties of rocks, which are essential for the interpretation of seismic images and acoustic logs, and the prediction of reservoir and rock mechanical characteristics (Lucia 1995, 1999; Mavko et al. 2009).

Depositional facies and (reservoir) rock types: The petrophysical characteristics are the end result of a process that starts with the primary depositional facies features (mineralogy, sediment sorting and texture), subsequently overprinted by early and burial diagenetic alterations, as outlined above. Carbonate sediment properties change fundamentally during this process. Distinction is, therefore, made between the primary sedimentary facies and the *rock types*, which represent the end product after diagenesis.

The changes that determine the rock type properties occur at different stages and at different scales as diagenetic processes partially overprint or obliterate the primary facies characteristics during fluid–rock interaction (Dunham 1962; Choquette and Pray 1970; Tucker and Wright 1990; Moore 2001; Fournier et al. 2005; Moore and Wade 2013; Brigaud et al. 2014; Worden et al. 2018; Teillet et al. 2019). Acoustic properties are sensitive to processes that infill (poronecrosis) or create (porogenesis) porosity, recrystallise the original matrix or even completely modify the texture (e.g. during dolomitisation) (Anselmetti and Eberli 1997; Kenter et al. 2002; Reijmer et al. 2022; Fournier and Borgomano 2007; Baechle et al. 2008a, b; Weger et al. 2009; Fournier et al. 2011; El Husseiny and Vanorio 2013, 2015; Kleipool et al. 2015, 2017; Soete et al. 2015; Bailly et al. 2019, 2022; Salih et al. 2023). Selective cementation may significantly reduce primary porosity in grain-supported lithologies by infilling the primary intergranular macropores. Early cements may act as 'early framework-stabilising cements' (Dewever et al. 2007) partly preventing compaction and partially preserving the primary porosity, while blocky calcite cement may reduce most of the remaining porosity (e.g. intergranular macropores, moulds and veins) (Van Geet et al. 2002). In addition, both early cementation and dissolution during succeeding stages of burial change the mechanical properties of the rock and their response to compressional and shear stress (tectonic and burial processes) and thus their brittle or ductile behaviour.

Grain size and cementation: In granular rocks, whether siliciclastic or carbonate, the grains are the principal medium of acoustic-wave propagation. In carbonates grain size and cementation and hence the porosity and pore type depend on (i) the depositional lithology, which includes (a) the carbonate mineralogy (aragonite, HMC, LMC and dolomite; sensitivity to diagenetic changes), (b) the bio-construction framework dominating the sedimentary system (e.g. coral reefs and microbial mats), or (c) the pore fabric that is linked to the origin of those grains (grain shape and internal fabric), as well as (d) sediment sorting, roundness and grain stacking, (ii) the grain versus matrix supported fabric and (iii) the diagenetic alterations controlling changes in pore type, porosity and permeability (Anselmetti and Eberli 2001; Eberli et al. 2003; Weger et al. 2009, 2023). In clastics and to some extent in mixed clastic-carbonate systems the velocity–density relationships depend on sediment sorting and composition (e.g. mineral composition and clay content) (Vernik and Nur 1992; Braaksma et al. 2003; Jafarian et al. 2017). In crystalline carbonates (dolomites) the principal medium of acoustic-wave propagation is the framework of interlocking crystals formed during diagenesis (Anselmetti and Eberli 1993; Verwer et al. 2008). An increase of carbonate cement, and infill of the pore space in grainy sediments, results in a decrease in porosity and a concommitant increase of acoustic velocity (Hossain and Zhou 2015).

Mineralogy: Both the mineralogy of the carbonate sediment and that of the admixed lithologies have an impact on the petrophysical characteristics. For a given mineral composition, velocities in carbonates display no clear trend with variations in porosity and for a given porosity may exhibit a wide scatter in velocities (Anselmetti and Eberli 1993; Eberli et al. 2003; Verwer et al. 2008). Different types of dolomitisation modify the original acoustic velocity differently: fabric-preserving dolomitisation has a limited impact on acoustic velocity, whereas non-fabric preserving dolomitisation leads to an increase of acoustic velocity (Eberli et al. 2003; Salih et al. 2021b, 2023). Admixed minerals that may influence the petrophysical properties of carbonate-dominated sediments are principally clays and siliciclastics. Clay contents of more than 12–15% may lead to a decrease in acoustic velocity (Braaksma et al. 2003; Hossain and Zhou 2015; Jafarian et al. 2017).

Porosity, permeability and acoustic properties: The adequate characterisation of the pore type network is of economic importance for reservoir evaluations. It is fundamental in determining the acoustic properties of carbonate rocks and is the basis for the definition of petrophysical rock type classes and porosity-permeability trends. The primary porosity distribution in carbonates is related first to the growth strategy of the carbonate-precipitating organisms, each of which produce an individual grain spectrum (shape and size) with variations also in the quantity of muds versus grains. Porosity will also depend on the sorting capacity of waves, currents and gravity-driven processes, but is mainly dependent on diagenetic modifications, for example, occlusion by carbonate cements, recrystallisation of the original (non-) skeletal grains and carbonate muds and dolomitisation processes during burial. These processes affect the initial porosity and permeability as they not only modify the pore shapes through cementation but also reduce pore throat sizes and impact permeability. Diagenetic alterations may also create secondary porosity (fabric selective dissolution; microporosity) or even completely alter the original fabric during dolomitisation (Eberli et al. 2003; Salih et al. 2021a).

The three most commonly used types of porosity classification are (Figure 11.10): matrix porosity versus visible porosity (Archie 1952), fabric-selective versus non-fabric-selective porosity (Choquette and Pray 1970) and interparticle porosity versus vuggy porosity (separate or touching pores; Lucia 1999). These classifications partly overlap and are complementary. They can, therefore, conveniently be combined in a subdivision of three main pore types (Figure 11.11):

1) *Inter-particle matrix pores*, which form the inter-particle pore space between grains and crystals representing the matrix interconnectivity. The diagenetic overprint can be fabric-selective or fabric-destructive. Four sub-pore types are distinguished: inter-grain, inter-crystal, mixed inter-grain and inter-crystal and microporosity.
2) *Separate vug pores*, which have a vug-matrix-vug connectivity. The diagenetic overprint is fabric selective. Five sub-pore types are distinguished: intra-particle, intra-growth framework, moldic, shelter and fenestral.
3) *Touching vug pores,* which have a vug-to-vug connectivity. The diagenetic overprint is fabric destructive. Five sub-pore types are distinguished: cavernous, breccia (large inter-particle porosity), fractures, solution enlarged fractures, and large fenestral pores.

The distinction between these three types is relevant for reservoir characterisation in order to allow a reasonable porosity–permeability relationship to be established and, for example, a correct evaluation of the potential reservoir capacity (Original Hydrocarbon In Place, OHIP; Lucia 1999).

One of the main challenges in carbonate reservoir characterisation is the definition of meaningful rock type

Petrophysical classifications of carbonate pore types

Archie (1952)	Matrix		
	Visible (A, B, C, and D)		
Choquette and Pray (1970)	Fabric selective		Non-fabric selective
Lucia (1983; 1999)	Inter-particle	Vuggy	
		Separate	Touching
		Vuggy porosity = within grains/crystals or larger (visible pore space)	
Summary	**Inter-particle matrix pores** Inter-grain Inter-crystal	**Separate vug pores** Intra-fossil Moldic Shelter Fenestral	**Touching vug pores** Cavernous Breccia Fractures (dissolution enlarged) Large fenestral

Figure 11.10 Synthesis of main pore type classes based on the integration of three commonly used petrophysical classifications (*Source:* Adapted from Lucia 1995). See also Figure 11.11.

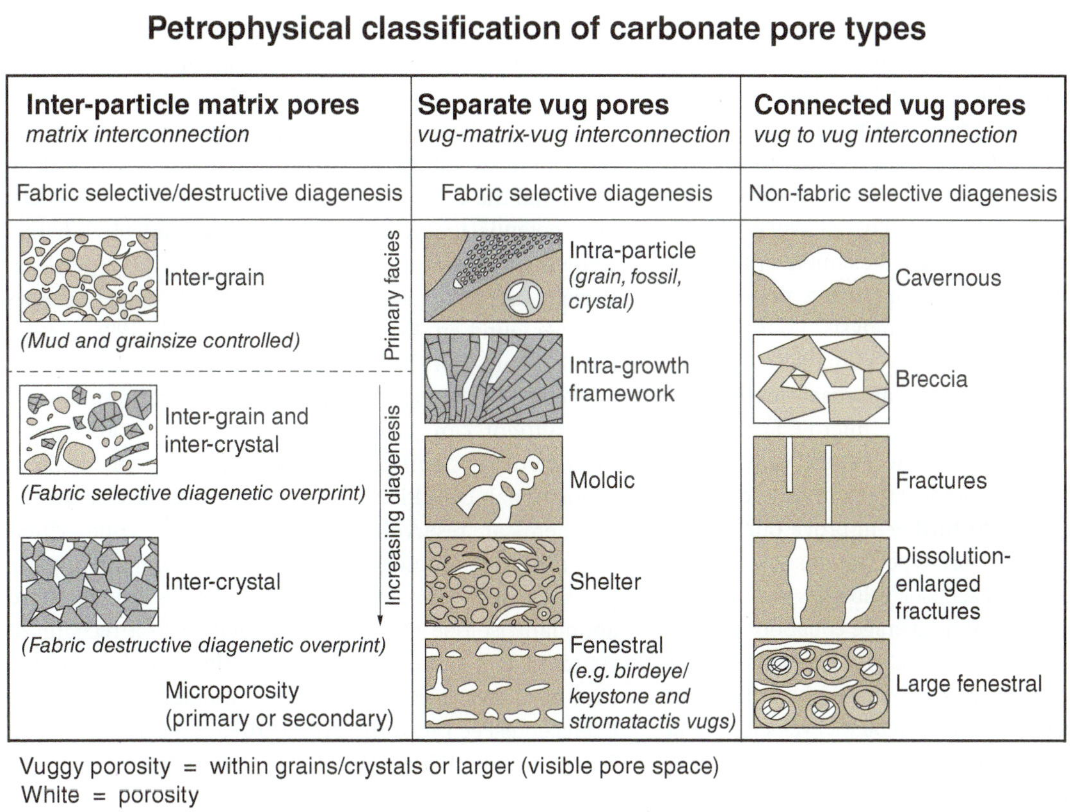

Figure 11.11 Synthesis of petrophysical classification of carbonate pore types based on the integration of Archie (1952), Choquette and Pray (1970) and Lucia (1983, 1999). See also Figure 11.10.

classes that have consistent pore type, porosity and permeability characteristics, and a predictable distribution in space to populate reservoir models. One often cited example of such a rock type classification is the work of Lucia (1999). For the class of matrix inter-particle pore systems (Figure 11.11), he established linear porosity–permeability relationships for three distinct petrophysical rock types (Figure 11.12a). These three sub-classes are essentially characterised by the pore throat sizes, which are linked to the size of the primary particles and the crystal size of the dolomitisation. To link these classes to the commonly used Dunham classification, Lucia (1999) introduced a matrix that added increasing levels of dolomite crystal size and made a further distinction between packstones with a mud-dominated matrix (P1), and packstones with a grain-dominated matrix (P2). Using the rock-type matrix of Figure 11.12b, the three petrophysical classes can now be visualised: Class 1 contains sediments with large (>100 μm) particles and crystals encountered in grainstones and in the most intensely dolomitised facies. Class 2 contains sediments with medium (20–100 μm) particles and crystals found in the P2 packstones and medium-grade dolomitised sediments. And Class 3 contains the sediments with the small (<20 μm) particles and crystals found in mudstones, wackestones and P1 packstones, and their low-grade dolomitised equivalents. An additional observation in this matrix is that the mud-dominated fabrics are sensitive to dolomitisation, which causes an increase in permeability, whereas the grain-dominated fabrics are much less so, to the extent that grainstones unaffected by diagenesis fall in the same class as fabric-destructive dolomite crystals (Figure 11.12b). Lucia's work was an important contribution and is a good example of an early systematic approach to rock typing using petrography (2D) and standard petrophysical analysis. Lønøy (2006) expanded further on this method, working with a worldwide database. A more modern approach uses high-resolution 3D digital imaging and micro-CT scans to define the pore-throat and volume system (e.g. Weger et al. 2009, 2023) and future applications of AI and digital innovations are expected to further improve our understanding of the carbonate pore-type system.

The relationship between acoustic velocities and porosity is expressed in the standard time average equations of Wyllie et al. (1956) and Raymer et al. (1980). Later studies demonstrated that porosity and, therefore, the bulk–grain density ratio, primarily controls acoustic velocities and impedance contrasts (Braaksma et al. 2003; Hossain and Zhou 2015), and that both P- and S-wave velocities have a high correlation with porosity (Anselmetti and Eberli 1993; Røgen et al. 2005; Fournier and Borgomano 2007; Fournier et al. 2011, 2018). Macro-scale porosity like fractures (of the touching vug pores class) can also significantly modify the elastic properties (Dewever et al. 2007; Xu and Paine 2009; Salih et al. 2021a).

In practice, acoustic properties vary depending on the original carbonate facies, with the grainy carbonate platform margin rocks displaying higher velocities, and the muddier slope to basin and lagoonal deposits having lower velocities (Brigaud et al. 2010; Regnet et al. 2015). The higher velocities for a given porosity are caused by the amount of inter-granular and moldic macro-pores present in the platform margin sediments. In contrast, basin deposits have a narrow acoustic velocity range, due to the non-porous mud-rich facies, for example, planktic foraminiferal mudstones and wackestones. Early/late cementation, neomorphism and mechanical compaction processes affect the initial porosity of the sediments and will reduce the porosity significantly as existing pores are partially to completely infilled by different types of cements, for example, dogtooth and syntaxial calcite cements, microsparitic crystals and blocky calcite.

11.3.4 Analytical Methods and Stable Isotopes

A wide range of tools is available to characterise the composition of carbonate rocks and their diagenetic transformations (Figure 11.13 and Table 11.2). Remarkable innovations and discoveries in isotope geochemistry have resulted in (i) the use of 'non-traditional' stable isotopes of metals (i.e. variations in isotopic compositions of Mg, Fe, Cu, etc.), (ii) a great expansion of mass-independent fractionation geochemistry, and (iii) the development of clumped isotope geochemistry (Eiler 2011; Eiler et al. 2014). Among these, new techniques combining use of the Δ_{47} geothermometer (Daëron et al. 2016) and the U–Pb chronometer (Roberts et al. 2020) have great potential to simultaneously (but independently) characterise the temperature and timing of diagenetic processes in ancient carbonate systems (Figure 11.14). Good reviews can be found in Kendall and McDonnell (1998), Swart (2015), Sharp (2017) and Stern and Wieman (2021).

The stable isotopes of carbon, oxygen, strontium, boron and sulphur can, under favourable preservation conditions, provide significant information about the primary depositional conditions such as climate, the states of geochemical reservoirs and atmospheric composition (Figure 11.15). These are expressed as differences in parts per thousand between the ratio of the less to the more abundant isotope of an element relative to a recognised standard using the conventional delta notation. In the case of C and O the standards are a belemnite from the Cretaceous Pee Dee formation (PDB) and *Vienna Standard Mean Ocean Water* (VSMOW), respectively. Dickson (1990) and Moore (2001) provide reviews for the use of the stable isotopes of C and O in carbonates, additional literature includes Banner (2004), Swart (2015) and Stern and Wieman (2021). Stable isotope geochemistry is treated further in Chapter 13.

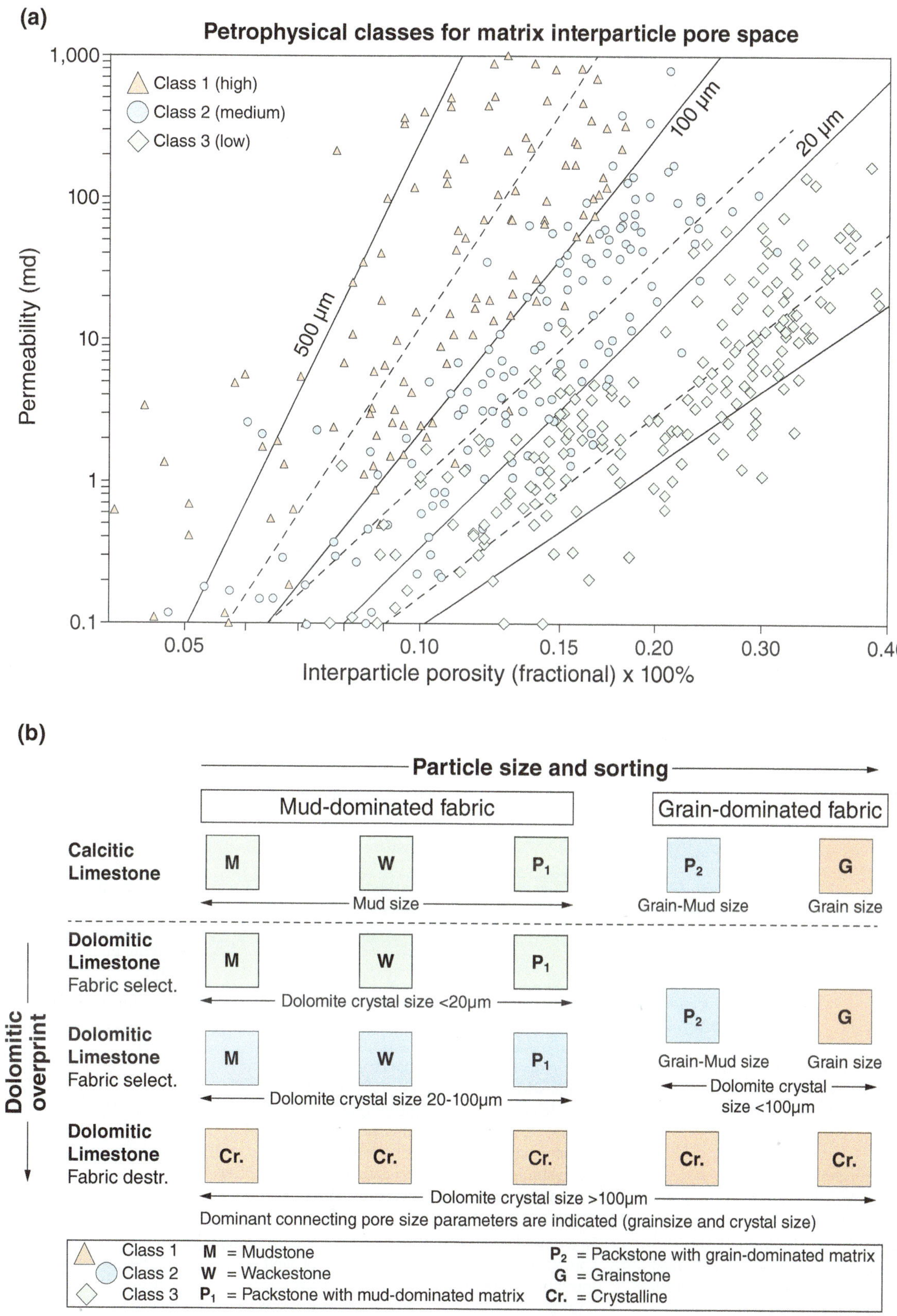

Figure 11.12 Rock type classification for the matrix inter-particle pore space system (*Source:* Lucia 1995 / GeoScienceWorld). (a) Permeability/porosity plot showing the distribution of three rock type classes based on the particle grain sizes. (b) Same three rock type classes (colour coding) and their relationship to Dunham facies classification, with the addition of a P1 and P2 packstone, and impact of dolomitisation. See text for further explanation.

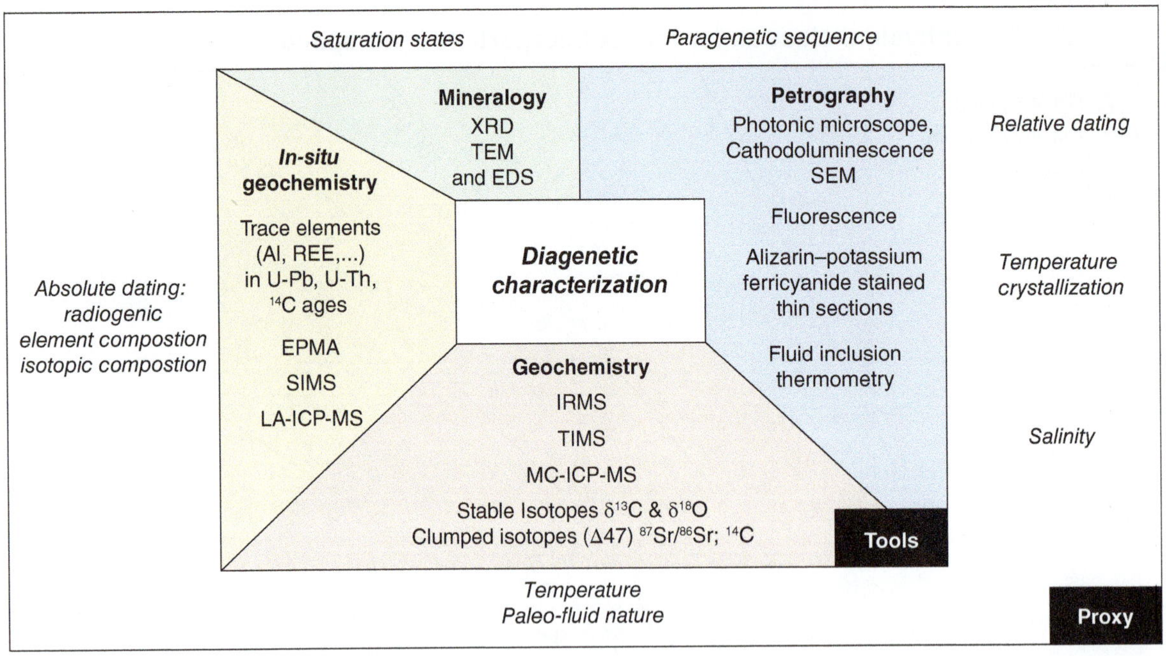

Figure 11.13 Schematic overview of geochemical, mineralogical and petrographical analytical tools available for carbonate diagenesis studies (*Source:* Adapted from Al-Aasm and Mansurberg 2021; Flügel 2004). See also accompanying Table 11.2.

Table 11.2 Summary of geochemical, mineralogical and petrographical analytical methods for carbonate analysis shown in Figure 11.14.

Tools for diagenetic characterisation	Observation	Characterisation	Results
Thin section petrography	Crystal morphology	Relative timing of diagenetic events through superposition or cross-cutting relationships; crystal shapes and distribution, petrophysics characteristics	Paragenetic sequence, relative dating, identification of diagenetic phases and domains
Alizarin–potassium ferricyanide stained thin sections	Crystal growth (Fe calcite, dolomite)	Rapid determination of carbonate (calcite, dolomite and siderite) cement stratigraphy, additional to cathodoluminesce microscopy	Mineralogy, paragenetic sequence, relative dating and reconstitution of the diagenetic phases and domains
Fluorescence and cathodoluminescence imaging	Chemistry, crystal structure, lattice defect Mn, Pb, Fe and other Earth elements	*Fluorescence*: identification of organic matter *Epifluorescence*: recognising hidden microfabrics, organic components and biomineralisation *Cathodoluminescence*: reconstructing diagenetic events which identify stages of cementation and provide cement-based stratigraphic correlations, change in the chemical environments, mineral transformations, recognising fabrics (recrystallisation and microcracks) and evaluating pore water chemistry reflected by sector zoning in crystals	Reconstitution of the diagenetic phases and domains
Fluid inclusion thermometry	Inclusion size: 10–30 μm: chemical composition, pressure and temperature	In a closed system fluid, inclusions have the same composition and volume at the time of their formation (engulfment with growing crystal or entrapment microfractures); provide information on the diagenetic environment of cements (vadose zone, low-temperature phreatic zone and high-temperature burial zone)	Thermometer (temperatures of the parent fluids) and reconstitution of the diagenetic domains

Table 11.2 (Continued)

Tools for diagenetic characterisation	Observation	Characterisation	Results
Scanning electron microscopy (SEM), transmission electron microscopy (TEM) and energy dispersive X-ray spectrometer (EDS)	Chemistry, crystal structure	*SEM*: mineralogy, morphology and spatial relations of carbonate cements, distribution and geometry of micropores *TEM*: microbial contribution and biomineralisation *EDS*: chemical compositions of carbonates expressed in oxide weight percentages	Reconstitution of the diagenetic phases and domains, identification of trace elements
X-ray microtomography (XRMT)	Pore structures	Characterisation of the hierarchy and variability of pore structure to complement the macroscopic properties of carbonates (porosity and permeability)	Porosity and permeability
X-ray diffractometry (XRD)	Mineralogy	Information on the chemical composition and proportion of different carbonate minerals, determine the ordering of dolomite crystals	Reconstitution of mineral and chemical crystal properties, nature of paleo-fluid (Sr isotopes, trace elements)
Electron probe micro analyser (EPMA)	*In situ* measurements of elemental composition	Major and minor element composition of carbonates (Ca, Fe, Mg, Mn, Al and Sr), measured on carbon-coated thin sections	Reconstitution of elemental composition
Spectrometry	Trace and Earth elements	Atomic adsorption spectro-photometer: identify minor amounts of elements in the carbonate lattice Inductively coupled plasma atomic emission spectrometry analysis (ICP-AES), neutron activation analysis (NAA), multicollector ICP-MS and thermal ionisation mass spectrometry (TIMS): contents of trace and rare earth elements in carbonate rocks Comparison of physical or chemical properties encoded in a mineral (e.g. formation temperature and/or $^{87}Sr/^{86}Sr$ composition) with regional or secular variations of these same parameters in reference curves (e.g. Sr isotopes)	Determination of original mineralogy of rock components; chemical gradients indicative of paleofluid flow direction; estimation of rock–water ratios during diagenesis as critical input for reactive transport modelling of diagenesis
Isotope ratio mass spectrometry	Isotope abundance for geochemical research	Stable isotope analyses of ancient carbonates provide proxies for paleoenvironmental constraints (e.g. temperature and salinity variations of sea water, terrigenous input), unravel changes in carbon cycling of ancient ocean-atmosphere systems and document global climate changes	Chemical composition of fluid inducing carbonate precipitation, thermometer ($^{18}O/^{16}O$ isotope ratio, ^{14}C), isotope curves supporting stratigraphic correlations
Clumped isotopes analyses Δ47	Temperature control on group ^{16}O, ^{17}O and ^{18}O isotopes; independent of isotopic fluid composition	Characterisation of formation temperature of diagenetic calcite and dolomite cements (in the range 60–100°C) and $\delta^{18}O$ water composition of diagenetic fluids that precipitated these carbonates	Paleoclimatology, paleoaltimetry, diagenetic conditions and water–rock interaction
Secondary ion mass spectrometry (SIMS)	*In situ* measurements of stable C and O isotopes (directly on polished thin sections)	Stable isotope analyses of ancient carbonates provide proxies for paleoenvironmental constraints (e.g. temperature and salinity variations of the sea water, terrigenous input), unravel changes in carbon cycling of ancient ocean-atmosphere systems and document global fluctuations in the climate	
Inductively coupled plasma mass spectrometry (ICP-MS) coupled to a laser ablation system	*In situ* (directly on thin section) measurements of element compositions (majors, minors and traces); Radiometric dating (U-Pb)	Geochemical analyses of carbonate: major elements (Ca, Mg, Fe and Mn), minor elements (Al, Sr, Y, REE, U, Pb and Th), Pb and U isotopes (^{206}Pb, ^{207}Pb and ^{238}U), *in situ* dating using isochron in a Tera-Wasserburg diagram	REE spectra, absolute age dating

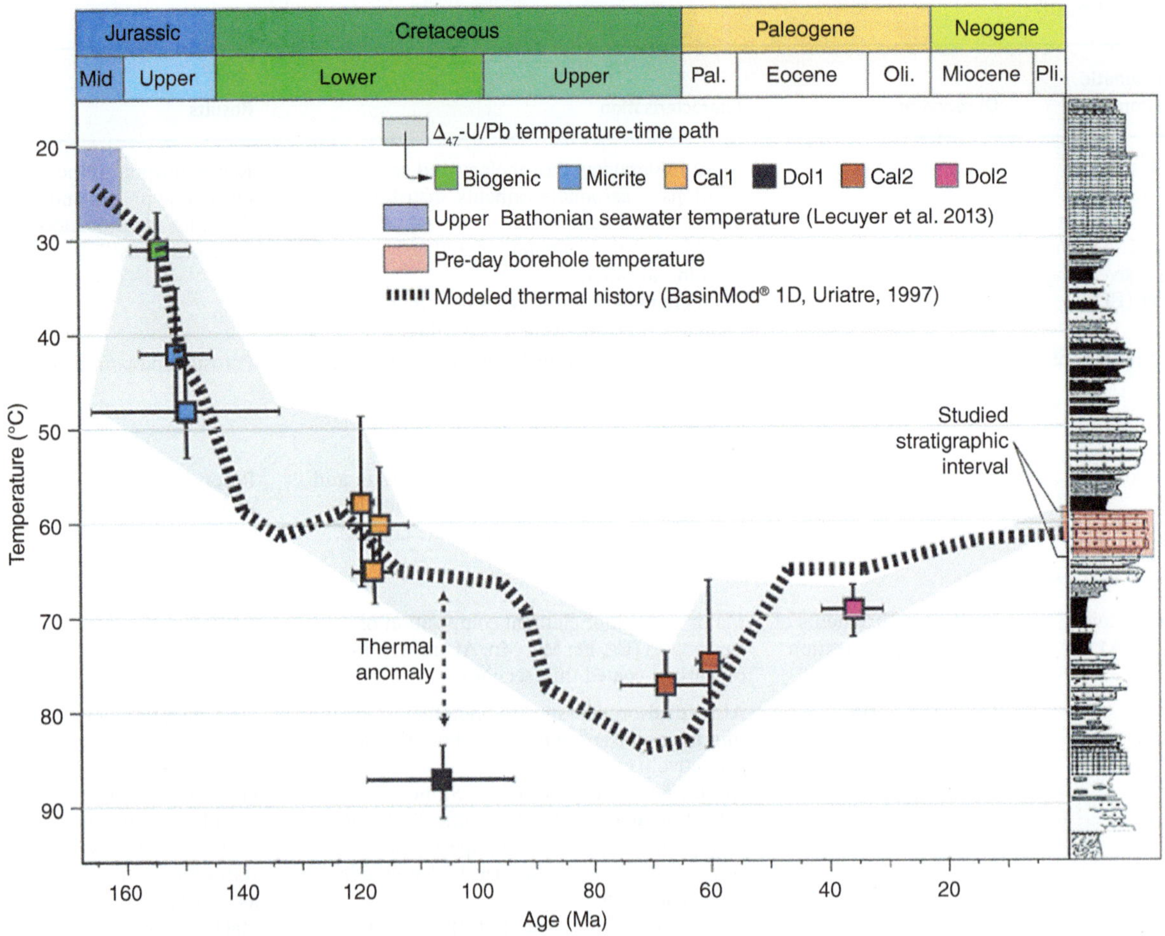

Figure 11.14 Example of thermo-chronometer for carbonate bearing rocks, showing the history of Middle Jurassic succession in the Paris Basin depocenter. The temperature–time data is deduced from Δ^{47} and U–Pb analyses (Mangenot et al. 2018). Cal 1, Cal2, Dol1 and Dol2 are analysed carbonate phases. Black dashed line represents thermal history modelled by Uriarte (1997) using conventional rock-based approaches (*Source:* Adapted from BasinMod 1D; https://www.platte.com/software/basinmod-2012.html). Pal.: Paleocene; Oli.: Oligocene; Pli.: Pliocene.

Oxygen isotopes, when preserved in a closed chemical system, can provide information about water temperatures, and as such climatic conditions and their evolution through time (Grossman and Joachimski 2020; Scotese et al. 2021; Figure 11.15a). Most oxygen in carbonate minerals derives from the fluid from which the crystal precipitated and undergoes temperature-dependent isotopic fractionation between liquid and mineral phase. For instance, increasingly more negative values are associated with decreasing salinity and increasing temperature (Hudson 1977). Oxygen isotope ratios of planktonic foraminifera provide the single best sea-level proxy for the Late Pliocene through Holocene carbonates and were instrumental in establishing the orbital forcing theory of the ice ages and associated rapid climate changes (Section 13.6.2; Emiliani 1955; Pearson 2012). More recently, carbonate clumped isotope thermometry has been applied to reconstruct mineral growth temperature in surface and subsurface environments independent of parent water isotopic composition (Eiler 2011; Mangenot et al. 2018). This method is particularly used for palaeoclimate reconstructions (Finnegan et al. 2011; Henkes et al. 2013). Oxygen isotopes can also be indicators of subaerial exposure, and early diagenetic overprint, thus providing important stratigraphic information. The oxygen stable isotope evolution is also used in dolomite undergoing burial and alteration to reconstruct the successive temperatures of precipitation and the isotopic compositions of the dolomitising fluids (see review in Warren 2000).

Carbon stable isotopes have multiple applications. The ratio $\delta^{12}C/\delta^{13}C$ in a marine carbonate can reflect the composition of past ocean water, which was then in equilibrium with the atmosphere. In this case, variations are caused by the relative accumulation of carbon in either the mineral carbon sink (carbonates) or the organic carbon sink (organic-matter-rich deposits) and are interpreted in terms of changes in primary production, burial and preservation of both organic matter and carbonates (Cramer and Jarvis 2020). The isotopic ratio $\delta^{13}C/\delta^{12}C$ is used as a proxy for the concentration of atmospheric CO_2 through geological time (Veizer and Prokoph 2015; Veizer et al. 2000;

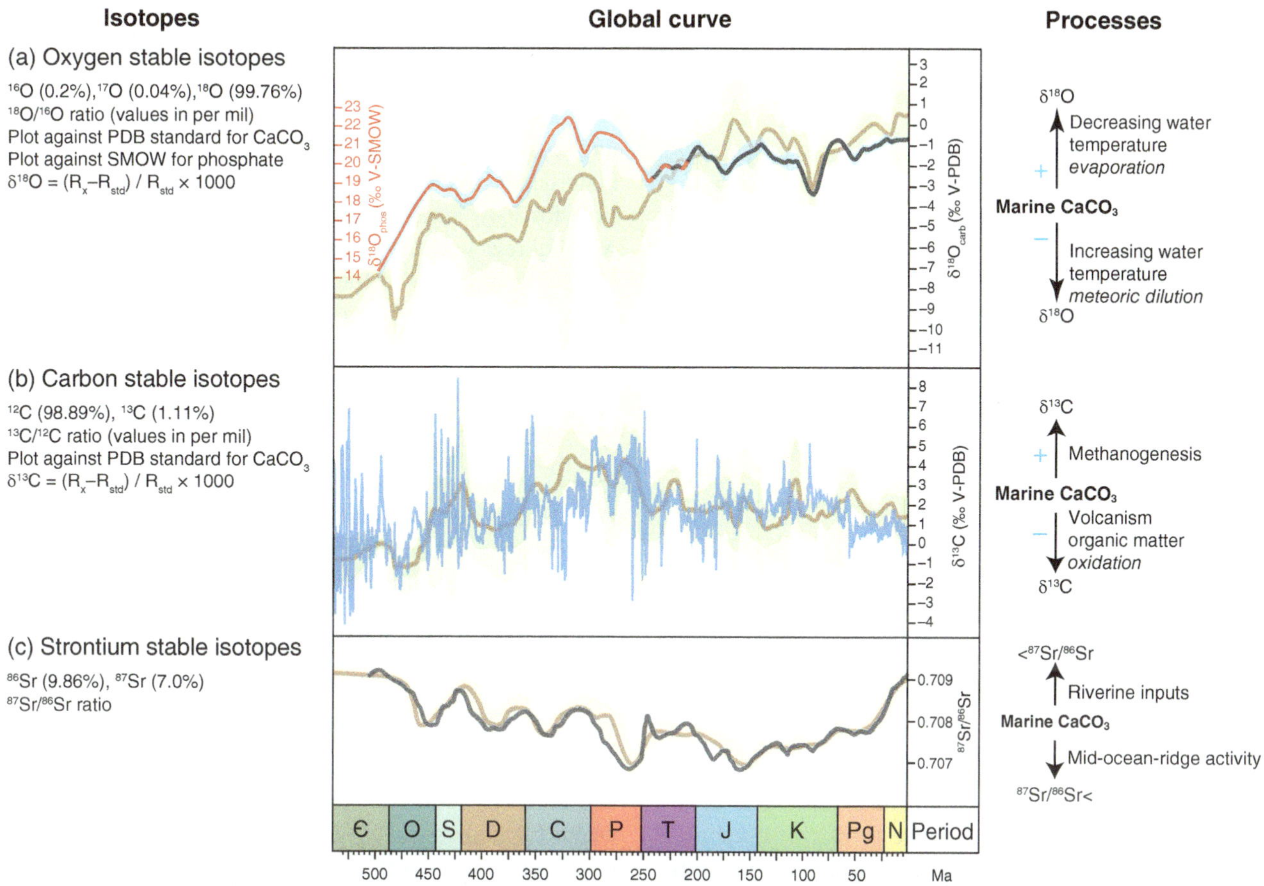

Figure 11.15 Phanerozoic oxygen, carbon and strontium isotope records showing variations at short- and long-time scales. A comparison is made between Veizer et al. (1999), brown line, and the latest dataset proposed in the literature, black-red line. The shaded area represents confidence intervals: 95%, yellow, and 68% green, for Veizer et al. (1999); and 95%, blue, for the latest datasets. (a) Smoothed $\delta^{18}O$ curves from phosphate (red) and carbonate (black) fossils modified after Song et al. (2019), and exclusively carbonate fossils for Veizer et al. (1999). $\delta^{18}O_{Phos}$ include phosphatic brachiopod, conodont and fish. $\delta^{18}O_{Carb}$ mostly include belemnite, bivalve, brachiopod and planktonic foraminifer. (b) Raw $\delta^{13}C$ curve from bulk and fossil carbonates (Cramer and Jarvis 2020), and smoothed $d^{13}C$ curve carbonate fossils from Veizer et al. (1999). Analysed carbonate fossils include belemnite, bivalve, brachiopod and planktonic foraminifer. (c) Smoothed $^{87}Sr/^{86}Sr$ ratios from carbonate fossils from Veizer et al. (1999) including brachiopod, belemnite and conodont fossils.

Figure 11.15b). With caution, the $\delta^{13}C$ values of bulk rocks can also be used as a stratigraphic correlation tool in pelagic deposits, and ancient shallow-water carbonate settings of ramps and carbonate platforms (e.g. Weissert 1989; Vahrenkamp 1996; Mutti et al. 1997; Grötsch et al. 1998; Wissler et al. 2003; Cozzi et al. 2004a, b; Swart and Eberli 2005; Swart 2008; Smith and Swart 2022).

The $\delta^{13}C$ values of precipitated carbonates primarily reflect the source(s) of bicarbonate dissolved in the waters, which can come from sea water ($\delta^{13}C$ near 0‰), the earth's mantle (−4‰) marine shell dissolution ($\delta^{13}C$ near 0‰), soil weathering processes ($\delta^{13}C$ near -10‰) bacterial oxidation or sulphate-reduction of organic matter ($\delta^{13}C$ near −25‰), bacterial methanogenic fermentation ($\delta\ ^{13}C$ near + 15‰), oxidation of methane ($\delta^{13}C$ from −50‰ to −80‰) or abiotic reactions associated with thermal cracking and decarboxylation ($\delta^{13}C$ from −10‰ to −25‰) (Hudson 1977; Irwin et al. 1977; Coleman 1993). The stable isotope composition is also strongly affected by biological processes, the 'vital effects' (Meibom et al. 2004; Colombié et al. 2011).

An effective way to evaluate the diagenetic overprint on carbonates is to use cross-plots of $\delta^{18}O$ and $\delta^{13}C$ values (Figure 11.16; see also Figure 4.2 in Tucker and Jones 2023). The isotopic values of present-day marine waters can vary several per mil depending on temperature, evaporation and dilution from melting ice (Anderson and Arthur 1983). The $\delta^{18}O$ value of meteoric waters becomes more negative with increasing distance from the ocean, altitude and latitude (Dickson 1990). Burial fluids are isotopically more positive in $\delta^{18}O$ values compared to shallower, earlier, marine and meteoric cements and the change in $\delta^{18}O$ is primarily a function of increasing water temperature (Swart 2015). In addition, the calcareous organisms show important deviations from stable isotope equilibrium (e.g. Milliman 1974a; Morrison and Brand 1986; Wefer and Berger 1991; Moore 2001; Swart et al. 2009; James and Jones 2016).

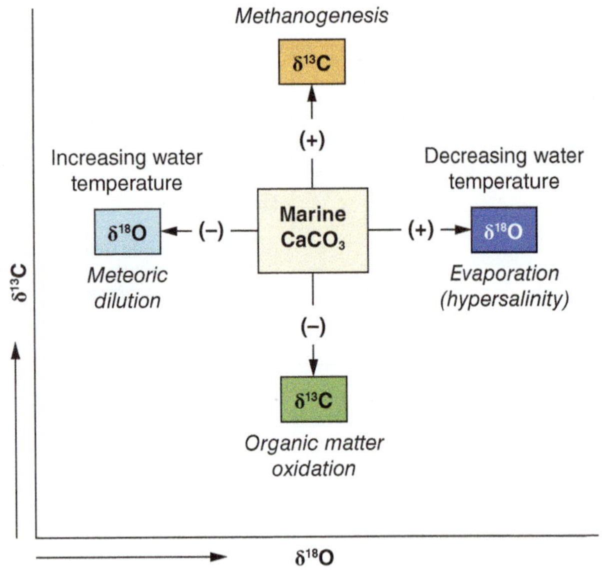

Figure 11.16 Carbon and oxygen stable isotopes plot and typical diagenetic modification pathways (after James and Jones 2016). See also further annotated version of this figure in Tucker and Jones (2023).

To establish $\delta^{34}S$ values of seawater through time, $\delta^{34}S$ in carbonate (Carbon Associated Sulphate, CAS; Gill et al. 2007) can be used as an alternative to marine evaporites (e.g. gypsum), which have limited geographic and stratigraphic distribution (Warren 2010; Rennie et al. 2018; Rose et al. 2019). This method is also used to constrain the source of diagenetic alteration (Gill et al. 2008; Smith et al. 2022). The palaeoenvironmental utility of $\delta^{34}S_{CAS}$ is encouragingly robust following several studies (Gill et al. 2008). However, a number of processes including bacterial/microbial sulphate reduction (BSR), pore-fluid variation during meteoric diagenesis and in marine sediments, changes in concentration of different carbonate polymorphs, can result in the fractionation of the $\delta^{34}S$ of sulphate inducing changes in the $\delta^{34}S_{(CAS)}$ value for diagenetic carbonate (see review in Smith et al. 2022).

Strontium (Sr) stable isotopes ($^{87}Sr/^{86}Sr$) are used to (i) trace the diagenetic evolution of sediments, such as to reveal mineral precursors and processes involved in the formation of dolomite (e.g. Saller and Koepnick 1990; Vahrenkamp et al. 1991); (ii) date carbonate rock successions using the empirical secular variation of Figure 11.16c (Smalley et al. 1994; Meredith et al. 2013; McArthur et al. 2020); and (iii) trace the continental weathering rate (Raymo et al. 1988).

Other useful elements found in calcite include Li, B, Na, Mg, Al, Mn, Fe, Zn, Sr, Cd, Ba and U, which provide, individually or in association, proxies for palaeoclimatic interpretations. For instance, the boron (B) concentration of B/Ca and its isotopes ($\delta^{11}B$) are much used proxies (Foster et al. 2013, 2016; Wu et al. 2017). The $\delta^{11}B$ value of unaltered carbonate provides information on both recent and ancient oceanic pH (Hönisch et al. 2008). B/Ca of benthic foraminifera helps in reconstructing the carbonate ion history (Yu et al. 2014) and past changes in oceanic circulation and surface productivity patterns. The use of ICP–MS methods allows to obtain a signal for small amounts of material and to discuss changes in ocean alkalinity responsible for P_{CO2} atmospheric variations (Misra et al. 2014). A useful review of the importance of trace elements during diagenesis and how to trace their association with carbonates is provided by Swart (2015).

11.4 Biological and Physical Aspects

In this section, we present (i) a selection of fundamental biological concepts which are important to understand the ecology and behaviour of carbonate-producing organisms and (ii) the main physical environmental parameters and oceanographic-atmospheric processes that control the rich variety of environments where carbonates precipitate.

11.4.1 Ecology

Ecology is the study of the relationship of organisms and their environment. A selection of ecological concepts useful for understanding biogenic carbonate precipitation in the present, as well as for ancient carbonate deposits (Schlager 2005), is presented below.

11.4.1.1 The Law of Sigmoidal Growth of Populations

The growth of a population of organisms generally follows a sigmoidal curve consisting of a slow starting phase, a phase of rapid growth and a final phase of slow down, where the population approaches a steady state in equilibrium with the carrying capacity of its habitat (Figure 11.17a). This curve can be described in a 'logistic equation', a mathematical expression that produces sigmoidal curves (Figure 11.17b). It describes the interaction between the intrinsic growth rate of a population and the limits set by the size of the living environment and other resources. At an early stage of growth, living space and resources are virtually unlimited, and lead to exponential growth (bold line in Figure 11.17b). Subsequently, the limitations on growth imposed by the finite size of the living space can be added to this equation by introducing the term N/K, the ratio of population size, N, and carrying capacity of the living space, K. N/K will eventually approach 1, that is, a constant population size when carrying capacity has been reached (the dotted line in Figure 11.17b). This principle of sigmoidal growth applies to carbonate production, whereby a distinction can be made between the adaptation and evolution of the individual species and the biotic assemblage as a whole, in their response to environment change (bathymetry, salinity, substratum, nutrient supply, light, etc.).

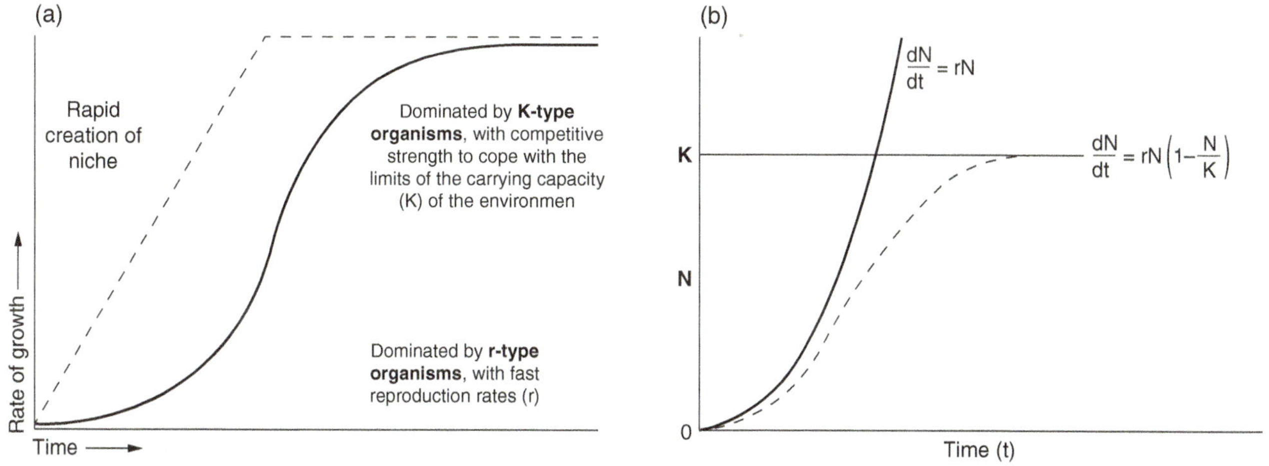

Figure 11.17 Growth curves of populations (*Source:* Adapted from Schlager 2005 and Townsend et al. 2003): (a) Sigmoidal growth curve – illustrating the response of a population of organisms to the opening up of new living space (niche – dotted line) in three steps: first, growth lags behind the creation of living space; second, population growth exceeds the rate of growth of living space; finally, population growth is limited by the growth in living space. (b) Mathematical representation of the sigmoidal growth curve, distinguishing between an exponential growth curve (bold line) and the sigmoidal growth curve governed by the logistic equation (dashed line). In natural systems, an initial phase of exponential growth usually gives way to decelerating growth that goes to zero as population size approaches the limits of space and other resources. N: number of individuals in the population, t: time, r: intrinsic reproduction rate of an individual and K: carrying capacity of the environment.

11.4.1.2 Life Strategies of Organisms

Inherent to this growth law are the life strategies of organisms. Since energy and other resources available to organisms are limited, they need to respond and either grow more and reproduce less or vice versa (Townsend et al. 2003). Two extreme life strategies to cope with this have been defined (MacArthur and Wilson 1967; Losos and Ricklefs 2009). One is a group of 'r-strategists', who spend most of their life as a population in the r-dominated (r = reproduction rate), nearly exponential, lower part of the population curve (Figure 11.17a). They are organisms that quickly appear in newly opened habitats, multiply rapidly, are short-lived and relatively small. Their abundance in carbonate settings can be such that they are rock-forming (e.g. large benthic foraminifera like nummulitids or orbitolinids; but also microbialites). The other group is called 'K-strategists', as most of their life is spent in the K-dominated (K = carrying capacity of the environment) upper part of the curve (Figure 11.17a). These are organisms that live in environments of intense competition for limited resources, where diversity and specialisation is a survival strategy, such as the extreme niche specialisation in barrier reefs. These adaptation strategies of individual species should be taken into consideration when interpreting the biotic evolution during the course of a depositional sequence (e.g. the Aptian Trilogy; Pittet et al. 2002).

11.4.1.3 Food Chains and Nutrient Levels

Most biotic interactions in a given ecosystem relate to the food web, which connects plant with grazer, predator with prey and so on (Townsend et al. 2003). The food chain normally starts with primary production, which in photic areas in most cases is photosynthesis. Alternatively, the food chain may start with chemosynthesis, bacterially mediated reactions such as oxidising methane, hydrogen sulphide or other substances. Biotic shallow-marine carbonate production nearly always depends on photosynthesis as a starting point (Schlager 2005). The organisms at the base of the food chain are called autotrophs (self-feeders), and organisms further up the food chain, which eat other organisms, are called heterotrophs (other-feeders). Carbonate-producing organisms belong to both types.

The rate of primary, photosynthetic production in the marine environment depends on the light intensity and the concentration of dissolved nutrients, such as phosphorus and nitrogen. Primary productivity in the world oceans shows a significant latitudinal variation, which is to a large extent controlled by ocean current patterns (Chapter 13, Figure 13.2). Shallow-marine carbonate systems have a tolerance window for nutrients, which can be too plentiful in zones with upwelling from the deeper ocean or areas with abundant sediment run-off from land masses (Mutti and Hallock 2003).

11.4.2 Environmental Controls on Marine Carbonate Precipitation

The main environmental parameters controlling biotically induced precipitation of carbonate include light, temperature, nutrients, salinity, substrate, carbonate saturation in the water column, and oceanographic and atmospheric processes such as currents, tides and winds.

11.4.2.1 Light

Light provides the essential energy for photosynthesis by autotrophic organisms at the base of the food chain. The maximum depth of the euphotic zone (that part of the sea water column in which sufficient light penetrates to enable the prolific growth of photosynthetic lifeforms) varies greatly. It is as shallow as 20 m for the Arabian/Persian Gulf, Singapore and W. Caribbean, 40 m for the Great Barrier Reef and the Yucatan shelf and reaches maximum values of almost 150 m for Pacific atolls (Figure 11.18). Turbidity is the main factor other than latitude that controls the depth of light penetration. Turbidity can be caused by mud in suspension, notably close to river mouths (Chapter 9, Figure 9.11; Morgan et al. 2016), where it varies due to seasonal and climatic factors (e.g. Piller and Pervesler 1989; Piller 1994). Turbidity can also be due to phytoplankton blooms.

Organisms have different strategies to deal with light reduction at depth. For example, corals adapt their growth towards a larger surface with depth to compensate for light diminution. Branching corals dominate the uppermost levels of the euphotic zone, head corals the intermediate levels and platy corals the deepest levels (Reiss and Hottinger 1984; Hallock and Schlager 1986; Hallock 2005; Einbinder et al. 2009). The growth rates of photosynthetic, carbonate-secreting benthos like large benthic foraminifera (Hallock 1985; Hohenegger 2005) and calcareous green algae like *Halimeda*, *Penicillus* and *Udotea* (Bosence 1989a,b) are also affected by water depth. In turbid waters, the decrease in light penetration combined with inherited seafloor morphology may set the threshold between a high and low coral cover at water depths as shallow as 2 m (Mozambique: Perry 2003; Great Barrier Reef: Morgan et al. 2016; Singapore: Morgan et al. 2020a,b).

11.4.2.2 Water Temperature

After light, water temperature is the determining factor for the type of calcifying biota that dominates any given environment. The critical temperature boundary is around 20°C, above which tropical species will thrive, and below which cooler water faunal assemblages take over (Lees and Buller 1972; Nelson 1988; Hayton et al. 1995). Each carbonate-secreting species has its own temperature window for optimum growth (Lees 1975). So, when moving from the equator to higher latitudes, the biotic associations and the type of bio-constructions change (Figure 11.19; Grigg 1982; Schlager 2005; Schlanger 1981). This can be an abrupt change between the end-members of the tropical

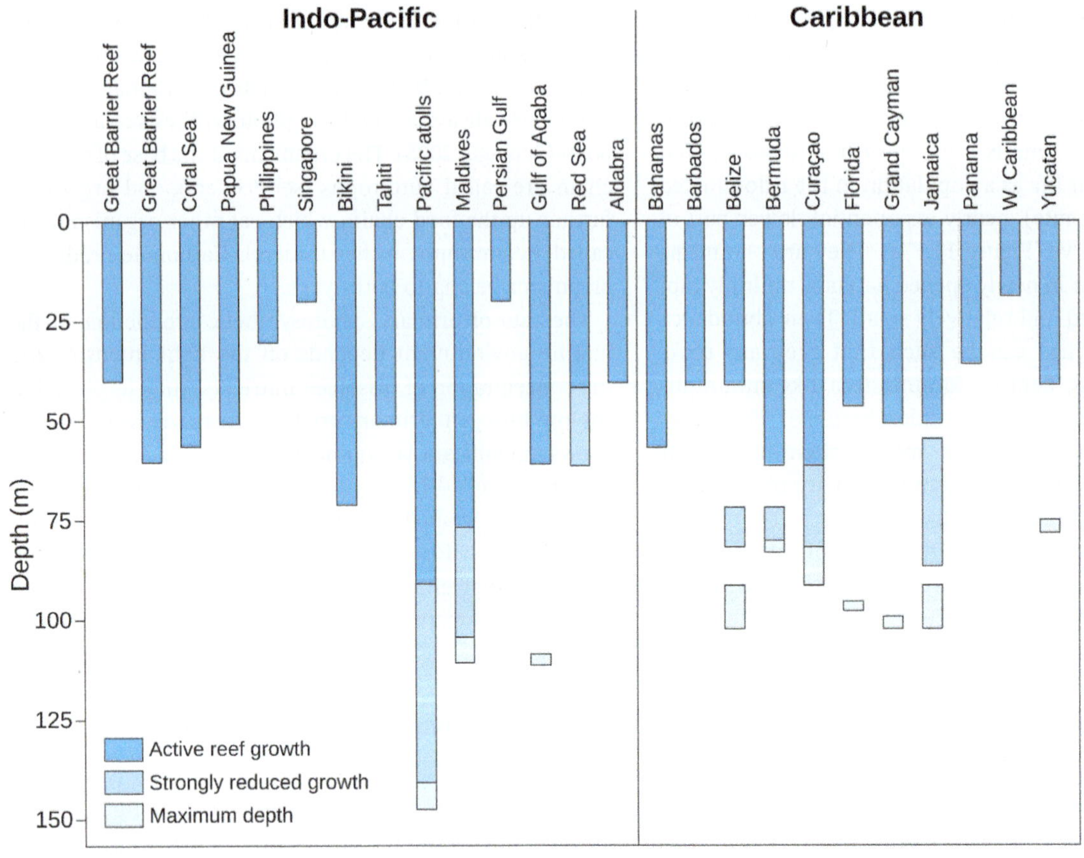

Figure 11.18 Comparison of depth of active reef growth in the Indo-Pacific and the Caribbean showing an optimum euphotic window of 0–40 m. Locally conditions may allow for a deeper or shallower penetration of light (*Source:* Schlager 2005 / Society for Sedimentary Geology).

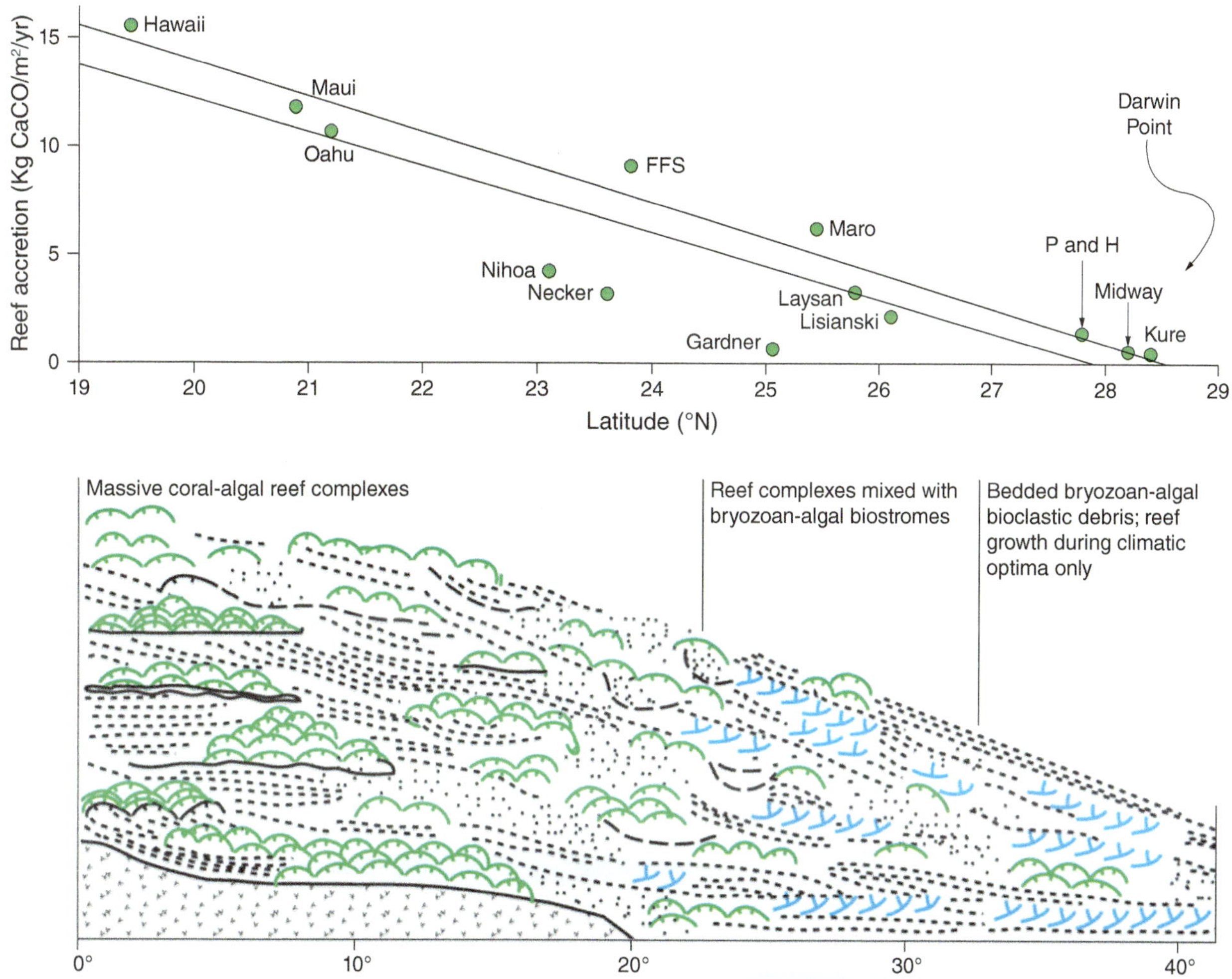

Figure 11.19 Marine carbonate reef growth as a function of latitude in the North Pacific (*Source:* Adapted from Compiled in Schlager 2005, based on Schlanger 1981 and Grigg 1982). Upper panel: decrease of the rate of reef growth towards the northern hemisphere. Darwin Point marks the northern limit of reef growth. Lower panel: latitudinal change from tropical to cool-water carbonate facies observed on the Hawaii-Emperor chain of islands and seamounts. Black dots and dashes: carbonate debris; green: coral reefs; blue: bryozoan biostromes.

and cool-water biological assemblages as on the Brazilian Shelf, in the Mediterranean, and along the Gulf of California (e.g. Betzler et al. 1997a; Carannante et al. 1988; Halfar et al. 2000, 2004, 2006). The latitudinal distribution of benthic assemblages is also influenced by ocean currents such as the Gulf Stream, introducing warmer waters at higher latitudes, and upwelling zones, such as along the west coast of Africa, which introduce colder waters to the surface in lower latitudes. Present-day global warming has also affected the ocean waters, and caused local coral mortality, a decrease in coral diversity and enhanced bioerosion (e.g. Glynn 2000; Evensen et al. 2022).

11.4.2.3 Nutrients

The third important factor is nutrient supply, which favours some carbonate-producing biota and inhibits others. For instance, high nutrient levels favour the development of filamentous algae, barnacles and certain taxa of bryozoans, and at the same time reduce calcification rates and cause the death of coral colonies (e.g. Halfar et al. 2004, 2006; Reijmer et al. 2012a). Four main categories of nutrient availability in surface waters are distinguished (Figure 11.20; Hallock 2001; Mutti and Hallock 2003).

Oligotrophic conditions (chlorophyll concentrations <0.1 mg m^{-3}): An environment of scarce nutrients where nutrient limitation is the primary control on biota development. Shallow benthic carbonate-producers like zooxanthellate corals thrive under these conditions in tropical/subtropical oceanic waters.

Mesotrophic conditions (chlorophyll concentrations of about 0.1–0.5 mg m^{-3}): An environment where some nutrients are available and light penetration is sufficient to support calcareous algal production besides coral growth (although the latter show lower growth rates than in the oligotrophic setting). Ample dissolved and particulate nutrients allows the development of a heterotrophic benthic community that includes algae, sponges, molluscs and echinoderms. These conditions are typical of upwelling

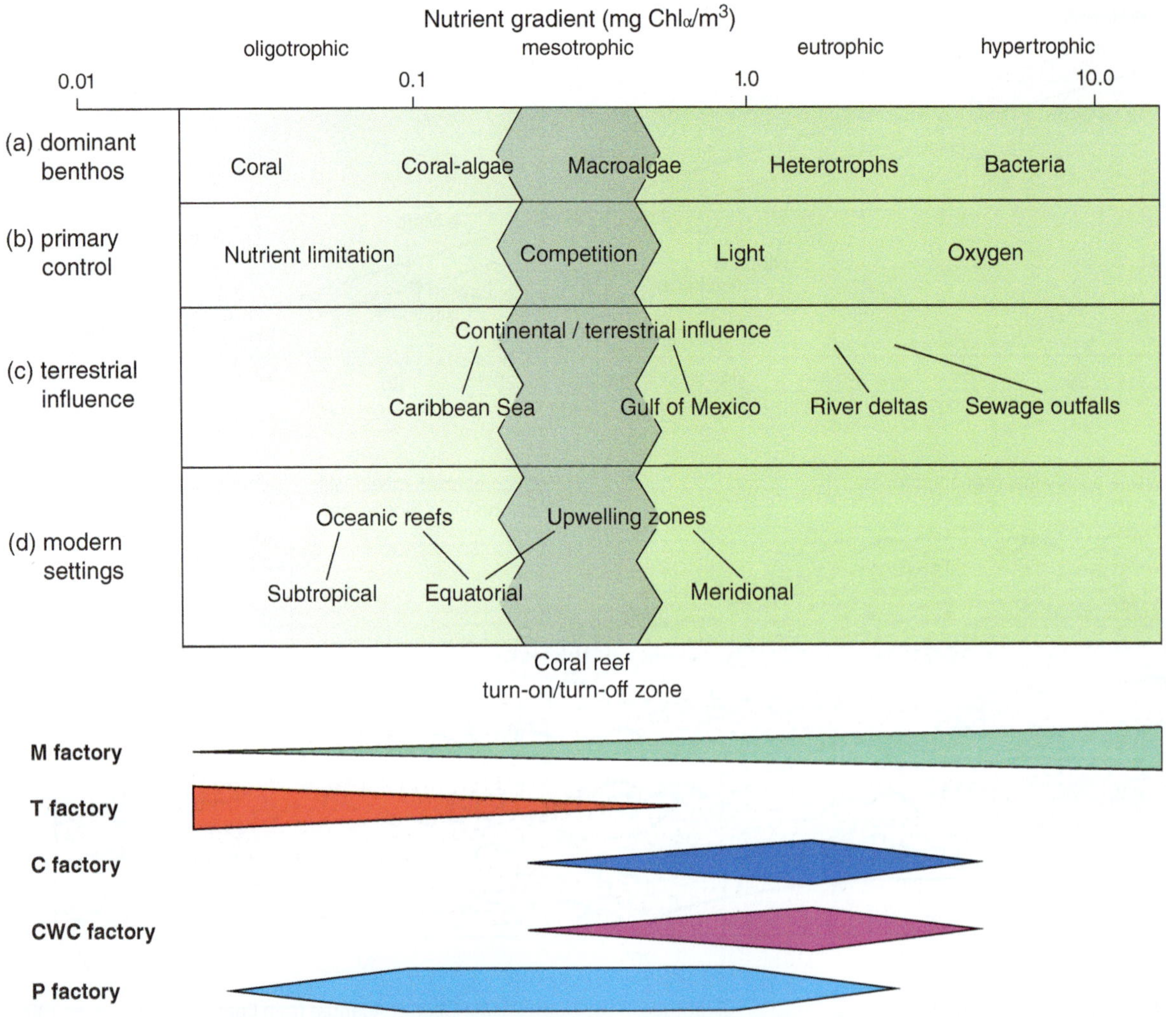

Figure 11.20 Nutrient concentration in surface waters of the ocean and its effect on marine biota (*Source:* Adapted from after Schlager 2005, modified after Hallock 2001). Nutrient levels are expressed in milligrams of chlorophyll per cubic meter of seawater on the horizontal axis. On the vertical axis, the impact of the nutrient gradient is shown on the dominant benthos and the primary limiting control on carbonate sedimentation. In addition, its relationship to terrestrial influx, the modern-day oceanographic settings and the carbonate factories are illustrated.

zones for instance. Inter-species competition regulates the diversity and abundance of biota.

Eutrophic conditions (chlorophyll concentrations $>1\ \mathrm{mg\ m^{-3}}$): An environment where abundant nutrients are available, and light is the limiting factor for benthic fauna development. Heterotrophic organisms, such as filter- and detritus-feeding benthos, including bivalves, sponges, ascidians (tunicates) and echinoderms thrive in these conditions.

Hypertrophic conditions: In these environments, oxygen supply to the benthic community is the restraining factor. The nutrient flux to surface waters may drive the development of dense phytoplankton populations, which may result in the development of episodic hypoxia or anoxia in the benthic community at the sediment–water interface, which is dominated by bacteria.

11.4.2.4 Salinity

The salinity of the sea water in the open oceans varies little from a mean of 35‰. Significant increases in salinity occur in restricted or isolated seaways and intra-shelf basins, or on large, flat and shallow carbonate shelf systems, with restricted water circulation and/or high evaporation rates. A study of biodiversity within environments with salinities ranging from brackish to hypersaline showed that only when salinities exceeded 40‰ did a sharp decrease in number of taxa occur (Figure 11.21; Heckel 1972). Coral reef communities do well in ocean environments with salinities ranging between 25‰ and 40‰ and can cope with both long- and short-term salinity variations (Muthiga and Szmant 1987; Coles and Jokiel 1992). However, values below 20‰ may result in rapid and severe decline in coral growth (Smith and Buddemeier 1992). In one ancient, restricted saline

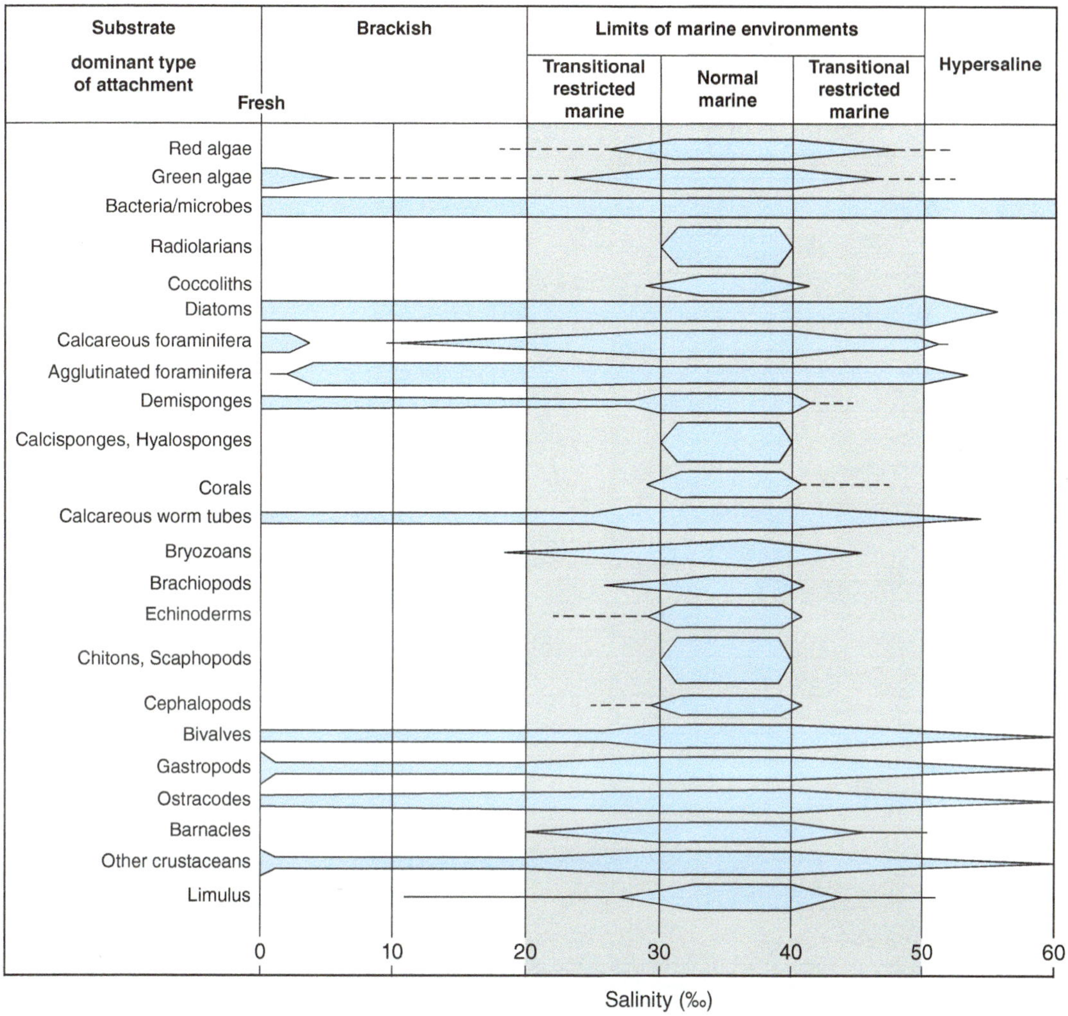

Figure 11.21 Salinity impact on carbonate producing marine biota (*Source:* Heckel 1972 / Society for Sedimentary Geology).

environment, the Miocene in SE Spain, only *Porites* sp. survived (e.g. Riding et al. 1991a, b; Braga and Martin 1996) or is replaced by microbialites (Bourillot et al. 2010a). Microbial communities may tolerate higher salinities and can populate large terminal lakes like the Great Salt Lake (Stephens 1990; Harris et al. 2013; Bouton et al. 2016; Vennin et al. 2019; Boussagol et al. 2023), hypersaline lakes (e.g. Eleuthera Island, Bahamas; Dupraz et al. 2004, 2009); and sabkha environments (e.g. Kenig et al. 1990; Duane and Al-Zamel 1999; Bontognali et al. 2010; Lokier et al. 2013).

11.4.2.5 Substrate

The substrate, the surface on which organisms, or often their larvae, settle, attracts particular biotic assemblages depending on its condition. Four types of sand and rock-related substrates are distinguished (Figure 11.22; Heckel 1972; Flügel 2010): (i) the *hard* substrate, which is preferred by colonial corals and certain bivalves (borers); (ii) the *firm* substrate, which is preferred by solitary corals, red algae and green algae, sponges, bryozoans, oysters and barnacles; (iii) the *soft* substrate, which is populated by foraminifera, microbial mats and brachiopods, among others; and (iv) the *mobile* substrate, which is preferred by starfish, echinoids, some clams (burrowers) and ostracods. Gastropods and microbial mats cover the entire hard to mobile substrate range (Heckel 1972), but preserved microbialites develop preferentially on hard-to-firm substrates (Roche et al. 2019; Vennin et al. 2021). In addition, epiphytic carbonate production occurs in *Thalassia* and other seagrass meadows, where epibionts or encrusters like crustose coralline red algae, small rotaliid and miliolid benthic foraminifera, rare encrusting acervulinid foraminifera, serpulids and bryozoans all live attached to the seagrass leaves (e.g. Nelson and Ginsburg 1986; Bosence 1989a; Perry and Beavington-Penney 2005). As the plants die, the carbonate fossil assemblage becomes part of the sediments in which the plants rooted.

11.4.2.6 Carbonate Saturation of the Water Column

The carbonate saturation of the water column influences the capacity and rate of biotic carbonate precipitation (e.g. Lowenstam 1981; Mann 1983; Pomar and Hallock

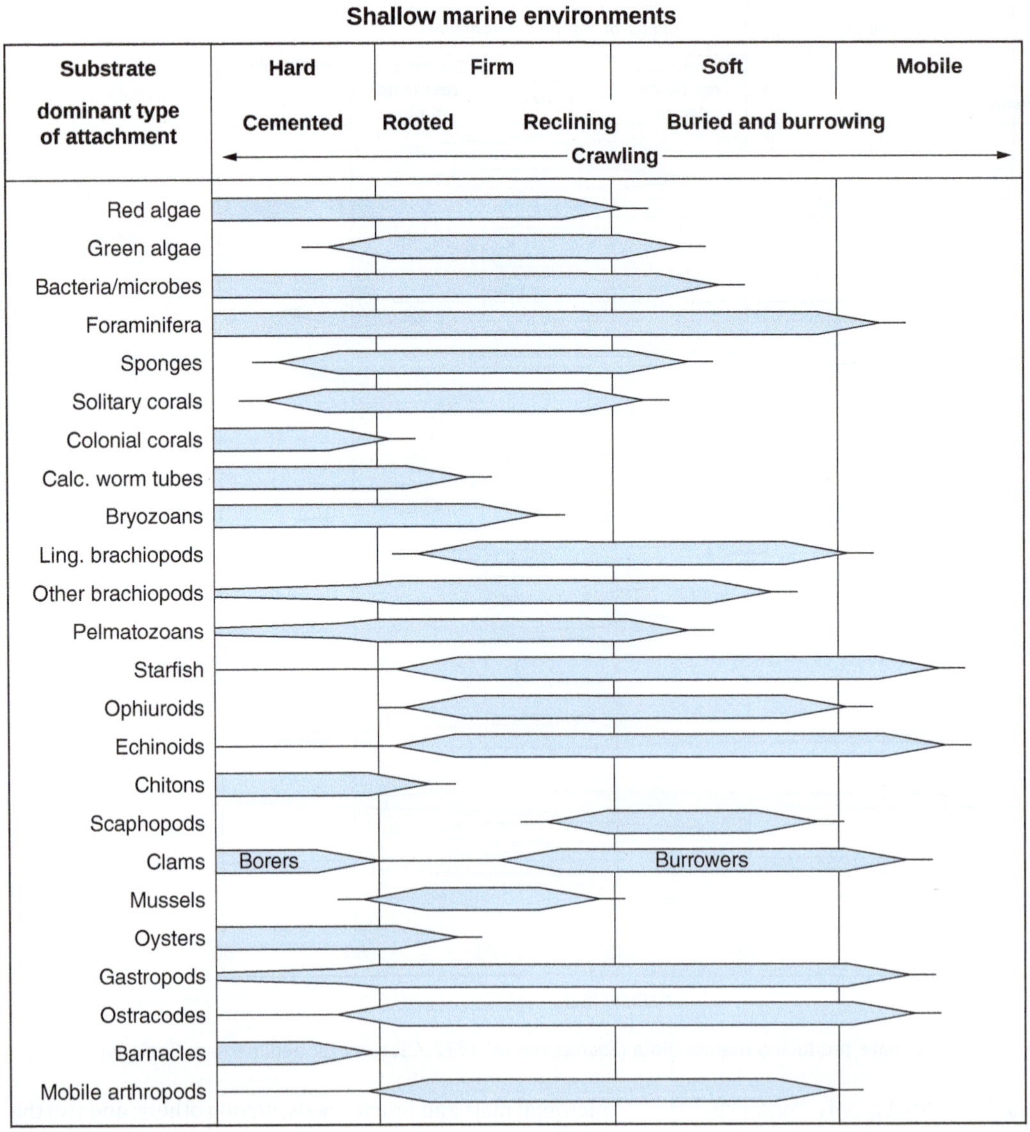

Figure 11.22 Substrate dependence of carbonate producing marine biota (after Heckel 1972).

2008). Different organisms require different saturation levels for prolific carbonate production (e.g. Broecker and Takahashi 1966; Opdyke and Wilkinson 1990, 1993; Smith and Buddemeier 1992; Gattuso et al. 1998; Weiner and Dove 2003; Riding and Liang 2005; Pomar and Hallock 2008; Michel et al. 2019). The saturation level of seawater depends largely on water temperature and shows significant global variation (Guan et al. 2015). Laugié et al. (2019) question the universality of the response of calcifiers to changing carbonate saturation based upon laboratory experiments demonstrating the varying response of marine calcifiers to reduced carbonate saturation and increase in atmospheric *pCO2* (Ries et al. 2009). Carbonate saturation may have been more relevant in the Precambrian when precipitation in microbial mats dominated (e.g. Visscher and Stolz 2005; Bontognali et al. 2008). The capacity of tropical zooxanthellate corals to respond to ocean acidification will be complicated by thermally induced loss of photosymbionts (Anthony and Fabricius 2000; Jokiel et al. 2008; Cameron et al. 2022).

11.4.2.7 Ocean Currents

Present-day ocean circulation patterns locally disrupt the temperature pattern that would be expected from a simple atmosphere-controlled latitudinal distribution of the main climate belts (Figure 11.23a, b; see also Chapter 13, Figures 13.2–13.5). For example, there is no large-scale tropical reef development at low latitudes off the west coasts of South America and Africa. These region-specific anomalies are related to ocean gyres transporting cold surface waters to low latitudes. Conversely, western boundary currents (e.g. the Gulf Stream bathing Bermuda, e.g.

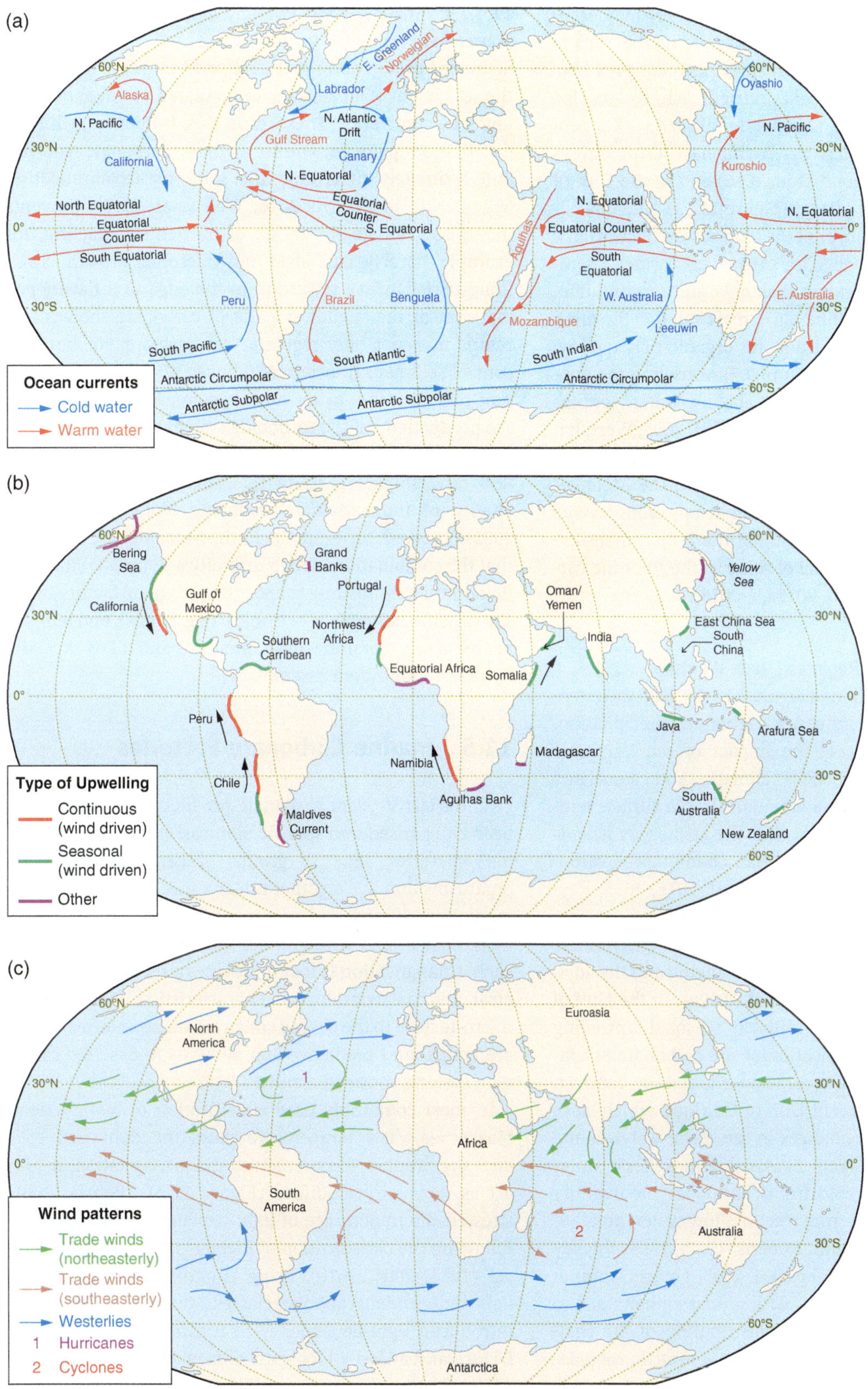

Figure 11.23 Global Ocean and Wind Circulation patterns. (a) Global Ocean currents. Note the transportation of tropical waters to higher latitudes, thus enabling tropical faunal associations to expand, for example, by the Agulhas Current along the east African coast, the Leeuwin Current with the Houtman Abrolhos islands (*Source:* Adapted from Crossland 1988; Collins et al. 1993, 2003) and the Kuroshio current reaching the Ryukyu Islands and Japan (e.g. Nagai et al. 2019), (b) global upwelling zones (Kâmpf and Chapman 2016) and (c) global wind patterns (*Source:* Adapted from Simmons and Alford 2012). See also Chapter 13.

Crossland 1988; Opdyke and Wilkinson 1990) may carry warm waters to temperate latitudes, expanding the tropical carbonate-production range. Ocean surface currents also cause local upwelling of cold nutrient-rich waters, for example, off Mauritania (Michel et al. 2011; Klicpera et al. 2015), in the Galapagos Archipelago (Humphreys et al. 2016, 2018; Reymond et al. 2016) and in the Gulf of Panama (Reijmer et al. 2012a) (Figure 11.23). The upwelling regime forces a turnover of the biotic assemblage from a tropical biotic assemblage to an adapted cool-water assemblage dominated by bivalves, balanids and foraminifera. The dominance of coralline-algal reefs on the Brazilian equatorial margin results from the organic-rich outflow of the Amazon (Chapter 9, Figure 9.7) restricting the population by coral larvae of the shallow-water settings (Jindrich 1983; Gherardi and Bosence 2001, 2005; Westphal et al. 2010; Amado-Filho et al. 2012). Seasonal changes in current pattern can also be significant. On the Maldives platform in the Indian Ocean seasonal upwelling events associated with the intensification of the Indian monsoon resulted in the partial demise of some of the Pliocene carbonate atolls (Betzler et al. 2013a,b, 2018).

11.4.2.8 Atmospheric Regimes (Trade Winds)

Dominant wind directions (Figure 11.23c) and their seasonal variations also influence carbonate precipitation. A present-day example is the impact of the El Niño–Southern Oscillation (ENSO) climatic and associated oceanic events. These cause variations in sea-surface temperature and nutrient content near the Galapagos Islands in the eastern tropical Pacific Ocean changing the carbonate-producing biota (Humphreys et al. 2016, 2018; Reymond et al. 2016; Riegl et al. 2019).

Such changes can be localised. Seasonal changes in trade wind patterns affect both the Gulf of Panama and the adjacent Gulf of Chiriquí (Reijmer et al. 2012a). In the Gulf of Panama, seasonal changes initiated seasonal upwelling that resulted in the modification of the carbonate assemblages from warm-water to cool-water fauna. By contrast, in the immediately adjacent Gulf of Chiriquí, which lacks seasonal upwelling, no changes in the warm-water faunal assemblage were observed. Seasonal atmospheric variations, such as passing cold fronts, influence the quantity of sediments exported from the shallow-water areas of the Bahamas (Wilson and Roberts 1992, 1995; Hickey et al. 2000; Roth and Reijmer 2004, 2005; Lopez-Gamundi et al. 2024). The trade winds, with a predominant east to west direction, determine the redistribution of the shallow-water carbonate sediments. The preferential progradation of the carbonate sedimentary system of the Bahamas and the infill of 'the Tongue of Ocean' seaway in Great Bahama Bank reflect this wind-driven sediment transport system (Eberli and Ginsburg 1989).

11.4.2.9 Shallow-Water Hydrodynamics – Tides, Waves and Freshwater Input

Sediment production rates, such as coral reef growth and the settlement of larvae, may be regulated by tides and shallow-water currents. These processes influence sediment distribution patterns within shallow carbonate shelves, and ultimately determine the infill of accommodation (e.g. Purkis and Harris 2016; Weij et al. 2019). Currents may also regulate the influx of nutrients impacting, for example, coral versus algal and microbial growth. Swell (long period) wave refraction at the edge of a flat-topped platform or the shallower parts of carbonate shelves and ramps may not only regulate reef development but also control sediment transport (Collins et al. 1993, 2003). Current-related influx of terrestrial sediments may influence the productivity of the carbonate system as it may, depending on the grain size, increase the turbidity of the water column and impact parts of the substrate, disturbing the vitality of the carbonate-producing biota. Perry (2003, 2005) observed, for example, in the Mozambique Channel, that the combination of high turbidity and mobile siliciclastic-dominated substrate shaped the carbonate factory into a site-specific assemblage of tropical (e.g. corals), and temperate biota (e.g. coralline algae, molluscs and a low diversity benthic foraminifera assemblage).

11.5 Marine Carbonate Factories

Over the last 50 years, carbonate-producing organisms have been categorised into specific biotic assemblages or 'carbonate factories', that is, groups of carbonate-producing organisms that live together under a specific set of environmental conditions, and as a result have a particular sedimentological and stratigraphic imprint. The advantage of such a classification is that it provides a set of sedimentological and palaeontological characteristics, ranging from micro-texture to the overall construction geometries, which are typical and predictive and allow us to describe carbonate systems more efficiently.

In most commonly used classification of biotic assemblages sea-water temperature plays the dominant role, and the trophic requirements and precipitation modes act as secondary modifiers (Figure 11.24). Pomar (2001) stressed the importance of light-dependency of organisms following the classification of Liebau (1980, 1984) and Flügel (1984, 2004, 2010), using euphotic, oligophotic and light-independent (aphotic) classes. Michel et al. (2020) also introduced light-dependency to further subdivide the C-factory and added a biochemical factory.

The frequently used classification by James (1997) is based on a nutrient/water temperature matrix and distinguishes between photozoan and heterozoan associations (Figure 11.25). Transitional biological associations, having

Authors	Tropical >22°C	Subtropical 22–18°C	Temperate 18–10°C	Cold 10–5°C	Polar <5°C	Classification criteria
Lees and Buller (1972)	Chlorozoan			Foramol		Temperature salinity latitude
Lees (1975)	Chlorozoan	Chloralgal		Foramol		
Schlanger (1981)		Coral/algal facies		Bryozoan/algal facies		
Carannante et al. (1988)	Chlorozoan	Chloralgal	Rhodalgal	Bryoalgal	Molechfor	
Nelson (1988)				Non-tropical		
Betzler et al. (1997b)			Warm temperate 20–11°C			
James (1997)	Photozoan			Heterozoan		Trophic requirements
Schlager (2005)		T-factory		C-factory		Trophic requirements and biotically induced/ controlled sediment
Reijmer (2021) and this book		T-factory		C-factory and CWC-factory		
				M- factory and P-factory		

Figure 11.24 Comparative table of different carbonate classification systems projected against a seawater temperature gradient (*Source:* Adapted from Mutti and Hallock 2003; Schlager 2005). Naming strategies for these biotic assemblages differ and include the dominant benthic grain types (Lees and Buller 1972; Lees 1975; Schlanger 1981), a more elaborate subdivision also using dominant grain type (Carannante et al. 1988), and the distinction between photozoan and heterozoan associations (James 1997). The contributions by Nelson (1988) and Betzler et al. (1997b) highlighted the presence of a transitional biotic association between the warm water and cool water endmembers. Schlager's (2005) and Reijmer's (2021) carbonate factories, used in this book chapter, have also been placed in this temperature-controlled scheme.

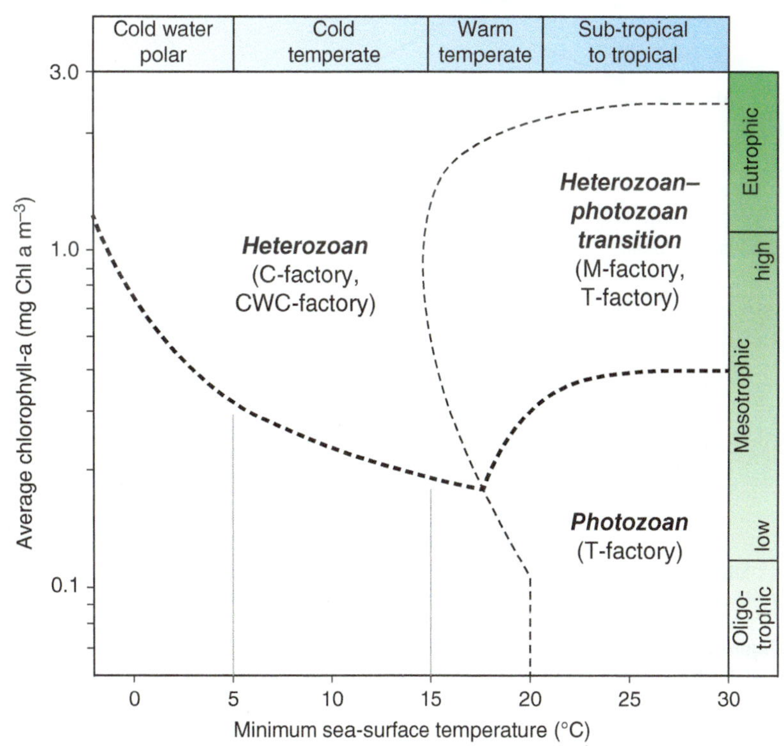

Figure 11.25 Impact of sea water temperatures and trophic resource levels on shallow water heterozoan and photozoan biotic associations from various carbonate environments (*Source:* Adapted from Halfar et al. 2006; James and Lukasik 2010). The *photozoan* association comprises carbonates deposited in euphotic, mostly oligotrophic settings located in warm-water environments, and characterised by green calcareous algae and corals; the association also incorporates ooids and inorganically produced muds and cements. The *heterozoan* association is typical for eutrophic to mesotrophic settings in cool to temperate-water conditions generally colder than 20°C. The position of the carbonate factories used in this book are also indicated.

characteristics of both end members, have also been reported. A good example is the work by Halfar et al. (2000, 2004, 2006) which showed that in the Gulf of California the boundaries between heterozoan and photozoan associations are gradual as a result of variations in temperature and nutrient levels.

Here we favour the carbonate factories of Schlager (2000, 2003) as recently refined by Reijmer (2021). This classification first distinguishes between different marine *carbonate precipitation modes*, which can be abiotic, biotically induced or biotically controlled (Figure 11.26). A further layer distinguishes between the autotrophic and heterotrophic nature of the organisms (Lowenstam 1981). Only a very small percentage of carbonate deposits is attributed to abiotic, purely chemical, precipitation (see Section 11.5.1: 'T factory', for Whiting Events). The vast majority of the shallow-water carbonate deposits, certainly in the Phanerozoic record, is thus of biological origin, with precipitation controlled by the catalysing effects of microbiotic biochemical reactions at the cell level (e.g. Dupraz et al. 2009). 'Biotically induced' refers to the non-skeletal elements or the soft body parts, such as microbialites and ooids, and peloids which are mostly of faecal pellet origin. By contrast, 'biotically controlled' precipitation refers to skeletal components (shells and other hard body parts). In the latter category a further subdivision is made into mostly sunlight-dependent organisms (phototrophic) and mostly nutrient-dependent organisms (heterotrophic). The resulting five factory types are the planktic P-factory, the tropical T-factory, the cool-water C-factory, the cold-water coral CWC-factory and the mud-mound M-factory (Figure 11.26).

The naming of the factories is one of convenience and helps to define theoretical end-members of specific biota classes. The role of controlling factors on these factories is subtle and composite, as is summarised in Table 11.3, and co-occurrences and transitions in space, down a depositional profile or along strike, from one to the other are common. In some cases, a strict attribution to one of the factories is debatable, such as for instance for the ooid-bearing or -dominated deposits (see below). Nevertheless, it is considered that the distinction of these five factories, their characteristics, and the combination with a measured evaluation of the controlling factors provide an effective way to advance the study of the wide variety of marine carbonate systems.

The five carbonate factories are summarised, with descriptions of each of the following: (i) main characteristics, present-day geographical distribution and history of research; (ii) the sediment composition, including biological origin, mineralogy and grain-size spectrum; (iii) the production profile (i.e. the growth rate of the factory plotted against depth) and the depositional profile. The depositional and production profiles for the different factories are summarised in Figure 11.27, and the mineralogical characteristics in Figure 11.28. The implications that different carbonate factory types have for sequence stratigraphic studies are treated in Section 11.7.

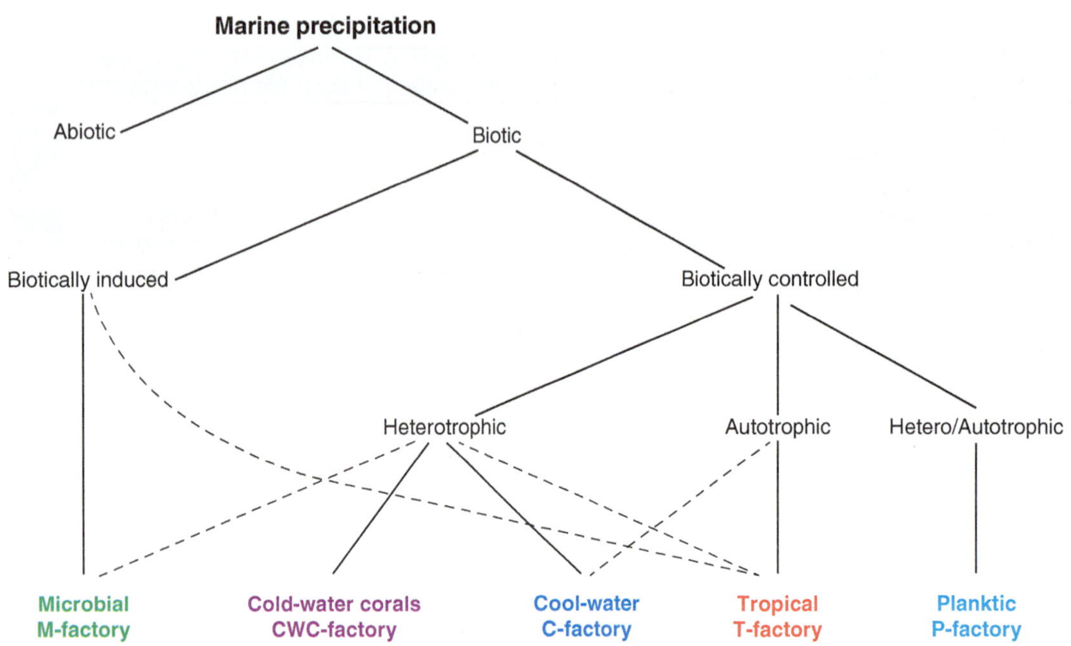

Figure 11.26 Marine carbonate precipitation modes and carbonate factories (modified after Schlager 2003, 2005, and Reijmer 2021). See text for further explanation.

Table 11.3 Summary of the relative impact of six environmental conditions on the five carbonate factories. The autotrophic T-factory stands out as sensitive to temperature, light (positive) and nutrients (negative). The C-, M- and CWC-factories are mostly heterotrophic, and sensitive to nutrient supply (positive), and C- and CWC-factories are sensitive to temperature (cold!). The M-factory appears to be the most adaptable. See text for further explanations.

Carbonate Factories and Impact of Environmental Conditions						
	Light	**Temperature**	**Nutrients**	**Salinity**	**Substrate**	**Carbonate saturation**
T-Factory	*Maximum within shallow-water depths: 0–40 m. Maximum water depth euphotic zone varies for ocean realms (Schlager 2005).*	*Warm: >23°C (min. 14°C)*	*Preferably low: <0.1 mg* Chl_α/m^3	*Normal to hypersaline*	*Needed for certain organisms*	*(A) Saturation index (SI) is defined as the decadic logarithm of the ratio of the ion activity product (IAP) over the solubility constant (Ks):* $SI = \log(IAP/Ks)$. *A solution is supersaturated with respect to a mineral phase when SI is positive. Laboratory experiments demonstrated that an SI > 0.8 is required before precipitation occurs (Kempe and Kazmierczak 1994)* *(B) Biota show varying response to SI variations (Ries et al. 2009)*
C-Factory		*Water temperature determines surface and deep-water occurrence (e.g. platform slopes); Cool: 13–19°C (min. 9–12°C)*	*High: >0.4–<5 mg* Chl_α/m^3	*Normal*		
M-Factory	*Depends on the type of microbes and their metabolism (Visscher and Stolz 2005; Dupraz et al. 2009)*	*Wide range of temperatures; from cool water to hot vents*	*Low to very high: >0.1 mg* Chl_α/m^3	*Normal to elevated saline*		
CWC-Factory		*Water temperature determines habitat together with nutrient input; cold: 4–13°C (min. 4°C)*	*High: >0.4–<5 mg* Chl_α/m^3	*Normal*	*Presence of suitable substrate important*	
P-factory	*Main production within shallow-water depth: 0–40 m, but also depends upon the organism; some foraminifera have algal symbionts*		*High: >0.4–<5 mg* Chl_α/m^3	*Normal*	*Not needed*	

Very positive	Negative
Positive	Moderately negative
Moderately positive	Neutral

11.5.1 T-Factory

The tropical (T) factory is the best documented of the factories, both in present-day environments and for Phanerozoic examples. The T-factory groups carbonate producing organisms that need sea-water temperatures for most of the year of 21.7 to 29.6°C, with lower and upper limits of 18.1 and 31.5°C, respectively (Guan et al. 2015). The T-factory biota only support low nutrient levels (oligotrophic) and are dependent on light (autotrophic) with an average minimum light intensity that should exceed 450 μmol photons m^{-2} s^{-1} (Geider and MacIntyre 2002; Bouman et al. 2018). These conditions limit the distribution of this carbonate factory to the tropical climate belt between 30°N and 30°S or the 18°C winter isotherm, as is clearly shown in Figure 11.29. Locally the T-factory can be suppressed within this optimum growth zone by the inhibiting effects of cold upwelling currents, an over-supply of nutrients favouring the proliferation of microbial organisms, or anomalously high salinities due to evaporation. The annual mean salinities tolerated by T-factory assemblages are between 28.7‰ and 40.4‰, with 23.3‰ and 41.8‰ as lower and upper limits (Flügel 2010).

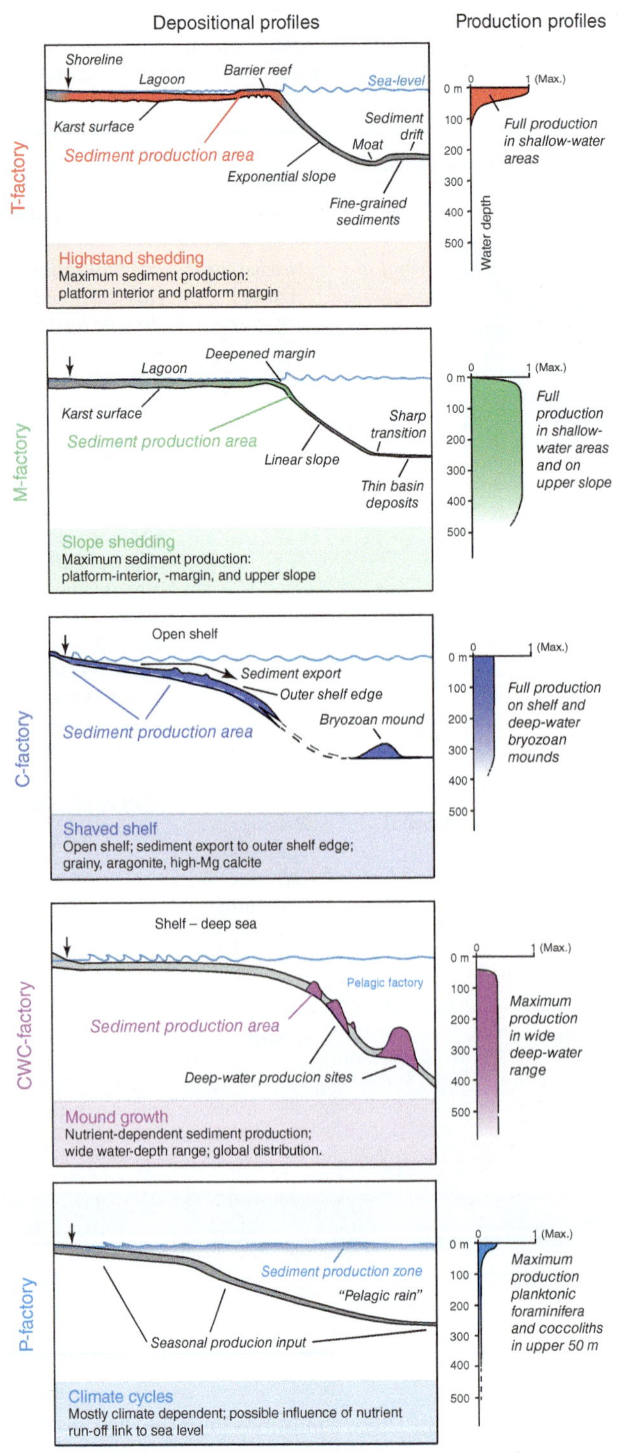

Figure 11.27 Carbonate factories and their typical depositional and production profiles (*Source:* Adapted from Reijmer 2021; Schlager 2003, 2005).

11.5.1.1 Sediment Composition

The T-factory biological assemblage represents the classical image of carbonate systems that we know today in the tropics, comprising zooxanthellate corals, calcareous green algae, calcareous red algae, benthic foraminifera and echinoderms (Figure 11.30). Bryozoans and bivalves are rare to common (Flügel 2010). The non-skeletal fraction includes ooids, peloids, aggregate grains and carbonate mud (Milliman et al. 1974; Flügel 2010). The sediments cover the full grain-size spectrum from muds to coarse-grained skeletal hash and reef boulders.

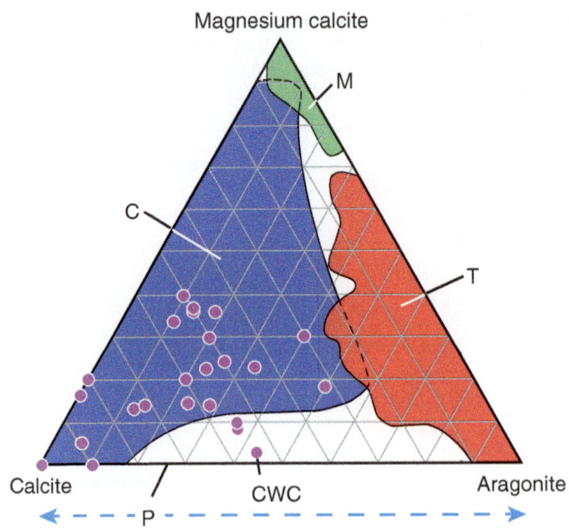

Figure 11.28 Carbonate mineralogy of the carbonate factories (*Source:* Adapted from Reijmer 2021; Schlager 2003, 2005). CWC-factory: purple dots, P-factory lower part of the triangle (ranging between calcite and aragonite).

Recent research into the formation of ooids, an important part of the T-factory, suggests that it is a dominantly biotically controlled process regulated by the interactions of redox microbial communities along with extra-cellular polymeric substance (EPS) biofilms (Pacton et al. 2012; Edgcomb et al. 2013; Diaz et al. 2014, 2015, 2017, 2022; Diaz and Eberli 2019). Similarly, the Bahamian Whiting Events originate from the interactions between (i) off-platform currents delivering carbonate supersaturated waters to the shallow platform top, (ii) the platform top topography with its shallow lagoon and (iii) the stagnant platform top waters which create an environment in which direct crystal precipitation initiated by cyanobacteria may occur (see more on the Bahama platform in Section 11.6).

The initial mineralogy of T-factory sediments is dominated by aragonite followed by HMC and then LMC (Schlager 2005) (Figure 11.28). The sediments on Great Bahama Bank, for example, contain between 77.7% and 100% aragonite (mean 93.3%). HMC content varies between 0% and 22.3% (mean 6.5%), and LMC from 0% to 3.9% (mean 0.2%) (Reijmer et al. 2009). The high aragonite content also makes these sediments very susceptible to early diagenetic processes such as dissolution during subaerial exposure (Section 11.3), and to cementation forming beachrock (Dickson et al. 2014) or cementation stabilising upper slope sediments (Grammer et al. 1993).

11.5.1.2 Sediment Production Profile and Depositional Profile

The sediment production profile of the T-factory, which depends primarily on the penetration of sunlight in the

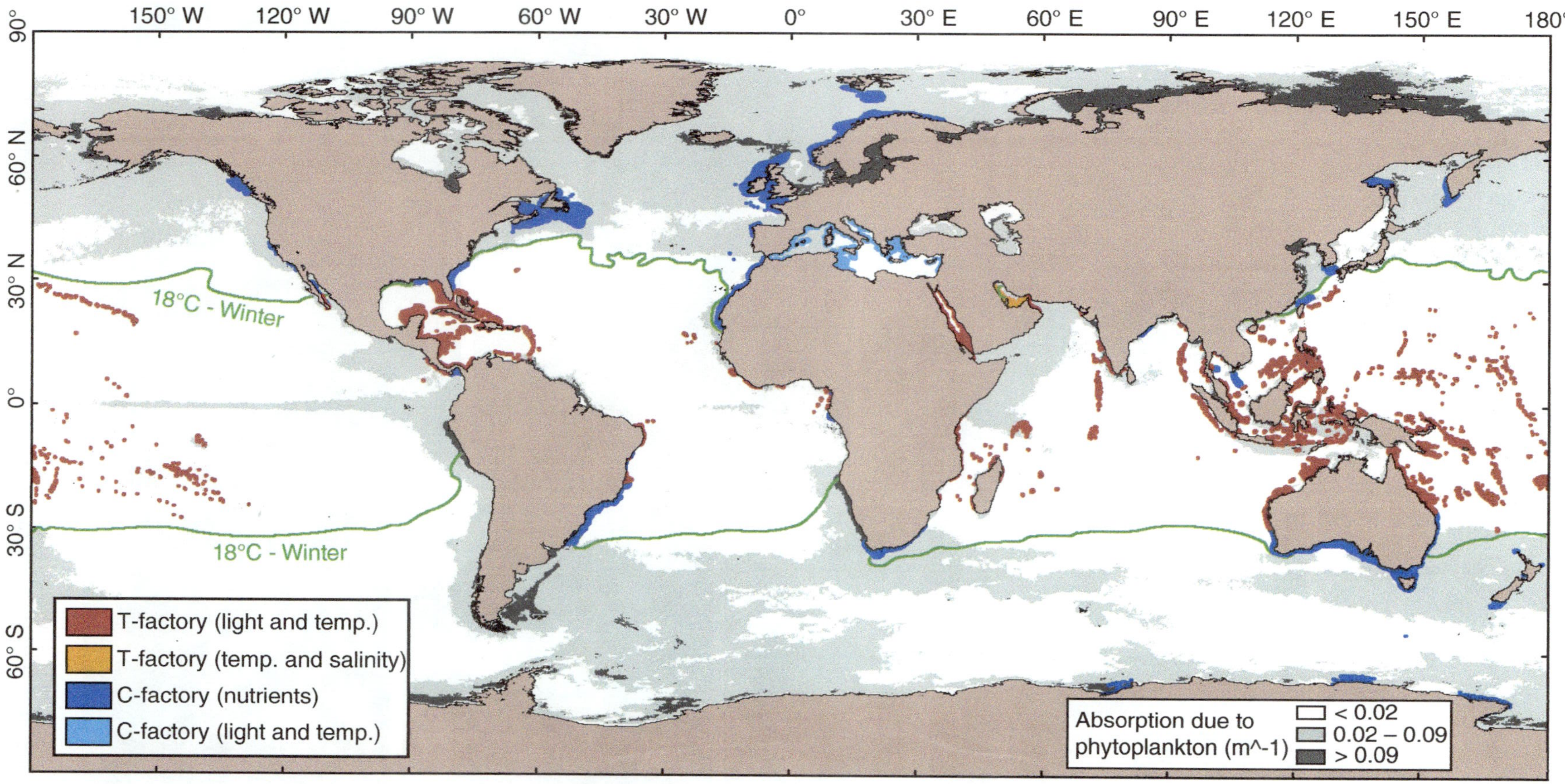

Figure 11.29 Present-day distribution map of the ocean areas where the T- (red; orange: Persian Gulf) and C-factories (dark blue; light blue: Mediterranean) are the dominant carbonate producers (*Source:* Adapted from Laugie et al. 2019; Reijmer 2021). Grey values: monthly mean sea surface diffuse attenuation coefficient (MODIS) representing light penetration in the water column at the blue-green wavelengths (c. 490 nm); represents combined effect of absorption and backscattering by water constituents, and structure surrounding light field (Werdell and Bailey 2005).

Figure 11.30 Examples of T-factory sediments; (a–c) coral reef constructions at the Maldives; (d, e) grainy, oolitic shoals and mangrove carbonate mudflats on the Bahamas; (f–i) *Halimeda* and coral reef constructions in Cuba (*Source:* Courtesy of Emmanuelle Vennin).

water column, was experimentally determined by measuring growth of corals in the Caribbean and shows an exponential decrease with water depth (Bosscher and Schlager 1992). Although light penetration varies substantially (Figure 11.18), a general optimum range of 0–40 m water depth is applied (Figure 11.27). Seagrass meadows provide a particular eco-system that produces substantial amounts of carbonate sediment in tropical waters. They too are restricted to the euphotic zone (e.g. Perry and Beavington-Penney 2005). It is the grazing food chain of the seagrass ecosystem that determines the type of carbonate sediment produced in this setting (Freiwald 1993).

In both the land-attached and land-detached T-factory carbonate systems, two types of depositional profile are found: the flat-topped shelf profile and the ramp profile (Figure 11.2). The former is the most common due to the tendency of the T-factory to build-up to sea level. The shelf profile is typified by the presence of bio-construction and/or grainy-shoal-dominated barriers separating the slope and deeper-water higher-energy setting from the protected, shallow water, lower-energy setting. The ramp profile is commonly a transient state of development towards a shelf profile (see Figure 11.5).

The composition and stability of the shelf barrier and the angle of the slope both vary depending on the grain-size spectrum of the marginal sediments, which is determined by the transportation efficiency and sediment production at the platform top. T-factory slopes tend to have a concave shape with a distal decrease in slope angle (for further details, see Section 11.6).

11.5.2 C-Factory

The biological assemblage of the cool-water (C) factory consists predominantly of heterotrophic organisms that live in sea water temperatures ranging from a few degrees (polar) to below 20°C, in meso- to eutrophic conditions, and in a water-depth range from 0 to 400 m (James 1997; Schlager 2005). In present-day surface waters, the C-factory operates at higher latitudes than the 18°C winter isotherm. The Great Australian Bight is a well-known example (Figure 11.29). C-factory assemblages also dominate in upwelling areas which are rich in nutrients, and sites below the thermocline, where ocean water temperatures are low. The delineation of the T-factory and the C-factory on the present-day map (Figure 11.29) is arbitrary since there are now well-documented examples where there is a gradual transition at 25–30°N and 25–30°S between these two, with a transitional, oligo-mesotrophic-eutrophic, warm-temperate system with red algal and seagrass-derived biota (Betzler et al. 1997a; Halfar et al. 2000, 2004; Michel et al. 2018). The Gulf of California (23°–30° N) is a good example with a coral-dominated, shallow-water T-factory in the south, an intermediate warm-temperate system with red algal-dominated, subtidal to inner-shelf carbonates, and a C-factory with molluscan bryozoan-rich inner to outer shelf carbonates in the northernmost part of the Gulf (Halfar et al. 2004, 2006).

The importance of the C-factory was only brought to the attention of the scientific community in the 1980s and 1990s and had a significant impact on the study of present-day and ancient carbonate systems (e.g. Carannante et al. 1988; Nelson 1988; James et al. 1994, 1999; Hayton et al. 1995; James and Bone 2011; Surlyk 1997; Betzler et al. 1997a,b; Braga et al. 2006; Pedley and Carannante 2006).

11.5.2.1 Sediment Composition

Carbonate production is largely from the skeletons of heterotrophic, calcareous grazers (gastropods and echinoids) and filter feeders (bryozoans, bivalves, serpulids, crinoids, barnacles and foraminifera) (Figure 11.31). The benthic phototrophs in this factory are restricted to macrophytes and coralline algae. In deeper environments, patches and mounds of bryozoans and sponges are built (James and Bone 2011; James et al. 2013). A useful subdivision of the C-factory proposed by James and Lukasik (2010) distinguishes cold, cool-water temperate and warm-water temperate associations (see Section 11.6). The biota decreases in diversity towards the higher, colder latitudes where kelp assemblages dominate (e.g. Leonard et al. 1981; Henrich et al. 1992, 1995; Surlyk 1997) as well as maerl-type carbonates (Brittany: Cabioch 1969; Wehrmann 1994, 1998; Western Ireland: Bosence 1976, 1980). Sea grass meadows form important C-factory production sites, hosting molluscs, coralline algae, and benthic foraminifera embedded in poorly sorted muds and sandy muds (Holocene, S. Australian Shelf; James and Bone 2011). They form the phaeophytic belt with its giant kelp forests where epiphytic carbonate sediment production depends on population densities of crustose coralline algae together with rotaliid and soritid benthic foraminifera, serpulids and bryozoans, but also bryozoans, balanids, serpulids and echinoderms that occupy the habitats in between the kelp forests (e.g. Henrich et al. 1995).

C-factory sediments may contain up to 50% aragonite, 40% intermediate-Mg calcite (4–12 mol% $MgCO_3$), HMC (>12 mol% $MgCO_3$) and up to 10% LMC (Figure 11.28; James and Bone 2011). Carbonate dissolution due to oxic/anoxic microbial degradation of sedimentary organic matter can result in the almost complete loss of aragonite in shallow sub-sea floor sediments. The shelf margin slope sediment may locally exhibit an alternation of (i) grainy, aragonitic and HMC-rich sediments with tunicate spicules, relict bioclasts and bryozoan detritus, and (ii) micritic LMC-rich sediments with large amounts of sponge spicules (Saxena and Betzler 2003).

Figure 11.31 Examples of C-factory sediments; (a) fenestrate bryozoans growing on macroalgae (branching brown biota) at upper left, with pink sponges at right and small encrusting grey bryozoan below, 12 m water depth, southern Victoria, Australia, width = 20 cm. (b) Close view of fenestrate bryozoan on gorgonian corals, 12 m water depth, southern Victoria, Australia, image view = 10 cm across. (c) Bedding plane view of bryozoan grainstone-rudstone, Oligocene, Pt. Willunga Formation, Pt. Willunga, South Australia, image = 10 cm across. (d) Bedding plane view of robust branching (centre) and fenestrate (right) bryozoans, Assistance Formation, Middle Permian, Ellesmere Island, northern Canada; image view c. 20 cm. (*Source:* Reproduced from Reijmer (2021) / John Wiley & Sons).

11.5.2.2 Sediment Production Profile and Depositional Profile

The sediment production profile of the C-factory shows relatively low production rates, about 25% that of the T-factory (James 1997; Schlager 2005), but covers a large depth range (0–400 m) (Figure 11.27). The heterotrophic, suspension-feeding, biota that dominate this assemblage is not light dependent and can thus cover a much wider depth range as long as there are nutrients available (Schlager 2005; James and Bone 2011).

The typical depositional profile of the grainy C-factory assemblages is the ramp. The absence of a shallow-water barrier combined with low cementation rates results in a wave and current-dominated sediment-distribution model, and hydrodynamic partitioning (e.g. Nelson 1988; Halfar et al. 2006). Sediments are deposited below wave base or transported to deeper waters (*shaved shelf* model of James et al. 1994), forming outer shelf sediment wedges (Saxena and Betzler 2003). This 'siliciclastic type' of sediment distribution is clearly shown in the Agua Amarga basin in SE Spain, where cool-water carbonates developed beach deposits, storm barriers and wash-over fans (Betzler et al. 1997b; Villalobos and Braga 2003). Slopes in C-factory systems have a sigmoidal or convex shape (Adams and Kenter 2014) and share process similarities with the continental slopes of siliciclastic passive margins in terms of erosion and redistribution of sediment by storm- and sea-level-induced fluctuations of base level (Schlager and Adams 2001; Adams and Kenter 2014). Within the ramp setting, specific faunal assemblages and production loci can be distinguished such as the seagrass, kelp communities and oyster banks (e.g. James and Bone 2011; Michel et al. 2018). In the outer ramp setting, mound-forming associations of bryozoan and scleractinian corals can occur (Surlyk 1997; Hvid et al. 2021); and in the offshore domains these are complemented by P-factory production of coccoliths (Surlyk 1997). Another setting where coarse-grained C-factory sediments are commonly found is in submarine canyon fills

(Oligocene-Holocene, Gippsland Basin, southeast Australia; Wallace et al. 2002; Mitchell et al. 2007) and tide-influenced incised-valley systems (Miocene, SE-France; Reynaud et al. 2006, 2012; Reynaud and James 2012; James et al. 2014) (see also Section 11.6). C-Factory sediments were also reported from Late Carboniferous tropical latitudes by Samankassou (2002).

11.5.3 CWC-Factory

The cold-water coral (CWC) factory consists of a heterotrophic faunal assemblage dominated by non-zooxanthellate hermatypic (reef-dwelling) corals. These corals grow at present in deep marine, meso- to eutrophic environments in water depths varying from 40 m to over 2000 m (average range of 200–1000 m), with water temperatures between 4 and 13°C (Kenyon et al. 2003; Freiwald et al. 2004, 2005; Hebbeln and Samankassou 2015; Hebbeln et al. 2019; Portilho-Ramos et al. 2022; Cordes and Mienis 2023).

The wide present-day global distribution clearly shows the relative importance of this system (Figure 11.32). CWC-factory systems build provinces of individual coral mounds, locally organised in large clusters, along the continental margins of the Atlantic Ocean, Mediterranean Sea and Gulf of Mexico as well as in the Indian and Pacific Oceans (Huvenne et al. 2007; Duineveld et al. 2007; Wienberg et al. 2018; Hebbeln et al. 2019; Hanz et al. 2019). Ocean bottom currents provide the main physical control of growth patterns. They shape the mounds, provide nutrients ('nutrient distribution window'; Kenyon et al. 2003) and are critical for reproduction strategies and colonisation of new sites (Rüggeberg et al. 2011; Henriet et al. 2014; Wienberg and Titschack 2017). Whether the triggering factors for mound locations are related to antecedent structural topography and fault systems or to regional productivity patterns and current systems is still debated (Hovland et al. 1994; Huvenne et al. 2007; Eisele et al. 2011; Hebbeln et al. 2014). The CWC-factory was only recently fully appreciated as an important carbonate factory, which is due to its hidden present-day location on the deep-sea floor below the photic zone. It has had limited recognition, so far, in the rock record.

11.5.3.1 Sediment Composition

CWC-factory mounds and banks usually consist of an alternation of coral-dominated intervals and lithified sediment intervals (e.g. van der Land et al. 2010; Mienis et al. 2009). Coral-dominated intervals consist of up to 82% hermatypic corals, for example, *Lophelia*, *Madrepora* (Teichert 1958), with minor sponge spicules, echinoderms, brachiopods and gastropods, embedded in a fine-grained matrix. The lithified intervals consist of coralline algal material (2%), coccoliths (68%) and planktic foraminifera (29%) (van der Land et al. 2010). The biodiversity is high (Freiwald and Roberts 2005; Cordes and Mienis 2023).

In the coral-dominated intervals 40% of the components are aragonitic, 51% LMC and less than 2% HMC (van der Land et al. 2010) (Figure 11.28). In the lithified intervals the sediment is dominantly LMC (94%), aragonite is reduced to 1–5%, and HMC varies between 1% and 5%. Consequently, because of the high aragonite content, the primary sediments are chemically unstable, and have thus a limited chance of 'un-recrystallised' preservation.

11.5.3.2 Sediment Production Profile and Depositional Profile

The sediment production rate of this factory is low (4–25 mm a^{-1}) compared to the other factories (T-factories reach up to 150 mm a^{-1}), but it has an exceptionally large depth range (Figure 11.27). The production rate depends on the influx of nutrients either through deep-water currents, pelagic input (Freiwald and Roberts 2005; Cordes and Mienis 2023) or from the sea floor through natural hydrocarbon seeps.

The typical depositional profile of the CWC-factory is one of positive mound-like structures, which vary in shape, from circular to arcuate and irregular with single or multiple peaks; in height, from several metres to over 300 m, and in length, up to several hundreds of metres (Freiwald et al. 2004; Mienis et al. 2007; Roberts et al. 2009). Frame-building corals predominate in this system, and sediment export is limited, only affecting the direct surroundings of the production sites. Belts of individual coral mounds display aligned, elongated morphologies with varying lengths (kms) but narrow widths (hundreds of metres) (e.g. Hebbeln et al. 2019). The mound slopes vary from 20°–30° to up to 80° (De Mol et al. 2002; de Haas et al. 2009). The angle of repose is a result of the combination of coral rubble itself and the stabilising effect of early diagenetic processes, forming hard- and firmgrounds and/or semi-lithified sands (Mienis et al. 2006; de Haas et al. 2009; Hebbeln et al. 2019).

11.5.4 M-Factory

The mud mound/microbial (M) factory is characterised by microbially induced sedimentation. In present-day environments, it is found in water depths ranging from the intertidal zone, down to 400 m along the slopes of isolated platforms, in individual mounds at varying water depths, and also near methane seeps at the bottom of the oceans. The source of energy for M-factory biota can be nutrients, light or methane (Dupraz et al. 2009). Microbial activity is ubiquitous in all present-day carbonate depositional environments, and yet hard to substantiate due to the limited direct preservation potential of the microbial organisms. Consequently, the present-day distribution map (Figure 11.32) is expected to undergo substantial updates in the coming years. The best-known markers of microbial activity are the stromatolites,

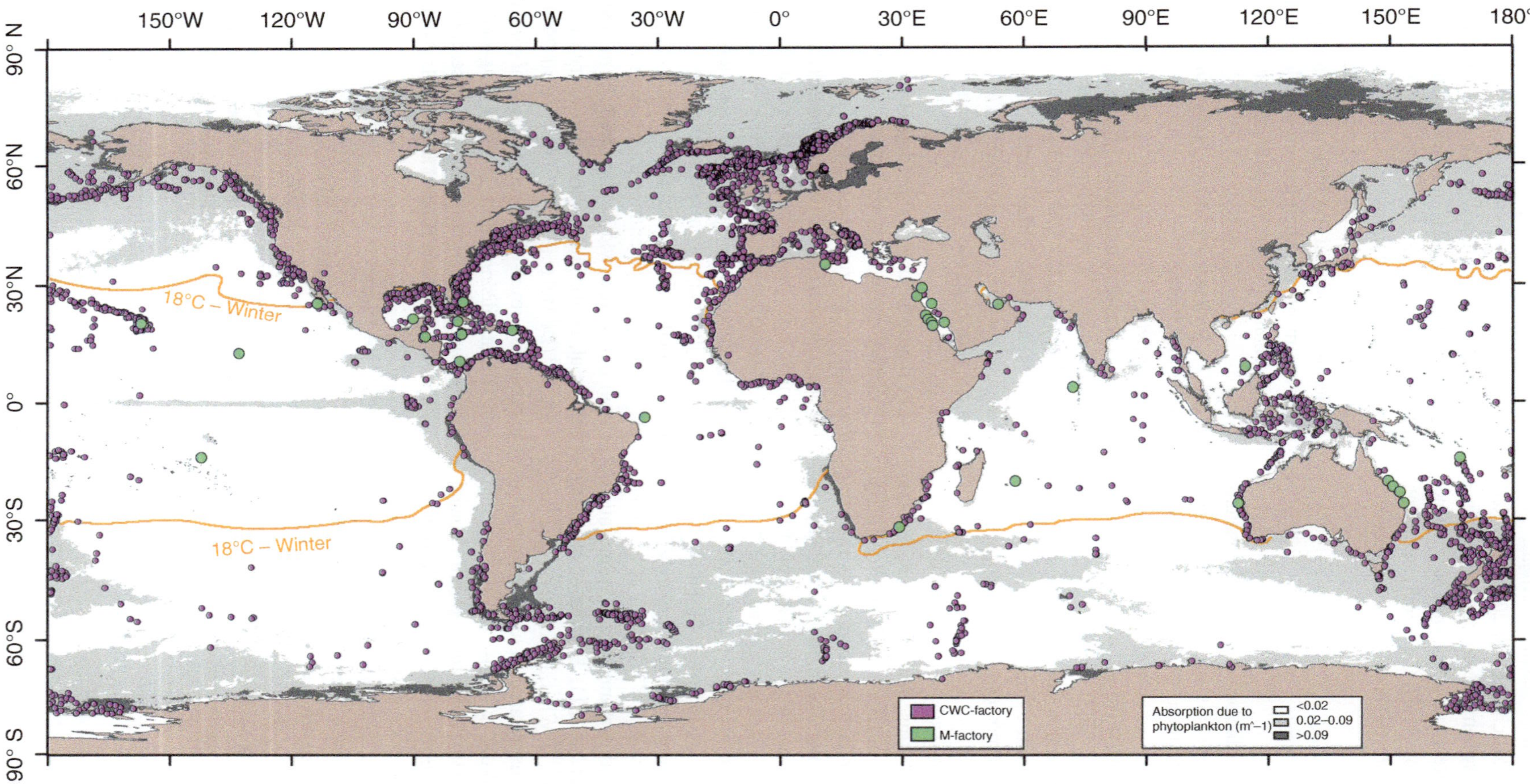

Figure 11.32 Present-day distribution map of CWC mounds (purple) and areas where M-factory sediments (green circles with numbers) dominate. CWC. Adapted from data are from Laugie et al. (2019) and Titschack et al. (2004) in Freiwald et al. (2004, 2021) UNEP report. Compilation made by Jacek Raddatz (Frankfurt, Germany). The M-factory is most likely present in many more places but has not been specifically recorded. Well-studied occurrences can be found at The Bahamas, Cuba, Red Sea, Shark Bay (Australia), and Tahiti (see text for references). Grey values: monthly mean sea surface diffuse attenuation coefficient (MODIS) representing light penetration in the water column at the blue-green wavelengths (ca. 490 nm); represents combined effect of absorption and backscattering by water constituents, and structure surrounding light field (Werdell and Bailey 2005).

which leave an imprint of a layered growth structure in the sediment, both recent and ancient (Figures 10.28 and 10.29). Good examples are the stromatolite growth forms found in the inter-tidal environment of Shark Bay in Australia (see Suosaari et al. 2016a, b for references), and in Peros Bay in Cuba (Bouton et al. 2016). Other bacteria-mediated precipitates are found in microbial mats as in Exumas, Bahamas (Reid et al. 1995; Dupraz et al. 2009), the Red Sea (e.g. Szilagyi et al. 2020; Vahrenkamp et al. 2024), Great Barrier Reef (Szilagyi et al. 2020) and Tahiti (e.g. Séard et al., 2011).

In the deep ocean M-factory, bathyal microbial mounds have been found, devoid of coral-rich frameworks and fragments, which are formed by microbially induced precipitation with sulphate reduction as the source of energy (Suess 2002; Teichert et al. 2005). These mounds are called 'chemoherms' (they were originally considered of abiotic origin; Aharon 1994). Chemoherms have now been described from the Gulf of Mexico (Roberts and Aharon 1994; Aharon et al. 1997), the Cascadia accretionary margin (Kulm et al. 1986; Bohrmann et al. 1998) and the Yaquina Basin (Hübscher and Kukowski 2003).

Although the M-factory is relatively poorly represented (or recognised?) in present-day marine settings, it was arguably the original Precambrian carbonate factory and has played a critical role in the evolution of the Earth (see Section 11.7). For this reason, modern microbial mats and microbialites represent an invaluable topic of research to study the origins of early life on earth, global biogeochemical cycles and the possibility of life elsewhere. Microorganisms are the life-support system of the biosphere and have had, and will continue to have major impact on climate and environment (Reinold et al. 2019).

11.5.4.1 Sediment Composition

The M-factory entraps carbonate sediment *in situ* by means of microbial mats, forming 'microbialites'. Both carbonate mineralisation and dissolution are facilitated by benthic microbial communities: primarily bacteria and sometimes also microeukaryotes such as diatoms. These microbial communities are organised in mats, organo-sedimentary structures that develop as benthic biofilms (Figures 11.33 and 11.34; Chapter 10, Figures 10.28, 10.29; van Gemerden 1993). They constitute robust ecosystems, semi-isolated from their immediate environment by a self-constructed polymer matrix (Decho 1990; Decho and Guiterrez 2017). The resulting microbialites are organo-sedimentary deposits which accreted through benthic microbial community trapping and binding of detrital sediment and/or mineral precipitation (Figure 11.35; Burne and Moore 1987; Riding 2011). Current research focusses on the large metabolic diversity of microbial mats, which is driven by the availability of various electron donors and acceptors, and mediated by environmental controls (Visscher and Stolz 2005). The metabolic reaction rates of all members of the microbial mat community collectively determine the overall change in alkalinity and consequently, the dynamic balance between mineral precipitation and dissolution (Visscher and Stolz 2005; Pace et al. 2018). Microbial mats are reported to facilitate carbonate cementation (Diaz and Eberli 2021; Medina Ferrer et al. 2021) and hence the stabilisation of sediments in various depositional environments. Microbial mats play an important role as they can be opportunistic, are very adaptable and hence can rapidly fill ecological niches. Microbial mats can be considered a type of biofilm which has become more developed and complex, hosting a more diverse microbial community, and having a relatively high degree of structure and permanence (Visscher et al. 2022). They also build a substrate on which other types of biota can settle.

In the fossil record, the biology of microbial organisms is not fully understood, and some have been affiliated to *Microproblematica* (Flügel 1982a, 2002). They are, however, remarkably abundant in reef communities, acting as encrusters as for instance the *Lithocodium-Baccinella* complex of the Mesozoic (Rameil et al. 2010; Schlagintweit et al. 2010; Schlagintweit and Bover-Arnal 2012; Esrafili-Dizaji et al. 2020) and the *Archaeolithoporella-Tubiphytes* of the Late Palaeozoic (Vennin et al. 1997). M-factory sediments vary from thick successions of stromatolites and thrombolites, deposited in shallow tidal flats (Riding 2011), to the blocky microbialite-encrusted platform slope debris deposits where microbialites cemented the continuous supply of crusts and peloidal mud clasts (see Section 11.6 for further details).

Early cementation is characteristic of the M-factory and occurs from the shelf-margin to the upper slope (van der Kooij et al. 2010), consolidating coarse-grained sediments, sands and large blocks on the slope (Kenter et al. 2005; Kenter and Schlager 2009; Schlager and Reijmer 2009; Brandner and Keim 2011; Della Porta et al. 2014; Playton and Kerans 2018). The deeper water mud-mounds represent the fine-grained end member of M-factory sediments. A peculiar characteristic of the M-factory is the locally abundant presence of *Stromatactis*, defined as a series of cavities filled with fibrous syn-sedimentary calcite cement and locally, internal sediment (Stenzel and James 1995; Flügel 2010; Hladil 2005; James and Jones 2016). The origin of the cavities remains a matter of debate.

The mineralogy of the M-factory biota is dominated by HMC (Russo et al. 1997; Neuweiler et al. 1999; Schlager 2005) (Figure 11.28). Other carbonate mineralogies occur but in minor quantities. Mineral precipitates in modern microbial mats exhibit a range of morphologies (e.g. micrite, rhombs, dumb-bells, needles and spherulites), and mineralogies (e.g. aragonite, calcite, monohydrocalcite, vaterite and HMC to Ca–dolomite).

Figure 11.33 Examples of M-factory sediments; (a) layered microbial mats consisting of several layers of different microbes; northern Belize (James and Jones 2016); (b) intertidal microbial mats and two hurricane layers of sediment deposited in 1960 and 1985; Bahamas (Photograph: courtesy H. Wanless); (c) microbial mat in growth position in shallow lagoonal setting, Perros Bay, Cuba (Photograph: Courtesy of Emmanuelle Vennin).

11.5.4.2 Sediment Production Profile and Depositional Profile

The production profile of this factory shows consistent production rates across a wide range of environments and water depths (0–400 m) and can reach tropical production levels (Figure 11.27; Kenter et al. 2005). In contrast to the T-factory, fine-grained sediment export towards the surrounding basins is restricted. The typical depositional profile for M-factory systems is twofold: flat-topped shelfs, either land-attached or land-detached, and deeper marine mud-mounds (Figure 11.27). In the platform settings, wave-sensitivity controls biological distribution and morphological expression: (i) at the platform top, in the shallow water tidal flats stromatolites dominate; (ii) in high turbulence zones along the margin, ooids are formed, and microbial filaments can grow in cavities between coarser-grained sediments; (iii) below the immediate influence of waves, in the upper slope of carbonate platform margins,

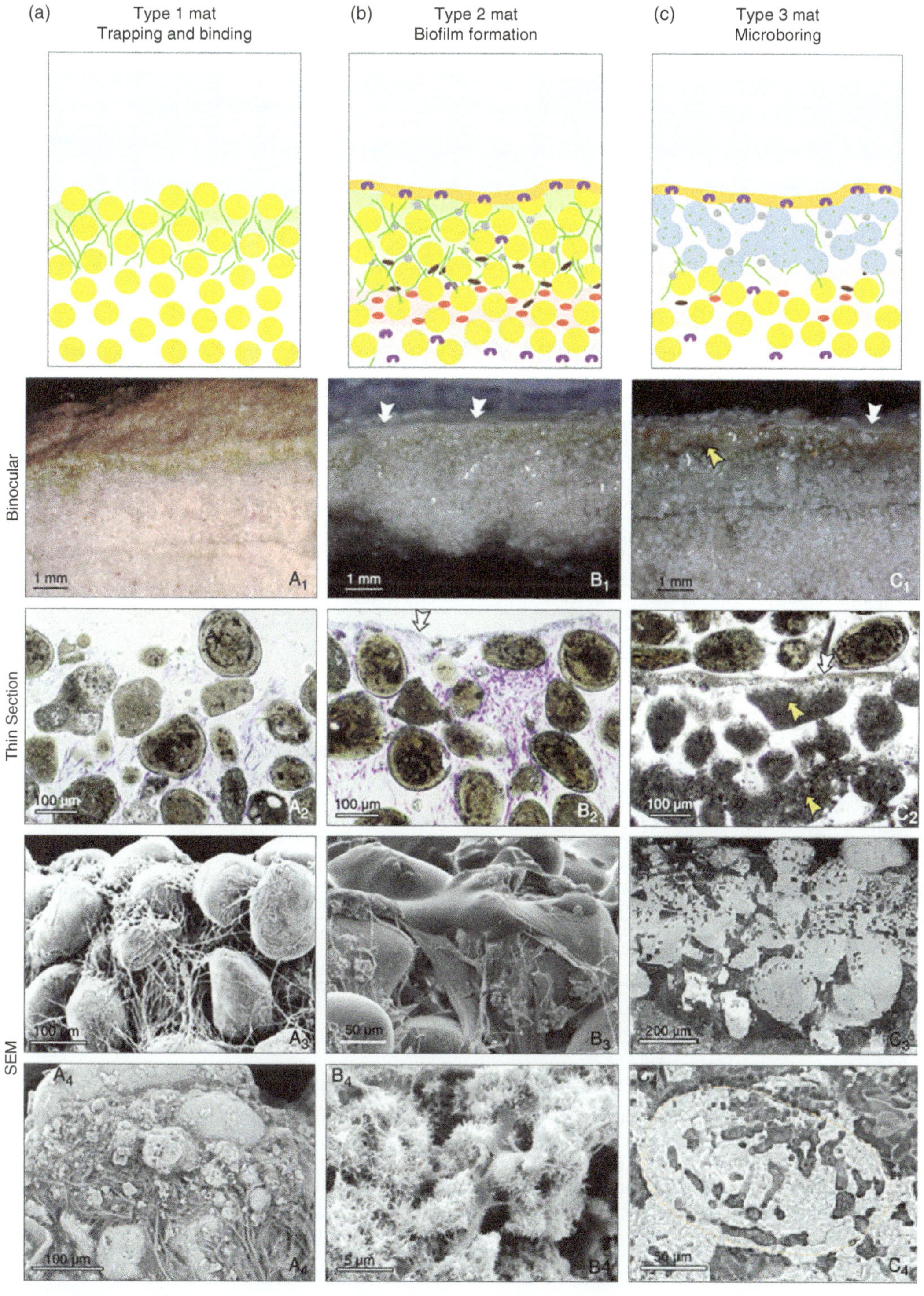

Figure 11.34 Micrographs illustrating the main features of the surface mats responsible for the formation of modern marine stromatolites (*Source:* Dupraz et al. (2011) / Springer Nature). Stromatolite lamination results from the iterative growth at the surface of the build-up of three mat types: (1) Type 1 mats are dominated by filamentous cyanobacteria, which trap carbonate sand grains through EPS stickiness (A1-4); (2) Type 2 mats consist of a continuous biofilm, draping the stromatolite surface (white arrows in B1-2); the biofilm is composed of extracellular polymeric substances (EPS, B3) containing numerous heterotrophic bacteria. This biofilm rapidly lithifies as a result of precipitation of aragonite needles within the biofilm (B4); (3) Type 3 mats are characterised by endolith-infested ooid grains (yellow arrows in C1-2), which appear grey and are fused together, below a surface biofilm. (A1-B1-C1) binocular micrographs, (A2-B2-C2) petrographic micrographs, (A3-B3-B4) high vacuum scanning electron microscope images (SEM; chemical drying), and (A4-C3-C4) cryo-SEM (frozen samples) images.

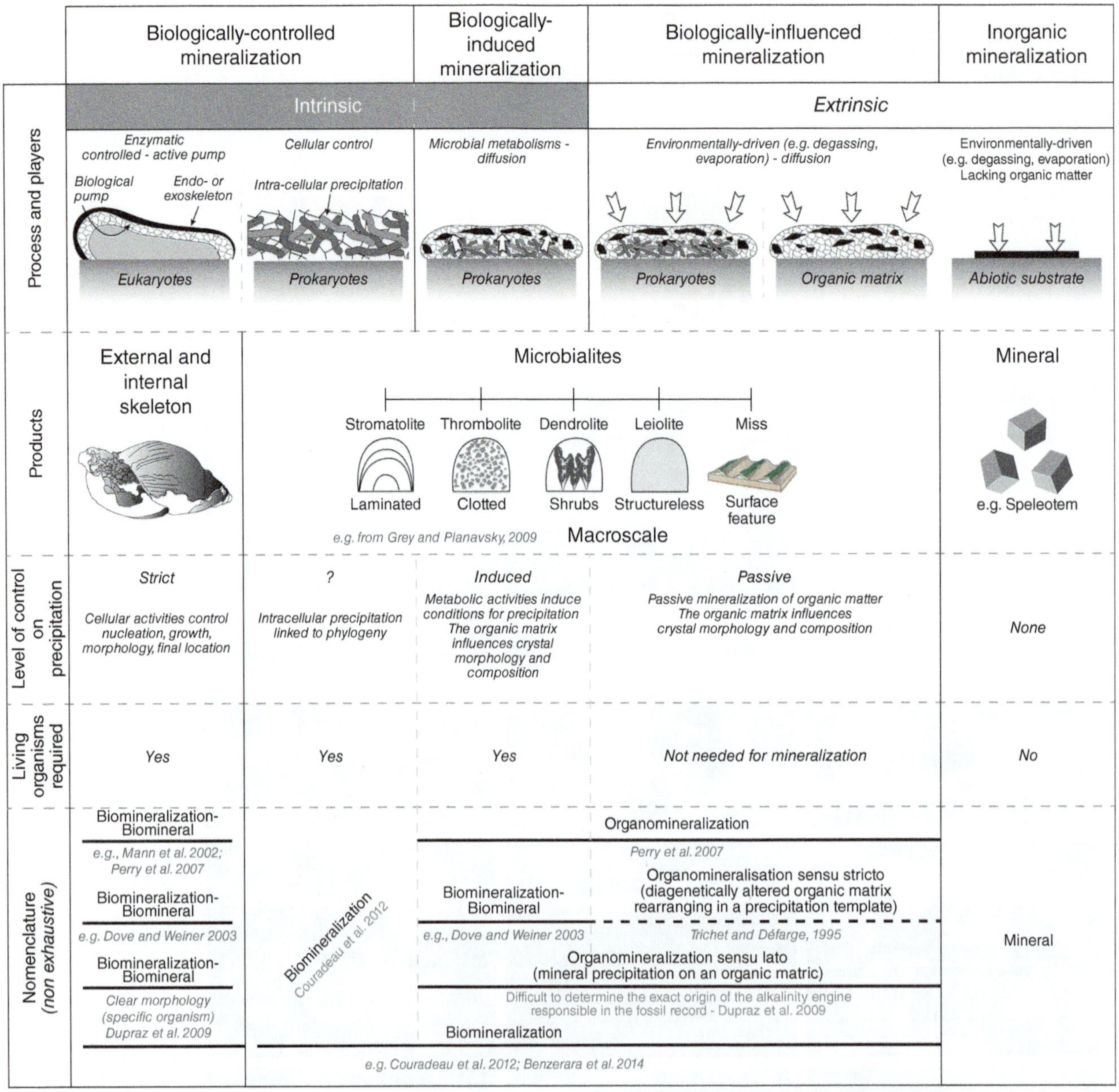

Figure 11.35 Classification of mineralisation types and processes as they relate to living (biotic) and non-living (abiotic) organic matter (modified from Dupraz et al. 2009). The biologically *controlled* mineralisation implies a stronger, possibly genetic control of the organism on precipitation (intracellular carbonates in some prokaryotes; Couradeau et al. 2012; Benzerara et al. 2014). The biologically *induced* mineralisation is characterised by microorganisms directly modulating the chemistry of their surrounding environment by their metabolic activity (Arp et al. 2001). In the biologically *influenced* mineralisation, the organic microbial mat and the extracellular polysaccharides (EPS) serve as a substrate and nucleation point for carbonate precipitation when adequate chemical conditions are provided by external factors (Dupraz et al. 2009).

microbial organisms form crusts and cements, and locally mud-mounds of varying sizes may form.

Depositional features typical for the M-factory include the following:

i) *Fall-in beds*: These arise from the absence of a rim in some of the steep-sided, flat-topped M-factory platforms, where the transition of the platform interior towards the slope at the platform edge shows a deepened margin (Della Porta et al. 2004; Reijmer 2021) (Figure 11.36; Kerans and Harris 1993; Tinker 1998). The basin-ward dipping beds at the edge of the platform have an angle of repose varying between 5° and 10° and connect the (sub)horizontal platform lagoon sediments, with the steeply inclined slope deposits, 30°–35°. Shelf margin profiles of the Jbel Bou Dahar (Scheibner and Reijmer 1999), the Sella carbonate platform by Keim and Schlager (2001), and the

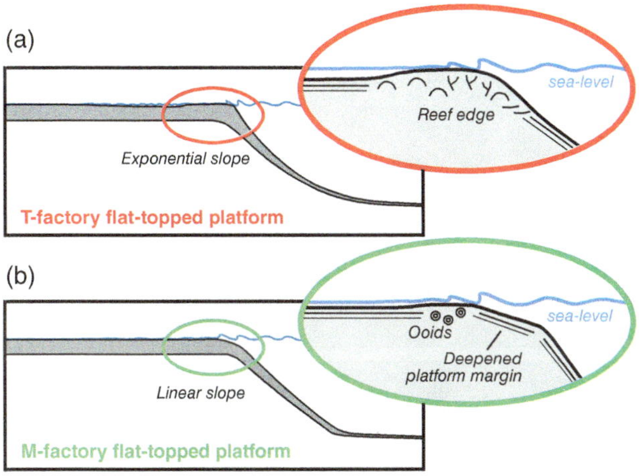

Figure 11.36 Comparison of platform margin constructions: (a) reefal margin of a T-factory platform, (b) fall-in beds typical for many M-factory platform margins (*Source:* Adapted from Kerans and Harris 1993 and Tinker 1998, in Reijmer 2021).

Sierra de la Cuera (Della Porta et al. 2003) all illustrate this pattern.

ii) *Slope shedding*: In contrast with the T-factory, the M-factory production has the capacity to develop high *in situ* production rates in the foreslope environment. This environment is fairly constant through time, independent of relative sea-level changes and can produce enough sediment to prograde irrespective of changes in accommodation or depth, leading to a phenomenon called 'slope shedding' (Kenter et al. 2005).

iii) *High-angle slopes*: M-factory slopes tend to develop high slope angles as microbial and other rapid cementation processes stabilize the upper parts of the slope (Keim and Schlager 2001; Della Porta et al. 2003; van der Kooij et al. 2010; Playton et al. 2010).

iv) *Mud mounds*: The mud mounds are a very typical feature of the M-factory, of which there are different types described notably from the Palaeozoic (see Section 11.6).

11.5.5 P-Factory (See Also Chapter 13)

The modern planktic (P-factory) biota comprises planktic foraminifera, coccolithophores and pteropods, which share their environment with other pelagic silicate producing organisms like radiolaria and diatoms (Figure 11.37). They occur mostly in the upper part of the water column in the open oceans, embayments, intra-cratonic basins and in intra-shelf basins. At present planktic foraminifera cover a wide latitudinal and temperature zone distribution. Those populating the photic zone often show a symbiotic relationship with photosynthesising algae, which is vital for species living in warm oligotrophic waters with ample light (Kucera 2007). Planktic foraminifera can be omnivores or selective carnivores (Hemleben et al. 1989). They show periodic pulses of primary productivity, linked to seasonal variations in sea-surface temperature. Planktonic foraminifera developed during the Late Triassic–Jurassic (Grigyalis and Gorbachik 1980; Kucera 2007). Another well-known group of marine zooplankton are pteropods, which are a group of heterobranch gastropods that originated in the Late Cretaceous.

The calcification of coccolithophores is driven by photosynthesis, which limits their occurrence to the photic zone (e.g. De Vargas et al. 2007). They occur in a range of trophic modes, from auto- to hetero- and mixotrophic (e.g. Müller 2019). They are known to occur from the Late Triassic onward and have become an important part of the oceanic phytoplankton system since then (Bown et al. 2014). At present, their diversity patterns vary and correlate well with climate tendencies; warmer climates show a higher diversity (Bown et al. 2004). Production rates vary at the seasonal scale, with surface nutrients and water stratification being important parameters (e.g. Poulton et al. 2014).

11.5.5.1 Sediment Composition

In the P-factory the grain-size of the carbonate fraction is dominated by the intact fossil skeletons or their components (e.g. coccoliths, the plates of coccolithophores). The grain size of the calcareous biota varies between 10 μm and 100 μm for the coccospheres and 2–20 μm for the coccolith plates, a maximum diameter of 600 μm for planktic foraminifera (Milliman 1974a), and up to 1–2 cm as maximum shell length for pteropods (Figure 11.37; Scholle and Ulmer-Scholle 2003). Important non-carbonate components are radiolarians and diatoms. Radiolarians range in size from 100 μm to more than 2 mm (Scholle and Ulmer-Scholle 2003). Diatoms vary between 20 and 200 μm, but may be up to 2 mm.

The mineralogy of P-factory carbonates is dominantly LMC as represented by the coccoliths and planktic foraminifera (Figure 11.28; Milliman 1974b). Pteropods are the main aragonite suppliers (Milliman 1974b; Berger 1978). Non-carbonate pelagic components like radiolaria and diatoms both have an amorphous (opaline) silica skeleton (Haq and Boersma 1978).

11.5.5.2 Sediment Production Profile and Depositional Profile

The productivity of the P-factory depends on seasonal changes in surface nutrient availability (nitrate, phosphate and silicate), which varies together with water stratification influencing nutrient positioning throughout the seasons (e.g. Poulton et al. 2014). For planktic foraminifera the seasons, water masses and water depths are important parameters regulating species abundance and associated sediment production (e.g. Schiebel and Hemleben 2005).

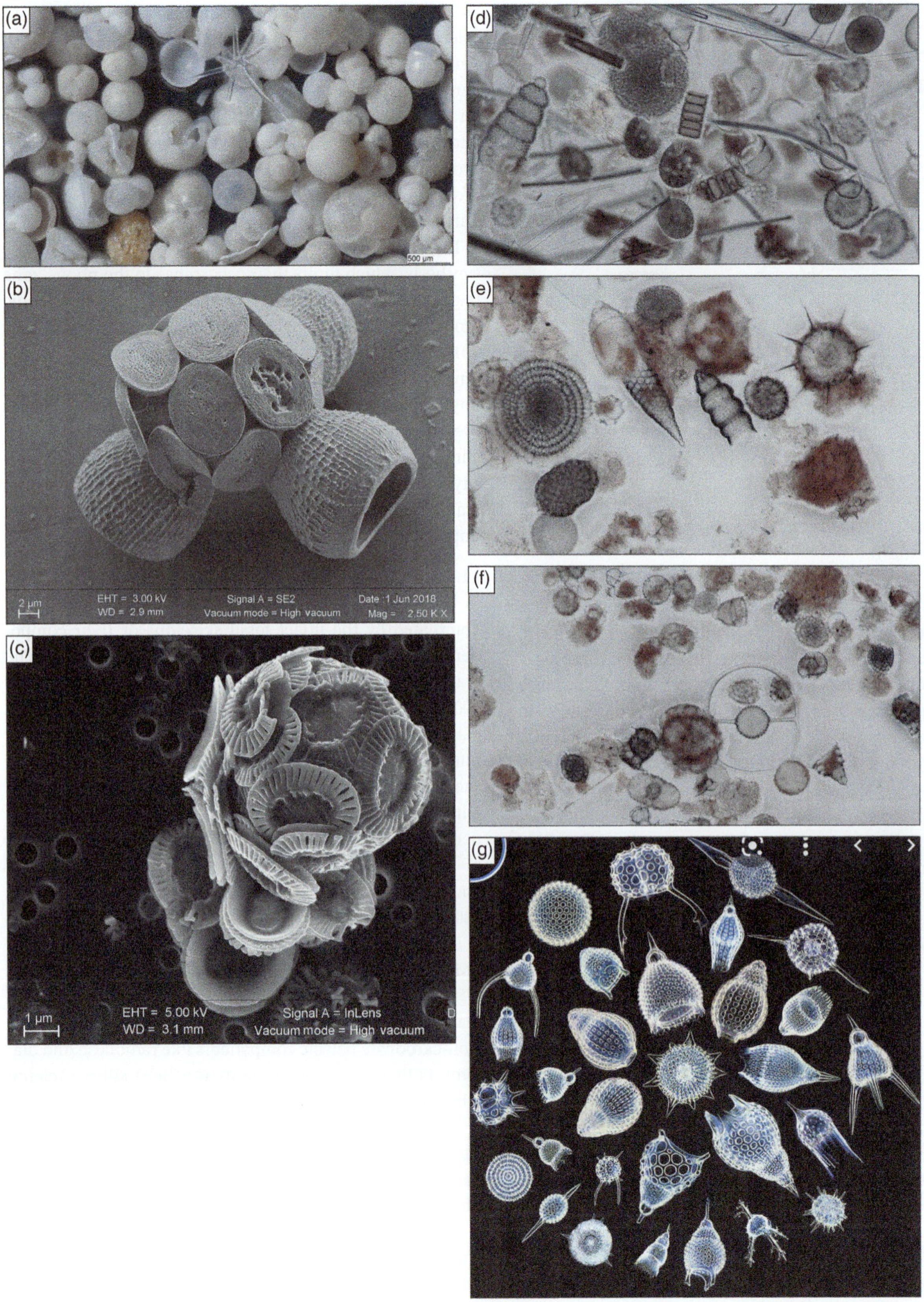

Figure 11.37 Biota components forming sediments of the P-factory; (a) planktic foraminifera; (b, c) Coccoliths (*Source:* Courtesy of Mickael Hermoso), (d–f) diatoms (*Source:* Courtesy of Emmanuelle Vennin) and (g) radiolaria.

The distribution and production of calcareous nannoplankton (coccolithophores) relates to zones with high organic productivity combined with quality and intensity of light, water temperature and latitudinal variations (Haq and Boersma 1978; Flügel 2010). Pteropods have a limited tolerance to changes in salinity and water temperature, and thrive in warm, tropical zones in the uppermost 500 m of the water column (Haq and Boersma 1978).

P-factory sediments, consisting of loose skeletons or parts thereof, may form sedimentary geometries in the deep sea such as mounded and sheeted drifts, valleys, moats, channels and sediment waves, as well as mass-transport deposits (as shown for the Cretaceous nannofossil oozes of the Chalk in the North Sea Basin; Surlyk and Lykke-Andersen 2007; Gale et al. 2015; Buls et al. 2017; van Buchem et al. 2018). Similar features are also known from the present-day Bahamas (Chabaud et al. 2016; Eberli and Betzler 2019; Mulder et al. 2019), and Maldives (e.g. Betzler et al. 2017a, b; Lüdmann et al. 2018a, b), where the deep-water deposits consist of a mix of P-factory and T-factory-derived sediments. P-factory slope steepness depends on the overall sand grain-size and mud content of the deposits and the general slope stability rules for loose granular materials.

11.6 Marine Carbonate Facies Models and Depositional Environments

Facies models are simplified communication and organisation tools to present the lateral relationships among contemporaneous sedimentary facies along a depositional profile in a consistent and logical manner (see also Chapter 2). As such they represent idealised or conceptual end-members that may serve as analogues when starting a study in a new area (e.g. Wilson 1975; Flügel 2010; James and Jones 2016). Refinements or simplifications of these schemes are logical results of new observations, be it at the microscopic, the macroscopic or regional context scale.

In modern environments, the study of carbonate systems along depositional profiles provides fundamental insights into the distribution of environmental factors and hydrodynamic regimes. Subsequent observations of geobodies and the biological and textural composition of facies, allow us to populate the depositional profiles with facies types, determine their affinity with the carbonate factories and provide a facies model. In the fossil record establishing the spatial variation at a given moment in time is more complicated but is achievable through a combination of 3D outcrop and subsurface information. A limitation of modern-day environments is, however, that they represent one snapshot in time during an interglacial stage in an ice-house period. Modern analogue carbonate systems were deposited during the eustatic sea-level rise of approximately 100/120 m amplitude, which started after the last glacial maximum at about 16,500 years ago (Smith et al. 2011; Törnqvist and Hijma, 2012; Lambeck et al. 2014). Although essential for understanding carbonate sedimentology, the modern-day observations should be treated with caution as indicators of the relative importance and nature of processes in the past.

The carbonate facies models presented below are illustrated where possible with examples from the present and complemented with selected case studies to document facies variability through time. This variability is a result of changing physico-chemical and biological conditions: climatic conditions (e.g. temperature fluctuations and rainfall-controlled siliciclastic influx), oceanographic conditions (nutrient-rich upwelling zones, currents and depth of thermocline) as well as the evolution of the carbonate-producing organisms (diversification and extinctions). When introducing each facies model, we refer to the different carbonate factories, and where appropriate provide a short overview of evolutionary trends. The impact of this evolution on the stratigraphic architecture of carbonate deposits through time is summarised in Section 11.7.

The facies models are subdivided into land-attached carbonate systems (shelf, ramp and open shelf), land-detached carbonate systems (isolated platforms and mounds) and seaway and valley fill carbonate systems. Apart from these purely carbonate-dominated systems, three types of mixed systems are briefly treated at the end of this section: carbonate-siliciclastic, carbonate-organic matter-rich and carbonate-evaporite systems (see also Section 9.2 and Chapter 10).

11.6.1 Land-Attached Carbonate Systems

11.6.1.1 Land-Attached Shelf Facies Model

The characteristic environments of the carbonate shelf depositional profile are the flat-top, the high angle slope, normally accompanied by a rim structure at the shelf-edge/slope break, and the sharp to gradual toe-of-slope to basin transition (Figure 11.38). Critically, the presence of the rim, as a reef, grainy shoal, or deepened margin, enhances the separation between the higher energy basin-facing side of the shelf at the shelf break, from the protected, lower-energy, flat-top of the shelf. Where the rim is absent, large ocean waves and swells can easily sweep across the whole shelf, which creates an open-shelf distally steepened ramp profile, which is treated as an independent facies model below.

The carbonate shelf profile hosts a large number of environments, where different physical, chemical and biological factors operate. This was captured in the rimmed shelf facies model of Wilson (1975), which remains a useful way to display all possible environments that can be encountered in a shelf setting, with the understanding that

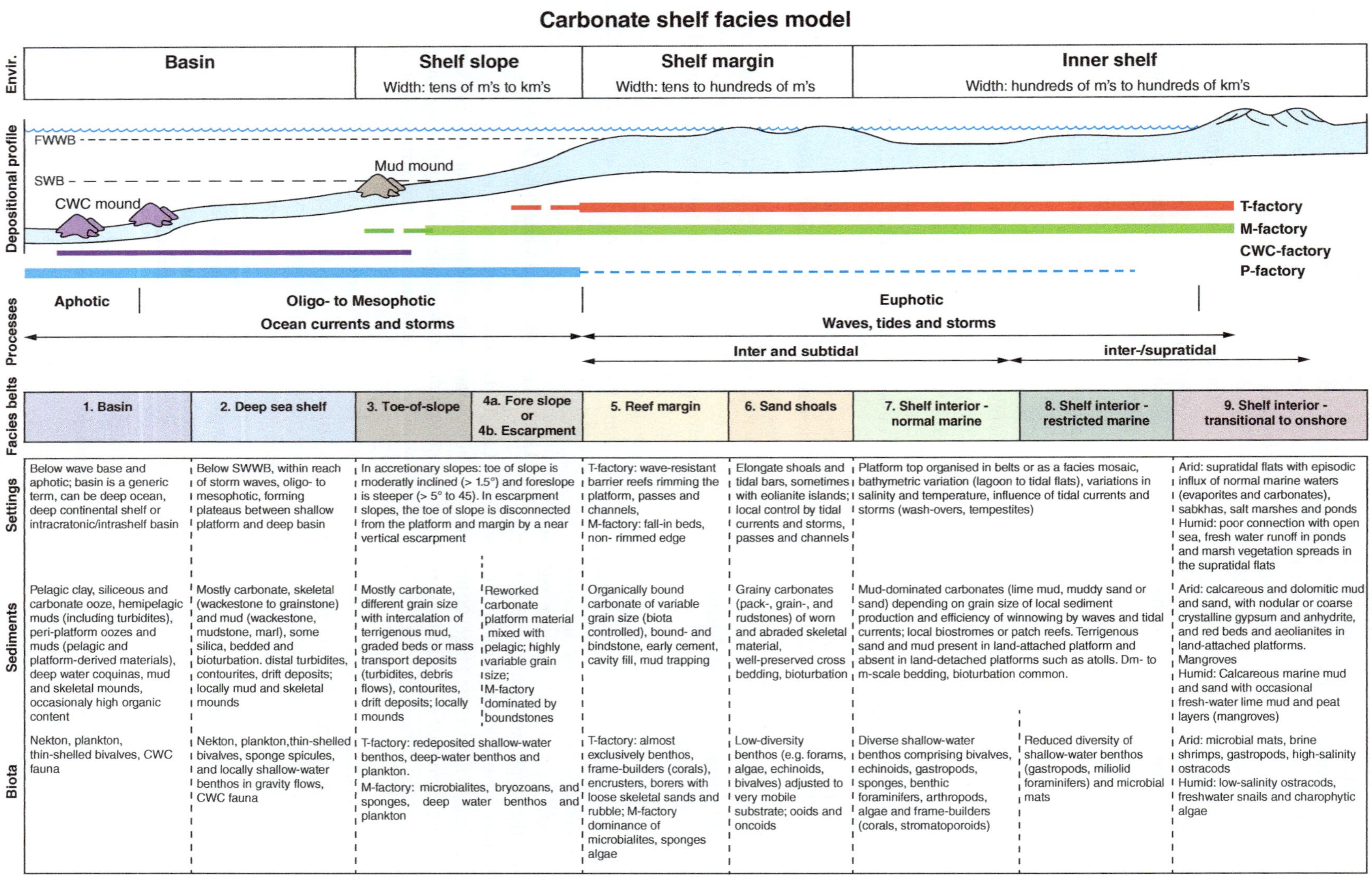

Figure 11.38 Conceptual carbonate shelf facies model (*Source:* Adapted from Wilson 1975; James and Jones 2016). Note the scale indication for the facies belts; the inner shelf facies belt is generally the place where the largest volumes of sediment are produced and preserved. The potential productive areas of four carbonate factories are indicated along this depositional profile, the C-factory is absent here. Note that biota from different carbonate factories may occur together. The preservation of the sediment can be at different places along the profile depending on sediment transportation by storms, ocean currents and tides and the available accommodation.

the size and number of facies belts present may differ from case to case and through time. Nine facies belts are distinguished, which can be grouped in (i) basin (facies belts 1 and 2), (ii) slope (3 and 4), (iii) platform margin (5 and 6) and (iv) inner platform (7–9) environments (Figure 11.38).

The potential distribution of the different carbonate factories is also shown in Figure 11.38. Their presence and distribution depends on the environmental conditions and coexistence of different factory associations along the same depositional profile may occur.

Note the indication of the width ranges for the different facies belts. The belt that covers by far the largest areal surface is the inner shelf, its width can be up to hundreds of kilometres, and it thus represents normally the most important part of the carbonate production in the shelf settings. The shelf margin, although important for the architecture, is normally of a reduced width, in the order of tens to hundreds of metres. The shelf slope also has a limited width (tens of metres to kilometres).

11.6.1.1.1 The Basin Environment

Two facies belts are distinguished: facies belt 1, the deeper and more distal basin, which is well below storm wave base and depends on settling from suspension and ocean floor currents for sediment input, and facies belt 2, the shallower deep sea shelf, which is indirectly within reach of storm activity through the sediment input by storm-generated density currents (Figure 11.38). The deep basin facies belt 1 occurs in water depths in excess of hundreds of metres, well below storm wave-base. Depending on the oceanographical conditions the water column may show a density and/or temperature stratification. The sediment consists mostly of fine-grained calcareous oozes produced by planktonic organisms (P-factory) with an admixture of fine argillaceous and siliceous material brought in either by wind or ocean currents. Closer to the platform, shallow-water derived carbonate muds might also be present (periplatform ooze; Schlager and James 1978). The calcitic sediments are only preserved above the CCD, whereas aragonite is only preserved above the shallower aragonite compensation depth (ACD) (Chapter 13, Figure 13.13).

The deep shelf facies association (facies belt 2) is characterised by a water column of tens to a few hundreds of metres, which is fully oxygenated, has a normal marine salinity, and active bottom currents. The association is located below normal storm wave base but occasionally receives storm deposits. The carbonates are derived from different sources: (i) *in situ* production by the cold-water coral mound fauna (CWC-factory), (ii) calcareous oozes (P-factory) and (iii) gravity flow deposits supplying bioclasts of benthic origin or muds from the shallow-water platform margin and top. These muds may be pure carbonates or siliciclastics direct from the hinterland or previously deposited in a marine environment. Bottom and slope currents (contourites) may further control sediment distribution as in the toe-of-slope setting described below.

11.6.1.1.2 The Slope Environment

The slope is an important environment in the depositional profiles of carbonate platforms, since it can act as a sink for excess sediment produced on the platform top, a locus of sediment production or as a zone of bypass, thus influencing the geometries of the platform margin. Three end members of slope morphology are distinguished (Figure 11.39): (i) the accretionary slope, where there is a gradual interfingering of platform, slope and basinal sediments (James and Mountjoy 1983), (ii) the bypass slope with neither deposition or erosion and (iii) the erosional slope, whereby the platform top and lower-slope and basinal sediments are physically detached. The erosional vigour, in particular of bottom currents (contourites), has a significant influence on the shaping of the erosional slope (e.g. Betzler et al. 2007). Both the neutral and erosional/escarpment slopes are conditions which, if they persist over a longer time, can lead to steep, locally vertically aggrading

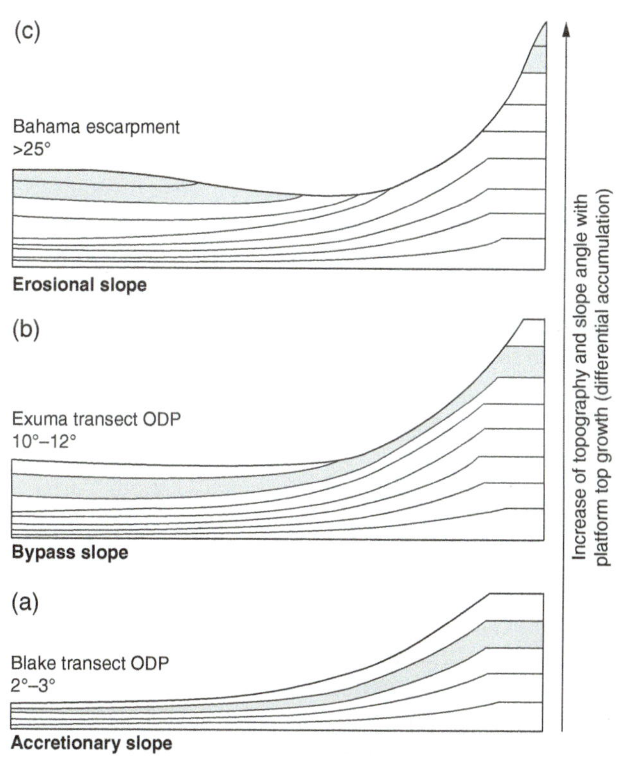

Figure 11.39 General slope morphologies because of the balance of erosion and deposition (*Source:* Adapted from Schlager and Camber 1986; Schlager 2005). Sediment supply is generally from the platform top, whereas erosion along the slope is in many localities controlled by contourites, vigorous long-shore currents that control erosion, transportation and deposition (e.g. Betzler et al. 2014, 2017).

carbonate platform margins. This feature is represented in the general facies model of Figure 11.38 as a modifier of facies belt 4, since it is a common feature in the geological record.

The curvatures of many accretionary slope profiles obey mathematical functions that provide clues to the transport regime on the slope and to sediment composition (Kenyon and Turcotte 1985; Adams and Schlager 2000; Schlager and Adams 2001). In a synthesis of commonly observed clinoform shapes in seismic profiles, Adams and Kenter (2014) distinguished three accretionary slope shapes. Although these are considered typical for particular carbonate factories (Figure 11.40), they cannot be considered as unique, due to the strong influence of the local hydrodynamics on the slope morphology.

The *sigmoidal slope profile* occurs in detrital accumulations, where reefs or early-lithified sands may occur as lenses, and commonly occurs in C-factory carbonate systems. The sigmoid profile has an upper convex part and a concave lower part and corresponds to a Gaussian mathematical expression (Adams and Schlager 2000). This profile resembles those of Holocene siliciclastic passive margin slopes but is the *least* common among carbonate margins (Adams and Kenter 2014). Sigmoidal margin profiles are dependent on mechanical reworking and a high debris-to-framework ratio (Pomar and Ward 1995). They are documented from the Great Australian Bight (Feary and James 1998; James et al. 2000; Saxena and Betzler 2003; James et al. 2004), and the cool-water carbonates of the southern Australia platform (James and von der Borch 1991).

The *concave slope profile* is typical for wave-resistant reef or lithified-sand-shoal rimmed shelves of the T-factory (Adams and Schlager 2000), such as part of the Tobacco Cay of Belize carbonate margin (James and Ginsburg 1979; Nolting et al. 2018). It results from the interplay between sediment input at the shelf edge and variably efficient downslope sediment transport. The slope pattern is caused by (i) a general fining of sediments downslope, which is associated with a decrease in slope angle, a trend which is often modified by erosional bottom currents (Rendle et al. 2000; Rendle and Reijmer 2002; Betzler et al. 2007; Adams and Kenter 2014); (ii) shallow longshore currents redistributing the sediment by removing the fines (Betzler et al. 2014); and (iii) the consolidation of the sediment by early cementation (Grammer et al. 1993, 1999) or microbial processes (Tahiti: Séard et al. 2011; Bahamas: Reolid et al. 2017a) which may occur at various positions along the depositional profile.

The *planar or linear slope profile* is controlled by lime sand and rubble at the angle of repose (Adams and Schlager 2000). It is typical for M-factory dominated systems where the microbial boundstone-dominated, deep and oligophotic sediment production sites generate rubble, boulders and sand along the slope itself (e.g. Keim and Schlager 2001). The planar slope has higher slope angles if microbial and rapid cementation processes stabilise the upper parts of the slope (Keim and Schlager 2001; Della Porta et al. 2003; van der Kooij et al. 2010; Playton et al. 2010). For the Sella platform (Triassic, Italy) the coarse-grained slopes are up to 35°; even 45° within the upper slopes where microbial binding prevails. Planar slopes are beautifully displayed as prograding clinoforms of the Triassic carbonate platforms in the Dolomites (e.g. Bosellini 1989; Kenter 1990).

The slope environment is divided into the toe-of-slope facies belt (3) and the fore-slope facies belt (4) (Figure 11.38). The toe-of-slope setting has similar water depth, wave base conditions and oxygen level to basin facies belt 2, with the main difference being its proximity to the platform and consequently to carbonate material derived from the platform margin and top. Once deposited, this sediment can be reworked by bottom currents, such as contour currents that locally winnow the fine sediments, reshaping the grain-size distribution such that coarse-grained sediments occur at the toe-of-slope instead of the more common fine muddy sediments (Betzler et al. 2014). Rebesco et al. (2014), following the definition of Stow et al. (2008), defined large accumulations of sediments constructed by bottom currents as 'contourite drifts' or simply 'drifts' and sediment deposited or substantially reworked by the persistent action of bottom currents as 'contourites' (Eberli and Betzler 2019).

Differentiation of contourites from turbidites, with which they are often interbedded, is addressed in Chapter 12, but helpful attributes include (James and Jones 2016; Reolid et al. 2020; Hüneke et al. 2021, 2023): (i) local gradational bases of ripple cross-stratified units; (ii) intercalated layers or drapes of mud within the cross-stratified unit; (iii) common internal erosion surfaces; (iv) lack of an ordered Bouma sequence like turbidites have; and (v) internal condensed horizons with Fe-Mn-PO_4 encrustations and hardground ichnofauna. In current-dominated areas, contourites are also often associated with cold-water coral mounds (CWC factory) as the bottom current provides food for the heterotrophic coral community and sediment to build the mounds (Hebbeln et al. 2016). Erosional features result both from surface and deep ocean currents. Erosion by surface currents produces abraded surfaces and terraces in 300- to 500-m water depth (Eberli and Betzler 2019). Large, linear erosional features from bottom currents are generally subdivided into three types: (i) moats, (ii) contourite channels and (iii) marginal valleys (Hernández-Molina et al. 2008; Rebesco et al. 2014). Under oxic conditions burrowed carbonates and red bioturbated shales accumulate, in dysoxic environments mainly structureless carbonates and green mudstones are found, and under anoxic conditions black laminated, organic-rich carbonates, and mudstones accumulate.

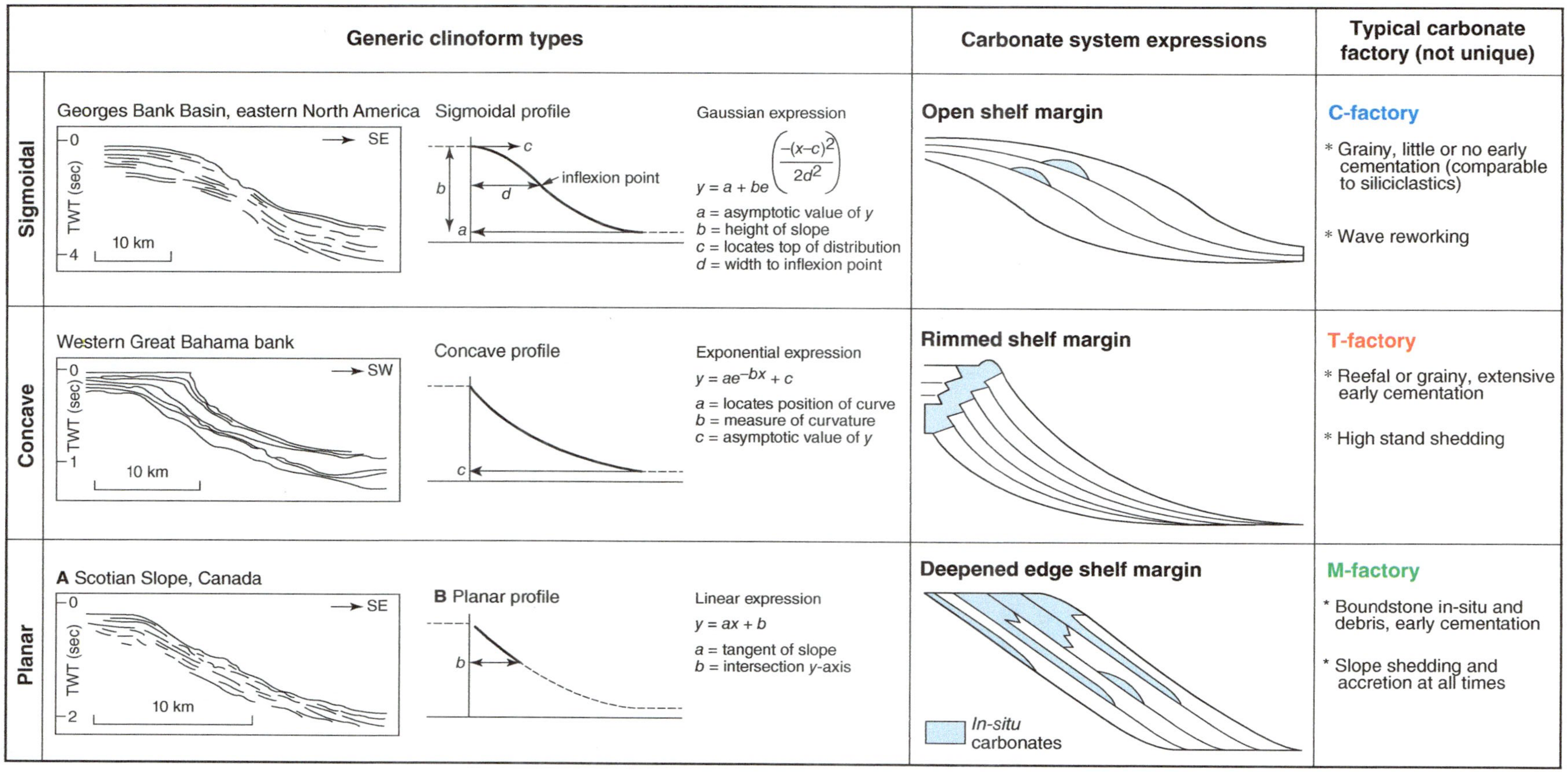

Figure 11.40 Three generic types of clinoform geometries in accretionary slopes are distinguished in both siliciclastic and carbonate systems. In carbonate systems, the sigmoidal slope is often observed in distally steepened ramp type settings typical for the C-factory, the concave slope is a typical feature of the rimmed carbonate shelves of the T-factory in tropical regions, and the planar slope with the deepened edge margin is a common feature of microbial dominated M-factory platforms (*Source:* Adapted from Adams and Kenter 2014; Reijmer 2021; Schlager 2005). Although these types are characteristic for certain factories, they are by no means unique.

The fore-slope facies belt (4) is positioned between storm and fair-weather wave base. Common sedimentary features observed are slumps, small bio-constructed mounds, wedge-shaped foresets, and debris of various sizes. Slope sediment accumulations of unstable debris can produce inclined surfaces as steep as 30°.

The composition of the sediment can vary greatly in size, and three classes are distinguished:

i) Fine-grained sediments (muddy carbonates) which consist of skeletal and non-skeletal sediments that are platform margin and platform lagoon derived (Perkins 1972; Droxler et al. 1983; Heath and Mullins 1984; Boardman et al. 1986) and include muds produced by whiting events (Shinn et al. 1989; Robbins and Blackwelder 1992; Robbins et al. 1997; Morse et al. 2003; Dierssen et al. 2009; Purkis et al. 2017) and the fall out of planktonic production (P-factory). The loci of planktonic production are determined by oceanographical conditions such as upwelling and seasonal blooms. Across large areas of shallow platforms copious quantities of carbonate muds are produced by the T- and M-factories, which are subsequently transported off-platform by tidal, wind-driven, and cascading density currents (Eberli and Betzler 2019). The amount of exported sediments decreases with increasing distance away from the platform margin. If the sediment–water suspension is dense enough, it can move downslope as a sediment gravity flow and spread out across the lowermost slope and adjacent basin as a mud turbidite. When lithified, the rocks consist of clay-sized and silt-sized carbonate mudstones, chalks and marls, with wackestone to muddy packstone textures.
ii) Medium to coarse-grained sediments (sand to coarse debris) result entirely from sediment gravity flows derived for the most part from the platform margin complex (Figure 11.41). Sand-to-mud ratios may change through time for instance between glacial and interglacial intervals, and in space, as an effect of leeward-windward grain-size sorting variations (Rendle-Bühring and Reijmer 2005; Reijmer and Andresen 2007). The material of the T-factory comes from ooid shoals, reefs, the upper slope and from the protected neritic environments behind them. It consists of ooid, skeletal and peloidal sands and reef clasts (Haak and Schlager 1989). This shallow-water sediment is swept outward by strong tidal currents flowing through reef passes and across the shoals. The material can reside for short periods on the uppermost slope and is then episodically transported into deeper water by grain flows or turbidity currents. The upper part of the T-factory slope may be stabilised by microbial crusts (Reolid et al. 2017b; Diaz and Eberli 2021) or early cementation processes (Grammer et al. 1993, 1999). Similar processes were also observed at the platform margins of Tahiti (Séard et al. 2011) where slopes may become over-steepened resulting in large-scale destabilisation processes, with margin collapse and slope failures (Jo et al. 2015; Principaud et al. 2015, 2017, 2018; Schnyder et al. 2016; Le Goff et al. 2020). The planar slope profiles of M-factory sediment production were described above. These slope sediments are dominated by micrite, carbonate sand and rubble (Keim and Schlager 2001; Della Porta et al. 2014). In contrast, in the T-factory the slopes are dominated by a coarser sandy grain size (Schlager and Reijmer 2009).
iii) Coarse (i.e. pebble grade and above) carbonate debris may contain breccias, olistoliths and conglomerates interbedded with, or in a matrix of, muds and sands. The depositional processes include coarse-grained turbidity currents and debris flows. Deposits are generally poorly sorted, massive, chaotic and ungraded sediments occurring as thick-bedded lenticular beds (e.g. Eberli 1991; Floquet and Hennuy 2003; Principaud et al. 2015; Wunsch et al. 2018; Le Goff et al. 2019, 2020). James and Jones (2016) classified the slope conglomerates depending on the origin of the clasts (margin/slope derived), their form (plate, chip and boulder) and their size, with fragments ranging from cobble-sized to tens of metres (the largest, up to >100 m in size, are called olistoliths) (Figure 11.41; e.g. Devonian, Canada: Mountjoy 1965a, b; Mountjoy and Playford 1972; Cretaceous, Albania: Le Goff et al. 2019, 2020).

11.6.1.1.3 The Shelf Margin Environment

The carbonate shelf margin or rim is a critical sedimentological feature. It consists either of a string of more- or less continuous carbonate bio-constructions (bioherms, reefs; facies belt 5) or bioclastic, grain-supported, early-cemented carbonate sand shoals (facies belt 6) (Figure 11.38). It absorbs most of the wave energy from the open sea and protects the much calmer environment of the inner platform.

The size and shape of a bioconstructed shelf margin (facies belt 5) (Figure 11.38) is dependent on water energy, steepness of the fore-slope, organic productivity, amount of frame construction, binding, or trapping, and the frequency of subaerial exposure events and consequent cementation. T-factory carbonate platforms often show a distinct rim formed by reefs and islands: the classical reef-rimmed margin of modern coral–algal assemblages. Well-known present-day, land-attached, reef-rimmed shelves dominated by the T-factory are the Great Barrier Reef, Australia (Webster and Davies 2003), New Caledonia (Montaggioni et al. 2011) and Belize (Purdy et al. 1975; Purdy and Gischler 2003), where they form coral ecosystems hundreds of kilometres long. Further examples of Quaternary coral reefs are given in Camoin and Hallmann (2022). Despite the large

Figure 11.41 Examples of carbonate debris flows: (a) internal organisation of beds (*Source:* James and Jones 2016 / John Wiley & Sons), (b) outcrop photo of metre scale beds of debris flows with clasts of cm to dm size (Upper Jurassic, Oman, W. Muyadin; Courtesy of Frans van Buchem) and (c) Olistolith in outcrop with person for scale (*Source:* James et al. (2015) / John Wiley & Sons).

variability of reef sub-environments, a broad pattern can be recognised in modern and fossil reefs using criteria such as the rock fabric, the organism types and morphologies, and early diagenesis patterns (reviews in Tucker and Wright 1990; Reading 2009). Barrier reef complexes are organised into reef 'crest' and 'flat' with back-reef lagoon and reef-front and fore-reef slope (Figure 11.42). Along the barrier reef, the

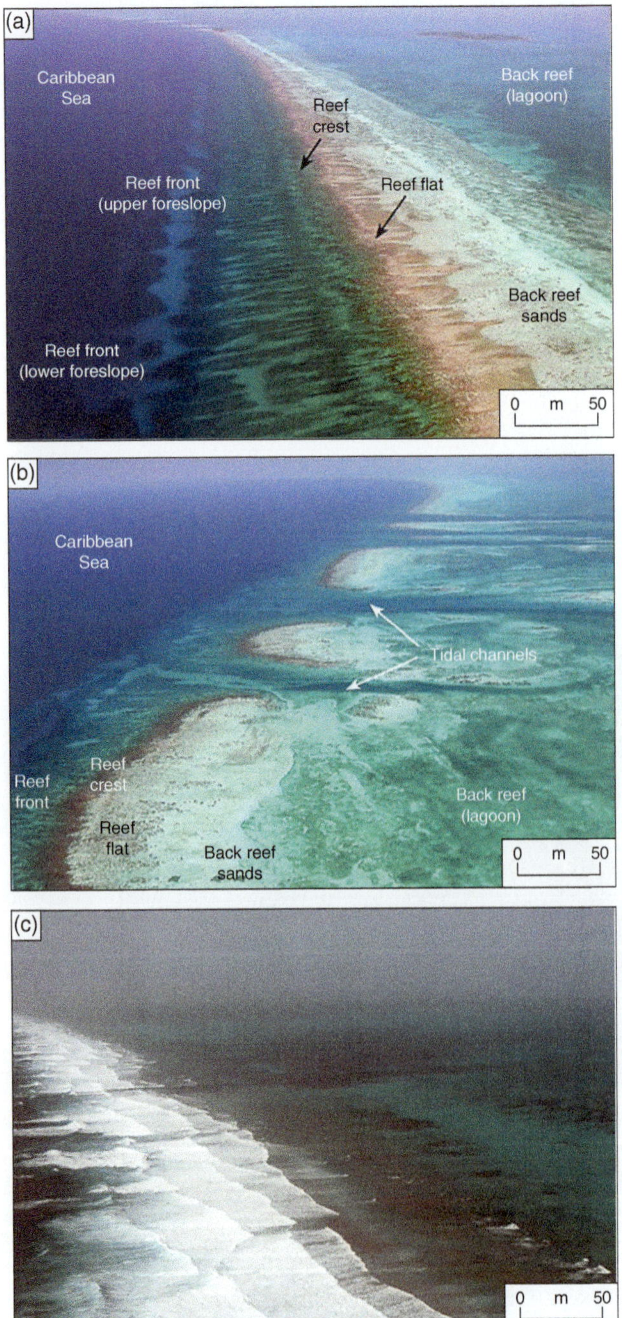

Figure 11.42 Carbonate shelf rims in present-day environments (*Source:* James et al. (2015) / John Wiley & Sons): (a) overview of the Belize shelf margin with a continuous coral reef rim, (b) close-up of the rimmed margin with tidal-inlet channels cross-cutting the rim and (c) same reef on a stormy day, showing how the interior part of the platform remains largely unaffected and tranquil.

edge position is favourable for carbonate-producing organisms because bottom shear is higher and the diffusive boundary layer between reef and water thinner. Thus, the reef edge has easier access to nutrients and is less likely to be buried by sediment (Rogers 1990). More specifically, corals grow faster in strong currents (Schutter et al. 2010) resulting in a higher carbonate production rate (Schlager 2005). Obviously bottom communities modify water circulation and bottom conditions through building barrier reefs (Ginsburg and Lowenstam 1958). Their growth forms and skeletal diversity are strongly controlled by hydrodynamic energy and available space: high energy favours bulbous forms and low energy favours platy and branching forms, and high diversity for the reef crest and deep water and low diversity for intermediate depths (Hallock 2005).

Vertical or steep reef walls and fringing reef slopes have been observed in both the recent and ancient (Betzler et al. 2007; Playton et al. 2010). Vertical accretion of coral reefs occurs during periods of fast, high magnitude sea-level rises, driven by glacio-eustasy, such as during the Holocene (Goreau 1969; Pirazzoli and Montaggioni 1988, Pirazzoli et al. 1990; Kennedy and Woodroffe 2002). Wave exposure was a shaping factor in this process (Rogers et al. 1993; Grigg 1998). Examples of steep reef-wall belts include the Pleistocene reef margin of Belize (James and Ginsburg 1979), the Great Barrier Reef (Davies et al. 1985), and St Croix US Virgin Island (Adey 1978; Hubbard et al. 1986). Microbialites can play an important role in sustaining vertical T-factory reef accretion during accelerated sea-level rise (Brachert and Dullo 1991, 1994; Reolid et al. 2017a; Yokoyama et al. 2018). Purely microbialite-dominated platform margins (M-factory) generally lack a raised rim (Harris 1993; Wood 1999; Keim and Schlager 2001; Webb 2001). The steep sided, flat-topped M-factory platforms with characteristic deepened margin ('fall-in beds') were described earlier (Figure 11.36).

Grainy carbonate shoals of facies belt 6 can also form wave-resistant barriers at or close to the platform margin. Winnowed platform-edge sands (5- to 10-m water depth to above sea level) develop as shoals, beaches and offshore tidal bars in fans or belts, or dune islands (James and Jones 2016). In Figure 11.43, an example is shown of a combination of facies belts 5 and 6, where the oolitic shoals sit just behind coral reefs towards the shelf edge (see also Rankey and Doolittle 2012). The environment is well oxygenated with a shifting grainy substrate. Early lithification is common and over time enhances the wave resistance of the shoal, reducing rates of lateral migration and erosion. The nature of the shoals depends on the topography (presence of reefs or islands), the orientation with respect to dominant winds and waves, and tidal ranges. Tucker and Wright (1990) provide an extensive review of carbonate shoreline sand systems, updated by James and Jones (2016). Barrier and beach sand bodies are observed in modern high-energy margin examples such as the Yucatan coastline in Mexico (Ward and Brady 1979),

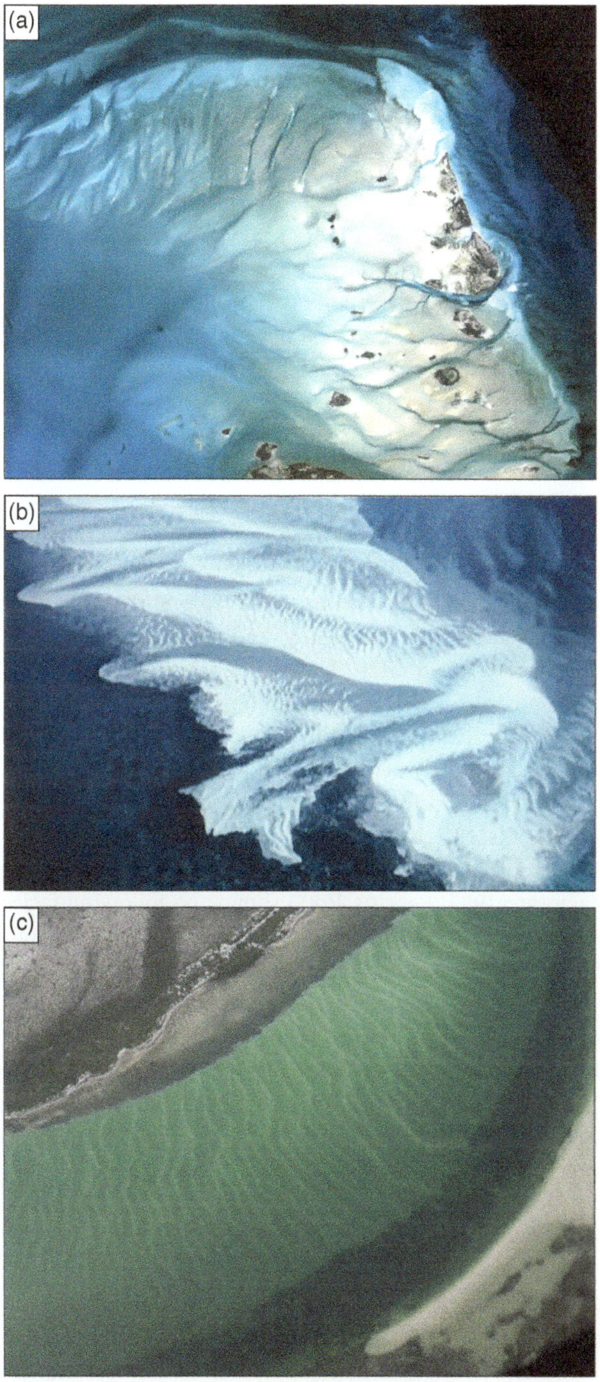

Figure 11.43 Carbonate sand shoal rim features in the Bahamas (from Eberli et al. 2007): (a) aerial photo of an oolitic shoal sitting just behind a coral reef shelf rim at Joulters Cay (Claudia Weinmann / Alamy Stock Photo) and (b) close-up of grainy tidal inlets spilling into the protected muddy lagoon (Joulters Cay). Note rapid lateral facies change, (c) large tidal channel with metre-scale sand waves at Caicos (*Source:* Eberli 2007 / Comparative Sedimentology Laboratory).

Shark Bay in Australia (Marshall and Davies 1975) and Belize coastline (Purdy and Gischler 2003). They develop where sediment supply exceeds removal by erosion and where adequate accommodation is supplied by platform topography.

11.6.1.1.4 The Inner Shelf Environment

Behind the shelf margin, three main environments can be distinguished (Figure 11.38): (i) the open–inner shelf domain of facies belt 7, a subtidal area of normal-marine open-water conditions; (ii) the more restricted, subtidal to intertidal inner shelf domain of facies belt 8; and (iii) the coastal, transitional marine domain of facies belt 9, a supratidal area with sporadic marine flooding. The relative areal extent of facies belts 7–9 varies. Two end-members are highlighted: a flat-topped scenario where restricted and transitional marine environments dominate (8 and 9), and an under-filled platform interior with a substantial lagoon (7), where the normal, open marine environment dominates.

The inner shelf (Figure 11.38) represents a zone of vigorous carbonate production. But paradoxically, although production rates are typically much higher than the rate of accommodation creation, unfilled accommodation is normal on the platform top, due to tidal and storm transport, notably of the finer-grained sediment fraction (e.g. Eberli 2013; Weij et al. 2019). The sediment is dominantly muddy, but grainy high-energy sedimentation also occurs as a result of wave, storm and tidal reworking, or fluvial discharge across the platform interior. These shallow-water areas are patchworks or mosaics of different sub-environments, characterised by different species, processes, grain sizes and carbonate production (Wright and Burgess 2005). Making a direct link between palaeo-water depth and depositional diversity of subtidal lithofacies in the reconstruction of depositional environments is, therefore, an oversimplification (Maloof and Grotzinger 2012; Purkis et al. 2015; Dyer et al. 2018; Weij et al. 2019). But, nonetheless, hydrodynamics processes are crucial in directing the distribution of facies in the shallow photic zone.

Extensive 'lagoonal' areas (up to 100 km wide) covering thousands or even tens of thousands of square kilometres are commonly observed in ancient carbonate shelf systems. Although some present-day lagoons are large and useful for our understanding of facies distribution (e.g. the lagoon and shelf behind the Great Barrier Reef), these are much smaller than the vast platforms that existed in past times. Modern lagoons are typically shallow (<10 m deep up to 20 m) and protected from strong wave action. But rarer deeper lagoons up to 80–140 m depth also exist (i.e. Belize margin: Purdy and Gischler 2003; Maldives: Betzler et al. 2016; Al Wajh platform lagoon, northern Red Sea, Saudi Arabia: Petrovic et al. 2022). A lagoonal area introduces larger facies variability in the platform interior. Sediments are derived from the platform margin reefs or shoal belts, as well as from the mudflats and channels in the restricted platform interior and marginal marine settings. Patch reefs can be common in deeper open lagoons, and their dimensions range from a few tens of metres to as much as 10 km (i.e. Great Barrier Reef; Marshall and Davies 1984). Their morphology, size and composition evolve in space reflecting a fractal distribution controlled by environmental and ecological parameters (Purkis et al. 2007; Schlager and Purkis 2015). Some reef mounds are built by organisms that lack significant frame-building potential (e.g. molluscs and *Halimeda*) and low relief mounds or ooid-shoal complexes have also been observed in Florida Bay, behind the Pleistocene Florida Keys. These mounds nucleate preferentially on antecedent topography (Bosence 1989b) and result from the alternation of mud deposition and erosion by storms with sea-grass colonisation and bioturbation (Wanless 1991). The sediment is mainly lime mud, muddy sand and clean sand, depending on the grain size of local sediment production and the efficiency of winnowing by waves, storms and tidal currents. The fine-grained sediments are mainly skeletal, including the products of skeletal abrasion, but also mud-sized skeletal constituents released through the decomposition of the organic tissues that bound them (Purdy and Gischler 2003). In addition, *in situ* deposition by planktonic organisms such as coccolithophores (Scholle and Kling 1972) and precipitation through whitings (Robbins et al. 1996; Yates and Robbins 2001; Reijmer et al. 2009; Harris et al. 2018) may contribute to mud production.

The shorelines of tropical to subtropical inner carbonate platforms (facies belts 8 and 9) are characterised by peritidal depositional environments, including nearshore, shallow subtidal zones, tidal flat and supratidal zones, and coastal marshes (Figures 11.38 and 11.44). The sediment is commonly affected by early diagenetic alteration by meteoric and dolomitisation processes. Three sub-environments are distinguished:

i) The subtidal zone, a permanently submerged area influenced by wave action and tides. The abundance and the diversity of the biota depend mainly on salinity and temperature controlled by the degree of tidal exchange (Reading 2009), but also any land drainage (Purdy and Gischler 2003).
ii) The intertidal zone is characterised by a mosaic of sedimentary and diagenetic features. The main criteria used for recognising intertidal deposits are related to exposure (desiccation cracks, fenestral porosity, evaporites and soils; Shinn 1969; Taylor and Illing 1969; Butler et al. 1982; Müller et al. 1990; Wilkinson and Drummond 2004; Gischler and Lomando 2007). Low-energy intertidal sediments are represented by mud, microbial flats and some mud-rich beaches, whereas spits and coarse-gained beaches indicate higher energy.
iii) The supratidal zone is occasionally flooded during high tides and storms. On low-relief prograding shorelines, it can be many kilometres wide. In arid and semi-arid settings exposed to wind deflation, extensive gypsum may precipitate (comparable to the coastal sabkhas of Chapter 10). In humid settings, large marshes and mangroves can develop as seen in the Florida Everglades, and along the Red Sea Coast.

Belize Platform

The present-day carbonate shelf complex offshore Belize in Central America is the largest in the Atlantic Ocean and a typical example of a T-factory system. The shelf complex is 250 km long and has a well-developed lagoon and bio-constructed barrier reef, which becomes an emergent fringing reef with aeolianites in northern Belize and adjoining Yucatán (Figure 11.45). The reef complex is an almost continuous barrier reef with numerous patch reefs and mangrove cays in its shoreward lagoon. The lagoon is 20–25 km wide and 5–20 m deep in the northern part, but where the southern third of the reef bends eastward, the lagoon becomes more than 40 km wide before opening into the Gulf of Honduras (Purdy and Gischler 2003). The distribution of modern facies, including the reefs, is strongly controlled by the underlying antecedent topography (Purdy 1974; Esker et al. 1998: Ferro et al. 1999; Purdy and Gischler 2003; Purdy et al. 2003). Three isolated buildups are positioned eastward of the shelf (Purdy and Gischler 2003).

A section across the Belize carbonate shelf (Rützler and Macintyre 1982; Perkins 1985) shows the following facies belt succession (using the scheme of Figure 11.38), from offshore to the coastline (Figure 11.45): (i) toe of slope facies belt (3), sits at the base of the escarpment, at about 100 m water depth, where blocks of limestone and finer sediment derived from the platform top constitute the talus slope (Miller and MacIntyre 1977); (ii) the fore-slope (4) is divided in an outer fore-slope of the reef (at least 12 m deep) and an inner fore-slope of the reef (1–12 m deep). The outer fore-slope is affected only by long-period storm waves. Towards the open ocean, the fore-slope reef passes over into a nearly vertical reef wall. The vertical reef wall is an important facies belt (Betzler et al. 2007; Playton et al. 2010) marking the platform margin destabilisation zone (Jo et al. 2015; Nolting et al. 2018, 2020), and not merely the transition zone between the outer fore-slope of the reef and the upper slope. The inner fore-slope is strongly affected by both storm waves (hurricanes) and waves related to normal trade wind conditions resulting in interdigitating coral buttresses and sandy troughs forming an extensive spur and groove system; (iii) the reef margin (5) is formed by a bio-constructed barrier reef crest in the high-energy surf zone consisting of a shallow rampart built of coral rubble and *Acropora palmata* (James and Ginsburg 1979); (iv) the back reef sand shoal (6) is 0.1–1.0 m deep and is subjected to strong currents and lagoon-ward transport of sediments. It displays a variety of sediment textures,

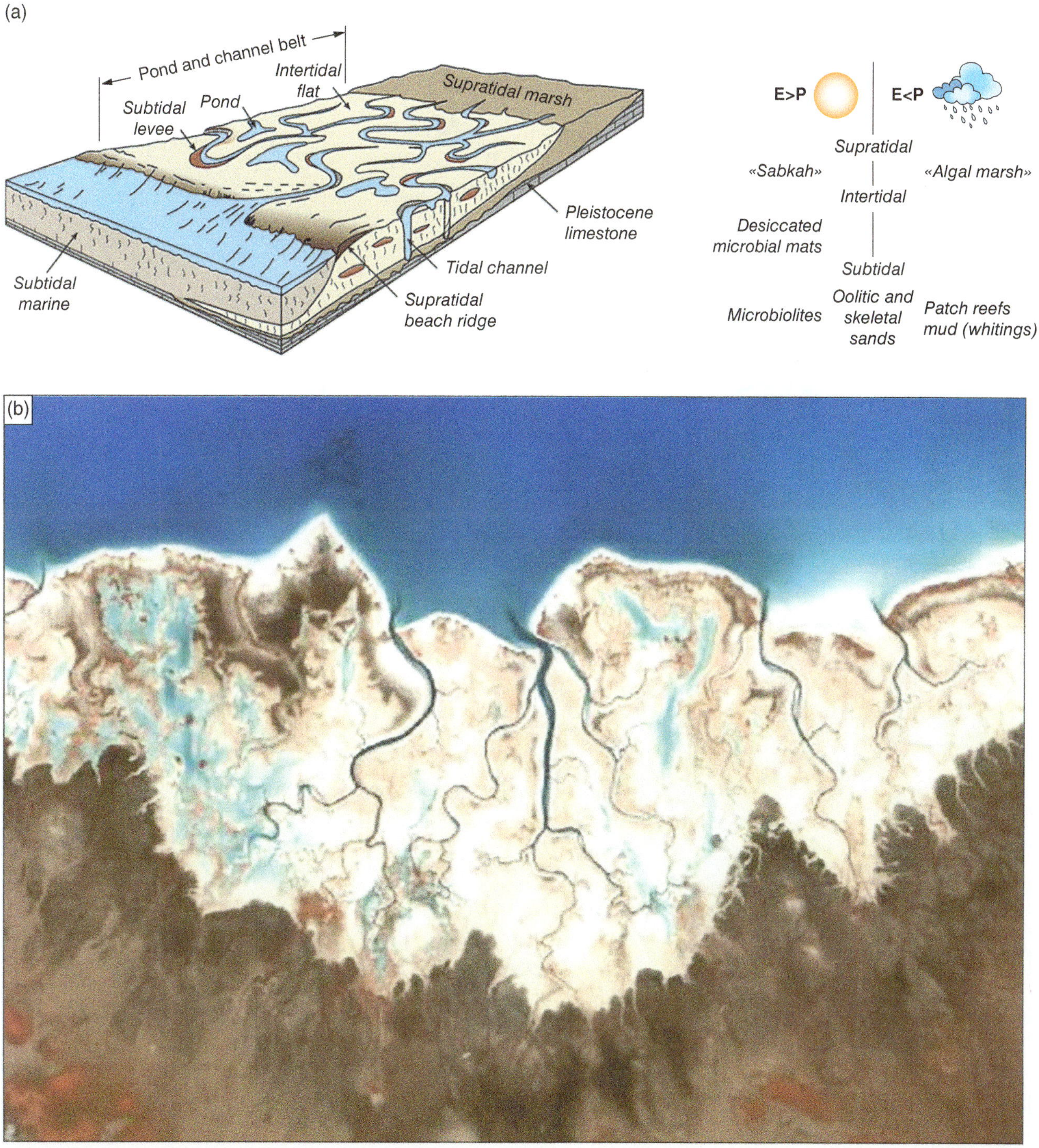

Figure 11.44 Carbonate tidal flat environment: (a) depositional elements of a platform interior tidal flat environment with facies characteristics for a dry (E > P) and humid (E < P) climatic setting, E = evaporation, P = precipitation (Shinn 1969); (b) aerial photo of tidal flat environment West of Andros Island (*Source:* Eberli (2007) / Comparative Sedimentology Laboratory).

grading from a rubble pavement near the reef crest to an apron of shifting sand abutting meadows of sea grass on the landward margin; (v) the inner shelf (7–9) is a calm-water lagoonal domain composed of extensive sea-grass meadows, numerous patch reefs of different sizes, a complex maze of 'Faroes' (circular bioherms, usually less than 3 or 4 km in diameter surrounding a shallow secondary lagoon with depths generally less than 20 m; Agassiz 1902), and vast mangrove swamps interfacing marine and terrestrial environments along virtually the entire coastline. This 1.5- to 2-m-deep, confined part of the shelf is marked by weak currents and a significant accumulation of fine sand and silt, it deepens up to 20 m in a northward direction.

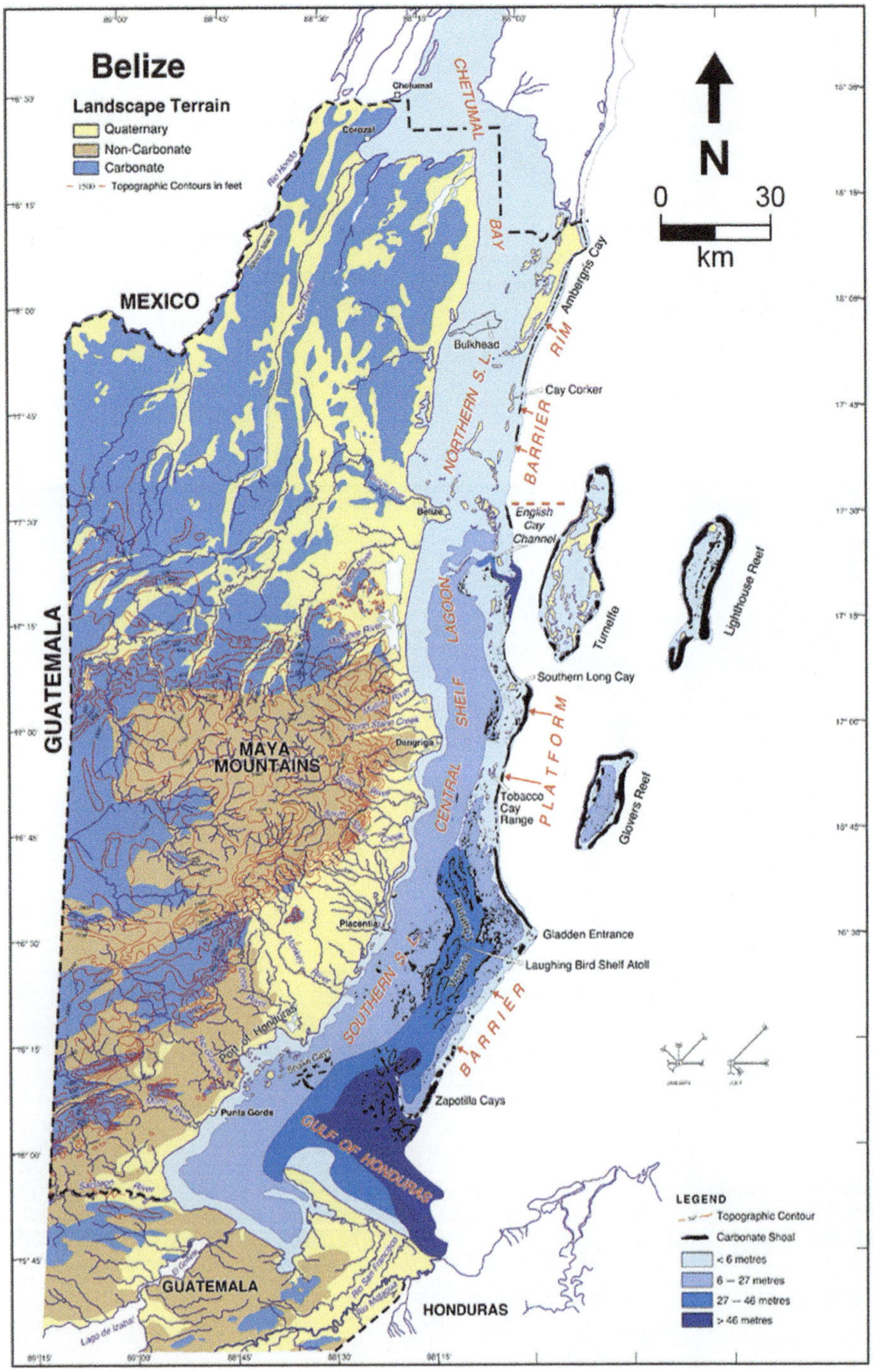

Figure 11.45 Geographical map of Belize, with the onshore and offshore distribution of carbonate deposits (*Source:* Purdy and Gischler 2003 / Springer Nature). This land-attached T-factory carbonate shelf is one of the largest in the world.

11.6.1.1.5 Present-Day Land-Attached Carbonate Shelf Facies Models

Two present-day examples of land-attached shelf systems are reviewed briefly below. These examples are all from the tropical to sub-tropical climatic belt, and in most the T-factory is dominant. Only a few present-day M-factory examples are currently documented.

Great Barrier Reef, Northern Australia

The Great Barrier Reef (GBR) is the largest coral reef system in the world. It extends from 9°30′S in the north to 24°30′S in the south, about 2300 km along the northeastern shelf of Australia (Figure 11.46). The spatial distribution of the present-day, Holocene reefs is largely controlled by ancestral, partly Pleistocene, topography and shows considerable geomorphologic complexity, with both latitudinal and cross-shelf depositional patterns (Davies and Kinsey 1977; Done 1982; Marshall and Davies, 1982, 1988; Hopley 1982; Davies and Hopley 1983; Drew and Abel 1988a,b; Hopley et al. 2007; Braithwaite and Montaggioni 2009; Davies 2011; Kench et al. 2012; Hinestrosa et al. 2014; Webster et al. 2018). In addition to the contribution of corals to the carbonate

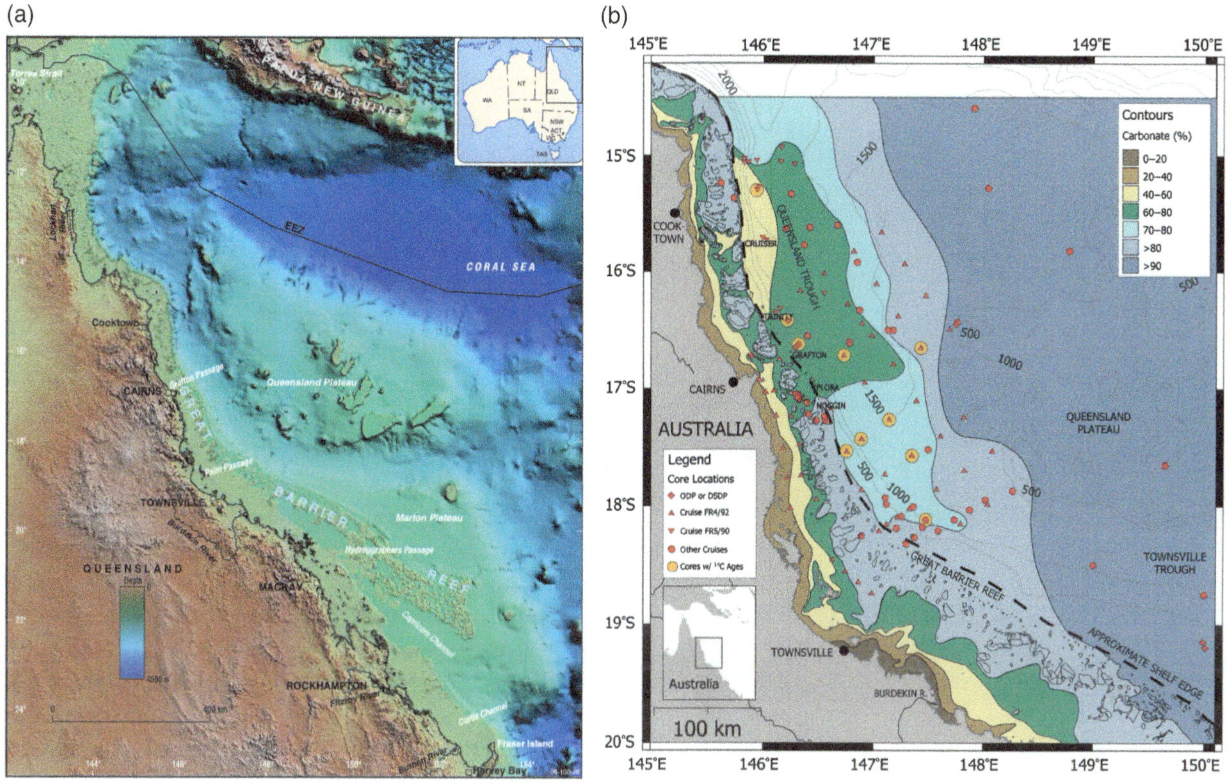

Figure 11.46 Great Barrier Reef (GBR) maps, northeastern Australian margin: (a) bathymetric map of the general setting of the GBR (Mathews et al. 2007) and (b) detailed map of the North Queensland Margin with contours of $CaCO_3$ content. Carbonate concentration was measured at each core top, depicted in red. Note offshore position of reefs (highest carbonate content) and decreasing carbonate content towards the land (*Source:* Francis et al. 2007 / Society for Sedimentary Geology (SEPM)).

sediments (T-factory), there is a very large area (6095 km^2) adjacent to the platform-edge barrier reef, which is covered by *Halimeda* bioherms (McNeil et al. 2016, 2021a, b; Nolting et al. 2020; Hinestrosa et al. 2022). The extended (6095 km^2) *Halimeda* mounds or bioherms exhibit a complex morphological pattern (Davies 2011; McNeil et al. 2016). The bioherm formation and distribution are controlled by the complex interaction of regional and local currents and outer-shelf geometry. In addition, the morphology and depth of continental slope submarine canyons determines the delivery of cool, nutrient-rich water upwelling through inter-reef passages (McNeil et al. 2016).

The land-attached systems of the Great Barrier Reef system and the Belize Shelf locally have mixed carbonate-siliciclastic facies (Figure 11.46b). For example, the depauperate coral assemblages and, in some cases, the complete inhibition of coral development along the Great Barrier Reef, have been attributed to the elevated turbidity associated with a proximity to river inflow and large tidal ranges (Kleypas 1996; van Woesik and Done 1997; Webster et al. 2018; see also Chapter 9: the Fly River Delta).

Shark Bay, Western Australia

Located on the west coast of Australia, Shark Bay with Hamelin Pool (Figure 11.47a) is the most notable present-day example of a restricted, hypersaline interior shelf (embayment) setting, with a particular ecosystem composed of microbial and sea-grass communities and coquina deposits (Playford et al. 2013). The stromatolites, exposed during low tide (Figure 11.47b), are world famous and are used as an analogue for Precambrian stromatolite systems. Hamelin Pool is a U-shaped embayment semi-closed to the north by a large seagrass bank cut by tidal channels (Logan and Cebulski 1970; Butcher et al. 1984). The embayment can be subdivided in a basinward direction into supratidal, intertidal, shallow subtidal and embayment plain (Logan et al. 1974). Several studies illustrate the distribution of microbial deposits along a platform-basin transect (Jahnert and Collins 2012, 2013). The extensive and prolific activity of bacteria during the last 2000 years resulted in a diverse assemblage of microbial deposits, morphologies and fabrics that is progressively expanding and colonising the subtidal zone due to sea-level fall (Figure 11.47c). Shark Bay is a rare present-day example of an inner shelf dominated by the M-factory, where the stromatolites are formed by microbial trapping and binding and microbial biomineralisation. Microbial trapping and binding are the primary accretionary mechanism in the *intertidal* zone, forming grainy, calcarenitic stromatolites. Microbial biomineralisation is the primary accretionary mechanism in the *subtidal* zone, forming muddy, micritic stromatolites (Reid et al. 2003;

Figure 11.47 Shark Bay and Hamlin Pool, W-Australia, represent a unique example of a M-factory dominated system (adapted from James and Jones 2016, Grey 1994). (a) Satellite map and main facies belts, (b) stromatolites exposed at low tide and (c) different shapes of stromatolites.

Suosaari et al. 2016a,b, 2019). Recent studies combining metatranscriptomic and organic geochemical techniques have shown that cyclonic events may increase the activity of some heterotrophic microbes leading to potential localised increase in eutrophication and increase in calcium carbonate precipitation, helping mat preservation (Campbell et al. 2022). The recognition of present-day micritic stromatolite build-ups is of great importance as many ancient stromatolites are micritic.

11.6.1.1.6 *Land-Attached Carbonate Shelf Facies Models in the Rock Record*

The shelf depositional profile is a consistent feature recognised in numerous outcrop and subsurface carbonate studies throughout the Phanerozoic (e.g. Wilson 1975; Read 1985; Tucker and Wright 1990; Handford and Loucks 1993; Pomar 2001; Williams et al. 2011). There are significant variations in the dominant carbonate-producing organisms that populated it, as well as in the grain-size distributions, textures, degrees of cementation and sedimentation rates. Here we provide a selection of land-attached carbonate shelf facies models to illustrate this variability. A selection of case studies is provided in Supplementary Information 11.S.1.

The first organisms capable of forming reef structures appeared already in the Precambrian in the form of stromatolites (Walter et al. 1980), made of cyanobacteria and other microbes (Grotzinger et al. 1989) which had a low level of ecological complexity. Although ramp profiles were the rule, the development into shelf morphologies also occurred (see also Section 11.7). Platform facies models for this interval are relatively simple. Ecologically complex reef structures appeared in the Ediacaran (microbilalites, *Cloudina, Namacalathus*) followed in the Early Cambrian by microbialites and archaeocyaths. A first acme of reef building was attained in the Silurian and Devonian with the proliferation of stromatoporoids, corals and algae, and the development of vast tropical carbonate shelves with continuous marginal reef belts (Copper 2002a). Well-known examples are the large Middle and Late Devonian carbonate platforms and build-ups of western Canada (Mountjoy 1965a,b; Geldsetzer et al. 1988; Wendte 1992; Whalen et al. 2000a,b; Weissenberger et al. 2017) and the Canning Basin in Australia (Figure 11.48a; Playford 1980, Playford et al. 2009; Playton and Kerans 2015a,b; Playton et al. 2017).

Typical for Late Palaeozoic and Triassic shelf margins is the importance of microbial boundstones (*Archaeolithoporella-Tubiphytes*; M-factory) in the construction of the margin and fore-slope. Good examples of these exposed in large-scale outcrops include the Dolomites (Bosellini 1984, 1989; Reijmer et al. 1991; Blendinger et al. 1997; Keim and Schlager 1999, 2001; Keim et al. 2001; Keim and Brandner 2002; Schlager and Reijmer 2009), Kazakhstan subsurface (Cook et al. 2002; Kenter et al. 2002; Zempolich et al. 2002), the Permian Capitan Reef margin in the Delaware basin of New Mexico and west Texas (Figure 11.48b; Ward et al. 1986; Wood et al. 1996; Tinker 1998; Harris and Saller 1999; Kerans et al. 2012; Playton and Kerans 2018) and the Pennsylvanian carbonate platform, Spain, Asturias (Bahamonde et al. 1988, 1997, 2004; Della Porta et al. 2003, 2004, 2017).

Figure 11.48 Outcrops of ancient carbonate shelf margins: (a) panoramic view (dip direction towards the left) of the Classic Face at Windjana Gorge, showing flat-bedded back reef and reef-flat limestones on the right, passing into massive reef-margin in the centre and steeply dipping marginal slope deposits on the left. (*Source:* Playford et al. 2016 / GeoScienceWorld); (b, c) composite panoramic view of the Capitan Reef escarpment showing low-angle prograding clinoforms (dip direction towards the right, starting at the left of figure b – Grayburg-Queen – and finishing at the escarpment margin at the right of figure c – Brushy Canyon) (Permian, Delaware Basin, New Mexico/West Texas, USA; Courtesy of C. Kerans).

In the Jurassic, corals were the dominant component of the T-factory platform profiles, although M-factory *Lithiotis* bioherms and sponge–microbial associations prevailed in the Pliensbachian–Toarcian period (Kenter and Campbell 1991; Scheibner and Reijmer 1999; Verwer et al. 2009a,b; Merino-Tomé et al. 2012; Della Porta et al. 2014). Siliceous sponges and microbes were locally important in the Middle, and especially Late, Jurassic (Brunton and Dixon 1994; Leinfelder et al. 2002). Typical for the large Jurassic and Early Cretaceous shelves are thick successions of peritidal cycles in the platform interior. The occurrences of these repetitive shallowing-upward patterns are interpreted by many authors to be allocyclic, driven by orbitally forced climatic and eustatic sea-level variations (Fischer 1964; Goodwin and Anderson 1985; Strasser 1988; Eriksson and Simpson 1990; Strasser et al. 1999; Hillgärtner and Strasser 2003).

In the Cretaceous, T-factory carbonate shelves were widespread in the tropical climate belt, but these had very different compositions than the Palaeozoic platforms. The contribution of microbialites was strongly reduced, to specific time intervals, such as the *Lithocodium-Baccinella* complex in the mid-Cretaceous Aptian stage (Schlagintweit et al. 2010 Schlagintweit and Bover-Arnal 2012; Rameil et al. 2010; Esrafili-Dizaji et al. 2020). On the shelf a prolific benthic biological association comprising large rudistic bivalves, benthic foraminifera, green algae, echinoderms and peloids dominated (e.g. Simmons and Hart 1987; Simo et al. 1993; Skelton and Gili 2012). Well-exposed examples are found worldwide (Supplementary Information 11.S.1).

It is only during the Cenozoic that the builders of the present-day reefs, the scleractinian corals, dominate reef buildups (Stanley 2003; Hallock 2005; Kiessling 2009). Since the Late Miocene, the reef corals have been able to grow up to sea level, as indicated by their association with green algae characteristic of the 'euphotic' zone (Stanley and Lipps 2011; Pomar et al. 2017; Stuhr et al. 2018). This evolution happened progressively through a gradually increasing abundance of modern reef-building organisms, changing the biological framework and associated geobodies. Typical examples include Cenozoic platforms exposed in the Mediterranean, such as the outcrops in the Mut Basin in S. Turkey (Bassant et al. 2005; Janson et al. 2010), and in the Sorbas Basin in S.E. Spain (Dabrio et al. 1981; Sánchez-Almazo et al. 2007), and subsurface examples from South-East Asia and Australia (Belde et al. 2017; and overview in Michel et al. 2020).

A particular feature of ancient, land-attached carbonate shelves are 'intrashelf basins' that tend to be even larger and deeper than the extensive lagoons mentioned earlier (Figure 11.3). These basins are not represented in the facies model (Figure 11.38) but is nevertheless a common feature of large ancient shelves (see Supplementary Information 11.S.1). The well-studied examples of the Mesozoic on the Arabian Plate had water depths typically ranging from 60 to 150 m and were surrounded by extensive shallow-water (<20 m) areas, with both low-angle or steep, inward-dipping margins, depending on the wind and current directions (Murris 1980; Droste 1990; Burchette 1993; van Buchem et al. 2002b, 2010b; Yose et al. 2006; Vahrenkamp et al. 2015). Organic-rich deposits formed in some of these intra-shelf basins in association with anoxic bottom water (Figure 11.49). Intra-shelf basins were widespread on the Cretaceous carbonate shelves around the Neo-Tethys Ocean, with reported occurrences in Croatia (Korbar et al. 2001), the southern United States (Phelps et al. 2014; Eldrett et al. 2017) and Venezuela (Vahrenkamp et al. 1993). A Plio-Quaternary example has been documented from the Bonaparte Shelf in NW Australia (Bourget et al. 2014). The origin of most of these intrashelf basins has a strong stratigraphic component, creating topographic relief through differential sedimentation rates of the carbonate system (see Section 11.7).

11.6.1.2 Land-Attached Carbonate Ramp Facies Model

The characteristic feature of the carbonate ramp facies model is the low-angle (<1.5°) homoclinal slope (Ahr 1973). The physical control on the facies distribution along this depositional profile is determined by the depth of the storm wave base, fair-weather wave base and the pycnocline, as well as the potential impact of tides (Burchette and Wright 1992). In this respect it resembles siliciclastic ramp models (e.g. Allen 1982; Hampson et al. 2008; see also Chapter 8). Although ramp depositional profiles are the rule for C-factories, they are not limited to these, and they may also occur in the T-factory (as transient features) and the M-factory.

A schematic representation of the typical facies distribution along a ramp depositional profile is shown in Figure 11.50. Four main environments are distinguished, of which only the inner ramp environment is further subdivided into facies belts due to its complexity:

a) The *basin* environment (facies belt 1) is the most distal and deepest setting, positioned where the sea bottom is below the pycnocline and out of reach of direct sediment supply. Sediments are dominated by planktic suspension fallout. Occasional turbidites, grain flows and debris flows might be present (Pasquier and Strasser 1997; Proust et al. 1998; Phelps et al. 2008).
b) The *outer ramp* environment (facies belt 2) covers the water depths from storm wave base to the basin floor. Depending on the extension of the ramp, storm beds might still occur with the preservation of graded beds or laminae. Sedimentation from suspension fallout derived from the inner and mid-ramp is the rule, and locally patch reefs and mud-mounds may occur (Burchette and Wright 1992). Although the deeper part of the ramp is a low-energy setting, geostrophic contour currents may induce zones of winnowed sediment at different depths (Mullins et al. 1988; Viana et al. 1998).
c) The *mid-ramp* environment (facies belt 3) is positioned between fair-weather and storm wave-base, where storm deposits are common, and coarser-grained, graded tempestites alternate with fine-grained sediments that settled out of suspension (e.g. Peters and Loss 2012). The abundance and degree of amalgamation of often-graded storm beds decrease systematically from proximal to distal (e.g. Aigner 1984).
d) The *inner ramp* environment is more complex than the other environments and is, therefore, subdivided into four facies belts (4, 5, 6 and 7). These cover the depth range from supratidal to fair-weather wave base, where wave and current activity move sediment almost continuously.

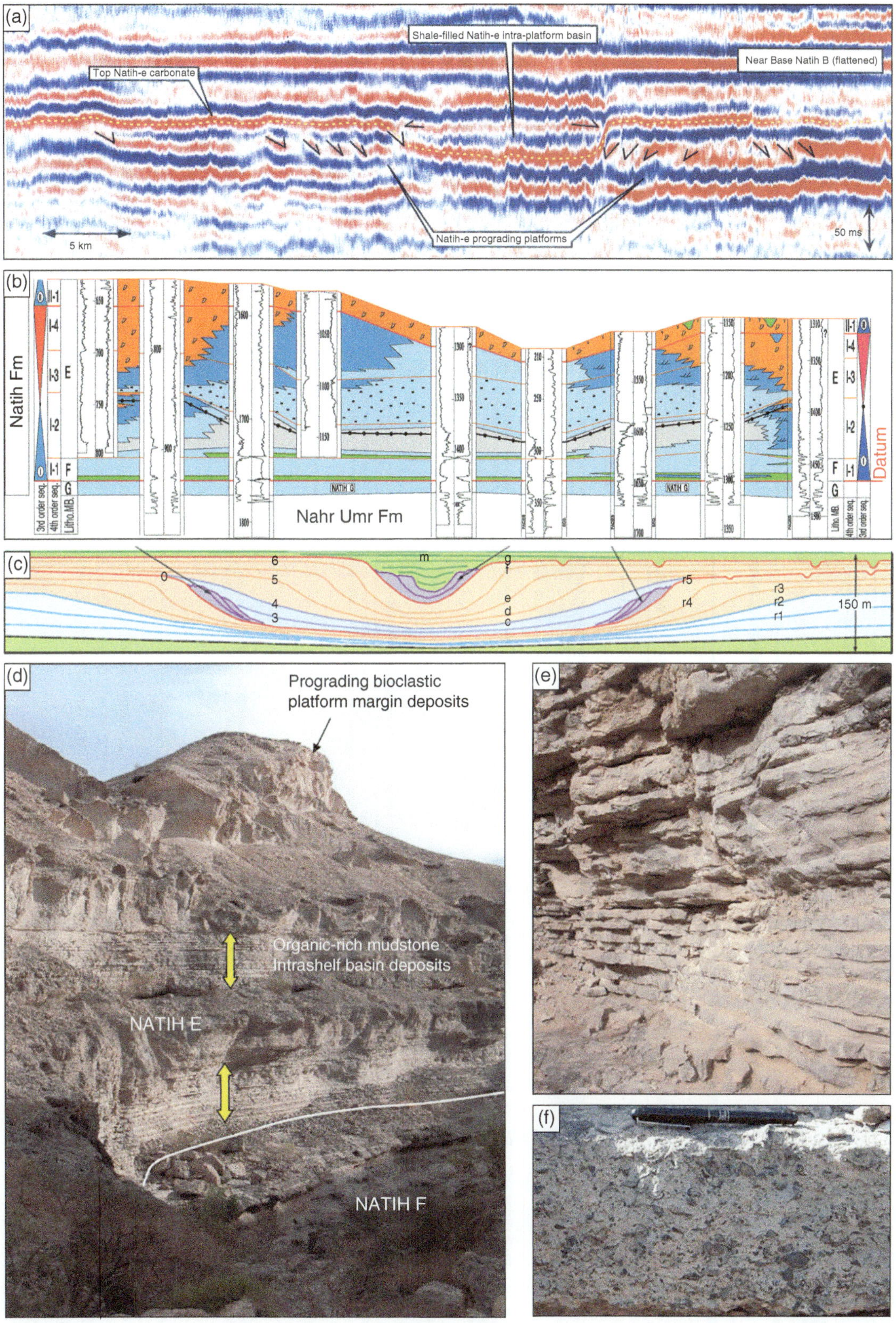

Figure 11.49 Cross sections and facies of the intrashelf basin in the Natih E Member (Lower Cenomanian) in Oman, Arabian Plate. (a) Seismic expression of an intrashelf seaway (*Source:* Droste and van Steenwinkel 2004 / GeoScienceWorld), (b) well log correlation roughly parallel to the seismic line shown in (a) (*Source:* van Buchem et al. 2002b / GeoScienceWorld), (c) sequence stratigraphic model of the same seaway based on seismic, well logs, core and outcrops (*Source:* Homewood et al. 2008 / GeoScienceWorld) and (d, e) intrashelf basinal facies of the Natih E Member in outcrop at Jebel Madmar (Oman). Note dm-scale bedding, TOC content reaches up to 8% (Razin et al. 2017; Wohlwend et al. 2016), (f) close-up of an Exogyra oyster layer which occurs in the basinal facies (*Source:* Razin et al. (2017) / American Association of Petroleum Geologists).

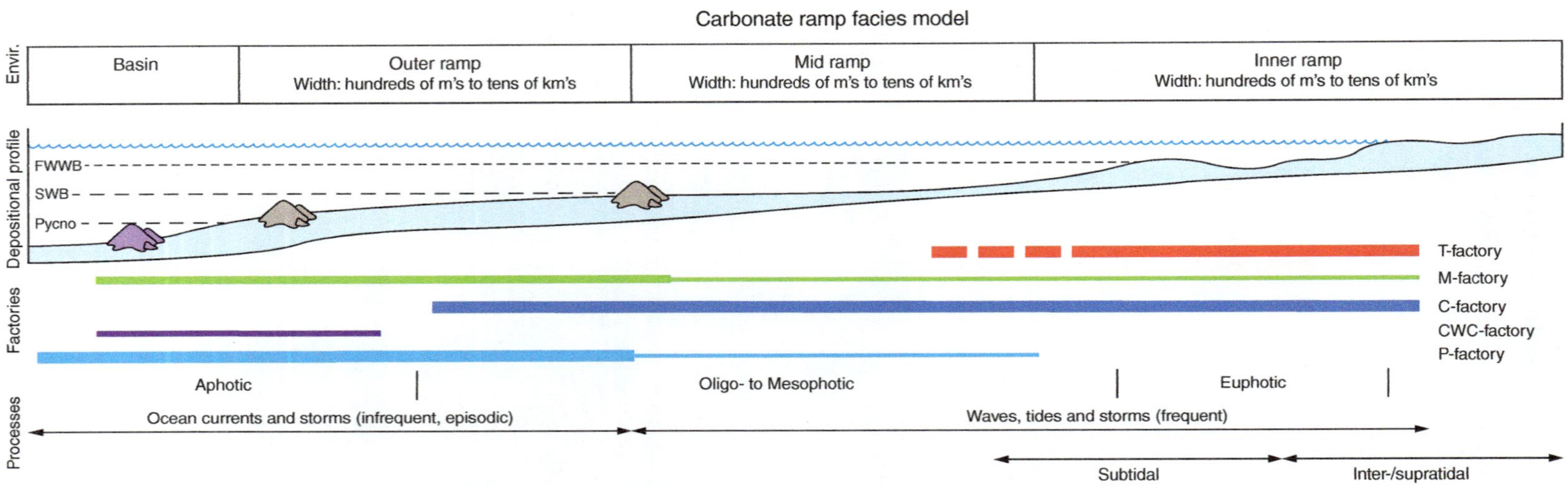

Facies belts	1. Basin	2. Outer ramp	3. Mid ramp	4. Inner ramp shoals / 5. Inner ramp backbarrier / 6. Inner ramp restricted marine	7. Inner ramp transitional to onshore
Settings	Below pycnocline and below SWB. Local distal turbidites. Anoxia may develop below pycnocline. Where outer ramp passes into a distally steepened ramp gravity transport is enhanced towards the basin	Below SWB to the basin floor; infrequent storm reworking but depending on the ramp extension and the water depth, various degrees of storm-induced deposits are present. Outer ramp to basin transition of intra-shelf basins may become suboxic to anoxic by restricted circulation	Between FWWB and SWB, storm wave and swell processes dominate and frequently rework sediment	Above FWWB, wave and current activity are continuous, lagoonal and tidal flat environments (channels and levees), depending on the profile morphology and the degree of restriction caused by the sand shoal, are well developed, small or absent.	*Beach and supra-tidal to onshore environments.* *Arid: sabkha and salinas, mangroves* *Humid: terrestrial influx, mangroves*
Sediments	Pelagic clay, siliceous and carbonate ooze, rare tsunami deposits. Locally organic-rich facies	Argilaceous lime mudstone and wackestone interbedded with marls, autochthonous and allochthonous carbonates and hemipelagic sediments; graded distal tempestite beds; common organic buildups (mud and skeletal mounds)	Grainy carbonates, graded packstone to grainstone, hummocky cross-stratification and tempestites; thick ooid or bioclastic sand shoals; lime mudstones and marls with fine grained terrigenous input; organic buildups are represented by patch reefs (bioherms) and mud mounds	Lime mud, muddy sand or sand depending of grain size of local sediment production and efficiency of winnowing by waves and tidal currents, local bioherms or biostromes; terrigenous sand and mud present but variable depending on the humid/arid conditions. Lagoon and tidal flat present the same characteristics as platform interior facies belts	Arid: calcareous and dolomitic mud and sand, with nodular or coarse crystalline gypsum and anhydrite. Mangroves. Humid: Calcareous marine mud and sand with occasional fresh-water lime mud and peat layers (mangroves)
Biota	Nekton, plankton, thin-shelled bivalves, sponges (spicules), CWC mounds	Diverse biota, with benthos, plankton and nekton; bryozoans, sponges, microbial biota in mudmounds	Large diversity of benthos (foraminifers, sponges, bryozoans, brachiopods, mollusks, red algae and echinoderms; C-factory dominated)	Depending on climatic conditions combinations of T-, M- and C-factory biota can occur; in addition ooids are common	Arid: microbial mats, gastropods, high-salinity ostracods. Humid: low-salinity ostracods freshwater snails and charophytic algae

Figure 11.50 Conceptual carbonate ramp facies model (*Source:* Adapted from Ahr 1973; Aigner 1984, 1985a; Burchette and Wright 1992). The distribution of carbonate factories along the depositional profile indicates their potential locus of production, the place of deposition and accumulation of the facies will depend on the intervention of sediment transportation by storms, ocean currents or tides. Biota from different carbonate factories may occur together along one depositional profile. This profile is typical for the C-factory but may also occur in T-factory-dominated systems, usually as a transient state.

As at siliciclastic ramp type shorelines, shoreface, beach, lagoon and tidal flat environments can be distinguished in carbonate-dominated ramps. In the inner ramp, carbonate sand shoals are common, shallow-water bio-construction builders are frequently present, and sea-grass meadows may develop creating a favourable environment for diverse marine life, including large benthic foraminifera. Tucker and Wright (1990) propose three different end-members for the inner ramp facies belt: (i) a beach barrier–tidal delta complex, with lagoons and a tidal flat behind it (equivalent to belts 4, 5, 6 and 7 of Figure 11.50; i.e. Trucial Coastline, Loreau and Purser 1973; Purser 1973). Lagoons in ramp systems range from narrow (a few kilometres) coastal lagoon to tens of kilometres wide shallow-water embayments. These may become highly restricted when the connection with the open sea is hampered through (i) the presence of inner ramp shoal belts; (ii) fringing shoal complexes with tidal flats but no lagoon (equivalent to belts 4 and 7 in Figure 11.50; (i.e. Shark Bay, Playford et al. 2013); and (iii) strand plains with linear beach ridges (equivalent to belt 4 in Figure 11.50) i.e. Yucatan coast, Ward and Halley 1985). The inner ramp facies complexity is well illustrated in Gischler and Lomando (2005, 2007) for Kuwait, Rivers et al. (2019, 2020, 2021) and Engel et al. (2022) for Qatar, and Paul and Lokier (2017), Lokier and Fiorini (2016) and Pedersen et al. (2021) for the UAE. For ancient examples, see Amour et al. (2013), Sequero et al. (2018) and Brandano et al. (2020).

11.6.1.2.1 Present-Day Carbonate Ramp Facies Models

At present, carbonate ramp systems occur in two geological settings: (i) open ocean facing continental shelves and (ii) interior basins, such as the Mediterranean, the Gulf of California and the Arabian/Persian Gulf. Ramp systems on ocean facing shelves are found in three different climatic zones: cold-water polar, cool temperate and warm temperate neritic domains, each characterised by different (heterozoan) biotic associations typical of the C-factory (Figure 11.51; James and Jones 2016). The interior basin examples occur in the warm temperate and arid tropical domains.

The *cold-water polar ramp systems* occur on high latitude continental shelves in water temperatures that rarely rise above 5°C (Figure 11.51a; James and Jones 2016). These deposits contain a small fraction of heterotrophic carbonate producers and are characterised by ice-rafted debris and a large bio-siliceous component composed of diatom frustules and sponge spicules. The shallowest part of the shelf, at 60 m water depth, is current swept with epibenthic suspension feeders growing on hard substrates. The carbonate sediments typically comprise debris from bivalves, bryozoans, echinoids, gastropods and benthic foraminifera. James et al. (2009), Frank et al. (2014, 2020) and Matheson and Frank (2020) provide detailed information on the polar low diversity benthic community comprising stylasterine hydrocorals, barnacles and bryozoans.

The *cool temperate ramp systems* are positioned on ocean-facing shelves at intermediate latitudes (Figure 11.51b; James and Jones 2016). Several of these were glaciated during the Neogene resulting in the deposition of glaciomarine sediments (notably diamictites), forming the substrate for the subsequent carbonate deposition. At present, many of these shelves are colonised by heterotrophic carbonate producers which live in seawater temperatures of 5–15°C. A typical facies zonation comprises an inner ramp with a water depth of approximately 30 m dominated by ocean-facing kelp forests which generate large amounts of barnacle-rich sediments. Protected areas, at the leeward side of islands, are dominated by coralline algae (maerl) whose detritus is swept into cross-bedded shoals and banks (Bosence 1976). The mid- to outer-ramp setting is sub-photic, densely colonised by bivalves, bryozoans, and serpulids (Light and Wilson 1998). The deep-water coral, *Lophelia pertusa*, is locally recorded from depths of 200–500 m. Their remains form rippled and burrowed carbonate sands that are habitats for both infaunal and epifaunal bivalves (James and Jones 2016). Enhanced aeolian transport and dune formation are observed during cold climatic intervals when sea level was low, sediment supply (exposed marine carbonate sediment) was high, winter windiness was high, and vegetation on coastal plains at a minimum (Fornós et al. 2009).

The low energy, warm temperate systems (Figure 11.51c) are tranquil, and so have a shallow fair-weather-wave-base (FWWB), and storms are generally local and of short duration. The facies spectrum is comparable to the high-energy settings (see open shelf below) but is generally finer-grained. Examples from east and west Australia show the following facies characteristics (Betzler et al. 1997b; James and Jones 2016): (i) the inner-ramp facies belt is diverse, including rocky shorelines and beaches, barrier islands and landward muddy lagoons with numerous bivalves, ostracods and gastropods. In more protected sites, mudflats can develop, seaward of which large muddy, biota-rich seagrass meadows exist next to grainy, cross-bedded sand shoals; (ii) the mid-ramp facies belt has sediments deposited below wave-base, with red algae in the form of encrusted rhodolith pavements, and articulate coralline algae associated with bryozoans, molluscs and benthic foraminifera. Deposits range from burrowed and thick-bedded, to cross-laminated packstone layers; (iii) the outer-ramp facies belt is muddy, and is inhabited by filter feeders, infaunal calcareous organisms and planktonic foraminifera (James and Jones 2016).

Other examples of warm-temperate carbonate ramp systems include coastal areas along the Mediterranean, where they are dominated by well-developed seagrass-derived sedimentation, rhodalgal production, locally in association

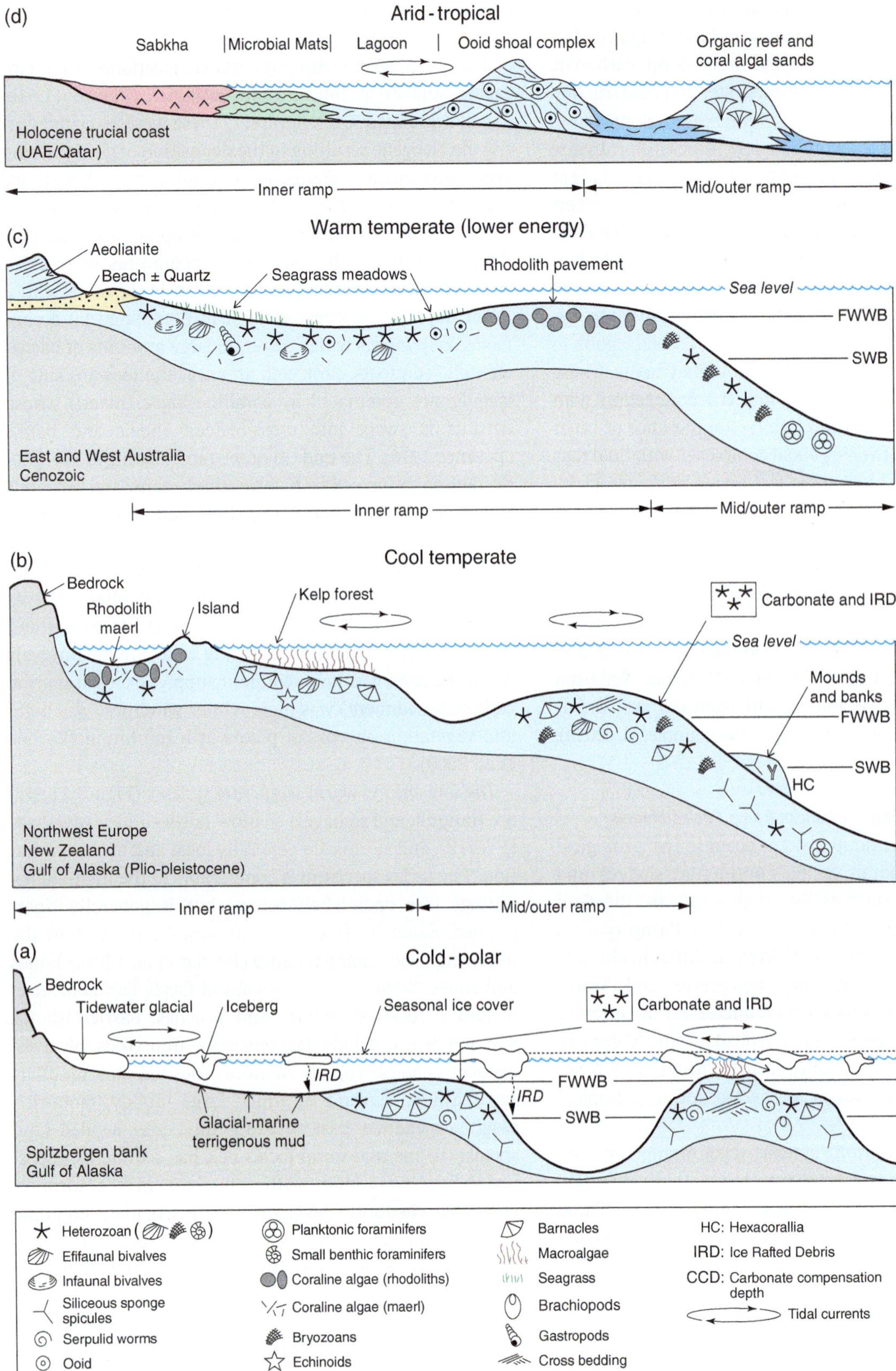

Figure 11.51 Present-day carbonate ramp models for different climatic zones, indicating biota and hydrodynamics (*Source:* Adapted from James and Jones 2016; James and Lukasik 2010): (a) cold water polar environment in the Northern hemisphere, (b) cool temperate environment, (c) warm temperate, lower energy environment and (d) arid, tropical environment (see also Figure 11.52).

with large benthic foraminifera and *Posidonia* (Betzler et al. 1997a,b; the photo-C factory in Michel et al. 2019) and where light is an important driving factor. In the subtropical settings of the middle part of the Gulf of California, rhodalgal-dominated sedimentation occurs which includes a few corals in proximal settings and bioclastic deposits in distal settings in response to changes in trophic conditions (Halfar et al. 2006). On the Mauritanian shelf, which is also a subtropical setting, carbonate deposition occurs along a ramp profile including phototrophic grain production by red algae in shallow water (a few metres of water depth) and a large production of suspension-feeding bivalves and barnacles in the mid-ramp environment. Conditions in this depositional setting are strongly influenced by a major upwelling system and desert dust input (Westphal et al. 2010; Klicpera et al. 2015).

The arid-tropical carbonate ramp system has been much studied along the coastlines of the present-day Arabian/Persian Gulf (Figures 11.51d and 11.52), which is mostly dominated by the T-factory, but also has an important M-factory component (e.g. Purser and Evans 1973; Reading 1996; Strohmenger et al. 2010; Kendall and Alsharhan 2011; Granier and Boichard 2017; Purkis et al. 2017; Rivers et al. 2018; Laugié et al. 2019; Rivers et al. 2020). The ramp slopes down to an intrashelf basin with a maximum water depth of approximately 80 m. Water circulation is fairly restricted, controlled by a mesotidal range (2.1 m), and influenced by NNE-directed seasonal winds (Shamals). Salinity increases towards the coast due to the hot, arid climate: 40–50‰ in the basin and up to 70‰ in the lagoons (Purser and Evans 1973; Brewer and Dyrssen 1985; Michel et al. 2019). The highest facies diversity is encountered in the inner-ramp domain where the following facies belts can be distinguished (Figure 11.51d): (i) supratidal and intertidal zones with sabkhas and microbial-mat-dominated tidal flats. The sabkha extends up to 10 km inland and was formed over the last 5000 years as a result of the seaward migration of the shoreline and infilling of lagoons (Patterson and Kinsman 1981). It is the location of widespread evaporitic sedimentation (gypsum, anhydrite and dolomite), and fringed on the lagoonal side by tidal flats that can be up to 5 km wide and are dominated by

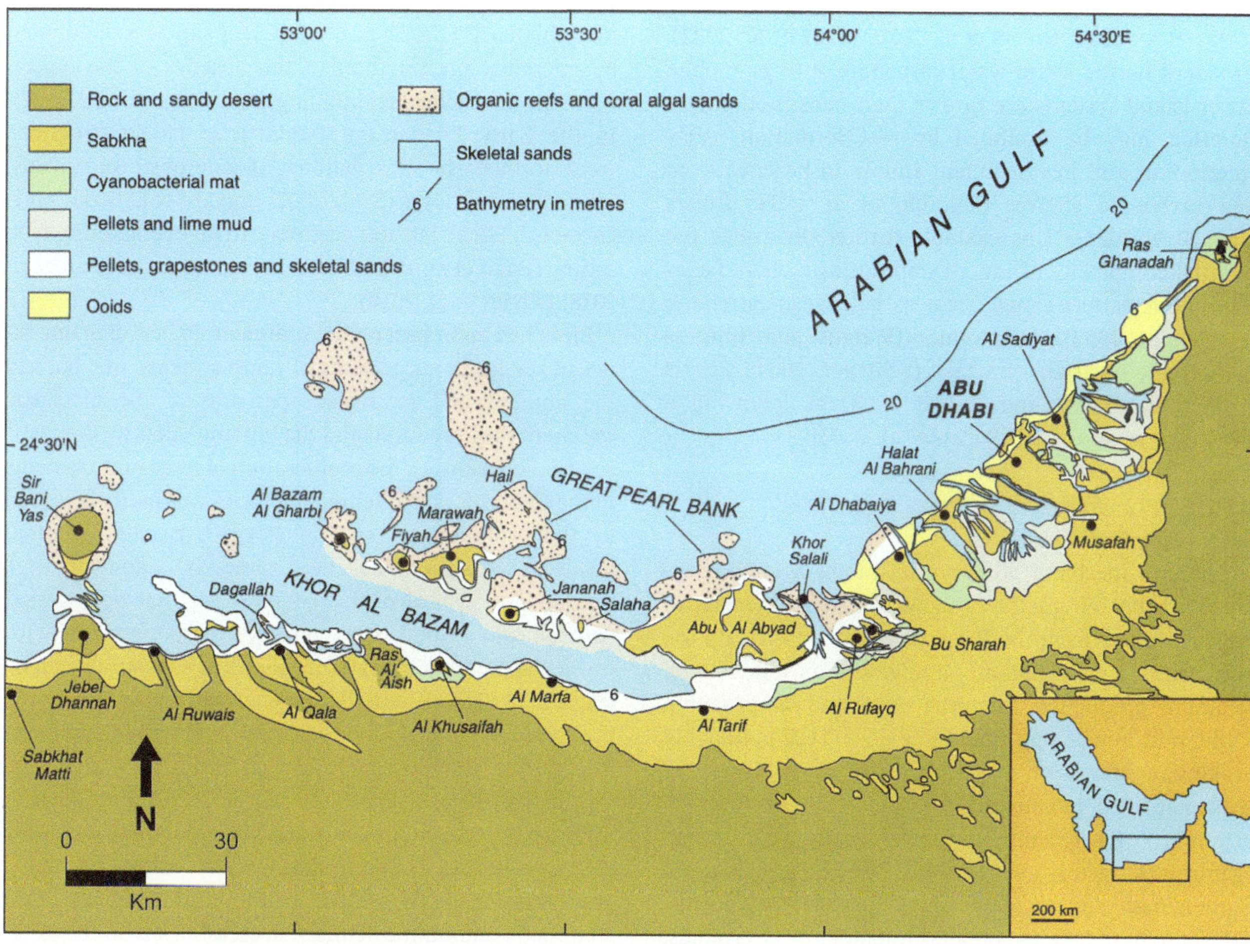

Figure 11.52 Facies map of an arid-tropical carbonate ramp system map (Trucial Coast, Abu Dhabi coastline; *Source:* Kendall and Alsharhan 2011 / John Wiley & Sons). Compare this map to the 2D transect of Figure 11.51d and note the variation in the spatial distribution of the facies belts. Cyanobacterial mat would now be classified as microbial mat.

microbial mats and pelleted lime-muds. (ii) These deposits border extensive coastal lagoons, a few metres deep and up to 30-km wide and characterised by muddy peloidal and bioclastic sediments. They have a low diversity biota due to the high salinity; (iii) subtidal barriers and shoals shelter the lagoons and are shaped by strong longshore currents induced by the seasonal 'Shamal' (northerly) winds. Small ebb-tidal deltas are sites of oolite production, developed between the barrier islands. These are seeded on Pleistocene erosional features at places where tidal inlets connect the lagoon with the distal Gulf open-water area; (iv) the mid- to outer-ramp (offshore zone) is characterised by skeletal sandy muds with bivalves and foraminiferal debris, and locally organic reefs with coralgal sands. Siliciclastic material is introduced mostly by the Tigris–Euphrates rivers and is reworked by currents and seasonal storms. Note the spatial variation of these facies belts in the regional map of Figure 11.52. For further details, see Chapter 10.

In the same region, the Al Ruwais area of northern Qatar is another example of an arid-tropical (inner) ramp type setting. This has been the site of shallow water carbonate sedimentation since the mid-Holocene and is characterised by the migration of a spectacular barrier island (Purkis et al. 2017; Rivers et al. 2020; Mauz et al. 2022). In modern neritic warm-water carbonate settings, mobile barrier island systems are rare as their construction and migration may be inhibited by reef formation, early cementation and low sediment supply to beaches from open carbonate shelves (Diloreto et al. 2019; Rivers et al. 2019, 2021). This example further illustrates the potential complexity of inner-ramp settings where facies distribution patterns may occur as belts (e.g. Burchette and Wright 1992), as mosaics (Wright and Burgess 2005), scale-invariant fractals (Schlager 2004; Purkis et al. 2005; Purkis and Kohler 2008) or following a power–law relationship (Purkis et al. 2005).

11.6.1.2.2 Land-Attached Carbonate Ramp Facies Models in the Rock Record

A review of the biota of carbonate ramps through time (Burchette and Wright 1992) shows an overall increase in species diversity from the Palaeozoic and Mesozoic to the Cenozoic; and concomitantly a migration of the locus of maximum production from the outer/mid-ramp setting to the mid/inner-ramp setting (Figure 11.53). Roughly speaking, a succession of six biological assemblages can be distinguished, punctuated by extinction events (see also Section 11.7): (i) Precambrian: almost exclusively microbial organisms (M-factory), with a broad morphological range of microbialites and widespread deposition of ooids, (ii) Cambrian to Late Devonian: stromatoporoids, crinoids, corals and archaeocyathids (Cambrian only), (iii) Late Devonian to Permian: bryozoa, phylloid algae, sponges and fusulinids, (iv) Triassic and Jurassic: corals, stromatoporoids and stromatolites, (v) Cretaceous: green algae, echinoderms, large benthic foraminifera, bivalves (rudists), pelagic foraminifera and nannoplankton and (vi) the Cenozoic: green algae, molluscs, large benthic foraminifera, corals, coralline algae and seagrass communities.

The major biological extinction events also affected the ramp system biota (Figure 11.53). When examined in more detail, it appears that just after these extinctions, ramp systems tend to dominate, probably due to the temporary disappearance of frame-building organisms populating the shallow-water environments, favouring a shift to massive skeletal – and ooid production, or extensive microbial activity (Burchette and Wright 1992). An example of the former is the mass extinction of corals and stromatoporids at the Late Devonian Frasnian–Famennian boundary, which put an end to rimmed platform constructions (Copper 2002a; Whalen et al. 2002; Playford et al. 2009), and led to a period dominated by the Mississippian ramp systems among which ooids are common (Handford 1988; Burchette et al. 1990). When rimmed platforms are preserved in the Early Famennian, they mainly result from microbialite developments (Stephens and Sumner 2003; Shen and Webb 2004). Also, following the end-Permian mass extinction, the Early Triassic post-crisis 'reef gap' was dominated by microbial precipitation in the absence of the typical Permian metazoan reef-building organisms (Flügel 2002). In the Early Triassic only small-scale bio-constructions were formed, by associations of microbial organisms, sponges and bivalves (Szulc 2007; Brayard et al. 2011; Marenco et al. 2012), but still occurring in an overall ramp type setting (Baud et al. 1997; Vennin et al. 2015; Grosjean et al. 2018; Olivier et al. 2018).

Biota had an impact on the sedimentological characteristics of the facies successions of ramp systems, due both to the amount and grain-size spectrum of the bioclastic sediment they produced and the binding-capacity they possessed. For example, the proliferation of large ramp systems in the Cenozoic warm-temperate zones was partly favoured by the evolution of the (C-factory) barnacles, infaunal bivalves, red algae (Neogene; Crame 2002) and the large foraminifera typical of the T-factory for that time period (Buxton and Pedley 1989), possibly in combination with the development of seagrass meadows as a favourable niche (Figure 11.51c). A remarkable example is the enormous proliferation of large benthic foraminifer genus *Nummulites* during the Eocene. These foraminifers were rock-forming along the coastlines of the Neo-Tethys Ocean. Their sediments deposited along wave-dominated distally steepened ramp profiles, display a proximal to distal grain-size sorting from intact Nummulitic grain- and rudstones, to Nummulitic debris of decreasing size in the distal direction (e.g. Aigner 1982, 1983; Hottinger 1983; Moody and Grant 1988; Jorry et al. 2003; Vennin et al. 2000, 2003; Rasser et al. 2005; Mateu-Vicens et al. 2012). The high

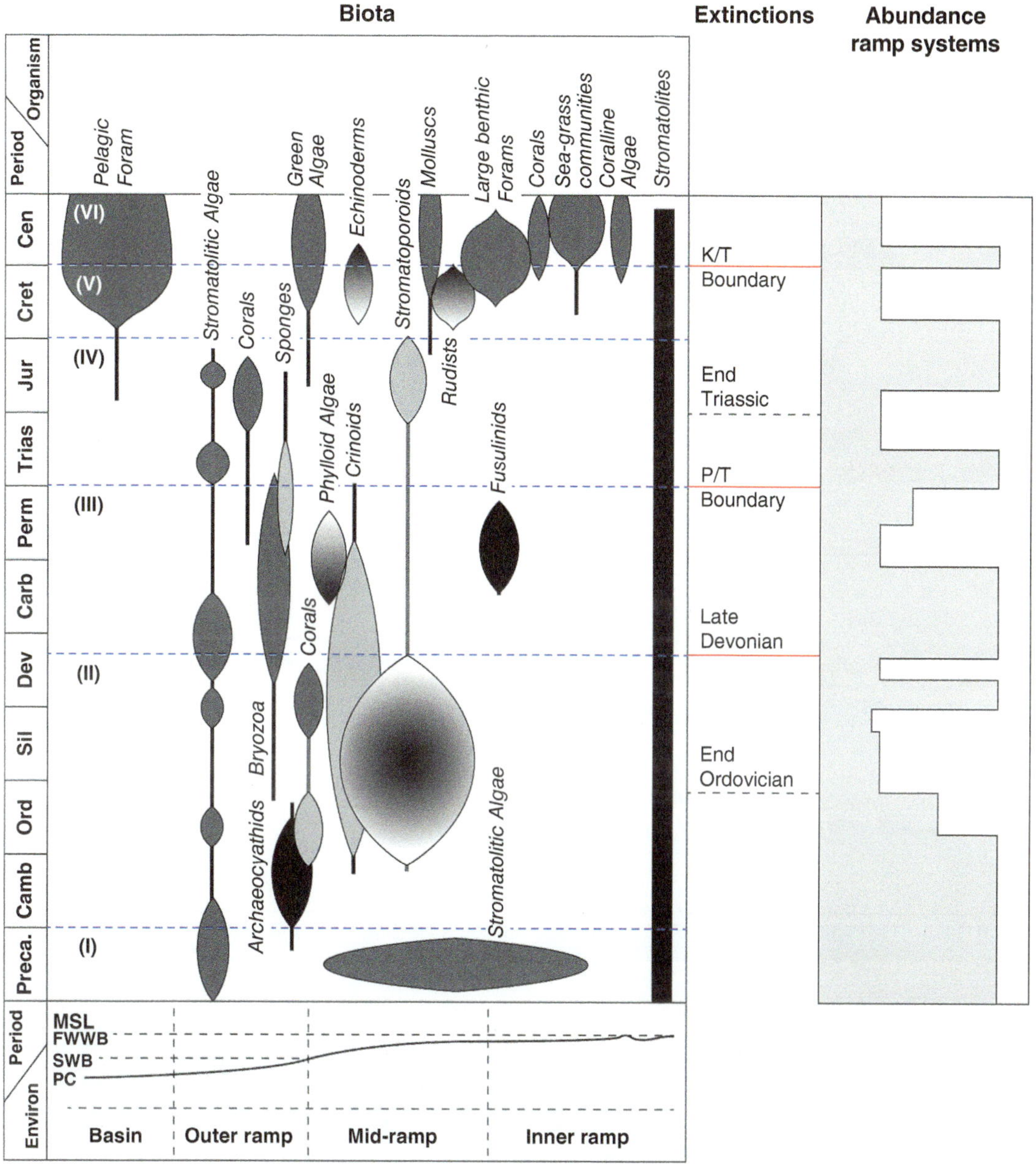

Figure 11.53 Dominant biota of ramp type carbonate systems through time and position along the depositional profile. Note migration of organisms towards the inner ramp domain in the Late Cretaceous and Cenozoic. Most of the major biological extinction events are followed by an increased presence of the ramp depositional systems. I–VI are stratigraphic intervals referred to in the text. *Source:* Adapted from Burchette and Wright 1992; Sepkoski 1981.

porosity and the low density of Nummulite tests can explain the broad range of depositional environments where they are found. Bottom and wave currents induced the formation of *in situ* winnowed bio-accumulations, whereas offshore or onshore transport far from the original biocoenosis also occurred (Jorry et al. 2006).

Climate change has also influenced the occurrence of ramp systems. During periods of cooling, such as in the Late Ordovician, Early Carboniferous, Early Permian and the Oligocene, the C-factory biological assemblages became dominant and led to the development of extensive ramp systems like the ones proposed for the Mississippian of Kazakhstan and the southern Urals (Russia, Zempolich et al. 2002), the Bashkirian fusulinid carbonate ramp in Russia (Proust et al. 1996) and the crinoid-rich Mississippian in North America, Europe and North Africa (Ausich 1997; Kammer and Ausich 2006). A particular cool- and temperate-water ramp setting is represented by the Late Cretaceous Chalks, which were dominated by coccoliths in the deeper part of the basin, by mound-forming bryozoans and hexacorallians above storm-wave base, and by oyster-dominated associations along rocky shorelines (Surlyk 1997). This setting would come close to the present-day Cool-temperate setting (Figure 11.51b).

Ramps that were deposited in the tropical climate belt include the Middle Jurassic (Bajocian/Bathonian) Amelago outcrops in the High-Atlas of Morocco, which display exceptional kms-scale, continuous, structurally undeformed ramp morphologies (Figure 11.54) composed of an alternation of ooid-dominated and micrite-rich, ooid-free deposits (Pierre et al. 2010; Bordenave et al. 2021). These compositional changes and the associated

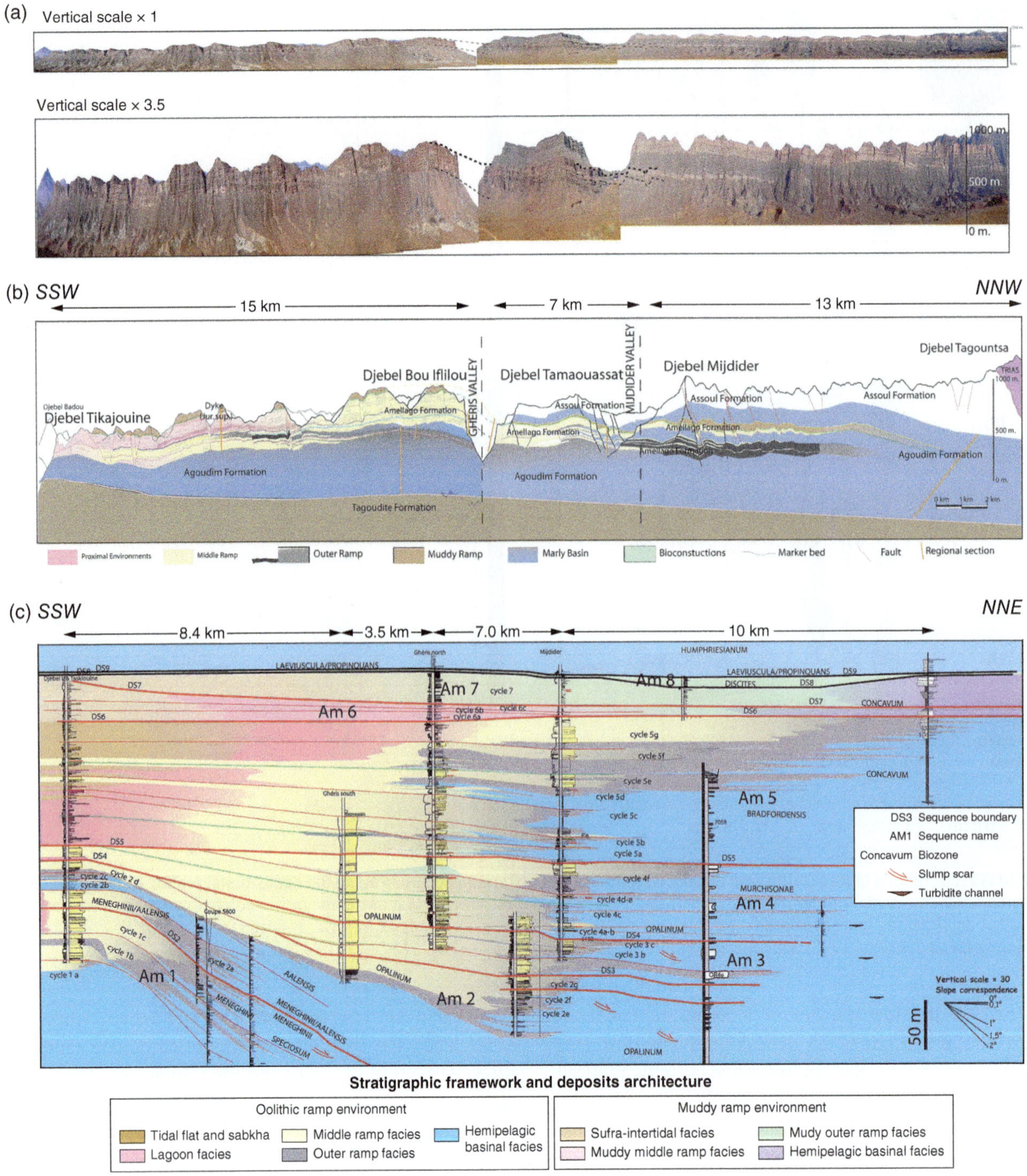

Figure 11.54 Unique example of a fully exposed ramp system along a 35-km-long outcrop, which allows to map out in detail lateral facies changes and stacking patterns (Amelago area, Morocco; *Source:* Pierre et al. 2010 / The Geological Society of London). (a) Outcrop photo at true vertical scale and same photo with a vertically exaggerated scale of 3.5; (b) stratigraphic correlations and facies distribution along the outcrop photo panel; (c) detailed stratigraphic correlation panel with measured sections, ammonite zones and different orders of depositional sequences.

differences in ramp geometries from low-angle muddy to a higher-angle grainy systems are interpreted to result from the combined effects of short-term eustatic sea-level changes and climatic variations inducing changes in carbonate production (change of the relative dominance of the T-factory and M-factory) under arid/humid conditions (Kenter et al. 2008; Pierre et al. 2010). Similarly, the Late Jurassic of the Northern Iberian Chain is consistently organised along ramp depositional profiles dominated by ooids (Bádenas and Aurell 2010; Aurell et al. 2010). The absence of rim-forming bio-constructing organisms in these successions, ensured that the ramp profile was maintained throughout their deposition.

Based on observations of the Late Miocene ramp-to-platform morphological evolution in the Western Mediterranean, Pomar (2020) pointed out that the observed sharp changes of skeletal associations (from C-factory to T-factory) were not only related to changes in temperature and climate (e.g. Nelson 1988; James and Bone 1989, 2004; James et al. 1992; Hayton et al. 1995; James and Clarke 1997) or light required for growth (photozoan/heterozoan; James 1997). Instead, the changes of skeletal association from the heterozoan or C-factory to photozoan or T-factory were coeval with sharp increases of the *thermal gradient* and the onset of a thermocline, which coincided with a sharp climate switch from humid to arid (Pomar et al. 2017). This model resembles the thermocline-stratified ramp model proposed for biotic changes in the Palaeozoic by Martindale and Boreen (1997). In addition, nutrient availability and trophic resources are key factors in controlling the dominant types of biotic communities in ramp systems (Brandano and Corda 2002; James et al. 2004; Pomar et al. 2004).

11.6.1.3 Land-Attached Carbonate Open Shelf Facies Model

The depositional profile of the open shelf facies model has elements from both the flat shelf with steep slope, or a slight seaward dipping ramp with distal steepening, the essential characteristic being that it has no elevated rim, which gives waves in ocean-facing settings free access to the platform top. Consequently, this model (Figure 11.55) has a limited range of deposits (5 facies belts), with the notable absence of a low-energy muddy lagoon, and with belts that have a smaller width than the models discussed above. Open shelves can occur along continental margins and on isolated platforms.

The open shelf facies model was defined based on the ocean-facing southern and western Australian coastlines, positioned adjacent to an arid hinterland of moderate to low relief (James et al. 1992; Collins et al. 1997). Cyclonic storms, long-period swells and large internal tides result here in preferential accumulation of coarse-grained sediments. The dominant C-factory is influenced by the combination of the south-flowing, low-salinity Leeuwin Current upwelling, the seaward-flowing saline bottom waters generated by seasonal evaporation, and flashy fluvial discharge. All these factors enhance mesotrophic-favouring biota, such as macroalgae and bryozoans, and sediment production by the P-factory (James et al. 2004; James and Clarke 1997; James and Bone 2010; James and Lukasik 2010; James and Jones 2016). The C-factory biotic assemblage thus dominates and mostly consists of heterotrophs, which predominantly produce calcitic bioclastic material that is broken down by bio-erosion, macerated and generally poorly cemented.

Carbonates are produced along the entire depositional profile, but due to the high energy (deeper fair-weather wave base level, c. 50–60 m) sediment accumulation is disjunct and occurs either onshore, in the form of aeolianites (Figure 7.1; James and von der Borch 1991; Betzler et al. 1997b; James and Bone 2015, 2017, 2022; Lipar et al. 2017) or offshore in prograding slope sediment wedges (e.g. ODP leg 182 results; Hine et al. 2004). The aeolianites consist of grains that were formed in the shallow-marine setting and were subsequently blown onshore from the carbonate beaches (Abegg et al. 2001). These deposits can reach a height of several tens of metres above sea level (Brooke et al. 2014; James and Bone 2015, 2022; Lipar et al. 2017). The inner shelf is further marked by a steep shoreface that descends to 10 m water depth, with the seafloor covered with rippled sand, bedrock with kelp, or seagrass. Bryozoans, molluscs, echinoids and small benthic foraminifera provide the material for the sands, and coralline red algae encrust most bedrock (James and Jones 2016). The mid-shelf, which is continuously affected by the large waves with their deep wave base, display wave-rippled sands that locally can develop into large subaqueous dunes. Sediment is produced in this zone, but little accumulates due to the wave erosion (James 1997; James et al. 2001). The outer shelf is a locus of sediment production and accumulation, with bryozoans, sponges and other heterotrophs. During Quaternary glacial lowstands, bryozoan mounds developed here (James et al. 2000) that were also interpreted as contourite-related sediment waves with a biogenic, bryozoan-dominated, cover (Anderskouv et al. 2010). The distally steepened slope of this profile becomes progressively muddier comprising less macro-biota, and receiving a pelagic fallout, especially of coccoliths.

11.6.2 Land-Detached Carbonate Systems

Land-detached carbonate systems are split into isolated platforms, which built up to sea level, and mounds, which

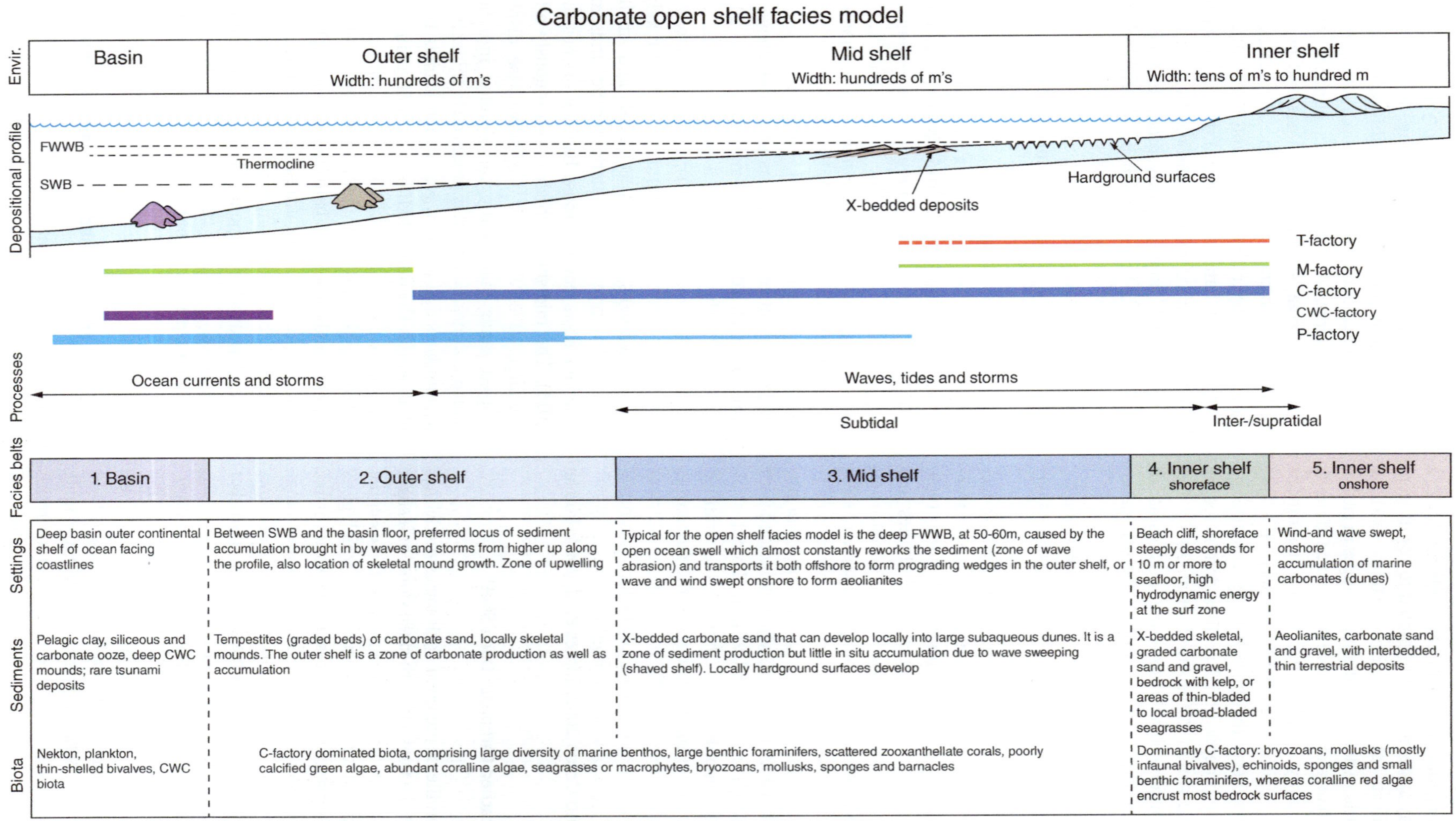

Figure 11.55 Conceptual carbonate open shelf facies model (Source: James and Jones 2016 / John Wiley & Sons). The distribution of carbonate factories along the depositional profile indicates their potential locus of production. Deposition is in this model preferably in the outer shelf domain, or onshore as aeolianites due to the reworking of the strong and deep fair weather wave swell. This high-energy open shelf model is typically characterised by C-factory dominated sediment.

have a rounded shape and most of which were deposited well below the sea surface.

11.6.2.1 Isolated Carbonate Platform Facies Model

Isolated platforms rise with steep slopes from the sea floor, building up close to sea level, with a slope break characterised by a bio-constructed reef margin, or by cemented sandy shoals (cf. Wilson 1975; Read 1985; Handford and Loucks 1993; Harris 1996). The modern isolated platforms occur, like their land-attached counterparts, mostly in the tropical climate belt and are dominated by the T-factory faunal association. The shallow-water production window makes them very sensitive to relative changes in sea level. A drop in sea level of just a few metres has the potential to eliminate most of the sediment production zone, expose the shallow-water sediments subaerially, and subject them to karstification (e.g. Purdy 1974; Schlager et al. 1994; Kievman 1998; Purdy and Waltham 1999; Purdy and Winterer 2001, 2006; Purdy and Gischler 2005; Purkis et al. 2014).

Specific characteristics of isolated carbonate platforms that justify their separate treatment are:

1) *Asymmetry:* the distribution of sediments on and around isolated platforms is controlled by wind and storms, waves, tides and ocean currents, often creating an asymmetrical, windward, high-energy and a leeward, low-energy side (e.g. Great Bahama Bank description below).
2) *Size and morphology:* many isolated platforms and atolls are built on an inherited topography (see Figure 11.3) such as fault-bounded syn-rift highs (Dorobek 2008), volcanic highs or salt diapirs (Bosence 2005). They range in size from a few to, more rarely, hundreds of kilometres across, with a sedimentary thickness of hundreds to over a thousand metres.
3) *Lithological composition:* because of the land-detached position, isolated from siliciclastic input, isolated platforms are entirely constructed of carbonate material, and production is not inhibited by siliciclastic material, with the notable exception of active volcanic atolls (Wilson and Lokier 2002).

11.6.2.1.1 Present-Day Isolated Carbonate Platforms

Isolated Flat-topped Carbonate Platform (Bahamas)

The Bahamas platform is one of the largest and best studied examples of an isolated, flat-topped carbonate platform. It comprises a group of barely submerged carbonate shoals with intervening deeply submerged embayments or channels (Providence Channel, Tongue of the Ocean, Exuma Sound, Columbus Basin). It lies off the Atlantic coast of Florida and enjoys a humid subtropical climate. In total, 96,000 km^2 of shallow-marine carbonate deposition is bounded on all sides by steep marginal slopes (often exceeding 40° in their upper parts) that terminate in water depths of hundreds or even thousands of metres (Figure 11.56a). Platform water

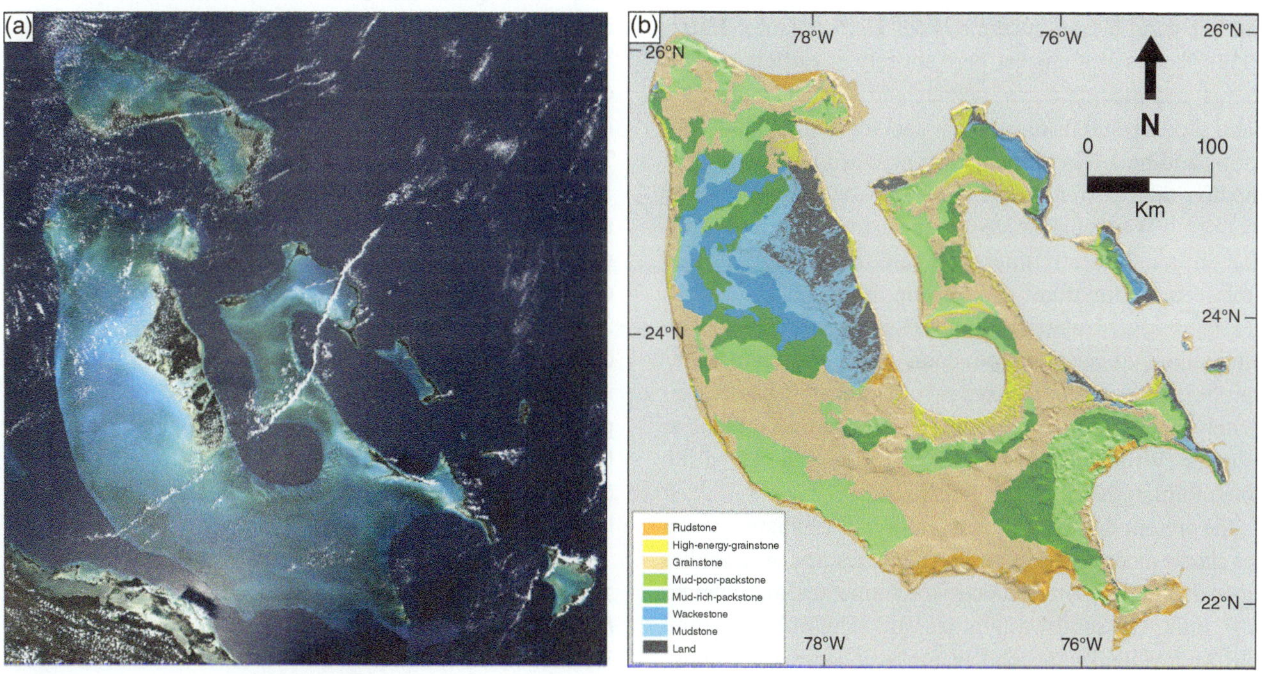

Figure 11.56 Bahamas isolated carbonate platforms. (a) Aerial view of the Bahamas archipelago (*Source:* Eberli (2007) / Comparative Sedimentology Laboratory; NASA / Wikimedia Commons / Public domain); (b) detailed sedimentary facies texture map of the Great Bahamas buildup (based on sampled database from Reijmer et al. 2009, and combined with satellite image observations in Harris et al. 2015).

depths seldom exceed 15 m and are normally less than 6 m. The eastern margins of the platform are rimmed by islands of cemented Pleistocene aeolianites. Many detailed facies models have been proposed (e.g. Purdy 1963a,b; Ginsburg and Shinn 1964; Harris 1979; Eberli and Ginsburg 1987; Reijmer et al. 2009; Bergman et al. 2010; Berkeley and Rankey 2012; Rankey and Doolittle 2012; Harris et al. 2015). A sedimentary texture map is presented in Figure 11.56b. The high-energy platform-top is made of sand bodies including a diversity of barriers and beaches (Figure 11.43; Wanless and Tedesco 1993; Harris et al. 2018). They develop where sediment supply exceeds removal by erosion and where adequate accommodation exists due to the platform topography. The shape and the extension of the carbonate sand bodies is highly variable and result from the platform geometry, bedrock topography, the presence of ridges or islands, the age of the shoal, orientation relative to predominant winds and waves and shelf break curvature. Off-bank transportation of neritic carbonate sands occurs in gravity flows (Perkins 1972; Droxler et al. 1983; Heath and Mullins 1984; Boardman et al. 1986), which alternate down-slope with the fine-grained winnowed aragonite fraction (whitings) that settled out of suspension (Shinn et al. 1989; Robbins and Blackwelder 1992; Robbins et al. 1997; Morse et al. 2003; Dierssen et al. 2009; Purkis et al. 2017; Jorry et al. 2020).

A predictive relationship between increasing peak current velocity and increasing area of a sand body is observed for the Bahama Cays where grainstone accumulation preferentially occurs in areas of high tidal exchange and intermittent wind-driven currents (such as hurricanes; Harris 1969). Tidal sands consist here entirely of carbonate sediments and include ooids, peloids and skeletal fragments. Sand shoals result from complex stacking of individual bedforms and bars. These features are shaped by, and in turn modify, tidal currents that move on and off the shallow platforms. Waves and wave-driven currents play a subordinate although locally important role in their genesis and architecture. Importantly, water depths, habitats and facies are not uniquely related or linked (Rankey 2014, 2021). The interactions of various production sites in different water depths in response to a mixed set of physical processes ultimately leads to a high diversity of facies belts and to facies mosaics (Wright and Burgess 2005; Strasser and Védrine 2009; Rankey 2016).

The present-day depositional profile of the Bahamas forms the classic model for steep carbonate slopes, and most studies agree that the slope morphology is asymmetrical and changes with time. The southwestern, leeward sides of both Bimini and Andros Banks are good examples of concave accretionary progradational wedges (Figure 11.57), whereas their northeastern, windward, sides are steeply aggrading escarpment margins (Eberli and Ginsburg 1987, 1989; Eberli 2000). Consequently, the Straits of Andros shows an asymmetrical infill pattern (Figure 11.57b, e). Leeward margins such as these are commonly characterised by 'highstand shedding' (Droxler et al. 1983; Schlager et al. 1994), facilitating progradation and platform expansion (Eberli and Ginsburg 1987, 1989). Recent studies on the 'wave-ward margins' of the Little Bahama Bank (Rankey and Doolittle 2012; Tournadour et al. 2015; Mulder et al. 2017), provide further detail of a complex mix of erosional (escarpment, gullies and canyons) and depositional processes and features such as several mass-transport complexes, a thick (up to 20 m) Holocene sediment sand wedge and rare slump features.

Atolls (Reunion, Mayotte, Mauritius, Maldives, Seychelles and Glorieuses Archipelago)

Atolls are carbonate structures that sit on a rocky pedestal, which can be volcanic, or basement rocks (Droxler and Jorry 2021). Although of relatively minor importance in terms of their rock volume, they are an emblematic example of isolated, carbonate platforms, ever since they were first described by Charles Darwin (1842). In contrast with the shallower lagoonal systems of present-day land-attached platforms, some atoll lagoons are notably deep: with the Maldives Archipelago around 40–60 m (Naseer 2003), and up to 76 m for some of the French Polynesia and Cook Islands atolls (Andréfouët et al. 2020). The morphology of atolls with a volcanic pedestal is initially controlled by the morphology of the volcanic edifice. Atolls vary from sub-circular (e.g. Anewetak; Quinn and Saller 2004; Reunion, Mayotte and Mauritius; Camoin et al. 1997) to elliptical (Bermuda; Vacher and Rowe 2004) to a complex of amalgamated volcanic pedestals and sub-circular volcanic pedestals with intervening grabens (Maldives; Purdy and Bertram 1993). The Seychelles have a granitic pedestal (Pirazzoli et al. 1990). Apart from the shape of the pedestal, the sediment distribution on atolls is strongly controlled by the prevailing wind and storm directions (Glorieuses Archipelago, Jorry et al. 2020; Maldives Archipelago, Belopolsky and Droxler 2004; Betzler et al. 2017a, b; Droxler and Jorry 2021; Seychelles, Braithwaite et al. 2000). For example, at the Glorieuses Archipelago about half of the carbonate platform production is shed at the leeward side of the platform, through deep-sea trough channels and deposited in base-of-slope lobes (Figure 11.58). Well-studied atolls are also located just offshore Belize (Gischler 1994, 2003, 2007).

11.6.2.1.2 Isolated Platforms and Atolls in the Rock Record

Isolated carbonate platforms are a common, well-documented feature of the Phanerozoic record. A selection of examples is provided in Supplementary Information

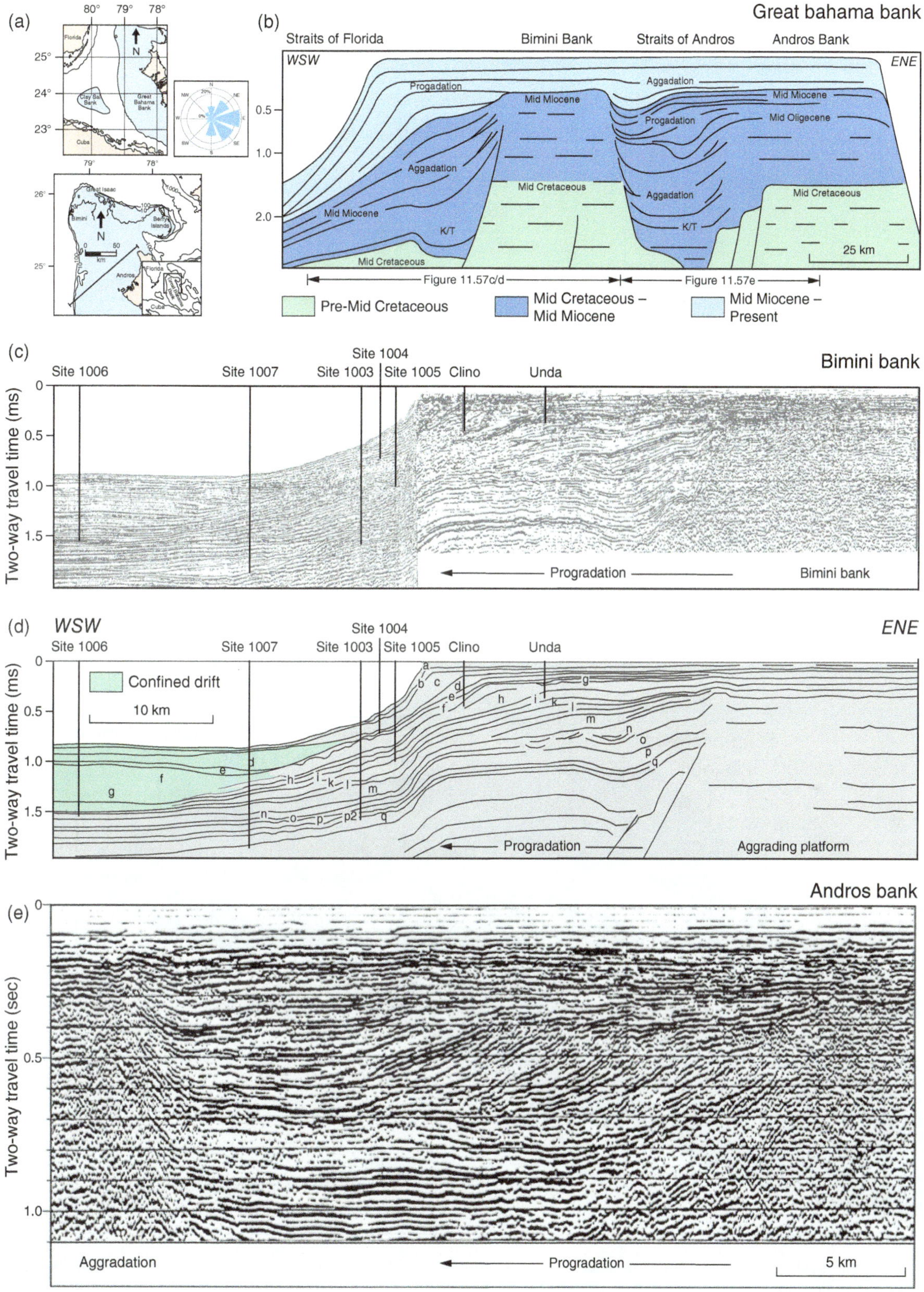

Figure 11.57 Seismic cross sections of the Bahamas carbonate buildup illustrating the evolution of the depositional profiles through time to the present day. (a) Location maps of Great Bahama Bank with location of seismic line and wind rose diagram showing prevailing wind direction from the east (*Source:* Adapted from Eberli and Ginsburg 1987); (b) Stratigraphic cross-section of the Great Bahama Bank showing is composite architecture with the amalgamation of the Andros and Bimini Banks. (c) Seismic line across Bimini Bank with position of the boreholes drilled along the line; (d) interpretation of seismic line showing the progradation of the accretionary slope (Eberli et al. 1996; Eberli 2000) and toe of slope deposits (Mulder et al. 2017); (e) seismic line across Andros Bank, with contrasting leeward progradation and windward aggradation (*Source:* Adapted from Eberli and Ginsburg 1989).

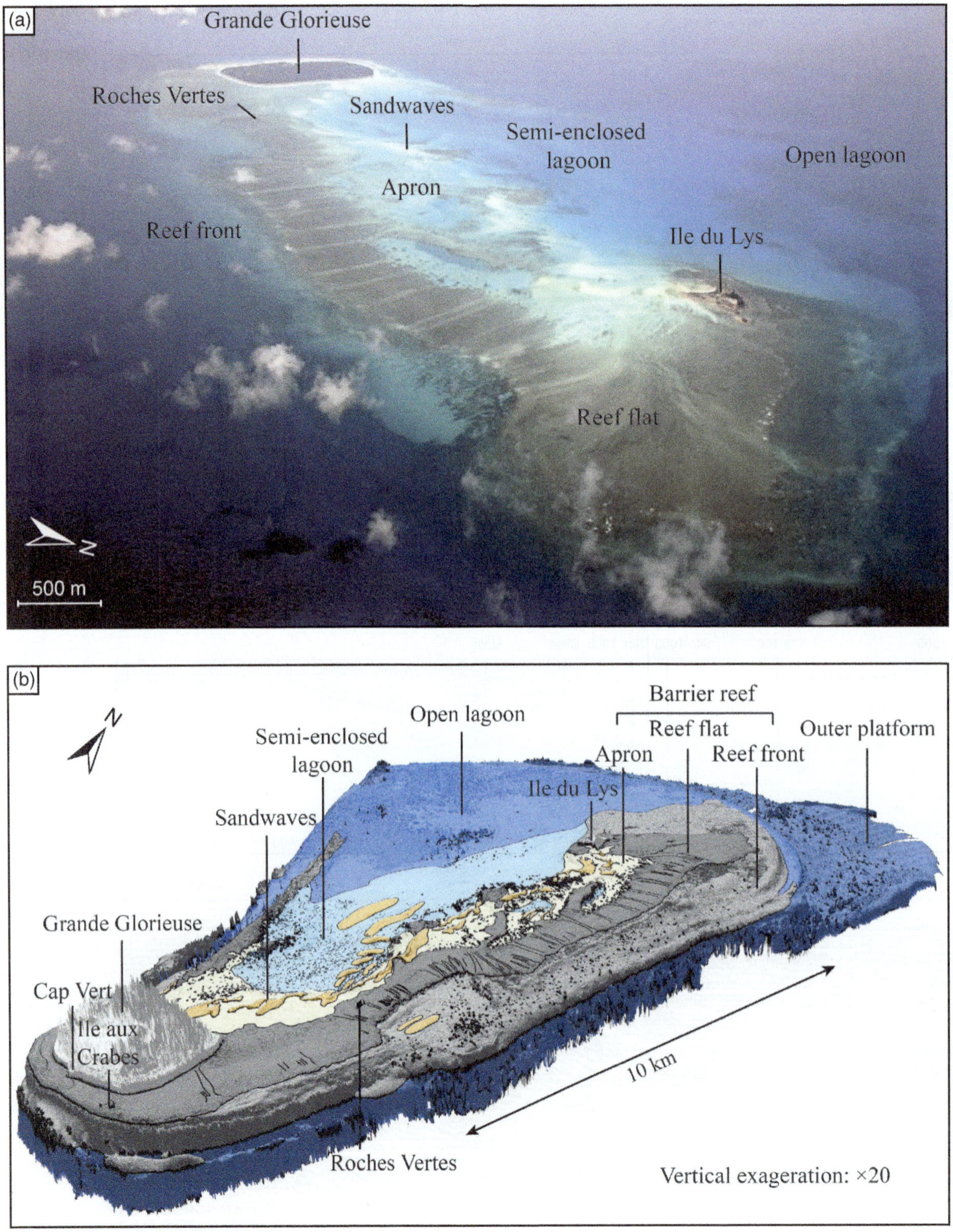

Figure 11.58 Atoll type isolated carbonate platform of Glorieuses Archipelago, Indian Ocean (*Source:* Prat et al. 2016 / Elsevier). (a) Satellite image and main depositional environments and (b) facies map.

11.S.2, and further references can be found in Gischler (1995), Kiessling et al. (2002) and Markello et al. (2008). An asymmetrical facies distribution and stratigraphic architecture is one of the most striking characteristics of many present-day and ancient isolated platforms; a feature that can be particularly well appreciated in high-resolution 3D seismic surveys such as shown in Figures 11.59 and 11.60.

One of the rare examples of a described ancient atoll is the El Joyazo atoll, which formed on top of a small elongate ridge of dacite close to an attached Miocene carbonate platform in southern Spain (Dabrio et al. 1981; Warrlich et al. 2005, 2008). The atoll displays the same sedimentation pattern as the well-developed coral and algal (*Halimeda*) reefs elsewhere in the Mediterranean (Pomar et al. 1996). The reef complex consists of a reef-core, constructed by a framework of vertical, branching colonies of *Porites* coated by laminated micrite, and probable submarine cements. The fore-reef deposits, consist of sloping (30°) beds of broken skeletal material and debris (Dabrio et al. 1981). In the northern Oman mountains, Triassic isolated platforms are seeded on Permian volcanics and tilted fault blocks and are now exposed in obducted tectonic units (Weidlich and Bernecker 2003).

11.6.2.2 Carbonate Mound Facies Models

The mound depositional profile consists of an elevated structure with a rounded, hill-like top that mostly formed in deeper waters (from several tens to hundreds of metres deep), which makes them less sensitive to wave and storm activity and allows a rounded shape. (When occurring in shallow water, the top can be somewhat flattened by waves and currents.) The deeper-water mound is a typical profile for the CWC-factory, and a common feature in both M- and T-factories in deep and shallow-waters. A variety of organisms has been involved in their formation since the Early Palaeozoic (Monty et al. 1995). Here, we distinguish between two types of mounds: skeletal frame-built mounds and microbial-mud mounds (Schlager 2005).

11.6.2.2.1 Skeletal, Frame-Built Mound Facies Models

Skeletal frame-built mounds are dominated by skeletal elements, such as corals, crinoids, bryozoans and phylloid algae, and occur throughout the Phanerozoic. The present-day cold-water coral mounds of the CWC-factory are the most important modern representative. They occur worldwide (Figure 11.32) at depths of 200–1000 m (or even deeper) and vary in size from several metres to over 300 m high (Figure 11.61a; Freiwald et al. 2004; Hebbeln and Samankassou 2015). They are composed of a mosaic of facies including cold-coral fragments, other biogenic components, and a variably biogenic (planktonic and benthic fragments) or terrigenous (siliciclastic, clays and silts) matrix (Figure 11.61c). For further information, see Section 11.5.

Ancient examples include Silurian reef mounds, Late Palaeozoic phylloid algae mounds, and Cenozoic bryozoan, scleractinian coral mounds (Copper and Brunton 1991; McLaughlin et al. 2019). The Silurian reef mounds of the Michigan and Illinois Basins occur in a wide range of settings from near-shore fringing structures, to mid-shelf or lagoonal patch reefs, shelf margin barrier and patch reefs, 'atolls', and spectacular mid- and deep-slope mounds (Figure 11.62a). The reef biotas varied from algae in restricted nearshore or embayment habitats, to impoverished patch reef faunas in marginal environmental settings, to thriving open marine coral-stromatoporoid (coralline sponges)-bryozoan communities, to sparsely fossiliferous lithistid or stromatactid mud-mounds in deeper water settings (Figure 11.62b, c). Rine et al. (2017) observed asymmetrical geometries in the Silurian reef complexes of the Michigan Basin, with internal facies distribution patterns influenced by a dominant easterly to southeasterly palaeowind direction. These pinnacles show a pulsed upward growth through successive evaporite cycles controlled by sea-level change.

In the Late Palaeozoic, phylloid algal mounds stand out as a dominant type of build-up (Wray 1968; Heckel 1974; Wilson 1975; Toomey 1980; Oborny et al. 2017). In particular, during the Pennsylvanian, they had a worldwide distribution (Canadian Arctic Archipelago: Davies et al. 2005 and Beauchamp et al. 1989; Arctic of Norway: Bruce and Toomey 1993; Russian Platform: Chuvashov 1983; Vennin et al. 2002b; northern Africa: Vachard et al. 1989; Toomey 1991; China: Fan and Rigby 1994; western United States: Heckel and Cocke 1969; Heckel 1974; Fagerstrom 1987; West 1988; Grammer et al. 2000; Samankassou and West 2002; Oborny et al. 2017). The phylloid algal mounds formed individual and amalgamated complexes, in various water depths (tens of metres to <5 m), resulting either from active constructional algal growth or passive accumulation of algal fragments within a mud matrix and/or early diagenetic marine cements. The shape of phylloid algae mounds in the Paradox Basin (USA) has been interpreted by Grammer and Ritter (2009) as the result of tidal currents based on a comparison with Miocene *Halimeda*-rich mounds (Braga et al. 1996).

Among the skeletal mounds, the Uralian Early Permian reef mounds show an upward transition from bryozoans, *Tubiphytes* and *Archaeolithoporella* to corals and *Palaeoaplisina* boundstones, the last facies being developed under wave action and bottom currents (Vennin et al. 2002b; Zempolich et al. 2002). Characteristic for these build-ups is the abundance of early marine cementation, which played a role in slope stabilisation.

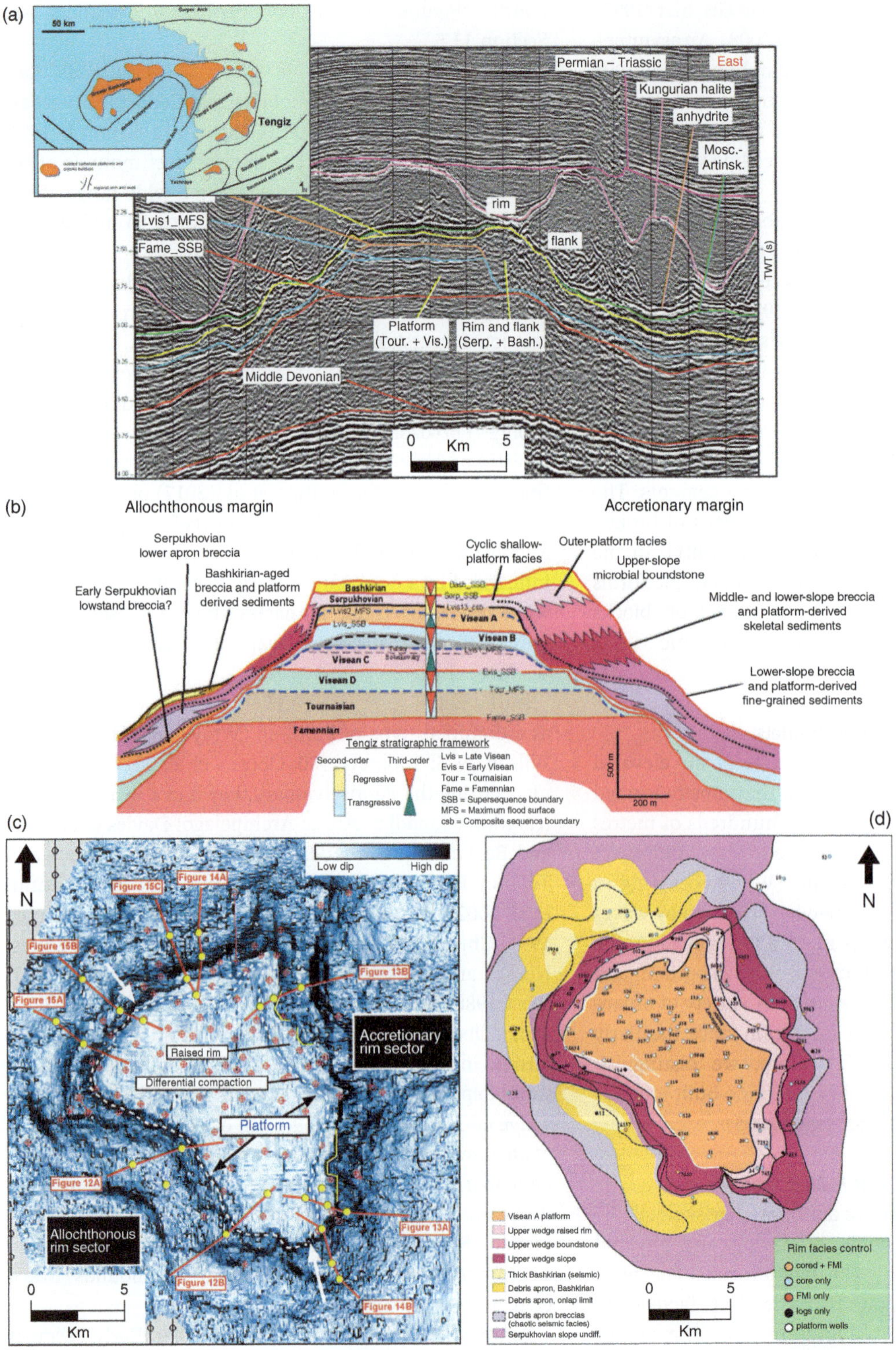

Figure 11.59 The Late Devonian/Early Carboniferous Tengiz platform (*Source:* Adapted from Kazakhstan; Collins et al. 2006; Weber et al. 2003) is a good example of a land-detached T-/M-factory dominated isolated carbonate platform. (a) Inset: palaeogeographical map showing the position of the Tengiz field as one of the isolated platforms in an archipelago in the i. PriCaspian Basin; the 2D Seismic cross section shows the complex internal architecture; (b) stratigraphic model based on core and seismic cross sections. The initial topography is created by a retrograding succession (Tournaisian-Visean), followed by the deposition of an early, lower debris apron that accumulated around the base of the isolated platform. Subsequently, asymmetrical progradation occurred along a bypass margin (West), and an accretionary margin (East) which has an upper wedge with a boundstone core; (c) dip magnitude map on the top of the Bashkirian surface measured from 3D seismic data, showing the extent of the accretionary and bypass (allochthonous) margins; (d) same map view as (c) with facies distribution in the rim and flank based on seismic and well log information.

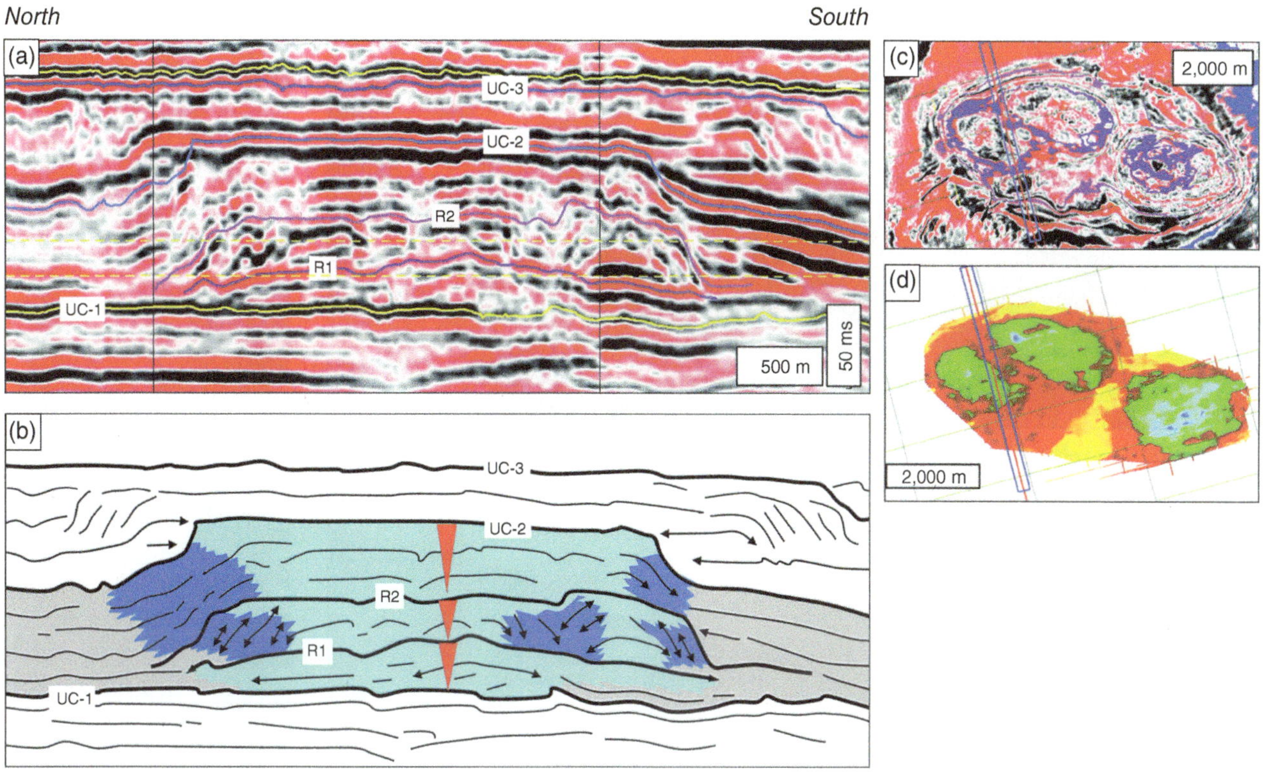

Figure 11.60 3D architecture of a Miocene isolated, T-factory carbonate platform (NorthWest Australian Shelf; *Source:* Rankey 2020 / GeoScienceWorld). (a) 2D seismic line illustrating complex internal architecture of the isolated platform; (b) stratigraphic interpretation of the seismic line in (a) with four development stages (R1, R2, UC2 and UC3); (c) horizon slice in the lower part of the isolated platform showing the presence of three individual mounds in the early development stage (R2); (d) thickness map of these mounds.

During the Late Jurassic, besides the deeper-water 'classical' mud-mound occurrences (see below), there are several examples of skeletal mound development in shallow-water environments (Nose 1995). These mounds are generally composed of scleractinian corals, stromatoporoids, detrital mud and minor amounts of binding microbial crusts. A well-documented example of cool-water skeletal mounds is that of the Cenozoic bryozoan and scleractinian coral mounds positioned in water depths of approximately 200–300 m in the Danian of Scandinavia (Surlyk 1997; Weidlich and Bernecker 2003; Hvid et al. 2021). These scleractinian coral mounds represent one of the few potential analogues of the present-day cool-water coral mounds (Hvid et al. 2021). Along the SW Australian margin, Pleistocene deposits contain bryozoan skeletal mounds, which were deposited during glacial lowstands, when these were sites of nutrient-rich coastal upwelling (Figure 11.83b; ODP Leg 182; Holbourn et al. 2002; James et al. 2004).

11.6.2.2.2 Microbial Framework Mud Mound Facies Models

The definition of a microbial mud-mound is based on ancient examples. It is a carbonate build-up with depositional relief, dominantly composed of carbonate mud, peloidal mud or micrite, either produced *in situ* or of detrital origin. Bacteria and cyanobacteria are the primary producers. These mounds contain an admixture of other biota, such as bryozoans, sponges, crinoids and corals (Monty 1995; Bosence and Bridges 1995).

Good reviews of ancient carbonate mud-mound systems and related processes are provided by Bourque et al. (2004) and Monty et al. (1995). Vennin et al. (2007b,c) provide a review of the Palaeozoic mud-mounds in a wider perspective. The terminology and classification of present-day and ancient microbial framework-built mud-mounds are a matter of active debate, due to the often partial or non-preservation of the biota that was involved in mound formation (Foubert and Henriet 2009, and Hebbeln and Samankassou 2015 for reviews).

Late Cambrian microbial mounds, exposed in Texas, USA, occur in clusters and show a three-phased evolution (Figure 11.63a, b). Detailed mapping of these in a horizontal bedding plane showed a scaled internal organisation (constructed of smaller mounds), ranging from the dm to the hundred-metre scale, with several amalgamation steps, to reach a total diameter ranging from 30–90 m and a

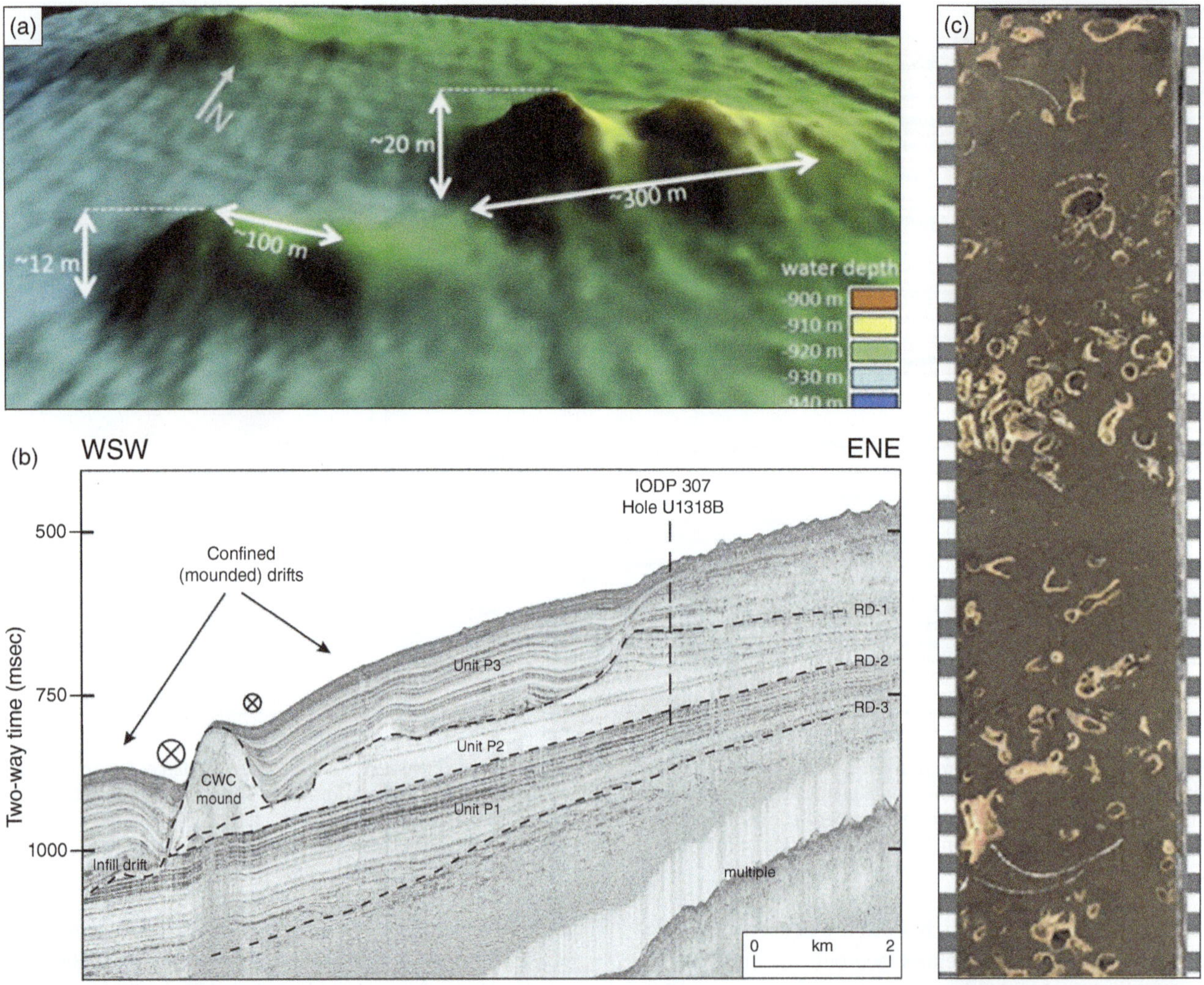

Figure 11.61 Cold Water Coral (CWC) mound architecture and facies. (a) CWC mounds on the sea floor offshore Morocco (Hebbeln and Samankassou 2015); (b) CWC mound surrounded by confined drifts (North Atlantic, seismic line, Ghent University) with location of IODP 307 site – Hole U1318B (*Source:* Hebbeln et al. 2016 / with permission of Elsevier); (c) core with typical cold-water coral mound facies of coral fragments (and other shells) embedded in a matrix of hemipelagic sediments (core GeoB14904-2, Mauritanian margin; Hebbeln et al. 2016).

thickness of 10–15 m (Khanna et al. 2020a,b; Lehrmann et al. 2020). The distribution and final shape of the build-ups was interpreted to be controlled by palaeo trade winds and tidal currents.

The Devonian carbonate mud-mounds (Figure 11.63c, f) are among the most spectacular of the various Palaeozoic carbonate mounds (e.g. Algeria: Wendt et al. 1997; Ireland: Lees 1964; Belgium: Boulvain 2001; Morocco: Brachert et al. 1992; Klug et al. 2018; Canada: Zhou and Pratt 2019a,b). In Belgium, their height varies from 30 to 80 m with diameters of 100–250 m, and they are embedded in mudstone, nodular mudstone and argillaceous limestone, and show a clear vertical succession of facies (Johnson et al. 1985; Boulvain 1993, 2001, 2007; Cooper 2002; Da Silva and Boulvain 2012). These mounds typically started growth below the euphotic zone and storm wave base and built up into shallow-water. The earliest carbonate mound facies rest on an argillaceous limestone substrate and consist of spiculitic wackestone with *Stromatactis* which becomes progressively enriched in crinoids, bryozoans and corals towards the top (Figure 11.63d). The mound resulted from the development of a community of sponges, below the photic zone, in a hypoxic environment, at depths of 100–150 m. The tops of the mounds then accreted into the wave-affected zone in less than 70–80 m water depth. Their morphology results from the difference between the accumulation rate of sediments in the off-mound environment and the mound's growth rate. Such mud-mounds have also been extensively studied in the 3D outcrops of northern Africa (Wendt et al. 1993, 1997, 2006; Wendt

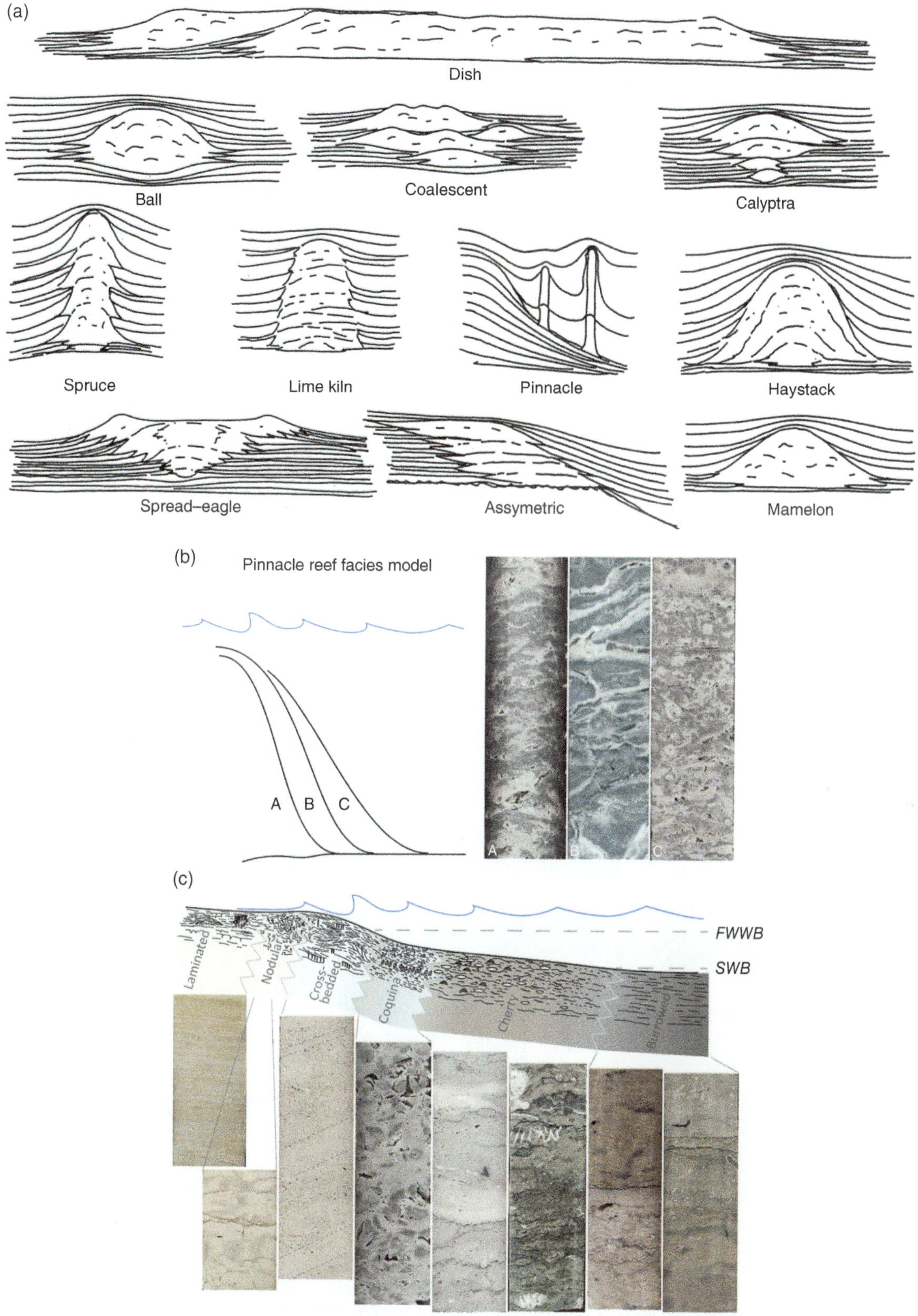

Figure 11.62 Silurian skeletal reef mound shapes and facies. (a) Idealised diagrams of the more common Silurian reef geometries, as seen in profile (*Source:* Adapted from Copper and Brunton 1991; expanded from Shaver 1977); (b) simplified facies model for Silurian pinnacle reefs. A: stromatactis core and flank facies, B: laminar stromatoporoid boundstone, C: crinoidal packstone to rudstone (McLaughlin et al. 2019); (c) depositional model depicting the relationship of facies (representative images of core for each) along an increasing water depth gradient from inter-tidal to deep subtidal (*Source:* McLaughlin et al. (2019) / Elsevier).

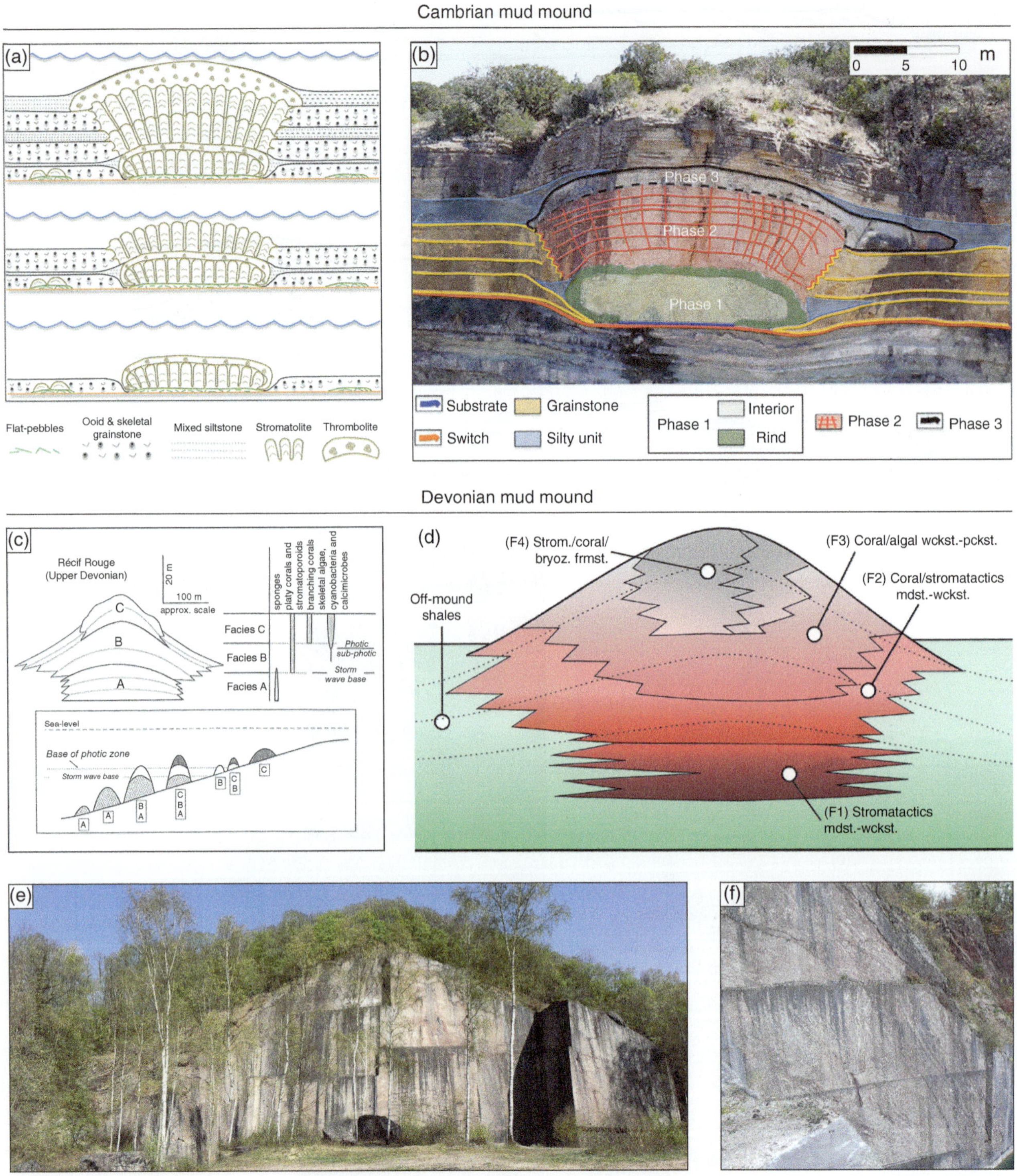

Figure 11.63 Palaeozoic mud mound examples, all showing a stepwise evolution of morphology and facies. (a,b) Cambrian mud mound, with stromatolites and thrombolites (Texas, USA; *Source:* Khanna et al. 2020a, b / John Wiley & Sons); (c–f) Devonian mud mound, with a facies succession of sponges, corals, and algae (*Source:* Adapted from Belgium; Boulvain 2001); (g,h) Carboniferous Waulsortian mud mound, with sponges, algae and crinoids (*Source:* Adapted from Belgium; Aretz and Chevalier 2007; Lees and Miller 1985); and (i–k) Permian mud mound, with *Tubiphytes-Archaeolithoporella, Palaeoaplisina* and bryozoa (Vennin et al. 2002a) (k Unknown Source / Wikimedia Commons / CC BY 4.0).

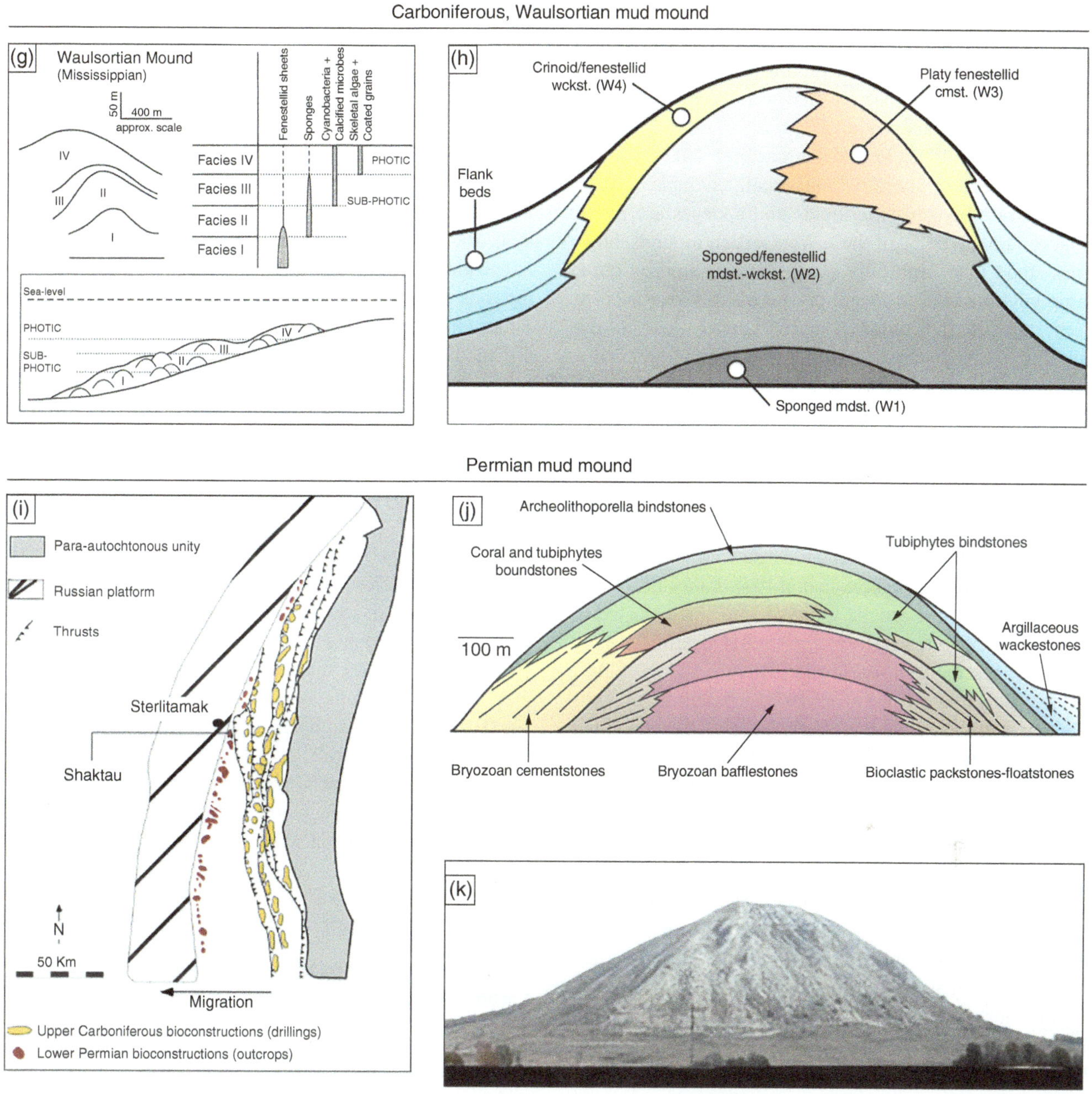

Figure 11.63 (Continued)

and Kaufmann 1998, 2006). Spectacular examples are in the Bechar Basin (Algeria), where several mounds developed in a distally steepened, open shelf ramp setting (Bourque et al. 1995). Another well-studied example is the Kess-Kess mountains (Hamar Laghdad, Morocco), which is considered as an oasis of biodiversity (e.g. Brachert et al. 1992; Joachimski and Buggisch 1999; Aitken et al. 2002; Jakubowicz et al. 2014; Klug et al. 2018; Berkowski et al. 2019). The location of these mounds is attributed to the presence of hydrothermal fluid circulation up faults, enabling carbonate precipitation and rapid lithification (Belka 1998; Mounji et al. 1998; Peckmann et al. 1999).

The Early Carboniferous seas were colonised by Waulsortian mud-mounds (Figure 11.63g, h), first comprehensively described by Lees (1964) in Belgium and later observed in other places in Europe, Asia and North America (Lees et al. 1985; Lees and Miller 1995; Monty et al. 1995; Pratt 1995). They are carbonate mud-rich build-ups that commonly occur isolated within the deeper parts of Lower Mississippian ramps and have facies, geometries and fossil communities distinct from those of the surrounding sea bottom (Figure 11.63h; Ahr and Stanton 1996; Jeffery and Stanton 1996). The carbonate mud is considered to be primarily microbial, with bacteria and cyanobacteria of the M-factory as the primary producers (Monty et al. 1995; Reitner and Neuweiler 1995; Rodriguez-Martinez 2011). Locally admixtures of low diversity, heterozoan assemblages such as fenestrate bryozoans, crinoids and sponges are found (Figure 11.63h). Some of these Waulsortian mounds, however, stood topographically above the surrounding outer ramp environment and were not necessarily restricted to deep, aphotic environments but occurred close to the euphotic zone (Kirkby and Hunt 1996; Bachtel and Dorobek 1998; Dorobek and Bachtel 2001).

The acme of mud-mounds was reached in the Late Palaeozoic, with the increase of skeletal components (bryozoan, sponges and others) and cements. Whereas bryozoans, sponges, bivalves and echinoderm-rich microbial mud-mounds generally developed in deeper water, the algal-rich and calcified sponge mounds are considered to have originated in shallow water (Beauchamp 1994; James 1997). Permian mound types can be grouped into microbial reefs, metazoan reefs and calcareous algae reefs (Flügel and Stanley 1984; Davies et al. 1989a–c; Flügel and Reinhardt 1989; Flügel 1994; Weidlich and Fagerstrom 1998). These mud-mounds are characterised by the rareness of large, well-skeletonised biota (e.g. Fagerstrom 1997). Many Permian mounds are dominated by microbially related bio-mineralization of the M-factory (Figure 11.63i–k; Webb 1996, 2001; Kirkland et al. 1998) and contain syn-sedimentary cements (Schmidt 1977; Grotzinger and Knoll 1995; Vennin et al. 2002b).

During the Late Jurassic, as during the Mid-Palaeozoic, microbial mud-mounds were widespread and occurred together with shallow-water reefs (James and Bourque 1992; Schmid et al. 2001; Neuweiler et al. 2001). Late Jurassic mounds were especially common along the northern Tethys margin and its adjacent shelf seas. Numerous studies have illustrated sponge–microbialite mud-mounds that developed within deeper ramp settings (Schmid et al. 2001 for review), whereas local to regional slope breaks were the sites for microbialite-rich mound growth. Tomás et al. (2019) documented an exceptional case of microbial-siliceous sponge mounds that lived in shallow water (Middle Jurassic, NE Spain). During the Cretaceous microbial mounds developed at specific times, such as in the early Aptian, when metre to decametre scale mounds and clusters of amalgamated mounds of *Lithocodium-Baccinella* occurred on the carbonate shelves bordering the Neo-Tethys Ocean (e.g. Pittet et al. 2002; Hillgärtner et al. 2003; Hillgärtner 2010; Rameil et al. 2010; Schlagintweit et al. 2010; Embry et al. 2010).

After the Mid-Cretaceous, mud-mounds became skeletal-rich (Monty 1995) and the role of the microbial organisms was reduced to locally stabilising the platform margin as in present-day T-factory platform margins (Grammer et al. 1991, 1993; Busson et al. 2019; Reolid et al. 2017a), tidal inlets (Reid and Brown 1991; Reid et al. 1995) and beaches (Bahamas, Reid et al. 1999). Microbial cements also infill the primary pore space between the coralgal framework (Camoin et al. 1999; Séard et al. 2011) within anoxic micro-environments (Heindel et al. 2010, 2012; Canet et al. 2003, 2005; Petrash et al. 2012).

The identification of modern equivalents of ancient mud mounds is still a controversial issue. Some authors (i.e. James and Bones 2015) state that there are no known modern analogues for ancient microbial mounds, whereas Hebbeln and Samankassou (2015) consider that the present-day deep-water carbonate mounds (CWC-factory) can serve as good analogues to ancient 'heterozoan' carbonate mounds (or skeletal frame-built mounds). Most commonly Miocene and present-day mounds are dominated by autotrophic organisms (T-factory) such as calcareous algae (e.g. *Halimeda*; Drew 1983; Drew and Abel 1988a, b; Marshall and Davies 1988; McNeil et al. 2021a,b). Several mounds have been described in modern shallow subtropical back reef and bay environments of southern Florida (Wanless et al. 1995; Bosence 1995) and compared with Cenozoic examples from the Eocene in Spain (Taberner and Bosence 1995). The Florida examples have more affinity with skeletal frame-built mounds, and they mainly differ from the Palaeozoic and Mesozoic mud-mounds in their low-relief and complex stratal geometries, and in their facies being mainly representative of seagrass breakdown and reworking (Bosence 1995).

11.6.3 Seaway, Incised Valley and Estuary Carbonate Systems

Bioclastic, carbonate-filled, incised valley systems are rarer than their siliciclastic counterparts. Where recorded, the carbonate fill is mostly skeletal or oolitic, well-sorted, transported by (tidal) currents and mostly of C-factory origin (Nelson 1988). Present-day examples of carbonate tide-dominated estuaries are described from Australia (Harris 1993), the Scottish and Celtic North Sea (Wilson 1988) and Canada (Carey et al. 1995). In shallow water coastal systems such as northern Brittany (Wehrmann 1998), macrophytes are the most efficient bioclastic carbonate producers. Their distribution is sensitive to the hydrodynamic regime with *Laminaria*-kelp colonising rocky substrates on open shelves episodically impacted by strong storms, and autochthonous coralline algae forming *maerl*-banks in the more sheltered coastal parts of the platform. The inner shelf deposits are predominantly 'bryomol' (bryozoan and mollusc) carbonates composed of bryozoans, diverse epifauna, brachiopods, balanids, ophiurids and benthic foraminifera.

Ancient examples of carbonate-filled incised valleys have been reported from the Cenozoic of France (Allen and Bass 1993), the Plio-Pleistocene of Cyprus (McCallum and Robertson 1995), the Middle Miocene of the Mut Basin of Turkey (Bassant 1999; Bassant et al. 2005); the Cretaceous of North America (Dennis et al. 2013) and the Aptian of Northern Spain (Bover-Arnal et al. 2009; Peropadre et al. 2013). Among the most spectacular examples are the complex Miocene palaeo-valleys exposed in the peri-Alpine terrane of southern France (Figure 11.64; Rubino 1988; Parize et al. 1997; Reynaud et al. 2006, 2012; James et al. 2014). These valleys correspond to a semi-enclosed depression that was connected to the open Mediterranean Sea by narrow seaways and filled with C-factory, heterozoan limestones (made of bryozoans, molluscs and red algae), in tide-dominated environments characterised by marginal conglomerates, tidal bars and channels, open-coast tidal deposits, subtidal dunes, basinal bioclastic beds and basinal muds. Increasing or decreasing tidal currents are related to changes in constriction of the tidal flow (Reynaud and James 2012). Whereas the cross-bedded grainy sediments were restricted to the narrow valley, once the valley became filled, the sediments onlapped across and overstepped the interfluves with reduced current speeds.

11.6.4 Mixed Systems

Mixed system facies models are only briefly reviewed here for completeness. Where relevant reference is made to overlap with other chapters of this book.

11.6.4.1 Mixed Carbonate – Siliciclastic Facies Models

At the basin scale, carbonates are frequently found adjacent to siliciclastics, deposited as separately operating depositional systems, and contracting and expanding as a result of sea-level fluctuations, climate and seasonal variations (e.g. Belize: Esker et al. 1998, Ferro et al. 1999; Great Barrier Reef: Dunbar and Dickens 2003; Gulf of Papua: Tcherepanov et al. 2008a,b, 2010, Jorry et al. 2010; Bonaparte Basin: Bourget et al. 2014). In these situations, the grain size of the supplied clastics is key to water turbidity, for instance, sand and gravel systems can coexist much better with carbonates (e.g. Gulf of Aqaba; Sinai coast, Strasser et al. 1992) than large tropical rivers with months-long flood stages (see Amazon and Papua New Guinea examples in Section 9.2). Chiarella et al. (2017) provide an overview of the different mixed carbonates and siliciclastic systems, from the build-ups in front of deltas down to the bed scale.

In Safaga Bay, on the Egyptian side of the Red Sea, a mixed system has been studied in detail with an 'actuo-palaeontological' approach providing insight into the relationship between clastic sediment dynamics and the response of the carbonate factories (Piller and Pervesler 1989; Piller and Mansour 1990, 1994; Piller 1994). The Great Barrier Reef system is also a mixed environment, where carbonate and siliciclastic sedimentation interfinger (Puga-Bernabeu et al. 2011; Daniell et al. 2020). Large submarine canyon systems have been mapped out as the bypass routes of siliciclastics during lowstands (Puga-Bernabeu et al. 2013, 2014).

The transportation of siliciclastics across carbonate shelves is often limited to bypass valley systems that formed during high magnitude sea-level falls: an example of a form of 'reciprocal sedimentation'. A good example is found in the Middle Cretaceous Shuaiba Formation carbonate platform which was incised by large valley systems locally up to 8 km wide and 60 m deep, transporting clastics from the exposed inboard Arabian Shield to the Arabian/Persian Gulf. These were back-filled during the subsequent transgression with clean sandstones and clays (Figure 11.81; Raven et al. 2010).

Many mixed settings have been reported from the Miocene, notably in the Mediterranean area and the Far East, which illustrate how the marine biota respond to clastic sediment flux (Spain: Braga et al. 1995; Vennin et al. 2002a, 2004; Bourillot et al. 2010a; Carpentier et al. 2020; Poyatos-Moré et al. 2022; and Turkey: e.g. Kelling et al. 2005; Bassant et al. 2005; Janson et al. 2007). Spectacular seismic examples of mixed systems have been reported from several continental shelves in the Far East, where isolated build-ups coexist with large delta systems. The Miocene build-ups offshore Sarawak (Malaysia) can be up to 2 km thick and reveal complex seismic geometries, including out-building, in-building, carbonate-siliciclastic intercalations and coalescence of neighbouring carbonate bodies (Figure 11.65; e.g. Epting 1987a,b; Burgess et al. 2013; Koša 2015; Koša et al. 2015). Similar examples of carbonate build-up and delta interaction occur in the Neogene of Indonesia (Wilson and Lokier 2002) and Papua New Guinea (Tcherepanov et al. 2008a,b, 2010; Mallarino et al. 2021). In the Indonesian Mahakam Delta, numerous, small Miocene patch reefs developed in areas with high siliciclastic input associated with the progradation of the delta; and the Quaternary deposits show *Halimeda* bioherms development during transgressions (e.g. Crumeyrolle et al. 2007; Santodomingo et al. 2015; Morgan et al. 2020a). Generally speaking, in these mixed settings, carbonates can mature and develop considerable topography close to the siliciclastic sources particularly during transgressions and (early) highstands when the clastics are trapped near the shoreline (e.g. Limon, Costa Rica: McNeill et al. 2000, Bauch et al. 2011; Kalimantan, Indonesia: Saller

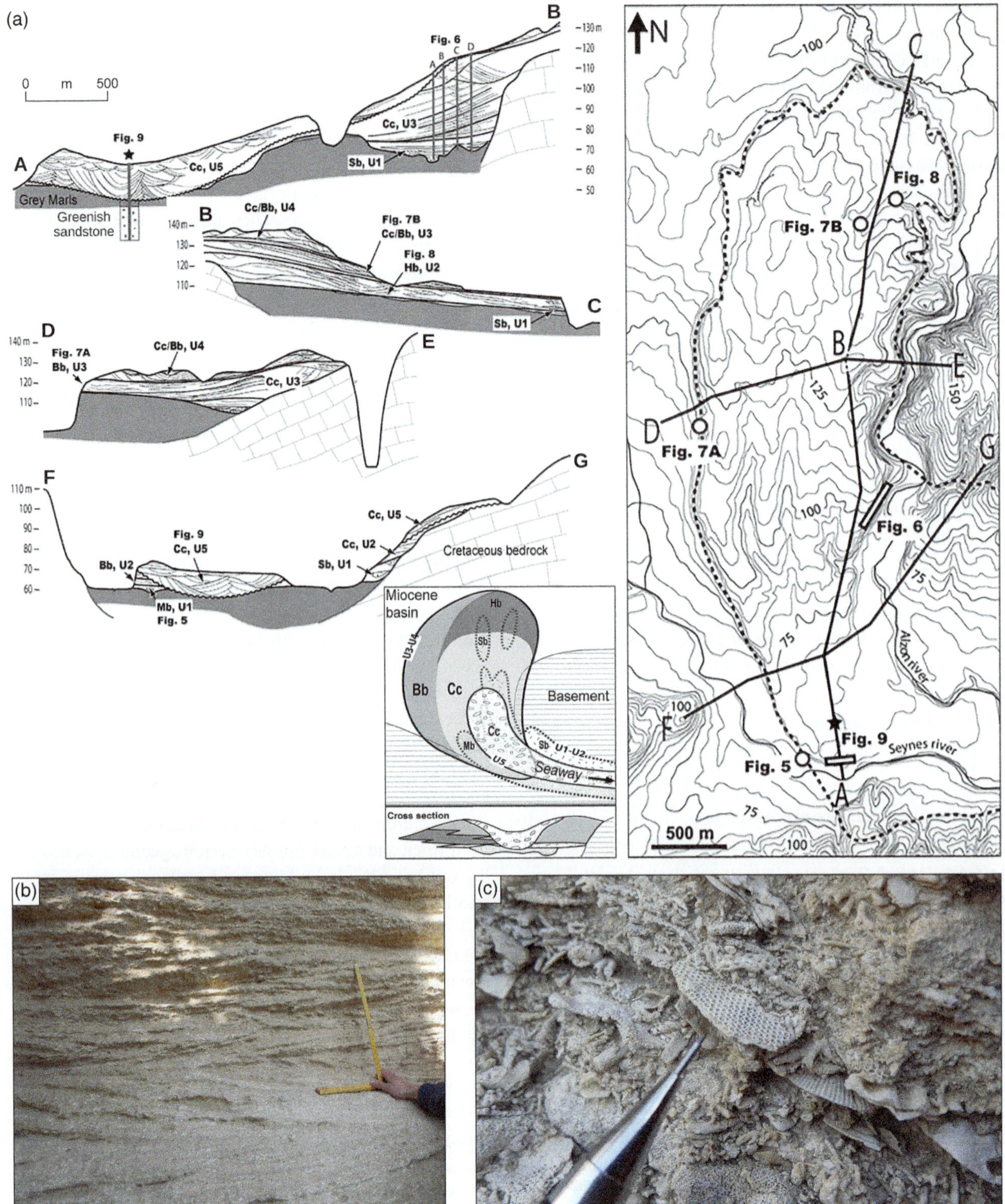

Figure 11.64 Complex architecture of Miocene incised valleys filled with estuary-type C-factory carbonate facies (Uzes Fm., S. France), *Source:* Reynaud et al. 2006 / With Permission of Society for Sedimentary Geology. (a) Indicated figure numbers refer to the original paper by Reynaud et al. (2006). Seaway reconstruction, with cross sections and depositional model; (b) bi-directional crossbedding structures; and (c) illustration of cool water carbonate facies in time equivalent incised valley systems in Sommière, France (Photographs: courtesy Emmanuelle Vennin).

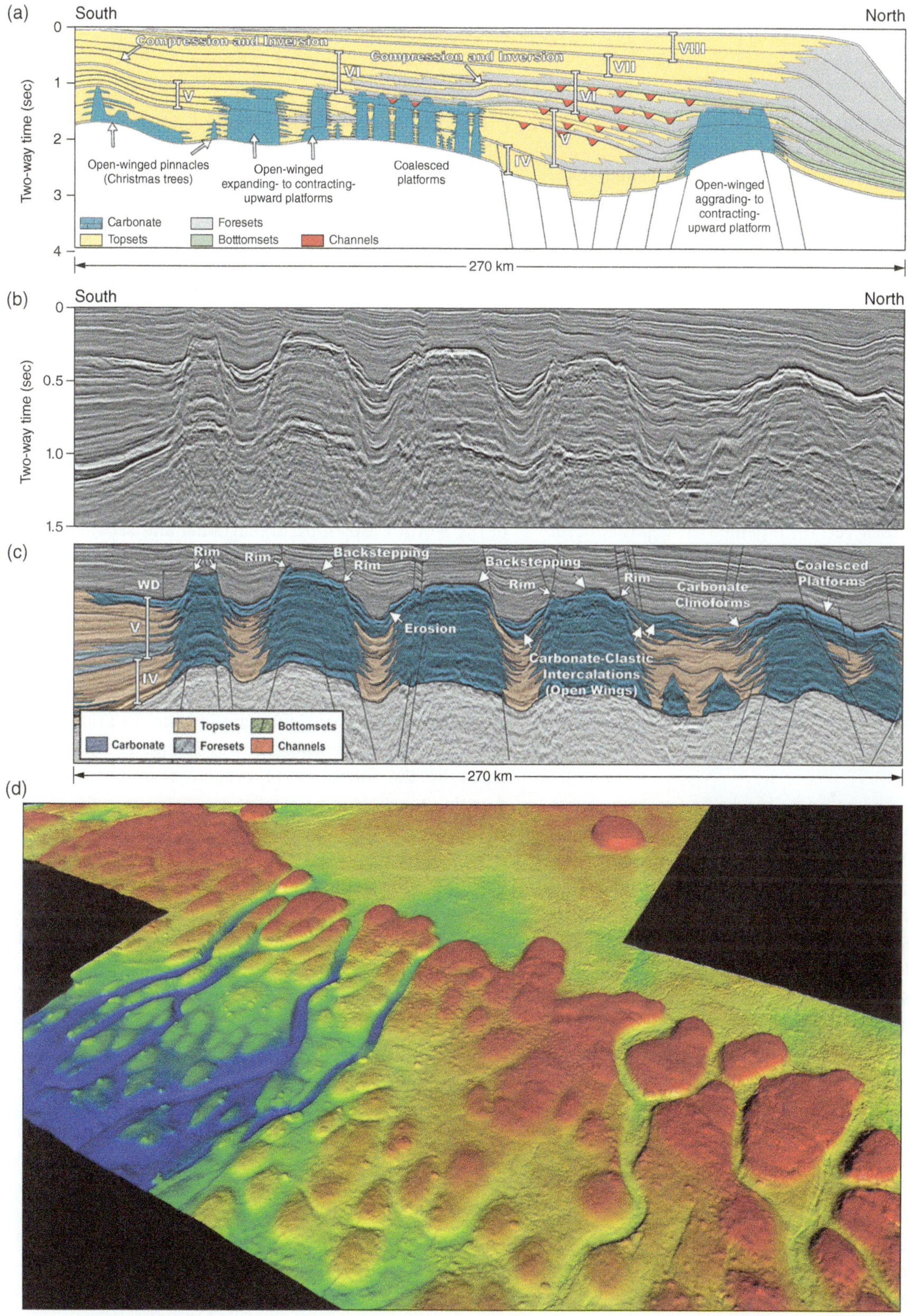

Figure 11.65 Miocene mixed carbonate-siliciclastic depositional systems in Indonesia and Malaysia. (a) Depositional model for carbonate buildups in prograding delta setting, Sarawak/Luconia (*Source:* Koša et al. 2015 / with permission of Elsevier); (b) 2D seismic line of the coalesced platforms in figure (a); (c) interpretation of seismic line in (b) (*Source:* Koša et al. 2015 / with permission of Elsevier); (d) seismic landscape image of Miocene reefs offshore Madura, crosscut by anastomosing tidal channels (*Source:* Posamentier et al. (2010) / Society for Sedimentary Geology (SEPM)).

et al. 2010, Santodomingo et al. 2015). In contrast, the Late Pliocene carbonate deposits on the Papua New Guinea shelf (Tcherepanov et al. 2008a,b, 2010) are much thinner, and do not create significant topography, but show up in seismic profiles as strong reflectors on top of the clinoforms of the prograding delta (see also Sections 2.5 and 9.2.4).

Mixed systems are also a common in the Quaternary (Francis et al. 2007; Gischler et al. 2010; Le Roy et al. 2019; Tournadour et al. 2021). A much subtler source of terrestrial material is windblown dust which easily crosses oceans (Garrison et al. 2003; Stuut et al. 2005; Lindhorst et al. 2019). In terms of sediment volume, it has a minimal impact, but by contributing nutrients, it is thought to be capable of triggering coral growth (Swart et al. 2014), or killing them by blowing in pathogens (Garrison et al. 2003).

11.6.4.2 Mixed Carbonate – Organic-Rich Facies Models

Throughout the rock record, organic-rich deposits occur adjacent to and interdigitating with carbonate deposits, both in shallow- and deep-water environments (see case studies in Brooks and Fleet 1986; Greenlee and Lehman 1993; Katz 1995a,b; Harris 2005). As in siliciclastic systems, the accumulation of organic-rich sediment depends on the primary production, the preservation potential (anoxic conditions) and absence of dilution by another sediment fraction (Figure 9.18; Calvert 1987; Ricken 1991; Calvert and Pedersen 1993).

Four general basin settings favourable for the accumulation of mixed carbonate-organic-rich deposits can be distinguished (Emery and Myers 1996): (i) *intra-shelf basins* in carbonate shelves created through differential accumulation, such as the Late Devonian in W. Canada (e.g. Stoakes 1980;), and on the Arabian plate the Jurassic Hanifa basin (Murris 1980; Droste 1990), the Late Jurassic Amu Darya Basin in Uzbekistan (Carmeille et al. 2020) and the Cenomanian Natih sub-basins (van Buchem et al. 2002b), the Eagle Ford Formation in the Western Interior Seaway in North America (Eldrett et al. 2015; Minisini et al. 2018); (ii) *intra-cratonic basins* where depressions are created through differential subsidence (Wang et al. 2017; Zheng et al. 2019; Zhai 2021); (iii) *unrestricted plate margins with open shelves* where ocean-current driven upwelling cells operated over long time intervals. Good examples include the La Luna Formation in Venezuela (Rangel et al. 2000; De Romero et al. 2003); the Tarfaya Formation in Morocco (Kuhnt et al. 2009), the Brown Limestone Formation in the Gulf of Suez (Emery and Myers 1996) and the Muwaqqar Formation in Jordan (Powell and Moh'd 2011); (iv) *deep ocean basins* where short periods of extraordinary widespread deposition of organic matter occurred as a result of palaeo-oceanographical conditions, which are called the Oceanic Anoxic Events (OAEs). These are particularly well investigated for the Mid Cretaceous OAE 1a and OAE 2 events (Schlanger et al. 1987; Trabucho et al. 2010). In addition, locally this kind of favourable conditions persisted for longer time intervals too (e.g. Katz 1995a,b). In shallow-water carbonates, OAEs are marked by carbonate platform 'drowning' in which deeper, organic-rich deposits replaced shallow-water carbonate facies (Arthur and Schlanger 1979; Schlager 1981; Eldrett et al. 2015; Phelps et al. 2015).

11.6.4.3 Mixed Carbonate – Evaporite Facies Models

A review of mixed carbonate-evaporite systems by Sarg (2001) includes good examples of giant evaporite basins (i.e. Permian/Delaware basins, West Texas; Middle Permian in the North Caspian basin), lagoonal and sabkha evaporites (Jurassic Smackover Formation, Gulf of Mexico), restricted basins in shelf areas (Triassic Zechstein Group in Europe) and m-scale succession of carbonates and evaporites in intrashelf basins (Permian San Andres Formation). As with siliciclastics the deposition of evaporites can either occur contemporaneously with active carbonate deposition, or they can alternate in time. In arid settings, the inner shelf or inner ramp environment can be dominated by the formation of evaporites in the inter-tidal domain. A well-studied present-day example is the sabkha environment along the Trucial Coastline (Figure 11.40) treated in more depth in Chapter 10. There are many good examples of ancient sabkha environments existing next to shallow water environments, such as in the Permian Khuff Formation of the Arabian Plate (Insalaco et al. 2006; Maurer et al. 2009; Koehrer et al. 2010), Permian San Andres Formation in Texas (Garcia-Fresca et al. 2012), the Cambrian Tarim basin in China (You et al. 2018) and the Precambrian Pilbara in Western Australia (Knoll 1989; Noffke et al. 2013).

Two good examples that illustrate the transition from a carbonate system to an evaporite (dominated) system include the Oligocene–Miocene evaporite deposits in Iraq and Iran, which were deposited during global sea-level lowstands in isolated, foreland basins surrounded by carbonate platforms (Aqrawi et al. 2010; van Buchem et al. 2010b; Daraei et al. 2015); and the Messinian salinity crisis, which is one of the greatest evaporitic events of Earth's history, considering its short duration, that is, 640 kyr (Krijgsman et al. 1999a,b, 2018; Rouchy and Caruso 2006; Borrelli et al. 2021; see Section 10.4.6 for more details).

11.7 Stratigraphic Architecture

11.7.1 Stratigraphic Principles of Marine Carbonate Systems

The most effective way to characterise the stratigraphic organisation of a sedimentary succession is the sequence stratigraphic methodology. The reader is referred to Chapter 4 for an explanation of the basic stratigraphic and sequence stratigraphic principles and the commonly used terminology. After some general reminders, the focus of this section is on carbonate system-specific aspects of sequence stratigraphy.

The application of sequence stratigraphic methodology in any depositional system needs to make the vital distinction between:

i) the description and interpretation of the sedimentary facies, special stratigraphic surfaces, depositional geometries and stacking patterns. This analysis is structured in terms of depositional environments and sequences organised at different process-linked time scales ('orders'). This approach is documented in a number of carbonate case study compilations (e.g. Crevello et al. 1989; Loucks and Sarg 1993; Lukasik and Simo 2008; van Buchem et al. 2010a); and
ii) the evaluation of the factors controlling sedimentation, distinguishing between the relative influence of local parameters (e.g. depositional events and autogenic processes) versus global parameters (e.g. geodynamic context, eustasy and long- and short-term cyclic climate change). The difference between local and global is best established through comparison of successions within and between sedimentary basins across the globe, rigorously constrained by age dating.

What makes the sequence stratigraphic approach widely applicable is the inference that, having addressed autogenic and local parameters, the fundamental control on stratigraphic architecture is the interplay of two key allogenic parameters: *Accommodation* (A), the rate of creation of space available to accumulate sediment, and *Sediment Supply* (S), the rate or flux of sediment produced at, or transported to, a given site (Jervey 1988). It is the balance between these two rates that shapes stratigraphy. Accommodation is controlled by the combination of tectonic subsidence or uplift, isostatic compensation of the sedimentary load and compaction, together with eustatic sea-level fluctuations and any inherited initial bathymetry. Accommodation is thus to some extent independent of the nature of the sedimentary system. The processes that control sediment flux, especially, as we have seen, in carbonate systems, vary significantly for different depositional systems. Sediment flux is depositional system specific. Siliciclastic systems are characterised as *source to sink systems* (Allen 2017). In contrast, many carbonate systems are better seen as *source and sink* systems because of their capacity to generate sediment *in situ*, as we have seen in the carbonate factory discussion (Section 11.4).

11.7.1.1 The Carbonate Factories

The carbonate factory concept is fundamental to clarifying the linkages between changes in accommodation and sediment supply. Due to the differences in dominant biota, chemical composition and stability, production profiles with water depth and depositional profiles, each factory has a unique response to changes in environmental conditions, and in particular to RSL (Relative Sea Level) fluctuations (Figure 11.66 and Section 11.5). Here we present end-member characteristics, with the understanding that the biotas of different factories can locally occur together.

T-Factory

The warm-tropical T-factory is normally considered the standard reference for carbonate systems (Section 11.5). Specific characteristics that influence stratigraphic architecture are (Figure 11.66) as follows:

i) a shallow optimum production profile from 0 to approximately 40-m water depth due to the light dependence of the dominantly autotrophic biota, and a consequent tendency to build up to sea level. For this reason, ramp geometries in T-factory systems form a transitional stage to a flat-topped or rimmed platform. The platforms are sensitive to even small changes in accommodation (metre-scale) and are therefore excellent recorders of both low and high magnitude RSL variations. Subaerial-exposure-induced karstification surfaces are common in T-factory carbonate platform deposits;
ii) high sensitivity to salinity, temperature and clay and nutrient supply, any of which may inhibit carbonate production and cause the system to back-step, mimicking the impact of a RSL rise (e.g. drowned platforms; Schlager and Camber 1986);
iii) early diagenetic cementation and microbial growth are common, reducing sediment mobility, thus allowing bio-constructed rims to be built and diminishing the impact of erosion during both sea-level fall and rise (Harris et al. 1985; Grammer and Ginsburg 1992; Grammer et al. 1993; Séard et al. 2011; Reolid et al. 2017b);
iv) with high sediment production occurring on the platform tops during sea-level highstands most of this sediment is exported to the slope and basin, a phenomenon

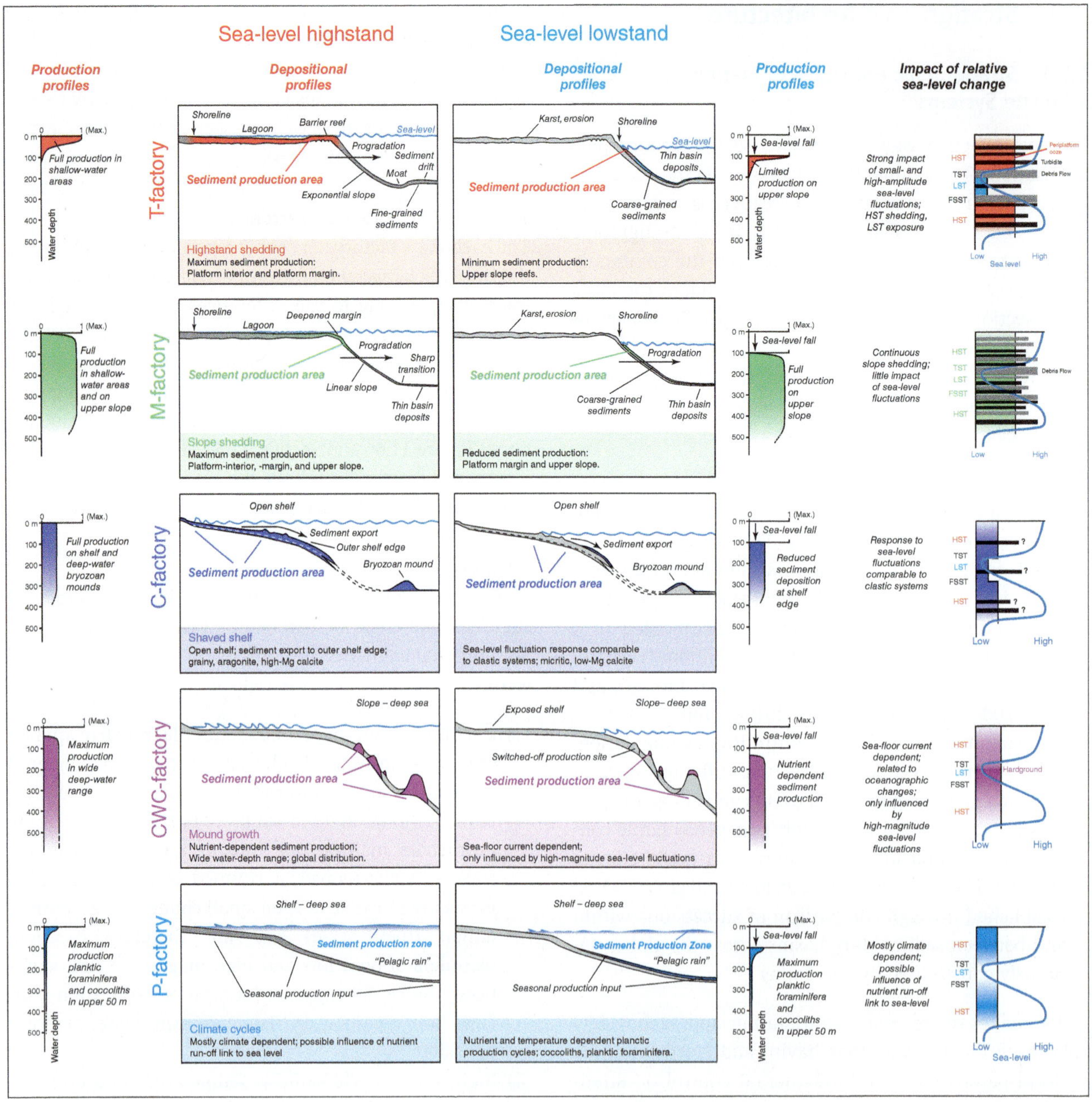

Figure 11.66 Carbonate factories production profile and depositional profiles with typical response to relative sea-level change (highstand and lowstand). Note the sharp platform sediment production differences for the T-factory between highstands and lowstands in sea level (highstand shedding) as the M-factory shows a consistent pattern (slope shedding). The C-factory shows sediment export variations like siliciclastic systems with sediment production variations depending on the size of the shelf area. The CWC-factory and the P-factory show a dependency on shallow- to deep-water current variations supplying nutrient-rich waters and/or waters with temperatures and salinities enabling the biota to flourish.

called *highstand shedding* (Droxler and Schlager 1985; Reijmer et al. 1988; Schlager et al. 1994; Lantzsch et al. 2007; Betzler et al. 2014; Webster et al. 2012; Jorry et al. 2020);

v) sensitivity of *land-detached*, isolated T-factory platforms to the influence of winds and currents on the sediment distribution. This may lead to the contemporaneous occurrence of retrograding, aggrading and prograding stacking patterns (Figure 11.57 and 11.67). In contrast, the typical expressions of the systems tracts in a *land-attached* T-factory platform setting are shown in Chapter 4 (Figure 4.33).

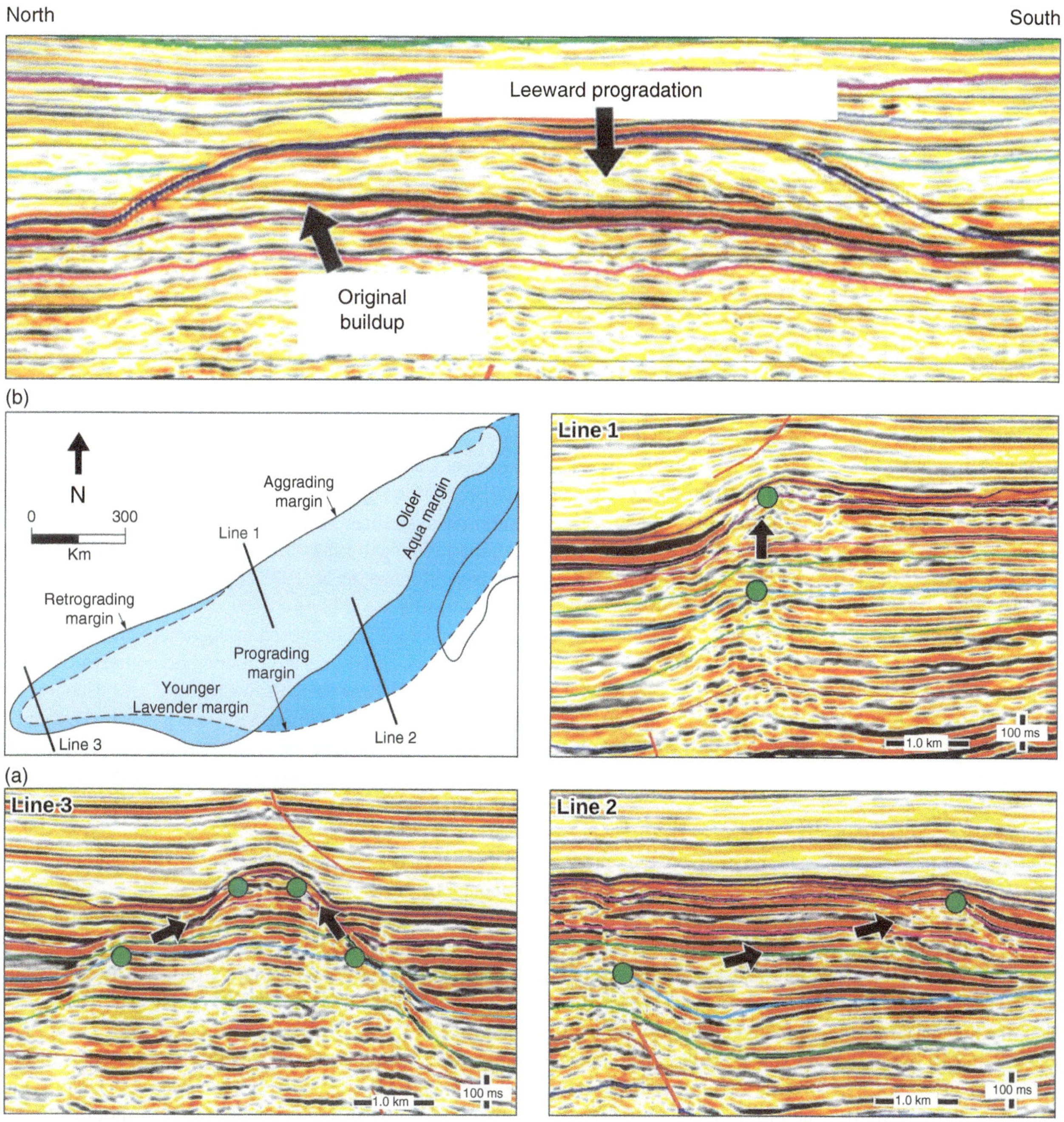

Figure 11.67 Land-detached carbonate platform displaying three different, contemporaneously formed, stacking patterns caused by the influence of dominant wind and ocean current directions (Segita platform, Lower Pliocene, Malaysia; Bachtel et al. 2004/American Association of Petroleum Geologists). (a) Location map with positions of three transects; (b) N-S cross section of the buildup; other figures show seismic lines with contrasting stacking patterns along the margin of the same buildup: Line 1 aggradation, Line 2 progradation and Line 3 retrogradation.

C-Factory

The C-factory is dominant in the cooler water continental shelf areas and carbonate dominated, tidal-driven estuaries (Section 11.5). Specific characteristics that influence stratigraphic architecture are (Figure 11.66):

i) a deep production profile (0–400 m) due to the dominantly heterotrophic biota, which are not dependent on light but rather on nutrient supply. The C-factory is characterised by the grainy, non-consolidated (not cemented) nature of its sediments. It thus comes closest to the physical characteristics of siliciclastic depositional systems, and consequently resembles them in its response to sea-level change, with low angle or distally steepened ramp profiles, and easily mobilised sediment;
ii) sediment is produced *in situ*, but at a fairly low pace, and is subjected to a large amount of remobilisation and reworking (James and Jones 2016). Depending

on the prevailing hydrodynamic conditions, sediment can be deposited in disconnected depocentres both above water as aeolianites, and offshore below storm wave base (Section 11.6);

iii) sub-assemblages of the C-factory, such as seagrass communities, kelp forests, oyster-rich beach settings and contributions from the cool-water planktonic factory, act at different water depths and each respond differently to relative sea-level change.

CWC-Factory

The predominantly deep-water (up to hundreds of metres) location of the CWC-factory implies that it is only indirectly influenced by sea-level fluctuations. Therefore, the classic sequence stratigraphic stratal patterns and systems tracts do not readily apply to this factory (Figure 11.66). Palaeoceanographic conditions such as ocean-bottom currents affect nutrient supply and high-magnitude RSL fluctuations may thus have an indirect impact (Fink et al. 2012; Hebbeln et al. 2019). Three typical responses of growth pattern adaptation to new oceanographic circumstances are observed: continued growth, the transposition of the system to another water depth or location (Foubert et al. 2008; Wienberg et al. 2009, 2022; Fink et al. 2013; Fentimen et al. 2022) or a temporary stop in sediment production and the formation of a hardground (e.g. van der Land et al. 2010).

M-Factory

The M-factory biota are diverse, found in different environments, locally in association with other carbonate factories and, therefore, have no single response to changing environments (Section 11.5; Figure 11.66). The M-factory has a large production profile, from 0–500 m depth, due to the heterotrophic biota, and a strong capacity to bind and consolidate sediment (Kenter and Campbell 1991; Vennin et al. 2002b; Schlager and Reijmer 2009; Hillgärtner 2010; Merino-Tomé et al. 2012; Della Porta et al. 2014). Its contributions are found (i) on the platform top tidal flats, where stromatolites can be rock-forming and resemble T-factory assemblages in accumulating up to sea level and being sensitive to relative sea-level fluctuations; (ii) as microbial mounds in the deeper lagoonal parts of the platform top settings, where they are preferentially deposited during early and late transgressive phases, such as the *Lithocodium-Baccinella* mounds in the early Aptian of the Arabian Plate (Pittet et al. 2002; Rameil et al. 2010); (iii) on the upper slope, below fair weather wave base, where microbial activity produces sediment *in situ*, and stabilises it by encrusting, and cavity in-fills. In these cases, the M-factory production on the slope is relatively little affected by high-frequency sea-level fluctuations (Della Porta et al. 2002; Kenter et al. 2005); (iv) in the deeper slope, where debris brought down from the upper slope can accumulate in a more chaotic fashion (Figure 11.68; Playton and Kerans 2018); and (v) in the more distal basinal settings where mud mounds of different scales typically occur (e.g. Figure 11.78, Cook et al. 2002).

P-Factory

The P-factory can contribute to sedimentation in the shallower water domains, but dominates in the deeper water, pelagic domains (Section 11.5 and Chapter 13). Where the P-factory dominates, the sedimentary record is commonly influenced by orbitally driven climatic variations in productivity, variously expressed as variations in carbonate, silicate, organic matter and clay content. The link between these productivity variations, climatic changes and orbital forcing has also been made in the shallow-water carbonates settings (Schwarzacher 1954; Fischer 1964; Goldhammer et al. 1987; Brack et al. 1996), and toe-of-slope deposits (Reijmer et al. 1991, 1994). Apart from the seasonal-to-Milanković-band-driven cyclicity, changes in the P-factory record are also used to document large-scale tectonic processes that affected productivity patterns (closure of the Central American Seaway: Haug and Tiedemann 1998; Reijmer et al. 2002; the SE Asian gateway: Holbourn et al. 2011), as well as variations in the carbon cycle and the acidification of the oceans (Hassenkam et al. 2011; Gattuso and Hansson 2011).

11.7.1.2 Carbonate Sediment Production and Building Topography

The capability of carbonate systems to produce sediment *in situ* means that they are capable of accumulating in both land-attached and land-detached settings such as (submarine) volcanoes and islands. A particular expression of *in situ* production is the creation of topography through differential accumulation/production rates. This is a well-understood and accepted concept used to explain the creation of modern, isolated platforms and mounds of the T-factory (e.g. the vertical aggrading carbonate systems of the Bahamian platforms and atolls and the Great Barrier Reef system), but also the deep-water mounds of the present-day CWC-factory, and multiple examples of the ancient mud-mounds of the M-factory (see Section 11.6).

A similar process of differential accumulation also operated on the large T-factory dominated carbonate shelves of the Palaeozoic and Mesozoic (Supplementary Information 11.S.2). The key message being that the gradual aggradation of part of the shelf and drowning and starvation of another part of the shelf, created topographic relief without any significant tectonic control. This stepwise process has

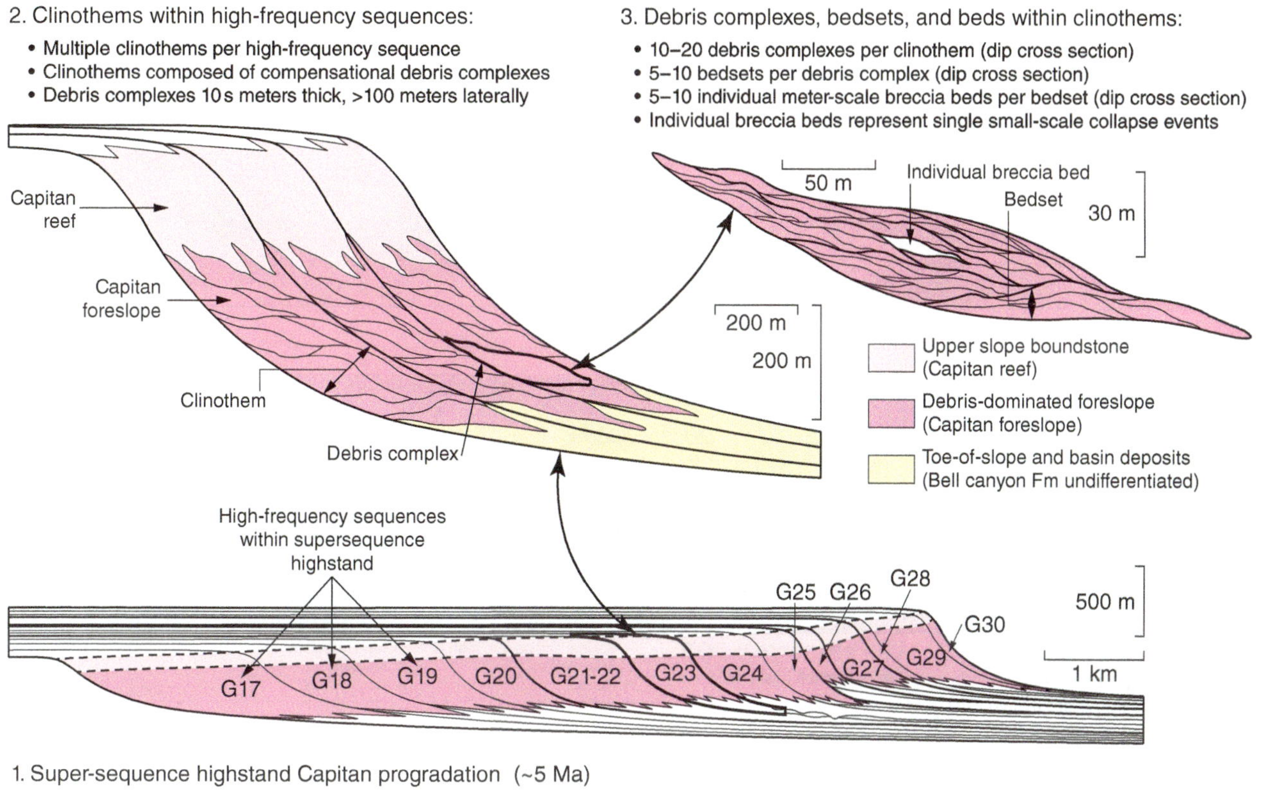

Figure 11.68 The stratigraphic organisation of a M-factory, land-attached carbonate shelf margin (Permian of Guadalupe Mountains, US; Playton and Kerans 2018 / with permission of Elsevier). A three-component hierarchy is defined of (1) the 'Supersequence' highstand, (2) the high-frequency sequences, and (3) the debris-complexes, bedsets and beds. The basic sedimentary process occurs on the foreslope through boundstone accretion and shedding, called foreslope failure and back-filling or 'slope shedding' (Kenter et al. 2005).

been summarised in conceptual models for the Palaeozoic (Schlager 1989; Playton et al. 2017), Mesozoic (Murris 1980; van Buchem et al. 2002a,b, 2010a; Razin et al. 2010, 2017; Wharton 2017; Gravestock et al. 2021) and Cenozoic (Bourget et al. 2014; Belde et al. 2017; Rankey et al. 2019). These models have a common theme: stepwise evolution from an initial flat ramp or platform to a pronounced platform-intrashelf basin topography, with subsequent (partial) infill (Figure 11.69). This pattern can be locally repeated, even within one formation or within subsequent formations, such as in the Mid-Cretaceous carbonate successions of the Arabian Plate (Murris 1980; van Buchem et al. 2002a,b, 2010a). This generally applicable model is here illustrated in more detail in three steps (Figure 11.70).

1) *Flat topography:* The starting point is a flat shelf, such as the top of a preceding platform phase, which can be capped by an exposure surface (type 1 Sequence Boundary (SB) or a submarine, type 2, SB: Figure 4.29). Contemporaneous lowstand deposits at this stage can potentially be found at the shelf margins. An initial slow rise of RSL (early Trangressive System Tract, TST) induces deposition of shallow-water sediments deposited along a sub-horizontal depositional profile. This is a low-energy ramp system, characterised by carbonate-mud rich sediments, which can locally be dominated by monospecific, r-type colonising fauna, such as benthic foraminifera. At this stage, the carbonate production rate follows the creation of accommodation, and the two rates are in balance (A = S) consequently an aggrading stacking pattern is formed. The slow overall increase in accommodation rate is expressed by a gradual thickening-upward trend in platform-top cycles.
2) *Creation of topography:* An acceleration in the rate of relative sea-level rise during late transgression (late TST) may cause, at the scale of the platform, a volumetric deficit in the carbonate production with respect to the creation of accommodation (A > S). When this deficit is homogeneous at the scale of the platform, a general deepening is observed and eventually drowning of the platform (Schlager 1989). When the production deficit is expressed by laterally varying sedimentation rates at the scale of the platform, it leads to a differentiation into shallow-water areas, where the carbonate production keeps up with the relative rise in sea level (A = S), and deeper-water areas, intra-shelf basins, where the

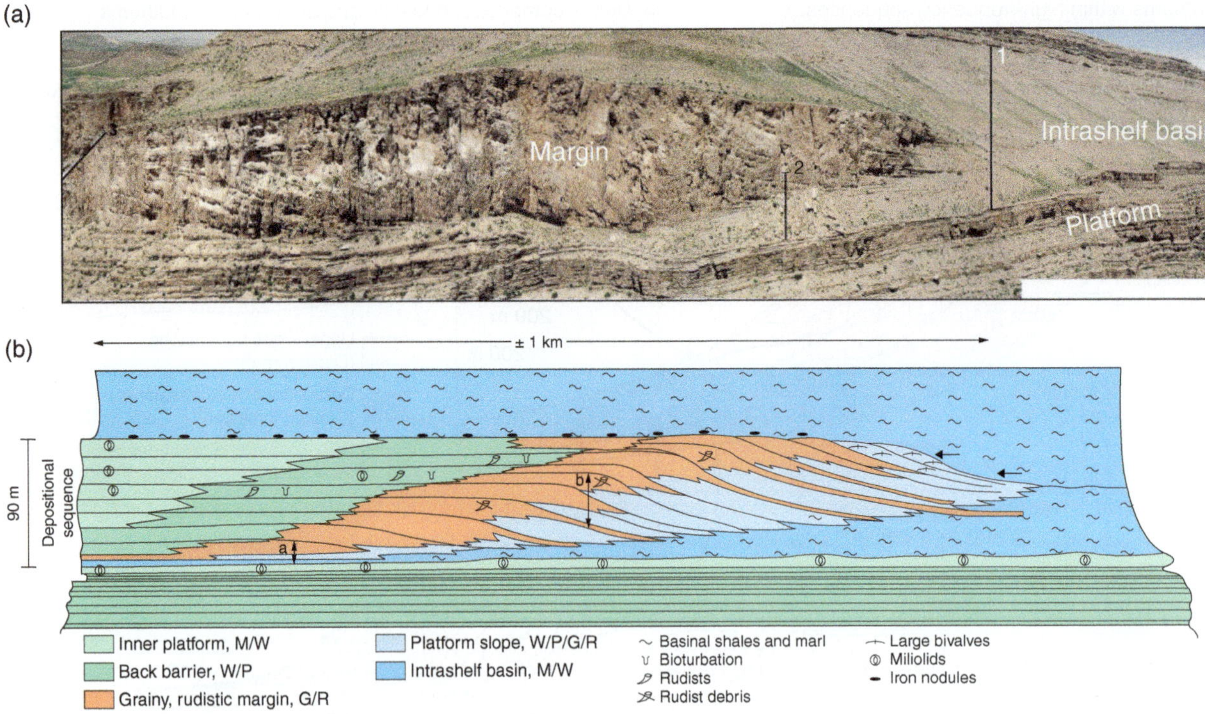

Figure 11.69 Outcrop of a T-factory, rudist-debris dominated land-attached, carbonate shelf margin (*Source:* Adapted from Sarvak Formation, Cenomanian High Zagros, Iran; Razin et al. 2010; Taati 2005). Upper part: outcrop picture showing the stepwise evolution of a full depositional sequence creating topographic relief through differential accumulation (Photograph courtesy Frans van Buchem). Note the evolution of the bedding pattern from very flat beds, with minimal topography (m-scale) at the base, to high angle clinoform geometries with 50–60 m relief towards the top of this shelf margin; lower part: depositional model illustrating the facies distribution. The margin and slope are mostly composed of rudist-rich grainstone and rudstone, the platform interior consists of a mud-dominated facies (M/W) with benthic foraminifers, including miliolids (indicative of restricted, shallow water carbonate top environments).

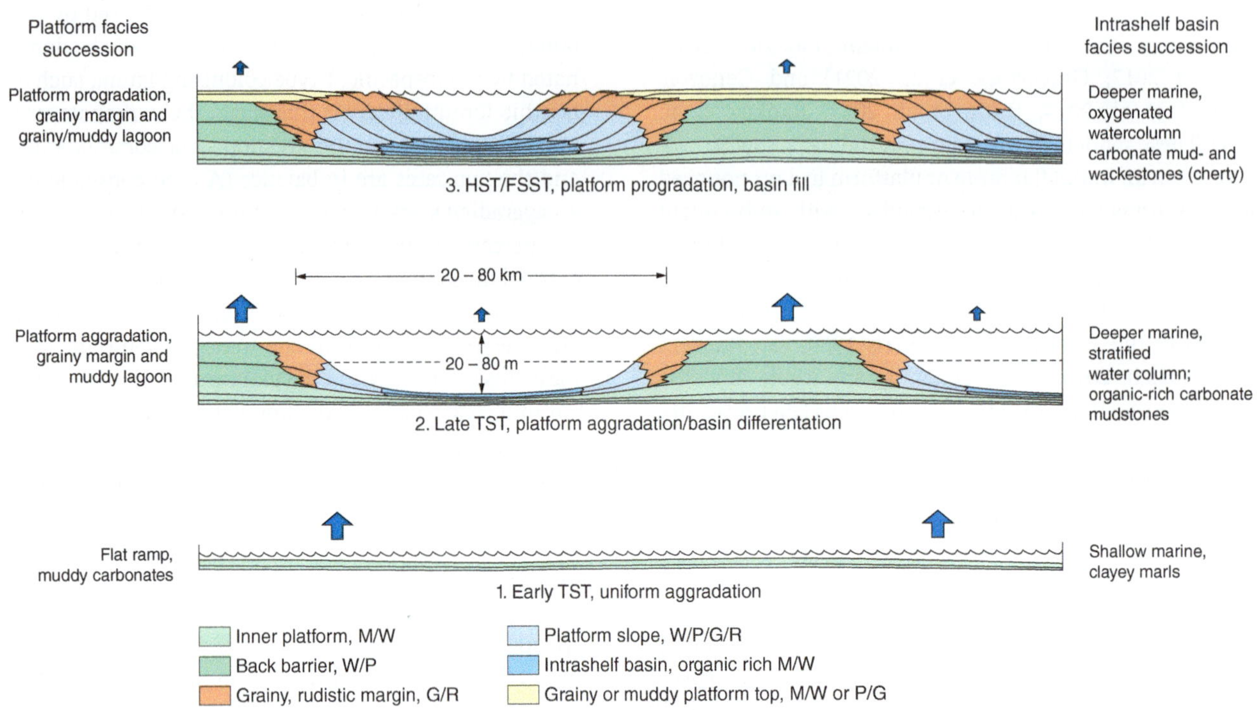

Figure 11.70 Conceptual model for intra-shelf basin formation through differential carbonate accumulation on a carbonate shelf setting (*Source:* Razin et al. 2010 / The Geological Society of London). Note that the MFS sits high in the carbonate platform, not in the marls at the base!

carbonate production does not keep up with the rising relative sea level and where local to regional bathymetric lows are formed (A > S). The change in topographical relief in turn causes a profound modification of the hydrodynamics on the platform and of the depositional systems. The increase in water depth, with intra-shelf basins varying from 40 to 150 m (or more in exceptional cases) enables high-energy sedimentary processes, such as waves (increased fetch) and tidal currents, to create higher energy settings where bioclastic shoals, barriers and channels may form, and storm deposits are laid down. A bioclastic barrier-rimmed platform and intra-shelf basin profile thus replaces the initial muddy ramp system of the early transgressive part of the sequence.

3) *Progradation and infill:* A slowing down and subsequent stop of the relative rise in sea level (early and late HST) is generally expressed as a progradational trend, gradually losing the aggradational component (A < S). This is accompanied by evolution from larger, lower angle sigmoidal clinoforms to smaller, steeper, clinoforms. The relative shallowing of the platform top causes an increase in winnowing of the finer grained fraction and its transport off-platform (highstand shedding). If this phase is followed by a gradual fall in RSL (Falling Stage Systems Tract), a forced regressive wedge develops, characterised by less inclined clinoforms and an upper toplap surface that dips gently seawards (Figure 11.69). If the RSL drop leads to subaerial exposure and karstification, it also causes a dramatic drop in carbonate production, and deposition will consequently be reduced to a volumetrically limited shelf-margin wedge. This can infill the remaining basinal topography, resulting in a new planar depositional surface such that the cycle can be repeated.

This conceptual depositional model has several important consequences for the interpretation of the depositional environments and the sequence stratigraphy, which are illustrated with examples below.

11.7.1.3 Identification of the maximum flooding surface (MFS)

The correct positioning of the MFS in carbonate platform successions can be problematic. Carbonate platform margins are in principle good settings for sequence stratigraphic interpretations, since shelf-break tracking methods can be applied as proxies for sea-level variations (Chapter 4, Figure 4.27). The challenge is that progradation may start well before the maximum rate of accommodation creation has been reached, due to the impact of increasing carbonate sediment production. An example of a Mid-Cretaceous carbonate platform on the Arabian Plate illustrates this well (Figure 11.69). The bedding pattern shows aggradation and progradation already during early sea-level rise, when there is still minimal topography along the depositional profile (at *a* in Figure 11.69b). With ongoing rise, progradation continues and the slope angle increases, till a maximum accommodation is reached, and RSL starts to fall (at *b* in Figure 11.69b). Following a geometry-based stacking pattern definition, as classically applied in siliciclastic systems, can thus be misleading and place the MFS too low in the succession, at MFS 1 in the A/S cycle, instead of at MFS 2 in the A cycle (Figure 11.71a). It is only in systems with a steady sediment supply, an assumption now increasingly challenged also for siliciclastic systems, that the retrogradation to progradation turnaround is in phase with the maximum rate of RSL rise (Figure 11.71b).

In the aggrading succession of the platform interior, overall facies variations may be minimal, and here increasing and decreasing thicknesses of high-frequency cycles (parasequences) can be used to interpret the evolution of accommodation and position the MFS (Figure 11.72). In the toe-of-slope setting, where accommodation is not a limiting factor, sediment supply variation is the clue to the interpretation, and caution is needed for a similar de-phasing of the MFS as observed in the marginal setting (Figure 11.71). The difficulties that can arise when correlating these domains with less ideal outcrop or subsurface data are obvious.

11.7.1.4 The Evolution of the Depositional Profile and Facies Model Succession

As illustrated above, the depositional profile within a single depositional sequence can evolve from a flat platform top/low angle ramp to an intrashelf basin/platform top topography with clinoforms along the margins. This has a profound impact on the facies composition and distribution (lithology, grain size and biological content). The increase in the angle of the depositional profile and the depth contrast between platform and intra-shelf basins favours the transition from a low-energy, muddy system to a high-energy, grainy and bioclastic system by increasing wave energy, possibly in combination with an increase in the tidal prism and tidal currents. This, in turn, has an impact on the biota that will dominate during the course of an individual depositional sequence. For instance, during early transgression r-strategy organisms tend to rapidly occupy new niches and can become rock-forming. In particular, benthic foraminifera have this capacity, as is observed throughout the rock record: fusulinids in the Permian (Barnaby and Ward 2007; Kolodka et al. 2012; Wahlman et al. 2013), the orbitolinids in the Mid Cretaceous (Vennin and Aurell 2001; Pittet et al. 2002; Embry et al. 2010; Schroeder et al. 2010); the nummulitids in the Eocene (Aigner 1982; Racey 2001; Vennin et al. 2003; Jorry et al. 2006; Beavington-Penney et al. 2004) and the lepidocyclinids in the Miocene (BouDagher-Fadel and Price 2010).

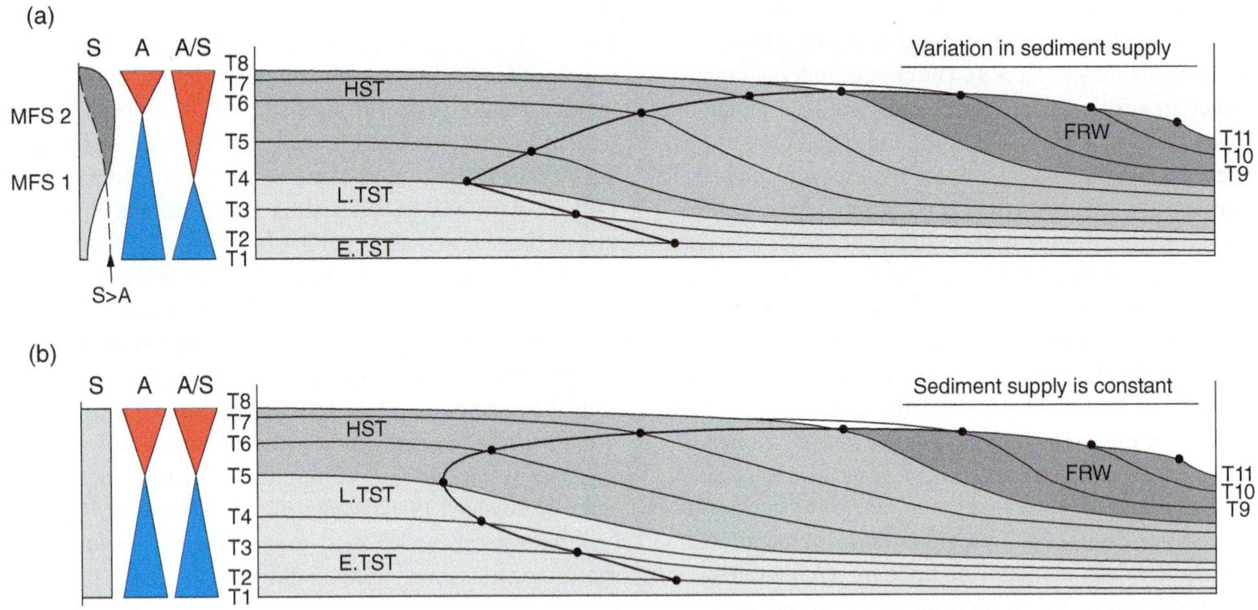

Figure 11.71 Timing of platform progradation in response to sediment supply and accommodation (*Source:* Razin et al. 2010 / The Geological Society of London). (a) Variable sediment supply: early increase in carbonate production causes the system to prograde before 'absolute' maximum water-depth is reached. The mfs of the A/S cycle occurs earlier (T4) than the maximum accommodation point (T6). This situation commonly occurs in carbonate systems. (b) Constant sediment supply: the mfs of the A/S cycle is in phase with the maximum accommodation point (T5). This situation commonly occurs in siliciclastic systems. See text for further explanation.

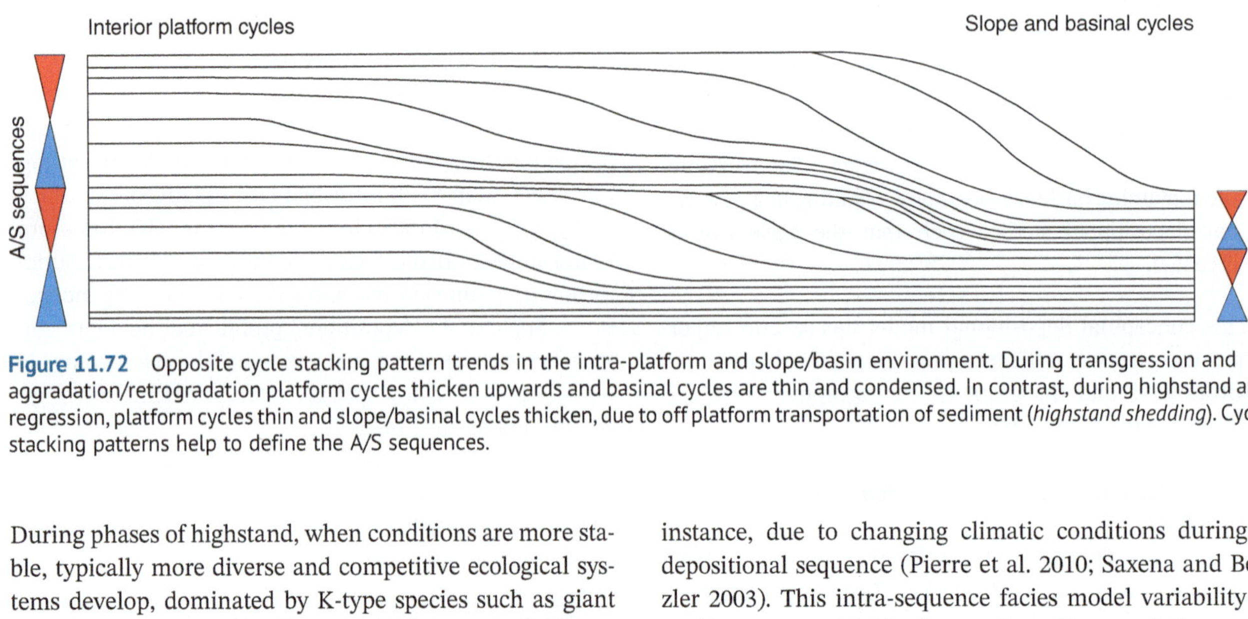

Figure 11.72 Opposite cycle stacking pattern trends in the intra-platform and slope/basin environment. During transgression and aggradation/retrogradation platform cycles thicken upwards and basinal cycles are thin and condensed. In contrast, during highstand and regression, platform cycles thin and slope/basinal cycles thicken, due to off platform transportation of sediment (*highstand shedding*). Cycle stacking patterns help to define the A/S sequences.

During phases of highstand, when conditions are more stable, typically more diverse and competitive ecological systems develop, dominated by K-type species such as giant gastropods or rudists (Kauffman and Johnson 1988). Therefore, it may be more appropriate to propose not one but a succession of facies models to characterise facies evolution during a depositional sequence, since at no point in time do all facies occur together. A good example of this is the Mid-Cretaceous of the Arabian Plate shown in Figure 11.73 (Yose et al. 2006, 2010). In other cases, a simpler succession between two alternating facies models can be proposed, for instance, due to changing climatic conditions during a depositional sequence (Pierre et al. 2010; Saxena and Betzler 2003). This intra-sequence facies model variability is an important principle for stratigraphic correlations and environmental interpretations in carbonate successions.

11.7.1.5 Biotic Distribution and Depositional Environment Interpretation

The presence of open marine organisms such as ammonites, brachiopods, planktonic foraminifera and nannofossils is a common feature in the shallow-water argillaceous

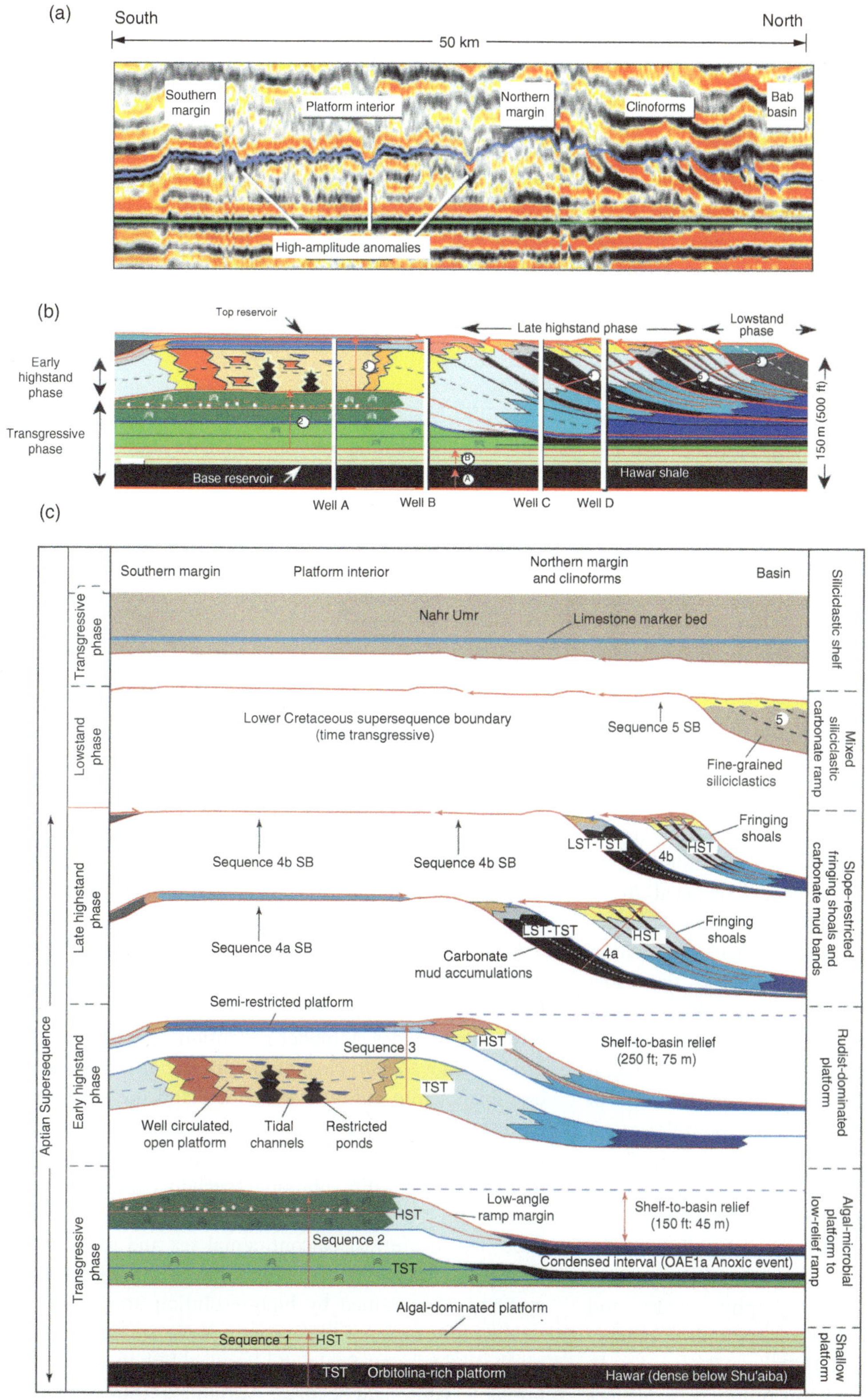

Figure 11.73 Carbonate facies and depositional profile evolution during a second-order depositional sequence in the Aptian age Shuaiba Formation (UAE; Yose et al. 2010). (a) Seismic line across the isolated platform; (b) sequence stratigraphic model, integrating seismic geometries and core derived sedimentary facies; (c) succession of facies models to capture the facies evolution during the second-order sequence: (i) in the transgressive phase (Sequences 1 and 2), a flat, muddy, ramp, dominated by benthic foraminifera *Orbitolina*, backstepped and aggraded, followed by microbial dominated, amalgamated mounds, consisting of the *Lithocodium-Baccinella* complex; (ii) during the early highstand phase (Sequence 3), the platform top was colonised by rudists, and an aggrading and prograding, coarse grained, rudist debris rim formed; (iii) during late highstand phase (Sequences 4a and 4b), the platform prograded with mixed carbonate-clay sequences; (iv) during the lowstand phase (Sequence 5) sedimentation was limited to the intrashelf basin. The lowstand was caused by a drop in sea level with a magnitude of approximately 60 m, and a duration of approximately 4 My, at which time the carbonate platform was exposed, and incised valleys formed (*Source:* Adapted from Maurer et al. 2013; Pierson et al. 2010; Raven et al. 2010; see also Figure 11.81). Finally, (v) during the subsequent transgressive phase, the platform was covered by shallow water shales.

carbonate facies of the early transgressive deposits at the base of depositional sequences in carbonate settings. If environmental interpretations are only based on the presence of these fossils, a common practice in borehole microfacies analysis, these units may be mistakenly interpreted as MFS, at the base of shallowing-up successions, an interpretation that can be contested on facies and geometrical arguments (Figure 11.70). Instead, the penetration of 'deeper-water' organisms into these shallow-water settings should be attributed to the extremely flat morphology and low energy of the depositional system in the early transgressive phase just after the initial flooding surface, before the development of a high energy barrier that would hinder such widespread distribution (Razin et al. 2010).

11.7.1.6 Climate and Oceanography

Longer-term changes in the depositional environment such as climatic and palaeoceanographic changes also have a dramatic impact on the biogenic sediment producing organisms. A striking example of this is the Carboniferous–Permian succession of the Sverdrup Basin (Figure 11.74) where during an overall cooling a stepwise transition occurred from a typical T-factory carbonate platform system (Bashkirian-Asselian) to a C-factory heterotrophic association dominated by sponges and silica in a mud ramp depositional setting (Artinskian–Kungurian), which caused a dramatic decrease in the sedimentation rate (Beauchamp and Desrochers 1997; Beauchamp et al. 2014). In addition to the progressive cooling, an additional component may have been the shoaling of the calcite saturation limit (lysocline) and ocean acidification at that time (Beauchamp and Grasby 2012). Other examples of longer term, climate-controlled changes in carbonate factories and geometries, include for instance the Jurassic succession in the Paris Basin (Brigaud et al. 2014), and the Cenozoic of the Mediterranean (Rouchy et al. 2007; Bourillot et al. 2010a,b; Baceta et al. 2011; Perrin and Kiessling 2012; Pomar et al. 2017).

The contemporaneous deposition of shallow water carbonates and organic-rich intra-shelf basinal deposits was discussed earlier. This commonly occurs during transgressive systems tracts when differential topography is created and conditions are favourable for the preservation of organic matter (Greenlee and Lehman 1993; van Buchem et al. 2005). This process acts at different orders of sequences and is a good example of how rate and magnitude of RSL fluctuations can shape stratigraphic architecture. For instance, (i) during the Carboniferous (Desmoinean) icehouse period, fast, high magnitude sea-level fluctuations at the fourth-order scale controlled the distribution of organic-rich deposits, (ii) during the Late Cretaceous (Cenomanian) a period of minimal presence of ice-caps, the third-order sequences controlled the deposition of organic-rich deposits, modulated at the dm-scale by the fourth-order orbital cyclicity, and (iii) during the Late Devonian (Frasnian), an intermediate case, the main control was sea-level change at the second-order sequence level (van Buchem et al. 2005). Climate thus seems to play an active role, both as a driver of sea-level changes and productivity cycles. In addition, the dilution of organic matter content through variations in carbonate productivity is demonstrated in these case studies (van Buchem et al. 2005).

11.7.1.7 Cyclostratigraphy and Sequence Stratigraphy

Rhythmic bedding at the dm and m scale is a common feature of many carbonate deposits, both in the shallow-water and pelagic domain (Chapter 13). In the shallow-water domain of the T- and M-factory platform tops, this is expressed as peri-tidal cycles: rhythmic deepening/shallowing trends in many cases capped by subaerial exposure surfaces. Most of these cycles are understood to reflect orbitally forced climatic changes in the Milanković band (see compilations in Goldhammer et al. 1993; De Boer and Smith 1994; House and Gale 1995; Hilgen and Krijgsman 1999; Weedon 2003; D'Argenio et al. 2004; Hilgen et al. 2004; Strasser et al. 2006; Hinnov et al. 2012; Laskar 2020), although in the classic shallow-water carbonates sections in the Triassic of the Dolomites it is now suggested that their duration may well be shorter, and their origin more complex (see Section 11.7.2 under Triassic).

A cyclostratigraphic component in sequence stratigraphic studies has several potential advantages. If orbital forcing can be demonstrated in a sufficiently long time series, it provides a higher-resolution age control than can be achieved with biostratigraphy, and thus a more precise insight into sedimentation rates and the presence of hiatuses. In principle, establishment of an orbital forcing mechanism also allows evaluation of the different processes that influenced sedimentation, especially climate-sensitive parameters. For example, the Jurassic T-factory platforms deposited in the vast epicontinental sea and rift domains of the European Western Tethys have a well-expressed cyclicity. Constrained by high-resolution ammonite biostratigraphy, regional studies in the Swiss and French Jura, southern Germany, Spain and Portugal link these cycles to orbitally controlled climate variations (Strasser 1988; Strasser et al. 1999, 2012; Pittet et al. 2000; Colombié et al. 2011; Hillgärtner and Strasser 2003). By linking the cyclostratigraphic time framework to biostratigraphy, a high-resolution chronostratigraphic model has been developed with a rare spatial coverage and insight. This enables the construction of super-regional facies maps documenting the evolution of shallow-water sedimentary systems and their sediment export patterns through time (e.g. Pittet

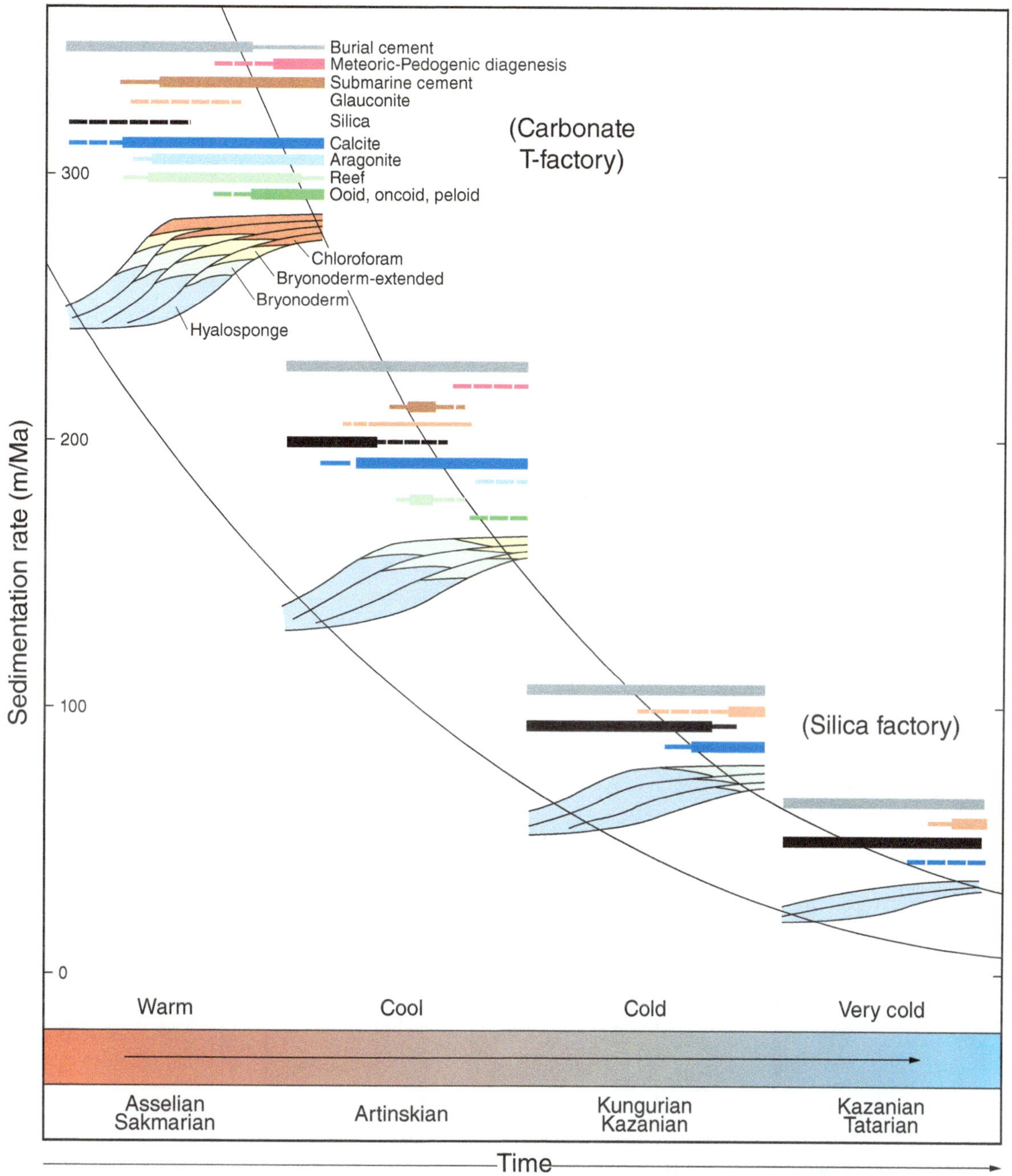

Figure 11.74 Evolution of the depositional system from a carbonate dominated T-factory shelf to a silica dominated ramp system because of a long-term cooling trend (*Source:* Beauchamp and Desrochers 1997 / Society for Sedimentary Geology).

and Strasser 1998a,b; Pittet et al. 2000; Colombié et al. 2011).

11.7.2 Marine Carbonate Depositional Patterns Through Time

The importance of the ecological and evolutionary history of carbonate-producing organisms for the stratigraphic architecture of marine carbonates has been highlighted by several authors, such as the historical reviews of reefs by Heckel (1974), of carbonate facies models by Wilson (1975), of dominant carbonate producing organisms by James (1984) and of ramp systems by Burchette and Wright (1992). More recent historical reviews, accompanied by plate-tectonic maps, include the synthesis of the Phanerozoic reef-building organisms and their global distribution by Kiessling et al. (2002; Figure 11.75a), and the synthesis of carbonate depositional patterns in relation to global

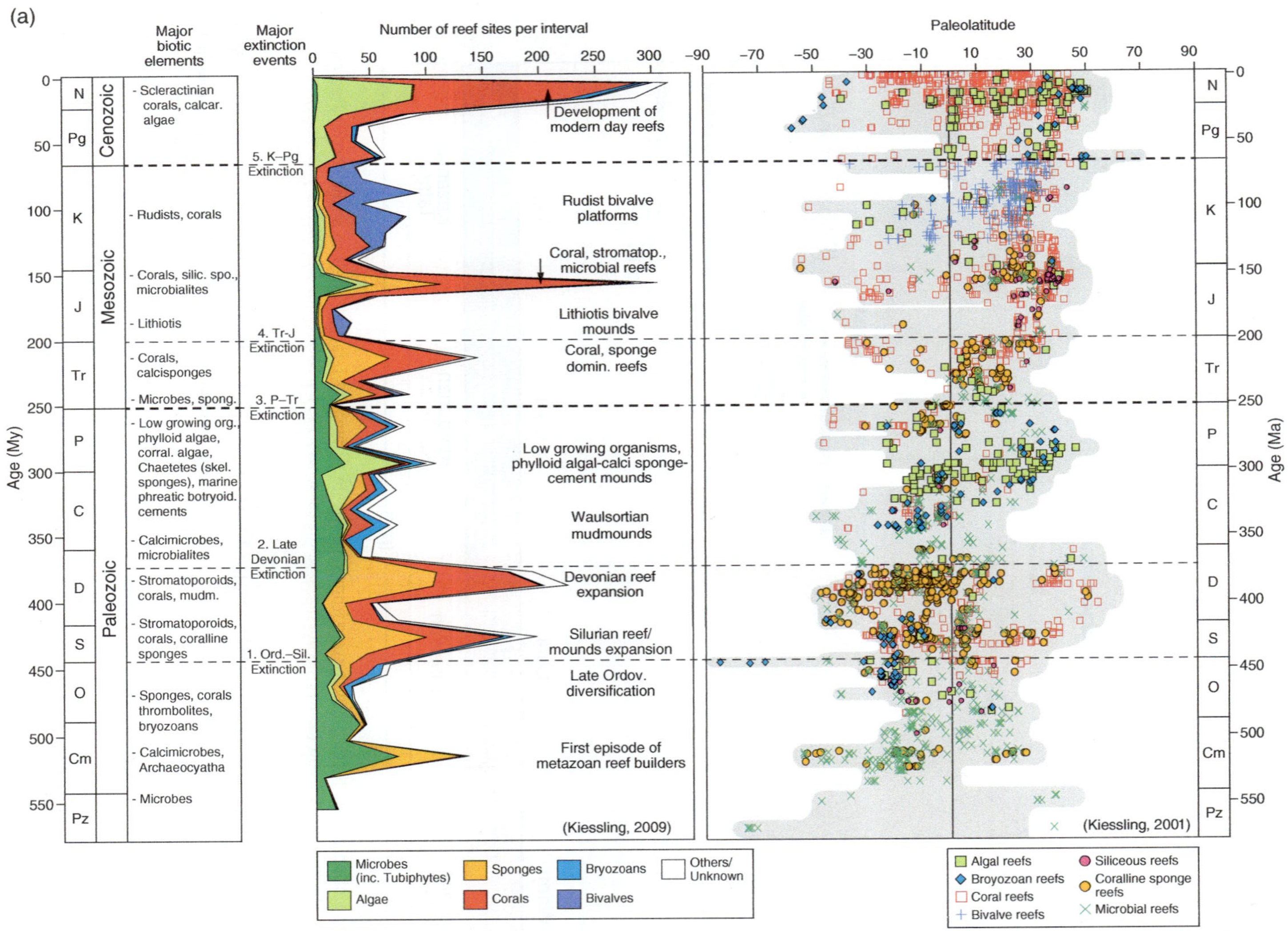

Figure 11.75a (a) Number of reef sites and their biotic composition plotted against time (10 Myr bins) and their palaeolatitudinal distribution (*Source:* Adapted from Kiessling 2001, 2009). The big five mass extinction episodes are indicated (Raup and Sepkoski 1982); note that extinctions 1 and 5 do not mark a global reef decline. See also discussion on reef crises in Kiessling and Simpson (2011). The present-day optimum zone for coral growth is between approximately 30° N and 30° S latitude (see Figure 11.29). The time scale is from GTS 2012 (Gradstein et al. 2012). Corraline sponge reefs (orange), including stromatoporoids, archaecyaths, chatetids, pharetronids; bivalves: rudists and other bivalves combined.

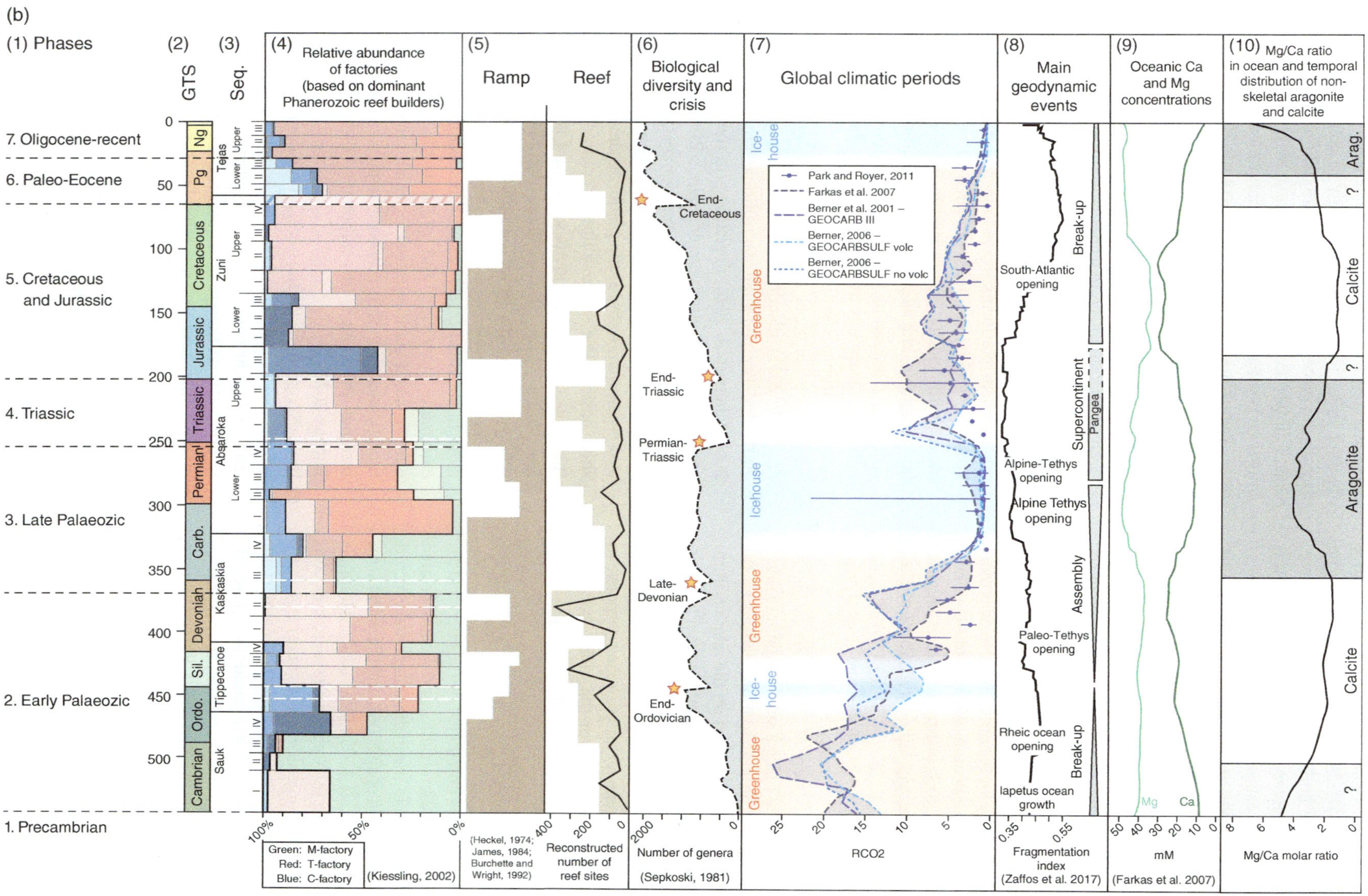

Figure 11.75b (Continued) (b) Phanerozoic summary of carbonate factory abundance and key physical controlling parameters based on the reefal database of Kiessling et al. (2002), of which the timescale is followed here. (1) Seven main evolutionary phases of the carbonate system are distinguished, punctuated by the extinction events; (2) Geological Time Scale 1996 with updates (Golonka and Kiessling 2002); (3) the time slices follow the second-order sequences from Sloss (1963, 1988) and the further subdivision by Haq et al. (1988), see Golonka and Kiessling (2002); (4) the relative abundance of the three main carbonate factories (T, C and M) based on our interpretation of the dataset of Kiessling (2002); (5) ramp and reef distribution from Burchette and Wright (1992), Heckel (1974), and James (1984); (6) biological diversity and crises from Sepkoski (1981 and Sepkoski et al. (1981)); (7) the references for the global climatic trends are provided in the figure; (8) main geodynamic events from Zaffos et al. (2017); (9) and (10) oceanic Ca and Mg concentrations and ratios from Farkaš et al. (2007). See text for discussion.

physical parameter change by Markello et al. (2008). The evolution and relative dominance of the main reef-building organisms during the Phanerozoic is punctuated by the 'big five' Phanerozoic crises in biodiversity which have had a dramatic impact on the evolution of life on earth, in the water and on land (Raup and Sepkoski 1982). Of these, the Permian–Triassic boundary extinction was the most dramatic of all, with an estimated extinction of 50% of the marine families, equating to 90–96% of the then existing species (Benton 2015). Second in importance is the K-Pg (Cretaceous–Palaeogene) boundary event. These two extinctions define the large-scale subdivision of the Phanerozoic in Palaeozoic ('old' life), Mesozoic ('middle' life) and Cenozoic ('new' life) systems (Figure 11.75a). The recovery phase after the Permo-Triassic extinction was the longest, with a full recovery of the pre-event diversity that was reached about 100 my later in the Early Cretaceous, whereas the recovery phase for the other 'big five' was more in the order of 5–10 My (Benton 2015, and references therein). To illustrate the link to the carbonate factories defined in this chapter, the reef-builders database of Kiessling et al. (2002) has been translated into a relative abundance curve of the T-, C- and M-factories and several other trends throughout the Phanerozoic (Figure 11.75b).

Below, the Phanerozoic history of the marine carbonate systems will be summarised using a subdivision into seven phases mostly guided by the punctuations caused by the major faunal extinctions and its effect on the relative abundance of the carbonate factories and ramp and reef type carbonate systems (Figure 11.75b). Although this is admittedly a simplification, it does provide a useful overall insight into the dynamics of the carbonate factories in their response to the changing physical conditions of the globe, including ocean chemistry, climate and geodynamic events.

11.7.2.1 Precambrian

Precambrian carbonates are dominated by the microbial M-factory biotic association. Microbialites and their associated microorganisms have been found in the sedimentary record since the Archean (Allwood et al. 2006), and the oldest stromatolites date back to about 3.5 Ga (Stal 2012). Although stromatolites are abundant in Archean deposits, they contain only rare fossil microorganisms, and their biogenicity was, therefore, difficult to demonstrate (Hofmann et al. 1999; Schopf et al. 2007). The first stromatolitic (layered) carbonates with organic laminae supporting benthic microbial mats are as old as 3.47 Ga (Lepot 2020). Photosynthetic production of O_2 most likely started between 3.2 and 2.8 Ga, that is, well before the Great Oxidation Event at 2.45–2.31 Ga (Figure 11.76. See Supplementary Information 11.S.3 for a selected case study list).

The Precambrian is characterised by a progressive decrease of abiotic factors in controlling microbialite growth and a concomitant increase in benthic microbial mats (Grotzinger and James 2000). The stromatolite diversity data show an increasing diversity of columnar forms through the Proterozoic and culminating in the Neoproterozoic, followed by a sharp decrease related to the radiation of organisms capable of disrupting mats and competing for space (Awramik 1971; Riding and Awramik 2000; Grey and Awramik 2020). The general pattern of evolution of the structure of the Precambrian platforms is summarised by Grotzinger and James (2000) in three steps (Figure 11.76):

a) Archean stromatolitic morphologies are generally considered as less complex than the Proterozoic ones (Hofmann 2000; Semikhatof and Raaben 2000). Late Archean and Palaeoproterozoic carbonates developed on rimmed platforms including major stromatolite buildups, grainstone belts, lagoonal to peritidal cyclic facies, with calcite pseudomorphs of giant aragonite fan structures indicating abundant sea-floor precipitation. The decrease in direct precipitation through the Palaeoproterozoic is attributed to a gradual change in oceanic carbonate saturation.
b) The Mesoproterozoic is characterised by ramp (Pratt and Rule 2021), distally steepened ramp (Bartley et al. 2007) and rimmed-shelf depositional profiles (Turner 2009; Chen et al. 2016) depending on the geodynamic context and the hydrodynamic regime. Rimmed shelves developed on sedimentary accretionary margins, but locally bypass margins occurred with both gullied slopes and escarpments, and some platforms developed intrashelf basins. The stromatolites presented greater textural diversity and an increase in mud is observed in association with the problematic but distinctive molar-tooth calcite. In the Mesoproterozoic (and likely the early Neoproterozoic) ocean, the saturation state lay close to a critical threshold with respect to the ease of carbonate nucleation and precipitation. The consequence was a marine system in which relatively minor modification of the physico-chemical environment resulted in changes in carbonate fabric (Kah and Bartley 2022).
c) In the Neoproterozoic, the ramp profile dominated, with limited sea-floor cementation, an increased mud content, and giant ooids and microbialites (Cozzi and Al-Siyabi 2004). Although stromatolites are still observed in reefs, the fabric of the microbialites evolved with thrombolites and calcified microbes associated with void spaces (loci of cementation, sand accumulation and coelobionts – cavity-dwelling organisms). The reason for the dominance of the ramps is unclear but they coincide

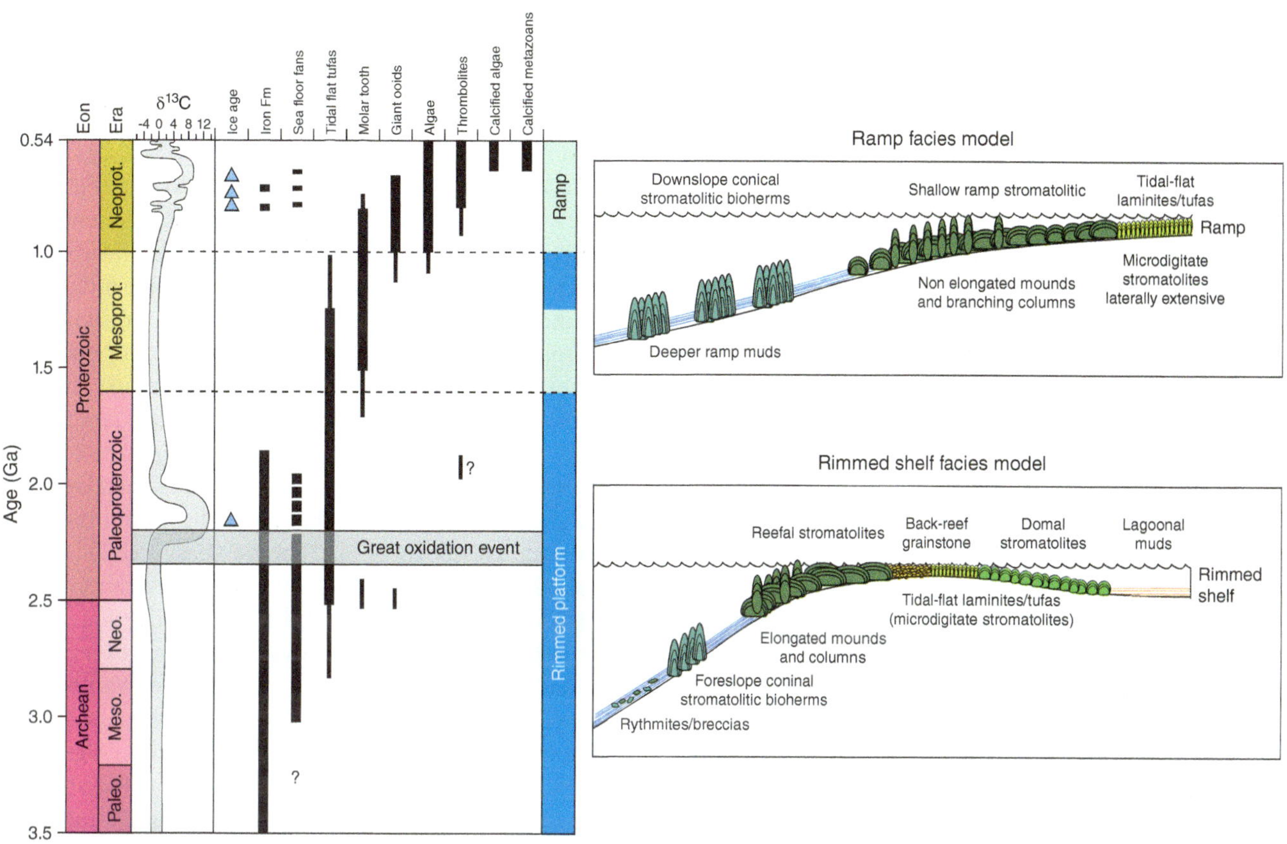

Figure 11.76 Pre-Cambrian carbonate depositional patterns. The generalised record of the $\delta^{13}C$ values of marine carbonates is based on Karhu and Holland (1996); Planavsky et al. (2011). The microbial-dominated (M-factory) carbonate system alternates between the ramp and platform depositional profiles (modified after Grotzinger 1989).

with abundant grainstones (Knoll 1989; Thorie et al. 2018), the decline of effective stromatolitic barriers (Grotzinger 1989), and the rise of other multicellular algae in competition for the substrate (Butterfield et al. 1988).

11.7.2.2 Early Palaeozoic (Cambrian to Devonian)

The Early Palaeozoic (Figure 11.75) was initially characterised by continued domination of microbial reefs (Riding 2006). High-relief bio-constructed build-ups first appeared in the Early to Middle Ordovician, with stromatactis-bearing mud mounds and the evolution of diverse metazoan reef assemblages, including for the first time an assemblage of corals, sponges and bryozoans (Webby 2002) that acted as important framework builders and sediment bafflers. This assemblage collapsed at the Late Ordovician extinction event (Copper 2001) and was replaced by the current and wave-resistant pinnacle reefal mounds of the T-factory which protruded tens to hundreds of metres above the Mid Silurian seafloor (Copper 2002a,b; McLaughlin et al. 2019). The two-pulse Late Ordovician extinction event is considered one of the most important step changes in the evolution of the marine biosphere (Flügel and Kiessling 2002; Kiessling et al. 2010). It coincided with the Late Ordovician glaciation (Brenchley et al. 2001; Pohl et al. 2016; Saupe et al. 2020), but an alternative hypothesis proposes volcanism, warming and anoxia as triggers to explain the mass extinction (Bond and Grasby 2020). The subsequent global development of large carbonate shelves and isolated buildups had two peaks, one in the Mid-Silurian and one in the Late Devonian (Figure 11.75). It was terminated by the major Late Devonian extinction event at the Frasnian-Fammenian boundary. See Supplementary Information 11.S.4 for a selected case study list that complements the examples cited in the text.

The Silurian strata of the North American midcontinent region represent one of the world's great Phanerozoic carbonate factories (Kiessling et al. 2003), with hundreds to thousands of pinnacle reefs recorded at different stratigraphic levels (Figure 11.62; Copper and Brunton 1991; Brunton et al. 1998; Copper 2002a,b). McLaughlin et al. (2019) demonstrated that most of the buildups originated just above sequence boundaries, during early and rapid phases of RSL rise. In addition, these seem to have coincided with positive excursions in the d^{13}C curve, suggesting a link to the carbon cycle ('calcification overshoot').

During the Middle and Late Devonian, a similar global development of carbonate shelf and isolated platform systems is observed in intra-cratonic basins, and along plate margins, as now exposed in world-class outcrops in Canada, Australia, Belgium and Morocco. The mud mounds of Belgium and Morocco were described above (Section 11.6.2.2, Figure 11.63c–f). The Middle and Late Devonian strata of the intra-cratonic Western Canadian Sedimentary Basin are organised in mega-cycles consisting of shallow water carbonate shelves and isolated buildups, basinal clays, locally enriched in organic-matter and evaporites (Stoakes 1980; Wendte et al. 1990; Wendte,1992; Whalen et al. 2000a,b; van Buchem et al. 2000, 2005; Weissenberger et al. 2017). The Late Devonian, Frasnian part of the succession displays a typical evolution from low-angle ramps to steep-sloped shelves and isolated flat-topped platforms surrounded by organic-rich intra-basinal shales and mudstones during overall transgression, followed by progradation and part clay-infill of the basin during regression (Figure 11.77a). Similarly insightful outcrops are present along the northern margin of the Canning Basin (Australia), where a 350-km-long Middle to Late Devonian barrier reef buildup complex exists. This carbonate shelf also developed from flat ramps to a reef-fringed, flat-topped isolated platforms standing tens to hundreds of metres above the surrounding seafloor and flanked by slope and basin deposits (Figures 11.48a and 11.77b; Kerans and Harris 2008; Playford et al. 2009). The striking stratigraphic similarity of these locations makes a strong case for potential global, eustatic control of the carbonate sedimentation patterns.

11.7.2.3 Late Palaeozoic (Latest Devonian, Carboniferous and Permian)

At the Famennian-Frasnian extinction event the typical Frasnian tabulate-stromatoporoid reef-builders of the T-factory disappeared and were initially replaced by an M-factory dominated assemblage of microbialites (thrombolites, stromatolites and oncoids), calci-microbes and siliceous sponges (Figure 11.75a; Fagerstrom 1987; Joachimski and Buggisch 1993, 1996; Joachimski et al. 1999; Wood 1999; Webb 2001). In the early Carboniferous, a major climatic change occurred (Figure 11.75b) with dramatic cooling and the beginning of an ice-dominated age with the associated glacio-eustatic sea-level changes having a magnitude of 40–120 m and frequencies in the Milanković band (Rygel et al. 2008). These conditions had a profound impact on the composition and architecture of the carbonate systems. They caused a shift from the shallow water T-factory to C- and M-factories, that cover a larger production profile, with the proliferation of the Waulsortian mud mounds in deeper foreslope and outer ramp, significant microbial contribution to the stabilisation and growth of the platforms in the mid to upper slope range, and the development of ooid and crinoid dominated ramps. The high magnitude of the sea-level changes profoundly affected sedimentation patterns, with locally (aeolian) clastics and carbonates commonly occurring within the same fourth-order sequences (e.g. Paradox basin; Goldhammer et al. 1991; Atchley and Loope 1993), or carbonates interdigitated with coal seams and deltaic clastics (Waters and Davies 2006). These successions were the focus of early definitions of depositional cycles (cyclothems; Wanless and Weller 1932; see review in Fielding 2021) and played a role in the establishment of Milanković cyclicity (Algeo and Wilkinson 1988; Goldhammer et al. 1994). Towards the Middle Permian, palaeoclimate became warmer and dryer again (Kiessling et al. 2002), and the T-factory re-established itself (Weidlich 2002). See Supplementary Information 11.S.4 for a selected case study list.

A characteristic Late Palaeozoic stratigraphic and faunal pattern is illustrated with outcrop and subsurface examples from Kazakhstan. In the Bolshoi-Karatau Mountains, the platform to basin transition of a large carbonate shelf is exposed, whereas in the nearby Pricaspian basin two time-equivalent isolated carbonate platforms are documented with high-resolution seismic and core (Figure 11.78). The outcrop sections show the platform to basin transition over 30 km in a 4500-m-thick succession of stacked carbonate platforms (shelves and ramps), with the occurrence of Waulsortian mounds on the outer slope of Latest Devonian (Famennian) to Middle Carboniferous (Pennsylvanian) age (Figure 11.78c; Cook et al. 2003). The seaward margin was probably structurally controlled by the rifted edges of the passive continental margin. The isolated, flat-topped, Karachaganak and Tengiz platforms in the subsurface of the North Pricaspian basin have a similar age as these outcrop sections, and in the case of the Tengiz buildup, also a comparable thickness (Cook et al. 2002; Weber et al. 2003; Collins et al. 2006, 2014; Katz et al. 2010). Both platforms show, after the initial establishment of the Famennian platform an overall backstepping trend up to the Late Visean, and in the

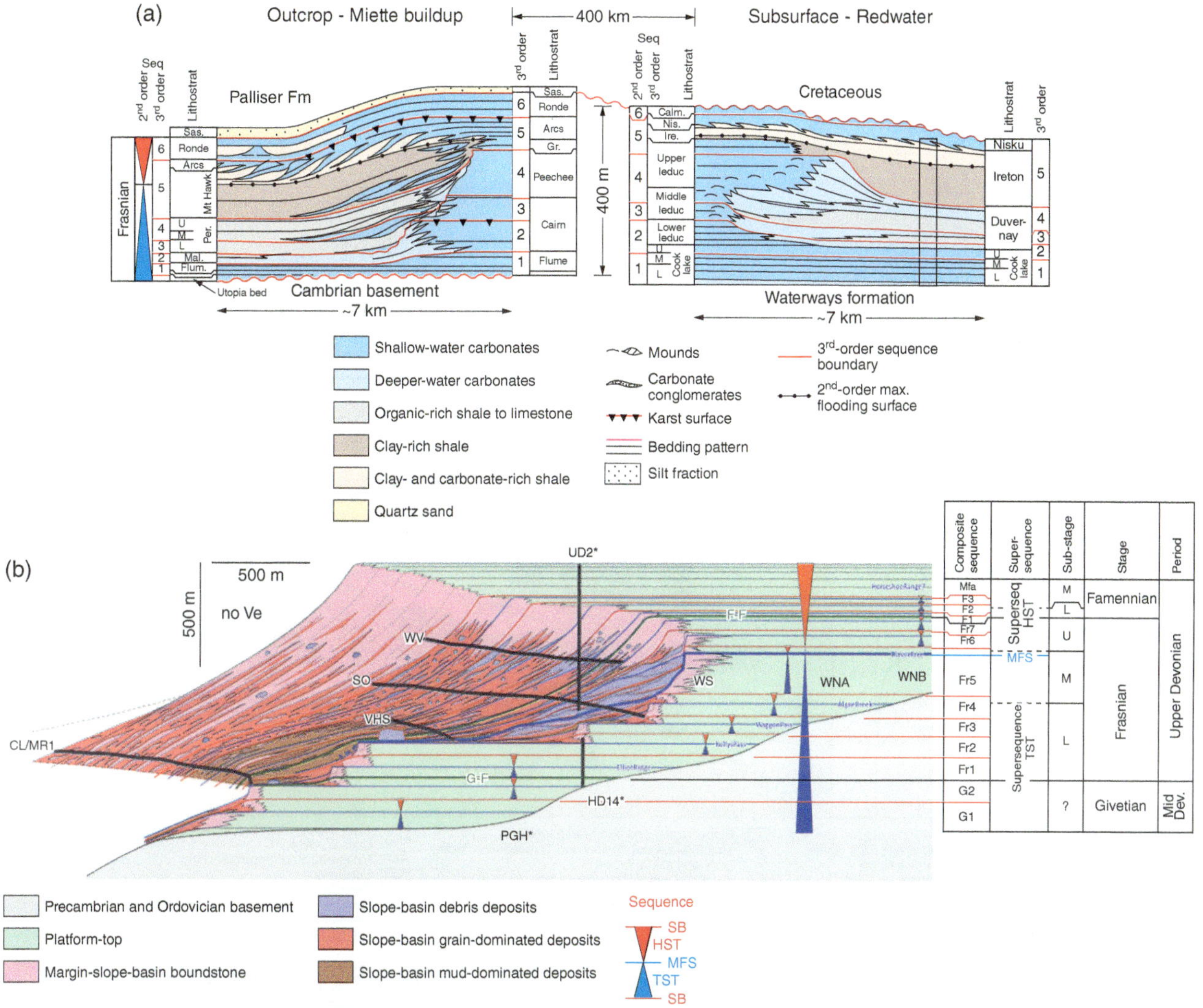

Figure 11.77 Comparison of time equivalent Upper Devonian (Frasnian) carbonate platforms in Canada and Australia. (a) Sequence stratigraphic cross section of two isolated Frasnian platforms in the Western Canadian Sedimentary Basin, which are approximately 400 km apart. Note the similarity in overall architecture and thickness, but contrasting margin geometries, with the interfingering of shallow water carbonate and basinal facies in the Redwater margin, and the highly erosional bypass margin in Miette, with onlapping slope deposits of olistoliths and debris flows (Whalen et al. 2000b; van Buchem et al. 2005; Wong et al. 2016); (b) sequence stratigraphic cross section of the Frasnian shelf margin in the Canning basin of W. Australia, displaying a bypass margin (Adapted from Playford et al. 2009; see also Figure 11.48a). The marginal-slope deposits include large amounts of coarse platform-derived debris, much of which was transported down the slopes as submarine debris flows, like what has been observed in the Canadian outcrops (Mountjoy and Playford 1972). The remarkably similar second-order, transgressive-regressive stratigraphic architecture in Canada and Australia is a strong argument for a global (eustatic) control on these sedimentation patterns.

Serpukhovian and Bashkirian the development of an asymmetrical prograding wedge of microbial boundstones, a stacking pattern also recognised in the outcrop sections, and thus likely reflecting a sea-level-ecosystem response (Figure 11.78c–d). The combined case studies from the Late Palaeozoic such as from Kazakhstan (Collins et al. 2006; Katz et al. 2010; Kenter et al. 2010a,b), northern Spain (Bahamonde et al. 2000; Della Porta et al. 2003, 2004; Merino-Tomé et al. 2007) and southern United States (Tinker 1998; Playton and Kerans 2018; Figures 11.54b and 11.79) provide good examples of the functioning of the M-factory, with the typical microbial boundstones growing on the upper slopes, and the platform-top derived sediments which largely bypassed to more distal environments.

A characteristic element of the Late Carboniferous (Pennsylvanian) are the phylloid algal mounds, which

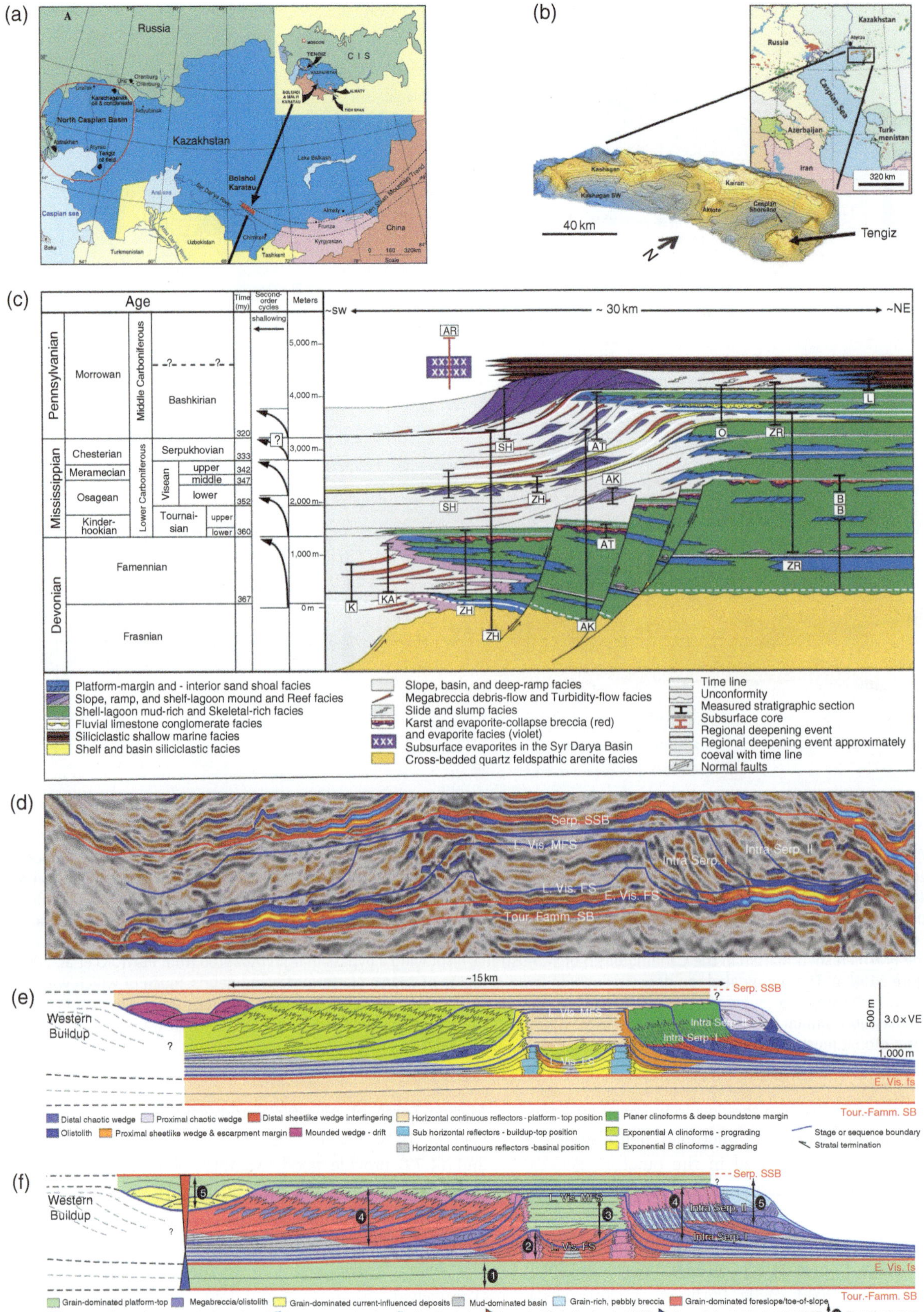
(a)
Russia
Kazakhstan
North Caspian Basin
Caspian sea
Uzbekistan
Turkmenistan
China
Bolshoi Karatau
C I S
(b)
Kazakhstan
Russia
Caspian Sea
Azerbaijan
Turkmenistan
Iran
320 km
Kashagan
Kashagan SW
Kairan
Aktote
Caspian Shoreline
Tengiz
40 km
N
(c)
Age
Time (my)
Second-order cycles
Meters
shallowing
~SW
~ 30 km
~NE
Pennsylvanian
Morrowan
Middle Carboniferous
Bashkirian
Mississippian
Chesterian
Meramecian
Osagean
Kinderhookian
Lower Carboniferous
Serpukhovian
Visean
upper
middle
lower
Tournaisian
upper
lower
Devonian
Famennian
Frasnian
320
333
342
347
352
360
367
5,000 m
4,000 m
3,000 m
2,000 m
1,000 m
0 m
AR
SH
AT
AK
ZH
O
ZR
L
B
K
KA
Platform-margin and - interior sand shoal facies
Slope, ramp, and shelf-lagoon mound and Reef facies
Shell-lagoon mud-rich and Skeletal-rich facies
Fluvial limestone conglomerate facies
Siliciclastic shallow marine facies
Shelf and basin siliciclastic facies
Slope, basin, and deep-ramp facies
Megabreccia debris-flow and Turbidity-flow facies
Slide and slump facies
Karst and evaporite-collapse breccia (red) and evaporite facies (violet)
Subsurface evaporites in the Syr Darya Basin
Cross-bedded quartz feldspathic arenite facies
Time line
Unconformity
Measured stratigraphic section
Subsurface core
Regional deepening event
Regional deepening event approximately coeval with time line
Normal faults
(d)
Serp. SSB
L. Vis. MFS
Intra Serp. I
Intra Serp. II
L. Vis. FS
E. Vis. FS
Tour. Famm. SB
(e)
~15 km
Western Buildup
Serp. SSB
500 m
3.0 x VE
1,000 m
L. Vis. MFS
Intra Serp. II
Intra Serp. I
L. Vis. FS
E. Vis. fs
Tour.-Famm. SB
Distal chaotic wedge
Proximal chaotic wedge
Distal sheetlike wedge interfingering
Horizontal continuous reflectors - platform-top position
Planer clinoforms & deep boundstone margin
Olistolith
Proximal sheetlike wedge & escarpment margin
Mounded wedge - drift
Sub horizontal reflectors - buildup-top position
Exponential A clinoforms - prograding
Stage or sequence boundary
Horizontal continuours reflectors -basinal position
Exponential B clinoforms - aggrading
Stratal termination
(f)
Western Buildup
Serp. SSB
L. Vis. MFS
Intra Serp. II
Intra Serp. I
L. Vis. FS
E. Vis. fs
Tour.-Famm. SB
Grain-dominated platform-top
Megabreccia/olistolith
Grain-dominated current-influenced deposits
Mud-dominated basin
Grain-rich, pebbly breccia
Grain-dominated foreslope/toe-of-slope
In situ reef
Margin collapse locally along strike
Stage or sequence boundary
Stratal termination
Low-order high stand systems tract (HST)
Low-order low stand systems tract (LST)
Depositional stages

thrived in tropical shallow water environments (Schatzinger 1983) and formed either organic frameworks on high-rising platforms or small clusters of decametre scale mounds (see Section 11.6).

11.7.2.4 Triassic

The mass extinction that occurred at the Permo-Triassic boundary is regarded as the most severe of the Phanerozoic. An estimated 90–96% of marine species disappeared including a large part of the carbonate producers (James 1984; Sepkoski 1996; Payne and Clapham 2012; Benton 2015). The aftermath of this event saw a two-step recovery in shallow water carbonate systems: first, during the 1.5 My after the event microbes and sponges dominated and reefs were absent (Figure 11.75; Schubert and Bottjer 1992; Woods et al. 1999; Flügel 1982a; Pruss et al. 2005; Payne et al. 2007; Woods 2009). In China, at this time, oolitic deposits are recorded that contain giant ooliths and large coated intraclasts, phenomena attributed to calcium carbonate saturated waters, due to the reduced biotic carbonate precipitation in the aftermath of the crisis (Lehrmann et al. 2012). The second step of the recovery was a migration of microbial organisms from the coastal to the more distal shelf environments, and the re-establishment of a more diverse assemblage in the coastal waters, consisting of echinoderms, ostracods, gastropods, bivalves, sponges, brachiopods and vertebrae remains (Brayard et al. 2011; Vennin et al. 2015; Olivier et al. 2016). See Supplementary Information 11.S.5 for a selected case study list.

The land-detached Triassic carbonate platforms spectacularly exposed in the Italian Dolomites region played an important role in our understanding of carbonate systems, since they were among the first for which large-scale steep-sloped margins were described (Bosellini 1984; Gischler et al. 2001). Most of these buildups evolved from an initial low relief ramp into flat-topped platform profiles that aggraded, prograded and retrograded, locally attaining a relief of 800–900 m with steep slope angles of up to 25°–35° (Biddle et al. 1992; Harris 1994; Maurer 2000; Keim et al. 2001; Brandner and Keim 2011; Gianolla et al. 2021). These platforms were dominated by the microbial activity of the M-factory, which helped to stabilise the steep slopes (Keim and Schlager 2001; Marangon et al. 2011; Adams and Kenter 2014). The platform tops show well-expressed cyclicity interpreted as deepening-shallowing trends (the Lofer cycles; Schwarzacher 1947, 1954; Fischer 1964; Haas 1982), either of orbitally forced Milanković cyclicity origin (Schwarzacher 1947, 1954, 2005; Fischer 1964; Goldhammer et al. 1987; Goldhammer and Harris 1989; Preto et al. 2001), or caused by shorter (superposed) climatic variations linked to solar and/or lunar forcing (Brack et al. 1996; Maurer 2000; Mundil et al. 2003; Zühlke et al. 2003; Roth and Reijmer 2005). Another characteristic feature of the Triassic are the large, land-attached shallow water carbonate shelves with a platform depositional profile that covered many of the tectonic plates at that time, such as the Arabian Plate (Maurer et al. 2008; Davies et al. 2019), Europe (Aigner and Bachmann 1992; Pöppelreiter et al. 2004; Warnecke and Aigner 2019) and China (Lehrmann et al. 2007).

Figure 11.78 Stratigraphic architecture of a Late Devonian/Carboniferous land-attached shelf margin exposed in outcrops and a time-equivalent isolated carbonate platform studied in the subsurface, both in Kazakhstan. (a) Location map of the Late Palaeozoic carbonate shelf exposed in outcrops of the Bolshoi-Karatau Mountains (*Source:* Cook et al. 2002 / With Permission of Society for Sedimentary Geology); (b) location and topographic map of the time-equivalent isolated platforms in the Pri-Caspian basin in the subsurface of the Caspian Sea (*Source:* Collins et al. 2013 / With Permission of Society for Sedimentary Geology); (c) stratigraphic cross section of the shelf margin where five second-order supersequences are defined (*Source:* Adapted from Cook et al. 2002; Zempolich et al. 2002): Supersequence 1, reef rimmed platform, M-factory dominated margin, platform interior of mud mounds, carbonate sands, cryptalgal and evaporitic laminites, capped by a karstification surface; Supersequence 2, ramp system with Waulsortian skeletal mud mounds and bioclastic turbidite aprons in the outer ramp, and C-factory biostromes and tidal flats in the mid and inner ramp; Supersequences 3 and 4, shoal-rimmed platforms with cross-bedded ooids and bioclastic sands, and Waulsortian skeletal slope mounds of up to 90 m high. The interior platform facies consist of a M-factory dominated production (with ooids and algae); Supersequence 5, reef-rimmed platform margin with Waulsortian skeletal mud mounds, and in the interior a deeper, muddy lagoon; (d–f) Karachaganak isolated platform (*Source:* Katz et al. (2010) / Society of Petroleum Engineers). An overall transgressive-regressive sequence is defined corresponding to super sequences 3 and 4 defined above. Five evolutionary steps are distinguished: (1) initial early Visean flat shelf; (2) late Visean, initiation of smaller mounds; (3) coalescence of mounds and development of steep margins of up to 35° with grain and debris-dominated slope sediments; (4) Serpukhovian progradation, distinctly asymmetrical and characterised by two end-member margin-to-slope types: escarpment margins with onlapping slopes, on the east side, and accretionary margins with interfingered slopes, on the west side, caused by prevailing wind directions. The interior platform is organised in meter to decametre scale shallowing-up cycles, capped by exposure surfaces in the Bashkirian during global ice-house conditions (Weber et al. 2003; Kenter et al. 2006); (5) Serpukhovian and Bashkirian, the upper-slope was the locus of in situ production – M-factory – with a lower growth limit down to ~300 m. These facies fed the middle and lower slope during periodic large-scale failure of the rim, triggered by overgrowth and the high magnitude glacio-eustatic sea-level fluctuations.

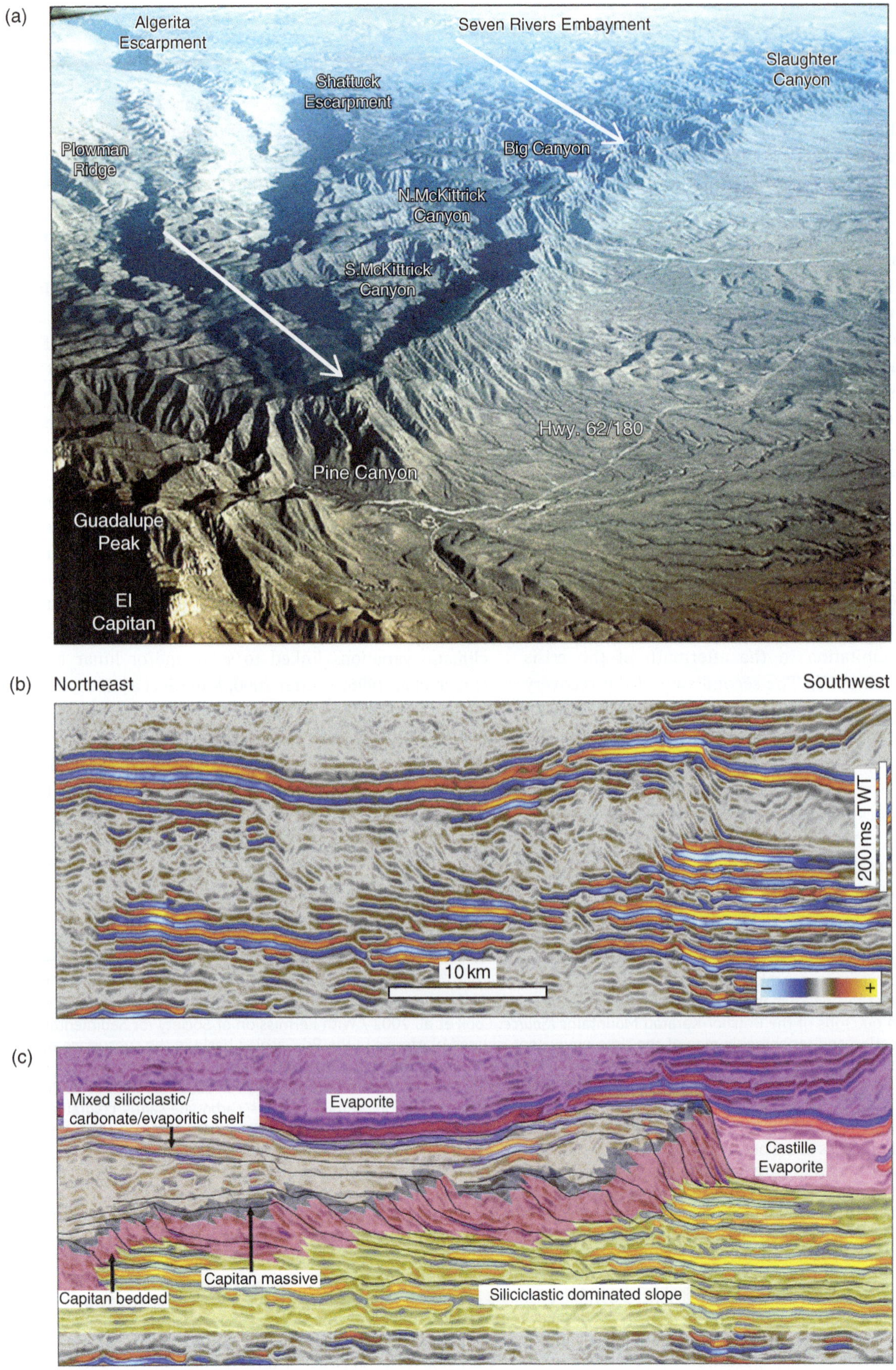

Figure 11.79 Permian carbonate shelf margin, Guadalupe Mountains Texas, USA. (a) Aerial view showing the fully exposed platform margin. Time-equivalent basinal facies are not only partly preserved but also extensively eroded away. See also Figure 11.48b for a perpendicular view of the platform margin. White arrows show platform progradation direction. *Source:* Bebout and Kerans (1993) / Bureau of Economic Geology - The University of Texas at Austin. (b) Seismic line through the Capitanian reefal margin. Courtesy of Fairfield Geotechnologies, (c) interpretation of seismic line showing seismic facies subdivision and the architecture of the prograding Capitan reef margin (*Source:* Janson et al. (2011) / American Association of Petroleum Geologists).

11.7.2.5 Jurassic and Cretaceous

The Jurassic period is marked by a profound plate tectonic reorganisation caused by the breakup of northern Pangea. This affected the geodynamic context along the NeoTethys margins, changed the major ocean current systems and was associated with an overall rise in sea level, which allowed carbonates to colonise broad shallow water shelves. After the end-Triassic extinction event, the Early Jurassic was dominated by the C-factory and ramp systems, due to the absence of reef builders (Figure 11.75). These returned, regaining a more widespread distribution in the Middle Jurassic (Leinfelder et al. 2002), when a richer variety of carbonate depositional systems includes large, grainy, homoclinal ramp settings and T- and M-factory flat-topped rimmed-shelves with steep slopes and different types of buildups (Leinfelder et al. 2002; Kiessling et al. 2003). Although climatic conditions generally are assumed to have been warm and equable, there is now also evidence for short periods of cooling notably during the Pliensbachian, Toarcian and the Callovian, which occurred in conjunction with intensive volcanism in large igneous provinces and in association with the development of OAEs. The resulting warming and cooling events were relatively rapid (0.5–1 Myr) and caused high magnitude sea-level fluctuations, most likely due to glacio-eustasy (Hallam 1988; Dromart et al. 2003a,b; Hesselbo et al. 2007; Nunn et al. 2009; Nunn and Price 2010; Dera et al. 2011; Donnadieu et al. 2011; Pellenard et al. 2014; Krencker et al. 2019). The end of the Jurassic saw a notable decrease in the shallow water carbonate deposition, with large shelves changing to arid evaporite-dominated deposition likely during a lowering of sea level (Leinfelder et al. 2002; Arabian Plate: Murris 1980; Alsharhan and Kendall 1994). See Supplementary Information 11.S.5 for a selected case study list.

The Jurassic T-factory land-attached platforms deposited in the vast epicontinental sea and rift domains of the European Western Tethys have been intensively studied and have benefitted from a high-resolution biostratigraphic framework to constrain regional correlations. In an exemplary regional synthesis of the Jurassic of the Paris Basin, Brigaud et al. (2014) proposed five carbonate systems, distinguished by factory and depositional profiles, that dominated in different time intervals. These authors proposed that the sensitivity of these epicontinental basin carbonate systems to climate change was such that during cool and humid periods the storm-dominated ramp or rimmed-ramp morphology dominated, whereas during warm and humid conditions, flat-topped, rimmed-shelf platforms prevailed. In Early Jurassic outcrops in the Djebel Bou Dahar mountain range in Morocco, a slower evolution is observed from a carbonate ramp (Hettangian-Sinemurian) to a flat-topped carbonate platform profile (Sinemurian-Pliensbachian; Merino-Tomé et al. 2012). The platform system in this case shows a typical M-factory morphology with platform lagoon sediments, hierarchical vertical facies stacking pattern, a deepened margin with ooid shoals and steep-sided slopes with linear clinoforms (Kenter and Campbell 1991; Scheibner and Reijmer 1999; Merino-Tomé et al. 2012, 2017). Long-lasting Jurassic oolitic-dominated ramp systems are reported from Morocco (Pierre et al. 2010; Figure 11.54), the Lusitanian Basin (Leinfelder 1986; Ellis et al. 1990) the Jurassic epeiric-sea carbonate ramp of the Iberian Basin in Spain (Aurell et al. 2003; Bádenas et al. 2005), as well as from the United States (e.g. Smackover Formation; Heydari and Moore 1994; Bhattacharjee et al. 2024).

Isolated carbonate platforms and deeper water mounds were both common during the Jurassic. In the shallow water domain, the large bivalve *Lithiotis* formed bioherms that prevailed in the Pliensbachian–Toarcian. These were succeeded by corals as the dominant reefal mound builders (Figure 11.75a). In the deeper water settings of the middle ramp (up to 70 m deep), the siliceous sponge and microbes associations are typical; they were present in the Early Jurassic (Pliensbachian, Morocco; Della Porta et al. 2014), but were more important in the Middle and Late Jurassic (Leinfelder et al. 1993; Reitner and Neuweiler 1995; Leinfelder and Schmid 2000; Leinfelder et al. 2002). Microbialites were associated with both these dominant reef builders. The abundant occurrence of deep-water siliceous sponge reefs in the Toarcian, Oxfordian and Kimmeridgian has been related to the unusual eutrophic oceanographic conditions, which also led to the accumulation of large amounts of organic matter in the sediment at these times (Leinfelder et al. 2005; Reolid 2011). A detailed study of the sponge and microbial-rich mounds occurrence and habitat along a proximal-distal transect of a distal to middle ramp setting in northeastern Spain showed a clear depth-related zonation of mound shape, spacing and composition (Tomás et al. 2019). The microbial and siliceous-sponge growth is governed by complex interactions between water depth, light, energy, sedimentation rate, substrate, oxygenation, nutrient availability and carbonate saturation (Brachert et al. 1992; Keupp et al. 1993; Leinfelder et al. 1993; Dupraz and Strasser 1999, 2002; Schmid et al. 2001; Riding and Liang 2005; Reolid 2011; Tomás et al. 2019).

During the Early Cretaceous shallow-water carbonate deposition expanded again as a result of long-term sea-level rise, re-creating large shallow-water shelves bordering the NeoTethys Ocean, a trend which reached its acme in the Late Cretaceous (Cenomanian-Turonian; Höfling and Scott

2002; Johnson et al. 2002). Carbonate systems worldwide became dominated by the T-factory (Figure 11.75), with a typical shallow-water benthic biota, including rudist bivalves, dictyoconid foraminifers, dasyclad algae, colonial corals and stromatoporoids (Masse 1992). A climatic optimum was reached during the Mid-Cretaceous. It is now recognised that this was punctuated by cold snaps, with rapid, high magnitude sea level drops probably caused by glacio-eustasy (Mutterlose et al. 2009; Maurer et al. 2013; Ray et al. 2019; Davies et al. 2020). These, in combination with several OAEs, influenced the palaeoceanographical conditions and caused extinction events in the shallow-water domain in the mid-Valanginian, late Aptian, and latest Cenomanian, affecting dominant carbonate producers such as the thick-shelled, rudistic bivalves and calcareous algae (Höfling and Scott 2002; Johnson et al. 2002; Skelton 2003; Skelton and Gili 2012). Notably OAE-1a also promoted the proliferation of microbial facies, as relief-forming microbial mounds (e.g. Rameil et al. 2010). A plate spreading rate switch influencing the Ca/Mg composition of the oceans is invoked for the Turonian switch from early Cretaceous aragonite-dominated coral–algal and coral–rudist associations to Turonian-Maastrichtian calcite-dominated rudist associations (Steuber 2002).

The stratigraphic succession of the Arabian Plate is a representative example of this overall Lower to Mid Cretaceous depositional trend. A 2-fold stratigraphic subdivision can be made (Figure 11.80a) in a lower part, of Berriasian to Barremian age, which consists of a large, land-attached, flat-topped, carbonate shelf that prograded over hundreds of kilometres (Figure 11.80b–d), and an upper part, of Aptian to Early Turonian age, when the large-scale stacking pattern changed to an aggradational mode, with the typical development of several generations of intrashelf basins creating substantial facies heterogeneity, with rudist dominating margins and locally organic-rich intra-shelf basin deposits as discussed above (Figures 11.49, 11.69, 11.73). The creation of one of the largest incised valley systems of the Cretaceous also occurred at this time, caused by the high magnitude, late Aptian (glacio-eustatic) sea-level drop (Figure 11.81).

Late Cretaceous carbonate sedimentation in the northern hemisphere was dominated by one of the longest lasting (40 My) and largest carbonate systems of the Phanerozoic: the Chalk system (Surlyk et al. 2003; Stanley et al. 2005; Mortimore 2011). It was dominated by the P-factory (coccoliths and nannofossils) and covered the temperate to cool-water boreal domain from the northern American continent into central Asia (Kauffmann and Caldwell 1993; Gale et al. 1999; Eldrett et al. 2017). Simultaneously, T-factory carbonate-dominated shelves became scarcer due to environmental changes affecting the Neo-Tethys Ocean. During the Campanian and Maastrichtian an increase in nutrients along the southern Neo-Tethys margin triggered high primary productivity with extensive chert, phosphate and oil shale deposition stretching from Jordan to Morocco (Notholt 1985; Notholt et al. 1989) and beyond to the northern part of Latin America (Föllmi et al. 1992). In Jordan, oyster mounds with a local relief of up to 40 m are associated with these deposits in a mid-ramp setting (Abed and Sadaqah 1998; Pufahl and James 2006 and Pufahl et al. 2003; Powell and Moh'd 2011). Their filter-feeding capacity allowed them to thrive in this highly eutrophic environment, contrary to the T-factory shallow-water biota. In a further example an early Cenomanian sea-level rise triggered stratification of the water column on the Texas shelf causing rapid facies change from an open marine, diverse carbonate platform facies (Buda Formation), to a dysaerobic, organic-rich setting (Eagle Ford Fm.) with a restricted planktonic carbonate production (Eldrett et al. 2017; Minisini et al. 2018). This planktonic dominated system subsequently returned to clean (non-organic rich) chalk sedimentation for the remainder of the Late Cretaceous when a connection through the Mid-Continent Seaway was established with cold, oxic seawater (Eldrett et al. 2017).

11.7.2.6 Paleocene–Eocene

The K/Pg boundary extinction event wiped out a significant part of the shallow-water carbonate producers: the rudists being a notable casualty (Figure 11.75; Sepkoski 1981). Their disappearance opened a niche in the T-factory, which was filled by corals and foraminifera (Baceta et al. 2005). A global inventory shows that the decline in carbonate depositional systems across the K/Pg boundary was, however, limited (Flügel and Kiessling 2002) and recovery occurred faster than previously thought (Fagerstrom 1987; Perrin 2002). In Spain, a series of Early Paleocene reef development phases of autotrophic reef ecosystems (T-factory) occurred (Baceta et al. 2005), as well as the development of calcareous red algae deposits (Aguirre et al. 2007), whereas in Denmark cool-water (C-factory) bryozoan mounds thrived (Bjerager and Surlyk 2007a,b; Bjerager et al. 2010). During the Paleocene–Eocene a new type of T-factory developed, which was dominated by large benthic foraminifera that thrived in seagrass meadows in wave-dominated ramp settings (see Section 11.6.1.2).

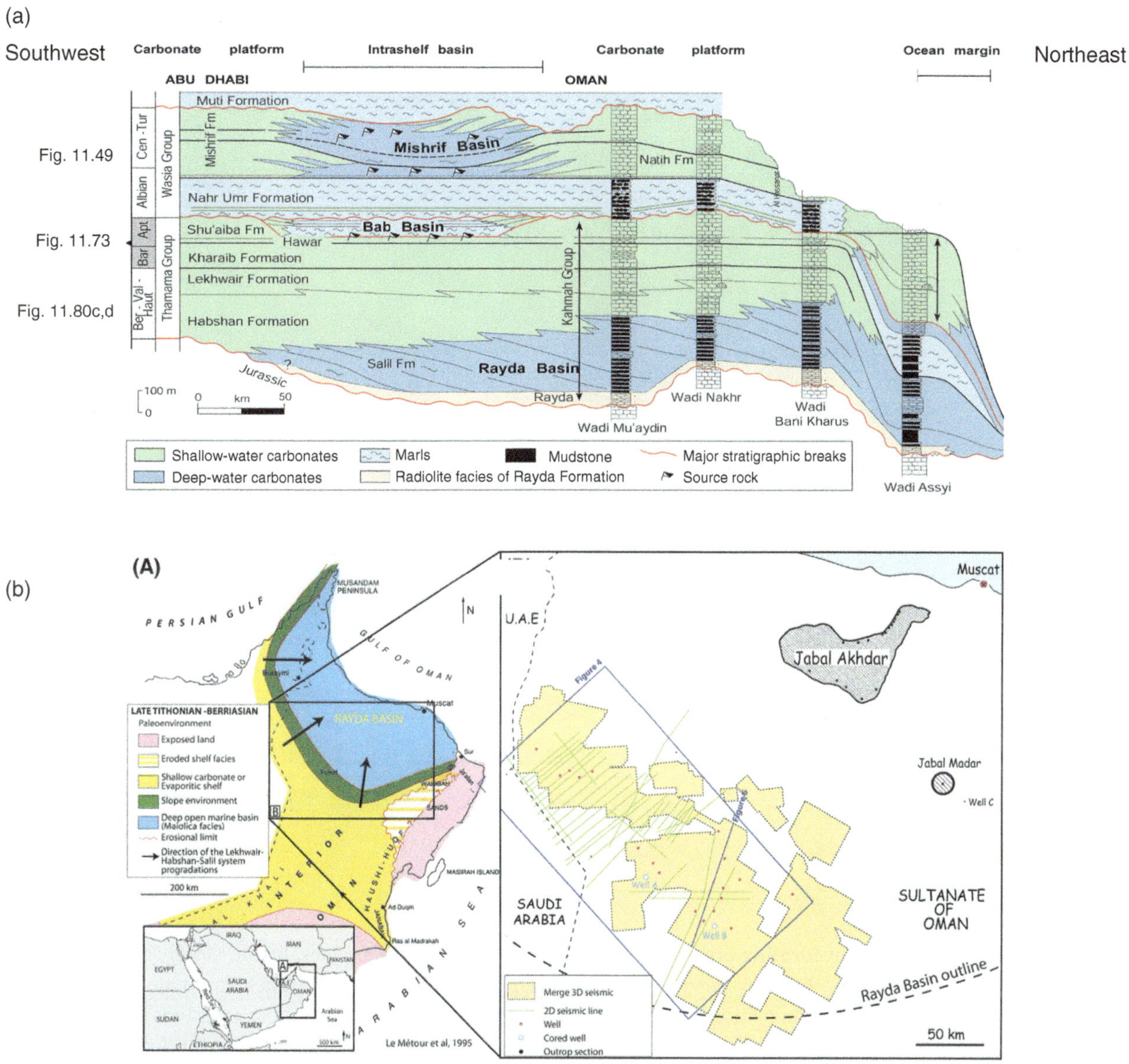

Figure 11.80 Lower Cretaceous depositional patterns on the Arabian Plate with contrasting depositional styles. (a) Schematic SW-NE stratigraphic cross section of the Eastern Arabian Plate, UAE and Oman (*Source:* Adapted from van Buchem et al. 2002b; Pratt and Smewing 1993), illustrating the overall evolution of a prograding, flat-topped, carbonate shelf (Salil, Habshan, Lekhwair and Kharaib formations), to an aggrading shelf with the presence of two generations of intrashelf basins, one in the Aptian Shuaiba Formation (see also Figure 11.73) and one in the Cenomanian Natih/Mishrif Formation (see also Figure 11.49); (b) location map of a mega-merge of seismic surveys in Oman covering an area of approximately 40,000 km^2. Location of seismic line of (d) is indicated (*Source:* Dujoncquoy et al. 2018 / GeoScienceWorld); (c) interpretation of a horizon slice from this mega-merge at the level of the Habshan Formation, based on the identification of 14 sequences identified in 2D lines of which one example is shown in (d). Note in (c) the lateral disappearance of some of the sequences (e.g. sequences IX and X), and the variations of the clinoform slope angles. The sediments consist of a T-factory assemblage dominated by rudistic bivalves on the top and margins, and oolitic facies forming the steep slopes with the admixture of abundant echinoderm debris. The low angle slopes have a muddier texture with common oncoids (Adapted from Lebec 2004; Droste and van Steenwinkel 2004; Dujoncquoy et al. 2018).

The success of larger foraminifera during the early Palaeogene appears to have been driven by the gradual rise in temperatures leading up to the Eocene Climatic Maximum. Because of the vulnerability of corals to high surface-water temperatures, this global warming may have favoured larger foraminifera at the expense of corals as the main carbonate-producing component on carbonate shelves at lower latitudes (Scheibner et al. 2005).

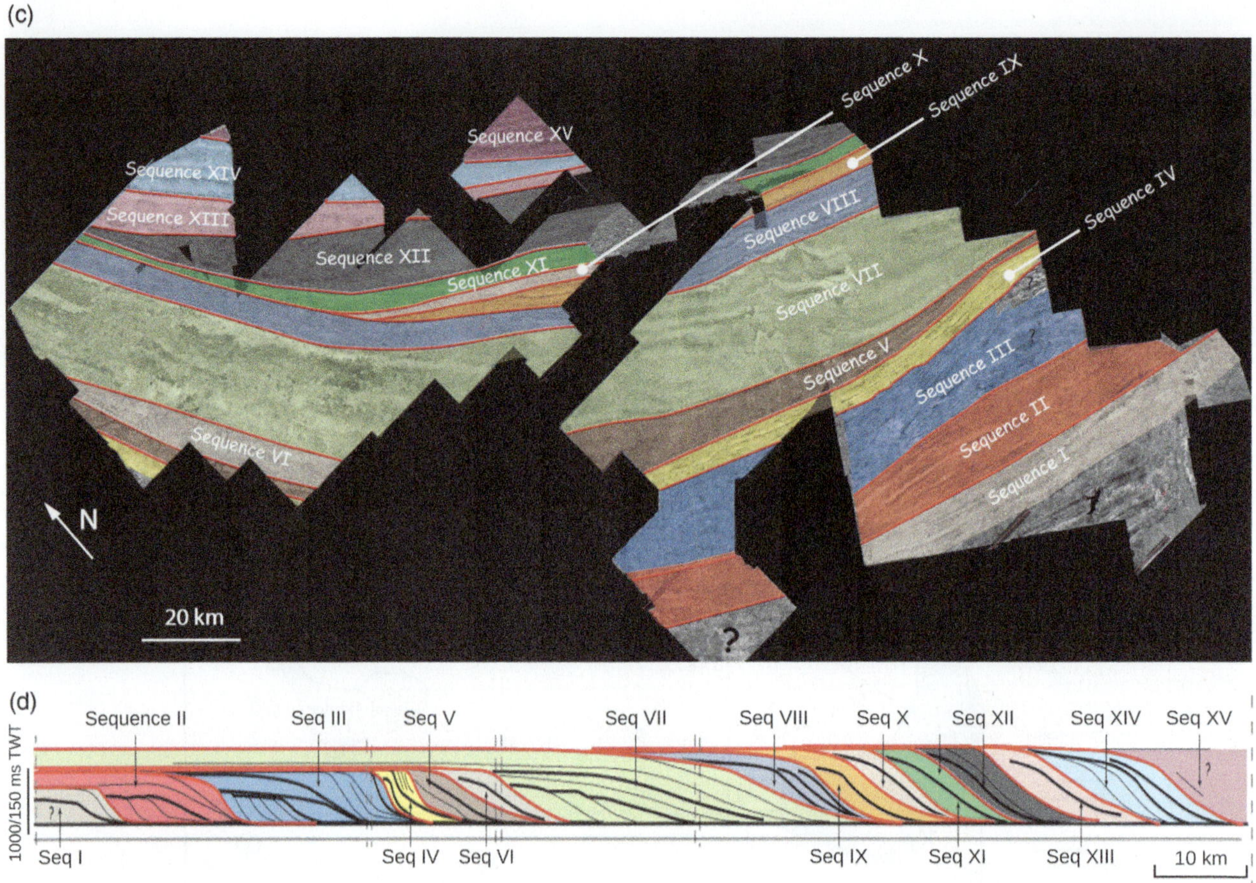

Figure 11.80 (Continued)

The resulting unavailability of rim-forming organisms at this time led to the dominance of ramp depositional profiles at the expense of platform depositional profiles (Figure 11.75).

The Paleocene–Eocene transition (Zeebe 2013) represents a defining step in this evolution of Palaeogene larger foraminifera with a rapid increase in species diversity, test size and adult dimorphism (Hottinger 1983; Scheibner et al. 2005). Enhanced oligotrophy from the late Paleocene onward favoured the diversification of K-strategist larger foraminifera. Scheibner et al. (2005) suggested that a short-term eutrophication during the Paleocene–Eocene Thermal Maximum (PETM) led to a temporary decline of extreme K-strategist larger-foraminifera species, providing opportunities for new taxa with different ecological strategies to develop. During post-PETM oligotrophic conditions, these new taxa were able to evolve rapidly and soon dominated early Eocene larger-foraminifera assemblages, whereas many Paleocene taxa gradually disappeared.

Outcrops in the Pyrenees, Spain, clearly show this transition from Paleocene limestones dominated by red algae, with minor benthic foraminifera (rotaliids) and corals, to Eocene limestones with larger foraminifera (alveolinids, nummulitids and orbitolitids) and minor corals and bivalves (Payros et al. 2000; Scheibner et al. 2005, 2007). The change in carbonate-producing biota, and their new living habitat in seagrass meadows and mangroves, also influenced the sediment accumulation mode. With reef builders virtually absent, the grainy, skeletal sediment sandbars and sand waves, displaying grain-size sorting along large-scale carbonate ramps became dominant. The global scale of this transition is testified by the many outcrop and subsurface case studies from the margins of the NeoTethys ocean that illustrate this type of sedimentation pattern, with the noted dominance of the large benthic Nummulitid foraminifera (Oman: Beavington-Penney et al. 2006; Egypt: Aigner 1982; NE-Italy: Bassi 2005; Bassi et al. 2013; Seddighi et al. 2015; Spanish Pyrenees:

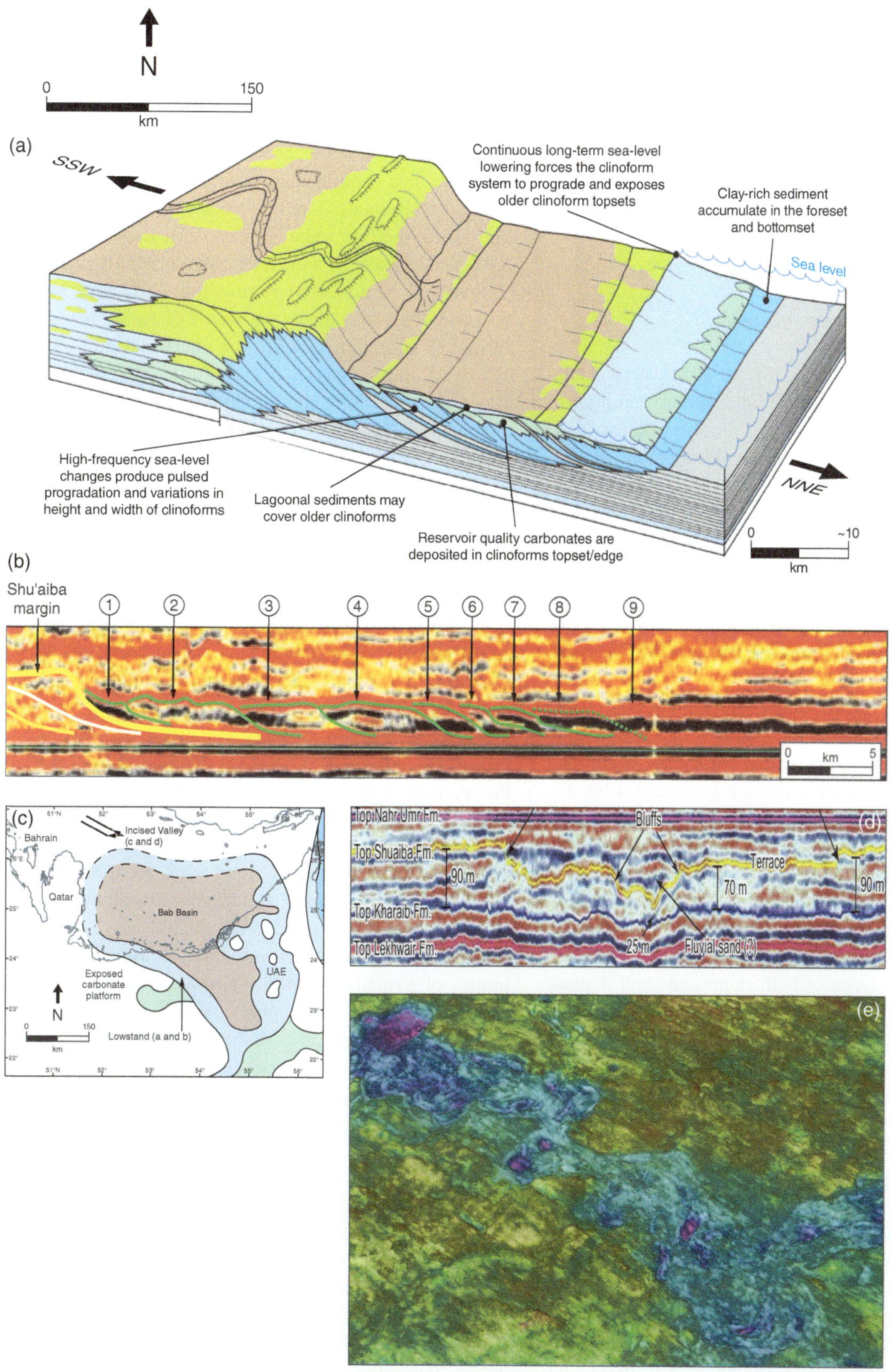

Figure 11.81 Seismic expression of Mid Cretaceous lowstand and incised valley incision (Late Aptian age Shuaiba Formation, Arabian Plate). (a) Sequence stratigraphic model for the Shuaiba Formation in the UAE (*Source:* Adapted from Pierson et al. 2010), see also Figure 11.73; (b) seismic line in the UAE illustrating onlap against the platform slope on the left, and progradation and further downstepping to the right as a result of a sea-level drop of approximately 60 m (*Source:* Pierson et al. (2010) / GeoScienceWorld); (c) palaeogeographical map of the Bab Basin during the Late Aptian sea-level lowstand, with the location of figures a,b and d,e (*Source:* Adapted from van Buchem et al. 2010a), (d) 2D seismic line illustrating an incised channel cutting into the Shuaiba Formation carbonate shelf in offshore Qatar (*Source:* Raven et al. 2010 / GeoScienceWorld); (e) horizon slice across the channel showing the incised valley system (in green) which can locally be 8 km wide and is filled with coarse grained siliciclastics (Source: Raven et al. 2010 / GeoScienceWorld). The Late Aptian sea-level fall has been observed worldwide (review in Maurer et al. 2013; Bover-Arnal et al. 2022), and the top Shuaiba exposure surface is thus a good example of a eustatically controlled sequence boundary.

Scheibner et al. 2007; Mateu-Vicens et al. 2012; Iran: Adabi et al. 2008; Tunisia: Moody 1987; Loucks et al. 1998; Racey 2001; Racey et al. 2001; Vennin et al. 2003; Beavington-Penney and Racey 2004; Jorry 2004; Jorry et al. 2006). See Supplementary Information 11.S.6 for a selected case study list.

11.7.2.7 Oligocene–Miocene–Pliocene–Pleistocene–Holocene

The Oligocene to recent interval was a time of overall global cooling marking the transition to full icehouse conditions. The cooling occurred stepwise, and a number of substantial and abrupt climatic changes occurred, notably at the beginning of the Oligocene and in the middle of the Miocene (Zachos et al. 2001; Mutti et al. 2010; Coxall and Wilson 2011; Reolid et al. 2019a; Waite et al. 2023). A combination of plate tectonic rearrangements and concomitant ocean circulation changes provided the conditions for Antarctic ice cap growth. The increased volumetric potential to store water on land caused glacio-eustatic sea-level fluctuations with a magnitude of up to 120 m and at a frequency in the Milanković band (Miller et al. 2005, 2020). These boundary conditions had a profound impact on the stratigraphic architecture and facies composition of the shallow water carbonate systems. Carbonate deposition at this time was characterised by relatively small land-attached carbonate shelves, and numerous isolated platforms, often in front of or interfingering with siliciclastic deltaic systems (Figure 11.65). A global synthesis of Oligocene and Miocene shelf and buildup geometries by Michel et al. (2020) shows the importance of antecedent topography, the presence of both ramp and shelf depositional profiles and the respective dominant carbonate factories (Figure 11.82). The carbonate shelves and isolated platforms in this glacio-eustasy dominated time-interval show a heavy imprint of diagenetic alterations and large-scale slope failures due to the frequent periods of subaerial exposure caused by fast and high magnitude sea-level fluctuations (Vahrenkamp et al. 2004; Bassant et al. 2004; Janson et al. 2011). It is only in this time interval that the T-factory as we know it today developed, comprising corals, benthic foraminifera and red algae (Scheibner and Speijer 2008). The dramatic climate variations caused rapid turnover between carbonate factories such as during the Middle Miocene Climate Optimum (MMCO; 15–17 Ma) the warmest time period of the Neogene (Holbourn et al. 2015). The MMCO saw a partial replacement of corals as the main carbonate producers by other organisms adapted to warmer and nutrient-richer waters, such as red algae and large benthic foraminifera (Halfar and Mutti 2005). The scleractinian corals diversified in the Eocene, but their productivity increased in the cooler Oligocene as did their reef-building potential during the Neogene (Perrin 2002).

During the Neogene global cooling, the benthic foraminifera were less prolific than during their Eocene heyday (Hallock et al. 2006). Coralline red algae increased in diversity during the Oligocene (Pedley 1998; Rasser and Piller 2004) and globally became dominant carbonate producers (Halfar and Mutti 2005) during the Early and Middle Miocene. This was likely the result of meso- to slightly eutrophic conditions caused by increased run-off and enhanced upwelling (Halfar and Mutti 2005). By the Late Miocene, corals were the dominant carbonate producers (Pomar and Hallock 2007). Their development has been attributed to their co-evolution with zooxanthellae (Pochon and Pawłowski 2006), and change in ocean geochemistry to support widespread aragonite precipitation through the Late Miocene (Pomar and Hallock 2008). In more general terms, these major changes in the skeletal factory that occurred throughout the Cenozoic are considered as adaptations to decreasing Ca^{2+} and pCO_2, rising Mg/Ca ratios, decreasing global temperature and increasing latitudinal and bathymetric temperature gradients (Scheibner and Speijer 2008; Stanley 2012).

Studies of Oligo-Miocene carbonates in the Mediterranean realm addressed three main themes: (i) an *ecological theme*, with a focus on the carbonate producing biota's capability to form carbonate ramps versus rimmed carbonate shelves (Franseen et al. 1991; Pomar 1991; Esteban 1996; Pomar 2001; Mateu-Vicens et al. 2008; Pomar and Kendall 2008; Bourillot et al. 2010a,b; Janson et al. 2010; Westphal et al. 2010; Brandano et al. 2017); (ii) a *combined structural-glacio-eustatic theme*, with a focus on the relative impact of structural control, substrate morphology, high magnitude sea-level fluctuations and climatic changes on the architecture of the carbonate systems (Sharp et al. 2000; Bosence 2005; Dorobek 2008; Bassant et al. 2005, 2017; Janson et al. 2010); and (iii) an *evaporite basin theme*, where the combination of basin configurations with narrow ocean gateways and the impact of high magnitude sea-level fluctuations led to the accumulation of large amounts of evaporites. The best-known example is the Messinian crisis, in the proto-Mediterranean Sea, which caused dramatic palaeoenvironmental changes (Roveri and Manzi 2006) and the abrupt transition from a coral dominated T-factory to a microbial dominated M-factory (Braga et al. 1995; Esteban et al. 1996; Bourillot et al. 2010a,b). Another example is from the Zagros Foreland basin of the Arabian Plate (Iraq and Iran), which became isolated from the ocean twice during the early Miocene (Aquitanian) due to high magnitude sea-level falls and became the locus of extensive evaporite deposition (Aqrawi et al. 2010; van Buchem et al. 2010b).

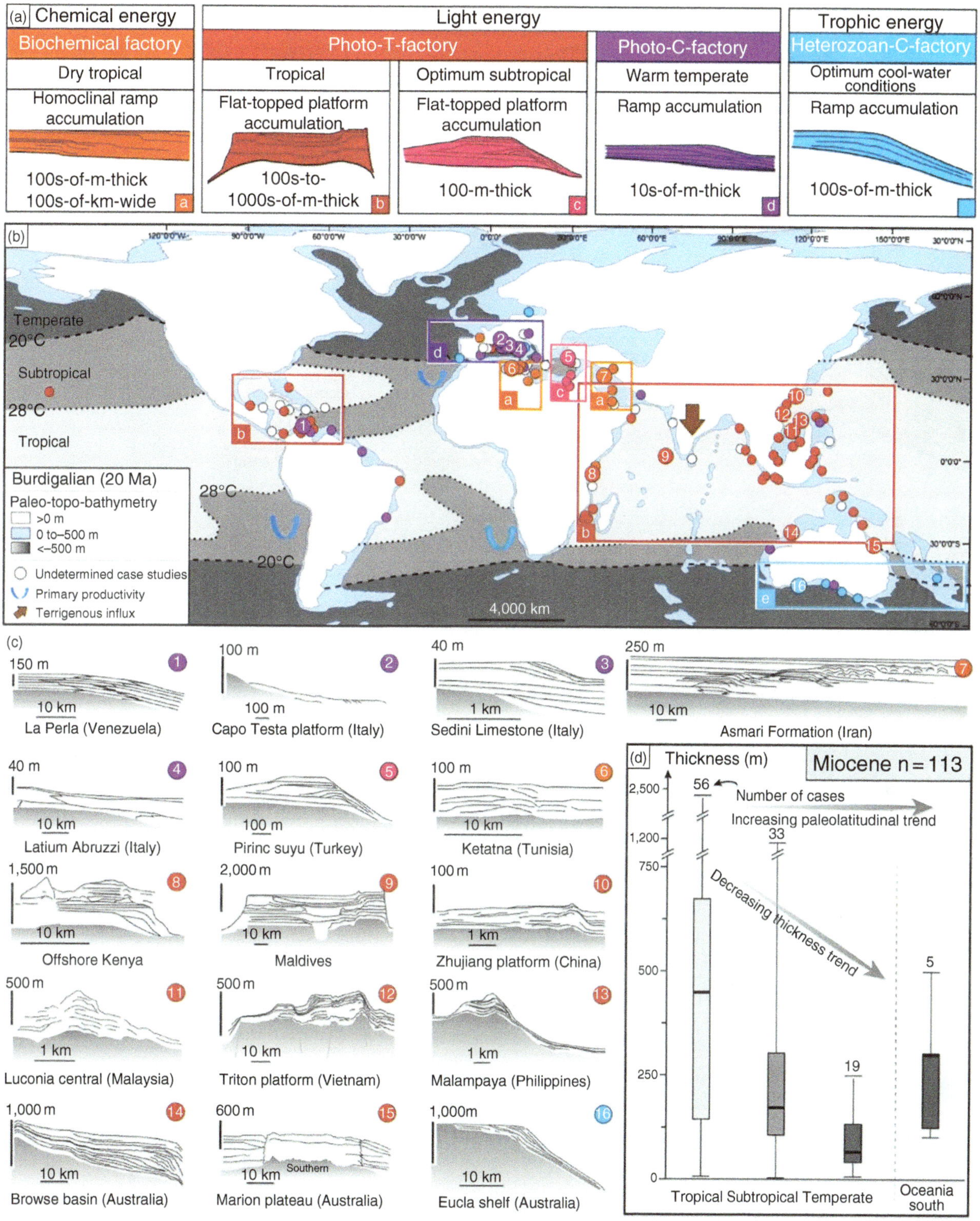

Figure 11.82 Global synthesis of Miocene land-attached and land-detached carbonate platforms (*Source:* Michel et al. 2021 / GeoScienceWorld). (a) Conceptual scheme of global trend of carbonate factories in terms of carbonate accumulation geometry and thickness; (b) case studies and main carbonate palaeoclimatic regions; (c) selected examples of carbonate stratigraphic architecture redrawn from the literature; (d) median, maximum, minimum and 75% and 25% quartile values of Miocene carbonate thickness through palaeolatitudes (see also Figure 11.19).

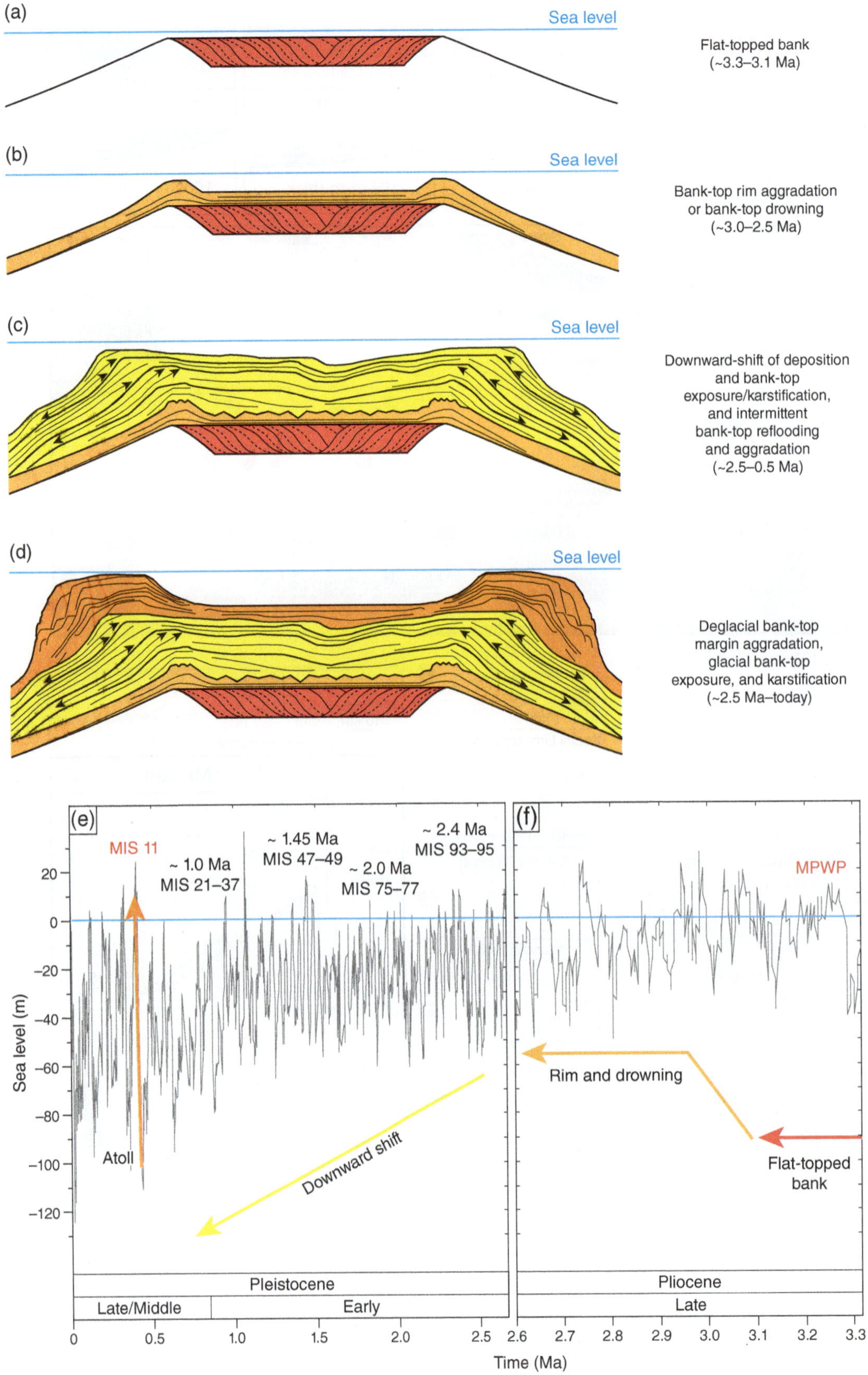

Figure 11.83 Antecedent karst model to explain origin of modern, T-factory atolls, based on current knowledge of well-established Plio-Quaternary sea-level fluctuations (Droxler and Jorry 2021). The model illustrates four different phases of modern atoll evolution (panels a–d) placed into the context of the Quaternary (panel e) and Late Pliocene (panel f) sea-level fluctuations (Miller et al. 2020). Abbreviations: Ma, million years ago; MIS, marine isotope stage; MPWP, mid-Pliocene warm period.

In the Far East, along Australia's north-western margin a 2000-km-long carbonate system existed during the middle to late Miocene. This giant T-factory reef system displays alternating phases of regional platforms and isolated pinnacle reefs controlled by the high magnitude sea-level changes (Figure 11.60; Rosleff-Soerensen et al. 2012, 2016; Belde et al. 2014, 2017; Rinke-Hardekopf et al. 2018; Rankey 2020; McCaffey et al. 2025).

In the Indo-Pacific, a great diversity of relatively small (km to deca-km scale) isolated carbonate platforms developed on antecedent topography within various passive and active tectonic settings (Fulthorpe and Schlanger 1989; Sales et al. 1997; Wilson 2002), and several occur in front of delta systems, or are stratigraphically interfingering with these (Figure 11.65; Dunn et al. 1996; Epting 1980; Saller and Vijaya 2002; Bachtel et al. 2004; Vahrenkamp et al. 2004; Fournier et al. 2005; Tcherepanov et al. 2010; Koša et al. 2015).

A recent synthesis by Droxler and Jorry (2021) of the Plio-Pleistocene history of the T-factory dominated modern atolls from the Maldives Archipelago, the tropical Pacific and southwest Indian Oceans provides a long-overdue update of Darwin's initial model (1842). These authors explain their origin as the interaction of an antecedent karst model and the well-established glacio-eustatic sea-level fluctuations during the last five million years, rooted on top of late Pliocene flat-topped isolated platforms (Figure 11.83). This interpretation highlights the stratigraphic and sedimentological value of atolls in their recording capacity of the Quaternary glacio-eustatic sea-level fluctuations, and where they have a longer sedimentary record, as valuable sea level and environmental control points in the middle of oceans. Contemporaneously, extensive C-factory deposits were laid down around the Australian continent with the typical occurrence of Bryozoan Mound complexes (Figure 11.84; James et al. 2000).

The present-day shallow water carbonate depositional environments were deposited during the Holocene sea-level rise of approximately 120 m. Their deposition on formerly exposed shelves during a transgression provides us with a snapshot in time of a system that has shown an enormous capacity to adapt and evolve ever since the first appearance of the stromatolites in the Archean, way back in the abyss of time.

11.8 Concluding Remarks

In conclusion, some reflections on outstanding research issues in marine carbonate studies:

- Even though the notion of carbonate factories can be further refined and has areas with 'shades of grey' that will require further work, we are now better equipped to map the variations of carbonate systems in space and time and establish links with changing environmental factors.
- Our understanding of the M-factory's biota, their functioning, depositional processes and present-day distribution has increased much over the last decades and highlights at the same time what still needs to be done to fully appreciate the importance and versatility of this carbonate factory.
- The relatively recent 'discovery' of a new carbonate factory (CWC factory) and its stunning global distribution begs the question whether we have missed their ancient forebears in our studies of the rock record.
- The improved understanding of the nature and operating conditions of different carbonate factories and their characteristic depositional geometries, combined with an improved age dating, has opened the way to the application of stratigraphic forward modelling. This powerful approach allows to test different scenarios of depositional conditions, carbonate factory response and impact on the 3D stratigraphic architecture (see Figures 4.40 and 4.41 for examples). A critical element in this approach is a proper understanding of the biological principles that control carbonate producing organisms. A further integration of these principles can only be beneficial, with the caution to be aware of the limits of using present-day environments and processes for the interpretation of the past.
- In this chapter, only a summary has been provided of the main evolutionary steps of the carbonate systems, and the impact this had on global (carbonate) sedimentation patterns. A far more in-depth documentation would be justified, and it is hoped that by introducing this element in this chapter it will draw further interest to this important stratigraphic feature of the carbonate systems.
- The evolution of carbonate systems is controlled and punctuated by the major earth processes and events, which is a larger picture that is slowly emerging. Further work is needed, for instance to establish the link of the carbonate factories with the global bio-geochemical cycles, and its support for the resilience of the living conditions on earth in the face of major tectonic and chemical change (Lenton 2016; Steffen et al. 2020). Such an event as the blooming of the Chalk in the late Cretaceous is one of the events alluded to in this way (Gaia Theory, Lovelock 1988).
- Finally, the marine carbonate system is, through its sensitivity to environmental change, both an accurate recorder of changes in the present-day, such as acidification, global warming, ocean current slowdown and nutrient pollution, and a valuable repository of information on the effects that similar changes have had on the marine environment in the past.

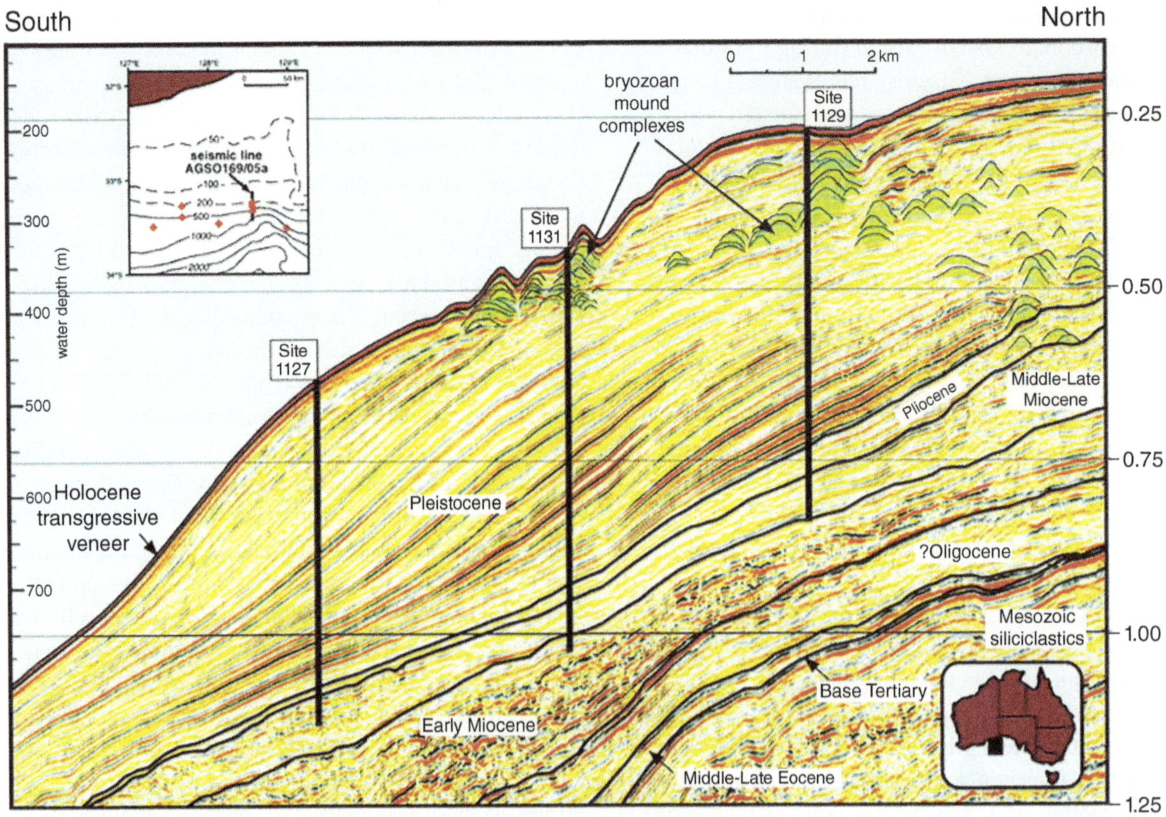

Figure 11.84 Pleistocene land-attached C-factory dominated open shelf deposits along the SW Australian margin, with bryozoan reef mounds down the slope, deposited during glacial lowstands when these were sites of nutrient rich coastal upwelling. Inset bottom right, map of Australia with study area; inset top left: detailed study area with Ocean Drilling Program Leg 182 with drill site and seismic line. The seismic image illustrates Cenozoic stratigraphy, a thick Pleistocene carbonate sediment wedge, bryozoan reef mounds and the location of the drill sites (James et al. 2000). See for a discussion on an alternative interpretation of the bryozoan reef mounds Anderskouv et al. (2010).

Further Reading

Ahr, W.M. (2008). *Geology of Carbonate Reservoirs: The Identification, Description, and Characterization of Hydrocarbon Reservoirs in Carbonate Rocks*. Hoboken, NJ: Wiley, 277 pp.

Burchette, T.P. and Wright, V.P. (1992). Carbonate ramp depositional systems. *Sedimentary Geology* 79: 3–57.

Cordes, E. and Mienis, F. (ed.) (2023). *Cold-Water Coral Reefs of the World*. Cham: Springer Nature Switzerland, 293 pp.

Dupraz, C., Reid, R.P., Braissant, O. et al. (2009). Processes of carbonate precipitation in modern microbial mats. *Earth Science Reviews* 96: 141–162.

Flügel, E. (2010). *Microfacies of Carbonate Rocks – Analysis, Interpretation and Application*. Heidelberg Dordrecht London New York: Springer, 976 pp.

Hendry, J., Burgess, P., Hunt, D. et al. (2021). *Seismic Characterization of Carbonate Platforms and Reservoirs*, Special Publications, vol. 509. London: Geological Society, London, 289 pp.

James, N.P. and Bone, Y. (2011). *Neritic Carbonate Sediments in a Temperate Realm – Southern Australia*. Dordrecht, Heidelberg/London/New York: Springer.

James, N.P. and Jones, B. (2016). *Origin of Carbonate Rocks*. AGU, Wiley, 464 pp.

Kiessling, W., Flügel, E., and Golonka, J. (ed.) (2002). *Phanerozoic Reef Patterns*. SEPM (Society for Sedimentary Geology), Special Publication no. 72, 775 pp.

Lucia, F.J. (2007). *Carbonate Reservoir Characterization*, 2e. Berlin Heidelberg New York: Springer.

Moore, C.H. (2001). *Carbonate Reservoirs: Porosity Evolution and Diagenesis in a Sequence Stratigraphic Framework*. Amsterdam, London, New York, Oxford, Paris, Shannon, Tokyo: Elsevier. 444 pp.

Sarg, J.F. (1988). Carbonate sequence stratigraphy. In: *Sea-Level Changes: An Integrated Approach*, Special Publication, vol. 42 (ed. C.K. Wilgus, B. S. Hastings, C.G.S.C. Kendall, et al.), 155–182.

Reijmer, J.J.G. (2021). Marine carbonate factories: review and update. *Sedimentology* 68: 1729–1796.

Schlager, W. (2005). *Carbonate Sedimentology and Sequence Stratigraphy*. Tulsa, Ok: SEPM (Society for Sedimentary Geology), 200 pp.

Scholle, P.A. and Ulmer-Scholle, D.S. (2003). *A Color Guide to the Petrography of Carbonate Rocks*. Tulsa, OK: American Association of Petroleum Geologists, 474 pp.

Tucker, M.E. and Jones, S.J. (2023). *Sediment Petrology*. Hoboken, NJ: Wiley, 448 pp.

Van Buchem, F.S.P., Al-Husseini, M.I., Maurer, F., and Droste, H.J. (ed.) (2010). *Barremian – Aptian Stratigraphy and Hydrocarbon Habitat of the Eastern Arabian Plate*, GeoArabia Special Publication 4. Manama, Bahrain: Gulf PetroLink (2 volumes), 614 pp.

Vennin, E., Aretz, M., Boulvain, F., and Munnecke, A. (ed.) (2007). *Facies from Palaeozoic Reefs and Bioaccumulations*. Paris: Muséum National d'Histoire Naturelle, 341 pp.

Verwer, K., Playton, T.E., and Harris, P.M. M. (ed.) (2014). *Deposits, Architecture, and Controls of Carbonate Margin, Slope, and Basinal Settings*, Special Publication, vol. 105. Tulsa, OK: SEPM (Society for Sedimentary Geology).

Wright, V.P. and Della Porta, G. (ed.) (2021). *Field Guides to Exceptionally Exposed Carbonate Outcrops*. Ghent: International Association of Sedimentologists, 792 pp.

References

Abed, A.M. and Sadaqah, R. (1998). Role of Upper Cretaceous oyster bioherms in the deposition and accumulation of high-grade phosphorites in central Jordan. *Journal of Sedimentary Research* 68: 1009–1020.

Abegg, R.E., Harris, P.M. and Loope, D.B. (Eds) (2001). *Modern and Ancient Carbonate Eolianites: Sedimentology, Sequence Stratigraphy and Diagenesis*. Tulsa: SEPM - Society for Sedimentary Geology, 207 pp.

Adabi, M., Zohdi, A., Ghabeishavi, A., and Amiri-Bakhtiyar, H. (2008). Applications of nummulitids and other larger benthic foraminifera in depositional environment and sequence stratigraphy: an example from the Eocene deposits in Zagros Basin, SW Iran. *Facies* 54: 499–512.

Adams, A. and Diamond, L.W. (2017). Early diagenesis driven by widespread meteoric infiltration of a Central European carbonate ramp: a reinterpretation of the Upper Muschelkalk. *Sedimentary Geology* 362: 37–52.

Adams, E.W. and Kenter, J.A.M. (2014). So different, yet so similar: comparing and contrasting siliciclastic and carbonate slopes. In: *Deposits, Architecture and Controls of Carbonate Margin, Slope and Basinal Settings*, SEPM Special Publication, vol. 105 (ed. K. Verwer, T.E. Playton, and P.M. Harris), 14–25. Tulsa, OK: SEPM (Society for Sedimentary Geology).

Adams, E.W. and Schlager, W. (2000). Basic types of submarine slope curvature. *Journal of Sedimentary Research* 70: 814–828.

Adey, W.H. (1978). Coral reef morphogenesis: a multidimensional model. *Science* 202: 831–837.

Agassiz, A. (1902). ART. XXIV. – an expedition to the Maldives. *American Journal of Science (1880–1910)* 13: 297.

Aguirre, J., Braga, J.C., and Martín, J.M. (2007). El Mioceno marino del Prebético occidental (Cordillera Bética, SE de España): historia del cierre del Estrecho Norbético. In: *XIII Jornadas de la Sociedad Española de Paleontología: Guía de Excursiones (Caravaca de la Cruz, 3-6 de Octubre de 2007)* (ed. J. Aguirre, M. Company, and F.J. Rodríguez-Tovar), 53–66. Granada: Instituto Geológico y Minero de España–Universidad de Granada.

Aharon, P. (1994). Geology and biology of modern and ancient submarine hydrocarbon seeps and vents: an introduction. In: *Hydrocarbon Seeps and Vents*, Geo-Marine Letters (ed. P. Aharon), 69–73. Berlin, Heidelberg: Springer Verlag.

Aharon, P., Schwarcz, H.P., and Roberts, H.H. (1997). Radiometric dating of submarine hydrocarbon seeps in the Gulf of Mexico. *Geological Society of America Bulletin* 109: 568–579.

Ahr, W.M. (1973). The carbonate ramp: an alternative to the shelf model. *Gulf Coast Association of Geological Societies* 23: 221–225.

Ahr, W.M. (2008). *Geology of Carbonate Reservoirs: The Identification, Description, and Characterization of Hydrocarbon Reservoirs in Carbonate Rocks*. Hoboken, NJ: Wiley, 277 pp.

Ahr, W.M. and Stanton, R.J. (1994). Comparative sedimentology and paleontology of waulsortian mounds and coeval level-bottom sediments of the lower lake valley formation (Lower Mississippian) in the Sacramento mountains (New Mexico, USA). *Abhandlungen der Geologischen Bundesanstalt* 50: 11–24.

Ahr, W.M. and Stanton, R.J. Jr. (1996). Constituent composition of Early Mississippian carbonate buildups and their level-bottom equivalents, Sacramento Mountains, New Mexico. *Geological Society, London, Special Publication* 107: 83–95.

Aigner, T. (1982). Calcareous tempestites: storm-dominated stratification in Upper Muschelkalk limestones (Middle Trias, SW-Germany). In: *Cyclic and Event Stratification* (ed. G. Einsele and A. Seilacher), 180–198. Berlin, Heidelberg: Springer.

Aigner, T. (1983). Storm-generated accumulation of nummulite banks in eocene of Cairo,Egypt. *American Association of Petroleum Geologists Bulletin* 67 (3): 411.

Aigner, T. (1984). Dynamic stratigraphy of epicontinental carbonates, Upper Muschelkalk limestones (Middle Triassic), South German Basin. *Neues Jahrbuch für Geologie und Paläontologie* 169: 127–159.

Aigner, T. (1985a). *Storm Depositional Systems – Dynamic Stratigraphy in*

Modern and Ancient Shallow-Marine Sequences. Berlin, Heidelberg: Springer, 174 pp.

Aigner, T. (1985b). Biofabrics as dynamic indicators in nummulite accumulations. *Journal of Sedimentary Research* 55: 131–134.

Aigner, T. and Bachmann, G.H. (1992). Sequence-stratigraphic framework of the German Triassic. *Sedimentary Geology* 80: 115–135.

Aissaoui, D.M. (1985). Botryoidal aragonite and its diagenesis. *Sedimentology* 32: 345–361.

Aitken, S.A., Collom, C.J., Henderson, C. M., and Johnston, P.A. (2002). Stratigraphy, paleoecology, and origin of Lower Devonian (Emsian) carbonate mud buildups, Hamar Laghdad, eastern Anti-Atlas, Morocco, Africa. *Bulletin of Canadian Petroleum Geology* 50: 217–243.

Al-Aasm, I. and Mansurberg, H. (ed.) (2021). *Minerals Chemical, Mineralogical and Isotopic Studies of Diagenesis of Carbonate and Clastic Sediments*. Basel: MDPI, 218 pp.

Alexeiev Dmitriy, V., Cook Harry, E., Djenchuraeva Alexandra, V., and Mikolaichuk Alexander, V. (2017). The stratigraphic, sedimentological and structural evolution of the southern margin of the Kazakhstan continent in the Tien Shan range during the Devonian to Permian. *Geological Society, London, Special Publication* 427: 231–269.

Algeo, T.J. and Wilkinson, B.H. (1988). Periodicity of mesoscale Phanerozoic sedimentary cycles and the role of Milankovitch orbital modulation. *Journal of Geology* 96: 313–322.

Algeo, T.J., Wilson, J.L., and Lohmann, K. C. (1991). Eustatic and tectonic controls on cyclic sediment accumulation patterns in lower-middle Pennsylvanian strata of the Orogrande Basin, New Mexico. In: *Geology of the Sierra Blanca, Sacramento, and Capitan Ranges, New Mexico* (ed. J.M. Barker, B.S. Kues, G.S. Austin, and S.G. Lucas), 203–212. New Mexico Geological Society 42-nd Annual Fall Field Conference Guidebook.

Allahkarampour Dill, M., Vaziri-Moghaddam, H., Seyrafian, A., and Behdad, A. (2018). Oligo-Miocene carbonate platform evolution in the northern margin of the Asmari intra-shelf basin, SW Iran. *Marine and Petroleum Geology* 92: 437–461.

Allan, J.R. and Matthews, R.K. (1982). Isotope signatures associated with early meteoric diagenesis. *Sedimentology* 29: 797–817.

Allen, J.R.L. (1982). *Sedimentary Structures, Their Character and Physical Basis – Volume 1*. Amsterdam: Elsevier Science, 592 pp.

Allen, P.A. (2017) *Sediment Routing Systems: The Fate of Sediment from Source to Sink*. Cambridge: Cambridge University Press, 407 pp.

Allen, P.A. and Bass, J.P. (1993) Sedimentology of the upper marine molasse of the Rhône-Alp Region, Eastern France: implications of basin evolution. *Eclogae Geologicae Helvetiae* 86: 121–172.

Allison, N., Finch, A.A., Webster, J.M., and Clague, D.A. (2007). Palaeoenvironmental records from fossil corals: the effects of submarine diagenesis on temperature and climate estimates. *Geochimica et Cosmochimica Acta* 71: 4693–4703.

Allwood, A.C., Walter, M.R., Kamber, B.S. et al. (2006). Stromatolite reef from the Early Archaean era of Australia. *Nature* 441: 714–718.

Al-Mojel, A., Razin, P., Le Nindre, Y.-M., and Dera, G. (2020). Shallow-marine depositional sequences in a transgressive mixed siliciclastic-carbonate system: the Early Jurassic Marrat formation from central Saudi Arabia. *Journal of African Earth Sciences* 167: 103429.

Alonso-Zarza, A.M. and Tanner, L.H. (Eds) (2006). *Paleoenvironmental Record and Applications of Calcretes and Palustrine Carbonates*. Boulder: Geological Society of America, 239 pp.

Alsharhan, A.S. and Kendall, C.G.S.C. (1994). Depositional setting of the Upper Jurassic Hith Anhydrite of the Arabian Gulf: an analog to Holocene evaporites of the United Arab Emirates and Lake MacLeod of Western Australia. *American Association of Petroleum Geologists Bulletin* 78: 1075–1096.

Amado-Filho, G.M., Moura, R.L., Bastos, A.C. et al. (2012). Rhodolith beds are major $CaCO_3$ bio-factories in the tropical South West Atlantic. *PLoS One* 7: e35171.

Amour, F., Mutti, M., Christ, N. et al. (2013). Outcrop analog for an oolitic carbonate ramp reservoir: A scaledependent geologic modeling approach based on stratigraphic hierarchy. *AAPG Bulletin* 97 (5): 845–871.

Anderskouv, K., Surlyk, F., Huuse, M. et al. (2010). Sediment waves with a biogenic twist in Pleistocene cool water carbonates, Great Australian Bight. *Marine Geology* 278: 122–139.

Anderson, T.F. and Arthur, M.A. (1983). Stable isotopes of oxygen and carbon and their application to sedimentologic and paleoenvironmental problems. In: *Stable Isotopes in Sedimentary Geology*, SEPM Short Course, vol. 10 (ed. M.A. Arthur, T.F. Anderson, I.R. Kaplan, et al.), 1–151. Tulsa, OK: Society of Economic Paleontologists and Mineralogists.

André, J.-P., Barthet, Y., Ferrandini, M. et al. (2011). The Bonifacio formation (Miocene of Corsica): transition from a wave- to tide-dominated coastal system in mixed carbonate-siliciclastic setting. *Bulletin de la Société Géologique de France* 182: 221–230.

Andréfouët, S., Genthon, P., Pelletier, B. et al. (2020). The lagoon geomorphology of pearl farming atolls in the Central Pacific Ocean revisited using detailed bathymetry data. *Marine Pollution Bulletin* 160: 111580.

Andreucci, S., Pistis, M., Funedda, A., and Loi, A. (2017). Semi-isolated, flat-topped carbonate platform (Oligo-Miocene, Sardinia, Italy): sedimentary architecture and processes. *Sedimentary Geology* 361: 64–81.

Andrieu, S., Brigaud, B., Barbarand, J. et al. (2016). Disentangling the control of tectonics, eustasy, trophic conditions and climate on shallow-marine carbonate production during the Aalenian–Oxfordian interval: from the western France platform to the western Tethyan domain. *Sedimentary Geology* 345: 54–84.

Andrieu, S., Brigaud, B., Barbarand, J., and Lasseur, E. (2017). Linking early diagenesis and sedimentary facies to sequence stratigraphy on a prograding oolitic wedge: the Bathonian of western France (Aquitaine Basin). *Marine and Petroleum Geology* 81: 169–195.

Andrieu, S., Brigaud, B., Barbarand, J. and Lasseur, E. (2018). The complex diagenetic history of discontinuities in shallow-marine carbonate rocks: New insights from high-resolution ion microprobe investigation of $\delta 18^{O}$ and $\delta 13^{C}$ of early cements. *Sedimentology* 65: 360–399.

Anselmetti, F.S. and Eberli, G.P. (1993). Controls on sonic velocity in carbonates. *Pure and Applied Geophysics PAGEOPH* 141: 287–323.

Anselmetti, F.S. and Eberli, G.P. (1997). Sonic velocity in carbonate sediments and rocks. In: *Carbonate Seismology*, SEG Geophysical Development Series, vol. 6 (ed. I. Palaz and K.J. Marfurt), 53–74.

Anselmetti, F.S. and Eberli, G.P. (2001). Sonic velocity in carbonates – a combined product of depositional lithology and diagenetic alterations. In: *Subsurface Geology of a Prograding Carbonate Platform Margin – Great Bahama Bank Results of the Bahamas Drilling Project*, SEPM Special Publication, vol. 70 (ed. R.N. Ginsburg), 193–216. Tulsa, OK: SEPM (Society for Sedimentary Geology).

Anthony, K.R.N. and Fabricius, K.E. (2000). Shifting roles of heterotrophy and autotrophy in coral energetics under varying turbidity. *Journal of Experimental Marine Biology and Ecology* 252: 221–253.

Aqrawi, A.A.M., Mahdi, T.A., Sherwani, G.H., and Horbury, A.D. (2010). Characterization of the Mid-Cretaceous Mishrif reservoir of the Southern Mesopotamian Basin, Iraq. *AAPG GEO 2010 Middle East Geoscience Conference & Exhibition*, vol. 7, 7–10, Manama, Bahrain.

Archie, G.E. (1952). Classification of carbonate reservoir rocks and petrophysical considerations. *American Association of Petroleum Geologists Bulletin* 36: 218–298.

Aretz, M. and Chevalier, E. (2007). After the collapse of stromatoporid–coral reefs – the Famennian and Dinantian reefs of Belgium: much more than Waulsortian mounds. *Geological Society, London, Special Publication* 275: 163–188.

Aretz, M., Poty, E., Devuyst, F.-X. et al. (2012). Late Tournaisian Waulsortian-like carbonate mud banks from South China (Longdianshan Hill, central Guangxi): preliminary investigations. *Geological Journal* 47: 450–461.

Arp, G., Wedemeyer, N. and Reitner, J. (2001). Fluvial Tufa Formation in a Hard-Water Creek (Deinschwanger Bach, Franconian Alb, Germany). *Facies* 44: 1–22.

Arthur, M.A. and Schlanger, S.O. (1979). Cretaceous "oceanic anoxic events" as causal factors in development of reef-reservoired giant oil fields. *American Association of Petroleum Geologists Bulletin* 63: 870–885.

Atchley, S.C. and Loope, D.B. (1993). Low-stand aeolian influence on stratigraphic completeness: Upper Member of the Hermosa Formation (Latest Carboniferous), Southeast Utah, USA. In: *Aeolian Sediments: Ancient and Modern*, IAS Special Publication, vol. 16 (ed. K. Pye and N. Lancaster), 127–149. Oxford: Blackwell Scientific Publishers/IAS.

Atchley, S. C., West, L. W., & Sluggett, J. R. (2006). Reserves growth in a mature oil field: The Devonian Leduc Formation at Innisfail field, south-central Alberta, *Canada. AAPG Bulletin* 90(8): 1153–1169.

Aubert, O. and Droxler, A.W. (1996). Seismic stratigraphy and depositional signatures of the Maldive carbonate system (Indian Ocean). *Marine and Petroleum Geology* 13: 503–536.

Aurell, M., Bádenas, B., Bosence, D.W.J., and Waltham, D.A. (1998). Carbonate production and offshore transport on a Late Jurassic carbonate ramp (Kimmeridgian, Iberian basin, NE Spain): evidence from outcrops and computer modelling. *Geological Society, London Special Publication*, (Eds V.P. Wright and T.P. Burchette) 149: 137–161.

Aurell, M., Robles, S., Bádenas, B. et al. (2003). Transgressive-regressive cycles and Jurassic paleogeography of northeast Iberia. *Sedimentary Geology* 162: 239–271.

Aurell, M., Bádenas, B., Ipas, J., and Ramajo, J. (2010). Sedimentary evolution of an Upper Jurassic epeiric carbonate ramp, Iberian Basin, NE Spain. In: *Mesozoic and Cenozoic Carbonate Systems of the Mediterranean and the Middle East: Stratigraphic and Diagenetic Reference Models*, Special Publication, vol. 329 (ed. F.S.P. Van Buchem, K.D. Gerdes, and M. Esteban), 89–111. London: Geological Society of London.

Aurell, M., Fregenal-Martínez, M., Bádenas, B. et al. (2019). Middle Jurassic–Early Cretaceous tectono-sedimentary evolution of the southwestern Iberian Basin (central Spain): major palaeogeographical changes in the geotectonic framework of the Western Tethys. *Earth-Science Reviews* 199: 102983.

Ausich, W.I. (1997). Regional encrinites: a vanished lithofacies. In: *Paleontological Events: Stratigraphic, Ecologic, and Evolutionary Implications* (ed. C.E. Brett and G.C. Baird), 509–519. NY: Columbia University Press.

Awramik, S.M. (1971). Precambrian columnar stromatolite diversity: reflection of metazoan appearance. *Science* 174: 825–827.

Bábek, O., Kalvoda, J., Cossey, P. et al. (2013). Facies and petrophysical signature of the Tournaisian/Viséan (Lower Carboniferous) sea-level cycle in carbonate ramp to basinal settings of the Wales-Brabant massif, British Isles. *Sedimentary Geology* 284–285: 197–213.

Baceta, J.I., Pujalte, V., and Bernaola, G. (2005). Paleocene coralgal reefs of the western Pyrenean basin, northern Spain: new evidence supporting an earliest Paleogene recovery of reefal ecosystems. *Palaeogeography, Palaeoclimatology, Palaeoecology* 224: 117–143.

Baceta, J., Pujalte, V., Wright, V.P., and Schmitz, B. (2011). Carbonate platform models, sea-level changes and extreme climatic events during the Paleocene-early Eocene greenhouse interval: a basin-platform-coastal plain transect across the southern Pyrenean basin. In: *Pre-Meeting Field trips Guidebook – 28th IAS Meeting. Zaragoza, Geo-Guias, vol. 7 (ed. C. Arenas, L. Pomar, and F. Colombo), 151–198. Zaragoza: Sociedad Geológica de España.*

Bachmann, M., Bassiouni, M.A.A., and Kuss, J. (2003). Timing of mid-Cretaceous carbonate platform depositional cycles, northern Sinai, Egypt. *Palaeogeography, Palaeoclimatology, Palaeoecology* 200: 131–162.

Bachtel, S.L. and Dorobek, S.L. (1998). Mississippian carbonate ramp-to-basin transitions in south-central New Mexico: sequence stratigraphic response to progressively steepening outer-ramp profiles. *Journal of Sedimentary Research* 68: 1189–1200.

Bachtel, S.L., Kissling, R.D., Martono, D. et al. (2004). Seismic stratigraphic evolution of the Miocene- Pliocene Segitiga Platform, East Natuna Sea, Indonesia: the origin, growth, and demise of an isolated carbonate platform. In: *Seismic Imaging of Carbonate Reservoirs and Systems*, AAPG Memoir, vol. 81 (ed. G.P. Eberli, J.L. Masaferro, and J.F.R. Sarg), 309–328. Tulsa, OK: American Association of Petroleum Geologists.

Back, W., Hanshaw, B.B., and Van Driel, J. N. (1984). Role of groundwater in shaping the eastern coastline of the Yucatan Peninsula, Mexico. In: *Groundwater as a Geomorphic Agent (Binghamton Symposium in Geomorphology No. 13)* (ed. R.G. LaFleur), 281–293. Boston: Allen and Unwin.

Bádenas, B. and Aurell, M. (2010). Facies models of a shallow-water carbonate ramp based on distribution of non-skeletal grains (Kimmeridgian, Spain). *Facies* 56: 89–110.

Bádenas, B. and Aurell, M. (2018). The Down-dip preferential sequence record of orbital cycles in greenhouse carbonate ramps: examples from the jurassic of the Iberian Basin (NE Spain), Chapter 7. In: *Stratigraphy & Timescales*, vol. 3 (ed. M. Montenari), 285–325. Cambridge, USA: Academic Press.

Bádenas, B., Aurell, M., and Gröcke, D.R. (2005). Facies analysis and correlation of high-order sequences in middle–outer ramp successions: variations in exported carbonate on basin-wide δ 13Ccarb (Kimmeridgian, NE Spain). *Sedimentology* 52: 1253–1275.

Bádenas, B., Pomar, L., Aurell, M., and Morsilli, M. (2012). A facies model for internalites (internal wave deposits) on a gently sloping carbonate ramp (Upper Jurassic, Ricla, NE Spain). *Sedimentary Geology* 271–272: 44–57.

Baechle, G.T., Colpaert, A., Eberli, G.P., and Weger, R.J. (2008a). Effects of microporosity on sonic velocity in carbonate rocks. *The Leading Edge* 27: 1012–1018.

Baechle, G.T., Eberli, G.P., Boyd, A. et al. (2008b). Oomoldic carbonates: pore structure and fluid effects on sonic velocity. *SEG Technical Program Expanded Abstracts 2008: SEG Technical Program Expanded Abstracts*, 1660–1664. Society of Exploration Geophysicists.

Bahamonde, J., Colmenero, J.R., and Heredia, N. (1988). Morfología de un margen de plataforma carbonatada en el Carbonífero Superior de la Zona Cantábrica. *Geogaceta* 5: 48–50.

Bahamonde, J.R., Colmenero, J.R., and Vera, C. (1997). Growth and demise of Late Carboniferous carbonate platforms in the eastern Cantabrian Zone, Asturias, northwestern Spain. *Sedimentary Geology* 110: 99–122.

Bahamonde, J.R., Vera, C., and Colmenero, J.R. (2000). A steep-fronted Carboniferous carbonate platform: clinoformal geometry and lithofacies (Picos de Europa, NW Spain). *Sedimentology* 47: 645–664.

Bahamonde, J.R., Kenter, J.A.M., Della Porta, G. et al. (2004). Lithofacies and depositional processes on a high, steep-margined Carboniferous (Bashkirian-Moscovian) carbonate platform slope, Sierra del Cuera, NW Spain. *Sedimentary Geology* 166: 145–156.

Bailly, C., Adelinet, M., Hamon, Y., and Fortin, J. (2019). Combined controls of sedimentology and diagenesis on seismic properties in lacustrine and palustrine carbonates (Upper Miocene, Samos Island, Greece). *Geophysical Journal International* 219: 1300–1315.

Bailly, C., Kernif, T., Hamon, Y. et al. (2022). Controlling factors of acoustic properties in continental carbonates: implications for high-resolution seismic imaging. *Marine and Petroleum Geology* 137: 105518.

Bally, A.W. (ed.) (1989). *Atlas of Seismic Stratigraphy*, vol. 3, 1–243. Tulsa, OK: American Association of Petroleum Geologists.

Banner, J.L. (2004). Radiogenic isotopes: systematics and applications to earth surface processes and chemical stratigraphy. *Earth-Science Reviews* 65: 141–194.

Barlow, E., Van Kranendonk, M.J., Yamaguchi, K.E. et al. (2016). Lithostratigraphic analysis of a new stromatolite–thrombolite reef from across the rise of atmospheric oxygen in the Paleoproterozoic Turee Creek Group, Western Australia. *Geobiology* 14: 317–343.

Barnaby, R.J. and Ward, W.B. (2007). Outcrop analog for mixed siliciclastic–carbonate ramp reservoirs – stratigraphic hierarchy, facies architecture, and geologic heterogeneity: Grayburg Formation, Permian Basin, U.S.A. *Journal of Sedimentary Research* 77: 34–58.

Barrett, S.J. and Webster, J.M. (2017). Reef Sedimentary Accretion Model (ReefSAM): Understanding coral reef evolution on Holocene time scales using 3D stratigraphic forward modelling. *Marine Geology* 391: 108–126.

Bartley, J.K., Knoll, A.H., Grotzinger, J.P. and Sergeev, V.N. (2000) Lithification and fabric genesis in precipitated stromatolites and associated peritidal carbonates, Mesoproterozoic Billyakh Group, Siberia. In: *Carbonate Sedimentation and Diagenesis in the Evolving Precambrian World* (Eds J.P. Grotzinger and N.P. James), SEPM Special Publication 67, pp. 60–73. SEPM (Society for Sedimentary Geology), Tulsa, OK

Bartley, J.K., Kah, L.C., McWilliams, J.L., and Stagner, A.F. (2007). Carbon isotope chemostratigraphy of the Middle Riphean type section (Avzyan Formation, Southern Urals, Russia): signal recovery in a fold-and-thrust belt. *Chemical Geology* 237: 211–232.

Bartley, J.K., Firmin, S.M., and Berger, J.T. (2022). Stromatolites and calcitized evaporite in a hypersaline playa lake: Rossport Formation (Mesoproterozoic, Ontario). *The Depositional Record* 8: 127–142.

Basilone, L. (2022). Jurassic–Cretaceous intraplatform basins from NW Sicily fold and thrust belt: implications for oblique rifting of the Southern Tethyan margin. *Sedimentary Geology* 440: 106255.

Basilone, L., Sulli, A., and Gasparo Morticelli, M. (2016). Integrating facies and structural analyses with subsidence history in a Jurassic–Cretaceous intraplatform basin: outcome for paleogeography of the Panormide Southern Tethyan margin (NW Sicily, Italy). *Sedimentary Geology* 339: 258–272.

Bassant, P. (1999). The high-resolution stratigraphic architecture and evolution of the Burdigalian carbonate-siliciclastic sedimentary systems of the Mut Basin, Turkey. In: *Doctor rerum naturalium, l'Université de Fribourg*. Fribourg: Suisse, 269 pp.

Bassant, P., Van Buchem, F., Strasser, A., and Lomando, A. (2004). A comparison of two Early Miocene carbonate margins: the Zhujiang Carbonate Platform (subsurface, South China Sea) and the Pirin Platform (outcrop, Southern Turkey). In: *Integration of Outcrop and Modern Analogs in Reservoir Modeling*, AAPG Memoir, vol. 80 (ed. G.M. Grammer, P.M. Harris, and G.P. Eberli), 153–170. Tulsa, OK: AAPG.

Bassant, P., van Buchem, F.S.P., Strasser, A., and Görür, N. (2005). The stratigraphic architecture and evolution of the Burdigalian carbonate–siliciclastic sedimentary systems of the Mut Basin, Turkey. *Sedimentary Geology* 173: 187–232.

Bassi, D. (2005). Larger foraminiferal and coralline algal facies in an Upper Eocene storm-influenced, shallow-water carbonate platform (Colli Berici, north-eastern Italy). *Palaeogeography, Palaeoclimatology, Palaeoecology* 226: 17–35.

Bassi, D., Nebelsick, J.H., Puga-Bernabéu, Á. And Luciani, V. (2013) Middle Eocene Nummulites and their offshore re-deposition: a case study from the Middle Eocene of the Venetian area, northeastern Italy. *Sedimentary Geology*, 297, 1–15.

Bathurst, R.G.C. (1975). *Carbonate Sediments and Their Diagenesis*. Amsterdam: Elsevier, 658 pp.

Bathurst, R.G.C. (1983). Early diagenesis of carbonate sediments. In: *Sediment Diagenesis* (ed. A. Parker and B.W. Sellwood), 349–377. Dordrecht: Springer Netherlands.

Bauch, T., Reijmer, J.J.G., McNeill, D.F., and Schäfer, P. (2011). Development of a Pliocene mixed-carbonate siliciclastic reef (Limon, Costa Rica). *Sedimentary Geology* 239: 37–47.

Baud, A., Cirilli, S., and Marcoux, J. (1997). Biotic response to mass extinction: the lowermost Triassic microbialites. *Facies* 36: 238–242.

Baud, A., Richoz, S., and Marcoux, J. (2005). Calcimicrobial cap rocks from the basal Triassic units: western Taurus occurrences (SW Turkey). *Comptes Rendus Palevol* 4: 569–582.

Baud, A., Richoz, S., and Pruss, S. (2007). The lower Triassic anachronistic carbonate facies in space and time. *Global and Planetary Change* 55: 81–89.

Baud, A., Richoz, S., Brandner, R. et al. (2021). Sponge takeover from end-Permian mass extinction to early Induan time: records in central Iran microbial buildups. *Frontiers in Earth Science* 9: https://doi.org/10.3389/feart.2021.586210.

Bayet-Goll, A., Daraei, M., and Seginsara, M.I. (2022). Palaeogeographic reconstruction and sequence architecture of the middle-upper Jurassic successions of Hawraman Basin (NW Iran): implications for tectono-depositional processes of the northeastern passive margin of the Arabian Plate. *Geological Journal* 57: 2058–2093.

Beauchamp, B. (1994). Permian climatic cooling in the Canadian Arctic. In: *Pangea: Paleoclimate, Tectonics, and Sedimentation During Accretion, Zenith, and Breakup of a Supercontinent*, vol. 288 (ed. G.O. Klein), 229–246. Boulder, CO: Geological Society of America.

Beauchamp, B. and Baud, A. (2002). Growth and demise of Permian biogenic chert along northwest Pangea: evidence for end-Permian collapse of thermohaline circulation. *Palaeogeography, Palaeoclimatology, Palaeoecology* 184: 37–63.

Beauchamp, B. and Desrochers, A. (1997). Permian warm- to very cold-water carbonates and cherts in Northwest Pangea. In: *Cool-Water Carbonates*, Society for Sedimentary Geology Special Publication, vol. 56 (ed. N.P. James and J.A.D. Clarke), 327–348. Tulsa, OK: SEPM (Society for Sedimentary Geology).

Beauchamp, B. and Grasby, S.E. (2012). Permian lysocline shoaling and ocean acidification along NW Pangea led to carbonate eradication and chert expansion. *Palaeogeography, Palaeoclimatology, Palaeoecology* 350–352: 73–90.

Beauchamp, E.G., Trevors, J.T., and Paul, J.W. (1989). Carbon sources for bacterial denitrification. In: *Advances in Soil Science*, vol. 10 (ed. B.A. Stewart), 113–142. NY: Springer.

Beauchamp, B., Shultz, C.V., and Anderson, K.D. (2014). Warm- vs. cool-water carbonate factories and adjacent slopes: Pennsylvanian-Early Permian Sverdrup Basin, Arctic Canada. In: *Deposits, Architecture, and Controls of Carbonate Margin, Slope and Basinal Settings*, vol. 105 (ed. K. Verwer, T.E. Playton, and P.M. Harris), 114–141.

Tulsa, OK: SEPM (Society for Sedimentary Geology).

Beavington-Penney, S.J. and Racey, A. (2004). Ecology of extant nummulitids and other larger benthic foraminifera: applications in palaeoenvironmental analysis. *Earth-Science Reviews* 67: 219–265.

Beavington-Penney, S.J., Paul Wright, V., and Woelkerling, W.J. (2004). Recognising macrophyte-vegetated environments in the rock record: a new criterion using 'hooked' forms of crustose coralline red algae. *Sedimentary Geology* 166: 1–9.

Beavington-Penney, S.J., Wright, V.P., and Racey, A. (2005). Sediment production and dispersal on foraminifera-dominated early Tertiary ramps: the Eocene El Garia Formation, Tunisia. *Sedimentology* 52: 537–569.

Beavington-Penney, S.J., Wright, V.P., and Racey, A. (2006). The middle eocene seeb formation of Oman: an investigation of acyclicity, stratigraphic completeness, and accumulation rates in shallow marine carbonate settings. *Journal of Sedimentary Research* 76: 1137–1161.

Beier, J.A. (1985). Diagenesis of Quaternary Bahamian beachrock; petrographic and isotopic evidence. *Journal of Sedimentary Research* 55: 755–761.

Belde, J., Back, S., and Reuning, L. (2014). Three-dimensional seismic analysis of sediment-waves and related geomorphological features on a carbonate shelf exposed to large amplitude internal waves, Browse Basin region, Australia. *Sedimentology* 67: 87–109.

Belde, J., Back, S., Bourget, J., and Reuning, L. (2017). Oligocene and Miocene carbonate platform development in the Browse Basin, Australian Northwest Shelf. *Journal of Sedimentary Research* 87: 795–816.

Belopolsky, A. and Droxler, A. (2003). Imaging Tertiary carbonate system – the Maldives, Indian Ocean: insights into carbonate sequence interpretation. *The Leading Edge* 2003: 646–652.

Belopolsky, A.V. and Droxler, A.W. (2004). Seismic expressions and interpretations of carbonate sequences: The Maldives carbonate platform, equatorial Indian Ocean. *American Association of Petroleum Geologists Studies in Geology 49*. Tulsa: American Association of Petroleum Geologists, 46 pp.

Belka, Z. (1998). Early Devonian Kess-Kess carbonate mud mounds of the eastern Anti-Atlas (Morocco), and their relation to submarine hydrothermal venting. *Journal of Sedimentary Research* 68: 368–377.

Benedetti, M.F., Menard, O., Noack, Y., Carvalho, A. and Nahon, D. (1994). Water-rock interactions in tropical catchments: field rates of weathering and biomass impact. *Chemical Geology* 118: 203–220.

Benton, M.J. (2015). *When Life Nearly Died: The Greatest Mass Extinction of All Time* (Revised edition). Thames & Hudson, 352 pp.

Benzerara, K., Skouri-Panet, F., Li, J. et al. (2014). Intracellular Ca-carbonate biomineralization is widespread in cyanobacteria. *Proceedings of the National Academy of Sciences* 111: 10933–10938.

Berger, W.H. (1978). Deep-sea carbonate: pteropod distribution and the aragonite compensation depth. *Deep Sea Research* 25: 447–452.

Bergman, K.L., Westphal, H., Janson, X. et al. (2010). Controlling parameters on facies geometries of the Bahamas, an isolated carbonate platform environment. In: *Carbonate Depositional Systems: Assessing Dimensions and Controlling Parameters – The Bahamas, Belize and Persian/Arabian Gulf* (ed. H. Westphal, B. Riegl, and G.P. Eberli), 5–80. Dordrecht, Heidelberg, London, New York: Springer Verlag.

Berkeley, A. and Rankey, E.C. (2012). Progradational Holocene carbonate tidal flats of Crooked Island, south-east Bahamas: an alternative to the humid channelled belt model. *Sedimentology* 59: 1902–1925.

Berkowski, B., Jakubowicz, M., Belka, Z., Król, J.J. and Zapalski, M.K. (2019). Recurring cryptic ecosystems in Lower to Middle Devonian carbonate mounds of Hamar Laghdad (Anti-Atlas, Morocco). *Palaeogeography, Palaeoclimatology, Palaeoecology* 523: 1–17.

Bernecker, M. and Weidlich, O. (1990). The Danian (Paleocene) coral limestone of Fakse, Denmark: a model for ancient aphotic, azooxanthellate coral mounds. *Facies*, 22, 103–138.

Berner, R. A. (1974). Kinetic models for the early diagenesis of nitrogen, sulfur, phosphorus, and silicon in anoxic marine sediments. Sea 5: 427–450.

Berner, R.A. (1975). The role of magnesium in the crystal growth of calcite and aragonite from sea water. *Geochimica et Cosmochimica Acta* 39: 489–504.

Berner, R. A. (2006). Inclusion of the weathering of volcanic rocks in the GEOCARBSULF model. *American Journal of Science* 306(5): 295–302.

Betzler, C., Brachert, T.C., Braga, J.C., and Martin, J.M. (1997a). Nearshore, temperate, carbonate depositional systems (lower Tortonian, Agua Amarga Basin, southern Spain): implications for carbonate sequence stratigraphy. *Sedimentary Geology* 113: 27–53.

Betzler, C., Brachert, T.C., and Nebelsick, J. (1997b). The warm temperate carbonate province: a review of the facies, zonations, and delimitations. *Courier Forschungsinstitut Senckenberg* 201: 83–99.

Betzler, C., Pawellek, T., Abdullah, M., and Kossler, A. (2007). Facies and stratigraphic architecture of the Korallenoolith Formation in North Germany (Lauensteiner Pass, Ith Mountains). *Sedimentary Geology* 194: 61–75.

Betzler, C., Fürstenau, J., Lüdmann, T. et al. (2013a). Sea-level and ocean-current control on carbonate-platform growth, Maldives, Indian Ocean. *Basin Research* 25: 172–196.

Betzler, C., Lüdmann, T., Hübscher, C., and Fürstenau, J. (2013b). Current and sea-level signals in periplatform ooze (Neogene, Maldives, Indian Ocean). *Sedimentary Geology* 290: 126–137.

Betzler, C., Lindhorst, S., Eberli, G.P. et al. (2014). Periplatform drift: the combined result of contour current and off-bank transport along carbonate platforms. *Geology* 42: 871–874.

Betzler, C., Hübscher, C., Lindhorst, S. et al. (2016). Lowstand wedges in carbonate platform slopes (Quaternary, Maldives, Indian Ocean). *The Depositional Record* 2: 196–207.

Betzler, C., Eberli, G., Alvarez Zarikian, C. et al. (2017a). Expedition 359 Summary. In: *Proceedings of the International Ocean Discovery Program, 359* (ed. C. Betzler, G. Eberli, C.A. Zarikian, and the Expedition 359 Scientists). College Station, TX, USA: International Ocean Discovery Program.

Betzler, C., Eberli, G.P., Alvarez Zarikian, C. A., and the Expedition 359 Scientists (2017b). *Maldives Monsoon and Sea Level*. College Station, TX, USA: International Ocean Discovery Program.

Betzler, C., Eberli, G.P., Lüdmann, T. et al. (2018). Refinement of Miocene sea level and monsoon events from the sedimentary archive of the Maldives (Indian Ocean). *Progress in Earth and Planetary Science* 5 (1): 1–18.

Beukes, N.J. (1980). Lithofacies and stratigraphy of the Kuruman and Griquatown iron-formations, Northern Cape Province, South Africa. *South African Journal of Geology* 83: 69–86.

Beukes, N.J. and Smit, C.A. (1987). New evidence for thrust faulting in Griqualand West, South Africa: implications for stratigraphy and the age of red beds. *South African Journal of Geology* 90: 378–394.

Bhattacharjee, S., Minzoni, M. and Prather, B.E. (2024). Investigating the controls on architecture and facies distribution of a carbonate ramp to shelf system: Insights from stratigraphic forward modeling of the Jurassic Smackover Formation, Gulf of Mexico. *Marine and Petroleum Geology* 160: 106622.

Biddle, K.T., Schlager, W., Rudolph, K.W., and Bush, T.L. (1992). Seismic model of a progradational carbonate platform, Picco di Vallandro, Dolomite Alps, Northern Italy. *American Association of Petroleum Geologists Bulletin* 76: 14–30.

Bischoff, J.L. (1969). Red Sea geothermal brine deposits: their mineralogy, chemistry, and genesis. In: *Hot Brines and Recent Heavy Metal Deposits in the Red Sea: A Geochemical and Geophysical Account* (ed. E.T. Degens and D.A. Ross), 368–401. Berlin, Heidelberg: Springer.

Bjerager, M. and Surlyk, F. (2007a). Danian cool-water bryozoan mounds at Stevns Klint, Denmark – a new class of non-cemented skeletal mounds. *Journal of Sedimentary Research* 77: 634–660.

Bjerager, M. and Surlyk, F. (2007b). Benthic palaeoecology of Danian deep-shelf bryozoan mounds in the Danish Basin. *Palaeogeography, Palaeoclimatology, Palaeoecology* 250: 184–215.

Bjerager, M., Surlyk, F., Lykke-Andersen, H. et al. (2010). Danian cool-water coral reefs in southern Scandinavia localised over seafloor highs. *Marine and Petroleum Geology* 27: 455–466.

Blendinger, W. (2001). Triassic carbonate buildup flanks in the Dolomites, northern Italy: breccias, boulder fabric and the importance of early diagenesis. *Sedimentology* 48: 919–933.

Blendinger, W., Bowlin, B., Zijp, F.R., and Ekroll, M. (1997). Carbonate buildup flank deposits: an example from the Permian (Barents Sea, northern Norway) challenges classical facies models. *Sedimentary Geology* 112: 89–103.

Boardman, M.R., Neumann, A.C., Baker, P.A. et al. (1986). Banktop responses to Quaternary fluctuations in sea level recorded in periplatform sediments. *Geology* 14: 28–31.

Boggs, S. (2001). *Principles of Sedimentology and Stratigraphy*. New Jersey: Pearson Prentice Hall, 726 pp.

Bohrmann, G., Greinert, J., Suess, E., and Torres, M. (1998). Authigenic carbonates from the Cascadia Subduction Zone and their relation to gas hydrate stability. *Geology* 26: 647–650.

Bond, D.P.G. and Grasby, S.E. (2020). Late Ordovician mass extinction caused by volcanism, warming, and anoxia, not cooling and glaciation. *Geology* 48: 777–781.

Bonin, A., Vennin, E., Pucéat, E. et al. (2012). Community replacement of neritic carbonate organisms during the late Valanginian platform demise: a new record from the Provence Platform. *Palaeogeography, Palaeoclimatology, Palaeoecology* 365–366: 57–80.

Bontognali, T.R.R., Vasconcelos, C., Warthmann, R.J. et al. (2008). Microbes produce nanobacteria-like structures, avoiding cell entombment. *Geology* 36: 663–666.

Bontognali, T.R.R., Vasconcelos, C., Warthmann, R.J. et al. (2010). Dolomite formation within microbial mats in the coastal sabkha of Abu Dhabi (United Arab Emirates). *Sedimentology* 57: 824–844.

Bordenave, A., Dujoncquoy, E., Razin, P. et al. (2021). drone-based photogrammetry: new insights for sequence stratigraphic interpretation of a Jurassic oolitic ramp (Amellago, Morocco). In: *35th Meeting of Sedimentology (Prague 2021)*, Contribution ID: 965. International Association of Sedimentologists, Prague, Czech Republic.

Borgomano, J.R.F. (2000). The Upper Cretaceous carbonates of the Gargano-Murge region, southern Italy: a model of platform-to-basin transition. *American Association of Petroleum Geologists Bulletin* 84: 1561–1588.

Borgomano, J.R., Fournier, F., Viseur, S., and Rijkels, L. (2008). Stratigraphic well correlations for 3-D static modeling of carbonate reservoirs. *American Association of Petroleum Geologists Bulletin* 92: 789–824.

Borgomano, J., Lanteaume, C., Léonide, P. et al. (2020). Quantitative carbonate sequence stratigraphy: Insights from stratigraphic forward models. *American Association of Petroleum Geologists Bulletin* 104: 1115–1142.

Borrelli, C., Katz, M.E., and Toggweiler, J. R. (2021). Middle to late eocene changes of the ocean carbonate cycle. *Paleoceanography and Paleoclimatology* 36 (12): e2020PA004168.

Bosellini, A. (1984). Progradation geometries of carbonate platforms: examples from the Triassic of the Dolomites, northern Italy. *Sedimentology* 31: 1–24.

Bosellini, A. (1989). Dynamics of Tethyan carbonate platforms. In: *Controls on Carbonate Platform and Basin Development*, SEPM Special Publication, vol. 44 (ed. P.D. Crevello, J.L. Wilson, J.F. Sarg, and J.F. Read), 3–13. Tulsa, OK:

Society of Economic Paleontologists and Mineralogists.

Bosence, D.W.J. (1976). Ecological studies on two unattached coralline algae from Western Ireland. *Palaeontology* 19: 365–395.

Bosence, D.W.J. (1980). Sedimentary facies, production rates and facies models for recent coralline algal gravels, Co. Galway, Ireland. *Geological Journal* 15: 91–111.

Bosence, D. (1989a). Surface sublittoral sediments of Florida Bay. *Bulletin of Marine Science* 44: 434–453.

Bosence, D. (1989b). Biogenic carbonate production in Florida Bay. *Bulletin of Marine Science* 44: 419–433.

Bosence, D.W.J. (1995). Anatomy of a recent biodetrital mud-mound, Florida Bay, USA. In: *Carbonate Mud-Mounds – Their Origin and Evolution*, Special Publication, vol. 23 (ed. C.L.V. Monty, D. W.J. Bosence, P.H. Bridges, and B.R. Pratt), 475–493. International Association of Sedimentologists.

Bosence, D. (2005). A genetic classification of carbonate platforms based on their basinal and tectonic settings in the Cenozoic. *Sedimentary Geology* 175: 49–72.

Bosence, D.W.J. and Bridges, P.H. (1995). A review of the origin and evolution of carbonate mud-mounds. In: *Carbonate mud-mounds - their origin and evolution* (ed. C.L.V. Monty, D.W.J. Bosence, P.H. Bridges, and B.R. Pratt), Special Publication Number 23, 3–9. Ghent: International Association of Sedimentologists.

Bosence, D. W. and Pedley, H. M. (1982). Sedimentology and palaeoecology of a Miocene coralline algal biostrome from the Maltese Islands. *Palaeogeography, Palaeoclimatology, Palaeoecology* 38(1–2): 9–43.

Bosscher, H. and Schlager, W. (1992). Computer simulation of reef growth. *Sedimentology* 39: 503–512.

BouDagher-Fadel, M.K. and Price, G.D. (2010). Evolution and paleogeographic distribution of the lepidocyclinids. *Journal of Foraminiferal Research* 40: 79–108.

Boulvain, F. (1993). Sédimentologie et diagenèse des monticules micritiques F2j du Frasnien de l'Ardenne, 2 fascicules. *Professional Paper-Geological Survey of Belgium*, 260(1, 2).

Boulvain, F. (2001). Facies architecture and diagenesis of Belgian Late Frasnian carbonate mounds. *Sedimentary Geology* 145: 269–294.

Boulvain, F. (2007). Frasnian carbonate mounds from Belgium: sedimentology and palaeoceanography. In: *Palaeozoic Reefs and Bioaccumulations: Climatic and Evolutionary Controls*, Special Publication, vol. 275 (ed. J.J. Álvaro, M. Aretz, F. Boulvain, et al.), 125–142. London: Geological Society, London.

Boulvain, F., Cornet, P., da Silva, A.-C. et al. (2004). Reconstructing atoll-like mounds from the Frasnian of Belgium. *Facies* 50: 313–326.

Bouman, H.A., Platt, T., Doblin, M. et al. (2018). Photosynthesis–irradiance parameters of marine phytoplankton: synthesis of a global data set. *Earth System Science Data* 10: 251–266.

Bourget, J., Ainsworth, R.B., and Nanson, R. (2014). Origin of mixed carbonate and siliciclastic sequences at the margin of a "giant" platform during the quaternary (Bonaparte Basin, NW Australia). In: *Deposits, Architecture, and Controls of Carbonate Margin, Slope and Basinal Settings*, SEPM Special Publication, vol. 105 (ed. K. Verwer, T.E. Playton, and P.M. Harris), 157–177. Tulsa, OK: SEPM (Society for Sedimentary Geology).

Bourillot, R., Vennin, E., Kolodka, C. et al. (2009). The role of topography and erosion in the development and architecture of shallow-water coral bioherms (Tortonian–Messinian, Cabo de Gata, SE Spain). *Palaeogeography, Palaeoclimatology, Palaeoecology* 281: 92–114.

Bourillot, R., Vennin, E., Rouchy, J.-M. et al. (2010a). Structure and evolution of a Messinian mixed carbonate-siliciclastic platform: the role of evaporites (Sorbas Basin, South-east Spain). *Sedimentology* 57: 477–512.

Bourillot, R., Vennin, E., Rouchy, J.M. et al. (2010b). The end of the Messinian Salinity Crisis in the western Mediterranean: Insights from the carbonate platforms of south-eastern Spain. *Sedimentary Geology* 229: 224–253.

Bourque, P.A. (2001). Mud-mounds: do they still constitute an enigma? In: *Anatomy of Carbonate Bodies/Anatomie des corps carbonates International Meeting*, Géologie Méditerranéenne, vol. 28 (ed. M. Floquet, J. Hennuy, and J.-P. Masse), 27–32. Marseille: Géologie Méditerranéenne.

Bourque, P.-A., Madi, A., and Mamet, B.L. (1995). Waulsortian-type bioherm development and response to sea-level fluctuations, upper Visean of Bechar Basin, western Algeria. *Journal of Sedimentary Research* 65: 80–95.

Bourque, P.A., Neuweiler, F., and Boulvain, F. (2004). The mud-mound system: products and processes. *32nd Intl. Geol. Congr.*, CD-ROM Abstracts, 2, 1079, Florence, Italy.

Boussagol, P., Vennin, E., Bouton, A., et al. (2023). Quaternary lacustrine carbonate deposits of the Great Basin, USA: Impact of climate, tectonics and substrate. *Sedimentology* 70(3): 969–1007.

Bouton, A., Vennin, E., Pace, A. et al. (2016). External controls on the distribution, fabrics and mineralization of modern microbial mats in a coastal hypersaline lagoon, Cayo Coco (Cuba). *Sedimentology* 63: 972–1016.

Bover-Arnal, T., Salas, R., Moreno-Bedmar, J.A., and Bitzer, K. (2009). Sequence stratigraphy and architecture of a late Early–Middle Aptian carbonate platform succession from the western Maestrat Basin (Iberian Chain, Spain). *Sedimentary Geology* 219: 280–301.

Bover-Arnal, T., Pascual-Cebrian, E., Skelton, P.W. et al. (2015). Patterns in the distribution of Aptian rudists and corals within a sequence-stratigraphic framework (Maestrat Basin, E Spain). *Sedimentary Geology* 321: 86–104.

Bover-Arnal, T., Salas, R., Guimerà, J., and Moreno-Bedmar, J.A. (2022). Eustasy in the Aptian world: a vision from the eastern margin of the Iberian Plate. *Global and Planetary Change* 214: 103849.

Bown, P.R., Lees, J.A., and Young, J.R. (2004). Calcareous nannoplankton evolution and diversity through time. In: *Coccolithophores: From Molecular Processes to Global Impact* (ed. H.R. Thierstein and J.R. Young), 481–508. Berlin, Heidelberg: Springer.

Bown, P.R., Gibbs, S.J., Sheward, R. et al. (2014). Searching for cells: the potential of fossil coccospheres in coccolithophore research. *Journal of Nannoplankton Research* 34: 5–21.

Braaksma, H., Kenter, J.A.M., Proust, J.N. et al. (2003). Controls on acoustic properties of Upper Jurassic siliciclastic rocks (Boulonnais, northern France). *Geophysics* 68: 58–69.

Brachert, T.C. and Dullo, W.-C. (1991). Laminar micrite crusts and associated foreslope processes, Red Sea. *Journal of Sedimentary Petrology* 61: 354–363.

Brachert, T.C. and Dullo, W.-C. (1994). Micrite crusts on Ladinian foreslopes of the Dolomites seen in the light of a modern scenario from the Red Sea. *Abhandlungen der Geologischen Bundesanstalt* 50: 57–78.

Brachert, T.C., Buggisch, W., Flügel, E. et al. (1992). Controls of mud mound formation: the Early Devonian Kess-Kess carbonates of the Hamar Laghdad, Anti-Atlas, Morocco. *Geologische Rundschau* 81: 15–44.

Brack, P., Mundil, R., Oberli, F. et al. (1996). Biostratigraphic and radiometric age data question the Milankovitch characteristics of the Latemar cycles (Southern Alps, Italy). *Geology* 24: 371–375.

Braga, J.C. and Martín, J.M. (1996). Geometries of reef advance in response to relative sea-level changes in a Messinian (uppermost Miocene) fringing reef (Cariatiz reef, Sorbas Basin, SE Spain). *Sedimentary Geology* 107: 61–81.

Braga, J.C., Martin, J.M., and Riding, R. (1995). Controls on microbial dome fabric development along a carbonate-siliciclastic shelf-basin transect, Miocene, SE Spain. *PALAIOS* 10: 347–361.

Braga, J.C., Jimenez, A.P., Martin, J.M., and Rivas, P. (1996). Middle Miocene coral-oyster reefs, Murchas, Granada, Southern Spain. *SEPM: Concepts in Sedimentology and Paleontology* 5: 131–139.

Braga, J.C., Martín, J.M., Betzler, C., and Aguirre, J. (2006). Models of temperate carbonate deposition in Neogene basins in SE Spain: a synthesis. In: *Cool-Water Carbonates: Depositional Systems and Palaeoenvironmental Controls*, Special Publication, vol. 255 (ed. H.M. Pedley and G. Carannante), 121–135. London, UK: Geological Society of London.

Braithwaite, C.J.R. and Montaggioni, L.F. (2009). The Great Barrier Reef: a 700 000 year diagenetic history. *Sedimentology* 56: 1591–1622.

Braithwaite, C.J.R., Montaggioni, L.F., Camoin, G.F. et al. (2000). Origins and development of Holocene coral reefs: a revisited model based on reef boreholes in the Seychelles, Indian Ocean. *International Journal of Earth Sciences* 89: 431–445.

Brandano, M. and Corda, L. (2002). Nutrients, sea level and tectonics: constrains for the facies architecture of a Miocene carbonate ramp in central Italy. *Terra Nova* 14: 257–262.

Brandano, M., Corda, L., Tomassetti, L., and Tagliavento, M. (2016). Frequency analysis across the drowning of a Lower Jurassic carbonate platform: the Calcare Massiccio Formation (Apennines, Italy). *Marine and Petroleum Geology* 78: 606–620.

Brandano, M., Cornacchia, I., and Tomassetti, L. (2017). Global versus regional influence on the carbonate factories of Oligo-Miocene carbonate platforms in the Mediterranean area. *Marine and Petroleum Geology* 87: 188–202.

Brandano, M., Tomassetti, L., Trippetta, F., and Ruggieri, R. (2020). Facies heterogeneities and 3D porosity modelling in an Oligocene (Upper Chattian) carbonate ramp, Salento Peninsula, Southern Italy. *Journal of Petroleum Geology* 43: 191–208.

Brandner, R. and Keim, L. (2011). A 4-day geological field trip in the Western dolomites. *Geo. Alp* 8: 76–118.

Brayard, A., Vennin, E., Olivier, N. et al. (2011). Transient metazoan reefs in the aftermath of the end-Permian mass extinction. *Nature Geoscience* 4: 693–697.

Brenchley, P.J., Marshall, J.D., and Underwood, C.J. (2001). Do all mass extinctions represent an ecological crisis? Evidence from the Late Ordovician. *Geological Journal* 36: 329–340.

Brewer, P.G. and Dyrssen, D. (1985). Chemical oceanography of the Persian Gulf. *Progress in Oceanography* 14: 41–55.

Bridges, P.H. and Chapman, A.J. (1988). The anatomy of a deep water mud-mound complex to the southwest of the Dinantian platform in Derbyshire, UK. *Sedimentology* 35: 139–162.

Bridges, P.H., Gutteridge, P., and Pickard, N.A.H. (1995). The environmental setting of Early Carboniferous mud-mounds. In: *Carbonate Mud-Mounds: Their Origin and Evolution*, IAS Special Publication, vol. 23 (ed. C.L.V. Monty, D.W.J. Bosence, P.H. Bridges, and B.R. Pratt), 171–190. Chichester, West Sussex: International Association of Sedimentologists–Blackwell.

Brigaud, B., Vincent, B., Durlet, C. et al. (2010). Acoustic properties of ancient shallow-marine carbonates: effects of depositional environments and diagenetic processes (Middle Jurassic, Paris Basin, France). *Journal of Sedimentary Research* 80: 791–807.

Brigaud, B., Vincent, B., Carpentier, C. et al. (2014). Growth and demise of the Jurassic carbonate platform in the intracratonic Paris Basin (France): interplay of climate change, eustasy and tectonics. *Marine and Petroleum Geology* 53: 3–29.

Broecker, W.S. and Takahashi, T. (1966). Calcium carbonate precipitation on the Bahama Banks. *Journal of Geophysical Research (1896–1977)* 71: 1575–1602.

Bromhead, A.D., van Buchem, F.S.P., Simmons, M.D., and Davies, R.B. (2022). Sequence stratigraphy, palaeogeography and petroleum plays of the Cenomanian – Turonian succession of the Arabian Plate: an updated synthesis. *Journal of Petroleum Geology* 45: 119–161.

Brooke, B.P., Olley, J.M., Pietsch, T. et al. (2014). Chronology of Quaternary coastal aeolianite deposition and the drowned shorelines of southwestern Western Australia – a reappraisal. *Quaternary Science Reviews* 93: 106–124.

Brooks, J. and Fleet, A. (1986). *Marine Petroleum Source Rocks*. Boston, MA: Blackwell Science Publishers.

Bruce, J.R. and Toomey, D.F. (1993). Late Palaeozoic bioherm occurrences of the Finnmark Shelf, Norwegian Barents Sea: analogues and regional significance. In: *Norwegian Petroleum Society Special Publication*, vol. 2 (ed. T. O. Vorren, E. Bergsager, Ø.A. Dahl-Stamnes, et al.), 377–392. Amsterdam: Elsevier.

Brunton, F.R. and Dixon, O.A. (1994). Siliceous sponge-microbe biotic associations and their recurrence through the phanerozoic as reef mound constructors. *PALAIOS* 9: 370–387.

Brunton, F.R., Smith, L., Dixon, O.A. et al. (1998). Silurian reef episodes, changing seascapes, and paleobiogeography. *New York State Museum Bulletin* 491: 265–282.

Budd, A.F. (1988). Large-scale evolutionary patterns in the reef-coral montastrea: the role of phenotypic plasticity. *Proceedings of the 6th International Coral Reef Symposium, Australia*, 3, 393–398.

Buls, T., Anderskouv, K., Friend, P.L. et al. (2017). Physical behaviour of Cretaceous calcareous nannofossil ooze: Insight from flume studies of disaggregated chalk. *Sedimentology* 64: 478–507.

Burchette, T.P. (1993). Mishrif formation (Cenomanian–Turonian), Southern Arabian Gulf: carbonate platform growth along a Cratonic Basin Margin. In: *Cretaceous Carbonate Platforms*, vol. 56 (ed. J.A.T. Simo, R.W. Scott, and J.-P. Masse), 185–199. Tulsa, OK: American Association of Petroleum Geologists.

Burchette, T.P. and Wright, V.P. (1992). Carbonate ramp depositional systems. *Sedimentary Geology* 79: 3–57.

Burchette, T.P., Paul Wright, V., and Faulkner, T.J. (1990). Oolitic sandbody depositional models and geometries, Mississippian of southwest Britain: implications for petroleum exploration in carbonate ramp settings. *Sedimentary Geology* 68: 87–115.

Burgess, P.M. and Wright, V.P. (2003). Numerical forward modeling of carbonate platform dynamics: an evaluation of complexity and completeness in carbonate strata. *Journal of Sedimentary Research* 73: 637–652.

Burgess, P.M., Winefield, P., Minzoni, M., and Elders, C. (2013). Methods for identification of isolated carbonate buildups from seismic reflection data. *American Association of Petroleum Geologists Bulletin* 97: 1071–1098.

Burne, R.V. and Moore, L.S. (1987). Microbialites: organosedimentary deposits of benthic microbial communities. *PALAIOS* 2: 241–254.

Busson, J., Joseph, P., Mulder, T. et al. (2019). High-resolution stratigraphic forward modeling of a Quaternary carbonate margin: Controls and dynamic of the progradation. *Sedimentary Geology* 379: 77–96.

Butcher, B.P., Van de Graaff, W.J.E., Hocking, R.M., and Australia, G.S.o.W. (1984). Shark Bay – Edel, Western Australia: sheets SG/49-8 and SG/49-12 International Index. Explanatory Notes, 21 p., scale 1:250,000. Geological Survey of Western Australia, Perth, W.A., Australia.

Butler, G.P., Kendall, C.G.S.C., and Harris, P.M. (1982). Recent evaporites from the Abu Dhabi coastal flats. In: *Depositional and Diagenetic Spectra of Evaporites*, SEPM Core Workshop, vol. 3 (ed. G.R. Handford, R.G. Loucks, and G.R. Davies), 33–64. Tulsa, OK: Society of Economic Paleontologists and Mineralogists.

Butterfield, N.J., Knoll, A.H., and Swett, K. (1988). Exceptional preservation of fossils in an Upper Proterozoic shale. *Nature* 334: 424–427.

Buxton, M.W.N. and Pedley, H.M. (1989). Short Paper: a standardized model for Tethyan Tertiary carbonate ramps. *Journal of the Geological Society* 146: 746.

Cabioch, J. (1969). *Les fonds de Maërl de la Baie de Morlaix et leur peuplement végétal*, 139–161. Tome X: Cahiers De Biologie Marine.

Cabioch, G., Montaggioni, L.F., and Faure, G. (1995). Holocene initiation and development of New Caledonian fringing reefs, SW Pacific. *Coral Reefs* 14: 131–140.

Calvert, S.E. (1987). *Oceanographic Controls on the Accumulation of Organic Matter in Marine Sediments*, Special Publication, vol. 26, 137–151. Geological Society, London.

Calvert, S.E. and Pedersen, T.F. (1993). Geochemistry of Recent oxic and anoxic marine sediments: Implications for the geological record. *Marine Geology* 113: 67–88.

Cameron, L.P., Reymond, C.E., Bijma, J., Büscher, J.V., De Beer, D., Guillermic, M., Eagle, R.A., Gunnell, J., Müller-Lundin, F., Schmidt-Grieb, G.M., Westfield, I., Westphal, H. and Ries, J.B. (2022) Impacts of warming and acidification on coral calcification linked to photosymbiont loss and deregulation of calcifying fluid pH. In: *Journal of Marine Science and Engineering*, 10(8), pp. https://doi.org/10.3390/jmse10081106.

Camoin, G.F. and Hallmann, N. (ed.) (2022). *Coral Reefs and Sea-Level Change: Quaternary Records and Modelling*. Gent, Belgium: IAS, 320 pp.

Camoin, G.F., Colonna, N., Montaggioni, L.F. et al. (1997). Holocene sea level changes and reef development in the southwestern Indian Ocean. *Coral Reefs* 16: 247–259.

Campbell, M.A., Bauersachs, T., Schwark, L. et al. (2022). Salinity-driven ecology and diversity changes of heterocytous cyanobacteria in Australian freshwater and coastal-marine microbial mats. *Environmental Microbiology* 24: 6493–6509.

Canet, C., Prol-Ledesma, R.M., Melgarejo, J.-C., and Reyes, A. (2003). Methane-related carbonates formed at submarine hydrothermal springs: a new setting for microbially-derived carbonates? *Marine Geology* 199: 245–261.

Canet, C., Prol-Ledesma, R.M., Torres-Alvarado, I. et al. (2005). Silica-carbonate stromatolites related to coastal hydrothermal venting in Bahía Concepción, Baja California Sur, Mexico. *Sedimentary Geology* 174: 97–113.

Carannante, G., Esteban, M., Milliman, J. D., and Simone, L. (1988). Carbonate lithofacies as paleolatitude indicators: problems and limitations. *Sedimentary Geology* 60: 333–346.

Carey, J.S., Moslow, T.F., and Barrie, J.V. (1995). Origin and distribution of Holocene temperate carbonates, Hecate Strait, western Canada continental shelf. *Journal of Sedimentary Research, A* 65: 185–194.

Carmeille, M., Bourillot, R., Pellenard, P. et al. (2020). Formation of microbial organic carbonates during the Late Jurassic from the Northern Tethys (Amu Darya Basin, Uzbekistan): implications for Jurassic anoxic events. *Global and Planetary Change* 186: 103127.

Carpentier, C., Lathuiliere, B., and Ferry, S. (2010). Sequential and climatic framework of the growth and demise of a carbonate platform: implications for the peritidal cycles (Late Jurassic, North-eastern France). *Sedimentology* 57: 985–1020.

Carpentier, C., Vennin, E., Rouchy, J.-M. et al. (2020). Ages and stratigraphical architecture of late Miocene deposits in the Lorca Basin (Betics, SE Spain): new insights for the salinity crisis in marginal basins. *Sedimentary Geology* 405: 105700.

Carter, J.G. (ed.) (1990). *Skeletal Biomineralization: Patterns, Processes and Evolutionary Trends*, vol. 1, 2. New York, USA/London, U.K.: Van Nostrand Reinhold – Chapman & Hall vii + 832 pp.v + 101 pp.

Chabaud, L., Ducassou, E., Tournadour, E. et al. (2016). Sedimentary processes determining the modern carbonate periplatform drift of Little Bahama Bank. *Marine Geology* 378: 213–229.

Chave, K.E. (1962). Factors influencing the mineralogy of carbonate sediments. *Limnology and Oceanography* 7: 218–223.

Chen, D., Tucker, M.E., Zhu, J., and Jiang, M. (2001). Carbonate sedimentation in a starved pull-apart basin, Middle to Late Devonian, southern Guilin, South China. *Basin Research* 13: 141–167.

Chen, Y., Fu, X., Xiao, A. et al. (2016). Type and evolution of carbonate platforms in Jixian Period Mesoproterozoic: southwestern margin of Ordos Basin. *Journal of Petroleum Exploration and Production Technology* 6: 555–568.

Chevalier, J.-P. (1961). Recherches sur les madréporaires et les formations récifales miocènes de la Méditerranée occidentale. Thèse, Paris, France, 562 pp.

Chiarella, D., Longhitano, S.G., and Tropeano, M. (2017). Types of mixing and heterogeneities in siliciclastic-carbonate sediments. *Marine and Petroleum Geology* 88: 617–627.

Choquette, P.W. and Pray, L.C. (1970). Geologic nomenclature and classification of porosity in sedimentary carbonates. *American Association of Petroleum Geologists Bulletin* 54: 207–250.

Christ, N., Immenhauser, A., Amour, F. et al. (2012). Characterization and interpretation of discontinuity surfaces in a Jurassic ramp setting (High Atlas, Morocco). *Sedimentology* 59: 249–290.

Chuvashov, B.I. (1983). Permian reefs of the urals. *Facies* 8: 191–211.

Clausen, O.R. and Huuse, M. (2002). Mid-Paleocene palaeogeography of the Danish area. *Bulletin of the Geological Society of Denmark* 49: 171–186.

Coleman, M.L. (1993). Microbial processes: controls on the shape and composition of carbonate concretions. *Marine Geology* 113: 127–140.

Coleman, M.L. and Raiswell, R. (1993). Microbial mineralization of organic matter: mechanisms of self-organization and inferred rates of precipitation of diagenetic minerals. *Philosophical Transactions of the Royal Society of London. Series A: Physical and Engineering Sciences* 344: 69–87.

Coles, S.L. and Jokiel, P.L. (1992). Effects of salinity on coral reefs. In: *Pollution in Tropical Aquatic Systems* (ed. D.W. Connell and D.W. Hawker), 147–166. London: CRC Press Inc.

Collins, J., Narr, W., Harris, P.M. et al. (2014). Lithofacies, depositional environments, burial diagenesis, and dynamic field behavior in a carboniferous slope reservoir, Tengiz Field (Republic of Kazakhstan), and comparison with outcrop analogs. In: *Deposits, Architecture, and Controls of Carbonate Margin, Slope and Basinal Settings*, SEPM Special Publication, vol. 105 (ed. K. Verwer, T.E. Playton, and P.M. Harris), 50–83. Tulsa, OK: SEPM (Society for Sedimentary Geology).

Collins, L.B., Zhu, Z.R., Wyrwoll, K.H. et al. (1993). Holocene growth history of reef complex on a cool-water carbonate margin: Easter Group of the Houtman Abrolhos, Eastern Indian Ocean. *Marine Geology* 115: 29–46.

Collins, L.B., France, R.E., Rong Zhu, Z., and Wyrwoll, K.-H. (ed.) (1997). *Warm-Water Platform and Cool-Water Shelf Carbonates of the Abrolhos Shelf, Southwest Australia*, 23–36. Tulsa, OK: SEPM (Society for Sedimentary Geology).

Collins, J.F., Kenter, J.A.M., Harris, P.M. et al. (2006). Facies and reservoir-quality variations in the Late Visean to Bashkirian Outer Platform, Rim, and Flank of the Tengiz Buildup, Precaspian Basin, Kazakhstan. In: *Giant Hydrocarbon Reservoirs of the World: From Rocksto Reservoir characterization and Modeling*, AAPG Memoir, vol. 88 (ed. P.M. Harris and L.J. Weber), 55–96.

Collins, L.B., Zhu, Z.R., Wyrwoll, K.-H., and Eisenhauer, A. (2003). Late Quaternary structure and development of the northern Ningaloo Reef, Australia. *Sedimentary Geology* 159: 81–94.

Collins, L.B., Read, J.F., Hogarth, J.W., and Coffey, B.P. (2006). Facies, outcrop gamma ray and C–O isotopic signature of exposed Miocene subtropical continental shelf carbonates, North West Cape, Western Australia. *Sedimentary Geology* 185: 1–19.

Colombié, C. and Strasser, A. (2005). Facies, cycles, and controls on the evolution of a keep-up carbonate platform (Kimmeridgian, Swiss Jura). *Sedimentology* 52: 1207–1227.

Colombié, C., Lécuyer, C., and Strasser, A. (2011). Carbon- and oxygen-isotope records of palaeoenvironmental and carbonate production changes in shallow-marine carbonates

(Kimmeridgian, Swiss Jura). *Geological Magazine* 148: 133–153.

Cook, H.E., Zhemchuzhnikov, V.G., Zempolich, W.G. et al. (2002). Devonian and Carboniferous Carbonate Platform Facies in the Bolshoi Karatau, Southern Kazakhstan: Outcrop Analogs for Coeval Carbonate Oil and Gas Fields in the North Caspian Basin, Western Kazakhstan. In: *Paleozoic Carbonates of the Commonwealth of Independent States (CIS): Subsurface Reservoirs and Outcrop Analogs*, SEPM Special Publication, vol. 74 (ed. W.G. Zempolich and H.E. Cook), 81–122. Tulsa, OK: SEPM (Society for Sedimentary Geology).

Cook, H.E., Zhemchuzhnikov, V.G., Zempolich, W.G. et al. (2003). Devonian and Carboniferous Carbonate Platform Facies in the Bolshoi Karatau, Southern Kazakhstan: Outcrop Analogs for Coeval Carbonate Oil and Gas Fields in the North Caspian Basin, Western Kazakhstan. In: *Paleozoic Carbonates of the Commonwealth of Independent States (CIS): Subsurface Reservoirs and Outcrop Analogs* (ed. W.G. Zempolich and H.E. Cook), SEPM Special Publication 74, 81–122. Tulsa: SEPM Society for Sedimentary Geology.

Cooper, J.A.G. (2002). The role of extreme floods in estuary-coastal behavior: contrasts between river- and tide-dominated microtidal estuaries. *Sedimentary Geology* 150: 123–137.

Copp, I.A. (2000). Subsurface facies analysis of Devonian reef complexes, Lennard Shelf, Canning Basin, Western. Report 58, Geological Survey of Western Australia, Perth, W.A, 200 pp.

Copper, P. (1994). Ancient reef ecosystem expansion and collapse. *Coral Reefs* 13: 3–11.

Copper, P. (2001). Reefs during the multiple crises towards the Ordovician-Silurian boundary: Anticosti Island, eastern Canada, and worldwide. *Canadian Journal of Earth Sciences* 38: 153–171.

Copper, P. (2002a). Silurian and Devonian Reefs: 80 Million Years of Global Greenhouse between two Ice Ages. In: *Phanerozoic Reef Patterns*, SEPM Special Publication, vol. 72 (ed. W. Kiessling, E. Flügel, and J. Golonka), 181–238. Tulsa, OK: SEPM (Society for Sedimentary Geology).

Copper, P. (2002b). Reef development at the Frasnian/Famennian mass extinction boundary. *Palaeogeography, Palaeoclimatology, Palaeoecology* 181: 27–65.

Copper, P. and Brunton, F. (1991). A global review of Silurian reefs. *Special Papers in Palaeontology* 44: 225–259.

Cordes, E. and Mienis, F. (ed.) (2023). *Cold-Water Coral Reefs of the World.* Cham: Springer Nature Switzerland, 293 pp.

Couradeau, E., Benzerara, K., Gérard, E., Moreira, D., Bernard, S., Brown, G.E. and López-García, P. (2012). An Early-Branching Microbialite Cyanobacterium Forms Intracellular Carbonates. *Science* 336: 459–462.

Courgeon, S., Bourget, J., and Jorry, S.J. (2016a). A Pliocene–Quaternary analogue for ancient epeiric carbonate settings: the Malita intrashelf basin (Bonaparte Basin, northwest Australia). *American Association of Petroleum Geologists Bulletin* 100: 565–595.

Courgeon, S., Jorry, S.J., Camoin, G.F. et al. (2016b). Growth and demise of Cenozoic isolated carbonate platforms: new insights from the Mozambique Channel seamounts (SW Indian Ocean). *Marine Geology* 380: 90–105.

Coxall, H.K. and Wilson, P.A. (2011). Early Oligocene glaciation and productivity in the eastern equatorial Pacific: insights into global carbon cycling. *Paleoceanography* 26 (2): PA2221.

Cozzi, A. and Al-Siyabi, H.A. (2004). Sedimentology and play potential of the late Neoproterozoic Buah Carbonates of Oman. *GeoArabia* 9: 11–36.

Cozzi, A., Allen, P.A., and Grotzinger, J.P. (2004a). Understanding carbonate ramp dynamics using $\delta^{13}C$ profiles: examples from the Neoproterozoic Buah Formation of Oman. *Terra Nova* 16: 62–67.

Cozzi, A., Grotzinger, J.P., and Allen, P.A. (2004b). Evolution of a terminal Neoproterozoic carbonate ramp system (Buah Formation, Sultanate of Oman): effects of basement paleotopography. *Geological Society of America Bulletin* 116: 1367–1384.

Crame, J.A. (2002). Evolution of taxonomic diversity gradients in the marine realm: A comparison of late jurassic and recent bivalve faunas. *Paleobiology* 28: 184–207.

Cramer, B.D. and Jarvis, I.J.G.T.S. (2020). Carbon isotope stratigraphy. In Geologic Time Scale. Elsevier, 309–343 pp.

Crevello, P.D., Wilson, J.L., Sarg, J.F., and Read, J.F. (ed.) (1989). *Controls on Carbonate Platforms and Basin Development*. Tulsa, OK: SEPM (Society for Sedimentary Geology), 409 pp.

Crossland, C.J. (1988). Latitudinal comparisons of coral reef structure and function. In: *Proceedings of the 6th International Coral Reef Symposium: Vol. 1: Plenary Addresses and Status Review (ed. J.H. Choat, D. Barnes, M. Borowitzka, et al.). Townsville, Australia: ICRS.*

Crumeyrolle, P., Renaud, I., and Suiter, J. (2007). The use of two- and three-dimensional seismic to understand sediment transfer from fluvial to deepwater via sinuous channels: example from the Mahakam shelf and comparison with outcrop data (south central Pyrenees). *Geological Society, London, Special Publication* 277: 85–103.

D'Argenio, B., Fischer, A.G., Premoli Silva, I. et al. (ed.) (2004). *Cyclostratigraphy: Approaches and Case Histories*, SEPM Special Publication, vol. 81. Tulsa, OK: SEPM (Society for Sedimentary Geology), 311 pp.

Da Silva, A.-C. and Boulvain, F. (2012). Analysis of the Devonian (Frasnian) platform from Belgium: a multi-faceted approach for basin evolution reconstruction. *Basin Research* 24: 338–356.

Da Silva, A.-C., Kershaw, S., and Boulvain, F. (2011). Sedimentology and stromatoporoid palaeoecology of Frasnian (Upper Devonian) carbonate mounds in southern Belgium. *Lethaia* 44: 255–274.

Dabrio, C.J., Esteban, M., and Martin, J.M. (1981). The coral reef of Nijar,

Messinian (uppermost Miocene), Almeria Province, S.E. Spain. *Journal of Sedimentary Research* 51: 521–539.

Daëron, M., Blamart, D., Peral, M., and Affek, H.P. (2016). Absolute isotopic abundance ratios and the accuracy of Δ47 measurements. *Chemical Geology* 442: 83–96.

Daniell, J., Manoy, T., Beaman, R.J. et al. (2020). Shelf-edge delta and reef development on a mixed siliciclastic–carbonate margin, central Great Barrier Reef. *Journal of Sedimentary Research* 90: 1286–1304.

Daraei, M., Amini, A., and Ansari, M. (2015). Facies analysis and depositional environment study of the mixed carbonate–evaporite Asmari Formation (Oligo-Miocene) in the sequence stratigraphic framework, NW Zagros Iran. *Carbonates and Evaporites* 30: 253–272.

Darwin, C. (1842). XXVII. — On the Distribution of the Erratic Boulders and on the Contemporaneous Unstratified Deposits of South America. *Transactions of the Geological Society of London* 6: 415–431.

Davies, A., Gréselle, B., Hunter, S.J. et al. (2020). Assessing the impact of aquifer-eustasy on short-term Cretaceous sea-level. *Cretaceous Research* 112: 104445.

Davies, A., Hunter, S.J., Gréselle, B., Haywood, A.M. and Robson, C. (2019). Evidence for seasonality in early Eocene high latitude sea-surface temperatures. *Earth and Planetary Science Letters* 519: 274–283.

Davies, D.J., Powell, E.N., and Stanton, R. J. Jr. (1989a). Relative rates of shell dissolution and net sediment accumula–ion - a commentary: can shell beds form by the gradual accumulation of biogenic debris on the sea floor? *Lethaia* 22: 207–212.

Davies, G. R. and Smith Jr, L. B. (2006). Structurally controlled hydrothermal dolomite reservoir facies: An overview. *AAPG Bulletin* 90(11): 1641–1690.

Davies, G.R., Richards, B.C., Beauchamp, B., and Nassichuk, W.W. (1989b). Carboniferous and Permian reefs in Canada and adjacent areas. In: *Reefs, Canada and Adjacent Area*, Canadian Society of Petroleum Geologists Memoir, vol. 13 (ed. H.H.J. Geldsetzer, N.P. James, and G.E. Tebbult), 565–574. Calgary, Canada: Canadian Society of Petroleum Geologists.

Davies, P.J. (2011). Halimeda bioherms. In: *Encyclopedia of Modern Coral Reefs: Structure, Form and Process* (ed. D. Hopley), 539–549. Netherlands, Dordrecht: Springer.

Davies, P.J. and Hopley, D. (1983). Growth fabrics and growth rates of Holocene reefs in the Great Barrier Reef. *BMR Journal of Australian Geology and Geophysics* 8: 237–251.

Davies, P.J. and Kinsey, D.W. (1977). Holocene reef growth – One Tree Island, Great Barrier Reef. *Marine Geology* 24 (1): M1–M11.

Davies, P.J., Symonds, P.A., Feary, D.A., and Pigram, C.J. (1989c). The evolution of the carbonate platforms of northeast Australia. In: *Controls on Carbonate Platform and Basin Development*, Special Publication No, vol. 44 (ed. P.D. Crevello, J.L. Wilson, J.F. Sarg, and J.F. Read), 233–258. SEPM (Society of Economic Paleontologists and Mineralogists).

Davies, R.B. and Simmons, M.D. (2018). Triassic sequence stratigraphy of the Arabian Plate, Chapter 4. In: *Lower Triassic to Middle Jurassic Sequence of the Arabian Plate* (ed. M.C. Poppelreiter), 101–162. Houten: EAGE.

Davies, R.B., Simmons, M.D., Jewell, T.O., and Collins, J. (2019). Regional controls on siliciclastic input into Mesozoic depositional systems of the Arabian plate and their petroleum significance. *AAPG Memoirs* 116: 103–140. https://doi.org/10.1306/13642165M 1183798.

Davies, R.B., Casey, D.M., Horbury, A.D. et al. (2002). Early to mid-Cretaceous mixed carbonate-clastic shelfal systems: examples, issues and models from the Arabian Plate. *GeoArabia* 7: 541–598.

de Haas, H., Mienis, F., Frank, N. et al. (2009). Morphology and sedimentology of (clustered) cold-water coral mounds at the south Rockall Trough margins, NE Atlantic Ocean. *Facies* 55: 1–26.

De Boer, P.L. and Smith, D.G. (1994). *Orbital Forcing and Cyclic Sequences.* Oxford: Blackwell Scientific Publications Ltd., 559 pp.

de Haas, H., Mienis, F., Frank, N. et al. (2009). Morphology and sedimentology of (clustered) cold-water coral mounds at the south Rockall Trough margins, NE Atlantic Ocean. *Facies* 55: 1–26.

De Mol, B., Van Rensbergen, P., Pillen, S. et al. (2002). Large deep-water coral banks in the Porcupine Basin, southwest of Ireland. *Marine Geology* 188: 193–231.

De Romero, L.M., Truskowski, I.M., Bralower, T.J. et al. (2003). An integrated calcareous microfossil biostratigraphic and carbon-isotope stratigraphic framework for the La Luna Formation, Western Venezuela. *PALAIOS* 18: 349–366.

De Vargas, C., Aubry, M.-P., Probert, I.A. N., and Young, J. (2007). Origin and evolution of coccolithophores: from coastal hunters to oceanic farmers, Chapter 12. In: *Evolution of Primary Producers in the Sea* (ed. P.G. Falkowski and A.H. Knoll), 251–285. Burlington: Academic Press.

Debrenne, F., Zhuravlev, A.Y., and Kruse, P.D. (2015). General features of the Archaeocyatha. In: *Treatise on invertebrate paleontology, Part E - Porifera revised, Hypercalcified Porifera*, vol. 5 (ed. P.A. Selden, F. Debrenne, W. D. Hartman, et al.), 845–1084. Lawrence, Kansas: University of Kansas Paleontological Institute.

Dechnik, B., Bastos, A.C., Vieira, L.S. et al. (2019). Holocene reef growth in the tropical southwestern Atlantic: evidence for sea level and climate instability. *Quaternary Science Reviews* 218: 365–377.

Decho, A.W. (1990). Microbial exopolymer secretions in ocean environments: their role(s) in food webs and marine processes. *Oceanography and Marine Biology* 28: 73–153.

Dennis, K.J., Cochran, J.K., Landman, N. H. and Schrag, D.P. (2013). The climate of the Late Cretaceous: New insights from the application of the carbonate clumped isotope thermometer to Western Interior Seaway macrofossil.

Earth and Planetary Science Letters 362: 51–65.

Decho, A.W. and Gutierrez, T. (2017). Microbial extracellular polymeric substances (EPSs) in ocean systems. *Frontiers in Microbiology* 8: https://doi.org/10.3389/fmicb.2017.00922.

Della Porta, G., Kenter, J.A.M., and Bahamonde, J.R. (2002). Microfacies and paleoenvironment of Donezella accumulations across an Upper Carboniferous high-rising carbonate platform (Asturias, NW Spain). *Facies* 46: 149–168.

Della Porta, G., Kenter, J.A.M., Bahamonde, J.R. et al. (2003). Microbial boundstone dominated carbonate slope (Upper Carboniferous, N Spain): microfacies, lithofacies distribution and stratal geometry. *Facies* 49: 175–208.

Della Porta, G., Kenter, J.A.M., and Bahamonde, J.R. (2004). Depositional facies and stratal geometry of an Upper Carboniferous prograding and aggrading high-relief carbonate platform (Cantabrian Mountains, N. Spain). *Sedimentology* 51: 267–295.

Della Porta, G., Merino-Tomé, O., Kenter, J.A.M., and Verwer, K. (2012). *Control of Lower Jurassic microbial Reef Communities on Carbonate Platform Geometry (Djebel Bou Dahar, High Atlas, Morocco).* AAPG Search and Discovery Article, 50654.

Della Porta, G., Merino-Tomé, O., Kenter, J.A.M., and Verwer, K. (2014). Lower Jurassic microbial and skeletal carbonate factories and platform geometry (Djebel Bou Dahar, High Atlas, Morocco). In: *Deposits, Architecture, and Controls of Carbonate Margin, Slope, and Basinal Settings*, SEPM Special Publication, vol. 105 (ed. K. Verwer, T.E. Playton, and P.M. Harris), 238–264. Tulsa, OK: SEPM (Society for Sedimentary Geology).

Della Porta, G., Bahamonde, J.R., Kenter, J.A.M., and Verwer, K. (2017). The Sierra del Cuera (Pennsylvanian microbial platform margin) in Asturias, north Spain. *American Association of Petroleum Geologists Bulletin* 101: 543–551.

Dera, G., Brigaud, B., Monna, F. et al. (2011). Climatic ups and downs in a disturbed Jurassic world. *Geology* 39: 215–218.

Devuyst, F.-X. and Lees, A. (2001). The initiation of Waulsortian buildups in Western Ireland. *Sedimentology* 48: 1121–1148.

Dewever, B., Breesch, L., Mezini, A., and Swennen, R. (2007). Sedimentological and marine eogenetic control on porosity distribution in Upper Cretaceous carbonate turbidites (central Albania). *Sedimentology* 54: 1–22.

Diaz, M.R. and Eberli, G.P. (2019). Decoding the mechanism of formation in marine ooids: a review. *Earth-Science Reviews* 190: 536–556.

Diaz, M.R. and Eberli, G.P. (2021). Microbial contribution to early marine cementation. *Sedimentology* 69: 798–822.

Diaz, M.R., Norstrand, J.D., Eberli, G.P. et al. (2014). Functional gene diversity of oolitic sands from Great Bahama Bank. *Geobiology* 12: 231–249.

Diaz, M.R., Swart, P.K., Eberli, G.P. et al. (2015). Geochemical evidence of microbial activity within ooids. *Sedimentology* 62: 2090–2112.

Diaz, M.R., Eberli, G.P., Blackwelder, P. et al. (2017). Microbially mediated organomineralization in the formation of ooids. *Geology* 45: 771–774.

Diaz, M.R., Eberli, G.P., and Weger, R.J. (2022). Indigenous microbial communities as catalysts for early marine cements: an in vitro study. *The Depositional Record* 9: 437–456.

Dickson, A.G. (1984). pH scales and proton-transfer reactions in saline media such as sea water. *Geochimica et Cosmochimica Acta* 48: 2299–2308.

Dickson, D.R., Yepsen, J.H., and Hales, J. V. (1965). Saturated vapor pressures over Great Salt Lake brine. *Journal of Geophysical Research* 70: 500–503.

Dickson, J.A.D. (1985). Diagenesis of shallow-marine carbonates. In: *Geological Society, London, Special Publication*, vol. 18 (ed. P.J. Brenchley and B.P.J. Williams), 173–188. London: The Geological Society of London.

Dickson, T. (1990). Carbonate mineralogy and chemistry. In: *Carbonate Sedimentology* (ed. M.E. Tucker and V. P. Wright), 284–313. Oxford: Wiley-Blackwell.

Dickson, J.A.D., Dickson, C.F., Dickson, W.M., et al. (2014) Marine beachrocks. In: Carbonate Rock Topics (ed. J.A.D. Dickson). Carbonate Diagenesis Limited, 41 pp.

Dierssen, H.M., Zimmerman, R.C., and Burdige, D.J. (2009). Optics and remote sensing of Bahamian carbonate sediment whitings and potential relationship to wind-driven Langmuir circulation. *Biogeosciences* 6: 487–500.

DiLoreto, Z.A., Bontognali, T.R.R., Al Disi, Z.A. et al. (2019). Microbial community composition and dolomite formation in the hypersaline microbial mats of the Khor Al-Adaid sabkhas, Qatar. *Extremophiles* 23: 201–218.

Done, T.J. (1982). Patterns in the distribution of coral communities across the central Great Barrier Reef. *Coral Reefs* 1: 95–107.

Donnadieu, Y., Dromart, G., Goddéris, Y. et al. (2011). A mechanism for brief glacial episodes in the Mesozoic greenhouse. *Paleoceanography* 26: PA3212.

Dorobek, S.L. (2008). Carbonate-platform facies in volcanic-arc settings: characteristics and controls on deposition and stratigraphic development. In: *Formation and Applications of the Sedimentary Record in Arc Collision Zones*, vol. 436 (ed. A.E. Draut, P.D. Clift, and D.W. Scholl), 55–90. Boulder, Colorado: Geological Society of America.

Dorobek, S.L. and Bachtel, S.L. (2001). Supply of allochthonous sediment and its effects on development of carbonate mud mounds, Mississippian Lake Valley Formation, Sacramento Mountains, South-Central New Mexico, U.S.A. *Journal of Sedimentary Research* 71: 1003–1016.

Dravis, J. (1979). Rapid and widespread generation of recent oolitic hardgrounds on a high energy Bahamian platform, Eleuthera Bank, Bahamas. *Journal of Sedimentary Petrology* 49: 195–208.

Dravis, J.J. (1996). Rapidity of freshwater calcite cementation - implications for carbonate diagenesis and sequence stratigraphy. *Sedimentary Geology* 107: 1–10.

Drew, E.A. (1983). Halimeda biomass, growth rates and sediment generation on reefs in the central Great Barrier Reef province. *Coral Reefs* 2: 101–110.

Drew, E.A. and Abel, K.M. (1988a). Studies on *Halimeda*. I. The distribution and species composition of *Halimeda* meadows throughout the Great Barrier Reef Province. *Coral Reefs* 6: 195–205.

Drew, E.A. and Abel, K.M. (1988b). Studies on *Halimeda*. II. Reproduction, particularly the seasonality of gametangia formation, in a number of species from the Great Barrier Reef Province. *Coral Reefs* 6: 207–218.

Dromart, G., Garcia, J.-P., Gaumet, F. et al. (2003a). Perturbation of the carbon cycle at the Middle/Late Jurassic transition: Geological and geochemical evidence. *American Journal of Science* 303: 667–707.

Dromart, G., Garcia, J.P., Picard, S. et al. (2003b). Ice age at the Middle–Late Jurassic transition? *Earth and Planetary Science Letters* 213: 205–220.

Droste, H. (1990). Depositional cycles and source rock development in an epeiric intra-platform basin: the Hanifa Formation of the Arabian Peninsula. *Sedimentary Geology* 69: 281–296.

Droste, H. and Van Steenwinkel, M. (2004). Stratal geometries and patterns of platform carbonates: the cretaceous of Oman. In: *Seismic imaging of carbonate reservoirs and systems*, AAPG Memoir, vol. 81 (ed. G.P. Eberli, J.L. Masaferro, and J.F. Sarg), 185–206. Tulsa, OK: American Association of Petroleum Geologists.

Droxler, A.W. and Jorry, S.J. (2021). The origin of modern atolls: challenging darwin's deeply ingrained theory. *Annual Review of Marine Science* 13: 537–573.

Droxler, A.W. and Schlager, W. (1985). Glacial versus interglacial sedimentation rates and turbidite frequency in the Bahamas. *Geology* 13: 799–802.

Droxler, A.W., Schlager, W., and Whallon, C.C. (1983). Quaternary aragonite cycles and oxygen-isotope record in Bahamian carbonate ooze. *Geology* 11: 235–239.

Duane, M.J. and Al-Zamel, A.Z. (1999). Syngenetic textural evolution of modern sabkha stromatolites (Kuwait). *Sedimentary Geology* 127: 237–245.

Duineveld, G.C.A., Lavaleye, M.S.S., Bergman, M.J.N. et al. (2007). Trophic structure of a cold-water coral mound community (Rockall Bank, NE Atlantic) in relation to the near-bottom particle supply and current regime. *Bulletin of Marine Science* 81: 449–467.

Dujoncquoy, E., Grélaud, C., Razin, P. et al. (2018). Seismic stratigraphy of a Lower Cretaceous prograding carbonate platform (Oman) and implications for the eustatic sea-level curve. *American Association of Petroleum Geologists Bulletin* 102: 509–543.

Dunbar, G.B. and Dickens, G.R. (2003). Massive siliciclastic discharge to slopes of the Great Barrier Reef Platform during sea-level transgression: constraints from sediment cores between 15°S and 16°S latitude and possible explanations. *Sedimentary Geology* 162: 141–158.

Dunham, R.J. (1962). Classification of carbonate rocks according to depositional texture. In: *Classification of Carbonate Rocks – A Symposium*, AAPG Memoir, vol. 1 (ed. W.E. Ham), 108–121. Tulsa; OK: American Association of Petroleum Geologists.

Dunham, J. (1970). Keystone vugs in carbonate beach deposits. *American Association of Petroleum Geologists Bulletin* 54: 845–845.

Dunn, C.E., Brooks, R.R., Edmondson, J. et al. (1996). Biogeochemical studies of metal-tolerant plants from southern Morocco. *Journal of Geochemical Exploration* 56: 13–22.

Dupraz, C. and Strasser, A. (1999). Microbialites and micro-encrusters in shallow coral bioherms (middle to late Oxfordian, Swiss Jura Mountains). *Facies* 40: 101–130.

Dupraz, C. and Strasser, A. (2002). Nutritional modes in coral—microbialite reefs (Jurassic, Oxfordian, Switzerland): evolution of trophic structure as a response to environmental change. *PALAIOS* 17: 449–471.

Dupraz, C., Visscher, P.T., Baumgartner, L.K., and Reid, R.P. (2004). Microbe-mineral interactions: early carbonate precipitation in a hypersaline lake (Eleuthera Island, Bahamas). *Sedimentology* 51: 745–765.

Dupraz, C., Reid, R.P., Braissant, O. et al. (2009). Processes of carbonate precipitation in modern microbial mats. *Earth Science Reviews* 96: 141–162.

Dupraz, C., Reid, R.P., and Visscher, P.T. (2011). Microbialites, Modern. In: *Encyclopedia of Geobiology* (ed. J. Reitner and V. Thiel), 617–635. Dordrecht: Springer Netherlands.

Dyer, B., Maloof, A.C., Purkis, S.J., and Harris, P.M. (2018). Quantifying the relationship between water depth and carbonate facies. *Sedimentary Geology* 373: 1–10.

Eberli, G.P. (1991). Calcareous turbidites and their relationship to sea-level fluctuations and tectonism. In: *Cycles and Events in Stratigraphy* (ed. G. Einsele, W. Ricken, and A. Seilacher), 340–359. Berlin: Springer-Verlag.

Eberli, G.P. (2000). The record of Neogene sea-level changes in the prograding carbonates along the Bahamas transect – Leg166 synthesis. In: *Proceedings of the Ocean Drilling Program, Scientific Results Leg 166*, vol. 166 (ed. P.K. Swart, G.P. Eberli, M.J. Malone, and J.F. Sarg), 167–177. TX: Ocean Drilling Program, College Station.

Eberli, G.P. (2013). The uncertainties involved in extracting amplitude and frequency of orbitally driven sea-level fluctuations from shallow-water carbonate cycles. *Sedimentology* 60: 64–84.

Eberli, G.P. and Betzler, C. (2019). Characteristics of modern carbonate contourite drifts. *Sedimentology* 66: 1163–1191.

Eberli, G.P. and Ginsburg, R.N. (1987). Segmentation and coalescence of platform, Tertiary, NW Great Bahama Bank. *Geology* 15: 75–79.

Eberli, G.P. and Ginsburg, R.N. (1989). Cenozoic progradation of northwestern Great Bahama Bank, a record of lateral platform growth and sea-level fluctuations. In: *Controls on Carbonate Platform and Basin Development*, SEPM Special Publication, vol. 44 (ed. P.D. Crevello, J.L. Wilson, J.F. Sarg, and J.F. Read), 339–351. Tulsa, OK: SEPM (Society of Economic Paleontologists and Mineralogists).

Eberli, G.P., Swart, P.K., Malone, M.J. et al. (1996). *Leg 166 Preliminary Report No. 166, The Bahamas Transect.* TX: Ocean Drilling Program, College Station www-odp.tamu.edu/publications/prelim/166_prel/166concl.html.

Eberli, G.P., Baechle, G.T., Anselmetti, F. S., and Incze, M.L. (2003). Factors controlling elastic properties in carbonate sediments and rocks. *The Leading Edge* 22: 654–660.

Eberli, G.P., Anselmetti, F.S., Betzler, C. et al. (2004a). Calibration of Seismic Data from the Western Margin of Great Bahama Bank with Exposed Strata in the Maiella Mountains (Italy). In: *Integration of Outcrop and Modern Analogs in Reservoir Modelling*, AAPG Memoir, vol. 81 (ed. G.M. Grammer, P. M. Harris, and G.P. Eberli), 207–250. Tulsa, OK: American Association of Petroleum Geologists.

Eberli, G.P., Anselmetti, F.S., Betzler, C. et al. (2004b). Carbonate platform to basin transitions on seismic data and in outcrops: Great Bahama Bank and the Maiella Platform margin, Italy. In: *Seismic Imaging of Carbonate Reservoirs And Systems*, AAPG Memoir, vol. 81, 207–250. Tulsa, OK: American Association of Petroleum Geologists.

Eberli, G.P., Masaferro, J.L., and Sarg, J.F. R. (ed.) (2004c). *Seismic Imaging of Carbonate Reservoirs and Systems.* Tulsa, OK: American Association of Petroleum Geologists, 376 pp.

Eberli, G.P., Harris, P.M., and Grammer, G.M. (2007). *Sequence Stratigraphy and Reservoir Distribution in a Modern Carbonate Platform, AAPG Core Workshop and Field Seminar Guide.* Tulsa, OK: American Association of Petroleum Geologists, 238 pp.

Eberli, G.P., Diaz, M.R., Weger, R.J. and Karaca, E. (2019). Early microbial cements – implications for marine diagenesis and rock physics. *GSA Annual Meeting 2019* (eds. M. Grammer, R. Gries, and L.S. Ruhl). *Geological Society of America Abstracts with Programs.* vol. 51, pp. Paper No. 176-7. Geological Society of America, Phoenix, Arizona.

Edgcomb, V.P., Bernhard, J.M., Beaudoin, D. et al. (2013). Molecular indicators of microbial diversity in oolitic sands of Highborne Cay, Bahamas. *Geobiology* 11: 234–251.

Eiler, J.M. (2011) Paleoclimate reconstruction using carbonate clumped isotope thermometry. *Quaternary Science Reviews* 30: 3575–3588.

Eiler, J.M., Bergquist, B., Bourg, I. et al. (2014). Frontiers of stable isotope geoscience. *Chemical Geology* 372: 119–143.

Einbinder, S., Mass, T., Brokovich, E. et al. (2009). Changes in morphology and diet of the coral Stylophora pistillata along a depth gradient. *Marine Ecology Progress Series* 381: 167–174.

Eisele, M., Frank, N., Wienberg, C. et al. (2011). Productivity controlled cold-water coral growth periods during the last glacial off Mauritania. *Marine Geology* 280: 143–149.

El Husseiny, A. and Vanorio, T. (2013). Synthesis of micritic carbonate analogs: effect on velocity-pressure sensitivity and dissolution. In: *Society of Exploration Geophysicists, 83rd Annual International Meeting, Expanded Abstracts, 83rd Annual International Meeting, Expanded Abstracts*, 2797–2801. Houston, TX: Society of Exploration Geophysicists.

El Husseiny, A. and Vanorio, T. (2015). The effect of micrite content on the acoustic velocity of carbonate rocks. *Geophysics* 80: L45–L55.

El Moursi, M. and Montaggioni, L.F. (1994). Diagenesis of Pleistocene reef-associated sediments from the Red Sea coastal plain, Egypt. *Sedimentary Geology* 90: 49–59.

Eldrett, J.S., Ma, C., Bergman, S.C. et al. (2015). An astronomically calibrated stratigraphy of the Cenomanian, Turonian and earliest Coniacian from the Cretaceous Western Interior Seaway, USA: implications for global chronostratigraphy. *Cretaceous Research* 56: 316–344.

Eldrett, J.S., Dodsworth, P., Bergman, S.C. et al. (2017). Water-mass evolution in the Cretaceous Western Interior Seaway of North America and equatorial Atlantic. *Climate of the Past* 13: 855–878.

Ellis, P.M., Wilson, R.C.L., and Leinfelder, R.R. (1990). Controls on Upper Jurassic carbonate buildup development in the Lusitanian Basin, Portugal. In: *Carbonate Platforms*, IAS Special Publication, vol. 9 (ed. M.E. Tucker, J.L. Wilson, P.D. Crevello, et al.), 169–202. Hoboken, NJ: Wiley-Blackwell.

Elrick, M. and Read, J.F. (1991). Cyclic ramp-to-basin carbonate deposits, Lower Mississippian, Wyoming and Montana: a combined field and computer modelling study. *Journal of Sedimentary Petrology* 61: 1194–1224.

Embry, A.F. and Klovan, J.E. (1971). A late Devonian reef tract on northeastern Banks Island Nordwest Territories. *Bulletin of Canadian Petroleum Geology* 19: 730–781.

Embry, J.-C., Vennin, E., van Buchem, F.S. P. et al. (2010). Sequence stratigraphy and carbon isotope stratigraphy of an Aptian mixed carbonate-siliciclastic platform to basin transition (Galve sub-basin, NE Spain). In: *Mesozoic and Cenozoic Carbonate Systems of the Mediterranean and the Middle East: Stratigraphic and Diagenetic Reference Models*, Geological Society, London, Special Publication, vol. 329 (ed. F.S.P. van Buchem, K.D. Gerdes, and M. Esteban), 113–143. London: Geological Society, London.

Emery, D. and Myers, K.J. (1996). *Sequence Stratigraphy*. Oxford: Blackwell Science, 297 pp.

Emiliani, C. (1955). Pleistocene temperatures. *Journal of Geology* 63: 538–578.

Engel, M., Strohmenger, C.J., Peis, K.T. et al. (2022). High-resolution facies analysis of a coastal sabkha in the eastern Gulf of Salwa (Qatar): A spatiotemporal reconstruction. *Sedimentology* 69 (3): 1119–1150.

Enos, P. (1977). Tamabra limestone of the Poza Rica trend, Cretaceous, Mexico. In: *Deep-Water Carbonate Environments*, SEPM Special Publication, vol. 25 (ed. H.E. Cook and P. Enos), 273–314. Tulsa, OK: SEPM (Society for Sedimentary Geology).

Epting, M. (1980). Sedimentology of miocene carbonate buildups, Central Luconia, offshore Sarawak. *Bulletin of the Geological Society of Malaysia* 12: 17–30.

Epting, M. (1987a). Active margin: miocene carbonate buildups of central Luconia. In: *Offshore Sarawak*, AAPG Studies in Geology v. 27, Atlas of Seismic Stratigraphy v. 3 (ed. A.W. Bally), 168–173. Tulsa, OK: AAPG.

Epting, M. (1987b). The Miocene carbonate buildups of Central Luconia, offshore Sarawak. In: *AAPG Studies in Geology #27, volume 3: Atlas of Seismic Stratigraphy*, AAPG Studies in Geology v. 27 (ed. A.W. Bally), 168–173. Tulsa, OK: American Association of Petroleum Geologists.

Eriksson, K.A. and Simpson, E.L. (1990). Recognition of high-frequency sea-level fluctuations in Proterozoic tidal deposits, Mount Isa, Australia. *Geology* 18: 474–477.

Ernst, A., Jiménez-Sánchez, A., and Baidder, L. (2015). Bryozoan fauna of the Upper Ordovician (Katian) of Alnif, Morocco. *Palaeobiodiversity and Palaeoenvironments* 95: 537–553.

Escudero-Mozo, M.J., Martín-Chivelet, J., Goy, A., and López-Gómez, J. (2014). Middle-Upper Triassic carbonate platforms in Minorca (Balearic Islands): implications for Western Tethys correlations. *Sedimentary Geology* 310: 41–58.

Esker, D., Eberli, G.P., and McNeill, D.F. (1998). The structural and sedimentological controls on the reoccupation of Quaternary incised valleys, Belize southern lagoon. *American Association of Petroleum Geologists Bulletin* 82: 2075–2109.

Esrafili-Dizaji, B., Hajikazemi, E., Dalvand, M., Hassanzadeh Nemati, M. and Swennen, R. (2020) Diagenesis and reservoir characteristics of the Lithocodium–Bacinella facies in a Lower Cretaceous reservoir, eastern Persian Gulf Basin. *Facies*, 66, 24.

Esteban, M. (1991). Palaeokarst: case histories. In: *Palaeokarsts and Palaeokarstic Reservoirs*, Postgraduate Research Institute for Sedimentology Occ. Publ. Series, vol. 2 (ed. V.P. Wright, M. Esteban, and P.L. Smart), 159. Reading, UK: University of Reading.

Esteban, M. (1996). An overview of Miocene reefs from Mediterranean areas: general trends and facies models. *SEPM: Concepts in Sedimentology and Paleontology* 5: 1–53.

Esteban, M. and Klappa, C.F. (1983). Subaerial exposure environment, Chapter 1. In: *Carbonate Depositional Environments*, vol. 33 (ed. P.A. Scholle, D.G. Bebout, and C.H. Moore), 1–54. Tulsa, OK: American Association of Petroleum Geologists.

Esteban, M. and Taberner, C. (2003). Secondary porosity development during late burial in carbonate reservoirs as a result of mixing and/or cooling of brines. *Journal of Geochemical Exploration* 78 (79): 355–359.

Esteban, M., Braga, J.C., Martín, J., and de Santisteban, C. (1996). Western mediterranean reef complex. *SEPM: Concepts in Sedimentology and Paleontology* 5: 55–72.

Evans, G., Murray, J.W., Biggs, H.E.J. et al. (1973). The oceanography, ecology, sedimentology and geomorphology of parts of the Trucial Coast barrier island complex, Persian Gulf. In: *The Persian Gulf: Holocene Carbonate Sedimentation and Diagenesis in a Shallow Epicontinental Sea* (ed. B.H. Purser), 233–277. Berlin, Heidelberg: Springer.

Evensen, N.R., Voolstra, C.R., Fine, M. et al. (2022). Empirically derived thermal thresholds of four coral species along the Red Sea using a portable and standardized experimental approach. *Coral Reefs* 41: 239–252.

Everts, A.-J.W. and Reijmer, J.J.G. (1995). Clinoform composition and margin geometries of a Lower Cretaceous carbonate platform (Vercors, SE France). *Palaeogeography, Palaeoclimatology, Palaeoecology* 119: 19–33.

Ezaki, Y., Liu, J.B., and Adachi, N. (2012). Lower Triassic stromatolites in Luodian County, Guizhou Province, South China: evidence for the protracted devastation of the marine environments. *Geobiology* 10: 48–59.

Fagerstrom, J.A. (1987). *The Evolution of Reef Communities*. New York, USA: Wiley, 600 pp.

Fan, J. S., & Rigby, J. K. (1994). Upper Carboniferous phylloid algal mounds in southern Guizhou, China. *Brigham Young University Geology Studies* 40: 17–24.

Farkaš, J., Buhl, D., Blenkinsop, J., and Veizer, J. (2007). Evolution of the oceanic calcium cycle during the late Mesozoic: evidence from $\delta^{44/40}$Ca of marine skeletal carbonates. *Earth and Planetary Science Letters* 253: 96–111.

Feary, D.A. and James, N.P. (1998). Seismic stratigraphy and geological evolution of the Cenozoic, cool-water, Eucla Platform, Great Australian Bight. *American Association of Petroleum Geologists Bulletin* 82: 792–816.

Fentimen, R., Feenstra, E., Rüggeberg, A. et al. (2022). A 300,000-year record of cold-water coral mound build-up at the East Melilla Coral Province (SE Alboran Sea, western Mediterranean). *Climate of the Past* 18 (8): 1915–1945.

Ferro, C.E., Droxler, A.W., Anderson, J.B., and Mucciarone, D. (1999). Late Quaternary shift of mixed siliciclastic-carbonate environments induced by glacial eustatic sea-level fluctuations in Belize. In: *Advances in Carbonate Sequence Stratigraphy: Application to Reservoirs, Outcrops and Models*, SEPM Special Publication, vol. 63 (ed. P.M. Harris, A.H. Saller, and T. Simo), 385–411. Tulsa, OK: SEPM (Society for Sedimentary Geology).

Fink, H.G., Wienberg, C., Hebbeln, D. et al. (2012). Oxygen control on Holocene cold-water coral development in the eastern Mediterranean Sea. *Deep Sea Research Part I: Oceanographic Research Papers* 62: 89–96.

Fink, H.G., Wienberg, C., De Pol-Holz, R. et al. (2013). Cold-water coral growth in the Alboran Sea related to high productivity during the Late Pleistocene and Holocene. *Marine Geology* 339: 71–82.

Finnegan, S., Bergmann, K., Eiler, J.M. et al. (2011). The magnitude and duration of late ordovician–early Silurian Glaciation. *Science* 331: 903–906.

Fischer, A.G. (1964). The lofer cyclothems of the Alpine triassic. *Kansas Geological Survey Bulletin* 169: 107–149.

Floquet, M. and Hennuy, J. (2003). Evolutionary gravity flow deposits in the Middle Turonian – Early Coniacian Southern Provence Basin (SE France): origins and depositional processes. In: *Submarine Mass Movements and Their Consequences*, Natural and Technical Hazard Research Series (ed. J. Locat and J. Mienert), 417–424. Dordrecht: Kluwer Academic Publishers.

Flügel, E. (1982a). Evolution of Triassic reefs: current concepts and problems. *Facies* 6: 297–328.

Flügel, E. (1982b). *Microfacies Analysis of Limestones*. Berlin: Springer Verlag, 375 pp.

Flügel, E. (1984). *Microfacies Analysis of Limestones*. Berlin-Heidelberg: Springer-Verlag, 633 pp.

Flügel, E. (1994). Pangean shelf carbonates: controls and paleoclimatic significance of Permian and Triassic reefs. In: *Pangea: Paleoclimate, Tectonics, and Sedimentation During Accretion, Zenith, and Breakup of a Supercontinent*, vol. 288 (ed. G.O. Klein), 247–266. Boulder, CO: Geological Society of America.

Flügel, E. (2002). Triassic reef patterns. In: *Phanerozoic Reef Patterns*, SEPM Special Publication, vol. 72 (ed. W. Kiessling, E. Flügel, and J. Golonka), 391–463. Tulsa, OK: SEPM (Society for Sedimentary Geology).

Flügel, E. (2004). *Microfacies of Carbonate Rocks*. Berlin: Springer, 976 pp.

Flügel, E. (2010). *Microfacies of Carbonate Rocks – Analysis, Interpretation and Application*. Heidelberg Dordrecht London New York: Springer, 1007 pp.

Flügel, E. and Bernecker, M. (1996). Upper Triassic reefs of the south Tethyan margin (Oman). In: *Global and Regional Controls on Biogenic Sedimentation 1. Reef-Evolution. Research Reports*, Göttinger Arbeiten zur Geologie und Paläontologie, Sonderband 2 (ed. J. Reitner, F. Neuweiler, and F. Gunkel). Göttingen: Geologisches Institut der Georg-August-Universität Göttingen.

Flügel, E. and Kiessling, W. (2002). A new look at ancient reefs. In: *Phanerozoic Reef Patterns*, SEPM Special Publication, vol. 72 (ed. W. Kiessling, E. Flügel, and J. Golonka), 3–10. Tulsa, OK: SEPM (Society for Sedimentary Geology).

Flügel, E. and Reinhardt, J. (1989). Uppermost Permian Reefs in Skyros (Greece) and Sichuan (China): Implications for the Late Permian Extinction Event. *PALAIOS* 4: 502–518.

Flügel, E. and Senowbari-Daryan, B. (2001). Triassic reefs of the Tethys. In: *The History and Sedimentology of Ancient Reef Systems* (ed. G.D. Stanley), 217–249. Boston, MA: Springer.

Flügel, E. and Stanley, G.D. (1984). Reorganization, development, and evolution of post-Permian reefs and reef organisms. *Bulletins of American Paleontology* 54: 177–186.

Folk, R.L. (1959). Practical petrographic classification of limestones. *American Association of Petroleum Geologists Bulletin* 43: 1–38.

Folk, R.L. (1962). Spectral subdivision of limestone types. In: *Classification of Carbonate Rocks – A Symposium*, AAPG Memoir, vol. 1 (ed. W.E. Ham), 62–84. Tulsa, OK: American Association of Petroleum Geologists.

Folk, R.L. (1974a). *Petrology of Sedimentary Rocks*. Austin, TX: Hemphill Publishing Company, 182 pp.

Folk, R.L. (1974b). The natural history of crystalline calcium carbonate: effect of magnesium content and salinity. *Journal of Sedimentary Petrology* 44: 40–53.

Föllmi, K.B. and Godet, A. (2013). Palaeoceanography of lower Cretaceous Alpine platform carbonates. *Sedimentology* 60: 131–151.

Föllmi, K.B., Garrison, R.E., Ramirez, P.C. et al. (1992). Cyclic phosphate-rich successions in the upper Cretaceous of Colombia. *Palaeogeography, Palaeoclimatology, Palaeoecology* 93: 151–182.

Ford, D.C. and Williams, P.W. (1989). *Karst Geomorphology and Hydrology*. London, U.K: Chapman & Hall, 562 pp.

Fornós, J.J., Clemmensen, L.B., Gómez-Pujol, L., and Murray, A.S. (2009). Late Pleistocene carbonate aeolianites on Mallorca, Western Mediterranean: a luminescence chronology. *Quaternary Science Reviews* 28: 2697–2709.

Foster, G.L., Hönisch, B., Paris, G. et al. (2013). Interlaboratory comparison of boron isotope analyses of boric acid, seawater and marine $CaCO_3$ by MC-ICPMS and NTIMS. *Chemical Geology* 358: 1–14.

Foster, G.L., Lécuyer, C., and Marschall, H.R. (2016). Boron stable isotopes. In: *Encyclopedia of Geochemistry: A Comprehensive Reference Source on the Chemistry of the Earth* (ed. W.M. White), 1–6. Cham: Springer International Publishing.

Foubert, A. and Henriet, J.-P. (2009). *Nature and Significance of the Recent Carbonate Mound Record*. Dordrecht Heidelberg London New York: Springer, 298 pp.

Foubert, A., Depreiter, D., Beck, T. et al. (2008). Carbonate mounds in a mud volcano province off north-west Morocco: key to processes and controls. *Marine Geology* 248: 74–96.

Fournier, F. and Borgomano, J. (2007). Geological significance of seismic reflections and imaging of the reservoir architecture in the Malampaya gas field (Philippines). *American Association of Petroleum Geologists Bulletin* 91: 235–258.

Fournier, F., Borgomano, J., and Montaggioni, L.F. (2005). Development patterns and controlling factors of tertiary carbonate buildups: insights from high-resolution 3D seismic and well data in the Malampaya gas field (Offshore Palawan, Philippines). *Sedimentary Geology* 175: 189–215.

Fournier, F., Leonide, P., Biscarrat, K. et al. (2011). Elastic properties of microporous cemented grainstones. *Geophysics* 76: E211–E226.

Fournier, F., Pellerin, M., Villeneuve, Q. et al. (2018). The equivalent pore aspect ratio as a tool for pore type prediction in carbonate reservoirs. *American Association of Petroleum Geologists Bulletin* 102: 1343–1377.

Franceschi, M., Corso, J.D., Cobianchi, M. et al. (2019). Tethyan carbonate platform transformations during the Early Jurassic (Sinemurian–Pliensbachian, Southern Alps): comparison with the Late Triassic Carnian Pluvial Episode. *Geological Society of America Bulletin* 131: 1255–1275.

Francis, J.M., Dunbar, G.B., Dickens, G.R. et al. (2007). Siliciclastic sediment across the North Queensland Margin (Australia): a holocene perspective on reciprocal versus coeval deposition in tropical mixed siliciclastic-carbonate systems. *Journal of Sedimentary Research* 77: 572–586.

Frank, T.D., Thomas, S.G., and Fielding, C.R. (2008). On using carbon and oxygen isotope data from glendonites as paleoenvironmental proxies: a case study from the permian system of Eastern Australia. *Journal of Sedimentary Research* 78: 713–723.

Frank, T.D., James, N.P., Bone, Y. et al. (2014). Late Quaternary carbonate deposition at the bottom of the world. *Sedimentary Geology* 305: 1–16.

Frank, T.D., James, N.P., and Shultis, A.I. (2020). Lack of synsedimentary chemical alteration in polar carbonates (Ross Sea, Antarctica): resolution of a conundrum. *Journal of Sedimentary Research* 90: 449–467.

Franklin, A. and Sarg, J.F. (2017). Sedimentology and ichnofacies, uppermost Three Forks Formation (Fammenian), Williston Basin, North Dakota and Montana – a storm dominated intrashelf Basin. *The Sedimentary Record* 15: 4–12.

Franklin, A. and Sarg, J.F. (2018). Storm-influenced intrashelf systems: sedimentological characterization of the Famennian Three Forks Formation, Williston Basin, U.S.A. *Journal of Sedimentary Research* 88: 583–612.

Franseen, E.K., Watney, W.L., Kendall, C. G.S.C., and Ross, W. (1991). *Sedimentary Modeling: Computer Simulations and Methods for Improved Parameter Definition.* Lawrence, Kansas: Kansas Geological Survey, 524 pp.

Freiwald, A. (1993). Coralline algal maerl frameworks - islands within the phaeophytic kelp belt. *Facies* 29: 133–148.

Freiwald, A. and Roberts, J.M. (ed.) (2005). *Cold-Water Corals and Ecosystems.* Berlin Heidelberg, Erlangen: Springer-Verlag, 1243 pp.

Freiwald, A., Fosså, J.H., Grehan, A. et al. (2004). *Cold-Water Coral Reefs.* Cambridge, UK: UNEP-WCMC.

Freiwald, A., Rogers, A., Hall-Spencer, J. et al. (2021). *Global Distribution of Cold-Water Corals (Version 5.1).* Fifth update to the dataset in Freiwald et al. (2004) by UNEP-WCMC, in collaboration with Andre Freiwald and John Guinotte. Cambridge: UN Environment Programme World Conservation Monitoring Centre.

Friedman, G.M. (1965). Terminology of crystallization textures and fabrics in sedimentary rocks. *Journal of Sedimentary Research* 35: 643–655.

Friesenbichler, E., Richoz, S., Baud, A. et al. (2018). Sponge-microbial build-ups from the lowermost Triassic Chanakhchi section in southern Armenia: microfacies and stable carbon isotopes. *Palaeogeography, Palaeoclimatology, Palaeoecology* 490: 653–672.

Fuller, M. and Jenkins, R. (2007). Reef corals from the Lower Cambrian of the Flinders Ranges, South Australia. *Palaeontology* 50: 961–980.

Fulthorpe, C.S. and Schlanger, S.O. (1989). Paleo-oceanographic and tectonic settings of early miocene reefs and associated carbonates of offshore Southeast Asia. *American Association of Petroleum Geologists Bulletin* 73: 729–756.

Gale, A., Hancock, J., and Kennedy, W. (1999). Biostratigraphical and sequence correlation of the Cenomanian successions in Mangyshlak (W. Kazakhstan) and Crimea (Ukraine) with those in southern England. *Bulletin de l'Institut Royal des Sciences Naturelles de Belgique - Bulletin van het Koninklijk Belgisch Instituut voor Natuurwetenschappen* 69 (Supplément A): 67–86.

Gale, A., Anderskouv, K., Surlyk, F., and Whalley, J. (2015). Slope failure of chalk channel margins: implications of an Upper Cretaceous mass transport complex, southern England. *Journal of the Geological Society* 172: 763.

Galloway, W.E. (2008). Depositional evolution of the Gulf of Mexico sedimentary basin, Chapter 15. In: *Sedimentary Basins of the World*, vol. 5 (ed. A.D. Miall), 505–549. Amsterdam: Elsevier.

Gamper, A., Heubeck, C., Demskec, D. and Hoehse, M. (2012). Composition and microfacies of archean microbial mats (Moodies Group, ca. 3.22 Ga, South Africa). In: *Microbial mats in siliciclastic depositional systems through time* (ed. N. Noffke and H. Chafetz), SEPM Special Publication No. 101, 65–74. Tulsa: Society for Sedimentary Geology.

Gao, Y., Fu, X., Li, Y. et al. (2020). Demise of the Middle Jurassic carbonate platform: Responses to sea-level fall in the Qiangtang Basin, Eastern Tethys. *Geological Journal* 55: 7935–7952.

Garcia-Fresca, B., Lucia, F.J., Sharp, J.M., and Kerans, C. (2012). Outcrop-constrained hydrogeological simulations of brine reflux and early dolomitization of the Permian San Andres Formation. *American Association of Petroleum Geologists Bulletin* 96: 757–1781.

Garcia-Mondéjar, J. and Fernández-Mendiola, P.A. (1995). *Albian carbonate mud-mounds: comparative study in the context of sea-level variations (Soba, northern Spain). In: Carbonate mud-mounds - their origin and evolution* (ed. C.L.V. Monty, D.W.J. Bosence, P.H.

Bridges and B.R. Pratt), Special Publication Number 23, 359–384. International Association of Sedimentologists.

Garrison, V.H., Shinn, E.A., Foreman, W. T. et al. (2003). African and Asian dust: from desert soils to coral reefs. *BioScience* 53: 469–480.

Garven, G. and Freeze, R.A. (1984). Theoretical analysis of the role of groundwater flow in the genesis of stratabound ore deposits; 2, Quantitative results. *American Journal of Science* 284: 1125–1174.

Gattuso, J.-P. and Hansson, L. (ed.) (2011). *Ocean Acidification*. Oxford: Oxford University Press, 326 pp.

Gattuso, J.-P., Frankignoulle, M., Bourge, I. et al. (1998). Effect of calcium carbonate saturation of seawater on coral calcification. *Global and Planetary Change* 18: 37–46.

Ge, Y., Della Porta, G., Pederson, C.L. et al. (2021). Botryoidal and spherulitic aragonite in carbonates associated with microbial mats: precipitation or diagenetic replacement product? *Frontiers in Earth Science* 9: 488.

Ge, Y., Pederson, C.L., Lokier, S.W. et al. (2022). Radiaxial fibrous calcite forms via early marine-diagenetic alteration of micritic magnesium calcite. *Sedimentology* 70: 434–450.

Geider, R.J. and MacIntyre, H.L. (2002). Physiology and biochemistry of photosynthesis and algal carbon acquisition. In: *Phytoplankton Productivity* (ed. P.J.l.B. Williams, D.N. Thomas, and C.S. Reynolds), 44–77. Wiley.

Geldsetzer, H.H.J., James, N.P., and Tebutt, G.E. (1988). *Reefs*. Calgary, Alberta: Canadian Society of Petroleum Geologists, 766 pp.

Gharbi, M., Boote, D., Sekatni Aich, N. et al. (2022). Aptian–Albian extension in Tethyan rifted continental passive margin of Tunisia inferred from sequential restorations of balanced cross-sections and expansion index. *Marine and Petroleum Geology* 143: 105818.

Gherardi, D.F.M. and Bosence, D.W.J. (2001). Composition and community structure of the coralline algal reefs from Atol das Rocas, South Atlantic, Brazil. *Coral Reefs* 19: 205–219.

Gherardi, D.F.M. and Bosence, D.W.J. (2005). Late Holocene reef growth and relative sea-level changes in Atol das Rocas, equatorial South Atlantic. *Coral Reefs* 24: 264–272.

Gianolla, P., Avanzini, M., Breda, A. et al. (2010). Fieldtrip to the World Heritage Site of the Tethyan Triassic - September 5-10, 2010 - Dolomites, Southern Alps, Italy. In: *Pan-European Correlation of the Triassic*, Triassic Field Workshops 2010, 122. Fondazione Dolomiti_Dolomiten-Dolomites-Dolomitis UNESCO, Dolomites, Southern Alps, Italy.

Gianolla, P., Caggiati, M., and Riva, A. (2021). The interplay of carbonate systems and volcanics: Cues from the 3D model of the Middle Triassic Sciliar/Schlern platform (Dolomites, Southern Alps). *Marine and Petroleum Geology* 124: 104794.

Gill, B.C., Lyons, T.W., and Saltzman, M. R. (2007). Parallel, high-resolution carbon and sulfur isotope records of the evolving Paleozoic marine sulfur reservoir. *Palaeogeography, Palaeoclimatology, Palaeoecology* 256: 156–173.

Gill, B.C., Lyons, T.W., and Frank, T.D. (2008). Behavior of carbonate-associated sulfate during meteoric diagenesis and implications for the sulfur isotope paleoproxy. *Geochimica et Cosmochimica Acta* 72: 4699–4711.

Ginsburg, R.N. (1956). Environmental relationships of grain size and constituent particles in some South Florida carbonate sediments. *American Association of Petroleum Geologists Bulletin* 40: 2384–2427.

Ginsburg, R.N. (1957). Early diagenesis and lithification of shallow-water carbonate sediments in South Florida. In: *Regional Aspects of Carbonate Deposition*, vol. 5 (ed. R.J. LeBlanc and J.G. Breeding), 80–99. Tulsa, OK: Society of Economic Paleontologists and Mineralogists.

Ginsburg, R.N. and James, N.P. (1974). Holocene carbonate sediments of continental shelves. In: *The Geology of Continental Margins* (ed. C.A. Burk and C.L. Drake), 137–155. Berlin, Heidelberg: Springer.

Ginsburg, R.N. and Lowenstam, H.A. (1958). The influence of marine bottom communities on the depositional environment of sediments. *Journal of Geology* 66: 310–318.

Ginsburg, R.N. and Shinn, E.A. (1964). Distribution of the reef-building community in Florida and the Bahamas. *American Association of Petroleum Geologists Bulletin* 48: 527–527.

Ginsburg, R.N., Lloyd, R.M., McCallum, J. S. et al. (1958). *Surface Sediments of Great Bahama Bank*. Houston, Texas: Shell Development Company.

Gischler, E. (1994). Sedimentation on three Caribbean atolls: Glover Reef, Lighthouse Reef and Turneffe Islands, Belize. *Facies* 31: 243–254.

Gischler, E. (1995). Current and wind induced facies patterns in a Devonian atoll: Iberg Reef, Harz Mts., Germany. *PALAIOS* 10: 180–189.

Gischler, E. (2003). Holocene lagoonal development in the isolated carbonate platforms off Belize. *Sedimentary Geology* 159: 113–132.

Gischler, E. (2007). Pleistocene facies of Belize barrier and atoll reefs. *Facies* 53: 27–41.

Gischler, E. (2008). Accretion patterns in Holocene tropical coral reefs: do massive coral reefs in deeper water with slowly growing corals accrete faster than shallower branched coral reefs with rapidly growing corals? *International Journal of Earth Sciences* 97: 851–859.

Gischler, E. and Lomando, A.J. (2005). Offshore sedimentary facies of a modern carbonate ramp, Kuwait, northwestern Arabian-Persian Gulf. *Facies* 50: 443–462.

Gischler, E. and Lomando, A.J. (2007). Offshore sedimentary facies of a modern carbonate ramp, Kuwait, northwestern Arabian-Persian Gulf. *Facies* 53: 317.

Gischler, E., Bosellini, A., Brack, P., and Homewood, P. (2001). *Mojsisovics on the Triassic reefs of the Dolomites (1879)*, 106. Miami, FL: Comparative

Sedimentology Laboratory, University of Miami.

Gischler, E., Ginsburg, R.N., Herrle, J.O., and Prasad, S. (2010). Mixed carbonates and siliciclastics in the Quaternary of southern Belize: Pleistocene turning points in reef development controlled by sea-level change. *Sedimentology* 57: 1049–1068.

Glynn, P.W. (2000). El Niño-Southern Oscillation mass mortalities of reef corals: a model of high temperature marine extinctions? In: *Carbonate Platform Systems: Components and Interactions*, Special Publication, vol. 178 (ed. E. Insalaco, P.W. Skelton, and T.J. Palmer), 117–133. London: Geological Society, London.

Goktas, P., Austin, J.A. Jr., Fulthorpe, C.S., and Gallagher, S.J. (2016). Morphologies and depositional/ erosional controls on evolution of Pliocene-Pleistocene carbonate platforms: Northern Carnarvon Basin, Northwest Shelf of Australia. *Continental Shelf Research* 124: 63–82.

Goldhammer, R.K. and Harris, M.T. (1989). Eustatic controls on the stratigraphy and geometry of the Latemar buildup (Middle Triassic), the Dolomites of northern Italy. In: *Controls on Carbonate Platform and Basin Development*, SEPM Special Publication, vol. 44 (ed. P.D. Crevello, J.L. Wilson, J.F. Sarg, and J.F. Read), 323–338. Tulsa, OK: SEPM (Society of Economic Paleontologists and Mineralogists).

Goldhammer, R.K., Dunn, P.A., and Hardie, L.A. (1987). High frequency glacio-eustatic sealevel oscillations with Milankovitch characteristics recorded in middle Triassic platform carbonates in Northern Italy. *American Journal of Science* 287: 853–892.

Goldhammer, R.K., Oswald, E.J., and Dunn, P.A. (1991). Hierarchy of stratigraphic forcing: example from Middle Pennsylvanian shelf carbonates of the Paradox basin. In: *Sedimentary Modeling: Computer Simulations and Methods for Improved Parameter Definition*, Kansas Geological Survey Bulletin, vol. 233 (ed. E.K. Franseen, W.L. Watney, C.G. S.C. Kendall, and W. Ross), 361–413. Kansas: Kansas Geological Survey Lawrence.

Goldhammer, R.K., Lehmann, P.J., and Dunn, P.A. (1993). The origin of high-frequency platform carbonate cycles and third-order sequences (Lower Ordovician El Paso, West Texas): constraints from outcrop data and stratigraphic modeling. *Journal of Sedimentary Petrology* 63: 318.

Goldhammer, R.K., Oswald, E.J., and Dunn, P.A. (1994). High frequency glacio-eustatic cyclicity in the Middle Pennsylvanian of the Paradox Basin: an evaluation of Milankovitch forcing. In: *Orbital Forcing and Cyclic Sequences*, IAS Special Publication, vol. 19 (ed. P.L. De Boer and D.G. Smith), 243–283. Oxford: International Association of Sedimentologists – Blackwell Scientific Publications.

Goldstein, R.H. (1991). Stable isotope signatures associated with palaeosols, Pennsylvanian Holder Formation, New Mexico. *Sedimentology* 38: 67–77.

Golonka, J. and Kiessling, W. (2002). Phanerozoic time scale and definition of time slices. In: *Phanerozoic Reef Patterns*, SEPM Special Publication, vol. 72 (ed. W. Kiessling, E. Flügel, and J. Golonka), 11–20. Tulsa, OK: SEPM (Society for Sedimentary Geology).

Goodwin, P.W. and Anderson, E.J. (1985). Punctuated aggradational cycles: a general hypothesis of episodic stratigraphic accumulation. *Journal of Geology* 93: 515–533.

Goreau, T. (1969). Post-Pleistocene urban renewal in coral reefs. *Micronesica* 5: 323–326.

Goudie, A.S. (1996). Organic agency in calcrete development. *Journal of Arid Environments* 32: 103–110.

Gradstein, F.M., Ogg, J.G., Schmitz, M.D., and Ogg, G.M. (2012). *The Geologic Time Scale 2012*. Boston: Elsevier.

Grammer, G.M. and Ginsburg, R.N. (1992). Highstand versus lowstand deposition on carbonate platform margins: insight from Quaternary foreslopes in the Bahamas. *Marine Geology* 103: 125–136.

Grammer, G.M. and Ritter, A.L. (2009). Phylloid Algal Mounds in the Paradox Basin, Southwestern USA: an alternative to the in situ constructional growth model? In: *Perspectives in Carbonate Geology*, IAS Special Publication, vol. 41 (ed. P.K. Swart, G.P. Eberli, J.A. McKenzie, et al.), 239–254. Chichester, West Sussex: Wiley-Blackwell/ International Association of Sedimentologists.

Grammer, G.M., Ginsburg, R.N., and McNeill, D.F. (1991). Morphology and development of modern carbonate foreslopes, Tongue of the Ocean, Bahamas. In: *Transactions of the 12th Caribbean Geological Conference* (ed. D. K. Larue and G. Draper), 27–32. St. Croix, U.S. Virgin Islands: Miami Geological Society.

Grammer, G.M., Ginsburg, R.N., Swart, P. K. et al. (1993). Rapid growth rates of syndepositional marine aragonite cements in steep marginal slope deposits, Bahamas and Belize. *Journal of Sedimentary Petrology* 63: 983–989.

Grammer, G.M., Crescini, C.M., McNeill, D.F., and Taylor, L.H. (1999). Quantifying rates of syndepositional marine cementation in deeper platform environments – new insight into a fundamental process. *Journal of Sedimentary Research* 69: 202–207.

Grammer, G.M., Eberli, G.P., van Buchem, F.S.P. et al. (2000). Application of high resolution sequence stratigraphy in developing an exploration and production strategy for a mixed carbonate/siliciclastic system (Carboniferous), Paradox basin, Utah, USA. In: *Genetic Stratigraphy on the Exploration and the Production Scales*, Bulletin Centre Recherche Elf Exploration Production Memoir, vol. 24 (ed. P.W. Homewood and G.P. Eberli), 29–69. Pau: Elf Exploration Production.

Granier, B. and Boichard, R. (2017). Sedimentological investigation on Holocene deposits in the Mussafah Channel (Abu Dhabi, United Arab Emirates). *Carnets de géologie (Notebooks on geology)* 17: 39–104.

Gravestock, C., Simmons, M., Van Buchem, F., and Davies, R. (2021). Challenging Middle Eastern Stratigraphic Misconceptions — An Example for Creation and Infill of the

Gotnia Basin. In: *82nd EAGE Annual Conference & Exhibition, 2021*, 1–5. Amsterdam: EAGE.

Greenlee, S.M. and Lehmann, P.J. (1993) Stratigraphic framework of productive carbonate buildups. In: *Carbonate Sequence Stratigraphy: Recent Developments and Applications* (Eds R. G. Loucks and J.F. Sarg), AAPG Memoir 57, pp. 43–62. American Association of Petroleum Geologists, Tulsa, OK.

Gregg, J.M., Bish, D.L., Kaczmarek, S.E., and Machel, H.G. (2015). Mineralogy, nucleation and growth of dolomite in the laboratory and sedimentary environment: a review. *Sedimentology* 62: 1749–1769.

Grélaud, C., Razin, P., Homewood, P.W., and Schwab, A.M. (2006). Development of incisions on a periodically emergent carbonate platform (Natih Formation, Late Cretaceous, Oman). *Journal of Sedimentary Research* 76: 647–669.

Grey, K. (1994). Stromatolites from the Palaeoproterozoic (Orosirian) Glengarry Group, Glengarry Basin, Western Australia. *Alcheringa: An Australasian Journal of Palaeontology* 18: 275–300.

Grey, K. and Awramik, S.M. (2020). *Handbook for the Study and Description of Microbialites*. Perth: Geological Survey of Western Australia, 290 pp.

Grey, K. and Planavsky, N.J. (2009). Microbialites of Lake Thetis, Cervantes, Western Australia: a fi Geological Survey of Western Australia, Record 2009/11, 21 pp.

Grigg, R.W. (1982). Darwin point: a threshold for atoll formation. *Coral Reefs* 1: 29–34.

Grigg, R.W. (1998). Holocene coral reef accretion in Hawaii: a function of wave exposure and sea level history. *Coral Reefs* 17: 263–272.

Grigyalis, A. and Gorbachik, T.N. (1980). Morphology and taxonomy of Jurassic and Early Cretaceous representatives of the superfamily Globigerinacea (Favusellidae). *Journal of Foraminiferal Research* 10: 180–190.

Grosjean, A.-S., Vennin, E., Olivier, N. et al. (2018). Early Triassic environmental dynamics and microbial development during the Smithian–Spathian transition (Lower Weber Canyon, Utah, USA). *Sedimentary Geology* 363: 136–151.

Grossman, E.L. and Joachimski, M.M. (2020). Oxygen isotope stratigraphy, Chapter 10. In: *Geologic Time Scale 2020* (ed. F.M. Gradstein, J.G. Ogg, M. D. Schmitz, and G.M. Ogg), 279–307. Amsterdam: Elsevier.

Grötsch, J., Billing, I., and Vahrenkamp, V. (1998). Carbon-isotope stratigraphy in shallow-water carbonates: Implications for Cretaceous black-shale deposition. *Sedimentology* 45: 623–634.

Grotzinger, J.P. (1989). Facies and evolution of Precambrian carbonate depositional systems: emergence of the modern platform archetype. In: *Controls on Carbonate Platform and Basin Development*, SEPM Special Publication, vol. 44 (ed. P.D. Crevello, J.L. Wilson, J.F. Sarg, and J.F. Read), 79–106. Tulsa, OK: Society of Economic Paleontologists and Mineralogists.

Grotzinger, J.P. and James, N.P. (2000). Precambrian carbonates: evolution of understanding. In: *Carbonate Sedimentation and Diagenesis in the Evolving Precambrian World*, SEPM Special Publication, vol. 67 (ed. J.P. Grotzinger and N.P. James), 1–19. Tulsa, OK: SEPM (Society for Sedimentary Geology).

Grotzinger, J.P. and Knoll, A.H. (1995). Anomalous carbonate precipitates: is the Precambrian the key to the Permian? *PALAIOS* 10: 578–596.

Grotzinger, J.P., Adams, R.D., McCormick, D.S., and Myrow, P. (1989). Sequence stratigraphy, correlations between Wopmay orogen and Kilohigok basin, and further investigations of the Bear Creek Group (Goulburn Supergroup), District of Mackenzie, NWT. *Geological Survey of Canada* 89: 107–119.

Guan, Y., Hohn, S., and Merico, A. (2015). Suitable environmental ranges for potential coral reef habitats in the tropical ocean. *PLoS One* 10 (6): e0128831.

Gunatilaka, A. (1991). Dolomite formation in coastal Al-Khiran, Kuwait Arabian Gulf: a re-examination of the sabkha model. *Sedimentary Geology* 72: 35–53.

Gutteridge, P. (1987). Dinantian sedimentation and the basement structure of the Derbyshire Dome. *Geological Journal* 22: 25–41.

Haak, A.B. and Schlager, W. (1989). Compositional variations in calciturbidites due to sea-level fluctuations, late Quaternary, Bahamas. *Geologische Rundschau* 78: 477–486.

Haas, J. (1982). Facies analysis of the cyclic Dachstein limestone formation (Upper Triassic) in the Bakony Mountains, Hungary. *Facies* 6: 75–84.

Halfar, J. and Mutti, M. (2005). Global dominance of coralline red-algal facies: a response to Miocene oceanographic events. *Geology* 33: 481–484.

Halfar, J., Gordinez-Orta, L., and Ingle, J. C. (2000). Microfacies analysis of Recent carbonate environments in the southern Gulf of California, Mexico – a model for warm-temperate to subtropical carbonate formation. *PALAIOS* 15: 323–342.

Halfar, J., Godinez-Orta, L., Mutti, M. et al. (2004). Nutrient and temperature controls on modern carbonate production. An example from the Gulf of California, Mexico. *Geology* 32: 213–216.

Halfar, J., Godinez-Orta, L., Mutti, M. et al. (2006). Carbonates calibrated against oceanographic parameters along a latitudinal transect in the Gulf of California, Mexico. *Sedimentology* 53: 297–320.

Hallam, A. (1988). The reevaluation of Jurassic eustasy in the light of new data and the revised Exxon curve. In: *Sea-Level Changes: An Integrated Approach*, SEPM Special Publication, vol. 42 (ed. C.K. Wilgus, B.S. Hastings, C.G.S.C. Kendall, et al.), 261–273. Tulsa, OK: SEPM (Society of Economic Paleontologists and Mineralogists).

Hallock, P. (1985). Why are larger foraminifera large? *Paleobiology* 11: 195–208.

Hallock, P. (2001). Coral reefs, carbonate sediments, nutrients, and global change. In: *The History and Sedimentology of Ancient Reef Systems* (ed. G.D. Stanley), 387–427. US, Boston, MA: Springer.

Hallock, P. (2005). Global change and modern coral reefs: new opportunities to understand shallow-water carbonate depositional processes. *Sedimentary Geology* 175: 19–33.

Hallock, P. and Schlager, W. (1986). Nutrient excess and the demise of coral reefs and carbonate platforms. *PALAIOS* 1: 389–398.

Hallock, P. and Seddighi, M. (2022). Why did some larger benthic foraminifera become so large and flat? *Sedimentology* 69: 74–87.

Hallock, P., Williams, D.E., Fisher, E.M., and Toler, S.K. (2006). Bleaching in foraminifera with algal sybionts: implications for reed monitoring and risk assessment. *Anuário do Instituto de Geociências* 29: 108–128.

Hampson, G.J., Steel, R.J., Burgess, P.M., and Dalrymple, R.W. (ed.) (2008). *Recent Advances in Models of Siliciclastic Shallow-Marine Stratigraphy*. Tulsa, OK: SEPM (Society for Sedimentary Geology), 497 pp.

Handford, C.R. (1988). Review of carbonate sand-belt deposition of ooid grainstones and application to Mississippian Reservoir, Damme Field, Southwestern Kansas. *American Association of Petroleum Geologists Bulletin* 72: 1184–1199.

Handford, C.R. and Loucks, R.G. (1993). Carbonate depositional sequences and system tracts - responses of carbonate platforms to relative sea-level changes. In: *Carbonate Sequence Stratigraphy: Recent Developments and Applications*, AAPG Memoir, vol. 57 (ed. R.G. Loucks and J.F. Sarg), 3–41. Tulsa, OK: American Association of Petroleum Geologists.

Hanz, U., Wienberg, C., Hebbeln, D. et al. (2019). Environmental factors influencing benthic communities in the oxygen minimum zones on the Angolan and Namibian margins. *Biogeosciences* 16: 4337–4356.

Haq, B.U. and Boersma, A. (1978). *Introduction to Marine Micropaleontology*. New York, NY, USA: Elsevier, 376 pp.

Haq, B. U., Hardenbol, J., and Vail, P. R. (1988). Mesozoic and Cenozoic Chronostratigraphy and Cycles of *Sea-Level Change. In: Sea-Level Changes: An Integrated Approach* (ed. C. K. Wilgus et al.). SEPM (Special Publication No. 42), 71–108 pp.

Hardie, L.A. (1996). Secular variation in seawater chemistry: An explanation for the coupled secular variation in the mineralogies of marine limestone and potash evaporites over the past 600 m.y. *Geology* 24: 279–283.

Hardie, L.A. (2003). Secular variations in Precambrian seawater chemistry and the timing of Precambrian aragonite seas and calcite seas. *Geology* 31: 785–788.

Harris, R.C. (1969). Boron regulation in the oceans. *Nature* 223: 290–291.

Harris, P.M. (1979). *Facies Anatomy and Diagenesis of a Bahamian Ooid Shoal*, 1–161. VII: Sedimenta.

Harris, M.T. (1988). Margin and foreslope deposits of the Latemar carbonate buildup (Middle Triassic), the Dolomites, Northern Italy. PhD, Johns Hopkins University. Baltimore, Maryland, USA, 488 pp.

Harris, M.T. (1993). Reef fabrics, biotic crusts and syndepositional cements of the Latemar reef margin (Middle Triassic), northern Italy. *Sedimentology* 40: 383–402.

Harris, M.T. (1994). The foreslope and toe-of slope facies of the Middle Triassic Latemar build up (Dolomites, Northern Italy). *Journal of Sedimentary Research* B64: 132–145.

Harris, M.T. and Sheehan, P.M. (1996). Upper Ordovician-Lower Silurian depositional sequences determined from middle shelf sections, Barn Hills and Lakeside Mountains, eastern Great Basin. In: *Paleozoic Sequence Stratigraphy: Views from the North American Craton*, Geological Society of America Special Paper, vol. 306 (ed. B.J. Witzke, G.A. Ludvigson, and J. Day), 161–176. Boulder, CO: Geological Society of America.

Harris, N.B., McMillan, J.M., Knapp, L.J., and Mastalerz, M. (2018). Organic matter accumulation in the Upper Devonian Duvernay Formation, Western Canada Sedimentary Basin, from sequence stratigraphic analysis and geochemical proxies. *Sedimentary Geology* 376: 185–203.

Harris, P.M. (1996). Reef styles of modern carbonate platforms. *Bulletin of Canadian Petroleum Geology* 44: 72–81.

Harris, P.M. and Saller, A.H. (1999). Subsurface expression of the Capitan depositional system and implications for hydrocarbon reservoirs, Northeastern Delaware Basin. In: *Geologic Framework of the Capitan Reef*, SEPM Special Publication, vol. 65 (ed. A.H. Saller, P.M.M. Harris, B.L. Kirkland, and S.J. Mazzullo), 37–49. Tulsa, OK: SEPM (Society for Sedimentary Geology).

Harris, P.M., Kendall, C.G.S.C., and Lerche, I. (1985). Carbonate cementation – a brief review. In: *Carbonate Cements*, SEPM Special Publication, vol. 36 (ed. N. Schneidermann and P.M. Harris), 79–95. Tulsa, OK: SEPM (Society of Economic Paleontologists and Mineralogists).

Harris, P.M., Ellis, J., and Purkis, S.J. (2013). Assessing the extent of carbonate deposition in early rift settings. *American Association of Petroleum Geologists Bulletin* 97: 27–60.

Harris, P.M., Purkis, S.J., Ellis, J. et al. (2015). Mapping bathymetry and depositional facies on Great Bahama Bank. *Sedimentology* 62: 566–589.

Harris, P.T. (1991). Reversal of subtidal dune asymmetries caused by seasonally reversing wind-driven currents in Torres Strait, northeastern Australia. *Continental Shelf Research* 11 (7): 655–662.

Hashim, M.S. and Kaczmarek, S.E. (2021). The transformation of aragonite to calcite in the presence of magnesium: implications for marine diagenesis. *Earth and Planetary Science Letters* 574: 117166.

Hassenkam, T., Pedersen, C.S., Dalby, K., Austad, T. and Stipp, S.L.S. (2011). Pore scale observation of low salinity effects on outcrop and oil reservoir sandstone.

Colloids and Surfaces A: Physicochemical and Engineering Aspects 390: 179–188.

Haug, G.H. and Tiedemann, R. (1998). Effect of the formation of the Isthmus of Panama on Atlantic Ocean thermohaline circulation. *Nature* 393: 673–676.

Hayton, S., Nelson, C.S., and Hood, S.D. (1995). A skeletal assemblage classification system for non-tropical carbonate deposits based on New Zealand Cenozoic limestones. *Sedimentary Geology* 100: 123–141.

Heath, K.C. and Mullins, H.T. (1984). Open-ocean, off-bank transport of fine-grained carbonate sediment in the northern Bahamas. In: *Fine-Grained Sediments: Deepwater Processes and Facies*, Special Publication, vol. 15 (ed. D.A.V. Stow and D.J.W. Piper), 199–208. London, UK: Geological Society of London.

Hebbeln, D. and Samankassou, E. (2015). Where did ancient carbonate mounds grow - in bathyal depths or in shallow shelf waters? *Earth-Science Reviews* 145: 56–65.

Hebbeln, D., Wienberg, C., Wintersteller, P. et al. (2014). Environmental forcing of the Campeche cold-water coral province, southern Gulf of Mexico. *Biogeosciences* 11: 1799–1815.

Hebbeln, D., Van Rooij, D., and Wienberg, C. (2016). Good neighbours shaped by vigorous currents: cold-water coral mounds and contourites in the North Atlantic. *Marine Geology* 378: 171–185.

Hebbeln, D., Bender, M., Gaide, S. et al. (2019). Thousands of cold-water coral mounds along the Moroccan Atlantic continental margin: distribution and morphometry. *Marine Geology* 411: 51–61.

Heckel, P.H. (1972). Recognition of ancient shallow marine environments. In: *Recognition of Ancient Sedimentary Environments*, SEPM Special Publication, vol. 16 (ed. J.K. Rigby and W.K. Hamblin), 226–286. Tulsa, OK: SEPM (Society for Sedimentary Geology).

Heckel, P.H. (1974). Carbonate buildups in the geologic record: a review. In: *Reefs in Time and Space – Selected Examples from the Recent and the Ancient*, SEPM Special Publication, vol. 18 (ed. L.F. Laporte), 90–154. Tulsa, OK: SEPM (Society of Economic Paleontologists and Mineralogists).

Heckel, P. H. (1983). Diagenetic model for carbonate rocks in Midcontinent Pennsylvanian eustatic cyclothems. *Journal of Sedimentary Petrology* 53: 733–759.

Heckel, P.H. and Cocke, J.M. (1969). Phylloid algal-mound complexes in outcropping upper pennsylvanian rocks of mid-continent. *American Association of Petroleum Geologists Bulletin* 53: 1058–1074.

Heidari, A., Shokri, N., Ghasemi-Nejad, E. et al. (2015). Application of petrography, major and trace elements, carbon and oxygen isotope geochemistry to reconstruction of diagenesis of carbonate rocks of the Sanganeh Formation (Lower Cretaceous), East Kopet-Dagh Basin, NE Iran. *Arabian Journal of Geosciences* 8: 4949–4967.

Heindel, K., Birgel, D., Peckmann, J. et al. (2010). Formation of deglacial microbialites in coral reefs off tahiti (IODP 310) involving sulfate-reducing bacteria. *PALAIOS* 25: 618–635.

Heindel, K., Birgel, D., Brunner, B. et al. (2012). Post-glacial microbialite formation in coral reefs of the Pacific, Atlantic, and Indian Oceans. *Chemical Geology* 304-305: 117–130.

Hemleben, C., Spindler, M., and Anderson, O.R. (1989). *Modern Planktonic Foraminifera*. New York: Springer, 363 pp.

Hendry, J., Burgess, P., Hunt, D. et al. (2021). Seismic characterization of carbonate platforms and reservoirs: an introduction and review. *Geological Society, London, Special Publication* 509: 1–28.

Henkes, G.A., Passey, B.H., Wanamaker, A.D. et al. (2013). Carbonate clumped isotope compositions of modern marine mollusk and brachiopod shells. *Geochimica et Cosmochimica Acta* 106: 307–325.

Henrich, R., Hartmann, M., Reitner, J. et al. (1992). Facies belts and communities of the Arctic Vesterisbanken Seamount (Central Greenland Sea). *Facies* 27: 71–104.

Henrich, R., Freiwald, A., Betzler, C. et al. (1995). Controls on modern carbonate sedimentation on warm-temperate to Arctic coasts, shelves and seamounts in the northern hemisphere: implications for fossil counterparts. *Facies* 32: 71–108.

Henriet, J.P., Hamoumi, N., Da Silva, A.C. et al. (2014). Carbonate mounds: from paradox to World Heritage. *Marine Geology* 352: 89–110.

Hernández-Molina, F.J., Llave, E. and Stow, D.A.V. (2008). Chapter 19 - Continental Slope Contourites, vol. 60. In: *Developments in Sedimentology* (ed. M. Rebesco and A. Camerlenghi), 379–408. Elsevier.

Hesselbo, S.P., Jenkyns, H.C., Duarte, L. V., and Oliveira, L.C.V. (2007). Carbon-isotope record of the Early Jurassic (Toarcian) Oceanic Anoxic Event from fossil wood and marine carbonate (Lusitanian Basin, Portugal). *Earth and Planetary Science Letters* 253: 455–470.

Heubeck, C. (2009). An early ecosystem of Archean tidal microbial mats (Moodies Group, South Africa, ca. 3.2 Ga). *Geology* 37: 931–934.

Heydari, E. and Moore, C.H. (1994) Paleoceanographic and paleoclimatic controls on ooid mineralogy of the Smackover Formation, Mississippi salt basin: implications for Late Jurassic seawater composition. *Journal of Sedimentary Research A* 64: 101–114.

Hickey, B.M., MacCready, P., Elliott, E., and Kachel, N.B. (2000). Dense saline plumes in Exuma Sound, Bahamas. *Journal of Geophysical Research* 105: 11,471–11,488.

Hilgen, F.J. and Krijgsman, W. (1999). Cyclostratigraphy and astrochronology of the Tripoli diatomite formation (pre-evaporite Messinian, Sicily, Italy). *Terra Nova* 11: 16–22.

Hilgen, F., Schwarzacher, W., and Strasser, A. (2004). Concept and definitions in cyclostratigraphy (Second

Report of the Cyclostratigraphy Working Group). In: *Cyclostratigraphy: Approaches and Case Histories*, SEPM Special Publication, vol. 81 (ed. B. D'Argenio, A.G. Fischer, I.P. Silva, et al.), 303–305. Tulsa, OK: SEPM (Society for Sedimentary Geology).

Hillgärtner, H. (2010). Anatomy of a microbially constructed, high-energy, ocean-facing carbonate platform margin (earliest Aptian, northern Oman Mountains). In: *Barremian – Aptian Stratigraphy and Hydrocarbon Habitat of the Eastern Arabian Plate (vol. 1)*, GeoArabia Special Publication 4, vol. 1 (ed. F.S.P. van Buchem, M.I. Al-Husseini, F. Maurer, and H.J. Droste), 285–300. Bahrain: Gulf PetroLink.

Hillgärtner, H. and Strasser, A. (2003). Quantification of high-frequency sea-level fluctuations in shallow-water carbonates: an example from the Berriasian–Valanginian (French Jura). *Palaeogeography, Palaeoclimatology, Palaeoecology* 200: 43–63.

Hillgärtner, H., Dupraz, C., and Hug, W. (2001). Microbially induced cementation of carbonate sands: are micritic meniscus cements good indicators of vadose diagenesis? *Sedimentology* 48: 117–131.

Hillgärtner, H., van Buchem, F.S.P., Gaumet, F. et al. (2003). The Barremian-Aptian evolution of the Eastern Arabian carbonate platform margin (Northern Oman). *Journal of Sedimentary Research* 73: 756–773.

Hine, A.C., Feary, D.A. and Malone, M.J. (Eds). (2004). *Proceedings ODP - Volume 182, Sci. Results - Great Australian Bight*. Ocean Drilling Program, College Station, TX, USA.

Hinestrosa, G., Webster, J.M., Beaman, R.J., and Anderson, L.M. (2014). Seismic stratigraphy and development of the shelf-edge reefs of the Great Barrier Reef, Australia. *Marine Geology* 353: 1–20.

Hinestrosa, G., Webster, J.M., and Beaman, R.J. (2022). New constraints on the postglacial shallow-water carbonate accumulation in the Great Barrier Reef. *Scientific Reports* 12: 924.

Hladil, J. (2005). The formation of stromatactis-type fenestral structures during the sedimentation of experimental slurries – a possible clue to a 120-year-old puzzle about stromatactis. *Bulletin of Geosciences* 80: 193–211.

Hönisch, B., Ridgwell, A., Schmidt, D.N. et al. (2012). The geological record of ocean acidification. *Science* 335: 1058–1063.

Höfling, R. and Scott, R.W. (2002). Early and Mid-Cretaceous buildups. In: *Phanerozoic Reef Patterns*, SEPM Special Publication, vol. 72 (ed. W. Kiessling, E. Flügel, and J. Golonka), 521–548. Tulsa, OK: SEPM (Society for Sedimentary Geology).

Hofmann, H.J. (2000). Archean stromatolites as microbial archives. In: *Microbial Sediments* (ed. R.E. Riding and S.M. Awramik), 315–327. Berlin Heidelberg: Springer.

Hofmann, H.J., Grey, K., Hickman, A.H., and Thorpe, R.I. (1999). Origin of 3.45 Ga coniform stromatolites in Warrawoona Group, Western Australia. *Geological Society of America Bulletin* 111: 1256–1262.

Hohenegger, J. (2005). Estimation of environmental paleogradient values based on presence/absence data: a case study using benthic foraminifera for paleodepth estimation. *Palaeogeography, Palaeoclimatology, Palaeoecology* 217: 115–130.

Holbourn, A., Kuhnt, W., and James, N. (2002). Late Pleistocene bryozoan reef mounds of the Great Australian Bight: isotope stratigraphy and benthic foraminiferal record. *Paleoceanography* 17: 1–13.

Holbourn, A., Kuhnt, W., and Xu, J. (2011). Indonesian throughflow variability during the last 140 ka: the Timor Sea outflow. In: *The SE Asian Gateway: History and Tectonics of the Australia–Asia Collision*, Special Publication, vol. 355 (ed. R. Hall, M.A. Cottam, and M.E.J. Wilson), 283–303. London: Geological Society of London.

Holbourn, A., Kuhnt, W., Kochhann, K.G. D. et al. (2015). Global perturbation of the carbon cycle at the onset of the Miocene Climatic Optimum. *Geology* 43: 123–126.

Homann, M., Heubeck, C., Airo, A., and Tice, M.M. (2015). Morphological adaptations of 3.22 Ga-old tufted microbial mats to Archean coastal habitats (Moodies Group, Barberton Greenstone Belt, South Africa). *Precambrian Research* 266: 47–64.

Homewood, P.W. and Eberli, G.P. (2000). Genetic stratigraphy on the exploration and production scales [Texte imprimé]: case studies from the upper Devonian of Alberta and the Pennsylvanian of the Paradox basin. *Bulletin des Centres de recherche exploration-production Elf-Aquitaine. Mémoire* 24: 290.

Homewood, P.W., Razin, P., Grelaud, C. et al. (2008). Outcrop sedimentology of the Natich Formation, northern Oman: a field guide to selected outcrops in the Adam Foothills and Al Jabal al Akhdar areas. *GeoArabia* 13: 39–120.

Hopley, D. (1982). *The Geomorphology of the Great Barrier Reef.* New York, USA: Wiley-Interscience, 453 pp.

Hopley, D., Parnell, K., and Smithers, S.G. (ed.) (2007). *The Geomorphology of the Great Barrier Reef: Development, Diversity and Change*. Cambridge: Cambridge University Press, 532 pp.

Hopson, H.H., Droxler, A.W., Khanna, P. et al. (2025). Controls on the evolution of large Upper Cambrian microbial buildups in Mason County, Texas, USA: Bridging the gaps between scales from microbes to large buildup architecture. *GSA Bulletin* 137: 5381–5402.

Horowitz, A.S. and Potter, P.E. (1971). *Introductory Petrography of Fossils*. Berlin Heidelberg New York: Springer Verlag, 296 pp.

Hossain, Z. and Zhou, Y. (2015). Petrophysics and rock physics modeling of diagenetically altered sandstone. *Interpretation* 3: SA107–SA120.

Hottinger, L. (1983). Processes determining the distribution of larger foraminifera in space and time. *Utrecht Micropaleontological Bulletins* 30: 239–253.

Hottinger, L. (2014). *Paleogene Larger Rotaliid Foraminifera from the Western and Central Neotethys*. Cham: Springer International Publishing.

House, M.R. and Gale, A.S. (ed.) (1995). *Orbital Forcing Timescales and Cyclostratigraphy*. London: Geological Society, London, 204 pp.

Hovland, M., Croker, P.F., and Martin, M. (1994). Fault-associated seabed mounds (carbonate knolls?) off western Ireland and north-western Australia. *Marine and Petroleum Geology* 11: 232–246.

Hubbard, D.K., Burke, R.B. and Gill, I.P. (1986). Styles of reef accretion along a steep, shelf-edge reef, St. Croix, U.S. Virgin Islands. *Journal of Sedimentary Research* 56: 848–861.

Hübscher, C. and Kukowski, N. (2003). Complex BSR pattern in the Yaquina Basin off Peru. *Geo-Marine Letters* 23: 91–101.

Hudson, J.D. (1977). Stable isotopes and limestone lithification. *Journal of the Geological Society* 133: 637–660.

Humphreys, A.F., Halfar, J., Rivera, F. et al. (2016). Variable El Niño-Southern Oscillation influence on biofacies dynamics of eastern Pacific shallow-water carbonate systems. *Geology* 44: 571–574.

Humphreys, A.F., Halfar, J., Ingle, J.C. et al. (2018). Effect of seawater temperature, pH, and nutrients on the distribution and character of low abundance shallow water benthic foraminifera in the Galápagos. *PLoS One* 13: e0202746.

Hüneke, H., Hernández-Molina, F.J., Rodríguez-Tovar, F.J. et al. (2021). Diagnostic criteria using microfacies for calcareous contourites, turbidites and pelagites in the Eocene–Miocene slope succession, southern Cyprus. *Sedimentology* 68: 557–592.

Hüneke, H., Gibb, M.A., Mayer, O. et al. (2023). Bioclastic bottom-current deposits of a Devonian contourite terrace: facies variability and depositional architecture (Tafilalt Platform, Morocco). *Sedimentology* 70: 1413–1471.

Husinec, A. and Jelaska, V. (2006). Relative sea-level changes recorded on an isolated carbonate platform: Tithonian to Cenomanian succession, Southern Croatia. *Journal of Sedimentary Research* 76: 1120–1136.

Husinec, A. and Read, J.F. (2011). Microbial laminite versus rooted and burrowed caps on peritidal cycles: Salinity control on parasequence development, Early Cretaceous isolated carbonate platform, Croatia. *Geological Society of America Bulletin* 123: 1896–1907.

Huvenne, V., Bailey, W., Shannon, P. et al. (2007). The Magellan mound province in the Porcupine Basin. *International Journal of Earth Sciences* 96: 85–101.

Hvid, J.M., van Buchem, F., Andreasen, F. et al. (2021). Stratigraphy and petrophysical characteristics of Lower Paleocene cool-water carbonates, Faxe quarry, Denmark. *Bulletin of the Geological Society of Denmark* 69: 97–121.

Immenhauser, A. (2022). On the delimitation of the carbonate burial realm. *The Depositional Record* 8: 524–574.

Insalaco, E., Virgone, A., Courme, B. et al. (2006). Upper Dalan Member and Kangan Formation between the Zagros Mountains and offshore Fars, Iran: depositional system, biostratigraphy and stratigraphic architecture. *GeoArabia* 11: 75–176.

Irwin, H., Curtis, C., and Coleman, M. (1977). Isotopic evidence for source of diagenetic carbonates formed during burial of organic-rich sediments. *Nature* 269: 209–213.

Jafarian, E., Kleipool, L.M., Scheibner, C. et al. (2017). Variations in petrophysical properties of Upper Palaeozoic mixed carbonate and non-carbonate deposits, Spitsbergen, Svalbard Archipelago. *Journal of Petroleum Geology* 40: 59–83.

Jahnert, R.J. and Collins, L.B. (2012). Characteristics, distribution and morphogenesis of subtidal microbial systems in Shark Bay, Australia. *Marine Geology* 303–306: 115–136.

Jahnert, R.J. and Collins, L.B. (2013). Controls on microbial activity and tidal flat evolution in Shark Bay, Western Australia. *Sedimentology* 60, 1071–1099.

Jakubowicz, M., Belka, Z. and Berkowski, B. (2014). Frutexites encrustations on rugose corals (Middle Devonian, southern Morocco): complex growth of microbial microstromatolites. *Facies* 60, 631–650

James, N.P. (1983). Depositional models for carbonate rocks. In: *Sediment Diagenesis*, NATO ASI Series, vol. 115 (ed. A. Parker and B.W. Sellwood), 289–348. Dordrecht: Springer.

James, N.P. (1984). Reefs. In: *Facies Models, Second Edition*, Geoscience Canada, Reprint Series 1 (ed. R.G. Walker), 229–244. Toronto: Geological Association of Canada.

James, N.P. (1997). The cool-water carbonate depositional realm. In: *Cool-Water Carbonates: Society for Sedimentary Geology Special Publication*, SEPM Special Publication no. 56 (ed. N.P. James and A.D. Clarke), 1–20. Tulsa, OK: SEPM (Society for Sedimentary Geology).

James, N.P. and Bone, Y. (1989). Petrogenesis of Cenozoic, temperature water calcarenites, South Australia: a model. *Journal of Sedimentary Petrology* 59: 191–203.

James, N.P. and Bone, Y. (2011). *Neritic Carbonate Sediments in a Temperate Realm – Southern Australia*. Dordrecht, Heidelberg, London, New York: Springer, 254 pp.

James, N.P. and Bone, Y. (2015). Pleistocene aeolianites at Cape Spencer, South Australia; record of a vanished inner neritic cool-water carbonate factory. *Sedimentology* 62: 2038–2059.

James, N.P. and Bone, Y. (2017). Quaternary aeolianites in south-east Australia – a conceptual linkage between marine source and terrestrial deposition. *Sedimentology* 64: 1005–1043.

James, N.P. and Bone, Y. (2022). Sedimentology and diagenesis of Pleistocene calcareous, non-dolomitic, interdune sediments, Coorong Coastal Plain, South Australia. *Sedimentology* 69: 2814–2843.

James, N.P. and Bourque, P.-A. (1992). Reefs and mounds. In: *Facies Models - Response to Sea Level Change* (ed. R.G. Walker and N.P. James), 323–347. St.

John's Newfoundland: Geological Association of Canada.

James, N.P. and Choquette, P.W. (1988). *Paleokarst*. NY: Springer-Verlag, 416 pp.

James, N.P. and Clarke, J.A.D. (1997). *Cool-Water Carbonates*. Tulsa, OK: Society for Sedimentary Geology, 440 pp.

James, N.P. and Ginsburg, R.N. (1979). *The Seaward Margin of Belize Barrier and Atoll Reefs*, IAS Special Publication, vol. 3. Oxford: International Association of Sedimentologists – Blackwell, 191 pp.

James, N.P. and Jones, B. (2016) *Origin of Carbonate Sedimentary Rocks*. Chichester: AGU – John Wiley & Sons, 446 pp.

James, N.P. and Lukasik, J.J. (2010). Cool- and cold-water carbonates. In: *Facies Models 4* (ed. N. James and R. Dalrymple), 369–398. St. John's Newfoundland: Geological Association of Canada.

James, N.P. and Mountjoy, E.W. (1983). Shelf-slope break in fossil carbonate platforms: an overview. In: *The Shelfbreak: Critical Interface on Continental Margins*, vol. 33 (ed. D.J. Stanley and G.T. Moore), 189–206. Tulsa, OK: SEPM (Society for Sedimentary Geology).

James, N.P. and Pufahl, P.K. (2026). Carbonate sedimentology: An evolved discipline. *The Depositional Record* 12: e70042.

James, N.P. and von der Borch, C. (1991). Carbonate shelf edge of Southern Australia: a prograding open-platform margin. *Geology* 19: 1005–1008.

James, N.P., Bone, Y., von der Borch, C.C., and Gostin, V.A. (1992). Modern carbonate and terrigenous sediments on a cool water, high energy, mid-latitude shelf: Lacepede, Southern Australia. *Sedimentology* 39: 877–905.

James, N.P., Bone, Y., Collins, L.B. and Kyser, T.K. (2001). Surficial sediments of the Great Australian Bight: facies dynamics and oceanography on a vast cool-water carbonate shelf. *Journal of Sedimentary Research* 71: 549–567.

James, N.P., Boreen, T.D., Bone, Y., and Feary, D.A. (1994). Holocene carbonate sedimentation on the west Eucla Shelf, Great Australian Bight: a shaved shelf. *Sedimentary Geology* 90: 161–177.

James, N.P., Collins, L.B., Bone, Y., and Hallock, P. (1999). Subtropical carbonates in a temperate realm: modern sediments on the southwest Australian Shelf. *Journal of Sedimentary Research* 69: 1297–1321.

James, N.P., Feary, D.A., Surlyk, F. et al. (2000). Quaternary bryozoan reef mounds in cool-water, upper slope environments: Great Australian Bight. *Geology* 28: 647–650.

James, N.P., Feary, D.A., Betzler, C. et al. (2004). Origin of Late Pleistocene Bryozoan reef mounds; Great Australian Bight. *Journal of Sedimentary Research* 74: 20–48.

James, N.P., Frank, T.D. and Fielding, C.R. (2009). Carbonate Sedimentation in a Permian High-Latitude, Subpolar Depositional Realm: Queensland, Australia. *Journal of Sedimentary Research* 79: 125–143.

James, N.P., Reid, C.M., Bone, Y. et al. (2013). The macroalgal carbonate factory at a cool-to-warm temperate marine transition, Southern Australia. *Sedimentary Geology* 291: 1–26.

James, N.P., Seibel, M.J., Dalrymple, R.W. et al. (2014). Warm-temperate, marine, carbonate sedimentation in an Early Miocene, tide-influenced, incised valley; Provence, south-east France. *Sedimentology* 61: 497–534.

Janson, X., Eberli, G.P., Bonnaffe, F. et al. (2007). Seismic expressions of a Miocene prograding carbonate margin, Mut Basin, Turkey. *American Association of Petroleum Geologists Bulletin* 91: 685–713.

Janson, X., van Buchem, F.S.P., Dromart, G. et al. (2010). Architecture and facies differentiation within a Middle Miocene carbonate platform, Ermenek, Mut Basin, southern Turkey. In: *Mesozoic and Cenozoic Carbonate Systems of the Mediterranean and the Middle East: Stratigraphic and Diagenetic Reference Models*, Special Publication, vol. 329 (ed. F.S.P. Van Buchem, K.D. Gerdes, and M. Esteban), 265–290. London: Geological Society of London.

Janson, X., Kerans, C., Loucks, R. et al. (2011). Seismic architecture of a Lower Cretaceous platform-to-slope system, Santa Agueda and Poza Rica fields, Mexico. *American Association of Petroleum Geologists Bulletin* 95: 105–146.

Jeffery, D.L. and Stanton, R.J. (1996). Growth history of Lower Mississippian Waulsortian Mounds: distribution, stratal patterns, and geometries, New Mexico. *Facies* 29–58.

Jell, J.S. (1984). Cambrian cnidarians with mineralized skeletons. *Palaeontographica Americana* 54: 105–109.

Jervey, M.T. (1988). Quantitative geological modeling of siliciclastic rock sequences and their seismic expression. In: *Sea-Level Changes: an Integrated Approach*, Special Publication, vol. 42 (ed. C.K. Wilgus, B. S. Hastings, C.G.S.C. Kendall, et al.), 47–69. Tulsa, OK: Society of Economic Paleontologists and Mineralogists.

Jiang, W., Yu, K., Fan, T. et al. (2019). Coral reef carbonate record of the Pliocene-Pleistocene climate transition from an atoll in the South China Sea. *Marine Geology* 411: 88–97.

Jiménez-Sánchez, A., Villas, E., and Vennin, E. (2015). New trepostomate bryozoans from the Upper Ordovician of Morocco and the temperature influence on zooid size. *Journal of Paleontology* 89: 385–404.

Jindrich, V. (1983). Structure and diagenesis of Recent algal-foraminifer reefs, Fernando de Noronha, Brazil. *Journal of Sedimentary Research* 53: 449–459.

Jo, A., Eberli, G.P., and Grasmueck, M. (2015). Margin collapse and slope failure along southwestern Great Bahama Bank. *Sedimentary Geology* 317: 43–52.

Joachimski, M.M. and Buggisch, W. (1993). Anoxic events in the late Frasnian—Causes of the

Frasnian-Famennian faunal crisis? *Geology* 21: 675–678.

Joachimski, M.M. and Buggisch, W. (1996). The Upper Devonian reef crisis - insights from the Carbon isotope record. In: *Global and Regional Controls on Biogenic Sedimentation 1. Reef-Evolution. Research Reports*, Göttinger Arbeiten zur Geologie und Paläontologie, Sonderband, vol. 2 (ed. J. Reitner, F. Neuweiler, and F. Gunkel), 365–370. Göttingen: Geologisches Institut der Georg-August-Universität Göttingen.

Joachimski, M.M. and Buggisch, W. (1999). Hydrothermal origin of Devonian conical mounds (Kess-Kess) of Hamar Laghdad Ridge, Anti-Atlas, Morocco. *Geology* 27: 863.

Joachimski, M.M., Buggisch, W., Mounji, D., Bourque, P.A. and Savard, M.M. (1999). Hydrothermal origin of Devonian conical mounds (Kess-Kess) of Hamar Lakhdad Ridge, Anti-Atlas, Morocco: Comment and Reply. *Geology* 27: 863–864.

Johnson, M.E. (1987). Extent and bathymetry of North American Platform Seas in the Early Silurian. *Paleoceanography* 2: 185–211.

Johnson, M.E. and Webb, G.E. (2007). Outer rocky shores of the Mowanbini Archipelago, Devonian Reef Complex, Canning Basin, Western Australia. *The Journal of Geology* 115: 583–600.

Johnson, J.G., Klapper, G., and Sandberg, C.A. (1985). Devonian eustatic fluctuations in Euramerica. *Geological Society of America Bulletin* 96: 567–587.

Johnson, C.C., Sanders, D., Kauffman, E. G., and Hay, W.W. (2002). Patterns and processes influencing Upper Cretaceous reefs. In: *Phanerozoic Reef Patterns*, SEPM Special Puclication, vol. 72 (ed. W. Kiessling, E. Flügel, and J. Golonka), 549–585. Tulsa, OK: SEPM (Society for Sedimentary Geology).

Jokiel, P.L., Rodgers, K.S., Kuffner, I.B. et al. (2008). Ocean acidification and calcifying reef organisms: a mesocosm investigation. *Coral Reefs* 27: 473–483.

Jones, B., Renaut, R.W., and Konhauser, K.O. (2005). Genesis of large siliceous stromatolites at Frying Pan Lake, Waimangu geothermal field, North Island, New Zealand. *Sedimentology* 52: 1229–1252.

Jorry, S. (2004). The Eocene nummulite carbonates (central Tunisia and NE Libya): sedimentology, depositional environments, and application to oil reservoirs. PhD thesis. Section des sciences de la Terre-Université de Genève, Genève, Switzerland, 206 pp.

Jorry, S., Davaud, E., and Caline, B. (2003). Controls on the distribution of nummulite facies: a case study from the late Ypresian El Garia Formation (Kesra Plateau, Central Tunisia). *Journal of Petroleum Geology* 26: 283–306.

Jorry, S.J., Hasler, C.-A., and Davaud, E. (2006). Hydrodynamic behaviour of *Nummulites*: implications for depositional models. *Facies* 52: 221–235.

Jorry, S.J., Droxler, A.W., and Francis, J.M. (2010). Deepwater carbonate deposition in response to re-flooding of carbonate bank and atoll-tops at glacial terminations. *Quaternary Science Reviews* 29: 2010–2026.

Jorry, S.J., Jouet, G., Edinger, E.N. et al. (2020). From platform top to adjacent deep sea: new source-to-sink insights into carbonate sediment production and transfer in the SW Indian Ocean (Glorieuses archipelago). *Marine Geology* 423: 106144.

Kah, L.C. and Bartley, J.K. (2022). Carbonate fabric diversity and environmental heterogeneity in the late Mesoproterozoic Era. *Geological Magazine* 159: 220–246.

Kammer, T.W. and Ausich, W.I. (2006). The "age of crinoids": a mississippian biodiversity spike coincident with widespread carbonate ramps. *PALAIOS* 21: 238–248.

Kämpf, J. and Chapman, P. (2016). The Benguela current upwelling system. In *Upwelling Systems of the World: A Scientific Journey to the Most Productive Marine Ecosystems*. Cham: Springer International Publishing.

Karhu, J.A. and Holland, H.D. (1996). Carbon isotopes and the rise of atmospheric oxygen. *Geology* 24: 867–870.

Katz, B. (ed.) (1995a). *Petroleum Source Rocks*. Berlin–Heidelberg: Springer-Verlag, 327 pp.

Katz, B.J. (1995b). The Green River Shale: an Eocene Carbonate Lacustrine Source Rock. In: *Petroleum Source Rocks* (ed. B. J. Katz), 309–324. Berlin Heidelberg, Berlin, Heidelberg: Springer.

Katz, D.A., Playton, T., Bellian, J. et al. (2010). Slope Heterogeneity and Production Results in a Steep-sided Upper Paleozoic Isolated Carbonate Platform Reservoir, Karachaganak Field, Kazakhstan. *SPE Caspian Carbonates Technology Conference*, pp. SPE-139960-MS, Atyrau, Kazakhstan.

Kauffman, E.G. and Caldwell, W.G.E. (1993). The Western Interior Basin in space and time. In: *Evolution of the Western Interior Basin*, Special Paper, vol. 39 (ed. W.G.E. Caldwell and E.G. Kauffman), 1–30. St. John's, Canada: Geological Association of Canada.

Kauffman, E.G. and Johnson, C.C. (1988). The morphological and ecological evolution of Middle and Upper Cretaceous reef-building Rudistids. *PALAIOS* 3: 194–216.

Kaufmann, B. (1998). Facies, stratigraphy and diagenesis of Middle Devonian reef- and mud-mounds in the Mader (eastern Anti-Atlas, Morocco). *Acta Geologica Polonica* 48: 43–106.

Kaufmann, B. and Wendt, J. (2000). Calcite cement successions in Middle Devonian (Givetian) carbonate mud buildups of the southern Ahnet basin (Algerian Sahara). *Carbonates and Evaporites* 15: 149–161.

Keim, L. and Brandner, R. (2002). Facies interfingering and synsedimentary tectonics on Late Ladinian-Early Carnian carbonate platforms (Dolomtes, Italy). *International Journal of Earth Sciences* 91: 555–556.

Keim, L. and Schlager, W. (1999). Automicrite facies on steep slopes (Triassic, Dolomites, Italy). *Facies* 41: 15–26.

Keim, L. and Schlager, W. (2001). Quantitative compositional analysis of a Triassic carbonate platform (Southern Alps, Italy). *Sedimentary Geology* 139: 261–283.

Keim, L., Brandner, R., Krystyn, L., and Mette, W. (2001). Termination of carbonate slope progradation: an example from the Carnian of the Dolomites, Northern Italy. *Sedimentary Geology* 143: 303–323.

Kelling, G., Robertson, A., and Van Buchem, F. (2005). Cenozoic sedimentary basins of southern Turkey: an introduction. *Sedimentary Geology* 173: 1–13.

Kempe, S. and Kazmierczak, J. (1994). The role of alkalinity in the evolution of ocean chemistry, organization of living systems, and biocalcification processes. In: *Past and present biomineralization processes – Considerations about the carbonate cycle* (ed. F. Doumenge), 61–117. Monaco: Bulletin de l'Institut oceanographique no. spec. 13.

Kench, P.S., Smithers, S.G., and McLean, R.F. (2012). Rapid reef island formation and stability over an emerging reef flat: Bewick Cay, northern Great Barrier Reef, Australia. *Geology* 40: 347–350.

Kendall, C.G.S.C. and Alsharhan, A.S. (ed.) (2011). *Quaternary Carbonate and Evaporite Sedimentary Facies and Their Ancient Analogues: A Tribute to Douglas James Shearman*. Oxford, UK: Wiley-Blackwell, 496 pp.

Kendall, C. and McDonnell, J.J. (ed.) (1998). *Isotope Tracers in Catchment Hydrology*. Amsterdam: Elsevier, 839 pp.

Kendall, C.G., Moore, P., Viiparelli, E. et al. (2014). Analysis of sequence stratigraphic models for the jurassic cretaceous sedimentary fill of the intrashelf basins of the Eastern Margin the Arabian Plate. *AAPG Annual Convention and Exhibition*, no. 30326, Houston, Texas, USA.

Kenig, F., Huc, A.Y., Purser, B.H., and Oudin, J.-L. (1990). Sedimentation, distribution and diagenesis of organic matter in a recent carbonate environment, Abu Dhabi, U.A.E. *Organic Geochemistry* 16: 735–747.

Kennedy, D.M. and Woodroffe, C.D. (2002). Fringing reef growth and morphology: a review. *Earth-Science Reviews* 57: 255–277.

Kenter, J.A.M. (1990). Carbonate platform flanks: slope angle and sediment fabric. *Sedimentology* 37: 777–794.

Kenter, J.A.M. and Campbell, A.E. (1991). Sedimentation on a Lower Jurassic carbonate platform flank: geometry, sediment fabric and related depositional structures (Djebel Bou Dahar, High Atlas, Morocco). *Sedimentary Geology* 72: 1–34.

Kenter, J.A.M. and Schlager, W. (2009). Slope angle and basin depth of the Triassic platform-basin transition at the Gosaukamm, Austria. *Austrian Journal of Earth Sciences* 102: 15–22.

Kenter, J.A.M., Anselmetti, F.S., Kramer, P.H. et al. (2002). Acoustic properties of "young" carbonate rocks, ODP Leg 166 and boreholes Clino and Unda, Western Great Bahama Bank. *Journal of Sedimentary Research* 72: 129–137.

Kenter, J.A.M., Harris, P.M., and Della Porta, G. (2005). Steep microbial boundstone-dominated platform margins—examples and implications. *Sedimentary Geology* 178: 5–30.

Kenter, J.A.M., Harris, P.M., Collins, J.F., Weber, L.J., Kuanysheva, G. and Fischer, D.J. (2006). Late Visean to Bashkirian platform cyclicity in the central Tengiz buildup, Precaspian Basin, Kazakhstan: Depositional evolution and reservoir development. In: *Giant hydrocarbon reservoirs of the world: From rocks to reservoir characterization and modeling* (ed. P.M. Harris and L.J. Weber), AAPG Memoir 88 / SEPM Special Publication, 7–54. Tulsa: AAPG and SEPM.

Kenter, J.A.M., Harris, P.M. and Collins, J. F. (2008). Facies and Reservoir Quality of the Tengiz Isolated Platform, Pricaspian Basin, Kazakhstan. *AAPG and AAPG European region Energy Conference & Exhibition*. American Association of Petroleum Geologists, Athens, Greece.

Kenter, J.A.M., Playton, T., Harries, P. et al. (2010a). Application of outcrop analogs to characterize carbonate reservoirs in the Pricaspian Basin. *SPE Caspian Carbonates Technology Conference*, pp. SPE-139934-MS, Atyrau, Kazakhstan.

Kenter, J.A.M., Tankersley, T., Skalinski, M. et al. (2010b). Tengiz field (Republic of Kazakhstan) Unit 1 Platform Static Model: Using a Hybrid Depositional – Diagenetic Approach. *SPE Caspian Carbonates Technology Conference*, pp. SPE-139935-MS.

Kenyon, P.M. and Turcotte, D.L. (1985). Morphology of a delta prograding by bulk sediment transport. *Geological Society of America Bulletin* 96: 1457–1465.

Kenyon, N.H., Akhmetzhanov, A.M., Wheeler, A.J. et al. (2003). Giant carbonate mud mounds in the southern Rockall Trough. *Marine Geology* 195: 5–30.

Kerans, C. 1985. Petrology of Devonian and Carboniferous carbonates of the Canning and Bonaparte basins. Western Australian Mining and Petroleum Research Institute, Report No. 12, Perth, Australia, 384 pp.

Kerans, C. and Fitchen, W.M. (1995). *Sequence Hierarchy and Facies Architecture of a Carbonate-Ramp System: San Andres Formation of Algerita Escarpment and Western Guadalupe Mountains, West Texas and New Mexico*, vol. 235. Austin, TX: The University of Texas at Austin, Bureau of Economic Geology, 86 pp.

Kerans, C. and Harris, P.M. (1993). Outer shelf and shelf crest. In: *Guide to the Permian Reef Geology Trail, McKittrick Canyon, Guadalupe Mountains National Park, Texas*, Guidebook 26 (ed. D.G. Bebout and C. Kerans), 32–43. Austin, Texas: Texas Bureau of Economic Ecology - Bureau of Economic Geology - University of Texas.

Kerans, C. and Harris, P.M.M. (2008). Devonian reefal platforms of the canning basin – lessons learned and value as analogs. *AAPG Annual Convention 2007*, pp. Search and

Discovery Article #40302. AAPG, Long Beach, CA, USA.

Kerans, C. and Kempter, K. (2002). *Hierarchical Stratigraphic Analysis of a Carbonate Platform, Permian of the Guadalupe Mountains, American Association of Petroleum Geologists/ Datapages Discovery Series No. 5, pp. CD-ROM*. Austin: The University of Texas at Austin, Bureau of Economic Geology.

Kerans, C., Bellian, J.A., Playton, T.E. et al. (2012). 3-D sequence framework of Leonardian through Guadalupian shelf-to-basin successions of the Guadalupe Mountains. In: *Digital Geospatial Context for 3-D Source-to-Sink Models: New Insights into the Classic Shelf to Basin System of the Guadalupe and Delaware Mountains*, SEPM Research Conference Extended Abstract (ed. C. Kerans and J.A. Bellian), 45–57, April 26–30, 2012, Carlsbad, NM, USA.

Kershaw, S. (1994). Classification and geological significance of biostromes. *Facies* 31(1): 81–91.

Kershaw, S., Zhang, T., and Li, Y. (2021). Calcilobes wangshenghaii n. gen., n. sp., microbial constructor of Permian–Triassic boundary microbialites of South China, and its place in microbialite classification. *Facies* 67(4): 28.

Keupp, H., Jenisch, A., Herrmann, R. et al. (1993). Microbial carbonate crusts – a key to the environmental analysis of fossil spongiolites? *Facies* 29: 41–54.

Khanna, P., Hopson, H.H., Droxler, A.W. et al. (2020a). Late Cambrian microbial build-ups, Llano Area, Central Texas: a three-phase morphological evolution. *Sedimentology* 67: 1135–1160.

Khanna, P., Pyrcz, M., Droxler, A.W. et al. (2020b). Implications for controls on Upper Cambrian microbial build-ups across multiple-scales, Mason County, Central Texas, USA. *Marine and Petroleum Geology* 121: 104590.

Kiessling, W. (2001). Phanerozoic reef trends based on the paleoreef database. In: *The History and Sedimentology of Ancient Reef Systems*, Topics in Geobiology (ed. G.D. Stanley Jr.), 41–88. New York: Kluwer Academic/Plenum Publishers.

Kiessling, W. (2002). Secular variations in the Phanerozoic reef ecosystem. In: *Phanerozoic Reef Patterns* (ed. W. Kiessling, E. Flügel and J. Golonka). Tulsa, OK: SEPM (Society for Sedimentary Geology), 775 pp.

Kiessling, W. (2009). Geologic and biologic controls on the evolution of reefs. *Annual Review of Ecology, Evolution, and Systematics* 40: 173–192.

Kiessling, W. and Simpson, C. (2011). On the potential for ocean acidification to be a general cause of ancient reef crises. *Global Change Biology* 17: 56–67.

Kiessling, W., Flügel, E., and Golonka, J. (1999). Paleoreef maps: evaluation of a comprehensive database on phanerozoic reefs. *American Association of Petroleum Geologists Bulletin* 83: 1552–1587.

Kiessling, W., Flügel, E., and Golonka, J. (ed.) (2002). *Phanerozoic Reef Patterns*. Tulsa, OK: SEPM (Society for Sedimentary Geology), 775 pp.

Kiessling, W., Flügel, E., and Golonka, J. (2003). Patterns of Phanerozoic carbonate platform sedimentation. *Lethaia* 36: 195–226.

Kiessling, W., Simpson, C., and Foote, M. (2010). Reefs as cradles of evolution and sources of biodiversity in the phanerozoic. *Science* 327: 196–198.

Kievman, C.M. (1998). Match between Late Pleistocene Great Bahama Bank and deep-sea oxygen isotope records of sea level. *Geology* 26: 635–638.

Kirkby, K.C. and Hunt, D. (1996). Episodic growth of a Waulsortian buildup: the Lower Carboniferous Muleshoe Mound, Sacramento Mountains, New Mexico, USA. *Geological Society, London, Special Publication* 107: 97–110.

Kirkland, B.L., Dickson, J.A.D., Wood, R. A., and Land, L.S. (1998). Microbialite and microstratigraphy: the origin of encrustations in the middle and Upper Capitan Formation, Guadalupe Mountains, Texas and New Mexico, U.S.A. *Journal of Sedimentary Research* 68: 956–969.

Kleipool, L.M., Reijmer, J.J.G., Bádenas, B., and Aurell, M. (2015). Variations in petrophysical properties along a mixed siliciclastic carbonate ramp (Upper Jurassic, Ricla, NE Spain). *Marine and Petroleum Geology* 68 (Part A): 158–177.

Kleipool, L.M., de Jong, K., de Vaal, E.L., and Reijmer, J.J.G. (2017). Seismic characterization of switching platform geometries and dominant carbonate producers (Miocene, Las Negras, Spain). *Sedimentology* 64: 1676–1707.

Kleypas, J.A. (1996). Coral reef development under naturally turbid conditions: fringing reefs near Broad Sound, Australia. *Coral Reefs* 15: 153–167.

Klicpera, A., Michel, J., and Westphal, H. (2015). Facies patterns of a tropical heterozoan carbonate platform under eutrophic conditions: the Banc d'Arguin, Mauritania. *Facies* 61: 421.

Klimchouk, A. (2017). Types and settings of hypogene karst. In: *Hypogene Karst Regions and Caves of the World* (ed. A. Klimchouk, A.N. Palmer, J.D. Waele, et al.), 1–39. Cham: Springer Nature.

Klug, C., Samankassou, E., Pohle, A. et al. (2018). Oases of biodiversity: Early Devonian palaeoecology at Hamar Laghdad, Morocco. *Neues Jahrbuch für Geologie und Paläontologie* 290: 9–48.

Knoll, A.H. (1989). The paleomicrobiological information in Proterozoic rocks. In: *Microbial Mats: Physiological Ecology of Benthic Microbial Communities* (ed. Y. Cohen and E. Rosenberg), 469–485. Washington DC: ASM Press.

Koehrer, B., Zeller, M., Aigner, T., Poeppelreiter, M., Milroy, P., Forke, H. and Al-Kindi, S. (2010). Facies and stratigraphic framework of a Khuff outcrop equivalent: Saiq and Mahil formations, Al Jabal al-Akhdar, Sultanate of Oman. *GeoArabia* 15: 91–156.

Kohout, F.A. (1967). Ground water flow and the geothermal regime of the Floridian Plateau. *Transactions, Gulf Coast Association of Geological Societies* 17: 339–354.

Kolodka, C., Vennin, E., Vachard, D. et al. (2012). Timing and progression of the end-Guadalupian crisis in the Fars province (Dalan Formation, Kuh-e Gakhum, Iran) constrained by foraminifers and other carbonate microfossils. *Facies* 58: 131–153.

Korbar, T., Fucek, L., Husinec, A. et al. (2001). Cenomanian carbonate facies and rudists along shallow intraplatform basin margin – The Island of Cres (Adriatic Sea, Croatia). *Facies* 45: 39–58.

Koša, E. (2015). Sea-level changes, shoreline journeys, and the seismic stratigraphy of Central Luconia, Miocene-present, offshore Sarawak, NW Borneo. *Marine and Petroleum Geology* 59: 35–55.

Koša, E., Warrlich, G.M.D., and Loftus, G. (2015). Wings, mushrooms, and Christmas trees: the carbonate seismic geomorphology of Central Luconia, Miocene–present, offshore Sarawak, northwest Borneo. *American Association of Petroleum Geologists Bulletin* 99: 2043–2075.

Kosnik, M.A., Kaufman, D.S., and Hua, Q. (2008). Identifying outliers and assessing the accuracy of amino acid racemization measurements for geochronology: I. Age calibration curves. *Quaternary Geochronology* 3: 308–327.

Kossovaya O. L. and Kropatcheva G., S. (2013). Extinction of Guadalupian rugose corals: an example of biotic response to the Kamura event (southern Primorye, Russia). *Geological Society, London, Special Publication* 376: 407–429.

Krencker, F.-N., Lindström, S., and Bodin, S. (2019). A major sea-level drop briefly precedes the Toarcian oceanic anoxic event: implication for Early Jurassic climate and carbon cycle. *Scientific Reports* 9: 12518.

Krijgsman, W., Hilgen, F.J., Raffi, I. et al. (1999a). Chronology, causes and progression of the Messinian salinity crisis. *Nature* 400: 652–655.

Krijgsman, W., Langereis, C.G., Zachariasse, W.J. et al. (1999b). Late Neogene evolution of the Taza–Guercif Basin (Rifian Corridor, Morocco) and implications for the Messinian salinity crisis. *Marine Geology* 153: 147–160.

Krijgsman, W., Capella, W., Simon, D. et al. (2018). The Gibraltar Corridor: Watergate of the Messinian Salinity Crisis. *Marine Geology* 403: 238–246.

Kruse, P.D., Andrey Yu, Z., and James, N. P. (1995). Primordial metazoan-calcimicrobial reefs: tommotian (Early Cambrian) of the Siberian Platform. *PALAIOS* 10: 291–321.

Kucera, M. (2007). Planktonic foraminifera as tracers of past oceanic environments. *Developments in Marine Geology* 1: 213–262.

Kuhnt, W., Holbourn, A., Gale, A. et al. (2009). Cenomanian sequence stratigraphy and sea-level fluctuations in the Tarfaya Basin (SW Morocco). *Geological Society of America Bulletin* 121: 1695–1710.

Kulm, L.D., Suess, E., Moore, J.C. et al. (1986). Oregon subduction zone: venting, fauna, and carbonates. *Science* 231: 561–566.

Lambeck, K., Rouby, H., Purcell, A. et al. (2014). Sea level and global ice volumes from the Last Glacial Maximum to the Holocene. *Proceedings of the National Academy of Sciences of the United States of America* 111: 15296–15303.

Lan, Z., Zhang, S., Tucker, M. et al. (2020). Evidence for microbes in early Neoproterozoic stromatolites. *Sedimentary Geology* 398: 105589.

Land, L.S. (1970). Carbonate mud: production by epibiont growth on *Thalassia testidunum*. *Journal of Sedimentary Petrology* 40: 1361–1363.

Lantzsch, H., Roth, S., Reijmer, J.J.G., and Kinkel, H. (2007). Sea-level related resedimentation processes on the northern slope of Little Bahama Bank (Middle Pleistocene to Holocene). *Sedimentology* 54: 1307–1322.

Lasemi, Y. and Jalilian, A.H. (2010). The Middle Jurassic basinal deposits of the Surmeh Formation in the Central Zagros Mountains, southwest Iran: facies, sequence stratigraphy, and controls. *Carbonates and Evaporites* 25: 283–295.

Laskar, J. (2020). Astrochronology. In: *Geologic Time Scale 2020*, 1e (2-volume set (ed. F.M. Gradstein, J.G. Ogg, M. Schmitz, and G. Ogg). Amsterdam: Elsevier, 1390 pp.

Laugié, M., Michel, J., Pohl, A. et al. (2019). Global distribution of modern shallow-water marine carbonate factories: a spatial model based on environmental parameters. *Scientific Reports* 9: 16432.

Le Goff, J., Reijmer, J.J.G., Cerepi, A. et al. (2019). The dismantling of the Apulian carbonate platform during the late Campanian – early Maastrichtian in Albania. *Cretaceous Research* 96: 83–106.

Le Goff, J., Slootman, A., Mulder, T. et al. (2020). On the architecture of intra-formational mass-transport deposits: insights from the carbonate slopes of Great Bahama Bank and the Apulian Carbonate Platform. *Marine Geology* 427: 106205.

Le Roy, P., Jorry, S., Jouet, G. et al. (2019). Late Pleistocene evolution of the mixed siliciclastic and carbonate southwestern New Caledonia continental shelf/ lagoon. *Palaeogeography, Palaeoclimatology, Palaeoecology* 514: 502–521.

Leadbeater, B.S.C. and Riding, R. (ed.) (1986). *Biomineralization in Lower Plants and Animals Systematics Association*. Oxford: Clarendon Press, 401 pp.

Lebec, A. (2004). *Architecture et dynamique des systèmes carbonatèes de la plate-forme crétacee inferieur du Sultanat d'Oman: Correlations entre donees de sismique et d'affleurements*. Pessac: Universite Michel de Montaigne, Bordeaux 3, 267 pp.

Lebrato, M., Andersson, A.J., Ries, J.B. et al. (2016). Benthic marine calcifiers coexist with $CaCO_3$-undersaturated seawater worldwide. *Global Biogeochemical Cycles* 30: 1038–1053.

Lee, J.-H. and Riding, R. (2018). Marine oxygenation, lithistid sponges, and the early history of Paleozoic skeletal reefs. *Earth-Science Reviews* 181: 98–121.

Lee, J.-H., Hong, J., Choh, S.-J. et al. (2016). Early recovery of sponge framework reefs after Cambrian archaeocyath extinction: Zhangxia Formation (early Cambrian Series 3), Shandong, North China.

Palaeogeography, Palaeoclimatology, Palaeoecology 457: 269–276.

Lees, A. (1964). The structure and origin of the waulsortian (Lower Carboniferous) 'reefs' of west-central eire. *Philosophical Transactions of the Royal Society of London, Series B, Biological Sciences* 247: 483–531.

Lees, A. (1975). Possible influence of salinity and temperatures on modern shelf carbonate sedimentation. *Marine Geology* 19: 159–198.

Lees, A. and Buller, A.T. (1972). Modern temperate-water and warm-water shelf carbonate sediments contrasted. *Marine Geology* 13: M67–M73.

Lees, A. and Miller, J. (1985). Facies variation in Waulsortian buildups. Part 2; Mid-Dinantian buildups from Europe and North America. *Geological Journal* 20: 159–180.

Lees, A. and Miller, J. (1995). Waulsortian banks. In: *Carbonate Mud-Mounds – Their Origin and Evolution*, Special Publication, vol. 23 (ed. C.L.V. Monty, D.W.J. Bosence, P.H. Bridges, and B.R. Pratt), 191–271. Oxford: International Association of Sedimentologists – Blackwell.

Lees, A., Hallet, V., and Hibo, D. (1985). Facies variation in Waulsortian buildups, Part 1; A model from Belgium. *Geological Journal* 20: 133–158.

Legrand, H.E. and Stringfield, V.T. (1973). Karst hydrology – a review. *Journal of Hydrology* 20: 97–120.

Lehrmann, D.J. (1999). Early Triassic calcimicrobial mounds and biostromes of the Nanpanjiang basin, south China. *Geology* 27: 359–362.

Lehrmann, D.J., Wei, J. and Enos, P. (1998). Controls on facies architecture of a large Triassic carbonate platform: the Great Bank of Guizhou, Nanpanjiang Basin, South China. *Journal of Sedimentary Research* 68: 311–326.

Lehrmann, D. J., Payne, J. L., Felix, S. V., Dillett, P. M., Wang, H., Yu, Y., and Wei, J. (2003). Permian–Triassic boundary sections from shallow-marine carbonate platforms of the Nanpanjiang Basin, South China: implications for oceanic conditions associated with the end-Permian extinction and its aftermath. *Palaios* 18(2): 138–152.

Lehrmann, D.J., Donghong, P., Enos, P. et al. (2007). Impact of differential tectonic subsidence on isolated carbonate-platform evolution: Triassic of the Nanpanjiang Basin, south China. *American Association of Petroleum Geologists Bulletin* 91: 287–320.

Lehrmann, D.J., Minzoni, M., Li, X. et al. (2012). Lower Triassic oolites of the Nanpanjiang Basin, south China: facies architecture, giant ooids, and diagenesis—implications for hydrocarbon reservoirs. *American Association of Petroleum Geologists Bulletin* 96: 1389–1414.

Lehrmann, D.J., Minzoni, M., Enos, P. et al. (2020). Giant sector-collapse structures (scalloped margins) of the Yangtze Platform and Great Bank of Guizhou, China: Implications for genesis of collapsed carbonate platform margin systems. *Sedimentology* 67: 3167–3198.

Leinfelder, R.R. (1986). Facies, Stratigraphy and Paleogeographic Analysis of Upper? Kimmeridgian to Upper Portlandian Sediments in the Environs of Arruda dos Vinhos, Estremadura, Portugal. PhD thesis. Johannes Gutenberg-Universität, Mainz, Germany, Münchner Geowissenschaftlicher Abhandlungen – Reihe A – Geologie und Paläontologie, 216 pp.

Leinfelder, R.R. and Schmid, D.U. (2000). Mesozoic reefal thrombolites and other microbolites. In: *Microbial Sediments* (ed. R.E. Riding and S.M. Awramik), 289–294. Berlin, Heidelberg: Springer.

Leinfelder, R.R., Nose, M., Schmid, D.U., and Werner, W. (1993). Microbial crusts of the Late Jurassic: composition, paleoecological significance and importance in reef construction. *Facies* 29: 195–203.

Leinfelder, R.R., Werner, W., Nose, M., Schmid, D.U., Krautter, M., Laternser, R., Takacs, M. and Hartmann, D. (1996). Paleoecology, growth parameters and dynamics of coral, sponge and microbolite reefs from the Late Jurassic. In: Global and Regional Controls on Biogenic Sedimentation 1. Reef-Evolution. Research Reports (ed. J. Reitner, F. Neuweiler, and F. Gunkel), Göttinger Arbeiten zur Geologie und Paläontologie, Sonderband 2, 227–248. Göttingen: Geologisches Institut der Georg-August-Universität Göttingen.

Leinfelder, R.R., Schmid, D.U., Nose, M., and Werner, W. (2002). Jurassic reef patterns – the expression of a changing globe. In: *Phanerozoic Reef Patterns*, SEPM Special Publication, vol. 72 (ed. W. Kiessling, E. Flügel, and J. Golonka), 465–520. Tulsa, OK: SEPM (Society for Sedimentary Geology).

Leinfelder, R.R., Schlagintweit, F., Werner, W. et al. (2005). Significance of stromatoporoids in Jurassic reefs and carbonate platforms – concepts and implications. *Facies* 51: 287–325.

Lenton, T. (2016). *Earth System Science, a Very Short Introduction*. Oxford: Oxford University Press, 153 pp.

Leonard, J.E., Cameron, B., Pilkey, O.H., and Friedman, G.M. (1981). Evaluation of cold-water carbonates as a possible paleoclimatic indicator. *Sedimentary Geology* 28: 1–28.

Leonard, N.D., Lepore, M.L., Zhao, J.-x. et al. (2020). Re-evaluating mid-Holocene reef "turn-off" on the inshore Southern Great Barrier Reef. *Quaternary Science Reviews* 244: 106518.

Lepot, K. (2020). Signatures of early microbial life from the Archean (4 to 2.5 Ga) eon. *Earth-Science Reviews* 209: 103296.

Li, Y., Kershaw, S., and Mu, X. (2004). Ordovician reef systems and settings in South China before the Late Ordovician mass extinction. *Palaeogeography, Palaeoclimatology, Palaeoecology* 205: 235–254.

Li, X., Yu, M., Lehrmann, D.J. et al. (2012). Factors controlling carbonate platform asymmetry: preliminary results from the Great Bank of Guizhou, an isolated Permian–Triassic Platform in the Nanpanjiang Basin, south China. *Palaeogeography,*

Palaeoclimatology, Palaeoecology 315–316: 158–171.

Li, F., Yan, J., Algeo, T., and Wu, X. (2013). Paleoceanographic conditions following the end-Permian mass extinction recorded by giant ooids (Moyang, South China). *Global and Planetary Change* 105: 102–120.

Liaghat, M., Adabi, M.H., Swennen, R. et al. (2021). An integrated facies, diagenesis and geochemical analysis along with sequence stratigraphy of the Lower Triassic Aghe-Darband basin (north-east Iran). *Journal of African Earth Sciences* 173: 103952.

Liebau, A. (1980). Paläobathymetrie und Ökofaktoren: Flachmeerzonierungen. *Neues Jahrbuch für Geologie und Paläontologie, Abhandlungen* 160: 173–216.

Liebau, A. (1984). Grundlagen der Ökobathymetrie. In: *Paläobathymetrie*, vol. 2 (ed. H. Luterbacher), 149–184. Paläontologische Kursbücher.

Light, J.M. and Wilson, J.B. (1998). Cool-water carbonate deposition on the West Shetland Shelf: a modern distally steepened ramp. In: *Carbonate Ramps: An Introduction*, vol. 149 (ed. V.P. Wright and T.P. Burchette) Special Publication, 73–105. Geological Society of London, London.

Lindhorst, S., Appoo, J., Artschwager, M., et al. 2019. Saya de Malha Carbonates, Oceanography and Biogeochemistry (Western Indian Ocean), Cruise No. SO270, 2019-09-06–2019-10-23, Hong Kong (China)–Port Louis (Mauritius), Gutachterpanel Forschungsschiffe, Bonn.

Lipar, M., Webb, J.A., Cupper, M.L., and Wang, N. (2017). Aeolianite, calcrete/ microbialite and karst in southwestern Australia as indicators of Middle to Late Quaternary palaeoclimates. *Palaeogeography, Palaeoclimatology, Palaeoecology* 470: 11–29.

Liu, J., Zhan, R., Dai, X. et al. (2011). Demise of Early Ordovician oolites in South China: evidence for paleoceanographic changes before the GOBE. In: *Ordovician of the World - Cuadernos del Museo Geominero 14*, Cuadernos del Museo Geominero, vol. 14 (ed. I.R.a.D.G.-B.e. J.C. Gutiérrez-Marco, Ordovician of the World. Cuadernos del Museo Geominero, 14. Instituto Geológico y Minero de España, Madrid. Instituto Geológico y Minero de España 2011), 309–317. Madrid: Instituto Geológico y Minero de España.

Logan, B.W. and Cebulski, D.E. (1970). Sedimentary environments of Shark Bay, Western Australia. In: *Carbonate Sedimentation and Environments, Shark Bay, Western Australia*, American Association of Petroleum Geologists, Memoir, vol. 13 (ed. B.W. Logan, G.R. Davies, J.F. Read, and D.E. Cebulski), 38–84. Tulsa, OK: American Association of Petroleum Geologists.

Logan, B.W., Hoffman, P., and Gebelein, C.D. (1974). Algal mats, cryptalgal fabrics, and structures, Hamelin Pool, Western Australia. In: *Evolution and Diagenesis of Quaternary Carbonate Sequences, Shark Bay, Western Australia*, American Association of Petroleum Geologists, Memoir, vol. 22 (ed. B.W. Logan, J.F. Read, G.M. Hagan, et al.), 140–194. Tulsa, OK: American Association of Petroleum Geologists.

Lokier, S.W. and Al Junaibi, M. (2016). The petrographic description of carbonate facies: are we all speaking the same language? *Sedimentology* 63: 1843–1885.

Lokier, S.W. and Fiorini, F. (2016). Temporal evolution of a carbonate coastal system, Abu Dhabi, United Arab Emirates. *Marine Geology* 381: 102–113.

Lokier, S.W., Knaf, A., and Kimiagar, S. (2013). A quantitative analysis of Recent arid coastal sedimentary facies from the Arabian Gulf Coastline of Abu Dhabi, United Arab Emirates. *Marine Geology* 346: 141–152.

Longman, M.W. (1980). Carbonate diagenetic textures from near surface diagenetic environments. *American Association of Petroleum Geologists Bulletin* 64: 461–487.

Longman, M.W. (1981). *Carbonate Diagenesis as a Control on Stratigraphic Traps (with examples from the Williston Basin)*, Educations Course Note Series, vol. 21. Tulsa, OK: American Association of Petroleum Geologists, 159 pp.

Lønøy, A. (2006). Making sense of carbonate pore systems. *AAPG Bulletin* 90: 1381–1405.

Lopez-Gamundi, C., Barnes, B.B., Bakker, A.C., Harris, P., Eberli, G.P. and Purkis, S.J. (2024). Spatial, seasonal and climatic drivers of suspended sediment atop Great Bahama Bank. *Sedimentology* 71: 769–792.

Loreau, J.P. and Purser, B.H. (1973). Distribution and ultrastructure of Holocene ooids in the Persian Gulf. In: *The Persian Gulf: Holocene Carbonate Sedimentation and Diagenesis in a Shallow Epicontinental Sea* (ed. B.H. Purser), 279–328. Berlin-Heidelberg: Springer-Verlag.

Losos, J.B. and Ricklefs, R.E. (2009). Adaptation and diversification on islands. *Nature* 457: 830–836.

Loucks, R.G. (1999). Paleocave carbonate reservoirs: origins, burial-depth modifications, spatial complexity, and reservoir implications. *American Association of Petroleum Geologists Bulletin* 83: 1795–1834.

Loucks, R.G. and Sarg, J.F. (ed.) (1993). *Carbonate Sequence Stratigraphy – Recent Developments and Applications*. Tulsa, OK: American Association of Petroleum Geologists, 545 pp.

Loucks, R.G., Moody, R.T.J., Bellis, J.K., and Brown, A.A. (1998). Regional depositional setting and pore network systems of the El Garia Formation (Metlaoui Group, Lower Eocene), offshore Tunisia. In: *Petroleum Geology of North Africa*, Geological Society of London Special Publication, vol. 132 (ed. D.S. McGregor, R.T.J. Moody, and D.D. Clark-Lowes), 355–374. London: Geological Society of London.

Loucks, R.G., Kerans, C., Zeng, H., and Sullivan, P.A. (2017). Documentation and characterization of the Lower Cretaceous (Valanginian) Calvin and Winn carbonate shelves and shelf margins, onshore northcentral Gulf of Mexico. *American Association of Petroleum Geologists Bulletin* 101: 119–142.

Lovelock, J.E. (1988). *The Ages of Gaia: A Biography of Our Living Earth*. Oxford University Press, 252 pp.

Lowenstam, H.A. (1950). Niagaran reefs of the Great Lakes area. *The Journal of Geology* 58: 430–487.

Lowenstam, H.A. (1981). Minerals formed by organisms. *Science* 211: 1126–1131.

Lowenstam, H.A. and Weiner, S. (1989). *On biomineralization*. New York, USA: Oxford University Press, 336 pp.

Lucia, F.J. (1995). Rock-fabric/petrophysical classification of carbonate pore space for reservoir characterization. *American Association of Petroleum Geologists Bulletin* 79: 1275–1300.

Lucia, F.J. (1999). Characterization of petrophysical flow units in carbonate reservoirs: discussion. *American Association of Petroleum Geologists Bulletin* 83: 1161–1163.

Lucia, F.J. (2007). *Carbonate Reservoir Characterization*, 2e. Berlin Heidelberg New York, Germany: Springer, 336 pp.

Lucia, F.J. (1983). Petrophysical parameters estimated from visual descriptions of carbonate rocks: A field classification of carbonate pore space. *Journal of Petroleum Technology* 35: 629–637.

Lüdmann, T., Kalvelage, C., Betzler, C. et al. (2013). The Maldives, a giant isolated carbonate platform dominated by bottom currents. *Marine and Petroleum Geology* 43: 326–340.

Lüdmann, T., Betzler, C., Eberli, G.P. et al. (2018a). Carbonate delta drift: a new sediment drift type. *Marine Geology* 401: 98–111.

Lüdmann, T., Betzler, C., Eberli, G.P. et al. (2018b). Corrigendum to "Carbonate delta drift: a new sediment drift type" [Mar. Geol. 401 (2018) 98–111]. *Marine Geology* 406: 214–215.

Lukasik, J. and Simo, J.A. (2008). Controls on development of phanerozoic carbonate platforms and reefs – introduction and synthesis. In: *Controls on Carbonate Platform and Reef Development*, Special Publication, vol. 89 (ed. J. Lukasik and J.A. Simo), 5–12. Tulsa, Oklahoma: SEPM Society for Sedimentary Geology.

Luo, M., Chen, J., Qie, W. et al. (2021). Microbially induced carbonate precipitation in a middle triassic microbial mat deposit from Southwestern China: new implications for the formational process of micrite. *Journal of Earth Science* 32: 633–645.

MacArthur, R.H. and Wilson, E.O. (1967). *The Theory of Island Biogeography*. Princeton, NJ: Princeton University Press, 224 pp.

Machel, H.G. (2004). Concepts and models of dolomitization: a critical reappraisal. In: *The Geometry and Petrogenesis of Dolomite Hydrocarbon Reservoirs*, vol. 235 (ed. C.J.R. Braithwaite, G. Rizzi, and G. Darke), 7–63. London: Geological Society of London.

Machel, H.-G. and Mountjoy, E.W. (1986). Chemistry and Environments of dolomitization —A reappraisal. *Earth-Science Reviews* 23: 175–222.

Machel, H.G., Mountjoy, E.W., Humphrey, J.D., and Quinn, T.M. (1990). Coastal mixing zone dolomite, forward modeling, and massive dolomitization of platform-margin carbonates; discussion and reply. *Journal of Sedimentary Research* 60: 1008–1016.

Machel, H.G., Krouse, H.R., and Sassen, R. (1995). Products and distinguishing criteria of bacterial and thermochemical sulfate reduction. *Applied Geochemistry* 10: 373–389.

Mackenzie, F.T. (ed.) (2005). *Sediments, Diagenesis, and Sedimentary Rocks*. Oxford: Elsevier, 425 pp.

Madden, R.H.C., Wilson, M.E.J., Mihaljević, M. et al. (2017). Unravelling the depositional origins and diagenetic alteration of carbonate breccias. *Sedimentary Geology* 357: 33–52.

Makhankova, A., Sautter, B., Mathew, M. et al. (2021). Seismic stratigraphy and sedimentology of a Miocene carbonate platform in Luconia, South China Sea. *Geological Journal* 56: 1–17.

Mallarino, G., Francis, J.M., Jorry, S.J. et al. (2021). Timescale dependent sedimentary record during the past 130 kyr from a tropical mixed siliciclastic–carbonate shelf edge and slope: Ashmore Trough (southern Gulf of Papua). *Sedimentology* 68: 2606–2648.

Maloof, A.C. and Grotzinger, J.P. (2012). The Holocene shallowing-upward parasequence of north-west Andros Island, Bahamas. *Sedimentology* 59: 1375–1407.

Mangenot, X., Gasparrini, M., Gerdes, A., Bonifacie, M. and Rouchon, V. (2018). An emerging thermochronometer for carbonate-bearing rocks: Δ47 /(U-Pb). *Geology* 46: 1067–1070.

Mann, S. (1983). Mineralization in biological systems. In: *Inorganic Elements in Biochemistry* (ed. P.H. Connett, H. Follmann, M. Lammers, et al.), 125–174. Berlin, Heidelberg: Springer.

Mann, S., Heywood, B.R., Rajam, S., and Birchall, J.D. (1989). Interfacial control of nucleation of calcium carbonate under organized stearic acid monolayers. *Proceedings of the Royal Society of London, A: Mathematical and Physical Sciences* 423: 457–471.

Mann, T., Wizemann, A., Stuhr, M. et al. (2022). Shallow-marine carbonate cementation in Holocene segments of the calcifying green alga Halimeda. *Sedimentology* 69: 282–300.

Mansouri-Daneshvar, P., Moussavi-Harami, R., Mahboubi, A. et al. (2015). Sequence stratigraphy of the petroliferous Dariyan Formation (Aptian) in Qeshm Island and offshore (southern Iran). *Petroleum Science* 12: 232–251.

Marangon, A., Gattolin, G., Della Porta, G., and Preto, N. (2011). The Latemar: a flat-topped, steep fronted platform dominated by microbialites and synsedimentary cements. *Sedimentary Geology* 240: 97–114.

Marenco, P.J., Griffin, J.M., Fraiser, M.L., and Clapham, M.E. (2012). Paleoecology and geochemistry of Early Triassic (Spathian) microbial mounds and implications for anoxia following the end-Permian mass extinction. *Geology* 40: 715–718.

Markello, J.R. and Read, J.F. (1981). Carbonate ramp-to-deeper shale shelf transitions of an Upper Cambrian intrashelf basin, Nolichuky Formation, Southwest Virginia Appalachians. *Sedimentology* 28: 573–597.

Markello, J., Koepnick, R., Waite, L., and Collins, J. (2008). The carbonate analogs through time (CATT) Hypothesis and the global atlas of carbonate fields — a systematic and predictive look at phanerozoic carbonate systems. In: *Controls on Carbonate Platform and Reef Development*, SEPM Special Publication, vol. 89 (ed. J. Lukasik and J.A. Simo), 15–45. Tulsa, OK: SEPM (Society for Sedimentary Geology).

Marshall, J.F. and Davies, P.J. (1975). High-magnesium calcite ooids from the Great Barrier Reef. *Journal of Sedimentary Research* 45: 285–291.

Marshall, J.F. and Davies, P.J. (1981). Submarine lithification on windward reef slopes; Capricorn-Bunker Group, southern Great Barrier Reef. *Journal of Sedimentary Research* 51: 953–960.

Marshall, J.F. and Davies, P.J. (1982). Internal structure and Holocene evolution of One Tree Reef, southern Great Barrier Reef. *Coral Reefs* 1: 21–28.

Marshall, J.F. and Davies, P.J. (1984). Last interglacial reef growth beneath modern reefs in the southern Great Barrier Reef. *Nature* 307: 44–46.

Marshall, J.F. and Davies, P.J. (1988). Halimeda bioherms of the northern Great Barrier Reef. *Coral Reefs* 6: 139–148.

Martindale, W. and Boreen, T.D. (ed.) (1997). *Temperature-Stratified Mississippian Carbonates as Hydrocarbon Reservoirs – Examples from the Foothils of the Canadian Rockies*, 391–410. Tulsa, OK: SEPM (Society for Sedimentary Geology).

Martindale, R.C., Corsetti, F.A., James, N. P., and Bottjer, D.J. (2015a). Paleogeographic trends in Late Triassic reef ecology from northeastern Panthalassa. *Earth Science Reviews* 142: 18–37.

Martindale, R.C., Strauss, J.V., Sperling, E. A. et al. (2015b). Sedimentology, chemostratigraphy, and stromatolites of lower Paleoproterozoic carbonates, Turee Creek Group, Western Australia. *Precambrian Research* 266: 194–211.

Martín-Martín, M., Guerrera, F., Tosquella, J., and Tramontana, M. (2020). Paleocene-Lower Eocene carbonate platforms of westernmost Tethys. *Sedimentary Geology* 404: 105674.

Masiero, I., Kozlowski, E., Antonatos, G. et al. (2020). Numerical stratigraphic forward models as conceptual knowledge repositories and experimental tools: an example using a new enhanced version of CarboCAT. *Computers and Geosciences* 138: 104453.

Masse, J.P. (1992). The Lower Cretaceous Mesogean benthic ecosystems: palaeoecologic aspects and palaeobiogeographic implications. *Palaeogeography, Palaeoclimatology, Palaeoecology* 91: 331–345.

Mateu-Vicens, G., Pomar, L., and Tropeano, M. (2008). Architectural complexity of a carbonate transgressive systems tract induced by basement physiography. *Sedimentology* 55: 1815–1848.

Mateu-Vicens, G., Pomar, L., and Ferràndez-Cañadel, C. (2012). Nummulitic banks in the upper Lutetian 'Buil level', Ainsa Basin, South Central Pyrenean Zone: the impact of internal waves. *Sedimentology* 59: 527–552.

Matheson, E.J. and Frank, T.D. (2020). An epeiric glass ramp: Permian low-latitude neritic siliceous sponge colonization and its novel preservation (Phosphoria Rock Complex). *Sedimentary Geology* 399: 105568.

Mathews, E., Heap, A., and Woods, M. (2007). *Inter-reefal Seabed Sediments and Geomorphology of the Great Barrier Reef, a Spatial Analysis*. Canberra, Australia: Geoscience Australia, Record 2007/9, 140 pp.

Matti, J.C. and McKee, E.H. (1977). Silurian and Lower Devonian paleogeography of the outer continental shelf of the Cordilleran miogeocline, central Nevada. In: *Paleozoic paleogeography of the western United States, Pacific Coast Paleogeography Symposium I: Pacific Section* (ed. J.H. Stewart, A.E. Fritsche, and C.H. Stevens), 181–215. Tulsa, OK: Society of Economic Paleontologists and Mineralogists.

Maurer, F. (2000). Growth mode of Middle Triassic carbonate platforms in the Western Dolomites (Southern Alps, Italy). *Sedimentary Geology* 134: 275–286.

Maurer, F., Rettori, R., and Martini, R. (2008). Triassic stratigraphy, facies and evolution of the Arabian shelf in the northern United Arab Emirates. *International Journal of Earth Sciences* 97: 765–784.

Maurer, F., Martini, R., Rettori, R., Hillgärtner, H. and Cirilli, S. (2009). The geology of Khuff outcrop analogues in the Musandam Peninsula, United Arab Emirates and Oman. *GeoArabia* 14: 125–158.

Maurer, F., van Buchem, F.S.P., Eberli, G. P. et al. (2013). Late Aptian long-lived glacio-eustatic lowstand recorded on the Arabian Plate. *Terra Nova* 25: 87–94.

Mauz, B., Shen, Z., Alsuwaidi, M. et al. (2022). The mid-Holocene sea-level change in the Arabian Gulf. *The Holocene* 32: 1173–1183.

Mavko, G., Mukerji, T., and Dvorkin, J. (2009). *The Rock Physics Handbook: Tools for Seismic Analysis of Porous Media*. New York, USA: Cambridge University Press, 511 pp.

Mazzullo, S.J., Mazzullo, J., and Harris, P. M. (1985). Eolian origin of quartzose sheet sands in Permian shelf facies, Guadalupe Mountains. In: *Permian Carbonate/Clastic Sedimentology, Guadalupe Mountains*, Permian Basin Section, Society of Economic Paleontologists and Mineralogists, Publication, vol. 85-24 (ed. B.K. Cunningham and C.L. Hedrick), 71. Midland, TX: Permian Basin Section, Society of Economic Paleontologists and Mineralogists.

McArthur, J.M., Howarth, R.J., Shields, G. A., and Zhou, Y. (2020). Strontium isotope stratigraphy, Chapter 7. In: *Geologic Time Scale 2020* (ed. F.M. Gradstein, J.G. Ogg, M.D. Schmitz, and G.M. Ogg), 211–238. Amsterdam, The Netherlands: Elsevier.

McCaffrey, J.C., Gallagher, S.J., Wallace, M.W. et al. (2025) The Rowley Shoals atolls: Remnants of a Miocene great barrier reef on the north-west Australian margin. *Global and Planetary Change* 245: 104688.

McCallum, J.E. and Robertson, A.H.F. (1995). Sedimentology of two fan-delta systems in the Pliocene-Pleistocene of the Mesaoria Basin, Cyprus. *Sedimentary Geology* 98: 215–244.

McKenzie, J., Bernoulli, D., and Schlanger, S.O. (1980). Shallow-water carbonate sediments from the Emperor Seamounts: their diagenesis and paleogeographic significance. In: *Initial Reports of the Deep Sea Drilling Project 55*, vol. 55 (ed. E.D. Jackson, I. Koisumi, et al.), 415–455. Washington: U.S. Govt. Printing Office.

McLaughlin, P.I., Emsbo, P., Brett, C.E. et al. (2019). The rise of pinnacle reefs: a step change in marine evolution triggered by perturbation of the global carbon cycle. *Earth and Planetary Science Letters* 515: 13–25.

McNeil, M.A., Webster, J.M., Beaman, R. J., and Graham, T.L. (2016). New constraints on the spatial distribution and morphology of the *Halimeda* bioherms of the Great Barrier Reef, Australia. *Coral Reefs* 35: 1343–1355.

McNeil, M.A., Firn, J., Nothdurft, L.D. et al. (2021a). Inter-reef Halimeda algal habitats within the Great Barrier Reef support a distinct biotic community and high biodiversity. *Nature Ecology and Evolution* 5: 647–655.

McNeil, M.A., Nothdurft, L.D., Dyriw, N.J. et al. (2021b). Morphotype differentiation in the Great Barrier Reef *Halimeda* bioherm carbonate factory: internal architecture and surface geomorphometrics. *The Depositional Record* 7: 176–199.

McNeill, D.F., Coates, A.G., Budd, A.F., and Borne, P.F. (2000). Integrated paleontologic and paleomagnetic stratigraphy of the upper Neogene deposits around Limon, Costa Rica: a coastal emergence record of the Central American Isthmus. *Geological Society of America Bulletin* 112: 963–981.

Medina Ferrer, F., Rosen, M.R., Feyhl-Buska, J. et al. (2021). Potential role for microbial ureolysis in the rapid formation of carbonate tufa mounds. *Geobiology* 20: 79–97.

Meibom, A., Cuif, J.-P., Hillion, F. et al. (2004). Distribution of magnesium in coral skeleton. *Geophysical Research Letters* 31: L23306.

Melim, L.A., Westphal, H., Swart, P.K. et al. (2002). Questioning carbonate diagenetic paradigms: evidence from the Neogene of the Bahamas. *Marine Geology* 185: 27–53.

Melim, L.A., Swart, P.K., and Eberli, G.P. (2004). Mixing-zone diagenesis in the subsurface of Florida and the Bahamas. *Journal of Sedimentary Research* 74: 904–913.

Menier, D., Pierson, B., Chalabi, A. et al. (2014). Morphological indicators of structural control, relative sea-level fluctuations and platform drowning on present-day and Miocene carbonate platforms. *Marine and Petroleum Geology* 58: 776–788.

Meredith, K., Moriguti, T., Tomascak, P. et al. (2013). The lithium, boron and strontium isotopic systematics of groundwaters from an arid aquifer system: implications for recharge and weathering processes. *Geochimica et Cosmochimica Acta* 112: 20–31.

Merino-Tomé, O.A., Della Porta, G., Verwer, K., and Kenter, J.A.M. (2007). Digital Outcrop Modelling of a Lower Jurassic Carbonate Platform (Djebel Bou Dahar, High Atlas, Morocco) – Part 1: Evolution of Platform-Wide Depositional Anatomy and Lithofacies, SIEP, Rijswijk.

Merino-Tomé, Ó., Della Porta, G., Kenter, J.A.M. et al. (2012). Sequence development in an isolated carbonate platform (Lower Jurassic, Djebel Bou Dahar, High Atlas, Morocco): influence of tectonics, eustacy and carbonate production. *Sedimentology* 59: 118–155.

Merino-Tomé, Ó., Della Porta, G., Pierre, A. et al. (2017). Intact seismic-scale platforms and ramps in the Lower to Middle Jurassic of Morocco: implications for stratal anatomy and lithofacies partitioning. *American Association of Petroleum Geologists Bulletin* 101: 505–513.

Meyers, J.H. (1987). Marine vadose beachrock cementation by cryptocrystalline magnesian calcite, Maui, Hawaii. *Journal of Sedimentary Research* 57: 558–570.

Michel, J., Mateu-Vicens, G., and Westphal, H. (2011). Modern heterozoan carbonates from a eutrophic tropical shelf (Mauritania). *Journal of Sedimentary Research* 81: 641–655.

Michel, J., Borgomano, J., and Reijmer, J.J. G. (2018). Heterozoan carbonates: when, where and why? A synthesis on parameters controlling carbonate production and occurrences. *Earth-Science Reviews* 182: 50–67.

Michel, J., Laugié, M., Pohl, A. et al. (2019). Marine carbonate factories: a global model of carbonate platform distribution. *International Journal of Earth Sciences* 108: 1773–1792.

Michel, J., Lanteaume, C., Lettéron, A. et al. (2020). Oligocene and miocene global spatial trends of shallow-marine carbonate architecture. *The Journal of Geology* 128: 563–570.

Mienis, F., Weering, T.v., de Haas, H. et al. (2006). Carbonate mound development at the SW Rockall Trough margin based on high resolution TOBI and seismic recording. *Marine Geology* 233: 1–19.

Mienis, F., De Stigter, H.C., White, M. et al. (2007). Hydrodynamic controls on cold-water coral growth and carbonate-mound development at the SW and SE Rockall Trough Margin, NE Atlantic Ocean. *Deep Sea Research Part I: Oceanographic Research Papers* 54: 1655–1674.

Mienis, F., de Stigter, H.C., de Haas, H., and van Weering, T.C.E. (2009). Near-bed particle deposition and resuspension in a cold-water coral mound area at the Southwest Rockall Trough margin, NE Atlantic. *Deep Sea Research I* 56: 1026–1038.

Mihaljevi, M., Renema, W., Welsh, K., and Pandolfi, J.M. (2014). Eocene-Miocene shallow-water carbonate platforms and increased habitat diversity in Sarawak, Malaysia. *PALAIOS* 29: 378–391.

Miller, J.A. and Macintyre, I.G. (1977). *Field guidebook to the Reefs of Belize*. The Atlantic Reef Committee: Third International Symposium on Coral Reefs, 36 pp.

Miller, E.L., Miller, M.M., Stevens, C.H. et al. (1992). Late Paleozoic paleogeographic and tectonic evolution of the western U.S. Cordillera. In: *The Cordilleran Orogen* (ed. B.C. Burchfiel, P.W. Lipman, and M.L. Zoback) G-3, 57–106. Boulder, CO: Geological Society of America.

Miller, K.G., Wright, J.D., and Browning, J.V. (2005). Visions of ice sheets in a greenhouse world. *Marine Geology* 217: 215–231.

Miller, K.G., Browning, J.V., Schmelz, W. J. et al. (2020). Cenozoic sea-level and cryospheric evolution from deep-sea geochemical and continental margin records. *Science Advances* 6: eaaz1346.

Milliman, J.D. (1974a). *Marine Carbonates*. Berlin, Heidelberg, New York: Springer Verlag, 375 pp.

Milliman, J.D. (1974b). Precipitation and cementation of deep-sea carbonate sediments. In: *Deep-Sea Sediments: Physical and Mechanical Properties* (ed. A.L. Inderbitzen), 463–476. US, Boston, MA: Springer.

Minisini, D., Eldrett, J., Bergman, S.C., and Forkner, R. (2018). Chronostratigraphic framework and depositional environments in the organic-rich, mudstone-dominated Eagle Ford Group, Texas, USA. *Sedimentology* 65: 1520–1557.

Misra, S., Greaves, M., Owen, R., Kerr, J., Elmore, A.C. and Elderfield, H. (2014). Determination of B/Ca of natural carbonates by HR-ICP-MS. *Geochemistry, Geophysics, Geosystems* 15: 1617–1628.

Mitchell Simon, F. (2019). The Chalk Group (Upper Cretaceous) of the Northern Province, eastern England – a review. *Proceedings of the Yorkshire Geological Society* 62: 153–177.

Mitchell, J.K., Holdgate, G.R., Wallace, M. W., and Gallagher, S.J. (2007). Marine geology of the Quaternary Bass Canyon system, southeast Australia: a cool-water carbonate system. *Marine Geology* 237: 71–96.

Montaggioni, L.F., Cabioch, G., Thouveny, N. et al. (2011). Revisiting the Quaternary development history of the western New Caledonian shelf system: From ramp to barrier reef. *Marine Geology* 280: 57–75.

Monty, C.L.V. (1995). The rise and nature of carbonate mud-mounds: an introductory actualistic approach. In: *Carbonate Mud-Mounds – Their Origin and Evolution*, IAS Special Publication, vol. 23 (ed. C.L.V. Monty, D.W.J. Bosence, P.H. Bridges, and B.R. Pratt), 11–48. Chichester, West Sussex: International Association of Sedimentologists–Blackwell.

Monty, C.L.V., Bosence, D.W.J., Bridges, P. H., and Pratt, B.R. (ed.) (1995). *Carbonate Mud-Mounds – Their Origin and Evolution*. Oxford, London: Blackwell Scientific Publications, 537 pp.

Moody, R.T.J. (1987). The Ypresian carbonates of Tunisia—a model of foraminiferal facies distribution. In: *Micropaleontology of Carbonate Environments* (ed. M.B. Hart), 82–92. UK: British Micropalaeontological Society – Ellis Horwood.

Moody, R.T.J. and Grant, G.G. (1988). Deposition of the Ypresian Carbonates of Tunisia, American Association of Petroleum Geologists Bulletin 72:8, Conference: Mediterranean Basins Conference and Exhibition, Nice (France), 25–28 September 1988. American Association of Petroleum Geologists, Tulsa, OK, USA.

Moore, C.H. (1989). *Carbonate Diagenesis and Porosity*. Elsevier, 338 pp.

Moore, C.H. (2001). *Carbonate Reservoirs: Porosity Evolution and Diagenesis in a Sequence Stratigraphic Framework*. Amsterdam, London, New York, Oxford, Paris, Shannon, Tokyo: Elsevier, 444 pp.

Moore, C.H. and Wade, W.J. (2013). *Carbonate Reservoirs: Porosity and Diagenesis in a Sequence Stratigraphic Framework*. Amsterdam, The Netherlands: Elsevier, 374 pp.

Morgan, K.M., Perry, C.T., Smithers, S.G. et al. (2016). Evidence of extensive reef development and high coral cover in nearshore environments: implications for understanding coral adaptation in turbid settings. *Scientific Reports* 6: 29616.

Morgan, K.M., Moynihan, M.A., Sanwlani, N., and Switzer, A.D. (2020a). Light limitation and depth-variable sedimentation drives vertical reef compression on turbid coral reefs. *Frontiers in Marine Science* 7: 931.

Morgan, K.M., Perry, C.T., Arthur, R. et al. (2020b). Projections of coral cover and habitat change on turbid reefs under future sea-level rise. *Proceedings of the Royal Society B: Biological Sciences* 287: 20200541.

Morrison, J.O. and Brand, U. (1986). PALEOSCENE #5. Geochemistry of Recent Marine Invertebrates. *Geoscience Canada* 13: 237–254.

Morse, J.W. and Mackenzie, F.T. (1990). *Geochemistry of Sedimentary Carbonates*. Amsterdam, The Netherlands: Elsevier, 707 pp.

Morse, J.W., Gledhill, D.K., and Millero, F. J. (2003). $CaCO_3$ precipitation kinetics in waters from the Great Bahama Bank: implications for the relationship between Bank hydrochemistry and whitings. *Geochimica et Cosmochimica Acta* 67: 2819–2826.

Mortimore, R. (2011). A chalk revolution: what have we done to the Chalk of England? *Proceedings of the Geologists' Association* 122: 232–297.

Mortimore, R. and Pomerol, B. (1997). Upper Cretaceous tectonic phases and end Cretaceous inversion in the Chalk of the Anglo-Paris Basin. *Proceedings of the Geologists' Association* 108: 231–255.

Mounji, D., Bourque, P.-A., and Savard, M. M. (1998). Hydrothermal origin of Devonian conical mounds (Kess-Kess) of Hamar Lakhdad Ridge, Anti-Atlas, Morocco. *Geology* 26: 1123–1126.

Mountjoy, E.W. (1965a). The Permo-Carboniferous succession of Mount Greenock and Northern Jasper Park: discussion. *Bulletin of Canadian Petroleum Geology* 13: 340–345.

Mountjoy, E.W. (1965b). *Stratigraphy of the Devonian Miette reef Complex and Associated Strata, Eastern Jasper National Park, Alberta*. Ottawa: Geological Survey of Canada.

Mountjoy, E.W. and Playford, P.E. (1972). Submarine megabreccia debris flows and slumped blocks of Devonian of Australia and Alberta—a comparison. *American Association of Petroleum Geologists Bulletin* 56: 641–641.

Muchez, P., Viaene, W., and Marshall, J.D. (1991). Origin of shallow burial cements in the late viséan of the Campine Basin, Belgium. *Sedimentary Geology* 73: 257–271.

Mulder, T., Joumes, M., Hanquiez, V. et al. (2017). Carbonate slope morphology revealing sediment transfer from bank-to-slope (Little Bahama Bank, Bahamas). *Marine and Petroleum Geology* 83: 26–34.

Mulder, T., Ducassou, E., Hanquiez, V. et al. (2019). Contour current imprints and contourite drifts in the Bahamian archipelago. *Sedimentology* 66: 1192–1221.

Müller, M.N. (2019). On the genesis and function of coccolithophore calcification. *Frontiers in Marine Science* 6: 49.

Müller, D.W., McKenzie, J.A., and Mueller, P.A. (1990). Abu Dhabi Sabkha, Persian Gulf, revisited: application of strontium isotopes to test an early dolomitization model. *Geology* 18: 618–621.

Mullins, H.T., Gardulski, A.F., Hine, A.C. et al. (1988). Three-dimensional sedimentary framework of the carbonate ramp slope of Central West Florida: a sequential seismic stratigraphic perspective. *Geological Society of America Bulletin* 100: 514–533.

Mundil, M., Zühlke, R., Bechstädt, T. et al. (2003). Cyclicities in Triassic platform carbonates: synchronizing radio-isotopic and orbital clocks. *Terra Nova* 15: 81–87.

Murris, R.J. (1980). Middle East: stratigraphic evolution and oil habitat. *American Association of Petroleum Geologists Bulletin* 64: 597–618.

Muthiga, N.A. and Szmant, A.M. (1987). The effect of salinity stress on the rates of aerobic respiration and photosynthesis in the hermatypic coral *Siderastrea Siderea*. *Biological Bulletin* 173: 539–551.

Mutterlose, J., Pauly, S. and Steuber, T. (2009). Temperature controlled deposition of early Cretaceous (Barremian–early Aptian) black shales in an epicontinental sea. *Palaeogeography, Palaeoclimatology, Palaeoecology* 273: 330–345.

Mutti, M. and Hallock, P. (2003). Carbonate systems along nutrient and temperature gradients: some sedimentological and geochemical constraints. *International Journal of Earth Sciences (Geologische Rundschau)* 92: 465–475.

Mutti, M., Bernoulli, D., and Stille, P. (1997). Temperate carbonate platform drowning linked to Miocene oceanographic events: Maiella platform margin, Italy. *Terra Nova* 9: 122–125.

Mutti, M., Pliller, W.E., and Betzler, C. (ed.) (2010). *Carbonate Systems During the Oligocene-Miocene Climate Transition*. Chichester (West Sussex): International Association of Sedimentologists/Wiley-Blackwell, 300 pp.

Nader, F.H. (2017). *Multi-Scale Quantitative Diagenesis and Impacts on Heterogeneity of Carbonate Reservoir Rocks*. Cham, Switzerland: Springer International Publishing, 146 pp.

Nagai, T., Durán, G.S., Otero, D.A. et al. (2019). How the Kuroshio current delivers nutrients to sunlit layers on the continental shelves with aid of near-inertial waves and turbulence. *Geophysical Research Letters* 46: 6726–6735.

Naseer, A. (2003). *The Integrated Growth Response of Coral Reefs to Environmental Forcing: Morphometric Analysis of Coral Reefs of the Maldives*. Halifax, Nova Scotia: Dalhousie University, 254 pp.

Neilson, J.E. (1990). Experimental investigation of controls on cementation in carbonates. *Journal of the Geological Society* 147: 949–958.

Nelson, C.S. (1988). An introductory perspective on non-tropical shelf carbonates. *Sedimentary Geology* 60: 3–12.

Nelson, J.E. and Ginsburg, R.N. (1986). Calcium carbonate production by epibionts on Thalassia in Florida Bay. *Journal of Sedimentary Petrology* 56: 622–628.

Neumeier, U. (1998). Tidal dunes and sand waves in deep outer-shelf enviroments, Bajocian, SE Jura, France. *Journal of Sedimentary Research* 68: 507–514.

Neuweiler, F., Gautret, P., Thiel, V. et al. (1999). Petrology of Lower Cretaceous carbonate mud mounds (Albian, N. Spain): insights into organomineralic deposits of the geological record. *Sedimentology* 46: 837–860.

Neuweiler, F., Mehdi, M., and Wilmsen, M. (2001). Facies of Liassic Sponge Mounds, Central High Atlas, Morocco. *Facies* 44: 243–264.

Neuweiler, F., Kershaw, S., Boulvain, F. et al. (2023). Keratose sponges in ancient carbonates - A problem of interpretation. *Sedimentology* 70: 927–968.

Neveux, L., Grgic, D., Carpentier, C. et al. (2014). Influence of hydrocarbon injection on the compaction by pressure-solution of a carbonate rock: an experimental study under triaxial stresses. *Marine and Petroleum Geology* 55: 282–294.

Noffke, N., Christian, D., Wacey, D., and Hazen, R.M. (2013). Microbially Induced Sedimentary Structures Recording an Ancient Ecosystem in the ca. 3.48 Billion-Year-Old Dresser Formation, Pilbara, Western Australia. *Astrobiology* 13: 1103–1124.

Nogues, J.P., Fitts, J.P., Celia, M.A., and Peters, C.A. (2013). Permeability evolution due to dissolution and precipitation of carbonates using reactive transport modeling in pore networks. *Water Resources Research* 49: 6006–6021.

Nolting, A., Zahm, C.K., Kerans, C., and Nikolinakou, M.A. (2018). Effect of carbonate platform morphology on syndepositional deformation: insights from numerical modeling. *Journal of Structural Geology* 115: 91–102.

Nolting, A., Zahm, C. K., Kerans, C., and Alzayer, Y. (2020). The influence of variable progradation to aggradation ratio and facies partitioning on the development of syndepositional deformation in steep-walled carbonate platforms. *Marine and Petroleum Geology* 114: 104171.

Nose, M. (1995). Vergleichende Faziesanalyse und Palökologie korallenreicher "shallowing-upwards" Abfolgen des iberischen Oberjura. *Profil* 8: 1–237.

Nose, M., Schmid, D.U., and Leinfelder, R. R. (2006). Significance of microbialites, calcimicrobes, and calcareous algae in reefal framework formation from the Silurian of Gotland, Sweden. *Sedimentary Geology* 192: 243–265.

Notholt, A.J.G. (1985). Phosphorite resources in the Mediterranean (Tethyan) phosphogenic province: a progress report. In: *Phosphorites. Sixth International Field-Workshop and Seminar on Phosphorites. I.G.C.P. 156. Maroc-Sénégal, Oct.–Nov. 1983.*, Sciences Géologiques, Mémoire 77, 9–17. Strasbourg, France: Strasbourg: Institut de Géologie - Université Louis-Pasteur.

Notholt, A.J.G., Sheldon, R., and Davidson, D.F. (1989). *Phosphate Deposits of the World, Vol. 2, Phosphate Rock Resources.* Cambridge: Cambridge University Press, 600 pp.

Nunn, E.V. and Price, G.D. (2010). Late Jurassic (Kimmeridgian–Tithonian) stable isotopes ($\delta^{18}O$, $\delta^{13}C$) and Mg/Ca ratios: new palaeoclimate data from Helmsdale, northeast Scotland. *Palaeogeography, Palaeoclimatology, Palaeoecology* 292: 325–335.

Nunn, E.V., Price, G.D., Hart, M.B. et al. (2009). Isotopic signals from Callovian–Kimmeridgian (Middle–Upper Jurassic) belemnites and bulk organic carbon, Staffin Bay, Isle of Skye, Scotland. *Journal of the Geological Society* 166: 633–641.

Oborny, S.C., Cramer, B.D., Heckel, P.H., Ludvigson, G.A. and Henthorne, R.W. (2017). Development of phylloid-algal carbonate mounds during regression: Expanding the build-and-fill model. *Journal of Sedimentary Research* 87: 688–706.

Olivier, N., Brayard, A., Vennin, E. et al. (2016). Evolution of depositional settings in the Torrey area during the Smithian (Early Triassic, Utah, USA) and their significance for the biotic recovery. *Geological Journal* 51: 600–626.

Olivier, N., Fara, E., Vennin, E. et al. (2018). Late Smithian microbial deposits and their lateral marine fossiliferous limestones (Early Triassic, Hurricane Cliffs, Utah, USA). *Facies* 64: 13.

Opdyke, B.N. and Wilkinson, B.H. (1990). Paleolatitude distribution of Phanerozoic marine ooids and cements. *Palaeogeography, Palaeoclimatology, Palaeoecology* 78: 135–148.

Opdyke, B.N. and Wilkinson, B.H. (1993). Carbonate mineral saturation state and cratonic limestone accumulation. *American Journal of Science* 293: 217–234.

O'Reilly, S.S., Mariotti, G., Winter, A.R. et al. (2017). Molecular biosignatures reveal common benthic microbial sources of organic matter in ooids and grapestones from Pigeon Cay, The Bahamas. *Geobiology* 15: 112–130.

Orpen, J.L. and Wilson, J.F. (1981). Stromatolites at ~3,500 Myr and a greenstone–granite unconformity in the Zimbabwean Archaean. *Nature* 291: 218–220.

Pace, A., Bourillot, R., Bouton, A. et al. (2018). Formation of stromatolite lamina at the interface of oxygenic–anoxygenic photosynthesis. *Geobiology* 16: 378–398.

Pacton, M., Ariztegui, D., Wacey, D. et al. (2012). Going nano: a new step toward understanding the processes governing freshwater ooid formation. *Geology* 40: 547–550.

Pagel, M., Barbarand, J., Beaufort, D. et al. (2014). *Bassins sédimentaires: Les marqueurs de leur histoire thermique*, 228. Paris, France: Edp Sciences, Société Géologique de France Géosphères.

Palmer, A.N. (1995). Geochemical models for the origin of macroscopic solution porosity in carbonate rocks. In: *Unconformities in Carbonate Strata: Their Recognition and the Significance of Associated Porosity*, Memoir 63 (ed. D. A. Budd, A.H. Saller, and P.M. Harris), 77–101. Tulsa, OK: American Association of Petroleum Geologists.

Parize, O., Rubino, J.-L., and Javaux, C. (1997). *Architecture et Géométrie des corps tidaux bioclastiques comblant les paléovallées miocènes au Sud-Est du bassin de Carpentras, Abstract Book: Publication ASF 27*, 215–216. Paris, France: Publication Association des Sedimentologists Francais.

Pasquier, J.B. and Strasser, A. (1997). Platform-to-basin correlation by high-resolution sequence stratigraphy and cyclostratigraphy (Berriasian, Switzerland and France). *Sedimentology* 44: 1071–1092.

Patterson, R.J. and Kinsman, D.J.J. (1981). Hydrologic framework of a sabkha along Arabian Gulf. *American Association of Petroleum Geologists Bulletin* 65: 1457–1475.

Paul, A. and Lokier, S. (2017). Holocene marine hardground formation in the Arabian Gulf: Shoreline stabilisation, sea level and early diagenesis in the coastal sabkha of Abu Dhabi. *Sedimentary Geology* 352: 1–13.

Paumard, V., Zuckmeyer, E., Boichard, R. et al. (2017). Evolution of Late Oligocene - Early Miocene attached and isolated carbonate platforms in a volcanic ridge context (Maldives type), Yadana field, offshore Myanmar. *Marine and Petroleum Geology* 81: 361–387.

Payne, J.L. and Clapham, M.E. (2012). End-Permian mass extinction in the oceans: an ancient analog for the twenty-first century? *Annual Review of Earth and Planetary Sciences* 40: 89–111.

Payne, J.L., Lehrmann, D.J., Christensen, S. et al. (2006). Environmental and biological controls on the initiation and growth of a Middle Triassic (Anisian) Reef Complex on the Great Bank of

Guizhou, Guizhou Province, China. *PALAIOS* 21: 325–343.

Payne, J.L., Lehrmann, D.J., Follett, D. et al. (2007). Erosional truncation of uppermost Permian shallow-marine carbonates and implications for Permian-Triassic boundary events. *Geological Society of America Bulletin* 119: 771–784.

Payros, A., Astibia, H., Cearreta, A. et al. (2000). The Upper Eocene South Pyrenean coastal deposits (Liedena Sandstone, Navarre): sedimentary facies, benthic foraminifera and Avian Ichnology. *Facies* 42: 107–132.

Payros, A., Pujalte, V., Tosquella, J., and Orue-Etxebarria, X. (2010). The Eocene storm-dominated foralgal ramp of the western Pyrenees (Urbasa–Andia Formation): an analogue of future shallow-marine carbonate systems? *Sedimentary Geology* 228: 184–204.

Payton, C.E. (1977). *Seismic Stratigraphy – Applications to Hydrocarbon Exploration*. Tulsa, OK: American Association of Petroleum Geologists, 516 pp.

Pearson, P.N. (2012). Oxygen Isotopes in Foraminifera: Overview and Historical Review. *The Paleontological Society Papers* 18: 1–38.

Peckmann, J., Walliser, O.H., Riegel, W., and Reitner, J. (1999). Signatures of hydrocarbon venting in a Middle Devonian Carbonate Mound (Hollard Mound) at the Hamar Laghdad (Antiatlas, Marocco). *Facies* 40: 281–296.

Pedersen, E., Rodriguez, L.L., Jones, G. and Qualman, H. (2021). Insights from Process-Based Models and Integration with Reservoir Characterization and Rock Typing Workflows for the Pre-Salt Lacustrine Carbonate Reservoirs. 2021 (1), 1-5.

Pedley, M. (1998). A review of sediment distributions and processes in Oligo-Miocene ramps of southern Italy and Malta (Mediterranean divide). In: *Carbonate Ramps: An Introduction*, Special Publication, vol. 149 (ed. V.P. Wright and T.P. Burchette), 163–179. London, UK: Geological Society of London.

Pedley, M. and Carannante, G. (2006). Cool-water carbonate ramps: a review. In: *Cool-Water Carbonates: Depositional Systems and Palaeoenvironmental Controls*, Special Publication, vol. 255 (ed. H.M. Pedley and G. Carannante), 1–9. UK: Geological Society, London.

Pellenard, P., Tramoy, R., Pucéat, E. et al. (2014). Carbon cycle and sea-water palaeotemperature evolution at the Middle–Late Jurassic transition, eastern Paris Basin (France). *Marine and Petroleum Geology* 53: 30–43.

Perkins, R.D. (1972). Aragonite crystals within codiacean algae: distinctive morphology and sedimentary implications. *Science* 175: 624–626.

Perkins, R.D. (1985). Timing of syntaxial cement, American Association of Petroleum Geologists Annual Meeting. American Association of Petroleum Geologists New Orleans, LA, USA. *American Association of Petroleum Geologists Bulletin* 69: 295–295.

Peropadre, C., Liesa, C.L., and Meléndez, N. (2013). High-frequency, moderate to high-amplitude sea-level oscillations during the late Early Aptian: insights into the Mid-Aptian event (Galve sub-basin, Spain). *Sedimentary Geology* 294: 233–250.

Perrin, C. (2002). Tertiary: the emergence of modern reef ecosystems. In: *Phanerozoic Reef Patterns*, SEPM Special Publication, vol. 72 (ed. W. Kiessling, E. Flügel, and J. Golonka), 587–621. Tulsa, OK: SEPM (Society for Sedimentary Geology).

Perrin, C. and Kiessling, W. (2012) Latitudinal trends in Cenozoic Reef patterns and their relationship to climate. In: *Carbonate Systems during the Oligocene–Miocene Climatic Transition* (ed M. Mutti, W. Piller and C. Betzler), IAS Special Publication no. 42, 17–33. Wiley-Blackwell.

Perry, C.T. (1999). Biofilm-related calcification, sediment trapping and constructive micrite envelopes: a criterion for the recognition of ancient grass-bed environments? *Sedimentology* 46: 33–45.

Perry, C.T. (2003). Coral reefs in a high-latitude, siliciclastic barrier island setting: reef framework and sediment production at Inhaca Island, southern Mozambique. *Coral Reefs* 22: 485–497.

Perry, C.T. (2005). Structure and development of detrital reef deposits in turbid nearshore environments, Inhaca Island, Mozambique. *Marine Geology* 214: 143–161.

Perry, C.T. and Beavington-Penney, S.J. (2005). Epiphytic calcium carbonate production and facies development within sub-tropical seagrass beds, Inhaca Island, Mozambique. *Sedimentary Geology* 174: 161–176.

Perry, R.S., Mcloughlin, N., Lynne, B.Y. et al. (2007). Defining biominerals and organominerals: Direct and indirect indicators of life. *Sedimentary Geology* 201: 157–179.

Peters, S.E. and Loss, D.P. (2012). Storm and fair-weather wave base: a relevant distinction? *Geology* 40: 511–514.

Petrash, D.A., Gingras, M.K., Lalonde, S.V. et al. (2012). Dynamic controls on accretion and lithification of modern gypsum-dominated thrombolites, Los Roques, Venezuela. *Sedimentary Geology* 245: 29–47.

Petrash, D.A., Bialik, O.M., Bontognali, T. R.R., Vasconcelos, C., Roberts, J.A., McKenzie, J.A. and Konhauser, K.O. (2017) Microbially catalyzed dolomite formation: from near-surface to burial. *Earth-Science Reviews*, 171, 558–582. Phelps, R. M., Kerans, C., Scott, S. Z., Janson, X., & Bellian, J. A. (2008). Three-dimensional modelling and sequence stratigraphy of a carbonate ramp-to-shelf transition, Permian Upper San Andres Formation. Sedimentology, 55(6), 1777-1813.

Petrovic, A., Ariza Fuentes, M., Putri, I. et al. (2022) Holocene sediment distribution in the Al Wajh platform lagoon (northern Red Sea, Saudi Arabia), a modern analogue for large rift basin carbonate platforms. *Sedimentology* 69: 1365–1398.

Phelps, R.M., Kerans, C., Scott, S.Z., Janson, X. and Bellian, J.A. (2008). Three-dimensional modelling and sequence stratigraphy of a carbonate ramp-to-shelf transition, Permian Upper San Andres formation. *Sedimentology* 55: 1777–1813.

Phelps, R.M., Kerans, C., Loucks, R.G. et al. (2014). Oceanographic and eustatic control of carbonate platform evolution and sequence stratigraphy on the Cretaceous (Valanginian–Campanian) passive margin, northern Gulf of Mexico. *Sedimentology* 61: 461–496.

Phelps, R.M., Kerans, C., Da-Gama, R.O.B. P. et al. (2015). Response and recovery of the Comanche carbonate platform surrounding multiple Cretaceous oceanic anoxic events, northern Gulf of Mexico. *Cretaceous Research* 54: 117–144.

Philip, J.M. and Gari, J. (2005). Late Cretaceous heterozoan carbonates: palaeoenvironmental setting, relationships with rudist carbonates (Provence, south-east France). *Sedimentary Geology* 175: 315–337.

Pierre, A., Durlet, C., Razin, P., and Chellai, E.H. (2010). Spatial and temporal distribution of ooids along a Jurassic carbonate ramp: Amellago outcrop transect, High-Atlas, Morocco. In: *Mesozoic and Cenozoic Carbonate Systems of the Mediterranean and the Middle East: Stratigraphic and Diagenetic Reference Models*, Special Publication, vol. 329, 65–88. Geological Society, London.

Pierson, B., Eberli, G., Al Mehsin, K. et al. (2010). Seismic stratigraphy and depositional history of the Upper Shu'aiba (Late Aptian) in the UAE and Oman. In: *Barremian – Aptian Stratigraphy and Petroleum Habitat of the Eastern Arabian Plate – Vol. 2*, GeoArabia Special Publication 4 - vol. 2 (ed. F.S.P. van Buchem, M.I. Al-Husseini, F. Maurer, and H.J. Droste), 411–444. Bahrain: Gulf-Petrolink.

Piller, W.E. (1994). The northern Bay of Safaga (Red Sea, Egypt): an actuopalaeontological approach. IV. Thin section analysis. *Beiträge zur Paläontologie von Österreich* 18: 1–73.

Piller, W.E. and Mansour, A.M. (1990). The northern Bay of Safaga (Red Sea, Egypt): an actuopaleontological approach. II: Sediment analysis and sedimentary facies. *Beiträge zur Paläontologie von Österreich* 16: 1–102.

Piller, W.E. and Mansour, A.M. (1994). Origin and transport mechanisms of non-carbonate sediments in a carbonate-dominated environment (northern Safaga Bay, Red Sea, Egypt). *Abhandlungen der Geologischen Bundesanstalt von Österreich* 50: 369–379.

Piller, W.E. and Pervesler, P. (1989). The northern Bay of Safaga (Red Sea, Egypt): an actuopaleontological approach. I. Topography and bottom facies. *Beiträge zur Paläontologie von Österreich* 15: 103–147.

Pirazzoli, P.A. and Montaggioni, L.F. (1988). Holocene sea-level changes in French Polynesia. *Palaeogeography, Palaeoclimatology, Palaeoecology* 68: 153–175.

Pirazzoli, P.A., Kaplin, P.A., and Montaggioni, L.F. (1990). Differential vertical crustal movements deduced from Late Holocene coral-rich conglomerates: Farquhar and St. Joseph Atolls (Seychelles, western Indian Ocean). *Journal of Coastal Research* 6: 381–389.

Piryaei, A., Reijmer, J.J.G., van Buchem, F. S.P. et al. (2010). The influence of Late Cretaceous tectonic processes on sedimentation patterns along the northeastern Arabian plate margin (Fars Province, SW Iran). In: *Tectonic and Stratigraphic Evolution of Zagros and Makran during the Mesozoic-Cenozoic*, Special Publication, vol. 330 (ed. P. Leturmy and C. Robin), 211–254. UK: Geological Society, London.

Piryaei, A., Reijmer, J.J.G., Borgomano, J., and van Buchem, F.S.P. (2011). Late Cretaceous tectonic and sedimentary evolution of the Bandar Abbas area, Fars region, Southern Iran. *Journal of Petroleum Geology* 34: 157–180.

Pittet, B. and Strasser, A. (1998a). Long-distance correlations by sequence stratigraphy and cyclostatigraphy: examples and implications (Oxfordian from the Swiss Jura, Spain, and Normandy). *Geologische Rundschau* 86: 852–874.

Pittet, B. and Strasser, A. (1998b). Depositional sequences in deep-shelf environments formed through carbonate-mud import from the shallow platform (Late Oxfordian, German Swabian Alb and eastern Swiss Jura). *Eclogae Geologicae Helvetiae* 91: 149–169.

Pittet, B., Strasser, A., and Mattioli, E. (2000). Depositional sequences in deep-shelf environments: a response to sea-level changes and shallow-platform carbonate productivity (Oxfordian, Germany and Spain). *Journal of Sedimentary Research* 70: 392–407.

Pittet, B., van Buchem, F.S.P., Hillgärtner, H. et al. (2002). Ecological succession, palaeoenvironmental change, and depositional sequences of Barremian-Aptian shallow-water carbonates in northern Oman. *Sedimentology* 49: 555–581.

Planavsky, N. and Grey, K. (2008). Stromatolite branching in the Neoproterozoic of the Centralian Superbasin, Australia: an investigation into sedimentary and microbial control of stromatolite morphology. *Geobiology* 6: 33–45.

Planavsky, N. J., McGoldrick, P., Scott, C. T. et al. (2011). Widespread iron-rich conditions in the mid-Proterozoic ocean. *Nature* 477(7365): 448–451.

Playford, P.E. (1980). Devonian "Great Barrier Reef" of Canning Basin, Western Australia. *American Association of Petroleum Geologists Bulletin* 64: 814–840.

Playford, P.E. (1984). Platform-Margin and Marginal-Slope Relationships in Devonian Reef Complexes of the Canning Basin. In: The Canning Basin, WA (Ed P.G. Purcell), *Proceedings Geological Society of Australia and Petroleum Exploration Society of Australia Symposium* Perth 1984, pp. 189–214. Australia.

Playford, P.E., Hocking, R.M., and Cockbain, A.E. (2009). *Devonian Reef Complexes of the Canning Basin, Western Australia*. Perth: Geological

Survey of Western Australia, 444 pp.

Playford, P.E., Cockbain, A.E., Berry, P.F. et al. (2013). *The Geology of Shark Bay*. Australia: Geological Survey of Western Australia, Perth, 281 pp.

Playton, T.E. and Kerans, C. (2015a). Late devonian carbonate margins and foreslopes of the Lennard Shelf, Canning Basin, Western Australia, Part A: development during backstepping and the aggradation-to-progradation transition. *Journal of Sedimentary Research* 85: 1334–1361.

Playton, T.E. and Kerans, C. (2015b). Late devonian carbonate margins and foreslopes of the Lennard Shelf, Canning Basin, Western Australia, Part B: development during progradation and across the Frasnian–Famennian Biotic crisis. *Journal of Sedimentary Research* 85: 1362–1392.

Playton, T.E. and Kerans, C. (2018). Architecture and genesis of prograding deep boundstone margins and debris-dominated carbonate slopes: examples from the Permian Capitan Formation, Southern Guadalupe Mountains, West Texas. *Sedimentary Geology* 370: 15–41.

Playton, T., Janson, X., and Kerans, C. (2010). Carbonate slopes. In: *Facies*, vol. 4 (ed. N.P. James and R.W. Dalrymple), 449–476. St. John's Newfoundland: Geological Association of Canada.

Playton, T.E., Weissenberger, J., and Kerans, C. (ed.) (2017). *New Advances in Devonian Carbonates: Outcrop Analogs, Reservoirs, and Chronostratigraphy*. Tulsa, OK: SEPM (Society for Sedimentary Geology), 408 pp.

Pochon, X. and Pawłowski, J. (2006). Evolution of the soritids-Symbiodinium symbiosis. *Symbiosis* 42: 77–88.

Pohl, A., Nardin, E., Vandenbroucke, T.R. A., and Donnadieu, Y. (2016). High dependence of Ordovician ocean surface circulation on atmospheric CO_2 levels. *Palaeogeography, Palaeoclimatology, Palaeoecology* 458: 39–51.

Pomar, L. (1991). Reef geometries, erosion surfaces and high-frequency sea-level changes, upper Miocene Reef Complex, Mallorca, Spain. *Sedimentology* 38: 243–269.

Pomar, L. (2001). Types of carbonate platforms: a genetic approach. *Basin Research* 13: 313–334.

Pomar, L. (2020). Carbonate systems. In: *Regional Geology and Tectonics* (ed. N. Scarselli, J. Adam, D. Chiarella, et al.), 235–311. Amsterdam: Elsevier.

Pomar, L. and Hallock, P. (2007). Changes in coral-reef structure through the Miocene in the Mediterranean province: adaptive versus environmental influence. *Geology* 35: 899–902.

Pomar, L. and Hallock, P. (2008). Carbonate factories: a conundrum in sedimentary geology. *Earth-Science Reviews* 87: 134–169.

Pomar, L. and Kendall, C.G.S.C. (2008). Architecture of carbonate platforms: A response to hydrodynamics and evolving ecology. In: *Controls on Carbonate Platform and Reef Development* (ed. J. Lukasik and J.A. Simo), SEPM Special Publication no. 89, 187–216. Tulsa: SEPM – Society for Sedimentary Geology.

Pomar, L. and Ward, W.C. (1995). Sea-level changes, carbonate production and platform architecture: the Llucmajor Platform, Mallorca, Spain. In: *Sequence Stratigraphy and Depositional Response to Eustatic, Tectonic and Climatic Forcing* (ed. E.B. U. Haq), 87–112. Kluwer Academic Publishers.

Pomar, L., Ward, W.C., and Green, D.G. (1996). Upper Miocene Reef complex of the Llucmajor Area, Mallorca, Spain. *SEPM: Concepts in Sedimentology and Paleontology* 5: 191–225.

Pomar, L., Obrador, A., and Westphal, H. (2002). Sub-wavebase cross-bedded grainstones on a distally steepened carbonate ramp, Upper Miocene, Menorca, Spain. *Sedimentology* 49: 139–169.

Pomar, L., Brandano, M., and Westphal, H. (2004). Environmental factors influencing skeletal grain sediment associations: a critical review of Miocene examples from the western Mediterranean. *Sedimentology* 51: 627–651.

Pomar, L., Mateu-Vicens, G., Morsilli, M., and Brandano, M. (2014). Carbonate ramp evolution during the Late Oligocene (Chattian), Salento Peninsula, southern Italy. *Palaeogeography, Palaeoclimatology, Palaeoecology* 404: 109–132.

Pomar, L., Baceta, J.I., Hallock, P. et al. (2017). Reef building and carbonate production modes in the west-central Tethys during the Cenozoic. *Marine and Petroleum Geology* 83: 261–304.

Popa, L., Panaiotu, C.E., and Grădinaru, E. (2014). An early Middle Anisian (Middle Triassic) Tubiphytes and cement crusts-dominated reef from North Dobrogea (Romania): facies, depositional environment and diagenesis. *Acta Geologica Polonica* 64: 189–206.

Pöppelreiter, M.C., Simone, A., and Hoetz, G. (2004). Reservoir characteristics of intracontinental carbonate ramp deposits-Upper Muschelkalk, Middle Triassic, NE Netherlands. *Netherlands Journal of Geosciences – Geologie en Mijnbouw* 83: 1–15.

Portilho-Ramos, R.d.C., Titschack, J., Wienberg, C. et al. (2022). Major environmental drivers determining life and death of cold-water corals through time. *PLOS Biology* 20: e3001628.

Posamentier, H.W., Laurin, P., Warmath, A. et al. (2010). Seismic stratigraphy and geomorphology of oligocene to miocene carbonate buildups, Offshore Madura, Indonesia. In: *Cenozoic Carbonate Systems of Australasia*, vol. 95 (ed. W.A. Morgan, A.D. George, P.M. Harris, et al.), 175–192. Tulsa, OK: SEPM (Society for Sedimentary Geology).

Poulton, A.J., Stinchcombe, M.C., Achterberg, E.P. et al. (2014). Coccolithophores on the north-west European shelf: calcification rates and environmental controls. *Biogeosciences* 11: 3919–3940.

Powell, J.H. and Moh'd, B.K. (2011). Evolution of Cretaceous to Eocene alluvial and carbonate platform sequences in central and south Jordan. *GeoArabia* 16: 29–82.

Poyatos-Moré, M., García-García, F., Rodríguez-Tovar, F.J. et al. (2022). Sharp-based, mixed carbonate–siliciclastic shallow-marine deposits (upper Miocene, Betic Cordillera, Spain): the record of ancient transgressive shelf ridges? *Sedimentary Geology* 429: 106077.

Prat, S., Jorry, S.J., Jouet, G. et al. (2016). Geomorphology and sedimentology of a modern isolated carbonate platform: The Glorieuses archipelago, *SW Indian Ocean. Marine Geology* 380: 272.

Pratt, B.R. (1995). The origin, biota and evolution of deep-water mud-mounds. In: *Carbonate Mud-Mounds – Their Origin and Evolution*, IAS Special Publication Number, vol. 23 (ed. C.L.V. Monty, D.W.J. Bosence, P.H. Bridges, and B.R. Pratt), 49–123. Chichester, West Sussex: International Association of Sedimentologists.

Pratt, B.R. and Rule, R.G. (2021). A Mesoproterozoic carbonate platform (lower Belt Supergroup of western North America): sediments, facies, tides, tsunamis and earthquakes in a tectonically active intracratonic basin. *Earth-Science Reviews* 217: 103626.

Pratt, B.R. and Smewing, J.D. (1993). Early Cretaceous platform margin configuration and evolution in the central Oman Mountains, Arabian Peninsula. *American Association of Petroleum Geologists Bulletin* 76: 225–244.

Preto, N., Hinnov, L.A., Hardie, L.A., and De Zanche, V. (2001). Middle Triassic orbital signature recorded in the shallow-marine Latemar carbonate buildup (Dolomites, Italy). *Geology* 29: 1123–1126.

Principaud, M., Mulder, T., Gillet, H., and Borgomano, J. (2015). Large-scale carbonate submarine mass-wasting along the northwestern slope of the Great Bahama Bank (Bahamas): morphology, architecture, and mechanisms. *Sedimentary Geology* 317: 27–42.

Principaud, M., Ponte, J.-P., Mulder, T. et al. (2017). Slope-to-basin stratigraphic evolution of the northwestern Great Bahama Bank (Bahamas) during the Neogene to Quaternary: interactions between downslope and bottom currents deposits. *Basin Research* 29: 699–724.

Principaud, M., Mulder, T., Hanquiez, V. et al. (2018). Recent morphology and sedimentary processes along the western slope of Great Bahama Bank (Bahamas). *Sedimentology* 65: 2088–2116.

Proust, J.N., Vennin, E., Vachard, D. et al. (1996). Étude sédimentologique et biostratigraphique du stratotype du Bashkirien (Oural du Sud, Russie). *Bulletin des centres de recherches exploration-Production Elf-Aquitaine* 20: 341–365.

Proust, J.N., Chuvashov, B.I., Vennin, E., and Boisseau, T. (1998). Carbonate platform drowning in a foreland setting: the mid-Carboniferous platform in western Urals (Russia). *Journal of Sedimentary Research* 68: 1175–1188.

Pruss, S.B., Corsetti, F.A., and Bottjer, D.J. (2005). The unusual sedimentary rock record of the Early Triassic: a case study from the southwestern United States. *Palaeogeography, alaeoclimatology, Palaeoecology* 222 (1-2): 33–52.

Pruss, S.B., Bottjer, D.J., Corsetti, F.A., and Baud, A. (2006). A global marine sedimentary response to the end-Permian mass extinction: examples from southern Turkey and the western United States. *Earth-Science Reviews* 78: 193–206.

Pufahl, P.K. and James, N.P. (2006). Monospecific Pliocene oyster buildups, Murray Basin, South Australia: Brackish water end member of the reef spectrum. *Palaeogeography, Palaeoclimatology, Palaeoecology* 233: 11–33.

Pufahl, P.K., Grimm, K.A., Abed, A.M., and Sadaqah, R.M.Y. (2003). Upper Cretaceous (Campanian) phosphorites in Jordan: implications for the formation of a south Tethyan phosphorite giant. *Sedimentary Geology* 161: 175–205.

Puga-Bernabéu, Á., Webster, J.M., Beaman, R.J., and Guilbaud, V. (2011). Morphology and controls on the evolution of a mixed carbonate–siliciclastic submarine canyon system, Great Barrier Reef margin, north-eastern Australia. *Marine Geology* 289: 100–116.

Puga-Bernabéu, Á., Webster, J.M., Beaman, R.J., and Guilbaud, V. (2013). Variation in canyon morphology on the Great Barrier Reef margin, north-eastern Australia: the influence of slope and barrier reefs. *Geomorphology* 191: 35–50.

Puga-Bernabéu, Á., Webster, J.M., Beaman, R.J. et al. (2014). Filling the gap: a 60-ky record of mixed carbonate-siliciclastic turbidite deposition from the Great Barrier Reef. *Marine and Petroleum Geology* 50: 40–50.

Purdy, E.G. (1963a). Recent calcium carbonate facies of the Great Bahama Bank. 1. Petrography and reaction groups. *Journal of Geology* 71: 334–355.

Purdy, E.G. (1963b). Recent calcium carbonate facies of the Great Bahama Bank. 2. Sedimentary facies. *Journal of Geology* 71: 472–497.

Purdy, E.G. (1974). Karst-determined facies patterns in British Honduras: Holocene carbonate sedimentation model. *American Association of Petroleum Geologists Bulletin* 58: 825–855.

Purdy, E.G. and Bertram, G.T. (1993). *Carbonate Concepts from the Maldives, Indian Ocean*, 1–56. American Association of Petroleum Geologists.

Purdy, E.G. and Gischler, E. (2003). The Belize margin revisited: 1. Holocene marine facies. *International Journal of Earth Sciences (Geologische Rundschau)* 92: 532–551.

Purdy, E.G. and Gischler, E. (2005). The transient nature of the empty bucket model of reef sedimentation. *Sedimentary Geology* 175: 35–47.

Purdy, E.G. and Waltham, D. (1999). Reservoir Implications of modern karst topography. *American Association of Petroleum Geologists Bulletin* 83: 1774–1794.

Purdy, E.G. and Winterer, E.L. (2001). Origin of atoll lagoons. *Geological*

Society of America Bulletin 113: 837–854.

Purdy, E.G. and Winterer, E.L. (2006). Contradicting Barrier Reef relationships for Darwin's Evolution of reef types. *International Journal of Earth Sciences* 95: 143–167.

Purdy, E.G., Pusey, W.C. III, and Wantlan, K.F. (1975). Continental shelf of belize-regional shelf attributes. In: *Belize Shelf—Carbonate Sediments, Clastic Sediments, and Ecology: Petrology and Diagenesis of Carbonate Eolianites of Northeastern Yucatán Peninsula, Mexico*, AAPG Studies in Geology, vol. 2 (ed. K.F. Wantland and W.C. Pusey III), 1–39. Tulsa, OK: American Association of Petroleum Geologists.

Purdy, E.G., Gischler, E., and Lomando, A. J. (2003). The Belize margin revisited. 2. Origin of Holocene antecedent topography. *International Journal of Earth Sciences (Geologische Rundschau)* 92: 552–572.

Purkis, S.J. and Harris, P.M. (2016). The extent and patterns of sediment filling of accommodation space on Great Bahama Bank. *Journal of Sedimentary Research* 86: 294–310.

Purkis, S. and Kohler, K. (2008). The role of topography in promoting fractal patchiness in a carbonate shelf landscape. *Coral Reefs* 27: 977–989.

Purkis, S.J., Kohler, K.E., Riegl, B.M., and Rohmann, S.O. (2007). The statistics of natural shapes in modern coral reef landscapes. *Journal of Geology* 115: 493–508.

Purkis, S., Kerr, J., Dempsey, A. et al. (2014). Large-scale carbonate platform development of Cay Sal Bank, Bahamas, and implications for associated reef geomorphology. *Geomorphology* 222: 25–38.

Purkis, S.J., Riegl, B.M. and Andrefouet, S. (2005). Remote sensing of geomorphology and facies patterns on a modern carbonate ramp (Arabian Gulf, Dubai, UAE). *Journal of Sedimentary Research* 75: 86–876.

Purkis, S.J., Rowlands, G.P., and Kerr, J.M. (2015). Unravelling the influence of water depth and wave energy on the facies diversity of shelf carbonates. *Sedimentology* 62: 541–565.

Purkis, S., Cavalcante, G., Rohtla, L. et al. (2017). Hydrodynamic control of whitings on Great Bahama Bank. *Geology* 45: 939–942.

Purser, B.H. (1973). *The Persian Gulf: Holocene Carbonate Sedimentation and Diagenesis in a Shallow Epicontinental Sea*. New York, Heidelberg, Berlin, Germany: Springer-Verlag, 471 pp.

Purser, B.H. and Evans, G. (1973). Regional sedimentation along the Trucial coast, SE Persian Gulf. In: *The Persian Gulf* (ed. B.H. Purser), 211–231. Berlin–Heidelberg: Springer.

Purser, B.H. and Loreau, J.P. (1973) Aragonitic, supratidal Encrustations on the Trucial Coast, Persian Gulf. In: *The Persian Gulf* (ed. B.H. Purser), 343–376. Berlin: Springer Berlin Heidelberg.

Quinn, T.M. and Saller, A.H. (2004). Geology of Anewetak Atoll, Republic of the Marshall Islands, Chapter 21. In: *Developments in Sedimentology*, vol. 54 (ed. H.L. Vacher and T.M. Quinn), 637–666. Amsterdam, The Netherlands: Elsevier.

Racey, A. (2001). A review of Eocene Nummulite accumulations: structure, formation and reservoir potential. *Journal of Petroleum Geology* 24: 79–100.

Racey, A., Bailey, H.W., Beckett, D. et al. (2001). The petroleum geology of the Early Eocene El Garia Formation, Hasdrubal Field, Offshore Tunisia. *Journal of Petroleum Geology* 24: 29–53.

Racki, G. (1998). Frasnian–Famennian biotic crisis: undervalued tectonic control? *Palaeogeography, Palaeoclimatology, Palaeoecology* 141: 177–198.

Radulović, M.M. (2013). A new view on karst genesis. *Carbonates and Evaporites* 28: 383–397.

Rameil, N., Immenhauser, A., Warrlich, G. et al. (2010). Morphological patterns of Aptian Lithocodium–Bacinella geobodies: relation to environment and scale. *Sedimentology* 57: 883–911.

Rangel, A., Parra, P., and Niño, C. (2000). The La Luna formation: chemostratigraphy and organic facies in the Middle Magdalena Basin. *Organic Geochemistry* 31: 1267–1284.

Rankey, E.C. (2014). Contrasts between wave- and tide-dominated oolitic systems: Holocene of Crooked–Acklins Platform, southern Bahamas. *Facies* 60: 405–428.

Rankey, E.C. (2016). On facies belts and facies mosaics: Holocene isolated platforms, South China Sea. *Sedimentology* 63: 2190–2216.

Rankey, E.C. (2020). Eustatic, climatic, and oceanographic influences on geomorphology and architecture of isolated carbonate platforms: Miocene, Northwest Shelf, Australia. *Lithosphere* 2020: 8844754.

Rankey, E.C. (2021). Platform-top reef sand apron morphodynamics and the half-empty bucket. *Sedimentary Geology* 412: 105825.

Rankey, E.C. and Doolittle, D.F. (2012). Geomorphology of carbonate platform-marginal uppermost slopes: insights from a Holocene analogue, Little Bahama Bank, Bahamas. *Sedimentology* 59: 2146–2171.

Rankey, E.C., Schlaich, M., Mokhtar, S. et al. (2019). Seismic architecture of a Miocene isolated carbonate platform and associated off-platform strata (Central Luconia Province, offshore Malaysia). *Marine and Petroleum Geology* 102: 477–495.

Rasser, M.W. and Piller, W.E. (2004). Crustose algal frameworks from the Eocene Alpine Foreland. *Palaeogeography, Palaeoclimatology, Palaeoecology* 206: 21–39.

Rasser, M.W., Scheibner, C., and Mutti, M. (2005). A paleoenvironmental standard section for Early Ilerdian tropical carbonate factories (Corbieres, France; Pyrenees, Spain). *Facies* 51: 217–232.

Raup, D.M. and Sepkoski, J.J. (1982). Mass extinctions in the Marine Fossil Record. *Science* 215: 1501–1503.

Raven, M.J., van Buchem, F.S.P., Larsen, P.-H. et al. (2010). Late Aptian incised valleys and siliciclastic infill at the top

Gotland (Sweden): facies interpretation based on stable isotopes in brachiopod shells. *Geologische Rundschau* 85: 278.

Sánchez-Almazo, I.M., Braga, J.C., Dinarès-Turell, J. et al. (2007). Palaeoceanographic controls on reef deposition: the Messinian Cariatiz reef (Sorbas Basin, Almería, SE Spain). *Sedimentology* 54: 637–660.

Sandberg, P.A. (1983). An oscillating trend in Phanerozoic non-skeletal carbonate mineralogy. *Nature* 305: 19–22.

Santodomingo, N., Novak, V., Pretković, V. et al. (2015). A diverse patch reef from turbid habitats in the Middle Miocene (East Kalimantan, Indonesia). *PALAIOS* 30: 128–149.

Sarg, J.F. (1982). Off-reef salina deposition (Silurian), Southern Michigan basin implications for reef genesis. In: *Depositional and Diagenetic Spectra of Evaporites – A Core Workshop*, Core Workshop, vol. 3 (ed. C.R. Handford, R. G. Loucks, and G.R. Davies), 354–384. Tulsa, OK: SEPM (Society for Sedimentary Geology).

Sarg, J.F. (1988). Carbonate sequence stratigraphy. In: *Sea-level changes: an integrated approach*, Special Publication, vol. 42 (ed. C.K. Wilgus, B.S. Hastings, C. G.S.C. Kendall, et al.), 155–182. Tulsa, OK: Society of Economic Paleontologists and Mineralogists.

Sarg, J.F. (2001). The sequence stratigraphy, sedimentology, and economic importance of evaporite-carbonate transitions: a review. *Sedimentary Geology* 140: 9–42.

Sattler, U., Immenhauser, A., Schlager, W., and Zampetti, V. (2009). Drowning history of a Miocene carbonate platform (Zhujiang Formation, South China Sea). *Sedimentary Geology* 219: 318–331.

Saupe, E.E., Qiao, H., Donnadieu, Y. et al. (2020). Extinction intensity during Ordovician and Cenozoic glaciations explained by cooling and palaeogeography. *Nature Geoscience* 13: 65–70.

Saxena, S. and Betzler, C. (2003). Genetic sequence stratigraphy of cool water slope carbonates (Pleistocene Eucla Shelf, southern Australia). *International Journal of Earth Sciences (Geologische Rundschau)* 92: 482–493.

Schatzinger, R.A. (1983). Phylloid algal and sponge-bryozoan mound-to-basin transition: a late paleozoic facies tract from the Kelly-Snyder Field, West Texas. In: *Carbonate Buildups – A Core Workshop*, SEPM Core Workshop, vol. 4 (ed. P.M. Harris), 197–211. Tulsa, OK: SEPM (Society for Sedimentary Geology).

Scheibner, C. and Reijmer, J.J.G. (1999). Facies patterns within a lower Jurassic upper slope to inner platform transect (Jbel Bou Dahar, Morocco). *Facies* 41: 55–80.

Scheibner, C. and Speijer, R.P. (2008). Late Paleocene–early Eocene Tethyan carbonate platform evolution – a response to long- and short-term paleoclimatic change. *Earth-Science Reviews* 90: 71–102.

Scheibner, C., Speijer, R.P., and Marzouk, A.M. (2005). Turnover of larger foraminifera during the Paleocene-Eocene Thermal Maximum and paleoclimatic control on the evolution of platform ecosystems. *Geology* 33: 493–496.

Scheibner, C., Rasser, M.W., and Mutti, M. (2007). The Campo section (Pyrenees, Spain) revisited: Implications for changing benthic carbonate assemblages across the Paleocene–Eocene boundary. *Palaeogeography Palaeoclimatology Palaeoecology* 248: 145–168.

Schiebel, R. and Hemleben, C. (2005). Modern planktic foraminifera. *Paläontologische Zeitschrift* 79: 135–148.

Schlager, W. (1981). The paradox of drowned reefs and carbonate platforms. *Bulletin of the Geological Society of America* 92: 197–211.

Schlager, W. (1989). Drowning unconformities on carbonate platforms. In: *Controls on Carbonate Platform and Basin Development* (ed. P. D. Crevello, J. L. Wilson, J. F. Sarg and J. F. Read), *Special Publication* vol. 44. Society of Economic Paleontologists and Mineralogists (SEPM), 15–25 pp.

Schlager, W. (2000). Sedimentation rates and growth potential of tropical, cool-water and mud-mound carbonate systems. In: *Carbonate Platform Systems: Components and Interactions*, Special Publication, vol. 178 (ed. E. Insalaco, P.W. Skelton, and Y.J. Palmer), 217–227. London: Geological Society.

Schlager, W. (2003). Benthic carbonate factories of the Phanerozoic. *International Journal of Earth Sciences* 92: 445–464.

Schlager, W. (2004). Fractal nature of stratigraphic sequences. *Geology* 32: 185–188.

Schlager, W. (2005). *Carbonate Sedimentology and Sequence Stratigraphy*. Tulsa, OK, USA: SEPM (Society for Sedimentary Geology), 200 pp.

Schlanger, S.O., Arthur, M.A., Jenkyns, H. C. and Scholle, P.A. (1987) The Cenomanian-Turonian Oceanic Anoxic Event, I. Stratigraphy and distribution of organic carbon-rich beds and the marine δ13C excursion. *Geological Society, London, Special Publications* 26: 371–399.

Schlager, W. and Adams, E.W. (2001). Model for the sigmoidal curvature of submarine slopes. *Geology* 29: 883–886.

Schlager, W. and Camber, O. (1986). Submarine slope angles, drowning unconformities and self-erosion of limestone escarpments. *Geology* 14: 762–765.

Schlager, W. and James, N.P. (1978). Low-magnesian calcite limestones forming at the deep-sea floor, Tongue of the Ocean, Bahamas. *Sedimentology* 25: 675–702.

Schlager, W. and Purkis, S. (2015). Reticulate reef patterns – antecedent karst versus self-organization. *Sedimentology* 62: 501–515.

Schlager, W. and Reijmer, J.J.G. (2009). Carbonate platform slopes of the Alpine Triassic and the Neogene – a comparison. *Austrian Journal of Earth Sciences* 102: 4–14.

Schlager, W., Reijmer, J.J.G., and Droxler, A. (1994). Highstand shedding of carbonate platforms. *Journal of Sedimentary Research; Section B – Stratigraphy and Global Studies* 64: 270–281.

Schlagintweit, F. and Bover-Arnal, T. (2012). The morphological adaptation of Lithocodium aggregatum Elliott (calcareous green alga) to cryptic microhabitats (Lower Aptian, Spain): an example of phenotypic plasticity. *Facies* 58: 37–55.

Schlagintweit, F., Bover-Arnal, T., and Salas, R. (2010). New insights into Lithocodium aggregatum Elliott 1956 and Bacinella irregularis Radoicic 1959 (Late Jurassic-Lower Cretaceous): two ulvophycean green algae (?Order Ulotrichales) with a heteromorphic life cycle (epilithic/euendolithic). *Facies* 56: 509–547.

Schlanger, S.O. (1981). Shallow-water limestones in oceanic basins as tectonic and paleoceanographic indicators. In: *The Deep Sea Drilling Project: A Decade of Progress*, Society Economic Paleontologists Mineralogists Special Publication, vol. 32 (ed. J.E. Warme, R. G. Douglas, and E.L. Winterer), 209–226. Tulsa, OK: SEPM (Society for Sedimentary Geology).

Schmid, D.U., Leinfelder, R.R., and Nose, M. (2001). Growth dynamics and ecology of Upper Jurassic mounds, with comparisons to Mid-Palaeozoic mounds. *Sedimentary Geology* 145: 343–376.

Schmidt, V. (1977). Inorganic and organic reef growth and subsequent diagenesis in the Permian Capitan reef complex, Guadalupe Mountains, Texas, New Mexico. In: *Upper Guadalupian facies, Permian reef complex, Guadalupe Mountains, New Mexico and West Texas*, Special Publication, vol. 77-16 (ed. M.E. Hileman and S.J. Mazzullo), 93–131. Midland, TX: SEPM (Society for Sedimentary Geology), Permian Basin Section.

Schnyder, J.S.D., Eberli, G.P., Kirby, J.T. et al. (2016). Tsunamis caused by submarine slope failures along western Great Bahama Bank. *Scientific Reports* 6: 35925.

Scholle, P.A. and Halley, R.B. (1985). Burial diagenesis: out of sight, out of mind! In: *Carbonate Cements: Based on a Symposium Sponsored by the Society of Economic Paleontologists and Mineralogists*, vol. 36 (ed. N. Schneidermann and P.M. Harris), 309–334. Tulsa, OK: SEPM (Society for Sedimentary Geology).

Scholle, P.A. and Kling, S.A. (1972). Southern British Honduras; lagoonal coccolith ooze. *Journal of Sedimentary Research* 42: 195–204.

Scholle, P.A. and Ulmer-Scholle, D.S. (2003). *A Color Guide to the Petrography of Carbonate Rocks*. Tulsa, OK: American Association of Petroleum Geologists, 474 pp.

Scholle, P.A., Bebout, D.G., and Moore, C.H. (1983). *Carbonate Depositional Environments*. Tulsa, OK: American Association of Petroleum Geologists, 708 pp.

Schopf, J.W., Kudryavtsev, A.B., Czaja, A. D., and Tripathi, A.B. (2007). Evidence of Archean life: stromatolites and microfossils. *Precambrian Research* 158: 141–155.

Schroeder, R., van Buchem, F.S.P., Cherchi, A. et al. (2010). Revised orbitolinid biostratigraphic zonation for the Barremian – Aptian of the eastern Arabian Plate and implications for regional stratigraphic correlations. In: *Barremian – Aptian Stratigraphy and Hydrocarbon Habitat of the Eastern Arabian Plate (vol. 1)*, GeoArabia Special Publication 4, vol. 1 (ed. F.S.P. van Buchem, M.I. Al-Husseini, F. Maurer, and H.J. Droste), 49–96. Bahrain: Gulf PetroLink.

Schubert, J.K. and Bottjer, D.J. (1992). Early Triassic stromatolites as post-mass extinction disaster forms. *Geology* 20: 883–886.

Schutter, M., Crocker, J., Paijmans, A. et al. (2010). The effect of different flow regimes on the growth and metabolic rates of the scleractinian coral Galaxea fascicularis. *Coral Reefs* 29: 737–748.

Schwarzacher, W. (1947). Über die sedimentäre Rhythmik des Dachsteinkalkes bei Lofer. *Verhandlungen der Geologischen Bundesanstalt* 10-12: 175–188.

Schwarzacher, W. (1954). Die Großryhtmik des Dachsteinkalkes von Lofer. *Tschermaks Mineralogische und Petrographische Mitteilungen* 4: 44–54.

Schwarzacher, W. (2005). The stratification and cyclicity of the Dachstein Limestone in Lofer, Leogang and Steinernes Meer (Northern Calcareous Alps, Austria). *Sedimentary Geology* 181: 93–106.

Scotese, C.R., Song, H., Mills, B.J.W., and van der Meer, D.G. (2021). Phanerozoic paleotemperatures: the earth's changing climate during the last 540 million years. *Earth-Science Reviews* 215: 103503.

Scott, R.W., Fernández-Mendiola, P.A.G., and E. and Simó, A. (1990). Persistence of coral-rudist reefs into the late cretaceous. *PALAIOS* 5: 98–110.

Séard, C., Camoin, G., Yokoyama, Y. et al. (2011). Microbialite development patterns in the last deglacial reefs from Tahiti (French Polynesia; IODP Expedition #310): implications on reef framework architecture. *Marine Geology* 279: 63–86.

Seddighi, M., Briguglio, A., Hohenegger, J., and Papazzoni, C.A. (2015). New results on the hydrodynamic behaviour of fossil Nummulites tests from two nummulite banks from the Bartonian and Priabonian of northern Italy. *Bollettino della Societa Paleontologica Italiana* 54: 103–116.

Semikhatov, M.A. and Raaben, M.E. (2000). Proterozoic stromatolite taxonomy and biostratigraphy. In: *Microbial Sediments* (ed. R.E. Riding and S.M. Awramik), 295–306. Berlin: Springer.

Sepkoski, J.J. (1981). A factor analytic description of the Phanerozoic marine fossil record. *Paleobiology* 7: 36–53.

Sepkoski, J.J. (1996). Patterns of Phanerozoic extinction: a perspective from global data bases. In: *Global Events and Event Stratigraphy in the Phanerozoic: Results of the International Interdisciplinary Cooperation in the IGCP-Project 216 "Global Biological Events in Earth History"* (ed. O.H. Walliser), 35–51. Berlin, Heidelberg: Springer.

Sepkoski, J.J., Bambach, R.K., Raup, D.M., and Valentine, J.W. (1981). Phanerozoic marine diversity and the fossil record. *Nature* 293: 435–437.

Sequero, C., Bádenas, B. and Aurell, M. (2018). Facies mosaic in the inner areas of a shallow carbonate ramp (Upper Jurassic, Higueruelas Fm, NE Spain). *Facies* 64: 9.

Sevillano, A., Rosales, I., Bádenas, B. et al. (2019). Spatial and temporal facies evolution of a Lower Jurassic carbonate platform, NW Tethyan margin (Mallorca, Spain). *Facies* 65 (3): 1–34.

Shalev, N., Bontognali, T.R.R., and Vance, D. (2021). Sabkha dolomite as an archive for the magnesium isotope composition of seawater. *Geology* 49: 253–257.

Shao, L., Li, Q., Zhu, W. et al. (2017). Neogene carbonate platform development in the NW South China Sea: Litho-, bio- and chemo-stratigraphic evidence. *Marine Geology* 385: 233–243.

Sharaf, E., Simo, J.A., Carroll, A.R., and Shields, M. (2005). Stratigraphic evolution of Oligocene–Miocene carbonates and siliciclastics, East Java basin, Indonesia. *American Association of Petroleum Geologists Bulletin* 89: 789–819.

Sharp, Z. (2017). *Principles of Stable Isotope Geochemistry*, 2e. Albuquerque, New Mexico: University of New Mexico, 416 pp.

Sharp, I.R., Gawthorpe, R.L., Underhill, J. R., and Gupta, S. (2000). Fault-propagation folding in extensional settings: examples of structural style and synrift sedimentary response from the Suez rift, Sinai, Egypt. *Geological Society of America Bulletin* 112: 1877–1899.

Shaver, R.H. (1977). Silurian reef geometry; new dimensions to explore, SEPM Presidential Address, Washington, D.C., June 13, 1977. *Journal of Sedimentary Research* 47: 1409–1424.

Shen, J.-W. and Webb, G.E. (2004). Famennian (Upper Devonian) calcimicrobial (Renalcis) reef at Miaomen, Guilin, Guangxi, South China. *Palaeogeography, Palaeoclimatology, Palaeoecology* 204: 373–394.

Shinn, E. (1963). Spur and groove formation on the Florida Reef Tract. *Journal of Sedimentary Research* 33: 291–303.

Shinn, E.A. (1969). Submarine lithification of Holocene carbonate sediments in the Persian Gulf. *Sedimentology* 12: 109–144.

Shinn, E.A. (2009). The mystique of beachrock. In: *Perspectives in Carbonate Geology: A Tribute to the Career of Robert Nathan Ginsburg*, IAS Special Publication, vol. 41 (ed. P.K. Swart, G.P. Eberli, J.A. McKenzie, et al.), 19–28. Oxford, UK: Wiley-Blackwell-IAS.

Shinn, E.A., Steinen, R.P., Lidz, B.H., and Swart, P.K. (1989). Whitings, a sedimentologic dilemma. *Journal of Sedimentary Petrology* 59: 147–161.

Simmons, H.L. and Alford, M.H. (2012). Simulating the long-range swell of internal waves generated by ocean storms. *Oceanography* 25 (2): 30–41.

Simmons, M.D. and Hart, M.B. (1987). The biostratigraphy and microfacies of the Early to mid-Cretaceous carbonates of Wadi Mi'aidin, central Oman Mountains. In: *Micropalaeontology of Carbonate Environments* (ed. M.B. Hart), 176–207. Chichester: Ellis Horwood.

Simo, J.A.T., Scott, R.W., and Masse, J.-P. (ed.) (1993). *Cretaceous Carbonate Platforms*. Tulsa, OK: American Association of Petroleum Geologists, 479 pp.

Skelton, P.W. (2003). Rudist evolution and extinction – a North African perspective. In: *North African Cretaceous Carbonate Platform Systems* (ed. E. Gili, M. El Hédi Negra, and P.W. Skelton), 215–227. Dordrecht, The Netherlands: Springer Netherlands.

Skelton, P.W. and Gili, E. (2012). Rudists and carbonate platforms in the Aptian: a case study on biotic interactions with ocean chemistry and climate. *Sedimentology* 59: 81–117.

Skelton Peter, W., Sano, S.-I., and Masse, J.-P. (2013). Rudist bivalves and the Pacific in the Late Jurassic and Early Cretaceous. *Journal of the Geological Society* 170: 513–526.

Skelton, P.W., Gili, E., Rosen, B.R., and Valldeperas, F.X. (1997). Corals and rudists in the late Cretaceous: a critique of the hypothesis of competitive displacement. *Boletín de la Sociedad Española de Historia Natural* 92: 225–239.

Sloss, L.L. (1963). Sequences in the Cratonic Interior of North America. *Geological Society of America Bulletin* 74: 93–114.

Sloss, L.L. (1988). Forty years of sequence stratigraphy. *Geological Society of America Bulletin* 100: 1661–1665.

Smalley, P.C., Higgins, A.C., Howarth, R.J. et al. (1994). Seawater Sr isotope variations through time: a procedure for constructing a reference curve to date and correlate marine sedimentary rocks. *Geology* 22: 431–434.

Smart, P.L., Dawans, J.M., and Whitaker, F. (1988). Carbonate dissolution in a modern mixing zone. *Nature* 335: 811–813.

Smit, F.W.H., Stemmerik, L., Smith, M.E. et al. (2022). The occurrence of a palaeo-lithification front in an overpressured basin (Chalk Group) – timing from clumped isotopes and seismic stratigraphic analysis. *Basin Research* 34: 1341–1373.

Smith, S.V. and Buddemeier, R.W. (1992). Global change and coral reef ecosystems. *Annual Review of Ecology and Systematics* 23: 89–118.

Smith, M.E. and Swart, P.K. (2022). The influence of diagenesis on carbon and oxygen isotope values in shallow water carbonates from the Atlantic and Pacific: implications for the interpretation of the global carbon cycle. *Sedimentary Geology* 434: 106147.

Smith, D.E., Harrison, S., Firth, C.R., and Jordan, J.T. (2011). The early Holocene sea level rise. *Quaternary Science Reviews* 30: 1846–1860.

Soete, J., Kleipool, L.M., Claes, H. et al. (2015). Acoustic properties in travertines and their relation to porosity and pore types. *Marine and Petroleum Geology* 59: 320–335.

Srinivasan, K. and Walker, K.R. (1993). Sequence stratigraphy of an intrashelf basin carbonate ramp to rimmed platform transition: Maryville Limestone (Middle Cambrian), southern Appalachians. *Geological Society of America Bulletin* 105: 883–896.

Stal, L.J. (2012). Cyanobacterial mats and stromatolites. In: *Ecology of Cyanobacteria II: Their Diversity in Space and Time* (ed. B.A. Whitton), 65–125. Dordrecht. The Netherlands: Springer Netherlands.

Stanley, G.D. Jr. (2003). The evolution of modern corals and their early history. *Earth-Science Reviews* 60: 195–225.

Stanley, G.D. and Lipps, J.H. (2011). Photosymbiosis: the driving force for reef success and failure. In: *Corals and Reef Crises, Collapse and Change*, Paleontological Society Short Course, vol. 17 (ed. J.G.D. Stanley), 33–60. Westminster, CO: Paleontological Society Short Course.

Stanley, S.M. (2006). Influence of seawater chemistry on biomineralization throughout phanerozoic time: paleontological and experimental evidence. *Palaeogeography, Palaeoclimatology, Palaeoecology* 232: 214–236.

Stanley, S.M. (2012). Geobiology of the phanerozoic. In: *Fundamentals of Geobiology* (ed. A.H. Knoll, D.E. Canfield, and K.O. Konhauser), 403–424. Oxford: Wiley-Blackwell/Wiley Online Library.

Stanley, S.M. and Hardie, L.A. (1998). Secular oscillations in the carbonate mineralogy of reef-building and sediment-producing organisms driven by tectonically forced shifts in seawater chemistry. *Palaeogeography, Palaeoclimatology, Palaeoecology* 144: 3–19.

Stanley, S.M., Ries, J.B., and Hardie, L.A. (2005). Seawater chemistry, coccolithophore population growth, and the origin of Cretaceous chalk. *Geology* 33: 593–596.

Steffen, W., Richardson, K., Rockstrom, J. et al. (2020). The emergence and evolution of Earth System Science. *Nature Reviews Earth and Enivronment* 1: 54–63.

Stemmerik, L., Bendix-Almgreen, S.E., and Piasecki, S. (2001). The Permian–Triassic boundary in central East Greenland: past and present views. *Bulletin of the Geological Society of Denmark* 48: 159–167.

Stenzel, S.R. and James, N.P. (1995). Shallow-water stromatactis mud-mounds on a Middle Ordovician foreland basin platform, western Newfoundland. In: *Carbonate Mud-Mounds – Their Origin and Evolution*, IAS Special Publication, vol. 23 (ed. C.L. V. Monty, D.W.J. Bosence, P.H. Bridges, and B.R. Pratt), 127–149. Chichester, West Sussex: International Association of Sedimentologists–Blackwell.

Stephens, D.W. (1990). Changes in lake levels, salinity and the biological community of Great Salt Lake (Utah, USA), 1847–1987. *Hydrobiologia* 197: 139–146.

Stephens, N. P., & Sumner, D. Y. (2003). Late Devonian carbon isotope stratigraphy and sea level fluctuations, Canning Basin, Western Australia. *Palaeogeography, Palaeoclimatology, Palaeoecology* 191(2): 203–219.

Stern, J.C. and Wieman, S.T. (2021). Traditional stable isotope geochemistry. In: *Encyclopedia of Geology*, 2e (ed. D. Alderton and S.A. Elias), 100–113. Oxford: Academic Press.

Steuber, T. (2002). Plate tectonic control on the evolution of Cretaceous platform-carbonate production. *Geology* 30: 259–262.

Stoakes, F.A. (1980). Nature and control of shale basin fill and its effect on reef growth and termination: Upper Devonian Duvernay and Ireton formations of Alberta, Canada. *Bulletin of Canadian Petroleum Geology* 28: 345–410.

Stow, D.A.V., Hunter, S., Wilkinson, D., and Hernández-Molina, F.J. (2008). The nature of contourite deposition, Chapter 9. In: *Developments in Sedimentology*, vol. 60 (ed. M. Rebesco and A. Camerlenghi), 143–156. Amsterdam, The Netherlands: Elsevier.

Strasser, A. (1988). Shallowing-upward sequences in Purbeckian peritidal carbonates (Lowermost Cretaceous, Swiss and French Jura Mountains). *Sedimentology* 35: 369–383.

Strasser, A. and Védrine, S. (2009). Controls on facies mosaics of carbonate platforms: a case study from the oxfordian of the Swiss Jura. In: *Perspectives in Carbonate Geology: a Tribute to the Career of Robert Nathan Ginsburg*, IAS Special Publication, vol. 41 (ed. P.K. Swart, G.P. Eberli, and J.A. McKenzie), 199–213. Chichester: Wiley-Blackwell.

Strasser, A., Davaud, E., and Jedoui, Y. (1989). Carbonate cements in Holocene beachrock: example from Bahiret el Biban, southeastern Tunisia. *Sedimentary Geology* 62: 89–100.

Strasser, A., Strohmenger, C., Davaud, E. and Bach, A. (1992) Sequential evolution and diagenesis of Pleistocene coral reefs (South Sinai, Egypt). *Sedimentary Geology* 78: 59–79.

Strasser, A., Pittet, B., Hillgärtner, H., and Pasquier, J.-B. (1999). Depositional sequences in shallow carbonate-dominated sedimentary systems: concepts for a high-resolution analysis. *Sedimentary Geology* 128: 201–221.

Strasser, A., Hilgen, F.J., and Heckel, P.H. (2006). Cyclostratigraphy-concepts, definitions, and applications. *Newsletters on Stratigraphy* 42: 75–114.

Strasser, A., Védrine, S., and Stienne, N. (2012). Rate and synchronicity of environmental changes on a shallow carbonate platform (Late Oxfordian, Swiss Jura Mountains). *Sedimentology* 59: 185–211.

Strohmenger, C.J., Steuber, T., Ghani, A. et al. (2010). Sedimentology and chemostratigraphy of the Hawar and Shu'aiba depositional sequences, Abu Dhabi, United Arab Emirates. In: *Barremian – Aptian Stratigraphy and Hydrocarbon Habitat of the Eastern Arabian Plate (vol. 2)*, GeoArabia Special Publication, vol. 4 (ed. F.S.P.

van Buchem, M.I. Al-Husseini, F. Maurer, and H.J. Droste), 341–365. Bahrain: Gulf PetroLink.

Stuhr, M., Blank-Landeshammer, B., Reymond, C.E. et al. (2018). Disentangling thermal stress responses in a reef-calcifier and its photosymbionts by shotgun proteomics. *Scientific Reports* 8: 3524.

Stuut, J.-B., Zabel, M., Ratmeyer, V. et al. (2005). Provenance of present-day eolian dust collected off NW Africa. *Journal of Geophysical Research: Atmospheres* 110: 14.

Suess, E. (2002). The evolution of an idea: from avoiding gas hydrates to actively drilling for them. In: *Achievements and Opportunities of Scientific Ocean Drilling – The Legacy of the Ocean Drilling Program* (ed. L.C. Peterson, H. Elderfield, C. Mevel and J. Tarduno), Joides Journal, 45–50. College Station: Ocean Drilling Program.

Suosaari, E.P., Reid, R.P., Araujo, T.A.A. et al. (2016a). Environmental pressures influencing living stromatolites in Hamelin Pool, Shark Bay, Western Australia. *PALAIOS* 31: 483–496.

Suosaari, E.P., Reid, R.P., Playford, P.E. et al. (2016b). New multi-scale perspectives on the stromatolites of Shark Bay, Western Australia. *Scientific Reports* 6: 20557.

Suosaari, E.P., Reid, R.P., Oehlert, A.M. et al. (2019). Stromatolite provinces of hamelin pool: physiographic controls on stromatolites and associated lithofacies. *Journal of Sedimentary Research* 89: 207–226.

Surlyk, F. (ed.) (1997). *A Cool-Water Carbonate Ramp with Bryozoan Mounds: Late Cretaceous-Danian of the Danish Basin*, 293–308. Tulsa, OK: SEPM (Society for Sedimentary Geology).

Surlyk, F. and Lykke-Andersen, H. (2007). Contourite drifts, moats and channels in the Upper Cretaceous chalk of the Danish Basin. *Sedimentology* 54: 405–422.

Surlyk, F., Dons, T., Clausen, C.K., and Higham, J. (2003). Upper Cretaceous. In: *The Millennium Atlas: Petroleum Geology of the Central and Northern North Sea* (ed. D. Evans, C. Graham, A. Armour, and P. Bathurst), 213–233. London: Geological Society of London.

Swart, P.K. (2008). Global synchronous changes in the carbon isotopic composition of carbonate sediments unrelated to changes in the global carbon cycle. *Proceedings of the National Academy of Sciences of the United States of America* 105: 13741.

Swart, P.K. (2015). The geochemistry of carbonate diagenesis: the past, present and future. *Sedimentology* 62: 1233–1304.

Swart, P.K. and Eberli, G. (2005). The nature of the $\delta^{13}C$ of periplatform sediments: implications for stratigraphy and the global carbon cycle. *Sedimentary Geology* 175: 115–129.

Swart, P.K., Reijmer, J.J.G., and Otto, R. (2009). A re-evaluation of facies on Great Bahama Bank II: variations in the $\delta^{13}C$, $\delta^{18}O$ and mineralogy of surface sediments. In: *Perspectives in Carbonate Geology: A Tribute to the Career of Robert Nathan Ginsburg*, IAS Special Publication, vol. 41 (ed. P.K. Swart, G.P. Eberli, J.A. McKenzie, et al.), 47–59. Oxford, UK: Wiley-Blackwell-IAS.

Swart, P.K., Oehlert, A.M., Mackenzie, G. J. et al. (2014). The fertilization of the Bahamas by Saharan dust: a trigger for carbonate precipitation? *Geology* 42: 671–674.

Szilagyi, Z., Webster, J.M., Patterson, M. A., et al. (2020). Controls on the spatio-temporal distribution of microbialite crusts on the Great Barrier Reef over the past 30,000 years. *Marine Geology* 429: 106312.

Szulc, J. (2007). Sponge-microbial stromatolites and coral-sponge reef recovery in the Triassic of the Western Tethys domain. *Museum of Natural History and Science Bulletin* 41: 402.

Szulczewski, M. (2013). Depositional evolution of the Holly Cross Mts. (Poland) in the Devonian and Carboniferous – a review. *Geological Quarterly* 39: 471–488.

Taati Qorayem, F. (2005). Stratigraphie sequentielle des iagene carbonates dans un contextede tectonique active: le groupe de Bangestan (Albien-Turonien) dans le Zagros Central (Iran) – Sequence stratigraphy of the Bangestan group (Albian – Turonian) in tectonically active setting (Zagros, Iran). Thèse doctorat, Universite de Bordeaux 3, Bordeaux, France, 266 pp.

Taberner, M.C. and Bosence, D.W.J. (1995). An Eocene biodetrital mud-mound from the southern Pyrenean foreland basin, Spain: an ancient analogue for Florida Bay mounds. In: *Carbonate Mud-Mounds – Their Origin and Evolution*, Special Publication Number, vol. 23 (ed. C.L.V. Monty, D. W.J. Bosence, P.H. Bridges, and B.R. Pratt), 423–437. Chichester, West Sussex: International Association of Sedimentologists.

Tassy, A., Crouzy, E., Gorini, C. et al. (2015). Egyptian Tethyan margin in the Mesozoic: evolution of a mixed carbonate-siliciclastic shelf edge (from Western Desert to Sinai). *Marine and Petroleum Geology* 68: 565–581.

Taylor, J.C.M. and Illing, L.V. (1969). Holocene intertidal calcium carbonate cementation, Qatar, Persian Gulf. *Sedimentology* 12: 69–107.

Tcherepanov, E.N., Droxler, A.W., Lapointe, P. et al. (2008a). Neogene evolution of the mixed carbonate-siliciclastic system in the Gulf of Papua, Papua New Guinea. *Journal of Geophysical Research: Earth Surface* 113: F01S21.

Tcherepanov, E.N., Droxler, A.W., Lapointe, P., and Mohn, K. (2008b). Carbonate seismic stratigraphy of the Gulf of Papua mixed depositional system: neogene stratigraphic signature and eustatic control. *Basin Research* 20: 185–209.

Tcherepanov, E.N., Droxler, A.W., Lapointe, P. et al. (2010). Siliciclastic influx and burial of the Cenozoic carbonate system in the Gulf of Papua. *Marine and Petroleum Geology* 27: 533–554.

Teichert, C. (1958). Cold- and deep-water coral banks. *American Association of Petroleum Geologists Bulletin* 42: 1064–1082.

Teichert, B.M.A., Bohrmann, G., and Suess, E. (2005). Chemoherms on Hydrate Ridge — Unique microbially-mediated carbonate build-ups growing into the water column. *Palaeogeography, Palaeoclimatology, Palaeoecology* 227: 67–85.

Teillet, T., Fournier, F., Gisquet, F. et al. (2019). Diagenetic history and porosity evolution of an Early Miocene carbonate buildup (Upper Burman Limestone), Yadana gas field, offshore Myanmar. *Marine and Petroleum Geology* 109: 589–606.

Thorie, A., Mukhopadhyay, A., Banerjee, T., and Mazumdar, P. (2018). Giant ooids in a Neoproterozoic carbonate shelf, Simla Group, Lesser Himalaya, India: an analogue related to Neoproterozoic glacial deposits. *Marine and Petroleum Geology* 98: 582–606.

Tian, L.I., Bottjer, D.J., Tong, J. et al. (2015). Distribution and size variation of ooids in the aftermath of the Permian–Triassic mass extinction. *PALAIOS* 30: 714–727.

Tinker, S.W. (1998). Shelf-to-basin facies distributions and sequence stratigraphy of a steep-rimmed carbonate margin: capitan depositional system, McKittrick Canyon, New Mexico and Texas. *Journal of Sedimentary Research* 68: 1146–1174.

Tomás, S., Aurell, M., Bádenas, B. et al. (2019). Architecture and paleoenvironment of mid-Jurassic microbial–siliceous sponge mounds, Northeastern Spain. *Journal of Sedimentary Research* 89: 110–134.

Toomey, D.F. (1980). History of a Late Carboniferous phylloid algal bank complex in northeastern New Mexico. *Lethaia* 13: 249–267.

Toomey, D.F. (1991). Late Permian reefs of southern Tunisia: facies patterns and comparison with the Capitan Reef, southwestern United States. *Facies* 25: 119–146.

Toomey, D.F. and Finks, R.M. (1969). Middle Ordovician (Chazyan) Mounds, Southern Quebec, Canada. In: *New York State Geological Association Guidebook to Field Excursion, 41st Annual Meeting*, 121–134. State University, Plattsburgh, New York, USA.

Törnqvist, T.E. and Hijma, M.P. (2012). Links between early Holocene ice-sheet decay, sea-level rise and abrupt climate change. *Nature Geoscience* 5: 601–606.

Tournadour, E., Mulder, T., Borgomano, J. et al. (2015). Origin and architecture of a Mass Transport Complex on the northwest slope of Little Bahama Bank (Bahamas): relations between off-bank transport, bottom current sedimentation and submarine landslides. *Sedimentary Geology* 317: 9–26.

Tournadour, E., Jorry, S.J., Etienne, S. et al. (2021). Neogene to Quaternary evolution of carbonate and mixed carbonate-siliciclastic systems along New Caledonia's eastern margin (SW Pacific). *Marine Geology* 438: 106524.

Townsend, A.R., Howarth, R.W., Bazzaz, F.A. et al. (2003). Human health effects of a changing global nitrogen cycle. *Frontiers in Ecology and the Environment* 1: 240–246.

Trabucho Alexandre, J., Tuenter, E., Henstra, G.A., van der Zwan, K.J., van de Wal, R.S.W., Dijkstra, H.A. and de Boer, P.L. (2010). The mid-Cretaceous North Atlantic nutrient trap: Black shales and OAEs. *Paleoceanography* 25: PA4201.

Trichet, J. and Defarge, C. (1995) Non-biologically supported organomineralization. *Bulletin de l'Institut Oceanographique (Monaco)* Special Issue 14 Part 2, 203–236.

Trudgill, S. (1985). *Limestone Geomorphology*. New York, NY: Wiley.

Tucker, M.E. and Bathurst, R.G.C. (ed.) (2009). *Carbonate Diagenesis*. Hoboken, NJ: Wiley-Blackwell Scientific Publications, 320 pp.

Tucker, M.E. and Jones, S.J. (2023). *Sediment Petrology*, 4e. Hoboken, NJ: Wiley, 448 pp.

Tucker, M.E. and Wright, V.P. (1990). *Carbonate Sedimentology*. Oxford, UK: Blackwell Scientific Publications, 482 pp.

Turner, E.C. (2009). Mesoproterozoic carbonate systems in the Borden Basin, Nunavut. *Canadian Journal of Earth Sciences* 46: 915–938.

Turner, E.C. and Long, D.G.F. (2008). Basin architecture and syndepositional fault activity during deposition of the Neoproterozoic Mackenzie Mountains supergroup, Northwest Territories, Canada. *Canadian Journal of Earth Sciences* 45: 1159–1184.

Turner, E.C., James, N.P., and Narbonne, G.M. (1997). Growth dynamics of Neoproterozoic calcimicrobial reefs, Mackenzie Mountains, northwestern Canada. *Journal of Sedimentary Research* 67: 437–450.

Uriarte Goti, J.A. (1997). *Maturité thermique des sédiments de la bordure sud-est du Bassin de Paris*. Genève: Université de Genève. Section des sciences de la terre, 146 pp.

Vachard, D., Perret, M.-F., and Delvolvé, J.-J. (1989). Algues, Pseudo-algues et Foraminifères des niveaux bachkiriens dans les secteurs d'escarra et aragon subordan (pyrénées aragonaises, espagne). *Geobios* 22: 697–723.

Vacher, H.L. and Rowe, M.P. (2004). Geology and hydrogeology of Bermuda, Chapter 2. In: *Developments in Sedimentology*, vol. 54 (ed. H.L. Vacher and T.M. Quinn), 35–90. Amsterdam, The Netherlands: Elsevier.

Vahrenkamp, V.C. (1996). Carbon isotope stratigraphy of the upper Kharaib and Shuaiba formations: implications for the early Cretaceous evolution of the Arabian Gulf region. *American Association of Petroleum Geologists Bulletin* 80: 647–662.

Vahrenkamp, V.C., David, F., Duijndam, P. et al. (2004). Growth architecture, faulting, and karstification of a middle miocene carbonate Platform, Luconia Province, Offshore Sarawak, Malaysia. In: *Seismic Imaging of Carbonate Reservoirs and Systems*, AAPG Memoir, vol. 81 (ed. G.P. Eberli, J.L. Masaferro, and J.F. Sarg), 329–350. Tulsa, OK, USA: American Association of Petroleum Geologists.

Vahrenkamp, V.C., Franssen, R.S.W.M., Grotsch, J., and Munoz, P.J. (1993). Maracaibo platform (Aptian—Albian), Northwestern Venezuela. In: *Cretaceous Carbonate Platforms*, AAPG

Memoir, vol. 56 (ed. J.A.T. Simo, R.W. Scott, and J.P. Masse), 25–33. Tulsa, OK: American Association of Petroleum Geologists.

Vahrenkamp, V.C., Swart, P. K. and Ruiz, J. (1991). Episodic dolomitization of late Cenozoic carbonates in the Bahamas; evidence from strontium isotopes. *Journal of Sedimentary Research* 61(6): 1002–1014.

Vahrenkamp, V., David, F., Duijndam, P. et al. (2004). Growth architecture, faulting, and karstification of a middle miocene carbonate Platform, Luconia Province, Offshore Sarawak, alaysia. In: *Seismic Imaging of Carbonate Reservoirs and Systems, AAPG Memoir*, vol. 81 (ed. G.P. Eberli, J.L. Masaferro, and J.F. Sarg), 329–350. Tulsa, OK, USA: American Association of Petroleum Geologists.

Vahrenkamp, V.C., Van Laer, P., Franco, B. et al. (2015). Late jurassic to Cretaceous source rock prone intra-shelf basins of the Eastern Arabian plate–interplay between tectonism, global anoxic events and carbonate platform dynamics. *International Petroleum Technology Conference*, pp. D021S002R003, Doha, Qatar.

Vahrenkamp, V., Chandra, V., Garuglieri, E., et al. (2024). Discovery of modern living intertidal stromatolites on Sheybarah Island, Red Sea, Saudi Arabia. *Geology* 52: 347–351.

Vail, P.R. and Mitchum, R.M. Jr. (1977). Seismic stratigraphy and global changes of sea level, part 1: overview. In: *Seismic Stratigraphy – Applications to Hydrocarbon Exploration*, vol. 26 (ed. C. E. Payton), 51–52. Tulsa, OK: American Association of Petroleum Geologists.

Van Buchem, F.S.P., Chaix, M., Eberli, G.P., et al. (2000) Outcrop to subsurface correlation of the upper Devonian (Frasnian) in the Alberta Basin (W. Canada) based on the comparison of Miette and Redwater carbonate buildup margins. In: *Genetic Stratigraphy on the Exploration and Production Scales – Case Studies from the Pennsylvanian of the Paradox Basin and the Upper Devonian of Alberta* (Eds P. Homewood and G.P. Eberli), Bull. Centre Rech. Elf Explor. Prod., Mémoir 24, pp. 225- 267. Elf Exploration and Production, Pau, France.

Van Buchem, F.S.P., Pittet, B., Hillgärtner, H. et al. (2002a). High-resolution sequence stratigraphic architecture of Barremian/Aptian carbonate systems in Northern Oman and the United Arab Emirates (Khairab and Shu'aiba Formations). *GeoArabia* 7: 461–500.

Van Buchem, F.S.P., Razin, P., Homewood, P.W. et al. (2002b). Stratigraphic organization of carbonate ramps and organic-rich intrashelf basins: Natih Formation (middle Cretaceous) of northern Oman. *American Association of Petroleum Geologists Bulletin* 86: 21–53.

Van Buchem, F.S.P., Huc, A., Pradier, B., and Stefani, M. (2005). Stratigraphic patterns in carbonate source rock distribution: 2nd to 4th order control and sediment flux. In: *The Deposition of Organic-Carbon-Rich Sediments: Models, Mechanisms and Consequences*, SEPM Special Publication, vol. 82 (ed. N.H. Harris), 191–221. Tulsa, OK: Society of Economic Paleontologists and Mineralogists.

Van Buchem, F.S.P., Al-Husseini, M.I., Maurer, F., and Droste, H.J. (2010a). *Barremian – Aptian stratigraphy and Hydrocarbon Habitat of the Eastern Arabian Plate*. Manama, Bahrain: Gulf PetroLink, 614 (2 volumes) pp.

Van Buchem, F.S.P., Allan, T.L., Laursen, G.V. et al. (2010b). Regional stratigraphic architecture and reservoir types of the Oligo-Miocene deposits in the Dezful Embayment (Asmari and Pabdeh Formations) SW Iran. In: *Mesozoic and Cenozoic Carbonate Systems of the Mediterranean and the Middle East: Stratigraphic and Diagenetic Reference Models*, Geological Society, London, Special Publication, vol. 329 (ed. F.S.P. Van Buchem, K.D. Gerdes, and M. Esteban), 219–263. London: Geological Society, London.

Van Buchem, F.S.P., Gerdes, K.D., and Esteban, M. (2010c). Mesozoic and Cenozoic carbonate systems of the Mediterranean and the Middle East: stratigraphic and diagenetic reference models – an introduction. In: *Mesozoic and Cenozoic Carbonate Systems of the Mediterranean and the Middle East: Stratigraphic and Diagenetic Reference Models*, Geological Society, London, Special Publication, vol. 329 (ed. F.S.P. Van Buchem, K.D. Gerdes, and M. Esteban), 1–7. London: Geological Society, London.

Van Buchem, F., Smit, F., Buijs, G. et al. (2018). Tectono-stratigraphic framework and depositional history of the Cretaceous-Danian succession of the Danish Central Graben (North Sea) – new light on a mature area. In: *Petroleum Geology of NW Europe: 50 Years of Learning – Proceedings of the 8th Petroleum Geological Conference, vol. 8 (ed. M. Bowman and B. Levell), 9–46. London, UK: The Geological Society of London.*

Van de Poel, H.M. and Schlager, W. (1994). Variations in Mesozoic-Cenozoic skeletal carbonate mineralogy. *Geologie en Mijnbouw* 73: 31–51.

Van der Kooij, B., Immenhauser, A., Csoma, A.J.B., and Steuber, T. (2009). Spatial geochemistry of a Carboniferous platform-margin-to-basin transect: balancing environmental and diagenetic factors. *Sedimentary Geology* 219: 136–150.

Van der Kooij, B., Immenhauser, A., Steuber, T. et al. (2010). Controlling factors of volumetrically important marine carbonate cementation in deep slope settings. *Sedimentology* 57: 1491–1525.

Van der Land, C., Mienis, F., de Haas, H. et al. (2010). Diagenetic processes in carbonate mound sediments at the south-west Rockall Trough margin. *Sedimentology* 57: 912–931.

Van Geet, M., Swennen, R., Durmishi, C. et al. (2002). Paragenesis of Cretaceous to Eocene carbonate reservoirs in the Ionian fold and thrust belt (Albania): relation between tectonism and fluid flow. *Sedimentology* 49: 697–718.

Van Gemerden, H. (1993). Microbial mats: a joint venture. *Marine Geology* 113: 3–25.

Van Kranendonk, M.J., Webb, G.E., and Kamber, B.S. (2003). Geological and trace element evidence for a marine sedimentary environment of deposition and biogenicity of 3.45 Ga stromatolitic carbonates in the Pilbara Craton, and support for a reducing Archaean ocean. *Geobiology* 1: 91–108.

Van Woesik, R. and Done, T.J. (1997). Coral communities and reef growth in the southern Great Barrier Reef. *Coral Reefs* 16: 103–115.

Vasconcelos, C., McKenzie, J.A., Bernasconi, S. et al. (1995). Microbial mediation as a possible mechanism for natural dolomite formation at low temperatures. *Nature* 377: 220–222.

Veizer, J. (1992). Depositional and iagenetic history of limestones: stable and radiogenic isotopes. *Lecture Notes Earth Science* 43: 13–48.

Veizer, J., Ala, D., Azmy, K. et al. (1999). $^{87}Sr/^{86}Sr$, $\delta^{13}C$ and $\delta^{18}O$ evolution of Phanerozoic seawater. *Chemical Geology* 161: 59–88.

Veizer, J., Godderis, Y., and François, L. M. (2000). Evidence for decoupling of atmospheric CO_2 and global climate during the Phanerozoic eon. *Nature* 408 (6813): 698–701.

Veizer, J. and Prokoph, A. (2015). Temperatures and oxygen isotopic composition of Phanerozoic oceans. *Earth-Science Reviews* 146: 92–104.

Vennin, E. (2007a). Coelobiontic communities in neptunian fissures of synsedimentary tectonic origin in Permian reef, southern Urals, Russia. *Geological Society, London, Special Publication* 275: 211–227.

Vennin, E. (2007b). Archaeolithoporella-Shamovella (= Tubiphytes)-Bryozoan reef-mound complexes, Russia. In: *Facies from Palaeozoic Reefs and Bioaccumulations*, Publications Scientifiques, vol. 195 (ed. E. Vennin, M. Aretz, F. Boulvain, and A. Munnecke), 295–297. Paris, France: Publications Scientifiques du Muséum.

Vennin, E. (2007c) Artinskian bryozoan-phylloid algal-Shamovella bioherms, Russia. In: *Facies from Palaeozoic reefs and Bioaccumulations* (Eds E. Vennin, M. Aretz, F. Boulvain and A. Munnecke), Publications Scientifiques 195, pp. 299–301. Publications Scientifiques du Muséum, Paris, France.

Vennin, E. and Aurell, M. (2001). Stratigraphie sequentielle de l'Aptien du sous-bassin de Galve (Province de Teruel, NE de l'Espagne). *Bulletin de la Société Géologique de France* 172: 397–410.

Vennin, E., Vachard, D., and Proust, J.-N. (1997). Taphonomie et synécologiedu "genre" Tubiphytes dans les bioconstructions de Tratau et de Nizhni-Irginsk (Permien inférieur de l'Oural, Russie). *Geobios* 30: 635–649.

Vennin, E., Álvaro, J.J., and Villas, E. (1998). High-latitude pelmatozoan–bryozoan mud-mounds from the late Ordovician northern Gondwana platform. *Geological Journal* 33: 121–140.

Vennin, E., Van Buchem, F., Sonnenfeld, M. et al. (2000). High Resolution Sequence Stratigraphy of a Nummulitic Carbonate System, Ypresian, Jebel Ousselat, Central Tunisia. *EAGE Conference on Geology and Petroleum Geology - St. Julians, Malta, 1 - 4 October 2000*. EAGE, Houten, The Netherlands.

Vennin, E., Rouchy, J.M., Chaix, C. et al. (2002a). Paléoécologie et morphologie des bioconstructions coralliennes du bassin de Lorca au Tortonien (SE Espagne). Travaux et Documents des Laboratoires de Géologie de Lyon – STRATI 2002. 3ème congrès français de stratigraphie. Lyon, 8-10 juillet 2002, 156, 233–233.

Vennin, E., Boisseau, T., Proust, J.N. et al. (2002b). Influence of sea level change on reef architecture in Early Permian buildup complexes, southern Urals, Russia. In: *Paleozoic Carbonates of the Commonwealth of Independent States (CIS): Subsurface Reservoirs and Outcrop Analogs*, SEPM Special Publication, vol. 74 (ed. W.G. Zempolich and H.E. Cook), 205–218. Tulsa, OK: SEPM (Society for Sedimentary Geology).

Vennin, E., van Buchem, F.S.P., Joseph, P. et al. (2003). A 3D outcrop analogue model for Ypresian nummulitic carbonate reservoirs: Jebel Ousselat, northern Tunisia. *Petroleum Geoscience* 9: 145–161.

Vennin, E., Rouchy, J.-M., Chaix, C. et al. (2004). Paleoecological constraints on reef-coral morphologies in the Tortonian–early Messinian of the Lorca Basin, SE Spain. *Palaeogeography, Palaeoclimatology, Palaeoecology* 213: 163–185.

Vennin, E., Olivier, N., Brayard, A. et al. (2015). Microbial deposits in the aftermath of the end-Permian mass extinction: a diverging case from the Mineral Mountains (Utah, USA). *Sedimentology* 62: 753–792.

Vennin, E., Bouton, A., Bourillot, R. et al. (2019). The lacustrine microbial carbonate factory of the successive Lake Bonneville and Great Salt Lake, Utah, USA. *Sedimentology* 66: 165–204.

Vennin, E., Bouton, A., Roche, A. et al. (2021). The Limagne Basin: a journey through modern and fossil microbial deposits. *Bulletin de la Société Géologique de France* 192 41: 1–35.

Vera, J.A., Molina, J.M., Montero, P., and Bea, F. (1997). Jurassic guyots on the Southern Iberian Continental Margin: a model of isolated carbonate platforms on volcanic submarine edifices. *Terra Nova* 9: 163–166.

Vernik, L. and Nur, A. (1992). Petrophysical classification of siliciclastics for lithology and porosity predictions from seismic velocities. *American Association of Petroleum Geologists Bulletin* 76: 1295–1309.

Verwer, K., Braaksma, H., and Kenter, J.A. M. (2008). Acoustic properties of carbonates: effects of rock texture and implications for fluid substitution. *Geophysics* 73: B51–B65.

Verwer, K., Della Porta, G., Merino-Tomé, O., and Kenter, J.A.M. (2009a). Controls and predictability of carbonate facies architecture in a Lower Jurassic three-dimensional barrier-shoal complex (Djebel Bou Dahar, High Atlas, Morocco). *Sedimentology* 56: 1801–1831.

Verwer, K., Merino-Tomé, O., Kenter, J.A. M., and Della Porta, G. (2009b). Evolution of a high-relief carbonate platform slope Using 3D digital outcrop models: lower jurassic Djebel Bou

Dahar, High Atlas, Morocco. *Journal of Sedimentary Research* 79: 416–439.

Verwer, K., Playton, T.E., and Harris, P.M. M. (ed.) (2014). *Deposits, Architecture, and Controls of Carbonate Margin, Slope, and Basinal Settings*. Tulsa, OK: SEPM (Society for Sedimentary Geology).

Viana, A.R., Faugeres, J.C., Kowsmann, R. O. et al. (1998). Hydrology, morphology and sedimentology of the Campos continental margin, offshore Brazil. *Sedimentary Geology* 115: 133–157.

Villalobos Megía, M. and Braga Alarcón, J. C. (ed.) (2003). *Geology of the Arid Zone of Almeria (Southeast Spain) – An Educational Field Guide*. State Water Company for the Southern Basin SA (ACUSUR) / Junta de Andalucia, 165 pp.

Vincent, B., van Buchem, F.S.P., Bulot, L. G. et al. (2010). Carbon-isotope stratigraphy, biostratigraphy and organic matter distribution in the Aptian – Lower Albian successions of southwest Iran (Dariyan and Kazhdumi formations). In: *Barremian – Aptian Stratigraphy and Hydrocarbon Habitat of the Eastern Arabian Plate (Vol. 1)*, GeoArabia Special Publication 4, vol. 1 (ed. F.S.P. van Buchem, M.I. Al-Husseini, F. Maurer, and H.J. Droste), 139–197. Manama, Bahrain: Gulf PetroLink.

Visscher, P.T. and Stolz, J.F. (2005). Microbial mats as bioreactors: populations, processes, and products. *Palaeogeography, Palaeoclimatology, Palaeoecology* 219: 87–100.

Visscher, P.T., Gallagher, K.L., Bouton, A., Vennin, E., Thomazo, C., White III, R. A. and Burns, B.P. (2022). Microbial Mats. In: Treatise Online, Number 163 – Part B, Volume 2, Chapter 3, pp. 163: 1–32. Kansas: KU – Paleontological Institute.

Wahlman, G.P., Orchard, D.M., and Buijs, G.J. (2013). Calcisponge-microbialite reef facies, middle Permian (lower Guadalupian), northwest shelf margin of Permian Basin, New Mexico. *American Association of Petroleum Geologists Bulletin* 97: 1895–1919.

Waite, L.E., Koepnick, R.B., and Markello, J.R. (ed.) (2023). *The Miocene World: A Brief Summary*, 32–48. Tulsa, OK: SEPM (Society for Sedimentary Geology).

Walker, R.G. and James, N.P. (1992). *Facies Models – Response to Sea Level Change*. St. John's Newfoundland: Geological Association of Canada, 409 pp.

Walker, K.R., Jernigan, D.G., and Weber, L.J. (1990). Petrographic criteria for the recognition of marine, syntaxial overgrowths, and their distribution in geologic time. *Carbonates and Evaporites* 5: 141–152.

Wallace, M.W. (1987). The role of internal erosion and sedimentation in the formation of stromatactis mudstones and associated lithologies. *Journal of Sedimentary Research* 57: 695–700.

Wallace, M.W., Holdgate, G.R., Daniels, J. et al. (2002). Sonic velocity, submarine canyons, and burial diagenesis in Oligocene-Holocene cool-water carbonates, Gippsland Basin, southeast Australia. *American Association of Petroleum Geologists Bulletin* 86: 1593–1607.

Walter, M.R., Buick, R., and Dunlop, J.S.R. (1980). Stromatolites 3,400–3,500 Myr old from the North Pole area, Western Australia. *Nature* 284: 443–445.

Wang, Z., Zhao, W., Hu, S. et al. (2017). Control of tectonic differentiation on the formation of large oil and gas fields in craton basins: A case study of Sinian–Triassic of the Sichuan Basin. *Natural Gas Industry B* 4: 141–155.

Wanless, H.R. (1983). Burial diagenesis in limestones. In: *Sediment Diagenesis* (ed. A. Parker and B.W. Sellwood), 379–417. Dordrecht, The Netherlands: Springer Netherlands.

Wanless, H.R. (1991). Observational foundation for sequence modeling. In: *Sedimentary Modeling: Computer Simulations and Methods for Improved Parameter Definition*, Bulletin, vol. 233 (ed. E.K. Franseen, W.L. Watney, C.G.S. C. Kendall, and W. Ross), 43–62. Lawrence, Kansas: Kansas Geological Survey.

Wanless, H.R. and Tedesco, L.P. (1993). Comparison of oolitic sand bodies generated by tidal vs. wind-wave agitation. In: *Mississippian Oolites and Modern Analogs*, AAPG Studies in Geology, vol. 35 (ed. B.D. Keith and C. W. Zuppann), 199–225. Tulsa, OK: American Association of Petroleum Geologists.

Wanless, H.R. and Weller, J.M. (1932). Correlation and Extent of Pennsylvanian Cyclothems 12. *Geological Society of America Bulletin* 43: 1003–1016.

Wanless, H.R., Cottrell, D.J., Tagett, M.G. et al. (1995). Origin and growth of carbonate banks in south Florida. In: *Carbonate Mud-Mounds – Their Origin and Evolution*, IAS Special Publication, vol. 23 (ed. C.L.V. Monty, D.W.J. Bosence, P.H. Bridges, and B.R. Pratt), 439–473. Chichester, West Sussex: International Association of Sedimentologists–Blackwell.

Ward, W.C. and Brady, M.J. (1979). Strandline sedimentation of carbonate grainstones, Upper Pleistocene, Yucatan Peninsula, Mexico1: GEOLOGIC NOTES. *American Association of Petroleum Geologists Bulletin* 63: 362–369.

Ward, W.C. and Halley, R.B. (1985) Dolomitization in a mixing zone of Near-Seawater Composition, Late Pleistocene, Northeastern Yucatan Peninsula. *Journal of Sedimentary Research* 55: 407–420.

Ward, R.F., Kendall, C.G.S.C., and Harris, P.M. (1986). Upper Permian (Guadalupian) facies and their association with hydrocarbons—Permian Basin, West Texas and New Mexico. *American Association of Petroleum Geologists Bulletin* 70: 239–262.

Warnecke, M. and Aigner, T. (2019). Influence of subtle paleo-tectonics on facies and reservoir distribution in epeiric carbonates: integrating stratigraphic analysis and modelling (U. Muschelkalk, SW Germany). *Sedimentary Geology* 383: 82–100.

Warren, J.K. (2000). Evaporites, brines and base metals: low-temperature ore emplacement controlled by evaporite diagenesis. *Australian Journal of Earth Sciences* 47: 179–208.

Warren, J.K. (2010). Evaporites through time: tectonic, climatic and eustatic controls in marine and nonmarine deposits. *Earth-Science Reviews* 98: 217–268.

Warren, J.K. and Kempton, R.H. (1997). Evaporite sedimentology and the origin of evaporite-associated Mississippi valley-type sulfides in the Cadjebut Mine Area, Lennard Shelf, Canning Basin, Western Australia. In: *Basin-Wide Diagenetic Patterns: Integrated Petrologic, Geochemical, and Hydrologic Considerations*, vol. 57 (ed. I.P. Montanez, J.M. Gregg, and K.L. Shelton), 183–205. Tulsa, OK: SEPM (Society for Sedimentary Geology).

Warrlich, G.M.D., Waltham, D.A., and Bosence, D.W.J. (2002). Quantifying the sequence stratigraphy and drowning mechanisms of atolls using a new 3-D forward stratigraphic modelling program (CARBONATE 3D). *Basin Research* 14: 379–400.

Warrlich, G., Bosence, D., and Waltham, D. (2005). 3D and 4D controls on carbonate depositional systems: sedimentological and sequence stratigraphic analysis of an attached carbonate platform and atoll (Miocene, Níjar Basin, SE Spain). *Sedimentology* 52: 363–389.

Warrlich, G., Bosence, D., Waltham, D. et al. (2008). 3D stratigraphic forward modelling for analysis and prediction of carbonate platform stratigraphies in exploration and production. *Marine and Petroleum Geology* 25: 35–58.

Waters, C.N. and Davies, S.J. (2006). Carboniferous: extensional basins, advancing deltas and coal swamps. In: *The Geology of England and Wales* (ed. P.J. Brenchley and P.F. Rawson), 173–223. London: Geological Society of London.

Watts, K.F. (1988). Triassic carbonate submarine fans along the Arabian platform margin, Sumeini Group, Oman. *Sedimentology* 35: 43–71.

Watts, K.F. and Garrison, R.E. (1986). Sumeini Group, Oman – Evolution of a Mesozoic carbonate slope on a South Tethyan continental margin. *Sedimentary Geology* 48: 107–168.

Watts, N.R. and Riding, R. (2000). Growth of rigid high-relief patch reefs, Mid-Silurian, Gotland, Sweden. *Sedimentology* 47: 979–994.

Webb, G.E. (1996). Was Phanerozoic reef history controlled by the distribution of non-enzymatically secreted reef carbonates (microbial carbonate biologically induced cement)? *Sedimentology* 43: 947–971.

Webb, G.E. (2001). Famennian mud-mounds in the proximal fore-reef slope, Canning Basin, Western Australia. *Sedimentary Geology* 145: 295–315.

Webby, B.D. (2002). Patterns of ordovician reef development. In: *Phanerozoic Reef Patterns*, SEPM Special Publication, vol. 72 (ed. W. Kiessling, E. Flügel, and J. Golonka), 129–179. Tulsa, OK: SEPM (Society for Sedimentary Geology).

Weber, L.J., Francis, B.P., Harris, P.M., and Clark, M. (2003). Stratigraphy, lithofacies, and reservoir distribution, Tengiz Field, Kazakhstan. In: *Permo-Carboniferous Carbonate Platforms and Reefs*, vol. 78 (ed. W.M. Ahr, P.M. Harris, W.A. Morgan, and I.D. Somerville), 351–394. Tulsa, OK: SEPM (Society for Sedimentary Geology).

Webster, J.M. and Davies, P.J. (2003). Coral variation in two deep drill cores: significance for the Pleistocene development of the Great Barrier Reef. *Sedimentary Geology* 159: 61–80.

Webster, J.M., Beaman, R.J., Puga-Bernabéu, Å. et al. (2012). Late Pleistocene history of turbidite sedimentation in a submarine canyon off the northern Great Barrier Reef, Australia. *Palaeogeography, Palaeoclimatology, Palaeoecology* 331-332: 75–89.

Webster, J.M., Braga, J.C., Humblet, M. et al. (2018). Response of the Great Barrier Reef to sea-level and environmental changes over the past 30,000 years. *Nature Geoscience* 11: 426–432.

Weedon, G.P. (2003). Time-series analysis and cyclostratigraphy – Examining stratigraphic records of environmental cycles. Press Syndicate of the University of Cambridge, Cambridge, UK, 259 pp.

Wefer, G. and Berger, W.H. (1991). Isotope palaeontology: growth and composition of extant calcareous species. *Marine Geology* 100: 207–248.

Weger, R.J., Eberli, G.P., Baechle, G.T. et al. (2009). Quantification of pore structure and its effect on sonic velocity and permeability in carbonates. *American Association of Petroleum Geologists Bulletin* 93: 1297–1317.

Weger, R.J., Eberli, G.P., Massaferro, J.L. et al. (2023). Theoretically derived pore geometry in carbonates using the extended biot theory. *Marine and Petroleum Geology* 155: 106359.

Wehrmann, A. (1994). Subboreale Flachwasserkarbonate des Holozäns in der Nordbretagne: Steuerungsmechanismen in karbonatproduzierenden Ökosystemen. Unpublished PhD thesis. Philipps-Universität Marburg – University of Marburg, Marburg, Germany, 191 pp.

Wehrmann, A. (1998). Modern cool-water carbonates on a coastal platform of Northern Brittany, France: carbonate production in macrophytic systems and sedimentary dynamics of bioclastic facies. *Senckenbergiana maritima* 28: 151–166.

Weidlich, O. (2002). Middle and Late Permian reefs – distributional patterns and reservoir potential. In: *Phanerozoic Reef Patterns*, SEPM Special Publication, 72 (ed. W. Kiessling, E. Flügel, and J. Golonka), 339–390. Tulsa, OK: SEPM (Society for Sedimentary Geology).

Weidlich, O. and Fagerstrom, J.A. (1998). Evolution of the Upper Capitan-Massive (Permian), Guadalupe Mountains, New Mexico. *Brigham Young University Geology Studies* 43: 167–187.

Weidlich, O. and Bernecker, M. (2003). Supersequence and composite sequence carbonate platform growth: Permian and Triassic outcrop data of the Arabian platform and Neo-Tethys. *Sedimentary Geology* 158: 87–116.

Weij, R., Reijmer, J.J.G., Eberli, G.P., and Swart, P.K. (2019). The limited link between accommodation space, sediment thickness, and inner platform facies distribution (Holocene–Pleistocene, Bahamas). *Depositional Record* 5: 400–420.

Weiner, S. and Dove, P.M. (2003). An overview of biomineralization processes and the problem of the vital effect. *Reviews in Mineralogy and Geochemistry* 54: 1–29.

Weissenberger, J.A.W., Wong, P.K., and Gilhooly, M.G. (2017). Stratigraphic architecture of the Frasnian South Jasper Basin, North-Central Alberta Front Ranges. In: *New advances in Devonian carbonates: outcrop analogs, reservoirs and chronostratigraphy*, SEPM Special Publication, vol. 107 (ed. T.E. Playton, C. Kerans, and J.A.W. Weissenberger), 80–112. Tulsa, OK: SEPM (Society for Sedimentary Geology).

Weissert, H. (1989). C-Isotope stratigraphy, a monitor of paleoenvironmental change: a case study from the early cretaceous. *Surveys in Geophysics* 10: 1–61.

Wendt, J. and Kaufmann, B. (1998). Mud buildups on a Middle Devonian carbonate ramp (Algerian Sahara). In: *Carbonate Ramps: an introduction*, Special Publications, vol. 149 (ed. V.P. Wright and T.P. Burchette), 397–415. London: Geological Society, London.

Wendt, J. and Kaufmann, B. (2006). Middle Devonian (Givetian) coral-stromatoporoid reefs in West Sahara (Morocco). *Journal of African Earth Sciences* 44: 339–350.

Wendt, J., Belka, Z., and Moussine-Pouchkine, A. (1993). New architecture of deep-water carbonate buildups: evolution of mud mounds into mud ridges (Middle Devonian, Algerian Sahara). *Geology* 21: 723–726.

Wendt, J., Belka, Z., Kaufmann, B. et al. (1997). The world's most spectacular carbonate mud mounds (Middle Devonian, Algerian Sahara). *Journal of Sedimentary Research* 67: 424–436.

Wendt, J., Kaufmann, B., Belka, Z. et al. (2006). Sedimentary evolution of a Palaeozoic basin and ridge system: the Middle and Upper Devonian of the Ahnet and Mouydir (Algerian Sahara). *Geological Magazine* 143: 269–299.

Wendte, J.C. (1992). Platform evolution and its control on reef inception and localization. In: *Devonian–Early Mississippian Carbonates of the Western Canada Sedimentary Basin: A Sequence Stratigraphic Framework*, SEPM - Short Course, vol. 28 (ed. J.C. Wendte, F.A. Stoakes, and C.V. Campbell), 41–88. Tulsa, OK: SEPM (Society for Sedimentary Geology).

Werdell, P.J. and Bailey, S.W. (2005). An improved in-situ bio-optical data set for ocean color algorithm development and satellite data product validation. *Remote Sensing of Environment* 98: 122–140.

West, R.R. (1988). Temporal changes in Carboniferous reef mound communities. *PALAIOS* 3: 152–169.

Westphal, H., Halfar, J., and Freiwald, A. (2010). Heterozoan carbonates in subtropical to tropical settings in the present and past. *International Journal of Earth Sciences* 99: 153–169.

Wetzel, A., Weissert, H., Schaub, M., and Voegelin, A.R. (2013). Sea-water circulation on an oolite-dominated carbonate system in an epeiric sea (Middle Jurassic, Switzerland). *Sedimentology* 60: 19–35.

Whalen, M.T., Eberli, G.P., van Buchem, F.S.P., and Mountjoy, E.W. (2000a). Facies models and architecture of Upper Devonian carbonate platforms (Miette and Ancient Wall), Alberta, Canada. In: *Genetic stratigraphy on the exploration and production scales - Case studies from the Pennsylvanian of the Paradox Basin and the Upper Devonian of Alberta*, Bulletin Centre Recherche Elf Exploration Production (ed. P.W. Homewood and G.P. Eberli), 139–178. Pau, France: Elf Exploration Production.

Whalen, M.T., Eberli, G.P., van Buchem, F.S.P. et al. (2000b). Bypass margins, basin-restricted wedges, and platform-to-basin correlation, Upper Devonian, Canadian Rocky Mountains: implications for sequence stratigraphy of carbonate platform systems. *Journal of Sedimentary Research* 70: 913–936.

Whalen, M.T., Day, J., Eberli, G.P., and Homewood, P.W. (2002). Microbial carbonates as indicators of environmental change and biotic crises in carbonate systems: Examples from the Late Devonian, Alberta basin, Canada. *Palaeogeography, Palaeoclimatology, Palaeoecology* 181: 127–151.

Wharton, S.R. (2017). The Rimthan Arch, basin architecture, and stratigraphic trap potential in Saudi Arabia. *Interpretation* 5: T563–T578.

Whitaker, F., Smart, P., Hague, Y. et al. (1997). Coupled two-dimensional diagenetic and sedimentological modeling of carbonate platform evolution. *Geology* 25: 175–178.

Wienberg, C. and Titschack, J. (2017) Framework-forming scleractinian cold-water corals through space and time: a late Quaternary North Atlantic perspective. In: *Marine Animal Forests: the Ecology of Benthic Biodiversity Hotspots* (Eds S. Rossi, L. Bramanti, A. Gori and C. Orejas Saco del Valle), pp. 699–732. Springer International Publishing Switzerland, Cham.

Wienberg, C., Hebbeln, D., Fink, H.G. et al. (2009). Scleractinian cold-water corals in the Gulf of Cadiz-First clues about their spatial and temporal distribution. *Deep-Sea Research Part I-Oceanographic Research Papers* 56: 1873–1893.

Wienberg, C., Titschack, J., Freiwald, A. et al. (2018). The giant Mauritanian cold-water coral mound province: Oxygen control on coral mound formation. *Quaternary Science Reviews* 185: 135–152.

Wienberg, C., Krengel, T., Frank, N. et al. (2022). Cold-water coral mounds in the western Mediterranean Sea: new insights into their initiation and development since the Mid-Pleistocene in response to changes of African hydroclimate. *Quaternary Science Reviews* 293: 107723.

Wilkinson, B.H. and Drummond, C.N. (2004). Facies mosaics across the Persian Gulf and around

Antigua – Stochastic and deterministic products of shallow-water sediment accumulation. *Journal of Sedimentary Research* 74: 513–526.

Williams, H.D., Burgess, P.M., Wright, V. P. et al. (2011). Investigating carbonate platform types: multiple controls and a continuum of geometries. *Journal of Sedimentary Research* 81: 18–37.

Wilson, J.L. (1975). *Carbonate Facies in Geological History*. Berlin: Springer Verlag, 471 pp.

Wilson, J.B. (1988). A model for temporal changes in the faunal composition of shell gravels during a transgression on the continental shelf around the British Isles. *Sedimentary Geology* 60 (1-4): 95–105.

Wilson, A.O. (2020a). Chapter 1 – Introduction to the Jurassic Arabian Intrashelf Basin. *Geological Society, London, Memoir* 53: 1–19.

Wilson, A.O. (2020b). Chapter 5 – Interpretation of the origin and evolution of the Arabian Intrashelf Basin and the development of the Dhruma Atash, Tuwaiq and Hanifa sequences. *Geological Society, London, Memoir* 53: 113–157.

Wilson, M.E.J. (2000). Tectonic and volcanic influences on the development and diachronous termination of a Tertiary tropical carbonate platform. *Journal of Sedimentary Research* 70: 310–324.

Wilson, M.E.J. (2002). Cenozoic carbonates in Southeast Asia: implications for equatorial carbonate development. *Sedimentary Geology* 147: 295–428.

Wilson, M.E.J. (2011). SE Asian carbonates: tools for evaluating environmental and climatic change in equatorial tropics over the last 50 million years. In: *The SE Asian Gateway: History and Tectonics of the Australia–Asia Collision*, Special Publications, vol. 355 (ed. R. Hall, M.A. Cottam, and M.E.J. Wilson), 347–372. London: Geological Society, London.

Wilson, M.E.J. and Lokier, S.W. (2002). Siliciclastic and volcaniclastic influences on equatorial carbonates: insights from the Neogene of Indonesia. *Sedimentology* 49: 583–601.

Wilson, P.A. and Roberts, H.H. (1992). Carbonate-periplatform sedimentation by density flows: a mechanism for rapid off-bank and vertical transport of shallow-water fines. *Geology* 20: 713–716.

Wilson, P.A. and Roberts, H.H. (1995). Density cascading: off-shelf sediment transport, evidence and implications, Bahama Banks. *Journal of Sedimentary Research* A65: 45–56.

Wilson, M.E.J. and Vecsei, A. (2005). The apparent paradox of abundant foramol facies in low latitudes: their environmental significance and effect on platform development. *Earth-Science Reviews* 69: 133–168.

Wilson, J.L. and Ward, W.C. (1993). Early cretaceous carbonate platforms of Northeastern and East-Central Mexico. In: *Cretaceous Carbonate Platforms*, vol. 56 (ed. J.A.T. Simo, R.W. Scott, and J.-P. Masse), 35–49. Tulsa, OK: American Association of Petroleum Geologists.

Wilson, P.A., Jenkyns, H.C., Elderfield, H., and Larson, R.L. (1998). The paradox of drowned carbonate platforms and the origin of Cretaceous Pacific guyots. *Nature* 392: 889–894.

Wilson, M.E.J., Bosence, D.W.J., and Limbong, A. (2000). Tertiary syntectonic carbonate platform development in Indonesia. *Sedimentology* 47: 395–419.

Wilson, E.N., Watney, W.L., and Grammer, G.M. (2019). An overview of the giant heterogeneous Mississippian carbonate system of the midcontinent: ancient structure, complex stratigraphy, conventional traps, and unconventional technology in a high fluid volume world. In: *Mississippian Reservoirs of the Midcontinent*, vol. 122 (ed. G.M. Grammer, J.M. Gregg, J. Puckette, et al.), 1–23. Tulsa, OK: American Association of Petroleum Geologists.

Wissler, L., Funk, H., and Weissert, H. (2003). Response of Early Cretaceous carbonate platforms to changes in atmospheric carbon dioxide levels. *Palaeogeography, Palaeoclimatology, Palaeoecology* 200: 187–205.

Wohlwend, S., Hart, M., and Weissert, H. (2016). Chemostratigraphy of the Upper Albian to mid-Turonian Natih Formation (Oman) – how authigenic carbonate changes a global pattern. *The Depositional Record* 2: 97–117.

Wong, P.K., Weissenberger, J.A.W., and Gilhooly, M.G. (2016). Revised regional Frasnian sequence stratigraphic framework, Alberta outcrop and subsurface. In: *New Advances in Devonian Carbonates: Outcrop Analogs, Reservoirs, and Chronostratigraphy*, SEPM Special Publication, vol. 107 (ed. T.E. Playton, C. Kerans, and J.A.W. Weissenberger), 16–79. Tulsa, OK: SEPM (Society for Sedimentary Geology).

Wood, R.A. (1999). *Reef Evolution*. Oxford: Oxford University Press, 414 pp.

Wood, R., Zhuravlev, A.Y., and Anaaz, C. T. (1993). The ecology of Lower Cambrian buildups from Zuune Arts, Mongolia: implications for early metazoan reef evolution. *Sedimentology* 40: 829–858.

Wood, R., Dickson, J.A.D., and Kirkland, B.L. (1996). New observations on the ecology of the Permian Capitan Reef, Texas and New Mexico. *Palaeontology* 39: 733–762.

Woods, A.D. (2009). Anatomy of an anachronistic carbonate platform: lower Triassic carbonates of the southwestern United States. *Australian Journal of Earth Sciences* 56: 825–839.

Woods, A.D., Bottjer, D.J., Mutti, M., and Morrison, J. (1999). Lower Triassic large sea-floor carbonate cements: their origin and a mechanism for the prolonged biotic recovery from the end-Permian mass extinction. *Geology* 27: 645–648.

Worden, R.H., Armitage, P.J., Butcher, A. R. et al. (2018). Petroleum reservoir quality prediction: overview and contrasting approaches from sandstone and carbonate communities. In: *Reservoir Quality of Clastic and Carbonate Rocks: Analysis, Modelling and Prediction*, Special Publication, vol. 435 (ed. P.J. Armitage, A.R.

Butcher, J.M. Churchill, et al.), 1–31. London: Geological Society.

Wray, J.L. (1968). Late paleozoic phylloid algal limestones in the United States. In: *Proceedings of the 23rd International Geological Congress, 8*, pp. 113–119, Prague, Czech Republic.

Wright, V.P. (1992). A revised classification of limestones. *Sedimentary Geology* 76: 177–185.

Wright, V.P. (1994). Early Carboniferous carbonate systems: an alternative to the Cainozoic paradigm. *Sedimentary Geology* 93: 1–5.

Wright, V.P. and Burchette, T.P. (1998). *Carbonate Ramps*, 465. London: The Geological Society.

Wright, V.P. and Burgess, P.M. (2005). The carbonate factory continuum, facies mosaics and microfacies: an appraisal of some of the key concepts underpinning carbonate sedimentology. *Facies* 51: 17–23.

Wright, V.P. and Della Porta, G. (Eds) (2021). *Field guides to exceptionally exposed carbonate outcrops*. Ghent: International Association of Sedimentologists, Ghent University, 792 pp.

Wright, V.P. and Tucker, M.E. (1991). *Calcretes*. Oxford: Wiley-Blackwell, 352 pp.

Wu, Y. and Chafetz, H.S. (2002). ^{13}C-Enriched Carbonate in Mississippian mud mounds: Alamogordo member, Lake Valley Formation, Sacramento Mountains, New Mexico, USA. *Journal of Sedimentary Research* 72: 138–145.

Wu, H.C., Dissard, D., Le Cornec, F. et al. (2017). Primary life stage boron isotope and trace elements incorporation in aposymbiotic Acropora millepora coral under ocean acidification and warming. Frontiers in Marine *Science* 4.

Wu, S., Zhang, X., Yang, Z. et al. (2016). Spatial and temporal evolution of Cenozoic carbonate platforms on the continental margins of the South China Sea: response to opening of the ocean basin. *Interpretation* 4: SP1–SP19.

Wunsch, M., Betzler, C., Eberli, G.P. et al. (2018). Sedimentary dynamics and high-frequency sequence stratigraphy of the southwestern slope of Great Bahama Bank. *Sedimentary Geology* 363: 96–117.

Wyllie, M.R.J., Gregory, A.R., and Gardner, L.W. (1956). Elastic wave velocities in heterogeneous and porous media. *Geophysics* 21: 41–70.

Xu, S. and Payne, M.A. (2009). Modeling elastic properties in carbonate rocks. *The Leading Edge* 28: 66–74.

Yates, K.K. and Robbins, L.L. (2001). Microbial lime-mud production and its relation to climate change. In: *Geological Perspectives of Global Climate Change*, AAPG Studies in Geology, vol. 47 (ed. E.L.C. Gerhard, W.E. Harrison, and B.M. Hanson), 267–283. Tulsa, OK: American Association of Petroleum Geologists.

Yin, S.-Z., Bai, G.-P., Gao, L., and Xu, C.-Y. (2018). A facies and palaeogeography-based approach for analysis of petroleum systems in United Arab Emirates. *Journal of Palaeogeography* 7: 168–178.

Yose, L.A., Ruf, A.S., Strohmenger, C.J. et al. (2006). Three-dimensional Characterization of a Heterogeneous Carbonate Reservoir, Lower Cretaceous, Abu Dhabi (United Arab Emirates). In: *Giant Hydrocarbon Reservoirs of the World: From Rocks to Reservoir Characterization and Modeling*, AAPG Memoir / SEPM Spec. Publ., vol. 88 (ed. P.M. Harris and L.J. Weber), 173–212. Tulsa, OK: American Association of Petroleum Geologists – Society of Economic Paleontologists and Mineralogists.

Yose, L.A., Strohmenger, C.J., Al-Hosani, I. et al. (2010). Sequence-stratigraphic evolution of an Aptian carbonate platform (Shu'aiba Formation), eastern Arabian Plate, onshore Abu Dhabi, United Arab Emirates. In: *Barremian – Aptian Stratigraphy and Hydrocarbon Habitat of the Eastern Arabian Plate (vol. 2)*, GeoArabia Special Publication 4, vol. 2 (ed. F.S.P. van Buchem, M.I. Al-Husseini, F. Maurer, and H.J. Droste), 309–340. Manama, Bahrain: Gulf PetroLink.

You, X.L., Sun, S., Lin, C.S. and Zhu, J.Q. (2018). Microbial dolomite in the sabkha environment of the middle Cambrian in the Tarim Basin, NW China. *Australian Journal of Earth Sciences* 65: 109–120.

Yu, J., Elderfield, H., Jin, Z., Tomascak, P. and Rohling, E.J. (2014). Controls on Sr/Ca in benthic foraminifera and implications for seawater Sr/Ca during the late Pleistocene. *Quaternary Science Reviews* 98: 1–6.

Zachos, J.C., Pagani, M., Sloan, L. et al. (2001). Trends, rhythms, and aberrations in global climate 65 Ma to Present. *Science* 292: 686–693.

Zaffos, A., Finnegan, S., and Peters, S.E. (2017). Plate tectonic regulation of global marine animal diversity. *Proceedings of the National Academy of Sciences of the United States of America* 114: 5653–5658.

Zampetti, V. (2010). Controlling factors of a Miocene Carbonate Platform: implications for platform architecture and off-platform reservoirs (Luconia Province, Malaysia). In: *Cenozoic Carbonate Systems of Australasia*, vol. 95 (ed. W.A. Morgan, A.D. George, P.M. Harris, et al.), 129–145. Tulsa, OK: SEPM (Society for Sedimentary Geology).

Zatoń, M., Niedźwiedzki, G., Rakociński, M. et al. (2018). Earliest Triassic metazoan bioconstructions from East Greenland reveal a pioneering benthic community in the immediate aftermath of the end-Permian mass extinction. *Global and Planetary Change* 167: 87–98.

Zeebe, R.E. (2013). Time-dependent climate sensitivity and the legacy of anthropogenic greenhouse gas emissions. *Proceedings of the National Academy of Sciences of the United States of America* 110: 13739–13744.

Zempolich, W.G., Cook, H.E., Zhemchuzhnikov, V.G., Zhaimina, V. Y., Zorin, A.Y., Buvtyshkin, V.M., Kotova, E.A., Golub, L.Y., Giovannelli, A., Viaggi, M., Lehmann, P.J., Fretwell, N., Alexeiev, D.V. and Lapointe, P. (2002) Biotic and abiotic influence on the stratigraphic architecture and diagenesis of Middle to Upper Paleozoic carbonates of the Bolshoi Karatau mountains, Kazakhstan and the

Southern Urals, Russia: implications for the distribution of early marine cements and reservoir quality in subsurface reservoirs. In: *Paleozoic Carbonates of the Commonwealth of Independent States (CIS): Subsurface Reservoirs and Outcrop Analogs* (Eds W. G. Zempolich and H.E. Cook), SEPM Special Publication 74, pp. 81–122. SEPM (Society for Sedimentary Geology), Tulsa, OK.

Zeng, H. and Kerans, C. (2003) Seismic frequency control on carbonate seismic stratigraphy: A case study of the Kingdom Abo sequence, west Texas. AAPG Bulletin 87(2): 273–293.

Zhai, M. (2021). Ordos Block (Basin) is a key to understand early continental evolution and tectonic regime of the North China Craton. *Chinese Science Bulletin.* https://doi.org/10.1360/tb-2021-0113

Zhang, K.-J., Xia, B., and Liang, X. (2002). Mesozoic–Paleogene sedimentary facies and paleogeography of Tibet, western China: tectonic implications. *Geological Journal* 37: 217–246.

Zhang, Y., Li, Y., and Munnecke, A. (2014). Late Ordovician microbial reefs in the Lianglitag Formation (Bachu, Tarim, NW China). *Facies* 60: 663–684.

Zhao, Y., Xiao, Z., Fan, C. et al. (2020). Comparative mechanical behaviors of four fiber-reinforced sand cemented by microbially induced carbonate precipitation. *Bulletin of Engineering Geology and the Environment* 79: 3075–3086.

Zheng, H., Ma, Y., Chi, G., Qing, H., Liu, B., Zhang, X., Shen, Y., Liu, J. and Wang, Y. (2019). Stratigraphic and structural control on hydrothermal dolomitization in the middle permian carbonates, Southwestern Sichuan Basin (China). *Minerals* 9: 32.

Zhou, K. and Pratt, B.R. (2019a). Upper Devonian (Frasnian) stromatactis-bearing mud mounds, western Alberta, Canada: reef framework dominated by peloidal microcrystalline calcite. *Journal of Sedimentary Research* 89: 833–848.

Zhou, K. and Pratt, B.R. (2019b). Composition and origin of stromatactis-bearing mud-mounds (Upper Devonian, Frasnian), southern Rocky Mountains, western Canada. *Sedimentology* 66: 2455–2489.

Zühlke, R., Bechstädt, T., and Mundil, R. (2003). Sub-Milankovitch and Milankovitch forcing on a model Mesozoic carbonate platform – the Latemar (Middle Triassic, Italy). *Terra Nova* 15: 69–80.

Supplementary Information 11.S.1 Ancient land-attached shelves and intrashelf basins.

Cambrian	N. American plate: Southern Virginia, Appalachians	Markello and Read (1981) and Srinivasan and Walker (1993)
Silurian	N. American plate: Michigan, Illinois	Harris and Sheehan 1996 (1996), Rine et al. (2017) and Sarg (1982)
Devonian	N. American plate: Williston Basin, Western Canadian Sedimentary basin; Canning Basin, Western Australia (cratonic basins); NW European shelf (Poland, Germany and Belgium); Central Asian plate: Kazakhstan, Russia and China	Franklin and Sarg (2017, 2018), Geldsetzer et al. (1988), Mountjoy (1965a, b), Wendte et al. (1990), Whalen et al. (2000), Kerans and Harris (2008), Playton et al. (2017), Szulczewski (2013), Bridges et al. (1995), Racki (1998) and Chen et al. (2001)
Early Carboniferous	N. American plate: Permian and Midland basins, Paradox basin and Sacramento Mt (cratonic basins); NW European plate	Algeo et al. (1991), Goldhammer et al. (1991), Gutteridge (1987) and Mazzullo et al. (1985)
Mesozoic	Mostly Jurassic and Cretaceous: shelf margins of the Neotethys domain: Afro-Arabian Plate (China, Egypt, Italy, Tunisia, Oman, UAE, Saudi Arabia and Iran; Gulf of Mexico, Uzbekistan)	Al-Mojel et al. (2020), Bachmann et al. (2003), Basilone et al. (2016), Basilone (2022), Bayet-Goll et al. (2022), Bromhead et al. (2022), Carmeille et al. (2020), Davies et al. (2002), Galloway (2008), Gao et al. (2020), Gharbi et al. (2022), Grélaud et al. (2006), Hillgärtner et al. (2003), Kendall et al. (2014), Lasemi and Jalilian (2010), Loucks et al. (2017), Mansouri-Daneshvar et al. (2015), Maurer et al. (2013), Phelps et al. (2014), Piryaei et al. (2010, 2011), Rousseau et al. (2005), Tassy et al. (2015), Van Buchem et al. (2000, 2002, 2010a, b), Vincent et al. (2010), Wilson (2020), Yin et al. (2018) and Zhang et al. (2002)

Supplementary Information 11.S.2 Ancient isolated buildups.

Silurian	USA	Johnson (1987)
Devonian	Central Nevada, USA	Matti and McKee (1977) and Miller et al. (1992)
	Rocky Mountains, Canada	Mountjoy (1965a, b), Wendte (1992), Whalen et al. (2000) and Atchley et al. (2006)
Carboniferous	West Texas, USA	Schatzinger (1983)
	Kazakhstan	Cook et al. (2002), Zempolich et al. (2002), Kenter et al. (2005), Alexeiev et al. (2017)
Permian	Oman	Weidlich (2002)
	China	Lehrmann et al. (2003) and Li et al. (2012)
Triassic	Dolomites, Italy	Bosselini (1984), Harris (1988), Maurer (2000), Blendinger (2001), Marango et al. (2011) and Gianolla et al. (2010)
	Oman	Watts and Garrison (1986), Watts (1988) and Flügel and Bernecker (1996)
Jurassic	Morocco	Merino-Tome et al. (2012) and Della Porta et al. (2014)
	Portugal and Spain	Ellis et al. (1990) and Vera et al. (1997)
	Others	Leinfelder et al. (1996, 2005)
Cretaceous	Mexico	Enos (1977), Wilson and Ward (1993) and Janson et al. (2011)
	Croatia	Husinec and Jelaska (2006) and Husinec and Read (2011)
	Pacific, atolls	Wilson et al. (1998)
	Mediterranean	
Oligo-Miocene	Central Luconia Province in offshore Malaysia	Epting (1987b), Rankey et al. (2019), Koša et al. (2015) and Zampetti (2010)
	Indonesia	Sharaf et al. (2005)
	Turkey, Mut Basin	Janson et al. (2007)
	Spain, El Hoyazo atoll	Dabrio et al. (1981) and Warrlich et al. (2005, 2008)

Supplementary Information 11.S.3 Precambrium case studies.

Age	Location	Carbonate production	References
Archean	Pilbara and Yilgarn ratons in Australia; Kaapvall and Zibabwe Cratons in Africa; North America; and Asia	Presentation of different stromatolite morphologies through times (conical, branching, oncoids)	Hofmann (2000)
Early Archean (Sebakwian Group-3.5 Ga)	Rhodesian Archaean craton	M-Factory: Stromatolites	Orpen and Wilson (1981)
Early Archean (Warrawoona Group-3.49 Ga; Kelly Group-3.43 Ga)	Pilbara craton (Dresser Formation and Strelley Pool Chert), Australia	M-Factory: Stromatolites (different morphologies from stratiform and domal (Dresser Fm.) to conical and pseudocolumnar (Strelley Chert)	Allwood et al. (2006, 2007) and Van Kranendonk et al. (2003)
Archean (Barberton Supergroup-3.22 Ga)	Barberton greenstone belt, South Africa and Swaziland	M-Factory: Planar, wavy-crinkly and tufted microbial mats and MISS (Microbially induced sedimentary structure)	Heubeck (2009), Gamper et al. (2012) and Homann et al. (2015)
Archean (Transvaal Supergroup-2.6–2.5 Ga)	Kaapvaal craton, South Africa, Griqualand West	M-Factory: Oolites and stromatolites	Beukes (1980) and Beukes and Smit (1987)
Paleoproterozoic (Glengarry group; Turee Creek Group)	Western Australia	M-Factory: Increasing diversity in stromatolites	Barlow et al. (2016), Grey (1994), and Martindale et al. (2015b)
Early Paleoproterozoic (Pethei Group)	Slave and Churchill Cratons, Canada	M-Factory: Stromatolites and early marine cements	Sami and James (1993)
Mesoproterozoic (Billyakh Group)	Siberia	M-Factory: stromatolites and peritidal carbonates	Bartley et al. (2000)
Mesoproterozoic (Sibley Group; Rossport Formation)	Ontario, Canada	M-Factory: Stromatolites and early marine cements	Bartley et al. (2022)
Mesoproterozoic (Itaiacoca Group)	São Francisco craton, Brazil	M-Factory: "restricted temporal distribution Conophyton" Stromatolites and ooids	Sallum Filho and Fairchild (2011) and Reis et al. (2017)
Mesoproterozoic (lower Belt Supergroup)	Western North America	M-Factory: Stromatolites and ooids	Pratt and Rule (2021)
Early Neoproterozoic (Jiuliqiao Formation)	North China Platform	M-Factory: Columnar stromatolites	Lan et al. (2020)
Neoproterozoic (Mackenzie Mountains Supergroup)	Paleolatitude (15°), North American craton	M-Factory: Calcimicrobes (filaments and clots) and thrombolites	Turner et al. (1997) and Turner and Long (2008)
Neoproterozoic	Central Basin, Australia	M-Factory: Hemispheroidal to columnar stromatolites	Planavsky and Grey (2008)
Late Neoproterozoic (Buah Formation)	Sultanate of Oman	M-Factory: Stromatolites and ooids	Cozzi et al. (2004a, b) and Cozzi and Al-Siyabi (2004)

Supplementary Information 11.S.4 Palaeozoic case studies.

		Early Paleozoic	
Age	**Location**	**Carbonate production**	**References**
Early and Late Cambrian	Nevada, Great Basin (USA), Iran, Korea, North China, Russia, Morocco	M-Factory: Earliest Cambrian: stromatolitic and calcimicrobial (Kruse et al., 1995) Early Cambrian: archeocyaths, microbialites (Wood et al., 1993; Wood, 1999; Debrenne et al., 2015), coralomorphs Jell, 1984; Fuller and Jenkins, 2007) Mid-late Cambrian: considered as reef gap (Fagerstrom, 1987; Kiessling, 2009); microbialites and lithistid sponges (Lee et al., 2016; Lee and Riding, 2018).	
Ordovician	N and S China	middle-late Tremadoc: Lithistid sponge–bryozoan–Calathium–calcimicrobiallymiddle Ashgill coral–stromatoporoid (Zhang et al., 2014; Webby, 2002)	Li et al. (2004), Liu et al. (2011), Weeby (2002) and Zhang et al. (2014)
	Gondwanan: Morocco and Spain	C-Factory: bryozoan and pelmatozoans	Ernst et al. (2015), Jiménez-Sánchez et al. (2015), Vennin et al. (1998) and Webby (2002).
Silurian	Wisconsin (North American midcontinent): Appalachian Basin and Algonquin Arch	T-Factory: Stromatoporoid, coral and pelmatozoan	McLaughlin et al. (2019) and Rine et al. (2020)
Middle Silurian (Wenlock, Ludlow)	Margin of the Baltic basin, Gotland, Sweden	T-Factory: Stromatoporoid, coral, calcimicrobial and algal	Copper and Brunton (1991), Kershaw (1994), Watts and Riding (2000), Nose et al. (2006) and Samtleben et al. (1996)
Early to Middle Devonian	Large latitudinal extension (up to 60° North and 55° south), extending from the Spanish Sahara to Novaya Zemlya in Russia	T-Factory: Stromatoporoid, corals, calcimicrobial and algal, calcimicrobial-lithistid	Kiessling et al. (1999) and Copper (2002a, b)
	Gondwana margin, Morocco	T-Factory: Stromatactis fabrics, Tabulate and rugose corals, brachiopods and crinoids	Brachert et al. (1992) and Kaufmann (1998)
Middle Devonian	Gondwana margin, Morocco	M-Factory: Clotted micrite and rimmed cement	Peckmann et al. (1999)
Late Devonian (Late Frasnian)	Laurussia (passive extensional margin affected by syn-sedimentary faults), Belgium	T- and M-Factories: stromatoporoid and corals to algal and microbes (Middle Frasnian)	Boulvain (2001, 2007), Bourque et al. (2004) and Boulvain et al. (2004)
		T-, C- and M- Factories: sponges and stromatactis, corals and stromatoporoids and microbial bindstones (Frasnian)	Da Silva and Boulvain (2012) and Da Silva et al. (2011)
Late Devonian (Givetian (*), Frasnian (**) to Famennian (***)	Continental margin of the Eastern Gondawana (between 10° and 20°), Australia	(*) T-Factory: stromatoporoid(**) T-, M-Factories: stromatoporoid and microbes (Renalcis) and cemented ooid shoals(***) M-Factory: microbes, calcareous algae, cements and scarce sponges	Playford (1984), Kerans (1985), Wallace (1987), Warren and Kempton (1997), Copp (2000), Johnson and Webb (2007)

(*Continued*)

Late Paleozoic			
Age	**Location**	**Carbonate production**	**References**
Carboniferous	Kazakhstan	M-Factory: Microbial boundstone	Collins et al. (2014), Kenter et al. (2002) and Kenter et al. (2010a, b)
Carboniferous (Late Tournaisian and Visean)	Laurussian continental margin, Western Europe (Ireland, Norther England, Belgium)	C-Factory: Bryozoans, crinoids or sponges, or foraminifera, or microbial and calcareous algae	Aretz et al. (2012), Bábek et al. (2013), Lees and Miller (1985), Bridges and Chapman (1988) and Wright (1994)
Carboniferous (Lower Mississipian)	Laurussian internal seaway, Sacramento Mountains	C-, M-Factories: Sponge, microbial carbonates	Elrick and Read (1991), Ahr and Stanton (1994, 1996) Wilson et al. (2019)
Upper Carboniferous	Gondwana: N Spain	M-Factory on carbonate slopes; ooid–pisoid shoals and algal packstone on platform top	Bahamonde et al. (1997), Bahamonde et al. (2000), Della Porta et al. (2002), Della Porta et al. (2003), Della Porta et al. (2004), Kenter et al. (2005) and Samankassou (2001)
Permian (Leonardian to Guadalupian)	Low latitude position 5° north of paleoequator, Guadalupe Mountains, Texas and New Mexico	C-, M-Factories (Ooids, fusuilinids, microbial carbonates) and C- Factory (Sponge, foraminifera)	Kerans and Fitchen (1995), Kerans and Kempter (2002) and Rush and Kerans (2010)
Permian (Asselian to Kungurian)	East European Platform, Russia	T-factory (phylloid algae, *Palaeoaplysina*, *Tubiphytes*, *Archaeolithoporella* and various calcisponges associated with submarine cements), ("photozoan") to C-factory (bryozoans, crinoids foraminifera and brachiopods), ("heterozoan")	Vennin et al. (2007a,b), Zempolich et al. (2002, 2003) and Kossovaya and Kropatcheva (2013)
Permian (Artinskian-Kazanian)	Northern Pangea, Greenland, Svalbard and Barents Sea	C-Factory: bryozoans/crinoids and bryozoan-Tubiphytes buildups and brachiopods/bryozoans	Stemmerik et al. (2001) and Vennin (2007c)
Permian (Asselian to Tatarian)	Northern Pangea, Sverdrup Basin to the Barents sea	T-Factory (phylloid algae, Palaeoaplysina, Tubiphytes, Archaeolithoporella and various calcisponges associated with submarine cements), ("chlorozoan" sensu Beauchamp, 1994) to C-Factory (bryozoans, echinoderms, sponges and brachiopods), ("heterozoan" or Bryonoderm association sensu Beauchamp, 1994)	Beauchamp and Desrochers (1997) Beauchamp and Baud (2002)

Supplementary Information 11.S.5 Mesozoic case studies.

Age	Location	Carbonate production	References
Early Triassic	South China	Ooliths, microbialites and sponges	Li et al. (2013), Payne et al. (2006) and Tian et al. (2015)
		Stromatolites	Ezaki et al. (2012), Kershaw et al. (2021), Lehrmann (1999) and Luo et al. (2021)
	Iran	Ooids, microbialites	Baud et al. (1997), Baud et al. (2007) and Liaghat et al. (2021)
	Turkey	Microbialites	Baud et al. (2005) and Pruss et al. (2006)
	Greenland	Microconchids	Zaton et al. (2018)
Triassic (Smithian, Spathian)	Western United States, Iran, Oman, South Armenia	Sponge and microbial reefs	Baud et al. (2021), Brayard et al. (2011), Friesenbichler et al. (2018), Martindale et al. (2015a), Vennin et al. (2015) and Olivier et al. (2018)
Triassic	Italy (Dolomites)	M- Factory: microbial deposits on slope	Bosellini (1984), Brandner and Keim (2011), Gianolla et al. (2021), Marangon et al. (2011), Maurer. (2000), Reijmer et al. (1991) and Reijmer et al. (1994)
	Arabian Plate	Ooids, microbialites	Davies and Simmons (2018), Maurer et al. (2008) and Weidlich and Bernecker (2003)
	China	*Tubiphytes* reefs	Lehrmann et al. (1998), Li et al. (2012) and Payne et al. (2006)
	Europe (Espagne, Romania) -tethyan realm	*Tubiphytes* reefs and mounds	Calvet and Tucker (1995), Escudero-Mozo et al. (2014), Flügel and Senowbari-Daryan (2001) and Popa et al. (2014)
Lower Jurassic	Morocco (High Atlas)	C-Factory: sponge, coral and microbial boundstone	Della Porta et al. (2012), Della Porta et al. (2014), Leinfelder et al. (2002) and Verwer et al. (2009a, b)
	Italy	M-Factory in shallower domain and C-Factory in deeper environments	Brandano et al. (2016) and Franceschi et al. (2019)
	Spain	M-Factory: peritidal and oolitic-peloidal inner platform	Sevillano et al. (2019)
Jurassic	France	Alternation of C- (bivalve/sponge) and T- Factories (ooid/coral) Early Jurassic: carbonate ramps Middle Jurassic: flat-topped platform with coral and stromatoporoid Late Jurassic: Carbonate rimmed shelf and ramp (Ooids to skeletal and mud ramp)	Andrieu et al. (2016), Brigaud et al. (2014) and Carpentier et al. (2010)
Middle and Late Jurassic	Spain	Ooids, microbial-siliceous sponge mounds	Aurell et al. (2019), Aurell et al. (1998), Bádenas et al. (2012), Bádenas and Aurell (2018) and Tomás et al. (2019)
	Switzerland	Ooids, microbial-siliceous sponge and coral mounds	Colombié and Strasser (2005), Dupraz and Strasser (2002), Strasser et al. (1999), Strasser et al. (2009) and Wetzel et al. (2013)
	Europe	Deep water mounds: sponge and microbialite	Leinfelder et al. (2005), Reolid (2011), and Reolid and Duarte (2014)

(*Continued*)

Age	Location	Carbonate production	References
Cretaceous	Western Europe (Denmark, Spain and France), West Africa (Gabon and Republic of Congo), South America (Peru and Brazil) and India	C-Factory: bryozoan, mollusc and echinoderm (inner platform and mid ramp); associated with T-factory (ooid, green algae, rudist)	James and Jones (2016), Michel et al. (2018), Philip and Gari (2005) and Wilson and Vecsei (2005)
Middle to Late Cretaceous	Tethyan domain	See intra-shelf: carbonate ramp and organic-rich basin	See table for references
Early Cretaceous	Europe, Middle East and Africa	T-Factory: (Rudist) and coral; and M-Factory during anoxic event (OAE-1)	Bonin et al. (2012), Bover-Arnal et al. (2009), Bover-Arnal et al. (2015), Embry et al. (2010), Everts and Reijmer (1995), Föllmi and Godet (2013) and Skelton et al. (2013)
Late Cretaceous	Europe, Middle East and Africa	T-Factory: Rudist and coral; and M-Factory during anoxic event	Borgomano (2000), Grélaud et al. (2006), Kauffman and Johnson (1988), Piryaei et al. (2010), Skelton et al. (1997) and Scott et al. (1990)
	North hemisphere	P-Factory	Mitchell (2019), Mortimore and Pomerol (1997), Mortimore (2011) and Stanley et al. (2005)

Supplementary Information 11.S.6 Cenozoic case studies.

Age	Location	Carbonate production	References
Paleocene-Eocene	Tethyan domain	T-Factory: coral, algal, foraminifera	Baceta et al. (2005), Baceta et al. (2011), Martin-Martin et al. (2020), Payros et al. (2010), Pedley (1998), Rasser and Piller (2004), Scheibner et al. (2005), Scheibner et al. (2007) and Scheibner and Speijer (2008)
		Specific carbonate Factory: Foraminifera	Adabi et al. (2008), Aigner (1985b), Bassi (2005), Beavington-Penney et al. (2005), Beavington-Penney et al. (2006), Hallock and Seddighi (2022), Hottinger and Bassi (2014), Jorry et al. (2006), Racey (2001) and Vennin et al. (2003)
	Denmark	C-Factory: bryozoan, aphotic coral	Bjerager and Surlyk (2007a, b), Bjerager et al. (2010), Bernecker and Weidlich (1990), Clausen and Huuse (2002) and Hvid et al. (2021)
Oligocene-Miocene	Myanmar, Malaysia, Indian Ocean, South oceania region; Australia	coral, coralline algae, foraminifera	Courgeon et al. (2016a, b), James and Bone (2011), Mihaljevi et al. (2014), Paumard et al. (2017) and Rankey et al. (2019)
	Mediterranean region, Middle East region	T-Factory: corals, red algae and C-Factory (mollusc, foraminifera)	Andreucci et al. (2017), Bourillot et al. (2009), Bourillot et al. (2010a, b), Brandano et al. (2017), Klicpera et al. (2015), Menier et al. (2014), Michel et al. (2020), Pomar et al. (2002), Pomar et al. (2014), Reolid et al. (2017) and Warrlich et al. (2005)
	Turkey, Iran, China	Coral, red algae and foraminifera	Dill et al. (2018), Janson et al. (2010), Makhankova et al. (2021), Sattler et al. (2009), Shao et al. (2017), Van Buchem et al. (2010c) and Wu et al. (2016)

Age	Location	Carbonate production	References
Miocene	Spain, France, Turkey	Mix terrigenous and carbonate grains (bryozoan, mollusc, echinoderm, red algae)	André et al. (2011), Bassant et al. (2005), Carpentier et al. (2020), Reynaud and James (2012) and Reynaud et al. (2012)
Plio-Pleistocene	Indian Ocean (modern Atolls), Pacific and Atlantic domains	T-Factory: coral, algae and foraminifers	Cabioch et al. (1995), Courgeon et al. (2016a, b), Davies and Kinsey (1977), Dechnik et al. (2019), Droxler and Jorry (2021), Gisler et al. (2010), Goktas et al. (2016), Jiang et al. (2019), Leonard et al. (2020), Roff (2020) Reid et al. (2003), Reijmer et al. (1992) and Weij et al. (2019)

12

Deep-Water Detrital Systems

David M. Hodgson[1], Elda Miramontes[2], Ian A. Kane[3], and Christopher A.-L. Jackson[4]

[1] School of Earth and Environment, University of Leeds, Leeds, UK
[2] MARUM-Center for Marine Environmental Sciences, University of Bremen, Bremen, Germany, and Faculty of Geosciences, University of Bremen, Bremen, Germany
[3] Department of Earth and Environmental Sciences, University of Manchester, Manchester, UK
[4] WSP UK Ltd, Manchester, UK, and Imperial College, London, UK

Summary

Imagine you were orbiting the Earth with the ability to see through our oceans. As a budding geoscientist, you would be struck by the major mountain ranges, such as the Andes, the Himalayas and the mid-Atlantic ridge. You might also note the vast and largely featureless plains with isolated highs in the Pacific and Indian Oceans. But possibly the most striking morphological features would be the heavily incised and degraded landscapes, ornamented with channels, in between the highest mountains and the flat plains. You would have to look more carefully to see the large fan-like features at the end of some channels, but if you could zoom-in on these features you would see wonderfully intricate leaf-like frondescent patterns, as in the Mississippi and Amazon Fans.

These deep-water detrital systems are the longest and largest sedimentary systems on Earth, but our blue planet obscures them. However, these are exciting times for deep-water detrital research as their secrets are being revealed at an astonishing rate. Major advances in monitoring of sediment gravity flows in the natural environment, increased resolution of seafloor imagery, detailed documentation of exhumed systems over large areas and integrated well and seismic reflection datasets from ancient buried systems are becoming publicly available. This richness in data is reflective of the diversity of physiographic configurations, and the range of external controls, which cautions against overly simple application of analogues and generalised deep-water detrital system classifications.

Cohesive mass flows, sediment gravity flows and bottom currents work both independently and together, to redistribute detrital sediment in deep-water settings. The amazing duration (days), volume (km^3) and transport distance (>1000 km) of individual sediment gravity flows has been revealed, and new flow structures have been proposed. The presence of elevated mud content in many deposits means that long-established process models of cohesionless flows are being re-evaluated, with the recognition that flow transformations in time and space help us to explain the wide range of bed types observed in deep-water detrital systems. New life has been breathed into the utility of sole structures preserved on the base of sandstone beds to be used as indicators of palaeohydraulic conditions and substrate rheology, as well as palaeoflow direction. The fine-scale depositional architecture of exhumed slope to basin-floor stratigraphy has shown that channel–levee system lengthening and basin-floor progradation is normal, as systems become more confined and efficient through time.

Therefore, the reader should be excited that, like an accelerating turbidity current, our understanding of deep-water detrital systems is rapidly moving from enigmatic to tractable, and even somewhat predictable. These are pertinent advances. Improved understanding of the processes and products in deep-water detrital settings is key to many contemporary societal challenges, including unlocking ancient records of environmental change, reducing uncertainty in the efficacy of carbon storage sites, predicting sites of destructive sediment gravity flows and landslides to protect infrastructure, and the identification of pollution hotspots in the modern oceans.

12.1 Introduction

12.1.1 What Are Deep Water Detrital Systems?

Sixty-three percent of the Earth's surface lies beyond the continental shelf-break, and although large parts of the ocean floor have very little siliciclastic sedimentation

Reading's Sedimentary Environments: Processes, Facies and Stratigraphy, Fourth Edition. Edited by Bruce Levell.

Companion website: www.wiley.com/go/levell4e

(Chapter 13), continental margins can receive huge volumes of detrital sediment (Dutkiewicz et al. 2015). Submarine fans, as depositional environments, contain by far the largest deep-water detrital accumulations on Earth (Curray et al. 2002). They form at the terminii of source-to-sink sediment transport systems, receiving sediment from continents, which are transported via erosional canyons that can incise deeply into the continental shelf, and which feed sinuous channels bounded by large levees that extend across the lower slope and continental rise (Figure 12.1). The sediments transported through and deposited within these conduits build submarine fan successions that can be several kilometres thick and provide long-lived archives of palaeoenvironmental change. For example, the largest submarine fan in the world, the Bengal Fan, which formed in response to India-Asia collision, is c. 3000 km long, c. 1000 km wide and, at its maximum beneath the Bay of Bengal, is up to 16.5 km thick.

Nonetheless, the detailed relief (bathymetry) of Earth's ocean floors is known in less detail than that of the Moon. A diverse range of flows transport sediment across the seafloor. These modify submarine seascapes through erosion and deposition developing complicated, and dynamic, patterns of seafloor relief. Although we have much to learn about deep-water detrital systems, there have been major advances in understanding the processes and products of these systems since the previous edition of this book. The main agents of detrital sediment transport in the oceans are a broad suite of particulate density currents that are driven by combinations of gravitational forces and density contrasts between the flow and the ambient fluid (seawater). For simplicity, these are collectively referred to as 'sediment gravity flows' (Middleton and Hampton 1973), which can range in rheology from dilute and fluidal to highly concentrated and cohesive. These flows interact with bottom currents related to oceanographic processes, which are sometimes powerful enough to affect sediment gravity flow behaviour, and which may also rework seafloor sediment to form contourites.

Submarine canyons and submarine fans are striking features of the modern seafloor (Figure 12.1), and many ancient examples have been recognised both in the subsurface and exhumed and exposed at the Earth's surface. The typically low resolution and patchy imaging of the seafloor, combined with the paucity of direct measurements of sediment gravity flow processes, means that sedimentologists remain unusually reliant on the stratigraphic record to understand the processes and products of deep-water detrital systems. Nonetheless, the last decade has seen a step-change in the number of direct measurements of sediment gravity flows, sediment coring of modern and recent systems, physical and numerical modelling and several large-scale (>1000 km^2) mapping projects of exhumed systems (Figure 12.2).

In part, the expansion of understanding of deep-water detrital systems has been driven by hydrocarbon exploration, as the sandstones within them can form large oil and gas reservoirs. In addition, submarine fans occurring at the end of source-to-sink transport systems can preserve relatively complete and datable archives of hinterland tectonic deformation, denudation and climate change. This potential to act as archives of environmental change has been exploited by collaborative international ocean drilling campaigns. For example, the interplay between atmospheric and oceanic circulation patterns and the Earth's surface can be recorded in submarine fan successions through sediment flux changes over millions of years (e.g. Clift and Gaedicke 2002). There are, however, complications; for example, the storage of large volumes of sediment in floodplains and on continental shelves can result in long lag times between tectonic events or changes in climate, and the signals of these perturbations being recorded in deep ocean sediments. Autogenic processes (e.g. channel avulsion, erosion) are also ubiquitous during the building of deep-water detrital successions, which makes reading stratigraphic records in terms of external forcings very challenging, especially when limited to using 1D data (e.g. single cores). Importantly, land area is a poor predictor of the volumes of sediment supplied to the deep sea. For example, oceanic islands, such as New Zealand and Taiwan, are estimated to contribute 17–35% of the particulate organic carbon to the world's oceans and submarine depositional systems (Lyons et al. 2002) despite comprising only a few percent of the Earth's total land area. The reason they punch above their weight in terms of sediment flux is because of the short transport distances and associated limited onshore or continental shelf sediment storage, their high and steep hinterland relief, and the extreme variability of rainfall on these islands (and typhoons in the case of Taiwan), which all enable rapid sediment transfer to the ocean (Milliman and Syvitski 1992; see also Figure 9.5).

The most widely recognised type of sediment gravity flow deposit is called a turbidite – the deposits of a turbidity current – and for this reason submarine fans are also referred to as turbidite systems (Bouma et al. 1985). There is a wide range of sediment gravity flow behaviours in time and space, and therefore an apparently bewildering assortment of deposit (bed) types. Among the more concentrated mass movements are some submarine landslides, which can be voluminous (several 100s km^3), and yet travel over very low gradients for long distances, and, unusually, can increase markedly in volume through erosion and entrainment of the substrate as they move into deeper water. The scars they leave behind, and the topography of their deposits, can strongly influence subsequent sediment dispersal patterns. Despite potentially generating tsunamis and

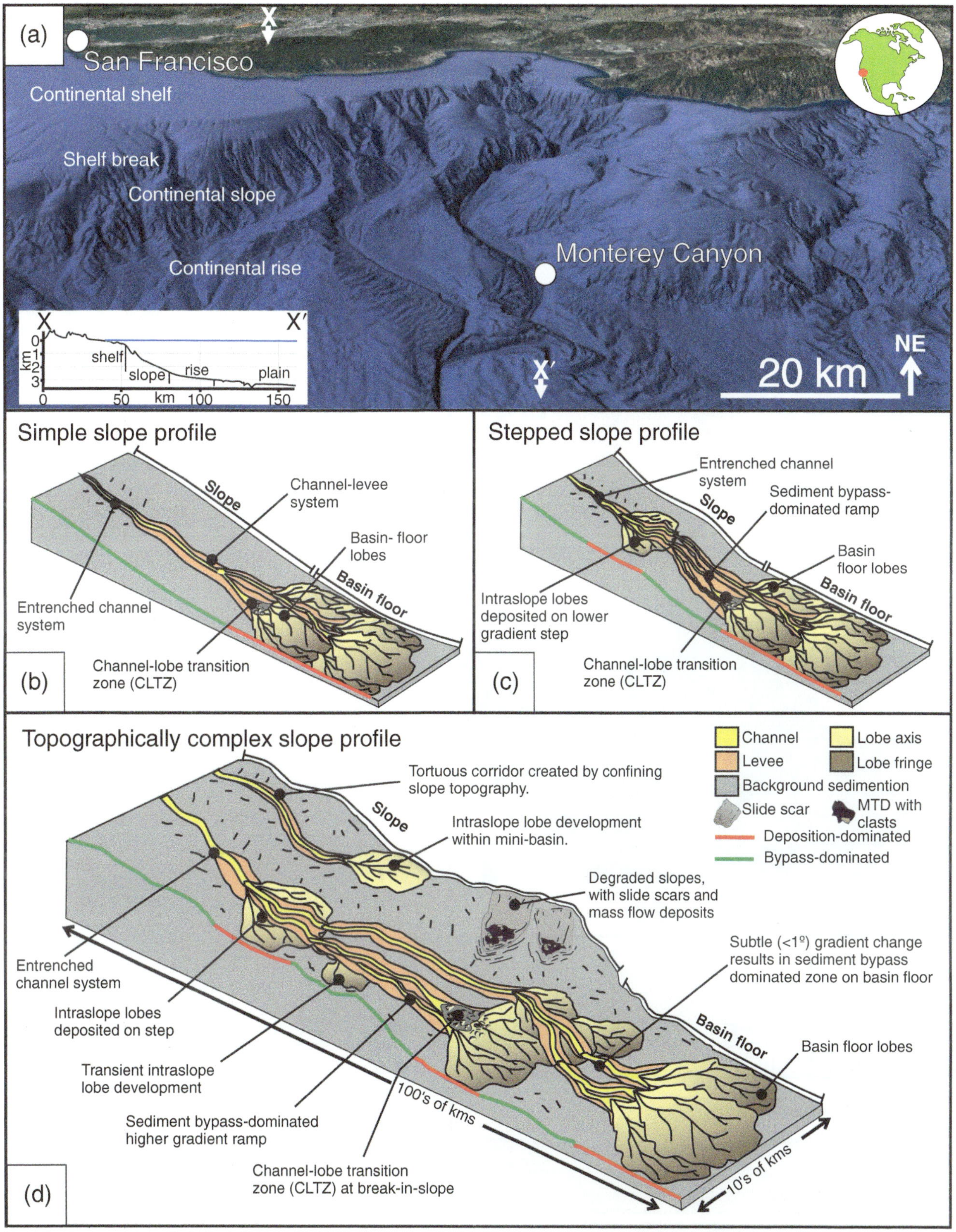

Figure 12.1 (a) Example of an incised continental margin seascape, offshore California, USA. The prominent valley feature is the canyon. Source: Google Earth. (b–d) Examples of submarine fans with different topographic configurations and depositional patterns. (b) Simple submarine slope profile with a single break-in-slope changing from a bypass-dominated channel–levee system to deposition-dominated basin-floor lobe complexes. (c) Stepped slope profile with incised higher gradient ramps linking lower gradient steps. (d) Topographically complex slope, encompassing different magnitudes and orientations of topography. MTD – mass transport deposit. *Source:* Adapted from Brooks et al. (2018).

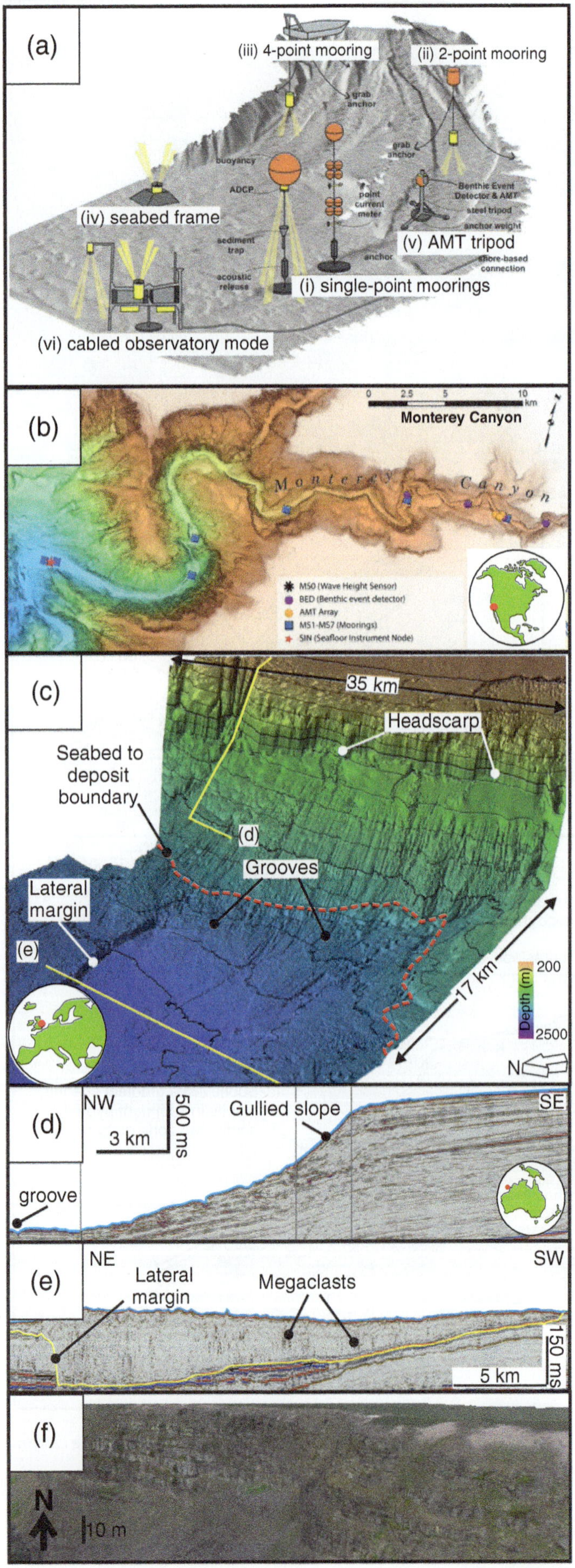

Figure 12.2 (a) Instrumentation to monitor modern deep-water detrital systems, including: (i) anchored single-point moorings (right: point current meters, left: Acoustic Doppler Current Profiler (ADCP) design); (ii) unanchored two-point mooring with suspended down-looking instrument; (iii) four-point mooring to deploy suspended instruments; (iv) seafloor frame to deploy an upward-facing instrument; (v) Acoustic Monitoring Transponder (AMT) tripod with Benthic Event Detector to track movement; (vi) platform connected to a seafloor cable network that may host many instruments with real-time communications and power. (b) Configuration of instrument deployment in the Monterey Canyon, offshore California, USA. (c) Seafloor and subsurface expression of the basal shear surface of a submarine landslide using seismic reflection data, offshore NW Australia. (d and e) Seismic reflection cross sections located on (c). (f) Image from a photogrammetric model of the Carboniferous Alport Castles outcrop, Derbyshire, UK, built using images collected by an uncrewed aerial vehicle. *Source:* (a, b) from Clare et al. (2020); (c–e) adapted from Nugraha et al. (2022); (f) photo credit: David Hodgson.

destroying seafloor infrastructure, they remain poorly understood and hard to forecast.

A single sediment gravity flow can transport over 100 km^3 of sediment (e.g. Rothwell et al. 1998; Talling et al. 2007), which is around ten times the annual sediment budget supplied to the oceans from all the world's rivers (Milliman and Syvitski 1992). This makes sediment gravity flows volumetrically the most significant particulate transport process across Earth's surface (Talling et al. 2012). The capability of sediment gravity flows to transport sediment long distances, across low gradients, and at high speeds, was first recognised following the 1929 Grand Banks earthquake. The earthquake triggered a submarine landslide that evolved into a turbidity current, which broke seafloor telecommunications cables. The timing of cable breaks was used to calculate a flow velocity of 9 $m s^{-1}$ down a 0.3° gradient (Heezen and Ewing 1952; Piper et al. 1999).

Deep-water detrital systems are of fundamental importance to many modern issues. For example, their large volumes mean they have a critical, yet poorly understood, role in modulating rates of organic carbon burial in the deep ocean, which potentially represents a key control on atmospheric CO_2 concentrations over millions of years (Galy et al. 2007). Burial of organic carbon in submarine-fan successions represents the second largest global sink of atmospheric carbon dioxide (after the combined process of silicate weathering and subsequent carbonate precipitation; Galy et al. 2007). The large volumes and porosities of subsurface turbidite systems make them attractive sites for carbon capture and storage projects (e.g. Ketzer et al. 2005). It is also apparent that large volumes of anthropogenic pollutants, such as microplastics, are being transported to the deep ocean, and reworked by bottom currents (e.g. Kane et al. 2020). Furthermore, around 95% of intercontinental global Internet traffic travels via undersea cables that run across the ocean floor; these cables are prone to breakage by sediment gravity flows (e.g. Clare et al. 2020).

12.1.2 History of Deep-Water Detrital Sedimentological Research

Conceptual models of deep-water detrital systems are necessarily founded on ancient successions, as only recently have there been advances in the documentation of active modern systems, which are largely inaccessible (and destructive). The reader is referred to Mutti et al. (2009), who provide a thorough historical review of the development of conceptual models of submarine fans and sediment gravity flows.

Oceanographers first documented the highly degraded nature of the world's submerged continental margins using marine geophysical data (see Normark et al. 2003). V-shaped submarine canyons were shown to extend for tens to hundreds of kilometres into the oceans, and to transition to U-shaped channel systems bounded by large sediment wedges, known as external levees. Other submarine channel types have been identified that are found in a wide range of physiographic settings and reflect different developmental controls, from continental slopes and basin plains to abyssal plains and ocean trenches (see Peakall and Sumner 2015, for a full discussion). Our focus is on submarine fans (Menard 1955) that comprise channel–levee systems that feed expansive depositional systems that form fan-like morphologies in map view across the seafloor. As multiple modern submarine fans were documented, so their irregular morphologies were highlighted. The variability was ascribed to differences in seafloor topography, feeder system, and sediment supply style from their terrestrial source areas (Reading and Richards 1994; Mattern 2005; Figure 12.1).

Contemporaneous with oceanographers recording the unexpectedly rugose bathymetry of the world's ocean floors, pioneering outcrop sedimentologists were documenting and discussing distinctive normally graded, and commonly tabular, sandstone beds associated with marine mud-rich successions. Kuenen and Migliorini (1950) proposed the concept of a normally graded sandstone bed being the product of resedimented fluvial and/or shallow-marine sediment transferred into deep-water environments by submarine turbidity currents. The recognition of turbidity currents, and related sediment gravity flows, as important agents of detrital sediment delivery to deep-water settings, drove a profound change in sedimentological thinking and palaeoenvironmental interpretations.

The turbidity current and turbidite concepts led to the development of sedimentary facies and architectural element schemes, and submarine fan models. The identification of architectural elements (Chapter 2) has proved to be a particularly effective descriptive tool to link ancient and modern deep-water detrital systems (e.g. Mutti and Normark 1987, 1991). Process-oriented sedimentary facies analysis introduced concepts such as sediment gravity flow efficiency, deflection, reflection, and ponding (e.g. Pickering and Hiscott 1985), and the recognition of bed types that strongly differed from the classic turbidite model (see Talling et al. 2012). The advent of large-scale mapping programmes of exhumed deep-water detrital systems that are at a scale comparable to modern systems has provided a missing scalar link (e.g. Gardner et al. 2003). Sedimentary facies and architectural elements can now be placed within submarine slope to basin-floor stratigraphic architectures. These have, for example, supported theoretical spatio-temporal relationships of bypass-dominated channel–levee systems and downdip lobe complexes (e.g. Walker 1978). In contrast, the links between bottom current deposits (contourites) identified on the ocean floor and outcrop-based facies and architectural schemes of these systems are still in their infancy.

12.1.3 Approaches to the Study of Deep-Water Detrital Sedimentary Systems

A continuing challenge is establishing rigorous links between flow processes and their products (deposits) and making observational connections across systems of different scales and physiographic configurations. Technological advances are helping to bridge these gaps.

Modern/recent systems: Detailed documentation of the seascapes of active, or recently active, deep-water detrital systems use side scan sonar and multibeam bathymetry data for surface characterisation, and sub-bottom profiler, multichannel high-resolution seismic reflection data and sediment cores for subsurface characterisation. As the resolution of these data improves, so does the complexity and richness of seafloor geomorphology, and the range of bedform types and scales, are revealed (e.g. Carvajal et al. 2017). The biggest step forward has been the (partially) successful instrumentation and monitoring of sediment gravity flows in the natural environment (Clare et al. 2020; Figure 12.2).

Ancient subsurface systems: Knowledge of the large-scale 3D stratigraphic architecture of deep-water detrital systems has advanced through data collection by the oil and gas industry, with extensive and integrated 3D seismic-reflection and well-log datasets. Well-preserved buried deep-water landscapes from Mesozoic and Cenozoic systems are widely documented (e.g. Fisher et al. 2021). These data permit an understanding of the impact of dynamic topography associated with salt and mud diapirism, and gravitational tectonics, on sediment dispersal patterns over millions of years and 1000s km^2 (e.g. Casagrande et al. 2022). Some of the most insightful imagery comes from shallow and non-prospective stratigraphy, such as the complicated patterns of migrating sinuous submarine channels (e.g. Janocko et al. 2013; Hansen et al. 2017), the intricate form of lobes (e.g. McHargue et al. 2021), and the kinematics of submarine landslides (e.g. Steventon et al. 2019; Figure 12.2).

Ancient exhumed systems: Field data collection from deep-water detrital systems uses exposed deposits to decipher transport and depositional processes within the context of sedimentary facies, or bed types (Section 12.4) and architectural element approaches (Section 12.5). A crucial advance has been the mapping and reconstruction of extensively exposed deep-water detrital systems (Figure 12.3), which comprise channel–levee and lobe systems of comparable scale to both modern and buried systems. Outcrop studies will remain the building blocks for the development of models that link sedimentary processes, deposits and stratigraphic evolution, as these are the only datasets with the spatial and temporal scales required for this type of multi-scale analysis.

Variable tectonic settings: Whereas elements of deep-water basins in different tectonic settings are comparable, there are distinct differences in terms of basin physiography, subsidence rates, sediment supply, and scale. Syn-extensional (syn- and post-rift) deep-water detrital systems are rarely well preserved at outcrop due to deformation during later basin inversion (Privat et al. 2021). Similarly, it is only recently that the sedimentology of exposed systems influenced by salt tectonics (e.g. Cumberpatch et al. 2021) has been documented. As a result, understanding of the processes and deposits in these tectonic settings lags behind understanding of convergent and oblique-slip basins, which have been documented in detail from many foredeep, wedge-top and strike-slip basin-fills, typically fed by small, high-gradient river systems (Mutti et al. 2009). How comparable these relatively small systems are to larger equivalents on passive margins fed by major rivers (Figure 12.4), deposited on slope profiles perturbed by gravitational and salt tectonics, and influenced by oceanic circulation, is questionable.

12.2 Environmental Configurations

In this section, we begin by looking at the challenge of identification of trigger mechanisms for sediment gravity flows. From here, the range of different flow types, and linkage of the main depositional environments, is covered with reference to present day systems where the configuration can be observed. Two case studies, the present-day Congo River-Canyon-Fan system and the exhumed Permian submarine fans of the Karoo Basin, South Africa, provide examples of modern and ancient detrital deep-water systems.

12.2.1 Introduction

Sediment gravity flows are notoriously difficult to monitor in the natural environment because they are rarely observed on historic time scales, and initiation mechanisms are poorly understood and unpredictable (Piper and Normark 2009). However, increasing numbers of sediment gravity flows are being monitored (Figure 12.2), and more rarely linked to their deposit (Maier et al. 2019), and in some cases erosion of the seafloor has been quantified (Figure 12.5). Many studies infer external trigger mechanisms for mass movements and sediment gravity flows. These external triggers are usually events, such as storms (e.g. Normandeau et al. 2020), river floods (e.g. Mulder and Syvitski 1995) and major earthquakes (e.g. Mountjoy et al. 2018; Figure 12.5). Indeed, sediment gravity flow deposits have been treated as records of such events with

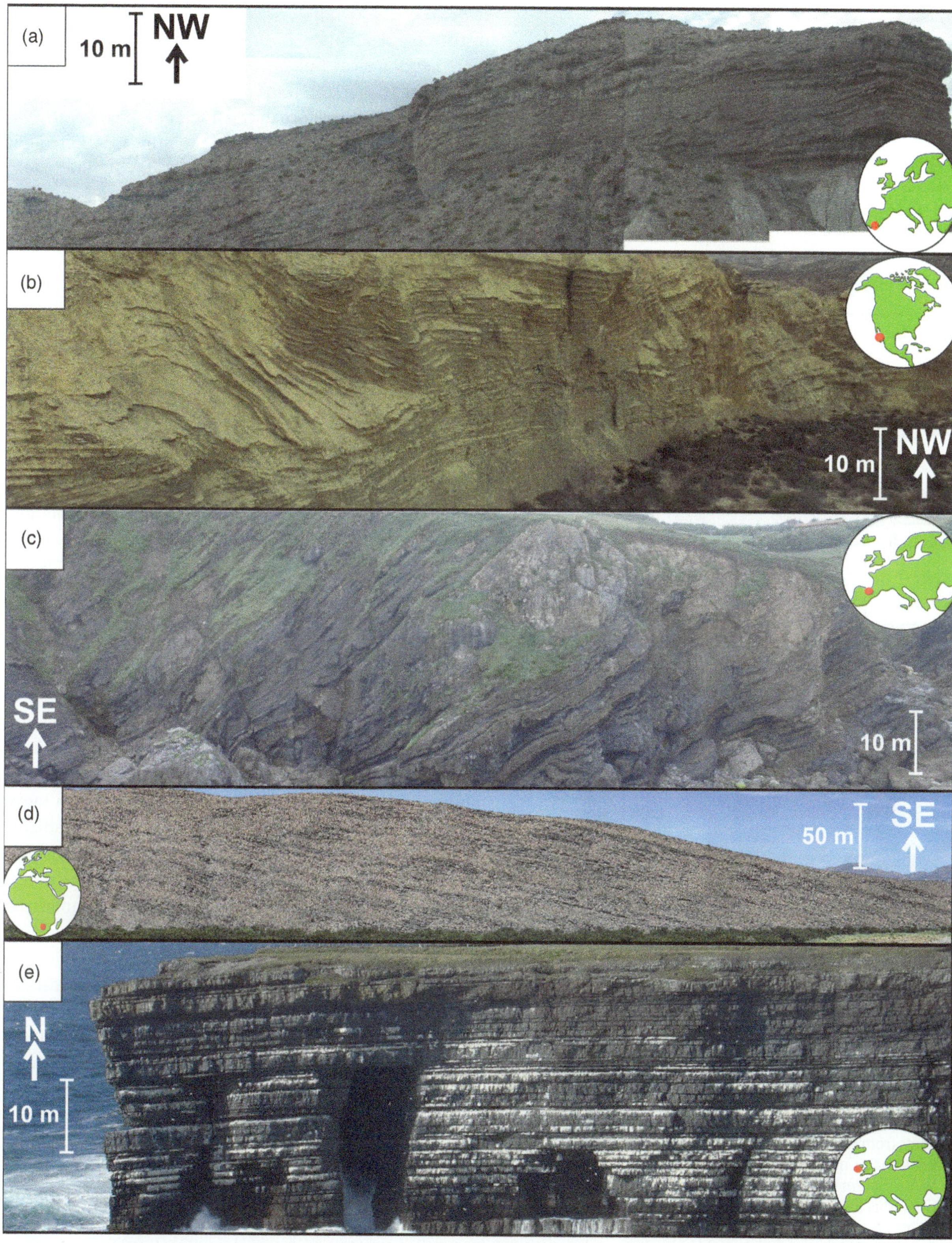

Figure 12.3 Examples of ancient exhumed deep-water detrital systems. (a) Dip-section through a Miocene submarine channel complex, palaeoflow from left to right, Tabernas-Sorbas Basin, southeastern Spain. (b) Strike section through a remobilised external levee succession, Cretaceous Rosario Formation, Baja California, Mexico. (c) Intercalated turbidites and debrites on the eastern flank of the Cretaceous Bakio salt diapir, northern Spain. (d) Extensive exposures to Permian submarine lobe complexes, Laingsburg depocentre, Karoo Basin, South Africa. (e) Stacked submarine lobes, Carboniferous Ross Formation, County Clare, western Ireland. *Source:* Photo credit: David Hodgson.

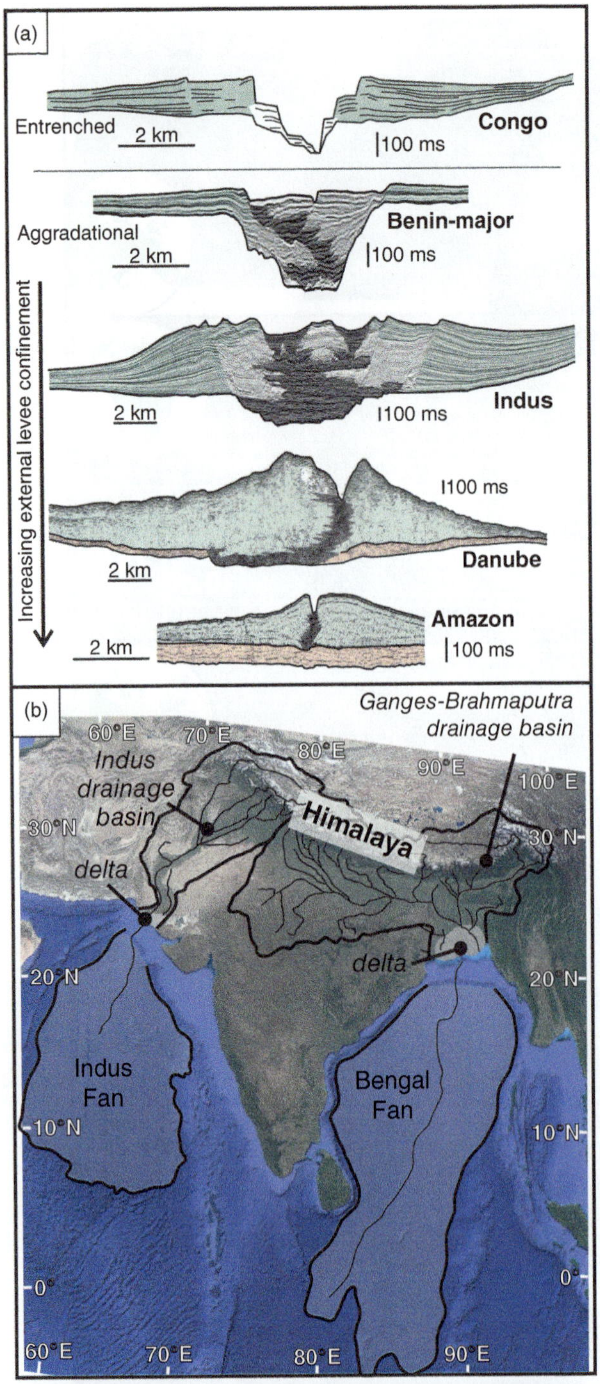

Figure 12.4 (a) Architectural variations in long-lived channel–levee systems from entrenched (top) to levee-confined (base). From top to base: Congo Channel (Babonneau et al. 2002); Benin-major channel levee system (CLS), western Niger Delta slope (Deptuck et al. 2012); Indus Fan CLS C (Deptuck et al. 2003); Danube CLS, Black Sea (Popescu et al. 2001); Amazon Channel (Amazon Fan; Pirmez et al. 2000). Light grey = internal levees/terrace deposits; green = external levee deposits; dark grey = channel-fills, and orange = basal lobe deposits. Adapted from Deptuck and Sylvester (2018). (b) South Asian source-to-sink sediment-routeing systems. The Ganges and Brahmaputra drainage basin provides sediment to the Bengal Fan, and the Indus drainage basin provides sediment to the Indus Fan. *Source:* Based on Covault (2011).

the aim of extending our knowledge beyond historical or instrumental archives (e.g. Goldfinger 2011). However, there is not a clear linkage between the nature or magnitude of the trigger, the velocity, volume or runout distance of the flow, and therefore the character of the deposit. Two similar magnitude earthquakes in the same location might trigger two very different sediment gravity flows and deposits, just one flow, or no sediment gravity flow at all. Therefore, the interpretation of turbiditic successions as a record of, for example, seismicity, risks circular reasoning.

Sediment gravity flows can occur when sedimentary accumulations exceed thresholds in fluid pore-pressure, loading of substrate, or gradient, without needing to invoke an external 'trigger'. For example, Bailey et al. (2021) deduced that rather than 'instantaneous trigger events', rapid seasonal supply of sediment to canyon heads or river mouths can result in elevated pore pressures, which may predispose slopes to failure and subsequent generation of sediment gravity flows. Recent monitoring, for example, has shown that the Whittard Canyon, which is located over 300 km from a terrestrial source, is active with six large-volume turbidity currents recorded in one year (Heijnen et al. 2022).

12.2.2 Submarine Landslides

Submarine landslides, like their subaerial counterparts, are mass movements that can transform fully or partially from one type of flow to another as they pass down slope. Unlike most subaerial landslides, submarine landslides can move for very long distances (100s km) over very low gradients (<1°). The Saharan debris flow (Gee et al. 1999) dated at about 60 ka travelled over 400 km as a single large debris flow of pelagic material travelling on basal volcaniclastic sand. They can 'bulk-up' during downslope movement through the entrainment of large volumes of substrate (Nugraha et al. 2022; Figure 12.2). Submarine landslides are well characterised using seafloor and shallow subsurface data (e.g. Steventon et al. 2019; Karstens et al. 2023). Despite the wide range of scales, and processes involved, submarine landslide systems can be broadly subdivided into: (i) a headwall domain, characterised by arcuate and coalesced scars, extensional faults and blocks; (ii) a translational domain, characterised by steep lateral margins and flat to stepped basal surface ornamented with grooves, with the landslide deposit having a complicated surface relief and protruding blocks (megaclasts); and (iii) a toe domain, characterised by verging folds and imbricate thrust faults, and outrunner blocks (Figure 12.6). The down-dip margins of submarine landslides may be frontally confined or frontally emergent (Frey-Martínez et al. 2006; Bull et al. 2009; Figure 12.6).

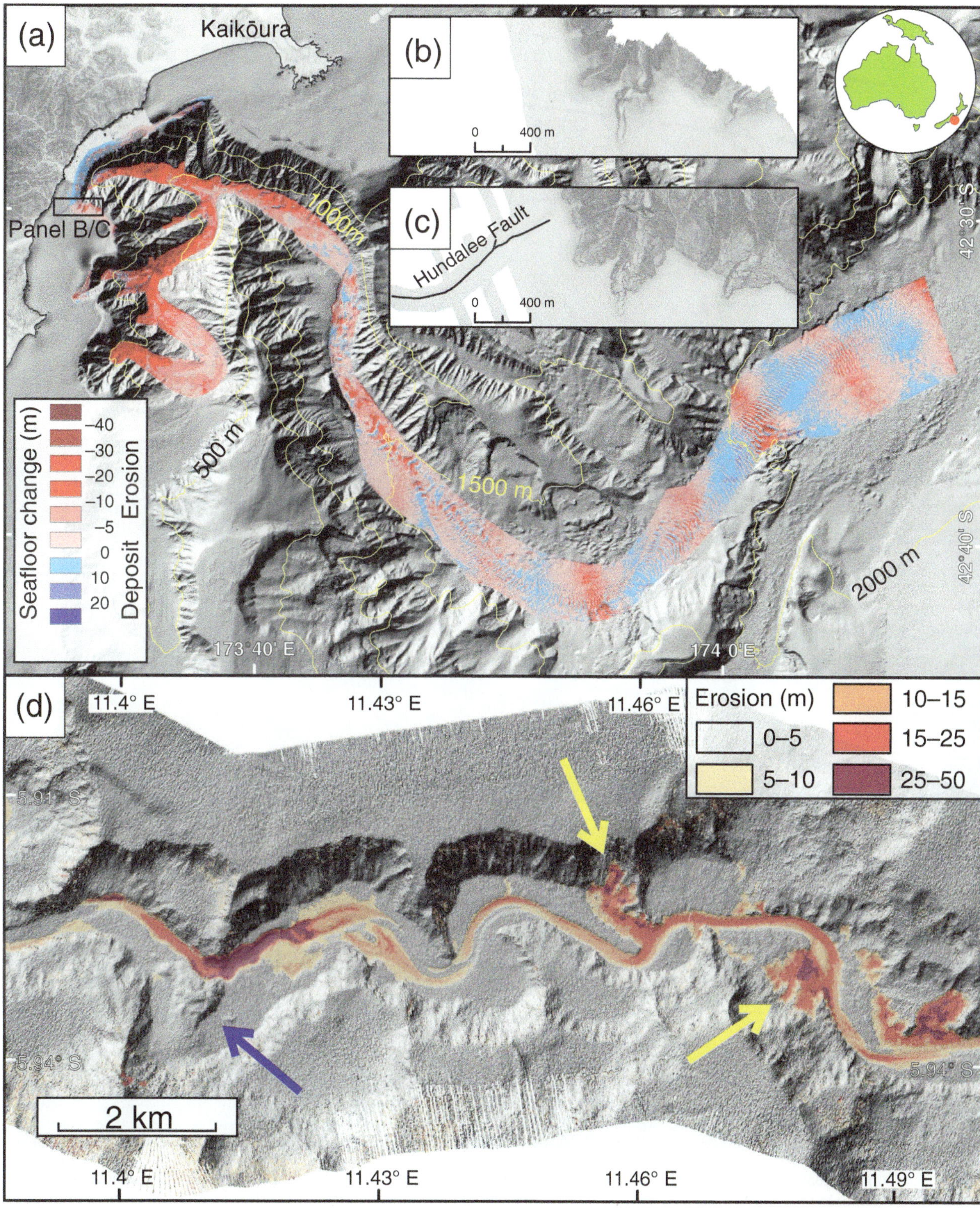

Figure 12.5 (a) Digital elevation model of the Kaikōura Canyon bathymetry, offshore Aotearoa/New Zealand (in grayscale) overlain by the magnitude of erosion and deposition within the canyon, measured by differencing the pre- and post-earthquake bathymetry datasets. (b and c) Show the pre-earthquake and post-earthquake bathymetry, respectively, where coseismic landslides have occurred at the canyon rim (location on the canyon rim indicated by the labelled black rectangle). (d) Changes in seafloor elevation along the upper Congo Canyon between September–October 2019 and October 2020. Patterns of localised seafloor erosion are associated with knickpoints (steeper zones in canyon long-profile). Depth of erosion is shown in red, with 2019 bathymetry in greyscale. Note the prominent canyon side-wall failures (yellow arrows), and an older (pre-2019) landslide-dam (blue arrow) that formed a major knickpoint. *Source:* (a–c) from Mountjoy et al. (2018), (d) from Talling et al. (2022).

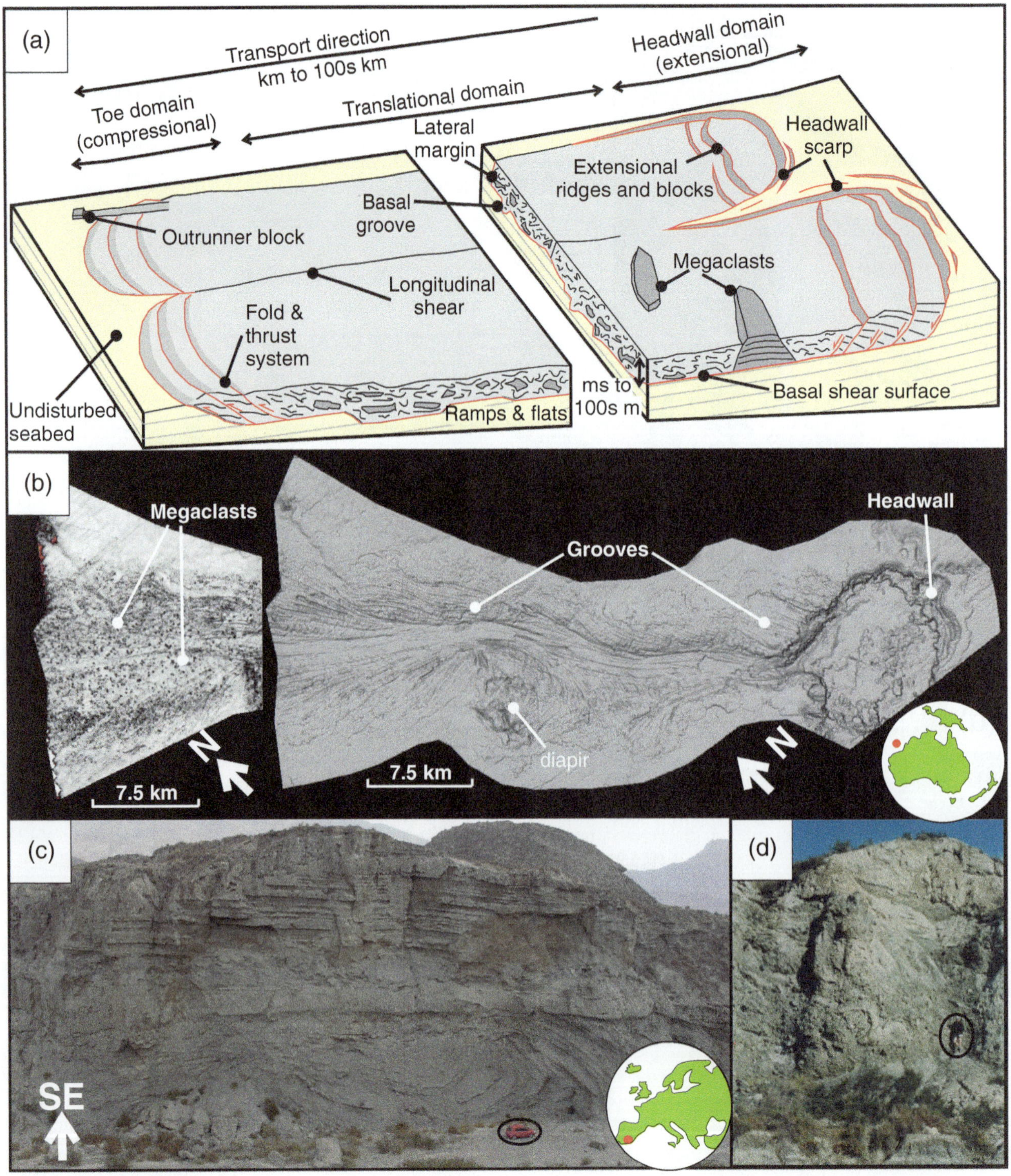

Figure 12.6 (a) Schematic illustration of the morphology, domains, and range of structures identified in submarine landslides. (b) Subsurface example of a submarine landslide, offshore NW Shelf Australia. Right: a variance map shows the distribution of grooves on the basal shear surface. Left: a variance map of the same area ~15 m below the top surface to show the distribution of megaclasts, which are a similar size to the grooves suggesting that these were the tools that made the grooves. (c) The folded structure is a megaclast transported in a debris flow, the Miocene Gordo Megabed, Tabernas-Sorbas Basin, SE Spain. Car for scale. (d) Close-up of the poorly sorted debrite. *Source:* (a) Adapted from Prior et al. (1984) and Bull et al. (2009). (b) Images from Yan Li. (c and d) Photo credit: David Hodgson.

12.2.3 Submarine Canyons

Submarine canyons are strongly incisional features that dissect the slope and shelf on all continental margins. Harris et al. (2014) estimated that there are 9477 submarine canyons on Earth, covering about 11% of the continental slope. Submarine canyons erode through all substrates, from unlithified sediments (e.g. the Congo Canyon; Figure 12.5) to crystalline bedrock (e.g. Monterey Canyon, Figure 12.1). Their character is related to tectonic setting. The narrowest continental shelves are associated with tectonically active and steep margins, where submarine canyons are more closely spaced and shorter compared to those on the gentler slopes found on passive margins (Harris and Whiteway 2011; Soutter et al. 2023). The walls of submarine canyons are generally very steep and can be near vertical and etched with scallop-shaped scars related to slope failure (Figure 12.5). Shelf-incising submarine canyons can incise the shelf to depths generally ranging from 0.1 to 3 km (Harris and Whiteway 2011). The canyon head can be located a few 10s metres from the shoreline (e.g. La Jolla, California) or even directly connected to the fluvial system (e.g. Congo Canyon), or >300 km from the coastline (e.g. Whittard Canyon). Some canyons are slope-confined (blind canyons *sensu* Harris and Whiteway 2011) and do not incise the shelf, but can reach depths comparable to shelf-incised canyons. At the present day, shelf-incising submarine canyons occur on flooded continental shelves, although during sea-level lowstand many canyon heads were exposed and directly connected to fluvial systems. Submarine canyons can also develop in areas remote from direct fluvial-deltaic input. For example, along glaciated margins, where the advance and retreat of shelf-crossing ice sheets, and associated meltwater events, transfer large quantities of poorly sorted glacial material onto the outer continental shelf and slope (e.g. Piper et al. 1985). Furthermore, the heads of several modern active canyons intercept sediment transported by longshore drift (e.g. Covault et al. 2007; Boyd et al. 2008), but were largely inactive during lower sea-levels due to sediment re-routeing.

12.2.4 Submarine Fans

Submarine fans are composite constructional features that evolve above the area of decreasing gradient connecting the continental slope, the gentler continental rise and onto the abyssal plain, and are normally fed by submarine canyons (Figure 12.1a). They can be hundreds of kilometres long and several kilometres thick (Figure 12.4) and are built from terrestrial and shallow marine detrital sediment that has been redistributed into deeper water via a range of sediment gravity flow types. Submarine fans are present on both active and passive continental margins, in relatively shallow (<500 m) to very deep (>4000 m) water, and at all latitudes. The best-known modern submarine fans, with the largest volumes, develop seaward of Earth's largest rivers or their deltas (Wetzel 1993), and include the Indus, Bengal, Amazon, Mississippi and Congo fans. Smaller submarine fans, on the order of 10s–100s km^2 develop seaward of high-gradient bedload-dominated rivers that cross narrow continental shelves, like the Golo fans off eastern Corsica (Gervais et al. 2006), and the Hueneme Fan offshore southern California (Piper et al. 1999).

Despite the wide range of attributes (e.g. size, longevity, grain-size range), most submarine fan systems comprise a combination of submarine channels that construct adjacent external levees that are associated with upper fan environments, and pass downstream into lobate deposits, 'lobes', which accumulate where sediment-gravity flows lose channel confinement and characterise the lower fan environments (e.g. Piper and Normark 1983). These basic building blocks of submarine fans can be intercalated with submarine landslides and pelagic or hemipelagic drapes (Chapter 13), which can all be modified by ocean bottom currents. In comparison to the erosional seascapes of the continental slope (kms), the surface relief of submarine fans is relatively subdued (10s to 100s m).

Submarine fans may contain a single channel–levee system, or form a distributive network that ornaments the fan surface and develops through channel avulsions to initiate new sites of lobe accumulation (Picot et al. 2016; Figure 12.7c). The active channel may be highly sinuous and narrow (100s m), flanked by relatively large symmetrical external levees (up to 150 m high, and extending >6 km away from the channel; Figure 12.4a), or may be relatively straight and wide, flanked by highly asymmetric external levees (e.g. Piper et al. 1985; Savoye et al. 1993). Typically, external levee height and thickness decrease laterally away from the channel to form a wedge geometry.

Despite gross morphological similarities, submarine channel–levee systems differ greatly from their riverine counterparts (Peakall et al. 2000). Submarine channels commonly attain much larger values of both external levee relief (10s metres) and channel belt aggradation prior to autogenic avulsion compared to rivers. Furthermore, they can aggrade to a superelevated position, i.e. the channel floor is perched many tens of metres above the surrounding slope. This is largely because the density contrast between sediment gravity flows and the ambient fluid is ~50 times lower in submarine channels than in rivers (Imran et al. 1998). The strong vertical density stratification in sediment gravity flows means that the velocity maximum is located much closer to the bed than in riverine flows (Sequeiros et al. 2010; Azpiroz-Zabala et al. 2017). External levees

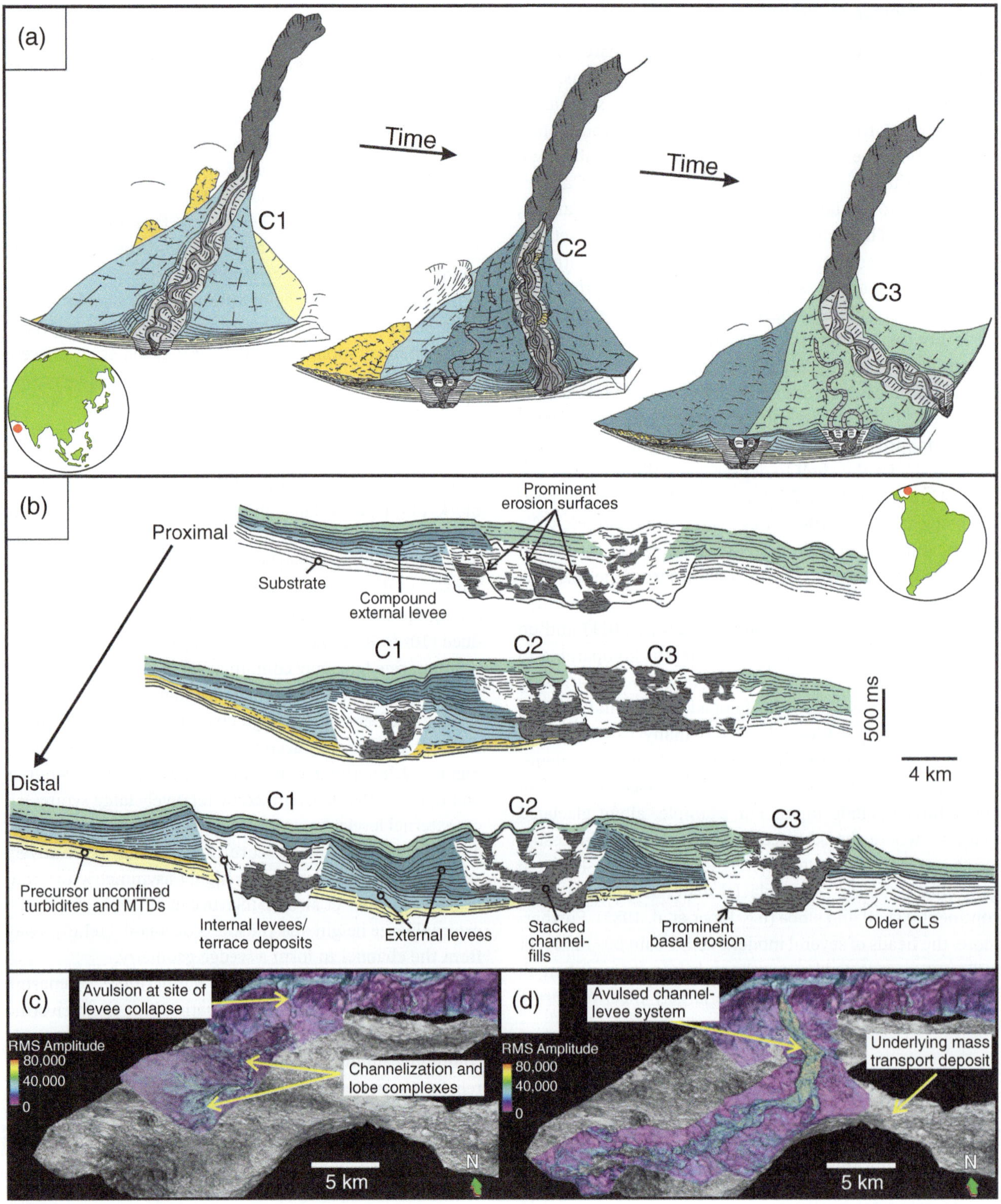

Figure 12.7 (a) Schematic illustrations showing a reconstruction of upper submarine fan avulsions and resulting architecture on the Indus fan (see Figure 12.4b). (b) Three transects across part of the Indus fan showing changes in fan architecture with increasing distance from the canyon-mouth. Highly self-cannibalised channel–levee systems (CLSs) with prominent erosive surfaces near the canyon-mouth pass down-system into well-preserved compensationally stacked CLSs, each bordered by prominent wedge-shaped external levees. (c and d) North-facing perspective views from seismic reflection data of amplitude extractions, overlain on variance extractions at the top surface of a submarine landslide on the Magdalena Fanoffshore Colombia. (c) A channelised segment develops downstream from the site of an external levee collapse. (d) A new submarine channel–levee system develops to the east of the lobe complex. *Source:* (a and b) Adapted from Deptuck and Sylvester (2018). (c and d) Adapted from Ortiz-Karpf et al. (2015).

are built by low-concentration turbidity currents that escape the confinement of their parent channel systems via flow stripping and overspill processes and then decelerate and deposit on the levee flanks (Peakall et al. 2000). Overspill occurs on both the outer and inner bends of the channel belt when the height of the sediment gravity flow is greater than the depth of the channel confinement. Flow stripping occurs preferentially on the outer bends of channel–levee systems, where the upper dilute and finer part of the density-stratified sediment gravity flows escape the channel confinement, and the denser, more concentrated and coarser part remains within the channel belt (Peakall et al. 2000). Because the more energetic and sediment-laden portion of the flow is nearer the channel base than the external levee crest (Hansen et al. 2015; Jobe et al. 2017), the potential for levee breaching and channel avulsion is limited (Jobe et al. 2020; Figure 12.7). Sediment waves can be well-developed on external levees at the outer bends of sinuous channels (Posamentier and Kolla 2003).

Submarine lobes form where sediment gravity flows exit channel confinement and lose their ability to carry some or all their load, and therefore deposit. The change from transport to deposition can happen suddenly where flow deceleration is greatest, such as at the channel-mouth (e.g. frontal lobes of Morris et al. 2014), where confining external levees are breached (e.g. avulsion lobes of Pirmez et al. 2000), or where there is a reduction in gradient on stepped slopes (e.g. transient fans of Adeogba et al. 2005). Typically, lobes are downstream elongate, high aspect ratio (width: thickness, 1000:1), and low-relief (a few metres) features with surface gradients <1°. The stacking pattern of lobe deposits at the distal ends of submarine channels is closely linked to the channel–levee avulsion history in both large (e.g. Jégou et al. 2008; Picot et al. 2016) and small (e.g. Piper and Normark 2001; Deptuck et al. 2008) fan systems.

12.2.5 Contourite Depositional Systems

In oceanic settings, contourites, "the sediments deposited or substantially reworked by the persistent action of bottom currents" (Stow et al. 2002; Rebesco et al. 2014), form a diverse, and volumetrically significant, component of deep-water detrital systems (Figure 12.8). Originally, the term contourite was applied to the deposits of geostrophic currents, flowing parallel to bathymetric contours (i.e. contour currents) that are part of the thermohaline circulation (Heezen et al. 1966). However, contourites are now considered to be deposits formed by any type of persistent current related to oceanographic processes flowing near the seafloor (i.e. bottom currents) including features dominated by erosion, winnowing and deposition. The term 'bottom currents' also comprises overflows, cascading of dense water, eddies, internal waves and other types of waves that can affect the seafloor in deep-water environments (Figure 12.8). Different processes may affect ocean currents at different spatial and time scales. For example, large-scale currents related to thermohaline circulation typically affect large areas (in the range of 1000s km^2) and show a long-term variability (hundreds to thousands of years) related to changes in climate. In contrast, internal waves related to tides have periods of 12–24 hours and usually affect smaller areas (0.1–100 km) (Talley et al. 2011). However, even major deep sea currents may be subject to tidal and seasonal variability (Bailey et al. 2024).

The association of different depositional and erosional contourite features in an area is called a 'contourite depositional system' (CDS), and the association of different adjacent CDSs, which may be stacked and/or emplaced beside one another, is named a 'Contourite Depositional Complex' (CDC) (Hernández-Molina et al. 2008). CDSs can cover >100 million km^2, and in some areas such as the eastern Argentine passive margin, can be found along the entire margin, from the shelf to the abyssal plain (Lantzsch et al. 2014; Hernández-Molina et al. 2010). Large-scale contourite features can be 10s to 100s km long, >100 km wide and up to 2 km thick, with reliefs of up to 1.5 km (Rebesco et al. 2014), with superimposed small-scale contourite features at sub-kilometre scale (Miramontes et al. 2021). The main depositional features are named 'contourite drifts' (Rebesco et al. 2014), which have various forms (Figure 12.8a), but are usually mounded muddy deposits where sediment accumulation is higher than in the surrounding environment.

Bottom currents can also affect the development of turbidite systems, forming mixed turbidite-contourite systems (Fuhrmann et al. 2020; see Section 12.5.4). This interaction can occur in different ways (Mulder et al. 2008; Fonnesu et al. 2020): (i) *Passive interaction*, where turbidite and contourite systems coexist in the same margin but develop independently; (ii) *Phased interaction*, where turbidity currents and bottom currents affect the same area but are not active at the same time, resulting in an alternation of sediment gravity-driven and bottom-current related deposits; and (iii) *Synchronous interaction*, where ocean currents interact with turbidity currents forming combined flows that have their own characteristics, distinct from both isolated bottom currents and sediment gravity flows.

12.2.6 Examples of Modern and Ancient Deep-Water Detrital Systems

12.2.6.1 The Congo Land-Attached Modern Canyon System

The Congo deep-water detrital system is located on the Congo-Angola passive margin (Figure 12.9). This system has been a focus for research because the incision of the

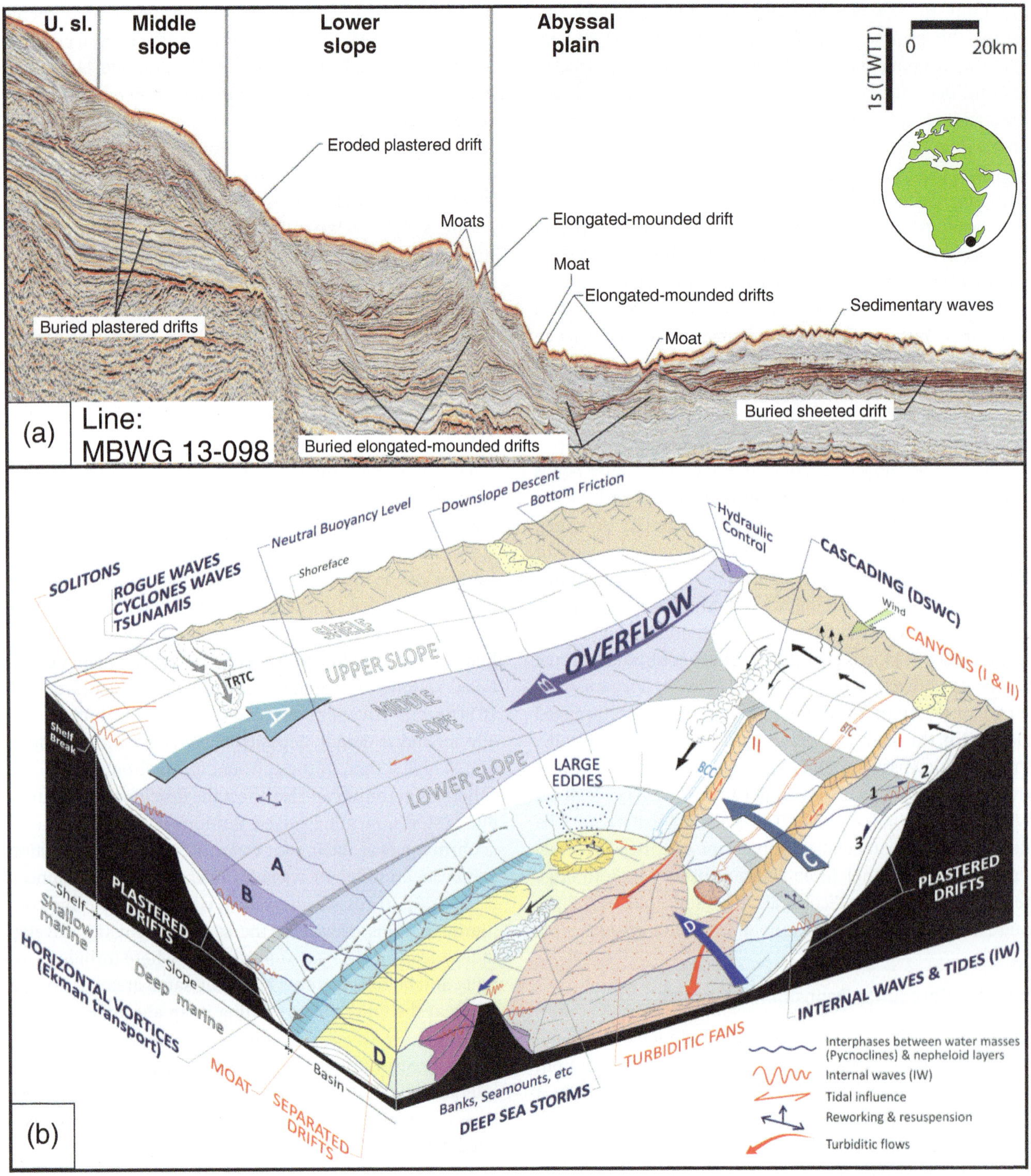

Figure 12.8 (a) An example of seismic reflection profiles of major morpho-sedimentary features of contourites, from the Mozambican continental margin. DSWC = dense shelf water cascade. U. sl. = upper slope; TWTT = two-way travel time. (b) 3D cartoon depicting the possible oceanographic processes in deep-water detrital environments. The seafloor current velocity can be affected by density currents and overflows, and by intermittent processes, such as cascading, eddies, and internal waves. *Source:* (a) is adapted from Thiéblemont et al. (2019) and (b) from Rebesco et al. (2014).

submarine canyon across the continental shelf and its direct physical connection with the Congo River means that it is very active. The submarine fan is also clearly imaged (Picot et al. 2016; Figure 12.9). The 4370 km-long Congo River is one of the largest rivers in the world in terms of fluid discharge (42,800 m^3 s^{-1}; Kinga-Mouzeo 1986). Annually, the river delivers ~43 Mt sediment and ~2 Mt of organic carbon into the canyon head (Baudin et al. 2020).

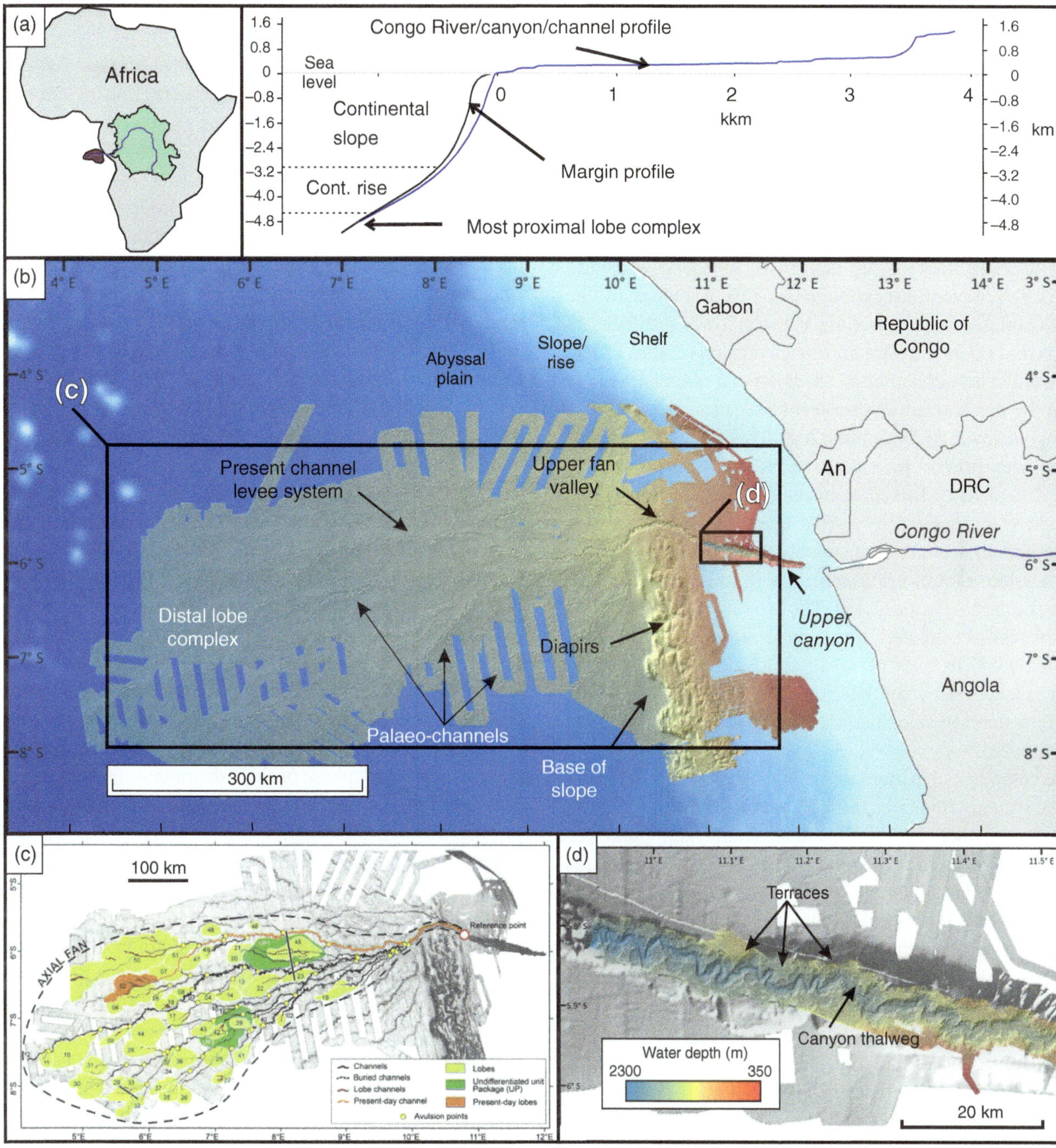

Figure 12.9 The Congo sedimentary system from source (Congo River and its catchment) to sink (Quaternary turbidite system). (a) Channel depth profile of the Congo River and the Congo Canyon and channel (blue). Vertical exaggeration x370. *Source:* Adapted from Picot et al. (2016). (b) Bathymetry map oceanward of the mouth of the Congo River shows the complicated pattern of sinuous submarine channel–levee systems emanating from the mouth of Congo Canyon, that together with lobes, build the Congo Fan. Congo Canyon/Fan. Basemap from GEBCO. An, Angola; DRC, Democratic Republic of Congo. *Source:* Adapted from Pope et al. (2022). (c) Map of the Late Quaternary Axial Congo Fan, with channels, lobes (numbered from oldest (1) to youngest (52), avulsions points and undifferentiated packages. Present-day channel and lobes are in orange. *Source:* From Picot et al. (2016). (d) Bathymetry map of the upper Congo Canyon. A sinuous canyon thalweg is present in the >800-m-deep canyon. Multiple terrace levels are visible along the canyon. *Source:* Adapted from Pope et al. (2022). Congo Canyon/Fan bathymetry Provided to Pope et al. (2022) by Ifremer.

The physical connection between the Congo River and canyon ensures a direct sediment conduit to the deep Atlantic Ocean at all sea levels (Babonneau et al. 2002). The canyon is >800 m deep, 6–12 km wide, and 130 km long (Figure 12.9). The canyon walls are marked by the headscarps of submarine landslides (Babonneau et al. 2010; Figure 12.5b). The canyon is characterised by numerous terrace surfaces, formed by a combination of erosion and sediment accumulation during overall incision of the canyon. The floor of the canyon is 50–800 m wide and marked by several knickpoints up to 10s m high that, like river rapids, are steep sections between lower gradient sections that migrate upstream though erosion (Figure 12.5b). One of the best-constrained examples of a sediment gravity flow in the natural environment was recorded from the Congo Canyon in December 2020. This flow is thought to have been primed by a surge in sediment transported by the Congo River linked to heavy rainfall and floods, and which was triggered by spring tides that increased pore pressures in accumulated sediments. This flow broke seafloor telecommunication cables as it travelled >1130 km into the deep Atlantic Ocean. The two-day 'canyon-flushing event' self-accelerated through entrainment of an estimated 1338–2675 Mt of substrate to increase the density contrast with seawater that helped drive the sediment gravity flow to reach velocities of 8 m s^{-1} (~29 km h^{-1}) (Talling et al. 2022).

A complicated pattern of sinuous submarine channel–levee systems emanates from the mouth of Congo Canyon. These, together with linked lobe complexes, build the Congo Fan which extends 800–1000 km westwards into the Atlantic Ocean, in 2000–5200 m water depth (Figure 12.9). Droz et al. (2003) and Marsset et al. (2009) showed that the architecture of the Congo Fan records successive major up-fan channel avulsions, and the channel–levee systems and lobe complexes are actually grouped into three fans; the Northern Fan (780–540 ka), the Southern Fan (540–210 ka), and the Axial Fan (210 ka to present). These three fans build a composite 400 km wide body. The present mean gradient of the seafloor of the Congo Fan decreases from a maximum of ~1.7° on the continental slope to <0.15° in the abyssal plain (Babonneau et al. 2002). Picot et al. (2016) identified 52 channel–levee systems in the Axial Fan alone, with channels that range from 300 to 1000 km in length. Only one channel–levee system of the Congo Fan is active at any given time (Droz et al. 1996), with 2–9 channel avulsions every 10 kyr for the last 38 kyr (Picot et al. 2019). Most of the channel–levees are documented to feed slightly elongated lobe complexes that show internal radial or progradational stacking patterns, which in cumulative volume reach 2600 km^3, and represent ~30% of the volume of the Axial Fan (8500 km^3).

12.2.6.2 Exhumed Submarine Fans of the Karoo Basin, South Africa

A major advance in understanding deep-water detrital systems has been the systematic documentation of ancient exhumed systems over large areas (>1000 km^2). In the Laingsburg depocentre, Karoo Basin, South Africa, several Permian sand-rich submarine fans (Units C-G), have been mapped over a 2500 km^2 area (Van der Merwe et al. 2014) from entrenched slope valleys through channel–levee systems to basin-floor lobe complexes. Each sand-rich submarine fan system is overlain by a 10–30 m thick regional mudstone unit, which comprises mud-rich sediment gravity flow deposits with minor hemipelagic and pelagic components (Boulesteix et al. 2022). The scale of the systems suggests there were discrete periods of coarse-grained sediment supply interspersed with periods of near-complete sand starvation.

In proximal locations, Units C and D are characterised by channel–levee systems and interpreted to represent lower to middle slope settings, respectively (Figure 12.10a). Unit C is a levee-confined channel system with partial preservation of frontal lobe deposits (Hodgson et al. 2011; Morris et al. 2014). Unit D represents an entrenched channel-system that cuts down into Unit C. Hodgson et al. (2011) identified a common horizontal stacking (lateral migration) followed by vertical stacking (aggradation) of component channel-fills. This channel stacking pattern was interpreted to record an initial phase of slope degradation, with formation of external levees and a composite bounding erosion surface, through a sediment bypass dominated period with horizontal stepping of channels and widening of the slope valley, to slope aggradation, with contemporaneous vertical stacking and internal levee development.

Each fan passes down-dip over 10s of km from channel–levee systems into sand-rich lobe-complexes, which each being several 10s km^3 in volume (Figure 12.10b). The stratigraphic architecture of Units C and D show a common pattern of channel lengthening and basinward growth, followed by backstepping and abandonment. This results in a typical stratigraphic expression of lobe complexes overlain by external levee deposits (Figure 12.10c). The Karoo Basin deep-water system has a strikingly narrow grain-size range – it is dominantly a lower fine sand to coarse silt system. The sediment gravity flows in the Karoo channel–levee systems were likely powerful enough to transport coarser material, given the size of the rip-up clasts in channel axis lag deposits, but the coarsest grain size observed is lower medium sand. Even so, there is a marked stratigraphic segregation of grain-sizes in the submarine fans, with mud-rich deposits preferentially in the external levee successions, and sand-rich deposits in the lobe complexes. This segregation is linked to sedimentary transport processes – the finest

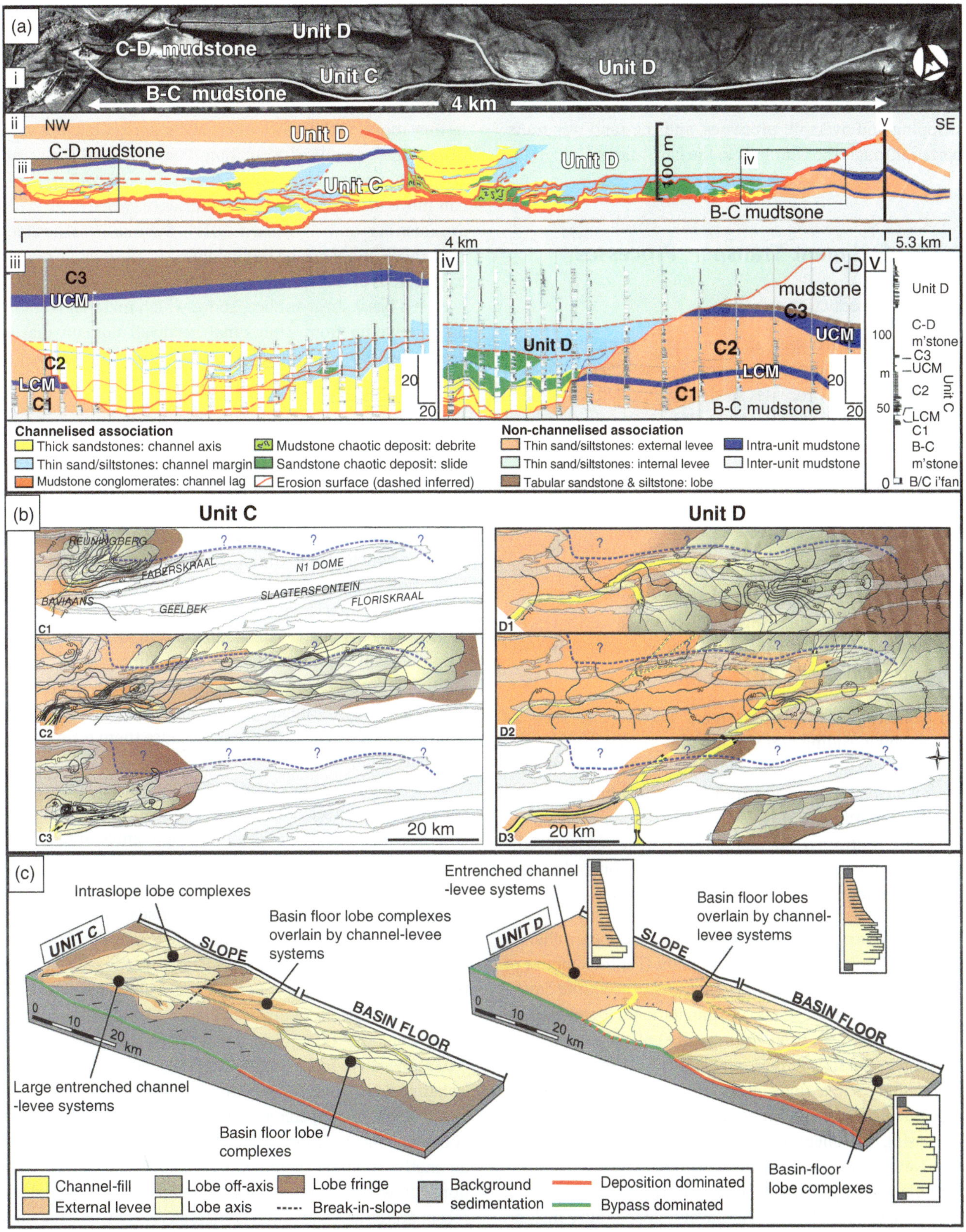

Figure 12.10 Compilation figure of two exhumed submarine fan systems, Units C and D, documented over a 4500 km^2 area in the Karoo Basin, South Africa. (a) (i) Aerial photograph and (ii) schematic correlation panel based on physical correlation of c. 200 closely spaced logged sections (see Hodgson et al. 2011 for detailed correlation panel). The panel demonstrates that an entrenched submarine channel–levee system (Unit D) partially removed an older levee-confined submarine channel–levee system (Unit C). (iii and iv) Insets of the NW and SE margins of Unit C with the detailed stratigraphy labelled (see (ii) for location). (v) A measured section from the study area (see (ii) for log location). UCM, upper C mudstone; LCM, lower C mudstone. (b) Threefold evolution for Units C (left) and D (right). Outcrop area in grey, and thickness is shown as isopachs with units in metres. Facies maps represent gross depositional environments for the given time intervals. Note the shared pattern of progradation and retreat. (c) Block model reconstructions for Units C and D at their most basinward position, with schematic logged sections (outside of channel axes) to illustrate typical stacking patterns of lobe and levee successions. *Source:* (a) from Hodgson et al. (2011) and (b) and (c) from Brooks et al. (2018).

particulate material was preferentially sequestered as muds (clay and silts) in external levee successions through flow stripping and overspill processes, and the sand was transported through the channels to the basin-floor.

12.3 Sediment Transport Processes

12.3.1 Introduction

The last decade has seen an unprecedented increase in the quantity and quality of monitoring of sediment transport and depositional processes in deep-water detrital settings (Clare et al. 2020; Figure 12.2). The data collected have revealed a wide range of flow structures and rheologies, and that flows can readily transform from one state to another. It is also clear that minor differences in the mix of grain-sizes and composition (a few %), and subtle changes in seafloor bed topography (fractions of a degree), can have profound impacts on flow behaviour, and therefore the patterns of erosion and deposition.

We consider the wide range of flow behaviours that transport sediment in deep-water detrital systems using the following subdivision:

i) Cohesive and coherent mass movements, such as creeping, sliding and slumping.
ii) Sediment gravity flows, such as debris flows and turbidity currents, and their transformations to transitional flows and hybrid flows.
iii) Bottom currents that include any type of persistent current related to oceanographic processes operating near the seafloor.

Cohesive and coherent mass movements are characterised by limited internal deformation (i.e. the amount of internal strain is relatively low) and occur when the downslope-oriented shear stress (the gravitational force) exceeds the shear strength (the resisting force) of the sediment(s), causing instability, failure, and sometimes movement, of sedimentary successions. Slope instability can be triggered by reducing the shear strength of the sediments or/and by increasing the shear stress. These mass movements are non-Newtonian – that is the moving sediment mass has variable viscosity when shear is applied. The preconditioning and triggering factors that favour slope failure include earthquakes, rapid sedimentation, oversteepening by basal erosion or diapirism and pore-pressure build-up, driven by, for example, the trapping of gas. Preconditioning factors influence the stresses and/or reduce the shear strength, while the triggering factor is the short-lived process that ultimately provokes destabilisation.

Mass movements are dominated by cohesive forces, they therefore maintain the coherence of the original stratigraphy as they move, with limited internal disaggregation, but with plastic and/or brittle deformation (Figure 12.11a). They can be quasi-instantaneous, or like their subaerial counterparts, can move incrementally over centuries (e.g. Gee et al. 2006; Posamentier and Martinsen 2011). Submarine landslides remobilise and displace sedimentary successions of metres to hundreds of metres thickness. They result in deposits that overlie a basal shear surface, above which material is translated downdip from a proximal headwall domain that is dominated by extensional deformation, to a distal toe domain that is dominated by compressional deformation (Figure 12.6a). The resulting range of deposits is referred to collectively in Section 12.4 as submarine landslides.

Sediment Gravity Flows are a type of underwater bottom-hugging particle-laden and gravity-driven current (Lowe 1982; Kneller et al. 1999; Mulder and Alexander 2001). These flows are particle (sediment grains) and water mixes in which, due to the excess density of the particulate-laden flow relative to the ambient fluid, the sediment component drags interstitial water down slope under the action of gravity (Bagnold 1962; Middleton and Hampton 1973). An important distinction between a sediment gravity flow and other turbulent unidirectional flows, such as riverine flows, is that the sediment gravity flow will stop once there is no more sediment to transport because there is no longer a density contrast.

The ambient fluid might be freshwater in lakes (Chapter 6) or saline water in seas and oceans. The sediment gravity flow will exchange ambient water through entrainment and interstitial water through detrainment. The sediment mass will also change through erosion and deposition, thereby modifying the unit weight, and the density contrast, of the flow as it moves. These exchanges between the flow, the ambient fluid, and the substrate result in a wide range of flow types, which change in both time and space, such that their depositional record is diverse (Figure 12.11d). To simplify these dynamic flows, we follow Pickering and Hiscott (2015) by subdividing sediment gravity flows into laminar '*cohesive flows*' (debris flows and mudflows) and turbulent '*frictional flows*' (hyperconcentrated density flows, and high- and low-concentration turbidity currents) based on their distinctly different rheological behaviours. In short, frictional flow behaviour is directly related to the turbulent mix of grains and water and is characterised by incremental (grain-by-grain) deposition, whereas cohesive flows have matrix strength arising from their high mud content, have laminar streamlines meaning that the particles are transported with little mixing, show a constant velocity profile, and deposit *en masse* (Figure 12.12a). Although debris flows can be highly mobile and travel long distances over low

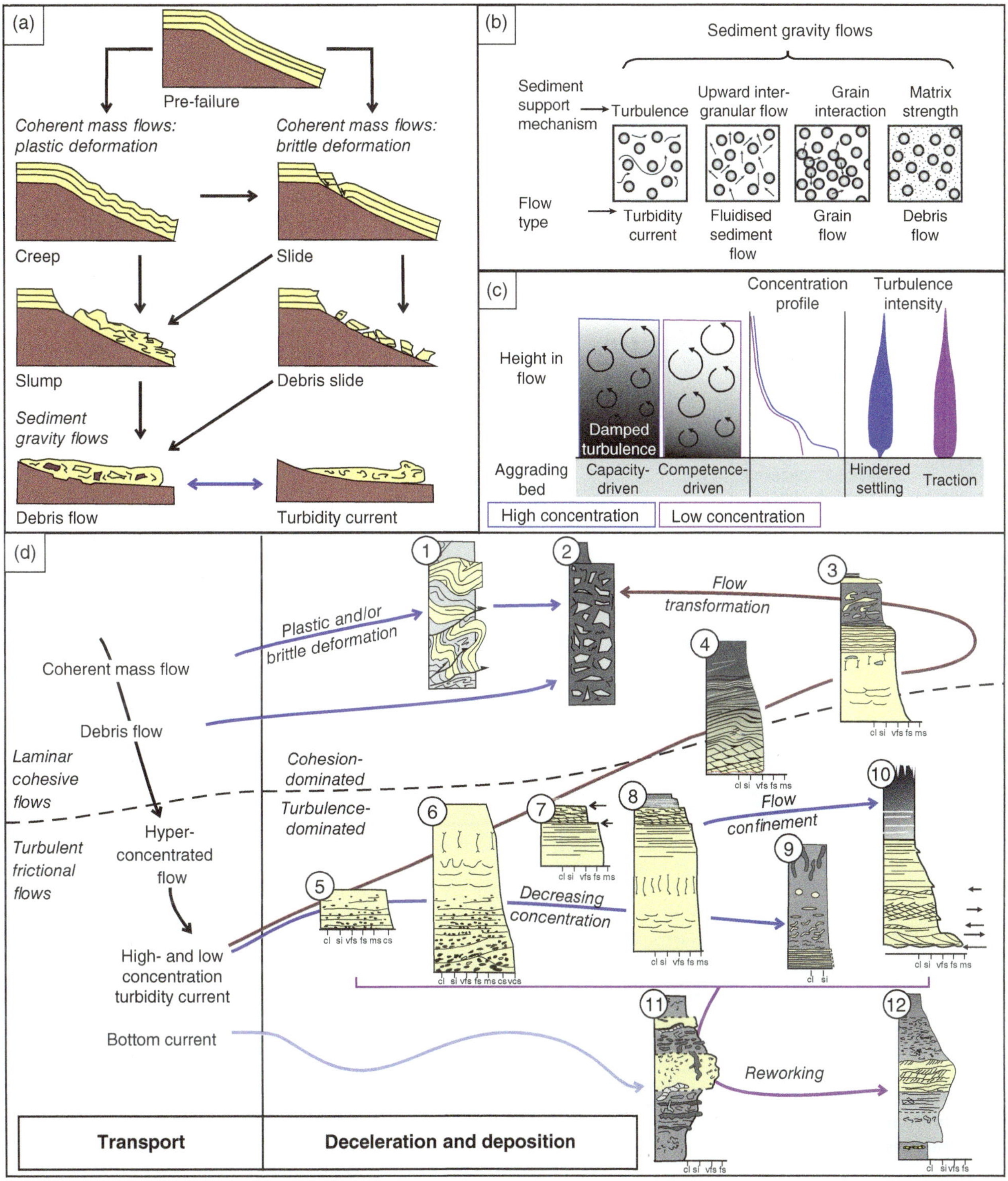

Figure 12.11 (a) Schematic diagram of end-member gravity-driven flows from coherent mass flows to turbulent frictional sediment gravity flows. *Source:* Modified from McHugh et al. (2002). (b) Sediment gravity flows differentiated according to dominant sediment support mechanism. *Source:* Modified from Middleton and Hampton (1973). (c) Contrasting near-bed grain support mechanisms between high (blue) and low (purple) concentration turbidity currents. Both types are turbulent away from the sedimentation surface (arrows are turbulent eddies). In high-concentration flows, grain interactions are strong at the aggrading bed, and turbulence damping inhibits the development of tractional structures that develop under low-concentration flows. *Source:* Adapted from Pickering and Hiscott (2015). (d) Summary of the evolution of sediment gravity flows, and other deep-marine transport processes, by their long-distance transport mechanism, and the flow transformations possible during deceleration and deposition. Bed types: 1 = slide/slump; 2 = debrite; 3 = hybrid event bed (Haughton et al. 2009); 4 = unidirectional transitional flow deposit (Taylor et al. 2024); 5 = 'top missing' turbidite suggesting sediment bypass dominated deposit (Stevenson et al. 2015); 6 = high-concentration turbidity current deposit (HCTCD; Lowe 1982); 7 = grain size breaks (arrows) in turbidite, suggesting sediment bypass-dominated deposit (Stevenson et al. 2015); 8 = low-concentration turbidity current deposit (LCTCD; Bouma 1962); 9 = mud turbidite (Piper 1978); 10 = reflected and ponded turbidite (Pickering and Hiscott 1985). Arrows indicate changing flow direction; 11 = Contourite bi-gradational sequence (Gonthier et al. 1984); 12 = bottom current reworked sands (De Castro et al. 2020).

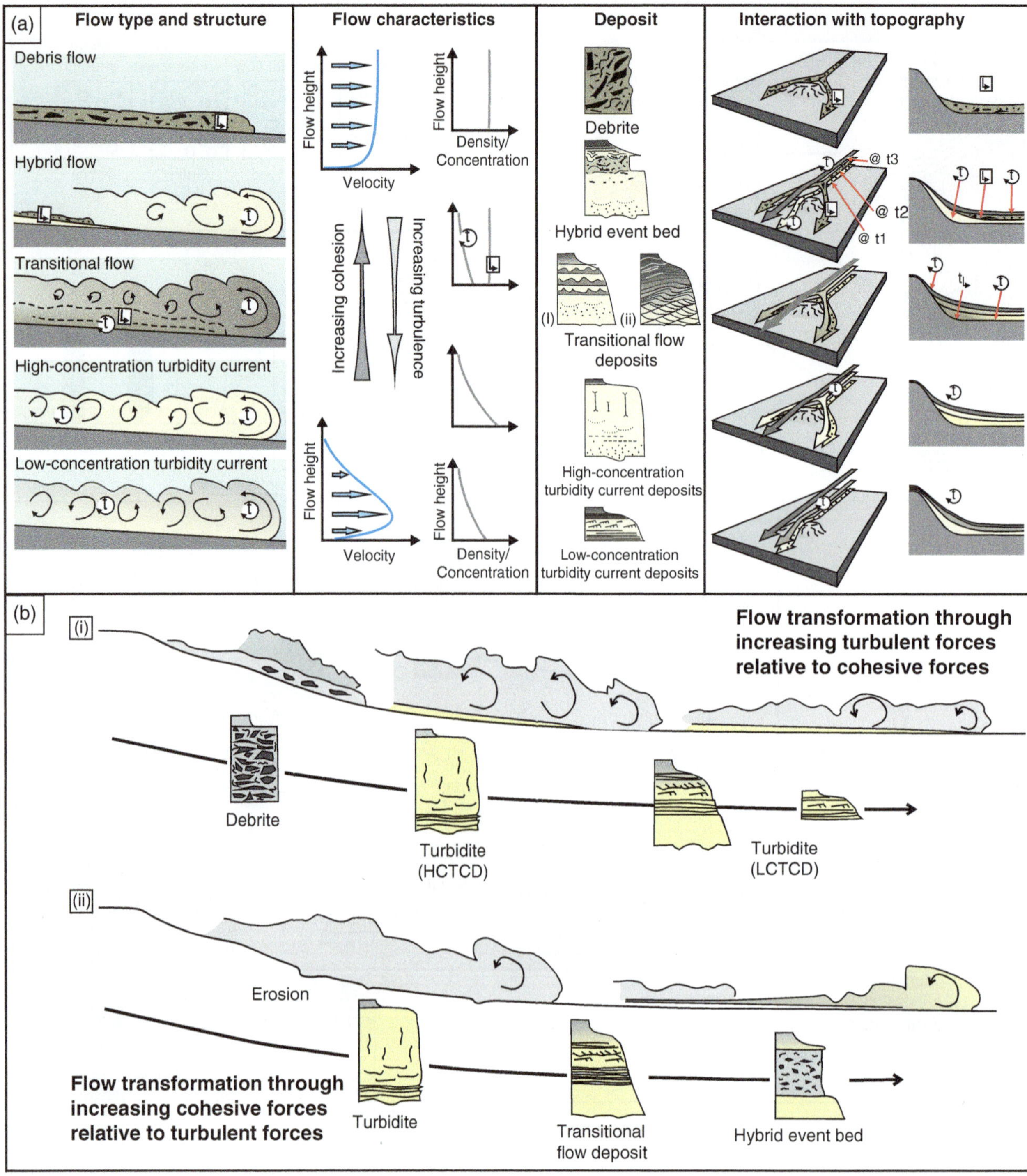

Figure 12.12 (a) Classification of sediment gravity flows based on their rheological properties, illustrating that different flow types create different deposits, and according to their velocity and density profiles may interact differently with seafloor topography. Laminar flows are ground-hugging and tend to follow topographic obstructions, whereas turbidity currents may be able to surmount topography. Well-stratified flows may be more inclined to follow topography, whereas well-mixed (less or unstratified) turbulent parts may surmount higher obstructions. In the case of cogenetic turbulent-laminar flows, the laminar or less turbulent parts of the flow may have a different onlap geometry to that of the associated turbidite sands. Figure from Bakke et al. (2013) based on the flow classification of Haughton et al. (2009) and the topographic interaction model of Al-Ja'aidi (2000). T = turbulent; L = laminar. Transitional flow deposits: (i) deposits of waning flow (little clock images) and (ii) more sustained flow (modified from Taylor et al. 2024). (b) Styles of flow transformation. (i) Increasing downstream flow dilution through entrainment of ambient water, leading to decreased cohesive forces relative to turbulent forces. HCTCD = High concentration turbidity current deposit, LCTCD = Low concentration turbidity current deposit. (ii) Increasing downstream flow concentration and decreased turbulence. *Source:* Adapted from Haughton et al. (2009).

gradients (<1°), they tend to come to an abrupt halt. Laminar flow conditions may also occur in cohesionless (i.e. mud-poor) flows, if sediment concentrations are sufficient for grains to interlock and give the flow frictional strength (Amy et al. 2005).

Submarine mass movements and sediment gravity flows can undergo significant flow transformations (Fisher 1983; Figures 12.11d and 12.12b). Laminar cohesive flows can transform into turbulent frictional flows, and vice versa, as well as turbulence-modulated transitional flows (Baas and Best 2002; Sumner et al. 2008; Baas et al. 2009, 2011, Figures 12.11 and 12.12). Therefore, a continuum of flow rheologies exists, and this is reflected in the range of deposits (Section 12.4, Figure 12.11d). The starting volumes, strengths, and grain-size mixes are inherently highly variable. Clay particles carried in sediment gravity flows can aggregate to form floccules and gels (Winterwerp and Van Kesteren 2004). These increase cohesive forces relative to turbulent forces during flow deceleration and can either enhance or suppress turbulence (Baas and Best 2002; Baas et al. 2009, 2011; Figure 12.12b). Nonetheless, the evolution of sediment gravity flows has been deduced from destruction of underwater infrastructure, such as submarine cable breaks which occurred during the 1929 Grand Banks event offshore eastern Canada (Heezen and Drake 1964), and the 1979 event offshore Nice, France (Mulder et al. 1997). Using these events, we can consider the initiation, evolution and deposition of a 'typical' sediment gravity flow. (i) Flow initiation, which may or may not require a trigger, where the flow may begin as a cohesive flow or a dilute turbulent suspension, or anything in-between. (ii) Early-stage flow transformation, during which ambient water is entrained and, if initiated as a cohesive flow, the original sedimentary pile disaggregates, and the flow accelerates to become a turbidity current. (iii) Flow stabilisation, if topography allows, when the flow has formed a coherent structure (head, body and tail), and might accelerate or decelerate, but can travel long distances. (iv) Late-stage flow transformation and deposition, when the flow may simply decelerate as a dilute turbidity current or may entrain clay-rich substrate and (partially) transform to a cohesive flow.

The term 'bottom current' has been used to define the ocean currents that flow near the seafloor (Rebesco et al. 2014; Figure 12.8). The circulation patterns of the deep ocean are driven by differences in water density, principally thermohaline effects controlled by temperature ('thermo') and salinity ('haline'). To build large-scale contourite drifts, it is necessary that the flow conditions that favour the localised accumulation of sediments in the contourite drift remain relatively stable over thousands to millions of years. This does not mean that the near-bottom flows are constant, and they can in fact be highly variable (e.g. Bailey et al. 2024). For example, periods of intense bottom current activity can be related to the passage of eddies and result in sediment resuspension. Such events are often called benthic storms (Gardner et al. 2017). Relatively short duration and extreme events can generate enough bed shear stress to induce sand transport. Separated mounded drifts parallel to the slope form from along slope currents, which have a secondary across-slope flow related to Ekman transport (Section 13.2.1; the result of the combined effects of Coriolis force and bottom friction) (Wilckens et al. 2023; Zhao et al. 2024).

12.3.2 Basic Process Characteristics of Sediment Gravity Flows

Cohesive mass flows, sediment gravity flows and bottom currents work both independently, and together, to redistribute detrital sediment in deep-water settings. We focus here on the key process characteristics of sediment gravity flows in more detail due to the variety of sediment gravity bed types identified in the rock record (Figure 12.11d), their great diversity when compared to mass flows, and the established process understanding and links to deposits when compared to less well understood bottom currents and contourites. These characteristics include flow structure, flow acceleration and deceleration in time and space (uniformity and steadiness, respectively), flow concentration and stratification, grain support mechanisms during transport and processes at the time of deposition (Figures 12.11–12.13).

Once a sediment gravity flow initiates, its evolution depends on the interplay between gravitational and frictional forces, which control whether the flow continues moving, or stops. We will consider debris flows and their deposits, debrites and turbidity currents, and their deposits, turbidites, separately as laminar cohesive flows and turbulent frictional flows, respectively. Within the process continuum between these endmembers there are several kinds of sediment gravity flows (Figures 12.11d and 12.12): cohesive and non-cohesive debris-flows (Talling et al. 2012), hyperconcentrated flows (hyperconcentrated density flows of Mulder and Alexander 2001), high-concentration turbidity currents (HCTC; Lowe 1982), low-concentration turbidity currents (LCTC; Bouma 1962), mud-rich flows (Piper 1978) and composite and transitional flows (Haughton et al. 2009; Taylor et al. 2024). This subdivision of flow types is used to link back to deposits in the rock record (Section 12.4).

12.3.2.1 Flow Structure

Experimental turbidity currents display grain-size segregation through the height of the flow due to density

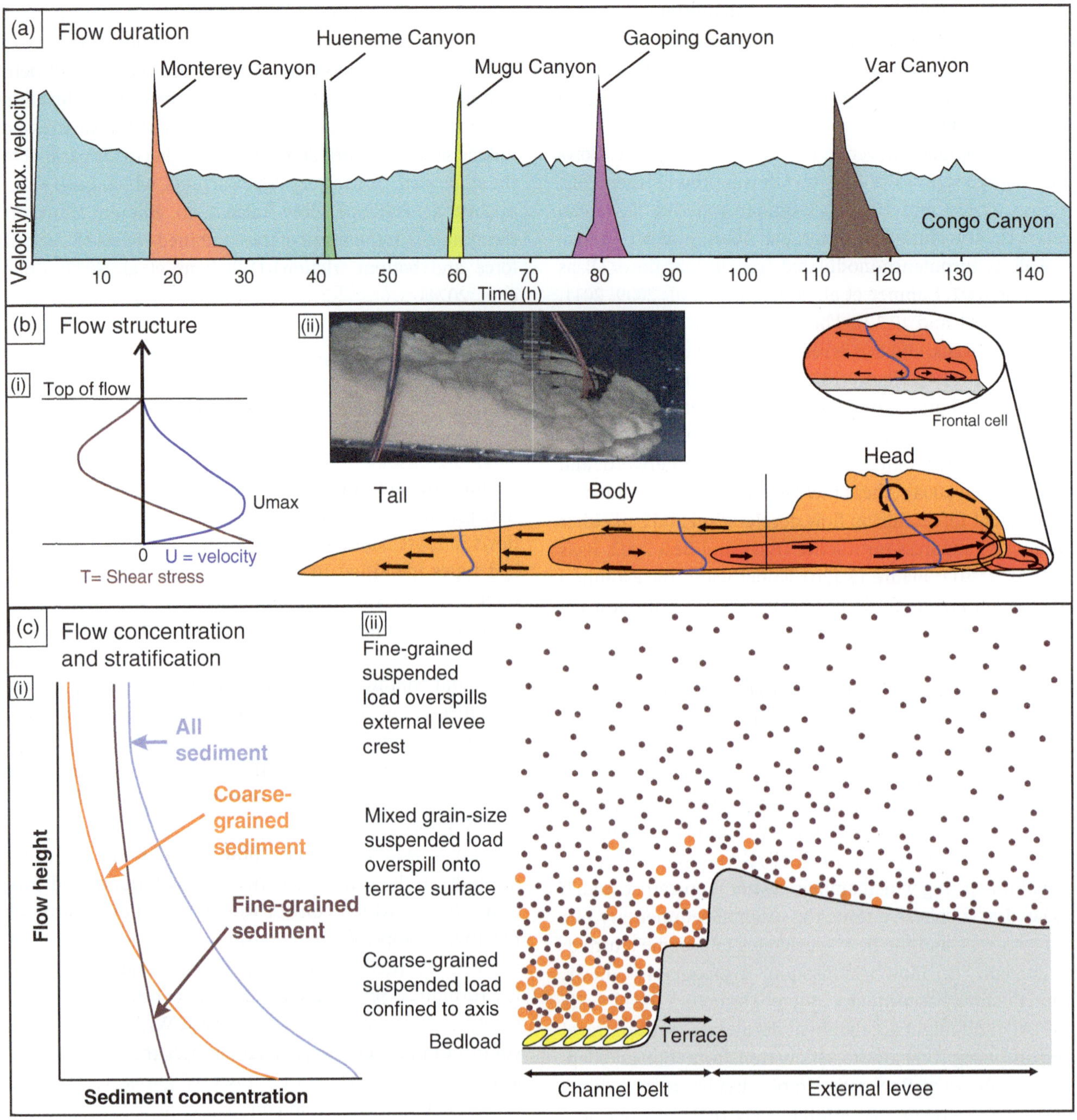

Figure 12.13 Summary figure of key attributes of fluidal sediment gravity flows. (a) Compilation of measured flow duration from natural sediment gravity flows. (b) Compilation of flow structure. (bi) Turbidity current structure with maximum shear stress on the upper body of the current and maximum velocity (Umax) of the current towards the base. (bii) Flow structure of a sediment gravity flow, comprising head, body and tail, with a finite-volume (surge-like) release. Black arrows denote the relative movement of sediment-laden fluid with respect to the flow front. Velocity profile shapes are shown by blue lines. Inset illustrates that the head may, in some cases, be the fastest part of the flow, shown by a high-concentration, fore-running basal flow cell. Photograph of an experimentally produced turbidity current. (c) Compilation of flow concentration and stratification. (ci) Schematic diagram of a bi-disperse sediment suspension illustrating the origin of flow stratification. (cii) Schematic diagram of channel belt, terrace and external levee showing the interaction of a stratified flow with topography. *Source:* (a bi) Adapted from Azpiroz-Zabala et al. (2017). (bi and c) Adapted from Hansen et al. (2015).

stratification, i.e. a vertical decrease in sediment concentration (Kneller 1995; Kneller and Buckee 2000; Figure 12.13). Theoretical and experimental research often simplifies the density stratification of turbidity currents into a two-layer flow with a density interface separating a lower dense layer that interacts with the substrate and an upper dilute layer (e.g. Middleton 1967; Kneller and Buckee 2000). Based on laboratory-scale experiments (Figure 12.13), models for the horizontal structure of sustained turbidity currents comprise an unsteady flow front (the head), followed by a steadier period of flow (the body) that finally wanes (the tail). The head displaces the surrounding ambient fluid meaning that it is slower and thicker than the body. The faster-moving body feeds the head with sediment-laden fluid. The highest sediment concentrations occur in the head and lower part of the body (Figure 12.13c). The sediment concentration and flow velocity and thickness all decline within the tail of the flow.

Monitoring of turbidity current behaviour in the Congo Canyon suggests some distinct differences in the internal structure of natural flows compared to laboratory flows. First, the Congo Canyon turbidity currents comprise a short-lived zone of faster, denser and coarser-grained flow, referred to as the 'frontal-cell' (Azpiroz-Zabala et al. 2017; Figure 12.13b). The frontal-cell is thinner than the trailing body, and velocity data suggest that the body does not feed the frontal-cell with sediment-laden fluid. Instead, sediment-laden fluid is shed from the frontal-cell into the trailing body implying that the frontal-cell must continuously entrain substrate to replenish the sediment lost into the trailing body. That the frontal-cell can erode and incorporate sediment from the canyon floor, thus becoming self-sustaining, was confirmed by the December 2020 Congo Canyon event where an estimated 2.68 km^3 of substrate was entrained (Talling et al. 2022). Whether or not these coarse-grained turbidity currents with high-concentration near-bed layers are restricted to submarine canyons needs to be tested.

12.3.2.2 Flow Uniformity and Steadiness

Sediment gravity flows can accelerate and decelerate in time and space through changes in various boundary conditions. Changing flow velocities control sites of erosion, sediment bypass (neither erosion nor deposition at a given point), and deposition, as well as the deposit character. A flow that changes velocity through space, along its flow path, is described as *non-uniform*. A flow that changes velocity through time, in reference to a fixed observational point, is *non-steady*. Theoretically, a flow that does not change velocity in time or space, a *uniform and steady* flow, is in 'bypass' mode (Stevenson et al. 2013).

Whether a flow accelerates (accumulative and waxing flows of Kneller 1995) or decelerates (depletive and waning flows of Kneller 1995) is controlled by several interacting factors. Increases in slope gradient and/or flow confinement will increase flow momentum and flow height, respectively, and lead to erosion and bypass (accumulative conditions). In contrast, decreases in slope gradient and/or flow confinement will lead to the flow being depositional (depletive conditions).

From a sedimentological point of view, a decelerating flow is important to understand as such flows are likely to deposit. Deposition can start through spatial and/or temporal slowing, or a combination. A deceleration through space along a flow path is described as a 'non-uniform and *depletive*' flow. A deceleration through time at a fixed observational point is a 'non-steady and *waning* flow'. A flow that is depletive and waning will deposit.

A flow that accelerates along its flow path (through space) is a non-uniform and *accumulative* flow, whereas a flow that accelerates through time is an unsteady and *waxing* flow. A flow that is accumulative and waxing will not deposit, and depending on the physical properties of the substrate, might sculpt an erosion surface. Identification of erosion surfaces recording the erosion and bypass of a single flow, or of many flows (a composite surface), means that the formative conditions of the surface are unrelated to the overlying deposit(s).

Sediment gravity flows can behave in a non-uniform manner through substrate interactions. For example, when turbidity currents erode sediment at a higher rate than they deposit sediment they increase in mass, become denser, and thus accelerate. The bulking-up of the flow can drive increased erosion, and further acceleration. This process is termed flow *ignition* (Parker et al. 1986). Alternatively, turbidity currents that deposit more sediment than they erode will decelerate, driving more deposition in a process termed *dissipation* (Parker et al. 1986). Where flows strike a balance between erosion and deposition, and therefore concentration is quasi-constant, they are said to be in equilibrium. This near uniform flow state has also been termed *autosuspension* (Pantin 1979) and has been invoked to explain the very long runout distances of frictional flows (100s km).

12.3.2.3 Sediment Concentration

The mixtures of sediment and water in a sediment gravity flow can only develop, and sustain, turbulence if the concentration is low enough. If sediment concentration is too high then turbulence is limited, or *dampened*, which means that particles with typical mineral densities cannot be suspended and transported far. Frictional flows like turbidity currents exchange particles with the substrate via deposition and erosion, and they exchange fluid principally with the ambient water (Parsons and Garcia 1998). Deposition of sediment decreases flow concentration, and therefore the excess density of the flow, resulting in deceleration. In contrast, erosion and entrainment of seafloor sediment by the flow leads to increased concentration and/or excess density, and acceleration. Ultimately, the flow could transform into a laminar cohesive flow (Figures 12.11a,d and 12.12b). Therefore, the concentration of a sediment gravity flow is a key parameter to consider.

Unfortunately, sediment concentration has never been measured directly in a long run-out flow beyond the continental slope. Although the bulk (depth-averaged) flow density of natural turbidity currents remains poorly known, the best estimates range from <0.1% to ~0.2% (Konsoer et al. 2013; Simmons et al. 2020). The depth-averaged sediment concentration of a turbidity current in the Congo Canyon that lasted one week in 2010 (Figure 12.13a) was estimated as just 0.02% (Azpiroz-Zabala et al. 2017).

Here, we follow the convention derived from bed characteristics in the field and subdivide flows into high-concentration (or -density) and low-concentration (or -density) turbidity currents (Figure 12.11d). Hyperconcentrated density (or gravity) flows (Mulder and Alexander 2001; Figure 12.11d), which are also referred to as inflated sandflows (Pickering and Hiscott 2015), liquefied flows (Middleton and Hampton 1973), cohesionless debris flows (Postma 1986) and density-modified grain flow (Lowe 1976) are sand-laden laminar or pseudo-laminar flow types that rapidly deposit. These flow types and their deposits remain poorly understood, but are thought not to have cohesive strength, are unlikely to have long runout distance and can 'freeze' as pore pressure dissipates at flow margins. The reader is pointed to Talling et al. (2012) for a comprehensive discussion of this flow type.

12.3.2.4 Grain Support Mechanisms

Sediment gravity flows can be differentiated by their dominant grain-support mechanism (Middleton and Hampton 1973; Figure 12.11b). Here, these processes are briefly outlined, working from more concentrated to more dilute flow types. Debris flows are at one end of the spectrum, where large grains and gravels are supported by the cohesive and frictional strength of the matrix comprising interstitial fluid and fine-grained sediment (Middleton and Hampton 1973). In turbidity currents, sediment is predominantly supported by the upward component of fluid turbulence (Middleton and Hampton 1973) and hindered settling (Richardson and Zaki 1954).

Matrix strength refers to a physical property of concentrated and poorly sorted sediment and water mixtures, where deformation is resisted by frictional strength between grains and cohesive strength from electrostatic attraction between clay and silt particles. Flows with matrix strength are non-Newtonian, which, in this case, means that a critical shear stress needs to be applied before they will flow, and that the flow will stop moving if the critical shear stress is higher than the downslope component of gravity. These flows are laminar, in that the particles are transported with limited collisions and little mixing. The matrix strength of a flow also means that larger particles will not settle as they must overcome frictional and cohesive strength to push material out of the way.

Grain interactions in high-concentration grain flows occur as faster moving particles collide with slower moving particles, leading to some of their downstream momentum being transferred into an upward dispersive pressure (Bagnold 1956). This results in an expanding, or dilating, mass of grains that reduce grain-to-grain friction.

Buoyancy is a force that provides support to a particle in a denser surrounding fluid. Positive buoyancy means a particle will float towards the upper surface of a flow. This can be an important factor in dense sediment and water mixtures where flows can carry large particles (Hampton 1979).

Excess pore pressures can occur when grains deposit rapidly to form a low permeability interface that hinders upward escape of interstitial pore space fluids. This creates a situation of excess fluid pressure in the pore space, which keeps the grains apart and reduces friction. If a fluid escape route is found through the low permeability interface, the finer and lower density particles are preferentially transported upwards towards sites of lower pressure.

Turbulence is a crucial, if complex, component of sediment transport in low-viscosity fluids such as turbidity currents (Mulder and Cochonat 1996; Kneller and Buckee 2000; Meiburg and Kneller 2010; Figure 12.11b) where inertial forces dominate viscous forces. Turbulence also imparts eddies and velocity fluctuations that disturb the downstream velocity of suspended particles. The settling velocity of individual non-cohesive particles depends on the particle size, shape and density distribution, the liquid density and the viscosity of the fluid (see Stoke's Law, Section 13.2.4). The presence of adjacent particles in a suspension results in 'hindered settling', in which flow around neighbouring falling particles causes a larger fluid drag than for a single falling particle. The effect is a lower fall velocity for individual particles in the suspension (Richardson and Zaki 1954; Baas et al. 2022). The hindered settling effect can reduce the settling velocity by up to several orders of magnitude for particles transported in dense suspensions (Richardson and Zaki 1954).

12.3.2.5 Deposition from Frictional Versus Cohesive Flows

In most sediment gravity flows, the particles are transported in suspension and spend very little time at the interface between the base of the flow and the bed. Particles in frictional flows are carried in a dilute turbulent suspension. Deposition from suspension is selective because the particles settle onto the bed according to their size, shape and density. This grain-by-grain settling out from the flow is referred to as *incremental deposition* (Talling et al. 2012). In contrast, in more cohesive and concentrated flows the particles are not free to move during transport. Deposition occurs *en masse*, that is the particles deposit all together (Talling et al. 2012), through increased resistance to flow during deceleration, which is driven by cohesive forces from mud, or increased friction between particles. Therefore, there is a fundamental difference in the way that concentrated non-turbulent (laminar) cohesive flows deposit particles and evolve compared to turbulent frictional flows.

Deposition from decelerating turbulent frictional flows is often described in terms of decreasing flow 'competence' and flow 'capacity'. Competence is a measure of the largest grain size that can be transported. Competence-driven deposition occurs when, as velocity decreases, the flow becomes progressively incompetent to carry its coarsest suspended sediment load, which results in grain-by-grain settling of that fraction. In progressively decelerating flows, this produces well-sorted normally graded deposits. A sediment gravity flow at full capacity carries as much suspended load as the turbulence intensity allows. Capacity-driven deposition occurs as turbulence intensity decreases and the suspended load at the base of the sediment gravity flow surpasses a flow's capacity. The excess suspended sediment load results in rapid (but not *en masse*) deposition (Allen 1991; Hiscott 1994a). The basal parts of sediment gravity flows will contain a wide spectrum of suspended grain sizes and densities (e.g. Garcia 1994), so the flux of material to the aggrading bed under capacity-drive sedimentation will yield a poorly sorted deposit that resembles the sediment composition at the base of the flow.

There is also a marked difference in where deposition starts. In turbulent frictional flows, the particles deposit on the bed, and aggrade from the base upwards, and the grains are available for tractional reworking by the overriding flow. This grain-by-grain deposition commonly results in a normally graded profile in grain-size as flow energy decreases. In contrast, in a cohesive flow the shear stress, which is responsible for internal deformation and particle transport, increases downwards and therefore the base of the flow stops moving last, i.e. deposition is from the top of the flow downwards. The upper part of the flow without internal deformation is commonly referred to as the rigid 'plug' (Johnson 1970). This simple subdivision is a useful starting point, as incremental deposition will tend to form laminated and graded beds, whereas *en masse* deposition will tend to form poorly sorted, structureless or deformed beds. In a continuum of sediment gravity flow behaviours, the balance between turbulent and cohesive forces is a key control of the deposit type. Crucially, this balance is dynamic, and flows can *transform* in time and space from one type to another, and some intermediate flow types, *transitional flows*, will have the characteristics of both incremental bottom-up and *en masse* top-down depositional processes.

12.3.3 Bottom Current Processes

A range of processes can cause bottom currents, and control sedimentation (Figure 12.8b), and sediment supply. The hydrodynamic implications of different sediment types, such as foraminiferal sand versus siliciclastic sand, are important considerations in the development and interpretation of contourites. At intermediate and abyssal ocean depths, circulation is associated with large-scale buoyancy forcing and weak diapycnal processes. The surface and deep layers of the ocean are connected through overturning or thermohaline circulation, driven by global density gradients related to fluxes in surface heat (e.g. export of heat from the tropics and subtropics) and freshwater (e.g. export of freshwater poleward from the rainy Intertropical Convergence Zone). The associated currents of this large-scale circulation are subject to long-term oscillations controlled by tectonic reconfigurations and climate changes (Talley et al. 2011). Deep ocean bottom currents are susceptible to Coriolis forcing, such that they are deflected to the left in the southern hemisphere and to the right in the northern hemisphere (Rebesco et al. 2014).

Bottom currents flowing along slope (i.e. contour currents) usually develop a secondary circulation-oriented up or down slope due to the combined effects of the Coriolis force and bottom friction. It is common to observe abraded surfaces and 'moats', elongated depressions, related to the core of the current (McCave and Tucholke 1986; Wilckens et al. 2021). Bottom friction induces Ekman transport in the bottom boundary layer, which is directed to the left of the flow in the northern hemisphere and to the right in the southern hemisphere (Wåhlin and Walin 2001; Rebesco et al. 2014).

Dense waters form over continental shelves by cooling and evaporation in several areas in the world ocean, such as in the Mediterranean Sea or in Antarctica (Ivanov et al. 2004; Amblas and Dowdeswell 2018). These dense waters sink across the continental slope, forming dense shelf water cascades, until they reach a level with the same density. This process is commonly focused inside submarine canyons but can also occur on open slopes (Palanques et al. 2012). Cascades of dense shelf water can form thick layers with high suspended sediment concentration (bottom nepheloid layers) in the water column (Puig et al. 2008; Foglini et al. 2016).

Open-ocean convection occurs when surface waters become denser, due to thermal or haline surface fluxes (Wadhams et al. 2002), and less dense deep-waters start to rise, inducing convection. Overflows can occur when denser water masses pass from one basin into another through a topographic barrier, as occurs in the Gulf of Cadiz, where the denser (warmer and saltier) Mediterranean waters flow through the Strait of Gibraltar into the

Atlantic Ocean and sink until they reach density equilibrium and flow along slope (Sánchez-Leal et al. 2017). Eddies are transient loops of swirling water of very different sizes and affect different portions of the water column. For instance, eddies in the Mozambique Channel (SW Indian Ocean) are >300 km diameter, affect the whole water column, strongly modifying the bottom current circulation and shaping the seafloor (Figure 12.8a; Miramontes et al. 2019, 2021).

The density of ocean water generally increases with depth, forming the internal stratification of the oceans. But when a parcel of water is displaced, it encounters water of different density, and it thus moves vertically, eventually becoming re-equilibrated. This process generates an oscillatory movement that results in the formation of internal gravity waves that either propagate horizontally or vertically. Internal waves related to tides are named 'internal tides' (Lamb 2014). Non-linear internal tides often propagate horizontally and form trains of short strongly non-linear internal solitary waves, which can be visible from satellite imagery or synthetic aperture radar (Jackson et al. 2012). Internal waves can be funnelled and amplified in canyons, generating upslope and downslope flows that can form nepheloid layers (Puig et al. 2014), and affect turbidity currents (Liu et al. 2023).

12.3.4 Post-depositional Remobilisation Processes and Products

Several physical processes can occur during and after deposition from sediment gravity flows; these collectively serve to disturb the original sedimentary structure. These processes commonly involve the movement of fluids and are often triggered by permanent or transient changes in pore pressure.

Liquefaction and fluidisation are two related processes whereby a granular material is transformed from a static solid-like (but not lithified) state to a dynamic fluid-like state. The difference is that in liquefied flows and beds, the grains settle downwards through the fluid, displacing the fluid upwards, whereas in fluidised beds, the fluid moves upwards through the bed between grains that are temporarily suspended, and potentially transported (Lowe 1976). Liquefaction involves a rapid and transient increase in pore fluid pressure and the consequent loss of shear resistance, which can result in the deformation, but not destruction, of sedimentary structures (i.e. convolute laminations). It can be triggered by cyclic loading at a relatively high frequency (e.g. seismic tremors; Craig 2004). Fluidisation occurs when the upward drag forces on grains exceed their submerged weight. Full fluidisation results in complete disruption of the primary sedimentary structures (Duranti and Hurst 2004).

Dish structures are concave-up, bowl-like sedimentary structures commonly developed in thick-bedded sand (-stones) (Lowe and LoPiccolo 1974). They form by liquefaction and fluidization of water-charged soft sediment either during or immediately following deposition. Water escaping upwards through the rapidly deposited sand encounters sub-horizontal lower permeability barriers, such as mud-rich laminae. These lower permeability barriers force the escaping water to flow laterally until a place is found where it can break through and flow upwards. The bounding surface of each dish (2–50 cm wide) is the low permeability barriers. The dishes are commonly separated by, or overlie, vertical streaks of massive sand called pillars. These vertical or subvertical features form during forceful water escape, which elutriates fine material, and often form sheets (rather than pipes).

Sand injection is a natural hydraulic fracturing processes (Lorenz and Finley 1991; Jolly and Lonergan 2002) requiring a pressure differential between the source of the injecting sediment–water mixture and the tip of a developing fracture to sustain propagation. For clastic injections to form the over- and underlying lithologies must form impermeable barriers that allow build-up of higher pore pressures in the parent unit than in the surrounding strata; such conditions are most commonly met during burial and compaction (Cosgrove 2001). The result is 'overpressure': the pore fluid pressures in the parent body become higher than the surrounding hydrostatic pressure (Maltman 1994). The overpressured and sealed sediment can remain unconsolidated, even at great burial depths (100s m). Finally, a trigger is needed to cause the sealing lithology to hydraulically fracture and the parent sand to liquefy and fluidise, subsequently migrating into the propagating fractures forming clastic injectites (Jolly and Lonergan 2002). For sediment to become fluidised, and remobilised, there must be a migrating interstitial fluid providing sufficient driving force (e.g. drag, lift) to overcome any forces resisting the movement of the particle (e.g. gravity, apparent cohesion or fluid–particle friction). The nature of flow during sand injection has been debated, with arguments for laminar (Dott 1966; Cobain et al. 2015) and turbulent (Obermeier 1996; Scott et al. 2009) flows.

Intrusion of mud as diapirs (Section 9.3.6; Figure 9.16) can also occur being characterised by the slow upward migration of clay-rich sediment and fluid bodies (Kopf 2002), which rise from a source layer and pierce overlying stratigraphy through differential loading, buoyancy forces (density contrast between the mud and surrounding sediments), or other driving mechanisms (i.e. overpressure and gas generation; Hedberg 1974). Mud has low mechanical yield strength when water-laden, unlithified and uncompacted and can flow plastically within the shallow subsurface (Bishop 1978; Maltman and Bolton 2003). The mobility of buried muds is caused by high initial porosity and fluid content, which can lead to overpressured conditions (Collinson 1994). If the diapir is connected to the seafloor then a mud volcano will form.

12.4 Bedforms and Bed Types

12.4.1 Introduction

Traditionally, a sedimentary facies and facies association approach, documenting grain-size and sedimentary structures, and their distributions, in combination with identification of architectural elements, has been employed to support palaeoenvironmental interpretations. However, sediment gravity flow transformations and interactions with topography result in a plethora of deposit types (e.g. Talling et al. 2012). This generates cumbersome facies schemes that struggle with textural transitions both between and within beds in deep-water detrital systems. Therefore, to provide a more accessible observational basis for documenting deep-water detrital systems we employ a bed-type approach (Section 12.4.3; Figures 12.11d and 12.12a), in which a bed can comprise one or more bedforms and their transitions (Figure 12.14b). Using idealised bed types is not without limitations. For example, sometimes it can be challenging to identify an individual bed due to erosion and amalgamation. Also, at a point, a bedform can be represented by a single lamina set, and a single bedform can represent a bed. Furthermore, in some cases, such as coherent mass flow deposits, the range of processes and scales mean that there is no simple set of characteristic bed types.

The following terms are used to describe component layers of beds: laminae and bands, laminaset and banded set, bedform sequence and bed (Figure 12.14b; e.g. Campbell 1967; Taylor et al. 2024). At the smallest-scale, laminae (1–5 mm thick) and bands (5–10 mm thick) refer to thin layers that have relatively sharp upper and lower boundaries, which represent a change in grain-size and/or sorting (Campbell 1967; Stevenson et al. 2020). Laminasets or banded sets are conformable laminae or bands with shared characteristics. One or more laminasets or banded sets with similar textural properties form a depositional bedform. Bedforms with different textural and morphological properties stack to form bedform sequences, which make up (parts of) beds (Figures 12.14 and 12.15). Therefore, we first

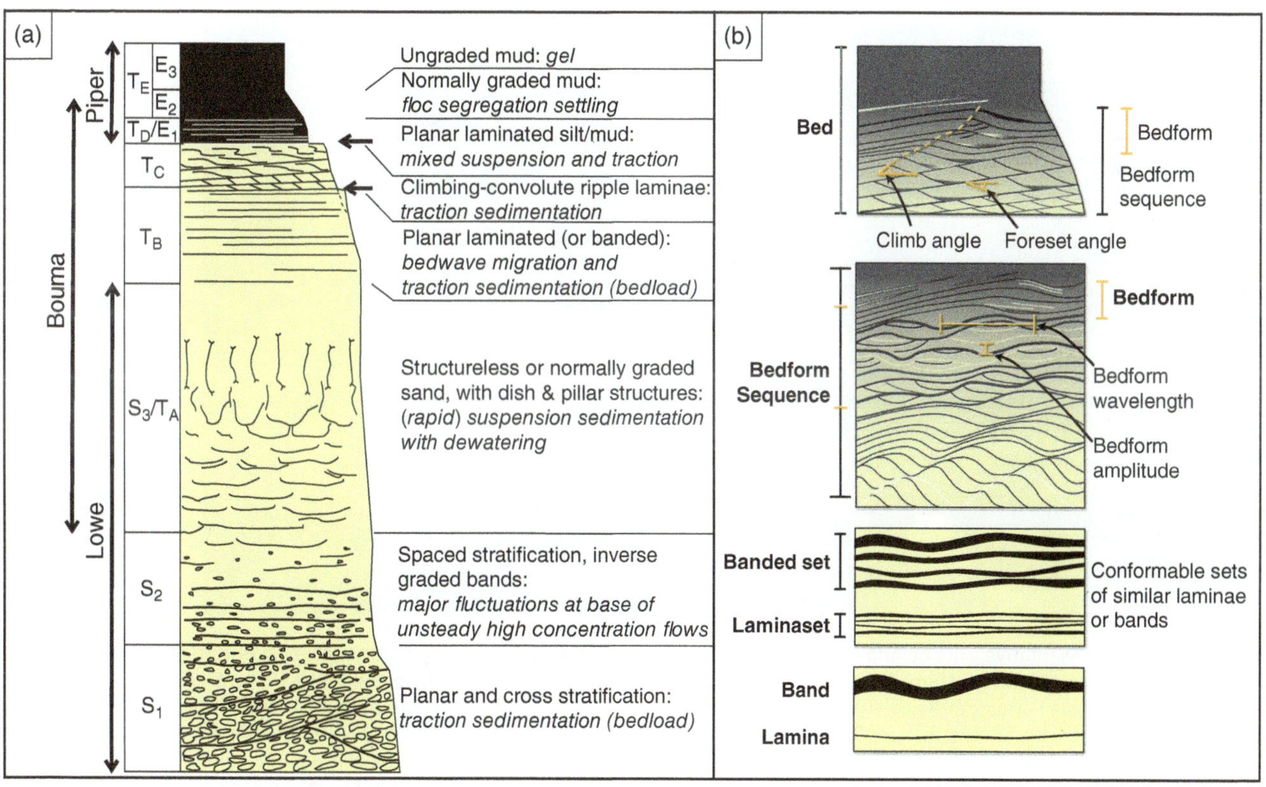

Figure 12.14 (a) Compilation of bed-type models for high-concentration (Lowe 1982), low-concentration (Bouma 1962) and mud-rich (Piper 1978) turbidity current deposits. Arrows point to grain-size breaks. *Source:* Adapted from Covault (2011). (b) Diagram highlighting the terminology used to describe the components of deep-water detrital beds. *Source:* Adapted from Taylor et al. (2024).

Figure 12.15 Representative outcrop and core photographs of deep-water bed types. (a) High-concentration turbidity current deposit (HCTCD) with dish and pillar structures (Cretaceous, Black Flysch Group, Cabo Matxitxako, Spain); (b) HCTCD with gravelly base (Cretaceous, Punta Baja Formation, Baja California, Mexico); (c) low-concentration turbidity current deposits (LCTCD) with abundant flame structures (Cretaceous, Rosario Formation, Baja California, Mexico); (d) LCTCD with climbing ripple cross-lamination (Permian, Fort Brown Formation, Karoo Basin, South Africa. Lens cap diameter 58 mm; (e) transitional flow deposit with banding and mud clasts (Carboniferous, Ross Formation, County Clare, Ireland); (f) transitional flow deposit with mixed grain size bedforms (Permian, Fort Brown Formation, Karoo Basin, South Africa); (g and h) hybrid event beds (Silurian, Aberystwyth Grits Group, Wales). Lens cap diameter 82 mm; (i) debrite with shell material (Cretaceous, Black Flysch Group, Cabo Matxitxako, Spain); (j) thick debrite with megaclasts (Black Flysch Group, Cabo Matxitxako, Spain). Coin is 24.25 mm; (k) thick graded HCTCD (Cretaceous, Springar Formation, Vøring Basin, Norway). *Source:* Photo credit (a–k): Ian Kane.

cover depositional and erosional bedforms (Section 12.4.2), followed by idealised bed types (Section 12.4.3) and contourite types (Section 12.4.4). We then consider the evidence at bed-scale for flow-topography interactions (Section 12.4.5). Finally, the expression of clastic injectites at outcrop is considered (Section 12.4.6), as these features are preferentially found where successions pinch-out against seabed topography.

Whether a sedimentary facies or bed-type approach is taken, our use of the rock record means we hold a descriptive and interpretative bias to late-stage near-bed processes operating at the time of deposition. The conditions at the base of many sediment gravity flows are markedly different to the transport conditions in the main part of the flow or during deposition (Figure 12.13c) meaning that there remains much to learn about sediment gravity flow transport mechanisms.

12.4.2 Bedforms in Deep-Water Detrital Systems

12.4.2.1 Depositional Bedforms

Flow criticality is a concept that links the flow process regime and underlying bedforms in deep-water detrital systems. It is tied to the Froude number (Fr), a dimensionless number that is defined as the ratio of the inertial to gravitational forces in the flow. A fundamental distinction is made between sediment gravity flows that are subcritical (Fr < 1) and supercritical (Fr > 1), with a critical flow being Fr = 1. Transitions from supercritical to subcritical states can occur through a reduction in velocity and/or gradient, and flows pass through a hydraulic jump, which is marked by an abrupt increase in turbulence, and represents a high energy loss and increased erosive potential. Subcritical flows are dominated by gravitational forces and behave in a slow or stable way and are linked to subcritical bedforms, such as lower-stage plane beds, ripples and dunes, all lower-regime bedforms. Supercritical flows are dominated by inertial forces and behave as rapid or unstable flows and are linked to supercritical bedforms, such as upper stage plane beds, antidunes and cyclic steps (Harms and Fahnestock 1965; Cartigny et al. 2011; Figure 12.16).

12.4.2.2 Subcritical Bedforms

Depositional bedforms that develop under subcritical turbidity currents in the lower flow regime, are ripples which are widely identified, and more rarely dunes. These downstream migrating bedforms are an important source of palaeocurrent information, and for providing information on hydraulic conditions at the time of deposition. Ripples and dunes are geometrically similar but hydrodynamically different and can be defined according to their dimensions

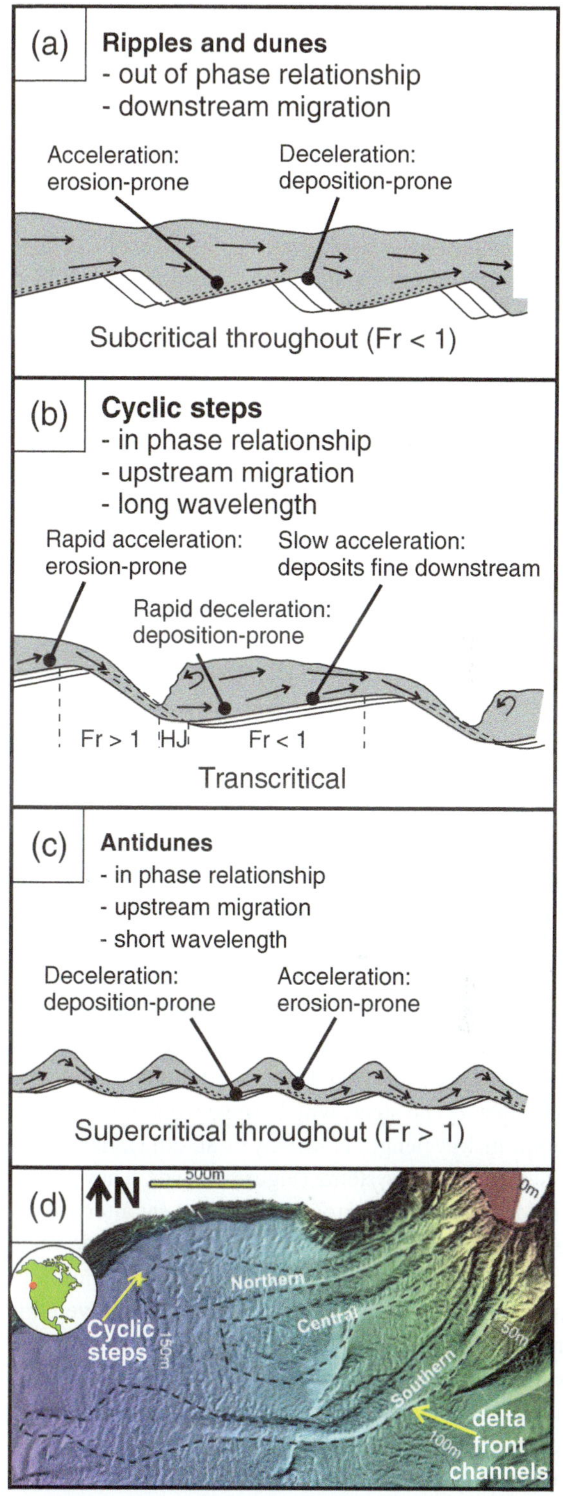

Figure 12.16 Summary cartoon of bedform types. (a) Ripples and dunes, formed in Froude-subcritical flow (Fr < 1). (b) Cyclic steps, characterised by transcritical flow (i.e. alternating Froude-subcritical and supercritical flow). (c) Antidunes, formed in Froude-supercritical flow (Fr > 1). HJ: hydraulic jump that marks the downstream transition from Froude-supercritical to Froude-subcritical flow. (d) Mulitbeam bathymetry of cyclic steps on the Squamish Delta, in Howe Sound, British Columbia, which is dissected by three active channels. The channelised margin and prodelta seafloor display upslope migrating cyclic steps. *Source:* (a–c) Adapted from Slootman and Cartigny (2020) and (d) Adapted from Clare et al. (2020).

(ripples are generally less than 3 cm high and 0.5 m wavelength; Collinson and Mountney 2019).

Climbing-ripple cross-lamination is structure formed by migration of a very fine-grained to fine-grained sand bedform (Figure 12.15d). This bedform develops during contemporaneous suspended load fallout and bedload transport under subcritical conditions (Sorby 1859; Jobe et al. 2012). The climb angle of the ripples reflects the ratio of the rates of suspended load fallout and bedload sedimentation. The stoss side of the ripple can be preserved at very high rates of suspended load fallout (Figure 12.15f). High rates of suspended rate fallout are associated with locations where flows expand rapidly due to abrupt loss of confinement and/or slope gradient decrease (Jobe et al. 2012; Morris et al. 2014). Confusingly, climbing-ripple cross-lamination is sometimes described as 'subcritical' or 'supercritical' based on the cross-sectional geometry and ripple climb angle (Hunter 1977), which has nothing to do with the identical terms used to describe flow criticality in relation to the Fr number.

12.4.2.3 Supercritical and Transcritical Bedforms

A suite of bedforms develops under entirely supercritical flow conditions (e.g. antidunes) or under alternating super- and subcritical conditions (e.g. cyclic steps). In shallow water, under supercritical flow conditions, antidune bedforms develop as the mobile substrate takes on the shape and amplitude of the surface waves, and may migrate upstream, downstream or remain stationary (Kennedy 1963; McLean 1990; Fourrière et al. 2010). Some short wavelength examples have been interpreted in the rock record, although unequivocal identification of antidunes remains rare (see Pickering and Hiscott 2015).

In contrast, cyclic steps, which range from tens of cm to many km in wavelength (Cartigny et al. 2011), have been identified in several systems (Figure 12.16d). They are an upper flow-regime bedform characterised by alternations of lee side Froude-supercritical flow and stoss side Froude-subcritical flow, separated by a hydraulic jump in the intervening bedform trough (Figure 12.16b). Given this alternation about $Fr = 1$, these bedforms are termed *transcritical* because lee-side erosion records supercritical flow conditions and stoss-side deposition records subcritical flow conditions. This spatial relationship leads to upstream migration of the transcritical bedform when the lee-side deposition is less than stoss-side deposition.

12.4.2.4 Bedforms Under Transitional Flows

The elevated mud content in transitional frictional flows leads to the development of a range of mixed-grain-size bedforms. The balance of cohesive and turbulent forces, and the rate of flow deceleration (aggradation rate), govern the mixed grain-size bedforms that form (Baas et al. 2016; Baker et al. 2017). *Banding* is distinct from lamination as it describes thick (a few mm to a few cm) alternations of subparallel mud- and sand-rich bands (Figure 12.14b). The juxtaposition of hydrodynamically different grain-sizes at the scale of bands contradicts the established models of bedform development under cohesionless flows (Stevenson et al. 2020). Banding encompasses a range of bedforms that likely form within, or at the boundaries of, the upper stage plane bed regime via turbulence suppression and tractional reworking beneath mud-rich decelerating transitional plug flows (Stevenson et al. 2020). Therefore, banding develops under flow conditions that are like those of the Bouma Tb division (planar-laminated sandstone, Section 12.4.3), the main difference being an elevated cohesive mud component in the flow.

If cohesive forces increase during tractional reworking, turbulence can be enhanced in the troughs of ripples, which increases ripple size (Baas and Best 2008), forming mud-rich current ripples and biconvex ripples. *Mud-rich current ripples* are asymmetrical, climbing laminasets that contain distinct, high-angle, concave or planar foreset laminae of mixed or alternating sandstone–mudstone (Figure 12.15f). *Rounded, biconvex ripples* are symmetrical to slightly asymmetrical with rounded profiles and sigmoidal-shaped foreset laminae of alternating sandstone–mudstone. Continued flow deceleration and increased mud concentration modulates turbulence leading to the thickening of the upper plug-flow region, which expands downward in the flow (Baas et al. 2009). This causes the flow to become cohesive and decelerate, which results in rapid aggradation of bedforms with high climb angles. These bedforms comprise gently dipping and alternating sandstone–mudstone foreset laminae and are referred to as *low amplitude bedwaves* (Baas et al. 2016), which are asymmetrical, long wavelength and low- to high-angle climbing laminasets with stoss-side preservation.

12.4.2.5 Where Are the Dunes?

Subaqueous dunes are one of the most readily recognised depositional bedforms in fluvial and shallow marine systems but are rarely identified in deep-water detrital successions. This is despite sediment gravity flows decelerating from upper-stage plane beds, through the dune bedform phase space (assuming grain-size is larger than 0.6 mm), to ripples (Arnott 2012), and the presence of sediment waves parallel and transverse to flow identified in many seafloor surveys (e.g. Wynn et al. 2002). The paucity of dune-scale bedforms has been explained by: (i) turbulent flows having deposited their coarse grains prior to entering the dune-forming velocity and grain-size phase (Walton 1967), (ii) the short duration that flows spend in the dune-forming phase (Walker 1965), (iii) clay content altering bed and flow rheology (Simons et al. 1963; Baas and Best 2002; Schindler et al. 2015), (iv) high flow-densities and sedimentation rates in the dune-forming velocity phase that suppresses dune formation (Lowe 1988), (v) the narrow grainsize range capable of generating steep enough density gradients to create the angular bed defects required for dune development (Tilston et al. 2015), and (vi) amplification of initial perturbations being restricted by very high bedload concentrations near the packing limit (Arnott 2012).

12.4.2.6 Erosional Bedforms

Sediment bypass is a condition where a sediment gravity flow neither erodes nor deposits. In the ancient record, we look for indicators where sediment bypass has dominated a stratigraphic package. Sediment bypass in deep-water detrital systems is closely associated with erosional bedforms. Stevenson et al. (2015) summarised the key physical processes that govern sediment bypass, and the rock record criteria for identification of sediment bypass in different deep-water detrital settings (Figure 12.17).

In common with sediment bypass-dominated surfaces, such as channel bases and scours, the flow(s) that formed the erosion surfaces will be markedly different in character from the flow(s) that overlie the surface. Distinguishing large scour-fills from channel-fills in two-dimensional data can be challenging. Scours span a wide range of scales, and can coalesce to form complicated planform patterns, but tend to have steep upstream headwalls, and shallow and widen downstream. Their presence in front of submarine channels means that they are incorporated into basal erosion surfaces as channels lengthen into a basin. However, they can be preserved lateral to channels and infilled by thin-bedded successions that contain limited evidence for sediment bypass (Hofstra et al. 2015).

Sole structures encompass a wide range of erosional bedforms of inorganic origin, which are found in heterolithic deep-water detrital successions on the bases of sandstone beds that overlie mudstones (Figure 12.18). Their position on the base of sandstone beds means they are widely utilised as way-up markers. In addition, they are valuable indicators of palaeocurrent direction. These structures are the infill of erosional bedforms generated in the underlying fine-grained substrate and therefore provide a record of the interactions between over-riding sediment gravity flow(s) and the muddy seafloor, and provide insights into the palaeohydraulic conditions during erosion, bypass and deposition (Peakall et al. 2020).

Typically, sole structures are subdivided into two types: (i) scour marks such as *flutes* formed by turbulent scour of the substrate; and (ii) tool marks formed by objects (tools) carried within the flow that sculpt the seafloor (Dżułyński and Sanders 1962; Collinson et al. 2006). Tool marks are subdivided further into continuous (grooves and chevron marks) and discontinuous types (prod, bounce, skip and roll marks) (Dżułyński and Sanders 1962).

'Flutes' are scour marks that occur individually or in clusters and are typically 1–20 cm wide, 5–50 cm long and a few cm deep (Collinson et al. 2006). They provide unidirectional palaeocurrent information with a narrower and deeper upstream 'nose' that widens and shallows downstream, and can be parabolic (wide), spindle- (long and thin) or comet-shaped (sinuous) (Figure 12.18a,b). 'Longitudinal ridge and furrows' are closely spaced (0.5–1 cm) parallel ridges and furrows (few mm relief; Figure 12.18f,g). Although elongate, they can form 'noses' like flutes, which point to the upstream direction (see Peakall et al. 2024 for details). 'Tool marks' are formed by objects being dragged through underlying cohesive substrate by sediment gravity flows. Finding the tool at the end of the mark is rare: mudclasts are the most likely tools given their ubiquity (Peakall et al. 2020). 'Grooves' are parallel-side tool marks that show flow trends, forming elongate ridges on bed bases that are remarkably constant in depth and width, and can present internal parallel longitudinal striae (Figure 12.18c). Grooves span a large range in dimensions from mm-40 m wide, up to 20 cm deep, and generally extend beyond the scale of outcrops (Peakall et al. 2020; Figure 12.6). Some grooves have adjacent closely spaced V-shaped or U-shaped ridges that close in the direction of travel, referred to as chevrons (Figure 12.18e), and their formation is linked to substrate rheology (McGowan et al. 2024). 'Skim', 'prod' and 'skip' marks are types of discontinuous tool marks, which are interpreted to form during transient impact of tools being carried in the flow with the cohesive substrate (Figure 12.18d). Typically, skip and bounce marks are symmetrical, and show a flow trend, but prod marks are asymmetric with a longer shallower upstream slope, and a shorter steeper downstream slope, and so are a directional palaeoflow indicator (Peakall et al. 2020).

Many depictions of Bouma bed types (see Section 12.4.3) have sole structures on the base of the Ta division, although both flutes and tool marks are found under Tb and Tc beds (e.g. Bouma 1962; Pett and Walker 1971). Indeed, sole structures on bed bases may have no genetic link to the overlying deposits (Peakall et al. 2020), which challenges the assumption that they provide the flow direction of those beds. In general, however, flutes are associated with more proximal locations, and tool marks with more distal locations (e.g. Walker 1967; Remacha and Fernández 2003; Figure 12.18h). Bed bases ornamented with both tool marks and flutes are rare (e.g. Wood and Smith 1958; Collinson et al. 2006), but where present commonly preserve cross-cutting relationships (Enos 1969; Draganits et al. 2008).

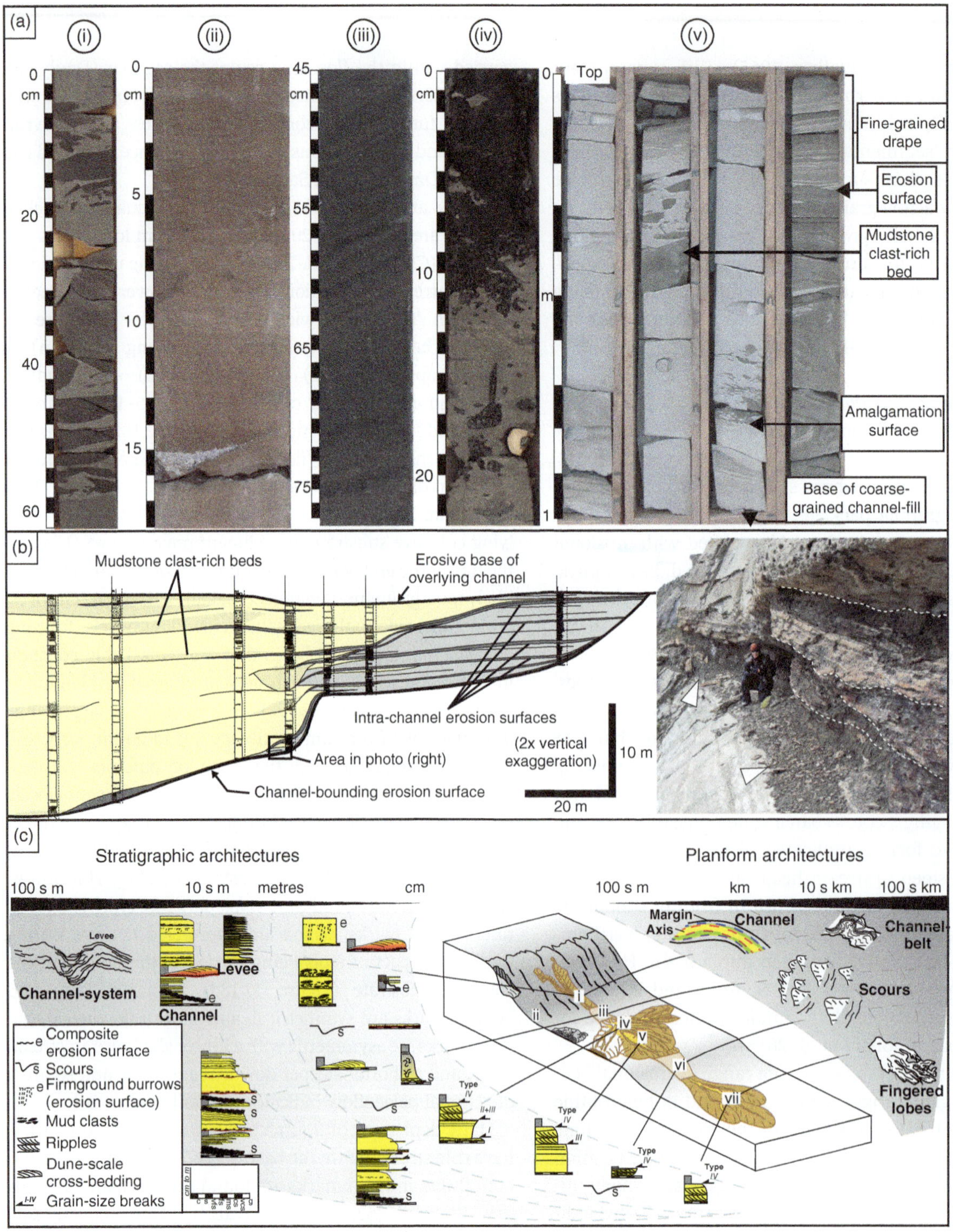

Figure 12.17 (a) Representative photographs of indicators of sediment bypass in core: (i) claystone and siltstone clasts in a fine-grained sandstone matrix (Fort Brown Formation, Karoo Basin, South Africa; (ii) thin coarse-grained lag draping an erosion surface, in turn overlain by fine silt and mud (Upper Puchkirchen Formation, Austria); (iii) moderately sorted cross-stratified glauconitic medium sand with clay-rich interlamination suggesting episodic bedform migration (IODP Expedition 313, offshore New Jersey); (iv) coarse-sand infilled Thalassinoides burrows indicating open burrows in firmground with bypass-dominated flows that infilled the exhumed burrows (IODP Expedition 313, offshore New Jersey); (v) interpreted channel-fill succession with an erosion surface draped by fine-grained beds overlain by progressively thicker sandstones (Upper Puchkirchen Formation, Austria). (b) Evidence for sediment bypass from the Tres Pasos Formation, Magallanes Basin, Chile. Left: Slope-channel cross section highlighting the axial to marginal shift from sandstone- to siltstone-dominated strata. The complicated intrachannel strata indicate a protracted history of erosion, sediment bypass, and deposition within the conduit. Right: Photograph of an erosional channel base (indicated by arrows) overlain by thinly interbedded siltstone and sandstone. The recessive, composite unit that drapes the channel base is characterised by numerous internal erosion surfaces (dashed lines), which are thought to have been incised when coarse-grained sediment was bypassed basinward. (c) Spatial variability of indicators of sediment bypass in an idealised deep-water clastic system (submarine fan). Areas of sediment bypass are highlighted with associated recognition criteria in both stratigraphic and planform perspectives. Areas of bypass include: (i) slope-channel bypass, (ii) slope bypass of mass-transport events, (iii) base-of-slope bypass, (iv, v, vii) basin-floor bypass and (vi) basin-floor bypass zones. *Source:* (a–c) From Stevenson et al. (2015).

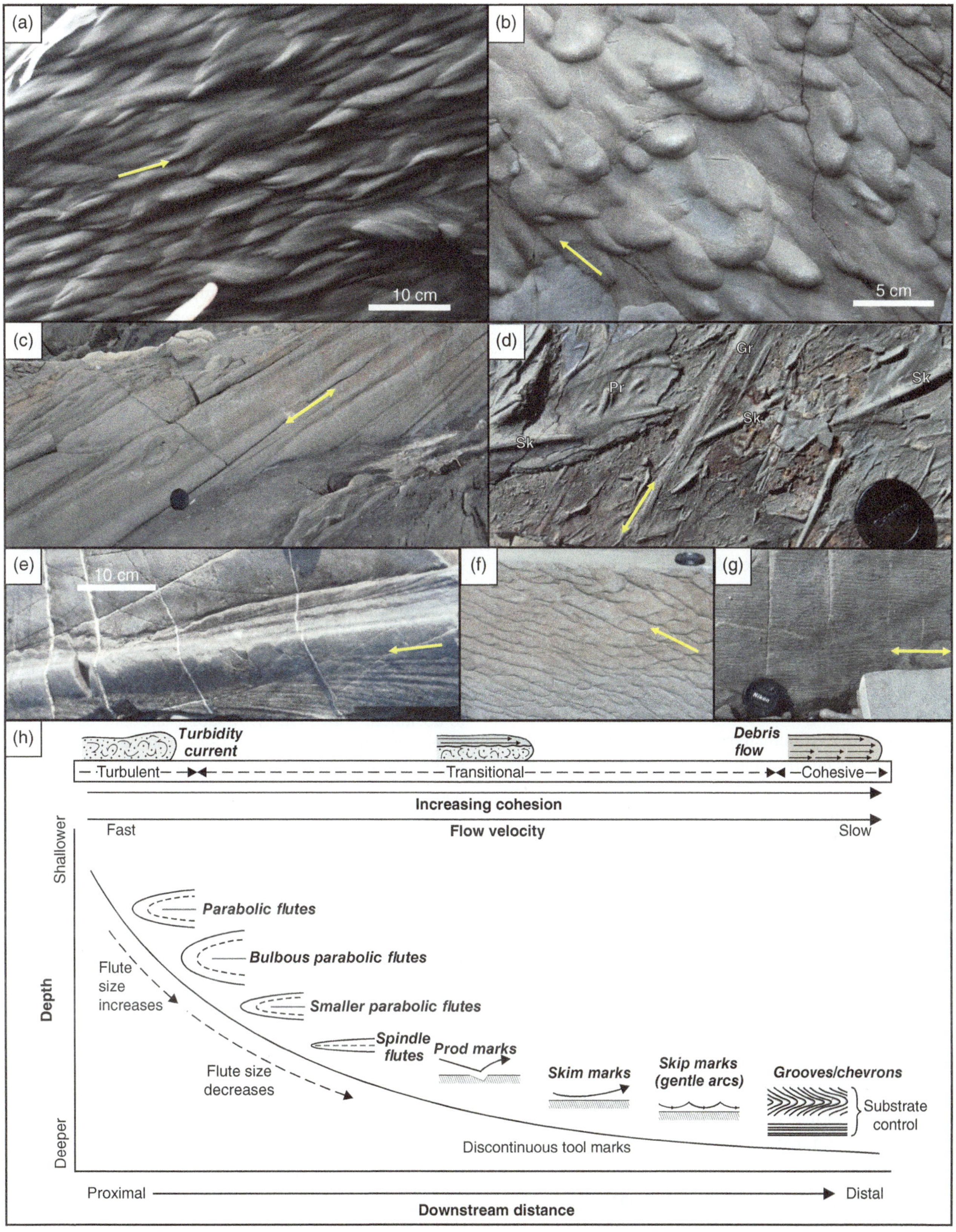

Figure 12.18 Photographs of sole structures. (a) Spindle flutes with some flutes exhibiting a pronounced spiralling pattern. (b) Parabolic flutes, with some exhibiting prominent median ridges, or simple smooth forms. (c) Parallel grooves. Lens cap diameter 58 mm. (d) Prod (Pr) and skim marks (Sk), with a large groove (Gr) displaying central striations. Lens cap diameter 58 mm. (e) Groove ornamented with interrupted chevrons, showing flow from right to left. (f) Longitudinal ridges and furrows. Flow to the top left. Lens cap diameter 77 mm. (g) Slightly sinuous and elongate ridges and furrows. Lens cap diameter 77 mm. (h) A process-orientated conceptual model for the longitudinal distribution of flutes and tool marks shown for a flow that increases in cohesive forces with longitudinal distance. Yellow arrow shows palaeoflow direction or trend. a, c, e, f, g from middle Ordovician Cloridorme Formation, Quebec, Canada, b from Aberystwyth Grits, Wales and d from middle Carboniferous Quebrada de las Lajas, Argentina. *Source:* (a–e and h) Adapted from Peakall et al. (2020) and (f, g) From Peakall et al. (2024).

Peakall et al. (2020) use recent advances in understanding of transitions between fully turbulent and laminar flows to propose that (i) flutes are not solely the product of turbulent flows, but can form under transitional flows, (ii) discontinuous tool marks (skim, prod and skip marks) form under transitional mud-rich flows, and (iii) grooves form under debris flows, slumps and slides, but not turbidity currents. Consequently, the observed spatial distribution of flutes in more proximal settings and tool marks in more distal settings is attributed to downstream increases in flow cohesion (Figures 12.11d, 12.12b, and 12.18h). There remains much potential in the use of sole structures to support interpretations of palaeohydraulic conditions, and flow–substrate interactions.

12.4.3 Bed Types (and Bed-Type Associations)

The bed types described below are linked to the flow types established in Section 12.3. Cohesive mass movements are grouped as *submarine landslide deposits* to acknowledge the transitions between creep, slide and slump deposits that make an unequivocal identification in the field or core challenging. Beds resulting from laminar flows dominated by cohesive forces are debrites and mudflows, and bed results from turbulent frictional flows are related to the deposition of a range of turbidites. We use the established idealised models for coarse-grained turbidites (Lowe-type), medium-grained turbidites (Bouma-type) and fine-grained turbidites (Piper-type). As illustrated in Section 12.3.2, sediment gravity flows can entrain enough clay and silt for cohesive forces to become important and establish a transitional state between fully turbulent and laminar states (Baas et al. 2011). The result is the deposition of bedforms with mixed grain-sizes, or beds comprising divisions with abrupt textural breaks. These bed types will be discussed using idealised bed models for transitional flow deposits (Baas et al. 2016; Taylor et al. 2024) and hybrid event beds (Haughton et al. 2009). The idealised bed types show characteristic sequences of sedimentary structures and grain-size grading patterns (Figure 12.11d). The different structures are referred to as divisions within a bed sequence. In all cases, it is rare to see complete examples of these bed sequences; in any one location it is normal to find base-missing or top-missing bed sequences and marked lateral variability.

12.4.3.1 Coherent Mass Flow Deposits: Submarine Landslide Deposits

The gravity-driven movement of submarine landslides into deeper water can be through incremental motion or in a single event. Deposits range from barely deformed, to intensely folded and faulted. Slides are distinguishable from debrites through their overall lack of disaggregation of the material in the flow. A continuum of movement types through increasing displacement, internal deformation and disaggregation is identified: from creep with no basal detachment, through slides with basal detachment but limited internal deformation, to slumps which are basally detached and internally deformed (Figure 12.11a).

Establishing strict criteria for defining the deposits of these different flow types would be of limited utility as a single coherent submarine landslide deposit can show characteristics of creep, slide and slump (and associated debris flow and turbidity current) processes. Furthermore, multiple failure events with similar or different types of movement can coexist within the same submarine landslide system (Mulder and Cochonat 1996; Scarselli 2020). The suite of deposits associated with coherent mass flows can range across several observational scales in volume (up to kms^3) and thickness (10s m), meaning that they can be much larger than typical outcrops, making their recognition as mass flow deposits, and their process interpretations difficult (Figure 12.6). These characteristics means that they do not lend themselves readily to a bed-type approach.

The simplest form of submarine landslide is submarine *creep*, which is a gravity-driven and very slow velocity mass movement without a basal detachment (Hill et al. 1982). Examples of submarine creep have not been identified at outcrop, and the lack of detachment means they are hard to document convincingly. A submarine *slide* deposit is a failed mass that moved across a basal detachment, with limited internal deformation. The coherent nature of a slide points to a simple basal detachment and often limited downslope movement. Downslope movement above a basal shear surface can lead to internal deformation during failure producing an internally contorted (folded) and displaced (faulted) deposit referred to as a *slump*. The contorted mass may present a strongly aligned structural fabric that can be used to infer transport direction (e.g. Steventon et al. 2019). In seismic reflection data, submarine landslide deposits are identified from a chaotic seismic facies in which coherent blocks and thrusts may be identified. These are referred to as mass transport deposits (MTDs) where just a single event is identified, or mass transport complexes (MTCs) where multiple coalesced events are recognised. Sustained down slope movement of a slide and/or slump can lead to mass disaggregation and sediment–water mixing causing debris flows to transform into turbidity currents (Talling et al. 2007; Figure 12.11a,d). Therefore, MTCs can comprise debrites and turbidites.

12.4.3.2 Cohesive Sediment Gravity Flow Deposits

12.4.3.2.1 Debris Flow Deposits: Debrites

Debris flows are ground-hugging, cohesive, laminar flows. Particles in these hyperconcentrated to concentrated flows are supported during transport by buoyancy and grain-to-grain support, which are 'held together' through electrostatic attraction between particles in the mud fraction that provides matrix strength. Their pseudo-plastic rheology means ambient fluid entrainment or particle loss, and therefore flow dilution, is limited. Their relatively high viscosity and density hinder incremental deposition and grain-size segregation, leading to *en masse* deposition, and therefore the development of beds, *debrites*, without organised sedimentary structures or grain-size distributions (Figure 12.16i,j). Therefore, their deposit character reveals details about the late-stage particulate transport processes (in contrast to more turbulent flows). Water can infiltrate between grains during transport, which reduces flow cohesion and matrix strength, and can ultimately result in their transformation into turbidity currents.

The 'freezing' and *en masse* deposition of debris flows results in deposits with relief, abruptly tapered frontal and lateral margins, and a deposit thickness that is closely related to flow thickness. Although debrites are characteristically poorly sorted, outsized clasts can be segregated towards the base or top of the debris flow, such that debrites can contain crude, normal and/or inverse grading. Large-volume cohesive debrites tend to occur in more proximal settings and extend immediately downdip from an initial slope failure (Figure 12.15j). These can be highly erosional and 'bulk-up' (i.e. increase in volume) through entrainment of substrate such that they are largely confined by their own erosion surface with groove marks at the base formed by dragging of megaclasts through the substrate (Figures 12.2c and 12.6b). These debrites, which can be 10s m in thickness, are commonly part of submarine landslide systems where an initial slide has disaggregated.

Thinner and highly mobile low-strength cohesive debris flows produce extensive deposits (Figure 12.15i). These low-strength debris flows travel long distances, or form from local flow transformation due to entrainment of cohesive mud that dampens turbulence (see hybrid event beds below). They can preserve weak grain-size grading, including (i) normal or inverse grading of outsized clasts, but not the finer matrix; (ii) normal grading in the uppermost part of the deposit due to reworking by an overriding turbulent flow; and (iii) thin basal layers with inverse grading due to 'kinetic sieving' (percolation of smaller grains through spaces between larger grains). Fluid mud flows (e.g. Bowles et al. 2003), comprising >95% clay and silt, can behave as cohesive or frictional flows, and transform from one to another, meaning that their position in a simple process classification is problematic. Laminar fluid muds can flow to low relief areas, where they will pond to form thick ungraded deposits.

12.4.3.3 Sediment Gravity Flow Deposits

Turbulent frictional flows sort the particles they carry vertically and laterally by size, shape and density, through fluid turbulence. During competence-driven deposition, the result is grain-by-grain (incremental) deposition, as evidenced by normal grading, and the presence of low-angle imbrication (e.g. Hiscott and Middleton 1980; Arnott and Hand 1989). During capacity-driven deposition, the result is relatively poorly sorted sands. In contrast to debrites, the thickness of a turbulent frictional flow deposit is unrelated to flow thickness. Bed types of turbulence-dominated (cohesionless) flows are covered first, followed by a suite of bed types where cohesive forces are considered important.

12.4.3.4 Turbidity Current Deposits: Turbidites

The deposits of sand-rich cohesionless turbulent frictional flows, turbidites, are exemplified in three widely used bed-type models, which each show a characteristic sequence of sedimentary structures (Figures 12.11d and 12.14a). These schemes are underpinned by an assumption of deposition under a single uniform turbidity current event in a decelerating (depletive) regime, where grain transport is dominated by turbulent forces with minor cohesive forces. These schemes loosely refer to different mean grain-sizes. The Lowe (1982) sequence is best applied to coarse-grained sandstones (very coarse sand to pebble grain-sizes) of interpreted high-concentration turbidity current deposits (e.g. Figure 12.15b). The Bouma (1962) sequence is the most widely used and is best applied to coarse- to very-fine-grained ('medium-grained') sandstones interpreted to be deposited by low-concentration turbidity currents. The Piper et al. (1978) sequence was developed for fine-grained (clay and silt) and thin-bedded low-concentration turbidity current deposits. These schemes are complementary and overlapping. For example, the Piper scheme is equivalent to the Td–Te divisions of the full Bouma sequence (Ta–Te), and the S3 division of the Lowe sequence is like the Ta division of the Bouma sequence (Figure 12.14a).

The 'Lowe (1982) bed sequence' (Figures 12.11d and 12.14a) describes a set of sedimentary structures in turbidite sandstone beds that are interpreted to be deposited by high-concentration turbidity currents (e.g. Figure 12.15b,k). Sediment transport within a high-concentration turbidity current, ~1–10% grains, is dominantly supported by fluid turbulence. Other mechanisms become increasingly important as the flow decelerates (e.g. Kneller and Branney 1995). At sediment volume concentrations of >10% hindered settling starts to occur (Richardson and Zaki 1954). Towards the base of the flow, grain-to-grain collisions, pore water escape, tractional reworking and frictional freezing are more pronounced than in the upper part of the flow where turbulence is more important. The rate of flow deceleration, and hence the rate of turbulence intensity decrease, will directly affect deposition. Rapid deceleration increases fallout rates of suspended load, which dampens turbulence and inhibits tractional or shear reworking of the aggrading bed (Figure 12.11c). More gradual deceleration may allow reworking and bedform development.

Description and interpretation (from base to top; Figure 12.14a)):

S1 division: Basal conglomerate or coarse-grained sandstone that display parallel- to cross-lamination. These features suggest tractional grain transport through bedload rolling and sliding.

S2 division: Stacked fine- to coarse-grained inverse graded stratification bands (5–10 cm thick). Burst-and-sweep cycles and major fluctuations in hydrodynamic conditions at the base of unsteady high-concentration turbidity currents. Inverse grading results in preferential settling out of smaller grains through voids between larger grains during progressive aggradation. This process can repeat multiple times to form spaced stratification (Hiscott 1994b).

S3 division: Structureless to normally graded, fine- to coarse-grained sandstones that represent deposition from a turbulent suspension. When the turbulence intensity at the base of the flow falls the suspended sediment load can no longer be sustained, resulting in rapid mass deposition, forming a massive to weakly normally graded division. As grains settle out, displaced water can move upwards to form dish and pillar structures (Figure 12.15a).

The 'Bouma (1962) bed sequence' (Figure 12.14a) is normally graded and comprises five divisions labelled from base to top Ta through Te, which are employed to describe the ideal vertical succession of sedimentary structures deposited by different processes under a low-concentration waning (decelerating) cohesionless turbidity current. The dominant sediment transport mechanism in these low-concentrations (<1% sediment) sand-rich turbidity currents is the suspension of particles by turbulence (Mulder and Cochonat 1996; Kneller and Buckee 2000; Meiburg and Kneller 2010).

Description and interpretation (from base to top):

Ta division: Structureless to normally graded sandstone division that sometimes contains basal extra- and/or intra-basinal clasts. In a similar manner to the S3 division of the Lowe sequence, the Ta division is deposited when the flow energy is low enough that fluid turbulence is unable to keep the coarsest grains in suspension, and the grains settle out rapidly to form a structureless bed (capacity-driven). If flow energy decreases more slowly, then the coarser grains will settle out first, leaving the finer grains still in suspension (competence-driven).

Tb division: Planar-laminated fine- to medium-grained sandstone. The Tb division is deposited via dilute flows under upper flow regime conditions where energy is high enough to carry sand grains by traction, wherein they slide and roll across the surface beneath the flow. Planar laminations can form through the gradual lateral migration and very slow aggradation of long wavelength (~1 m) and low amplitude (<5 mm) bedwaves (Best and Bridge 1992). The planar (or wavy) laminae develop through sediment sorting on the lee side of the bedwave.

Tc division: Ripple-laminated fine-grained sandstone that is deposited under lower flow regime conditions where there is enough energy for the flow to carry fine sand by bouncing (saltating) across the bed surface. As grains settle out, current ripples develop. Climbing ripple lamination will develop if sediment fallout rates are high enough. If the overrunning flow imposes enough shear on the aggrading ripples, then convolute laminations can develop (e.g. Figure 12.15d).

Td division: Parallel-laminated siltstone deposited by suspension settling where a minor current exists. Minor energy fluctuations cause settling out of alternating coarser and finer silt grains to form laminations.

Te division: Massive, ungraded mudstone, sometimes with evidence of bioturbation. The Te division results from suspension settling where essentially no current exists but is difficult to differentiate from the Td division below and is easily eroded by subsequent turbidity currents, so is often not present.

The 'Piper (1978) bed sequence' (Figures 12.11d and 12.14a) forms the uppermost (Td and Te) divisions of the Bouma (1962) bed sequence and needs to be differentiated from pelagic and hemipelagic mudstones (Chapter 13) that settle out from the water column between sediment gravity flow events. Piper et al. (1978) subdivided these mud divisions into three divisions (Te1–Te3, Figure 12.14a) and differentiated the laminated mud (Te1 division) from the laminated silt of the Bouma Td division based on the finer modal grain-size. Mud-rich turbidites have been differentiated into nine subdivisions (T0–T8; Stow and Shanmugam (1980). As argued by Talling et al. (2012), this subdivision is not practical when working at outcrop, and we follow the Piper scheme for mud-rich turbidites.

Description and interpretation (from base to top):

Te1 division: Laminated mud divisions (0.1–0.25 m thick) that are better developed in more proximal locations (Piper 1978; McCave and Jones 1988). Typically, there is a gradual upwards transition from Td or Tc divisions to Te1 divisions, and a sharper transition to Te2. The coarser-grained laminae comprise (non-cohesive) silt grains, with irregular lenticular bedding. Consecutive silt laminae typically fine upwards, whereas the intervening more poorly sorted mud laminae show less well-developed trends, showing formation under the same overall waning flow event (Stow and Bowen 1978).

Te2 division: Massive and normally graded mud division, which is typically 0.1–0.2 m thick but can form >1 m thick 'megabeds' (Jones et al. 1992). The formation is via segregation settling of silt without reworking into well-defined laminae. Segregation settling may occur when a low-strength gel (a dense cohesive mud suspension) develops that supports the preferential settling out of coarser silt grains.

Te3 division: Massive ungraded mud division that is best developed in basin plains or in confined intra-slope depocentres (up to 20 m thick; McCave and Jones 1988; Jones et al. 1992). Talling et al. (2012) invoke a model where these deposits originate from very dilute and turbulent suspensions that became more concentrated as they settle out, until they consolidate *en masse*. This is because thick ungraded deposits do not support deposition from dilute suspensions. Gels or fluid mud layers with higher concentrations ('densite mud' of Talling et al. 2012) can flow for very long distances across low gradient sea floor before ponding in the lowest point of a depocentre.

12.4.3.5 Transitional and Compound Flow Deposits

The concept of flows transforming between turbulent, transitional and laminar behaviours is crucial for interpreting the depositional record of deep-water detrital systems in general and is invoked in particular for the interpretation of argillaceous sandstones (Figure 12.11d). These flow transformations are strongly dependent on the volume fraction of cohesive mud within a flow, which controls the balance between turbulent and cohesive forces. The clay particles aggregate to form floccules and gels and thus enhance or suppress the shear-generated turbulence and can form laminar plug flows (Baas et al. 2009; Baker et al. 2017). Small volumes of clay added to a turbulent, sand-laden flow produce distinctive mixed grain-size bedforms with a variety of different textures, shapes and sizes when compared with bedforms associated with cohesionless flows (i.e. clean sands; Baas et al. 2011, 2013; Schindler et al. 2015).

Argillaceous bed types that are products of flow transformation processes are diverse in character, but can be simplified into two categories: (i) transitional flow deposits (TFDs; Baas et al. 2011, 2021; Taylor et al. 2024) that contain mixed grain-size (mud-rich) bedforms, and (ii) hybrid event beds (HEBs; Haughton et al. 2009) that juxtapose clean and argillaceous sandstone divisions. This subdivision captures the difference that HEBs have a debrite component which TFDs lack. A HEB sequence can contain a transitional flow division. The uppermost division of both TFDs and HEBs may represent dilute low-concentration deposition from the tail of the flow. TFDs, which record slow flow deceleration, and HEBs, which record rapid flow deceleration, are distributed preferentially in proximal and distal settings, respectively (Stevenson et al. 2020).

Transitional flow deposits (TFDs) consist of various mixed grain-size sedimentary structures and bedforms deposited by mud-rich transitional flows (Baas et al. 2009, 2016; Kane and Pontén 2012; Stevenson et al. 2020) including banding, biconvex ripple lamination, and low amplitude bedwaves (see '*Bedforms under transitional flows*' above; e.g. Figure 12.15e,f). These structures record decelerating flows undergoing a transformation from fully turbulent to increasingly laminar flow conditions, but without formation of a clear cohesive flow component at the point of deposition. Taylor et al. (2024) propose idealised bed types for unidirectional transitional flows (Figure 12.11d), and in the presence of seafloor topography, combined transitional flows.

Hybrid event beds (HEBs) are composite beds, with abrupt textural breaks of contrasting mud content indicating deposition from two or more very different flow rheologies (e.g. Figure 12.15g,h). Megabeds (10s m thick), with a basal debrite overlain by a turbidite (e.g. Labaume et al. 1987; Figure 12.6c) are also considered hybrid event beds by Haughton et al. (2009). Here, the focus is on the idealised fivefold (H1–5, H = hybrid) hybrid event bed sequence that was defined by Haughton et al. (2009; Figures 12.11d and 12.12b). From the base, H1 is a structureless sandstone deposited by high-concentration turbidity currents, H2 is a banded sandstone division deposited by a transitional flow, H3 is argillaceous and commonly mud clast-bearing deposited by a laminar flow (debrite), H4 is thin fine-grained sand deposited by the trailing low-concentration turbidity current, and H5 is a graded mudstone recording late-stage suspension fallout.

Several mechanisms have been proposed to explain the juxtaposition of different flow rheologies in HEBs, each involving an increase in flow strength (or cohesion) either throughout or in parts of the flow:

i) The 'longitudinal segregation' model (Haughton et al. 2009) proposes that the cohesive, clay-rich and hydrodynamically 'lighter' components of a flow, such as mud clasts and organic material, are lofted within the flow, forming a clay-rich trailing tail to the leading turbulent head. Progressively, the trailing part of the flow develops a higher mud concentration, and owing to this cohesive fraction, forms a debris flow, which ultimately deposits on top of the sand-rich turbidite deposited by the head and body of the flow.

ii) The vertical transformation model (Baas et al. 2011; Kane et al. 2017) in which turbulence is modulated at the base of the flow by increased mud content during rapid deceleration, and rapidly passes through transitional to laminar behaviour, to develop a plug-like upper part.

iii) The late-stage settling model (Sumner et al. 2009), where coarser/denser grains settle through a decelerating high-concentration muddy flow to form a deposit with a relatively cleaner lower division and a muddier upper division.

iv) The combined segregation model (Baas et al. 2021), which invokes transformation of the flow head into a debris flow, and longitudinal segregation of bedload and vertical segregation of suspension load.

12.4.4 Contourite Deposits

Contourite successions have been intersected with cores to develop sedimentological models. Two main contourite deposit types have been described, both comprising mixes of biogenic and terrigenous material (Stow and Lovell 1979): (i) muddy contourites that comprise poorly sorted and thoroughly bioturbated mixes of biogenic and siliciclastic grains, which range from homogenous and structureless mud to mottled silt and mud with irregular lamination, and (ii) sandy contourites that comprise irregular, bioturbated lag deposits, or more rarely clean, cross-laminated beds,produced by winnowing of fines under stronger flows (Driscoll et al. 1985). These deposit types were arranged in a contourite bed-type model, which is commonly used for muddy contourite drifts (Gonthier et al. 1984; Figure 12.11d). The idealised bed-type coarsens upwards from homogenous mud, through mottled silt and mud, to a fine-grained sand, which is overlain by a fining-upwards pattern back to homogenous mud. The bigradational bed sequence is interpreted to record long-term increases and then decreases in bottom current velocity. Deposition rates are typically relatively low (commonly 10–100 cm ka^{-1}), which allows for pervasive bioturbation to take place. The contourite bed type has been refined with variants that emphasise the presence of omission surfaces that form during periods of maximum velocity, as well as partial bed sequences (i.e. top- or base-absent; Stow and Faugères 2008).

Coarser-grained contourites can be found in environments with strong bottom currents, such as moats, contourite channels or contourite terraces. Commonly, silt- to fine-grained sand-rich contourites are poorly to moderately sorted and show a very fine to fine skew, while medium- to coarse-grained sand-rich contourites present a coarse skew (Brackenridge et al. 2018). Gravel lags and erosion surfaces are identified in contourite channel-fills (Brackenridge et al. 2018). In locations far from terrigenous sources, sandy contourites may be found in condensed layers with low sedimentation rates, primarily composed of planktonic foraminifera (Lopes et al. 2023). Recently, bed-type schemes have been developed for bottom-current reworked sands (BCRS; e.g. De Castro et al. 2020; Figure 12.11d). They may consist of discretely rippled and well-sorted, very fine- to fine-sand or heterolithic deposits with minor to moderate bioturbation (De Castro et al. 2020). These deposits are often found in environments that are close to the source of the sediment such as at the channel/moat-drift transition (where turbidity currents may arrive from the adjacent slope), or in mixed turbidite–contourite systems (see Section 12.5.4).

12.4.5 Depositional Record of Flow-Topography Interactions

Seafloor topography can influence sediment gravity flow behaviour and transformations, and associated processes of erosion and deposition, above even subtle changes in slope (Kneller 1995; Kneller and McCaffrey 1999). Modern systems can switch from depositional to erosional regimes over gradient breaks of just a fraction of a degree (e.g. Mignard et al. 2019). The different topographic configurations (i.e. incidence angle of flows with respect to seafloor slopes, slope gradient and rugosity of relief) and the range of sediment gravity flow characteristics (flow thickness, velocity, concentration) result in a diverse range of process responses, which control the resulting direction and behaviour of flows, and affect the location and character of their deposits (Kneller 1995; Soutter et al. 2019; Figure 12.12a).

Palaeocurrent measurements from the rock record have shown that turbidity currents are highly susceptible to being deflected and reflected off seafloor slopes (e.g. Pickering and Hiscott 1985) and can become partially to completely ponded (i.e. trapped) in topographic lows (e.g. Haughton 1994; Figure 12.11d). Different flow rheologies have markedly different responses when interacting with seafloor topography, impacting the vertical and lateral distribution of their deposits (Figure 12.12a). For example, debris flows and mudflows are ground-hugging, and will not run far up topography, meaning their deposits are concentrated in topographic lows. Low concentration turbidity currents, with lower basal friction and density stratification, in contrast, result in thicker flows with much of the sediment load carried above the bed allowing them to run-up and override topography (e.g. Muck and Underwood 1990; Figure 12.13c). High-concentration turbidity currents are more strongly affected by topographic interaction than low-concentration turbidity currents (Al-Ja'aidi 2000; Gladstone et al. 2004). The interaction of strongly density-stratified flows with a slope may result in the high-concentration lower part of the flow being deflected and/or abruptly deposited, whereas the more dilute, well-mixed and finer-grained upper part may override the topography (Figure 12.12a). In 'flow stripping', this process separates flow in the submarine channel from that over the adjacent terrace or levee (Piper and Normark 1983; Figure 12.13c). Transitional and composite flow types will have complicated interactions with topography (Figure 12.12a), and, as we have seen, flow transformations can occur as flows, or parts of flows, increase cohesion through entrainment on mud-draped slopes.

Turbidity currents are said to be 'ponded' when the flow becomes fully trapped by seafloor topography to produce a flat-topped suspension cloud that spreads across the depocentre (e.g. Toniolo et al. 2006; Figure 12.11d). The orthogonal or oblique interaction of stratified sand-rich turbidity currents with an intrabasinal slope may result in flow 'reflection' or 'deflection', respectively. In both cases, the high-concentration lower part of the flow will deposit abruptly, whereas the more dilute, well-mixed and finer-grained upper part may override the topography. Flow deflection can also result in (i) downslope acceleration resulting in enhanced bypass and erosion (Kneller 1995; Kneller and McCaffrey 1999; Jobe et al. 2017), (ii) remobilisation and redeposition of sand higher-up on confining slopes (McCaffrey and Kneller 2001; Puigdefàbregas et al. 2004), and (iii) run-up and collapse of the high-velocity base of a flow into its more dilute cloud (Haughton 1994).

An idealised bed type deposited by large turbidity currents that deflected and reflected multiple times off topography in a constricted foreland basin (the Cloridorme Formation, Quebec, Canada) is presented by Hiscott and Pickering (1984; Figure 12.11d). Notable characteristics of deflected, reflected and ponded turbidites include: (i) sole structures indicting primary flow direction at a high angle to divisions of cross-ripple lamination with diverse palaeocurrent directions recording flow deflection and reflection; (ii) complicated grading patterns, with multiple subtle grain-size breaks with an overall normal graded profile suggesting multiple flow reversals; (iii) complicated sequences of sedimentary structures within an individual bed, with alternations of parallel, ripple, wavy and convolute lamination in the upper sandstone division, suggested highly unsteady flows; and (iv) a thick, normally graded to texturally homogeneous silty mudstone cap, suggesting the ponding of a large portion of the flow.

When flows ride obliquely up and down slopes they generate combined flow fields, with interacting unidirectional and oscillatory flow components. Flow deflection and reflection can result in flow transformations from unidirectional to 'combined flows'. Deposits of combined flows can be recognised by the formation of distinctive depositional bedforms (Harms 1969; Tinterri 2011; Taylor et al. 2024) and reworking of their parent deposit into hummock-like structures (Pickering and Hiscott 1985; Tinterri 2011; Privat et al. 2021). This reworking during flow-topography interactions can also result in palaeocurrent divergence adjacent to topographic obstacles, which can be used to reconstruct the physiography at the time of deposition (Kneller et al. 1991).

Documenting the termination, or pinch-out, of sandstone beds (and packages) in deep-water detrital systems can help constrain the physiography of the basin during deposition, and can indicate whether seafloor topography is static or dynamic. Large volume sediment gravity flows, relative to the size of the receiving basin, will interact with bounding slopes and promote onlap at basin margins and/or remobilisation (Haughton 1994). The flow direction relative to the orientation of the intrabasinal slope and the flow rheology, velocity and thickness are factors to consider as they influence run-up heights and directions, and flow transformation processes (e.g. Soutter et al. 2019). The different parts of density-stratified flows will interact differently with topography, with the high-concentration lower portion of the flow decelerating rapidly resulting in an abrupt pinch-out, and the low-concentration upper portion of the flow able to drape sediment onto the topography (Kneller and McCaffrey 1999; Figure 12.12a). This results in grain-size segregation, with the coarser fraction preferentially infilling the floor of the confined basin and the finer fraction of the density-stratified flow being able to deposit on the topography. Therefore, draping onlap and heterolithic pinch-out terminations deposited by thick low-concentration turbidity currents develop above low-gradient confining slopes with, whereas abrupt onlap of thick sandstones deposited by ground-hugging high-concentration turbidity currents typically occurs against higher gradient confining slopes (Gardiner 2006).

Topography on the seafloor also steers and modifies bottom currents and thus controls the development of contourites (Figure 12.8). Through time, the growth of a contourite drift increasingly affects bottom currents, resulting in positive feedback and further contourite drift growth under similar general circulation patterns (Miramontes et al. 2016). Different contourite types (Section 12.5.4) are influenced by seafloor topography in different ways. Separated mounded drifts and their related moats (moat-drift systems) or contourite channels are commonly found where there is a strong decrease in slope gradient, such as the base of continental slopes or escarpments formed by slope failure or faults (Rebesco et al. 2014; Wilckens et al. 2023). The steeper the slope the more intense the secondary flow across the slope, resulting in the formation of larger and steeper drifts compared to locations with more gentle slopes at the same current velocities (Wilckens et al. 2023). In contrast, plastered drifts are commonly found on planar low relief slopes (Hernández-Molina et al. 2009; Miramontes et al. 2021).

Seafloor topography also affects the generation and propagation of internal waves, which can be generated in straits, at ridges, or at the shelf-edge break, and then propagate onshore and offshore until they finally break on the slope or dissipate in the open ocean (Lamb 2014). The interaction of internal waves with the slope can drive sediment resuspension and transport processes, and the formation of a 200–1000 m thick layer of water with increased suspended sediment concentration, known as a nepheloid layer (e.g. Bourgault et al. 2014). Internal waves being reflected or breaking on slopes can also induce the formation of (i) sand dunes (e.g. Droghei et al. 2016; Miramontes et al. 2020); (ii) sediment waves with crests oriented parallel to the slope contours (e.g. Ribó et al. 2016); (iii) channels oriented parallel to the slope; and (iv) erosional surfaces (Miramontes et al. 2020).

12.4.6 Recognising Clastic Injectites at Outcrop

Sand injectite networks have four components: (i) the parent sandbody, such as a channel-fill, which is partially or completely liquified and fluidised; (ii) sills, which are roughly concordant with bedding with steps up and down across the host strata (e.g. Truswell 1972); (iii) dykes, which are discordant with stratigraphy, and commonly bifurcate, taper and/or are ptygmatically folded (e.g. Jolly and Lonergan 2002); and (iv) seafloor extrusions (e.g. Vigorito et al. 2008), such as sand volcanoes in 'open systems'. Where the host rock is cohesive, sheet intrusions occur in the form of dykes, sills or cones, and where the host sediments are cohesionless, pipes form that feed surface extrusions (Maltman 1994; Ross et al. 2011). Injectites have been documented across a wide range of scales (Figure 12.19) from kms in length and 10s–100s m thick, most often recognised in seismic reflection data (e.g. Dixon et al. 1995; Cobain et al. 2021) to cm scale examples documented in core and at outcrop (e.g. Scott et al. 2009).

At outcrop or in core, internal and external structures help distinguish clastic injectites from normal sandstone bedding and the interaction between the injecting sand flow and the host muds/mudstones. External features are preserved at the sharp contacts between injectites and host strata and are either caused by the flow of fluidised sand (scours, scallops, flutes,) or record the fracture morphology of the surrounding mudstone (plumose marks). Most injectites appear structureless but internal structures can include laminations, banding, clasts and grain-size grading (Scott et al. 2009). Grading is seen perpendicular to the walls/margins of the injection and may occur as a single or multiple normal or inverse graded layers (Hubbard et al. 2007). Clasts are commonly angular, locally derived and tend to be concentrated at injectite margins. Laminar flows will preserve structures created on the margins of clastic injectites during initial fracture, like plumose features (Figure 12.19c; Cobain et al. 2015), whereas turbulent flows will cause erosion of the host lithology along the margins of clastic injectites, consequently producing features such as flutes and scours.

12.5 Architectural Elements and Hierarchy

One approach to understanding deep-water detrital systems is to view them as composite bodies formed of building blocks, or 'architectural elements' (e.g. Mutti and Normark 1987, 1991; Ghosh and Lowe 1993). Architectural elements (Section 2.1) show a progressive increase in scale, or a hierarchy, from the deposit of a single sediment gravity flow (a bed) through bedsets (or storeys), elements, complexes, to the accumulated deposits that comprise entire slope or basin floor successions (complex sets). Collectively, these stratigraphic building blocks stack to build the sedimentary architecture of submarine fans. Their identification is based solely on their physical stratigraphic characteristics: architectural elements are defined using their geometries and vertical profiles, their bounding surfaces and their assemblages of bed types. Adopting a hierarchical approach means that, for example, the different types of architectural elements common to channel–levee systems and lobe systems can be compared across different deep-water detrital systems. Architectural elements identified at outcrop can be usefully compared to those seen in seismic reflection profiles and horizon slices, and vice versa (e.g. Deptuck et al. 2003; Posamentier and Kolla 2003; Figure 12.20). We characterise the dominant architectural elements that comprise deep-water detrital systems in proximal to distal order across a simple slope profile (Figure 12.1), from channel–levee systems (Section 12.5.1), with channel-fills and adjacent overbank environments (external and internal levees and terrace deposits), through channel mouth settings (Section 12.5.2), submarine lobe systems (Section 12.5.3) and contourite environments (Section 12.5.4).

12.5.1 Channel–Levee Systems

Channel–levee systems are confined by a composite bounding surface that forms through a combination of substrate incision (channel-belt erosion surface) and adjacent

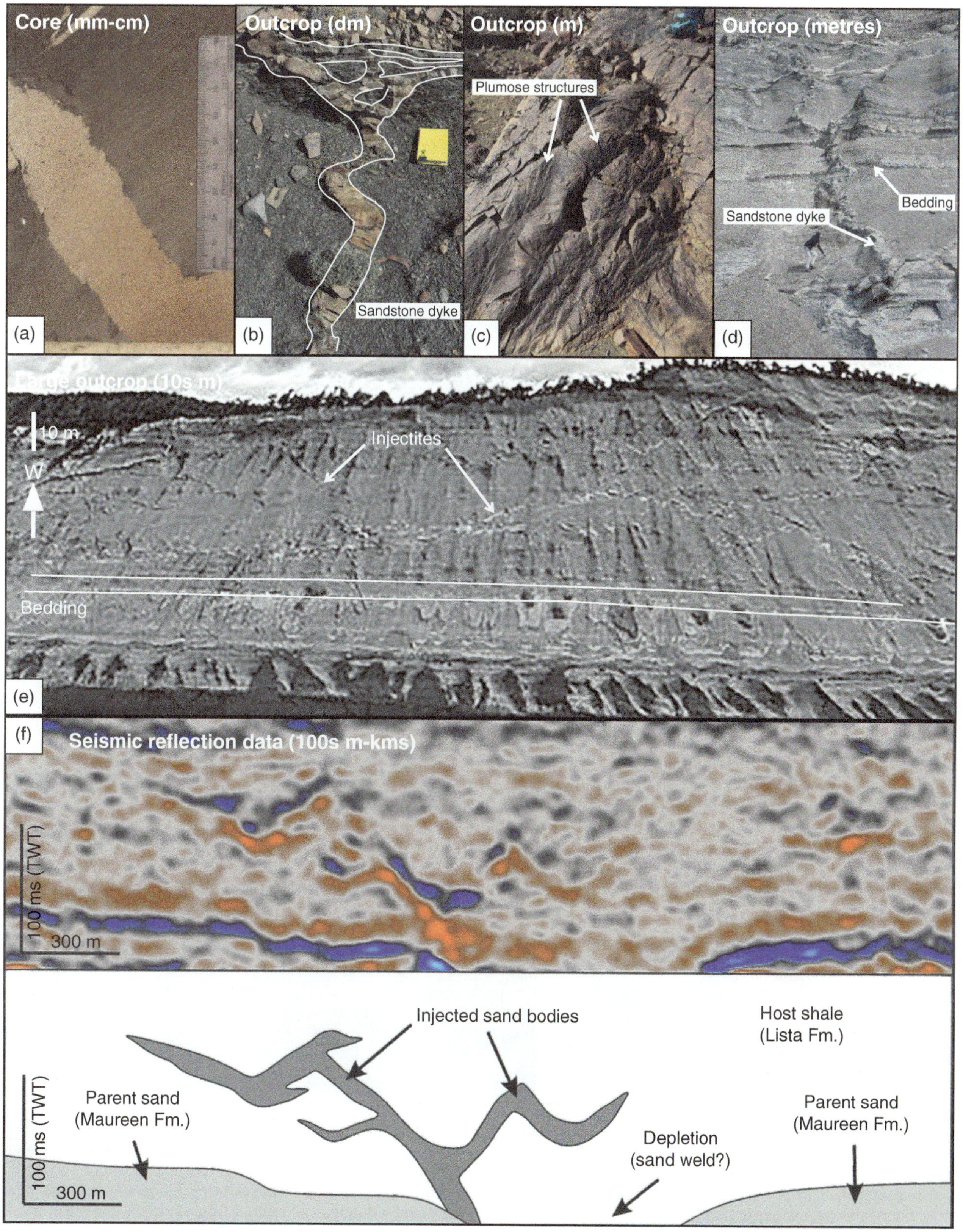

Figure 12.19 Photographs and images of clastic injectites across a wide range of scales. (a) Core photograph from a Paleocene succession, North Sea. (b) Sandstone dykes, Fort Brown Formation, Laingsburg depocentre, Karoo Basin, South Africa. (c) Top surface of injectite showing plumose structures, Skoorsteenberg Formation, Tanqua depocentre, Karoo Basin, South Africa. (d and e) Injectite network in Miocene deep-water sediments, Austral Basin, Patagonia, Argentina. (f) Top: representative reflection seismic section of injectites in the Palaeogene North Sea UK. Bottom: simplified interpretation of depleted parent sand (Maureen Fm.) and resultant injected sand. *Source:* (a, b, e, and f) Adapted from Cobain et al. (2021).

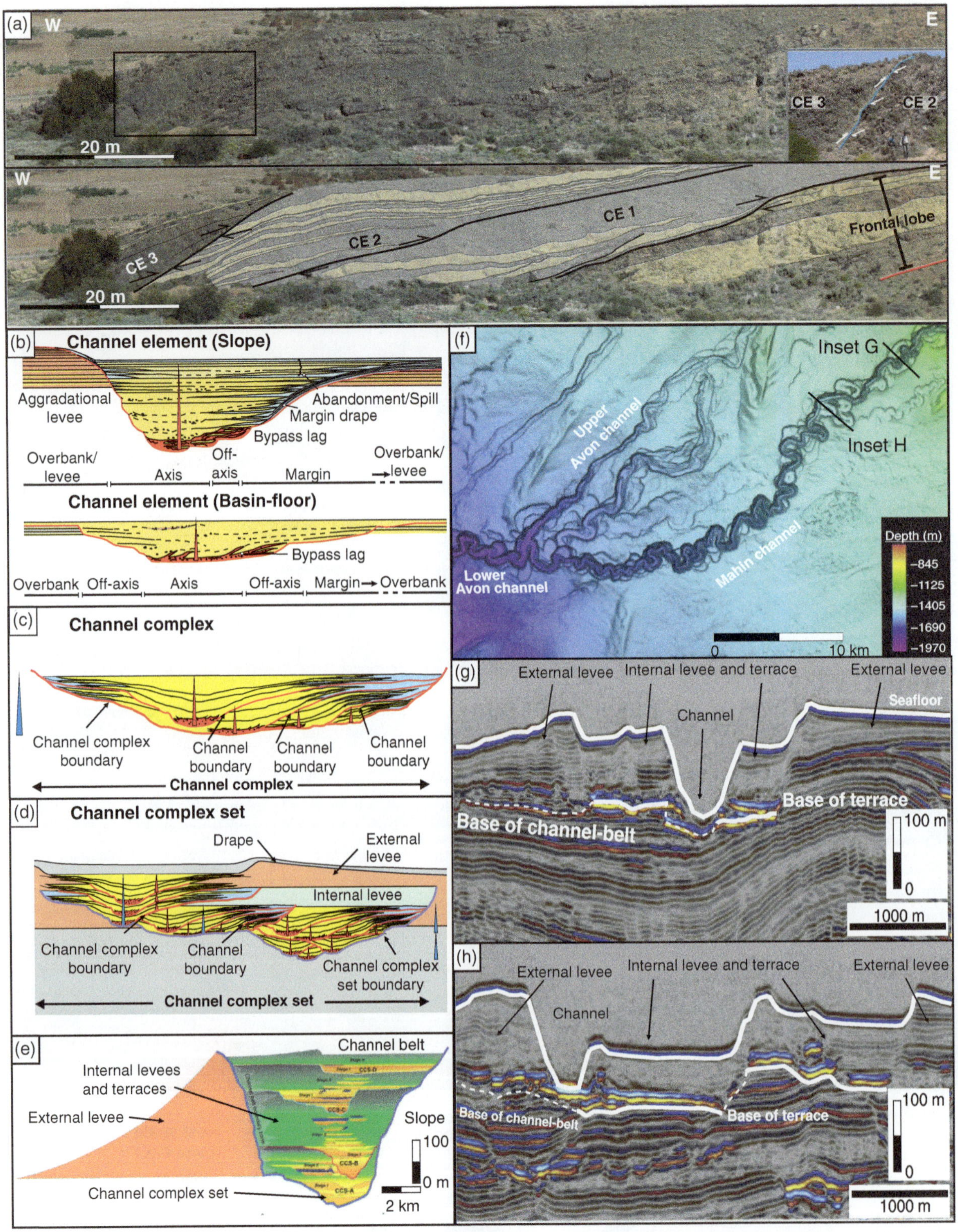

Figure 12.20 (a) Outcrop example of a submarine channel complex, from Unit C, Fort Brown Fm., Karoo Basin, South Africa. Note the three channel-fill elements (CE) that stack laterally to form a channel complex, and the high proportion of thin beds. Photo credit: David Hodgson. (b) Schematic representations of submarine channel-fill element types. Top, channel-fill element in a slope setting with coarse-grained amalgamated sandstone beds in the axis that transition laterally through an off-axis to a thinner bedded sandstone and siltstone channel margin. Bottom, an isolated basin-floor channel-fill element with a less heterolithic fill, and more symmetrical cross-sectional form. Typical scales of channel elements are 15–25 m thick and 150–300 m wide. (c) Several channel-fill elements stack to form a channel complex. (d) Several channel complexes stack to form a channel complex set. (e) Stacked channel complex sets (CCS a–d) forming a channel belt from the Rosario Formation, Mexico. (b–e) Yellow: sand-rich amalgamated channel-fill sediments, blue: mud-rich channel-fill sediments, red: mud-clast-rich channel-fill sediments, gray: mud-rich pre-existing sediments. (f) Seafloor image of the Avon and Mahon submarine channels, offshore Nigeria. (g and h) Cross-sectional geometry of Mahin channel, showing flat-lying terraces, wedge-shaped levees and high-amplitude (yellow and bright blue) reflectors marking the channel base. *Source:* (b) Modified from McHargue et al. (2011). (c and d) Modified from Di Celma et al. (2011). (e) Modified after Kneller et al. (2020). (f–h) Modified after Hansen et al. (2017).

construction of external levees (*sensu* Kane and Hodgson 2011) by flow stripping and overspill. The shape of these composite erosional and constructional surfaces resembles large-scale submarine slope valleys. Above simple submarine slope profiles, they pass downstream from more erosionally confined (entrenched) to more levee-confined architectures (e.g. Deptuck et al. 2008; Janocko et al. 2013; Figure 12.4). The bounding surface is strongly diachronous, and in ancient systems, it should be noted that their preserved composite 3D forms likely never existed as single bathymetric features on the seafloor (Sylvester et al. 2011; Hodgson et al. 2016).

12.5.1.1 Channel-Fills

A threefold hierarchical scheme describes submarine channel features (Figure 12.20). The elementary channel-fill, the 'channel element', defines the depositional body formed within a discrete conduit for confined flow. The fill of a channel element will comprise beds and bedsets, collectively referred to as storeys (as in fluvial systems). Stacking of two or more channel elements of similar grain-size, sedimentary facies and architectural style form a 'channel complex' (Figure 12.20a), and two or more channel complexes stack in consistent patterns to form 'channel complex sets'. 'Channel-belts' comprise a channelised axis component (channel complex [sets]) and a margin component comprising terrace deposits and internal levees (Kane and Hodgson 2011; Hansen et al. 2017; Kneller et al. 2020).

At outcrop, a channel-fill element is defined by a basal erosion surface that truncates underlying bedding, and/or has a constructional relationship with adjacent and contemporaneous levee/overbank deposits. The sediments that comprise channel elements are usually organised into a predictable pattern of bed types and degree of bed amalgamation. The 'axis' of a channel element is located at the thickest part of the channel-fill, it displays the greatest amount of bed amalgamation, has the thickest beds, the highest concentration of coarse sediment and often includes a basal unit, comprising intra- and/or extraformational clasts in a sand matrix, interpreted as a lag deposit recording sediment bypass dominated periods (Stevenson et al. 2015). The 'channel margin' of a channel-element is located towards the lateral edges of the channel fill, and beds are thinner, display less amalgamation and are finer grained. A channel 'off-axis' association is sometimes defined to describe the transition between channel axial and channel marginal environments (e.g. McHargue et al. 2011, Figure 12.20).

In terms of scale, channel elements are typically 10–30 m thick and 100–400 m wide (~10:1, width: thickness; McHargue et al. 2011). Commonly, in channel-belts only part of a channel element is preserved due to the repeated phases of erosion and deposition. Where a full cross section is preserved the channel elements tend to be asymmetric in their geometry and pattern of infill. This is attributed to a planform sinuosity with the steeper margin, representing the outer bend, being overlain by amalgamated sandstones and the shallower margin, representing the inner bend, overlain by thin beds. Channel-fills can extend beyond the slope onto the basin floor to form distributive channel networks where they are often isolated (i.e. do not stack to form channel complexes). Typically, these channel-fills cut underlying lobe deposits and are more symmetrical in cross-section as they have low sinuosity and their infill is sand-rich to the margins (Figure 12.20).

The stratigraphic evolution of submarine channel–levee systems generally results in the formation of a large-scale, composite, erosional bounding surface that forms during incision and lateral migration of the active channel floor during early channel-system evolution (Deptuck et al. 2003; McHargue et al. 2011). This dominantly degradational phase is followed by increased preservation of the component channel elements, which may stack laterally, obliquely and vertically. The channel elements stack in a range of patterns, and trajectories, from lateral- through oblique-, to vertical-stacking. Lateral-stacking of channel fills is widely reported from the base of channel-belts, with repeated cycles of cutting, filling and stepping of channel elements (Figure 12.20a). The lateral stepping widens the composite basal surface, and thin-bedded channel margin deposits are preferentially preserved (Hodgson et al. 2011). In geophysical data, this stepped migration could be interpreted as lateral accretion (i.e. lateral migration of an open and sinuous channel). Commonly, a change from lateral- to vertical-stacking of channel elements, and aggradation of adjacent terrace deposits and internal levees, is observed in the later stages of channel–levee system evolution prior to abandonment (see Section 12.6.2 for an explanation).

12.5.1.2 External and Internal Levees, and Terrace Deposits

The main diagnostic criteria for the identification of 'external levees' are a constructional wedge-shaped body of sediment (10s of metres thick) that thins for up to several kilometres away from the channel-belt, with basal downlap (Figure 12.21). They are readily identified in subsurface seismic reflection and seafloor datasets (e.g. Deptuck et al. 2003; Posamentier and Kolla 2003). They comprise substantial volumes of fine-grained thin-bedded sediment (Figure 12.21e,f) and fine away from the channel belt, which means that their full extent is often challenging to constrain (e.g. Posamentier and Kolla 2003). Commonly, stratigraphic fining- and thinning-upward trends are

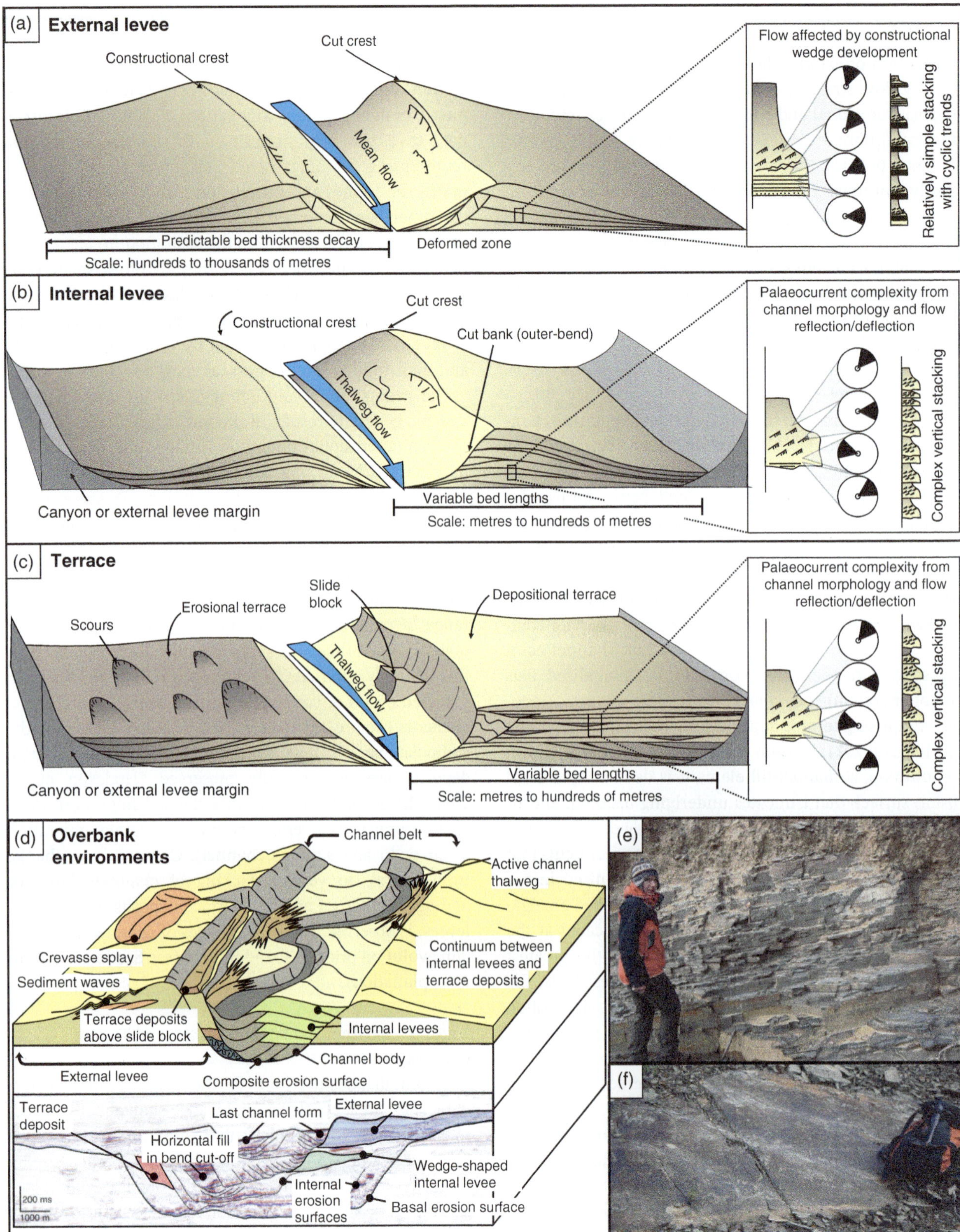

Figure 12.21 Channel bounding sedimentary deposits. (a) External levees bound an entire channel belt; (b) internal levees develop a distinctive wedge shape geometry but are confined to a wider channel belt by external levees or canyon walls; (c) terraces are flat-lying and may be depositional, where flows are relatively large within a confined conduit, or may be erosional and marked by scours and bedforms. A continuity between terraces and internal levees is common and is determined by the magnitude and grain-size of the overspilling flows. A levee geometry can only form if the overspilling flows are relatively small compared to the main channel confining surface. (d) Top: Schematic diagram of depositional environments within an aggradational submarine channel belt. Base: Subsurface seismic reflection image from the Indus channel. (e) Aggradational thin-bedded siltstones in external levee, and (f) large-scale asymmetric bedforms in the upper part of a distal external levee (Unit D, Fort Brown Fm., Karoo Basin, South Africa, see Figure 12.10). *Source:* (a–c) Modified from Kane and Hodgson et al. (2011). (d) Modified from Allen (2020) and Sylvester et al. (2011). (e and f) Photo credit: David Hodgson.

reported in external levee successions from core and outcrop (Kane and Hodgson 2011). This pattern is attributed to increasing flow confinement through time as the levee is constructed and the vertical distance between the channel thalweg and the levee crest increases, resulting in increasingly dilute parts of turbidity currents being flow-stripped.

Large volumes of channel-belt fills are not composed of the channel complex sets, but instead comprise a combination of 'internal levees', 'terrace deposits' and slide blocks from the steep margins (Hansen et al. 2015; Figure 12.21). The term terrace has been used in deep-water environments to describe areas of subdued topography adjacent to a channel (Babonneau et al. 2004) and terraces can be of erosional and/or depositional origin (Hansen et al. 2015; Figures 12.9 and 12.21). Internal levees are defined as channel element and complex flanking features within a channel belt, which have a wedge shape in cross section and thin away from the channel but that still lie within the broader confines of the channel belt (Kane and Hodgson 2011, Figure 12.21).

The differentiation of internal levees and terrace deposits is largely semantic given that both are supplied by flows from the adjacent channel via flow-stripping and overspill but remain confined by the channel-belt surface. Also, both settings show complicated palaeocurrent patterns with multiple flow directions due to flow deflection and reflection (Kane and Hodgson 2011; Figure 12.21), and both are best preserved during the later stage fill of a channel system. More crucial are the concepts that: (i) volumetrically significant thin-bedded and fine-grained successions can accumulate within large-scale confining surfaces of channel-belts, and (ii) these settings, together with external levees, are part of an effective grain-size segregation process where finer grained parts of density-stratified flows are sequestered on the submarine slope. This means that the flows exiting the mouths of submarine channels can be relatively enriched in sand.

12.5.2 Channel-Mouth Settings

Submarine channel mouth settings, widely referred to as submarine 'channel-lobe transition zones' (CLTZs), are areas between channel–levee systems and lobes (Mutti and Normark 1987, 1991; Wynn et al. 2002) and are associated with a distinctive assemblage of erosional and depositional bedforms (e.g. scours and scour fields, sediment waves, incipient channels). Typically, channel mouth settings are dominated by sediment bypass (Stevenson et al. 2015; Brooks et al. 2018), and rapid flow deceleration and expansion.

A wide range of submarine channel mouth configurations has been imaged in modern systems (Figure 12.22), which provides important understanding of the transfer, and preservation, of channel mouth settings into the stratigraphic record. The dominance of erosional bedforms exacerbates the difficulty in their identification in outcrop. However, a relative abundance of supercritical and transcritical bedforms, including cyclic steps, sediment waves and scour-fills is predicted. Interpreted examples from outcrops range from single surfaces that separate lobe elements from overlying channel–levee systems (e.g. Gardner et al. 2003; Hodgson et al. 2016) to thick successions of aggradational beds in close association with scour-fill features (e.g. Pemberton et al. 2016). Although criteria for the stratigraphic identification of channel mouth settings are still evolving, treating deep-water detrital successions as being either channel–levee- or lobe-dominated settings is too simplistic.

12.5.3 Submarine Lobe Systems

In many submarine fan successions, the dominant architectural elements by volume are laterally extensive but thin (high aspect ratio [width: thickness]: 1000:1) lobate, sand-rich accumulations referred to as 'submarine lobes'. In unconfined settings, submarine lobes form down-dip of submarine channel mouths (e.g. Normark 1970; Deptuck et al. 2008), with elongate elliptical to equi-dimensional planforms (e.g. Jégou et al. 2008; Saller et al. 2008) and subtle convex-up profiles. Originally, lobe deposits were conceived as simple radial (lobate) bodies that thin and fine from an apex (e.g. Normark 1970; Mutti 1977). It is now recognised that submarine lobes are more variable and complicated in terms of morphology, sub-environments, grain-size distributions and stacking patterns (e.g. Prélat et al. 2009; Etienne et al. 2012; Fildani et al. 2018; McHargue et al. 2021; Figure 12.23). Documented dimensions of lobes range between 4 and 50 km long, 4–18 km wide and 2–47 m thick, and fall into two broad populations: thin unconfined lobes and thick confined lobes (Prélat et al. 2010).

An architectural hierarchy has been established for submarine lobe deposits (Gervais et al. 2006; Deptuck et al. 2008; Prélat et al. 2009; Spychala et al. 2017), which is adapted here to align with the channel hierarchy (Figure 12.23b): (i) one or more beds stack to form a 'lobe bedset' (or storey); (ii) several bedsets form a 'lobe element' (the 'lobe' of Prélat and Hodgson 2013); (iii) several related lobes stack to form a 'lobe complex'; and (iv) multiple lobe complexes stack to form a 'lobe-complex set'. These bodies systematically increase in thickness and lateral extent with increasing hierarchical rank. Each higher rank within the sandstone hierarchy is separated by a siltstone-rich unit that correspondingly increases in thickness (Prélat et al. 2009), which represents the fine-grained fringe of a body

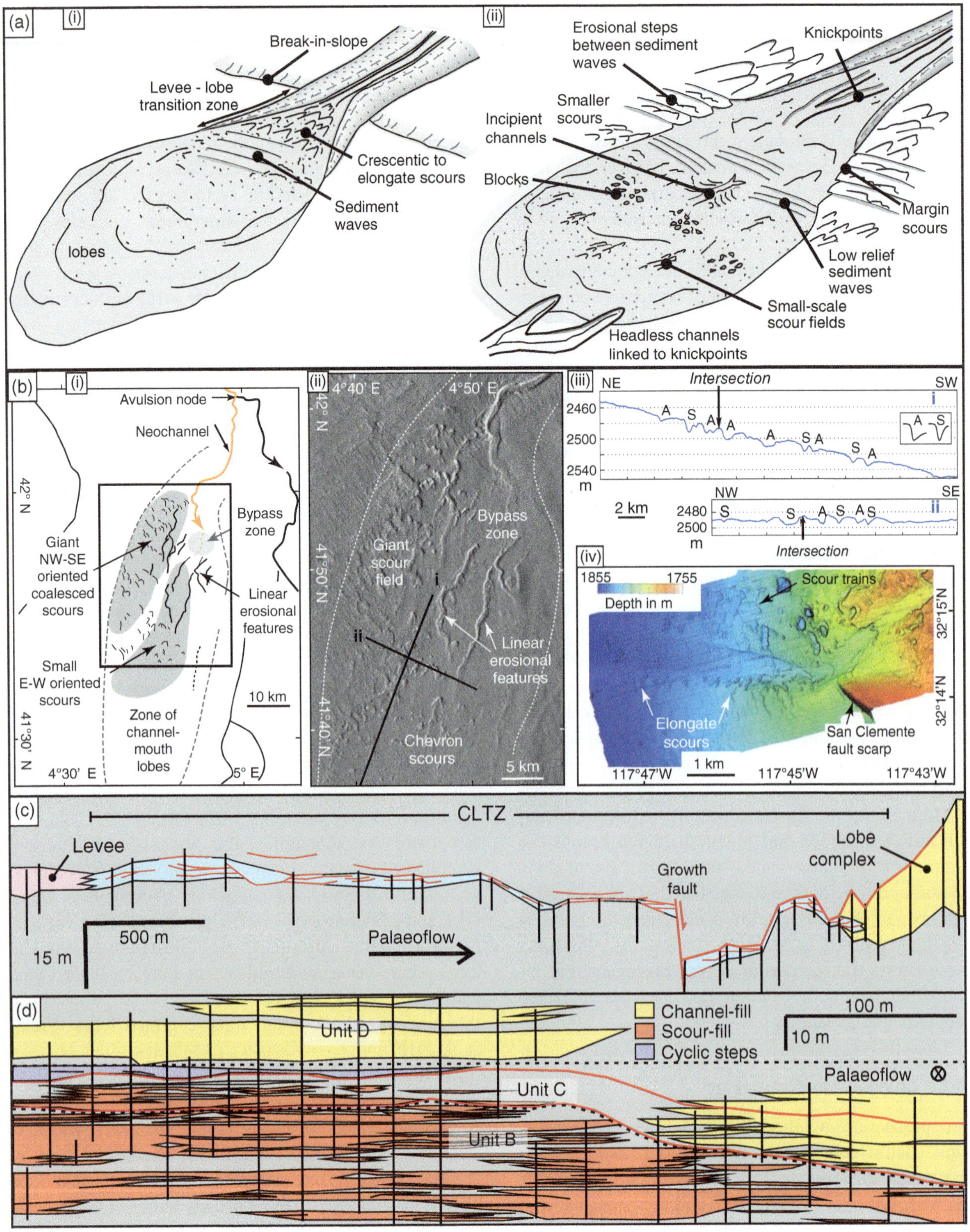

Figure 12.22 (a) Cartoons of end member channel mouth settings. (i) A channel-lobe transition zone (CLTZ), based on Wynn et al. (2002) and Brooks et al. (2018). Note the transition from scours to sediment waves downstream of the channel mouth. (ii) A channel mouth expansion zone (CMEZ) based on Maier et al. (2020). Note the scours and sediment waves inside and outside the area where the channel mouth flares downstream. (b) Erosional CLTZ seafloor features. Schematic interpretive map (i) shaded multibeam bathymetry (ii) and seafloor relief profiles (iii) of the Rhône neofan CLTZ). Scours are designated as asymmetrical (A) and symmetrical (S). (iv) High-resolution bathymetry of a channel mouth setting on the Navy Fan, offshore California, showing large sectors of the seafloor sculpted by elaborate bedforms and scours. *Source:* Adapted from Carvajal et al. (2017). (c) A thin CLTZ in Unit E3, Fort Brown Formation, Laingsburg depocentre, Karoo Basin, South Africa, in a base-of-slope location. *Source:* Adapted from Brooks et al. (2018). Red lines are scour surfaces, and the aquamarine unit is the interpreted CLTZ, which is <10 m thick. Note the growth fault, and the abrupt thickening downdip of the sand-rich lobe complex (yellow). (d) A thicker succession recording the overall progradation of a lower slope succession with many scour-fills, and local cyclic steps, below and adjacent to stacking channel elements from the Tres Pasos Formation, Magallanes retroarc foreland basin in southern Chile. *Source:* Adapted from Pemberton et al. (2016). All adapted from Hodgson et al. (2022).

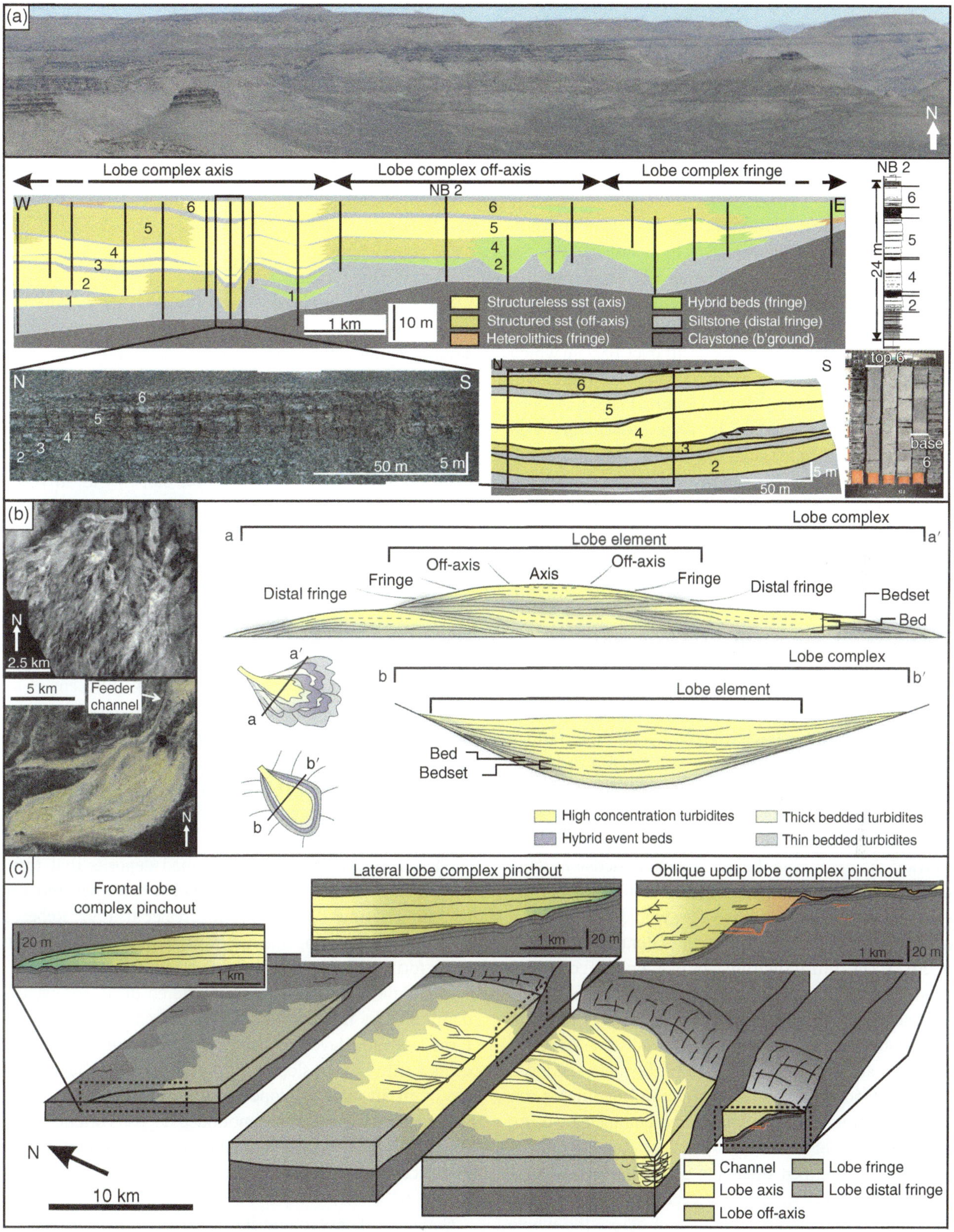

Figure 12.23 (a) Top, several exhumed Permian submarine fans, Tanqua depocentre, Karoo Basin, South Africa, where individual lobe elements have been mapped. Field of view ~3.5 km. Base, strike correlation panel of lobe elements in a lobe complex (Fan 3). Numbers refer to lobe elements documented by Prélat et al. (2009). Outcrop photograph and inset panel is a dip section showing amalgamation of lobe elements. Colours related to lobe sub-environments, and grey units are thin-bedded distal fringes of lobe elements. Log and photograph are from a research borehole (NB2) located on the correlation panel. (b) Left, planform images of submarine lobes, both Root Mean Squared RMS) amplitude extractions from 3D seismic reflection volume on the middle slope, offshore Nigeria. Right: (a–a′) Cartoon of a lobe complex showing compensational stacking and facies distribution of lobe elements in an unconfined setting. (b–b′) Lobe complex in a confined setting dominated by aggradation and complex onlap styles. (c) Block diagram based on the Fan 3 lobe complex to illustrate different lobe element environments and oblique up-dip, lateral and frontal pinchouts of Fan 3. (a and c) Adapted from Hansen et al. (2019), (b) adapted from McHargue et al. (2021) and Fuhrmann et al. (2022).

at the same hierarchical rank (Prélat and Hodgson 2013). Lobe elements are defined by their laterally extensive bounding surfaces and marked by abrupt stratigraphic facies changes. The bounding surfaces between sandstone and siltstone units are interpreted to record avulsion of feeder channels.

Prélat et al. (2009) identified four lobe element sub-environments that are characterised by sedimentary facies associations and thicknesses (Figure 12.23b). These are termed: lobe element axis, lobe element off-axis, lobe element fringe and lobe element distal fringe. The transitions from one environment to another occur over 100s m, which means that definitions of lobe element sub-environments of deposition are necessarily broad (Figure 12.23a). A 'lobe element axis' is characterised by thick-bedded sand-rich deposits (>85% sandstone) that commonly show bed amalgamation and scouring. The deposits of the lobe axis are laterally extensive down-dip and across-strike for several hundred metres. Amalgamated lobe element axes can be traced into non-amalgamated medium-bedded units of the 'lobe element off-axis' (50–85% sandstone), which is characterised by medium-bedded sandstones with abundant sedimentary structures (planar lamination, banding and ripple lamination). The lobe off-axis transitions over several hundred metres in both dip and strike directions to 'lobe element fringe' settings (20–50% sandstone), which display a heterolithic character. Typically, the lateral fringe of lobe elements is characterised by thin-bedded turbidites and the frontal fringe of lobe elements by HEBs (Spychala et al. 2017). This difference is attributed to more erosion and entrainment of fines by turbidity currents with a high basinward momentum in the lobe axis that increases cohesive forces to drive flow transformation, compared to the laterally spreading parts of flows that would deposit cohesionless Bouma-type turbidites in the lateral fringe (Kane et al. 2017). The transition from lobe element fringe to 'lobe element distal fringe' (<20% sandstone) settings marks the sand pinch-out (Figure 12.23c). The distal fringe environment of lobe elements is dominated by siltstone deposits, which form an extensive 'halo' around the main sand-rich body and extend for several kilometres, and, at outcrop can be hard to distinguish from pelagic and hemipelagic deposition (Boulesteix et al. 2020).

The stratigraphic bed thickness pattern preserved in a lobe element reflects the internal organisation of bedsets, rather than initiation and progradation across the whole lobe element as implied by a thickening-upward only pattern. Three broad types of stacking patterns in lobe systems can be considered: compensational, aggradational and longitudinal stacking, which occur across different hierarchical scales (Deptuck et al. 2008; Prélat and Hodgson 2013). 'Compensational stacking' describes the tendency of sediment to fill topographic lows created by preceding deposits (Mutti and Sonnino 1981; Straub et al. 2009). 'Aggradational stacking' occurs in highly confined settings where avulsion is restricted due to the scale of confinement (e.g. Marini et al. 2015). 'Longitudinal stacking' patterns occur where there is lateral confinement.

12.5.4 Contourite Systems

Contourites can be classified according to their orientation and shape, which are in turn controlled by sediment deposition, winnowing and erosion and non-deposition (Miramontes et al. 2020; Figure 12.24). Small-scale contourites (<1 km length) can be found superimposed on large-scale contourites (10s km in length). 'Separated mounded drifts' are slope parallel elongated mounded sediment bodies. They are separated from the slope by an incision, which is called a 'moat' when the drift has a pronounced crest, or 'contourite channel' when no evident crest is present and the channel is dominated by erosion forming truncations on the drift flank (Rebesco et al. 2014; Wilckens et al. 2023). Separated mounded drifts, and their related moats (moat-drift systems) or contourite channels, are commonly located at the foot of continental slopes and any type of obstacle (Rebesco et al. 2014; Wilckens et al. 2023). Typically, separated mounded drifts are composed of muddy sediments, but they can also contain bi-gradational beds, which coarsen and thicken towards the moat (Rebesco et al. 2014). Moats and contourite channels are usually dominated by relatively coarser sediments (García et al. 2009).

'Plastered drifts' are convex-shaped deposits that form depocentres with a contour-parallel orientation and are commonly found on slopes (Faugères et al. 1999; Rebesco et al. 2014). The upper (shallower) part of the plastered drifts forms a flat area referred to as a 'contourite terrace'. The middle and lower part of plastered drifts are usually dominated by muddy sediments while contourite terraces form alternate layers of sand and mud. Since plastered drifts are found on slopes and have a convex shape, they are often affected by slope instability (Miramontes et al. 2019). 'Detached drifts' are mounded deposits oriented perpendicular or oblique to bathymetric contours found at the intersection of two bottom currents flowing in opposite directions.

Generally, large contourite drifts are muddy. However, when bottom current velocities are high enough, they can generate intense winnowing and reworking of sediments (Figure 12.24d), forming 'sandy contourites' and building large-scale features: (i) contourite terraces, typically associated with plastered drifts on slopes; (ii) sandbanks, common on seamounts and drowned platforms; (iii) condensation

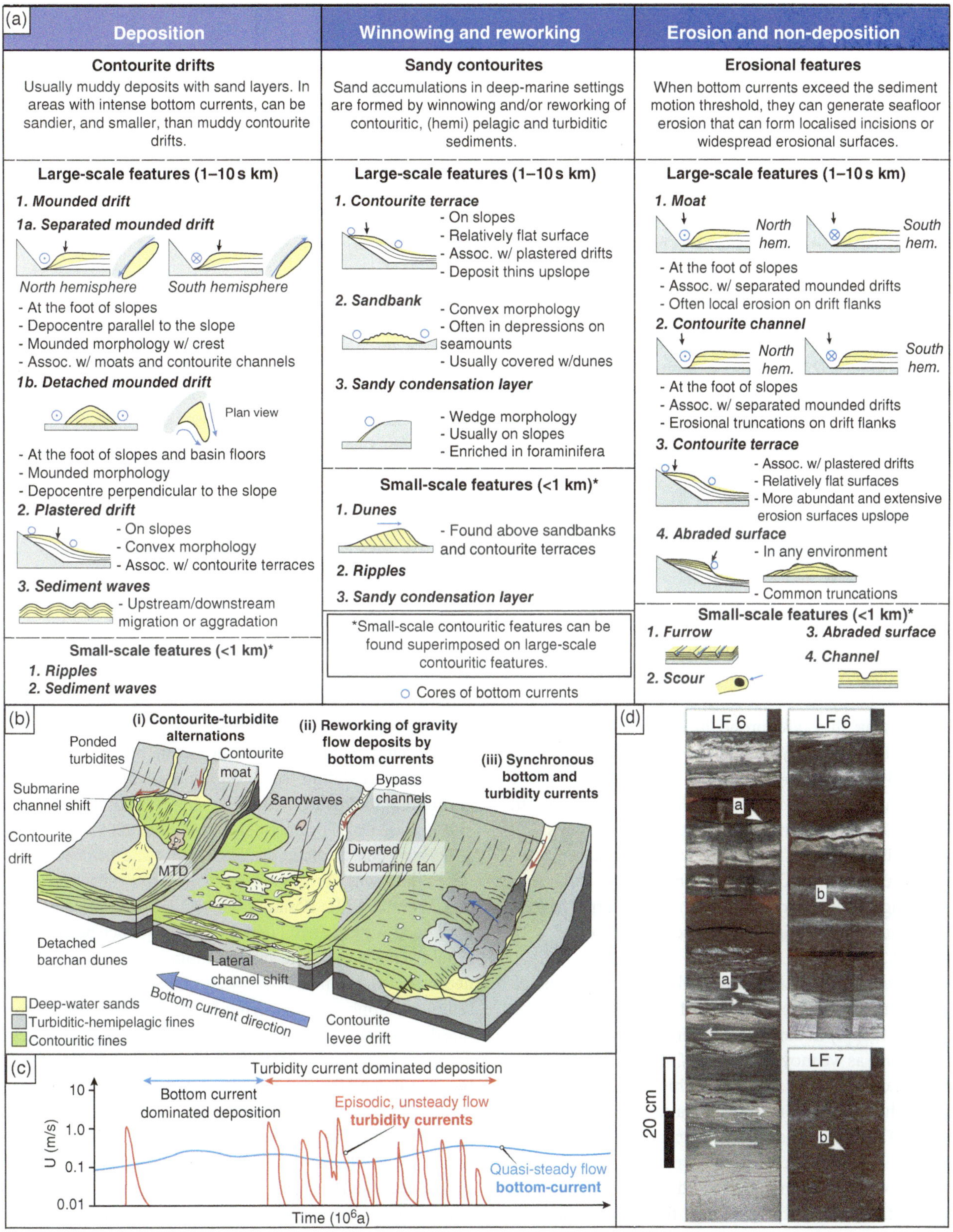

Figure 12.24 (a) Classification of contourites based on their general geometry and internal architecture, and the related bottom current intensity increasing from deposition to erosion and non-deposition. (b) Conceptual deep-water depositional and process models for the interaction of sediment gravity flows and bottom currents: (i) contourite/turbidite alternations; (ii) reworking of sediment gravity flow deposits by contour current processes; (iii) interaction of bottom currents and sediment gravity flows. The red arrows indicate the sediment gravity flow pathways, the blue arrows indicate the effect and direction of bottom currents. (c) Conceptual velocity (U) profiles illustrating the superimposition of short-lived turbidity currents (red) on a background quasi-steady bottom current (blue). (d) Bottom current reworked sandstones above lobe fringe turbidites showing changing palaeoflow directions (a) with low sedimentation rates that allowed for pervasive bioturbation (b; lithofacies [LF]6). Drift deposit comprises mottled, red to green mud-rich siltstones (LF 7). *Source:* (a) Adapted from Miramontes et al. (2021). (b) Adapted from Fonnesu et al. 2020. (c and d)Adapted from Fuhrmann et al. (2022).

layers, which in distal areas sometimes are formed by foraminiferal sand. Small-scale (<1 km) sandy contourites, like ripples and dunes, can be superimposed on large-scale features. In areas affected by intense bottom currents, dominated by erosion and non-deposition, it is common to find: (i) moats and contourite channels that parallel the slope, associated with a slope break; (ii) contourite terraces, commonly associated with plastered drifts; and (iii) widespread erosion surfaces (abraded surfaces). Increasingly, systems where the interaction of sediment gravity flows and bottom currents are identified, and conceptual depositional and process models have been developed (Fonnesu et al. 2020); Fuhrmann et al. 2022). A range of architectural configurations can develop through interactions between quasi-steady bottom currents and episodic and unsteady turbidity currents (Figure 12.24b,c).

12.6 Stratigraphic Controls

Disentangling allogenic and autogenic signals is challenging, particularly in deep-time stratigraphy and using core/well data without good 3D constraints for correlation. External (allogenic) signals of palaeoenvironmental change may not be faithfully recorded in the deep-water detrital successions as they are prone to being overprinted or shredded by internally generated (autogenic) controls (Chapter 4) by the long timescales and travel distances involved in many systems. Nonetheless, when changes in grain-size and provenance, in architectural elements and their stacking patterns and in palaeocurrents and processes are observed, we should consider whether the balance of these changes is being driven by external or internal factors, even if the record is inevitably a combination of both (Figure 12.25).

Sequence stratigraphic frameworks have emphasised allogenic drivers such as relative sea-level change and rates of sediment supply as dominating over autogenic drivers. In simple terms, as we move our scale of observation from the individual event deposit, the bed, to thick sedimentary successions, the main drivers shift from autogenic to allogenic controls. That said, deep-water detrital systems are the most distal part of siliciclastic sediment supply systems, and so should a stratigraphic record only attributable to allogenic factors be expected? We first cover allogenic (Section 12.6.1) and autogenic (Section 12.6.2) controls in deep-water detrital systems, then consider the stratigraphic evolution of intraslope and confined basins (Section 12.6.3), which are commonly preserved in the rock record. Section 12.4.54 presents two subsurface case studies (, both compiled by Mike Mayall, Imperial College, UK); first, a passive margin system with salt diapirism (offshore Angola), followed by a shallow marine to deepwater system (Cretaceous foreland basin, North Slope Alaska, USA). This chapter concludes with Section 12.6.45, which discusses the record and influence of anthropogenic activity on deep-water detrital systems, which means that the interplay of controls on their evolution will continue to be important to unravel.

12.6.1 Allogenic Controls on the Deep-Water Detrital Stratigraphic Record

The distal parts of deep-water detrital successions have been identified as attractive palaeoenvironmental archives due to their thickness and relative completeness. For example, submarine fan successions have been used to investigate the interplay between orogenic processes, the denudation and erosion of the crust, the transfer of siliciclastic sediments to the deep oceans and the long-term feedback on Earth's climate (e.g. Hessler and Fildani 2019). To support interpretations of the stratigraphic organisation of deep-water systems being a record of 'allogenic' controls, an intimate relationship between relative sea-level change and the waxing-to-waning sediment supply cycles to the deep ocean is invoked. Indeed, a widely applied sequence stratigraphic framework is underpinned by the idea that times of low shelf accommodation (sea level lowstands) record periods of abundant siliciclastic sediment flux to the deep ocean, and times of high shelf accommodation (sea level highstands) record periods of limited basinward sediment supply (e.g. Posamentier and Kolla 2003 ; Posamentier and Walker 2006).

This proposed allogenically driven cycle of deep-water deposition was developed from seismic reflection data (e.g. Weimer 1991; Posamentier and Kolla 2003). An idealised deep-water detrital depositional sequence begins with a basal chaotic seismic facies (submarine landslides and debris flows) that mark the initiation of deep-water detrital sediment supply. The overlying distributive sand-rich turbidite deposits (lobe complexes) and overlying slope channel–levee complexes record the time when the submarine fan is being built, and the slope is dominated by sediment bypass. This idealised cycle is completed by filling of the slope conduits and the whole system is overlain by a hemipelagic drape. In this framework, the waxing deep-water detrital supply cycle starts during a fall in relative sea-level, when a subaerial unconformity develops on the shelf and the shoreline and river mouths are at their most basinward positions close to, or even beyond, the shelf-edge. The deep-water expression of this regressive phase is the formation of a correlative conformity above which there is a marked increase in the supply of coarse-grained sediment. Deep-water detrital systems evolve through time to become

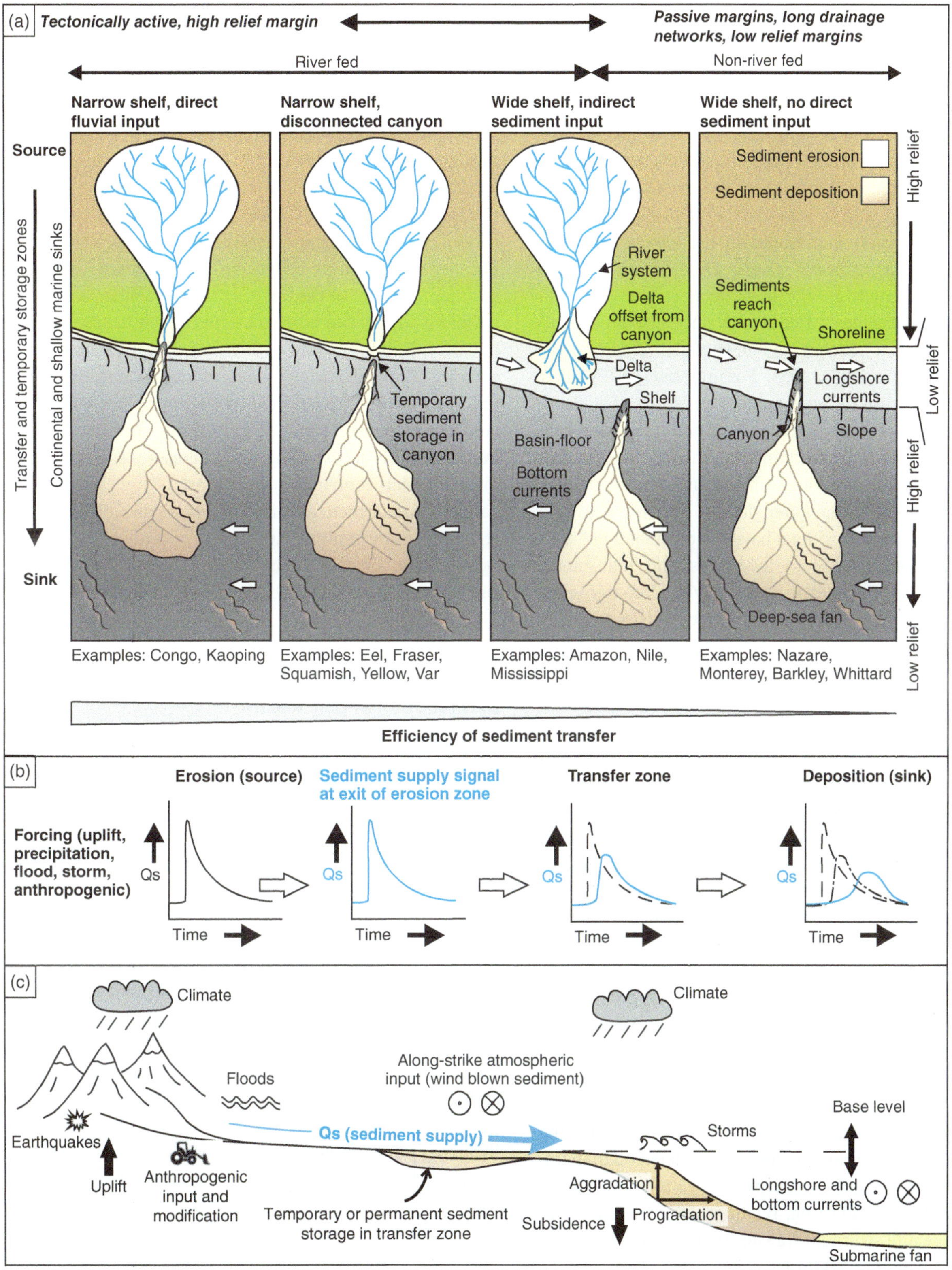

Figure 12.25 (a) The efficiency of transfer of sediment from the terrestrial to deep-marine detrital systems is dependent on the duration of onshore sediment storage and the proximity of the canyon head to the principal terrestrial sources of sediment, which is influenced primarily by the tectonic setting. Longshore currents as well as the volume and density of fluvial outflow will also have a strong influence on the likelihood and timing of sediment reaching the canyon head. (b) Schematic portrayal of a sediment supply (Qs) signal from the erosion zone to the deep-marine sink. The leftmost signal represents Qs as measured at the exit of the erosion zone. The transfer zone Qs signal is measured within the transfer zone at some distance from exit of the erosion zone and the rightmost signal represents that which reaches the accumulation zone and is an input for the stratigraphic record. Dashed lines refer to Qs signal(s) in up-system segment(s) to illustrate that a signal can be modified during propagation. (c) 2D profile of a generic sediment-routeing system emphasising erosion, transfer and accumulation and important controls of tectonics (including earthquakes), climate (including storms), base-level and anthropogenic factors. *Source:* (a) Modified from Sømme et al. (2009). (b and c) Modified from Romans et al. (2016).

more efficient in the transport of sediment, meaning that channel lengthening, and submarine fan progradation, is commonplace, and that the basin-floor correlative conformity as a lithological boundary, is likely highly time transgressive. The waning phase of a deepwater sediment supply cycle begins when relative sea-level rise outpaces sediment supply, and a landward shift of environments during transgression that results in progressive deactivation of shelf-edge sediment staging areas. Ultimately, the delivery of sediment gravity flows to deep-water detrital environments becomes infrequent as only suspension fallout from plumes reaches the deep-water environment (Posamentier and Kolla 2003; Posamentier and Walker 2006). The idealised stratigraphic expression of a deepwater detrital depositional sequence is therefore a sand-rich lowstand systems tract overlain by mud-rich deposits that represent the transgressive and highstand systems tracts.

Several river-fed deep-water detrital systems active in mid- to low-latitude positions have a close relationship in their growth and decay to lowstands and highstands, respectively (e.g. Catuneanu et al. 2009). However, even in these situations, multiple factors hinder the simple translation of sequence stratigraphic terms and principles defined for shallow water systems to the interpretation of deepwater detrital systems. These factors include: (i) the position of deepwater deposits at the distal end of the sediment transport pathway, which means that the interplay of many allogenic and autogenic factors influences the timing and location of delivery of sediment to deep-water environments (Posamentier and Allen 1999; Sømme et al. 2009) (Fig 12.25); (ii) a physical disconnect between coeval shallow- and deep-water deposits due to the position of the shoreline, the instability and remobilisation of sediment at the shelf-edge break and the bypass of sediment in steeper portions of the submarine slope (Catuneanu et al. 2009); (iii) a lack of direct evidence of relative sea-level change in deep-water sediments, in contrast to shallow-water systems (e.g. shoreline position, ravinement surfaces); (iv) an insufficient temporal resolution to correlate shallow- and deepwater deposits (Catuneanu 2006); and (v) inaccessibility and rarely monitored activity of modern systems, meaning that deep-water processes are less well understood than fluvial and shallow-water counterparts.

The simple, even alluring, relationships of relative sea-level change and sediment transfer to the deep-water have been challenged (e.g. Bourget et al. 2010; Covault et al. 2011; Heijnen et al. 2022). In tectonically active settings, and/or systems with narrow shelves, sediment supply might be high at all stages of a base-level cycle (Porebski and Steel 2006), although a (glacio-) eustatic signal may still be apparent in the volume and grain-size of sediment supplied (Pickering et al. 1999; Catuneanu et al. 2009). Furthermore, the temporal relationship of sediment generation and supply, and fluvial transport to the shelf involves many factors other than the state of accommodation on the shelf. For example, some systems increase in sediment flux during rising sea-levels with material generated during lowstand periods but not transported until relative sea level rise (e.g. Ducassou et al. 2009). Also, isostatic responses in areas close to waxing and waning ice sheets result in complicated relationships between shelf accommodation and sediment transfer to the deep ocean. Finally, strong littoral sediment supply can intersect with canyon heads meaning that supply is high during highstand periods and might even be curtailed during lowstand periods if no upstream drainage basin is connected (Covault et al. 2007; Boyd et al. 2008; Figure 12.25). Deep-water detrital systems that more faithfully record allogenic processes, i.e. those with a direct fluvial connection and short sediment transport pathway, are termed 'reactive', whereas typically larger systems with long sediment transport pathways and temporary storage of sediments tend to shred the allogenic signal and are termed 'buffered' (Romans et al. 2016; Figure 12.25).

12.6.2 Autogenic Controls on the Deep-Water Detrital Stratigraphic Record

'Autogenic', or internal, controls on the deep-water detrital stratigraphic records are relatively poorly understood. In part, this relates back to the difficulty in observing active systems. Nonetheless, the influence of channel avulsion and compensational stacking of lobes, the role of inherited seafloor topography of difference scales and configurations, the dynamic relief above mobile substrates and flow–substrate–deposit interactions are all important controls to consider when interpreting deep-water detrital successions. The range of configurations, and process-relief interactions, is too broad to consider in detail, but the fact that sediment gravity flows are sensitive to subtle changes in seafloor topography means that the role of topography on deep-water detrital stratigraphy should always be considered. Seafloor topography is generated by a wide variety of geological processes, including salt diapirism (e.g. Kane et al. 2012; Cumberpatch et al. 2021), relief above submarine landslides (e.g. Ortiz-Karpf et al. 2015), syn-depositional tectonic deformation (e.g. Hodgson and Haughton 2004; Kane et al. 2010) and differential compaction (e.g. Privat et al. 2021). The key autogenic controls are outlined below.

12.6.2.1 Salt Diapirism

The density of rock salt is less than most other rocks at depths of 1.3–1.5 km below the surface or more, which establishes a density inversion meaning that salt has positive buoyancy and rises in response to sediment loading (Warren 2006). As salt rises it intrudes into the overburden and deforms the seafloor to generate complicated topographic configurations that exert substantial control on the location, pathway and architecture of many deep-water detrital systems (e.g. Mayall et al. 2010; Casagrande et al. 2022). Successions of genetically related growth strata influenced by near-surface diapiric or extrusive salt form unconformity-bounded packages of thinned and folded strata termed halokinetic sequences (Giles and Rowan 2012). Exposed examples from deep-water detrital successions are rare but show the sub-seismic scale intercalations of cohesive mass flows and sediment gravity flows that change in relation to the evolving salt structures (Cumberpatch et al. 2021; Figure 12.3c).

12.6.2.2 Channel Avulsion

The architecture of submarine fans is strongly influenced by channel avulsion processes, especially in distributary channel systems (Picot et al. 2016). Trigger mechanisms for channel avulsion include flow-stripping and overspill (Piper and Normark 1983; Fildani et al. 2006), channel-floor aggradation (Armitage et al. 2012), levee failure by mass-transport emplacement (Ortiz-Karpf et al. 2015) and climate cyclicity (Picot et al. 2019).

12.6.2.3 Compensational Stacking

When an avulsion occurs, the new channel will feed sediment via sediment gravity flows to new sites of deposition, which will be topographic lows. Because turbidity currents are so susceptible to the influence of seafloor topography even the subtle relief generated by high aspect ratio bodies, like lobe elements, can deflect flows and control the site of deposition. The stratigraphic result is the vertical alternation of lobe element axes and off-axes with lobe element fringes and distal fringes (Figure 12.23a). Without three-dimensional control, the vertical representation of compensationally stacked high-aspect ratio (1000:1, length: thickness) km-scale bodies can be misinterpreted as externally driven changes in sediment supply.

12.6.2.4 Self-confinement

Progressive (self) confinement is an autogenic control that can explain the widely documented stratigraphic relationship in deep-water detrital systems of submarine channel-fills that have incised into underlying lobe-dominated successions. Hodgson et al. (2016) show several examples of frontal lobes abruptly overlain by external levee successions separated by a subtle scoured surface. The juxtaposition of channels cutting into lobes points to lengthening of channel–levee systems. Channel–levee systems are very effective at increasing their confinement at a point along the sediment transport route through a combination of building relief (external levees) and incision. This progressive confinement at a point means that successive sediment gravity flows of similar magnitude will maintain their downslope energy and travel longer distances into the basin. Evidence for time transgressive confinement of submarine channel–levee systems include composite basal erosion surfaces, and cut-off bends and hanging valleys (Figure 12.26a). The basinward response to time transgressive development of flow confinement on the slope is growth, and progradation, of the linked basin floor fan, until a maximum basinward extent is reached, which corresponds to the time of maximum through-channel sediment bypass (Figure 12.26b,c B, C). Theoretically, this coupled progressive slope confinement and basin-floor fan growth could be autogenic, but would, for example, be amplified by an allogenic waxing-then-waning sediment supply cycle.

12.6.3 Stratigraphic Evolution of Deep-Water Detrital Systems in Confined Basins

Many of the well-known exhumed deep-water detrital successions are from confined basin-fills. Sinclair and Tomasso (2002) describe a four-stage evolution of a confined basin-fill (Figure 12.27a). The evolving physiography and natural variability in flow magnitude and type means that these stages overlap in time and space. *Flow ponding* occurs where incoming flows are trapped on all sides, and the confined suspension clouds deposit anomalously thick normally graded mud caps (Sinclair and Tomasso 2002). In successions with ponded turbidites, there may be a dominance of thick beds meaning that architectural elements are challenging to identify. The bounding relief will reduce as a confined basin fills up. This marks a transition to healed slope accommodation with a stepped slope profile. This open or partial downflow-confined slope configuration leads to the development of aggradationally stacked sand-rich intraslope lobe complexes, and downstream 'flow stripping' of the finer grained dilute flow component. The stripped parts of the flows can form laterally extensive fine-grained aprons downstream. When the step is sufficiently filled with sediment, the break of slope that defined the margin of the upper basin is healed (Prather et al. 1998) and the opposing intrabasinal high is predominantly buried. Sediment bypass occurs where accommodation is limited, and flows traversing over the filled basin lip result in incision and headward erosion at the next break of slope leading to channels progressively incising through the step (Figure 12.27a).

Many systems have stepped slopes (Figures 12.1 and 12.28), and these form transient intraslope depocentres, which can retain large proportions of sand due to the flow

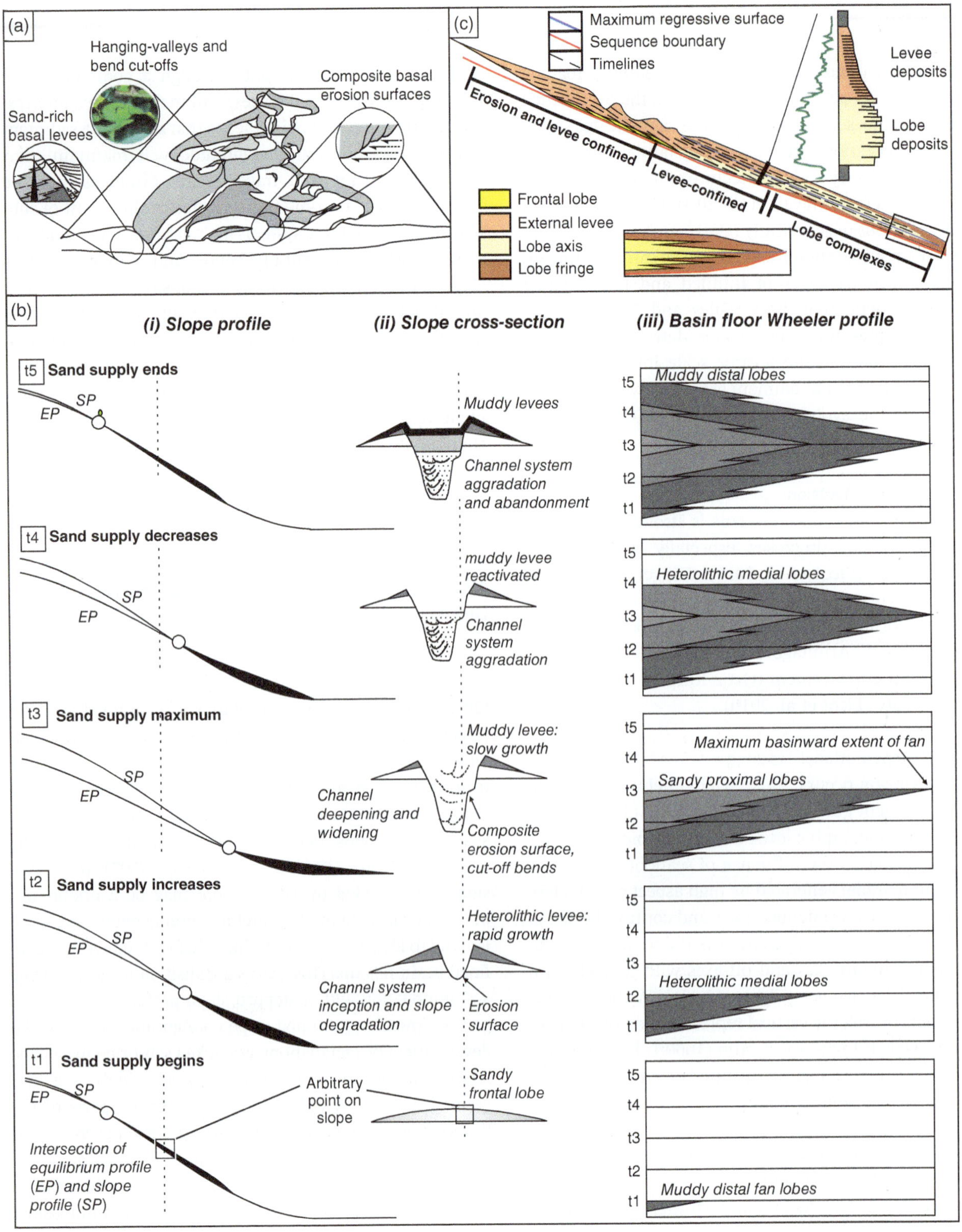

Figure 12.26 Evolution of deep-water detrital systems – through time increased confinement on the slope drives channel lengthening and basin-floor progradation. (a) Cartoon of a channel–levee system illustrating evidence for the progressive confinement of submarine channel–levee systems through a combination of downcutting and construction, including a frontal lobe, overlain by an external levee, a composite erosion surface and cut-off bends and hanging valleys. (b) Schematic diagram illustrating the response on the slope and basin floor to an evolving equilibrium profile and progressive slope confinement (t1-to-t5). As the profile lowers, in response to relative sea-level fall and/or waxing sediment supply, flows at an arbitrary point of the slope become more confined and efficient through a combination of construction and erosion. Gravity flows can supply sediment farther into the basin, and thus the maximum basinward extent of the basin-floor fan, illustrated in a Wheeler diagram, corresponds to a composite erosion surface. (c) Cartoon dip section parallel to an external levee crest that shows the levee–lobe transition, indicating a contemporaneous propagation of the channel–levee system and progradation of a coeval basin-floor fan. Inset schematic vertical section with external levee deposits overlying lobe deposits is a common stratigraphic configuration observed in progradational deep-water systems. *Source:* All modified from Hodgson et al. (2016).

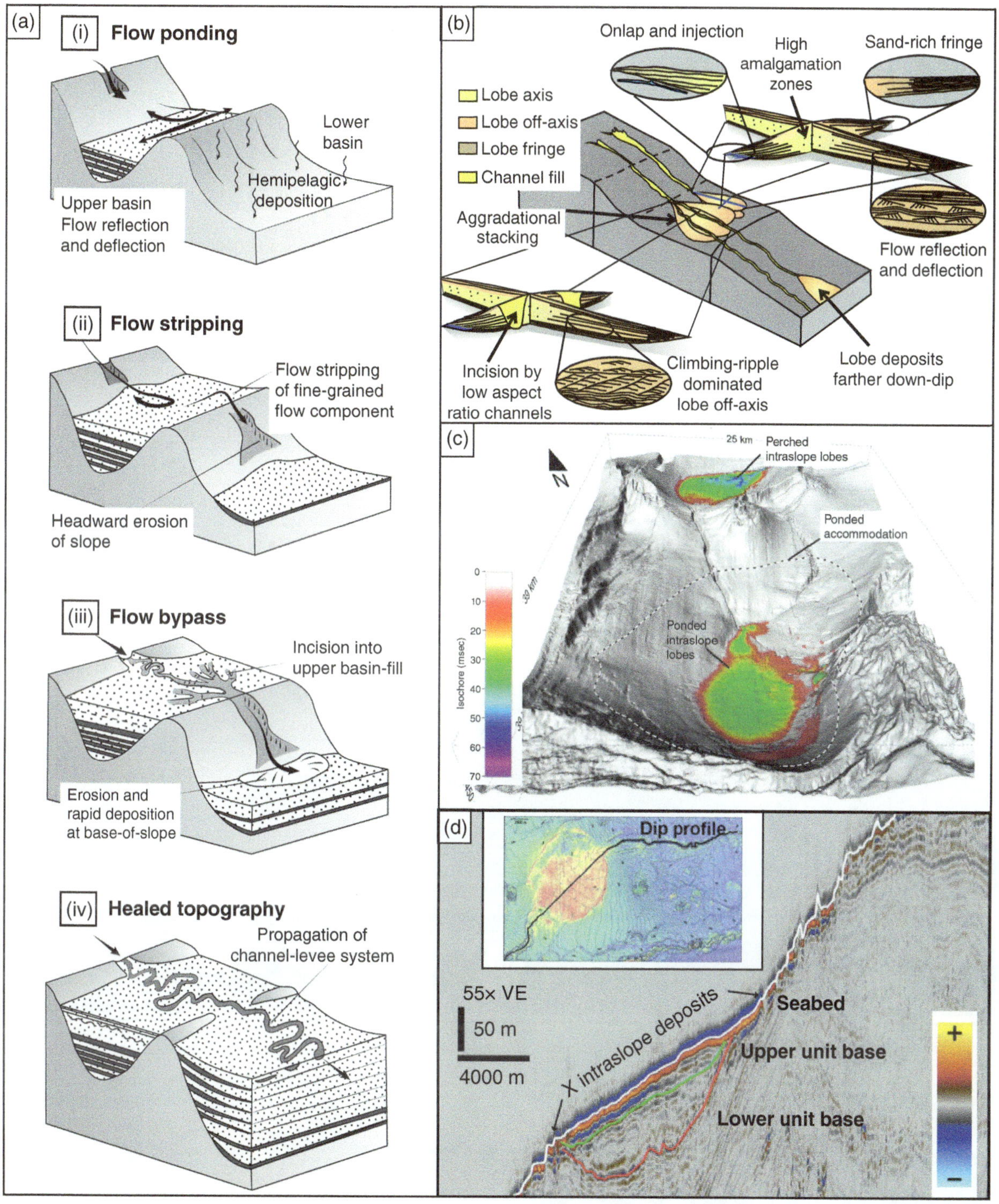

Figure 12.27 Intraslope depocentres range from highly confined to open steps. (a) Four-stage depositional model for the progressive infill of a confined turbidite basin and associated basinward deposits: (i) flow ponding, when sediment gravity flows are fully confined by depocentre topography; (ii) flow stripping, where the finer, more dilute portion of the flow escape over the confining topography to be deposited basinward, and erosion of slope initiates; (iii) Flow bypass, either by flows traversing over the filled depocentre than is connected basinward through channelisation, and (iv) healing of topography and propagation of channel–levee system. (b) Block diagram showing the key recognition criteria of intraslope lobe complexes on open stepped slopes (i.e. no flow ponding). (c) Perspective view of partially (perched) and fully (ponded) confined intraslope depocentres linked by a higher gradient area dominated by sediment bypass from the Brazos-Trinity system, Gulf of Mexico, USA. (d) Thickness map and dip seismic profile of intraslope basin fill from the Western Niger Delta slope, demonstrating the formation and healing of an intraslope depocentre above mobile shale. *Source:* (a) Modified from Sinclair and Tomasso (2002). (b) Modified from Spychala et al. (2015). (c) Modified from Prather et al. (2017). (d) Modified from Jobe et al. (2017).

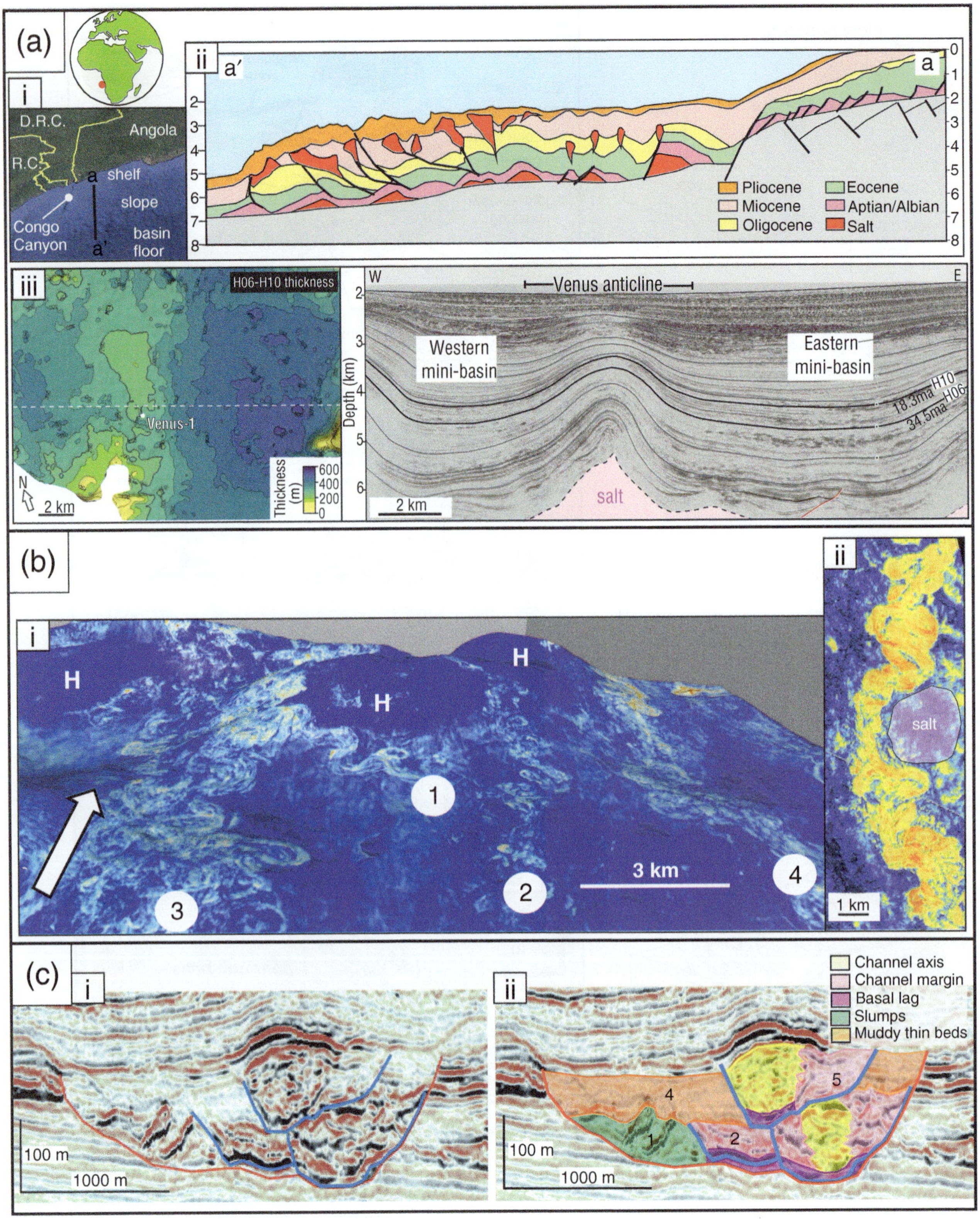

Figure 12.28 (a) (i) Seafloor map of part of West Africa. The large present-day drainage basin of the Congo River can be seen feeding into the Congo Canyon. showing the smooth shelf areas, the slope which has complex irregular topography due to underlying salt diapirs. At the base of the irregular slope is the basin floor. (ii) Cross section (line on map) shows the general stratigraphy and structure across the slope. (iii) Complex spatial and temporal changes of stratigraphic thickness record how sedimentation responds to the movement of the salt diapirs.

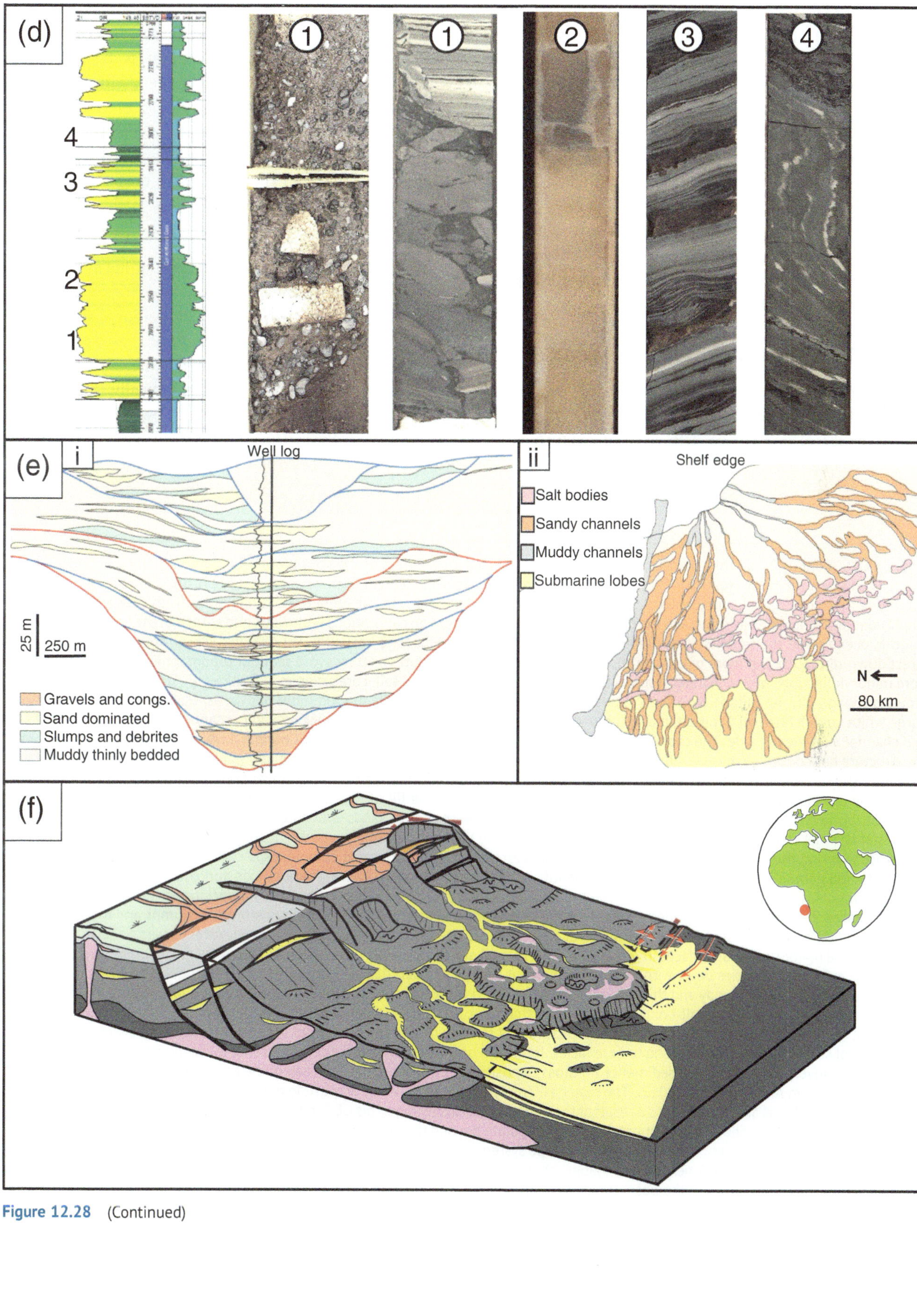

Figure 12.28 (Continued)

stripping effect described above (Spychala et al. 2015; Figure 12.27b). The progressive healing of topography applies to deep-water detrital systems with static seafloor relief, or where sedimentation rate outpaces the deformation rate. Tectonic and/or diapiric processes can deform the seafloor at a higher rate than the sedimentation rate, which can increase the scale of seafloor topography (Figure 12.28; see Hodgson and Haughton 2004; Brooks et al. 2018). Commonly, this dynamic topography is considered mainly in the vertical sense. However, any spatial difference in deformation rates will result in a component of lateral tilting and shifts in the sites of erosion and deposition (Casagrande et al. 2022).

12.6.4 Subsurface Case Studies

To conclude, two subsurface case studies are presented, compiled by Mike Mayall (London Imperial College). The large number of subsurface datasets that capture deep-water detrital systems have driven much of the advances in understanding the linkage and large-scale stratigraphic evolution of these systems. The first is from offshore Angola, which is a passive margin system with widespread salt diapirism that drives deformation of the sea floor (Figure 12.28). The second is from the Cretaceous foreland basin, North Slope Alaska, USA, which allows a source-to-sink assessment from the shallow- to deep-water detrital system (Figure 12.29).

12.6.4.1 Offshore Angola Deep-Water Detrital System Impacted by Salt Tectonics

This example demonstrates how depositional models and palaeogeographic maps can be developed based on data sets dominated by extensive 3D seismic calibrated with well logs and cores. The example is graphically illustrated in Figure 12.28 and b and is from the Lower Miocene of the Lower Congo Basin (LCB) offshore Angola and focusses

Figure 12.28 (Cont'd) A seismic line across part of the slope shows an example of the stratigraphy variably thickening between diapirs and thinning over the crest. Isopach is from the H06–H10 Interval. *Source:* Modified from Lewis et al. (2023). (b) Root mean square (RMS) amplitude maps of slope channel systems showing interaction of channels with salt-generated topography. In this data, the bright amplitudes are interpreted as sands and the dim amplitudes as muds based on calibrations with wells and cores. On the Lower Congo Basin Oligo-Miocene slope, the sediments are dominated by large channel systems (1–3 km wide, 100–200 m thick). In some areas, they show a direct down-dip course but more often they show complex and contorted routes as they avoid seafloor topographic obstacles. This varies from minor deflections of the channel systems to extensive diversions around salt-cored diapirs and ridges. (i) Channel systems avoiding growing topography. Amplitude map superimposed on present-day structure, salt-formed highs shown as 'H'. Flow away from the view (large white arrow). Channel 1 (oldest) passes around the front of the structure and continues downdip, channel 2 passes on the right side and channel 3 on the left side. The channels are also avoiding the course of earlier channels. Channel 4 has moved further to the right as the structure has continued to grow. (ii) Channel system with sinuosity deflecting around a salt diapir (pink). *Source:* Modified from Mayall et al. (2010). (c) Seismic line, uninterpreted and interpreted, from a slope channel system in the Lower Congo Basin. In seismic sections, the base of the channel systems can be readily identified and mapped (red). The fill of the channel systems shows a complex association of cross-cutting stratigraphic surfaces and bright and dim amplitudes interpreted as sands and mudstones, respectively. With careful analysis, a series of channel complexes (blue) within the channel system can be identified and mapped reflecting multiple phases of sedimentation and erosion during the depositional history of the channel system. Within the channel complexes, a combination of amplitude strength, continuity and geometries can be used to further identify conglomeratic channel bases, channel axis and channel margins. *Source:* Modified from Mayall and Kneller (2021). (d) Well log through a typical channel system. Key facies associations: 1. exotic and mudclast conglomerates, 2. structureless to graded sands, 3. thinly interbedded sands (Tc and Td) with mudstones, 4. slumps and muddy debrites. *Source:* Modified from Porter et al. (2006) and Mayall et al. (2006). (e) (i) Depositional model for a channel system based on seismic, logs and cores. Note the major erosional surfaces in red and lesser erosional surfaces in blue. Conglomerate facies are most common near the base of the system. Sandstone facies occur within each complex but slumps and debrites are very common. The model shows that much of the channel system is filled by muddy and thin-bedded facies (modified from Porter et al. 2006). Overall, throughout the channel system, the well logs and cores show an overall fining upwards reflecting finer grain sizes, thinner sands and thicker mudstones. The model is a conceptual representation of the stratigraphic architecture and facies within a slope channel system, incorporating observations over a wide range of scales from regional setting to core facies. It forms a powerful framework for characterising reservoirs and building models to study fluid movement within these sediments. (ii) Gross Depositional Environment (GDE) map of the Lower Miocene in the Lower Congo Basin, Modified from Anderson et al. (2012). The map has largely been generated by amplitude maps from extensive 3D seismic data and calibrated with well logs and core data. As a result of the extensive data coverage, this GDE map is deterministic, i.e. largely based on direct observations without the need to interpret between or extrapolate beyond the data. As such, it provides a very powerful tool for predicting the distribution of depositional bodies throughout the slope system. (f) General palaeogeographic model for slope systems with deforming seafloor topography showing sediment supply system, shelf and shelf edge, channels passing through or being ponded in the slope and deposition of depositional lobes at the base of slope beyond the deformed slope. *Source:* Modified from Mayall et al. (2010). 3D block diagrams of depositional systems provide very graphic representations, particularly to show the relationship between slope channels and developing seafloor topography. The models may be based on the observations at one particular location and time interval or can synthesis observations from a number of similar settings. The incorporation of observations from ancient and modern analogues provides the basis for broader generic facies models.

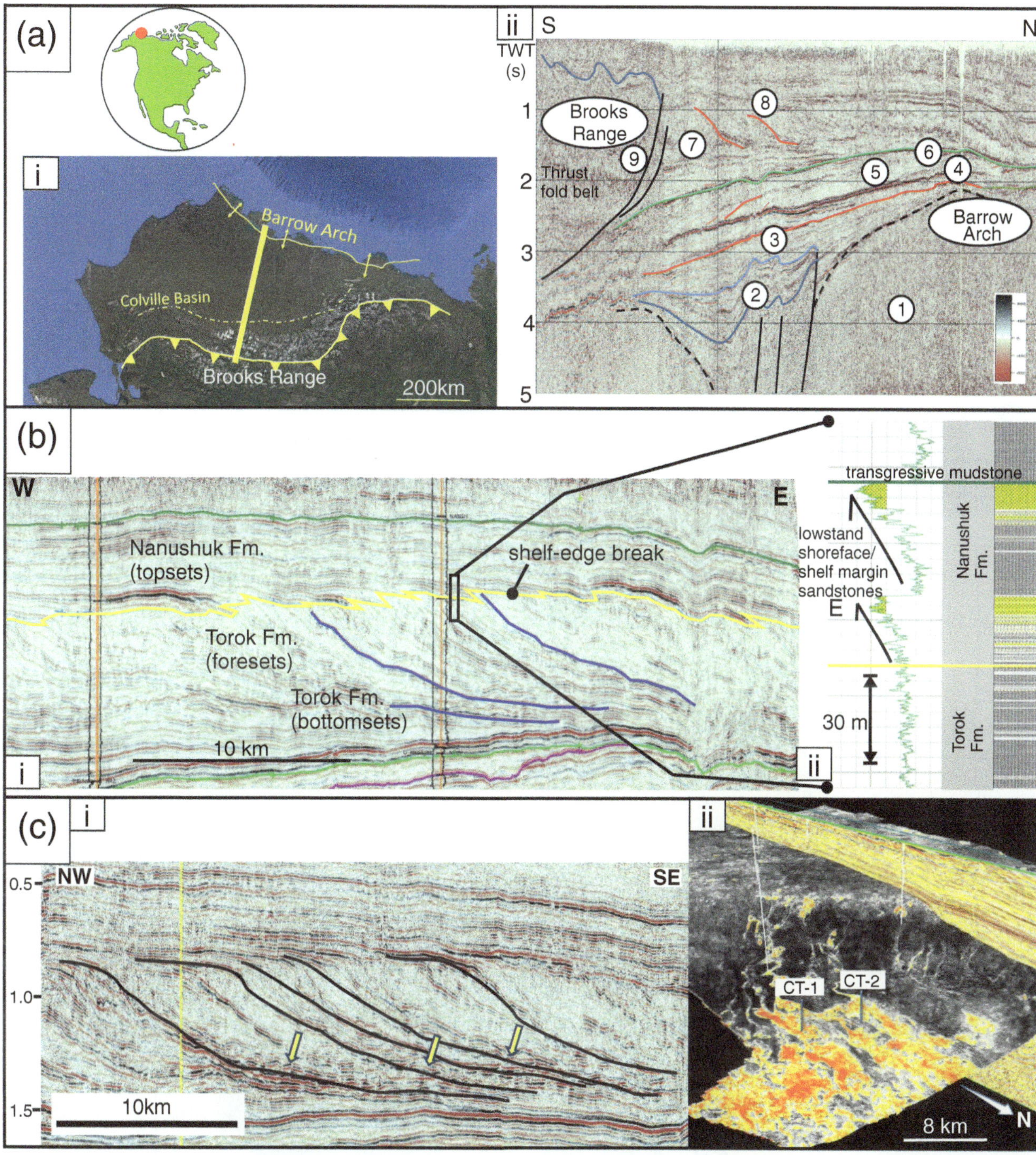

Figure 12.29 (a) Location map of Alaska showing the main structural features and regional seismic line (R8 USGS 2d survey) across the North Slope. From Bird and Houseknecht (2011). 1. Basement. 2. Synrift megasequence. 3. Sag megasequence. 4. Passive margin megasequence prograding from north. 5. Barrow Arch. 6. Start of foreland basin megasequence onlapping older sequences. 7. Prograding sequence from south (Torok). 8. Topsets of clinoforms

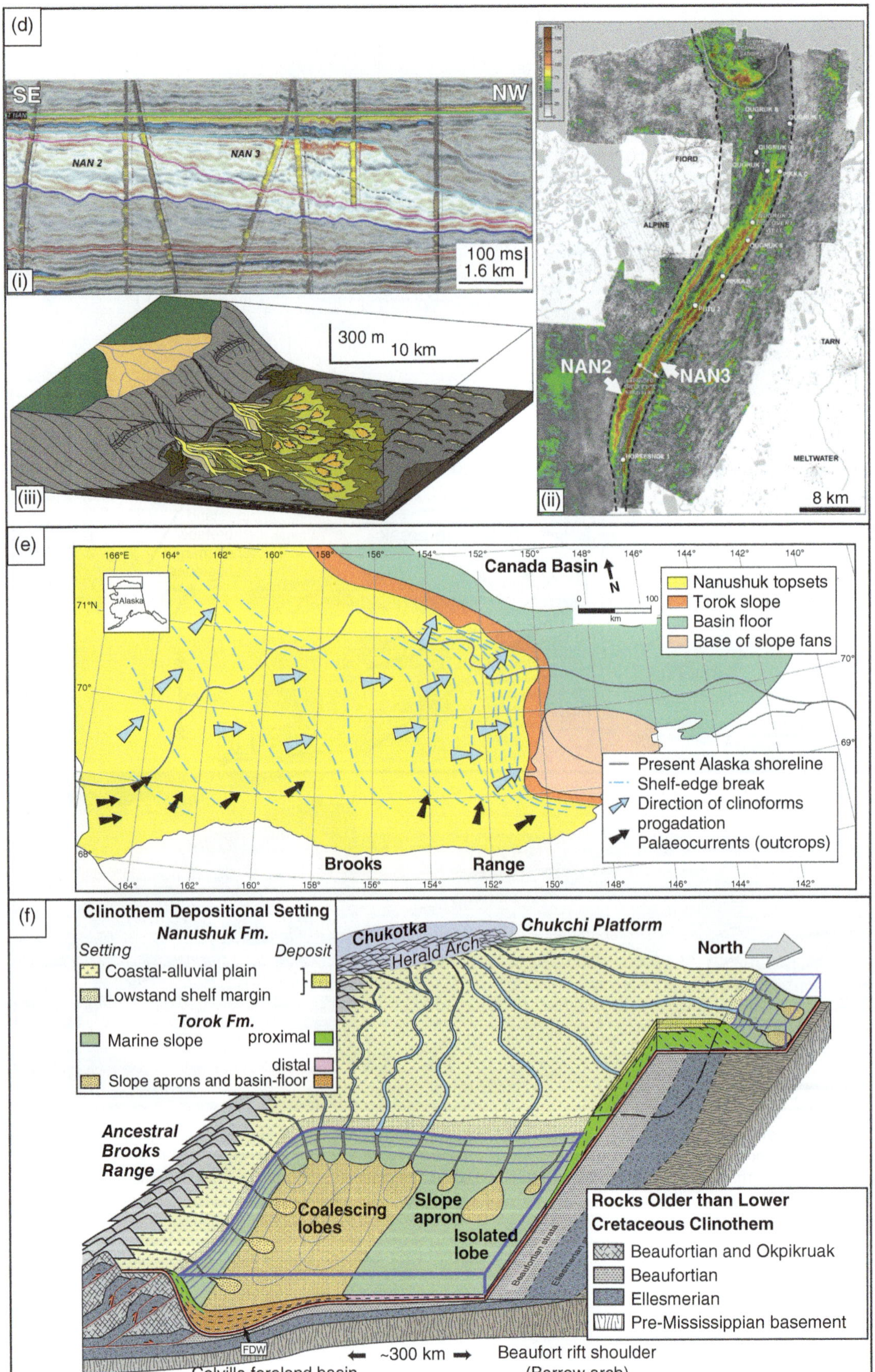

Figure 12.29 (Continued)

on the interaction between mobile substrates, cause by salt tectonics, and deep-water sandstone deposition on the continental slope between the shelf edge and basin floor. In this example, the extensive 3D seismic data allow for a robust tectonostratigraphic framework to be established within which detailed facies and depositional interpretation can be made.

12.6.4.2 Summary

Datasets that contain extensive 3D seismic, calibrated with wells and cores, can provide unique insights into depositional systems. They capture the true distribution and scale of sedimentary bodies and structures. The orientation and geometry of depositional elements can be imaged, mapped and measured. The combination of the seismic reflection data with well logs and cores allows the sedimentary systems to be understood at a range of scales from regional to detailed facies and facies associations. Hence, the stratigraphic geometries from seismic data are linked to the depositional processes observed in cores. The amplitude maps demonstrate spectacular images of deep-water channel systems across the slope and have completely changed our understanding of the development of deep-water detrital systems.

12.6.4.3 Shallow- to Deep-Water Detrital System, Cretaceous Foreland Basin, North Slope Alaska, USA

This case study demonstrates the integration of 2D seismic data and well logs, with sequence stratigraphic concepts, including a prograding shallow marine sequence, slope sequences and basin floor submarine fans. The work has allowed detailed palaeogeographic maps and depositional models to be generated. The example is from the Albian-Cenomanian of the North Slope of Alaska. A foreland basin sequence was generated during the deformation of the Brooks Range and the sedimentary sequences infilled the elongate Colville Basin.

12.6.4.4 Summary

Understanding the contact between the muddy Torok and sandy Nanushuk formations in a vertical section (well logs/cores/outcrops) is critical to recognising the stratigraphic geometries. The gradational nature of the contact suggests that the two lithostratigraphic units are laterally equivalent to each other which is confirmed by the seismic reflection data. An alternative interpretation could be that there is a sharp contact which may represent a sequence boundary caused by moderate sea-level fall and an abrupt basinward shift of coastal sands onto slope mudstones, i.e. a forced regression. The example clearly shows how the underlying

Figure 12.29 (Cont'd) (Nanushuk). 9. Fold and thrust belt of Brooks Range. The interval of interest is within the prograding Torok Fm and the time-equivalent top-sets of the Nanushuk Fm. (b) Seismic line and well log showing the transition from the Nanushuk top sets and the Torok clinoforms (foresets and bottom sets). Outcrop studies from the Brooks Range and initial wells drilled in the basin recognised a lithostratigraphy based on an upper sandy interval: the Nanushuk Formation, and a lower muddy interval: the Torok Formation. As more seismic data were acquired it was recognised, helped by the identification of clinoforms, that the two formations are laterally equivalent with the Nanushuk representing the shallow marine topsets and the Torok the laterally muddy slope equivalent. Gamma logs from a well show well-developed coarsening-upward profiles representing prograding shallow marine sequences interrupted by flooding surface. *Source:* From Decker (2018) and Houseknecht (2019). (c) (i) Seismic line with the clinoforms showing a succession of discrete basinward steps, typically of c. 1–2 km with a series of interpreted sequence boundaries. The position of the shelf margin break can readily be mapped on the seismic data and shows the lateral progradation of the systems. At the base of some of the clinoforms are brighter amplitude mounded features interpreted as submarine fans (arrowed) generated during lowstands. *Source:* From Decker (2018). (ii) Amplitude map superimposed on present-day topography showing channels from the shelf edge feeding base of slope submarine fans. This image beautifully shows the link from the shelf edge, through channels across the slope supplying the basin-floor fans. *Source:* From From Bhattacharya and Verma 2020. (d) (i) Seismic line highlighting two of the clinoform packages and ii. thickness map of the bright amplitudes, calibrated by logs and cores as shallow-marine to shoreface sands, shows that the sand bodies are developed as a series of linear bodies with abrupt terminations both landward and basinward (from Sommer et al. 2021). Linear sandbodies such as these are characteristic of wave-dominated shorelines where sand is distributed alongshore from the point of fluvial input. The map of the shelf-margin breaks. (e) shows an eastward bulge in the centre of the area suggesting this is where the fluvial input was more dominant. (iii) Depositional model for the Torok and Nanushuk Formations. A block-diagram geological model combines the observations from above and illustrates the shelf edge deltas, channels transporting sediment downslope and the submarine fans composed of channels and depositional lobes (Jones 2023). The model focusses on presenting the linkage between the systems and some of the complexities of the stratigraphic architecture in the submarine fan. *Source:* From Jones (2023). (e and f). General depositional environment (GDE) map and depositional model for the Nanushuk-Torok Formations. The figures show the integration of palaeocurrent data from outcrops in the Brooks Range, the history of generally west-to-east clinoform progradation along the axis of the eastward plunging Colville Basin (FDW = foredeep wedge) and submarine fans deposited at the base of slope. The palaeocurrents show that sediments were derived from the west and south but that the overall progradation of the system was controlled by the Barrow Arch and the Brooks Range. The map depicts a time of the most eastward extension of the Nanushuk foresets and as such demonstrates the maximum extent of this unit. A submarine fan is shown at the base of slope at this time which in this case has been identified in outcrops from the Brooks Range and subsurface well data. *Source:* From Houseknecht et al. (2009) and Houseknecht (2019).

basin morphology, in this case an elongate foreland basin controls the direction of progradation. Palaeocurrents indicate that at least some sediments were derived from the Brooks Range to the south, but the basin shape has promoted overall progradation to the east.

The stratigraphic geometries, combined with isopach and amplitude maps and some core observations, demonstrate that the Nanushuk is part of a wave-dominated shoreline. This in turn has generated a novel stratigraphic trapping mechanism for hydrocarbons as the shoreface sand pinch-out unusually abruptly landwards into lagoonal or marsh mudstones. Recognising sequence boundaries from the seismic data and distinctive seismic facies geometries has successfully predicted the timing and location of submarine fans at the base of the slope.

12.6.5 Anthropogenic Influences

Human influence on deep marine detrital systems is manifested in several ways. For example, human activity has influenced sediment supply rates to deep-water settings for thousands of years (e.g. Mohamed et al. 2024). Sediment supply to the oceans is affected by human-induced erosion of the landscape and interception of sediment pathways, leading to increases or decreases in sediment yield. Factors that affect sediment yield include mining (both waste from quarrying and riverbed/coastal sand extraction), deforestation, farming, urbanisation and damming of rivers (Syvitski and Kettner 2011). All these affect the volume, composition and calibre of sediment delivered to deep-water detrital systems.

Many pollutants are delivered to the coast by rivers, meaning they enter deep-marine detrital systems in the same way as detrital sediments. Large volumes of marine 'litter' are transferred to submarine canyons by rivers, as has been documented in the Messina Strait, where rivers are common sites of waste dumping. This waste is picked up by flash floods and flushed into submarine canyons and ranges in scales from microplastics to chairs, tables and even cars (Pierdomenico et al. 2019). Microplastics have been found at thousands of metres water depth and are transported in turbidity currents and bottom currents (e.g. Kane et al. 2020). Many harmful pollutants can sorb to mineral grains. These include dichloro-diphenyl-trichloroethane (DDT, an insecticide), which has been found in the Monterey Canyon at thousands of metres water depth, associated with clay grains transported by turbidity currents (Hartwell 2004), and perfluoroalkyl substances (PFASs), used in surface coatings, which are transported with fine particulate matter by dense water cascades into the Cap de Creus Canyon, offshore Spain (Sanchez-Vidal et al. 2015). The marked increase in the volume and variety of pollutants being carried by sediment gravity flows means that understanding the transport processes and pathways are critical to quantify in terms of the global budgets and the way these relate to delicate benthic ecosystems.

The deep-water seafloor is increasingly being used to host critical infrastructure, which is at great risk from large destructive sediment gravity flows. Indeed, the first known record of a turbidity current in nature was provided by the successive breaking of telecommunication cables by a turbidity current triggered by the 1929 Grand Banks earthquake (Heezen et al. 1954). Today, ~99% of all intercontinental digital data traffic is carried by cables on the seafloor, and this is at great risk from sediment gravity flows. For example, debris flows generated by the 2022 Hunga Tonga–Hunga Ha'apai eruption, in Tonga, destroyed domestic and international telecommunication cables (Clare et al. 2023). The deep-water stratigraphic record therefore provides important information on past destructive events that can be used to guide the installation of seabed infrastructure.

In summary, whether predicting geohazards, quantifying terrestrial carbon flux, constraining pollutant dispersal in the oceans, using ancient successions to interpret rates and controls of environmental changes or reducing uncertainty of geological carbon storage projects, it is more crucial than ever to investigate and improve understanding of the processes and products that operate in deep-water detrital systems.

Further Reading

Bowman, M., Cobain, S., Hurst, A. et al. (2021). *Subsurface Sand Remobilization and Injection*. Geological Society of London Special Publication, 493.

Deptuck, M.E. and Sylvester, Z. (2018). Submarine fans and their channels, levees, and lobes. In: *Submarine Geomorphology* (ed. A. Micallef, S. Krastel, and A. Savini). Springer, Cham: Springer Geology ISBN 978-3-319-57851-4.

Georgiopoulou, A., Amy, L.A., Benetti, S., Chaytor, J.D., Clare, M.A., Gamboa, D., Haughton, P.D.W., Moernaut, J. and Mountjoy, J.J. eds., 2020, Subaqueous mass movements and their consequences: advances in process understanding, monitoring and hazard assessments. Geological Society, London, Special Publications, 500.

Hodgson, D.M. and Flint, S.S. eds., 2005. Submarine slope systems: processes and products. *Geological Society of London Special Publication* 244.

Kneller, B., Martinsen, O.J., and McCaffrey, W. (ed.) (2009). *External Controls on Deep-Water Depositional Systems*, vol. 92, 402. Socio Economic Paleontologists and Mineralogists Special Publication.

Lintern, D.G., Mosher, D.C., Moscardelli, L. G., Bobrowsky, P.T., Campbell, C., Chaytor, J.D., Clague, J.J., Georgiopoulou, A., Lajeunesse, P., Normandeau, A. and Piper, D.J.W. eds., 2019, Subaqueous mass movements and their consequences: assessing geohazards, environmental implications and economic significance of subaqueous landslides. *Geological Society of London Special Publication* 477.

Lomas, S.A. and Joseph, P., 2004. Confined turbidite systems. *Geological Society, London, Special Publications*, 222.

McKie, T., Rose, P.T., Hartley, A.J. et al. (2015). *Tertiary Deep-Marine Reservoirs of the North Sea region*. London: Geological Society of London Special Publication 403.

Nilsen, T.H., Shew, R.D., Steffens, G.S., and Studlick, J.R. (2008). *Atlas of Deep-Water outcrops. AAAPG Studies in Geology*, vol. 56. American Association of Petroleum Geologists.

Pickering, K.T. and Hiscott, R.N. (2015). *Deep Marine Systems: Processes, Deposits, Environments, Tectonics and Sedimentation*. Wiley 978-1-405-12578-9.

Rotzien, J.R., Yeilding, C.A., Sears, R.A. et al. (ed.) (2022). *Deepwater Sedimentary Systems: Science, Discovery, and Applications*. Elsevier ISBN: 9780323919180.

Scarselli, N. (2020). Submarine landslides, architecture, controlling factors and environments. A summary. In: *Regional Geology and Tectonics: Principles of Geologic Analysis: Volume 1: Principles of Geologic Analysis* (ed. N. Scarselli, J. Adam, D. Chiarella, et al.). Elsevier.

References

Adeogba, A.A., McHargue, T.R., and Graham, S.A. (2005). Transient fan architecture and depositional controls from near-surface 3-D seismic data, Niger Delta continental slope. *AAPG Bulletin* 89 (5): 627–643.

Al-Ja'aidi, O.S.S. (2000). The influence of topography and flow efficiency on the deposition of turbidites. PhD thesis. University of Leeds, Leeds, United Kingdom, 162 p.

Allen, C.D. (2020). The evolution, processes, and deposits of high latitude submarine channel systems. PhD thesis. University of Leeds, Leeds, United Kingdom.

Allen, J.R.L. (1991). The Bouma division A and the possible duration of turbidity currents. *Journal of Sedimentary Research* 61 (2): 291–295.

Amblas, D. and Dowdeswell, J.A. (2018). Physiographic influences on dense shelf-water cascading down the Antarctic continental slope. *Earth-Science Reviews* 185: 887–900.

Amy, L.A., Talling, P.J., Peakall, J. et al. (2005). Bed geometry used to test recognition criteria of turbidites and (sandy) debrites. *Sedimentary Geology* 179 (1–2): 163–174.

Anderson, A.V., Sickafoose, D.K., Fahrer, T.R., and Gottschalk, R.R. (2012). Interaction of Oligocene–Miocene deep-water depositional systems with actively evolving structures: The Lower Congo Basin, offshore Angola. In: *Tectonics and sedimentation*, Implications for petroleum systems. *AAPG Memoir*, vol. 100 (ed. D. Gao), 291–313.

Armitage, D.A., McHargue, T., Fildani, A., and Graham, S.A. (2012). Postavulsion channel evolution: Niger Delta continental slope. *AAPG Bulletin* 96 (5): 823–843.

Arnott, R.W.C. (2012). Turbidites, and the case of the missing dunes. *Journal of Sedimentary Research* 82 (6): 379–384.

Arnott, R.W.C. and Hand, B.M. (1989). Bedforms, primary structures and grain fabric in the presence of suspended sediment rain. *Journal of Sedimentary Research* 59 (6): 1062–1069.

Azpiroz-Zabala, M., Cartigny, M.J., Talling, P.J. et al. (2017). Newly recognized turbidity current structure can explain prolonged flushing of submarine canyons. *Science Advances* 3 (10): e1700200.

Baas, J.H. and Best, J.L. (2002). Turbulence modulation in clay-rich sediment-laden flows and some implications for sediment deposition. *Journal of Sedimentary Research* 72 (3): 336–340.

Baas, J.H. and Best, J.L. (2008). The dynamics of turbulent, transitional and laminar clay-laden flow over a fixed current ripple. *Sedimentology* 55 (3): 635–666.

Baas, J.H., Best, J.L., Peakall, J., and Wang, M. (2009). A phase diagram for turbulent, transitional, and laminar clay suspension flows. *Journal of Sedimentary Research* 79 (4): 162–183.

Baas, J.H., Best, J.L., and Peakall, J. (2011). Depositional processes, bedform development and hybrid bed formation in rapidly decelerated cohesive (mud–sand) sediment flows. *Sedimentology* 58 (7): 1953–1987.

Baas, J.H., Davies, A.G., and Malarkey, J. (2013). Bedform development in mixed sand–mud: the contrasting role of cohesive forces in flow and bed. *Geomorphology* 182: 19–32.

Baas, J.H., Best, J.L., and Peakall, J. (2016). Predicting bedforms and primary current stratification in cohesive mixtures of mud and sand. *Journal of the Geological Society* 173 (1): 12–45.

Baas, J.H., Tracey, N.D., and Peakall, J. (2021). Sole marks reveal deep-marine depositional process and environment: implications for flow transformation and hybrid-event-bed models. *Journal of Sedimentary Research* 91 (9): 986–1009.

Baas, J.H., Baker, M.L., Buffon, P. et al. (2022). Blood, lead and spheres: a hindered settling equation for sedimentologists based on metadata analysis. *The Depositional Record* 8 (2): 603–615.

Babonneau, N., Savoye, B., Cremer, M., and Klein, B. (2002). Morphology and architecture of the present canyon and channel system of the Zaire deep-sea fan. *Marine and Petroleum Geology* 19: 445–467.

Babonneau, N., Savoye, B., Cremer, M., and Bez, M. (2004). Multiple terraces within the deep incised Zaire Valley (ZaïAngo Project): are they confined levees? *Geological Society, London, Special Publications* 222: 91–114.

Babonneau, N., Savoye, B., Cremer, M., and Bez, M. (2010). Sedimentary architecture in meanders of a submarine channel: detailed study of the present Congo turbidite channel (Zaiango project). *Journal of Sedimentary Research* 80: 852–866.

Bagnold, R.A. (1956). The flow of cohesionless grains in fluids. *Philosophical Transactions of the Royal Society of London. Series A, Mathematical and Physical Sciences* 249 (964): 235–297.

Bagnold, R.A. (1962). Auto-suspension of transported sediment; turbidity currents. *Proceedings of the Royal Society of London. Series A. Mathematical and Physical Sciences* 265 (1322): 315–319.

Bailey, L.P., Clare, M.A., Rosenberger, K.J. et al. (2021). Preconditioning by sediment accumulation can produce powerful turbidity currents without major external triggers. *Earth and Planetary Science Letters* 562: 116845.

Bailey, L.P., Clare, M.A., Hunt, J.E. et al. (2024). Highly variable deep-sea currents over tidal and seasonal timescales. *Nature Geoscience* 17 (8): 787–794.

Baker, M.L., Baas, J.H., Malarkey, J. et al. (2017). The effect of clay type on the properties of cohesive sediment gravity flows and their deposits. *Journal of Sedimentary Research* 87 (11): 1176–1195.

Bakke, K., Kane, I.A., Martinsen, O.J. et al. (2013). Seismic modeling in the analysis of deep-water sandstone termination styles. *AAPG Bulletin* 97 (9): 1395–1419.

Baudin, F., Rabouille, C., and Dennielou, B. (2020) Routing of terrestrial organic matter from the Congo River to the ultimate sink in the abyss: a mass balance approach (André Dumont medallist lecture 2017). Geol. Belgica https://doi.org/10.20341/gb.2020.004

Best, J.I.M. and Bridge, J. (1992). The morphology and dynamics of low amplitude bedwaves upon upper stage plane beds and the preservation of planar laminae. *Sedimentology* 39 (5): 737–752.

Bhattacharya, S. and Verma, S. (2020). Seismic attribute and petrophysics-assisted interpretation of the Nanushuk and Torok Formations on the North Slope, Alaska. *Interpretation* 8 (2): SJ17–SJ34.

Bird, K.J. and Houseknecht, D.W. (2011). Geology and petroleum potential of the Arctic Alaska petroleum province. Chapter 32. In: *Arctic Petroleum Geology*, Geological Society, London, Memoirs, vol. 35 (ed. A.M. Spencer, A. F. Embry, D.L. Gautier, et al.), 485–499. Geological Society.

Bishop, R.S. (1978). Mechanism for emplacement of piercement diapirs. *AAPG Bulletin* 62 (9): 1561–1583.

Boulesteix, K., Poyatos-Moré, M., Hodgson, D.M. et al. (2020). Fringe or background: characterizing deep-water mudstones beyond the basin-floor fan sandstone pinchout. *Journal of Sedimentary Research* 90 (12): 1678–1705.

Boulesteix, K., Poyatos-Moré, M., Flint, S. S. et al. (2022). Sedimentologic and stratigraphic criteria to distinguish between basin-floor and slope mudstones: implications for the delivery of mud to deep-water environments. *The Depositional Record* 8 (2): 958–988.

Bouma, A.H. (1962). *Sedimentology of Some Flysch Deposits: A Graphic Approach To Facies Interpretation, 168*. Elsevier Publishing Company.

Bouma, A.H., Stelting, C.E., and Coleman, J.M. (1985). Mississippi Fan, Gulf of Mexico. In: *Submarine Fans and Related Turbidite Systems* (ed. A.H. Bouma, W.R. Normark, and N.E. Barnes), 143–150. New York, NY: Springer New York.

Bourgault, D., Morsilli, M., Richards, C. et al. (2014). Sediment resuspension and nepheloid layers induced by long internal solitary waves shoaling orthogonally on uniform slopes. *Continental Shelf Research* 72: 21–33.

Bourget, J., Zaragosi, S., Mulder, T. et al. (2010). Hyperpycnal-fed turbidite lobe architecture and recent sedimentary processes: A case study from the Al Batha turbidite system. Oman margin. *Sedimentary Geology* 229 (3): 144–159.

Bowles, F.A., Faas, R.W., Vogt, P.R. et al. (2003). Sediment properties, flow characteristics, and depositional environment of submarine mudflows, Bear Island Fan. *Marine Geology* 197 (1–4): 63–74.

Boyd, R., Ruming, K., Goodwin, I. et al. (2008). Highstand transport of coastal sand to the deep ocean: a case study from Fraser Island, southeast Australia. *Geology* 36 (1): 15–18.

Brackenridge, R.E., Stow, D.A., Hernández-Molina, F.J. et al. (2018). Textural characteristics and facies of sand-rich contourite depositional systems. *Sedimentology* 65 (7): 2223–2252.

Brooks, H.L., Hodgson, D.M., Brunt, R.L. et al. (2018). Deep-water channel-lobe transition zone dynamics: processes and depositional architecture, an example from the Karoo Basin, South Africa. *GSA Bulletin* 130 (9–10): 1723–1746.

Bull, S., Cartwright, J., and Huuse, M. (2009). A review of kinematic indicators from mass-transport complexes using 3D seismic data. *Marine and Petroleum Geology* 26 (7): 1132–1151.

Campbell, C.V. (1967). Lamina, laminaset, bed and bedset. *Sedimentology* 8 (1): 7–26.

Cartigny, M.J., Postma, G., Van den Berg, J.H., and Mastbergen, D.R. (2011). A comparative study of sediment waves and cyclic steps based on geometries, internal structures and numerical modeling. *Marine Geology* 280 (1–4): 40–56.

Carvajal, C., Paull, C.K., Caress, D.W. et al. (2017). Unraveling the channel–lobe transition zone with high-resolution AUV bathymetry: Navy Fan, offshore Baja California, Mexico. *Journal of Sedimentary Research* 87 (10): 1049–1059.

Casagrande, J., Hodgson, D.M., Peakall, J., and Benac, P.M. (2022). Fill-and-Spill, Tilt-and-Repeat (FaSTaR) cycles: stratigraphic evolution above a dynamic submarine stepped slope. *Basin Research* 34 (6): 2162–2188.

Catuneanu, O. (2006). *Principles of Sequence Stratigraphy*. Elsevier.

Catuneanu, O., Abreu, V., Bhattacharya, J. P. et al. (2009). Towards the standardization of sequence stratigraphy. *Earth-Science Reviews* 92 (1–2): 1–33.

Clare, M., Lintern, D.G., Rosenberger, K. et al. (2020). Lessons learned from the monitoring of turbidity currents and guidance for future platform designs. In: *Subaqueous Mass Movements and their Consequences: Advances in Process Understanding, Monitoring and Hazard Assessments*, Geological Society, London, Special Publications, vol. 500 (ed. A. Georgiopoulou, L.A. Amy, S. Benetti, et al.), 605–634. Geological Society, London.

Clare, M.A., Yeo, I.A., Watson, S. et al. (2023). Fast and destructive density currents created by ocean-entering volcanic eruptions. *Science* 381 (6662): 1085–1092.

Clift, P. and Gaedicke, C. (2002). Accelerated mass flux to the Arabian Sea during the middle to late Miocene. *Geology* 30 (3): 207–210.

Cobain, S.L., Peakall, J., and Hodgson, D. M. (2015). Indicators of propagation direction and relative depth in clastic injectites: implications for laminar versus turbulent flow processes. *Bulletin of Geological Society of America* 127 (11–12): 1816–1830.

Cobain, S.L., Hodgson, D.M., Peakall, J., and Silcock, S.Y. (2021). Relationship between bowl-shaped clastic injectites and parent sand depletion: implications for their scale-invariant morphology and composition. In: *Subsurface Sand Remobilization and Injection*, Geological Society Special Publications, vol. 493 (ed. S. Silcock, M. Huuse, M. Bowman, et al.), 135–149. Geological Society.

Collinson, J. (1994). Sedimentary deformational structures. In: *The Geological Deformation of Sediments* (ed. A. Maltman), 95–125. Dordrecht: Springer Netherlands.

Collinson, J. and Mountney, N. (2019). *Sedimentary Structures*, 4e. Liverpool University Press.

Collinson, J., Mountney, N., and Thompson, D. (2006). *Sedimentary Structures*, 3e, 292. Harpenden, England: Terra Publishing.

Cosgrove, J.W. (2001). Hydraulic fracturing during the formation and deformation of a basin: a factor in the dewatering of low-permeability sediments. *AAPG Bulletin* 85 (4): 737–748.

Covault, J.A. (2011). Submarine fans and canyon-channel systems: a review of processes, products, and models. *Nature Education Knowledge* 3 (10): 4.

Covault, J.A., Normark, W.R., Romans, B. W., and Graham, S.A. (2007). Highstand fans in the California borderland: the overlooked deep-water depositional systems. *Geology* 35 (9): 783–786.

Covault, J.A., Romans, B.W., Graham, S. A. et al. (2011). Terrestrial source to deep-sea sink sediment budgets at high and low sea levels: Insights from tectonically active Southern California. *Geology* 39 (7): 619–622.

Craig, W. (2004). Centrifuge model tests for offshore structures research and design. In: *Cyclic Behaviour of Soils and Liquefaction Phenomena*, 355–362. Bochum, Germany/Balkema, Netherland.

Cumberpatch, Z.A., Kane, I.A., Soutter, E. L. et al. (2021). Interactions between deep-water gravity flows and active salt tectonics. *Journal of Sedimentary Research* 91 (1): 34–65.

Curray, J.R., Emmel, F.J., and Moore, D.G. (2002). The Bengal Fan: morphology, geometry, stratigraphy, history and processes. *Marine and Petroleum Geology* 19 (10): 1191–1223.

De Castro, S., Hernández-Molina, F.J., Rodríguez-Tovar, F.J. et al. (2020). Contourites and bottom current reworked sands: bed facies model and implications. *Marine Geology* 428: 106267.

Decker, P. (2018). Nanushuk Formation discoveries confirm world-class exploration potential in a newly proven stratigraphic play, Alaska North Slope: Abstract only. *Presented at the Annual Convention and Exhibition, 2018. American Association Petroleum Geologists*.

Deptuck, M.E. and Sylvester, Z. (2018). Submarine fans and their channels, levees, and lobes. In: *Submarine Geomorphology* (ed. A. Micallef, S. Krastel, and A. Savini). Springer, Cham: Springer Geology.

Deptuck, M.E., Steffens, G.S., Barton, M., and Pirmez, C. (2003). Architecture and evolution of upper fan channel-belts on the Niger Delta slope and in the Arabian Sea. *Marine and Petroleum Geology* 20 (6–8): 649–676.

Deptuck, M.E., Piper, D.J., Savoye, B., and Gervais, A. (2008). Dimensions and architecture of late Pleistocene submarine lobes off the northern margin of East Corsica. *Sedimentology* 55 (4): 869–898.

Deptuck, M.E., Sylvester, Z., and O'Byrne, C. (2012). Pleistocene seascape evolution above a "simple" stepped slope—Western Niger Delta. In: *Application of the Principles of Seismic Geomorphology to Continental Slope and Base-of-Slope Systems: Case Studies from Seafloor and Near-Seafloor Analogues* (ed. B.E. Prather, M.E. Deptuck, D. Mohrig, et al.), 99. SEPM Special Publications.

Di Celma, C.N., Brunt, R.L., Hodgson, D. M. et al. (2011). Spatial and temporal evolution of a Permian submarine slope channel–levee system, Karoo Basin, South Africa. *Journal of Sedimentary Research* 81 (8): 579–599.

Dixon, R.J., Schofield, K., Anderton, R. et al. (1995). Sandstone diapirism and clastic intrusion in the tertiary submarine fans of the Bruce-Beryl Embayment, Quadrant 9, UKCS. *Geological Society, London, Special Publications* 94 (1): 77–94.

Dott, R.H. (1966). Cohesion and flow phenomena in clastic intrusions. *AAPG Bulletin* 50 (3): 610–611.

Draganits, E., Schlaf, J., Grasemann, B., and Argles, T. (2008). Giant submarine landslide grooves in the Neoproterozoic/Lower Cambrian Phe Formation, northwest Himalaya: mechanisms of formation and palaeogeographic implications. *Sedimentary Geology* 205 (3–4): 126–141.

Driscoll, M.L., Tucholke, B.E., and McCave, I.N. (1985). Seafloor zonation in sediment texture on the Nova Scotian lower continental rise. *Marine Geology* 66 (1–4): 25–41.

Droghei, R., Falcini, F., Casalbore, D. et al. (2016). The role of Internal Solitary Waves on deep-water sedimentary processes: the case of up-slope migrating sediment waves off the Messina Strait. *Scientific Reports* 6 (1): 36376.

Droz, L., Rigaut, F., Cochonat, P., and Tofani, R. (1996). Morphology and recent evolution of the Zaire turbidite system (Gulf of Guinea). *Geological Society of America Bulletin* 108 (3): 253–269.

Droz, L., Marsset, T., Ondreas, H. et al. (2003). Architecture of an active mud-rich turbidite system: The Zaire Fan (Congo–Angola margin southeast Atlantic) Results from ZaïAngo 1 and 2 cruises. *AAPG Bulletin* 87 (7): 1145–1168.

Ducassou, E., Migeon, S., Mulder, T. et al. (2009). Evolution of the Nile deep-sea turbidite system during the Late Quaternary: influence of climate change on fan sedimentation. *Sedimentology* 56 (7): 2061–2090.

Duranti, D. and Hurst, A. (2004). Fluidization and injection in the deep-water sandstones of the Eocene Alba Formation (UK North Sea). *Sedimentology* 51 (3): 503–529.

Dutkiewicz, A., Müller, R.D., O'Callaghan, S., and Jónasson, H. (2015). Census of seafloor sediments in the world's ocean. *Geology* 43 (9): 795–798.

Dżułyński, S. and Sanders, J.E. (1962). Current marks on firm mud bottoms. *Connecticut Academy of Science Transactions* 42: 57–96.

Enos, P. (1969). Anatomy of a flysch. *Journal of Sedimentary Research* 39 (2): 680–723.

Etienne, S., Mulder, T., Bez, M. et al. (2012). Multiple scale characterization of sand-rich distal lobe deposit variability: examples from the Annot Sandstones Formation, Eocene–Oligocene, SE France. *Sedimentary Geology* 273: 1–18.

Faugères, J.C., Stow, D.A., Imbert, P., and Viana, A. (1999). Seismic features diagnostic of contourite drifts. *Marine Geology* 162 (1): 1–38.

Fildani, A., Normark, W.R., Kostic, S., and Parker, G. (2006). Channel formation by flow stripping: large-scale scour features along the Monterey East Channel and their relation to sediment waves. *Sedimentology* 53 (6): 1265–1287.

Fildani, A., Clark, J., Covault, J.A. et al. (2018). Muddy sand and sandy mud on the distal Mississippi fan: implications for lobe depositional processes. *Geosphere* 14 (3): 1051–1066.

Fisher, R.V. (1983). Flow transformations in sediment gravity flows. *Geology* 11 (5): 273–274.

Fisher, W.L., Galloway, W.E., Steel, R.J. et al. (2021). Deep-water depositional systems supplied by shelf-incising submarine canyons: recognition and significance in the geologic record. *Earth-Science Reviews* 214: 103531.

Foglini, F., Campiani, E., and Trincardi, F. (2016). The reshaping of the South West Adriatic Margin by cascading of dense shelf waters. *Marine Geology* 375: 64–81.

Fonnesu, M., Palermo, D., Galbiati, M. et al. (2020). A new world-class deep-water play-type, deposited by the syndepositional interaction of turbidity flows and bottom currents: the giant Eocene Coral Field in northern Mozambique. *Marine and Petroleum Geology* 111: 179–201.

Fourrière, A., Claudin, P., and Andreotti, B. (2010). Bedforms in a turbulent stream: formation of ripples by primary linear instability and of dunes by nonlinear pattern coarsening. *Journal of Fluid Mechanics* 649: 287–328.

Frey-Martínez, J., Cartwright, J., and James, D. (2006). Frontally confined versus frontally emergent submarine landslides: a 3D seismic characterisation. *Marine and Petroleum Geology* 23: 585604.

Fuhrmann, A., Kane, I.A., Clare, M.A. et al. (2020). Hybrid turbidite-drift channel complexes: an integrated multiscale model. *Geology* 48 (6): 562–568.

Fuhrmann, A., Kane, I.A., Schomacker, E. et al. (2022). Bottom current modification of turbidite lobe complexes. *Frontiers in Earth Science* 9: 752066.

Galy, V., France-Lanord, C., Beyssac, O. et al. (2007). Efficient organic carbon burial in the Bengal fan sustained by the Himalayan erosional system. *Nature* 450 (7168): 407–410.

Garcia, M.H. (1994). Depositional turbidity currents laden with poorly sorted sediment. *Journal of Hydraulic Engineering* 120 (11): 1240–1263.

García, M., Hernández-Molina, F.J., Llave, E. et al. (2009). Contourite erosive features caused by the Mediterranean Outflow Water in the Gulf of Cadiz: Quaternary tectonic and oceanographic implications. *Marine Geology* 257 (1-4): 24–40.

Gardiner, R.G. (2006). The variability of turbidite sand body pinchout and its impact on hydrocarbon recovery in stratigraphically trapped fields: the deliberate search for the stratigraphic trap. *Geological Society (London) Special Publication* 254: 267–287.

Gardner, M.H., Borer, J.M., Melick, J.J. et al. (2003). Stratigraphic process-response model for submarine channels and related features from studies of Permian Brushy Canyon outcrops. *West Texas. Marine and Petroleum Geology* 20 (6–8): 757–787.

Gardner, W.D., Tucholke, B.E., Richardson, M.J., and Biscaye, P.E.

(2017). Benthic storms, nepheloid layers, and linkage with upper ocean dynamics in the western North Atlantic. *Marine Geology* 385: 304–327.

Gee, M.J.R., Masson, D.G., Watts, A.B., and Allen, P.A. (1999). The Saharan debris flow: an insight into the mechanics of long runout submarine debris flows. *Sedimentology* 46: 317–335.

Gee, M.J.R., Gawthorpe, R.L., and Friedmann, S.J. (2006). Triggering and evolution of a giant submarine landslide, offshore Angola, revealed by 3D seismic stratigraphy and geomorphology. *Journal of Sedimentary Research* 76 (1): 9–19.

Gervais, A., Savoye, B., Mulder, T., and Gonthier, E. (2006). Sandy modern turbidite lobes: a new insight from high resolution seismic data. *Marine and Petroleum Geology* 23 (4): 485–502.

Ghosh, B. and Lowe, D.R. (1993). The architecture of deep-water channel complexes, Cretaceous Venado sandstone member, Sacramento Valley, California. In: Advances in the Sedimentary Geology of the Great Valley Group, Sacramento Valley, California. p. 51–65. Pacific Section SEPM.

Giles, K.A. and Rowan, M.G. (2012). Concepts in halokinetic-sequence deformation and stratigraphy. *Geological Society, London, Special Publications* 363 (1): 7–31.

Gladstone, C., Ritchie, L.J., Sparks, R.S.J., and Woods, A.W. (2004). An experimental investigation of density-stratified inertial gravity currents. *Sedimentology* 51 (4): 767–789.

Goldfinger, C. (2011). Submarine paleoseismology based on turbidite records. *Annual Review of Marine Science* 3 (1): 35–66.

Gonthier, E.G., Faugères, J.C., and Stow, D.A.V. (1984). Contourite facies of the Faro drift, Gulf of Cadiz. *Geological Society, London, Special Publications* 15 (1): 275–292.

Hampton, M.A. (1979). Buoyancy in debris flows. *Journal of Sedimentary Research* 49 (3): 753–758.

Hansen, L.A., Callow, R.H., Kane, I.A. et al. (2015). Genesis and character of thin-bedded turbidites associated with submarine channels. *Marine and Petroleum Geology* 67: 852–879.

Hansen, L., Janocko, M., Kane, I., and Kneller, B. (2017). Submarine channel evolution, terrace development, and preservation of intra-channel thin-bedded turbidites: Mahin and Avon channels, offshore Nigeria. *Marine Geology* 383: 146–167.

Hansen, L.A.S., Hodgson, D.M., Pontén, A. et al. (2019). Quantification of basin-floor fan pinchouts: examples from the Karoo Basin. *South Africa. Frontiers in Earth Science* 7: 12.

Harms, J.C. (1969). Hydraulic significance of some sand ripples. *Geological Society of America Bulletin* 80: 363–396.

Harms, J. and Fahnestock, R.K. (1965). Stratification, bed forms, and flow phenomena (with an example from the Rio Grande). In: *Primary Sedimentary Structures and Their Hydrodynamic Interpretation, Society for Sedimentary Geology Special Publication 12* (ed. G.V. Middleton), 84–115. Okla: Tulsa.

Harris, P.T. and Whiteway, T. (2011). Global distribution of large submarine canyons: geomorphic differences between active and passive continental margins. *Marine Geology* 285 (1–4): 69–86.

Harris, P.T., Macmillan-Lawler, M., Rupp, J., and Baker, E.K. (2014). Geomorphology of the oceans. *Marine Geology* 352: 4–24.

Hartwell, S.I. (2004). Distribution of DDT in sediments off the central California coast. *Marine Pollution Bulletin* 49 (4): 299–305.

Haughton, P.D. (1994). Deposits of deflected and ponded turbidity currents, Sorbas Basin, Southeast Spain. *Journal of Sedimentary Research* 64 (2a): 233–246.

Haughton, P., Davis, C., McCaffrey, W., and Barker, S. (2009). Hybrid sediment gravity flow deposits–classification, origin and significance. *Marine and Petroleum Geology* 26 (10): 1900–1918.

Hedberg, H.D. (1974). Relation of methane generation to undercompacted shales, shale diapirs, and mud volcanoes. *AAPG Bulletin* 58 (4): 661–673.

Heezen, B.C. and Drake, C.L. (1964). Grand Banks slump. *AAPG Bulletin* 48 (2): 221–225.

Heezen, B.C. and Ewing, W.M. (1952). Turbidity currents and submarine slumps, and the 1929 Grand Banks earthquake. *American Journal of Science* 250: 849–873.

Heezen, B.C., Ericson, D.B., and Ewing, M. (1954). Further evidence for a turbidity current following the 1929 Grand Banks earthquake. *Deep Sea Research (1953)* 1 (4): 193–202.

Heezen, B.C., Hollister, C.D., and Ruddiman, W.F. (1966). Shaping of the continental rise by deep geostrophic contour currents. *Science* 152 (3721): 502–508.

Heijnen, M.S., Mienis, F., Gates, A.R. et al. (2022). Challenging the highstand-dormant paradigm for land-detached submarine canyons. *Nature Communications* 13 (1): 3448.

Hernández-Molina, F.J., Maldonado, A., and Stow, D.A.V. (2008). Abyssal plain contourites. *Developments in Sedimentology* 60: 345–378.

Hernández-Molina, F.J., Paterlini, M., Violante, R. et al. (2009). Contourite depositional system on the Argentine Slope: an exceptional record of the influence of Antarctic water masses. *Geology* 37 (6): 507–510.

Hernández-Molina, F.J., Paterlini, M., Somoza, L. et al. (2010). Giant mounded drifts in the Argentine Continental Margin: origins, and global implications for the history of thermohaline circulation. *Marine and Petroleum Geology* 27 (7): 1508–1530.

Hessler, A.M. and Fildani, A. (2019). Deep-sea fans: tapping into Earth's changing landscapes. *Journal of Sedimentary Research* 89 (11): 1171–1179.

Hill, P.R., Moran, K.M., and Blasco, S.M. (1982). Creep deformation of slope sediments in the Canadian Beaufort Sea. *Geo-Marine Letters* 2: 163–170.

Hiscott, R.N. (1994a). Loss of capacity, not competence, as the fundamental process governing deposition from turbidity currents. *Journal of Sedimentary Research* 64 (2a): 209–214.

Hiscott, R.N. (1994b). Traction-carpet stratification in turbidites; fact or fiction? *Journal of Sedimentary Research* 64 (2a): 204–208.

Hiscott, R.N. and Middleton, G.V. (1980). Fabric of coarse deep-water sandstones, Tourelle Formation, Quebec, Canada. *Journal of Sedimentary Research* 50 (3): 703–721.

Hiscott, R.N. and Pickering, K.T. (1984). Reflected turbidity currents on an Ordovician basin floor, Canadian Appalachians. *Nature* 311 (5982): 143–145.

Hodgson, D.M. and Haughton, P.D. (2004). Impact of syndepositional faulting on gravity current behaviour and deep-water stratigraphy: Tabernas-Sorbas Basin, SE Spain. In: *Geological Society*, vol. 222 (ed. S.A. Lomas and P. Joseph), 135–158. London: Special Publications.

Hodgson, D.M., Di Celma, C.N., Brunt, R. L., and Flint, S.S. (2011). Submarine slope degradation and aggradation and the stratigraphic evolution of channel–levee systems. *Journal of the Geological Society* 168 (3): 625–628.

Hodgson, D.M., Kane, I.A., Flint, S.S. et al. (2016). Time-transgressive confinement on the slope and the progradation of basin-floor fans: implications for the sequence stratigraphy of deep-water deposits. *Journal of Sedimentary Research* 86 (1): 73–86.

Hodgson, D.M., Peakall, J., and Maier, K. L. (2022). Submarine channel mouth settings: processes, geomorphology, and deposits. *Frontiers in Earth Science* 10: 790320.

Hofstra, M., Hodgson, D.M., Peakall, J., and Flint, S.S. (2015). Giant scour-fills in ancient channel-lobe transition zones: formative processes and depositional architecture. *Sedimentary Geology* 329: 98–114.

Houseknecht, D.W. (2019). Petroleum systems framework of significant new oil discoveries in a giant Cretaceous (Aptian–Cenomanian) clinothem in Arctic Alaska. *AAPG Bulletin* 103 (3): 619–652.

Houseknecht, D.W., Bird, K.J., and Schenk, C.J. (2009). Seismic analysis of clinoform depositional sequences and shelf-margin trajectories in Lower Cretaceous (Albian) strata, Alaska North Slope. *Basin Research* 21 (5): 644–654.

Hubbard, S.M., Romans, B.W., and Graham, S.A. (2007). An outcrop example of large-scale conglomeratic intrusions sourced from deep-water channel deposits, Cerro Toro Formation, Magallanes basin, southern Chile. In: *Sand Injectites*, Implications for Hydrocarbon Exploration and Production. AAPG Memoir, vol. 87 (ed. A. Hurst and J. Cartwright), 199–208.

Hunter, R.E. (1977). Terminology of cross-stratified sedimentary layers and climbing-ripple structures. *Journal of Sedimentary Research* 47: 697–706.

Imran, J., Parker, G., and Katopodes, N. (1998). A numerical model of channel inception on submarine fans. *Journal of Geophysical Research: Oceans* 103 (C1): 1219–1238.

Ivanov, V.V., Shapiro, G.I., Huthnance, J.M. et al. (2004). Cascades of dense water around the world ocean. *Progress in Oceanography* 60 (1): 47–98.

Jackson, C.R., Da Silva, J.C., and Jeans, G. (2012). The generation of nonlinear internal waves. *Oceanography* 25 (2): 108–123.

Janocko, M.N.W.H.S.W.M., Nemec, W., Henriksen, S., and Warchoł, M. (2013). The diversity of deep-water sinuous channel belts and slope valley-fill complexes. *Marine and Petroleum Geology* 41: 7–34.

Jégou, I., Savoye, B., Pirmez, C., and Droz, L. (2008). Channel-mouth lobe complex of the recent Amazon Fan: the missing piece. *Marine Geology* 252 (1–2): 62–77.

Jobe, Z.R., Lowe, D.R., and Morris, W.R. (2012). Climbing-ripple successions in turbidite systems: depositional environments, sedimentation rates and accumulation times. *Sedimentology* 59 (3): 867–898.

Jobe, Z.R., Sylvester, Z., Howes, N. et al. (2017). High-resolution, millennial-scale patterns of bed compensation on a sand-rich intraslope submarine fan, western Niger Delta slope. *GSA Bulletin* 129 (1–2): 23–37.

Jobe, Z.R., Howes, N.C., Straub, K.M. et al. (2020). Comparing aggradation, superelevation, and avulsion frequency of submarine and fluvial channels. *Frontiers in Earth Science* 8: 53.

Johnson, A.M. (1970). *Physical Processes in Geology*. San Francisco: Freeman, Cooper & Co.

Jolly, R.J. and Lonergan, L. (2002). Mechanisms and controls on the formation of sand intrusions. *Journal of the Geological Society* 159 (5): 605–617.

Jones, V.L. (2023). Sedimentary processes and depositional model for the Badami Reservoir, North Slope Alaska. MSc thesis. University of Anchorage, Alaska.

Jones, K.P.N., McCave, I.N., and Weaver, P. P.E. (1992). Textural and dispersal patterns of thick mud turbidites from the Madeira Abyssal Plain. *Marine Geology* 107 (3): 149–173.

Kane, I.A. and Hodgson, D.M. (2011). Sedimentological criteria to differentiate submarine channel levee subenvironments: exhumed examples from the Rosario Fm. (Upper Cretaceous) of Baja California, Mexico, and the Fort Brown Fm. (Permian), Karoo basin, S. Africa. *Marine and Petroleum Geology* 28 (3): 807–823.

Kane, I.A., McGee, D.T., and Jobe, Z.R. (2012). Halokinetic effects on submarine channel equilibrium profiles and implications for facies architecture: Conceptual model illustrated with a case study from Magnolia Field, Gulf of Mexico. In: *Salt Tectonics, Sediments and Prospectivity*, vol. 363 (ed. G. I. Alsop, S. G. Archer, A. J. Hartley, N. T. Grant, R. Hodgkinson), 289–302. London: Geological Society, Special Publications.

Kane, I.A. and Pontén, A.S. (2012). Submarine transitional flow deposits in the Paleogene Gulf of Mexico. *Geology* 40 (12): 1119–1122.

Kane, I.A., Catterall, V., McCaffrey, W.D., and Martinsen, O.J. (2010). Submarine channel response to intrabasinal tectonics: The influence of lateral tilt. *AAPG Bulletin* 94 (2): 189–219.

Kane, I.A., Pontén, A.S., Vangdal, B. et al. (2017). The stratigraphic record and processes of turbidity current transformation across deep-marine lobes. *Sedimentology* 64 (5): 1236–1273.

Kane, I.A., Clare, M.A., Miramontes, E. et al. (2020). Seafloor microplastic hotspots controlled by deep-sea circulation. *Science* 368 (6495): 1140–1145.

Karstens, J., Haflidason, H., Berndt, C., and Crutchley, G.J. (2023). Revised Storegga Slide reconstruction reveals two major submarine landslides 12,000 years apart. *Communications Earth & Environment* 4 (1): 55.

Kennedy, J.F. (1963). The mechanics of dunes and antidunes in erodible-bed channels. *Journal of Fluid mechanics* 16 (4): 521–544.

Ketzer, J.M., Carpentier, B., Le Gallo, Y., and Le Thiez, P. (2005). Geological sequestration of CO2 in mature hydrocarbon fields. Basin and reservoir numerical modelling of the Forties Field, North Sea. *Oil & Gas Science and Technology* 60 (2): 259–273.

Kinga-Mouzeo (1986). Transport particulaire actuel du fleuve Congo et de quelques affluents; enregistrement quaternaire dans l'éventail détritique profond (sédimentologie, minéralogie et géochimie). Univ de Papipan, 251 p.

Kneller, B. (1995). Beyond the turbidite paradigm: physical models for deposition of turbidites and their implications for reservoir prediction. *Geological Society, London, Special Publications* 94 (1): 31–49.

Kneller, B.C. and Branney, M.J. (1995). Sustained high-density turbidity currents and the deposition of thick massive sands. *Sedimentology* 42 (4): 607–616.

Kneller, B. and Buckee, C. (2000). The structure and fluid mechanics of turbidity currents: a review of some recent studies and their geological implications. *Sedimentology* 47: 62–94.

Kneller, B. and McCaffrey, W. (1999). Depositional effects of flow nonuniformity and stratification within turbidity currents approaching a bounding slope; deflection, reflection, and facies variation. *Journal of Sedimentary Research* 69 (5): 980–991.

Kneller, B., Edwards, D., McCaffrey, W., and Moore, R. (1991). Oblique reflection of turbidity currents. *Geology* 19 (3): 250–252.

Kneller, B.C., Bennett, S.J., and McCaffrey, W.D. (1999). Velocity structure, turbulence and fluid stresses in experimental gravity currents. *Journal of Geophysical Research: Oceans* 104 (C3): 5381–5391.

Kneller, B., Bozetti, G., Callow, R. et al. (2020). Architecture, process, and environmental diversity in a late Cretaceous slope channel system. *Journal of Sedimentary Research* 90 (1): 1–26.

Konsoer, K., Zinger, J., and Parker, G. (2013). Bankfull hydraulic geometry of submarine channels created by turbidity currents: relations between bankfull channel characteristics and formative flow discharge. *Journal of Geophysical Research: Earth Surface* 118 (1): 216–228.

Kopf, A.J. (2002). Significance of mud volcanism. *Reviews of Geophysics* 40 (2): 2–1.

Kuenen, P.H. and Migliorini, C.I. (1950). Turbidity currents as a cause of graded bedding. *The Journal of Geology* 58 (2): 91–127.

Labaume, P., Mutti, E., and Seguret, M. (1987). Megaturbidites: a depositional model from the Eocene of the SW-Pyrenean foreland basin, Spain. *Geo-Marine Letters* 7: 91–101.

Lamb, K.G. (2014). Internal wave breaking and dissipation mechanisms on the continental slope/shelf. *Annual Review of Fluid Mechanics* 46 (1): 231–254.

Lantzsch, H., Hanebuth, T.J., Chiessi, C. M. et al. (2014). The high-supply, current-dominated continental margin of southeastern South America during the late Quaternary. *Quaternary Research* 81 (2): 339–354.

Lewis, M.M., Mugwanya, K.E., and Mayall, M. (2023). Quantitative analysis of the structural evolution of salt diapirs and their impact on sediment routing and the architecture of deep-water channel reservoirs, Venus field offshore Angola. *Marine and Petroleum Geology* 155: 106380.

Liu, M., Wang, Z., Yu, K., and Xu, J. (2023). Two distinct types of turbidity currents observed in the Manila Trench, South China Sea. *Communications Earth & Environment* 4 (1): 108.

Lomas, S.A. and Joseph, P. (2004). Confined turbidite systems. *Geological Society, London, Special Publications* 222.

Lopes, U., Babonneau, N., Fierens, R. et al. (2023). Foraminiferal sandy contourite of the Limpopo Corridor (Mozambique margin): facies characterization and paleoceanographic record. *Marine Geology* 459: 107031.

Lorenz, J.C. and Finley, S.J. (1991). Regional fractures II: fracturing of Mesaverde reservoirs in the Piceance basin, Colorado. *AAPG Bulletin* 75 (11): 1738–1757.

Lowe, D.R. (1976). Grain flow and grain flow deposits. *Journal of Sedimentary Research* 46 (1): 188–199.

Lowe, D.R. (1982). Sediment gravity flows; II, Depositional models with special reference to the deposits of high-density turbidity currents. *Journal of Sedimentary Research* 52 (1): 279–297.

Lowe, D.R. (1988). Suspended-load fallout rate as an independent variable in the analysis of current structures. *Sedimentology* 35 (5): 765–776.

Lowe, D.R. and LoPiccolo, R.D. (1974). The characteristics and origins of dish and pillar structures. *Journal of Sedimentary Research* 44 (2): 484–501.

Lyons, W.B., Nezat, C.A., Carey, A.E., and Hicks, D.M. (2002). Organic carbon fluxes to the ocean from high-standing islands. *Geology* 30 (5): 443–446.

Maier, K.L., Gales, J.A., Paull, C.K. et al. (2019). Linking direct measurements of turbidity currents to submarine canyon-floor deposits. *Frontiers in Earth Science* 7: 144.

Maier, K.L., Paull, C.K., Caress, D.W. et al. (2020). Submarine-fan development revealed by integrated high-resolution datasets from La Jolla Fan, offshore California, USA. *Journal of Sedimentary Research* 90 (5): 468–479.

Maltman, A. (1994). Deformation structures preserved in rocks. In: *The Geological Deformation of Sediments* (ed. A. Maltman), 261–307. Dordrecht: Springer Netherlands.

Maltman, A.J. and Bolton, A. (2003). How sediments become mobilized. *Geological Society, London, Special Publications* 216 (1): 9–20.

Marini, M., Milli, S., Ravnås, R., and Moscatelli, M. (2015). A comparative study of confined vs. semi-confined turbidite lobes from the Lower Messinian Laga Basin (Central Apennines, Italy): implications for assessment of reservoir architecture. *Marine and Petroleum Geology* 63: 142–165.

Marsset, T., Droz, L., Dennielou, B., and Pichon, E. (2009). Cycles in the architecture of the Quaternary Zaire turbidite system: a possible link with climate. In: *External Controls on Deep-Water Depositional Systems* (ed. B. Kneller, O.J. Martinsen, and B. McCaffrey). SEPM Society for Sedimentary Geology.

Mattern, F. (2005). Ancient sand-rich submarine fans: depositional systems, models, identification, and analysis. *Earth-Science Reviews* 70 (3–4): 167–202.

Mayall, M. and Kneller, B. (2021). Seismic interpretation workflows for deep-water systems: a practical guide for the subsurface. *AAPG Bulletin* 105 (11): 2127–2157.

Mayall, M., Jones, E., and Casey, M. (2006). Turbidite channel reservoirs—Key elements in facies prediction and effective development. *Marine and Petroleum Geology* 23 (8): 821–841.

Mayall, M., Lonergan, L., Bowman, A. et al. (2010). The response of turbidite slope channels to growth-induced seabed topography. *AAPG Bulletin* 94 (7): 1011–1030.

McCaffrey, W. and Kneller, B. (2001). Process controls on the development of stratigraphic trap potential on the margins of confined turbidite systems and aids to reservoir evaluation. *AAPG bulletin* 85 (6): 971–988.

McCave, I.N. and Jones, K.P.N. (1988). Deposition of ungraded muds from high-density non-turbulent turbidity currents. *Nature* 333 (6170): 250–252.

McCave, I.N. and Tucholke, B.E. (1986). Deep current-controlled sedimentation in the western North Atlantic. In: *The Western North Atlantic Region* (ed. P.R. Vogt and B.E. Tucholke). Geology of North America.

McGowan, D., Salian, A., Baas, J.H. et al. (2024). On the origin of chevron marks and striated grooves, and their use in predicting mud bed rheology. *Sedimentology* 71 (2): 687–708.

McHargue, T.R., Hodgson, D.M., and Shelef, E. (2021). Architectural diversity of submarine lobate deposits. *Frontiers in Earth Science* 9: 697170.

McHargue, T., Pyrcz, M.J., Sullivan, M.D. et al. (2011). Architecture of turbidite channel systems on the continental slope: patterns and predictions. *Marine and Petroleum Geology* 28 (3): 728–743.

McHugh, C.M., Damuth, J.E. and Mountain, G.S., 2002. Cenozoic mass-transport facies and their correlation with relative sea-level change, *New Jersey continental margin. Marine Geology*, 184(3-4), pp.295-334.

McLean, S.R. (1990). The stability of ripples and dunes. *Earth-Science Reviews* 29 (1–4): 131–144.

Meiburg, E. and Kneller, B. (2010). Turbidity currents and their deposits. *Annual review of Fluid Mechanics* 42 (1): 135–156.

Menard, H.W. (1955). Deformation of the northeastern Pacific basin and the west coast of North America. *Geological Society of America Bulletin* 66 (9): 1149–1198.

Middleton, G.V. (1967). Experiments on density and turbidity currents: III. Deposition of sediment. *Canadian Journal of Earth Sciences* 4 (3): 475–505.

Middleton, G.V. and Hampton, M.A. (1973). Part I. Sediment gravity flows: mechanics of flow and deposition. Pacific Section SEPM. *Geology, Environmental Science* 1–38.

Mignard, S., Mulder, T., Martinez, P., and Garlan, T. (2019). The Ogooue Fan (offshore Gabon): a modern example of deep-sea fan on a complex slope profile. *Solid Earth* 10 (3): 851–869.

Milliman, J.D. and Syvitski, J.P. (1992). Geomorphic/tectonic control of sediment discharge to the ocean: the importance of small mountainous rivers. *The Journal of Geology* 100 (5): 525–544.

Miramontes, E., Cattaneo, A., Jouet, G. et al. (2016). The Pianosa contourite depositional system (northern Tyrrhenian Sea): drift morphology and Plio-Quaternary stratigraphic evolution. *Marine Geology* 378: 20–42.

Miramontes, E., Penven, P., Fierens, R. et al. (2019). The influence of bottom currents on the Zambezi Valley morphology (Mozambique Channel, SW Indian Ocean): in situ current observations and hydrodynamic modelling. *Marine Geology* 410: 42–55.

Miramontes, E., Jouet, G., Thereau, E. et al. (2020). The impact of internal waves on upper continental slopes: insights from the Mozambican margin (southwest Indian Ocean). *Earth Surface Processes and Landforms* 45 (6): 1469–1482.

Miramontes, E., Thiéblemont, A., Babonneau, N. et al. (2021). Contourite and mixed turbidite-contourite systems in the Mozambique Channel (SW Indian Ocean): link between geometry, sediment characteristics and modelled bottom currents. *Marine Geology* 437: 106502.

Mohamed, M.A., Collier, R.E., Hodgson, D.M., Gawthorpe, R.L., Shillington, D. J., Muravchik, M. and Sakellariou, D., 2024. Sediment flux variation as a record of climate change in the Late Quaternary deep-water active Corinth Rift, *Greece. Basin Research*, 36(5), p. e12896.

Morris, E.A., Hodgson, D.M., Brunt, R.L., and Flint, S.S. (2014). Origin, evolution and anatomy of silt-prone submarine external levées. *Sedimentology* 61 (6): 1734–1763.

Mountjoy, J.J., Howarth, J.D., Orpin, A.R. et al. (2018). Earthquakes drive large-scale submarine canyon development and sediment supply to deep-ocean basins. *Science Advances* 4 (3): eaar3748.

Muck, M.T. and Underwood, M.B. (1990). Upslope flow of turbidity currents: a comparison among field observations, theory, and laboratory models. *Geology* 18 (1): 54–57.

Mulder, T. and Alexander, J. (2001). The physical character of subaqueous sedimentary density flows and their deposits. *Sedimentology* 48 (2): 269–299.

Mulder, T. and Cochonat, P. (1996). Classification of offshore mass movements. *Journal of Sedimentary Research* 66 (1): 43–57.

Mulder, T. and Syvitski, J.P. (1995). Turbidity currents generated at river mouths during exceptional discharges to the world oceans. *The Journal of Geology* 103 (3): 285–299.

Mulder, T., Savoye, B., and Syvitski, J.P.M. (1997). Numerical modelling of a mid-sized gravity flow: the 1979 Nice turbidity current (dynamics, processes, sediment budget and seafloor impact). *Sedimentology* 44 (2): 305–326.

Mulder, T., Faugères, J.C., and Gonthier, E. (2008). Mixed turbidite–contourite systems. *Developments in Sedimentology* 60: 435–456.

Mutti, E. (1977). Distinctive thin-bedded turbidite facies and related depositional environments in the Eocene Hecho Group (south-central Pyrenees, Spain). *Sedimentology* 24: 107–131.

Mutti, E. and Normark, W.R. (1987). Comparing examples of modern and ancient turbidite systems: problems and concepts. In: *Marine Clastic Sedimentology: Concepts and Case studies* (ed. J.K. Leggett and G.G. Zuffa), 1–38. Graham & Tortman.

Mutti, E. and Normark, W.R. (1991). An integrated approach to the study of turbidite systems. In: *Seismic Facies and Sedimentary Processes of Submarine Fans and Turbidite Systems* (ed. P. Weimer and M.H. Link), 75–106. New York, NY: Springer New York.

Mutti, E. and Sonnino, M. (1981). Compensation cycles: a diagnostic feature of turbidite sandstone lobes. In: *International Association of Sedimentologists—European Regional Meeting*, vol. 2, 120–123.

Mutti, E., Bernoulli, D., Lucchi, F.R., and Tinterri, R. (2009). Turbidites and turbidity currents from Alpine 'flysch' to the exploration of continental margins. *Sedimentology* 56 (1): 267–318.

Normandeau, A., Bourgault, D., Neumeier, U. et al. (2020). Storm-induced turbidity currents on a sediment-starved shelf: insight from direct monitoring and repeat seabed mapping of upslope migrating bedforms. *Sedimentology* 67 (2): 1045–1068.

Normark, W.R. (1970). Growth patterns of deep-sea fans. *AAPG Bulletin* 54 (11): 2170–2195.

Normark, W.R., Carlson, P.R., Chan, M. A., and Archer, A.W. (2003). Giant submarine canyons: is size any clue to their importance in the rock record? *Special Papers-Geological Society of America* 175–190.

Nugraha, H.D., Jackson, C.A.L., Johnson, H.D. et al. (2022). Extreme erosion by submarine slides. *Geology* 50 (10): 1130–1134.

Obermeier, S.F. (1996). Use of liquefaction-induced features for paleoseismic analysis—an overview of how seismic liquefaction features can be distinguished from other features and how their regional distribution and properties of source sediment can be used to infer the location and strength of Holocene paleo-earthquakes. *Engineering Geology* 44 (1–4): 1–76.

Ortiz-Karpf, A., Hodgson, D.M., and McCaffrey, W.D. (2015). The role of mass-transport complexes in controlling channel avulsion and the subsequent sediment dispersal patterns on an active margin: the Magdalena Fan, offshore Colombia. *Marine and Petroleum Geology* 64: 58–75.

Palanques, A., Puig, P., de Madron, X.D. et al. (2012). Sediment transport to the deep canyons and open-slope of the western Gulf of Lions during the 2006 intense cascading and open-sea convection period. *Progress in Oceanography* 106: 1–15.

Pantin, H.M. (1979). Interaction between velocity and effective density in turbidity flow: phase-plane analysis, with criteria for autosuspension. *Marine Geology* 31 (1–2): 59–99.

Parker, G., Fukushima, Y., and Pantin, H. M. (1986). Self-accelerating turbidity currents. *Journal of Fluid Mechanics* 171: 145–181.

Parsons, J.D. and Garcia, M.H. (1998). Similarity of gravity current fronts. *Physics of Fluids* 10 (12): 3209–3213.

Peakall, J. and Sumner, E.J. (2015). Submarine channel flow processes and deposits: a process-product perspective. *Geomorphology* 244: 95–120.

Peakall, J., McCaffrey, B., and Kneller, B. (2000). A process model for the evolution, morphology, and architecture of sinuous submarine channels. *Journal of Sedimentary Research* 70 (3): 434–448.

Peakall, J., Best, J., Baas, J.H. et al. (2020). An integrated process-based model of flutes and tool marks in deep-water environments: implications for palaeohydraulics, the Bouma sequence and hybrid event beds. *Sedimentology* 67 (4): 1601–1666.

Peakall, J., Best, J.L., Baas, J.H. et al. (2024). Flow-induced interfacial deformation structures (FIDS): implications for the interpretation of palaeocurrents, flow dynamics and substrate rheology. *Sedimentology* 71 (6): 1709–1743.

Pemberton, E.A., Hubbard, S.M., Fildani, A. et al. (2016). The stratigraphic expression of decreasing confinement along a deep-water sediment routing system: outcrop example from southern Chile. *Geosphere* 12 (1): 114–134.

Pett, J.W. and Walker, R.G. (1971). Relationship of flute cast morphology to internal sedimentary structures in turbidites. *Journal of Sedimentary Research* 41 (1): 114–128.

Pickering, K.T. and Hiscott, R.N. (1985). Contained (reflected) turbidity currents from the Middle Ordovician Cloridorme Formation, Quebec, Canada: an alternative to the antidune hypothesis. *Sedimentology* 32: 373–394.

Pickering, K.T. and Hiscott, R.N. (2015). *Deep Marine Systems: Processes, Deposits, Environments, Tectonics and Sedimentation*. John Wiley & Sons.

Pickering, K.T., Souter, C., Oba, T. et al. (1999). Glacio-eustatic control on deep-marine clastic forearc sedimentation, Pliocene-mid-Pleistocene (c. 1180–600 ka) Kazusa Group, SE Japan. *Journal of the Geological Society* 156 (1): 125–136.

Picot, M., Droz, L., Marsset, T. et al. (2016). Controls on turbidite sedimentation: insights from a quantitative approach of submarine channel and lobe architecture (Late Quaternary Congo Fan). *Marine and Petroleum Geology* 72: 423–446.

Picot, M., Marsset, T., Droz, L. et al. (2019). Monsoon control on channel avulsions in the Late Quaternary Congo Fan.

Quaternary Science Reviews 204: 149–171.

Pierdomenico, M., Casalbore, D., and Chiocci, F.L. (2019). Massive benthic litter funnelled to deep sea by flash-flood generated hyperpycnal flows. *Scientific reports* 9 (1): 5330.

Piper, D.J. (1978). Turbidite muds and silts on deep-sea fans and abyssal plains. In: *Sedimentation in Submarine Canyons, Fans, and Trenches* (ed. D.J. Stanley and G. Kelling), 163–176. Stroudsburg, PA: Dowden, Hutchison & Ross.

Piper, D.J. and Normark, W.R. (1983). Turbidite depositional patterns and flow characteristics, Navy submarine fan. California Borderland. *Sedimentology* 30 (5): 681–694.

Piper, D.J. and Normark, W.R. (2001). Sandy fans-from Amazon to Hueneme and beyond. *AAPG Bulletin* 85 (8): 1407–1438.

Piper, D.J. and Normark, W.R. (2009). Processes that initiate turbidity currents and their influence on turbidites: a marine geology perspective. *Journal of Sedimentary Research* 79 (6): 347–362.

Piper, D.J., Panagos, A.G., and Pe, G.G. (1978). Conglomeratic Miocene flysch, western Greece. *Journal of Sedimentary Research* 48 (1): 117–125.

Piper, D.J., Shor, A.N., Farre, J.A. et al. (1985). Sediment slides and turbidity currents on the Laurentian Fan: sidescan sonar investigations near the epicenter of the 1929 Grand Banks earthquake. *Geology* 13 (8): 538–541.

Piper, D.J., Hiscott, R., and Normark, W.R. (1999). Outcrop-scale acoustic facies analysis and latest Quaternary development of Hueneme and Dume submarine fans, offshore California. *Sedimentology* 46 (1): 47–78.

Pirmez, C., Beaubouef, R.T., Friedmann, S.J., and Mohrig, D.C. (2000). Equilibrium profile and baselevel in submarine channels: examples from Late Pleistocene systems and implications for the architecture of deepwater reservoirs. In: *Deep-Water Reservoirs of the World: 20th Annual* (ed. P. Weimer), 782–805. SEPM Society for Sedimentary Geology.

Pope, E.L., Heijnen, M.S., Talling, P.J. et al. (2022). Carbon and sediment fluxes inhibited in the submarine Congo Canyon by landslide-damming. *Nature Geoscience* 15 (10): 845–853.

Popescu, I., Lericolais, G., Panin, N. et al. (2001). Late Quaternary channel avulsions on the Danube deep-sea fan, Black Sea. *Marine Geology* 179 (1–2): 25–37.

Porebski, S.J. and Steel, R.J. (2006). Deltas and sea-level change. *Journal of Sedimentary Research* 76 (3): 390–403.

Porter, M., Sprague, A., Sullivan, M. et al. (2006). Stratigraphic organization and predictability of mixed coarse- and fine-grained lithofacies successions in a Lower Miocene deep-water slope-channel system, Angola Block15. In: *Giant Hydrocarbon Reservoirs of the World: From Rocks to 843 Reservoir Characterization and Modeling*, AAPG Memoir 88 (ed. P.M. Harris and L.J. Weber), 281–305. SEPM Special Publication.

Posamentier, H.W. and Allen, G.P. (1999). Siliciclastic sequence stratigraphy-concepts and applications. In: *SEPM Concepts in Sedimentology and Paleontology Series 7*, 210. Tulsa Oklahoma: Society for Sedimentary Geology.

Posamentier, H.W. and Kolla, V. (2003). Seismic geomorphology and stratigraphy of depositional elements in deep-water settings. *Journal of Sedimentary Research* 73: 367–388.

Posamentier, H.W. and Martinsen, O.J. (2011). The character and genesis of submarine mass-transport deposits: insights from outcrop and 3D seismic data. *Special Publications series of the Society for Sedimentary Geology* 96: 7–38.

Posamentier, H.W. and Walker, R.G. (2006). Deep-water turbidites and submarine fans. In: *Facies Models Revisited* (ed. H.W. Posamentier and R. G. Walker), 86. SEPM Special Publications.

Postma, G. (1986). Classification for sediment gravity-flow deposits based on flow conditions during sedimentation. *Geology* 14 (4): 291–294.

Prather, B.E., Booth, J.R., Steffens, G.S., and Craig, P.A. (1998). Classification, lithologic calibration, and stratigraphic succession of seismic facies of intraslope basins, deep-water Gulf of Mexico. *AAPG Bulletin* 82 (5): 701–728.

Prather, B.E., O'Byrne, C., Pirmez, C., and Sylvester, Z. (2017). Sediment partitioning, continental slopes and base-of-slope systems. *Basin Research* 29 (3): 394–416.

Prélat, A. and Hodgson, D.M. (2013). The full range of turbidite bed thickness patterns in submarine lobes: controls and implications. *Journal of the Geological Society* 170 (1): 209–214.

Prélat, A., Hodgson, D.M., and Flint, S.S. (2009). Evolution, architecture and hierarchy of distributary deep-water deposits: a high-resolution outcrop investigation from the Permian Karoo Basin, South Africa. *Sedimentology* 56 (7): 2132–2154.

Prélat, A., Covault, J.A., Hodgson, D.M. et al. (2010). Intrinsic controls on the range of volumes, morphologies, and dimensions of submarine lobes. *Sedimentary Geology* 232 (1–2): 66–76.

Prior, D.B., Bornhold, B.D., and Johns, M. W. (1984). Depositional characteristics of a submarine debris flow. *Journal of Geology* 92: 707–727.

Privat, A.M.L., Hodgson, D.M., Jackson, C. A.L. et al. (2021). Evolution from syn-rift carbonates to early post-rift deep-marine intraslope lobes: the role of rift basin physiography on sedimentation patterns. *Sedimentology* 68 (6): 2563–2605.

Puig, P., Palanques, A., Orange, D.L. et al. (2008). Dense shelf water cascades and sedimentary furrow formation in the Cap de Creus Canyon, northwestern Mediterranean Sea. *Continental Shelf Research* 28 (15): 2017–2030.

Puig, P., Palanques, A., and Martín, J. (2014). Contemporary sediment-transport processes in submarine canyons. *Annual Review of Marine Science* 6: 53–77.

Puigdefàbregas, C., Gjelberg, J., and Vaksdal, M. (2004). The Grès d'Annot in the Annot syncline: outer basin-margin onlap and associated soft-sediment deformation. *Geological Society, London, Special Publications* 221 (1): 367–388.

Reading, H.G. and Richards, M. (1994). Turbidite systems in deep-water basin margins classified by grain size and feeder system. *AAPG Bulletin* 78 (5): 792–822.

Rebesco, M., Hernández-Molina, F.J., Van Rooij, D., and Wåhlin, A. (2014). Contourites and associated sediments controlled by deep-water circulation processes: state-of-the-art and future considerations. *Marine Geology* 352: 111–154.

Remacha, E. and Fernández, L.P. (2003). High-resolution correlation patterns in the turbidite systems of the Hecho Group (South-Central Pyrenees, Spain). *Marine and Petroleum Geology* 20 (6–8): 711–726.

Ribó, M., Puig, P., Muñoz, A. et al. (2016). Morphobathymetric analysis of the large fine-grained sediment waves over the Gulf of Valencia continental slope (NW Mediterranean). *Geomorphology* 253: 22–37.

Richardson, J.F. and Zaki, W.N. (1954). The sedimentation of a suspension of uniform spheres under conditions of viscous flow. *Chemical Engineering Science* 3 (2): 65–73.

Romans, B.W., Castelltort, S., Covault, J.A. et al. (2016). Environmental signal propagation in sedimentary systems across timescales. *Earth-Science Reviews* 153: 7–29.

Ross, J.A., Peakall, J., and Keevil, G.M. (2011). An integrated model of extrusive sand injectites in cohesionless sediments. *Sedimentology* 58 (7): 1693–1715.

Rothwell, R.G., Thomson, J., and Kähler, G. (1998). Low-sea-level emplacement of a very large Late Pleistocene 'megaturbidite' in the western Mediterranean Sea. *Nature* 392 (6674): 377–380.

Rotzien, J.R., Yeilding, C.A., Sears, R.A. et al. (2022). *Deepwater Sedimentary Systems: Science, Discovery, and Applications*. Elsevier ISBN: 9780323919180.

Saller, A., Werner, K., Sugiaman, F. et al. (2008). Characteristics of Pleistocene deep-water fan lobes and their application to an upper Miocene reservoir model, offshore East Kalimantan, Indonesia. *AAPG Bulletin* 92 (7): 919–949.

Sánchez-Leal, R.F., Bellanco, M.J., Fernández-Salas, L.M. et al. (2017). The Mediterranean overflow in the Gulf of Cadiz: a rugged journey. *Science Advances* 3 (11): eaao0609.

Sanchez-Vidal, A., Llorca, M., Farré, M. et al. (2015). Delivery of unprecedented amounts of perfluoroalkyl substances towards the deep-sea. *Science of the Total Environment* 526: 41–48.

Savoye, B., Piper, D.J., and Droz, L. (1993). Plio-Pleistocene evolution of the Var deep-sea fan off the French Riviera. *Marine and Petroleum Geology* 10 (6): 550–571.

Scarselli, N. (2020). Submarine landslides, architecture, controlling factors and environments. A summary. In: *Regional Geology and Tectonics: Principles of Geologic Analysis, Principles of Geologic Analysis*, vol. 1 (ed. N. Scarselli, J. Adam, D. Chiarella, et al.). Elsevier.

Schindler, R.J., Parsons, D.R., Ye, L. et al. (2015). Sticky stuff: redefining bedform prediction in modern and ancient environments. *Geology* 43 (5): 399–402.

Scott, A., Vigorito, M., and Hurst, A. (2009). The process of sand injection: internal structures and relationships with host strata (Yellowbank Creek Injectite Complex, California, USA). *Journal of Sedimentary Research* 79 (8): 568–583.

Sequeiros, O.E., Spinewine, B., Beaubouef, R.T. et al. (2010). Characteristics of velocity and excess density profiles of saline underflows and turbidity currents flowing over a mobile bed. *Journal of Hydraulic Engineering* 136 (7): 412–433.

Simmons, S.M., Azpiroz-Zabala, M., Cartigny, M.J.B. et al. (2020). Novel acoustic method provides first detailed measurements of sediment concentration structure within submarine turbidity currents. *Journal of Geophysical Research: Oceans* 125 (5): e2019JC015904.

Simons, D.B., Richardson, E.V., and Haushild, W.L. (1963). Some effects of fine sediments on flow phenomena, Water-Supply Paper, No. 1498G, United States Geological Survey, Washington, D.C., 1963.

Sinclair, H.D. and Tomasso, M. (2002). Depositional evolution of confined turbidite basins. *Journal of Sedimentary Research* 72 (4): 451–456.

Slootman, A. and Cartigny, M.J. (2020). Cyclic steps: review and aggradation-based classification. *Earth-Science Reviews* 201: 102949.

Sømme, T.O., Helland-Hansen, W., Martinsen, O.J., and Thurmond, J.B. (2009). Relationships between morphological and sedimentological parameters in source-to-sink systems: a basis for predicting semi-quantitative characteristics in subsurface systems. *Basin Research* 21 (4): 361–387.

Sommer, J.V., Dorrington, K., Van Denburg, C. et al. (2021). Pikka Field and the opening of the Nanushuk play, North Slope, Alaska. In: *Giant fields of the decade: 2010–2020*, vol. 125 (ed. C.A. Sternbach, R.K. Merrill, and J.C. Dolson), 301–338. Tulsa Oklahoma: American Association of Petroleum Geologists.

Sorby, H.C. (1859). On the structures produced by the currents present during the deposition of stratified rocks. *The Geologist* 2: 137–147.

Soutter, E.L., Kane, I.A., Fuhrmann, A. et al. (2019). The stratigraphic evolution of onlap in siliciclastic deep-water systems: autogenic modulation of allogenic signals. *Journal of Sedimentary Research* 89 (10): 890–917.

Soutter, E.L., Kane, I.A., Hodgson, D.M., and Flint, S.S. (2023). Controls on continental shelf width: a machine learning approach. *Geomorphology* 436: 108729.

Spychala, Y.T., Hodgson, D.M., Flint, S.S., and Mountney, N.P. (2015). Constraining the sedimentology and stratigraphy of submarine intraslope lobe deposits using exhumed examples from the Karoo Basin, South Africa. *Sedimentary Geology* 322: 67–81.

Spychala, Y.T., Hodgson, D.M., Prélat, A. et al. (2017). Frontal and lateral submarine lobe fringes: comparing sedimentary facies, architecture and flow processes. *Journal of Sedimentary Research* 87 (1): 75–96.

Stevenson, C.J., Talling, P.J., Wynn, R.B. et al. (2013). The flows that left no trace: very large-volume turbidity currents that bypassed sediment through submarine channels without eroding the sea floor. *Marine and Petroleum Geology* 41: 186–205.

Stevenson, C.J., Jackson, C.A.L., Hodgson, D.M. et al. (2015). Deep-water sediment bypass. *Journal of Sedimentary Research* 85 (9): 1058–1081.

Stevenson, C.J., Peakall, J., Hodgson, D.M. et al. (2020). TB or not TB: banding in turbidite sandstones. *Journal of Sedimentary Research* 90 (8): 821–842.

Steventon, M.J., Jackson, C.A.L., Hodgson, D.M., and Johnson, H.D. (2019). Strain analysis of a seismically imaged mass-transport complex, offshore Uruguay. *Basin Research* 31 (3): 600–620.

Stow, D.A. and Bowen, A.J. (1978). Origin of lamination in deep sea, fine-grained sediments. *Nature* 274 (5669): 324–328.

Stow, D.A.V. and Faugères, J.C. (2008). Contourite facies and the facies model. In: *Contourites, Developments in Sedimentology*, vol. 60 (ed. M. Rebesco and A. Camerlenghi), 223–256. Amsterdam: Elsevier.

Stow, D. and Lovell, J.P.B. (1979). Contourites – their recognition in modern and ancient sediments. *Earth-Science Reviews* 14: 251–291.

Stow, D.A. and Shanmugam, G. (1980). Sequence of structures in fine-grained turbidites: comparison of recent deep-sea and ancient flysch sediments. *Sedimentary Geology* 25 (1–2): 23–42.

Stow, D.A., Faugères, J.C., Howe, J.A. et al. (2002). Bottom currents, contourites and deep-sea sediment drifts: current state-of-the-art. *Geological Society, London, Memoirs* 22 (1): 7–20.

Straub, K.M., Paola, C., Mohrig, D. et al. (2009). Compensational stacking of channelized sedimentary deposits. *Journal of Sedimentary Research* 79 (9): 673–688.

Sumner, E.J., Amy, L.A., and Talling, P.J. (2008). Deposit structure and processes of sand deposition from decelerating sediment suspensions. *Journal of Sedimentary Research* 78 (8): 529–547.

Sumner, E.J., Talling, P.J., and Amy, L.A. (2009). Deposits of flows transitional between turbidity current and debris flow. *Geology* 37 (11): 991–994.

Sylvester, Z., Pirmez, C., and Cantelli, A. (2011). A model of submarine channel-levee evolution based on channel trajectories: implications for stratigraphic architecture. *Marine and Petroleum Geology* 28 (3): 716–727.

Syvitski, J.P. and Kettner, A. (2011). Sediment flux and the Anthropocene. *Philosophical Transactions of the Royal Society A: Mathematical, Physical and Engineering Sciences* 369 (1938): 957–975.

Talley, L.D., Pickard, G.L., Emery, W.J., and Swift, J.H. (2011). *Descriptive Physical Oceanography: An Introduction*, 6e, 560. Boston: Elsevier.

Talling, P.J., Amy, L.A., and Wynn, R.B. (2007). New insight into the evolution of large-volume turbidity currents: comparison of turbidite shape and previous modelling results. *Sedimentology* 54 (4): 737–769.

Talling, P.J., Masson, D.G., Sumner, E.J., and Malgesini, G. (2012). Subaqueous sediment density flows: depositional processes and deposit types. *Sedimentology* 59 (7): 1937–2003.

Talling, P.J., Baker, M.L., Pope, E.L. et al. (2022). Longest sediment flows yet measured show how major rivers connect efficiently to deep sea. *Nature Communications* 13 (1): 4193.

Taylor, W.J., Hodgson, D.M., Peakall, J. et al. (2024). Unidirectional and combined transitional flow bedforms: Controls on process and distribution in submarine slope settings. *Sedimentology* 71: 1329–1362.

Thiéblemont, A., Hernández-Molina, F.J., Miramontes, E. et al. (2019). Contourite depositional systems along the Mozambique channel: the interplay between bottom currents and sedimentary processes. *Deep Sea Research Part I: Oceanographic Research Papers* 147: 79–99.

Tilston, M., Arnott, R.W., Rennie, C.D., and Long, B. (2015). The influence of grain size on the velocity and sediment concentration profiles and depositional record of turbidity currents. *Geology* 43 (9): 839–842.

Tinterri, R. (2011). Combined flow sedimentary structures and the genetic link between sigmoidal-and hummocky-cross stratification. *GeoActa* 10 (4): 1–43.

Toniolo, H., Parker, G., Voller, V., and Beaubouef, R.T. (2006). Depositional turbidity currents in diapiric minibasins on the continental slope: experiments—numerical simulation and upscaling. *Journal of Sedimentary Research* 76 (5): 798–818.

Truswell, J.F. (1972). Sandstone sheets and related intrusions from Coffee bay, Transkei, South Africa. *Journal of Sedimentary Research* 42 (3): 578–583.

Van der Merwe, W.C., Hodgson, D.M., Brunt, R.L., and Flint, S.S. (2014). Depositional architecture of sand-attached and sand-detached channel-lobe transition zones on an exhumed stepped slope mapped over a 2500 km^2 area. *Geosphere* 10 (6): 1076–1093.

Vigorito, M., Hurst, A., Cartwright, J., and Scott, A. (2008). Regional-scale subsurface sand remobilization: geometry and architecture. *Journal of the Geological Society* 165 (3): 609–612.

Wadhams, P., Holfort, J., Hansen, E., and Wilkinson, J.P. (2002). A deep convective chimney in the winter Greenland Sea. *Geophysical research letters* 29 (10): 76–71.

Wåhlin, A.K. and Walin, G. (2001). Downward migration of dense bottom currents. *Environmental Fluid Mechanics* 1: 257–279.

Walker, R.G. (1965). The origin and significance of the internal sedimentary structures of turbidites. *Proceedings of the Yorkshire Geological Society* 35 (1): 1–32.

Walker, R.G. (1967). Turbidite sedimentary structures and their relationship to proximal and distal depositional environments. *Journal of Sedimentary Research* 37 (1): 25–43.

Walker, R.G. (1978). Deep-water sandstone facies and ancient submarine fans: models for exploration for stratigraphic traps. *AAPG Bulletin* 62 (6): 932–966.

Walton, E.K. (1967). The sequence of internal structures in turbidites. *Scottish Journal of Geology* 3 (2): 306–317.

Warren, J.K. (2006). Salt tectonics. In: *Evaporites: Sediments, Resources and Hydrocarbons* (ed. J.K. Warren), 375–453. Springer.

Weimer, P. (1991). Seismic facies, characteristics, and variations in channel evolution, Mississippi Fan (Plio-Pleistocene), Gulf of Mexico. In: *Seismic Facies and Sedimentary Processes Of Submarine Fans and Turbidite Systems*, 323–347. New York, NY: Springer New York.

Wetzel, A. (1993). The transfer of river load to deep-sea fans: a quantitative approach. *AAPG Bulletin* 77 (10): 1679–1692.

Wilckens, H., Miramontes, E., Schwenk, T. et al. (2021). The erosive power of the Malvinas Current: Influence of bottom currents on morpho-sedimentary features along the northern Argentine margin (SW Atlantic Ocean). *Marine Geology* 439: 106539.

Wilckens, H., Schwenk, T., Lüdmann, T. et al. (2023). Factors controlling the morphology and internal sediment architecture of moats and their associated contourite drifts. *Sedimentology* 70 (5): 1472–1495.

Winterwerp, J.C. and Van Kesteren, W.G. (2004). *Introduction to the Physics of Cohesive Sediment Dynamics in the Marine Environment*. Elsevier.

Wood, P.A. and Smith, A.J. (1958). The sedimentation and sedimentary history of the Aberystwyth Grits (Upper Llandoverian). *Quarterly Journal of the Geological Society* 114 (1–4): 163–195.

Wynn, R.B., Kenyon, N.H., Masson, D.G. et al. (2002). Characterization and recognition of deep-water channel-lobe transition zones. *AAPG Bulletin* 86 (8): 1441–1462.

Zhao, Y., Liu, Z., Zhang, Y. et al. (2024). Formation mechanism of drift-moat contourite systems revealed by in-situ observations in the South China Sea. *Earth and Planetary Science Letters* 628: 118585.

13

Pelagic Environments

Stuart A. Robinson

Department of Earth Sciences, University of Oxford, Oxford, UK

Summary

About 60% of the surface area of our planet is covered by water that is more than 1600 m deep and, consequently, the pelagic depositional environments of the deep-sea cover more of our planet than all other environments combined. The oceans, and the biology they host, play fundamental roles in the Earth system, influencing climate and biogeochemical cycles. It is the rich time-series of data from pelagic sediments that allow us to understand the temperature, chemistry, fauna and flora of modern and past oceans. These insights into the interactions between oceanography, biology and geology provide the context for anthropogenically driven scenarios for our future oceans. However, these pelagic archives need to be read correctly, and in this chapter, we therefore explore the processes of sediment generation, accumulation, erosion and preservation in the deep-sea and the controls on their variable facies distributions in space and time.

13.1 Introduction

Throughout much of the global open-ocean, there is only a relatively thin sedimentary cover above basement rocks (typically hundreds of metres), compared to the many kilometres that have accumulated in the subsiding sedimentary basins on continental crust, passive continental margins and subduction zones (Figure 13.1a). The sediments comprising this thin deep-sea cover are generally all referred to as 'pelagic', yet their origins are varied. The term 'pelagic' is used differently by different communities of marine scientists but common to all is the reference to environments that are far from land and the influence of shallow water hydrodynamics. In ecology, 'pelagic' denotes that part of the water column that is close to the sea surface (in contrast to 'benthic' environments on the seafloor). In sedimentology, 'pelagic sediments' have been variously defined over time but, here, they are defined as those deposited or forming in 'deep' marine environments, typically from below the storm-wave base to the abyssal plain. Pelagic sediments are typically sourced from material generated by biological production within the water column or particulate material (largely atmospheric dust and volcanic ash) that has been transported to the ocean surface through the atmosphere, before sinking to the seafloor. Although not strictly a pelagic process, authigenic and hydrothermal mineral precipitation at the seafloor is included here as the third process by which sediments can accumulate in deep-marine environments. These three sources generate three major classes of sediments in the open-ocean: *biogenic pelagic sediment, pelagic clays and authigenic precipitates.* Pelagic sediments are generally extremely fine-grained and dominated by mud to medium sand-sized components. In the modern ocean, pelagic sediments are found predominantly in deep-water environments, away from the influence of terrigenous input, off the continental shelves. The definition of pelagic sediments used here therefore specifically excludes so-called 'hemipelagic' deposits containing a high proportion of silicate mud derived ultimately from rivers (Chapters 9 and 12), clastic sediments in the deep-sea transported via submarine canyons (Chapter 12) or lime muds derived from relatively shallow-water benthic carbonate factories (Chapter 11). The division between pelagic settings and these other environments is not, however, distinct and, closer to the margins and on the shelves (Figure 13.1b) pelagic sediments arefound interbedded with, or as a component of, deposits from other sources that were deposited by processes other than settling.

Pelagic environments may appear as simple sedimentary systems, remote from everyday experience and lacking the exciting macroscopic fossils or sediment dynamics present in other settings. The reality though is more interesting and deserving of attention! Many pelagic sedimentary records are rich in microfossils and accumulated continuously for long intervals of geological time (millions of years), providing

Reading's Sedimentary Environments: Processes, Facies and Stratigraphy, Fourth Edition. Edited by Bruce Levell.

Published 2026 by John Wiley & Sons Ltd.
Companion website: www.wiley.com/go/levell4e

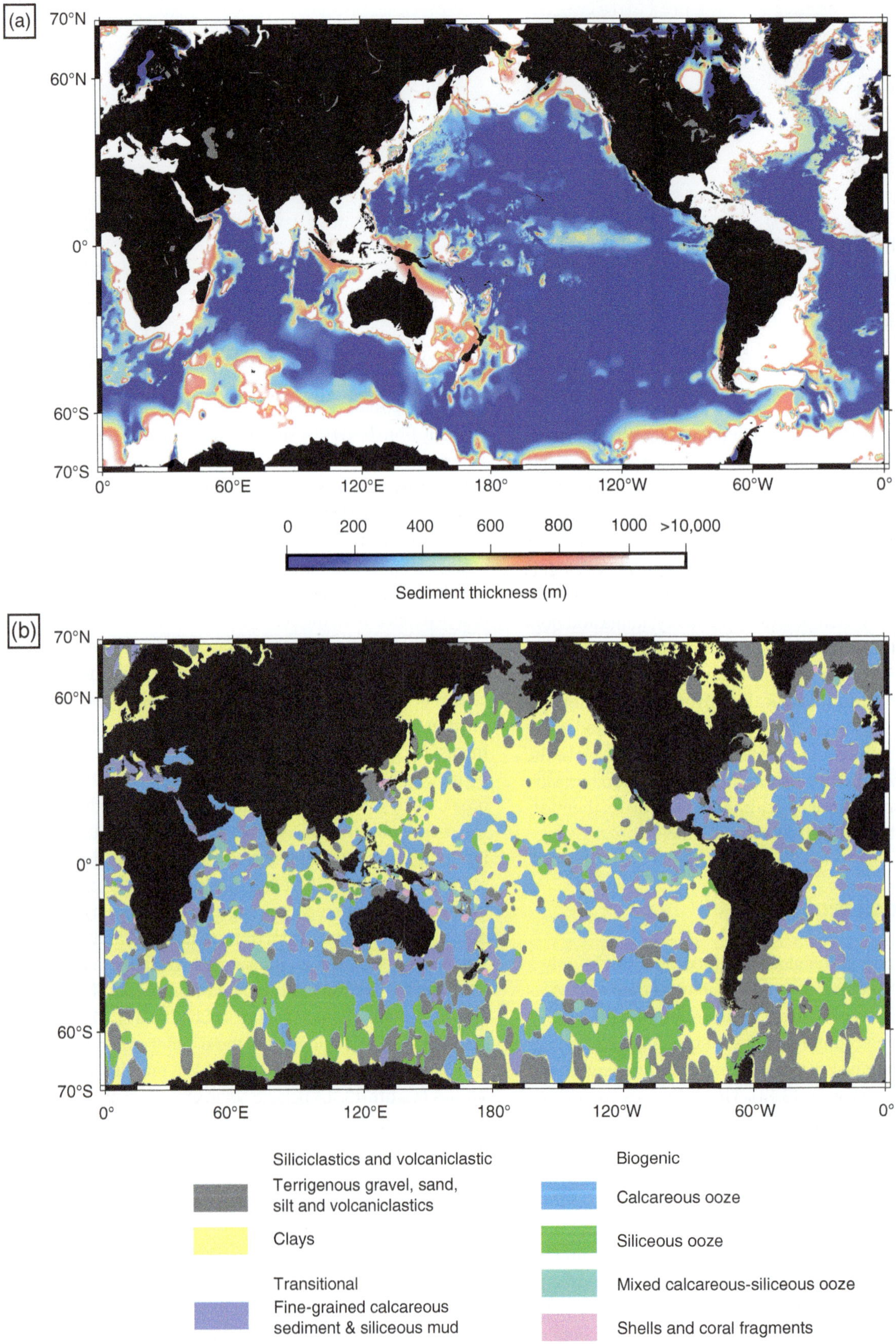

Figure 13.1 (a) Map of global sediment thickness (data from Straume et al. 2019), scaled to highlight differences between 0 and 1000 m. White areas have sediment thicknesses of 1000 to >10,000 m. (b) Distribution of sediment types in the ocean (data from Dutkiewicz et al. 2015). Both maps are drawn using GMT6 (Wessel et al. 2019) and a Mercator projection.

unique perspectives on biotic evolutionary processes (e.g. Aze et al. 2011). The microfossils provide a superb archive of geochemical data from readily identifiable, distinct and globally distributed, species. Their stable isotope compositions form the basis for palaeoceanographic and palaeoclimatological reconstructions of the last 100 million years (e.g. Tierney et al. 2020; Westerhold et al. 2020) and provide an important archive for reconstructing atmospheric carbon dioxide levels (e.g. Rae et al. 2021; CenCO_2PIP consortium 2023), thereby contextualising ongoing anthropogenic climate change. Pelagic biogenic materials play key roles in many interlinked biogeochemical cycles (e.g. C, S, O, N and P), including the transport and storage of carbon in the deep ocean which, in turn, helps regulate climate (e.g. Armstrong et al. 2001; Klaas and Archer 2002; Jin et al. 2006). Over long timescales, the global climatic and environmental changes driven by periodic changes in Earth's orbital parameters (Milanković cycles) have been recorded best by pelagic lithologies. Pelagic sediments enable refinement of the geological timescale through 'astrochronology' (e.g. Hinnov and Hilgen 2012). Recognition of Milanković cycles in pelagic sediments revolutionised our understanding of Quaternary glaciation (Hays et al. 1976), and of how the physics and chemistry of the ocean are interwoven with biogeochemical cycles, climate and the cryosphere. Pelagic sediments provided key evidence for plate tectonics, through dating the ocean floor away from mid-ocean ridges (Hsü 1992). They also established the existence of past ocean basins, through their association with obducted ocean crust (ophiolite complexes) found in collisional mountain belts (e.g. Bernoulli and Jenkyns 2009). Finally, pelagic sediments have provided resources required by society, including hydrocarbons from organic-rich black mudstones. Looking to the future, they are regarded by some as important sources of rare Earth elements (Hein et al. 2020), albeit with potentially significant detrimental impacts on benthic environments and ecosystems that are not yet fully understood (e.g. Washburn et al. 2023). In summary, pelagic sediments are fundamental components of the Earth system and are essential for understanding its future.

13.2 Modern Pelagic Sedimentation: Processes and Products

13.2.1 Fundamental Oceanographic Processes

The distribution and volume of pelagic sediment types in the ocean (Figure 13.1) are determined by several major oceanographic processes that control sediment production and preservation and, to a lesser degree, lateral transport. The open-ocean is not a single, homogeneous environment. Important variations in physical and chemical conditions occur both in depth and with latitude and longitude. A brief overview of the fundamental, and most pertinent, tenets of physical, chemical and biological oceanography is provided here. For in-depth discussion, see one of the many introductory textbooks to oceanography that are available, for example, the Open University series (e.g. Open University 1989a, 1989b, 1995; Lalli and Parsons 1997; Rothery 1998), Sverdrup and Armbrust (2008) or Trujillo and Thurman (2020).

The low latitudes receive more annual solar radiation than higher latitudes due to the low angle of incidence of the Sun's rays at high latitudes (and longer path length through the atmosphere). Heating of the low latitudes causes convection in the atmosphere with vertical movement of moist airmasses, which, at altitude, lose moisture, cool, move poleward and sink at around 30° N and S of the equator. This movement of air masses is known as the Hadley Cell of atmospheric circulation. The Coriolis force modifies this pattern to create spirals of atmospheric circulation, which, at Earth's surface we recognise as the trade winds. Over the oceans, the trade winds transfer energy to the surface of the ocean – creating wind-driven waves and, importantly, driving surface currents (Figure 13.2). At the sea surface, the direction of the surface current is at an angle to the wind direction (typically less than 45°) due, again, to the Coriolis Force. If the upper ocean is considered as a series of horizontal layers, then the frictional coupling between layers should mean that the direction of movement in the next lower layer is further deflected by the Coriolis Force, creating a spiral of transport directions in the uppermost ocean down to a depth below which the wind energy can no longer drive a current. This theoretical construction is known as the 'Ekman Spiral' and the depth interval over which it occurs (<100 m) is the 'Ekman layer'. The average motion of water in the Ekman layer defines the net transport of the surface ocean and is approximately at 90° to the wind direction. ('Ekman Transport': 90° to the right of the wind direction in the Northern Hemisphere and 90° to the left in the Southern; Figure 13.3). Ekman transport is key to understanding major features of surface ocean circulation, in particular the major ocean gyres and open-ocean upwelling regions. Wind-driven circulation can lead to areas of the ocean where Ekman transport is moving water towards a convergence zone with an elevated water surface. Gravity acts to return the slope of the sea-surface height to the horizontal, thereby counteracting the effects of the Coriolis Force, which (under steady-state conditions) creates a geostrophic flow of the upper ocean in the same direction as the wind direction, such as occurs in the North and South Pacific gyres (Figure 13.2) (Box 13.1). Conversely, where Ekman transport

Box 13.1 History of Pelagic Sedimentological Research

Pelagic sediment research began in the 19th century as a consequence of oceanographic expeditions by the navies of countries such as Britain, the United States and Prussia that sought (variously) to gain scientific knowledge, to maintain their global influence and power, and to better understand trade routes.

In the 1840s and 1850s, deep-sea sediment samples collected by vessels of these navies led to the recognition of the fine-grained nature of pelagic sediments. Subsequent microscopic investigation of these sediments by scientists such as Christian Gottfried Ehrenberg and Thomas Henry Huxley led to the description of their key biogenic and non-biogenic constituents (even if the origin of some, such as the coccoliths, was then not understood). In 1872, the British naval vessel H.M.S. *Challenger* set sail on a four-year global voyage sponsored by the Royal Society in London that provided the first detailed, systematic study of the oceans and their sediments (e.g. Murray and Renard 1891). The *Challenger* expedition laid the foundations of scientific oceanography by mapping the distribution of pelagic sediment types with remarkable accuracy; recognising the importance of water depth in the preservation of calcium carbonate; providing evidence for the patterns of ocean circulation; and discovering abundant iron-manganese nodules on the abyssal plains. The results from *Challenger* and other expeditions also provided a conceptual framework with which to interpret 'unusual' sediments exposed on land, such as the widespread Mesozoic red and white limestones and radiolarites of the Alps (Bernoulli and Jenkyns 2009). The similarities between these Alpine rocks and modern deep-sea sediments led workers in the 19th and early 20th centuries to suggest the existence of former ocean basins and the possibility of disappearing oceans and great vertical movements of the crust, albeit at a time when convincing explanations of such dynamics were lacking.

In the early 20th century, the techniques and disciplines needed to make sense of deep-sea sediments started to develop. A pioneering study of Cenozoic foraminifera from Gulf Coast oil wells by Esther Richards Applin, Alva Ellisor and Hedwig Kniker (Applin et al. 1925), demonstrated the importance of microfossils for biostratigraphy and correlation, although the contribution of these women was overlooked by many of the men that followed (see Gries 2018). Around the same time, a number of oceanographic research institutes were being founded (e.g. in 1930, the Woods Hole Oceanographic Institution in Massachusetts) or refocused (in 1925, the Scripps Biological Research Station became the Scripps Institute of Oceanography), while new techniques were being developed to map the seafloor and image the internal structure of the sediments and crust that lay below (e.g. Ewing and Vine 1938). After the Second World War, further advances in marine geophysics, and the volume and wealth of data being collected (driven, in large part, by the Cold War), provided completely new perspectives on the deep-ocean seafloor and crust, including, for example, the famous bathymetric maps drawn by Marie Tharp (see Higgs 2021). These new insights from mapping and advances in marine geophysics led to the plate tectonic revolution of the 1960s, which finally explained the tectonic mechanisms needed to uplift deep-sea sediments that had been hinted at by the earlier geological studies of the Alps.

Prior to the 1960s, the age and lithologies of in situ deep-sea sediments were known almost entirely from dredges and shallow (m-scale) cores. As zonation schemes based on microfossils continued to be developed from marine sediments on land, it became possible to use these shallow deep-sea samples to constrain the age of the ocean basins (which, at that time, was contentious, with views ranging from the Precambrian to the Mesozoic). For example, in 1966, Early Cretaceous sediments were recovered from a depth of 3500 m below sea level (mbsl) in a 73 cm-long core on the Shatsky Rise in the NW Pacific Ocean (Ewing et al. 1966). This core contained planktonic foraminifera identical to those in Early Cretaceous pelagic sediments of the Alps. The ratio of planktonic to benthic foraminifera suggested deposition in water depths on the order of 1.8 km, thereby confirming that pelagic environments had existed in the Pacific Ocean since at least the Early Cretaceous. However, only with the advent of deep-ocean drilling did it become possible to drill hundreds of metres into the seafloor to directly sample both ocean crust and the overlying sediments in order to provide more definitive evidence of the history of the ocean basins.

The first attempt at deep-sea drilling in 1961 was the US-funded Project Mohole that was conceived and led by the intriguingly named American Miscellaneous Society, an informal group of eminent US scientists and engineers, formed by the US Office of Naval Research. The aim of Project Mohole was to drill through the ocean crust and sample the mantle (thereby crossing the seismically defined Mohorovičić Discontinuity, or 'Moho'). The project failed in this regard, but, by drilling and recovering Miocene sediments and basalt from 180 m below the sea floor, from a platform floating in water depths of 3.6 km, it succeeded in establishing a future for scientific ocean drilling (Hsü 1992). The technological challenge of keeping a free-floating vessel stationary on the ocean surface above a fixed point on the seafloor had been solved. This success led directly to the next US-funded program – the Deep-Sea Drilling Project

(DSDP: 1968–1985), using the purpose-built *Glomar Challenger*. The cores that this ship recovered revealed many fundamental aspects of ocean structure and history including confirming the hypothesis of seafloor spreading (thereby effectively ending all debate about the existence of plate tectonics), demonstrating that the Mediterranean Sea had dried out in the Miocene and discovering evidence for widespread anoxia in the oceans during the Cretaceous (Hsü 1992). The DSDP eventually became an international scientific program, featuring funding, leadership and scientists from many nations. This model continued in four successor programs – the Ocean Drilling Project (ODP: 1985–2002), the Integrated Ocean Drilling Program (IODP: 2003–2013) and the International Ocean Discovery Program (IODP: 2013–2024), with the latest iteration, the International Ocean Drilling Programme (IODP[3]) beginning in January 2025. The *Glomar Challenger* was replaced in 1985 by the US-operated *JOIDES Resolution* (affectionately known as '*the JR*'). The *JR* operated almost continuously from 1985 to 2024 when it was retired. It was joined in 2013 by the Japanese-operated *D/V Chikyu* (meaning 'Earth') and various 'mission-specific' vessels organised and operated by a consortium of nations, the majority of which are in Europe. These three platforms – the *JR*, the *Chikyu* and the mission-specific platforms – allowed ODP and IODP to drill and core in a wide variety of water depths, environments and geological conditions. For pelagic depositional environments, the *JR* has been the most suitable vessel to use and, consequently (and by virtue of its long service) much of what we know about the geological record of pelagic sedimentation in the deep-ocean comes from cores recovered by this vessel and its predecessor, the *Glomar Challenger*.

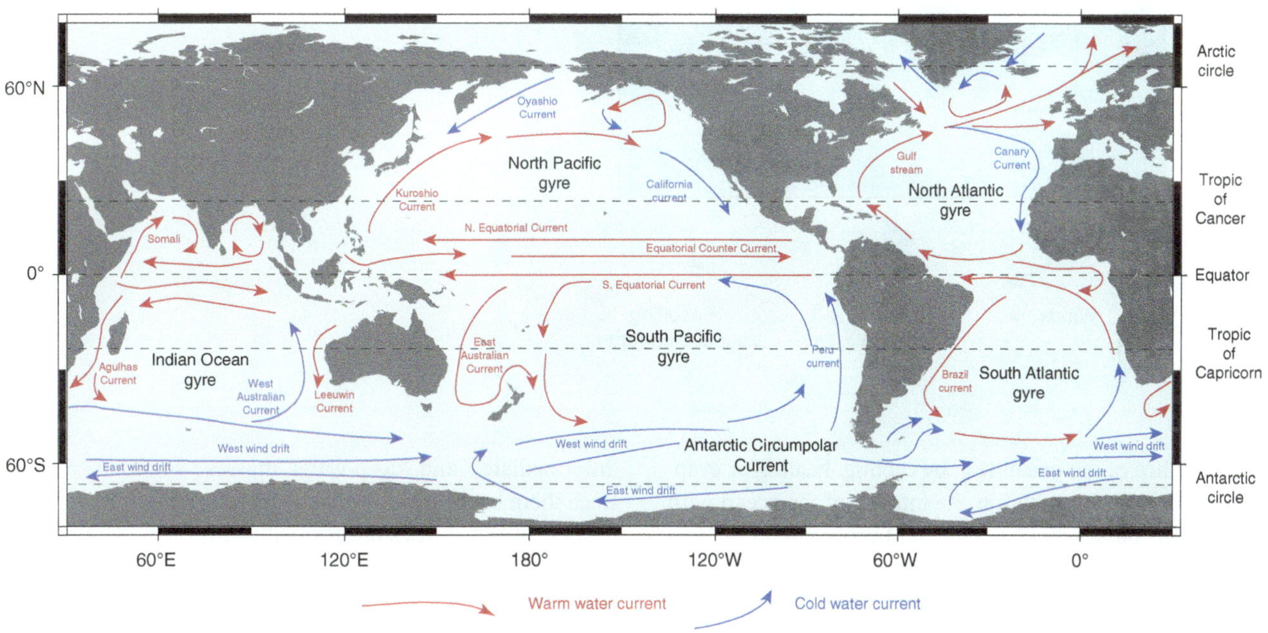

Figure 13.2 The generalised pattern of wind-driven surface ocean circulation and the major gyres and currents. Map generated using GMT 6 (Wessel et al. 2019). Position of currents after Thurman and Burton (2001).

leads to divergent movement of surface waters *upwelling* of colder, subsurface waters results, bringing waters from below to replace those moved away at the surface (Figure 13.4). This phenomenon occurs, for example, around the equator where the trade winds in both hemispheres drive surface circulation away from the equator. Upwelling can also occur along continental margins due to Ekman transport moving surface water in an offshore direction (common along the western coasts of continents). As will become clear, upwelling is a key process for biology and the production of sediments in the pelagic realm.

Wind-driven circulation is only capable of directly influencing the Ekman Layer, with a maximum depth of ~100 m. To understand the ocean interior and movement of waters at greater depths, the density of seawater, determined by temperature and salinity, must be considered. A water mass can be defined by its temperature and salinity as these are conservative properties that are modified only by processes

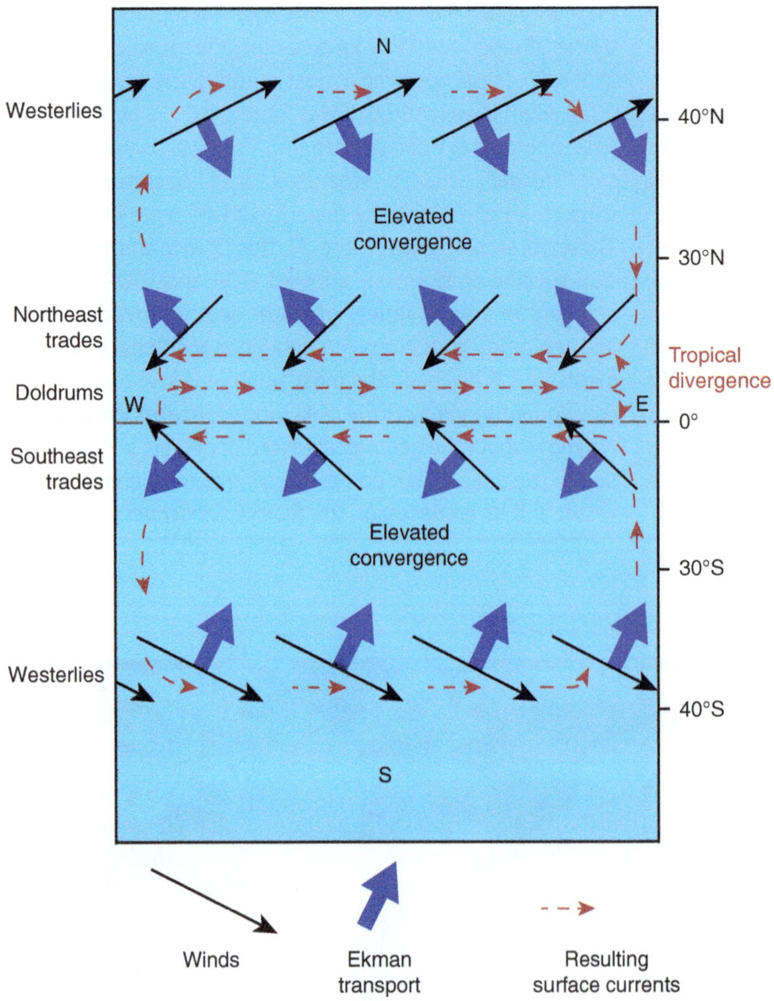

Figure 13.3 Schematic illustration of interactions between winds, surface-water currents and Ekman transport of the upper (<100 m) part of the water column in the Northern and Southern Hemispheres. Ekman transport can lead to accumulation of water (and consequent elevation of sea-level) in the centre of ocean gyres. Adapted from Sverdrup and Armbrust (2008).

at the surface of the ocean (e.g. by cooling, heating or evaporation) or by mixing within the interior of the ocean with other water masses that have different properties. If the surface layer of the ocean has a lower density than those below it, the water column is said to be stable and stratified. Examples of stable water columns come from deep lakes that have cold bottom waters underlying warmer surface layers (Section 6.2.2), from silled marine basins with abundant riverine input (like the Black Sea) with low salinity freshwater 'lids' overlying more saline deep-waters, and from low-latitude oceans with very warm surface waters. In these settings, the density stratification limits vertical movements in the water column. In contrast, processes such as intense evaporation (as seen in the Eastern Mediterranean; Section 10.6.2), or formation of sea-ice, can elevate surface water salinities (and, consequently, densities), leading to unstable water columns that can overturn, sink and move along surfaces of equal density (isopycnals) within the ocean interior. Using conserved temperature and salinity, it is possible to define and map water masses (the so-called 'intermediate'- and 'deep-water masses'; Figure 13.5) and trace them back to their surface source regions. The densest, deepest, water masses are formed north of the Greenland-Iceland-Faeroe Ridge in the Greenland-Nordic Seas (North Atlantic Deep Water, NADW) and in the Ross and Weddell Seas around Antarctica (Antarctic Bottom Water, AABW). These locations are known as areas of 'deep water formation'. In the North Atlantic, overturning is driven by the extreme cooling of surface waters that become turbulently mixed with saline waters below. These saline waters at depth are sourced from the Arctic Ocean, where sea-ice formation causes the formation of salty 'brines' (as sea-ice contains very little salt compared to seawater). In combination with the very cold temperatures, the elevated salinity increases the density of the seawater in the Greenland-Nordic Seas sufficiently that it can sink and overflow the Greenland-Iceland-Faeroe Ridge and move southwards into the deep Atlantic basins. Around Antarctica, areas of open water within the sea-ice ('polynyas') form where heat is lost from seawater, either by sea-ice formation or by conduction and

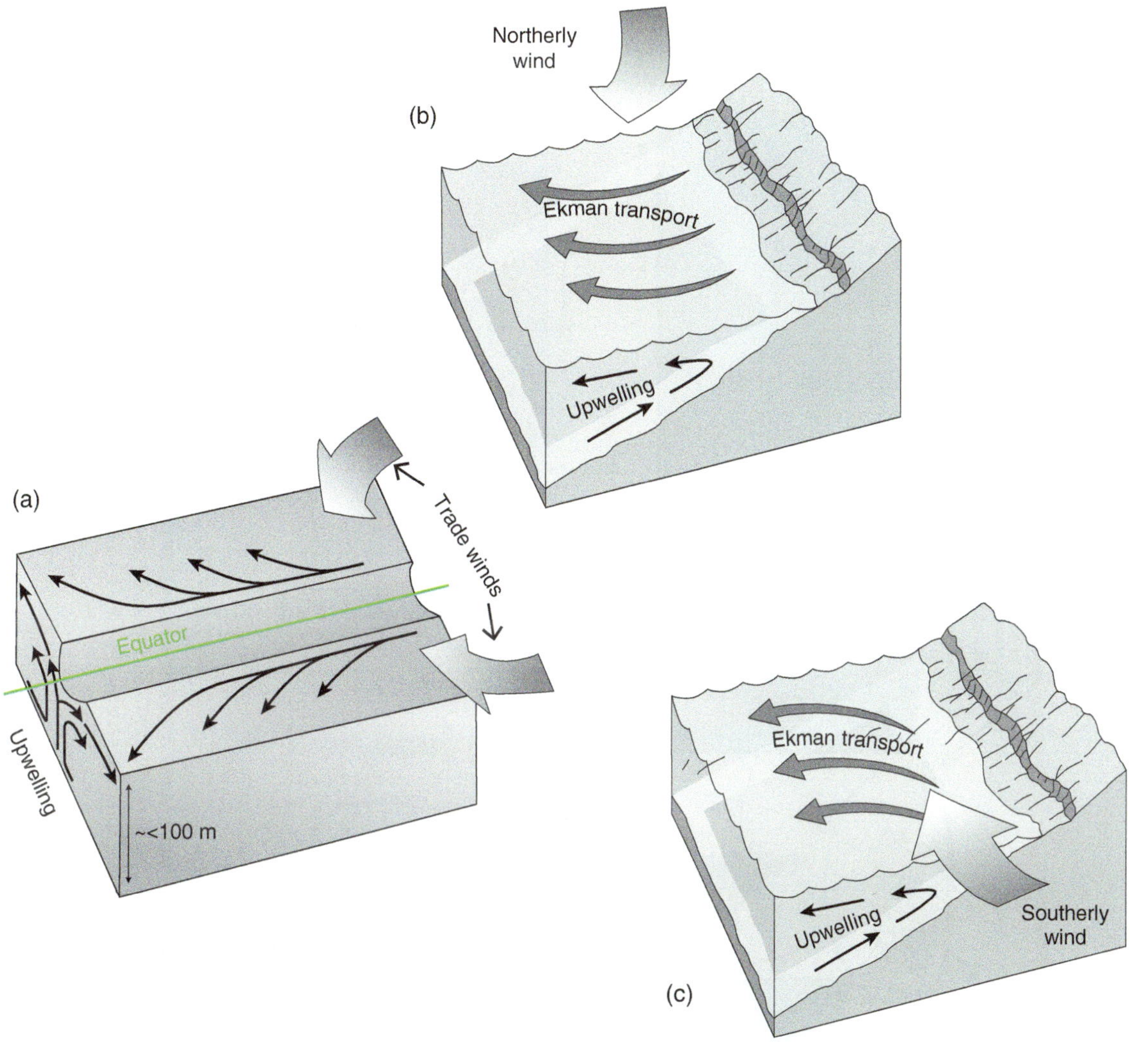

Figure 13.4 Styles of upwelling associated with Ekman transport along the equator (a) and along the western margin of continents in the Northern (b) and Southern (c) Hemispheres (redrafted from Emerson and Hedges 2008, after Thurman 1994).

convection. The cool surface waters can then sink to be replaced by relatively warmer waters from below. In the Atlantic Ocean, NADW flows southwards and AABW flows northwards (Figure 13.5). In the Southern Ocean, NADW and AABW become partially mixed (due to the wind-driven Antarctic Circumpolar Current) and ultimately exported into the Indian and Pacific Oceans (e.g. Talley 2013 and Figure 13.5) where deep water eventually returns to the surface to become entrained once more in surface ocean circulation. As a consequence of this 'ocean conveyor belt' (Broecker 1987, 1991), the time elapsed since a discrete volume of deep-water was last at the surface is relatively short in the Atlantic Ocean (<400 years based on radiocarbon ages) but considerably more in the Pacific (up to about 1360 years; Khatiwala et al. 2012). The relative 'age' of the water mass (young water masses being those that most recently were at the surface) is an important determinant of its chemistry, with implications for the types of sediment preserved at any given location. Furthermore, the movement of deep-ocean water masses, particularly as boundary currents adjacent to the continental margins, can be at sufficiently high velocity to transport sediment laterally along the ocean floor (see sediment drifts in Chapter 12).

13.2.2 Biogenic Pelagic Sediment Production Processes and Controls

13.2.2.1 Photosynthesis and Respiration

The majority of modern pelagic sediments are made of mineralised materials (calcium carbonate or silica) produced by organisms in the upper 200 m of the water column (Figure 13.6). The organisms producing these 'biogenic pelagic sediments' are planktonic and are autotrophs (e.g.

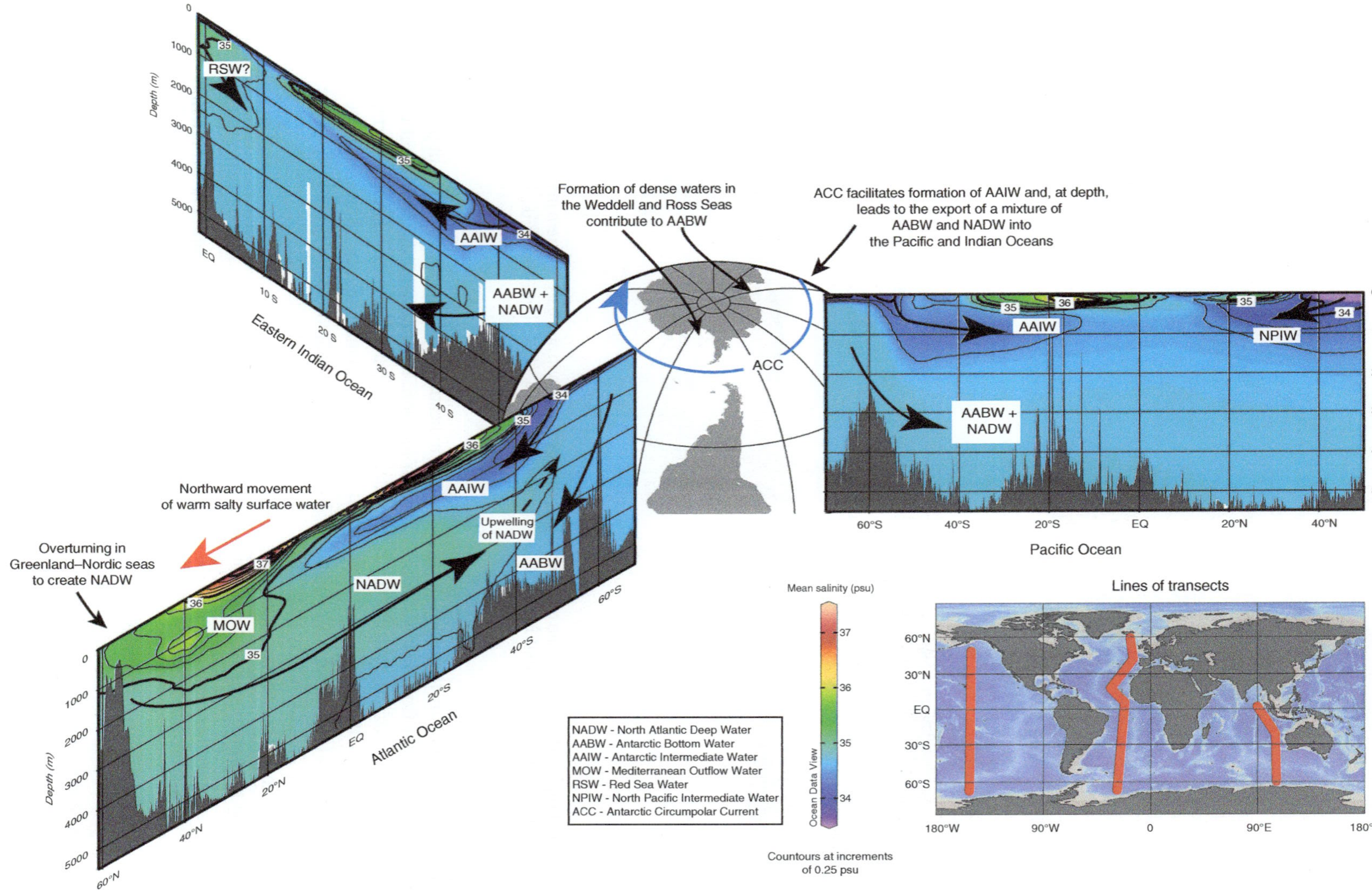

Figure 13.5 North-South sections of mean ocean salinity (in psu) in the Atlantic, Pacific and Indian Oceans, illustrating the origin and flow direction of the major intermediate and deep-water masses in each basin (inspired by Figure I-90 in Schmitz 1996). Salinity data from the World Ocean Atlas 2018 dataset (Boyer et al. 2018), plotted using Ocean Data View (Schlitzer, Reiner, Ocean Data View, odv.awi.de Schlitzer 2021). Map of Southern Hemisphere created using GMT6 (Wessel et al. 2019).

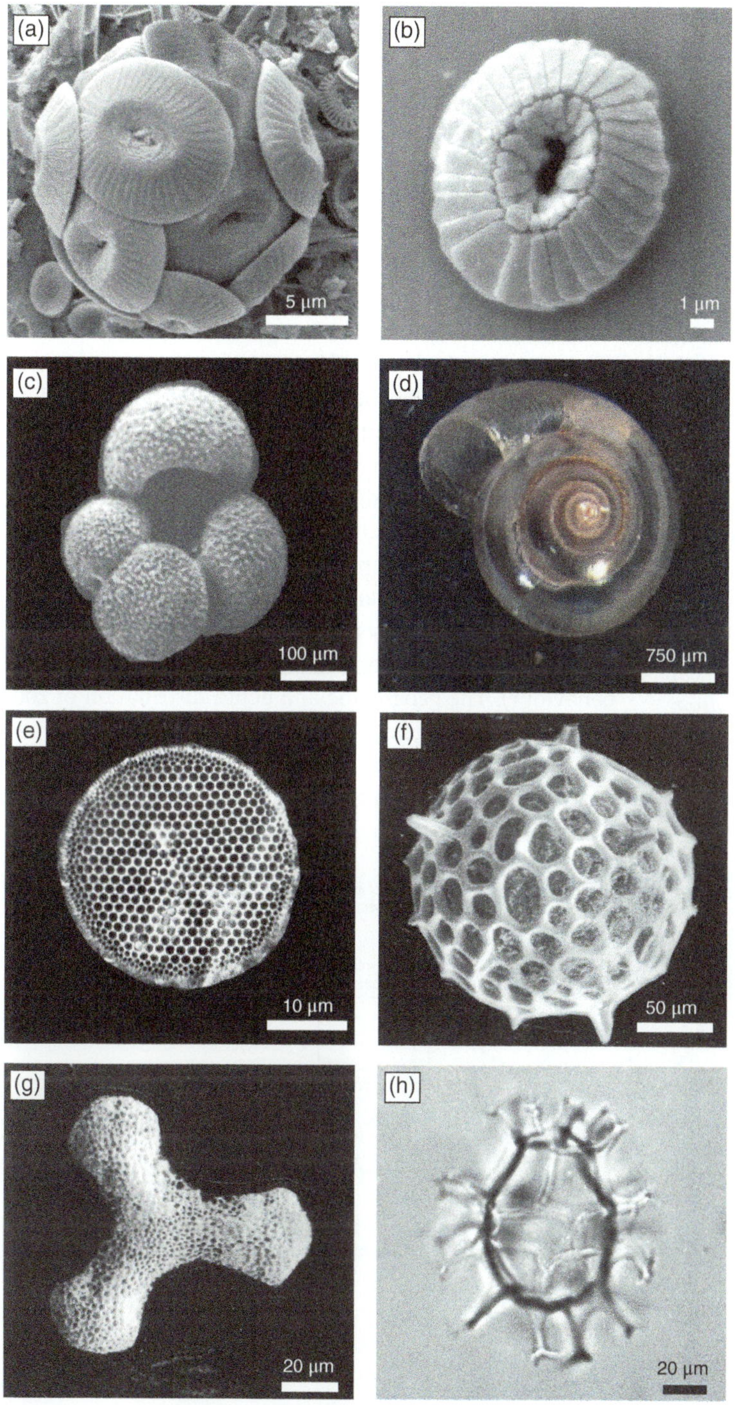

Figure 13.6 Examples of modern sediment-producing plankton. (a) A coccosphere of *Coccolithus pelagicus*; SEM image. (b) An individual coccolith of *Coccolithus pelagicus*; SEM image. (c) The test of the planktonic foraminifera, *Globigerina bulloides*; SEM image. (d) A shell of the planktonic gastropod (pteropod), *Limacina helicina*; light microscope. (e) The frustule of a diatom *Thalassiosire ferelineata*; SEM image. (f) The test of a radiolaria *Actinomma sp.;* SEM image. (g) The test of a radiolaria *Dictyocoryne truncatum*; SEM image (h) The dinoflagellate *Spiniferites bentori;* light microscope. All scale bars approximate. All images (except d) from the University College London Department of Earth Sciences Micropalaeontological Unit Miracle website (https://www.ucl.ac.uk/GeolSci/micropal/welcome.html), Courtesy of Professor Paul Bown. Image d from Miller et al. (2023) / Oxford University Press / CC BY 4.0.

photosynthesisers), heterotrophs (consumers) or mixotrophs (capable of either photosynthesising or consumption). Photosynthesisers utilise dissolved CO_2, macro and micro-nutrients (e.g. P, N and Fe) and light to produce their organic matter (Eq. 13.1) and are also known as *primary producers*.

$$6CO_2 + 6H_2O + \text{nutrients} + \text{light} \rightarrow C_6H_{12}O_6 + 6\,O_2 \quad (13.1)$$

Planktonic photosynthesisers range in size from the picoplankton (~10^{-9} m), dominated by bacteria, to the microplankton (<10^{-3} m). Their distribution in the surface ocean is controlled by nutrient availability and light, and so varies with depth, latitude, ocean currents and ocean turbidity. In most settings, the highest rates of primary production occur within the upper few metres of the water column. Primary productivity is often limited in these settings by nutrient availability, and as a consequence, rates of biological utilisation are high and concentrations of key nutrients in surface seawater are near zero (Figure 13.7).

Unlike photosynthetic primary production, which can only occur within the photic zone, *aerobic respiration*, the biochemical process of oxidising organic matter (Eq. 13.2), can, in principle, occur throughout the water column, wherever oxygen and organic matter exist.

$$C_6H_{12}O_6 + 6O_2 \rightarrow 6CO_2 + 6H_2O + \text{biochemical energy} + \text{nutrients} \quad (13.2)$$

Respiration releases nutrients back into seawater (Figure 13.7), which, if in the photic zone, can be utilised again by phytoplankton. However, much respiration occurs at deeper levels below the photic zone, thereby preventing immediate reutilisation of nutrients. An exception occurs in areas of upwelling, where continual nutrient resupply from deeper in the water column can sustain higher rates of primary productivity (Figure 13.8). Different groups of phytoplankton are adapted to different conditions of light and nutrient availability and, consequently, their geographic distribution in the ocean, and, as sediments on the seafloor, varies predictably with oceanographic setting (Figure 13.1b).

The *zooplankton* are planktonic organisms that are heterotrophs and consumers of primary producers and their derivatives, although some zooplankton have photosymbionts supplying them directly with food (mixotrophic behaviour). The geographic and depth distribution of zooplankton is controlled by the availability of their food sources, and in some cases, the availability of key nutrients, water temperature and turbidity.

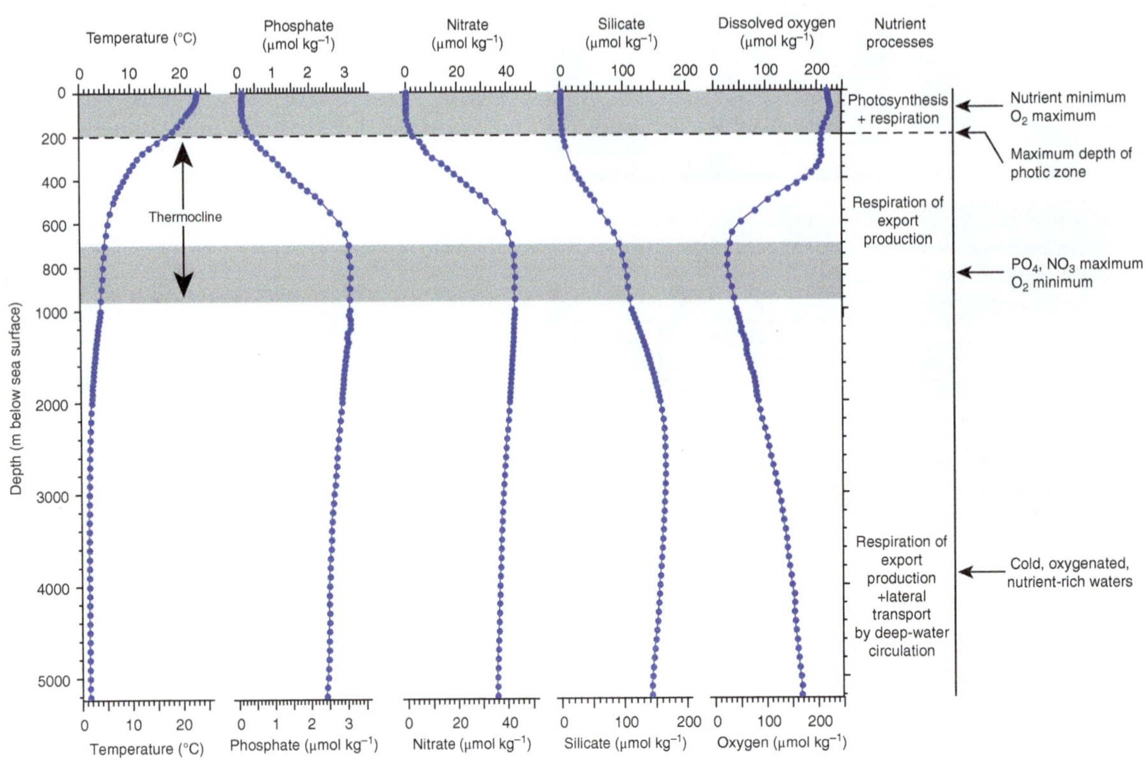

Figure 13.7 A typical open-ocean water column depth profile of temperature, key nutrient concentrations (phosphate, nitrate and silicate) and dissolved oxygen concentrations in the northern Pacific Ocean (145.5°W, 24.5°N; Station 28530 (B) in the World Ocean Atlas 2018 dataset; Boyer et al. 2018 / National Centers for Environmental Information / Public domain).

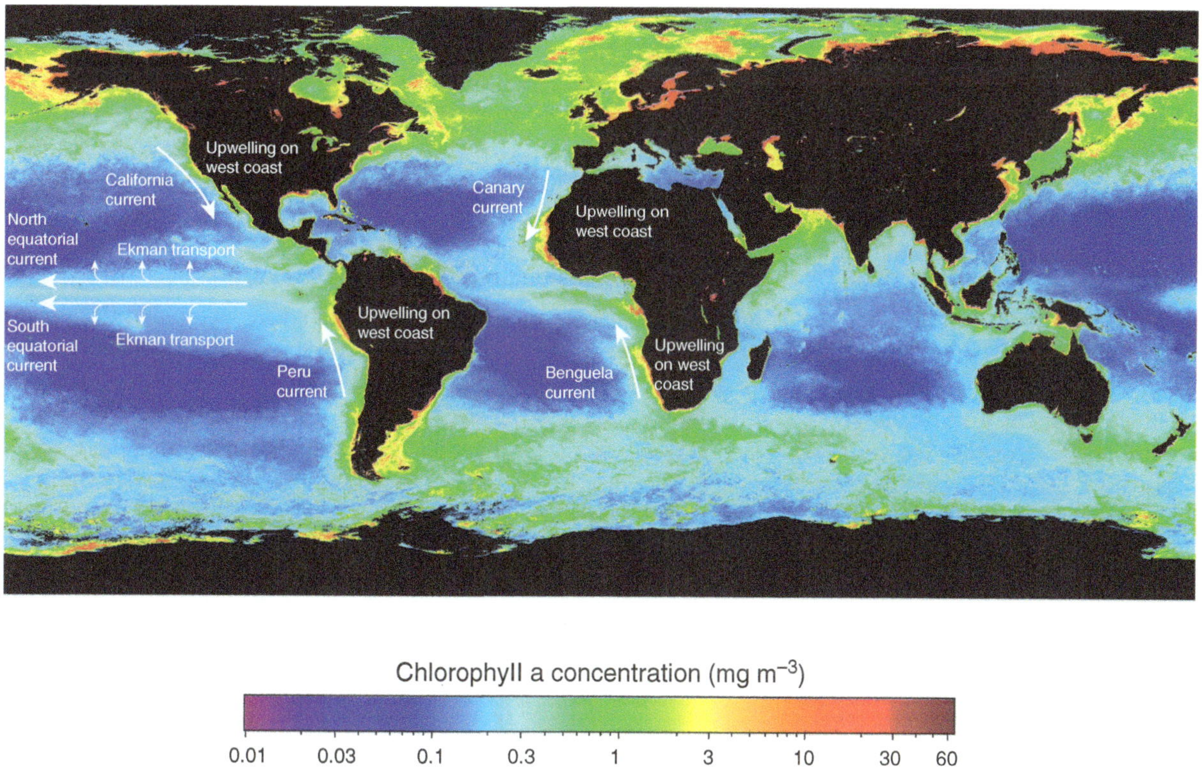

Figure 13.8 Satellite image of annual chlorophyll-a concentrations in surface waters for 2018 from the MODIS-Aqua instrument and schematic position of major surface water currents that lead to significant upwelling zones in equatorial latitudes and along the western margin of continents. NASA / Goddard Space Flight Center / Ocean Ecology Laboratory / Ocean Biology Processing Group / Public domain / Link - (https://oceancolor.gsfc.nasa.gov/l3/).

Although the phytoplankton and zooplankton include an extremely wide range of organisms, only some produce mineralised components in sufficient quantities to be sediment forming. Consequently, the sedimentary record of the deep-sea only provides a partial view of coeval pelagic ecosystems. Calcareous planktonic organisms form structures using aragonite and calcite, two polymorphs of calcium carbonate ($CaCO_3$). Siliceous planktonic organisms use hydrated amorphous silica (opaline silica, $SiO_2.nH_2O$). Due to their high porosity and interstitial water content, the very fine-grained biogenic sediments that are produced from these materials are known as calcareous or siliceous 'ooze'. Modern calcareous ooze is dominantly sourced from coccolithophores (Haptophyte algae), planktonic foraminifera (Rhizarian protists) calcifying pteropods (planktonic gastropods), or some mixture of these components. Siliceous ooze is sourced from diatoms (Stramenopile protists) and/or radiolaria (Rhizarian protists).

13.2.2.2 Modern Pelagic Carbonate Producers

The *coccolithophores* are unicellular photosynthesisers (although some species have been shown to be capable of heterotrophic and mixotrophic behaviour), that produce micron-scale plates of calcite (known as 'coccoliths'; less than 30 μm, but typically 5–10 μm, in diameter; Figure 13.6b). These are used by some forms to cover the external surface of their cell, creating a rigid 'coccosphere' (typically a few tens of microns in diameter; Figure 13.6a). Over their life cycle, coccolithophores shed old coccoliths and replace them with new coccoliths produced within the cell. The microfossil record of coccolithophores is dominated by 'heterococcoliths' that are typically circular to elliptical and made up of calcite plates, rods and grains. In contrast, smaller, less rigid structures, made of many tiny euhedral crystallites of calcite (~0.1 μm in size), 'holococcoliths', have poor preservation potential, even in modern deep-sea sediments, and so are rarely reported in the geological record (e.g. Bown et al. 2008). As photosynthesisers, coccolithophores can be found throughout the photic zone (e.g. Young 1994), occupying different ecological niches in different oceanographic settings (e.g. Balch 2018). Coccolithophore abundance is generally highest in warm, stratified, oligotrophic (Section 11.4) tropical and subtropical waters, where carbonate saturation is greatest, but they can be found in sediment-producing quantities across all latitudes up to the sub-polar regions of the North Atlantic (below 70°N; Young 1994). Holococcoliths tend to dominate in low-nutrient settings, whereas heterococcoliths are more prevalent in lower-light, higher nutrient settings. Many modern coccolithophores produce these two morphological forms at different stages of their life cycle, perhaps in response to seasonally changing light levels and nutrient availability. Consequently, in coastal and upwelling settings coccolithophores can form large seasonal blooms of heterococcoliths capable of producing a disproportionate amount of sediment in a relatively short period (reaching concentrations of >1,000,000 cells per litre of seawater, e.g. Tyrrell and Merico 2004). Although major and trace nutrients are a requirement for coccolithophore growth (Young 1994), they may require less iron than other groups of phytoplankton (e.g. diatoms), which allows them to maintain growth in oceanographic settings where iron availability is limited (Balch 2018). Calcareous oozes dominated by coccolithophores are called 'coccolith ooze'.

The *planktonic foraminifera* are heterotrophic single-celled protozoans that feed on bacteria, phytoplankton and small zooplankton, and, in some cases, can also host photosymbionts (Armstrong and Brasier 2005; Schiebel and Hemleben 2017). They produce small porous tests, usually of low magnesium calcite, that typically range from about 30 to 400 μm in diameter (Figure 13.6c). Individual tests typically comprise a small number of chambers that are fused together and formed progressively through the lifetime of the individual. Their lives tend to be short (~1 month) and start in the surface layer of the ocean. They often move deeper in the water column during adulthood. During reproduction, adult planktonic foraminifera can add a thick layer of gametogenic calcite. The tests of planktonic foraminifera exhibit adaptations that allow different species to occupy different niches by maintaining buoyancy at an optimal depth (Armstrong and Brasier 2005). These adaptations include variations in the porosity of the test, differences in the number and shape of spines and the addition of keels to reduce sinking rates. Those with photosymbionts are restricted to the photic zone, but other species can inhabit water depths down to several kilometres (albeit in relatively low abundance). In the modern ocean, high abundances of planktonic foraminifera are typically found in surface waters that are relatively nutrient-rich whereas lower abundances exist in oligotrophic gyres (e.g. Bé and Tolderlund 1971; Schiebel and Hemleben 2017). Assemblages show a strong biogeographic provinciality from the equatorial to polar regions, with individual species being useful indicators of particular temperature ranges (e.g. Schiebel and Hemleben 2017). Greatest diversity is found in the subtropics and temperate regions. In areas of high planktonic foraminiferal productivity, they are deposited in sufficient quantities to form planktonic foraminiferal ooze.

The final important pelagic carbonate producer in the modern oceans are planktonic calcifying gastropods, commonly referred to as the *pteropods* (although the order 'Pteropoda' includes non-calcifying forms). The calcifying pteropods are typically small (<30 mm) and produce a thin shell of aragonite (Figure 13.6d). Such pteropods are an important contributor to the zooplankton and can produce large mucous webs which trap a variety of detritus, phytoplankton and small zooplankton for consumption (Gilmer and Harbison 1986). They are common in the upper 500 m of the water column (Chen 1971) and, although widely distributed, are generally restricted to tropical and sub-tropical waters in the modern. Like the planktonic foraminifera, they are generally found where productivity is relatively high, that is, their source of food is abundant (Bé and Gilmer 1977). Despite the importance of pteropods in modern ecosystems and a likely origination in the Cretaceous, their preservation potential in the sediment is relatively poor due to their thin aragonitic shells, and so they may be under-represented in the geological record.

13.2.2.3 Modern Pelagic Silica Producers

Modern siliceous oozes are commonly dominated by the remains of *diatoms*. Diatoms are a very abundant and diverse group of unicellular photosynthesisers (reviewed in, e.g. Armstrong and Brasier 2005) found in most aquatic environments that are exposed to light (including freshwater). In the modern oceans, they are the dominant primary producer and, thus, play a significant role in global biogeochemical cycles. Diatoms are typically 10–100 μm in size and produce, within their cell wall, a frustule of opaline silica that comprises two overlapping valves (Figure 13.6e). Some diatoms living in shelf-sea environments also produce a heavily silicified resting cyst that is used to protect the diatom on the seafloor when temperature and nutrient conditions are unfavourable. In the open-ocean, diatoms are restricted to the photic zone with different species being adapted to exploit variations in, for example, light, salinity, temperature and nutrient availability. In addition to the standard nutrient requirements of phosphate, nitrate and iron, diatoms also require silica. They are abundant in regions of upwelling where availability of nutrients is high, such as the Southern Ocean and the equatorial regions of the Indian and Pacific Oceans. Conversely, in regions of low nutrient supply, such as within the oceanic gyres, concentrations of diatoms are extremely low. Sediments rich in diatoms are referred to as 'diatomaceous ooze'.

Radiolaria are single-celled protists that consume a variety of food including phytoplankton and small zooplankton, but can also harbour photosymbionts (e.g. Biard

2022). As a consequence of their zooplanktonic lifestyle, radiolaria are typically found in greatest abundance where food is readily available, namely regions of upwelling nutrient-rich waters. Like the diatoms, they construct siliceous tests. These are typically 50–300 μm in size (although some are larger). The tests tend to be spherical, discoidal or conical, with additional radial spines in some species (Figure 13.6f,g). Radiolaria are particularly abundant in warm low-latitude areas (e.g. Anderson 1983), where they can occur in sufficient quantities to create 'radiolarian ooze'. They are also found in high-latitude regions alongside diatoms. Like planktonic foraminifera, radiolaria are found across a wide range of depths from the sea-surface to below 1000 m and exhibit morphological adaptations to maintain buoyancy, such as the number of holes in the test and the length of their spines.

13.2.2.4 Other Biogenic Components

It is possible to find a range of other biogenic sedimentary grains within modern deep-sea pelagic sediments, although these are rarely the dominant component. Such grains can be produced by organisms living within the water column or at the seafloor. *Dinoflagellates*, inhabiting the water column, are the second most abundant primary producer in the oceans after the diatoms (although not all species are autotrophic). They produce spectacular 'red-tides' when they bloom in huge numbers. During a resting-phase of their life cycle, some dinoflagellates produce an organic-walled cyst (Figure 13.6h) made of the robust biopolymer 'dinosporin'. This is not easily destroyed by bacteria and can be preserved in sediments ranging from coastal settings to the continental slope and rise. A number of dinoflagellate species produce calcareous cysts, tens of microns in diameter, which have been reported from neritic and open-ocean settings (Vink 2004; Zonneveld et al. 2005). Other carbonate-producing organisms that make a minor contribution to modern pelagic carbonate sediments include benthic calcareous foraminifera, deep-sea ostracods, echinoderms, and molluscs. The phosphatic skeletal remains of fish can also be found in pelagic sediments, at all depths, particularly where concentrated due to slow background sedimentation rates.

13.2.3 Non-biogenic Detrital Sediment Sources

Clastic sediment, sourced from the land or from shallow-water carbonate platforms, is an important additional source of sediment in some pelagic settings. Terrigenous and shallow-water carbonate sediments are transported to the deep-sea by mass transport processes, most notably turbidity flows (Chapter 12), while glacial sediments can also be transported by ice-rafting (Section 14.5.10). However, the abundance of such clastic deposits decreases significantly with distance from the relatively shallow-water settings of continental margins and carbonate platforms. Far from the influence of ice sheets and the continental shelves, terrigenous sediment is supplied to the open-ocean largely by wind and is, consequently, extremely fine-grained dust (<20 μm) sourced from arid areas of the continents, and volcanic ash (Figure 13.9). Most of the wind-blown dust comprises silt-sized quartz grains and clay minerals dominated by kaolinite, chlorite or illite (reviewed in Jenkyns 1986; Open University 1989a). The clay mineralogy is determined largely by the climatic regime of the source areas (Section 9.1.4) and, so, kaolinite tends to dominate in low-latitude regions, where the continents undergo extensive chemical weathering, whereas chlorite is sourced from regions dominated by physical weathering. Illite can be deposited almost everywhere, although it is in greatest abundance in the North Atlantic (Open University 1989a). Although wind-blown and volcanogenic dust can be deposited across much of the ocean surface, this sediment type is only rock-forming in regions where biogenic pelagic sediments are not preserved at the seafloor. In such areas, the fine-grained deposits produce pelagic clays, often referred to as deep-sea clays. The presence of oxidised iron in these deposits (due to the low organic matter content) gives them, typically, a brown or red-brown colour. Novel additional detrital components that occur in many modern and ancient deep-sea sediments are interplanetary dust particles (IDPs). These are constantly settling through Earth's atmosphere and, in slowly accumulating pelagic sediments, can be sufficiently concentrated to be observed directly or inferred by their elevated $^3He/^4He$ ratio relative to the bulk Earth (e.g. Farley 1995).

13.2.4 Sediment Transport and Deposition

Although relatively large particles (e.g. planktonic foraminifera: up to 400 μm) may sink readily, finer-grained sediments (whether biogenic or not) may remain in suspension. The balance between particle weight and particle buoyancy determines whether any given particle will float or sink. Particles that are sinking through a fluid also experience a drag force that, for small particles, can be described by Stoke's Law. Combining gravitational, buoyancy and drag effects allows the terminal velocity of a small (spherical) particle sinking in a fluid to be expressed as:

$$v = \frac{2}{9}\frac{(\rho_p - \rho_f)}{\mu} g R^2 \qquad (13.3)$$

where v is the terminal velocity (m s^{-1}), ρ_p is the density of the particle, ρ_f is the density of the fluid (kg m^{-3}) μ is the dynamic viscosity (kg m^{-1}s^{-1}), g is the acceleration due to

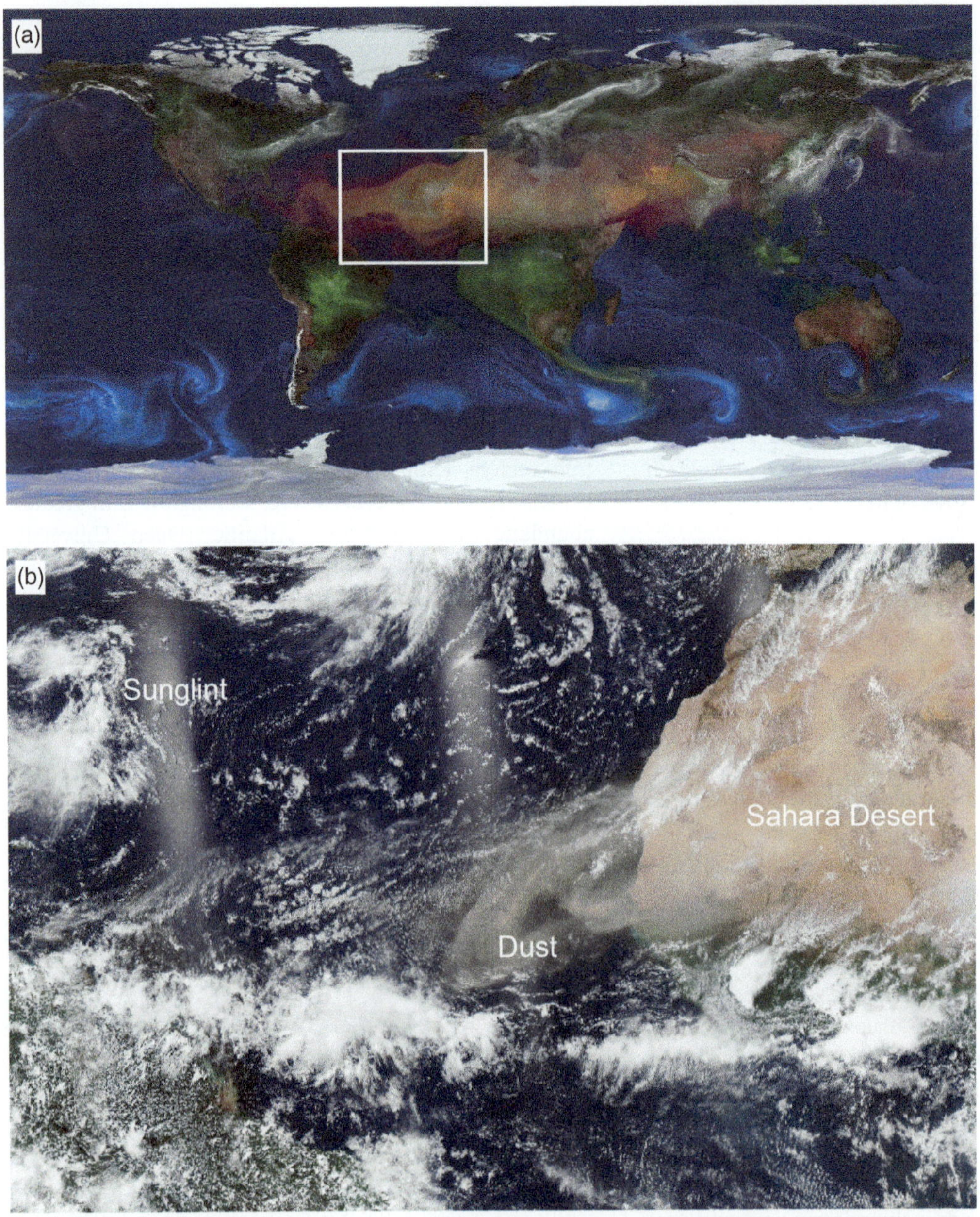

Figure 13.9 (a) A simulation of atmospheric dust (red), sea salt (blue), organic/black carbon (green) and sulphates (white) based upon satellite data (GEOS-5) and the Goddard Chemistry Aerosol Radiation and Transport (GOCART) Model (NASA / Goddard Space Flight Center / Public Domain / Link - https://svs.gsfc.nasa.gov/30017). Approximate area covered by image b shown. (b) Visual image of a dust plume transporting material from the Sahara desert across the equatorial Atlantic Ocean (NASA Earth Observatory / Public Domain / Link - https://earthobservatory.nasa.gov/images/149918/a-burst-of-saharan-dust).

gravity ($m\ s^{-2}$) and R is the radius of the sphere (m). On this basis, a sand-sized solid calcite sphere with a radius of 0.1 mm (100 μm) would take about 20 days to settle through 3.8 km of seawater, the average depth of the deep ocean. In contrast, a mud-sized solid calcite sphere with a radius of 5 μm would take more than 22 years! Such simple calculations make no allowance for the real shapes of particles, their internal porosity nor for turbulence in the water column, all of which can be expected to reduce the sinking rate further. Yet, there is evidence from sediment traps and seafloor imaging that very-fine grained particles, for example, coccoliths, can be transported from the surface ocean to the seafloor on timescales of just days to months (Figure 13.10). The apparent paradox is resolved by considering the

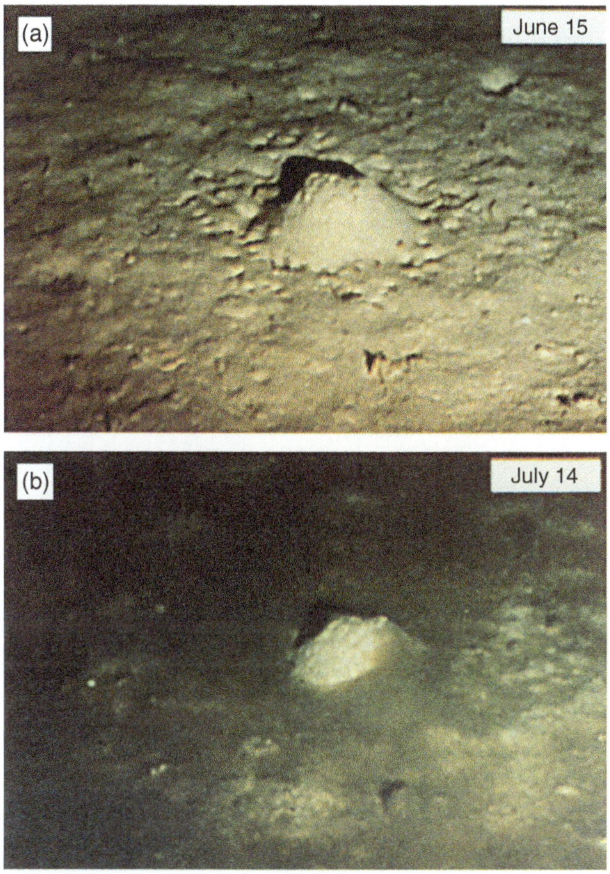

Figure 13.10 These sequential images show deposition from a phytoplankton bloom that occurred in 1983 in the north-east Atlantic. They show an $\sim$1.5 m^2 area of seafloor at a water depth of 4000 m. A small mound clearly visible on June 15th (a) has been partially buried by marine snow by the time of the next image was taken on July 14th (b). Courtesy of Professor R. Lampitt, National Oceanography Centre, UK.

importance of biological processes and the interactions between particles within the water column. Faecal pellets produced predominantly by zooplankton provide the most important mechanism of packaging finer-grained material such that it can settle relatively rapidly from the surface ocean to the deep ocean interior, and, ultimately, the seafloor. Another pathway for sedimentation results from Brownian motion that leads to mud-sized particles being concentrated together in the upper water column where they can coagulate or flocculate. Dissolved organic matter and Extra-cellular Polymeric Substances (EPS), produced for example by diatoms, are significant binding agents for larger particles (Figure 13.11). This macroscopic material, known as 'marine snow', gains density through these accretionary processes, and consequently, can transport the aggregates of fine-grained particles to the seafloor much more rapidly than the sinking of individual micron-sized grains (e.g. Alldredge and Silver 1988). Marine snow also provides, along with plankton, an important food source for many organisms in the ocean, and, thus engenders further faecal pellets. Due to currents within the water column and at the seafloor, some lateral transport of sinking particles (whether individual grains, pellets or marine snow) can occur during transit to their ultimate site of deposition. In some areas, the sediments on the seafloor are therefore not necessarily sourced directly from the overlying water column (e.g. Honjo et al. 1982; Neuer et al. 1997; Fahl and Nöthig 2007). Deposition onto a sloping seafloor can also lead to lateral re-deposition of pelagic sediments through gravitational processes, such as slumping, sliding and turbidity flows.

13.2.5 Biogenic Pelagic Sediment Preservation

The processes of primary production, consumption and the formation of marine snow and faecal pellets are key parts of the 'biological pump' (Figure 13.12), the linked set of processes by which carbon is transferred from CO_2 dissolved in the surface ocean to the ocean interior (for more details see, e.g. Open University 1989a; Emerson and Hedges 2008; Hüneke and Henrich 2011). Approximately, 10% of the organic carbon produced in the photic zone is exported to deeper levels ('export production'). Most of this export production is recycled and respired in the upper 500–1000 m of the ocean by consumers through aerobic respiration and released back into seawater as CO_2. Only about 1% of the organic carbon produced in the modern surface ocean ultimately reaches the seafloor, where well-oxygenated deep-waters and further respiration combine to reduce the organic carbon content of modern deep ocean sediments even more. Consequently, at present, biogenic pelagic sediments on the seafloor are low in total organic carbon content (TOC) and are dominated by carbonate minerals (calcite or aragonite) or opaline silica. The preservation of these biogenic minerals, both during transit to the seafloor and at the sea floor, is controlled by solubility, particularly the ambient saturation state of seawater with respect to the solid mineral phase. For carbonate minerals, the saturation state in seawater depends upon the concentration of the carbonate (CO_3^{2-}) and bicarbonate (HCO_3^-) ions, temperature and pressure, all of which vary with depth and water mass. In warm waters, typical of the surface ocean at lower latitudes, seawater is supersaturated with respect to calcite and aragonite, meaning that particles of these minerals will not dissolve. As depth and pressure increase, however, carbonate ion concentrations decrease, as do temperatures, leading to a decrease in saturation state. At the aragonite and calcite lysoclines, seawater becomes undersaturated with respect to each carbonate

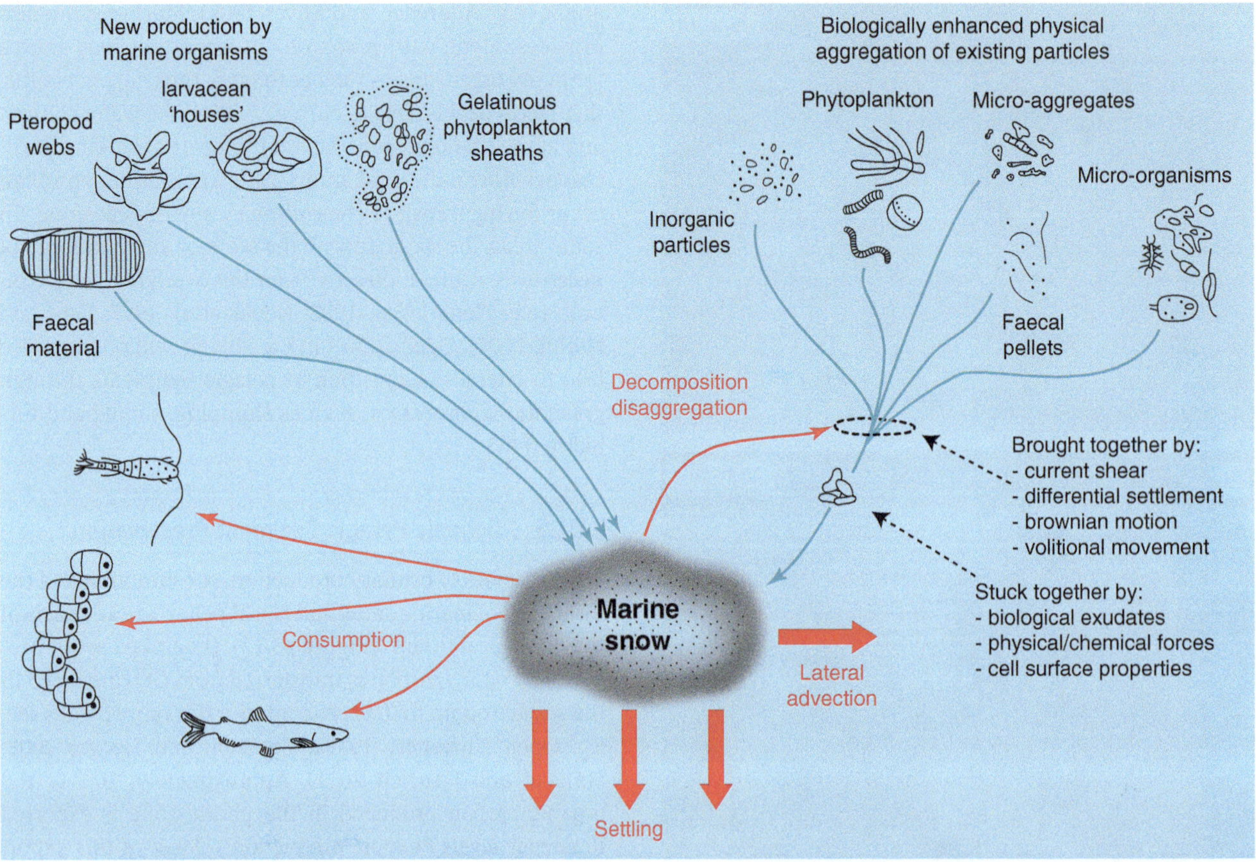

Figure 13.11 A cartoon summarising the components, formation and fate of marine snow formed in the pelagic zone of the ocean. Primary biologically produced materials (left-hand side) and biological and inorganic materials are aggregated through a number of chemical and physical processes (right-hand side). The fate of marine snow is shown by the red arrows, with much being exported to deeper levels in the ocean and, ultimately, the seafloor. Modified from Alldredge and Silver (1988) and Open University (1989a).

mineral and carbonate dissolution rates increase noticeably, leading to an observable decrease in carbonate mineral preservation in sediments (Figure 13.13). Planktonic foraminifera deposited below the lysocline may show evidence of incomplete dissolution in the form of corrosion, test-wall thinning or fragmentation of chambers. Below the lysocline is an empirical level (known as the 'compensation depth') below which no carbonate will be preserved at all, as the rate of dissolution equals the rate of sediment supply (Figure 13.13). The compensation depth for aragonite is shallower ($\sim$>3 km in the modern) than for calcite (typically $\sim$4 to 5 km) reflecting the higher solubility of aragonite compared to calcite; these depths are known as the 'Aragonite Compensation Depth' (ACD) and 'Calcite Compensation Depth' (CCD; note that 'carbonate compensation depth' is also used to refer to the latter). In areas of high carbonate supply, depositional rates of carbonate are sufficiently high to allow preservation, despite the presence of undersaturated waters. Consequently, below the high-productivity equatorial upwelling region the ACD and CCD are relatively deep due to the high rates of carbonate supply, most clearly seen in the CCD of the central equatorial Pacific Ocean (Figure 13.14). The importance of the CCD for pelagic sediments is that carbonate sediments can only be preserved where the seafloor is shallower than the local CCD; where the seafloor is deeper than the CCD biogenic pelagic carbonate settling from the surface ocean will be entirely dissolved within the water column before reaching the sediment–water interface. The depths of the ACD and CCD are not uniform globally (e.g. Figure 13.14 for the CCD) due to varying rates of carbonate supply and because deep-ocean circulation also affects the saturation state of bottom-waters. Young deep-water masses, such as NADW in the North Atlantic Basin, are less corrosive than older water masses which have accumulated additional CO_2 from respiration of organic carbon over their longer sub-surface transport histories (e.g. Pacific and Indian Oceans; Figure 13.14).

The preservation of opaline silica in deep-ocean sediments is also determined by seawater chemistry (e.g. Open University 1989a; Emerson and Hedges 2008; Hüneke and Henrich 2011) but, in contrast to carbonate, seawater is

Figure 13.12 Cartoon of the biological and solubility pumps that are key in transporting carbon into the deep ocean interior. The 'biological pump' describes the movement of a proportion of organic carbon from the shallow ocean (photosynthetic assimilation and respiration by heterotrophs), and its export into the deep-ocean, where most of it is respired into CO_2. Very little of this organic carbon is deposited on the ocean floor in the modern. The 'solubility pump' describes the physical/chemical transfer of carbon in the form of CO_2 between the atmosphere, surface ocean and deep ocean. Adapted from Chisholm (2000).

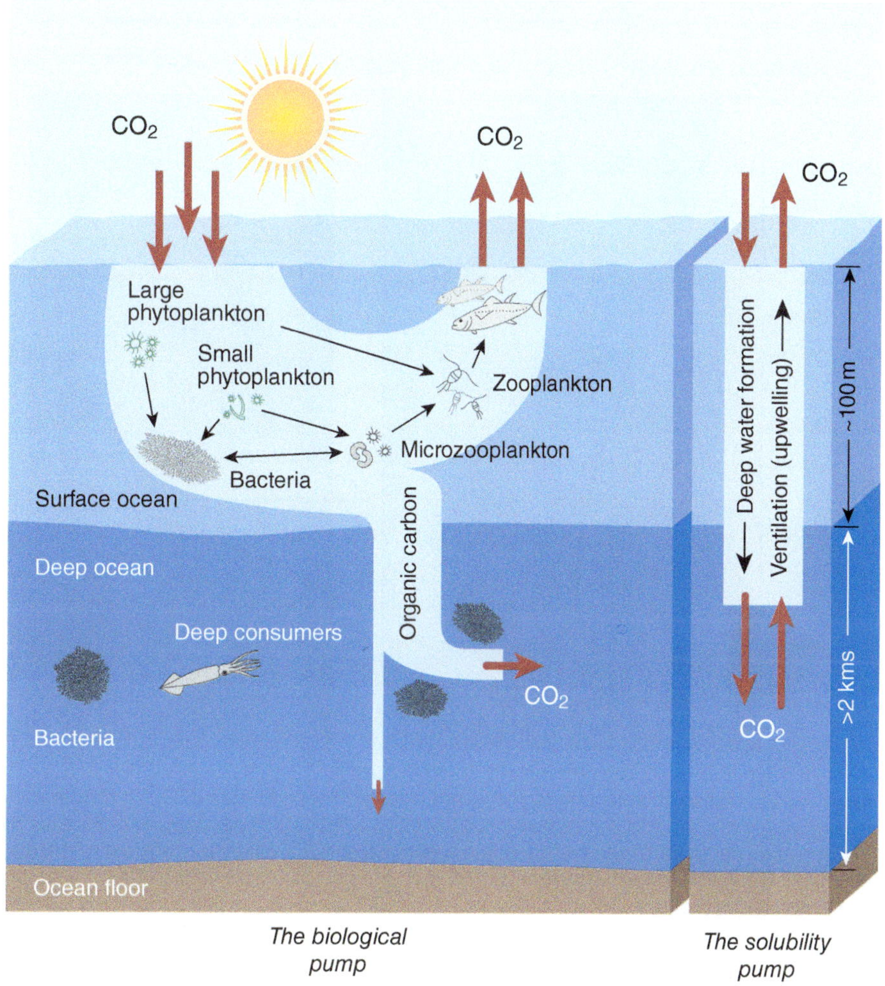

undersaturated with respect to opaline silica throughout the ocean everywhere. The degree of undersaturation is, however, most pronounced in the surface ocean due to the rapid biological uptake of silica by siliceous plankton (Figure 13.15). With increasing depth and, consequently, decreasing temperature, seawater moves towards saturation, but never reaches it. Particles of opaline silica, thus, dissolve rapidly in seawater and can only be preserved if silica supply rates to the seafloor are sufficiently high to exceed the rate of dissolution and allow a proportion of silica particles to enter the sediment pile (Figure 13.15). In this context, the inclusion of siliceous particles in marine snow and faecal pellets is important in facilitating rapid transport to the seafloor.

13.2.6 Other Sediment Generating Processes, Controls and Products

13.2.6.1 Authigenic Ferro-Manganese Nodules and Crusts

In most pelagic depositional environments, the rate of supply of biogenic particles is sufficiently high to prevent substantial accumulation of relatively slow-growing authigenic minerals at the seafloor (sometimes referred to as 'chemogenic' sediments). However, in areas where sedimentation rates are very low (or even negative), ferro-manganese nodules and crusts can accumulate (Figure 13.16). As a consequence, fields of nodules are commonly associated with deposition of deep-sea clays on the abyssal plains, below the CCD, where surface water productivity and terrigenous input are low. These authigenic accumulations of manganese and iron oxides and hydroxides can range in size from a few microns to more than 10 cm in diameter and, internally, possess concentric (or quasi-concentric) laminations typically surrounding a nucleus, which can be either fragments of rock or fragments of older nodules (Hein 2016). Detrital and biogenic (e.g. microfossils) particles may also be present trapped within the laminae. The iron and manganese minerals forming crusts and nodules grow very slowly from cold seawater, sediment pore waters or a combination of the two (Hein 2016). Although nodule formation has been traditionally viewed as the consequence of abiogenic, physical and chemical conditions, microbiological

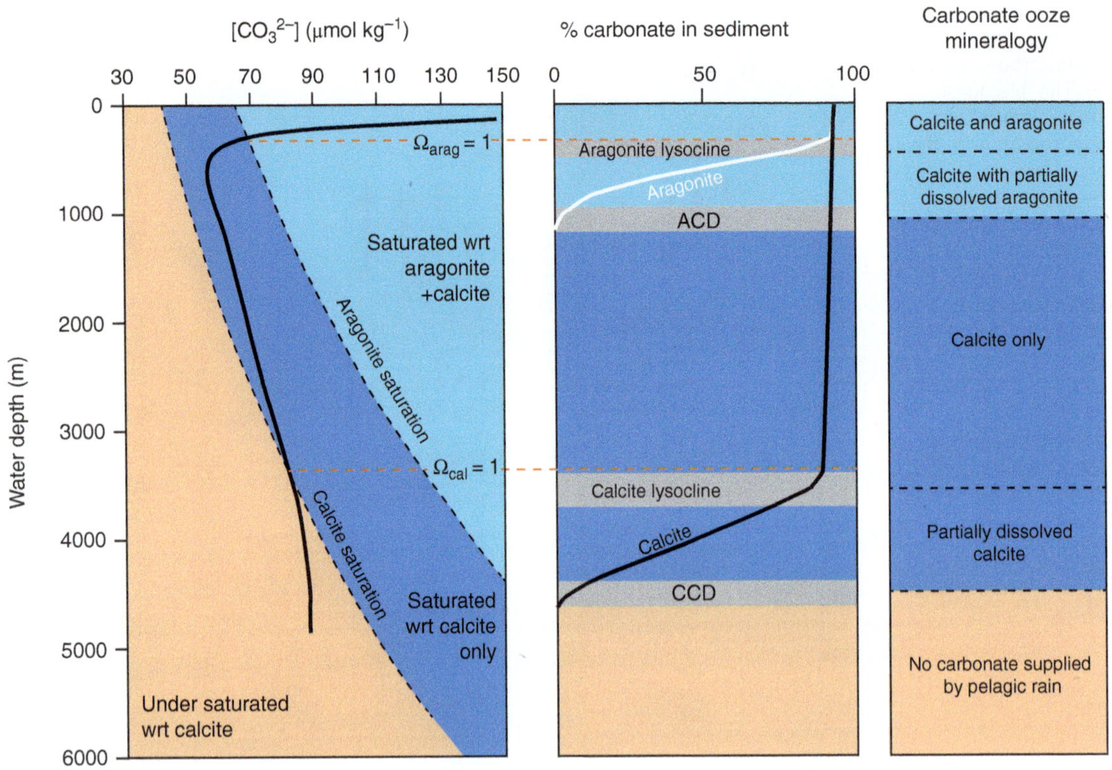

Figure 13.13 Schematic diagram to show carbonate ion concentration ($[CO_3^{2-}]$) the saturation state (Ω) of calcite and aragonite in an oceanic water column, and the carbonate content and carbonate mineralogy of calcareous biogenic pelagic sediments. The aragonite compensation depth (ACD) and calcite compensation depth (CCD) are empirically determined by either the point at which the % carbonate in sediments falls below a fixed value (e.g. 10%) or the point at which carbonate mass accumulation rates are zero. The carbonate ion concentration profile and saturation state calculations are from the equatorial Pacific Ocean (redrawn and modified from Emerson and Hamme 2022). In the Atlantic Ocean, the carbonate ion concentrations are higher and, consequently, the ACD and CCD are deeper. The generalised profiles of sediment composition and mineralogy are based on James and Jones (2016).

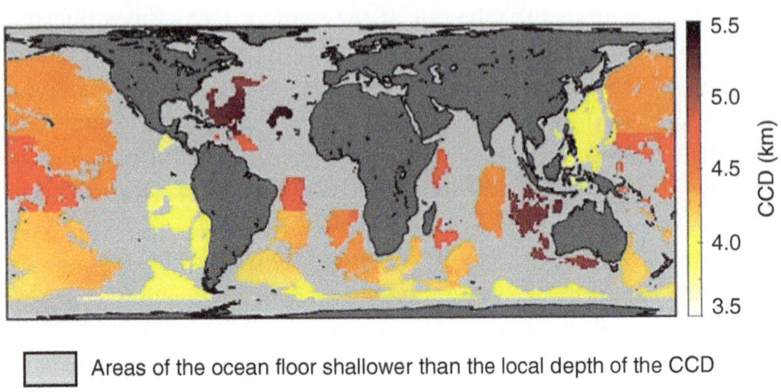

Figure 13.14 Present-day calcite compensation depth (CCD) reconstructed from in situ measurements of surface sediment $CaCO_3$ content. The CCD was not reconstructed for the Arctic and Southern Ocean south of 60°S. Continental shelf and near-shore areas were also not included in the database used. Modified from Sulpis et al. (2018).

processes may also be very important in the precipitation of the minerals (e.g. Tully and Heidelberg 2013; Shiraishi et al. 2016). In addition to manganese and iron, these nodules and crusts can contain relatively high concentrations of critical metals, such as Ni, Cu, Co and Rare Earth Elements, at levels that are leading to speculative attempts to exploit these resources, despite the poor understanding of their environments of formation and relationship with benthic ecology (reviewed in Hein et al. 2020).

Deep-sea ferro-manganese nodules grow very slowly (1–2 cm Ma^{-1}) yet remain at (or very close to) the seafloor for millions of years (Hein 2016). Many modern nodules are estimated to have nucleated around 12 Ma

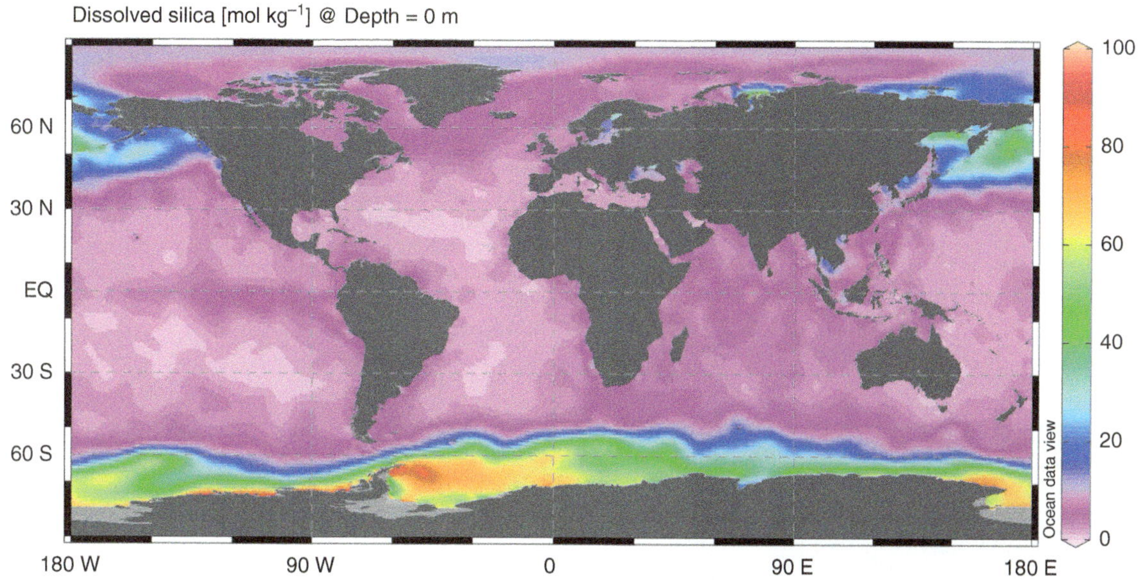

Figure 13.15 Surface seawater silica mean annual concentrations at 0 m depth from the World Ocean Atlas 2018 dataset (Boyer et al. 2018) plotted using Ocean Data View (Schlitzer, Reiner, Ocean Data View, odv.awi.de Schlitzer 2021). Note that surface seawater is undersaturated everywhere with respect to opaline silica by up to two orders of magnitude.

(see Dutkiewicz et al. 2020). This continuity of growth and maintenance of position at the sediment–water interface, occurs despite ambient sedimentation rates that, although slow (100–500 cm Ma^{-1} for pelagic clay), are still much greater than the rates of nodule growth. It has been hypothesised that the action of bioturbation by benthic echinoderms, arthropods and molluscs lifts nodules and keeps them free of sediment, allowing the nodules to be maintained within the bottom-water environment at the seafloor (Dutkiewicz et al. 2020). Although bottom-water currents may also be important in removing sediment in some settings (and perhaps for the development of crusts, as opposed to nodules), generally the speed of such currents in areas of Fe-Mn nodule growth is too slow to transport clay and silt-sized particles. In addition to low-sedimentation rates and bioturbation (elaborated on in Section 13.4.4), nodules are also associated with moderately high oxygen concentrations, which are required to allow the formation of oxide and hydroxide minerals, and low amounts of organic carbon in the surrounding sediments.

13.2.6.2 Volcaniclastics and Hydrothermal Deposits

Volcanic ash is transported to the pelagic environment via explosive eruption into the atmosphere and subsequent settling to the seafloor (e.g. Carey and Schneider 2011). As a consequence, volcanic ash layers can often be found interbedded with biogenic and terrigenous pelagic sediments across very large areas. Similarly, low-density pumice can float for several years and be dispersed over a very wide area before sinking into deep-sea depositional environments. Other processes can transport volcaniclastic material into pelagic environments in localities that are proximal to submarine and sub-aerial volcanoes, such as gravity-driven turbidity flows, debris flows and large-scale (>10^6 m^3) volcanic debris avalanches (VDAs), triggered initially by collapse on the flanks of volcanic edifices (e.g. Carey and Schneider 2011). Once in the deep-sea, volcaniclastic materials (irrespective of origin) can be physically and biologically reworked and mixed with other sediments, which may obscure their presence. In areas of submarine volcanism, such as the mid-ocean ridges, hyaloclastites and volcanic breccias are common as the hot basaltic lavas interact with cold seawater. Repeated explosive submarine eruptions associated with the mid-Atlantic ridge have been inferred from volcaniclastic deposits interbedded with calcareous pelagic sediments (Eissen et al. 2003).

The circulation of seawater through newly erupted (hot) igneous rocks leaches metals into solution and can lead to the precipitation of hydrothermal minerals and metalliferous sediments, around the seafloor vents of these metal-bearing hydrothermal fluids. At the vent, there may be significant quantities of mineral precipitation, in the form of, for example, black- (Fe, S) and white- (Ba, Ca, Si) smoker chimneys or mounds of hydrothermal minerals. In some settings, where bottom-water chemistry allows, hydrothermal minerals can form within the water column or be deposited by suspension from a hydrothermal plume (reviewed in Barrett et al. 2021). Crusts of hydrothermal minerals can form directly onto volcanic rocks on both seamounts and spreading ridges. Depending on the chemistry of the bottom-waters and

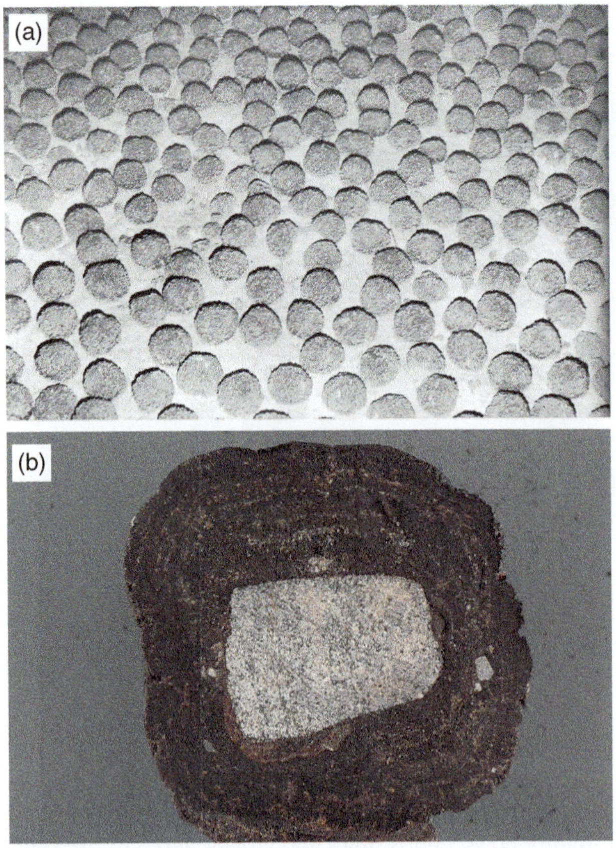

Figure 13.16 (a) A field of ferromanganese nodules from the Pacific Ocean at a water depth of 5292 m from Heezen and Hollister (1971). Images of this type typically imaged an area of 2–10 m^2 with an average of about 6 m^2. GeoMapApp / Marine Geoscience Data System / www.geomapapp.org. Location of image: 42°31′S 145°06′W. (b) A view of the internal structures of a ferromanganese nodule showing the concentric laminations around a central nucleus – in this case (and rather unusually) a dropstone. Other occluded grains can also be seen trapped in the laminae. The diameter of this sample is ~5 cm. Collected using a Scripps dredge in the Scotia Sea (56°36.6′S, 42°44′W; 2907 m water depth) on expedition NBP08-05. Photo courtesy of Professor L. Robinson and Professor R. Waller.

the hydrothermal fluids, the mineralogy of hydrothermal deposits can range from sulphides, to silicates to oxyhydroxides, with iron and/or manganese often being the major metals present. Metalliferous sediments associated with Mid-ocean ridges tend to be poor in Al and Ti compared to deep-sea clays but enriched in Fe and Mn (Jenkyns 1986). As with the ferromanganese nodules described previously, these hydrothermally sourced deposits (known as 'umbers') can also contain ore-grade concentrations of critical metals and Rare Earth Elements which are increasingly being considered for economic extraction (Box 13.2).

Box 13.2 Approaches to Studying and Sampling Pelagic Sediments

Oceanographic processes have been discovered and observed via a variety of direct and indirect techniques. Many of the tools used in oceanography have, since the days of HMS *Challenger*, been deployed and recovered during the same expedition, directly from vessels. From the late 20th century onwards, the use of satellites, autonomous floats and (semi-) permanent monitoring stations has provided spatial and temporal coverage in real-time which would have been impossible to achieve with ships alone (Summerhayes and Thorpe 1996). Satellites provide detailed data about surface ocean productivity, temperatures and dust flux, while autonomous floats measure the density of the upper 2000 m by moving through the water column over time, recording temperature and salinity (e.g. ARGO floats). These remote autonomous observations constrain many of the major physical and biological characteristics of the ocean in near-real time. Nonetheless, direct observations by sea-going expeditions are still critical for many aspects of oceanography relevant to pelagic sedimentation, including the taxonomy and ecology of planktonic organisms and communities (sampled via towing plankton nets or filtered from seawater samples), depth profiles of ocean chemistry (through sea-water sampling at fixed depths in the water column) and particle abundance and composition (via particle traps within the water column, that can be anchored to the seafloor or suspended from a vessel).

The seafloor itself can be sampled very crudely by dredging or, more precisely, by coring (Mulder et al. 2011). Shallow- and deep-penetrating seismic reflection surveys are used to image the subsurface to depths of metres to kilometres allowing the identification of target sites for coring and drilling. Various coring techniques with different capabilities allow sampling to different depths and have different applications to sedimentology depending on the objectives (Table B13.1 and Figure B13.1). For cored sediments,

Table B13.1 A summary of methods of sampling deep-sea sediment samples (Data from Mulder et al. 2011).

Coring technique	Basic principle	Depth of penetration	Applications and limitations
Box coring	A 'box' with a large area is driven into sediments and then pulled out. A mechanism closes a door to prevent material loss	Typically <0.5 m	Relatively quick sampling of the sediment–water interface, including in relatively coarse-grained deposits. Good for sampling benthic fauna.
Interface corer	Gravity-driven penetration of one or more tubes suspended from a frame lying on the seafloor	Typically <1 m	Precise sampling of sediment–water interface, including the overlying seawater
Gravity corer	Free-fall, gravity-driven penetration of a tube suspended from a wireline; a core-catcher at the base of the tube closes to prevent loss of core when the tube is pulled up to the ship	Typically a few metres	Good for recovering short cores that penetrate deeper than box or interface cores across a wide range of ocean depths, but gravity cores tend to destroy the sediment–water interface and do not always penetrate the seafloor vertically
Piston corer/giant piston corer	Like the gravity corer but with a piston attached to the wireline that is triggered when the seafloor is reached, firing the corer into the sediments (see Figure B13.1)	Up to 70 m	Like standard gravity cores, piston cores provide excellent archives for palaeoceanography but can experience problems with core deformation and flow
Seafloor rock drills	A rock-drill mounted to a frame that is lowered onto the seafloor	Up to 200 m	Initially developed for hard-rock drilling (e.g. directly into ocean crust) but now modified for soft-sediment settings
Wells	Used by the IODP (and its predecessors), drilling wells at sea requires a free-floating vessel to remain in a near stationary position above the borehole (achieved by a seafloor beacon, GPS and the deployment of multiple computer-controlled thrusters on the vessel). A drill-pipe is assembled by screwing together sections of pipe that are lowered through a Moon-pool on the vessel and then suspended below the vessel until the seafloor is reached. At the bottom of the pipe is a cylindrical drill bit that allows core to pass through the centre of it into the pipe. Coring is achieved by lowering a core tube and coring mechanism through the pipe, which is then returned to the ship and replaced by an empty core tube for the next core. In this way, cores can be taken continuously until a target depth is achieved. For softer lithologies, a piston coring mechanism is used that fires through the drill bit into the sediments below, before the drill pipe is advanced over it (see Figure B13.1), which may recover sediment to over 300 m below sea floor. This method will usually produce high-quality recovery and preservation of the sediment. For firmer lithologies, the rotation of the drill pipe provides the method of penetration with core being collected inside the drill string and can be used to recover cores from >1000 m below sea floor. This latter technique can result in poorer recovery and preservation of sedimentary material	Up to several km (although typically <1.5 km in sediments and sedimentary rocks)	The only method suitable for recovering long, continuous archives of the sedimentary history of the ocean basins. The sediment–water interface is easily lost or missed, while coring in intervals with mixed lithologies of varying duration (e.g. ooze interbedded with chert) leads to deformation and loss of soft materials and fracturing and break-up of hard materials

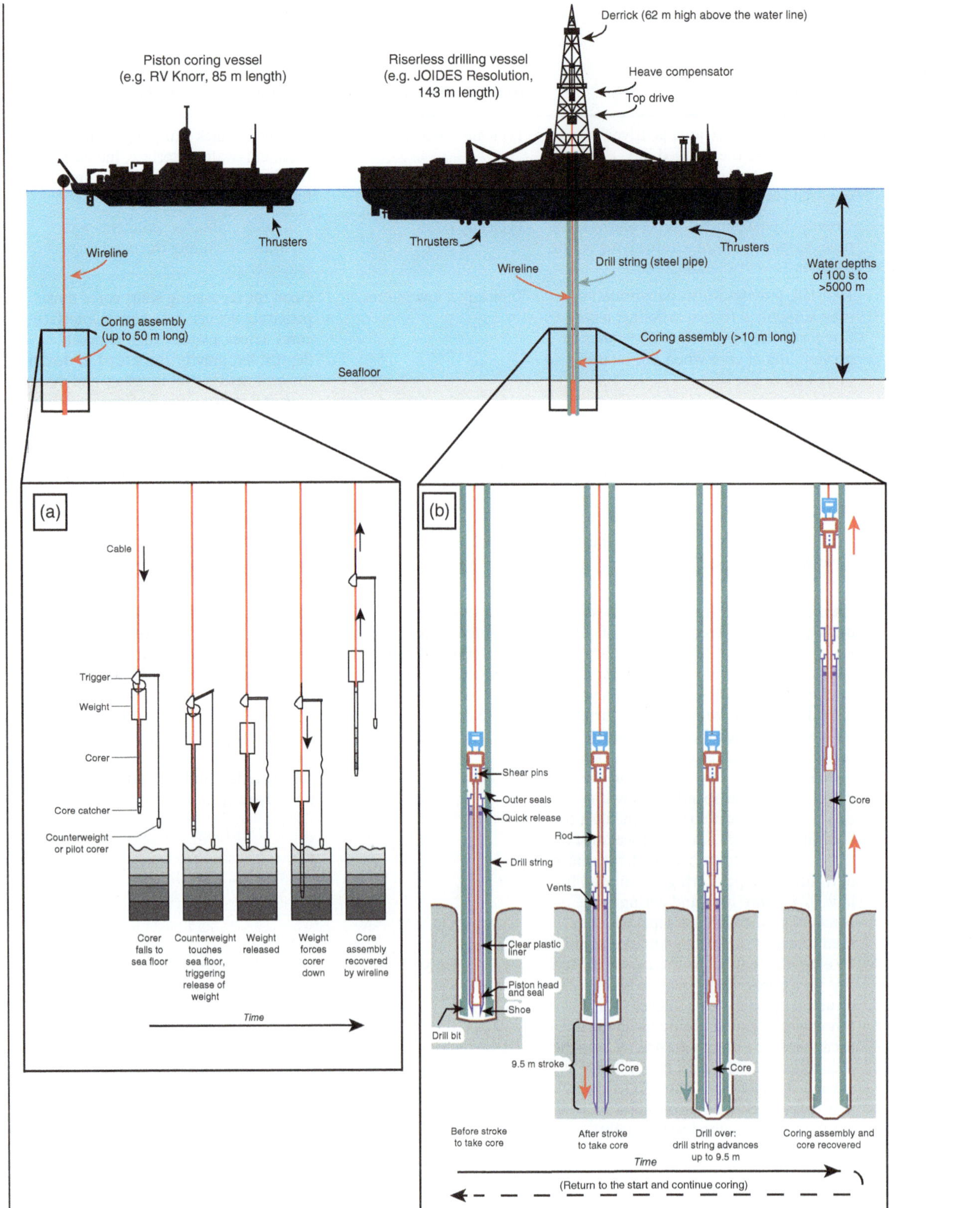

Figure B13.1 Examples of coring methods used to sample pelagic sediments in the ocean. (a) Piston and gravity coring, allowing relatively shallow penetration into the seafloor (metres to 70 m). (b) Coring and drilling by vessels such as the JOIDES Resolution allow deeper penetration into the subsurface (typically hundreds of metres to kilometres). Shown here is the Advanced Piston Corer (APC), which shoots the core barrel into the sediments ahead of the drill bit. (a) Mulder et al. (2011) / with permission of Elsevier; (b) Graber et al. 2002 / IODP publications / Public Domain.

ephemeral (i.e. wet bulk density, thermal conductivity, interstitial water and gas chemistry) and other chemical and physical properties (e.g. radioactivity, magnetic susceptibility, p-wave velocity) can be measured shortly after sampling (onboard the ship in the case of the *JOIDES Resolution* or the *Chikyu*) – such data have value for local and regional stratigraphy as well as sedimentology. Cores are often split along their length (into 'archive' and 'working' halves) for description and sampling. The working half is used for destructive measurements and sampling, while the archive half is commonly preserved for further description or non-destructive measurements. A fairly common set of methods exists for the description and dating of pelagic sequences. Macroscopic observations are combined with microscopic observations of smear-slides (for soft lithologies) or thin sections (for harder lithologies) and supplemented by digital core scanning, colour data and core-scanning XRF (X-ray fluorescence) measurements.

13.3 Modern Pelagic Depositional Environments and Sediment Distribution

Modern pelagic sediments are found in a wide range of depositional settings (Figure 13.17), across a variety of tectonic contexts, latitudes and water depths. In each of these settings, the relative importance of the major controls on sediment type (i.e. biological productivity, preservation and siliciclastic dilution) varies both in space and time, leading to significant differences in lithologies, accumulation rate and, over time, the generation of recognisable stratigraphic patterns (reviewed in detail in Jenkyns 1986). Pelagic sediment thicknesses can vary from being almost completely absent in some areas to accumulations in excess of a kilometre-thick where the oceanic basement is sufficiently old, and sediment supply has been sufficiently constant (Figure 13.1a). Typical thicknesses of biogenic pelagic sediment bodies are on the order of hundreds of metres and broadly increase with the age of the ocean crust (Olson et al. 2016; Straume et al. 2019).

13.3.1 Mature Ocean Basins

In the mature ocean basins of the Atlantic and Pacific Oceans, the type of pelagic sedimentation at low to mid-latitudes is largely governed by water depth and, therefore, by proximity to the mid-ocean ridges, the crests of which are generally at depths of around 2.5 km. As ocean crust

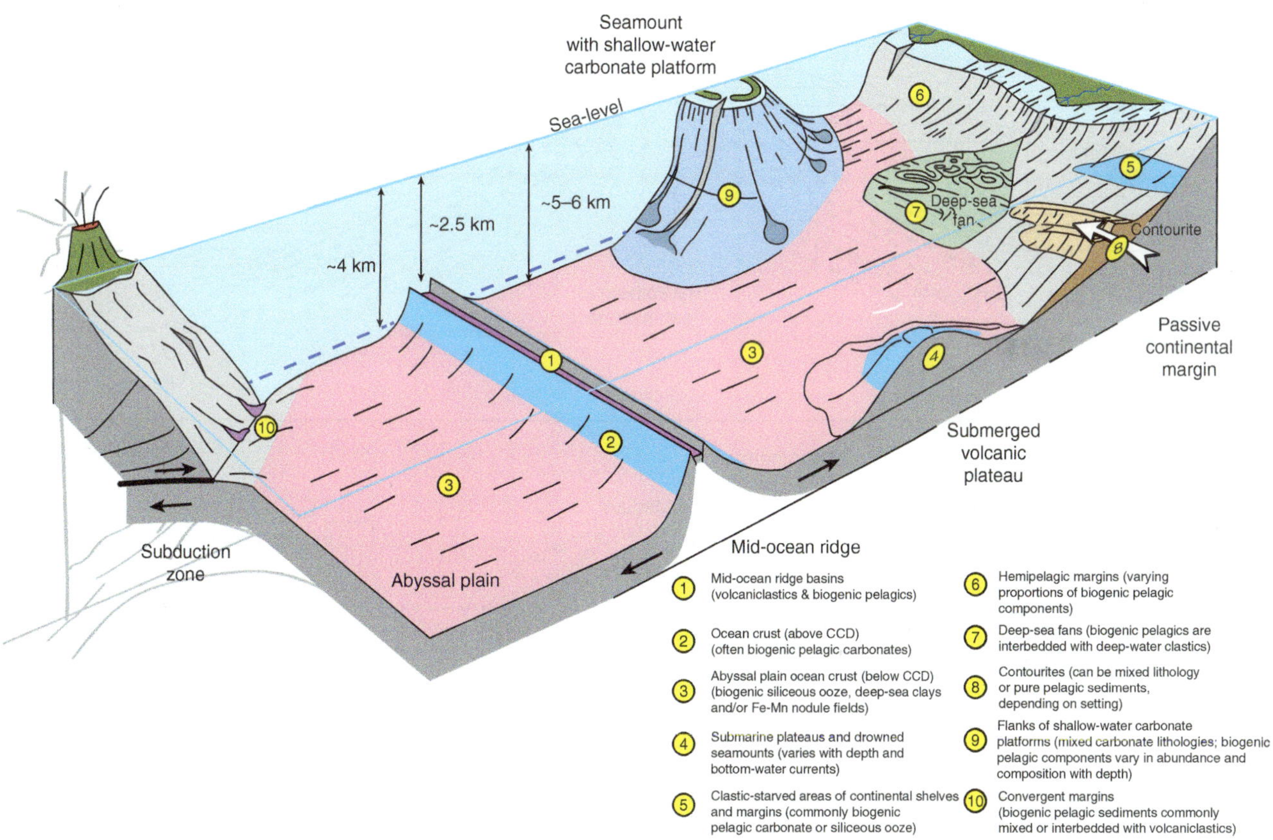

Figure 13.17 Schematic of a mature ocean basin showing the range of possible pelagic sediment depositional environments and common lithologies (and mixtures of lithologies). Adapted from Mulder et al. (2011).

cools, increases in density and subsides away from the ridge crests, water depths increase, with the square root of time, down to the abyssal plains at more than 5 km. Concurrently, the influence of volcanic and hydrothermal activity decreases. Consequently, on a transect from ridge axis to abyssal plain, pelagic sediments transition laterally from dominantly volcaniclastic and hydrothermal at the ridge axis, to biogenic pelagic carbonate (containing calcifying pteropods, foraminifera and/or coccoliths) and, finally, below the CCD to pelagic clays and/or siliceous oozes. This lateral succession becomes a vertical succession as the ocean crust moves away from the ridge axis. The intersection between subsiding ocean crust and the CCD provides the major control on carbonate preservation and largely explains the geographic distribution of carbonate sediment in the modern deep ocean (Figure 13.1b). Sedimentation rates in areas below the CCD are generally low and so authigenic ferromanganese nodule fields are common in these regions. Where surface-water environments are dominated by high rates of siliceous production and/or the CCD is locally shallow (due to upwelling of undersaturated waters and/or high rates of organic carbon rain), biosiliceous sediments, or mixed calcareous-siliceous lithologies, can accumulate; for example, below the equatorial upwelling zone in the Pacific Ocean and at high-latitudes (Figure 13.1b; Dutkiewicz et al. 2015).

Spreading rate exerts a secondary control on the detailed sediment distribution in mature ocean basins, particular around the ridge axis (Figure 13.18; Jenkyns 1986). On the fast-spreading East Pacific Rise, the high heat flow and, consequently, extensive hydrothermal activity leads to relatively thick accumulations of metalliferous sediments close to the ridge as coatings on basaltic breccias and around active vent systems. Further away, dilute hydrothermal plumes lead to the deposition of fine-particulates rich in Fe-Mn oxyhydroxides together with biogenic pelagic sediments. In contrast, on the slower-spreading North Atlantic ridge there is considerably less hydrothermal activity, more rugged topography, and, locally, exposure of both mafic and ultramafic rocks. These local bathymetric highs of exposed basement can become sites for authigenic formation of ferromanganese crusts, whereas the deeper areas between can become sites of sediment accumulation. Although sedimentary lithologies in

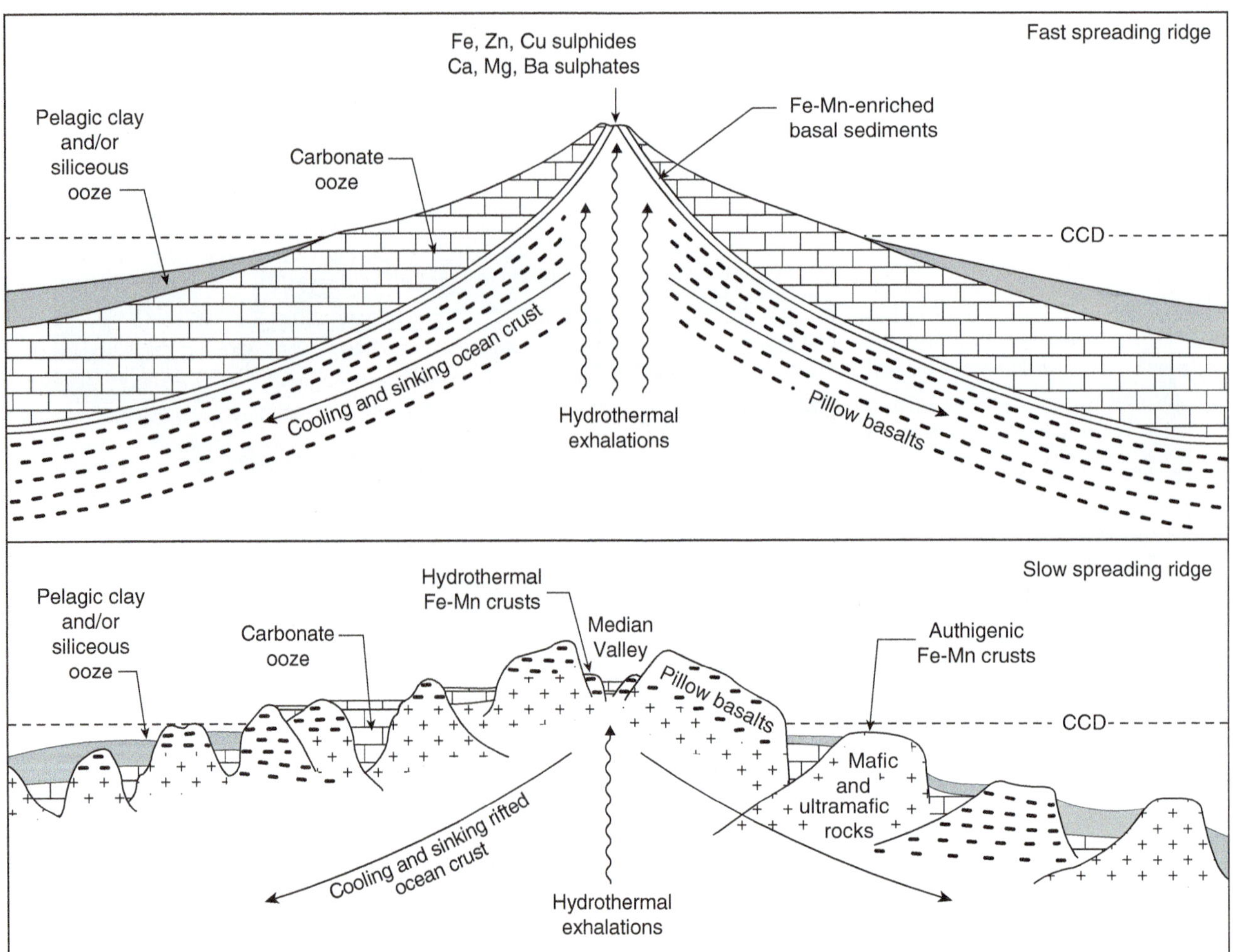

Figure 13.18 Comparison of pelagic sedimentation associated with fast and slow spreading ridges (Adapted from Jenkyns 1986).

these 'valleys' are dominated by pelagic carbonate, there is evidence for re-deposition of sediment by turbidity currents and mixing of carbonate with volcaniclastic and hydrothermal debris.

In addition to mid-ocean ridges, volcanic seamounts, volcanic plateaus and aseismic ridges also provide sites for pelagic sediment accumulation, most notably by elevating the local seafloor above the CCD and allowing pelagic carbonates to be deposited atop these structures (e.g. on the Rio Grande Rise and Walvis Ridge in the South Atlantic). Sediment type and thickness associated with such edifices depends on a number of factors including the depth change during growth of the structure and the length of time since volcanic activity ceased. These factors together determine the uplift and subsidence history (i.e. water depth over time) and, hence, the stratigraphy of any sediment accumulation. The morphology and topography of the structures may create local basins within the structures that are sites of deposition and may focus and accelerate bottom-water currents such that erosion and re-deposition are common. Locally, carbonate hardgrounds and ferromanganese crusts can be present. Volcanic features that do not grow sufficiently to rise above the CCD will not possess a pelagic carbonate cap, whereas large features that become subaerial may possess a more complex sedimentary history recording the erosion and drowning of the volcanic edifice.

On the margins of mature ocean basins, whether passive or active, pelagic sediments become increasingly associated with terrigenous sediments, thereby diluting the pelagic biogenic carbonate and silica. This dilution is expressed as interbedded pelagic and turbiditic sediments or as increasing amounts of clay associated with deposition of pelagic components (i.e. hemipelagic sedimentation, see Chapter 9). In the high-latitude regions, addition of terrigenous material can also be through deposition from icebergs, leading to the occurrence of dropstones within predominantly diatomaceous oozes. Approaching a passive margin from the mid-ocean ridge, an increase in terrigenous sediment is often encountered on the abyssal plain and continental rise as a consequence of the deposition of turbidites and, in some areas, the development of deep-sea fans (see Chapter 12). Along active margins, in both fore-arc and back-arc settings, volcaniclastics commonly dilute pelagic sedimentation by both ash fall and gravity-driven re-deposition from shallower water.

On passive continental margins, there is commonly topography associated with seamounts, banks and basins. These reflect structural (e.g. fault blocks and grabens) and depositional (e.g. carbonate build-ups and depositional slopes) features created during the rifting and thinning of the continental margin. Pelagic sediments can predominate in such settings only if the terrigenous sediment supply is low (e.g. due to either supply *per se* or local routing). For example, deposition on the modern Blake Plateau (south-eastern margin of North America) in water depths averaging 850 m, includes carbonate oozes, winnowed by bottom-water currents and pavements of phosphatic and ferromanganese nodules overlying older cemented pelagic carbonate deposits (reviewed in Jenkyns 1986). To the south of the Blake Plateau, lie the Bahama Banks, an extensive shallow-water carbonate platform dissected by deeper water channels (Section 11.6.2). The dominantly pelagic sediments of the Blake Plateau transition southwards to 'peri-platform ooze' containing inter-mixed pelagic and shallow-water components transported off the platform by storms (aragonite needles, carbonate mud and sand). This mixed material also occurs in the deep channels that lie between the banks, where graded beds of shallow-water carbonate debris provide evidence for episodic carbonate turbidity flows from the platform margin.

13.3.2 Small Oceanic Basins

Pelagic sedimentation patterns in smaller oceanic basins are typically highly sensitive to changes in tectonics, volcanism and/or sea-level, through controls on, for example, gateways, current patterns, siliciclastic dilution and nutrient supply. Examples of such basins where pelagic sediments accumulate in the modern include incipient ocean basins, remnant basins due to incomplete ocean closure and strike-slip basins associated with transform faults between plates.

The Red Sea provides the classic example of an incipient ocean basin resulting from continental rifting (Figure 13.19), with a spreading centre that has been active for just the last four million years separating the African and Arabian plates (e.g. Bosworth 2015). Water depths in the axial trough, the site of seafloor spreading, are typically 600–1000 m, reaching 2000 m in the south. The relatively shallow-water margins on either side of the trough are tens of kms in width and, in the warm, arid climate of the region, are characterised by deposition of biogenic carbonate, dominated by coccolithophores, planktonic foraminifera and calcifying pteropods (e.g. Berggren and Boersma 1969; Almogi-Labin et al. 1991; Chernihovsky et al. 2018). Surface waters in the Red Sea possess a slightly elevated salinity (relative to the open-ocean), due to the 'anti-estuarine' circulation pattern of surface waters flowing into the basin, experiencing evaporation and forming out-flowing deep-water. This circulation pattern is very sensitive to the depth of water over 'sills' (areas of relatively shallow water between the basins and the open-ocean). If tectonics or eustatic sea-level were to alter the depth of these sills, then the Red Sea could easily become dominated by evaporation, and pelagic deposition would be replaced by salts, as indeed occurred in the Miocene prior to the development of the axial trough (e.g. Hughes and Beydoun 1992). These Miocene salts now contribute

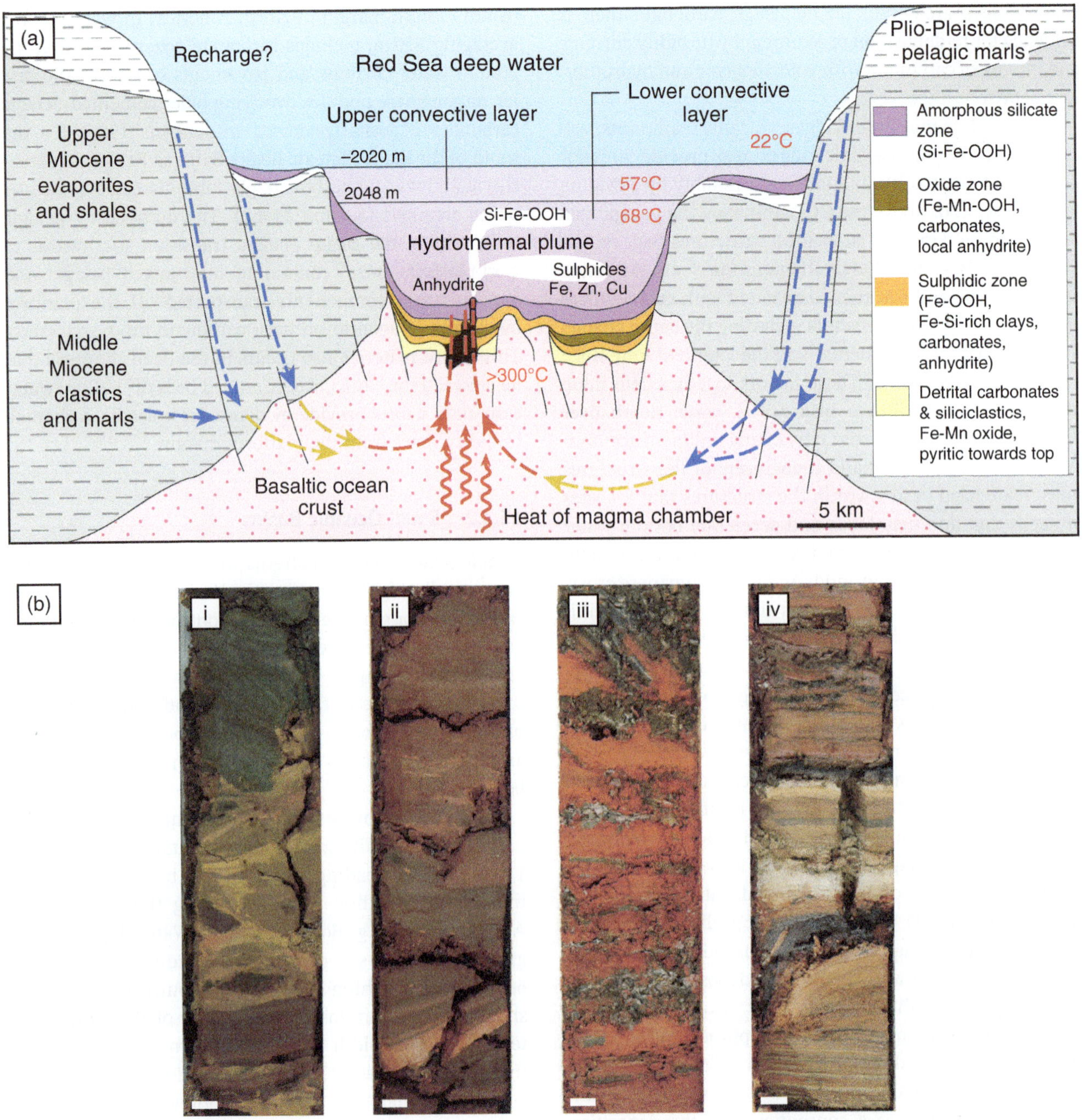

Figure 13.19 (a) Schematic cross section of a Red Sea spreading centre (Atlantis II Deep) showing the relationship between ocean crust, rift-related sedimentation (Miocene clastics, marls and evaporites), biogenic pelagic sediments and hydrothermal deposits. Barrett et al. (2021) / with permission of Elsevier. (b) Cores of hydrothermal sediments from Atlantis II Deep, adapted from Laurila et al. (2015). Scale bars are 2 cm. (i) Brecciated mud dominated by green clays at the top and Fe-(oxy)hydroxides in the lower part; (ii) gradational and diffuse laminations of hydrothermal muds; (iii) red Fe-(oxy)hydroxides and grey anhydrite-rich layers from a core situated close to a hydrothermal vent; (iv) finely laminated sediments composed of Fe-(oxy)hydroxides (red and brown), hydrothermal carbonates and Mn-(oxy)hydroxides (dark grey). The white layer contains detrital silicates and biogenic calcite.

to the generation of saline brines in the deepest parts of the axial trough which play an important role in the deposition of the abundant metalliferous hydrothermal sediments found there (Figure 13.19; Barrett et al. 2021).

The Gulf of California lies at the southern end of the San Andreas fault zone where the East Pacific Rise meets the North American plate. Water depths span from 600 m in the north to 3000 m to the south; the deepest waters occur in troughs that are associated with incipient spreading centres, where some hydrothermal deposits occur (e.g. Lonsdale et al. 1980). Sedimentation in the centre of the Gulf is dominated by diatoms (although radiolaria are also present), with, in places, a very significant clay content (e.g. Calvert 1966; Thunell et al. 1994). Unlike the Red Sea, circulation in the Gulf of California is predominantly 'estuarine', with deep-waters flowing in and surface waters flowing out due to offshore winds. However, significant seasonal variations affect the strength and direction of surface water flow, which in turn determines the timing and abundance of primary production. Strong seasonal upwelling fuels high productivity which leads to varved alternations of diatom and clay-rich laminae being deposited (e.g. Calvert 1966; Pike and Kemp 1999). At depths between 500 and 1000 m, a well-developed oxygen-minimum zone exists, in part due to the inflowing and poorly oxygenated Pacific Intermediate Water, but also because of the high fluxes of organic carbon and high levels of respiration in the water column (Thunell 1998). The combination of high primary productivity and significant clay input and deposition means that sedimentation rates in the Gulf of California can be very high; for example >40 to >200 cm ka^{-1} during the Pleistocene (e.g. Dean et al. 2004; Teske et al. 2021).

Both the Red Sea and the Gulf of California represent proto-oceanic basins that may or may not become mature ocean basins in the future. Although mature ocean basins may persist for hundreds of millions of years, subduction will, eventually, lead to the destruction of extant ocean crust. In some cases, plate tectonics will close a mature ocean basin by bringing two continents together. Prior to complete continent-continent collision, deep remanent basins largely surrounded by land may occur, as is the case for large parts of the Mediterranean (water depths up to 5.3 km and, in parts of the eastern Mediterranean, underlain by, purportedly, fragments of Tethyn Ocean crust as old as Carboniferous). The hydrology of these basins is key to determining the depositional environment. In the case of the Mediterranean, circulation is anti-estuarine (like the Red Sea) with the connection to the open-ocean via the Straits of Gibraltar (Chapter 10: reviewed in Rohling et al. 2015). Away from areas of detrital input, low-productivity calcareous pelagic sediments dominated by coccoliths and planktonic foraminifera are widespread. Ultimately, as continental convergence progresses, ocean basins trapped during collision will become dominated by extremely high-rates of clastic input ('flysch' type deposits) and the conditions conducive to pelagic sedimentation will disappear.

Areas with predominantly horizontal tectonic movements along *en echelon* strike-slip faults can produce significant amounts of both crustal uplift and subsidence, often in adjacent areas. Subsidence rates can be extremely high and the pull-apart basins so produced can be very deep even though their areal extent may be small. Such basins are found in a variety of plate tectonic settings where pelagic (and hemipelagic) sediments may accumulate, including along continental margins where two plates are moving laterally past each other (e.g. Cariaco Basin offshore Venezuela) or along transform faults in oceanic crust (e.g. the Cayman Trench/Trough in the Caribbean). The Cariaco Basin lies along the transform-faulted plate boundary between the South American and Caribbean plates. Hemipelagic Holocene sedimentation in the Cariaco Basin is dominated by laminated, silty clays containing biogenic carbonate, opal and organic carbon that accumulated at sedimentation rates of approximately 30 cm ka^{-1} (e.g. Peterson et al. 1991, 2000). Seasonal upwelling and proximity to the South American continent impart a strong control on biogenic and terrigenous sedimentation and are expressed as annual cycles of light-dark laminae in the varved Holocene sequence (e.g. Hughen et al. 1996).

13.4 The Geological Record of Pelagic Sedimentation: Processes and Products

The mass of preserved biogenic pelagic sediment in today's oceans is estimated to be about 140×10^{18} kg (Hay et al. 1988), of which approximately 50% is younger than 30 Ma and only about 10% is Cretaceous or older (>66 Ma). The sources, environments and external controls of modern pelagic sedimentation provide the starting point for interpreting these deposits. However, to fully understand the pelagic rock record, fundamental long-term processes also need to be included. Over geological time, plate tectonics drives the changes in geographic location and water depth that control sediment production and preservation as well as destroying large parts of the record through subduction. Evolution has determined the nature of biogenic pelagic sediment producers and consumers. Furthermore, during and after deposition, all pelagic sediments can be modified by physical, biological and chemical processes. In this section, we explore the tectonic, evolutionary and post-depositional processes that determine the distribution, lithologies, bedforms and sedimentary structures observed in the pelagic rock record (Box 13.3).

Box 13.3 Classification of Pelagic Lithologies and Bedforms

Pelagic sediments either contain varying proportions of grains of biogenic (e.g. skeletal) and detrital (e.g. aeolian or volcanic) origin or are sourced directly from seawater through chemical precipitation (e.g. Fe-Mn nodules). For those sediments of a biogenic or detrital origin, the proportions of different grain types are used to classify unconsolidated lithologies. For example, guidelines for scientists aboard the *JOIDES Resolution* define 'biogenic pelagic sediments' as those where the combined biogenic sediment and neritic carbonate (i.e. coarse bioclasts, ooids, non-pelagic micrite, etc.) component is more than 60% of the whole *and* where the pelagic sediment component is greater than that of the neritic carbonates (Figure B13.2; Mazzullo et al. 1988). Unconsolidated biogenic pelagic sediments, irrespective of composition, tend to be soft at, or close to, the seafloor: 'oozes'. With increased burial depth, oozes and clays become more indurated (due to compaction and cementation [Section 13.4.5]). Classification of consolidated materials is based on the mineralogy, degree of induration and dominant constituent particles. Carbonate oozes become chalks or limestones; siliceous oozes become radiolarites, diatomites or cherts; and pelagic clays become pelagic claystones. Lithologies with an unquantified mixture of biogenic carbonate material with terrigenous fine-grained material can be referred to as *marls/marlstones*. Analogous, albeit rarely used, terms exists for silica-bearing deposits – *sarls/sarlstones* and *smarls/smarlstones* (siliceous calcareous mud). These terms, particularly 'marls', are useful informal terms for describing such mixtures in the field or core. Where biogenic pelagic sediments are exposed on land, they have generally experienced diagenesis during burial and/or uplift and so are more indurated lithologies than oozes. Fe-Mn deposits are classified based upon their mineralogy and morphology (e.g. nodules versus crusts).

Pelagic sediments exhibit a range of bedforms, sedimentary structures and fabrics, as a consequence of factors such as changing sediment sources, physical and chemical bottom-water conditions and diagenesis (Figure B13.2). Beds (typically tens of cms thick) are generally tabular but may exhibit sharp or gradational bedding contacts, frequently either obscured or enhanced by diagenetic

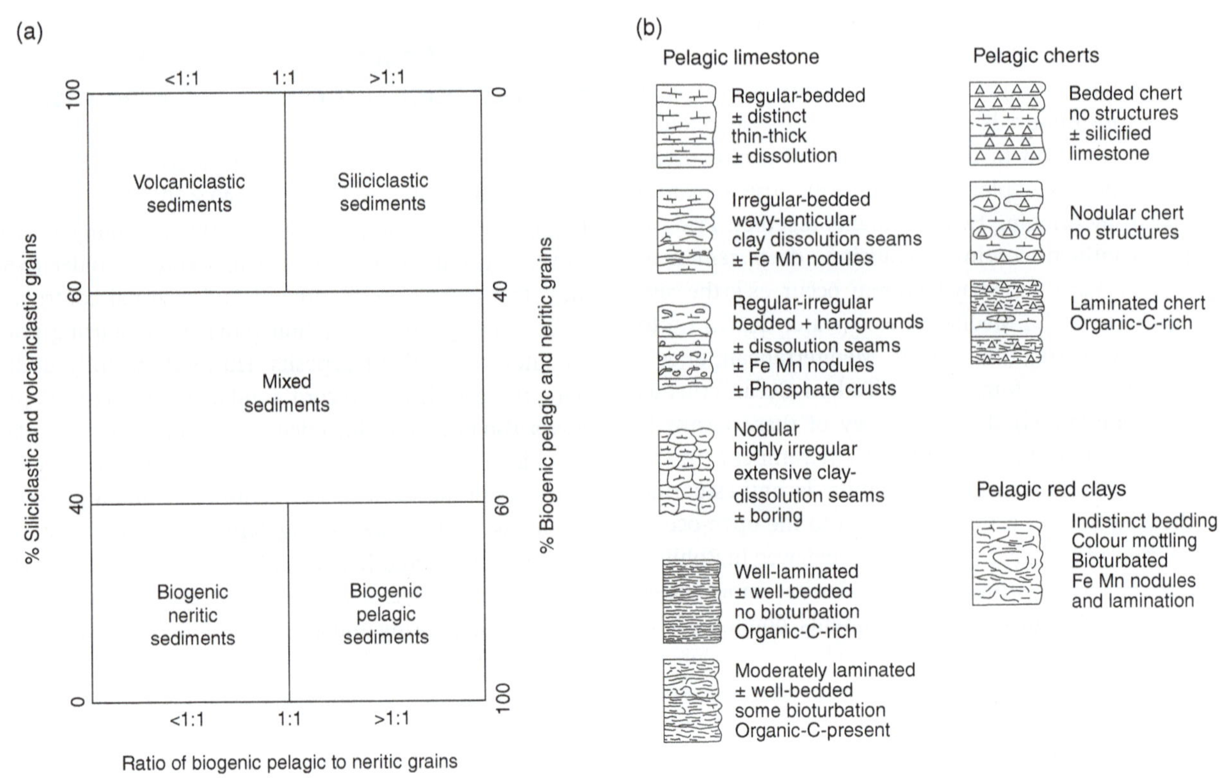

Figure B13.2 (a) Classification scheme for sediments used in scientific ocean drilling (ODP, IODP) adapted from Mazzullo et al. (1988). (b) Facies models for pelagic sedimentary rocks adapted from Stow et al. (1996).

changes. Where beds (or packages of beds) are seen or traced (e.g. at outcrop or on seismic profiles), they are often continuous over many km^2. Where suspension sedimentation is dominant and there is minimal remobilisation of sediment at the seafloor, laminae (often due to variation in the amount of organic carbon) may develop, and, in the absence of bioturbation, be preserved. Where bottom-water currents exist, sediments can be transported by unidirectional flow, leading to the development of asymmetric ripples and ripple cross-lamination within beds. At a larger scale (typically >tens to hundreds of metres of sediment thickness), vigorous and consistent bottom currents, driven by deep-ocean circulation, lead to the development of pelagic sediment drifts and contourites. Nodular textures are common in pelagic lithologies but have various origins, often associated with bioturbation. Slow sedimentation of pelagic carbonates promotes early cementation (which is often nodular), hardground formation and the development of Fe-Mn nodules. Deep burial of biogenic sediments can lead to extensive dissolution seams and the development of false bedding and nodules. In silica-bearing sediments, burrows can act as sites for nucleation of chert nodules. Deep-sea red clays are generally structureless (due to bioturbation under slow sedimentation rates) but can contain Fe-Mn nodules.

13.4.1 Tectonic Processes

Thermal subsidence of ocean crust moves a site vertically with respect to the CCD, controlling carbonate preservation, and lateral movement of a plate moves a site horizontally relative to the oceanographic and climatic conditions, thereby controlling sediment production. For example, the northward movement of a site in the Pacific Ocean from the Southern to the Northern Hemisphere means that the site passes beneath the equatorial divergence zone and there records an increased flux of sediment due to high-productivity driven by upwelling (Figure 13.20). This leads to the concept of 'plate stratigraphy' – stratigraphic variations in lithology that can be related to the tectonic evolution of a site in both water depth and geographic location (Berger and Winterer 1974).

Tectonics also determines the long-term preservation of pelagic sediments. Subduction consumes ocean crust normally together with its blanket of pelagic sediment, such

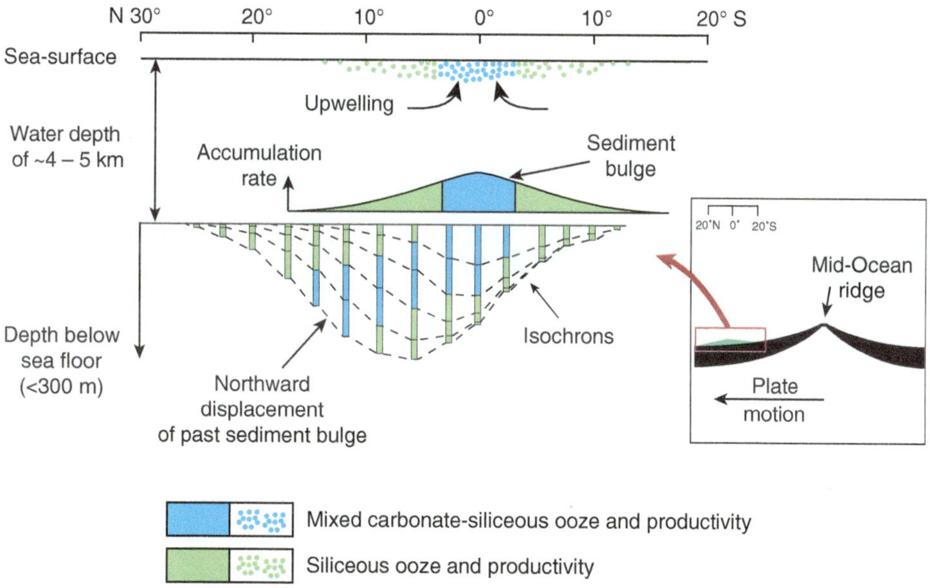

Figure 13.20 A cartoon to illustrate the concept of 'plate stratigraphy' (adapted from Berger and Winterer 1974) and the effects on sedimentation rates and facies successions of the movement of a plate beneath a zone of high-sediment production. The vertical lines below the seafloor represent the stratigraphy at a series of sites in a north-south transect across an equatorial upwelling zone of high productivity, such as seen in the modern Pacific Ocean, that leads to high rates of sediment production and a depression of the CCD. These processes allow enhanced preservation of carbonate and siliceous ooze on the ocean floor. In the subsurface, expanded sequences, containing mixed carbonate-siliceous ooze, occur to the north of the modern equatorial upwelling zone due to the northward displacement of the sediment-thick bulge by plate motion, relative to the zone of highest productivity. This model assumes that the zone of sediment accumulation remains fixed with respect to latitude.

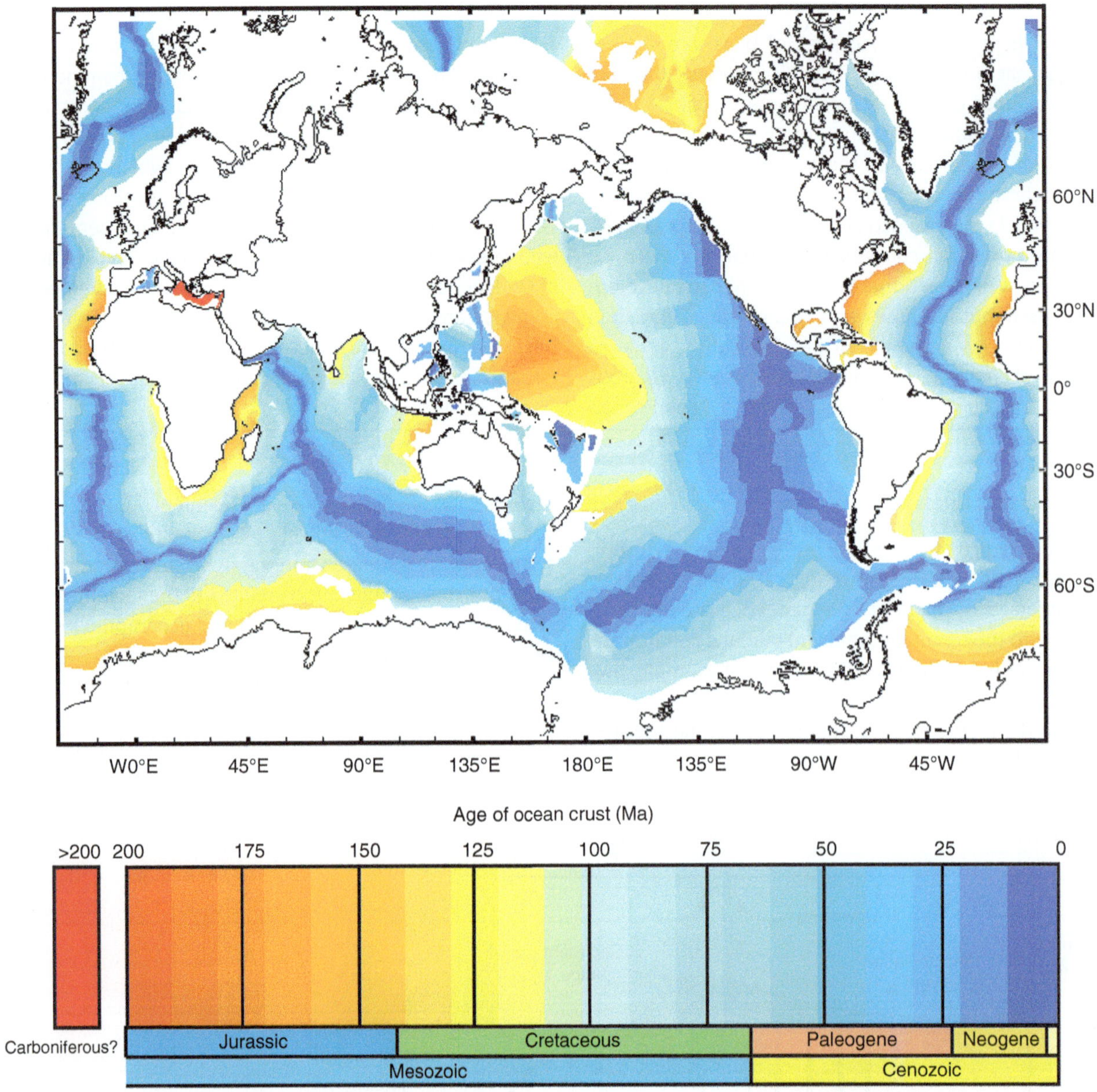

Figure 13.21 Map of the age of the oceanic lithosphere (crustal age data from Müller et al. 2008; figure made using GeoMapApp; www.geomapapp.org).

that the ocean crust and pelagic records are predominantly Cenozoic in age (Figure 13.21), with very little older than the Cretaceous. Although ophiolites and accretionary complexes can contain pelagic sediments from now-subducted oceanic areas (e.g. Devonian radiolarian cherts associated with the Lizard Ophiolite in SW England; Jurassic and Cretaceous pelagic cherts and limestones associated with basalts and serpentinites in the Franciscan Complex of California; Late Cretaceous umbers, radiolarian cherts and chalks associated with the Troodos Ophiolite in Cyprus), these archives have typically experienced tectonic deformation and/or metamorphism. Nonetheless, such fragments provide important insights into the spatial and temporal history of pelagic sedimentation that would otherwise not be available.

13.4.2 Evolution and the Composition of Biogenic Pelagic Sediment

Of the modern sediment-generating mineralised plankton, only the radiolaria possess a fossil record stretching back into the Palaeozoic (Figure 13.22). They appear to have been the *only* part of the plankton then capable of producing mineralised tests in sufficient quantities to be

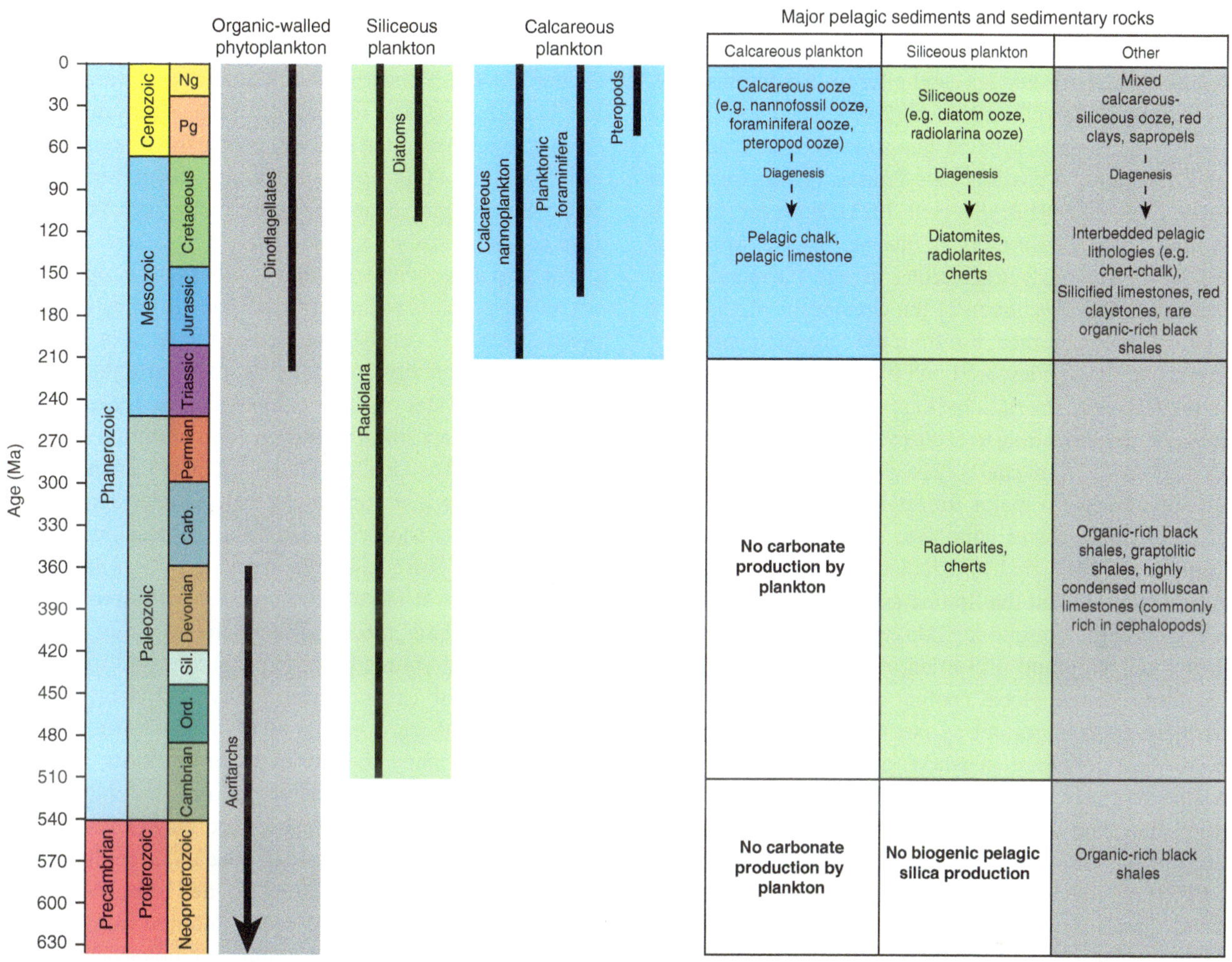

Figure 13.22 The evolution of major sediment-producing planktonic organisms and the associated types of pelagic sediments and sedimentary rocks commonly found during the Late Precambrian to Recent.

rock-forming. In addition, small volume, but dense, accumulations of nektonic organisms, such as cephalopods (e.g. nautiloids, ammonoids and belemnoids), have been interpreted to be examples of Palaeozoic and Early Mesozoic pelagic carbonate sedimentation.

Prior to the Late Triassic, there were apparently *no* major calcareous pelagic sediment producers. During the mid-Mesozoic, there were major changes in marine ecosystems and ocean chemistry, that are known collectively as the Mid-Mesozoic Revolution (MMR). A feature of the MMR is the appearance in the fossil record of the major modern planktonic sediment-generating organisms, starting with the calcareous nannofossils and dinoflagellates in the Late Triassic and the planktonic foraminifera in the Middle Jurassic (reviewed in Hull 2017). The term 'calcareous nannofossils' is used to describe both coccoliths (from coccolithophores) and nannoliths of unknown affinity, but similar size, that are assumed to be from extinct relatives of coccolithophores (Bown and Young 1998). Nannofossil-rich limestones become increasingly common through the Jurassic whereas planktonic foraminiferal-rich sediments gain in prominence through the Cretaceous. The calpionellids (~40 to 150 μm long; single-celled, now extinct, eukaryotes), nannoconids (<10 s μm; a type of calcareous nannofossil also now extinct) and calcispheres (<400 μm in diameter) made a significant contribution to pelagic carbonate production during some time intervals of the Jurassic and Cretaceous although their biological affinities are somewhat uncertain (e.g. Bown and Young 1998; Remane 1998; Armstrong and Brasier 2005). Most of the Mesozoic–Cenozoic calcispheres were likely dinoflagellate cysts. (Note that Palaeozoic calcispheres may have had a different biological affinity and do not appear to be associated with planktonic environments; e.g. Wray 1998, Versteegh et al. 2009.) The evolutionary events of the MMR allowed carbonate oozes to be deposited in the open-ocean on a large scale for the first time and likely drove major changes in ocean carbonate chemistry by providing an additional sink for carbonate that could act as a buffer to changes in atmospheric CO_2 (Ridgwell and Zeebe 2005). The evolutionary origin of diatoms is debated due to molecular clock evidence indicating a possible Late Triassic or Jurassic origin (e.g. Sims et al. 2006; Bryłka et al. 2023). However, the oldest undisputed well-preserved, diverse, marine, fossil diatom floras are Early Cretaceous in age (Gersonde and Harwood 1990). For much of the Cretaceous diatoms are not commonly observed, but from the Paleogene onwards, diatoms become widespread and highly diverse. Throughout the Mesozoic, the radiolaria appear to have been diverse and widespread, leading to extensive deposition of radiolarites and radiolarian cherts. The proliferation of diatoms may have caused a major transition in the oceanic silica cycle towards lower dissolved silica concentrations (Conley et al. 2017) that contributed to the marginalisation of the radiolaria and the limited extent of radiolarian-rich sediments at the present-day (Maliva et al. 1989). Although silica availability and utilisation are likely factors in this long-term change, global cooling over the Cenozoic may also have been important as modern radiolaria are most diverse and abundant in warm, nutrient-rich conditions (Anderson 1983); perhaps as these are the conditions to which they had adapted in the past (Racki and Cordey 2000). It has also been argued that the Cenozoic timing of the rise of the diatoms may be more apparent than real, due to diagenetic destruction of the frustules rather than absence in the ecosystem (Westacott et al. 2021). If correct, then any effect of diatoms on seawater silica concentrations and cycling would also be earlier in the Cenozoic, or even Mesozoic, than the diatom fossil record appears to suggest (e.g. Conley et al. 2017).

13.4.3 Physical Processes

Settling from suspension provides the primary physical process of transporting pelagic sediment from the upper ocean to the seafloor. In areas of extremely low-energy conditions at the seafloor (i.e. far from the influence of bottom-water currents), this produces regularly parallel-bedded deposits which drape seafloor topography. Sedimentation rates in such settings are thus a simple function of sediment supply, and, if applicable, dissolution in the water column. In the case of biogenic sediments, this means sedimentation rates are limited by rates of biological production. Beds in unconsolidated sediments (i.e. oozes) in such settings are recognisable due to subtle compositional differences caused by changes in sediment production and preservation and are likely to have gradational boundaries. If the bottom-waters are well-oxygenated then the sediments are likely to be structureless due to extensive bioturbation (see Section 13.4.4). In settings with low oxygen (and hence low bioturbation) and regularly varying conditions of sediment production (often seasonal productivity blooms, such as in the Gulf of California and Cariaco Basin) laminations reflecting changing fluxes of organic matter, silica or carbonate deposition may be preserved.

Not all pelagic sediments, however, are the simple product of suspension sedimentation. Both composition and bedding features can be modified at the seafloor through the action of gravity and bottom-water currents. Due to high primary water contents of biogenic ooze (porosity typically >80%), deposition onto even gently inclined slopes of less than 2 to 3° can lead to failure and mass-wasting. Mass-wasting of slowly accumulating pelagic deposits may be assisted by other triggering factors such as local seismicity, oceanographic currents and/or the escape of gas and fluids (e.g. Rodriguez et al. 2012). Downslope transport occurs through the same mechanisms found in other settings and lithologies – namely slumping, sliding and mass flow of the material (including turbidity and debris flows: Chapter 12). Modern examples of mass-wasting can be found around many oceanic highs in the form of slump scars, glide blocks, gully networks and undulating seafloors. Examples from the rock record may exhibit slump folds from core to outcrop scale (Figure 13.23) or classic turbidite sedimentary structures (e.g. graded beds, sole marks and current lamination) and can be inferred from seismic profiles using the same criteria and terminology as applied to continental margins.

Bottom-water currents are generated by thermohaline circulation, particularly along continental margins, and by smaller-scale, local physical oceanographic processes, such as internal waves caused, for example, by the movement of tidal currents over submarine ridges and seamounts (Simmons 2008; Mitchell et al. 2015). Experimental results, corrected to represent deep-sea conditions, suggest that calcareous biogenic sediments can be eroded by current velocities on the order to 15–35 cm s^{-1} – depending on the degree of consolidation and water content (Southard et al. 1971). Although mean current velocities in much of the ocean are typically less than these critical erosion thresholds, some areas, particularly where internal tides are intense or significant thermohaline boundary currents exist, can experience higher current velocities (reviewed in Shanmugam 2021). When local flow rates exceed the critical threshold for erosion, pelagic sediment will be eroded and transported. Current action is recognised by features such as hiatuses, winnowed planktonic-foraminiferal-rich sands and unidirectional ripples and dunes (Figure 13.24). Where deep-sea thermohaline currents are at their strongest a turbid layer

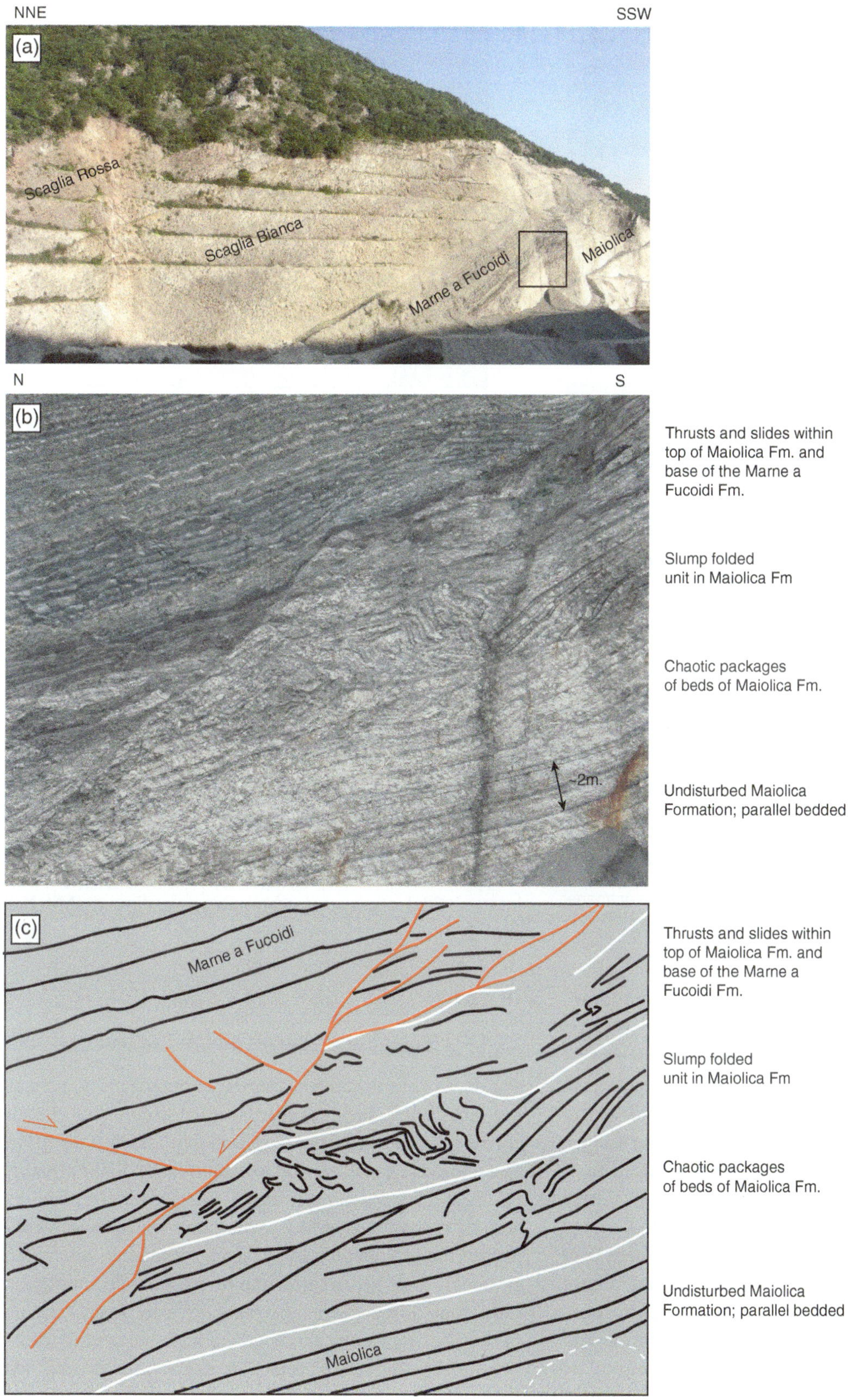

Figure 13.23 Slumping (due to syn-sedimentary faulting or to deposition on a slope) can occur within parallel-bedded pelagic sediments. (a) Cretaceous pelagic stratigraphy of Vispi Quarry near Gubbio (Italy) looking to the south-east, showing the generally parallel-bedded nature of the deposits. Formations labelled. Black rectangle indicates area of the outcrop shown in b. (b, c): a close-up of part of the Vispi Quarry (view approximately eastwards) with an interpretative sketch in c of synsedimentary deformation, guided by Menichetti (2016). Photo by S. Robinson (2010).

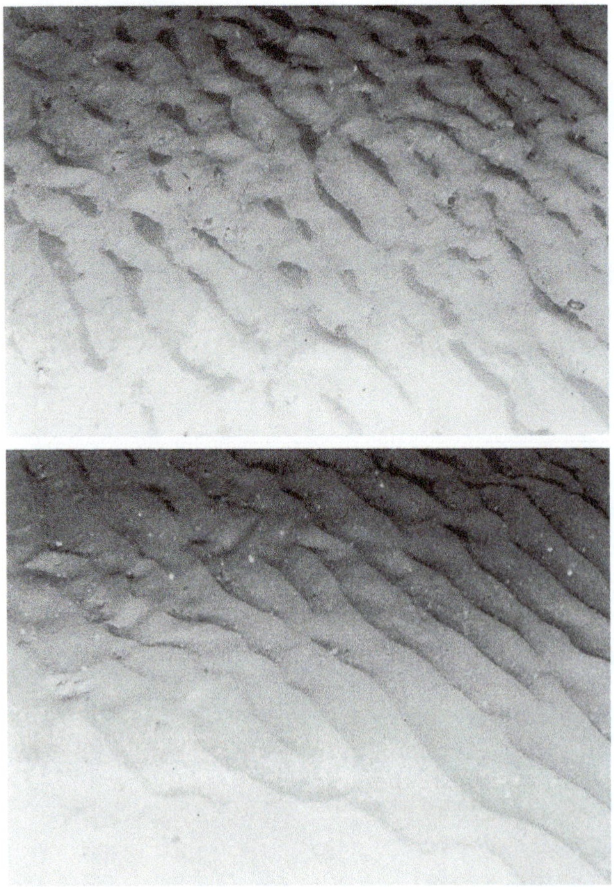

Figure 13.24 Images of deep-sea ripples in (probably calcareous) ooze on the mid-ocean ridge in the South Pacific at a water depth of 3347 m, from Heezen and Hollister (1971). Suspended particles can be seen in the lower image. Images of this type typically imaged an area of 2–10 m^2 with an average of about 6 m^2. GeoMapApp / www.geomapapp.org / CC BY 4.0. Location of image: 56°53′S 135°04′W.

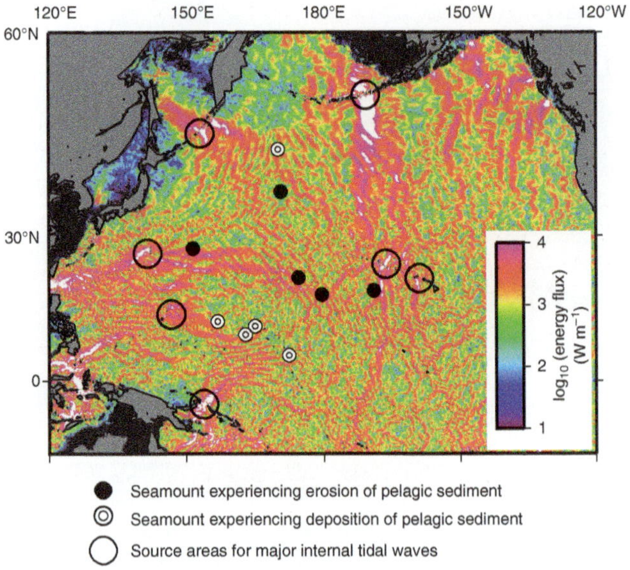

Figure 13.25 Map of the modelled internal tidal wave field in the Pacific Ocean (from Simmons 2008) showing selected seamount sites that are either currently experiencing deposition (white circles) or erosion (filled circles) of pelagic sediments. Large open circles are areas where major internal tidal waves are generated. The areas characterised by the highest energy flux probably experience the largest amplitude tidal waves and the greatest current velocities. Adapted from Mitchell et al. (2015).

of re-suspended fine-grained material (a 'nepheloid layer') can develop. As these currents expand and decelerate after flowing past topography on the seafloor, they can deposit their sedimentary load into thick drifts (contourites), which can have relatively high sedimentation rates (20 to 100 mm a^{-1}). Although in many continental margin settings contourites comprise a mixture of sediment sourced from (hemi-) pelagic and turbiditic sources (see Chapter 12), drifts can also occur in the abyssal ocean, far from land, where pelagic sediments are the only source, such as the Nazca Drift system in the Pacific Ocean which occurs at water depths of ~2 to 5 km and is dominated by calcareous ooze (Calvès et al. 2022). Pelagic sediments deposited on top of seamounts and plateaus are not free from physical erosion and transport by currents, either. Internal tidal waves in the Pacific Ocean, in particular, generate specific areas of high energy flux (Figure 13.25). Seamounts within these areas of high energy do not appear to be accumulating pelagic sediment, due to removal at the seafloor, while seamounts in areas of lower internal tidal energy are accumulating sediment (Mitchell et al. 2015).

13.4.4 Biological Processes on the Seafloor

The deep ocean is home to a variety of organisms that create trace fossils in pelagic sediments and in so doing affect their chemistry and diagenesis (e.g. Uchman and Wetzel 2011). The relatively small size and low abundance of bioturbators in these deep ocean settings is strongly controlled by organic carbon availability and bottom-water oxygen concentrations. Nonetheless, pelagic sediments are often extremely well-bioturbated due to the slow accumulation rates that allow deep burrowers to extensively explore the sub-seafloor environment. The high burrowing intensity means that the spectacular surface traces seen on

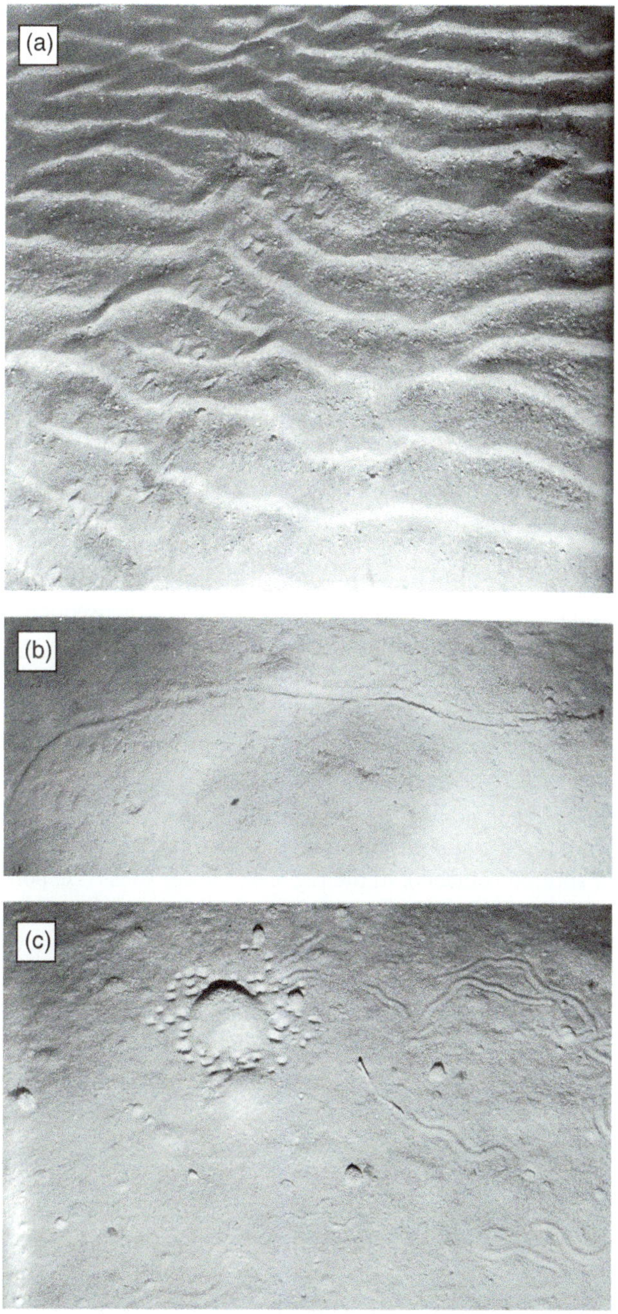

Figure 13.26 Bioturbation of the modern seafloor from Heezen and Hollister (1971). (a) Decapod crustacean (shrimp) footprints on a sandy rippled bottom, Blake Plateau, western North Atlantic. Water depth of 789 m (30°59′N 79°00′W); (b) ploughed seafloor from a water depth of 4165 m off Central Chile (34°48′S 74°54′W); (c) cones, rings and plough marks (probably made by echinoids) on the mid-ocean ridge, SW Pacific Ocean at a water depth of 3464 m (48°59′S 120°00′W). Images of this type typically imaged an area of 2–10 m^2 with an average of about 6 m^2. GeoMapApp / Marine Geoscience Data System / www.geomapapp.org.

deep-sea photographs (Figure 13.26) and the shallowest burrows (<5–8 cm below the seafloor) are rarely preserved. In modern sediments, the uppermost layer (5–8 cm below the seafloor) is extremely well-mixed by surface trails and burrows (the 'mixed layer'; Uchman and Wetzel 2011). This mixing is evidenced from surface sediments by the co-occurrence of 'dead' and 'living' benthic foraminifera (identified by using 'Rose Bengal' a chemical stain that detects living foraminiferal protoplasm) and by a range of radiocarbon ages from foraminifera in core-top samples (e.g. Peng and Broecker 1984; Barker et al. 2007). Below the mixed layer, a 'transition zone' occurs with partial sediment mixing by infaunal organisms. This transition zone can extend 20–35 cm below the seafloor, below which depth there are no active burrows. The preserved fabric in pelagic sediments is normally overprinted by this transition zone and, as a consequence, deep-sea pelagic ichnofacies contain the deep-burrowing elements of both the *Zoophycos* and *Nereites* ichnofacies (Chapter 3; Figures 3.18 and 3.19). Burrows such as *Skolithos, Planolites, Chondrites, Teichichnus* and *Zoophycos* have been observed in box cores of modern carbonate ooze (e.g. Ekdale and Berger 1978). Lower diversity assemblages dominated by *Planolites* with few *Chondrites, Teichichnus* and *Zoophycos* have been recorded in pelagic clays below the CCD (Ekdale et al. 1984). Although oxygen and organic carbon content may play a role in the different assemblages associated with carbonate ooze and pelagic clays, the strength of the sediment may also play a role. Increasing clay content reduces the strength of the sediment preventing larger burrows from remaining open and, consequently, *Skolithos* is not present in clay-rich sediments with <60% carbonate (Ekdale and Berger 1978; Ekdale et al. 1984). It is worth noting that the record of bioturbation in pelagic sediments and sedimentary rocks may be biased by coring due to the lower probability of recovering vertical burrows (such as *Skolithos* and *Teichichnus*) in narrow-diameter cores compared to the higher chances of intersecting horizontal structures. Bioturbation provides conduits for water and oxygen to enter the sediment pile, thereby 'irrigating' the environment allowing better-oxygenated waters to penetrate the sediment and promoting the re-mineralization of organic matter (e.g. Aller 2014). However, the relationships between macrofauna, redox and organic carbon re-mineralisation are complex. For example, the production of faecal pellets in burrows can create microchemical redox gradients and mm-scale areas of more reduced conditions within otherwise well-oxygenated sediments.

13.4.5 Chemical Processes

As discussed in Section 13.2.5, organic and inorganic materials produced by plankton are subject to respiration and dissolution in the water column. As a consequence, the rock record of pelagic sediment production is an inherently biased record of pelagic biology and of the potential sediment that is generated in the upper ocean. The degree of bias is a function of the preservational potential of particles in the water column and through both early and late diagenesis.

During early diagenesis, any labile organic matter reaching the seafloor is subject to aerobic and anaerobic respiration, leading to yet more recycling of organic carbon and release of CO_2 and CH_4 into solution (e.g. Aller 2014). Within well-oxygenated sediments below the seafloor aerobic bacteria use oxygen to respire organic carbon, thereby consuming oxygen. At some depth, determined by sedimentation rate, organic carbon flux and bottom-water oxygen concentrations, pore-water oxygen concentrations fall to zero and other, less energetically favourable, bacterially mediated pathways for organic carbon respiration take over. As the key reductants are consumed with increasing depth in the sediment profile, different reactions become dominant in turn, each less energetically favourable than the one before, as summarised in the context of the 'redox tower' (Figure 9.17, Section 9.3.10). In settings with abundant organic carbon (e.g. restricted basins, high-productivity settings), these reactions are important for the genesis of early diagenetic mineral phases, in particular authigenic carbonate cements and the formation of sulphides, like pyrite. Rates of organic carbon oxidation reduce dramatically with each step of the redox tower but nonetheless, these reactions (and the availability of reductants in porewaters) are important for determining the amount of organic carbon entering the rock record in pelagic settings.

Pelagic carbonate sediments begin to experience changes within the water column, even before deposition, due to increasing undersaturation at depth (particularly below the lysocline) that leads, for example, to corrosion and fragmentation of planktonic foraminiferal tests (e.g. Berger 1971; Schiebel and Hemleben 2017). At or just below the seafloor, planktonic foraminifera are susceptible to micron-scale recrystallisation and formation of infilling diagenetic calcite from pore-waters, particularly in carbonate-rich sediments (see Pearson 2012). These processes normally retain the overall morphology of the original test but can lead to significant changes to the microscopic test wall structure and geochemistry (Figure 13.27). In particular, the stable oxygen-isotopes ($\delta^{18}O$) of planktonic foraminifera in pelagic sediments can be offset towards bottom-water conditions, compared to co-eval well-preserved specimens with 'glassy' preservation (e.g. Sexton et al. 2006; Pearson 2012; Figure 13.27e). With increasing burial compaction, soft carbonate oozes begin to lose volume and porosity and, as a consequence, density increases (Figure 13.28). Deep-sea drilling has found that at depths of around 200 m or more, carbonate oozes transition into 'chalk' (Box 13.2; Schlanger and Douglas 1974). (Note an important distinction must be made between the use of 'chalk' as a lithological description and 'Chalk' in the context of Cretaceous lithostratigraphy, see also Section 13.5.2.). Further burial of chalk leads to yet more loss of volume and porosity and, with cementation, leads to a transition to hard pelagic limestones. The cements in pelagic limestones largely come from pressure-solution along grain boundaries, a process by which grain dissolution occurs as a consequence of compressional stress at the contact point between two grains. Dissolved ions are transported laterally away from the point of contact in a very thin film of fluid through solution transfer or diffusion down a chemical gradient to an area of lower stress and sometimes lower solubility. Over time, the pressure can lead to the partial dissolution of one or both grains (visible in thin sections as concavo-convex or sutured contacts). In the regions of lower stress (i.e. pore spaces between grains), cements can be precipitated. Under burial, dissolution in carbonates can occur along extensive horizons throughout the sediment and pelagic limestones commonly develop pressure dissolution seams and stylolites observable at all scales from thin section to outcrop (Figure 13.29).

Silica occurs in pelagic sediments in one of three polymorphs: opal-A (a disordered form of opaline silica containing up to 10–14 weight % water), opal-CT (a disordered layered mixture of cristobalite and tridymite, with significant water content) or quartz (highly ordered silica tetrahedra with trace amounts of water). Upon deposition, siliceous oozes are dominated by opal-A but with increasing burial temperature, silica solubility increases leading to dissolution of opal-A and an increase in pore-water saturation with respect to opal-CT (Figure 13.30; also see review in Hesse and Schacht 2011). Opal-CT can replace biogenic structures (e.g. undissolved radiolaria or diatoms) or precipitate a variety of cement morphologies (including 'lepispheres': rosette-like spherical aggregates of bladed opal-CT). With further burial and increased temperatures, opal-CT itself begins to dissolve and pore waters become

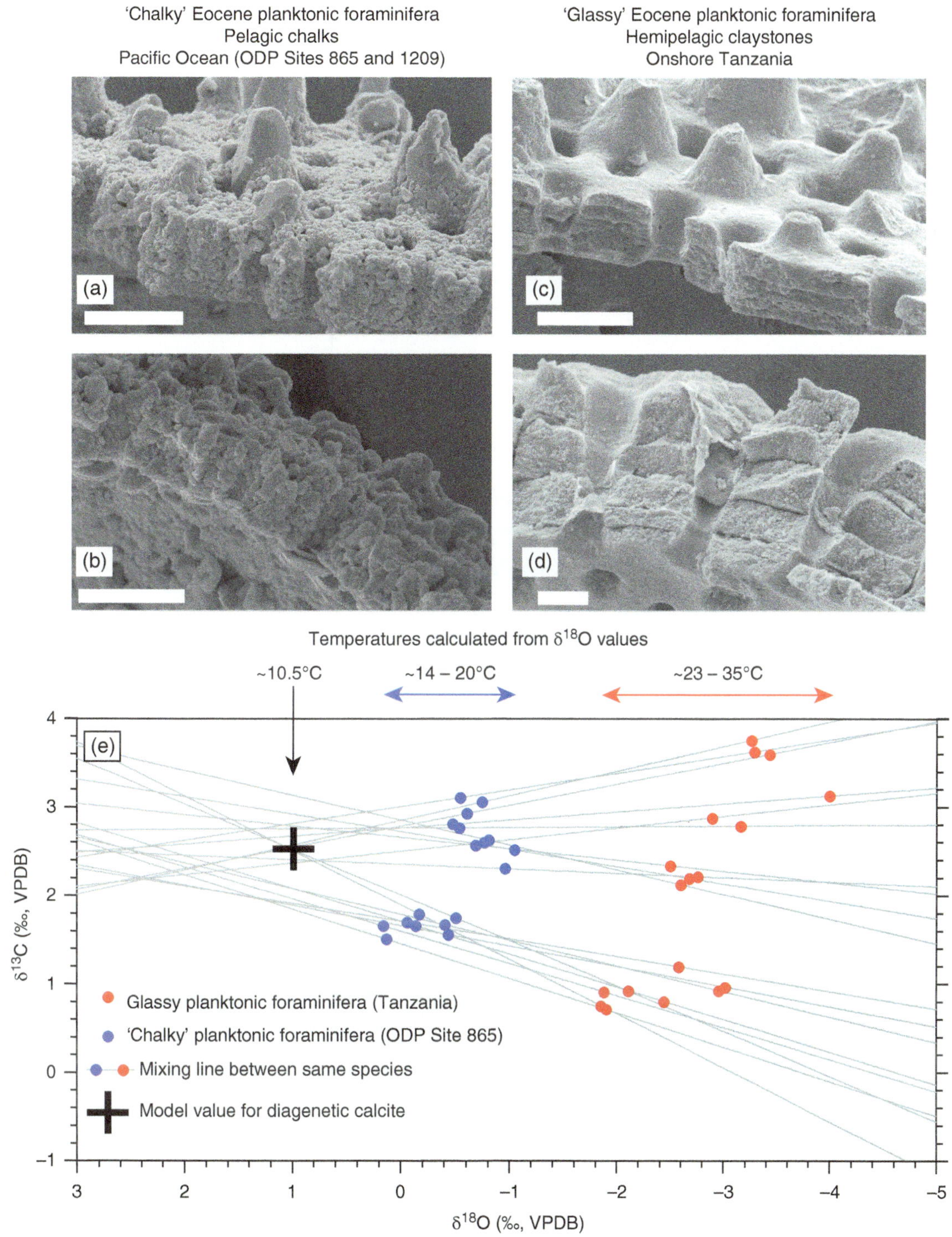

Figure 13.27 Comparison of Eocene planktonic foraminifera that display relatively poor ('chalky') and relatively good ('glassy') preservation using scanning electron microscope images (a–d) and stable isotope data (e). All of these foraminifera are of the same age and were deposited at low palaeolatitude. Images (a) and (b) show planktonic foraminifera from pelagic chalk recovered at ODP Sites 865 and 1209 in the Pacific Ocean. Similar age specimens shown in images (c) and (d) are from hemipelagic claystones, recovered from Tanzania. Panel (e) shows the effect of preservation on the stable-isotope geochemistry of samples from these different host lithologies. Scale bar for SEM images is 10 μm. Data and images from Pearson et al. (2007) and Pearson (2012) and supplied courtesy of Paul Pearson.

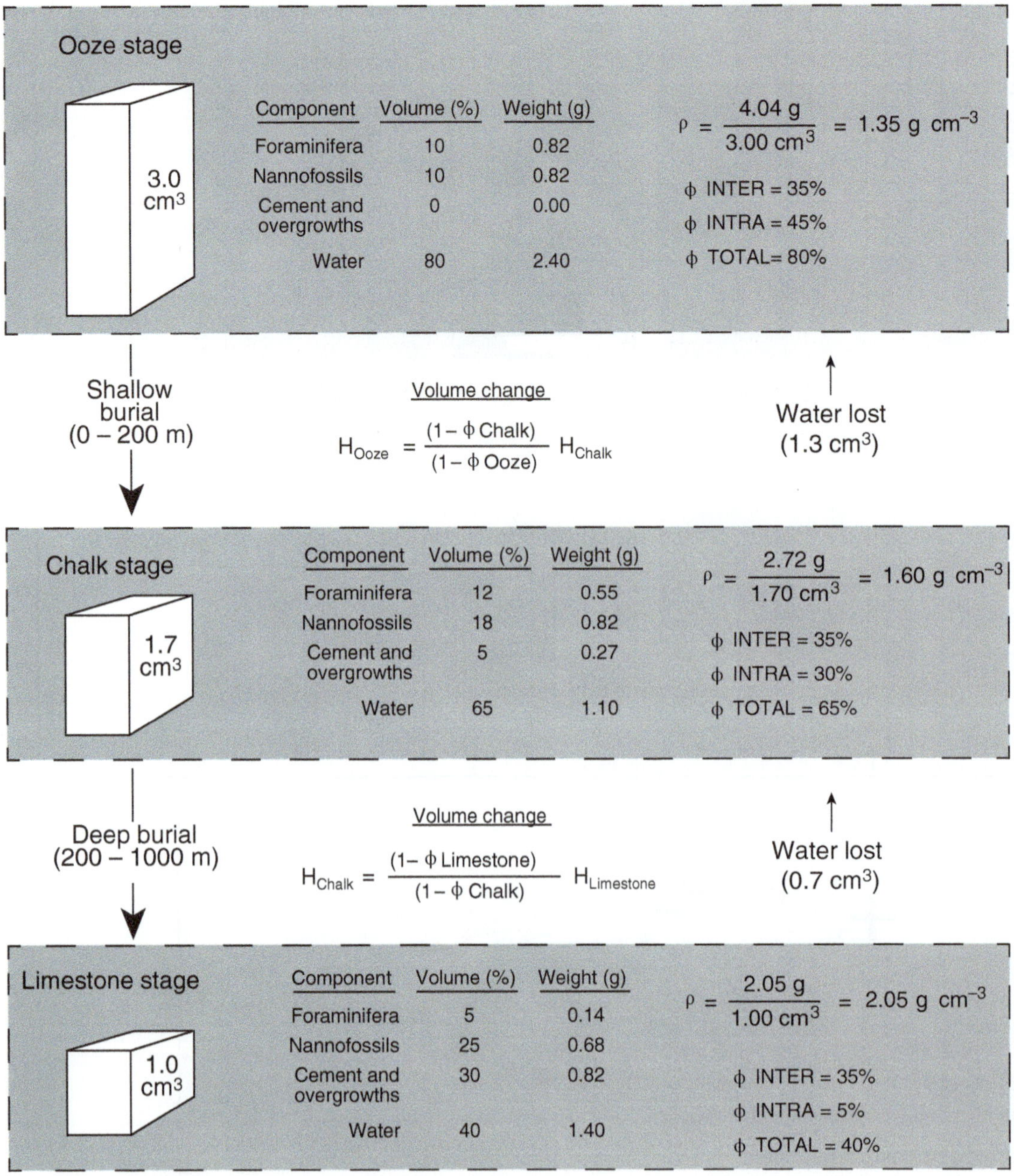

Figure 13.28 A model of diagenesis for carbonate ooze during burial, illustrating the volumetric, component and cement changes that occur in the transition from soft ooze through chalk to hard limestone. ϕINTER = porosity between particles; ϕINTRA = porosity within particles (e.g. within chambers of foraminifera). Adapted from Schlanger and Douglas (1974).

saturated with respect to quartz, leading to the precipitation of microquartz, with coarser forms of megaquartz and chalcedony occurring within pore spaces, such as the interior of radiolarian tests (Figure 13.31). The diatomaceous oozes of the Miocene Monterey Formation of California provide clear examples of these diagenetic transformations in a setting where the same lithostratigraphic unit has been buried to varying depths (Behl 1999). Estimates of burial temperature based upon oxygen-isotope analysis of the silica polymorphs suggest that, in this setting, the opal-A to opal-CT and opal-CT to quartz transitions occur at temperatures of around 40–50°C (~0.5 to 2 km depth) and 65–80°C (~1.5 to 3 km depth), respectively (Figure 13.32; Keller and Isaacs 1985; Behl 1999). For very pure diatomites (<10% non-diatomaceous material), relatively low burial temperatures are needed to convert opal-A to opal-CT or

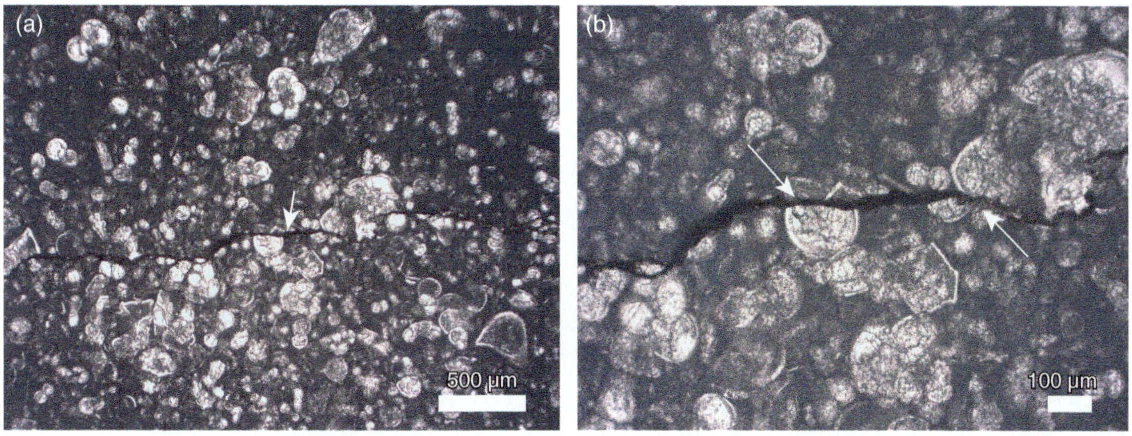

Figure 13.29 (a) A planktonic foraminiferal-rich pelagic limestone with dissolution seams (stylolites, arrowed) rich in, most likely, insoluble oxides, clays and organic matter. Such surfaces develop as the consequence of chemical compaction during burial diagenesis and the dissolution and loss of carbonate. (b) A close-up of (a), showing the stylolite cross cutting foraminiferal tests both above and below the surface (arrowed). Both images in plane polarised light. Sample is from Cretaceous pelagic limestones located in the Indo-Burma Range (Indo-Myanmar Range) of western Myanmar (Burma). Sample collected by Tin Tin Naing and described in Naing et al. (2023). Photo by S. Robinson.

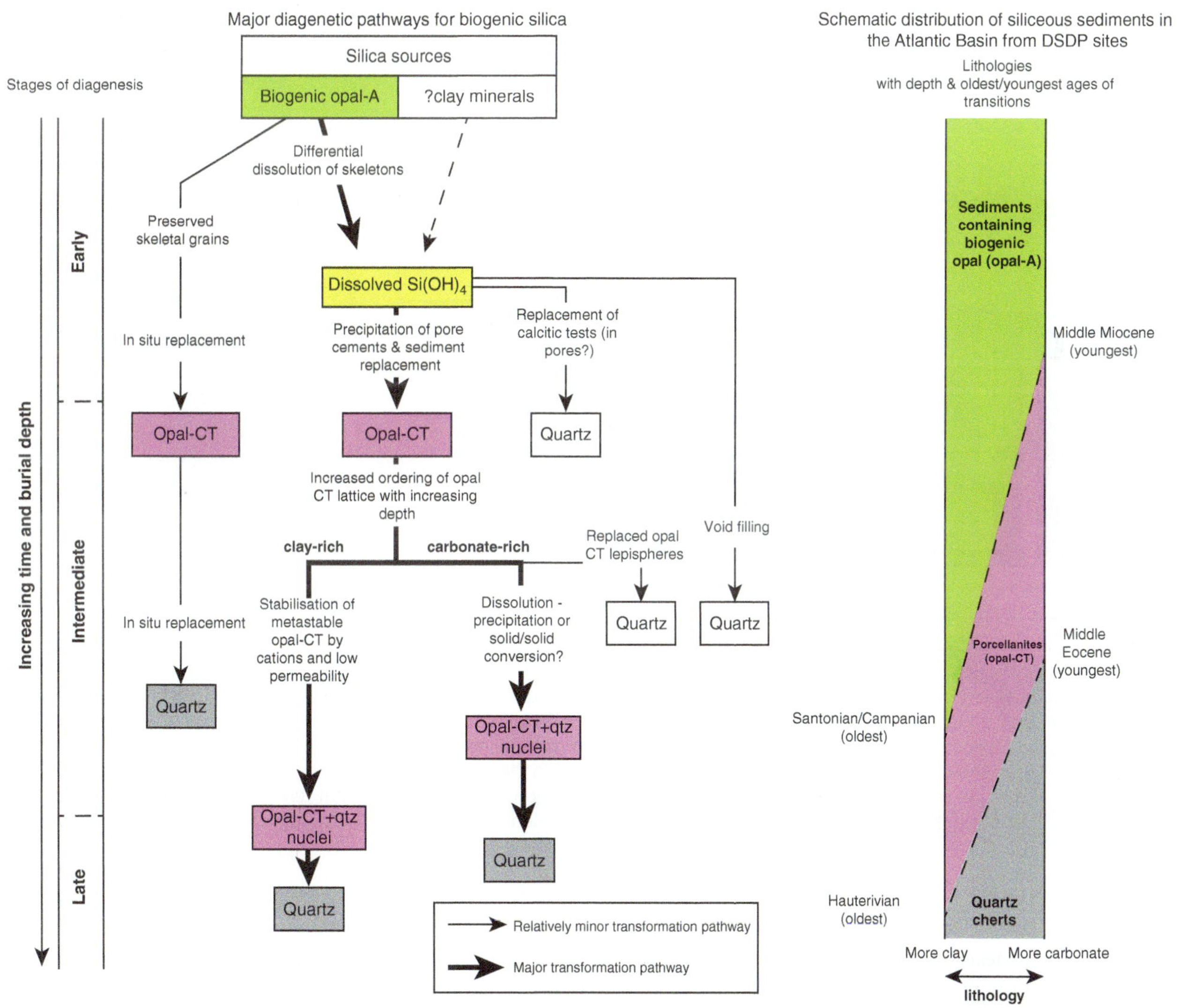

Figure 13.30 Flow chart to illustrate the possible diagenetic pathways for biogenic silica in pelagic sediments with increasing burial depth, age and different associated sediments (e.g. carbonates and clays). Also shown is the relative distribution of siliceous lithologies with time based on North Atlantic DSDP sites. Adapted from Riech and von Rad (1979).

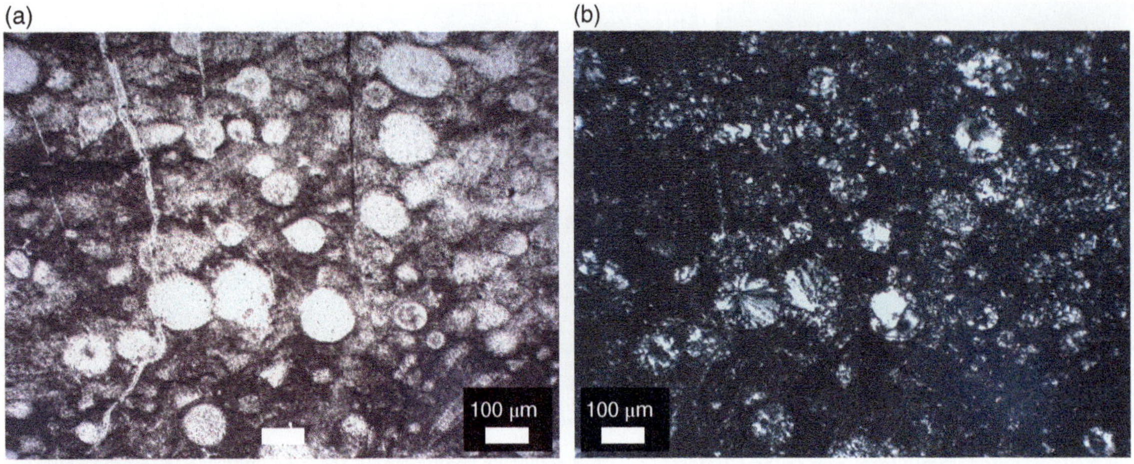

Figure 13.31 An example of a radiolarian-rich chert in thin section in plane-polarised (a) and cross-polarised light (b). In (a), the colourless grains are radiolaria. In (b), radiolaria are infilled by fibrous chalcedonic quartz or megaquartz crystals. The matrix is a mixture of microcrystalline quartz and iron-oxides that give the rock a red to brown colour in hand specimen. Sample is from Jurassic radiolarian cherts associated with the Kaleymo Ophiolites in Myanmar (Burma) described in Naing et al. (2023). Photo by S. Robinson.

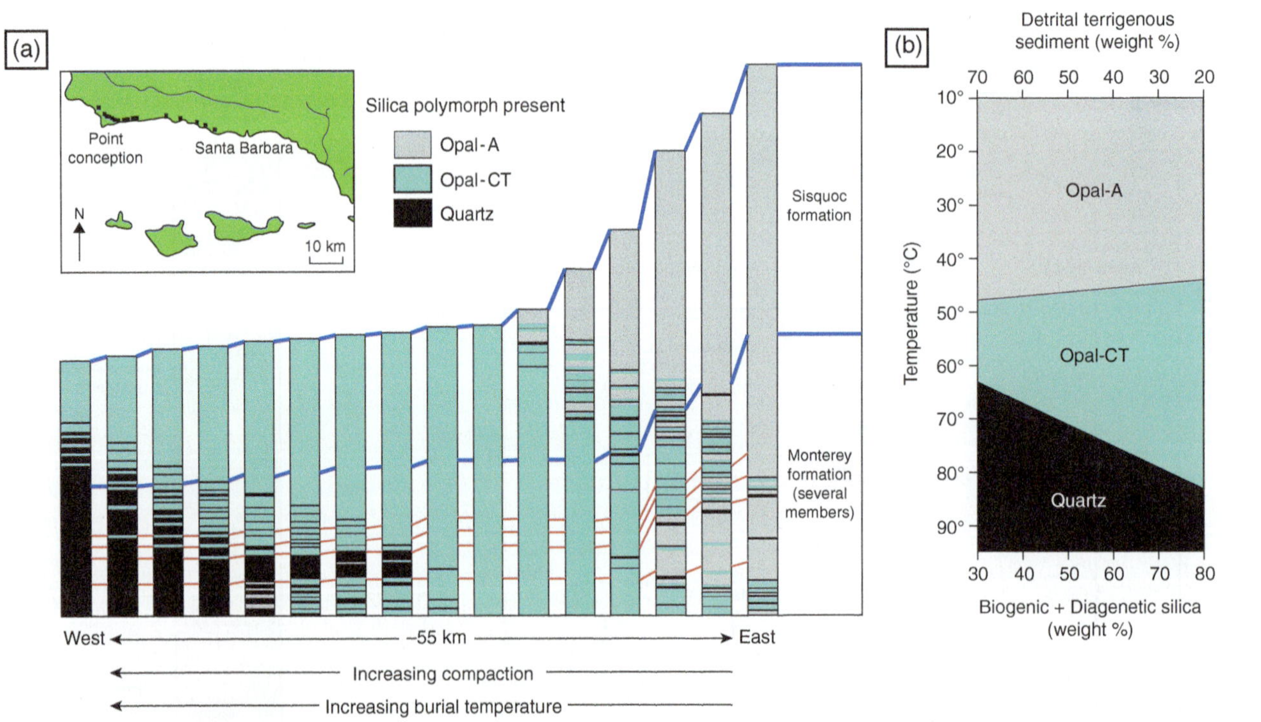

Figure 13.32 (a) Cross section illustrating changes in dominant silica polymorphs in the Monterey and Sisquoc Formations along a west to east transect of the coastline between Point Conception and Santa Barbara (southern California, USA). Cross-cutting relationships between silica mineralogy and lithostratigraphic boundaries are evident both at formation and member level. Blue lines show correlation of formations and red lines show correlation of members (not named here) within the Monterey Formation. Modified from Behl (1999). (b) The temperatures of silica phase changes and the importance of the relative proportion of silica to detrital terrigenous sediment in the initial sediment based on data from the Monterey Formation at Point Conception. Modified from Keller and Isaacs (1985) and Behl (1999).

opal-CT to quartz (2–33°C and 36–76°C, respectively; summarised in Behl 1999). The relationship between temperature and silica diagenesis is not, therefore, simple: sediment composition and time play significant roles (Figure 13.30). For example, the opal-CT to quartz transition can be enhanced by increased alkalinity in porewaters (i.e. in carbonate-rich sediments) or retarded in clay-rich sediments. The diagenetic changes among silica polymorphs are reflected in the types of silica-rich pelagic lithologies found in the rock-record. The grains of siliceous oozes and mudstones dominated by opal-A or opal-CT will generally retain a high proportion of recognisable grains recording the biogenic source (e.g. diatoms, radiolaria). Upon transition to quartz the lithology is (usually) termed chert and the original siliceous grains become less common as most have been dissolved. The primary source of the silica can sometimes be inferred from the presence of replaced siliceous microfossils, which can often be seen in thin sections of deep-sea cherts (Figure 13.31). The porosity change associated with the opal-A to opal-C/T transition is sufficient to cause a reflection on conventional marine seismic data, which, being thermally influenced, often crosses lithostratigraphic boundaries (e.g. Davies and Cartwright 2002).

In settings where both carbonate and silica have been deposited, whether simultaneously or in alternation, interbedded or mixed lithologies are found. As the burial transformations that carbonate and silica experience do not occur at the same depths, it is possible to find sequences of, for example, soft chalks with hard cherts, which makes core recovery very difficult (e.g. Shipboard Scientific Party 2002). With sufficient burial, limestone-chert sequences can be generated although these are rarely sampled by ocean drilling (due to depth). However, such sequences are commonly seen in collisional tectonic settings where they have been uplifted and exposed (e.g. the Cretaceous Calera Limestone of California; Wachs and Hein 1975; Robinson et al. 2008; Figure 13.33).

Sequences of repetitively alternating lithologies or variations in sediment composition in relatively soft, non-diagenetically altered pelagic sediments typically reflect primary variations in depositional processes (e.g. climatic or oceanographic conditions; see Section 13.5.1). In contrast, cyclic alternations in pelagic and hemipelagic sedimentary rocks that have undergone some degree of diagenesis (e.g. chert and limestone, claystone and limestone, marls and black mudstones) may reflect the operation of both depositional and diagenetic processes. One extreme end-member interpretation of such alternations is that each lithology (or lithofacies) represents purely the *in situ* diagenesis of the original deposits; a sequence of alternating thin chert and limestone beds would thus represent periods of silica production alternating with periods of carbonate production. The opposing view is that the original components were well-mixed upon deposition and that the cyclic bedding is the result of chemical segregation during diagenesis, termed 'diagenetic un-mixing' (e.g. Hallam 1986). In this view, chemical components are mobilised vertically during diagenesis to reprecipitate in homogeneous beds. For example, it could be imagined that biogenic opal, originally uniformly distributed within a calcareous ooze, might dissolve during diagenesis and precipitate out as opal-CT or quartz along a particular horizon in which the pore-water conditions were favourable (e.g. Murray et al. 1992). Continued migration of silica to this horizon would deplete the adjacent beds and lead to a new 'bed' of silica, most likely as a chert. The more likely model for many cyclic pelagic and hemipelagic sedimentary rocks is that some degree of diagenetic un-mixing occurs (as in concretionary growth) and enhances the primary differences in sediment composition that were the consequence of cyclic variations in deposition (e.g. Research on Cretaceous Cycles (R.O.C.C.) Group 1986; Trabucho-Alexandre et al. 2011). The presence of different syn-depositional sedimentary structures and/or styles of bioturbation in alternating lithologies with different mineralogies provides good evidence for primary sedimentological variations (e.g. Frank

Figure 13.33 Views of the Early Cretaceous Calera Limestone exposed in the Permanente Quarry, California. (a) Part of the stratigraphy of the Barremian–Aptian part of the Calera Limestone showing the alternations of thin bedded limestones (white and light grey) with cherts (generally dark brown, dark grey and black). The prominent thicker dark band in the middle of the outcrop is likely the local expression of Oceanic Anoxic Event 1a and has slightly elevated organic carbon contents above background (Robinson et al. 2008). (b) Close-up of alternating limestone and chert beds, cross-cut by later veins related to the later tectonic history of these rocks. Photo by S. Robinson.

et al. 1999; Gambacorta et al. 2014). And so, despite diagenesis, such sequences can still inform our understanding of past sediment supply and depositional processes.

13.5 Facies Associations and Successions of Pelagic Sediments and Sedimentary Rocks from Cores and Outcrops

An understanding of modern pelagic sedimentary processes and products, and the possible diagenetic pathways, provides a means with which to interpret facies associations and successions from the Late Mesozoic to Recent in terms of biogenic sediment production, dilution, preservation, lateral transport, diagenesis and/or tectonics. However, due to the changing biogenic sources of sediment and the destruction of ocean crust a strictly uniformitarian approach becomes less valid with increasing geological age. The division between 'analogous' and 'non-analogous' conditions is not sharp but occurs over the course of the Mesozoic (in particular the Jurassic and Cretaceous: Section 13.4). In the Palaeozoic, the record attests to styles of pelagic sedimentation for which no modern analogue exists, therefore requiring considered extrapolation away from direct observations of oceanography, biology and geochemical cycling in the modern ocean.

13.5.1 Facies of the Late Mesozoic and Cenozoic – Analogous to the Modern

13.5.1.1 Late Pleistocene Pelagic Facies in the Atlantic and Pacific

The modern distribution of pelagic sediments (Figure 13.1b) provides a geological snapshot of the balances of production, preservation and dilution that control lithologies in the deep-ocean. The Quaternary record of these sediments reveals how rapidly (and dramatically) these balances are altered through regional and global climatic change. During the Quaternary, global climate fluctuated between glacial and inter-glacial states in response to changing orbital parameters (Box 13.4), marked most significantly by the growth and collapse of large Northern Hemisphere continental ice sheets. These orbitally driven changes in the Earth system were felt in the oceans through changes in surface temperatures, circulation patterns and biological productivity, which impacted the characteristics of pelagic sedimentation through time via a set of complex Earth system feedbacks (Figure 13.34).

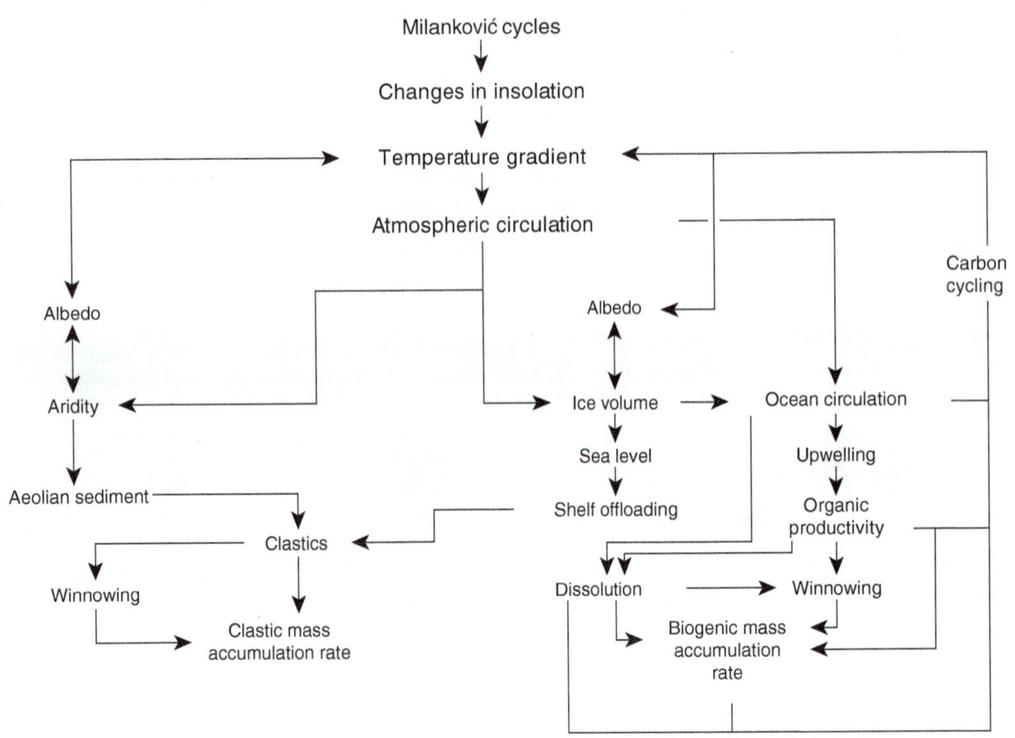

Figure 13.34 Simplified flow chart demonstrating some of the difficulties in connecting an observed change in pelagic sedimentation (e.g. changes in clastic or biogenic mass accumulation) to a specific Milanković -driven set of processes (Dean and Gardner 1986 / John Wiley & Sons).

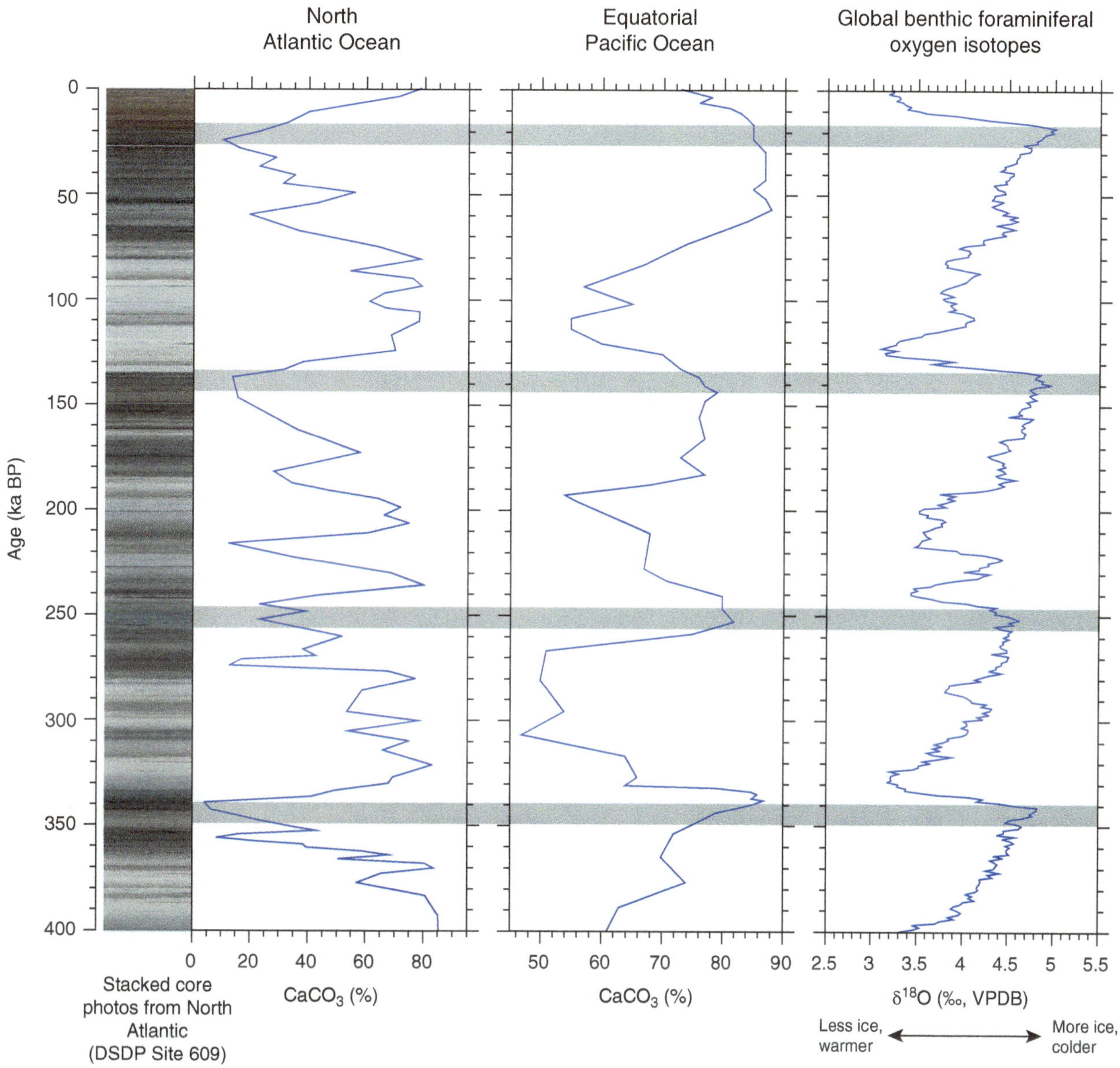

Figure 13.35 Late Quaternary cycles of carbonate content in pelagic sediments from the North Atlantic and equatorial Pacific Oceans showing the anti-correlation between the two basins. The composite global benthic foraminiferal oxygen-isotope curve from Lisiecki and Raymo (2005) is also shown. Peak glacial conditions (grey bars) coincide with high carbonate contents in the Pacific and low carbonate contents in the Atlantic. Carbonate data originally from Ruddiman et al. (1989) and LaMontagne et al. (1996), plotted using age models from Sexton and Barker (2012). Data kindly supplied by P. Sexton. Note that the Pacific Ocean $CaCO_3$ record has a lower sampling resolution and lower absolute variability than the North Atlantic record. Core photos for the North Atlantic site from Ruddiman et al. (1987) have been scaled to age using the average sedimentation rate over the last 400 ka. Light colours represent carbonate-rich oozes; dark colours and carbonate-poor and relatively rich in clays. Photo by Ocean Drilling Program.

Many pelagic sedimentary archives of the Late Pleistocene from the Atlantic and Pacific Oceans show dramatic cycles in carbonate content (Figure 13.35) that, in the equatorial Pacific at least, can be correlated over thousands of kms. Over the last ~700 kyrs these carbonate cycles are broadly aligned with eccentricity-paced glacial–interglacial cycles (as defined by oxygen-isotope stratigraphy, Lisiecki and Raymo 2005). The origin of these carbonate cycles, and their apparent anti-correlation between the Atlantic and Pacific over the Late Pleistocene (Sexton and Barker 2012), has caused significant debate (e.g. Berger 1992; Anderson et al. 2008). Were elevated carbonate contents caused by increased carbonate preservation alone or did carbonate production and preservation both increase? What role, if any, did dilution from changing dust or silica fluxes play? What is the significance of the anti-correlation for understanding the role of ocean circulation and carbon storage in glacial–interglacial climate changes? In the equatorial Pacific Ocean (and Indian and Southern Oceans too), carbonate minima occur during interglacials ('Pacific-type cycles'), while in the Atlantic the opposite is true – interglacials are times of carbonate maxima ('Atlantic-type cycles'). For cycles in the equatorial Pacific Ocean, some workers argued on the basis of sedimentation rate estimates that periods of increased carbonate deposition were also characterised by increased fluxes of organic carbon and opal, leading to the suggestion that the carbonate cycles had a productivity origin (e.g. Arrhenius 1988; Lyle et al. 1988; Archer 1991). However, more refined (and accurate) estimates of accumulation rates of biogenic materials (based on a geochemical proxy for sedimentation accumulation called 'the 230-Th normalised method') and a proxy directly related to primary production (biogenic barite accumulation) suggests little change in productivity between glacials and interglacials (see Anderson et al. 2008). In contrast to the productivity model, a number of studies have used quantitative evidence for carbonate dissolution (such as measures of planktonic foraminiferal fragmentation, or the relative abundance of dissolution-resistant versus dissolution-prone species) to demonstrate that, in most cases, carbonate-poor sediments correspond to poor carbonate preservation (e.g. Berger 1973, 1992; Lalicata and Lea 2011). These observations strongly suggest that such cycles in the equatorial Pacific Ocean are determined by lower carbonate saturation states during interglacials driven by changes in the carbon cycle and strength of overturning circulation (i.e. implicating older more-corrosive bottom waters). In the Northwest Pacific, similar carbonate cycles have also been recorded and interpreted as dissolution cycles, albeit in lithologies with less carbonate on average than sites in the equatorial region (e.g. Hovan et al. 1991). The remaining sediment consists of opaline silica and continentally derived aeolian material. Analysis of the dust component reveals changes in aeolian dust flux related to more arid climatic conditions in eastern Asia (the major source of wind-blown dust) during glacial times (evidenced by the well-dated sequence of East Asian loess deposits). In the deep Equatorial and North Atlantic Ocean, glacials are associated with the minima in carbonate preservation in the deep-sea (e.g. Berger 1973), which has been generally ascribed to increased dissolution caused by a weakening (and shallowing) of North Atlantic Deep Water (e.g. Howard and Prell 1994), leading to the ingress of more corrosive Antarctic Bottom Water. Nonetheless, here too changes in productivity and dilution have been invoked as important factors in some settings. Dilution may be particularly important for locations that are more proximal to the continents.

Orbital forcing was not the only climatic driver of the pelagic sedimentary record during the Late Pleistocene. Millennial-scale climate changes during glacial periods have been reconstructed from ice core records from Greenland (e.g. Dansgaard et al. 1993), that attest to abrupt atmospheric warming events (up to 15°C), termed 'interstadials', followed by a more gradual return to cooler conditions ('stadials'). These climatic perturbations are known as Dansgaard-Oeschger (D-O) events (or cycles) but do not show a regular periodicity, nor a constant duration, lasting typically between 500 and 2500 years. D-O events have been recognised in the marine sedimentary record of the North Atlantic as transitory changes to warm-water planktonic foraminiferal assemblages during interstadials (e.g. Rasmussen et al. 2016). More significantly, D-O cycles are also implicated in the occurrence of widespread layers of ice-rafted debris (IRD) in North Atlantic sediment cores. The larger of these IRD layers are found every ~10,000 years or so during the last glacial and are known as 'Heinrich Events' (Bond et al. 1992, 1993). The thickness of individual Heinrich Event layers varies from 1 m to a few cms with increasing distance from the source. The detrital grains are used to determine the precise location of the source and, consequently, aspects of past ice sheet dynamics. At some sites, Heinrich layers can be associated with an increased abundance of cool-water planktonic foraminiferal assemblages (e.g. Bond et al. 1993). Although Heinrich Events correspond in time to the longest D-O cycles, the exact mechanisms connecting the two phenomena, and the ultimate driver of both, are still debated. One suggestion is surging behaviour of the Laurentide Ice Sheet. During these surge events, large quantities of ice are hypothesised to have been discharged into the North Atlantic. For each Heinrich event, it is estimated that $\sim 5 \times 10^{15}$ kg of IRD was deposited across an area of $\sim 5 \times 10^{6}$ km^2 (Cronin 2010) extending far away from the ice sheet (Figure 13.36).

13.5.1.2 Neogene Cyclic Sedimentation in the Eastern Mediterranean Basin

The effect of orbital forcing of climate can be seen on pelagic sedimentation in much smaller basins than the Atlantic and Pacific, such as the Mediterranean. Here the sedimentological signal and temporal patterns are different again. The small size of the Mediterranean means that surface water conditions and circulation are particularly sensitive to changes in hydrology on the surrounding land masses. In the modern, evaporation over the Eastern Mediterranean drives the formation of a density-driven

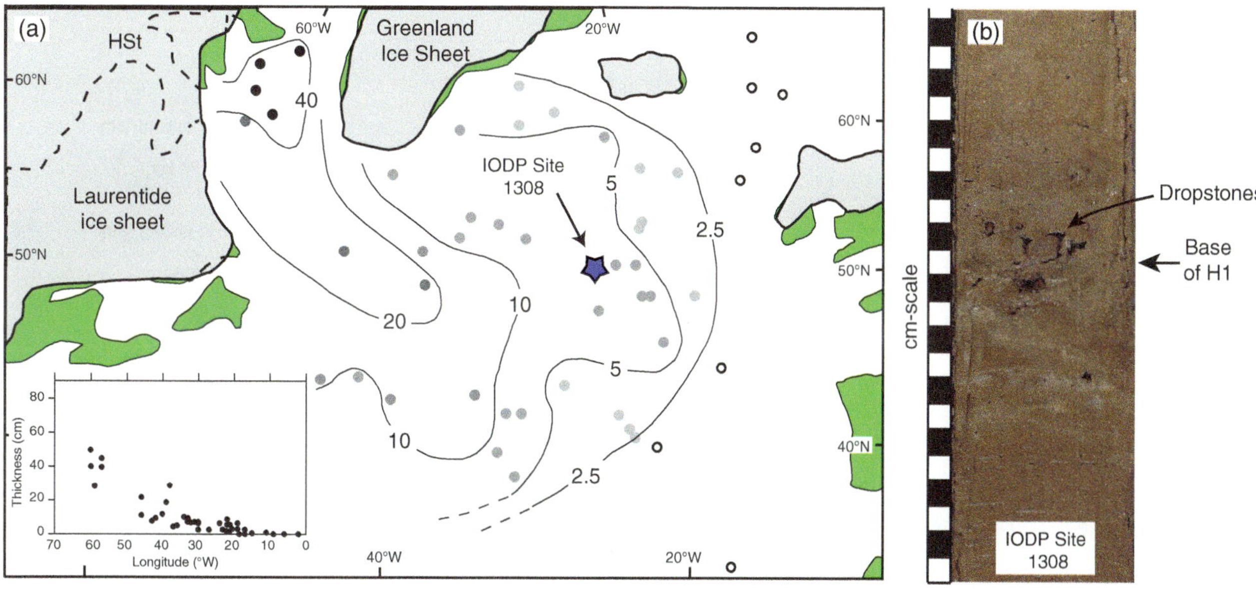

Figure 13.36 (a) Isopach map (in cm) of Heinrich Event 1 layer (H1; ~16–15 ka BP) in North Atlantic cores. The inset graph shows the relationship between longitude and thickness, clearly demonstrating that the thickest deposits lay closest to the Laurentide Ice Sheet. It has been shown that much of the terrigenous sediment carried by icebergs and deposited during Heinrich events was sourced in the Hudson Strait region (HSt). Dowdeswell et al. (1995) / GeoScienceWorld. (b) Close-up photograph of Heinrich event 1 from IODP Site 1308 (Hole U1308B Core 1H Section 1, Depth interval 79–99 cm). Depth of H1 based on Hodell et al. (2017) / John Wiley & Sons.

Box 13.4 Milanković Cycles and Orbital Forcing

The shape of Earth's orbit around the Sun (eccentricity), the tilt of the Earth's axis of rotation relative to the orbital plane (obliquity), the movement of that axis relative to the stars (axial precession) and the timing of the perihelion (apsidal precession) vary cyclically over periods of 10 to 100 s of kyrs (Figure B13.3). The behaviour of these so-called 'orbital cycles' and their impact on climate was first identified by a Serbian mathematician, Milutin Milanković (commonly anglicised to Milankovitch), who, in 1941, proposed that they were drivers of glacial–interglacial climate change (Milanković 1941). Since then, sedimentary signals of orbital variations in climate forcing have been recognised throughout much of the Phanerozoic geological record. Consequently, orbital cycles are now considered a fundamental driver of climate change over geological timescales, irrespective of the mean climate state and the existence, or not, of large polar ice caps at any given time. More detailed overviews of orbital climate change and cyclostratigraphy can be found in Cronin (2010), Weedon (2003), Hinnov and Hilgen (2012), Laskar (2020) and Berger and Yin (2021).

Eccentricity is the measure of how elliptical Earth's orbit is; higher values indicate a more elliptical orbit whereas zero indicates a circular shape. The gravitational interactions between the Earth and the large gas giants, Jupiter and Saturn, result in an orbital path that varies in shape cyclically with dominant periodicities of 95, 124 (short eccentricity) and 405 kyr (long eccentricity). These periodicities are thought to have been constant over much of the Phanerozoic. Eccentricity determines the seasonal changes in distance between the Sun and Earth and therefore causes some change in the length of the seasons and the seasonal contrasts in incoming solar radiation. However, the annual net global effect of varying eccentricity on total solar insolation received annually by Earth is very small (Berger and Loutre 1991).

Currently the axis of Earth's rotation is orientated at ~23.44° from the normal to the orbital plane (the ecliptic) but over timescales of ~41 kyr, the axial tilt, or obliquity, changes between 22.1° and 24.5°. Variations in Earth's obliquity are caused by interference between precession and the gravitational interactions with the other planets of the solar system. The tilt determines the magnitude of seasonal contrast within each hemisphere and, critically, determines how much solar insolation reaches the high latitudes. As with eccentricity, the effect on total global insolation of varying obliquity is small but the effect on seasonal distribution is significant in driving the behaviour of the climate system, particularly in polar regions (Figure B13.3), and obliquity has, thus, been implicated as a major control on the growth and collapse of Northern Hemisphere icesheets during the Quaternary.

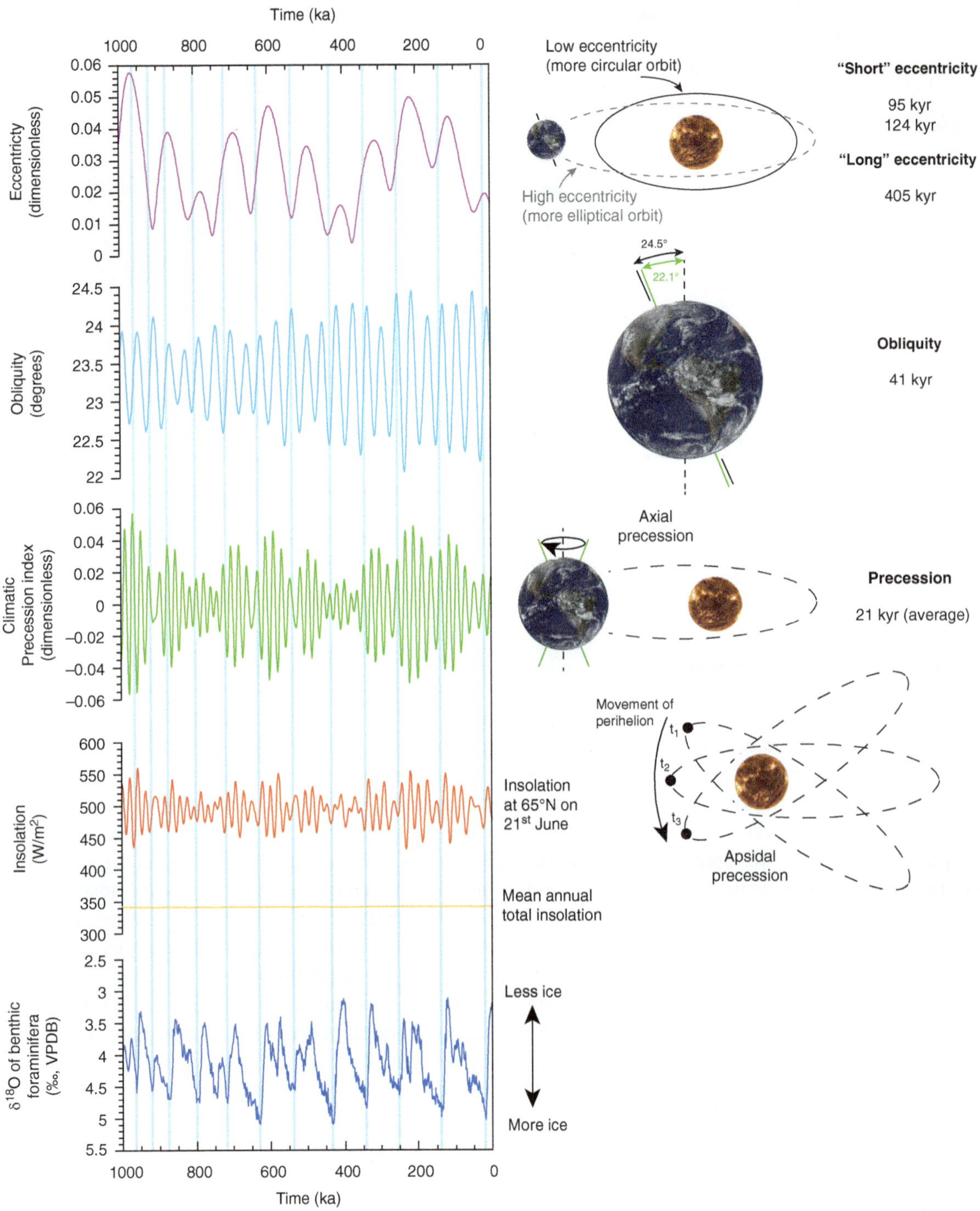

Figure B13.3 A summary of Earth's orbital parameters calculated over the last million years. The graphs illustrate how the major orbital parameters of eccentricity, obliquity and climatic precession have varied over the last million years and how these changes are manifested in the seasonal insolation received at 65°N and total annual insolation (Laskar et al. 2004; downloaded from http://vo.imcce.fr/insola/earth/online/earth/online/index.php). The combined effect of these periodicities is shown by the benthic foraminiferal oxygen isotope ($\delta^{18}O$) that, largely, records changes in continental ice volume (data from Lisiecki and Raymo 2005). Peak glacial conditions shown by vertical blue lines. Images of Earth and the Sun from NASA / Public Domain.

In addition to variations in the angle of tilt, the rotation axis also 'wobbles', relative to the stars, defining a circle over a 26 kyr period, known as axial precession. This form of precession is a consequence of the torque imposed on Earth's equatorial bulge by the Sun and Moon; consequently, over geological timescales, the periodicity of axial precession has gradually increased by a few per cent per 100 million years as the Moon has moved further away from the Earth. As obliquity is, in part, determined by precession, then it follows that the period of the obliquity cycle has also increased over geological time. Axial precession determines when in the year Earth is closest to the Sun (the perihelion), thereby affecting the amount of incoming radiation received during that season and the seasonal contrast. At present, perihelion occurs in Southern Hemisphere summer. In addition to axial precession, apsidal precession describes the rotation of the line of apsides (the line connecting the perihelion and aphelion, i.e. the long axis of the orbital ellipse) around the Sun, with a periodicity of about 112 kyrs. The combination of axial and apsidal precession causes climatic precession cycles with an average periodicity of ~21 kyrs (although these can vary between ~19 and 23 kyrs). Eccentricity determines the contrast in Earth–Sun distance between perihelion and aphelion (the time when Earth is farthest from the Sun) and therefore acts to modulate precession cycles.

None of the orbital parameters has much effect on total mean annual insolation (Figure B13.3) and, thus, they do not of themselves make the planet significantly warmer or colder. However, through their impact on the latitudinal and temporal distribution of incoming solar radiation, changes in the orbital parameters do drive major changes in the physical climate system that, through feedbacks in the carbon cycle, oceans and cryosphere can cause global changes in climate. Such regional and global changes in climate can, ultimately, cause changes in sedimentation that can be recognised in the geological record (e.g. Figure B13.3 and Figure 13.34).

overturning circulation that ventilates the deep basins (Section 10.6.2). However, the presence throughout much of the Neogene record (from at least the middle Miocene to Holocene) of thin organic-carbon-rich shales (typically with 1–10% TOC), known as sapropels, rhythmically interbedded with normal, low TOC pelagic and hemipelagic sediments (Figure 4.20), suggests that very different oceanographic conditions occurred episodically in the past (e.g. Rohling et al. 2015; Grant et al. 2022). Sapropel horizons are often laminated and exhibit a lack of benthic faunas and bioturbation, suggesting low-oxygen bottomwaters. The TOC content of the sediments does not, however, provide a complete record of the duration of low-oxygen conditions as some sapropels show evidence (in the form of other geochemical or palaeontological indicators) for 'burn down' – loss of organic carbon from the uppermost part of the sapropel layer due to re-oxygenation of the seafloor. The timing of sapropel depositional events seems to align with increasing strength of African monsoons, driven by Northern Hemisphere summer insolation maxima, and thus is determined by precessional forcing (Rossignol-Strick et al. 1982; Rossignol-Strick 1983). However, not all insolation maxima are associated with sapropel deposition, which is attributed to the modulation of precession cycles by varying eccentricity. Increased monsoonal rainfall over northern Africa during high insolation conditions is thought to have increased freshwater fluxes to the eastern Mediterranean, in particular via the River Nile, leading to stable water stratification and a shut-down of overturning circulation in the Eastern Mediterranean. The stable, stratified water column led to the deoxygenation of bottom waters and an increase in organic carbon preservation in the sapropels, possibly accompanied by increased productivity during some events, although this remains an area of debate (e.g. Rohling et al. 2015; Grant et al. 2022).

13.5.1.3 Cenozoic Cycles, Deep-Sea Currents and Pelagic Sedimentation

The Pleistocene pelagic sedimentary record provides a detailed archive of how orbitally-driven insolation changes impact on the climate system, carbon cycle and cryosphere which, in-turn, drive changes in oceanography and pelagic sedimentation. Inherent to our understanding, and interpretation, of Pleistocene pelagic sedimentation and the Earth system, is the existence of major ice sheets in both the Northern and Southern Hemispheres with major consequences for the link between insolation and climate through albedo effects. However, this configuration of the cryosphere is relatively recent. Persistent, intense Northern Hemisphere glaciations are thought to have developed from around 2.7 Ma (e.g. Shackleton et al. 1984; Lear et al. 2000; Lisiecki and Raymo 2005), although the transition is likely to have been more complex and less globally synchronous than previously thought (McClymont et al. 2023). Looking further back, the first large Antarctic ice sheet likely appeared at the Eocene–Oligocene boundary, ~34 Ma (Zachos et al. 2001). Development of large ice sheets was probably controlled by long-term changes in climate and ocean circulation driven both by changing CO_2

and plate tectonics (Zachos et al. 2001; DeConto and Pollard 2003). Given the central role that ice sheet/climate feedbacks play in the Pleistocene, it is not unreasonable to ask whether orbital forcing produces similar records in pelagic sedimentation in periods of Earth history characterised by one or no ice sheet. The cyclicity of the Cenozoic pelagic sedimentary record clearly demonstrates that orbitally forced climate changes are a ubiquitous feature of pelagic sedimentary archives (e.g. Herbert 1997; Shackleton et al. 1999; Westerhold and Röhl 2009), irrespective of the state of the cryosphere. Sedimentary cycles throughout the Cenozoic can be interpreted as a consequence of cyclical changes in the balance of productivity, preservation and/or dilution, but the precise processes connecting these parameters to climate change driven by orbital forcing can be both difficult to establish and varies through time and space. The sedimentological variations will depend to a large degree on the state, and response, of ocean circulation, which can be further modified by opening and closing of tectonic gateways between oceans. Similar-looking lithological cycles to those observed in the Pleistocene can be found in other parts of the Cenozoic record although the explanations of these cycles may be very different. For example, middle Miocene carbonate dissolution cycles recovered from the equatorial Pacific show maxima of dissolution during warm phases. One interpretation of these cycles, on the basis of carbon-isotope data, is that CO_2 was released from high-latitude soils during recession of ice sheets driven by orbitally forced warming (Kochhann et al. 2016). This mechanism stands in contrast to one of changing Southern Ocean circulation used to explain similar lithological variations in the Late Pleistocene.

In addition to modifying the relationship between orbital forcing and sedimentary cycles, the slow, long-term Cenozoic changes in geography (through tectonics) and decline of atmospheric CO_2, had profound effects on the spatial pattern and intensity of deep-sea circulation. Hiatuses are evidenced in many late Cenozoic abyssal deep-sea pelagic sedimentary cores by missing biostratigraphic zones (e.g. Moore et al. 1978; Barron and Keller 1982; Keller and Barron 1983; Dutkiewicz and Müller 2022). Analysis of the spatial and temporal distribution of hiatuses in deep-sea cores shows that some periods of time, such as the Miocene, are characterised by extremely widespread and frequent hiatuses occurring in palaeowater depths >2 km, that can be traced across multiple ocean basins (Figure 13.37). In the early Miocene, such deep-water hiatuses occur across low and mid-latitudes in all ocean basins. They broadly correspond in time to an interval of cooling and Antarctic ice-sheet growth, a strengthening of deep-water flow and increased erosiveness and corrosiveness of deep waters. These changes may have been initiated by changes in critical oceanic gateways (Dutkiewicz and Müller 2022). The complete opening of the Drake Passage allowed a full Antarctic Circumpolar Current (ACC) to develop, and the gradual closing of the Tethys gateway (Figure 13.37) eventually led to the termination of long-standing circum-equatorial connections between the Indian and Atlantic Oceans. Around the early to middle Miocene boundary (~16 Ma), changes in the spatial distribution of pelagic sedimentation occurred: with silica deposition reducing in the North Atlantic and, perhaps, becoming more dominant in the Pacific between ~16 and 13 Ma (Keller and Barron 1983; Cortese et al. 2004). This interval of time is also marked by increased pelagic organic carbon deposition, such as in the siliceous and organic-rich Monterey Formation of California. The same period broadly corresponds to a peak in total hiatus frequency across all water depths and ocean basins (Dutkiewicz and Müller 2022). These changes in sedimentation and the hiatuses have been attributed to a reorganisation of deep-water circulation due either to the shoaling of the Caribbean gateway and/or the onset of deep-water formation in the North Atlantic (due to tectonic gateway changes in the Fram Strait and between Greenland and Iceland). The intensification of North Atlantic deep-waters is likely to have led to less corrosive waters in the Atlantic as a whole, permitting more carbonate deposition and diluting silica. To further complicate matters, the middle Miocene was also marked by warming, known as the Miocene Climatic Optimum (MCO), caused by CO_2 most likely emitted during the eruption around 17 to 15 Ma of the Columbia River Flood Basalts (e.g. Longman et al. 2022). This was followed by an interval of ice sheet expansion on East Antarctica and significant global climate cooling, the Middle Miocene Climatic Transition (MMCT; ~14 Ma; Figure 13.37). Similar to early transitions, this event appears to have been associated with changes in ocean circulation (e.g. Holbourn et al. 2013) but by the end middle Miocene, around 12 Ma, the modern pattern of sediment distribution in the Pacific and Atlantic Oceans appears to have been largely established. The frequency of hiatuses since the late Miocene has gradually declined, perhaps due to an overall deepening of the CCD and/or a reduction in deep-water flow velocities caused by a slowing of ocean circulation as global climates have cooled over the last ~10 million years (Dutkiewicz and Müller 2022).

Similar widespread hiatuses are also known from around the Eocene–Oligocene boundary (Moore et al. 1978; Moore Jr. 2013) and even the Cretaceous (e.g. Sliter 1992). The Eocene–Oligocene boundary events are possibly due to the opening of the Tasman Gateway and Drake Passage and the onset of Antarctic glaciation during the Eocene–Oligocene Transition (Dutkiewicz and Müller 2022; Figure 13.37). However, interpretation of the origin of

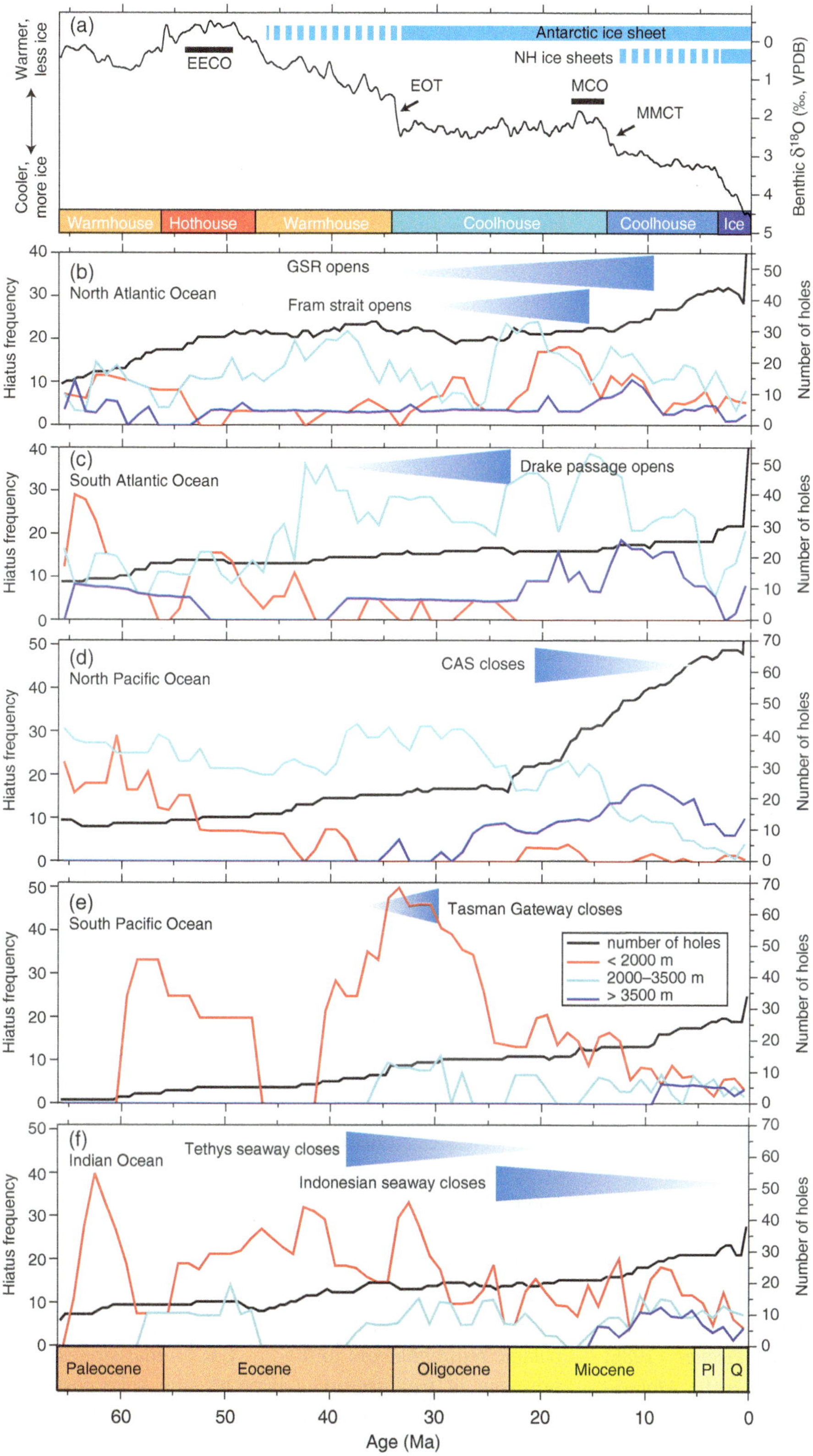

Figure 13.37 A comparison of Cenozoic climate (a) (as recorded by benthic foraminiferal $\delta^{18}O$) with frequency distributions of Cenozoic hiatuses in different ocean basins (b–f) at palaeowater depths of <2000 m (red), 2000–3500 m (cyan) and >3500 m (blue), based on deep-sea drill holes in 0.5 Ma bins. The black line is the number of drilled holes that sampled each 0.5 Ma bin, and the coloured lines are the proportion of holes at that water depth characterised by missing section within each bin. NH: Northern Hemisphere; EECO: Early Eocene Climatic Optimum; EOT: Eocene–Oligocene Transition; MCO: Miocene Climatic Optimum; MMCT: Middle Miocene Climatic Transition; GSR: Greenland-Scotland Ridge; CAS: Central American Seaway; Pl: Pliocene; Q: Quaternary; Adapted from Dutkiewicz and Müller (2022).

hiatuses becomes increasingly difficult with age due to the reduced number of sites available for study. A salutary lesson is provided by Cenozoic hiatuses found in pelagic sediments deposited atop Pacific Ocean guyots that do not appear to correlate with the timing of hiatuses from abyssal sedimentary records (Mitchell et al. 2015), such as those discussed above in the Miocene. In the case of the guyots the presence of winnowed sands between the hiatuses and the varying duration of hiatuses at different sites, all suggests physical erosion of sediment, most likely due to internal tidal waves (e.g. Figure 13.24). These internal tidal wave fields in turn depend upon the bathymetry of submarine ridges and rises that are controlled by plate movement and subsidence. Reconstructing such changes, even in the Neogene, is problematic, and thus the significance of older hiatuses present in pelagic sediments on top of submarine highs is likely to remain highly speculative.

13.5.1.4 The Cretaceous Ocean – Pelagic Sedimentation in an Extreme Greenhouse

The limited area of extant ocean crust of Cretaceous, or older, age constrains the number of ocean drilling sites that have penetrated and recovered open-ocean pelagic sediment of this age. In the Pacific, much of the ocean crust that existed during the Cretaceous has now been subducted, while the North Atlantic and South Atlantic were relatively small ocean basins at the time, and so pelagic sediments were deposited in relatively proximal areas and, in some cases, are now covered by thick post-Cretaceous passive margin sequences. The record of Cretaceous deep-ocean pelagic sedimentation is thus supplemented extensively by sequences exposed on-land that were, primarily, deposited in the Tethys Ocean and subsequently uplifted and deformed during the Cenozoic Alpine-Himalayan orogenies. Despite the complications of burial diagenesis, it is possible to recognise, and interpret, many aspects of the Cretaceous pelagic sedimentary record through a broadly uniformitarian approach. (Pelagic facies of the Cretaceous that are not readily comparable with the modern are addressed in Section 13.5.2.)

As in the Cenozoic, cyclic sedimentation is common in the Cretaceous pelagic record, but, unlike many Cenozoic examples, these records can comprise multiple, and varied, lithologies as a consequence of both oceanographic and diagenetic variations (e.g. Herbert and Fischer 1986; Batenburg et al. 2016). One such example is the Albian–Cenomanian (Cretaceous) Scaglia Bianca (e.g. Gambacorta et al. 2014), deposited in the Tethys Ocean and now exposed in the Appenines of central Italy (Figure 13.38). The Scaglia Bianca is dominated by pelagic limestones that vary from homogeneous to laminated, with varying amounts of clay, silica (both radiolarites and cherts) and organic matter (often in black mudstones, known as 'black shales'). Critically, the Scaglia Bianca shows that the sedimentological record of suspension sedimentation can be modified by bottom-water currents even in a deep-water setting that would seem to be very low energy. These weak bottom-water currents are evidenced by erosion, winnowing of fine grains and deposition of coarser grained (foraminiferal and intraclastic) lags and wavy- and parallel-laminated limestones with small-scale, low-amplitude bed forms (Figure 13.39). Graded limestones suggest the existence of contour currents or pelagic turbidites. The association of these different facies in vertical successions point to the systematic redistribution of carbonate pelagic sediment at the sea floor. Further evidence of a dynamic seafloor is provided by the siliceous sediments – radiolarites show subtle variations from homogeneous beds (interpreted as representing suspension sedimentation) to sharp-based, laminated beds (interpreted as representing winnowing and/or re-deposition of radiolaria).

Diagenesis of pelagic sediments can significantly impact their appearance and interpretation at the bed-scale. In the Scaglia Bianca the formation of cherts, for example, can obscure sedimentary structures seen in the radiolarites. Perhaps more significantly, however, the formation of indurated, cemented lithologies during diagenesis can lead to the development of beds dominated by one mineralogy through diagenetic un-mixing. This can be demonstrated through joint petrographic and geochemical studies. Early Cretaceous-age pelagic limestone-marl cycles recovered from a number of North Atlantic DSDP sites provide a good example (Frank et al. 1999). Bioturbated white and light grey limestones are interbedded with laminated to micro-burrowed marlstones, homogeneous claystones and graded sandstones. The claystones and sandstones represent deposition as turbidites. The cyclic pelagic limestones and marls alternate on a dm to m-scale and exhibit primary sedimentological variability (i.e. variations in the extent of bioturbation and lamination) that indicates clearly that these sediments were not homogeneous upon deposition. At all sites studied by Frank et al. (1999), the limestones are characterised by oxygen-isotope ($\delta^{18}O$) and Sr/Ca values that are consistent with significant addition of diagenetic carbonate (i.e. relatively low $\delta^{18}O$ values and low Sr concentrations, compared to both primary pelagic sediments and to the low $CaCO_3$ marlstones). The presence of well-preserved burrows and microfossils in the limestones suggests that they were well-consolidated (and potentially even partially cemented) prior to deep burial. In contrast, the marlstones at the studied sites contain dissolution seams and fragmented, etched coccoliths, suggesting that they have lost carbonate. All of this suggests that the availability of primary $CaCO_3$ was sufficient to

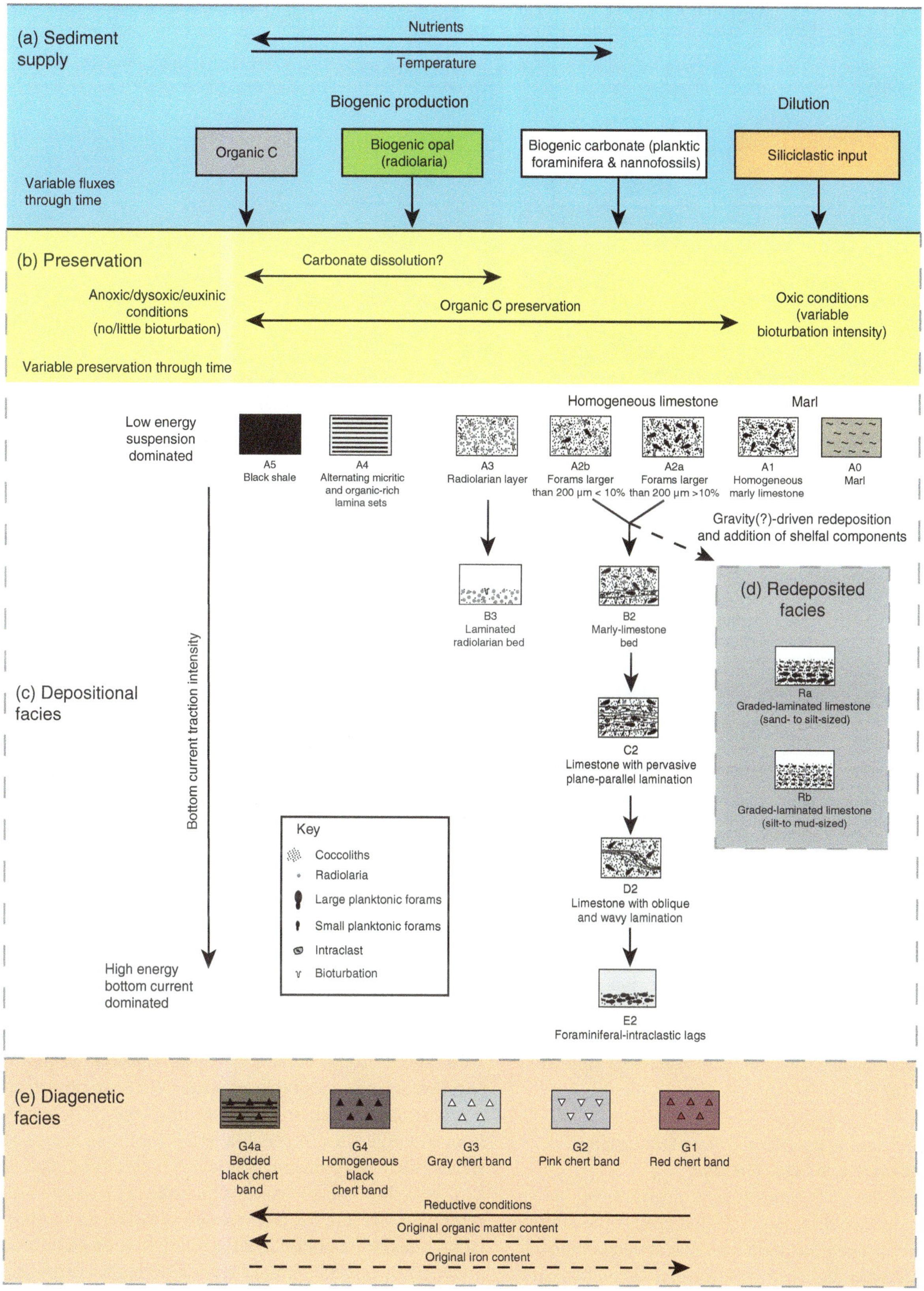

Figure 13.38 An example of a depositional and diagenetic facies scheme for pelagic sediments, and the associated processes, based upon observations of the Scaglia Bianca and Scaglia Variegata in the Umbria-Marche and Belluno Basins of Italy. Oceanographic conditions determined the production and mixture of sediment supplied (a), which, during deposition, passed through a preservational filter (b) within the water column and at the seafloor. Redeposition of sediments occurred under both bottom-water currents (c) and through gravity-driven processes (d). Finally, some lithofacies resulted from the chemical diagenetic alteration of primary lithologies (e). Adapted from Gambacorta et al. (2014).

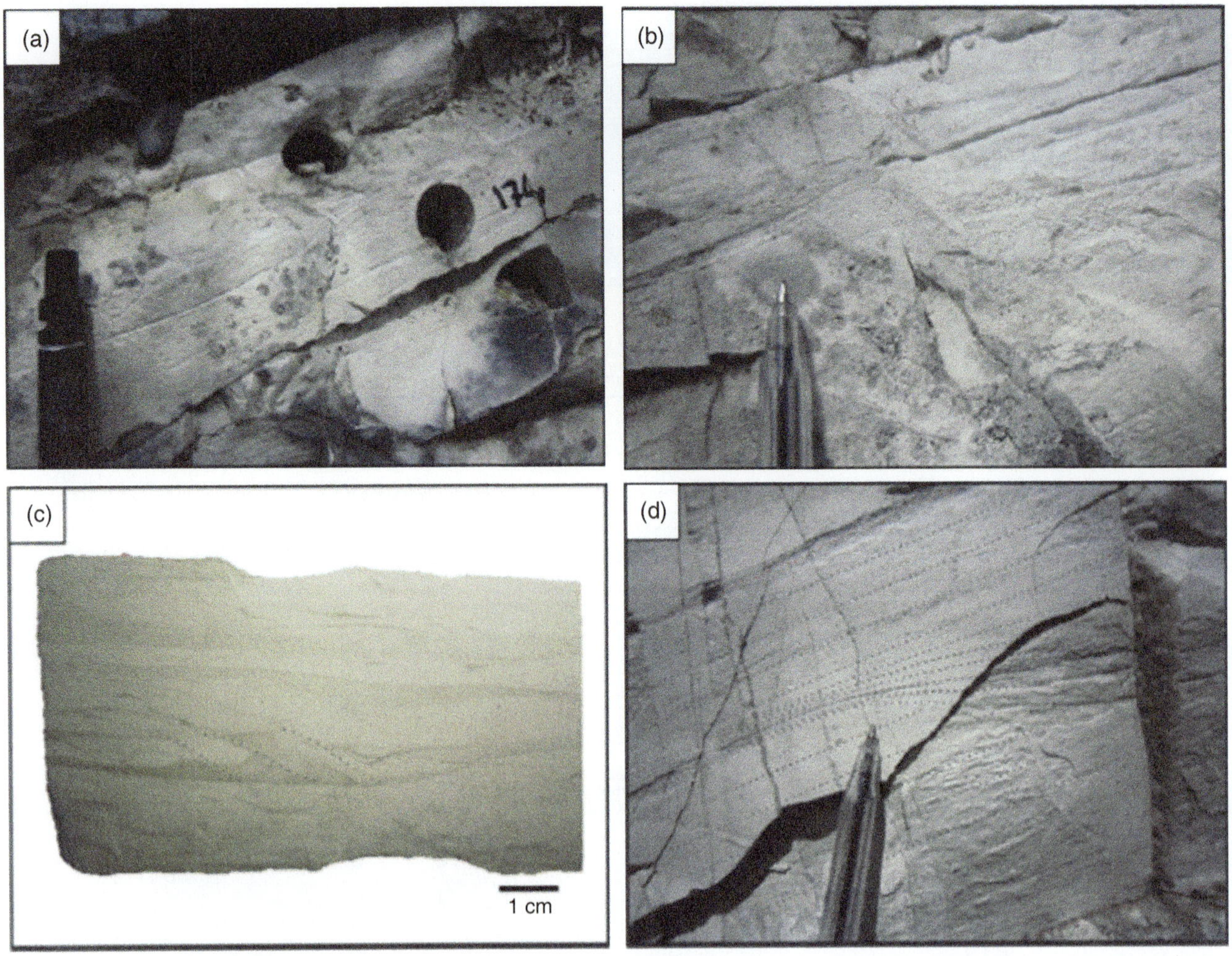

Figure 13.39 Examples of oblique and wavy lamination interpreted as evidence for bottom-water currents in the pelagic limestones of the Cretaceous Scaglia Bianca. (a), (b) and (d) are all images from outcrops with a pen for scale; (c) is an image of a polished block. Reproduced from Gambacorta et al. (2014) / Elsevier.

allow the addition of diagenetic cement to the limestone intervals over distances of tens of cm during burial compaction and pressure solution. In this example, it is suggested that diagenetic un-mixing results in the near-reversal of the primary geochemical differences (Sr/Ca and $\delta^{18}O$) between originally carbonate-rich and carbonate-poor intervals (Figure 13.40).

13.5.1.5 Changes in the CCD and Plate Stratigraphy During the Cenozoic and Late Mesozoic

The depth of the CCD varies over geological time due to changes in the global carbon cycle that change the saturation state of the deep-ocean through absorption of more or less CO_2. Changes in ocean temperature, circulation and local biogenic carbonate sediment supply also impact the depth of the CCD, which can be reconstructed from pelagic facies variations in deep-sea cores. The time at which the ocean-floor subsided below the CCD is marked by the total loss of carbonate in the stratigraphic record at that site, thereby providing the position of the CCD at that time and place. If the palaeo-water depth can be constrained, for example by back-tracking the subsidence of the oceanic crust, and if the age of the lithological transition is known (e.g. through biostratigraphy or magnetostratigraphy), then it is possible to reconstruct the history of the CCD, ideally from multiple sites (e.g. Palike et al. 2012). Figure 13.41 shows such a reconstruction of the equatorial Pacific CCD over the Cenozoic based upon detailed carbonate concentration measurements (expressed as carbonate accumulation rates) from multiple sedimentary records on normal ocean crust. In general, the sites exhibit higher carbonate accumulation rates in the oldest parts of the record, overlying basement (at least where basement was recovered). This represents deposition on young hot ocean crust at water

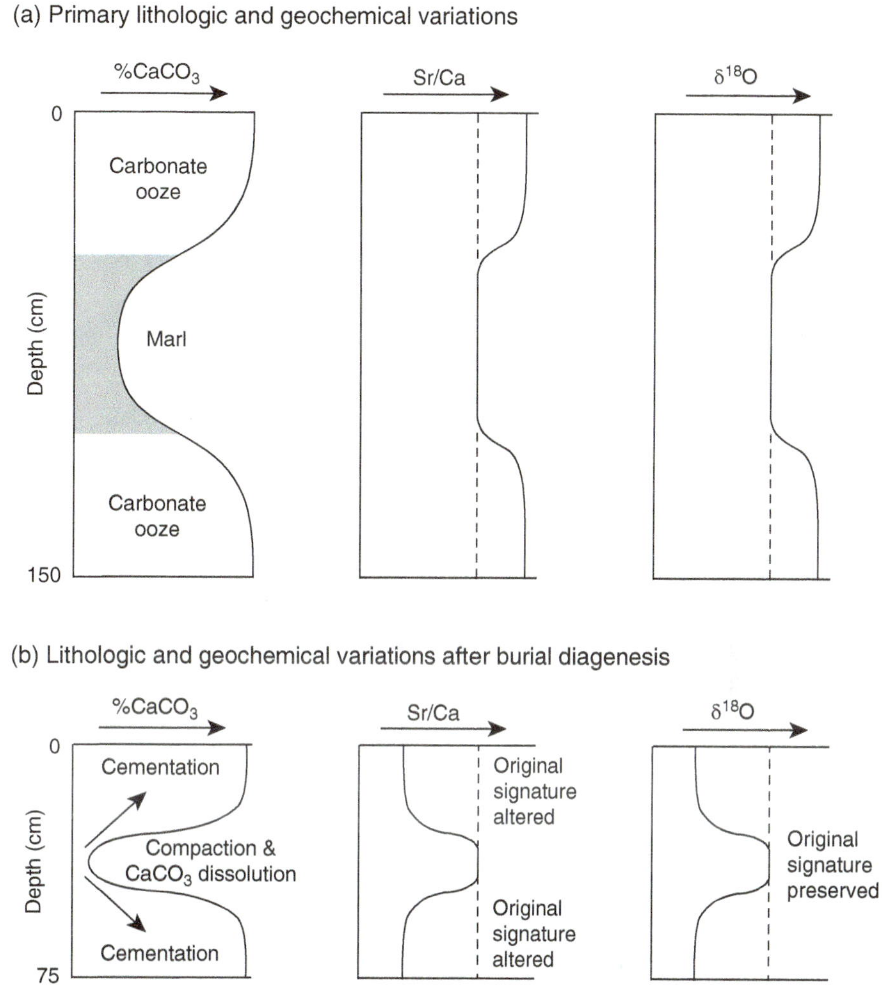

Figure 13.40 Schematic diagrams to illustrate the effects of burial diagenesis on the geochemical composition of interbedded marls and carbonate oozes. In (a), stratigraphic profiles of $\%CaCO_3$, bulk Sr/Ca and bulk carbonate $\delta^{18}O$ are shown; (b) shows the same datasets following compaction and the transfer of $CaCO_3$ from marls into carbonate oozes. These diagenetic processes lead to exaggeration of primary differences in $\%CaCO_3$ (through dissolution in marls and cementation in oozes) and the reversal of primary differences in Sr/Ca and $\delta^{18}O$. Redrawn from Frank et al. (1999) / The University of Nebraska - Lincoln / Public Domain.

depths of ~2.5 km. As the sites moved away from the spreading centre and began to subside below the lysocline, carbonate accumulation rates declined, until the CCD was reached. Intriguingly, Figure 13.41 shows that at some sites (e.g. U1333, DSDP 70), carbonate deposition does not simply decline and disappear – instead carbonate deposition came and went, as a consequence of the level of the CCD moving up and down through time. Those particular sites that are subsiding slowly and are at water depths close to the CCD when such changes occur can therefore record multiple facies changes and give information on the fluctuations of the CCD itself.

Purely lateral movements of the plates can also create a stratigraphy that appears similar to that produced by cooling subsidence below the CCD or by vertical movements of the CCD. For CCD reconstructions, such as shown in Figure 13.41, all sites chosen to reconstruct the CCD should be selected from a narrow palaeolatitudinal range, that is: with similar productivity. An example of how latitudinal change effects sedimentation has been proposed from the Mesozoic Tethyan deep-water succession now exposed in the southern Alps (Figure 13.42). In the Lombardian Basin, Early Jurassic pelagic limestones are overlain by Middle to Late Jurassic green and red radiolarites, before carbonate returns in the later Jurassic and Early Cretaceous (Figure 13.42). This succession has been interpreted as recording changes in the level of the CCD during the Jurassic and Cretaceous (Jenkyns 1986). However, palaeomagnetic reconstructions of palaeolatitude suggest an alternative interpretation (Muttoni et al. 2005). These data suggest that Early Jurassic pelagic carbonates were deposited at palaeolatitudes >15°N, the Middle and Late Jurassic radiolarites at palaeolatitudes of ~5 to 15°N, and the overlying Early Cretaceous pelagic limestones also south of 15°N. The depositional sequence is therefore interpreted as resulting from southward drift of the basin starting in the Early Jurassic into a zone of near-equatorial upwelling with abundant radiolarian

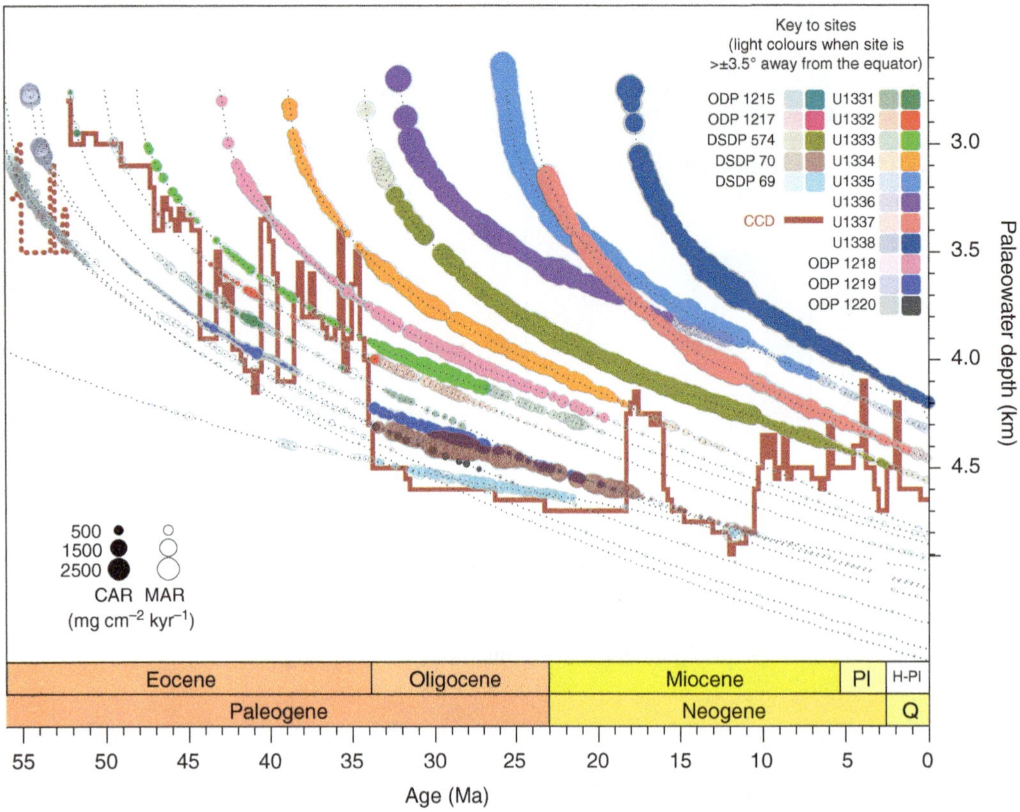

Figure 13.41 A reconstruction of the Cenozoic history of the calcite compensation depth (CCD) in the equatorial Pacific Ocean based upon carbonate accumulation rates (CAR; filled circles) at a number of deep-water sites overlying ocean crust. The thermal subsidence curve for each site is shown, upon which is superimposed the CAR and MAR (mass accumulation rates; empty circles) data for that site. The CCD is constrained by calculating the gradient in water depth for a 250 kyr window of CAR values and projecting it to the depth at which CAR is zero (i.e. where carbonate supply = carbonate dissolution and net preservation is zero). For CAR, light colours are used when the site was more than ±3.5° north or south of the equator. Palike et al. (2012) / Springer Nature.

productivity by the Middle to Late Jurassic. This southward plate movement was likely caused by the opening of the seaway between the North Atlantic and Tethys. In the latest Jurassic and earliest Cretaceous, the site reversed its motion and gradually moved northwards back into a latitudinal belt of carbonate production. This particular interpretation is not without criticism, regarding the size of the errors on the palaeomagnetic data (Baumgartner 2013), but similar latitudinally caused changes in pelagic sedimentation have also been reconstructed from the Early Cretaceous of the Pacific Ocean, where changes in calcareous nannoplankton towards high nutrient assemblages and increased biosiliceous sedimentation have been used to infer the existence of a latitudinal band of high productivity corresponding to the so-called 'equatorial sediment bulge' (Ogg et al. 1992, Figure 13.43). Together, these examples from the Tethyan and Pacific Oceans demonstrate the idea that plate stratigraphies involve changes both through time and through space.

13.5.2 Facies of the Mesozoic and Palaeozoic with No Direct Modern Analogues

13.5.2.1 Epicontinental Pelagic Facies of the Late Cretaceous

Late Cretaceous sea-levels were among the highest of the Phanerozoic and led to the extensive development of epicontinental seas and seaways, with water depths of, at most, a few 100 m. In such settings, fine-grained chalks and limestones, dominated by nannofossils, comprising mainly coccoliths, were deposited across northwest Europe, North America and the Middle East. This lithological unit is normally referred to as 'The Chalk'. The (generally) parallel-bedded sediments contain >75% planktonic components, are, in that sense, pelagic and, in many cases, are almost pure carbonates. Like deep-sea pelagic oozes, the Chalk often contains a minor non-carbonate fraction (that commonly includes clay minerals, silt-sized quartz grains and chert nodules, (known in the Chalk as 'flints'). Cyclic bedding can be recognised through variations in carbonate content,

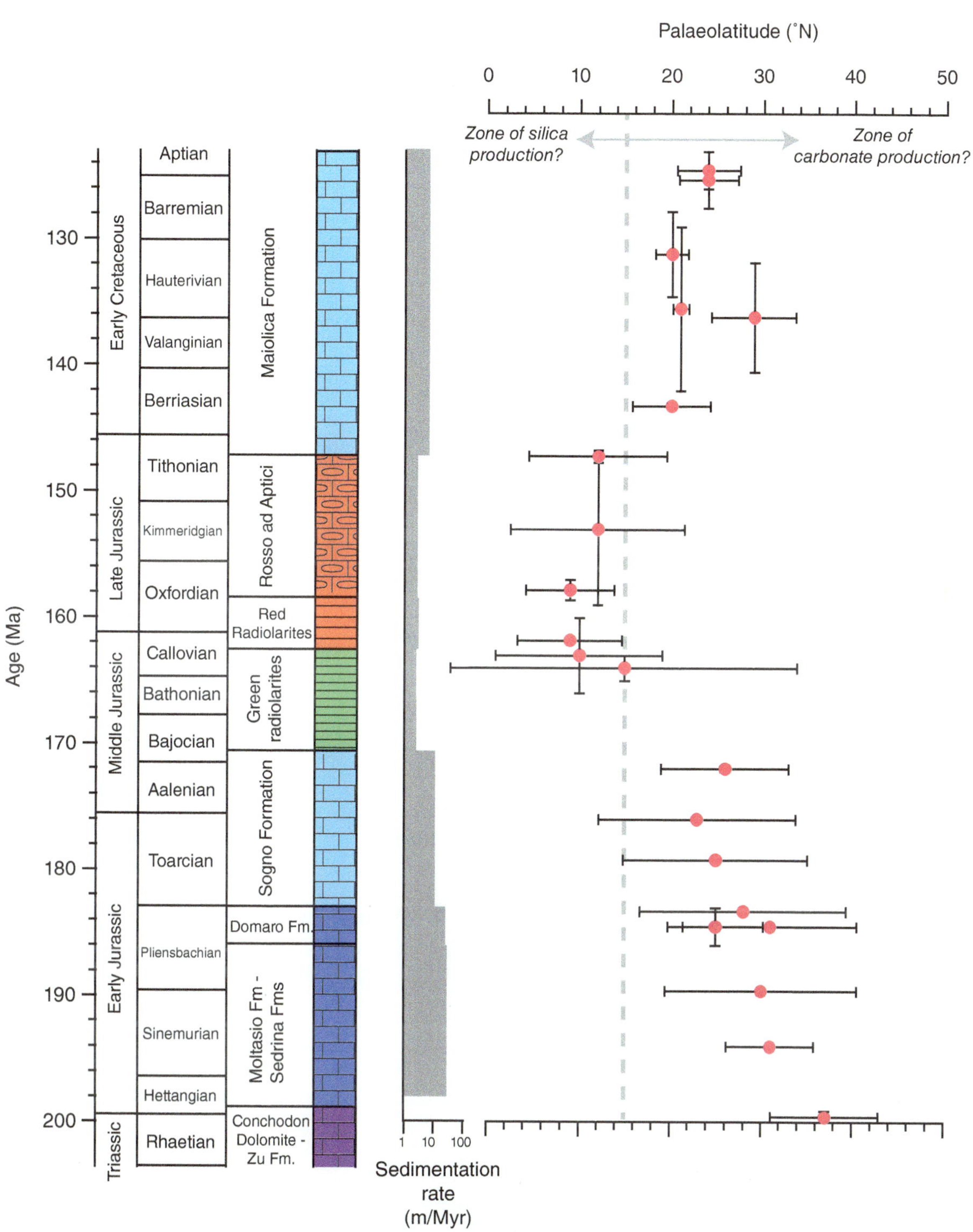

Figure 13.42 Mesozoic stratigraphy of the Lombardian Basin (Southern Alps of northern Italy) from Collo Sogno and reconstructed palaeolatitudes. Pelagic sedimentation dominated in this basinal setting from the late Early Jurassic (Toarcian Sogno Formation) to the Early Cretaceous (Aptian Maiolica Formation). The pelagic siliceous content increased when palaeolatitudes were below ~15°N, which has been suggested as evidence of latitudinal facies belts controlled by biogenic sediment production in the Tethyan Ocean. Adapted from Muttoni et al. (2005).

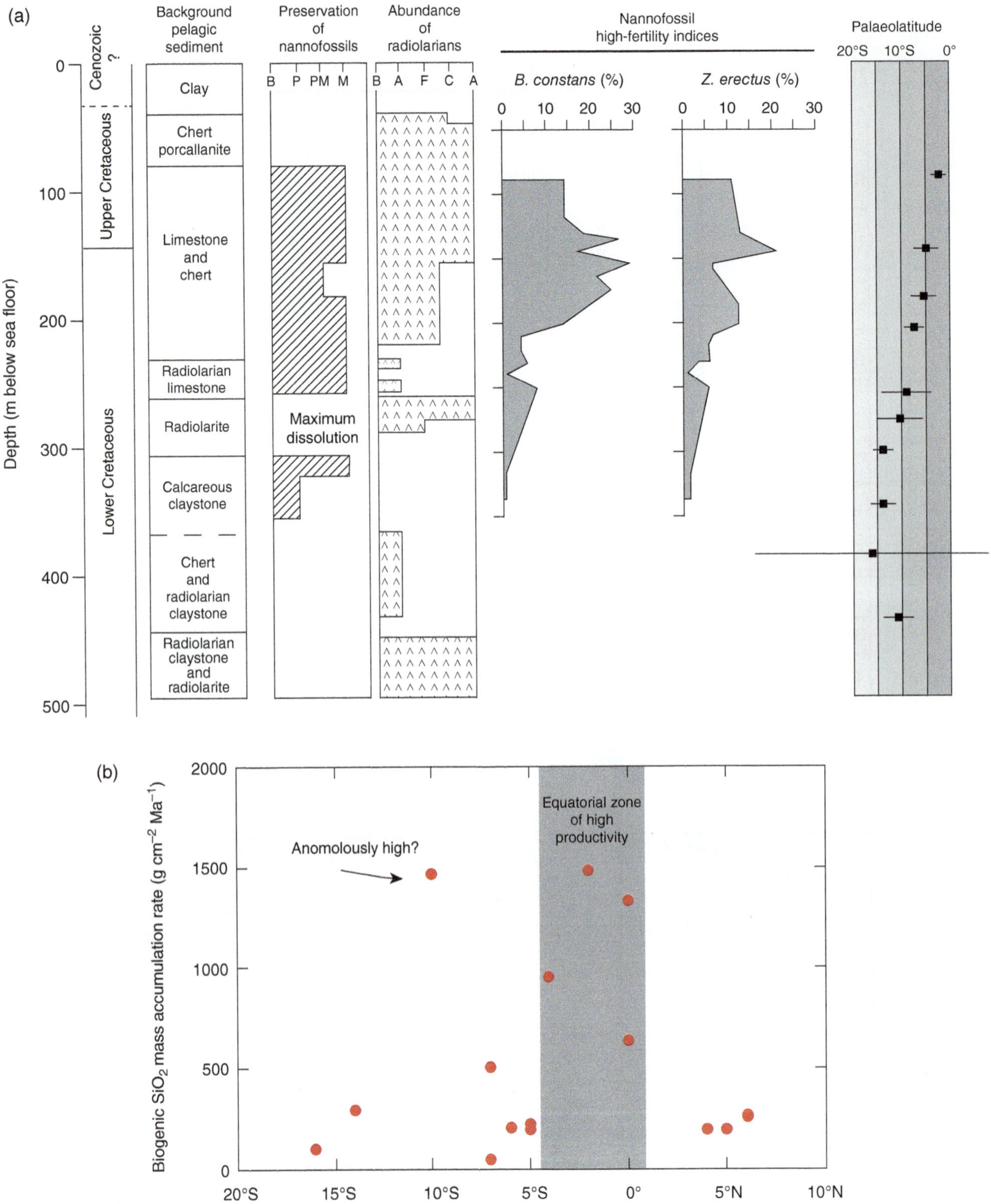

Figure 13.43 (a) Lithological, micropalaeontological and palaeolatitudinal data from ODP Site 801, illustrating the changes associated with the northward movement of the Pacific plate under a near-equatorial zone of high productivity. This zone is recognised by the enhanced carbonate preservation, an increase in radiolaria abundance and a change in nannofossil assemblage towards species associated with high nutrient (fertility) conditions. Erba (1992) / Proceedings of the Ocean Drilling Program / Public Domain. (b) Biogenic silica mass accumulation rates of Middle Jurassic to Early Cretaceous radiolarian-rich sediments plotted against palaeolatitude at the time of deposition, revealing high-rates of silica accumulation within ~5° of the equator. The high point at ~10°S may be anomalous for various reasons including the quality of the age model used, enhanced silica preservation, redeposition or more vigorous gyre circulation. Adapted from Ogg et al. (1992).

although some of the more clay-rich horizons are thought to be bentonites (e.g. Gale et al. 1999; Wray and Gale 2006). Aspects of the depositional processes and environments of the Chalk, however, differ significantly from true deep-sea pelagic deposits. The Chalk of northwest Europe provides good examples of differences between pelagic deposition in a relatively shallow epeiric sea and the deep-ocean. Florally, there are nannofossils in the Chalk that are more common in epeiric shelf settings than coeval open-ocean records (e.g. *Braarudosphaera bigelowii*; Fonseca et al. 2019). Faunally, a diversity of nektonic and benthic organisms inhabited the Chalk seas including ammonites, belemnites, inoceramid bivalves, echinoids, sponges, bryozoa and brachiopods (e.g. Owen and Smith 1987) – all components that are relatively rare to absent in most Late Cretaceous deep-water pelagic sediments. Similarly, the presence of abundant *Thalassanoides* in the Chalk contrasts with coeval deep-sea pelagic deposits, that rarely contain this trace fossil (Ekdale and Bromley 1984). Mineralogically, parts of the Chalk contain abundant flints with silica probably sourced from siliceous sponge spicules (e.g. Maliva et al. 1989; Madsen and Stemmerik 2010). Silica cementation often appears to be concentrated around macrofossils and burrows, commonly creating nodular cherts (in contrast to deep-sea bedded-cherts, although the nodules themselves are often aligned along bedding). Sedimentation rates of the Chalk were spatially and temporally highly variable (e.g. from 2.5 cm ka^{-1} up to >25 cm ka^{-1}; Wray and Gale 2006), compared to the rates typical of deep-sea carbonate ooze accumulation (1–4 cm ka^{-1}). Condensation caused by slow accumulation rates, erosion and/or discontinuous deposition is marked, as in the deep-sea, by nodules and/or hardgrounds, but these horizons also commonly contain the shallow marine mineral, glauconite. As a consequence of the spatial and temporal variability of sedimentation rates, there are significant lateral variations in Chalk thickness (e.g. Mortimore 2011). Additionally, there are many documented examples of metre to tens-of-metres scale channel structures and cross-cutting bedding relations, both at outcrop (e.g. Figure 13.44) and in seismic data. All these observations are inconsistent with simple pelagic settling from a surface

Figure 13.44 An example of erosional surfaces and channel fills in the Late Cretaceous Chalk in northern France (Pays de Caux). Surfaces A, B and C are major hardgrounds associated with erosion. D and E represent channel fill facies. F represents cyclic, parallel bedded pelagic deposition. Mortimore (2011) / Permission of Elsevier.

ocean with regionally uniform carbonate productivity. The thickness variations are due to a combination of active syn-depositional tectonics, creating basins, 'swells' and fault blocks, and a relatively dynamic depositional environment in which sediment was at times removed and subsequently re-deposited due to changes in sea-level and/or seawater circulation (Mortimore 2011). Hardgrounds likely formed during both sea-level fall and rise – the former are found at basin margins associated with erosive channels and surfaces, the latter onlap onto raised areas of seafloor topography (i.e. the swells). In areas of relatively shallow-water the channels can be tens of metres deep (e.g. Gale et al. 2013), while in the deep-water areas of the North Sea and Anglo-Paris Basin, channel depths are shown to be 100 m or more (e.g. Esmerode and Surlyk 2009). In this latter context, slumping and re-deposition of pelagic carbonate as mass-flows (e.g. Arfai et al. 2016), drifts and mounds were common and help explain the observed dramatic variations in sedimentation rates. See Section 11.6.1 and Figure 11.49 for more discussion of the Chalk in the context of cold-water carbonate ramp settings.

13.5.2.2 Pelagic Black Shales of the Jurassic and Cretaceous

The common observation that most Mesozoic and Cenozoic pelagic deep-sea sediments are well-bioturbated with low organic carbon contents indicates that the ancient deep-sea environment at those times was likely broadly analogous to the modern, at least with respect to oxygenation. However, there is clear evidence for multiple, discrete, geologically brief periods of widespread ocean-bottom anoxia that were not analogous to the modern ocean. These are evidenced by sediments rich in total organic carbon (TOC >1% wt%; known collectively as 'black shales') that are generally fine-grained. Deposition and preservation of black shales require that the rates of production and supply of marine organic carbon to the deep ocean exceed the rates of respiration in the water column and at the seafloor. Tipping this balance towards the deposition of black shales has classically been viewed either as a result of increased primary production (e.g. through increased upwelling or terrigenous nutrient supply) or increased preservation (e.g. through reduced bottom-water oxygenation caused by stratified water columns). However, high primary production can itself lead to low oxygen concentrations at the seafloor as is seen in modern oxygen minimum zones. Equally, low bottom-water oxygenation does not necessarily translate to organic carbon-rich sediments if there is insufficient export production (e.g. the modern Black Sea). With these general considerations in mind, the pelagic black shales of the Late Mesozoic probably relate to combinations of palaeogeography and palaeoclimate that together determined the locations of productive surface waters and areas of bottom-water mixing and ventilation. Even so, the specific reasons why pelagic black shales occur in individual basins and particular intervals of time clearly varies. In the case of the Cretaceous Scaglia Bianca (Figure 13.38), thin black shales occur interbedded rhythmically with pelagic limestones and cherts (e.g. Gambacorta et al. 2014). The black shales and associated black cherts likely represent increased productivity caused by orbitally forced climate change perhaps through enhanced upwelling or increased riverine nutrient input (e.g. Batenburg et al. 2016). The thin, cyclically bedded black shales and cherts of the Scaglia Bianca can be found over tens of kms in central Italy but appear to be constrained to the NW Tethys, implying that they represent the consequences of a very specific combination of regional factors. In contrast, other black shales of the Jurassic and Cretaceous are far more widespread (Figure 13.45) and can be stratigraphically correlated across different depositional settings, water depths and even ocean basins (Schlanger and Jenkyns 1976; Jenkyns 1980). These black shales are frequently laminated, containing abundant fish debris, thin interbeds of radiolaria and extremely high TOC values (up to 35 wt%). They were deposited on timescales of hundreds of kyrs and represent deposition during intervals of time known as 'Oceanic Anoxic Events' or OAEs (Schlanger and Jenkyns 1976; Jenkyns 1980, 2010). The black shales deposited in pelagic environments during OAEs are often bounded by typical pelagic carbonate- and silica-bearing sediments (Figure 13.45), that represent deposition in well-oxygenated, 'normal' conditions. Clearly, during OAEs, pelagic environments were significantly perturbed from an oxic state analogous to the modern. Evidence from various geochemical palaeoclimatic and palaeoceanographic proxies (Jenkyns 2010) indicates that OAEs were associated with high temperatures, high productivity, changes in weathering and volcanism, and major changes in water column redox, including photic-zone euxinia in some areas (H_2S dissolved within the water column). The organic-carbon deposited during OAEs is thought to have been a negative feedback response to greenhouse warming caused, most likely, by CO_2 outgassing associated with large igneous province volcanism (Jenkyns 2010).

13.5.2.3 Pelagic Environments of the Tethys Ocean in the Jurassic

Pelagic carbonate production prior to the evolution of the modern calcareous phytoplankton and zooplankton was largely due to nektonic organisms such as cephalopods, although, in some examples, benthic carbonate sources (e.g. bivalves) also contributed significantly to sediment generation. Sedimentation rates were generally low

Figure 13.45 Photographs of the Livello Bonarelli, a ~1-m thick black shale interval, deposited during the latest Cenomanian OAE2. (a) The black shale occurs within a thick sequence of pelagic carbonates and cherts representing normal, well-oxygenated sedimentation. (b) The black shale itself is heterogenous on the cm-scale, with beds of laminated, organic-rich black shale interbedded with organic-poor radiolarian-rich sands (light coloured layers). (c) Map showing Late Cenomanian palaeogeography and sites where TOC-rich sediments deposited during OAE2 have been documented. All figures modified from Robinson et al. (2017) / John Wiley & Sons.

(mm ka^{-1}) rather than cm ka^{-1}), and condensed sedimentation, hardground development and discontinuities are common. A number of insightful examples of such deposits occur within the Jurassic deep-water sequences of the northwest Tethyan Ocean (summarised in Jenkyns 1986; Weissert 2011; Baumgartner 2013), often in stratigraphic association with demonstrably pelagic radiolarian-rich sediments.

The Mesozoic sediments of the Southern Alps in northern Italy were deposited on the southern edge of the Tethyan Ocean (including the Lombardian Basin succession shown in Figure 13.42), overlying the rifted continental crust of the African passive margin (Bernoulli and Jenkyns 1974; Winterer and Bosellini 1981). The Triassic comprises continental and shallow-water deposits laid down during the earliest phase of rifting, which continued into the Early Jurassic. Facies distributions here, as elsewhere in the Tethys (e.g. as represented by sequences in Sicily and the Appenines of central Italy), were strongly controlled by faulting and a series of shallow-water carbonate platforms, submarine plateaus and intervening faulted basins developed (Figure 13.46). Re-deposited material from relatively shallower-water settings is common proximal to the margins of these platforms and plateaus. Many of the shallow-water carbonate platforms drowned in the Early to Middle Jurassic, creating submerged bathymetric highs, such as the Trento Plateau, upon which a variety of condensed 'pelagic' limestones were deposited at various times (and likely in different water depths). Some of these pelagic limestones, containing ammonites, ferromanganese crusts and nodules, stromatolites, boring algae and extensive encrusters were likely associated with the initial drowning of the shallow-water carbonate platforms and might even have been deposited in the photic zone, swept by bottom-water currents. On submerged platforms and their margins, red-coloured condensed limestones commonly occur, often referred to as 'Rosso Ammonitico'-type facies, although the specific lithologies, faunas, depositional-water depth and age can vary throughout the Tethyan region. The Middle Jurassic Rosso Ammonitico Inferiore Formation in the Southern Alps comprises red, marly calcareous, nodular sediments and is typically thin (0–10 m). Calcareous nannofossils are generally rare and other faunal elements, such as encrusters and stromatolites, are rare to absent (in contrast to some other shallower-water condensed limestone facies). There is also a notable lack of radiolaria. The nodular appearance is most likely to have developed due to aragonite dissolution and calcite cementation during early diagenesis. Sedimentation rates of the Rosso Ammonitico Inferiore were exceedingly slow (<1 cm ka^{-1}; Winterer and Bosellini 1981; Jenkyns 1986) and water depths of deposition are estimated as several hundred metres to a kilometre. In adjacent basinal areas, re-deposited pelagic limestones, radiolarites and black shales accumulated that are much thicker than the coeval condensed Rosso Ammonitico-type facies. The re-deposited carbonates contain pelagic bivalves, crinoid ossicles and radiolaria and show evidence of downslope transport such as occasional sole marks (e.g. flute casts), slump folds and pebbly mudstone breccias. The basinal radiolarites exhibit red, green or grey, cm-thick, regularly bedded layers of chert separated by mm-thick clay partings. In addition to radiolaria, these sediments contain a few sponge spicules and very-rare, poorly preserved nannofossils in more calcareous lithologies. Although it has been suggested that these basinal sediments were deposited below the CCD, the absence of radiolaria in the Rosso Ammonitico compared with the high sedimentation rates of the basinal radiolarites suggests that deposition of the radiolaria was not just by simple suspension settling (Baumgartner 2013). Currents sweeping the submerged plateaus may have effectively removed any radiolaria (and other fine-grained detritus) and re-deposited them into the lower energy basinal regions, thereby increasing the apparent sedimentation rates there. Thus, the condensation of the Rosso Ammonitico may have been largely due to winnowing of fine-grained material, although deposition below the aragonite lysocline may also have limited carbonate accumulation. By the Early Cretaceous, the bathymetry of the region was much more uniform, and the entire region was draped by the pelagic cherty limestones of the *Maiolica* (also known as the *Biancone)* facies, rich in calcareous nannofossils, with rare ammonites, belemnites and radiolaria. Such deposits clearly represent sedimentation above the CCD in a manner analogous to modern nannofossil-rich carbonate ooze.

13.5.2.4 Mesozoic Radiolarites in the Panthalassic Ocean (Palaeo-Pacific)

The Pacific Ocean today represents the modern manifestation of the Panthalassic Ocean, which once surrounded Pangaea. The Mesozoic (and, indeed, Palaeozoic) geologic record of this ocean, from both deep ocean drilling sites and accreted fragments in the entire circum-Pacific region, point to major differences in the distribution and importance of radiolarian-rich sediments compared to the modern (e.g. Racki and Cordey 2000 ; Baumgartner 2013). At the present-day, radiolaria are most abundant in warm, low latitude divergence (i.e. upwelling) zones. However, evidence from, for example, the mid-Cretaceous of Shatsky Rise shows the presence of chert within calcareous oozes throughout the Aptian–Campanian, over which time the sites drifted from equatorial latitudes to ~15 to 20°N, well outside the range of the modern equatorial divergence (Shipboard Scientific Party 2002). The broader distribution of radiolaria could represent a difference in surface water

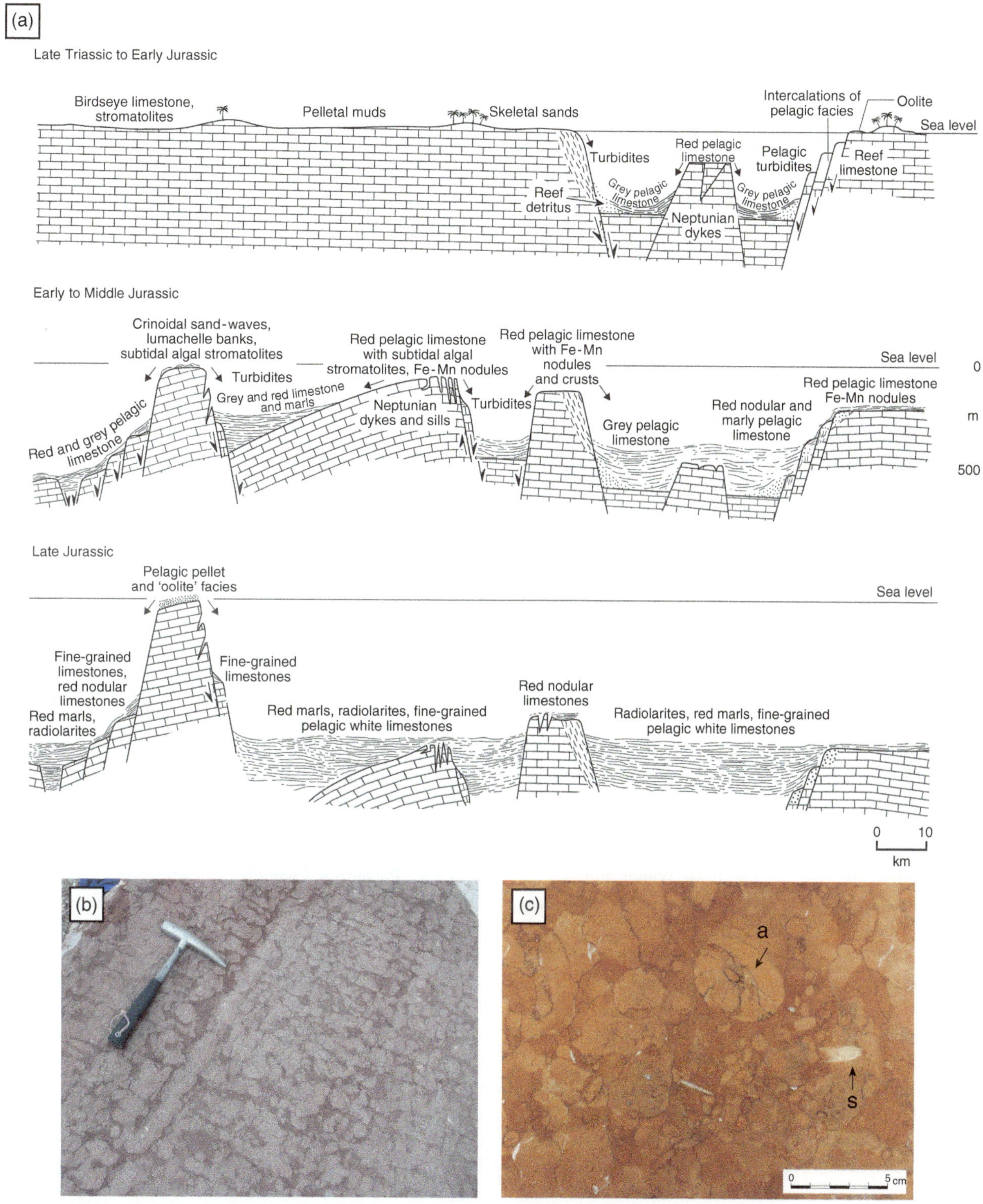

Figure 13.46 (a) Schematic cross sections of the southern Tethyan margin from the Late Triassic to Late Jurassic, showing the relationship between fault-bounded basins, facies development and variable sediment thicknesses. By the end of the Jurassic much of the faulting had ceased and the margin subsided thermally, becoming blanketed in the Early Cretaceous by a much more uniform thickness of biogenic pelagic sediment, with the exception of a few persistent areas of shallow-water on isolated seamounts and plataeus. Modified from Bernoulli and Jenkyns (1974) / Society for Sedimentary Geology (SEPM). (b) Cut block of Ammonitico Rosso from a quarry near Monte Verzegnis, Italy showing a view perpendicular to bedding. Courtesy of H.C. Jenkyns. (c): Small polished block of Ammonitico Rosso from Monte Inici, Sicily showing nodular texture, a partially dissolved ammonite labelled (a) and a structure interpreted as a stromatolite head labelled (s). Courtesy of H.C. Jenkyns and originally figured in Jenkyns (1974).

circulation patterns, or a difference in the ecological niches occupied by the radiolaria. Evidence for the latter is provided by the extremely widespread occurrence of Jurassic (and even older – back to the Devonian) radiolarites and bedded-cherts in accreted terranes around the Pacific Ocean (see Baumgartner 2013). Although some early workers thought these deposits originated in small, marginal basins (analogous to the diatomites of the Gulf of California; Jenkyns and Winterer 1982), it is now thought, on the basis of palaeomagnetic evidence, that many of the accreted chert units were likely deposited in open-ocean settings, far from terrigenous nutrient or sediment supply (Racki and Cordey 2000; Baumgartner 2013). The persistence of radiolarite deposition for >100 Myr in some accreted terranes (e.g. the Bridge River complex of Canada; Cordey and Schiarizza 1993) is highly suggestive of extensive areas of persistent radiolarian deposition for long periods of time – open-ocean basins are the obvious contender. The view that radiolarians were once prevalent across a wide range of palaeolatitudes, palaeolongitudes and depositional settings is further supported by the presence of radiolarians within a wide variety of other sedimentary deposits found in Palaeozoic-Mesozoic accreted terranes (Racki and Cordey 2000).

13.5.2.5 Cephalopod Limestones of the Palaeozoic Continental Margins

Condensed, cephalopod limestones, similar to the Jurassic Rosso Ammonitico described above but lacking any components from planktonic carbonate producers, are well-known from the Late Palaeozoic (Devonian–Carboniferous) of Europe and North Africa and referred to as the 'Griotte' or 'Cephalopodenkalk' (e.g. Tucker 1974; Wendt and Aigner 1985; Hüneke et al. 2023). Similar deposits are also known from the Ordovician and Silurian (e.g. Histon 2012; Bábek et al. 2021). The faunas of the Late Palaeozoic limestones are 'pelagic', dominated by cephalopods but also including conodonts, rare radiolaria and, small, thin-shelled bivalves and gastropods (Tucker 1974) that are, purportedly, nektonic or nektoplanktonic. Other potential micro-palaeontological indicators of the depositional environment include *Tentaculites* (a group of small conical fossils of uncertain affinity) and entomozoacean ostracods, both of which have been interpreted by some authors as pelagic indicators, although the ecology of both remains uncertain. Benthic carbonate sources, including larger bivalves and crinoids are secondary in significance but can be locally abundant. Unlike many Late Mesozoic–Recent pelagic deposits containing abundant coccolith-sourced fine-grained carbonate, the micrite present in the Griotte and similar deposits most likely relates either to re-deposition of peri-platform ooze from shallow-water settings or the mechanical breakdown of macrofossils. The cephalopod limestones were deposited in a wide range of continental margin settings: typically, shallow platforms or ridges, shelves, slopes and some starved basins (e.g. Wendt and Aigner 1985). On the bathymetric highs, the limestones are typically highly condensed, with well-developed hardgrounds, nodularity and encrustation by ferromanganese minerals, leading to a characteristic 'variegated' appearance. Sedimentation rates are estimated at 1–5 mm kyr^{-1}. On the Devonian Tafilalt Platform of North Africa, at least, bottom-water currents were a significant control on facies variations and depositional architecture on the platforms, leading to the development of condensed limestones and contourite deposits (Hüneke et al. 2023). In basinal areas, the lateral equivalents of the Griotte-type limestones are typically thicker and comprise nodular, marly limestones and shales (Figure 13.47). Evidence for re-deposition in the form of slumps and debris flows can be locally abundant. With increasing distance from the platform regions, clays become more dominant. Condensed carbonate facies do however also occur in basinal settings, where they may be rhythmically interbedded with turbidites. Estimated water depths for the Griotte facies deposition vary widely, from tens to thousands of metres. These limestones are found in association with both shallow-water (i.e. peri-tidal) and deeper-water (i.e. turbidite) indicators. The wide variety of water depth estimates probably reflects the importance of clastic sediment starvation in allowing pelagic carbonate facies to develop, particularly in those periods lacking abundant calcareous plankton.

13.5.2.6 Palaeozoic Ophiolites and Accretionary Complexes

In the case of the Palaeozoic and Early Mesozoic, ophiolites and accretionary complexes provide the only available direct insight into sedimentation in the deep-ocean prior to the evolution of the modern plankton. Interpretation of sediments associated with ophiolites is not without its difficulties. The geochemistry of ophiolitic basalts often suggests that they formed above subduction zones and not as 'normal' oceanic crust at mature mid-ocean ridges (summarised in Dilek and Furnes 2014). Obduction and later tectonism or preservation in complex tectonic melanges mean that many ophiolitic bodies are highly fragmented and original vertical or lateral stratigraphic relationships can be hard to discern. The Late Cambrian to Ordovician Ballantrae Complex and Southern Upland Belt of Scotland are classic examples of an ophiolitic fragment and an accretionary prism, respectively, that both

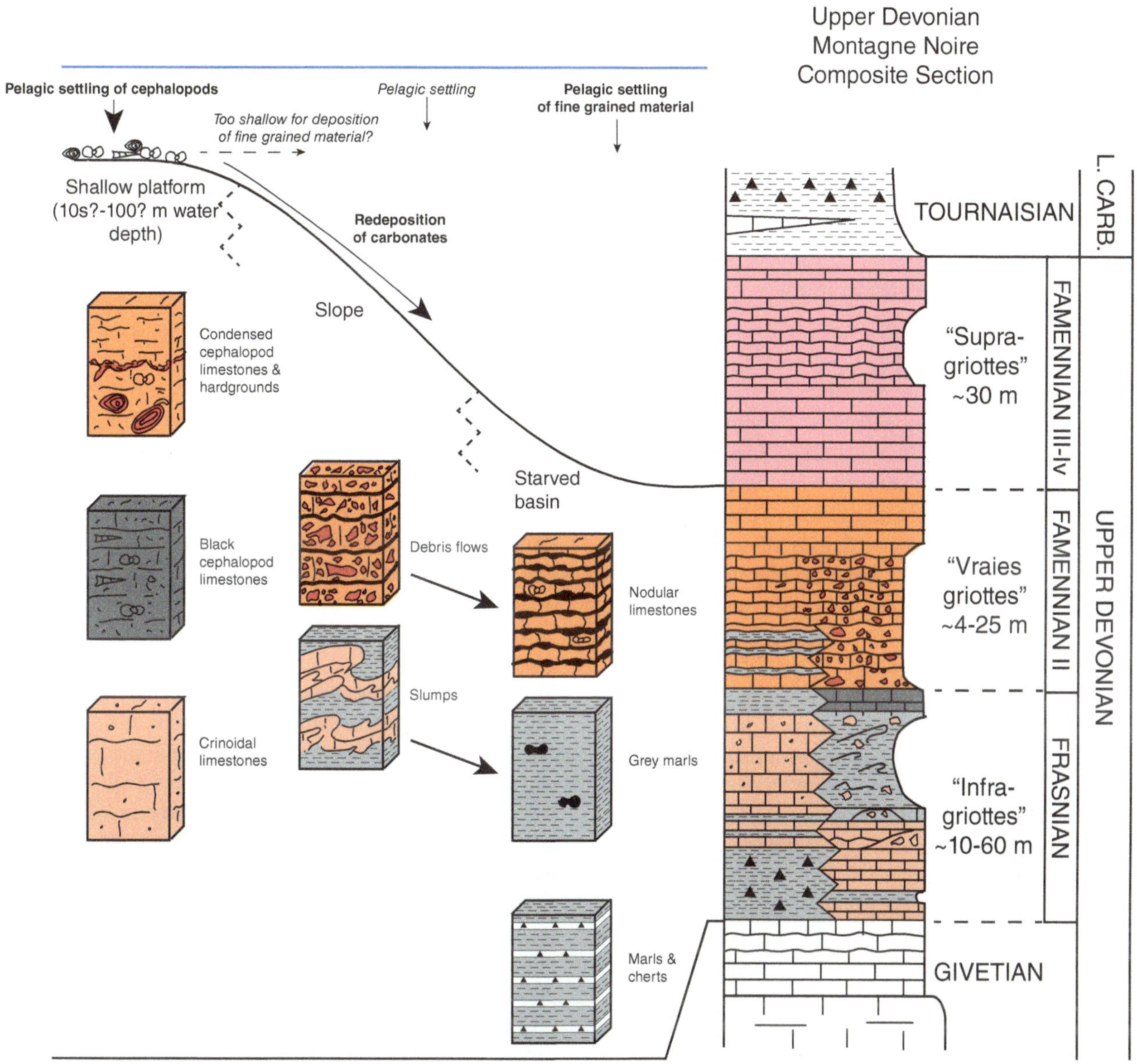

Figure 13.47 A schematic representation of the facies and key depositional processes of the Upper Devonian pelagic cephalopod limestones ('Griotte') and associated sediments from the Montagne Noire region of France. The arrows show how the slope and basinal facies transition into each other. The colours shown are largely schematic as these units can be variegated. Modified from Wendt and Aigner (1985) / With permission of Elsevier.

formed during the closure of the Iapetus Ocean (e.g. Stone 2014). Truly pelagic sediments are associated with both (e.g. Leggett 1979; Leggett et al. 1979; Aitchison 1998; Stone 2014). The basaltic crustal rocks of the Ballantrae Complex are variously massive, brecciated or in pillows and are associated with biogenic pelagic sediments including radiolarian cherts (Aitchison 1998) that contain conodonts. In the Southern Uplands, basalts associated with cherts and metalliferous muds, potentially analogous to modern hydrothermal deposits, have been reported (Leggett and Smith 1980). The cherts of the Southern Uplands are commonly interbedded on the scale of metres with mudstones, that also contain radiolaria. Differences in the proportions of radiolaria and mud in cm-scale thin beds have been attributed to transport and re-deposition of the radiolaria by bottom-water currents or downslope movement (e.g. Leggett 1979). Within the cherts and mudstones, trace fossils similar to *Planolites and Teichichnus* have been reported, consistent with a deep-water setting (Kakuwa and Floyd 2016). Arc-derived, altered bentonites provide clear evidence for the subduction-related depositional environment associated with the sediments of the Southern Uplands. The radiolaria-bearing sediments are predominantly Early to Middle Ordovician in age and are overlain by Late Ordovician organic-rich graptolitic shales. These likely represent (hemi) pelagic deposition in low-oxygen conditions, perhaps induced by high-productivity and an expanded oxygen minimum zone (Leggett 1979).

13.5.3 Pelagic Sedimentation in the Precambrian?

Although there are a number of purported ophiolites and accretionary complexes of Precambrian age (e.g. Moores 2002), these are often very difficult to interpret due to subsequent tectonism and, in some cases, fundamental disputes regarding the tectonic origins of the mafic igneous rocks (similar to the discussions about Palaeozoic ophiolites). The sediments associated with these Precambrian mafic and ultramafic rocks range widely in composition and mode of deposition. Some are clearly not pelagic (e.g. turbidites; Moores 2002) or are unlikely to be globally widespread for most of the Precambrian (e.g. diamictites of glacial origin; Stern 1981; Stern et al. 2004). Other lithologies, including cherts, dolomites and banded iron formations (chemogenic sediments commonly comprising alternating layers of iron-oxides and cherts), may be more representative of Precambrian pelagic sedimentation. However, interpreting the origins of these 'pelagic' Precambrian sediments is also not straightforward given an absence of primary biogenic sediment producers, uncertainties over basinal settings and seawater chemistry, and the potential for a local hydrothermal origin for some lithologies (e.g. cherts and banded iron formations).

Because of the difficulties inherent in working with Precambrian ophiolites, our understanding of the Precambrian oceans is mainly reliant on sediments that were deposited in sedimentary basins on continental crust. Like the ophiolites, these sediments have also commonly experienced significant diagenesis and even metamorphism in some cases. Estimating depositional water depths for sediments deposited below the storm-wave base is, in the absence of depth-related ichnofacies, difficult for Precambrian sedimentary rocks (infaunal bioturbating organisms only became widespread from the Cambrian onwards). Convincingly identifying facies as 'deep-water pelagic' can, therefore, be problematic. Furthermore, arguing from the basis of a record preserved in sedimentary basins on continental crust what the geochemistry of the contemporary oceans were like, requires an understanding of the nature of oceanographic connections between these, potentially, more marginal basins and the open, pelagic Precambrian oceans. Geochemical extrapolations to the open-ocean must be approached cautiously.

Despite all these difficulties, however, it is certainly interesting to speculate on what open-ocean pelagic sedimentation was possibly like in a world so very different from the modern: characterised, at times, by very low levels of oxygen, high levels of CO_2 and, throughout, by the absence of any biogenic pelagic carbonate or silica production. Following the evolution of oxygenic photosynthesis and the ensuing Great Oxidation Event (~2400 to 2000 Ma), cyanobacteria likely occupied the phytoplanktonic ecologic niche. Some Proterozoic acritarchs (organic-walled unicellular microfossils of uncertain origin) may have been phytoplankton (Armstrong and Brasier 2005; Butterfield 2015), although it has also been suggested that some at least were heterotrophic and benthic (Butterfield 1997, 2007). As in the modern ocean, much of the primary production in the Proterozoic was likely undertaken by the picoplankton (0.2–2 μm in diameter) that would have been consumed by microzooplankton in the upper water column (either heterotrophic bacteria or early eukaryotes). The microzooplankton are not capable of producing faecal pellets, so there would have been very little export production (e.g. Butterfield 1997). Through the Proterozoic and into the Early Phanerozoic, the size of planktonic organisms and the complexity of pelagic ecosystems evolved. It has been hypothesised that these evolutionary and ecological changes caused changes in export production, through the production of larger-sized particles and faecal pellets, more capable of sinking into the ocean interior. However, modelling work strongly suggests that despite these biological changes, temperature and oxygen concentrations played more critical roles in determining the strength of the biological pump and the efficiency with which organic carbon could be transferred to the seafloor (Fakhraee et al. 2020). As a consequence of this interplay between ecological and environmental factors, sediments deposited in an ocean with low primary production, but a highly efficient biological pump, could have had similar TOCs to sediments deposited during times of higher productivity, larger cell-sizes and more zooplankton but with a more inefficient biological pump.

The absence of CO_2-buffering pelagic carbonate producers in the Precambrian may have resulted in an unstable carbon cycle (e.g. Ridgwell and Zeebe 2005). This was likely more significant for shallow-water environments than for the deep-ocean since the deep ocean apparently lacked a source of carbonate sediment anyway. The absence of silica producers in the Precambrian will have meant higher concentrations of SiO_2 in seawater (Conley et al. 2017) which may have

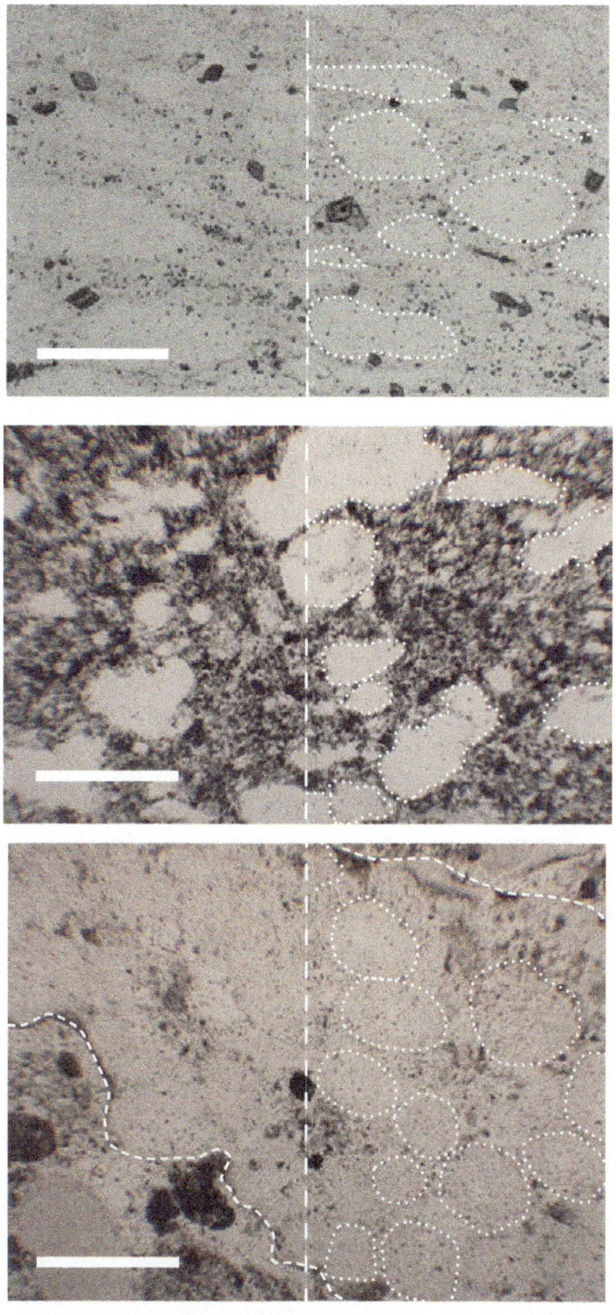

Figure 13.48 Examples of silica granules (outlined by dotted white lines on the right-hand side of each image), preserved as quartz, that are thought to be evidence for direct precipitation of silica within Archaean seawater. Scale bars are 500 μm. These examples are from Archaean sedimentary rocks exposed in South Africa dating to ~3.3 Ga that were deposited in relatively low-energy environments below the storm-wave base. Reproduced from Stefurak et al. (2015) / Geological Society of America.

permitted easier production of authigenic clays, Fe(II) silicates, and/or amorphous silica compared to the modern (e.g. Siever 1992). In the Archaean, intriguing granules of silica have been observed in shallow to basinal settings (Figure 13.48) and it has been argued that these granules were precipitated directly from seawater and then deposited through suspension (Stefurak et al. 2015). If so, this clearly underlines how our knowledge of pelagic sedimentation based upon the modern ocean is incomplete when viewed through the lens of the entirety of Earth history!

13.6 Significance and Applications of Pelagic Sediments and Environments

13.6.1 Pelagic Sediments as a Fundamental Component of the Earth System

As has been discussed, biogenic pelagic sediments play a fundamental role in regulating ocean chemistry and climate through their role in the carbon cycle (and other linked biogeochemical processes). Furthermore, the deposition of biogenic and terrigenous pelagic sediments upon ocean crust, means that they (and the pore fluids they entrain), play a role in the subduction process, the chemistry of arc volcanism and of the mantle, and elemental cycling on plate-tectonic timescales (e.g. Plank and Langmuir 1998). Pelagic sediments are a critical element of the Earth system that connects the surficial environment with solid Earth processes. The nature and significance of the role they have played over time has varied with evolution of different planktonic groups, such that it is difficult to generalise. In the modern world, the existence of pelagic carbonate producers and sediment deposition provides an important buffer on ocean chemistry and climate that allows CO_2 to be dissolved in the deep-ocean and stored as sediment. Prior to the evolution of diverse pelagic calcifiers in the mid-Mesozoic, the oceanic carbonate saturation state was highly susceptible to changes in the extent and abundance of shallow-water carbonate factories that are very sensitive to changes in sea-level, weathering rates and oceanic calcium concentrations (Section 11.3, Ridgwell and Zeebe 2005; Ridgwell 2005). Although carbonate-rich pelagic sediments are estimated to be a relatively small component of the total sediment subducted at present (approximately 20%; e.g. Plank and Langmuir 1998), that may well reflect the current configuration of subduction zones in relation to terrigenous sediment supply. It is reasonable to ask how the processes and products of subduction and mantle chemistry were affected when terrigenous sediment supply, plate tectonic configurations, and the composition and volumetric contributions of pelagic sediments to subduction were different.

13.6.2 The Pelagic Sediment Record – An Archive of Earth History

Pelagic sediments are in principle a key archive of Earth's climatic and oceanographic history, but in order to use this record to reconstruct past climates and oceanography, we need to convert the *thickness* of sediment in an archive into the *time* it represents. Where sediment redistribution is minor or absent a simplistic view of steady and continuous (Chapter 4) pelagic sediment, accumulation is a plausible assumption over relatively long intervals of geological time (say tens of thousands to millions of years). However, even in such settings with typical sedimentation rates of 1–10 cm ka^{-1}, caution must be exercised with relatively brief intervals of time (e.g. <10 kyrs) and thin beds, as bioturbation and potentially diagenetic un-mixing will result in averaging effects. For example, with a depth of bioturbation of 10 cm and a sedimentation rate of 10 cm ka^{-1}, 1000 years of sedimentation will, potentially, be mixed together. At 1 cm ka^{-1}, the same depth of bioturbation could mix together sedimentation that occurred over 10,000 years. The implication is that high-temporal resolution records can only be found in high-sedimentation rate environments where the averaging effect of mixing integrates a smaller interval of time, or in low O_2 environments that exclude benthic organisms and hence reduce bioturbation. The relationship between thickness and time is called an 'age model'. This is built by identifying marker horizons for which the age is known, either absolutely through radiometric ages (e.g. ^{14}C of organic matter or carbonate, $^{40}Ar/^{39}Ar$ ages of tephras) or relatively through biostratigraphic events (e.g. Base or Top occurrence of a particular microfossil), magnetic polarity reversals or geochemical anomalies. With a set of age-calibrated tie points, it is then possible to model continuous sediment accumulation by, for example, assuming constant sedimentation rates between pairs of tie points or by fitting a linear regression to a set of tie points. The age model allows ages to be assigned to depths and samples between the tie points.

A more detailed approach to constructing an age model relies on recognising orbital cycles within a sedimentary sequence, alongside bio and magnetostratigraphic markers ('astrochronology'; Weedon 2003, Hinnov and Hilgen 2012). Models of solar system dynamics, taking into account the orbits and gravitational interactions between the planets, their Moons and large asteroids, can be used to calculate Earth's precession, obliquity and eccentricity over the last 50 Ma (Box 13.4; Laskar et al. 2004, 2011). For older time intervals, this is more problematic due to uncertainties in orbital parameters and the somewhat chaotic behaviour of large asteroids. Nonetheless, for much of the Cenozoic, it is possible to 'tune' those cyclic records that are demonstrably orbitally forced to the astronomical solution and thus determine the age-thickness relationship of a sedimentary sequence with a high degree of precision. The chain of reasoning is long. Cyclicity is first established by measuring the physical or chemical properties (e.g. colour, magnetic susceptibility, gamma radiation, carbonate content, stable oxygen or carbon isotopes) of closely spaced samples. Samples need to be closely and, ideally, evenly spaced such that the shortest cycle present is adequately sampled (e.g. Weedon 2003). Many palaeoceanographers would recommend at least 8–10 samples per shortest cycle (i.e. to recognise a 24 kyr precession cycle, sample spacing would need to equate to <3000 years). The cyclical properties, after rigorous testing for statistical significance (e.g. Weedon 2003, 2022), are then linked by the available (geochemical) evidence to climatic/oceanographic processes that are themselves responses to changes in orbital forcing. In different settings and time periods, different properties will exhibit a variety of responses to orbital forcing (and these are almost always indirect responses (Box 13.4).

With a robust chronology, it becomes possible to compare pelagic records from many different sites and, therefore, reconstruct spatial and temporal variations in pelagic environments. As discussed in Section 13.5, the pelagic sedimentary record has revealed the palaeoceanographic history of the last 100 Myrs in detail, providing sedimentological evidence for events such as the OAEs, glacial–interglacial cycles and millennial-scale climate perturbations. Other events also stand out in the pelagic sedimentary record: the end-Cretaceous mass extinction is typically identified by the appearance of a thin layer of clay and impact spherules at the boundary sourced from the Chicxulub impact (Schulte et al. 2010; Hull et al. 2020) whereas the Paleocene–Eocene Thermal Maximum, a period of rapid warming and increased CO_2, is marked in many deep-sea records by a transient decline in carbonate content caused by a rapid shoaling of the CCD (Zachos et al. 2005). However, in order to develop a detailed understanding of such events, and the intervening periods of background oceanographic and climatic conditions, quantified reconstructions of the Earth system are needed that utilise proxies for climatic and oceanographic variables. Given their wide geographic distribution and relative continuity of deposition over long periods of time, pelagic sediments are excellent archives for this purpose. Proxies commonly used in palaeoceanography and palaeoclimatology (e.g. Table 13.2) all share some common aspects. All proxies are grounded in an empirical understanding of how the proxy measurement relates to the variable of interest either through laboratory or field studies (note, though, that for some proxies, multiple processes may influence the measured value, and in some cases, it might not yet be known what all those processes are!). Refinement

Table 13.2 Examples of proxies commonly used in pelagic settings to reconstruct past oceanic, climatic and environmental variability (Cronin 2010 / Columbia University Press).

Parameter of interest	Proxy name	Common timescale of application	Comments	Further reading
Sea-surface temperature (SSTs)	Faunal/floral assemblages	Cenozoic	Common approaches use the distribution of modern planktonic foraminifera or to reconstruct past temperature distributions. Was critical in many early reconstructions of Last Glacial Maximum temperature estimates	MARGO project members (2009)
	Alkenones (U_k^{37})	Cenozoic	Long-chain hydrocarbon molecules (biomarkers) produced by coccolithophores that change in relative abundance due to growth temperature	Herbert (2014)
	Mg/Ca of planktonic foraminifera	Cenozoic	Mg incorporation into carbonate increases at higher temperature; Mg/Ca composition of seawater needs to be estimated	Lea (2014)
	TEX_{86}	Jurassic–Recent	Biomarkers produced by marine archaea that change in relative abundance due to growth temperature (and other factors)	Inglis and Tierney (2020)
	$\delta^{18}O$ in planktonic foraminifera	Cretaceous and Early Cenozoic	Can be used as an SST proxy if $\delta^{18}O$ of local seawater can be estimated (i.e. ice volume and local salinity are known) and if the preservation is suitable	Pearson (2012)
Sea surface salinity	$\delta^{18}O$ in planktonic foraminifera	Cenozoic	If $\delta^{18}O$ is used in combination with an independent palaeothermometer (e.g. Mg/Ca), then local seawater $\delta^{18}O$ and salinity can be reconstructed	Pearson (2012)
Bottom water temperatures (BWTs)	$\delta^{18}O$ in benthic foraminifera	Cretaceous and Early Cenozoic	In an ice-free world, and assuming an isotopically well-mixed deep ocean, benthic $\delta^{18}O$ reflects temperature	Pearson (2012)
	Mg/Ca or Δ_{47} (clumped isotopes) in benthic foraminifera	Cenozoic	These proxies for temperature are independent of salinity and ice volume. Mg/Ca requires estimation of seawater Mg/Ca	Lea (2014); Meckler et al. (2022)
Ocean productivity	Micropalaeontological assemblages	Mesozoic–Cenozoic	Use inferred ecology of planktonic species to infer nutrient levels	Kucera (2007)
	Biogenic opal, organic carbon and carbonate accumulation rates	Cenozoic	Use accumulation rates to infer levels of surface water productivity	Ragueneau et al. (2000)
	Biogenic barite accumulation rates and Ba/Al	Cretaceous and Cenozoic	Biogenic barite ($BaSO_4$) forms within reducing microenvironments present in marine snow. Areas of high productivity thus have high rates of biogenic barite accumulation. Productivity can also be inferred from bulk sediment Ba/Al ratios (normalised to Al to account for changes in terrigenous sources of Ba)	Paytan and Griffith (2007)
Oceanic oxygenation	Biomarkers (e.g. isorenieratene)	Phanerozoic	Some biomarkers are produced by photosynthetic green sulphur bacteria, that require both light and free H_2S in the water column (i.e. euxinic conditions)	Grice and Eiserbeck (2014)
	Redox-sensitive trace metals (e.g. Cr, V and Mo)	Proterozoic–Phanerozoic	Changes in redox can affect the solubility or species of some metals, which can lead to changes in their abundance or isotopic composition in seawater	Tribovillard et al. (2006)
Deep ocean circulation	Radiogenic isotopes of Nd, Pb	Cretaceous to Cenozoic (Nd), Cenozoic (Pb)	Measured in biogenic apatite (Nd only) and ferromanganese crusts and coatings. Isotopes of Nd and Pb vary in different rock types and water	Goldstein and Hemming (2014)

(Continued)

Table 13.2 (Continued)

Parameter of interest	Proxy name	Common timescale of application	Comments	Further reading
			masses, so potentially provide information about sediment sources and water mass mixing	
	$\delta^{13}C$ in benthic foraminifera Cd/Ca ratios in foraminifera	Cenozoic Quaternary	Carbon-isotope values of deep-waters reflect the source and age of the water. Older waters are more enriched in ^{12}C and Cd due to more organic carbon respiration	Lynch-Stieglitz and Marchitto (2014)
	Sortable silt	Late Quaternary	Higher bottom-water current velocities can transport higher abundances of relatively coarse silt	McCave et al. (2017)
Atmospheric CO_2	$\delta^{13}C$ of alkenones, $\delta^{13}C$ of phytane	Cenozoic Palaeozoic to recent	Alkenones and phytane are organic molecules produced by photosynthesising plankton. The amount of stable carbon-isotope fractionation during photosynthesis is, in part, controlled by the concentration of dissolved CO_2, which can be related to atmospheric CO_2	Pagani (2014); Witkowski et al. (2018)
	$\delta^{11}B$ of planktonic foraminifera	Cenozoic	The stable boron-isotopic composition of seawater, recorded by carbonates, is sensitive to the pH of the water, which, in the surface ocean, can be related to atmospheric CO_2	Foster and Rae (2016)
Ice volume	$\delta^{18}O$ of benthic foraminifera	Cenozoic	When used in combination with a benthic temperature proxy (e.g. Mg/Ca, Δ_{47} ratios), seawater $\delta^{18}O$ can be reconstructed and related to changes in continental ice volume	Lear et al. (2000); Meckler et al. (2022)
Continental weathering	Strontium isotopes ($^{87}Sr/^{86}Sr$) Osmium isotopes ($^{187}Os/^{186}Os$)	Phanerozoic	In seawater, these isotopic ratios reflect (mainly) a balance between weathering of continental and ocean crust, which changes over long intervals (millions of years) and can be recorded by sediments	Ravizza and Zachos (2014)

and more detailed understanding of how an individual proxy works, and the limitations of each proxy, are a normal part of proxy method development and a continual process, even for the most established approaches. A useful proxy is one which can be applied in multiple settings across a wide expanse of space and time and, ideally, is relatively unsusceptible to diagenesis (although this is often impossible to avoid if samples have been buried sufficiently).

Of particular importance to palaeoceanographic studies are the oxygen and carbon-isotopic compositions of benthic and planktonic foraminifera (Pearson 2012). The oxygen-isotopic composition of deep-sea benthic foraminifera is determined largely, by the oxygen-isotopic composition of seawater, controlled by continental ice volume and the temperature of the deep-sea. Given the relative uniformity of global deep-sea temperatures and $\delta^{18}O$ values, global compilations of benthic foraminiferal $\delta^{18}O$ from the Cenozoic provide a high-resolution, continuous record of deep-sea ocean temperature and ice volume (Zachos et al. 2001, 2008; Westerhold et al. 2020; Figure 13.49). For the Cenozoic, this record provides a global view of climatic and environmental change against which many other studies from marine and terrestrial environments are compared. Overall, the record shows increasing $\delta^{18}O$ values towards the present day, representing long-term global cooling and the establishment of continental ice sheets. Variations in benthic $\delta^{13}C$ values can be used to understand changes in the long-term global carbon cycle and their connection to the major climatic events evident in the $\delta^{18}O$ record. When the benthic $\delta^{18}O$ and $\delta^{13}C$ records are compared with independent evidence from sedimentology, tectonics and palaeontology (Figure 13.49), it becomes possible to interpret the short- and long-term variations in the Earth system over the course of the Cenozoic and understanding how climate forcings have varied over time.

13.6.3 What Does the Future Hold for Pelagic Sedimentary Environments?

Although far from land, the open-ocean and pelagic marine environments are by no means immune to the activities of humanity. Changes in surface ocean temperatures and pH

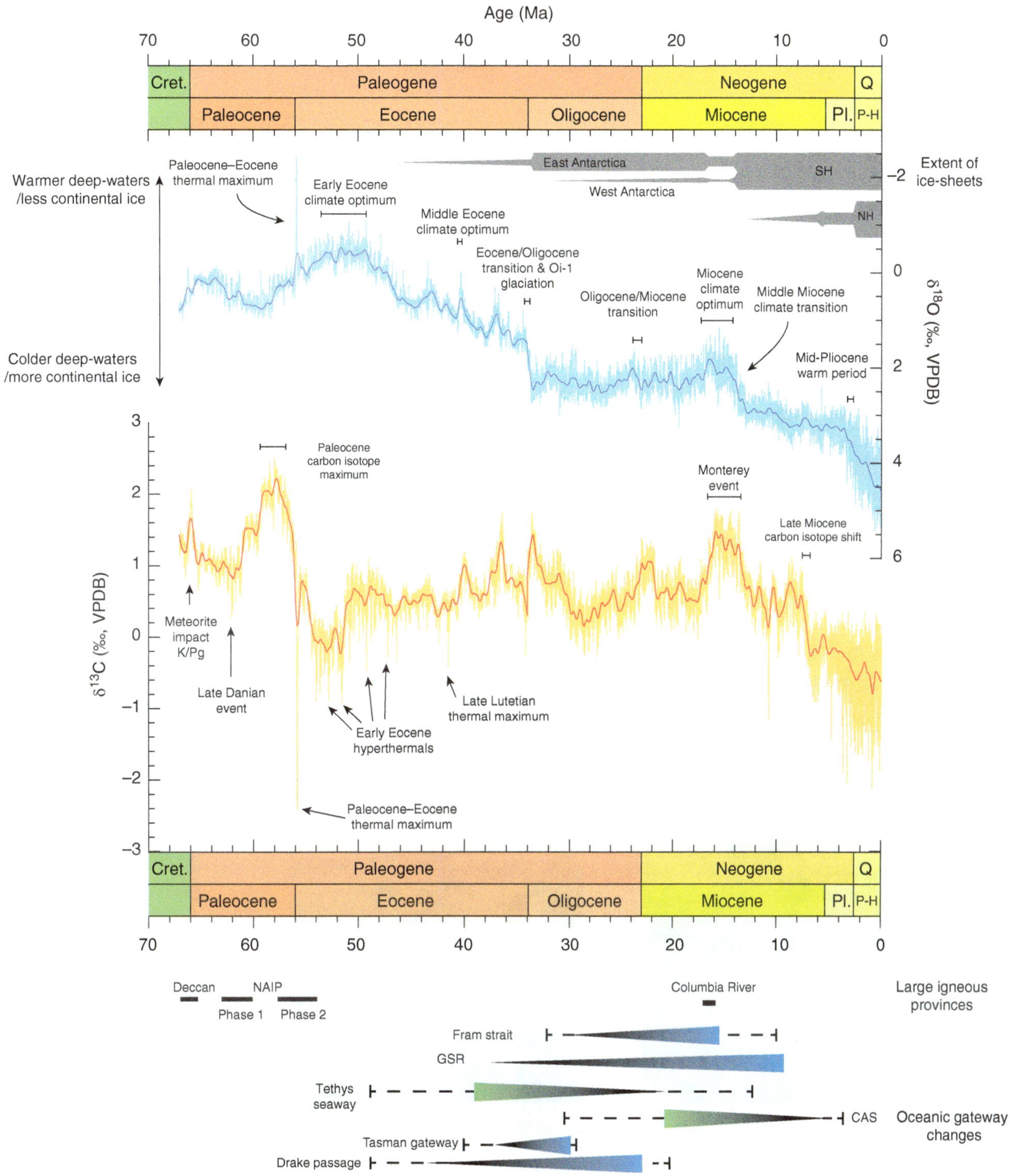

Figure 13.49 A compilation of globally sourced stable oxygen- and carbon-isotope data ($\delta^{18}O$, $\delta^{13}C$) from benthic foraminiferal carbonate, sampled from pelagic settings with water depths >1000 m (CENOGRID; adapted from Westerhold et al. 2020). Also shown are the timings of major climatic and carbon cycle events and the extent of northern (NH) and southern hemisphere (SH) icesheets. Some of these changes may have been caused by large igneous province eruptions and opening (blue) and closing (green) of oceanic gateways (the latter taken from Straume et al. 2020). The dashed lines on the gateways show the range of estimate ages for changes in each gateway, whereas the triangles show the modelled gateway evolution of Straume et al. (2020). The solid lines shown for the isotope records are the result of smoothing the original data (shown as the higher-frequency background) over 1 Ma. NAIP = North Atlantic Igneous Province; GSR = Greenland-Scotland Ridge; CAS = Central American Seaway.

are already evident in response to increased atmospheric CO_2 and there is a growing body of evidence and models that suggest that these changes will be significant for planktonic ecosystems, and, by extension, pelagic sediment production through, for example, reduced carbonate production (e.g. Grigoratou et al. 2022). Much of the anthropogenically sourced CO_2 has been absorbed by the ocean, which, over centennial to millennial timescales, will likely lead to a decrease in deep-ocean pH, carbonate saturation and a shoaling of the ACD and CCD (Ridgwell and Schmidt 2010; Figure 13.50), thereby impacting carbonate preservation: it has been estimated that the CCD has already risen by ~300 m in the western North Atlantic since the industrial revolution (Sulpis et al. 2018; Figure 13.51).

Over the last 50 years, there has been a significant decline in dissolved oxygen concentrations in both coastal settings and the upper open-ocean (Breitburg et al. 2018). The former can be attributed largely to eutrophication associated with nutrient run-off (largely derived from agriculture) while the latter is ultimately caused by a global warming-driven decrease in oxygen solubility and, in places, ocean stratification. Coupled with the potential for increased wind-driven upwelling in a warmer world (wind-driven increases in current velocities have already been observed and modelled; e.g. Hu et al. 2020), there may be future local increases in organic-carbon burial rates in pelagic settings. The coeval changes in nutrient availability and productivity in the surface ocean may also drive changes in sediment production and preservation. However, the extent of deoxygenation is unlikely to reach that attained in the Mesozoic when ocean basin configurations and global climate were very different.

The modern and future changes in production and preservation of organic carbon and carbonate may impart a signature on the future pelagic sedimentary record, depending on how long carbon dioxide emissions continue to rise and the full magnitude of global warming. Eventually, once emissions cease, negative, relatively slow feedbacks in the carbon cycle (including further deep-ocean carbon storage, increased silicate weathering, and/or increased organic carbon burial) will reduce atmospheric CO_2. These feedbacks will return the carbon-cycle and global climate to a state of quasi-equilibrium over time, albeit one that may be different from the pre-industrial world. However, models suggest that this may take thousands of years to achieve, particularly in the case of deep-ocean carbonate saturation (possibly more than 10,000 years in some models, e.g. Figure 13.50; Ridgwell and Schmidt 2010). If the latter bears out as predicted, then a dissolution horizon coupled with negative carbon and oxygen-isotope excursions, analogous to past hyperthermal events such as the Paleocene–Eocene Thermal Maximum (Zachos et al. 2008; Foster et al. 2018),

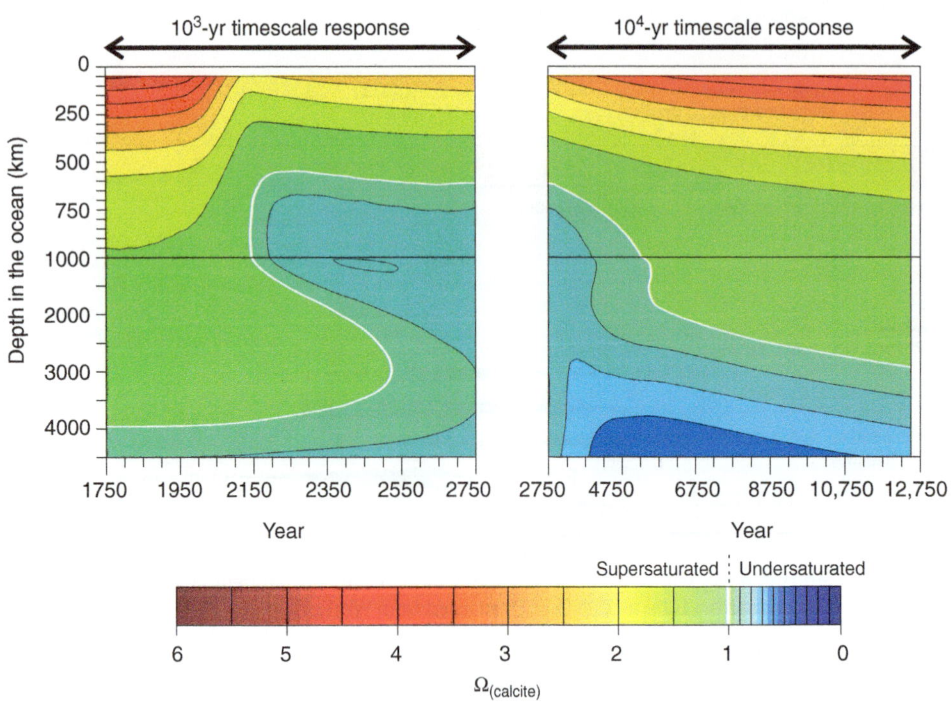

Figure 13.50 A model output of future ocean calcite saturation in response to increased CO_2 levels, using the historical emissions record to 2000 and then continued emissions to 2100, resulting in a total cumulative release of 2180 PgC. Adapted from Ridgwell and Schmidt (2010).

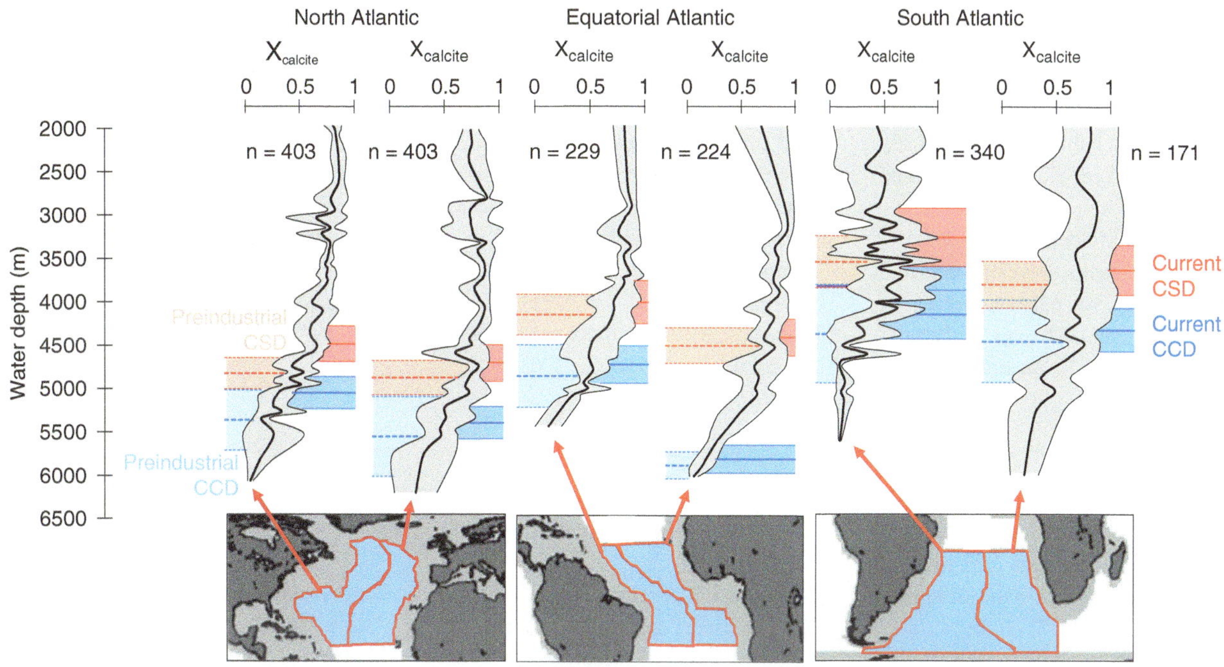

Figure 13.51 Proportion of calcite ($X_{calcite}$) in sediments at different depths in the Atlantic Ocean compared with the depths of the Preindustrial (dashed lines) and current (solid lines) CCD (Calcite Compensation Depth) and CSD (carbonate saturation depth). The Atlantic is divided into three different regions (North, Equatorial and South), each of which is split into a western (left) and eastern (right) part, allowing regional variability in the $X_{calcite}$, CCD and CSD to be observed. Note the significant difference (by about ~300 m) between the Preindustrial and Current depths of the CCD and CSD in the western North Atlantic. Error bars on the $X_{calcite}$ values, CCD and CSD are ±1 standard deviation. Sulpis et al. (2018) / Proceedings of the National Academy of Sciences PNAS.

may be evident in pelagic sediments of the future, depending on the duration of events and the homogenising effects of bioturbation. The horizon is also likely to be characterised by other signatures of human activity too, such as microplastics which have already been shown to be transported to deep-sea sediments (Woodall et al. 2014) via suspension settling and turbidity currents and may have long-term preservational potential (Kane and Clare 2019). The impact of these physical and chemical changes on benthic ecosystems and organisms is, as yet, unknown.

The concern for the future of pelagic environments is not just restricted to those characterised by biogenic sedimentation. The high concentrations of rare Earth metals in ferromanganese nodules have made them a target for deep-sea mining (Hein et al. 2020), but with a very incomplete knowledge of the geographic distribution of nodules and the associated ecosystems, there is considerable potential for harm, aspects of which, such as the effect of plumes of sediment, are only just starting to be fully appreciated (e.g. Washburn et al. 2023; Crane et al. 2024).

Further Reading

Bernoulli, D. and Jenkyns, H.C. (2009). Ancient oceans and continental margins of the Alpine-Mediterranean Tethys: deciphering clues from Mesozoic pelagic sediments and ophiolites. *Sedimentology* 56: 145–190. https://doi.org/10.1111/j.1365-3091.2008.01017.x.

Broecker, W.S. (1991). The great ocean conveyor. *Oceanography* 4: 79–89.

Dutkiewicz, A., Judge, A., and Müller, R.D. (2020). Environmental predictors of deep-sea polymetallic nodule occurrence in the global ocean. *Geology* 48: 293–297. https://doi.org/10.1130/G46836.1.

Dutkiewicz, A., Müller, R.D., O'Callaghan, S.O., and Jonasson, H. (2015). Census of seafloor sediments in the world's ocean. *Geology* 43: 795–798. https://doi.org/10.1130/G36883.1.

Ekdale, A.A. and Berger, W.H. (1978). Deep-sea ichnofacies: modern organism traces on and in pelagic carbonates of the western equatorial Pacific. *Palaeogeography, Palaeoclimatology, Palaeoecology* 23: 263–278. https://doi.org/10.1016/0031-0182(78)90096-2.

Hsü, K.J. (1992). *Challenger at Sea: A Ship That Revolutionized Earth Science*, 416.

Princeton, NJ: Princeton University Press.

Hsü, K.J. and Jenkyns, H.C. (ed.) (1974). *Pelagic Sediments: on Land and Under the Sea*, Special Publications of the IAS, vol. 1, 448. Oxford: Blackwell Scientific Publishing.

Hüneke, H. and Mulder, T. (ed.) (2011). *Deep-Sea Sediments*, Developments in Sedimentology, vol. 63, 849. Elsevier.

Jenkyns, H.C. (1986). Chapter 11: Pelagic environments. In: *Sedimentary Environments and Facies*, 2e (ed. H.G. Reading). Oxford: Blackwell Scientific Publishing.

Open University (1989a). *Ocean Chemistry and Deep-Sea sediments*, 134. Oxford: The Open University, Pergamon Press.

Palike, H. et al. (2012). A Cenozoic record of the equatorial Pacific carbonate compensation depth. *Nature* 488: 609–614.

Tierney, J.E. et al. (2020). Past climates inform our future. *Science* 370: eaay3701. https://doi.org/10.1126/science.aay3701.

Trujillo, A.P. and Thurman, H.V. (2020). *Essentials of Oceanography*, 13e, 626. Hoboken, NJ: Pearson.

Westerhold, T., Marwan, N., Drury, A.J. et al. (2020). An astronomically dated record of Earth's climate and its predictability over the last 66 million years. *Science* 369: 1383–1387. https://doi.org/10.1126/science.aba6853.

References

Aitchison, J.C. (1998). A lower Ordovician (Arenig) radiolarian fauna from the Ballantrae Complex, Scotland. *Scottish Journal of Geology* 34: 73–81. https://doi.org/10.1144/sjg34010073.

Alldredge, A.L. and Silver, M.W. (1988). Characteristics, dynamics and significance of marine Snow. *Progress in Oceanography* 20: 41–82.

Aller, R.C. (2014). 8.11 - Sedimentary diagenesis, depositional environments, and benthic fluxes. In: *Treatise on Geochemistry*, 2e (ed. H.D. Holland and K.K. Turekian), 293–334. https://doi.org/10.1016/B978-0-08-095975-7.00611-2.

Almogi-Labin, A., Hemleben, C., Meischner, D., and Erlenkeuser, H. (1991). Paleoenvironmental events during the last 13,000 years in the central Red Sea as recorded by pteropoda. *Paleoceanography* 6 (1): 83–98. https://doi.org/10.1029/90PA01881.

Anderson, O.R. (1983). *Radiolaria*, 355. New York: Springer-Verlag.

Anderson, R.F., Fleisher, M.Q., Lao, Y., and Winckler, G. (2008). Modern $CaCO_3$ preservation in equatorial Pacific sediments in the context of late-Pleistocene glacial cycles. *Marine Chemistry* 111: 30–46. https://doi.org/10.1016/j.marchem.2007.11.011.

Applin, E., Richards, A.E.E., and Kniker, H.T. (1925). Subsurface stratigraphy of the Coastal Plain of Texas and Louisiana. *AAPG Bulletin* 9: 79–122.

Archer, D.E. (1991). Equatorial Pacific calcite preservation cycles: production or dissolution? *Paleoceanography* 6 (5): 561–571. https://doi.org/10.1029/91PA01630.

Arfai, J., Lutz, R., Franke, D. et al. (2016). Mass-transport deposits and reservoir quality of upper cretaceous chalk within the German Central Graben, North Sea. *International Journal of Earth Sciences* 105: 797–818.

Armstrong, H.A. and Brasier, M.D. (2005). *Microfossils*, 2e, 296. Blackwell Publishing.

Armstrong, R.A., Lee, C., Hedges, J.I. et al. (2001). A new, mechanistic model for organic carbon fluxes in the ocean based on the quantitative association of POC with ballast minerals. *Deep Sea Research Part II: Topical Studies in Oceanography* 49: 219–236. https://doi.org/10.1016/S0967-0645(01)00101-1.

Arrhenius, G. (1988). Rate of production, dissolution and accumulation of biogenic solids in the ocean. *Palaeogeography Palaeoclimatology Palaeoecology* 67: 119–146.

Aze, T., Ezard, T.H.G., Purvis, A. et al. (2011). A phylogeny of Cenozoic macroperforate planktonic foraminifera from fossil data. *Biological Reviews* 86: 900–927. https://doi.org/10.1111/j.1469-185X.2011.00178.x.

Bábek, O., Kumpan, T., Calner, M. et al. (2021). Redox geochemistry of the red 'orthoceratite limestone' of Baltoscandia: possible linkage to mid-Ordovician palaeoceanographic changes. *Sedimentary Geology* 420: 105934. https://doi.org/10.1016/j.sedgeo.2021.105934.

Balch, W.M. (2018). The ecology, biogeochemistry and optical properties of coccolithophores. *Annual Review of Marine Science* 10: 71–98. https://doi.org/10.1146/annurev-marine-121916-063319.

Barker, S., Broecker, W., Clark, E., and Hajdas, I. (2007). Radiocarbon age offsets of foraminifera resulting from differential dissolution and fragmentation within the sedimentary bioturbated zone. *Paleoceanography* 22: PA2205. https://doi.org/10.1029/2006PA001354.

Barrett, T.J., Jarvis, I., Hannington, M.D., and Thirwall, M.D. (2021). Chemical characteristics of modern deep-sea metalliferous sediments in closed versus open basins, with emphasis on rare-earth elements and Nd isotopes. *Earth Science Reviews* 222: 103801. https://doi.org/10.1016/j.earscirev.2021.103801.

Barron, J.A. and Keller, G. (1982). Widespread Miocene deep-sea hiatuses: coincidence with periods of global cooling. *Geology* 10: 577–581. https://doi.org/10.1130/0091-7613(1982)10<577:WMDHCW>2.0.CO;2.

Batenburg, S.J., De Vleeschouwer, D., Sprovieri, M. et al. (2016). Orbital control on the timing of oceanic anoxia in the Late cretaceous. *Climate of the Past* 12: 1995–2009. https://doi.org/10.5194/cp-12-1995-2016.

Baumgartner, P.O. (2013). Mesozoic radiolarites – accumulation as a function of sea surface fertility on

Tethyan margins and in ocean basins. *Sedimentology* 60: 292–318. https://doi.org/10.1111/sed.12022.

Bé, A.W.H. and Gilmer, R.W. (1977). A zographic and taxonomic review of euthescosomatous Pteropoda. In: *Oceanic micropalaeontology*, vol. 1 (ed. A.T.S. Ramsay), 733–808. London: Academic Press.

Bé, A.W.H. and Tolderlund, D.S. (1971). Distribution and ecology of living planktonic Foraminifera in surface waters of the Atlantic and Indian Oceans. In: *The Micropalaeontology of Oceans* (ed. B.M. Funnell and W.R. Riedel), 105–150. Cambridge University Press.

Behl, R.J. (1999). Since Bramlette (1946): the Miocene Monterey Formation of California revisited. In: *Classic Cordilleran Concepts: A View from California*, vol. 338 (ed. E.M. Moores, D. Sloan, and D.L. Stout), 301–313. Boulder, CO: Geological Society of America Special Paper.

Berger, W.H. (1971). Sedimentation of planktonic foraminifera. *Marine Geology* 11: 325–358. https://doi.org/10.1016/0025-3227(71)90035-1.

Berger, W.H. (1973). Deep-sea carbonates: Pleistocene dissolution cycles. *Journal of Foraminiferal Research* 3: 187–195.

Berger, W.H. (1992). Pacific carbonate cycles revisited: arguments for and against productivity controls. In: *Centenary of Japanese Micropaleontology* (ed. K. Ishizaki and T. Saito), 15–25. Tokyo: Terra Scientific Publishing Company.

Berger, A. and Loutre, M.F. (1991). Insolation values for the climate of the last 10 million years. *Quaternary Science Reviews* 10: 297–317. https://doi.org/10.1016/0277-3791(91)90033-Q.

Berger, W.H. and Winterer, E.L. (1974). Plate stratigraphy and the fluctuating carbonate line. In: *Pelagic Sediments: On Land and Under the Sea*, vol. 1 (ed. K.J. Hsü and H.C. Jenkyns), 11–48. Special Publications of the IAS.

Berger, A. and Yin, Q. (2021). Orbital forcing (astronomical theory of paleoclimates). In: *Encyclopedia of Geology*, 2e (ed. D. Alderton and S.A. Elias), 435–443. Academic Press https://doi.org/10.1016/B978-0-12-409548-9.11990-6.

Berggren, W.A. and Boersma, A. (1969). Late Pleistocene and Holocene planktonic foraminifera from the Red Sea. In: *Hot Brines and Recent Heavy Metal Deposits in the Red Sea* (ed. E.T. Degens and D.A. Ross), 282–298. New York: Springer-Verlag.

Bernoulli, D. and Jenkyns, H.C. (1974). Alpine, Mediterranean, and Central Atlantic Mesozoic facies in relation to the early evolution of the Tethys. In: *Modern and Ancient Geosynclinal Sedimentation* (ed. R.H. Dott and R.H. Shaver), 129–160. Tulsa, OK: Special Publications of the SEPM, 19.

Bernoulli, D. and Jenkyns, H.C. (2009). Ancient oceans and continental margins of the Alpine-Mediterranean Tethys: deciphering clues from Mesozoic pelagic sediments and ophiolites. *Sedimentology* 56: 145–190. https://doi.org/10.1111/j.1365-3091.2008.01017.x.

Biard, T. (2022). Diversity and ecology of Radiolaria in modern oceans. *Environmental Microbiology* 24: 2179–2200. https://doi.org/10.1111/1462-2920.16004.

Bond, G., Broecker, W., Johnsen, S. et al. (1993). Correlations between climate records from North Atlantic sediments and Greenland ice. *Nature* 365: 143–147. https://doi.org/10.1038/365143a0.

Bond, G., Heinrich, H., Broecker, W. et al. (1992). Evidence for massive discharges of icebergs into the North Atlantic ocean during the last glacial period. *Nature* 360: 245–249. https://doi.org/10.1038/360245a0.

Bosworth, W. (2015). Geological evolution of the Red Sea: historical background, review, and synthesis. In: *The Red Sea: The Formation, Morphology, Oceanography and Environment of a Young Ocean Basin* (ed. N.M.A. Rasul and I.C.F. Stewart), 45–78. Berlin: Springer Earth System Sciences.

Bown, P.R., Dunkley Jones, T., Lees, J.A. et al. (2008). A Paleogene calcareous microfossil Konservat-Lagerstätte from the Kilwa Group of coastal Tanzania. *GSA Bulletin* 120: 3–12. https://doi.org/10.1130/B26261.1.

Bown, P.R. and Young, J.R. (1998). Introduction. In: *Calcareous Nannofossil Biostratigraphy* (ed. P.R. Bown), 1–15. Kluwer Academic Publishers.

Boyer, T.P., Garcia, H.E., Locarnini, R.A. et al. (2018). World Ocean Atlas 2018. NOAA National Centers for Environmental Information. Dataset. Schlitzer, Reiner, Ocean Data View, odv.awi.de, 2021. https://www.ncei.noaa.gov/archive/accession/NCEI-WOA18.

Breitburg, D. et al. (2018). Declining oxygen in the global ocean and coastal waters. *Science* 359: eaam7240. https://doi.org/10.1126/science.aam7240.

Broecker, W.S. (1987). The biggest chill. *Natural History Magazine* 97: 74–82.

Broecker, W.S. (1991). The great ocean conveyor. *Oceanography* 4: 79–89.

Bryłka, K., Alverson, A.J., Pickering, R.A. et al. (2023). Uncertainties surrounding the oldest fossil record of diatoms. *Scientific Reports* 13: 8047. https://doi.org/10.1038/s41598-023-35078-8.

Butterfield, N. (1997). Plankton ecology and the Proterozoic-Phanerozoic transition. *Paleobiology* 23 (2): 247–262. https://doi.org/10.1017/S009483730001681X.

Butterfield, N.J. (2007). Macroevolution and macroecology through deep time. *Palaeontology* 50: 41–55. https://doi.org/10.1111/j.1475-4983.2006.00613.x.

Butterfield, N.J. (2015). Early evolution of the Eukaryota. *Palaeontology* 58: 5–17. https://doi.org/10.1111/pala.12139.

Calvert, S.E. (1966). Origin of diatom-rich, varved sediments from the Gulf of California. *The Journal of Geology* 74 (5): 546–565. http://www.jstor.org/stable/30059298.

Calvès, G., Mix, A., Giosan, L. et al. (2022). The Nazca Drift System – palaeoceanographic significance of a giant sleeping on the SE Pacific Ocean floor. *Geological Magazine* 159: 322–336. https://doi.org/10.1017/S0016756821000960.

Carey, S.N. and Schneider, J.-L. (2011). Volcaniclastic processes and deposits in the Deep-Sea. In: *Deep-Sea Sediments*, Developments in Sedimentology, vol. 63 (ed. H. Hüneke and T. Mulder), 457–516.

https://doi.org/10.1016/B978-0-444-53000-4.00007-X.

CenCO2PIP Consortium, (The Cenozoic CO2 Proxy Integration Project (CenCO2PIP) Consortium) (2023). Toward a Cenozoic history of atmospheric CO_2. *Science* 382: eadi5177. https://doi.org/10.1126/science.adi5177.

Chen, C. (1971). Distribution of shell-bearing pteropods in the oceans (abstract). In: *The Micropalaeontology of Oceans* (ed. N.M. Funnell and W.M. Riedel), 161–162. Cambridge University Press.

Chernihovsky, N., Torfstein, A., and Almogi-Labin, A. (2018). Seasonal flux patterns of planktonic foraminifera in a deep, oligotrophic, marginal sea: sediment trap time series from the Gulf of Aqaba, northern Red Sea. *Deep Sea Research Part I: Oceanographic Research Papers* 140: 78–94. https://doi.org/10.1016/j.dsr.2018.08.003.

Chisholm, S.W. (2000). Stirring times in the Southern Ocean. *Nature* 407: 685–687.

Conley, D.J., Frings, P.J., Fontorbe, G. et al. (2017). Biosilicification drives a decline of dissolved Si in the oceans through geologic time. *Frontiers in Marine Science* 4: 397. https://doi.org/10.3389/fmars.2017.00397.

Cordey, F. and Schiarizza, P. (1993). Long-lived panthalassic remnant: the Bridge River accretionary complex, Canadian Cordillera. *Geology* 21: 263–266. https://doi.org/10.1130/0091-7613(1993)021<0263:LLPRTB>2.3.CO;2.

Cortese, C., Gersonde, R., Hillenbrand, C.-D., and Kuhn, G. (2004). Opal sedimentation shifts in the World Ocean over the last 15 Myr. *Earth and Planetary Science Letters* 224: 509–527. https://doi.org/10.1016/j.epsl.2004.05.035.

Crane, R., Laing, C., Littler, K. et al. (2024). Deep-sea mining poses an unjustifiable environmental risk. *Nature Sustainability* 7: 836–838. https://doi.org/10.1038/s41893-024-01326-6.

Cronin, T.M. (2010). *Paleoclimates: Understanding Climate Change Past and Present*, 441. New York: Columbia University Press.

Dansgaard, W., Johnsen, S., Clausen, H. et al. (1993). Evidence for general instability of past climate from a 250-kyr ice-core record. *Nature* 364: 218–220. https://doi.org/10.1038/364218a0.

Davies, R.J. and Cartwright, J. (2002). A fossilized Opal A to Opal C/T transformation on the northeast Atlantic margin: support for a significantly elevated Palaeogeothermal gradient during the Neogene? *Basin Research* 14: 467–486. https://doi.org/10.1046/j.1365-2117.2002.00184.x.

Dean, W.E. and Gardner, J.V. (1986). Milankovitch cycles in Neogene deep-sea sediment. *Paleoceanography* 1 (4): 539–553. https://doi.org/10.1029/PA001i004p00539.

Dean, W., Pride, C., and Thunell, R. (2004). Geochemical cycles in sediments deposited on the slopes of the Guaymas and Carmen Basins of the Gulf of California over the last 180 years. *Quaternary Science Reviews* 23: 1817–1833. https://doi.org/10.1016/j.quascirev.2004.03.010.

DeConto, R. and Pollard, D. (2003). Rapid Cenozoic glaciation of Antarctica induced by declining atmospheric CO_2. *Nature* 421: 245–249. https://doi.org/10.1038/nature01290.

Dilek, Y. and Furnes, H. (2014). Ophiolites and their origins. *Elements* 10 (2): 93–100. https://doi.org/10.2113/gselements.10.2.93.

Dowdeswell, J.A., Maslin, M.A., Andrews, J.T., and McCave, I.N. (1995). Iceberg production, debris rafting, and the extent and thickness of Heinrich layers (H-1, H-2) in North Atlantic sediments. *Geology* 23: 301–304.

Dutkiewicz, A., Judge, A., and Müller, R.D. (2020). Environmental predictors of deep-sea polymetallic nodule occurrence in the global ocean. *Geology* 48: 293–297. https://doi.org/10.1130/G46836.1.

Dutkiewicz, A. and Müller, R.D. (2022). Deep-sea hiatuses track the vigor of Cenozoic ocean bottom currents. *Geology* 50: 710–715. https://doi.org/10.1130/G49810.1.

Dutkiewicz, A., Müller, R.D., O'Callaghan, S.O., and Jonasson, H. (2015). Census of seafloor sediments in the world's ocean. *Geology* 43: 795–798. https://doi.org/10.1130/G36883.1.

Eissen, J.-P., Fouquet, Y., Hardy, D., and Ondréas, H. (2003). Recent MORB volcaniclastic explosive deposits formed between 500 and 1750 m.b.s.l. on the axis of the Mid-Atlantic Ridge, South of the Azores. In: *Explosive Subaqueous Volcanism* (ed. J.D.L. White, J.L. Smellie, and D.A. Clague). https://doi.org/10.1029/140GM09.

Ekdale, A.A. and Berger, W.H. (1978). Deep-sea ichnofacies: modern organism traces on and in pelagic carbonates of the western equatorial Pacific. *Palaeogeography, Palaeoclimatology, Palaeoecology* 23: 263–278. https://doi.org/10.1016/0031-0182(78)90096-2.

Ekdale, A.A. and Bromley, R.G. (1984). Comparative ichnology of shelf-sea and deep-sea chalk. *Journal of Paleontology* 58: 322–332.

Ekdale, A.A., Bromley, R.G., and Pemberton, G.S. (1984). *Ichnology: The Use of Trace Fossils in Sedimentology and Stratigraphy, SEPM Short Course 15*, 1–317. Society of Economic Geologists and Paleontologists.

Emerson, S.R. and Hamme, R.C. (2022). *Chemical Oceanography*. Cambridge University Press.

Emerson, S.R. and Hedges, J.I. (2008). *Chemical Oceanography and the Marine Carbon Cycle*. Cambridge University Press.

Erba, E. (1992). Middle Cretaceous calcareous nannofossils from the western Pacific (Leg 129): evidence for paleoequatorial crossings. In: *Proceedings of the Ocean Drilling Program, Scientific Results*, vol. 129 (ed. R.L. Larson and T. Lancelot), 189–201. College Station, TX: Ocean Drilling Program.

Esmerode, E.V. and Surlyk, F. (2009). Origin of channel systems in the Upper Cretaceous Chalk Group of the Paris Basin. *Marine and Petroleum Geology* 26: 1338–1349. https://doi.org/10.1016/j.marpetgeo.2009.05.001.

Ewing, M., Saito, T., Ewing, J.I., and Burckle, L.H. (1966). Lower Cretaceous sediments from the Northwest Pacific. *Science* 152: 751–755. https://doi.org/10.1126/science.152.3723.751.

Ewing, M. and Vine, A. (1938). Deep-sea measurements without wires or cables. *Eos, Transactions American Geophysical Union* 19 (1): 248–251. https://doi.org/10.1029/TR019i001p00248.

Fahl, K. and Nöthig, E.-M. (2007). Lithogenic and biogenic particle fluxes on the Lomonosov Ridge (central Arctic Ocean) and their relevance for sediment accumulation: vertical vs. lateral transport. *Deep Sea Research Part I: Oceanographic Research Papers* 54: 1256–1272. https://doi.org/10.1016/j.dsr.2007.04.014.

Fakhraee, M., Planavsky, N.J., and Reinhard, C.T. (2020). The role of environmental factors in the long-term evolution of the marine biological pump. *Nature Geoscience* 13: 812–816 (2020). https://doi.org/10.1038/s41561-020-00660-6.

Farley, K.A. (1995). Cenozoic variations in the flux of interplanetary dust recorded by ^{3}He in a deep sea sediment. *Nature* 376: 153–156.

Fonseca, M.D.L., Scotese, C.R., and Mário Cachão, M. (2019). Late Cretaceous paleobiogeography of Braarudosphaera bigelowii. *Marine Micropaleontology* 152: 101738. https://doi.org/10.1016/j.marmicro.2019.03.010.

Foster, G.L., Hull, P., Lunt, D.J., and Zachos, J.C. (2018). Placing our current 'hyperthermal' in the context of rapid climate change in our geological past. *Philosophical Transactions of the Royal Society A* 376: 20170086. https://doi.org/10.1098/rsta.2017.0086.

Foster, G.L. and Rae, J.W.B. (2016). Reconstructing ocean pH with boron isotopes in foraminifera. *Annual Review of Earth and Planetary Sciences* 44: 207–237. https://doi.org/10.1146/annurev-earth-060115-012226.

Frank, T.D., Arthur, M.A., and Dean, W.A. (1999). Diagenesis of lower Cretaceous pelagic carbonates, North Atlantic: paleoceanographic signals obscured. *Journal of Foraminiferal Research* 29: 340–351.

Gale, A.S., Surlyk, F., and Anderskouv, K. (2013). Channelling versus inversion: origin of condensed Upper Cretaceous chalks, eastern Isle of Wight, UK. *Journal of the Geological Society, London* 170: 281–290. https://doi.org/10.1144/jgs2012-044.

Gale, A.S., Young, J.R., Shackleton, N.J. et al. (1999). Orbital tuning of Cenomanian marly chalk successions: towards a Milankovitch time-scale for the Late Cretaceous. *Philosophical Transactions of the Royal Society of London. Series A* 357: 1815–1829. https://doi.org/10.1098/rsta.1999.0402.

Gambacorta, G., Bersezio, R., and Erba, E. (2014). Sedimentation in the Tethyan pelagic realm during the Cenomanian: monotonous settling or active redistribution? *Palaeogeography, Palaeoclimatology, Palaeoecology* 409: 301–319. https://doi.org/10.1016/j.palaeo.2014.05.011.

Gersonde, R. and Harwood, D.M. (1990). Lower Cretaceous diatoms from ODP Leg 113 site 693 (Weddell Sea) Part 1: vegetative cells. *Proceedings of the Ocean Drilling Program, Scientific Results* 113: 365–402.

Gilmer, R.W. and Harbison, G.R. (1986). Morphology and field behaviour of pteropod molluscs: feeding methods in the families Cavoliniidae Limacinidae and Peraclididae (Gastropoda: Thecosomata). *Marine Biology* 91: 47–57.

Goldstein, S.L. and Hemming, S.R. (2014). 8.17 - Long-lived Isotopic tracers in oceanography, paleoceanography, and ice-sheet dynamics. In: *Treatise on Geochemistry*, 2e (ed. H.D. Holland and K.K. Turekian), 453–483. Elsevier https://doi.org/10.1016/B978-0-08-095975-7.00617-3.

Graber, K.K., Pollard, E., Jonasson, B., and Schulte, E. (ed.) (2002). Technical note 31: overview of ocean drilling program engineering tools and hardware. Ocean Drilling Program. https://doi.org/10.2973/odp.tn.31.2002

Grant, K.M., Amarathunga, U., Amies, J. D. et al. (2022). Organic carbon burial in Mediterranean sapropels intensified during Green Sahara Periods since 3.2 Myr ago. *Communications Earth & Environment* 3: 11. https://doi.org/10.1038/s43247-021-00339-9.

Grice, K., and Eiserbeck, C., (2014), 12.3 - The analysis and application of biomarkers, in Editor(s): Heinrich D., Holland, H.D., Turekian, K.K., *Treatise on Geochemistry* (2), Elsevier, 47-78, https://doi.org/10.1016/B978-0-08-095975-7.01006-8

Gries, R. (2018). Three women, one breakthrough: unsung heroines who transformed oil exploration, search and discover article, #70319. https://www.searchanddiscovery.com/pdfz/documents/2018/70319gries/ndx_gries.pdf.html (retrieved 18 October 2024).

Grigoratou, M., Monteiro, F.M., Wilson, J. D. et al. (2022). Exploring the impact of climate change on the global distribution of non-spinose planktonic foraminifera using a trait-based ecosystem model. *Global Change Biology* 28: 1063–1076. https://doi.org/10.1111/gcb.15964.

Hallam, A. (1986). Origin of minor limestone–shale cycles: climatically induced or diagenetic? *Geology* 14: 609–612.

Hay, W.W., Sloan, J.L. II, and Wold, C.N. (1988). Mass/age distribution and composition of sediments on the ocean floor and the global rate of sediment subduction. *Journal of Geophysical Research* 93 (B12): 14933–14940. https://doi.org/10.1029/JB093iB12p14933.

Hays, J.D., Imbrie, J., and Shackleton, N.J. (1976). Variations in the Earth's orbit: pacemaker of the ice ages. *Science* 194: 1121–1132. https://doi.org/10.1126/science.194.4270.1121.

Heezen, B.C. and Hollister, C.D. (1971). *The Face of the Deep*, 659. New York, London, and Toronto: Oxford University Press.

Hein, J.R. (2016). Manganese nodules. In: *Encyclopedia of Marine Geosciences* (ed. J. Harff et al.), 408–412. Dordrecht: Springer https://doi.org/10.1007/978-94-007-6238-1.

Hein, J.R., Koschinsky, A., and Kuhn, T. (2020). Deep-ocean polymetallic nodules as a resource for critical materials. *Nature Reviews Earth and Environment* 1: 158–169. https://doi.org/10.1038/s43017-020-0027-0.

Herbert, T.D. (1997). A long marine history of carbon cycle modulation by orbital-climatic changes. *Proceedings of the National Academy of Sciences* 94: 8362–8369. https://doi.org/10.1073/pnas.94.16.8362.

Herbert, T.D. (2014). 8.15 - Alkenone paleotemperature determinations. In: *Treatise on Geochemistry*, 2e (ed. H.D. Holland and K.K. Turekian), 399–433. Elsevier https://doi.org/10.1016/B978-0-08-095975-7.00615-X.

Herbert, T. and Fischer, A. (1986). Milankovitch climatic origin of mid-Cretaceous black shale rhythms in central Italy. *Nature* 321: 739–743. https://doi.org/10.1038/321739a0.

Hesse, R. and Schacht, U. (2011). Chapter 9: Early diagenesis of deep-sea sediments. In: *Deep-Sea Sediments*, Developments in Sedimentology, vol. 63 (ed. H. Hüneke and T. Mulder), 557–573. Amsterdam, The Netherlands: Elsevier https://doi.org/10.1016/B978-0-444-53000-4.00009-3.

Higgs, B.M. (2021). Understanding the Earth: the contribution of Marie Tharp. In: *Celebrating 100 Years of Female Fellowship of the Geological Society: Discovering Forgotten Histories*, vol. 506 (ed. C.V. Burek and B.M. Higgs), 231–243. Geological Society, London, Special Publications https://doi.org/10.1144/SP506-2019-248.

Hinnov, L.A. and Hilgen, F.J. (2012). Chapter 4: Cyclostratigraphy and astrochronology. In: *The Geologic Time Scale*, vol. 2012 (ed. F.M. Gradstein, J.G. Ogg, M.D. Schmitz, and G.M. Ogg), 63–83. https://doi.org/10.1016/B978-0-444-59425-9.00004-4.

Histon, K. (2012). Paleoenvironmental and temporal significance of variably colored Paleozoic orthoconic nautiloid cephalopod accumulations. *Palaeogeography, Palaeoclimatology, Palaeoecology* 367–368: 193–208. https://doi.org/10.1016/j.palaeo.2012.07.008.

Hodell, D.A. et al. (2017). Anatomy of Heinrich Layer 1 and its role in the last deglaciation. *Paleoceanography* 32: 284–303. https://doi.org/10.1002/2016PA003028.

Holbourn, A., Kuhnt, W., Frank, M., and Haley, B.A. (2013). Changes in Pacific Ocean circulation following the Miocene onset of permanent Antarctic ice cover. *Earth and Planetary Science Letters* 365: 38–50. https://doi.org/10.1016/j.epsl.2013.01.020.

Honjo, S. et al. (1982). Deep advective transport of lithogenic particles in Panama Basin. *Science* 216: 516–518. https://doi.org/10.1126/science.216.4545.516.

Hovan, S.A., Rea, D.K., and Pisias, N.G. (1991). Late Pleistocene continental climate and oceanic variability recorded in northwest Pacific sediments. *Paleoceanography* 6 (3): 349–370. https://doi.org/10.1029/91PA00559.

Howard, W.R. and Prell, W.L. (1994). Late quaternary $CaCO_3$ production and preservation in the Southern Ocean: implications for oceanic and atmospheric carbon cycling. *Paleoceanography* 9 (3): 453–482. https://doi.org/10.1029/93PA03524.

Hsü, K.J. (1992). *Challenger at Sea: A Ship That Revolutionized Earth Science*, 416. Princeton, NJ: Princeton University Press.

Hu, S., Sprintall, J., Guan, C. et al. (2020). Deep-reaching acceleration of global mean ocean circulation over the past two decades. *Science Advances* 6 (6): eaax7727. https://doi.org/10.1126/sciadv.aax7727.

Hughen, K.A., Overpeck, J.T., Peterson, L. C., and Anderson, R.F. (1996). The nature of varved sedimentation in the Cariaco Basin, Venezuela and its palaeoclimatic significance. In: *Palaeoclimatology and Palaeoceanography from Laminated Sediments*, vol. 116 (ed. A.E.S. Kemp), 171–183. Geological Society Special Publication https://doi.org/10.1144/GSL.SP.1996.116.01.15.

Hughes, G.W. and Beydoun, Z.R. (1992). The Red Sea — Gulf of Aden: biostratigraphy, lithostratigraphy and palaeoenvironments. *Journal of Petroleum Geology* 15: 135–156. https://doi.org/10.1111/j.1747-5457.1992.tb00959.x.

Hull, P.M. (2017). Emergence of modern marine ecosystems. *Current Biology* 27: R466–R469. https://doi.org/10.1016/j.cub.2017.04.041.

Hull, P.M. et al. (2020). On impact and volcanism across the Cretaceous-Paleogene boundary. *Science* 367: 266–272. https://doi.org/10.1126/science.aay5055.

Hüneke, H., Gibb, M.A., Mayer, O. et al. (2023). Bioclastic bottom-current deposits of a Devonian contourite terrace: facies variability and depositional architecture (Tafilalt Platform, Morocco). *Sedimentology* 70: 1413–1471. https://doi.org/10.1111/sed.13089.

Hüneke, H. and Henrich, R. (2011). Pelagic sedimentation in modern and ancient oceans. In: *Deep-Sea Sediments*, Developments in Sedimentology, vol. 63 (ed. H. Hüneke and T. Mulder), 215–352. https://doi.org/10.1016/B978-0-444-53000-4.00004-4.

Inglis, G.N. and Tierney, J.E. (2020). *The TEX86 Paleotemperature Proxy*. Cambridge: Cambridge University Press (Elements in Geochemical Tracers in Earth System Science).

James, N.P. and Jones, B. (2016). *Origin of Carbonate Sedimentary Rocks*. Wiley.

Jenkyns, H.C. (1980). Cretaceous anoxic events: from continents to oceans. *Journal of the Geological Society* 137: 171–188. https://doi.org/10.1144/gsjgs.137.2.0171.

Jenkyns, H.C. (1986). Chapter 11: Pelagic environments. In: *Sedimentary Environments and Facies*, 2e (ed. H.G. Reading). Oxford: Blackwell Scientific Publishing.

Jenkyns, H.C. (2010). Geochemistry of oceanic anoxic events. *Geochemistry, Geophysics, Geosystems* 11: Q03004. https://doi.org/10.1029/2009GC002788.

Jenkyns, H.C. and Winterer, E.L. (1982). Palaeoceanography of Mesozoic ribbon radiolarites. *Earth and Planetary Science Letters* 60: 351–375. https://doi.org/10.1016/0012-821X(82)90073-5.

Jin, X., Gruber, N., Dunne, J.P. et al. (2006). Diagnosing the contribution of phytoplankton functional groups to the production and export of particulate organic carbon, $CaCO_3$, and opal from global nutrient and alkalinity distributions. *Global Biogeochemical Cycles* 20: https://doi.org/10.1029/2005GB002532.

Kakuwa, Y. and Floyd, J.D. (2016). Trace fossils in Ordovician radiolarian chert successions in the Southern Uplands, Scotland. *Earth and Environmental Science Transactions of the Royal Society of Edinburgh* 107 (1): 13–22. https://doi.org/10.1017/S1755691017000044.

Kane, I.A. and Clare, M.A. (2019). Dispersion, accumulation, and the ultimate fate of microplastics in deep-marine environments: a review and future directions. *Frontiers in Earth Science* 7: 80. https://doi.org/10.3389/feart.2019.00080.

Keller, G. and Barron, J.A. (1983). Paleoceanographic implications of Miocene deep-sea hiatuses. *GSA Bulletin* 94: 590–613. https://doi.org/10.1130/0016-7606(1983)94<590:PIOMDH>2.0.CO;2.

Keller, M.A. and Isaacs, C.M. (1985). An evaluation of temperature scales for silica diagenesis in diatomaceous sequence including a new approach based on the Miocene Monterey Formation, California. *Geo-Marine Letters* 5: 31–35.

Khatiwala, S., Primeau, F., and Holzer, M. (2012). Ventilation of the deep ocean constrained with tracer observations and implications for radiocarbon estimates of ideal mean age. *Earth and Planetary Science Letters* 325–326: 116–125. https://doi.org/10.1016/j.epsl.2012.01.038.

Klaas, C. and Archer, D.E. (2002). Association of sinking organic matter with various types of mineral ballast in the deep sea: implications for the rain ratio. *Global Biogeochemical Cycles* 16 (4): 1116. https://doi.org/10.1029/2001GB001765.

Kochhann, K.G.D., Holbourn, A., Kuhnt, W. et al. (2016). Eccentricity pacing of eastern equatorial Pacific carbonate dissolution cycles during the Miocene Climatic Optimum. *Paleoceanography* 31: 1176–1192. https://doi.org/10.1002/2016PA002988.

Kucera, M. (2007). Chapter 6: Planktonic foraminifera as tracers of past oceanic environments. In: *Developments in Marine Geology*, vol. 1 (ed. C. Hillaire-Marcel and A. De Vernal), 213–262. Elsevier https://doi.org/10.1016/S1572-5480(07)01011-1.

Lalicata, J.L. and Lea, D.W. (2011). Pleistocene carbonate dissolution fluctuations in the eastern equatorial Pacific on glacial timescales: evidence from ODP Hole 1241. *Marine Micropaleontology* 79: 41–51. https://doi.org/10.1016/j.marmicro.2011.01.002.

Lalli, C.M. and Parsons, T.R. (1997). *Biological Oceanography: An introduction*, 2e, 314. Oxford, UK: The Open University, Pergamon Press.

LaMontagne, R.W., Murray, R.W., Wei, K. Y. et al. (1996). Decoupling of carbonate preservation, carbonate concentration, and biogenic accumulation: a 400-kyr record from the central equatorial Pacific Ocean. *Paleoceanography* 11: 553–562.

Laskar, J. (2020). Chapter 4: Astrochronology. In: *Geologic Time Scale 2020* (ed. F.M. Gradstein, J.G. Ogg, M.D. Schmitz, and G.M. Ogg), 139–158. Elsevier https://doi.org/10.1016/B978-0-12-824360-2.00004-8.

Laskar, J., Fienga, A., Gastineau, M., and Manche, H. (2011). La2010: a new orbital solution for the long-term motion of the Earth. *Astronomy and Astrophysics* 532: A89. https://doi.org/10.1051/0004-6361/201116836.

Laskar, J., Robutel, P., Joutel, F. et al. (2004). A long-term numerical solution for the insolation quantities of the Earth. *Astronomy and Astrophysics* 428: 261–285. https://doi.org/10.1051/0004-6361:20041335.

Laurila, T.E., Hannington, M.D., Leybourne, M. et al. (2015). New insights into the mineralogy of the Atlantis II Deep metalliferous sediments, Red Sea. *Geochemistry, Geophysics, Geosystems* 16: 4449–4478. https://doi.org/10.1002/2015GC006010.

Lea, D.W. (2014). 8.14 - Elemental and isotopic proxies of past ocean temperatures. In: *Treatise on Geochemistry*, 2e (ed. H.D. Holland and K.K. Turekian), 373–397. Elsevier https://doi.org/10.1016/B978-0-08-095975-7.00614-8.

Lear, C.H. et al. (2000). Cenozoic deep-sea temperatures and global ice volumes from Mg/Ca in benthic foraminiferal calcite. *Science* 287: 269–272. https://doi.org/10.1126/science.287.5451.269.

Leggett, J.K. (1979). Oceanic sediments from the Ordovician of the Southern Uplands. *Geological Society Special Publication* 8: 495–498. https://doi.org/10.1144/GSL.SP.1979.008.01.59.

Leggett, J.K., McKerrow, W.S., Morris, J. H. et al. (1979). The north-western margin of the Iapetus Ocean. *Geological Society Special Publication* 8: 499–512. https://doi.org/10.1144/GSL.SP.1979.008.01.60.

Leggett, J.K. and Smith, T.K. (1980). Fe-rich deposits associated with Ordovician basalts in the Southern Uplands of Scotland: possible lower Palaeozoic equivalents of modern active ridge sediments. *Earth and Planetary Science Letters* 47: 431–440. https://doi.org/10.1016/0012-821X(80)90031-X.

Lisiecki, L.E. and Raymo, M.E. (2005). A Pliocene–Pleistocene stack of 57 globally distributed benthic delta O-18 records. *Paleoceanography* 20 (1): PA2007.

Longman, J., Mills, B.J.W., Donnadieu, Y., and Goddéris, Y. (2022). Assessing volcanic controls on Miocene climate change. *Geophysical Research Letters* 49: e2021GL096519. https://doi.org/10.1029/2021GL096519.

Lonsdale, P.F., Bischoff, J.L., Burns, V.M. et al. (1980). A high-temperature hydrothermal deposit on the seabed at a gulf of California spreading center. *Earth and Planetary Science Letters* 49: 8–20. https://doi.org/10.1016/0012-821X(80)90144-2.

Lyle, M., Murray, D.W., Finney, B.P. et al. (1988). The record of late Pleistocene biogenic sedimentation in the eastern tropical Pacific Ocean. *Paleoceanography* 3 (1): 39–59. https://doi.org/10.1029/PA003i001p00039.

Lynch-Stieglitz, J. and Marchitto, T.M. (2014). 8.16 - Tracers of past ocean circulation. In: *Treatise on*

Geochemistry, 2e (ed. H.D. Holland and K.K. Turekian), 435–451. Elsevier https://doi.org/10.1016/B978-0-08-095975-7.00616-1.

Madsen, H.B. and Stemmerik, L. (2010). Diagenesis of flint and porcellanite in the Maastrichtian chalk at Stevns Klint, Denmark. *Journal of Sedimentary Research* 80: 578–588. https://doi.org/10.2110/jsr.2010.052.

Maliva, R.G., Knoll, A.H., and Siever, R. (1989). Secular change in chert distribution: a reflection of evolving biological participation in the silica cycle. *PALAIOS* 4: 519–532.

MARGO Project Members (2009). Constraints on the magnitude and patterns of ocean cooling at the Last Glacial Maximum. *Nature Geoscience* 2: 127–132. https://doi.org/10.1038/ngeo411.

Mazzullo, J.M., Meyer, A., and Kidd, R.B. (1988). New sediment classification scheme for the Ocean Drilling Program. In: *Technical Note 8: Handbook for Shipboard Sedimentologists* (ed. J. Mazzullo and A.G. Graham), 44–67. Ocean Drilling Program https://doi.org/10.2973/odp.tn.8.1988.

McCave, I.N., Thornalley, D.J.R., and Hall, I.R. (2017). Relation of sortable silt grain-size to deep-sea current speeds: calibration of the 'mud current meter'. *Deep Sea Research Part I: Oceanographic Research Papers* 127: 1–12. https://doi.org/10.1016/j.dsr.2017.07.003.

McClymont, E.L., Ho, S.L., Ford, H.L. et al. (2023). Climate evolution through the onset and intensification of Northern Hemisphere Glaciation. *Reviews of Geophysics* 61: e2022RG000793. https://doi.org/10.1029/2022RG000793.

Meckler, A.M. et al. (2022). Cenozoic evolution of deep ocean temperature from clumped isotope thermometry. *Science* 377: 86–90. https://doi.org/10.1126/science.abk0604.

Menichetti, M. (2016). Early Cretaceous tectonic event in the Adria: insight from Umbria-Marche pelagic basin (Italy). In: *The Stratigraphic Record of Gubbio: Integrated Stratigraphy of the Late Cretaceous–Paleogene Umbria-Marche Pelagic Basin*, vol. 524 (ed. M. Menichetti, R. Coccioni, and A. Montanari), 35–55. Geological Society of America, Special Publication https://doi.org/10.1130/2016.2524(04).

Milanković, M.M. (1941). *Canon of Insolation and the Ice-Age Problem*. Israel Program for Scientific Translations.

Miller, M.R., Oakes, R.L., Covert, P.A. et al. (2023). Evidence for an effective defence against ocean acidification in the key bioindicator pteropod *Limacina helicina*. *ICES Journal of Marine Science* 80: 1329–1341. https://doi.org/10.1093/icesjms/fsad059.

Mitchell, N.C., Simmons, H.L., and Lear, C.H. (2015). Modern and ancient hiatuses in the pelagic caps of Pacific guyots and seamounts and internal tides. *Geosphere* 11: 1590–1606. https://doi.org/10.1130/GES00999.1.

Moore, T.C. Jr. (2013). Erosion and reworking of Pacific sediments near the Eocene-Oligocene boundary. *Paleoceanography* 28: 263–273. https://doi.org/10.1002/palo.20027.

Moore, T.C., van Andel, T., and H., Sancetta, C., & Pisias, N. (1978). Cenozoic hiatuses in pelagic sediments. *Micropaleontology* 24 (2): 113–138. https://doi.org/10.2307/1485246.

Moores, E.M. (2002). Pre–1 Ga (pre-Rodinian) ophiolites: their tectonic and environmental implications. *GSA Bulletin* 114: 80–95. https://doi.org/10.1130/0016-7606(2002)114<0080:PGPROT>2.0.CO;2.

Mortimore, R. (2011). A chalk revolution: what have we done to the Chalk of England? *Proceedings of the Geologists' Association* 122: 232–297. https://doi.org/10.1016/j.pgeola.2010.09.001.

Mulder, T., Hüneke, H., and Van Loon, A. J. (2011). Progress in deep-sea sedimentology. In: *Deep-Sea Sediments*, Developments in Sedimentology, vol. 63 (ed. H. Hüneke and T. Mulder), 1–24. Amsterdam, The Netherlands: Elsevier.

Müller, R.D., Sdrolias, M., Gaina, C., and Roest, W.R. (2008). Age, spreading rates and spreading symmetry of the world's ocean crust. *Geochemistry, Geophysics, Geosystems* 9: Q04006. https://doi.org/10.1029/2007GC001743.

Murray, R.W., Buchholtz, M.R., Brink, T. et al. (1992). Rare earth, major, and trace element composition of Monterey and DSDP chert and associated host sediment: assessing the influence of chemical fractionation during diagenesis. *Geochimica et Cosmochimica Acta* 56: 2657–2671. https://doi.org/10.1016/0016-7037(92)90351-I.

Murray, J. and Renard, A.F. (1891). Report on deep-sea deposits based on specimens collected during the voyage of H.M.S. 'Challenger' in the years 1873–1876. In: *Challenger Reports*, 525. Edinburgh: HMSO.

Muttoni, G., Erba, E., Kent, D.V., and Bachtadse, V. (2005). Mesozoic Alpine facies deposition as a result of past latitudinal plate motion. *Nature* 434: 59–63.

Naing, T.T., Robinson, S.A., Searle, M.P. et al. (2023). Age, depositional history and tectonics of the Indo-Myanmar Ranges, Myanmar. *Journal of the Geological Society, London* 180: https://doi.org/10.1144/jgs2022-091.

Neuer, S., Ratmeyer, V., Davenport, R. et al. (1997). Deep water particle flux in the Canary Island region: seasonal trends in relation to long-term satellite derived pigment data and lateral sources. *Deep Sea Research Part I: Oceanographic Research Papers* 44: 1451–1466. https://doi.org/10.1016/S0967-0637(97)00034-4.

Ogg, J.G., Karl, S.M., and Behl, R.J. (1992). Jurassic through Early Cretaceous sedimentation history of the central equatorial Pacific and of Sites 800 and 801. In: *Proceedings of the Ocean Drilling Program, Scientific Results*, vol. 129 (ed. R.L. Larson and T. Lancelot), 571–613. College Station, TX: Ocean Drilling Program.

Olson, P., Reynolds, E., Hinnov, L., and Goswami, A. (2016). Variation of ocean sediment thickness with crustal age. *Geochemistry, Geophysics, Geosystems* 17: 1349–1369. https://doi.org/10.1002/2015GC006143.

Open University (1989a). *Ocean Chemistry and Deep-Sea Sediments*, 134. Oxford, UK: The Open University, Pergamon Press.

Open University (1989b). *Ocean Circulation*, 238. Oxford, UK: The Open University, Pergamon Press.

Open University (1995). *Seawater: Its Composition, Properties and Behaviour*, 2e, 168. Oxford, UK: The Open University, Pergamon Press.

Owen, E.F. and Smith, A.B. (1987). *Fossils of the Chalk*. London: Palaeontological Association. Print. Field Guides to Fossils; No. 2.

Pagani, M. (2014). 12.13 - Biomarker-based inferences of past climate: the alkenone pCO_2 proxy. In: *Treatise on Geochemistry*, 2e (ed. H.D. Holland and K.K. Turekian), 361–378. Elsevier https://doi.org/10.1016/B978-0-08-095975-7.01027-5.

Palike, H. et al. (2012). A Cenozoic record of the equatorial Pacific carbonate compensation depth. *Nature* 488: 609–614.

Paytan, A. and Griffith, E.M. (2007). Marine barite: recorder of variations in ocean export productivity. *Deep Sea Research Part II: Topical Studies in Oceanography* 54: 687–705. https://doi.org/10.1016/j.dsr2.2007.01.007.

Pearson, P.N. (2012). Oxygen isotopes in formainifera: overview and historical review. In: *Reconstructing Earth's Deep-Time Climate—The State of the Art in 2012*, Paleontological Society Short Course, November 3, 2012. The Paleontological Society Papers, vol. 18 (ed. L.C. Ivany and B.T. Huber), 1–38. Paleontological Society.

Pearson, P.N., van Dongen, B.E., Nicholas, C.J. et al. (2007). Stable warm tropical climate through the Eocene Epoch. *Geology* 35: 211–214. https://doi.org/10.1130/G23175A.1.

Peng, T.H. and Broecker, W.S. (1984). The impacts of bioturbation on the age difference between benthic and planktonic foraminifera in deep-sea sediments. *Nuclear Instruments and Methods in Physics Research Section B: Beam Interactions with Materials and Atoms* 233: 346–352. https://doi.org/10.1016/0168-583X(84)90540-8.

Peterson, L.C., Haug, G.H., Murray, R.W. et al. (2000). Late quaternary stratigraphy and sedimentation at site 1002, Cariaco Basin (Venezuela). In: *Proceedings of the Ocean Drilling Program, Scientific Results*, vol. 165 (ed. R.M. Leckie, H. Sigurdsson, G.D. Acton, and G. Draper), 85–99. College Station, TX: Ocean Drilling Program https://doi.org/10.2973/odp.proc.sr.165.017.2000.

Peterson, L.C., Overpeck, J.T., Kipp, N.G., and Imbrie, J. (1991). A high-resolution Late Quaternary upwelling record from the anoxic Cariaco Basin. *Venezuela, Paleoceanography* 6 (1): 99–119. https://doi.org/10.1029/90PA02497.

Pike, P. and Kemp, A.E.S. (1999). Diatom mats in Gulf of California sediments: implications for the paleoenvironmental interpretation of laminated sediments and silica burial. *Geology* 27: 311–314. https://doi.org/10.1130/0091-7613(1999)027<0311:DMIGOC>2.3.CO;2.

Plank, T. and Langmuir, C.H. (1998). The chemical composition of subducting sediment and its consequences for the crust and mantle. *Chemical Geology* 145: 325–394. https://doi.org/10.1016/S0009-2541(97)00150-2.

Racki, G. and Cordey, F. (2000). Radiolarian palaeoecology and radiolarites: is the present the key to the past? *Earth-Science Reviews* 52: 83–120.

Rae, J.W.B., Zhang, Y.G., Liu, X. et al. (2021). Atmospheric CO_2 over the past 66 million years from marine archives. *Annual Review of Earth and Planetary Sciences* 49: 609–641. https://doi.org/10.1146/annurev-earth-082420-063026.

Ragueneau, O., Tréguer, P., Leynaert, A. et al. (2000). A review of the Si cycle in the modern ocean: recent progress and missing gaps in the application of biogenic opal as a paleoproductivity proxy. *Global and Planetary Change* 26: 317–365. https://doi.org/10.1016/S0921-8181(00)00052-7.

Rasmussen, T., Thomsen, E., and Moros, M. (2016). North Atlantic warming during Dansgaard-Oeschger events synchronous with Antarctic warming and out-of-phase with Greenland climate. *Scientific Reports* 6: 20535. https://doi.org/10.1038/srep20535.

Ravizza, G.E. and Zachos, J.C. (2014). 8.20 - Records of Cenozoic ocean chemistry. In: *Treatise on Geochemistry*, 2e (ed. H.D. Holland and K.K. Turekian), 543–568. Elsevier https://doi.org/10.1016/B978-0-08-095975-7.00620-3.

Remane, J. (1998). 6 - Calpionellids. In: *Introduction to Marine Micropaleontology*, 2e (ed. B.U. Haq and A. Anne Boersma), 161–170. Elsevier Science B.V https://doi.org/10.1016/B978-044482672-5/50006-4.

Research on Cretaceous Cycles (R.O.C.C.) Group (1986). Rhythmic bedding in Upper Cretaceous pelagic carbonate sequences: varying sedimentary response to climatic forcing. *Geology* 14: 153–156. https://doi.org/10.1130/0091-7613(1986)14<153:RBIUCP>2.0.CO;2.

Ridgwell, A. (2005). A mid Mesozoic revolution in the regulation of ocean chemistry. *Marine Geology* 217: 339–358.

Ridgwell, A. and Schmidt, D. (2010). Past constraints on the vulnerability of marine calcifiers to massive carbon dioxide release. *Nature Geoscience* 3: 196–200. https://doi.org/10.1038/ngeo755.

Ridgwell, A. and Zeebe, R.E. (2005). The role of the global carbonate cycle in the regulation and evolution of the Earth system. *Earth and Planetary Science Letters* 234: 299–315. https://doi.org/10.1016/j.epsl.2005.03.006.

Riech, V. and von Rad, U. (1979). Silica diagenesis in the Atlantic Ocean: diagenetic potential and transformations. In: *Deep Drilling in the Atlantic Ocean: Continental Margins and Paleoenvironments*, Maurice Ewing Series 3 (ed. M. Talwani et al.), 315–340. Washington, DC: American Geophysical Union.

Robinson, S.A., Clark, L.J., Nederbragt, A., and Wood, I.G. (2008). Mid-Cretaceous oceanic anoxic events in the Pacific Ocean revealed by carbon-isotope stratigraphy of the Calera Limestone, California, USA. *Geological Society of America Bulletin* 120: 1416–1427.

Robinson, S.A., Heimhofer, U., Hesselbo, S.P., and Petrizzo, M.R. (2017). Mesozoic climates and oceans – a tribute to Hugh Jenkyns and Helmut Weissert. *Sedimentology* 64: 1–15.

Rodriguez, M., Fournier, M., Chamot-Rooke, N. et al. (2012). Mass wasting processes along the Owen Ridge

(Northwest Indian Ocean). *Marine Geology* 326–328: 80–100. https://doi.org/10.1016/j.margeo.2012.08.008.

Rohling, E.J., Marino, G., and Grant, K.M. (2015). Mediterranean climate and oceanography, and the periodic development of anoxic events (sapropels). *Earth-Science Reviews* 143: 62–97. https://doi.org/10.1016/j.earscirev.2015.01.008.

Rossignol-Strick, M. (1983). African monsoons, an immediate climate response to orbital insolation. *Nature* 30: 446–449.

Rossignol-Strick, M., Nesteroff, V., Olive, P., and Vergnaud-Grazzini, C. (1982). After the deluge; Mediterranean stagnation and sapropel formation. *Nature* 295: 105–110.

Rothery, D.A. (ed.) (1998). *The Ocean Basins: Their Structure and Evolution*, 2e, 185. Oxford, UK: The Open University, Pergamon Press.

Ruddiman, W.F., Kidd, R.B., Thomas, E. et al. (1987). *Deep Sea Drilling Project Initial Reports Volume 94*. Washington, DC: U.S. Govt. Printing Office.

Ruddiman, W.F., Raymo, M.E., Martinson, D.G. et al. (1989). Pleistocene evolution: Northern hemisphere ice sheets and North Atlantic Ocean. *Paleoceanography* 4: 353–412.

Schiebel, R. and Hemleben, C. (2017). *Planktic Foraminifers in the Modern Ocean*, 358. Berlin: Springer.

Schlanger, S.O. and Douglas, R.G. (1974). The pelagic ooze-chalk-limestone transition and its implications for marine stratigraphy. In: *Pelagic Sediments: On Land and Under the Sea*, vol. 1 (ed. K.J. Hsü and H.C. Jenkyns), 117–148. Special Publications of the IAS.

Schlanger, S.O. and Jenkyns, H.C. (1976). Cretaceous oceanic anoxic events: causes and consequences. *Geologie en Mijnbouw* 55: 179–184.

Schlitzer, R. (2021). Ocean data view, odv.awi.de.

Schmitz, W.J. (1996). On the world ocean circulation: volume I. Some global features/North Atlantic circulation. *Woods Hole Oceanographic Institution Technical Report*. WHOI-96-03, Woods Hole, MA, 141.

Schulte, P. et al. (2010). The Chicxulub asteroid impact and mass extinction at the Cretaceous-Paleogene boundary. *Science* 327: 1214–1218. https://doi.org/10.1126/science.1177265.

Sexton, P.F. and Barker, S. (2012). Onset of 'Pacific-style' deep-sea sedimentary carbonate cycles at the mid-Pleistocene transition. *Earth and Planetary Science Letters* 321–322: 81–94.

Sexton, P.F., Wilson, P.A., and Pearson, P. N. (2006). Microstructural and geochemical perspectives on planktic foraminiferal preservation: "Glassy" versus "Frosty". *Geochemistry, Geophysics, Geosystems* 7: Q12P19. https://doi.org/10.1029/2006GC001291.

Shackleton, N., Backman, J., Zimmerman, H. et al. (1984). Oxygen isotope calibration of the onset of ice-rafting and history of glaciation in the North Atlantic region. *Nature* 307: 620–623. https://doi.org/10.1038/307620a0.

Shackleton, N.J., Crowhurstm, S.J., Weedon, G.P., and Laskar, J. (1999). Astronomical calibration of Oligocene–Miocene time. *Philosophical Transactions of the Royal Society A* 357: 1907–1929. https://doi.org/10.1098/rsta.1999.0407.

Shanmugam, G. (2021). Chapter 8: Bottom currents. In: *Mass Transport, Gravity Flows, and Bottom Currents* (ed. G. Shanmugam), 309–375. https://doi.org/10.1016/B978-0-12-822576-9.00008-4.

Shipboard Scientific Party (2002). Leg 198 summary. In: *Proceedings of the Ocean Drilling Program, Initial Reports*, vol. 198 (ed. T.J. Bralower, I. Premoli Silva, M.J. Malone, et al.), 1–148. College Station, TX: Ocean Drilling Program https://doi.org/10.2973/odp.proc.ir.198.101.2002.

Shiraishi, F., Mitsunobu, S., Suzuki, K. et al. (2016). Dense microbial community on a ferromanganese nodule from the ultra-oligotrophic South Pacific Gyre: implications for biogeochemical cycles. *Earth and Planetary Science Letters* 447: 10–20. https://doi.org/10.1016/j.epsl.2016.04.021.

Siever, R. (1992). The silica cycle in the Precambrian. *Geochimica et Cosmochimica Acta* 56: 3265–3272. https://doi.org/10.1016/0016-7037(92)90303-Z.

Simmons, H.L. (2008). Spectral modification and geographic redistribution of the semi-diurnal internal tide. *Ocean Modelling* 21: 126–138.

Sims, P.A., Mann, D.G., and Medlin, L.K. (2006). Evolution of the diatoms: insights from fossil, biological and molecular data. *Phycologia* 45 (4): 361–402. https://doi.org/10.2216/05-22.1.

Sliter, W.V. (1992). Cretaceous planktonic foraminiferal biostratigraphy and paleoceanographic events in the Pacific Ocean with emphasis on indurated sediment. In: *Centenary of Japanese Micropaleontology* (ed. K. Ishizaki and T. Saito), 281–299. Tokyo: Terra Scientific Publishing Company.

Southard, J.B., Young, R.A., and Hollister, C.D. (1971). Experimental erosion of calcareous ooze. *Journal of Geophysical Research* 76: 5903–5909. https://doi.org/10.1029/JC076i024p05903.

Stefurak, E.J.T., Lowe, D.R., Zentner, D., and Fischer, W.W. (2015). Primary silica granules—A new mode of Paleoarchean sedimentation. *Geology* 42: 283–286. https://doi.org/10.1130/G35187.1.

Stern, R.J. (1981). Petrogenesis and tectonic setting of late Precambrian ensimatic volcanic rocks, central eastern desert of Egypt. *Precambrian Research* 16: 195–230. https://doi.org/10.1016/0301-9268(81)90013-9.

Stern, R.J., Johnson, P.R., Kröner, A., and Yibas, B. (2004). Neoproterozoic ophiolites of the Arabian-Nubian shield. In: *Developments in Precambrian Geology*, vol. 13, 95–128. Elsevier https://doi.org/10.1016/S0166-2635(04)13003-X.

Stone, P. (2014). A review of the geological origins and relationships in the Ballantrae Complex. *SW Scotland* 50: 1–25. https://doi.org/10.1144/sjg2013-012.

Stow, D.A.V., Reading, H.G., and Collinson, J.D. (1996). Chapter 10: Deep seas. In: *Sedimentary Environments: Processes, Facies and Stratigraphy*, 3e (ed. H.G. Reading). Oxford: Blackwell Science Ltd.

Straume, E.O., Gaina, C., Medvedev, S. et al. (2019). GlobSed: updated total sediment thickness in the world's oceans. *Geochemistry, Geophysics, Geosystems* 20: 1756–1772. https://doi.org/10.1029/2018GC008115.

Straume, E.O., Gaina, C., Medvedev, S., and Nisancioglu, K.H. (2020). Global Cenozoic paleobathymetry with a focus on the Northern Hemisphere oceanic gateways. *Gondwana Research* 86: 126–143. https://doi.org/10.1016/j.gr.2020.05.011.

Sulpis, O., Boudreau, B.P., Mucci, A. et al. (2018). Current $CaCO_3$ dissolution at the seafloor caused by anthropogenic CO_2. *PNAS* 115: 11700–11705. https://doi.org/10.1073/pnas.1804250115.

Summerhayes, C.P. and Thorpe, S.A. (1996). *Oceanography an Illustrated Guide*, 352. London: Manson Publishing.

Sverdrup, K.A. and Armbrust, E.V. (2008). *An introduction to the World's Oceans*, 10e. McGraw-Hill.

Talley, L.D. (2013). Closure of the global overturning circulation through the Indian, Pacific, and Southern Oceans: schematics and transports. *Oceanography* 26 (1): 80–97. https://doi.org/10.5670/oceanog.2013.07.

Teske, A., Lizarralde, D., Höfig, T.W. et al. (2021). Expedition 385 summary. In: *Guaymas Basin Tectonics and Biosphere*. Proceedings of the International Ocean Discovery Program, vol. 385 (ed. A. Teske, D. Lizarralde, T.W. Höfig, and the Expedition 385 Scientists). College Station, TX: International Ocean Discovery Program https://doi.org/10.14379/iodp.proc.385.101.2021.

Thunell, R.C. (1998). Seasonal and annual variability in particle fluxes in the Gulf of California: a response to climate forcing. *Deep Sea Research Part I: Oceanographic Research Papers* 45: 2059–2083. https://doi.org/10.1016/S0967-0637(98)00053-3.

Thunell, R.C., Pride, C.J., Tappa, E., and Muller-Karger, F.E. (1994). Biogenic silica fluxes and accumulation rates in the Gulf of California. *Geology* 22: 303–306. https://doi.org/10.1130/0091-7613(1994)022<0303:BSFAAR>2.3.CO;2.

Thurman, H. (1994). *Essentials of Oceanography*, 7e. New York, NY: MacMillan.

Thurman, H. and Burton, E.A. (2001). *Essentials of Oceanography*, 9e. Upper Saddle River, NJ: Prentice-Hall Inc., Pearson Education.

Tierney, J.E. et al. (2020). Past climates inform our future. *Science* 370: eaay3701. https://doi.org/10.1126/science.aay3701.

Trabucho-Alexandre, J., Negri, A., and de Boer, P.L. (2011). Early Turonian pelagic sedimentation at Moria (Umbria-Marche, Italy): primary and diagenetic controls on lithological oscillations. *Palaeogeography, Palaeoclimatology, Palaeoecology* 311: 200–214. https://doi.org/10.1016/j.palaeo.2011.08.021.

Tribovillard, N., Algeo, T.J., Lyons, T., and Riboulleau, A. (2006). Trace metals as paleoredox and paleoproductivity proxies: an update. *Chemical Geology* 232: 12–32. https://doi.org/10.1016/j.chemgeo.2006.02.012.

Trujillo, A.P. and Thurman, H.V. (2020). *Essentials of Oceanography*, 13e, 626. Hoboken, NJ: Pearson.

Tucker, M.E., (1974), Sedimentology of Palaeozoic pelagic limestones: the Devonian Griotte (Southern France) and Cephalopodenkalk (Germany) *in* Hsü, K.J. and Jenkyns, H.C., (eds), *Pelagic Sediments: on Land and Under the Sea*. Special Publication of the IAS, 1, 71-92.

Tully, B.J. and Heidelberg, J.F. (2013). Microbial communities associated with ferromanganese nodules and the surrounding sediments. *Frontiers in Microbiology* 4: https://doi.org/10.3389/fmicb.2013.00161.

Tyrrell, T. and Merico, A. (2004). *Emiliania huxleyi*: bloom observations and the conditions that induce them. In: *Coccolithophores: From Molecular Processes to Global Impact* (ed. H.R. Thierstein and J.R. Young), 75–97. Berlin: Springer-Verlag.

Uchman, A. and Wetzel, A. (2011). Chapter 8: Deep-sea ichnology: the relationships between depositional environment and endobenthic organisms. In: *Deep-Sea Sediments*, Developments in Sedimentology, vol. 63 (ed. H. Hüneke and T. Mulder), 517–556. https://doi.org/10.1016/B978-0-444-53000-4.00008-1.

Versteegh, G.J.M., Servais, T., Streng, M. et al. (2009). A discussion and proposal concerning the use of the term calcispheres. *Palaeontology* 52: 343–348. https://doi.org/10.1111/j.1475-4983.2009.00854.x.

Vink, A. (2004). Calcareous dinoflagellate cysts in South and equatorial Atlantic surface sediments: diversity, distribution, ecology and potential for palaeoenvironmental reconstruction. *Marine Micropaleontology* 50: 43–88. https://doi.org/10.1016/S0377-8398(03)00067-7.

Wachs, D. and Hein, J.R. (1975). Franciscan limestones and their environments of deposition. *Geology* 3: 29–33. https://doi.org/10.1130/0091-7613(1975)3<29:FLATEO>2.0.CO;2.

Washburn, T.W., Simon-Lledó, E., Yee Soong, G., and Suzuki, A. (2023). Seamount mining test provides evidence of ecological impacts beyond deposition. *Current Biology* 33: 1–7. https://doi.org/10.1016/j.cub.2023.06.032.

Weedon, G.P. (2003). *Time-Series Analysis and Cyclostratigraphy: Examining Stratigraphic Records of Environmental Cycles*. Cambridge: Cambridge University Press https://doi.org/10.1017/CBO9780511535482.

Weedon, G.P. (2022). Problems with the current practice of spectral analysis in cyclostratigraphy: Avoiding false detection of regular cyclicity. *Earth-Science Reviews* 235: 104261. https://doi.org/10.1016/j.earscirev.2022.104261.

Weissert, H. (2011). Chapter 11: Mesozoic pelagic sediments: archives for ocean and climate history during green-house conditions. In: *Deep-Sea Sediments*, Developments in Sedimentology, vol. 63 (ed. H. Heiko Hüneke and T. Thierry Mulder), 765–792. https://doi.org/10.1016/B978-0-444-53000-4.00011-1.

Wendt, J. and Aigner, T. (1985). Facies patterns and depositional environments of Paleozoic cephalopod limestones. *Sedimentary Geology* 44: 263–300.

Wessel, P., Luis, J.F., Uieda, L. et al. (2019). The generic mapping tools version 6.

Geochemistry, Geophysics, Geosystems 20: 5556–5564. https://doi.org/10.1029/2019GC008515.

Westacott, S., Planavsky, N.J., Zhao, M.-Y., and Hull, P.M. (2021). Revisiting the sedimentary record of the rise of the diatoms. *Proceedings of the National Academy of Sciences* 118: e2103517118. https://doi.org/10.1073/pnas.2103517118.

Westerhold, T., Marwan, N., Drury, A.J. et al. (2020). An astronomically dated record of Earth's climate and its predictability over the last 66 million years. *Science* 369: 1383–1387. https://doi.org/10.1126/science.aba6853.

Westerhold, T. and Röhl, U. (2009). High resolution cyclostratigraphy of the early Eocene – new insights into the origin of the Cenozoic cooling trend. *Climate of the Past* 5: 309–327. https://doi.org/10.5194/cp-5-309-2009.

Winterer, E.L. and Bosellini, A. (1981). Subsidence and sedimentation on Jurassic passive continental Margin, Southern Alps, Italy. *AAPG Bulletin* 65: 394–421. https://doi.org/10.1306/2F9197E2-16CE-11D7-8645000102C1865D.

Witkowski, C.R. et al. (2018). Molecular fossils from phytoplankton reveal secular Pco_2 trend over the Phanerozoic. *Science Advances* 4: eaat4556. https://doi.org/10.1126/sciadv.aat4556.

Woodall, L.C., Sanchez-Vidal, A., Canals, M. et al. (2014). The deep sea is a major sink for microplastic debris. *Royal Society Open Science* 1: https://doi.org/10.1098/rsos.140317#d1e333.

Wray, J.L. (1998). 7 - Calcareous Algae. In: *Introduction to Marine Micropaleontology*, 2e (ed. B.U. Haq and A. Anne Boersma), 171–187. Elsevier Science B.V https://doi.org/10.1016/B978-044482672-5/50007-6.

Wray, D.S. and Gale, A.S. (2006). The palaeoenvironment and stratigraphy of Late Cretaceous Chalks. *Proceedings of the Geologists' Association* 117: 145–162. https://doi.org/10.1016/S0016-7878(06)80006-4.

Young, J.R. (1994). Functions of coccoliths. In: *Coccolithophores* (ed. A. Winter and W.G. Siesser), 63–82. Cambridge, UK: Cambridge University Press.

Zachos, J.C., Dickens, G.R., and Zeebe, R. E. (2008). An early Cenozoic perspective on greenhouse warming and carbon-cycle dynamics. *Nature* 451: 279–283. https://doi.org/10.1038/nature06588.

Zachos, J.C., Pagani, M., Sloan, L. et al. (2001). Trends, rhythms, and aberrations in global climate 65 Ma to present. *Science* 292: 686–693. https://doi.org/10.1126/science.1059412.

Zachos, J.C., Röhl, U., Schellenberg, S.A. et al. (2005). Rapid acidification of the ocean during the Paleocene–Eocene thermal maximum. *Science* 308: 1611–1615. https://doi.org/10.1126/science.1109004.

Zonneveld, K.A.F., Sebastian Meier, K.J., Esper, O. et al. (2005). The (palaeo) environmental significance of modern calcareous dinoflagellate cysts: a review. *Paläontologische Zeitschrift* 79: 61–77. https://doi.org/10.1007/BF03021754.

14

Glacial Landsystems

Bethan Davies[1], *Clare Boston*[2], *and Daniel Paul Le Heron*[3]

[1] *School of Geography, Politics and Sociology, Newcastle University, Newcastle upon Tyne, UK*
[2] *School of the Environment and Life Sciences, University of Portsmouth, Portsmouth, UK*
[3] *Department of Geology, University of Vienna, Vienna, Austria*

Summary

Glaciers are significant landscape modifying agents; they carve huge valleys, produce vast quantities of sediment through erosion, and transport it for hundreds of kilometres. Over the Earth's history, glaciers and ice sheets have grown and shrunk repeatedly. At times, almost the entire Earth was covered by ice, and at other times, there was hardly any frozen ground at all. Today, there are large polar ice sheets in Greenland and Antarctica, and smaller glaciers, icefields and ice caps at high latitudes and on mountain ranges across the globe. Glaciers are sensitive to climate change and are therefore indicators of changing temperature or precipitation. At present, global glacier and ice sheet recession is currently driving sea-level rise, reducing freshwater resources and changing hazard potential in mountain regions worldwide. In this chapter, we first discuss different types of ice mass, consider the relationships between glaciers and climate, review how glaciers gain and lose mass, and how glaciers transfer this mass from their upper to lower regions. Second, we address glaciers as powerful erosive agents that generate vast volumes of sediment. This sediment is transported towards the glacier margins and ultimately to large depocentres beyond the ice margin. Here, we consider characteristic lithofacies associated with different glaciogenic settings. Finally, this chapter introduces the fundamental concept of 'Glacial Landsystems'. This theory rests on the idea that ice masses in different climates, topographic settings and terminal environments (e.g. terrestrial, lacustrine, marine) behave differently and have different impacts on the landscape and depositional record. This results in a characteristic suite of sediment-landform assemblages that can be collated to identify key processes at the ice-bed interface, at the ice margin, and within and on top of the ice. We include examples of some, but by no means all, glacial landsystems, from a variety of different environments, and for each one discuss the key processes at play, the resultant lithofacies, and the diagnostic assemblages of sediments and landforms. This gives a glimpse into the astonishingly wide array of glacial landscapes that can be formed, and preserved, in different environments and climates across our globe.

14.1 Introduction

The Cryosphere includes all forms of ice, including sea ice, lake ice, river ice, glaciers and ice sheets, snow cover, ground ice and permafrost. Mass and energy are constantly exchanged between the Cryosphere and other elements of the Earth System. In this chapter, we are primarily concerned with glaciers and ice sheets, sometimes called 'land ice'.

A 'glacier' is fundamentally a mass of water ice moving under its own mass, either through internal deformation of the ice, or, for temperate glaciers, basal slippage (Cuffey and Paterson 2010). Movement is driven by the accumulation of mass, mainly from snow, and the loss of mass (ablation) through the processes of melting, calving and sublimation. Glaciers can form anywhere where land lies above the limit of the lowest snowline (i.e. where snow persists all year around) (Hambrey and Glasser 2012).

Glaciers are sensitive to environmental change, since they fundamentally rely on the balance between temperature, which controls ice melt, and precipitation, which governs the accumulation of ice mass. They grow or shrink in response to changes in climatic forcing. Currently, glaciers and ice sheets cover ~10% of the Earth's surface, equating to over 15 million km^2. Ninety-nine per cent of this ice volume is found in the ice sheets of Greenland and Antarctica.

Reading's Sedimentary Environments: Processes, Facies and Stratigraphy, Fourth Edition. Edited by Bruce Levell.
 Published 2026 by John Wiley & Sons Ltd.
Companion website: www.wiley.com/go/levell4e

The 'sea-level equivalent' (i.e. the amount that global sea levels would rise on average if the ice melted) is a total of 65.64 m (Morlighem et al. 2017, 2019; Farinotti et al. 2019). In recent decades, the reduction in the size of ice masses globally has raised concerns about global water supply (Immerzeel et al. 2020) and increased hazards from outburst floods (Carrivick and Tweed 2016), glacier surges (Kääb et al. 2018; Bhambri et al. 2020) and ice avalanches (Margreth et al. 2017; Jacquemart et al. 2020), alongside sea-level rise from melting of land ice (Howard et al. 2014; Bamber et al. 2018; Frederikse et al. 2020). Quantifying how our world may be impacted as global land ice responds to climate change, and how glacier and ice sheet fluctuations may have influenced past humans and perhaps even the origin of animals, is fundamental to understanding our world. The sediments deposited by former ice masses also present geotechnical challenges to urban infrastructure, coastal defences and, for example, offshore wind farms (Martin et al. 2017), as well as influencing groundwater and pollution pathways in landfill sites.

The main types of land ice covered in this chapter are ice sheets, ice caps, mountain glaciers and icefields. These ice masses can terminate on land, in the ocean (glaciomarine), or in proglacial lakes (glaciolacustrine). The wide variety of climatic, topographic and geographic settings for land ice results in a distinctive array of different assemblages of glacial sediments and associated landforms that can be used to understand glacial environments at the time of deposition. Here, we use a 'glacial landsystems' approach, whereby the assemblages of glacial sediments and landforms associated with different glacierised regimes are described (Evans 2003). The suite of sediments and landforms is principally controlled by climate, type of ice mass (ice sheet, mountain glacier, icefield etc.), topography, terminal environment (terrestrial, marine, lacustrine) and geographic position, including latitude.

14.2 Characteristics of Glaciers

In this section, we describe the broad types of glaciers and environments in which ice masses are found. More detailed accounts can be found in Benn and Evans (2010), Bennett and Glasser (2009). Extensive photographs and brief descriptions of all glacial phenomena can be found in Hambrey and Alean (2016).

14.2.1 Types of Ice Mass

14.2.1.1 Ice Sheets

Ice sheets are by far the largest masses of land ice: by definition, they are over 50,000 km^2 in area. Today, there are two ice sheets, the Greenland and Antarctic ice sheets. Ice sheets bury the landscape, with only the summits of the largest mountains (nunataks) emerging through the ice. The ice in Antarctica is on average over 2000 m thick and reaches over 4700 m at its thickest. Ice flow at this scale is not constrained by bed topography, especially in the ice sheet's central portions. The land surface underneath may be a topographic high, or, as for West Antarctica and parts of East Antarctica and the former Laurentide Ice Sheet, a vast lowland (Benn and Evans 2010). The central parts of ice sheets are typically characterised by an ice dome, which acts as an ice-dispersal centre, from which the ice mass is dispersed radially (Figure 14.1). 'Ice divides' occur where ice flows away from ice dispersal centres in opposing directions. Ice sheets usually contain multiple ice domes or ice dome complexes, and ice divides associated with these domes migrate as an ice sheet evolves through time. These shifting ice divides cause changes in ice flow direction in the central parts of the ice sheet, leading to overprinted glacial landforms (Hughes et al. 2014). Most of the ice is discharged through 'ice streams' to the periphery and into the ocean. Ice streams are defined as corridors of fast-flowing ice within an ice sheet, flanked by slower-moving ice or by topography (Bennett 2003) (Section 14.5.5).

During the global Last Glacial Maximum (Marine Isotope Stage (MIS) 2, ~21,000–27,000 years ago) (Clark et al. 2009), ice sheets expanded, resulting in global sea levels that were 130 m lower than today (Simms et al. 2019). For example, a huge Laurentide Ice Sheet, with a sea-level equivalent of 76.0 $\pm$ 6.7 m, was present over North America (Simms et al. 2019; Dalton et al. 2020), and a large Eurasian Ice Sheet with a sea-level equivalent of 18.4 $\pm$ 4.9 m expanded in Europe and Russia (Hughes et al. 2016; Simms et al. 2019). In the Southern Hemisphere, the growth of ice sheets was restricted to an additional 9.9 m sea-level equivalent for the Antarctic Ice Sheet (Bentley et al. 2014; Simms et al. 2019), and 1.5 m for the Patagonian Ice Sheet (Davies et al. 2020), compared to present day. These large ice sheets expanded during each cold marine isotope stage throughout the Quaternary, with particularly large advances after the Mid-Pleistocene Revolution (1.25–0.7 Ma) and during MIS 16, 12, 10, 8 and 6 (Batchelor et al. 2019).

Significant pre-Quaternary glaciations occurred during the Cryogenian Period (c. 720–635 Ma) (Long et al. 2019), including the Sturtian interval (c. 715–680 Ma) and the Marinoan (c. 650–635 Ma). These periods were associated with probably the most extensive glaciations of any period on Earth (*ibid.*). Significant ice ages also occurred during the Gaskiers Glaciation (Ediacaran period: c. 580 Ma) (Kendall et al. 2004), and during the Late Ordovician (445.6–443.7 Ma), Late Devonian to Early Carboniferous (361–349 Ma) and Permo-Carboniferous (326–267 Ma) as well as other intervals.

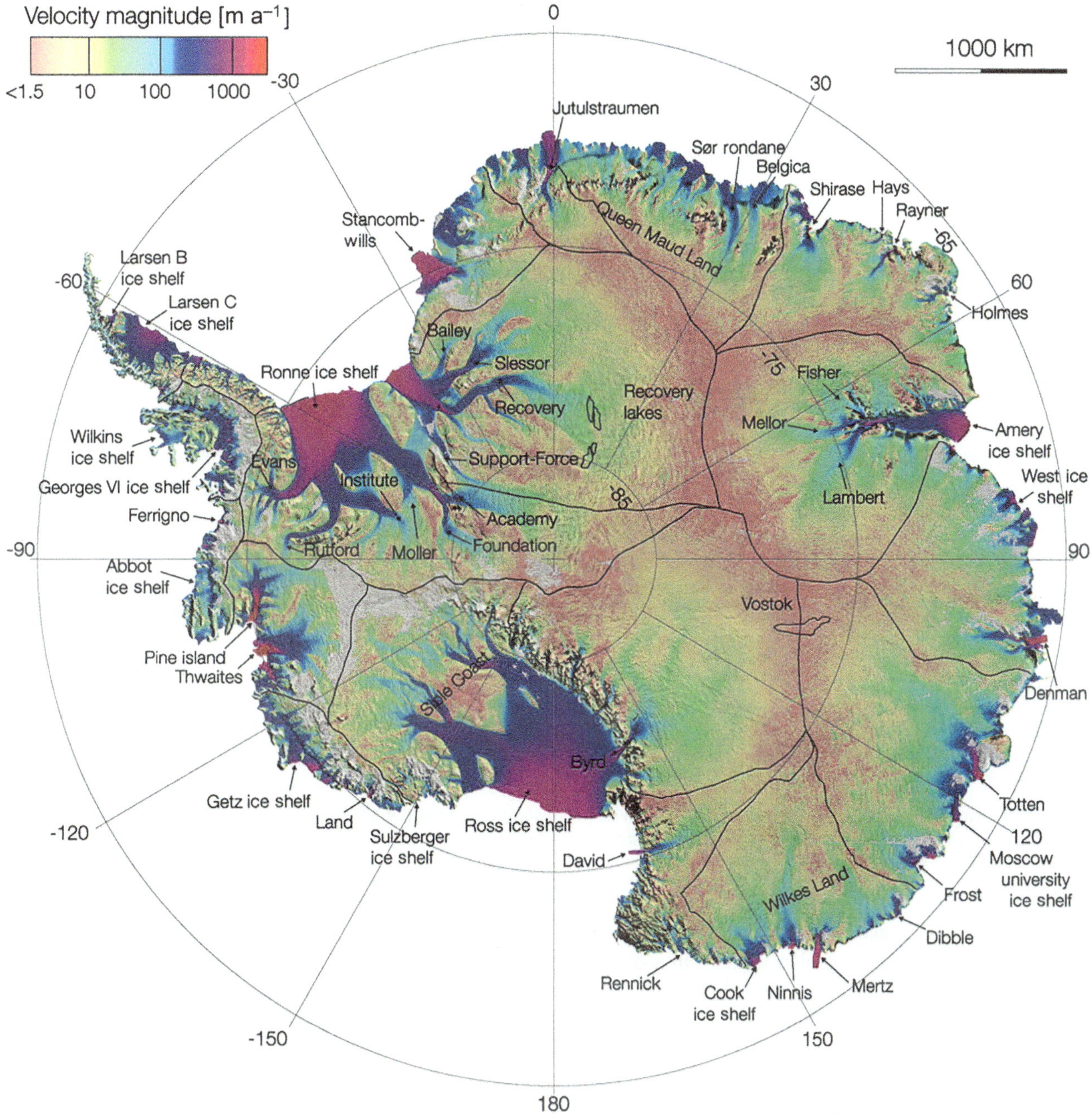

Figure 14.1 Ice flow velocities (m a^{-1}) within the Antarctic Ice Sheet, showing the ice divides (black lines), surrounded by areas of slower moving ice, and ice streams as areas of faster-flowing ice. From Rignot et al. (2011, p. 1428) / American Association for the Advancement of Science.

14.2.1.2 Ice Caps and Icefields

Similar to ice sheets, ice caps are unconstrained by bed topography and have a domed surface topography, but are smaller than 50,000 km^2. Today, their location is restricted to the high latitudes, where they exist at a low elevation. Ice caps occur today in Iceland, Arctic Canada and Alaska, among other places. Ice caps typically form on topographic highs and are drained by outlet glaciers flowing radially outwards from the centre (e.g. Hannesdóttir et al. 2015). As an example, the Vatnajökull Ice Cap, Iceland, is one of Europe's largest ice caps at 7800 km^2 (Schmidt et al. 2020) with an average ice thickness of ~400 m (Marshall et al. 2005) (Figure 14.2).

Icefields, by contrast, are influenced by the underlying topography. They too are drained by outlet glaciers and, for alpine icefields, ice divides are clearly demarcated by nunataks. Examples include the Columbia, St Elias and Juneau icefields in Canada and Alaska, plateau icefields

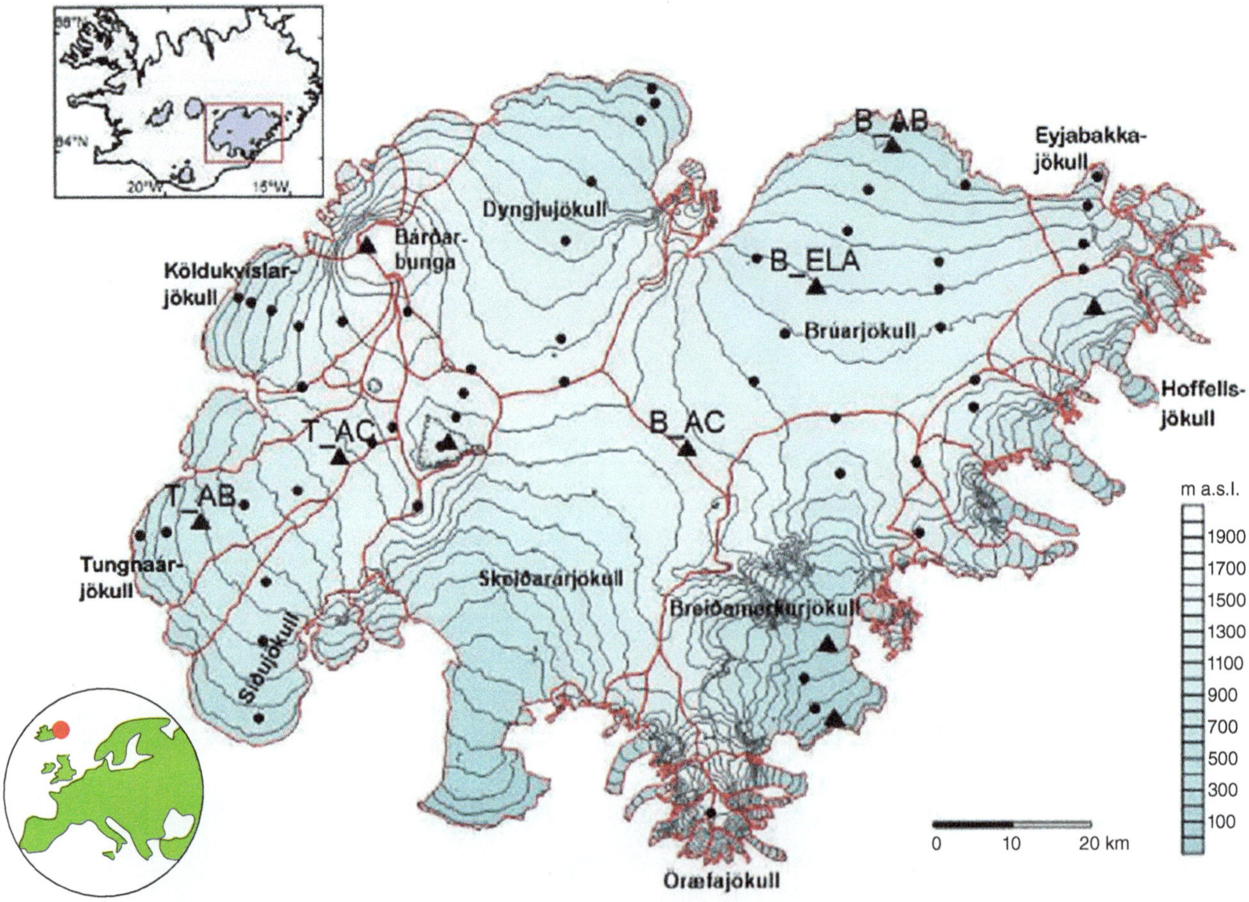

Figure 14.2 Vatnajökull Ice Cap, Iceland. Ice divides are in red, with some named outlet glaciers. Ice surface contours are shown. Black triangles are automatic weather stations, and circles are sites where surface mass balance and velocity are monitored in the field. From Schmidt et al. (2020, p. 98) / Cambridge University Press.

in Norway, transection glacier icefields on Spitsbergen, the Northern and Southern Patagonian icefields in southernmost South America (Figure 14.3), and the Tien Shan and Kunlan Shan icefields in China.

14.2.1.3 Mountain Glaciers

Mountain glaciers fringe larger ice masses and are many and varied in form. They include valley glaciers (Section 14.5.2), where ice is discharged from a cirque or icefield; transection glaciers, that cover deeply dissected mountainous landscapes and form interconnected systems of valley glaciers; piedmont glaciers, that spill out as they reach unconfined lowlands; and smaller cirque glaciers that form armchair-shaped bedrock hollows on the sides of mountains (Benn and Evans 2010). Smaller still are the thin niche glaciers and 'glacierets' that form on steep mountain sides or are left behind as glaciers shrink. They are differentiated from snow patches by movement due to the internal deformation of ice.

Today, mountain glaciers occur in many mountainous regions worldwide, including around the margins of the Greenland and Antarctic ice sheets (e.g. Figure 14.4). Worldwide, 215,547 glaciers are included in Version 6 of the Randolph Glacier Inventory (Pfeffer et al. 2014; Randolph Glacier Inventory Consortium 2017), with a global glacierised area of 705,738.79 km^2 (this includes icefields and icecaps; all ice outside the two ice sheets). These ice masses have a total volume of $158 \pm 41 \times 10^3$ km^3 (Farinotti et al. 2019), which is equivalent to 0.32 ± 0.08 m sea level.

14.2.1.4 Glaciomarine and Glaciolacustrine Environments

The term 'glaciomarine' refers to the marine environment that is influenced by glaciers and ice sheets, including where marine water and glacier ice are in contact (Powell and Molnia 1989; Ó Cofaigh 2013; Domack and Powell 2018). Glaciomarine environments may be proximal to the ice margin, or very distal, and receive sediments via suspension settling from long-distance sediment plumes or rafting by debris-laden icebergs (see Section 14.5.10). This environment therefore has a wide variety of different

Figure 14.3 Landsat 8 image composite of the Southern Patagonian Icefield, acquired April 29 to May 24th, 2016. *Source:* NASA Earth Observatory / Public domain.

processes and sediment sources, with complex interactions between glacier ice, sea ice (frozen sea water), icebergs, seawater, the actions of wind and waves, tides and gravity flows. The long-term preservation bias towards subsiding continental margins means that the pre-Cenozoic glaciomarine stratigraphic successions are dominated by glaciomarine sequences (Vandyk et al. 2019).

Glaciers that reach the ocean and calve icebergs are termed 'tidewater glaciers' (van der Veen 2002; Benn et al. 2007). Calving accounts for 70% of the world's annual mass transfer from glaciers and ice sheets to the ocean (Vieli 2011). Calving from marine-terminating outlets is the predominant way in which the Antarctic Ice Sheet loses mass (Shepherd et al. 2018). Tidewater glaciers may be grounded at their terminus, where they remain in contact with the seafloor and form a vertical ice cliff in the ocean at their terminus (e.g. Figure 14.5). This is usually the case for tidewater glaciers in temperate regions, such as Alaska, where the local climate exceeds the thermal limit of ice-shelf viability. In these more temperate regions, the ice is insufficiently stiff to form an ice shelf when the water depth is sufficient to float the ice. Here, the position of the tidewater glacier terminus is instead determined by water depth. The calving of icebergs occurs where the water depth increases and is sufficient to float the ice. In colder polar regions, tidewater glaciers, such as outlet glaciers and ice streams in Greenland and Antarctica are often floating at their terminus (van der Veen 2002). As the ice is stiffer, the floating ice from several glaciers may coalesce into floating ice shelves, as for example, around the Antarctic Peninsula (e.g. Cook et al. 2014; see Section 14.5.11 and Cuffey and Paterson 2010, for further details).

'Glaciolacustrine' refers to the processes and sediments involving a lake that receives meltwater from a glacier (Carrivick and Tweed 2013). Lakes frequently form in over-deepenings in valley floors scoured out by glaciers, forming deep basins infilled with a sedimentary record that can yield information on ice dynamics (e.g. Figure 14.6) (Bendle et al. 2017a). Commonly, glaciolacustrine environments are separated into 'ice-contact' environments, where the lake is in direct contact with a glacier, and 'ice-distal' environments, where the lake is not directly influenced by the ice front (Fitzsimons and Howarth 2018). Ice-contact lakes may be ephemeral, and drain once glaciers recede (e.g. Thorndycraft et al. 2019). Their sedimentary record may therefore relate to only the most recent glacial advance. The environment is characterised by energetic processes, changes in lake level, rapid accumulation of sediment and rapidly changing sub-environments, leading to abrupt changes in sedimentary facies (Section 14.5.9). Ice-distal lakes may persist for multiple glacial cycles,

Figure 14.4 The Alpine glacier Gepatschferner, Kaunertal, Austrian Alps, in 2021. Photo by Bethan Davies.

and contain a long sedimentary record (Fitzsimons and Howarth 2018). They are influenced by glacial meltwater, which introduces sands and gravels and fine-grained glacial sediment in suspension but lack more energetic processes associated with the ice margin.

14.2.2 Changes in Global Ice Cover

The proportion of the Earth's land surface covered by ice (presently ~10%) has undergone dramatic fluctuations throughout Earth's history. On timescales of hundreds of millions of years, the Earth System has tipped between periods with high average ice cover (icehouse, like today, with polar ice sheets) and little or no ice cover (greenhouse) (Benn and Evans 2010). These longer-term cycles are overprinted by shorter-term cycles on the scale of thousands of years, resulting in constantly changing global ice volumes. The occurrence of glaciations on all timescales is modulated by ice-albedo feedbacks which can result in sudden 'tipping points' between ice accumulation and recession.

On 10 My and longer time scales plate tectonics, through changing ocean gateways and positioning of high latitude land masses, has probably been a fundamental control of glaciation. Globally significant glaciations in the Late Ordovician and Permo-Carboniferous occurred when parts of Gondwana occupied the southern high-latitude regions during the slow assembly of the supercontinent Pangea. Global icehouse intervals in the Phanerozoic are also associated with high atmospheric O_2, low atmospheric pCO_2 and relatively high magnitude and high-frequency changes in eustasy (Montañez and Poulsen 2013; Isbell et al. 2021). During the severest Sturtian and Marinoan glaciations in the Cryogenian Period in the Neoproterozoic (740–630 Ma), glacier ice reached equatorial latitudes, vastly diminishing the global hydrological cycle (Hoffman et al. 2017a). However, Sturtian glaciation followed the break-up of a tropical supercontinent with no clear polar land mass, and the emplacement of a large igneous province with CO_2 release into what was already a much higher CO_2 atmosphere than present.

Figure 14.5 Point Wild, Elephant Island, Antarctic Peninsula. This grounded tidewater outlet glacier makes a steep vertical ice cliff ('ice wall') where it meets the ocean. Photo by Bethan Davies.

Extensive southern hemisphere Cenozoic glaciation began to leave a sedimentary record around 34 Ma, when Antarctica rifted away from South America and became isolated from oceanic heat transfer from the equator by the formation of the Antarctic Circumpolar Current, following the development of Drake Passage (Siegert 2008; Anderson et al. 2011). The uplift of the Tibetan Plateau and the North and South American Cordillera in the Mid-Cenozoic may have increased CO_2 drawdown by weathering, encouraging intensification of glaciation (Ruddiman et al. 1997), and the higher landmasses may have favoured snow persistence and glacier formation.

Shorter-term climatic oscillations are superimposed on both ice- and greenhouse periods. Quaternary climates have cycled between ice ages and interglacials, linked to cyclical changes in the Earth's orbit around the sun (Denton and Hughes 1983; Imbrie 1984; Chapter 13). Key cyclical orbital variations include the eccentricity (becoming more or less elliptical), the obliquity or tilt of the Earth's axis relative to the orbital plane, and the direction of tilt of the Earth's axis (precession). These cycles together cause variations in the amount and timing (seasonality) of solar radiation received in the Northern Hemisphere, driving a series of responses and feedbacks in the Earth System (see box on Milankovitch cyclicity in Chapter 13; Box 13.3).

At present, the Greenland and Antarctic ice sheets, and glaciers and icefields worldwide, are shrinking, at an accelerating rate (Hanna et al. 2020). Between 2006 and 2016, glaciers and ice caps contributed 335 ± 144 gigatonnes (Gt) of ice to the ocean per annum, leading to 0.92 ± 0.39 mm a^{-1} sea-level rise (Zemp et al. 2019). This accounts for 25–30% of observed sea-level rise over this period. The contribution to sea-level rise from the Greenland Ice Sheet has accelerated sharply since 1995 (Bamber et al. 2018). From 2012 to 2016, the Greenland Ice Sheet lost 247 ± 15 Gt of ice per year, equivalent to 0.69 ± 0.04 mm a^{-1} sea-level rise. In this same period, the West Antarctic Ice Sheet (including the Antarctic Peninsula)

Figure 14.6 Calluqueo Glacier, Monte San Lorenzo, Patagonia, Chile. Mountain glaciers frequently have lakes forming in front of them. Large trimlines are visible higher up the glacier, and in the foreground, large lateral moraines are prominent. Medial and lateral supraglacial debris is visible on the glacier surface. Photo by Bethan Davies.

lost $172 \pm 27\,\mathrm{Gt\,a^{-1}}$ ($0.48 \pm 0.08\,\mathrm{mm\,a^{-1}}$ sea-level rise) (Bamber et al. 2018). Rates of early 21st-century mass loss from glaciers are without historical precedent on a global scale, and mass loss rates will continue even if climate stabilises (Zemp et al. 2015) because glaciers and ice caps are now in disequilibrium with climate.

14.3 Processes

14.3.1 Mass Balance

Ice masses are systems with inputs and outputs. The main input to the system is snowfall which builds up in layers on the glacier surface. Over time the snow compresses, becoming denser and air pockets become smaller. Snow that is in the process of transforming to ice is called 'firn' (Cuffey and Paterson 2010). This process of mass gain, known as 'accumulation', occurs over a number of years until the interconnected air spaces are completely sealed-off, resulting in trapped air bubbles within the ice. While snowfall may occur over the whole glacier surface in winter, net accumulation over the course of a year is confined to the upper part of a glacier ('accumulation zone'), where temperatures are lower and snow survives the summer melt season. Ice mass is lost through melting, evaporation, sublimation, wind scour, calving of icebergs into water, and, in very cold terrestrial conditions, dry calving (avalanching) of ice blocks (Benn and Evans 2010). Collectively, this loss of mass is known as 'ablation', and the lower part of the glacier, where this is dominant, is the 'ablation zone'. Accumulation and ablation zones are bounded by the equilibrium line, where ice mass is more or less constant over time.

Ice flow (see Section 14.3.2) moves ice from the accumulation zone to the ablation zone. In order for a glacier to remain in equilibrium, and the glacier snout to remain at the same location, the mass lost in the ablation zone must be replaced with ice from the accumulation zone. Glaciers that are in equilibrium with climate and have a high mass turnover have a faster 'balance velocity' (Sugden and John 1976) that redistributes this mass, while glaciers with a smaller mass balance gradient and therefore a smaller mass turnover have a lower balance velocity, as less mass is transferred annually through the system.

Whether a glacier terminus advances or recedes is controlled by the balance between mass gains (accumulation) and mass losses (ablation). This is referred to as glacier 'mass balance': the change in mass of a glacier (or part of a glacier)

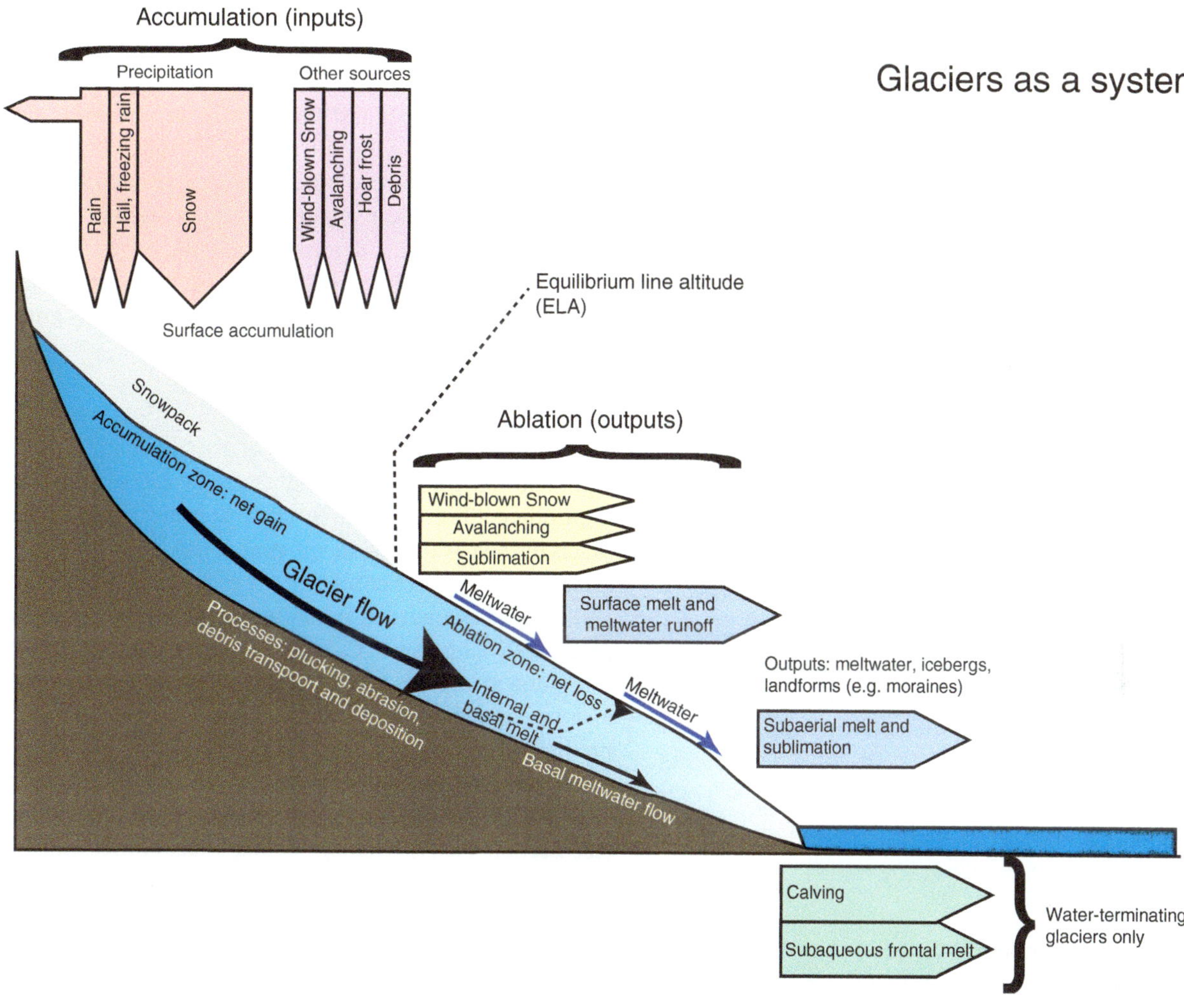

Figure 14.7 The glacier as a system. Ice mass gain is largely from precipitation, but also from wind-blown snow and avalanches. The glacier loses mass (ablates) mainly through calving and surface and subaqueous melt. The Equilibrium Line is the point on the glacier where net accumulation and net ablation are equal. Above the equilibrium line altitude (ELA) there is more accumulation; below it, there is more ablation. The net gain above the ELA means that the snowpack here is retained over summer.

over a specific time period (Cogley et al. 2011). Most glaciers worldwide today have negative mass balances because ablation outweighs accumulation, and as a result, most are receding (Zemp et al. 2019). A glacier with a mass balance of zero is considered to be in equilibrium with the climate.

Glacier mass balance varies seasonally as well as spatially. Over the area of a typical clean-ice valley glacier (see Section 14.5.3 for debris-covered glaciers), ablation and accumulation may both occur across the whole ice surface. Glacier mass balance is usually evaluated over the course of a year because accumulation and ablation vary seasonally. At mid- or high-latitude glaciers, winter is dominated by accumulation across the glacier surface, while summer is characterised by ablation, particularly in the lower part of the glacier. Over the course of one year, the 'net mass balance' at any given point is the total accumulation minus the total ablation.

Ice mass balance can also be considered at specific locations along the long profile of a glacier or ice stream. Over the course of one year, total ablation is greatest at the snout and therefore mass balance at this location is the most negative (Figure 14.7). Generally speaking, annual mass balance becomes less negative with altitude further up-glacier, as ablation reduces, and accumulation increases with increased snowfall and colder temperatures at higher altitudes.

At an altitude known as the 'equilibrium line altitude' (ELA), the net mass balance is zero because, over the course of the mass balance year, net accumulation balances net ablation. Up-glacier of this point net mass balance becomes increasingly positive with altitude as accumulation increases and ablation decreases. Up-glacier of the ELA is known as the 'accumulation zone'. Down-glacier of the ELA, net ablation increases and this forms the 'ablation zone'. The rates at which annual accumulation and ablation change with altitude are known as the 'accumulation gradient' and 'ablation gradient', and the sum is the 'mass balance gradient'. Accumulation and ablation gradients on the same glacier are usually different, with the ablation gradient steeper than the accumulation gradient. The ratio between the two gradients is referred to as the 'balance ratio' (Furbish and Andrews 1984; Rea 2009), and this is increasingly used to estimate the ELAs of palaeo-glaciers.

14.3.2 Glacier Flow

Glacier movement is driven by gravitational stresses acting on the ice. These driving stresses are balanced, or opposed, by resistive stresses, such as basal and lateral drag, which act where the ice meets the surrounding terrain or less mobile ice. The viscosity of the ice will also resist the driving stress. Additionally, longitudinal stress within the ice mass, which changes spatially due to variations in compressive and tensile forces, can act as either a resistive or driving stress. Glacier ice deforms in response to all these stresses, through processes of creep and fracture. Creep is the most important in terms of glacier flow and results from movement within and between individual ice crystals (Benn and Evans 2010). The process is described by Glen's Flow Law (Glen 1955; Nye 1957) which, in its simplest form where basal shear stress is assumed to balance the driving stress, can be expressed as:

$$\varepsilon^{\cdot} = A\tau^{n}$$

where $\varepsilon^{\cdot}$ is the strain rate, τ is the basal shear stress, A is a parameter that increases exponentially with temperature, and the exponent n is the flow law component, which is often around 3.

Glen's Flow Law shows that the amount of ice deformation is dependent on the basal shear stress and the temperature of the ice; warmer ice deforms more readily. Strain rates are highest at the bed of the glacier where the shear stresses are high, but velocity may be zero here due to resistive forces (Benn and Evans 2010). Velocity thus increases both away from the bed and away from lateral resistive stresses at the sides, meaning that ice velocity is highest towards the central area of the glacier surface as a result of the cumulative effect of strain rates in the surrounding ice.

Where there is water at the glacier bed, resistive stresses are reduced, and ice masses can move through a second mechanism: sliding. Water occurs at the glacier bed when ice is at a pressure melting point (Weertman 1957), since both pressure and temperature affect the melting point of ice (see Section 14.3.3). Water at the bed may flow within channels, as a thin film, or a braided or networked cavity system of streams or as groundwater flow. Frictional stresses limit the amount of sliding that can occur, and sliding is spatially and temporally variable across a glacier bed.

In addition to ice deformation and sliding, glacier movement is also caused by the deformation of soft sediments at the glacier bed. This third mechanism is thought to be the primary contributor to fast flow within ice streams (Alley et al. 1986, 1987; Blankenship et al. 1986, 1987). Bed deformation occurs when pore water pressure within a soft sediment bed increases, reducing the yield strength of the sediment and allowing deformation (Tulaczyk et al. 2000; Boulton et al. 2001; Iverson 2010). Strain rates are highest at the top of the sediment pile, where shear stress is greatest, and bed deformation decreases with depth.

14.3.3 Thermal Regime

Although 'all glaciers are cold', some glaciers are colder than others. There are three main thermal regimes. The first is termed 'warm-based' or 'temperate', which describes ice masses with bases entirely at the pressure melting point. Under high overburden pressure, ice melts, even at temperatures well below 0°C. Liquid water is then stable as a thin film at the ice-bed interface, between ice crystals, and in channelised systems. Because more meltwater is produced, more erosion can occur, and for this reason warm thermal regimes are important in the production of glacial sediments and for the glacial sedimentary record (see Section 14.3.4). The lubricating effect of meltwater at the bed also plays a vital role in terms of glacier flow, allowing sliding, increasing the velocity of ice masses.

The second type are 'cold-based' ice masses, sometimes referred to as 'polar', which are glaciers and ice sheets that are below pressure melting point and therefore frozen to their beds and produce relatively less meltwater. Cold-based glaciers have no free-flowing basal meltwater (Hambrey and Glasser 2012), because surface, englacial and basal heating is too limited to raise the temperature to pressure melting point. Since there is no melting at the glacier bed, forward momentum is by ice deformation only (Section 14.3.2). As a result, generally, cold-based glaciers are considered to have a greatly reduced potential for erosion and debris entrainment (Section 14.3.4). This means that they can preserve relict landscapes and protect them from both erosion and deposition (Hättestrand and Stroeven 2002). However, the underlying substrate and proglacial area can be modified by glaciotectonic deformation particularly when the glacier flows over permafrost (e.g. in the Dry Valleys, Antarctica; Fitzsimons and Howarth 2020) (Section 14.5.7).

Cold-based glaciers are found in cold, arid environments, where mean annual air temperatures are well below 0°C, with limited snow accumulation and strong winds (Fitzsimons 2004). However, the thick ice sheets of Antarctica and Greenland have extensive areas of ice that reach pressure melting point at their bed, with free-flowing basal water, due to the effects of geothermal heat flow, the poor heat conductivity of ice (Cuffey and Paterson 2010), and the overlying weight of the ice (Hambrey and Glasser 2012). Hence, cold-based glaciers dominate in polar areas only where ice is thinner, and at the margins of ice sheets.

The third type, 'polythermal' ice masses, are a complex, patchwork mosaic of warm- and cold-based areas. Polythermal regimes can be in glaciers of any scale. At the large scale, the Antarctic ice sheet has a polythermal structure (warm-based zones include subglacial lakes and ice stream beds; cold-based areas correspond to thinner zones of ice including some margin areas). Other areas where polythermal glaciers dominate include Arctic regions such as Svalbard, the Canadian Arctic and mountain areas around the periphery of the Greenland Ice Sheet. Polythermal glacier structures can include (i) predominantly cold ice with a zone of warm ice at the bed where ice is thickest and/or caused by strain heating, or (ii) predominantly warm ice with a zone of cold ice at higher altitudes, or at the surface caused by cold winter air temperatures. Recently, as glaciers thin in a warming climate, the proportion of cold-based ice within polythermal (and some temperate) glaciers is increasing (e.g. Reinardy et al. 2019; Carrivick et al. 2023).

14.3.4 Debris Erosion, Entrainment, Transportation and Deposition

14.3.4.1 Erosion and Entrainment

Debris is incorporated into the underside of a glacier via a range of processes including quarrying, abrasion, net adhesion, regelation and freeze-on. Quarrying (or plucking) is the process by which larger fragments of rock are removed from a hard bedrock surface (Hallet 1996). Quarrying occurs because the normal force (weight) of a glacier is not evenly distributed across its bed and the underlying bedrock surface is not flat. These pressure differences induce stress fields in the subglacial bedrock that may be sufficient to induce new or expand pre-existing fractures (Iverson 2002; Glasser and Bennett 2004; Hooyer et al. 2012). Fluctuating water pressures at the ice-bed interface have also been shown in experiments to create and expand cracks (Iverson 1991; Cohen et al. 2006). This eventually leads to detachment of rock fragments, on scales of centimetres to metres, which become entrained in the basal ice.

Abrasion occurs as particles already incorporated into basal ice scrape against a bedrock surface. The bedrock becomes smoothed and polished as small-scale protuberances are removed. Particles in the basal ice may also produce narrow scratches in the bedrock surface, known as 'striae' (singular: 'striation'), or, at a larger scale, grooves. These striae and grooves are produced as a result of a series of brittle failure events as the particle moves in a jerky motion across the rock surface (Drewry 1986). The process of abrasion leads to edge-rounding of clasts towards sub-angular and sub-rounded, and towards a blockier clast shape (Benn and Ballantyne 1994; see Section 14.3.4.2). This process, along with quarrying, is a major contributor to the creation of sediment in subglacial systems and is the primary source of silt-sized sediment in glacial systems, often referred to as 'glacial flour' (Alley et al. 2019). Because they tend to have more debris entrained in their basal ice, warm-based glaciers often produce impressive striated and polished bedrock surfaces, which can be recognised in both the modern and ancient records (Figure 14.8).

Macro-scale landforms associated with glacial abrasion and quarrying include roches moutonées and whalebacks. Roches moutonées are asymmetric bedrock bumps that are characterised by an abraded stoss side and plucked lee side. This shape is caused by an initial increase in pressure at the glacier bed as it flows over a bedrock bump causing abrasion. The development of a cavity on the lee side promotes quarrying, resulting in an asymmetric shape. In contrast, whalebacks form when ice remains in contact with the bedrock bump, resulting in abraded stoss and lee sides and a symmetrical shape. Roberts and Long (2005) argue that roches moutonées are associated with thinner ice, which promotes cavity formation due to lower overburden pressures, while whalebacks form under thicker ice.

Although, in general, cold-based glaciers are considered to flow only through ice deformation, sliding has been observed at the bed of some cold-based glaciers. For example, basal sliding has been observed at the Urumqi Glacier Number 1 in China (Echelmeyer and Zhongxiang 1987) and at −17°C basal ice temperatures beneath the Meserve Glacier, Antarctica (Cuffey et al. 1999). While it is possible for clean ice to cause some abrasion, particularly where there are higher stress concentrations on bedrock bumps, it is significantly less effective as an erosive agent than ice containing entrained debris. Conditions at the bed of cold-based glaciers are, however, thought to be conducive to plucking due to adhesion (Boulton 1972), and Fitzsimons et al. (1999) demonstrated entrainment and deformation of debris beneath the Suess Glacier in Antarctica.

Debris can be entrained into basal ice by a process called 'regelation' (Weertman 1957). Regelation is a process where ice melts under relatively high pressure and subsequently refreezes under relatively low pressures. The process is associated with warm-based glaciers because basal ice has to be close to the pressure melting point. As ice flows around an obstacle at the glacier bed, pressure increases on the stoss (up-ice) side, which causes melting. The meltwater migrates to lower pressure on the lee (down-glacier) side of the obstacle, whereupon it refreezes to the base of the glacier,

Figure 14.8 Striated surfaces through space and time. (a) Polished bedrock (paragneiss) exposed in recent decades following the recession of the Gepatsch Glacier in Kaunertal, Austria, see also Figure 14.4 for a neighbouring glacier. (b) Polished bedrock (orthogneiss) in northern Namibia, which was glaciated approximately 300 million years ago (Dietrich et al., in press). (c) Early Permian striated surface developed on Neoproterozoic siltstone at Hallett Cove, South Australia. Note crosscutting striations and chatter marks. (d) Surface sheen (glacial polish) at Pingdingshan, central China, reveals evidence for two contrasting flow directions beneath an Ediacaran ice sheet. (e) Interaction of a modern creek with an Early Permian grooved and ridged surface at Glacier Rock, South Australia. (f) The peeling crust of glacial varnish, preserving striations, weathers away from paragneiss bedrock in Ontario, Canada. This surface is of Pleistocene age and is associated with Mega-Scale Glacial Lineations (MSGL) developed in bedrock (cf. Krabbendam et al. 2016). (g) Probable Late Cryogenian striated pavement in the Tarim Craton, NW China, beneath the Yuermeinak Formation (see Vandyk et al. 2019). (h) Example of a peeling and highly weathered soft-sediment striated surface (e.g. Le Heron et al. 2020) from the Late Ordovician of Saudi Arabia. Photo by D.P. Le Heron.

incorporating sediment to create a debris-rich ice layer. Regelation ice is therefore often laminated, consisting of fine layers of clean ice and debris-rich ice (Hubbard and Sharp 1995). Sediment can also be entrained by regelation of ice into a porous substrate, which is known as 'regelation infiltration' (Alley et al. 1997). This process can occur where a glacier is underlain by soft sediments. Layers of subglacial debris entrained in this manner can be decimetres to metres thick (Alley et al. 1997; Rempel 2008). Net adfreezing can also incorporate debris into basal ice in areas where freezing dominates. This process can occur when meltwater flows to colder areas of the glacier bed, for example, at thin glacier margins when the winter cold wave is able to penetrate into the bed, lowering the freeze-front. Debris is also entrained into, or moved within, moving ice, by mechanical deformation processes such as folding, thrusting and fracturing (Jennings and Hambrey 2021).

At subglacial overdeepenings close to the ice margin, meltwater is forced to flow upslope out of the overdeepening by the pressure gradient. As the water ascends, the pressure drops, and therefore the pressure melting point rises. If the pressure melting point rises quicker than the ascending water can warm, the water becomes 'supercooled', temporarily existing as a liquid below pressure melting point (Alley et al. 1998; Lawson et al. 1998). As the water freezes to the glacier bed, it captures debris carried by the meltwater. Ice formed from supercooled water freezes as 'frazil' and 'anchor ice'. This ice has distinctive characteristics that include layering from successive freezing events, sediment concentrations of about 30% by volume, high proportions of silt (because frazil ice growth preferentially selects silt), and a high proportion of edge-rounded fluvial clasts indicating glaciofluvial rather than subglacial sediment transport (Cook et al. 2011).

Debris also becomes incorporated into the glacier system supraglacially by material deposited onto the glacier surface via rock falls, snow avalanching, debris flows and meltwater streams from the valley sides. Some of this debris is then incorporated into the glacier via crevasses and through snow accumulation (Jennings and Hambrey 2021; see Sections 14.5.2 and 14.5.3). Edge rounding of clasts does not occur within debris that is transported passively on the ice surface, except where the debris is moved by supraglacial meltwater streams. The differences between the degree of edge rounding on clasts that have been transported subglacially and supraglacially enable the mode of debris transport to be identified in glaciogenic sediments in the glacier foreland (Lukas et al. 2013).

14.3.4.2 Transportation

The concept of the 'glacial debris cascade' summarises the complex transport pathways of sediments within a glacier (Benn and Evans 1998; Evans 2017b) (Figure 14.9). In mountain glaciers, debris can be sourced from the valley sides and the subglacial environment, including plucked and abraded bedrock or overridden sediments. Once debris is entrained supraglacially or subglacially, it is carried along a debris transport path (supraglacially, englacially and subglacially) (Figure 14.10). Debris transported in supraglacial pathways will undergo little erosion or modification, while subglacially transported debris is subjected to abrasion (Benn and Ballantyne 1994; Benn 2004; Goodsell et al. 2005; Hambrey and Glasser 2012; Lukas et al. 2013). Debris may also be transported via glaciofluvial pathways in temperate glaciers, whereby the debris will undergo further modification and edge rounding (Benn et al. 2003).

The debris transportation pathway can be reflected in the shape of clasts deposited in ice-marginal environments. Clast shape can be explored in an RA-C_{40} plot (Figure 14.11), where materials with a longer transport history with more abrasion (beach cobbles, fluvial cobbles) will trend towards the bottom left of the plot, as they become more edge-rounded and more blocky/spherical in shape. Materials with a shorter transport history (such as rock fall or scree) have sharper edges and a more slabby or elongated shape, and will therefore trend towards the top right (Benn and Ballantyne 1994; Lukas et al. 2013). Clasts transported subglacially will tend to be somewhat blocky, edge-rounded and frequently faceted and striated.

14.3.4.3 Deposition

Deposition of glacially transported material occurs predominantly in the sub-marginal and ice-marginal zones (Boulton 1996). Debris deposited by a glacier often consists of an unsorted mixture of a wide range of particle sizes from clay to large boulders. The term 'diamicton' is used to describe any poorly sorted sediment (not just glacially derived), and 'diamictite' is used for the lithified equivalent (Flint et al. 1960). The term 'subglacial traction till' is used for sediment that has been directly deposited at the bed of a glacier and largely homogenised (Evans et al. 2006a; Evans 2017a), and is therefore interpretive of this process. This term incorporates all processes that occur subglacially and characterise debris deposition: lodgement, ploughing, bed deformation and meltout (Benn and Evans 2010). Tills can also include evidence of water flow, giving rise to stratified sediments (Evans et al. 2006a). Previously, tills were considered as end-members of one of these processes, and were assigned to lodgement till, deformation till or meltout till. Evans et al. (2006a) argued that a range of subglacial processes is likely to occur at the glacier bed at any given time, and therefore any subglacial till is a hybrid product of these processes. Consequently, a subglacial traction till is likely to contain evidence for a range of processes that are difficult to separate into clear end-members.

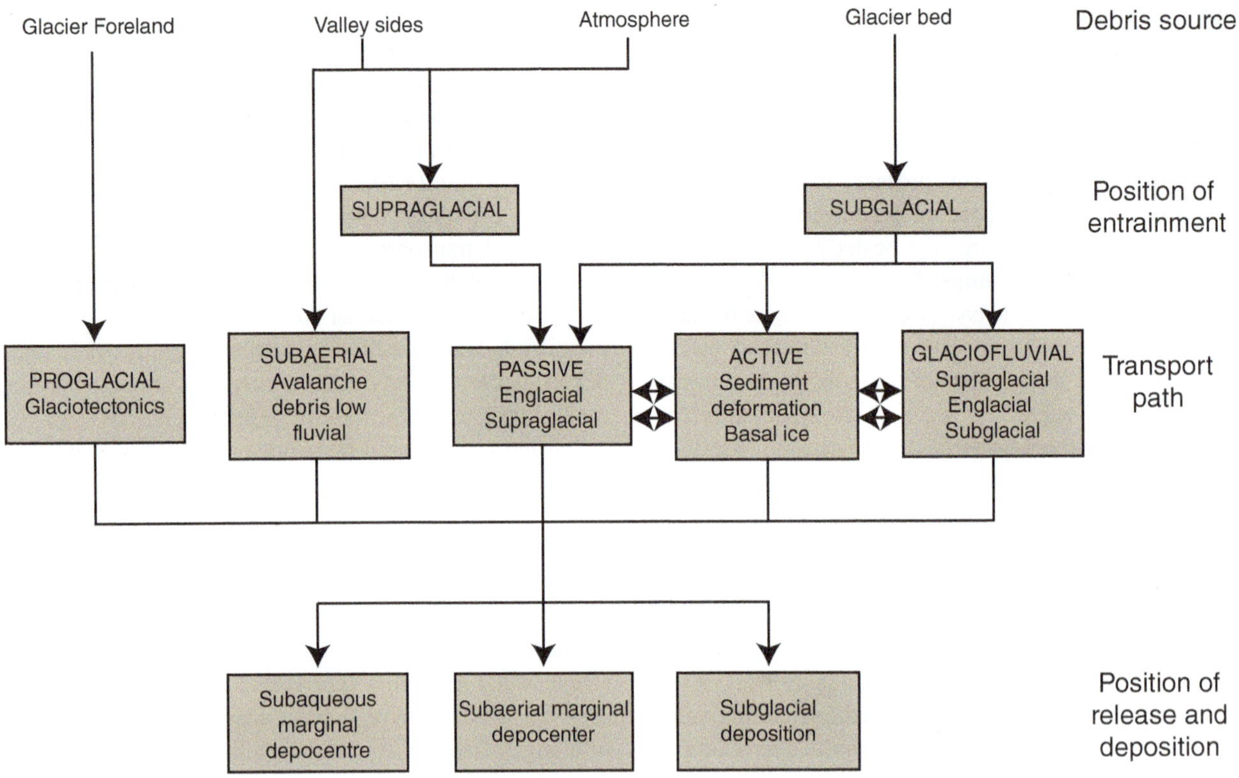

Figure 14.9 The glacial debris cascade. These transport pathways are relevant for all glaciers, but are particularly relevant for the valley glacier landsystem. Modified from Benn and Evans (2010, p. 334).

A two-tiered till structure, consisting of an A and B horizon (Evans et al. 2006a; Phillips et al. 2018), has been identified in sections at many Icelandic glacier forelands (e.g. Breiðamerkurjökull; Benn 1995) (Figure 14.12). The upper A horizon is a low-strength, dilatant diamicton that in a subglacial position would have been the actively deforming layer with high pore water pressures and freely rotating clasts. By contrast, the lower B horizon is a higher-density, compact diamicton, with clast a-axis (longest axis) fabrics organised parallel to ice flow, and a fissility that develops through brittle shear (Benn 1995; Evans et al. 2006a; Evans 2017a). In sections, the A horizon is often missing however, due to either (i) subsequent partial or total erosion, (ii) overprinting by repeated reworking of sub-marginal till or (iii) by compaction and shearing during till accretion (Evans and Twigg 2002; Evans and Hiemstra 2005).

Subglacial traction tills can sometimes be difficult to distinguish from other ice-marginal deposits, which can also consist of massive and stratified diamictons. Compaction and evidence for shearing, sliding or deformation are used as defining criteria (Evans 2017a), alongside edge-rounded, faceted, striated, blocky-shaped clasts (Lukas et al. 2013) (Figure 14.11). Strong a-axis clast fabrics have also been used to identify subglacial traction tills (Hooyer and Iverson 2000), although there have also been many cases of weak or variable clast fabrics in tills (Evans 2017a).

As a glacier or ice sheet flows, subglacial sediment may be organised into subglacial bedforms. At the smaller scale, 'flutes' (or 'flutings') are produced beneath cirque or (outlet) valley glaciers, forming elongate ridges of sediment that are orientated parallel to ice flow and are tens of centimetres to a few metres in height, with widths typically around 1–2 m. Flutes develop as the subglacial deforming layer is squeezed into cavities in the lee of lodged boulders (Boulton 1976; Benn 1994; Evans et al. 2010; Ely et al. 2017; Ives and Iverson 2019) or is ploughed by ice keels or boulders embedded in the basal ice (Evans and Rea 2003; Eyles et al. 2015). They are composed of subglacial traction till, and since they generally have grain size distributions and massive structures that are the same as the adjacent subglacial till, they are difficult to identify in outcrop.

Beneath faster-flowing ice (large valley glaciers, e.g. Johnson et al. 2010, and ice streams, e.g. King et al. 2009, 2016; Barcheck et al. 2020), 'drumlins' form. Drumlins are egg-shaped mounds that can be both asymmetric and symmetric in shape and can be 0.5–40 m in height, although almost half of drumlins measured in Britain have a relief of <5 m (Spagnolo et al. 2011, 2012). The composition of Quaternary drumlins is highly variable, including

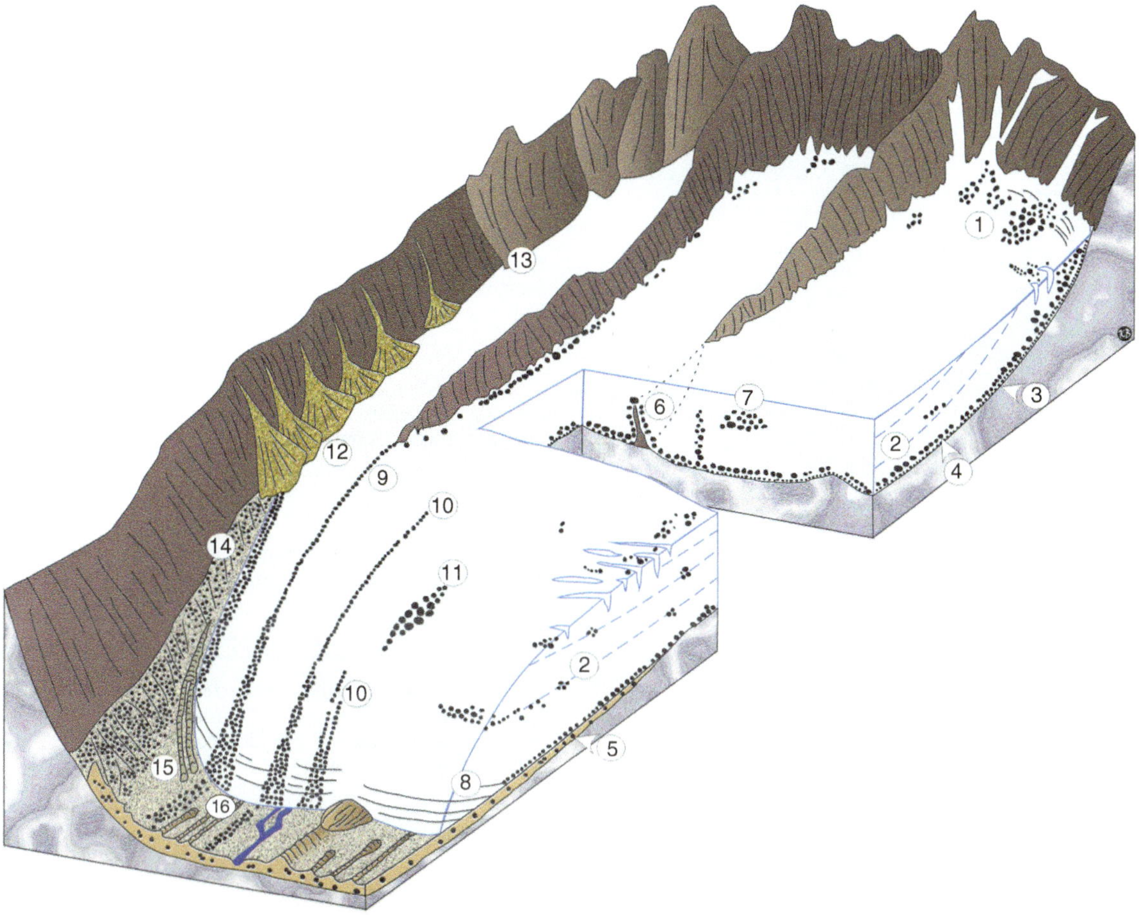

Figure 14.10 Debris transport pathways in a typical clean-ice mountain glacier: (1) Rockfall debris entrained supraglacially and buried by snow or fallen into crevasses. (2) Ice flow transports debris away from the surface in the accumulation area, and towards the surface in the ablation area (dashed lines). (3) Subglacial traction zone. (4) Suspension zone. (5) Subglacial traction till. (6) Debris septum elevated from the bed below a confluence. (7) Diffuse septum and cluster of rockfall debris. (8) Debris elevated from the bed by shearing. (9) Ice-flow interaction medial moraine. (10) Ablation-dominant medial moraines. (11) Avalanche-type medial moraines. (12) Supraglacial lateral moraines. (13) Periglacial trimline. (14) Lateral moraine. (15) Latero-frontal moraines. (16) Flutes. Modified from Benn and Evans (2010, p. 339).

Figure 14.11 A RA-C_{40} plot. This shows modification of clasts during passage through a glacier, based upon a co-variance plot of roundness (RA: % clasts that are angular or very angular) against C_{40} (percentage of clasts with a C/A axial rations $\leq$0.4), from Storbreen, Norway. Here, the 'subglacial' sample is from an elevated debris band and therefore differentiated from 'till'. Benn and Ballantyne (1994, p. 221) / with permission of Elsevier.

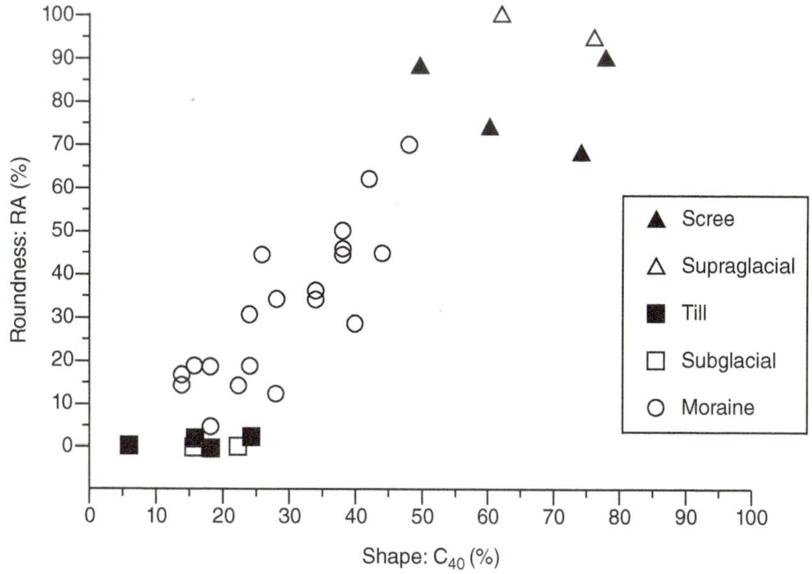

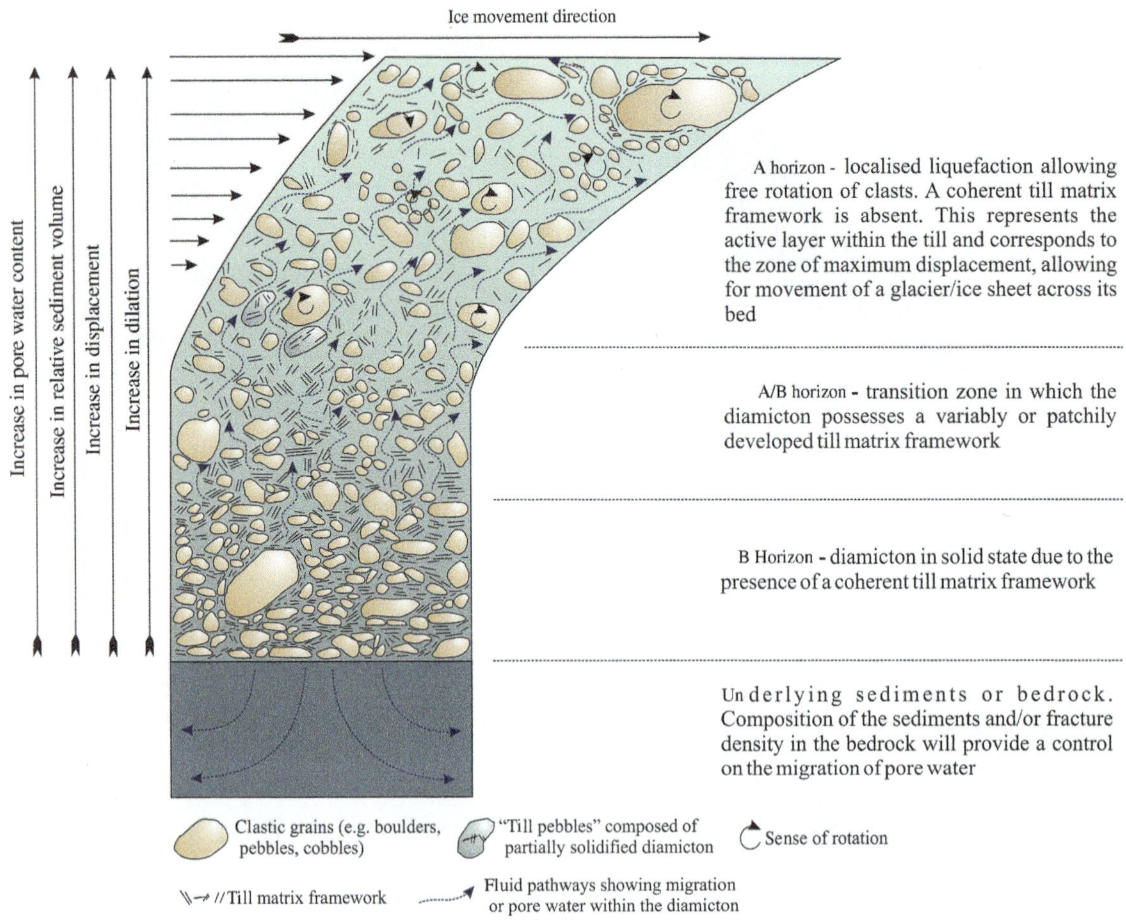

Figure 14.12 A and B horizons of subglacial till. Diagram shows the zonation of a relatively homogenous subglacially deforming till with dilation, displacement, sediment volume, shear strength, connectivity and pore water pressure. Adapted from Phillips et al. (2018, p. 124); after Evans et al. (2006a).

various combinations of subglacial till, glaciofluvial and glaciolacustrine sediments, and bedrock, which has made it difficult to identify a unifying theory for drumlin formation (Stokes et al. 2011). There has been a significant amount of research focussed on explaining the development of subglacial bedforms, particularly the formation of drumlins (e.g. Ely et al. 2016; Iverson et al. 2017; Clark et al. 2018; Zoet et al. 2021). Recent work has argued that subglacial bedforms grow and evolve as a result of instabilities at the ice-bed interface (e.g. Stokes et al. 2013; Ely et al. 2018). Furthermore, numerical modelling by Ely et al. (2023) shows that over time, 'subglacial ribs' (ridges perpendicular to ice flow) may evolve into drumlins which may then evolve into 'megascale glacial lineations' (MSGL). MSGL are highly elongated ridges of subglacial sediment that are a few kilometres long, a few hundred metres in width and lateral spacing, and form beneath fast flowing zones within ice sheets (Spagnolo et al. 2014).

On a microscale, a wafer-thin layer of chemical precipitate can form directly beneath an ice mass at the interface with hard bedrock (Siman-Tov et al. 2017). This silica gel is responsible for the 'glacial polish' seen on land surfaces of modern to Ediacaran (e.g. Le Heron et al. 2019b) age. The ultrastructure of glacial polish developed in Neoproterozoic glacial pavements of Norway led Bestmann et al. (2006) to propose that flash temperatures produced in response to subglacial shearing may even exceed 1000°C.

At the ice margin, sediment is deposited through a range of processes, which are dependent on the location of sediment within a glacier, glacier thermal regime and dynamics. The efficiency and location of sub-, en- and supraglacial and ice-marginal fluvial systems will also affect sediment deposition in this zone (Benn et al. 2003). This interplay between glacial, glaciofluvial and ice melt-out processes makes the ice-marginal zone a complex mix of subglacial traction till, debris flow deposits and sorted material with a wide range of grain sizes (e.g. Lawson 1982; Lukas 2012).

The most distinctive landform associated with the ice-marginal zone is a moraine ridge. While the term 'moraine' has been used with a variety of connotations, here it is used

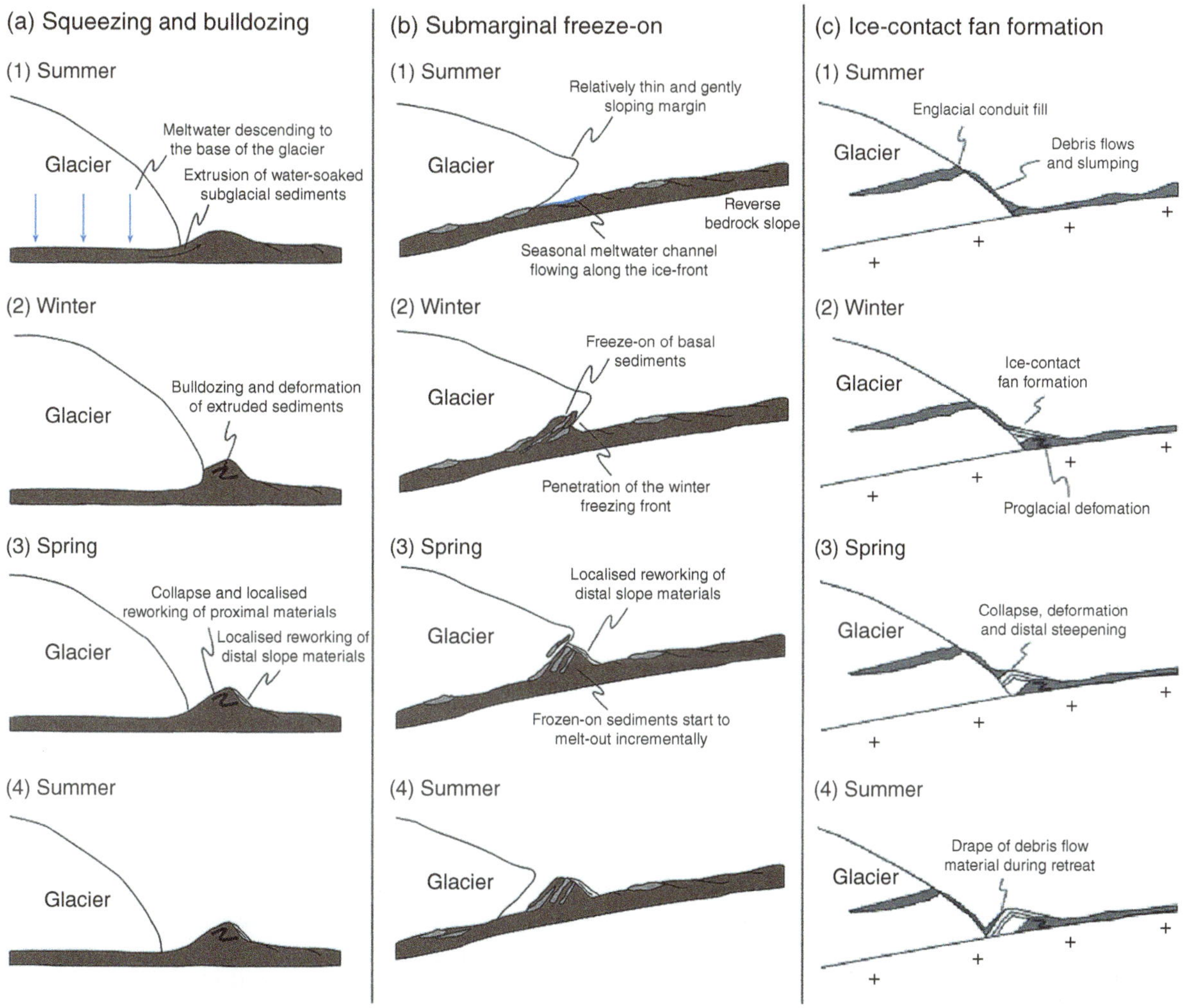

Figure 14.13 Formation of ice-marginal moraines: (a) squeezing and bulldozing, (b) submarginal freeze-on, modified after Chandler et al. (2016a, p. 194) and (c) ice-contact fan formation, Lukas (2012, p. 475) / with permission of Elsevier.

simply for a ridge or mound that is formed at the ice margin. Moraines derived from subglacial sediment can form through squeezing and extrusion of saturated subglacial sediments by ice overburden pressures followed by ice push (bulldozing) (Price 1970; Sharp 1984) (Figure 14.13a). This results in small (<1 m high) ridges of often massive diamicton that mimic the shape of the ice margin and are common, for example, on recently deglaciated Icelandic forelands (e.g. Evans and Twigg 2002; Chandler et al. 2016a, 2020a). Similarly, bulldozing of proglacial sediment immediately in front of an ice margin during advance will produce moraines consisting of a combination of subglacial traction till and glaciofluvial, glaciolacustrine and debris flow material (Lukas 2012; Chandler et al. 2016a). Subglacial sediment can also be emplaced in a moraine through the freeze-on of slabs of subglacial material to the glacier bed and subsequent forward motion (Krüger 1996; Reinardy et al. 2013) (Figure 14.13b). Ice margins may remain at or revisit a particular position for a number of years, creating a larger composite moraine (e.g. Evans and Hiemstra 2005).

Moraines can also form through the 'dumping' of supra- or englacial debris when an ice margin remains stationary. This process is often associated with valley or cirque glaciers, particularly at lateral margins, where there is an abundant source of supraglacial material from the valley sides (e.g. Boulton and Eyles 1979; Benn and Owen 2002; Lukas et al. 2012; Martin et al. 2019), and can occur in combination with ice push (e.g. Krüger et al. 2002). Very large lateral and frontal moraines can be deposited if a glacier remains stationary or the same moraine is revisited multiple times (e.g. Lukas et al. 2012; Hornsey et al. 2022), and are often associated with high sediment supply and low meltwater discharges (Benn et al. 2003). Delivery of sediment from the glacier surface can also be via subaerial debris flows (e.g. Zieliński and Van Loon 2000; Lukas

2005, 2012; Chandler et al. 2020b). Such moraines consist of dipping beds of diamicton with thin intervening layers or lenses of laminated sands (Figure 14.13c), documenting episodic debris flows and sheet wash. In a lacustrine or marine setting, moraines form as De Geer moraines, morainal banks, ice-contact fans and composite grounding line fans. Sediment can be bulldozed at the ice margin and/or deposited subaqueously by meltwater exiting the glacier subglacially, englacially and as supraglacial runoff. Further details are discussed in Sections 14.5.9, 14.5.10 and 14.5.11.

14.3.5 Glaciotectonism

Glaciotectionism is the deformation of sediment and bedrock by glaciers, under a combination of vertical load and lateral (push) processes (Aber et al. 1989). Glaciotectonism can occur both subglacially and proglacially and involves ductile and/or brittle deformation, predominantly within soft sediments. This process is particularly important at polythermal ice margins, where the thinner ice near the terminus is not at a pressure melting point, and remains frozen to the substrate, or where permafrost in the proglacial area leads to folding and thrusting as ice margins advance (Waller et al. 2012). The analogy of orogen development, characterised by in-sequence and occasionally out-of-sequence foreland propagating thrusts, with limited back-thrusts, is commonly invoked to understand glaciotectonic ridges (Schack Pedersen 2005; Figure 14.14).

The products of this range of processes are called 'glaciotectonites', defined by Benn and Evans (1998, p. 387) as 'rock or sediment that has been deformed by subglacial shearing, but retains some of the structural characteristics of the parent material', thus differentiating them from subglacial traction tills (Evans et al. 2006a). Glaciotectonites present a huge spectrum of products, the precise style at any one location representing the sum of the substrate composition, ice mass behaviour and size, and thermal regime. A range of glaciotectonic landforms can be produced, such as glaciotectonic composite ridges or thrust-block moraines (Figure 14.14), hill-hole pairs, cupola-hills and megarafts. Glaciotectonic moraines are sometimes referred to as 'push moraines' (e.g. van der Wateren 1995; Hart and Watts 1997; Boulton et al. 1999), but 'glaciotectonic composite ridge' or 'thrust-block moraine' is preferred to differentiate these large ridges in which >25% of the moraine is composed of glaciotectonites (cf. Benn and Evans 2010) from smaller push ridges.

Within glaciotectonic ridges, a range of folds and smaller-scale fractures is commonly encountered between propagating thrust slices. Deformation styles are largely lithology-dependent. In homogenous gravels, diamictons or sands, deformation may be expressed as kink bands or chevron folds. Shear zones can occur at the boundary between thrust units in ice-marginal deformation structures as well as within deformed subglacial substrates. In ancient records, an example of subglacial shear zones occurred in South Africa as the so-called Fold Zone within the Ordovician Peninsula Formation (Blignault and Theron 2017).

14.3.6 Glacial Hydrology

Meltwater flows supraglacially, englacially and subglacially, and as groundwater within and beneath a temperate glacier. In the supraglacial pathway, surface meltwater reworks supraglacial debris, and cuts down into the glacier surface, releasing englacial sediment as it does. Moulins are vertical channels where supraglacial streams plunge downwards into the glacier, sometimes following fractures and heterogeneities (Gulley et al. 2012), in a manner similar to sinkholes in a karstified landscape. These downward-directed meltwater routes may either continue within the glacier as englacial conduits or extend to the glacier bed (Gulley et al. 2012). Here, water may pond, forming a subglacial lake, or flow towards the glacier margin.

Subglacial meltwater can flow through either 'channelised systems' (conduits) or 'distributed systems' (across/within the bed) (Benn and Evans 2010). Channelised systems include 'Röthlisberger' or 'Hooke' channels, which are incised upwards into the ice, 'Nye' channels that are incised downwards into underlying rock or sediment, and 'tunnel valleys', which are large-scale channels formed when glaciers advance over a poorly consolidated substrate (Kirkham et al. 2024) (Section 14.5.4). Distributed systems include thin water films, linked cavity systems and braided canal networks that occur between the ice and bedrock or soft substrate and groundwater flow within subglacial sediment or bedrock. It should be noted that large hydraulic heads and therefore large erosive power are often characteristic of subglacial meltwater due to both loading by ice and topographic gradients.

Subglacial landforms associated with glacial hydrology include tunnel valleys, meltwater channels, eskers (which can also be englacial) and accumulations of sorted, bedded sediments (Section 14.5.4). Proglacial or ice-marginal landforms associated with flowing meltwater include kames, which are accumulations of sorted sediments, which can form flat-topped terraces between the glacier and valley sides, proglacial meltwater channels and outwash plains or sandar systems (Section 14.5.1). Landforms associated with glacial meltwater entering lakes are covered in Section 14.5.9.

14.3.7 Periglaciation

Periglacial environments are those dominated by frost-action and permafrost-related processes (Ballantyne 2018b). French (2018) defines the periglacial domain as

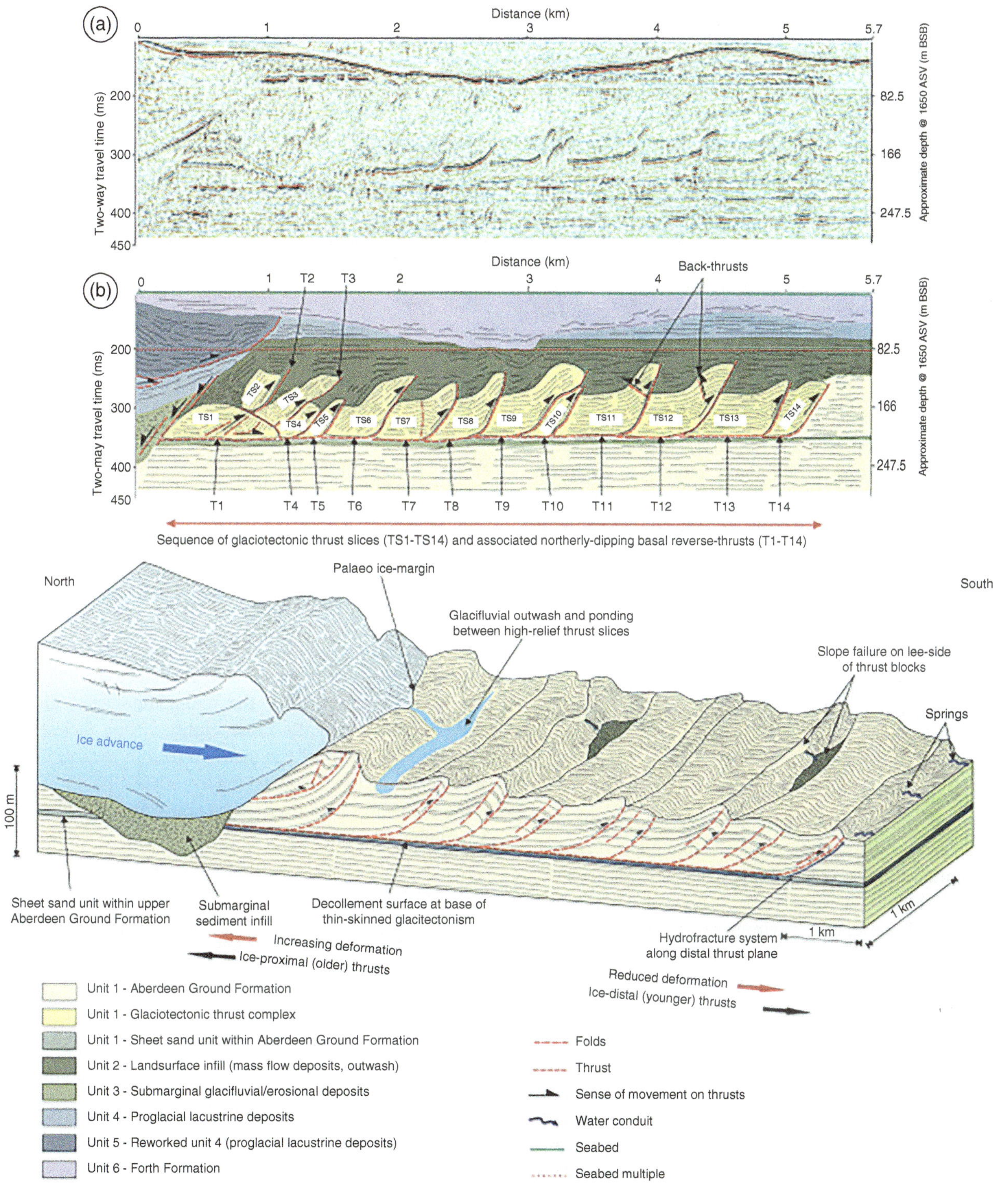

Figure 14.14 The products of large-scale pro-glacial to ice-marginal glaciotectonic deformation as imaged in the North Sea on seismic reflection data (a), together with their interpretation (b) and 3D reconstruction (c). The deformed blocks T1-14 are composed of Middle Pleistocene clays. The decollement surface is a sheet-sand. From Phillips (2018, p. 492) / with permission of Elsevier.

areas with a mean annual temperature of less than +3°C, although frost action processes do not dominate until the mean annual temperature is below −2°C. Periglacial environments are wide-ranging and include high arctic climates (weak diurnal, strong seasonal pattern), continental climates (weak diurnal, extreme seasonal pattern), subarctic island climates (low annual temperature range), mid-latitude alpine climates (well-developed diurnal and seasonal patterns), low latitude mountain climates (strong diurnal pattern, low annual temperature range) and the Tibetan Plateau (high insolation, strong diurnal pattern, moderate seasonal pattern) (French 2018). A wide range of periglacial landforms occur as a result of permafrost (e.g. palsas, lithalsas, pingos, ice wedges, thermokarst, rock glaciers) and frost-action in the active layer (the near surface layer subject to alternate freeze and thaw) (e.g. patterned ground: sorted and non-sorted circles and stripes, solifluction lobes, talus slopes). Further information on these landforms relating to processes of formation, relict examples and engineering implications can be found in Griffiths and Martin et al. (2017), French (2018) and Ballantyne (2018a). Periglacial features on Mars are discussed in Section 15.3.3).

Periglacial environments occur immediately beyond glacier limits. Here, they are particularly relevant because frost-related processes can influence, modify, or overprint freshly deposited glacial material. They also directly contribute to sediment. For example, by the addition of debris to the glacier surface by freeze-thaw weathering on valley sides and cirque backwalls (Section 14.5.2). Since glaciers broadly protect the underlying landscape from the effects of frost-action (because thick glacier ice insulates the ground from cold air temperatures and basal processes generate heat), the limits of periglacial landforms can be used to reconstruct the extent of former glaciers (e.g. Lukas 2006). The most common example of this is on valley sides where a periglacial trimline separates glacially eroded and smoothed bedrock below from frost-shattered bedrock above (e.g. Ballantyne 1997), although some trimlines may actually represent an englacial boundary between warm- and cold-based ice (Ballantyne and Stone 2015; McCarroll 2016).

Glacial and periglacial environments are however not mutually exclusive. A growing body of research has demonstrated that glacier-permafrost interactions are common at cold-based glacier margins (Echelmeyer and Zhongxiang 1987; Fitzsimons et al. 1999; Bennett et al. 2003; Haeberli 2005; Harris and Murton 2005; Waller et al. 2012). The majority of these interactions are likely to occur during ice advances over frozen terrain. This can result in compression of the permafrost, leading to the development of thrust-block moraines (Aber et al. 1989; Evans and England 1991) (Section 14.3.5). The base of the permafrost is a natural plane of weakness and is therefore thought to act as a décollement surface (Le Heron et al. 2005). The presence of permafrost may also act as a barrier to groundwater flow from the ice margin, resulting in high porewater pressures at the base of the permafrost, and facilitating décollement (Waller et al. 2012). Further examples of glacier-permafrost interactions are in Section 14.5.7.

14.4 Glacial Facies Analysis

14.4.1 Facies Associations

14.4.1.1 Approaches and Methods

In glacial sedimentology, facies analysis aims to recognise a discrete assemblage of lithologies and sedimentary structures that first demonstrate a glacial origin and secondly record a range of subglacial, ice-marginal and proglacial processes (Figure 14.15) (Griffiths and Martin, 2017). A particular facies (e.g. a stratified diamicton (if lithified, 'diamictite')) will often have characteristics that allow the processes of final deposition to be deduced, but usually not the wider depositional environment. For example, a stratified diamicton may be deposited in a subglacial environment, or alternatively in an ice-marginal environment (e.g. a lake, with abundant ice-rafted debris), and so more information is required to determine the precise depositional setting. This is typically obtained by virtue of context, commonly through 'facies associations' (Figure 14.15). For example, our stratified diamicton may be found interbedded with cross-stratified sand facies, which we interpret as recording deposition by flowing water. Noting that, in the vertical sequence, these facies are duplicated multiple times and cut through by multiple thrusts, we may propose that this facies association records glaciotectonic deformation. Careful sampling of the diamicton facies and its faulted contact with the cross-stratified sand facies may allow detailed study of the material in thin section in the laboratory, and to further test the hypothesis that the layers in the diamicton was acquired by shearing in a subglacial environment.

In the Quaternary record, particular facies codes are widely used. Specific lithologies (e.g. a diamicton) assume a primary descriptor: for matrix-supported diamictons, this is 'Dm' (Table 14.1). Noting the presence of stratification in the diamicton, we would add a secondary descriptor: 's' for 'stratified', thus, 'Dms'. Likewise, a massive, matrix-supported diamicton would be described as 'Dmm'. The facies code approach was first established for glacial deposits by Eyles et al. (1983). The advantage of this method is that sections from different locations can be compared and contrasted, and the vertical and spatial distribution of facies can be documented in a semi-quantitative way. An up-to-date summary of the facies code approach can be found in Lee (2018) and Evans and Benn (2021) and is summarised in Table 14.1.

Due to the variability and importance of grain shape, texture and fabric, there are some techniques which are more widely used in glacial sedimentology. The Udden-Wentworth scheme is modified to include clasts larger than boulders (Terry and Goff 2014). Clast roundness (~35–125 mm) is

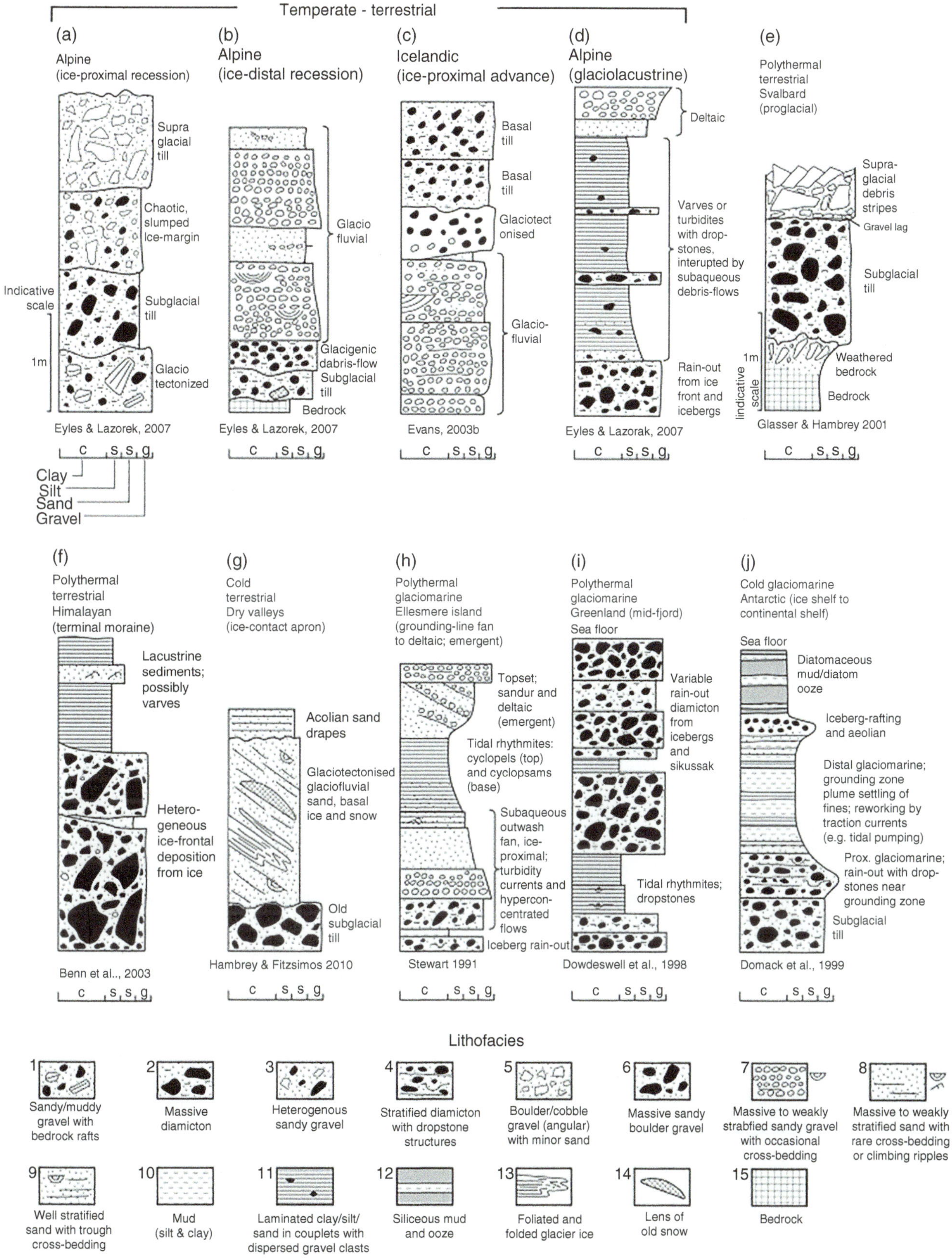

Figure 14.15 Idealised facies associations for different thermal regimes in terrestrial and marine settings. From Hambrey and Glasser (2012, p. 13) / with permission of Elsevier. (a–d) Refer to temperate terrestrial environments. (e–i) Are from polythermal environments. (j) Is from a cold glaciomarine environment.

Table 14.1 Facies codes commonly used in descriptions of Quaternary sediments.

Diamicton	Fine gravel (2–8 mm)
Dm Matrix-supported	**GRcl** Massive with clay laminae
Dmm Matrix-supported, massive,	**GRch** Massive and infilling channels
Dms Matrix-supported, stratified	**GRh** Horizontally bedded
Dcm Clast-supported	**GRm** Massive and homogenous
Dmg Matrix-supported, graded	**GRmb** Massive and pseudo-bedded
Dml Matrix-supported, laminated	**GRmc** Massive with isolated outsize clasts
— **(p)** Includes clast pavement	**GRmi** Massive with isolated, imbricated clasts
— **(g)** Graded diamicton	**GRmp** Massive with clast stringers
— **(b/s)** Banded/sheared	**GRo** Openwork structure
	GRruc Repeated upward-coarsening cycles
Silts and Clays (<0.063 mm)	**GRruf** Repeated upward-fining cycles
Fm Fines, massive	**GRt** Trough cross-stratified gravel
Fl Fines, laminated	**GRcu** Upward-coarsening (inverse grading)
Flv Fine lamination with rhythmites or varves	**GRfu** Upward-fining (normal)
Frg Graded or climbing-ripple cross-lamination	**GRp** Cross-stratified
Fcpl Cyclopels (couplets of silt & mud)	**GRfo** Deltaic foresets
Fp Intraclast or lens	
—**(d)** with dropstones	**Coarse Gravel (8–256 mm)**
—**(w)** with dewatering	**Gms** Matrix-supported, massive
	Gm Clast-supported, massive
Sands (0.063–2 mm)	**Gsi** Matrix-supported, imbricated
Sm Massive	**Gmi** Clast-supported, massive, imbricated
St (A) Ripple cross-laminated (Type A)	**Gfo** Deltaic foresets
St (B) Ripple cross-laminated (Type B)	**Gh** Horizontally stratified
St (S) Ripple cross-laminated (Type S)	**Gt** Trough cross-stratified
Scr Climbing ripples	**Gp** Planar cross-stratified
Ssr Starved ripples	**Gfu** Upward-fining (normal grading)
Sr Ripple cross-laminated	**Gcu** Upward-coarsening (inverse grading)
Sh Very fine to very coarse and horizontally/planar-bedded or low-angle cross-lamination	**Go** Open framework
Sd Deformed bedding	**Gd** Deformed bedding
St Medium to very coarse trough cross-stratified	**Glg** Palimpsest (marine) or bedload lag
Sp Medium to very coarse planar cross-stratified	
Sl Horizontal or draped lamination	**Boulders (>256 mm)**
Sh Sheared	**B** Boulders
Sfo Deltaic foresets	**Bh** Horizontally bedded
Sfl Flasar-bedded	**Bms** Matrix-supported, massive
Se Erosional scours with intraclasts and crudely cross-stratified	**Bmg** Matrix-supported, graded
Su Fine to coarse with broad shallow scours and cross-stratification	**Bcg** Clast-supported, graded
Sc Steeply dipping planar cross-bedding	**BL** Boulder lag or pavement
Suc Upward-coarsening	**Bfo** Deltaic foresets
Suf Upward-fining	
Srg Graded cross-lamination	**Structure**
SB Bouma sequence	**Bo** Boudinage
Scps Cyclopsams (couplets of sand & mud)	**Be** Bedding
— **(d)** with dropstones	**Ba** Banding
— **(w)** with dewatering	

Adapted from Evans and Benn (2021), Krüger and Kjær (1999).

typically determined using the Powers (1953) scale (rating clasts as either very angular, angular, sub-angular, sub-rounded, rounded or well-rounded) to differentiate between supraglacial (very angular to angular), subglacial (sub-angular to sub-rounded) and fluvial (rounded to well-rounded) clasts. This can be complemented by measuring the a, b and c axes of each clast (usually of 50 clasts per sample) to examine clast shape ('blocky', 'platy', 'elongate'; subglacial clasts tend to be blockier). Thus, the shape-roundness data can be used to make inferences about sediment transfer pathways and erosional processes (cf. Benn and Ballantyne 1994; Lukas et al. 2013; Evans and Benn 2021).

14.4.1.2 Challenges in Glacial Facies Analysis

Modern terrestrial glacial environments provide a 'snapshot' of the land surface where processes and deposits are studied together to understand the ancient record. The glaciomarine environment, with a considerably higher long-term preservation potential, is understood inadequately. For example, the delivery of large volumes of sediment to the continental margins by ice streams favours the accumulation of trough-mouth fans in modern high-latitude settings (e.g. Dowdeswell et al. 2016). Although the internal architecture of these fans remains little studied, they are known to consist of a mixture of diamicton, glaciomarine mud and graded sandstones that imply debris flow deposition, background sedimentation and turbidites, respectively (e.g. Le Heron et al. 2013; Gales et al. 2019). The background glaciomarine muds may contain ice-rafted debris and iceberg grounding structures. The diamictons may include inherited striated, faceted, clasts, but much of the evidence for a glacial origin may be lost.

Even in the terrestrial environment gravitational reworking of modern glacier forefields also presents challenges to facies interpretation. Gravitational instability is common in upland glacial environments, and buried ice melt, rockfalls and rockslides rework sediments that were originally glacial in origin (e.g. Lawson 1982; Vehling et al. 2017). For example, Ballantyne and Benn (1994) document how rapidly oversteepened slopes readjust after glacial recession towards a stable angle of repose. Similarly, Rose (1991) documented the rapid reworking of fluted till in just a few decades after deposition, particularly by water erosion. These examples illustrate the role of reworking of earlier-deposited glacial sediments whereby the original textural characteristics (sorting, shape) are modified or destroyed.

Other changes occur on burial. Compaction (either through a glacial re-advance over pre-existing sediments, or burial by successive layers) favours grain and clast re-alignment as pore space is reduced. This effect will be more pronounced in mud-rich sediments (e.g. clast-poor diamictons) and less pronounced in sand- or gravel-rich sediments (clast-supported deposits). The compaction is often accompanied by expulsion of fluids to produce a range of features including dewatering structures, mud volcanoes or sedimentary dykes (e.g. Ravier et al. 2014). If the buried sediment is sufficiently mud-rich, under overburdens of kms, a bedding-parallel foliation can be produced by secondary clay mineral growth which can be confused with 'bedding'. This complicates the determination of whether diamictons are originally stratified or later foliated. Fabric analysis also needs to consider that significant re-orientation of clasts and sedimentary structures can occur in tectonic folding (Spencer 1971).

14.4.2 Facies Sequences and Preservation

Vertical and lateral relationships of facies and facies associations are particularly complex in terrestrial glacial successions. This is largely due to the prevalence of erosional contacts, which preclude the establishment of facies relationships. Additional complexity arises from the uncertain nature of the formation of the accommodation space essential for preservation: by erosion, base-level change or subsidence. The guiding principle of Walther's law (Sections 2.1.2 and 4.2) is a useful thought experiment. This states that only deposits of those environments that are adjacent can become vertically superimposed with a gradational contact over time: for example, through glacial recession, sub-environments will become spatially re-arranged (i.e. switch position to reflect the recessional process). If we consider the spatial relationships between, for example, moraines, subglacial traction till and glaciofluvial sands and gravels. This lateral sequence may, for example, result in a vertical sequence of proglacial sands and gravels on top of, but possibly cutting down into, moraines deposited at a receding ice margin. If that same ice margin re-advances, this process may result in the deposition of an overlying second layer of subglacial traction till (e.g. Boulton 1996). Thus, the vertical succession of facies (recording each neighbouring sub-environment) charts the oscillatory behaviour of the ice margin over time (Figure 14.16). On the other hand, the lateral distribution of sub-environments may change purely as a result of re-organisation of the hydrology and dynamics of the ice mass, for example, the impact of the switching position of meltwater drainage pathways or a change in ice-marginal depositional processes (autogenic processes).

A further issue is the accommodation space available for the sediment to fill, that is, the potential of sediment deposition in the first place and then to be preserved from subsequent erosion (compare an exposed gravel bar in a proglacial river to a deep ice-marginal lake) (Ghienne et al. 2007). Finally, classic river terrace evolution produces

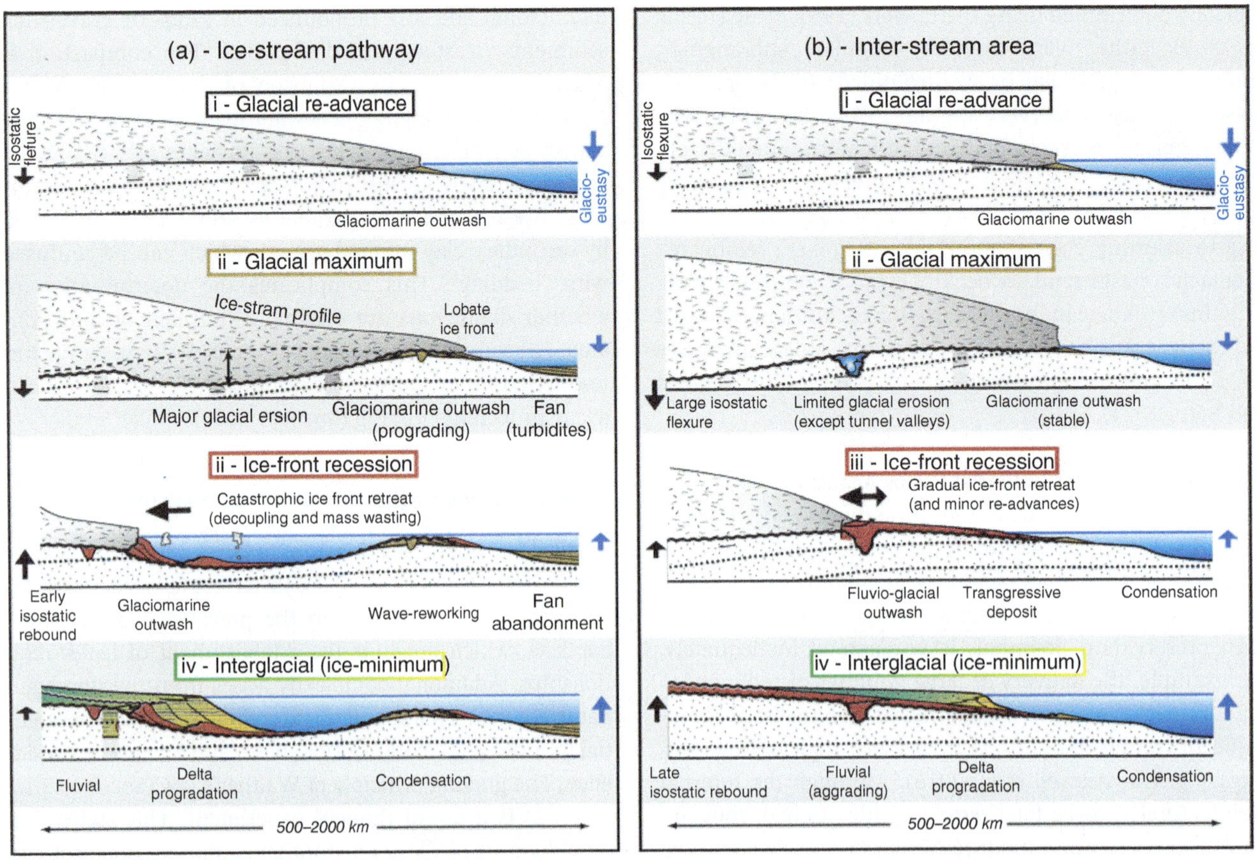

Figure 14.16 Advance (brown) and recession (red) phase till sequences deposited during ice sheet advance and recession, from an ice stream pathway (a) and the inter ice stream area (b). This model derives from detailed outcrop work in the Late Ordovician (Hirnantian) of Libya where multiple glacial cycles occur, stacked vertically and exposed over hundreds of kilometres. The roles of sediment input from the ice margin together with global sea-level changes (glacioeustacy), and the complex influence of isostatic flexure of the crust and subsequent rebound combine to control the amount of accommodation space available in the basin at a specific point in the glacial cycle. Black arrows indicate isostasy and blue arrows indicate sea level. From Ghienne et al. (2007, p. 313) / International Association of Sedimentologists.

a stratigraphy where the oldest deposits are more elevated, and the youngest deposits occur at lower levels. Consider the example of 'Kaunteral', Austria, where the 1850 lateral moraines occur at hundreds of metres elevation above the modern valley floor, with the most recent tills at the lower elevations in the valley. Thus, for stratigraphy to be unravelled in a glacial setting, it is vital to know the landsystem context before attempting to use the superposition of strata to interpret ancient glacial processes.

Although much is to be learned from the terrestrial sedimentary record of land-terminating present-day glaciers, and from the recent Quaternary record, the long-term preservation potential of sediments in these settings is slight (Eyles 1993, 2008). Their preservation is dependent on sediments avoiding subsequent erosion via meltwater and wind erosion immediately following deposition (e.g. Rose 1991). Preservation can occur by virtue of (i) being deposited in areas of deep erosion below base level (accounting for differential preservation of tunnel valleys for example (LeHeron et al. 2010), (ii) rapid subsidence (Fielding et al. 2009; Ali et al. 2018) or (iii) non-erosional marine transgression (Qsaiba Formation Magreb). Long-term preservation is inextricably linked to subsidence (Section 4.5.5) and therefore the sedimentary record is biased towards glaciomarine deposition. In the same vein, Kirkbride and Winkler (2012) discuss the potential for the misinterpreting the climatic significance of glacial advances even in the Quaternary based on the moraine record due to erosion censoring and a complex glacier response to climate change.

The records of some glaciations, such as the Cryogenian examples of the Neoproterozoic, are dominantly of a proglacial character. In rocks of this age, the evidence for subglacial deposits is rare. Notable exceptions are in northern Namibia where, despite reworking processes, excellent examples of terrestrially deposited subglacial tillites are found plastered against the walls of deeply incised Late Palaeozoic Ice Age (LPIA) (~300 Ma) topography (Dietrich et al. 2021), adjacent to glacially sculpted bedrock and roches moutonnées (Figure 14.17a,b) (Fedorchuck et al. 2023). Following deposition of the subglacial tillites, glacier recession led to the

Figure 14.17 Examples of preservation of both terrestrial and marine glacial deposits in the deep-time sedimentary record. (a) Well-expressed roche moutonnée (ice flow from the left) on Precambrian granites of northern Namibia, cut during the Late Palaeozoic Ice Age (Dietrich et al. 2021). (b) Thin patches of diamictite plastered to the same bedrock granite surface shown in A. (c) Upper surface of a diamictite in the Neoproterozoic (Sturtian) Port Askaig Formation. The deposits here are continental diamictites (Ali et al. 2018), and the surface shows spectacular sand wedge polygons (interpreted as an ice wedge pseudomorph) recording periglacial conditions. (d) Neoproterozoic diamictites (glaciomarine environment) of the Kingston Peak Formation (foreground), Beck Spring Canyon, California. (e) Large dropstone within turbidites of the Kingston Peak Formation, Sperry Wash, California. (f) Multiple dropstones within the Luoquan Formation, central China Orogen (described in Le Heron et al. 2018). Photo by D.P. Le Heron.

establishment of terrestrial sandur systems, prior to marine incursion that transformed them into fjords and deposited mudstone. Away from the fjord systems, in central to southern Namibia, extensive fluvial sands and gravels, intercalated with tillites, are well developed (Le Heron et al. 2021). Other examples of terrestrial preservation of ancient glacial facies include the c. 700 Ma Port Askaig Formation of the Garvellachs, NW Scotland, which preserves multiple levels of frost-wedge polygons (Figure 14.17c) with many terrestrial diamictites (Ali et al. 2018).

The record of the Late Ordovician glaciation in North Africa and Arabia is dominantly subaqueous (Ghienne et al. 2007). For the Late Palaeozoic Ice Age (LPIA) deposits, a much more diverse setting is probably recorded, including fluvial to lacustrine deposits in Oman (Braakman et al. 1982; Levell et al. 1988; Martin et al. 2012) and Namibia (Dietrich et al. 2021). There are also spectacular dropstone-bearing turbidites of the Kingston Peak Formation of Death Valley, and the Chuos Formation of Namibia (e.g. Le Heron et al. 2013, Le Heron et al. 2014; Ali et al. 2018) (Figure 14.17d,e).

The preservation bias towards marine sediments deposited in depocentres implies that it is not always easy to demonstrate that diamictites are of direct glacial origin (cf. Busfield and Le Heron 2016; Kennedy and Eyles 2021). Textures at the fine (thin section) scale are often identical, making the differentiation of subglacial versus mass-flow diamict(ites) sometimes impossible (Phillips and Kearsey 2020; Phillips et al. 2022). Conclusive evidence of ice rafting comes from marine and lake basins which show 'lonestones' (that may be interpreted as 'dropstones' provided that deflected and punctured underlying laminae and undistorted overlying layers) can be identified (Figure 14.17f) (Le Heron 2015).

Taking a different perspective, the wafer-thin layer of chemical precipitate on a landscape (Siman-Tov et al. 2017) may be enough to at least archive evidence for the former presence of ice masses in the ancient record. The presence of glacial polish is critical in recognizing striated pavements in deep time, which are often very fragmentary and localised. A mm-thick deposit has a stratigraphic significance totally disproportionate to its thickness!

14.4.3 Sediment-Landform Assemblages

The concept of 'sediment-landform assemblages' rests on the assumption that, at the landscape scale, it is possible to identify landscape features with common attributes that can be related to processes involved in their development (Evans 2003; Benn and Evans 2010; Hambrey and Glasser 2012). The sedimentology of specific landforms is used to elucidate processes of formation and post-depositional modification. Thus, sediment-landform assemblages emphasise process-form relationships, specifying the interrelationships of specific sediment-landform associations at local and regional scales.

'Landsystems' are areas of common terrain attributes, reflecting underlying geology, glacier thermal regime and ice-flow history, previous erosional and depositional processes, and climate (Benn and Evans 2010; Knight et al. 2011). They comprise suites of sediment-landform assemblages and are based on the recognition of a continuum of glacial styles and dynamics, controlled by topography, climate, environmental setting and scale. The expanding database of sediment-landform assemblages with detailed, high-resolution surveys facilitates a quantitative approach to palaeoglaciology, which, applying the principle of uniformitarianism, allows the inversion of landscapes to reconstruct palaeo-ice dynamics (Benn and Evans 2010). The process-based glacial landsystems approach has its roots in modern analogues and provides a broader context and methodological approach for interpreting glacial landscapes. This approach is required in order to determine larger-scale ice dynamics, advance and recession cycles, and the manner and style of glaciation.

To date, a wide range of different landsystems models have been proposed for different glacierised and glaciated regions. These landsystems may evolve through time and overprint one another if the dominant processes change. For example, during the recession, glaciers may thin and no longer reach pressure melting point throughout, becoming polythermal. They may stagnate and slow down, accumulating increased debris cover on their surface (cf. Davies et al. 2022). Large proglacial lakes may develop in front of previously land-terminating glaciers, or alternatively, ice-dammed lakes may drain during recession. Outlet glaciers terminating in marine fjords may become terrestrial during recession and local sea-level fall. During glacier advance or recession cycles, local icefields may interact with other ice masses, such as larger outlet glaciers from regional ice sheets. During the recession, icefields may fragment, and ice divides can migrate, driving variations in ice-flow direction.

14.5 Glacial Landsystems

Here, we discuss some, but by no means all, examples of different glacial landsystems, with a focus on first, process, second on typical lithofacies, and finally, case studies and exemplar landsystems. We present some of the more common landsystems, but the reader is advised that they have a huge diversity and important differences can be caused by subtle changes in climate, topography, or geology. This provides a simple and powerful framework to understand the key processes controlling sediment-landform assemblages and lithofacies development in different parts of the world and under different climatic and topographic regimes.

14.5.1 Active-Temperate Glacial Landsystems (Lowland)

14.5.1.1 Processes

In an active-temperate (warm-based) glacier landsystem, basal ice is at pressure melting point throughout the year.

The presence of water at the bed allows temperate glaciers to flow not only through ice deformation, but also by sliding and soft-sediment deformation of the subglacial bed (Section 14.3.2). These flow mechanisms enable a temperate glacier to erode its bed and lateral margins, abrading material that is subsequently deposited either subglacially or at the ice margin. Glacial meltwater plays a key role in transporting and eroding debris supra-, en- and subglacially, as well as moving material away from the ice margin. An efficient glaciofluvial system can result in very few moraines being either deposited at the margin or preserved in the foreland, giving way to a proglacial zone dominated by proglacial outwash sediments (Benn et al. 2003).

The term 'active' is used to describe glaciers that are actively flowing and are eroding, transporting and depositing material, even during overall recession. 'Active recession' refers to glacier recession that is punctuated by small advances that produce recessional moraines, usually during a winter stabilisation. This, combined with the bed being at pressure-melting point, results in a wide range of landforms and sediments associated with active-temperate glaciers. This section focuses on processes of sediment deposition and landforms associated with ice margins located in lowland terrain, but there are many crossovers with the temperate valley glacier landsystem (Section 14.5.2) and ice sheet landsystem (Section 14.5.4). The landsystem and lithofacies described below, based on modern ice caps, may provide a useful analogue for the margins of former Pleistocene or older ice sheets, for which there are currently no modern equivalents (Evans and Twigg 2002).

14.5.1.2 Lithofacies and Sediment-Landform Assemblages

The forelands of receding active-temperate glaciers contain a wide range of subglacial bedforms and ice-marginal landforms (Figure 14.18). While polished and striated bedrock can be exposed (e.g. at Skálafellsjökull, Iceland; Chandler et al. 2016b), the proglacial area is often dominated by sediments of subglacial, glaciofluvial and glaciolacustrine origin. Subglacial traction till can be widespread and thicknesses of up to 5 m have been observed in exposures at Flaájökull and Fjallsjökull, Iceland (Evans and Hiemstra 2005). However, since active deformation of the subglacial bed is restricted to depths of a few metres, thicker homogenised till sequences are formed through advection and stacking (Evans and Hiemstra 2005; Evans 2017a). Flutes are commonly found in areas of exposed subglacial traction till. A range of moraine types can be found in active temperate glacier forelands, including annual push moraines formed through either squeezing of subglacial material, bulldozing

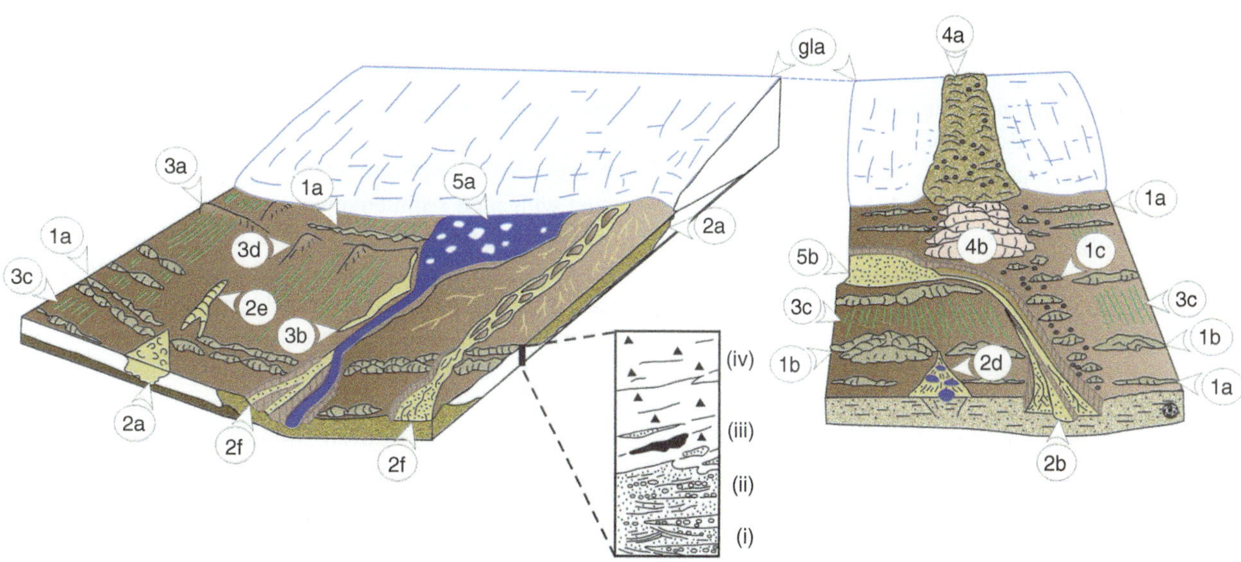

Figure 14.18 Conceptual landsystem model for lowland active-temperate glacial environments. The model is split into two segments, each a few kms wide, to accommodate the range of landforms associated with the active-temperate landsystem. Landform numbers relate to their domain: 1 = morainic domain, 2 = glaciofluvial domain, 3 = subglacial domain, 4 = supraglacial domain, 5 = glaciolacustrine domain. (1a) Small, often annual, push moraines, (1b) superimposed push moraines, (1c) hummocky moraine, (2a) ice-contact sandur fans, (2b) spillway-fed sandur fan, (2c) ice margin parallel outwash tract/kame terrace, (2d) pitted sandur, (2e) eskers, (2f) entrenched ice-contact outwash fans, (3a) overridden (fluted) push moraines, (3b) overridden, pre-advance ice-contact outwash fan, (3c) flutes, (3d) drumlins, (4a) supraglacial debris/medial moraine, (4b) hummocky terrain with buried ice beneath, (5a) proglacial lake and (5b) drained proglacial lake. The section log shows an idealised depositional sequence recording glacier advance over glaciofluvial sediments, comprising: (i) = undeformed outwash; (ii) = glaciotectonised outwash/glaciotectonite; (iii) = massive, sheared till with basal inclusions of pre-advance peat and glaciofluvial sediment; (iv) = massive sheared till with basal erosional contact. Modified from Evans and Twigg (2002, p. 2173).

of proglacial sediments or winter-freeze on of subglacial sediment slabs (e.g. Krüger 1996; Chandler et al. 2016a). Evans and Twigg (2002) observed that flutes often terminate at push moraines, suggesting that the two landforms are genetically linked, whereby the flutes appear to record the flow events that deposited the push moraines.

Glaciofluvial landforms are often well-developed at active temperate glacier margins forming extensive sequences of proglacial outwash sands and gravels, often referred to using the Icelandic term 'sandar' (singular: 'sandur'). Outwash plains that are laterally constricted within valleys are known as 'valley trains' or 'valley sandar'. In laterally unrestricted terrain, such as in southern Iceland, sandur fans emanate from en- and subglacial meltwater portals. Depositional environments are similar to those of non-glacial braided fluvial systems and alluvial fans (see Chapter 5), but a key difference is that glaciofluvial systems (see also Section 5.3.1) are subject to large fluctuations in meltwater discharge and debris supply due to daily and seasonal cycles, flood events and changes in the location of the ice margin and meltwater portals. These dynamic factors, and in particular topographic forcing caused by glacier recession, control channel migration and aggradation, and incision on the outwash plain (Marren and Toomath 2014).

Proglacial outwash (sandur) fans in currently glaciated environments (Figure 14.19), such as Iceland and Alaska, are typically characterised by three zones; ice-proximal, middle and distal; which reflect a progressive reduction in meltwater stream energy away from the ice margin. The ice-proximal zone is dominated by a small number of high-energy channels in which massive gravels are deposited to form longitudinal bars. As the channels diverge, a braided channel network becomes established in the middle part of the fan, typically depositing transverse gravel foresets and planar cross-stratified sand (Zielinski and Van Loon 2002), reflecting fluctuating run-off conditions and lateral channel movement (Church and Gilbert 1975; Maizels 1993). In the distal zone, sandur can be composed of either sheet wash (Krigström 1962), braided sand-bed channel networks (Boothroyd and Nummedal 1978), or sinuous channels with point-bar successions (Boothroyd and Ashley 1975; Zielinski and Van Loon 2002). However, in more restricted topographic settings (e.g. valleys) such a well-defined progressive transition in meltwater stream energy is often absent.

Progradation and coalescence of fans can lead to extensive sandur, such as the 20 km wide Breiðamerkursandur in Iceland (Evans and Twigg 2002). As ice recession progresses, changes in the position of the ice margin and meltwater portal lead to outwash fan incision and channel abandonment. Proglacial lake drainage or episodic high discharge events lead to the development of spillway-fed proglacial outwash fans, which cause incision and terrace development into earlier deposited outwash fans (Evans and Twigg 2002; Chandler et al. 2020c).

If meltwater drainage is constrained between a lateral ice margin and the valley side, a 'kame terrace' will be deposited. This is a gently sloping terrace deposited on the valley side, consisting of sands and gravels, and sometimes lacustrine sediments deposited in ephemeral ponds (e.g. Pisarska-Jamroży et al. 2010). As ice recession continues, a staircase of kame terraces can develop reflecting progressive ice margin lowering. These terraces are inherently unstable and are likely to degrade through debris flows and landsliding (Benn and Evans 2010). A large number of Icelandic temperate glaciers (e.g. Schomacker 2010; Chandler et al. 2020c; Guðmundsson and Evans 2022) and temperate glaciers elsewhere (e.g. Glasser and Hambrey 2002; Boston et al. 2023) are now receding into their overdeepenings, resulting in lake development around the ice margin.

'Eskers' (sinuous ridges of sand and gravel) form as meltwater channels flowing subglacially, englacially and more rarely supraglacially infilled with sediment (Figure 14.19b). They occur in a range of forms including single continuous ridges (tunnel fills), segmented ridges, flat-topped eskers (ice-channel fills), beaded eskers and anastomosing ridge networks (Warren and Ashley 1994; Brennand 2000; Benn and Evans 2010). Eskers typically form parallel to ice flow reflecting major drainage routes within the ice and can be found at outwash fan apexes, marking a former transition from glacial to proglacial drainage. However, complex esker networks have also been recorded parallel to the former ice margin, documenting en- and subglacial meltwater flow between active and stagnant ice at the glacier margin, associated with high sedimentation rates and frequent drainage reorganisation events (Storrar et al. 2020). In Iceland, many eskers appear to have formed englacially. Here, delivery of debris through englacial channels progressively buries the glacier margin, which eventually becomes detached from the active ice margin (Evans et al. 2023). This results in large volumes of buried ice within the immediate glacier foreland beneath glaciofluvial sediments, including within recently formed eskers (Chandler et al. 2020b). Subsequent downwasting of this buried ice leads to large areas of emerging kame and kettle topography and pitted outwash (Figure 14.19a).

14.5.1.3 Example of a Lowland Active-Temperate Glacier Landsystem

The active-temperate landsystem of Fjallsjökull is described by Chandler et al. (2020c) (Figure 14.20). Three landform zones are recognised based on changes in moraine characteristics. The outer part of the foreland (zone 1) is dominated by subglacially streamlined overridden moraines, broadly

Figure 14.19 Active-temperate landsystem examples from Iceland. Ice advance direction is from top to bottom on both images. (a) Outwash (sandur) deposits around the margin of Skeiðarárjökull, southern Iceland. The many pits and pockmarks that break up the outwash surface form by the melting of buried ice over time. Image: Google Earth. (b) Sinuous esker ridges on the foreland of the Breiðamerkurjökull glacier, southern Iceland. Google Earth. Image credit: Jacob Bendle.

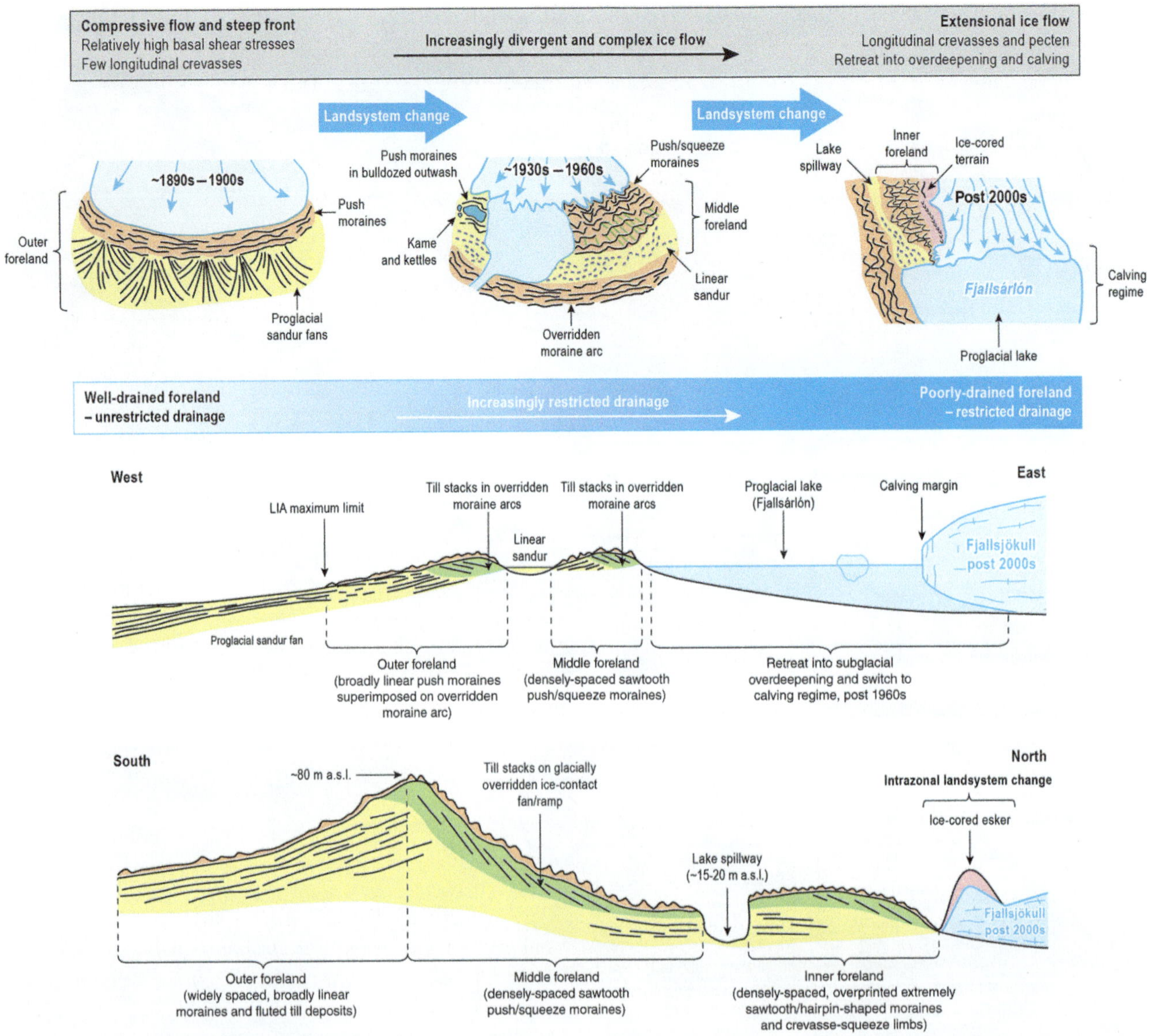

Figure 14.20 Conceptual diagram to show the evolution of the ice margin and sediment-landform assemblages produced as a result of ice recession at Fjallsjökull into an overdeepening (From Chandler et al. 2020c, p. 26) / Elsevier / CC BY 4.0.

linear recessional moraines and proglacial outwash/sandur fans. This zone is argued to represent a stage of early recession from the Little Ice Age maximum (~1890 CE; Hannesdóttir et al. 2015) when proglacial drainage was relatively unrestricted (allowing the development of outwash fans) and ice flow was compressive (leading to a steep ice front and linear/arcuate moraines). The middle part of the foreland (zone 2) is characterised by closely spaced 'sawtooth' moraines, concentrated on topographic high points, and channelised ice-margin-parallel outwash pathways. This zone represents a transition from compressive to extensional ice flow, associated with a change to longitudinal crevasses at the ice margin, which creates a sawtooth-shaped ice margin (leading to sawtooth moraines), and a switch in proglacial drainage characteristics (Figure 14.20). Zone 3 (inner foreland), is characterised by an overdeepening close to the present-day ice margin. It contains extremely sawtooth/hairpin-shaped moraines and crevasse-squeeze ridges. Chandler et al. (2020c) suggest the change in sawtooth moraine shape is a result of increased air temperatures and high surface meltwater fluxes to the bed, combined with poor drainage caused by the overdeepening. These conditions result in saturated ice-marginal sediments that are conducive to ice-marginal and longitudinal crevasse squeezing.

14.5.2 Temperate Valley Glacier Landsystem (Clean Ice)

14.5.2.1 Processes

The valley glacier landsystem describes the characteristic sediments and landforms associated with upland or mountain valley glaciers (e.g. Benn et al. 2003). This landsystem may relate to temperate outlet glaciers from high-altitude icefields such as in Patagonia and Alaska, or to independent mountain glaciers, such as those in the European Alps or New Zealand. It also occurs in low-altitude, high-latitude environments such as Svalbard and Iceland. In all cases, the glacier is laterally confined within a valley, with the valley sides being an important debris source (Benn and Evans 2010). Alpine glaciers therefore form a continuum across clean-ice glaciers, debris-covered glaciers and rock glaciers (Anderson et al. 2018). Geology is a key control on the amount of rockfall. For example, in High-Mountain Asia, extreme topography and fissile sedimentary rocks promote rockfall. Conversely, in Norway, crystalline rocks yield low rockfall rates and glaciers have little surface cover (Benn et al. 2003). Surface debris imparts a strong control on ablation (Benn et al. 2012), resulting in very different processes and sediment-landform signatures for clean-ice and debris-covered end members of this spectrum. The 'glaciated valley' landsystem can therefore be regarded as a family of landsystems, with considerable variety (Benn et al. 2003). Glaciated valleys across this broad spectrum have two recurrent features: the importance of supraglacial debris and the strong influence of topography on glacier morphology, sediment transport and depositional basins (Benn et al. 2003) (see debris transport pathways in Section 14.3.4). In this chapter, we focus on two end-members: temperate clean-ice valley glaciers (this section) and debris-covered glaciers (Section 14.5.3).

14.5.2.2 Lithofacies and Sediment-Landform Assemblages

Temperate valley glaciers are associated with classic parabolic (U-shaped) valleys created through subglacial erosion, large often multi-crested, lateral moraines, ice-marginal recessional moraines and lateral-terminal moraine complexes (Figure 14.21a). Generally, preservation of subglacial sediments is rare but large accumulations of subglacial till can occur (Figure 14.21b), sometimes fashioned into flutes and ice-marginal moraines. Roches moutonnées and striated and polished bedrock, sometimes with perched erratics provide evidence for erosion (Figure 14.21c). The distribution of subglacial sediment over bedrock is typically uneven, with till accumulations pinching out laterally against topographic highs (Figure 14.21d). Older lateral moraines may be recognised high on the valley slopes by semi-parallel ridges of diamicton (Figure 14.21e). Thinning and stagnation of a glacier in a temperate valley setting may be accompanied by the transition to a rock glacier due to large quantities of debris within the ice (Figure 14.21f). Other typical sediment-landform assemblages within the glaciated valley landsystem include: outwash fans and sandur, meltwater channels, and ice and moraine-dammed lakes (Małecki et al. 2018; Martin et al. 2019), talus, ice-moulded bedrock and scarps. The principal sedimentary lithofacies associated with temperate valley glaciers are listed in Table 14.2.

Lateral moraines are a ubiquitous feature in alpine glacier settings, reflecting a previous ice surface (Figure 14.22). They are largely derived from supraglacial and ice-marginal sediments (Hambrey and Glasser 2012) and may contain matrix-supported stratified diamictons with sandy layers and lenses, large boulders and clasts with faceting and striae. Other lithofacies include horizontally bedded and massive granule lenses, laminated or massive sand layers and lenses, and thinner sorted sediment layers (Lukas and Sass 2011). Lukas and Sass (2011) interpret these lithofacies to indicate lateral moraine formation from cohesive debris flows of supraglacial or valley-side material. Lukas et al. (2012) describe lateral moraines at Findelengletscher composed of (i) loose, stratified, matrix-supported diamictons, intercalated with sorted sediments, near the crestline and on the distal slope, interpreted as supraglacial debris flows with fluvial 'wash' horizons, (ii) over-consolidated diamicton with sorted sediment lenses lower in the moraine on the proximal side, interpreted as glaciotectonite and subglacial traction till. Both diamictons contain clasts of predominantly subglacial origin, suggesting elevation of debris to the surface via englacial debris bands, prior to deposition as supraglacial debris flows. The moraines also contain a number of intercalated palaeosols, indicating formation during multiple glacial advances that overrode the pre-existing lateral moraine and added new debris flow deposits on top.

In Alpine regions, these lateral moraines curve into end-moraine complexes. Lithofacies here may include sandy gravel (glaciofluvially derived), stratified sand, mud and laminites (glaciolacustrine). In regions with more meltwater, such as Patagonia, terminal moraines may be destroyed (J. R. V. Martin et al. 2019). In tropical regions with less meltwater, moraines may be larger, and hold back moraine-dammed lakes (Hubbard et al. 2005, e.g. Emmer and Vilímek 2013). Subglacial traction till is relatively uncommon in glaciated valleys in alpine environments, occurring in lateral moraines as described above or overlying ice-abraded bedrock (Evans et al. 2006a; Benn and Evans 2010; Hambrey and Glasser 2012; Lukas et al. 2012). Lithofacies deposited by temperate valley glaciers are frequently reworked by subglacial and proglacial streams.

Figure 14.21 Typical landforms and lithofacies in a temperate glaciated valley, exemplified by the Kaunertal of Tirol, Austria. (a) View taken from the Gepatsch Glacier looking towards a large lateral moraine plastered onto crystalline bedrock. This rises approximately 60 m above the glacier surface. (b) View of the Gepatsch Glacier margin in July 2021, showing a thick deposit of diamicton, approximately 15 m thick, directly overlain by the glacier. The main meltwater stream is to the left. (c) An erratic boulder, approximately 2 m wide, resting on striated bedrock. With time, winnowing through a combination of processes leaves large blocks such as these as the only sedimentary deposits in large parts of the valley. (d) A view under the ice of the Gepatsch Glacier margin, taken in July 2019. A thin veneer of glaciofluvial gravel rests upon striated bedrock here. (e) Metre-scale lateral moraines in the wider 'Kaunertal' area. (f) A rock glacier in the Riffeltal hanging valley, Kaunertal, Austria. In the image, the deposits are about 30 m thick, and are made almost entirely of boulders. Photos by D.P. Le Heron.

Table 14.2 Typical lithofacies, interpretation and relative abundance in temperate valley glaciers.

Lithofacies	Interpretation	Abundance	% striated clasts	% faceted clasts	% clasts	Matrix			Sorting coefficient	Sorting category
						Sand	Silt	Clay		
Clast-rich muddy or sandy diamicton	Basal glacial	*	6	12	5–30	50	40	10	1.05–1.8	Poorly sorted
Sandy boulder-gravel	Ice marginal	****	2	8	50–70	60	30	10	1.3–1.7	Poorly sorted
Rounded gravel	Glaciofluvial	**	0	0	90–100	90	10	0	0.35	Well sorted
Angular gravel	Supraglacial	*****	0	0	90–100	95	5	0	0.45	Well sorted
Sandy gravel	Reworked glaciofluvial	*****	2	4	80–100	70	30	0	0.5–0.6	Moderate sorted
Sand	Reworked glaciolacustrine	**	—	—	—	10	90	0	0.37–0.5	Well sorted
Laminite	Reworked ice-proximal glaciolacustrine	**	—	—	—	10	10	80	0.4	Well sorted
Silt	Glaciolacustrin	**	—	—	—	0	100	0	0.34–0.48	Well sorted

Key to relative abundance: * rare; to ***** dominant. Hambrey and Glasser (2012, p. 9) / with permission of Elsevier.

Figure 14.22 A set of lateral moraines at Findelengletscher, Switzerland. Photo by C.M. Boston.

14.5.2.3 Example of an Upland Clean Ice Temperate Valley Glacier Landsystem

During the Younger Dryas in Britain (12.9–11.7 ka), glaciers expanded in upland areas, forming a large icefield in western Scotland, and numerous satellite icefields and glaciers (Bickerdike et al. 2018, 2016). Coire Ardair is a deeply incised valley in Creag Meagaidh (Central Scottish Highlands), with a lake in the cirque at the head of the valley (Figure 14.23). High ground above the cirque contains frost-shattered, angular detritus, resulting from periglacial activity. Gullied glaciogenic sediments, comprising sub-angular to sub-rounded cobbles in a sandy matrix, occur on both sides of the valley. Closely spaced laterofrontal moraines occur throughout the lower part of the valley. Jones et al. (2017) recognise two types of moraines (Figure 14.23): type 1 moraines are most widespread, consisting of large ridges and mounds, strewn with sub-rounded boulders, that are interpreted as Younger Dryas in age; type 2 moraines occur only at the valley mouth and are more subdued, with heavily fractured and weathered boulders. These moraines are argued to be older (Jones et al. 2017), relating to ice sheet deglaciation ~16–15 ka.

Type 1 moraines are formed from a matrix-supported diamicton, with sub-angular to sub-rounded cobbles to boulders, some of which are striated (Jones et al. 2017). Numerous discontinuous silty-sand lenses within the diamicton suggest deposition as an ice-contact fan (cf. Lukas 2005). Gravel at the top of a moraine forms an erosional contact with the underlying diamicton, and is interpreted as glaciofluvial outwash at the glacier margin following moraine deposition.

14.5.3 Temperate Valley Glacier Landsystem (Debris-Covered)

14.5.3.1 Processes

The debris-covered valley glacier landsystem has many similarities with the clean-ice valley glacier landsystem, but increased rockfall from surrounding topography means that the ablation zone is covered in a thick layer of debris (Figure 14.24). Debris-covered glaciers are more common in areas of very steep or high topography, such as the Himalayas (Benn et al. 2012; Scherler et al. 2018). Around 7.3% of the global mountain glacier area is debris-covered,

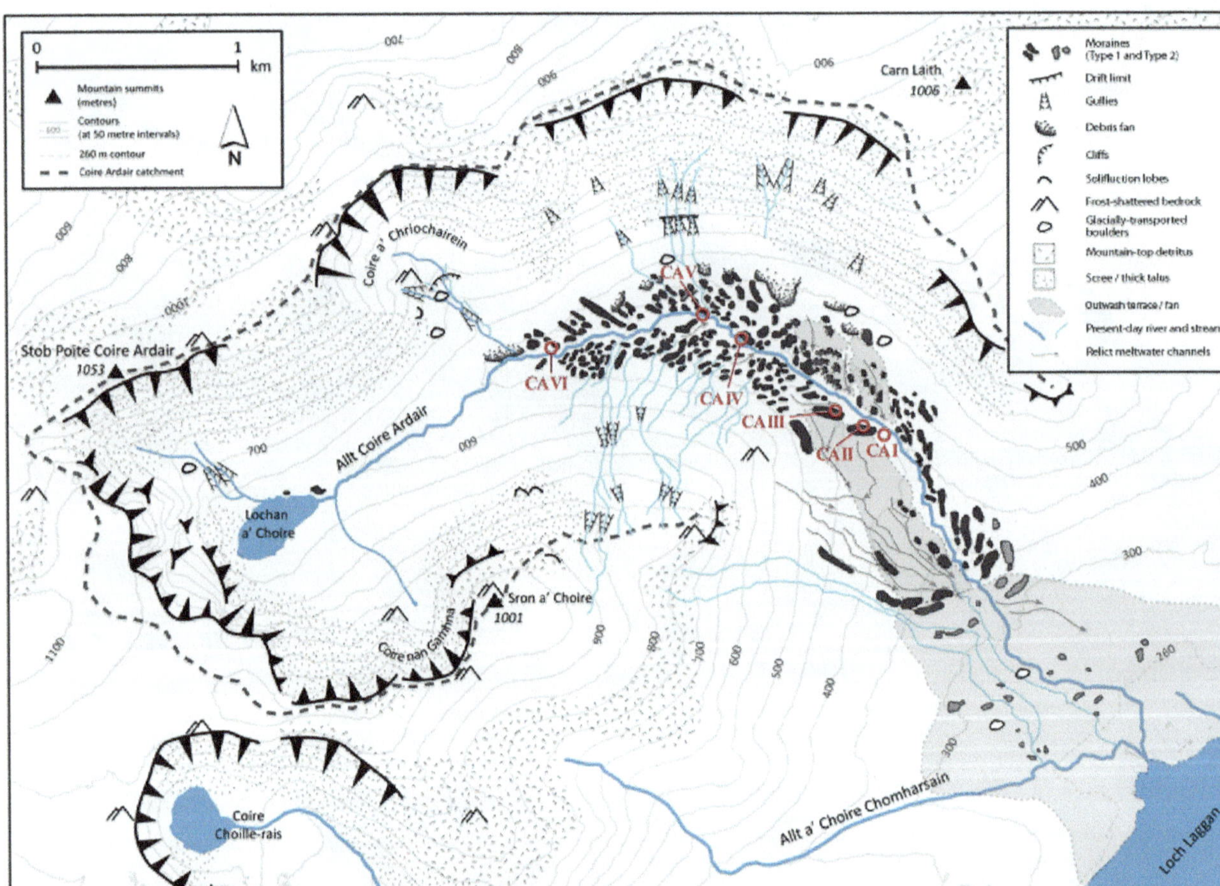

Figure 14.23 Glacial geomorphological map of Coire Ardair. From Jones et al. (2017, p. 57) / with permission of Elsevier.

Figure 14.24 The surface of the debris-covered Pasterze Glacier, Austria. (a) View of the glacier margin in July 2018, showing a debris-covered glacier and a fluvial system feeding an ice-marginal delta. (b) Dirt cone derived from debris melting out onto the surface of the glacier in August 2021. Dirt accumulates in hollows on the ice. This material thermally insulates the glacier, promoting slow downwasting rather than recession, and ultimately results in a topographic inversion, with a cone of dirt-covered ice as the surrounding ice melts faster or recedes away from the glacier. Photo by D.P. Le Heron.

with half of the world's debris-covered glaciers concentrated in Alaska, Southwest Asia and Greenland (Herreid and Pellicciotti 2020).

A schematic model illustrating a debris-covered glacier landsystem is shown in Figure 14.25. Sediment sources for debris-covered glaciers include debris flows onto the ice surface, snow avalanches, landslides, rockfalls and rock avalanches. Thin debris cover (under 5 cm) enhances ablation since it reduces albedo and increases the absorption of short and longwave radiation (Østrem 1959). However, thicker debris insulates underlying ice, reducing ablation (Benn et al. 2003; Gibson et al. 2017). Debris-covered glaciers are defined as those where at least 50% of the ablation zone is debris-covered (Kirkbride 2011). This protection against ablation allows the debris-rich ice tongue to extend much further downvalley than a clean glacier (Anderson et al. 2018). This results in a larger ablation area, making the ratio between the accumulation and ablation areas (accumulation area ratio: AAR) lower (<0.5).

Due to the thick debris insulating the underlying ice, debris-covered glaciers tend to thin and downwaste first, rather than immediately recede via a change in length. This reduces the ice surface slope and velocity, leading to stagnant glacier tongues (Quincey and Glasser 2009; Benn et al. 2012). These debris-mantled tongues often feature supraglacial ponds and ice cliffs, which can be steep-sided and expose bare ice, enhancing ablation in these areas (Herreid and Pellicciotti 2018). Supraglacial ponds also increase ablation by absorbing and transmitting thermal energy, which promotes ice-cliff calving (Watson et al. 2017). This process can lead to pond amalgamation and the expansion of a large supraglacial lake on the low-gradient glacier tongue (Benn et al. 2012), dammed by a terminal moraine at the downglacier end. If the moraine dam fails, a glacial lake outburst flood (GLOF) may occur (Westoby et al. 2014).

14.5.3.2 Lithofacies and Sediment-Landform Assemblages

Debris-covered mountain glacier sediment-landform assemblages (Figure 14.25) are in many ways similar to those associated with a clean-ice valley glacier. This is because debris transport pathways are similar, and similar assemblages of medial, lateral and terminal moraines are produced. They are also associated with similar erosional features in their upper reaches, including cirque headwalls, steep arêtes, ice-scoured bedrock and over-deepened valleys. The main difference is in the volume of supraglacial debris, particularly in the ablation areas. This debris significantly impacts glacier ablation, leading to fundamentally different dynamics in the mass balance processes than in clean-ice glacier (Benn et al. 2003; Gibson et al. 2017).

In Himalayan examples (e.g. Figure 14.26), supraglacial debris can include scree, water-dominated debris fans and avalanche fans. Debris on the ice surface is typically a sandy boulder-gravel, with up to 30% sand matrix. Some debris bears hallmarks of active glacial transport in the basal traction zone of the glacier, with striations, edge-rounding and faceting. This debris may be mixed with areas of highly angular, mono-lithology blocks, derived from rockfalls and transported passively on the ice surface (Benn and Owen 2002; Hambrey et al. 2008).

14.5.3.3 Example of an Upland Debris-Covered Temperate Valley Glacier Landsystem

An example of a debris-covered glacier landsystem is the Ghulkin Glacier, in northern Pakistan (Figure 14.25). Large volumes of debris are delivered to the ice surface by avalanching and supraglacial debris transport is dominant. Thick sediment occurs in the ablation zone, forming a large, steep, end-moraine complex and latero-frontal moraines form a continuous rampart around the glacier margin, high above the valley floor. This is because meltwater discharge is low compared with debris flux. The moraines form from mass movements and debris flows, originating from the ice surface depositing massive to stratified diamictons, with rounded to angular boulders that are rarely faceted or striated. These thick sediment accumulations are continually reworked by meltwater erosion and slope failure (Benn and Owen 2002). As ice support is removed during recession, these landforms are susceptible to collapse and reworking (Benn and Owen 2002). The ice-marginal landforms may be underlain by subglacial traction tills and melt-out tills, which occur at Ghulkin Glacier as metre-thick massive or stratified matrix-supported bouldery diamictons. The tills tend to include more finer-grained material than the supraglacial and mass-movement sediments, and the clasts are more edge-rounded, with striae and faceting. These diamictons are interbedded with supraglacial glaciolacustrine and glaciofluvial sediments, forming metre-scale channel infills.

14.5.4 Ice Sheet Landsystem

14.5.4.1 Processes

In this section, we discuss the ice-sheet landsystem and consider it separately from the ice-stream landsystem. This is because although an ice stream is a zone of fast flow within an ice sheet, different dynamics result in different sediment-landform assemblages. Ice sheets are topographically unconstrained, which means that ice flow direction is controlled by the ice surface profile rather than the underlying topography. The present-day Antarctic and Greenland ice sheets predominantly terminate in water, with extensive floating ice shelves at the margins of many

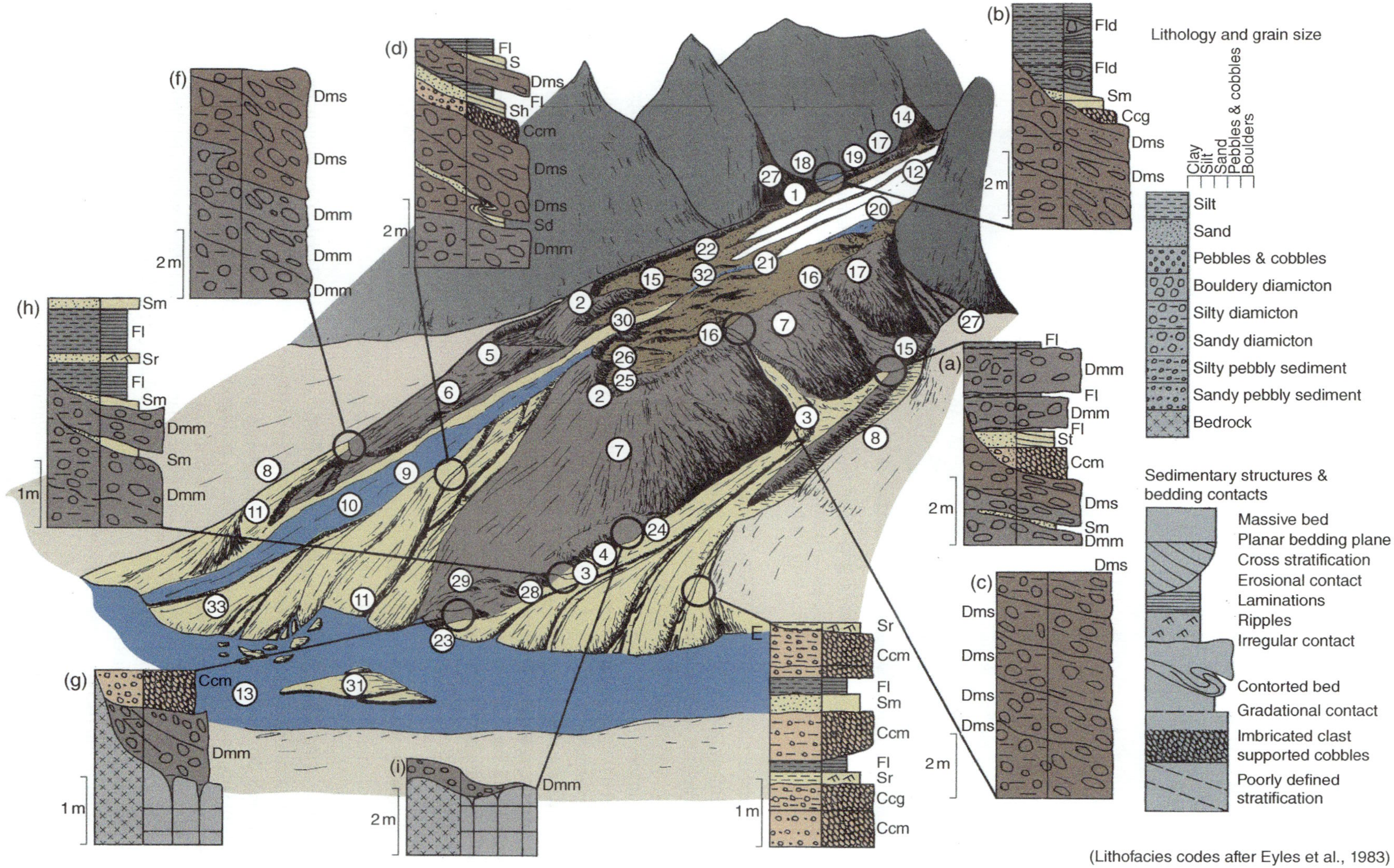

Figure 14.25 Debris-covered mountain glacier, based on Ghulkin Glacier, Hunza Valley, Karakorum Mountains, northern Pakistan. The major landforms are listed 1–33. (1) Truncated scree, (2) latero-terminal dump moraines, (3) lateral outwash channels, (4) glaciofluvial outwash fan, (5) slide moraine, (6) slide and debris flow cone, (7) slide-modified lateral moraine, (8) abandoned lateral outwash fan, (9) meltwater channel, (10) meltwater fan, (11) abandoned meltwater fan, (12) bare ice areas, (13) trunk-valley river, (14) debris flow, (15) flowslide, (16) gullied lateral moraine, (17) lateral moraine, (18) ablation valley lake, (19) ablation valley, (20) supraglacial lake, (21) supraglacial stream, (22) ice-contact terrace, (23) lateral lodgement till, (24) roche moutonnée, (25) fluted moraine, (26) collapsed latero-frontal moraine, (27) high-level till remnant, (28) bedrock knoll, (29) hummocky moraine, (30) ice-cored moraines, (31) river terraces, (32) supraglacial debris and (33) dead ice. The lithofacies associations are labelled (a) to (h). (a) outwash channel sediments and fan associated with a lateral moraine, (b) lateral moraine and ablation valley sediments, (c) lateral moraine sediments, (d) latero-frontal moraine and glaciofluvial outwash fan sediments, (e) proglacial outwash fan sediments, (f) frontal moraine sediments, (g) bedrock, subglacial till and outwash, (h) frontal till and proglacial lacustrine and (i) exposed bedrock and subglacial till. Modified from Benn and Owen (2002, p. 11).

Figure 14.26 Debris cover on the ablation area of Khumbu Glacier, Nepal, looking down-glacier from Kala Pathar showing the elevation difference between the Little Ice Age (LIA) lateral moraine crests and the glacier surface, and relict landslide material that has remained in situ from at least 2003–2014. Rowan et al. (2015, p. 428) / with permission of Elsevier.

Antarctic glaciers. By comparison, during the Late Pleistocene, the southern margins of many large ice sheets, including the Laurentide, Fennoscandian and British-Irish ice sheets were land terminating, a scenario for which there are currently no modern analogues.

Ice sheets are several kilometres thick and therefore tend to be temperate due to high pressures at the bed, geothermal heating at the ice-bed interface, and the bed being insulated by ice from cold air temperatures. However, cold-based ice may be present where ice sheets overlay mountainous topography and ice is thinner or where geothermal heat flux is low. The central zone of the Late Pleistocene Fennoscandian Ice Sheet is also thought to have been cold-based, resulting in preserved central areas (Klemen et al. 1997, 1999a; Hättestrand and Stroeven 2002). Conversely, if ice is at pressure melting point (warm-based), it can erode and entrain debris. This results in deposition and accretion of thick till sequences, often through glaciotectonic folding and stacking (van der Wateren 1995; Boston et al. 2010), and moulding of subglacial sediment into elongate forms such as ribbed moraines, drumlins and megascale glacial lineations (Ely et al. 2023) (Section 14.3.4).

Meltwater drainage within ice sheets strongly influences ice-flow dynamics due to its effect on pore water pressures in the subglacial deforming layer and frictional drag at the ice-bed interface (Boulton et al. 2001, 2009; Evans et al. 2006a). Where temperate ice sheets rest on unconsolidated materials or soft sedimentary rocks, meltwater can drain through 'tunnel valleys', particularly towards ice sheet margins. These channel systems can be tens of kilometres in length, several kilometres in width and hundreds of metres deep (Bradwell et al. 2008; Moreau and Huuse 2013), and are thought to provide major drainage routes for large volumes of subglacial water. They are well documented in Quaternary and Late Ordovician records, either as individual valleys or anastomosing networks. A clue that they form subglacially is the fact that the base of the valleys often shows considerable undulation, including reverse bedrock slopes (Greenwood et al. 2007; Bradwell et al. 2008), which implies formation under the confining pressure of an overlying ice sheet (Van der Vegt et al. 2012). The precise mechanisms of tunnel valley formation remain hotly debated, despite the progress from analogue experiments (Lelandais et al. 2016). It has been attributed to outburst floods from supra- or subglacial lakes, which may have formed from trapped meltwater upglacier of a frozen margin (e.g. Hooke and Jennings 2006; Zoet et al. 2019). Alternatively, Livingstone and Clark (2016) suggest that tunnel valleys primarily form gradually through headward erosion, with secondary contributions from subglacial flood events.

Subglacial eskers (see also Section 14.5.1) record the fill of tunnels cut up into the ice, and in ice sheets typically occur over hard bedrock (e.g. shield) areas (Clark and Walder 1994; Boulton et al. 2009; Storrar et al. 2013). This is because of the relative ease with which meltwater can carve upwards into the ice, rather than downwards into rock.

Despite eskers having previously been used as a record of former drainage routes (e.g. Boulton et al. 2009), Livingstone et al. (2015) surprisingly show that eskers tended to be located away from modelled subglacial drainage routes of the former Laurentide Ice Sheet. These authors suggest that the eskers document drainage routes close to the ice margin immediately prior to deglaciation. This association of eskers with deglaciation is well-known and has been documented elsewhere. For example, within the Brampton Kame Belt, which was deposited during deglaciation of the last British-Irish Ice Sheet as part of a glacier karst system of ice-walled lake plains and migrating drainage channels (Livingstone et al. 2010; Lovell et al. 2019).

14.5.4.2 Lithofacies and Sediment-Landform Assemblages

The ice sheet landsystem is characterised by an inner zone of erosional landforms and streamlined features, such as crag and tails, and an outer depositional zone of ribbed moraines, drumlins and megascale glacial lineations (Boulton 1996) (Sections 14.3.4 and 14.5.5). Landforms may be discordant with topography as a result of ice flow direction being controlled by the ice sheet surface profile. Glaciofluvial landforms, such as eskers, tunnel valleys and kames (Section 14.5.1), are widespread as a result of a well-developed meltwater drainage network (Section 14.3.6).

As waterlain deposits, eskers typically show internal organisation and share many characteristics of fluvial sediments deposited in other settings. They consist of massive, stratified or cross-stratified sands and gravels (Section 14.5.1). Tunnel valleys show a great range of sedimentary fills. In the North European Plain, multiple generations of tunnel valley incisions are found. Most of these are known from the subsurface, and thus a detailed impression of their infill can only be gained through borehole investigations including wireline log analyses (Figure 14.27) (Janszen et al. 2013). Elsewhere, on a similar but much older glaciated plain (North Africa-Arabia during the Late Ordovician), tunnel valleys are spectacularly exposed (Figure 14.28). The tunnel valleys almost exclusively cut down into shallow marine and fluvial sandstones and evidence for gravitational loading at their margins suggests that these substrates were unconsolidated during deposition (Le Heron et al. 2004). Despite almost 100% exposure in the desert, it is often impossible to objectively correlate the sedimentary facies, with the fill locally being chaotic in nature (Tofaif et al. 2019). Despite the heterogeneity, tunnel valleys often contain a significant sand or sandstone component. This makes them excellent reservoirs for oil and gas or water (Le Heron et al. 2009). In both the Quaternary and deep time records, temperate ice sheet landsystems and ice-stream glacial landsystems may represent different phases of an ice sheet's evolution. This can make depositional interpretations difficult, but both are clearly expressed in the Late Ordovician record whereby regional belts of ice streams over the Tassili N'Ajjers plateau lie in clear contrast to swarms of tunnel valleys in central Libya (Ghienne et al. 2003; Moreau et al. 2005; Le Heron and Craig 2008).

14.5.4.3 Example of an Ice Sheet Landsystem

The Late Devensian (MIS 2; 33–15 ka) British-Irish Ice Sheet (BIIS) (Figure 14.29) is one of the most studied former ice sheets. Evidence for its extent and deglacial history includes moraines and grounding zone complexes on the continental shelf to the west of Scotland and Ireland and within the Irish Sea and North Sea basins, tunnel valleys in the North Sea Basin, terrestrial moraines, drumlins, megascale glacial lineations, glaciofluvial deposits and evidence of former ice-dammed lakes (Clark et al. 2018, 2022). In the central sector of the ice sheet, there is evidence for cross-cutting ice flow pathways that document changes in ice divide location as the ice sheet evolved (Livingstone et al. 2012; Hughes et al. 2014).

The timing of ice sheet growth and recession is constrained by geochronometric ages that show that the ice sheet reached its maximum positions asynchronously (Clark et al. 2022) (Figure 14.29). The northwestern margin of the BIIS advanced first, reaching a maximum position on the continental shelf edge west of Scotland by around 30 ka BP (Bradwell et al. 2021). West of Ireland the ice sheet reached the continental shelf edge around 27 ka BP (Ó Cofaigh et al. 2021). To the south, a rapid, but short-lived, advance of the Irish Sea Ice Stream to the shelf break south of Ireland occurred around 25.5 ka BP (Scourse et al. 2021). In the east, the North Sea lobe reached its maximum extent around 25.8 ka BP, impinging on the present-day north Norfolk coast (Evans et al. 2021). These maximum positions were transitory, with asynchronous recession commencing from ~29 to 20.8 ka BP. By ~15 ka BP, ice was limited to higher ground in Scotland and Ireland (Clark et al. 2022).

The asynchronous advance of the ice sheet is attributed to differences in the geography of the areas surrounding the upland ice sheet source areas (Clark et al. 2022). In west Scotland, mountainous terrain is close to the coastline and so the advancing ice quickly reached the continental shelf edge and deep water, which then limited further advance. Conversely, there was more accommodation space in the Irish Sea and North Sea basins, which were also further away from the ice sheet source areas, meaning that ice reached its maximum extent in these areas much later. The constant reconfiguration of the ice sheet margins led to ice divide migration and changes in the internal flow direction, as documented by the drumlin record (e.g. Livingstone et al. 2012).

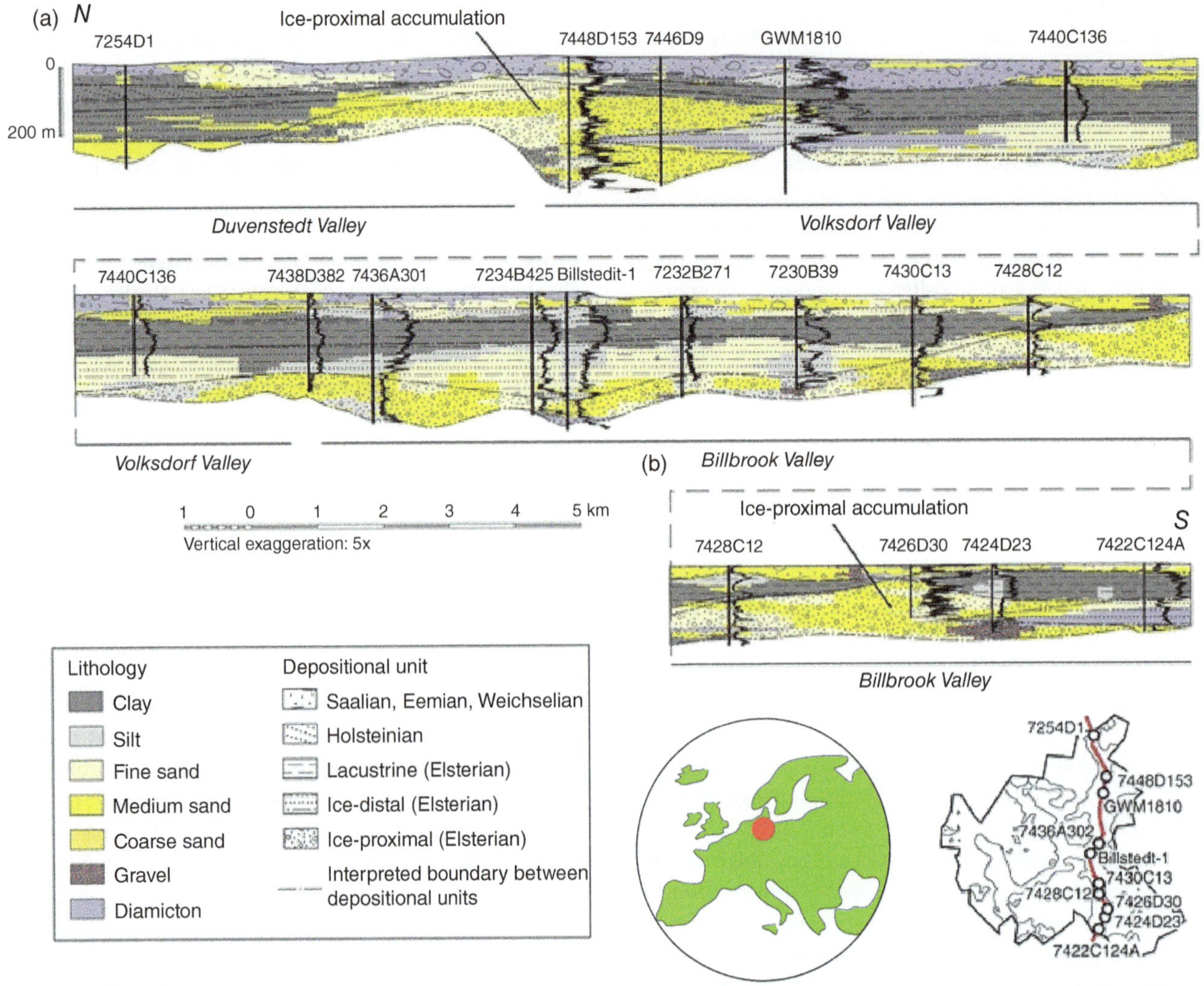

Figure 14.27 The sedimentary architecture of three N-S Pleistocene tunnel valleys beneath Hamburg (NW Germany). Correlation between boreholes along a single depositional dip section incorporates kriged data from a 3D dataset based on 1057 wells and illustrates the complexity of the lateral lithofacies changes. Janszen et al. (2013, p. 704) / John Wiley & Sons.

Figure 14.28 Margin of a partially exhumed tunnel valley at Tabuk, Saudi Arabia, of Late Ordovician age (Tofaif et al. 2019). The thin, horizontal beds on the skyline to the left of the photo are pre-glacial and of mid-Ordovician age. The tunnel valley cuts down into these towards the right, and is filled with highly complex sandstones (Tofaif et al. 2019). Photo by D.P. Le Heron.

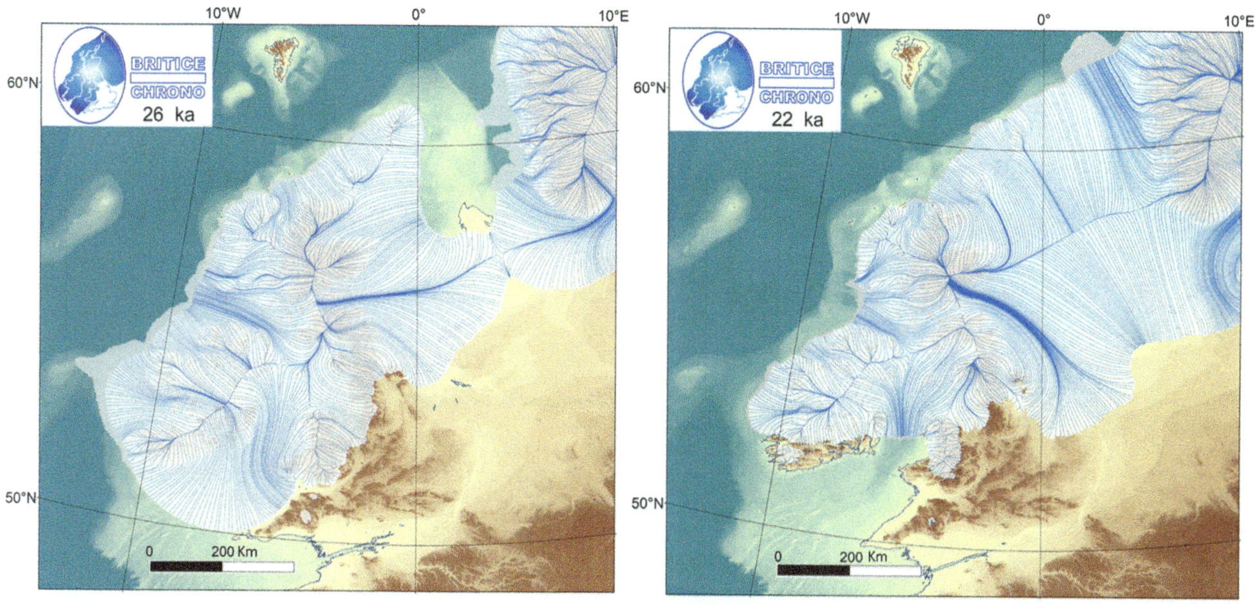

Figure 14.29 Extent and flow geometry of the Late Devensian (MIS 2) British-Irish Ice Sheet (BIIS) at 26 and 22 ka BP, showing maximum areal extent at 26 ka BP, and coalescence and growth of ice from the British and Scandinavian lobes in the North Sea Basin at 22 ka BP. Ice divides and flow directions are in blue and ice shelves are marked in grey (Clark et al. 2022, p. 734) / John Wiley & Sons / CC BY 3.0.

14.5.5 Ice-Stream Landsystem

14.5.5.1 Processes

Ice streams are fast-flowing corridors of ice within an ice sheet. They are major agents of ice flux, for example, accounting for ~90% of the drainage of the East Antarctic Ice Sheet (Bennett 2003) (Figure 14.1). They also contribute significantly to bedrock erosion and sediment transport. The scale of ice streams is highly variable, but is related to the size of the host ice mass. The Lambert Glacier in East Antarctica, for example, measures more than 500 km in length and 40 km in width, with a velocity of 400–800 m a^{-1}. At the opposite end of the spectrum, a recently formed ice stream within the Vavilov Ice Cap in Russia measured around 35 km long and 5–10 km wide in 2018, but reached velocities of around 1800 m a^{-1} (Zheng et al. 2019). Ice stream location can be controlled through the preferential excavation of inherited structural depressions (e.g. the Lambert Graben), which results in topographic focusing, as well as bed roughness, subglacial geology, geothermal heat flux and subglacial meltwater pathways (Winsborrow et al. 2010). Ice streams can be 'pure ice streams' (Stokes and Clark 1999), with defined lateral boundaries surrounded by slower moving ice, or can be constrained by topography; 'topographic ice streams' (Stokes 2018).

Ice streams within the Antarctic Ice Sheet, that are not topographically constrained, have low surface profiles and driving stresses (~10 ka). Their fast flow is attributed to deformable sediments at the bed (Alley et al. 1986, 1987). These ice streams are known to switch on and off, and water piracy, where water is diverted from one ice stream to another, is thought to be a key mechanism for this (e.g. Alley et al. 1994; Anandakrishnan and Alley 1997). Lelandais et al. (2018) argued that ice streams switch on and suddenly accelerate as a result of catastrophic outburst floods at the ice margin, which promote fast ice flow. Ice stream deceleration then occurs when the water 'self-organises' (develops) into channels, creating tunnel valleys.

14.5.5.2 Lithofacies and Sediment-Landform Assemblages

Beneath an ice stream, sediment is generated, comminuted, transported, deposited and possibly cannibalised. This means that in terms of a 'predictable' lithology, the type of sediment generated is dependent on the material that is being eroded, and thus the pre-existing geological history of the area (cf. Boulton 1996). This sediment is moulded and streamlined into subglacial ribs (ribbed moraines), drumlins and megascale glacial lineations (MSGL), which are thought to be characteristic of fast ice stream flow (Section 14.3.4.). This ice stream 'imprint' is abundant in the Quaternary geomorphological record (e.g. Livingstone et al. 2008; Margold et al. 2015; Stokes 2018) and can show a convergent pattern of streamlined landforms in the ice stream onset zone and well-defined lateral boundaries (Figure 14.30).

Arguably, some of the best-preserved lithofacies of any ice stream glacial-landsystem are not of Pleistocene, but rather of Late Ordovician age. An extensive suite of

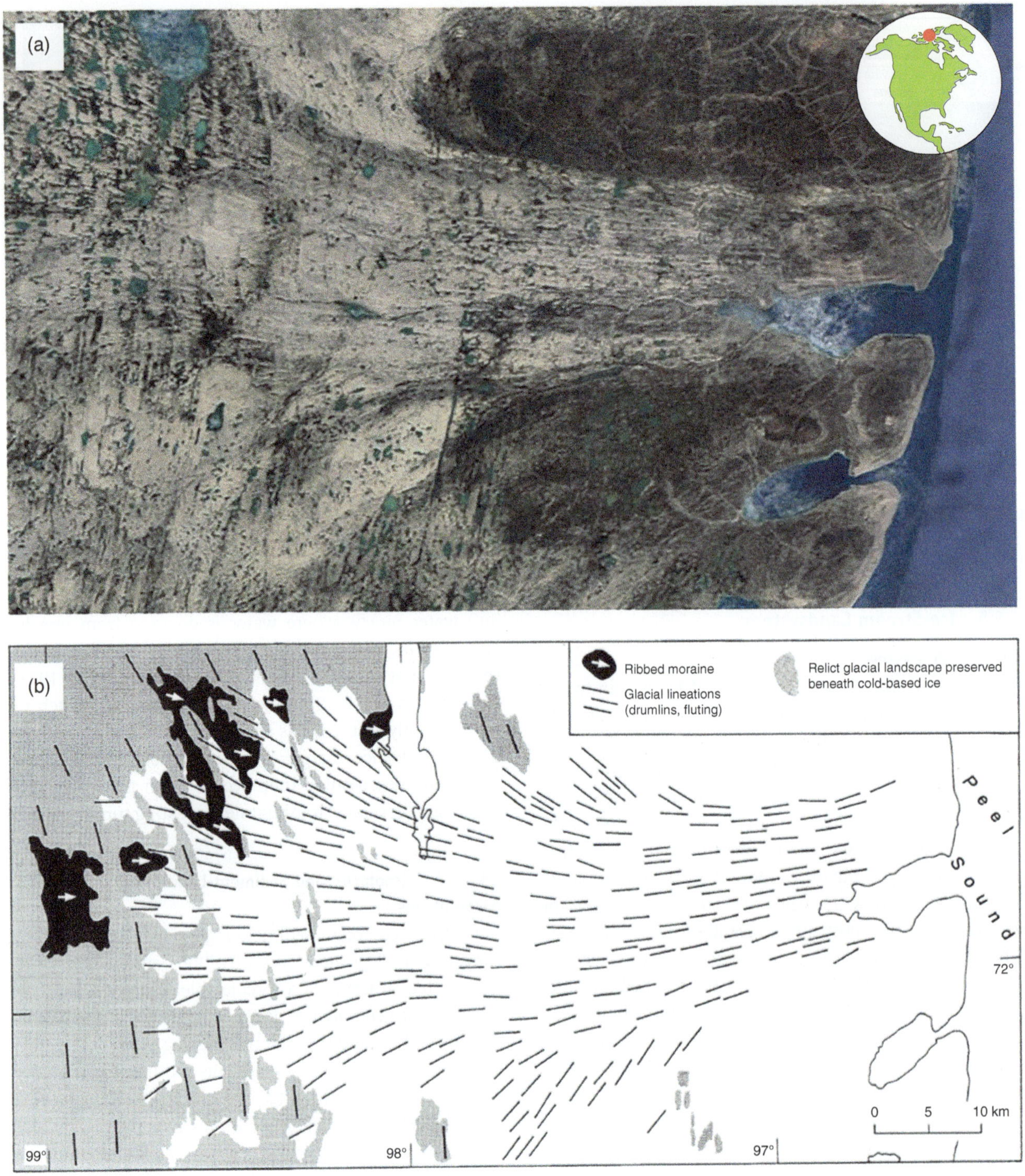

Figure 14.30 Ice stream geomorphological imprint. (a) Google Earth image of an erratic dispersal train and subglacial bedforms on Prince of Wales Island, Canadian Arctic, originally described by Dyke and Morris (1988) and Dyke et al. (1992). Note the distinct lateral margins. (b) Simplified map of the glacial geomorphology, showing convergent drumlins leading into the dispersal train (Adapted from Hättestrand and Kleman 1999, redrawn from Dyke et al. 1992). Stokes (2018, p. 94) / John Wiley & Sons / CC BY 4.0.

sandstone-dominated glacial rocks extends from Mauritania to Saudi Arabia and were deposited on an extensive, shallow-dipping platform 443 million years ago when this region was at high southern latitudes. The rocks are sandstone because repeated sediment cannibalisation for around 100 Myr before the ice sheets caused the development of thick quartz arenite deposits, and then the ice sheets entrained and deposited this sand (Le Heron et al. 2018). Consequently, the dominant lithofacies within the MSGL is deformed sandstone: a range of shear zones at different levels in the stratigraphy testify to competing ductile and brittle processes. Evidence for ice loading of the substrate is commonly found, including transverse box folds at the flanks of mega-scale glacial lineations with a wavelength of many tens of metres.

14.5.5.3 Example of an Ice Stream Landsystem

A Quaternary example of the ice stream landsystem is the former Laurentide Ice Sheet in North America during the Last Glacial Maximum (Margold et al. 2015, 2018; Dalton et al. 2020, 2022). Ice streams have long been recognised as part of this ice sheet (Margold et al. 2018), and their southerly terrestrial-terminating margins provide excellent evidence for ice dynamics during glaciation and deglaciation. The majority of the ice streams were marine-terminating and occupied large bathymetric troughs. The terrestrial margin in the south was lobate and was drained by large ice streams which were only loosely constrained by topography (Figure 14.31). These ice streams were transient, with the low-relief and soft-bedded topography allowing dynamic

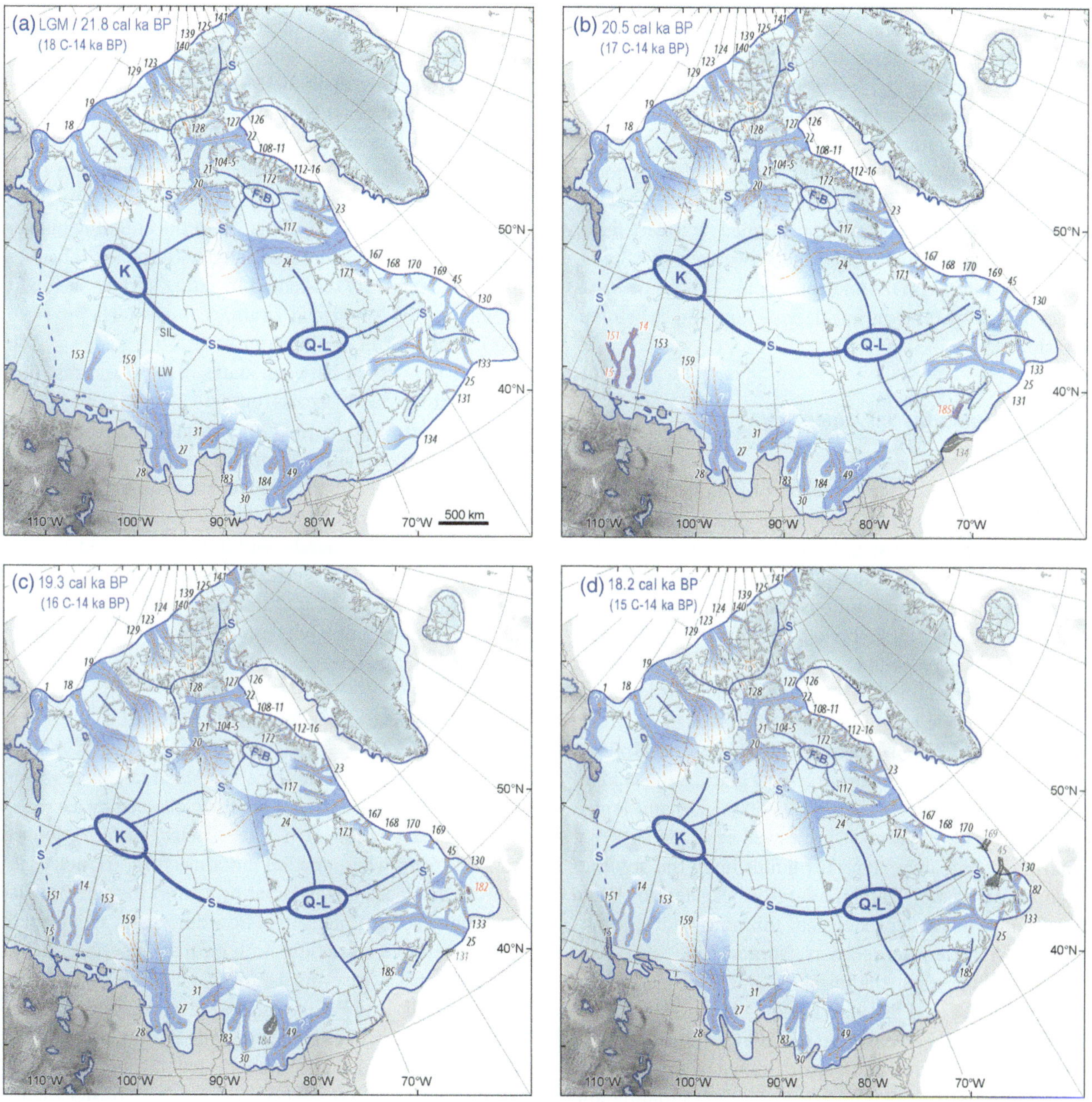

Figure 14.31 Ice streams of the Laurentide Ice Sheet during the Last Glacial Maximum, showing the time transgressive nature of the ice streams and the ice margin during the LGM (a) and deglaciation (b–d). Ice divides and ice domes ([K] Keewatin, [Q-L] Quebec-Labrador, [F-B] Foxe-Baffin, and [S] saddles) are shown in blue. In panel (a): [SIL] Southern Indian Lake, [LW] Lake Winnipeg. Central Alberta Ice Stream is ice stream number 14 on the southern margin of the Laurentide Ice Sheet (panel b). From Margold et al. (2018, p. 7) / Elsevier / CC BY 4.0.

changes in the sheet configuration during deglaciation (Margold et al. 2018). The terrestrial lobes were likely also characterised by slower-flowing ice and higher ablation than their marine-terminating counterparts.

The ice streams of southern Alberta, Canada (number 14 on Figure 14.31) deposited a glacial landform assemblage typical of terrestrial-terminating ice streams with a lobate glacier snout (Evans et al. 2014). This Central Alberta Ice Stream contains flowsets of elongate and streamlined landforms, such as MSGL, that were formed at the bed of fast-flowing ice. These typically have a fan shape, reflecting the lobate ice margin. The shape of the ice margin is clearly demarcated by large transverse terminal moraine ridges that are frequently constructed by glaciotectonic thrusting of bedrock and have often been overridden by ice (Evans et al. 2014). Nested behind these larger terminal moraines are sequences of smaller, closely spaced recessional push moraines and arcs of hummocky moraines. These landforms indicate that the thermal regime of the glacier snout transitioned between faster steady-state temperate ice flow and slower velocities relating to cold/polythermal snout conditions. This process was particularly apparent during deglaciation and ice-stream recession.

14.5.6 Surging Glacier Landsystem

14.5.6.1 Processes

Surging glaciers experience a cycle of high velocities and significant terminus advance (active phase) followed by reduced velocities, ice stagnation and terminus recession (quiescent phase). The active phase is shorter than the quiescent phase, and their durations are proportional; glaciers with a short overall surge cycle have shorter active phases compared to glaciers with longer surge cycles (Sevestre and Benn 2015). For example, Variegated Glacier in Alaska has a short surge cycle length of between 13 and 18 years and active phases of only 1–1.5 years (Kamb et al. 1985; Eisen et al. 2001). By comparison, surge-type glaciers in Svalbard have longer surge cycle lengths (50–500 years) and longer active phases (3–10 years) (Dowdeswell et al. 1991).

During a surge, mass typically builds up in the accumulation zone (reservoir), and the surge propagates down glacier, leading to glacier extension, thinning and terminus advance. At tidewater glaciers, surges have also been observed to initiate at the terminus and to propagate up-glacier (Sevestre et al. 2018). While only around 1% of modern glaciers have been observed to surge (Sevestre and Benn 2015), surging occurs within a wide range of glacier types, from cirques to ice streams, at both land-terminating and tidewater glaciers, and within temperate and polythermal regimes (Benn et al. 2019). Surging glaciers are located within distinct clusters in Svalbard, Alaska, Arctic Canada, parts of East and West Greenland, Iceland, Novaya Zemlya, the Karakoram, Pamirs and western Tien Shan. These clusters have been shown to exist within a well-defined climatic envelope, bounded by precipitation and temperature thresholds (Sevestre and Benn 2015).

The cause of surge initiation is not fully understood, but a change in internal glacier dynamics is thought to be central. An enthalpy model has been proposed as a single, overarching model to explain controls on surging (Sevestre and Benn 2015, see references therein for previous models; Benn et al. 2019). In this context, enthalpy is defined as the internal energy of a glacier system, which is controlled by ice temperature and water content (Aschwanden et al. 2012; Sevestre and Benn 2015). Changes in enthalpy will occur as a result of (i) energy exchange at the ice surface through heat loss and surface melt, (ii) frictional heating due to ice flow and (iii) evacuation of excess energy by basal or englacial meltwater drainage (Sevestre and Benn 2015). Surging results from an imbalance in the enthalpy budget, where the enthalpy gains associated with frictional heating cannot be efficiently dissipated by either surface heat conduction or the subglacial drainage system (Benn et al. 2019). This lack of equilibrium within the glacier system causes it to oscillate between states of slow and fast flow in order to remove excess energy (Sevestre and Benn 2015). Sevestre and Benn (2015) argue that this occurs under certain climatic (temperature and precipitation) regimes, and therefore the primary control on surging is climate. These authors also show that geometry is a secondary control on surging, whereby within these specified climatic boundaries, surge-type glaciers tend to be longer and have greater complexity owing to more tributaries than normal glaciers. This explains why not all glaciers surge within the same climatic envelope (Benn et al. 2019).

14.5.6.2 Lithofacies and Sediment-Landform Assemblages

Surge-type glaciers produce a distinct landsystem (Figure 14.32) a result of fast ice flow followed by ice recession and stagnation. On the glacier surface, looped moraines form where medial moraines from tributary glaciers are displaced downglacier by a surge in the main glacier (Lønne 2016). Extensional flow during the active phase causes extensive crevassing both on the ice surface and at the glacier base (Rea and Evans 2011; Flink et al. 2015). Subglacial till at the bed is squeezed into basal crevasses to form a network of cross-cutting ridges sub-perpendicular and sub-parallel to the ice flow direction. These are commonly known as 'crevasse-squeeze ridges' (CSR) (Sharp 1985; Rea and Evans 2011), but are often referred to using the broader term 'geometrical ridge networks' that also

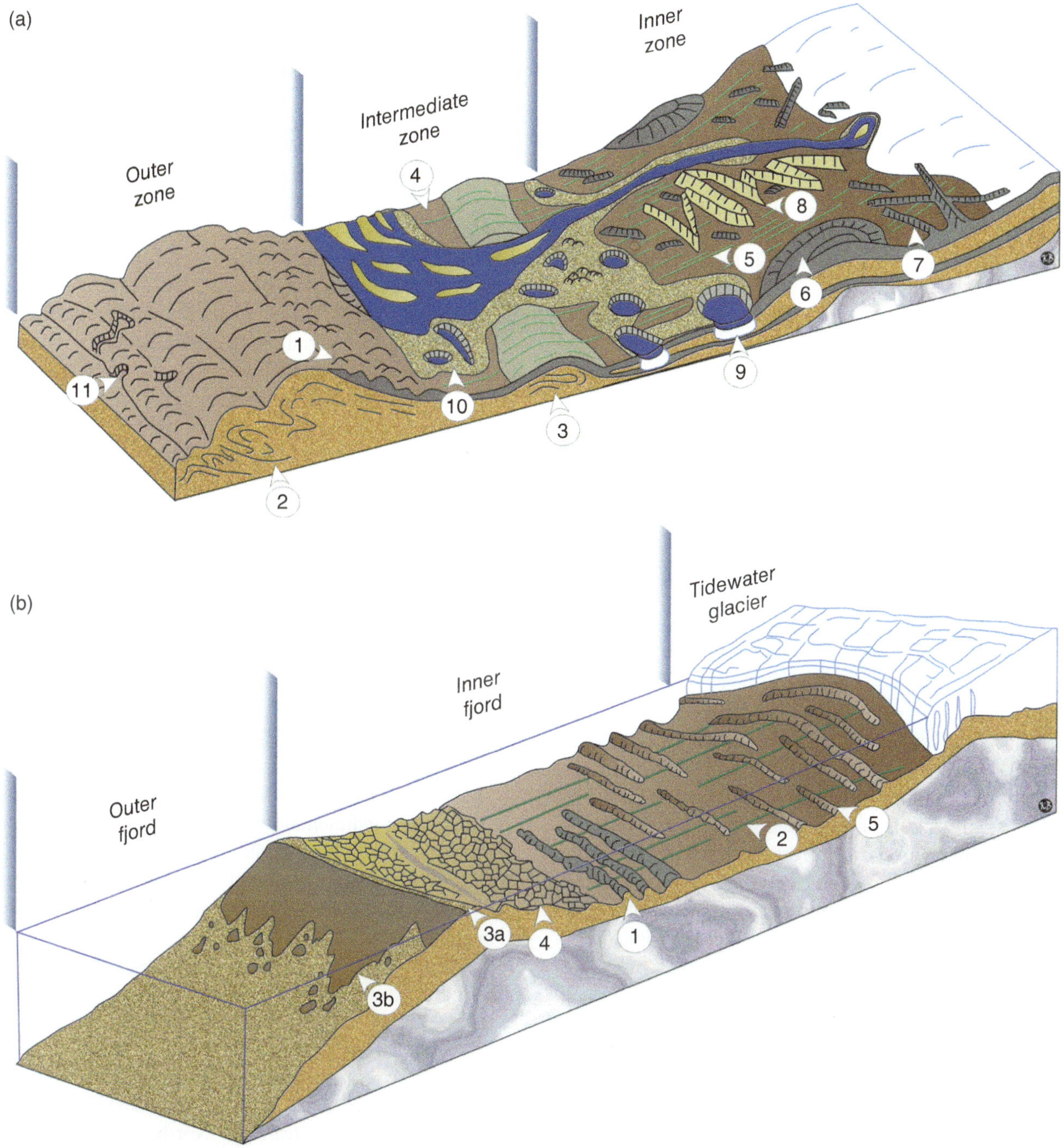

Figure 14.32 (a) A model of the terrestrial surging glacier landsystem after a glacial surge, showing the expected features. (1) Hummocky moraine, (2) Glaciotectonic end moraine, (3) Overridden end moraine, (4) Subglacial till, (5) Flutes, (6) Drumlin, (7) Crevasse squeeze ridges, (8) Concertina esker, (9) Dead-ice, (10) Pitted outwash and (11) Blow-out structure. Modified from Schomacker et al. (2014, p. 106). (b) Landsystem model for surging tidewater glaciers after a surge. (1) Overridden recessional moraines, (2) Streamlined glacial lineations, (3a) Terminal moraine of glacial advance, (3b) Lobe-shaped debris flows, (4) Rhombohedral crevasse-fill ridges and (5) Transverse annual recessional moraines. Modified from Ottesen and Dowdeswell (2006, p. 14).

encompasses other subglacial processes such as hydrofracturing and squeezing of sediment into thrust faults (e.g. Bennett et al. 1996; Lovell et al. 2015). As ice melts around crevasse-squeeze ridges, the removal of lateral support may cause them to collapse and become remobilised as debris flows (Rea and Evans 2011). However, in contrast to active recession at non-surging glaciers, which is likely to remove crevasse-squeeze ridges, ice stagnation favours preservation. Crevasse-squeeze ridges have therefore been used as an identifying landform for surging (e.g. Farnsworth et al. 2016), although they are not fully diagnostic and need to be considered alongside other landforms (Evans and Rea 1999).

Crevasse-squeeze ridges are often superimposed on top of flutings, which form due to the fast glacier flow and a saturated, mobile substrate. Based on Sharp's (1985) detailed observations of crevasse-squeeze ridge and fluting intersections at Eyjabakkajökul, Iceland, it is thought that the two landforms form simultaneously, with the final phase of emplacement at the start of the quiescent phase. Concertina (zig-zag shaped) eskers, composed of sands and gravels, have also been reported from a number of surging glacier forelands (e.g. Schomacker et al. 2014). These eskers are thought to form towards the end of a surge when high water discharges exploit the existing crevasse network (Evans and Rea 1999), leading to their zig-zag shape.

The rapid advance of the glacier terminus on land causes proglacial stacking and folding of soft sediment to produce a 'glaciotectonic composite ridge system' (Aber et al. 1989) (Figure 14.32) that marks the maximum surge extent. This is a multi-crested landform that can be up to 100 m high. Glaciotectonic composite ridge systems predominantly comprise sorted and fine-grained sediment, although there is significant variability in internal structure and proportions of mud, sand and gravel between ridge systems (Lovell and Boston 2017). Ductile deformation results in folding and contortion, while brittle deformation causes faulting and stacking of large blocks of proglacial material. The latter are referred to as 'thrust-block moraines', and the term has been used to describe glaciotectonic moraines composed of proglacially thrust blocks of permafrozen sediment (Evans and England 1991; Evans and Rea 1999). Composite ridges have also been referred to as 'push moraines' or 'push moraine complexes' (e.g. Boulton et al. 1999), but this terminology is problematic as 'push moraine' is also used to refer to smaller single ridge landforms formed through bulldozing at temperate non-surging glacier margins (e.g. Chandler et al. 2016a).

After surging ceases, the glacier tongue stagnates and small melt ponds form on the low-profile ice surface. The ice is debris-rich and continued melt can lead to a debris-covered surface which becomes pitted as the dead-ice continues to melt out. This debris melt-out may lead to debris flows of supra- and englacial sediment, and ponding will deposit isolated areas of glaciolacustrine (fine-grained) sediment. Outwash from the active glacier margin may rework sediments and remove landforms within the dead-ice zone, leading to a braided channel network and deposition of glaciofluvial sediments. This meltwater typically forms a deep gorge through the outer composite ridge system as it exits the immediate glacier foreland (Lovell and Boston 2017).

At tidewater glacier margins, glaciotectonic composite ridges may form above water at the lateral margins, composed of elevated fjord sediments (e.g. Bennett et al. 1999). The development of mud aprons due to bulldozing and tectonic thickening of seafloor sediments in front of a surging tidewater glacier margin has also been observed (e.g. Lovell et al. 2018a). The maximum position of a tidewater glacier surge is marked by a terminal moraine, from which debris flow lobes of glaciomarine silts and clays emanate on the distal side (Figure 14.32). Streamlined bedforms (e.g. MSGL) that are parallel to ice flow are found inside the terminal moraine. Crevasse-squeeze ridges can be found on the proximal slope of the terminal moraine often in a rhombohedral pattern, but typically do not occur over such large areas as for a terrestrial surge landsystem. This is because the recession of surging tidewater glaciers is predominantly characterised by mass loss through calving, rather than ice stagnation. As a result, streamlined bedforms are often cross-cut by small transverse ridges perpendicular to ice flow (annual push moraines) (Ottesen and Dowdeswell 2006), that mark minor winter glacier advances due to reduced calving losses. Down-fjord from the terminus, fine-grained sediments rain out from meltwater plumes exiting the glacier, draping and eventually burying glacial landforms on the seafloor (Ottesen and Dowdeswell 2006; Ottesen et al. 2017).

Multiple glacier surges will modify pre-existing surge landforms and sediments, leading to overridden moraines, often with streamlined bedforms superimposed. Lovell et al. (2018b) demonstrated that some glaciotectonic composite ridges formed by multiple glacial advances. In a glaciomarine setting, sedimentation rates may be high enough to bury the geomorphological imprint of surging within a matter of centuries (Ottesen and Dowdeswell 2006). Within the Quaternary landform and sedimentary record, evidence for surging has been difficult to identify. The presence of thick stacked till sequences at several locations has been tentatively attributed to surging. For example, surging has been suggested within the North Sea lobe of the British-Irish Ice Sheet (Boston et al. 2010; Evans and Thompson 2010). Extensive areas of hummocky terrain, megaflutings, crevasse-squeeze ridges and thrust-block moraines at the southwest margin of the former Laurentide ice sheet have also been associated with a surging margin (Clayton et al. 1985; Evans et al. 2014; Ankerstjerne et al. 2015). Recently, Benn (2021) suggested that some glaciers may have surged during the Younger Dryas (Loch Lomond Stadial) in Scotland, based on geomorphological, geometric and climatic criteria established by Sevestre and Benn (2015).

14.5.6.3 Example of a Surging Glacier Landsystem

Nathorstbreen is a surging tidewater glacier in Svalbard. Its most recent surge occurred between 2008 and 2016, advancing over 15 km and reaching a maximum velocity of 25 m d^{-1} (Sund et al. 2014). Previous surges are thought to have occurred during the Little Ice Age ~1890, ~1160 CE, and twice between 2.61 and 2.79 cal. ka BP (Kempf et al. 2013; Lovell et al. 2018a). During its recent advance, an extensive mud apron developed in front of the ice margin, formed through bulldozing and tectonic thickening of marine sediments from the fjord floor. The apron surface consisted of saturated flat areas with numerous surface ponds and areas of flowing water, active flow structures indicating the continuous formation of the apron, and small (~1 m high) transverse ridges that were interpreted as minor push moraines (Lovell et al. 2018a). At the lateral margins to the north and south, Nathorstbreen has advanced onto terrestrial areas on multiple occasions leading to terrain characterised by an outer glaciotectonic composite ridge system (Figure 14.32), inside of which an extensive zone of hummocky terrain with intervening pools occurs, indicating glacier stagnation and melting of buried ice. Crevasse-squeeze ridges are common within this zone, becoming exposed and preserved as the ice stagnates (Lovell et al. 2018a) (Figure 14.32).

14.5.7 Polar Continental (Cold-Based) Glacial Landsystem

14.5.7.1 Processes

Polar deserts were common in the Pleistocene in high latitude continental interiors, and are widespread today in the Arctic (Svalbard, Canadian High Arctic) and Antarctic (e.g. Vestfold Hills, Allan Hills, Alexander Island; James Ross Island) (Davies et al. 2013). While cold-based glaciers are known to protect the landscape, there is subtle evidence in some areas that they have some ability to erode, entrain and transport sediment. In these circumstances, basal processes remain active at sub-zero temperatures, resulting in glacier motion (Waller 2001) (Section 14.3.2).

At glaciers where there is a warm-based interior and cold-based margin, entrained basal debris can be elevated to the glacier surface by englacial thrusting at the thermal boundary due to compressive flow (Hambrey and Huddart 1995). This leads to a debris-covered ice margin and the development of controlled moraine (hummocky moraine) as the ice margin becomes buried (Evans 2009). In cold environments, this ice becomes aggraded into the permafrost. Remnants of Late Pleistocene ice masses are still present in northern Russia, western Canadian Arctic and East Antarctica (Waller et al. 2012).

14.5.7.2 Lithofacies and Sediment-Landform Assemblages

While some minor erosional, depositional and deformational features are associated with cold-based glacier flow (Lloyd Davies et al. 2009), the potential for fine-grained diamicton/diamictite generation is low (Figure 14.33). Evidence of glacial erosion in the Allan Hills (Victoria Land, Antarctica) includes broad scrapes, striae and grooves on boulders, and ridge and groove lineations (Atkins et al. 2002; Lloyd Davies et al. 2009). Here, deposits from cold-based outlet glaciers of the East Antarctic Ice Sheet are dominated by thin accumulations of poorly sorted sandstone and siltstone breccia with a friable, structureless matrix (Atkins et al. 2002; Lloyd Davies et al. 2009; Hambrey and Glasser 2012). Sand, gravel and residual ice dominate, and diamicton is rare (Hambrey and Glasser 2012). Isolated boulders are scattered over the foreland, some forming poorly defined boulder trains (Atkins et al. 2002). Clasts tend towards angular to sub-rounded, with RA values lower than those of rockfall. Ice-cored debris cones exist in the Allan Hills, up to 1 km inland of the present-day glacier snout (Lloyd Davies et al. 2009).

Landforms associated with cold-based glaciers include thrust-block moraines formed of sand, gravel and silt, ice-contact sediment ramps comprising foliated ice, and tilted masses of glaciofluvial sand and gravel, covered with aeolian sediment (Hambrey and Glasser 2012) (Figure 14.15g). In the McMurdo Dry Valleys, cold-based glaciers of the East Antarctic Ice Sheet have created ice-marginal ridges comprising a sandy boulder/cobble-gravel. Preservation potential of the sediment-landform assemblages is high, as landforms and sediments are frozen in permafrost, and fluvial and slope processes are limited. This results in a terrain that is little modified on timescales of millions of years, especially in the McMurdo Dry Valleys (Figure 14.33).

14.5.7.3 Example of a Polar Continental (Cold-Based) Landsystem

The classic example of the cold-based glacial landsystem is the glaciers of the Dry Valleys, Victoria Land, Antarctica (Hambrey and Fitzsimons 2010; Atkins 2013). Basal glacier temperatures here range from −18°C at the margin to −15°C at the centre (Hambrey and Glasser 2012). Any basal meltwater is refrozen. Flow is predominantly by internal deformation of ice in the lower metres of the ice mass. Cold-based glaciers often have thick sequences of stratified debris-rich basal ice, as a result of overriding and incorporating proglacial sediments, regelation and stacking of debris sequences. In these cold environments, ablation is dominated by dry calving and sublimation, with some minor meltwater (Hambrey and Glasser 2012). Deposition is dominated by the removal of ice by sublimation, and the

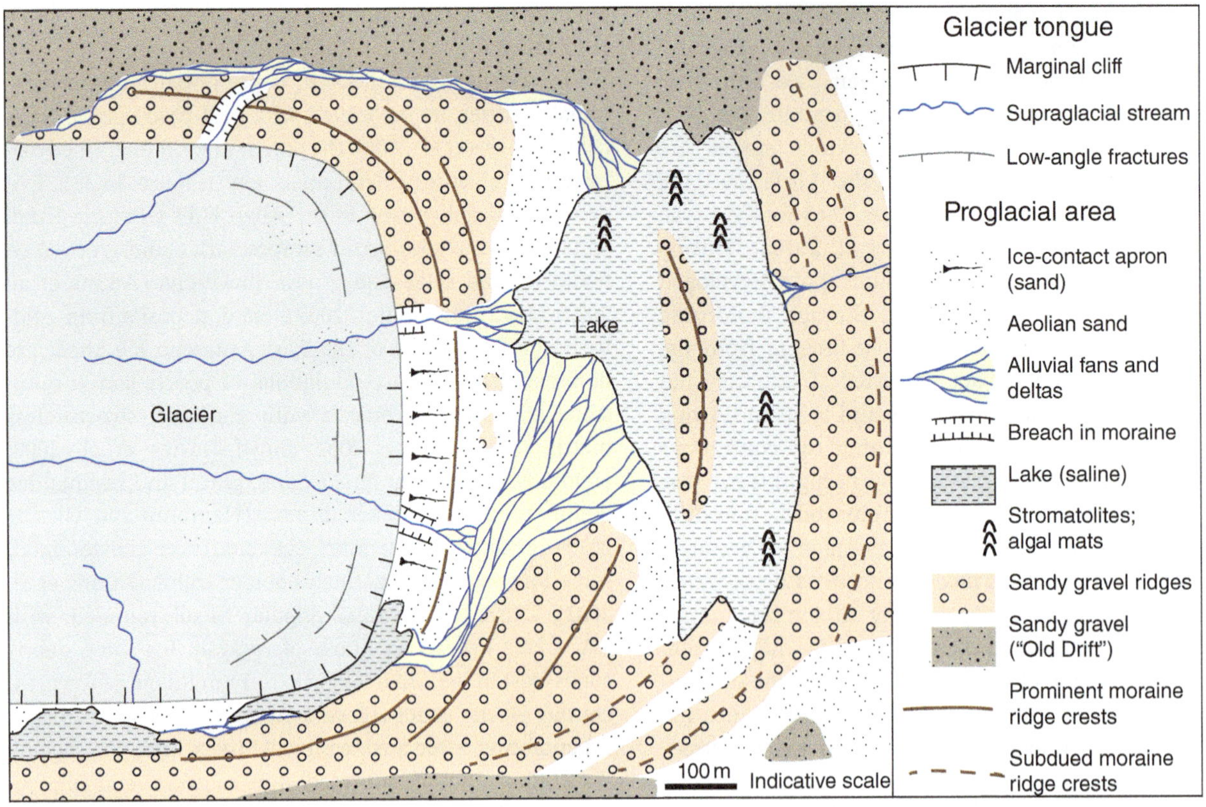

Figure 14.33 Conceptual model for cold-based polar terrestrial glacial landsystem based on valley glaciers in the Dry Valleys of Antarctica. Hambrey and Glasser (2012, p. 17) / with permission of Elsevier.

structure of the ice, including fissility and foliation, may be well preserved.

14.5.8 Plateau Icefield Landsystem

14.5.8.1 Processes

Plateau icefields are topographically constrained ice masses that form in upland terrain characterised by plateaux and rounded summits. These icefields are independent ice dispersal centres that have existed both during glacial periods and at present, and therefore can be considered as long-term glacial landsystems representative of average glaciation conditions (Porter 1989; Rea and Evans 2003; Evans 2016). Plateau icefields often occur close to the margins of ice sheets, both in current (e.g. Greenland) and past settings (e.g. Ellesmere Island and Norway), and therefore also have the potential to influence the configuration of initial ice sheet advance into a region and patterns of ice sheet recession (Rea and Evans 2003; Regnéll et al. 2022).

Plateau icefields consist of a flat to gently sloping central zone from which outlet glaciers flow down-valley (cf. Davies et al. 2022; Chandler et al., 2019). There are thus stark topographic differences between the central accumulation area and downvalley ablation areas, which are often reflected in contrasting glacier dynamics between these zones (Rea and Evans, 2003). When the ELA is situated on the plateau, plateau icefields are highly sensitive to changes in climate, because a small increase or decrease in the ELA will lead to a large increase in the size of the ablation or accumulation area, respectively (Oerlemans 2012; Boston and Lukas 2019). Therefore, plateau icefield response to climate fluctuations during the Holocene has been complex (Nesje et al. 2008; Levy et al. 2014; Margreth et al. 2014). Additionally, this can result in rapid deglaciation on flatter areas and could cause the ice to stagnate rather than to recede actively (Giesen and Oerlemans 2010).

Due to icefield hypsometry, ice on plateau areas is usually relatively thin. This, combined with low flow velocities, and therefore limited strain heating, means that plateau ice is often (although not always) cold-based (Rea and Evans 2003). This means that there may be little erosion or debris transport on the plateau, resulting in limited geomorphological and sedimentological evidence to indicate the presence of former ice. Indeed, blockfields, felsenmeer, weathered bedrock and periglacial features have been observed to have survived intact beneath cold-based plateau ice in Norway (Dahl 1966; Whalley et al. 1981; Gellatly et al. 1988; Rea et al. 1996). There is increasing support for former plateau icefields in areas previously thought to be devoid of ice (e.g. McDougall 2001 McDougall 2013; Evans et al. 2012; Boston et al. 2015). However, there are icefields where the plateau ice is thick enough to be temperate, for example, Hardangerjokulen, Norway (Reinardy et al. 2019) and the Juneau Icefield, Alaska (Davies et al. 2022), and moraines have been documented on the plateau in formerly glaciated areas, such as the Monadhliath, Scotland (Boston 2012). Nunataks may provide a source of supraglacial debris for ice on the plateau, but in general plateau ice is likely to have little supraglacial debris.

In contrast to the plateau, outlet glaciers are often temperate as a result of thicker ice within valleys and strain heating as the ice flows over the plateau edge (Rea et al. 1998; Rea and Evans 2003). These outlet glaciers therefore behave similarly to active temperate valley glaciers and produce similar sediment-landform assemblages (see Section 14.5.2). The connection between the plateau ice and outlet glaciers is variable depending on the underlying topography; in some cases, the plateau slopes gently into outlet valleys, in contrast to a steep backwall at other locations (Figures 14.34 and 14.35). In areas where there is a steep backwall, there is likely to be an icefall (cf. Davies et al. 2022). Reconstituted glaciers can also occur, where ice and snow avalanche down vertical slopes to reform at the bottom. Increasingly, as current plateau icefields continue to recede, disconnections between the plateau accumulation area and the valley glacier tongues across these very steep slopes and in icefalls (e.g. Figure 14.34b) are becoming more frequent (e.g. Rippin et al. 2020; Carrivick et al. 2022; Davies et al. 2022).

14.5.8.2 Lithofacies and Sediment-Landform Assemblages

On plateaux, cold-based ice may leave little trace, resulting in plateau areas being dominated by periglacial landforms such as tors and blockfields (e.g. Dahl 1966; Whalley et al. 1981; Gellatly et al. 1988; Rea et al. 1996) (Figure 14.35). In these circumstances, key features to help identify a former plateau icefield include lateral meltwater channels on the plateau and plateau edge that depict former ice margins, alongside glacially smoothed bedrock at the plateau edge, marking the transition to a warm-based thermal regime in the outlet glacier. At temperate plateau ice margins, there is the potential for moraines to form and till may be deposited. The low slope gradient of the plateau may cause ice stagnation during recession if the ELA rises above the plateau surface (Giesen and Oerlemans 2010), resulting in ice melt-out and debris flow deposits, alongside sorted glaciofluvial sediments. Locally sourced erratics on the plateau may also indicate the presence of plateau ice, otherwise, more far-travelled erratics may record submersion of the plateau surface beneath an ice sheet.

In the outlet valleys, the lithofacies are similar to either that of a cold-based or temperate valley glacier landsystem (Sections 14.5.2 and 14.5.7). Supraglacial debris cover will increase due to input from the valley sides. Within a temperate system, moraines can form at the ice front and lateral margins, and glaciofluvial sediments, for example, eskers, kames and outwash plains may also be deposited (e.g. Reinardy et al. 2013; Weber et al. 2019; Davies et al. 2022; Boston et al. 2023). Polythermal icefield margins may be characterised by ice-cored hummocky moraine pitted with melt-out hollows, subglacial till and flutings, glaciofluvial and

Figure 14.34 (a) Ice flowing from the plateau into the outlet glacier Midtdalsbreen at Hardangerjøkulen, Norway. (b) A glacier disconnection (sensu Davies et al. 2022) near Østre Svartisen, Norway, where the glacier has receded into the cirque and is no longer receiving substantial nourishment from ice at higher elevations. Photos by C.M. Boston.

Figure 14.35 Conceptual diagram of a polythermal [P] plateau icefield landsystem, with different types of geomorphological signature due to a spatially variable thermal regime. (a) Close to the maximum extent, warm-based [W] outlet glaciers drain from the plateau, depositing recessional moraines. Cold-based [C] outlet glaciers will produce a network of ice-marginal meltwater channels with sparse moraines. Former ice margin positions depict a range of recession styles controlled by valley morphology (see Boston and Lukas 2019 for further details). (b) during recession, the plateau icefield will thin and in some places break up into smaller ice masses, with disconnection of ice at the valley heads possible. Modified from Boston and Lukas (2019, p. 446) / John Wiley & Sons.

glaciolacustrine deposits (e.g. Evans et al. 2006b, Evans et al. 2016; Evans 2010). Some outlet valley glaciers (e.g. in Norway) rest on hard bedrock, typically where the slope gradient is steeper. In these areas, there may only be a thin, patchy, till veneer or collection of boulders on the bedrock surface, as the majority of glaciogenic debris is efficiently removed by meltwater streams (Weber et al. 2019).

14.5.8.3 Example of a Plateau Icefield Landsystem

A contemporary example of a plateau icefield is Hardangerjøkulen in Norway. Hardangerjøkulen is predominantly temperate, although recent geophysical surveys indicate that in areas of thinner ice such as around nunataks and at the margin of the outlet glacier, Midtdalsbreen, the ice is cold-based (Etzelmüller and Hagen 2005; Reinardy et al. 2019). Reinardy et al. (2013) demonstrated that annual moraines at Midtdalsbreen are formed by winter freeze-on of subglacial slabs to the glacier bed (e.g. Krüger 1996). The icefield reached a maximum size of 109.7 km^2 during the Little Ice Age (~1750 CE) and the limit is marked by bouldery moraines, drift limits and erosional and vegetational trimlines (Weber et al. 2019). The recession of the outlet glaciers has revealed moraines, flutings, eskers and lakes. Weber et al. (2019) show that Hardangejøkulen lost 37% of its area up to 2010. Due to its hypsometry, the icefield is very sensitive to surface mass balance changes (Åkesson et al. 2017). It is predicted to catastrophically recede over the next few decades and may have disappeared entirely by 2100 (Giesen and Oerlemans 2010).

14.5.9 Glaciolacustrine Landsystem: Large Proglacial Lakes

14.5.9.1 Processes

Proglacial lakes are ubiquitous in temperate terrestrial glaciated environments, forming in both mountainous environments and lowlands. Glacial lakes today are growing as glaciers globally recede (Wilson et al. 2018; Shugar et al. 2020). During Pleistocene glacial intervals, lakes grew large enough to alter regional oceanic and atmospheric circulation upon sudden drainage (Teller 2013; Glasser et al. 2016). Glacial lakes are pervasive in worldwide records of Quaternary glaciation (Carrivick and Tweed 2013) and encompass a wide variety of environments, including proximal and ice-contact lakes, subglacial and supraglacial and ice-distal (Fitzsimons and Howarth 2018). Different glacier-lake settings are associated with diverse and varied sediment-landform assemblages. Proglacial lakes may form where outflows are impounded by ice, moraines, landslides, or bedrock. Moraine-dammed lakes are associated with the debris-covered glacier landsystem (Section 14.5.3). The focus of this section is large proglacial ice-contact lakes and their associated landsystems. Key limnological processes and terminology are covered in Chapter 6 (see for example Section 6.2.2).

Proglacial lakes may impart a significant control on glacier behaviour, by inducing calving and decoupling glacier dynamics from climate (Thorndycraft et al. 2019; Sutherland et al. 2020). The growth of an ice-marginal lake may affect the longitudinal stress balance of a glacier, affecting glacier mass balance and velocity (Carrivick and Tweed 2013). Proglacial lakes can increase the ice-surface gradient, promote flotation, raise subglacial water pressure, promote calving and cause intense ice flexure and fracture if they drain and refill rapidly. Water at the base of a glacier can increase basal water pressure and enhance sliding, and if basal water pressures approach the ice overburden pressure, may cause basal crevasse development and hydrofracturing (Figure 14.36) (Carrivick and Tweed 2013).

Thermal stratification in lakes results in changes in water density associated with temperature change (Fitzsimons and Howarth 2018). The maximum water density of 1000 kg m^{-3} is reached at −4°C, with a systematic decrease in density both below (until freezing point is reached) and above this point. In temperate latitudes in the summer, surface waters are warmed, producing a warm surface layer (epilimnion). This rests above cooler water, the hypolimnion (Figure 14.37). High suspended sediment concentrations can enhance surface warming by altering the lake's albedo (Fitzsimons and Howarth 2018). Between the epilimnion and hypolimnion is the thermocline, where temperature changes rapidly with depth. In summer, there is limited mixing and the lake is stable. Wind may cause thermal mixing, by inducing circulation within the lake. Thermal stratification may break down in winter, with surface waters cooling. If the surface water cools more than the deep water, the dense surface water overturns and moves downward. Ice may form on the surface, and the densest water (4°C) sinks to the bottom of the lake. Thermal stratification with a new, deeper thermocline may develop, before breaking down again in the spring. These 'monomictic' lakes therefore experience stratification in the winter, with the lake surface freezing, isothermal water temperatures in spring and summer, and a thermocline developing in the summer (Palmer et al. 2019).

This thermal stratification is key for controlling the distribution of sediment that flows into a proglacial lake, resulting in overflows, interflows and underflows. Overflows occur where the inflowing water has a lower density than the surface water (Fitzsimons and Howarth 2018). A surface plume spreads across the lake surface, and sediment is deposited by settling from the suspension of fine particles. This is more common in marine settings, where the density of seawater is higher than the inflowing fresh,

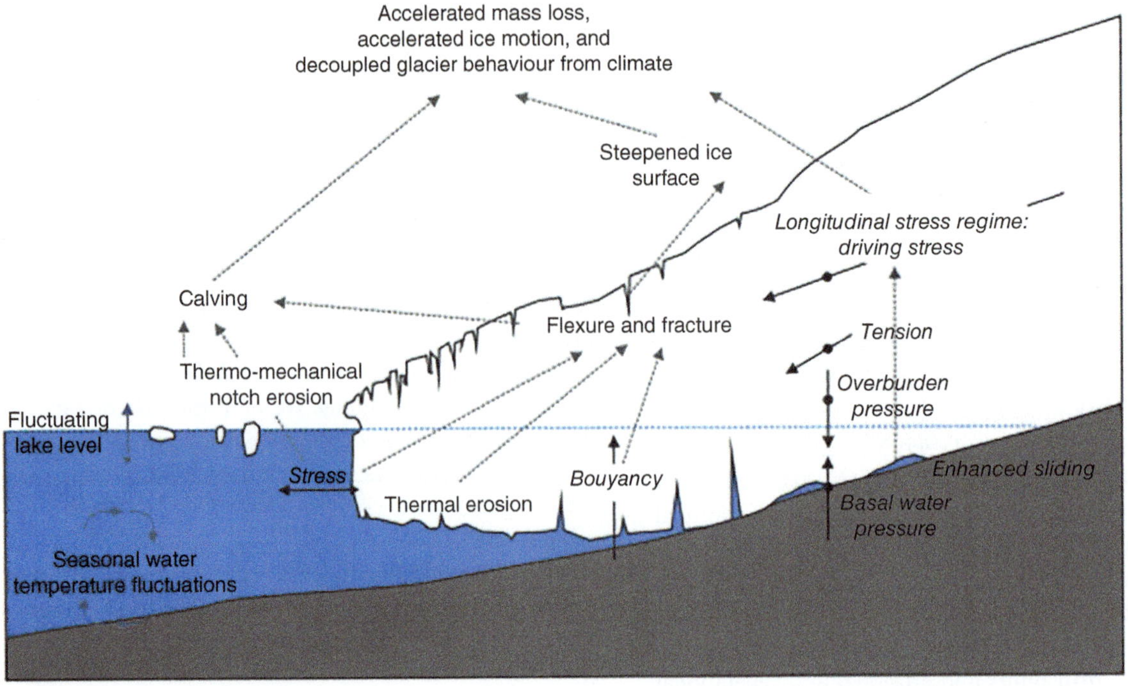

Figure 14.36 Summary schematic diagram of the influence of an ice-marginal lake on glacier dynamics. From Carrivick and Tweed (2013, p. 45) / with permission of Elsevier.

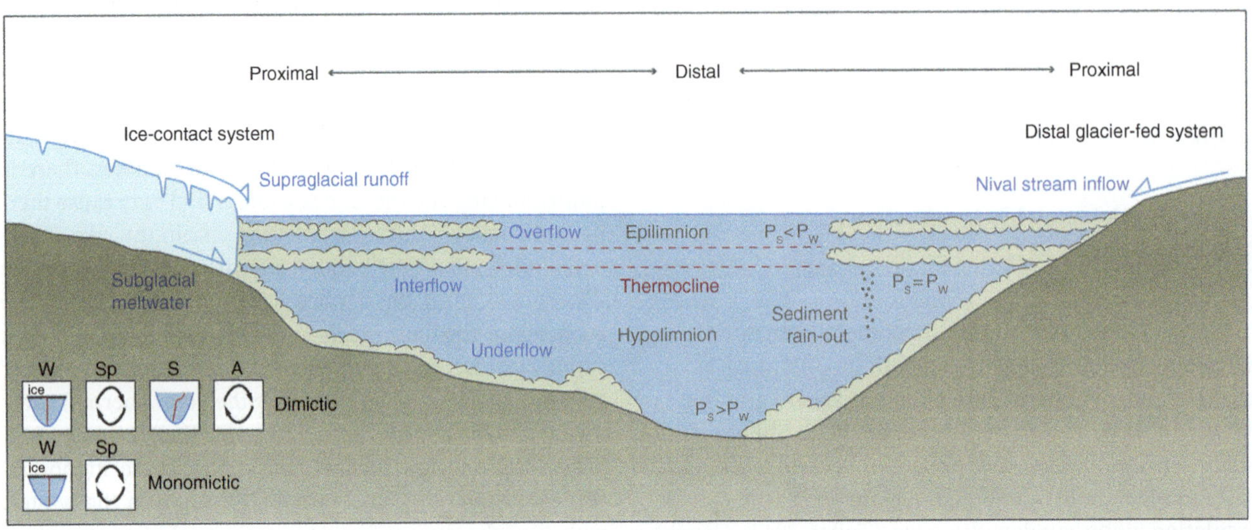

Figure 14.37 The nature of flows within an ice-contact lake and distal glaciolacustrine systems. This example lake is thermally stratified in the summer (S) and overturning in spring (Sp) and autumn (A). The position of the sediment flows is influenced by the seasonally changing comparative density of the sediment inflows and the lake water. Palmer et al. (2019, p. 4) / Elsevier / CC BY 4.0.

cold water from glacier termini. In glacial lakes, a relatively small proportion of lake sediment is derived from overflows, as inflowing water is usually more dense than lake water because glacier-derived water is typically colder and has a higher sediment load.

Interflows (Figure 14.37) occur where the inflowing glacial meltwater has a higher density than the epilimnion but less dense than the hypolimnion. The water plume, and any suspended sediment, flows along the thermocline and sediment is again deposited by suspension settling. Inflows with lower sediment concentrations within glacial meltwater may form interflows (Palmer et al. 2019), while denser, higher sediment concentrations are associated with underflows. Interflows with suspended sediment can be widely distributed across lakes, producing distinctive couplets or rhythmites of sediment.

Underflows form where inflowing water is densest, and the sediment is transported along the lake floor (Figure 14.37) (Gilbert and Crookshanks 2009). They are volumetrically important and comprise gravity, turbidity or density currents, especially in more proximal locations. These underflows may be interbedded with rhythmites derived from interflow or overflows and may form discontinuous beds of stratified diamictons.

14.5.9.2 Lithofacies and Sediment-Landform Assemblages

Sediments within glaciolacustrine environments may be sourced directly from the glacier (ice-contact), via meltwater streams (ice-distal), or via nival meltwater streams. In ice-proximal or ice-contact settings, the sediment is released directly from subglacial and supraglacial transport networks, iceberg rainout and supraglacial slumping (Bennett et al. 2002; Palmer et al. 2019). Depositional processes include jets emerging from subaqueous tunnels, flows, slides and slumps, aqueous underflows of various sorts, and suspension settling (Figure 14.38). Lithofacies include bedded silt and clay (mud-rich) facies, mainly derived from flow, slides and slumps, and sand and gravel facies, which are derived from quasi-continuous and surge-like flows. The larger grain sizes often form fans along the ice front, which may include grains up to boulder size that fall into the water and accumulate at the terminus of the glacier (Teller 2003). This results in massive, matrix-supported diamictons, in both proximal locations, especially where a slope or flows propagate finer materials further into the basin, or more distal locations where transport has removed the coarser fraction (Palmer et al. 2019). At the ice margin, sorted sand and gravel facies are more common (Figure 14.38).

In the distal parts of the lake basin, particle size decreases, depositing laminated silts and clays (Figure 14.38). These may form rhythmites or, if they are annually laminated, varves (Bennett et al. 2002; Palmer et al. 2019). These sediments are deposited by hyperconcentrated to concentrated flows (30–90% sediment concentration by volume) (see Section 12.3.1). Rainout of larger material from icebergs

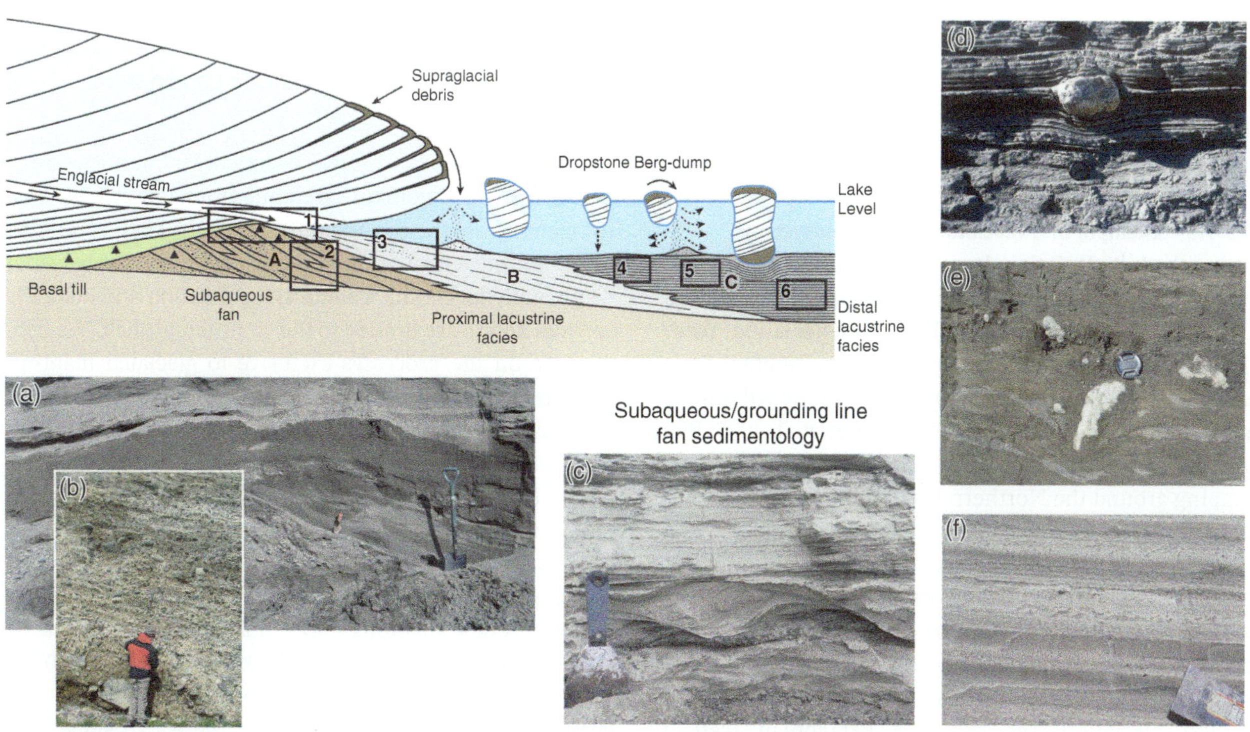

Figure 14.38 The range of depositional environments within a glaciolacustrine setting. (1) Gravel and sand clinoforms with glaciotectonised carapace produced by glacier overriding of stream portal/grounding line fan apex. Sands drape the depo-centre and relate to ice recession, (2) clinoforms composed of interbedded stratified diamictons and poorly sorted gravels deposited as sediment gravity flows on proximal fan surface; (3) alternating sequence of planar bedded sand, climbing sand ripples and planar cross-stratified fine gravels, with localised scour and fill structures, deposited by traction currents on the distal slope of the fan; (4) dropstone in distal glaciolacustrine facies; (5) dropstone and iceberg dump structure in distal glaciolacustrine facies; (6) distal glaciolacustrine rhythmites composed of alternating clay and silt/fine sand laminae with small dropstones, the combined product of suspension sedimentation, turbidites and iceberg rafting. From Thomas (1984, p. 18) / Geological Society of London, and Evans and Benn (2021, p. 355) / Taylor and Francis.

floating above can result in larger clasts, or 'dropstones', dropping through the water column to be deposited within these laminated sediments, deforming the bedding below and draped by the overlying bedding (Figures 14.17f and 14.38). Eventually, sedimentation is dominated by underflows and turbidity current processes (1–10% sediment concentration by volume) in the more distal parts of the basin (Palmer et al. 2019). Suspended sediment settling out from overflows and interflows forms a blanket-like drape across the basin, with proximal to distal fining and thinning.

Varves form in glacial lakes where a lighter silt to sandy layer is deposited in spring and summer by sediment-laden glacial meltwater inflowing into the lake. These fining-up sequences record coarse sediment input from overflow and interflow plumes during the glacial ablation season. In the winter, a darker, clay-rich layer is deposited by gradual suspension settling as the lake surface freezes (Palmer et al. 2008; Fitzsimons and Howarth 2018). Clastic varve sequences have been linked to climatic and meteorological events (Bendle et al. 2019) and can provide a precise chronological control for deglaciation (Palmer et al. 2020).

Large proglacial lake landsystems can also include lithofacies produced by other processes, including alluvial fans and deltas, subaqueous fans and inputs from the valley sides. Rainout from icebergs can result in material being transported into the middle of the lake system, depositing iceberg-rafted debris and dropstones within lacustrine sediments (Thomas and Connell 1985; Palmer et al. 2019).

14.5.9.3 Example of a Glaciolacustrine Landsystem

During the last glaciation, large proglacial lakes formed in front of the former Patagonian Ice Sheet (Thorndycraft et al. 2019; Davies et al. 2020). During deglaciation, these lakes grew, evolved and drained, with a continental-scale reorganisation of meltwater drainage from the Atlantic to the Pacific Ocean (Caldenius 1932; Turner et al. 2005; Bell 2008; Glasser et al. 2016). Systematic geomorphological mapping around the Northern Patagonian Icefield reveals a detailed dataset of regional glacial and palaeolake geomorphology, that can be used to identify ice margins and palaeolake levels (Bendle et al. 2017b). East of the Northern Patagonian Icefield, at ~47°S, Río Baker drains Lago General Carrera/Buenos Aires and Lago Cochrane/Pueyrredón and flows south into the Pacific Ocean. During the last glaciation, the Northern Patagonian Icefield dammed the Baker Valley, resulting in large ice-dammed lakes to the east of the icefield (Davies et al. 2018; Thorndycraft et al. 2019). These palaeolakes occupied the overdeepened basins in front of outlet glaciers and grew in size during deglaciation. As the ice receded past key cols and drainage pathways, the palaeolakes evolved in size and shape, and eventually drainage changed from eastwards, towards the Atlantic, to westwards, towards the Pacific (Turner et al. 2005; Bell 2008; Glasser et al. 2016; Thorndycraft et al. 2019).

In the Salto valley, which was previously occupied by the Calluqueo Glacier, a series of glaciolacustrine landforms are associated with an ice advance dated to the Antarctic Cold Reversal (Davies et al. 2018). The glacier terminated in a glaciolacustrine environment, forming subaqueous, asymmetric morainal banks and sharp-crested, subaerial moraines above the water level (Figure 14.39). The glaciolacustrine landsystem includes raised deltas and palaeoshorelines that formed along the margins of the palaeolake, and subaqueous fans on the valley floor, which formed during temporary pauses in recession. Well-sorted varved sediments are found on the valley floor beyond the moraines. The subaqueous moraines are composed of stratified sandy gravels and stratified diamicton (Davies et al. 2018). These proximal sediments represent sediments deposited subaqueously in front of the ice margin by underflows and turbidity currents (Figure 14.37). This association of ice-contact fans and morainal banks is common in temperate glacial environments with abundant meltwater and erosive valley glaciers (Dowdeswell et al. 2016).

14.5.10 Temperate Glaciomarine Landsystem

14.5.10.1 Processes

The glaciomarine landsystem is actually a 'marine system' as it consists of a domain extending from a tidewater environment across the continental shelf to the shelf break and slope. Marine terminating glaciers are constrained by water depth, with deepening water driving flotation and inducing calving, and so are limited to the continental shelf. We concentrate on the processes exclusive to glaciated margins. The processes and products of gravity remobilisation are covered in Chapter 12. The glaciomarine landsystem records the interplay between ice advection (e.g. through grounded ice), sediment deformation and cannibalisation (grounding, reworking) and non-glacial reworking by processes such as waves, tides or currents.

The 'grounding line' is the point at which grounded ice terminates or transitions into a floating glacier tongue or an ice shelf in polar regions (Fricker and Padman 2006). This area is often referred to as the 'grounding zone' to depict the transition zone between these two states and can be affected by water depths, tides and the ice thickness. Typical landforms at this site include grounding zone wedges and asymmetric sedimentary depocentres that form through the rapid accumulation of rapidly deforming subglacial sediments at the grounding zone (Batchelor and Dowdeswell 2015; Batchelor et al. 2018, 2020). Moraines and ice-proximal fans may build up during retreat during

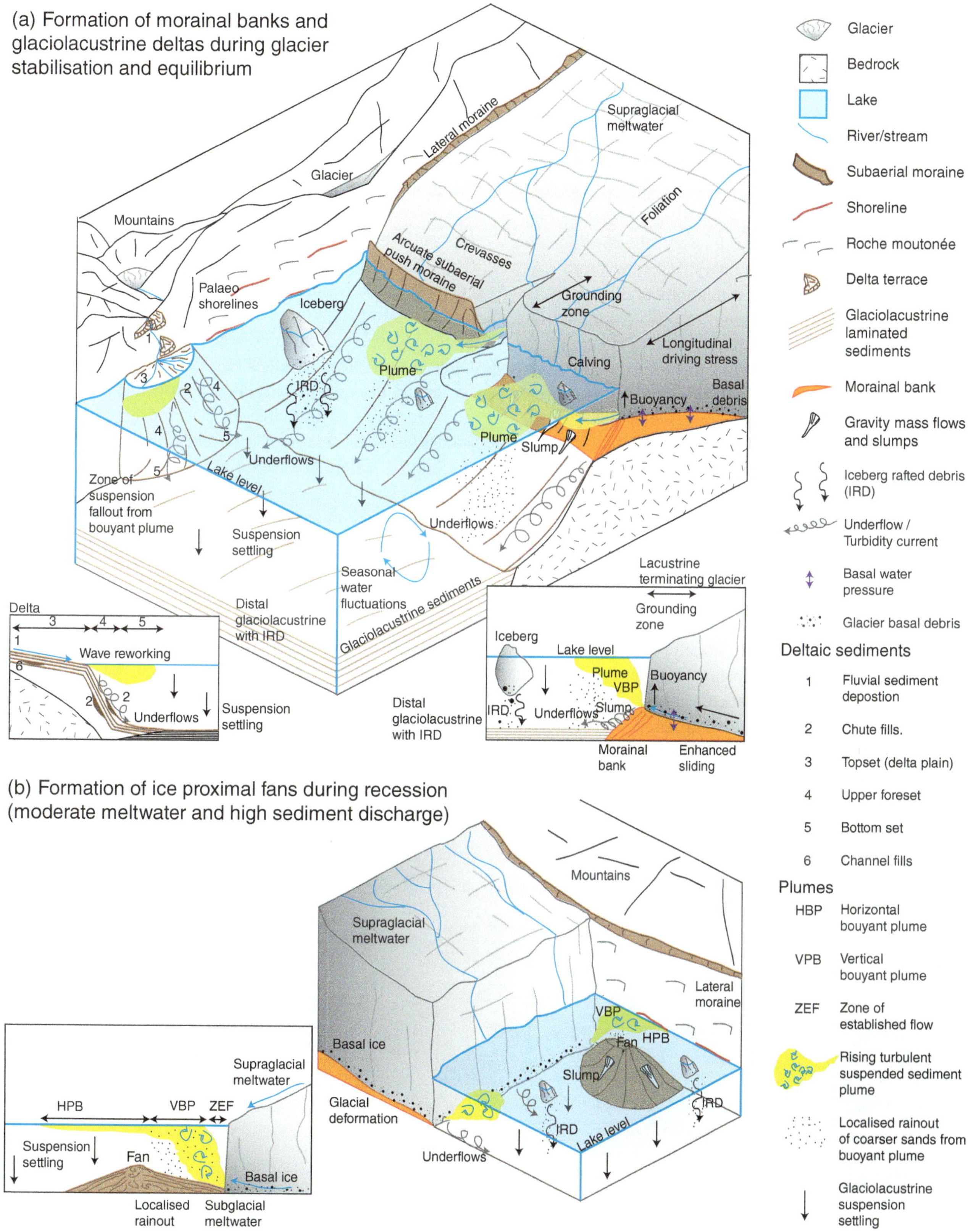

Figure 14.39 (a) 3D plan view and along-profile 2D views illustrating glaciolacustrine sedimentary processes in Patagonia. (b) 3D plan view and 2D along-profile views illustrating formation of an ice-proximal fan. Diagram is not to scale but the lakes in this example were up to 300 m deep. From Davies et al. (2018, p. 307) / Elsevier / CC BY 4.0.

temporary still stands in the ice margin, with sediment delivered from point-sourced subglacial meltwater streams.

14.5.10.2 Lithofacies and Sediment-Landform Assemblages

Modern high-latitude continental shelves have been intensively investigated using geophysical techniques such as multibeam echo-sounding and 3D seismic data, with the focus resting on glacial sedimentology and geomorphology (Dowdeswell et al. 2016), as well as coring expeditions led by the IODP and ODP. Sedimentary facies are variable, and likely to consist of thick accumulations of diamicton, glaciomarine mud, sand and gravel. At the grounding line, for example, a complex combination of mud-bearing ice-rafted debris (dropstones), diamicton and sediments apparently devoid of any clear glacial influence can be expected (Domack and Harris 1998). Where gravitational reworking is important, often in more distal settings at the continental slope, intercalations of diamicton (typically gravity flow reworked glacial deposits), sandstone (turbidites) and hemipelagic (glacio)marine muds are characteristic. Well-documented examples of the latter are found on the Wilke's Land continental margin, Antarctica (Escutia et al. 2000), as well as in the Neoproterozoic record (Le Heron et al. 2013).

The glaciomarine landsystem exhibits a series of unique features (Figure 14.40). Unfortunately, on account of their scale, some of these are challenging to adequately recognise in the ancient record. Cross-shelf troughs are produced by ice streams incising the continental shelf (Batchelor and Dowdeswell 2014). Cross-shelf troughs have considerable size variations, from about 35 to 1400 km long, 12–260 km wide, and are cut in modern water depths of 200–1000 m (Batchelor and Dowdeswell 2014). They are oriented length-parallel to the ice flow direction, and their orientation and dimensions are considered to reflect the shape of the ice stream that formed them. As such, these elongated basins may contain an array of subglacial bedforms including mega-scale glacial lineations and drumlins, some stacked at multiple stratigraphic levels and separated by tens to hundreds of metres of stratigraphy (Kurjanski et al. 2019).

Trough-mouth fan systems are thick accumulations of material dumped leeward of the shelf break by ice streams flowing down the troughs (e.g. Dowdeswell et al. 2019; Ottesen and Dowdeswell 2022). Outcrops in the ancient record are dominated by boulder-rich conglomerates, diamictites probably ultimately of glacial origin, and graded sandstone beds organised into lobes like any other deep marine fan system (Le Heron et al. 2013). A distinction is observed between fans that are dominated by glaciogenic debris flows (produced by a collapsing 'wall' of diamicton at the ice margin) and those where more sustained meltwater contributes to fan growth (Bellwald et al. 2020).

Grounding zone wedges are stacked accumulations of diamicton and record ice margin stabilisation (Batchelor and Dowdeswell 2015; Kowal and Worster 2020). For example, they have been found extensively on the continental shelf off the west coast of Ireland, and document both the maximum extent of the British-Irish Ice Sheet and periods of stabilisation during ice-sheet recession (Ó Cofaigh

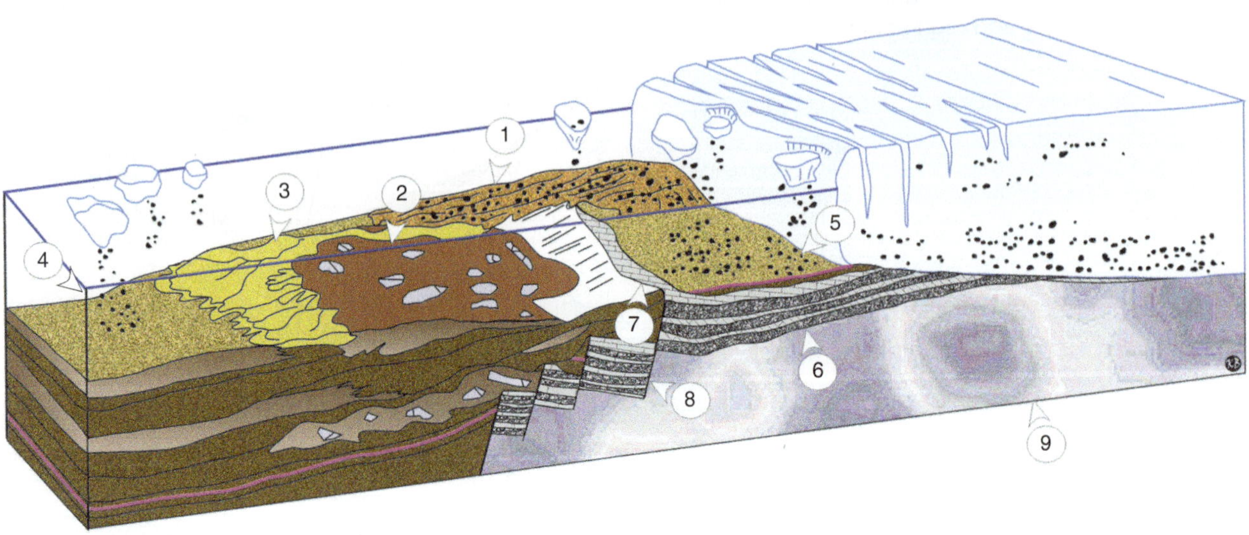

Figure 14.40 The glaciomarine landsystem as a model for the Cryogenian of Death Valley. In this diagram, a complex interplay of glacial sedimentation and extensional tectonics leads to an admixture of glacially sourced deposits and gravitational collapse structures. (1) Glaciogenic debris flows, (2) Olistostrome, (3) Turbidites, (4) Ice-rafted debris (IRD) (distal), (5) IRD (proximal), (6) Preglacial stratigraphy, (7) Fault scarp (exposing Beck Spring Dolomite), (8) Syn-depositional faulting and (9) Crystalline basement (gneiss, schist, granitoid). Adapted from Le Heron et al. (2017, p. 33).

et al. 2021). Here, moraines on the Porcupine Bank transition laterally into grounding zone wedges in the Slyne Trough due to increasing water depths.

Finally, glaciomarine landforms include glacial lineations of different orders of magnitude, produced by ice plastically moulding sea floor sediment or through the passage of icebergs (Zecchin et al. 2016) and iceberg ploughmarks (Dowdeswell and Ottesen 2013; Linch and Dowdeswell 2016). In their distal reaches, large volumes of sediment are dumped as trough-mouth fans at the margins of the last British-Irish Ice Sheet (Ó Cofaigh et al. 2021; Clark et al. 2022).

14.5.10.3 Example of a Temperate Glaciomarine Landsystem

During the Last Glacial Maximum, outlet glaciers draining from the Patagonian Ice Sheet flowed westwards onto the continental shelf (Davies et al. 2020). The continental shelf and western Patagonia comprises hundreds of channels, islands and fjords, cut by repeated Cenozoic glaciations. These fjords have an over-deepened sea-floor topography excavated by glaciers during full-glacial conditions. An off-shore temperate glaciomarine landsystem comprises fjords, offshore moraine ridges, streamlined sedimentary lineations, turbidity-current channels, raised fluvial deltas and slope failures (Dowdeswell and Vasquez 2013; Dowdeswell et al. 2016). This environment is the mildest climatic and oceanographic end-member of glacially influenced tidewater environments.

The western, Pacific, continental shelf of Patagonia (42°–55°S) includes 12 deep cross-shelf troughs up to 300 km long. These troughs originate in narrow, deep, inter-island fjords characteristic of the western Patagonian coastline. These inner fjords reach up to 1000 m below sea level, and are 3–10 km wide (Davies et al. 2020). On the continental shelf, the troughs are shallower (~200 m below sea level) and broader (11–38 km wide). They form a dendritic pattern, with fjords coalescing into a single trough on the middle to outer continental shelf. These cross-shelf troughs are interpreted to have been cut by palaeo ice streams during full glacial conditions, with over-deepening by glacial erosion on the inner shelf and depths decreasing seaward, similar to those observed on other glaciated continental shelves (Batchelor and Dowdeswell 2014).

High-resolution swath bathymetry reveals a temperate, meltwater-dominated inner fjord setting, characterised by the rapid delivery of fine-grained sorted sediments from meltwater plumes and glaciofluvial systems for land-terminating glaciers (Dowdeswell and Vasquez 2013). Fjord floors are characterised by transverse moraine ridges, turbidity current channels and Holocene basin infill. There are prominent perched glaciofluvial and fluvial deltas on valley sides, reflecting meltwater input during a period of higher relative sea level. The transverse moraine ridges demarcate stillstands during grounded glacier recession. Several are located at fjord mouths, representing pinning points before further deepening of the fjord floor (Dowdeswell and Vasquez 2013). Streamlined features are rare, likely because they have been buried by fine-grained Holocene infill. Turbidity current channels emanate from steep delta fronts in several fjords (Dowdeswell and Vasquez 2013). Irregular sea floor depressions attributed to ploughing by the keels of icebergs are absent from these Patagonian fjords, despite the presence of calving tidewater glaciers, because the icebergs produced tend to be small and shallow, and rarely grounded.

14.5.11 Ice-Shelf Glaciomarine Landsystem

14.5.11.1 Processes

Ice shelves are the floating extensions of land ice that extend beyond the grounding line. They form in polar regions at the margins of an ice sheet, where the ice becomes thin enough to float (Jenkins 2011) but remains stiff enough to resist the lateral extensional forces promoting thinning. Today, they almost completely surround West Antarctica. They can be fed by multiple tributary glaciers, which coalesce to form a larger ice shelf. Ice shelves can be 100 m to over 2000 m thick. In Antarctica, they comprise 14% of the total glacierised area. The majority of grounded ice in Antarctica today drains through peripheral ice shelves (Smith et al. 2019). Calving and melting from Antarctic ice shelves is a crucial part of Antarctic mass balance, and these processes dominate ablation in Antarctica (Shepherd et al. 2012; Rignot et al. 2019). Ice shelves are changing rapidly in Antarctica, particularly in the Antarctic Peninsula and West Antarctica (Cook and Vaughan 2010), and these changes are important for influencing glacier change far up-glacier of the grounding line (Pattyn et al. 2006; Schoof 2007).

14.5.11.2 Lithofacies and Sediment-Landform Assemblages

Where the ice streams or outlet glaciers flow into ice shelves, a diamicton is produced at the grounding line, often as part of a feature such as a grounding-zone wedge or morainal bank (Hambrey and Glasser 2012). Sediments here comprise discontinuous layers of clast-rich mud or diamicton, interlayered with ice (Smith et al. 2019). This region is also commonly associated with landforms such as MSGL and drumlins (Smith et al. 2011; Smith et al. 2019).

Analysis of marine sediment cores taken from the Antarctic Peninsula continental shelf shows a succession of sedimentary sequences that evolve as the environment transitions from subglacial to sub-ice shelf to open glaciomarine conditions (Pudsey and Evans 2001; Domack et al. 2005; Pudsey et al. 2006) (Figure 14.41). Since ice shelves accrete marine ice on their base through basal freeze-on, sub-ice-shelf sediments are characterised by low depositional rates and low biogenic matter, and any remaining sediment is trapped within the ice shelf (Hambrey and Glasser 2012). This sediment is released when the ice calves icebergs, which float away and melt, depositing the debris as iceberg-rafted debris. Sedimentation increases with proximity to sediment sources (grounding line and calving line) and is also strongly influenced and modulated by marine currents (Smith et al. 2019). These sediments may typically be stratified diamictons, massive to stratified, and well-sorted glaciomarine sands and muds. These sediments fine laterally into a 'null zone' (Figure 14.41). Towards the calving line sediments become coarser again. This coarsening indicates proximity to the calving line and comprises biogenic matter and silty to sandy material advected under the ice shelf from the open ocean. Beyond the calving line, sediments comprise an iceberg-rafted diamicton deposited at the ice-shelf front, in turn, overlain by a dropstone-bearing biogenic mud deposited in seasonally open marine conditions (Smith et al. 2019).

Sediment cores taken from the Larsen shelf (eastern Antarctic Peninsula), including from areas formerly underneath ice shelves, record this transition from sub-ice stream to sub-ice shelf and then open glaciomarine conditions. The subglacial unit comprises a soft to stiff dark grey diamicton and gravelly mud (Reinardy et al. 2009). This unit has a high shear strength, related to compaction and dewatering and a high overburden pressure (Evans et al. 2005). Overlying units, interpreted to originate from a proximal, sub-ice shelf environment, comprise normally consolidated diamicton, gravelly mud, muddy gravel and sandy gravel, with frequent stratification and lamination (Reinardy et al. 2009). Over this are normally consolidated sandy muds and diamictons with variable amounts of clasts. These sub-shelf units are interpreted as being deposited by melt-out debris, subaqueous debris flows and current winnowing (Evans and Pudsey 2002; Reinardy et al. 2009). Sediment plumes emanating from the grounding line can drape the sea floor under the ice shelf with finely laminated muds, deposited through suspension settling (Smith et al. 2019; Figure 14.41). The distal sub-shelf environment is characterised by a low shear strength, sparse ice-rafted debris and an increase in diatom content (Figure 14.41).

14.5.11.3 Example of an Ice-Shelf Glaciomarine Landsystem

During the Last Glacial Maximum, two ice streams, Pine Island Glacier and Thwaites Glacier, that drain the West Antarctic Ice Sheet into the Amundsen Sea Embayment,

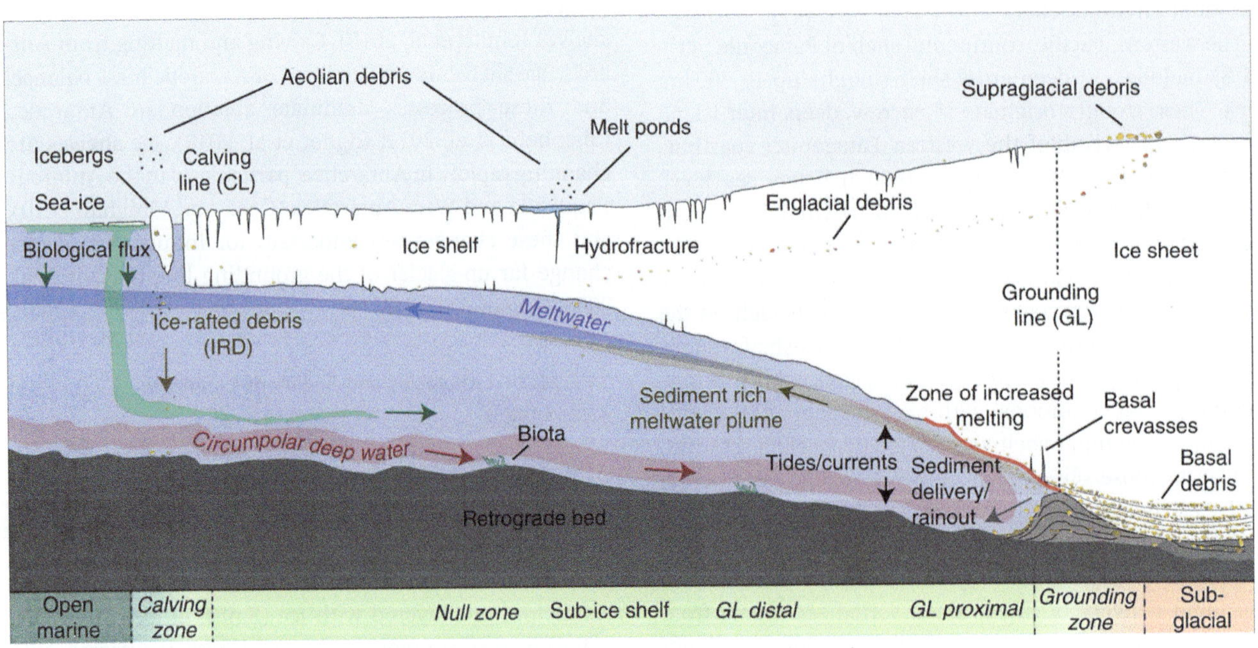

Figure 14.41 Conceptual landsystem for an Antarctic continental shelf a few hundred metres deep, with an ice shelf extending past the grounding line. GL: Grounding Line. Note that in reality the ice shelf would normally be proportionately much longer than shown. This model is for a temperate ice stream on the Antarctic continental shelf. The ice stream is temperate due to geothermal heat flux and ice thickness, but the polar waters support the ice shelf. Ice shelves do not form in warmer waters at lower latitudes due to increased surface melt, and the increased deformability and decreased stiffness of the warmer ice, resulting in thinning and calving. From Smith et al. (2019, p. 3) / Springer Nature / CC BY 4.0.

reached the continental shelf edge. Their recession across the continental shelf deposited a well-preserved sediment-landform record across a 500 km trough (Graham et al. 2016). These ice streams are thought to have terminated in ice shelves, leaving a sub-ice shelf landsystem superimposed on an ice stream landsystem as they receded, which is also shown in Autonomous Underwater Vehicle (AUV) data of seafloor bathymetry beneath the present-day Pine Island Glacier ice shelf.

The landsystem includes parallel, streamlined MSGL, with identical alignment to modern ice-shelf flow. These MSGL most likely formed during the most recent period of grounded ice flow over this site (Graham et al. 2016). These lineations have an amplitude of 2–7 m and a wavelength of c. 130–400 m, with ridge lengths of >1 km, and are focused on the back slope of a bedrock ridge under the ice shelf. This bedrock ridge has a 3-km wide planed-off crest, with 24 regularly spaced low-amplitude transverse ridges ('corrugations') between the mega-scale glacial lineations. These regularly spaced ridges are inferred to have been formed by sub-ice-shelf keels (Graham et al. 2016) as a result of tidally induced grounding line recession across fairly flat slopes (Batchelor et al. 2023). This buoyancy-driven ice sheet recession can occur at hundreds of metres per day.

On the mid-shelf trough in front of Pine Island Glacier, the trough floor is characterised by isolated bedrock sills and dominated by highly streamlined MSGL. Wedge-shaped asymmetrical landforms interpreted as grounding-zone wedges extend across the axis of the bathymetric trough. Characteristic thicknesses are up to 55 m, with a width of ~20 km. The grounding-zone wedges are cross-cut by sinuous and randomly aligned V-shaped iceberg ploughmarks. In deeper water below 750 m below sea level, the furrows are replaced by parallel MSGL (Graham et al. 2016).

The inner shelf is also characterised by a series of smooth, sedimentary basins with late Holocene glacial infill (Graham et al. 2016). Crystalline bedrock, polished into crag-and-tails, drumlinoid ridges and streamlined grooves, crops out adjacent to these basins. Broad tunnel valleys incise through the landforms, indicative of strong meltwater input.

14.5.12 Glaciovolcanic Landsystem

14.5.12.1 Processes

Glacier-capped volcanoes can be divided into two types: (i) ice-clad volcanoes that support icefields and at which the volcano is clearly visible; and (ii) ice-covered volcanoes, which exist beneath extensive ice sheets. Volcanic activity directly affects glaciers through subglacial heating, subglacial eruptions and dome growth, lava flows, floods and lahars, and supraglacial deposition of eruption material (Barr et al. 2018) (Figure 14.42). Subglacial heating due to increased geothermal heat flux will lead to ice melt at the glacier bed, which may cause enhanced sliding and glacier velocity increases (e.g. Sturm et al. 1991). This mechanism has been recognised as a cause for concern for the stability of the West Antarctic Ice Sheet, which overlies the West Antarctic Rift System (Blankenship et al. 1993; Corr and Vaughan 2008; Van Wyk de Vries et al. 2018). Enhanced melt can also lead to the formation of subglacial lakes or cavities, alongside pitting on the ice surface and the development of lake-filled melt pits (cauldrons) (e.g. Bleick et al. 2013). Drainage of these lakes has led to outburst floods (jökulhlaups in Iceland), which can also increase glacier velocity due to enhanced sliding and glacier lift from high subglacial water pressures (Einarsson et al. 2016; Magnússon et al. 2012).

Substantial ice melt also occurs as a result of subglacial eruptions and dome growth, in addition to potential losses in glacier size or changes in shape. For example, the 1980 eruption at Mount St Helens, USA, destroyed the Loowit and Leschi glaciers and beheaded others (Brugman and Post 1981). Subsequent subglacial dome growth at Mount St Helens during the 2004–2006 eruption caused the newly formed Crater Glacier to be split in two through the extrusion of a silicic lava dome (Walder et al. 2007). Lava flows produced during eruptions can flow either subglacially or supraglacially, often confined or diverted by ice barriers. Intense steaming occurs during lava-ice contact resulting in substantial ice melt, which may form deep channels or pits on the ice surface or enlarge existing subglacial tunnels (Lescinsky and Fink 2000; Barr et al. 2018). Pyroclastic density currents, formed with hot volcanic debris and gas, will also cut channels into the ice surface through melting and abrasion (Barr et al. 2018). These currents can be incredibly destructive and have been documented as scouring a section of the Drift Glacier on Mount Redoubt to bedrock during the 1966–1968 and 1989–1990 eruptions (Trabant et al. 1994). Pyroclastic density currents can also entrain ice and produce lahars as they progress downvalley. Tephra deposited on the glacier surface during eruptions can increase glacier melt by reducing the ice surface albedo. However, if the tephra layer is thick enough, the underlying ice will be protected and enable glacier advance (Kirkbride and Dugmore 2003; Nield et al. 2013).

The primary impact of volcanic activity on glaciers is melting, and therefore floods (including jökulhlaups) and lahars are common. Meltwater can drain supraglacially, eroding deep surface channels, or subglacially by either distributed or channelised flow (Smellie 2023). Outburst floods or jökulhlaups result from the release of meltwater

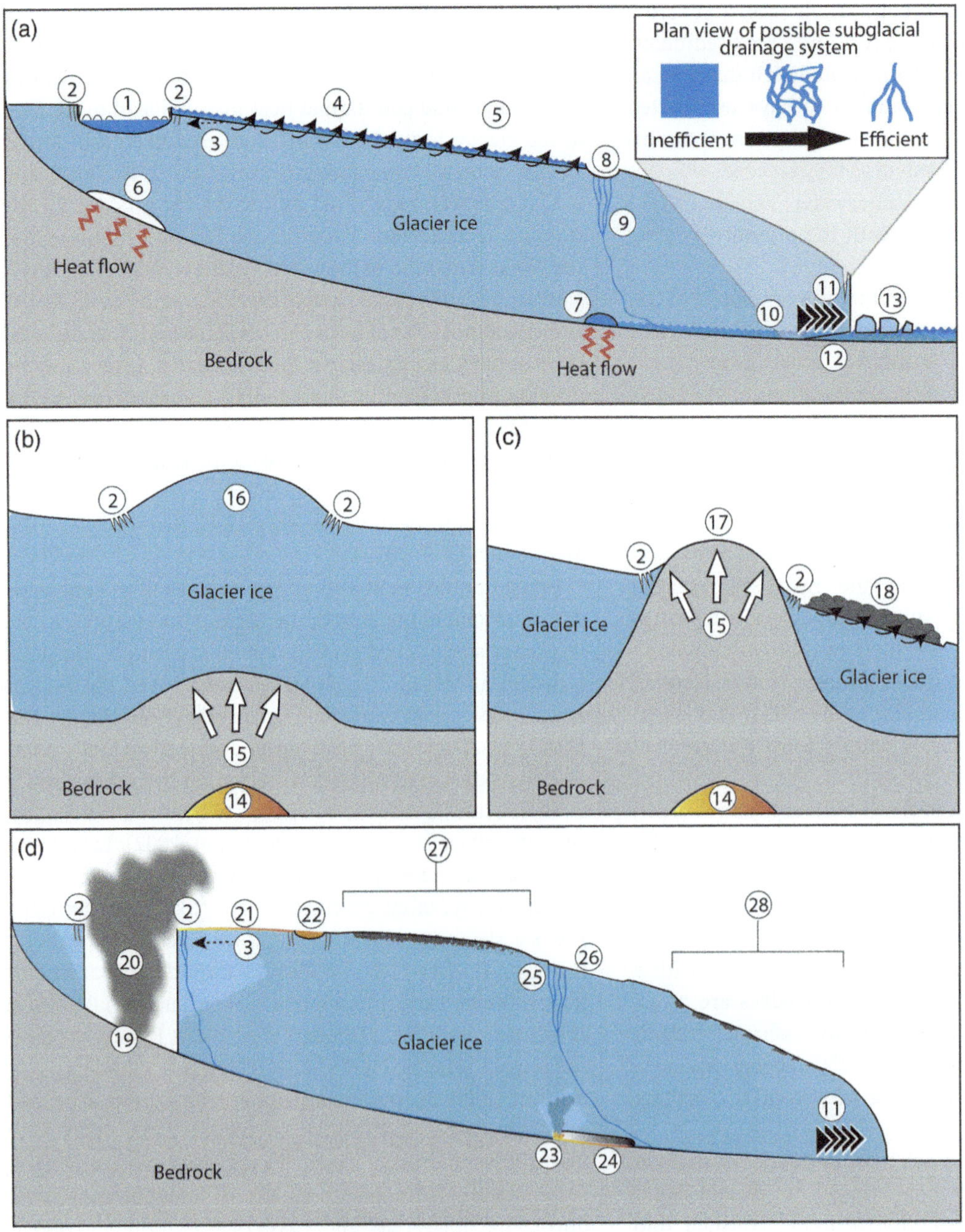

Figure 14.42 Schematic figure showing the impacts of different forms of volcanic activity on glacier behaviour. Forms of volcanic activity include (a) enhanced subglacial heat flow and flood/lahars; (b) subglacial volcanic dome growth; (c) volcanic dome extrusion and a pyroclastic density current; and (d) a subglacial volcanic eruption, lava flows and supraglacial tephra deposition. Numbers in the figure refer to different processes or features. (1) Lake-filled supraglacial cauldron, with ice breaking from vertical cauldron walls. (2) Zones of crevassing and fracturing. (3) Local reversal in ice flow direction. (4) Supraglacial flood (formed by lake drainage) transitioning to a lahar. (5) Supraglacial channel/canyon formed by ice melt, erosion and entrainment (by a flood/lahar). (6) Subglacial cavity caused by enhanced heat flow. (7) Subglacial lake formed by enhanced heat flow. (8) Supraglacial depression formed above a subglacial lake. (9) Crevasses (above a subglacial lake) acting as a route for supraglacial meltwater drainage to the bed. (10) Subglacial meltwater drainage with inset plan view of different possible drainage styles. (11) Glacier advance/acceleration due to subglacial meltwater drainage (more widespread if drainage is inefficient). (12) Glacier front uplifted by subglacial flooding. (13) Ice blocks torn from a glacier front by subglacial flooding. (14) Magma upwelling. (15) Subglacial dome growth. (16) Glacier doming. (17) Dome extrusion through a glacier. (18) Pyroclastic density current (due to dome collapse) and associated supraglacial channel (formed by ice melt, erosion and entrainment). (19) Subglacial eruption. (20) Ice crater. (21) Supraglacial lava flow and associated melt channel. (22) Supraglacial lava ponding and associated crevasse bounded melt pit. (23) Site of small and/or phreatic eruption (with lava fountain). (24) Subglacial lava flow and associated melt cavity. (25) Crevasses (above the site of a small subglacial eruption) acting as a route for supraglacial meltwater drainage to the bed. (26) Supraglacial depression formed above a subglacial lava flow melt cavity. (27) Reduced melt under thick and/or continuous supraglacial tephra. (28) Increased melt under thin and/or discontinuous supraglacial tephra. From Barr et al. (2018, p. 189) / with permission of Elsevier.

within glacial lakes and have the potential to cause significant landscape change, eroding large canyons and depositing vast quantities of glaciogenic debris and ice across the glacier foreland (Björnsson 2003). Conversely, lahars contain higher debris concentrations and are usually triggered by more continuous ice or snow melt. They are primarily caused by pyroclastic density currents during eruptions, and lava dome collapse which will trigger extensive melting as the hot rock comes into contact with snow and ice (Smellie 2023). Outburst floods can also evolve into lahars as debris content increases (Barr et al. 2018).

14.5.12.2 Lithofacies and Sediment-Landform Assemblages

Stratigraphic sequences within a glaciovolcanic landsystem reflect the interplay between glaciogenic and volcanic deposition. Due to the distinctive nature of glaciovolcanic deposits, they can be used to (i) establish whether glacier ice was present during an eruption, (ii) estimate the thickness of overlying ice and (iii) date the timing of glaciation by directly dating the volcanic rocks (Wilson et al. 2019; e.g. Hambrey et al. 2008; Smellie et al. 2008; Edwards et al. 2011; Russell et al. 2014; Wilson and Russell 2018; Smellie 2018).

At ice-clad volcanoes, explosive eruptions are largely unimpeded and meltwater may be able to drain freely, resulting in the formation of pyroclastic cones (e.g. Magnússon et al. 2012). In contrast, beneath ice masses that are over 200 m thick, explosive eruptions are mainly suppressed and melting creates a subglacial water reservoir (Smellie 2023). Lava is therefore deposited subaqueously. Hydrovolcanic activity results in the construction of tuff cones, tindar (ridges of volcanic rock), subaqueous lava effusions, hyaloclastites, pillow mounds and tuyas (Smellie 2023). In Iceland, many plateau icefields now exist on tuyas (flat-topped volcanoes) that were formed beneath the Icelandic Ice Sheet during the last glaciation (Evans et al. 2006b). In Antarctica, Mount Haddington on James Ross Island is an example of a polygenetic tuya, constructed during numerous eruptions (Smellie et al. 2008).

If subaerial lava flows have been in direct contact with ice, evidence for excessive cooling should be expected. This includes a glassy outer shell containing polygonal (column-forming), sheet-like, psuedopillow, or hackly fractures, in contrast to a crystalline flow interior with platy or broad polygonal fractures (Lescinsky and Fink 2000; Conway et al. 2015). Thin pyroclastic deposits such as palagonite breccia and glassy vesicular cinder would also be expected below basal lavas of a volcanic group, caused by explosive contact between the lava and ice (Downie 1964).

Stratigraphic sequences containing glacial tills, outwash sands and gravels and lava flows have been documented at a number of locations (e.g. Downie 1964; Carrivick et al. 2004a). For example, an exposed section at Kverkfjallarani, Iceland, shows jökulhlaup deposits beneath predominantly subaerial lava flow deposits (Carrivick et al. 2004a). Where lava flows are found intercalated with morainic deposits or subglacial tills, dating of lava flows can enable the timing of glacial events to be constrained, for example, at Kilimanjaro (Downie 1964). Tephra layers overlying, within, or between moraines also provide valuable age markers for constraining the age of glacial advances. For example, the use of tephrostratigraphy at two northern outlets of the Eyjafjallajökull ice cap, Iceland, enabled Kirkbride and Dugmore (2008) to identify multiple glacial advances over the last 2000 years.

Glacial lake outburst floods deliver large quantities of debris to the glacier foreland, and jökulhlaup deposits on Icelandic sandar have been studied extensively (Russell and Knudsen 1999; Roberts et al. 2001; Carrivick et al. 2004a, 2004b; Duller et al. 2008; Harrison et al. 2022). The sediments deposited are usually polymict by nature, reflecting the range of debris that is entrained en route (Smellie 2023). A distinctive facies succession also develops that is linked to the jökulhlaup hydrograph shape and available sediment supply (Harrison et al. 2022). This makes it possible to distinguish jökulhlaup-dominated sandar from sandar that develop predominantly from lower magnitude seasonal meltwater (e.g. Marren and Schuh 2009; Section 14.5.1). Key sedimentary facies associated with jökulhlaups include hyperconcentrated flow deposits, which contain too high a volume of sediment to be deposited by 'normal' flow, thick upwards coarsening sediment sequences accreted during the rising stage of an outburst flood, and large-scale cross-stratified gravel deposited within dunes and pendant bars (Maizels 1993; Marren 2002; Marren and Schuh 2009). Other less distinctive indicators include cross-stratified gravel with reactivation surfaces, laterally extensive erosion surfaces and consistent palaeoflow indicators (Marren 2002; Marren and Schuh 2009). Ice block obstacle marks can also indicate a previous jökulhlaup if they are of sufficient quantity and size (Maizels 1992; Marren and Schuh 2009).

14.5.12.3 Example of a Glaciovolcanic Landsystem

Many glaciers in Iceland are underlain by volcanoes. The largest ice cap, Vatnajökull, sits on top of five volcanic systems (Björnsson 2003) and high geothermal heat fluxes from these volcanoes lead to continuous melt at the glacier bed, producing subglacial lakes (Benn and Evans 2010). The best known of these lakes is Grímsvötn, which empties relatively regularly at intervals of 1–10 years (Björnsson 2003) by jökulhlaup. The water drains beneath Skeiðarárjökull and emerges onto Skeiðarársandar (Gudmundsson et al. 1997). The impact of the 1996 jökulhlaup from Grímsvötn on Skeiðarársandar has been particularly well-studied (e.g. Russell et al. 1999; Russell and Knudsen 1999; Roberts et al. 2001). Following this jökulhlaup, Russell and Knudsen (1999) recorded upward coarsening sediment successions, composed of sands, gravels, boulders and blocks of ice, within sections through ice-proximal fans on Skeiðarársandar. The upwards coarsening sequence indicates the rising limb of the flood hydrograph. Evidence for hyperconcentrated flow and river-bed scour was also identified. Waning stage deposits were characterised by sandy climbing-ripple sequences, suggesting rapid sedimentation. Russell and Knudsen (1999) argued that the sedimentological characteristics of the 1996 jökulhlaup deposits were controlled by high sediment concentrations within the flood waters, sediment reworking during the waning stage, large-scale channel geometry, the distribution of ice blocks brought down by the flood and the flood duration.

14.6 Conclusions

Glacial landscapes are some of the most diverse on Earth. Glacier ice behaves very differently in different climates or environments since this affects the glacier thermal regime. The thermal regime, or the temperature of the ice, is a first-order control on ice dynamics, including glacier erosion, glacier flow, and depositional processes and landforms. Second-order controls include topography, which can control ice thickness and direction of flow, and bedrock geology, which controls how easily rocks are eroded. Because glaciers occur in a wide variety of climates and topographies, this results in a diverse range of glacial landforms. We can group together associations or assemblages of landforms, and their sediments, into characteristic packages, or 'landsystems'. In this chapter, we introduce some key landsystems, but there are far more than can be described here. By linking diagnostic sediment-landform assemblages into characteristic landsystems, we can say something about the style and manner of glaciation, even if the glaciers are no longer extant.

Regardless of this diversity, some key features occur across multiple landsystems. Characteristic indications of glaciation in sediments include diamictons with embedded faceted, edge-rounded and striated clasts, with 'erratics' that have been transported beyond their catchment. Where erosion and basal sliding are active, polished and faceted bedrock and boulders are typical, along with scratches, or striations, that can differentiate from slickensides. For large ice sheets or in polar regions, (outlet) glaciers may reach the glaciomarine realm. Here, typical indicative glaciomarine sediments and lithofacies include sedimentary laminations, dropstones, iceberg dump structures and associations of cold-water micro and macro fauna. In contrast, glaciolacustrine settings are typically barren and may exhibit annual laminations. A suite of sediments is usually insufficient on its own to diagnose a landsystem however; typically this must also include reference to the wider scale geomorphology and topographic context, if known (e.g. for Late Pleistocene deposits).

Further Reading

Barr, I.D., Lynch, C.M., Mullan, D. et al. (2018). Volcanic impacts on modern glaciers: a global synthesis. *Earth-Science Reviews* 182: 186–203.

Benn, D.I. and Evans, D.J.A. (2010). *Glaciers & Glaciation*. London: Hodder Education.

Benn, D.I. and Owen, L.A. (2002). Himalayan glacial sedimentary environments: a framework for reconstructing and dating the former extent of glaciers in high mountains. *Quaternary International* 97–98: 3–25.

Chandler, B.M., Chandler, S.J., Evans, D.J. et al. (2020a). Sub-annual moraine formation at an active temperate Icelandic glacier. *Earth Surface Processes and Landforms* 45 (7): 1622–1643.

Clark, C.D., Ely, J.C., Hindmarsh, R.C. et al. (2022). Growth and retreat of the last British–Irish Ice Sheet, 31000 to 15000 years ago: the BRITICE-CHRONO reconstruction. *Boreas* 51 (4): 699–758.

Ó Cofaigh, O.C., Callard, S.L., Roberts, D. H. et al. (2021). Timing and pace of ice-sheet withdrawal across the marine–terrestrial transition west of Ireland during the last glaciation. *Journal of Quaternary Science* 36: 805–832.

Cuffey, K.M. and Paterson, W.S.B. (2010). *The Physics of Glaciers*, 4e. Academic Press.

Davies, B.J., Darvill, C.M., Lovell, H. et al. (2020). The evolution of the Patagonian Ice Sheet from 35 ka to the present day (PATICE). *Earth-Science Reviews* 204: 103152.

Evans, D.J.A. (ed.) (2003). *Glacial Landsystems*. Arnold.

Evans, D.J.A., Phillips, E.R., Hiemstra, J.F., and Auton, C.A. (2006a). Subglacial till: formation, sedimentary characteristics and classification. *Earth-Science Reviews* 78: 115–176.

Graham, A.G.C., Jakobsson, M., Nitsche, F.O. et al. (2016). Submarine glacial-landform distribution across the West Antarctic margin, from grounding line to slope: the Pine Island–Thwaites ice-stream system. *Geological Society, London, Memoirs* 46: 493–500.

Hambrey, M.J. and Fitzsimons, S.J. (2010). Development of sediment-landform associations at cold glacier margins, Dry Valleys, Antarctica. *Sedimentology* 57: 857–882.

Hambrey, M.J. and Glasser, N.F. (2012). Discriminating glacier thermal and dynamic regimes in the sedimentary record. *Sedimentary Geology* 251–252: 1–33.

Jones, R.S., Lowe, J.J., Palmer, A.P. et al. (2017). Dynamics and palaeoclimatic significance of a Loch Lomond Stadial glacier: Coire Ardair, Creag Meagaidh, Western Highlands, Scotland. *Proceedings of the Geologists Association* 128: 54–66. https://doi.org/10.1016/j.pgeola.2015.11.004.

Lovell, H., Benn, D.I., Lukas, S. et al. (2018a). Multiple Late Holocene surges of a High-Arctic tidewater glacier system in Svalbard. *Quaternary Science Reviews* 201: 162–185.

Margold, M., Stokes, C.R., and Clark, C.D. (2018). Reconciling records of ice streaming and ice margin retreat to produce a palaeogeographic reconstruction of the deglaciation of the Laurentide Ice Sheet. *Quaternary Science Reviews* 189: 1–30.

Menzies, J. and van der Meer, J.J.M. (ed.) (2018). *Past Glacial Environments*, 2e. Elsevier.

Thorndycraft, V.R., Bendle, J.M., Benito, G. et al. (2019). Glacial lake evolution and Atlantic-Pacific drainage reversals during deglaciation of the Patagonian Ice Sheet. *Quaternary Science Reviews* 203: https://doi.org/10.1016/j.quascirev.2018.10.036.

Waller, R.I., Murton, J.B., and Kristensen, L. (2012). Glacier–permafrost interactions: processes, products and glaciological implications. *Sedimentary Geology* 255–256: 1–28.

Weber, P., Boston, C.M., Lovell, H., and Andreassen, L.M. (2019). Evolution of the Norwegian plateau icefield Hardangerjøkulen since the 'Little Ice Age'. *The Holocene* 29 (12): 1885–1905.

References

Aber, J.S., Croot, D.G., and Fenton, M.M. (1989). *Glaciotectonic Landforms and Structures*, 201. Boston: Kluwer Academic.

Åkesson, H., Nisancioglu, K.H., Giesen, R.H., and Morlighem, M. (2017). Simulating the evolution of Hardangerjøkulen ice cap in southern Norway since the mid-Holocene and its sensitivity to climate change. *The Cryosphere* 11 (1): 281–302.

Ali, D.O., Spencer, A.M., Fairchild, I.J. et al. (2018). Indicators of relative completeness of the glacial record of the Port Askaig Formation, Garvellach Islands, Scotland. *Precambrian Research* 319: 69–78. https://doi.org/10.1016/j.precamres.2017.12.005.

Alley, R. et al. (2019). Glacial erosion: status and outlook. *Annals of Glaciology* 60 (80): 1–13.

Alley, R.B., Anandakrishnan, S., Bentley, C.R., and Lord, N. (1994). A water-piracy hypothesis for the stagnation of Ice Stream C, Antarctica. *Annals of Glaciology* 20: 187–194.

Alley, R.B., Blankenship, D.D., Bentley, C.R., and Rooney, S.T. (1986). Deformation of till beneath ice stream B West Antarctica. *Nature* 322 (6074): 57–59.

Alley, R.B., Blankenship, D.D., Bentley, C.R., and Rooney, S.T. (1987). Till beneath ice stream B: 3. Till deformation: evidence and implications. *Journal of Geophysical Research: Solid Earth* 92 (B9): 8921–8929.

Alley, R.B., Cuffey, K.M., Evenson, E.B. et al. (1997). How glaciers entrain and transport basal sediment: physical constraints. *Quaternary Science Reviews* 16 (9): 1017–1038.

Alley, R.B., Lawson, D.E., Evenson, E.B. et al. (1998). Glaciohydraulic supercooling: a freeze-on mechanism to create stratified, debris-rich basal ice: II. Theory. *Journal of Glaciology* 44 (148): 563–569.

Anandakrishnan, S. and Alley, R.B. (1997). Stagnation of ice stream C, West Antarctica by water piracy. *Geophysical Research Letters* 24 (3): 265–268.

Anderson, J.B., Wamy, S., Askain, R.A. et al. (2011). Progressive Cenozoic cooling and the demise of Antarctica's last refugium. *Proceedings of the National Academy of Sciences* 108: 11356–11360.

Anderson, R.S., Anderson, L.S., Armstrong, W.H. et al. (2018). Glaciation of alpine valleys: the glacier – debris-covered glacier – rock glacier continuum. *Geomorphology* 311: 127–142. https://doi.org/10.1016/j.geomorph.2018.03.015.

Ankerstjerne, S., Iverson, N.R., and Lagroix, F. (2015). Origin of a washboard moraine of the Des Moines Lobe inferred from sediment properties. *Geomorphology* 248: 452–463.

Aschwanden, A. et al. (2012). An enthalpy formulation for glaciers and ice sheets. *Journal of Glaciology* 58 (209): 441–457.

Atkins, C.B. (2013). Geomorphological evidence of cold-based glacier activity in South Victoria Land, Antarctica. *Geological Society, London, Special Publications* 381: 299–318.

Atkins, C.B., Barrett, P.J., and Hicock, S.R. (2002). Cold glaciers erode and deposit: evidence from Allan Hills, Antarctica. *Geology* 30: 659–662.

Ballantyne, C.K. (1997). Periglacial trimlines in the Scottish Highlands. *Quaternary International* 38: 119–136.

Ballantyne, C.K. (2018a). After the ice: Lateglacial and Holocene landforms and landscape evolution in Scotland. *Earth and Environmental Science Transactions of the Royal Society of Edinburgh* 1–39. https://doi.org/10.1017/S175569101800004X.

Ballantyne, C.K. (2018b). *Periglacial Geomorphology*. Chichester: Wiley.

Ballantyne, C.K. and Benn, D.I. (1994). Paraglacial slope adjustment and resedimentation following recent glacier retreat Fåbergstølsdalen. *Arctic and Alpine Research* 26: 255–269. https://doi.org/10.2307/1551938.

Ballantyne, C.K. and Stone, J.O. (2015). Trimlines, blockfields and the vertical extent of the last ice sheet in southern Ireland. *Boreas* 44 (2): 277–287.

Bamber, J.L., Westaway, R.M., Marzeion, B., and Wouters, B. (2018). The land ice contribution to sea level during the satellite era. *Environmental Research Letters* 13: 99502.

Barcheck, C.G., Schwartz, S.Y., and Tulaczyk, S. (2020). Icequake streaks linked to potential mega-scale glacial lineations beneath an Antarctic ice stream. *Geology* 48 (2): 99–102.

Barr, I.D., Lynch, C.M., Mullan, D. et al. (2018). Volcanic impacts on modern glaciers: a global synthesis. *Earth-Science Reviews* 182: 186–203.

Batchelor, C.L., Christie, F.D.W., Ottesen, D. et al. (2023). Rapid, buoyancy-driven ice-sheet retreat of hundreds of metres per day. *Nature* 617: 105–110.

Batchelor, C.L. and Dowdeswell, J.A. (2014). The physiography of High Arctic cross-shelf troughs. *Quaternary Science Reviews* 92: 68–96.

Batchelor, C.L. and Dowdeswell, J.A. (2015). Ice-sheet grounding-zone wedges (GZWs) on high-latitude continental margins. *Marine Geology* 363: 65–92. https://doi.org/10.1016/j.margeo.2015.02.001.

Batchelor, C.L., Dowdeswell, J.A., and Ottesen, D. (2018). Submarine glacial landforms. In: *Submarine Geomorphology* (ed. A. Micallef, S. Krastel, and A. Savini), 207–234. Cham: Springer International Publishing https://doi.org/10.1007/978-3-319-57852-1_12.

Batchelor, C.L., Margold, M., Krapp, M. et al. (2019). The configuration of Northern Hemisphere ice sheets through the Quaternary. *Nature Communications* 10: 3713. https://doi.org/10.1038/s41467-019-11601-2.

Batchelor, C.L., Montelli, A., Ottesen, D. et al. (2020). New insights into the formation of submarine glacial landforms from high-resolution Autonomous Underwater Vehicle data. *Geomorphology* 107396. https://doi.org/10.1016/j.geomorph.2020.107396.

Bell, C.M. (2008). Punctuated drainage of an ice-dammed Quaternary lake in Southern South America. *Geografiska Annaler Series A Physical Geography* 90: 1–17.

Bellwald, B., Planke, S., Becker, L.W.M., and Myklebust, R. (2020). Meltwater sediment transport as the dominating process in mid-latitude trough mouth fan formation. *Nature Communications* 11: https://doi.org/10.1038/s41467-020-18337-4.

Bendle, J.M., Palmer, A.P., Thorndycraft, V.R., and Matthews, I.P. (2017a). High-resolution chronology for deglaciation of the Patagonian Ice Sheet at Lago Buenos Aires (46.5°S) revealed through varve chronology and Bayesian age modelling. *Quaternary Science Reviews* 177: 314–339. https://doi.org/10.1016/j.quascirev.2017.10.013.

Bendle, J.M., Palmer, A.P., Thorndycraft, V.R., and Matthews, I.P. (2019). Phased Patagonian Ice Sheet response to Southern Hemisphere atmospheric and oceanic warming between 18 and 17 ka. *Scientific Reports* 9: 4133. https://doi.org/10.1038/s41598-019-39750-w.

Bendle, J.M., Thorndycraft, V.R., and Palmer, A.P. (2017b). The glacial geomorphology of the Lago Buenos Aires and Lago Pueyrredón ice lobes of central Patagonia. *Journal of Maps* 13: 654–673.

Benn, D.I. (1994). Fluted moraine formation and till genesis below a temperate valley glacier: Slettmarkbreen, Jotunheimen, southern Norway. *Sedimentology* 41 (2): 279–292.

Benn, D.I. (1995). Fabric signature of subglacial till deformation, Breidamerkurjökull, Iceland. *Sedimentology* 42 (5): 735–747.

Benn, D.I. (2004). Clast morphology. In: *A Practical Guide to the Study of Glacial Sediments* (ed. D.J.A. Evans and D.I. Benn), 77–92. London: Arnold.

Benn, D.I. (2021). Surging glaciers in Scotland. *Scottish Geographical Journal* 137 (1–4): 1–40.

Benn, D.I. and Ballantyne, C.K. (1994). Reconstructing the transport history of glaciogenic sediments: a new approach based on the co-variance of clast form indices. *Sedimentary Geology* 91: 215–227.

Benn, D.I., Bolch, T., Hands, K. et al. (2012). Response of debris-covered glaciers in the Mount Everest region to recent warming, and implications for outburst flood hazards. *Earth-Science Reviews* 114: 156–174. https://doi.org/10.1016/j.earscirev.2012.03.008.

Benn, D.I. and Evans, D.J.A. (1998). *Glaciers and Glaciation*. London: Arnold.

Benn, D.I. and Evans, D.J.A. (2010). *Glaciers & Glaciation*. London: Hodder Education.

Benn, D.I., Fowler, A.C., Hewitt, I., and Sevestre, H. (2019). A general theory of glacier surges. *Journal of Glaciology* 65 (253): 701–716.

Benn, D.I., Hulton, N.R.J., and Mottram, R.H. (2007). "Calving laws", "sliding laws" and the stability of tidewater glaciers. *Annals of Glaciology* 46: 123–130.

Benn, D.I., Kirkbride, M.P., Owen, L.A., and Brazier, V. (2003). Glaciated valley landsystems. In: *Glacial Landsystems* (ed. D.J.A. Evans), 372–406. London: Arnold.

Benn, D.I. and Owen, L.A. (2002). Himalayan glacial sedimentary environments: a framework for reconstructing and dating the former extent of glaciers in high mountains. *Quaternary International* 97–98: 3–25.

Bennett, M.R. (2003). Ice streams as the arteries of an ice sheet: their mechanics, stability and significance. *Earth-Science Reviews* 61: 309–339. https://doi.org/10.1016/S0012-8252(02)00130-7.

Bennett, M.R. and Glasser, N.F. (2009). *Glacial Geology: Ice sheets and landforms*, 2nde. Wiley-Blackwell: Chichester.

Bennett, M.R., Hambrey, M.J., Huddart, D., and Ghienne, J.F. (1996). The formation of a geometrical ridge network by the surge-type glacier Kongsvegen, Svalbard. *Journal of Quaternary Science: Published for the Quaternary Research Association* 11 (6): 437–449.

Bennett, M.R., Hambrey, M.J., Huddart, D. et al. (1999). The landform and sediment assemblage produced by a tidewater glacier surge in Kongsfjorden,

Svalbard. *Quaternary Science Reviews* 18 (10–11): 1213–1246.

Bennett, M.R., Huddart, D., and Thomas, G.S.P. (2002). Facies architecture within a regional glaciolacustrine basin: Copper River, Alaska. *Quaternary Science Reviews* 21: 2237–2279.

Bennett, M.R., Waller, R.I., Midgley, N.G. et al. (2003). Subglacial deformation at sub-freezing temperatures? Evidence from Hagafellsjökull-Eystri, Iceland. *Quaternary Science Reviews* 22: 915–923.

Bentley, M.J., Cofaigh, O.C., Anderson, J.B. et al. (2014). A community-based geological reconstruction of Antarctic Ice Sheet deglaciation since the Last Glacial Maximum. *Quaternary Science Reviews* 100: https://doi.org/10.1016/j.quascirev.2014.06.025.

Bestmann, M., Rice, A.H.N., Langenhorst, F. et al. (2006). Subglacial bedrock welding associated with glacial earthquakes. *Journal of the Geological Society* 163: 417–420.

Bhambri, R., Watson, C.S., Hewitt, K. et al. (2020). The hazardous 2017–2019 surge and river damming by Shispare Glacier, Karakoram. *Scientific Reports* 10: 1–14.

Bickerdike, H.L., Cofaigh, O.C., Evans, D.J.A., and Stokes, C.R. (2018). Glacial landsystems, retreat dynamics and controls on Loch Lomond Stadial (Younger Dryas) glaciation in Britain. *Boreas* 47: 202–224. https://doi.org/10.1111/bor.12259.

Bickerdike, H.L., Evans, D.J.A., Cofaigh, Ó.C., and Stokes, C.R. (2016). The glacial geomorphology of the Loch Lomond Stadial in Britain: a map and geographic information system resource of published evidence. *Journal of Maps* 12: 1178–1186. https://doi.org/10.1080/17445647.2016.1145149.

Björnsson, H. (1992). Jökulhlaups in Iceland: prediction, characteristics and simulation. *Annals of Glaciology* 16: 95–106.

Björnsson, H. (2003). Subglacial lakes and jökulhlaups in Iceland. *Global and Planetary Change* 35 (3–4): 255–271.

Blankenship, D.D., Bell, R.E., Hodge, S.M. et al. (1993). Active volcanism beneath the West Antarctic ice sheet and implications for ice-sheet stability. *Nature* 361 (6412): 526–529.

Blankenship, D.D., Bentley, C.R., Rooney, S.T., and Alley, R.B. (1986). Seismic measurements reveal a saturated porous layer beneath an active Antarctic ice stream. *Nature* 322 (6074): 54–57.

Blankenship, D.D. et al. (1987). Till beneath Ice Stream B: 1. Properties derived from seismic travel times. *Journal of Geophysical Research: Solid Earth* 92 (B9): 8903–8911.

Bleick, H.A., Coombs, M.L., Cervelli, P.F. et al. (2013). Volcano–ice interactions precursory to the 2009 eruption of Redoubt Volcano, Alaska. *Journal of Volcanology and Geothermal Research* 259: 373–388.

Blignault, H.J. and Theron, J.N. (2017). Diapirism and the Fold Zone controversy of the Ordovician glaciomarine Pakhuis Formation, South Africa. *South African Journal of Geology* 120: 209–222.

Boothroyd, J.C. and Ashley, G.M. (1975). Processes, bar morphology, and sedimentary structures on braided outwash fans, NE Gulf of Alaska. In: *Glaciofluvial and Glaciolacustrine Sedimentation*, vol. 23 (ed. A.V. Jopling and B.C. McDonald), 193–222. Society of Economic Paleontologists and Mineralogists Special Publication.

Boothroyd, J.C. and Nummedal, D. (1978). Proglacial braided outwash: a model for humid alluvial-fan deposits. *Memoir - Canadian Society of Petroleum Geologists* 5: 641–668.

Boston, C.M. (2012). A glacial geomorphological map of the Monadhliath Mountains, Central Scottish Highlands. *Journal of Maps* 8 (4): 437–444.

Boston, C.M., Chandler, B.M., Lovell, H. et al. (2023). The role of topography in landform development at an active temperate glacier in Arctic Norway. *Earth Surface Processes and Landforms* .

Boston, C.M., Evans, D.J., and Cofaigh, C. Ó. (2010). Styles of till deposition at the margin of the Last Glacial Maximum North Sea lobe of the British–Irish Ice Sheet: an assessment based on geochemical properties of glacigenic deposits in eastern England. *Quaternary Science Reviews* 29 (23–24): 3184–3211.

Boston, C.M. and Lukas, S. (2019). Topographic controls on plateau icefield recession: insights from the Younger Dryas Monadhliath Icefield, Scotland. *Journal of Quaternary Science* 34: 433–451.

Boston, C.M., Lukas, S., and Carr, S.J. (2015). A Younger Dryas plateau icefield in the Monadhliath, Scotland, and implications for regional palaeoclimate. *Quaternary Science Reviews* 108: 139–162.

Boulton, G.S. (1972). The role of thermal régime in glacial sedimentation. In: *Polar Geomorphology*, vol. 4 (ed. R.J. Price and D.E. Sugden), 1–19. IBG Special Publication.

Boulton, G.S. (1976). The origin of glacially fluted surfaces-observations and theory. *Journal of Glaciology* 17 (76): 287–309.

Boulton, G.S. (1996). Theory of glacial erosion, transport and deposition as a consequence of subglacial sediment deformation. *Journal of Glaciology* 42 (140): 43–62.

Boulton, G.S., Dobbie, K.E., and Zatsepin, S. (2001). Sediment deformation beneath glaciers and its coupling to the subglacial hydraulic system. *Quaternary International* 86 (1): 3–28.

Boulton, G.S. and Eyles, N. (1979). Sedimentation by valley glaciers: a model and genetic classification. In: *Moraines and Varves* (ed. C. Schluchter), 11–23. Rotterdam: Balkema.

Boulton, G.S., Hagdorn, M., Maillot, P.B., and Zatsepin, S. (2009). Drainage beneath ice sheets: groundwater–channel coupling, and the origin of esker systems from former ice sheets. *Quaternary Science Reviews* 28 (7–8): 621–638.

Boulton, G.S., van der Meer, J.J.M., Beets, D.J. et al. (1999). The sedimentary and structural evolution of a recent push moraine complex: Holmstrømbreen, Spitsbergen. *Quaternary Science Reviews* 18: 339–371.

Braakman, J.H., Levell, B.K., Rutten, K.W. et al. (1982). Late Palaeozoic Gondwana glaciation in Oman. *Nature* 299: 48–50.

Bradwell, T., Stoker, M.S., Golledge, N.R. et al. (2008). The northern sector of the last British Ice Sheet: maximum extent and demise. *Earth Science Reviews* 88: 207–226.

Bradwell, T. et al. (2021). Pattern, style and timing of British–Irish Ice Sheet retreat: Shetland and northern North Sea sector. *Journal of Quaternary Science* 36 (5): 681–722.

Brennand, T.A. (2000). Deglacial meltwater drainage and glaciodynamics: inferences from Laurentide eskers, Canada. *Geomorphology* 32 (3-4): 263–293.

Brugman, Melinda M., and Austin Post (1981). *Effects of volcanism on the glaciers of Mount St. Helens*. No. 850. US Department of the Interior, Geological Survey.

Busfield, M.E., Le Heron, D.P. (2016). A Neoproterozoic ice advance sequence, Sperry Wash, California. *Sedimentology* 63: 307–330.

Caldenius, C.C. (1932). Las glaciaciones cuaternarias en la Patagonia y Tierra del Fuego. *Geografiska Annaler* 14: 1–164.

Carrivick, J.L., Andreassen, L.M., Nesje, A., and Yde, J.C. (2022). A reconstruction of Jostedalsbreen during the Little Ice Age and geometric changes to outlet glaciers since then. *Quaternary Science Reviews* 284: 107501.

Carrivick, J.L., Russell, A.J., and Tweed, F.S. (2004b). Geomorphological evidence for jökulhlaups from Kverkfjöll volcano, Iceland. *Geomorphology* 63 (1–2, 102): 81.

Carrivick, J.L., Russell, A.J., Tweed, F.S., and Twigg, D. (2004a). Palaeohydrology and sedimentary impacts of jökulhlaups from Kverkfjöll, Iceland. *Sedimentary Geology* 172 (1–2): 19–40.

Carrivick, J.L. and Tweed, F.S. (2013). Proglacial lakes: character, behaviour and geological importance. *Quaternary Science Reviews* 78: 34–52. https://doi.org/10.1016/j.quascirev.2013.07.028.

Carrivick, J.L. and Tweed, F.S. (2016). A global assessment of the societal impacts of glacier outburst floods. *Global and Planetary Change* 144: 1–16.

Carrivick, J.L. et al. (2023). Cooling glaciers in a warming climate since the Little Ice Age at Qaanaaq, northwest Kalaallit Nunaat (Greenland). *Earth Surface Processes and Landforms* 48 (13): 2446–2462.

Chandler, B.M., Chandler, S.J., Evans, D.J. et al. (2020a). Sub-annual moraine formation at an active temperate Icelandic glacier. *Earth Surface Processes and Landforms* 45 (7): 1622–1643.

Chandler, B.M., Evans, D.J., Roberts, D.H. et al. (2016b). Glacial geomorphology of the Skálafellsjökull foreland, Iceland: a case study of 'annual' moraines. *Journal of Maps* 12 (5): 904–916.

Chandler, B.M., Evans, D.J.A., and Roberts, D.H. (2016a). Characteristics of recessional moraines at a temperate glacier in SE Iceland: insights into patterns, rates and drivers of glacier retreat. *Quaternary Science Reviews* 135: 171–205.

Chandler, B.M., Lukas, S., and Boston, C.M. (2020b). Processes of 'hummocky moraine' formation in the Gaick, Scotland: insights into the ice-marginal dynamics of a Younger Dryas plateau icefield. *Boreas* 49 (2): 248–268.

Chandler, B.M.P., Boston, C.M., and Lukas, S. (2019). A spatially-restricted Younger Dryas plateau icefield in the Gaick, Scotland: reconstruction and palaeoclimatic implications. *Quaternary Science Reviews* 211: 107–135.

Chandler, B.M.P., Evans, D.J.A., Chandler, S.J.P. et al. (2020c). The glacial landsystem of Fjallsjökull, Iceland: spatial and temporal evolution of process-form regimes at an active temperate glacier. *Geomorphology* 361: 107192.

Church, M. and R. Gilbert (1975). Proglacial fluvial and lacustrine environments.

Clark, C.D., Ely, J.C., Hindmarsh, R.C. et al. (2022). Growth and retreat of the last British–Irish Ice Sheet, 31000 to 15000 years ago: the BRITICE-CHRONO reconstruction. *Boreas* 51 (4): 699–758.

Clark, C.D., Ely, J.C., Spagnolo, M. et al. (2018). Spatial organization of drumlins. *Earth Surface Processes and Landforms* 43 (2): 499–513.

Clark, P.U., Dyke, A.S., Shakun, J.D. et al. (2009). The last glacial maximum. *Science* 325: 710–714.

Clark, P.U. and Walder, J.S. (1994). Subglacial drainage, eskers, and deforming beds beneath the Laurentide and Eurasian ice sheets. *Geological Society of America Bulletin* 106 (2): 304–314.

Clayton, L., Teller, J.T., and Attig, J.W. (1985). Surging of the southwestern part of the Laurentide Ice Sheet. *Boreas* 14 (3): 235–241.

Ó Cofaigh, Callard, S.L., Roberts, D.H. et al. (2021). Timing and pace of ice-sheet withdrawal across the marine–terrestrial transition west of Ireland during the last glaciation. *Journal of Quaternary Science* 36: 805–832.

Cogley, J.G., Hock, R., Rasmussen, L.A. et al. (2011). *Glossary of Glacier Mass Balance and Related Terms*, IHP-VII Technical Documents in Hydrology No. 86, IACS Contribution No. 2. Paris: UNESCO-IHP.

Cohen, D., Hooyer, T.S., Iverson, N.R. et al. (2006). Role of transient water pressure in quarrying: A subglacial experiment using acoustic emissions. *Journal of Geophysical Research: Earth Surface* 111: https://doi.org/10.1029/2005JF000439.

Conway, C.E., Townsend, D.B., Leonard, G.S. et al. (2015). Lava-ice interaction on a large composite volcano: a case study from Ruapehu, New Zealand. *Bulletin of Volcanology* 77: 1–18.

Cook, A.J. and Vaughan, D.G. (2010). Overview of areal changes of the ice shelves on the Antarctic Peninsula over the past 50 years. *The Cryosphere* 4: 77–98. https://doi.org/10.5194/tc-4-77-2010.

Cook, A.J., Vaughan, D.G., Luckman, A.J., and Murray, T. (2014). A new Antarctic Peninsula glacier basin inventory and observed area changes since the 1940s.

Antarctic Science 26: 614–624. https://doi.org/10.1017/S0954102014000200.

Cook, S.J., Graham, D.J., Swift, D.A. et al. (2011). Sedimentary signatures of basal ice formation and their preservation in ice-marginal sediments. *Geomorphology* 125 (1): 122–131.

Corr, H.F. and Vaughan, D.G. (2008). A recent volcanic eruption beneath the West Antarctic ice sheet. *Nature Geoscience* 1 (2): 122–125.

Cuffey, K.M., Conway, H., Hallet, B. et al. (1999). Interfacial water in polar glaciers and glacier sliding at −17°C. *Geophysical Research Letters* 26: 751–754. https://doi.org/10.1029/1999gl900096.

Cuffey, K.M. and Paterson, W.S.B. (2010). *The Physics of Glaciers*, 4e. Academic Press.

Dahl, R. (1966). Block fields, weathering pits and tor-like forms in the Narvik Mountains, Nordland, Norway. *Geografiska Annaler* 46A: 55–85.

Dalton, A.S., Margold, M., Stokes, C.R. et al. (2020). An updated radiocarbon-based ice margin chronology for the last deglaciation of the North American Ice Sheet Complex. *Quaternary Science Reviews* 234: 106223. https://doi.org/10.1016/j.quascirev.2020.106223.

Dalton, A.S., Stokes, C.R., and Batchelor, C.L. (2022). Evolution of the Laurentide and Innuitian ice sheets prior to the Last Glacial Maximum (115 ka to 25 ka). *Earth-Science Reviews* 224: 103875. https://doi.org/10.1016/j.earscirev.2021.103875.

Davies, B., Bendle, J., Carrivick, J. et al. (2022). Topographic controls on ice flow and recession for Juneau Icefield (Alaska/British Columbia). *Earth Surface Processes and Landforms* 47: 2357–2390.

Davies, B.J., Darvill, C.M., Lovell, H. et al. (2020). The evolution of the Patagonian Ice Sheet from 35 ka to the present day (PATICE). *Earth-Science Reviews* 204: 103152.

Davies, B.J., Glasser, N.F., Carrivick, J.L. et al. (2013). Landscape evolution and ice-sheet behaviour in a semi-arid polar environment: James Ross Island, NE Antarctic Peninsula. *Geological Society Special Publication*. https://doi.org/10.1144/SP381.1.

Davies, B.J., Thorndycraft, V.R., Fabel, D., and Martin, J.R.V. (2018). Asynchronous glacier dynamics during the Antarctic Cold Reversal in central Patagonia. *Quaternary Science Reviews* 200: https://doi.org/10.1016/j.quascirev.2018.09.025.

Denton, G.H. and Hughes, T.J. (1983). Milankovitch theory of ice ages: hypothesis of ice-sheet linkage between regional insolation and global climate. *Quaternary Research* 20: 125–144. https://doi.org/10.1016/0033-5894(83)90073-X.

Dietrich, P., Griffis, N.P., Le Heron, D.P. et al. (2021). Fjord network in Namibia: A snapshot into the dynamics of the late Paleozoic glaciation. *Geology* 49: 1521–1526. https://doi.org/10.1130/G49067.1.

Domack, E., Duran, D., Leventer, A. et al. (2005). Stability of the Larsen B ice shelf on the Antarctic Peninsula during the Holocene epoch. *Nature* 436: 681–685.

Domack, E. and Harris, P.T. (1998). A new depositional model for ice shelves, based upon sediment cores from the Ross Sea and the MacRobertson shelf, Antarctica. *Annals of Glaciology* 27: 281–284.

Domack, E.W. and Powell, R. (2018. Chapter 7 -). Modern glaciomarine environments and sediments: an Antarctic perspective. In: *Past Glacial Environments*, 2e (ed. J. Menzies and J.J.M. van der Meer), 181–272. Elsevier https://doi.org/10.1016/B978-0-08-100524-8.00030-0.

Dowdeswell, J.A., Dowdeswell, E.K., Rodrigo, C., and Diaz, J. (2016). Assemblage of glacial and related landforms in the fjords of southern Chile. *Geological Society London, Memoirs* 46: 131–134. https://doi.org/10.1144/m46.139.

Dowdeswell, J.A., Hamilton, G.S., and Hagen, J.O. (1991). The duration of the active phase on surge-type glaciers: contrasts between Svalbard and other regions. *Journal of Glaciology* 37 (127): 388–400.

Dowdeswell, J.A., Hogan, K.A., and Le Heron, D.P. (2019). The glacier-influenced marine record on high-latitude continental margins: synergies between modern, Quaternary and ancient evidence. In: *Glaciated Margins: The Sedimentary and Geophysical Archive* (ed. D.P. Le Heron, K.A. Hogan, E.R. Phillips, et al.). London: Geological Society Special Publications, 475, https://doi.org/10.1144/SP475.13.

Dowdeswell, J.A. and Ottesen, D. (2013). Buried iceberg ploughmarks in the Early Quaternary sediments of the central North Sea: a two-million year record of glacial influence from 3D seismic data. *Marine Geology* 344: 1–9.

Dowdeswell, J.A. and Vasquez, M. (2013). Submarine landforms in the fjords of southern Chile: implications for glacimarine processes and sedimentation in a mild glacier-influenced environment. *Quaternary Science Reviews* 64: 1–19.

Downie, C. (1964). Glaciations of Mount Kilimanjaro, Northeast Tanganyika. *Geological Society of America Bulletin* 75: 1–16.

Drewry, D.J. (1986). *Glacial Geologic Processes*. London: Edward Arnold.

Duller, R.A., Mountney, N.P., Russell, A.J., and Cassidy, N.C. (2008). Architectural analysis of a volcaniclastic jökulhlaup deposit, southern Iceland: sedimentary evidence for supercritical flow. *Sedimentology* 55 (4): 939–964.

Dyke, A.S. and Morris, T.F. (1988). Drumlin fields, dispersal trains, and ice streams in Arctic Canada. *Canadian Geographer/Le Géographe Canadien* 32 (1): 86–90.

Dyke, A.S., Morris, T.F., Green, D.E.C., and England, J. (1992). Quaternary geology of Prince of Wales Island, Arctic Canada. *Geological Survey of Canada Memoir* 433: 142.

Echelmeyer, K. and Zhongxiang, W. (1987). Direct observations of basal sliding and deformation of basal drift at subfreezing temperatures. *Journal of Glaciology* 33: 83–98.

Edwards, B.R., Russell, J.K., and Simpson, K. (2011). Volcanology and petrology of Mathews Tuya, northern British

Columbia, Canada: glaciovolcanic constraints on interpretations of the 0.730 Ma Cordilleran paleoclimate. *Bulletin of Volcanology* 73 (5): 479–496.

Einarsson, B., Magnússon, E., Roberts, M. J., Pálsson, F., Thorsteinsson, T., & Jóhannesson, T. (2016). A spectrum of jökulhlaup dynamics revealed by GPS measurements of glacier surface motion. *Annals of Glaciology*, 57 (72), 47–61

Eisen, O., Harrison, W.D., and Raymond, C.F. (2001). The surges of Variegated glacier, Alaska, U.S.A., and their connection to climate and mass balance. *Journal of Glaciology* 47 (158): 351–358.

Ely, J.C., Clark, C.D., Spagnolo, M. et al. (2018). Using the size and position of drumlins to understand how they grow, interact and evolve. *Earth Surface Processes and Landforms* 43 (5): 1073–1087.

Ely, J.C., Clark, C.D., Spagnolo, M. et al. (2016). Do subglacial bedforms comprise a size and shape continuum? *Geomorphology* 257: 108–119.

Ely, J.C., Graham, C., Barr, I.D. et al. (2017). Using UAV acquired photography and structure from motion techniques for studying glacier landforms: application to the glacial flutes at Isfallsglaciären. *Earth Surface Processes and Landforms* 42 (6): 877–888.

Ely, J.C., Stevens, D., Clark, C.D., and Butcher, F.E.G. (2023). Numerical modelling of subglacial ribs, drumlins, herringbones, and mega-scale glacial lineations reveals their developmental trajectories and transitions. *Earth Surface Processes and Landforms* 48: 956–978.

Emmer, A. and Vilímek, V. (2013). Lake and breach hazard assessment for moraine-dammed lakes: an example from the Cordillera Blanca (Peru). *Natural Hazards and Earth System Sciences* 13 (6): 1551–1565.

Escutia, C., Eittreim, S.L., Cooper, A.K., and Nelson, C.H. (2000). Morphology and acoustic character of the Antarctic Wilkes Land turbidite systems: Ice-sheet sourced versus river sourced fans. *Journal of Sedimentary Research* 70 (1): 84–93.

Etzelmüller, B. and Hagen, J.O. (2005). Glacier–permafrost interaction in Arctic and alpine mountain environments with examples from southern Norway and Svalbard. In: *Cryospheric Systems: Glaciers and Permafrost* (ed. C. Harris and J.B. Murton). London: Geological Society, London, Special Publications.

Evans, D.J.A., Ewertowski, M., Orton, C., (2016). Eiríksjökull plateau icefield landsystem, Iceland. *Journal of Maps* 12: 747–756.

Evans, D.J., Nelson, C.D., and Webb, C. (2010). An assessment of fluting and "till esker" formation on the foreland of Sandfellsjökull, Iceland. *Geomorphology* 114 (3): 453–465.

Evans, D.J. and Thomson, S.A. (2010). Glacial sediments and landforms of Holderness, eastern England: a glacial depositional model for the North Sea Lobe of the British–Irish Ice Sheet. *Earth-Science Reviews* 101: 147–189.

Evans, D.J.A. (2003). Introduction to glacial landsystems. In: *Glacial Landsystems* (ed. D.J.A. Evans), 1–11. New York: Arnold.

Evans, D.J.A. (2009). Controlled moraines: origins, characteristics and palaeoglaciological implications. *Quaternary Science Reviews* 28: 183–208.

Evans, D.J.A. (2010). Controlled moraine development and debris transport pathways in polythermal plateau icefields: examples from Tungnafellsjökull, Iceland. *Earth Surface Processes and Landforms* 35 (12): 1430–1444.

Evans, D.J.A. (2016). Landscapes at the periphery of glacierization – retrospect and prospect. *Scottish Geographical Journal* 132: 140–163.

Evans, D.J.A. (2017a). *Till: A Glacial Process Sedimentology*. Wiley.

Evans, D.J.A. (2017b). Conceptual glacial ground models: British and Irish case studies. *Engineering Geology Special Publication* 28: 369–500.

Evans, D.J.A. and Benn, D.I. (ed.) (2021). *A Practical Guide to the Study of Glacial Sediments*, 2e. London: Quaternary Research Association.

Evans, D.J.A. and England, J. (1991). Canadian landform examples 19, high arctic thrust block moraines. *Canadian Geographer* 35: 93–97.

Evans, D.J.A., Ewertowski, M.W., Tomczyk, A., and Chandler, B.M.P. (2023). Active temperate glacial landsystem evolution in association with outwash head/depositional overdeepenings. *Earth Surface Processes and Landforms* 48 (8): 1573–1598.

Evans, D.J.A. and Hiemstra, J.F. (2005). Till deposition by glacier submarginal, incremental thickening. *Earth Surface Processes and Landforms* 30 (13): 1633–1662.

Evans, D.J.A., Phillips, E.R., Hiemstra, J.F., and Auton, C.A. (2006a). Subglacial till: Formation, sedimentary characteristics and classification. *Earth-Science Reviews* 78: 115–176.

Evans, D.J.A. and Rea, B.R. (1999). Geomorphology and sedimentology of surging glaciers: a land-systems approach. *Annals of Glaciology* 28: 75–82.

Evans, D.J.A. and Rea, B.R. (2003). Surging glacier landsystem. In: *Glacial Landsystems* (ed. D.J.A. Evans), 259–288. London. Chapter 11: Hodder Arnold.

Evans, D.J.A., Roberts, D.H., Bateman, M. D. et al. (2021). Retreat dynamics of the eastern sector of the British–Irish Ice Sheet during the last glaciation. *Journal of Quaternary Science* 36 (5): 723–751.

Evans, D.J.A., Strzelecki, M., Milledge, D.G., and Orton, C. (2012). Hørbyebreen polythermal glacial landsystem, Svalbard. *Journal of Maps* 8: 146–156.

Evans, D.J.A. and Twigg, D.R. (2002). The active temperate glacial landsystem: a model based on Breiethamerkurjokull and Fjallsjokull, Iceland. *Quaternary Science Review* 21: 2143–2177.

Evans, D.J.A., Twigg, D.R., and Shand, M. (2006b). Surficial geology and geomorphology of the þórisjökull plateau icefield, west-central Iceland. *Journal of Maps* 2 (1): 17–29.

Evans, D.J.A., Young, N.J.P., and O Cofaigh, C. (2014). Glacial geomorphology of terrestrial-terminating fast flow lobes/ice stream margins in the southwest Laurentide Ice Sheet. *Geomorphology* 204: 86–113. 10.1016/j.geomorph.2013.07.031.

Evans, J. and Pudsey, C.J. (2002). Sedimentation associated with Antarctic Peninsula ice shelves:

implications for palaeoenvironmental reconstructions of glacimarine sediments. *Journal of the Geological Society* 159 (3): 233–237.

Evans, J., Pudsey, C.J., Cofaigh, Ó.C. et al. (2005). Late Quaternary glacial history, flow dynamics and sedimentation along the eastern margin of the Antarctic Peninsula Ice Sheet. *Quaternary Science Reviews* 24: 741–774. https://doi.org/10.1016/j.quascirev.2004.10.007.

Eyles, N. (1993). Earth's glacial record and its tectonic setting. *Earth Science Reviews* 35: 1–248.

Eyles, N. (2008). Glacio-epochs and the supercontinent cycle after∼ 3.0 Ga: tectonic boundary conditions for glaciation. *Palaeogeography, Palaeoclimatology, Palaeoecology* 258 (1–2): 89–129.

Eyles, N., Boyce, J.I., and Putkinen, N. (2015). Neoglacial (< 3000 years) till and flutes at Saskatchewan Glacier, Canadian Rocky Mountains, formed by subglacial deformation of a soft bed. *Sedimentology* 62 (1): 182–203.

Eyles, N., Eyles, C.H., and Miall, A.D. (1983). Lithofacies types and vertical profile models; an alternative approach to the description and environmental interpretation of glacial diamict and diamictite sequences. *Sedimentology* 30: 393–410. https://doi.org/10.1111/j.1365-3091.1983.tb00679.x.

Farinotti, D., Huss, M., Fürst, J.J. et al. (2019). A consensus estimate for the ice thickness distribution of all glaciers on Earth. *Nature Geoscience* https://doi.org/10.1038/s41561-019-0300-3.

Farnsworth, W.R., Ingólfsson, Ó., Retelle, M., and Schomacker, A. (2016). Over 400 previously undocumented Svalbard surge-type glaciers identified. *Geomorphology* 264: 52–60.

Fedorchuk, N.D., Isbell, J.L., Rosa, E.L. et al. (2023). Reappraisal of exceptionally preserved s-forms, striae, and fractures from late Paleozoic subglacial surfaces in paleofjords. *NW Namibia. Sedimentary Geology* 456: 106498.

Fielding, C., Atkins, C., Bassett, K. et al. (2009). Sedimentology and stratigraphy of the AND-2A core, ANDRILL Southern McMurdo Sound project, Antarctica. *Terra Antarctica* 15: 77–112.

Fitzsimons, S. and Howarth, J. (2018). Chapter 9: Glaciolacustrine processes. In: *Past Glacial Environments*, 2e (ed. J. Menzies and J.J.M. van der Meer), 309–334. Elsevier https://doi.org/10.1016/B978-0-08-100524-8.00009-9.

Fitzsimons, S. and Howarth, J. (2020). Development of push moraines in deeply frozen sediment adjacent to a cold-based glacier in the McMurdo Dry Valleys, Antarctica. *Earth Surface Processes and Landforms* 45 (3): 622–637.

Fitzsimons, S.J. (2004). Ice-marginal terrestrial landsystems: polar continental glacier margins. In: *Glacier Landsystems* (ed. D.J.A. Evans), 89–110. London: Hodder Arnold.

Fitzsimons, S.J., McManus, K.J., and Lorrain, R.D. (1999). Structure and strength of basal ice and substrate of a dry-based glacier: evidence for substrate deformation at subfreezing temperatures. *Annals of Glaciology* 28: 236–240.

Flink, A.E., Noormets, R., Kirchner, N. et al. (2015). The evolution of a submarine landform record following recent and multiple surges of Tunabreen glacier, Svalbard. *Quaternary Science Reviews* 108: 37–50.

Flint, R.F., Sanders, J.E., and Rodgers, J. (1960). Diamictite, a substitute term for symmictite. *Geological Society of America Bulletin* 71 (12): 1809–1810.

Frederikse, T., Landerer, F., Caron, L. et al. (2020). The causes of sea-level rise since 1900. *Nature* 584: 393–397. https://doi.org/10.1038/s41586-020-2591-3.

French, H.M. (2018). *The Periglacial Environment.* Hoboken, NJ: Wiley.

Fricker, H.A. and Padman, L. (2006). Ice shelf grounding zone structure from ICESat laser altimetry. *Geophysical Research Letters* 33: 1–5.

Furbish, D.J. and Andrews, J.T. (1984). The use of hypsometry to indicate long-term stability and response of valley glaciers to changes in mass transfer. *Journal of Glaciology* 30 (105): 199–211.

Gales, J., Hillenbrand, C.D., Larter, R. et al. (2019). Processes influencing differences in Arctic and Antarctictrough mouth fan sedimentology. *Geological Society, London, Special Publications* 475 (1): 203–221.

Gellatly, A.F., Gordon, J.E., Whalley, W.B., and Hansom, J.D. (1988). Thermal regime and geomorphology of plateau ice caps in northern Norway: observations and implications. *Geology* 16: 983–986.

Ghienne, J.-F., Deynoux, M., Manatschal, G., and Rubino, J.-L. (2003). Paléovallées et dépocentres sur failles dans les séries glaciaires fini-ordoviciennes du bassin de Murzuq (Libye centrale). *Comptes Rendus Geoscience* 335: 1091–1100. https://doi.org/10.1016/j.crte.2003.09.010.

Ghienne, J.-F., Le Heron, D.P., Moreau, J. et al. (2007). The Late Ordovician glacial sedimentary system of the North Gondwana Platform. In: *Glacial Sedimentary Processes and Products* (ed. P. Christoffersen, N.F. Glasser, B. Hubbard, and M.J. Hambrey), 295–319. Blackwell Publishing Ltd.

Gibson, M.J., Glasser, N.F., Quincey, D.J. et al. (2017). Temporal variations in supraglacial debris distribution on Baltoro Glacier, Karakoram between 2001 and 2012. *Geomorphology* 295: 572–585. 10.1016/j.geomorph.2017.08.012.

Giesen, R.H. and Oerlemans, J. (2010). Response of the ice cap Hardangerjøkulen in southern Norway to the 20th and 21st century climates. *The Cryosphere* 4 (2): 191–213.

Gilbert, R. and Crookshanks, S. (2009). Sediment waves in a modern high-energy glacilacustrine environment. *Sedimentology* 56: 645–659. 10.1111/j.1365-3091.2008.00990.x.

Glasser, N.F. and Bennett, M.R. (2004). Glacial erosional landforms: origins and significance for palaeoglaciology. *Progress in Physical Geography* 28 (1): 43–75.

Glasser, N.F. and Hambrey, M.J. (2002). Sedimentary facies and landform genesis at a temperate outlet glacier: Soler Glacier, North Patagonian Icefield. *Sedimentology* 49 (1): 43–64.

Glasser, N.F., Jansson, K.N., Duller, G.A. T. et al. (2016). Glacial lake drainage in Patagonia (13-8 kyr) and response of the adjacent Pacific Ocean. *Scientific*

Reports 6: 21064. https://doi.org/10.1038/srep21064 http://www.nature.com/articles/srep21064#supplementary-information.

Glen, J.W. (1955). The creep of polycrystalline ice. *Proceedings of the Royal Society of London. Series A. Mathematical and Physical Sciences* 228 (1175): 519–538.

Goodsell, B., Hambrey, M.J., and Glasser, N.F. (2005). Debris transport in a temperate valley glacier: Haut Glacier d'Arolla, Valais, Switzerland. *Journal of Glaciology* 51: 139–146.

Graham, A.G.C., Jakobsson, M., Nitsche, F.O. et al. (2016). Submarine glacial-landform distribution across the West Antarctic margin, from grounding line to slope: the Pine Island–Thwaites ice-stream system. *Geological Society, London, Memoirs* 46: 493–500.

Greenwood, S.L., Clark, C.D., Hughes, A. L.C., (2007). Formalising an inversion methodology for reconstructing ice-sheet retreat patterns from meltwater channels: applications to the British ice sheet. *Journal of Quaternary Science* 22: 637–645.

Griffiths, J.S. and Martin, C.J. (2017). *Engineering Geology and Geomorphology of Glaciated and Periglaciated Terrains*. Engineering Group Working Party Report. Engineering Geology Special Publications, 28. London: Geological Society of London.

Gudmundsson, M.T., Sigmundsson, F., and Björnsson, H. (1997). Ice–volcano interaction of the 1996 Gjálp subglacial eruption, Vatnajökull, Iceland. *Nature* 389 (6654): 954–957.

Guðmundsson, S. and Evans, D.J.A. (2022). Geomorphological map of Breiðamerkursandur 2018: the historical evolution of an active temperate glacier foreland. *Geografiska Annaler: Series A, Physical Geography* 104 (4): 298–332. https://doi.org/10.1080/04353676.2022.2148083.

Gulley, J.D., Grabiec, M., Martin, J.B. et al. (2012). The effect of discrete recharge by moulins and heterogeneity in flow-path efficiency at glacier beds on subglacial hydrology. *Journal of Glaciology* 58 (211): 926–940.

Haeberli, W. (2005). Investigating glacier–permafrost relationships in high-mountain areas: historical background, selected examples and research needs. In: *Cryospheric Systems: Glaciers and Permafrost: Geological Society*, vol. 242 (ed. C. Harris and J.B. Murton), 29–37. London: Special Publications.

Hallet, B. (1996). Glacial quarrying: a simple theoretical model. *Annals of Glaciology* 22: 1–8.

Hambrey, M.J. and Alean, J.C. (2016). *Colour atlas of glacial phenomena*. CRC Press.

Hambrey, M.J. and Fitzsimons, S.J. (2010). Development of sediment-landform associations at cold glacier margins, Dry Valleys, Antarctica. *Sedimentology* 57: 857–882.

Hambrey, M.J. and Glasser, N.F. (2012). Discriminating glacier thermal and dynamic regimes in the sedimentary record. *Sedimentary Geology* 251–252: 1–33.

Hambrey, M.J. and Huddart, D. (1995). Englacial and proglacial glaciotectonic processes at the snout of a thermally complex glacier in Svalbard. *Journal of Quaternary Science* 10: 313–326.

Hambrey, M.J., Smellie, J., Nelson, A.E., and Johnson, J.S. (2008). Late Cenozoic glacier-volcano interaction on James Ross Island and adjacent areas, Antarctic Peninsula region. *Geological Society of America Bulletin* 120 (5-6): 709–731.

Hanna, E., Pattyn, F., Navarro, F. et al. (2020). Mass balance of the ice sheets and glaciers–progress since AR5 and challenges. *Earth-Science Reviews* 201: 102976.

Hannesdóttir, H., Björnsson, H., Pálsson, F. et al. (2015). Changes in the southeast Vatnajökull ice cap, Iceland, between ~ 1890 and 2010. *The Cryosphere* 9: 565–585. https://doi.org/10.5194/tc-9-565-2015.

Harris, C. and Murton, J.B. (2005). Interactions between glaciers and permafrost: an introduction. In: *Cryospheric Systems: Glaciers and Permafrost*, vol. 242 (ed. C. Harris and J.B. Murton), 1–9. Geological Society, London, Special Publications.

Harrison, D., Ross, N., Russell, A.J., and Jones, S.J. (2022). Ground-penetrating radar (GPR) investigations of a large-scale buried ice-marginal landsystem, Skeiðarársandur, SE Iceland. *Boreas* 51 (4): 824–846.

Hart, J.K. and Watts, R.J. (1997). A comparison of the styles of deformation associated with two recent push moraines, south Van Keulenfjorden, Svalbard. *Earth Surface Processes and Landforms* 22: 1089–1107.

Hättestrand, C. and Kleman, J. (1999). Ribbed moraine formation. *Quaternary Science Reviews* 18 (1): 43–61.

Hättestrand, C. and Stroeven, A.P. (2002). A relict landscape in the centre of Fennoscandian glaciation: geomorphological evidence of minimal Quaternary glacial erosion. *Geomorphology* 44: 127–143.

Herreid, S. and Pellicciotti, F. (2018). Automated detection of ice cliffs within supraglacial debris cover. *The Cryosphere* 12: 1811–1829.

Herreid, S. and Pellicciotti, F. (2020). The state of rock debris covering Earth's glaciers. *Nature Geoscience* https://doi.org/10.1038/s41561-020-0615-0.

Hoffman, P.F., Abbot, D.S., Ashkenazy, Y. et al. (2017a). Snowball Earth climate dynamics and Cryogenian geology-geobiology. *Science Advances* 3: e1600983. https://doi.org/10.1126/sciadv.1600983.

Hooke, R.L. and Jennings, C.E. (2006). On the formation of the tunnel valleys of the southern Laurentide ice sheet. *Quaternary Science Reviews* 25 (11–12): 1364–1372.

Hooyer, T.S., Cohen, D., and Iverson, N.R. (2012). Control of glacial quarrying by bedrock joints. *Geomorphology* 153: 91–101.

Hooyer, T.S. and Iverson, N.R. (2000). Diffusive mixing between shearing granular layers: constraints on bed deformation from till contacts. *Journal of Glaciology* 46: 641–651.

Hornsey, J., Rowan, A.V., Kirkbride, M.P. et al. (2022). Be-10 dating of ice-marginal moraines in the Khumbu Valley, Nepal, Central Himalaya, reveals the response of monsoon-influenced glaciers to Holocene climate change. *Journal of Geophysical Research: Earth Surface* 127 (8): e2022JF006645.

Howard, T., Ridley, J., Pardaens, A.K., et al. (2014). The land-ice contribution to 21st-century dynamic sea level rise. *Ocean Science* 10: 485–500.

Hubbard, B., Heald, A., Reynolds, J.M. et al. (2005). Impact of a rock avalanche on a moraine-dammed proglacial lake: Laguna Safuna Alta, Cordillera Blanca, Peru. *Earth Surface Processes and Landforms* 30: 1251–1264. https://doi.org/10.1002/esp.1198.

Hubbard, B. and Sharp, M. (1995). Basal ice fades and their formation in the western Alps. *Arctic and Alpine Research* 27 (4): 301–310.

Hughes, A.L.C., Clark, C.D., and Jordan, C.J. (2014). Flow-pattern evolution of the last British Ice Sheet. *Quaternary Science Reviews* 89: 148–168.

Hughes, A.L.C., Gyllencreutz, R., Lohne, Ø.S. et al. (2016). The last Eurasian ice sheets–a chronological database and time-slice reconstruction, DATED-1. *Boreas* 45: 1–45.

Imbrie, J. (1984). The orbital theory of Pleistocene climate: support from a revised chronology of the marine 18O record. In: *Milankovitch and Climate: Understanding the Response of Astronomical Forcing* (ed. A.L. Berger), 269–305. Dordrecht: Reidel.

Immerzeel, W.W., Lutz, A.F., Andrade, M. et al. (2020). Importance and vulnerability of the world's water towers. *Nature* 577: 364–369. https://doi.org/10.1038/s41586-019-1822-y.

Isbell, J.L., Vesely, F.F., Rosa, E.L.M. et al. (2021). Evaluation of physical and chemical proxies used to interpret past glaciations with a focus on the late Paleozoic Ice Age. *Earth-Science Reviews* 221: 103756.

Iverson, N.R. (1991). Potential effects of subglacial water-pressure fluctuations on quarrying. *Journal of Glaciology* 37 (125): 27–36.

Iverson, N.R. (2002). Processes of glacial erosion. In: *Modern and Past Glacial Environments*, 131–145. Butterworth-Heinemann.

Iverson, N.R. (2010). Shear resistance and continuity of subglacial till: hydrology rules. *Journal of Glaciology* 56 (200): 1104–1114. https://doi.org/10.3189/002214311796406220.

Iverson, N.R., McCracken, R.G., Zoet, L.K. et al. (2017). A theoretical model of drumlin formation based on observations at Múlajökull, Iceland. *Journal of Geophysical Research: Earth Surface* 122 (12): 2302–2323.

Ives, L.R.W. and Iverson, N.R. (2019). Genesis of glacial flutes inferred from observations at Múlajökull, Iceland. *Geology* 47 (5): 387–390.

Jacquemart, M., Loso, M., Leopold, M. et al. (2020). What drives large-scale glacier detachments? Insights from Flat Creek glacier, St. Elias Mountains, Alaska. *Geology* https://doi.org/10.1130/G47211.1.

Janszen, A., Moreau, J., Moscariello, A. et al. (2013). Time-transgressive tunnel-valley infill revealed by a three-dimensional sedimentary model, Hamburg, north-west Germany. *Sedimentology* 60 (3): 693–719.

Jenkins, A. (2011). Ice shelf. In: *Encyclopedia of Snow, Ice and Glaciers* (ed. V.P. Singh, P. Singh, and U.K. Haritashya), 613–616. Netherlands, Dordrecht: Springer https://doi.org/10.1007/978-90-481-2642-2_278.

Jennings, S.J. and Hambrey, M.J. (2021). Structures and deformation in glaciers and ice sheets. *Reviews of Geophysics* 59 (3): e2021RG000743.

Johnson, M.D., Schomacker, A., Benediktsson, Í.Ö., et al. (2010). Active drumlin field revealed at the margin of Múlajökull, Iceland: a surge-type glacier. *Geology* 38: 943–946.

Jones, R.S., Lowe, J.J., Palmer, A.P. et al. (2017). Dynamics and palaeoclimatic significance of a Loch Lomond Stadial glacier: Coire Ardair, Creag Meagaidh, Western Highlands, Scotland. *Proceedings of the Geologists Association* 128: 54–66. https://doi.org/10.1016/j.pgeola.2015.11.004.

Kääb, A., Leinss, S., Gilbert, A. et al. (2018). Massive collapse of two glaciers in western Tibet in 2016 after surge-like instability. *Nature Geoscience* 11: 114–120. https://doi.org/10.1038/s41561-017-0039-7.

Kamb, B., Raymond, C.F., Harrison, W.D. et al. (1985). Glacier surge mechanism: 1982-1983 surge of Variegated Glacier, Alaska. *Science* 227 (4686): 469–479.

Kempf, P., Forwick, M., Laberg, J.S., and Vorren, T.O. (2013). Late Weichselian and Holocene sedimentary palaeoenvironment and glacial activity in the high-arctic van Keulenfjorden, Spitsbergen. *The Holocene* 23 (11): 1607–1618. https://doi.org/10.1177/0959683613499055.

Kendall, B.S., Creaser, R.A., Ross, G.M., and Selby, D. (2004). Constraints on the timing of Marinoan "Snowball Earth" glaciation by 187Re–187Os dating of a Neoproterozoic, post-glacial black shale in Western Canada. *Earth and Planetary Science Letters* 222: 729–740. https://doi.org/10.1016/j.epsl.2004.04.004.

Kennedy, K. and Eyles, N. (2021). Syn-rift mass flow generated 'tectonofacies' and 'tectonosequences' of the Kingston Peak Formation, Death Valley, California, and their bearing on supposed Neoproterozoic panglacial climates. *Sedimentology* 68: 352–381. https://doi.org/10.1111/sed.12781.

Kennedy, K. and Eyles, N. (2021). Syn-rift mass flow generated 'tectonofacies' and 'tectonosequences' of the Kingston Peak Formation, Death Valley, California, and their bearing on supposed Neoproterozoic panglacial climates. *Sedimentology* 68 (1): 352–381.

King, E.C., Hindmarsh, R.C., and Stokes, C.R. (2009). Formation of mega-scale glacial lineations observed beneath a West Antarctic ice stream. *Nature Geoscience* 2 (8): 585–588.

King, E.C., Pritchard, H.D., and Smith, A.M. (2016). Subglacial landforms beneath Rutford Ice Stream, Antarctica: detailed bed topography from ice-penetrating radar. *Earth System Science Data* 8 (1): 151–158.

Kirkbride, M.P. (2011). Debris-covered glaciers. In: *Encyclopedia of Snow, Ice and Glaciers* (ed. V.P. Singh, P. Singh, and U.K. Haritashya), 180–182. Netherlands, Dordrecht: Springer https://doi.org/10.1007/978-90-481-2642-2_622.

Kirkbride, M.P. and Dugmore, A.J. (2003). Glaciological response to distal tephra fallout from the 1947 eruption of Hekla, south Iceland. *Journal of Glaciology* 49 (166): 420–428.

Kirkbride, M.P. and Dugmore, A.J. (2008). Two millennia of glacier advances from southern Iceland dated by tephrochronology. *Quaternary Research* 70 (3): 398–411.

Kirkbride, M.P. and Winkler, S. (2012). Correlation of Late Quaternary moraines: impact of climate variability, glacier response, and chronological resolution. *Quaternary Science Reviews* 46: 1–29.

Kirkham, J.D., Hogan, K.A., Larter, R.D. et al. (2024). Tunnel valley formation beneath deglaciating mid-latitude ice sheets: Observations and modelling. *Quaternary Science Reviews* 323: 107680.

Kleman, J., Hättestrand, C., and Clarhäll, A. (1999a). Zooming in on frozen-bed patches: scale-dependent controls on Fennoscandian ice sheet basal thermal zonation. *Annals of Glaciology* 28: 189–194.

Knight, J., Mitchell, W.A., and Rose, J. (2011). Chapter 6: Geomorphological field mapping. In: *Developments in Earth Surface Processes* (ed. M.J. Smith, P. Paron, and J.S. Griffiths), 151–187. Elsevier.

Kowal, K.N. and Worster, M.G. (2020). The formation of grounding zone wedges: theory and experiments. *Journal of Fluid Mechanics* 898: https://doi.org/10.1017/jfm.2020.393.

Krabbendam, M., Eyles, N., Putkinen, N. et al. (2016). Streamlined hard beds formed by palaeo–ice streams: a review. *Sedimentary Geology* 338: 24–50. https://doi.org/10.1016/j.sedgeo.2015.12.007.

Krigström, A. (1962). Geomorphological studies of sandur plains and their braided rivers in Iceland. *Geografiska Annaler* 44 (3-4): 328–346.

Krüger, J. (1996). Moraine ridges formed from subglacial frozen-on sediment slabs and their differentiation from push moraines. *Boreas* 25 (1): 57–64.

Krüger, J. and Kjær, K.H. (1999). A data chart for field description and genetic interpretation of glacial diamicts and associated sediments... with examples from Greenland, Iceland, and Denmark. *Boreas* 28 (3): 386–402.

Krüger, J., Kjaer, K.H., and Van Der Meer, J.J. (2002). From push moraine to single-crested dump moraine during a sustained glacier advance. *Norsk Geografisk Tidsskrift-Norwegian Journal of Geography* 56 (2): 87–95.

Kurjanski, B., Rea, B.R., Spagnolo, M. et al. (2019). Morphological evidence for marine ice stream shutdown, central Barents Sea. *Marine Geology* 414: 64–76. https://doi.org/10.1016/j.margeo.2019.05.001.

Lawson, D.E. (1982). Mobilization, movement and deposition of active subaerial sediment flows, Matanuska Glacier, Alaska. *The Journal of Geology* 90 (3): 279–300.

Lawson, D.E., Strasser, J.C., Evenson, E.B. et al. (1998). Glaciohydraulic supercooling: a freeze-on mechanism to create stratified, debris-rich basal ice: I. Field evidence. *Journal of Glaciology* 44 (148): 547–562.

Le Heron, D.P. (2015). The significance of ice-rafted debris in Sturtian glacial successions. *Sedimentary Geology* 322: 19–33. 10.1016/j.sedgeo.2015.04.001.

Le Heron, D.P., Busfield, M.E., and Kamona, F. (2013). An interglacial on snowball Earth? Dynamic ice behaviour revealed in the Chuos Formation, Namibia. *Sedimentology* 60: 411–427.

Le Heron, D.P., Busfield, M.E., and Kettler, C. (2020). Ice-rafted dropstones in "postglacial" Cryogenian cap carbonates. *Geology* 49 (3): 263–267.

Le Heron, D.P., Busfield, M.E., Prave, A.R. (2014). Neoproterozoic ice sheets and olistoliths: multiple glacial cycles in the Kingston Peak Formation, California. *Journal of the Geological Society* 171: 525–538.

Le Heron, D.P. and Craig, J. (2008). First-order reconstructions of a Late Ordovician Saharan ice sheet. *Journal of the Geological Society* 165: https://doi.org/10.1144/0016-76492007-002.

Le Heron, D.P., Craig, J., and Etienne, J.L. (2009). The relationship between glaciation and hydrocarbon accumulations across North Africa and the Middle East. *Earth Science Reviews* 93: 47–76.

Le Heron, D.P., Kettler, C., Davies, B. et al. (2021). Rapid geomorphological and sedimentological changes at a modern Alpine ice margin: lessons from the Gepatsch Glacier, Tirol, Austria. *Journal of the Geological Society, London* 179: https://doi.org/10.1144/jgs2021-052.

Le Heron, D.P., Sutcliffe, O.E., Bourgig, K. et al. (2004). Sedimentary architecture of Upper Ordovician tunnel valleys, Gargaf Arch, Libya: implications for the genesis of a hydrocarbon reservoir. *GeoArabia* 9: 137–160.

Le Heron, D.P., Sutcliffe, O.E., Whittington, R.J., and Craig, J. (2005). The origins of glacially related soft-sediment deformation structures in Upper Ordovician glaciogenic rocks: implication for ice-sheet dynamics. *Palaeogeography, Palaeoclimatology, Palaeoecology* 218 (1-2): 75–103.

Le Heron, D.P., Tofaif, S., and Melvin, J. (2018). The Early Palaeozoic glacial deposits of Gondwana: overview, chronology and controversies. In: *Past Glacial Environments: A New and Revised Edition* (ed. J. Menzies and J. van der Meer), 47–73. Elsevier https://doi.org/10.1016/B978-0-08-100524-8.00002-6.

Le Heron, D.P., Tofaif, S., Vandyk, T., and Ali, D.O. (2017). A diamictite dichotomy: Glacial conveyor belts and olistostromes in the Neoproterozoic of Death Valley, California, USA. *Geology* 45: 31–34. https://doi.org/10.1130/G38460.1.

Lee, J. (2018). Chapter 11: Glacial lithofacies and stratigraphy. In: *Past Glacial Environments*, 2e (ed. J. Menzies and J.J.M. van der Meer), 377–429. Elsevier https://doi.org/10.1016/B978-0-08-100524-8.00011-7.

Lelandais, T., Mourgues, R., Ravier, É. et al. (2016). Experimental modeling of pressurized subglacial water flow: implications for tunnel valley formation. *Journal of Geophysical Research: Earth Surface* 121 (11): 2022–2041. https://doi.org/10.1002/2016JF003957.

Lelandais, T., Ravier, É., Pochat, S. et al. (2018). Modelled subglacial floods and tunnel valleys control the life cycle of transitory ice streams. *The Cryosphere* 12 (8): 2759–2772. https://doi.org/10.5194/tc-12-2759-2018.

Lescinsky, D.T. and Fink, J.H. (2000). Lava and ice interaction at stratovolcanoes: use of characteristic features to determine past glacial extents and future volcanic hazards. *Journal of Geophysical Research: Solid Earth* 105 (B10): 23711–23726.

Levell, B.K., Braakman, J.H., and Rutten, K.W. (1988). Oil-bearing sediments of Gondwana glaciation in Oman. *AAPG Bulletin* 72 (7): 775–796.

Levy, L.B., Kelly, M.A., Lowell, T.V. et al. (2014). Holocene fluctuations of Bregne ice cap, Scoresby Sund, east Greenland: a proxy for climate along the Greenland Ice Sheet margin. *Quaternary Science Reviews* 92: 357–368.

Linch, L.D. and Dowdeswell, J.A. (2016). Micromorphology of diamicton affected by iceberg-keel scouring, Scoresby Sund, East Greenland. *Quaternary Science Reviews* 152: 169–196.

Livingstone, S.J. and Clark, C.D. (2016). Morphological properties of tunnel valleys of the southern sector of the Laurentide Ice Sheet and implications for their formation. *Earth Surface Dynamics* 4: 567–589. https://doi.org/10.5194/esurf-4-567-2016.

Livingstone, S.J., Cofaigh, C.Ó., and Evans, D.J. (2008). Glacial geomorphology of the central sector of the last British-Irish Ice Sheet. *Journal of Maps* 4 (1): 358–377.

Livingstone, S.J., Evans, D.J., Cofaigh, C.Ó. et al. (2012). Glaciodynamics of the central sector of the last British-Irish Ice Sheet in Northern England. *Earth-Science Reviews* 111 (1–2): 25–55.

Livingstone, S.J., Evans, D.J., Cofaigh, C.Ó., and Hopkins, J. (2010). The Brampton kame belt and Pennine escarpment meltwater channel system (Cumbria, UK): morphology, sedimentology and formation. *Proceedings of the Geologists' Association* 121 (4): 423–443.

Livingstone, S.J., Storrar, R.D., Hillier, J.K. et al. (2015). An ice-sheet scale comparison of eskers with modelled subglacial drainage routes. *Geomorphology* 246: 104–112.

Lloyd Davies, M.T., Atkins, C.B., van der Meer, J.J.M. et al. (2009). Evidence for cold-based glacial activity in the Allan Hills, Antarctica. *Quaternary Science Reviews* 28: 3124–3137. https://doi.org/10.1016/j.quascirev.2009.08.002.

Long, J., Zhang, S., and Luo, K. (2019). Cryogenian magmatic activity and early life evolution. *Scientific Reports* 9: 6586. https://doi.org/10.1038/s41598-019-43177-8.

Lønne, I. (2016). A new concept for glacial geological investigations of surges, based on High-Arctic examples (Svalbard). *Quaternary Science Reviews* 132: 74–100.

Lovell, H., Benn, D.I., Lukas, S. et al. (2018a). Multiple Late Holocene surges of a High-Arctic tidewater glacier system in Svalbard. *Quaternary Science Reviews* 201: 162–185.

Lovell, H., Benn, D.I., Lukas, S. et al. (2018b). Geomorphological investigation of multiphase glacitectonic composite ridge systems in Svalbard. *Geomorphology* 300: 176–188.

Lovell, H. and Boston, C.M. (2017). Glacitectonic composite ridge systems and surge-type glaciers: an updated correlation based on Svalbard, Norway. *Arktos* 3: 1–16.

Lovell, H., Fleming, E.J., Benn, D.I. et al. (2015). Debris entrainment and landform genesis during tidewater glacier surges. *Journal of Geophysical Research: Earth Surface* 120 (8): 1574–1595.

Lovell, H., Livingstone, S.J., Boston, C.M. et al. (2019). Complex kame belt morphology, stratigraphy and architecture. *Earth Surface Processes and Landforms* 44 (13): 2685–2702.

Lukas, S. (2005). A test of the englacial thrusting hypothesis of 'hummocky' moraine formation: case studies from the northwest Highlands, Scotland. *Boreas* 34 (3): 287–307.

Lukas, S. (2006). Morphostratigraphic principles in glacier reconstruction-a perspective from the British Younger Dryas. *Progress in Physical Geography* 30 (6): 719–736.

Lukas, S. (2012). Processes of annual moraine formation at a temperate alpine valley glacier: insights into glacier dynamics and climatic controls. *Boreas* 41 (3): 463–480.

Lukas, S., Benn, D.I., Boston, C.M. et al. (2013). Clast shape analysis and clast transport paths in glacial environments: a critical review of methods and the role of lithology. *Earth-Science Reviews* 121: 96–116.

Lukas, S., Graf, A., Coray, S., and Schlüchter, C. (2012). Genesis, stability and preservation potential of large lateral moraines of Alpine valley glaciers–towards a unifying theory based on Findelengletscher, Switzerland. *Quaternary Science Reviews* 38: 27–48.

Lukas, S. and Sass, O. (2011). The formation of Alpine lateral moraines inferred from sedimentology and radar reflection patterns: a case study from Gornergletscher, Switzerland. In: *Ice Marginal and Periglacial Processes and Sediments*, vol. 354 (ed. I.P. Martini, A.P.R. Alberti, and H.M. French), 77–92. Geological Society London, Special Publication.

Magnússon, E., Gudmundsson, M.T., Roberts, M.J. et al. (2012). Ice-volcano interactions during the 2010 Eyfjallajökull eruption, as revealed by airborne imaging radar. *Journal of Geophysical Research* 117: B07405.

Maizels, J. (1992). Boulder ring structures produced during jökulhlaup flows: origin and hydraulic significance. *Geografiska Annaler: Series A, Physical Geography* 74 (1): 21–33.

Maizels, J. (1993). Lithofacies variations within sandur deposits: the role of runoff regime, flow dynamics and sediment supply characteristics.

Sedimentary Geology 85 (1–4): 299–325.

Małecki, J., Lovell, H., Ewertowski, W. et al. (2018). The glacial landsystem of a tropical glacier: Charquini Sur, Bolivian Andes. *Earth Surface Processes and Landforms* 43: 2584–2602.

Margold, M., Stokes, C.R., and Clark, C.D. (2015). Ice streams in the Laurentide Ice Sheet: identification, characteristics and comparison to modern ice sheets. *Earth-Science Reviews* 143: 117–146.

Margold, M., Stokes, C.R., and Clark, C.D. (2018). Reconciling records of ice streaming and ice margin retreat to produce a palaeogeographic reconstruction of the deglaciation of the Laurentide Ice Sheet. *Quaternary Science Reviews* 189: 1–30.

Margreth, A., Dyke, A.S., Gosse, J.C., and Telka, A.M. (2014). Neoglacial ice expansion and late Holocene cold-based ice cap dynamics on Cumberland Peninsula, Baffin Island, Arctic Canada. *Quaternary Science Reviews* 91: 242–256.

Margreth, S., Funk, M., Tobler, D. et al. (2017). Analysis of the hazard caused by ice avalanches from the hanging glacier on the Eiger west face. *Cold Regions Science and Technology* 144: 63–72.

Marren, P.M. (2002). Glacier margin fluctuations, Skaftafellsjökull, Iceland: implications for sandur evolution. *Boreas* 31 (1): 75–81.

Marren, P.M. and Schuh, M. (2009). Criteria for identifying jokulhlaup deposits in the sedimentary record. In: *Megaflooding on Earth and Mars* (ed. D.M. Burr, P.A. Carling, and V.R. Baker), 225–242. Cambridge University Press.

Marren, P.M. and Toomath, S.C. (2014). Channel pattern of proglacial rivers: topographic forcing due to glacier retreat. *Earth Surface Processes and Landforms* 39 (7): 943–951.

Marshall, S.J., Björnsson, H., Flowers, G. E., and Clarke, G.K.C. (2005). Simulation of Vatnajökull ice cap dynamics. *Journal of Geophysical Research - Earth Surface* 110: https://doi.org/10.1029/2004JF000262.

Martin, C.J., Morley, A.L., and Griffiths, J.S. (2017). Introduction to engineering geology and geomorphology of glaciated and periglaciated terrains. In: *Geological Society London Engineering Geology Special Publications*, vol. 28, 1–30. https://doi.org/10.1144/EGSP28.1.

Martin, J.R., Redfern, J., and Williams, B. P.J. (2012). Evidence for multiple ice centres during the late Palaeozoic ice age in Oman: outcrop sedimentology and provenance of the Late Carboniferous–Early Permian Al Khlata Formation. In: *Glaciogenic Reservoirs and Hydrocarbon Systems*, vol. 368 (ed. M. Huuse, D.P. Le Heron, R. Dixon, et al.), 229–256. Geological Society, London, Special Publications https://doi.org/10.1144/SP368.18.

Martin, J.R.V., Davies, B.J., and Thorndycraft, V.R. (2019). Glacier dynamics during a phase of Late Quaternary warming in Patagonia reconstructed from sediment-landform associations. *Geomorphology* 337: 111–133.

McCarroll, D. (2016). Trimline trauma: the wider implications of a paradigm shift in recognising and interpreting glacial limits. *Scottish Geographical Journal* 132 (2): 130–139.

McDougall, D. (2013). Glaciation style and the geomorphological record: evidence for Younger Dryas glaciers in the eastern Lake District, northwest England. *Quaternary Science Reviews* 73: 48–58. https://doi.org/10.1016/j.quascirev.2013.05.002.

Montañez, I.P. and Poulsen, C.J. (2013). The Late Paleozoic ice age: an evolving paradigm. *Annual Review of Earth and Planetary Sciences* 41: 629–656.

Moreau, J., Ghienne, J.F., Le Heron, D.P. et al. (2005). 440 Ma ice stream in North Africa. *Geology* 33: 753–756. https://doi.org/10.1130/G21782.1.

Moreau, J. and House, M. (2013). Infill of tunnel valleys associated with landward-flowing ice sheets: the missing Middle Pleistocene record of the NW European rivers? *Geochemistry, Geophysics, Geosystems* 14: https://doi.org/10.1002/2013GC005007.

Morlighem, M., Rignot, E., Binder, T. et al. (2019). Deep glacial troughs and stabilizing ridges unveiled beneath the margins of the Antarctic ice sheet. *Nature Geoscience* https://doi.org/10.1038/s41561-019-0510-8.

Morlighem, M., Williams, C.N., Rignot, E. et al. (2017). BedMachine v3: complete bed topography and ocean bathymetry mapping of Greenland from multibeam echo sounding combined with mass conservation. *Geophysical Research Letters* 44: 11051–11061. https://doi.org/10.1002/2017GL074954.

Nesje, A., Bakke, J., Dahl, S.O. et al. (2008). Norwegian mountain glaciers in the past, present and future. *Global and Planetary Change* 60 (1–2): 10–27.

Nield, J.M., Chiverrell, R.C., Darby, S.E. et al. (2013). Complex spatial feedbacks of tephra redistribution, ice melt and surface roughness modulate ablation on tephra covered glaciers. *Earth Surface Processes and Landforms* 38 (1): 95–102.

Nye, J.F. (1957). The distribution of stress and velocity in glaciers and ice-sheets. *Proceedings of the Royal Society of London. Series A. Mathematical and Physical Sciences* 239 (1216): 113–133.

Ó Cofaigh, C. (2013). GLACIAL LANDFORMS, SEDIMENTS: Glaciomarine sediments and ice-rafted debris. In: *Encyclopedia of Quaternary Science*, 2e (ed. S.A. Elias and C.J. Mock), 30–42. Amsterdam: Elsevier https://doi.org/10.1016/B978-0-444-53643-3.00085-6.

Oerlemans, J. (2012). Linear modelling of glacier length fluctuations. *Geografiska Annaler: Series A, Physical Geography* 94: 183–194.

Østrem, G. (1959). Ice melting under a thin layer of moraine, and the existence of ice cores in moraine ridges. *Geografiska Annaler* 41: 228–230.

Ottesen, D. and Dowdeswell, J.A. (2006). Assemblages of submarine landforms produced by tidewater glaciers in Svalbard. *Journal of Geophysical Research: Earth Surface* 111 (F1): 1–16.

Ottesen, D. and Dowdeswell, J.A. (2022). Distinctive iceberg ploughmarks on the mid-Norwegian margin: tidally influenced chains of pits with implications for iceberg drift. *Arctic, Antarctic, and Alpine Research* 54 (1): 163–175.

Ottesen, D., Dowdeswell, J.A., Bellec, V. K., and Bjarnadóttir, L.R. (2017). The geomorphic imprint of glacier surges into open-marine waters: examples from eastern Svalbard. *Marine Geology* 392: 1–29.

Palmer, A.P., Bendle, J.M., MacLeod, A. et al. (2019). The micromorphology of glaciolacustrine varve sediments and their use for reconstructing palaeoglaciological and palaeoenvironmental change. *Quaternary Science Reviews* 226: 105964. https://doi.org/10.1016/j.quascirev.2019.105964.

Palmer, A.P., Matthews, I.P., Lowe, J.J. et al. (2020). A revised chronology for the growth and demise of Loch Lomond Readvance ('Younger Dryas') ice lobes in the Lochaber area, Scotland. *Quaternary Science Reviews* 248: 106548. https://doi.org/10.1016/j.quascirev.2020.106548.

Palmer, A.P., Rose, J., Lowe, J.J., and Walker, M.J.C. (2008). Annually laminated Late Pleistocene sediments from Llangorse Lake, South Wales, UK: a chronology for the pattern of ice wastage. *Proceedings of the Geologists Association* 119: 245–258. https://doi.org/10.1016/S0016-7878(08)80304-5.

Pattyn, F., Huyghe, A., De Brabander, S., and De Smedt, B. (2006). Role of transition zones in marine ice sheet dynamics. *Journal of Geophysical Research* 111: F02004. https://doi.org/10.1029/2005jf000394.

Pedersen, S.A.S. (2005). Structural analysis of the Rubjerg Knude Glaciotectonic Complex, Vendsyssel, northern Denmark. *Geological Survey of Denmark and Greenland Bulletin* 8: 192.

Pfeffer, W.T., Arendt, A.A., Bliss, A. et al. (2014). The Randolph Glacier Inventory: a globally complete inventory of glaciers. *Journal of Glaciology* 60: https://doi.org/10.3189/2014JoG13J176.

Phillips, E., Carter, G., and Teasdale, T. (2022). Microscale evidence of the role of water during the emplacement of mass flows in glacial environments. *Quarterly Journal of Engineering Geology and Hydrogeology* 55: 1–20. https://doi.org/10.1144/qjegh2021-026.

Phillips, E. and Kearsey, T.I. (2020). Mass flow and hydrofracturing during Late Devensian moraine emplacement, NE Scotland. *Proceedings of the Geologists Association* 131: 730–750. https://doi.org/10.1016/j.pgeola.2020.08.002.

Phillips, E.R., Evans, D.J., van der Meer, J.J., and Lee, J.R. (2018). Microscale evidence of liquefaction and its potential triggers during soft-bed deformation within subglacial traction tills. *Quaternary Science Reviews* 181: 123–143.

Pisarska-Jamroży, M., Machowiak, K., and Krzyszkowski, D. (2010). Sedimentation style of a Pleistocene kame terrace from the Western Sudety Mountains, S Poland. *Geologos* 16 (2): 101–110.

Porter, S. (1989). Some geological implications of average quaternary glacial conditions. *Quaternary Research* 32: 245–261.

Powell, R.D. and Molnia, B.F. (1989). Glacimarine sedimentary processes, facies and morphology of the south-southeast Alaska shelf and fjords. *Marine Geology* 85: 359–390.

Powers, M.C. (1953). A new roundness scale for sedimentary particles. *Journal of Sedimentary Research* 23: 117–119.

Price, R.J. (1970). Moraines at Fjallsjokull, Iceland. *Arctic and Alpine Research* 2: 27–42.

Pudsey, C.J. and Evans, J. (2001). First survey of Antarctic sub-ice shelf sediments reveals Mid-Holocene ice shelf retreat. *Geology* 29: 787–790. https://doi.org/10.1130/0091-7613(2001)029<0787:FSOASI>2.0.CO;2.

Pudsey, C.J., Murray, J.W., Appleby, P., and Evans, J. (2006). Ice shelf history from petrographic and foraminiferal evidence, Northeast Antarctic Peninsula. *Quaternary Science Reviews* 25: 2357–2379. https://doi.org/10.1016/j.quascirev.2006.01.029.

Quincey, D.J. and Glasser, N.F. (2009). Morphological and ice-dynamical changes on the Tasman Glacier, New Zealand, 1990–2007. *Global and Planetary Change* 68: 185–197. https://doi.org/10.1016/j.gloplacha.2009.05.003.

Randolph Glacier Inventory Consortium (2017). Randolph glacier inventory–a dataset of global glacier outlines: Version 6.0: technical report, global land ice measurements from space, Colorado, USA. Digit. Media.

Ravier, E., Buoncristiani, J.-F., Guiraud, M. et al. (2014). Porewater pressure control on subglacial soft sediment remobilization and tunnel valley formation: a case study from the Alnif tunnel valley (Morocco). *Sedimentary Geology* 304: 7195.

Rea, B.R. (2009). Defining modern day Area-Altitude Balance Ratios (AABRs) and their use in glacier-climate reconstructions. *Quaternary Science Reviews* 28 (3–4): 237–248.

Rea, B.R. and Evans, D.J. (2011). An assessment of surge-induced crevassing and the formation of crevasse squeeze ridges. *Journal of Geophysical Research: Earth Surface* 116 (F4).

Rea, B.R. and Evans, D.J.A. (2003). Plateau icefield landsystems. In: *Glacial Landsystems* (ed. D.J.A. Evans), 407–431. Oxford: Arnold.

Rea, B.R., Walley, B.W., Rainey, M.M., and Gordon, J.E. (1996). Blockfields, old or new? Evidence and implications from some plateaus in northern Norway. *Geomorphology* 15: 109–121.

Rea, B.R., Whalley, W.B., Evens, D.J.A. et al. (1998). Plateau icefields: geomorphology and dynamics. *Journal of Quaternary Science* 13 (6): 35–54.

Regnéll, C., Briner, J.P., Haflidason, H. et al. (2022). Deglaciation of the Scandinavian Ice Sheet and a Younger Dryas ice cap in the outer Hardangerfjorden area, southwestern Norway. *Boreas* https://doi.org/10.1111/bor.12568.

Reinardy, B.T., Booth, A.D., Hughes, A.L. et al. (2019). Pervasive cold ice within a temperate glacier–implications for glacier thermal regimes, sediment transport and foreland geomorphology. *The Cryosphere* 13 (3): 827–843.

Reinardy, B.T., Leighton, I., and Marx, P.J. (2013). Glacier thermal regime linked to processes of annual moraine formation at Midtdalsbreen, southern Norway. *Boreas* 42 (4): 896–911.

Reinardy, B.T.I., Pudsey, C.J., Hillenbrand, C.-D. et al. (2009). Contrasting sources for glacial and interglacial shelf sediments used to interpret changing ice flow directions in the Larsen Basin, Northern Antarctic Peninsula. *Marine Geology* 266: 156–171. https://doi.org/10.1016/j.margeo.2009.08.003.

Rempel, A.W. (2008). A theory for ice-till interactions and sediment entrainment beneath glaciers. *Journal of Geophysical Research: Earth Surface* 113 (F1): 1–20.

Rignot, E., Mouginot, J., and Scheuchl, B. (2011). Ice flow of the Antarctic ice sheet. *Science* 333 (6048): 1427–1430. https://doi.org/10.1126/science.1208336.

Rignot, E., Mouginot, J., Scheuchl, B. et al. (2019). Four decades of Antarctic Ice Sheet mass balance from 1979–2017. *Proceedings of the National Academy of Sciences* 116: 1095–1103. https://doi.org/10.1073/pnas.1812883116.

Rippin, D.M., Sharp, M., Van Wychen, W., and Zubot, D. (2020). 'Detachment' of icefield outlet glaciers: catastrophic thinning and retreat of the Columbia Glacier (Canada). *Earth Surface Processes and Landforms* 45: 459–472.

Roberts, D.H. and Long, A.J. (2005). Streamlined bedrock terrain and fast ice flow, Jakobshavns Isbrae, West Greenland: implications for ice stream and ice sheet dynamics. *Boreas* 34 (1): 25–42.

Roberts, M.J., Russell, A.J., Tweed, F.S., and Knudsen, Ó. (2001). Controls on englacial sediment deposition during the November 1996 jökulhlaup, Skeiðarárjökull, Iceland. *Earth Surface Processes and Landforms* 26: 935–952.

Rose, J. (1991). Subaerial modification of glacier bedforms immediately following ice wastage. *Norsk Geografisk Tidsskrift* 45: 143–153. https://doi.org/10.1080/00291959108552267.

Rowan, A.V., Egholm, D.L., Quincey, D.J., and Glasser, N.F. (2015). Modelling the feedbacks between mass balance, ice flow and debris transport to predict the response to climate change of debris-covered glaciers in the Himalaya. *Earth and Planetary Science Letters* 430: 427–438.

Ruddiman, W.F., Raymo, M.E., Prell, W.L., and Kutzbach, J.E. (1997). The uplift-climate connection: a synthesis. *Tectonic Uplift and Climate Change. Springer* 471–515.

Russell, A.J. and Knudsen, O. (1999). Controls on the sedimentology of the November 1996 jökulhlaup deposits, Skeiðarársandur, Iceland. *Fluvial Sedimentology VI* 315–329.

Russell, A.J., Knudsen, Ó., Maizels, J.K., and Marren, P.M. (1999). Channel cross-sectional area changes and peak discharge calculation on the Gígjukvísl during the November 1996 jökulhlaup, Skeiðarársandur, Iceland. *Jökull* 47: 1–14.

Russell, J.K., Edwards, B.R., Porritt, L., and Ryane, C. (2014). Tuyas: a descriptive genetic classification. *Quaternary Science Reviews* 87: 70–81.

Scherler, D., Wulf, H., and Gorelick, N. (2018). Global assessment of supraglacial debris-cover extents. *Geophysical Research Letters* 45: 11,798–11,805. https://doi.org/10.1029/2018GL080158.

Schmidt, L.S., Aðalgeirsdóttir, G., Pálsson, F. et al. (2020). Dynamic simulations of Vatnajökull ice cap from 1980 to 2300. *Journal of Glaciology* 66: 97–112. https://doi.org/10.1017/jog.2019.90.

Schomacker, A. (2010). Expansion of ice-marginal lakes at the Vatnajökull ice cap, Iceland, from 1999 to 2009. *Geomorphology* 119 (3-4): 232–236. https://doi.org/10.1016/j.geomorph.2010.03.022.

Schomacker, A., Benediktsson, Í.Ö., and Ingólfsson, Ó. (2014). The Eyjabakkajökull glacial landsystem, Iceland: geomorphic impact of multiple surges. *Geomorphology* 218: 98–107.

Schoof, C. (2007). Ice sheet grounding line dynamics: steady states, stability, and hysteresis. *Journal of Geophysical Research: Earth Surface* 112: https://doi.org/10.1029/2006jf000664.

Scourse, J.D., Chiverrell, R.C., Smedley, R. K. et al. (2021). Maximum extent and readvance dynamics of the Irish sea ice stream and Irish Sea Glacier since the last glacial maximum. *Journal of Quaternary Science* 36 (5): 780–804.

Sevestre, H. and Benn, D.I. (2015). Climatic and geometric controls on the global distribution of surge-type glaciers: implications for a unifying model of surging. *Journal of Glaciology* 61 (228): 646–662.

Sevestre, H., Benn, D.I., Luckman, A. et al. (2018). Tidewater glacier surges initiated at the terminus. *Journal of Geophysical Research: Earth Surface* 123 (5): 1035–1051.

Sharp, M. (1984). Annual moraine ridges at Skálafellsjökull, south-east Iceland. *Journal of Glaciology* 30 (104): 82–93.

Sharp, M. (1985). "Crevasse-fill" ridges—a landform type characteristic of surging glaciers? *Geografiska Annaler: Series A, Physical Geography* 67 (3–4): 213–220.

Shepherd, A., Ivins, E., Rignot, E. et al. (2018). Mass balance of the Antarctic Ice Sheet from 1992 to 2017. *Nature* 558: 219–222. https://doi.org/10.1038/s41586-018-0179-y.

Shepherd, A., Ivins, E.R., Barletta, A.G.V. R. et al. (2012). A reconciled estimate of ice-sheet mass balance. *Science* 338: 1183–1189. https://doi.org/10.1126/science.1228102.

Shugar, D.H., Burr, A., Haritashya, U.K. et al. (2020). Rapid worldwide growth of glacial lakes since 1990. *Nature Climate Change* https://doi.org/10.1038/s41558-020-0855-4.

Siegert, M.J. (2008). Antarctic subglacial topography and ice-sheet evolution. *Earth Surface Processes and Landforms* 33: 646–660.

Siman-Tov, S., Stock, G.M., Brodsky, E.E., and White, J.C. (2017). The coating layer of glacial polish. *Geology* 45: 987–990. https://doi.org/10.1130/G39281.1.

Simms, A.R., Lisiecki, L., Gebbie, G. et al. (2019). Balancing the last glacial maximum (LGM) sea-level budget. *Quaternary Science Reviews* 205: 143–153.

Smellie, J.L. (2018). Glaciovolcanism: A 21st century proxy for palaeo-ice. In: *Past glacial environments*, 2nde (ed. J. Menzies, v. der Meer, and J.M. Jaap), 335–375. Elsevier.

Smellie, J.L. (2023). Sedimentation associated with glaciovolcanism: a review. In: *Volcanic Processes in the Sedimentary Record: When Volcanoes Meet the Environment*, vol. 520 (ed. A. Di Capua, R. De Rosa, G. Kereszturi, et al.), 121–164. Geological Society of London, Special Publications.

Smellie, J.L., Johnson, J.S., McIntosh, W. C. et al. (2008). Six million years of glacial history recorded in volcanic lithofacies of the James Ross Island Volcanic Group, Antarctic Peninsula. *Palaeogeography, Palaeoclimatology, Palaeoecology* 260 (1–2): 122–148.

Smith, J.A., Graham, A.G.C., Post, A.L. et al. (2019). The marine geological imprint of Antarctic ice shelves. *Nature Communications* 10: 1–16.

Smith, J.A., Hillenbrand, C.-D., Kuhn, G. et al. (2011). Deglacial history of the West Antarctic Ice Sheet in the western Amundsen Sea Embayment. *Quaternary Science Reviews* 30: 488–505. https://doi.org/10.1016/j.quascirev.2010.11.020.

Spagnolo, M., Clark, C.D., Ely, J.C. et al. (2014). Size, shape and spatial arrangement of mega-scale glacial lineations from a large and diverse dataset. *Earth Surface Processes and Landforms* 39 (11): 1432–1448.

Spagnolo, M., Clark, C.D., and Hughes, A. L. (2012). Drumlin relief. *Geomorphology* 153: 179–191.

Spagnolo, M., Clark, C.D., Hughes, A.L., and Dunlop, P. (2011). The topography of drumlins; assessing their long profile shape. *Earth Surface Processes and Landforms* 36 (6): 790–804.

Spencer, A.M. (1971). Late pre-Cambrian glaciation in Scotland. *Geological Society of London Memoirs* 6: 98. https://doi.org/10.1144/GSL.MEM.1971.006.01.02.

Stokes, C.R. (2018). Geomorphology under ice streams: moving from form to process. *Earth Surface Processes and Landforms* 43 (1): 85–123.

Stokes, C.R. and Clark, C.D. (1999). Geomorphological criteria for identifying Pleistocene ice streams. *Annals of Glaciology* 28: 67–74.

Stokes, C.R., Fowler, A.C., Clark, C.D. et al. (2013). The instability theory of drumlin formation and its explanation of their varied composition and internal structure. *Quaternary Science Reviews* 62: 77–96.

Stokes, C.R., Spagnolo, M., and Clark, C.D. (2011). The composition and internal structure of drumlins: complexity, commonality, and implications for a unifying theory of their formation. *Earth-Science Reviews* 107 (3–4): 398–422.

Storrar, R.D., Ewertowski, M., Tomczyk, A.M. et al. (2020). Equifinality and preservation potential of complex eskers. *Boreas* 49 (1): 211–231.

Storrar, R.D., Stokes, C.R., and Evans, D.J. (2013). A map of large Canadian eskers from Landsat satellite imagery. *Journal of Maps* 9 (3): 456–473.

Sturm, M., Hall, D.K., Benson, C.S., and Field, W.O. (1991). Non-climatic control of glacier-terminus fluctuations in the Wrangell and Chugach Mountains, Alaska, USA. *Journal of Glaciology* 37 (127): 348–356.

Sugden, D.E. and John, B.S. (1976). *Glaciers and landscape: a geomorphological approach*, vol. 365. London: Edward Arnold.

Sund, M., Lauknes, T.R., and Eiken, T. (2014). Surge dynamics in the Nathorstbreen glacier system, Svalbard. *The Cryosphere* 8: 623–638.

Sutherland, J.L., Carrivick, J.L., Gandy, N. et al. (2020). Proglacial lakes control glacier geometry and behavior during recession. *Geophysical Research Letters* 47: e2020GL088865.

Teller, J.T. (2003). Subaquatic landsystems: large proglacial lakes. *Glacial Landsystems* 348–371.

Teller, J.T. (2013). Lake Agassiz during the Younger Dryas. *Quaternary Research* 80: 361–369. https://doi.org/10.1016/j.yqres.2013.06.011.

Terry, J.P. and Goff, J. (2014). Megaclasts: Proposed revised nomenclature at the coarse end of the Udden-Wentworth grain-size scale for sedimentary particles. *Journal of Sedimentary Research* 84: 192–197. https://doi.org/10.2110/jsr.201419.

Thomas, G.S.P. (1984). Sedimentation of a sub-aqueous esker-delta at Strabathie, Aberdeenshire. *Scottish Journal of Geology* 20: 9–20. https://doi.org/10.1144/sjg20010009.

Thomas, G.S.P. and Connell, R.J. (1985). Iceberg drop, dump, and grounding structures from Pleistocene glacio-lacustrine sediments, Scotland. *Journal of Sedimentary Research* 55: 243–249.

Thorndycraft, V.R., Bendle, J.M., Benito, G. et al. (2019). Glacial lake evolution and Atlantic-Pacific drainage reversals during deglaciation of the Patagonian Ice Sheet. *Quaternary Science Reviews* 203: https://doi.org/10.1016/j.quascirev.2018.10.036.

Tofaif, S., Le Heron, D.P., and Melvin, J. (2019). Development of a Palaeovalley complex on a late Ordovician glaciated margin in NW Saudi Arabia. *Geological Society Special Publication* https://doi.org/10.1144/SP475.8.

Trabant, D.C., Waitt, R.B., and Major, J.J. (1994). Disruption of Drift glacier and origin of floods during the 1989–1990 eruptions of Redoubt Volcano, Alaska. *Journal of volcanology and geothermal research* 62 (1–4): 369–385.

Tulaczyk, S., Kamb, W.B., and Engelhardt, H.F. (2000). Basal mechanics of Ice Stream B, west Antarctica: 1. Till mechanics. *Journal of Geophysical Research* 105 (B1): 463–481. https://doi.org/10.1029/1999JB900329.

Turner, K.J., Fogwill, C.J., McCulloch, R. D., and Sugden, D.E. (2005). Deglaciation of the eastern flank of the North Patagonian Icefield and associated continental-scale lake diversions. *Geografiska Annaler: Series A, Physical Geography* 87: 363–374.

van der Veen, C.J. (2002). Calving glaciers. *Progress in Physical Geography* 26: 96–122. https://doi.org/10.1191/0309133302pp327ra.

Van der Vegt, P., Janszen, A., and Moscariello, A. (2012). Tunnel valleys: current knowledge and future perspectives. *Geological Society, London, Special Publications* 368: https://doi.org/10.1144/SP368.13.

van der Wateren, F.M. (1995). Structural geology and sedimentology of push moraines: processes of soft sediment deformation in a glacial environment and the distribution of glaciotectonic styles. *Mededelingen Rijks Geologische* 54: 1–168.

Van Wyk de Vries, M., Bingham, R.G., and Hein, A.S. (2018). A new volcanic province: an inventory of subglacial volcanoes in West Antarctica. In: *Exploration of Subsurface Antarctica: Uncovering Past Changes and Modern Processes*, vol. 461 (ed. M.J. Siegert, S.S. R. Jamieson, and D.A. White), 231–248. Geological Society of London, Special Publications.

Vandyk, T.M., Wu, G., Davies, B.J. et al. (2019). Temperate glaciation on a Snowball Earth: Glaciological and palaeogeographic insights from the Cryogenian Yuermeinak Formation of NW China. *Precambrian Research* 331: https://doi.org/10.1016/j.precamres.2019.105362.

Vehling, L., Baewert, H., Glira, P. et al. (2017). Quantification of sediment transport by rockfall and rockslide processes on a proglacial rock slope (Kaunertal, Austria). *Geomorphology* 287: 46–57. https://doi.org/10.1016/j.geomorph.2016.10.032.

Vieli, A. (2011). Tidewater glaciers. In: *Encyclopedia of Snow, Ice and Glaciers* (ed. V.P. Singh, P. Singh, and U.K. Haritashya), 1175–1179. Springer.

Walder, J.S., LaHusen, R.G., Vallance, J.W., and Schilling, S.P. (2007). Emplacement of a silicic lava dome through a crater glacier: Mount St Helens, 2004–06. *Annals of Glaciology* 45: 14–20.

Waller, R.I. (2001). The influence of basal processes on the dynamic behaviour of cold-based glaciers. *Quaternary International* 86: 117–128. https://doi.org/10.1016/s1040-6182(01)00054-4.

Waller, R.I., Murton, J.B., and Kristensen, L. (2012). Glacier–permafrost interactions: processes, products and glaciological implications. *Sedimentary Geology* 255–256: 1–28.

Warren, W.P. and Ashley, G.M. (1994). Origins of the ice-contact stratified ridges (eskers) of Ireland. *Journal of Sedimentary Research* 64 (3a): 433–449.

Watson, C.S., Quincey, D.J., Carrivick, J. L., and Smith, M.W. (2017). Ice cliff dynamics in the Everest region of the Central Himalaya. *Geomorphology* 278: 238–251.

Weber, P., Boston, C.M., Lovell, H., and Andreassen, L.M. (2019). Evolution of the Norwegian plateau icefield Hardangerjøkulen since the 'Little Ice Age'. *The Holocene* 29 (12): 1885–1905.

Weertman, J. (1957). On the sliding of glaciers. *Journal of Glaciology* 3 (21): 33–38.

Westoby, M.J., Glasser, N.F., Brasington, J. et al. (2014). Modelling outburst floods from moraine-dammed glacial lakes. *Earth-Science Reviews* 134: 137–159.

Whalley, W.B., Gordon, J.E., and Thompson, D.L. (1981). Periglacial features on the margins of a receding plateau ice cap, Lyngen, North Norway. *Journal of Glaciology* 27: 492–496.

Wilson, A.M. and Russell, J.K. (2018). Quaternary glaciovolcanism in the Canadian Cascade volcanic arc–Paleoenvironmental implications. In: *Field Volcanology: A Tribute to the Distinguished Career of Don Swanson*, vol. 538 (ed. M. Poland, M. Garcia, V. Camp, and A. Grunder), 133–157. Geological Society of America Special Papers.

Wilson, A.M., Russell, J.K., and Ward, B.C. (2019). Paleo-glacier reconstruction in southwestern British Columbia, Canada: a glaciovolcanic model. *Quaternary Science Reviews* 218: 178–188.

Wilson, R., Glasser, N.F., Reynolds, J.M. et al. (2018). Glacial lakes of the Central and Patagonian Andes. *Global and Planetary Change* 162: 275–291. https://doi.org/10.1016/j.gloplacha.2018.01.004.

Winsborrow, M.C.M., Clark, C.D., and Stokes, C.R. (2010). What controls the location of ice streams? *Earth-Science Reviews* 103: 45–59.

Zecchin, M., Catuneanu, O., and Rebesco, M. (2015). High-resolution sequence stratigraphy of clastic shelves IV: high-latitude settings. *Marine and Petroleum Geology* 68: 427–437.

Zecchin, M., Rebesco, M., Lucchi, R.G. et al. (2016). Buried iceberg-keel scouring on the southern Spitsbergenbanken, NW Barents Sea. *Marine Geology* 382: 68–79.

Zemp, M., Frey, H., Gärtner-Roer, I. et al. (2015). Historically unprecedented global glacier decline in the early 21st century. *Journal of Glaciology* 61: 745–762. https://doi.org/10.3189/2015JoG15J017.

Zemp, M., Huss, M., Thibert, E. et al. (2019). Global glacier mass changes and their contributions to sea-level rise from 1961 to 2016. *Nature* 568: 382–386.

Zheng, W., Pritchard, M.E., Willis, M.J., and Stearns, L.A. (2019). The possible transition from glacial surge to ice stream on Vavilov Ice Cap. *Geophysical Research Letters* 46 (23): 13892–13902.

Zieliński, T. and Van Loon, A.J. (2000). Subaerial terminoglacial fans III: overview of sedimentary characteristics and depositional model. *Netherlands Journal of Geosciences* 79 (1): 93–107.

Zielinski, T. and Van Loon, A.J. (2002). Present-day sandurs are not representative of the geological record. *Sedimentary Geology* 152 (1–2): 1–5.

Zoet, L.K., Muto, A., Rawling, J.E., and Attig, J.W., (2019). The effects of tunnel channel formation on the Green Bay Lobe, Wisconsin, USA. *Geomorphology* 324: 36–47.

Zoet, L.K., Rawling, J.E. III, Woodard, J.B. et al. (2021). Factors that contribute to the elongation of drumlins beneath the Green Bay Lobe, Laurentide Ice Sheet. *Earth Surface Processes and Landforms* 46 (13): 2540–2550.

15

Extraterrestrial Sedimentology

Mathieu Lapôtre[1], *Lauren Edgar*[2,3], *and Sanjeev Gupta*[4]

[1] *Department of Earth & Planetary Sciences, Stanford University, Stanford, CA, USA*
[2] *U.S. Geological Survey, Astrogeology Science Center, Flagstaff, AZ, USA*
[3] *NASA Johnson Space Center, Houston, TX, USA*
[4] *Department of Earth Science and Engineering, Imperial College London, London, UK*

Summary

The past 50 years of planetary exploration have revealed that other bodies in our Solar System have extensive sedimentary environments and records that are in many ways like those found on Earth. These "extraterrestrial" sedimentary environments are shaped by a diverse array of aeolian, fluvial, deltaic, glacial and mass-wasting processes. In particular, the surfaces of Mars, Saturn's moon Titan and Venus suggest that these planetary bodies host sedimentary processes that are also known to occur on Earth. However, while other planetary bodies share similar sedimentary features, there are body-specific environmental parameters, such as gravity, surface temperature, surface composition and atmospheric density, which result in unique and distinct sedimentary systems. This chapter largely focuses on planetary bodies with well-characterised sedimentary systems, including Mars, Venus, and Titan, with additional discussions of sedimentary processes on Pluto, comet 67P/Churyumov-Gerasimenko and elsewhere in our Solar System. Although we review the state-of-the-art knowledge and understanding of planetary sedimentary processes, environments and records, it is important to recognise that the field of extraterrestrial sedimentology is very active and is continually updated as new data become available. We first present a brief history of planetary sedimentological research through data obtained from orbiters, landers, rovers and helicopters. Second, we review sediment transport processes and key differences between planets. Third, we review current knowledge about the sedimentary geology of Mars, including the controls on sediment production, accumulation and diagenesis, the distribution of modern and ancient sedimentary systems, the planet's stratigraphy as observed by rovers and orbiters and sedimentary facies observed by rovers to date. Next, we focus on the sedimentary systems and geology of Venus and Titan. Finally, we turn an eye towards the future of the emerging field of planetary sedimentology.

15.1 Introduction

15.1.1 A Brief History of Planetary Sedimentological Research

One of the key questions driving planetary exploration is to determine when and where life arose in the Solar System. Sedimentary environments on planetary surfaces record important interactions among the planet's solid surface, hydrosphere and atmosphere and are thus natural targets for studies of past habitability. The field of extraterrestrial sedimentology has its roots in early observations of changing dust deposition on Mars based on telescope observations (Carpenter 1948; McLaughlin 1954; Sagan and Pollack 1969). Early applications also included a sedimentological approach to evaluating lunar clastic rock samples (Lindsay 1972), but subsequent analyses focused more on volcanic and impact-related processes. With the increased resolution provided by the Mariner 9 and Viking orbiters as they flew around Mars in the 1970s, observations of the sedimentologic and stratigraphic records of Mars began to expand (Figure 15.1). Data from these spacecraft revealed valley networks and outflow channels (Sharp and Malin 1975; Carr 1987, 1996). And yet, for many decades, Mars was still regarded as a dominantly volcanic planet, with later aeolian modification (Masursky 1973; Cutts and Smith 1973; McCauley 1973; Edgett and

Reading's Sedimentary Environments: Processes, Facies and Stratigraphy, Fourth Edition. Edited by Bruce Levell.

Companion website: www.wiley.com/go/levell4e

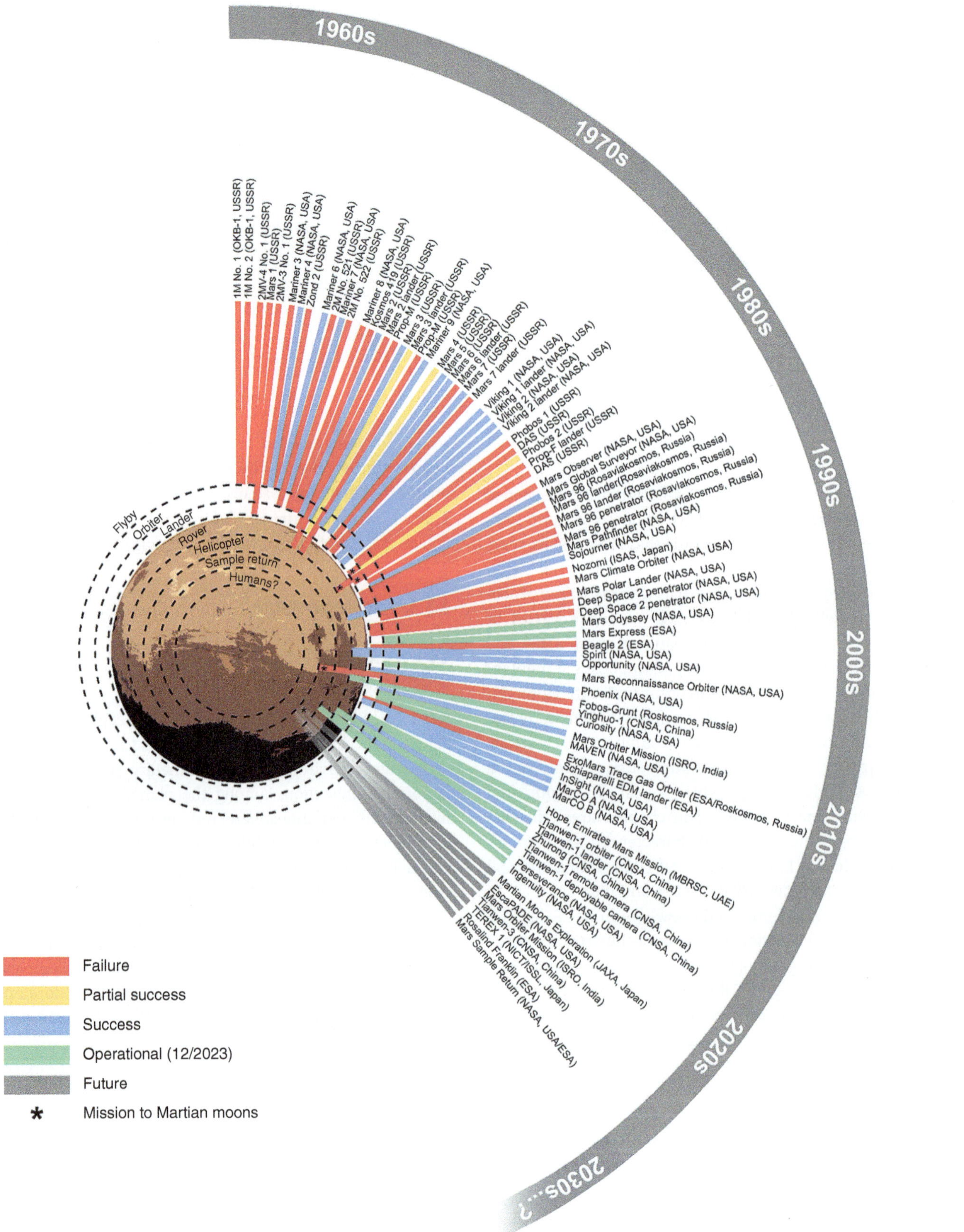

Figure 15.1 Timeline of past, present and planned missions to Mars (CNSA, China National Space Administration; ESA, European Space Agency; ISAS, Institute of Space and Astronautical Science; ISRO, Indian Space Research Organisation; ISSL, Intelligent Space Systems Laboratory; JAXA, Japanese Aerospace Exploration Agency; NASA, National Aeronautics and Space Administration; NICT, National Institute of Information and Communications Technology; MBRSC, Mohamed Bin Rashid Space Center).

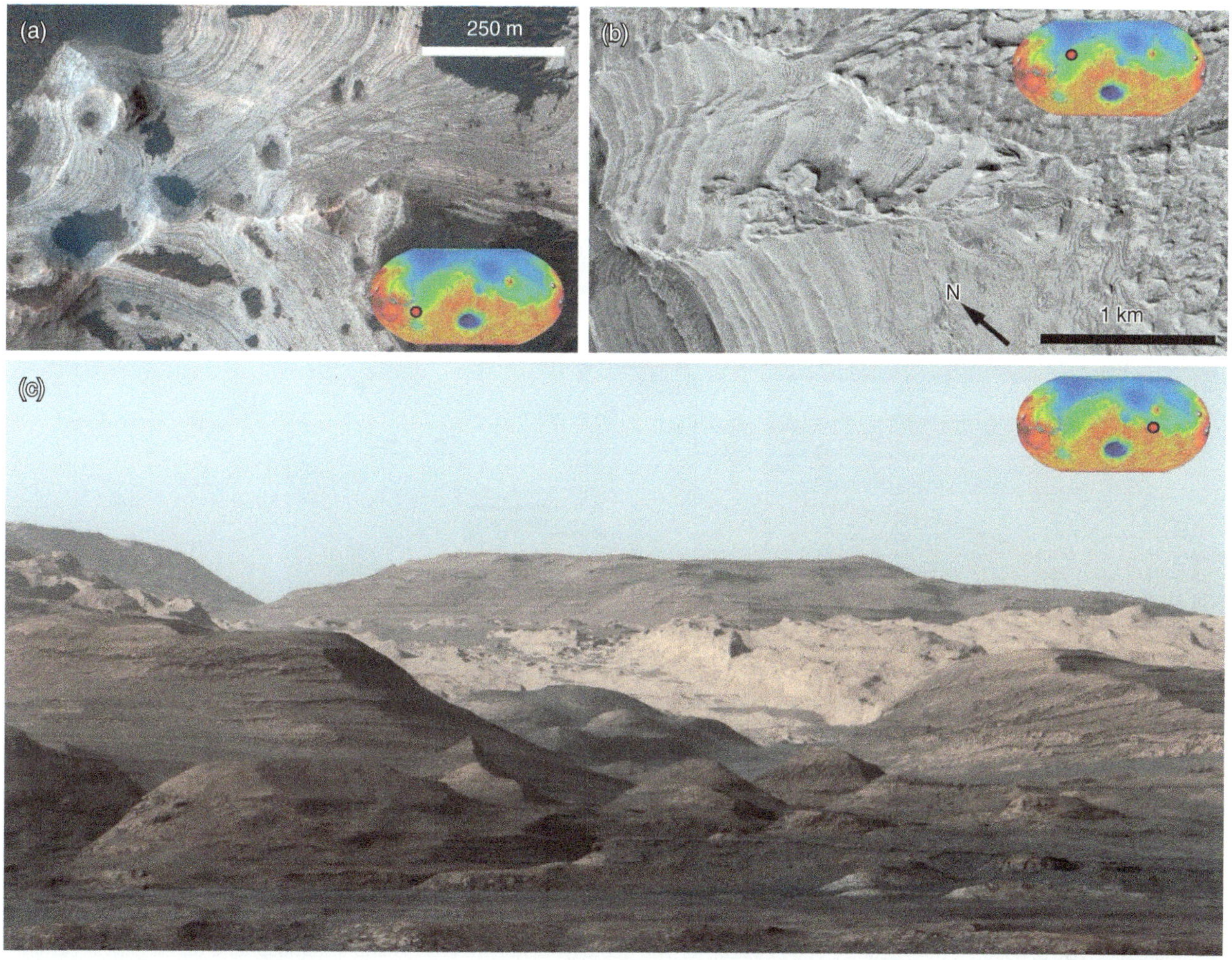

Figure 15.2 Thick sequences of layered sedimentary deposits: (a) Near Holden crater (HiRISE ESP_064288_1525), (b) in Becquerel crater (HiRISE PSP_001546_2015) and (c) in *Mount Sharp* (formally, Aeolis Mons) as seen by NASA's Curiosity rover in Gale crater, Mars (mosaic of Mastcam images; credit: Photos by NASA/JPL- Caltech/MSSS / Public domain). The Mastcam mosaic in (c) was acquired by Curiosity over 3 km away from the ridge (*Vera Rubin ridge*; Edgar et al. 2020; Fraeman et al. 2020) visible in the foreground. The summit of Mount Sharp rises over 4 km above that ridge.

Lancaster 1993). It was the arrival of the Mars Global Surveyor orbiter in 1997 and its subsequent decade of observations that revealed that exposures of sedimentary rocks were widespread on Mars (Malin and Edgett 2000a). Evidence for up to km-thick, cliff-forming stratigraphic sequences (Figure 15.2) suggested that early Mars had an extensive sedimentary history involving both subaqueous and subaerial processes (Box 15.1). In turn, the arrival of the Mars Odyssey spacecraft in 2001 provided the first direct and definitive identification of vast amounts of water ice in the polar regions and subsurface.

The turning point in the field of extraterrestrial sedimentology was the arrival of the Mars Exploration Rovers in early 2004 (Squyres et al. 2004a,b). The ability to conduct ground-based investigations at the level of detail available in terrestrial geology greatly advanced the field. The twin rovers, known as Spirit and Opportunity, each contained a suite of instruments capable of acquiring full colour, panoramic and stereoscopic images at the outcrop and hand-lens scales, along with related geochemical instruments. Observations from the Opportunity rover led to the discovery of ancient aeolian strata with intermittent subaqueous deposition (Squyres et al. 2004b; Grotzinger et al. 2005; Metz et al. 2009a; Squyres et al. 2009; Hayes et al. 2011; Edgar et al. 2012, 2014) and clay minerals suggestive of prolonged contact with groundwater (Arvidson et al. 2014) at Meridiani Planum (Box 15.1; see also Box 15.1 for Martian toponyms). Meanwhile, the Spirit rover uncovered evidence for pyroclastic sedimentation and subsequent aeolian reworking (Squyres et al. 2004a; Squyres et al. 2007; Lewis et al. 2008a). Both rovers confirmed an important role of sedimentary processes on Mars. These ground-based observations were paired with the arrival of the Mars Reconnaissance Orbiter (MRO) in 2006, including the High Resolution Imaging Science Experiment (HiRISE) camera, which is capable of capturing images ~5–6 km wide at 25 cm/pixel (McEwen et al. 2007, 2010). This unprecedented level of detail revealed a complex sedimentary history and rich stratigraphic record. Also onboard the MRO spacecraft is the Context Camera (CTX), which acquires

Box 15.1 Mars: a brief geological primer

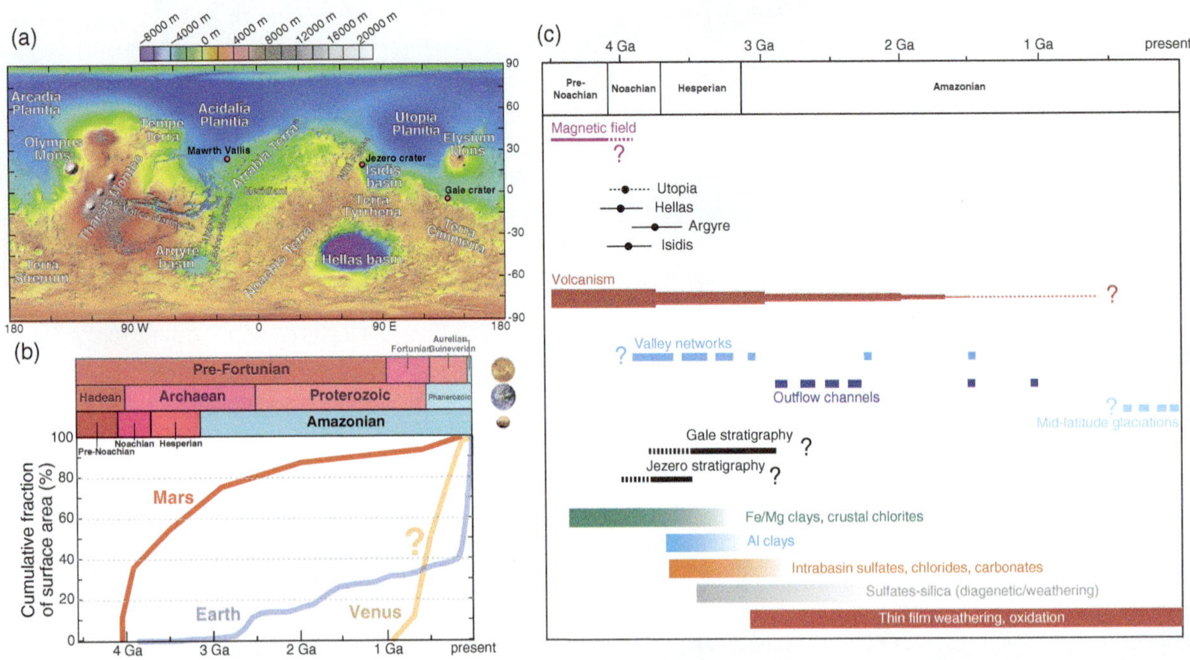

Figure 15.1 (a) Martian topography is dominated by a hemispheric crustal dichotomy, separating more heavily cratered (older) terrains to the south – the southern highlands – and smoother (younger) surfaces in the north – the northern lowlands. Second-order topographic features include large impact basins, volcanic edifices and Valles Marineris – a ~4000 km-long canyon. The map shows common toponyms used in this chapter. (b) Cumulative surface area of surfaces of different ages across Mars, Earth (both continental and oceanic crust) and Venus. The surface of Mars (red line) preserves a record of some of the oldest sedimentary environments in the Solar System. The Martian geologic timescale is divided into four epochs – the pre-Noachian, Noachian, Hesperian and Amazonian – the last three of which are defined from the spatial density of impact craters in terrains of corresponding ages. This comparison of crater-retention ages of Martian terrains with those of Venus and terrestrial crustal ages highlights the uniqueness of the Martian geologic record (a,b: adapted from Lapôtre et al. 2022a). (c) The geologic history of Mars is one of atmospheric loss and desiccation. Upon cessation of the Martian dynamo and formation of large impact basins, Mars hosted an active surface hydrological cycle, forming valley networks under a climate that was conducive to clay formation, depositing sediments in terminal basins. As liquid water became more scarce at the surface, with most flows occurring in the form of catastrophic outburst floods, evaporitic environments became common. By the early Amazonian, liquid water was still stable in the subsurface, facilitating diagenesis of previously deposited sediments, whereas Martian surface rocks were largely left to oxidise. Intermittently, orbital cycles have permitted the migration of water ice from the poles to lower latitudes, creating glaciers at mid-latitudes and on Tharsis in the late Amazonian (and possibly earlier). Compiled and adapted from Ehlmann et al. 2011a, 2016; Ehlmann and Edwards 2014; Hepburn et al. 2019.

~30 km-wide images at a spatial resolution of ~6 m/pixel (Malin et al. 2007), providing lower resolution but greater spatial coverage and important context for the HiRISE images.

The picture of Mars that emerged from the 2000s was one of a planet with a long history of active sedimentary processes, right up to the present day. This wealth of data changed the Mars exploration strategy from "follow the water" to "identify past habitable environments", which ultimately shaped the design of the Mars Science Laboratory "Curiosity" and Mars 2020 "Perseverance" rovers by the National Aeronautics and Space Administration (or NASA). These mobile laboratories are actively exploring ancient fluvial, deltaic, lacustrine and aeolian records in Gale and Jezero craters (e.g. Vasavada 2022; Mangold et al. 2021) (Box 15.1). The Perseverance rover is conducting the most complex drilling activities ever attempted on another planetary body (Beaty et al. 2019). It is now widely recognised that Mars has a rich sedimentary record (Grotzinger and Milliken 2012; McLennan et al. 2019; Edgett and Sarkar 2021; Lapôtre et al. 2022b). Current and future missions will continue to refine our understanding of the physical and chemical processes recorded on its surface and in its atmosphere (Nielsen 2004; Saunders et al. 2004; Zurek and Smrekar 2007; Grotzinger et al. 2012; Jakosky et al. 2015; López-Valverde et al. 2018; Farley et al. 2020; Sharaf et al. 2020; Balaram et al. 2021; Tian et al. 2021; Yongliao et al. 2021).

Meanwhile, planetary missions to other bodies in the Solar System have shown that Mars is not our only neighbour likely to have a rich sedimentary history. On Titan, the Cassini spacecraft flybys and Huygens descent probe provided tantalising evidence for surprisingly Earth-like, active rivers, lakes and sand dunes, but likely made of hydrocarbons (Lorenz et al. 2006, 2008; Stofan et al. 2007; Lopes et al. 2019). In contrast to Mars and Titan, Venus remains largely unexplored, mainly due to its 93-bar atmosphere, with just a few successful flybys and landings by the USSR and the United States between the 1960s and 1980s. Most of what is known about Venus' surface resulted from the Soviet Venera missions in the 1970s and 1980s (Florensky et al. 1977; Florensky et al. 1983; Garvin et al. 1984; Surkov and Barsukov 1985), and NASA's Magellan mission in the early 1990s (Saunders et al. 1992), with episodic new observations through flybys and orbiters by various space agencies in subsequent decades. Sediments on Venus are thought to be largely derived from impacts (Carter et al. 2023). The Venusian surface hosts limited aeolian deposits (Greeley et al. 1992, 1995) and possible yardangs and landslides (Carter et al. 2023). Although no unequivocal geomorphic evidence for past liquid flows has been identified to date, lava-flow embayment along the margins of tesserae (heavily deformed terrains characterized by high radar backscatter) has been hypothesized to highlight possible remnants of water-carved valleys (Khawja et al. 2020).

Surfaces of other planetary bodies, imaged during relatively brief flybys or orbiting missions, also display a range of sedimentary processes. For example, NASA's New Horizons spacecraft demonstrated in 2015 that Pluto hosts a variety of glacial processes, including likely nitrogen glaciers grinding over water-ice bedrock (Grundy et al. 2016) with wind-blown dunes of methane ice (Telfer et al. 2018). Comet 67P/Churyumov-Gerasimenko is a small body (acceleration of gravity, $g \sim 0.75 \times 10^{-4}$ m/s^2; Shi et al. 2024) with no atmosphere. The European Space Agency (ESA)'s Rosetta spacecraft revealed episodic transient surface flows driven by the sublimation of near-surface ices and dynamic erosion and deposition, for example, through airfall and mass wasting processes (Jindal et al. 2024), as well as the presence (Thomas et al. 2015; Jia et al. 2017) and evolution of ripples (El-Maarry et al. 2019; Barrington et al. 2023). Finally, dune-like ridge features were identified at the surface of Io by Galileo near an active volcanic edifice (Williams et al. 2002; Bart et al. 2004) and were proposed to form in places where sublimation vapour from lava-SO_2 frost interactions can drive saltation of SO_2 grains (Diniega et al. 2017; McDonald et al. 2022).

The abundance of extraterrestrial sedimentary systems has highlighted the value of comparative planetology in deciphering the history of our Solar System. This extraterrestrial evidence for weathering, erosion, transport and deposition of sediment suggests that our understanding of sedimentary systems on Earth, and perhaps even a "source-to-sink" framework (Section 4.2.1), could be applied to interpreting processes on other planetary bodies (Grotzinger et al. 2013). Not only can Earth guide our understanding of other planetary bodies, but observations from other planetary bodies can, in turn, teach us about the Earth (Lapôtre et al. 2020). For example, studies of other planets can help us to better understand controls on atmospheric evolution (e.g. when and how did Mars lose most of its atmosphere?) and even the mechanics of sedimentary processes. Finally, much of Earth's sedimentary record has been lost to plate tectonics or altered by deformation and metamorphism. In contrast, Mars has preserved a much more complete history of the Solar System's early period (Lapôtre et al. 2022a) (Box 15.1).

15.1.2 Studying Extraterrestrial Sedimentary Systems

15.1.2.1 Flybys, Orbiters, Landers, Rovers and the Helicopter

Observations of sedimentary systems on other planetary bodies have primarily come from orbiting, landed and roving spacecraft and (to date) a helicopter. Each platform type provides a unique perspective on its target. Combined, this range of spacecraft provides images and geochemical data from sub-millimetre to multi-kilometre scales (e.g. Bridges et al. 2016; Stack et al. 2016).

However, there are trade-offs between the spatial (and spectral) resolution provided by these various datasets and their spatial coverage. On Mars, for example, the HiRISE camera onboard the MRO enables imaging of the surface of Mars at a resolution of down to ~25 cm/pixel, but it had only imaged less than 4% of the planet's surface between 2006 and 2024 (McEwen et al. 2024). Another important consideration is the duration of exploration missions in relation to the time scales of active sedimentary processes. To date, "long-term" monitoring has only been possible on Mars over timescales of years to decades. The trade-off of acquiring additional spatial coverage versus repeated imaging of the same locations to monitor changes on shorter timescales requires careful consideration by mission teams.

Mars has been the target of more missions than any other planetary body (Figure 15.1), and therefore our knowledge of its extensive sedimentary record and processes is drawn from a larger dataset than anywhere else but Earth. In select places, sedimentological studies on Mars have reached a level of detail that is comparable to that of terrestrial studies (Section 15.3). To further advance our knowledge of sedimentary environments on Mars, the Mars 2020 Perseverance rover is actively collecting samples that are intended for return to Earth (Kminek et al. 2022; Simon et al. 2023). The Mars 2020 mission delivered the Ingenuity helicopter to the surface of Mars to serve as a technology demonstration (Balaram et al. 2021). Since accomplishing the first powered flight on another planetary body, the Ingenuity helicopter reached many new milestones while also providing a unique vantage point for linking orbital and ground-based observations in Jezero crater (e.g. Sun et al. 2023). While Ingenuity is the only helicopter to have been flown on another planetary body to date, NASA's Dragonfly octocopter is bound to explore Titan's surface in the mid-2030s.

15.1.2.2 Data Types and Methods of Study

Orbiter-based remote sensing enables a multitude of investigations of planetary surfaces, including (but not limited to) visible, spectral and radar observations. Optically thick atmospheres, such as those of Venus and Titan, render observations of planetary surfaces challenging, leaving radar as the instrument of choice to image the surfaces of such bodies. Synthetic-aperture radar (SAR) images of Venus and Titan have resolutions as fine as ~150 m/pixel and ~300 m/pixel, respectively (though noise increases the minimum size of resolvable objects). Radar altimetry also enabled the characterisation of Venus' and Titan's topography (Saunders et al. 1990; Elachi et al. 2004). In contrast, the thin atmosphere of Mars has enabled detailed imaging and compositional studies of the surface from orbiters. Visible images exist across the planet at ~5–6 m/pixel (CTX), enabling the identification of modern and ancient sedimentary landforms and thick exposures of sedimentary rocks at scales greater than a decametre (Dickson et al. 2023a). Stereo imaging also enables the production of digital elevation models (DEMs; e.g. Gwinner et al. 2010) over smaller spatial scales but at finer spatial and vertical resolutions than through laser altimetry (Zuber et al. 1992). From DEMs, detailed geomorphic properties can be measured, including, for example, slopes, channel dimensions and long profiles, sediment volumes, bedform properties and bed thickness distributions. MRO has been observing Mars for over 10 Martian years (or over 19 Earth years). Repeated imaging of specific targets from orbit has also enabled change-detection campaigns to investigate, for example, the formation of gullies, landslides, wind ripple migration and new impact craters (e.g. Dundas et al. 2021; Sutton et al. 2022; Rangarajan et al. 2023). Finally, a variety of spectrometers have been flown to planetary surfaces to better understand their chemical and mineral composition at a variety of spatial and spectral scales (e.g. at visible-shortwave infrared wavelengths; Bibring et al. 2005; Murchie et al. 2007; Barnes et al. 2007), as well as their physical properties (e.g. at thermal infrared wavelengths; Christensen et al. 2003, 2004).

At the other end of the spectrum, studies of planetary sedimentary systems have been greatly enhanced by instruments onboard landed missions, especially on Mars through observations made by the Spirit, Opportunity, Curiosity, Zhurong and Perseverance rovers. Landed remote imagers, including the Panoramic Cameras (Pancam; Bell et al. 2003), Mast Camera (Mastcam; Malin et al. 2017) and Mast Camera Zoom (Mastcam-Z; Bell et al. 2021) and various Navigation Cameras (Navcams), have enabled the identification of sedimentary facies and facies associations. These cameras are capable of stereo imaging, which has, in turn, been used to constrain bedding orientations, measure palaeoflow directions and draw stratigraphic correlations (Hayes et al. 2011; Barnes et al. 2018). The multispectral and panoramic capabilities of these cameras also provide information about the textural and spectral variability of surface materials. By tracking different lithologies and facies along rover traverses, scientists also could construct detailed stratigraphic sections (e.g. Grotzinger et al. 2005, 2015). At the hand-lens scale, instruments mounted at the end of the rover arms, such as the Microscopic Imagers (MI; Herkenhoff et al. 2003), Mars Hand Lens Imager (MAHLI; Edgett et al. 2012) and Wide-Angle Topographic Sensor for Operations and eNgineering (WATSON; Bhartia et al. 2021) have enabled studies of grain size, sorting, rounding and detailed analyses of sedimentary structures. These studies are enhanced by data from the Remote Micro Imagers (RMI) associated with the Chemistry Camera (ChemCam; Maurice et al. 2012) and SuperCam (Maurice et al. 2021) onboard the MSL and Mars 2020 rovers, which provide very high-resolution context images and mosaics and are also used for the identification of grain sizes and sedimentary structures. Finally, in addition to detailed physical sedimentology, these rover missions have collected vast compositional datasets used for mineralogy, chemostratigraphic analyses, diagenetic features characterisation and the search for organic matter. For further suggested reading on sedimentary geochemistry on Mars, see "Further Reading".

15.1.2.3 Naming Conventions

Mars missions use unofficial names (not approved by the International Astronomical Union or the International Commission on Stratigraphy) to keep track of the many individual small-scale geologic features and rock targets and to define informal stratigraphic nomenclature. Throughout this chapter, italics denote all first mentions of such informal names. Since these are informal names, it is convention in the planetary sedimentology literature to lowercase stratigraphic units, that is, "group, formation, member" instead of "Group, Formation, Member". This convention is adopted throughout the chapter.

15.2 Sediment Transport Processes in the Solar System

15.2.1 Modes of Sediment Transport

Deciphering the sedimentary records of other planets requires a quantitative understanding of morphodynamics, that is, of how surface fluid flows, sediments and topography co-evolve. Mechanistic (rather than descriptive) models are particularly important on other planetary bodies because variables such as the acceleration of gravity, sediment density or flow density may differ by orders of magnitude from more familiar terrestrial values (Table 15.1). Furthermore, sedimentary processes that may be rare or absent on Earth may play a significant role in shaping the landscapes of other planetary bodies; for example, frost sublimation is thought to generate sediment transport in Martian gullies (e.g. Dundas et al. 2019a) and on comets (Belton and Melosh 2009; Thomas et al. 2015; Jia et al. 2017). Here, we discuss how planet-specific environmental parameters may affect sediment transport and bedforms with illustrations drawn from extraterrestrial fluvial and aeolian environments. To learn more about how planetary regolith forms in the first place, see Sections 15.3.1 (Mars) and Further Reading.

Fluvial environments are here defined as surface liquid streams eroding landscapes, transporting and depositing sediments along their course and in terminal depositional sinks. They are known to have shaped landscapes on Earth, Mars and Titan (Figure 15.3) and have been hypothesised on Venus. Aeolian environments, in turn, are shaped by gas flows (typically winds) transporting sediment grains, sometimes collecting them into migrating bedforms – ripples and dunes. They have been reported on Venus, Earth, Mars, Titan, Pluto, comet 67P/Churyumov-Gerasimenko (Figure 15.4) and possibly Io (Williams et al. 2002; Bart et al. 2004).

On Earth, fluvial and aeolian environments are characterised by distinct modes of sediment transport (Figure 15.5). Typical terrestrial rivers transport coarser, heavier grains (gravel) in *bedload* – a combination of grain rolling and sliding and ballistic hopping (or *saltation*). Finer particles (sand and mud aggregates or flocs) are transported in *suspension*, spending significant time in the water column but episodically interacting with the sediment bed. Even finer grains (clay particles and colloids) are transported in *washload*, virtually behaving as flow tracers and rarely interacting with the riverbed. Finally, some sediments may also dissolve in water and get transported to depositional sinks in the form of *dissolved load*.

On Earth, wind-blown sediments follow similar transport patterns as fluvial sediments, with coarser grains transported closer to the sediment bed and finer particles travelling higher up in the air column, with the notable difference that saltation-driven grain-splash plays a significant role in mobilising sediments (due to the lesser viscous dampening upon grain–grain collisions). As sand-sized grains are transported through *saltation*, their impact onto the sediment bed triggers the ejection of impacted grains. Through splash, ejected dust-sized grains may become *suspended*, sand-sized grains may saltate and coarser grains may roll or slide. The impact-driven mobilisation of grains into near-bed transport trajectories is often referred to as *reptation*. Together with wind-drag-driven grain rolling and sliding, impact-driven reptation mobilises the *creep* population. For example, grains up to 3 mm in diameter were observed to be actively transported in creep (drag-induced rolling or reptation creep; Baker et al. 2018a) in Gale crater, Mars (see also Section 7.3.2).

The phase (liquid or gas) of the sediment-transporting fluid defines the distinction between fluvial and aeolian environments. It is important to note, however, that the relationship between the transport environment and the modes of sediment transport therein is non-unique. Notably, grain-splash (from grain-to-grain collisions) is known to play a significant role in aeolian saltation but not in fluvial bedload on Earth. However, this dichotomy in the importance of grain-splash does not directly translate to all aeolian and fluvial environments across the Solar System. To a first order, the ratio of sediment to fluid density, $\frac{\rho_s}{\rho_f}$, can be used as a proxy to evaluate whether grain-splash is a significant driver of transport in loose sediments.

Table 15.1 Flow, sediment and planet-specific properties for fluvial and aeolian environments on Venus, Earth, Mars and Titan. See also Grotzinger et al. (2013).

	Venus		Earth		Mars		Titan	
	Flows	Winds	Flows	Winds	Flows	Winds	Flows	Winds
Fluid density, ρ_f (kg/m^3)	n/a	67	1000	1.27	1000 (freshwater) 1300 (brines)	0.02	520–670	5.1
Fluid kinematic viscosity, ν (m^2/s)	n/a	5×10^{-7}	1×10^{-6}	1.4×10^{-5}	1×10^{-6} (freshwater) 4×10^{-5} (brines)	5.4×10^{-4}	4.8×10^{-7}	1.7×10^{-6}
Sediment density, ρ_s (kg/m^3)	2900 (basalt)		2650 (quartz)		2900 (basalt)		940 (ice) 600–1250[a] (organics)	
Acceleration of gravity, g (m/s^2)	8.87		9.81		3.71		1.35	

[a] Most estimates within 600–1000 kg/m^3.
Adapted from Grotzinger et al. (2013).

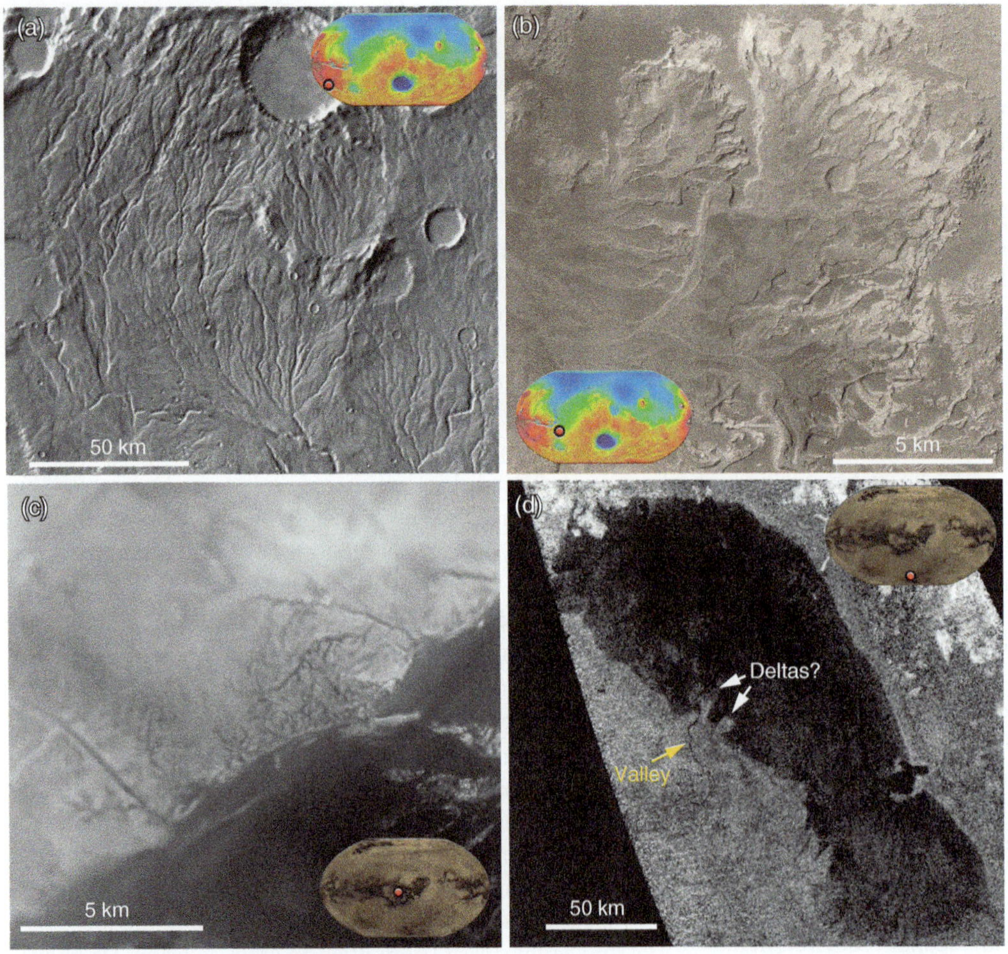

Figure 15.3 Fluvio-lacustrine landforms in the Solar System. (a) Dendritic valley network of Warrego Valles, Mars (THEMIS daytime; credit: Photo by NASA/ JPL-Caltech/ASU / Public domain). (b) Exhumed remnants of the Eberswalde delta deposit, Mars (MGS MOC mosaic; credit: Photos by NASA/JPL- Caltech/MSSS / Public domain). (c) Area surrounding the Huygens landing site during descent, showing dendritic valley networks (Descent Imager; credit: Photo by ESA/NASA/JPL-Caltech/U. of A / Public domain). (d) Ontario Lacus (dark), a hydrocarbon lake near Titan's south pole, with fluvial valleys and putative deltas (Cassini RADAR; credit: Photo by NASA/JPL-Caltech / Public domain).

On Earth, fluvial and aeolian environments are characterised by $\frac{\rho_s}{\rho_f}$ of ~2.5 and 2100, respectively, such that grain-splash must become significant for $\frac{\rho_s}{\rho_f}$ between those values. On planetary bodies with relatively thin atmospheres and dense sediments (e.g. Mars), grain-splash can be expected to contribute significantly to aeolian sediment transport. But what about Venus and Titan? On Venus, $\frac{\rho_s}{\rho_f} \sim 40$ in aeolian environments, suggesting that aeolian sediment transport may be more akin to fluvial bedload transport on Earth, with minimal importance of grain-splash (e.g. Kok et al. 2012; Gunn and Jerolmack 2022). On Titan, the composition and porosity of wind-blown sand are so poorly constrained (e.g. Hörst and Tolbert 2010; Imanaka et al. 2012) that uncertainty in $\frac{\rho_s}{\rho_f}\left(\frac{\rho_s}{\rho_f} \sim 120\text{-}250\right)$ allows for either scenario, such that the importance of grain-splash on Titan remains to be determined.

15.2.2 Fluvial Environments

15.2.2.1 Sediment Transport

Under liquid flows, grains of a given size transition from no motion to bedload, suspension and washload as flow strength increases. Conversely, for a given flow strength, coarser grains are transported in bedload, whereas finer grains are transported in suspension and washload. Quantitatively, the onset of fluvial transport modes in a given fluid and on a given planet can be defined as a function of median grain diameter (d) and the shear stress (τ) exerted by the flow onto the sediment bed (Figure 15.6a,b).

Incipient motion of loose sediments into bedload occurs when destabilising forces exerted by the flow onto sediment grains (drag and lift) exceed stabilising forces (weight and friction) or when τ exceeds some critical value (τ_c). Because τ and τ_c are affected by fluid and sediment properties as well as gravity (g), it is convenient to express the incipient motion criterion in a dimensionless form that can be easily exported to extraterrestrial environments.

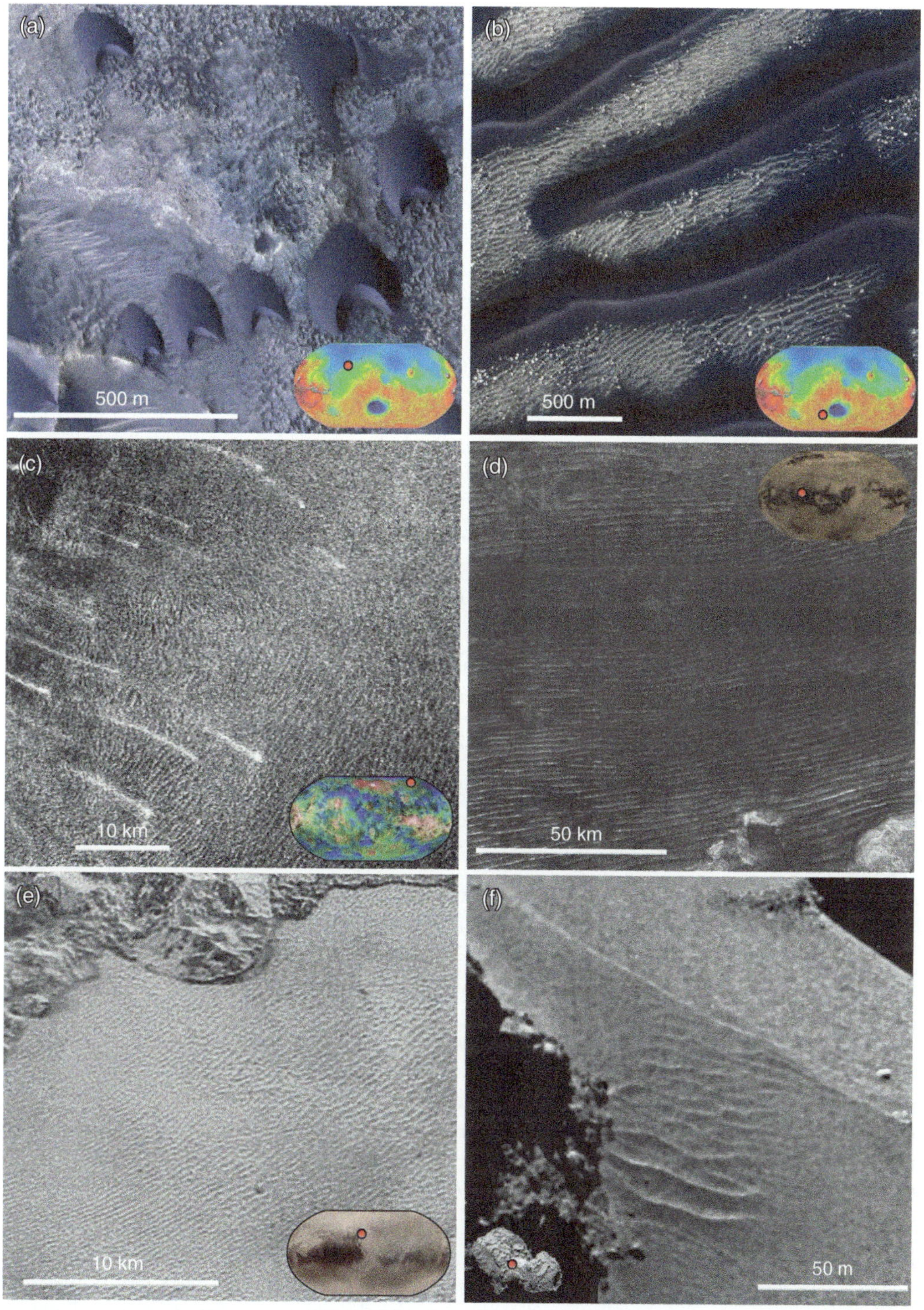

Figure 15.4 Wind-blown dunes in (a) Arabia Terra, Mars (HiRISE), and (b) Proctor crater, Mars (HiRISE). (c) Putative dune field near Al-Uzza Undae, Venus (Magellan radar; credit: Photo by NASA/JPL-Caltech / Public domain). NE–SW linear features are interpreted as dunes; bright SE–NW linear features are interpreted as wind streaks (north is up). (d) Linear dunes in the Belet dune field of Titan (Cassini RADAR, denoised; Charnay et al. 2015 / Springer Nature). (e) Regularly spaced linear ridges on Pluto (LORRI, New Horizons; credit: Photo by NASA/JHU-APL/SWRI / Public domain). (f) Ripple-like features on comet 67P/Churyumov-Gerasimenko (OSIRIS narrow-angle camera, Rosetta; credit: Photo by ESA/Rosetta/MPS / CC BY-SA 3.0 IGO).

To that end, dimensional analysis can provide a set of independent dimensionless numbers to describe sediment transport. It is important to note that the choice of these dimensionless quantities is non-unique, and while we illustrate sediment transport with one set of such dimensionless numbers, other equivalent sets of dimensionless numbers may be found elsewhere in the literature.

A widely used criterion to describe incipient transport in bedload is the so-called Shields criterion, stating that incipient motion occurs when $\tau_* > \tau_{*c}$, where

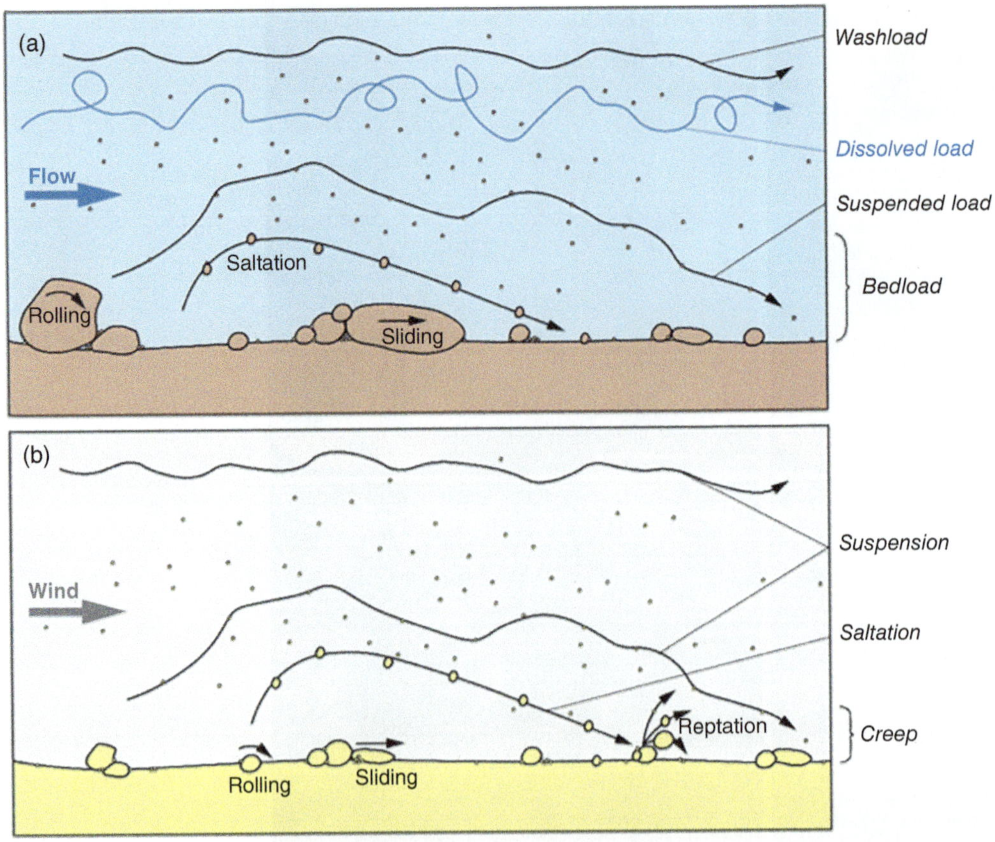

Figure 15.5 Modes of (a) fluvial and (b) aeolian sediment transport on Earth and Mars. On Venus, aeolian sediment transport may be more akin to fluvial sediment transport on Earth (a), with a minimal importance of grain-splash.

$\tau_* = \frac{\tau}{(\rho_s - \rho_f)gd}$ and τ_{*c} is the Shields number, which is a function of the particle Reynolds number, $\mathrm{Re_p} = \frac{u_* d}{\nu}$ (where $u_* = \sqrt{\frac{\tau}{\rho_f}}$ is the flow shear velocity and ν is the kinematic viscosity of the fluid). The dependence of τ_{*c} on $\mathrm{Re_p}$ is shown as the boundary separating the "no motion" and "bedload" fields in Figure 15.6a.

Intuitively, significant suspension should occur when the typical upward velocity of turbulent eddies (u_*) exceeds the settling velocity of sediment grains (w_s). Quantitatively, suspension is expected to occur when $\frac{u_*}{w_s}$ exceeds a critical value, which also varies with $\mathrm{Re_p}$:

$$\left.\frac{u_*}{w_s}\right|_{\mathrm{critical}} = \begin{cases} 21.2\,\mathrm{Re_p}^{-1.2} \text{ for } 1 < \mathrm{Re_p} < 27.3 \\ 0.4 \text{ for } \mathrm{Re_p} \geq 27.3 \end{cases}$$

The low $\mathrm{Re_p}$ formulation arises because under low $\mathrm{Re_p}$, grains sitting on the bed are shielded from turbulent eddies by the laminar viscous sublayer, the thickness of which increases with decreasing $\mathrm{Re_p}$ (Niño et al. 2003). This criterion, in turn, can be rewritten as a function of τ_* and $\mathrm{Re_p}$ (or in dimensional form, τ and d). The onset of suspension is shown in Figure 15.6a as the line separating the "bedload" and "suspension" fields.

Finally, the onset of transport into washload is often assumed to occur when $\frac{u_*}{w_s}$ is about 3-fold that required for suspension (e.g. Lamb and Venditti 2016). Figure 15.6a shows the onset of washload transport under that assumption.

Because it explicitly accounts for sediment and fluid properties as well as gravity, the transport diagram of Figure 15.6a should remain unchanged on other planetary bodies for loose sediment (although note that the effect of gravity on sediment flux is grain-size dependent such that, for example, Earth-based empirical laws describing total sediment flux should not be applied to other planetary bodies; Braat et al. 2024). It can then be recast in dimensional form as a function of bed shear stress, τ, and median grain diameter, d, for specific transport environments, as illustrated in Figure 15.6b for siliciclastic sediments transported by freshwater on Earth. For example, a 1-mm quartz grain in a terrestrial river will transition from no motion to bedload, from bedload to suspension and from suspension to washload transport at bed stresses of ~0.4, 6 and 50 Pa, respectively.

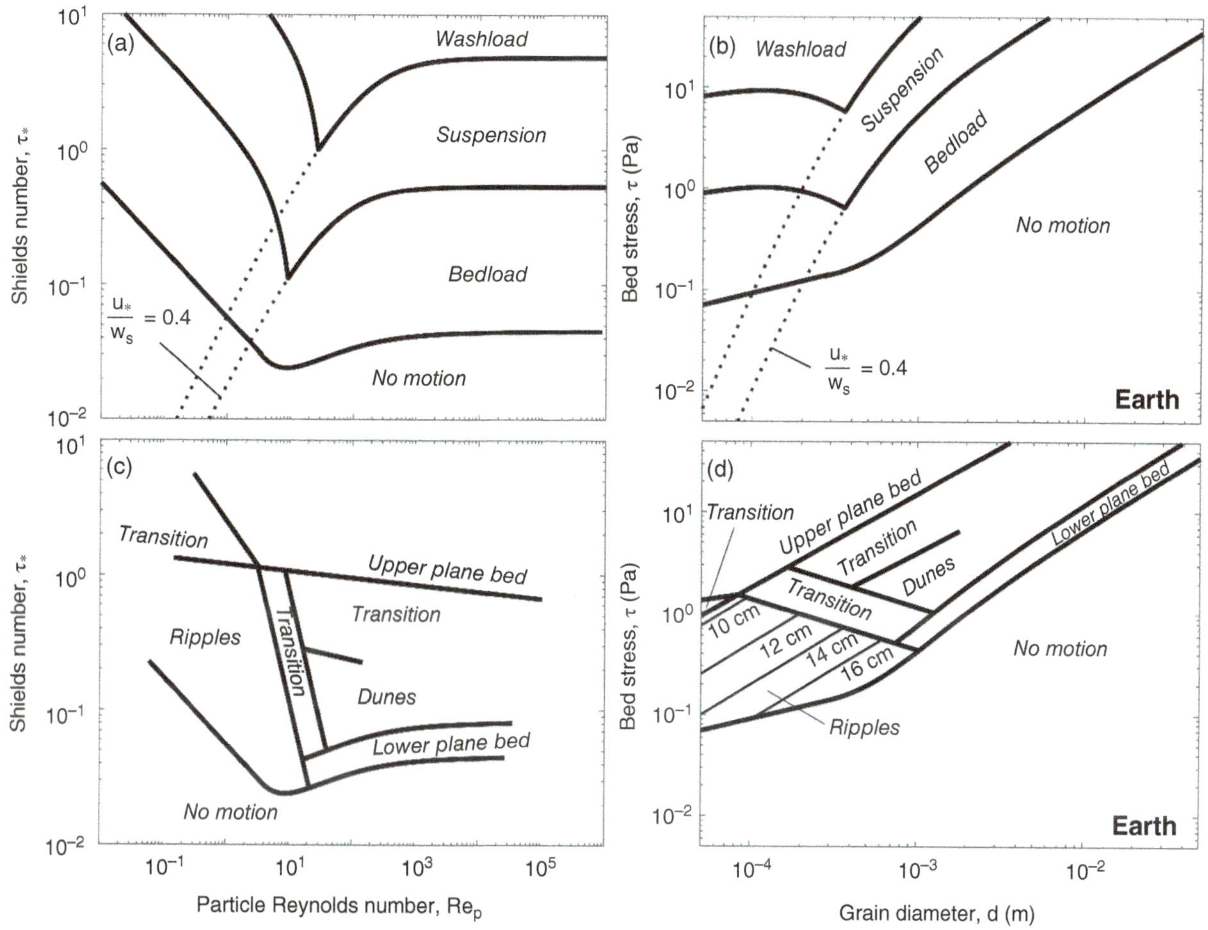

Figure 15.6 (a) Phase diagram of fluvial modes of transport, expressed as a function of flow Shields number and particle Reynolds number (adapted from Lamb and Venditti 2016). (b) Same as (a) but in dimensional form, expressed as a function of bed stress and sediment grain diameter, for quartz grains in freshwater on Earth. In (a) and (b), dotted lines that deviate from the solid boundary at low Re_p represent predictions in the case where the suspension threshold were independent of Re_p. (c) Dimensionless bedform stability diagram for Froude subcritical flows, expressed as a function of flow Shields number and particle Reynolds number (adapted from Lamb et al. 2012, and Lapôtre et al. 2017). (d) Same as (c) but in dimensional form, expressed as a function of bed stress and sediment grain diameter, for quartz grains in freshwater on Earth. Thin contours within the ripple stability field indicate predicted ripple wavelengths (adapted from Lapôtre et al. 2017).

Ancient Martian rivers largely transported basaltic sediments in fresh to moderately saline aqueous flows (e.g. Lamb et al. 2012). Rivers on Titan are thought to be dominated by methane, with some dissolved nitrogen (~80% methane, 20% nitrogen; Hayes 2016; Malaska et al. 2017; Steckloff et al. 2020). Figure 15.7 shows the same transport diagram as predicted for the transport of basaltic sediment by freshwater and briny flows on Mars (Figure 15.7a,b, respectively), icy sediment by methane–nitrogen flows on Titan (Figure 15.7c) and organic sediment of variable density by methane–nitrogen flows on Titan (Figure 15.7d). Because Mars has a lower acceleration of gravity, weaker freshwater flows are capable of mobilising sediments of a given size relative to terrestrial flows (Figure 15.7a). Conversely, briny flows on Mars could have a significantly higher kinematic viscosity than freshwater flows (Table 15.1), which for a given grain size and flow strength, would decrease the particle Reynolds number (and thus, e.g. w_s), shifting the respective fields on the transport diagram to overall coarser grain sizes (Figure 15.7b). Finally, the radically different sediment and fluid types on Titan are thought to be characterised by $\frac{\rho_s}{\rho_f} \sim 1-2$ (possibly up to 3), that is, fluvial sediments are likely more buoyant on Titan than in typical terrestrial and Martian fluvial environments. The low kinematic viscosity of methane–nitrogen mixtures and low acceleration of gravity on Titan (Table 15.1) further result in an overall decrease in the transport threshold bed stresses for a given grain size (Figure 15.7c,d). Similar dimensional analyses can be used to predict other bankfull properties of streams on Mars and Titan (e.g. Birch et al. 2023). Box 15.2 illustrates how sediment transport mechanics can be applied to palaeohydraulic reconstructions on other planetary bodies.

15.2.2.2 Bedforms

As a liquid flows over a granular bed, it may generate a variety of different *bed states*, including certain types of *bedforms*. In a similar way to the transport diagram

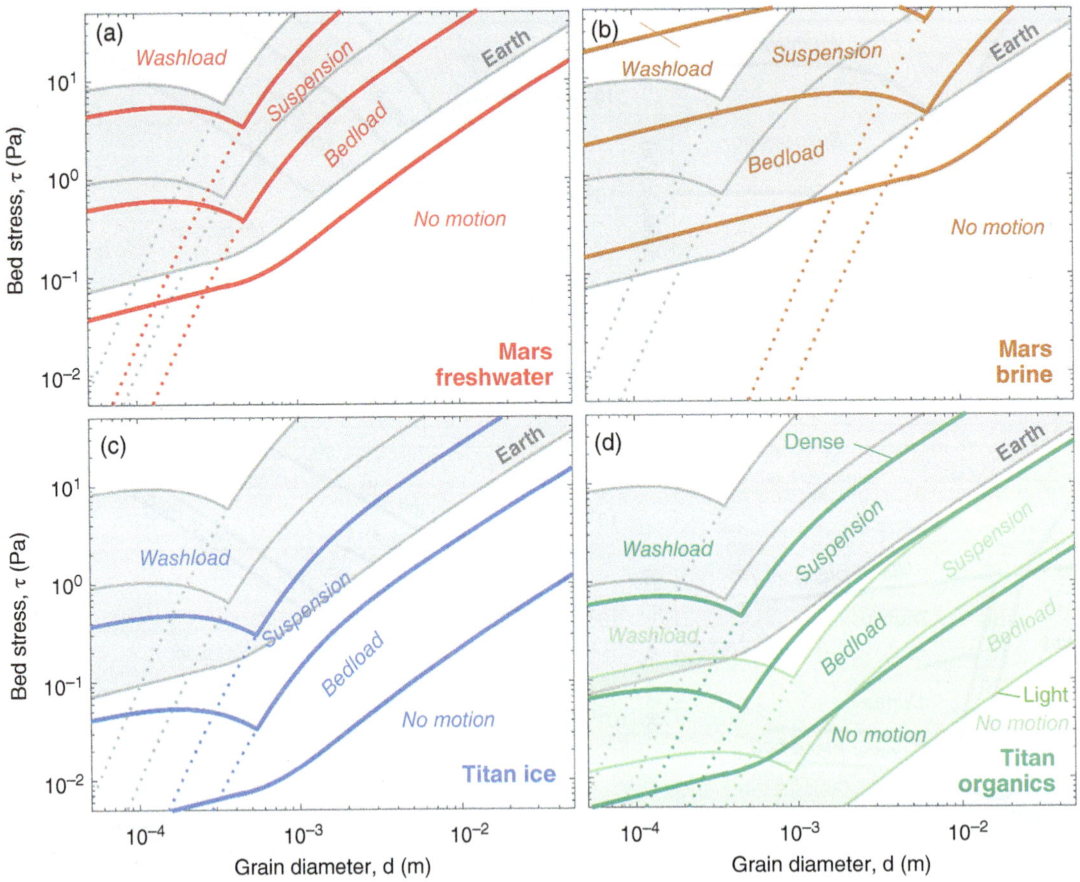

Figure 15.7 Modes of fluvial sediment transport predicted as a function of bed stress and grain diameter for (a) basaltic sediments under freshwater flows on Mars, (b) basaltic sediments under concentrated briny flows on Mars, (c) icy sediments under methane–nitrogen flows on Titan (consistent with Perron et al. 2006) and (d) organic sediments under methane–nitrogen flows on Titan (dark green lines: dense organics, ρ_s = 1250 kg/m^3; shaded light green field: light organics, ρ_s = 600 kg/m^3; consistent with Burr et al. 2006). Fluvial transport on Titan (c, d) is illustrated using a fluid density ρ_f = 520 kg/m^3, though it is known to vary with composition from ~520 to 670 kg/m^3. For non-buoyant grains ($\frac{\rho_s}{\rho_f} > 1$, e.g. for ice and dense organics), transport thresholds in (c, d) decrease by ~30% from ρ_f = 520 to 670 kg/m^3. Conversely, buoyant particles would be transported in washload exclusively. Dotted lines that deviate from the solid boundary at low grain diameters represent predictions in the case where the suspension threshold was independent of Re_p.

Box 15.2 Palaeohydraulic reconstruction on Mars and Titan

Bedload-dominated fluvial flows adjust their geometry to bypass their bed materials. As a result, bed stresses in bedload-dominated rivers are close to the threshold of motion of bedload sediment (e.g. Parker 1978). Thus, in dimensionless form:

$$\tau_* \sim \tau_{*c},$$

where $\tau_* = \frac{\tau}{(\rho_s - \rho_f)gd}$ is the Shields stress (with τ being the bed stress, ρ_s and ρ_f being the sediment and fluid densities, g being the acceleration of gravity, and d being the diameter of bedload sediment grains), and τ_{*c} is the Shields number (which varies with particle Reynolds number and thus, with d). Assuming steady uniform flow conditions in a wide, gently sloping rectangular channel, conservation of fluid momentum dictates that:

$$\tau = \rho_f g h S,$$

where h is the flow depth, and S is the bed slope. Substituting for τ and solving for flow depth yield

$$h \approx \frac{Rd}{S}\tau_{*c},$$

where $R = \frac{(\rho_s - \rho_f)}{\rho_f}$ is the specific submerged density of bed sediments. Thus, one can estimate flow depth from the size of bedload clasts found in fluvial and flood deposits and estimates of bed slope. We illustrate this palaeohydraulic reconstruction with three planetary examples: a pebble conglomerate imaged by Curiosity in Gale crater, Mars, a cobble and boulder-rich conglomerate imaged by Perseverance at Jezero crater, Mars, and rounded ice clasts found on a dry riverbed at the Huygens landing site on Titan.

Grain sizes were measured from Mastcam images at Gale crater (Williams et al. 2013b), SuperCam RMI images at Jezero crater (Mangold et al. 2024) and Descent Imager images on Titan (Tomasko et al. 2005; Perron et al. 2006). Shields numbers were estimated from grain sizes (Perron et al. 2006; Williams et al. 2013b; Mangold et al. 2021). Bed slopes were constrained from terrestrial analogues and a nearby alluvial fan at Gale crater (Williams et al. 2013b), basal and top deposit slopes at Jezero crater (Mangold et al. 2024) and a DEM produced from Descent Imager images during Huygens' descent into the atmosphere of Titan (Daudon et al. 2020). The density of sediment and fluid are given in Table 15.1 – for Mars, we considered basaltic clasts transported by freshwater; on Titan, water ice clasts transported by a methane flow. In all three cases, $Re_p > 100$. We thus assume a value of $\tau_{*c} \sim 0.045$ (Figure 15.6a) in Gale crater; that value is adjusted to 0.1 in Jezero crater and on Titan to account for the large grain size relative to flow depth (Perron et al. 2006) and steep bed slopes (Lamb et al. 2008).

We chose this approach to illustrate applications of palaeohydraulics to planetary sedimentary systems, but calculations could be refined to adopt more realistic assumptions. For example, it is unlikely that the floods that deposited Jezero's boulder-rich conglomerate were steady and uniform, and a high sediment concentration could have further affected boulder transport. Yet, this simple approach yields flow depths in relatively good agreement with those obtained using an independent approach by Mangold et al. (2024) ($h \approx 1.87$-2.53 m).

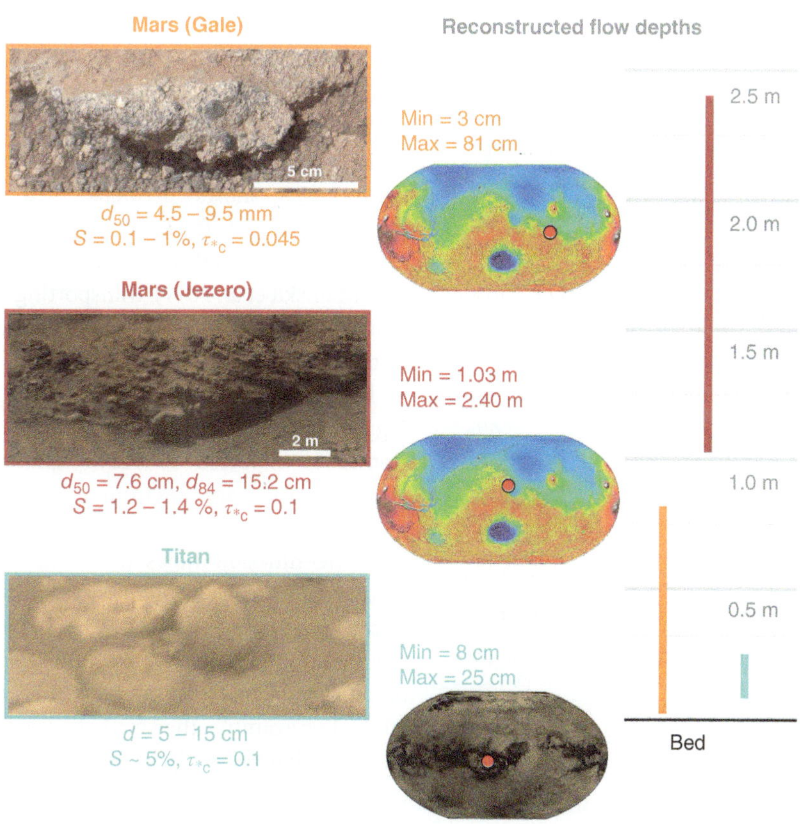

(Figure 15.6a,b), a bedform stability diagram can be represented as a function of τ_* and Re_p (or in dimensional form of τ and d; Figure 15.6c,d). On Earth, *current ripples* tend to emerge on a sandy bed as flow speed increases above the threshold of motion and bedload transport initiates; further increasing flow strength leads to a transition to an *upper plane bed* regime. In contrast, coarser sediment will first form a *lower plane bed* and develop *dunes* before transitioning to an upper plane bed regime as flow strength increases (Figure 15.6d). As flow strength is increased further, the flow Froude number,

$$\mathrm{Fr} = \frac{\overline{U}}{\sqrt{gh}},$$

where $\overline{U}$ is the depth-averaged flow velocity and h is the flow depth, exceeds a value of 1, and flow becomes *supercritical*. Supercritical flows host a variety of other bedform types, including *antidunes*, *chutes-and-pools*, *cyclic steps* and *chevrons* (not shown in Figure 15.6).

Ripples and dunes both arise in unidirectional currents from a hydrodynamic instability that creates a spatial lag between shear stresses exerted by the flow onto the bed and sediment flux and promotes sediment transport upstream and deposition downstream of bedform crests (e.g. Smith 1970; Durán Vinent et al. 2019). On Earth's riverbeds, current ripples (or *drag ripples*) are smaller than dunes, and contrary to dunes, their equilibrium wavelength does not depend on flow depth (e.g. Lapôtre et al. 2017). It was shown that the transition from ripples to dunes is relatively abrupt and occurs for values of the Yalin number, $\chi = \mathrm{Re}_p \tau_*^{1/2}$, between 4 and 9 (Lapôtre et al. 2017). Durán Vinent et al. (2019) proposed that this abrupt transition may be caused by a hydrodynamic anomaly that leads to a shift of the locus of maximum sediment flux downstream of bedform crests for intermediate values of $\mathrm{Re}_\lambda = \frac{\lambda u_*}{\nu}$, where λ is the bedform wavelength (or, conversely, for intermediate values of $\frac{l_{\mathrm{sat}} u_*}{\nu}$, where l_{sat} is the sediment saturation length scale which varies with the mode of sediment transport and which, for bedload transport, is proportional to $\frac{\nu}{u_*}\chi$). Such a shift would preclude the formation of bedforms with wavelengths intermediate to those of ripples and dunes on Earth's riverbeds (i.e. it generates a "bedform gap"; Courrech du Pont et al. 2024). That hydrodynamic anomaly, referred to as the *Hanratty anomaly* (first described in Zilker et al. 1977; Abrams and Hanratty 1985; and Frederick and Hanratty 1988), only occurs under hydrodynamically smooth flow conditions, that is, at low roughness Reynolds numbers, $\mathrm{Re}_k = \frac{k u_*}{\nu} < \sim 20$ (where k is the roughness length scale) as is typical of, for example, fluvial environments on Earth. This anomaly is thought to result from a laminar-to-turbulent transition, triggered by the bedform, resulting in a modification of the structure of the inner boundary layer (e.g. Durán Vinent et al. 2019).

Following the model of Durán Vinent et al. (2019), the equilibrium wavelength of current ripples is limited by the smallest ripple size at which the Hanratty anomaly arises, whereas the equilibrium wavelength of dunes is ultimately limited by flow depth. Using an extensive data compilation from flume experiments and natural rivers, Lapôtre et al. (2017) developed a scaling relationship between Re_λ and χ, which allows the prediction of the equilibrium wavelength of current ripples as a function of sediment and fluid properties, as well as gravity,

$$\mathrm{Re}_\lambda \approx 2500\chi^{1/3},$$

or, in the dimensional form,

$$\lambda \approx 2500\frac{\nu^{2/3} d^{1/6}}{(Rg)^{1/6}\, u_*^{1/3}},$$

where $R = \frac{\rho_s - \rho_f}{\rho_f}$ is the specific submerged density of sediment grains. Predicted ripple wavelengths in siliciclastic grains under freshwater flows on Earth are shown as thin contour lines in Figure 15.6d.

Similar to the transport diagrams of Figure 15.6, Figure 15.8 shows predicted bed states for basaltic sediments under freshwater and briny flows on Mars (Figure 15.8a,b) and methane flows (with dissolved nitrogen; Malaska et al. 2017), transporting icy and organic sediments on Titan (Figure 15.8c,d). In addition to overall shifts in the diagram boundaries that follow similar trajectories as those observed by the transport diagrams shown in Figure 15.7, it is notable that the wavelength of current ripples is expected to vary significantly between the different planetary environments – from decimetre to multi-decimetre wavelengths in freshwater on Mars and methane–nitrogen flows on Titan to potentially multi-metre scale ripples in highly concentrated brines on Mars. Because ripple size (and thus, the thickness of any resulting stratification) is sensitive to fluid properties, they can be used as palaeo-environmental indicators (Lapôtre et al. 2017). For example, they could be used to detect small temperature changes on Titan, where fluid properties are especially sensitive to temperature (Steckloff et al. 2020).

Owing to Mars' lower gravity, it was hypothesised that ancient alluvial rivers might develop steeper bed slopes than their terrestrial counterparts for a given channel width, such that Froude supercritical flows – and resulting bedforms – might have been more common on Mars than they are in Earth's lowland rivers (Konsoer et al. 2018).

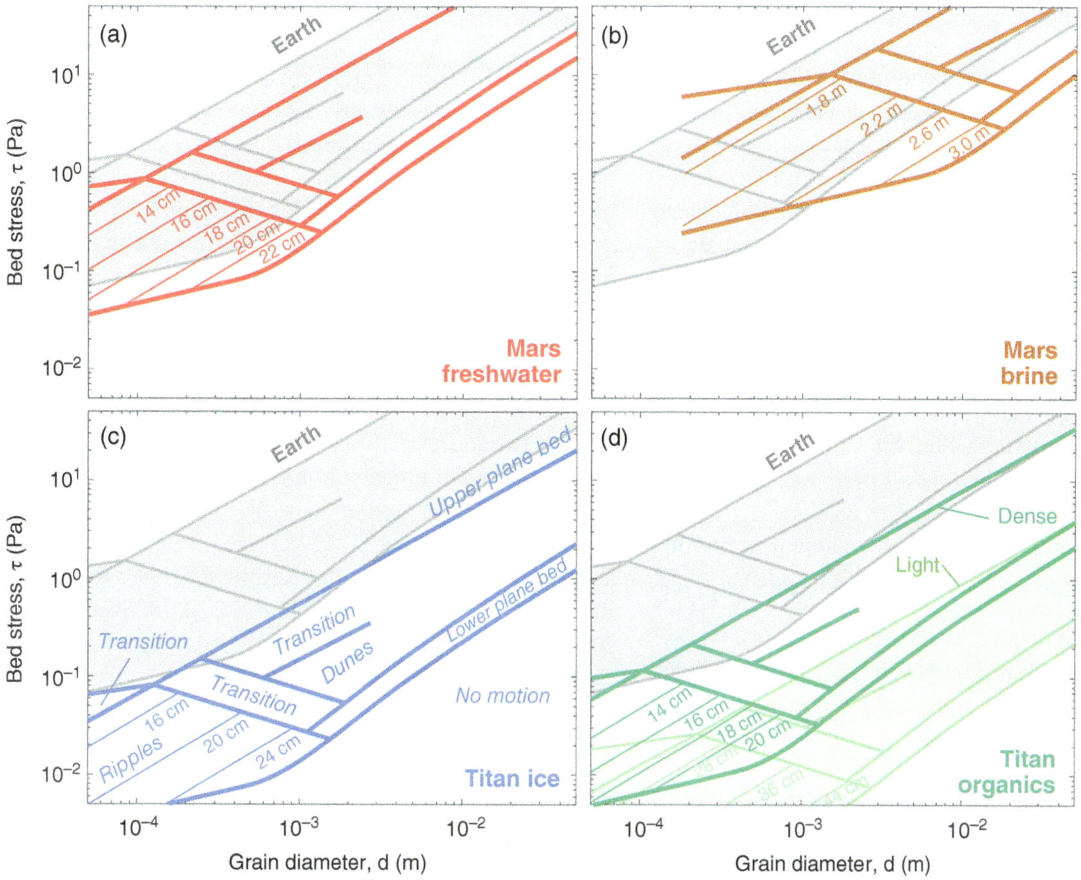

Figure 15.8 Bedform stability diagrams as a function of bed stress and grain diameter for (a) basaltic sediments under freshwater flows on Mars, (b) basaltic sediments under concentrated briny flows on Mars, (c) icy sediments under methane–nitrogen flows on Titan and (d) organic sediments under methane–nitrogen flows on Titan (dark green lines: dense organics, ρ_s = 1250 kg/m^3; shaded light-green field: light organics, ρ_s = 600 kg/m^3). Fluvial bedforms on Titan (c, d) are illustrated using a fluid density ρ_f = 520 kg/m^3, though it is known to vary with composition from ~520 to 670 kg/m^3. For non-buoyant grains ($\frac{\rho_s}{\rho_f} > 1$, e.g. for ice and dense organics), domain boundaries in (c, d) decrease by ~30% (in bed stress) from ρ_f = 520 to 670 kg/m^3. Conversely, buoyant particles would be transported in washload exclusively and thus not contribute to the formation of bedforms. Bedform fields are only indicated in (c) for clarity; they are the same as in Figure 15.6c,d. Thin contours within ripple stability fields indicate predicted ripple wavelengths.

However, this effect could be counteracted, for example, by lower sediment settling velocities (Braat et al., 2024), and others have predicted the bed slopes of Martian rivers to be similar to those of their terrestrial counterparts (e.g. Birch et al., 2023).

Another important difference between Martian fluvial systems and those on the modern Earth is the absence of stabilising overbank vegetation. Without land plants, the lateral migration of meandering channels on Mars was likely relatively rapid (Matsubara et al. 2015; Ielpi and Lapôtre 2020a), possibly leading to extensive reworking of floodplain deposits as is observed in unvegetated basins on Earth today (e.g. Ielpi and Lapôtre 2018, 2019, 2020b; Hasson et al. 2023). These differences affect the resulting facies and stratigraphic architecture of fluvial deposits (Hasson et al., 2025), and the deposits of ancient Martian rivers may be more akin to those of Precambrian rivers than of modern terrestrial streams. Titan's active streams might also offer a window into stream dynamics in unvegetated landscapes, though our knowledge of bank-material properties is exceedingly poor at this time, preventing any analogies to the early Earth until further Titan data are collected.

15.2.3 Aeolian Environments

15.2.3.1 Sediment Transport

As for the incipient motion of sediment grains in rivers, the threshold wind speed for initiation of saltation – or *fluid threshold* – is a function of sediment and fluid properties, as well as gravity and thus can be expected to vary significantly across planetary bodies. Interestingly, on Earth, saltation may be sustained (once initiated) by wind speeds that are ∼20% lower than the fluid threshold – a threshold commonly referred to as the *impact threshold*. This phenomenon is known as *transport hysteresis*. The difference between the fluid and impact thresholds is caused by the relative efficiency of grain-splash upon impact, which on Earth permits saltation to be sustained through grain-splash below the fluid threshold once initiated. The impact threshold is expected to vary significantly across planetary bodies in a similar way to the fluid threshold as the efficiency of atmospheres of variable densities and viscosities, for example, may significantly alter saltation trajectories and grain-impact velocities (e.g. Kok et al. 2012; Gunn and Jerolmack 2022). The impact threshold is thought to be significantly lower than the fluid threshold on Mars (perhaps up to an order of magnitude lower; e.g. Kok 2010; Gunn and Jerolmack 2022).

In contrast, the transport of loose sediment grains in aeolian environments with $\frac{\rho_s}{\rho_f}$ < a few 10s–100s (e.g. on Venus) is more analogous to subaqueous transport as described for fluvial environments above, with wind drag rather than impact splash controlling sediment motion. In these environments, the impact threshold is similar to (or even higher than) the fluid threshold velocity, minimising the importance of grain-splash in triggering and sustaining sediment transport. Significant uncertainty exists, however, about saltation mechanics on Titan, and grain-splash could play an important role if, for example, wind-blown particles on Titan displayed significant cohesion (e.g. Lorenz 2017; Comola et al. 2022). It was also proposed that triboelectrification (electric charge transfer) of wind-blown organic particles could lead to particle aggregation and raise the saltation threshold on Titan (e.g. Mendéz Harper et al. 2017).

Current models for the initiation and cessation of saltation on extraterrestrial planetary bodies are fully derived from field measurements on Earth or wind-tunnel experiments, theoretical and numerical models (e.g. Bagnold 1937, 1941; Greeley et al. 1974, 1976, 1980, 1984; Iversen and White 1982; Iversen and Rasmussen 1999; Kok 2010; Burr et al. 2015, 2020; Martin and Kok 2018; Swann et al. 2020; Andreotti et al. 2021; Gunn and Jerolmack 2022). Figure 15.9 shows example threshold models (from Gunn and Jerolmack 2022) for Venus, Mars and Titan compared

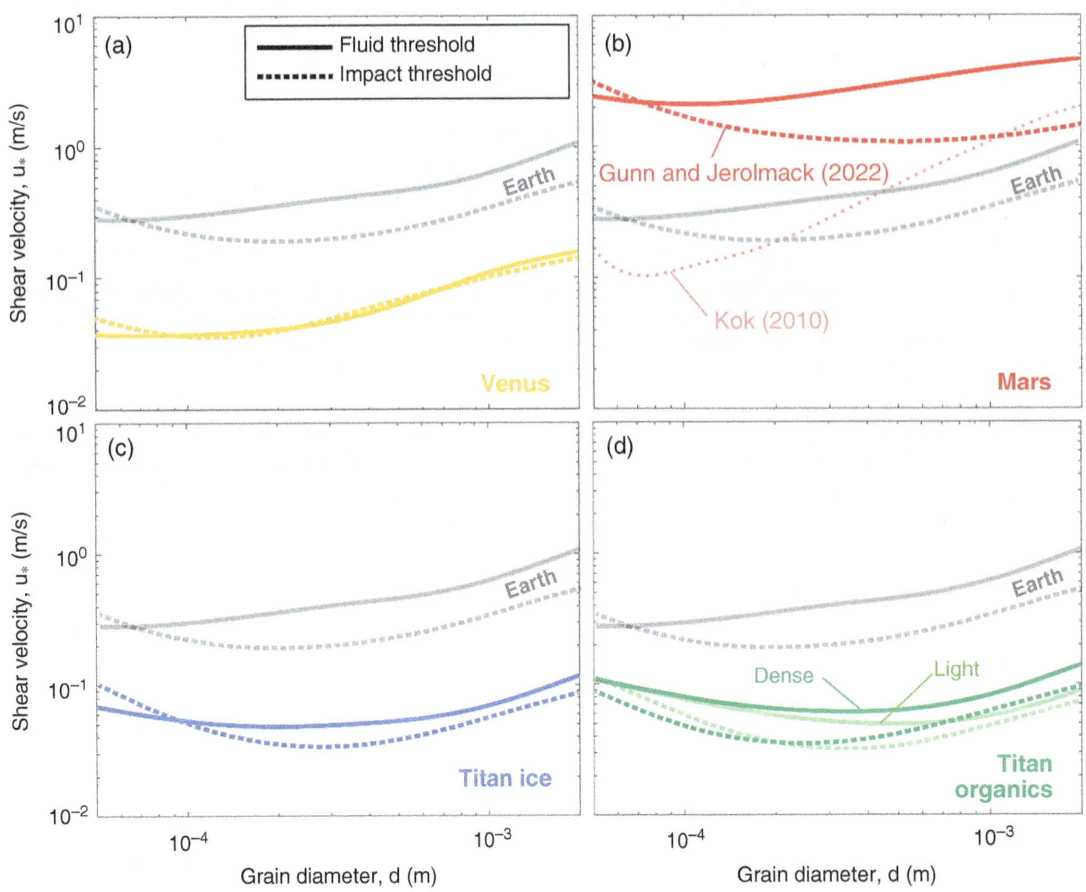

Figure 15.9 Fluid and impact threshold for saltation on (a) Venus, (b) Earth, (c) Mars and (d) Titan (adapted from Gunn and Jerolmack 2022). Two different models are shown for the impact threshold on Mars (adapted from Kok 2010; Gunn and Jerolmack 2022).

to calibrated saltation fluid and impact thresholds for Earth. On Venus, the impact threshold is not expected to be lower than the fluid threshold, suggesting that fluid drag from the thick Venusian atmosphere is the most efficient driver of sediment motion. Conversely, the impact threshold on Mars is thought to be significantly lower than the fluid threshold (although specific predictions vary between models).

Owing to Mars' low atmospheric density and despite the lower acceleration of gravity, the wind speeds required to initiate saltation are thought to be much higher on Mars than on Earth and are typically higher than the speeds of surface winds measured on Mars by landed spacecraft. The prediction of such high wind speeds to initiate sand transport is difficult to reconcile with unambiguous evidence, from orbit and on the ground, that active sand transport occurs globally on Mars (e.g. Baker et al. 2018a,b, 2021). Unfortunately, no in situ or orbiter-based observations have fully succeeded at testing or calibrating transport threshold models on Mars to date (for attempts, see e.g. Ayoub et al. 2014; Baker et al. 2018a,b, 2021).

As a possible solution to the apparent paradox of widespread sand motion on Mars, it was proposed that sand transport might initiate below the fluid threshold (but above the impact threshold) in the form of low-flux saltation clusters (Sullivan and Kok 2017). Specifically, Sullivan and Kok (2017) noted that, although the onset of widespread and continuous sand motion is not expected for winds below the fluid threshold, sporadic motion of individual grains may occur. Such grains, in turn, may develop saltation trajectories prone to splashing other grains into saltation. These cascades of grains travel together in the form of saltation clusters, resulting in relatively low but non-zero sand fluxes.

On Titan, significant uncertainty exists about the composition and density of sediments. Based on existing constraints (Table 15.1), $\frac{\rho_s}{\rho_f}$ is ~120–250 in Titan's aeolian environments. This value makes the efficiency of grain-splash unclear. Assuming that wind-blown sediments on Titan have low cohesion, predicted impact thresholds might be slightly lower than fluid thresholds, leaving open the possibility of transport hysteresis on Titan. However, it is not known whether wind-blown particles are cohesive on Titan, which would also affect the propensity of grain-splash to mobilise sediment particles (e.g. Comola et al. 2022).

15.2.3.2 Bedforms

On Earth, wind-blown deserts primarily host two scales of bedforms – decimetre-scale ripples and tens-of-metres to kilometre-scale dunes. Although these two scales of bedform mirror the ripple and dune populations found on Earth's riverbeds, current and aeolian ripples are thought to form through distinct mechanisms (Anderson 1987). Unlike current ripples, the formation of aeolian ripples is intimately tied to grain-splash (Chapter 7.3.5). As grains saltate above a granular bed, grains sitting on upwind-facing surfaces of bed defects are preferentially exposed to impacts from incoming saltating particles and, as a result, will preferentially be splashed into reptation. Conversely, grains located on downwind-facing surfaces of bed defects tend to be shielded from incoming saltators, leading to a net accumulation of sand grains in those areas. This pattern of erosion and deposition leads to the formation of so-called *impact ripples* (e.g. Sharp 1963), the wavelength of which is controlled by the reptation length scale. Thus, for impact ripples to form, grain-splash needs to be an efficient mechanism of sediment mobilisation. As a result, impact ripples are only expected to form in environments with $\frac{\rho_s}{\rho_f}$ < a few 10s–100s, hence their absence on Earth's riverbeds. *Anti-ripples*, or impact ripples migrating upwind, have recently been predicted to form on Venus and, perhaps, Titan; Lester et al. 2025).

In contrast with impact ripples, wind-blown sand dunes arise from an aerodynamic instability that is analogous to that responsible for the formation of current ripples and subaqueous dunes. However, whereas the roughness Reynolds number is relatively low in Earth's rivers, such that both drag ripples and dunes may coexist (through Hanratty's hydrodynamic anomaly; Section 15.2.2; Durán Vinent et al. 2019), winds in Earth's sandy deserts are typically aerodynamically rough. Accordingly, aeolian ripples are thought to overwhelmingly consist of impact ripples in Earth's wind-blown sandy deserts, though recent experiments suggest that aeolian drag ripples could also form on Earth under specific conditions favouring low roughness Reynolds numbers (Alvarez et al., 2025a,b).

Planetary bodies of the Solar System span a wide range in atmospheric properties, allowing for the possibilities that (i) impact ripples may not exist in all wind-blown landscapes (and anti-ripples might exist instead), and (ii) drag ripples (here defined as the smaller mode of drag bedforms forming under low Re_k) may be formed by winds in some of them. For example, bedforms formed in wind tunnel experiments conducted under Venus-like atmospheric pressure were interpreted as drag bedforms referred to as *microdunes* (Greeley et al. 1984). Existing radar observations of the Venusian surface revealed the presence of dunes but were not able to resolve any smaller, superimposed features (Figure 15.4c).

Unlike Earth, Martian dune fields host three scales of bedforms – decimetre-scale impact ripples (e.g. Yizhaq et al. 2014), metre-scale ripples (here referred to as *large ripples*), and decametre- to kilometre-scale dunes (Lapôtre et al. 2016a; Ewing et al. 2017). These are all forming concurrently in well-sorted, relatively fine-grained sand (Figure 15.10a–c; Weitz et al. 2018). Small impact ripples on Mars are similar to their terrestrial counterparts (e.g. Lapôtre et al. 2016a), and Martian dunes are remarkably similar to terrestrial dunes morphologically, even following similar morphometric scaling relationships (Rubanenko et al. 2022). Large ripples and dunes display similar morphologies, with highly sinuous to straight crestlines oriented transversely to longitudinally to the net sand-flux direction (Lapôtre et al. 2016a, 2018; Silvestro et al. 2016; Vaz et al. 2017), and in some cases, display grainflow and grainfall deposits (Sections 7.4.3 and 7.4.4) along their slipfaces (Lapôtre et al. 2016a; Ewing et al. 2017).

Orbiter-based measurements of large ripple wavelengths globally revealed that wavelength decreases with increasing atmospheric density (Figure 15.10b; Lorenz et al. 2014; Lapôtre et al. 2016a; Vaz et al. 2023), following a trend that is quantitatively consistent with expectations from drag bedforms (such as current ripples; Figure 15.10b). This relationship was interpreted as evidence for distinct formation mechanics between large Martian ripples and smaller impact ripples, and large Martian ripples could be drag ripples enabled by Mars' low-density atmosphere (Lapôtre et al. 2016a). Under that scenario, large ripples are in essence aeolian dunes that form below the "dune gap" (i.e. their wavelength is smaller than that at which the Hanratty anomaly arises), a phenomenon that could occur on Mars as the low atmospheric density gives rise to the Hanratty anomaly under aeolian transport conditions (Durán Vinent et al. 2019; Alvarez et al. 2025b). Also consistent with large Martian ripples being drag ripples, the size of Mars' smallest dunes decreases with increasing atmospheric density, following a quantitative relationship predicted by theory (Rubanenko et al. 2022). In contrast, Sullivan et al. (2020) proposed that large Martian ripples are impact ripples, which can grow larger on Mars owing to low dynamic wind pressures. Under this mechanism, incipient impact ripples would form with decimetre wavelengths and grow larger over time. However, no bedforms with wavelengths in the ~20–80 cm range have been found to date in well-sorted, relatively fine sand within dune fields (e.g. Lapôtre et al. 2016a, 2018, 2021; Weitz et al. 2018, 2022), and such intermediately sized bedforms do not form, even transiently, in low-pressure wind-tunnel experiments (Yizhaq et al. 2024; Alvarez et al. 2025a) – an observation that only the drag-ripple model can explain. Furthermore, the size of impact ripples was recently shown to be controlled by properties of the sediment bed rather than by saltation trajectories, resulting in similar predicted impact-ripple sizes on Earth and Mars (Lester et al. 2025). These theoretical predictions are supported by observations in low-pressure wind-tunnel experiments that impact-ripple size varies little with atmospheric pressure within the Earth-to-Mars range (Miller et al. 1987; Andreotti et al. 2021; Yizhaq et al. 2024; Alvarez et al. 2025a).

In addition to impact and large ripples, which can form and migrate on top of dunes, a variety of other aeolian bedforms are found on flatter surfaces and surrounding dune fields on Mars, including the so-called *megaripples* and *transverse aeolian ridges* (or TARs; Figure 15.11a). The term 'megaripple' has been used to refer to a variety of bedform types over time. On Earth, the term is sometimes used to refer to *coarse-grained ripples* (i.e. ripples forming in polydisperse beds, with a concentration of coarse grains along their crests). On Mars, 'megaripple' was initially used to describe all metre-scale bedforms regardless of their dynamics, but some more recent studies have excluded the large Martian ripples discussed above from the definition of megaripples. Here, we use the term 'megaripples' to refer to metre-scale bedforms that typically display brighter crests and longer wavelengths than neighbouring large ripples in orbiter-based images (Figure 15.11a) without implications for formation mechanism. Such megaripples migrate at a much slower rate than neighbouring large ripples (Silvestro et al. 2020; Chojnacki et al. 2021), suggesting that they may be coarse-grained ripples. Mars rovers have observed coarse-grained ripples from the ground (Figure 15.11b; e.g. Sullivan et al. 2005; Jerolmack et al. 2006; Lapôtre et al. 2018), with grain sizes along the ripples' crests sometimes coarser than bulk sand by a factor of 5 or more (e.g. 500–>1000 μm along the crests versus ~100–200 μm for bulk sand; Weitz et al. 2018, 2022).

TARs are yet another type of aeolian bedforms on Mars, characterised by their multi-metre to decametre wavelengths and overall bright tone (likely due to dust accumulation; Balme et al. 2008; Zimbelman 2010). Although TARs likely migrated in the past (Day 2021), their migration has not been detected to date, such that they are largely thought to be inactive or reworked by smaller bedforms today (Lu et al. 2022; Gou et al. 2022). Their formation mechanism is uncertain; it was proposed that TARs could have formed from the deposition, induration and erosion of fine, dust-sized grains (Geissler 2014), or more likely, that they are the dusty remnants (Lu et al. 2022) of past bedforms that formed through surface creep of coarse grains (e.g. Zimbelman 2010; Zimbelman and Foroutan 2020). A candidate TAR was crossed by the Curiosity rover at *Dingo Gap* in Gale crater, Mars (Figure 15.11c; Zimbelman and Foroutan 2020), and an extensive field of linear to crescentic TARs was explored by the Chinese Zhurong rover in Utopia Planitia (Ding et al. 2022; Gou et al. 2022; Lu et al. 2022; Wang et al. 2023). The term 'TAR' here is used as

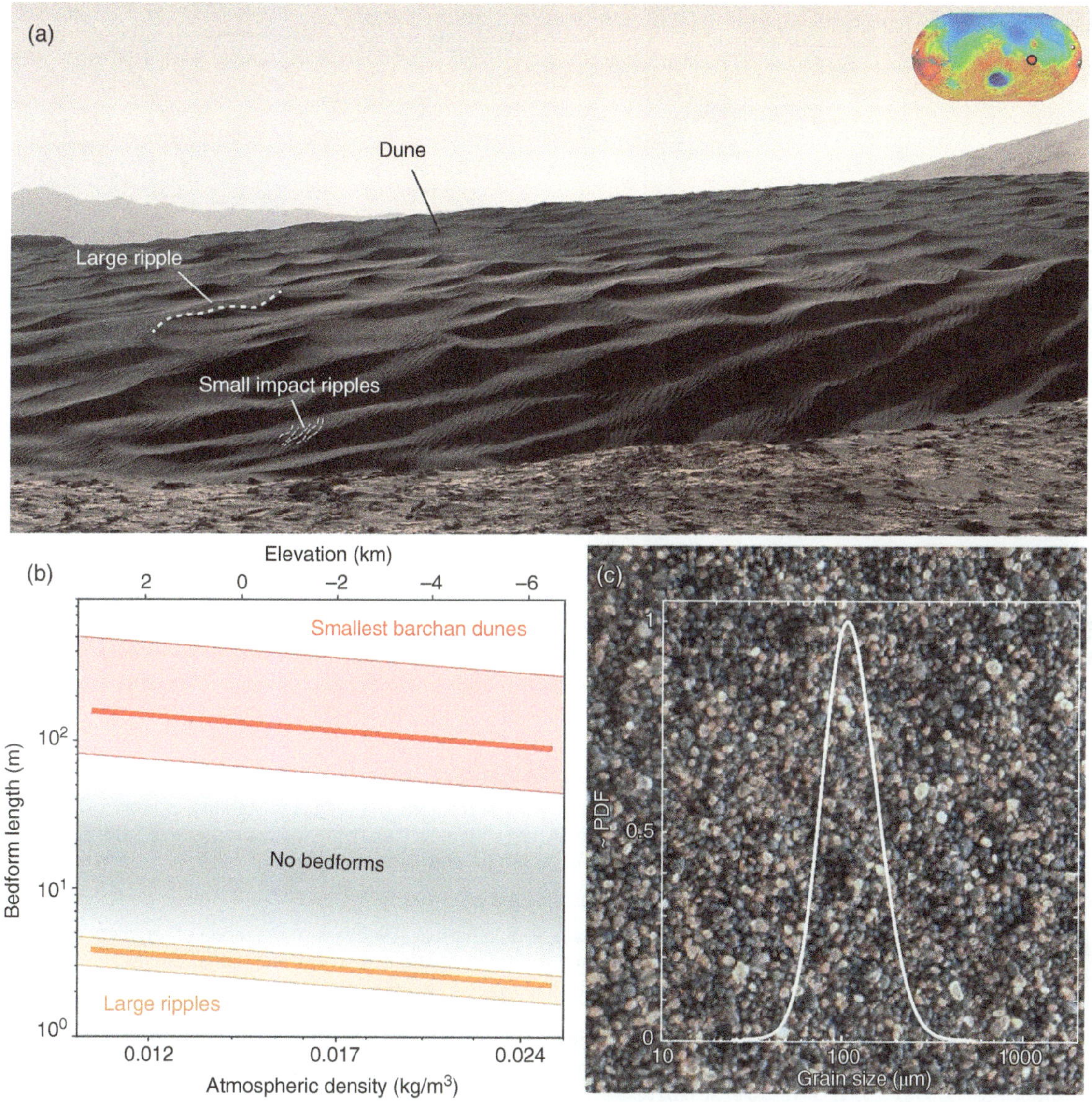

Figure 15.10 (a) Large Martian ripples migrating on top of *Namib Dune* within the *Bagnold dune field* of Gale crater, Mars (Mastcam mosaic; credit: Photo by NASA/JPL-Caltech/MSSS/Thomas Appéré / Public domain). (b) Wavelength of large Martian ripples (data from Lapôtre et al. 2016a and Vaz et al. 2023) and length of Mars' smallest barchan dunes (data from Rubanenko et al. 2022) as a function of atmospheric density (or elevation, see top x-axis). The observed gap in bedform length on Mars matches predictions from the drag-ripple theory (Durán Vinent et al. 2019) for ~125 μm sand and shear velocities near the fluid threshold (Rubanenko et al. 2022). (c) Portion of MAHLI image of the *Otavi* sand target, along the crest of a decimetre-scale impact ripple within the Bagnold dune field, Gale crater (image 1242MH0005740000403707R00, sol 1242), with overlain kernel density of the grain size (Lapôtre et al. 2018 / John Wiley & Sons / CC BY-NC-ND 4.0).

legacy terminology for consistency with the published literature, though we contend that the term has been used to refer to bedforms that might not have migrated transversely to their formative winds (e.g. some of them form networks; Balme et al., 2008).

Extensive fields of kilometre-scale dunes (in spacing) with linear crests are found all around the equatorial regions of Titan (Lorenz et al. 2006). Candidate barchans and star dunes were also identified in sediment-limited regions (Ewing et al. 2015). As for Venus, radar data of Titan's surface does not permit the resolution of any smaller bedforms that may be superimposed on the moon's equatorial dunes. Uncertainty related to sediment properties limits the ability to predict whether impact or drag ripples should form on Titan. NASA's upcoming Dragonfly mission, due to launch in the mid-2030s, will provide definitive in situ, ground and airborne observations of aeolian landscapes on Titan, and in doing so, possibly answer

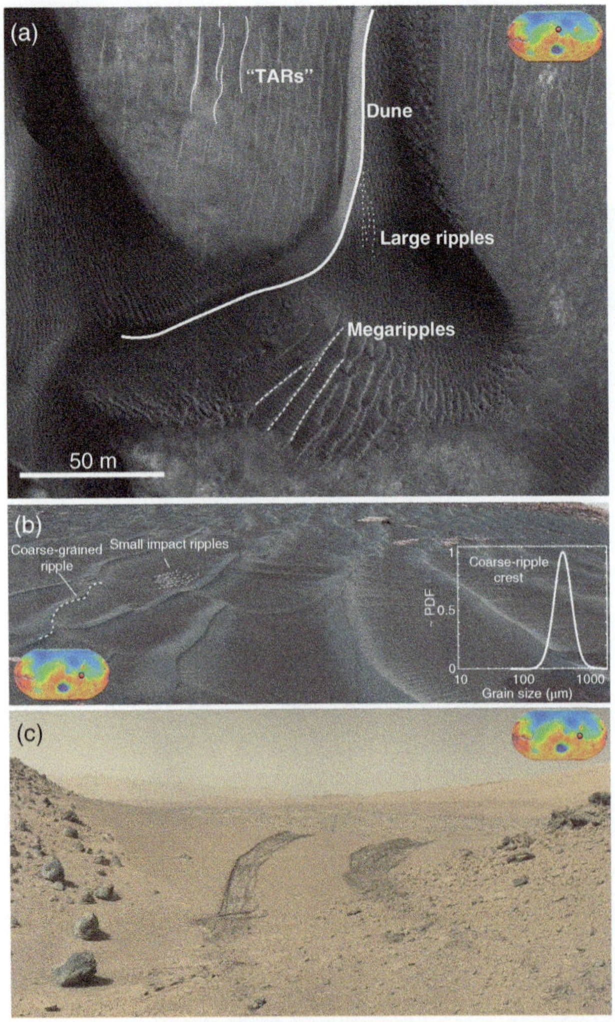

Figure 15.11 Other aeolian bedforms on Mars. (a) Satellite image of dunes, megaripples (candidate coarse-grained ripples), large ripples (aeolian drag ripples) and transverse aeolian ridges (TARs) near Nili Fossae (HiRISE PSP_003587_2015). Silvestro et al. (2020) detected migration of all bedform scales in this image, except for TARs, over a time span of 9 Earth years. (b) Coarse-crested ripples in an isolated ripple field in Gale crater, observed by the Curiosity rover (Mastcam mosaic; credit: Photos by NASA/JPL- Caltech/MSSS / Public domain). (c) A small TAR, marking the entrance to a location called *Dingo Gap*, traversed by the Curiosity rover in Gale crater (Mastcam mosaic; credit: Photo by NASA/JPL-Caltech/MSSS/ Thomas Appéré / Public domain).

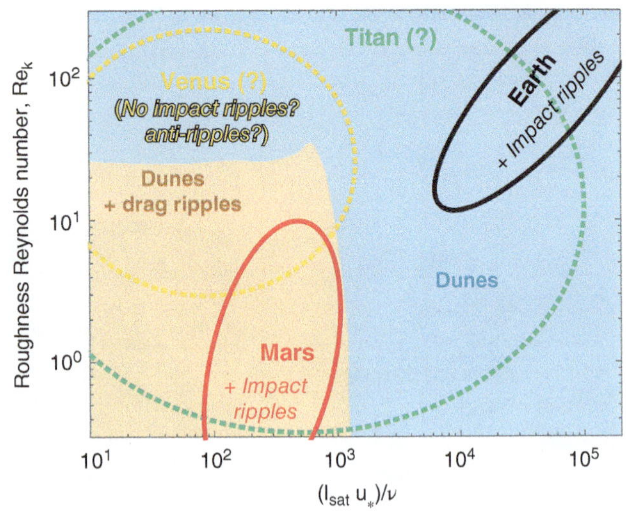

Figure 15.12 Stability of drag ripples and dunes on Venus, Earth, Mars and Titan (adapted from Durán Vinent et al. 2019). Drag ripples and dunes coexist under aerodynamically smooth conditions (low Re_k) when saturation length (l_{sat}) is small compared to that of the "dune gap" (i.e. the bedform wavelength at which the Hanratty anomaly arises). In contrast, no dune gap forms under aerodynamically rough conditions or for large enough saturation lengths. The formation of impact ripples, in contrast, is controlled by the sediment-to-fluid density ratio, $\frac{\rho_s}{\rho_f}$ (i.e. it adds a third dimension to this diagram). Rather than showing a three-dimensional diagram, the predicted stability of impact ripples is indicated within each planetary body's respective field. The larger size of Venus' and Titan's fields reflects the large uncertainties in parameters that control Re_k and $\frac{l_{sat}u_*}{\nu}$. Notably, transport saturation length (l_{sat}) varies with the mode of sediment transport and is poorly constrained on extraterrestrial bodies.

remaining questions regarding the efficiency of grain-splash and stability of ripple-scale bedforms. Figure 15.12 summarises our current understanding and predictions of aeolian-bedform stability (impact ripples, drag ripples and dunes) on Venus, Earth, Mars and Titan.

Finally, putative aeolian bedforms have been reported at the surface of many other planetary bodies. For example, regularly spaced linear ridges on Pluto have been interpreted as wind-blown dunes made of methane-ice grains (Figure 15.4e; Telfer et al. 2018). Upon exploring comet 67P/Churyumov-Gerasimenko, the Rosetta spacecraft discovered ripple-like bedforms on the comet's surface (Figure 15.4f; Thomas et al. 2015). The presence of such bedforms was initially surprising as drag from episodic outgassing flows (driven by sublimation of near-surface ices and subsequent gas expansion) is typically too low to initiate motion of the comet's gravel-sized grains (Thomas et al. 2015). Reworking of the ripples by a bright, scarp-forming circular feature was observed near the perihelion, after which ripples reformed in the same area (El-Maarry et al. 2019; Barrington et al. 2023). As was the case for large Martian ripples, two formation mechanisms have been proposed to explain the comet's bedforms – one that is analogous to the formation of impact ripples but for which initial motion is triggered by airfall of non-escaping particles (Thomas et al. 2015) and one in which drag from thermal winds near the perihelion is strong enough to mobilise grains and initiate the formation of drag bedforms analogous to current ripples (or aeolian dunes; Jia et al. 2017). For an in-depth discussion of the mechanistic controls on dune sizes across the Solar System, we refer the reader to the review of Courrech du Pont et al. (2024).

15.3 Sedimentary Geology of Mars

15.3.1 Sediment Production

Crustal materials on Mars are predominantly basaltic to basaltic andesitic (Christensen et al. 2005; McSween et al. 2009) though some degree of magmatic evolution has been recognised (McCubbin et al. 2008; Udry et al. 2018; Payré et al. 2020). In addition to primary igneous rocks, a significant fraction of bedrock is thought to be of clastic origin (including volcaniclastic, impact-generated and detrital sedimentary rocks; e.g. Figure 15.13a) (Malin and Edgett 2000a; Rogers et al. 2018), and evidence for extra-formational recycling in ancient sediments was observed by the Curiosity rover (Edgett et al. 2020) (Figure 15.13b).

Erosion of parent rocks, like on Earth, has occurred through a combination of physical and chemical processes, with their relative importance varying through space and time (Grotzinger et al. 2011, 2013). Fragmentation of bedrock by meteor impacts has contributed to the continuous formation and mobilisation of the regolith (*impact gardening*; Hartmann et al. 2001) (Figure 15.13c). In addition, solar-induced thermal fracturing (Eppes et al. 2015), glacial erosion (Levy et al., 2016), mass wasting (Diniega et al. 2021; Dundas et al. 2021) fluvial erosion, and aeolian abrasion (Ward 1979; Bridges et al. 1999; Szabó et al. 2015; Herkenhoff et al. 2023) (Figure 15.13d) have contributed to the formation and comminution of sediments.

The importance of chemical weathering changed dramatically over time as near-surface liquid water became scarce. Evidence for limited chemical weathering includes possible weathering rinds in rocks of Gusev crater (Hurowitz and McLennan 2007). Overall low chemical indices of alteration (CIA) in sedimentary rocks of Gale crater suggest cold or arid surface environments during deposition in the Hesperian (McLennan et al. 2014; Siebach et al. 2017; Mangold et al. 2019; Thorpe et al. 2021). In contrast, spectroscopic analyses of Noachian-aged exposures revealed compositional layering, with Al-rich clays overlying Fe/Mg-clay-rich layers (Figure 15.13e), which has been interpreted as evidence for more pronounced chemical weathering earlier in Mars' history (Bishop et al. 2008, 2013; Lowe et al. 2020).

15.3.2 Sediment Accumulation, Preservation and Diagenesis

Mars did not develop an Earth-like plate-tectonic regime. Thus, crustal deformation largely occurred through a combination of secular cooling and crustal loading by large, isolated mantle plumes (e.g. Golombek and Phillips 2010), leading to overall limited subsidence (Grotzinger and Milliken 2012; Davis et al. 2021). In contrast, the formation of impact-generated craters and basins has generated sediment traps across Mars (McLennan et al. 2019; Day et al. 2019; Day and Catling 2020; Roback et al. 2020; Dorn and Day 2020; Gunn et al. 2022) (Figure 15.14). Other sediment sinks include tectonic depressions (Davis et al. 2021) and the northern lowlands (Tsoar et al. 1979; Lucchitta et al. 1986; Lancaster and Greeley 1990; Tanaka 1997; Tanaka et al. 2003; Skinner et al. 2012; Salvatore and Christensen 2014a,b; Ivanov et al. 2015) (Figure 15.15).

Without plate tectonics, sediments deposited billions of years ago remained largely undeformed, with little evidence for structural disturbance or metamorphism other than through hydrothermal activity near impact structures (e.g. McSween et al. 2015). Perhaps surprisingly, thick sedimentary successions have been preserved, revealing burial depths up to several kilometres in places (Cosgrove et al. 2022; De Hon 1987; Pain et al. 2007; Milliken et al. 2010; Zabrusky et al. 2012; Bennett and Bell 2016; Caswell and Milliken 2017; Schieber et al. 2017; Day and Catling 2020).

Diagenesis occurred throughout Mars history (in the Noachian, Hesperian and Amazonian) in the presence of groundwater (Okubo and McEwen 2007; Okubo et al. 2009; Grotzinger and Milliken 2012; Martin et al. 2017). Evidence for significant diagenesis includes abundant concretions of varied compositions (Squyres et al. 2004a; Arvidson et al. 2014), pore-filling cements, dissolution features, cement overgrowths and lenticular crystal moulds (Grotzinger et al. 2005, 2014; McLennan et al. 2005; Stack et al. 2014), cement-filled raised ridges and mineralised veins (Siebach et al. 2014; Nachon et al. 2014; Kah et al. 2018; Kronyak et al. 2019), dendrites (Sun et al. 2019; Yingst et al. 2023), erosion-resistant haematite-rich strata (Fraeman et al. 2020) and X-ray amorphous materials in sediments (Smith et al. 2021) (Figure 15.16).

15.3.3 Distribution of Sedimentary Environments on Modern Mars

When considering a planetary body as a whole, it may be helpful to assess the global distribution of sedimentary environments to evaluate factors associated with the prevalence of fluid-driven and atmospheric processes through space and time. Modern-day sedimentary activity on Mars is dominated by winds, frost action and mass wasting (e.g. Diniega et al. 2021; Dundas et al. 2021). Active aeolian sand transport was reported globally (with high-flux regions on par with wind-blown sediment fluxes measured in Victoria Valley, Antarctica; Bridges et al. 2012a,b; Chojnacki et al., 2019), and fields of sand dunes are distributed all around

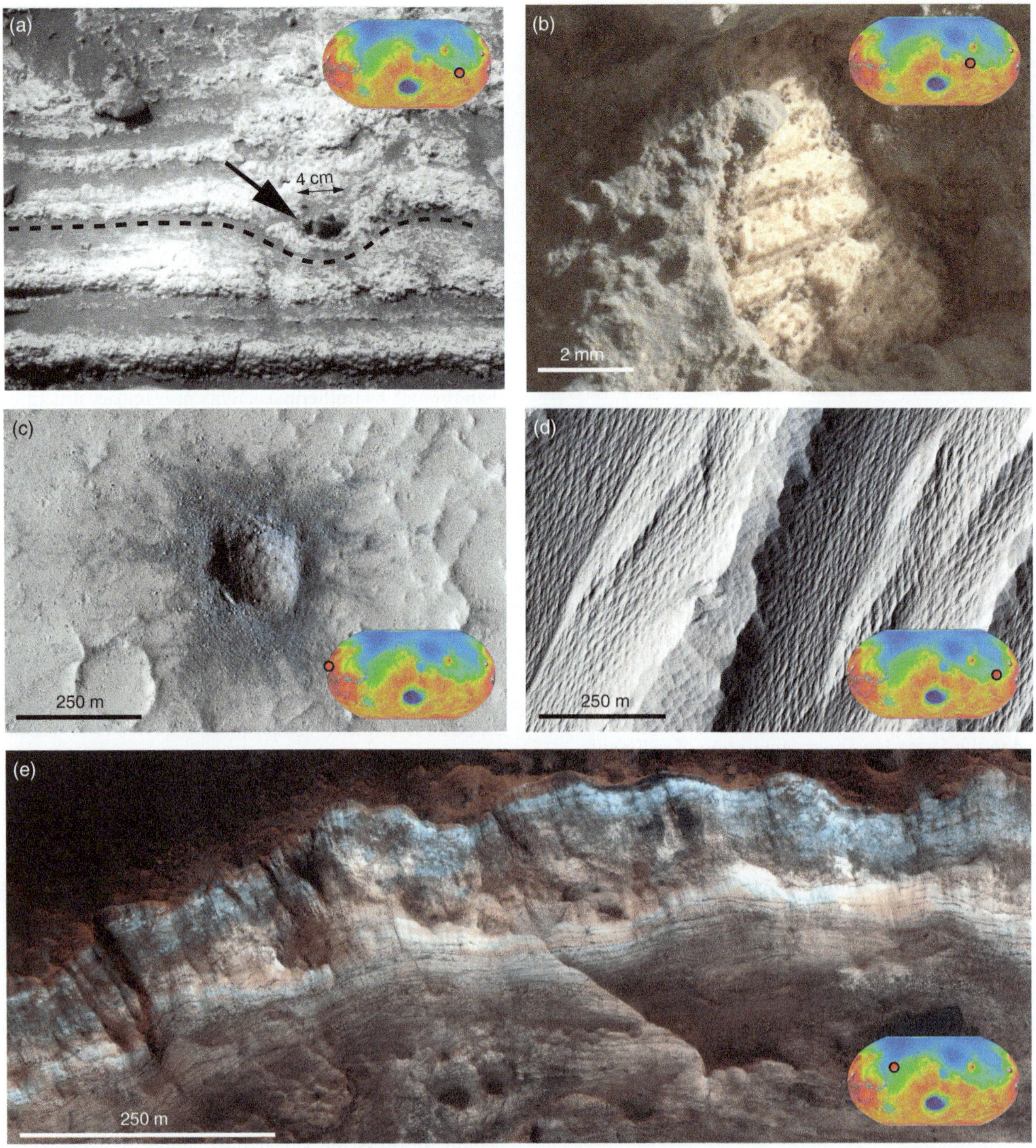

Figure 15.13 Sediment production on Mars. (a) Possible bomb sag as evidence for volcaniclastic sediment production (NASA/JPL-Caltech/Cornell Univ / Public domain; Squyres et al. 2007; Lewis et al. 2008a), imaged by the Spirit rover at *Home Plate*, Gusev crater (Pancam 2P193033032RADA055P2276R1C1). (b) Sedimentary lithoclast in extraformational conglomerate, *Tumba-Funda* boulder, NASA/JPL-Caltech/MSSS / Public domain). (c) Field of impact generated boulders near Tharsis (NASA/JPL-Caltech/MSSS / Public domain). (d) Abraded yardang of the Medusa Fossae formation (Photo by NASA/JPL-Caltech/Univ. of Arizona / Public domain). (e) Possible weathering profile evidenced by compositional layering (here visible through vertical colour variations), exposed in a crater wall near Mawrth Vallis (Photo by NASA/JPL-Caltech/Univ. of Arizona / Public domain).

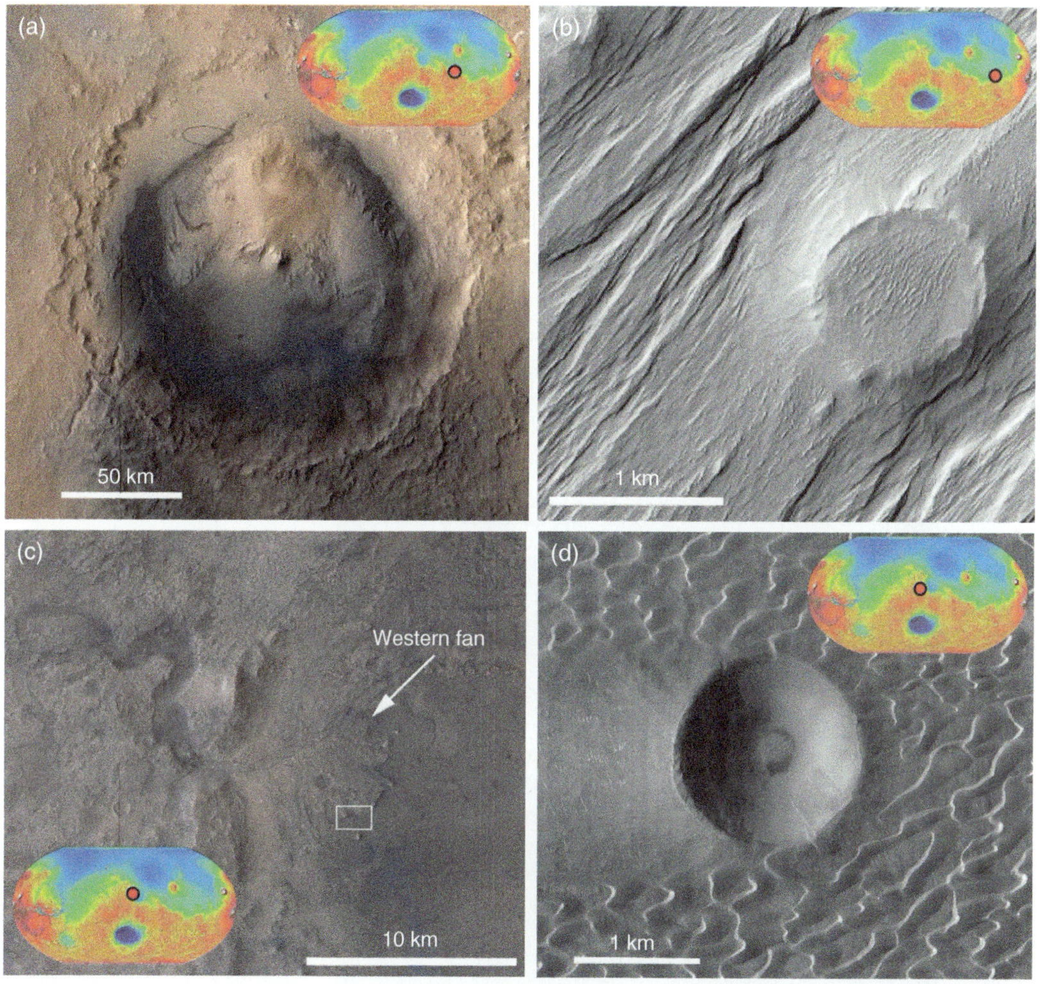

Figure 15.14 Evidence for sediment accumulation and preservation on Mars. (a) Mount Sharp (Aeolis Mons) within Gale crater, Mars, with Curiosity's landing ellipse (HRSC/CTX/Viking composite; credit: Photo by NASA/JPL-Caltech/ESA/DLR/FU Berlin/MSSS / Public domain). Note the dune fields of dark, mafic sand blown by modern winds around Mount Sharp. (b) Exhumed impact crater near Memnonia Sulci, Mars (Photo by NASA/JPL/Malin Space Science Systems / Public domain). (c) Remnants of delta deposits within Jezero crater, Mars (HRSC/CTX/CRISM composite; credit: Photo by NASA/JPL-Caltech/ESA/DLR/FU Berlin/MSSS/JHUAPL/Aster Cowart / Public domain). The rectangle indicates the overall location of stratigraphic sections summarised in Figure 15.23. (d) Modern deposition of wind-blown sand in barchanoid transverse dunes around and inside a small impact crater near Meroe Patera, Mars (NASA/JPL-Caltech/ Public domain).

the planet (Hayward et al. 2007, 2014; Fenton 2020; Rubanenko et al. 2021, 2023). The morphology of modern aeolian dunes on Mars is similar to that of Earth's dunes (see Chapter 7), spanning the full range of dune types observed on Earth (Hayward et al. 2007; Diniega et al. 2021; Rubanenko et al. 2021; Courrech du Pont et al. 2024). Like on Earth, the patterns formed by dune crestlines on Mars also offer a record of environmental conditions (Ewing et al. 2010; Kocurek and Ewing, 2012; Marvin et al. 2023).

Observations of modern sand transport on Mars have been used to infer the direction (Tsoar et al. 1979; Hayward et al. 2007, 2014; Rubanenko et al. 2021) and seasonality (Ayoub et al. 2014; Baker et al. 2018a,b; Bridges et al. 2017) of surface winds on Mars. Notably, the orientation of barchan dunes, globally, appears to follow continuous pathways over thousands of kilometres (Rubanenko et al. 2023). Inferred sand transport pathways are largely aligned with southern summer winds – dominantly westerly (i.e. to the east) polewards of 45° and northerly (i.e. to the south) at mid-latitudes (Figure 15.15a; Rubanenko et al. 2023). They also correlate spatially with dust-storm tracks (Battalio and Wang 2021), consistent with the hypothesis that aeolian saltation contributes to lofting dust and triggering dust storms (Pollack et al. 1976; Greeley 2002; Newman et al. 2002). Around the northern polar ice cap, barchan dunes generally follow Mars' polar vortex circulation (Lancaster and Greeley 1990).

Inferred wind convergence zones correlate with significant accumulations of wind-blown sand, for example, at

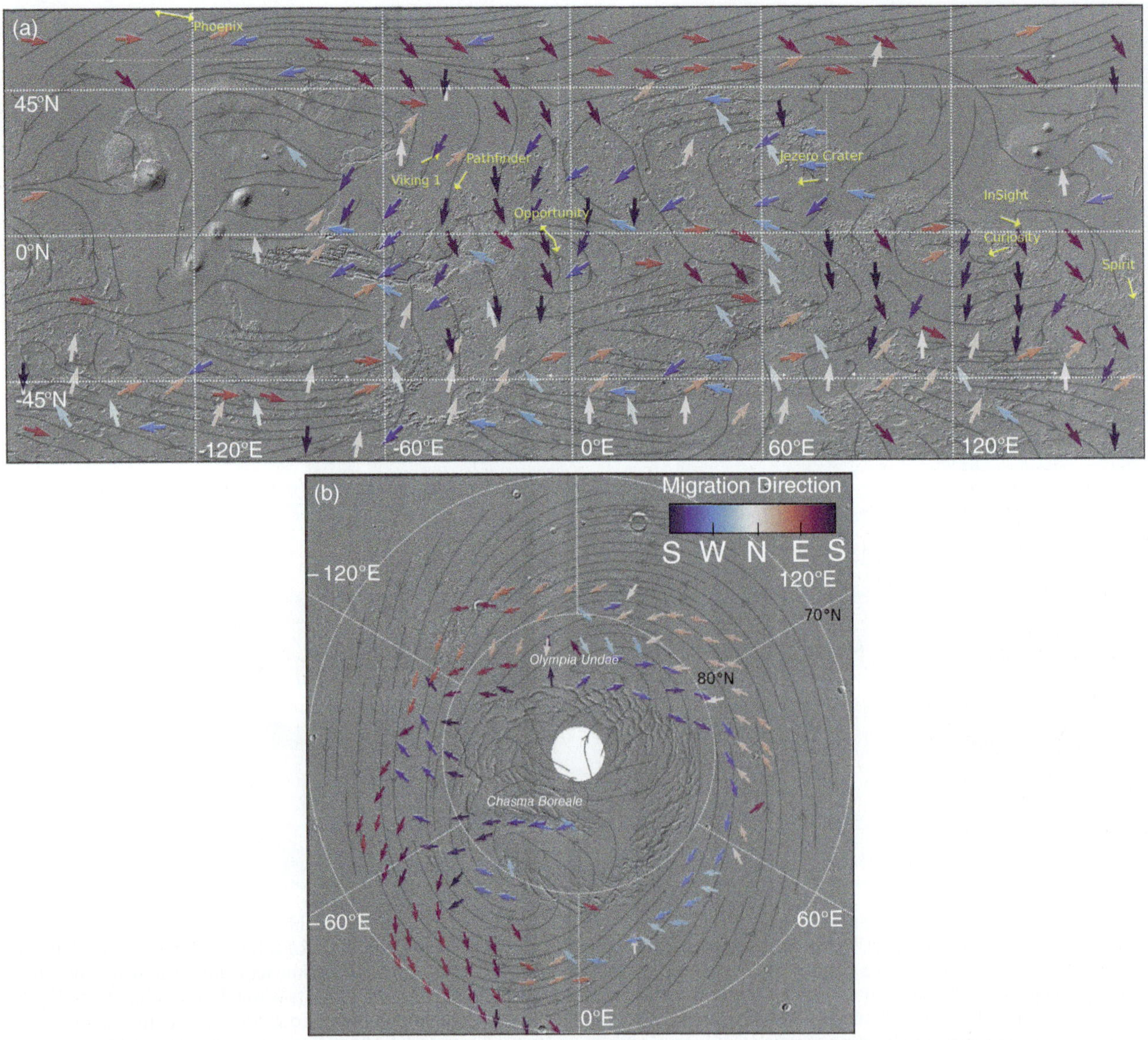

Figure 15.15 Global transport pathways of wind-blown sand on modern Mars overlain on MOLA-shaded relief (a) between −70°(S) and 70°N and (b) north polar view at latitudes >70°N. Sand transport directions were derived from the orientation of barchan dunes by Rubanenko et al. (2023). Gray streamlines indicate annually averaged net sand-flux direction as calculated from a general circulation model (Forget et al. 1999). Dashed white lines highlight dust-storm paths (from Battalio and Wang 2021). Yellow arrows illustrate local, ground or orbital constraints on wind direction or transport direction, as compiled by Rubanenko et al. (2023); they do not necessarily reflect the regional net sand-flux directions. Rubanenko et al. (2023) / John Wiley & Sons / CC BY-NC 4.0.

the southern end of Chasma Boreale and between cyclonic and anticyclonic polar flows near Olympia Undae (Figure 15.15b; Rubanenko et al., 2023). Besides the northern lowlands, impact craters constitute a significant sink for aeolian sediments, where wind-blown sand accumulates over time (Roback et al. 2020; Gunn et al. 2022). The Curiosity rover explored a dune field in Gale crater, reporting on the first in situ exploration of an active extraterrestrial dune field named the *Bagnold dune field* (Bridges and Ehlmann 2018; Lapôtre and Rampe 2018).

Finally, a variety of slope and periglacial processes (encompassing mass wasting and, possibly, liquid flow events) have been observed to occur on Mars today (Figure 15.17; reviewed in more detail by Diniega et al. 2021; Dundas et al. 2021). Martian slope processes have been classified based on the morphology, distribution and seasonality of the landforms they create. At low- to mid-latitudes, common slope processes and features include gullies, slope lineae (some of which recur and are referred to as Recurring Slope Lineae, or RSL; McEwen et al. 2011,

Figure 15.16 An extensive record of diagenesis revealed by Mars rover missions. (a) Haematite-rich spherule in evaporite-cemented sandstone of the *Burns formation*, Eagle crater, Mars (NASA/JPL-Caltech/Cornell University/ASU / Public domain). (b) Likely diagenetic concretions with concentric internal structure in smectite-rich sandstone of the *Matijevic formation*, Endeavour crater (NASA/JPL-Caltech/Cornell University/ASU / Public domain). (c) Diagenetic concretions preserving primary laminations in fine-grained sediments of the *Murray formation*, Mars (NASA/JPL-Caltech/Cornell University/ASU / Public domain). (d) Crystal-mould porosity in evaporite-cemented sandstone of the Burns formation, Eagle crater, Mars (mosaic of MER Opportunity Microscopic Imager images 1 M130670922 and 1 M130671284). (e) Diagenetic dendrite in fine-grained sediments of the *Pahrump Hills member* of the Murray formation, Gale crater, Mars (NASA/JPL-Caltech/MSSS / Public domain). (f) Multiple generations of veins in fine-grained sediments at *Garden City*, Pahrump Hills member, Murray formation, Gale crater, Mars (NASA/JPL-Caltech/MSSS / Public domain).

2014), slumps, slope streaks and other mass wasting events (Figures 15.17 and 15.18; Diniega et al. 2010; Dundas et al. 2019a, 2022; Pasquon et al. 2016; Ohja et al. 2017; Becerra et al. 2020; Thomas et al. 2020; Stillman et al. 2020).

Martian gullies form on crater walls and other steep slopes and are characterised by erosional alcoves, channels and depositional aprons downslope (Malin and Edgett, 2000b). Their activity is seasonal and coincides with the formation of CO_2 frost during the cold season (Diniega et al. 2010; Dundas et al. 2010, 2012, 2015, 2019a; Raack et al. 2015, 2020). Gullies appear to be shaped by multiple flows. It is unclear whether they were

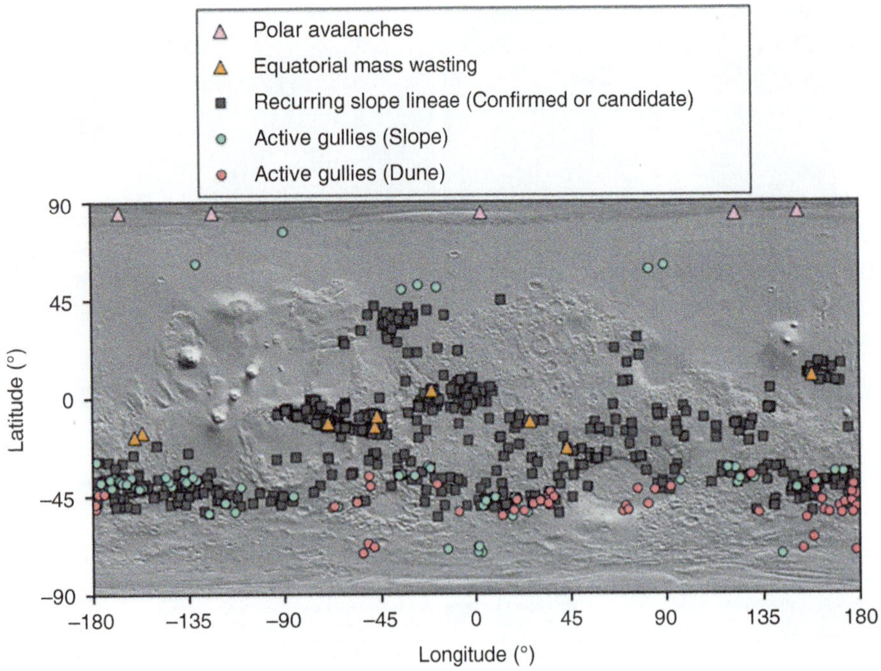

Figure 15.17 Distribution of detected slope processes on modern Mars overlain on MOLA-shaded relief as compiled by Diniega et al. (2021) from Becerra et al. (2020); Thomas et al. (2020); Stillman et al. (2020); Dundas et al. (2019a); Diniega et al. (2010); Pasquon et al. (2016) (see also Dundas et al. 2022). Adapted from Diniega et al. (2021). Photo by MOLA Science Team / NASA / Public domain.

first created under past climate conditions, for example, from snow or water ice melt, and are only being reworked by modern CO_2 frost processes (Christensen 2003; Dickson et al. 2023b), or whether modern processes are fully responsible for their formation (Dundas et al. 2022). Gully activity includes changes in albedo, erosion and deposition (sometimes leading to changes in channel position and sinuosity; Pasquon et al. 2019a,b). The distribution of active gullies is highly asymmetric, with significantly more activity detected in the southern hemisphere (Figure 15.17), likely due to Mars' current orbital eccentricity (Dundas et al. 2015, 2019a). Gully activity is thought to recur over a period of decades to centuries in individual gullies (Dundas et al. 2015, 2022).

Unlike gullies, RSL are superficial flows when seen at orbital scales; they are characterised by their dark albedo and gradual downslope propagation (McEwen et al. 2011, 2014). Most RSL are found at southern mid-latitudes (Figure 15.17). Their formation is seasonal, typically during the warm season (especially after global dust storms; Stillman 2018; Stillman and Grimm 2018; Stillman et al. 2014, 2016, 2017; McEwen et al. 2021, Ohja et al. 2014), with growth and fading observed annually. Originally thought to require flows of water or brines (McEwen et al. 2011, 2014; Stillman et al. 2014, 2016, 2017), several formation mechanisms have been proposed that require little to no liquid (Dundas et al. 2017; Schaefer et al. 2019, Vincendon et al. 2019; Dundas 2020a). Their formation remains ambiguous to date (Stillman and Grimm 2018; Stillman et al. 2020).

Other types of slope lineae have been observed that do not share the same seasonality. Like slope lineae, slope streaks are mostly superficial, but they persist for many years without progressive growth and are largely found in dusty regions of Mars (Sullivan et al. 2001); they are thought to be formed by dry dust avalanches (Sullivan et al. 2001; Dundas 2020b).

At high latitudes, slope processes are most often associated with frost activity (Figure 15.18). Alcoves form seasonally on Mars' circumpolar dunes (Hansen et al. 2011; Diniega et al. 2019) and are often associated with changes in albedo (appearance of dark spots and flows), triggered by sublimation of surface frost (Gardin et al. 2010). However, some dune alcoves are thought to form as early as mid-to-late northern summer, which is inconsistent with the hypothesis that defrosting drives alcove formation (Horgan and Bell 2012). It was thus proposed that winds initiate mass wasting and that alcoves may form due to enhanced cohesion conferred by near-surface frost (Horgan and Bell 2012; Hansen et al. 2015; Diniega et al. 2019). Other active processes observed at high latitudes include CO_2 sublimation features such as "spiders" (or araneiforms; Piqueux et al. 2003; Schwamb et al. 2018) and fan-shaped deposits (Kieffer et al. 2006; Hansen et al. 2013), rock shifts (Dundas et al. 2019b) and polar avalanches (Russell et al. 2008; Becerra et al. 2020).

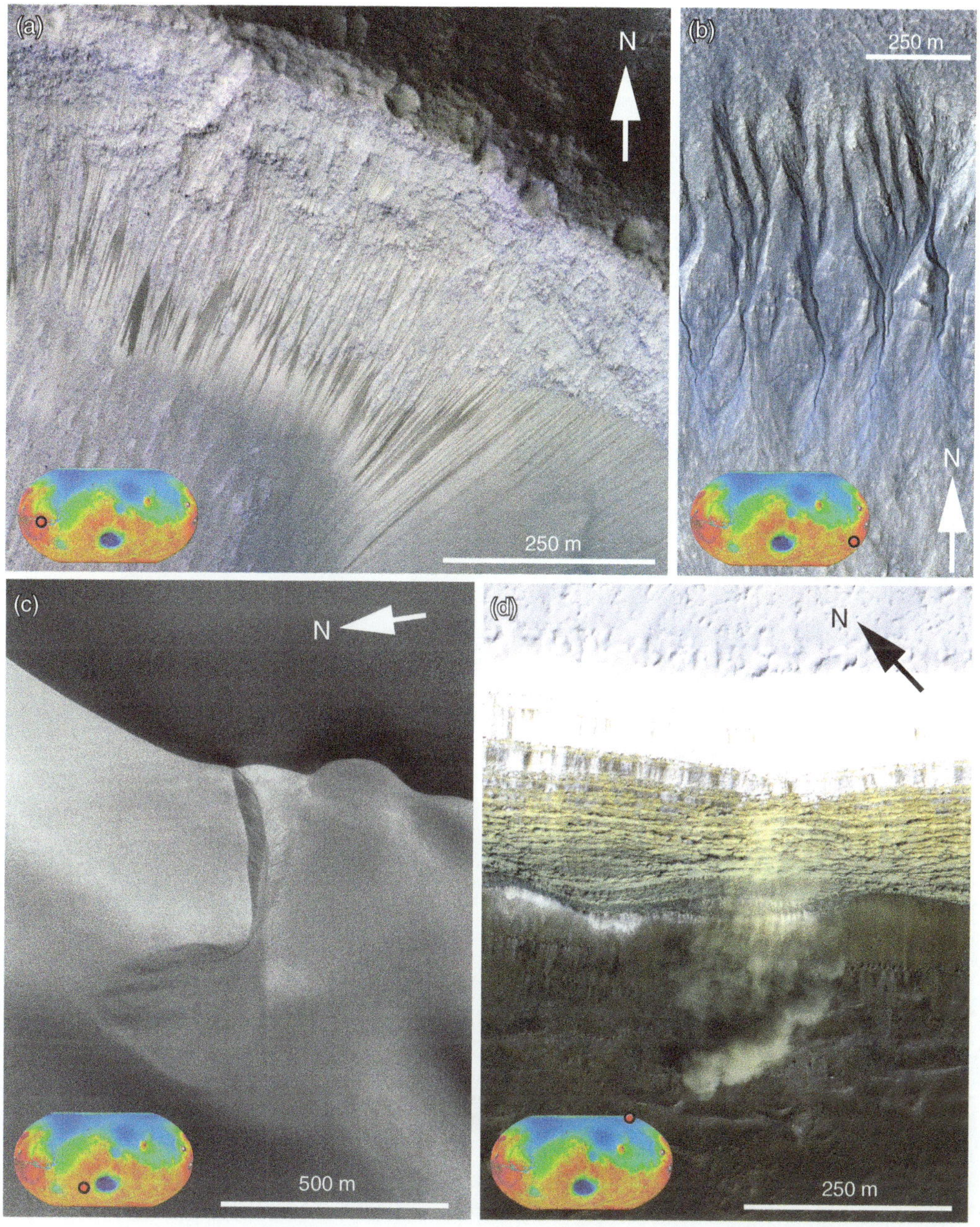

Figure 15.18 Example modern slope processes. (a) Recurring slope lineae on a crater wall within Valles Marineris (Photo by MOLA Science Team / NASA / Public domain), (b) (Terra Sirenum; NASA/JPL-Caltech/Uarizona / Public domain), (c) dune gully (Kaiser crater; NASA/JPL-Caltech/Uarizona / Public domain) and (d) polar avalanche (NASA/JPL-Caltech/Uarizona / Public domain).

15.3.4 Distribution of Sedimentary Environments on Early Mars

While modern Mars is cold and dry, aqueous depositional systems are known to have existed in the past from orbital data (Figure 15.3a,b). There is abundant evidence including bedrock-incised valley networks (Irwin et al. 2008), crater-lake basins (Cabrol and Grin 1999, 2001; Fassett and Head 2008a; Goudge et al. 2012, 2015a; Stucky de Quay et al. 2020), exhumed fluvial channel belts (Dickson et al. 2021) and possible delta deposits (Di Achille and Hynek 2010). These systems are found in the equatorial, mid-latitude and southern highlands of Mars.

Dendritic valley networks, interpreted to be carved by fluvial flows, are concentrated in the southern highlands (Figure 15.19a; Pieri 1976; Baker et al. 1992; Carr 1996; Irwin et al. 2005, 2008; Lasue et al. 2019) and attest to a

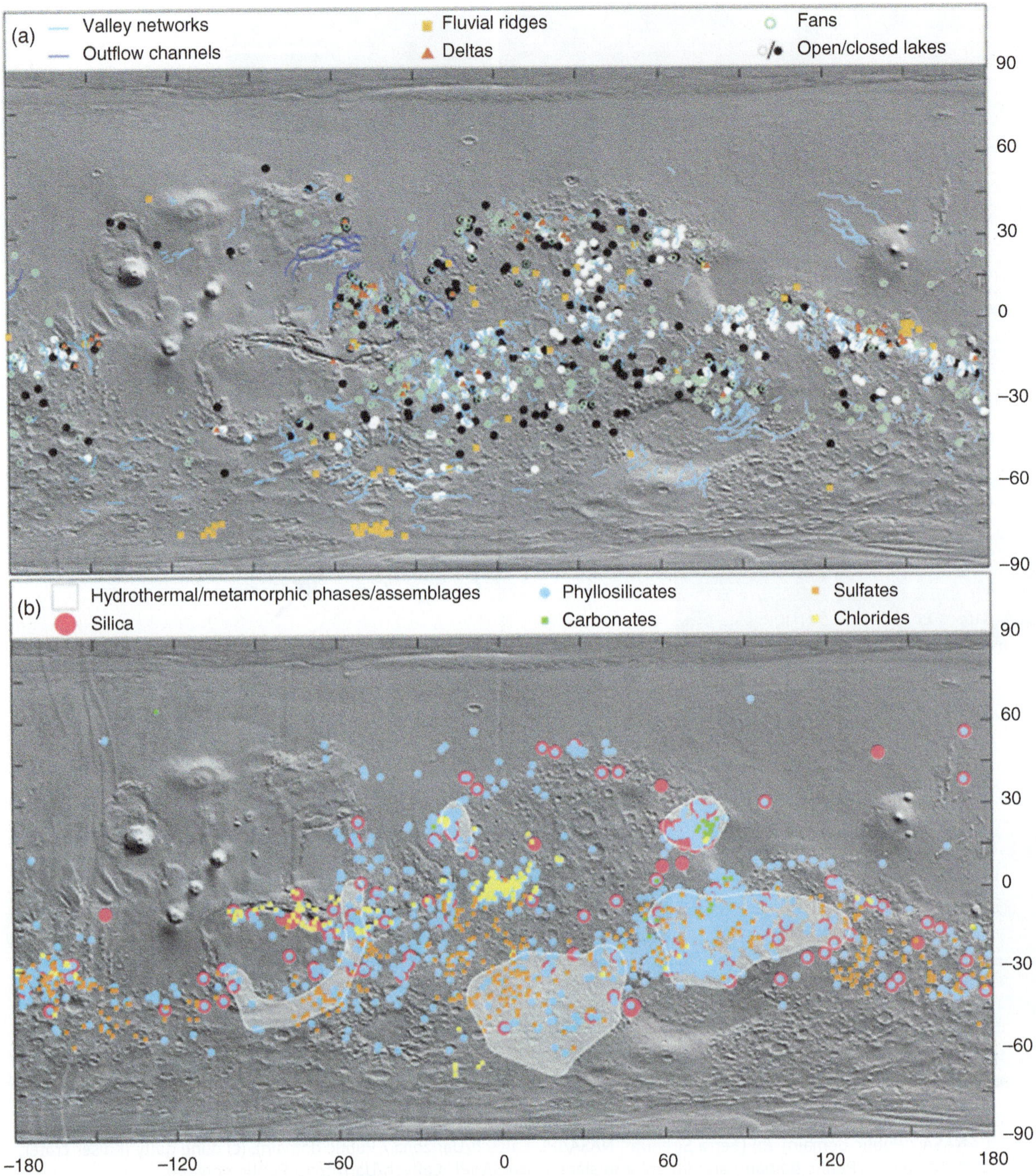

Figure 15.19 Global distributions of (a) ancient alluvial, fluvial, deltaic, and lacustrine landforms (compiled from Di Achille and Hynek 2010; Fassett and Head 2008a; Tanaka et al. 2014a,b; Goudge et al. 2015a; Dickson et al. 2021; and Wilson et al. 2021), and (b) mineral detections from orbiting spectrometers (compiled from Ehlmann et al. 2010, 2011a,b; Ehlmann and Edwards 2014; Bultel et al. 2015; Viviano-Beck et al. 2017; and Amador et al. 2018). Their concentration in the southern highlands highlights the role of surface liquid water on early Mars. Adapted from Lapôtre et al. (2022a). Photo by MOLA Science Team / NASA / Public domain.

well-integrated hydrologic cycle in the Noachian, fed by both precipitation and groundwater sources (Carr and Malin 2000; Mangold et al. 2004; Seybold et al. 2018; Lapôtre and Lamb 2018). Downstream from valley networks, orbital observations reveal sinuous ridges, which are interpreted to be the eroded remnants of amalgamated channel bodies from rivers that once flowed over hundreds of kilometres across alluvial plains (Figure 15.19a; Burr et al. 2010; DiBiase et al. 2013; Williams et al. 2013a; Kite et al. 2015; Davis et al. 2016, 2019; Hayden et al. 2019;

Balme et al. 2020; Dickson et al. 2021; Zaki et al. 2021; Cardenas et al. 2022a). These fluvial ridges formed through preferential erosion of floodplain materials, likely muddy sediments (Lapôtre et al. 2019). Channelised fluvial deposits are also hypothesised in outcrops from orbit (Salese et al. 2020) and observed on the ground by rovers (Section 15.3.6).

Whereas many fluvial systems terminated in the northern lowlands, many also flowed through or into open- or closed-basin lakes within impact craters (Figure 15.19a; Cabrol and Grin 1999, 2001; Fassett and Head 2008a; Goudge et al. 2012, 2015a; Stucky de Quay et al. 2020). Accordingly, delta and sublacustrine fan deposits are found along the hemispheric dichotomy boundary, in impact craters and in fault-bounded basins (Metz et al. 2009b; Wood 2006; Di Achille and Hynek 2010; Goudge et al. 2018; Lapôtre and Ielpi 2020; Rivera-Hernández and Palucis 2019; Palucis et al. 2016, Fawdon et al. 2018; Davis et al. 2018; De Toffoli et al. 2021). Lacustrine deposits have been observed by rovers in impact craters, including possible evidence for wave-worked lacustrine shorelines (Caravaca et al. 2022; Rubin et al. 2022; Section 15.3.6). Alluvial fans are also commonly found along the walls of impact craters and grabens (Figure 15.19a; Moore and Howard 2005; Palucis et al. 2014; Wilson et al. 2021; Davis et al. 2021).

In addition to sustained surface flows, very large (up to thousands of kilometres long and hundreds of metres wide) outflow channels were carved by episodic, catastrophic megafloods, largely into the northern lowlands (Figure 15.19a; Baker and Milton 1974; Komatsu and Baker 1997; Pacifici et al. 2009; Warner et al. 2009; Robinson and Tanaka 1990; Williams et al. 2000; Williams and Malin 2004; Lapôtre et al. 2016b). Preserved palaeobedforms are associated with outflow channels in places (Chojnacki et al. 2024). Whereas some of those floods were sourced from outbursts from rift-basin lake systems (Warner et al. 2013), many have clear subsurface water sources (Sharp and Malin 1975; Carr 1979; Head et al. 2003).

A transition from wet surface conditions, with well-integrated surface hydrology, to a more arid surface, with mostly episodic outbursts of flood waters and rare smaller surface flows, is hinted at by the younger age of Mars' outflow channels (Fassett and Head 2008b; Warner et al. 2009) and alluvial fans (Grant and Wilson 2011, 2019; Holo et al. 2021) relative to older dendritic valley networks. The hypothesis of a (possibly protracted and non-monotonic) wet-to-arid climate transition is also supported by the orbiter-based mineralogy of Mars' surface (Bibring et al. 2006; Ehlmann and Edwards 2014), with abundant Fe–Mg clays and some carbonates in Noachian terrains (Poulet et al. 2005; Ehlmann et al. 2008; Mustard et al. 2008) and sulphates, chlorides and oxides in younger terrains (Bibring et al. 2006; Ehlmann and Edwards 2014) (Figure 15.19b; Section 15.3.5).

Besides lakes, several lines of geological evidence have been used to argue that large, long-lived oceans once existed on Mars and served as terminal basins for water and sediments. Multiple scenarios have been proposed for the formation of such oceans, including a perennial valley-network-fed ocean in the Noachian and ephemeral Hesperian oceans fed by Mars' episodic outburst floods (Kreslavsky and Head 2002; Baker et al. 1991; Clifford and Parker 2001; Carr and Head 2003; Fairén et al. 2003).

Two sets of curvilinear geologic contacts, roughly concentric to Mars' dichotomy boundary, were first reported from Viking Orbiter images and proposed to represent the remnants of ancient wave-eroded shorelines from one or multiple unfrozen oceans (Parker et al. 1989, 1993). These two proposed palaeoshorelines are referred to as the Arabia (outer) and Deuteronilus (inner) levels. The Deuteronilus level roughly approaches an equipotential contour (with deviations observed in places attributed to post-formation geological activity; Head et al. 1998, 1999). Long-wavelength topography along both the Arabia and Deuteronilus levels was proposed to be consistent with deformation due to true polar wander (i.e. the changed figure of the planet following the emplacement of the Tharsis volcanic edifice; Perron et al. 2007) or from crustal loading during the emplacement of Tharsis (Citron et al. 2018).

Ancient delta deposits were shown to roughly follow the Arabia level, possibly outlining the extent of a ~ 3.5 Ga ocean (Di Achille and Hynek 2010). Valley networks also terminate near the Arabia level with little evidence for significant fluvial incision in the northern lowlands (Hynek et al. 2010; Chan et al. 2018), consistent with the hypothesis that a large body of water was present at the time of, or after, valley network formation. Inverted fluvio-deltaic deposits near Aeolis Dorsa are also consistent with formation along a fluctuating coastline (Cardenas et al. 2018; Hughes et al. 2019; Cardenas and Lamb 2022).

Three other pieces of evidence for ancient oceans include: (1) the Vastitas Borealis formation (a sediment- and possibly ice-rich deposit whose boundaries largely coincide with the Deuteronilus level and thought to have formed in the final stages of aqueous activity in the northern lowlands; Kreslavsky and Head 2002; Campbell et al. 2008; Mouginot et al. 2012); (2) possible evidence for ancient tsunami deposits (Rodriguez et al. 2016; Costard et al. 2017, 2019; Di Pietro et al. 2021; De Blasio 2020; Rodriguez et al. 2022); and (3) atmospheric deuterium-to-hydrogen (D/H) ratio measurements and geochemical modelling (Salvatore and Christensen 2014a; Villanueva et al. 2015; Scheller et al. 2021). Modern D/H ratios are 5–10 times

SMOW (standard mean ocean water on Earth; Donahue 1995; Webster et al. 2013; Villanueva et al. 2015) and have been constrained to 2–4 times SMOW about 4 billion years ago from Martian meteorites (Boctor et al. 2003; Greenwood et al. 2008). These measurements, combined with geochemical modelling of atmospheric escape and crustal hydration suggest that Noachian Mars hosted a global equivalent layer (GEL) of water hundreds to over 1500 m deep (Kurokawa et al. 2014; Villanueva et al. 2015; Lammer et al. 2003; Alsaeed and Jakosky 2019; Scheller et al. 2021).

In strong contrast with observations made from ~50 to 70 m/pixel Viking Orbiter images, high-resolution images from the ~2 to 20 m/pixel MGS Mars Orbiter Camera (MOC) revealed a general paucity of evidence for strandlines and other coastal landforms along the previously proposed ocean levels (Malin and Edgett 1999). Subsequent tests of the Martian palaeoshoreline hypothesis that combined even higher-resolution topographic data, geomorphology and spectroscopy showed that curvilinear features previously mapped as shorelines are inconsistent with the wave-generated shoreline hypothesis (Sholes et al. 2019, 2021; Sholes and Rivera-Hernández 2022). Potential terraces do exist, but they are discontinuous, appear to tilt in different directions and do not follow equipotential surfaces, suggesting they may instead represent locally exposed lithological boundaries. Correcting for deformation (by either true polar wander or Tharsis loading) of the various mapped Arabia levels still yields multi-kilometre variations in shoreline elevation.

An analysis of the elevation of delta fronts in the greater Gale crater region varies by ~2400 m and cannot be explained by either true polar wander or Tharsis loading (Rivera-Hernández and Palucis 2019). Rather, observations are consistent with delta formation in neighbouring but discontinuous palaeolakes. Furthermore, an exhaustive survey of exposed mineral phases using visible-shortwave infrared spectral data in the walls and ejecta of impact craters within the northern lowlands found no evidence of concentrated salt deposits that would be consistent with the evaporation of a long-lived ocean (Pan et al. 2017). Finally, the formation of long-lived, ice-free oceans is difficult to reconcile with predictions from 3D global circulation models (Forget et al. 2013; Wordsworth et al. 2015). In summary, whether the Martian climate ever allowed an ocean to be stable over geologically relevant timescales remains a subject of considerable debate.

Beyond aqueous processes, Mars' early surface is thought to have experienced a wide range of subaerial sedimentary processes (e.g. Grotzinger et al. 2013; McLennan et al. 2019), including aeolian processes and mass wasting for most of Mars' history and, at least more recently, glaciers (e.g. Hubbard et al. 2014; Levy et al. 2021). Landslide and rockfall deposits are widespread on Mars (Section 15.3.3), with large concentrations in the Valles Marineris region (Crosta et al. 2018). Their formation is thought to have spanned the past ~3.5 billion years (Quantin et al. 2004; Grindrod and Warner 2014). Aeolian processes, in turn, have left a significant imprint on Martian landscapes over time. Numerous lithified aeolian deposits have been identified, both from orbit (Milliken et al. 2014; Day et al. 2019) and in situ by rovers (Grotzinger et al. 2005; Banham et al. 2018, 2021, 2022; Rubin et al. 2022) (Section 15.3.6), including aeolian strata and exhumed lithified dune fields, termed "palaeodunes" (Chojnacki et al. 2020, 2024; Hunt et al. 2022). "Ghost" dune fields are found where dunes were once engulfed and buried by a rapid flow (of unconstrained origin) before differential erosion preferentially removed dune sand leaving behind crescent-shaped pits (Day and Catling 2018). Finally, aeolian erosional features such as yardangs (Ward, 1979; Zimbelman and Griffin 2010) and periodic bedrock ridges (Montgomery et al. 2012; Hugenholtz et al. 2015; Silvestro et al. 2021; Stack et al. 2022; Bretzfelder et al. 2024; Favaro et al. 2024) are also common on Mars and are likely still evolving to this day.

15.3.5 Global Stratigraphic Trends and Key Reference Sections

15.3.5.1 Global Stratigraphic Trends

Observations by orbiters offer a unique opportunity to constrain stratigraphic and compositional trends at the global scale, albeit lacking sedimentary facies information. For example, orbiting spectrometers have revealed a strong correlation between terrain age and mineralogy, with phyllosilicate-bearing Pre-Noachian to Late Noachian rocks, sulphate-bearing Early Hesperian to Early Amazonian terrains and oxide-dominated Early Amazonian and younger surfaces, interpreted as the result of Mars' loss of liquid water at its surface (Bibring et al. 2006) (Box 15.1). A similar, "drying-upward" trend is reflected in stratigraphic exposures across Mars. Grotzinger and Milliken (2012) broadly classified sedimentary successions across Mars into six "orbital facies" based on depositional framework, bed-scale textural and morphologic attributes and mineralogy: (1) massive breccia, (2) complexly stratified clay-bearing, (3) laterally continuous sulphate-bearing, (4) laterally continuous heterolithic, (5) distributary network and (6) rhythmite facies.

Massive breccias contain 1–100 m blocks, display poorly defined layers tens of metres thick and are laterally discontinuous. Massive breccias can form through a variety of processes, including brecciation by meteor impact (e.g. Scheller and Ehlmann 2020). The complexly stratified clay-bearing facies is defined by metre-scale or thinner light-toned clay-bearing layers, often displaying complex stratal geometries. Although the facies can be laterally extensive, individual layers are laterally discontinuous when traced over hundreds of metres to kilometres. Clay mineralogy can vary within and across sites. Fe/Mg clays are the most common, though Al-rich clays are also found. A defining feature of the laterally continuous sulphate-bearing facies is the great lateral (up to hundreds of kilometres) and sometimes vertical (up to several kilometres) extent of sulphate-bearing strata. This facies is also often associated with detections of Fe-oxides. Bed thicknesses are highly variable (~1–100 m). Rocks of the laterally continuous sulphate-bearing facies are typically light-toned and wind-sculpted. The laterally continuous heterolithic facies consists of metre-thick, cliff-forming strata, dominated by planar bedding, that are laterally continuous over tens to hundreds of kilometres. Rock colour varies vertically, sometimes from one bed to another. Rocks of the laterally continuous heterolithic facies do not exhibit consistent mineralogy but rather can contain opal, sulphates, clay minerals or even lack spectral signatures of hydrated minerals. The distributary network facies was defined by the presence of depositional bodies associated with distributary channel systems, such as alluvial fans, deltas and sublacustrine fans. Observed stratal geometries include clinoforms (e.g. Dromart et al. 2007; Ansan et al. 2011). These depositional bodies are typically metre- to decametre-thick and can contain a variety of hydrated minerals and carbonates. Finally, the rhythmite facies is characterised by the cyclic nature of often planar, metre-scale beds over relatively thick sequences (Lewis et al. 2008b). Visible-shortwave infrared spectra of rhythmites resemble those of Martian dust (Grotzinger and Milliken 2012).

Grotzinger and Milliken (2012) illustrated these orbital facies through seven orbiter-derived type sections across Mars: Nili Fossae, Mawrth Vallis, the Uzboi-Ladon-Morava outflow system (including Eberswalde and Holden craters), Gale crater, Valles Marineris, Sinus Meridiani and Arabia Terra (Figure 15.20). The Nili Fossae and Mawrth Vallis sections are the most ancient (Noachian) and contain abundant massive breccia and complexly stratified Fe/Mg clay-bearing rocks. At Mawrth Vallis, which was a candidate landing site for the Mars Science Laboratory, Mars 2020 and ExoMars, Al-rich clays overlie Fe/Mg clays (Bishop et al. 2008; Loizeau et al. 2012). The younger type section of the Uzboi-Ladon-Morava system displays finely stratified Fe/Mg clays of the laterally continuous heterolithic facies, overlain by alluvial and fluvio-deltaic deposits of the distributary network facies (Grant et al. 2008). In turn, the interior layered deposits of Valles Marineris and lower-to-middle stratigraphy of Sinus Meridiani are dominated by the laterally continuous sulphate-bearing facies (Figure 15.20). The upper stratigraphy of Sinus Meridiani consists of rhythmites. Similarly, thick rhythmites are pervasive in the Hesperian to Amazonian stratigraphy of Arabia Terra (Lewis et al. 2008b; Figure 15.20). Finally, Gale crater displays a transition up-section from complexly stratified clay-bearing units to laterally continuous sulphate-bearing strata, with rhythmites lying unconformably in the uppermost section of *Mount Sharp*, which is as yet unexplored (Milliken et al. 2010; Figure 15.20).

Overall, the global stratigraphy of Mars suggests a widespread but possibly complex desiccation history of the planet's surface (Grotzinger et al. 2012). However, caution and scepticism should always be applied when considering palaeo-environmental interpretations from orbiter-based data alone, which cannot resolve fine-scale but critical features such as individual sand grains and cross-stratification.

15.3.5.2 Type Section of the Burns Formation at Meridiani Planum

Unlike orbiters, rovers have the capability to resolve fine-scale sedimentary features at the outcrop level and provide a high-resolution picture of local stratigraphy. The first ground-based stratigraphic column describing a sedimentary succession on Mars was built from observations by the Opportunity rover of rocks exposed by a series of impact craters (notably Eagle and Endurance craters) in Meridiani Planum (Grotzinger et al. 2005). At Eagle and Endurance craters, a ~7 m thick section was identified, consisting of genetically related sandstone-dominated facies (Section 15.3.6), informally called the *Burns formation* (Figure 15.21). The grain size of these sandstones is dominantly medium sand and appears to be derived from siliciclastic-rich evaporites, though pervasive recrystallisation of evaporite minerals occurred, coarsening and obscuring most primary aeolian textures. The evaporitic nature of the sediments may indicate a playa-like sand source (McLennan et al. 2005; Grotzinger et al. 2005). Haematite-rich concretions are also found throughout the section (Chan et al. 2004; Christensen and Ruff 2004; Squyres et al. 2004a). The Burns formation is divided into three units.

As described in Grotzinger et al. (2005), the lower unit of the Burns formation consists of ~2.5 m of metre-scale cross-stratified sandstone (Section 15.3.6), interpreted as the result of the migration of large aeolian dunes in a dry setting. The contact between the lower and middle units is unconformable and was interpreted as a significant deflation surface. This contact is irregular, displaying multiple scours that resemble those formed by wind erosion of moist sand and was thus interpreted as a water-table controlled

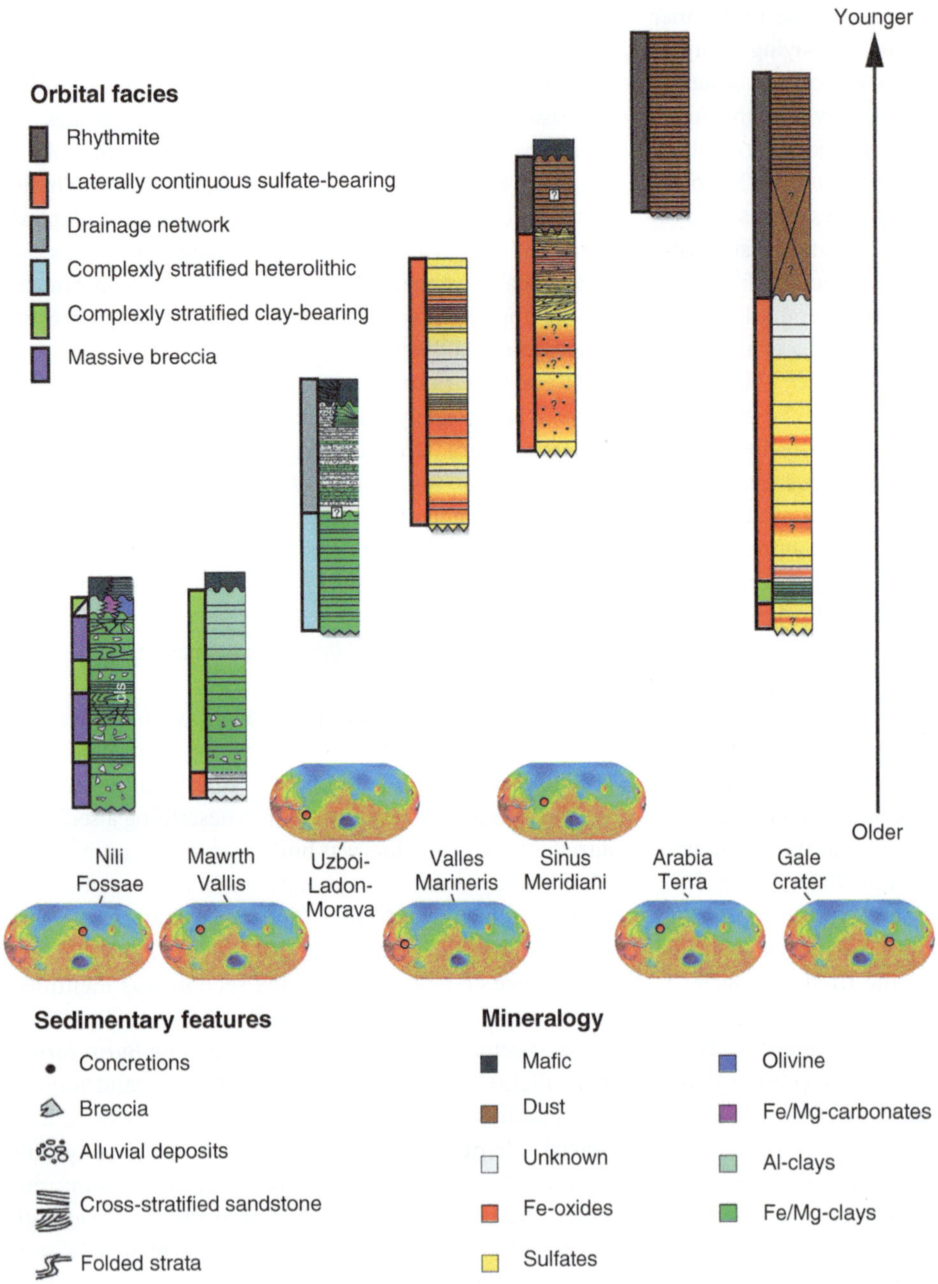

Figure 15.20 Type stratigraphic sections and orbital facies on Mars (adapted from Grotzinger and Milliken 2012).

transition to a damp interdune. The middle unit is a ~ 3-m-thick sandstone dominated by planar to low-angle cross-stratified sandstone (Section 15.3.6). Trough cross-lamination from rare subaqueous ripples (Section 15.3.6) is observed in places, indicating episodic wet conditions. A 30–40 cm-thick diagenetic zone, characterised by significant recrystallisation, defines the contact between the middle and upper units; it is thought to have resulted from early diagenesis in the capillary fringe of a stagnant, shallow water table. The middle unit thus appears to have recorded a transition from a damp to a wet aeolian setting. Finally, the upper unit consists of both sand sheet and wet interdune facies, including trough cross-lamination from subaqueous ripples, soft-sediment deformation and possible small tepee or salt-ridge structures, consistent with sand sheet deposition in a persistently wet aeolian system.

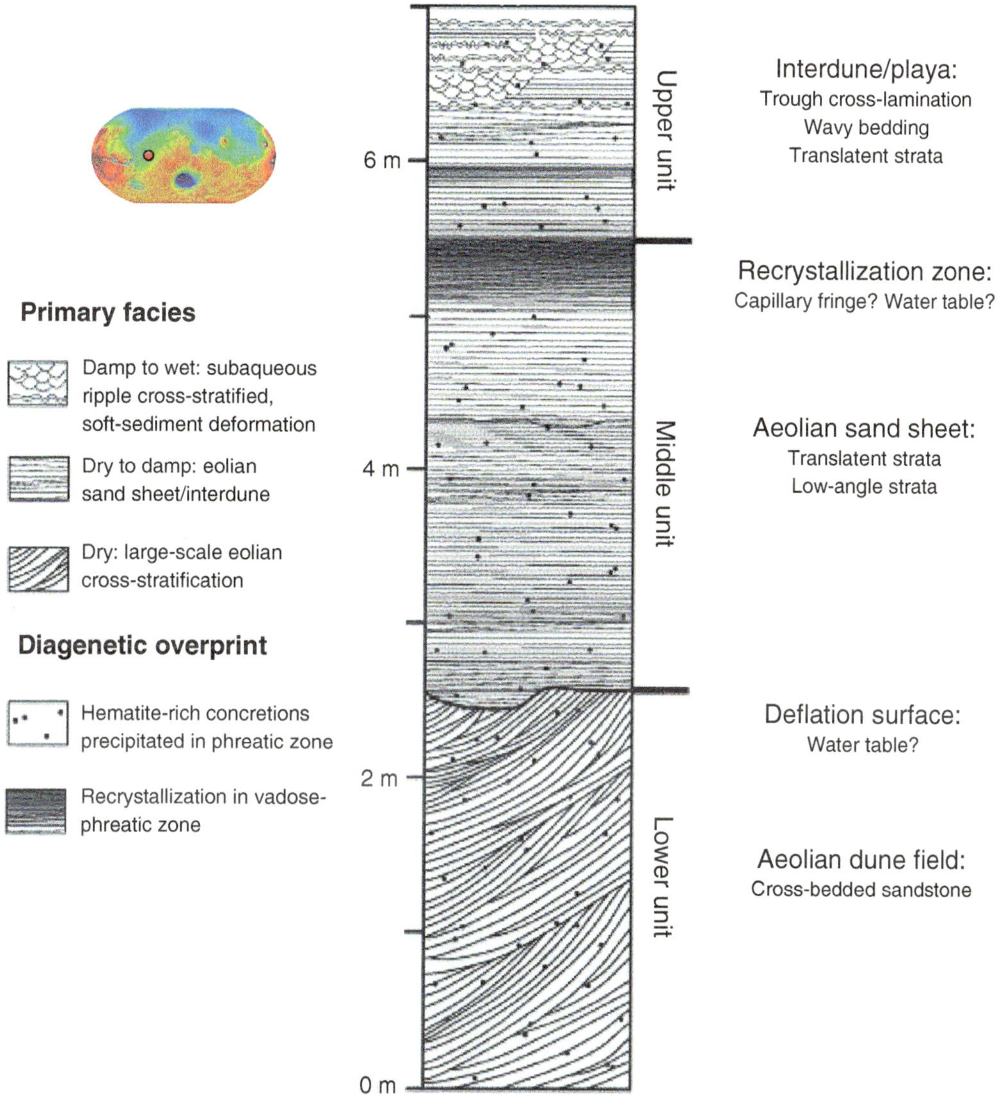

Figure 15.21 Type section of the Burns formation at Meridiani Planum, Mars (Grotzinger et al. 2005 / with permission of Elsevier).

More wetting and drying episodes of an overall aeolian environment were later documented higher up in the stratigraphy through observations at Erebus and Victoria craters further along Opportunity's traverse (Metz et al. 2009a; Hayes et al. 2011; Edgar et al. 2012). Finally, on its way to Endeavour crater, Opportunity inspected individual Burns formation blocks excavated by the impact that created the Santa Maria crater, which occurred ~100 m lower in elevation than Victoria crater. Although many blocks consisted of aeolian sandstone, several appeared more massive, fine-grained and nodular. Edgar et al. (2014) interpreted these unique blocks as mudstones formed in an interdune lake.

The stratigraphic description of the Burns formation marked a significant turning point in planetary geology, demonstrating how principles of sedimentology and stratigraphy could be applied to decipher the history of other planetary bodies. Specifically, it painted a picture of an ancient, dynamic aeolian environment, fed by playa-derived, sulphate-rich sediments and controlled by a fluctuating groundwater table that at least episodically broke the surface, creating wet surface conditions. Inspired by Opportunity's success, NASA sent its next two rovers, Curiosity and Perseverance, to locations where thick sedimentary successions had been identified from orbit – Gale and Jezero craters.

15.3.5.3 Gale Crater Stratigraphy

In 2012, Curiosity landed on the floor of Gale crater (a ~ 150 km diameter, ~4.5 km deep impact crater formed between ~3.8 and ~3.5 Ga), with the goal of investigating strata exposed in the lower slopes of Aeolis Mons (informally known as Mount Sharp), the crater's ~ 5 km high, layered central mound (Figure 15.14a; Grotzinger et al. 2012). To date, the strata explored by Curiosity have been divided into three lithostratigraphic groups, thought to consist of ~3.5–2.9 Ga deposits (e.g. Farley et al. 2014) (Figure 15.22): the *Bradbury group*, comprising fluvial, deltaic, and lacustrine sedimentary rocks near the rover's landing site; the *Mount Sharp group*, comprising lacustrine mudstones with some fluvial sandstones which pass upwards into aeolian sulphate-bearing strata; and the *Siccar Point group* that unconformably overlies rocks of the Mount Sharp group.

The Bradbury group is ~60 m thick and composed of mudstones, sandstones and pebble conglomerates (Section 15.3.6). It is divided into four intervals, two of which are named. The first interval, known as the *Yellowknife Bay formation*, is dominantly composed of lacustrine mudstone and fluvial sandstone (Grotzinger et al. 2014; Edgar et al. 2018). Notably, the *Sheepbed member* of the Yellowknife Bay formation was the first lacustrine mudstone to be described from the ground on Mars (Grotzinger et al. 2014; Schieber et al. 2017). Above the Yellowknife Bay formation lies ~25 m of interbedded conglomerates and sandstones, some pebbly and cross-stratified (Section 15.3.6). The conglomerates are typically weakly stratified and poorly sorted, with angular to subrounded clasts (Section 15.3.6), and are thought to represent fluvial deposition in an alluvial fan setting (Williams et al. 2013b) (Box 15.2). The *Kimberley formation* is ~5–10 m thick and exposes clinoform sandstones at its base, which are characterised by gently-sloping topsets that gradually transition into steeper foresets, which in turn transition into gently sloping bottomsets (Section 15.3.6). The clinoform sandstones are overlain by cross-stratified sandstone and lacustrine mudstone, interpreted to record the progradation of a river delta into a lake (Grotzinger et al. 2015; Rice et al. 2017). Above the Kimberley formation, similar sequences of interbedded clinoform sandstone and cross-stratified sandstone, with minor mudstone and conglomerate, record a dynamic fluviolacustrine system. At their contact, rocks of the Bradbury and Mount Sharp groups are interpreted to interfinger laterally.

To date, the Mount Sharp group is divided into three formations – the *Murray*, *Carolyn Shoemaker* and *Mirador formations*. Both the Murray and Carolyn Shoemaker formations belong to the "complexly stratified clay-bearing

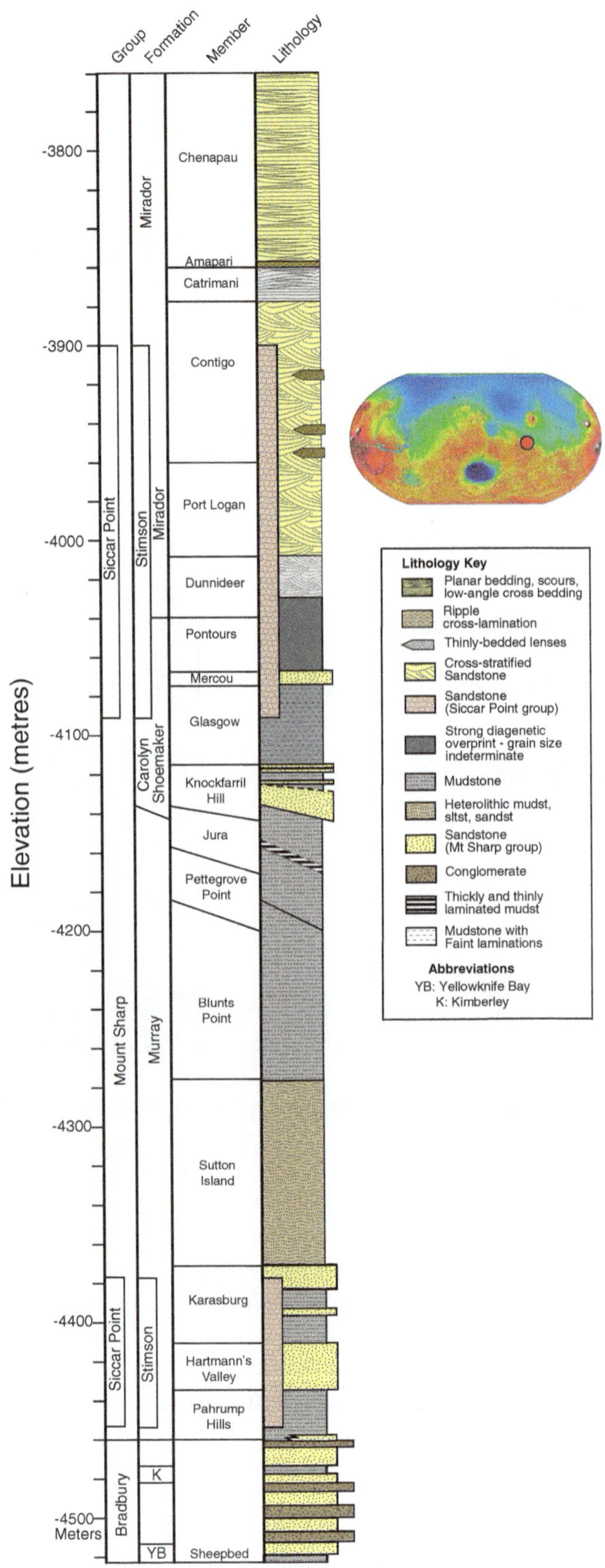

Figure 15.22 Stratigraphic column constructed from Curiosity rover observations during its ascent from the Gale crater floor up Mount Sharp (Aeolis Mons).

units" mapped from orbit, while the "laterally continuous sulphate-bearing strata" largely consist of the Mirador formation (Figure 15.20). However, the clay-to-sulphate transition mapped from orbit is not a perfect match to changes in facies observed on the ground, and some of the sulphate detections proved to be of diagenetic origin rather than a primary signal (Meyer et al. 2024).

The Murray formation is ~330 m thick and is overwhelmingly composed of planar-laminated lacustrine mudstones, with some sandstones and heterolithic mudstone, siltstone and sandstones. The Murray formation is divided into seven stratigraphic members: the *Pahrump Hills* (Stack et al. 2019; Schieber et al. 2022), *Hartmann's Valley* (Gwizd et al. 2022), *Karasburg* (Gwizd et al. 2022), *Sutton Island* (Edgar et al. 2020; Stein et al. 2018), *Blunts Point* (Edgar et al. 2020), *Pettegrove Point* (Edgar et al. 2020) and *Jura* (Edgar et al. 2020; Fedo et al. 2022; Caravaca et al. 2022) *members*. Collectively, deposits of the Murray formation suggest the past existence of a long-lived and dynamic lacustrine system, bordered and fed by fluvial and aeolian systems. Derived from erosion of the crater rim, the fluvial-deltaic strata of the Bradbury group are interpreted to pass laterally by facies transition into lacustrine strata of the Murray formation that are deposited towards the centre of the crater, thus defining a Martian source-to-sink sediment dispersal system. It should be noted that for most of the Murray formation succession, the coarser-grained proximal facies are not preserved.

The contact between mudstones of the Jura member and the lowermost units of the Carolyn Shoemaker formation is sharp and planar. This abrupt transition was interpreted as an environmental transition, from lacustrine to fluvial erosion and deposition in the *Glen Torridon* region of Gale crater (Bennett et al. 2023; Fedo et al. 2022; Caravaca et al. 2022). The Carolyn Shoemaker formation is ~105 m thick and is composed of interstratified mudstones and sandstones. It is divided into the *Knockfarril Hill* (Fedo et al. 2022; Caravaca et al. 2022), *Glasgow* (Fedo et al. 2022; Cardenas et al. 2022b), *Mercou* (Cardenas et al. 2022b) and *Pontours* (Rapin et al. 2023) members. Notably, river bar deposits were identified in the Glasgow and Mercou members of the Carolyn Shoemaker formation, consistent with an overall transition from a lacustrine to an alluvial setting, possibly along the lake shoreline. The Pontours member is the uppermost member of the Carolyn Shoemaker formation. It is severely altered by diagenetic overprints, including abundant nodules. It displays polygonal ridges with Y-shaped triple junctions, which have been interpreted as evidence for sustained, repeated wetting and drying cycles (Rapin et al. 2023).

Finally, the Mirador formation overlies the Carolyn Shoemaker formation and is ~275 m thick (Meyer et al. 2024). Curiosity is still exploring the Mirador formation as of March 2025. To date, it is divided into the *Dunnideer*, *Port Logan*, *Contigo*, *Catrimani*, *Amapari* and *Chenapau* members (Gupta et al. 2022; Roberts et al. 2023; Edgar et al. 2024). The Mirador formation is dominated by planar, low-angle and cross-stratified sandstone facies (Section 15.3.6), interpreted as deposited by aeolian sand dunes and sand sheets. Distinct from these aeolian facies, two stratigraphic intervals (meter-scale sandstone lenses interstratified within the Contigo member, and the laterally continuous *Amapari Marker Band* of the Amapari member) contain current and wave-ripple cross-lamination with mud drapes (Figure 15.26b), indicating deposition in a wave-influenced shallow water body (Gupta et a. 2022; Mondro et al. 2025a,b).

Finally, the Siccar Point group consists of the *Stimson formation* at its base, which unconformably overlies the Murray, Carolyn Shoemaker and Mirador formations of the Mount Sharp group (Watkins et al. 2022). The Stimson formation consists of varied aeolian sandstone facies (Section 15.3.6), dominated by very fine to medium sand (Preston et al. 2024). It is interpreted to have formed by the migration of simple crescentic to oblique compound dunes in a dry aeolian system (Box 15.3) (Banham et al. 2018, 2021, 2022), similar to the modern Bagnold dune field but with opposite net sand-flux direction (Bridges and Ehlmann 2018; Lapôtre and Rampe 2018). The Stimson formation is overlain by boulder-bearing deposits exposed in the *Gediz Vallis* ridge that are interpreted to represent deposition in a debris flow fan deposit (Bryk et al. 2024).

Box 15.3 A snapshot of Mars' Hesperian landscapes from rover observations of multibillion-year-old aeolian sandstones

During its ascent of Mount Sharp, Curiosity encountered cliff-forming mesas known as the *Murray Buttes*. The Murray Buttes consist of over a dozen multi-metre tall outcrops of aeolian sandstone (*Stimson formation, Siccar Point group*), lying unconformably over lacustrine mudstones of the Murray formation (Figure 15.22). Here, we illustrate how observations made by Curiosity at the Murray Buttes helped reconstruct the palaeo-environment in which the sediments were deposited from observations at one of these mesas – butte M1b (Banham et al. 2021).

The M1b butte is an ∼8 m tall by ∼45 m long outcrop of largely planar cross-stratified sandstone. Cross-lamination is pervasive and characterised by inclined and curved laminae that downlap asymptotically onto more horizontal to sub-horizontal surfaces and are truncated by upper bounding surfaces (white). Two orders of bounding surfaces are observed in the M1b butte sandstone based on their lateral continuity. Bounding surfaces that can be traced laterally for tens of metres (orange) are largely horizontal. Within the cosets (or groups of associated cross-sets) formed by those bounding surfaces, individual sets are separated by inclined bounding surfaces (yellow), which themselves can be truncated by upper bounding surfaces and downlap onto lower ones, and thus display lower lateral continuity (1–2 m or less). Banham et al. (2021) referred to highly laterally continuous bounding surfaces as compound-set (or coset) bounding surfaces (orange), and the other ones as subset bounding surfaces (yellow). They interpreted coset bounding surfaces as scouring surfaces from trough migration of large dunes (or draas). Cosets could have formed from the migration of smaller superimposed bedforms or reversing flows (reactivation surfaces). However, a detailed analysis of dip azimuth revealed variations between sets within individual cosets, and no simple sets were observed, so the hypothesis of superimposed bedforms was favoured.

Combining observations by the Curiosity rover from different vantage points, Banham et al. (2021) were able to reconstruct the three-dimensional geometry and dip orientations of different surfaces. Cumulatively, their observations led to the conclusion that the Stimson formation sandstone at the Murray Buttes recorded the migration of large (10s of metres tall, 100s of metres long), relatively linear-crested dunes to the north-northwest, with smaller (likely ∼5 m tall) superimposed linear dunes migrating across the draa's lee slope and obliquely to the draa's crestline along the stoss towards the northeast. Thus, net sand flux was roughly to the north at the time of deposition – opposite to the modern transport direction in the Bagnold dunes.

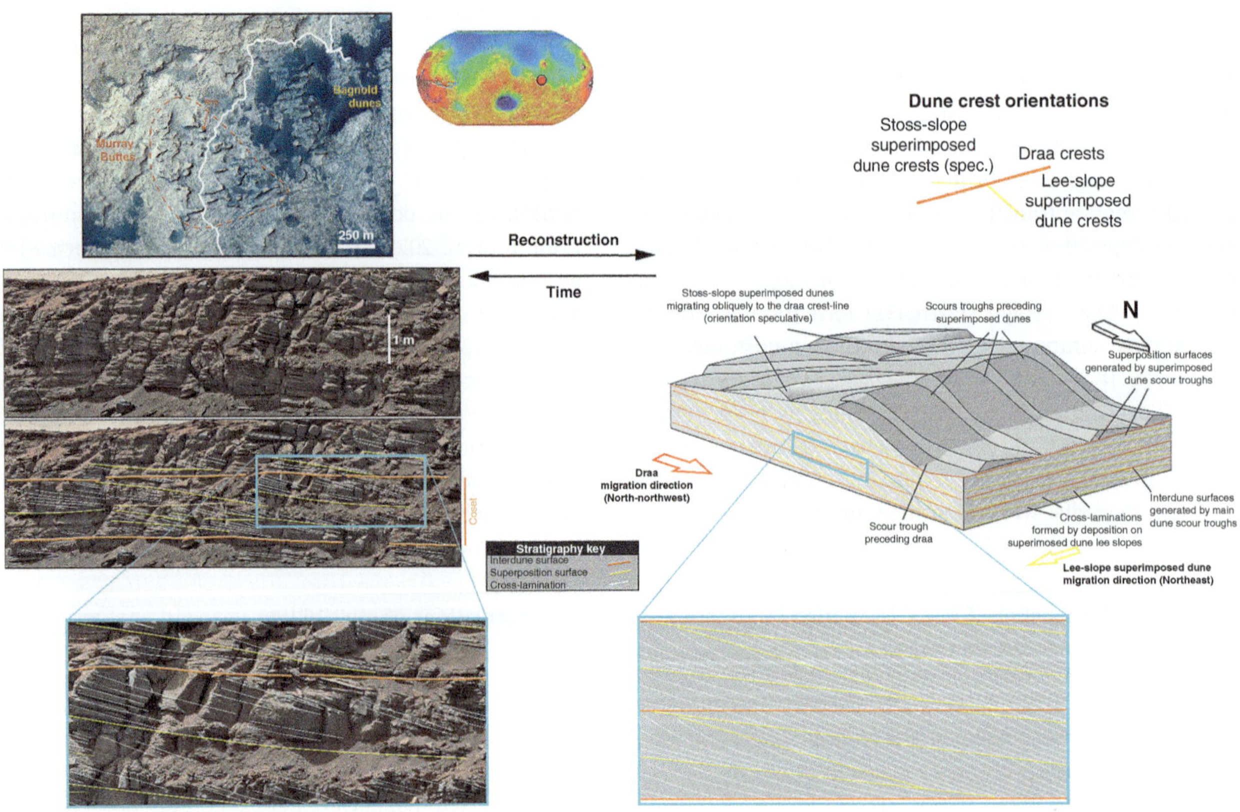

Curiosity's exploration of Gale crater extended the snapshot provided by the Opportunity rover further back in time, covering a remarkably thick sedimentary section. It revealed an important environmental transition, from sustained wet, clay-forming, habitable environments to sulphate-bearing strata, characterised by episodic aqueous activity under otherwise drier surface conditions. Interestingly, alternating wet and dry episodes in the sulphate-rich aeolian environments of the Mirador formation are reminiscent of those inferred from the Burns formation at Meridiani Planum. Could they be the record of Mars' global drying?

15.3.5.4 Western Fan Stratigraphy, Jezero Crater

Even older than Gale crater, the ~45 km diameter, ~1 km deep Jezero crater contains a sedimentary archive deposited ~3.8–3.2 Ga (Stack et al., 2024) of possibly habitable ancient environments, including the eroded remnants of fan-shaped deposits, formed where rivers derived from outside the crater debouched into a series of lakes. Specifically, its western fan is thought to host deposits of the fluvio-deltaic origin, making it an ideal astrobiological target as deltas concentrate sediments (and any organics), gathered over large drainages by rivers, into a single, well-defined location. Furthermore, fine-grained facies typical of distal delta deposits are known for their high organic preservation potential (McMahon et al. 2018). With a goal to determine whether Mars was inhabited prior to or during the Late Noachian-Early Hesperian, the Perseverance rover landed on the Jezero crater floor in 2021 to collect what will be the first set of rock samples to be returned to Earth (by the Mars Sample Return mission; Farley et al. 2020).

As of July 2024, Perseverance has investigated crater floor units on which it first landed, fan-forming sedimentary facies and possible lake-marginal facies near the Jezero Lake palaeoshoreline. For the first time in Mars exploration history, Perseverance has deployed a helicopter (Balaram et al. 2021) to aid with reconnaissance and has used a ground-based ground penetrating radar (GPR) to shed light onto subsurface stratigraphy (Hamran et al. 2022; Paige et al. 2024). Here we briefly describe the distal stratigraphy of the fan (Stack et al. 2024).

A significant unknown before landing was the stratigraphic position of the fan deposits relative to orbitally mapped crater floor units (e.g. Goudge et al. 2015b; Grant et al. 2018; Stack et al. 2020). Importantly, the surface area of the fan deposits is too small for dating through crater counting, whereas a surface age could be derived for the crater floor units (e.g. Shahrzad et al. 2019). Upon landing, crater floor units were divided into two formations – the *Séítah* and *Máaz formations*. The Séítah formation was determined to be an altered olivine cumulate (Liu et al. 2022; Farley et al. 2022). Basaltic lava flows overlying the Séítah formation near the fan's eastern scarp constitute the Máaz formation (Farley et al. 2022). Ground-based observations, and specifically, GPR imaging, demonstrated that fan deposits overlie both the Séítah and Máaz formations (Paige et al. 2024).

Unlike the Gale crater mound, which displays laterally continuous stratigraphy over kilometre scales, Jezero's western fan stratigraphy is laterally heterogenous (in both downstream and cross-stream directions) over decametre length scales (with the exception of the *Hogwallow Flats member*, which is laterally continuous over kilometre scales; Figure 15.23). As a result, more than one stratigraphic column is necessary to capture the complex three-dimensional distributions of various sedimentary facies.

The lower ~25 m of sedimentary deposits observed above the Séítah and Máaz formations by Perseverance where it climbed onto the fan belong to the *Shenandoah formation* (Stack et al. 2024; Figure 15.23). Perseverance investigated the Shenandoah formation in three locations along the fan's eastern scarp, yielding three stratigraphic sections (Stack et al. 2024). Although different member names were assigned at different locations, three facies associations could be spatially correlated across all three locations (Figure 15.23).

The lowermost facies association is composed of laminated fine- to coarse-grained sandstone, with soft-sediment deformation in places, and is interbedded with thin lenses of granule-to-pebble sandstone (Section 15.3.6). It directly overlays the crater floor units and was interpreted to have resulted from deposition in a distal alluvial fan or pro-delta setting (Stack et al. 2024). The second facies association is dominated by laminated mudstone to very fine-grained sandstone, with interbedded low-angle laminated sandstone (Section 15.3.6). This association was interpreted as lacustrine settling deposition of fine suspended sediments. Finally, the uppermost facies association of the Shenandoah formation consists of cross-stratified to sub-horizontal plane parallel, medium sandstone to granule conglomerate, with evidence for clast imbrication (Section 15.3.6). It was interpreted as reflecting the progradation of a fluvial delta into the lake (Stack et al. 2024).

Overlying these basal deposits is a succession that was observed remotely by the Perseverance rover through high-resolution imaging by the Mastcam-Z cameras, including clinoform features in the cliff faces of a butte named *Kodiak* (Section 15.3.6) (Mangold et al. 2021). This succession records large-scale foreset beds up to ~20 m thick and dipping basinwards at ~20–30°; these

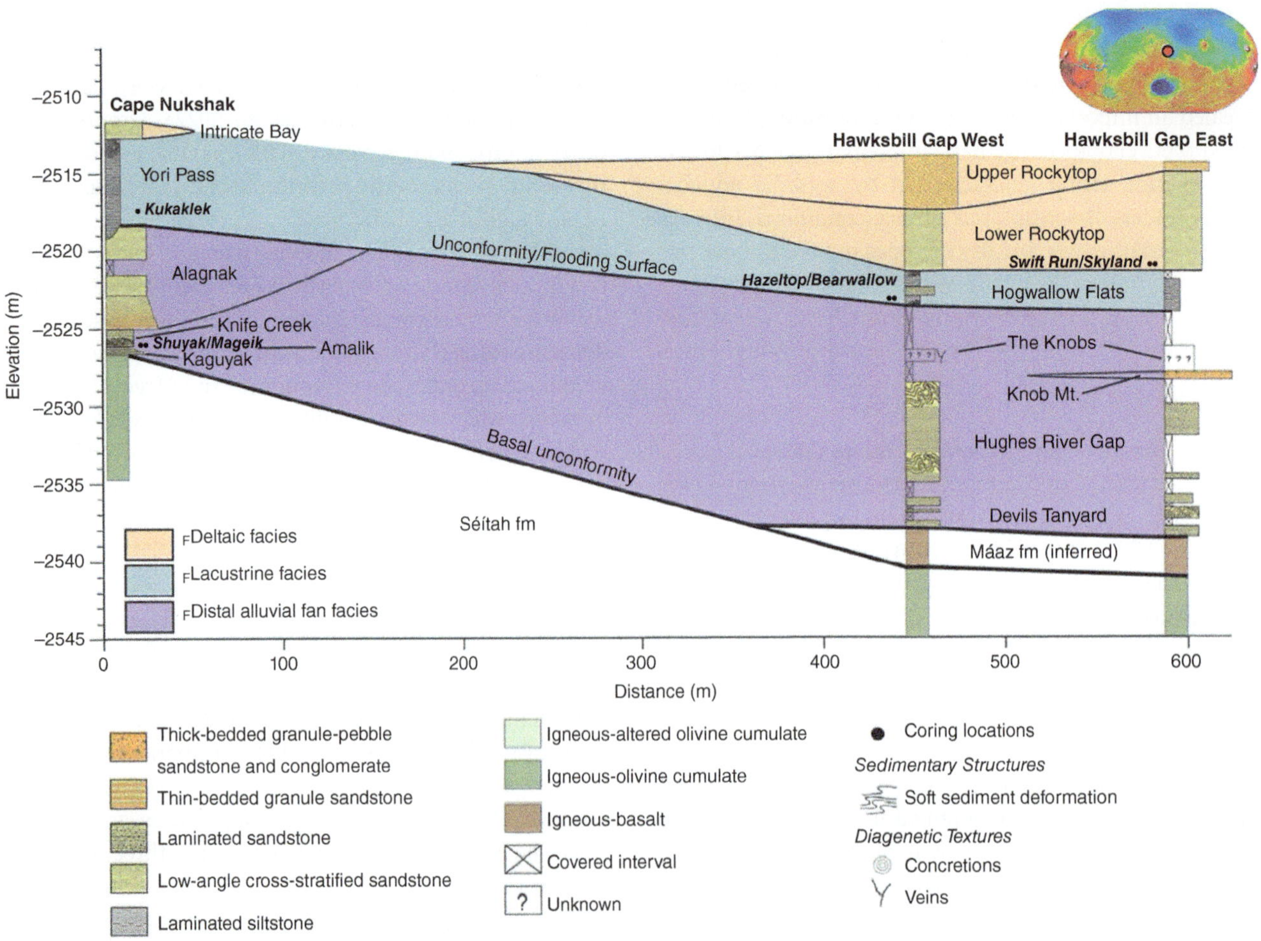

Figure 15.23 Stratigraphic columns constructed from Perseverance rover observations during its ascent from the Jezero crater floor up the western fan (adapted from Stack et al. 2024). Location of stratigraphic sections summarised here is shown in Figure 15.14c.

are sharply overlain by sub-horizontal topset strata comprising horizontally stratified sandstones and conglomerates (Caravaca et al. 2024) (Section 15.3.6). The foreset beds are interpreted as Gilbert-type delta front avalanche strata, whilst the sub-horizontal topset beds represent delta-top fluvial deposits (Mangold et al. 2021, 2024, Gupta et al. 2024).

Thus, like the deposits of Gale crater, sedimentary units explored in Jezero crater attest to a protracted and dynamic history of wet, habitable surface environments early in Mars' history. In particular, the deposits of the western fan point to sustained water and sediment transport by fluvio-deltaic processes, providing clear evidence of at least intermittently high-water discharge on the early Martian surface. The Perseverance rover's investigation of the fan's upper stratigraphy and possible lake-marginal deposits promises to deliver many more insights into Mars' ancient sedimentary environments and habitability.

15.3.5.5 Complementarity of Orbiter- and Rover-Based Observations for Planetary Stratigraphy

Rovers provide a great level of detail, but the spatial extent of their observations is limited. In turn, orbiters provide global observations but are limited in their spatial resolution and therefore our ability to robustly interpret sedimentary facies and processes. For example, orbiter-based spectral detections of sulphate minerals have been interpreted as a signature of desiccation at Gale crater; however, ground-truth from the Curiosity rover revealed that at least some of those detections reflect a later, diagenetic overprint rather than a primary environmental signal (Meyer et al. 2024). Such critical observations of how mineralogy relates to facies could not be performed from orbit. Similarly, a property as fundamental as the stratigraphic position of the fan deposits relative to crater floor units could not be determined with confidence from orbit prior to landing at Jezero crater (e.g. Grant et al. 2018). Thus, solving fundamental mysteries, such as the timing, pace and processes that drove Mars' atmospheric loss and surface drying, will require more orbiter missions along with ground-truthing by rovers, coupled with radiometric dating to develop a robust stratigraphic chronology.

15.3.6 Ancient Depositional Environments and Associated Facies: A Rover Perspective

A wide range of stratigraphic geometries are observed on Mars from orbital images, suggesting that a variety of depositional processes operated at the Martian surface (Stack et al. 2013). Nevertheless, it remains difficult to make robust interpretations of Martian palaeo-environments from orbital images alone. Here, we focus on rover-derived observations of sedimentary facies and facies associations that have been interpreted to reflect deposition in aeolian, alluvial/fluvial, shoreline/deltaic and lacustrine environments. References to remote-sensing studies of glacial/periglacial, mass wasting, volcaniclastic and impact-generated deposits are provided in "Further Reading".

15.3.6.1 Aeolian Deposition and Facies Associations

Evidence for aeolian deposition and erosion has been identified at every Mars landing site, and aeolian strata have been explored in detail at the landing sites for the Spirit, Opportunity and Curiosity Mars rovers. These sites form the basis for the following facies descriptions. For more details on modern aeolian processes, see Sections 15.2.3 and 15.3.3. Aeolian deposits, dominated by sandstones, were primarily identified from three facies associations – corresponding to dunes, sand sheets and interdunes – in the Burns formation at Meridiani Planum by Opportunity (Grotzinger et al. 2005; Metz et al. 2009a; Squyres et al. 2009; Hayes et al. 2011; Edgar et al. 2012; Edgar et al. 2014) and in the Stimson formation (Banham et al. 2018, 2021, 2022) (Box 15.3), and Hartmann's Valley and Karasburg members of the Murray formation (Gwizd et al. 2022) in Gale crater by Curiosity.

Similar to terrestrial systems, four basic types of stratification are recognised in Martian aeolian deposits: wind-ripple strata, grainflow strata, grainfall strata and planar-laminated pinstripe strata (see Section 7.4 for more detail). Aeolian strata on Mars were first recognised via landed missions by the observation of large-scale planar and trough cross-bedding in fine-to-medium-grained sandstones (Grotzinger et al. 2005). Dune facies include planar and trough cross-bedded sandstones, avalanche deposits and wind-ripple lamination. Troughs occur at a variety of scales, from the decimetre- to metre-scale (Grotzinger et al. 2005; Banham et al. 2018) (Figure 15.24a,c). Martian dune deposits are dominated by wind-ripple lamination (or pinstripe lamination), which produces thin, resistant strata. Evidence for pinstripe lamination was also observed in the Burns formation, though fine-scale primary textures were largely obscured by significant recrystallisation (Grotzinger et al. 2005). Pinstripe lamination is the dominant style of stratification in dune foreset strata of the Stimson formation. By contrast to dune deposits on Earth, avalanche deposits or grainflows (characterised by their wedge shape, relatively poor internal structure and recessive nature) have been rarely observed, although they have been found together in alternating sequences with wind-ripple lamination in the Stimson formation (Banham et al. 2022) (Figure 15.24b) and are observed in Mars' modern aeolian environments (e.g. Ewing et al. 2017).

Sand sheet facies are dominated by planar laminated sandstone, with some low-angle cross-stratified sandstone (Figure 15.24d). Both dune and sand sheet facies associations host decimetre-scale trough cross-stratified sandstones, indicative of the migration of large sinuous ripples, likely analogous to Mars' modern metre-scale "large ripples". Exhumed and fully preserved metre-scale bedforms, observed in the Kimberley formation by Curiosity, have also been interpreted as possible aeolian ripples (Rubin et al. 2022).

Finally, facies associated with both wet and dry interdune areas have been observed on Mars. Wet interdune facies observed to date include subaqueous ripple cross-lamination (Grotzinger et al. 2006; Lamb et al. 2012) (Figure 15.24e), wavy-to-irregular bedded sandstone, convolute bedding and possible salt-ridge or small tepee structures in the Burns formation (McLennan et al. 2005; Metz et al. 2009a). Wet interdunes were also inferred on the basis of preserved thicknesses and stacking patterns (Gwizd et al. 2022). Observations of soft-sediment deformation structures in aeolian dune strata of the Stimson formation have been interpreted as evidence of shallow groundwater within the dune deposits (Banham et al. 2024).

15.3.6.2 Fluvial and Alluvial Fan Deposition and Facies Associations

A diverse array of fluvial- and alluvial fan deposits has been observed on Mars, notably by the Curiosity and Perseverance rovers, and includes a variety of sandstone and conglomerate facies. Alluvial deposits were identified, for example, in the Bradbury group (Williams et al. 2013b; Grotzinger et al. 2015), the Yellowknife Bay formation (Grotzinger et al. 2014; Edgar et al. 2018), and the Carolyn Shoemaker formation (Caravaca et al. 2022; Cardenas et al. 2022b; Fedo et al. 2022) of Gale crater and the Shenandoah formation of Jezero crater (Stack et al. 2024).

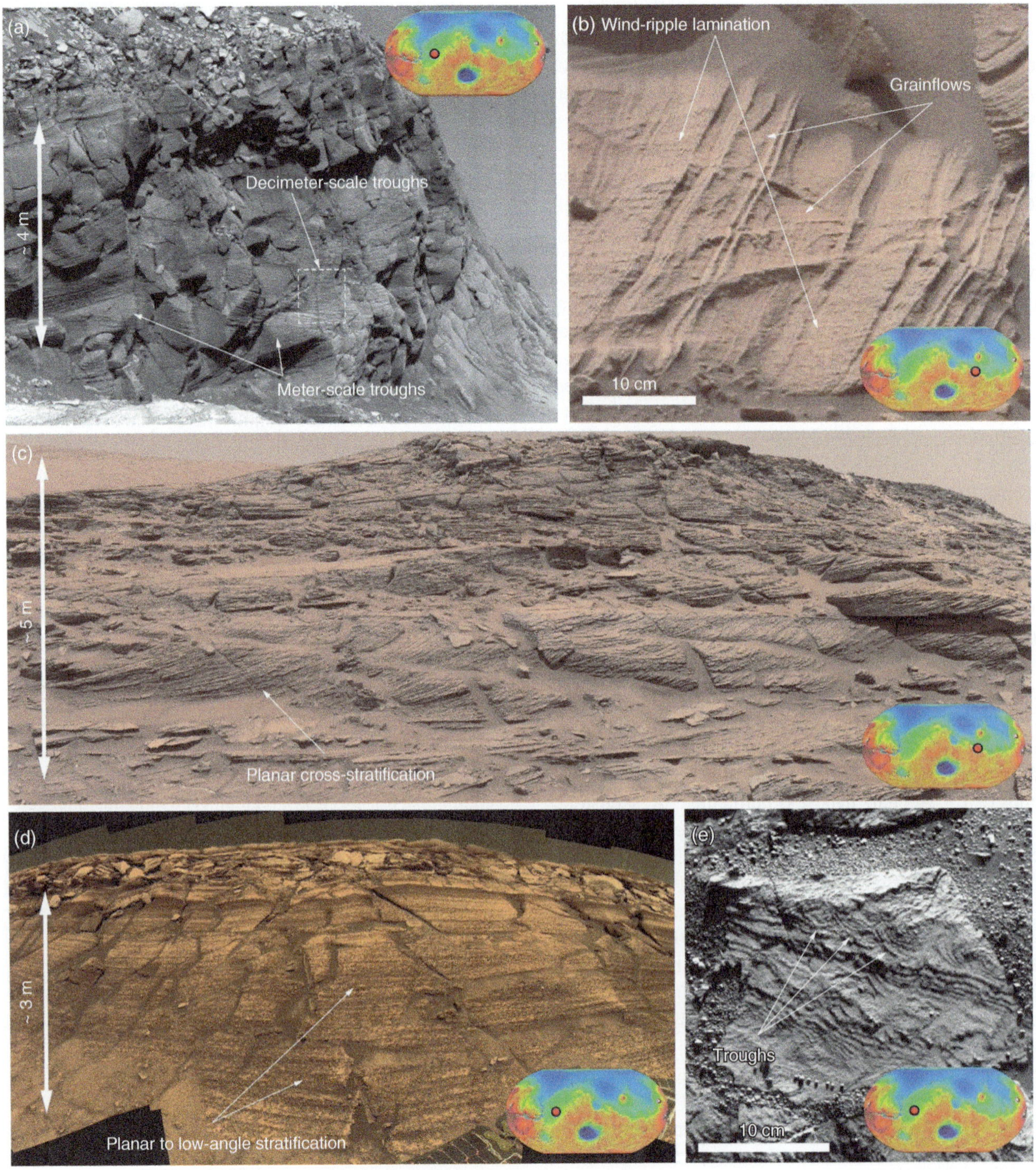

Figure 15.24 Examples of aeolian facies on Mars. (a) Trough cross-stratified sandstone of the Burns formation, Victoria crater, showing both metre- and decimetre-scale troughs (Photo by NASA/JPL-Caltech/Cornell University / Public domain). (b) Alternating grainflow deposits (relatively structureless and recessive) and wind-ripple lamination at the *Ladder* outcrop of the *Stimson formation*, Gale crater (NASA/JPL-Caltech/MSSS / Public domain). (c) Trough cross-stratified and planar laminated sandstone at the *Williams* outcrop of the Stimson formation, Gale crater (NASA/JPL-Caltech/MSSS / Public domain). (d) Planar to low-angle stratification in Burns formation sandstone at *Burns Cliff*, Endurance crater (Photo by NASA/JPL/Cornell / Public domain). (e) Centimetre-scale trough cross-stratification from subaqueous ripples in interdune deposits, *Overgaard* outcrop, Burns formation, Erebus crater (NASA/JPL-Caltech/Cornell Univ / Public domain).

Poorly sorted pebbly sandstones (Figure 15.25a) and conglomerates (Figure 15.25b,d) of the Bradbury group display angular to subrounded clasts, interpreted as bedload sediment transported over relatively short distances in an alluvial fan setting (Williams et al. 2013b; Grotzinger et al. 2015).

Channel facies include poorly to moderately well-sorted, crudely stratified to cross-stratified, fine- to coarse-grained sandstones with decimetre-scale trough cross-sets, for example in the *Shaler* outcrop of the Yellowknife Bay formation of Gale crater (Edgar et al. 2018) and in the Shenandoah formation of Jezero crater (Stack et al. 2024). Cross-stratified sandstones of the Knockfarril Hill member (Carolyn Shoemaker formation), which lack clear channels and display minimal interbedded mud, were interpreted as stacked barforms within a sand-bedded braided stream (Caravaca et al. 2022; Fedo et al. 2022) (Figure 15.25e). Barforms were also identified in the Glasgow member of the Carolyn Shoemaker formation, e.g. at *Mont Mercou* (Cardenas et al. 2022b) (Figure 15.25f).

Finally, boulder conglomerates in the upper stratigraphy of the western Jezero crater fan have been interpreted as indicative of episodic, high-energy floods (Mangold et al. 2021) (Figure 15.25c; Box 15.2).

15.3.6.3 Shore-Zone, Deltaic and Lacustrine Facies Associations and Depositional Environments

Both Gale and Jezero craters hosted lakes fed by rivers. Shore-zone facies include mixed mudstone, sandstone and conglomerates. Some mudstones of the Murray formation in Gale crater display desiccation cracks, formed from repeated wetting and drying along the lake margins (Stein et al. 2018; Gwizd et al. 2024) (Figure 15.26a). Candidate wave ripples were also identified in several outcrops of the Murray formation, possibly indicating proximity to the lake margin (Rubin et al. 2022), and in the Contigo and Amapari members of the Mirador formation (Figure 15.26b), reflecting deposition in shallow lakes coeval with aeolian dunes (Mondro et al. 2025a,b).

At Gale crater, delta-front facies are dominantly composed of fine-grained clinoform sandstones (Figure 15.26c), with planar to sigmoidal geometry and consistent dip direction over kilometre length scales (Grotzinger et al. 2015). At *Kodiak Butte* and in the eroded edge of the western fan in Jezero crater, Perseverance documented a coarse-grained Gilbert-type delta (Figure 15.26d), with steep foreset characteristic of progradation into a deep lake system (Mangold et al. 2021; Caravaca et al. 2024; Mangold et al. 2024).

In both craters, lacustrine facies are overwhelmingly composed of laminated mudstone, with some sandstones. Lacustrine mudstones were observed in the Shenandoah formation of Jezero crater (Stack et al. 2024) (Figure 15.27a). In Gale crater, facies associated with plunging river plumes include thickly (proximal; Figure 15.27b) and thinly (distal; Figure 15.27c) laminated mudstones (Grotzinger et al. 2015; Stack et al. 2019; Edgar et al. 2020). Other lacustrine facies include cross-stratified sandstones (Figure 15.27d) formed from the migration of climbing dunes or ripples in a proximal sublacustrine channel (Grotzinger et al. 2015; Stack et al. 2019; Edgar et al. 2020) and low-angle cross-stratified mudstone (Figure 15.27e) formed from migration of supercritical bedforms or bars in sublacustrine channels (Stack et al. 2019) at Gale crater.

15.4 Other Sedimentary Systems in the Solar System

15.4.1 Sedimentary Environments on Venus

Little is known about the sedimentary environments of Venus. At present, sediments are thought to largely consist of reworked ejecta materials generated by meteor impacts (Carter et al. 2023), with some possible generation by mass wasting (e.g. Malin 1992; Bulmer and Guest 1996), but minimal contribution from chemical weathering (Dyar et al. 2021). Estimates of sediment volumes from impact deposits imply an average sediment layer 8–63 cm thick if it were distributed globally (Ganey et al. 2023). The Soviet Venera landers revealed a diversity of surfaces, with exposures of bare bedrock surfaces overlain by scattered gravel and boulders, and either patchy veneers or deeper accumulations of finer sediment (e.g. Florensky et al. 1977; Florensky et al. 1983; Garvin et al. 1984; Surkov and Barsukov 1985) (Figure 15.28).

Due to the planet's high atmospheric density ($\rho_f \approx 65$ kg/m^3), even relatively weak winds like those measured by the Venera landers should mobilise dust and sand-sized particles (Lorenz 2016), possibly explaining detected sediment removal after landing (Selivanov et al. 1982). Though dust particles could not be resolved directly in Venera images, photometric measurements revealed settling of dust plumes shortly after the landing of Venera 9 and 10 (Moshkin et al. 1979; Garvin et al. 1984), thus confirming the existence of finer-grained sediment. Together, these observations suggest that impact-derived, sand-sized sediments are likely reworked by winds and possibly accumulate in discrete aeolian sand sheets or dune fields.

Only two dune fields have been recognised on Venus – Al-Uzza (Figure 15.4) and Menat Undae (previously known as the *Fortuna-Meshkent* and *Aglaonice* dune fields, respectively; Greeley et al. 1992, 1995; Weitz et al. 1994; Kreslavsky

Figure 15.25 Examples of fluvial facies on Mars. (a) Pebbly sandstone of the *Bradbury group* at *Cooperstown*, NASA/JPL-Caltech/MSSS / Public domain. (b) Well-rounded pebble conglomerate of the Bradbury group at *Hottah* outcrop, Gale crater (Photo by NASA/JPL-Caltech/MSSS / Public domain). (c) Boulder conglomerate within western fan scarp at Jezero crater as seen from the Perseverance rover landing site (sequence of Mars 2020 RMI images, sol 397; NASA/JPL-Caltech/LANL/CNES/CNRS / Public domain). (d) Poorly stratified, poorly to moderately well-sorted coarse pebble conglomerate of the Bradbury group at *Kylie*, Gale crater (Photo by NASA/JPL/Thomas Appéré / Public domain). (e) Cross-stratified sandstone at *Knockfarril Hill, Carolyn Shoemaker formation*, NASA/JPL-Caltech/MSSS / Public domain. (f) Cross-stratified sandstone of the Carolyn Shoemaker formation at *Mont Mercou*, NASA/JPL-Caltech/MSSS / Public domain.

and Bondarenko 2017). Both dune fields are thought to consist of transverse dunes, as evidenced by the bright radar backscatter in Magellan images and orientation perpendicular to wind streaks; detected dunes have spacings over 500 m (Greeley et al. 1992). Four additional fields of microdunes have been hypothesised based on backscattering properties of the surface (Weitz et al. 1994) and their formation in pressurised wind-tunnel experiments (Greeley et al. 1984). In addition to depositional aeolian landforms, putative yardangs were also identified (Greeley et al. 1992). Interestingly, the relative paucity of aeolian deposits on Venus leaves large volumes of impact-derived sediments not accounted for (Ganey et al. 2023), suggesting that other, yet undetected sediment sinks likely exist.

Very little is known about the sedimentary past of Venus. It is thought that its surface could have hosted liquid water, prior to the runaway greenhouse warming of the atmosphere as the Sun became brighter earlier in Solar System

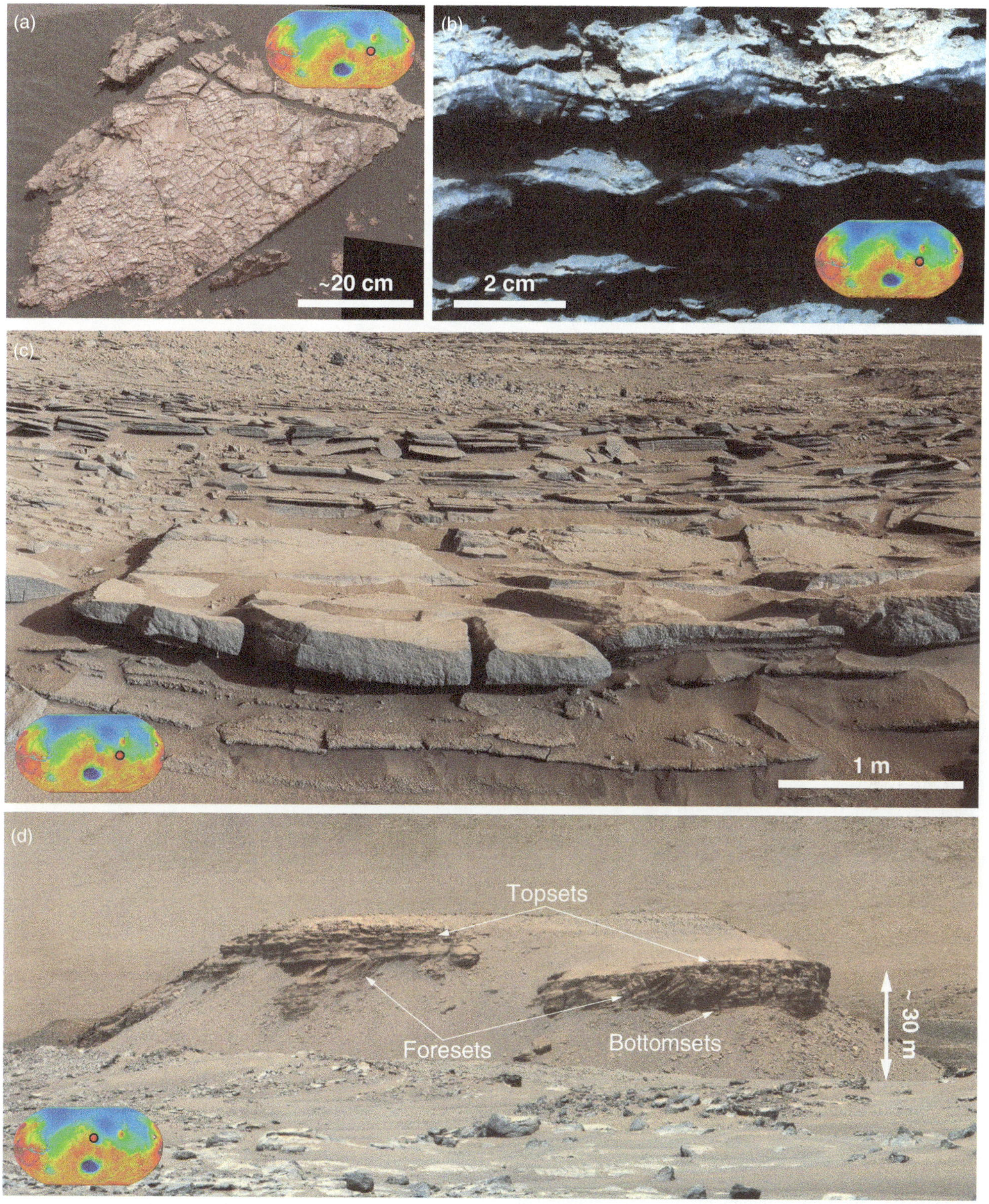

Figure 15.26 Examples of shore-zone and deltaic facies on Mars. (a) Desiccation cracks in the *Sutton Island member* of the Murray formation at *Old Soaker*, Gale crater (Photo by NASA/JPL-Caltech/MSSS / Public domain). (b) Wave ripples in the *Amapari Marker Band* outcrop of the *Amapari member* of the *Mirador formation, Marker Band Valley*, Gale crater (Photo by NASA/JPL-Caltech/MSSS / Public domain). (c) South-dipping (into the page) clinoform sandstone at *Kimberley*, Gale crater (Photo by NASA/JPL-Caltech/MSSS / Public domain). (d) Gilbert-type delta deposits with steep foresets at *Kodiak Butte*, Jezero crater (Photo by NASA/JPL-Caltech / Public domain).

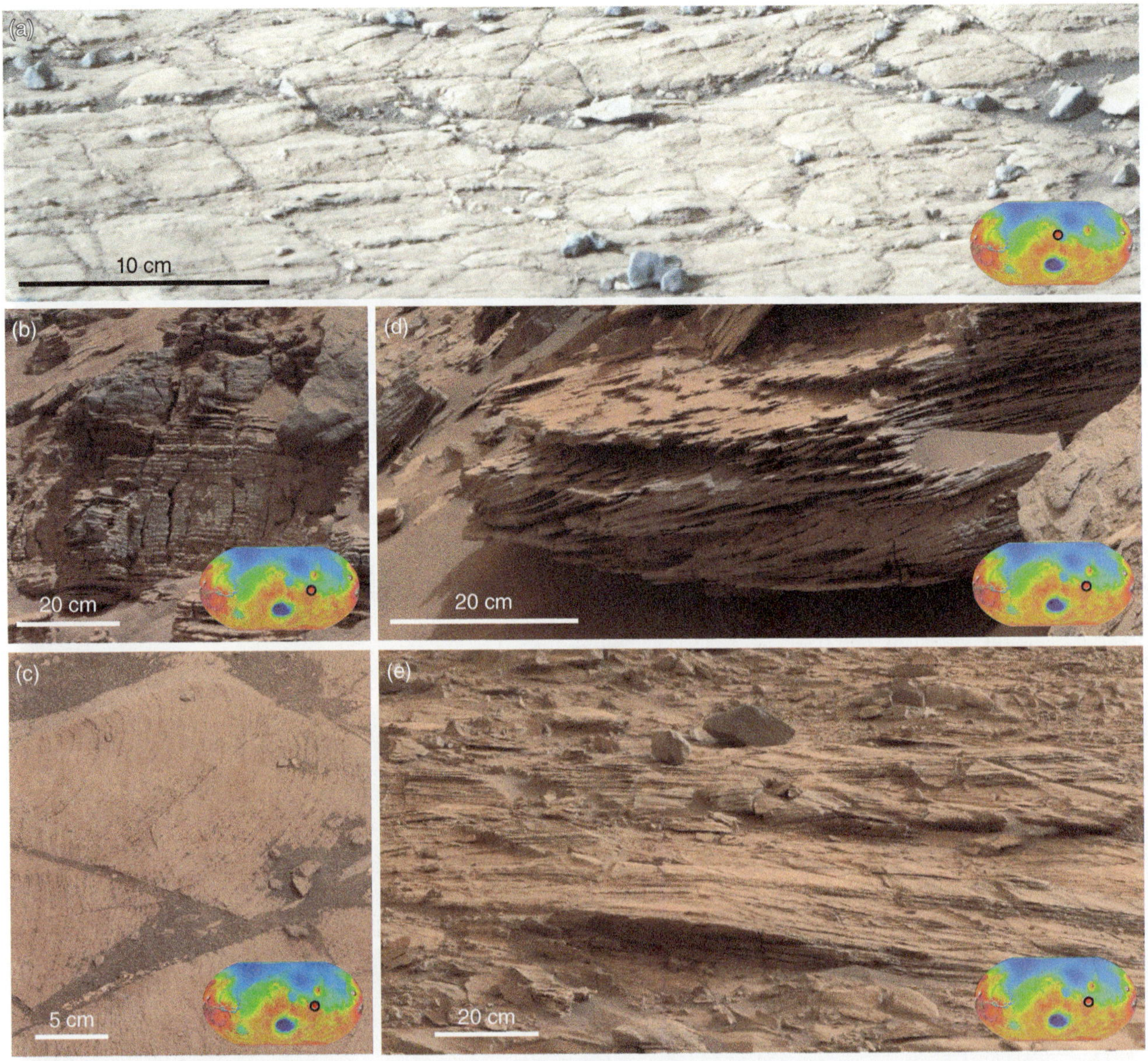

Figure 15.27 Example of lacustrine facies on Mars. (a) Laminated mudstone of the *Shenandoah formation* at *Yori Pass*, Jezero crater (NASA/JPL-Caltech/MSSS / Public domain). (b) Thickly laminated mudstone of the Pahrump Hills member of the Murray formation, Gale crater (Photo by NASA/JPL-Caltech/Imperial College / Public domain). (c) Thinly laminated mudstone of the Pahrump Hills member of the Murray formation at *Chinle*, Gale crater (NASA/JPL-Caltech/MSSS / Public domain). (d) Cross-stratified sandstone of the Pahrump Hills member of the Murray formation at *Whale Rock*, Gale crater (NASA/JPL-Caltech/MSSS / Public domain). (e) Planar to low-angle stratified mudstone showing variably dipping lamination in the Pahrump Hills member of the Murray formation at Chinle, Gale crater (NASA/JPL-Caltech/MSSS / Public domain).

history (e.g. Ingersoll 1969; Donahue et al. 1982). If so, ancient terrains such as the Venusian tesserae could have preserved a record of past liquid flows. Possibly supporting this scenario, lava-flow embayments along the margins of tesserae have been proposed to represent the remnants of ancient, water-carved valleys forming a diversity of valley network geometries (Khawja et al. 2020). Finally, the layered nature of the bedrock at the Venera landing sites (Figure 15.28) has led to the speculation that it could consist of sedimentary rocks (Florensky et al. 1983), though a lava-flow interpretation was later favoured (Garvin et al. 1984).

15.4.2 Sedimentary Environments on Titan

Titan's landscapes host rivers thought to be currently active (Figure 15.3c,d). However, its frigid surface temperature (~90 K) renders water ice as hard as silicate rocks on Earth,

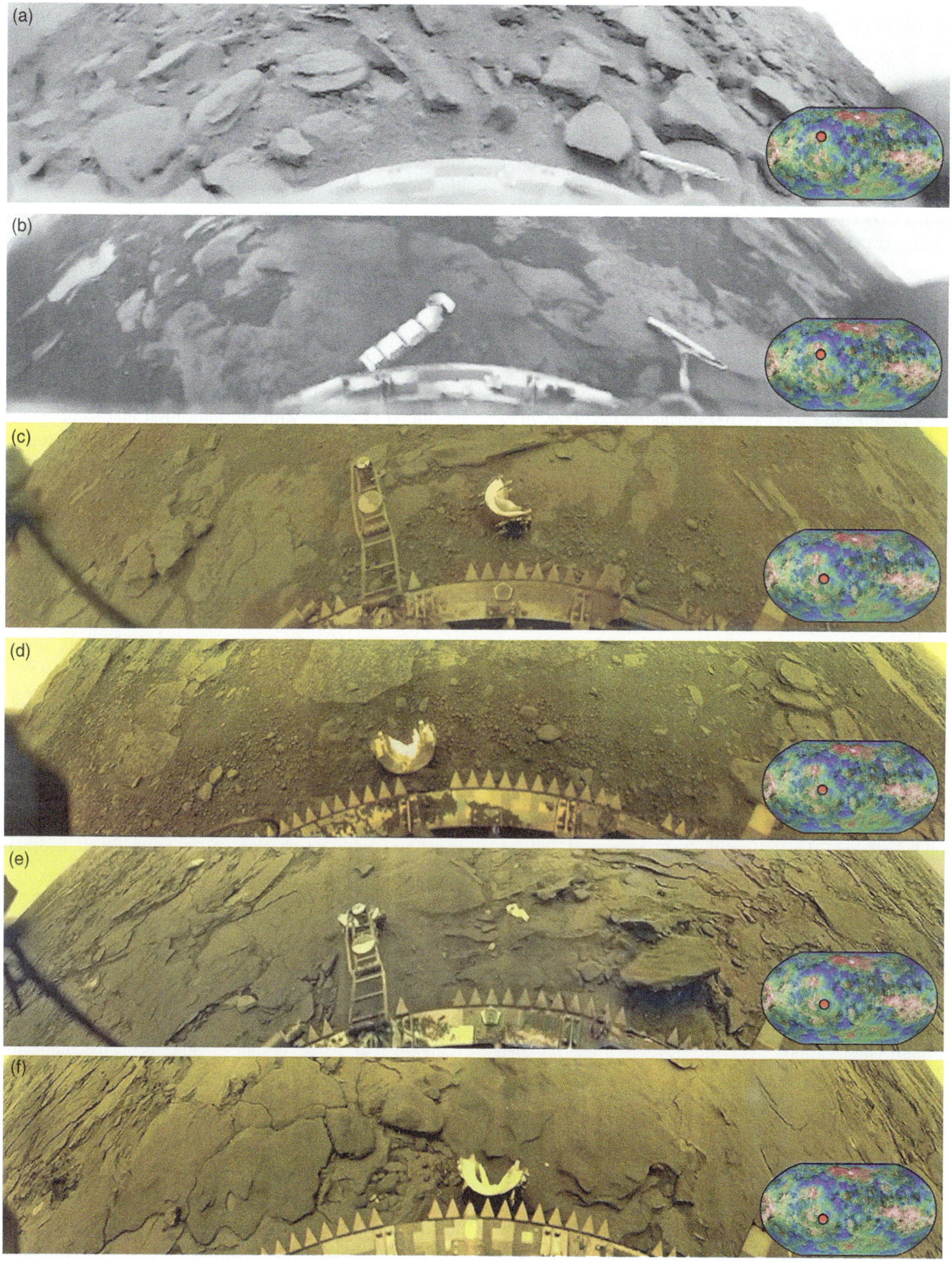

Figure 15.28 Surface panoramas of the Venusian surface at the (a) Venera 9, (b) Venera 10, (c, d) Venera 13 (front and rear cameras, respectively) and (e, f) Venera 14 (front and rear cameras, respectively) landing sites (credit: Photo by Russian Academy of Sciences/ Ted Stryk).

whereas methane, which is typically found in its gaseous form on Earth, is liquid on Titan. Thus, Titan's hydrologic cycle is methane-based (Hayes et al. 2018), with globally distributed river valleys (e.g. Langhans et al. 2012; Burr et al. 2013a,b; Miller et al. 2021), putative eroded canyonlands (Poggiali et al. 2016), alluvial fans largely concentrated in the mid-latitudes (Birch et al. 2016), potential polar deltas (Wall et al. 2010), polar lakes and seas (Hayes 2016) forming over water-ice bedrock and several impact craters that could serve as sediment sinks (Figure 15.29a) (Hedgepeth et al. 2020).

A year on Titan lasts ~29 Earth years, such that the Cassini spacecraft, which orbited Saturn for about 13 Earth years, has provided observations of Titan's surface spread over about half of the moon's year, revealing seasonal changes at its surface. Methane rainstorms are more frequent near Titan's poles (~100 storms/season) relative to the equator (<1 storm/season; Hayes et al. 2018). Equatorial storms, which may lead to episodic inundation, are thought to occur at equinox (e.g. Turtle et al. 2011), likely driving fluvial transport at least seasonally even at low latitudes (Figure 15.29a).

Whereas ~97% (by area) of liquid-filled lakes and seas are currently in the north polar regions (mostly in Kraken, Ligeia and Pungas Maria; Hayes et al. 2018), dry lakes and seas are found in the south polar and, possibly, equatorial regions (Moore and Howard 2010; Birch et al. 2018) (Figure 15.29). Ontario Lacus is the only large liquid-filled lake in the south polar region and the only one known to host potential deltas (Wall et al. 2010) (Figure 15.3). Similarly to lakes and seas, not all fluvial valleys appear to be liquid-filled; radar-bright valleys have been interpreted as either dry or hosting sub-resolution braided streams (Le Gall et al. 2010; Burr et al. 2013b). Most valley networks are dendritic, though more stubby valleys are also observed (Langhans et al. 2012). Although valleys span all latitudes, drainage density appears to be enhanced at polar latitudes and around Xanadu (Miller et al. 2021) – an enigmatic, bright, high-relief terrain of likely tectonic origin (Brown et al. 2011; Radebaugh et al. 2011; Langhans et al. 2013). During its descent through Titan's atmosphere, the Huygens probe revealed the presence of fine-scale (i.e. below Cassini radar resolution) dissected fluvial valleys at equatorial latitudes, adjacent to wind-blown dune fields (Soderblom et al. 2007). Upon landing on a dry riverbed, Huygens imaged rounded bright-toned cobbles, possibly made of water ice, and finer (< a few millimetres) dark grains, possibly made of organics (Tomasko et al. 2005; Zarnecki et al. 2005) (Figure 15.29b,c).

Extensive wind-blown dune fields (Senkyo, Belet, Shangri-La, Fensal and Aztlan) also occupy a near-continuous band between ±30° around Titan's equator (Lorenz et al. 2006; Radebaugh et al. 2008; Lorenz and Radebaugh 2009; Radebaugh 2013), only interrupted by Xanadu (Marvin et al. 2025). Dunes are thought to form near-longitudinally to westerly winds (as determined from dune interactions with topography) all around the moon's equatorial belt (Lorenz et al. 2006) (Figure 15.29a). This observation is difficult to reconcile with meteorological constraints, suggesting that dominant average winds are easterlies unless the transport threshold of dune particles is relatively high and most transport occurs during westerly equinox storms (Tokano 2008, 2010; Charnay et al. 2015). Titan's dunes are thought to have been active for at least tens of thousands to hundreds of thousands of years (Ewing et al. 2015), with the indication that some wind-blown sediments might be redistributed from equatorial and polar to mid-latitudes (Malaska et al. 2016). Dust-lifting events are thought to occur on Titan, especially during equinoctial storms that are strong enough to generate dust plumes (Rodriguez et al. 2018).

Several mechanisms have been proposed to generate saltatable sand from fine haze particles that settle onto Titan's surface (e.g. Hörst 2017): (1) the flocculation of very fine grains into coarser aggregates when they are transported by a liquid; (2) weathering of an evaporitic crust that precipitated after some liquid (with dissolved organics) evaporated, forming sand-sized grains; (3) sintering, whereby small grains coalesce into larger grains over time through largely diffusive processes; (4) lithification and subsequent weathering and erosion into sand-sized materials (Barnes et al. 2015). Spectral data over dune materials are possibly consistent with weathering of an organic-rich bedrock (Neish et al. 2015).

It is thought that wind-blown sand on Titan is ~150–300 μm in diameter, though the grain size could be either finer or coarser if sandy materials were cohesive (Lorenz 2017; Comola et al. 2022). A combination of abrasion and sintering, leading to equilibrium sand sizes, was proposed to sustain Titan's dune fields with saltatable sand (Lapôtre et al. 2022b) despite organics being mechanically weak and prone to comminution upon collisions (Yu et al. 2018). Consistent with those hypotheses, spectral observations suggest that dune materials are likely made of organics and are compositionally distinct from interdune materials (Bonnefoy et al. 2016).

Finally, very little is known about diagenesis and the formation of sedimentary "rocks" on Titan. Consolidation of loose sediment could occur through pressure-driven phase

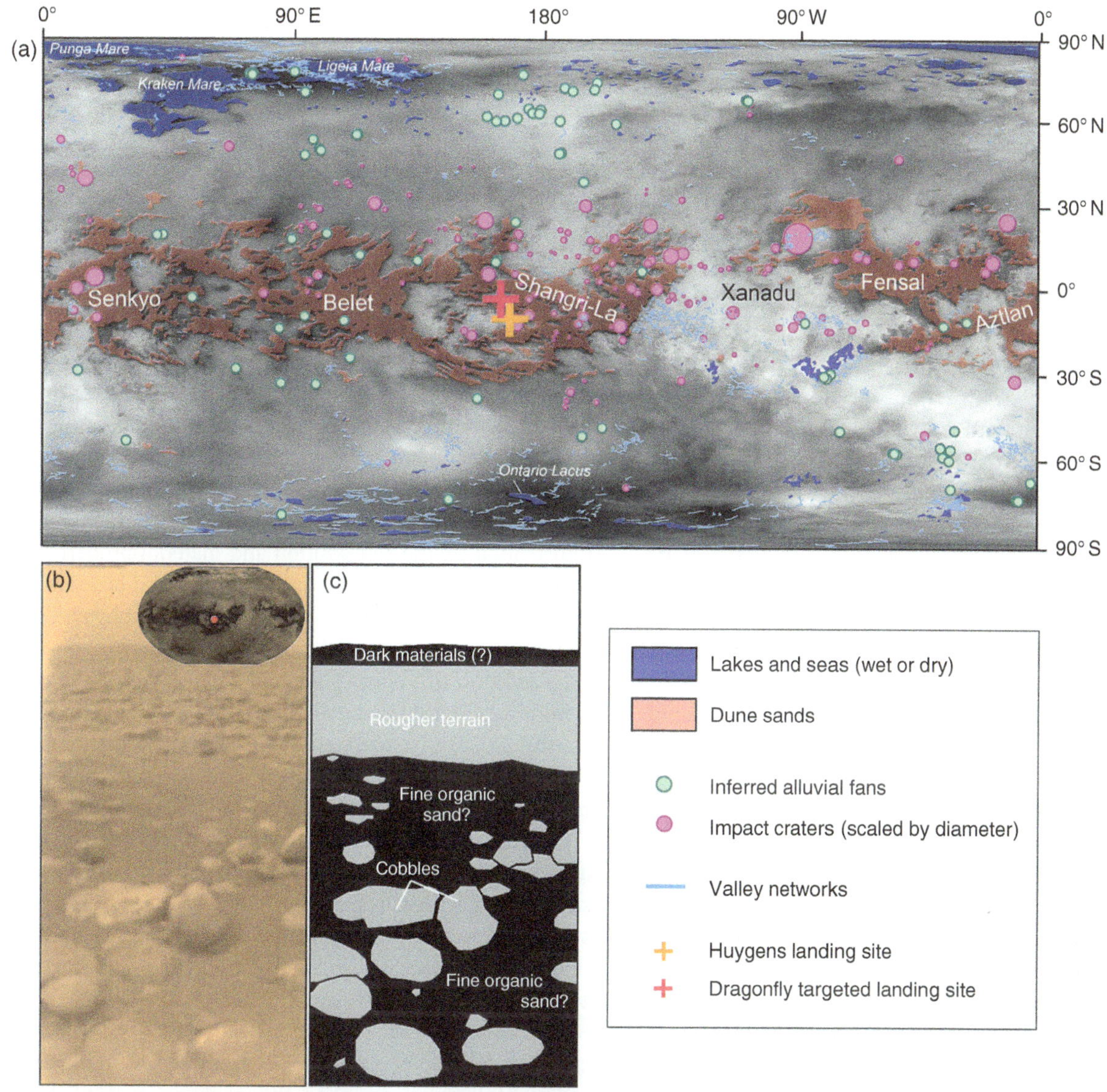

Figure 15.29 (a) Modern sedimentary environments on Titan (compiled from Lopes et al. 2020; Birch et al. 2016; Hedgepeth et al. 2020; Miller et al. 2021) over Cassini Imaging Science Subsystem (ISS) global mosaic of surface brightness at 938 nm (credit: Photo by NASA/JPL-Caltech/University of Arizona / Public domain). (b) Landing site of the European Space Agency's Huygens probe (credit: Photo by NASA/JPL/ESA/University of Arizona / Public domain) and (c) corresponding interpretation of a dry riverbed covered in fluvial sediments (Tomasko et al. 2005). Lapôtre et al. (2022b) / John Wiley & Sons.

changes and co-crystal formation (e.g. Maynard-Casely et al. 2018; Cable et al. 2021), sintering (Barnes et al. 2015; Lapôtre et al. 2022b) or cementation, e.g. from precipitation of methane–ethane soluble organic species, such as butane or acetylene (e.g. Barnes et al. 2011; Cordier et al. 2013; MacKenzie et al. 2014; Cordier et al. 2016; MacKenzie and Barnes 2016). Evidence for sedimentary-rock formation includes possible detections of organic evaporites on dry lake beds (Barnes et al. 2011; MacKenzie et al. 2014; MacKenzie and Barnes 2016) and karst-like terrains (Malaska et al. 2020).

15.5 The Future of Planetary Sedimentology: From Earth Outwards and Back

With a wealth of data returned by planetary missions led by space agencies around the world, the field of planetary sedimentology is simultaneously booming and in its infancy. Upcoming missions are bound to provide a wealth of new observations of sedimentary processes on Venus. NASA's VERITAS orbiter mission (Venus Emissivity,

Radio Science, InSAR, Topography and Spectroscopy; Smrekar et al. 2022), with a tentative launch date in the late 2020s–early 2030s, was designed to answer fundamental questions about the planet's geological and climate history by producing global, high-resolution maps of surface emissivity, topography and SAR images with a 15–30 m/pixel resolution. In turn, NASA's DAVINCI mission (Deep Atmosphere Venus Investigation of Noble gases, Chemistry and Imaging; Garvin et al. 2022) is anticipated to fly by and send a descent probe to uncover the planet's atmospheric past, and notably, whether it ever hosted a surface ocean. Its descent probe will also deliver high-resolution, high-contrast images of tessera terrains at the descent location. Tesserae are thought to be Venus' oldest terrains and may hold clues about surface processes when the planet might have been habitable (e.g. Khawja et al. 2020). ESA's EnVision mission (Ghail et al. 2011), also planned for a launch in the early 2030s, will seek to answer similarly profound questions regarding Venus' current activity, geological history and past habitability. Together, these missions are bound to revolutionise our understanding of modern sedimentary processes and may disclose the existence of a yet undiscovered but accessible Venusian stratigraphic record.

Whereas much has been learned about Mars' geological history and modern sedimentary processes over the past decades, many new questions have arisen that could only be answered through the acquisition of more ground-based data and absolute stratigraphic tie-points. The return of samples to the Earth will enable complex laboratory studies using instrumentation that has thus far been impossible to fly to Mars (McSween et al. 2025; Herd et al. 2025; Beaty et al. 2019). To date, three space agencies (NASA, ESA and China National Space Administration, or CNSA) have planned to return samples from the surface of Mars back to Earth in the 2030s. Meanwhile, similar platforms to the Ingenuity helicopter are being considered for relatively low-cost missions or even sample return on Mars (Mier-Hicks et al. 2023; Mittelholz et al. 2023; Zappek et al. 2023; Zhu et al. 2024). ESA's ExoMars program is expected to deliver the Rosalind Franklin rover to Oxia Planum, along the eastern border of Chryse Planitia to investigate ancient clay-bearing rocks by drilling up to 3 m deep (Vago et al. 2017; Quantin-Nataf et al. 2021). Finally, the development of significant interest by private companies in the exploration of Mars in recent years has boosted technological development to send both cargo and crewed missions. By essentially enabling both field- and laboratory-based sedimentology on Mars, human exploration would mark the beginning of a new era in planetary sedimentology.

Titan will be the target of NASA's upcoming Dragonfly mission, set to launch in 2028. Dragonfly will consist of an eight-rotored lander, with the ability to perform short flights and conduct studies of prebiotic chemistry and sedimentology upon landing in the mid-2030s (Barnes et al. 2021). To date, fundamental properties of both the sediments and fluids are very poorly constrained. The composition of sediments is unknown and could include any combinations of water ice, clathrates, complex hydrocarbon compounds such as tholins and cryo-minerals or organic co-crystals (e.g. Cable et al. 2021). Whereas surface liquids consist of methane–ethane–nitrogen mixtures, the relative proportions of each component are not known, and their material properties (density and viscosity) vary greatly with temperature (e.g. Steckloff et al. 2020). Dragonfly will provide crucial information on the composition and material properties of sediments and surface fluids and enable ground-based investigations of modern surface processes, stratigraphy and diagenesis in an icy, hydrocarbon-rich world.

In addition to Venus, Mars and Titan, other space missions are expected to take flight in the next decade that will contribute to our understanding of planetary sedimentary processes. For example, a wealth of missions to the Moon by the international community will deliver orbiters, landers, rovers and even crews to the lunar surface over the next decades. Although the Moon does not have an atmosphere, such missions may deepen our understanding of processes, such as impact-driven sediment production and accumulation, regolith processes and stratigraphy and mass wasting on planetary surfaces.

Last but not least, we demonstrated over the past several decades that we can use the concepts and methods of sedimentology to decipher the geologic past of other planetary bodies. In turn, we are now beginning to realise that other planets can teach us about the early Earth. Other planetary bodies provide full-scale experiments of sedimentary processes under very different boundary conditions to those of the modern Earth. Moreover, they provide archives of the Solar System's early history that are long-lost in Earth's geological record and can also be used as analogues to the early Earth, before life (micro- or macroscopic) or even plate tectonics developed (e.g. Sasselov et al. 2020; Lapôtre et al. 2020, 2022a). Thus, the emerging field of planetary sedimentology is poised to deliver some of the most fundamental advances in many sub-fields of the geo- and planetary sciences, including in Precambrian sedimentology, for decades to come.

Further Reading

Regolith Production and Space Weathering

Christensen, P.R. and Moore, H.J. (1992). The martian surface layer. In: *Mars* (ed. H.H. Kieffer, B.M. Jakovsky, C.W. Snyder, and M.S. Matthews), 686–727. Tucson: University of Arizona Press.

Pieters, C.M. and Noble, S.K. (2016). Space weathering on airless bodies. *Journal of Geophysical Research: Planets* 12 (10): 1865–1884. https://doi.org/10.1002/2016JE005128.

Speyerer, E.J., Povilaitis, R.Z., Robinson, M.S. et al. (2016). Quantifying crater production and regolith overturn on the Moon with temporal imaging. *Nature* 538: 215–218. https://doi.org/10.1038/nature19829.

Impact-Generated and Volcaniclastic Deposits on Mars

Hynek, B.M., Phillips, R.J., and Arvidson, R.E. (2003). Explosive volcanism in the Tharsis region: Global evidence in the martian geologic record. *Journal of Geophysical Research: Planets* 108 (E9): https://doi.org/10.1029/2003JE002062.

Kerber, L., Head, J.W., Madeleine, J.B. et al. (2012). The dispersal of pyroclasts from ancient explosive volcanoes on Mars: Implications for the friable layered deposits. *Icarus* 219 (1): 358–381. https://doi.org/10.1016/j.icarus.2012.03.016.

Koeberl, C. and Martinez-Ruiz, F. (2003). The stratigraphic record of impact events: A short overview. In: *Impact Markers in the Stratigraphic Record. Impact Studies* (ed. C. Koeberl and F. Martinez-Ruiz). Berlin, Heidelberg: Springer https://doi.org/10.1007/978-3-642-55463-6_1.

Mass-wasting Deposits on Mars

Roback, K.P. and Ehlmann, B.L. (2021). Controls on the global distribution of martian landslides. *Journal of Geophysical Research: Planets*, 126(5), https://doi.org/10.1029/2020JE006675

Watkins, J.A., Ehlmann, B.L., Yin, A. et al. (2020). Spatiotemporal evolution, mineralogical composition, and transport mechanisms of long-runout landslides in Valles Marineris. *Mars. Icarus* 350: https://doi.org/10.1016/j.icarus.2020.113836.

Periglacial and Glacial Deposits on Mars

Balme, M.R. and Gallagher, C. (2009). An equatorial periglacial landscape on Mars. *Earth and Planetary Science Letters* 285 (1-2): 1–15. https://doi.org/10.1016/j.epsl.2009.05.031.

Conway, S.J. and Balme, M.R. (2014). Decameter thick remnant glacial ice deposits on Mars. *Geophysical Research Letters* 41 (15): 5402–5409. https://doi.org/10.1002/2014GL060314.

Garvin, J.B., Head, J.W., Marchant, D.R. et al. (2006). High-latitude cold-based glacial deposits on Mars: Multiple superposed drop moraines in a crater interior at 70°N latitude. *Meteoritics and Planetary Science* 41 (10): 1659–1674. https://doi.org/10.1111/j.1945-5100.2006.tb00443.x.

Head, J.W., Marchant, D.R., Agnew, M.C. et al. (2006). Extensive valley glacier deposits in the northern mid-latitudes of Mars: Evidence for Late Amazonian obliquity-driven climate change. *Earth and Planetary Science Letters* 241 (3-4): 663–671. https://doi.org/10.1016/j.epsl.2005.11.016.

Sedimentary Geochemistry and Mineralogy on Mars

Lapôtre, M.G.A. (2017). Compositional variations in sands of the Bagnold Dunes, Gale crater, Mars, from visible-shortwave infrared spectroscopy and comparison with ground truth from the Curiosity rover. *Journal of Geophysical Research: Planets* 122 (12): 2489–2509. https://doi.org/10.1002/2016JE005133.

McLennan, S.M. (2003). Sedimentary silica on Mars. *Geology* 31 (4): 315–318.

McLennan, S.M. (2012). Geochemistry of sedimentary processes on Mars. In: *Sedimentary Geology of Mars* (ed. J.P. Grotzinger and R.E. Milliken). SEPM Special Publication No. 102 https://doi.org/10.2110/pec.12.102.0119.

Rampe, E.B., Blake, D.F., Bristow, T.F. et al. (2020). Mineralogy and geochemistry of sedimentary rocks and eolian sediments in Gale crater, Mars: A review after six Earth years of exploration with Curiosity. *Geochemistry* 80 (2): https://doi.org/10.1016/j.chemer.2020.125605.

Other Topics of Interest

Howard, A.D., Moore, J.M., Umurhan, O. M. et al. (2017). Present and past glaciation on Pluto. *Icarus* 287: 287–300. https://doi.org/10.1016/j.icarus.2016.07.006.

Lucas, A., Mangeney, A., Ampuero, J.P. et al. (2014). Frictional velocity-weakening in landslides on Earth and on other planetary bodies. *Nature Communications* 5: 3417. https://doi.org/10.1038/ncomss4417.

Miyamoto, H., Yano, H., Scheeres, D.J. et al. (2007). Regolith migration and sorting on asteroid Itokawa. *Science* 316 (5827): 1011–1014. https://doi.org/10.1126/science.1134390.

White, O.L., Moore, J.M., Howard, A.D. et al. (2018). Washboard and fluted terrains on Pluto as evidence for ancient glaciation. *Nature Astronomy* 3: 62–68. https://doi.org/10.1038/s41550-018-0592-z.

References

Abrams, J. and Hanratty, T. (1985). Relaxation effects observed for turbulent flow over a wavy surface. *Journal of Fluid Mechanics* 151: 443–455. https://doi.org/10.1017/S0022112085001045.

Alsaeed, N.R. and Jakosky, B.M. (2019). Mars water and D/H evolution from 3.3 Ga to present. *Journal of Geophysical Research: Planets* 124 (12): 3344–3353. https://doi.org/10.1029/2019JE006066.

Alvarez, C.A., Lapôtre, M.G.A., Swann, C., and Ewing, R.C. (2025a). Ripples formed in low-pressure wind tunnels suggest Mars' large ripples are not impact ripples. *Nature Communications* https://doi.org/10.1038/s41467-025-58140-7.

Alvarez, C.A., Lapôtre, M.G.A., Swann, C., Ewing, R.C., Jia, P, Claudin, P. (2025b). Aerodynamic roughness of rippled beds under active saltation at Earth-to-Mars atmospheric pressures. *Nature Communications* 16: 5113. https://doi.org/10.1038/s41467-025-60212-7.

Amador, E.S., Bandfield, J.L., Thomas, N.H. et al. (2018). A search for minerals associated with serpentinization across Mars using CRISM spectral data. *Icarus* 311: 113–114. https://doi.org/10.1016/j.icarus.2018.03.021.

Anderson, R.S. (1987). A theoretical model for aeolian impact ripples. *Sedimentology* 34 (5): 943–956.

Andreotti, B., Claudin, P., Iversen, J.J. et al. (2021). A lower-than-expected saltation threshold at Martian pressure and below. *Proceedings of the National Academy of Sciences of the United States of America* 118 (5): e2012386118. https://doi.org/10.1073/pnas.2012386118.

Ansan, V., Loizeau, D., Mangold, N. et al. (2011). Stratigraphy, mineralogy, and origin of layered deposits inside Terby crater, Mars. *Icarus* 211 (1): 273–304. https://doi.org/10.1016/j.icarus.2010.09.011.

Arvidson, R.E., Squyres, S.W., Bell, J.F. III et al. (2014). Ancient aqueous environments at Endeavour crater, Mars. *Science* 343 (6169). https://doi.org/10.1126/science.1248097.

Ayoub, F., Avouac, J.P., Newman, C.E. et al. (2014). Threshold for sand mobility on Mars calibrated from seasonal variations of sand flux. *Nature Communications* 5: 5096. https://doi.org/10.1038/ncomm6096.

Bagnold, R.A. (1937). The size-grading of sand by wind. *Proceedings of the Royal Society of London Series A–Mathematical and Physical Sciences* 163: 250–264.

Bagnold, R.A. (1941). *The Physics of Blown Sand and Desert Dunes*. Courier Corporation.

Baker, M.M., Lapôtre, M.G.A., Minitti, M.E. et al. (2018b). The Bagnold Dunes in southern summer: active sediment transport on Mars observed by the Curiosity rover. *Geophysical Research Letters* 45 (17): 8853–8863. https://doi.org/10.1029/2018GL079040.

Baker, M.M., Newman, C., Charalambous, C. et al. (2021). Vortex-dominated aeolian activity at Insight's landing site, Part 2: Local meteorology, transport dynamics, and model analysis. *Journal of Geophysical Research: Planets* 126 (4): e2020JE006514. https://doi.org/10.1029/2020JE006514.

Baker, M.M., Newman, C.E., Lapôtre, M.G.A. et al. (2018a). Coarse sediment transport in the modern martian environment. *Journal of Geophysical Research: Planets* 123 (6): 1380–1394. https://doi.org/10.1002/2017JE005513.

Baker, V.R., Carr, M.H., Gulick, V.C. et al. (1992). Channels and valley networks. In: *Mars*, 493–522.

Baker, V.R. and Milton, D.J. (1974). Erosion by catastrophic floods on Mars and Earth. *Icarus* 23 (1): 27–41.

Baker, V.R., Strom, R.G., Gulick, V.C. et al. (1991). Ancient oceans, ice sheets and the hydrological cycle on Mars. *Nature* 352: 589–594.

Balaram, J., Aung, M.M., Golombek, M.P. et al. (2021). The Ingenuity helicopter on the perseverance rover. *Space Science Reviews* 217 (56). https://doi.org/10.1007/s11214-021-00815-w.

Balme, M., Berman, D.C., Bourke, M.C. et al. (2008). Transverse aeolian ridges (TARs) on Mars. *Geomorphology* 101 (4): 703–720. https://doi.org/10.1016/j.geomorph.2008.03.011.

Balme, M.R., Gupta, S., Davis, J.M. et al. (2020). Aram Dorsum: an extensive mid-Noachian age fluvial depositional system in Arabia Terra. *Journal of Geophysical Research: Planets* 125 (5). https://doi.org/10.1029/e2019JE006244.

Banham, S.G., Gupta, S., Rubin, D.M. et al. (2018). Ancient Martian aeolian processes and palaeogeomorphology reconstructed from the Stimson formation on the lower slope of Aeolis Mons, Gale crater, Mars. *Sedimentology* 65: 993–1042. https://doi.org/10.1111/sed.12469.

Banham, S.G., Gupta, S., Rubin, D.M. et al. (2021). A rock record of complex aeolian bedforms in a Hesperian desert landscape: the Stimson formation as exposed in the Murray Buttes, Gale crater, Mars. *Journal of Geophysical Research: Planets* 126 (4): e2020JE006554. https://doi.org/10.1029/2020JE006554.

Banham, S.G., Gupta, S., Rubin, D.M. et al. (2022). Evidence for fluctuating wind in shaping an ancient martian dune field: the Stimson formation at the Greenheugh pediment, Gale crater. *Journal of Geophysical Research: Planets* 127 (9). https://doi.org/10.1029/2021JE007023.

Banham, S.G., Roberts, A.L., Gupta, S. et al. (2024). Ice? Salt? Pressure? Sediment deformation structures as evidence of late-stage shallow groundwater in Gale crater, Mars. *Geology* 52. https://doi.org/10.1130/G51849.1.

Barnes, J.W., Bow, J., Schwartz, J. et al. (2011). Organic sedimentary deposits in Titan's dry lakebeds: probable evaporite. *Icarus* 216 (1): 136–140. https://doi.org/10.1016/j.icarus.2011.08.022.

Barnes, J.W., Brown, R.H., Soderblom, L. et al. (2007). Global-scale surface spectral variations on Titan seen from Cassini/VIMS. *Icarus* 186 (1): 242–258. https://doi.org/10.1016/j.icarus.2006.08.021.

Barnes, J.W., Lorenz, R.D., Radebaugh, J. et al. (2015). Production and global transport of Titan's sand and particles. *Planetary Science* 4 (1). https://doi.org/10.1186/s13535-015-0004-y.

Barnes, R., Gupta, S., Traxler, C. et al. (2018). Geological analysis of Martian rover-derived digital outcrop models using the 3-D visualization tool, Planetary Robotics 3-D Viewer—Pro3D. *Earth and Space Science* 5 (7): 285–307.

Barnes, J.W., Turtle, E.P., Trainer, M.G. et al. (2021). Science goals and objectives for the Dragonfly Titan rotorcraft relocatable lander. *The Planetary Science Journal* 2 (4). https://doi.org/10.3847/PSJ/abfdcf.

Barrington, M.N., Birch, S.P.D., Jindal, A. et al. (2023). Quantifying morphological changes and sediment transport pathways on comet 67P/Churyumov-Gerasimenko. *Journal of Geophysical Research: Planets* 128 (7). https://doi.org/10.1029/2022JE007723.

Bart, G.D., Turtle, E.P., Jaeger, W.L. et al. (2004). Ridges and tidal stress on Io. *Icarus* 169 (1): 111–126. https://doi.org/10.1016/j.icarus.2004.01.003.

Battalio, M. and Wang, H. (2021). The mars dust activity database (MDAD): a comprehensive statistical study of dust storm sequences. *Icarus* 354. https://doi.org/10.1016/j.icarus.2020.114059.

Beaty, D.W., Grady, M.M., McSween, H.Y. et al. (2019). The potential science and engineering value of samples delivered to Earth by Mars sample return. *Meteoritics and Planetary Science* 54 (S1): S3–S125. https://doi.org/10.1111/maps.13242.

Becerra, P., Herny, C., Valantinas, A. et al. (2020). Avalanches of the martian north polar cap. Europlanet Science Congress 2020, online, 21 September 2020, EPSC2020-1093, 10.5194/epsc2020-1093.

Bell, J.F. III, Maki, J.N., Mehall, G.L. et al. (2021). The Mars 2020 perseverance rover mast camera zoom (Mastcam-Z) multispectral, stereoscopic imaging investigation. *Space Science Reviews* 217: 24. https://doi.org/10.1007/s11214-020-00755-x.

Bell, J.F. III, Squyres, S.W., Herkenhoff, K. E. et al. (2003). Mars exploration rover Athena panoramic camera (Pancam) investigation. *Journal of Geophysical Research: Planets* 108 (E12): https://doi.org/10.1029/2003JE002070.

Belton, M.J.S. and Melosh, J. (2009). Fluidization and multiphase transport of particulate cometary materials as explanation of the smooth terrains and repetitive outbursts on 9P/Tempel 1. *Icarus* 200 (1): 280–291. https://doi.org/10.1016/j.icarus.2008.11.012.

Bennett, K.A. and Bell, J.F. (2016). A global survey of martian central mounds: Central mounds as remnants of previously more extensive large-scale sedimentary deposits. *Icarus* 264: 331–341. https://doi.org/10.1016/j.icarus.2015.09.041.

Bennett, K.A., Fox, V.K., Bryk, A. et al. (2023). The Curiosity rover's exploration of Glen Torridon, Gale crater, Mars: An overview of the campaign and scientific results. *Journal of Geophysical Research: Planets* 128 (1): e2022JE007185.

Bhartia, R., Beegle, L.W., DeFlores, L. et al. (2021). Perseverance's scanning habitable environments with Raman and luminescence for organics and chemicals (SHERLOC) investigation. *Space Science Reviews* 217: 58. https://doi.org/10.1007/s11214-021-00812-z.

Bibring, J.-P., Langevin, Y., Gendrin, A. et al. (2005). Mars surface diversity as revealed by the OMEGA/Mars Express observations. *Science* 307 (5715): 1576–1581. https://doi.org/10.1126/science.1108806.

Bibring, J.-P., Langevin, Y., Mustard, J.F. et al. (2006). Global mineralogical and aqueous Mars history derived from OMEGA/Mars Express data. *Science* 312: 5772. https://doi.org/10.1126/science.1122659.

Birch, S.P.D., Hayes, A.G., Corlies, P. et al. (2018). Morphological evidence that Titan's southern hemisphere basins are paleoseas. *Icarus* 310: 140–148. https://doi.org/10.1016/j.icarus.2017.12.016.

Birch, S.P.D., Hayes, A.G., Howard, A.D. et al. (2016). Alluvial fan morphology, distribution and formation on Titan. *Icarus* 270: 238–247. https://doi.org/10.1016/j.icarus.2016.02.013.

Birch, S.P.D., Parker, G., Corlies, P. et al. (2023). Reconstructing river flows remotely on Earth, Titan, and Mars. *Proceedings of the National Academy of Sciences of the United States of America* 120 (29). https://doi.org/10.1073/pnas.2206837120.

Bishop, J.L., Dobrea, E.Z.N., McKeown, N.K. et al. (2008). Phyllosilicate diversity and past aqueous activity at Mawrth Vallis, Mars. *Science* 321: 830–833. https://doi.org/10.1126/science.1159699.

Bishop, J.L., Loizeau, D., McKeown, N.K. et al. (2013). What the ancient phyllosilicates at Mawrth Vallis can tell us about possible habitability on early Mars. *Planetary and Space Science* 86: 130–149. https://doi.org/10.1016/j.pss.2013.05.006.

Boctor, N.Z., Alexander, C.M.O.D., Wang, J. et al. (2003). The sources of water in martian meteorites: Clues from hydrogen isotopes. *Geochimica Cosmochimica Acta* 67: 3971–3989. https://doi.org/10.1016/S0016-7037(03)00234-5.

Bonnefoy, L.E., Hayes, A.G., Hayne, P.O. et al. (2016). Compositional and spatial variations in Titan dune and interdune regions from Cassini VIMS and RADAR. *Icarus* 270: 222–237. https://doi.org/10.1016/j.icarus.2015.09.014.

Braat, L., Brückner, M.Z.M., Sefton-Nash, E. et al. (2024). Gravity-driven differences in fluvial sediment transport on Mars and Earth. *Journal of Geophysical Research: Planets* 129 (2): https://doi.org/10.1029/2023JE007788.

Bretzfelder, J.M., Stack, K.M., Fraeman, A.A. et al. (2024). Aeolian bedrock ridges in Gale crater, Mars. *Icarus* 408: 115855. https://doi.org/10.1016/j.icaru.2023.115855.

Bridges, N.T., Mary C. Bourke, Paul E. Geissler, Maria E. Banks, Cindy Colon, Serina Diniega, Matthew P. Golombek, Candice J. Hansen, Sarah Mattson, Alfred S. McEwen, Michael T. Mellon, Nicholas Stantzos, Bradley J. Thomson; Planet-wide sand motion on Mars. *Geology* 2012a; 40 (1): 31–34. https://doi.org/10.1130/G32373.1

Bridges, N.T., Ayoub, F., Avouac, J.P. et al. (2012b). Earth-like sand fluxes on Mars. *Nature* 485: 339–342. https://doi.org/10.1038/nature11022.

Bridges, N.T., Dundas, C.M., Edgar, L.A. et al. (2016). Special section introduction on MicroMars to MegaMars. *Icarus* 280: 1–2. https://doi.org/10.1016/j.icarus.2016.08.021.

Bridges, N.T. and Ehlmann, B.L. (2018). The Mars Science Laboratory (MSL)

Bagnold Dunes campaign phase I: overview and introduction to the special issue. *Journal of Geophysical Research: Planets* 123 (1). https://doi.org/10.1002/2017JE005401.

Bridges, N.T., Greeley, R., Haldemann, A.F.C. et al. (1999). Ventifacts at the Pathfinder landing site. *Journal of Geophysical Research: Planets* 104 (E4): 8595–8615. https://doi.org/10.1029/98JE02550.

Bridges, N.T., Sullivan, R., Newman, C.E. et al. (2017). Martian aeolian activity at the Bagnold Dunes, Gale crater: the view from the surface and orbit. *Journal of Geophysical Research: Planets* 122 (10): 2077–2110. https://doi.org/10.1002/2017JE005263.

Brown, R.H., Barnes, J.W., Melosh, H.J. et al. (2011). On Titan's Xanadu region. *Icarus* 214 (2): 556–560. https://doi.org/10.1016/j.icarus.2011.03.018.

Bryk, A.B., Dietrich, WE, Fedo, C, et al. (2024). In situ investigation of the Gediz Vallis ridge: a remnant late-stage debris flow dominated fan deposit in Gale crater. 55th Lunar and Planetary Science Conference, Lunar and Planetary Science Institute (March 2024). Texas, USA: The Woodlands. 1770.

Bulmer, M.H. and Guest, J.E. (1996). Modified volcanic domes and associated debris aprons on Venus. *Special Publications of the Geological Society of London* 110: 349–371.

Bultel, Q., Quantin-Nataf, C., Andréani, M. et al. (2015). Deep alteration between Hellas and Isidis Basins. *Icarus* 260: 141–160. https://doi.org/10.1016/j.icarus.2015.06.037.

Burr, D.M., Bridges, N.T., Marshall, J.R. et al. (2015). Higher-sand predicted saltation threshold wind speeds on Titan. *Nature* 517: 60–63. https://doi.org/10.1038/nature14088.

Burr, D.M., Drummond, S.A., Cartwright, R. et al. (2013a). Morphology of fluvial networks on Titan: evidence for structural control. *Icarus* 226 (1): 742–759. https://doi.org/10.1016/j.iacrus.2005.11.012.

Burr, D.M., Emery, J.P., Lorenz, R.D. et al. (2006). Sediment transport by liquid surficial flow: application to Titan. *Icarus* 181 (1): 235–242. https://doi.org/10.1016/j.icarus.2005.11.012.

Burr, D.M., Perron, J.T., Lamb, M.P. et al. (2013b). Fluvial features on Titan: insights from morphology and modeling. *Bulletin of the Geological Society of America* 125 (3-4): 299–321. https://doi.org/10.1130/B30612.1.

Burr, D.M., Sutton, S.L.F., Emery, J.P. et al. (2020). A wind tunnel study of the effect of intermediate density ration on saltation threshold. *Aeolian Research* 45: 100601. https://doi.org/10.1016/j.aeolia.2020.100601.

Burr, D.M., Williams, R.M.E., Wendell, K.D. et al. (2010). Inverted fluvial features in the Aeolis/Zephyria Plana region, Mars: formation mechanism and initial paleodischarge estimates. *Journal of Geophysical Research: Planets* 115: E7. https://doi.org/10.1029/2009JE003496.

Cable, M.L., Runčevski, T., Maynard-Casely, H.E. et al. (2021). Titan in a test tube: organic co-crystals and implications for Titan mineralogy. *Accounts of Chemical Research* 54 (15): 3003–3124. https://doi.org/10.1021/acs.accounts.1c00250.

Cabrol, N.A. and Grin, E.A. (1999). Distribution, classification, and ages of Martian impact crater lakes. *Icarus* 142 (1): 160–172.

Cabrol, N.A. and Grin, E.A. (2001). The evolution of lacustrine environments on Mars: Is Mars only hydrologically dormant? *Icarus* 149 (2): 291–328.

Campbell, B., Carter, L., Phillips, R. et al. (2008). SHARAD radar sounding of the Vastitas Borealis Formation in Amazonis Planitia. *Journal of Geophysical Research Planets* 113 (E12). https://doi.org/10.1029/2008JE0003177.

Caravaca, G., Dromart, G., Mangold, N. et al. (2024). Depositional facies and sequence stratigraphy of Kodiak Butte, western delta of Jezero crater, Mars. *Journal of Geophysical Research: Planets* 129 (4). https://doi.org/10.1029/2023JE008205.

Caravaca, G., Mangold, N., Dehouck, E. et al. (2022). From lake to river: documenting an environmental transition across the Jura/Knockfarril Hill members boundary in the Glen Torridon region of Gale crater (Mars). *Journal of Geophysical Research: Planets* 127 (9): https://doi.org/10.1029/2021JE007093.

Cardenas, B.T., Grotzinger, J.P., Lamb, M.P. et al. (2022b). Barform deposits of the Carolyn Shoemaker formation, Gale crater, Mars. *Journal of Sedimentary Research* 92 (12): 1071–1092. https://doi.org/10.2110/jsr.2022.032.

Cardenas, B.T. and Lamb, M.P. (2022). Paleogeographic reconstruction of an ocean margin on Mars based on deltaic sedimentology at Aeolis Dorsa. *Journal of Geophysical Research Planets* 127 (10). https://doi.org/10.1029/2022JE007390.

Cardenas, B.T., Lamb, M.P., Grotzinger, J.P. et al. (2022a). Martian landscapes of fluvial ridges carved from ancient sedimentary basin fill. *Nature Geoscience* 15: 871–877. https://doi.org/10.1038/s41561-022-01058-2.

Cardenas, B.T., Mohrig, D., Goudge, T.A. et al. (2018). Fluvial stratigraphy of valley fills at Aeolis Dorsa, Mars: evidence for base-level fluctuations controlled by a downstream water body. *Bulletin of the Geological Society of America* 130 (3–4): 484–498. https://doi.org/10.1130/B31567.1.

Carpenter, A.H. (1948). Principles of historical geology applied to neighboring planets and life on Mars. *Popular Astronomy* 56: 233.

Carr, M.H. (1979). Formation of martian flood features by release of water from confined aquifers. *Journal of Geophysical Research: Solid Earth* 84 (B6): 2995–3007.

Carr, M.H. (1987). Water on Mars. *Nature* 326 (6108): 30–35. https://doi.org/10.1038/326030a0.

Carr, M.H. (1996). *Water on Mars*. Oxford, U.K.: Oxford University Press.

Carr, M.H. and Head, J.W. (2003). Oceans on Mars: an assessment of the observational evidence and possible fate. *Journal of Geophysical Research Planets* 108 (E5). https://doi.org/10.1029/2002JE001963.

Carr, M.H. and Malin, M.C. (2000). Meter-scale characteristics of Martian channels and valleys. *Icarus* 146 (2): 366–386. https://doi.org/10.1006/icar.2000.6428.

Carter, L.M., Gilmore, M.S., Ghail, R.C. et al. (2023). Sedimentary processes on

Venus. *Space Science Reviews* 219 (85). https://doi.org/10.1007/s11214-023-01033-2.

Caswell, T.E. and Milliken, R.E. (2017). Evidence for hydraulic fracturing at Gale crater, Mars: implications for burial depth of the Yellowknife Bay formation. *Earth and Planetary Science Letters* 468: 72–84. https://doi.org/10.1016/j.epsl.2017.03.033.

Chan, M.A., Beitler, B., Parry, W.T. et al. (2004). A possible terrestrial analogue for haematite concretions on Mars. *Nature* 429: 731–734. https://doi.org/10.1038/nature02600.

Chan, N.-H., Perron, J.T., Mitrovica, J.X. et al. (2018). New evidence of an ancient martian ocean from the global distribution of valley networks. *Journal of Geophysical Research Planets* 123 (8): 2138–2150. https://doi.org/10.1029/2018JE005536.

Charnay, B., Barth, E., Rafkin, S. et al. (2015). Methane storms as a driver of Titan's dune orientation. *Nature Geoscience* 8: 362–366. https://doi.org/10.1038/ngeo2406.

Chojnacki, M., Banks, M.E., Fenton, L.K. et al. (2019). Boundary condition controls on the high-sand-flux regions of Mars. *Geology* 47 (5): 427–430. https://doi.org/10.1130/G45793.1.

Chojnacki, M., Fenton, L.K., Edgar, L.A. et al. (2024). Global survey of paelo-bedforms on Mars. *Geomorphology* 466. https://doi.org/10.1016/j.geomorph.2024.109428.

Chojnacki, M., Fenton, L.K., Weintraub, A.R. et al. (2020). Ancient martian aeolian sand dune deposits recorded in the stratigraphy of Valles Marineris and implications for past climates. *Journal of Geophysical Research: Planets* 125 (9): e2020JE006510. https://doi.org/10.1029/2020JE006510.

Chojnacki, M., Vaz, D.A., Silvestro, S. et al. (2021). Widespread megaripple activity across the north polar ergs of Mars. *Journal of Geophysical Research - Planets* 126: 12. https://doi.org/10.1029/2021JE006970.

Christensen, P.R. (2003). Formation of recent martian gullies through melting of extensive water-rich snow deposits. *Nature* 422: 45–48. https://doi.org/10.1038/nature01436.

Christensen, P.R., Bandfield, J.L., Bell, J.F. III et al. (2003). Morphology and composition of the surface of Mars: Mars Odyssey THEMIS results. *Science* 300 (5628): 2056–2061.

Christensen, P.R., Jakosky, B.M., Kieffer, H.H. et al. (2004). The Thermal Emission Imaging System (THEMIS) for the Mars 2001 Odyssey mission. *Space Science Reviews* 110 (85): 85–130. https://doi.org/10.1023/B:SPAC.0000021008.16305.94.

Christensen, P.R., McSween, H.Y., Bandfield, J.L. et al. (2005). Evidence for magmatic evolution and diversity on Mars from infrared observations. *Nature* 436: 504–509. https://doi.org/10.1038/nature03639.

Christensen, P.R. and Ruff, S.W. (2004). Formation of the hematite-bearing unit in Meridiani Planum: evidence for deposition in standing water. *Journal of Geophysical Research: Planets* 109: E8. https://doi.org/10.1029/2003JE002233.

Citron, R.I., Manga, M., Hemingway, D.J. et al. (2018). Timing of oceans on Mars from shoreline deformation. *Nature* 555: 643–646. https://doi.org/10.1038/nature26144.

Clifford, S.M. and Parker, T.J. (2001). The evolution of the Martian hydrosphere: implications for the fate of a primordial ocean and the current state of the northern plains. *Icarus* 154 (1): 40–79. https://doi.org/10.1006/icar.2001.6671.

Comola, F., Kok, J.F., Lora, J.M. et al. (2022). Titan's prevailing circulation might drive highly intermittent, yet significant sediment transport. *Geophysical Research Letters* 49: 7. https://doi.org/10.1029/2022GL097913.

Cordier, D., Barnes, J.W., Ferreira, A.G. et al. (2013). On the chemical composition of Titan's dry lakebed evaporites. *Icarus* 224 (2): 1431–1437. https://doi.org/10.1016/j.icarus.2013.07.026.

Cordier, D., Cornet, T., Barnes, J.W. et al. (2016). Structure of Titan's evaporites. *Icarus* 270: 41–56. https://doi.org/10.1016/j.icarus.2015.12.034.

Cosgrove, J.W., Banham, S.G., Gupta, S., and Barnes, R. (2022). The origin of the fracture networks in the mudstones of Gale crater Mars; their implications regarding the state of stress and fluid pressure during their formation and the depth to which they were buried. *Journal of Geophysical Research: Planets* 127 (12): e2022JE007313.

Costard, F., Séjourné, A., Kelfoun, K. et al. (2017). Modeling tsunami propagation and the emplacement of thumbprint terrain in an early Mars ocean. *Journal of Geophysical Research Planets* 122 (3): 633–649. https://doi.org/10.1002/2016JE005230.

Costard, F., Séjourné, A., Lagain, A. et al. (2019). The Lomonosov crater impact event: a possible mega-tsunami source on Mars. *Journal of Geophysical Research Planets* 124 (7): 1840–1851. https://doi.org/10.1029/2019JE006008.

Courrech du Pont, S., Rubin, D.M., Narteau, C. et al. (2024). Complementary classifications of aeolian dunes based on morphology, dynamic, and fluid mechanics. *Earth-Science Reviews* https://doi.org/10.1016/j.earscirev.2024.104772.

Crosta, G.B., Frattini, P., Valbuzzi, E. et al. (2018). Introducing a new inventory of large martian landslides. *Earth and Space Science* 5 (4): 89–119. https://doi.org/10.1002/2017JE005398.

Cutts, J.A. and Smith, R.S.U. (1973). Eolian deposits and dunes on Mars. *Journal of Geophysical Research* 78 (20): 4139–4154.

Daudon, C., Lucas, A., Rodriguez, S. et al. (2020). A new digital terrain model of the Huygens landing site on Saturn's largest moon, Titan. *Earth and Space Science* 7 (12). https://doi.org/10.1029/2020EA001127.

Davis, J.M., Balme, M., Grindrod, P.M. et al. (2016). Extensive Noachian fluvial systems in Arabia Terra: implications for early Martian climate. *Geology* 44 (10): 847–850. https://doi.org/10.1130/G38247.1.

Davis, J.M., PM Grindrod, SG Banham et al. (2021). A record of syn-tectonic sedimentation revealed by perched alluvial fan deposits in Valles Marineris, Mars. *Geology*, 49. https://doi.org/10.1130/G48971.1

Davis, J.M., Grindrod, P.M., Fawdon, P. et al. (2018). Episodic and declining fluvial processes in southwest Melas Chasma, Valles Marineris, Mars. *Journal of Geophysical Research: Planets* 123 (10): 2527–2549. https://doi.org/10.1029/2018JE005710.

Davis, J.M., Gupta, S., Balme, M. et al. (2019). A diverse array of fluvial depositional systems in Arabia Terra: Evidence for mid-Noachian to early Hesperian rivers on Mars. *Journal of Geophysical Research: Planets* 124 (7): 1913–1934. https://doi.org/10.1029/2019JE005976.

Day, M. (2021). Interaction bounding surfaces exposed in migrating transverse aeolian ridges on Mar. *Geology* 49 (12): 1527–1530. https://doi.org/10.1130/G49373.1.

Day, M.D. and Catling, D.C. (2018). Dune casts preserved by partial burial: the first identification of ghost dune pits on Mars. *Journal of Geophysical Research: Planets* 123 (6): 1431–1448. https://doi.org/10.1029/2018JE005613.

Day, M.D. and Catling, D.C. (2020). Potential aeolian deposition of intra-crater layering: a case study of Henry crater, Mars. *GSA Bulletin* 132 (3–4): 608–616. https://doi.org/10.1130/B35230.1.

Day, M.D., Edgett, K.S., Stumbaugh, D. et al. (2019). Ancient stratigraphy preserving a wet-to-dry, fluvio-lacustrine to aeolian transition near Barth crater, Arabia Terra, Mars. *Journal of Geophysical Research: Planets* 124 (12): 3402–3421. https://doi.org/10.1029/2019JE006226.

De Blasio, F.V. (2020). Frontal Aureole Deposit on Acheron Fossae ridge as evidence for landslide-generated tsunami on Mars. *Planetary and Space Science* 187 (104911). https://doi.org/10.1016/j.pss.2020.104911.

De Hon, R.A. (1987). Ring furrows: inversion of topography in martian highland terrains. *Icarus* 71: 287–297.

De Toffoli, B., Plesa, A.C., Hauber, E. et al. (2021). Delta deposits on Mars: a global perspective. *Geophysical Research Letters* 48 (17). https://doi.org/10.1029/2021GL094271.

Di Achille, G. and Hynek, B.M. (2010). Ancient ocean on Mars supported by global distribution of deltas and valleys. *Nature Geoscience* 3 (7): 459–463. https://doi.org/10.1038/ngeo891.

Di Pietro, I., Séjourné, A., Costard, F. et al. (2021). Evidence of mud volcanism due to the rapid compaction of martian tsunami deposits in southeastern Acidalia Planitia, Mars. *Icarus* 354 (114096). https://doi.org/10.1016/j.icarus.2020.114096.

DiBiase, R.A., Limaye, A.B., Scheingross, J.S. et al. (2013). Deltaic deposits at Aeolis Dorsa: sedimentary evidence for a standing body of water on the northern plains of Mars. *Journal of Geophysical Research: Planets* 118 (6): 1285–1302. https://doi.org/10.1002/jgre.20100.

Dickson, J.L., Ehlmann, B.L., Kerber, L.H. et al. (2023a). Release of the global CTX mosaic of Mars: an experiment in information-preserving image data processing. *LPI Contributions* 2806: 2353.

Dickson, J.L., Lamb, M.P., Williams, R.M.E. et al. (2021). The global distribution of depositional rivers on early Mars. *Geology* 49 (5): 504–509. https://doi.org/10.1130/G48457.1.

Dickson, J.L., Palumbo, A.M., Head, J.W. et al. (2023b). Gullies on Mars could have formed by melting of water ice during periods of high obliquity. *Science* 380 (6652): 1363–1367. https://doi.org/10.1126/science.abk2464.

Ding, L., Zhou, R., Yu, T. et al. (2022). Surface characteristics of the Zhurong Mars rover traverse at Utopia Planitia. *Nature Geoscience* 15: 171–176. https://doi.org/10.1038/s41561-022-00905-6.

Diniega, S., Bramson, A.M., Buratti, B. et al. (2021). Modern Mars' geomorphological activity, driven by winds, frost, and gravity. *Geomorphology* 380: 107627. https://doi.org/10.1016/j.geonorph.2021.107627.

Diniega, S., Byrne, S., Bridges, N.T. et al. (2010). Seasonality of present-day martian dune-gully activity. *Geology* 38 (11): 1047–1050. https://doi.org/10.1130/G31287.1.

Diniega, S., Hansen, C.J., Allen, A. et al. (2019). Dune-slope activity due to frost and wind throughout the north polar erg, Mars. *Special Publications of the Geological Society of London* 467: 95–114. https://doi.org/10.1144/SP467.6.

Diniega, S., Kreslavsky, M., Radebaugh, J. et al. (2017). Our evolving understanding of aeolian bedforms, based on observation of dunes on different worlds. *Aeolian Research* 26: 5–27. https://doi.org/10.1016/j.aeolia.2016.10.001.

Donahue, T.M. (1995). Evolution of water reservoirs on Mars from D/H ratios in the atmosphere and crust. *Nature* 374: 432–434. https://doi.org/10.1038/374432a0.

Donahue, T.M. et al. (1982). Venus was wet: a measurement of the ratio of deuterium to hydrogen. *Science* 216: 630–633.

Dorn, T. and Day, M.D. (2020). Intracrater sediment trapping and transport in Arabia Terra, Mars. *Journal of Geophysical Research: Planets* 125. https://doi.org/10.1029/2020JE006581.

Dromart, G., Quantin, C., Broucke, O. et al. (2007). Stratigraphic architectures spotted in southern Melas Chasma, Valles Marineris, Mars. *Geology* 35 (4): 363–366. https://doi.org/10.1130/G23350A.1.

Dundas, C.M. (2020a). An aeolian grainflow model for martian recurring slope lineae. *Icarus* 343. https://doi.org/10.1016/j.icarus.2020.113681.

Dundas, C.M. (2020b). Geomorphological evidence for a dry dust avalanche origin of slope streaks on Mars. *Nature Geoscience* 13: 473–476. https://doi.org/10.1038/s41561-020-0598-x.

Dundas, C.M., Becerra, P., Byrne, S. et al. (2021). Active Mars: a dynamic world. *Journal of Geophysical Research: Planets* 126 (8). https://doi.org/10.1029/2021JE006876.

Dundas, C.M., Conway, S.J., Cushing, G.E. et al. (2022). Martian gully activity and the gully sediment transport system. *Icarus* 386: https://doi.org/10.1016/j.icarus.2022.115133.

Dundas, C.M., Diniega, S., Hansen, C.J. et al. (2012). Seasonal activity and morphological changes in martian gullies. *Icarus* 220 (1): 124–143. https://doi.org/10.1016/j.icarus.2012.04.005.

Dundas, C.M., Diniega, S., McEwen, A.S. et al. (2015). Long-term monitoring of martian gully formation and evolution with MRO/HiRISE. *Icarus*

251: 244–263. https://doi.org/10.1016/j.icarus.2014.05.013.

Dundas, C.M., McEwen, A.S., Chojnacki, M. et al. (2017). Granular flows at recurring slope lineae on Mars indicate a limited role for liquid water. *Nature Geoscience* 10: 903–907. https://doi.org/10.1038/s41561-017-0012-5.

Dundas, C.M., McEwen, A.S., Diniega, S. et al. (2010). New and recent gully activity on Mars as seen by HiRISE. *Geophysical Research Letters* 37 (7). https://doi.org/10.1029/2009GL041351.

Dundas, C.M., McEwen, A.S., Diniega, S. et al. (2019a). The formation of gullies on Mars today. *Special Publications of the Geological Society of London* 467: 67–94. https://doi.org/10.1144/SP467.5.

Dundas, C.M., Mellon, M.T., Conway, S.J. et al. (2019b). Active boulder movement at high martian latitudes. *Geophysical Research Letters* 46 (10): 5075–5082. https://doi.org/10.1029/2016GL082293.

Durán Vinent, O., Andreotti, B., Claudin, P. et al. (2019). A unified model of ripples and dunes in water and planetary environments. *Nature Geoscience* 12: 345–350. https://doi.org/10.1038/s41561-019-0336-4.

Dyar, M.D., Helbert, J., Cooper, R.F. et al. (2021). Surface weathering on Venus: Constraints from kinetic, spectroscopic, and geochemical data. *Icarus* 358: 114139. https://doi.org/10.1016/j.icarus.2020.114139.

Edgar, L.A., Fedo, C.M., Gupta, S. et al. (2020). A lacustrine paleoenvironment recorded at Vera Rubin Ridge, Gale crater: overview of the sedimentology and stratigraphy observed by the Mars Science Laboratory Curiosity rover. *Journal of Geophysical Research: Planets* 125 (3). https://doi.org/10.1029/2019JE006307.

Edgar, L.A., Grotzinger, J., Fedo, C.M. et al. (2024). Wet to dry depositional environments recorded in the clay-sulfate transition region of Gale crater, Mars: overview and stratigraphic context for Curiosity's exploration campaign. In: *55th Lunar and Planetary Science Conference*, 1016. The Woodlands, Texas, USA: Lunar and Planetary Science Institute.

Edgar, L.A., Grotzinger, J.P., Bell, J.F. III et al. (2014). Hypotheses for the origin of fine-grained sedimentary rocks at Santa Maria crater, Meridiani Planum. *Icarus* 234: 36–44. https://doi.org/10.1016/j.icarus.2014.02.019.

Edgar, L.A., Grotzinger, J.P., Hayes, A.G. et al. (2012). Stratigraphic architecture of bedrock reference section, Victoria crater, Meridiani Planum, Mars. In: *Sedimentary Geology of Mars*, SEPM Special Publication No. 102, 195–209. https://doi.org/10.2110/pec.12.102.0195.

Edgar, L.A., Gupta, S., Rubin, D.M. et al. (2018). Shaler: In situ analysis of a fluvial sedimentary deposit on Mars. *Sedimentology* 65 (1): 96–122. https://doi.org/10.1111/sed.12370.

Edgett, K.S., Banham, S.G., Bennett, K.A. et al. (2020). Extraformational sediment recycling on Mars. *Geosphere* 16 (6): 1508–1537. https://doi.org/10.1130/GES02244.1.

Edgett, K.S. and Lancaster, N. (1993). Volcaniclastic aeolian dunes: terrestrial examples and applications to martian sands. *Journal of Arid Environments* 25 (3): 271–297.

Edgett, K.S. and Sarkar, R. (2021). Recognition of sedimentary rock occurrences in satellite and aerial images of other worlds—insights from Mars. *Remote Sensing* 13 (21). https://doi.org/10.3390/rs13214296.

Edgett, K.S., Yingst, R.A., Ravine, M.A. et al. (2012). Curiosity's Mars Hand Lens Imager (MAHLI) investigation. *Space Science Reviews* 170: 259–317. https://doi.org/10.1007/s11214-012-9910-4.

Ehlmann, B.L., Anderson, F.S., Andrews-Hanna, J. et al. (2016). The sustainability of habitability on terrestrial planets: insights, questions, and needed measurements from Mars for understanding the evolution of Earth-like worlds. *Journal of Geophysical Research: Planets* 121 (10): 1927–1961. https://doi.org/10.1002/2016JE005134.

Ehlmann, B.L. and Edwards, C.S. (2014). Mineralogy of the Martian surface. *Annual Review of Earth and Planetary Sciences* 42: 291–315. https://doi.org/10.1146/annurev-earth-060313-055024.

Ehlmann, B.L., Mustard, J.F., Clark, R.N. et al. (2011b). Evidence for low-grade metamorphism, hydrothermal alteration, and diagenesis on Mars from phyllosilicate mineral assemblages. *Clay and Clay Minerals* 59: 359–377. https://doi.org/10.1346/CCMN.2011.0590402.

Ehlmann, B.L., Mustard, J.F., Murchie, S. L. et al. (2008). Orbital identification of carbonate-bearing rocks on Mars. *Science* 322 (5909): 1828–1832. https://doi.org/10.1126/science.1164759.

Ehlmann, B.L., Mustard, J.F., Murchie, S.L. et al. (2010). Geologic setting of serpentine deposits on Mars. *Geophysical Research Letters* 37. https://doi.org/10.1029/2010GL042596.

Ehlmann, B.L., Mustard, J.F., Murchie, S. L. et al. (2011a). Subsurface water and clay mineral formation during the early history of Mars. *Nature* 479: 53–60. https://doi.org/10.1038/nature10582.

Elachi, C., Allison, M.D., Borgarelli, L. et al. (2004). Radar: the Cassini Titan radar mapper. In: *The Cassini-Huygens Mission* (ed. C.T. Russel). Dordrecht: Springer.

El-Maarry, M.R., Groussin, O., Keller, H.U. et al. (2019). Review of Rosetta's observations of 67P/Churyumov-Gerasimenko. *Space Science Reviews* 216: 36. https://doi.org/10.1007/s11214-019-0602-1.

Eppes, M.-C. et al. (2015). Cracks in martian boulders exhibit preferred orientations that point to solar-induced thermal stresses. *Nature Communications* 6: https://doi.org/10.1038/ncomms7712.

Ewing, R.C., Hayes, A.G., Lucas, A. et al. (2015). Sand dune patterns on Titan controlled by long-term climate cycles. *Nature Geoscience* 8: 15–19. https://doi.org/10.1038/ngeo2323.

Ewing, R.C., Lapôtre, M.G.A., Lewis, K.W. et al. (2017). Sedimentary processes of the Bagnold Dunes: implications for the eolian rock record of Mars. *Journal of Geophysical Research: Planets* 122 (12): 2544–2573. https://doi.org/10.1002/2017JE005324.

Ewing, R.C., Peyret, A.P.B., Kocurek, G. et al. (2010). Dune field pattern formation and recent transporting winds in the Olympia Undae dune field, north polar region of Mars. *Journal of Geophysical Research: Planets* 115 (E8): https://doi.org/10.1029/2009JE003526.

Fairén, A., Dohm, J.M., Baker, V.R. et al. (2003). Episodic flood inundations of the northern plains of Mars. *Icarus* 165 (1): 53–67. https://doi.org/10.1016/S0019-1035(03)00144-1.

Farley, K.A., Malespin, C., Mahaffy, P. et al. (2014). In situ radiometric and exposure age dating of the martian surface. *Science* 343 (6169). https://doi.org/10.1126/science.1247166.

Farley, K.A., Stack, K.M., Shuster, D.L. et al. (2022). Aqueously altered igneous rocks sampled on the floor of Jezero crater, Mars. *Science* 377 (6614). https://doi.org/10.1126/science.abo2196.

Farley, K.A., Williford, K.H., Stack, K.M. et al. (2020). Mars 2020 mission overview. *Space Science Reviews* 216: 142. https://doi.org/10.1007/s11214-020-00762-y.

Fassett, C.I. and Head, J.W. (2008a). Valley network-fed, open-basin lakes on Mars: distribution and implications for Noachian surface and subsurface hydrology. *Icarus* 198 (1): 37–56. https://doi.org/10.1016/j.icarus.2008.06.016.

Fassett, C.I. and Head, J.W. (2008b). The timing of martian valley network activity: constraints from buffered crater counting. *Icarus* 195 (1): 61–89. https://doi.org/10.1016/j.icarus.2007.12.009.

Favaro, E.A., Balme, M.R., McNeil, J.D. et al. (2024). Periodic bedrock ridges at Oxia Planum and Chryse Planitia, Mars: evidence for widespread aeolian erosion of an ancient surface by regional paleowinds. *Icarus* 626: 118522. https://doi.org/10.1016/j.epsl.2023.118522.

Fawdon, P., Gupta, S., Davis, J.M. et al. (2018). The Hypanis Valles delta: the last highstand of a sea on early Mars? *Earth and Planetary Science Letters* 500: 225–241. https://doi.org/10.1016/j.epsl.2018.07.040.

Fedo, C.M., Bryk, A.B., Edgar, L.A. et al. (2022). Geology and stratigraphic correlation of the Murray and Carolyn Shoemaker formations across the Glen Torridon region, Gale crater, Mars. *Journal of Geophysical Research: Planets* 127 (9). https://doi.org/10.1029/2022JE007408.

Fenton, L.K. (2020). Updating the global inventory of dune fields on Mars and identification of many small dune fields. *Icarus* 352: 114018. https://doi.org/10.1016/j.icarus.2020.114018.

Florensky, C.P., Basilevsky, A.T., Kryuchkov, V.P. et al. (1983). Venera 13 and Venera 14: Sedimentary rocks on Venus? *Science* 221 (4605): 57–59. https://doi.org/10.1126/science.4605.57.

Florensky, C.P., Ronca, L.B., Basilevsky, A.T. et al. (1977). Surface of Venus as revealed by Soviet Venera 9 and 10. *Bulletin of the Geological Society of America* 88 (11): 1537–1545.

Forget, F., Hourdin, F., Fournier, R. et al. (1999). Improved general circulation models of the martian atmosphere from the surface to above 80 km. *Journal of Geophysical Research* 104 (E10): 24155–24175. https://doi.org/10.1029/1999JE001025.

Forget, F., Wordsworth, R., Millour, E. et al. (2013). 3D modelling of the early martian climate under a denser CO_2 atmosphere: temperatures and CO_2 ice clouds. *Icarus* 222 (1): 81–99. https://doi.org/10.1016/j.icarus.2012.10.019.

Fraeman, A.A., Edgar, L.A., Rampe, E.B. et al. (2020). Evidence for a diagenetic origin of Vera Rubin Ridge, Gale crater, Mars: summary and synthesis of Curiosity's exploration campaign. *Journal of Geophysical Research: Planets* 125 (12). https://doi.org/10.1029/2020JE006527.

Frederick, K.A. and Hanratty, T.J. (1988). Velocity measurements for a turbulent nonseparated flow over solid waves. *Experiments in Fluids* 6: 447–486.

Ganey, T.M., Gilmore, M.S., Brossier, J. et al. (2023). Reassessment of the volumes of sediment sources and sinks on Venus. *The Planetary Science Journal* 4 (1). https://doi.org/10.3847/PSJ/aca521.

Gardin, E., Allemand, P., Quantin, C. et al. (2010). Defrosting, dark flow features, and dune activity on Mars: example in Russell crater. *Journal of Geophysical Research: Planets* 115 (E6). https://doi.org/10.1029/2009JE003515.

Garvin, J.B., Getty, S.A., Arney, G.N. et al. (2022). Revealing the mysteries of Venus: the DAVINCI mission. *The Planetary Science Journal* 3 (5). https://doi.org/10.3847/PSJ/ac63c2.

Garvin, J.B., Head, J.W., Zuber, M.T. et al. (1984). Venus: the nature of the surface from Venera panoramas. *Journal of Geophysical Research: Solid Earth* 89 (B5): 3381–3399.

Geissler, P.E. (2014). The birth and death of transverse aeolian ridges on Mars. *Journal of Geophysical Research: Planets* 119 (12): 2583–2599. https://doi.org/10.1002/2014JE004633.

Ghail, R.C., C Wilson, M Galand et al. (2011). EnVision: taking the pulse of our twin planet. *Experimental Astronomy* 33: 337–363. https://doi.org/10.1007/s1068-011-9244-3

Golombek, M.P. and Phillips, R.J. (2010). Mars Tectonics. In: *Planetary Tectonics* (ed. T.R. Watters and R.A. Schultz), 183–232. Cambridge: Cambridge University Press.

Gou, S., Yue, Z., Di, K. et al. (2022). Transverse aeolian ridges in the landing area of the Tianwen-1 Zhurong rover on Utopia Planitia, Mars. *Earth and Planetary Science Letters* 595: 117764. https://doi.org/10.1016/j.epsl.2022.117764.

Goudge, T.A., Aureli, K.L., Head, J.W. et al. (2015a). Classification and analysis of candidate impact crater-hosted closed-basin lakes on Mars. *Icarus* 260: 346–367. https://doi.org/10.1016/j.icarus.2015.07.026.

Goudge, T.A., Head, J.W., Mustard, J.F. et al. (2012). An analysis of open-basin lake deposits on Mars: Evidence for the nature of associated lacustrine deposits and post-lacustrine modification processes. *Icarus* 219 (1): 211–229. https://doi.org/10.1016/j.icarus.2012.02.027.

Goudge, T.A., Mohrig, D., Cardenas, B.T. et al. (2018). Stratigraphy and paleohydrology of delta channel deposits, Jezero crater, Mars. *Icarus* 301: 58–75. https://doi.org/10.1016/j.icarus.2017.09.034.

Goudge, T.A., Mustard, J.F., Head, J.W. et al. (2015b). Assessing the mineralogy of the watershed and fan deposits of the Jezero crater paleolake system, Mars. *Journal of Geophysical Research: Planets* 120 (4). https://doi.org/10.1002/2014JE004782.

Grant, J.A., Golombek, M.P., Wilson, S.A. et al. (2018). The science process for selecting the landing site for the 2020 Mars rover. *Planetary and Space Science* 164: 106–126. https://doi.org/10.1016/j.pss.2018.07.001.

Grant, J.A., Irwin, R.P. III, Grotzinger, J.P. et al. (2008). HiRISE imaging of impact megabreccia and sub-meter aqueous strata in Holden crater, Mars. *Geology* 36 (3): 195–198. https://doi.org/10.1130/G24340A.1.

Grant, J.A. and Wilson, S.A. (2011). Late alluvial fan formation in southern Margaritifer Terra, Mars. *Geophysical Research Letters* 38 (8). https://doi.org/10.1029/2011GL046844.

Grant, J.A. and Wilson, S.A. (2019). Evidence for late alluvial activity in Gale crater. *Mars. Geophysical Research Letters* 46 (13): 7287–7294. https://doi.org/10.1029/2019GL083444.

Greeley, R. (2002). Saltation impact as a mean for raising dust on Mars. *Planetary and Space Science* 50 (2): 151–155.

Greeley, R., Arvidson, R.E., Elachi, C. et al. (1992). Aeolian features on Venus: preliminary Magellan results. *Journal of Geophysical Research* 97 (E8): 13319–13345.

Greeley, R., Bender, K., Thomas, P.E. et al. (1995). Wind-related features and processes on Venus: summary of Magellan results. *Icarus* 115 (2): 399–420.

Greeley, R., Iversen, J., Leach, R. et al. (1984). Windblown sand on Venus: Preliminary results of laboratory simulations. *Icarus* 57: 112–124.

Greeley, R., Iversen, J.D., Pollack, J.B. et al. (1974). Wind tunnel studies of martian aeolian processes. *Proceedings of the Royal Society of London, Series A–Mathematical and Physical Sciences* 341: 331–360.

Greeley, R., Leach, R., White, B. et al. (1980). Threshold windspeeds for sand on Mars: Wind tunnel simulations. *Geophysical Research Letters* 7: 121–124.

Greeley, R., White, B., Leach, R. et al. (1976). Mars: Wind friction speeds for particle movement. *Geophysical Research Letters* 3: 417–420.

Greenwood, J.P., Itoh, S., Sakamoto, N. et al. (2008). Hydrogen isotope evidence for loss of water from Mars through time. *Geophysical Research Letters* 35 (5). https://doi.org/10.1029/2007GL032721.

Grindrod, P.M. and Warner, N.H. (2014). Erosion rate and previous extent of interior layered deposits on Mars revealed by obstructed landslides. *Geology* 42 (9): 795–798. https://doi.org/10.1130/G35790.1.

Grotzinger, J.P., Arvidson, R.E., Calvin, W. et al. (2005). Stratigraphy and sedimentology of a dry to wet eolian depositional system, Burns formation, Meridiani Planum, Mars. *Earth and Planetary Science Letters* 240 (1): 11–72. https://doi.org/10.1016/j.epsl.2005.09.039.

Grotzinger, J.P., Beaty, D., Dromart, G. et al. (2011). Mars sedimentary geology: key concepts and outstanding questions. *Astrobiology* 11 (1): 77–87. https://doi.org/10.1089/ast.2010.0571.

Grotzinger, J.P., Bell, J. III, Herkenhoff, K. et al. (2006). Sedimentary textures formed by aqueous processes, Erebus crater, Meridiani Planum, Mars. *Geology* 34 (12): 1085–1088. https://doi.org/10.1130/G22985A.1.

Grotzinger, J.P., Crisp, J., Vasavada, A.R. et al. (2012). Mars Science Laboratory mission and science investigation. *Space Science Reviews* 170: 5–56. https://doi.org/10.1007/s11214-012-9892-2.

Grotzinger, J.P., Gupta, S., Malin, M.C. et al. (2015). Deposition, exhumation, and paleoclimate of an ancient lake deposit, Gale crater, Mars. *Science* 350 (6257). https://doi.org/10.1126/science.aac7575.

Grotzinger, J.P., Hayes, A.G., Lamb, M.P. et al. (2013). Sedimentary processes on Earth, Mars, Titan, and Venus. In: *Comparative Climatology of Terrestrial Planets* (ed. S.J. Mackwell, A.A. Simon-Miller, J.W. Harder, et al.), 439–472. Tucson: University of Arizona.

Grotzinger, J.P. and Milliken, R.E. (2012). The sedimentary rock record of Mars: Distribution, origins, and global stratigraphy. In: *Sedimentary Geology of Mars*, SEPM special paper 102, 1–48. https://doi.org/10.2110/pec.12.102.0001.

Grotzinger, J.P., Sumner, D.Y., Kah, L.C. et al. (2014). A habitable fluvio-lacustrine environment at Yellowknife Bay, Gale crater, Mars. *Science* 343 (6169). https://doi.org/10.1126/science.1242777.

Grundy, W.M., Binzel, R.P., Buratti, B.J. et al. (2016). Surface compositions across Pluto and Charon. *Science* 351: 6279. https://doi.org/10.1126/science.aad9189.

Gunn, A. and Jerolmack, D.J. (2022). Conditions for aeolian transport in the Solar System. *Nature Astronomy* 6: 923–929. https://doi.org/10.1038/s41550-022-01669-0.

Gunn, A., Rubanenko, L., Lapôtre, M.G.A. et al. (2022). Accumulation of windblown sand in impact craters on Mars. *Geology* 50 (9): 981–985. https://doi.org/10.1130/G49936.1.

Gupta, S., Edgar, L., Yingst, R.A. et al. (2022). Episodic aqueous conditions punctuated dominantly aeolian deposition within the layered sulphate-bearing unit, Gale crater (Mars). In: *Europlanet Science Congress 2022*, vol. 16, EPSC2022–EPSC2963.

Gupta, S., K. Stack Morgan, N. Mangold, et al. (2024). Going With the Flow: Sedimentary Evolution of the Jezero Western Fan, Mars. In 55th Lunar and Planetary Science Conference, 2607.

Gwinner, K., Scholten, F., Preusker, F. et al. (2010). Topography of Mars from global mapping by HRSC high-resolution digital terrain models and orthoimages: characteristics and performance. *Earth and Planetary Science Letters* 294 (3–4): 506–519. https://doi.org/10.1016/j.epsl.2009.11.007.

Gwizd, S., Fedo, C., Grotzinger, J. et al. (2022). Sedimentological and geochemical perspectives on a marginal lake environment recorded in the

Hartmann's Valley and Karasburg members of the Murray formation, Gale crater, Mars. *Journal of Geophysical Research: Planets* 127 (8). https://doi.org/10.1029/2022JE007280.

Gwizd, S., C Fedo, J Grotzinger et al. (2024). Evolution of a lake margin recorded in the Sutton Island member of the Murray formation, Gale crater, Mars. *Journal of Geophysical Research: Planets*, 129(1). https://doi.org/10.1029/2023JE007919

Hamran, S. E., DA Paige, A Allwood et al. (2022). Ground penetrating radar observations of subsurface structures in the floor of Jezero crater, Mars. Science *Advances*, 8(34). https://doi.org/10.1126/sciadv.abp8564

Hansen, C.J., Bourke, M., Bridges, N.T. et al. (2011). Seasonal erosion and restoration of Mars' northern polar dunes. *Science* 331 (6017): 575–578. https://doi.org/10.1126/science.1197636.

Hansen, C.J., Byrne, S., Portyankina, G. et al. (2013). Observations of the northern seasonal polar cap on Mars: I. Spring sublimation activity and processes. *Icarus* 225 (2): 881–897. https://doi.org/10.1016/j.icarus.2012.09.024.

Hansen, C.J., Diniega, S., Bridges, N. et al. (2015). Agents of change on Mars' northern dunes: CO_2 ice and wind. *Icarus* 251: 264–274. https://doi.org/10.1016/j.icarus.2014.11.015.

Hartmann, W.K., Anguita, J., de la Casa, M.A. et al. (2001). Martian cratering 7: The role of impact gardening. *Icarus* 149 (1): 37–53. https://doi.org/10.1006/icar.2000.6532.

Hasson, M., Finotello, A., Ielpi, A., & Lapôtre, M. G. A. (2025). Vegetation changes the trajectory of river bends. *Science* 389: (6763): 915–920. https://doi.org/10.1126/science.adv4939

Hasson, M., MC Marvin, A Gunn et al. (2023). A depositional model for meandering rivers without land plants. *Sedimentology* 70. https://doi.org/10.1111/sed.13121

Hayden, A.T., Lamb, M.P., Fischer, W.W. et al. (2019). Formation of sinuous ridges by inversion of river-channel belts in Utah, USA, with implications for Mars. *Icarus* 332: 92–110. https://doi.org/10.1016/j.icarus.2019.04.019.

Hayes, A.G. (2016). The lakes and seas of Titan. *Annual Reviews of Earth and Planetary Sciences* 44: 57–83. https://doi.org/10.1146/annurev-eart-060115-012247.

Hayes, A.G., Grotzinger, J.P., Edgar, L.A. et al. (2011). Reconstruction of eolian bed forms and paleocurrents from cross-bedded strata at Victoria crater, Meridiani Planum, Mars. *Journal of Geophysical Research: Planets* 116 (E7). https://doi.org/10.1029/2010JE003688.

Hayes, A.G., Lorenz, R.D., Lunine, J.I. et al. (2018). A post-Cassini view of Titan's methane-based hydrologic cycle. *Nature Geoscience* 11: 306–313. https://doi.org/10.1038/s41561-018-0103-y.

Hayward, R.K., Fenton, L.K., Titus, T.N. et al. (2014). Mars global digital dune database (MGD^3): Global dune distribution and wind pattern observations. *Icarus* 230: 34–46. https://doi.org/10.1016/j.icarus/j.icarus.2013.04.011.

Hayward, R.K., Mullins, K.F., Fenton, L.K. et al. (2007). Mars global digital dune database and initial science results. *Journal of Geophysical Research* 112 (E11). https://doi.org/10.1029/2007JE002943.

Head, J.W., Hiesinger, H., Ivanov, M.A. et al. (1999). Possible ancient oceans on Mars: evidence from Mars orbiter laser altimeter data. *Science* 286 (5447): 2134–2137.

Head, J.W., Kreslavsky, M., Hiesinger, H. et al. (1998). Oceans in the past history of Mars: tests for their presence using Mars orbiter laser altimeter (MOLA) data. *Geophysical Research Letters* 25 (24): 4401–4404.

Head, J.W., Wilson, L., Mitchell, K.L. et al. (2003). Generation of recent massive water floods at Cerberus Fossae, Mars by dyke emplacement, cryospheric cracking, and confined aquifer groundwater release. *Geophysical Research Letters* 30 (11). https://doi.org/10.1029/2003GL017135.

Hedgepeth, J.E., Neish, C.D., Turtle, E.P. et al. (2020). Titan's impact crater population after Cassini. *Icarus* 344: 113664. https://doi.org/10.1016/j.icarus.2020.113664.

Hepburn, A.J., Ng, F.S.L., Livingstone, S.J. et al. (2019). Polyphase mid-latitude glaciation on Mars: chronology of the formation of superposed glacier-like forms from crater-count dating. *Journal of Geophysical Research: Planets* 125 (2). https://doi.org/10.1029/2019JE006102.

Herd, C.D., Bosak, T., Hausrath, E.M. et al. (2025). Sampling Mars: Geologic context and preliminary characterization of samples collected by the NASA Mars 2020 Perseverance Rover Mission. *Proceedings of the National Academy of Sciences* 122 (2). e2404255121.

Herkenhoff, K.E., Squyres, S.W., Bell, J.F. III et al. (2003). Athena microscopic imager investigation. *Journal of Geophysical Research: Planets* 108 (E12). https://doi.org/10.1029/2003JE002076.

Herkenhoff, K.E., Sullivan, R.J., Newman, C.E. et al. (2023). Comparison of ventifact orientations and recent wind direction indicators on the floor of Jezero crater, Mars. *Journal of Geophysical Research: Planets* 128 (3). https://doi.org/10.1029/2022JE007599.

Holo, S.J., Kite, E.S., Wilson, S.A. et al. (2021). The timing of alluvial fan formation on Mars. *The Planetary Science Journal* 2 (5). https://doi.org/10.3847/PSJ/ac25ed.

Horgan, B.H. and Bell, J.F. (2012). Seasonally active slipface avalanches in the northern polar sand sea of Mars: evidence for a wind-related origin. *Geophysical Research Letters* 39 (9). https://doi.org/10.1029/2012GL051329.

Hörst, S.M. (2017). Titan's atmosphere and climate. *Journal of Geophysical Research: Planets* 122 (3): 432–482. https://doi.org/10.1002/2016JE005240.

Hörst, S.M. and Tolbert, M.A. (2010). In situ measurements of the size and density of Titan aerosol analogs. *The Astrophysical Journal* 770 (1). https://doi.org/10.1088/2041-8205/770/1/L10.

Hubbard, B., Souness, C., Brough, S. et al. (2014). Glacier-like forms on Mars. *The Cryosphere* 8: 2047–2061. https://doi.org/10.5194/tc-8-2047-2014.

Hugenholtz, C.H., Barchyn, T.E., Favaro, E.A. et al. (2015). Formation of periodic bedrock ridges on Earth. *Aeolian Research* 18: 135–144. https://doi.org/10.1016/j.aeolia.2015.07.002.

Hughes, C.M., Cardenas, B.T., Goudge, T.A. et al. (2019). Deltaic deposits indicative of a paleo-coastline at Aeolis Dorsa, Mars. *Icarus* 317: 442–453. https://doi.org/10.1016/j.icarus.2018.08.009.

Hunt, A.R.G., Day, M., Edgett, K.S. et al. (2022). The lithified aeolian dune field adjacent to the Apollinaris Sulci, Mars: geological history and paleo-wind record. *Icarus* 373. https://doi.org/10.1016/j.icarus.2021.114788.

Hurowitz, J.A. and McLennan, S.M. (2007). A ~3.5 Ga record of water-limited, acidic weathering conditions on Mars. *Earth and Planetary Science Letters* 260 (3-4): 432–443. https://doi.org/10.1016/j.epsl.2007.05.043.

Hynek, B.M., Beach, M., Hoke, M.R.T. et al. (2010). Updated global map of Martian valley networks and implications for climate and hydrological processes. *Journal of Geophysical Research Planets* 115 (E9). https://doi.org/10.1029/2009JE003548.

Ielpi, A. and Lapôtre, M.G.A. (2018). Biotic forcing militates against river meandering in the modern Bonneville Basin of Utah. *Sedimentology* 66 (5): 1896–1929. https://doi.org/10.1111/sed.12562.

Ielpi, A. and Lapôtre, M.G.A. (2019). Barren meandering streams in the modern Toiyabe Basin of Nevada, USA, and their relevance to the study of the pre-vegetation rock record. *Journal of Sedimentary Research* 89 (5): 399415. https://doi.org/10.2110/jsr.2019.25.

Ielpi, A. and Lapôtre, M.G.A. (2020a). A tenfold slowdown in river meander migration driven by plant life. *Nature Geoscience* 13. https://doi.org/10.1038/s41561-019-0491-7.

Ielpi, A. and Lapôtre, M.G.A. (2020b). Channel mobility drives a diverse stratigraphic architecture in the dryland Mojave River (California, USA). *Earth Surface Processes and Landforms* 45 (8): 1717–1731. https://doi.org/10.1002/esp.4841.

Imanaka, H., Cruikshank, D.P., Khare, B. N. et al. (2012). Optical constants of Titan tholins at mid-infrared wavelengths (2.5–25 μm) and the possible chemical nature of Titan's haze particles. *Icarus* 218 (1): 247–261. https://doi.org/10.1016/j.icarus.2011.11.018.

Ingersoll, A.P. (1969). The runaway greenhouse: a history of water on Venus. *Journal of Atmospheric Sciences* 26 (6): 1191–1198.

Irwin, R.P., Craddock, R.A., Howard, A.D. et al. (2005). Interior channels in Martian valley networks: discharge and runoff production. *Geology* 33 (6): 489–492. https://doi.org/10.1130/G21333.1.

Irwin, R.P., Howard, A.D., Craddock, R.A. et al. (2008). Fluvial valley networks on Mars. In: *River Confluences, Tributaries and Fluvial Network* (ed. S.P. Rice, A.G. Roy, and B.L. Rhodes). Wiley.

Ivanov, M.A., Hiesinger, H., Erkeling, G. et al. (2015). Evidence for large reservoirs of water/mud in Utopia and Acidalia Planitiae on Mars. *Icarus* 248: 383–391. https://doi.org/10.1016/j.icarus.2014.11.013.

Iversen, J.D. and Rasmussen, K.R. (1999). The effect of wind speed and bed slope on sand transport. *Sedimentology* 46: 723–731.

Iversen, J.D. and White, B.R. (1982). Saltation threshold on Earth, Mars and Venus. *Sedimentology* 29 (1): 111–119. https://doi.org/10.1111/j.1365-3091.1982.tb01713.x.

Jakosky, B., Lin, R.P., Grebowsky, J.M. et al. (2015). The mars atmosphere and volatile evolution (MAVEN) mission. *Space Science Reviews* 195: 3–48. https://doi.org/10.1007/s11214-015-0139-x.

Jerolmack, D.J., Mohrig, D., Grotzinger, J.P. et al. (2006). Spatial grain size sorting in eolian ripples and estimation of wind conditions on planetary surfaces: application to Meridiani Planum, Mars. *Journal of Geophysical Research: Planets* 111 (E12). https://doi.org/10.1029/2005JE002544.

Jia, P., Andreotti, B., Claudin, P. et al. (2017). Giant ripples on comet 67P/Churyumov-Gerasimenko sculpted by sunset thermal wind. *Proceedings of the National Academy of Science of the United States of America* 114 (20): 2509–2514. https://doi.org/10.1073/pnas.1612176114.

Jindal, A.S., Birch, S.P.D., Hayes, A.G. et al. (2024). Measuring erosional and depositional patterns across comet 67P's Imhotep region. *Journal of Geophysical Research: Planets* 129 (2): https://doi.org/10.1029/2023JE008089.

Kah, L.C., Stack, K.M., Eigenbrode, J.L. et al. (2018). Syndepositional precipitation of calcium sulfate in Gale crater, Mars. *Terra Nova* 30 (6): 431–439. https://doi.org/10.1111/ter.12359.

Khawja, S., Ernst, R.E., Samson, C. et al. (2020). Tesserae on Venus may preserve evidence of fluvial erosion. *Nature Communications* 11 (5789): https://doi.org/10.1038/s41467-020-19336-1.

Kieffer, H.H., Christensen, P.R., Titus, T. N. et al. (2006). CO_2 jets formed by sublimation beneath translucent slab ice in Mars' seasonal south polar ice cap. *Nature* 442: 793–796. https://doi.org/10.1038/nature04945.

Kite, E.S., Howard, A.D., Lucas, A.S. et al. (2015). Stratigraphy of Aeolis Dorsa, Mars: stratigraphic context of the great river deposits. *Icarus* 253: 223–242. https://doi.org/10.1016/j.icarus.2015.03.007.

Kminek, G., Meyer, M.A., Beaty, D.W. et al. (2022). Mars sample return (MSR): planning for returned sample science. *Astrobiology* 22 (S1). https://doi.org/10.1089/ast.2021.0198.

Kocurek, G. and Ewing, R.C. (2012). Source-to-Sink: an Earth/Mars comparison of boundary conditions for Eolian dune systems. In: *Sedimentary Geology of Mars* (ed. J.P. Grotzinger and R.E. Milliken), 151–168. Tulsa, OK: SEPM (Society for Sedimentary Geology). https://doi.org/10.2110/pec.12.102.0151.

Kok, J.F. (2010). Difference in the wind speeds required for initiation versus continuation of sand transport on Mars: Implications for dunes and dust storms. *Physical Review Letters* 104: 0745502. https://doi.org/10.1103/PhysRevLett.104.074502.

Kok, J.F., Parteli, E.J.R., Michaels, T.I. et al. (2012). The physics of wind-blown sand and dust. *Reports on Progress in Physics* 75 (10): 106901. https://doi.org/10.1088/0034-4885/75/10/106901.

Komatsu, G. and Baker, V.G. (1997). Paleohydrology and flood geomorphology of Ares Vallis. *Journal of Geophysical Research: Planets* 102 (E2): 4151–4160.

Konsoer, K., LeRoy, J., Burr, D. et al. (2018). Channel slope adjustment in reduced gravity environments and implications for martian channels. *Geology* 46 (2): 183–186. https://doi.org/10.1130/G39666.1.

Kreslavsky, M.A. and Bondarenko, N.V. (2017). Aeolian sand transport and aeolian deposits on Venus: a review. *Aeolian Research* 26: 29–46. https://doi.org/10.1016/j.aeolia.2016.06.001.

Kreslavsky, M.A. and Head, J.W. (2002). Fate of outflow channel effluents in the northern lowlands of mars: the Vastitas Borealis formation as a sublimation residue from frozen ponded bodies of water. *Journal of Geophysical Research Planets* 107 (E12). https://doi.org/10.1029/2001JE001831.

Kronyak, R.E., Kah, L.C., Edgett, K.S. et al. (2019). Mineral-filled fractures as indicators of multigenerational fluid flow in the Pahrump Hills member of the Murray formation, Gale crater, Mars. *Earth and Space Science* 6 (2): 238–265. https://doi.org/10.1029/2018EA000482.

Kurokawa, H., Sato, M., Ushioda, M. et al. (2014). Evolution of water reservoirs on mars: constraints from hydrogen isotopes in martian meteorites. *Earth and Planetary Science Letters* 394: 179–185. https://doi.org/10.1016/j.epsl.2014.03.027.

Lamb, M.P., Dietrich, W.E., Venditti, J.G. et al. (2008). Is the critical shields stress for incipient sediment motion dependent on channel-bed slope? *Journal of Geophysical Research: Earth Surface* 113: F2. https://doi.org/10.1029/2007JF000831.

Lamb, M.P., Grotzinger, J.P., Southard, J.B. et al. (2012). Were aqueous ripples on Mars formed by flowing brines? In: *Sedimentary Geology of Mars*. Special Publication (SPEM), 102, 139–150. Tulsa, OK: Society for Sedimentary Geology.

Lamb, M.P. and Venditti, J.G. (2016). The grain size gap and abrupt gravel-sand transitions in rivers due to the suspension fallout. *Geophysical Research Letters* 43 (8): 3777–3785. https://doi.org/10.1002/2016GL068713.

Lammer, H., Kolb, C., Penz, T. et al. (2003). Estimation of the past and present Martian water-ice reservoirs by isotopic constraints on exchange between the atmosphere and the surface. *International Journal of Astrobiology* 2 (3): 195–202. https://doi.org/10.1017/S1473550403001605.

Lancaster, N. and Greeley, R. (1990). Sediment volume in the north polar sand seas of Mars. *Journal of Geophysical Research: Solid Earth* 95 (B7): 10921–10927.

Langhans, M., Lunine, J.I., Mitri, G. et al. (2013). Titan's Xanadu region: geomorphology and formation scenario. *Icarus* 223 (2): 796–803. https://doi.org/10.1016/j.icarus.2013.01.016.

Langhans, M.H., Jaumann, R., Stephan, K. et al. (2012). Titan's fluvial valleys: morphology, distribution, and spectral properties. *Planetary and Space Science* 60 (1): 34–51. https://doi.org/10.1016/j.pss.2011.01.020.

Lapôtre, M.G.A., Bishop, J.L., Ielpi, A. et al. (2022a). Mars as a time machine to Precambrian Earth. *Journal of the Geological Society* 179 (5). https://doi.org/10.1144/jgs2022-047.

Lapôtre, M.G.A., Ewing, R.C., Lamb, M.P. et al. (2016a). Large wind ripples on mars: a record of atmospheric evolution. *Science* 353 (6294): 55–58. https://doi.org/10.1126/science.aaf3206.

Lapôtre, M.G.A., Ewing, R.C., Lamb, M.P. et al. (2021). An evolving understanding of enigmatic large ripples on Mars. *Journal of Geophysical Research: Planets* 126 (2). https://doi.org/10.1029/2020JE006729.

Lapôtre, M.G.A., Ewing, R.C., Weitz, C.M. et al. (2018). Morphologic diversity of Martian ripples: implications for large-ripple formation. *Geophysical Research Letters* 45 (19): 10,229–10,239. https://doi.org/10.1029/2018GL079029.

Lapôtre, M.G.A. and Ielpi, A. (2020). The pace of fluvial meanders on Mars and implications for the western delta deposits of Jezero crater. *AGU Advances* 1 (2). https://doi.org/10.1029/2019AV000141.

Lapôtre, M.G.A., Ielpi, A., Lamb, M.P. et al. (2019). Model for the formation of single-thread rivers in barren landscapes and implications for pre-Silurian and martian fluvial deposits. *Journal of Geophysical Research: Earth Surface* 124 (12): 2757–2777. https://doi.org/10.1029/2019JF005156.

Lapôtre, M.G.A. and Lamb, M.P. (2018). Substrate controls on valley formation by groundwater on Earth and Mars. *Geology* 46 (6): 531–534. https://doi.org/10.1130/G40007.1.

Lapôtre, M.G.A., Lamb, M.P., McElroy, B. et al. (2017). What sets the size of current ripples? *Geology* 45 (3): 243–246. https://doi.org/10.1130/G38598.1.

Lapôtre, M.G.A., Lamb, M.P., Williams, R.M. et al. (2016b). Canyon formation constraints on the discharge of catastrophic outburst floods on Earth and Mars. *Journal of Geophysical Research: Planets* 121 (7): 1232–1263. https://doi.org/10.1002/2016JE005061.

Lapôtre, M.G.A., Malaska, M.J., Cable, M.L. et al. (2022b). The role of seasonal sediment transport and sintering in shaping Titan's landscapes: a hypothesis. *Geophysical Research Letters* 49: 8. https://doi.org/10.1029/2021GL097605.

Lapôtre, M.G.A., O'Rourke, J.G., Schaefer, L.K. et al. (2020). Probing space to understand Earth. *Nature Reviews Earth and Environment* 1: 170–181. https://doi.org/10.1038/s43017-020-0029-y.

Lapôtre, M.G.A. and Rampe, E.B. (2018). Curiosity's investigation of the Bagnold Dunes, Gale crater: overview of the two-phase scientific campaign and introduction to the special collection. *Geophysical Research Letters* 45 (19): 10200–10210. https://doi.org/10.1029/2018GL079032.

Lasue, J., Clifford, S.M., Conway, S.J. et al. (2019). The hydrology of Mars Including a potential cryosphere.

In: *Volatiles in the Martian Crust* (ed. J. Filiberto and S.P. Schwenzer), 185–246. https://doi.org/10.1016/B978-0-12-804191-8.00007-6.

Le Gall, A., Janssen, M.A., Paillou, P. et al. (2010). Radar-bright channels on Titan. *Icarus* 207 (2): 948–958. https://doi.org/10.1016/j.icarus.2009.12.027.

Lester, C.W., Murray, A.B., Orencio, D., et al. (2025). Emergence of wind ripples controlled by mechanics of grain–bed impacts. *Nature Geoscience* 18 (4): 344–350. https://doi.org/10.1038/s41561-025-01672-w.

Levy, J.S., Fassett, C.I., Head, J.W. et al. (2016). Enhanced erosion rates on mars during Amazonian glaciation. *Icarus* 264: 213–219. https://doi.org/10.1016/j.icarus.2015.09.037.

Levy, J.S., Fassett, C.I., Holt, J.W. et al. (2021). Surface boulder banding indicates Martian debris-covered glaciers formed over multiple glaciations. *Proceedings of the National Academy of Sciences of the United States of America* 118 (4). https://doi.org/10.1079/pnas.2015971118.

Lewis, K.W., Aharonson, O., Grotzinger, J.P. et al. (2008a). Structure and stratigraphy of Home Plate from the spirit Mars exploration rover. *Journal of Geophysical Research: Planets* 113 (E12). https://doi.org/10.1029/2007JE003025.

Lewis, K.W., Aharonson, O., Grotzinger, J.P. et al. (2008b). Quasi-periodic bedding in the sedimentary rock record of Mars. *Science* 332 (5907): 1532–1535. https://doi.org/10.1126/science.1161870.

Lindsay, J. (1972). Sedimentology of clastic rocks returned from the Moon by Apollo 15. *Bulletin of the Geological Society of America* 83 (10): 2957–2970.

Liu, Y. et al. (2022). An olivine cumulate outcrop on the floor of Jezero crater, Mars. *Science* 377 (6614): 1513–1519. https://doi.org/10.1126/science.abo2756.

Loizeau, D., Werner, S.C., Mangold, N. et al. (2012). Chronology of deposition and alteration in the Mawrth Vallis region, Mars. *Icarus* 72 (1): 31–43. https://doi.org/10.1016/j.pss.2012.06.023.

Lopes, R.M.C., Malaska, M.J., Schoenfeld, A.M. et al. (2020). A global geomorphic map of Saturn's moon Titan. *Nature Astronomy* 4: 228–233. https://doi.org/10.1038/s41550-019-0917-6.

Lopes, R.M.C., Wall, S.D., Elachi, C. et al. (2019). Titan as revealed by the Cassini radar. *Space Science Reviews* 215 (33). https://doi.org/10.1007/s11214-019-0598-6.

López-Valverde, M.A., Gerard, J.C., González-Galindo, F. et al. (2018). Investigation of the Mars upper atmosphere with ExoMars trace gas orbiter. *Space Science Reviews* 214 (29). https://doi.org/10.1007/s11214-017-0463-4.

Lorenz, R.D. (2016). Surface winds on Venus: probability distribution from in-situ measurements. *Icarus* 264: 311–315. https://doi.org/10.1016/j.icarus.2015.09.036.

Lorenz, R.D. (2017). Physics of saltation and sand transport on Titan: a brief review. *Icarus* 230: 162–167. https://doi.org/10.1013/j.icarus.2013.06.023.

Lorenz, R.D., Bridges, N.T., Rosenthal, A.A. et al. (2014). Elevation dependence of bedform wavelength on Tharsis Montes, Mars: Atmospheric density as a controlling parameter. *Icarus* 230: 77–80. https://doi.org/10.1016/j.icarus.2013.10.026.

Lorenz, R.D., Lopes, R.M., Paganelli, F. et al. (2008). Fluvial channels on Titan: initial Cassini RADAR observations. *Planetary and Space Science* 56 (8): 1132–1144. https://doi.org/10.1016/j.pss.2008.02.009.

Lorenz, R.D. and Radebaugh, J. (2009). Global pattern of Titan's dunes: Radar survey from Cassini prime mission. *Geophysical Research Letters* 36 (3): https://doi.org/10.1029/2008GL036850.

Lorenz, R.D., Wall, S., Radebaugh, J. et al. (2006). The sand seas of Titan: Cassini RADAR observations of longitudinal dunes. *Science* 312 (5774): 724–727. https://doi.org/10.1126/science.1123257.

Lowe, D.R., Bishop, J.L., Loizeau, D. et al. (2020). Deposition of >3.7 Ga clay-rich strata of the Mawrth Vallis group, Mars, in lacustrine, alluvial, and aeolian environments. *Bulletin of the Geological Society of America* 132: 17–30. https://doi.org/10.1130/B35185.1.

Lu, Y., Edgett, K.S., Wu, B. et al. (2022). Aeolian disruption and reworking of TARs at the Zhurong rover field site, southern Utopia Planitia, Mars. *Earth and Planetary Science Letters* 595: 117785. https://doi.org/10.1016/j.epsl.2022.117785.

Lucchitta, B.K., Ferguson, H.M., and Summers, C. (1986). Sedimentary deposits in the northern lowland plains, Mars. *Journal of Geophysical Research: Solid Earth* 91 (B13): E166–E174.

MacKenzie, S.M. and Barnes, J.W. (2016). Compositional similarities and distinctions between Titan's evaporitic terrains. *The Astrophysical Journal* 821 (1). https://doi.org/10.3847/0004-637X/821/1/17.

MacKenzie, S.M., Barnes, J.W., Sotin, C. et al. (2014). Evidence of Titan's climate history from evaporite distribution. *Icarus* 243: 191–207. https://doi.org/10.1016/j.icarus.2014.08.022.

Malaska, M.J., Hodyss, R., Lunine, J.I. et al. (2017). Laboratory measurements of nitrogen dissolution in Titan lake fluids. *Icarus* 289: 94–105. https://doi.org/10.1016/j.icarus.2017.01.033.

Malaska, M.J., Lopes, R.M., Hayes, A.G. et al. (2016). Material transport map of Titan: the fate of dunes. *Icarus* 270: 183–196. https://doi.org/10.1016/j.icarus.2015.09.029.

Malaska, M.J., Radebaugh, J., Lopes, R.M.C. et al. (2020). Labyrinth terrain on Titan. *Icarus* 344: 113764. https://doi.org/10.1016/j.icarus.2020.113764.

Malin, M.C. (1992). Mass movements on Venus: preliminary results from Magellan cycle 1 observations. *Journal of Geophysical Research: Planets* 97 (E10): 16337–17352.

Malin, M.C., Bell, J.F. III, Cantor, B.A. et al. (2007). Context camera investigation on board the Mars Reconnaissance Orbiter. *Journal of Geophysical Research: Planets* 112 (E5): https://doi.org/10.1029/2006JE002808.

Malin, M.C. and Edgett, K.S. (1999). Oceans or seas in the Martian northern lowlands: High resolution imaging tests of proposed coastlines. *Geophysical Research Letters* 26 (19): 3049–3052.

Malin, M.C. and Edgett, K.S. (2000a). Sedimentary rocks of early mars. *Science* 290 (5498): 1927–1937. https://doi.org/10.1126/science.290.5498.1927.

Malin, M.C. and Edgett, K.S. (2000b). Evidence for recent groundwater seepage and surface runoff on Mars. *Science* 288 (5475): 2330–2335. https://doi.org/10.1126/science.288.5478.2330.

Malin, M.C., Ravine, M.A., Caplinger, M.A. et al. (2017). The mars science laboratory (MSL) mast cameras and descent imager: investigation and instrument description. *Earth and Space Science* 4 (8): 506–539. https://doi.org/10.1002/2016EA000252.

Mangold, N., Caravaca, G., Gupta, S. et al. (2024). Architecture of fluvial and deltaic deposits exposed along the eastern edge of the western fan of Jezero crater, Mars. *Journal of Geophysical Research: Planets* 129 (3). https://doi.org/10.1029/2023JE008204.

Mangold, N., Dehouck, E., Fedo, C. et al. (2019). Chemical alteration of fine-grained sedimentary rocks at Gale crater. *Icarus* 321: 619–631. https://doi.org/10.1016/j.icarus.2018.11.004.

Mangold, N., Gupta, S., Gasnault, O. et al. (2021). Perseverance rover reveals an ancient delta-lake system and flood deposits at Jezero crater, Mars. *Science* 374 (6568): 711–717. https://doi.org/10.1126/science.abl4051.

Mangold, N., Quantin, C., Ansan, V. et al. (2004). Evidence for precipitation on Mars from dendritic valleys in the Valles Marineris area. *Science* 305 (5680): 78–81. https://doi.org/10.1126/science.1097549.

Martin, P.E., Farley, K.A., Baker, M.B. et al. (2017). A two-step K-Ar experiment on Mars: Dating the diagenetic formation of jarosite from Amazonian groundwaters. *Journal of Geophysical Research: Planets* 122 (12): 2803–2818. https://doi.org/10.1002/2017JE005445.

Martin, R.L. and Kok, J.F. (2018). Distinct thresholds for the initiation and cessation of aeolian saltation from field measurements. *Journal of Geophysical Research: Earth Surface* 123 (7): 1546–1565. https://doi.org/10.1029/2017JF004416.

Marvin, M.C., Lapôtre, M.G.A., Gunn, A. et al. (2023). Dune interactions record changes in boundary conditions. *Geology*. https://doi.org/10.1130/G51264.1.

Marvin, M.C., Lapôtre, M.G.A., Radebaugh, J., and Bo, W. (2025). From Xanadu around and back: A ca. 11,000 km journey of windblown sand revealed by global dune patterns on Titan. *Geophysical Research Letters* 52 (5). https://doi.org/10.1029/2024GL112760.

Masursky, H. (1973). An overview of geological results from Mariner 9. *Journal of Geophysical Research* 78 (20): 4009–4030.

Matsubara, Y., Howard, A.D., Burr, D.M. et al. (2015). River meandering on Earth and Mars: a comparative study of Aeolis Dorsa meanders, Mars and possible terrestrial analogs of the Usuktuk River, AK, and the Quinn River, NV. *Geomorphology* 240: 102–120. https://doi.org/10.1016/j.geomorp.2014.08.031.

Maurice, S., Wiens, R.C., Bernardi, P. et al. (2021). The SuperCam instrument suite on the Mars 2020 rover: Science objectives and mast-unit description. *Space Science Reviews* 27: 47. https://doi.org/10.1007/s11214-021-00807-w.

Maurice, S., Wiens, R.C., Saccoccio, M. et al. (2012). The ChemCam instrument suite on the Mars Science Laboratory (MSL) rover: Science objectives and mast unit description. *Space Science Reviews* 170: 95–166. https://doi.org/10.1007/s11214-012-9912-2.

Maynard-Casely, H.E., Cable, M.L., Malaska, M.J. et al. (2018). Prospects for mineralogy on Titan. *American Mineralogist* 103 (3). https://doi.org/10.2138/am-2018-6259.

McCauley, J.F. (1973). Mariner 9 evidence for wind erosion in the equatorial and mid-latitude regions of Mars. *Journal of Geophysical Research* 78 (20): 4123–4137.

McCubbin, F.M., Nekvasil, H., Harrington, A.D. et al. (2008). Compositional diversity and stratification of the martian crust: Inferences from crystallization experiments on the picrobasalt Humphrey from Gusev crater, Mars. *Journal of Geophysical Research: Planets* 113 (E11). https://doi.org/10.1029/2008JE003165.

McDonald, G.D., Harper, J.M., Ojha, L. et al. (2022). Aeolian sediment transport on Io from lava-frost interactions. *Nature Communications* 13: 2076. https://doi.org/10.1038/s41496-022-29682-x.

McEwen, A.S., Banks, M.E., Baugh, N. et al. (2010). The high resolution imaging science experiment (HiRISE) during MRO's primary science phase (PSP). *Icarus* 205 (1): 2–37. https://doi.org/10.1016/j.icarus.2009.04.023.

McEwen, A.S., Dundas, C.M., Mattson, S. S. et al. (2014). Recurring slope lineae in equatorial regions of Mars. *Nature Geoscience* 7: 53–58. https://doi.org/10.1038/ngeo2014.

McEwen, A.S., Eliason, E.M., Bergstrom, J.W. et al. (2007). Mars reconnaissance orbiter's high resolution imaging science experiment (HiRISE). *Journal of Geophysical Research: Planets* 112 (E5). https://doi.org/10.1029/2005JE002605.

McEwen, A.S., Ojha, L., Dundas, C.M. et al. (2011). Seasonal flows on warm martian slopes. *Science* 333 (6043): 740–743. https://doi.org/10.1126/science.1204816.

McEwen, A.S., Schaefer, E.I., Dundas, C.M. et al. (2021). Mars: Abundant recurring slope lineae (RSL) following the planet-encircling dust event (PEDE) of 2018. *Journal of Geophysical Research: Planets* 126 (4). https://doi.org/10.1029/2020JE006575.

McEwen, A. S., S. Byrne, and the HiRISE Team (2024). The future of high-resolution orbital imaging of Mars. Tenth International Conference on Mars, Abstract #3150, Pasadena, California, USA

McSween, H.Y. Jr., Hamilton, V.E., and Farley, K.A. (2025). Perspectives on Mars Sample Return: A critical resource for planetary science and exploration. *Proceedings of the National Academy of Sciences* 122 (2): e2404248121.

McLaughlin, B. (1954). Interpretation of some martian features. *Publications of the Astronomical Society of the Pacific* 66 (391): 161–170.

McLennan, S.M., Anderson, R.B., Bell, J.F. III et al. (2014). Elemental geochemistry of sedimentary rocks at Yellowknife Bay, Gale crater, Mars. *Science* 343: 1244734. https://doi.org/10.1126/science.1244734.

McLennan, S.M., Bell, J.F. Iii, Calvin, W.M. et al. (2005). Provenance and diagenesis of the evaporite-bearing Burns formation, Meridiani Planum, Mars. *Earth and Planetary Science Letters* 240(1: 95–121. https://doi.org/10.1016/j.epsl.2005.09.041.

McLennan, S.M., Grotzinger, J.P., Hurowitz, J.A. et al. (2019). The sedimentary cycle on early Mars. *Annual Review of Earth and Planetary Sciences* 47: 91–118. https://doi.org/10.1149/annurev-earth-053018-060332.

McMahon, S., Bosak, T., Grotzinger, J.P. et al. (2018). A field guide to finding fossils on Mars. *Journal of Geophysical Research: Planets* 123 (5): 1012–1040. https://doi.org/10.1029/2017JE005478.

McSween, H.Y., Labotka, T.C., and Viviano-Beck, C.E. (2015). Metamorphism in the martian crust. *Meteoritics and Planetary Science* 50: 590–603. https://doi.org/10.1111/maps.12330.

McSween, H.Y., Taylor, G.J., Wyatt, M.B. et al. (2009). Elemental composition of the martian crust. *Science* 324 (5928): 736–739. https://doi.org/10.1126/science.1165871.

Mendéz Harper, J.S., McDonald, G.D., Dufek, J. et al. (2017). Electrification of sand on Titan and its influence on sediment transport. *Nature Geoscience* 10: 260–265. https://doi.org/10.1038/neo2921.

Metz, J.M., Grotzinger, J.P., Mohrig, D. et al. (2009b). Sublacustrine depositional fans in southwest Melas Chasma. *Journal of Geophysical Research: Planets* 114: E10. https://doi.org/10.1029/2009JE003365.

Metz, J.M., Grotzinger, J.P., Rubin, D.M. et al. (2009a). Sulfate-rich eolian and wet interdune deposits, Erebus crater, Meridiani Planum, Mars. *Journal of Sedimentary Research* 79 (5): 247–264. https://doi.org/10.2110/jsr.2009.033.

Meyer, M.J., Milliken, R.E., Stack, K.M. et al. (2024). Geological context and significance of the clay-sulfate transition region of Mount Sharp, Gale crater, Mars: an integrated assessment based on orbiter and rover data. *Bulletin of the Geological Society of America* https://doi.org/10.1130/B37355.1.

Mier-Hicks, F., HF Grip, A Kalantari et al. (2023). Sample recovery helicopter. 2023 IEEE Aerospace Conference, https://doi.org/10.1109/AERO55745.2023.10115951

Miller, J., Marshall, J., Greeley, R. et al. (1987). Wind ripples in low density atmospheres. *NASA Technical Memorandum* 89810: 268–270.

Miller, J.W., Birch, S.P.D., Hayes, A.G. et al. (2021). Fluvial features on Titan and Earth: Lessons from planform images in low-resolution SAR. *The Planetary Science Journal* 2 (4). https://doi.org/10.3847/PSJ/ac0245.

Milliken, R.E., Ewing, R.C., Fischer, W.W. et al. (2014). Wind-blown sandstones cemented by sulfate and clay minerals in Gale crater, Mars. *Geophysical Research Letters* 41 (4): 1149–1154. https://doi.org/10.1002/2013GL059097.

Milliken, R.E., Grotzinger, J.P., Thomson, B.J. et al. (2010). Paleoclimate of Mars as captured by the stratigraphic record in Gale crater. *Geophysical Research Letters* 37. https://doi.org/10.1029/2009GL041870.

Mittelholz, A., Heagy, L., Johnson, C.L. et al. (2023). Exploring martian magnetic fields with a helicopter. *The Planetary Science Journal* 4 (8): https://doi.org/10.3847/PSJ/ace9c1.

Mondro, C. A., Fedo, C. M., Grotzinger, J. P., Lamb, M. P., Gupta, S., Dietrich, W. E., Banham, S., Wetiz, C. M., Gasda, P., Edgar, L. A., Rubin, D., Bryk, A. B., Kite, E. S., Caravaca, G., Schieber, J. & Vasavada, A. R. (2025a). Wave ripples formed in ancient, ice-free lakes in Gale crater, Mars. *Science Advances*, 11(3), eadr0010.

Mondro, C.A., Grotzinger, J., Fedo, C.M. et al. (2025b). Depositional environment of the Amapari Marker Band: rising water levels formed kilometer-scale lake in Gale Crater, Mars. *Journal of Geophysical Research: Planets* 130 (3): e2024JE008606.

Montgomery, D.R., Bandfield, J.L., and Becker, S.K. (2012). Periodic bedrock ridges on Mars. *Journal of Geophysical Research: Planets* 117: E3. https://doi.org/10.1029/2011JE003970.

Moore, J.M. and Howard, A.D. (2005). Large alluvial fans on Mars. *Journal of Geophysical Research: Planets* 110 (E4). https://doi.org/10.1029/2004JE002352.

Moore, J.M. and Howard, A.D. (2010). Are the basins of Titan's Hotei Regio and Tui Regio sites of former low latitude seas? *Geophysical Research Letters* 37 (22). https://doi.org/10.1029/2010GL045234.

Moshkin, B.E., Ekonomov, A.P., and Golovin, I.M. (1979). Dust on the surface of Venus. *Cosmic Research* 17 (2): 280–285.

Mouginot, J., Pommerol, A., Beck, P. et al. (2012). Dielectric map of the Martian northern hemisphere and the nature of plain filling materials. *Geophysical Research Letters* 39 (2). https://doi.org/10.1029/2011GL050286.

Murchie, S., Arvidson, R., Bedini, P. et al. (2007). Compact reconnaissance imaging spectrometer for mars (CRISM) on mars reconnaissance orbiter (MRO). *Journal of Geophysical Research: Planets* 112 (E5). https://doi.org/10.1029/2006JE002682.

Mustard, J.F., Murchie, S.L., Pelkey, S.M. et al. (2008). Hydrated silicate minerals on Mars observed by the Mars Reconnaissance Orbiter CRISM instrument. *Nature* 454: 305–309. https://doi.org/10.1038/nature07097.

Nachon, M., Clegg, S.M., Mangold, N. et al. (2014). Calcium sulfate veins characterized by ChemCam/Curiosity at Gale crater. *Journal of Geophysical Research: Planets* 119 (9): 1991–2016. https://doi.org/10.1002/2013JE004588.

Neish, C.D., Barnes, J.W., Sotin, C. et al. (2015). Spectral properties of Titan's impact craters imply chemical weathering of its surface. *Geophysical Research Letters* 42 (10): 3746–3754. https://doi.org/10.1002/2015GL063824.

Newman, C.E., Lewis, S.R., Read, P.L. et al. (2002). Modeling the martian dust cycle, 1. Representations of dust transport processes. *Journal of Geophysical Research: Planets* 107 (E12). https://doi.org/10.1029/2002JE001910.

Nielsen, E. (2004). Mars express and MARSIS. *Space Science Reviews* 111: 245–262. https://doi.org/10.1023/B:SPAC.0000032712.05204.5e.

Niño, Y., Lopez, F., Garcia, M. et al. (2003). Threshold for particle entrainment into suspension. *Sedimentology* 50 (2): 247–263. https://doi.org/10.1046/j.1365-3091.2003.00551.x.

Ohja, L., Chojnacki, M., McDonald, G.D. et al. (2017). Seasonal slumps in Juventae Chasma, Mars. *Journal of Geophysical Research: Planets* 122 (10): 2193–2214. https://doi.org/10.1002/2017JE005375.

Ohja, L., McEwen, A., Dundas, C. et al. (2014). HiRISE observations of recurring slope lineae (RSL) during southern summer on Mars. *Icarus* 231: 365–376. https://doi.org/10.1016/j.icarus.2013.12.021.

Okubo, C.H. and McEwen, A.S. (2007). Fracture-controlled paleo-fluid flow in Cando Chasma, Mars. *Science* 315: 983–985. https://doi.org/10.1126/science.1136855.

Okubo, C.H., Schultz, R.A., Chan, M.A. et al. (2009). Deformation band clusters on Mars and implications for subsurface fluid flow. *Bulletin of the Geological Society of America* 121: 474–482. https://doi.org/10.1130/B26421.1.

Pacifici, A., Komatsu, G., Pondrelli, M. et al. (2009). Geological evolution of Ares Vallis on Mars: formation by multiple events of catastrophic flooding, glacial and periglacial processes. *Icarus* 202 (1): 60–77. https://doi.org/10.1016/j.icarus.2009.02.029.

Paige, D. A., SE Hamran, HEF Amundsen et al. (2024). Ground penetrating radar observations of the contact between the western delta and the crater floor of Jezero crater, Mars. *Science Advances*, 10(4). https://doi.org/10.1126/sciadv.adi8339

Pain, R.M., Clarke, J.D.A., Thomas, M. et al. (2007). Inversion of relief on Mars. *Icarus* 190: 478–491. https://doi.org/10.1016/j.icarus.2007.03.017.

Palucis, M.C., Dietrich, W.E., Hayes, A.G. et al. (2014). The origin and evolution of the Peace Vallis fan system that drains to the Curiosity landing site area, Gale crater, Mars. *Journal of Geophysical Research: Planets* 119 (4): 705–728. https://doi.org/10.1002/2013JE004583.

Palucis, M.C., Dietrich, W.E., Williams, R. M.E. et al. (2016). Sequence and relative timing of large lakes in Gale crater (Mars) after the formation of Mount Sharp. *Journal of Geophysical Research: Planets* 12 (3): 472–496. https://doi.org/10.1002/2015JE004905.

Pan, L., Ehlmann, B.L., Carter, J. et al. (2017). The stratigraphy and history of Mars' northern lowlands through mineralogy of impact craters: a comprehensive survey. *Journal of Geophysical Research: Planets* 122 (9): 1824–1854. https://doi.org/10.1002/2017JE005276.

Parker, G. (1978). Self-formed straight rivers with equilibrium banks and mobile bed. Part 2. The gravel river. *Journal of Fluid Mechanics* 89 (1): 127–146.

Parker, T.J., Gorsline, D.S., Saunders, R.S. et al. (1993). Coastal geomorphology of the Martian northern plains. *Journal of Geophysical Research Planets* 98 (E6): 11061–11078.

Parker, T.J., Saunders, R.S., Schneeberger, D.M. et al. (1989). Transitional morphology in west Deuteronilus Mensae, Mars: implications for modification of the lowland/upland boundary. *Icarus* 82: 111–145.

Pasquon, K., Gargani, J., Massé, M. et al. (2016). Present-day formation and seasonal evolution of linear dune gullies on Mars. *Icarus* 274: 195–210. https://doi.org/10.1016/j.icarus.2016.03.024.

Pasquon, K., Gargani, J., Massé, M. et al. (2019b). Present-day development of gully-channel sinuosity by carbon dioxide gas supported flows on Mars. *Icarus* 329: 269–313. https://doi.org/10.1016/j.icarus.2019.03.034.

Pasquon, K., Gargani, J., Nachon, M. et al. (2019a). Are different martian gully morphologies due to different processes on the Kaiser dune field? *Special Publications of the Geological Society of London* 467: 145–164. https://doi.org/10.1144/SP467.13.

Payré, V., Siebach, K.L., Dasgupta, R. et al. (2020). Constraining ancient magmatic evolution on Mars using crystal chemistry of detrital igneous minerals in the sedimentary Bradbury group, Gale crater, Mars. *Journal of Geophysical Research: Planets* 125 (8): https://doi.org/10.1029/2020JE006467.

Perron, J.T., Lamb, M.P., Koven, C.D. et al. (2006). Valley formation and methane precipitation rates on Titan. *Journal of Geophysical Research: Planets* 111 (E11): https://doi.org/10.1029/2005JE002602.

Perron, J.T., Mitrovica, J.X., Manga, M. et al. (2007). Evidence for an ancient martian ocean in the topography of deformed shorelines. *Nature* 447: 840–843. https://doi.org/10.1038/nature05873.

Pieri, D. (1976). Distribution of small channels on the Martian surface. *Icarus* 27 (1): 25–50.

Piqueux, S., Byrne, S., and Richardson, M. I. (2003). Sublimation of Mars' southern seasonal CO_2 ice cap and the formation of spiders. *Journal of Geophysical Research: Planets* 108 (E8). https://doi.org/10.1029/2002JE002007.

Poggiali, V., Mastrogiuseppe, M., Hayes, A.G. et al. (2016). Liquid-filled canyons on Titan. *Geophysical Research Letters* 43 (15): 7887–7894. https://doi.org/10.1002/2016GL069679.

Pollack, J.B., Haberle, R., Greeley, R. et al. (1976). Estimates of the wind speeds required for particle motion on Mars. *Icarus* 29 (3): 395–417.

Poulet, F., Bibring, J.P., Mustard, J.F. et al. (2005). Phyllosilicates on Mars and implications for early martian climate. *Nature* 438: 623–627. https://doi.org/10.1038/nature04274.

Preston, S.L., Siebach, K.L., Lapôtre, M.G.A., and Banham, S.G. (2024). Grain size measurements of the eolian Stimson formation, Gale crater, Mars and implications for sand provenance and paleoatmospheric conditions. *Journal of Geophysical Research: Planets* 129 (11). https://doi.org/10.1029/2024JE008369.

Quantin, C., Allemand, P., Mangold, N. et al. (2004). Ages of Valles Marineris

(Mars) landslides and implications for canyon history. *Icarus* 172: 555–572. https://doi.org/10.1016/j.icarus.2004.06.013.

Quantin-Nataf, C., Carter, J., Mandon, L. et al. (2021). Oxia Planum: the landing site for the ExoMars "Rosalind Franklin" rover mission: geological context and prelanding interpretation. *Astrobiology* 21 (3): 345–366.

Raack, J., Conway, S.J., Heyer, T. et al. (2020). Present-day gully activity in Sisyphi Cavi, Mars - Flow-like features and block movement. *Icarus* 350. https://doi.org/10.1016/j.icarus.2020.113899.

Raack, J., Reiss, D., Appéré, T. et al. (2015). Present-day seasonal gully activity in a south polar pit (Sisyphi Cavi) on Mars. *Icarus* 251: 226–243. https://doi.org/10.1016/j.icarus.2014.03.040.

Radebaugh, J. (2013). Dunes on Saturn's moon Titan as revealed by the Cassini mission. *Aeolian Research* 11: 23–41. https://doi.org/10.1016/j.aeolia.2013.07.001.

Radebaugh, J., Lorenz, R.D., Lunine, J.I. et al. (2008). Dunes on Titan observed by Cassini radar. *Icarus* 194 (2): 690–703. https://doi.org/10.1016/j.icarus.2007.10.015.

Radebaugh, J., Lorenz, R.D., Wall, S.D. et al. (2011). Regional geomorphology and history of Titan's Xanadu province. *Icarus* 211 (1): 672–685. https://doi.org/10.1016/j.icarus.2010.07.022.

Rangarajan, V.G., Tornabene, L.L., Osinski, G.R. et al. (2023). Change detection and monitoring of active martian surface phenomena with the Colour and Stereo Surface Imaging System (CaSSIS) onboard the ExoMars Trace Gas Orbiter (TGO). *Icarus* 394. https://doi.org/10.1016/j.icarus.2023.115443.

Rapin, W., Dromart, G., Clark, B.C. et al. (2023). Sustained wet-dry cycling on early Mars. *Nature* 620: 299–302. https://doi.org/10.1038/s41586-023-06220-3.

Rice, M.S., Gupta, S., Treiman, A.H. et al. (2017). Geologic overview of the Mars Science Laboratory rover mission at the Kimberley, Gale crater, Mars. *Journal of Geophysical Research: Planets* 122 (1): 2–20. https://doi.org/10.1002/2016JE005200.

Rivera-Hernández, F. and Palucis, M.C. (2019). Do deltas along the crustal dichotomy boundary of Mars in the Gale crater region record a northern ocean? *Geophysical Research Letters* 46 (15): 8689–8699. https://doi.org/10.1029/2019GL083046.

Roback, K.P., Runyon, K.D., Avouac, J.P. et al. (2020). Craters as sand traps: dynamics, history, and morphology of modern sand transport in an active martian dune field. *Icarus* 342: 113642. https://doi.org/10.1016/j.icarus.2020.113642.

Roberts, A.L., S Gupta, WE Dietrich et al. (2023). What Depositional Processes and Paleoenvironments Formed the Layered Sulfate Unit in Gale Crater, Mars? Insights from Marker Band Valley. LPI Contributions, 2806, p. 2945.

Robinson, M.S. and Tanaka, K.L. (1990). Magnitude of a catastrophic flood event at Kasei Valles, Mars. *Geology* 18 (9): 902–905.

Rodriguez, J.A.P., Fairén, A.G., Tanaka, K.L. et al. (2016). Tsunami waves extensively resurfaced the shorelines of an early Martian ocean. *Scientific Reports* 6 (25106): https://doi.org/10.1038/srep25106.

Rodriguez, J.A.P., Robertson, D.K., Kargel, J.S. et al. (2022). Evidence of an oceanic impact and megatsunami sedimentation in Chryse Planitia, Mars. *Scientific Reports* 12: 19589. https://doi.org/10.1038/s41598-022-18082-2.

Rodriguez, S., Le Mouélic, S., Barnes, J.W. et al. (2018). Observational evidence for active dust storms on Titan at equinox. *Nature Geoscience* 11: 727–732. https://doi.org/10.1038/s41561-018-0233-2.

Rogers, A.D., Warner, N.H., Golombek, M.P. et al. (2018). Areally extensive surface bedrock exposures on Mars: many are clastic rocks, not lava. *Geophysical Research Letters* 45 (4): 1767–1777. https://doi.org/10.1002/2018GL077030.

Rubanenko, L., Gunn, A., Pérez-López, S. et al. (2023). Global surface winds and aeolian sediment pathways on Mars from the morphology of barchan dunes. *Geophysical Research Letters* https://doi.org/10.1029/2022GL102610.

Rubanenko, L., Lapôtre, M.G.A., Ewing, R.C. et al. (2022). A distinct ripple-formation regime on Mars revealed by the morphometrics of barchan dunes. *Nature Communications* 13: 7156. https://doi.org/10.1038/s41467-022-39974-3.

Rubanenko, L., Pérez-López, S., Schull, J. et al. (2021). Automatic detection and segmentation of barchan dunes on Mars and Earth using a convolutional neural network. *IEEE Journal of Selected Topics in Applied Earth Observations and Remote Sensing* 14: 9364–9371. https://doi.org/10.1109/JSTARS.2021.3109900.

Rubin, D.M., Lapôtre, M.A.G., Stevens, A.W. et al. (2022). Ancient winds, waves, and atmosphere in Gale crater, Mars, inferred from sedimentary structures and wave modeling. *Journal of Geophysical Research: Planets* 127 (4). https://doi.org/10.1029/2021JE007162.

Russell, P., Thomas, N., Byrne, S. et al. (2008). Seasonally active frost-dust avalanches on a north polar scarp of Mars captured by HiRISE. *Geophysical Research Letters* 35 (23). https://doi.org/10.1029/2008GL035790.

Sagan, C. and Pollack, J.B. (1969). Windblown dust on Mars. *Nature* 223: 791–794.

Salese, F., McMahon, W.J., Balme, M.R. et al. (2020). Sustained fluvial deposition recorded in Mars' Noachian stratigraphic record. *Nature Communications* 11: 2067. https://doi.org/10.1038/s41467-020-15622-0.

Salvatore, M.R. and Christensen, P.R. (2014a). On the origin of the Vastitas Borealis Formation in Chryse and Acidalia Planitiae, Mars. *Journal of Geophysical Research Planets* 119 (12): 2437–2456. https://doi.org/10.1002/2014JE004682.

Salvatore, M.R. and Christensen, P.R. (2014b). Evidence for widespread aqueous sedimentation in the northern plains of Mars. *Geology* 42 (5): 423–426. https://doi.org/10.1130/G35319.1.

Sasselov, D.D., Grotzinger, J.P., Sutherland, J.D. et al. (2020). The origin of life as a planetary phenomenon.

Science Advances 6 (6). https://doi.org/10.1126/sciadv.aax3419.

Saunders, R.S., Arvidson, R.E., Badhwar, G.D. et al. (2004). 2001 Mars Odyssey mission summary. *Space Science Reviews* 110m: 2–36. https://doi.org/10.1023/B:SPAC.0000021006.84299.18.

Saunders, R.S., Pettengill, G.H., Arvidson, R.E. et al. (1990). The Magellan Venus radar mapping mission. *Journal of Geophysical Research: Solid Earth* 95 (B6): 8339–8355.

Saunders, R.S., Spear, A.J., Allin, P.C. et al. (1992). Magellan mission summary. *Journal of Geophysical Research: Planets* 97 (E8): 13067–13090.

Schaefer, E.I., McEwen, A.S., Sutton, S.S. et al. (2019). A case study of recurring slope lineae (RSL) at Tivat crater: Implications for RSL origins. *Icarus* 317: 621–648. https://doi.org/10.1016/j.icarus.2018.07.014.

Scheller, E.L. and Ehlmann, B.L. (2020). Composition, stratigraphy, and geological history of the Noachian basement surrounding the Isidis impact basin. *Journal of Geophysical Research: Planets* 125 (7): https://doi.org/10.1029/2019JE006190.

Scheller, E.L., Ehlmann, B.L., Hu, R. et al. (2021). Long-term drying of Mars by sequestration of ocean-scale volumes of water in the crust. *Science* 372 (6537): 56–62. https://doi.org/10.1126/science.abs7717.

Schieber, J., Bish, D., Coleman, M. et al. (2017). Encounters with an unearthly mudstone: Understanding the first mudstone found on Mars. *Sedimentology* 64: 311–358. https://doi.org/10.1111/sed.12318.

Schieber, J., Bohacs, K.M., Coleman, M. et al. (2022). Mars is a mirror – Understanding the Pahrump Hills mudstones from a perspective of Earth analogues. *Sedimentology* 69 (6): 2371–2435. https://doi.org/10.1111/sed.13024.

Schwamb, M.E., Aye, K.M., Portyankina, G. et al. (2018). Planet four: Terrains - Discovery of araneiforms outside of the South Polar layered deposits. *Icarus* 308: 148–187. https://doi.org/10.1016/j.icarus.2017.06.017.

Selivanov, A.S., Gektin, Y.M., Naraeva, M. K. et al. (1982). Evolution of the VENERA-13 imagery. *Soviet Astronomy Letters* 8: 235–236.

Seybold, H.J., Kite, E., Kirchner, J.W. et al. (2018). Branching geometry of valley networks on Mars and Earth and its implications for early Martian climate. *Science Advances* 4 (6): eaar6692. https://doi.org/10.1126/sciadv.aar6692.

Shahrzad, S., Kinch, K.M., Goudge, T.A. et al. (2019). Crater statistics on the dark-toned, mafic floor unit in Jezero crater, Mars. *Geophysical Research Letters* 46 (5): 2408–2416. https://doi.org/10.1029/2018GL081402.

Sharaf, O., Amiri, S., AlDhafri, S. et al. (2020). Sending hope to Mars. *Nature Astronomy* 4 (722). https://doi.org/10.1038/S41550-020-1151-y.

Sharp, R.P. (1963). Wind ripples. *The Journal of Geology* 71 (5): 617–636.

Sharp, R.P. and Malin, M.C. (1975). Channels on Mars. *Bulletin of the Geological Society of America* 86 (5): 593–609.

Shi, X., Hu, X., Agarwal, J. et al. (2024). Diurnal ejection of boulder clusters on comet 67P lasting beyond 3 AU. *The Astrophysical Journal Letters* 961 (1). https://doi.org/10.3847/2041-8213/ad18d9.

Sholes, S.F., Dickeson, Z.I., Montgomery, D.R. et al. (2021). Where are Mars' hypothesized ocean shorelines? Large lateral and topographic offsets between different versions of paleoshorelines maps. *Journal of Geophysical Research Planets* 126 (5). https://doi.org/10.1029/2020JE006486.

Sholes, S.F., Montgomery, D.R., and Catling, D.C. (2019). Quantitative high-resolution reexamination of a hypothesized ocean shoreline in Cydonia Mensae on Mars. *Journal of Geophysical Research Planets* 124 (2): 316–336. https://doi.org/10.1029/2018JE005837.

Sholes, S.F. and Rivera-Hernández, F. (2022). Constraints on the uncertainty, timing, and magnitude of potential Mars oceans from topographic deformation models. *Icarus* 378 (114934): https://doi.org/10.1016/j.icarus.2022.114934.

Siebach, K.L., Baker, M.B., Grotzinger, J.P. et al. (2017). Sorting out compositional trends in sedimentary rocks of the Bradbury group (Aeolis Palus), Gale crater, Mars. *Journal of Geophysical Research: Planets* 122: 295–328. https://doi.org/10.1002/2016JE005195.

Siebach, K.L., Grotzinger, J.P., Kah, L.C. et al. (2014). Subaqueous shrinkage cracks in the Sheepbed mudstone: implications for early fluid diagenesis, Gale crater, Mars. *Journal of Geophysical Research: Planets* 119 (7): 1597–1613. https://doi.org/10.1002/2014JE004623.

Silvestro, S., Chojnacki, M., Vaz, D.A. et al. (2020). Megaripple migration on Mars. *Journal of Geophysical Research: Planets* 125 (8). https://doi.org/10.1029/2020JE006446.

Silvestro, S., Pacifici, A., Salese, F. et al. (2021). Periodic bedrock ridges at the ExoMars 2022 landing site: evidence for a changing wind regime. *Geophysical Research Letters* 48 (4). https://doi.org/10.1029/2020GL091651.

Silvestro, S., Vaz, D.A., Yizhaq, H. et al. (2016). Dune-like dynamic of martian aeolian large ripples. *Geophysical Research Letters* 43 (16): 8384–8389. https://doi.org/10.1002/2016GL070014.

Simon, J.I., Hickman-Lewis, K., Cohen, B.A. et al. (2023). Samples collected from the floor of Jezero crater with the Mars 2020 Perseverance rover. *Journal of Geophysical Research: Planets* 128 (6). https://doi.org/10.1029/2022JE007474.

Skinner, J.A., Tanaka, K.L., Platz, T. et al. (2012). Widespread loess-like deposit in the Martian northern lowlands identifies Middle Amazonian climate change. *Geology* 40 (12): 1127–1130. https://doi.org/10.1130/G33513.1.

Smith, J.D. (1970). Stability of a sand bed subjected to a shear flow of low Froude number. *Journal of Geophysical Research* 75: 5928–5940. https://doi.org/10.1029/JC075i030p05928.

Smith, R.J., McLennan, S.M., Achilles, C. N. et al. (2021). X-ray amorphous components in sedimentary rocks of Gale crater, Mars: Evidence for ancient formation and long-lived aqueous activity. *Journal of Geophysical*

Research: Planets 126 (3): https://doi.org/10.1029/2020JE006782.

Smrekar, S., S Hensley, R Nybakken et al. (2022). VERITAS (Venus Emissivity, Radio Science, InSAR, Topography, and Spectroscopy): A Discovery mission. 2022 IEEE Aerospace Conference, https://doi.org/10.1109/AERO53065.2022.9843269

Soderblom, L.A., Tomasko, M.G., Archinal, B.A. et al. (2007). Topography and geomorphology of the Huygens landing site on Titan. *Planetary and Space Science* 55 (13): 2015–2024. https://doi.org/10.1016/j.pss.2007.04.015.

Squyres, S.W., Aharonson, O., Clark, B.C. et al. (2007). Pyroclastic activity at Home Plante in Gusev crater, Mars. *Science* 316 (5825): 738–742. https://doi.org/10.1126/science.1139045.

Squyres, S.W., Arvidson, R.E., Bell, J.F. III et al. (2004a). The Spirit rover's Athena science investigation at Gusev crater, Mars. *Science* 305 (5685): 794–799. https://doi.org/10.1126/science.3050794.

Squyres, S.W., Grotzinger, J.P., Arvidson, R.E. et al. (2004b). In situ evidence for an ancient aqueous environment at Meridiani Planum, Mars. *Science* 306 (5702): 1709–1714. https://doi.org/10.1126/science.1104559.

Squyres, S.W., Knoll, A.H., Arvidson, R.E. et al. (2009). Exploration of Victoria crater by the Mars rover opportunity. *Science* 324 (5930): 1058–1061. https://doi.org/10.1126/science.1170355.

Stack, K.M., Dietrich, W.E., Lamb, M.P. et al. (2022). Orbital and in-situ investigation of periodic bedrock ridges in Glen Torridon, Gale crater, Mars. *Journal of Geophysical Research: Planets* 127 (6). https://doi.org/10.1029/2021JE007096.

Stack, K.M., Edwards, C.S., Grotzinger, J.P. et al. (2016). Comparing orbiter and rover image-based mapping of an ancient sedimentary environment, Aeolis Palus, Gale crater, Mars. *Icarus* 280: 3–21. https://doi.org/10.1016/j.icarus.2016.02.024.

Stack, K.M., Grotzinger, J.P., Kah, L.C. et al. (2014). Diagenetic origin of nodules in the Sheepbed member, Yellowknife Bay formation, Gale crater, Mars. *Journal of Geophysical Research: Planets* 119 (7): 1639–1664. https://doi.org/10.1002/2014JE004617.

Stack, K.M., Grotzinger, J.P., Lamb, M.P. et al. (2019). Evidence for plunging river plume deposits in the Pahrump Hills member of the Murray formation, Gale crater, Mars. *Sedimentology* 66 (5): 1768–1802. https://doi.org/10.1111/sed.12558.

Stack, K.M., Grotzinger, J.P., and Milliken, R.E. (2013). Bed thickness distributions on Mars: An orbital perspective. *Journal of Geophysical Research: Planets* 118 (6): 1323–1349. https://doi.org/10.0102/jgre.20092.

Stack, K.M., Ives, L.R.W., Gupta, S. et al. (2024). Sedimentology and stratigraphy of the Shenandoah formation, western fan, Jezero crater, Mars. *Journal of Geophysical Research: Planets* 129 (2): https://doi.org/10.1029/2023JE008187.

Stack, K.M., Williams, N.R., Calef, F. et al. (2020). Photogeologic map of the Perseverance rover field site in Jezero crater constructed by the Mars 2020 Science Team. *Space Science Reviews* 216: 127. https://doi.org/10.1007/s11214-020-00739-x.

Steckloff, J.K., Soderblom, J.M., Farnsworth, K.K. et al. (2020). Stratification dynamics of Titan's lakes via methane evaporation. *The Planetary Science Journal* 1: 26. https://doi.org/10.3847/PSJ/ab974e.

Stein, N., Grotzinger, J.P., Schieber, J. et al. (2018). Desiccation cracks provide evidence of lake drying on Mars, Sutton Island member, Murray formation, Gale crater. *Geology* 46 (6): 515–518. https://doi.org/10.1130/G40005.1.

Stillman, D.E. (2018). Unraveling the mysteries of recurring slope lineae. In *Dynamic Mars: Recent and Current Landscape Evolution of the Red Planet* (eds. R.J. Soare, S.J. Conway, S.M. Clifford), 51–85.

Stillman, D.E., Bue, B.D., Wagstaff, K.L. et al. (2020). Evaluation of wet and dry recurring slope lineae (RSL) formation mechanisms based on quantitative mapping of RSL in Garni crater, Valles Marineris, Mars. *Icarus* 335: https://doi.org/10.1016/j.icarus.2019.113420.

Stillman, D.E. and Grimm, R.E. (2018). Two pulses of seasonal activity in martian southern mid-latitude recurring slope lineae (RSL). *Icarus* 302: 126–133. https://doi.org/10.1016/j.icarus.2017.10.026.

Stillman, D.E., Michaels, T.I., Grimm, R.E. et al. (2014). New observations of martian southern mid-latitude recurring slope lineae (RSL) imply formation by freshwater subsurface flows. *Icarus* 233: 328–341. https://doi.org/10.1016/j.icarus.2014.01.017.

Stillman, D.E., Michaels, T.I., Grimm, R.E. et al. (2016). Observations and modeling of northern mid-latitude recurring slope lineae (RSL) suggest recharge by a present-day martian briny aquifer. *Icarus* 265: 125–138. https://doi.org/10.1016/j.icarus.2015.10.007.

Stillman, D.E., Michaels, T.I., Grimm, R.E. et al. (2017). Characteristics of the numerous and widespread recurring slope lineae (RSL) in Valles Mariners, Mars. *Icarus* 285: 195–210. https://doi.org/10.1016/j.icarus.2016.10.025.

Stofan, E.R., Elachi, C., Lunine, J.I. et al. (2007). The lakes of Titan. *Nature* 445: 61–64. https://doi.org/10.1038/nature05438.

Stucky de Quay, G., Goudge, T.A., Fassett, C.I. et al. (2020). Precipitation and aridity constraints from paleolakes on early Mars. *Geology* 48: 1189–1193. https://doi.org/10.1130/G47886.1.

Sullivan, R., Banfield, D., Bell, J.F. et al. (2005). Aeolian processes at the Mars Exploration Rover Meridiani Planum landing site. *Nature* 436 (7047): 58–61. https://doi.org/10.1038/nature03641.

Sullivan, R. and Kok, J.F. (2017). Aeolian saltation on Mars at low wind speeds. *Journal of Geophysical Research: Planets* 122 (10): 2111–2143. https://doi.org/10.1002/2017JE005275.

Sullivan, R., Kok, J.F., Katra, I. et al. (2020). A broad continuum of aeolian impact ripple morphologies on Mars is enabled by /low wind dynamic pressures. *Journal of Geophysical Research: Planets* 125 (10). https://doi.org/10.1029/2020JE006485.

Sullivan, R., Thomas, P., Veverka, J. et al. (2001). Mass movement slope streaks imaged by the Mars Orbiter Camera. *Journal of Geophysical Research: Planets* 106 (E10): 23607–23633. https://doi.org/10.1029/2000JE001296.

Sun, V.A., Stack, K.M., Kah, L.C. et al. (2019). Late-stage diagenetic concretions in the Murray formation, Gale crater, Mars. *Icarus* 321: 866–890. https://doi.org/10.1016/j.icarus.2018.12.030.

Sun, V.Z., Hand, K.P., Stack, K.M. et al. (2023). Overview and results from the Mars 2020 Perseverance rover's first science campaign on the Jezero crater floor. *Journal of Geophysical Research: Planets* 128 (6). https://doi.org/10.1029/2022JE007613.

Surkov, Y.A. and Barsukov, V.L. (1985). Composition, structure and properties of Venus rocks. *Advances in Space Research* 5 (8): 17–29.

Sutton, S.S., Chojnacki, M., McEwen, A.S. et al. (2022). Revealing active Mars with HiRISE digital terrain models. *Remote Sensing* 14 (10): 2403. https://doi.org/10.3390/rs14102403.

Swann, C., Sherman, D.J., and Ewing, R.C. (2020). Experimentally derived thresholds for windblown sand on Mars. *Geophysical Research Letters* 47 (3): e2019GL084484. https://doi.org/10.1029/2019GL084484.

Szabó, T., Domokos, G., Grotzinger, J.P. et al. (2015). Reconstructing the transport history of pebbles on Mars. *Nature Communications* 6: https://doi.org/10.1038/ncomms9366.

Tanaka, K.L. (1997). Sedimentary history and mass flow structures of Chryse and Acidalia Planitiae, Mars. *Journal of Geophysical Research: Planets* 102 (E2): 4131–4149. https://doi.org/10.1029/96JE02862.

Tanaka, K.L., Robbins, S.J., Fortezzo, C.M. et al. (2014a). The digital global geologic map of Mars: Chronostratigraphic ages, topographic and crater morphologic characteristics, and updated resurfacing history. *Planetary and Space Science* 95: 11–24. https://doi.org/10.1016/j.pss.2013.03.006.

Tanaka, K.L., JA Skinner, JM Dohm et al. (2014b). Geologic map of Mars. U.S. Geological Survey Scientific Investigations Map, 3292, scale 1:20,000,000, pamphlet, https://doi.org/10.3133/sim3292

Tanaka, K.L., Skinner, J.A., Hare, T.M. et al. (2003). Resurfacing history of the northern plains of Mars based on geologic mapping of Mars Global Surveyor data. *Journal of Geophysical Research: Planets* 108 (E4). https://doi.org/10.1029/2002JE001908.

Telfer, M.W., Parteli, E.J.R., Radebaugh, J. et al. (2018). Dunes on Pluto. *Science* 360 (6392): 992–997. https://doi.org/10.1126/science.aao2975.

Thomas, N., Davidsson, B., El-Maarry, M.R. et al. (2015). Redistribution of particles across the nucleus of comet 67P/Churyumov-Gerasimenko. *Astronomy and Astrophysics* 583: A17. https://doi.org/10.1051/0004-6361/201526049.

Thomas, N., McEwen, A.S., Dundas, C.M. et al. (2020). Present-day mass wasting in sulfate-rich sediments in the equatorial regions of Mars. *Icarus* 342. https://doi.org/10.1016/j.icarus.2019.113566.

Thorpe, M.T., Hurowitz, J.A., and Siebach, K.L. (2021). Source-to-sink terrestrial analogs for the paleoenvironment of Gale crater, Mars. *Journal of Geophysical Research: Planets* 126. https://doi.org/10.1029/2020JE006530.

Tian, H., Zhang, T., Jia, Y. et al. (2021). Zhurong: Features and mission of China's first Mars rover. *The Innovation* 2: 100121.

Tokano, T. (2008). Dune-forming winds on Titan and the influence of topography. *Icarus* 194 (1): 243–262. https://doi.org/10.1016/j.icarus.2007.10.007.

Tokano, T. (2010). Relevance of fast westerlies at equinox for the eastward elongation of Titan's dunes. *Aeolian Research* 2: 113–127. https://doi.org/10.1016/j.aeolia.2010.04.003.

Tomasko, M.G., Perry, J.E., Hayes, A.G. et al. (2005). Rain, winds and haze during Huygens probe's decent to Titan's surface. *Nature* 438: 765–778. https://doi.org/10.1126/science.1201063.

Tsoar, H., Greeley, R., Peterfreund, A.R. et al. (1979). Mars: The North Polar Sand Sea and related wind patterns. *Journal of Geophysical Research: Solid Earth* 84 (B14): 8167–8180.

Turtle, E.P., Perry, J.E., Hayes, A.G. et al. (2011). Rapid and extensive surface changes near Titan's equator: Evidence of April showers. *Science* 331 (6023): 1414–1417. https://doi.org/10.1126/science.1201063.

Udry, A., Gazel, E., McSween, H.Y. et al. (2018). Formation of evolved rocks at Gale crater by crystal fractionation and implications for Mars crustal composition. *Journal of Geophysical Research: Planets* 123 (6): 1525–1540. https://doi.org/10.1029/2018JE005602.

Vago, J.L., Westall, F., Coates, A.J. et al. (2017). Habitability on early Mars and the search for biosignatures with the ExoMars Rover. *Astrobiology* 17 (6-7): 471–510.

Vasavada, A.R. (2022). Mission overview and scientific contributions from the Mars Science Laboratory Curiosity rover after eight years of surface operations. *Space Science Reviews* 218 (14). https://doi.org/10.1007/s11214-022-00882-7.

Vaz, D.A., Silvestro, S., Chojnacki, M. et al. (2023). Constraining the mechanisms of aeolian bedform formation on Mars through a global morphometric survey. *Earth and Planetary Science Letters* 614. https://doi.org/10.1016/j.epsl.2023.118196.

Vaz, D.A., Silvestro, S., Sarmento, P.T.K. et al. (2017). Migrating meter-scale bedforms on Martian dark dunes: Are terrestrial aeolian ripples good analogues? *Aeolian Research* 26: 101–116. https://doi.org/10.1016/j.aeolia.2016.08.003.

Villanueva, G.L., Mumma, M.J., Novak, R.E. et al. (2015). Strong water isotopic anomalies in the martian atmosphere: Probing current and ancient reservoirs. *Science* 348 (6231): 218–221. https://doi.org/10.1126/science.aaa3630.

Vincendon, M., Pilorget, C., Carter, J. et al. (2019). Observational evidence for a dry dust-wind origin of Mars seasonal dark flows. *Icarus* 325: 115–127. https://doi.org/10.1016/j.icarus.2019.02.024.

Viviano-Beck, C.E., Murchie, S.L., Beck, A.W. et al. (2017). Compositional and structural constraints on the geologic history of eastern Tharsis Rise, Mars. *Icarus* 284: 43–58. https://doi.org/10.1016/j.icarus.2016.09.005.

Wall, S., Hayes, A., Bristow, C. et al. (2010). Active shoreline of Ontario Lacus, Titan: A morphological study of the lake and its surroundings. *Geophysical Research Letters* 37 (5): https://doi.org/10.1029/GL041821.

Wang, J., Zhao, J., Xiao, L. et al. (2023). Recent aqueous activity on Mars evidenced by transverse aeolian ridges in the Zhurong exploration region of Utopia Planitia. *Geophysical Research Letters* 50: 6. https://doi.org/10.1029/2022GL101650.

Ward, A.W. (1979). Yardangs on Mars: Evidence of recent wind erosion. *Journal of Geophysical Research: Solid Earth* 84 (B14): 8147–8166.

Warner, N., Gupta, S., Muller, J.P. et al. (2009). A refined chronology of catastrophic outflow events in Ares Vallis, Mars. *Earth and Planetary Science Letters* 288 (1-2): 58–69. https://doi.org/10.1016/j.epsl.2009.09.008.

Warner, N., Sowe, M., Gupta, S. et al. (2013). Fill and spill of giant lakes in the eastern Valles Marineris region of Mars. *Geology* 41 (6): 675–678. https://doi.org/10.1130/G34172.1.

Watkins, J.A., Grotzinger, J.P., Stein, N.T. et al. (2022). Burial and exhumation of sedimentary rocks revealed by the base Stimson erosional unconformity, Gale crater, Mars. *Journal of Geophysical Research: Planets* 127 (7). https://doi.org/10.1029/2022JE007293.

Webster, C.R., Mahaffy, P.R., Flesch, G.J. et al. (2013). Isotope ratios of H, C, and O in CO_2 and H_2O of the martian atmosphere. *Science* 341: 260–263. https://doi.org/10.1126/science.1237961.

Weitz, C.M., O'Connell-Cooper, C., Thompson, L. et al. (2022). The physical properties and geochemistry of grains in aeolian bedforms at Gale crater, Mars. *Journal of Geophysical Research: Planets* 127 (11). https://doi.org/10.1029/2021JE007061.

Weitz, C.M., Plaut, J.J., Greeley, R. et al. (1994). Dunes and microdunes on Venus: Why were so few found in the Magellan data? *Icarus* 112 (1): 282–295.

Weitz, C.M., Sullivan, R.J., Lapôtre, M.G. A. et al. (2018). Sand grain sizes and shapes in eolian bedforms at Gale crater. *Mars. Geophysical Research Letters* 45 (18): 9471–9479. https://doi.org/10.1029/2018GL078972.

Williams, D.A., Radebaugh, J., Keszthelyi, L.P. et al. (2002). Geologic mapping of the Chaac-Camaxtli region of Io from Galileo imaging data. *Journal of Geophysical Research* 107 (E9): https://doi.org/10.1029/2001JE001821.

Williams, R.M.E., Grotzinger, J.P., Dietrich, W.E. et al. (2013b). Martian fluvial conglomerates at Gale crater. *Science* 340 (6136): 1068–1072. https://doi.org/10.1126/science.1237317.

Williams, R.M.E., Irwin, R.P. III, Burr, D. M. et al. (2013a). Variability in martian sinuous ridges form: Case study of Aeolis Serpens in the Aeolis Dorsa, Mars, and insight from the Mirackina paleoriver, South Australia. *Icarus* 225 (1): 308–324. https://doi.org/10.1016/j.icarus.2013.03.016.

Williams, R.M.E. and Malin, M.C. (2004). Evidence for late stage fluvial activity in Kasei Valles, Mars. *Journal of Geophysical Research: Planets* 109 (E6). https://doi.org/10.1029/20063JE002178.

Williams, R.M.E., Phillips, R.J., Malin, M. C. et al. (2000). Flow rates and duration within Kasei Valles, Mars: Implications for the formation of a martian ocean. *Geophysical Research Letters* 27 (7): 1073–1076. https://doi.org/10.1029/1999GL010957.

Wilson, S.A., Morgan, A.M., Howard, A.D. et al. (2021). The global distribution of craters with alluvial fans and deltas on Mars. *Geophysical Research Letters, in press* https://doi.org/10.1029/2020GL091653.

Wood, L.J. (2006). Quantitative geomorphology of the Mars Eberswalde delta. *Bulletin of the Geological Society of America* 118 (5–6): 557–566. https://doi.org/10.1130/B25822.1.

Wordsworth, R., Kerber, L., Pierrehumbert, R.T. et al. (2015). Comparison of "warm and wet" and "cold and icy" scenarios for early Mars in a 3-D climate model. *Journal of Geophysical Research Planets* 120 (6): 1201–1219. https://doi.org/10.1002/2015JE004787.

Yingst, R.A., Cowart, A.C., Kah, L.C. et al. (2023). Depositional and diagenetic processes of martin lacustrine sediments as revealed at Pahrump Hills by the Mars Hand Lens Imager, Gale crater, Mars. *Journal of Geophysical Research: Planets* 128 (5): https://doi.org/10.1029/2022JE007394.

Yizhaq, H., Kok, J.F., Katra, I. et al. (2014). Basaltic sand ripples at Eagle Crater as indirect evidence for the hysteresis effect in martian saltation. *Icarus* 230: 143–150. https://doi.org/10.1016/j.icarus.2013.08.006.

Yizhaq, H., Tholen, K., Saban, L. et al. (2024). Coevolving aerodynamic and impact ripples on Earth. *Nature Geoscience* 17: 66–72. https://doi.org/10.1038/s41561-023-01348-3.

Yongliao, Z., Zhu, Y., Bai, Y. et al. (2021). Scientific objectives and payloads of Tianwen-1, China's first Mars exploration mission. *Advances in Space Research* 67 (2): 812–823. https://doi.org/10.1016/j.asr.2020.11.005.

Yu, X., Hörst, S.M., He, C. et al. (2018). Where does Titan sand come from: Insights from mechanical properties of Titan sand candidates. *Journal of Geophysical Research: Planets* 123 (9): 2310–2321. https://doi.org/10.1029/2018JE005651.

Zabrusky, K., Andrews-Hanna, J.C., Wiseman, S.M. et al. (2012). Reconstructing the distribution and depositional history of the sedimentary deposits in Arabia Terra, Mars. *Icarus* 220: 311–330. https://doi.org/10.1016/j.icarus.2012.05.007.

Zaki, A.S., Pain, C.F., Edgett, K.S. et al. (2021). Global inventory of fluvial ridges on Earth and lessons applicable to Mars. *Earth-Science Reviews* 103561. https://doi.org/10.1016/j.earscirev.2021.103561.

Zappek, V., Rinker, M., Daxer, L. et al. (2023). Evaluation of aerial vehicle configurations for high-range Mars missions. *CEAS Space Journal* 15: 761–775. https://doi.org/10.1007/s12567-022-00479-4.

Zarnecki, J.C., Leese, M.R., Hathi, B. et al. (2005). A soft solid surface on Titan as revealed by the Huygens Surface Science Package. *Nature* 438: 792–795. https://doi.org/10.1038/nature04211.

Zhu, K., Quan, Q., Tang, D. et al. (2024). A Mars quadcopter capable of autonomous flight and sample collection: Structure and avionics. *Acta Astronautica* 214: 712–721. https://doi.org/10.1016/j.actaastro.2023.11.034.

Zilker, D.P., Cook, G.W., Hanratty, T.J. et al. (1977). Influence of the amplitude of a solid wavy wall on a turbulent flow. Part 1. Non-separated flows. *Journal of Fluid Mechanics* 82 (1): 29–51. 10.1017/S0022112077000524.

Zimbelman, J.R. (2010). Transverse aeolian ridges on Mars: First results from HiRISE images. *Geomorphology* 121 (1–2): 22–29. https://doi.org/10.1016/j.geomorph.2009.5.012.

Zimbelman, J.R. and Foroutan, M. (2020). Dingo Gap: Curiosity went up a small transverse aeolian ridge and came down a megaripple. *Journal of Geophysical Research: Planets* 125 (12): e2020JE006489. https://doi.org/10.1029/2020JE006489.

Zimbelman, J.R. and Griffin, L.J. (2010). HiRISE images of yardangs and sinuous ridges in the lower member of the Medusae Fossae Formation, Mars. *Icarus* 205 (1): 198–210. https://doi.org/10.1016/j.icarus.2009.04.003.

Zuber, M.T., Smith, D.E., Solomon, S.C. et al. (1992). The Mars Observer laser altimeter investigation. *Journal of Geophysical Research: Planets* 97 (E5): 7781–7797.

Zurek, R.W. and Smrekar, S.E. (2007). An overview of the Mars Reconnaissance Orbiter (MRO) science mission. *Journal of Geophysical Research: Planets* 112 (E5). https://doi.org/10.1029/2006JE002701.

Index

Reading's Sedimentary Environments: Processes, Facies and Stratigraphy, Fourth Edition. Edited by Bruce Levell.

Companion website: www.wiley.com/go/levell4e

b

C

q

r

Afterword

Howard Reading with students in the field in July 1968 on the Digermul Peninsula, Finnmark, Northern Norway. The outcrop geology is the Lower Cambrian Duolbasgaissa Formation. Photo credit Nigel Banks.

As Harold taught me:

"Bruce, the first thing that you do when you get to an outcrop is to walk away from it. Take in the overview, ask yourself where and why you are going to make observations and then test your hypotheses against those observations"

Good advice for life in general.

Reading's Sedimentary Environments: Processes, Facies and Stratigraphy, Fourth Edition. Edited by Bruce Levell.

Published 2026 by John Wiley & Sons Ltd.
Companion website: www.wiley.com/go/levell4e